PHYSICS

Consulting editor

Paul Davies	University of Adelaide

Reviewers

Colin Bates	University of Nottingham
M. Bister	Helsinki University of Technology
Tomas Bohr	Niels Bohr Institute, Copenhagen
Ronal Brown	University of Strathclyde
Jay Burns	Florida Institute of Technology
R. H. Dean	University of Southampton
Dominic P. E. Dickson	University of Liverpool
Lorella M. Jones	University of Illinois at Urbana–Champaign
W. G. Jones	Imperial College, University of London
Karl Lüchner	University of Munich
D. G. McCartan	University of Newcastle upon Tyne
W. O'Reilly	University of Newcastle upon Tyne
W. Poon	University of Edinburgh
Edward W. Thomas	Georgia Institute of Technology
Alma C. Zook	Pamona State College, California

PHYSICS

Marcelo Alonso
Florida Institute of Technology

Edward J. Finn
Georgetown University

Addison-Wesley Publishing Company
Wokingham, England • Reading, Massachusetts • Menlo Park, California • New York
Don Mills, Ontario • Amsterdam • Bonn • Sydney • Singapore • Tokyo • Madrid
San Juan • Milan • Paris • Mexico City • Seoul • Taipei

Developmental editor: Jane Hogg
Production editor: Sheila Chatten
Production controller: Jim Allman
Development assistant: Davina Arkell
Picture research: Barbara Johnson
Copy-editor: Ian Kingston
Text designer: Lesley Stewart
Illustrators: PanTek Arts
Typesetter: P & R Typesetters Ltd (Salisbury, UK)
Cover designer: Designers & Partners of Oxford. Photo courtesy of CERN.
Printed in the United States of America.

First printed 1992.

Acknowledgments
Permission to reproduce photographs and material from other sources is gratefully acknowledged. The publisher wishes to thank the following for permission to reproduce chapter opening photographs: AIP, Niels Bohr Library (Chapters 3, 6 (Physics Today Collection), 17, 19 (Burndy Library), 23 (W. F. Meggers Collection), 27, 30, 31, 34 and 35 (W. F. Meggers Collection), 36 (photo by Friedrich Hund), 37 (photo by Francis Simon), 38, 39, 40); The Royal Society (Chapters 9, 15, 25, 26, 29); The Royal Institution (Chapters 16, 22); The Bettmann Archive (Chapter 11); The Danmarks Tekniske Museum (Chapter 24); The UKAEA (Chapter 39).

British Library Cataloguing in Publication Data
A catalogue record for this book is available from the British Library.

Library of Congress Cataloging in Publication Data
Alonso, Marcelo, 1921–
Physics/Marcelo Alonso, Edward J. Finn.
p. cm.
Includes bibliographical references and index.
ISBN 0-201-56518-8
1. Physics. I. Finn, Edward J. II. Title.
QC21.2.A56 1992
530–dc20 91-41987
CIP

Preface

This book is designed for the introductory calculus physics course for engineering and science students. It addresses the subject from a contemporary, coherent point of view, integrating, to the extent possible, the Newtonian, relativistic and quantum descriptions of nature. In that sense the approach can be called 'modern'.

The primary goal of this text is to provide students with a sound understanding of *how* physical phenomena are analyzed, illustrated with applications to specific situations. The overall theme is to show how the macroscopic properties of matter can be related to its microscopic structure. At every possible opportunity the molecular and atomic properties of matter are brought into focus, without necessarily waiting for a formal treatment of their experimental foundation and of quantum theory, which come in later parts of the book. We hope that this methodology will offer students an integrated and coherent overview of physics that should be a solid foundation for the more advanced courses to follow.

It is expected that when students complete the course they will be able to recognize that there are:

(1) Two levels of the description of nature. One is **macro** and thus is global and phenomenological; it corresponds to the world we perceive directly. The other is **micro**, that is, structural; it is the domain of atoms and quantum theory.
(2) Two complementary descriptions of natural phenomena. One description employs **particles**; that is, balls, molecules, atoms etc. The other employs **fields** (gravitational, electromagnetic etc.), including **waves**.
(3) Two broad energy levels. One level is **low**, which corresponds to the world we normally deal with, described by Newtonian–Maxwellian physics. The other is **high**, which is the world of Einstein's relativity and nuclear forces.
(4) Two kinds of physical laws. One kind is **fundamental**, such as the conservation principles, the laws of gravitation and electromagnetism, the law of entropy etc. The other is **statistical**, corresponding to the laws of friction and viscosity, the gas laws, Ohm's law etc.

By logical necessity the text begins by developing the basic conceptual foundations and vocabulary of physics, for which Newtonian mechanics provides the natural framework. However, non-Newtonian notions, such as relativity and the quantization of energy and angular momentum, are introduced on several occasions in the proper context (electronic motion in atoms, simple harmonic oscillator and molecular vibrations, rigid body and molecular rotations, high energy processes etc.). Quantum mechanics is developed in a graphical and intuitive way that minimizes the mathematical requirements and highlights the physical content. This method shows that quantum mechanics is not an abstract theory but a theoretical framework that explains the properties of matter in a way that is complementary to the 'classical' particle picture. 'Modern' topics are presented whenever the occasion is appropriate.

The book is organized in a way that offers great flexibility to meet the needs of the instructor of a calculus general physics course. For example, those instructors who prefer to teach waves early in the course can use Chapter 28 immediately after either Chapter 10 or Chapter 17. Each chapter consists of the *main text*, (in which the fundamental concepts and experimental results are discussed), *examples* (which are applications of the theory or simply manipulations of data) and *notes* (which, in general, are more detailed elaborations or extensions of the text, and may be included or omitted at the discretion of the instructor). It is expected that many students will be interested in going beyond the material in the main text and will read the notes. Those proofs that are not an essential part of the analysis of a topic have been set aside from the main text; this will make it easier for the student to review a topic whenever needed. There are the usual sets of questions and problems at the end of each chapter.

The first 23 chapters are devoted primarily to the 'particle' aspects of physics, covering single particle and many particle systems, thermodynamics and statistical mechanics, relativity, Bohr's theory of atomic structure and the gravitational and electromagnetic interactions. The next 12 chapters deal with the 'field' or 'wave' aspects of physics, with an emphasis on the electromagnetic field. The remaining six chapters address the particle–field interface, with special consideration of quantum mechanics and its application to the structure of matter, nuclear phenomena and elementary particles. We have adhered to the recommendations of the Commission on Symbols, Units, and Nomenclature of IUPAP and have consistently used the SI units. All physical constants are expressed up to four decimal figures.

The book has been conceived for a course of about 100 lecture hours. For a shorter course of about 80 lecture hours, the instructor may use only Chapters 1 through 19 and Chapters 21 through 36, leaving some sections out. Some chapters (for example, Chapters 1, 2 and 12) should take up, at the most, one lecture since each can easily be studied independently by the student. Other chapters will require three or four lectures, while the majority can be comfortably discussed in two lectures.

It is assumed that the students have taken a physics course in high school or, at least, a general science course, which is the general rule. In any case, it is safe to assume that the students are not totally illiterate in science and are acquainted to some extent with many physical concepts. The mathematical requirements are well within the capabilities of most students and are standard to most calculus general physics courses. They correspond basically to elementary calculus and algebraic and trigonometric manipulations. Appendix A provides the necessary information for students not acquainted with vector algebra. Some simple differential equations are used to emphasize the basic features of certain phenomena (e.g. SHM and waves). It is not expected that students find a formal solution to these equations, since that is unnecessary and, most probably, beyond their mathematical sophistication. Rather, the solutions (which are normally sine and cosine or exponential functions) are given and the only requirement is to verify that they, indeed, satisfy the equation.

The motivation for producing this text came from the authors' participation in meetings sponsored by the IUPP/AIP. As a result, special attention has been paid to the recommendations of the IUPP Working Groups, without necessarily adhering to each recommendation. However, this project has no formal sponsorship of the IUPP and the authors are solely responsible for the content and organization of the text.

We want to express our gratitude to all those who, through their assistance, encouragement, comments and criticisms have made the completion of this book possible, helping us eliminate errors and improving the presentation of several topics. In particular, we wish to thank Tomas Bohr (Niels Bohr Institute, Copenhagen), Jay Burns (Florida Institute of Technology, Melbourne) and Karl Lüchner (Ludwig-Maximilians University, Munich).

M. Alonso
E. J. Finn

February, 1992

Contents

Appendices

Introduction

Studying physics is an exciting and challenging adventure, even if it may seem difficult at times. To be a professional physicist is even more exciting. It is a pleasing activity and, in the authors' opinion, nothing is more appealing than to learn about the world we live in and to help unravel the secrets of nature.

It may seem unnecessary at this point to discuss what physics is about, why it is so challenging and interesting, or what its methods are, since most probably the student already has some familiarity with this science. However, precisely because of this familiarity with many of the notions of physics, it is desirable to analyze and review its objectives and methods before embarking on its detailed study.

What is physics?

The word **physics** comes from a Greek term meaning **nature**, and historically became the term used to designate the study of natural phenomena. Until early in the nineteenth century, the term 'natural philosophy' was also used.

People, having inquiring minds, have always had a great curiosity about how nature works. As a result physics has evolved as more was learned about nature. At the beginning the only sources of information were our senses, and therefore observed phenomena were classified according to the way they were sensed. **Light** was related to the act of vision and **optics** was developed as a more or less independent science associated with this act. **Sound** was related to the act of hearing and **acoustics** developed as a correlative science. **Heat** was related to another kind of physical sensation, which we designate as hot or cold, and for many years the study of heat (called **thermodynamics**) was yet another autonomous branch of physics. **Motion**, of course, is the most common of all directly observed phenomena and the science of motion, **mechanics**, developed earlier than any other branch of physics. The motion of the planets caused by their gravitational interactions, as well as the free fall of bodies, was very nicely explained by the laws of mechanics; therefore **gravitation** was traditionally discussed as a chapter of mechanics. **Electromagnetism**, not being directly related to any sensory experience – in spite of being responsible for most of them – did not appear as an organized branch of physics until the nineteenth century.

So physics in the nineteenth century appeared to be divided into a few (called *classical*) sciences or branches: mechanics, thermodynamics, acoustics, optics and electromagnetism, with little or no connection between them, although mechanics was, quite properly, the guiding principle for all.

However, since the end of the nineteenth century a profound conceptual revolution, assisted by a refinement in experimental and observational methods, has taken place. This revolution, whose leaders were Max Planck and Albert Einstein, has modified our points of view and methods of attack on physical problems as well as our understanding of natural phenomena, particularly the structure of matter, and constitute the theories of **relativity** and **quantum mechanics**. These new theories have provided a more unified view of natural phenomena and have evolved into what has been called 'modern' physics, as well as requiring a reappraisal of the 'classical' branches of physics. However, 'modern' physics is not a new branch of physics. Rather, it is a 'modern', or new, approach to the analysis of natural phenomena, based on a deeper understanding of the structure of matter and the interactions among its components. In that sense, then, there will always be a **modern physics** based on contemporary physics being developed in one's time. This modern physics will require, at each instant, a revision and a re-evaluation of previous ideas and principles. But physics will always be a whole that must be considered in a unified, consistent and logical way. It is our aim in this text to present physics in such a way. Accordingly, we may say that

> *physics is a science whose objective is to study the components of matter and their mutual interactions. In terms of these components and interactions, the scientist tries to explain the properties of matter in bulk, as well as the other natural phenomena we observe.*

The student, while progressing through this course, will see how this program is developed through an analysis of a large variety of phenomena, apparently unrelated but obeying the same fundamental laws.

The relation of physics to other sciences

We have indicated that the objective of physics is to enable us to understand the basic components of matter and their mutual interactions, and thus to explain all natural phenomena, including the properties of matter in bulk. From this statement, which may seem a bit ambitious, we can see that physics is the most fundamental of all natural sciences. Chemistry deals basically with one particular aspect of this program: the application of the laws of physics to the structure and formation of molecules and their interactions as well as to the different practical means of transforming certain molecules into others. And biology must lean very heavily on physics and chemistry to explain the processes that occur in complex living systems. The application of the principles of physics and chemistry to practical problems, in research and technical development as well as in professional practice, has given rise to the different branches of engineering. Modern engineering practice and research would be impossible without a sound understanding of the fundamental ideas of physics.

But physics is important not only because it provides the basic conceptual and theoretical framework on which the other natural sciences are founded. From the practical point of view, it is important because it provides techniques that can be used in almost any area of pure or applied research. The astronomer requires optical, spectroscopic and radio techniques. The geologist uses gravimetric, acoustic, nuclear and mechanical methods. The same may be said of the oceanographer, the meteorologist, the seismologist etc. A modern hospital is equipped with laboratories in which the most sophisticated physical techniques are used. In medicine, ultrasonics, lasers, nuclear magnetic resonance and radio-isotopes are used routinely. In summary, hardly any activity of research, including such fields as archaeology, paleontology, history and art, can proceed without the use of modern physical techniques. This gives the physicist a gratifying feeling to advance not only our body of knowledge about nature, but to contribute to the social progress of mankind.

Another important point is that although physics, like chemistry, biology, psychology, sociology and other sciences, is a well-defined area of scientific specialization, most current scientific research is of interdisciplinary nature involving scientists with diverse but complementary interests. In this respect it is important to recognize the major areas of interdisciplinary research where physicists are involved in different degrees:

(1) *Understanding the universe.* This area includes research in cosmology, astrophysics, planetary and galactic exploration, space travel, space habitats, gravitation and relativity.

(2) *Understanding matter and energy.* As an illustration, this area includes research on the properties of the basic components of matter (particles, nuclei, atoms and molecules) and their interactions, states of matter (solids, liquids, gases and plasmas), properties of materials (superconductors, semiconductors etc.), behavior of matter under extreme conditions (very high or very low pressure and temperature, intense radiation, strong electromagnetic fields) etc.

(3) *Understanding the biosphere,* a thin peel about 15 km thick surrounding both sides of the Earth's surface. Critical areas of research comprise the dynamics of the biosphere (flow of matter, energy and life), effects of processes and changes (pollution, depletion and extinction) introduced by mankind and new habitats.

(4) *Understanding life.* This area of active research includes the flow of energy and matter in living systems, and above all bioengineering and genetic research (gene transplant, plant regeneration, genome mapping etc.).

(5) *Understanding the human body,* considered as a complex system. A branch of active research closely connected to physics is that of *medical technology,* which comprises many fields for analyzing the functioning of the human body. Another is related to genetics (gene therapy, disease control, aging, gene splicing etc.).

(6) *Understanding intelligence.* Some areas of research are brain dynamics (consciousness, perception etc.), neural nets, information processing, artificial intelligence, computers, robotics etc.

(7) *Understanding human relations*, which comprises research on population dynamics, communication, transportation, education, and the quality of life.

Even these seven major areas of interdisciplinary research are not completely independent of each other. But the important point is that physicists have a role to play in all of them, sometimes providing the theoretical insight, while at other times designing the experimental techniques and sometimes simply operating the scientific equipment.

The experimental method

In order to fulfill its objective, physics, as well as all natural sciences, both pure and applied, depends on **observation** and **experimentation**. Observation consists in the careful and critical examination of a phenomenon by noting, measuring and analyzing the different factors and circumstances that appear to influence it. Unfortunately, the conditions under which phenomena occur naturally rarely offer enough variation and flexibility. In some cases they occur only infrequently so that analyzing them is a difficult and slow process. For that reason experimentation is necessary. Experimentation consists in the observation of a phenomenon under prearranged and carefully controlled conditions. Thus the scientist can vary the conditions at will, making it easier to disclose how they affect the process. Without experimentation and measurement modern science would never have achieved the advances it has. This is why laboratories are so essential to the scientist.

Of course, experimentation is not the only tool a physicist has. From the known facts a scientist may infer new knowledge in a **theoretical** way. By theoretical we mean that the physicist proposes a **model** of the physical situation under study. Using relations previously established, logical and deductive reasoning is applied to the model. Ordinarily the scientist works out this reasoning by means of mathematical techniques. The end result may be the prediction of some phenomenon not yet observed or the verification of the relations among several processes. The knowledge a physicist acquires by theoretical means is in turn used by other scientists to perform new experiments for checking the model itself, or to determine its limitations and failures. The theoretician then revises and modifies the model so that it will agree with the new information. It is this interwoven relation between experimentation and theory that allows science to progress steadily and on solid ground. This means that physics, like most sciences, is a dynamic subject where nothing is taken for granted or is a dogma.

1 The structure of matter

Great Nebula in Andromeda also called M-31. The nearest of the large regular galaxies, it is still about 2 500 000 light years, or 2.4×10^{22} m, from the solar system. Its diameter is about 125 000 light years, or 10^{21} m, and it contains more than 10^{11} stars. (Photograph courtesy of California Institute of Technology.)

1.1 Introduction

In the introduction to this text it was said that **physics** is a science whose objective is the study of the components of matter and their mutual interactions. Therefore, we begin with a brief review of the current ideas about the structure of matter. As this course develops, the topics considered here will be re-examined in more detail as the occasion presents itself.

The question of the structure of matter has preoccupied philosophers and scientists since the dawn of civilization. The evolution of our ideas about matter is an exciting story, which we do not have time or space to consider here. One of the great achievements of modern physics is that we have formulated a rather coherent, although still incomplete, picture of the make-up of the bodies we are able to see and touch.

1.2 Particles

Matter appears to our senses in a diversity of forms, textures, colors etc. and seems to have a continuous structure. Actually, it is composed of distinct units grouped in many different ways. The basic units or building blocks of matter are,

Table 1.1 Some fundamental particles

Particle		Mass*	Charge
Neutrino	(ν)	~ 0	0
Electron	(e^-)	1	$-e$
Positron	(e^+)	1	$+e$
Muon	(μ)	~ 207	$+e, -e$
Pion	(π)	~ 270	$+e, -e, 0$
Proton	(p)	1836.2	$+e$
Neutron	(n)	1838.7	0

* The mass is expressed relative to the electron mass.

for convenience, called **particles**. This term does not imply that they are like little specks of dust or sand or that they do not have some kind of internal structure. Rather, by particles we understand entities with well-defined properties, such as mass, charge, etc. At present we consider that matter is composed of a few fundamental (or **elementary**) particles. We infer that all bodies, both living and inert, are made up of different groupings or arrangements of such particles. The three fundamental particles that are most important for our understanding of the structure and properties of matter are **electrons** (e), **protons** (p) and **neutrons** (n). We also must add the **photon** to these three particles. The photon (the symbol for which is γ) is a special kind of particle associated with electromagnetic radiation. Most of the phenomena we observe in nature, and which are considered in this text, may be explained in terms of **interactions** among these particles. There are a few other fundamental particles that manifest themselves in special phenomena. Some of these particles are listed in Table 1.1. The existence of these particles is made manifest only by means of rather elaborate observational techniques.

Actually, the *very* fundamental particles are the **leptons** (which, in Greek, means 'light') and the **quarks** (a term suggested by Murray Gell-Mann). **Electrons, neutrinos** (ν) and **muons** (μ) are kinds of leptons. There are several types of quarks (each of which comes in several kinds, called 'flavors') and are the basic components of other particles called **hadrons**, such as **pions** (π), **protons** and **neutrons**. Protons and neutrons are supposed to be composed of three quarks, while pions, which belong to a group of particles called **mesons**, are composed of two quarks or, rather, a quark and an antiquark. However, these features, which will be analyzed in more detail in Chapters 6 and 41, are not required for an analysis of most physical phenomena. In many instances, protons and neutrons can be considered as fundamental particles, and their internal structure may be ignored.

Particles are distinguished by their physical properties, such as mass and electric charge. Electrons are negatively charged, protons are positively charged, and neutrons are uncharged, or neutral, particles. In fact, electrons and protons carry equal charges, of opposite sign, designated $-e$ and $+e$, respectively. Protons and neutrons have about the same mass, which is about 1840 times greater than the mass of the electron. Also, it appears that for each particle, there is an **antiparticle**, which has some opposite properties. The antiparticle of the electron (e^-) is the **positron** (e^+). The antiparticle of the **neutrino** (ν), represented by $\bar{\nu}$, is

Figure 1.1 Aerial view of the Fermi National Accelerator Laboratory, Batavia, Illinois, which operates the Tevatron. It is the world's most powerful accelerator (1992). Protons and antiprotons are accelerated up to 900 GeV each, circulating in opposite directions in an underground ring of 1 km radius. When protons and antiprotons collide head-on, with a total energy of 1800 GeV, they produce many new particles.

called an **antineutrino**. (In actuality, there are several kinds of neutrinos; this will be discussed in Chapter 41.)

There are two interesting features that distinguish particles. First, many particles, like the muons and pions, have a transient life and decay in a very short time into other particles, while a few, like protons, electrons and neutrinos, are stable and do not decay. Second, when two particles collide they may be destroyed and new particles may be created. This means that particles can be transformed into others either through spontaneous decay or through violent collisions. Large machines, called **particle accelerators** (Figure 1.1), have been built to energize particles and produce new ones through collisions. By analyzing these processes we learn about the behavior of particles and determine their properties.

The roles many of the particles play in nature are not yet fully understood. Active research is being conducted to attain a better understanding of the reasons for such a variety of particles and the relations among them.

1.3 Atoms

In an oversimplified language, we may say that the three fundamental particles, electron, proton and neutron, are present in all matter in well-defined groups called **atoms**. Protons and neutrons are clustered in a very small central region, called the **nucleus**, which has a size of the order of 10^{-14} m. Protons and neutrons are held together in a nucleus by **strong nuclear forces**. The electrons move about the

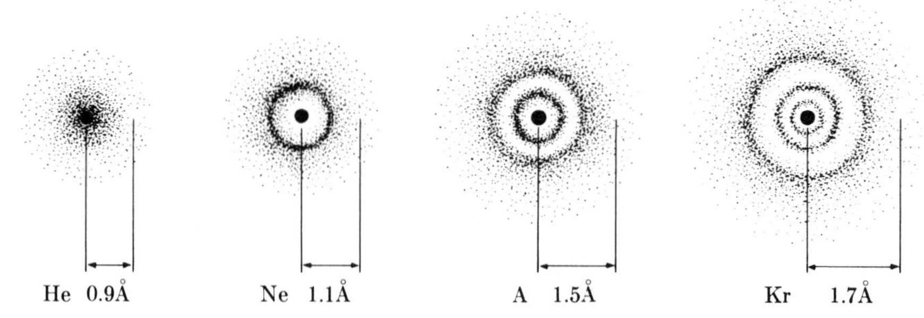

Figure 1.2 Schematic arrangement of electrons around the nucleus in some simple atoms (He = helium, Ne = neon, Ar = argon, Kr = krypton). Since electrons do not follow well-defined paths, the dark regions are those which are more likely to be occupied by electrons (1 Å = 1 angstrom = 10^{-10} m).

nucleus in a region with a diameter of the order of 10^{-10} m (Figure 1.2) or about 10^4 times larger than the nucleus. So, it is truly correct to think of the nucleus as a tiny speck at the center of the atom. Most of the mass of the atom is concentrated in the nucleus. The simplest of all atoms is hydrogen, which consists of one electron orbiting about a proton. The **electromagnetic interaction** between the oppositely charged orbiting electrons and nuclear protons holds an atom together. Electrons tend to move in certain regions around the nucleus, called **electron shells**, as shown schematically in Figure 1.2.

Atoms have the same number of protons as electrons, which is called the **atomic number** and designated by Z. Hence, the nucleus is positively charged with a charge $+Ze$. The total charge of the electrons is $-Ze$ and thus atoms have a zero net charge; that is, all atoms are electrically neutral. The total number of protons and neutrons in the nucleus of an atom is called its **mass number**, designated by A. Therefore, the number of neutrons in an atom is $A - Z$. All atoms with the same atomic number (same number of protons) belong to the same atomic 'species', or, as we usually say, to the same **chemical element**. Thus all atoms with $Z = 1$ are hydrogen atoms, all atoms with $Z = 6$ are carbon atoms, and all uranium atoms have $Z = 92$. All atoms with the same Z are designated by the symbol of the chemical element to which they belong.

Atoms with the same atomic number may have different mass numbers. Atoms of the same chemical element (same Z) but with a different mass number (different A) – that is, a different number of neutrons – are called **isotopes**. For most practical purposes the isotopes of a given element behave the same chemically. To distinguish the different isotopes of a chemical element, the value of A is shown as a superscript to the left of the chemical element. For example, the three isotopes of hydrogen ($Z = 1$) are ^1H, ^2H and ^3H; all have $Z = 1$ (one proton) but, as shown schematically in Figure 1.3, $A = 1$ (no neutron), $A = 2$ (one neutron), and $A = 3$ (two neutrons), respectively. Atoms of ^2H are called **deuterium** (double) and those of ^3H are called **tritium** (triple). In Figure 1.3, two isotopes of helium ($Z = 2$) are also shown, one with $A = 3$ and the other with $A = 4$, which is the more abundant isotope of helium. In fact, ^3He does not exist naturally on Earth; however, there is plenty in the Moon's crust as well as in space. About 109 different

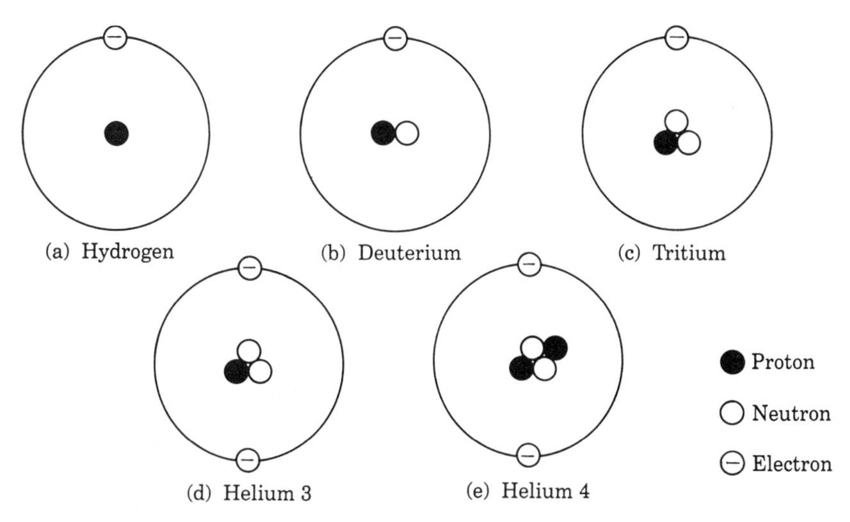

Figure 1.3 Simplified representation of the isotopes of hydrogen and helium.

chemical elements have been recognized (see Table A.1), but there are more than 1300 different isotopes.

Sometimes an atom may gain or lose some electrons, becoming negatively or positively charged. Such atoms are called **ions**. For example, a hydrogen atom may lose its only electron and be reduced to a proton or **hydrogen ion**, which is designated by H^+ or p. An ion is indicated by the symbol of the chemical element with a superscript to the right showing the net charge of the ion. For example, Cu^{2+} is an atom of copper that has lost two electrons while P^{3-} is a phosphorus atom that has gained three electrons. Most chemical processes correspond to an exchange of electrons between atoms.

1.4 Molecules

It is very rare to find isolated atoms. Rather, atoms form aggregates called **molecules**, of which many thousand different kinds are known to exist. Some molecules contain just a few atoms. Hydrochloric acid molecules are formed from one atom of hydrogen and one atom of chlorine. Carbon dioxide molecules are composed of one carbon atom and two oxygen atoms, and water molecules are composed of one oxygen atom and two hydrogen atoms (Figure 1.4). Other molecules, such as the proteins, enzymes and nucleic acids (DNA and RNA, Figure 1.5), may have several hundreds or thousands of atoms. Organic polymers, such as polyethylene or polyvinylchloride (PVC), which are commonly used plastics, are also very large molecules.

When a molecule is formed, the atoms (to a certain extent) lose their identity. We could say instead that a molecule is a system composed of several nuclei and a group of electrons moving about the nuclei in such a way that a stable configuration results. The forces that hold the molecule together are also of electromagnetic origin.

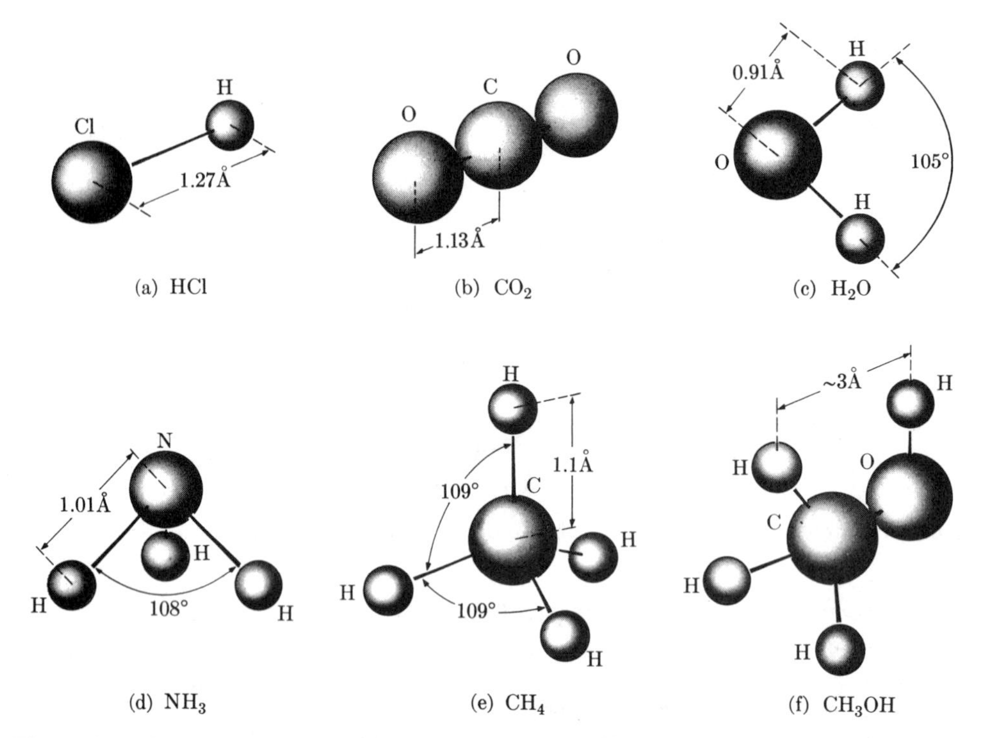

Figure 1.4 Simplified representation of some molecules: (a) hydrochloric acid, (b) carbon dioxide, (c) water, (d) ammonia, (e) methane, (f) methanol. The atoms in a molecule are arranged in well-defined geometric patterns. The inner electrons remain attached to the respective atoms, but the outer ones either move in the space between two atoms or move more or less freely over the molecule.

The simplest of all molecules is the **hydrogen molecule ion** H_2^+, consisting of two protons and one electron (Figure 1.6). Chemists write the formation of this molecule as:

$$H + H^+ \rightarrow H_2^+,$$

where H^+ is just a proton. In other words, the H_2^+ molecule is formed when a hydrogen atom captures a proton. But once the H_2^+ molecule is formed, it is no longer possible to tell which is the hydrogen atom and which is the proton. The electron, of course, is not at rest but moves around the protons, most of the time being in the region between them.

The next molecule in order of complexity is the **hydrogen molecule**, H_2, composed of two hydrogen atoms, or rather, two protons and two electrons (Figure 1.7). The two electrons move preferentially in the region between the protons. When two atoms more complex than hydrogen combine to form a molecule, the more tightly bound, or inner, electrons of each atom are practically undisturbed, remaining attached to their original nuclei. Only the outermost (also called **valence**) electrons are affected, and they move under the resultant force of

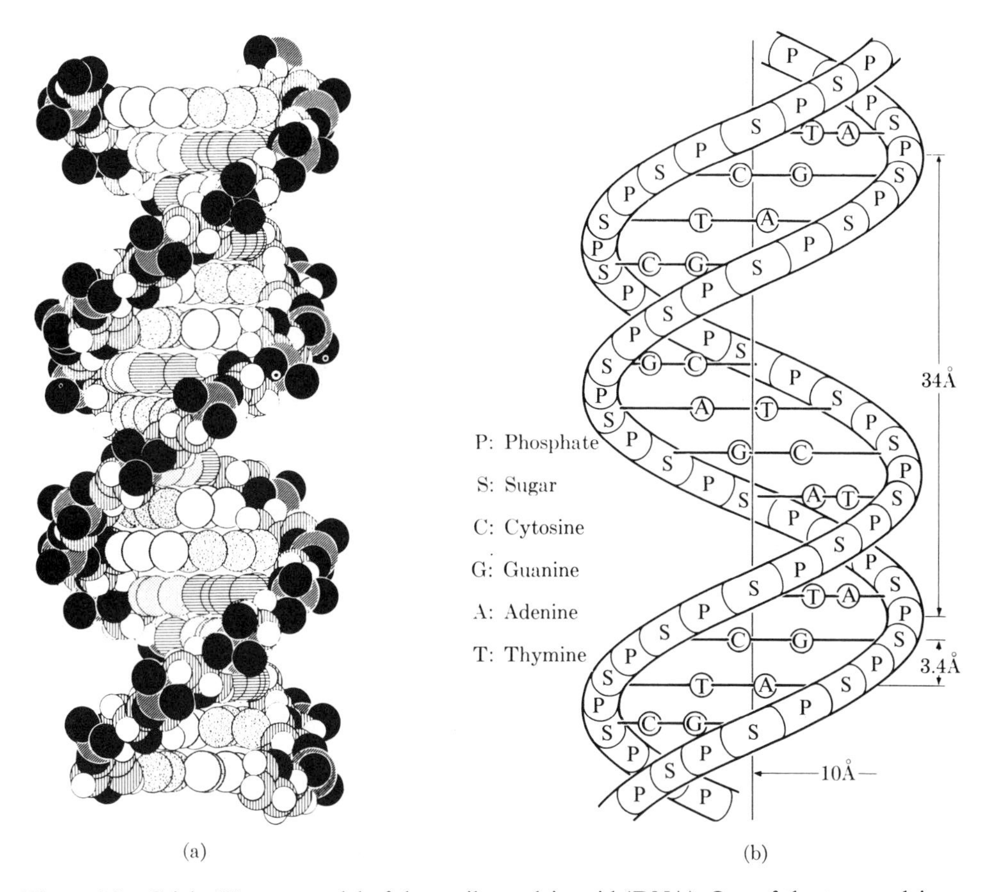

Figure 1.5 Crick–Watson model of deoxyribonucleic acid (DNA). One of the two nucleic acids involved in the composition of a chromosome, DNA carries genetic information and is one of the best studied biomolecules. X-ray diffraction has shown that it consists of two antiparallel helices composed of a sequence of sugar (S) and phosphate (P) groups. The sugar, called deoxyribose, contains five carbon atoms. The two helices are interlocked by pairs of hydrogen-bonded base groups. One pair is formed by two substances called adenine and thymine (A–T) and the other by cytosine and guanine (C–G). The genetic code of the DNA molecule depends on the sequence or ordering of each base pair. These base pairs are like rungs along a helical stepladder, each rung being about 11 Å long. The pitch of each helix is about 34Å and its overall diameter is about 18 Å (1 Å = 10^{-10} m).

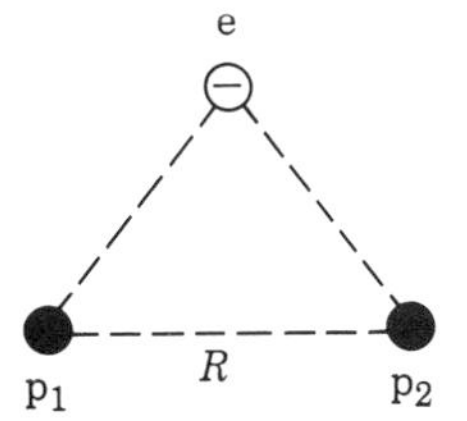

Figure 1.6 The hydrogen molecule ion.

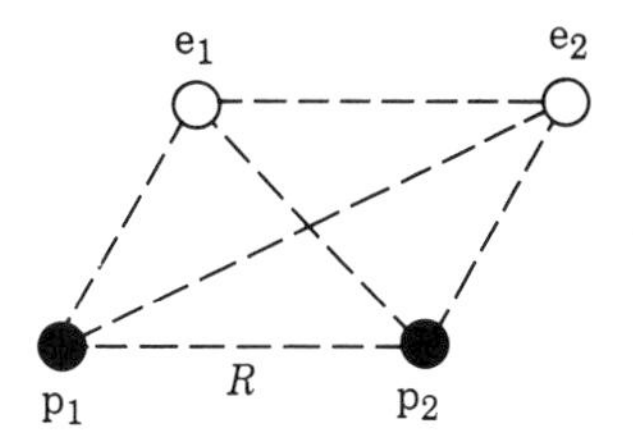

Figure 1.7 The hydrogen molecule.

the ions, composed of the nuclei and the inner electrons, as well as their mutual repulsion. These valence electrons are responsible for chemical bonding and for most physical properties of the molecule. Also, the nuclei are not fixed but tend to vibrate about their equilibrium positions, and the molecules as a whole may have rotational motion. Thus a molecule is a dynamic structure.

Atoms (or nuclei) in molecules are arranged in regular patterns characteristic of each molecule, as shown in Figure 1.4. In the carbon dioxide molecule, CO_2, the three nuclei are in a straight line with the carbon nucleus in the middle. The water molecule, H_2O, has the three nuclei at the vertices of a triangle. The ammonia molecule, NH_3, is a pyramid with the nitrogen nucleus at the summit. The methane molecule, CH_4, is a tetrahedron with the carbon nucleus at the center and the hydrogen nuclei at the vertices. Among the biomolecules, one of the most interesting is that of DNA (Figure 1.5), which has the shape of two intertwined helices. Modern experimental techniques (such as X-rays and electron beams) have provided invaluable information about the structure of molecules. In addition, it is possible today to 'manufacture' molecules to meet specific needs.

1.5 Matter in bulk

Matter in bulk, in the way it affects our senses, is an aggregate of a very large number of atoms or molecules, held together by electric forces. Grossly speaking, these aggregates appear to be in one of three physical states, or **phases**, designated as gases, liquids and solids. It is necessary to have a few thousand atoms in a relatively small volume in order to display the properties of matter in bulk.

In **gases** the average distance between molecules is much greater than the size of the molecules. Thus molecules in gases retain their individuality. Under normal conditions, that is, room pressure and temperature, the intermolecular distance is of the order of 3×10^{-9} m, or 10 times molecular dimensions. As a result, the intermolecular electric forces are much weaker than the electric forces that hold the molecule together. Also, it is relatively easy to change the volume occupied by a gas. Molecules in a gas are moving continuously throughout all the space occupied by the gas, colliding among themselves and with the surfaces of the bodies with which they are in contact. This molecular mobility explains why gases diffuse so easily. For example, if we have two vessels containing two different gases A and B (Figure 1.8) and the stopper K is opened, in a short time both gases are thoroughly mixed in both vessels.

In a **solid** the atoms (or molecules) are tightly packed. They are held in more or less fixed positions by forces of electromagnetic origin. These forces have

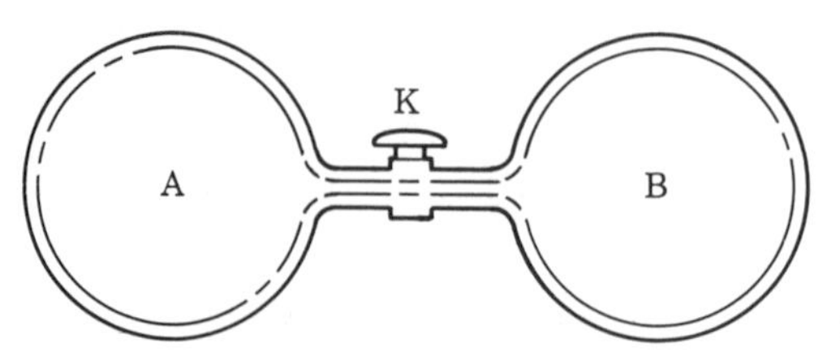

Figure 1.8 Gases in A and B diffuse into each other through K.

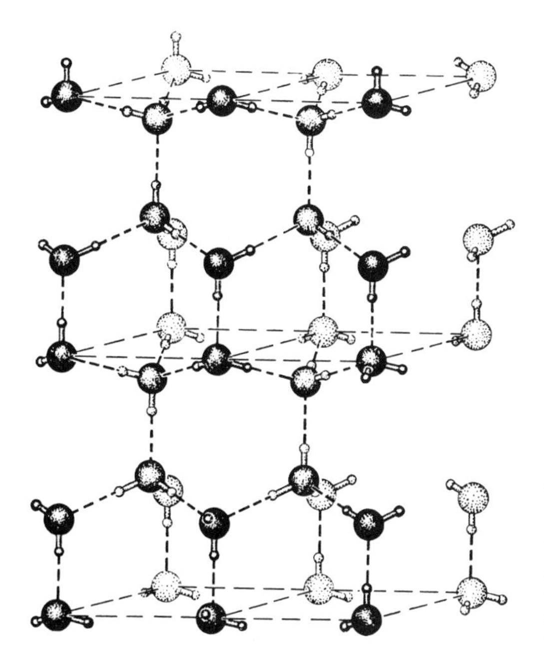

Figure 1.9 Arrangement of water molecules in ice (from L. Pauling, *The Nature of the Chemical Bond.* Ithaca, NY: Cornell University Press, 1960; by permission of the publisher). The oxygen atoms (large black balls) interact electrically with the hydrogen atoms (small open balls) of different molecules. This is called a 'hydrogen bond' and is indicated in the figure by the dashed lines.

the same order of magnitude as those involved in molecular binding. Thus the shape and volume of a solid remain essentially constant so long as the physical conditions, such as pressure and temperature, do not undergo an appreciable change. The average interatomic (or intermolecular) separation in solids is about 10^{-10} m, which is comparable to atomic and molecular dimensions.

In most solids, the atoms (or molecules) do not exist as isolated entities; rather their properties are modified by the nearby atoms. The regular arrangement of the atoms or groups of atoms is one of the most important features of solids. That is, the structure of solids exhibits a regularity or **periodicity** that results in what is called a **crystal lattice**. In certain solids the basic units making up the crystal lattice are the molecules themselves. This is the case for ice. The structure of one of the states of ice is shown in Figure 1.9. The water molecules are held together by electric forces between the oxygen atom of a molecule and a hydrogen atom of another molecule. In other cases the lattice is made up of oppositely charged ions, such as in the NaCl and CsCl lattices shown in Figure 1.10. In solids composed of a single class of atoms, the lattice elements are the atoms themselves. Figures 1.11 and 1.12 show the graphite and diamond lattices; the elements in each case are carbon atoms. The lines represent the regions where valence electrons from two adjacent carbon atoms tend to concentrate. In some cases, such as metals, the lattice is composed of positive ions, with electrons moving more or less freely in the space between them, constituting a sort of electron gas.

The physical properties of solids are directly related to the nature and geometrical arrangement of the units that compose the lattice. Some substances that appear as solids do not show this regular arrangement of atoms or molecules. This is the case, for example, with glass. These solids are called **amorphous**, and their physical properties are very different from those of crystalline solids. In

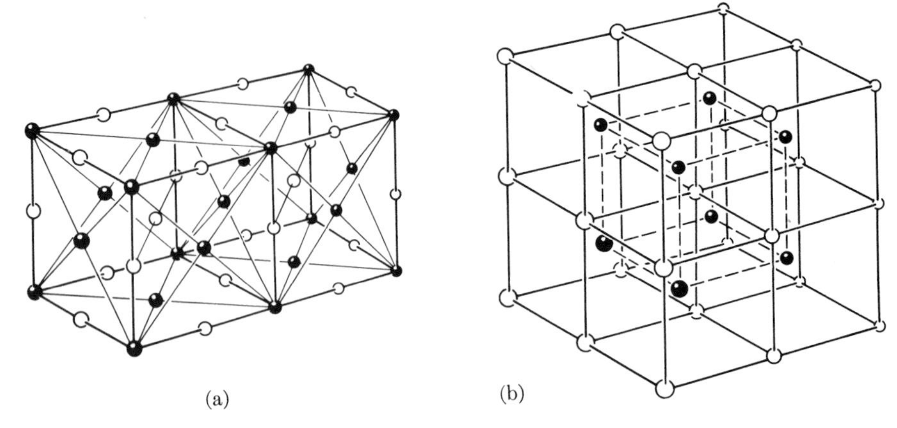

Figure 1.10 (a) NaCl lattice; (b) CsCl lattice.

general they are easily deformable and show some plasticity under pressure or when heated.

The properties of **liquids** fall between those of gases and solids. Molecules in liquids are separated by distances of the order of molecular dimensions and are held by electric forces stronger than in gases. Molecules in liquids have great mobility, although not as much as molecules in gases, and during their motion maintain a constant average distance from their closest neighbors. As a result, liquids have a fixed volume and a low compressibility. One consequence is that liquids do not have a shape of their own, but conform to the shape of a containing vessel. They yield to small shear stresses, a property called **fluidity**. Another property that identifies liquids is that they do not have the orderly arrangement found in solid lattices. However, it has been found that liquid molecules tend to group into small clusters, having sizes of the order of several intermolecular distances. Each cluster may last for a short time; the molecules continually disband and then regroup into new clusters. Liquids vary tremendously in their properties and

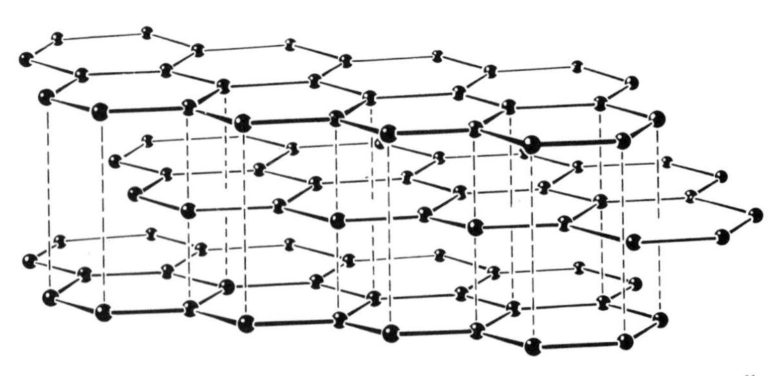

Figure 1.11 Graphite lattice. The carbon atoms are arranged hexagonally in parallel layers bonded by electrons localized along the solid lines. The bonding between atoms in adjacent layers is much weaker.

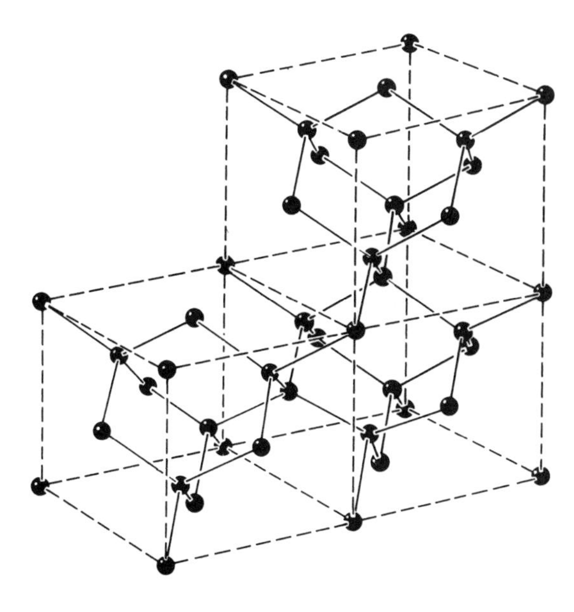

Figure 1.12 Diamond lattice. The solid lines represent the regions where two electrons from two carbon atoms are localized. The dashed lines represent the units of the crystal lattice.

behavior, depending on the composition, size and structure of the molecules and on the intermolecular forces.

A fourth state of matter, **plasma**, consists of a gaseous mixture of positively and negatively charged particles or ions which are at very high temperature, so that the ions move with very large velocities. The temperature must be great enough so that oppositely charged particles do not stick together when they collide. A gas heated to a very high temperature becomes a plasma. The vast majority of the matter in the universe, particularly in stars, is in the form of plasma. However, on Earth a plasma can be produced only in the laboratory by heating a gas to a very high temperature of a few million degrees or by sending an electric discharge through the gas.

From the above discussion, we see that the properties of matter in bulk depend directly on its atomic or molecular structure as well as on the pressure, temperature and other external factors.

1.6 Living systems

Physical laws apply not only to inert matter but also to the processes that occur in living systems. We do not yet fully understand all the details of the functioning of living systems. We do know, however, that living systems are extremely complex and are characterized by *functional* organization; that is, the different components of a living system must complement and support each other in the functioning of the whole system. Nevertheless, we know that living systems are composed of atoms and molecules, like the rest of matter in the universe. Some molecules in living systems are rather simple, such as water (H_2O), sodium chloride (NaCl), hydrochloric acid (HCl) and carbon dioxide (CO_2). Other molecules are rather complicated and are composed of many atoms, such as sugars (sucrose, lactose, fructose), urea, creatine etc. But the most important molecules associated with life, such as enzymes, proteins and nucleic acids, are composed of numerous atoms arranged in different patterns (Figure 1.5). The formation of these molecules can be explained in terms of the properties of the atoms constituting the molecules. Many of the processes in living systems can be explained by invoking well-known physical laws. Consequently, the complex functional organization of living systems can be described in great detail. But how these systems originated and evolved in such complexity and with such diversity is still a subject open to discussion and further research. In any case, physics is essential for the study of living systems. In fact, the specialization called **biophysics** is dedicated to the application of physics to the study of the phenomenon of life.

1.7 Interactions

Some questions naturally come to mind when we think about the particles that make up our world. Why and how are electrons, protons and neutrons bound together to form atoms? Why and how are atoms bound together to form molecules? Why and how are molecules bound together to form bodies? How does it happen that matter aggregates itself in size from small dust particles to huge planets, from bacteria to this marvelous creature called *Homo sapiens*? Why do all bodies fall toward the Earth, while planets revolve around the Sun? We may answer these fundamental questions, in principle, by introducing the notion of **interactions**. We say that the particles in an atom interact among themselves such that they produce a stable configuration. Atoms in turn interact to produce molecules, and molecules interact to form bodies. Matter in bulk also exhibits certain obvious interactions, such as gravitation, manifested for example in what we call **weight**.

One of the primary objectives of the physicist is to disclose the various interactions of matter and to express them in a quantitative way, for which mathematics is required. Finally, an attempt is made to formulate general rules about the behavior of matter in bulk which result from these fundamental interactions.

So far, four kinds of fundamental interaction have been recognized: **gravitational**, **electromagnetic**, **weak** and **strong** or **nuclear**. Each is related to a particular set of properties of matter and phenomena. The *gravitational* interaction is the weakest of the four. However, it is the most easily recognized force since it is manifested by an attraction between all matter. It is responsible for the existence of stars, planetary systems, galaxies and in general all large structures in the universe; it is related to that property of matter called **mass**. The *electromagnetic* interaction is related to a property of matter called **electric charge**. As we indicated before, it is responsible for holding atoms, molecules and matter in bulk together. The *weak* interaction manifests itself through certain processes, such as some kinds of **radioactive disintegration** or **decay**. It is related to a property called 'weak' charge. The *strong* or *nuclear* interaction holds protons and neutrons together in atomic nuclei as well as quarks inside protons, neutrons and pions. It is related to a property of matter called **color** charge (which has *nothing* to do with what we call color in daily life). See Note 6.1 for a further discussion of this.

In later chapters the structure of matter and the fundamental interactions will be discussed in greater detail. What has been said in this chapter is sufficient to give an appreciation of the physicist's view of the universe. On the other hand, many properties of matter in bulk (i.e. bodies composed of a large number of atoms or molecules) can be analyzed without explicit reference to the fundamental interactions. In general, this makes the analysis somewhat simpler. For this reason we begin by developing some methods for the study of matter in bulk. However, reference to atomic structure will be made whenever appropriate.

QUESTIONS

1.1 How can you explain the greater compressibility of a gas in comparison with that of a solid or a liquid?

1.2 If a plasma is a very hot gas composed of positive and negative ions, why doesn't it collapse under their mutual electrical attractions?

1.3 What interactions hold the electrons in an atom together? What interactions hold the protons and neutrons in a nucleus together?

1.4 Is the picture of an atom as a billiard ball an adequate representation?

1.5 Explain what we mean when we say that an atom is an electrically neutral system of particles. What happens when we ionize an atom?

1.6 If we could see a water molecule, would it be possible to recognize the two hydrogen atoms and the oxygen atom as separate entities?

1.7 Which do you think is more 'empty' (that is, has fewer particles per unit volume): a one cubic meter box containing 4×10^{26} gas molecules or an atom with a volume of 10^{-30} m^3 containing 25 orbiting electrons?

1.8 Investigate the structure of the following molecules: (a) carbon monoxide, CO; (b) hydrogen sulfide, H_2S; (c) phosphamine, PH_3; (d) chloroform, $ClCH_3$; (e) ethane, C_2H_6.

1.9 What is a particle accelerator?

1.10 Do we get a new chemical element when the number of protons in a nucleus is changed? What happens if the change is in the number of neutrons?

1.11 What important difference distinguishes living systems from inert matter?

1.12 Consider why it is necessary to have at least a minimum number of atoms, say 2500, in a small volume before bulk properties of matter become apparent.

2 Measurement and units

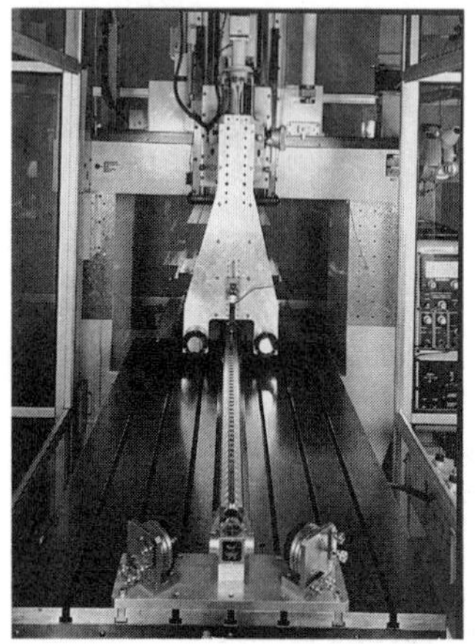

Helium–neon laser interferometer used to measure the distance between successive steps of a step gauge. By using two beams the effects of angular tilts in the machine are compensated for. The total accuracy of the measurement is of the order of 1 μm per meter. (Crown copyright – The National Physical Laboratory.)

Notes

2.1 Introduction

Before we embark on the study of physical phenomena, it is useful to see how physicists communicate their observations to each other and the world at large. An observation is generally incomplete unless it results in **quantitative** information. To get such information, a **measurement** of a physical property must be obtained. And so, measurement makes up a good part of the daily routine of the experimental physicist. Lord Kelvin (1824–1907) said that our knowledge is satisfactory only when we can express it in terms of numbers. Although this assertion is perhaps exaggerated, it expresses a philosophy of research that a physicist must keep in mind at all times. But the expression of a physical property in terms of numbers is not enough. Physicists always look for relations between the quantities they measure when investigating a particular phenomenon or process. These relations are usually expressed in mathematical form by means of formulas and equations.

Then they try to extrapolate from them or try to predict the results of other measurements, for which new measurements might be necessary.

2.2 Measurement

Measurement is a technique used to determine the magnitude of a physical property by comparing it with a similar, standard quantity that has been adopted as a **unit**. Most measurements performed in the laboratory reduce essentially to the measurement of quantities such as a length, an angle or a voltage. By using these measurements (and certain conventions expressed by formulas), we can obtain the desired quantity.

When a measurement is taken, the physicist must take great care to produce the minimum possible disturbance of the system that is under observation. For example, when we measure the temperature of a body, we place it in contact with a thermometer. But when we place the two together, some energy or 'heat' is exchanged between the body and the thermometer. This results in a slight change in the temperature of the body and thus affects the very quantity we wanted to measure. In addition, all measurements are influenced by some degree of **experimental error** due to the inevitable imperfections in the measuring device. The limitations imposed by our senses (vision and hearing), which record the information, also introduce some experimental error. Therefore the measuring technique is designed so that the disturbance of the quantity measured is smaller than the experimental error. In general, this is always possible when we are measuring quantities in the macroscopic range (i.e. in bodies composed of a large number of molecules or matter in bulk). All we have to do is to use a measuring device that produces a disturbance smaller, by several orders of magnitude, than the quantity measured. In other cases the amount of disturbance can be estimated and the measured value corrected.

The situation, however, is quite different when we are measuring atomic properties, such as the motion of an electron. Now we do not have the option of using a measuring device that produces an interaction smaller than the quantity to be measured, because we do not have a device that small. The disturbance introduced is of the same order of magnitude as the quantity to be measured, and it may not be possible even to estimate or account for it. Therefore a distinction must be made between the measurement of macroscopic quantities and that of atomic quantities.

Another important requirement is that the definitions of physical quantities must be **operational**. What we mean is that the definition must indicate, explicitly or implicitly, how to measure the quantity that is defined. This operation may be either direct or through the measurement of other quantities with which it is related. For example, to say that velocity is an expression of the rate at which a body moves is not an operational definition of velocity. But to say that *velocity is the distance moved divided by the elapsed time* is an operational definition of velocity. This definition implies the measurement of a distance and a time in order to measure the velocity.

2.3 Fundamental quantities

Before we measure something, we must first select a unit for each quantity to be measured. For purposes of measurement, there are fundamental and derived quantities and units. The physicist recognizes four fundamental independent quantities: **length**, **time**, **mass** and **electric charge**.

Length is a primary concept and is a notion we all acquire naturally. Although length is directly related to the notion of distance, it is useless to attempt to give a definition of it. We may say the same about time, even if the meaning of time as an ordering parameter is not fully understood. All physical phenomena occur in space (requiring the notion of length) and at a certain time. Mass and charge, however, are not that intuitive. The concept of mass will be analyzed in detail in Chapters 6 and 11. Let us say here only that mass is a coefficient, characteristic of each particle, that determines the strength of its gravitational interaction with other particles as well as the particle's behavior when subject to other forces.

Similarly, electric charge is another coefficient, characteristic of each particle, that determines the strength of its electromagnetic interaction with other particles. There may exist other coefficients characterizing other interactions between particles, but we do not need to be concerned with these now.

At this stage we can define mass operationally using the principle of the equal-arm balance (Figure 2.1); that is, a symmetric balance supported at its center O. Two bodies C and C' are said to have equal masses when, with one body placed on each pan, the balance remains in equilibrium. Experiments have verified that if, at some time, a balance is in equilibrium at one place on the Earth, it remains in equilibrium when placed anywhere else at any other time, as long as the bodies have not suffered any change. Therefore, the equality of mass is a property of the bodies, independent of the place where (and the time when) they are compared. If C' is composed of standard units, the mass of C can be obtained as a multiple of the standard mass. The mass obtained this way is really the **gravitational mass** (Chapter 11); but in Chapter 6 we shall see how to compare masses dynamically. Mass obtained dynamically is called **inertial mass**. No difference has been found between the results obtained by the two methods of measuring mass.

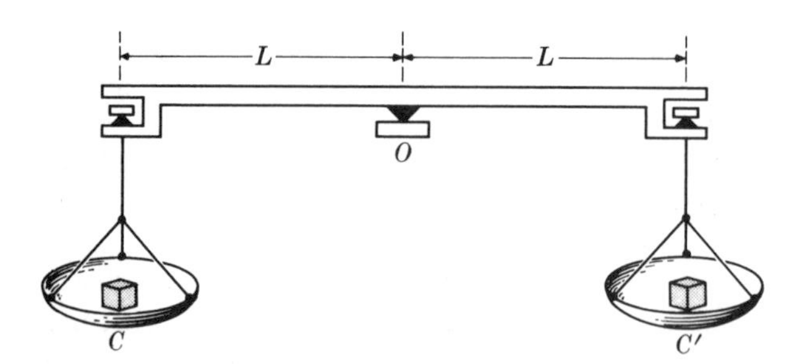

Figure 2.1 Equal-arm balance for comparing the masses of two bodies.

2.4 Fundamental units

With few exceptions, all quantities used in physics can be related to the four quantities length, mass, time and charge. Their definitions may be expressed as mathematical relations involving these quantities. The units of all derived quantities are in turn expressed in terms of the units of the four fundamental quantities by means of their defining relations. Therefore it is only necessary to agree on the units for the four fundamental quantities in order to have a consistent system of units. Physicists have agreed (at the Eleventh General Conference on Weights and Measures, held in Paris in 1960) to use the MKSC system of units, and this is what we shall adhere to in this book. The initials stand for meter, kilogram, second and coulomb. Their definitions are as follows.

The **meter** (abbreviated m) is the unit of length. It is equal to the distance traveled by light in vacuum in $3.335\,640\,952 \times 10^{-9}$ s. This is equivalent to fixing the velocity of light at $2.997\,924\,58 \times 10^{8}\ \mathrm{m\,s^{-1}}$, which is reasonable since the velocity of light in vacuum is one of the constants of nature. In practice, it is also equal to 1 650 763.73 wavelengths in a vacuum of a certain electromagnetic radiation emitted by the isotope ^{86}Kr. The radiation appears as a red line in a spectroscope. Historically, the meter was defined as 1/10 000 000 of the distance from the equator to either pole (Figure 2.2).

The **kilogram** (abbreviated kg) is the unit of mass. It is defined as the mass of Prototype Kilogram No. 1 or **international kilogram**, a platinum block kept at the International Bureau of Weights and Measures in Sèvres, near Paris. For practical purposes, it is equal to the mass of $10^{-3}\ \mathrm{m^3}$ of distilled water at 4 °C. The mass of 1 $\mathrm{m^3}$ of water is thus 10^3 kg. A volume of $10^{-3}\ \mathrm{m^3}$ is called one **liter**.

We can also associate the kilogram with an atomic property by saying that it is equal to the mass of a certain number of atoms. For example, the atoms of the isotope ^{12}C have a mass of $1.992\,648\,24 \times 10^{-26}$ kg. Then we could instead say that the kilogram is equal to the mass of $1/1.992\,648\,24 \times 10^{-26}$ or $5.018\,447\,21 \times 10^{25}$ atoms of ^{12}C.

In fact, this is the criterion adopted in defining the international scale of atomic masses. The **atomic mass unit** (abbreviated amu, although it is sometimes designated by u) is defined as

$$1\ \text{amu} = \tfrac{1}{12}\ \text{of the mass of a }^{12}\text{C atom}$$

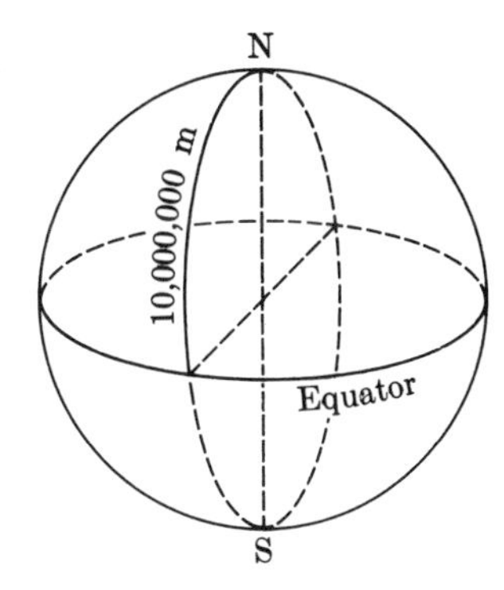

Figure 2.2 Historical definition of the meter.

This unit is commonly used in dealing with atomic and nuclear processes. The equivalence of the amu with the kilogram is:

$$1 \text{ amu} = \tfrac{1}{12} \times 1.992\,648\,24 \times 10^{-26}\,\text{kg}$$
$$= 1.660\,540\,2 \times 10^{-27}\,\text{kg}$$

The atomic masses in Table A.1 are expressed in amu and correspond to the *average* mass of the isotopes of the element, in the proportion they occur naturally.

The **second** (abbreviated s) is the unit of time. The second is defined as the duration of 9 192 631 770 periods of the radiation corresponding to a certain transition of the ^{133}Cs atom. Historically, the second was defined as 1/86 400 of the **mean solar day**, the time interval between two successive passages of a point on the Earth in front of the Sun, averaged over one year. But this definition has a drawback. The period of the Earth's rotation is decreasing gradually because of tidal action, and therefore this unit also changes gradually.

Atomic properties have been used to measure time with great precision. For example, the molecule of ammonia (NH_3) has a pyramidal structure, with the three H atoms at the base and the N atom at the vertex (Figure 2.3). There is a symmetric position, N′, for the nitrogen atom at the same distance from the H–H–H plane but on the opposite side. The N atom oscillates between these two positions of equilibrium at the rate of $2.387\,013 \times 10^{16}$ oscillations per second. The first atomic clock, based on this principle, was built at the US National Bureau of Standards (now the National Institute of Standards and Technology (NIST)) in 1948. Since then other substances have been tried as atomic clocks.

The **coulomb** (abbreviated C) is the unit of electric charge. Its precise definition will be given in Chapter 21. At this time we may say that it is equal, in absolute value, to the negative charge contained in $6.241\,508 \times 10^{18}$ electrons, or to the positive charge of an equal number of protons. The absolute value of the electron or proton charge is called the fundamental charge, designated ***e***. Clearly

$$e = \frac{1}{6.241\,508 \times 10^{18}}\,\text{C} = 1.602\,177 \times 10^{-19}\,\text{C}$$

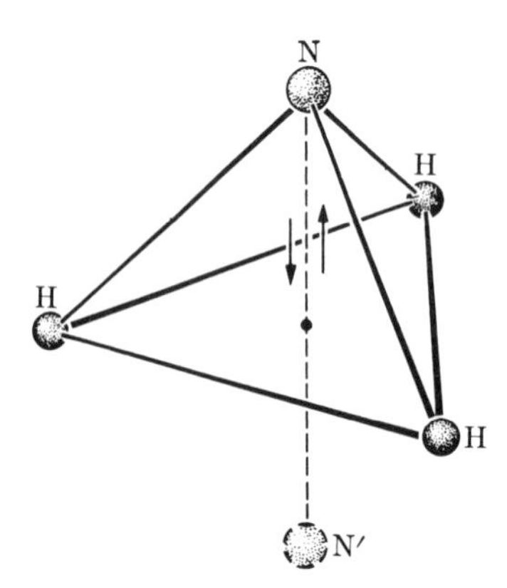

Figure 2.3 Oscillation of the nitrogen atom between two symmetric positions in the ammonia molecule.

Table 2.1 Prefixes for powers of ten

Magnitude	Prefix	Symbol
10^{-18}	atto-	a
10^{-15}	femto-	f
10^{-12}	pico-	p
10^{-9}	nano-	n
10^{-6}	micro-	μ
10^{-3}	milli-	m
10^{-2}	centi-	c
10^{-1}	deci-	d
$10^0 = 1$	Fundamental unit	
10	deca-	D
10^2	hecto-	H
10^3	kilo-	k (or K)
10^6	mega-	M
10^9	giga-	G
10^{12}	tera-	T
10^{15}	penta-	P
10^{18}	exa-	E

For practical reasons multiples and submultiples of the fundamental and derived units have been introduced as powers of ten. They are designated with a prefix, according to the scheme given in Table 2.1.

Strictly speaking, in addition to the meter, the kilogram and the second, the **ampere** (instead of the coulomb) was adopted at the Eleventh Conference as the unit of **electric current**. The ampere will be defined in Section 24.14. The coulomb is thus officially defined as the amount of electric charge that passes through a section of a conductor during one second when the current is one ampere. The reason for choosing the ampere is that a current is more easily established as a standard. Our decision to use the coulomb is based mainly on our wish to express the more fundamental character of electric charge, without departing essentially from the recommendations of the Eleventh Conference. The MKSA (where A stands for ampere) is the **International System** of units, designated by the symbol **SI**.

Another unit incorporated in the SI is the **mole**, defined as the amount of any substance that contains as many entities (atoms, molecules, ions, electrons or any other particles) as there are atoms in 0.012 kilograms of ^{12}C. Since, as indicated previously, the mass of $5.018\,447\,21 \times 10^{25}$ atoms of ^{12}C is equal to one kilogram, we conclude that one mole of any substance contains $0.012 \times 5.018\,447\,21 \times 10^{25}$ or $6.022\,136\,65 \times 10^{23}$ units or particles. This amount is called **Avogadro's number** and is designated by N_A.

Note 2.1 Historical basis for the fundamental units

The meter and the kilogram are units originally introduced during the French revolution. At that time, the French government, upon a recommendation from the Academy of Sciences made in 1790, decided to establish a rational system of units, known since then as the metric system, to supplant the chaotic and varied units in use at that time. The system was adopted officially by France in 1799. The meter was at first defined as 'the ten-millionth (10^{-7}) part of a quadrant of a terrestrial meridian'. For that purpose an arc of a meridian passing through Barcelona and Dunkirk was carefully measured, an operation that took from 1792 to 1798. Later on, in 1879, a standard platinum bar measuring one meter was fabricated and kept under controlled conditions at 0 °C at the International Bureau of Weights and Measures at Sèvres. Later measurements indicated that the standard bar was shorter by 1.8×10^{-4} m than the ten-millionth part of the quadrant of a meridian. It was then decided in 1889, at an International Conference, to adopt the length of the bar as the standard meter, without further reference to the Earth meridian. Duplicates of the standard meter exist in many countries. However, the convenience of having a standard of more permanent character and easy availability at any laboratory was recognized. For that reason, in 1960 the wavelength of the red line of ^{86}Kr was chosen and is still a practical secondary standard, even though in 1983 the present standard, based on the defined value for the speed of light, has been adopted.

For mass, the unit chosen by the French was the **gram** (abbreviated g) defined as the mass of one cubic centimeter (1 cm = 10^{-2} m and 1 cm^3 = 10^{-6} m^3) of distilled water at 4 °C. This temperature was chosen because it is the temperature at which the density of water is a maximum. The kilogram is then equal to 10^3 grams. A platinum block with a mass of one kilogram was built in 1879. Later on, at the International Conference in 1889, this block was adopted as the standard kilogram without further reference to water. Modern definitions of the meter and second have been adopted to be as close as possible to the historical definitions, while making them highly accurate and reproducible in laboratories throughout the world.

In the **cgs system**, used before the SI was officially adopted, the unit of length is the **centimeter**, the unit of mass is the **gram** and the unit of time is the **second**.

In the United States, another system of units is still used for some practical and engineering applications. The unit of length is the **foot** (abbreviated ft), the unit of mass is the **pound** (abbreviated lb) and the unit of time is again the **second**. The equivalent metric units are:

1 foot = 0.3048 m 1 m = 3.281 ft

1 pound = 0.4536 kg 1 kg = 2.205 lb

2.5 Derived units and dimensions

Most of the quantities used in physics, or for that matter in our daily life, cannot be expressed in just one of the fundamental units; rather, they require some combination of them. For example, the velocity of a body is expressed in meters per second or m s^{-1}, because it is obtained by dividing a distance (m) by a time interval (s). The area of a rectangular table is measured in square meters (m^2) because it is obtained by multiplying the length (m) by the width (m). Similarly, the volume of a box is obtained by multiplying the length by the product of the width and the height; the result is expressed in cubic meters (m^3). Of course, all areas are measured in m^2 and all volumes in m^3, regardless of their shape, although other rules are applied to obtain the numerical results. For example, the area of a circle is obtained using the relation $A = \pi r^2$, and the volume of a cylinder is found using $V = \pi r^2 h$. To measure the rate at which water flows in a pipe, the volume (m^3) of water discharged in a certain time interval (s) is measured. The rate of flow is then m^3 s^{-1}. Similarly, the **density** of a body, defined as its mass

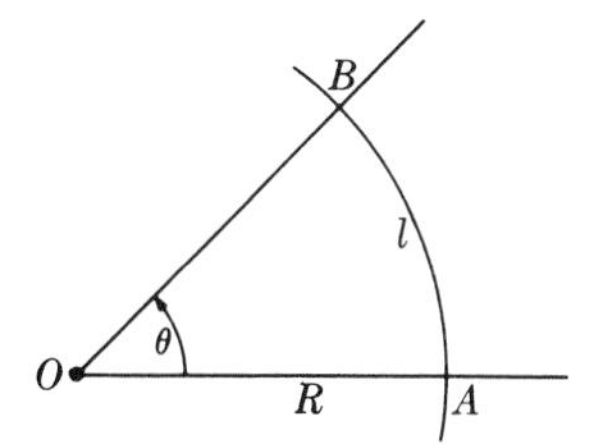

Figure 2.4 Measurement of an angle in radians.

per unit volume, is expressed by

$$\rho = \frac{m}{V} \tag{2.1}$$

and is measured in $kg\,m^{-3}$ (or $g\,cm^{-3}$). For most practical purposes the density of water is

$$\rho = 10^3\,kg\,m^{-3} \quad \text{or} \quad 1\,g\,cm^{-3} \tag{2.2}$$

From Equation 2.2 we see that to change from a density expressed in $g\,cm^{-3}$ to the same density expressed in $kg\,m^{-3}$, multiplication by 10^3 is required.

The expression of a physical quantity in terms of the fundamental units is called a **dimensional equation**. Some quantities are independent of the units and are called dimensionless. An example of a dimensionless unit is the measurement of an angle in **radians**. To express a plane angle in radians, an arc AB, of radius R (Figure 2.4), is drawn. The center of the arc is set at the vertex O of the angle. Then, the measure of θ in radians (abbreviated rad) is

$$\theta(\text{rad}) = \frac{l}{R} \tag{2.3}$$

where l is the length of the arc AB. This method is based on the fact that, for a given angle, the ratio l/R is constant and independent of the radius and may thus be used as a measure of the angle. The radian is a dimensionless unit because it is the ratio of two lengths both measured in the same unit. From Equation 2.3 we have

$$l = R\theta \tag{2.4}$$

so that an arc length may be found, given the angle subtended by the arc and the arc's radius.

Table 2.2 Densities of various materials in $kg\,m^{-3}$.

Solids		Liquids		Gases (at STP)	
Ice	9.17×10^2	Gasoline	6.70×10^2	Hydrogen	8.988×10^{-2}
Magnesium	1.74×10^3	Ethyl alcohol	7.91×10^2	Helium	1.7847×10^{-1}
Aluminum	2.70×10^3	Air (−147°C)	9.2×10^2	Nitrogen	1.250 55
Iron	7.86×10^3	Water (4°C)	1.000×10^3	Air	1.2922
Uranium	1.87×10^4	Mercury	1.359×10^4	Oxygen	1.429 04

Note 2.2 Space, time and matter

In the analysis of the physical universe, scientists are basically interested in what are called **phenomena**, which is a generic term for referring to observed facts, events, processes and changes that occur naturally or are induced in the laboratory. Examples of phenomena are the motion of bodies (cars, planes, electrons in TV tubes or in metallic wires, planets circling the Sun etc.), the collision of gas molecules with the walls of a container, the evaporation of a liquid, the fission of a uranium nucleus, the vibrations of a string in a musical instrument, and so on. In describing phenomena, scientists invariably consider *where* they occur, *when* they occur and *what* is involved.

We may say that all phenomena occur at a particular place in **space**, at a specific **time** or during a certain **time interval** and involve **matter** in one way or another. This is why the primary notions physicists deal with are space, time and matter. Space is considered to be three-dimensional, in the sense that three distances or coordinates must be measured to locate a place relative to a body chosen as reference, such as the surface of the Earth, the walls of the laboratory or the Sun. The time of occurrence of a phenomenon, or its duration, is determined by correlating it with some standard phenomenon that repeats itself with regularity, such as the rotation of the Earth or the oscillation of a pendulum. Matter is expressed quantitatively by using the concept of mass. For example, we buy many products, such as food, according to their mass content, commonly referred to as weight. For these reasons it is important to define at the outset the units for measuring lengths, time intervals and masses, as we explained in Sections 2.3 and 2.4. (Electric charge was added because it refers to an important property of matter.)

Traditionally space, time and matter have been considered independent notions. This assumption results from our own perceptions and is one that is satisfactory for analyzing most phenomena. However, as a result of some observations and experiments, it has been found that space and time are not truly independent, but coupled. It is more appropriate to talk about space–time. This coupling of space and time becomes important only for processes where bodies move with large velocities, comparable to that of light, as well as when large energies are involved, as in the fission of uranium. In our daily life the coupling of space and time may be ignored for all practical purposes. Other experiments and observations indicate that space–time is affected by the presence of matter. To put it in other terms, the *local* properties of space–time are determined by the quantity of matter in the locality and its effects appear in what we call gravitation. The effect of matter on space–time is more noticeable near large concentrations of matter, such as the Sun, a star or a black hole. However, near the Earth's surface, the coupling between matter and space–time is noticeable only through very delicate experiments and precise measurements.

The coupling of space, time and matter is the subject of the theories of special and general relativity, developed by Albert Einstein. An interesting aspect of these theories is that they also provide a coupling between mass and energy. For explaining most of the phenomena discussed in this book, space, time and matter will be treated as independent, unless otherwise stated.

QUESTIONS

2.1 What are the inconveniences in defining the meter as one ten-millionth of a quadrant of an Earth meridian?

2.2 If people on the Moon had used the same criteria for defining a unit length as was done on Earth in the 19th century for defining the meter, what would be the ratio between the 'Earth' meter and the 'Moon' meter?

2.3 What are the disadvantages in defining the second as one 86 400th part of a mean solar day?

2.4 The **relative density** of two substances is the ratio of the two densities, i.e. $\rho_{12} = \rho_1/\rho_2$. Is the relative density independent of the units chosen for measuring the two densities that are compared? In what units should the relative density be measured? Transform Table 2.2 into a table of densities relative to water.

2.5 Explain why it is convenient in physics to measure angles in radians.

2.6 Noting that the circumference of a circle is $2\pi R$, verify that a complete plane angle around a point, measured in radians, is 2π rad. So, 2π radians are equivalent (not equal!) to 360°.

2.7 Verify that the mass of one mole of a chemical element or of a compound, expressed in grams, is equal to the atomic mass of the element or compound expressed in amu.

2.8 Explain why the number of molecules in 1 cm^3 of water is equal to $N_A/18$.

2.9 How long, in years, would it take a light signal to go from the Sun to the center of the galaxy (see Figure 6.2)?

2.10 What would happen to the surface area, the volume and the density of a planet if the radius is (a) doubled, (b) halved without changing the mass?

2.11 What would happen to the mass of a planet if its radius is (a) doubled, (b) halved, while keeping the density constant?

PROBLEMS

2.1 Since one amu is equal to 1.6605×10^{-27} kg, express the masses of one atom of (a) hydrogen and (b) oxygen in kilograms. (c) How many atoms of (i) H, and (ii) O are there in one kg of each substance? Use Table A.1 for the atomic masses of hydrogen and oxygen.

2.2 (a) How many water molecules, each composed of one atom of oxygen and two atoms of hydrogen, are there in one gram? (b) In 18 grams? (c) In one cubic centimeter? (Use Table A.1.)

2.3 Consider molecules of hydrogen, of oxygen and of nitrogen, each composed of two identical atoms. (a) Calculate the number of molecules of each of these gases (at STP) in 1 m^3. Use the values of densities given in Table 2.2. (b) Extend your calculation to other gases. What general conclusion can you draw from this result?

2.4 (a) Assuming that air is composed of 20% oxygen and 80% nitrogen and that these gases have molecules each comprising two atoms, obtain the 'effective' molecular mass of air. (b) Estimate the number of molecules in one cubic centimeter of air at STP. How many molecules are oxygen, and how many are nitrogen?

2.5 The density of interstellar matter in our galaxy is estimated to be about 10^{-21} kg m^{-3}. (a) Assuming that the matter is mainly hydrogen, estimate the number of hydrogen atoms per cubic centimeter. (b) Compare the result with air at STP (Problem 2.4).

2.6 A glass containing water has a radius of 2 cm. In 2 hours the water level drops 1 mm. (a) Estimate, in grams per hour, the rate at which water is evaporating. (b) How many water molecules are evaporating per second from each square centimeter of water surface? (We suggest that the student perform this experiment and obtain personal data. Why do you get different results on different days?)

2.7 Using the data in Tables 2.2 and A.1, estimate the average separation between molecules in (a) hydrogen at STP (gas), (b) in water (liquid), and (c) in iron (solid). (Hint: Obtain the number of molecules per m^3. Assume that the volume occupied by each molecule is a cube and find its edge.)

2.8 The mass of an atom is practically all in its nucleus. The radius of the nucleus of uranium is 8.68×10^{-15} m. (a) Using the atomic mass of uranium given in Table A.1, obtain the density of 'nuclear matter'. (b) This nucleus contains 238 nucleons. Estimate the average separation between nucleons. (c) From your result and those of Problem 2.7, would you conclude that it is reasonable to treat nuclear matter in the same manner as matter in bulk, i.e. solids, liquids and gases?

2.9 (a) Obtain the average density of the Earth, the planets and the Sun. (Use the data in Table 11.1.) (b) When you compare these values with the data in Table 2.2, what do you conclude about the structure of these bodies?

2.10 The speed of light in vacuum is 2.9979×10^8 m s^{-1}. (a) Express it in kilometers per hour. (b) How many times could a light ray travel around the Earth in one second? The radius of the Earth is 6.37×10^6 m. (c) What distance would it travel in one year? This distance is called a **light year** (Lyr).

2.11 The radius of the Earth's orbit about the Sun is 1.49×10^{11} m. This length is called an **astronomical unit**. Express a light year in astronomical units (see Problem 2.10).

2.12 **Parallax** is the difference in the apparent direction of an object due to a change in the position of the observer. (Hold a pencil in front of you and cover first the right and then the left eye. Note that in each case the pencil appears against a different background.) **Stellar parallax** is the change in the apparent position of a star as a result of the Earth's orbital motion around the Sun. It is expressed quantitatively by one-half the angle subtended by the Earth's diameter E_1E_2 perpendicular to the line joining the star and the Sun (see Figure 2.5). It is given by $\theta = \frac{1}{2}(180° - \alpha - \beta)$, where the angles α and β are measured at the two positions E_1 and E_2 separated by six months. The distance r from the star to the Sun can be obtained from $a = r\theta$, where a is the radius of the Earth's orbit and θ is expressed in radians. The closest star, Alpha Centauri, has the largest parallax of 0.76″. Find its distance from the Sun expressed in (a) meters, (b) light years and (c) astronomical units.

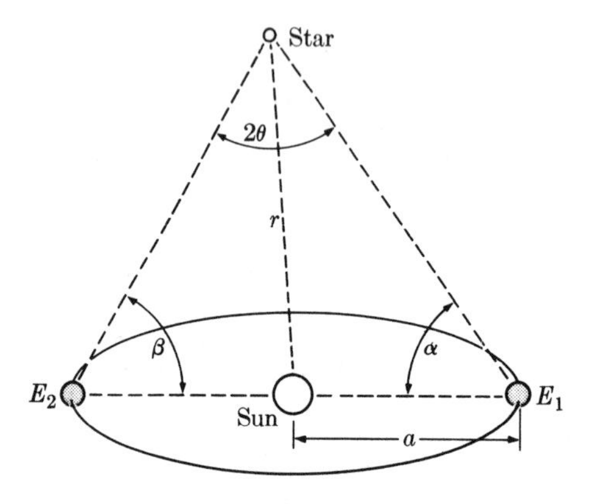

Figure 2.5

2.13 A **parsec** is equal to the distance from the Sun to a star whose parallax is 1″. Express the parsec in (a) meters, (b) light years and (c) astronomical units.

2.14 The distance between San Francisco and New York, measured along the great circle passing through these two cities, is 4137 km. Compute the angle between the verticals at the two cities. (The radius of the Earth is 6.37×10^6 m.)

2.15 By looking at a table of trigonometric functions or using your calculator, find the angle for which (a) $\sin\theta$ is less than $\tan\theta$ by (i) 10%, (ii) 1%, (iii) 0.1%. (b) Do the same for $\sin\theta$ and θ, and for θ and $\tan\theta$, where θ must be expressed in radians. (c) What conclusion can you draw from your result?

2.16 An automobile describes a circular curve with a radius of 100 m. The length of the curve is 60 m. (a) What is the angle through which the automobile has turned? (b) What distance has it traveled when it has turned through an angle of 20°?

2.17 The distance from the Moon to the Earth is 3.84×10^8 m. What distance has the Moon moved when its position relative to the Earth changes by 30°?

2.18 (a) Find the distance that a point, located on the Earth's equator, moves in one hour due to the Earth's rotation. (b) What is the distance for a point at 30° latitude? (c) And at 80° latitude? (The radius of the Earth is 6.37×10^6 m.)

2.19 (a) What is the angle subtended by the Moon as seen from the Earth? (b) And the angle subtended by the Sun? (Use data of Table 11.1.)

2.20 (a) Assuming that the galaxy can be considered as two rather flat cones, of height 10^{20} m and radius 5×10^{20} m, attached by their common bases (see Figure 6.2), verify that the volume of the galaxy is of the order of 6×10^{61} m^3. (b) If the total number of stars in the galaxy is about 10^{11}, verify that their average separation is 10 Lyr.

3 Rectilinear motion

Galileo Galilei was an Italian scientist who, among other discoveries, demonstrated that all bodies fall with constant acceleration that is independent of their mass. Legend has it that he dropped cannon-balls and bullets from the Leaning Tower of Pisa to show that they would hit the ground at the same time, although his writings contain no record of this experiment. He also invented the telescope with which he discovered the moons of Jupiter, craters on the Moon, moving spots on the Sun and the phases of Venus.

Notes

3.1 Mechanics

The most fundamental and obvious phenomenon we observe around us is **motion**. Blowing air, waves in the ocean, flying birds, running animals, falling leaves – all these are examples of motion. Practically all imaginable processes can be traced back to the motion of certain particles or objects. The Earth and the other planets move around the Sun. The Sun in turn carries the solar system around the center of the galaxy. Electrons move within atoms, giving rise to the absorption and emission of light. Electrons move inside a metal, producing an electric current. Gas molecules move in a random fashion, giving rise to pressure and diffusion processes. Our everyday experience tells us that the motion of a body is influenced by the bodies that surround it; that is, by its **interactions** with them. In a television tube, or the monitor of a computer system, the electron beam must move in a certain fashion to produce a pattern on the screen. In a thermal engine, the molecules of the burnt fuel must move in a particular way so that a piston or a turbine, in turn, moves in a desired direction. A chemical reaction is the consequence of certain atomic interactions resulting in a new arrangement, forming new classes of molecules.

One role of physicists and engineers is to discover the relation between motions and the interactions responsible for them and to arrange things so that useful motions are produced.

To analyze and predict the motion of particles resulting from the different kinds of interactions some important concepts have been invented, such as **momentum**, **force** and **energy**. Momentum, force and energy are so important that we rarely analyze a process without expressing it in terms of them. The rules or principles that apply to the analysis of all kinds of motion are called **mechanics**.

Mechanics is one of the fundamental areas of physics and must be understood thoroughly before beginning a consideration of particular interactions. Galileo recognized this basic role of mechanics in the statement '*Ignorato motu, ignoratur natura*'. ['If we do not understand motion, we remain ignorant about nature'.]

The science of mechanics as we understand it today is mainly the result of the genius of Sir Isaac Newton (1642–1727) who, in the 17th century, produced the great synthesis called Newton's laws of motion. However, more people have contributed to its advance over the centuries. Some of the more illustrious names are Archimedes (287?–212 BC), Galileo Galilei (1564–1642), Johannes Kepler (1571–1630), René Descartes (1596–1650), Christiaan Huygens (1629–1695), Joseph Louis Lagrange (1736–1813), William R. Hamilton (1788–1856), Ernst Mach (1838–1916) and, more recently, Albert Einstein (1879–1955).

To carry out a program in mechanics, we must begin by learning how to *describe* the motions we observe. The description of motion is called **kinematics**, a word derived from the Greek word *kinema*, meaning 'motion'.

3.2 Frames of reference

An object is in motion relative to another when its position, measured relative to the second body, is changing with time. On the other hand, if this relative position does not change with time, the object is at relative rest. Both rest and motion are *relative* concepts; that is, they depend on the condition of the object relative to the body that serves as reference. A tree and a house are at rest relative to the Earth, but in motion relative to the Sun. When a train passes a station we say that the train is in motion relative to the station. A passenger on the train might as well say that the station is in motion relative to the train, but moving in the opposite direction (Figure 3.1).

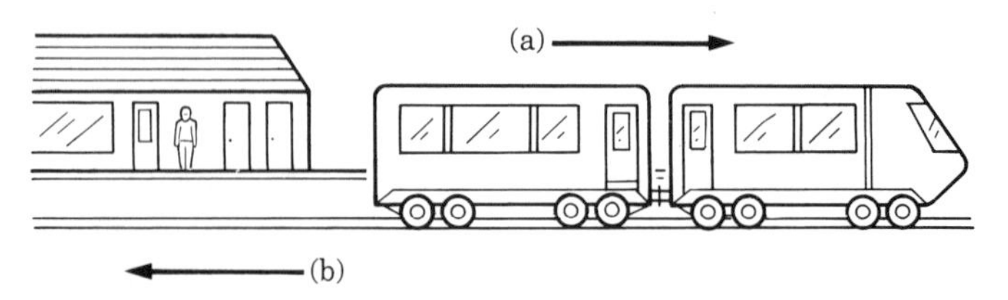

Figure 3.1 Motion is relative. (a) View of train from station; (b) view of station from train.

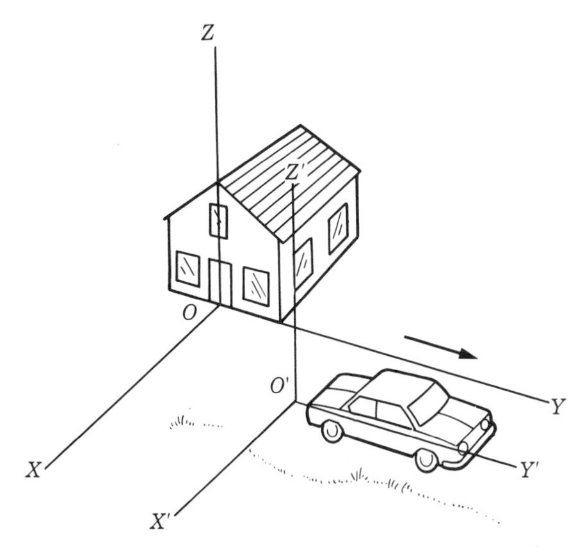

Figure 3.2 Frames of reference in relative motion.

To describe motion, the observer must define a **frame of reference** relative to which the motion is analyzed. A frame of reference may be considered as an object or set of objects at rest relative to the observer. To be more precise, the observer attaches a set of coordinate axes to the object(s) at rest relative to the observer, as we customarily do in geometry. In Figure 3.2 we have indicated two observers O and O', one at the house and the other in an automobile. These observers use the frames of reference XYZ and $X'Y'Z'$ that are attached to the house and the automobile, respectively. If O and O' are at rest relative to each other, they will observe the same motion of any body. But if O and O' are in relative motion, their observations of the motion of a third body will be different. For example, ancient astronomers used a geocentric frame of reference attached to the Earth to study planetary motion, obtaining very complicated orbits. When Nicolaus Copernicus (1473–1543) and Kepler referred the motion of the planets to a heliocentric frame of reference (that is, one attached to the Sun), planetary motion became quite simple.

3.3 Rectilinear motion: velocity

The simplest motion to analyze is rectilinear motion. The motion of a body is **rectilinear** when its trajectory is a straight line. This is, for example, the motion of a body falling vertically near the surface of the Earth. Take our frame of reference so that the X-axis (Figure 3.3) coincides with the trajectory. The position of the object is defined by its x coordinate or distance from an arbitrary point O, chosen as the origin of the coordinate system. In principle, the coordinate x can be correlated with the time by means of a functional relation $x = f(t)$, as represented in Figure 3.4 for an arbitrary case. Obviously, x may be positive or negative, depending on the position of the body relative to O.

To express, in a quantitative way, how fast a body is moving, we use the concept of **velocity**. To determine the velocity of a body, the distance moved is

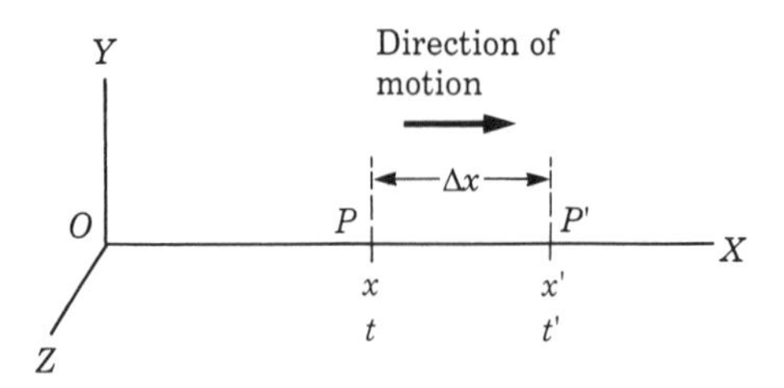

Figure 3.3 Rectilinear motion.

divided by the time elapsed. Thus, if an automobile moves 150 km in 3 hours, the average velocity of the automobile has been 50 km h^{-1}. To be more precise, suppose that at time t the object is at position P, with $OP = x$ (Figure 3.3). At a later time t', it is at P', with $OP' = x'$. Thus, the body has experienced a **displacement** $PP' = x' - x$ during the time interval $t' - t$. The displacement is positive or negative, depending on the direction of motion of the body.

The **average velocity** of the body when it moves from P to P' is then defined by means of the relation

$$v_{\text{ave}} = \frac{x' - x}{t' - t} = \frac{\Delta x}{\Delta t} \tag{3.1}$$

where $\Delta x = x' - x$ is the displacement of the body and $\Delta t = t' - t$ is the elapsed time. Thus

the average velocity during a certain time interval is given by the ratio of the displacement to the time interval.

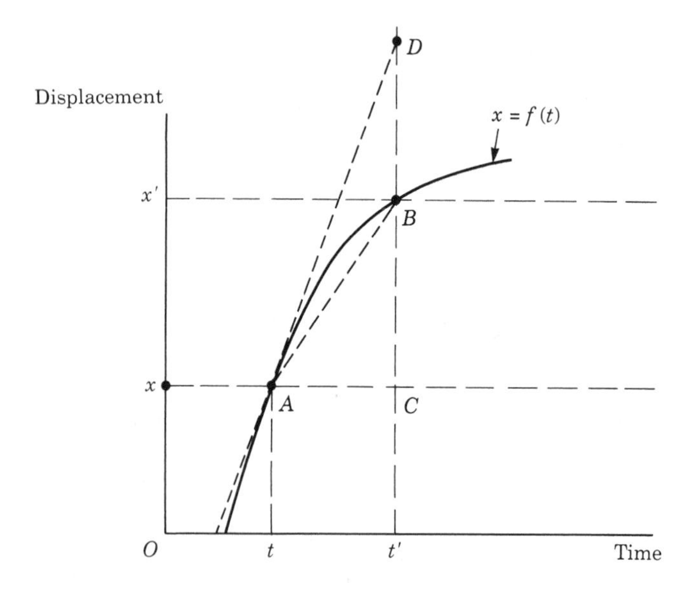

Figure 3.4 Graph of displacement as a function of time.

However, during a time interval a body may move faster at some times and slower at others.

To determine the **instantaneous velocity** at a certain time when the body passes a given point, such as P, we must make the time interval Δt as small as possible. In mathematical language this is equivalent to computing the limiting value of the ratio appearing in Equation 3.1 as the denominator Δt approaches zero. This is written in the form

$$v = \lim_{\Delta t \to 0} v_{\text{ave}} = \lim_{\Delta t \to 0} \frac{\Delta x}{\Delta t}$$

But this is the mathematical definition of the time derivative of x; that is,

$$v = \frac{dx}{dt} \tag{3.2}$$

so that

the instantaneous velocity is equal to the time rate of change of the displacement.

Operationally, the instantaneous velocity is found by observing the moving body at two very close positions separated by the small distance dx, and measuring the small time interval dt required to go from one position to the other. In an automobile this is done continuously by the speedometer, using an electric device. In this book the term 'velocity' will always refer to instantaneous velocity.

Generally, the velocity of a body varies during its motion. We say that it is a function of time; that is, $v = f(t)$. If the velocity remains constant, the motion is called **uniform**.

The velocity can also be easily computed from a graph that gives the position as a function of time; that is, $x = f(t)$. Referring to Figure 3.4, the average velocity between t and t' is obtained by measuring $\Delta t = AC$ and $\Delta x = BC$ in the triangle ABC. Then,

$$v_{\text{ave}} = \frac{\Delta x}{\Delta t} = \frac{BC}{AC} = \text{slope of line } AB$$

If we want to measure the instantaneous velocity at time t, we have to find the slope of the tangent to the curve at A, which, for example, is given by DC/AC.

If we know how the velocity varies with time, we may calculate the displacement $x - x_0$ of the body during any time interval $t - t_0$ by the following method. First, we divide the time interval $t - t_0$ into successive small intervals dt_0, dt_1, dt_2 etc. Next we calculate the average velocity v_0, v_1, v_2 etc. in each small time interval. The corresponding displacements, in accordance with Equation 3.2, are $v_0\,dt_0$, $v_1\,dt_1$, $v_2\,dt_2$ etc. Then the total displacement of the body is the sum of each small displacement:

$$\text{Displacement} = x - x_0 = v_0\,dt_0 + v_1\,dt_1 + v_2\,dt_2 + \cdots = \sum_i v_i\,dt_i$$

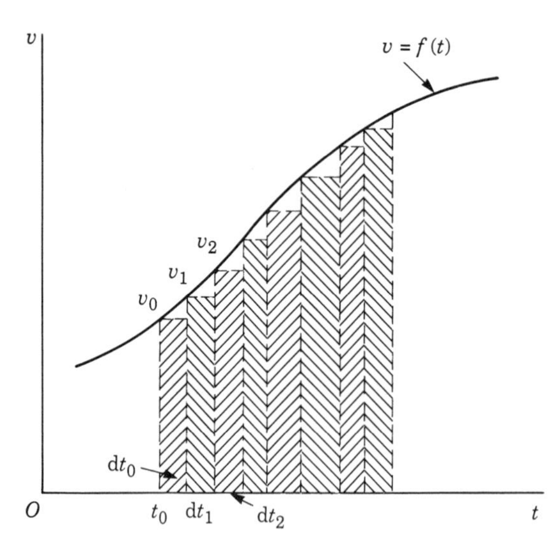

Figure 3.5 Graph of velocity as a function of time.

When the dts are very small, the summation $\sum v_i \, dt_i$ is equivalent to the mathematical operation called **integration** and is designated by $\int_{t_0}^{t} v \, dt$. Therefore, the displacement of the body is given by

$$x - x_0 = \int_{t_0}^{t} v \, dt \qquad \text{or} \qquad x = x_0 + \int_{t_0}^{t} v \, dt \tag{3.3}$$

This mathematical expression is useful whenever it is possible to calculate the integral. Otherwise, a graphical method may be used. Referring to Figure 3.5, the sum $\sum_i v_i \, dt_i$ is equal to the area under the curve. This provides a simple geometrical way for calculating the displacement when the graph of the velocity, as a function of time, is known.

We must remind the student that the displacement Δx (or dx) may be positive or negative, depending on whether the motion of the particle is to the right or to the left. The result is a positive or negative sign for the velocity. Thus,

the sign of the velocity in rectilinear motion indicates the direction of motion.

The direction is along $+OX$ if the velocity is positive and along $-OX$ if it is negative. Sometimes, in colloquial language, when referring only to the magnitude of the velocity, the word **speed** is used.

Units of velocity. In the SI system of units, velocity is expressed in meters per second, or m s^{-1}. This is the velocity of a body moving uniformly through one meter in one second. Of course, the velocity can also be expressed in any combination of space and time units, such as kilometers per hour, meters per minute etc.

EXAMPLE 3.1

Calculate the average velocity of a car, traveling in a straight line (i.e. rectilinearly), whose odometer readings at certain times are given in the following chart:

Time (h:min)	03:02	03:06	03:12	03:15	03:20	03:24
Odometer (km)	1582.6	1586.8	1593.4	1598.2	1606.4	1613.1

▷ The average velocity between the first and last observations is

$$v_{\text{ave}} = \frac{\Delta x}{\Delta t} = \frac{1613.1\text{ km} - 1582.6\text{ km}}{3\text{ h }24\text{ min} - 3\text{ h }2\text{ min}} = \frac{30.5\text{ km}}{22\text{ min}} = 1.38\,\frac{\text{km}}{\text{min}}$$

Similarly, it is easy to verify that the average velocity between the first two observations is 1.05 km min^{-1}. We leave the task of calculating the average velocity for each of the other intervals to the student.

EXAMPLE 3.2

Displacement of a car whose velocity, recorded at several times, is given in the chart below:

Time (h:min)	09:06	09:10	09:13	09:18	09:20	09:24
Velocity (km h^{-1})	38	44	48	50	42	24

▷ The velocity, as a function of time, has been plotted in Figure 3.6. To estimate the total distance moved between the first and last observation, we can determine the area under

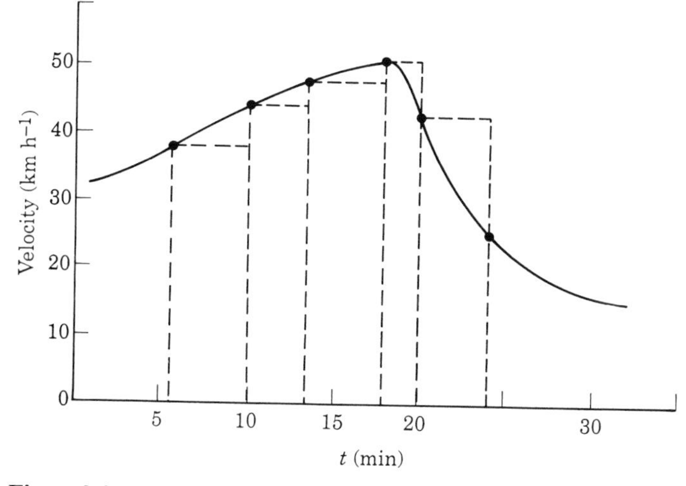

Figure 3.6

the curve. As a first approximation, we can measure the area of each rectangle shown in Figure 3.6 and add them up. For example, the area of the first rectangle is

$$(38\ \mathrm{km\,h^{-1}}) \times (4\ \mathrm{min}) = (38\ \mathrm{km}/60\ \mathrm{min}) \times (4\ \mathrm{min}) = 2.53\ \mathrm{km}$$

Repeating this procedure for the other rectangles gives a value of $2.53 + 2.20 + 4.00 + 1.67 + 2.80 = 13.20$ km. The calculation can be improved if we use the average velocity during each interval, which for small intervals can be approximated by finding the mean value of successive measurements. Thus, for the first interval, the average velocity is $\frac{1}{2}(38 + 44)\ \mathrm{km\,h^{-1}} = 41\ \mathrm{km\,h^{-1}}$. Then, the distance for the first interval is

$$(41\ \mathrm{km\,h^{-1}}) \times (4\ \mathrm{min}) = (41\ \mathrm{km}/60\ \mathrm{min}) \times (4\ \mathrm{min}) = 2.73\ \mathrm{km}$$

Repeating this procedure for the other intervals gives $2.73 + 2.30 + 4.08 + 1.53 + 2.20 = 12.84$ km for the distance. Other approximations may suggest themselves to the student. The techniques for numerically solving for areas under curves is a special subject in mathematics and have many applications. A computer is an especially useful tool for doing such repetitious tasks as those suggested by this and the previous example.

EXAMPLE 3.3

A disk, sliding along an inclined plane, is found to have its position (measured from the top of the plane) at any instant given by $x = 3t^2 + 1$, where x is in meters and t is in seconds. Compute its average velocity in the time interval between (a) 2 s and 3 s, (b) 2 s and 2.1 s, (c) 2 s and 2.001 s, (d) 2 s and 2.000 01 s. (e) Also find the instantaneous velocity at 2 s.

▷ We call $t_0 = 2$ s, which is common for the entire problem. Using $x = 3t^2 + 1$, we have $x_0 = 3(2)^2 + 1 = 13$ m. Therefore, for each question,

$$\Delta x = x - x_0 = x - 13 \qquad \text{and} \qquad \Delta t = t - t_0 = t - 2$$

(a) For $t = 3$ s, we have $\Delta t = 1$ s,

$$x = 3(3)^2 + 1 = 28\ \mathrm{m} \qquad \text{and} \qquad \Delta x = 28\ \mathrm{m} - 13\ \mathrm{m} = 15\ \mathrm{m}$$

Thus $v_{\mathrm{ave}} = \Delta x/\Delta t = 15\ \mathrm{m}/1\ \mathrm{s} = 15\ \mathrm{m\,s^{-1}}$.

(b) For $t = 2.1$ s, we have $\Delta t = 0.1$ s,

$$x = 3(2.1)^2 + 1 = 14.23\ \mathrm{m} \qquad \text{and} \qquad \Delta x = 1.23\ \mathrm{m}$$

Thus $v_{\mathrm{ave}} = \Delta x/\Delta t = 1.23\ \mathrm{m}/0.1\ \mathrm{s} = 12.3\ \mathrm{m\,s^{-1}}$.

(c) For $t = 2.001$ s, we have $\Delta t = 0.001$ s,

$$x = 3(2.001)^2 + 1 = 13.012003\ \mathrm{m} \qquad \text{and} \qquad \Delta x = 0.012\,003\ \mathrm{m}$$

Thus $v_{\mathrm{ave}} = \Delta x/\Delta t = 0.012\,003\ \mathrm{m}/0.001\ \mathrm{s} = 12.003\ \mathrm{m\,s^{-1}}$.

(d) The student may verify that for $t = 2.000\,01$ s,

$$v_{\mathrm{ave}} = 12.000\,03\ \mathrm{m\,s^{-1}}$$

(e) We note then that as Δt becomes smaller the velocity approaches the value of $12\,\mathrm{m\,s^{-1}}$. We may thus expect that this is the instantaneous velocity at $t = 2$ s. In fact,

$$v = \frac{\mathrm{d}x}{\mathrm{d}t} = \frac{\mathrm{d}}{\mathrm{d}t}(3t^2 + 1) = 6t$$

When we set $t = 2$ s, then we obtain $v = 12\,\mathrm{m\,s^{-1}}$.

3.4 Rectilinear motion: acceleration

In general, when a body is moving, its velocity is not always the same (watch the speedometer of an automobile). At some time, the body may go faster than at others. To measure how the velocity of a body is changing, the concept of acceleration has been introduced. Referring to Figure 3.7, suppose that at time t there is an object at P with velocity v, and at time t' it is at P' with velocity v'. The **average acceleration** between P and P' is calculated using the relation

$$a_{\mathrm{ave}} = \frac{v' - v}{t' - t} = \frac{\Delta v}{\Delta t} \tag{3.4}$$

where $\Delta v = v' - v$ is the change in velocity and, as before, $\Delta t = t' - t$ is the elapsed time. Thus

the average acceleration during a certain time interval is given by the ratio of the change in velocity to the time interval.

The **instantaneous acceleration** is the limiting value of the average acceleration when the time interval Δt becomes very small. That is,

$$a = \lim_{\Delta t \to 0} a_{\mathrm{ave}} = \lim_{\Delta t \to 0} \frac{\Delta v}{\Delta t}$$

resulting in

$$a = \frac{\mathrm{d}v}{\mathrm{d}t} \tag{3.5}$$

or, since $v = \dfrac{\mathrm{d}x}{\mathrm{d}t}$, then $a = \dfrac{\mathrm{d}^2x}{\mathrm{d}t^2}$.

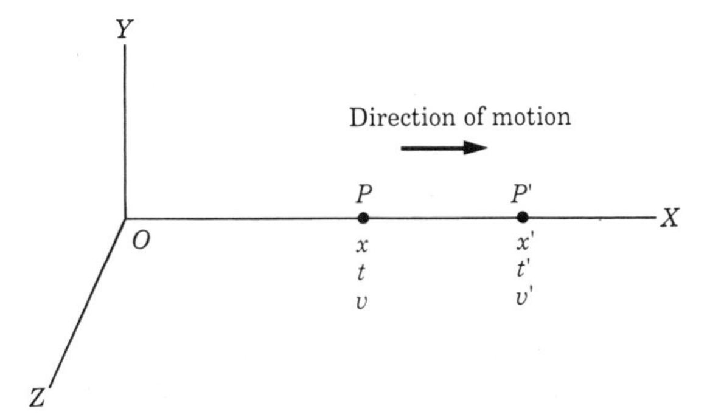

Figure 3.7

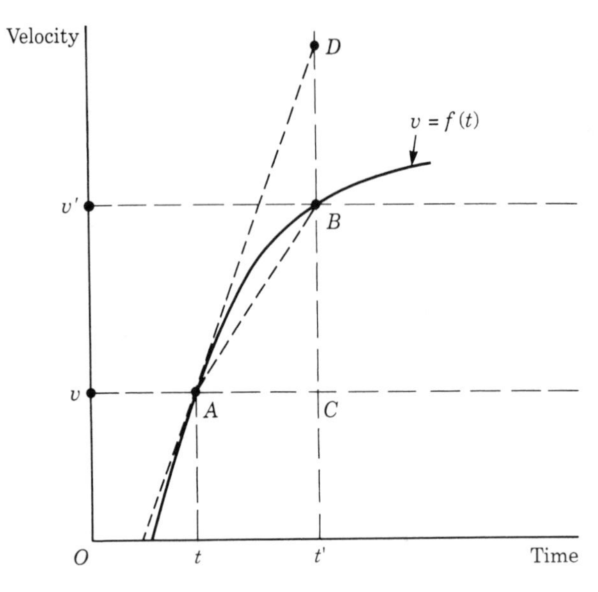

Figure 3.8 Graph of velocity as a function of time.

so that

the instantaneous acceleration is equal to the time rate of change of the velocity.

Operationally, the instantaneous acceleration is found by observing the small change of velocity $\mathrm{d}v$ that takes place in a very small time interval $\mathrm{d}t$. In the future, whenever we say 'acceleration' we shall mean the instantaneous acceleration. Note that the sign of the acceleration depends on whether $\mathrm{d}v/\mathrm{d}t$ is positive (v increases) or negative (v decreases).

In general, the acceleration varies during the motion. If the rectilinear motion has constant acceleration, the motion is said to be **uniformly accelerated**. If the velocity increases in absolute value with time, as happens when the accelerator pedal in an automobile is pressed, the motion is said to be 'accelerated'; but if the velocity decreases in absolute value with time, the motion is termed retarded or 'decelerated', as happens when the brakes in an automobile are applied.

The acceleration can also be computed using a graphic representation of $v(t)$. Referring to Figure 3.8, the average acceleration between t and t' is obtained by measuring $\Delta t = AC$ and $\Delta v = CB$ in the triangle ABC. Then

$$a_{\text{ave}} = \frac{\Delta v}{\Delta t} = \frac{BC}{AC} = \text{slope of line segment } AB$$

If we want to measure the acceleration at time t, we must find the slope of the tangent to the curve at A, which, for example, is given by CD/AC.

If we know the acceleration as a function of time, we may compute the change in velocity $v - v_0$ at any time by a procedure similar to that used for the displacement. First, we divide the time interval $t - t_0$ into successive small time intervals $\mathrm{d}t_0$, $\mathrm{d}t_1$, $\mathrm{d}t_2$ etc. Next, we calculate the average acceleration a_0, a_1, a_2 etc. in each interval. Then, in accordance with Equation 3.5, the corresponding changes

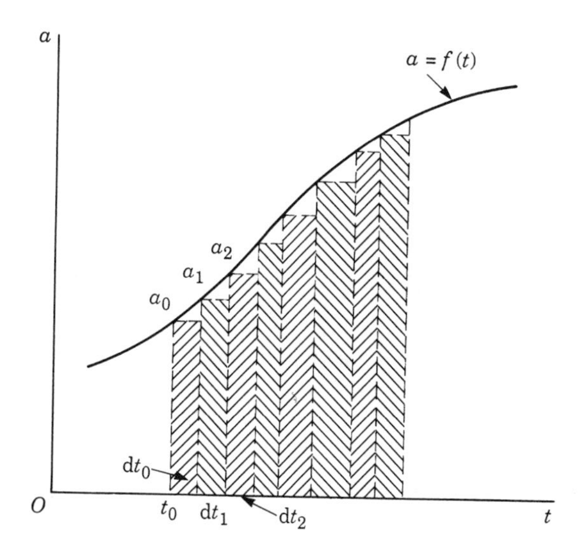

Figure 3.9 Graph of acceleration as a function of time.

in velocity are $a_0\,\mathrm{d}t_0$, $a_1\,\mathrm{d}t_1$, $a_2\,\mathrm{d}t_2$ etc. Therefore, the total change in velocity is

$$\text{Change in velocity} = v - v_0 = a_0\,\mathrm{d}t_0 + a_1\,\mathrm{d}t_1 + a_2\,\mathrm{d}t_2 + \cdots = \sum_i a_i\,\mathrm{d}t_i$$

When the time intervals $\mathrm{d}t_i$ are very small, the summation $\sum a_i\,\mathrm{d}t_i$ is equivalent to the integral $\int_{t_0}^{t} a\,\mathrm{d}t$. Therefore, the change in velocity of the body is

$$v - v_0 = \int_{t_0}^{t} a\,\mathrm{d}t \qquad \text{or} \qquad v = v_0 + \int_{t_0}^{t} a\,\mathrm{d}t \tag{3.6}$$

This expression can be used whenever it is possible to carry out the integration. Otherwise, we may use the same graphical method explained for the displacement. Referring to Figure 3.9, which is a graph of acceleration as a function of time, we note that the change in velocity between t_0 and t is equal to the area under the line representing $a(t)$.

Nothing prevents us from introducing another concept related to the time rate of change of the acceleration and usually called 'jolt' or 'jerk'. This might be useful for cases where the acceleration varies rapidly, such as in the take-off of a rocket ship. However, for most applications of mechanics the concepts of velocity and acceleration are sufficient.

Units of acceleration. In the SI system, acceleration is expressed in meters per second per second, or $(\mathrm{m\,s^{-1}})\mathrm{s^{-1}} = \mathrm{m\,s^{-2}}$. This is the acceleration of a body whose velocity increases one meter per second in one second, with constant acceleration. But the acceleration may also be expressed in other units, such as $(\mathrm{km\,h^{-1}})\mathrm{s^{-1}}$, which gives the change in velocity, expressed in $\mathrm{km\,h^{-1}}$, in one second.

EXAMPLE 3.4

Referring to the data given in Example 3.2, calculate the average acceleration.

▷ The average acceleration during the entire interval is

$$a_{\text{ave}} = \frac{\Delta v}{\Delta t} = \frac{(24\,\text{km}\,\text{h}^{-1} - 38\,\text{km}\,\text{h}^{-1})}{(9\,\text{h}\,24\,\text{min} - 9\,\text{h}\,6\,\text{min})} = \frac{-14\,\text{km}\,\text{h}^{-1}}{18\,\text{min}} = -0.778\,\frac{\text{km}}{\text{h}\,\text{min}}$$

$$= -\frac{0.78\,\text{km}}{(60\,\text{min})\text{min}} = -0.013\ \text{km}\ \text{min}^{-2}$$

The negative sign indicates that, over the whole interval, the motion was decelerated. By the same method of calculation, the acceleration was $+1.5\,\text{km}\,\text{h}^{-1}\,\text{min}^{-1}$ or $+0.025\,\text{km}\,\text{min}^{-2}$ during the first interval and the motion was accelerated, while the acceleration was $-4.5\,\text{km}\,\text{h}^{-1}\,\text{min}^{-1}$ or $-0.075\,\text{km}\,\text{min}^{-2}$ in the last interval and the motion was decelerated.

EXAMPLE 3.5

A car is accelerated for a few seconds so that its position along the road as a function of time is given by

$$x = 0.2t^3 + 0.5t^2 + 0.5$$

where x is in meters and t is in seconds. Find (a) the velocity and the acceleration at any time; (b) the position, velocity and acceleration at $t = 2$ s and 3 s; and (c) the average velocity and acceleration between $t = 2$ s and $t = 3$ s.

▷ (a) Using Equations 3.2 and 3.5, we may write

$$v = \frac{\text{d}x}{\text{d}t} = \frac{\text{d}}{\text{d}t}(0.2t^3 + 0.5t^2 + 0.5)$$

$$= (0.6t^2 + 1.0t)\,\text{m}\,\text{s}^{-1}$$

$$a = \frac{\text{d}v}{\text{d}t} = \frac{\text{d}}{\text{d}t}(0.6t^2 + 1.0t) = (1.2t + 1.0)\,\text{m}\,\text{s}^{-2}$$

(b) At $t = 2$ s, using the respective expressions, we have

$$x = 4.1\,\text{m},\ v = 4.4\,\text{m}\,\text{s}^{-1},\ a = 3.4\,\text{m}\,\text{s}^{-2}$$

Similarly, for $t = 3$ s,

$$x = 10.4\,\text{m},\ v = 8.4\,\text{m}\,\text{s}^{-1},\ a = 4.6\,\text{m}\,\text{s}^{-2}$$

(c) To find the average velocity and acceleration between $t = 2$ s and $t = 3$ s, we have $\Delta t = 1$ s, and from (b) we have $\Delta x = 6.3$ m and $\Delta v = 4.0\,\text{m}\,\text{s}^{-1}$. Thus

$$v_{\text{ave}} = \frac{\Delta x}{\Delta t} = \frac{6.3\,\text{m}}{1\,\text{s}} = 6.3\,\text{m}\,\text{s}^{-1}$$

$$a_{\text{ave}} = \frac{\Delta v}{\Delta t} = \frac{4.0\,\text{m}\,\text{s}^{-1}}{1\,\text{s}} = 4.0\,\text{m}\,\text{s}^{-2}$$

3.5 Some special motions

We now consider two important types of motion. The relations derived in this section will be used many times throughout the text.

(a) **Uniform rectilinear motion.** As we shall see in Chapter 6, this motion is produced when the forces acting on a body add to zero. When a body is in uniform rectilinear motion, its velocity v is constant. Therefore

$$a = \frac{\mathrm{d}v}{\mathrm{d}t} = 0$$

that is, there is no acceleration. From Equation 3.3, when v is constant, we have

$$x = x_0 + \int_{t_0}^{t} v \,\mathrm{d}t = x_0 + v \int_{t_0}^{t} \mathrm{d}t$$

or

$$x = x_0 + v(t - t_0) \tag{3.7}$$

In Figure 3.10(a), v is plotted as a function of t and in Figure 3.10(b) x is plotted as a function of t for uniform rectilinear motion.

(b) **Uniformly accelerated rectilinear motion.** This motion is produced when the net force acting on a body is constant. When a body is in uniformly accelerated rectilinear motion, its acceleration a is constant. Therefore, from Equation 3.6, we have

$$v = v_0 + \int_{t_0}^{t} a \,\mathrm{d}t = v_0 + a \int_{t_0}^{t} \mathrm{d}t$$

or

$$v = v_0 + a(t - t_0) \tag{3.8}$$

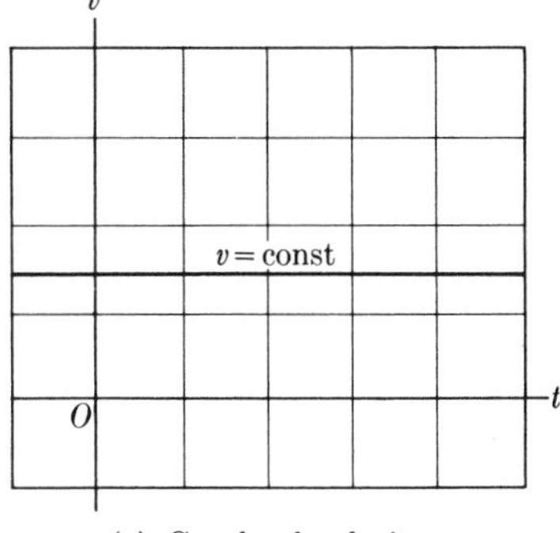

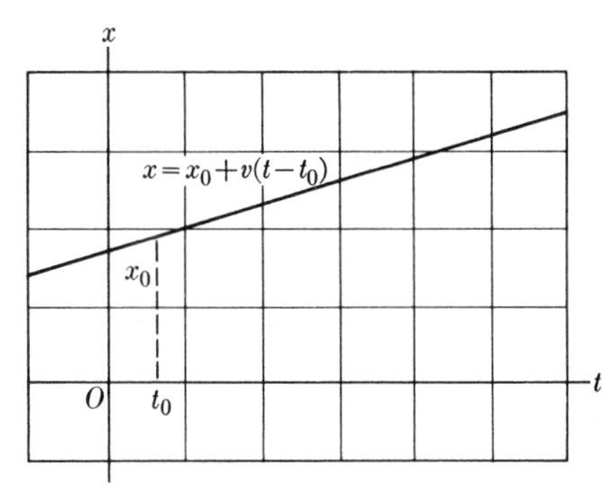

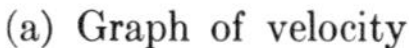
(a) Graph of velocity

(b) Graph of displacement

Figure 3.10 Graph of velocity and displacement in uniform motion.

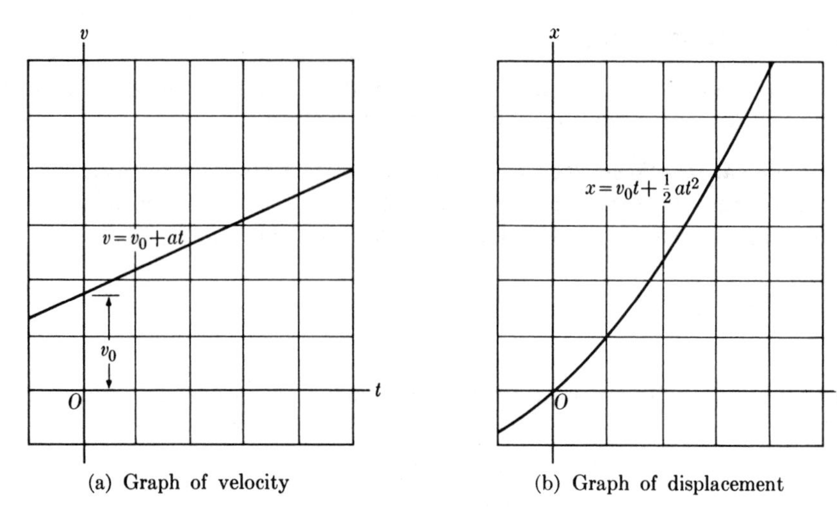

Figure 3.11 Graph of velocity and displacement in uniformly accelerated motion.

Also, from Equation 3.3 we have

$$x = x_0 + \int_{t_0}^{t} [v_0 + a(t - t_0)]\, dt$$

$$= x_0 + v_0 \int_{t_0}^{t} dt + a \int_{t_0}^{t} (t - t_0)\, dt$$

or

$$x = x_0 + v_0(t - t_0) + \tfrac{1}{2}a(t - t_0)^2 \tag{3.9}$$

Further, when we eliminate $t - t_0$ between Equations 3.8 and 3.9, we obtain the useful relation

$$v^2 = v_0^2 + 2a(x - x_0) \tag{3.10}$$

When, for simplicity, we set $t_0 = 0$ and $x_0 = 0$, Equations 3.8 and 3.9 become

$$v = v_0 + at \qquad \text{and} \qquad x = v_0 t + \tfrac{1}{2}at^2$$

Both equations have been plotted in Figure 3.11. In the special case when we also let $v_0 = 0$, Equations 3.8, 3.9 and 3.10 become

$$v = at, \qquad x = \tfrac{1}{2}at^2 \qquad \text{and} \qquad v^2 = 2ax \tag{3.11}$$

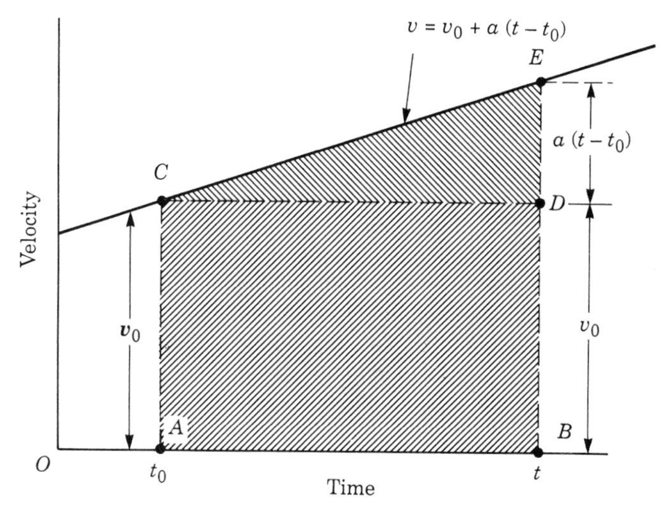

Figure 3.12 Calculation of displacement in uniformly accelerated motion.

Equations 3.11 show that the displacement of a uniformly accelerated body varies as the square of the time and that the velocity varies as the square root of the displacement. These are interesting features that serve to identify uniformly accelerated motion. In fact, using the proportionality relation $x \propto t^2$, Galileo demonstrated that freely falling bodies, as well as bodies sliding or rolling down an inclined plane, were uniformly accelerated.

Relation 3.9 can also be derived very easily by calculating the area under the graph of the velocity of a body undergoing uniform acceleration (Figure 3.12); that is,

$$\text{displacement} = x - x_0 = \text{area } ABEC = \text{area } ABCD + \text{area } CDE$$

But area $ABCD = AC \times AB = v_0(t - t_0)$ and area $CDE = \frac{1}{2}CD \times DE = \frac{1}{2}a(t - t_0)^2$. Therefore,

$$x - x_0 = v_0(t - t_0) + \tfrac{1}{2}a(t - t_0)^2$$

Table 3.1 Summary of relations for rectilinear motion

Uniform motion	$a = 0$ $v = \text{const.}$ $x = x_0 + v(t - t_0)$
Uniformly accelerated motion	$a = \text{const.}$ $v = v_0 + a(t - t_0)$ $x = x_0 + v_0(t - t_0) + \frac{1}{2}a(t - t_0)^2$ $v^2 = v_0^2 + 2a(x - x_0)$

EXAMPLE 3.6

A train is moving along a straight section of track with a velocity of $180\,\text{km}\,\text{h}^{-1}$. The braking deceleration is $2\,\text{m}\,\text{s}^{-2}$. Assuming the deceleration remains constant, at what distance from a train station should the engineer apply the brakes so that the train stops at the station? Also, how long will it take to bring the train to a halt?

▷ We are given $v_0 = 180\,\text{km}\,\text{h}^{-1} = 50\,\text{m}\,\text{s}^{-1}$ and $a = -2\,\text{m}\,\text{s}^{-2}$. Also, since the train stops, $v = 0$. Then, using Equation 3.10,

$$0 = (50\,\text{m}\,\text{s}^{-1})^2 + 2(-2.0\,\text{m}\,\text{s}^{-2})(x - x_0)$$

so that

$$x - x_0 = \frac{2500\,\text{m}^2\,\text{s}^{-2}}{4\,\text{m}\,\text{s}^{-2}} = 625\text{ m}$$

The time required to stop the train may be found by using Equation 3.8 with $v = 0$:

$$0 = 50\,\text{m}\,\text{s}^{-1} + (-2\,\text{m}\,\text{s}^{-2})(t - t_0) \qquad \text{or} \qquad t - t_0 = 25\,\text{s}$$

3.6 Free vertical motion under the action of gravity

It has been verified experimentally that when a body falls under the action of gravity through a relatively short distance of a few meters, the motion is uniformly accelerated. This acceleration is the same for all bodies, is designated by g and is called the **acceleration of gravity**. Of course, if the motion is upward, the acceleration is the same (i.e. downward) but the motion is decelerated. These statements are correct as long as we can neglect the effects due to air resistance and therefore apply to compact bodies when they move vertically through distances not larger than a few hundred meters.

When discussing vertical motion, we may take either the upward or downward direction as positive. If the upward direction is chosen as positive, we define $a = -g$, where the minus sign is due to the fact that the gravitational acceleration is always downward. Also, the velocity of the body is positive if it is moving upward and negative if it is moving downward. The value of g varies from one place on the Earth's surface to another, but it is always close to $g = 9.8\,\text{m}\,\text{s}^{-2}$. As we will see in Chapter 11, the acceleration of gravity decreases as we go above or below the Earth's surface. However, the change is very small for changes of even a few thousand meters. For example, at the top of a mountain 1000 m high, the value of g decreases by only 0.03%.

For vertical motion, with $a = -g = -9.8\,\text{m}\,\text{s}^{-2}$, the relations from Table 3.1 are

$$\text{Positive direction up} \begin{cases} v = v_0 - g(t - t_0) \\ y = y_0 + v_0(t - t_0) - \frac{1}{2}g(t - t_0)^2 \\ v^2 = v_0^2 - 2g(y - y_0) \end{cases} \qquad \textbf{(3.12)}$$

It may be verified, using Equations 3.12, that the maximum height reached by a projectile is $v_0^2/2g$ and the time required to reach that height is v_0/g.

In those cases where only downward motion has to be considered, we may prefer to choose down as the positive direction. In that case, the acceleration of a freely falling body is positive ($a = +g$). The velocity is positive as the body is falling but negative if it is going up. With this choice, the relations are

$$\text{Positive direction down} \begin{cases} v = v_0 + g(t - t_0) \\ y = y_0 + v_0(t - t_0) + \tfrac{1}{2}g(t - t_0)^2 \\ v^2 = v_0^2 + 2g(y - y_0) \end{cases} \quad \textbf{(3.13)}$$

Therefore, in discussing the vertical motion of a body, it is important to agree first on the positive direction. The decision depends on the specific situation.

EXAMPLE 3.7

A ball is thrown straight upward with a velocity of 98 m s^{-1} from the top of a building 100 m high (see Figure 3.13). Find (a) its maximum height above the ground, (b) the time required to reach it, (c) the velocity it has when it reaches the ground, and (d) the total elapsed time before the ball reaches the ground.

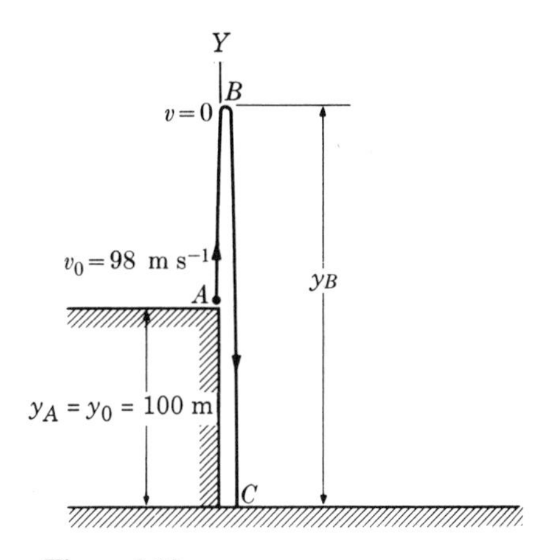

Figure 3.13

▷ Using Equations 3.12, with $t_0 = 0$, $v_0 = 98\,\text{m}\,\text{s}^{-1}$, $y_0 = y_A = 100\,\text{m}$ (the origin of coordinates C has been placed at street level), and $g = 9.8\,\text{m}\,\text{s}^{-2}$, we have at any time t,

$$v = 98\,\text{m}\,\text{s}^{-1} - (9.8\,\text{m}\,\text{s}^{-2})t$$

$$y = 100 + 98t - 4.9t^2$$

(a) At the point of maximum height $v = 0$. Thus

$$98 - 9.8t = 0 \quad \text{or} \quad t = 10\,\text{s}$$

(b) Placing this value of t in the expression for y, we have

$$y_B = 100 + 98(10) - 4.9(10)^2 = 590\,\text{m}$$

(c) To obtain the time required to reach the ground (that is, point C), we set $y_C = 0$, since C is our origin of coordinates. Then

$$0 = 100\,\text{m} + (98\,\text{m}\,\text{s}^{-1})t - (4.9\,\text{m}\,\text{s}^{-2})t^2$$

This is a second-degree (or quadratic) equation in t, whose roots are

$$t = -0.96\,\text{s} \quad \text{and} \quad t = 20.96\,\text{s}$$

The negative answer corresponds to a time previous to the throw ($t = 0$) and must be discarded, since it has no physical meaning in this problem (it may have in others). To obtain the velocity at C, we introduce the value $t = 20.96$ s in the expression for v_C, obtaining

$$v_C = 98 \text{ m s}^{-1} - (9.8 \text{ m s}^{-2}) \times (20.96 \text{ s}) = -107.41 \text{ m s}^{-1}$$

The negative sign means that the ball is moving downward. It is suggested that the student verify the results for y_B and v_C by using the third of Equations 3.12, which for this problem reads

$$v^2 = 9604 - 19.6(y - 100)$$

Also the student should solve the problem by placing the origin of coordinates at A. Then $y_0 = y_A = 0$ and $y_C = -100$ m.

EXAMPLE 3.8

A ball is thrown vertically upward with an initial velocity of 50 m s^{-1}. Two seconds later, another ball is thrown upward with the same velocity. Where and when do they meet? What is their velocity when they meet?

▷ Take the origin at the initial position of the bodies and the upward direction as positive. For the first ball, where $t_0 = 0$, we have

$$v = v_0 - gt \qquad \text{and} \qquad y = y_0 t - \tfrac{1}{2}gt^2$$

Since the second ball is thrown up at $t'_0 = 2$ s, we must write its equations as

$$v' = v_0 - g(t-2) \qquad \text{and} \qquad y' = v_0(t-2) - \tfrac{1}{2}g(t-2)^2$$

When they meet, $y' = y$ giving

$$v_0 t - \tfrac{1}{2}gt^2 = v_0(t-2) - \tfrac{1}{2}g(t-2)^2$$

Canceling common terms on both sides, we get

$$0 = -2v_0 - 2g + 2gt = 2(-v_0 - g + gt)$$

Therefore $t = v_0/g + 1 = 50/9.8 + 1 = 6.10$ s. The height at which they meet can now be found using either of the expressions for y or y'. The result is 122.7 m.

The respective velocities, found by using the first equation of each set, are $v = -9.78$ m s^{-1} and $v' = +9.82$ m s^{-1}. In other words, within the accuracy of our calculations, *both* bodies are moving with the same speed (9.8 m s^{-1}) but in opposite directions. The first body has already reached its maximum height, $y_{max} = v_0^2/2g = 127.6$ m, at the time $t = v_0/g = 5.10$ s and is returning while the second body is still going up.

This example shows the importance in maintaining a consistent orientation of the axes throughout a problem.

3.7 Vector representation of velocity and acceleration in rectilinear motion

When a particle moves in a straight line, its position can be expressed by a vector $\boldsymbol{x} = \boldsymbol{OP}$. This vector gives the displacement of the particle relative to the origin O. Similarly, velocity in rectilinear motion is represented by a vector whose length is given by Equation 3.2 and whose direction coincides with that of the motion (Figure 3.14). Since rectilinear motion can only be in either of two opposite directions, the velocity vector may be in the direction $+OX$ or $-OX$. This may also be determined by the sign of $\mathrm{d}x/\mathrm{d}t$.

Acceleration is represented by a vector whose magnitude is given by Equation 3.5. The vector points in the direction of OX or the opposite, depending on the direction of *change* of the velocity, that is, whether $\mathrm{d}v/\mathrm{d}t$ is positive or negative.

If $\boldsymbol{i}$ is a unit vector in the positive or $+OX$ direction of the X-axis, we may write in vector form

$$\boldsymbol{x} = \boldsymbol{i}x, \qquad \boldsymbol{v} = \boldsymbol{i}v = \boldsymbol{i}\frac{\mathrm{d}x}{\mathrm{d}t} \qquad \text{and} \qquad \boldsymbol{a} = \boldsymbol{i}a = \boldsymbol{i}\frac{\mathrm{d}v}{\mathrm{d}t} \tag{3.14}$$

The vectors $\boldsymbol{x}$, $\boldsymbol{v}$ and $\boldsymbol{a}$ point along the direction of $\boldsymbol{i}$ or in the opposite direction, depending on the signs of x, $\mathrm{d}x/\mathrm{d}t$ and $\mathrm{d}v/\mathrm{d}t$, respectively. The motion is accelerated or retarded, depending on whether $\boldsymbol{v}$ and $\boldsymbol{a}$ point in the same direction or opposite directions (Figure 3.14). Therefore, if v and a have the same sign, the motion is accelerated; if the signs are opposite, the motion is retarded.

For the case of vertical motion of a body under the action of gravity, the acceleration $\boldsymbol{g}$ is always downward (Figure 3.15). The velocity is a vector pointing either up or down, depending on the direction in which the body is moving. Thus, *free upward motion is always decelerated* while *free downward motion is accelerated.*

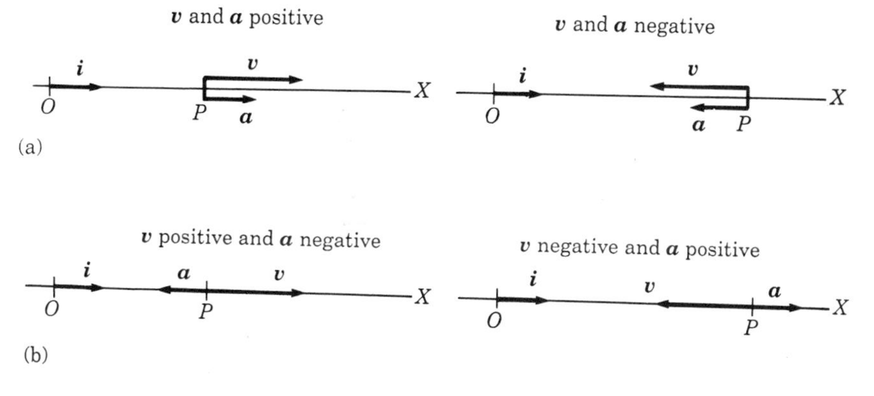

Figure 3.14 Vector relation between velocity and acceleration in rectilinear motion. The position of the particle, in each case, is given by the vector $\boldsymbol{x} = \boldsymbol{OP}$. (a) Accelerated motion: $\boldsymbol{v}$ and $\boldsymbol{a}$ are in the same direction. (b) Retarded motion: $\boldsymbol{v}$ and $\boldsymbol{a}$ are in opposite directions.

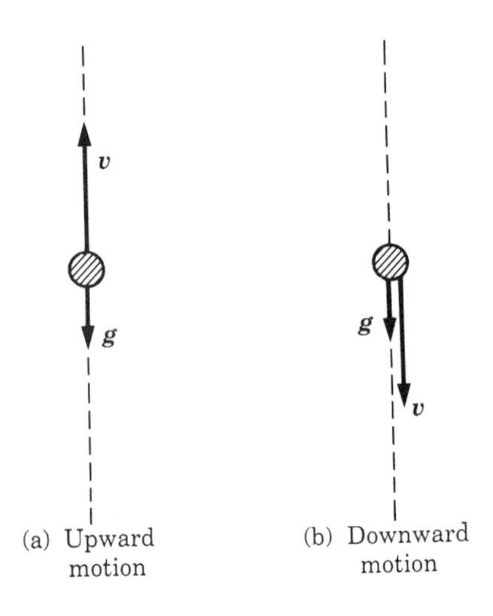

Figure 3.15 (a) Upward motion: the velocity and acceleration are in opposite directions; (b) downward motion: the velocity and acceleration are in the same direction.

3.8 Composition of velocities and accelerations

In many instances a body is subject to different agents, each giving the body a different velocity and/or acceleration. The resulting motion is obtained by combining the respective velocities and accelerations, according to the rules of vector addition.

Suppose, for example, that there is a boat moving with a velocity $\boldsymbol{V}_B$ relative to the water (Figure 3.16). If the water is still, $\boldsymbol{V}_B$ is also the velocity of the boat

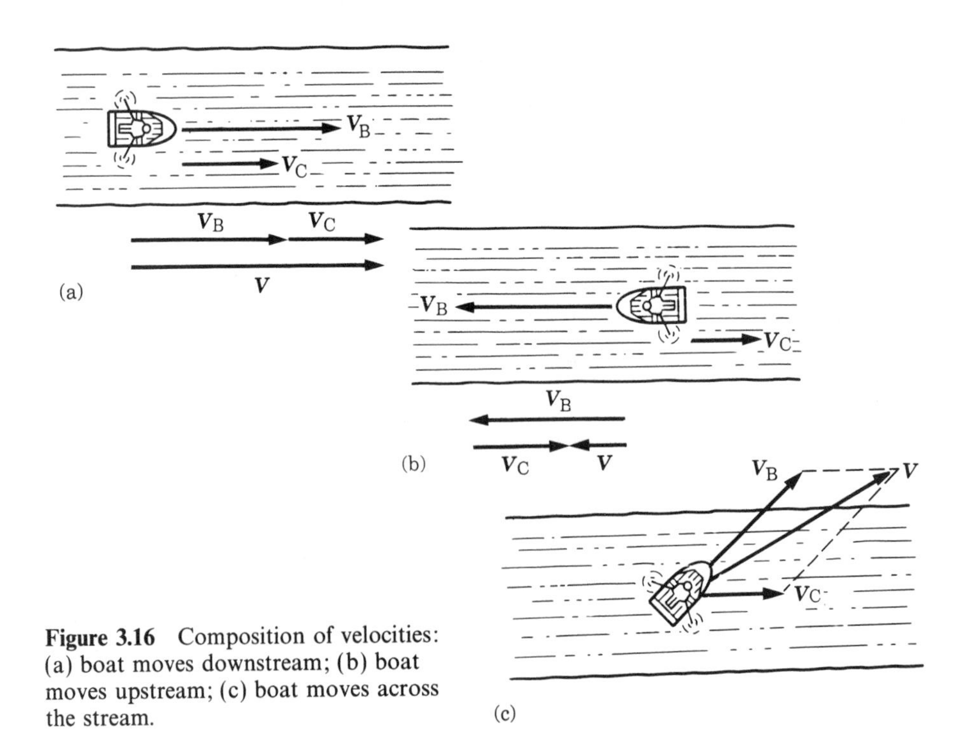

Figure 3.16 Composition of velocities: (a) boat moves downstream; (b) boat moves upstream; (c) boat moves across the stream.

as measured by an observer on the shore. But if the water is flowing at a certain rate, this introduces a drift factor that must be added to the boat's velocity. Thus the resultant velocity of the boat, as measured by an observer on the shore, is the vector sum of the velocity of the boat V_B relative to the water and the drift velocity V_C due to the water current; that is,

$$V = V_B + V_C$$

Similar logic applies to objects moving through the air, such as airplanes.

EXAMPLE 3.9

A motorboat is heading due north at $15\,\text{km}\,\text{h}^{-1}$ in a place where the current is $5\,\text{km}\,\text{h}^{-1}$ in the direction S 70° E. Find the resultant velocity of the boat relative to the shore.

▷ This problem is solved graphically in Figure 3.17, where V_B is the boat velocity relative to the water, V_C the current or drift velocity, and V the resultant velocity relative to the shore obtained from

$$V = V_B + V_C$$

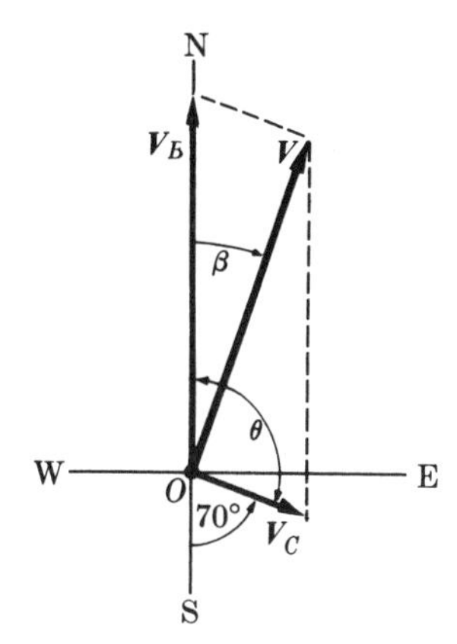

Figure 3.17

Analytically, since $\theta = 180° - 70° = 110°$, the law of cosines, Equation A.3, lets us find the magnitude of V as

$$V = [15^2 + 5^2 + 2(15)(5)\cos 110°]^{1/2} = 14.1\,\text{km}\,\text{h}^{-1}$$

To obtain its direction, we apply the law of sines, Equation A.4,

$$\frac{V}{\sin\theta} = \frac{V_C}{\sin\beta} \quad \text{or} \quad \sin\beta = \frac{V_C \sin\theta}{V} = 0.332$$

or $\beta = 19.4°$ and the resultant motion is in the direction N 19.4°E.

EXAMPLE 3.10

A racing boat is headed N 30° E at $25\,\text{km}\,\text{h}^{-1}$. The boat is in a place where the current gives it a resultant motion of $30\,\text{km}\,\text{h}^{-1}$ in the direction N 50° E. Find the velocity of the current.

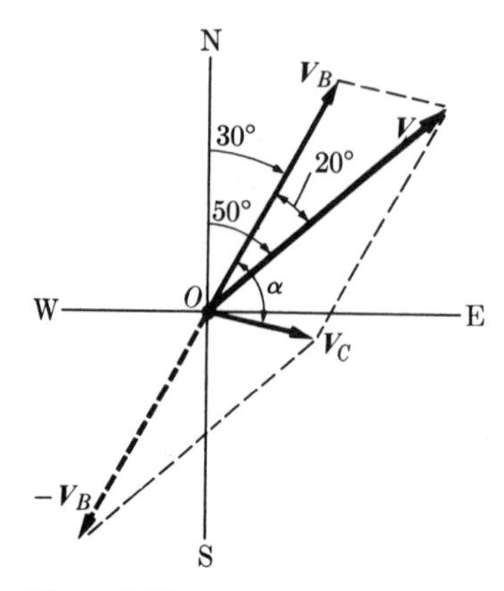

Figure 3.18

▷ Designating the velocity of the boat by V_B, the velocity of the current by V_C, and the resultant velocity by V, we have $V = V_B + V_C$, so that $V_C = V - V_B$. The vectors V and V_B have been drawn in Figure 3.18, as well as the difference between them, which is V_C. To compute V_C, we note that the angle between V and V_B is 20°. Thus, using Equation A.6,

$$\begin{aligned} V_C &= [30^2 + 25^2 - 2(30)(25)\cos 20°]^{1/2} \\ &= 10.74 \text{ km h}^{-1} \end{aligned}$$

To obtain the direction of V_C, we first obtain the angle α between V_C and V_B, using Equation A.4

$$\frac{V}{\sin\alpha} = \frac{V_C}{\sin 20°} \quad \text{or} \quad \sin\alpha = \frac{V\sin 20°}{V_C} = 0.955$$

giving $\alpha = 72°45'$. Therefore the angle with the NS-axis is $72°45' + 30° = 102°45'$, and the direction of V_C is S 77°15′ E.

EXAMPLE 3.11

Acceleration of a body sliding along an inclined plane.

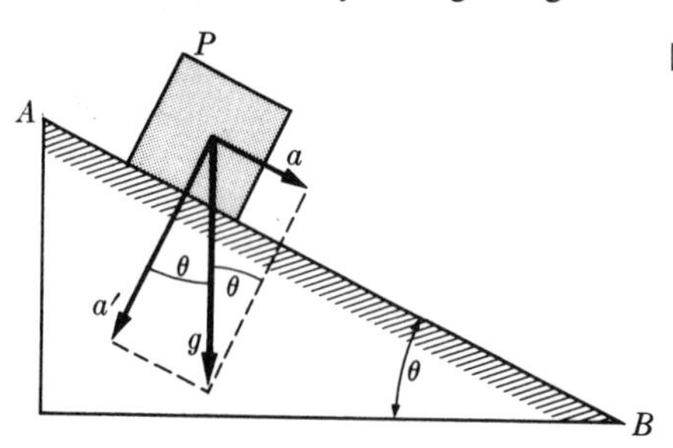

Figure 3.19 Acceleration along an inclined plane.

▷ Let P (Figure 3.19) be a body sliding down plane AB without friction. The plane AB is inclined at an angle θ. If the plane were not there, the body would fall freely along the vertical with the acceleration of gravity $g = 9.8 \text{ m s}^{-2}$. The components of $\boldsymbol{g}$ parallel and perpendicular to the plane are $a = g \sin\theta$ and $a' = g\cos\theta$. The component a gives the acceleration of the body along the plane. The component a' does not result in any motion because the plane resists motion perpendicular to it (i.e. the sum of the forces normal to the plane is zero, as we shall see in Chapter 7). Note that the smaller the inclination of the plane from the horizontal, the smaller the acceleration along the plane.

3.9 Relative motion

Consider two objects A and B and an observer O, using as frame of reference the axes XYZ (Figure 3.20). The velocities of A and B relative to O are V_A and V_B. Then the velocity of B relative to A is given by

$$V_{BA} = V_B - V_A \tag{3.15}$$

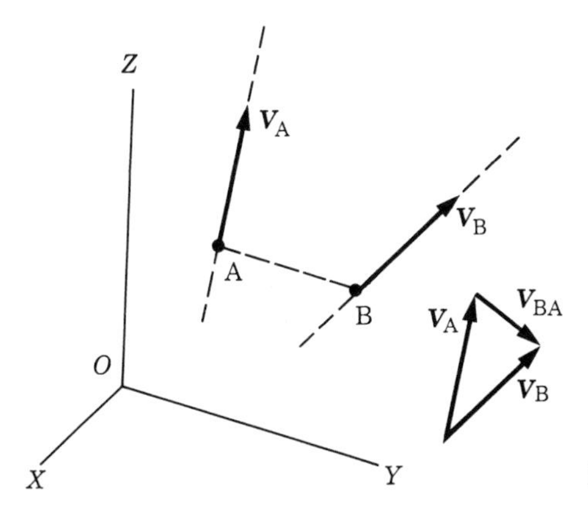

Figure 3.20 Definition of relative velocity.

and the velocity of A relative to B is given by

$$V_{AB} = V_A - V_B \tag{3.16}$$

Therefore, to obtain the **relative velocity** of two bodies, their velocities, relative to the observer, are subtracted. Note that

$$V_{BA} = -V_{AB} \tag{3.17}$$

In other words, the velocity of B relative to A is equal and opposite to the velocity of A relative to B.

Similarly, if $\boldsymbol{a}_A$ and $\boldsymbol{a}_B$ are, respectively, the accelerations of B and A relative to O, then

$$\boldsymbol{a}_{BA} = \boldsymbol{a}_B - \boldsymbol{a}_A \qquad \text{and} \qquad \boldsymbol{a}_{AB} = \boldsymbol{a}_A - \boldsymbol{a}_B \tag{3.18}$$

are, respectively, the accelerations of B relative to A and of A relative to B. Therefore, to get the **relative acceleration** of two bodies, their accelerations, relative to the observer, are subtracted.

EXAMPLE 3.12

An airplane A (Figure 3.21) flies due North at 300 km h^{-1} relative to the ground. At the same time another plane B flies in the direction N 60° W at 200 km h^{-1} relative to the ground. Find the velocity of A relative to B and of B relative to A.

▷ In Figure 3.21(a) and (b), the velocities of planes A and B relative to the ground are represented. On the right, in Figure 3.21(c) we have the velocity of A relative to B, that is, $V_{AB} = V_A - V_B$, and of B relative to A, that is, $V_{BA} = V_B - V_A$. We may note that $V_{AB} = -V_{BA}$, as it should be according to Equation 3.17.

To compute V_{AB}, we use Equation A.6, noting that the angle θ between V_A and V_B is 60°. Thus

$$V_{AB} = [300^2 + 200^2 - 2(300)(200)\cos 60°]^{1/2} = 264.6 \text{ km h}^{-1}.$$

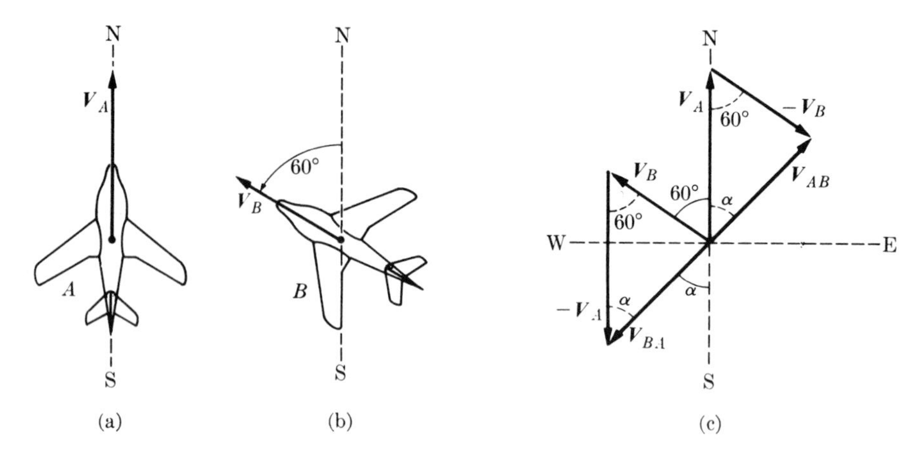

Figure 3.21

To obtain the direction of V_{AB}, we use the law of sines, Equation A.4:

$$\frac{V_B}{\sin\alpha} = \frac{V_{AB}}{\sin 60^\circ} \quad \text{or} \quad \sin\alpha = \frac{V_B \sin 60^\circ}{V_{AB}} = 0.654$$

giving $\alpha = 40.7^\circ$. Therefore, to a passenger in plane B it seems as if plane A moves at 264.6 km h^{-1} in the direction N 40.7° E. The relative velocity V_{BA} has the same magnitude, 264.6 km h^{-1}, but the opposite direction, S 40.7° W.

Note 3.1 The age of the universe

A problem that has always fascinated philosophers and scientists is how the universe began, if it had a beginning, and how long ago it happened. Much has been learned in the last few decades about the conditions of the early universe. Some aspects will be examined later on in this book (see Sections 11.12 and 41.9). At this time it is enough to say that current thinking assumes the universe began as a singularity in space and time. The singularity has been named the **Big Bang**, at which time everything was condensed into a very small space where there were both extremely high densities and exceptionally high temperatures (actually, the singularity was the beginning of space and time as we know them). Since that time the universe has been in continuous expansion, so that the average density and temperature have been decreasing continuously.

The rate of expansion of the universe (that is, the rate at which galaxies are receding from each other) is given by **Hubble's Law**, discovered in 1929 by the astronomer Edwin Hubble (1889–1953) by analyzing the spectra of distant galaxies (see Section 28.14). This law states that the rate of separation of any two galaxies in the universe is directly proportional to their separation. Thus, if we have two galaxies a distance R apart, their *present* relative velocity of separation is

$$v = HR$$

where H is a proportionality factor called **Hubble's parameter**. The currently accepted value for Hubble's parameter (with an error of $\pm 30\%$) is

$$H = 22 \text{ km s}^{-1} \text{ MLyr}^{-1} = 2.32 \times 10^{-18} \text{ s}^{-1}$$

One MLyr (or mega-light year) is 9.46×10^{21} m (See Problem 2.10).

We are not sure whether H is a constant or is changing with time. For sake of argument, let us assume that it *is* constant. Then we may define a time t_H equal to R/v, which corresponds to the time when two galaxies have reached a separation of R. Using Hubble's law, we find that

$$t_H = \frac{R}{v} = \frac{1}{H} = 4.3 \times 10^{17} \text{ s} = 1.36 \times 10^{10} \text{ years}$$

This time, which is called **Hubble's time**, is considered to be a fair estimate of the order of magnitude of the time that has elapsed since the Big Bang and, thus, of the age of the universe. Other estimates, ranging between 1.0 and 2.0×10^{10} years have been made. It is most probably about 1.5×10^{10} years. We return to this question in more detail in Chapters 11 and 41.

From the Hubble time we can estimate the 'size' of the observable universe. Since light (i.e. electromagnetic radiation) travels with a velocity of $c = 3 \times 10^8 \text{ m s}^{-1}$, the farthest distance, called **Hubble's distance**, that can be observed from Earth is

$$d_H = ct_H = (3 \times 10^8 \text{ m s}^{-1}) \times (4.3 \times 10^{17} \text{ s}) = 1.3 \times 10^{26} \text{ m}$$

or about 1.3×10^4 MLyr. This distance is also called the **horizon** of the universe.

The farthest, and oldest, object (observed in 1989) is near the constellation of Ursa Major. It is a star-like object, called a **quasar**, designated PC-1158 + 4635, and is estimated to be at a distance of 1.4×10^4 MLyr. Therefore it is very close to the horizon of the universe. Its light has taken about 1.4×10^{10} years to reach us and therefore we see it as it was about 0.1×10^{10} years after the Big Bang. Information obtained from its study is expected to provide some clues about the early universe. The term 'quasar' is short for *quas*i-stell*ar* radio source.

QUESTIONS

3.1 What is meant by the expression 'motion is relative'?

3.2 Why must an observer define a frame of reference for analyzing the motion of bodies?

3.3 A body is thrown vertically upward with a speed v_0. What are the velocity and acceleration when (a) the body reaches its highest point and (b) the body returns to the Earth's surface?

3.4 Draw the vectors representing the velocity and the acceleration of (a) a freely falling body and (b) a body moving vertically upward.

3.5 How do the velocity and the acceleration of a body change when it moves freely in the (a) upward and (b) downward directions?

3.6 How can one experimentally differentiate uniform motion from accelerated motion?

3.7 A body is thrown vertically upward. After it reaches a certain height, it begins to fall down. Make a diagram showing the velocity and the acceleration when it passes through the same point going up and down.

3.8 Why are velocity and acceleration vector quantities?

3.9 A bird is flying horizontally and in a straight line with constant velocity relative to the ground. Under what conditions would the bird appear stationary with respect to an observer in a car speeding along a road? How could the bird seem to fly backwards?

3.10 Explain why the velocity of A relative to B and the velocity of B relative to A have the same magnitude but opposite directions.

3.11 Meteorites arrive at the top of the Earth's atmosphere with a velocity of the order of 10^4 m s^{-1}. What is their velocity in km h^{-1}?

PROBLEMS

3.1 A body starts moving with an initial velocity of 3 m s^{-1} and a constant acceleration of 4 m s^{-2} in the same direction as the velocity. (a) What is the velocity of the body and the distance covered at the end of 7 s? (b) Solve the same problem for a body whose acceleration is in the direction opposite to that of the velocity. (c) In each case write the expression for the velocity and the displacement as a function of time.

3.2 An airplane, in taking off, covers a 600 m path in 15 s. (a) Assuming a constant acceleration, calculate the take-off velocity. (b) Also calculate the acceleration in m s^{-2}.

3.3 An automobile, starting from rest, reaches 60 km h^{-1} in 15 seconds. (a) Calculate the average acceleration in m min^{-2} and the distance moved. (b) Assuming that the acceleration is constant, how many more seconds will it take for the car to reach 80 km h^{-1}? (c) What has been the total distance covered?

3.4 A car starts from rest and moves with an acceleration of 1 m s^{-2} for 10 s. The motor is then turned off and the car is allowed to decelerate, due to friction and air drag, for 10 s at a rate of 5×10^{-2} m s^{-2}. Then the brakes are applied and the car is brought to rest in 5 more seconds. (a) Make a plot of a, v and x versus t. (b) Calculate the total distance traveled by the car.

3.5 A car starts from rest with an acceleration of 4 m s^{-2} for 4 s. During the next 10 s it moves with uniform motion. The brakes are then applied and the car decelerates at a rate of 8 m s^{-2} until it stops. (a) Make a plot of the velocity versus time. (b) Verify that the area bounded by the velocity curve and the time axis measures the total distance traveled.

3.6 An automobile is moving at the rate of 45 km h^{-1} when a red light flashes on at an intersection. If the reaction time of the driver is 0.7 s, and the car decelerates at the rate of 2 m s^{-2} as soon as the driver applies the brakes, calculate how far the car travels from the time the driver notices the red light until the car is brought to a stop. ('Reaction time' is the interval between the time the driver notices the light and the time the brakes are applied.)

3.7 Two cars, A and B, are traveling in the same direction with velocities v_A and v_B, respectively. When car A is a distance d behind car B, the brakes on A are applied, causing a deceleration a. Verify that to prevent a collision between A and B, it is necessary that $v_A - v_B < (2ad)^{1/2}$. (Hint: use the initial position of A as the origin of coordinates.)

3.8 Two bodies, A and B, are moving in the same direction. When $t = 0$, their respective velocities are 1 m s^{-1} and 3 m s^{-1}, and their respective accelerations are 2 m s^{-2} and 1 m s^{-2}. If car A is 1.5 m ahead of car B at $t = 0$, calculate when and where they will be side by side.

3.9 A body is moving along a straight line according to the law $x = 16t - 6t^2$, where x is measured in meters and t in seconds. (a) Find the position of the body at $t = 1$ s. (b) At what times does the body pass the origin? (c) Calculate the average velocity for the time interval $0 < t < 2$ s. (d) Find the general expression for the average velocity for the interval $t_0 < t < (t_0 + \Delta t)$. (e) Calculate the instantaneous velocity at any given time. (f) Calculate the velocity at $t = 0$. (g) At what times and position(s) will the body be at rest? (h) Find the general expression for the average acceleration for the time interval $t_0 < t < (t_0 + \Delta t)$. (i) Find the general expression for the acceleration at any time. (j) At what time(s) is the acceleration zero? (k) Plot, on a single set of axes, x versus t, v versus t, and a versus t. (l) At what time(s) is the motion accelerated and at what time(s) is it retarded?

3.10 The acceleration of a body moving along a straight line is given by $a = 4 - t^2$, where a is in

$m\,s^{-2}$ and t is in seconds. (a) Find the expressions for the velocity and displacement as functions of time, given that when $t = 3$ s, $v = 2\ m\,s^{-1}$ and $x = 9$ m. (b) Plot a, v and x as functions of time. (c) When is the motion accelerated and when is it retarded?

3.11 A body is in rectilinear motion with an acceleration given by $a = 32 - 4v$. The initial conditions are $x = 0$ and $v = 4$ at $t = 0$. Find (a) v as a function of t, (b) x as a function of t, and (c) x as a function of v.

3.12 The position of a moving body in terms of time is given in Figure 3.22. Indicate (a) where the motion is in the positive or negative x-direction, (b) when the motion is accelerated or retarded, (c) when the body passes through the origin, and (d) when the velocity is zero. Also make a sketch of the velocity as a function of time. Estimate from the graph the average velocity between (e) $t = 1$ s and $t = 3$ s, (f) $t = 1$ s and $t = 2.2$ s, (g) $t = 1$ s and $t = 1.8$ s.

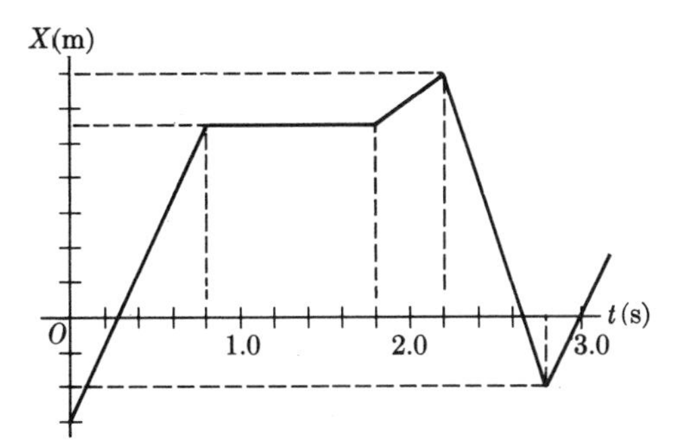

Figure 3.22

3.13 A stone falls from a balloon that is descending at a uniform rate of $12\ m\,s^{-1}$. (a) Calculate the velocity and the distance traveled by the stone after 10 s. (b) Solve the same problem for the case of a balloon rising uniformly at $12\ m\,s^{-1}$.

3.14 A stone is thrown upward from the bottom of a well 48 m deep with an initial velocity of $73.5\ m\,s^{-1}$. (a) Calculate the amount of time it will take the stone to reach the edge of the well, and its velocity. (b) Discuss the possible answers.

3.15 A man standing at the top of a building throws a ball vertically upward with a velocity of $12.25\ m\,s^{-1}$. The ball reaches the ground 4.25 s later. (a) What is the maximum height reached by the ball? (b) How high is the building? (c) With what velocity does the ball reach the ground?

3.16 A falling body travels 65.1 m in the last second of its motion. Assuming that the body started from rest, determine (a) the altitude from which the body fell and (b) how long it took to reach the ground.

3.17 One body is dropped while a second body, at the same instant, is thrown downward with an initial velocity of $1\ m\,s^{-1}$. When will the distance between them be 18 m?

3.18 Two bodies are thrown vertically upward, with the same initial velocity of $98\ m\,s^{-1}$, but 4 s apart. (a) How long after the first one is thrown will they meet? (b) How far above (below) the initial position will they meet?

3.19 A stone is dropped from the top of a building. The sound of the stone hitting the ground is heard 6.5 s later. If the velocity of sound is $340\ m\,s^{-1}$, calculate the height of the building.

3.20 Two trains, A and B, are running on parallel tracks at $70\ km\,h^{-1}$ and $90\ km\,h^{-1}$, respectively. Calculate the relative velocity of B with respect to A, when: (a) they move in the same direction, (b) they move in opposite directions.

3.21 Solve the previous problem if the tracks are set at an angle of 60° with respect to each other.

3.22 A train leaves a city A at 12 noon for city B 400 km away, and maintains a constant velocity of $100\ km\,h^{-1}$. Another train leaves city B at 2:00 p.m. and maintains a constant speed of $70\ km\,h^{-1}$. Determine the time at which the trains pass and the distance to city A if (a) the second train heads toward city A, and (b) the second train heads away from city A.

3.23 A person driving through a rainstorm at $80\ km\,h^{-1}$ observes that the raindrops make tracks on the side windows that have an angle of 80° with the vertical. When the person stops the car, the rain is observed to fall vertically. Calculate the relative velocity of the rain with respect to the car (a) when it is still, and (b) when it is moving at $80\ km\,h^{-1}$.

3.24 Two cars moving along perpendicular roads are traveling north and east, respectively. (a) If their velocities with respect to the ground are $60\ km\,h^{-1}$ and $80\ km\,h^{-1}$, calculate their relative velocity. (b) Does the relative velocity depend on the position of the cars on their respective roads? (c) Repeat the problem, assuming that the second car moves west.

3.25 A boat is moving in the direction N 60° W at $4.0\ km\,h^{-1}$ relative to the water. The current is in such a direction that the resultant motion relative to the Earth is due west at $5.0\,\text{km}\,\text{h}^{-1}$. Calculate the velocity and direction of the current with respect to the Earth.

3.26 The velocity of a speedboat in still water is $55\,\text{km}\,\text{h}^{-1}$. The driver wants to go to a point located 80 km away, at S 20° E. The current is very strong at $20\,\text{km}\,\text{h}^{-1}$ in the direction S 70° W. (a) Draw vectors that show the current direction and magnitude, the speedboat's relative velocity and direction and the resultant velocity and direction of S 20° E. (b) Calculate which direction the speedboat should be headed so that it travels in a straight line. (c) Determine the total time for the trip.

3.27 Two places, *A* and *B*, are 1 km apart and located on the bank of a rectilinear (perfectly straight) river section. A man goes from *A* to *B* and back to *A* in a rowboat at $4\,\text{km}\,\text{h}^{-1}$ relative to the river. Another man walks along the bank from *A* to *B* and back again at $4\,\text{km}\,\text{h}^{-1}$. If the river flows at $2\,\text{km}\,\text{h}^{-1}$, calculate the time taken by each man to make the round trip.

3.28 Using the data of the previous problem, determine the speed of the river so that the time difference for the two round trips is 6 min.

3.29 A river flows due north with a velocity of $3\,\text{km}\,\text{h}^{-1}$. A boat is going east with a velocity relative to the water of $4\,\text{km}\,\text{h}^{-1}$. (a) Calculate the velocity of the boat relative to the Earth. (b) If the river is 1 km wide, calculate the time necessary for a crossing. (c) What is the northward deviation of the boat when it reaches the other side of the river?

3.30 A river is 1 km wide. The current is $2\,\text{km}\,\text{h}^{-1}$. Determine the time it would take to row a boat directly across the river and back again. Compare this time with the time it would take to row 1 km upstream and back again. The rowboat moves with a velocity of $4\,\text{km}\,\text{h}^{-1}$ with respect to the water.

3.31 Consider a bullet that has a muzzle velocity of $245\,\text{m}\,\text{s}^{-1}$ fired from the tail gun of an airplane moving horizontally with a velocity of $215\,\text{m}\,\text{s}^{-1}$. Describe the motion of the bullet (a) in the coordinate system attached to the Earth, (b) in the coordinate system attached to the plane. (c) Calculate the angle at which the gunner must point the gun so that the velocity of the bullet has no horizontal component in the Earth's coordinate system. (Assume for simplicity that the Earth is flat and 'motionless'.)

3.32 A train passes through a station at $30\,\text{m}\,\text{s}^{-1}$. A ball is rolling along the floor of the train with a velocity of $15\,\text{m}\,\text{s}^{-1}$ directed (a) along the direction of the train's motion, (b) in the opposite direction and (c) perpendicular to the motion. Find, in each case, the velocity of the ball relative to an observer standing on the station platform.

4 Curvilinear motion

Stroboscopic photograph of a bouncing golf ball, showing parabolic trajectories after each bounce. Successive images are separated by equal time intervals. Each peak in the trajectories is lower than the preceding one because of energy loss during the 'bounce' or collision with the horizontal surface. (© Estate of Harold Edgerton, Courtesy of Palm Press, Inc.)

4.1 Introduction

Most bodies move along curved lines describing the most diverse paths or trajectories rather than along a straight line. A ball, after being hit by a bat or kicked with a foot, follows a path that is nearly parabolic. An artificial satellite describes an almost circular (or, rather, elliptical) path around the Earth. A train and an airplane often change direction during their motion. And birds and insects follow seemingly unpredictable paths.

In curvilinear motion we must first decide how to determine the position of the moving body. If XYZ is the frame of reference used by observer O (Figure 4.1), the position of P is determined by its coordinates x, y, z. If we introduce the unit vectors $\boldsymbol{i}$, $\boldsymbol{j}$, $\boldsymbol{k}$ along the three coordinate axes, we define the **position vector** of P as

$$\boldsymbol{r} = \boldsymbol{OP}$$

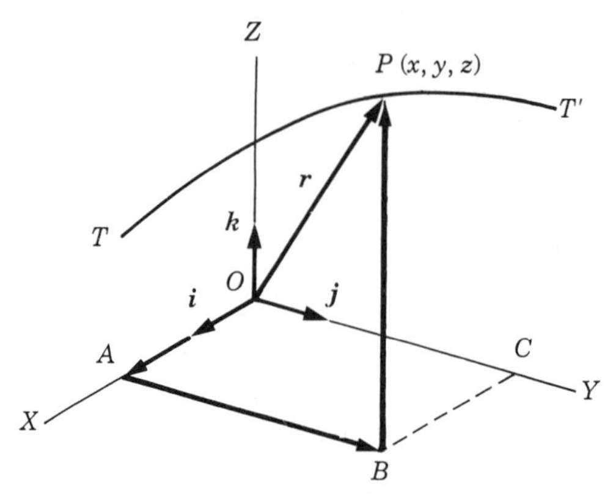

Figure 4.1 Position vector in space.

Vector $\boldsymbol{OP}$ is the sum of vectors $\boldsymbol{OA} = \boldsymbol{i}x$, $\boldsymbol{AB} = \boldsymbol{j}y$ and $\boldsymbol{BP} = \boldsymbol{k}z$. Therefore, the position vector of P can be written as

$$\boldsymbol{r} = \boldsymbol{i}x + \boldsymbol{j}y + \boldsymbol{k}z \tag{4.1}$$

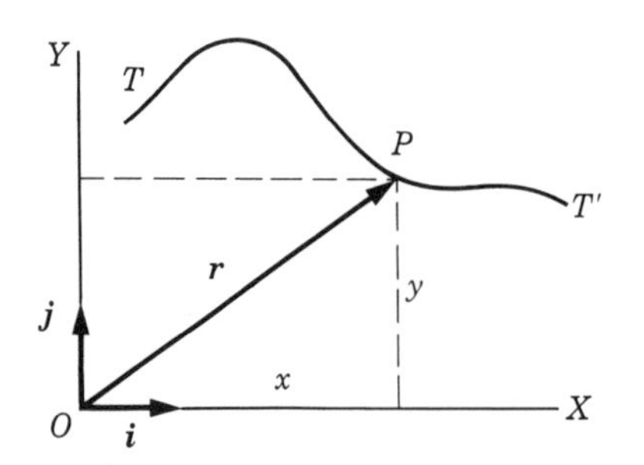

Figure 4.2 Position vector in a plane.

It is clear that as P moves, the position vector (as well as the three coordinates) changes. Thus, we say that $\boldsymbol{r}$ is a vector function of time, $\boldsymbol{r}(t)$.

When the motion is in a plane, we can place the X- and Y-axes on the plane so that only two coordinates, x and y, are necessary to fix the position of P (Figure 4.2). Clearly, in this case

$$\boldsymbol{r} = \boldsymbol{i}x + \boldsymbol{j}y \tag{4.2}$$

4.2 Curvilinear motion: velocity

Consider a particle describing a curvilinear path C, as illustrated in Figure 4.3. At time t the particle is at point P, given by position vector $\boldsymbol{r} = \boldsymbol{OP}$. At a later time, t', the particle will be at P' with $\boldsymbol{r}' = \boldsymbol{OP}'$. Although the particle has moved

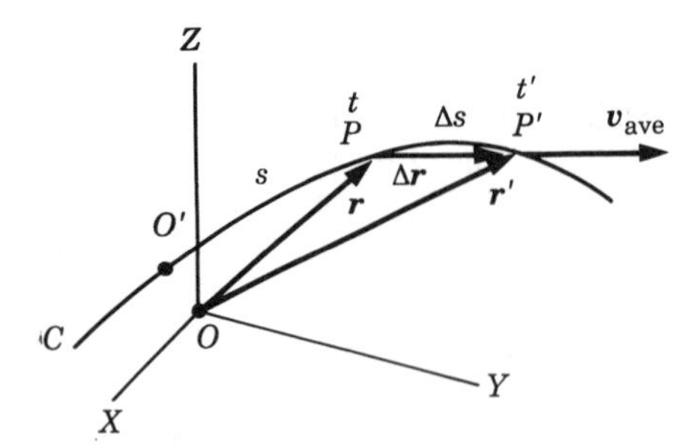

Figure 4.3 Displacement and average velocity in curvilinear motion.

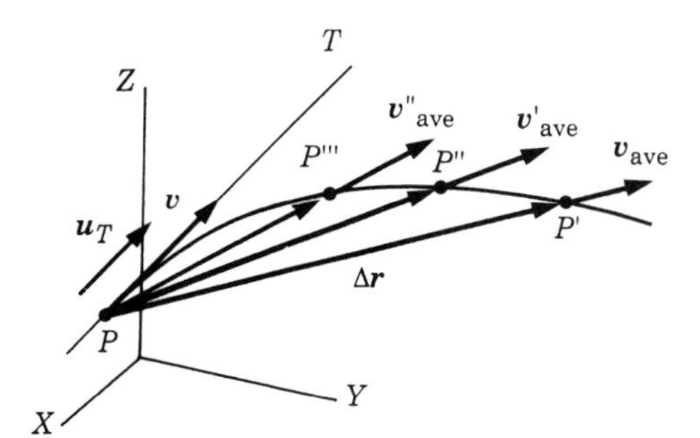

Figure 4.4 The velocity is tangent to the path in curvilinear motion.

along the arc $PP' = \Delta s$, the **displacement** in the time interval $\Delta t = t' - t$ is the vector

$$\boldsymbol{PP'} = \Delta \boldsymbol{r} = \boldsymbol{r}' - \boldsymbol{r}$$

The average velocity, also a vector, is defined by

$$\boldsymbol{v}_{ave} = \frac{\Delta \boldsymbol{r}}{\Delta t} \tag{4.3}$$

The average velocity is represented by a vector parallel to the displacement $\Delta \boldsymbol{r}$, that is, along the line PP'. To compute the instantaneous velocity we must, as in the case of rectilinear motion, make Δt very small. That is,

$$\boldsymbol{v} = \lim_{\Delta t \to 0} \boldsymbol{v}_{ave} = \lim_{\Delta t \to 0} \frac{\Delta \boldsymbol{r}}{\Delta t} \tag{4.4}$$

Now, as Δt approaches zero, point P' approaches point P, as indicated by the points P'', P''', ... in Figure 4.4. During this process the displacement $\Delta \boldsymbol{r}$ changes continuously in magnitude *and* direction, and so does the average velocity. In the limit, when P' is very close to P, the displacement practically coincides in direction with the tangent, T, to the path at P. In curvilinear motion, therefore, the instantaneous velocity is a vector *tangent* to the path, and is given by

$$\boldsymbol{v} = \frac{d\boldsymbol{r}}{dt} \tag{4.5}$$

where $d\boldsymbol{r}$ is the small displacement along the tangent during the small interval dt. Relation 4.5 shows that:

> *the velocity of a particle in curvilinear motion is equal to the time rate of change of the position vector of the particle.*

and is represented by a vector tangent at each point of the path. In Figure 4.5 the velocity vector is shown at different positions along the path.

Since $\boldsymbol{r} = \boldsymbol{i}x + \boldsymbol{j}y + \boldsymbol{k}z$, where x, y and z are the coordinates of the moving particle, the velocity is the vector

$$\boldsymbol{v} = \boldsymbol{i}\frac{dx}{dt} + \boldsymbol{j}\frac{dy}{dt} + \boldsymbol{k}\frac{dz}{dt} \tag{4.6}$$

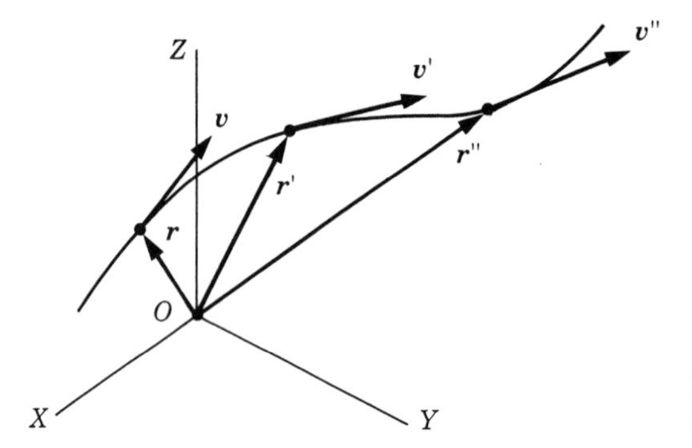

Figure 4.5 Velocity at three different positions.

The components of the velocity along the X-, Y- and Z-axes are

$$v_x = \frac{dx}{dt}, \qquad v_y = \frac{dy}{dt}, \qquad v_z = \frac{dz}{dt}$$

Let O' (Figure 4.3) be an arbitrary reference point on the path. Then $s = O'P$ gives the position of the particle as measured by the displacement *along the curve.* As in the rectilinear case, s may be positive or negative, depending on which side of O' the particle is on. When the particle moves from P to P', the displacement Δs along the curve is given by the length of the arc PP'. Then

$$v = \lim_{\Delta t \to 0} \frac{\Delta s}{\Delta t} = \frac{ds}{dt} \tag{4.7}$$

is the magnitude of the velocity, in accordance with our previous definition of velocity given in Equation 3.2. In Equation 4.7, ds is the displacement along the curvilinear path in the time dt. So, ds plays the same role in curvilinear motion as dx does in rectilinear motion.

If we introduce the unit vector $\boldsymbol{u}_T$, tangent to the path, then $d\boldsymbol{r} = \boldsymbol{u}_T \, ds$ and we can write the velocity $\boldsymbol{v}$ in the vector form

$$\boldsymbol{v} = \boldsymbol{u}_T \frac{ds}{dt} \tag{4.8}$$

Equation 4.8 for curvilinear motion is equivalent to Equation 3.14 for rectilinear motion.

4.3 Curvilinear motion: acceleration

In curvilinear motion, the velocity, in general, changes both in magnitude and in direction. This is, for example, what happens if we accelerate an automobile while going around a curve. The magnitude of the velocity changes because the particle may speed up or slow down. The direction of the velocity changes because the velocity is tangent to the path and the path bends continuously. Figure 4.6 indicates the velocity at times t and t', when the particle is at P and P', respectively. The vector change in velocity in going from P to P' is indicated by $\Delta \boldsymbol{v}$ in the vector triangle, drawn off to the side in Figure 4.6. The average acceleration in the time interval Δt is defined by

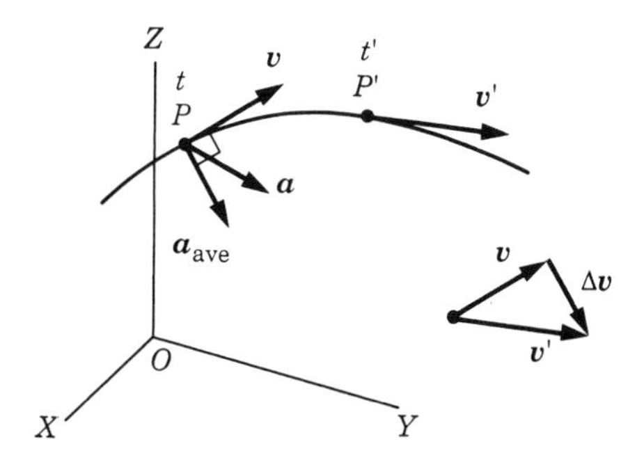

Figure 4.6 Acceleration in curvilinear motion. The acceleration always points towards the concave side.

$$\boldsymbol{a}_{\text{ave}} = \frac{\Delta \boldsymbol{v}}{\Delta t} \tag{4.9}$$

which is a vector parallel to $\Delta \boldsymbol{v}$.

The **instantaneous acceleration** is defined by

$$\boldsymbol{a} = \lim_{\Delta t \to 0} \boldsymbol{a}_{\text{ave}} = \lim_{\Delta t \to 0} \frac{\Delta \boldsymbol{v}}{\Delta t}$$

or

$$\boldsymbol{a} = \frac{\mathrm{d}\boldsymbol{v}}{\mathrm{d}t} \tag{4.10}$$

Since $\boldsymbol{v} = \boldsymbol{i}v_x + \boldsymbol{j}v_y + \boldsymbol{k}v_z$, the acceleration is the vector

$$\boldsymbol{a} = \boldsymbol{i}\frac{\mathrm{d}v_x}{\mathrm{d}t} + \boldsymbol{j}\frac{\mathrm{d}v_y}{\mathrm{d}t} + \boldsymbol{k}\frac{\mathrm{d}v_z}{\mathrm{d}t} \tag{4.11}$$

The components of the acceleration along the X-, Y- and Z-axes are

$$a_x = \frac{\mathrm{d}v_x}{\mathrm{d}t}, \qquad a_y = \frac{\mathrm{d}v_y}{\mathrm{d}t}, \qquad a_z = \frac{\mathrm{d}v_z}{\mathrm{d}t}$$

The acceleration is a vector that has the same direction as the instantaneous change in velocity. Since the direction of the velocity changes in the direction in which the curve bends, the acceleration in curvilinear motion is always pointing toward the concavity of the curve. In general, the acceleration is neither tangential nor perpendicular to the path. Figure 4.7 shows the acceleration for several positions.

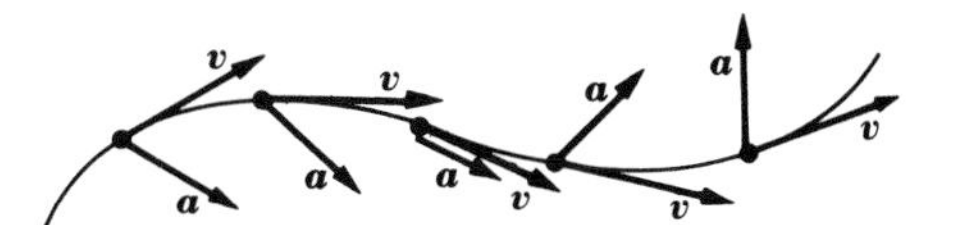

Figure 4.7 Vector relation between velocity and acceleration in curvilinear motion.

4.4 Tangential and normal acceleration

Since the acceleration points *toward* the concavity of the path, we can break the acceleration into two components, one tangent to the path and another perpendicular to the path. The tangential component $\boldsymbol{a}_T$, parallel to the tangent PT (Figure 4.8), is called the **tangential acceleration**. The normal component $\boldsymbol{a}_N$, parallel to the normal PN is called the **normal** or **centripetal acceleration**. That is, we may write

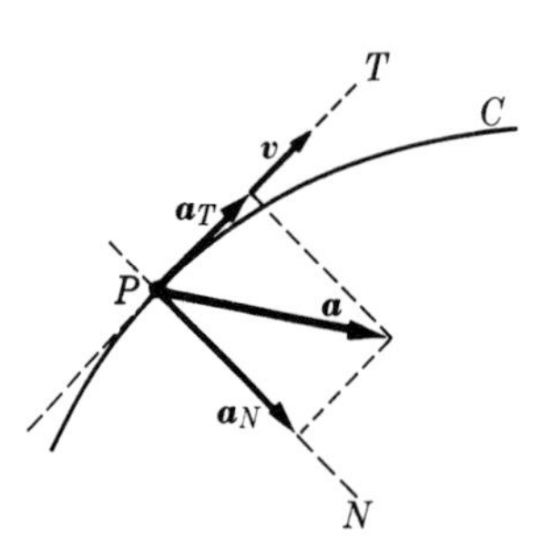

Figure 4.8 Tangential and normal accelerations in curvilinear motion.

$$\boldsymbol{a} = \boldsymbol{a}_T + \boldsymbol{a}_N$$

Each component has a well-defined meaning. When the particle moves it may go faster or slower. That is, the *magnitude* of the velocity may change and this change is related to the tangential acceleration. Also, the *direction* of the velocity changes, and this change is related to the normal acceleration. That is:

Change in magnitude of velocity → tangential acceleration

Change in direction of velocity → normal acceleration

As shown below, the magnitude of the tangential acceleration is

$$a_T = \frac{dv}{dt} \tag{4.12}$$

and the magnitude of the normal or centripetal acceleration is

$$a_N = \frac{v^2}{R} \tag{4.13}$$

where R is the radius of curvature of the path. Equation 4.12 is equivalent to Equation 3.5 for rectilinear motion. Equation 4.13 is easy to understand. The smaller the radius R and the larger the velocity, the more rapidly the velocity changes direction, as can be seen from Figure 4.9. Also, the larger the velocity, the larger the change in velocity in a perpendicular direction for a given change in direction. Therefore, the centripetal acceleration, a_N, must be proportional to $v(v/R)$. In vector form we can write

$$\boldsymbol{a} = \boldsymbol{u}_T \frac{dv}{dt} + \boldsymbol{u}_N \frac{v^2}{R}$$

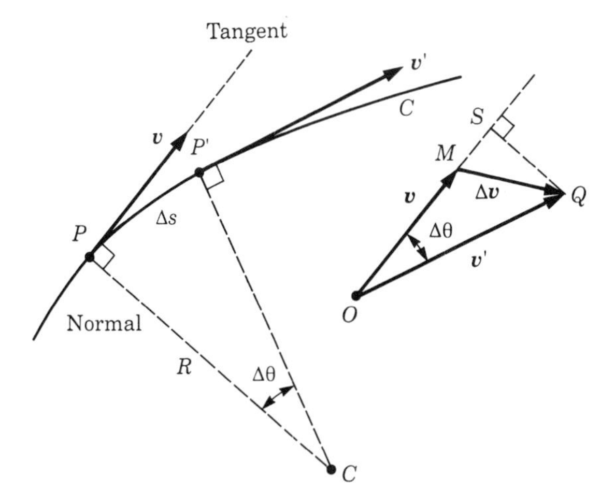

Figure 4.9 Calculation of the tangential and normal accelerations.

If the motion is *uniform* (i.e. the *magnitude* of the velocity remains constant), $v = const.$ and $a_T = 0$, so that there is no tangential acceleration, but there is a centripetal acceleration if the motion is curvilinear. On the other hand, if the motion is *rectilinear* (i.e. the *direction* of the velocity does not change), the radius of curvature is infinite ($R = \infty$), so that $a_N = 0$ and there is no normal acceleration; that is, the acceleration is parallel to the velocity.

Derivation of the expressions for the tangential and normal components of acceleration

Consider a particle moving along a curved path C (Figure 4.9). At time t the particle is at P with velocity $\boldsymbol{v}$, and at time $t + \Delta t$ the particle is at P' with velocity $\boldsymbol{v}'$. The change in velocity is

$$\Delta \boldsymbol{v} = \boldsymbol{v}' - \boldsymbol{v}$$

given by the vector $\boldsymbol{MQ}$ as shown in the auxiliary diagram. We designate the angle between $\boldsymbol{v}$ and $\boldsymbol{v}'$ by $\Delta\theta$. Then we have

Tangential component of $\Delta\boldsymbol{v}$: $\quad MS = v' \cos \Delta\theta - v$

Normal component of $\Delta\boldsymbol{v}$: $\quad SQ = v' \sin \Delta\theta$

Hence the components of the average acceleration in the interval Δt are

$$\begin{pmatrix}\text{Tangential component of} \\ \text{the average acceleration}\end{pmatrix} = \frac{v' \cos \Delta\theta - v}{\Delta t}$$

$$\begin{pmatrix}\text{Normal component of} \\ \text{the average acceleration}\end{pmatrix} = \frac{v' \sin \Delta\theta}{\Delta t}$$

The components of the acceleration at P are then given by

$$a_T = \lim_{\Delta t \to 0} \frac{v' \cos \Delta\theta - v}{\Delta t} \qquad \text{and} \qquad a_N = \lim_{\Delta t \to 0} \frac{v' \sin \Delta\theta}{\Delta t}$$

Now, when $\Delta t \to 0$, then $\Delta\theta \to 0$ also and

$$\lim_{\Delta t \to 0} \cos \Delta\theta = 1$$

Thus we may write

$$a_T = \lim_{\Delta t \to 0} \frac{v' \cos \Delta\theta - v}{\Delta t} = \lim_{\Delta t \to 0} \frac{v' - v}{\Delta t}$$

But $\Delta v = v' - v$ is the change in the magnitude of the velocity, which must not be confused with the vector difference $\Delta \boldsymbol{v} = \boldsymbol{v}' - \boldsymbol{v}$. Therefore

$$a_T = \lim_{\Delta t \to 0} \frac{v' - v}{\Delta t} = \lim_{\Delta t \to 0} \frac{\Delta v}{\Delta t} = \frac{dv}{dt}$$

which is Equation 4.12.

Again when $\Delta t \to 0$ we have that $\sin \Delta\theta$ can be replaced by $\Delta\theta$ and

$$\lim_{\Delta t \to 0} v' = v$$

Hence

$$a_N = \lim_{\Delta t \to 0} \frac{v' \Delta\theta}{\Delta t} = v \lim_{\Delta t \to 0} \frac{\Delta\theta}{\Delta t}$$

Calling R the radius of curvature of the path at P and making $\Delta s = \text{arc } PP'$, we have from the figure that

$$\Delta\theta = \frac{\Delta s}{R}$$

Hence, recalling that $v = ds/dt$, we can write

$$a_N = v \lim_{\Delta t \to 0} \frac{1}{R} \frac{\Delta s}{\Delta t} = v\left(\frac{1}{R} \frac{ds}{dt}\right) = v\left(\frac{v}{R}\right) = \frac{v^2}{R}$$

which is Equation 4.13.

4.5 Curvilinear motion with constant acceleration

Curvilinear motion with constant acceleration occurs wherever the motion is due to constant forces, as will be explained in Chapter 7. Two examples are the motion of bodies near the surface of the Earth under the action of the gravitational attraction and the motion of charged particles in a uniform electric field. For that reason it is important to understand this kind of motion.

When the acceleration in curvilinear motion is constant, both in magnitude and direction, we may integrate Equation 4.10 and get

$$\int_{\boldsymbol{v}_0}^{\boldsymbol{v}} \mathrm{d}\boldsymbol{v} = \int_{t_0}^{t} \boldsymbol{a}\, \mathrm{d}t = \boldsymbol{a} \int_{t_0}^{t} \mathrm{d}t$$

where $\boldsymbol{v}_0$ is the velocity at time t_0. Then the change in velocity during the time interval $t - t_0$ is

$$\boldsymbol{v} - \boldsymbol{v}_0 = \boldsymbol{a}(t - t_0)$$

and the velocity at any time is

$$\boldsymbol{v} = \boldsymbol{v}_0 + \boldsymbol{a}(t - t_0) \tag{4.14}$$

Recalling from Equation 4.5 that $\boldsymbol{v} = \mathrm{d}\boldsymbol{r}/\mathrm{d}t$, $\mathrm{d}\boldsymbol{r} = \boldsymbol{v}\, \mathrm{d}t$, and integrating again, we have

$$\int_{\boldsymbol{r}_0}^{\boldsymbol{r}} \mathrm{d}\boldsymbol{r} = \int_{t_0}^{t} \boldsymbol{v}\, \mathrm{d}t$$

Introducing Equation 4.14, we can write

$$\begin{aligned} \boldsymbol{r} - \boldsymbol{r}_0 &= \int_{t_0}^{t} [\boldsymbol{v}_0 + \boldsymbol{a}(t - t_0)]\, \mathrm{d}t \\ &= \boldsymbol{v}_0 \int_{t_0}^{t} \mathrm{d}t + \boldsymbol{a} \int_{t_0}^{t} (t - t_0)\, \mathrm{d}t \end{aligned}$$

where $\boldsymbol{r}_0$ gives the position at time t_0. Then

$$\boldsymbol{r} - \boldsymbol{r}_0 = \boldsymbol{v}_0(t - t_0) + \tfrac{1}{2}\boldsymbol{a}(t - t_0)^2 \tag{4.15}$$

which gives the displacement of the particle at any time. Note that the motion can be considered as the superposition of a uniform motion in the direction of $\boldsymbol{v}_0$ and a uniformly accelerated motion in the direction of $\boldsymbol{a}$. These results must be compared with those (Equations 3.8 and 3.9) obtained for rectilinear motion under constant acceleration.

In rectilinear motion, both the velocity and the acceleration have the same (or opposite) direction. However, in the more general case we are now discussing, $\boldsymbol{v}_0$ and $\boldsymbol{a}$ may have different directions. Therefore $\boldsymbol{v}$ as given by Equation 4.14 is not parallel to $\boldsymbol{a}$, but is always in the plane defined by $\boldsymbol{v}_0$ and $\boldsymbol{a}$. However, the change in velocity, $\boldsymbol{v} - \boldsymbol{v}_0$, is always parallel to the acceleration, as shown in Figure 4.10. Also, from Equation 4.15, we see that the displacement $\boldsymbol{r} - \boldsymbol{r}_0$ is a combination of two vectors, one parallel to $\boldsymbol{v}_0$ and the other parallel to $\boldsymbol{a}$. Therefore, the end-point

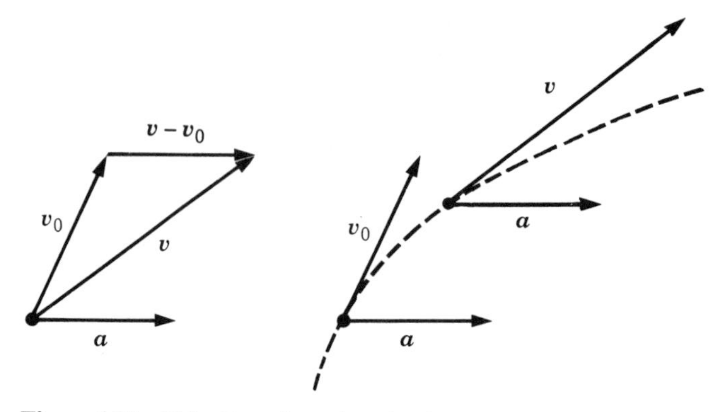

Figure 4.10 Velocity and acceleration in uniform curvilinear motion.

of the vector $\boldsymbol{r}$ is always in the plane, parallel to $\boldsymbol{v}_0$ and $\boldsymbol{a}$, that passes through the point defined by $\boldsymbol{r}_0$. We conclude then that:

motion under constant acceleration is always in a plane; the trajectory of the motion, described by Equation 4.15, is a parabola.

EXAMPLE 4.1
Graphical representation of motion with constant acceleration (Figure 4.11).

▷ For this case, according to Equation 4.14, the velocity at each instant can be considered as the sum of two vectors. One vector represents the initial velocity $\boldsymbol{v}_0$ and is constant. The other vector is $\boldsymbol{a}(t - t_0)$ and increases with time. Figure 4.11(a) shows the velocity at various times, corresponding to $t - t_0 = 1$ s, 2 s, 3 s, 4 s etc.

Similarly, the displacement of the particle, $\boldsymbol{r} - \boldsymbol{r}_0$, is the sum of two vectors. One represents $\boldsymbol{v}_0(t - t_0)$ while the other is $\frac{1}{2}\boldsymbol{a}(t - t_0)^2$. Using the same values of $t - t_0$ as before, the path of the particle when $\boldsymbol{v}_0$ and $\boldsymbol{a}$ have different directions can be plotted, as shown in Figure 4.11(b). As may be seen, the path is a parabola. This, for example, is the motion of a projectile initially thrown at an angle with the vertical from the top of a building or of an electron thrown at an angle with a uniform electric field.

It can be verified that the velocity is tangent at each point A, B, C,... of the trajectory by drawing tangents to the graph of Figure 4.11(b) at A, B etc. and observing that these tangents are parallel to the velocities that have been plotted in Figure 4.11(a).

Note that when $\boldsymbol{v}_0$ and $\boldsymbol{a}$ have the same direction, the motion can only be in a straight line. Therefore, for given values of the magnitudes $\boldsymbol{v}_0$ and $\boldsymbol{a}$, the resultant motion depends on the angle of the initial velocity with the acceleration.

EXAMPLE 4.2
General projectile motion in rectangular coordinates.

▷ This example is a special case of Example 4.1 when $\boldsymbol{a} = \boldsymbol{g}$, the acceleration of gravity, and where the initial velocity $\boldsymbol{v}_0$ forms an arbitrary angle α with the horizontal. Choose the XY-plane coincident with the vertical plane defined by $\boldsymbol{v}_0$ and $\boldsymbol{g}$, with the X-axis horizontal and the Y-axis directed upward so that $\boldsymbol{g}$ points in the $-Y$ direction, as shown in Figure 4.12. Place the origin O coincident with $\boldsymbol{r}_0$ so that $\boldsymbol{r}_0 = 0$. The motion will then

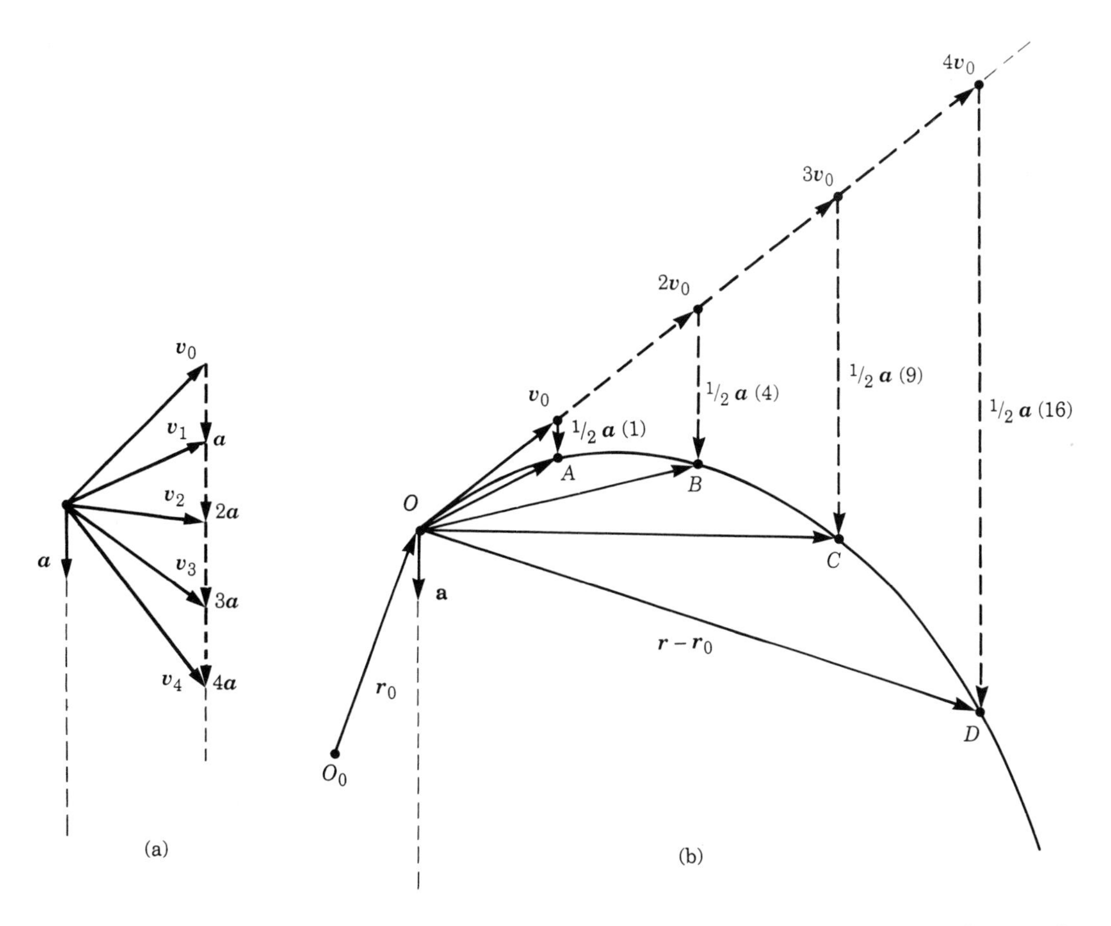

Figure 4.11 Vector representation of the relations $\boldsymbol{v} = \boldsymbol{v}_0 + \boldsymbol{a}(t - t_0)$ and $\boldsymbol{r} - \boldsymbol{r}_0 = \boldsymbol{v}_0(t - t_0) + \frac{1}{2}\boldsymbol{a}(t - t_0)^2$. When the acceleration is constant the path is a parabola.

be in the vertical XY-plane. If α is the angle of $\boldsymbol{v}_0$ with the horizontal,

$$v_{0x} = v_0 \cos \alpha \quad \text{and} \quad v_{0y} = v_0 \sin \alpha$$

Equation 4.14 can be separated into its components (setting $t_0 = 0$) by writing

$$v_x = v_{0x} \qquad \text{and} \qquad v_y = v_{0y} - gt$$

indicating that the X-component of $\boldsymbol{v}$ remains constant, as it should, since there is no acceleration in that direction. That is, we can say that the motion results from combining a uniform horizontal motion and a uniformly accelerated vertical motion with an acceleration $\boldsymbol{g} = -\boldsymbol{j}g$.

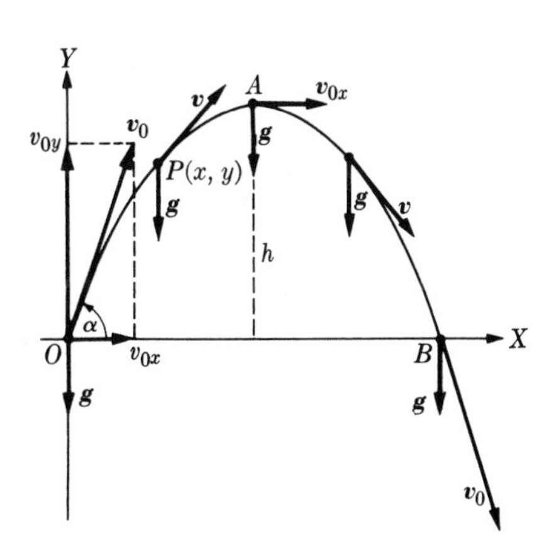

Figure 4.12 General motion of a projectile.

The magnitude of the velocity at any time is $v = (v_x^2 + v_y^2)^{1/2}$. Similarly, Equation 4.15, with $\boldsymbol{r}_0 = 0$ and $t_0 = 0$, when separated into its components, becomes

$$x = v_{0x}t \qquad y = v_{0y}t - \tfrac{1}{2}gt^2$$

which gives the coordinates of the particle as functions of time. As in the previous example, these two equations show that the motion corresponds to uniform motion in the X-direction and uniformly accelerated motion in the Y-direction.

By eliminating time between the two equations, the equation of the path may be found as

$$y = \frac{v_{0y}}{v_{0x}}\,x - \frac{g}{2v_{0x}^2}\,x^2$$

which is a parabola. The **range**, R, can be obtained by letting $y = 0$ and solving for x. This gives

$$R = \frac{2v_{0y}v_{0x}}{g} = \frac{v_0^2 \sin 2\alpha}{g}$$

The horizontal range is greatest for $\sin 2\alpha = 1$ or a launch angle of 45° because then $2\alpha = 90°$ and $\sin 2\alpha = 1$. The maximum height, H, may be found by setting $x = \frac{1}{2}R$ (symmetry dictates that the maximum height is attained midway through the flight), which gives

$$H = \frac{v_{0y}^2}{2g} = \frac{v_0^2 \sin^2 \alpha}{2g}$$

These results are valid when the initial velocity is small enough so that: (1) the curvature of the Earth, (2) the variation of gravity with height, and (3) the air resistance may all be neglected. If we take into account the resistance of the air, the path departs from a parabola, as shown in Figure 4.13, and the maximum height and range are diminished (see Section 7.5).

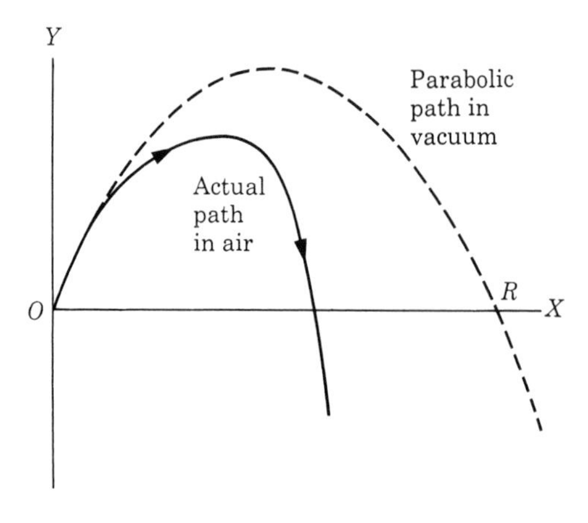

Figure 4.13 Effect of air resistance on the motion of a projectile.

For a long-range projectile (such as an intercontinental ballistic missle or ICBM), the situation is as shown in Figure 4.14, where all g vectors point toward the center of the Earth and their magnitudes vary with height. The path is, in this case, an arc of an ellipse, as will be discussed in Chapter 11. In long range projectile motion, it is also necessary to take into account the Coriolis acceleration, discussed in Section 5.5.

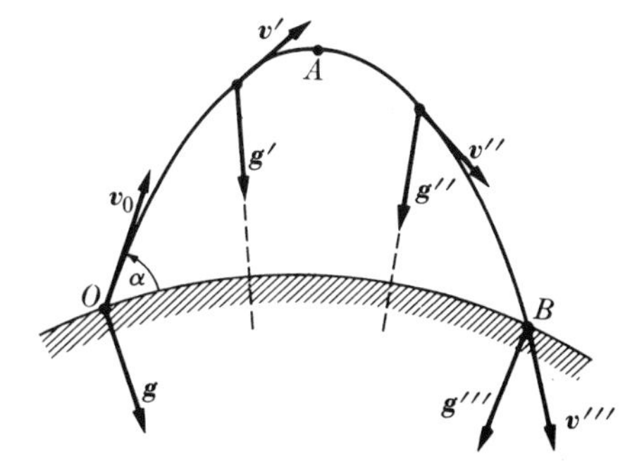

Figure 4.14 The path of the long-range projectile is not a parabola, but an arc of an ellipse.

Projectile motion is found in many sporting activities (golf, tennis, cricket, football, baseball etc.), but in those cases the motion is affected by the shape of the ball and its rotational movement, as well as by air resistance. It also applies to a skier jumping from a ramp and a swimmer diving from a board.

EXAMPLE 4.3

The electron gun in a TV tube produces electrons with a velocity of about 6×10^7 m s^{-1}. The distance from the gun to the screen is 0.40 m. If the electrons are initially thrown horizontally, calculate the vertical deviation from a straight line due to the action of gravity.

▷ The time it takes the electrons to reach the screen is

$$t = \frac{x}{v} = \frac{0.4\ \text{m}}{6 \times 10^7\ \text{m s}^{-1}} = 0.67 \times 10^{-8}\ \text{s}$$

The vertical distance fallen in this time is

$$y = \tfrac{1}{2}gt^2 = 2.17 \times 10^{-16}\ \text{m}$$

which is undetectable. Therefore, we may ignore gravitational effects in the motion of electrons in TV tubes.

4.6 Relative translational motion: the Galilean transformation

As we have said, motion is a relative concept in that it must always be referred to a particular frame of reference, chosen by the observer. Since different observers may use different frames of reference, it is important to know how observations made by different observers are related. For example, most of the observations made on Earth are related to a frame of reference attached to it, and therefore moving with the Earth. In atomic physics the motion of the electrons is determined relative to the nucleus, which may also be in motion relative to the observer. An experimenter usually chooses a frame of reference where data-taking and analysis are most easily accomplished. Since two observers may be in relative motion themselves, it is important to see how they can correlate their measurements. This is the purpose of this section.

Consider two observers O and O' who move, relative to each other, with rectilinear motion. That is, the observers do not rotate relative to each other. Therefore, when observer O sees observer O' moving with velocity $\boldsymbol{v}$, O' sees O moving with velocity $-\boldsymbol{v}$ (recall Section 3.9). We are interested in comparing their descriptions of the motion of an object. For example, one observer may be on the platform of a railroad station, another in a passing train moving in a straight line and both observers watching the flight of a plane overhead.

For simplicity we choose the X- and X'-axes along the line of the relative motion (Figure 4.15). We also choose the YZ- and $Y'Z'$-axes parallel to each other. The coordinate axes remain parallel because of the absence of relative rotation. Then $\boldsymbol{v}$ is parallel to the X-axis. Designate the position vector of O' relative to O by $\boldsymbol{R} = \boldsymbol{OO'}$, which is along the common X-axis. From the figure note that if $\boldsymbol{r} = \boldsymbol{OP}$ and $\boldsymbol{r'} = \boldsymbol{O'P}$ are the position vectors of a particle P relative to observers O and O', the vector relation:

$$\boldsymbol{OP} = \boldsymbol{OO'} + \boldsymbol{O'P} \qquad \text{or} \qquad \boldsymbol{O'P} = \boldsymbol{OP} - \boldsymbol{OO'}$$

holds and may be rewritten as

$$\boldsymbol{r'} = \boldsymbol{r} - \boldsymbol{R} \tag{4.16}$$

This expression gives the rule for comparing the position of P as measured by O and O'. We shall assume that observers O and O' use the same time so that $t = t'$; that is, we assume that time measurements are independent of the relative motion of the observers. This seems very reasonable, but it is only an assumption, which may be disproved by experiment.

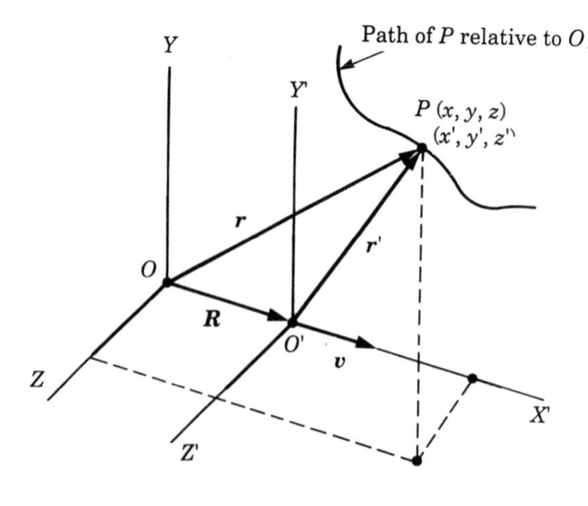

Figure 4.15 Frames of reference in uniform relative translational motion. The description of the motion of P is different for observers O and O'.

Taking the time derivative of Equation 4.16 we obtain

$$\frac{\mathrm{d}\boldsymbol{r}'}{\mathrm{d}t} = \frac{\mathrm{d}\boldsymbol{r}}{\mathrm{d}t} - \frac{\mathrm{d}\boldsymbol{R}}{\mathrm{d}t}$$

But $\boldsymbol{V} = \mathrm{d}\boldsymbol{r}/\mathrm{d}t$ and $\boldsymbol{V}' = \mathrm{d}\boldsymbol{r}'/\mathrm{d}t$ are the velocities of P as measured by observers O and O'. Also, $\boldsymbol{v} = \mathrm{d}\boldsymbol{R}/\mathrm{d}t$ is the velocity of O' relative to O. Therefore, we may write

$$\boldsymbol{V}' = \boldsymbol{V} - \boldsymbol{v} \tag{4.17}$$

This is the rule for comparing the velocity of a body as measured by two observers in relative translation motion.

If we take the time derivative of Equation 4.17, we obtain

$$\frac{\mathrm{d}\boldsymbol{V}'}{\mathrm{d}t} = \frac{\mathrm{d}\boldsymbol{V}}{\mathrm{d}t} - \frac{\mathrm{d}\boldsymbol{v}}{\mathrm{d}t}$$

But, $\boldsymbol{a} = \mathrm{d}\boldsymbol{V}/\mathrm{d}t$ and $\boldsymbol{a}' = \mathrm{d}\boldsymbol{V}'/\mathrm{d}t$ are the accelerations of P as measured by the two observers. And $\boldsymbol{a}_\mathrm{r} = \mathrm{d}\boldsymbol{v}/\mathrm{d}t$ is their relative acceleration. Therefore,

$$\boldsymbol{a}' = \boldsymbol{a} - \boldsymbol{a}_\mathrm{r} \tag{4.18}$$

This is the rule for comparing the acceleration of a body measured by two observers in relative translational motion.

Of particular interest is the case when the two observers are in *uniform* relative translational motion. For this case, $\boldsymbol{v}$ is a constant vector and, if we set $t = 0$ when the observers O and O' are coincident, we may write $\boldsymbol{OO'} = \boldsymbol{R} = \boldsymbol{v}t$. Also, $\boldsymbol{a}_\mathrm{r} = \mathrm{d}\boldsymbol{v}/\mathrm{d}t = 0$; that is, the relative acceleration is zero. Then we may write the previous set of equations as

$$\begin{aligned} \boldsymbol{r}' &= \boldsymbol{r} - \boldsymbol{v}t \\ \boldsymbol{V}' &= \boldsymbol{V} - \boldsymbol{v} \\ \boldsymbol{a}' &= \boldsymbol{a} \\ t' &= t \end{aligned} \tag{4.19}$$

This set of relations constitutes the **Galilean transformation** of coordinates, velocities and accelerations. (This name was given to this set of equations by Albert Einstein to distinguish it from the Lorentz transformation, to be discussed in Chapter 19 in connection with Einstein's theory of relativity, which must be used when the relative velocity of the observers is close to the velocity of light.)

An important feature of the Galilean transformation is that both observers measure the same acceleration. That is:

according to the Galilean transformation the acceleration of a particle is the same for all observers in uniform relative translational motion.

This result offers us an example of a physical quantity – the acceleration of a particle – that appears to be independent of the motion of an observer. In other words, we have found that:

the acceleration of a body remains invariant when passing from one frame of reference to another if the two frames are in uniform relative translational motion.

The invariance of acceleration to the Galilean transformation is the first example we encounter of a physical quantity that remains invariant under a transformation. Later on we will find other physical quantities that behave in the same manner. This result, as we shall see, has had a profound influence on the formulation of laws in physics.

Also, the Galilean transformation shows that the states of rest or of rectilinear motion are not intrinsic properties of a body but depend on its relation to the observer.

EXAMPLE 4.4

Expression of the Galilean transformation in terms of the rectangular components of the vectors.

▷ Referring to Figure 4.15, where the coordinate axes have been oriented with the X- and X'-axes parallel to the relative velocity $\boldsymbol{v}$, we have that the term $\boldsymbol{v}t$ in the first relation of Equation 4.19 affects only the x-coordinate. Therefore, the Galilean transformation for the coordinates becomes

$$x' = x - \boldsymbol{v}t, \quad y' = y, \quad z' = z, \quad t' = t \tag{4.20}$$

We have added $t' = t$ to the three space equations to emphasize that we are assuming that the two observers are using the same time. Similarly, the Galilean transformation of velocities becomes

$$V'_x = V_x - v, \quad V'_y = V_y, \quad V'_z = V_z \tag{4.21}$$

The transformation of the acceleration components is given by

$$a'_x = a_x, \quad a'_y = a_y, \quad a'_z = a_z \tag{4.22}$$

which expresses the invariance of the acceleration under the Galilean transformation.

EXAMPLE 4.5

The velocity of sound in still air at 25°C is 358 $\mathrm{m\,s^{-1}}$. Find the velocity measured by an observer moving with a velocity of 90 $\mathrm{km\,h^{-1}}$ (a) away from the source, (b) toward the source, (c) perpendicular to the direction of propagation in air, and (d) in a direction such that the sound appears to propagate perpendicularly relative to the moving observer. Assume that the source is at rest relative to the ground.

▷ Choose a frame of reference XYZ (Figure 4.16(a)) fixed on the ground, and thus at rest relative to the air, and a frame $X'Y'Z'$ moving with the observer, with the X- and X'-axes parallel to the velocity of the observer. Relative to XYZ, the sound source is at O, the velocity of the observer O' is $v = 90\ \mathrm{km\,h^{-1}} = 25\ \mathrm{m\,s^{-1}}$, and the velocity of sound is $V = 358\ \mathrm{m\,s^{-1}}$. The velocity of sound, relative to $X'Y'Z'$, as recorded by the moving observer O', is V'. Applying Equation 4.21, we have for case (a)

$$V' = V - v = 333\ \mathrm{m\,s^{-1}}$$

In case (b), we note that O' moves along the negative direction of the X-axis. Thus, replacing v by $-v$ in Equation 4.21, we have

$$V' = V + v = 383\ \mathrm{m\,s^{-1}}$$

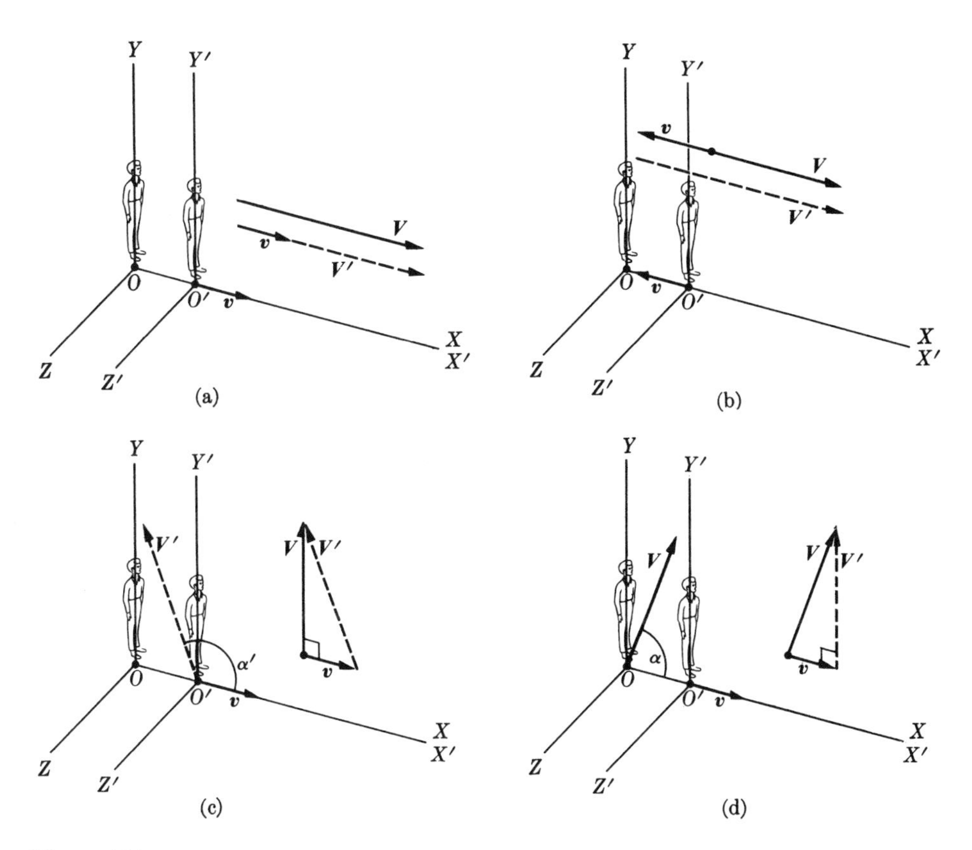

Figure 4.16

For situation (c) we note that V and v are perpendicular, so

$$V' = (V^2 + v^2)^{1/2} = 358.9\ \mathrm{m\,s^{-1}}$$

To the moving observer, the sound appears to propagate in a direction that makes an angle α' with the X'-axis such that

$$\tan\alpha' = \frac{V'_y}{V'_x} = \frac{V}{-v} = -15.32 \qquad \text{or} \qquad \alpha' = 93.7^\circ$$

Finally, in case (d), the direction of propagation of the sound in air is such that it appears to O' to be moving in the Y' direction. Thus V' is perpendicular to v. Therefore,

$$V^2 = v^2 + V'^2, \qquad \text{or} \qquad V' = (V^2 - v^2)^{1/2} = 357.1\ \mathrm{m\,s^{-1}}$$

In this case sound propagates through the still air in a direction making an angle α with the X-axis such that

$$\tan\alpha = \frac{V_y}{V_x} = \frac{V'}{v} = 14.385 \qquad \text{or} \qquad \alpha = 86.0^\circ$$

QUESTIONS

4.1 How does the velocity change in (a) uniform rectilinear motion, (b) uniformly accelerated rectilinear motion, and (c) uniformly decelerated rectilinear motion? Explain each case.

4.2 Which aspect of the velocity remains constant, and which changes, in uniform curvilinear motion?

4.3 Why is the velocity in curvilinear motion always a vector tangent to the path?

4.4 Why must the acceleration in curvilinear motion always point toward the concave side of the path?

4.5 Verify that the units of both tangential and normal acceleration are those of acceleration, i.e. $\mathrm{m\,s^{-2}}$.

4.6 What motion results when (a) the centripetal acceleration is zero, (b) the tangential acceleration is zero?

4.7 Can you explain why curvilinear motion with constant acceleration must be in a plane? Which vectors determine the plane?

4.8 Verify that the same range of a projectile is obtained for the angles α and $90^\circ - \alpha$. Is the time of flight the same in both cases?

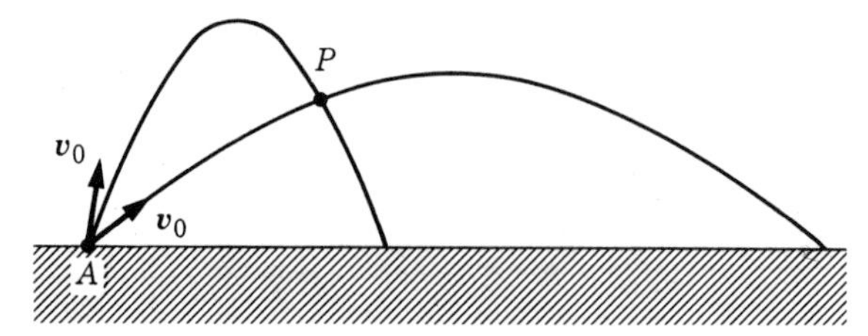

Figure 4.17

4.9 What physical result is meant by the statement that the acceleration of a body remains invariant under a Galilean transformation?

4.10 Two projectiles are launched at A with the same initial velocity but different angles (Figure 4.17). Compare their velocities and accelerations when they pass by point P.

4.11 Compare the velocities and accelerations of a projectile when it passes through points A and B that are at the same height (Figure 4.18).

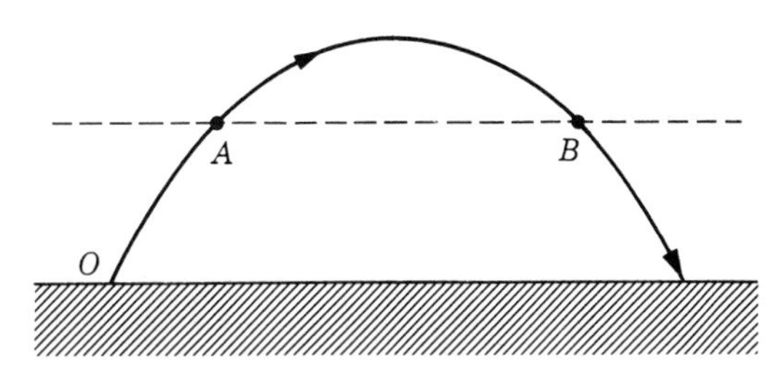

Figure 4.18

PROBLEMS

4.1 A body is traveling along a plane curve such that its rectangular coordinates, as a function of time, are given by

$$x = 2t^3 - 3t^2, \qquad y = t^2 - 2t + 1$$

Assuming that t is given in seconds and the coordinates in meters, calculate (a) the position of the body when $t = 1$ s, (b) the rectangular components of the velocity at any time, (c) the rectangular components of the velocity when $t = 1$ s, (d) the velocity at any time (write the value as a vector), (e) the velocity when $t = 0$ s, (f) the time(s) when the velocity is zero, (g) the rectangular components of the acceleration when $t = 1$ s, (h) the rectangular components of the acceleration at any time, (i) the acceleration at any time (write the acceleration as a vector), (j) the acceleration when $t = 0$ s, (k) the time(s) at which the acceleration is parallel to the y-axis.

4.2 The coordinates of a moving particle are given by

$$x = t^2, \quad y = (t - 1)^2$$

(a) Find the rectangular components of its average velocity and acceleration in the time interval between t and $t + \Delta t$. (b) Apply the results to the case when $t = 2$ s and $\Delta t = 1$ s. (c) Compare the results of (b) with the values of the rectangular components of the velocity and acceleration at $t = 2$ s.

4.3 (a) Referring to Problem 4.1, express the magnitudes of the velocity and the acceleration of the body as a function of time. (b) Obtain the tangential and the centripetal accelerations as well as the radius of curvature.

4.4 Repeat the previous problem for a particle moving according to the relations given in Problem 4.2.

4.5 Repeat Problem 4.2 using vector notation rather than writing the results in rectangular components.

4.6 A projectile is shot with a velocity of $100\ \mathrm{m\,s^{-1}}$ at an angle of 60° with the horizontal. Calculate (a) the horizontal range, (b) the maximum height, (c) the time of flight, (d) the velocity and height after 10 s.

4.7 (a) Referring to the projectile of Problem 4.6, express the coordinates and the components of the velocity and of the acceleration as a function of time. (b) Obtain the magnitude of the velocity, the tangential acceleration and the centripetal acceleration as functions of time.

4.8 A bomber plane is flying horizontally at an altitude of 1.2 km with a velocity of $360\ \mathrm{km\,h^{-1}}$. (a) How long before the plane is over its target should it drop a bomb? (b) What is the velocity of the bomb when it reaches the ground? (c) What is the horizontal distance covered by the bomb?

4.9 A projectile is shot up at an angle of 35°. It strikes the ground at a horizontal distance of 4 km. Calculate (a) the initial velocity, (b) the time of flight, (c) the maximum altitude, (d) the velocity at the point of maximum altitude.

4.10 A gun is placed at the base of a hill whose slope makes an angle ϕ with the horizontal. If the gun is set at an angle α with the horizontal and has a muzzle velocity v_0, find the distance, measured *along the hill*, at which the shell will hit.

4.11 An airplane is flying horizontally at an altitude of 1 km with a velocity of 200 km h^{-1}. It drops a bomb which is meant to hit a ship moving in the same direction at a velocity of 20 km h^{-1}. Verify that the bomb should be dropped when the horizontal distance between the plane and the ship is 715 m.

4.12 A baseball player hits the ball so that it has a velocity of 14.5 m s^{-1} and an angle of 30° above the horizontal. A second player, standing 30.5 m from the batter and in the same plane as the ball's trajectory, begins to run the instant the ball is struck. (a) Calculate his minimum velocity if he is to catch the ball if he can reach up to 2.4 m above the ground and the ball was 0.92 m high when it was struck. (b) How far does the second player have to run?

4.13 The position of a particle in a coordinate system O is measured in m as

$$\boldsymbol{r} = \boldsymbol{i}(6t^2 - 4t) + \boldsymbol{j}(-3t^3) + \boldsymbol{k}3$$

(a) Determine the relative velocity of system O' with respect to O if the position of the particle relative to O' is measured in m as

$$\boldsymbol{r}' = \boldsymbol{i}(6t^2 + 3t) + \boldsymbol{j}(-3t^3) + \boldsymbol{k}3$$

(b) Show that the acceleration of the particle is the same in both systems.

4.14 A ball is moving due north at 3.00 m s^{-1} when a force is applied for 40 s, causing an acceleration of 0.10 m s^{-2} due east, after which the force is removed. Determine (a) the magnitude and direction of the ball's final velocity, (b) the equation of its path, (c) its displacement from the starting point.

4.15 A particle is moving along a parabola $y = x^2$ so that at any time $v_x = 3$ m s^{-1}. Calculate the magnitude and direction of the velocity and the acceleration of the particle at the point $x = 2/3$ m.

4.16 A point is moving in the XY-plane such that $v_x = 4t^3 + 4t$ m s^{-1} and $v_y = 4t$ m s^{-1}. If the position of the point is (1, 2) when $t = 0$, find the cartesian equation of the trajectory and the components of the acceleration.

4.17 If the coordinates of a moving body are $x = at$ and $y = b \sin at$, show that the value of the acceleration is proportional to the distance of the body to the X-axis. Make a plot of the path.

4.18 Find the radius of curvature at the highest point of the path of a projectile fired at an initial angle α with the horizontal. (Hint: At the highest point, the velocity is horizontal and the acceleration is vertical.)

5 Circular motion

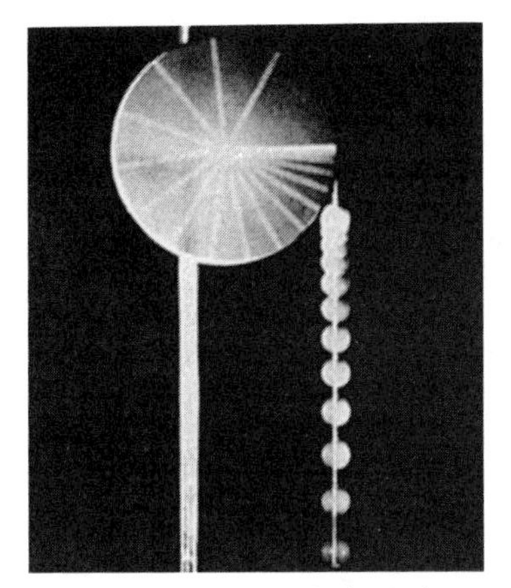

The multiflash photograph shows that the mass falls with uniformly accelerated motion. (This may be verified by taking actual measurements on the photograph.)

5.1 Introduction

Circular motion is a curvilinear motion whose path is a circle. It is, for example, the motion of any point in a rotating disk or wheel as well as the points on the hands of a clock. As a first approximation, it is the motion of the Moon around the Earth and of the electron around the proton in a hydrogen atom. Because of the daily rotation of the Earth, all bodies on the Earth's surface have circular motion relative to the Earth's axis of rotation.

5.2 Circular motion: angular velocity

In circular motion the velocity $\boldsymbol{v}$, which is tangential to the circle, is perpendicular to the radius $R = CA$ (Figure 5.1). When we measure distances along the circumference of the circle from the center O, we have that $s = R\theta$, where the

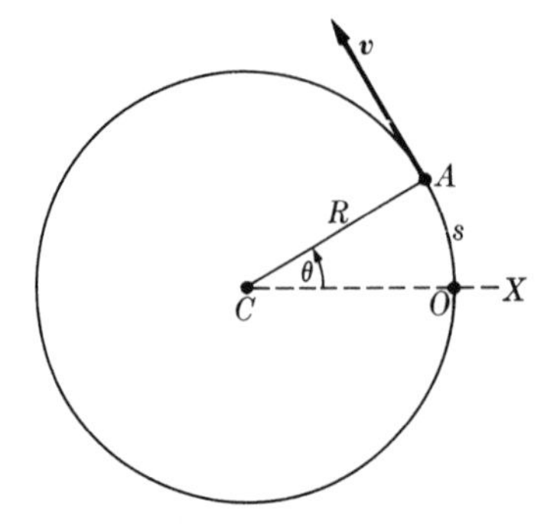

Figure 5.1 Circular motion.

angle θ is measured in radians. Applying Equation 4.7 and considering that R remains constant, the magnitude of the velocity is given by

$$v = \frac{ds}{dt} = R\frac{d\theta}{dt}$$

The quantity

$$\omega = \frac{d\theta}{dt} \tag{5.1}$$

is called **angular velocity**, and is equal to the time rate of change of the angle swept by the radius during the motion. Angular velocity is expressed in radians per second, rad s^{-1}, although it must be kept in mind that the radian is a dimensionless unit. For that reason, when there is no danger of confusion, angular velocity is often expressed simply as s^{-1}. Then

$$v = \omega R \tag{5.2}$$

indicating that in circular motion the velocity is directly proportional to the radius for a given angular velocity.

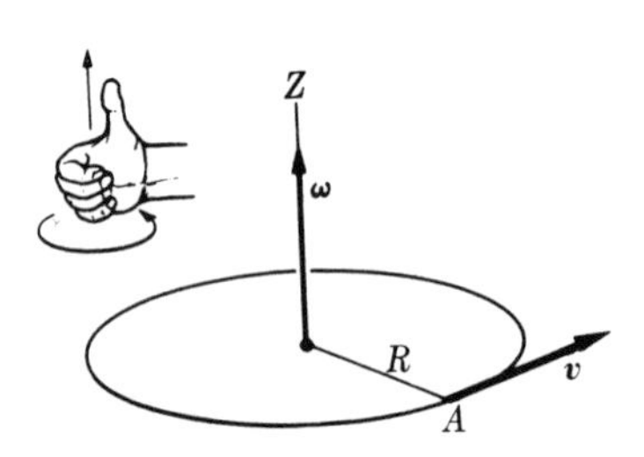

Figure 5.2 Vector relation between angular velocity, linear velocity and position vector in circular motion.

Angular velocity may be represented as a vector quantity whose direction is perpendicular to the plane of motion in the sense given by the right-hand rule (Figure 5.2).

Of special interest is the case of **uniform circular motion**, i.e. motion with constant angular velocity ω. In this case, the motion is periodic and the particle passes through each point of the circle at regular intervals of time.

The **period,** P, when a body is in uniform circular motion, is the time required for a complete turn or revolution, and the **frequency** ν is the number of revolutions per unit time. So if in time t the particle makes n revolutions, the period is $P = t/n$ and the frequency is $\nu = n/t$. Note that the frequency and period are the reciprocal of each other; that is,

$$\nu = \frac{1}{P} \tag{5.3}$$

When the period is expressed in seconds, the frequency must be expressed in $(\text{seconds})^{-1}$ or s^{-1}. This unit is called **hertz**, abbreviated Hz, after Heinrich Hertz

(1857–1894), who experimentally proved the existence of electromagnetic waves. The colloquial term is revolutions (or cycles) per second (rps) instead of s^{-1} or Hz. Thus, a frequency of one hertz corresponds to one revolution (or cycle) per second. Sometimes the frequency of a motion is expressed in revolutions per minute (rpm), which is the same as saying $(\text{minute})^{-1}$. Thus $1\ \text{min}^{-1} = \frac{1}{60}\text{Hz}$.

The concepts of period and frequency are applicable to all processes that occur in periodic or cyclic form. For example, the motion of the Earth around the Sun is neither circular not uniform, but periodic to a very good approximation. It is a motion that repeats itself every time the Earth completes one orbit. Therefore, in general, the *period* is the time required for a process to complete one cycle, and the *frequency* is the number of cycles per second, with one hertz corresponding to one cycle per second.

If ω is constant, we have, integrating Equation 5.1,

$$\int_{\theta_0}^{\theta} d\theta = \int_{t_0}^{t} \omega\ dt = \omega \int_{t_0}^{t} dt$$

or

$$\theta = \theta_0 + \omega(t - t_0)$$

This relation, which is valid only for uniform circular motion, should be compared with the equivalent expression 3.7 for uniform rectilinear motion. Usually the initial values of θ_0 and t_0 are set to zero ($\theta_0 = 0$, $t_0 = 0$), giving

$$\theta = \omega t \qquad \text{or} \qquad \omega = \frac{\theta}{t} \tag{5.4}$$

For a complete revolution, $t = P$ and $\theta = 2\pi$, resulting in

$$\omega = \frac{2\pi}{P} = 2\pi\nu \tag{5.5}$$

EXAMPLE 5.1

Angular velocity (spin) of the Earth about its axis.

▷ Our first impulse would be to use Equation 5.5, $\omega = 2\pi/P$, and use the value of 8.640×10^4 s for the period P, corresponding to one mean **solar** day. However, this is incorrect. Consider a point A on the Earth's surface that is at 'high noon', as shown in Figure 5.3 (not drawn to scale). After the Earth completes one revolution about its polar axis (called a **sidereal** day), it will be at E' due to its translational motion and our point will be at A'. But to complete one day and again be at 'high noon', the Earth must still rotate through the angle γ until the point is at A'', again facing the Sun. The period of revolution of the Earth (sidereal day) is then slightly *less* than 8.640×10^4 s. Its measured value is $P = 8.616 \times 10^4$ s

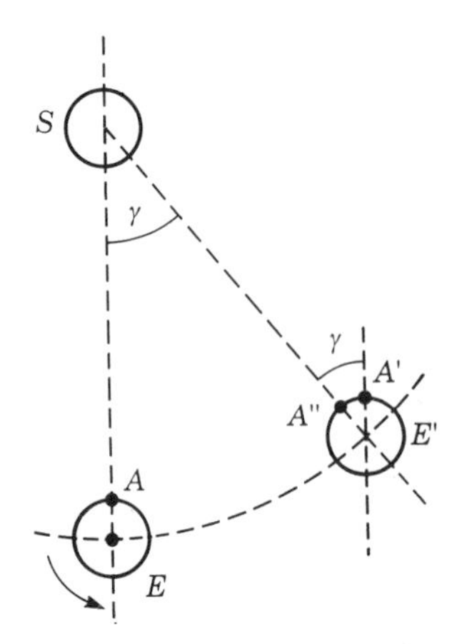

Figure 5.3 Sidereal and solar days.

or about 240 s shorter than the mean solar day. The angular velocity of the Earth is then

$$\omega = 2\pi/P = 7.292 \times 10^{-5} \text{ rad s}^{-1}$$

It is relatively simple to estimate this difference of 240 s. The Earth covers its complete orbit around the Sun in 365 days, which means that the angle γ corresponding to one day is slightly less than 1° or 0.01745 rad. The time required to move through this angle with the angular velocity given above is, by Equation 5.4,

$$t = \frac{\theta}{\omega} = \frac{1.745 \times 10^{-2} \text{ rad}}{7.292 \times 10^{-5} \text{ rad s}^{-1}} = 239 \text{ s}$$

which is in excellent agreement with the measured value.

For comparison, the angular velocity of the Earth about the Sun is approximately $2\pi/(1 \text{ year}) = 2\pi/(365 \times 24 \times 60 \times 60 \text{ s}) = 1.99 \times 10^{-7} \text{ rad s}^{-1}$ while the angular velocity of the electron in a hydrogen atom is $4.13 \times 10^{16} \text{ rad s}^{-1}$, or 10^{22} times larger!

Note 5.1 Radial and transverse velocity in plane curvilinear motion

When a particle moves in a plane, it is often convenient to consider the motion as a combination of a **radial** motion (since the distance of the particle from the origin changes with time) and an **angular** motion around the origin (since the direction of the position vector $\boldsymbol{r}$ also changes with time). Then it is possible to break the velocity into a component along the position vector of the particle relative to the origin of coordinates and another along a perpendicular direction. This is particularly useful when dealing with the motion of a planet about the Sun or of an electron about a nucleus.

It is easy to obtain the components of the velocity. From Figure 5.4(a), we note that when the particle moves from P to P', its displacement can be decomposed into a component of length dr along the radius and a component $r\,d\theta$ perpendicular to it; that is, the motion can be considered to be composed of a **radial displacement** and a **transverse angular displacement**. If the displacement takes place in the time interval dt, we see that the velocity, $\boldsymbol{v} = d\boldsymbol{r}/dt$, has radial and transverse components (Figure 5.4(b)) given by

$$v = \frac{dr}{dt} \quad \text{and} \quad v_\theta = r\frac{d\theta}{dt} \tag{5.6}$$

If the motion is circular, and the origin of coordinates coincides with the center of the circle, then r is constant so that $v_r = 0$. Also, for circular motion, $d\theta/dt$ coincides with the angular velocity ω, so that v_θ is equivalent to $r\omega$ (Equation 5.2).

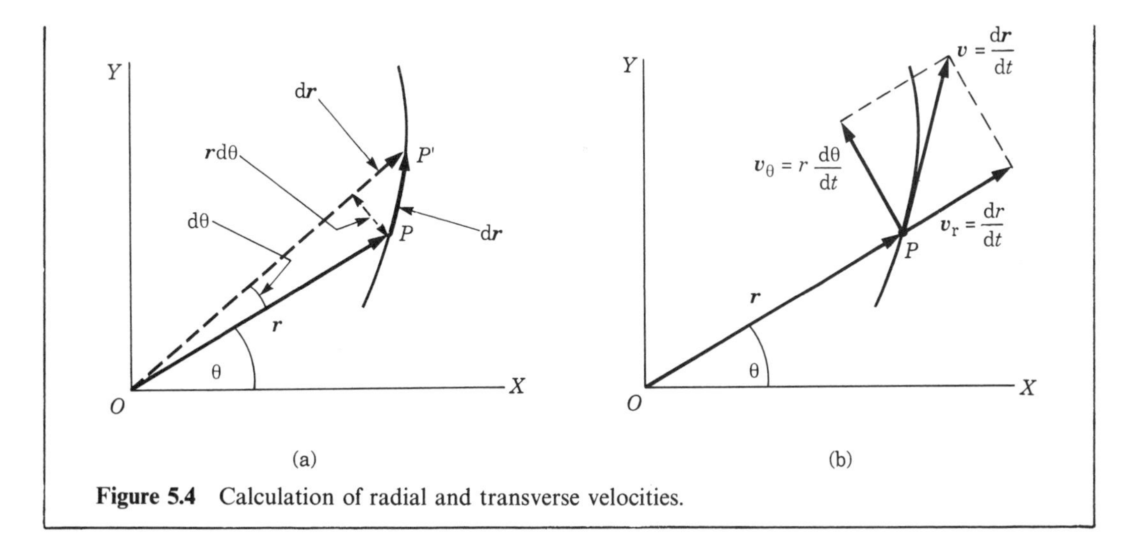

Figure 5.4 Calculation of radial and transverse velocities.

5.3 Circular motion: angular acceleration

When the angular velocity of a particle moving with circular motion changes with time, the **angular acceleration** is defined by

$$\alpha = \frac{\mathrm{d}\omega}{\mathrm{d}t} = \frac{\mathrm{d}^2\theta}{\mathrm{d}t^2} \qquad \textbf{(5.7)}$$

Angular acceleration is expressed in rad s^{-2}, or simply s^{-2}. When the angular acceleration is constant (i.e. when the circular motion is uniformly accelerated), we have, by integration of Equation 5.7,

$$\int_{\omega_0}^{\omega} \mathrm{d}\omega = \int_{t_0}^{t} \alpha\ \mathrm{d}t = \alpha \int_{t_0}^{t} \mathrm{d}t$$

or

$$\omega = \omega_0 + \alpha(t - t_0) \qquad \textbf{(5.8)}$$

where ω_0 is the value of ω at time t_0. Integrating again, recalling that $\omega = \mathrm{d}\theta/\mathrm{d}t$ or $\mathrm{d}\theta = \omega\ \mathrm{d}t$, we obtain

$$\int_{\theta_0}^{\theta} \mathrm{d}\theta = \int_{t_0}^{t} \omega\ \mathrm{d}t = \int_{t_0}^{t} \omega_0\ \mathrm{d}t + \alpha \int_{t_0}^{t} (t - t_0)\ \mathrm{d}t$$

so that

$$\theta = \theta_0 + \omega_0(t - t_0) + \tfrac{1}{2}\alpha(t - t_0)^2 \qquad \textbf{(5.9)}$$

This gives the angular position at any time. When $t_0 = 0$ and $\theta_0 = 0$, we have

$$\theta = \omega_0 t + \tfrac{1}{2}\alpha t^2$$

The student should compare these relations with the similar equations for uniformly accelerated rectilinear motion.

5.4 Vector relations in circular motion

Consider a particle A (Figure 5.5) moving in circular motion with center at C and radius R. From Figure 5.5, $R = r \sin \gamma$, where $\boldsymbol{r}$ is the position vector relative to O, chosen as the origin of coordinates. Therefore we may write the velocity $v = \omega R$ as $v = \omega r \sin \gamma$. Since $\boldsymbol{v}$ is perpendicular to $\boldsymbol{\omega}$ and $\boldsymbol{r}$, the following vector relation between $\boldsymbol{v}$, $\boldsymbol{\omega}$ and $\boldsymbol{r}$ holds, both in magnitude and direction (see Equation A.21):

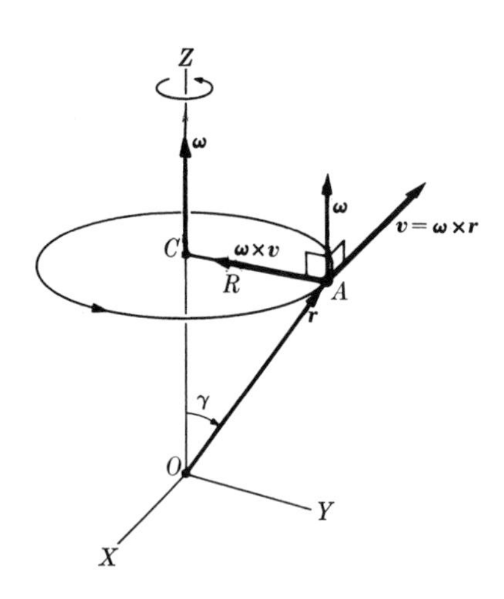

Figure 5.5 Centripetal acceleration in uniform circular motion.

$$\boldsymbol{v} = \boldsymbol{\omega} \times \boldsymbol{r} \qquad \textbf{(5.10)}$$

The acceleration of the particle is $\boldsymbol{a} = \mathrm{d}\boldsymbol{v}/\mathrm{d}t$. When the circular motion is uniform, the angular velocity $\boldsymbol{\omega}$ is constant. The acceleration, however, is not zero since the vector $\boldsymbol{v}$ is changing. Therefore, recalling that $\boldsymbol{v} = \mathrm{d}\boldsymbol{r}/\mathrm{d}t$

$$\boldsymbol{a} = \frac{\mathrm{d}\boldsymbol{v}}{\mathrm{d}t} = \boldsymbol{\omega} \times \frac{\mathrm{d}\boldsymbol{r}}{\mathrm{d}t} = \boldsymbol{\omega} \times \boldsymbol{v} \qquad \textbf{(5.11)}$$

From Figure 5.5 we see that the vector $\boldsymbol{\omega} \times \boldsymbol{v}$ points toward the center C of the circle. Therefore *in uniform circular motion the acceleration is perpendicular to the velocity and points radially inward.*

Thus, the acceleration given by Equation 5.11 is the *normal* or *centripetal* acceleration a_N referred to in Section 4.4. Further, if we use Equation 5.10 we may also write the acceleration as

$$\boldsymbol{a} = \boldsymbol{\omega} \times (\boldsymbol{\omega} \times \boldsymbol{r}) \qquad \textbf{(5.12)}$$

Since $\boldsymbol{\omega}$ is perpendicular to $\boldsymbol{v}$, the magnitude of $\boldsymbol{\omega} \times \boldsymbol{v}$ is ωv. Using Equation 5.2, $v = R\omega$, we may express the centripetal or normal acceleration in circular motion in either of the two forms:

$$a_N = \omega^2 R \qquad \text{or} \qquad a_N = \frac{v^2}{R} \tag{5.13}$$

The latter is the expression we gave in Equation 4.13.

In uniform circular motion the *magnitude* of the velocity remains constant. Thus the acceleration arises from the change in *direction* of the velocity during the motion.

In the general case of non-uniform circular motion, both the magnitude and the direction of the velocity change. The change in magnitude of the velocity gives rise to a *tangential* acceleration, a_T. Recalling Equation 4.12, $a_T = dv/dt$, and knowing that $v = \omega R$, we have that in circular motion

$$a_T = R\frac{d\omega}{dt} = R\alpha \tag{5.14}$$

The total acceleration of the particle, $\boldsymbol{a} = \boldsymbol{a}_N + \boldsymbol{a}_T$, has been shown in Figure 5.6.

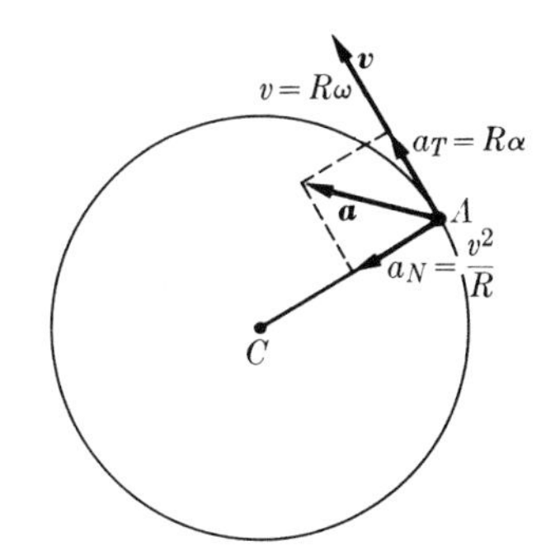

Figure 5.6 Tangential and normal acceleration in circular motion.

EXAMPLE 5.2

Expression of the velocity and acceleration as functions of time in uniform circular motion.

▷ Referring to Figure 5.7, the coordinates of a particle at point A are $x = R\cos\theta$ and $y = R\sin\theta$. But if the circular motion is uniform, $\theta = \omega t$. Thus

$$x = R\cos\omega t, \quad y = R\sin\omega t \tag{5.15}$$

The components of the velocity are then

$$v_x = \frac{dx}{dt} = -\omega R\sin\omega t$$
$$v_y = \frac{dy}{dt} = \omega R\cos\omega t \tag{5.16}$$

These results for the velocity could have been obtained directly from the geometry of Figure 5.7.

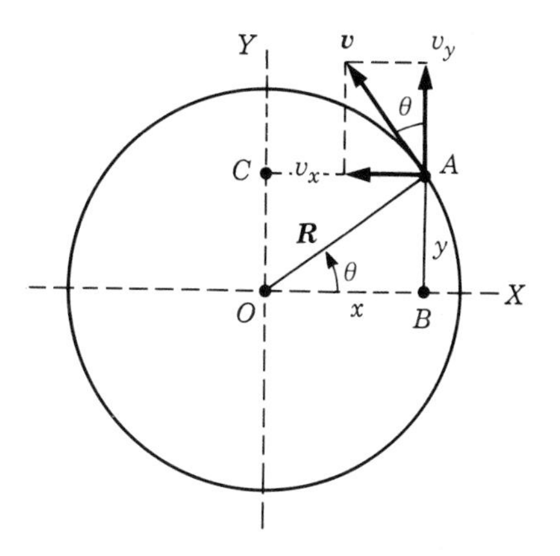

Figure 5.7 Rectangular components of the velocity in circular motion.

The rectangular components of the acceleration can be obtained from Equation 5.16 as follows:

$$a_x = \frac{dv_x}{dt} = -\omega^2 R \cos \omega t \qquad a_y = \frac{dv_y}{dt} = -\omega^2 R \sin \omega t \tag{5.17}$$

or $a_x = -\omega^2 x$ and $a_y = -\omega^2 y$, which means that

$$\boldsymbol{a} = -\omega^2 \boldsymbol{R} \tag{5.18}$$

confirming that the acceleration has a direction opposite to that of $\boldsymbol{R}$ and therefore points *toward* the center of the circle.

EXAMPLE 5.3

Velocity and acceleration of a point on the Earth's surface.

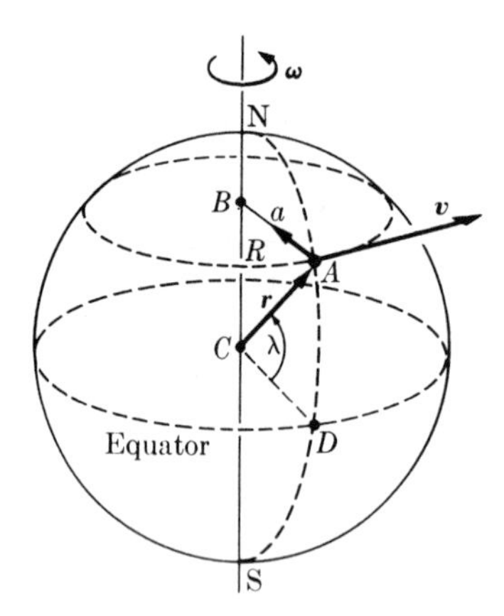

Figure 5.8 Velocity and acceleration of a point on the Earth's surface.

▷ Due to the rotational motion of the Earth, all points on its surface move at the same angular velocity (Example 5.1). The **latitude** of point A (Figure 5.8) is defined as the angle λ between the radius of the Earth $r = CA$ and the radius CD lying on the equator. When the Earth rotates around its axis, point A describes a circle (center at B) of radius $R = AB = r \cos \lambda$.

The velocity of any point on the Earth's surface is tangential to a circle that is parallel to the equator. Its magnitude, by Equation 5.2, is $v = \omega R = \omega r \cos \lambda$. The acceleration is centripetal, because the circular motion is uniform, and is directed toward B. Its magnitude, by Equation 5.13, is

$$a = \omega^2 R = \omega^2 r \cos \lambda \tag{5.19}$$

Introducing the values of the angular velocity of the Earth, $\omega = 7.292 \times 10^{-5}\,\text{s}^{-1}$ and its radius ($r = 6.35 \times 10^6$ m), we have $v = 459 \cos \lambda\ \text{m s}^{-1}$ and $a = 3.34 \times 10^{-2} \cos \lambda\ \text{m s}^{-2}$. The velocity has its maximum value at the equator ($\lambda = 0°$), where $v = 459\ \text{m s}^{-1}$ or $1652\ \text{km h}^{-1}$. We do not feel the effects of the Earth's rotation because our bodies and senses are accustomed to it. But we would immediately notice a change in it. The maximum value of the acceleration is $3.34 \times 10^{-2}\ \text{m s}^{-2}$, which is about 0.3% of the acceleration due to gravity; this value is so small relative to g that we do not feel any change in centripetal acceleration as we move away from the equator. In working out this example, we have assumed that the Earth has a perfect spherical shape, which is not true. The Earth is slightly flattened at the poles. However, this deviation from sphericity can be ignored in most instances.

EXAMPLE 5.4

A disk D (Figure 5.9) can rotate freely about its horizontal axis. A cord is wrapped around the outer circumference of the disk and a body A, attached to the cord, falls under the action of gravity. The motion of A is uniformly accelerated but, as will be seen in Chapter 13, its acceleration is less than that due to gravity. Suppose that at $t = 0$ the velocity of body A is 0.04 m s^{-1}, and 2 s later it has fallen 0.2 m. Find the tangential and normal accelerations, at any instant, of any point on the rim of the disk.

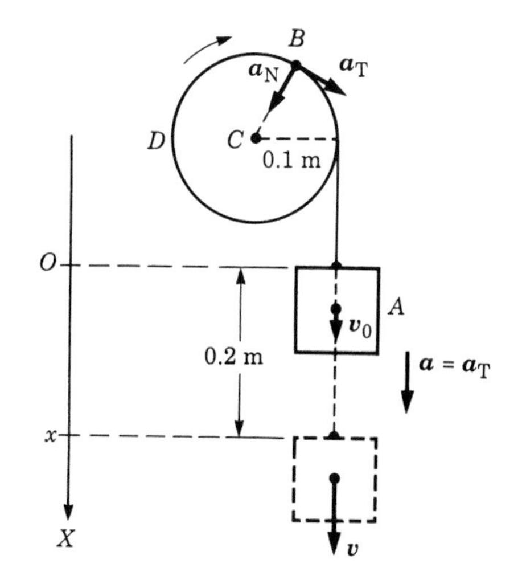

Figure 5.9

▷ The origin of coordinates is at the position O when $t_0 = 0$. The equation of the uniformly accelerated motion of A is

$$x = v_0 t + \tfrac{1}{2}at^2$$

where x is the distance fallen in time t. But we know that $v_0 = 0.04$ m s^{-1}. Thus

$$x = (0.04t + \tfrac{1}{2}at^2)\,\text{m}$$

When $t = 2$ s, $x = 0.2$ m. By substituting these values into the above equation, we find $a = 0.06$ m s^{-2}. We can now write the general equation as

$$x = (0.04t + 0.03t^2)\,\text{m}$$

By differentiation, the velocity of A at any time t is

$$v = \frac{dx}{dt} = (0.04 + 0.06t)\,\text{m s}^{-1}$$

This equation also gives the velocity of any point on the rim of the disk. The tangential acceleration of the points on the rim is thus the same as the acceleration of A:

$$a_T = \frac{dv}{dt} = (0.06)\,\text{m s}^{-2}$$

and, since $R = 0.1$ m, the normal acceleration is

$$a_N = \frac{v^2}{R} = \frac{(0.04 + 0.06t)^2}{0.1} = (0.016 + 0.048t + 0.036t^2)\,\text{m s}^{-2}$$

5.5 Relative rotational motion

Consider two observers O and O' rotating relative to each other but with no relative translational motion. For simplicity, assume that observers O and O'

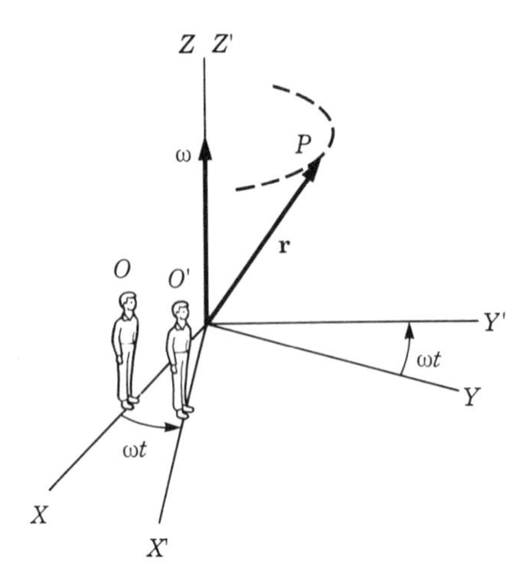

Figure 5.10 Frames of reference in uniform relative rotational motion.

use their own frame of reference attached to each of them, but with a common origin. In this case, observer O, who uses frame XYZ (Figure 5.10), notes that the frame $X'Y'Z'$ attached to O' is rotating with angular velocity $\boldsymbol{\omega}$. For observer O', the reverse occurs; that is, O' sees the frame XYZ (and O) rotating with angular velocity $-\boldsymbol{\omega}$. Let $\boldsymbol{r}$ be the position vector of a particle P relative to the common origin. The velocity of P as measured by O (relative to O's frame of reference XYZ) is

$$\boldsymbol{V} = \left(\frac{d\boldsymbol{r}}{dt}\right)_O \tag{5.20}$$

where the subscript O has been used to emphasize that it is the velocity of P measured by O. If particle P is at rest relative to the rotating observer O', P appears to move in a circle with an angular velocity $\boldsymbol{\omega}$ when observed by O. Therefore, it has a velocity relative to O given by $\boldsymbol{\omega} \times \boldsymbol{r}$. If, instead, particle P is observed by O' to move with a velocity $\boldsymbol{V}'$ (that is $\boldsymbol{V}' = (d\boldsymbol{r}/dt)_{O'}$), then the velocity of particle P *relative to* O must be

$$\boldsymbol{V} = \boldsymbol{V}' + \boldsymbol{\omega} \times \boldsymbol{r} \tag{5.21}$$

This expression relates the velocities $\boldsymbol{V}$ and $\boldsymbol{V}'$ of the same particle as measured by observers O and O' in relative rotational motion with angular velocity $\boldsymbol{\omega}$. It would apply, for example, to the case of an observer rotating with a carrousel and a second observer standing beside the carrousel, both observers using the center of the carrousel as the origin of coordinates. The relation between the acceleration of P measured by O and O' is a little more complicated and, as shown in the derivation below, is given by

$$\boldsymbol{a} = \boldsymbol{a}' + 2\boldsymbol{\omega} \times \boldsymbol{V}' + \boldsymbol{\omega} \times (\boldsymbol{\omega} \times \boldsymbol{r}) \tag{5.22}$$

or

$$\boldsymbol{a}' = \boldsymbol{a} - 2\boldsymbol{\omega} \times \boldsymbol{V}' - 2\boldsymbol{\omega} \times (\boldsymbol{\omega} \times \boldsymbol{r}) \tag{5.23}$$

Equation 5.23 relates the accelerations $\boldsymbol{a}$ and $\boldsymbol{a}'$ of the particle P as measured by observers O and O' who are in uniform relative rotational motion. The second term is called the **Coriolis acceleration**, after Gaspard G. de Coriolis (1792–1843), who described this effect related to rotational motion. The third term corresponds to a **centrifugal acceleration**.

Both the Coriolis and centrifugal accelerations are the result of the relative rotational motion of the observers. They are not accelerations due to any specific action applied to the particle.

Derivation of the relation between accelerations measured by observers in uniform relative rotational motion

To compare the acceleration of P measured by O and O', suppose again that P appears at rest relative to O'. Then, relative to O, the particle moves with uniform circular motion with the centripetal acceleration $\boldsymbol{\omega} \times (\boldsymbol{\omega} \times \boldsymbol{r})$. If, however, P is moving relative to O' with an acceleration $\boldsymbol{a}' = (\mathrm{d}\boldsymbol{V}'/\mathrm{d}t)_{O'}$, we might be tempted to say that the acceleration relative to O is $\boldsymbol{a} = \boldsymbol{a}' + \boldsymbol{\omega} \times (\boldsymbol{\omega} \times \boldsymbol{r})$. This is incorrect. To calculate the acceleration of P relative to O when P is in motion relative to O' we must take into account the variation with respect to time of *both* terms in Equation 5.21. Thus the acceleration of particle P as measured by O is

$$\boldsymbol{a} = \left(\frac{\mathrm{d}\boldsymbol{V}}{\mathrm{d}t}\right)_O = \left(\frac{\mathrm{d}\boldsymbol{V}'}{\mathrm{d}t}\right)_O + \boldsymbol{\omega} \times \left(\frac{\mathrm{d}\boldsymbol{r}}{\mathrm{d}t}\right)_O \tag{5.24}$$

The last term can be simplified by using Equation 5.21 in the following way:

$$\begin{aligned}\boldsymbol{\omega} \times \left(\frac{\mathrm{d}\boldsymbol{r}}{\mathrm{d}t}\right)_O &= \boldsymbol{\omega} \times \boldsymbol{V} = \boldsymbol{\omega} \times (\boldsymbol{V}' + \boldsymbol{\omega} \times \boldsymbol{r}) \\ &= \boldsymbol{\omega} \times \boldsymbol{V}' + \boldsymbol{\omega} \times (\boldsymbol{\omega} \times \boldsymbol{r})\end{aligned} \tag{5.25}$$

The term $(\mathrm{d}\boldsymbol{V}'/\mathrm{d}t)_O$ of Equation 5.24 is the time rate of change of the vector $\boldsymbol{V}'$ as measured by observer O. This is not the same as the time rate of change of $\boldsymbol{V}'$ relative to O', which is $(\mathrm{d}\boldsymbol{V}'/\mathrm{d}t)_{O'}$ or $\boldsymbol{a}'$. For example, suppose P moves relative to O' in a straight line with *constant* velocity so that $(\mathrm{d}\boldsymbol{V}'/\mathrm{d}t)_{O'} = 0$. Then, relative to O, the particle describes a helical path and the direction of $\boldsymbol{V}'$ changes, rotating with angular velocity $\boldsymbol{\omega}$ so that

$$(\mathrm{d}\boldsymbol{V}'/\mathrm{d}t)_O = \boldsymbol{\omega} \times \boldsymbol{V}'$$

Therefore, in the general case, when O' observes particle P moving with an acceleration $\boldsymbol{a}' = (\mathrm{d}\boldsymbol{V}'/\mathrm{d}t)_{O'}$, we must write

$$\left(\frac{\mathrm{d}\boldsymbol{V}'}{\mathrm{d}t}\right)_O = \left(\frac{\mathrm{d}\boldsymbol{V}'}{\mathrm{d}t}\right)_{O'} + \boldsymbol{\omega} \times \boldsymbol{V}' = \boldsymbol{a}' + \boldsymbol{\omega} \times \boldsymbol{V}' \tag{5.26}$$

Using Equations 5.25 and 5.26 in Equation 5.24 we can then write

$$\boldsymbol{a} = \boldsymbol{a}' + 2\boldsymbol{\omega} \times \boldsymbol{V}' + \boldsymbol{\omega} \times (\boldsymbol{\omega} \times \boldsymbol{r})$$

which is the expression given in Equation 5.22.

5.6 Motion relative to the Earth

One of the most interesting applications of Equations 5.21 and 5.22 is the study of a body's motion relative to the Earth. It is also important for the analysis of hurricanes, as well as for calculating the motion of ballistic missiles and artificial satellites. As indicated in Example 5.1, the angular velocity of the Earth is $\omega = 7.292 \times 10^{-5}\ \mathrm{rad\,s^{-1}}$. Its direction is that of the North–South axis of rotation of the Earth. Consider a point A on the Earth's surface and call $\boldsymbol{g}_0$ the acceleration of gravity measured by a non-rotating observer. Then $\boldsymbol{g}_0$ corresponds to $\boldsymbol{a}$ in Equation 5.22. Therefore the acceleration measured by an observer rotating with the Earth is

$$\boldsymbol{a}' = \boldsymbol{g}_0 - \boldsymbol{\omega} \times (\boldsymbol{\omega} \times \boldsymbol{r}) - 2\boldsymbol{\omega} \times \boldsymbol{V}' \tag{5.27}$$

This relation shows that the acceleration of a body relative to the Earth depends on the velocity $\boldsymbol{V}'$ relative to the Earth and on the position $\boldsymbol{r}$ of the body. We shall now examine the effect of the centrifugal and Coriolis accelerations separately. Note that the centrifugal term is of the order of 3.3×10^{-2} m s^{-2}, while the Coriolis term is of the order of $7.3 \times 10^{-5} V'$ m s^{-2}. Thus, for velocities smaller than 400 m s^{-1} or 1500 km h^{-1}, the Coriolis acceleration can be neglected relative to the centripetal acceleration. However, as we shall see, it has an important directional effect.

(a) Centrifugal acceleration Consider a body initially at rest, or moving very slowly, so that the Coriolis term $2\boldsymbol{\omega} \times \boldsymbol{V}'$ is zero or negligible when compared with the centrifugal term $-\boldsymbol{\omega} \times (\boldsymbol{\omega} \times \boldsymbol{r})$. The acceleration $\boldsymbol{a}'$ measured in this case is called the **effective acceleration** of gravity, and is designated by $\boldsymbol{g}$. Thus

$$\boldsymbol{g} = \boldsymbol{g}_0 - \boldsymbol{\omega} \times (\boldsymbol{\omega} \times \boldsymbol{r}) \tag{5.28}$$

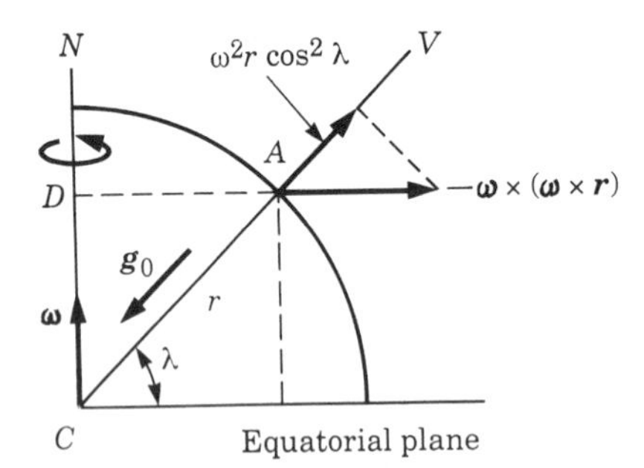

Figure 5.11 Centrifugal acceleration due to the rotation of the Earth.

The acceleration $-\boldsymbol{\omega} \times (\boldsymbol{\omega} \times \boldsymbol{r})$ points in the outward direction DA as shown in Figure 5.11.

Assuming that the Earth is spherical (actually it departs slightly from this shape) and that there are no local anomalies, we may consider that $\boldsymbol{g}_0$ points toward the center of the Earth along the radial direction. Due to the second term in Equation 5.28, the direction of $\boldsymbol{g}$ (called the **vertical**) deviates slightly from the radial direction; it is determined by a plumb line. Liquids always rest in equilibrium, with their surface perpendicular to $\boldsymbol{g}$. However, for practical purposes, and in the absence of local disturbances, the vertical may be assumed to coincide with the radial direction. The magnitude of $\boldsymbol{g}$ is slightly less than that of $\boldsymbol{g}_0$ and can be expressed approximately as (see the derivation below):

$$g = g_0 - \omega^2 r \cos^2 \lambda \tag{5.29}$$

Although the last term is very small (about 0.3%) compared to g_0, it accounts for most of the observed variations of the acceleration of gravity with latitude (see Table 5.1).

Table 5.1 Experimental values of the acceleration of gravity

Location	Latitude	$g\,(\mathrm{m\,s^{-2}})$
North Pole	90°00′	9.8321
Anchorage, Alaska	61°10′	9.8218
Greenwich, England	51°29′	9.8119
Paris, France	48°50′	9.8094
Washington DC, USA	38°53′	9.8011
Key West, Florida	24°34′	9.7897
Panama City, Panama	8°55′	9.7822
Equator	0°00′	9.7799

Variation of gravity with latitude

From Figure 5.11 we see that while the vector $\boldsymbol{g}_0$ points toward the center C of the Earth, the vector $-\boldsymbol{\omega} \times (\boldsymbol{\omega} \times \boldsymbol{r})$ is parallel to the equatorial plane. The magnitude of this vector is

$$|\boldsymbol{\omega} \times (\boldsymbol{\omega} \times \boldsymbol{r})| = \omega^2 r \cos \lambda = 3.34 \times 10^{-2} \cos \lambda \ \mathrm{m\,s^{-2}}$$

which is very small in comparison with $g_0 \simeq 9.8\ \mathrm{m\,s^{-2}}$. Therefore in computing the magnitude of $\boldsymbol{g} = \boldsymbol{g}_0 - \boldsymbol{\omega} \times (\boldsymbol{\omega} \times \boldsymbol{r})$, we may assume that $\boldsymbol{g}$ departs so little from the vertical direction AV that it is sufficient to subtract (from g_0) the component of $-\boldsymbol{\omega} \times (\boldsymbol{\omega} \times \boldsymbol{r})$ along AV. The magnitude of this component is

$$|\boldsymbol{\omega} \times (\boldsymbol{\omega} \times \boldsymbol{r})| \cos \lambda = \omega^2 r \cos^2 \lambda$$

Therefore $g = g_0 - \omega^2 r \cos^2 \lambda$, which is Equation 5.29.

(b) Coriolis effect The Coriolis acceleration $-2\boldsymbol{\omega} \times \boldsymbol{V}'$ is perpendicular to the velocity $\boldsymbol{V}'$. Therefore, its effect is to deviate the particle in a direction perpendicular to its velocity. For example, it can be seen from Figure 5.12 that the Coriolis effect on a freely falling body is to deviate its path slightly from a straight line so that the body strikes the Earth at a point due east of the point directly below the initial position.

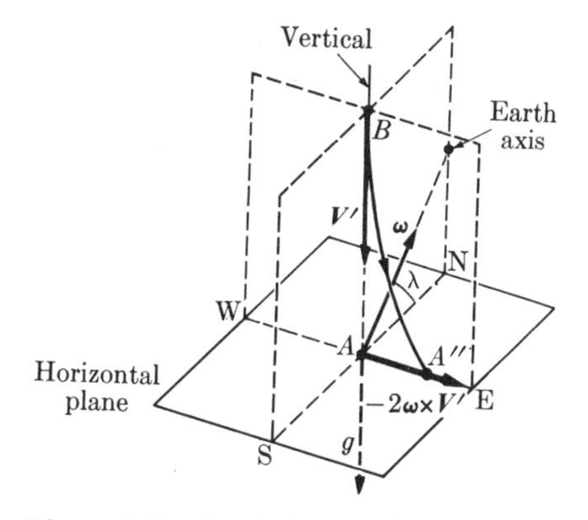

Figure 5.12 Deviation to the east of a falling body due to Coriolis acceleration.

If the particle moves in a horizontal plane, the Coriolis acceleration tends to deviate the path toward the right in the Northern hemisphere (and toward the left in the Southern hemisphere) as can be seen from Figure 5.13, in which a_{H} is the horizontal component of $-2\boldsymbol{\omega} \times \boldsymbol{V}'$.

The Coriolis effect may be seen in two common phenomena: the rotational motion of storms around low-pressure centers and the rotation of the plane of oscillation of a pendulum.

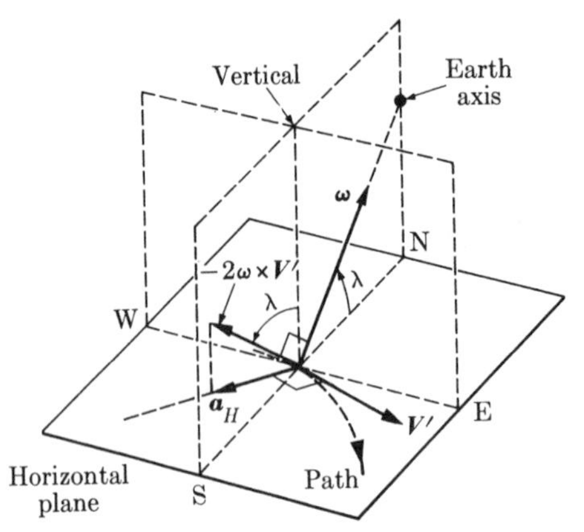

Figure 5.13 Deviation to the right of a body moving horizontally in the Northern hemisphere.

If a low-pressure center develops in the atmosphere, the wind will flow radially toward the center (Figure 5.14). However, the Coriolis acceleration deviates the air molecules toward the right of their paths in Northern latitudes, resulting in a counter-clockwise motion. In the Southern hemisphere the rotation is clockwise.

As a second example of the Coriolis effect, consider a pendulum. When the amplitude of oscillations is small, we can assume that the motion of the bob is along a horizontal path. If the pendulum were initially set to oscillate in the East–West direction and were released at A (see Figure 5.15), it would continue oscillating between A and B if the Earth were not rotating. Because of the Coriolis acceleration due to the Earth's rotation however, the path of the pendulum is deflected continuously to the right in the Northern hemisphere and to the left in the Southern hemisphere. Therefore, at the end of the first oscillation, it reaches B' instead of B. On its return, it goes to A' and not to A. In successive complete oscillations, it arrives at A'', A''', etc. In other words, the plane of oscillation of the pendulum rotates clockwise in the Northern hemisphere and counter-clockwise

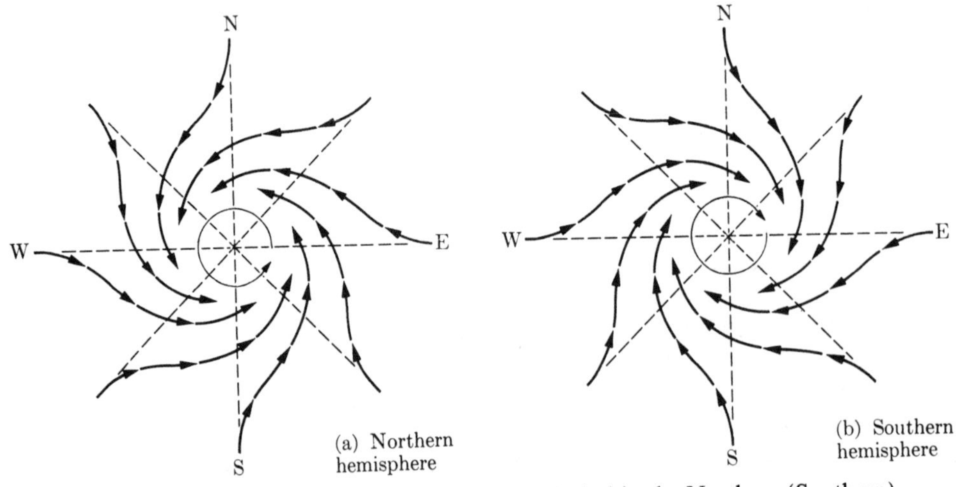

Figure 5.14 Counter-clockwise (clockwise) whirling of wind in the Northern (Southern) hemisphere resulting from a low-pressure center combined with Coriolis acceleration.

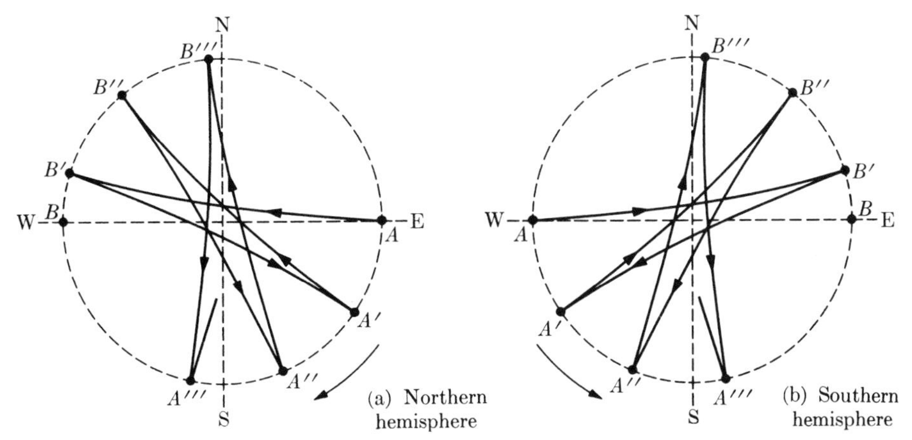

Figure 5.15 Rotation of plane of oscillation of a pendulum in the Southern hemisphere is in the opposite direction to that in the Northern hemisphere.)

in the Southern hemisphere. The Coriolis effect on a pendulum is a maximum at the poles and zero at the equator.

This effect was spectacularly demonstrated by Jean Léon Foucault in 1851. From the dome of Les Invalides, in Paris, he hung a pendulum 67 m long. During each oscillation, the pendulum's bob dropped sand on a circle, experimentally demonstrating that its plane of oscillation rotated at the rate of $11°15'$ each hour. Foucault's experiment is an effective proof of the rotation of the Earth. Even if the Earth had always been covered by clouds, this experiment would have told scientists that the Earth was rotating and would have allowed them to find the period of rotation.

QUESTIONS

5.1 What is meant by a periodic process? Give examples of periodic processes.

5.2 Why is the acceleration in uniform circular motion called 'centripetal'? (Look the word up in a dictionary).

5.3 Is it possible for a body to have centripetal acceleration but no tangential acceleration? Or to have tangential acceleration but no centripetal acceleration?

5.4 Why is the angular velocity represented by a vector perpendicular to the plane of the circular motion?

5.5 Verify that in accelerated circular motion the acceleration may be written as $\boldsymbol{a} = \boldsymbol{\alpha} \times \boldsymbol{r} + \boldsymbol{\omega} \times (\boldsymbol{\omega} \times \boldsymbol{r})$, where $\boldsymbol{\alpha} = \mathrm{d}\boldsymbol{\omega}/\mathrm{d}t$ is the vector angular acceleration.

5.6 Explain why the effective acceleration of gravity increases as the latitude increases.

5.7 Discuss the effect of the Coriolis acceleration on a body moving in a horizontal plane in the Southern hemisphere.

5.8 Verify that, in the case of a falling body in the Northern hemisphere, the Coriolis acceleration points east and has a magnitude $2\omega V' \cos \lambda$, where λ is the latitude. Verify that if the body moves vertically upward, the Coriolis acceleration points west and has the same magnitude given above.

5.9 Verify that, in the case of a body in the Northern hemisphere moving north, the Coriolis acceleration points east and has a magnitude $2\omega V' \sin \lambda$. Show that if the body is moving south, the Coriolis acceleration points west and has the same magnitude given above.

5.10 Discuss the effect of the Coriolis acceleration on a falling body in the Southern hemisphere.

PROBLEMS

5.1 (a) Calculate the angular velocity of a disk that rotates with uniform motion 13.2 rad every 6 s. (b) Calculate the period and frequency of rotation. How long will it take the disk (c) to rotate through an angle of 780°, and (d) to make 12 revolutions?

5.2 Calculate the angular velocity of each of the three hands of a clock.

5.3 Calculate (a) the angular velocity, (b) the linear velocity and (c) the centripetal acceleration of the Moon. The Moon makes a complete revolution in 28 days and the average distance from the Earth to the Moon is 3.84×10^8 m.

5.4 Find (a) the magnitude of the velocity and (b) the centripetal acceleration of the Earth in its motion around the Sun. The radius of the Earth's orbit is 1.49×10^{11} m and its period of revolution around the Sun is one year (3.16×10^7 s).

5.5 Find (a) the magnitude of the velocity and (b) the centripetal acceleration of the Sun in its motion through the Milky Way. The radius of the Sun's orbit is 2.4×10^{20} m and its period of revolution is 6.3×10^{15} s.

5.6 Find (a) the velocity and (b) the centripetal acceleration of the electron in a hydrogen atom, assuming the orbit is a circle of radius 5×10^{-11} m and the period of the motion is 1.5×10^{-16} s.

5.7 A flywheel whose diameter is 3 m is rotating at 120 rpm. Calculate: (a) its frequency, (b) the period, (c) the angular velocity, and (d) the linear velocity of a point on the rim.

5.8 The angular velocity of a flywheel increases uniformly from 20 rad s^{-1} to 30 rad s^{-1} in 5 s. Calculate (a) the angular acceleration and (b) the total angle through which it has rotated.

5.9 A body, initially at rest ($\theta = 0$ and $\omega = 0$ at $t = 0$) is accelerated in a circular path of radius 1.3 m according to the equation $\alpha = 120t^2 - 48t + 16$. Find (a) the angular position and (b) the angular velocity of the body as functions of time, and (c) the tangential and (d) the centripetal components of its acceleration.

5.10 A particle is moving in a circle according to the law $\theta = 3t^2 + 2t$ where θ is measured in radians and t in seconds. Calculate (a) the angular velocity and (b) the angular acceleration after 4 s. (c) Calculate the centripetal acceleration as a function of time if the radius of the circle is 2 m.

5.11 A wheel starts from rest and accelerates in such a manner that its angular velocity increases uniformly to 200 rpm in 6 s. After it has been rotating for some time at this speed, brakes are applied, stopping the wheel in 5 min. The total number of revolutions of the wheel is 3100. (a) Plot the angular velocity as a function of time. Calculate (b) the total time of rotation and (c) the total angle the wheel has rotated.

5.12 A flywheel 1.6 m in radius is rotating about a horizontal axis by means of a rope wound about its rim and having a weight at its end. If the vertical distance traveled by the weight is given by the equation $x = 10t^2$, where x is measured in meters and t in seconds, calculate (a) the angular velocity and (b) angular acceleration of the flywheel at any time.

5.13 A wheel of radius R rolls with constant velocity v_0 along a horizontal plane (Figure 5.16). (a) Verify that the position of a point on its edge initially at O is given by the equations $x = R(\omega t - \sin \omega t)$ and $y = R(1 - \cos \omega t)$, where $\omega = v_0/R$ is the angular velocity of the wheel and t is measured from the instant when the point is initially in contact with the plane. (b) Find the components of the velocity and the acceleration of the point. (c) Draw the velocity and the acceleration of the point. (d) Draw the paths of a point on the rim of the wheel and at a distance 2/3 the radius from the axis.

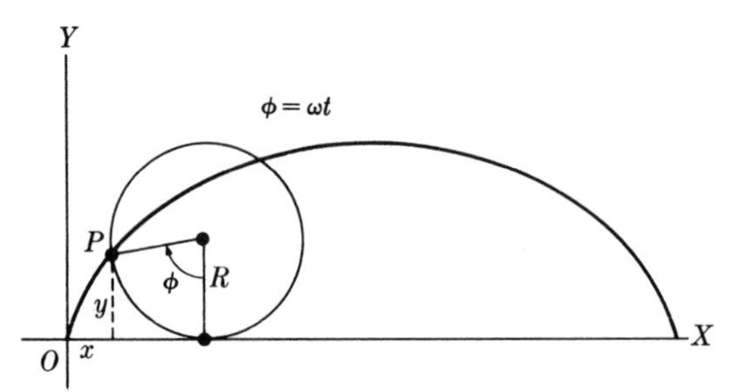

Figure 5.16

5.14 A particle with a velocity of 500 m s^{-1} relative to the Earth is heading due south at latitude 45° N. Compute (a) the centrifugal acceleration and (b) the Coriolis acceleration relative to the Earth of the particle. (c) Repeat the problem for a position at latitude 45° S.

5.15 A river flows (i) due N, (ii) due S, (iii) due E at 9 km h^{-1} at 45° latitude N. (a) Find the Coriolis acceleration. (b) Which side does the water push against, producing a larger erosion? (c) Repeat when the river is at 45°S latitude.

6 Force and momentum

Sir Isaac Newton was the author of *Principia Mathematica*, published in 1687, which contains the three laws of motion that are the foundation of **classical dynamics**. Newton also derived the law of universal gravitation in the *Principia*. This demonstrated that the same law governs the motion of bodies on Earth and the motion of the planets and other bodies in the universe. He discovered many optical phenomena as well, including the dispersion of light. Newton was a coinventor of the calculus.

6.1 Introduction

In the previous three chapters we discussed the elements that enter into the 'description' of the motion of a particle. We will now investigate the reasons *why* particles move the way they do. Why do bodies near the surface of the Earth fall with constant acceleration? Why does the Earth move around the Sun in an elliptical orbit? Why do atoms bind together to form molecules? Why does a spring oscillate when it is stretched? Understanding these motions and many others is important not only to our basic knowledge of nature, but also for engineering and practical applications. When we just 'describe' motions, we only learn about the specific situation we are describing. But when we understand how motions in general are produced, we are able to design machines and other practical devices that move as we desire. The study of the relationship between the motion of a body and the causes for this motion is called **dynamics**.

From daily experience we know that

the motion of a body is a direct result of its interactions with the other bodies around it.

When a person strikes a ball, there is an interaction with the ball, modifying its motion. The path of a projectile is but a result of its gravitational interaction with the Earth. The Earth moves around the Sun because there is a mutual gravitational interaction. The motion of an electron around a nucleus is the result of its electric interaction with the nucleus and perhaps with other electrons. Protons and neutrons are held together in a nucleus through their mutual nuclear or strong interaction.

Interactions are often expressed quantitatively in terms of a concept called **force**. We have an intuitive notion of force and of its strength in terms of a push or a pull. We say the weight of a body is the result of the Earth's gravitational pull on it. Other similar situations are the pull of a magnet on iron filings and the pull of a rubbed glass rod on small pieces of paper. For the concept of force to be useful to the physicist, however, we must be able to express force in a precise and quantitative way. To be more specific, a force must be expressed in terms of the parameters that describe the physical system, such as the distance between the particles, their masses, their charges etc. This important question will be developed as we proceed further in this text, when the different interactions found in nature will be analyzed in more detail. In the meantime, we shall assume that we have an experimental means for measuring forces and for determining how they depend on the relative position of the interacting particles. For example, by hanging several equal masses from a spring it is found that the force required to stretch a spring is proportional to the amount of stretching (Figure 6.1). This is the principle of the spring balance.

This chapter describes the methods of determining the motion of a body when the forces acting on it are known. Conversely, we discuss methods for determining the forces acting on a body when the motion is known. The study of dynamics is basically the analysis of the relation between force and the motion of a body.

The laws of motion we present here are generalizations. They arise from a careful analysis of motions we observe around us. Observations are extrapolated

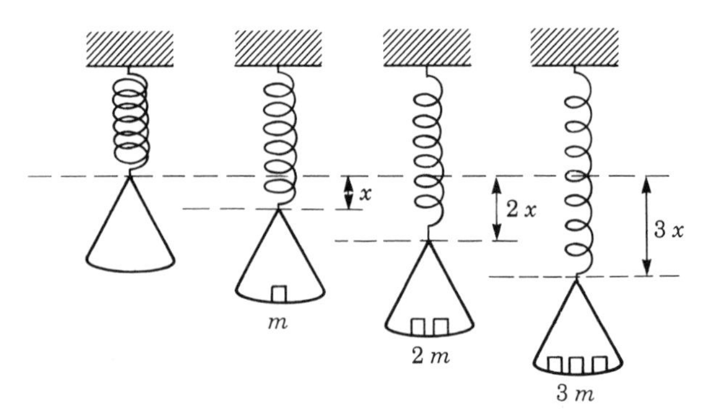

Figure 6.1 Spring balance for measuring forces.

to certain ideal or simplified experiments. These laws were first stated in a formal way by Sir Isaac Newton (1642–1727), but prior to his work they were suggested in a diverse form by Galileo Galilei. These laws are valid for particles moving with velocities small compared with that of light. In Chapter 19 we will see how they have to be modified when the particles move with very large velocities.

6.2 The law of inertia

A **free particle** is one that is not subject to any interaction. Strictly speaking, there is no such thing, because each particle is subject to interactions with the rest of the particles in the world. In practice, however, there are some particles that may be considered free. For example, if they are sufficiently far away from others, their interactions may be negligible or, because the interactions with the other particles cancel, give a zero net interaction.

The **law of inertia** states that

> *a free particle always moves with constant velocity, or (which amounts to the same thing) without acceleration.*

That is, a free particle either moves in a straight line with constant velocity or is at rest (zero velocity). This statement is also called **Newton's first law**.

Since motion is relative, when we state the law of inertia we must indicate to whom or what the motion of the free particle is referred. We assume that the motion of the particle is relative to an observer who may also be considered a free particle or system, i.e. who is not subject to interactions with the rest of the world. Such an observer is called an **inertial observer**, and the frame of reference used is called an **inertial frame of reference**. Different inertial observers may be in uniform relative motion. Therefore, a free particle at rest relative to an inertial observer may appear in motion with constant velocity relative to other inertial observers.

We assume that inertial frames of reference are not rotating. Rotation (a change in velocity due to a change in direction) implies acceleration. This is contrary to our definition of the inertial observer as a 'free particle', that is, one without acceleration.

We shall assume that inertial observers correlate their observations through the Galilean transformation (Section 4.6). Later on (Chapter 19) we will learn that when the relative velocity of the two observers is very large, they must correlate their observations by using the Lorentz transformation. For most practical situations, the Galilean transformation is adequate.

Because of its daily rotation and elliptical orbital path, the Earth is not an inertial frame of reference. In many cases, however, the effect of the Earth's motion is negligible. Therefore, in most cases, the frames of reference attached to our terrestrial laboratories can be considered inertial without too much error. Nor is the Sun an inertial frame of reference. Because of its interactions with the other bodies in our galaxy, it describes a curved orbit about the center of the galaxy (Figure 6.2). However, the orbital acceleration of the Sun is 150 million times

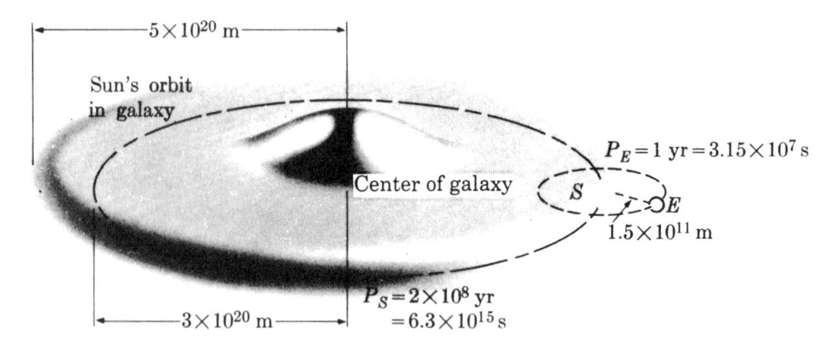

Figure 6.2 A coordinate system attached to the Earth is not inertial because of its daily rotation and its accelerated motion around the Sun; nor is the Sun an inertial frame, because of its motion about the center of the galaxy. However, for practical purposes, either of these two bodies may be used to define an inertial frame.

smaller than that of the Earth. The Sun's resemblance to an inertial frame is therefore much greater.

Many experiments performed in our terrestial laboratories support the law of inertia. A ball resting on a smooth horizontal surface will remain at rest unless acted upon. That is, its velocity relative to the surface remains constant, with value equal to zero. We assume that the surface on which the ball is resting balances the interaction between the ball and the Earth. Hence the ball is essentially free of interactions as long as it moves on the surface. When the ball is hit, as in billiards, it momentarily suffers an interaction and gains velocity. Immediately afterward it is free again, moving in a straight line along the surface with the velocity it acquired when it was struck. If the ball were rigid and perfectly spherical, and the surface perfectly horizontal and smooth, we may assume that the ball would continue to move that way indefinitely. In practice this is not the case, for the ball slows down and eventually stops. We say that there has been an additional unbalanced interaction between the ball and the surface. This interaction, called **friction**, will be discussed in Section 7.4.

6.3 Mass

In Section 2.3 we gave an operational definition of **mass** as a number attached to each particle or body obtained by comparison with a standard body, whose mass is defined as unity, by use of an equal arm balance. Such a procedure is based on the downward pull exerted by the Earth on all bodies; that is, the Earth's gravitational attraction. For that reason, mass so measured is designated as the **gravitational mass** of the body.

The above definition of mass, although of great practical value for comparing the masses of small bodies, has some inconveniences. First, not all bodies can be placed on a balance to find their masses. Second, aside from the gravitational attraction, bodies may (indeed, do) experience other types of interactions. Therefore, we must explore whether or not the concept of mass applies to those cases as well.

Finally, the operational definition of mass is based on the assumption that the body is at rest; we do not know from that definition whether or not mass will be the same when the particle is in motion.

One immediate consequence of the law of inertia is that an inertial observer recognizes that a particle is *not* free (i.e. it is interacting with other particles) when it is observed that the velocity of the particle fails to remain constant.

Suppose that, instead of observing one isolated particle in the universe as we did for the law of inertia, we observe two isolated particles that are subject only to their mutual interaction and are otherwise isolated from the rest of the world. As a result of their interaction, their individual velocities are not constant, but change with time, and their paths, in general, are curved. As a further simplification, assume that the particles do not interact until they are very close, inside the darkened region in Figure 6.3(a). In practice, this is what happens in collision experiments. We observe the particles before and after their interaction. Their respective velocities before the interaction are $\boldsymbol{v}_1$ and $\boldsymbol{v}_2$; after the interaction their velocities are $\boldsymbol{v}'_1$ and $\boldsymbol{v}'_2$, respectively. The change in the velocity of particle 1 as a result of the interaction is:

$$\Delta \boldsymbol{v}_1 = \boldsymbol{v}'_1 - \boldsymbol{v}_1$$

and the change in the velocity of particle 2 is

$$\Delta \boldsymbol{v}_2 = \boldsymbol{v}'_2 - \boldsymbol{v}_2$$

The first experimental result is that the *changes* in velocity, $\Delta \boldsymbol{v}_1$ and $\Delta \boldsymbol{v}_2$, produced by the interaction are always in *opposite* directions (Figure 6.3(b)). The second experimental result is that the ratio of the magnitudes of the changes in velocity is *constant*:

$$\frac{|\Delta \boldsymbol{v}_1|}{|\Delta \boldsymbol{v}_2|} = const.$$

In addition, when the masses of the particles are known, it is found that the particle with the larger mass experiences a smaller change in velocity as a result of the

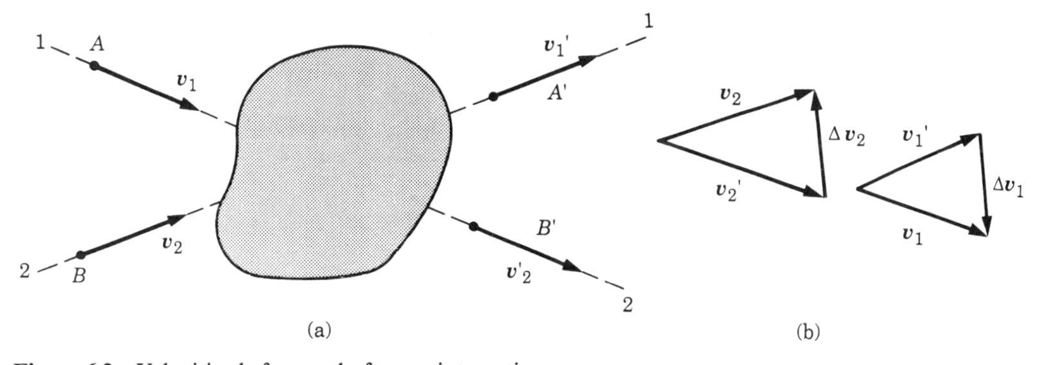

Figure 6.3 Velocities before and after an interaction.

interaction. More precisely, the changes in velocity are always in inverse proportion to the masses m_1 and m_2 of the particles; that is

$$\frac{|\Delta \boldsymbol{v}_1|}{|\Delta \boldsymbol{v}_2|} = \frac{m_2}{m_1} \tag{6.1}$$

Equation 6.1 also serves to compare the masses of two particles. So, if we know the mass m_1 of particle 1, we can determine the mass m_2 of particle 2 by letting particle 2 interact with particle 1 and measuring the respective changes in velocity. The mass obtained in this way is called the **inertial mass**. For example, the mass of the neutron was first determined based on this method (see Example 14.7).

Experiments show that the inertial mass obtained using Equation 6.1 coincides with the gravitational mass obtained using the balance method, wherever that method can be used. Thus, as far as our experimental measurements are concerned, gravitational and inertial masses are identical, which shows that the mass of a particle is a property independent of the interaction to which it is subject and the particles with which it interacts. This fact will be considered in more detail in Section 11.4. Therefore, we conclude that

the inertial mass of a particle is a property that determines how its velocity changes when it interacts with other bodies.

In that sense, mass is a fundamental property of matter.

We may write Equation 6.1 in a more general vector form as

$$m_1 \Delta \boldsymbol{v}_1 = -m_2 \Delta \boldsymbol{v}_2 \tag{6.2}$$

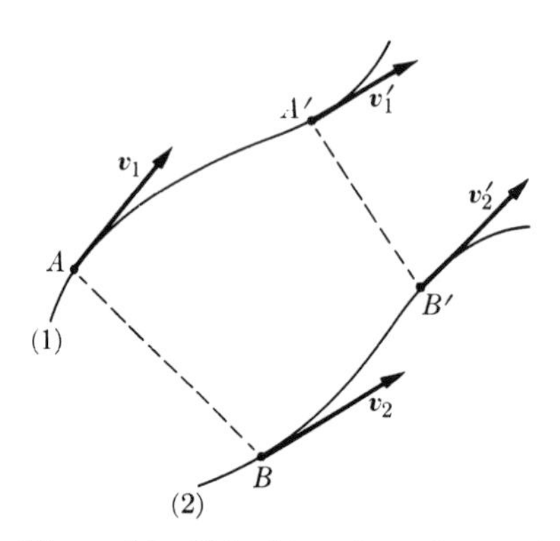

Figure 6.4 Velocities of two interacting particles.

where the negative sign is because $\Delta \boldsymbol{v}_1$ and $\Delta \boldsymbol{v}_2$ have opposite directions. Suppose now that we are able to observe the particles *while* they are interacting (Figure 6.4) and we measure the velocities of the particles at any two times t and t'. We find that, as long as the particles are not too far apart and move slowly in comparison with the velocity of light, relation 6.2 remains valid. Finally, Equation 6.2 has been found to be valid for *all* interactions, within the limitations stated.

EXAMPLE 6.1

The diagram shown in Figure 6.5 illustrates a simple experiment for comparing masses dynamically. Carts 1 and 2 are joined by a compressed spring and the system is held together by a string tied to the two carts. When the string is cut the compressed spring will press on both carts and they acquire velocities v_1 and v_2. In an experiment the velocities

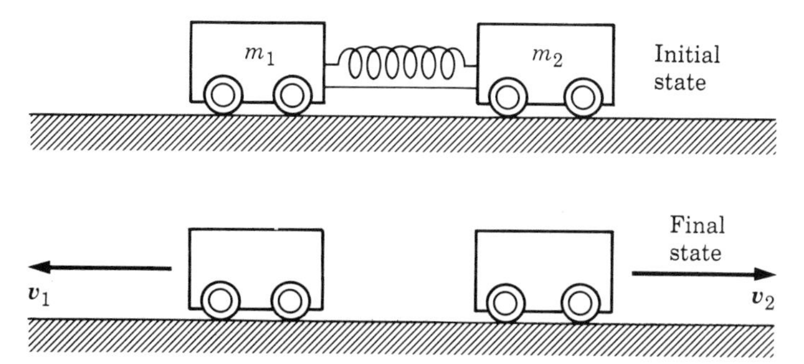

Figure 6.5

of carts 1 and 2 are found to be 0.6 m s^{-1} and 0.3 m s^{-1}, respectively. When the experiment is repeated using carts 1 and 3, the velocities are 0.4 m s^{-1} and 0.5 m s^{-1}. When the experiment is done a third time using carts 2 and 3, cart 2 has a velocity of 0.8 m s^{-1}. What is the velocity of cart 3? If the mass of cart 1 is 2 kg, what are the masses of carts 2 and 3?

▷ Let m_1, m_2 and m_3 be the respective masses of the carts. The initial velocities of the carts are zero so that $\Delta v_1 = v_1$ and $\Delta v_2 = v_2$. Then, using magnitudes only, we can write:

$$m_1 v_1 = m_2 v_2 \qquad \text{or} \qquad \frac{m_2}{m_1} = \frac{v_1}{v_2} = \frac{0.6}{0.3} = 2.0$$

and

$$m_1 v_1 = m_3 v_3 \qquad \text{or} \qquad \frac{m_3}{m_1} = \frac{v_1}{v_3} = \frac{0.4}{0.5} = 0.8$$

Finally

$$m_2 v_2 = m_3 v_3 \qquad \text{or} \qquad v_3 = \frac{m_2}{m_3} v_2 = 0.8 \frac{m_2}{m_3}$$

However, from the first two experiments we see that

$$\frac{m_2}{m_3} = \frac{m_2/m_1}{m_3/m_1} = \frac{2.0}{0.8} = 2.5$$

Therefore, $v_3 = 0.8 \times 2.5 = 2$ m s^{-1}. In this way we have obtained the velocity of cart 3 without finding its mass. When $m_1 = 2$ kg it is easily seen that $m_2 = 4$ kg and $m_3 = 1.6$ kg.

6.4 Linear momentum

The way mass appears in the relation given by Equation 6.2 suggests that a new physical quantity be introduced. This quantity is called the **linear momentum** of a particle and is defined as the product of its mass and its velocity. Designating the linear momentum by $\boldsymbol{p}$, we write

$$\boldsymbol{p} = m\boldsymbol{v} \tag{6.3}$$

Linear momentum is a vector quantity, and it has the same direction as the velocity. It is a very important physical concept because it combines two elements that characterize the dynamical state of a particle: its mass and its velocity. From here on we shall use the word **momentum** instead of 'linear momentum'. In the SI system, momentum is expressed in kg m s^{-1} (no special name has been given to this unit).

That momentum is a more informative dynamical quantity than velocity alone can be seen from several simple experiments. For example, a loaded truck moving with velocity $\boldsymbol{v}$ is more difficult to stop or speed up than when empty because the momentum of the loaded truck is greater.

We may now restate the law of inertia by saying that:

a free particle always moves with constant momentum relative to an inertial frame of reference:

$$\boldsymbol{p} = \textit{const.}$$

On the other hand, if the particle is not free and its velocity changes during an interval Δt, its change in momentum is

$$\Delta \boldsymbol{p} = \Delta(m\boldsymbol{v}) = m\Delta\boldsymbol{v} \tag{6.4}$$

as long as the mass does not change. Later on we will consider the case of a rocket burning fuel (see Example 7.6) where the mass does change. The definition (Equation 6.3) of momentum is valid only for velocities small compared with the velocity of light. In Chapter 19 we will modify the definition for fast-moving particles.

6.5 Principle of conservation of momentum

Consider two particles with masses m_1 and m_2 interacting with each other (Figure 6.6(a)) so that Equation 6.2 holds (i.e. $m_1\Delta\boldsymbol{v}_1 = -m_2\Delta\boldsymbol{v}_2$). Then, using Equation 6.4, we may rewrite Equation 6.2 as

$$\Delta \boldsymbol{p}_1 = -\Delta \boldsymbol{p}_2 \tag{6.5}$$

This vector relation indicates that, for two interacting particles, the change in momentum of one particle in a certain time interval is equal in magnitude and opposite in direction to the change in momentum of the other during the same time interval (Figure 6.6(b)). The above result may be expressed by saying

an interaction produces an exchange of momentum.

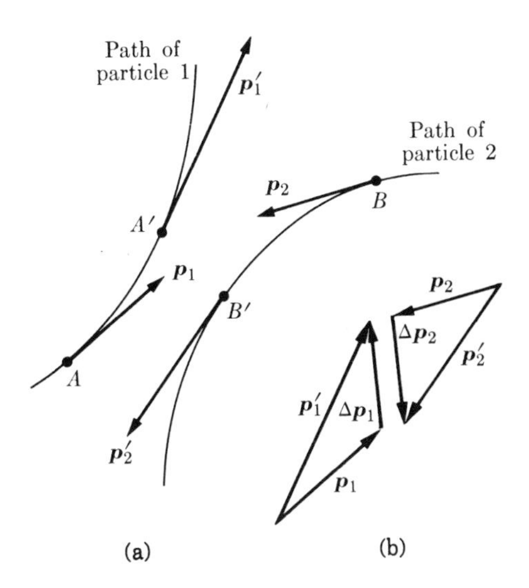

Figure 6.6 Momentum exchange as a result of the interaction between two particles.

That is, the momentum 'lost' by one of the interacting particles is equal to the momentum 'gained' by the other.

The change in momentum of particle 1 in the time interval $\Delta t = t' - t$ is

$$\Delta \boldsymbol{p}_1 = \boldsymbol{p}_1' - \boldsymbol{p}_1$$

The corresponding change of momentum of particle 2 in the same time interval is

$$\Delta \boldsymbol{p}_2 = \boldsymbol{p}_2' - \boldsymbol{p}_2$$

Therefore, we may rewrite Equation 6.5 as

$$\boldsymbol{p}_1' - \boldsymbol{p}_1 = -(\boldsymbol{p}_2' - \boldsymbol{p}_2) = -\boldsymbol{p}_2' + \boldsymbol{p}_2$$

By rearranging terms, we can write

$$\boldsymbol{p}_1' + \boldsymbol{p}_2' = \boldsymbol{p}_1 + \boldsymbol{p}_2 \tag{6.6}$$

We shall define

$$\boldsymbol{P} = \boldsymbol{p}_1 + \boldsymbol{p}_2 = m_1\boldsymbol{v}_1 + m_2\boldsymbol{v}_2 \tag{6.7}$$

as the **total momentum** of the two particles at time t. Similarly, $\boldsymbol{P}' = \boldsymbol{p}_1' + \boldsymbol{p}_2'$ is the total momentum of the two particles at time t'. The left-hand side of Equation 6.6 is the total momentum of the system of two particles at t'. The right-hand side is the total momentum of the same system at time t. So we have $\boldsymbol{P}' = \boldsymbol{P}$. Since this is true, no matter what t and t' are, we conclude that the total momentum of a system of two particles remains the same. In other words:

the total momentum of a system composed of two particles subject only to their mutual interaction remains constant:

$$\boldsymbol{P} = \boldsymbol{p}_1 + \boldsymbol{p}_2 = const. \tag{6.8}$$

This result constitutes the **principle of the conservation of momentum**, one of the most fundamental and universal principles of physics. For example, a hydrogen atom can be considered composed of an electron revolving around a proton. Assume that it is isolated so that only the interaction between the electron and the proton has to be considered. Then the sum of the momenta of the electron and the proton relative to an inertial frame of reference is constant. Similarly, consider the system composed of the Earth and the Moon. Neglecting the interactions of the Sun and the other bodies of our solar system, the sum of the momenta of the Earth and the Moon, relative to an inertial frame of reference, is constant. The principle of conservation of momentum was first recognized by Christiaan Huygens after analyzing collisions between balls.

The above-stated principle of the conservation of momentum considers only two particles. Nevertheless, this principle holds for any number of particles forming an isolated system, i.e. particles that are subject only to their own mutual interactions and not to interactions with other parts of the world. The principle of conservation of momentum in its general form says that

the total momentum of an isolated system of particles is constant:

$$\boldsymbol{P} = \sum \boldsymbol{p}_i = \boldsymbol{p}_1 + \boldsymbol{p}_2 + \boldsymbol{p}_3 + \cdots = const. \tag{6.9}$$

For example, consider a hydrogen molecule composed of two hydrogen atoms (therefore of two electrons and two protons). If the molecule is isolated, only the interactions among these four particles have to be considered. Then the sum of their momenta relative to an inertial frame of reference will be constant. Similarly, consider our planetary system, composed of the Sun, the planets, and their satellites. If we could neglect the interactions with all other heavenly bodies, the total momentum of the planetary system relative to an inertial frame of reference would be constant.

No exceptions to this general principle of conservation of momentum are known. In fact, whenever this principle seems to be violated in an experiment, the physicist immediately looks for some unknown or hidden particle that went unnoticed in order to make up for the apparent lack of conservation of momentum. It is this search that has led physicists to identify the neutron, the neutrino, the photon, and many other elementary particles.

The conservation of momentum implies that, in an isolated system, the change of momentum of a particle during a particular interval of time is equal and opposite to the change of momentum of the rest of the system during the same time interval. So, in the case of an isolated hydrogen molecule, the change of momentum of one of the electrons is equal and opposite to the sum of the changes of momenta of the other electron and the two protons.

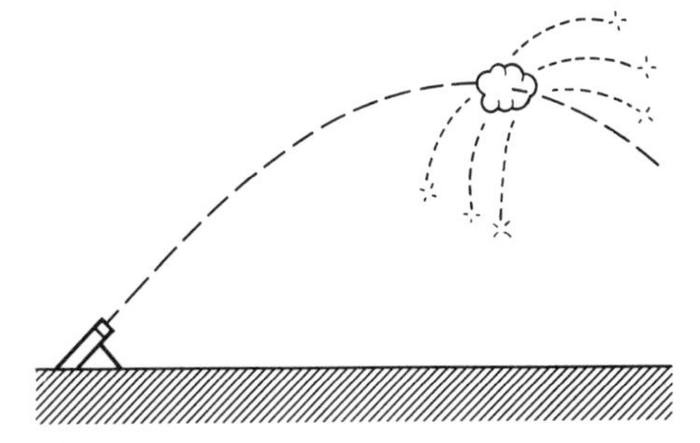

Figure 6.7 An exploding firework. Momentum is conserved at the explosion.

The law of inertia stated in Section 6.2 is one particular case of the principle of conservation of momentum. If we have only one isolated particle instead of several, Equation 6.9 has only one term and becomes $\boldsymbol{P} = m\boldsymbol{v} = const.$ For a constant mass this is equivalent to $\boldsymbol{v} = const.$, which is the law of inertia.

The recoil of a gun is an example of the principle of conservation of momentum. Initially the system of gun + bullet is at rest, and the total momentum is zero. When the gun is fired, it recoils to compensate for the forward momentum gained by the bullet. (Usually the recoil momentum is absorbed by the person holding the gun.) Similarly, when a nucleus (initially at rest) disintegrates, emitting an electron and a neutrino, the total momentum of the electron, the neutrino, and the resultant nucleus must add to zero. Finally, consider a firecracker that explodes in flight. The total momentum of all the fragments immediately after the explosion must add to a value equal to the momentum of the firecracker immediately before exploding (Figure 6.7). This is why the fragments all tend to move in the forward direction.

The form in which Equations 6.5 and 6.8 are written presupposes that the interaction between two particles is instantaneous. However, interactions propagate with a finite velocity, presumably equal to that of light. In order to take into account the retardation in the interaction due to the finite velocity of propagation, an additional term must be incorporated into Equations 6.5 and 6.8. Nevertheless, as previously stated, so long as: (1) the particles move slowly compared with the velocity of light and (2) are not very far from each other, the relation $\boldsymbol{p}_1 + \boldsymbol{p}_2 =$ *constant* (and the theory developed from it) constitutes an excellent approximation for describing the physical situation.

EXAMPLE 6.2

A gun whose mass is 0.80 kg fires a bullet whose mass is 0.016 kg with a velocity of 700 m s^{-1}. Compute the velocity of the gun's recoil.

▷ Initially both the gun and the bullet are at rest and their total momentum is zero. After the explosion the bullet is moving forward with a momentum

$$p_1 = m_1 v_1 = (0.016 \text{ kg})(700 \text{ m s}^{-1}) = 11.20 \text{ kg m s}^{-1}$$

The gun must then recoil with an equal but opposite momentum, so $\mathbf{p}_2 = -\mathbf{p}_1$. Therefore we must also have $p_2 = p_1$ so that

$$p_2 = 11.20 \text{ kg m s}^{-1} = m_2 v_2$$

Then, since $m_2 = 0.80$ kg,

$$v_2 = \frac{11.20 \text{ kg m s}^{-1}}{0.80 \text{ kg}} = 14.0 \text{ m s}^{-1}, \text{ about } 50 \text{ km h}^{-1}$$

The same method may be applied to get the relation between the velocities of the fragments when a grenade explodes into two fragments. Similarly, if a nucleus at rest emits an α-particle, then the resulting nucleus N recoils with a momentum such that $\mathbf{p}_N + \mathbf{p}_\alpha = 0$. Since $\mathbf{p} = m\mathbf{v}$, in magnitude we see that $v_\alpha / v_N = m_N / m_\alpha$, so that the velocities are in inverse proportion to the masses.

EXAMPLE 6.3

Analysis of conservation of momentum in interactions between atomic particles.

▷ The cloud chamber photograph in Figure 6.8(a) shows an incoming α-particle (or helium nucleus). It collides with an atom of hydrogen that was initially at rest, and which was part of the gas in the chamber. The α-particle is deflected from its original direction and the atom of hydrogen is set in motion. The relative masses of helium and hydrogen are in the ratio of 4 to 1. If we measure their velocities (by special techniques devised to analyze cloud and bubble chamber photographs), we can draw the momentum diagram of Figure 6.8(b). When, after the collision, the two momenta are added, the result is equal to the momentum of the incoming α-particle; that is,

$$\mathbf{p}_\alpha = \mathbf{p}'_\alpha + \mathbf{p}_H$$

So far conservation of momentum has been observed to hold in all atomic and nuclear interactions. In fact, momentum conservation is always used when analyzing processes

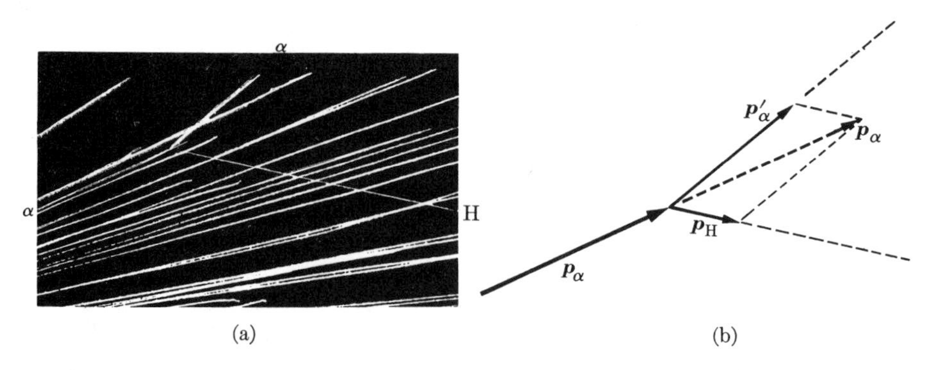

Figure 6.8 Momentum conservation in the collision of an α-particle (helium nucleus) and a proton (hydrogen nucleus).

involving fundamental particles, such as those shown in Figures 20.5 and 20.7. This problem will be discussed in more detail in Chapters 20, 40 and 41.

EXAMPLE 6.4

Momentum exchanges in an encounter with a planet.

▷ When a space probe swings around a planet, there is an exchange of momentum between the planet and the probe. If the proper approach is arranged, the probe gains velocity relative to the Sun, which helps the probe to continue its exploration of the planetary system.

Consider, for simplicity, the special case of a space probe that approaches a planet P almost directly opposite to the planet's motion relative to the Sun, so that it is a close encounter, as depicted in Figure 6.9. After passing around the planet further assume it is sent back almost in the same direction as the planet's orbital velocity. Relative to the Sun, the velocity of the planet is v, the probe's velocity on approach is V and the velocity of the probe as it leaves the planet is V'. An observer on the planet observes the probe at point A with an approach velocity of $V + v$ and a recession velocity of $V' - v$ when it is at point B. Also, relative to the planet, the motion of the probe appears to be symmetric. Therefore, relative to the planet, the probe's velocity at A and B must be the same (but in the opposite direction) because symmetry requires that the velocity gained approaching the planet is lost as the probe moves away. Therefore,

$$V' - v = V + v \qquad \text{or} \qquad V' = V + 2v$$

This means that, relative to the Sun, the probe has *gained* a velocity $2v$ and therefore the velocity and momentum of the probe has increased. The increase must come from the momentum of the planet relative to the Sun, which must lose some velocity (and momentum) because of the principle of conservation of momentum. However, since the mass of the planet is many orders of magnitude larger than the mass of the probe, its change in velocity (according to Equation 6.2) is so small that it can be regarded as zero.

The situation we have considered in this example is the most favorable for a gain of velocity (and momentum) by the probe. For other angles of approach, the gain in velocity is less. Such a gain in momentum has been used in propelling space probes through the solar system (see Section 12.4). This principle has been used with the space probe Galileo, launched in October 1989 to reach Jupiter in December 1995, and to study the Jovian atmosphere and its major moons for two years. The path of Galileo (Figure 6.10) involves a pass close to Venus and two passes near the Earth, until it gains enough momentum to reach Jupiter. For that reason the path is known as Venus–Earth–Earth Gravity Assist, or VEEGA. This path considerably reduces the initial momentum required for launch from the space shuttle. A reliable solid fuel booster, called the Inertial Upper Stage, was the source of momentum. An alternative path could have taken only two and a half years to

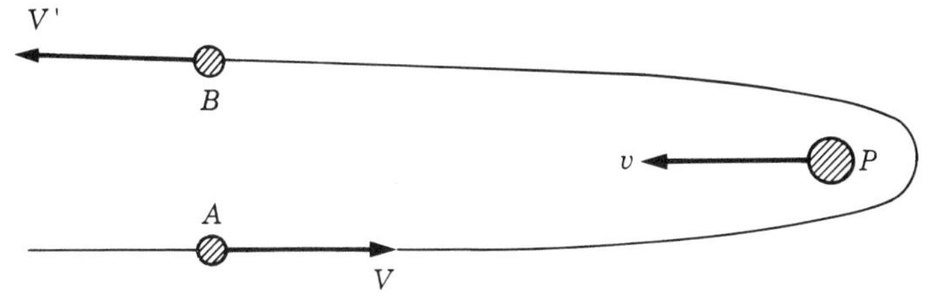

Figure 6.9

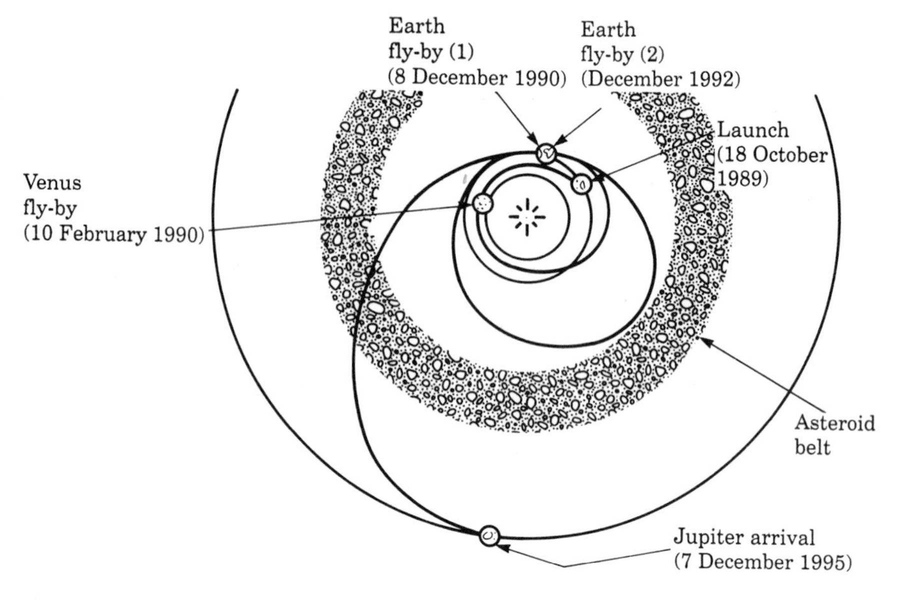

Figure 6.10 Galileo flight path.

reach Jupiter. However, it would have required a much more powerful booster, using liquid hydrogen and oxygen, thus posing a risk to the shuttle. The space probe Ulysses, launched in October 1990, uses the same principle (see Chapter 12).

The student should consider the case where the probe approaches the planet 'from the rear'. For this case the two are moving in the same direction relative to the Sun and so there is a *loss* of velocity (and momentum) by the probe relative to the Sun.

6.6 Newton's second and third laws

In many cases we observe the motion of one particle without reference to the other particles with which it interacts, either because we have no way of observing the other particles or because we deliberately ignore them. In this situation it is rather difficult to use the principle of conservation of momentum. However, there is a practical way of circumventing this difficulty. The corresponding theory is called the **dynamics of a particle**.

Equation 6.5, $\Delta \boldsymbol{p}_1 = -\Delta \boldsymbol{p}_2$, relates the momentum changes of particles 1 and 2 during the time interval $\Delta t = t' - t$. Dividing both sides of this equation by Δt, we may write

$$\frac{\Delta \boldsymbol{p}_1}{\Delta t} = -\frac{\Delta \boldsymbol{p}_2}{\Delta t} \tag{6.10}$$

which indicates that the average (vector) rates of change of momentum of the particles in a time interval Δt are equal in magnitude and opposite in direction.

If we make Δt very small, i.e. if we find the limit of Equation 6.10 as $\Delta t \to 0$, we get

$$\frac{d\boldsymbol{p}_1}{dt} = -\frac{d\boldsymbol{p}_2}{dt} \quad (6.11)$$

so that the instantaneous (vector) rates of change of the momentum of two interacting particles, at any time t, are equal and opposite. For example, the rate of change of momentum of the electron in an isolated hydrogen atom is equal and opposite to the rate of change of momentum of the proton, relative to an inertial frame. Or, assuming that the Earth and the Moon constitute an isolated system, the rate of change of momentum of the Earth is equal and opposite to the rate of change of momentum of the Moon, relative to an inertial frame.

We must emphasize that the change in momentum of a particle is due to its interactions with other particles. Interactions are expressed quantitatively in terms of the notion of **force**. Thus we shall now, following Newton, introduce force as a dynamical quantity related to the rate of change of momentum by

$$\boldsymbol{F} = \frac{d\boldsymbol{p}}{dt} \quad (6.12)$$

This relation constitutes **Newton's second law of motion**:

the time rate of change of momentum of a particle is equal to the force acting on the particle.

Expressed in another way, Newton's second law says that the force on a particle determines the rate of change of its momentum.

If the particle is free,

$$\boldsymbol{p} = const. \quad \text{and} \quad \boldsymbol{F} = \frac{d\boldsymbol{p}}{dt} = 0$$

Hence we can say that no force acts on a free particle, which means that the particle is either at rest or moves with constant velocity with respect to any inertial frame, which is Newton's first law.

Using the concept of force, we can write Equation 6.11 in the form

$$\boldsymbol{F}_1 = -\boldsymbol{F}_2 \quad (6.13)$$

where $\boldsymbol{F}_1 = d\boldsymbol{p}_1/dt$ is the force on particle 1 due to its interaction with particle 2 and $\boldsymbol{F}_2 = d\boldsymbol{p}_2/dt$ is the force on particle 2 due to its interaction with particle 1. Then we conclude that:

when two particles interact, the force on the first particle exerted by the second particle is equal and opposite to the force on the second particle exerted by the first particle.

This is **Newton's third law of motion**, a consequence of Newton's second law of motion and the principle of the conservation of momentum. It is also called the **law of action and reaction** (Figure 6.11).

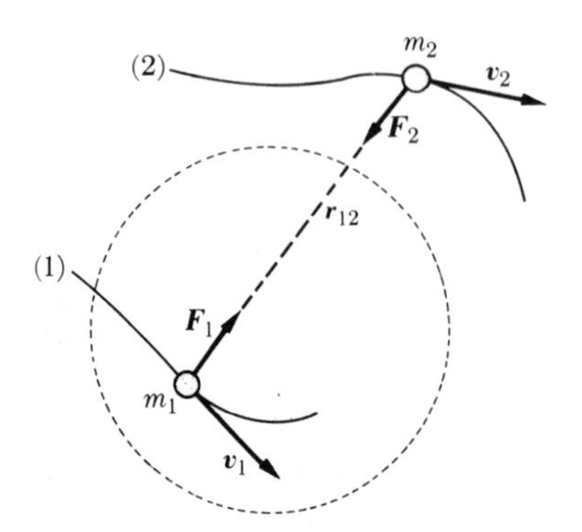

Figure 6.11 As a result of momentum conservation, action and reaction are equal and opposite.

Newton's second law of motion, Equation 6.12, is not just a definition of force. The fact is that in numerous experiments $\boldsymbol{F}_1$ (and of course $\boldsymbol{F}_2$ also), or the corresponding rates of exchange of momentum, can be expressed as a function of the relative position vector of the two particles, $\boldsymbol{r}_{12}$, and perhaps also as a function of their relative velocity. It is also usually expressed in terms of other parameters (mass, charge etc.) characteristic of the particles. Also, it has been found that this expression for the force applies not only to two specific particles, but to all particles subject to the same kind of interaction. The determination of $\boldsymbol{F}(\boldsymbol{r}_{12})$ for the many interactions found in nature is one of the most important problems of physics. It is precisely because physicists have been able to associate specific functional forms of $\boldsymbol{F}(\boldsymbol{r}_{12})$ with different interactions observed in nature that the concept of force has been so useful in analyzing the motion of particles under given interactions.

For example, the Earth–Moon system and the Sun–Earth system each move under their mutual gravitational interaction. Similarly, an electron moves relative to a nucleus in an atom under their mutual electrical interaction. As we shall see later, by measuring their rates of change of momentum, it has been found that the forces in both cases are inversely proportional to the square of the distance r between the two interacting bodies; that is, $F \propto 1/r^2$. However, in the case of gravitation, the force also depends on the masses of the particles while in the case of the electric interaction, the force depends on the charges on the particles (see Note 6.1). The constant of proportionality is determined experimentally in each case.

Unlike the simple two-body cases mentioned above, there are instances where a particle is interacting with many others, so that the force cannot be easily expressed in terms of a few parameters. However, in those cases the force can be obtained empirically by carrying out a series of measurements and then used in Equation 6.13 without any reference to the basic interaction(s). We may say that they are 'statistical' forces. This is the case, for example, with elastic, frictional and viscous forces, which in the final analysis are all of electrical origin but involve very large numbers of atoms or molecules.

6.7 Relationship between force and acceleration

Remembering the definition 6.3 of momentum, we may write Equation 6.12, $\boldsymbol{F} = \mathrm{d}\boldsymbol{p}/\mathrm{d}t$, in the form

$$\boldsymbol{F} = \frac{\mathrm{d}(m\boldsymbol{v})}{\mathrm{d}t} \tag{6.14}$$

In the case where m is constant we have

$$\boldsymbol{F} = m\frac{\mathrm{d}\boldsymbol{v}}{\mathrm{d}t} \quad \text{or} \quad \boldsymbol{F} = m\boldsymbol{a} \tag{6.15}$$

This is another version of Newton's second law of motion; but it is more restrictive because of the assumption that the mass remains constant. Also it is applicable only to a particle whose velocity is small compared with the velocity of light (see Section 19.9). Note that the acceleration has the same direction as the force (Figure 6.12).

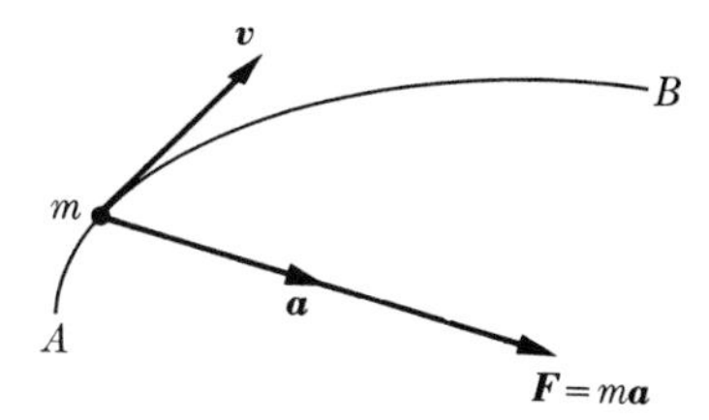

Figure 6.12 The force has the same direction as the acceleration.

From Equation 6.15 we see that if the force is constant, the acceleration $\boldsymbol{a} = \boldsymbol{F}/m$ is also constant and the motion is uniformly accelerated. This is what happens to bodies falling near the Earth's surface. It is a well known experimental fact that all bodies near the surface of the Earth fall toward the Earth with the same acceleration $\boldsymbol{g}$. Thus the force of gravitational attraction of the Earth, called **weight**, is

$$\boldsymbol{W} = m\boldsymbol{g} \tag{6.16}$$

Since the acceleration of gravity is the same for all bodies, we conclude that the weight of a body is proportional to its mass. This has given rise in our daily experience to considering weight and mass as equivalent, which is not correct.

Equation 6.15 can be used to determine the mass of a particle by measuring the acceleration produced by a known force and applying $m = F/a$. This, for example, may be done for a charged particle moving in uniform electric or magnetic fields. In fact this was how the electron mass was first measured (see Chapter 22). If the masses of two interacting particles are m_1 and m_2, their respective accelerations $\boldsymbol{a}_1$ and $\boldsymbol{a}_2$ are such that the forces are $\boldsymbol{F}_1 = m_1\boldsymbol{a}_1$ and $\boldsymbol{F}_2 = m_2\boldsymbol{a}_2$. Since we have shown that $\boldsymbol{F}_1 = -\boldsymbol{F}_2$, we also have $m_1\boldsymbol{a}_1 = -m_2\boldsymbol{a}_2$ or, using magnitudes, $a_1/a_2 = m_2/m_1$. This relation shows that when two bodies interact, their respective accelerations are inversely proportional to their masses, which is equivalent to Equation 6.1, where the same relation is expressed in terms of changes in velocity. The result is logical because the more massive body is more difficult to accelerate *with the same force*. If one of the bodies is much more massive than the other (e.g. $m_1 \gg m_2$), its acceleration is very small (i.e. $a_1 \ll a_2$) or even negligible. In such a case, we may assume that the massive body is (practically) at rest in an inertial system and we only have to consider the motion of the lighter body. Some examples where this approximation can be used include the motion of the Earth around the Sun or the Moon around the Earth, a ball dropped near the Earth's surface and the motion of an electron around the proton in a hydrogen atom. Examples 6.5 and 6.6 will discuss two of these cases.

6.8 Units of force

From Equations 6.12 or 6.15, we see that the unit of force must be expressed in terms of the units of mass (kg) and acceleration ($\mathrm{m\,s^{-2}}$). Thus in the SI system the force is measured in $\mathrm{kg\,m\,s^{-2}}$, a unit called the **newton** and denoted by N; that is,

$$\mathrm{N} = \mathrm{kg\,m\,s^{-2}}$$

Accordingly, we define the newton as the force that is applied to a body whose mass is one kg to produce an acceleration of one $\mathrm{m\,s^{-2}}$.

Another unit, used frequently by engineers, is based on Equation 6.16, which defines the weight of a body. The **kilogram force**, abbreviated kgf, is defined as a force equal to the weight of a mass equal to one kilogram. That is, setting $m = 1$ kg in Equation 6.16, we have

$$1\ \mathrm{kgf} = g\ \mathrm{N} \simeq 9.81\ \mathrm{N}$$

The student should be aware that the equivalence between kgf and N involves the gravitational acceleration, g, which varies slightly from place to place on the surface of the Earth. Therefore, 1 kgf is a different amount of force depending on the place. Note that mass measured in kilograms and weight measured in kilograms force are expressed by the same number. So a mass of 7.24 kg weighs 7.24 kgf (70.95 N). Although weight, being a force, should be expressed in N, it is customary, especially in engineering and household uses, to express it in kilograms force. In practice, one often speaks of a force of so many kilograms, but this is *not* good practice.

EXAMPLE 6.5

The force of gravitational attraction between the Earth and the Moon, as will be shown in Chapter 11, is $F = 1.985 \times 10^{17}$ N. The mass of the Earth is 5.97×10^{24} kg and that of the Moon is 7.35×10^{22} kg. Calculate the acceleration of each body due to that force.

▷ From the law of the conservation of momentum (or the law of action and reaction), the gravitational force with which the Earth attracts the Moon is equal to the force with which the Moon attracts the Earth. This is indicated in Figure 6.13. This force results in a different acceleration on each body because their masses are different. Thus for the Moon the acceleration is

$$a_{\mathrm{M}} = \frac{F}{m_{\mathrm{M}}} = \frac{1.985 \times 10^{17}\ \mathrm{N}}{7.35 \times 10^{22}\ \mathrm{kg}} = 2.70 \times 10^{-6}\ \mathrm{m\,s^{-2}}$$

This is the centripetal acceleration of the orbital motion of the Moon. For the Earth, the acceleration is

$$a_{\mathrm{E}} = \frac{F}{m_{\mathrm{E}}} = \frac{1.985 \times 10^{17}\ \mathrm{N}}{5.97 \times 10^{24}\ \mathrm{kg}} = 3.32 \times 10^{-8}\ \mathrm{m\,s^{-2}}$$

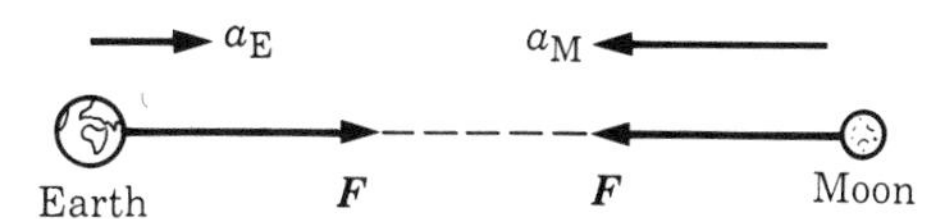

Figure 6.13 Gravitational forces in the Earth–Moon system.

or about 100 times less. For this reason, when we discuss the motion of the Earth–Moon system we often assume that the Earth remains at rest and the Moon moves around it. Actually, as we shall discuss in Chapter 13, the two bodies move around their center of mass.

EXAMPLE 6.6

The force of electric attraction between the electron and the proton in a hydrogen atom, as will be shown in Example 21.2, is $F = 8.20 \times 10^{-8}$ N. The mass of the electron is 9.109×10^{-31} kg and that of the proton is 1.672×10^{-27} kg. Calculate the acceleration of each particle due to their mutual interaction.

▷ This example is similar to the preceding example, except that the force is of electrical origin, rather than gravitational (Figure 6.14). The acceleration of the electron due to this force is

$$a_e = \frac{F}{m_e} = \frac{8.20 \times 10^{-8}\ \mathrm{N}}{9.109 \times 10^{-31}\ \mathrm{kg}} = 9.02 \times 10^{22}\ \mathrm{m\,s^{-2}}$$

which is about 10^{22} times the acceleration of gravity! The acceleration of the proton due to the same force is

$$a_p = \frac{F}{m_p} = \frac{8.20 \times 10^{-8}\ \mathrm{N}}{1.672 \times 10^{-27}\ \mathrm{kg}} = 4.91 \times 10^{19}\ \mathrm{m\,s^{-2}}$$

or about 1840 times smaller. For this reason, when we discuss the motion of the electron–proton system we normally assume (at least to a good first approximation) that the proton in a hydrogen atom is at rest and the electron moves around it. However, the fact that the proton *is not* at rest produces certain effects that can be detected with modern measuring techniques. We will discuss these effects in Chapter 23.

Figure 6.14 Electrical forces in a proton–electron system.

EXAMPLE 6.7

Calculate the force required to deviate an electron in a TV tube (see Example 4.3) by 0.1 m.

▷ Since the electron in Example 4.3 reaches the screen in 6.7×10^{-9} s, using $y = \frac{1}{2}at^2$ with $y = 0.1$ m, the necessary acceleration of the electron is

$$a = \frac{2y}{t^2} = \frac{2(0.1\ \mathrm{m})}{(6.7 \times 10^{-9}\ \mathrm{s})^2} = 4.46 \times 10^{15}\ \mathrm{m\,s^{-2}}$$

or about 5×10^{14} times the acceleration of gravity! The force required is

$$F = ma = (9.109 \times 10^{-31}\ \text{kg})(4.46 \times 10^{15}\ \text{m s}^{-2}) = 4.06 \times 10^{-15}\ \text{N}$$

which is about 5×10^{14} times the weight of the electron but 4.4×10^{-8} times the force on an electron in a hydrogen atom.

6.9 Classical principle of relativity

In Section 4.6, we derived the Galilean transformation of velocities and accelerations measured by two observers in uniform relative translational motion. Also, we have stated that the laws of motion must be considered relative to an inertial observer. Therefore, we must now verify that if the laws of motion hold for one inertial observer they also hold for all inertial observers.

Consider two different inertial observers O and O' moving with a constant relative velocity $\boldsymbol{v}$. Let us assume that the two observers correlate their respective observations of the same phenomenon by the Galilean transformation, Equation 4.19

$$\boldsymbol{r}' = \boldsymbol{r} - \boldsymbol{v}t, \quad \boldsymbol{V}' = \boldsymbol{V} - \boldsymbol{v}, \quad \boldsymbol{a}' = \boldsymbol{a}, \quad t' = t$$

Consider two particles, of masses m_1 and m_2, and call $\boldsymbol{V}_1$ and $\boldsymbol{V}_2$ their velocities as measured by inertial observer O. If no external forces act on the particles, the principle of conservation of momentum requires that

$$m_1 \boldsymbol{V}_1 + m_2 \boldsymbol{V}_2 = const. \tag{6.17}$$

For another inertial observer O', moving relative to O with the constant velocity $\boldsymbol{v}$, the velocities of m_1 and m_2 are

$$\boldsymbol{V}'_1 = \boldsymbol{V}_1 - \boldsymbol{v} \quad \text{and} \quad \boldsymbol{V}'_2 = \boldsymbol{V}_2 - \boldsymbol{v}$$

Substituting these values into Equation 6.17, we have

$$m_1(\boldsymbol{V}'_1 + \boldsymbol{v}) + m_2(\boldsymbol{V}'_2 + \boldsymbol{v}) = const.$$

or, rearranging terms,

$$m_1 \boldsymbol{V}'_1 + m_2 \boldsymbol{V}'_2 = const. - (m_1 + m_2)\boldsymbol{v}$$

But if the relative velocity $\boldsymbol{v}$ is constant, the right-hand side of the above equation is also constant and observer O' can write

$$m_1 \boldsymbol{V}'_1 + m_2 \boldsymbol{V}'_2 = const. \tag{6.18}$$

Note that this result is valid only if $\boldsymbol{v}$ is also constant; that is, if O' is another inertial observer. Equation 6.18 is entirely similar to Equation 6.17 and therefore *both inertial observers verify the principle of conservation of momentum.* We are assuming that both observers measure the same mass, an assumption substantiated by experience.

Let us next discuss the relation between the force measured by two inertial observers O and O' moving with a constant relative velocity $\boldsymbol{v}$ so that both inertial observers measure the same acceleration of the particle (that is, $\boldsymbol{a}' = \boldsymbol{a}$). According to Newton's second law of motion, the force measured by observer O is

$$\boldsymbol{F} = \frac{\mathrm{d}\boldsymbol{p}}{\mathrm{d}t} = m\frac{\mathrm{d}\boldsymbol{V}}{\mathrm{d}t} = m\mathbf{a}$$

and the force measured by observer O' is

$$\boldsymbol{F}' = \frac{\mathrm{d}\boldsymbol{p}'}{\mathrm{d}t} = m\frac{\mathrm{d}\boldsymbol{V}'}{\mathrm{d}t} = m\boldsymbol{a}'$$

In view of the fact that $\boldsymbol{a} = \boldsymbol{a}'$, we conclude that

$$\boldsymbol{F} = \boldsymbol{F}' \tag{6.19}$$

Therefore *both inertial observers measure the same force on the particle* when they compare their measurements using the Galilean transformation. The fact that both inertial observers must use the same laws of motion suggests that:

> *all laws of motion are the same for all inertial observers moving with constant velocity relative to each other, when the observers compare their observations using the Galilean transformation.*

This statement constitutes the **classical principle of relativity**. Our proof is valid as long as the particles move with velocities much smaller than that of light.

To illustrate this principle, consider an observer O at a railroad station and an observer O' on a train passing by with uniform rectilinear motion. Let O and O' perform the same experiment; they each play billiards. Both must use the same laws for discussing the motion of the balls and the balls will move in exactly the same way at the station as they do on the train.

EXAMPLE 6.8

Equation of motion with reference to a non-inertial observer: **inertial forces**.

▷ If an observer O' is non-inertial, this means that the velocity of O', $\boldsymbol{v}$, relative to an inertial observer O, is not constant in time. Thus if $\boldsymbol{a}_r = \mathrm{d}\boldsymbol{v}/\mathrm{d}t$ is the acceleration of O' relative to O, according to Equation 4.18 we have that

$$\boldsymbol{a}' = \boldsymbol{a} - \boldsymbol{a}_r$$

The force on a mass m measured by the inertial observer O is

$$\boldsymbol{F} = m\boldsymbol{a}$$

Then, if (non-inertial) observer O' uses the same definition of force, the force on mass m as measured by O' is

$$\boldsymbol{F}' = m\boldsymbol{a}'$$

Therefore, in view of the relation between $\boldsymbol{a}$ and $\boldsymbol{a}'$,

$$\boldsymbol{F}' = \boldsymbol{F} - m\boldsymbol{a}_r \qquad \textbf{(6.20)}$$

Thus:

the non-inertial observer measures a force different from that of the inertial observer.

In other words, the non-inertial observer O' considers that there is another force $\boldsymbol{F}_i$ acting on the particle given by

$$\boldsymbol{F}_i = -m\boldsymbol{a}_r \qquad \textbf{(6.21)}$$

This force is in addition to the force $\boldsymbol{F}$ measured by the inertial observer O (which includes all interactions to which the particle is subject). Therefore, the resultant force on the particle as measured by O' is $\boldsymbol{F} + \boldsymbol{F}_i$. The fictitious force $\boldsymbol{F}_i$ is called an **inertial force**.

When we want to describe the motion of a particle relative to a rotating frame, such as the Earth (which is *not* an inertial frame of reference), we sometimes use this kind of logic. In this case $\boldsymbol{a}_r$ is given by the last two terms in Equation 5.23. Therefore, the inertial force on a particle when the observer is attached to the Earth is

$$\boldsymbol{F}_i = -m\boldsymbol{\omega} \times (\boldsymbol{\omega} \times \mathbf{r}) - 2m\boldsymbol{\omega} \times \boldsymbol{V}' \qquad \textbf{(6.22)}$$

This corresponds to a centrifugal force and a Coriolis force acting on the particle in addition to $\boldsymbol{F}$.

Note 6.1 The forces we know

In this chapter we have become acquainted with the general concept of force as equal to the time rate of change of momentum. But an important question that has not been formally addressed is: 'where do these forces come from?'.

When we look around, we see many different 'kinds' of force. We apply a force on the ground when we walk. We push and lift objects. To stretch a string, we must apply a force. The wind upsets a tree or pushes a sailboat by applying a force. The expansion of gases in internal combustion engines produces a force that causes an automobile, a boat or an airplane to move. Electric motors produce a force that moves objects. We can think of numerous instances where forces are generated and applied.

The above examples of physical forces occur in complicated systems, such as our bodies, a machine, the Earth's atmosphere and on wires carrying electric current. We may say that they are *statistical* forces because, if we analyze the situations where these forces appear, we notice that they involve bodies composed of large numbers of atoms. The question is, then, whether we can break these phenomena down into microscopic components, namely molecules, atoms, electrons and their interactions. One of the great achievements of the last few decades has been to reduce all forces observed in nature to a very few *basic* or *fundamental* interactions between the basic components of matter. This means that statistical forces are simply manifestations of the fundamental forces when a very large number of particles are involved. In Section 1.7 we said that we currently recognize four fundamental forces or interactions between elementary particles. These forces are: gravitational, electromagnetic, strong and weak. We will now review the general characteristics of these basic forces. They have been recognized as a result of the analysis of a large number of experiments.

Fundamental forces or interactions are associated with **sources**. These entities can be considered as both the origin of the forces or the subjects on which the forces act. This follows directly from the law of conservation of momentum, or the equivalent Newton's third law of 'action' and 'reaction', since any source that is the origin of a force is also an entity that can be acted upon by such a force. In other words, a source is something that *grabs* and also something that is *grabbed*. This is natural because forces are expressions of the momentum exchange between interacting bodies. As explained in Section 1.7, different types of force have different kinds of sources associated with different particle properties: **mass** is the source of the gravitational force; **electric charge** is the source of the electric, or rather electromagnetic, force; the **color charge**, which is an attribute of the particles called quarks, is the source of the strong force; and the weak interaction has its source in the **weak charge**. Not all particles have these four attributes and this is why particles are affected differently by the four interactions.

Sources may come in one or different kinds. The source of gravitational force, mass, comes in only one kind. Electric charge, on the other hand, comes in two kinds. Color charge comes in six different kinds, as will be explained in Chapter 41. This feature has an effect on the direction of the force; that is, whether the force is always attractive or may be either attractive or repulsive.

Sources may also be **point-like** or **extended**. In the first case they are just points, without structure, while in the second case they have some sort of internal structure. Our present view of nature considers point-like sources as fundamental. Extended sources are thought of as secondary, most probably composite systems of point-like sources. For example, electrons are considered point-like sources, while protons and neutrons are apparently composite, consisting of three quarks each. The fundamental laws of force are generally formulated in terms of point-like sources. Perhaps there are no genuinely point-like sources. Nevertheless, the abstraction of a point-like source provides a simple description that can then be used to analyze the systems we actually see in nature. In fact, some composite sources are considered point-like under certain circumstances or approximations.

In a similar attempt to achieve simplicity, physicists try to reduce forces among many sources to forces between two sources; that is, *two-body* forces or interactions. In many instances, such an attempt has been fully successful. However, we are not certain that *no* force exists that is genuinely a many-body (many source) force and thus cannot be reduced to some superposition of two-body forces. In such two-body interactions, the magnitude of the force is proportional to a symmetric combination, such as the product of the two quantities that describe the 'strength' of the two sources. For example, the gravitational force is proportional to the product of the masses of the two interacting bodies ($F \propto mm'$) and the electric force is proportional to the product of the charges on the two bodies ($F \propto qq'$). The proportionality constant depends on the type of force and often is called the **coupling constant**. It also seems reasonable to assume that the magnitude of the force between two sources depends on the distance, r, between the two sources. This distance, or **radial dependence**, is a very important property of the force and plays a decisive role in determining the basic characteristics of the world we live in. For two basic forces (gravitational and electric), the force decreases as the distance apart increases, varying as $1/r^2$. Therefore the gravitational force is of the form (Chapter 11):

$$F \propto \frac{mm'}{r^2} \qquad \text{or} \qquad F = G\frac{mm'}{r^2}$$

and the electric force has the form (Chapter 21):

$$F \propto \frac{qq'}{r^2} \qquad \text{or} \qquad F = K_e \frac{qq'}{r^2}$$

where G and K_e are the coupling constants. In the case of the electric force, the expression is correct as long as the particles are at rest or move with a very small relative velocity. This is why we sometimes use the expression 'electrostatic' for the expression of the electric force given above. The nuclear force between protons and neutrons is a manifestation of the *strong* interaction. It decreases much faster than $1/r^2$ with increasing r. Its actual dependence on the distance is not yet well known (Chapters 39 and 41). The same may be said for the weak interaction.

The gravitational and electric forces are **long-range** forces, their influence extending from one side of the universe to the other. The strong force, on the other hand, is a **short-range** force, of the order of 10^{-15} m. The weak force is apparently of even shorter range than the strong force (Table 6.1).

Whether a force is of short or long (finite or infinite) range has important consequences. For example, in a molecule or any piece of matter in bulk, the atomic nuclei are separated by distances of the order of 10^{-10} m. Therefore the strong interactions among them are practically zero. As a consequence, atoms in matter in bulk are held together exclusively by electric forces (as we will see later, the gravitational force between atoms is *much* too small to even be considered). Thus strong forces do not produce measurable effects in ordinary everyday experiments. This accounts for the fact that the strong force was not discovered until the 1930s, when special machines were able to detect effects at distances as small as 10^{-14} m. In contrast, the gravitational and electric forces have been known for over two centuries. In turn, the weak force is responsible for some radiative processes that can be very easily observed. Actually, because of their long range, gravitational and electrical forces account for the vast majority of the phenomena we observe in the world. Although oversimplified, we can say that gravitational forces are responsible for the extremely large structures in the universe (galaxies, planetary systems, stars), while the strong and weak interactions operate at the other extreme (nuclei, elementary particles). In between, electric forces are responsible for most of the properties of matter in bulk (atoms, molecules, liquids, solids) as well as chemical reactions. Actually, most of the statistical forces of matter in bulk are electric or gravitational in origin. In fact, many of the statistical forces we observe in our daily life, such as friction, viscosity, gas pressure etc. are essentially the result of electric interactions between large numbers of atoms and molecules.

The direction of a force must somehow be determined by the properties of the sources and their relative positions. If the sources have no directional properties attached to them, then the only possible direction is the one defined by the line connecting the two sources. Experiment has shown that the force *is* in this direction. Such forces are called **central** because,

Table 6.1 Fundamental interactions

Interaction	Relative strength	Range	Matter property	Carrier boson
strong	1	10^{-15} m	'color'* charge	gluon
weak	10^{-14}	10^{-18} m	'weak' charge	weak boson
electromagnetic	10^{-2}	∞	electric charge	photon
gravitation	10^{-38}	∞	mass	graviton

* Not related to the colors of objects we see

if we regard one of the sources as the center of our coordinate system, the force on the other source always points toward or away from this center. The gravitational and electric forces are central forces as long as we consider interactions between point sources. However, if scalar sources aggregate into a composite system (atoms, molecules etc.) that, by its shape and symmetry, defines an additional direction, the resultant force may not be central. This is the case for intermolecular forces, forces between small magnets (or magnetic 'dipoles'), as well as between electric currents. However, for arrangements of masses or electric charges that exhibit spherical symmetry the force remains central. This occurs, to a very good approximation, with the gravitational force between the Earth and the Sun. There are other properties that determine the direction of the forces between elementary particles. One of these properties is called **intrinsic angular momentum** or **spin** (discussed in Chapter 13). In this case, the force may not be central.

Another important feature is the relative strengths of the four forces. These are given in Table 6.1, where the strength of the strong force has been taken as unity.

An important question that must be considered is *how* forces operate between sources at a distance, exchanging momentum. The four types of forces can be given a common basis in terms of an **exchange model** of forces. According to this model, a force between two sources is generated by the exchange of some entity or 'particle', called **carrier boson**, that plays the role of a **momentum carrier**. If the carrier boson has a non-zero mass, then it can be shown that the resulting force has a finite range. In particular, the larger the mass of the exchanged particle, the shorter the range of the resulting force. Electromagnetic forces, of infinite range, are said to be generated by the exchange of zero mass particles or **photons**, designated by the symbol γ. In weak interactions the exchange is due to particles of mass such that they produce a force with a very short range. These particles, called **intermediate** or **weak** bosons (designated by the symbols $W^{\pm}$ and Z^0) were first identified in the 1980s. Strong interactions are said to come from the exchange, between quarks, of certain particles called gluons. Gluons are considered to be massless, but other properties of the gluon, to be examined in Chapter 41, make the range of the strong interaction about 10^{-15} m. These carrier particles are found in nature only *within* protons, neutrons and mesons. The carrier of the gravitational force is supposed to be the **graviton**, a zero mass particle that, so far, has not been detected experimentally.

The particle-exchange model is currently the basis of all considerations about fundamental interactions. The validity of the model has been verified experimentally in a general sense. The details that are necessary in order to make precise calculations have not yet been fully developed (except for the electric force) and many questions still remain unanswered. The implementation of the model remains at the center of current research in theoretical and experimental physics.

A final remark is in order: conservation of momentum applies to all interactions. Newton's law of motion, $\boldsymbol{F} = \mathrm{d}\boldsymbol{p}/\mathrm{d}t$, seems to apply to all 'statistical' forces. However, the 'fundamental' forces or interactions cannot (except the long range gravitational and electromagnetic forces in some special circumstances) be expressed in the same terms. One of the reasons is the need to take the exchange phenomenon, by which momentum is carried from one source to another, into account. We will return to this matter in Chapter 41.

QUESTIONS

6.1 What is a 'free' particle? How do you recognize (experimentally) that a particle is 'free'?

6.2 Discuss the concept of an inertial observer. How is it possible to differentiate an inertial observer from a non-inertial observer?

6.3 State the principle of conservation of momentum when it is applied to (a) an isolated hydrogen molecule and (b) to the solar system.

6.4 Why can we say that 'mass' is a property of each particle independent of the forces acting on it?

6.5 State the law of action and reaction when applied (a) to the Earth–Moon system, (b) to the electron–proton system of a hydrogen atom.

6.6 What is the relation between the direction of a force and (a) the change in momentum and (b) the acceleration of a particle?

6.7 At a given time the force acting on a particle is F. At a later time the force is twice as large. What is the relation between the time rates of change of the momentum of the particle at the two times?

6.8 How do we recognize that the Earth exerts a force on all bodies near its surface? How do we measure the force? What do we call this force?

6.9 What physical quantities are 'invariant' under a Galilean transformation?

6.10 The **impulse** of a force $\boldsymbol{F}$ acting on a particle during time t_0 to time t is defined by

$$\boldsymbol{I} = \int_{t_0}^{t} \boldsymbol{F}\, \mathrm{d}t$$

Show that the impulse is equal to the change of momentum: $\Delta \boldsymbol{p} = \boldsymbol{I}$. Verify that when the force is constant $\boldsymbol{I} = \boldsymbol{F}\Delta t$, where $\Delta t = t - t_0$.

6.11 Plot the force $\boldsymbol{F}$ acting on a body as a function of time. Verify that the area under the curve corresponding to the interval $t - t_0$ is equal to the impulse.

6.12 Under what conditions does a very strong force acting during a very short time produce the same change in momentum as a weak force acting during a long time? Make a diagram of the forces as a function of time to explain your answer.

6.13 Discuss the relation between the concepts of 'interaction' and 'force'.

6.14 Designate the force exerted by the Earth on a body located at its surface by F and the radius of the Earth by R. Plot the points that correspond to the force on the body when it is at distances $2R$, $3R$ and $4R$ from the Earth's center if the force varies in inverse proportion to the square of the distance. Join the points with a smooth line. If the acceleration for free fall near the surface is $9.8\ \mathrm{m\,s^{-2}}$, what is the acceleration at distances $2R$, $3R$ and $4R$?

6.15 What relation(s) is(are) the same for all inertial observers according to the classical principle of relativity?

6.16 Discuss the origin of 'inertial forces' and give some examples.

6.17 An object moving with velocity $\boldsymbol{V}$ relative to an inertial observer O hits a wall that moves in the opposite direction with velocity $\boldsymbol{v}$ relative to O. What is the velocity of the object relative to the wall before and after the collision? What is the velocity of the object relative to O after the collision? (Note: recall Example 6.4.)

6.18 Write Equation 6.2 when particles 1 and 3 interact and also for particles 2 and 3.

PROBLEMS

6.1 A wagon at rest on a railway track is hit by another wagon moving at $0.04\ \mathrm{m\,s^{-1}}$. After the collision the first wagon is set in motion with a velocity of $0.053\ \mathrm{m\,s^{-1}}$, while the velocity of the second wagon drops to $0.015\ \mathrm{m\,s^{-1}}$. (a) Calculate the ratio of the masses of the two wagons. (b) Draw the vectors representing the velocity of each wagon before and after the collision and the vector *change* in their velocities.

6.2 Repeat the preceding problem if the first wagon has a final velocity of $0.01\ \mathrm{m\,s^{-1}}$ and (a) continues in the same direction as the second, (b) is moving in the opposite direction after the collision.

6.3 Two carts, A and B, are pushed toward each other. Initially B is at rest, while A moves to the right at $0.5\ \mathrm{m\,s^{-1}}$. After they collide, A rebounds at $0.1\ \mathrm{m\,s^{-1}}$, while B moves to the right at $0.3\ \mathrm{m\,s^{-1}}$. In a second experiment, A is loaded with a mass of 1 kg and pushed against B with a velocity of $0.5\ \mathrm{m\,s^{-1}}$. After the collision, A remains at rest, while B moves to the right at $0.5\ \mathrm{m\,s^{-1}}$. Find the mass of each cart.

6.4 (a) Find the momentum acquired by a mass of 1 g, 1 kg and 10^3 kg when each falls through a distance of 100 m. (b) Since the momentum acquired by the Earth is equal and opposite, determine the velocity (upward) acquired by the Earth. The mass of the Earth is 5.98×10^{24} kg. What do you conclude?

6.5 A particle whose mass is 0.2 kg is moving at $0.4\ \mathrm{m\,s^{-1}}$ along the X-axis when it collides with another particle, of mass 0.3 kg, which is at rest. After the collision the first particle moves at $0.2\ \mathrm{m\,s^{-1}}$ in a direction making an angle of $+40°$ with the X-axis. Determine (a) the magnitude and direction of the velocity of the second particle after the collision, and (b) the change in the velocity and the momentum of each particle.

6.6 A particle of mass 3.2 kg is moving due west with a velocity of 6.0 m s^{-1}. Another particle of mass 1.6 kg is moving due north with a velocity of 5.0 m s^{-1}. The two particles interact and, after 2 s, the first particle is moving in the direction N 30° E with a velocity of 3.0 m s^{-1}. Find: (a) the magnitude and direction of the velocity of the other particle, (b) the total momentum of the two particles, both at the beginning and after the 2 s have elapsed, (c) the change in momentum of each particle, (d) the change in velocity of each particle, and (e) the magnitudes of these changes in velocity.

6.7 In the chemical reaction H + Cl → HCl the H atom was initially moving in the positive X-direction with a velocity of 1.57×10^5 m s^{-1}, while the Cl atom was moving in the positive Y-direction with a velocity of 3.4×10^4 m s^{-1}. Find the magnitude and direction (relative to the original motion of the H atom) of the velocity of the resulting HCl molecule. Use the atomic masses of Table A.1.

6.8 In 28 days, the Moon (whose mass is 7.34×10^{22} kg) rotates about the Earth in a circle of radius 3.84×10^8 m. (a) What is the change in the momentum of the Moon in 14 days? (b) What must be the change in momentum of the Earth in 14 days? (c) Is the Earth stationary in the Earth–Moon system? (d) The mass of the Earth is 80 times that of the Moon. What is the change in velocity of the Earth in 14 days? (e) Repeat for 7 days.

6.9 A disc sliding over a smooth surface at 0.025 m s^{-1} hits another disc at rest. After the collision the first disc moves at 0.015 m s^{-1} in a direction making 30° with the initial direction. (a) What is the direction of motion of the second disc? (b) If the second disc recoils with a velocity of 0.20 m s^{-1}, what is the ratio of the masses of the discs?

6.10 A grenade moving horizontally at 8 km s^{-1} relative to the Earth explodes into three equal fragments. One continues to move horizontally at 16 km s^{-1}, another moves upward at an angle of 45°, and the third moves at an angle of 45° below the horizontal. Find the magnitude of the velocities of the second and third fragments.

6.11 A cart having a mass of 1.5 kg moves along its track at 0.20 m s^{-1} until it runs into a fixed bumper at the end of the track. What is its change in momentum and the average force exerted on the cart if, in 0.1 s, it (a) is brought to rest, or (b) rebounds with a velocity of 0.10 m s^{-1}? (c) Discuss the conservation of momentum in the collision.

6.12 An automobile has a mass of 1500 kg and its initial speed is 60 km h^{-1}. When the brakes are applied to produce a constant deceleration, the car stops in 1.2 min. Determine the force applied to the car.

6.13 (a) Calculate the time needed for a constant force of 80 N to act on a body of 12.5 kg in order to take it from rest to a velocity of 72 km h^{-1}. (b) How long, and in what direction, must the same constant force act on the body to return it to rest?

6.14 A body with a mass of 10 g falls from a height of 3 m on to a pile of sand. The body penetrates the sand a distance of 3 cm before stopping. What force was exerted on the body?

6.15 A body whose mass is 2 kg is moving on a smooth horizontal surface under the action of a horizontal force $F = 55 + t^2$ where F is in newtons and t in seconds. Calculate the velocity of the body when $t = 5$ s (the body was at rest when $t = 0$).

6.16 The force on an object of mass m is $F = F_0 - kt$ where F_0 and k are constants and t is the time. (a) Find the acceleration. (b) Find also the velocity and position at any time.

6.17 A particle of mass m, initially at rest, is acted on by a force $F = F_0[1 - (2t - T)^2/T^2]$ during the interval $0 \geqslant t \geqslant T$. (a) Verify that the velocity of the particle at the end of the interval is $2F_0T/3m$. Note that it depends only on the product F_0T and, if T is made smaller, the same velocity is attained by making F_0 proportionately larger. (b) Make a plot of F and v against t. (c) Describe some physical situation for which this problem could be an adequate description.

6.18 A force F that lasts 20 s is applied to a body of mass 500 kg. The body, which is initially at rest, is given a velocity of 0.5 m s^{-1} as a result of the force. If the force increases from zero linearly with time for 15 s and then decreases to zero linearly for 5 s, (a) find the impulse (see Question 6.10) on the body caused by the force, (b) find the maximum force exerted on the body and (c) make a graph of F versus t, and find the area under the curve. Does this area agree with the result of (a)?

7 Applications of the laws of motion

A parachutist falls with constant velocity (uniform motion) when the upward force due to the air drag of the parachute balances the weight of the parachutist. To increase air drag, parachutes have a large area and a special shape. (Courtesy of Airborne Forces Museum, Aldershot.)

7.1 Introduction

When the force acting on a body is known, it is possible to determine the motion of the body by application of Newton's second law. This law is expressed by the relations $\boldsymbol{F} = \mathrm{d}\boldsymbol{p}/\mathrm{d}t$ or $\boldsymbol{F} = m\boldsymbol{a}$. In this chapter we are going to illustrate some simple but important applications of these relations.

7.2 Motion under a constant force

When a particle is subject to a constant force $\boldsymbol{F}$, the acceleration is also constant and given by

$$\boldsymbol{a} = \frac{\boldsymbol{F}}{m} \qquad \textbf{(7.1)}$$

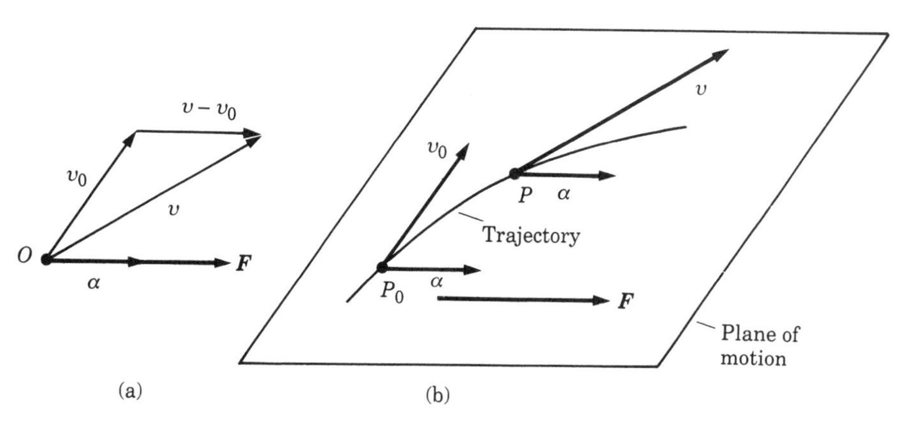

Figure 7.1

In this case we may apply the relations obtained in Section 4.4 and write

$$\boldsymbol{v} - \boldsymbol{v}_0 = \frac{\boldsymbol{F}}{m}(t - t_0) \tag{7.2}$$

Equation 7.2 shows that the velocity always changes in a direction parallel to the applied force (Figure 7.1(a)). Therefore, the trajectory tends, asymptotically, toward the direction of the force (Figure 7.1(b)).

The vector equation for the trajectory is

$$\boldsymbol{r} - \boldsymbol{r}_0 = \boldsymbol{v}_0(t - t_0) + \frac{1}{2}\frac{\boldsymbol{F}}{m}(t - t_0)^2 \tag{7.3}$$

From Equation 7.3 we see that when the force is constant the displacement is a combination of two vectors. One vector is in the direction of the initial velocity $\boldsymbol{v}_0$. The other is in the direction of the applied force $\boldsymbol{F}$. If $\boldsymbol{v}_0$ is parallel to $\boldsymbol{F}$, the motion is rectilinear. When $\boldsymbol{v}_0$ and $\boldsymbol{F}$ are not parallel, then the motion will be in the plane determined by the directions of $\boldsymbol{v}_0$ and $\boldsymbol{F}$.

This situation corresponds, for example, to the case of a projectile where the force is its weight, which is a constant force pointing downward. We have already discussed this case in Section 4.5. The same analysis applies when a charged particle, such as an electron, passes through a region of uniform electric field. The force on the charged particle is also a constant and in the direction of the electric field. In Example 6.7 the acceleration and force on an electron when moving in such a field was calculated.

7.3 Resultant force

In writing Equation 6.12, $\boldsymbol{F} = \mathrm{d}\boldsymbol{p}/\mathrm{d}t$, we have implicitly assumed that the particle interacts with only one other particle. However, if particle m interacts with particles $m_1, m_2, m_3, \ldots$ (Figure 7.2), each one produces a change in the

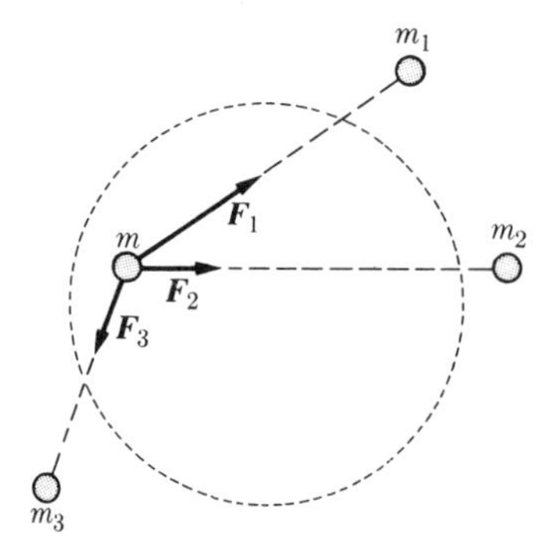

Figure 7.2 The resultant force on particle m is $\boldsymbol{F} = \boldsymbol{F}_1 + \boldsymbol{F}_2 + \boldsymbol{F}_3$.

momentum of m that is characterized by the respective forces $\boldsymbol{F}_1, \boldsymbol{F}_2, \boldsymbol{F}_3, \ldots$. Then the *total* rate of change of momentum of particle m is

$$\frac{\mathrm{d}\boldsymbol{p}}{\mathrm{d}t} = \boldsymbol{F}_1 + \boldsymbol{F}_2 + \boldsymbol{F}_3 + \cdots = \boldsymbol{F} \quad \textbf{(7.4)}$$

The vector sum on the right-hand side is called the **resultant** force $\boldsymbol{F}$ acting on m. Thus the resultant force $\boldsymbol{F}$ gives the net rate of change of momentum of a particle as a result of its momentum exchanges with the other particles with which it interacts.

For simplicity, we have not indicated in Figure 7.2 the possible interactions between m_1 and m_2, m_1 and m_3, m_2 and m_3 etc. These interactions are irrelevant to our present purpose. Also we have implicitly assumed that the interaction between m and m_1, for example, is not altered by the presence of $m_2, m_3, \ldots$; in other words, we have assumed that there are no interference effects.

As an example, consider a hydrogen molecule ion, composed of two protons and one electron (Figure 7.3). Each of the particles moves under the resultant force produced by each of the other two particles. In the figure, only the forces on the electron are shown.

As a second example, if a body of mass m and weight $\boldsymbol{W} = m\boldsymbol{g}$ is subject to another force $\boldsymbol{F}$, the resultant force has a magnitude $\boldsymbol{F} + \boldsymbol{W}$ or $\boldsymbol{F} + m\boldsymbol{g}$ (Figure 7.4). Therefore, the acceleration of the body is found from the equation of motion

$$\boldsymbol{F} + m\boldsymbol{g} = m\boldsymbol{a}$$

If the force $\boldsymbol{F}$ acts downward, the relation between the *magnitudes* is

$$F + mg = ma$$

and the downward acceleration is greater than that due to gravity. If $\boldsymbol{F}$ acts upward, the relation between the *magnitudes* is

$$F - mg = ma$$

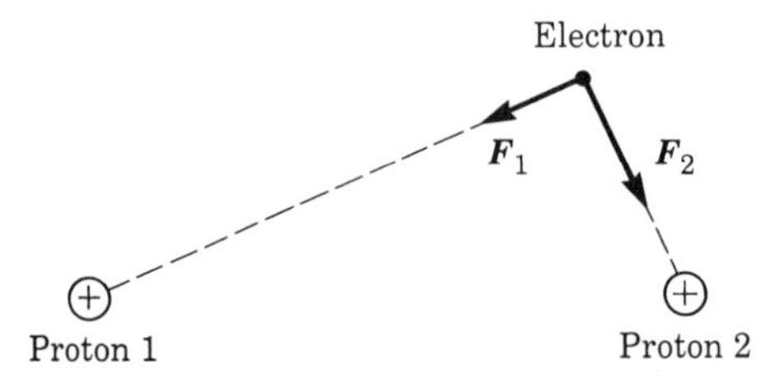

Figure 7.3 Forces on the electron in a hydrogen molecule ion, H_2^+.

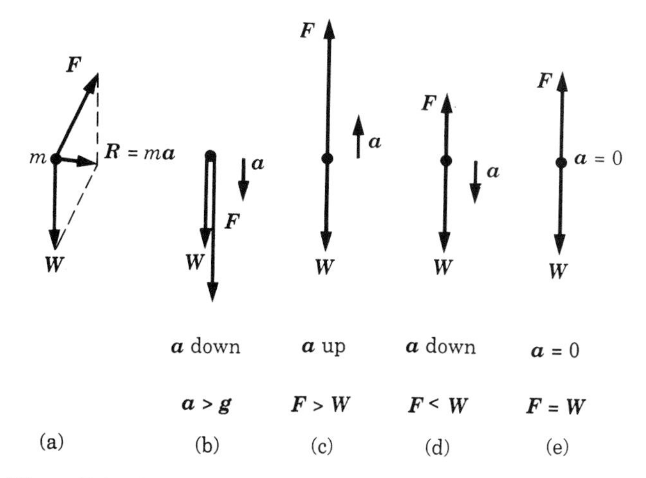

Figure 7.4 Acceleration of a body subject to its weight $\boldsymbol{W}$ and a force $\boldsymbol{F}$.

If the force F is larger than the weight, the vertical motion is accelerated upward. This is, for example, the case of an ascending balloon or a rocket. However, if F is smaller than the weight, the acceleration is downward, but less than that of gravity. This is the case for a body falling through air or another fluid.

EXAMPLE 7.1

Acceleration of masses m and m' of Figure 7.5. Assume that the wheel can rotate freely around O and disregard any possible effects due to the mass of the wheel (these effects will be considered later, in Chapters 13 and 14) as well as the mass of the string.

▷ Suppose that the motion is in the direction shown by the arrow at the top of the wheel, so that mass m is falling and mass m' rising. Both masses move with the same acceleration a. The masses interact through the string. Equal and opposite forces, designated F, are exerted on each other. Then the downward motion of m with acceleration a is given by

$$mg - F = ma$$

and the upward motion of m' with the same acceleration is

$$F - m'g = m'a$$

By adding the two equations we eliminate F and obtain

$$a = \frac{m - m'}{m + m'} g$$

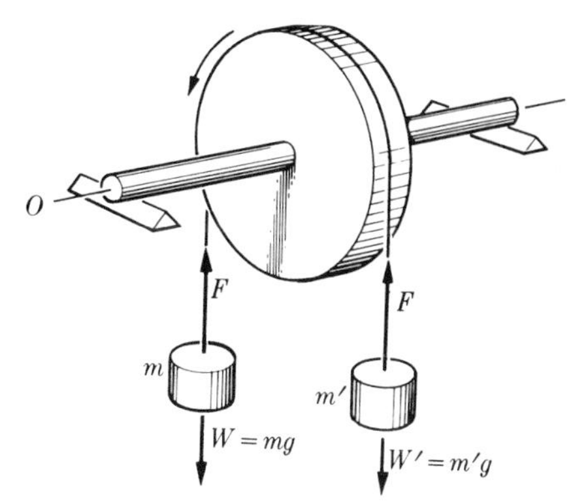

Figure 7.5 Atwood's machine.

for their common acceleration. By inserting this value of the acceleration into either of the two equations, the tension in the string is found to be

$$F = \frac{2mm'}{m + m'} g$$

A device similar to the arrangement of Figure 7.5, which is called **Atwood's machine**, is used to study the laws of uniformly accelerated motion. One advantage of this arrangement is that, by setting m very close to m', we can make the acceleration a very small, which makes it easier to observe the motion and make measurements.

EXAMPLE 7.2

An automobile whose mass is 1000 kg moves uphill along a street inclined 20° with the horizontal. Determine the force the motor must produce if the car is to move (a) with uniform motion, (b) with an acceleration of 0.2 m s^{-2}. Find also, for each case, the force exerted on the automobile by the street. All friction effects are to be ignored.

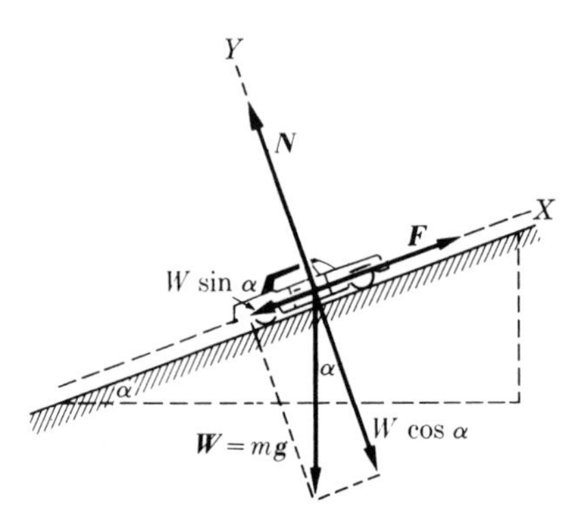

Figure 7.6 Motion on an inclined plane.

▷ Designate the mass of the automobile by m; the forces acting on it are illustrated in Figure 7.6. They are: its weight $\boldsymbol{W} = m\boldsymbol{g}$, pointing downward, the force $\boldsymbol{F}$ due to the motor, pointing uphill, and the force $\boldsymbol{N}$ due to the street and perpendicular to it. Using a set of axes as indicated in the figure, and employing Equation 6.15, we find that the motion along the X direction satisfies the equation

$$F - mg \sin \alpha = ma$$

or

$$F = m(a + g \sin \alpha)$$

The car has no motion along the Y-axis, and thus

$$N - mg \cos \alpha = 0 \qquad \text{or} \qquad N = mg \cos \alpha$$

We note that the force N due to the street is independent of the acceleration of the car and, introducing numerical values, is equal to 9210 N. But the force F due to the motor *does* depend on the acceleration of the car. When the car moves with uniform motion, $a = 0$ and

$$F = mg \sin \alpha$$

In our example F is 3350 N. When the car moves with an acceleration of 0.2 m s^{-2}, then $F = 3550$ N.

We suggest that the student solve the problem again, this time for a car moving downhill both with a constant velocity and under acceleration.

7.4 Equilibrium of a particle

Referring to Figure 7.4(e), when the upward force $\boldsymbol{F}$ is equal to the weight, the resultant force is zero and the acceleration of the particle is also zero. The particle is then at rest or in uniform motion. We say that the particle is in **equilibrium**, for example, a balloon whose total weight is just balanced by the buoyancy of the air it displaces is in equilibrium. Similarly, a body floating on water is in equilibrium because its weight is balanced by the buoyancy of water displaced. An airplane flying horizontally with a constant velocity is in equilibrium, even though it is not at rest, because the thrust of the engines is balanced by the drag on the plane and the weight of the plane is balanced by the lift of the wings. In general:

> *a particle is in equilibrium if the sum of all the forces acting on the particle is zero and, therefore, its acceleration is zero:*

$$\boldsymbol{F}_1 + \boldsymbol{F}_2 + \boldsymbol{F}_3 + \boldsymbol{F}_4 + \cdots = 0 \qquad \text{or} \qquad \sum_i \boldsymbol{F}_i = 0$$

For example, a body (Figure 7.7) on an inclined plane is in equilibrium only if the force of the pull $\boldsymbol{F}$, its weight $\boldsymbol{W}$ and the normal force $\boldsymbol{N}$ exerted by the plane add to zero. That is, the body is in equilibrium if

$$\boldsymbol{F} + \boldsymbol{W} + \boldsymbol{N} = 0$$

For this to be true, it is necessary that the resultant of forces $\boldsymbol{N}$ and $\boldsymbol{F}$ be equal and opposite to the weight $\boldsymbol{W}$.

In terms of the rectangular components of the forces the equilibrium conditions are expressed by

$$\sum \boldsymbol{F}_{ix} = 0, \qquad \sum \boldsymbol{F}_{iy} = 0 \qquad \text{and} \qquad \sum \boldsymbol{F}_{iz} = 0$$

For example, referring to Figure 7.7 and using the X- and Y-axes as shown, F is along the X-axis, N is along the Y-axis and the components of W are $-W \sin \alpha$ along the X-axis and $-W \cos \alpha$ along the Y-axis. Therefore the condition of equilibrium gives the two equations

$$F - W \sin \alpha = 0 \text{ or } F = W \sin \alpha$$

$$N - W \cos \alpha = 0 \text{ or } N = W \cos \alpha$$

which give the relations between the magnitudes of the three forces.

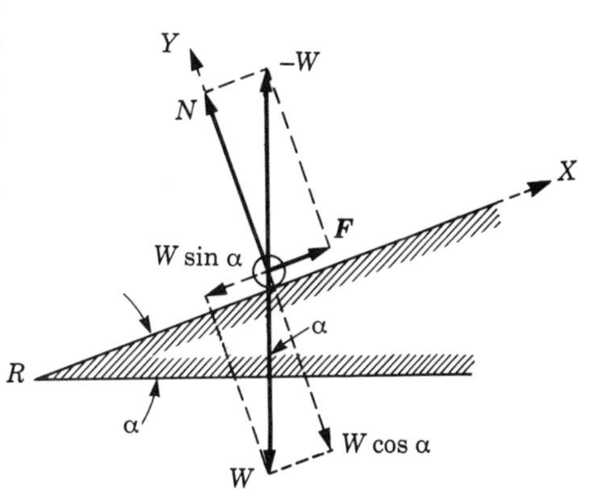

Figure 7.7 Equilibrium of a body.

EXAMPLE 7.3

Equilibrium of three forces.

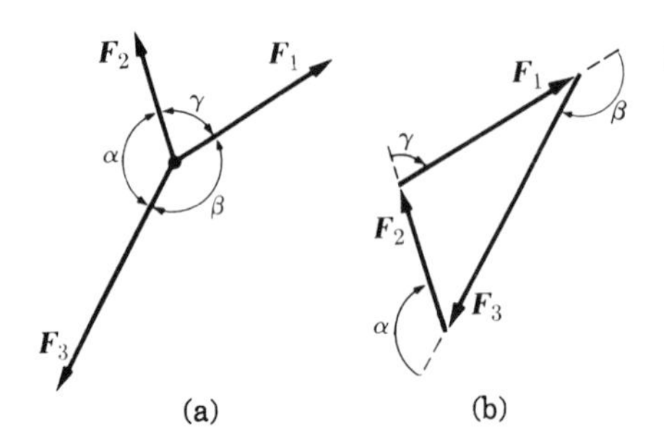

Figure 7.8 Equilibrium of three forces: $\boldsymbol{F}_1 + \boldsymbol{F}_2 + \boldsymbol{F}_3 = 0$.

▷ For the case of three forces $\boldsymbol{F}_1$, $\boldsymbol{F}_2$, and $\boldsymbol{F}_3$ (Figure 7.8(a)) that act on the same point and are in equilibrium, we have

$$\boldsymbol{F}_1 + \boldsymbol{F}_2 + \boldsymbol{F}_3 = 0$$

So, if we draw a polygon with the three forces we get a triangle, as shown in Figure 7.8(b). Thus when three concurrent forces are in equilibrium, they must be in one plane, and also the resultant of $\boldsymbol{F}_1$ and $\boldsymbol{F}_2$ must be equal and opposite to the third force; that is, $\boldsymbol{F}_1 + \boldsymbol{F}_2 = -\boldsymbol{F}_3$.

Applying the law of sines (Equation A.5) to the triangle whose sides are the three forces, we get

$$\frac{F_1}{\sin\alpha} = \frac{F_2}{\sin\beta} = \frac{F_3}{\sin\gamma}$$

which is a very useful formula relating the magnitudes of the forces and the angles between them when the forces are in equilibrium.

EXAMPLE 7.4

Equilibrium of a particle on a smooth inclined plane.

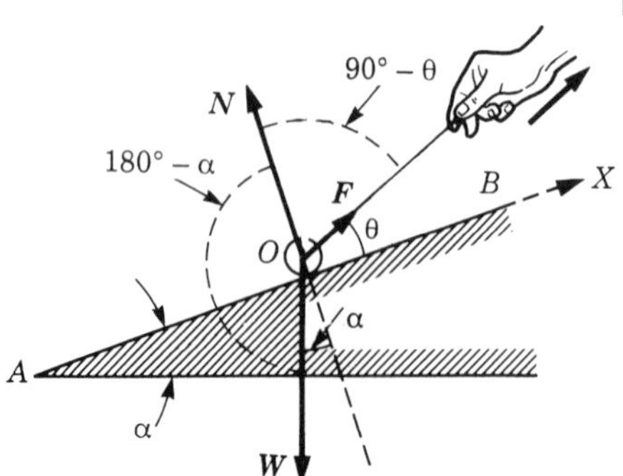

Figure 7.9 Equilibrium of a body on an inclined plane.

▷ Consider the particle O, resting on a smooth (frictionless) inclined plane AB (Figure 7.9). It is subject to the following forces: its weight $\boldsymbol{W}$, the pull $\boldsymbol{F}$, and the normal reaction of the plane $\boldsymbol{N}$. We wish to express $\boldsymbol{F}$ and $\boldsymbol{N}$ in terms of $\boldsymbol{W}$, α and θ. Using the law of sines and considering the geometry of Figure 7.9, we have

$$\frac{F}{\sin(180^\circ - \alpha)} = \frac{N}{\sin(90^\circ + \alpha + \theta)}$$

$$= \frac{W}{\sin(90^\circ - \theta)}$$

or

$$\frac{F}{\sin\alpha} = \frac{N}{\cos(\alpha + \theta)} = \frac{W}{\cos\theta}$$

Solving for F and N gives

$$F = \frac{W\sin\alpha}{\cos\theta}, \quad N = \frac{W\cos(\alpha + \theta)}{\cos\theta}$$

Alternatively, we may require that the forces along the X- and Y-axes, shown in Figure 7.9, each be zero. The result, of course, is the same.

When the plane is not smooth, we must include a frictional force parallel to the plane in a direction opposite to that in which the body tends to slide (see the next section).

7.5 Frictional forces

Whenever there are two bodies in contact, such as in the case of a book resting on a table, there is a resistance that opposes the relative motion of the two bodies. For example, suppose we push a book along the table, giving it some velocity. After we release it, the book slows down and eventually stops. This loss of momentum is indicative of a force opposing the motion; the force is called **sliding friction**. The force of friction is due to the interaction between the molecules of the two bodies. It is referred to as **cohesion** when the two bodies are of the same material or **adhesion** when the two bodies are of different materials. Thus friction involves an extremely large number of molecules from each of the two bodies in contact, which makes it very difficult to calculate it from first principles. Therefore, frictional forces must be determined experimentally. In fact, the phenomenon is rather complex and depends on many factors, such as the condition and nature of the surfaces, the relative velocity etc. We can experimentally verify that the magnitude of the force of friction, for most practical purposes, is proportional to the normal force N pressing one body against the other (Figure 7.10). The constant of proportionality is called the **coefficient of friction**, and is designated by μ. That is, in magnitude,

$$\text{Force of sliding friction} = \mu N \quad \textbf{(7.5)}$$

The force of sliding friction always opposes the relative motion of the bodies, and so has a direction opposite to the relative velocity.

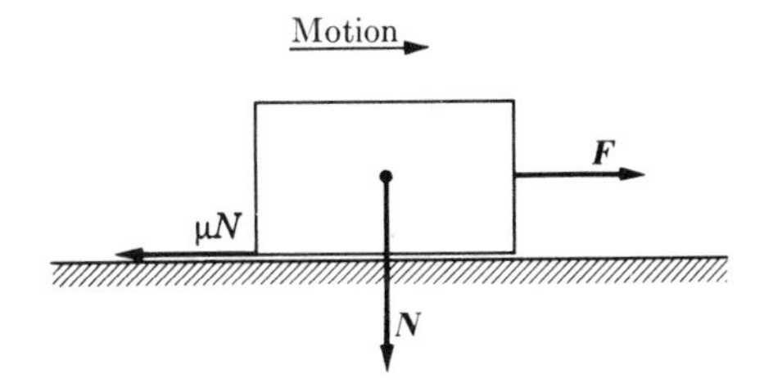

Figure 7.10 The force of friction opposes the motion and depends on the normal force (The upward force exerted by the surface on the body has not been drawn. It is equal and opposite to N.)

For example, when a force $\boldsymbol{F}$ is applied to the body of Figure 7.10 so that it moves to the right, the frictional force μN is to the left and the resultant force on the body is $F - \mu N$. If m is the mass of the body and a its acceleration, the equation of motion of the body is

$$F - \mu N = ma \tag{7.6}$$

In the case where the normal force N is the weight of the body, which is true when the motion is along a horizontal surface, we may write $N = W = mg$. The equation of motion becomes

$$F - \mu mg = ma \qquad \text{or} \qquad F = m(a + \mu g)$$

which gives the force required to slide a body along a horizontal plane with a certain acceleration. When the motion is uniform ($a = 0$), the force is $F = \mu N$; i.e. the applied force is equal to the force of friction.

In general there are two kinds of coefficient of friction. The *static* coefficient of friction, μ_s, when multiplied by the normal force, gives the minimum force needed to set two bodies, initially in contact and at relative rest, into relative motion. The *kinetic* coefficient of friction, μ_k, when multiplied by the normal force, gives the force necessary to maintain uniform relative motion between the two bodies. It has been found experimentally that μ_s is larger than μ_k for most materials. Table 7.1 lists representative values of μ_s and μ_k for several materials.

We want to emphasize that the force of friction, expressed as μN, is a macroscopic force of a *statistical* nature because it represents the sum of a very large number of interactions between the molecules of the two bodies in contact. It is impossible to take into account individual molecular interactions; hence they are determined in a collective way by some experimental method and represented approximately by the coefficient of friction. Therefore, it must be recognized that Equation 7.5 is not a law of physics in the same sense as Newton's laws. It is rather a convenient empirical approximation to the actual force between two sliding bodies. Such a description in terms of empirical concepts and parameters

Table 7.1 Coefficients of friction (all surfaces dry)*

Material	μ_s	μ_k
Steel on steel (hard)	0.78	0.42
Steel on steel (mild)	0.74	0.57
Lead on steel (mild)	0.95	0.95
Copper on steel (mild)	0.53	0.36
Nickel on nickel	1.10	0.53
Cast iron on cast iron	1.10	0.15
Teflon on teflon (or on steel)	0.04	0.04

* These values must be considered as only indicative, since the coefficients of friction are macroscopic quantities that depend on microscopic properties of both materials and fluctuate greatly.

is called, in physics, a **phenomenological** description. It serves to create some order among observations, even in the absence of a full understanding of these phenomena.

EXAMPLE 7.5

A body whose mass is 0.80 kg is on a plane inclined 30° with the horizontal. What force must be applied to the body so that it slides (a) uphill and (b) downhill? In both cases assume that the body moves with: (i) uniform motion and (ii) an acceleration of $0.10\ \mathrm{m\,s^{-2}}$. The coefficient of sliding friction with the plane is 0.30.

▷ (a) Consider first the body moving uphill. The forces acting on the body are illustrated in Figure 7.11(a). They are its weight $W = mg$, pointing downward, the applied force F (which we assume is uphill), the normal force N exerted by the plane on the body and the force of friction μN, which is always *against* the motion and thus in this case must be downhill. When we separate the weight into its component along the plane and its component perpendicular to the plane, the motion of the body along the plane, using Equation 7.6, is

$$F - W \sin \alpha - \mu N = ma$$

But from Figure 7.11(a) we see that the normal force pressing the body against the plane is $N = W \cos \alpha$. Thus the equation of motion becomes

$$F - W(\sin \alpha + \mu \cos \alpha) = ma$$

This equation serves two purposes. If we know the acceleration a, we can find the applied force F. Conversely, if we know the force F we can find the acceleration. In the first case, we have, when we replace W by mg,

$$F = m[a + g(\sin \alpha + \mu \cos \alpha)]$$

For example, if the motion is uniform, $a = 0$, and when we insert the corresponding numerical values, $F = 5.95$ N. When the body is moving uphill, with an acceleration of $0.10\ \mathrm{m\,s^{-2}}$, we obtain $F = 6.03$ N.

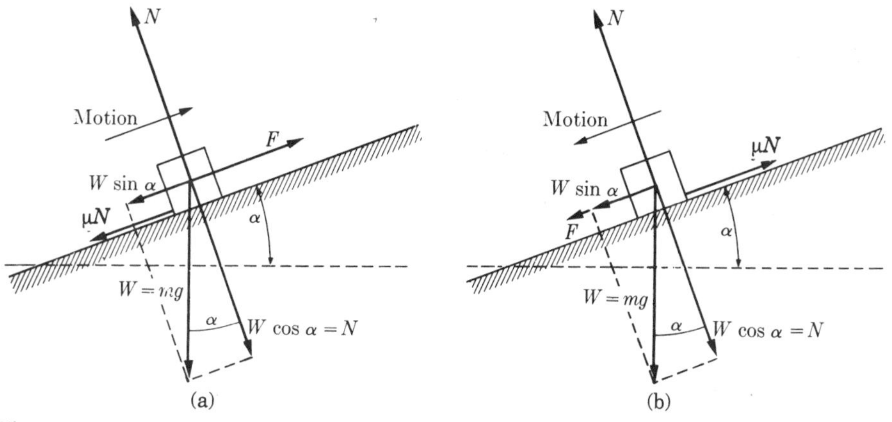

Figure 7.11 (a) Body moving uphill. (b) Body moving downhill.

(b) When the body moves downhill, the forces are as illustrated in Figure 7.11(b). Now we have assumed that F is downhill, but we could also have made the opposite assumption. However, the force of friction must be uphill to oppose the motion. Taking downhill as the positive direction, the student may verify that the equation of motion is now

$$F + W \sin \alpha - \mu N = ma$$

or, since $N = W \cos \alpha$,

$$F + W(\sin \alpha - \mu \cos \alpha) = ma$$

Replacing W by mg, we may write

$$F = m[a - g(\sin \alpha - \mu \cos \alpha)]$$

If the motion is uniform ($a = 0$), when we insert numerical values we obtain $F = -1.88$ N. If the block slides down with an acceleration of 0.10 m s^{-2}, we get $F = -1.80$ N. The negative sign in each case means that the force F is uphill instead of downhill as we had assumed.

We suggest the student determine the motion of the body if no force F is applied and, in view of the result obtained, justify the negative sign for F obtained in part (b).

7.6 Frictional forces in fluids

When a body moves through a fluid (such as a gas or a liquid) it experiences a frictional force (or **drag**) which increases as the velocity of the body, relative to the fluid, increases. This is because when a body moves through the fluid it has to push the molecules of the fluid aside, transferring momentum to them. The faster the body moves, the greater the rate of transfer of momentum; i.e. the larger the frictional force or drag. In addition, there is a frictional force between different layers of a fluid that are moving with different velocities; this will be discussed in Chapter 18. Frictional forces in fluids are called **viscous** forces. They are also macroscopic statistical forces because they involve the action of a very large number of molecules. Therefore, viscous forces must be determined experimentally. At a relatively low velocity, the force of friction is approximately proportional to the velocity, and opposed to it. We therefore write

$$\text{Force of fluid friction} = -K\eta v \qquad \textbf{(7.7)}$$

where η is the **coefficient of viscosity** of the fluid and K is the **drag coefficient**.

The coefficient of viscosity η depends on the molecular properties of the fluid. The drag coefficient K depends on the shape and size of the body and must be obtained experimentally, although in some cases it can also be calculated. For example, in the case of a slowly moving sphere of radius R, laborious calculation indicates that

$$K = 6\pi R \qquad \textbf{(7.8)}$$

a relation known as **Stokes' law**.

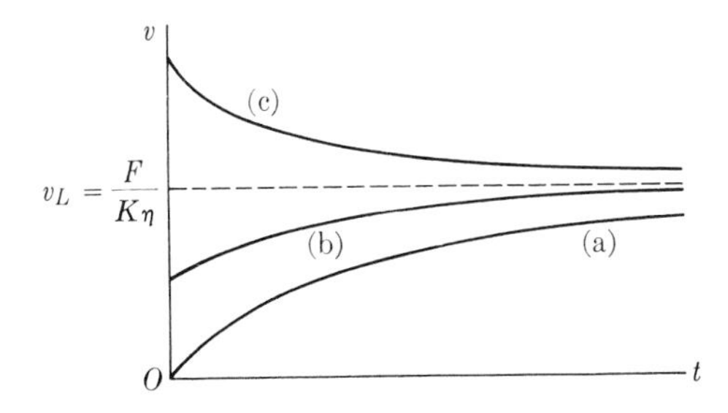

Figure 7.12 Velocity as a function of time for a body falling through a viscous fluid. (a) $v_0 = 0$, (b) $v_0 < v_L$, (c) $v_0 > v_L$.

When a body moves through a viscous fluid under the action of a force $\boldsymbol{F}$, the resultant force is $\boldsymbol{F} - K\eta\boldsymbol{v}$ and the equation of motion is

$$m\boldsymbol{a} = \boldsymbol{F} - K\eta\boldsymbol{v} \tag{7.9}$$

Assuming a constant force $\boldsymbol{F}$, the acceleration $\boldsymbol{a}$ produces an increase in $\boldsymbol{v}$ and a corresponding increase in fluid friction. This, in turn, results in a decrease of the acceleration. Eventually, when a certain velocity is reached, the right-hand side of Equation 7.9 becomes zero. At that moment the acceleration is also zero and there is no further increase in velocity, the applied force being exactly balanced by the fluid friction. The particle continues moving in the direction of the force with a constant velocity, called the **limiting** or **terminal** velocity, which is obtained by making $\boldsymbol{a} = 0$ in Equation 7.9. Thus

$$\boldsymbol{v}_L = \frac{\boldsymbol{F}}{K\eta} \tag{7.10}$$

Therefore the limiting velocity depends on η and K (that is, on the viscosity of the fluid and the shape and size of the body) but is independent of the initial velocity. The variation of the velocity with time, for different initial velocities, is shown in Figure 7.12.

In free fall under the influence of gravity, $\boldsymbol{F} = \boldsymbol{W} = m\boldsymbol{g}$ and Equation 7.10 becomes

$$\boldsymbol{v}_L = \frac{m\boldsymbol{g}}{K\eta} \tag{7.11}$$

This result applies not only to vertical motion but to the motion of a projectile thrown in any direction in the air, or in any fluid. The velocity of the projectile approaches a direction parallel to $\boldsymbol{g}$ and eventually falls vertically, regardless of the direction of the initial velocity, because the viscosity tends to reduce the horizontal velocity to zero, since no other horizontal force acts on the particle. Figure 7.13 illustrates the path taken by a body falling under the action ot its own weight $\boldsymbol{F} = \boldsymbol{W} = m\boldsymbol{g}$ and the viscous force $\boldsymbol{F}' = -K\eta\boldsymbol{v}$.

Equation 7.11 must be corrected for the buoyant force exerted by the fluid, which, according to **Archimedes' principle** (see Section 14.10), is equal to the weight

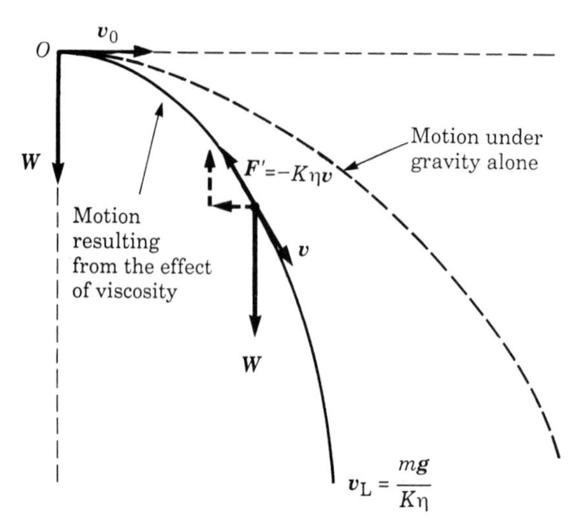

Figure 7.13 Motion under the action of gravity and a viscous force. The graph shows the horizontal and vertical components of the viscous force.

of the fluid displaced by the body. Buoyancy is accounted for by subtracting the mass m' of fluid displaced by the body from m, the mass of the body. Hence Equation 7.11 must be replaced by

$$\boldsymbol{v}_L = \frac{(m - m')\boldsymbol{g}}{K\eta}$$

Let V be the volume of the body and ρ and ρ' the density of the body and of the fluid, respectively. Then $m = V\rho$ and $m' = V\rho'$ and we may write the preceding equation in the alternative form

$$\boldsymbol{v}_L = \frac{(\rho - \rho')V\boldsymbol{g}}{K\eta} \tag{7.12}$$

When the density of the body is much larger than the density of the fluid, such as a solid body falling through air, the buoyancy effect may be ignored. If the density of the body is *less* than that of the fluid, such as a balloon filled with helium, the body will *rise* and reach a terminal velocity as well.

The expression $\boldsymbol{F} = -K\eta\boldsymbol{v}$ for viscous forces is another example of a phenomenological description of *statistical* forces. It does not express a fundamental law; rather it is a convenient approximation when the relative velocity of the body and the fluid is small. For large relative velocities, we need to use other expressions for the viscous force, such as one proportional to the square of the velocity.

Units of viscosity. The coefficient of viscosity in the SI system is expressed in $\mathrm{N\,s\,m^{-2}}$. This can be seen as follows. From Stokes' law, Equation 7.8, we see that K is expressed in meters (the same unit applies to bodies of different shapes). Thus, according to Equation 7.7, η must be expressed in $\mathrm{N/m(m\,s^{-1})}$, which is the same as the unit indicated above. Remembering $\mathrm{N = kg\,m\,s^{-2}}$, we may also express viscosity in $\mathrm{kg\,m^{-1}\,s^{-1}}$. Viscosity may also be expressed by a unit called the **poise**,

Table 7.2 Coefficients of viscosity, in poises*

Liquids	η	Gases	η
Water (0°C)	1.792×10^{-2}	Air (0°C)	1.71×10^{-4}
Water	1.005×10^{-2}	Air	1.81×10^{-4}
Water (40°C)	0.656×10^{-2}	Air (40°C)	1.90×10^{-4}
Alcohol	0.367×10^{-2}	Hydrogen	0.89×10^{-4}
Glycerine	8.33	Ammonia	0.97×10^{-4}
Castor oil	9.86×10^{-2}	Carbon dioxide	1.46×10^{-4}

* All at 20°C, except where noted

and abbreviated P. The poise is equal to one-tenth of the SI unit for viscosity; that is,

$$1 \text{ kg m}^{-1} \text{s}^{-1} = 10 \text{ P}$$

Table 7.2 lists the coefficients of viscosity of several fluids.

EXAMPLE 7.6

Limiting velocity of a steel ball, with a radius of 2 mm, falling through glycerine.

▷ Since the density of steel is 7.9×10^3 kg m^{-3} and that of glycerine is 1.3×10^3 kg m^{-3}, we cannot neglect the effect of buoyancy. Using Equation 7.8, $K = 6\pi r$, and applying Equation 7.12 with $V = (4/3)\pi r^3$, we find that the limiting velocity is given by

$$v_L = \frac{2(\rho - \rho')r^2 g}{9\eta}$$

Substituting numerical values, we find that

$$v_L = 6.9 \times 10^{-2} \text{ m s}^{-1}$$

7.7 Systems with variable mass

The great majority of systems we encounter in physics may be considered as having constant mass. However, in certain cases the mass of a part of the system is variable. The simplest example is that of a raindrop. While it falls, moisture may condense on its surface or water may evaporate, resulting in a change of mass for the drop. Suppose that the mass of the drop is m when it is moving with velocity $\boldsymbol{v}$. Let moisture, with velocity $\boldsymbol{v}_0$, condense on the drop at the rate $\mathrm{d}m/\mathrm{d}t$. The total time rate of change of momentum of the drop is the sum of $m\,\mathrm{d}\boldsymbol{v}/\mathrm{d}t$, due to the acceleration of the drop, and $(\mathrm{d}m/\mathrm{d}t)(\boldsymbol{v} - \boldsymbol{v}_0)$, corresponding to the gain of mass by condensation. Thus the equation of motion of the drop, using Equation 6.14, is

$$\boldsymbol{F} = \frac{\mathrm{d}\boldsymbol{p}}{\mathrm{d}t} = m\frac{\mathrm{d}\boldsymbol{v}}{\mathrm{d}t} + \frac{\mathrm{d}m}{\mathrm{d}t}(\boldsymbol{v} - \boldsymbol{v}_0) \quad (7.13)$$

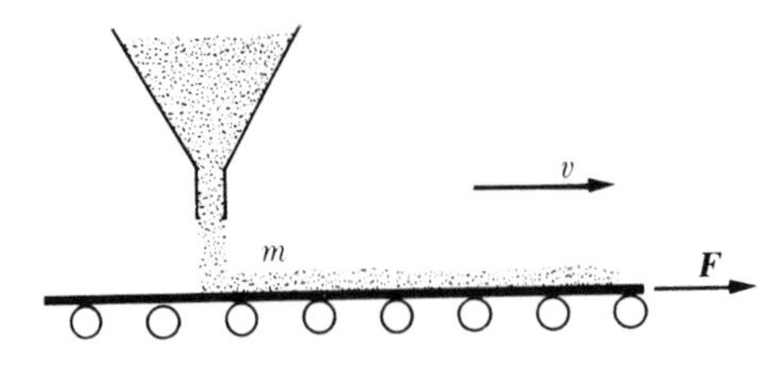

Figure 7.14

To solve this equation it is necessary to make some assumptions about how the mass varies with time.

Another example of a system of variable mass is a conveyor on which material is dropped at one end (Figure 7.14) and/or discharged at the other end. Suppose that material is dropped continuously on the moving belt at the rate of $\mathrm{d}m/\mathrm{d}t$. The conveyor is moving at the constant velocity $\mathbf{v}$ and a force $\boldsymbol{F}$ is applied to keep the belt moving at the constant velocity. If M is the mass of the belt and m is the mass of the material already dropped at time t, the total momentum of the system at that time is $\boldsymbol{P} = (m + M)\boldsymbol{v}$. Therefore, the force that must be applied to the belt to keep it moving with a constant velocity is

$$\boldsymbol{F} = \frac{\mathrm{d}\boldsymbol{P}}{\mathrm{d}t} = \boldsymbol{v}\frac{\mathrm{d}m}{\mathrm{d}t}$$

Note that the force in this case is related to the change in mass and not to the change in velocity.

Perhaps the most interesting example is that of a rocket, whose mass decreases because it expels the fuel it consumes. In the following example the dynamics of a rocket is analyzed.

EXAMPLE 7.7

Motion of a rocket.

▷ When a projectile is thrown into the air, it must receive an initial impulse. For example, baseball and golf players must hit the ball hard for a very short time. Similarly, in a rifle or cannon, the missile (bullet or shell) receives an initial impulse for a very short time from the expansion of gases in the gun barrel. On the other hand, a rocket is a missile that is acted on by a continuous force derived from the exhaust of gases produced in the combustion chamber within the missile itself (Figure 7.15). The rocket at take-off has a certain amount of fuel that is used gradually, and therefore the mass of the rocket decreases gradually.

Call $\boldsymbol{v}$ the velocity of the rocket relative to an inertial system, which we shall assume to be the Earth, and $\boldsymbol{v}'$ the velocity of the exhaust gases, also relative to the Earth. Then the exhaust velocity of the gases *relative to the rocket* is

$$\boldsymbol{v}_{\mathrm{e}} = \boldsymbol{v}' - \boldsymbol{v}$$

This velocity is opposed to $\boldsymbol{v}$. Let m be the mass of the rocket, including its fuel, at any time. During a very small time interval $\mathrm{d}t$, the mass of the rocket experiences a small change $\mathrm{d}m$, which is negative because the mass decreases. In the same time interval the velocity of the rocket changes by $\mathrm{d}\boldsymbol{v}$. The momentum of the system (the rocket, including the unexpended fuel) relative to the Earth at time t is $\boldsymbol{p} = m\boldsymbol{v}$. The momentum of the system (rocket plus exhausted gases) at time $t + \mathrm{d}t$ when the mass of the rocket is $m + \mathrm{d}m$ (where $\mathrm{d}m$ is quite

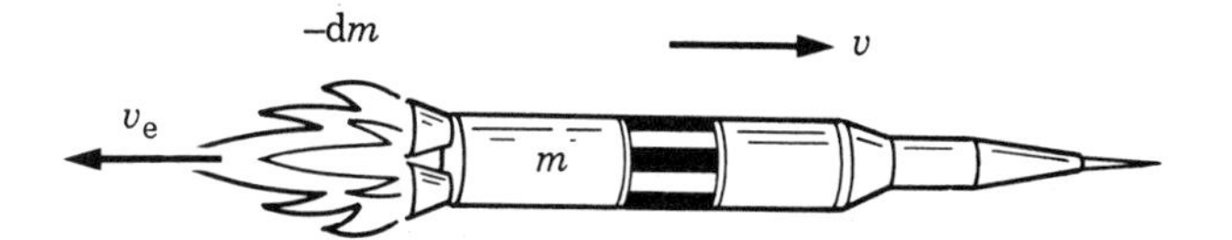

Figure 7.15

small and negative) is

$$\boldsymbol{p}' = \underbrace{(m + \mathrm{d}m)(\boldsymbol{v} + \mathrm{d}\boldsymbol{v})}_{\text{Rocket}} + \underbrace{(-\mathrm{d}m)\boldsymbol{v}'}_{\text{Gases}}$$

or

$$\boldsymbol{p}' = m\boldsymbol{v} + m\,\mathrm{d}\boldsymbol{v} - (\boldsymbol{v}' - \boldsymbol{v})\,\mathrm{d}m$$

where $-\mathrm{d}m$ is the positive value of the mass of the expelled gases and we have neglected the second-order term $\mathrm{d}m\,\mathrm{d}\boldsymbol{v}$. Since $\boldsymbol{v}_e = \boldsymbol{v}' - \boldsymbol{v}$, we can rewrite the above equation as

$$\boldsymbol{p}' = m\boldsymbol{v} + m\,\mathrm{d}\boldsymbol{v} - \boldsymbol{v}_e\,\mathrm{d}m$$

Since $\boldsymbol{p} = m\boldsymbol{v}$, the change in momentum in the time $\mathrm{d}t$ is

$$\mathrm{d}\boldsymbol{p} = \boldsymbol{p}' - \boldsymbol{p} = m\,\mathrm{d}\boldsymbol{v} - \boldsymbol{v}_e\,\mathrm{d}m$$

and the change of momentum of the system per unit time is

$$\frac{\mathrm{d}\boldsymbol{p}}{\mathrm{d}t} = m\frac{\mathrm{d}\boldsymbol{v}}{\mathrm{d}t} - \boldsymbol{v}_e\frac{\mathrm{d}m}{\mathrm{d}t}$$

When there are no external forces acting on the system, $\mathrm{d}\boldsymbol{p}/\mathrm{d}t = 0$ and the 'force' $m(\mathrm{d}\boldsymbol{v}/\mathrm{d}t)$ experienced by the rocket must be equal to the 'force' due to the escaping gases, given by $-\boldsymbol{v}_e(\mathrm{d}m/\mathrm{d}t)$ and designated as the **thrust** of the rocket.

If there is an external force $\boldsymbol{F}$ acting on the rocket, the equation of motion, according to Newton's second law, Equation 6.12, is $\boldsymbol{F} = \mathrm{d}\boldsymbol{p}/\mathrm{d}t$; then for our case we have

$$\boldsymbol{F} = m\frac{\mathrm{d}\boldsymbol{v}}{\mathrm{d}t} - \boldsymbol{v}_e\frac{\mathrm{d}m}{\mathrm{d}t} \quad \text{or} \quad m\frac{\mathrm{d}\boldsymbol{v}}{\mathrm{d}t} = \boldsymbol{F} + \boldsymbol{v}_e\frac{\mathrm{d}m}{\mathrm{d}t} \tag{7.14}$$

To solve this equation we must make some assumption about $\boldsymbol{v}_e$. In general it is assumed that $\boldsymbol{v}_e$ is constant. Also, neglecting air resistance and the variation of gravity with altitude, we may write $\boldsymbol{F} = m\boldsymbol{g}$, so that Equation 7.14 becomes

$$\frac{\mathrm{d}\boldsymbol{v}}{\mathrm{d}t} = \boldsymbol{g} + \frac{\boldsymbol{v}_e}{m}\frac{\mathrm{d}m}{\mathrm{d}t} \tag{7.15}$$

Multiplying by $\mathrm{d}t$ and integrating from the beginning of the motion ($t = 0$), when the velocity is $\boldsymbol{v}_0$ and the mass is m_0, up to an arbitrary time t, we have

$$\int_{\boldsymbol{v}_0}^{\boldsymbol{v}} \mathrm{d}\boldsymbol{v} = \boldsymbol{g}\int_0^t \mathrm{d}t + \boldsymbol{v}_e\int_{m_0}^{m} \frac{\mathrm{d}m}{m}$$

Calculation of each integral gives

$$\boldsymbol{v} - \boldsymbol{v}_0 = \boldsymbol{g}t + \boldsymbol{v}_e \ln \frac{m}{m_0}$$

For the special case where the motion is vertically upward, $\boldsymbol{v}$ is directed upward, while $\boldsymbol{g}$ and $\boldsymbol{v}_e$ are directed downward. The motion is then rectilinear and the above equation can be written in scalar form as

$$v = v_0 - gt + v_e \ln \frac{m_0}{m} \tag{7.16}$$

Let t be the time required for burning *all* the fuel. Then, in Equation 7.16, m is the final mass and v is the maximum velocity attained by the rocket. For example, if a rocket has an initial mass of 2.72×10^6 kg, a final mass of 2.52×10^6 kg after the fuel is burned, and the gases are expelled at a rate of 1290 kg s^{-1}, the total burn time is 155 s. If we assume an exhaust velocity of 55 000 m s^{-1} and $v_0 = 0$, the maximum velocity of the rocket, assuming it is shot vertically upward, will be

$$v = 55\,000 \ln \frac{2.72 \times 10^6}{2.52 \times 10^6} - (9.8)(155) = 2681 \text{ m s}^{-1}$$

This speed is more than 9600 km h^{-1}.

Rockets are used for the propulsion of vehicles for several different purposes. They are essential for the propulsion of vehicles in space, as will be discussed in Chapter 12.

QUESTIONS

7.1 Explain why, when the force is constant, the motion is in the plane determined by the direction of the force and the direction of the initial velocity.

7.2 Draw, in Figure 7.2, the forces on m_1 due to the other masses. Assume the forces are attractive.

7.3 A body is hanging from a string. A person pulls upward on the other end of the string. Is the force of the pull larger than, equal to, or smaller than the weight of the body when the upward motion is (a) uniform, (b) accelerated, (c) decelerated?

7.4 Explain what is meant by the statement that frictional and viscous forces are statistical in nature.

7.5 Is there a specific relation between the direction of the force of friction and the direction of the velocity of a body? Is the direction of the force of friction of a body related to the direction of its acceleration?

7.6 A body moves across a horizontal surface under an applied force. What kind of motion results when the applied force is (a) larger than, (b) equal to and (c) smaller than the force of friction? What kind of motion results if the force suddenly becomes zero?

7.7 Explain why a body falling through a viscous fluid attains a constant or limiting velocity. What is the relation between the weight of the body and the viscous force when the body reaches the limiting velocity? Does the shape of the body affect the limiting velocity?

7.8 Can a body be in equilibrium and in motion? What kind of motion? Can a body be at rest but not be in equilibrium?

7.9 Is there a specific relation between the direction of the resultant force on a particle and the directions of the acceleration of the particle and of its velocity?

7.10 Aristotle said different bodies fell at different rates. Galileo proved that all bodies fall with the same acceleration. Under what conditions might both assertions be right?

7.11 Equation 7.9 for motion in a viscous fluid may be written as $m\,d\boldsymbol{v}/dt = \boldsymbol{F} - K\eta\boldsymbol{v}$. When $\boldsymbol{F}$ is the weight $m\boldsymbol{g}$ of the body, this equation reduces to

$$\frac{d\boldsymbol{v}}{dt} = \boldsymbol{g} - \frac{K\eta}{m}\boldsymbol{v}$$

(a) Verify by direct substitution that the solution of this equation is

$$\boldsymbol{v} = \frac{m\boldsymbol{g}}{K\eta}(1 - e^{-K\eta t/m}) + \boldsymbol{v}_0 e^{-K\eta t/m}$$

where $\boldsymbol{v}_0$ is the initial velocity ($t = 0$). (b) Confirm that the limiting velocity ($t \to \infty$) is given by Equation 7.11 regardless of the value of $\boldsymbol{v}_0$. Note that the contribution of the initial velocity decreases exponentially with time. Recalling that $\boldsymbol{v} = d\boldsymbol{r}/dt$, obtain $\boldsymbol{r}$ as a function of t. This gives the vector equation of the path described by a projectile. (c) By making $K\eta/m \ll 1$, find the asymptotic values and compare with the results of Section 7.2.

7.12 Is it possible for three forces to be in equilibrium and not be in the same plane?

PROBLEMS

7.1 A person is standing on the flat bed of a truck moving at 36 km h^{-1}. At what angle, and in which direction, should the person lean to avoid falling if, in 2 s, the velocity of the truck changes to (a) 45 km h^{-1}, (b) 9 km h^{-1}?

7.2 An elevator whose mass is 1500 kg is carrying three persons whose masses are 60 kg, 80 kg and 100 kg. The initial force exerted by the motor is 20 000 N. (a) With what acceleration will the elevator rise? (b) Starting from rest, how high will it go in 5 s? (c) What should be the force exerted by the motor to move the elevator with constant velocity?

7.3 A person whose mass is 60 kg is in an elevator. Determine the force the floor exerts on the person when: (a) the elevator goes up with uniform motion, (b) the elevator goes down with uniform motion, (c) the elevator accelerates upward at 3 m s^{-2}, (d) the elevator accelerates downward at 3 m s^{-2}, and (e) the cable breaks and the elevator falls freely.

7.4 A body with a mass of 1.0 kg is on a smooth plane inclined at an angle of 30° with the horizontal. With what acceleration and in what direction will the body move if there is a force of 8.0 N applied parallel to the plane and directed (a) upward, (b) downward?

7.5 The bodies A, B and C in Figure 7.16 have masses 10 kg, 15 kg and 20 kg, respectively. A force F, equal to 50 N, is applied to C. (a) Find the acceleration of the system, assuming no friction, and (b) the tensions in each cable. (c) Discuss the problem when the system moves up a plane inclined 20° with the horizontal.

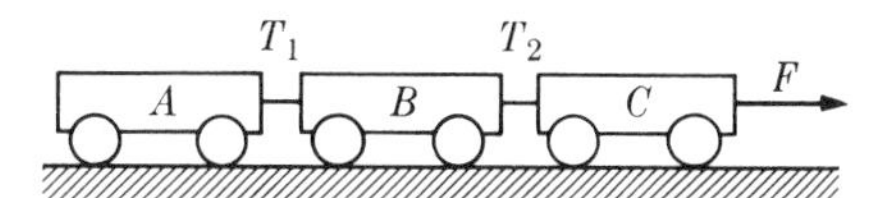

Figure 7.16

7.6 Repeat parts (a) and (b) of Problem 7.5 when the coefficient of friction of the bodies with the surface is 0.1.

7.7 Calculate the acceleration of the bodies in Figure 7.17 and the tension in the string. First solve the problem algebraically and apply to the case $m_1 = 50$ kg, $m_2 = 80$ kg and $F = 1000$ N.

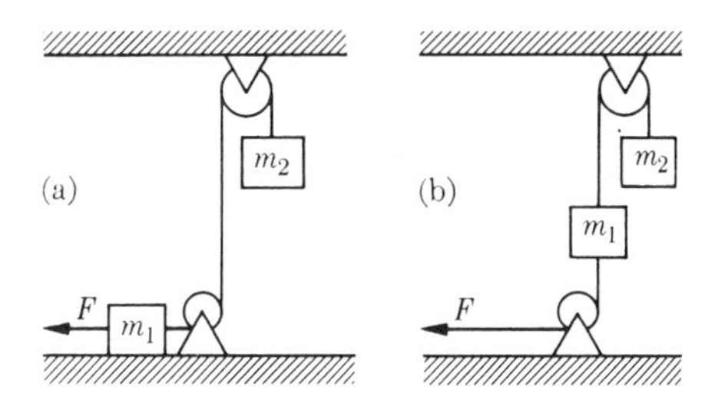

Figure 7.17

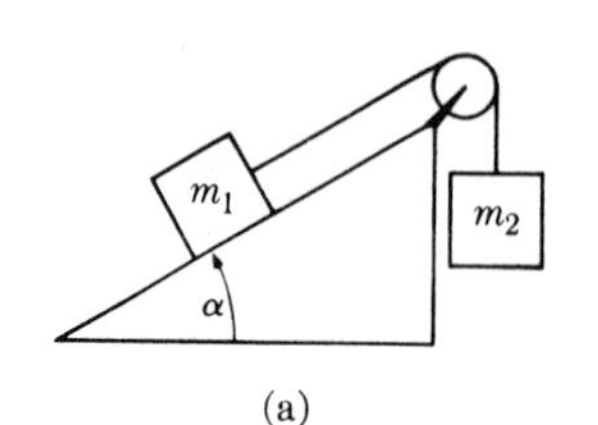

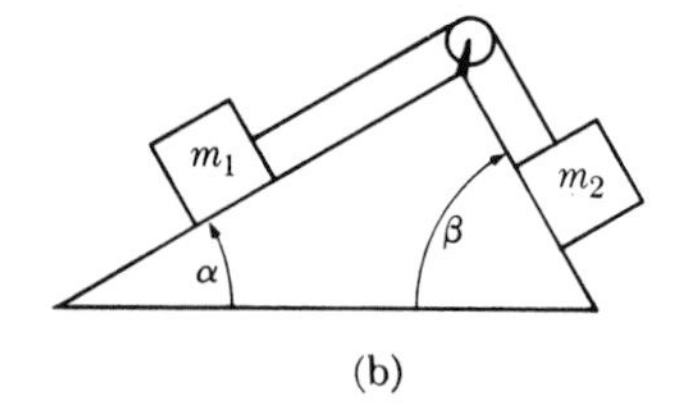

Figure 7.18

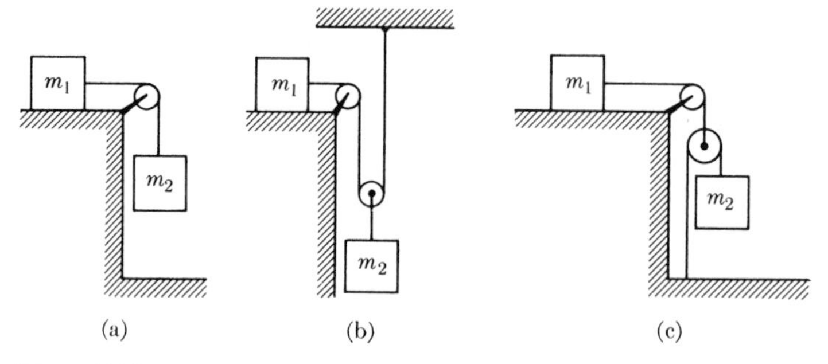

Figure 7.19

7.8 Determine the acceleration with which the bodies in Figure 7.18(a) and (b) move, and also the tensions in the strings. Assume that the bodies slide without friction. First solve the problem generally, and then apply to the case $m_1 = 20$ kg, $m_2 = 18$ kg, $\alpha = 30°$, $\beta = 60°$.

7.9 Repeat case (b) in the previous problem when there is friction, with coefficients 0.12 on the first surface and 0.10 on the second. Discuss all possible motions.

7.10 Calculate the acceleration of the bodies m_1 and m_2 and the tension in the ropes (Figure 7.19). All pulleys are weightless and frictionless and the bodies slide without friction. Solve algebraically; then apply to the case $m_1 = 4$ kg, $m_2 = 6$ kg.

7.11 The masses of A and B in Figure 7.20 are, respectively, 10 kg and 5 kg. The coefficient of friction of A with the table is 0.20. (a) Find the minimum mass of C that will prevent A from moving. (b) Compute the acceleration of the system if C is removed.

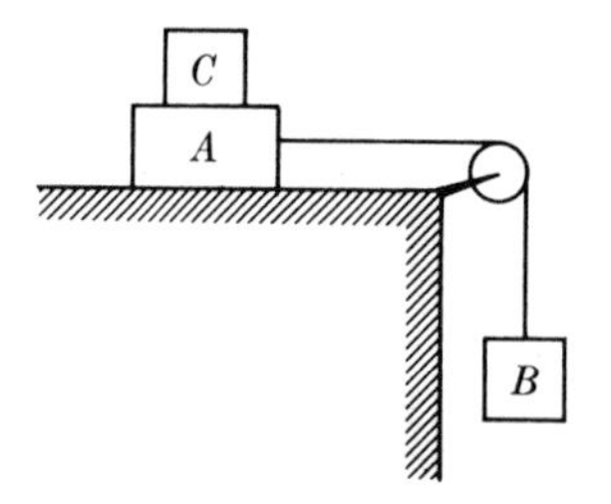

Figure 7.20

7.12 (a) Repeat Example 7.5 for a case where there is no applied force. The initial velocity of the body is $2\ \mathrm{m\,s^{-1}}$ up the plane. (b) How far up the plane will the body move before it stops?
(c) What is the *least* value for the coefficient of friction so that the body, once stopped, will not come back down?

7.13 A block of mass 0.2 kg starts up a plane inclined 30° with the horizontal with a velocity of $12\ \mathrm{m\,s^{-1}}$. (a) If the coefficient of sliding friction is 0.16, determine how far up the plane the block travels before stopping. (b) What is the block's speed when (if) it returns to the bottom of the plane?

7.14 Determine the frictional force exerted by the air on a body whose mass is 0.4 kg when it is falling with an acceleration of $9.0\ \mathrm{m\,s^{-2}}$.

7.15 (a) By integrating Equation 7.9, with $F = mg$, show that the velocity of a body falling freely from rest through a viscous medium is given by

$$v = v_L(1 - e^{-Pt}),$$

where $P = m/K\eta$. (b) Compare the velocity of the body at the time $t = P$ from the above equation with that of a freely falling body.

7.16 A 10^3 kg rocket is set vertically on its launching pad. The propellant is expelled at the rate of $2\ \text{kg}\,\text{s}^{-1}$. (a) Find the minimum velocity of the exhaust gases so that the rocket just begins to rise. (b) Find the rocket's velocity 10 s after ignition, assuming the minimum exhaust velocity.

7.17 A rocket, launched vertically, expels mass at a constant rate equal to $5 \times 10^{-2}\ m_0\ \text{kg}\,\text{s}^{-1}$, where m_0 is its initial mass. The exhaust velocity of the gases relative to the rocket is $5 \times 10^3\ \text{m}\,\text{s}^{-1}$. Find (a) the velocity and (b) the height of the rocket after 10 s.

7.18 A railroad tank car with a volume of $10\ \text{m}^3$ is full of water. The tank gets a leak in the bottom so that it loses water at the rate of $100\ \text{cm}^3\,\text{s}^{-1}$. (a) How does the acceleration of the car change if a constant force F is applied to the car? (b) How should the force vary to maintain a constant acceleration?

7.19 An empty railroad car of mass 10^5 kg coasts with a velocity of $0.5\ \text{m}\,\text{s}^{-1}$ beneath a stationary coal hopper. If 2×10^5 kg of coal are dumped into the car as it passes beneath the hopper, then: (a) What is the car's final velocity? (b) What will be the car's velocity if, instead, the coal is allowed to leave the car through bottom hoppers falling straight down relative to the car?

7.20 A body moves under the action of a constant force F through a fluid that opposes the motion with a force proportional to the square of the velocity; that is, $-kv^2$. Show that the limiting velocity is $v_L = (F/k)^{1/2}$.

7.21 A telephone pole is held in a vertical position by means of a cable that is fixed on the pole at a height of 10 m and also fixed to the ground 7 m from the base of the pole. If the tension in the cable is 5000 N, what are the horizontal and vertical forces exerted on the pole by the cable?

7.22 An inclined plane is 2 m high and 5 m long. There is a stone block (weight 100 N) on the plane, held in place by a fixed obstacle. Find the force exerted by the block (a) on the plane and (b) on the obstacle.

7.23 Find the magnitude and direction of the resultant of the systems of forces represented in Figure 7.21.

7.24 Determine the tensions on the ropes AC and BC (Figure 7.22) if M has a mass of 40 kg.

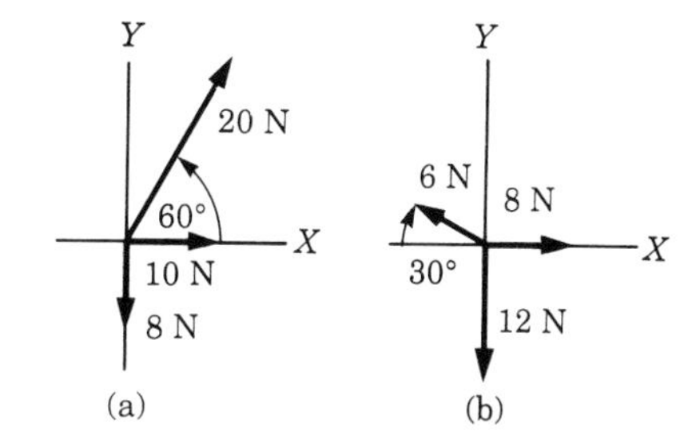

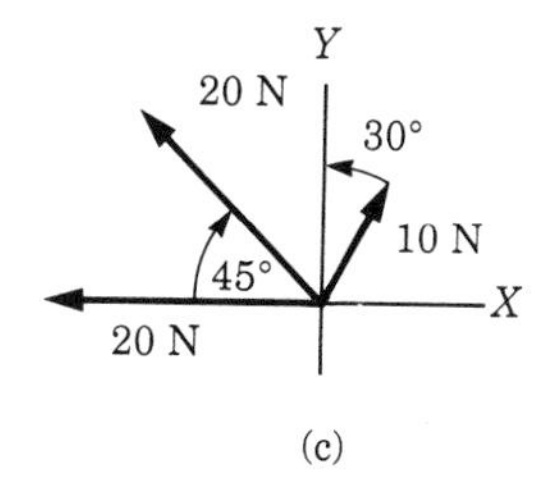

Figure 7.21

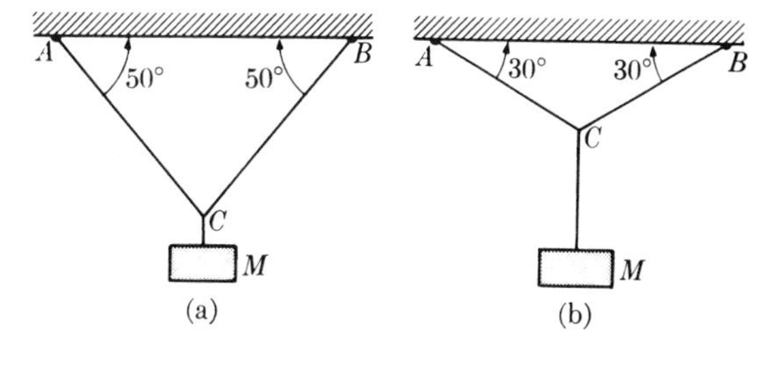

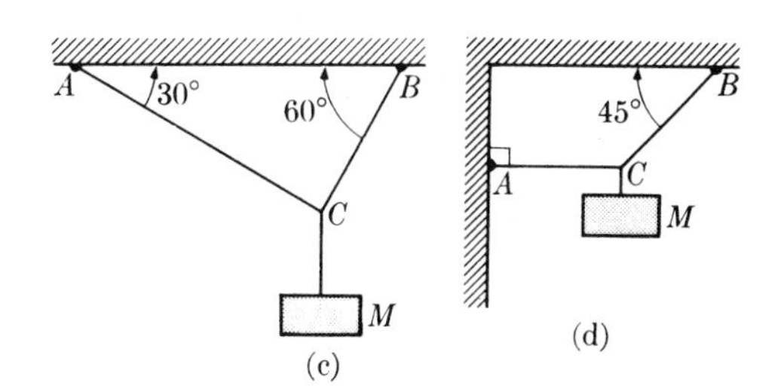

Figure 7.22

7.25 The body represented in Figure 7.23 weighs 400 N. It is held in equilibrium by means of the tension on the rope AB and the horizontal force $\boldsymbol{F}$. Given that $AB = 1.50$ m and the distance between the wall and the body is 0.90 m, calculate (a) the value of the force $\boldsymbol{F}$ and (b) the tension in the rope.

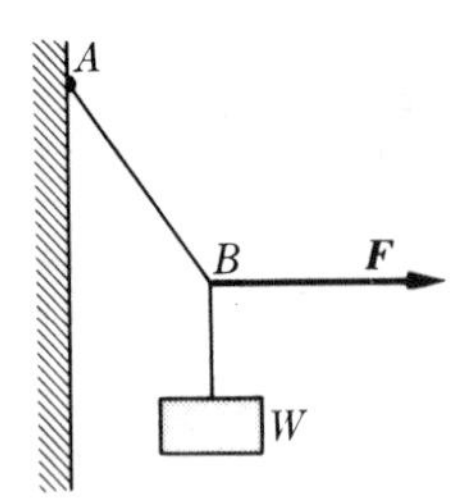

Figure 7.23

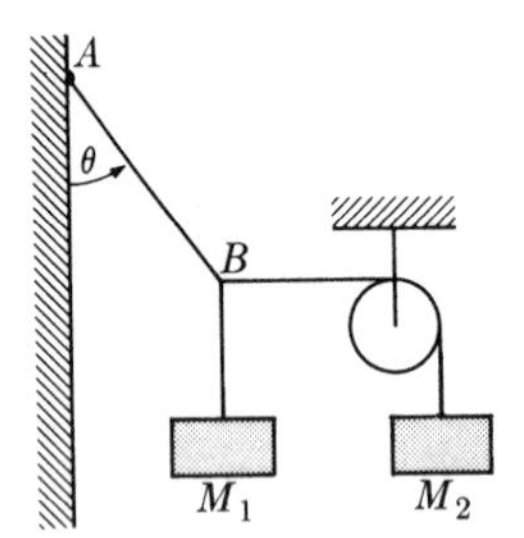

Figure 7.24

7.26 For Figure 7.24, calculate the angle θ and the tension in the rope AB if $M_1 = 300$ kg and $M_2 = 400$ kg.

7.27 A person with mass of 45 kg is holding onto a chinning bar. What force does each of the person's arms exert on the bar when (a) the arms are parallel to each other and (b) each arm makes an angle of 30° with the vertical? (c) Plot the force as a function of the angle. (d) What do you conclude from the graph?

7.28 A rope is hanging from the fixed points A and D (Figure 7.25). At point B there is a mass of 12 kg and at C an unknown weight. If the angle of AB with the horizontal is 60°, BC is horizontal, and CD forms an angle of 30° with the horizontal, calculate the value of W in order for the system to be in equilibrium.

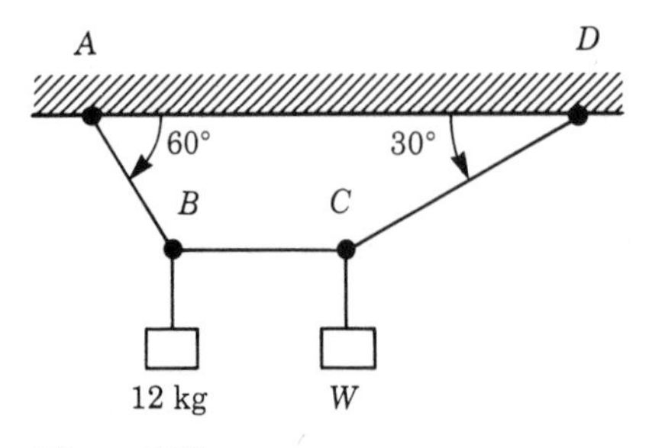

Figure 7.25

7.29 A sphere whose mass is 50 kg is leaning on two smooth planes, respectively inclined 30° and 45° with respect to the horizontal. Calculate the forces of the two planes on the sphere.

7.30 Determine the forces (Figures 7.26(a), (b) and (c)) on the beam BA by the mass M and the cable AC, assuming that M has a mass of 40 kg and the weight of the cable and the beam may be neglected.

7.31 Calculate the weight P needed to maintain equilibrium in the system shown in Figure 7.27, when A is 100 kg and Q is 10 kg. The plane and the pulleys are all smooth. Rope AC is horizontal and rope AB is parallel to the plane.

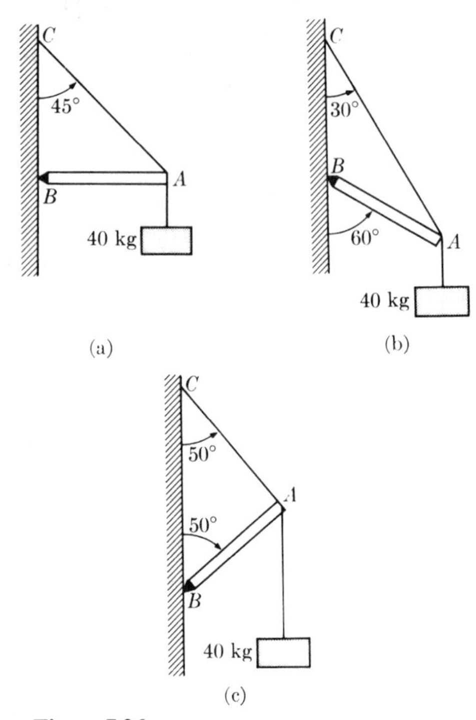

Figure 7.26

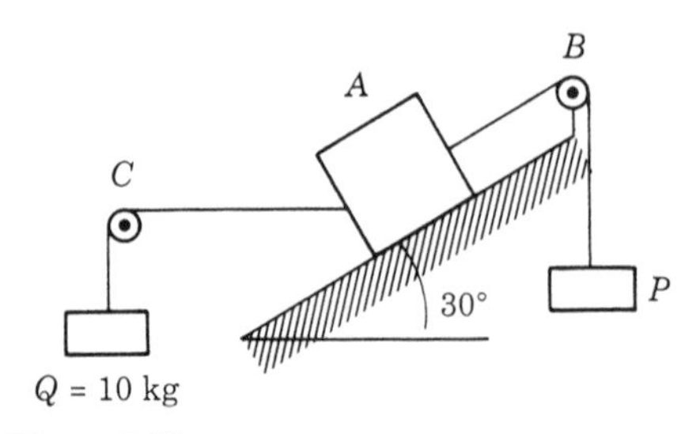

Figure 7.27

8 Torque and angular momentum

The force exerted by wind on the sails of a windmill causes them to rotate, that is it results in angular movement of the wheel. The measure of how effectively a force causes rotation of the body on which it acts is called **torque**. (Courtesy of Canadian Pacific Limited M8646.)

Notes

8.1 Introduction

It is very seldom that a body moves in a straight line or under a constant force. This is so because the resultant force on a body generally makes an angle with the velocity of the body and varies from point to point. To address such cases, it has been found useful to introduce new concepts, such as **torque** and **angular momentum**.

8.2 Curvilinear motion

If the force acting on a particle has the same direction as the velocity, the motion is in a straight line. To produce curvilinear motion, the resultant force must be at an angle with the velocity as already discussed in Section 7.2. Recall that the acceleration is always parallel to the force. Then the acceleration will have a component parallel to the velocity that changes its magnitude and another component perpendicular to the velocity that accounts for its change in direction

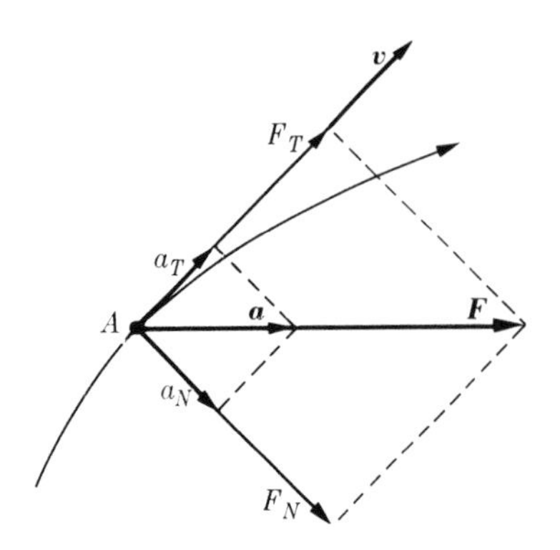

Figure 8.1 Relationship between the components of the force and the acceleration in curvilinear motion.

of the motion. The relationship between all these vectors in curvilinear motion is illustrated in Figure 8.1.

From the relation $\boldsymbol{F} = m\boldsymbol{a}$ and Equation 4.12, that gives the tangential acceleration as $a_T = \mathrm{d}v/\mathrm{d}t$, we conclude that the component of the force tangent to the path, or the **tangential force**, is

$$F_T = ma_T \quad \text{or} \quad F_T = m\frac{\mathrm{d}v}{\mathrm{d}t} \tag{8.1}$$

Using Equation 4.13 for the normal acceleration, $a_N = v^2/R$, the component of the force perpendicular to the path, called the **normal** or **centripetal force**, is

$$F_N = ma_N \quad \text{or} \quad F_N = \frac{mv^2}{R} \tag{8.2}$$

where R is the radius of curvature of the path. The centripetal force is always pointing toward the center of curvature at each point along the trajectory. The tangential force is responsible for the change in the magnitude of the velocity while the centripetal force is responsible for the change in the direction of the velocity. When the tangential force is zero, there is no tangential acceleration, which results in uniform curvilinear motion. If, in addition, the centripetal force is constant, the motion is circular. When the centripetal force is zero, there is no normal acceleration and the motion is rectilinear.

In the particular case of circular motion, $v = \omega R$, so that the centripetal force can also be written as

$$F_N = m\omega^2 R \tag{8.3}$$

where R is now the radius of the circle. In uniform circular motion the only force is F_N.

EXAMPLE 8.1

Banking of curves.

▷ Railroad tracks and highways are banked at curves to produce the centripetal force required by a vehicle moving along a curve. If curves were not banked, the centripetal force would have to be provided by a sidewise force on the track or the wheels or of the friction of the road on the tires. Figure 8.2 illustrates banking, although the angle has been exaggerated. The forces acting on the car are its weight $W = mg$ and the normal force N due to the tracks. To go around the curve of radius R at a velocity v such that there is no force applied to the wheels by the track along the direction of the axle, the resultant, F_N, of W and N must be enough to produce the centripetal force given in Equation 8.2. Thus $F_N = mv^2/R$,

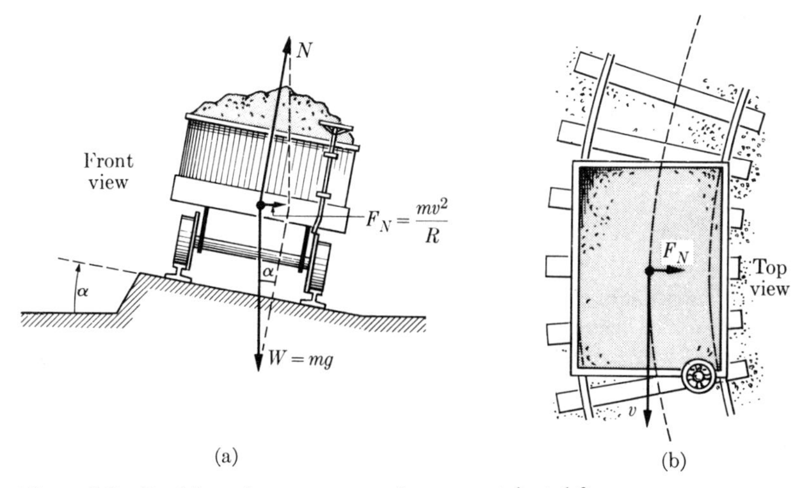

Figure 8.2 Banking of curves to produce a centripetal force.

where R is the radius of the curve. Then, from the figure we see that

$$\tan \alpha = \frac{F_N}{W} = \frac{v^2}{Rg}$$

The banking angle is thus independent of the mass of the body. Since α is fixed once the tracks have been laid or the roadway built, this formula gives the correct speed to traverse the curve so that there will be no sidewise forces acting on the vehicle. For smaller or slightly larger speeds, there is no great problem with the curve because the tracks or tire friction provide the balancing sidewise force needed. However, for large speeds the car will tend to skid off the curve.

EXAMPLE 8.2

A mass suspended from a fixed point by a string of length L is made to rotate around the vertical with angular velocity ω. Find the angle of the string with the vertical. This device is called a **conical pendulum**.

▷ The system has been illustrated in Figure 8.3. The body A moves around the vertical OC, describing a circle of radius $R = CA = OC \sin \alpha = L \sin \alpha$. The forces acting on A are its weight $W = mg$ and the tension F of the string. Their resultant F_N must be just the centripetal force required to describe the circle. Thus, using Equation 8.3, we have $F_N = m\omega^2 R = m\omega^2 L \sin \alpha$. From the figure, we see that

$$\tan \alpha = \frac{F_N}{W} = \frac{\omega^2 L \sin \alpha}{g}$$

or, since $\tan \alpha = \sin \alpha / \cos \alpha$,

$$\cos \alpha = g/\omega^2 L$$

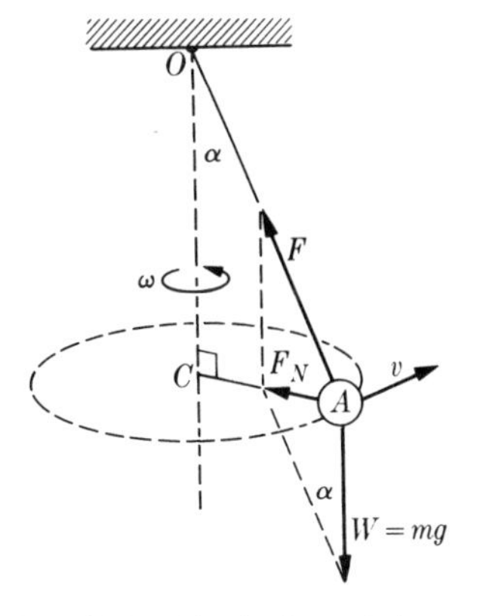

Figure 8.3 Conical pendulum.

Therefore the larger the angular velocity ω, the larger the angle α. For this reason the conical pendulum has long been used as a speed regulator (or 'governor') for steam engines; it closes the steam intake valve when the velocity goes above a predetermined limit and opens it when the velocity falls below.

EXAMPLE 8.3

Helical motion under an **axial force**. The case of $\boldsymbol{F} = \boldsymbol{A} \times \boldsymbol{v}$, where $\boldsymbol{A}$ is a constant vector.

▷ An important variation of circular motion is **helical** motion, i.e. motion in a helix. Helical motion results when a circular motion is combined with uniform motion along a direction perpendicular to the plane of the circle. Consider a particle P subject to a force $\boldsymbol{F}$ that is constant in magnitude and always perpendicular to an axis ZZ'. This is an example of what is called an **axial** force (Figure 8.4). If the initial velocity of the particle is in a plane perpendicular to the axis, say $\boldsymbol{v}$, the particle will move in a circle of radius $R = mv^2/F$. If, however, the initial velocity $\boldsymbol{v}$ makes an angle with the plane, it thus has components $\boldsymbol{v}_1$ in the plane and $\boldsymbol{v}_2$ perpendicular to the plane. Then the particle will move around ZZ' with the velocity $\boldsymbol{v}_1$ while it also moves parallel to the ZZ' axis with uniform motion with velocity $\boldsymbol{v}_2$. The resultant trajectory is a helix.

Of special interest is the case when the axial force is given by the vector product $\boldsymbol{F} = \boldsymbol{A} \times \boldsymbol{v}$, where $\boldsymbol{A}$ is a constant vector and $\boldsymbol{v}$ is the velocity of the particle. This force occurs, for example when electrons, or any charged particle, move in a magnetic field (see Chapter 22). It is also the force that generates hurricanes and tornados because of the Coriolis acceleration (recall Section 5.6). From the properties of the vector product, we see that the force $\boldsymbol{F}$ is in the direction perpendicular to both $\boldsymbol{A}$ and $\boldsymbol{v}$ (Figure 8.5) and its magnitude is proportional to the velocity. Place the Z-axis in the direction of $\boldsymbol{A}$. Then the equation of motion is $\boldsymbol{F} = m\boldsymbol{a}$ which, when we use Equation 5.11, $\boldsymbol{a} = \boldsymbol{\omega} \times \boldsymbol{v}$, for the centripetal acceleration gives $m\boldsymbol{\omega} \times \boldsymbol{v} = \boldsymbol{A} \times \boldsymbol{v}$, which shows that the (constant) angular velocity of the particle is

$$\boldsymbol{\omega} = \frac{\boldsymbol{A}}{m} \tag{8.4}$$

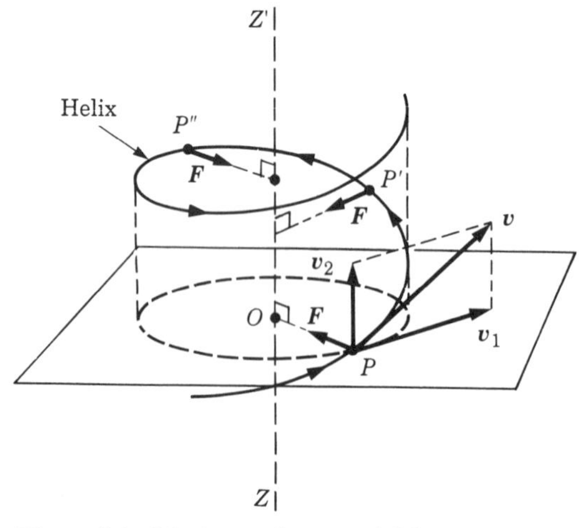

Figure 8.4 Motion under an axial force.

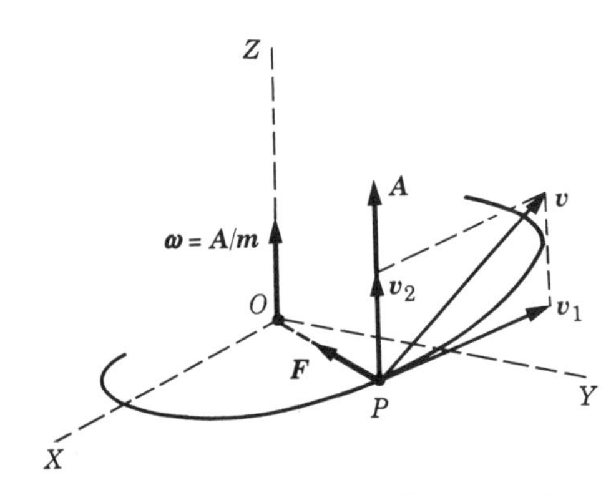

Figure 8.5 Motion under a force $\boldsymbol{F} = \boldsymbol{A} \times \boldsymbol{v}$.

a value independent of the velocity but inversely proportional to the mass of the particle. This result was to be expected, since the larger the mass the more difficult it is for the force $\boldsymbol{F} = \boldsymbol{A} \times \boldsymbol{v}$ to turn the direction of the velocity and maintain a circular path.

When the velocity is *not* perpendicular to $\boldsymbol{A}$, we may separate the velocity into a component $\boldsymbol{v}_2$ parallel to $\boldsymbol{A}$ and another $\boldsymbol{v}_1$ perpendicular to $\boldsymbol{A}$. The perpendicular velocity results in circular motion with an angular velocity given by Equation 8.4, while the parallel velocity $\boldsymbol{v}_1$ is not affected by the applied force since $\boldsymbol{A} \times \boldsymbol{v}_2 = 0$. Therefore the resultant motion is the helix.

8.3 Torque

When a force acts on a body, the body does not merely move in the direction of the force but it also usually turns about some point. Consider a force $\boldsymbol{F}$ acting on a particle A (Figure 8.6). Suppose that the effect of the force is to move the particle around O. Our daily experience suggests that the rotating effectiveness of F increases with the perpendicular distance (called **lever arm**) $b = OB$ from the center of rotation O to the **line of action** of the force. For example, when we open a door, we always push or pull as far as possible from the hinges and attempt to keep the direction of our push or pull perpendicular to the door. This common experience therefore suggests the convenience of defining a physical quantity τ that will be called **torque**, according to

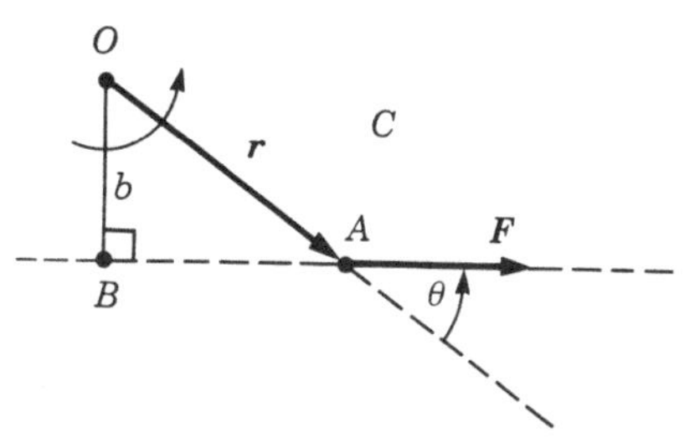

Figure 8.6 Torque of a force.

$$\tau = Fb \qquad \text{or} \qquad \text{torque} = \text{force} \times \text{lever arm}$$

Torque must be expressed as the product of a *unit of force* and a *unit of distance*. Thus in the SI system, torque is expressed in newton meters or N m.

Noting from the figure that $b = r \sin \theta$, we also may write

$$\tau = Fr \sin \theta \tag{8.5}$$

Comparing this equation with Equation A.21 for the vector product, we conclude that the torque may be considered as a vector quantity given by the vector product

$$\boldsymbol{\tau} = \boldsymbol{r} \times \boldsymbol{F} \tag{8.6}$$

where $\boldsymbol{r}$ is the position vector, relative to O, of the point A on which the force is acting.

According to the properties of the vector product, the torque is represented by a vector perpendicular to both $\boldsymbol{r}$ and $\boldsymbol{F}$. That is, the torque is perpendicular to the plane drawn through both $\boldsymbol{r}$ and $\boldsymbol{F}$, and directed according to the right-hand rule (Figure 8.7).

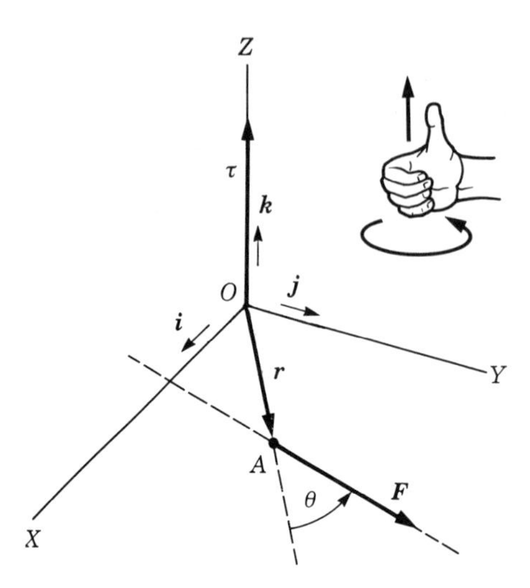

Figure 8.7 The torque is perpendicular to the plane determined by $\boldsymbol{r}$ and $\boldsymbol{F}$.

EXAMPLE 8.4

Expression of the torque in terms of the components of $\boldsymbol{r}$ and $\boldsymbol{F}$.

▷ Placing the X- and Y-axes in the plane determined by $\boldsymbol{r}$ and $\boldsymbol{F}$ (Figure 8.7), we have that

$$\boldsymbol{r} = \boldsymbol{i}x + \boldsymbol{j}y \qquad \text{and} \qquad \boldsymbol{F} = \boldsymbol{i}F_x + \boldsymbol{j}F_y$$

where (x, y) are the coordinates of A and (F_x, F_y) are the components of $\boldsymbol{F}$. Hence, $\boldsymbol{\tau} = (\boldsymbol{i}x + \boldsymbol{j}y) \times (\boldsymbol{i}F_x + \boldsymbol{j}F_y)$, and by application of Equation A.24 we obtain

$$\boldsymbol{\tau} = \boldsymbol{k}(xF_y - yF_x) \tag{8.7}$$

which is a vector parallel to the Z-axis, as illustrated in Figure 8.7. The magnitude of the torque is

$$\tau = xF_y - yF_x \tag{8.8}$$

The torque is positive or negative depending on the sense of rotation around the Z-axis.

8.4 Angular momentum

The **angular momentum** with respect to a point O (Figure 8.8) of a particle of mass m moving with velocity $\boldsymbol{v}$ (and therefore having momentum $\boldsymbol{p} = m\boldsymbol{v}$) is defined by the vector product

$$\boldsymbol{L} = \boldsymbol{r} \times \boldsymbol{p} \qquad \text{or} \qquad \boldsymbol{L} = m\boldsymbol{r} \times \boldsymbol{v} \tag{8.9}$$

where $\boldsymbol{r}$ is the position vector of the particle relative to O. The angular momentum is therefore a vector perpendicular to the plane determined by $\boldsymbol{r}$ and $\boldsymbol{v}$. In addition, its magnitude depends on the position of the point O and is given by

$$L = mrv \sin \theta$$

where θ is the angle between $\boldsymbol{r}$ and $\boldsymbol{v}$. The angular momentum of the particle in general changes in magnitude and direction while the particle moves. However, for a particle moving in a plane that contains the point O, the direction of the angular momentum remains the same (that is, perpendicular to the plane) since both $\boldsymbol{r}$ and $\boldsymbol{v}$ are in the plane. For the special case of circular motion (Figure 8.9), when the angular momentum is calculated relative to the center of the circle, vectors $\boldsymbol{r}$ and $\boldsymbol{v}$ are perpendicular ($\theta = 90°$) and we also have the relation $v = \omega r$, so that we can write

$$L = mrv = mr^2\omega \tag{8.10}$$

The direction of $\boldsymbol{L}$ is the same as that of $\boldsymbol{\omega}$, so that Equation 8.10 can be written in vector form as

$$\boldsymbol{L} = mr^2\boldsymbol{\omega} \tag{8.11}$$

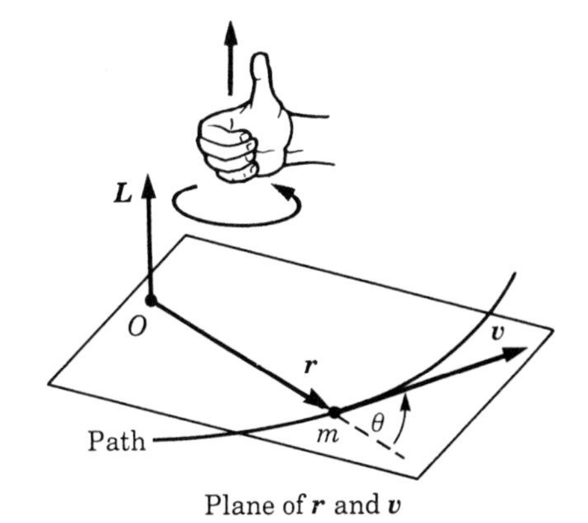

Figure 8.8 Angular momentum of a particle relative to O.

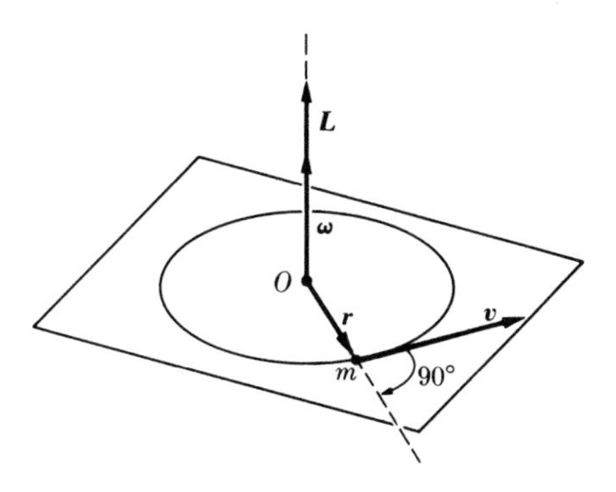

Figure 8.9 Vector relation between angular velocity and angular momentum in circular motion.

The orbital angular momentum and the torque of the forces acting on a particle, given by $\boldsymbol{\tau} = \boldsymbol{r} \times \boldsymbol{F}$ (both quantities evaluated with respect to the same point), have a very important relation. As shown below:

the time rate of change of the angular momentum of a particle is equal to the torque of the force applied to the particle.

That is,

$$\frac{\mathrm{d}\boldsymbol{L}}{\mathrm{d}t} = \boldsymbol{\tau} \tag{8.12}$$

We emphasize that this equation is correct only if $\boldsymbol{L}$ and $\boldsymbol{\tau}$ are measured relative to the same point.

Equation 8.12 bears a great resemblance to Equation 6.12, $d\boldsymbol{p}/dt = \boldsymbol{F}$, with the linear momentum $\boldsymbol{p}$ replaced by the angular momentum $\boldsymbol{L}$, and the force $\boldsymbol{F}$ replaced by the torque $\boldsymbol{\tau}$. This relation is fundamental to the discussion of rotational motion. As we shall see later on, this relation is also important when the angular momentum is the known physical quantity. This is particularly the case in scattering experiments, as well as in the analysis of the motion of electrons in atoms.

Derivation of the relation between torque and angular momentum

Taking the time derivative of Equation 8.9 gives

$$\frac{d\boldsymbol{L}}{dt} = \frac{d\boldsymbol{r}}{dt} \times \boldsymbol{p} + \boldsymbol{r} \times \frac{d\boldsymbol{p}}{dt} \tag{8.13}$$

But $d\boldsymbol{r}/dt = \boldsymbol{v}$ and $\boldsymbol{p} = m\boldsymbol{v}$ is always parallel to $\boldsymbol{v}$. Then

$$\frac{d\boldsymbol{r}}{dt} \times \boldsymbol{p} = \boldsymbol{v} \times \boldsymbol{p} = m\boldsymbol{v} \times \boldsymbol{v} = 0$$

On the other hand, $d\boldsymbol{p}/dt = \boldsymbol{F}$ according to Equation 6.12. Therefore, Equation 8.13, with $d\boldsymbol{p}/dt$ replaced by $\boldsymbol{F}$, becomes

$$\frac{d\boldsymbol{L}}{dt} = \boldsymbol{r} \times \boldsymbol{F} = \boldsymbol{\tau}$$

which is Equation 8.12. Note that while $d\boldsymbol{p}/dt = \boldsymbol{F}$ is a fundamental law of motion, $d\boldsymbol{L}/dt = \boldsymbol{\tau}$ is a consequence of the law of motion.

8.5 Central forces

If the torque on a particle is zero ($\boldsymbol{\tau} = \boldsymbol{r} \times \boldsymbol{F} = 0$), then according to Equation 8.12, we must have

$$\frac{d\boldsymbol{L}}{dt} = 0 \qquad \text{or} \qquad \boldsymbol{L} = \text{constant vector}$$

Thus the angular momentum of a particle relative to a point is constant in magnitude and direction if the torque relative to the same point is zero. This condition is fulfilled if $\boldsymbol{F} = 0$; that is, if the particle is free and thus moves with constant velocity. From Figure 8.10, we have

$$L = mvr \sin\theta = mvd$$

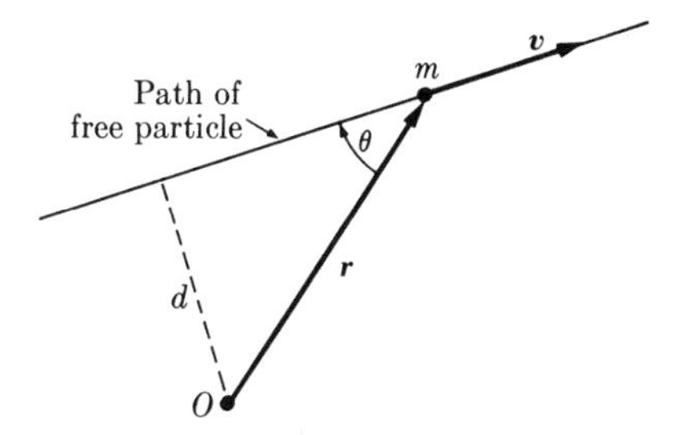

Figure 8.10 Angular momentum is constant for a free particle.

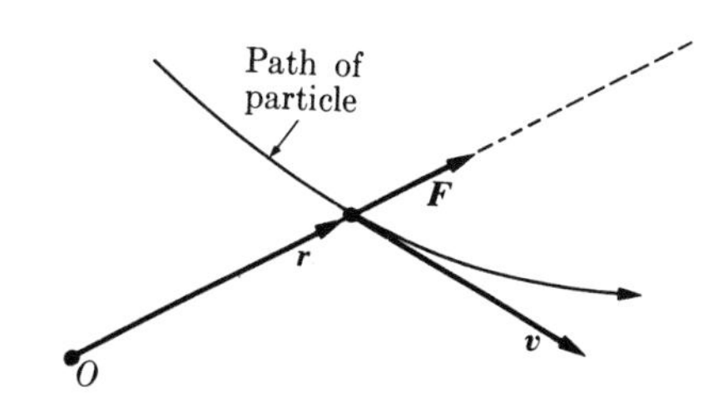

Figure 8.11 Angular momentum is constant for motion under a central force.

where $d = r \sin\theta$. The path of the free particle is in a straight line and the velocity does not change. Therefore, the quantity mvd remains constant because all factors involved are also constants.

The condition $\boldsymbol{\tau} = \boldsymbol{r} \times \boldsymbol{F} = 0$ is also fulfilled if $\boldsymbol{F}$ is parallel to $\boldsymbol{r}$; in other words, the direction of $\boldsymbol{F}$ passes through the point O. Then the torque due to that force, relative to O, is zero. A force whose direction always passes through a fixed point is called a **central force** (Figure 8.11). The fixed point is called the **center of force**. Therefore:

when a body moves under the action of a central force, the angular momentum relative to the center of force is a constant of motion, and conversely.

This result is very important because the forces appearing in many systems in nature are central. For example, the Earth moves around the Sun under the influence of a central force whose direction is always through the center of the Sun (Figure 8.12(a)). The Earth's angular momentum relative to the Sun is thus constant. The electron in a hydrogen atom essentially moves under the central force due to the electric interaction with the nucleus, which always points toward the nucleus (Figure 8.12(b)). Thus the angular momentum of the electron relative to the nucleus is constant. If a proton is thrown, with great velocity, against a nucleus, it is subject to a central repulsive force (Figure 8.12(c)). Therefore, while the proton is deviated (or **scattered**) from its original path, its angular momentum remains constant.

In atoms having many electrons, the resultant force on each electron is not rigorously central. In addition to the central interaction with the nucleus, there

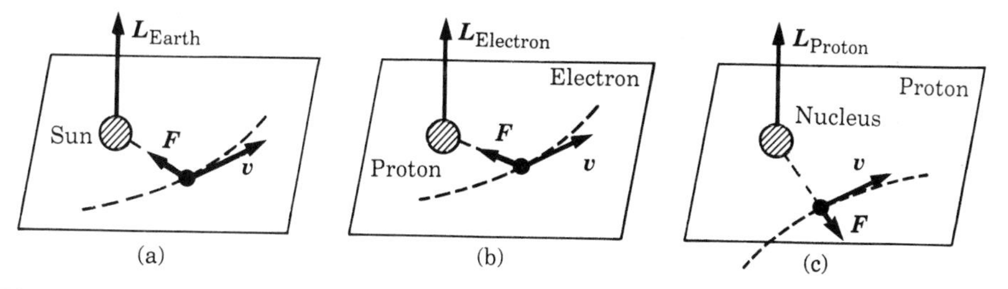

Figure 8.12 Angular momentum of (a) the Earth moving around the Sun, (b) an electron in a hydrogen atom, (c) a proton scattered by a nucleus.

are also the interactions with the other electrons. However, in many situations the average force on the electron can be considered as central to a satisfactory approximation. Therefore, we can consider that each electron moves with a constant angular momentum relative to the nucleus. Also, in nuclei we may assume, as a first approximation, that their components (protons and neutrons) move under average central forces and thus each nucleon has a constant angular momentum. However, in most nuclei this is not a very satisfactory approximation.

EXAMPLE 8.5

Angular momentum of the Earth around the Sun and of the Sun around the galaxy.

▷ To a good approximation, the Earth's orbit can be assumed circular. Then the relations shown in Figure 8.9 apply. The mass of the Earth is 5.98×10^{24} kg and its mean distance from the Sun is 1.49×10^{11} m. Also, from our definition of the second given in Section 2.4, we conclude that the period of revolution of the Earth around the Sun is 3.16×10^{7} s. Thus the average angular velocity of the Earth around the Sun is

$$\omega = \frac{2\pi}{P} = \frac{2\pi}{3.16 \times 10^{7}\ \mathrm{s}} = 1.98 \times 10^{-7}\ \mathrm{s}^{-1}$$

Therefore, from Equation 8.10, the angular momentum of the Earth relative to the Sun is

$$\begin{aligned} L &= mr^2\omega \\ &= (5.98 \times 10^{24}\ \mathrm{kg})(1.49 \times 10^{11}\ \mathrm{m})^2(1.98 \times 10^{-7}\ \mathrm{s}^{-1}) \\ &= 2.67 \times 10^{40}\ \mathrm{kg\ m^2\ s^{-1}} \end{aligned}$$

Similarly, the Sun moves around the center of the galaxy describing a path that is approximately a circle of radius 3×10^{20} m with an angular velocity of $10^{-15}\ \mathrm{s}^{-1}$. We leave the calculation of the angular momentum of the Sun about the center of our galaxy as an exercise for the student. It is of the order of $10^{55}\ \mathrm{kg\ m^2\ s^{-1}}$.

EXAMPLE 8.6

Angular momentum of an electron around the nucleus in a hydrogen atom.

▷ In the previous example we considered an extreme macroscopic situation, with planetary distances and masses. We now turn to the opposite extreme and consider an atomic situation. Consider the electron in a hydrogen atom: mass 9.11×10^{-31} kg, mean distance to the nucleus of 5.29×10^{-11} m and angular velocity of $4.13 \times 10^{16}\ \mathrm{s}^{-1}$ (recall Example 5.1). Thus, again assuming a circular orbit and using Equation 8.10, we find the angular momentum of the electron around the nucleus to be

$$\begin{aligned} L &= mr^2\omega \\ &= (9.11 \times 10^{-31}\ \mathrm{kg})(5.29 \times 10^{-11}\ \mathrm{m})^2(4.13 \times 10^{16}\ \mathrm{s}^{-1}) \\ &= 1.05 \times 10^{-34}\ \mathrm{kg\ m^2\ s^{-1}} \end{aligned}$$

This numerical value constitutes one of the most important constants in physics, and is designated by the symbol $\hbar$, read h-bar. The angular momentum of atomic particles is usually expressed in units of $\hbar$. The quantity $h = 2\pi\hbar$ is called **Planck's constant**.

Although there is a tremendous disparity between the values of the angular momentum of the Earth relative to the Sun and of the electron relative to a nucleus, in both cases the forces are central and the angular momentum is constant. However, in dealing with the electron, an atomic particle, a certain modification of our methods is required, using the technique called **quantum mechanics**, a subject we will take up in Chapter 36. Nevertheless, one result is that the angular momentum of electrons in atoms cannot have any arbitrary value; its value is restricted to those given by

$$L^2 = \hbar^2 l(l+1) \qquad \textbf{(8.14)}$$

where l is a positive integer (0, 1, 2, 3, ...) (see the discussion in Section 23.5). This property is called the **quantization** of angular momentum. In addition, the angular momentum of the electron can have only certain orientations in space. This restriction is called **space quantization**. As a consequence, the component of the angular momentum of an atomic electron in a given direction is limited to certain values.

EXAMPLE 8.7

Angular momentum for motion under an axial force.

▷ In Example 8.3 an axial force was defined as one whose direction always passes through a fixed axis, which we took (and will take again) as the Z-axis (Figure 8.13). The angular momentum of particle P is $\boldsymbol{L} = \boldsymbol{r} \times \boldsymbol{p}$ and the torque of the axial force is $\boldsymbol{\tau} = \boldsymbol{r} \times \boldsymbol{F}$, where both $\boldsymbol{L}$ and $\boldsymbol{\tau}$ are measured with respect to the origin O.

The torque $\boldsymbol{\tau}$ is perpendicular to the plane OPQ defined by $\boldsymbol{r}$ and $\boldsymbol{F}$. Therefore, it must be in the XY-plane, so that the Z-component of $\boldsymbol{\tau}$ is always zero. Thus, the Z-component of the equation $\mathrm{d}\boldsymbol{L}/\mathrm{d}t = \boldsymbol{\tau}$ becomes, in this case,

$$\frac{\mathrm{d}L_z}{\mathrm{d}t} = \tau_z = 0 \qquad \text{or} \qquad L_z = const. \qquad \textbf{(8.15)}$$

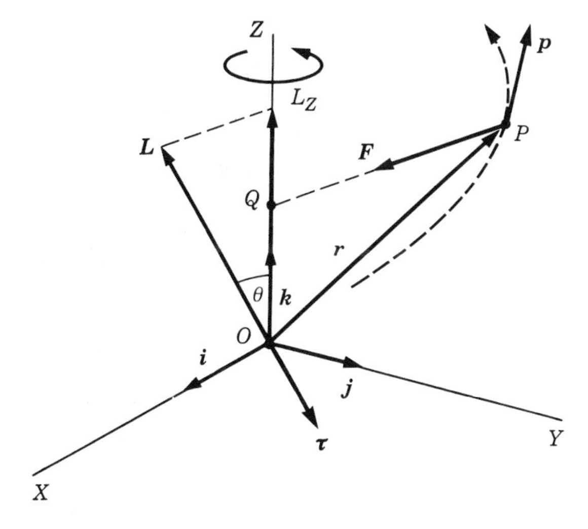

Figure 8.13 Angular momentum of a particle moving under an axial force.

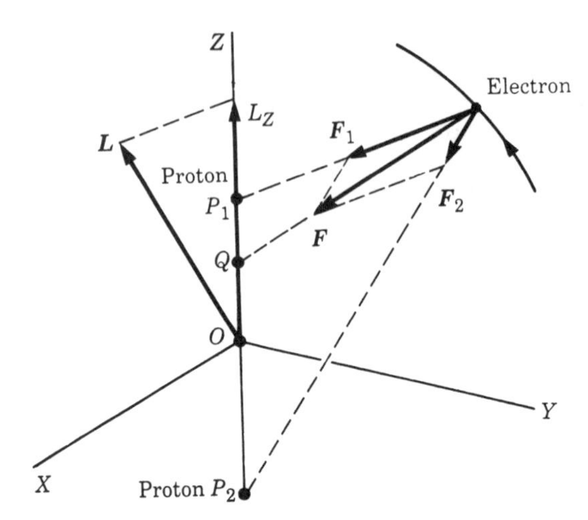

Figure 8.14 Angular momentum of the electron in the hydrogen molecule ion, H_2^+. The component L_Z remains constant.

Thus, when a particle moves under an axial force, the Z-component of the angular momentum is constant. However, the total angular momentum is not constant and precesses around the Z-axis as the particle moves, but its angle θ with the Z-axis varies in such a way that L_z remains constant.

This type of motion, for example, is that of the electron in the H_2^+ molecule ion, which is composed of two protons and only one electron (Figure 8.14). The resultant force on the electron always passes through the line P_1P_2 that joins the two protons.

EXAMPLE 8.8

Scattering of a particle by a central force.

▷ This problem is of special interest because of its application in atomic and nuclear physics. For example, consider a proton accelerated by a machine such as a cyclotron or a synchrotron. When it passes near a nucleus of a target material, it is scattered (or deflected) under the action of the force due to the electric repulsion of the nucleus. Similarly, comets coming from outside the planetary system and planetary space probes launched from the Earth are deflected, or scattered, by the gravitational attraction of the Sun (see Example 6.4) or by a planet, respectively. For this example, however, we consider only a *repulsive* central force.

Let O be the scatterer, or center of force, and A a particle thrown against O from a great distance with velocity $\boldsymbol{v}_0$ (Figure 8.15). The distance b, called the **impact parameter,** is the perpendicular distance from A to line OX drawn through O parallel to $\boldsymbol{v}_0$. Assuming that the force between A and O is repulsive and central (a similar logic can be used for the case where the force is attractive), the particle will follow the path AMB. The form of the path depends on how the force varies with the distance. When a particle is at A its angular momentum, L, relative to O is $L = mv_0b$. At any position such as M, its angular momentum still has the same value.

As the particle recedes from O the force decreases. At a point B, very far from O, the magnitude of the velocity is again $\boldsymbol{v}_0$ because, by symmetry, the velocity lost when the particle approaches O must be regained when it recedes from O. (The principle of conservation of energy, to be discussed in the next chapter, also supports this statement). However, the velocity at B is in a direction different from that of the initial velocity. The conservation of angular momentum under the action of a central force requires that the distance from B to OX' also be the impact parameter b.

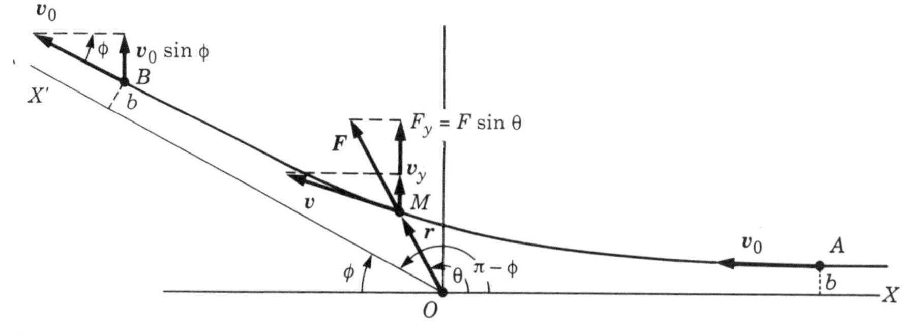

Figure 8.15 Scattering of a particle under a repulsive central force.

The angle ϕ by which the direction of the velocity has changed is called the **scattering angle**. Clearly, the stronger the repulsive force, the larger the scattering angle. On the other hand, the larger the angular momentum of the particle (large m, v_0 or b), the smaller the scattering angle because the force is less effective in changing the direction of the motion.

Since the scattering angle is determined by how the force depends on the distance, scattering experiments are very useful in determining the law of force in interactions between particles, or the size and structure of the scatterer. For example, by analyzing the scattering of α-particles obtained from a sample of radium, an experiment first performed by Hans Geiger (1882–1945) and Edward Marsden (1889–1970) in 1911 at the suggestion of Ernest Rutherford (1871–1937), it was possible to prove that atoms have an extremely small positively charged nucleus with a size about 10 000 times smaller than the atom itself. In Note 8.1 and Example 9.10 a quantitative analysis of scattering by an inverse square force is given.

Note 8.1 Scattering of a particle by a central repulsive inverse square force

The problem of scattering of a particle subject to a force inversely proportional to the square of the distance is of special interest because of its application to scattering by gravitational and electric forces.

Let O be the center of force (or scattering center) and A a particle thrown against O from a great distance with initial velocity v_0 (Figure 8.15). We set the X-axis so that it passes through O and is parallel to v_0, so that the distance b is the impact parameter. When the force is inversely proportional to the square of the distance, we have $F = k/r^2$ and it can be shown that the path is a hyperbola (see Note 11.1). The angular momentum relative to O of the particle at A is mv_0b. At any other position, such as M, its angular momentum, according to Equation 8.11, is $mr^2(\mathrm{d}\theta/\mathrm{d}t)$. The angular momentum must remain constant because the force is central. Therefore,

$$mr^2 \frac{\mathrm{d}\theta}{\mathrm{d}t} = mv_0 b \tag{8.16}$$

The equation of motion in the Y-direction is obtained by applying Newton's second law. That is,

$$m \frac{\mathrm{d}v_y}{\mathrm{d}t} = F_y = F \sin \theta = \frac{k}{r^2} \sin \theta$$

Eliminating r^2 by using Equation 8.16, we may write

$$\frac{dv_y}{dt} = \frac{k}{mv_0 b} \sin\theta \frac{d\theta}{dt}$$

To find the deflection of the particle, we must integrate this equation from one extreme of the path to the other. At A the value of v_y is zero (because the X-axis was chosen parallel to the initial motion) and also $\theta = 0$. At B we have $v_y = v_0 \sin\phi$ and $\theta = \pi - \phi$. Then

$$\int_0^{v_0 \sin\phi} dv_y = \frac{k}{mv_0 b} \int_0^{\pi - \phi} \sin\theta \, d\theta$$

Integrating, we get

$$v_0 \sin\phi = \frac{k}{mv_0 b}(1 + \cos\phi)$$

Remembering that $\cot \frac{1}{2}\phi = (1 + \cos\phi)/\sin\phi$, we finally get

$$\cot \tfrac{1}{2}\phi = \frac{mv_0^2 b}{k} \qquad \textbf{(8.17)}$$

This relation gives the scattering angle ϕ in terms of the impact parameter b and the force strength k and is valid only for an inverse square force.

QUESTIONS

8.1 A body moves along a curved path. Is the force tangent to the path, toward the concave side of the path, or toward the convex side of the path?

8.2 If the magnitude of the velocity remains constant in uniform circular motion, why is a centripetal force necessary?

8.3 What happens to the angular velocity and to the radius if the tangential force is zero and the centripetal force increases steadily?

8.4 Answer the previous question if the tangential force increases steadily while the centripetal force remains constant.

8.5 Why is the banking of curves desirable in highways and railroads?

8.6 What is the effect on the velocity of (a) a centripetal force and (b) a tangential force?

8.7 What happens to the magnitude of the velocity under an axial force $\boldsymbol{F} = \boldsymbol{A} \times \boldsymbol{v}$? What happens to the direction?

8.8 Verify that torque can also be defined as the product of the length of the position vector $\boldsymbol{r}$ and the component of the force $\boldsymbol{F}$ perpendicular to $\boldsymbol{r}$.

8.9 What is the relation between the direction of the angular momentum of a body and its velocity?

8.10 Is there a relation between the direction of the torque and that of (a) the force, (b) the angular momentum, (c) the rate of change of angular momentum?

8.11 Which equation is more fundamental: $\boldsymbol{F} = d\boldsymbol{p}/dt$ or $\boldsymbol{\tau} = d\boldsymbol{L}/dt$? Why?

8.12 What quantity is a constant of the motion when the force is central? What quantity remains constant if the force is axial?

8.13 What is meant by 'quantization' of the angular momentum of an electron in an atom?

8.14 Verify that Planck's constant $h = 2\pi\hbar$ is equal to 6.6256×10^{-34} kg m^2 s^{-1}.

8.15 How does the scattering angle vary as the impact parameter decreases? What is the scattering angle when the impact parameter is zero (i.e. a head-on collision)?

8.16 Since the angular momentum remains constant in a scattering experiment, how is the velocity of the particle at the distance of closest approach related to the initial velocity?

8.17 Define the concept of 'impact parameter'. Consider why such a term is useful.

8.18 Discuss the importance of scattering experiments for determining the law of force between particles.

8.19 How does the distance of closest approach vary as the initial velocity v_0 decreases? How does the distance of closest approach vary as the coupling constant k decreases? See Example 8.8 and Note 8.1.

PROBLEMS

8.1 A stone, whose mass is 0.4 kg, is tied to the end of a rope and the rope is held in one's hand. (a) If the stone is forced to move in a horizontal circle of 0.8 m radius at the rate of 80 rev/min, what is the magnitude of the force the rope exerts on the stone? (b) If the rope breaks when the tension is greater than 500 N, what is the largest possible angular velocity of the stone? (c) Can we ignore the weight of the stone?

8.2 A small block with a mass of 1 kg is tied to a rope 0.6 m long and is spinning at 60 rpm in a vertical circle. Calculate the tension in the rope when the block is (a) at the highest point of the circle, (b) at the lowest point, (c) when the rope is horizontal. (d) Calculate the linear velocity the block must have at the highest point in order for the tension in the rope to be zero.

8.3 The electron in a hydrogen atom revolves around the proton, following an almost circular path of radius 0.5×10^{-10} m with a velocity estimated to be about 2.2×10^6 m s^{-1}. Estimate the magnitude of the force between the electron and the proton.

8.4 A train travels around a banked curve at 63 km h^{-1}. The radius of the curve is 300 m. Calculate: (a) the degree of banking the curve must have in order that the wheels will experience no sidewise force, (b) the angle a chain hanging from the ceiling of one of the cars makes with the vertical.

8.5 A highway is 13.6 m wide. Calculate the difference in level between the external and internal edges of the road in order for a car to be able to travel at 60 m s^{-1} (without experiencing sidewise forces) around a curve whose radius is 600 m.

8.6 A highway curve whose radius is 300 m is *not* banked. Assume that the coefficient of friction between the tires and dry asphalt is 0.75, between the tires and wet asphalt is 0.50, and between the tires and ice is 0.25. Determine the maximum safe speed for traversing the curve on (a) dry days, (b) rainy days, (c) icy days.

8.7 A body D which has a mass of 12 kg (Figure 8.16) hangs from point E' and rests on a conical surface ABC. The body and cone are spinning about the axis EE' with an angular velocity of 10 rev/min. Calculate: (a) the linear velocity of the body, (b) the reaction of the surface on the body, (c) the tension in the string, and (d) the angular velocity necessary to reduce the reaction of the plane to zero.

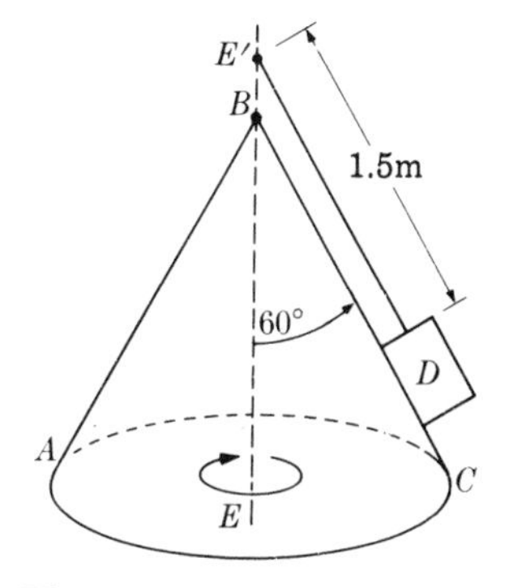

Figure 8.16

8.8 A small ball of mass m, initially at A, slides on the smooth circular surface ADB (Figure 8.17). When the ball is at the point C, show that the angular velocity and the force exerted by the surface are $\omega = [2g \sin(\alpha/r)]^{1/2}$ and $F = 3mg \sin \alpha$.

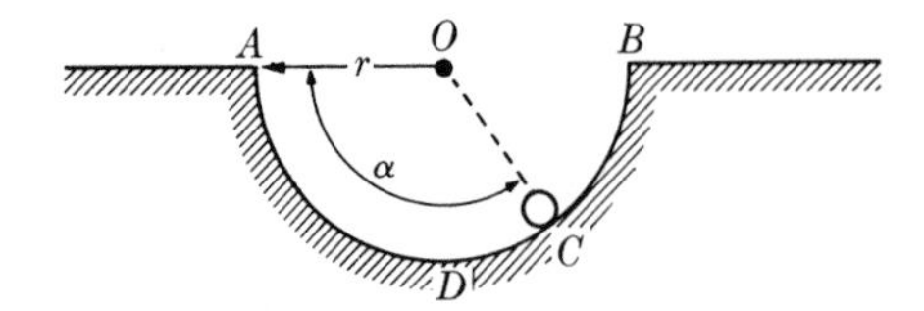

Figure 8.17

8.9 The position vector of a body of mass 6 kg is given as $\boldsymbol{r} = \boldsymbol{i}(3t^2 - 6t) + \boldsymbol{j}(-4t^3)$ m. Find: (a) the force acting on the particle, (b) the torque, with respect to the origin, acting on the particle, and (c) the momentum and angular momentum of the particle with respect to the origin.

8.10 At $t = 0$ s, a 3 kg mass is located at $\boldsymbol{r} = \boldsymbol{i}5$ (m) and has a velocity of $\boldsymbol{j}10$ (m s^{-1}). There are no forces acting on the mass. Determine the angular momentum of the mass with respect to the origin at (a) $t = 0$ s and (b) $t = 12$ s.

8.11 When the Earth is at *aphelion* (the position at which it is farthest away from the Sun) on June 21, its distance from the Sun is 1.52×10^{11} m and its orbital velocity is 2.93×10^4 m s^{-1}. (a) Find its orbital velocity at *perihelion* (the position at which it is closest to the Sun) about six months later, when its distance from the Sun is 1.47×10^{11} m. (b) Find the angular velocity of the Earth about the Sun for both cases. (Hint: both at aphelion and at perihelion the velocity is perpendicular to the radius vector.)

8.12 (a) Referring to Problem 8.1, if the stone is allowed to slide away while rotating by slowly letting the rope get longer, what happens to the angular velocity of the rope? (b) What happens if, instead, one pulls the rope in? (c) Is the angular momentum conserved in the process? (d) Obtain a relation between the rate of change of the angular velocity of the stone and the rate of change of the length of the rope.

8.13 A projectile (mass m) is fired from point O with an initial velocity of v_0 at an angle α above the horizontal. (a) Calculate the X- and Y-components of the angular momentum of the projectile as functions of time. (b) Determine the time rate of change of the angular momentum of the projectile. (c) Calculate the torque of the weight of the projectile and compare with the result of part (b). Compute both torque and angular momentum with respect to point O.

9 Work and energy

The connection between heat flow and work was conclusively demonstrated in 1843 by James Prescott Joule. He made the first measurement of the direct conversion of mechanical energy into heat. In honor of his research into heat and energy his name was given to the common unit of work, the **joule**.

Notes

9.1 Introduction

In this chapter we are going to consider some new concepts that do not correspond to direct experience but have become extremely important. They apply to a large number of phenomena and have been incorporated into our daily language. These are the concepts of **work**, **power** and **energy**. For example, we say we get tired when we work hard. We compensate a 'worker' according to the work done. We pay our monthly electricity bill based on the amount of energy we use. An automobile consumes gasoline because it needs energy to keep moving. And everyone needs a daily intake of energy, built into the food consumed, to stay alive.

The ideas of work and energy were developed in the form we use them today in physics mostly during the 19th century. This was accomplished more than a hundred years after Newton formulated his laws of motion. Among the people who contributed significantly to the development of these concepts we mention James P. Joule (1818–1889), Robert Mayer (1814–1878) and Lord Kelvin (William Thompson) (1824–1907). Today these basic notions have been extended to a broader range of phenomena, and we talk about electric energy, nuclear energy, radiation energy etc.

9.2 Work

It is a common experience that when pushing an object on a horizontal surface we get more tired the harder we push and the longer the distance through which we push. Similarly, to lift a body, the heavier the body and the higher we raise the body the more tired we get. Or, in the case of a diesel locomotive pulling a train, it uses more fuel the larger the force it applies to pull the train and the longer it moves.

All these examples, and many others, suggest that we introduce a concept that combines force and distance. Thus if we have a body (Figure 9.1) that moves a distance $s = AB$ under a constant force F, we define the **work** done by the expression

$$\text{Work} = \text{force} \times \text{distance}$$

$$W = Fs \tag{9.1}$$

That is, the work done by a constant force is given by the product of force and displacement. If the force acts at an angle with the direction of the displacement (Figure 9.2), then the work done is calculated by using the component of the force, F_s, along the direction of the displacement; that is

$$W = F_s s$$

Thus:

work is equal to the product of the displacement and the component of the force along the displacement.

Since $F_s = F \cos \theta$,

$$W = Fs \cos \theta \tag{9.2}$$

Using the concept of the scalar product, Equation A.16, and introducing the displacement vector $\boldsymbol{s} = \boldsymbol{AB}$, we may write work as the scalar product

$$W = \boldsymbol{F} \cdot \boldsymbol{s} \tag{9.3}$$

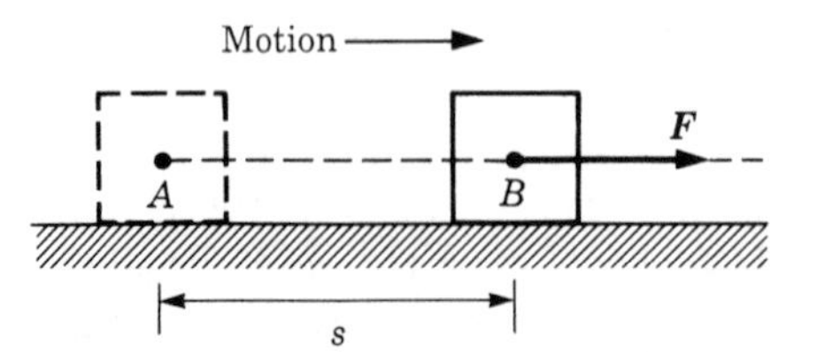

Figure 9.1 Work of a constant force parallel to the displacement.

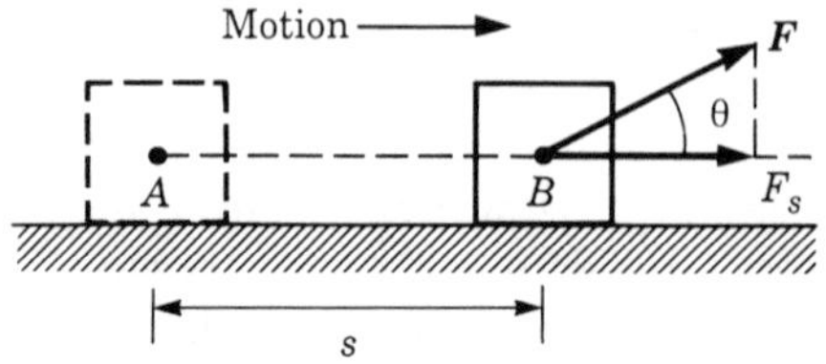

Figure 9.2 Work by a constant force at an angle with the displacement.

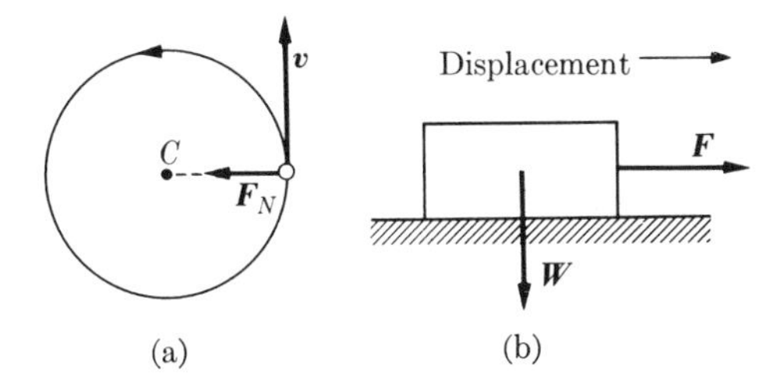

Figure 9.3 Forces that do no work: (a) $\boldsymbol{F}_N$, (b) $\boldsymbol{W}$.

Note that if a force is perpendicular to the displacement, so that $\theta = 90°$, the work done by the force is zero. For example, this is the case of the centripetal force $\boldsymbol{F}_N$ in circular motion (Figure 9.3(a)) or the weight $\boldsymbol{W}$ of a body when the body moves on a horizontal surface (Figure 9.3(b)). Also, if θ is larger than 90°, then $\cos\theta$ is negative and the work is negative. This is, for example, the case for frictional and viscous forces, which always act in the direction opposite to the direction of motion. In the case of a falling body, the weight does positive work, while if the body moves upward, the work of the weight is negative.

Equation 9.3 for work is valid when the force is constant and the body moves in a straight line. Let us now consider a more general case. Suppose a particle A is moving along a curve C under the action of a variable force $\boldsymbol{F}$ (Figure 9.4). In a very short time $\mathrm{d}t$ the particle moves from A to B. The displacement, which also is very small, may then be written as

$$\boldsymbol{AB} = \mathrm{d}\boldsymbol{r}$$

Using relation 9.3, the small amount of work done by the force $\boldsymbol{F}$ during that displacement is the scalar product

$$\mathrm{d}W = \boldsymbol{F}\cdot\mathrm{d}\boldsymbol{r} \tag{9.4}$$

Designating the magnitude of $\mathrm{d}\boldsymbol{r}$ (that is, the distance moved along the curve) by $\mathrm{d}s$, we may also write Equation 9.4 in the form

$$\mathrm{d}W = F\,\mathrm{d}s\cos\theta \tag{9.5}$$

where θ is the angle between the direction of the force $\boldsymbol{F}$ and the displacement $\mathrm{d}\boldsymbol{r}$, which is tangent to the path. Since $F\cos\theta$ is the component F_T of the force along the tangent to the path, we have that

$$\mathrm{d}W = F_T\,\mathrm{d}s \tag{9.6}$$

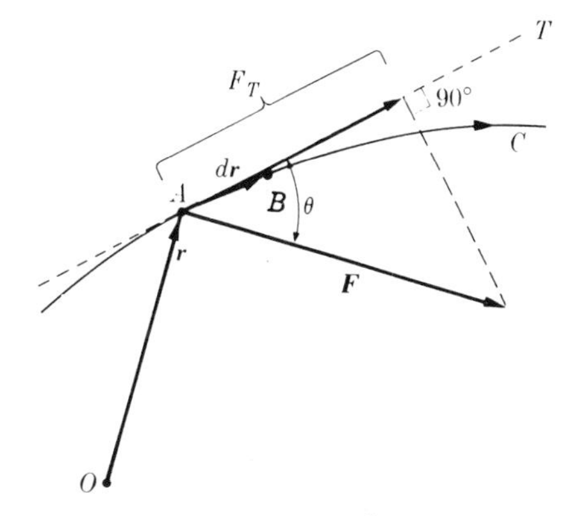

Figure 9.4 Work is equal to displacement multiplied by the component of the force along the displacement.

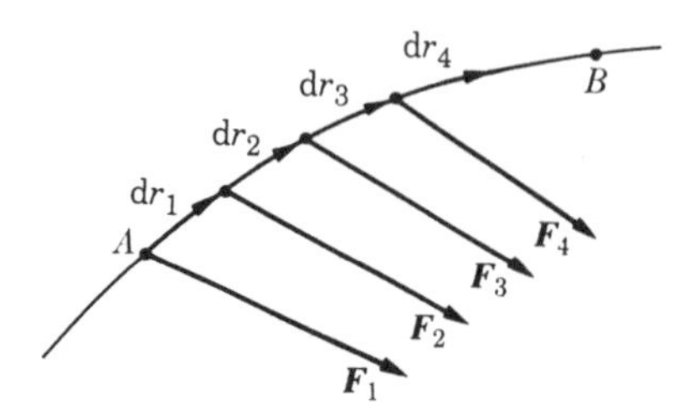

Figure 9.5 The total work is the sum of many infinitesimal works.

Thus, in curvilinear motion the work is also equal to the displacement times the component of the force along the tangent to the path, except that the direction of the tangent varies along the path.

Equation 9.4 gives the work for an infinitesimal displacement. The total work done on the particle when moving from A to B (Figure 9.5) is the sum of all the work done during successive small displacements along the path. That is,

$$W = \boldsymbol{F}_1 \cdot \mathrm{d}\boldsymbol{r}_1 + \boldsymbol{F}_2 \cdot \mathrm{d}\boldsymbol{r}_2 + \boldsymbol{F}_3 \cdot \mathrm{d}\boldsymbol{r}_3 + \cdots = \sum \boldsymbol{F}_i \cdot \mathrm{d}\boldsymbol{r}_i$$

When the displacements $\mathrm{d}\boldsymbol{r}_i$ are very small, the summation may be replaced by an integral. Therefore,

$$W = \int_A^B \boldsymbol{F} \cdot \mathrm{d}\boldsymbol{r} = \int_A^B F_T \, \mathrm{d}s \tag{9.7}$$

In order to perform the integral that appears in Equation 9.7, we must know $\boldsymbol{F}$ as a function of $\boldsymbol{r}$. Also, in general, we must know the equation of the path along which the particle moves. Alternatively, we must know $\boldsymbol{F}$ and $\boldsymbol{r}$ as functions of time or some other variable. The integral that appears in Equation 9.7 is called a **line integral** because it is calculated along a certain path (see Appendix A.10).

EXAMPLE 9.1

Work as the area under a curve.

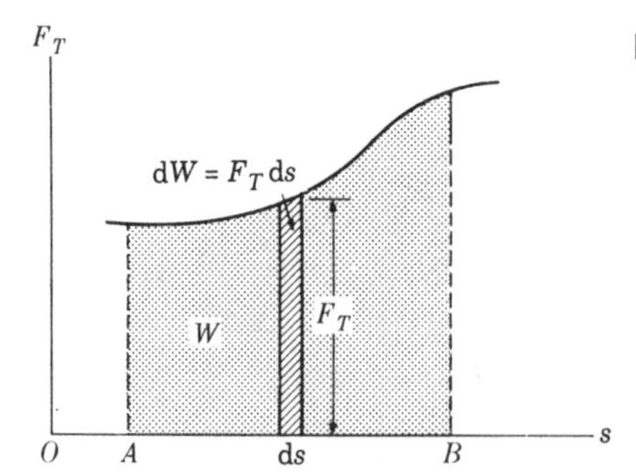

Figure 9.6 The total work done in going from A to B is equal to the area under the curve.

▷ As a body moves, the tangential component of the force F_T acting on the body in general varies. In Figure 9.6 we have plotted F_T as a function of the distance s measured along the path. The work $\mathrm{d}W = F_T \, \mathrm{d}s$ done during a small displacement $\mathrm{d}s$ corresponds to the area of the narrow rectangle. And so we can find the total work done on the particle in Figure 9.5 to move it from A to B by first dividing the whole shaded area in Figure 9.6 into narrow rectangles and then adding their areas. That is, *the total work done is given by the shaded area* in Figure 9.6. This result is important, from the practical point of view, for calculating the work done by various machines.

9.3 Power

In many practical applications it is important to know the rate at which work is done. The **power** is defined as the work done per unit time. Thus, if dW is the work done in the small time interval dt, the power is

$$P = \frac{dW}{dt} \tag{9.8}$$

Using Equation 9.4, $dW = \boldsymbol{F} \cdot d\boldsymbol{r}$, and recalling that $\boldsymbol{v} = d\boldsymbol{r}/dt$, we may also write

$$P = \boldsymbol{F} \cdot \frac{d\boldsymbol{r}}{dt} = \boldsymbol{F} \cdot \boldsymbol{v} \tag{9.9}$$

Therefore, power can also be defined as the scalar product of force and velocity. Equation 9.8 gives the **instantaneous power**. The **average power** during a time interval $\Delta t = t - t_0$ is obtained by dividing the total work W, as given in Equation 9.7, by the time interval Δt, yielding

$$P_{ave} = \frac{W}{\Delta t} \tag{9.10}$$

From the engineering point of view, the concept of power is very important. When an engineer designs a machine, it is generally the *rate* at which the machine can do work that matters, rather than the total amount of work the machine can do, although this is also important.

9.4 Units of work and power

From Equations 9.1 or 9.4 we see that work must be expressed as the product of a unit of force and a unit of distance. In the SI system, work is expressed in **newton meters**, a unit called a **joule**, abbreviated J. Thus one joule is the work done by a force of one newton when it moves a particle one meter in the same direction as the force. Recall that $N = kg\,m\,s^{-2}$. Therefore, we have that

$$J = N\,m = kg\,m^2\,s^{-2}$$

The name joule was chosen to honor James P. Joule, famous for his research on the concepts of heat and energy.

According to definition 9.8, power must be expressed as the ratio between a unit of work and a unit of time. In the SI system, power is expressed in joules per second, a unit called a **watt**, abbreviated W. One watt is the power of a machine that does work at the rate of one joule every second. Recalling that $J = kg\,m^2\,s^{-2}$, we have that

$$W = J\,s^{-1} = kg\,m^2\,s^{-3}$$

The name watt was chosen to honor James Watt (1736–1819), who improved the steam engine with his inventions. Three multiples of the watt in general use are the **kilowatt** (kW), the **megawatt** (MW) and the **gigawatt** (GW) defined by

$$1\ \text{kW} = 10^3\ \text{W}, \quad 1\ \text{MW} = 10^6\ \text{W}, \quad 1\ \text{GW} = 10^9\ \text{W}$$

Engineers use still another unit of power, called **horsepower**, which is equal to 745.7 W.

Another unit used to express work is the **kilowatt hour** (kWh). The kilowatt hour is equal to the work done during one hour by an engine having a power of one kilowatt. Thus

$$1\ \text{kilowatt hour} = (10^3\ \text{W})(3.6 \times 10^3\ \text{s}) = 3.6 \times 10^6\ \text{J}$$

We normally pay our electricity bills according to the number of kWh consumed.

EXAMPLE 9.2

An automobile, with a mass of 1200 kg, moves up a long hill inclined 5° with a constant velocity of 36 km h^{-1}. Calculate: (a) the work the engine does in five minutes and (b) the power developed by it. Neglect all frictional effects.

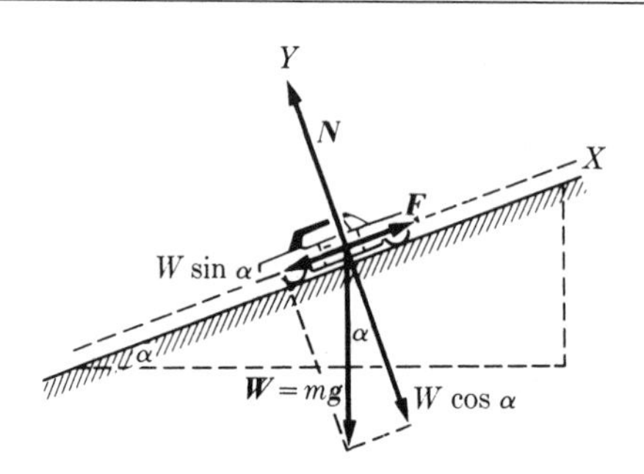

Figure 9.7

▷ The motion of the automobile along the hill is the result of the force F exerted by the engine and the force $W \sin \alpha$, which is the component of the weight of the automobile along the hill (Figure 9.7). Thus we write, using $W = mg$,

$$F - mg \sin \alpha = ma$$

Since the motion is uniform, $a = 0$, and

$$F = mg \sin \alpha = (1.2 \times 10^3\ \text{kg})(9.8\ \text{m s}^{-2}) \sin 5^\circ = 1.023 \times 10^3\ \text{N}.$$

The velocity of the automobile is $v = 36\ \text{km h}^{-1} = 36(10^3\ \text{m})/(3.6 \times 10^3\ \text{s}) = 10\ \text{m s}^{-1}$. In five minutes (or 300 s) it moves the distance $s = v\Delta t = (10\ \text{m s}^{-1})(300\ \text{s}) = 3 \times 10^3\ \text{m}$.

(a) Therefore, using Equation 9.8, the work done by the engine is

$$W = Fs = (1.023 \times 10^3\ \text{N})(3 \times 10^3\ \text{m}) = 3.069 \times 10^6\ \text{J}$$

(b) The average power can be computed in two different ways. First we may say that

$$P = \frac{W}{\Delta t} = \frac{3.069 \times 10^6\ \text{J}}{3 \times 10^2\ \text{s}} = 1.023 \times 10^4\ \text{W} = 10.23\ \text{kW}$$

Alternatively, using Equation 9.9, we may say that

$$P = Fv = (1.023 \times 10^3 \text{ N})(10 \text{ m s}^{-1}) = 1.023 \times 10^4 \text{ W}$$

EXAMPLE 9.3

Work required to expand a spring a distance x without acceleration.

▷ It is easy to verify experimentally that when a spring is extended a small distance x (Figure 9.8), the spring exerts a force proportional and opposite to the displacement, that is, $-kx$. If the spring is extended without acceleration it is necessary to apply an equal and opposite force, $F = +kx$. The variation of the applied force with the displacement is shown in Figure 9.9.

To find the work done by the applied force to extend the spring a distance x without acceleration, we must use Equation 9.7, which now becomes

$$W = \int_0^x F \, dx = \int_0^x kx \, dx = k \int_0^x x \, dx = \tfrac{1}{2}kx^2 \qquad \mathbf{(9.11)}$$

This result can also be obtained by computing the area of the shaded triangle in Figure 9.9; that is

$$W = \tfrac{1}{2}(\text{base} \times \text{height}) = \tfrac{1}{2}(x)(kx) = \tfrac{1}{2}kx^2$$

When a body is hung from a spring, exerting the force of its weight (that is, $F = mg$), the spring expands the distance x, so that $mg = kx$. Hence if we know k and measure x, we can obtain the weight of the body. Normally, k is obtained by hanging a body of known weight from the spring. This is the principle of the **spring balance** (or **dynamometer**), commonly used for measuring forces. We mentioned this device earlier in Section 6.2 as a means for comparing masses. For example, suppose that the spring extends 1.5×10^{-2} m when $m = 4$ kg. Then

$$k = \frac{mg}{x} = \frac{39.2 \text{ N}}{1.5 \times 10^{-2} \text{ m}} = 2.61 \times 10^3 \text{ N m}^{-1}$$

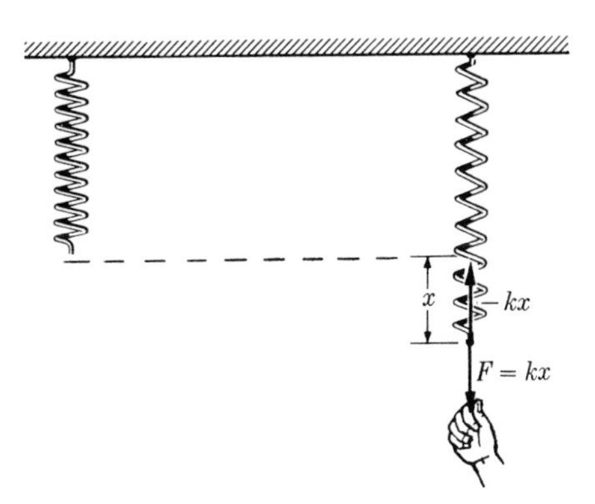

Figure 9.8 Work done in stretching a spring.

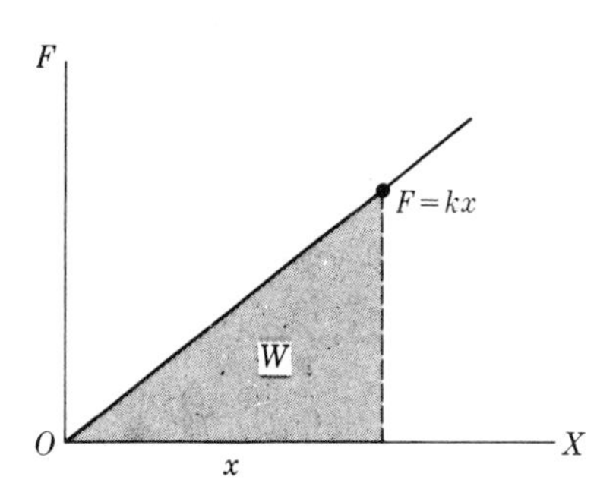

Figure 9.9 Work done by a force proportional to the displacement.

If we introduce this numerical value for k in Equation 9.11, we find that the work required to extend the spring by a distance $x = 0.02$ m is

$$W = \tfrac{1}{2}(2.61 \times 10^3\ \text{N m}^{-1})(0.02\ \text{m})^2 = 0.522\ \text{J}$$

9.5 Kinetic energy

Consider a body moving along a curve C under the action of a force $\boldsymbol{F}$ (Figure 9.10). The tangential force at P is

$$F_T = m(\mathrm{d}v/\mathrm{d}t)$$

(recall Equation 6.15). Therefore, the work done by F_T in a displacement $\mathrm{d}s$ is

$$\mathrm{d}W = F_T\ \mathrm{d}s = m\frac{\mathrm{d}v}{\mathrm{d}t}\ \mathrm{d}s = m\frac{\mathrm{d}s}{\mathrm{d}t}\ \mathrm{d}v = mv\ \mathrm{d}v \tag{9.12}$$

since $v = \mathrm{d}s/\mathrm{d}t$, according to Equation 4.7. The total work done in moving the body from A to B is then

$$W = \int_A^B F_T\ \mathrm{d}s = \int_A^B mv\ \mathrm{d}v = \tfrac{1}{2}mv_B^2 - \tfrac{1}{2}mv_A^2 \tag{9.13}$$

where v_B is the particle's velocity at B and v_A is the particle's velocity at A. The result 9.13 indicates that the work W done on a particle is equal to the difference of the quantity $\frac{1}{2}mv^2$ evaluated at the end and at the beginning of the path. The quantity $\frac{1}{2}mv^2$ is called the **kinetic energy** of the particle and is designated by E_k. Therefore

$$E_k = \tfrac{1}{2}mv^2 \qquad \text{or} \qquad E_k = \frac{p^2}{2m} \tag{9.14}$$

since $p = mv$. Equation 9.13 can then be expressed as

$$W = E_{k,B} - E_{k,A} \qquad \text{or} \qquad W = \Delta E_k \tag{9.15}$$

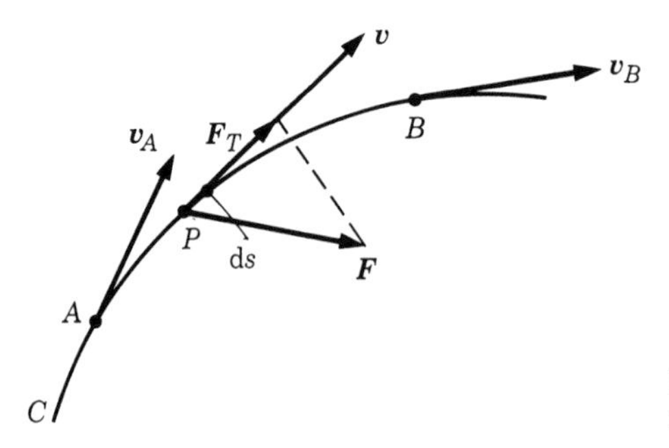

Figure 9.10 Work in curvilinear motion.

Equation 9.15 can be stated as:

> *the work done by the force acting on a particle is equal to the change in its kinetic energy.*

This is a result of general validity whatever the motion of the particles. However, when this relation is applied to a body composed of many particles, it might need some changes, as will be discussed in Chapter 14.

The relation $W = \Delta E_k$ does not involve any new law of nature, but is simply a consequence of the definition of work and Newton's law of motion $\boldsymbol{F} = m(\mathrm{d}\boldsymbol{v}/\mathrm{d}t)$ applied to a particle of mass m. However, it is of great practical application. For example, if we know the initial velocity, and therefore the initial kinetic energy, of a body, we can easily calculate the final velocity in terms of the work done on the body.

The kinetic energy of a particle is a quantity that can be calculated at any position during the motion if the velocity is known. However, the work done is a quantity associated with a force and a displacement and, in general, depends on the path followed. Thus one interesting aspect of Equation 9.15 is that it relates a particle characteristic (kinetic energy) with a quantity that depends on the path taken (work done).

Also, if the resultant force on a particle is zero, its velocity is constant and thus its kinetic energy remains constant. The total work done is also zero even though each of the forces applied to the body does some work. For example, if an automobile moves with constant velocity, its kinetic energy does not change and the net work done on the automobile is zero. The motor exerts a forward force and does positive work. But the force of the motor is opposed by frictional forces and air resistance, both of which are opposite to the motion and do negative work. Thus the net work is zero. However, the driver must pay for the gasoline required to do the positive part of the work done. Unfortunately, the driver does not receive any bonus for the negative work done by the frictional forces. For that reason, to reduce the cost of operating an automobile, one must try to minimize the frictional and viscous forces.

The term 'energy' is derived from the Greek by combining the word '*ergon*', which means 'work' and the prefix '*en*', which means 'in' or 'content'.

EXAMPLE 9.4

Kinetic energy of a particle attached to a spring. The spring of Example 9.3 has a mass m attached to its end, as shown in Figure 9.11. The mass is moved to the right a distance a, and then released. Calculate its kinetic energy when it is at distance x from the equilibrium position.

▷ According to our explanation in Example 9.3, the spring will exert a force $F = -kx$ on the mass m when it is at a distance x from the unextended position as long as the displacement x is not too large. (The minus sign indicates that the force produced by the spring is to the left when the body is displaced to the right and vice versa.) The equilibrium position is where $F = 0$ and therefore corresponds to the position $x = 0$.

When the mass is about to be released (Figure 9.11(b)), $x = a$, and $F = -ka$, and the initial velocity is zero ($v_0 = 0$), resulting in an initial kinetic energy of zero. Let us call

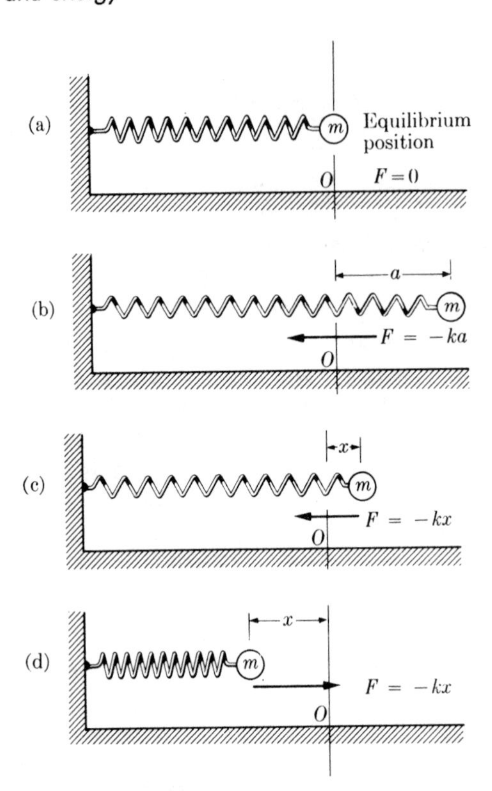

Figure 9.11 Motion of a body attached to a spring.

v the velocity at the intermediate position x. Then, using Equation 9.13, we have that $W = \frac{1}{2}mv^2$. But

$$W = \int_a^x F\,\mathrm{d}x = \int_a^x (-kx)\,\mathrm{d}x = \tfrac{1}{2}k(a^2 - x^2)$$

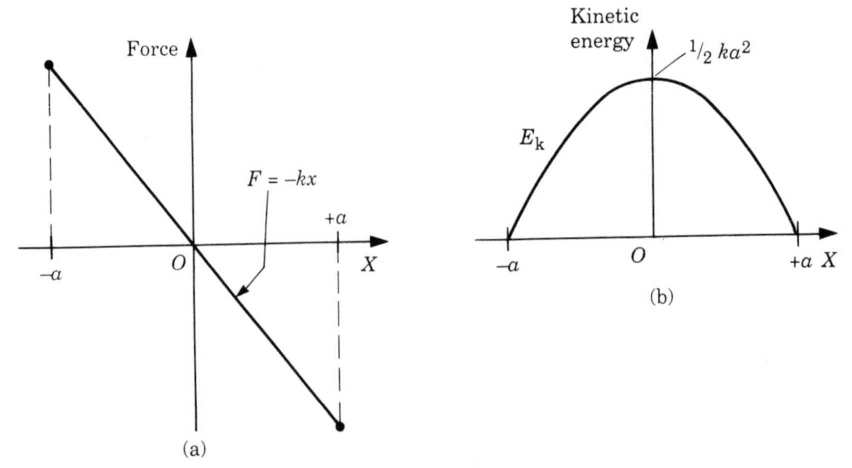

Figure 9.12 Force and kinetic energy when the force is proportional to the displacement.

Therefore

$$E_k = \tfrac{1}{2}mv^2 = \tfrac{1}{2}k(a^2 - x^2) \qquad \textbf{(9.16)}$$

This expression gives the kinetic energy of the particle as a function of position (Figure 9.12(b)) when $F = -kx$ (Figure 9.12(a)). Note that the kinetic energy is symmetric with respect to $x = 0$ and has a zero value for $x = \pm a$.

Since the kinetic energy cannot be negative (look what would happen to the velocity!), the motion of the particle is confined between $x = +a$ and $x = -a$. From Equation 9.16 we see that the velocity of the particle is

$$v = \pm[(k/m)(a^2 - x^2)]^{1/2}$$

which indicates that the velocity has the same magnitude for the same positive or negative values of x; i.e. the motion is symmetrical with respect to O. Also the double sign indicates that the velocity at x is the same whether the particle moves to the right or to the left. Finally, the radical $(a^2 - x^2)^{1/2}$ confirms the fact that the motion of the particle is confined to the region between $x = +a$ and $x = -a$.

9.6 Units of energy

In view of Equation 9.15, we can see that kinetic energy is measured in the same units as work; i.e. in joules in the SI system. This is easily verified by noting, from Equation 9.14, that E_k in the SI system must be expressed in $\text{kg m}^2\,\text{s}^{-2}$, which is the dimensional expression for joules in terms of the fundamental units.

Another unit of energy widely used by physicists to describe chemical and nuclear processes is the **electronvolt**, abbreviated eV, whose precise definition will be given in Chapter 21. Its equivalence is

$$1\ \text{eV} = 1.60218 \times 10^{-19}\ \text{J}$$

A useful multiple of the electron volt is the megaelectronvolt, MeV:

$$1\ \text{MeV} = 10^6\ \text{eV} = 1.60218 \times 10^{-13}\ \text{J}$$

EXAMPLE 9.5

An electron accelerated in a TV tube reaches the screen with a kinetic energy of 10 000 eV. Find the velocity of the electron.

▷ We must express the kinetic energy of the electron in J. Thus

$$E_k = 10\,000\ \text{eV} = (1.0 \times 10^4)(1.602 \times 10^{-19})\ \text{J} = 1.602 \times 10^{-15}\ \text{J}$$

From $E_k = \tfrac{1}{2}mv^2$, we get $v = (2E_k/m)^{1/2}$. Inserting the above value of E_k and the mass of the electron, 9.109×10^{-31} kg, we obtain

$$v = 5.931 \times 10^7\ \text{m s}^{-1}$$

or about $14 \times 10^{10}\ \text{km h}^{-1}$!

9.7 Work of a constant force

Consider a particle m moving under the action of a force $\boldsymbol{F}$ that is constant in magnitude and direction (Figure 9.13). There may be other forces acting on the particle that may or may not be constant, but we do not wish to be concerned with them now. The work of $\boldsymbol{F}$, when the particle moves from A to B is

$$W = \int_A^B \boldsymbol{F}\cdot \mathrm{d}\boldsymbol{r} = \boldsymbol{F}\cdot \int_A^B \mathrm{d}\boldsymbol{r} = \boldsymbol{F}\cdot(\boldsymbol{r}_B - \boldsymbol{r}_A) \tag{9.17}$$

One important conclusion of Equation 9.17 is that the work in this case is independent of the path joining points A and B. The work only depends on the resultant displacement $\boldsymbol{r}_B - \boldsymbol{r}_A = \boldsymbol{AB}$ and the component of the force in the direction of this displacement. For example, if instead of moving along path (1), the particle moves along path (2), which also joins A and B, the work of $\boldsymbol{F}$ will be the same because the displacement $\boldsymbol{r}_B - \boldsymbol{r}_A = \boldsymbol{AB}$ is the same. Note that Equation 9.17 can also be written in the form

$$W = \underbrace{\boldsymbol{F}\cdot\boldsymbol{r}_B}_{\text{final}} - \underbrace{\boldsymbol{F}\cdot\boldsymbol{r}_A}_{\text{initial}} \tag{9.18}$$

and therefore W is equal to the difference between $\boldsymbol{F}\cdot\boldsymbol{r}$ evaluated at the end of the path and the value at the beginning of the path.

An important application of Equation 9.17 is the work done by the force of gravity (Figure 9.14). Choosing the positive Y-axis vertically up, the force is $\boldsymbol{F} = m\boldsymbol{g} = -\boldsymbol{j}mg$. Also $\boldsymbol{r} = \boldsymbol{i}x + \boldsymbol{j}y$. Therefore, $\boldsymbol{F}\cdot\boldsymbol{r} = -mgy$. Substituting this value in Equation 9.18 we have $W = -mgy_B - (-mgy_A)$ or

$$W = \underbrace{mgy_A}_{\text{initial}} - \underbrace{mgy_B}_{\text{final}} \tag{9.19}$$

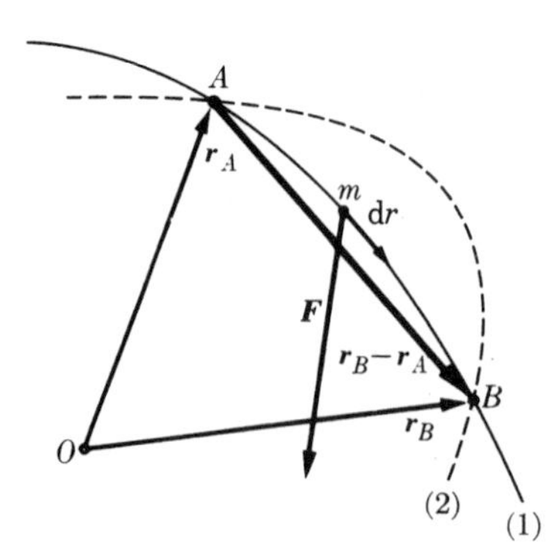

Figure 9.13 The work done by a constant force is independent of the path.

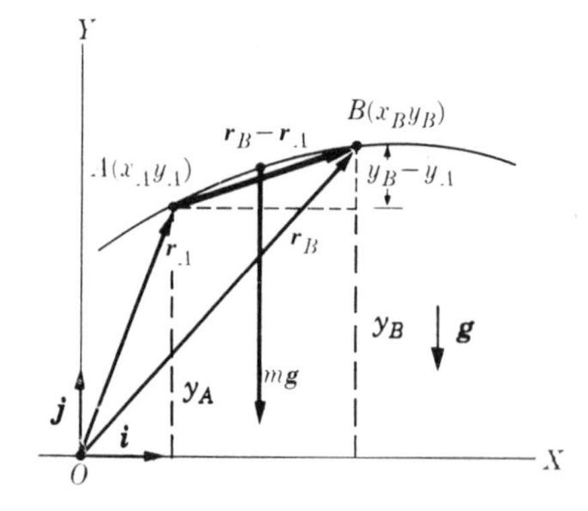

Figure 9.14 The work done by gravity depends on the height through which the body moves.

Again in Equation 9.19 there is no reference to the path, and the work depends only on the difference of the quantity mgy evaluated at the initial and final points.

EXAMPLE 9.6

Change in kinetic energy of a body moving under the action of gravity (free fall).

▷ Consider a freely falling body that has a velocity v_A at height y_A and a velocity v_B when at height y_B. Combining Equations 9.13 and 9.19 we find the change in kinetic energy to be

$$\tfrac{1}{2}mv_B^2 - \tfrac{1}{2}mv_A^2 = mgy_A - mgy_B$$

This relation holds whether the particle moves vertically or along any curved path joining points A and B. Note that the above equation can be written in the form $v_B^2 = v_A^2 - 2g(y_B - y_A)$, which is identical to Equation 3.12. Alternatively, we may write

$$v_B^2 + 2gy_B = v_A^2 + 2gy_A$$

which suggests that the quantity $v^2 + 2gy$ remains constant during the motion. In the next section we will see the implications of this result.

9.8 Potential energy

The situation illustrated in Section 9.7 is just one example of a large and important class of forces, called **conservative forces**. A force is conservative if its dependence on the position vector $\boldsymbol{r}$ of the particle is such that the work W done by the force can be expressed as the difference between a quantity $E_p(\boldsymbol{r})$ evaluated at the *initial* and at the *final* points (Figure 9.15), regardless of the path followed by the particle. The quantity $E_p(\boldsymbol{r})$ is called the **potential energy** of the particle associated with the applied conservative force, and is a function only of the position of the particle. Then, if $\boldsymbol{F}$ is a conservative force,

$$W = \int_A^B \boldsymbol{F} \cdot \mathrm{d}\boldsymbol{r} = \underbrace{E_{p(A)}}_{\text{initial}} - \underbrace{E_{p(B)}}_{\text{final}} \tag{9.20}$$

Note that we write $E_{p(A)} - E_{p(B)}$ and not $E_{p(B)} - E_{p(A)}$; that is, the work done is equal to E_p at the initial point minus E_p at the end-point. Once we know the expression for the potential energy, the work of a conservative force can be calculated without any reference to the path followed. That is,

$$W = E_{p(\text{initial})} - E_{p(\text{final})}$$

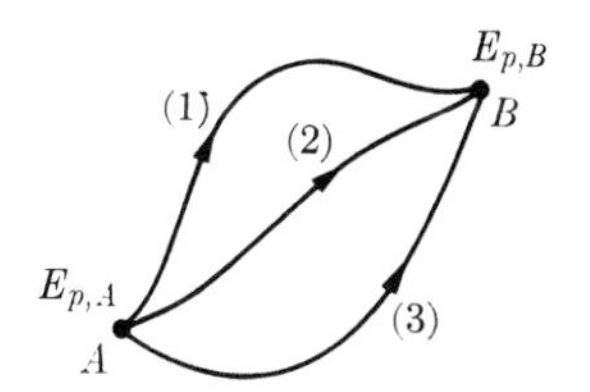

Figure 9.15 The work done by a conservative force is independent of the path.

or, since $\Delta E_p = \Delta E_{p(final)} - E_{p(initial)}$ is the change in potential energy,

$$W = -\Delta E_p \tag{9.21}$$

It is important to realize that no matter what the force $\boldsymbol{F}$ may be, it is always true that the kinetic energy is defined as $\frac{1}{2}mv^2$ and that Equation 9.15, $W = \Delta E_k$ always holds. On the other hand, in Equation 9.21 the form of the potential energy $E_p(\boldsymbol{r})$ depends on the nature of the force $\boldsymbol{F}$. Not all forces satisfy the condition set by Equation 9.20, only those called conservative. For example, comparing Equation 9.20 with Equation 9.18, we see that the potential energy corresponding to a constant force is

$$E_p = -\boldsymbol{F}\cdot\boldsymbol{r} \tag{9.22}$$

Similarly, from Equation 9.19 we note that the force of gravity is a conservative force and the potential energy due to gravity, when we take the Y-axis in the vertical direction, is defined as

$$E_p = mgy \tag{9.23}$$

Potential energy is always defined within an arbitrary constant, because, for example, if we write $mgy + C$ instead of mgy, Equation 9.21 still remains the same, since the constant C, appearing in the two terms, cancels out.

Because of this arbitrariness, the zero or **reference level** of potential energy may be defined wherever it best suits us. For example, for falling bodies the Earth's surface is the most convenient reference level. The potential energy due to gravity is therefore chosen as zero at the Earth's surface. For an Earth satellite, either natural or man-made, the zero of potential energy is usually defined at an infinite distance from the Earth. For an electron in an atom, the zero of potential energy is chosen at an infinite distance from the nucleus.

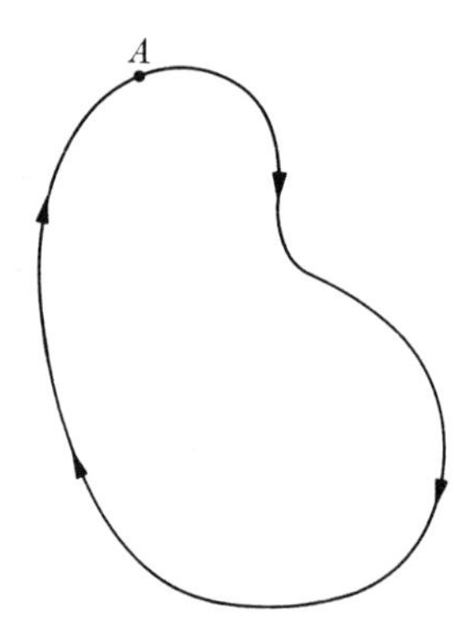

Figure 9.16 The work done by a conservative force along a closed path is zero.

It is important to emphasize that the work done by conservative forces is independent of the path joining points A and B; the difference $E_{p(A)} - E_{p(B)}$ is always the same because it depends only on the coordinates of A and B. In particular, if the path is *closed* (Figure 9.16), so that the final point coincides with the initial point (that is, A and B are the same point), then $E_{p(A)} = E_{p(B)}$ and the work is zero ($W = 0$). This means that during part of the path the work of the conservative force is positive and during the other part it is negative by exactly the same amount, giving a zero net result. For example, in climbing a staircase, the potential energy due to gravity increases; but returning to the ground floor decreases the potential energy by the same amount so that the net change in potential energy is zero. Similarly, for astronauts orbiting the Earth, their gravitational potential energy changes with their altitude; but every time they pass the same point in their orbit, their potential energy has the same value. Therefore, the change in potential energy, whenever they complete an orbit, is zero.

When the path is closed, it is customary to write the integral appearing in Equation 9.20 as $\oint$. (The circle on the integral sign indicates that the path is closed and the line integral is called the **circulation**. See Appendix A.10.) Therefore, for conservative forces,

$$W_{\text{(closed path)}} = \oint \boldsymbol{F} \cdot \mathrm{d}\boldsymbol{r} = 0 \qquad \textbf{(9.24)}$$

which means that: *the work of a conservative force along any closed path is zero.*

Conversely, it can be proved that the condition expressed by Equation 9.24 may be adopted as the definition of a conservative force. In other words, if a force $\boldsymbol{F}$ satisfies Equation 9.24 for any closed path arbitrarily chosen, then it can be proved that Equation 9.20 is valid.

9.9 Relation between force and potential energy

To satisfy Equation 9.21, $W = -\Delta E_p$, it is necessary that for each small displacement the work done is related to the change in potential energy by

$$\mathrm{d}W = \boldsymbol{F} \cdot \mathrm{d}\boldsymbol{r} = -\mathrm{d}E_p \qquad \textbf{(9.25)}$$

Since $\boldsymbol{F} \cdot \mathrm{d}\boldsymbol{r} = F_s\,\mathrm{d}s$, where F_s is the component of $\boldsymbol{F}$ in the direction of the displacement $\mathrm{d}\boldsymbol{r}$, and $\mathrm{d}s$ is the magnitude of the displacement $\mathrm{d}\boldsymbol{r}$, we may write, instead of Equation 9.25, $F_s\,\mathrm{d}s = -\mathrm{d}E_p$, or

$$F_s = -\frac{\mathrm{d}E_p}{\mathrm{d}s} \qquad \textbf{(9.26)}$$

Therefore, if we know $E_p(\boldsymbol{r})$ we may obtain the component of the force $\boldsymbol{F}$ in any direction by computing the quantity $-\mathrm{d}E_p/\mathrm{d}s$, where $\mathrm{d}s$ corresponds to a displacement in that direction. Thus, the components of $\boldsymbol{F}$ along the X-, Y- and Z-axes of coordinates are related to the potential energy by

$$F_x = -\frac{\mathrm{d}E_p}{\mathrm{d}x}, \quad F_y = -\frac{\mathrm{d}E_p}{\mathrm{d}y}, \quad F_z = -\frac{\mathrm{d}E_p}{\mathrm{d}z} \qquad \textbf{(9.27)}$$

The quantity dE_p/ds is equal to the change in potential energy per unit displacement and is called the *space rate of change* of E_p in the *direction* associated with ds. For that reason it is also called the **directional derivative** of E_p (see Appendix A.9).

When a vector is such that its component in any direction is equal to the derivative of a function in that direction, the vector is called the **gradient** of the function. Thus we say that a *conservative force* $\boldsymbol{F}$ *is the negative of the gradient of the potential energy* E_p.

When the potential energy E_p depends only on the distance r from a fixed point O but not on the direction of $\boldsymbol{r}$, we can write $E_p(r)$. Then, the potential energy only varies along the radial direction and so the only component of the force is along the direction in which r increases or decreases; that is, the direction of the radius vector $\boldsymbol{r}$, so that the force is central and its line of action always passes through the point O. Replacing ds by dr in Equation 9.26, the magnitude of the central force is then

$$F = -\frac{dE_p}{dr} \tag{9.28}$$

and conversely, it can be shown (see Note 9.1) that

> *if the force is central, the potential energy depends only on the distance from the center.*

For example, the potential energy of the Earth in its motion around the Sun depends only on its distance to the Sun. The same applies to an astronaut moving around the Earth and to an electron moving around the proton in a hydrogen atom.

EXAMPLE 9.7

Potential energy of a particle associated with the following central forces: (a) $F = -kr$, and (b) $F = -k/r^2$, where r is the distance from the center to the particle. The negative sign in both cases indicates that the force is attractive with respect to the center. If a positive sign is used, the force is repulsive.

▷ Using Equation 9.28, for case (a), we have

$$F = -dE_p/dr = -kr \qquad \text{or} \qquad dE_p = kr\,dr.$$

Integrating, we obtain

$$E_p = \int kr\,dr = \tfrac{1}{2}kr^2 + C \tag{9.29}$$

The constant of integration C is determined by assigning a value of E_p to a given position. In this case it is customary to make $E_p = 0$ at $r = 0$, so that $C = 0$ and $E_p = \frac{1}{2}kr^2$. This expression is useful in discussing plane oscillatory motion, as we will see in Chapter 10.

For case (b), which, as we shall see in Chapters 11 and 21, corresponds to the gravitational and electric interactions (see Section 8.5), we have

$$F = -\frac{dE_p}{dr} = -\frac{k}{r^2} \quad \text{or} \quad dE_p = k\frac{dr}{r^2}$$

Integrating, we have

$$E_p = \int k\frac{dr}{r^2} = -\frac{k}{r} + C \tag{9.30}$$

For inverse r forces, it is customary to determine C by making $E_p = 0$ at $r = \infty$, so that $C = 0$ and thus we have for our case

$$E_p = -k/r$$

This expression is used in the study of planetary motion (Chapter 11) and of atomic structure (Chapter 23). If the force is repulsive, $F = k/r^2$, the potential energy is $E_p = k/r$. This is the case, for example, when a proton is scattered by a nucleus.

Note 9.1 Relation between force, torque and potential energy in plane curvilinear motion

Consider a particle moving in a plane under the action of a conservative force $\boldsymbol{F}$ (Figure 9.17). For motion in a plane, when polar coordinates r, θ (Figure 9.18) are used, we may decompose a displacement $d\boldsymbol{r}$ into a radial component dr along the position vector $\boldsymbol{r}$ and a transverse component $r\,d\theta$ in a direction perpendicular to $\boldsymbol{r}$. Then, the **radial** and **transverse components** of $\boldsymbol{F}$ are related to the potential energy by

$$F_r = -\frac{dE_p}{dr}, \quad F_\theta = -\frac{1}{r}\frac{dE_p}{d\theta} \tag{9.31}$$

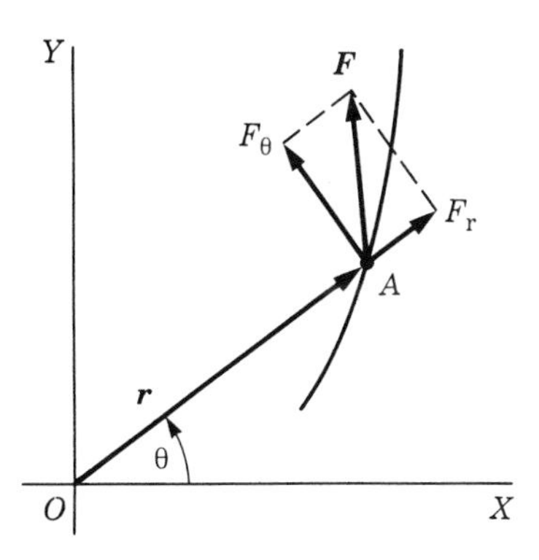

Figure 9.17 Radial and transverse components of force.

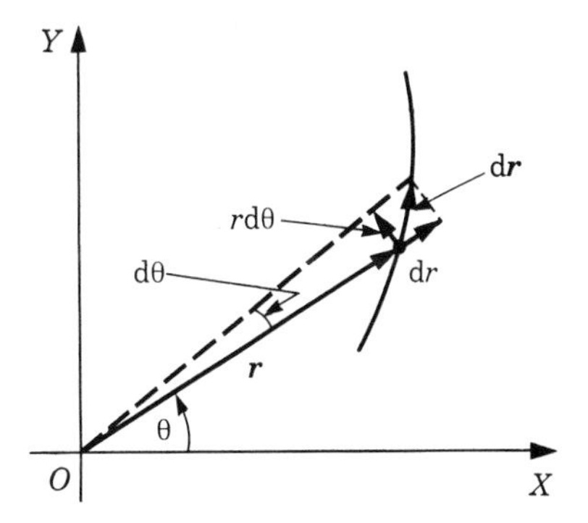

Figure 9.18 Radial and transverse displacements.

The torque of $\boldsymbol{F}$ with respect to O is the sum of the torques of the components F_r and F_θ. But the torque of F_r is zero because its direction passes through O. Thus the torque of $\boldsymbol{F}$ is

$$\tau = F_\theta r = -\left(\frac{1}{r}\frac{dE_p}{d\theta}\right)r = -\frac{dE_p}{d\theta} \quad \textbf{(9.32)}$$

that is, the torque of a force is related to the angular dependence of the potential energy E_p. For the case of a central force, E_p depends only on r so that $dE_p/d\theta = 0$. Therefore the torque relative to the center of force is zero ($\tau = 0$), a fact we already mentioned in Section 8.5. The converse is also true. If the torque of the force relative to a given point is zero no matter where the moving particle is, then, according to Equation 9.32, the potential energy is independent of the angle and depends only on the distance to the given point. The only component of the force is F_r so that the force is central.

9.10 Conservation of energy of a particle

When the force acting on a particle is conservative, we may combine Equation 9.21, $W = -\Delta E_p$, with the general relation $W = \Delta E_k$, which gives

$$\Delta E_k = -\Delta E_p \quad \textbf{(9.33)}$$

indicating that the changes of E_k and E_p are equal and opposite. This expression can also be written as

$$\Delta(E_k + E_p) = 0 \quad \textbf{(9.34)}$$

The quantity $E_k + E_p$ is called the **total energy** of the particle and is designated by E. That is, the total energy of a particle is equal to the sum of its kinetic energy and its potential energy, or

$$E = E_k + E_p = \tfrac{1}{2}mv^2 + E_p \quad \textbf{(9.35)}$$

Equation 9.34 indicates that the change in total energy is zero. Therefore,

when the force is conservative the total energy E of the particle remains constant.

In other words, *the total energy of a particle is conserved.* Thus we may write for any position of the particle,

$$E = E_k + E_p = const. \qquad \text{or} \qquad (E_k + E_p)_A = (E_k + E_p)_B$$

where A and B are any two positions of the particle. Recalling that $E_k = \frac{1}{2}mv^2$, we may write

$$E = \tfrac{1}{2}mv^2 + E_p = const. \quad \textbf{(9.36)}$$

This is the reason why we say that when there is a potential energy, the forces are conservative. During motion under conservative forces, both the kinetic and potential energies may vary, but always in such a way that their sum remains constant. Therefore, if the kinetic energy increases, the potential energy must decrease by the same amount, and conversely. We say that during the motion there is a continuous exchange of kinetic energy into potential energy and vice versa. Actually, nothing is exchanged but this is a convenient way of referring to how nature operates in the case of conservative forces.

For example, in the case of a body moving under the influence of gravity we have seen (Equation 9.23) that the potential energy is $E_p = mgy$, and the conservation of energy becomes

$$E = \tfrac{1}{2}mv^2 + mgy = const. \tag{9.37}$$

which is equivalent to $v^2 + 2gy = const.$ (see Example 9.5) if we multiply through by $\frac{1}{2}m$. As the body moves up or down and the height y increases or decreases, the velocity of the body decreases or increases so that Equation 9.37 holds. If the particle is initially at height y_0 and its velocity is zero, the total energy is mgy_0 and we have

$$\tfrac{1}{2}mv^2 + mgy = mgy_0, \qquad \text{or} \qquad v^2 = 2g(y_0 - y) = 2gh$$

where $h = y_0 - y$ is the height through which it has fallen. This result is the well-known formula for the velocity acquired by a body in free fall through a height h. We must note, however, that Equation 9.37 is not restricted to vertical motion; it is equally valid for the motion of any projectile moving at an angle with the vertical.

As a second example, consider a particle subject to a force $F = -kx$, a case we have already analyzed in Example 9.4. From Equation 9.29, with r replaced by x, we conclude that the potential energy of the particle is

$$E_p = \tfrac{1}{2}kx^2$$

Then the total energy of the particle is

$$E = \tfrac{1}{2}mv^2 + \tfrac{1}{2}kx^2 = const.$$

This should be compared with Equation 9.16, from which we conclude that $E = \frac{1}{2}ka^2$.

At first sight, the conservation of energy of a particle may seem to be simply a mathematical result based on the definition of potential energy and the relation between work and kinetic energy (which, as we mentioned, is in turn a consequence of Newton's law of motion). However, the fact that there exist forces in nature to which certain forms of potential energy can be associated is an indication of some profound relation operating at a fundamental level. This matter will be explored later on.

EXAMPLE 9.8

Minimum height at which a disk should start sliding down a ramp in order to complete successfully the loop shown in Figure 9.19.

▷ Assume that the disk slides without friction and that it is released at point A at a height h above the base of the circle in Figure 9.19. The disk gains velocity while moving down and loses velocity when moving up the circle. At any point on the track, the forces acting on the disk are: (i) its weight mg and (ii) the force F due to the track. (The force F points toward the center of the loop at all times, since the track 'pushes' but does not 'pull'.) At the highest point on the loop, both mg and F point toward the center O, and according to Equation 8.2 we must have

$$F + mg = mv^2/R$$

where R is the radius of the track. Since F cannot be negative, the minimum velocity of the disk at B (if it is to describe the loop) must correspond to $F = 0$ or $mg = mv^2/R$, which gives

$$v^2 = gR \qquad \text{or} \qquad v = (gR)^{1/2}$$

If the velocity is less than $(gR)^{1/2}$, the downward pull of the weight is larger than the required centripetal force, and the disk will separate from the loop before it reaches point B.

To obtain the corresponding height h, we note that the total energy at any point is $E = \frac{1}{2}mv^2 + mgy$. At A, where $v = 0$, the total energy is

$$E_A = (E_k + E_p)_A = mgh$$

At B, where $y = 2R$ and $v^2 = gR$,

$$E_B = (E_k + E_p)_B = \tfrac{1}{2}m(gR) + mg(2R) = \tfrac{5}{2}mgR$$

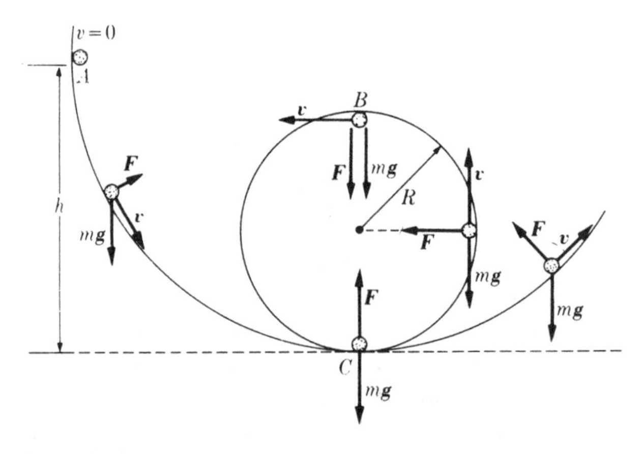

Figure 9.19

Thus, equating the values of E_A and E_B, we get $h = (\frac{5}{2})R$, which is the minimum height of the starting point of the disk if it is to successfully describe the loop. The total energy during the motion is $E = mgh = (\frac{5}{2})mgR$. This result is correct so long as we can neglect frictional forces. If instead of a disk that slides we have a ball that rolls, we must take into account the energy of rotation. Under those conditions the methods to be introduced in Chapter 14 must be used. Note also that since the force F does no work, it does not contribute to the energy of the particle.

At point C, where $y = 0$, the total energy is all kinetic and so $E = \frac{1}{2}mv^2$. But, the total energy is $(\frac{5}{2})mgR$. And so, $v^2 = 5gR$. Since at point C, forces F and mg have opposite directions, we must have $F - mg = mv^2/R$. Therefore, $F = 6mg$, or six times the weight of the disk. The disk, using aviators' jargon, 'feels 6 gees'.

9.11 Discussion of potential energy curves

The graphs representing $E_p(\mathbf{r})$ versus $\mathbf{r}$ are very useful in understanding the motion of a particle, even without solving the equation of motion. For simplicity we shall consider first the case of rectilinear or one-dimensional motion, so that the potential energy depends only on a variable x; that is, $E_p(x)$. The same analysis will apply to central forces if we replace x by r. In Figure 9.20 a possible potential energy curve for one-dimensional motion is illustrated. When we use Equation 9.26, the force on the particle for any value of x is given by

$$F = -\frac{dE_p}{dx}$$

Now dE_p/dx is the slope of the curve $E_p(x)$. The slope is positive whenever the curve is increasing or upward, and negative whenever the curve is decreasing or downward. Therefore, the force F (i.e. the negative of the slope) is negative, or directed to the left, whenever the potential energy is increasing and positive, or

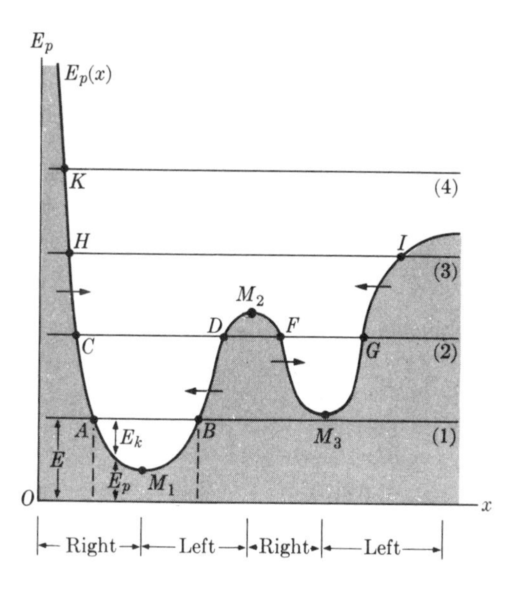

Figure 9.20 Relation between motion in a straight line and potential energy. The direction of the force in each region is shown by the horizontal arrows.

directed to the right whenever the potential energy is decreasing. This situation has been indicated in Figure 9.20 by the horizontal arrows in the different regions and marked below the figure.

At the points where the potential energy is minimum or maximum, such as M_1, M_2 and M_3, we have $dE_p/dx = 0$. Therefore at those points $F = 0$; that is, they are positions of equilibrium. Those positions where $E_p(x)$ is a *minimum* are of **stable** equilibrium because, when the particle is displaced slightly from its equilibrium position, it is acted on by a force that tends to restore it to that position. Where $E_p(x)$ is *maximum*, the equilibrium is **unstable**, since a slight displacement from the equilibrium position causes the particle to experience a force that tends to move it even farther away.

Consider now a particle having total energy E, as indicated by the horizontal line (1) of Figure 9.20. At any position x the potential energy E_p corresponds to the ordinate of the curve and the kinetic energy, $E_k = E - E_p$, is given by the distance from curve $E_p(x)$ to the E line. Now the E line intersects curve $E_p(x)$ at points A and B. To the left of A and to the right of B the energy E is less than the potential energy $E_p(x)$, and therefore in that region the kinetic energy would be negative. But that is impossible because $E_k = \frac{1}{2}mv^2$ is necessarily positive. Therefore the motion of the particle is limited to the interval AB, and the particle oscillates between A and B. At these points the velocity becomes zero and the particle reverses its motion. These points are called **turning points.**

If the particle has a higher energy, such as that corresponding to line (2), it has two possible regions of motion. These regions are called **potential wells.** In one region, the particle oscillates between C and D; in the other region the particle oscillates between F and G. Also, if the particle is in one region it can never jump to the other, because that would require passing through the region DF where the kinetic energy would be negative and is therefore forbidden. We say that the two regions, or potential wells, where the motion is allowed are separated by a **potential barrier**.

At energy level (3), the motion is oscillatory between H and I. Finally at energy level (4) the motion is no longer oscillatory and the particle moves between K and infinity. For example, if the particle is moving initially to the left, when it reaches K it 'bounces' back, receding toward the right and never returning.

When we consider the motion of atomic particles, so that quantum mechanics applies, the description we have given requires some modification. There is a possibility that the particles move a little beyond the turning points, or even 'tunnel' through a potential barrier (Section 37.8).

From the above discussion, we can see that potential energy curves can be very useful in analyzing the possible motions of a particle. In Note 9.2, energy relations in plane curvilinear motion are analyzed in more detail.

EXAMPLE 9.9

Potential energy for the interaction between two atoms in a diatomic molecule.

▷ A diatomic molecule is a stable system composed of two atoms whose nuclei are maintained at a certain equilibrium separation. The intermolecular interaction is electrical in nature;

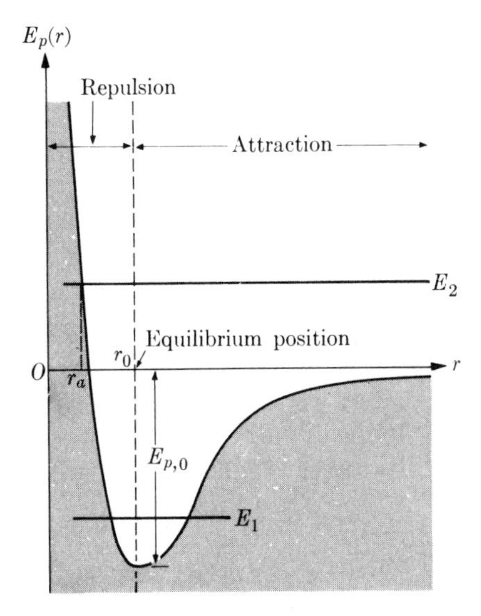

Figure 9.21 Intermolecular potential energy.

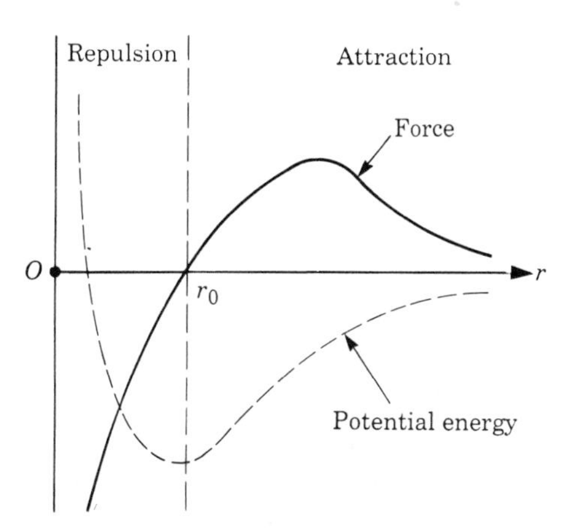

Figure 9.22 The force is related to the potential energy by $F = -dE_p/dr$.

however, because atoms are more or less complex structures composed of several charged particles, the determination of $E_p(r)$ can only be carried out in an approximate way. In any case, the potential energy can be represented by the curve of Figure 9.21, where r is the separation between the atoms. The equilibrium separation r_0 corresponds to the minimum, $-E_{p,0}$, of the potential energy of the molecule. For distances less than r_0, the interatomic force is repulsive ($E_p(r)$ is a decreasing function) and for distances larger than r_0 it is attractive ($E_p(r)$ is an increasing function). As r increases beyond r_0, the force $F = -dE_p/dr$ decreases because the slope of $E_p(r)$ decreases. However, for $r < r_0$, the repulsive force is very strong because the slope of $E_p(r)$ is very large.

If the total energy of the two atoms is negative, such as E_1, the two atoms oscillate between minimum and maximum values of the interatomic separation, forming a bound system or molecule. If the energy is positive, such as E_2, the atoms approach each other until their distance has a minimum value. Then they separate and no bound system is formed unless the excess energy is released in some way, as happens in some chemical reactions.

When a molecule is in equilibrium, it is necessary to supply the molecule with at least the energy $E_{p,0}$ to separate the two atoms. For that reason $E_{p,0}$ is called the **binding energy** of the molecule. (Actually, the binding energy is slightly less for reasons to be explained in Chapters 10 and 37.) The graph of the force $F = -dE_p/dr$ as a function of the distance is given in Figure 9.22. The graph of $E_p(r)$ has been superimposed for comparison. Note that the force is independent of the total energy of the particle. The potential energy for the interaction between two gas molecules is also very similar to that of Figure 9.21, except that the total energy of the molecules is positive, like for energy E_2.

EXAMPLE 9.10

Calculation of the distance of closest approach in the scattering of a particle by a repulsive inverse square central force.

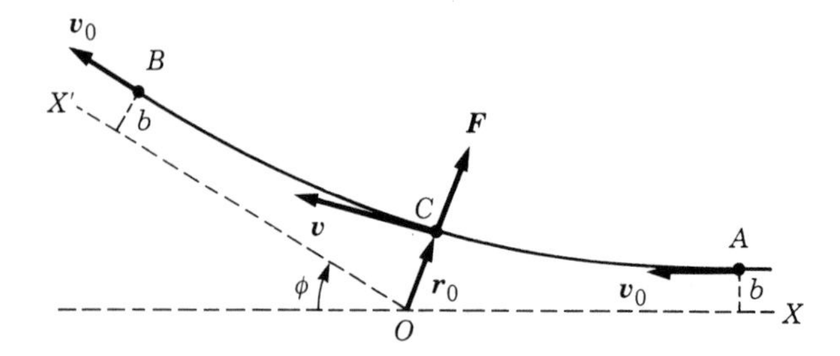

Figure 9.23 Scattering of a particle under a repulsive inverse square central force.

▷ In Example 8.7, the scattering of a particle was discussed. When the force is repulsive there is a position C at which the particle is closest to the scattering center O (Figure 9.23). At this position the velocity is perpendicular to the force because at C the radial velocity is zero. The distance $r_0 = OC$ depends on v_0, b and the strength of the force. Assuming an inverse square repulsive force $F = k/r^2$, it is clear that the larger the coupling constant k or the smaller the velocity v_0, the larger r_0 will be (since it will be easier to turn the particle away under either of these conditions). This can be verified in the following manner: at A, which is very far from O, the total energy of the particle is only kinetic energy; that is, $E = E_k + E_p = \frac{1}{2}mv_0^2$. At the point C, however, where the velocity has a value v and the potential energy is $+k/r_0$ (recall Example 9.7), the total energy is $E = \frac{1}{2}mv^2 + k/r_0$. Equating these expressions for the total energy gives

$$\tfrac{1}{2}mv^2 + \frac{k}{r_0} = \tfrac{1}{2}mv_0^2$$

On the other hand, the angular momentum of the particle, relative to O, when it is at A is $mv_0 b$. At the position of closest approach, where the velocity is normal to the radial line, the angular momentum is mvr_0. Equating these two values (since angular momentum is conserved for a central force), we have

$$mvr_0 = mv_0 b \qquad \text{or} \qquad v = v_0\left(\frac{b}{r_0}\right)$$

Therefore, using this value for the velocity, the energy relation becomes

$$\tfrac{1}{2}mv_0^2\left(\frac{b}{r_0}\right)^2 + \frac{k}{r_0} = \tfrac{1}{2}mv_0^2$$

This is a quadratic equation in r_0, the distance of closest approach. For a head-on collision, $b = 0$ and the equation becomes $k/r_0 = \frac{1}{2}mv_0^2$ or $r_0 = k/(\frac{1}{2}mv_0^2)$. The particle stops momentarily at $r = r_0$ before being sent back. In the general case, with some manipulation of the previous equation, we find that

$$r_0 = \frac{k}{mv_0^2} + \left(\frac{k^2}{m^2v_0^4} + b^2\right)^{1/2} \qquad \textbf{(9.38)}$$

It may be verified that if v_0 or b are very large, the value of r_0 approaches b, and the particle passes by the force center O without appreciable deviation. Also, for b or v_0 very small or k very large, r_0 approaches a value of $k/\frac{1}{2}mv_0^2$, which corresponds to a direct or head-on collision.

EXAMPLE 9.11

Potential energy of a pendulum.

▷ Consider a mass m suspended from a point O by a thin rod of negligible mass and length l (Figure 9.24). This system constitutes a **pendulum**. When the mass is displaced to the side or given an initial velocity, the pendulum moves along the arc of a vertical circle whose radius is l under the action of its weight and the tension on the rod. We shall define the (vertical) angle θ as positive when the pendulum is to the right of the lowest point, C, and negative when it is to the left. Therefore, θ may vary from $-\pi$ to $+\pi$ at D and is zero at C.

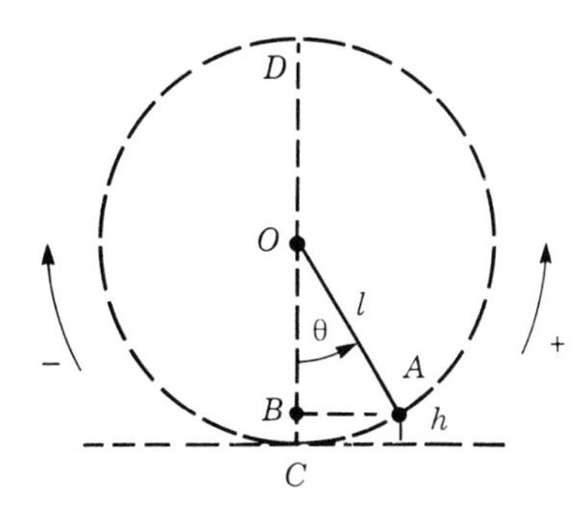

Figure 9.24 Motion of a pendulum of length l.

The potential energy of the pendulum at the arbitrary point A (when we take the zero of potential energy at the lowest point C) is

$$E_p = mgh$$

But

$$h = BC = OC - OB = l - l\cos\theta = l(1 - \cos\theta)$$

Therefore,

$$E_p = mgl(1 - \cos\theta)$$

If we plot this potential energy, we see that it corresponds to an inverted cosine curve as shown in Figure 9.25. E_p is a minimum ($E_p = 0$) for $\theta = 0$, at point C, and maximum (equal to $E_0 = 2mgl$) for $\theta = \pm\pi$, at point D. Therefore, the bottom ($\theta = 0$) is a position of stable equilibrium and the top ($\theta = \pm\pi$) is a position of unstable equilibrium. (It is very difficult to balance a rod with a mass at the upper end; try it!)

For a total energy E smaller than E_0, the pendulum oscillates within the limits P and Q given by the intersection of the line $P'Q'$ with the E_p curve in Figure 9.25. When E is larger than E_0, the motion of the pendulum is a rotation around O. The velocity decreases as the pendulum approaches D and increases as it moves towards C. We say that the motion is a modulated rotation. The motion is clockwise or counter-clockwise, depending on the initial conditions. For $E = E_0$, the pendulum reaches D with a zero velocity. However, it is very difficult to give a pendulum exactly that amount of energy because that would require giving the pendulum a velocity of exactly $(2gl)^{1/2}$ and neither g nor l are known exactly.

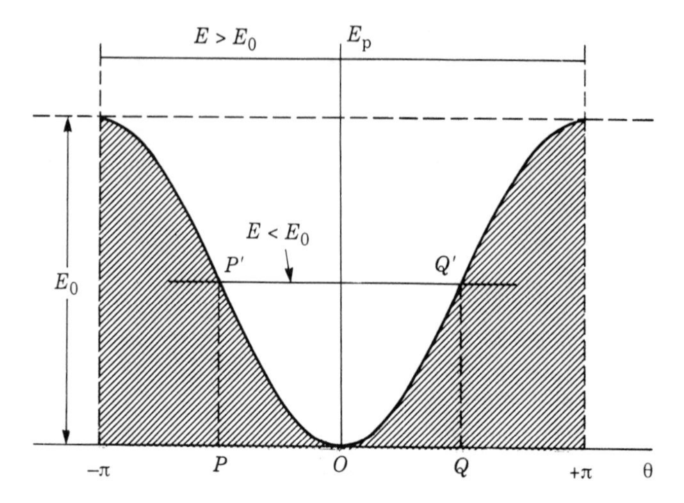

Figure 9.25 Potential energy of a pendulum. In the figure, $E_0 = 2mgl$.

Note 9.2 Energy in plane curvilinear motion

Consider a particle moving in a plane under the action of a conservative force whose potential energy is E_p. The total energy of the particle is then

$$E = \tfrac{1}{2}mv^2 + E_p$$

We may express the velocity $\boldsymbol{v}$ in terms of radial and transverse velocities (Figure 9.26). Using Equations 5.6 (that is $v_r = dr/dt$ and $v_\theta = r(d\theta/dt)$), we can write

$$v^2 = v_r^2 + v_\theta^2 = \left(\frac{dr}{dt}\right)^2 + r^2\left(\frac{d\theta}{dt}\right)^2$$

Recalling Equation 8.11, $\boldsymbol{L} = mr^2\omega$, where the angular velocity is $\omega = d\theta/dt$, we can write

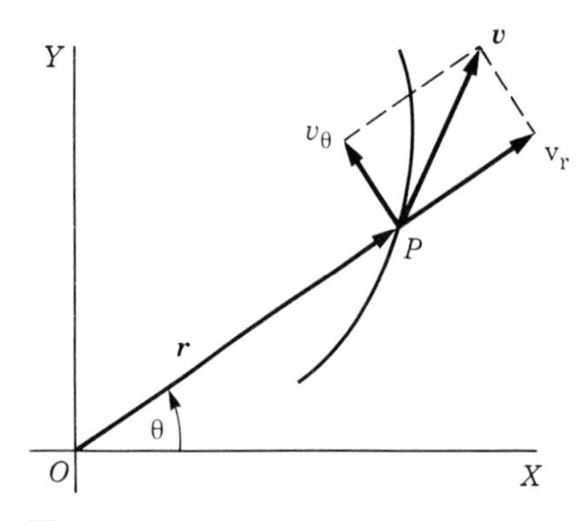

Figure 9.26 Radial and transverse components of the velocity.

$$\omega^2 = \left(\frac{d\theta}{dt}\right)^2 = \frac{L^2}{m^2r^4}$$

Then, the second term of the above equation may be written as

$$r^2\left(\frac{d\theta}{dt}\right)^2 = \frac{L^2}{m^2r^2}$$

The total energy of a particle in plane curvilinear motion can then be written as

$$E = \tfrac{1}{2}m\left(\frac{dr}{dt}\right)^2 + \frac{L^2}{2mr^2} + E_p \qquad \textbf{(9.39)}$$

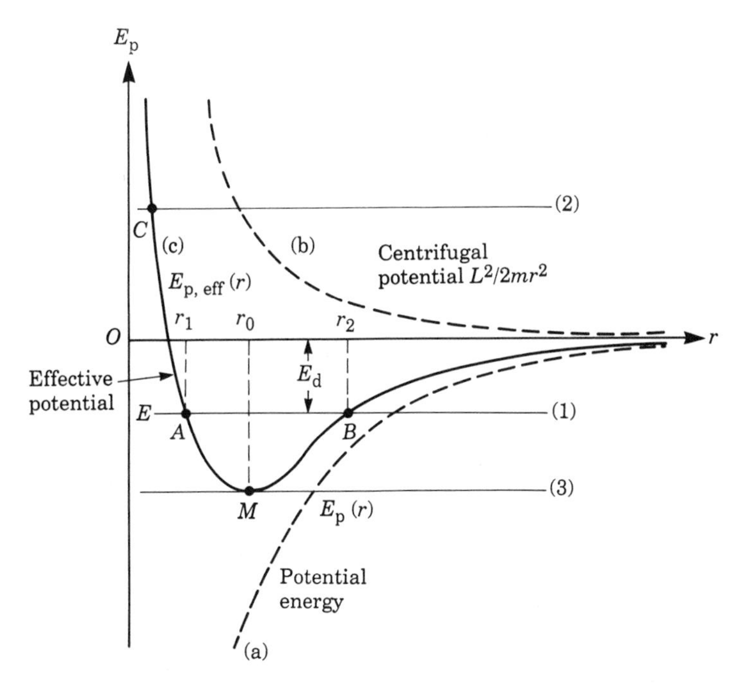

Figure 9.27 Energy relations for motion under central forces.

The first term of Equation 9.39 represents the kinetic energy associated with radial motion; the second term is the kinetic energy associated with the angular motion; the third term is the potential energy of the particle.

Now consider the case of motion under the action of a central force with potential energy $E_p(r)$. Then the angular momentum is constant and the second term in Equation 9.39 depends only on the radial distance and so it may be formally considered equivalent to a potential energy. We then say that the radial motion under a central force can be assumed to be due to an *effective* potential energy

$$E_{eff} = \frac{L^2}{2mr^2} + E_p(r) \qquad \textbf{(9.40)}$$

The term $L^2/2mr^2$ is always positive and increases very rapidly as the radial distance decreases, since it varies inversely as the square of the distance. This term has been represented in Figure 9.27 by the dashed curve (b). It can be considered equivalent to the potential energy of a fictitious centrifugal force and is called **centrifugal potential**.

Let us assume that the actual potential energy corresponds to an attractive potential energy at all distances. This is the situation found in planetary motion as well as in the motion of electrons in atoms. Then, if we take the zero of potential energy at great distances, the actual potential energy $E_p(r)$ is always negative and has the general shape of curve (a) in Figure 9.27.

When we combine the two terms in Equation 9.40, the general shape of the effective potential energy corresponds to curve (c) in Figure 9.27. In many cases of physical interest, the centrifugal potential energy is the dominant term at small distances, resulting in an effective potential energy $E_{p,\,eff}$ with the shape shown as curve (c). For the case of an electron in an atom, we have seen in Example 8.6 that the angular momentum is quantized according to Equation 8.14. Therefore there are several effective potential energy curves, depending on the value of L^2. If the total energy E of the particle is negative and corresponds to the horizontal

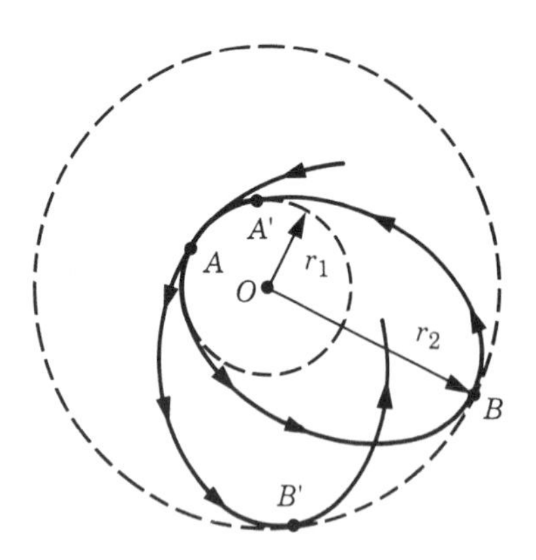

Figure 9.28 General shape of path for motion under central forces when $E < 0$.

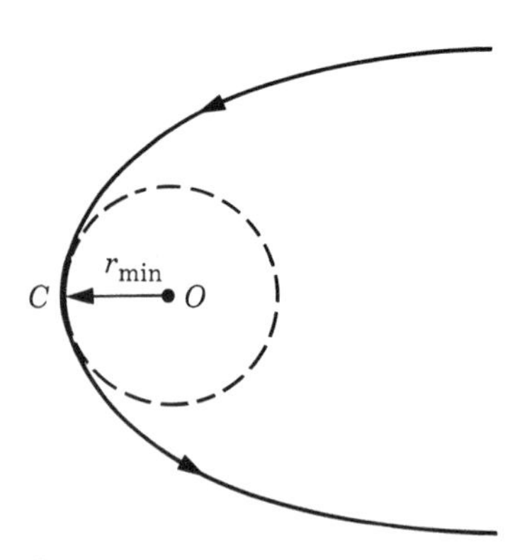

Figure 9.29 Distance of closest approach when $E > 0$.

line (1), the radius of the orbit will oscillate between the minimum and maximum values r_1 and r_2; the orbit will have the shape illustrated in Figure 9.28. But, if the energy is positive, corresponding to a value such as that given by line (2) of Figure 9.27, the orbit is not bound. The particle comes from infinity to the **point of closest approach**, C, at the distance r_{min} (Figure 9.29) and then recedes again without ever returning. If the energy corresponds to the minimum M of $E_{p,eff}$, shown in Figure 9.27 by line (3), there is only one intersection and the distance to the center remains constant; the particle describes a circular path of radius r_0. The distance of closest approach increases with increasing angular momentum, due to the effect of the centrifugal potential energy.

If, by some mechanism, a particle that has an energy equal to level (1) (Figure 9.27) can absorb energy and thereby 'jump' to energy level (2), the particle will fly away from the center of force; that is, the particle **dissociates** from the center of force. The minimum energy a particle requires to dissociate from energy level (1) is indicated in Figure 9.27 by E_d. On the other hand, if the particle is initially in energy level (2) and by some process loses energy when it passes near the center of force, the particle may jump into energy level (1) and then will remain in a bound orbit. We say that the particle has been **captured** by the center of force. Both of these situations are found, for example, in molecular formation and dissociation.

9.12 Non-conservative forces and energy dissipation

At first sight we find forces in nature that are not conservative. One example is friction. Sliding friction always opposes the displacement. Its work depends on the path followed, and although the path may be closed, the work is not zero, so that Equation 9.24 does not hold. Similarly, fluid friction opposes the velocity, and depends on velocity but not on position. A body may thus be subject to conservative and to non-conservative forces at the same time.

For example, a body falling through a fluid is subject to the conservative gravitational force and to the non-conservative fluid friction. Call E_p the potential energy corresponding to the conservative forces and W' the work done by the non-conservative forces (work which, in general, is negative because frictional forces oppose the motion). The total work done on the body when moving from

A to B is found by adding W' to the right-hand side of Equation 9.21; that is,

$$W = -\Delta E_p + W'$$

Using Equation 9.15, $W = \Delta E_k$, we then write

$$\Delta E_k = -\Delta E_p + W'$$

or

$$\Delta(E_k + E_p) = W' \qquad \textbf{(9.41)}$$

In this case the quantity $E_k + E_p$ does not remain constant but decreases (increases) if W' is negative (positive). On the other hand, we cannot call $E_k + E_p$ the total energy of the body, because this concept is not applicable in this case since it does not include all the forces present. The concept of total energy of a body is meaningful only when all the forces are conservative. However, Equation 9.41 is useful when we wish to make a comparison between the case where only conservative forces act (so that $E_k + E_p$ is the total energy) and the case where there are additional non-conservative forces. Then we say that Equation 9.41 gives the gain or loss of energy due to the non-conservative forces.

The existence of non-conservative forces, such as friction, must not be considered as necessarily implying that there may exist non-conservative interactions between elementary constituents of matter and that energy might disappear or be created. We must recall that frictional forces do not correspond to an interaction between two particles. They are really forces of a statistical nature involving interactions between large numbers of molecules (recall the discussion of Section 7.4). Sliding friction, for example, is the result of a large number of energy and momentum exchanges between the molecules of the two bodies in contact. Energy is conserved in each of these individual interactions. However, the macroscopic effect is not conservative because the energy is redistributed (or dissipated) among a very large number of molecules in a form that is very difficult to recover. That means that, although the bodies may be returned to the same *macro*scopic state through some external action, there remain changes at the *micro*scopic level. Thus, the final state is not microscopically identical to the initial one, nor is it equivalent to the initial state in a statistical sense. We say that friction results in an irreversible dissipation of energy.

EXAMPLE 9.12

A body falls through a viscous fluid starting from rest at a height y_0. Calculate the rate of dissipation of its kinetic and gravitational potential energies after the limiting velocity has been reached.

▷ When the body is at height y falling with velocity v, its kinetic plus its gravitational potential energy is $\frac{1}{2}mv^2 + mgy$. The rate of dissipation of energy (or energy lost per unit time) due to the action of the non-conservative viscous forces is, using Equation 9.41,

$$\frac{\mathrm{d}}{\mathrm{d}t}(E_k + E_p) = \frac{\mathrm{d}W'}{\mathrm{d}t} = F'\frac{\mathrm{d}y}{\mathrm{d}t}$$

where F' is the non-conservative force. In this example F' is due to the fluid friction, and has the form $F' = -K\eta v$ given in Equation 7.7. Thus

$$\frac{\mathrm{d}}{\mathrm{d}t}(E_k + E_p) = F'\frac{\mathrm{d}y}{\mathrm{d}t} = (-K\eta v)v = -K\eta v^2$$

We saw in Equation 7.11 that after a long time the velocity becomes constant and equal to $mg/K\eta$. Therefore, replacing v by $mg/K\eta$ in the above equation, we obtain the rate of energy dissipation, at the limiting velocity, as

$$\frac{\mathrm{d}}{\mathrm{d}t}(E_k + E_p) = -\frac{m^2g^2}{K\eta}$$

The negative sign shows that energy is lost by the body. Since all quantities appearing in the expression are constants, the energy decreases at a constant rate. However, this energy is not really 'lost', but is transferred to the molecules of the fluid in a form that is practically impossible to recover. It is an irreversible dissipative process.

Actually, since the kinetic energy of the body remains constant once the limiting velocity is reached, it is the gravitational potential energy lost by the body that is dissipated into molecular agitation of the fluid. This is a different way of saying that the downward force of gravity is balanced by the opposing force due to the viscosity of the fluid.

QUESTIONS

9.1 State the conditions under which the work done by a force is (a) zero, (b) positive and (c) negative.

9.2 How do you measure the average power of a machine during a certain time interval?

9.3 Refer to Figure 9.7 and assume the car moves uphill. Indicate which forces do positive work, which do negative work and which do no work at all. Repeat if the car moves downhill.

9.4 What happens to the kinetic energy of a particle when the work of the applied force is positive? And when it is negative?

9.5 What happens to the potential energy of a particle when the work of the applied force is positive? And when it is negative?

9.6 What is the relation between the changes in kinetic and potential energy of a particle when the forces are conservative?

9.7 A particle of mass m falls vertically through a height h. Write an equation relating the change in kinetic energy of the particle with the work done by the weight of the particle.

9.8 Refer to Example 9.4. What physical meaning can be associated with the fact that the kinetic energy depends on the negative of the square of the displacement, x, of the particle? Is there a limitation to the possible values of x?

9.9 A particle of mass m is attached to a spring of elastic constant k. The spring is extended the distance a and released. Relate the potential energy of the particle at $x = a$ to its kinetic energy at $x = 0$. What is the velocity of the particle at $x = 0$? Recall Example 9.4.

9.10 Refer to Example 9.8. Why can we ignore the force $\boldsymbol{F}$ in writing the conservation of energy?

9.11 What is 'conserved' by a conservative force?

9.12 What is the physical meaning of dissipative forces?

9.13 Explain how, from the slope of the graph of $E_p(r)$ versus r, we can determine whether a central force is attractive or repulsive. Plot a graph of $E_p(r)$ corresponding to a central force that is repulsive at short distances and attractive at large distances. Repeat for a central force attractive at short distances and repulsive at large distances.

9.14 A particle moves under the influence of a force whose potential energy may be described as a **potential well**, with depth E_0 and width b. Draw the potential energy curve, with one edge of the well at $x = 0$ and the other at $x = b$. When is the motion bound? Unbound?

9.15 What three pairs of related concepts have been introduced so far to discuss the motion of a particle?

PROBLEMS

9.1 Calculate the work of a constant force of 12 N when the particle on which it acts moves 7 m if the angle between the directions of the force and the displacement is (a) 0°, (b) 60°, (c) 90°, (d) 145°, (e) 180°.

9.2 Calculate the work done by a person who drags a 65 kg bushel of flour 10 m along the floor with a force of 250 N and then lifts it up to a truck whose platform is 75 cm above the floor. What is the average power developed by the person if it took 40 s for the whole process?

9.3 A body with a mass of 4 kg moves upward on a plane inclined 20° to the horizontal. Under the forces shown in Figure 9.30, the body slides 20 m on the plane in the upward direction. Calculate the total work done by the system of forces acting on the body.

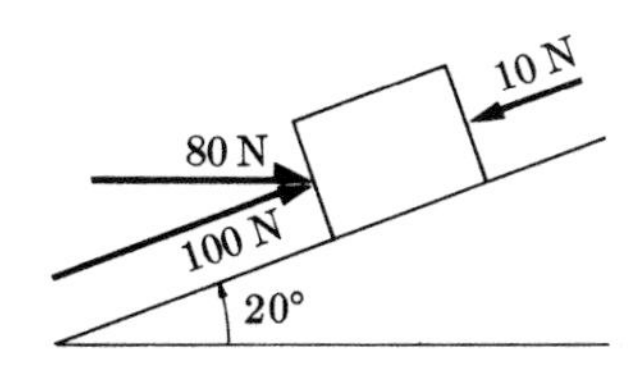

Figure 9.30

9.4 A man whose mass is 80 kg goes up an inclined plane forming a 10° angle with the horizontal at a velocity of 6 km h^{-1}. Calculate the power developed.

9.5 An automobile goes up a road inclined 3° with a constant velocity of 45 km h^{-1}. The mass of the automobile is 1600 kg. (a) What is the power developed by the motor? (b) What is the work done by the motor in 10 s? Ignore friction.

9.6 An electric cart weighs 1000 N and is moving on a horizontal path. It reaches a maximum velocity of 25 m s^{-1} when the motor develops its maximum power, 42 kW. Calculate the maximum velocity of the cart when it is going up a hill whose slope is 5%. Assume that the resistance of the air is constant. (Note: a 1% slope has an angle α such that $\tan \alpha = 0.01$.)

9.7 (a) What constant force must be exerted by the motor of an automobile whose mass is 1500 kg in order to increase the velocity of the automobile from 4.0 km h^{-1} to 40 km h^{-1} in 8 s? (b) Determine the variation of the momentum and kinetic energy. (c) Determine the work done by the force. (d) Compute the average power of the motor.

9.8 A small steel ball of mass 1 kg is tied to the end of a wire 1 m long spinning in a vertical circle about the other end with a constant angular velocity of 120 rad s^{-1}. (a) Calculate the kinetic energy. (b) What is the difference in potential energy at the top and the bottom of the circle?

9.9 A body of mass m is moving with velocity $\boldsymbol{V}$ relative to an observer O and with velocity $\boldsymbol{V}'$ relative to O'. The velocity of O' relative to O is $\boldsymbol{v}$. Find the relation between the kinetic energies E_k and E'_k of the particle as measured by O and O'.

9.10 (a) Express, in eV, the kinetic energy of an electron (mass $= 9.109 \times 10^{-31}$ kg) moving at a velocity of 10^6 m s^{-1}. (b) Repeat for a proton (mass $= 1.675 \times 10^{-27}$ kg) at the same velocity.

9.11 Find the velocity of (a) an electron in a television tube which hits the screen with an energy of 2.5×10^4 eV, and (b) a proton which emerges from a particle accelerator with an energy of 3×10^5 eV.

9.12 Given $\boldsymbol{F} = \boldsymbol{i}(7\,\text{N}) - \boldsymbol{j}(6\,\text{N})$, (a) compute the work done when a particle goes from the origin to $\boldsymbol{r} = \boldsymbol{i}(-3\,\text{m}) + \boldsymbol{j}(4\,\text{m})$. Is it necessary to specify the path followed by the particle? (b) Compute the average power if it took 0.6 s to go from one place to the other. Express your answer in watts. (c) If the mass of the particle is 1.0 kg, calculate the change in kinetic energy.

9.13 The force in the previous problem is conservative, since it is constant. (a) Calculate the potential energy difference between the origin and the point $(-3, 4)$. (b) Determine the potential energy at the point $\boldsymbol{r} = \boldsymbol{i}(7\,\text{m}) + \boldsymbol{j}(16\,\text{m})$ if the potential energy is zero at the origin.

9.14 A particle moves under an attractive inverse square force, $F = -k/r^2$. The path is a circle of radius r. Show that (a) the total energy is $E = -k/2r$, (b) the velocity is $v = (k/mr)^{1/2}$ and (c) the angular momentum is $L = (mkr)^{1/2}$.

9.15 Plot the kinetic and potential energies as a function of (a) time and (b) height, as a body falls from rest from a height h. Verify that the curves in each case always add to the same constant value.

9.16 A body of mass 20 kg is launched vertically upward with an initial velocity of 50 m s^{-1}. Calculate (a) the initial E_k, E_p and E, (b) E_k and E_p after 3 s, (c) E_k and E_p at 100 m altitude, and (d) the body's altitude when E_k is reduced to 80% of the initial value. Take the zero of potential energy at the Earth's surface.

9.17 Solve Problem 9.16 for a case in which the body is launched at a 70° angle above the horizontal.

9.18 A 0.40 kg ball is launched horizontally from the top of a hill 120 m high with a velocity of 6 m s^{-1}. Calculate (a) the ball's initial kinetic energy, (b) its initial potential energy, (c) its kinetic energy when it hits the ground and (d) its velocity when it hits the ground.

9.19 A 0.5 kg mass is dropped from a height of 1 m on to a small vertical spring which is in an upright position on the floor and becomes attached to it. The constant of the spring is $k = 2000$ N m^{-1}. Calculate the maximum deformation of the spring.

9.20 In the NH_3 molecule the N atom occupies the vertex of a tetrahedron with the three H atoms at the base (see Figure 2.3). It is clear that the N atom has two symmetric stable equilibrium positions. Draw a schematic potential energy curve for the N atom as a function of its distance to the base of the tetrahedron and discuss its possible motion in terms of its total energy.

9.21 In the ethane molecule (C_2H_6), the two CH_3 groups are tetrahedral with the C atom at one vertex (Figure 9.31). The two CH_3 groups may rotate relative to each other around the line joining the two carbon atoms. Symmetry suggests that there are two sets of positions of equilibrium for this motion; one set consists of stable positions and the other consists of unstable ones. (a) Determine these positions and make a schematic plot of the potential energy as a function of the angle ϕ between 0 and 2π. (b) Discuss the possible motions for different values of the total energy.

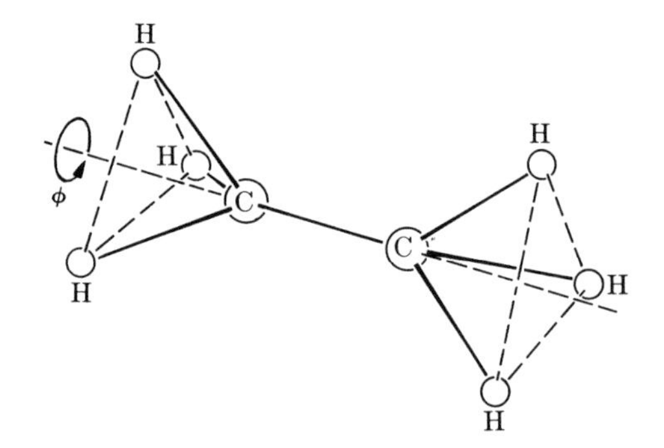

Figure 9.31

9.22 A 20 kg sled slides on a hill, starting at an altitude of 20 m. The sled starts from rest and has a velocity of 16 m s^{-1} when it reaches the bottom of the slope. Calculate the loss of energy due to friction.

9.23 A ball of mass 0.5 kg, which is launched vertically upward with an initial velocity of 20 m s^{-1}, reaches an altitude of 15 m. Calculate the loss of energy due to air resistance.

9.24 A car starting from rest travels 300 m down a 1% slope. With the impetus thus acquired, it goes 60 m up a 2% slope and comes to rest. Calculate the force of resistance to the motion of the car. (Note: a 1% slope has an angle α such that $\tan\alpha = 0.01$.)

9.25 A body of mass m slides downward along a plane inclined at an angle α. The coefficient of friction is μ. Find the rate at which kinetic plus gravitational potential energy is dissipated.

9.26 A particle is subject to a force associated with the potential energy $E_p(x) = 3x^2 - x^3$. (a) Make a plot of $E_p(x)$. (b) Determine the direction of the force in each appropriate range of the variable x. (c) Discuss the possible motions of the particle for different values of its total energy. (d) Find its positions of equilibrium (stable and unstable).

9.27 The interaction between two nucleons can be represented by the **Yukawa potential**: $E_p(r) = -V_0(r_0/r)e^{-r/r_0}$, where V_0 and r_0 are two constants. (a) Find the force between the two particles as a function of their separation. (b) Find the value of the force at $r = r_0$. (c) Find the value of r where the force is 1% of its value at $r = r_0$. (d) Can you justify why r_0 is called the **range** of the force?

9.28 In scattering under a repulsive inverse square central force, it can be shown that the distance of closest approach is related to the scattering angle ϕ (see Figure 9.23) by $r_0 = (k/mv_0^2)(1 + \csc\frac{1}{2}\phi)$. Using Equation 8.17, verify that this expression is identical to Equation 9.38. (Hint: use relation B.4.)

9.29 Verify that the quadratic equation for the distance of closest approach in Example 9.10 can be written as $r_0^2 - (2k/mv_0^2)r_0 - b^2 = 0$. By solving this equation, obtain Equation 9.38.

10 Oscillatory motion

Every musical instrument in an orchestra is a source of oscillatory motion, consisting of a fundamental frequency and several harmonics. The oscillatory motion is transmitted to the surrounding air and the sound perceived by the audience is the superposition of the oscillations produced by each instrument. (Courtesy of the Guildford Philharmonic Orchestra conducted by Vernon Handley.)

Notes

10.1 Introduction

A particle has oscillatory (vibrational) motion when it moves periodically about an equilibrium position. The motion of a pendulum is oscillatory. A weight attached to a stretched spring, once it is released, starts oscillating. The atoms in a solid and in a molecule are vibrating relative to each other. The electrons in a radiating or receiving antenna are in rapid oscillation. An understanding of vibrational motion is also essential for the discussion of wave phenomena, which are related to sound and light.

Of all the oscillatory motions, the most important is called **simple harmonic motion** (SHM). Besides being the simplest motion to describe and analyze, it constitutes a rather accurate description of many oscillations found in nature. Most of our discussion in this chapter will concentrate on this kind of motion. However, not all oscillatory motions are harmonic.

10.2 Kinematics of simple harmonic motion

A particle has simple harmonic motion along an axis OX when its displacement x relative to the origin of the coordinate system is given as a function of time by the relation

$$x = A \cos(\omega t + \alpha) \tag{10.1}$$

The quantity $(\omega t + \alpha)$ is called the **phase angle** or simply the phase of the SHM, and α is the initial phase, i.e. the phase at $t = 0$. Although we have defined simple harmonic motion in terms of a cosine function, it may just as well be expressed in terms of a sine function. The only difference between the two forms is an initial phase difference of $\pi/2$. Since the cosine (or sine) function varies between a value of -1 and $+1$, the displacement of the particle varies between $x = -A$ and $x = +A$. The maximum displacement from the origin, A, is the **amplitude** of the simple harmonic motion. The cosine function repeats itself every time the angle ωt increases by 2π. Thus the displacement of the particle repeats itself after a time interval of $2\pi/\omega$. Therefore, simple harmonic motion is periodic, and its **period** is

$$P = 2\pi/\omega$$

The frequency ν of a simple harmonic motion is equal to the number of complete oscillations per unit time; thus

$$\nu = 1/P$$

and is measured in hertz. The quantity ω, called the **angular frequency** of the oscillating particle, is related to the frequency by a relation similar to Equation 5.5 for a circular motion, namely

$$\omega = 2\pi/P = 2\pi\nu \tag{10.2}$$

The velocity of the particle is

$$v = \frac{dx}{dt} = -\omega A \sin(\omega t + \alpha) \tag{10.3}$$

which varies periodically between the values $+\omega A$ and $-\omega A$. Similarly, the acceleration is given by

$$a = \frac{dv}{dt} = -\omega^2 A \cos(\omega t + \alpha) = -\omega^2 x \tag{10.4}$$

and therefore varies periodically between the values $+\omega^2 A$ and $-\omega^2 A$. This expression also indicates that:

> *in simple harmonic motion the acceleration is proportional and opposite to the displacement.*

In Figure 10.1, x, v and a as functions of time are illustrated.

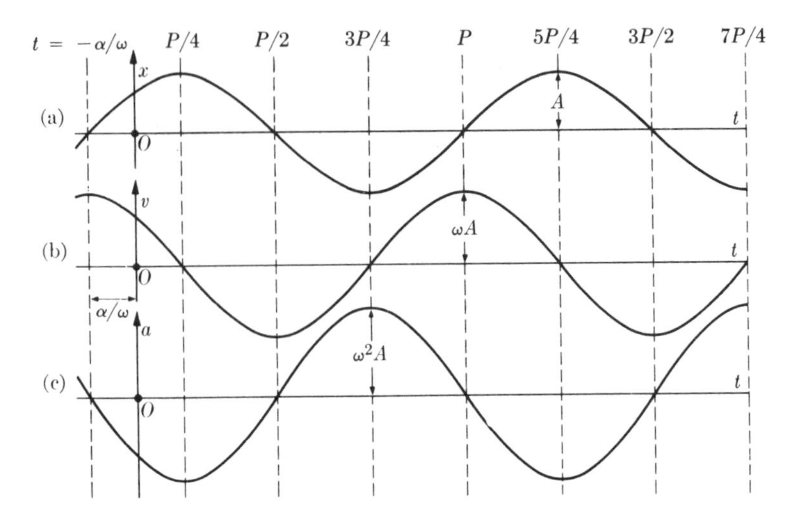

Figure 10.1 Graphs of (a) displacement, (b) velocity and (c) acceleration vs. time in SHM.

10.3 Rotating vectors or phasors

The displacement of a particle moving with SHM can also be considered as the X-component of a vector $\boldsymbol{OP'}$, with $OP' = A$. The vector rotates counter-clockwise around O with angular velocity ω and at each instant makes an angle $(\omega t + \alpha)$ with the X-axis. In Figure 10.2 we have represented the vector $\boldsymbol{OP'}$ in several positions. The student may verify that at any time the X-component of $\boldsymbol{OP'}$ is given by

$$x = OP = OP' \cos(\omega t + \alpha)$$

Equations 10.3 and 10.4 can be written as

$$v = \omega A \cos(\omega t + \alpha + \pi/2)$$

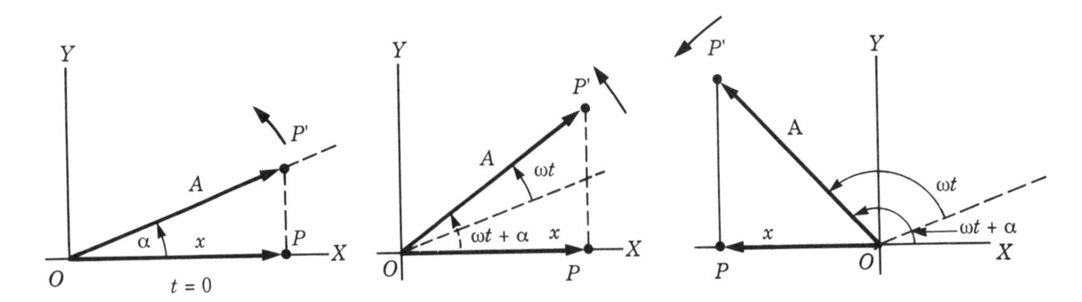

Figure 10.2 Rotating vector for displacement in SHM.

and

$$a = \omega^2 A \cos(\omega t + \alpha + \pi)$$

These forms show that v and a differ in phase from x by the amounts $\pi/2$ and π respectively. Therefore, the velocity and acceleration of the particle can also be represented by rotating vectors $\boldsymbol{OV'}$ and $\boldsymbol{OA'}$, having lengths ωA and $\omega^2 A$, respectively, with $\boldsymbol{OV'}$ an angle $\pi/2$ and $\boldsymbol{OA'}$ an angle π ahead of the rotating vector $\boldsymbol{OP'}$, as shown in Figure 10.3. The components of $\boldsymbol{OV'}$ and $\boldsymbol{OA'}$ along the X-axis give the velocity $\boldsymbol{v}$ and the acceleration $\boldsymbol{a}$ of the particle that is moving with SHM. Rotating vectors are also called **phasors**.

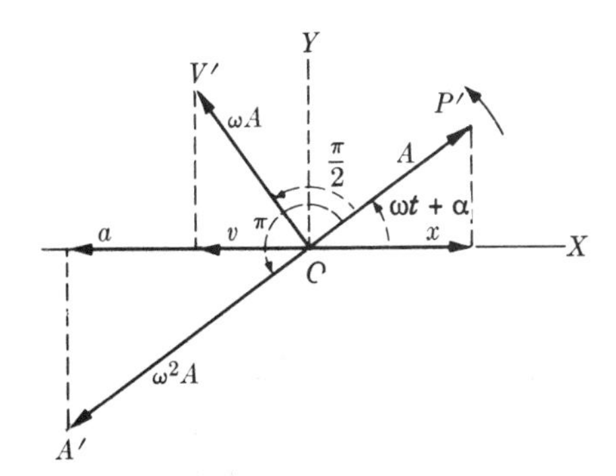

Figure 10.3 Rotating vectors for displacement, velocity and acceleration in SHM.

EXAMPLE 10.1

Relation between simple harmonic motion and uniform circular motion.

▷ Consider a particle Q (Figure 10.4), moving on a circle of radius A with constant angular velocity. The projection of Q on a diameter BC is P. It is clear from the figure that as Q moves around the circle the projection P moves back and forth (oscillates) between B and C. The angle that the radius OQ makes with the X-axis is $\theta = \omega t + \alpha$ and we have $OP = OQ \cos\theta$, or $x = A\cos(\omega t + \alpha)$. In other words, P moves with SHM. That is:

> *when a particle moves with uniform circular motion, its projection on a diameter moves with SHM.*

This kinematical relation is very useful in the design of certain mechanisms.

The velocity of Q is perpendicular to OQ and, according to Equation 5.2, has the magnitude of $v' = \omega A$. The component of v' along the X-axis is

$$v = -v' \sin\theta, \qquad \text{or} \qquad v = -\omega A \sin(\omega t + \alpha)$$

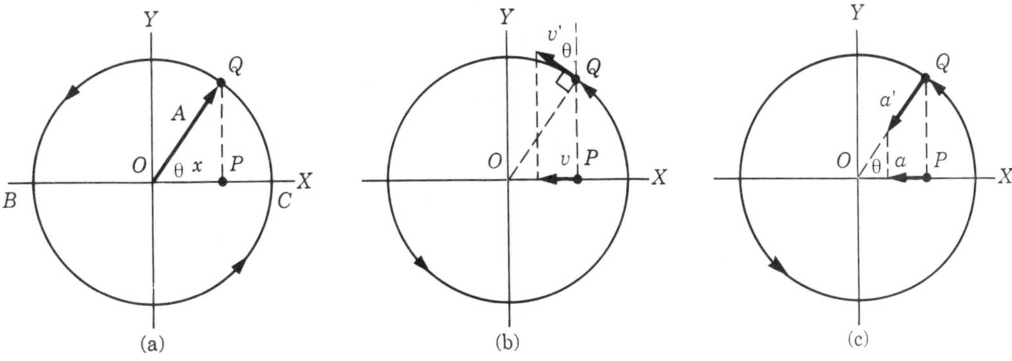

Figure 10.4 Relation between SHM and uniform circular motion. (a) Position, (b) velocity and (c) acceleration.

which is also the velocity of P. The acceleration of Q is centripetal and, according to Equation 5.13, has the magnitude $a' = \omega^2 A$. The component of a' along the X-axis is

$$a = -a' \cos\theta, \qquad \text{or} \qquad a = -\omega^2 A \cos(\omega t + \alpha)$$

which again coincides with the acceleration of P. Therefore, the velocity and the acceleration of the projection P are equal to the components along the diameter BC of the velocity and acceleration of Q.

10.4 Force and energy in simple harmonic motion

In Equation 10.4, we found that the acceleration of a body in SHM is $a = -\omega^2 x$. Applying the equation of motion $\boldsymbol{F} = m\boldsymbol{a}$, we can compute the force that must act on a particle of mass m in order for it to oscillate with simple harmonic motion. This gives

$$F = -m\omega^2 x = -kx \tag{10.5}$$

where we have set

$$k = m\omega^2 \qquad \text{or} \qquad \omega = (k/m)^{1/2} \tag{10.6}$$

Equation 10.5 indicates that

in simple harmonic motion the force is proportional and opposite to the displacement.

That is, when the displacement is to the right (positive), the force points to the left, and when the displacement is to the left (negative), the force points to the right. Thus the force is always pointing toward the origin O. This is the point of equilibrium, since at the origin $F = 0$ because $x = 0$. We may also say that the force F is central and attractive, where the center of attraction is the point O. The force given by Equation 10.5 is the type of force that appears when an elastic body, such as a spring, is deformed. The constant $k = m\omega^2$ is sometimes called the **elastic constant**. It represents the force required to displace the particle one unit of distance. Combining Equations 10.2 and 10.6, we can write

$$P = 2\pi\left(\frac{m}{k}\right)^{1/2} \qquad \nu = \frac{1}{2\pi}\left(\frac{k}{m}\right)^{1/2} \tag{10.7}$$

Equation 10.7 expresses the period and the frequency of a simple harmonic motion in terms of the mass of the particle and the elastic constant of the applied force. The kinetic energy of the particle is

$$E_k = \tfrac{1}{2}mv^2 = \tfrac{1}{2}m\omega^2 A^2 \sin^2(\omega t + \alpha) \tag{10.8}$$

Since $\sin^2\theta = 1 - \cos^2\theta$, and using $x = A\cos(\omega t + \alpha)$ for the displacement, we can also express the kinetic energy as

$$E_k = \tfrac{1}{2}m\omega^2 A^2[1 - \cos^2(\omega t + \alpha)]$$

which can be written as

$$E_k = \tfrac{1}{2}m\omega^2(A^2 - x^2)$$

or (10.9)

$$E_k = \tfrac{1}{2}k(A^2 - x^2)$$

This is the same result we obtained in Example 9.4. Note that in SHM:

the kinetic energy is maximum at the center ($x = 0$) *and zero at the extremes of oscillation* ($x = \pm A$).

To obtain the potential energy we remember Equation 9.27, $F_x = -\mathrm{d}E_p/\mathrm{d}x$. Using Equation 10.5 for the force, we can write $\mathrm{d}E_p/\mathrm{d}x = kx$. Integrating (choosing the zero of the potential energy at the origin or equilibium position), we obtain (recall Example 9.7)

$$\int_0^{E_p} \mathrm{d}E_p = \int_0^x kx\,\mathrm{d}x$$

or

$$E_p = \tfrac{1}{2}kx^2 = \tfrac{1}{2}m\omega^2 x^2 \tag{10.10}$$

which is the result obtained in Example 9.3. Thus in SHM:

the potential energy has a minimum value (zero) at the center ($x = 0$) *and increases as the particle approaches either extreme of the oscillation* ($x = \pm A$).

Adding Equations 10.9 and 10.10, we obtain the total energy of the simple harmonic oscillator as

$$E = E_k + E_p = \tfrac{1}{2}m\omega^2 A^2 = \tfrac{1}{2}kA^2 \tag{10.11}$$

which is a constant quantity. This was to be expected since the force is conservative. Therefore we may say that, during an oscillation, there is a continuous exchange of kinetic and potential energies. While moving away from the equilibrium position, the potential energy increases at the expense of the kinetic energy. When the particle moves toward the equilibrium position, the reverse happens. Note that the total energy of an oscillator is proportional to the square of the amplitude. This has several important practical consequences as will be seen later on.

Figure 10.5 shows the potential energy $E_p = \tfrac{1}{2}kx^2$ is represented by a parabola. For a given total energy E, indicated by the horizontal line, the limits

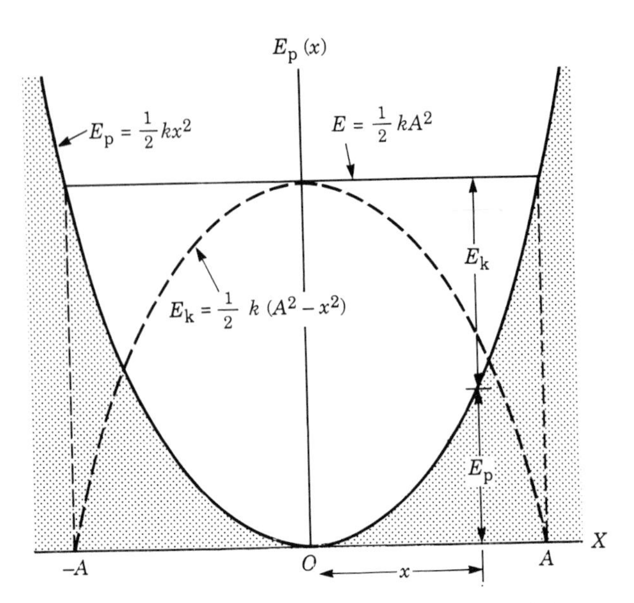

Figure 10.5 Energy relations in SHM. (a) Potential energy, E_p. (b) Kinetic energy, E_k. Total energy, $E = E_p + E_k$.

of oscillation are determined by the intersections of this line with the potential energy curve, as explained in Section 9.11. Since the parabola E_p is symmetric, the limits of oscillation are at equal distances $\pm A$ from O. At any point x the kinetic energy E_k is given by the distance between the curve $E_p(x)$ and the line E. It is represented by the dashed line, which is also a parabola.

In Figure 10.1 the displacement x and the velocity $v = \mathrm{d}x/\mathrm{d}t$ are represented as functions of time. However, in some cases it is important to represent v as a function of x using Equation 10.9. This is explained in Note 10.4.

10.5 Basic equation of simple harmonic motion

In Section 10.2 we defined simple harmonic motion by means of its kinematic properties, as expressed by Equation 10.1. At a later stage we discussed the kind of force required to produce such a motion (given by Equation 10.5). That is, given an attractive force proportional to the displacement,

$$F = -kx$$

the resulting motion is simple harmonic.

From the equation of motion, $F = ma$, with $F = -kx$, and, remembering that in rectilinear motion $a = \mathrm{d}^2x/\mathrm{d}t^2$, we may write the equation

$$m\frac{\mathrm{d}^2x}{\mathrm{d}t^2} = -kx \qquad \text{or} \qquad m\frac{\mathrm{d}^2x}{\mathrm{d}t^2} + kx = 0$$

Setting $\omega^2 = k/m$, we have

$$\frac{d^2x}{dt^2} + \omega^2 x = 0 \tag{10.12}$$

This is an equation relating acceleration and displacement whose solutions are known to be cosine or sine functions of ωt. For instance, substituting $x = A\cos(\omega t + \alpha)$ and its second time derivative into Equation 10.12, we can see that this cosine expression for x satisfies Equation 10.12, regardless of the values of A and α. Thus

$$x = A\cos(\omega t + \alpha)$$

is a general solution of Equation 10.12 because it has two arbitrary constants, the amplitude A and the initial phase α. Similarly, by the same process, we can see that $x = A\sin(\omega t + \alpha)$ and $x = A\sin\omega t + B\cos\omega t$ are also general solutions of Equation 10.12. We therefore say that Equation 10.12 is the basic equation of simple harmonic motion.

Equation 10.12 appears in many different situations in physics. Whenever it is found, it indicates that the corresponding phenomenon is oscillatory according to the law $x = A\cos(\omega t + \alpha)$ or one of the other general solutions. The phenomenon is oscillatory whether x describes a linear or angular displacement of a particle, a current in an electric circuit, the ion concentration in a plasma, the temperature of a body, or any of a multitude of other physical situations.

10.6 The simple pendulum

An example of simple harmonic motion is the motion of a pendulum. A simple pendulum is defined as a particle of mass m suspended from a point O by a string of length l and of negligible mass (Figure 10.6). When the particle is pulled aside to position B, so that the string makes an angle θ_0 with the vertical OC, and then released, the pendulum will oscillate between B and the symmetric position B'. The oscillatory motion is due to the tangential component F_T of the weight mg of the particle. Note that F_T is maximum at B and B', is zero at C and always points toward C, as can be seen by repeating the diagram for several positions of the pendulum. Although the angular motion is oscillatory with amplitude θ_0, it is not SHM unless the amplitude of the oscillations is very small. In such a case, as shown in the

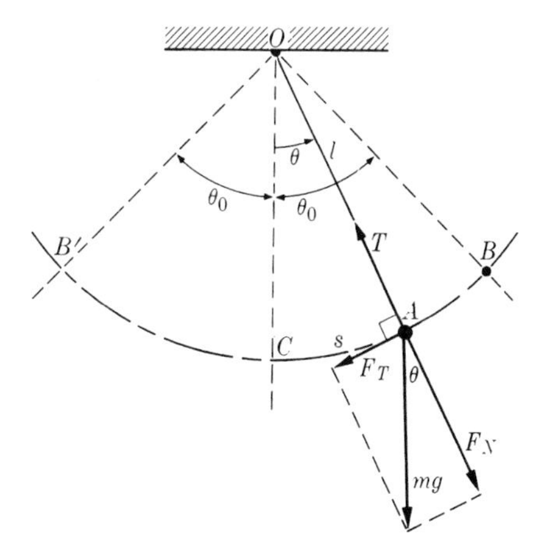

Figure 10.6 Oscillatory motion of a pendulum.

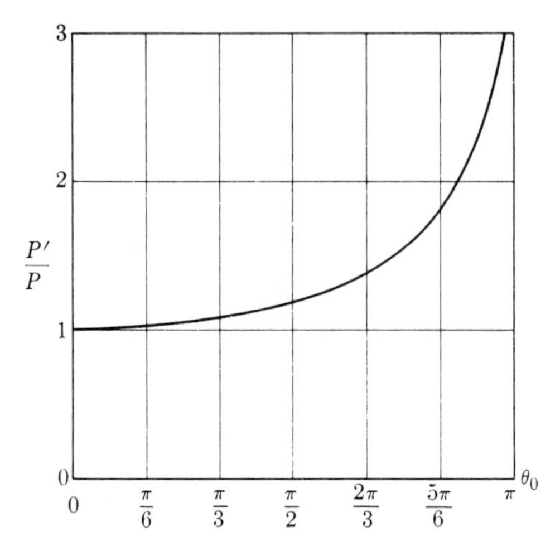

Figure 10.7 Variation of the period of a pendulum with the amplitude.

derivation below, the period of the oscillations is given by

$$P = 2\pi\sqrt{\frac{l}{g}} \tag{10.13}$$

Note that the period is independent of the mass of the pendulum and of the amplitude.

For larger amplitudes, the period of a pendulum, P', does depend on the amplitude θ_0. The variation of the period P' with the amplitude θ_0, expressed in terms of the period P corresponding to very small amplitudes, is illustrated in Figure 10.7. Note that only for very large amplitudes does the period P' differ appreciably from P (see Equation 10.16).

Derivation of the expression for the period of a simple pendulum

The particle in a simple pendulum moves in an arc of a circle of radius $l = OA$ (see Figure 10.6). The forces acting on the particle are its weight mg and the tension T along the string. The tangential component of the resultant force is, from the figure,

$$F_T = -mg \sin\theta$$

where the minus sign appears because it is opposed to the displacement $s = CA$. The equation for the tangential motion is

$$F_T = ma_T$$

Since the particle moves along a circle of radius l, we may use Equations 5.7 and 5.14, $a_T = R(\mathrm{d}^2\theta/\mathrm{d}t^2)$, with R replaced by l, to express the tangential acceleration. That is,

$$a_T = l\frac{\mathrm{d}^2\theta}{\mathrm{d}t^2}$$

The equation for the tangential motion is thus

$$ml \frac{d^2\theta}{dt^2} = -mg \sin \theta$$

or

$$\frac{d^2\theta}{dt^2} + \frac{g}{l} \sin \theta = 0 \qquad \textbf{(10.14)}$$

This equation is not of the same type as Equation 10.12 because of the presence of sin θ instead of just θ. However, if the angle θ is small, which is true if the amplitude of the oscillations is very small, we may use the series expansion of sin θ (see Appendix, Equation B.25) and keep only the first term. So we write $\sin \theta \simeq \theta$ in Equation 10.14 for the motion of the pendulum, which becomes

$$\frac{d^2\theta}{dt^2} + \frac{g}{l} \theta = 0 \qquad \textbf{(10.15)}$$

This is an equation identical to Equation 10.12, with x replaced by θ, this time referring to angular rather than linear motion. Thus we may conclude that, within our approximation, the angular motion of the pendulum is simple harmonic, with

$$\omega^2 = \frac{g}{l}$$

The angle θ can thus be expressed in the form

$$\theta = \theta_0 \cos(\omega t + \alpha)$$

Using Equation 10.2, $P = 2\pi/\omega$, we can express the period of oscillation of the pendulum as

$$P = 2\pi \sqrt{\frac{l}{g}}$$

which is Equation 10.13. For large amplitudes the approximation $\sin \theta \simeq \theta$ is not valid and the calculation of the period is more complex.

It is relatively easy and instructive to obtain a better approximation for P. Again using the series expansion of sin θ (Appendix, Equation B.25), keeping only the first two terms, we may write

$$\sin \theta = \theta - \frac{\theta^3}{3!} + \cdots \approx \theta\left(1 - \frac{\theta^2}{6}\right)$$

which when inserted in Equation 10.15 gives a second-order approximation to the equation of motion,

$$\frac{d^2\theta}{dt^2} + \frac{g}{l}\left(1 - \frac{\theta^2}{6}\right)\theta = 0$$

We may make another approximation by replacing θ^2 in the parentheses by $(\theta^2)_{av} = \frac{1}{2}\theta_0^2$ (see Problem 10.16). Then

$$\frac{d^2\theta}{dt^2} + \frac{g}{l}\left(1 - \frac{\theta_0^2}{12}\right)\theta = 0$$

which again is an equation of type 10.12. Therefore, up to this order of approximation

$$\omega^2 = \frac{g}{l}\left(1 - \frac{\theta_0^2}{12}\right)$$

Then

$$P' = \frac{2\pi}{\omega} = 2\pi\sqrt{\frac{l}{g}}\left(1 - \frac{\theta_0^2}{12}\right)^{-1/2}$$

$$\approx 2\pi\sqrt{\frac{l}{g}}\left(1 + \frac{\theta_0^2}{24}\right)$$

A more precise approximation gives

$$P' = 2\pi\sqrt{\frac{l}{g}}\left(1 + \frac{\theta_0^2}{16} + \cdots\right) \qquad \textbf{(10.16)}$$

where the amplitude θ_0 must be expressed in radians. This is a sufficient approximation for most practical situations. In fact, the corrective term $\theta_0^2/16$ amounts to less than 1% for amplitudes less than 23° or 0.40 radians.

10.7 Superposition of two SHMs of the same direction and frequency

When a particle is subject to more than one harmonic force, each trying to move the particle in its own direction with SHM, we say that there is an **interference** of SHMs. Interference effects are easy to observe on the surface of a lake when two stones are thrown into the water. Interference is also important in optics and acoustics, as will be discussed in Chapter 34. It also is manifested in radio signals.

Consider first the superposition of two simple harmonic motions that produce a displacement of the particle along the same line. Let us begin with the case when both have the same frequency. The displacement of the particle produced by each simple harmonic motion is given by

$$x_1 = A_1 \cos \omega t$$

$$x_2 = A_2 \cos(\omega t + \delta)$$

In writing these equations we have assumed that the initial phase of x_1 is zero, and that of x_2 is δ. This can always be done by a proper choice of the origin of time. The quantity δ is called the **phase difference**, and is the parameter that really matters.

The resulting displacement of the particle is given by the linear combination

$$x = x_1 + x_2 = A_1 \cos \omega t + A_2 \cos(\omega t + \delta)$$

and is periodic with a period $P = 2\pi/\omega$, since both terms have the same period.

Consider two important special cases. If $\delta = 0$ we say that the two motions are **in phase**. Then the resultant motion is

$$x = A_1 \cos \omega t + A_2 \cos \omega t = (A_1 + A_2)\cos \omega t \tag{10.17}$$

Equation 10.17 shows that the resultant motion is also SHM with the same angular frequency. The motion has an amplitude equal to the *sum* of the amplitudes of the two motions; that is,

$$A = A_1 + A_2$$

The graphs of the two motions and their resultant have been shown in Figure 10.8. The two rotating vectors OP'_1 and OP'_2 corresponding to the two motions are also represented. The rotating vector corresponding to the resultant motion is the sum of OP'_1 and OP'_2; that is,

$$OP' = OP'_1 + OP'_2$$

As a second case, when $\delta = \pi$ we have

$$x_2 = A_2 \cos(\omega t + \pi) = -A_2 \cos \omega t$$

Then the resultant motion is

$$x = A_1 \cos \omega t - A_2 \cos \omega t = (A_1 - A_2)\cos \omega t \tag{10.18}$$

which shows that the resultant motion is SHM with the same angular frequency and an amplitude equal to the *difference* of the amplitudes of the two motions; that is,

$$A = A_1 - A_2$$

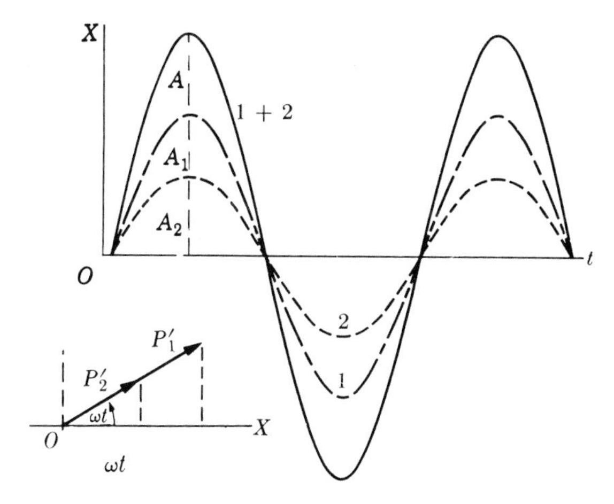

Figure 10.8 Composition of two SHMs in phase. $A = OP'_1 + OP'_2 = A_1 + A_2$.

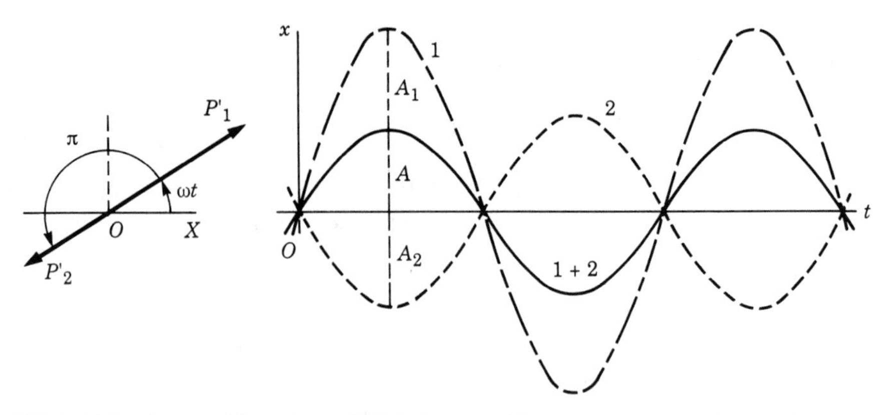

Figure 10.9 Composition of two SHMs in opposition. $A = OP'_1 - OP'_2 = A_1 - A_2$.

For that reason we say that the motions are **in opposition**. In Figure 10.9 we indicate, for $\delta = \pi$, the different motions considered and their rotating vectors.

In the general case, where the phase difference is arbitrary, the resultant motion is also SHM with the same angular frequency ω and an amplitude given by

$$A = (A_1^2 + A_2^2 + 2A_1A_2 \cos \delta)^{1/2} \quad \textbf{(10.19)}$$

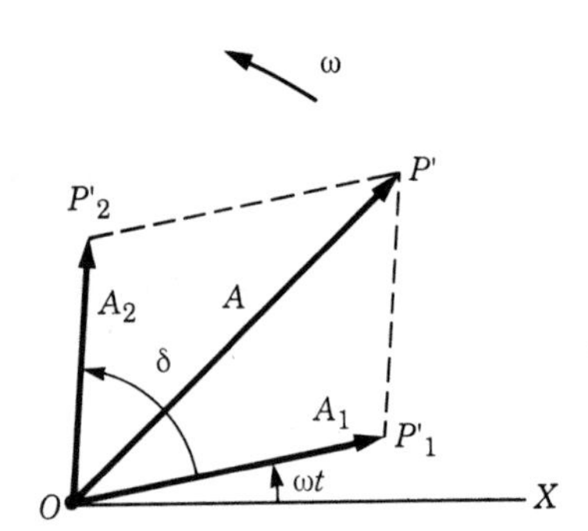

Figure 10.10 Composition of two SHMs of same frequency and a phase difference of δ.

This is easy to understand from examining Figure 10.10, where the rotating vectors $\mathbf{OP}'_1$ and $\mathbf{OP}'_2$ corresponding to x_1 and x_2 are drawn. It can be seen that $\mathbf{OP}'_1$ and $\mathbf{OP}'_2$ form a fixed angle δ. Their resultant $\mathbf{OP}'$ is the rotating vector corresponding to x. Using the rule of vector addition, the magnitude of $\mathbf{OP}'$ is given by Equation 10.19. Note that when in Equation 10.19 we make $\delta = 0$ or $\delta = \pi$, we obtain the two cases considered previously.

EXAMPLE 10.2

A particle is subjected, simultaneously, to two simple harmonic motions of the same frequency and direction. Find the resultant motion when their equations are

$$x_1 = 10 \cos 2t \qquad \text{and} \qquad x_2 = 6 \cos\left(2t + \frac{5}{12}\pi\right)$$

▷ The phase difference is $\delta = 5\pi/12$ radians or 75°. Therefore, since the amplitudes are $A_1 = 10$ and $A_2 = 6$, the resultant amplitude is

$$A = [10^2 + 6^2 + 2(10)(6)\cos(5\pi/12)]^{1/2} = 12.925$$

The equation expressing the resultant motion is then

$$x = 12.925 \cos(2t + \alpha)$$

where α is the phase difference between x and x_1. To find α, we see that at $t = 0$,

$$x_{t=0} = 12.925 \cos \alpha$$

Also

$$x_{t=0} = (x_1 + x_2)_{t=0} = 10 + 6 \cos(5\pi/12) = 11.553$$

Therefore $\cos \alpha = 11.553/12.925 = 0.8938$, or

$$\alpha = 26.64° = 0.148\pi \text{ rad}$$

Hence, $x = 12.925 \cos(2t + 0.148\pi)$, from which the resultant velocity and acceleration can be found.

10.8 Superposition of two SHMs with the same direction but different frequency

The case where two interfering simple harmonic motions in the same direction have different frequencies is also of importance. This is the kind of interference that results when two radio signals are broadcast with frequencies that are close but not the same. Consider the case of two oscillatory motions described by the equations

$$x_1 = A_1 \cos \omega_1 t \qquad \text{and} \qquad x_2 = A_2 \cos \omega_2 t$$

For simplicity the initial phases have been assumed to be zero. The angle between the rotating vectors $\boldsymbol{OP}'_2$ and $\boldsymbol{OP}'_1$ (Figure 10.11) is now $\omega_1 t - \omega_2 t = (\omega_1 - \omega_2)t$ and is not constant. Therefore the resultant vector $\boldsymbol{OP}'$ does not have constant length and does not rotate with constant angular velocity. In consequence, the resultant motion,

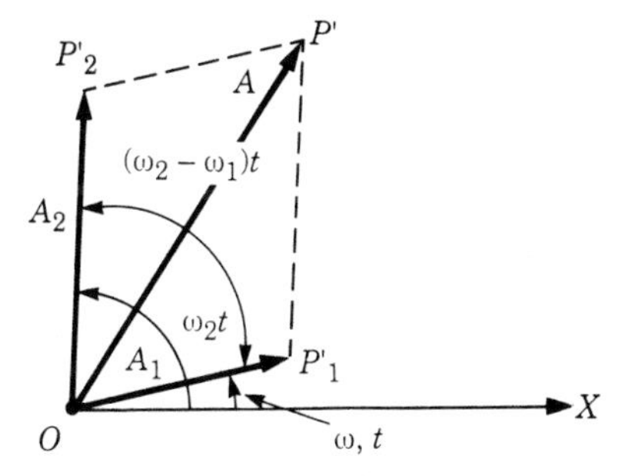

Figure 10.11 Composition of two SHMs of different frequencies.

$$x = x_1 + x_2$$

is not simple harmonic. From Figure 10.11, we see the 'amplitude' of the motion is

$$A = [A_1^2 + A_2^2 + 2A_1A_2 \cos(\omega_1 - \omega_2)t]^{1/2} \quad \textbf{(10.20)}$$

The amplitude 'oscillates' between the values

$$A = A_1 + A_2 \qquad \text{when} \qquad (\omega_1 - \omega_2)t = 2n\pi$$

and

$$A = |A_1 - A_2| \qquad \text{when} \qquad (\omega_1 - \omega_2)t = 2n\pi + \pi$$

It is then said that the amplitude is **modulated**. The frequency of the amplitude oscillation is expressed by

$$\nu = (\omega_1 - \omega_2)/2\pi = \nu_1 - \nu_2 \quad \textbf{(10.21)}$$

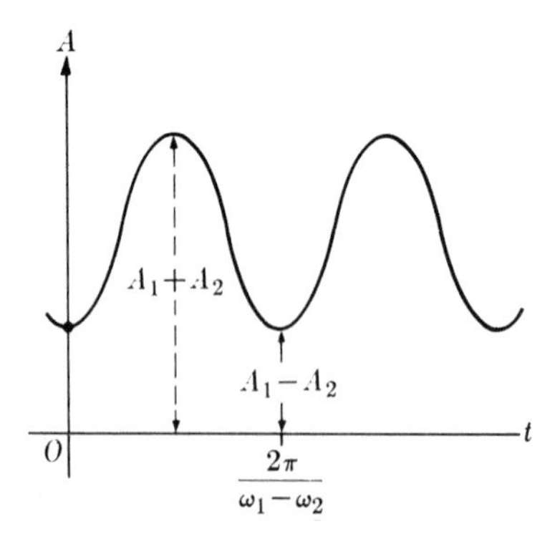

Figure 10.12 Amplitude fluctuation or beats.

and equals the difference of the frequencies of the two interfering motions. Figure 10.12 shows the variation of A with t.

The situation described arises when two sound sources (for example, two tuning forks), of close but different frequencies are vibrating simultaneously at nearby places. A listener observes a fluctuation in the intensity of the sound, called **beats**, which is due to the change in amplitude. Beats are important in tuning pianos, organs and other non-percussion instruments.

An interesting case occurs when $A_1 = A_2$; that is, when the two amplitudes are equal. Then, Equation 10.20 gives

$$A = A_1\{2[1 + \cos(\omega_1 - \omega_2)t]\}^{1/2}$$

Using $1 + \cos\theta = 2\cos^2 \frac{1}{2}\theta$ (see Appendix, Equation B.14), we obtain an amplitude

$$A = 2A_1 \cos \tfrac{1}{2}(\omega_1 - \omega_2)t \quad \textbf{(10.22)}$$

which oscillates between zero and $2A_1$. The resultant motion when $A_1 = A_2$ is given by

$$\begin{aligned} x = x_1 + x_2 &= A_1 \cos \omega_1 t + A_1 \cos \omega_2 t \\ &= A_1(\cos \omega_1 t + \cos \omega_2 t) \end{aligned}$$

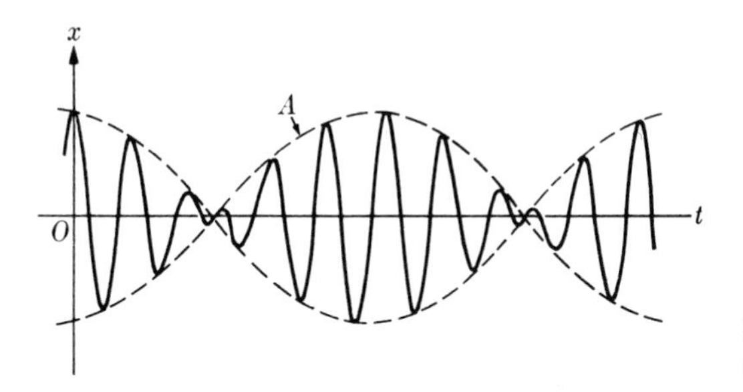

Figure 10.13 Beats when the two amplitudes are the same.

Using Equation B.8, we may combine the two cosine functions and write

$$x = [2A_1 \cos \tfrac{1}{2}(\omega_1 - \omega_2)t] \cos \tfrac{1}{2}(\omega_1 + \omega_2)t = A \cos \tfrac{1}{2}(\omega_1 + \omega_2)t$$

where A is given by Equation 10.22. The motion can then be interpreted as a harmonic motion with a frequency of $\omega = \frac{1}{2}(\omega_1 + \omega_2)$ and an amplitude modulated according to Equation 10.22. The plot of x against t is illustrated in Figure 10.13, where the dashed line shows the modulation of the amplitude.

10.9 Superposition of two SHMs in perpendicular directions

As a third example of superposition of SHM, consider the case where a particle moves in a plane such that its X- and Y-coordinates oscillate with simple harmonic motion. We first examine a case where the two motions have the same frequency. The motion along the X-axis is given by

$$x = A \cos \omega t \tag{10.23}$$

and the motion along the Y-axis is described by

$$y = B \cos(\omega + \delta) \tag{10.24}$$

where δ is the phase difference between the X- and Y-oscillations. We have assumed that the amplitudes A and B are different. The path of the particle is obviously limited to the region defined by the lines

$$x = \pm A \qquad \text{and} \qquad y = \pm B$$

We shall now consider some special cases. If the two motions are in phase, then $\delta = 0$ and

$$x = A \cos \omega t \qquad \text{and} \qquad y = B \cos \omega t$$

The two equations may be combined (by eliminating $\cos \omega t$) to yield

$$y = \frac{B}{A} x$$

This is the equation of the straight line PQ in Figure 10.14. The displacement along the line PQ is

$$r = (x^2 + y^2)^{1/2} = (A^2 + B^2)^{1/2} \cos \omega t \tag{10.25}$$

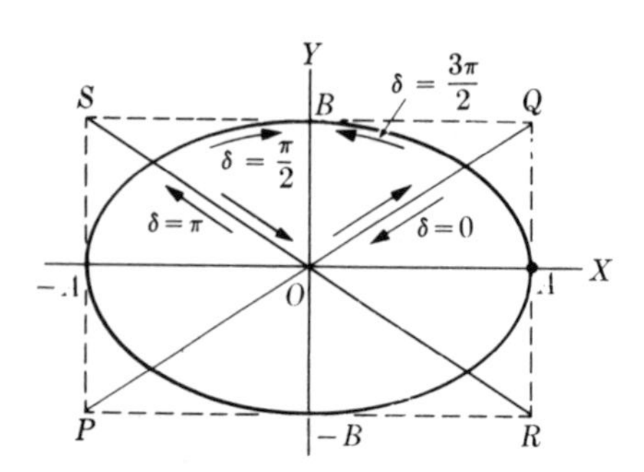

Figure 10.14 Composition of two SHMs of the same frequency but in perpendicular directions. The path depends on the phase difference.

The motion that results is simple harmonic, with amplitude $OQ = (A^2 + B^2)^{1/2}$.

If the two motions are in opposition, $\delta = \pi$ and

$$x = A \cos \omega t, \quad y = -B \cos \omega t$$

Combined, these give

$$y = -\frac{B}{A} x$$

which is the equation of the straight line RS. The motion is again simple harmonic, with amplitude $OS = (A^2 + B^2)^{1/2}$. Therefore we say that when $\delta = 0$ or π, the superposition of two perpendicular simple harmonic motions of the same frequency results in **rectilinear harmonic motion**.

When $\delta = \pi/2$, it is said that the motions along the X- and Y-axes are in **quadrature**. The equations now read

$$x = A \cos \omega t$$

$$y = B \cos(\omega t + \pi/2) = -B \sin \omega t$$

These relations, recalling that $\sin^2 \omega t + \cos^2 \omega t = 1$, may be combined by squaring and adding to give

$$\frac{x^2}{A^2} + \frac{y^2}{B^2} = 1$$

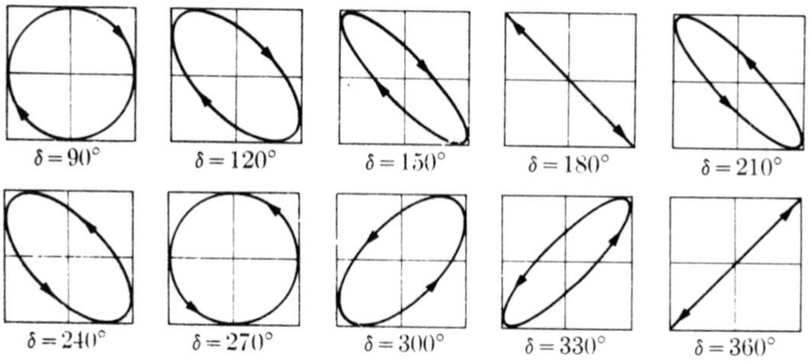

$\delta = 90°$ $\delta = 120°$ $\delta = 150°$ $\delta = 180°$ $\delta = 210°$

$\delta = 240°$ $\delta = 270°$ $\delta = 300°$ $\delta = 330°$ $\delta = 360°$

Figure 10.15 Paths for selected phase differences when $A = B$ and $\omega_1 = \omega_2$.

which is the equation of the ellipse illustrated in Figure 10.14. The ellipse is traversed in a clockwise sense. This may be verified by finding the velocity of the particle at $x = +A$. When $x = +A$, from Equation 10.23, then we must have that $\cos \omega t = 1$ (or $\omega t = 0$ or 2π rad). The X-component of the velocity is $v_x = \mathrm{d}x/\mathrm{d}t = -\omega A \sin \omega t = 0$. The Y-component of the velocity is $v_y = \mathrm{d}y/\mathrm{d}t = -\omega B \cos \omega t$ so that, in this case $v_y = -\omega B$ and the velocity is parallel to the Y-axis. Since it is negative, the point passes through A moving downward, corresponding to a clockwise sense of rotation.

The same ellipse is obtained if $\delta = 3\pi/2$ or $-\pi/2$, but then the motion is counter-clockwise. Thus we may say that when the phase difference δ is $\pm\pi/2$,

Figure 10.16 Lissajous figures. They depend on the ratio $\omega_2:\omega_1$ and on the phase difference of the motions along the X- and Y-directions.

the superposition of two simple harmonic motions of the same frequency results in **elliptical motion**. The axes of the ellipse are parallel to the directions of the two motions. When $A = B$, the ellipse transforms into a circle and we have **circular motion**. That is, circular motion can be generated by combining two oscillatory motions of the same frequency and amplitude along perpendicular directions but with a phase difference of $\pm\pi/2$.

For an arbitrary value of the phase difference δ, the path is still an ellipse, but its axes are rotated relative to the coordinate axes. The paths for selected phase differences are shown in Figure 10.15. One way of observing this motion is to use a simple pendulum that is displaced from the vertical. Instead of simply letting go, the pendulum is kicked in a direction perpendicular to the plane defined by the vertical and the pendulum.

Another interesting situation is the superposition of two perpendicular oscillatory motions of different frequencies; that is,

$$x = A_1 \cos \omega_1 t, \qquad y = A_2 \cos(\omega_2 t + \delta) \tag{10.26}$$

The resultant path depends on the ratio $\omega_2{:}\omega_1$ and on the phase difference δ. These paths are called **Lissajous figures**, and are illustrated in Figure 10.16 for several values of the ratio ω_2/ω_1 and several phase differences in each case. Of course, for $\omega_1 = \omega_2$, we obtain the paths illustrated in Figure 10.15.

10.10 Coupled oscillations

A situation frequently encountered is that of **coupled** oscillators. Two possible cases are illustrated in Figure 10.17. In (a), we have two masses m_1 and m_2 attached to springs k_1 and k_2 and coupled by spring k, so that the motions of m_1 and m_2 are not independent. In (b), we have two pendulums coupled by string AB. We shall encounter a similar case in Chapter 27 when we discuss coupled oscillating electric circuits. Coupled oscillations also occur in polyatomic molecules; we will discuss them in the next section. The net effect of the coupling of two or more oscillators can be described as an exchange of energy between them.

A particular motion of coupled oscillators is called a **normal mode** of oscillation. Normal modes correspond to the case where the two particles move with the same frequency and maintain a constant phase difference. In one normal

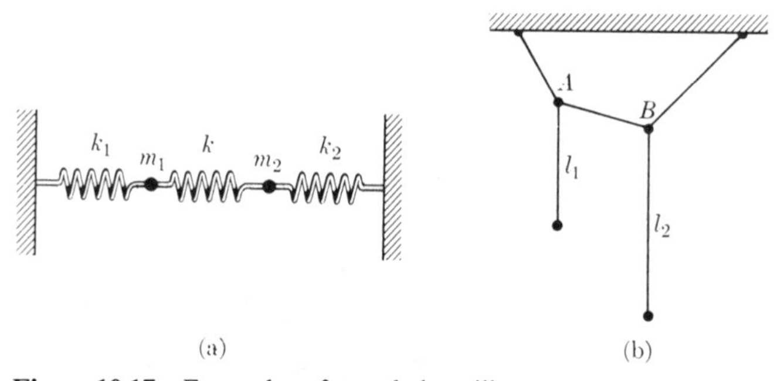

Figure 10.17 Examples of coupled oscillators.

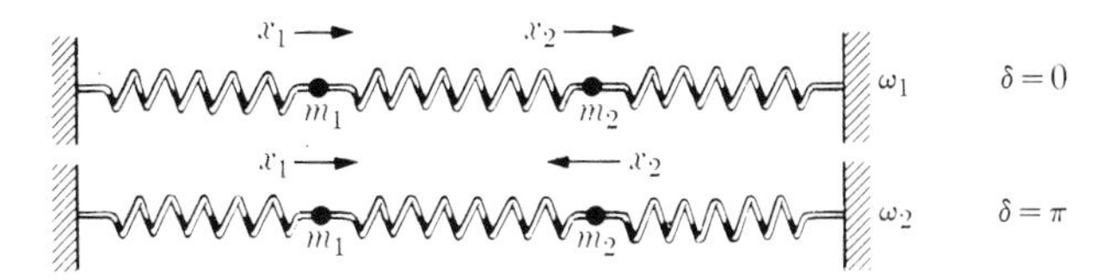

Figure 10.18 Normal modes of oscillation of two coupled oscillators.

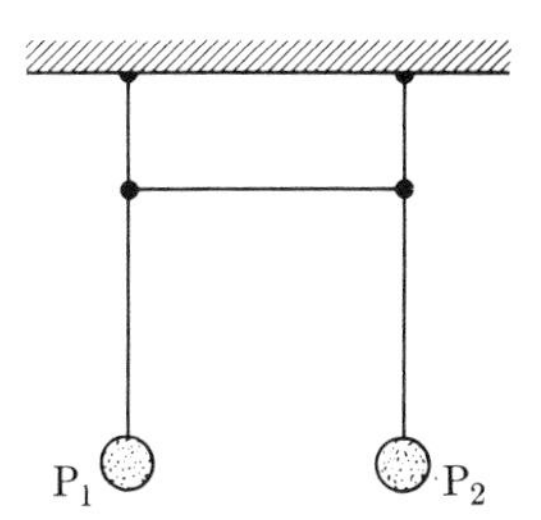

Figure 10.19 Coupled pendulums of the same length.

mode, the two oscillators move in phase ($\delta = 0$). In the second normal mode, the two oscillators move in opposition ($\delta = \pi$). These two normal modes of oscillation are represented schematically in Figure 10.18. The two particles simultaneously pass through their equilibrium positions and reach their maximum displacements simultaneously. It can be proved that the general motion of two coupled oscillators is a linear combination or superposition of normal modes. In this case the analysis of Sections 10.8 and 10.9 applies.

When the motion of coupled oscillators can be described by only one of the normal modes, their amplitudes remain constant and therefore their energies remain constant, i.e. there is no energy exchange. In the general case, however, the amplitude of each oscillator does not remain constant and it may happen that the amplitude of one oscillator increases as the amplitude of the other decreases. Shortly the situation is reversed. For example, consider the two coupled pendulums of Figure 10.19. If we displace only P_2 and release it, we note that its amplitude begins to decrease while P_1 starts oscillating with an increasing amplitude. When P_2 stops, P_1 reaches maximum amplitude. Then the trend reverses and the amplitude of P_1 decreases while that of P_2 increases.

The displacement of the coupled oscillators corresponding to the case we have just described is shown in Figure 10.20. The periodic variation of the amplitude

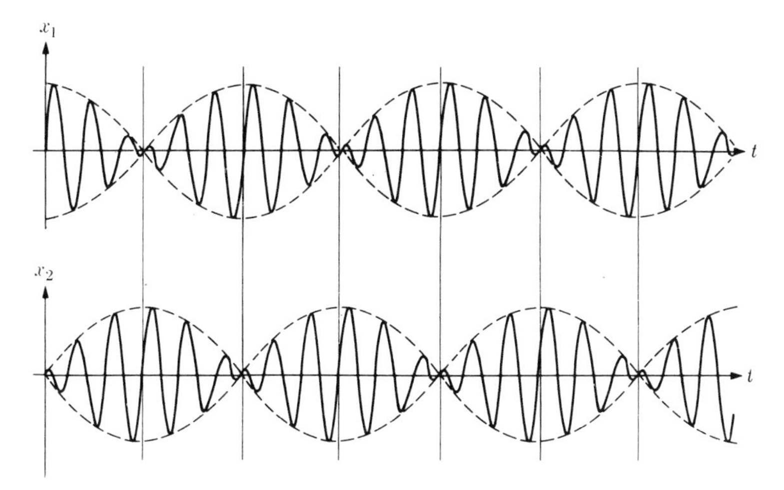

Figure 10.20 Displacement vs. time for each of two identical coupled oscillators having the same amplitude.

shows that there is an exchange of energy between the two oscillators. During the interval when the amplitude of oscillator 1 decreases and that of 2 increases, there is a transfer of energy from oscillator 1 to oscillator 2. During the next interval, there is a reversal and energy flows in the opposite direction. The process repeats itself continuously.

10.11 Molecular vibrations

An example of coupled oscillators is the vibration of atoms in a molecule. A molecule is not a rigid structure and the atoms oscillate about their equilibrium positions. For small energies, the atoms may be assumed to oscillate with SHM. In a diatomic molecule, such as CO or HCl, the atoms simply oscillate relative to each other with a certain frequency. The vibration frequency for CO is $4.08 \times 10^{14}\,s^{-1}$ and for HCl it is $5.40 \times 10^{14}\,s^{-1}$.

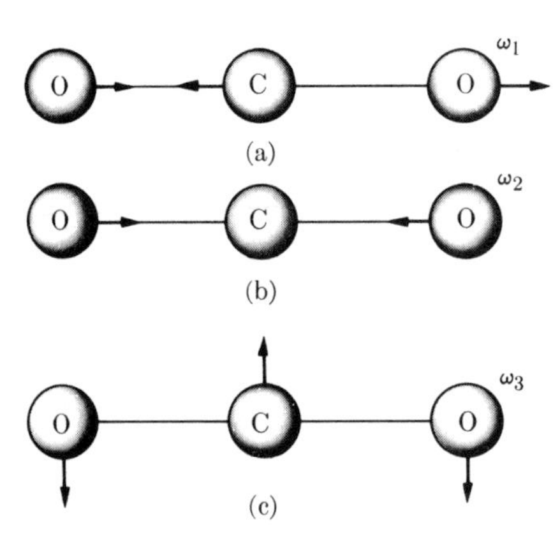

Figure 10.21 Normal vibrations of the CO_2 molecule.

For the case of a polyatomic molecule however, the oscillation of each atom affects its interaction with the others, and therefore they form a system of coupled oscillators. Consider, for example, the case of a linear triatomic molecule such as CO_2. Geometrically this molecule has the array O=C=O, as indicated in Figure 10.21, and is similar to the oscillators in Figure 10.17(a). The relative motion of the three atoms can be described in terms of normal oscillations, where the atoms vibrate while maintaining a constant phase difference of zero or π. In Figure 10.21(a), the oxygen atoms oscillate in phase, with the carbon atom moving in the opposite direction. This mode corresponds to oscillation ω_1 of Figure 10.18. In Figure 10.21(b), the two oxygen atoms move in opposite directions relative to the carbon atom, which remains fixed. This mode corresponds to oscillation ω_2 of Figure 10.18. The mode shown in Figure 10.21(c) corresponds to a motion perpendicular to the line joining the atoms with an angular frequency ω_3, resulting in a bending of the molecule. For the CO_2 molecule, the values of the three angular frequencies are

$$\omega_1 = 1.261 \times 10^{14}\,s^{-1}$$

$$\omega_2 = 2.529 \times 10^{14}\,s^{-1}$$

$$\omega_3 = 4.443 \times 10^{14}\,s^{-1}$$

If the molecule is not linear or if it has more than three atoms, the analysis of the normal oscillations becomes more complicated, but the physical considerations essentially remain the same. For example, for the water molecule, where the O atom is at the vertex of an angle of 105° and the H atoms are on each side, the

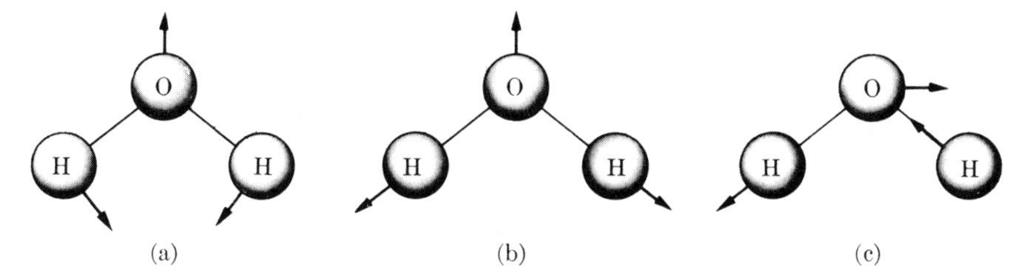

Figure 10.22 Normal vibrations of the H_2O molecule.

normal vibrations are as illustrated in Figure 10.22. Their frequencies are

$$\omega_1 = 3.017 \times 10^{14}\,\mathrm{s}^{-1}$$

$$\omega_2 = 6.908 \times 10^{14}\,\mathrm{s}^{-1}$$

$$\omega_3 = 7.104 \times 10^{14}\,\mathrm{s}^{-1}$$

These frequencies have been measured by analyzing the vibrational spectra of the molecules, which will be considered in Chapter 31.

An interesting feature of molecular vibrations is that the energy of vibration is restricted to certain values given by

$$E_n = (n + \tfrac{1}{2})h\nu \qquad \textbf{(10.27)}$$

where h is Planck's constant (introduced in Example 7.9 in connection with the angular momentum of the electron) and n is a positive integer, including zero. That is, n may take on only the values 0, 1, 2, 3, ... and no fractional values. We say that the vibrational energy of the molecule is **quantized** since it may have only *discrete* values. Since $\hbar = h/2\pi$ and $\omega = 2\pi\nu$, we can write the energy as

$$E_n = (n + \tfrac{1}{2})\hbar\omega \qquad \textbf{(10.28)}$$

This result can be explained only by using quantum mechanics (see Example 37.1).

Note that there is a minimum vibrational energy for each normal vibration, given by $\frac{1}{2}\hbar\omega$, which is called the **zero point energy**. The difference in the energy of consecutive levels is $\hbar\omega$. Since the frequencies are of the order of $10^{14}\,\mathrm{s}^{-1}$, the energy, $\hbar\omega$, is of the order of 10^{-20} J or 0.2 eV. This is the order of magnitude of the energy required to excite molecular vibrations.

EXAMPLE 10.3

Vibration of the carbon monoxide (CO) molecule.

▷ The carbon monoxide molecule vibrates with a frequency of $4.08 \times 10^{14}\,\mathrm{s}^{-1}$. Therefore, the separation between two energy levels is

$$\hbar\omega = (1.05 \times 10^{-34}\,\mathrm{J\,s})(4.08 \times 10^{14}\,\mathrm{s}^{-1})$$

$$= 4.28 \times 10^{-20}\,\mathrm{J} = 0.268\,\mathrm{eV}$$

The zero point energy is one-half of this amount or 0.134 eV, which is the minimum vibrational energy for this molecule.

If we treat the CO vibrations as an elastic oscillator, we can estimate the elastic constant of the C=O bond by using Equation 10.6, $k = m\omega^2$. However, in this case we must use the **reduced mass** of the molecule, a concept that will be introduced in Section 13.4. The reduced mass of the CO molecule is calculated in Example 13.4 and is equal to 11.38×10^{-27} kg. Therefore,

$$\begin{aligned} k &= (11.38 \times 10^{-27}\,\text{kg})(4.08 \times 10^{14}\,\text{s}^{-1})^2 \\ &= 1.89 \times 10^3\,\text{kg}\,\text{s}^{-2} \end{aligned}$$

Using Equation 10.11 for the energy of a harmonic oscillator, we may calculate the amplitude of the molecular oscillations that correspond to the zero point energy in CO. That is, we equate $E = \frac{1}{2}kA^2$ to the zero point energy $\frac{1}{2}\hbar\omega$ and find

$$A = \left(\frac{2E}{k}\right)^{1/2} = \left(\frac{2(2.14 \times 10^{-20}\,\text{J})}{1.89 \times 10^3\,\text{kg}\,\text{s}^{-2}}\right)^{1/2} = 4.75 \times 10^{-12}\,\text{m}$$

Therefore, the amplitude of the oscillation is about 4% of the CO bond length, which is 1.138×10^{-10} m. The results of this example should be taken only as an indication of the order of magnitude of values since atomic oscillations must be analyzed using quantum mechanics (Chapter 37).

10.12 Anharmonic oscillations

Simple harmonic motion is generated by a force proportional to the displacement, $F = -kx$. The potential energy is proportional to the square of the displacement, $E_p = \frac{1}{2}kx^2$. For both force and potential energy, the displacement is measured from the equilibrium position O. When the equilibrium position is at x_0 instead of the origin, as in Figure 10.23, then we must write

$$F = -k(x - x_0) \qquad \text{and} \qquad E_p = \tfrac{1}{2}k(x - x_0)^2$$

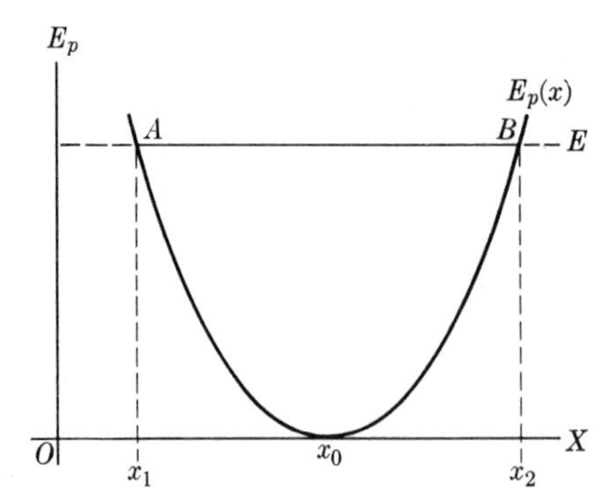

Figure 10.23 Harmonic oscillator with equilibrium position at x_0.

The graph of E_p is a parabola with its lowest point at x_0. If the total energy is E (represented by the horizontal line intersecting E_p at A and at B), the particle oscillates with SHM between positions x_1 and x_2. These points are symmetrically located with respect to x_0.

Now consider a potential energy that is *not* a parabola but has a well-defined minimum at x_0, the equilibrium position (Figure 10.24). This is the situation more often found in physical systems and results in **anharmonic oscillatory**

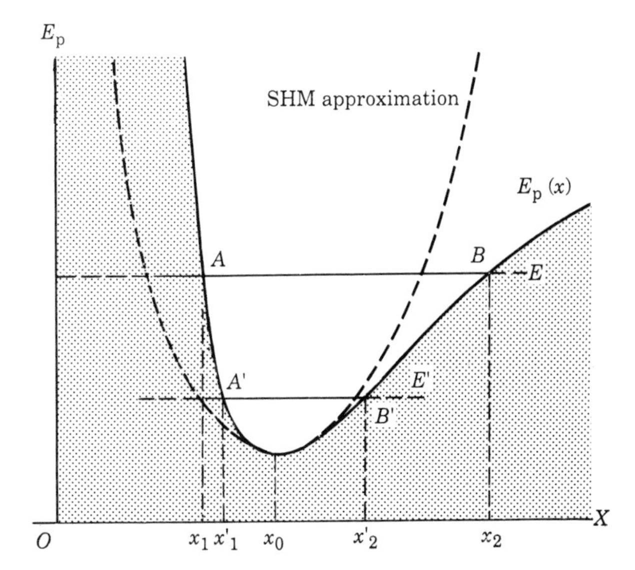

Figure 10.24 Anharmonic oscillator with equilibrium position at x_0. The limits of oscillations are shown for two energies E and E'.

motion. If the total energy is E, the particle will oscillate between positions x_1 and x_2, which, in general, are asymmetric with respect to the equilibrium position x_0. The frequency of the oscillation now depends on the energy and therefore Equation 10.6 does not apply. This is the case of a pendulum when the amplitude is large.

As mentioned in Example 9.10, the potential energy of two atoms in a diatomic molecule very closely resembles that of Figure 10.24. Therefore, the two atoms are in relative oscillatory motion, which can be considered as SHM only if the energy is very low, such as E', so that at low energies we may approximate the potential energy curve by a harmonic potential energy (dashed parabola in Figure 10.24). In other words, at low energies vibrational motions can be approximated by SHM. As the vibrational energy of the molecule increases, the motion becomes more and more markedly anharmonic and the frequency of the vibrations changes.

10.13 Damped oscillations

In simple harmonic motion the amplitude is a constant and the energy of the oscillator also remains constant. However, we know from experience that the amplitude of a vibrating body, such as a spring or a pendulum, gradually decreases, indicating that the oscillator gradually loses energy. We say that the oscillatory motion is damped.

To explain the damping dynamically, we may assume that, in addition to the elastic force $\boldsymbol{F} = -k\boldsymbol{x}$, there is a force that is opposed to the velocity. In Section 6.10, we considered a force of this kind, due to the viscosity of the medium in which the motion takes place. Then, we write the damping force as

$$\boldsymbol{F}' = -\lambda \boldsymbol{v}$$

where λ is a constant that depends on the medium and the shape of the body and $\boldsymbol{v}$ is the velocity of the body. The negative sign makes sure $\boldsymbol{F}'$ is opposed to $\boldsymbol{v}$. Other types of damping forces – proportional to higher powers of the velocity, or having other, different, physical relationships – may also be present in actual physical situations. However, a linear damping force is sufficient to analyze most damped oscillations.

The resultant force on the body is

$$\boldsymbol{F} + \boldsymbol{F}' = -k\boldsymbol{x} - \lambda \boldsymbol{v}$$

and the equation of motion is

$$m\boldsymbol{a} = -k\boldsymbol{x} - \lambda \boldsymbol{v}$$

Assuming that the motion is rectilinear, namely the motion is along the X-direction, the equation of motion may be written in scalar form as

$$ma + \lambda v + kx = 0 \tag{10.29}$$

Remembering that $v = \mathrm{d}x/\mathrm{d}t$ and $a = \mathrm{d}^2x/\mathrm{d}t^2$, we may also write

$$m\frac{\mathrm{d}^2x}{\mathrm{d}t^2} + \lambda\frac{\mathrm{d}x}{\mathrm{d}t} + kx = 0 \tag{10.30}$$

This is an equation that differs from Equation 10.12 for SHM in that it contains an additional term proportional to $\mathrm{d}x/\mathrm{d}t$. The general solution of Equation 10.30 can be obtained by the application of techniques learned in a calculus course.

One effect of damping is to decrease the frequency of the oscillations, which is given by

$$\omega = (\omega_0^2 - \gamma^2)^{1/2} \tag{10.31}$$

where $\omega_0 = (k/m)^{1/2}$ is the natural frequency of the oscillator without damping and $\gamma = \lambda/2m$. The relation between ω_0, ω and γ can be represented graphically by the triangle in Figure 10.25.

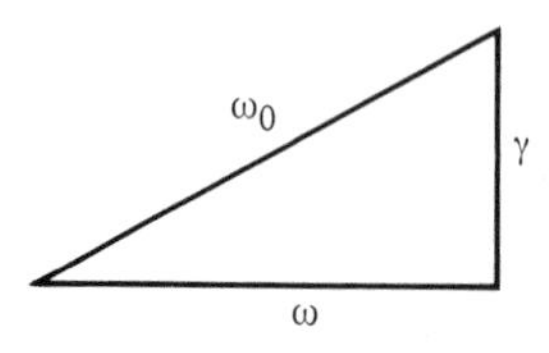

Figure 10.25 Relation between ω, ω_0 and γ.

Another effect attributable to the damping term in Equation 10.30 is that the amplitude of the oscillations is not constant but decreases exponentially as $A\mathrm{e}^{-\gamma t}$ with increasing time. Therefore the displacement of the particle is given by an expression of the form

$$x = A\mathrm{e}^{-\gamma t}\cos(\omega t + \alpha) \tag{10.32}$$

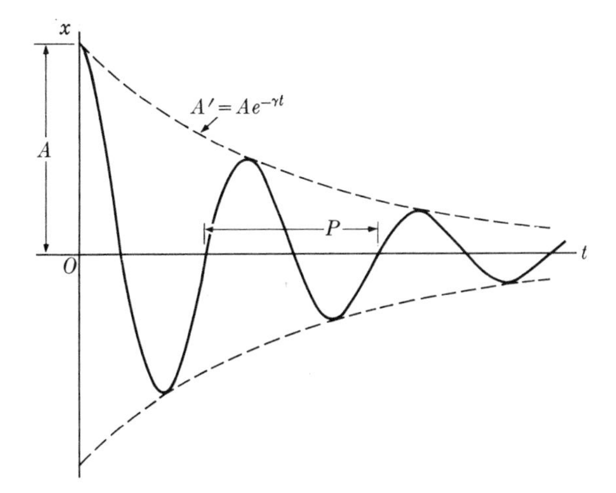

Figure 10.26 Displacement vs. time for damped oscillations.

Figure 10.26 shows how x changes with t. Although the motion is oscillatory it is not strictly periodic due to the decrease in the amplitude.

Since the amplitude of the damped oscillator decreases with time, the energy of the particle also decreases. The energy lost by the particle in damped oscillations is absorbed by the surrounding medium or radiated in some form.

EXAMPLE 10.4

A pendulum consists of an aluminum sphere suspended from a string 1 m long. It is found that in 27 minutes the amplitude decreases from 6.00° to 5.40°. Determine the coefficient γ and discuss how the viscosity of the air affects the period of the pendulum.

▷ Since the amplitude of damped oscillations decreases exponentially, we may write

$$A' = Ae^{-\gamma t}$$

To find γ we take logarithms,

$$\ln A' = \ln A - \gamma t \qquad \text{or} \qquad \gamma = \frac{1}{t}\ln\left(\frac{A}{A'}\right)$$

Introducing numerical values $t = 27\text{ min} = 1.62 \times 10^3\text{ s}$, $A = 6.00°$, and $A' = 5.40°$, we have $\gamma = 6.50 \times 10^{-5}\text{ s}^{-1}$.

In the case of a pendulum we must replace k/m by g/l in Equation 10.30 and we may then write the angular frequency of the pendulum as $\omega = (g/l - \gamma^2)^{1/2}$ so that the period is

$$P = \frac{2\pi}{\omega} = \frac{2\pi}{(g/l - \gamma^2)^{1/2}}$$

In our example $g/l = 9.80\,\mathrm{s}^{-2}$, while $\gamma^2 = 4.23 \times 10^{-9}\,\mathrm{s}^{-2}$, a value negligible compared to g/l. We may then write with sufficient accuracy that

$$P = 2\pi\left(\frac{l}{g}\right)^{1/2} = 2.01\ \mathrm{s}$$

and $\omega = (g/l)^{1/2} \approx 3.13\,\mathrm{s}^{-1}$. Accordingly, we conclude that the viscosity of the air practically does not affect the period of the pendulum considered in this example, although it does affect its amplitude, which decreases with time, as may be observed experimentally. Therefore we can describe the oscillatory motion of the pendulum by the expression

$$\theta = 6.00^\circ\ \mathrm{e}^{-(6.50 \times 10^{-5}t)} \cos(3.13t)$$

10.14 Forced oscillations

A problem of great practical importance is that of the forced motion of an oscillator; that is, the oscillations that result when an external oscillatory force is applied to a particle subject also to an elastic force. For example, when a vibrating tuning fork is placed on a resonating box, the walls of the box (and the air inside) are forced to oscillate. Also, when electromagnetic waves, absorbed by an antenna, act on the electric circuit of our radio or television set, forced electric oscillations are produced.

For simplicity, consider the rectilinear case where all motion and forces are along the X-axis. Let $F = F_0 \cos \omega_f t$ be the oscillating applied force, where its angular frequency is given by ω_f. Assuming that the particle is subject also to an elastic force $-kx$ and a damping force $-\lambda v$, the resulting force is

$$F = -kx - \lambda v + F_0 \cos \omega_f t$$

The equation of motion, $F = ma$, is then

$$ma = -kx - \lambda v + F_0 \cos \omega_f t$$

which may be rewritten as

$$ma + \lambda v + kx = F_0 \cos \omega_f t \tag{10.33}$$

Making the substitutions $v = \mathrm{d}x/\mathrm{d}t$, $a = \mathrm{d}^2x/\mathrm{d}t^2$, we also have

$$m\frac{\mathrm{d}^2x}{\mathrm{d}t^2} + \lambda\frac{\mathrm{d}x}{\mathrm{d}t} + kx = F_0 \cos \omega_f t \tag{10.34}$$

This is an equation similar to Equation 10.20, differing only in that the right-hand side is not zero. It can be solved by standard techniques. Instead of using these techniques, let us use our physical intuition for guidance. It seem logical that in this case the particle will oscillate with neither its free undamped angular frequency $\omega_0 = (k/m)^{1/2}$ nor the damped angular frequency $\omega = (\omega_0^2 - \gamma^2)^{1/2}$ but rather the particle is forced to oscillate with the angular frequency ω_f of the applied force. Thus the solution of Equation 10.34 can be written as

$$x = A \sin(\omega_f t - \alpha) \tag{10.35}$$

where A is the amplitude of the forced oscillations and a negative sign has been assigned to the initial phase α for reasons to be given later. We also have used the sine instead of the cosine function to simplify calculations. This amounts to a change of phase of $\pi/2$. Both the amplitude A and the initial phase α are no longer arbitrary, but are fixed quantities that depend on the frequency ω_f of the applied force. The larger the difference between ω_f and the natural frequency ω_0, the smaller the amplitude of the forced oscillations because it is more difficult for the oscillator to respond to the applied force when the forcing frequency is not near its natural frequency.

As shown in Note 10.1, the amplitude of the forced oscillations is

$$A = \frac{F_0/\omega_f}{[(m\omega_f - k/\omega_f)^2 + \lambda^2]^{1/2}} \tag{10.36}$$

which is a maximum for $\omega_f = (\omega_0^2 - 2\gamma^2)^{1/2}$. The variation of the amplitude with ω_f and λ is shown in Figure 10.27.

Equation 10.35 indicates that the forced oscillations are not damped, but are of constant amplitude. This means that the applied force overcomes the damping forces, and thus provides the energy necessary to maintain the oscillations. Actually, when a damped oscillator is set in forced motion, the initial motion is a combination of damped oscillations and forced oscillations. To be precise, we must add to the expression 10.35 for the forced oscillations Equation 10.32, corresponding to free

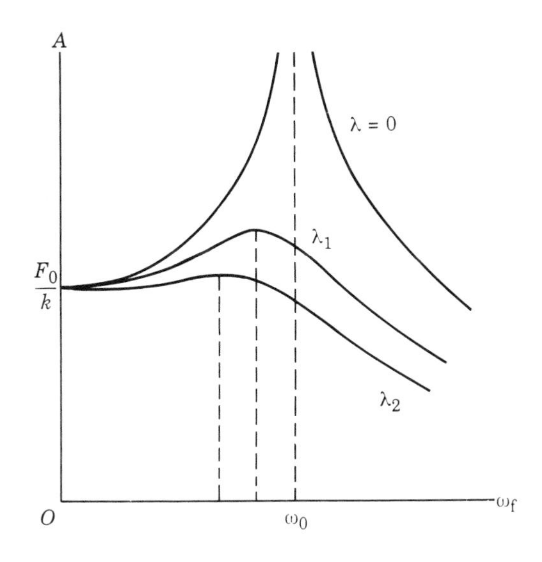

Figure 10.27 Variation of the amplitude of the forced oscillations with the damping (in the figure, λ_2 is larger than λ_1).

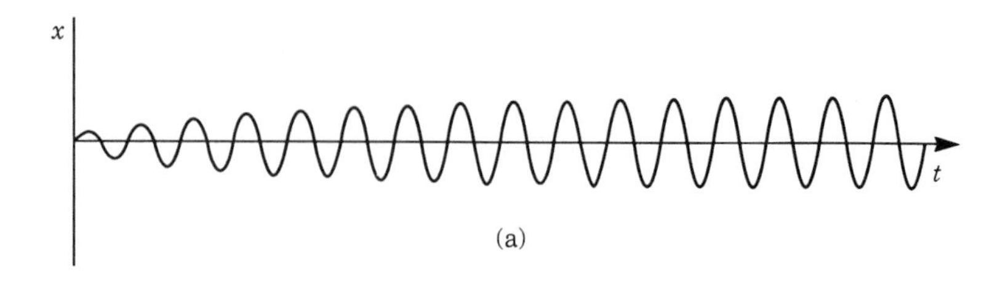

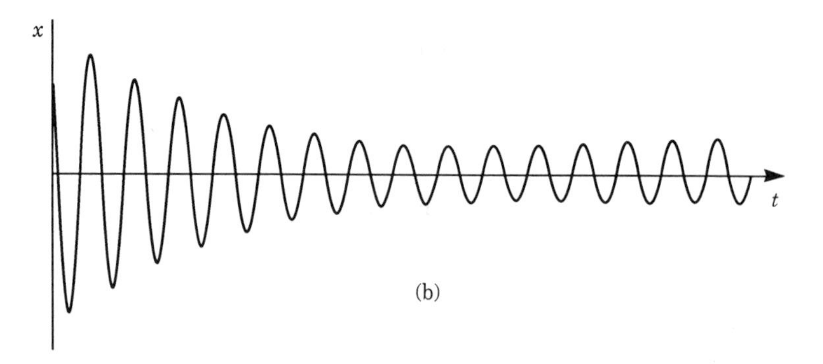

Figure 10.28 Superposition of forced oscillations and transients. Initial amplitude (a) smaller, (b) larger than the forced oscillation.

damped oscillations. However, after a certain time the amplitude of the damped oscillations becomes so small that they can be ignored and only the forced oscillations remain (see Figure 10.28). We then say the motion has reached its **steady state**.

The velocity of the forced oscillator is

$$v = \mathrm{d}x/\mathrm{d}t = \omega_\mathrm{f} A \cos(\omega_\mathrm{f} t - \alpha) \tag{10.37}$$

The **velocity amplitude** is

$$v_0 = \omega_\mathrm{f} A = \frac{F_0}{[(m\omega_\mathrm{f} - k/\omega_\mathrm{f})^2 + \lambda^2]^{1/2}}$$

and α is the phase angle by which the velocity *lags* the force (because of the negative sign). The variation of the velocity amplitude v_0 with the frequency ω_f has been illustrated in Figure 10.29. From the expression for v_0, it can be seen that when the frequency of the driving force is equal to the natural frequency, that is, when

$$\omega_\mathrm{f} = \omega_0 = \left(\frac{k}{m}\right)^{1/2} \tag{10.38}$$

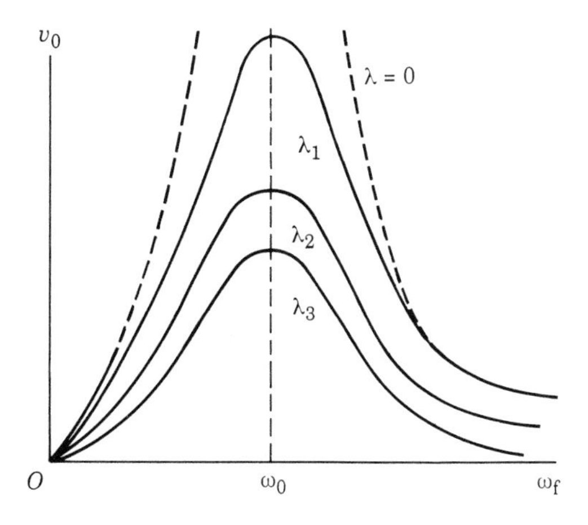

Figure 10.29 Variation of the velocity amplitude of forced oscillations with the frequency of the applied force. $\lambda_3 > \lambda_2 > \lambda_1$.

the velocity amplitude, and therefore also the kinetic energy, of the oscillator are maximum. This fact can be verified experimentally. Under this condition, it is said that there is **energy resonance**. Therefore

> *energy resonance occurs when the frequency of the applied force ω_f is equal to the natural frequency ω_0 of the oscillator without damping.*

At resonance the velocity is in phase with the applied force ($\alpha = 0$). Since the rate of work done on the oscillator by the applied force is Fv, this quantity is always positive when F and v are in phase ($\alpha = 0$). These are the most favorable conditions for transfer of energy to the oscillator. Therefore

> *at energy resonance the applied force and the velocity are in phase and the rate of energy or power transfer from the applied force to the forced oscillator is maximum.*

Resonance occurs whenever a system is subject to an external action that varies periodically with time and with the proper frequency. For example, if a gas is placed in a region where an oscillatory electric field exists (such as in an electromagnetic wave), forced oscillations will be induced in the atoms of the molecules of the gas. Since, as we explained in Section 10.11, gas molecules have well-defined natural vibration frequencies, energy absorption will be at a maximum when the frequency of the applied electric field coincides with one of the natural frequencies of the molecules. We can then obtain the **vibrational spectrum** of molecules by varying the frequency of the electric field and observing the reasonances. Similarly, we can think of the electrons in an atom as being oscillators with certain natural frequencies. The energy an atom absorbs from an oscillating electric field is a maximum when the frequency of the field coincides with one of the natural frequencies of the atom. This will be discussed in Chapter 31. Some crystals, such as sodium chloride, are composed of positively and negatively charged particles (called **ions**). If the crystal is subject to an external oscillating electric field, the positive and negative ions will oscillate relative to each other. Energy absorption by the crystal will be greatest when the frequency of the electric field matches the natural frequency of the relative oscillations of the ions. In the case of sodium chloride, this is found to be approximately $5 \times 10^{12}\,s^{-1}$.

Perhaps the most familiar example of resonance is what happens when we tune a radio or a TV receiver to a broadcasting station. All broadcasting stations are producing forced oscillations on the circuit of the receiver at all times. But, to each setting of the tuner, there corresponds a natural frequency of oscillation of the electric circuit of the receiver. When this frequency coincides with that of a broadcasting station, the energy absorption is a maximum, and hence this is the only station we hear.

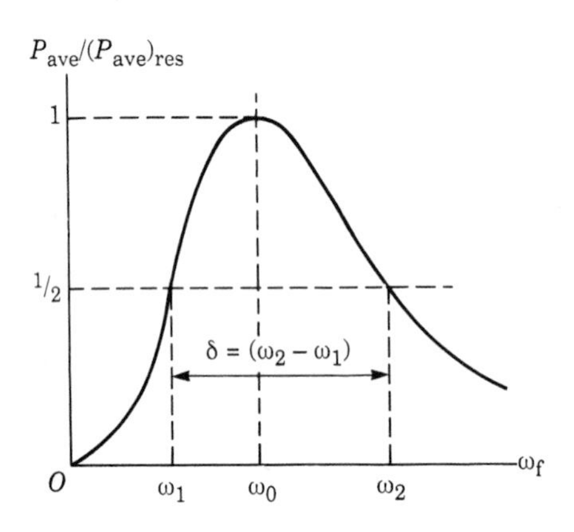

Figure 10.30 Relation between P_{ave} and $(P_{ave})_{res}$. The bandwidth is $\delta = \omega_2 - \omega_1$.

The power transmitted to an oscillator by the driving force varies with the frequency of the driving force and is a maximum at resonance. Figure 10.30 shows the variation of the average power P_{ave} with the frequency ω_f, in relation to the average power at resonance $(P_{ave})_{res}$. The faster P_{ave} falls to either side of ω_o, the sharper the resonance. It can be seen in Figure 10.30 that for each value of P_{ave} there are two possible frequencies of the driving force. The difference $\delta = \omega_2 - \omega_1$ between the frequencies, for which $P_{ave} = \frac{1}{2}(P_{ave})_{res}$ is called the **bandwidth** of the resonance. Thus, the smaller δ the sharper the resonance. For example, when two stations have broadcast frequencies very close together we sometimes hear both signals at the same time. The overlapping of signals results in an undesirable **interference**. A circuit with sharper resonance (smaller bandwidth) will eliminate such interference.

We can extend the concept of resonance to many processes where there are favorable conditions for transfer of energy from one system to another, even if we cannot describe the process in terms of forced oscillations. In this sense it is possible to talk about resonances in nuclear reactions and in processes that take place between fundamental particles. In this extended sense, the concept of energy resonance plays an important role in the description of many phenomena.

When an anharmonic oscillator is forced into motion, a new feature appears. As mentioned in Section 10.12, the frequency of oscillation for an anharmonic oscillator depends on its amplitude and therefore on its energy. Suppose the frequency of the driving force is adjusted for resonance at the natural frequency of the anharmonic oscillator. As the amplitude and energy of the forced oscillations increase under the action of the driving force, the natural frequency of the oscillator changes and resonance disappears. This results in a decrease of the amplitude, and thus also of the energy, of the oscillator, which tends to return to the frequency at which the driving force was in resonance.

Note 10.1 Impedance of an oscillator

The technique of rotating vectors or phasors is very useful in analyzing the forced vibrations of a damped oscillator.

Referring to Equation 10.33,

$$ma + \lambda v + kx = F_0 \cos \omega_f t \qquad \textbf{(10.33)}$$

we have the following quantities that can be represented by rotating vectors:

Quantity	Expression	Amplitude	Phase relative to F_0
Force	$F = F_0 \cos \omega_f t$	F_0	0
Displacement	$x = A \sin(\omega_f t - \alpha)$	A	$-\frac{1}{2}\pi - \alpha$
Velocity	$v = \omega_f A \cos(\omega_f t - \alpha)$	$\omega_f A$	$-\alpha$
Acceleration	$a = -\omega_f^2 A \sin(\omega_f t - \alpha)$	$\omega_f^2 A$	$+\frac{1}{2}\pi - \alpha$

We may represent the force by a rotating vector of length $\boldsymbol{F}_0$, making an angle $\omega_f t$ with the X-axis (Figure 10.31). Then it is clear from the above list that the velocity is represented by a rotating vector of length $\omega_f \boldsymbol{A}$ that *lags* $\boldsymbol{F}_0$ by the angle α. (If α is negative, then $\boldsymbol{v}_0$ actually leads $\boldsymbol{F}_0$). Recalling Figure 10.3, we represent the displacement by a rotating vector of length $\boldsymbol{A}$ lagging the velocity by an angle of $\frac{1}{2}\pi$. Similarly, the acceleration is represented by a rotating vector of length $\omega_f^2 \boldsymbol{A}$ that *leads* the velocity by an angle of $\frac{1}{2}\pi$. These vectors are represented in Figure 10.31.

Equation 10.33 implies that the sum of the three rotating vectors corresponding to the quantities on the left-hand side of the equation, that is ma, λv and kx, must be equal to the rotating vector corresponding to $\boldsymbol{F}$. All these rotating vectors are represented in Figure 10.32. They are the same as the vectors shown in Figure 10.31, except that they have been multiplied by the appropriate constants according to Equation 10.33.

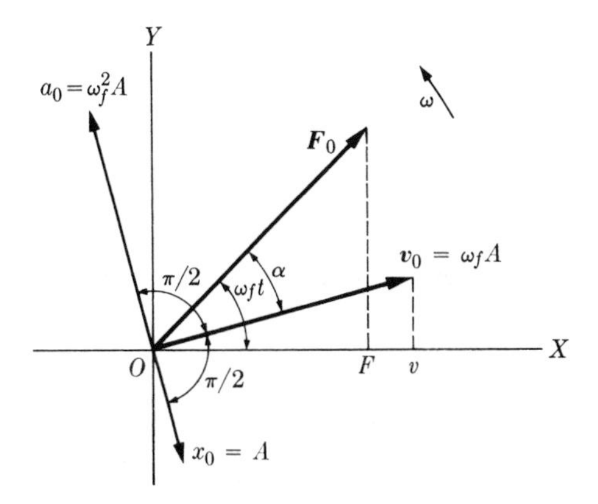

Figure 10.31 Relation between the force and displacement, velocity and acceleration rotating vectors in forced oscillators.

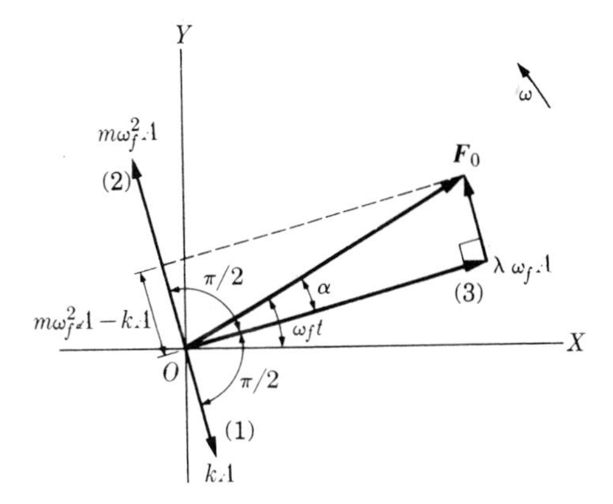

Figure 10.32 Composition of the rotating vectors appearing in Equation 10.33.

Vectors (1) and (2) are in opposite directions and perpendicular to vector (3). The resultant of vectors (1) and (2) has a magnitude

$$m\omega_f^2 A - kA.$$

This result, combined with vector (3) of length $\lambda\omega_f A$, gives

$$F_0 = [(m\omega_f^2 A - kA)^2 + (\lambda\omega_f A)^2]^{1/2}$$

We can write this equation in the alternative form

$$F_0 = \omega_f A[(m\omega_f - k/\omega_f)^2 + \lambda^2]^{1/2}$$

or

$$A = \frac{F_0/\omega_f}{[(m\omega_f - k/\omega_f)^2 + \lambda^2]^{1/2}}$$

which is the expression quoted in Equation 10.36. The velocity amplitude is then

$$v_0 = \omega_f A = \frac{F_0}{[(m\omega_f - k/\omega_f)^2 + \lambda^2]^{1/2}} \quad \textbf{(10.39)}$$

The quantity appearing in the denominator of Equation 10.39 is called the **impedance** of the oscillator, and is designated by Z. Then

$$Z = [(m\omega_f - k/\omega_f)^2 + \lambda^2]^{1/2} \quad \textbf{(10.40)}$$

This allows us to write

$$v_0 = \frac{F_0}{Z} \quad \text{or} \quad Z = \frac{F_0}{v_0}$$

Therefore, the impedance of an oscillator is the ratio of the amplitude of the driving force to that of the velocity amplitude. Similarly, the **reactance** X and the **resistance** R of the oscillator are defined by

$$X = m\omega_f - k/\omega_f \quad \text{and} \quad R \equiv \lambda \quad \textbf{(10.41)}$$

Therefore

$$Z = (X^2 + R^2)^{1/2} \quad \textbf{(10.42)}$$

The relationship between Z, X and R is indicated in Figure 10.33, which makes it easy to remember the above formulas.

From Figure 10.33 we see that

$$\tan\alpha = \frac{X}{R} = \frac{m\omega_f - k/\omega_f}{\lambda} \quad \textbf{(10.43)}$$

In the figure we note that v_0 is a maximum, and accordingly the kinetic energy of the oscillator is maximum, when Z is minimum. Since λ is a constant, the minimum of Z occurs when

$$X = m\omega_f - \frac{k}{\omega_f} = 0 \quad \text{or} \quad \omega_f = \left(\frac{k}{m}\right)^{1/2} = \omega_0$$

This is the condition given in Equation 10.38 for energy resonance. Also, when this condition is fulfilled

$$\tan \alpha = 0 \quad \text{or} \quad \alpha = 0$$

Therefore, the force and the velocity are in phase at energy resonance, as previously explained, while the force and the displacement are in quadrature. The plot of Equation 10.39 was shown in Figure 10.29.

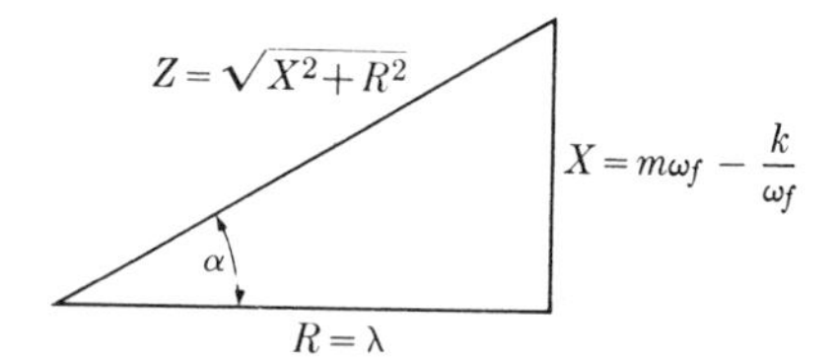

Figure 10.33 Relation between impedance, resistance and reactance in forced oscillations.

The power transferred to the oscillator by the force is

$$P = Fv = \frac{F_0^2}{Z} \cos \omega_f t \cos(\omega_f t - \alpha)$$

Expanding the second cosine and multiplying out, we have

$$P = \frac{F_0^2}{Z} (\cos^2 \omega_f t \cos \alpha - \cos \omega_f t \sin \omega_f t \sin \alpha)$$

We are more interested in the average power, P_{ave}, since this is what counts when we are computing the energy absorbed by the oscillator in a certain time. Now

$$(\cos^2 \omega_f t)_{ave} = \tfrac{1}{2} \quad \text{and} \quad (\cos \omega_f t \sin \omega_f t)_{ave} = 0$$

resulting in

$$P_{ave} = \frac{1}{2}\frac{F_0^2}{Z} \cos \alpha = \tfrac{1}{2} F_0 v_0 \cos \alpha = \frac{F_0^2 R}{2Z^2} = \tfrac{1}{2} R v_0^2 \tag{10.44}$$

This verifies that the maximum transfer of energy occurs when v_0 is a maximum, since R is fixed. At energy resonance, $\alpha = 0$ and $Z = R$, resulting in

$$(P_{ave})_{res} = \frac{F_0^2}{2R}$$

Using Equation 10.39 in Equation 10.44, it can be seen that

$$\frac{P_{ave}}{(P_{ave})_{res}} = \frac{\lambda^2}{[m\omega_f - (k/\omega_f)]^2 + \lambda^2}$$

The ratio of P_{ave} to $(P_{ave})_{res}$ is shown in Figure 10.34 for several values of km/λ. Recalling Figure 10.30, for each value of P_{ave} there are two possible frequencies of the driving force. As indicated before, the difference $\delta = \omega_2 - \omega_1$ between the frequencies where $P_{ave} = \frac{1}{2}(P_{ave})_{res}$ is the bandwidth of the oscillator. It is a measure of the **sharpness** of the resonance, since the smaller δ, the narrower the resonance peak.

To obtain the bandwidth we equate the expression for $P_{ave}/(P_{ave})_{res}$ to $\frac{1}{2}$, resulting in

$$\left(m\omega_f - \frac{k}{\omega_f}\right)^2 = \lambda^2 \quad \text{or} \quad \frac{\omega_f}{\omega_0} - \frac{\omega_0}{\omega_f} = \pm \frac{\lambda}{m\omega_0}$$

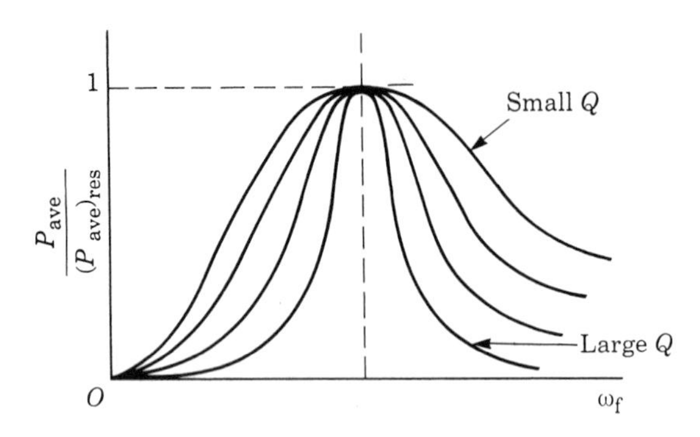

Figure 10.34 Average power for different values of Q.

where the relation $\omega_0 = (k/m)^{1/2}$ has been used. Therefore, the two frequencies ω_1 and ω_2, corresponding to each sign, are given by

$$\frac{\omega_1}{\omega_0} - \frac{\omega_0}{\omega_1} = -\frac{\lambda}{m\omega_0} \quad \text{and} \quad \frac{\omega_2}{\omega_0} - \frac{\omega_0}{\omega_2} = +\frac{\lambda}{m\omega_0}$$

By reversing the order of terms in the first relation and combining with the second, we can see that

$$\frac{\omega_1}{\omega_0} = \frac{\omega_0}{\omega_2}$$

Therefore the bandwidth satisfies the relation

$$\frac{\delta}{\omega_0} = \frac{\omega_2 - \omega_1}{\omega_0} = \frac{\omega_2}{\omega_0} - \frac{\omega_1}{\omega_0} = \frac{\omega_2}{\omega_0} - \frac{\omega_0}{\omega_2} = \frac{\lambda}{m\omega_0} = \frac{1}{Q} \qquad \textbf{(10.45)}$$

where

$$Q = \frac{m\omega_0}{\lambda} = \frac{(km)^{1/2}}{\lambda} \qquad \textbf{(10.46)}$$

This quantity is called the **figure of merit** of the oscillator. The larger Q ($km \gg \lambda^2$), the smaller the bandwidth and the sharper the resonance or the more sensitive the oscillator is to the frequency of the driving force (Figure 10.34). Therefore, increasing the elastic constant or the mass of the oscillator reduces the bandwidth, while increasing the damping broadens the bandwidth, a result in agreement with physical intuition, since damping makes the response of the oscillator more sluggish. This has important practical consequences, particularly in oscillatory electric circuits (Chapter 27).

Note 10.2 Fourier analysis of periodic motion

Simple harmonic motion is just one specific case of periodic or oscillatory motion. A general periodic motion of period P must be described by a function $x = f(t)$ that repeats itself at intervals equal to P. That is, the function must have the property that

$$f(t) = f(t + P) \qquad \textbf{(10.47)}$$

as shown in Figure 10.35.

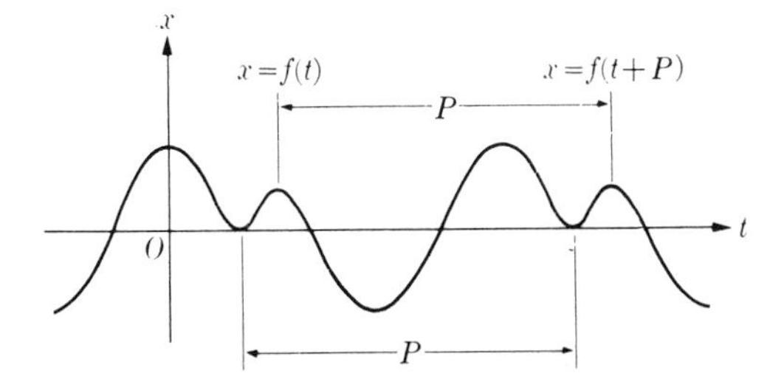

Figure 10.35 A periodic function of time.

Consider a motion whose displacement is described by

$$x = A \sin \omega t + B \sin 2\omega t \tag{10.48}$$

This represents the superposition of two simple harmonic motions of angular frequencies ω and 2ω or periods P and $\frac{1}{2}P$. Then x is also periodic, and its period will be P. This is seen in the graph of Figure 10.36 where curve (a) corresponds to $\sin \omega t$ and curve (b) to $\sin 2\omega t$. Although x is periodic, it is not simple harmonic. We can add terms of the form

$$\sin 3\omega t, \sin 4\omega t, \ldots, \sin n\omega t, \ldots$$

that have angular frequencies $3\omega, 4\omega, \ldots, n\omega, \ldots$ and periods $P/3, P/4, \ldots, P/n, \ldots$ (or cosine functions of the same frequencies), and still get a displacement x that is periodic with period P. Its exact form depends on the number of sine and cosine functions we add, and their relative amplitudes.

Thus we see that, by adding simple harmonic motions whose frequencies are multiples of a fundamental frequency and whose amplitudes are properly selected, we may obtain almost any periodic function. The reverse is also true. **Fourier's theorem** asserts that a periodic function $f(t)$ of period $P = 2\pi/\omega$ can be expressed as a superposition of simple harmonic terms written as follows:

$$\begin{aligned} f(t) = a_0 + a_1 \cos \omega t + a_2 \cos 2\omega t + \cdots + a_n \cos n\omega t + \cdots \\ + b_1 \sin \omega t + b_2 \sin 2\omega t + \cdots + b_n \sin n\omega t + \cdots \end{aligned} \tag{10.49}$$

This is known as a **Fourier series**. The frequency ω is called the **fundamental** and the frequencies $2\omega, 3\omega, \ldots, n\omega, \ldots$ are the **harmonics** or overtones.

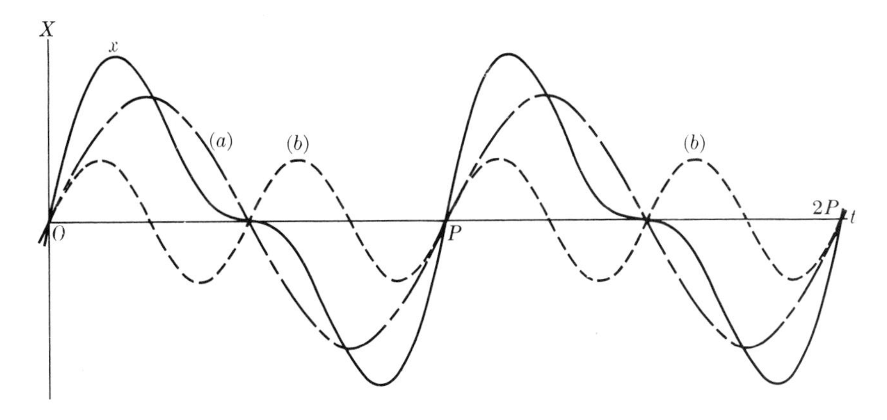

Figure 10.36 Superposition of two SHMs of frequencies ω and 2ω.

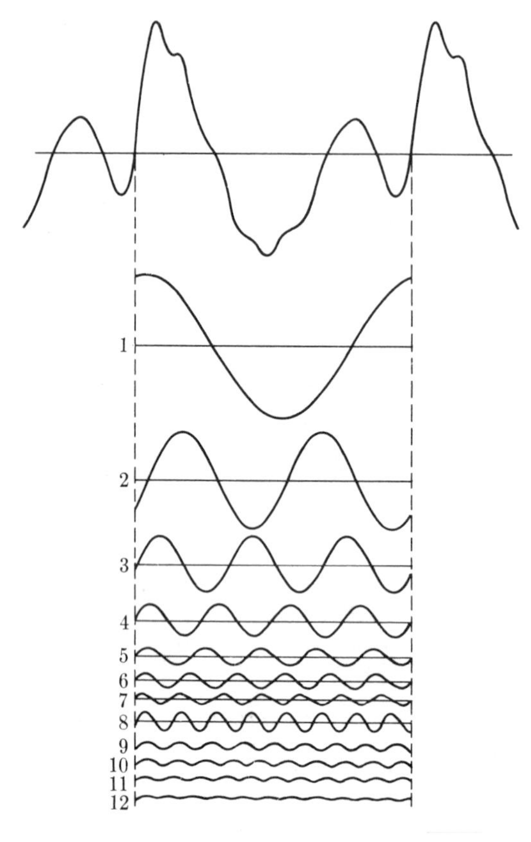

Figure 10.37 Fourier analysis of the periodic function shown at top.

By an application of Fourier's theorem, any periodic motion can be considered as the superposition of simple harmonic motions. In Figure 10.37 the periodic motion shown at the top is analyzed into its Fourier components. The first 12 harmonics are shown. Fourier's theorem, for example, helps to explain the different **quality** of sound produced by different musical instruments. The same note or musical tone produced by a piano, a guitar, and an oboe sounds different to our ears, in spite of the fact that the tones have the same fundamental frequency. The difference is due to the presence of the harmonics or overtones with different relative amplitudes. In other words, the Fourier analysis of the sound is different for each instrument even if all produce the same fundamental note.

Note 10.3 Representation of oscillatory motion in phase space

In Figure 10.1, the displacement, x, and the velocity, $v = \mathrm{d}x/\mathrm{d}t$, for SHM were represented as functions of time. There is another representation for oscillatory motion that is very useful for a number of situations. It consists in plotting x along the abscissa of a two-dimensional coordinate system and the *simultaneous* value of $v = \mathrm{d}x/\mathrm{d}t$ along the ordinate. This representation is called **phase space**.

To obtain the phase space representation of oscillatory motion, we note, from Equation 10.9, that the kinetic energy of a linear oscillator in SHM, $E_k = \frac{1}{2}mv^2$, and its potential energy, $E_p = \frac{1}{2}kx^2$, are related by

$$E_p + E_k = \tfrac{1}{2}kx^2 + \tfrac{1}{2}mv^2 = E$$

where $E = \frac{1}{2}kA^2$ is the total energy and A is the amplitude of the oscillation. This expression can be written in the alternative form

$$\left(\frac{k}{2E}\right)x^2 + \left(\frac{m}{2E}\right)v^2 = 1$$

Recalling the equation of an ellipse, $x^2/a^2 + y^2/b^2 = 1$, the preceding expression can be represented in phase space by an ellipse, whose semi-axes are $a = (2E/k)^{1/2}$ (or $a = A$) and $b = (2E/m)^{1/2}$ (or $b = \omega_0 A$, where $\omega_0 = (k/m)^{1/2}$) as shown in Figure 10.38. The size of the ellipse, for given m and k, depends on the energy E (or amplitude A) of the SHM. The larger the energy (or amplitude), the larger the ellipse (Figure 10.39). The diagram in phase space is a closed curve because, in SHM, the values of x and dx/dt keep repeating each period. We emphasize that the points of the ellipse correspond to simultaneous values of x and v and not to the actual trajectory of the oscillator in physical space.

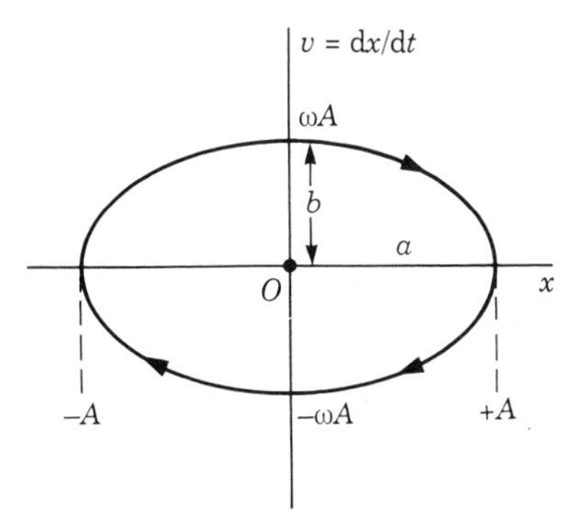

Figure 10.38 Phase space representation of SHM. $a = (2E/k)^{1/2} = A$, $b = (2E/m)^{1/2} = \omega A$, $\omega = (k/m)^{1/2} = b/a$.

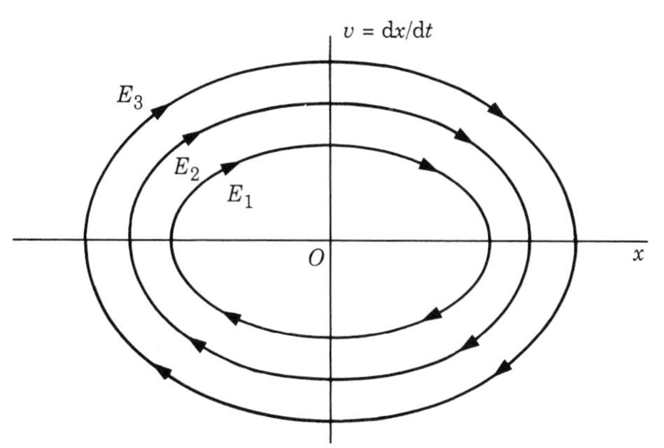

Figure 10.39 Phase space representation of SHM for different energies. The arrows show the sense in which the representative point describes the ellipse during an oscillation of the particle.

Note from these diagrams that $x = 0$ corresponds to $v = +\omega_0 A$ or $-\omega_0 A$ and $v = 0$ corresponds to $x = +A$ or $-A$, as we also saw in Figure 10.1. The *motion* of the oscillator is described in phase space by a representative point that traces out the ellipse in a clockwise sense. Why?

Of particular interest is the case of a pendulum. As discussed in Example 9.11, for small amplitudes (or energy) the equation of motion can be approximated by Equation 10.15 and the pendulum has SHM (Figure 10.40). Therefore, when we take the angular displacement θ and the angular velocity $d\theta/dt$ as the variables in phase space, the diagram is still an ellipse (the central curve in Figure 10.41).

However, as the amplitude (energy) increases, the difference between $\sin\theta$ and θ becomes appreciable and we must use Equation 10.14 to describe the motion. The motion is no longer SHM but is still periodic so that the phase space diagrams are closed curves resembling ellipses, marked (a) in Figure 10.41.

In Example 9.11 it was shown that when the pendulum has enough energy so that it rises all the way to the top of the circle and still has some kinetic energy, it crosses over and continues down on the other side. In this case the motion of the pendulum is a periodically modulated rotation instead of an oscillation. The maximum value of the angular velocity, $d\theta/dt$, occurs when the pendulum is at the bottom ($\theta = 0$) and minimum when it is at the top ($\theta = \pm\pi$). Note also that at the top, point D, θ changes from π to $-\pi$, or conversely, depending on the sense of rotation. Thus,

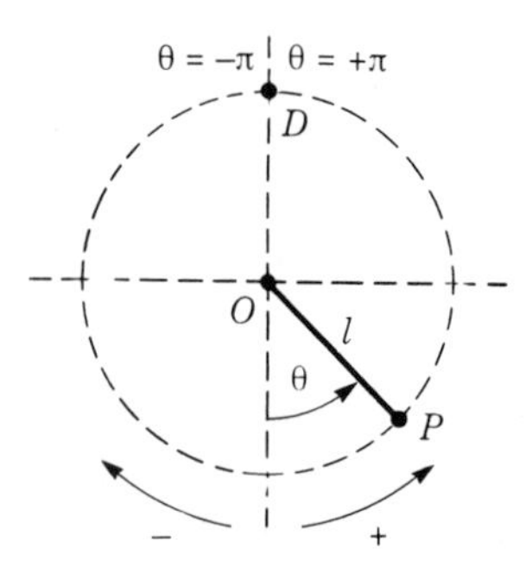

Figure 10.40 Pendulum motion.

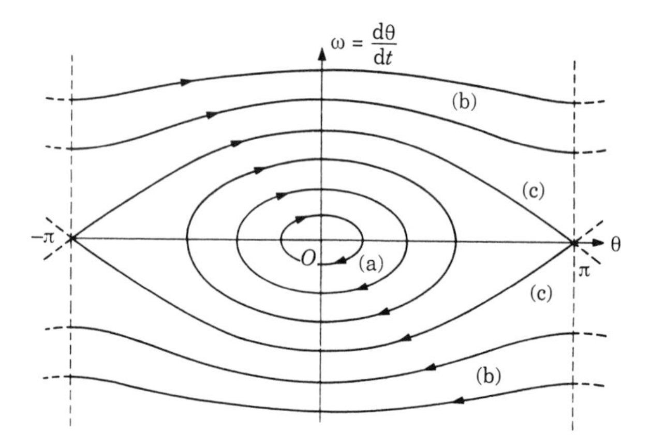

Figure 10.41 Phase space diagram of pendular motion. (a) Oscillatory motion $E < E_0$, (b) modulated rotation $E > E_0$, (c) $E = E_0$.

we get the lines marked (b) in Figure 10.41. The lower curves correspond to clockwise rotation of the pendulum and the upper ones to counter-clockwise. The representative point repeatedly describes only one of the lines, depending on the initial conditions.

A critical case occurs when the kinetic energy of the pendulum at the bottom is equal to the potential energy difference between top and bottom. That is, when the energy of the pendulum is $E_0 = 2mgl$ (recall Example 9.11). In this case the angular velocity of the pendulum at the top ($\theta = \pm\pi$) is zero ($d\theta/dt = 0$) because all the kinetic energy at the bottom has been transformed into potential energy at the top. (This is similar to the motion of a body thrown vertically upward; the maximum height reached is determined by the initial kinetic energy.) The two lines in phase space corresponding to $E = E_0$ are designated by (c) in Figure 10.41. They are called the **separatrix** because they are the boundary between the diagrams for oscillatory motion with $E < E_0$ and for modulated rotational motion with $E > E_0$. The pendulum can describe either branch of the separatrix, depending on the sense of motion, determined by the initial conditions.

When $E = E_0$, once the pendulum reaches the top it should remain there. However, since the top is a position of unstable equilibrium, because it is a maximum of potential energy, the pendulum can easily move one way or the other under the effect of a perturbation, no matter how small. This is why the two branches of the separatrix meet at $\theta = \pm\pi$, which are called **singular points**.

It is a well known experimental fact that it is extremely difficult to set the energy of the pendulum exactly to the value $E = E_0$ so that the pendulum stays at rest at the top. That is because the initial conditions of the motion of the pendulum must be *exactly* those that give $E = E_0$. But this is practically impossible. In the first place, we cannot know *exactly* the values of the three factors that enter into $E_0 = 2mgl$. In the second place, even if we knew them, we cannot experimentally reproduce them exactly by giving an initial push to the pendulum.

For small deviations from the energy E_0, the pendulum approaches the top very slowly and it can either just make it across or come back, depending on whether the energy is slightly larger or smaller than E_0, which experimentally requires almost identical initial conditions. In other words, although the motion of the pendulum is *determined* by Equation 10.14, exactly what the motion will be is hard to predict when $E \approx E_0$ because of the sensitivity to a change in the initial conditions.

A very different phase space diagram results in the case of a damped oscillator, discussed in Section 10.13. When we add a damping term to the 'elastic' force acting on an oscillator, the equation of motion becomes Equation 10.29. In this case the motion is no longer periodic and the displacement and the velocity decrease continuously in time. The phase space diagram is no longer a closed curve, as in SHM, since the values of the displacement and the velocity are never repeated.

For weak damping the phase space diagram will look like the spirals shown in Figure 10.42. Curve (a) corresponds to the case where the oscillator is released on the right (positive

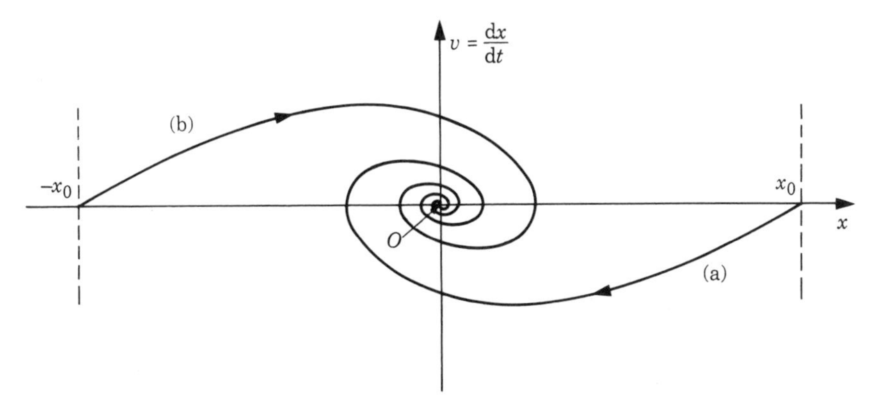

Figure 10.42 Phase space diagram of a weakly damped oscillator. The attractor is the origin.

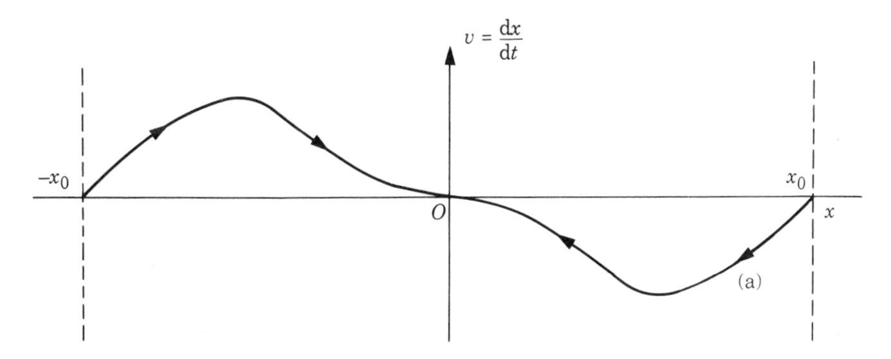

Figure 10.43 Phase space diagram of a strongly damped oscillator.

x) side of the equilibrium position and curve (b) when it is released on the left. Note that the spiral converges toward the point O, where, finally, $x = 0$ and $v = 0$, although it might take a long time to reach that point. Such a point in phase space is called an **attractor**. For strong damping, when the oscillator never goes beyond the equilibrium position, the phase space diagram is shown in Figure 10.43. The attractor is still O.

For two- and three-dimensional oscillators, phase space has four or six dimensions, making it impossible to represent these motions graphically.

Note 10.4 Non-linear oscillations and dynamical chaos

For most physical problems, Equation 10.12

$$m\frac{d^2x}{dt^2} + kx = 0 \tag{10.12}$$

for undamped oscillations and Equation 10.30

$$m\frac{d^2x}{dt^2} + \lambda\frac{dx}{dt} + kx = 0 \tag{10.30}$$

for damped oscillations are only a first approximation, valid for small oscillations only. The same small amplitude requirement applies to Equation 10.34,

$$m\frac{d^2x}{dt^2} + \lambda\frac{dx}{dt} + kx = F_0 \cos \omega_f t \tag{10.34}$$

which describes the forced motion of a damped oscillator. That is because, in those equations, the 'elastic' force has been assumed proportional to the displacement, $F = -kx$. But, as explained in Section 10.12, for large displacements the 'elastic' force is generally expressed by a more complex relation, involving higher powers of x.

The three equations have another element in common: they are **linear equations**; that is, the unknown variable x and its derivatives appear in no higher powers than the first (i.e. there are no terms like x^2 or $(dx/dt)^2$). Physically, this means that if, for example, the solution of Equation 10.34 for a driving force of amplitude F_0 is $x = f(t)$, then when the amplitude of the driving force is doubled to $2F_0$, the solution to the new problem is $x = 2f(t)$. This is clear from Note 10.1, where the amplitude A of the oscillation is directly proportional to F_0 (see Equation 10.36). Linearity is also exemplified by the way we treated the superposition of two SHMs in Sections 10.7 and 10.8. In actual physical systems, this linear property cannot hold forever. For example, in the case of a spring, as we continue to increase the driving force its amplitude will become so large that the spring eventually breaks! But before that happens, other terms enter into the equation and make it **non-linear**. One important feature of non-linear systems is their sensitivity to initial conditions or to any change in their parameters such as damping and the amplitude of the driving force.

The effect of non-linearity becomes particularly clear in the case of an undamped pendulum, which was discussed in Section 10.6. The equation of motion of the pendulum, given by Equation 10.14,

$$\frac{d^2\theta}{dt^2} + k \sin \theta = 0 \quad (k \equiv g/l) \tag{10.50}$$

is not linear because the sin θ term is not linear. Equation 10.50 reduces to the linear Equation 10.15 only for small amplitudes when sin θ can be approximated by θ and the motion can be considered SHM.

As explained in Example 9.11 and Note 10.3, the motion is oscillatory when $E < E_0$, where $E_0 = 2mgl$, but when the pendulum has enough energy ($E > E_0$), it can go all the way to the top of the circle, cross over and come down the other side. In this case, we have a periodic rotation modulated by an oscillation of the velocity instead of just an oscillation. But for energies very close to E_0 the motion of the pendulum becomes extremely sensitive to the initial conditions.

The situation becomes even more dramatic for the case of the forced oscillations of a damped pendulum, although this situation occurs in many other physical systems that show oscillatory properties. The equation that describes forced oscillations of a damped pendulum is obtained by adding a damping term proportional to $d\theta/dt$ to Equation 10.50 as well as a term corresponding to the driving force; that is

$$\frac{d^2\theta}{dt^2} + \Lambda \frac{d\theta}{dt} + k \sin \theta = f_0 \cos \omega_f t \tag{10.51}$$

where $\Lambda = \lambda/ml$, $k = g/l$, and $f_0 = F_0/ml$. This expression differs from Equation 10.34 by the presence of the non-linear sin θ term. However, for small amplitudes, we may replace sin θ by θ and the equation is of the same type as Equation 10.34.

Therefore, for small amplitudes, the forced oscillations of a damped pendulum are given by the expression obtained in Section 10.14 for a linear damped oscillator. These forced oscillations correspond to the asymptotic, or steady state, motion after the damping effects have been overcome by the driving force. Therefore, the motion evolves until it reaches a steady oscillation, called its **limit cycle**, as shown in Figure 10.44. In Figure 10.28(a), the initial conditions are such that the initial amplitude is smaller than that of the forced oscillations. Then the oscillations increase in amplitude under the action of the driving force until they finally reach the steady state value corresponding to the forced oscillations. The phase space diagram then takes the form shown in Figure 10.44(a), spiralling asymptotically out to become the ellipse corresponding to forced SHM. Another possibility is to have initial conditions such that the motion starts with a larger amplitude than that of the steady state forced oscillations. Then, the damping force reduces the amplitude until the pendulum settles into the steady value required by the driving force (Figure 10.28(b)). The corresponding phase space diagram is shown in Figure 10.44(b). In either case, we find that the **attractor** or limit cycle is an ellipse instead of a point, as in the case of a damped oscillator (recall Figure 10.43). If the driving force is not harmonic, the attractor is a closed curve, but not necessarily an ellipse. This is the case of a clock pendulum that is driven by a periodic but not harmonic force.

Actually, the phase space diagram of forced oscillations is three-dimensional, where the third axis corresponds to the phase $\phi = \omega_f t$ of the driving force (Figure 10.45). To select a trajectory in phase space uniquely, θ, $d\theta/dt$ and the phase of the driving force must each be

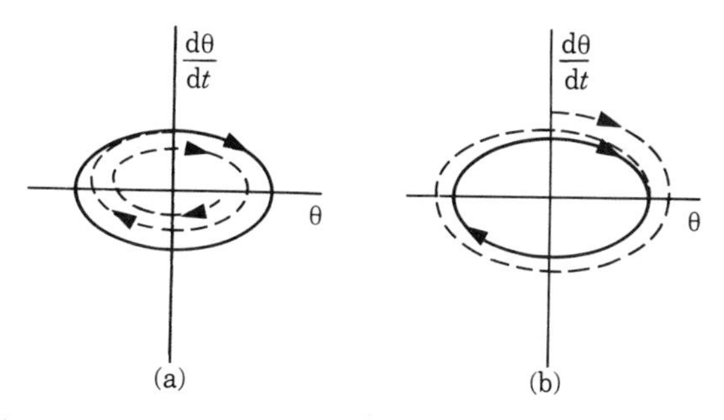

Figure 10.44 Phase space diagram of a forced pendulum.

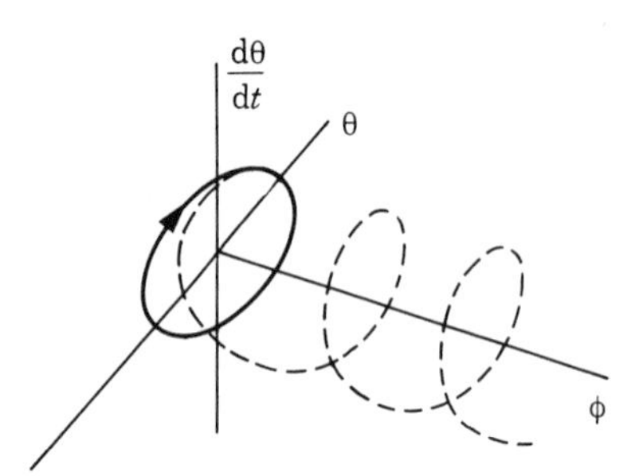

Figure 10.45 Three dimensional phase space diagram of a forced oscillator.

specified. The representation of the motion in three-dimensional phase space is a spiral whose axis coincides with the ϕ-axis and has a pitch equal to 2π. The projection of the spiral on the θ–$d\theta/dt$ plane are the curves shown in Figure 10.44.

For larger displacements, we must use the non-linear Equation 10.51. Unfortunately, non-linear equations generally do not have solutions that can be simply written down, as we did for Equations 10.32 and 10.34. Nevertheless, we can determine some properties of the solutions either by analytical methods we cannot describe here or by solving the equations numerically with a computer. The computer will only tell us the behavior of the system for concrete values of the parameters and in the limited interval of time that the computer can manage. It won't tell us the general behavior as a function of the parameters.

The motion can become very sensitive to the frequency ω_f of the driving force, since the non-linear terms create many new **resonances**. In Section 10.14 very strong resonant response was seen when the forcing frequency ω_f is close to the frequency ω_0 of the undamped linear oscillator. However, Equation 10.51 has resonances every time ω_f/ω_0 is a rational number (i.e. the ratio of two integers). Under normal conditions not too many resonances are seen because they tend to become very sharp (small bandwidth), so that the ratio ω_f/ω_0 must be tuned very accurately, particularly when the integers defining the resonance become large.

For certain values of ω_f and of the damping, the non-linear forced motion is no longer periodic and becomes very irregular or *chaotic*. Periodic motion is clearly recognizable by closed orbits in phase space. Chaotic motion is recognizable by irregular, fuzzy looking trajectories in phase space. In Figures 10.46(a)–(d) we show a sequence of asymptotic trajectories (i.e. after initial transients) of a forced pendulum. Here $\omega_f = 2\omega_0$, which is a resonant condition, and only the damping was varied between cases. Figures 10.46(a) and 10.46(b) show periodic motion – closed curves – of which the second is a bifurcated version of the first, with approximately twice the period. This type of bifurcation, where a closed curve in phase space changes into a 'double orbit' that doesn't quite close the first time around, plays an important role in the onset of chaos. The last two cases are **chaotic**. In (c) the motion is weakly chaotic and the pendulum continues to oscillate back and forth, as in (a) and (b), but its phase space path never closes. In case (d), where the damping has been reduced sharply, the motion is strongly chaotic and the pendulum winds back and forth across the top in a very irregular manner. Compare these with the phase–space trajectories of Figure 10.41.

Sensitivity to initial conditions means that a chaotic system started with two different, but very close, initial conditions will, after some time, end up in quite different states and the phase space trajectories diverge from each other exponentially in time. This is illustrated in Figure 10.47(a), where the angular displacement of the forced pendulum in the chaotic state corresponding to the phase diagram of Figure 10.46(d) is shown as a function of time for two slightly different initial conditions (θ and $d\theta/dt$ for $t = 0$). At the beginning of each motion the angular displacements are practically the same, but gradually a marked difference appears, rapidly growing with time. If we plot both motions of the pendulum in the same phase space, we find that the separation of points in each diagram corresponding to the same elapsed time increases exponentially. This is shown in Figure 10.47(b), where the logarithm of the separation d in phase space approximately increases linearly with time; that is, $\ln d \simeq \lambda t$ or $d \simeq e^{\lambda t}$. The positive quantity λ is called a **Lyapunov exponent**. Its value is 0.13 for the case under consideration. For the weakly chaotic pendulum corresponding to the phase space diagram of Figure 10.46(c), the Lyapunov exponent is 0.033. The more sensitive the system is to initial conditions and the more chaotic the motion, the larger the Lyapunov exponent. Therefore, if we use Equation 10.51 to predict where the pendulum will be at some time, such a prediction

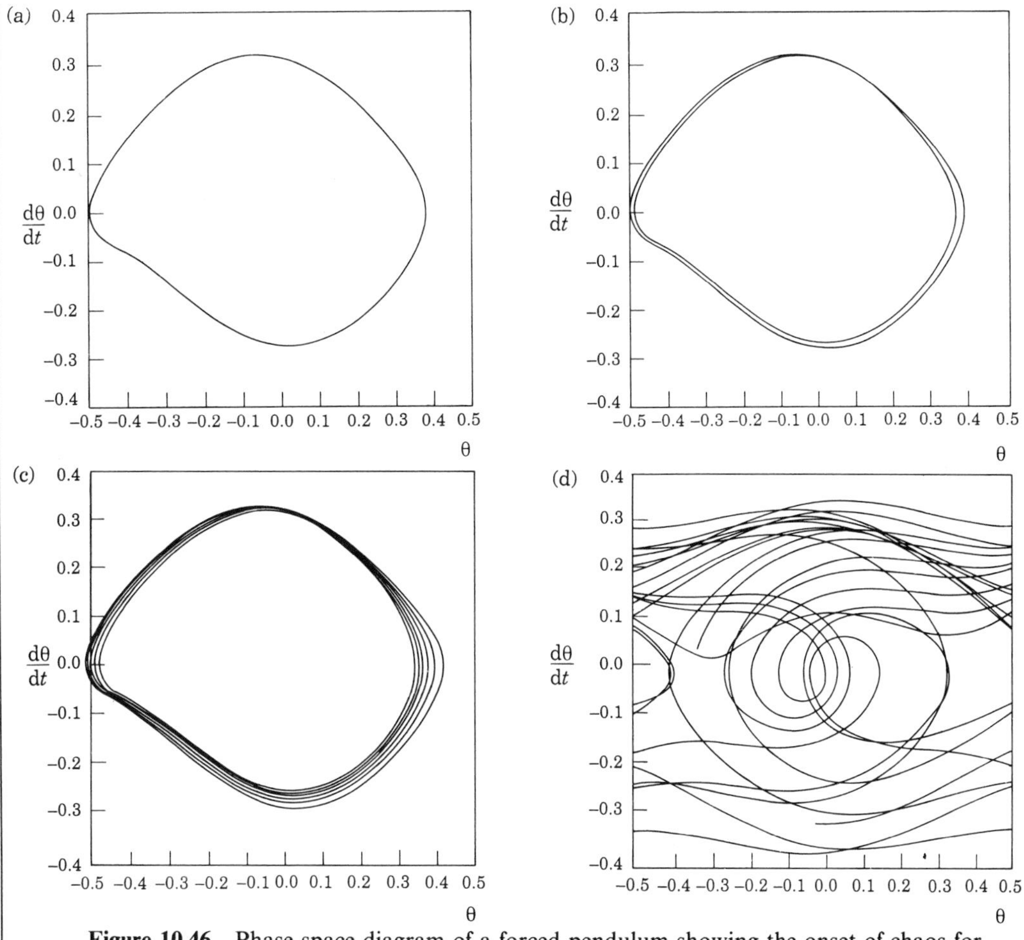

Figure 10.46 Phase space diagram of a forced pendulum showing the onset of chaos for $\omega_f = 2\omega_0$ and different dampings. (Courtesy of T. Bohr and F. Christiansen, Niels Bohr Institute.)

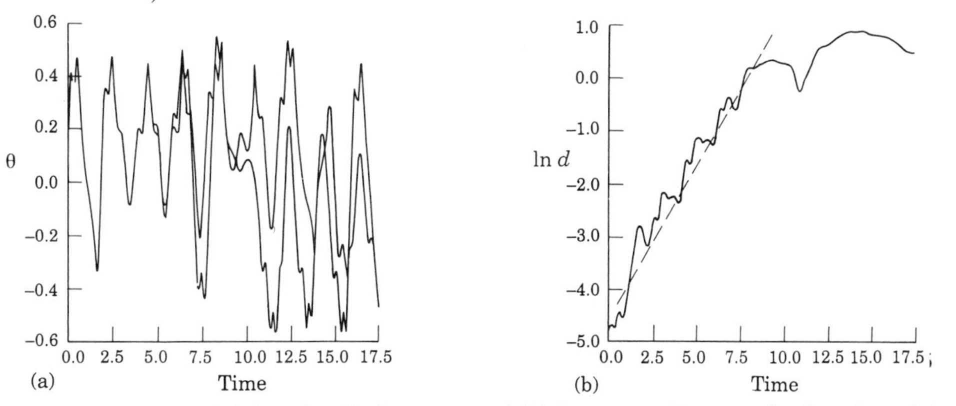

Figure 10.47 (a) Angular displacement and (b) Lyapunov diagram of a forced pendulum in the chaotic state corresponding to Figure 10.46(d). (Courtesy of T. Bohr and F. Christiansen, Niels Bohr Institute.)

must be done with great care. Since the initial conditions are known only to some limited accuracy, there is a limit to how far in the future any prediction can be trusted. With a Lyapunov exponent of 0.5, the initial error is amplified by a factor of 10 every 5 seconds!

Although the projection of the path of non-periodic and chaotic motion on the two-dimensional θ–$d\theta/dt$ plane of phase space is a line that intersects itself several times, the actual representation in three-dimensional phase space is a line that progresses along the ϕ-axis without ever intersecting itself. In fact, chaotic motion can only exist when the system has at least three dimensions in phase space, since that is a minimum to make sure that the path in phase space converges toward the attractor without ever crossing itself.

In summary, we call **chaos** the irregular and unpredictable evolution in time of non-linear, dynamical systems. Chaotic motion is aperiodic, never repeating itself, and is extremely sensitive to initial conditions and other parameters of the system. Slightly different initial conditions may result in quite different motions and their paths in phase space diverge from each other exponentially in time, as quantified by the Lyapunov exponent. We conclude then that although chaotic non-linear systems are described by deterministic equations, as illustrated for the case of an oscillator, they are unpredictable in the long term. It was Henri Poincaré (1854–1912) who first pointed out this property of non-linear systems.

Chaotic systems are not restricted to mechanical oscillations and many complex systems described by non-linear equations also show chaotic behavior. See Notes 11.5, 17.6 and 18.2.

QUESTIONS

10.1 Give a kinematical definition of SHM. Give a dynamical definition of SHM. Are the two definitions fully equivalent?

10.2 How is the period of SHM changed when (a) the mass of the particle is increased without changing the elastic constant, (b) when the elastic constant is increased without changing the mass, (c) when the mass and the elastic constant are changed by the same ratio?

10.3 A particle moves according to the equation $x = A \sin(\omega t + \alpha)$. Write equations for the velocity and acceleration of the particle. Does it move with SHM? What is the phase difference with respect to $x = A \cos(\omega t + \alpha)$?

10.4 Given a displacement $x = A \sin(\omega t + \alpha)$, plot the rotating vectors for displacement, velocity and acceleration; indicate how the phase angle $(\omega t + \alpha)$ should be measured.

10.5 Repeat Figure 10.6 for several positions of the pendulum on either side of C and determine F_T for each position. What do you conclude about the magnitude and direction of F_T?

10.6 Under what conditions does a pendulum move with (a) oscillatory motion, (b) SHM, (c) modulated circular motion?

10.7 The length of the pendulum of a clock is adjusted so that it gives the correct time when the amplitude of the oscillations is very small. If the pendulum is inadvertently set into oscillation with a large amplitude, does the clock go too fast or too slow?

10.8 The pendulum of a clock is adjusted to give the correct time at 40° latitude. What will happen to the clock if it is moved to the equator and if it is moved to a place at 80° latitude? What adjustment will be necessary in each case?

10.9 Express the conservation of energy in SHM in terms of x and v. From this expression obtain v in terms of x and compare with the result of Example 9.4.

10.10 Two SHMs of the same frequency and direction are superposed. What kind of motion results? Which properties of the resulting motion depend on the phase difference and which do not? What would happen if the oscillatory motions are anharmonic?

10.11 Two SHMs of the same frequency and perpendicular directions are superposed. Under what conditions is the resultant motion (a) SHM, (b) uniform circular motion? Is the resultant motion always periodic?

10.12 Under what conditions are beats produced?

10.13 What are the main features of the normal modes of coupled oscillators?

10.14 When is oscillatory motion anharmonic? Are anharmonic oscillations symmetric relative to the equilibrium position?

10.15 Why does the addition of a force $-\lambda v$ to an elastic force $-kx$ result in damped oscillatory motion?

10.16 What energy exchanges occur in damped oscillatory motion?

10.17 Why are the forced oscillations of a damped oscillator not damped?

10.18 Why must the force and the velocity be in phase at energy resonance?

10.19 Represent the relation $P_{ave}/(P_{ave})_{res}$ (Figure 10.30) for several values of Q (see Note 10.2).

10.20 Analyze the physical meaning of Fourier's theorem. In terms of this theorem, explain the difference in quality of the same musical note produced by different instruments (see Note 10.3).

10.21 A particle moves under the action of a force $\boldsymbol{F} = -k\boldsymbol{r}$. Write the equation of motion. Is the motion in a plane? What determines the plane of the motion? Separate the equation into X- and Y-components. Find the corresponding solutions. What kind of motion results? What must be the initial conditions so that the resultant motion is rectilinear? Is the angular momentum constant?

10.22 Substitute Equation 10.32 for the displacement of a damped oscillator into Equation 10.30 and verify that it is a satisfactory solution for arbitrary A and α if ω is given by Equation 10.31.

10.23 Verify that the tension T of the string of a pendulum is given by $T = mg(3 \cos \theta - 2 \cos \theta)$.

PROBLEMS

10.1 A wheel of 30 cm radius is provided with a handle at its edge. The wheel is rotating at 0.5 rev s^{-1} with its axis in a horizontal position. Assuming that the Sun's rays fall vertically on the Earth, the shadow of the handle will describe simple harmonic motion. Find (a) the period of the motion of the shadow, (b) its frequency, and (c) its amplitude. (d) Write the equation that expresses its displacement as a function of time. Assume zero initial phase.

10.2 A particle is moving with simple harmonic motion of 0.10 m amplitude and a period of 2 s. (a) Make a table indicating the values of the elongation, the velocity, and the acceleration at the following times: $t = 0$, $P/8$, $P/4$, $3P/8$, $P/2$, $5P/8$, $3P/4$, $7P/8$ and P. Plot the curves for (b) elongation, (c) velocity, and (d) acceleration, each as a function of time. Assume zero initial phase.

10.3 A simple harmonic oscillator is described by the equation $x = 0.4 \sin(0.1t + 0.5)$, where x and t are expressed in m and s, respectively. Find (a) the amplitude, period, frequency and initial phase of the motion, (b) the general expression for the velocity and acceleration, (c) the initial conditions, (d) the position, velocity and acceleration for $t = 5$ s. Represent (e) position, velocity and acceleration as functions of time.

10.4 A particle situated at the end of one arm of a tuning fork passes through its equilibrium position with a velocity of 2 m s^{-1}. The amplitude is 10^{-3} m. (a) Determine the frequency and period of the tuning fork. (b) Write the equations expressing its displacement and velocity as a function of time.

10.5 A particle of mass 1 g is vibrating with a simple harmonic motion of 2 mm amplitude. Its acceleration at the end of the trajectory is 8.0×10^3 m s^{-2}. Calculate (a) the frequency of the motion and (b) the velocity of the particle when it passes through the equilibrium point and when the elongation is 1.2 mm. (c) Write the equation expressing the force acting on the particle as a function of position and as a function of time.

10.6 A particle is vibrating with a frequency of 100 Hz and an amplitude of 3 mm. (a) Calculate its velocity and acceleration at the middle and at the extremes of the trajectory. (b) Write the equation expressing the elongation as a function of time. Assume zero initial phase.

10.7 A particle moving with simple harmonic motion of 0.15 m amplitude is vibrating 100 times per second. What is its angular frequency? Calculate (a) its velocity, (b) its acceleration, and (c) its phase, when its displacement is 0.075 m.

10.8 The motion of the needle in a sewing machine is practically simple harmonic. If the amplitude is 0.3 cm and the frequency is 600 vib min^{-1}, what will be the (i) elongation, (ii) velocity, and (iii) acceleration one-thirtieth of a second after the needle passes through the center of the trajectory (a) in the upward or positive sense, (b) in the downward or negative sense?

10.9 A particular simple harmonic motion has an amplitude of 8 cm and a period of 4 s. Calculate the velocity and acceleration (a) 0.5 s after the particle passes through the extreme of the trajectory and (b) when the particle passes through the center.

10.10 (a) Using the data of Problem 10.2, calculate the kinetic, potential and total energy at each time, assuming the particle has a mass of 0.5 kg. Verify that the total energy remains constant. (b) Plot the kinetic and potential energy (i) as functions of time and (ii) as functions of position. What are your conclusions?

10.11 A 0.50 kg particle is moving with simple harmonic motion. Its period is 0.1 s and its amplitude is 10 cm. Calculate the acceleration, the force, the potential energy and the kinetic energy when the particle is 5 cm from the equilibrium position.

10.12 A 4 kg particle is moving along the X-axis under the action of the force

$$F = -\left(\frac{\pi^2}{16}\right)x \text{ N}$$

When $t = 2$ s the particle passes through the origin, and when $t = 4$ s its velocity is 4 m s^{-1}. (a) Find the equation of the elongation. (b) Show that the amplitude of the motion is $32(2)^{1/2}/\pi$ m.

10.13 When a person of mass 60 kg gets into a car, the center of gravity of the car lowers 0.3 cm. (a) What is the elastic constant of the springs of the car? (b) Given that the mass of the car is 500 kg, what is its period of vibration when it is empty and when the person is inside?

10.14 A wooden block whose density relative to water is ρ has dimensions a, b and c. While it is floating in water with side a vertical, it is pushed down and released. Find the period of the resulting oscillations. (Recall that the buoyant force is equal to the weight of the fluid displaced.)

10.15 A particle moves so that its coordinates as functions of time are given by $x = v_0 t$ and $y = y_0 \sin \omega t$. (a) Plot x and y as functions of t. (b) Plot the path of the particle. (c) What force is required to produce this motion? (d) Find the magnitudes of its velocity and acceleration as functions of time.

10.16 Find, for simple harmonic motion, the values of $(x)_{ave}$ and $(x^2)_{ave}$, where the averages refer to time.

10.17 Find the average values of the kinetic and potential energies in simple harmonic motion relative to (a) time, (b) position.

10.18 The period of a pendulum is 3 s. What will be its period if its length is (a) increased, (b) decreased by 60%?

10.19 What should be the percentage change of length of a pendulum in order that a clock have the same period when moved from a place where $g = 9.80$ m s^{-2} to another where $g = 9.81$ m s^{-2}?

10.20 A simple pendulum whose length is 2 m is in a place where $g = 9.80$ m s^{-2}. The pendulum oscillates with an amplitude of 2°. Express, as a function of time, (a) its angular displacement, (b) its angular velocity, (c) its angular acceleration, (d) its linear velocity, (e) its centripetal acceleration and (f) the tension on the string if the mass of the bob is 1 kg.

10.21 A pendulum 1.00 m long and having a bob of 0.60 kg is raised along an arc so that it is 4 cm above its equilibrium height. (a) Find its angular amplitude. Express, as a function of the pendulum's height, (b) the force tangent to its path, (c) its tangential acceleration, (d) its velocity and (e) its angular displacement when allowed to swing. Find the values of these quantities at (f) the point of maximum amplitude and (g) the lowest point of the pendulum's path.

10.22 Estimate the relative order of magnitude of the corrective term in Equation 10.16 for the period of a simple pendulum if the amplitude is (a) 3°, (b) 10°, (b) 30°.

10.23 (a) Find the equation of motion resulting from superposing two parallel simple harmonic motions whose equations are $x_1 = 6 \sin 2t$ and $x_2 = 8 \sin(2t + \alpha)$, if $\alpha = 0, \pi/2$ and π. (b) Make a plot of the resultant motion in each case.

10.24 (a) Find the equation of motion resulting from superposing two parallel simple harmonic motions whose equations are $x_1 = 2 \sin(\omega t + \pi/3)$ and $x_2 = 3 \sin(\omega t + \pi/2)$. (b) Make a plot of the resultant motion.

10.25 (a) Find the equation of the path of the resulting motion of two perpendicular simple harmonic motions whose equations are: $x = 4 \sin \omega t$ and $y = 3 \sin(\omega t + \alpha)$, when $\alpha = 0$, $\pi/2$ and π. (b) In each case plot the path of the particle and show the sense in which the particle traverses it.

10.26 (a) Find the equation of the path of a particle resulting from the application of two perpendicular simple harmonic motions, given that

$$\frac{\omega_1}{\omega_2} = \frac{1}{2}$$

and $\alpha = 0$, $\pi/3$ and $\pi/2$. (b) In each case, plot the path and show the sense in which it is traversed.

10.27 A simple pendulum has a period of 2 s and an amplitude of 2°. After 10 complete oscillations its amplitude has been reduced to 1.5°. Find the damping constant γ.

10.28 In the case of a damped oscillator, the quantity $\tau = 1/2\gamma$ is called the **relaxation time** where γ is $\lambda/2m$ (see Section 10.13). (a) Verify that it is expressed in units of time. (b) How much has the amplitude of the oscillator changed after a time τ? (c) Express, as a function of τ, the time required for the amplitude to reduce to one-half its initial value. (d) What are the values of the amplitude after times equal to twice, three times, etc., the relaxation time?

10.29 (a) Write the expression for the velocity and the acceleration of a damped oscillator whose displacement is given by $x = A\,\mathrm{e}^{-\gamma t} \sin \omega t$. (b) Find the limiting values when γ is much smaller than ω_0.

10.30 Assume that, for a damped oscillator, γ is much smaller than ω_0 so that the amplitude remains essentially constant during one oscillation. Verify (a) that the kinetic energy of the damped oscillator can be written in the form $E_k = \frac{1}{2} m\omega_0^2 A^2 \mathrm{e}^{-2\gamma t} \cos^2(\omega_0 t + \alpha)$ and (b) that the total energy is $E = \frac{1}{2} m\omega_0^2 A^2 \mathrm{e}^{-2\gamma t}$. (Hint: recall Problem 10.29.)

10.31 Find the limiting values of the amplitude and the phase of a forced damped oscillator when (a) ω_f is much smaller than ω_0, and (b) ω_f is much larger than ω_0. Determine the dominant factors in each case.

10.32 Verify that for the forced oscillations of a damped oscillator, the average power of the applied force is equal to the average power dissipated by the damping force.

10.33 (a) Write the equation of motion of a simple harmonic undamped oscillator to which a force $F = F_0 \cos \omega_f t$ is applied. (b) Verify that its solution is

$$x = \frac{F_0}{m(\omega_0^2 - \omega_f^2)} \cos \omega_f t$$

Discuss resonance in this case.

10.34 A particle slides back and forth between two inclined frictionless planes joined smoothly at the bottom (Figure 10.48). (a) Find the period of the motion if h is the initial height. (b) Is the motion oscillatory? Is it simple harmonic?

Figure 10.48

10.35 A particle of mass m placed on a horizontal frictionless table (Figure 10.49) is held by two identical springs of unstretched length l_0; the other ends of the springs are fixed at points P_1 and P_2, a distance $2l_0$ apart. (a) If the particle is displaced sideways an amount x_0 (which is small compared with the length of the springs) and then released, determine its subsequent motion. (b) Find its frequency of oscillation and write the equation of motion.

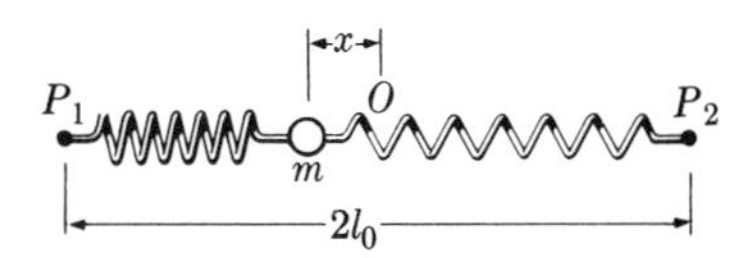

Figure 10.49

10.36 A particle of mass m is subject to the force shown in Figure 10.50, called a **square wave**; i.e. the force is constant in magnitude but reverses direction at regular time intervals of π/ω. This force can be represented by the Fourier series:

$$F = F_0(4/\pi) \times (\sin \omega t + 1/3 \sin 3\omega t + 1/5 \sin 5\omega t + \cdots)$$

(a) Write the equation of motion of the particle. (b) Verify, by direct substitution, that its solution

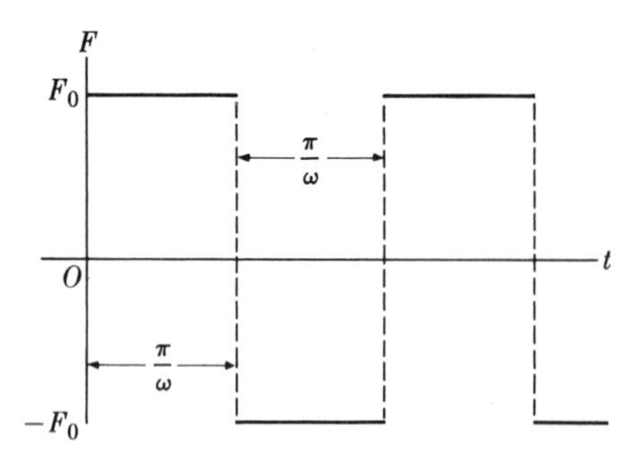

Figure 10.50

can be written as

$$x = a + bt + A \sin \omega t + B \sin 3\omega t + C \sin 5\omega t + \cdots$$

where a and b are arbitrary constants. (c) Determine the values of the coefficients A, B, $C, \ldots$, so that the equation of motion is satisfied.

10.37 Consider a particle oscillating under the influence of the anharmonic potential $E_p(x) = \frac{1}{2}kx^2 - \frac{1}{3}ax^3$, where a is positive and much smaller than k. (a) Make a schematic plot of $E_p(x)$. Is the curve symmetric around the value $x = 0$? (b) In view of your previous answer, in what direction is the center of oscillation displaced as the energy is increased? Do you expect x_{ave} to be zero? (c) Obtain the force as a function of x and make a schematic plot. (d) Describe the effect of the anharmonic term on the force.

10.38 Repeat Problem 10.37, assuming that the potential energy is $E_p(x) = \frac{1}{2}kx^2 - \frac{1}{4}ax^4$. As before, a is much smaller than k.

11 Gravitational interaction

Johannes Kepler, astronomer and mathematician of the 17th century, adopted the heliocentric planetary system, proposed in the 16th century by Nicolaus Copernicus, as superior to the Ptolemaic geocentric system, which was the accepted model of the solar system of his time. In his search for regularities in planetary motion, Kepler carefully analyzed, over more than ten years, the astronomical observations of Tycho Brahe, discovering the three laws that are followed by the planets in their motion around the Sun. Kepler's laws served as the kinematical foundation for Newton's law of gravitation.

Notes

11.1 Introduction

The motion of heavenly bodies has intrigued mankind since the dawn of civilization. Perhaps one of the most interesting processes in the history of science has been the evolution of our understanding of planetary motion.

The Greeks liked to consider the Earth as the center of the universe. They assumed that the Earth was the geometric center and that all 'heavenly bodies' moved around the Earth. Those bodies known at that time were placed according to their average distance from the Earth. This order was Moon, Mercury, Venus, Sun, Mars, Jupiter and Saturn, with the fixed stars scattered over an outer sphere.

The first hypothesis about planetary motion had these planets describe circles concentric with the Earth. This assumption, however, did not fit the observed motion of the planets relative to the Earth (Figure 11.1). And so the geometry of planetary motion became more and more complex to account for astronomical

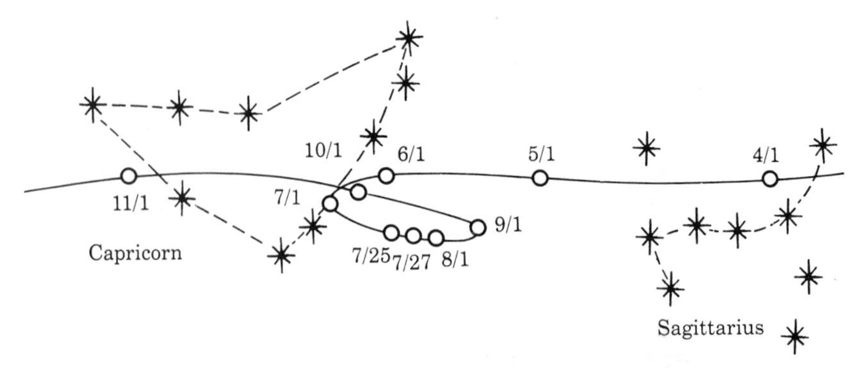

Figure 11.1 Position of Mars with respect to the fixed stars from 1 April to 1 November in 1985.

observations. In the second century AD the astronomer Ptolemy of Alexandria developed his theory of epicycles to explain this *geocentric* motion. In the simplest case the planet was assumed to move uniformly on a circle called an **epicycle**. The center of the epicycle, in turn, moved on a larger circle, called the **deferent**, that was concentric with the Earth. The resulting path of the planet was an **epicycloid** (Figure 11.2). Some planets needed an even more complicated arrangement. Using present language, what the Greeks did was to describe planetary motion relative to a frame of reference attached to the Earth.

This description was accepted as correct until, in the sixteenth century, Nicolaus Copernicus (1473–1543) developed a different method. He was looking for a simpler explanation and so proposed a description of the motion of all planets, including the Earth, relative to the Sun, which would be at the center. This *heliocentric* model was not new; it had been proposed by the astronomer Aristarchus about the third century BC. According to Copernicus, the orbits of the planets were placed in the following order with respect to the Sun: Mercury, Venus, Earth, Mars, Jupiter and Saturn, with the Moon revolving around the Earth. Essentially what Copernicus proposed was another frame of reference, attached to the Sun. In such a frame the motion of the planets had a simpler description.

The Sun, the largest body in our planetary system, is practically coincident with the center of mass of the system. This justifies its choice as center of reference,

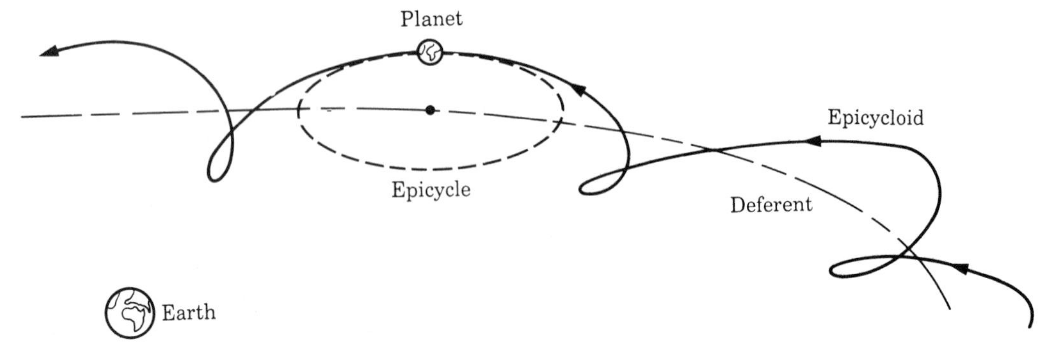

Figure 11.2 Epicycle model for planetary motion referred to the Earth.

Table 11.1 Basic data of the solar system

Body	Equatorial radius (m)	Mass (kg)	Period of rotation (s)	Semi-major axis of orbit (m)	Period of orbital motion (s)	Eccentricity of orbit
Sun	6.69×10^{8}	1.99×10^{30}	2.4×10^{6}	—	—	—
Mercury	2.44×10^{6}	3.30×10^{23}	5.07×10^{6}	5.79×10^{10}	7.60×10^{6}	0.2056
Venus	6.05×10^{6}	4.87×10^{24}	2.10×10^{7}†	1.08×10^{11}	1.94×10^{7}	0.0068
Earth	6.38×10^{6}	5.97×10^{24}	8.62×10^{4}	1.50×10^{11}	3.16×10^{7}	0.0167
Mars	3.39×10^{6}	6.42×10^{23}	8.86×10^{4}	2.28×10^{11}	5.94×10^{7}	0.0934
Jupiter	7.14×10^{7}	1.90×10^{27}	3.54×10^{4}	7.78×10^{11}	3.75×10^{8}	0.0483
Saturn	6.00×10^{7}	5.69×10^{26}	3.84×10^{4}	1.43×10^{12}	9.30×10^{8}	0.0560
Uranus	2.61×10^{7}	8.70×10^{25}	6.20×10^{4}	2.87×10^{12}	2.65×10^{9}	0.0461
Neptune	2.43×10^{7}	1.03×10^{26}	6.48×10^{4}	4.59×10^{12}	5.20×10^{9}	0.0100
Pluto	1.14×10^{6}	1.20×10^{22}	5.52×10^{5}	5.91×10^{12}	7.84×10^{9}	0.2484
Moon*	1.74×10^{6}	7.35×10^{22}	2.36×10^{6}	3.84×10^{8}	2.36×10^{6}	0.0550

* Orbital data of the Moon are relative to the Earth
† Venus exhibits retrograde motion

since it is, practically, an inertial frame, except for its motion relative to the center of the galaxy. Copernicus' proposal helped the astronomer Johannes Kepler (1571–1630) to discover the laws of planetary motion. An analysis of the astronomical measurements of Tycho Brahe (1546–1601) led Kepler to formulate his laws, which are a kinematical description of planetary motion. Kepler's laws are stated as follows:

1. *The planets describe elliptical orbits, with the Sun at one focus.*
2. *The position vector of any planet relative to the Sun sweeps out equal areas of its ellipse in equal times.*
3. *The squares of the periods of revolution are proportional to the cubes of the average distances of the planets from the Sun.*

The second law is called the **law of areas**. The third law may be stated by the equation $P^2 = kr^3_{ave}$, where k is a proportionality constant.

Undoubtedly the discovery of these laws by Kepler was an astonishing analytical effort, particularly in view of the limited accuracy of the data he had at his disposal. There is evidence to suggest that the experimental data were insufficient to 'prove' his laws because of their imprecision. Kepler may have adjusted numbers for planetary positions to agree with his laws, but even if this is true, that does not detract from his achievement.

The more important data about the solar system have been collected in Table 11.1.

11.2 The law of gravitation

Kepler's laws provide a description of how the planets move, but do not provide any clue as to why the planets move in that way and no other.

Next to his statement of the laws of motion (Chapter 6), Newton's second, and perhaps greatest, contribution to physics was the discovery of the **law of**

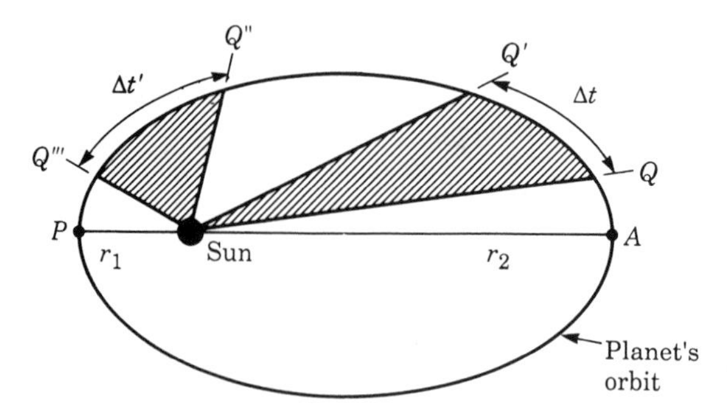

Figure 11.3 Law of areas.

universal gravitation. It predicts the attractive interaction between two bodies, either planets or small particles, which produces a motion that fits the description given by Kepler's laws. This law was formulated by Newton in 1666, but it was not published until 1687, when it appeared as a chapter in his monumental work *Principia Mathematica Philosophiae Naturalis.*

The fact that the planets describe a closed orbit and do not fly away from the Sun indicates that the gravitational force is attractive. (Of course an attractive force may also produce an open orbit, but a repulsive force cannot produce a closed orbit; see Note 9.2). The distance of a planet from the Sun fluctuates from a minimum r_1 at the **perigee** P (Figure 11.3) to a maximum r_2 at **apogee** A. If $r_1 = r_2$ the path is a circle. The larger the difference between r_1 and r_2 the more elongated or eccentric the ellipse. Actually, the **eccentricity** of the ellipse is measured by the ratio

$$\varepsilon = \frac{r_2 - r_1}{r_2 + r_1} = \frac{r_2 - r_1}{2a} \tag{11.1}$$

where $r_2 + r_1 = 2a$ is the length of the major axis of the ellipse. Table 11.1 gives the eccentricity for each planet and for the Moon. The eccentricity of a circle is zero ($r_1 = r_2$). Thus we note that except for Mercury and Pluto, the orbits of all planets and of the Moon are almost circles.

Suppose in the time interval Δt a planet moves from Q to Q' (Figure 11.3) so that its radius vector sweeps out an area ΔA and that later on, during a time interval $\Delta t'$, it moves from Q'' to Q''', sweeping out an area $\Delta A'$. The law of areas (or Kepler's second law) means that the areas swept per unit time must be the same in both cases. That is

$$\frac{\Delta A}{\Delta t} = \frac{\Delta A'}{\Delta t'} = const.$$

As shown in the derivation below, the law of areas implies that the angular momentum of a planet relative to the Sun is constant. A constant angular momentum means that

the force associated with the gravitational interaction is central.

Therefore the gravitational force exerted by the Sun on a planet acts along the line joining the two bodies and points toward the Sun (Figure 11.4). By the law of action and reaction the planets exert an equal and opposite force on the Sun. The same considerations apply to the gravitational interaction between any two masses M and m.

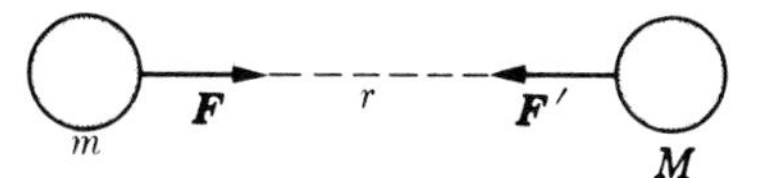

Figure 11.4 Gravitational interaction between two masses. The forces on M and m are equal in magnitude but opposite in direction.

If we assume that the gravitational interaction is a *universal* property of all matter, it seems plausible to assume also that the force $\boldsymbol{F}$ associated with the interaction is proportional to the 'amount of matter' in each body, i.e. to their respective masses M and m, and also depends on their distance apart, r. Thus we may write

$$\boldsymbol{F} = Mmf(\boldsymbol{r})$$

To decide upon the dependence of the force $\boldsymbol{F}$ on the distance r, expressed by $f(\boldsymbol{r})$, is a more difficult problem. Through a mathematical analysis, too complex to be carried out here, we may determine the form of $f(\boldsymbol{r})$ so that the orbit required by Kepler's laws results. In fact that was Newton's procedure, as will be explained in Section 11.3 for circular orbits. Alternatively, we could determine this dependence experimentally by measuring the force between masses M and m at several separations and deducing from our observations the relation between $\boldsymbol{F}$ and $\boldsymbol{r}$. Then we could extend our results to the planets. The experiment requires a sensitive set-up because the gravitational force is very small unless the two masses are very large or the distance r is very small. Also, at very short distances other interactions, stronger than the gravitational one, may enter into play and mask gravitational effects.

The results of these calculations and many experiments that have been performed allow us to conclude that

the gravitational interaction is always attractive and varies inversely with the square of the distance between the two bodies.

That is, $f(\boldsymbol{r}) \propto 1/r^2$. Therefore we express the force of gravitation as

$$F = G\frac{Mm}{r^2} \tag{11.2}$$

where the proportionality constant G depends on the units used for the other quantities. The constant G may be determined experimentally by measuring the force $\boldsymbol{F}$ between two known masses M and m at a known distance $\boldsymbol{r}$ (Figure 11.5). The value of G in MKSC units is

$$G = 6.67 \times 10^{-11}\,\mathrm{N\,m^2\,kg^{-2}} \quad (\text{or } \mathrm{m^3\,kg^{-1}\,s^{-2}})$$

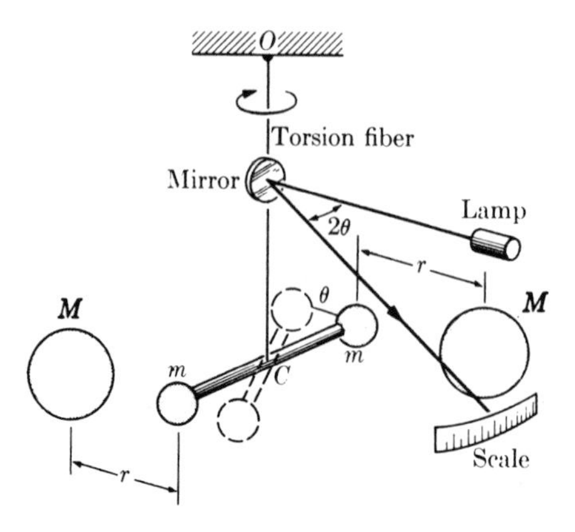

Figure 11.5 Cavendish torsion balance. When the masses M are placed close to the masses m, their gravitational attraction produces a torque on the horizontal rod that results in a torsion of the fiber OC. Equilibrium is established when the gravitational and torsional torques are equal. The torsional torque is proportional to the angle θ, which is measured by the deflection of a ray reflected from a mirror attached to the fiber. By repeating the experiment at several distances r, and using different masses m and M, we can verify law 11.1.

We may state that

the gravitational interaction between two bodies is expressed by an attractive central force proportional to the masses of the bodies and inversely proportional to the square of the distance between them.

This is **Newton's law of universal gravitation.**

Relation between the law of areas and angular momentum

Consider a particle describing a curved path as shown in Figure 11.6. In a short time interval $\mathrm{d}t$, the particle moves from P to P' and the radius vector $\boldsymbol{r}$ sweeps out the shaded area OPP' whose value is approximately $\mathrm{d}A = \frac{1}{2}$ base $\times$ height $= \frac{1}{2}r(r\,\mathrm{d}\theta) = \frac{1}{2}r^2\,\mathrm{d}\theta$. The area swept out by the particle per unit time is then

$$\frac{\mathrm{d}A}{\mathrm{d}t} = \frac{1}{2}r^2\frac{\mathrm{d}\theta}{\mathrm{d}t} \tag{11.3}$$

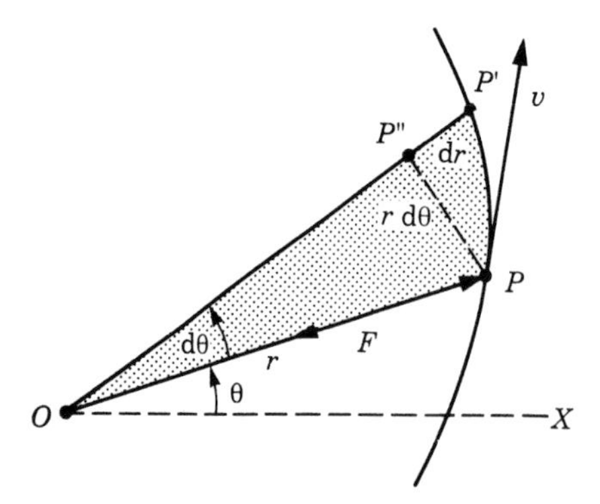

Figure 11.6 Under central forces, the position vector sweeps out equal areas in equal times.

Kepler's law of areas requires that $\mathrm{d}A/\mathrm{d}t$ is constant. According to Equation 8.11, the angular momentum, with $\omega = \mathrm{d}\theta/\mathrm{d}t$, can be expressed as $L = mr^2\,\mathrm{d}\theta/\mathrm{d}t$. By comparison of this equation with Equation 11.3 we may write

$$L = 2m\frac{\mathrm{d}A}{\mathrm{d}t} = const. \tag{11.4}$$

Therefore, Kepler's law of areas implies the constancy of the angular momentum of the planet, and hence that the force applied to the planet is central.

11.3 Newton's derivation of the law of force

Kepler's first law states that the orbit of a planet is an ellipse. A particular case of an ellipse is the circle, where the two foci coincide with the center. In this case, according to the second law, the force $\boldsymbol{F}$, being attractive, points toward the center of the circle. Thus, referring the motion of m relative to a frame of reference attached to M (Figure 11.7) and using Equation 8.2 for the centripetal force in circular motion, we may express the gravitational force on m as $F = mv^2/r$. Strictly speaking, this is valid only if M is at rest in an inertial system, as explained in Section 6.7, which is correct only if M is much larger than m (see Section 13.2). In the case of the Sun and the planets it is approximately true. Remembering that $v = 2\pi r/P$, we have

$$F = 4\pi^2 \frac{mr}{P^2}$$

For the special case of a circular orbit when the distance between M and m is the radius of the circle, the third law of Kepler becomes $P^2 = kr^3$. Therefore, cancelling an r factor,

$$F = 4\pi^2 \frac{m}{kr^2} \propto \frac{1}{r^2}$$

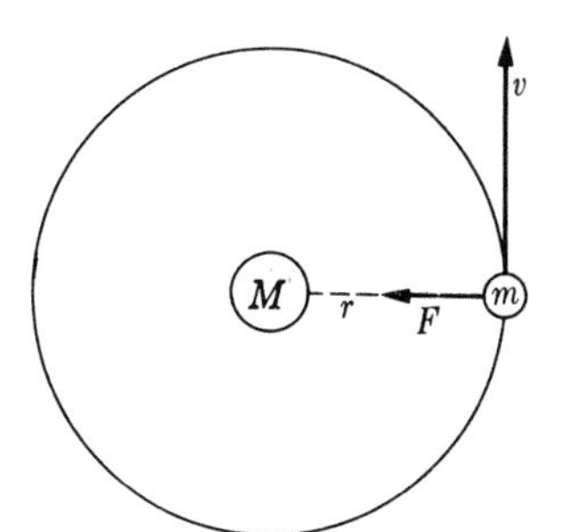

Figure 11.7 Motion of particle m under its gravitational interaction with M.

Then, to satisfy Kepler's laws for a circular orbit, the gravitational interaction must be both central and inversely proportional to the square of the distance. The same result is obtained for an elliptical orbit, but the calculation is more elaborate.

Newton checked that his law was correct by comparing the centripetal acceleration of the Moon with the acceleration of gravity, $g = 9.80\ \text{m s}^{-2}$. Since both accelerations are due, respectively, to the gravitational attraction of the Earth on the Moon and on any body close to the Earth's surface, their values should be related to the distance in the same way as the force. The centripetal acceleration of the Moon is

$$a_M = \frac{v^2}{r} = 4\pi^2 \frac{r}{P^2}$$

Using the values for the Moon's orbit given in Table 11.1, the acceleration is $a_M = 2.72 \times 10^{-3}\ \text{m s}^{-2}$. Therefore,

$$\frac{g}{a_M} = 3602 \sim (60)^2$$

Since the radius of the Earth is $R = 6.37 \times 10^6$ m and $r = 384 \times 10^6$ m,

$$\left(\frac{r}{R}\right)^2 = \left(\frac{384}{6.37}\right)^2 \sim (60)^2.$$

Therefore $g/a_M = (r/R)^2$ within the accuracy of his rough calculation. Since the two accelerations are in inverse proportion to the square of the distances from the center of the Earth, the same relation applies to the respective forces.

EXAMPLE 11.1

Relation between the acceleration of gravity and the mass of the Earth. Calculation of the mass of the Earth.

▷ Consider a particle of mass m on the Earth's surface. Its distance from the center of the Earth is equal to the Earth's radius R. Thus, if we denote the mass of the Earth by M, the gravitational force on the body is

$$F = G\frac{Mm}{R^2}$$

This force is the weight of the body, and therefore must be equal to mg (Equation 6.16), where g is the acceleration of gravity. Therefore

$$mg = G\frac{Mm}{R^2}$$

or, canceling the common factor m, we have

$$g = \frac{GM}{R^2} \tag{11.5}$$

Equation 11.5 gives the acceleration of gravity in terms of the mass and radius of the Earth. Note that the mass of the body does not appear in this expression, and thus (when we neglect air resistance) all bodies should fall with the same acceleration, in agreement with observation.

Solving for the mass of the Earth M, we obtain

$$M = \frac{gR^2}{G}.$$

Introducing the proper numerical values $g = 9.80\ \mathrm{m\,s^{-2}}$, $R = 6.37 \times 10^6$ m, $G = 6.67 \times 10^{-11}\ \mathrm{m^3\,kg^{-1}\,s^{-2}}$, gives

$$M = 5.98 \times 10^{24}\ \mathrm{kg}$$

The student must note that in working this example we used the distance of the mass m from the *center* of the Earth. In other words, we are implicitly assuming that the force on m is the same as if all the mass of the Earth were concentrated at its center. This assumption will be justified in Section 11.9.

EXAMPLE 11.2

Computation of the mass of a planet that has a satellite.

▷ Suppose that a satellite of mass m describes, with a period P, a circular orbit of radius r around a planet of mass M. The force of attraction between the planet and the satellite is $F = GMm/r^2$. This force must be equal to m times the centripetal acceleration $v^2/r = 4\pi^2 r/P^2$ of the satellite. Thus,

$$\frac{4\pi^2 mr}{P^2} = \frac{GMm}{R^2}$$

Canceling the common factor m and solving for M gives

$$M = 4\pi^2 r^3/GP^2.$$

We suggest that the student use this expression to re-evaluate the mass of the Earth, using the data for the Moon, $r = 3.84 \times 10^8$ m and $P = 2.36 \times 10^6$ s. Agreement with the result of Example 11.1 is a proof of the consistency of the theory. This formula can also be used for obtaining the mass of the Sun, using the data for the different planets.

11.4 Inertial and gravitational mass

In Chapter 6 we introduced the concept of **inertial mass** and called it simply **mass**. We also assumed that the laws of motion are of universal validity. Therefore they are the same for all kinds of matter, whether they be electrons, protons, neutrons, or groups of these particles. On the other hand, to characterize the strength of the gravitational interaction, we should have attached to each portion of matter a **gravitational mass** designated by m_g. We should then have written Equation 11.2 in the form

$$F = G\,\frac{M_g m_g}{r^2}$$

However, if we assume that gravitation is a universal property of all kinds of matter, it seems reasonable to consider that gravitational mass is proportional to inertial mass. Therefore, the ratio

$$K = \frac{\text{gravitational mass, } m_g}{\text{inertial mass, } m}$$

must be the same for all bodies. By a proper choice of units for m_g, we can make this ratio equal to one. Then the same number may be used for the gravitational mass as for the inertial mass. This has been done implicitly by our selection of the value of the constant G. The constancy of K, which is equivalent to the constancy of G, has been verified experimentally for all kinds of bodies with a precision better than one part in 10^{11}, and can be considered as a fundamental property of matter. The well-proven fact that all bodies near the Earth fall with the same acceleration is an indication of the fact that inertial and gravitational mass are the same. The acceleration of gravity is, according to Equation 11.5,

$$g = \frac{GM}{R^2}$$

where g is independent of the mass of the falling body. If m_g were different from m we would write instead $mg = Gm_gM/R^2$ or

$$g = \frac{m_g}{m}\frac{GM}{R^2}$$

If the ratio m_g/m is not the same for all bodies, the acceleration g would be different for each body, contrary to experience. Therefore, in what follows we shall use the term 'mass' to refer to either the inertial or the gravitational mass, since the two are indistinguishable, within the precision of our measurements.

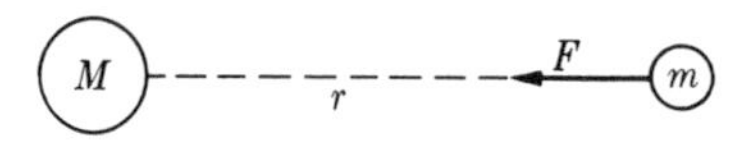

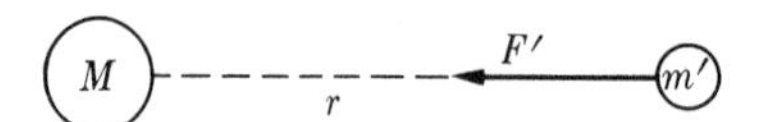

Figure 11.8 Method for comparing two masses m and m' by means of their gravitational interaction with a third mass M.

A way of measuring or comparing the masses of two bodies is by using a third body as reference. Consider two masses m and m' placed at the same distance r from a third mass M (Figure 11.8). Then, according to Equation 11.2, the forces on m and m' are

$$F = \frac{GMm}{r^2} \quad \text{and} \quad F' = \frac{GMm'}{r^2}$$

The ratio of these two forces is $F/F' = m/m'$. Therefore, if we have a method of comparing forces without necessarily measuring each one, the preceding relation provides a method for comparing and measuring masses. The principle of the balance allows us to use this method when the reference body M is the Earth. The balance achieves equilibrium when the two forces are equal, and therefore the masses are equal. We have now justified the method indicated in Section 2.3 for measuring mass by means of a balance.

Around 1909 Roland Eötvös (1848–1919) demonstrated experimentally, using a very sensitive double torsion balance, that all bodies with the same mass are subject to the same gravitational force, regardless of their chemical composition. His experiments confirmed the proportionality of the gravitational force to the mass of a body and the equivalence of gravitational and inertial mass. More recently (1964), Roll, Krotkov and Dicke compared the gravitational attraction of the Sun on different materials having the same inertial mass. They used a highly sensitive torsion balance, similar to that of Figure 11.5 where the masses m were of gold and copper respectively, and observed the position of the rod as the Earth rotated around its axis and moved along its orbit. They could not detect any change in the position of the rod due to a fluctuation in the Sun's attraction on the two masses as the Earth rotated, confirming with a precision of the order of 10^{-11} the equivalency of the gravitational and inertial masses and the proportionality between the gravitational force and the mass of a body.

EXAMPLE 11.3

Accelerations of two bodies subject to their mutual gravitational interaction.

▷ One important consequence of the identity between gravitational and inertial mass is that when two bodies exert gravitational forces on each other, the acceleration of each body is proportional to the mass of the other. Consider two bodies of masses m_1 and m_2 separated by a distance r. The magnitudes of their accelerations, using Equation 11.2 for the force, are

$$m_1 a_1 = G\frac{m_1 m_2}{r^2} \quad \text{or} \quad a_1 = \frac{Gm_2}{r^2}$$

$$m_2 a_2 = G\frac{m_1 m_2}{r^2} \quad \text{or} \quad a_2 = \frac{Gm_1}{r^2}$$

which may be combined to give the relation

$$\frac{a_1}{a_2} = \frac{m_2}{m_1}$$

In the particular case where m_1 is much larger than m_2, as is the case of a body falling toward the Earth or the Earth moving around the Sun, the acceleration a_1 is very small compared to the acceleration of m_2. In a first approximation, the motion of the more massive body can be ignored.

11.5 Gravitational potential energy

Because the gravitational interaction given by Equation 11.2 is central and depends only on the distance, it corresponds to a conservative force. We may therefore associate with it a **gravitational potential energy**. As shown below, the gravitational potential energy of the system composed of two particles whose masses are M and m and are separated by a distance r is

$$E_p = -\frac{GMm}{r} \tag{11.6}$$

If the particles move with velocities V and v relative to an inertial frame and are subject only to their gravitational interaction, their total energy in that frame is

$$E = E_k + E_p = \tfrac{1}{2}MV^2 + \tfrac{1}{2}mv^2 - \frac{GMm}{r} \tag{11.7}$$

However, when M is much larger than m ($M \gg m$), as in the case of the Sun and the planets or the Earth–Moon system, we may assume that M is at rest in an inertial system (Section 6.7) and write the total energy in that inertial system as

$$E = \tfrac{1}{2}mv^2 - \frac{GMm}{r} \tag{11.8}$$

In the rest of this chapter we shall assume that one of the masses is always much larger than the other, unless specified otherwise.

Derivation of gravitational potential energy

Take the origin of coordinates at M and consider only the force acting upon m (Figure 11.9). We note that $\boldsymbol{F}$, since it is attractive, is in the direction opposed to the vector $\boldsymbol{r} = \boldsymbol{OA}$. Therefore, instead of Equation 11.2, we may write the force on m in vector form as

$$\boldsymbol{F} = -\frac{GMm}{r^2}\boldsymbol{u} \tag{11.9}$$

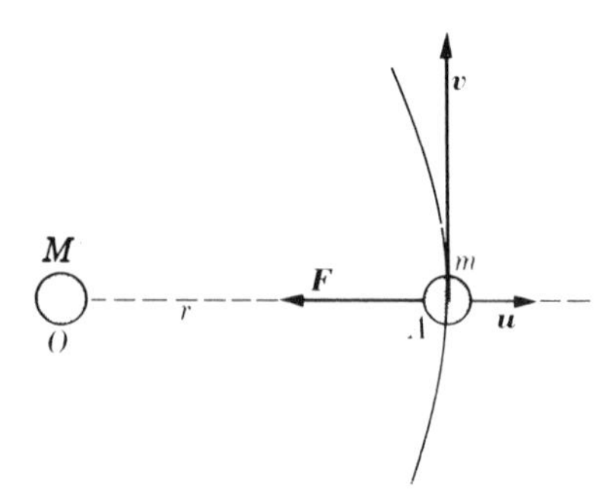

Figure 11.9 The gravitational attraction of M is opposed to the vector $\boldsymbol{r} = \boldsymbol{OA}$ directed away from M.

where the negative sign reflects the fact that $\boldsymbol{F}$ and $\boldsymbol{r}$ have opposite directions and $\boldsymbol{u}$ is a unit vector in the direction of $\boldsymbol{r}$.

The gravitational force 11.9 is equal to the negative of the gradient of the potential energy. In our case, since the force is central and acts along the radius, the potential energy depends only on r. Therefore, using Equation 9.28,

$$F = -\frac{dE_p}{dr}$$

From Equation 11.9, we have for this case that $F = -GMm/r^2$. Thus we get

$$\frac{dE_p}{dr} = G\frac{Mm}{r^2}$$

Integrating, and assigning the value zero to the potential energy at a very large distance ($r = \infty$), we obtain

$$E_p = \int_0^{E_p} dE_p = GMm\int_\infty^r \frac{dr}{r^2} = GMm\left(-\frac{1}{r}\right)_\infty^r = -\frac{GMm}{r}$$

for the gravitational potential energy of the system composed of masses M and m.

11.6 Relation between energy and orbital motion

Consider two particles of masses M and m, where particle M is very massive compared with m ($M \gg m$), so that M practically is at rest in an inertial system; we may then express the energy of the system in that inertial system as (recall Equation 11.8)

$$E = \tfrac{1}{2}mv^2 - \frac{GMm}{r}$$

If particle m moves in a circular orbit around M, the centripetal force acting on the mass is given by Equation 8.2, $F_N = mv^2/r$. Replacing F_N by the gravitational force of Equation 11.2, we have $mv^2/r = GMm/r^2$. Therefore the orbital kinetic energy is

$$\tfrac{1}{2}mv^2 = \frac{GMm}{2r}$$

and the **orbital velocity** is

$$v = \left(\frac{GM}{r}\right)^{\frac{1}{2}}.$$

Then the total energy of the circular orbit reduces to

$$E = \frac{GMm}{2r} - \frac{GMm}{r}$$

or

$$E = -\frac{GMm}{2r} \tag{11.10}$$

indicating that the total energy of the system is negative and equal in absolute value to the kinetic energy. This result is more general than our proof may suggest; all *elliptical* (or bound) orbits have a negative total energy ($E < 0$) when the gravitational potential energy is set to zero for infinite separation (Figure 11.10). The energy in this case is given by Equation 11.10 with r replaced by the semi-major axis a of the ellipse, $E = -GMm/2a$.

The bound nature of the orbit means that the kinetic energy is not enough at any point in the orbit to take the particle to infinity. This can be seen because, at an infinite distance the second term in Equation 11.8 is zero and we have $E = \tfrac{1}{2}mv^2$, a relation impossible to satisfy if E is negative.

Consider a particle whose energy is positive, that is, the particle has a velocity v when it is at a distance r such that $E > 0$. The particle can then reach infinity and still have some kinetic energy left. From Equation 11.8, if we set

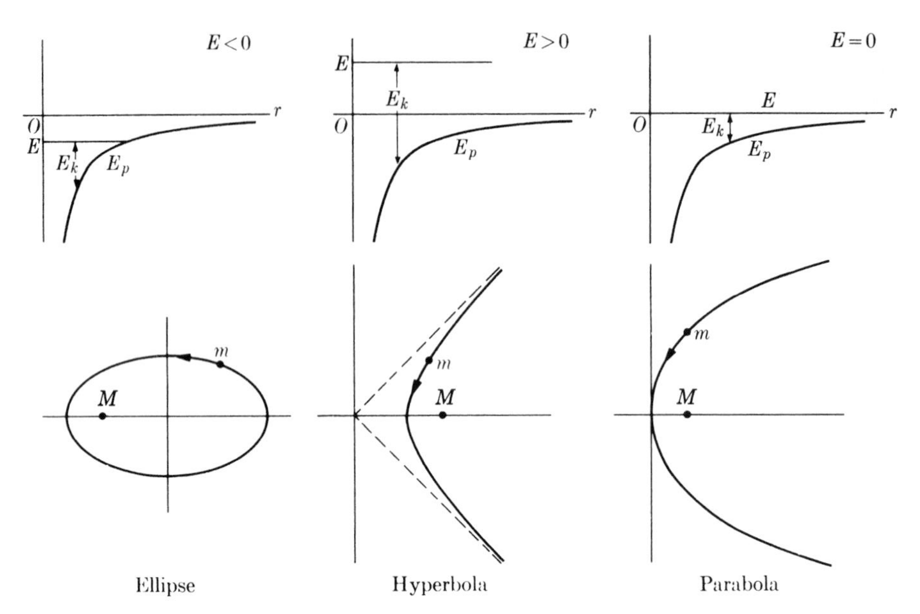

Figure 11.10 Relation between total energy and path for motion under an attractive inverse square force. In all cases, $E_p = -GMm/r$.

$r = \infty$, and designate the velocity at infinity by v_∞, then

$$E = \tfrac{1}{2}mv_\infty^2 \qquad \text{or} \qquad v_\infty = \left(\frac{2E}{m}\right)^{\frac{1}{2}} \tag{11.11}$$

Suppose now that a particle m is initially at a very large distance from M and is thrown toward it with velocity v_∞, called the **approach velocity**, so that the total energy is thus determined by Equation 11.11. As the particle m approaches M, its potential energy decreases (becoming more negative). The kinetic energy increases until it reaches its maximum value at the point of closest approach, which depends on the angular momentum of the particle (recall Figure 9.23). After that the particle begins to recede, losing kinetic energy until eventually, at large distances, its velocity is again v_∞. The path is an open curve, and it can be proved to be a *hyperbola*. This is the case for some comets as they approach a solar system as well as for some of the probes that have been launched to explore outer space. If the particle m is aimed directly at M, so that its angular momentum relative to M is zero, it eventually hits M in a head-on collision.

The particular case of zero total energy ($E = 0$) is interesting because then the particle, according to Equation 11.11, is at rest at infinity ($v_\infty = 0$). The orbit is still open but instead of being a hyperbola, it is now a *parabola*. Physically it corresponds to the situation where particle m is released at a distance from M with an initial velocity that makes its kinetic energy equal to its potential energy.

Figure 11.10 shows the three possible cases we have discussed, indicating in each one the total energy, the potential energy, the kinetic energy and the type of orbit.

These results are very important in placing an artificial satellite in orbit. Suppose that a satellite is launched from the Earth. After reaching its maximum height h, the satellite receives a final thrust at point A, producing a horizontal velocity v_0 (Figure 11.11), called the **insertion velocity**. The total energy of the satellite at A relative to the Earth is thus

$$E = \tfrac{1}{2}mv_0^2 - \frac{GMm}{R+h} \qquad \textbf{(11.12)}$$

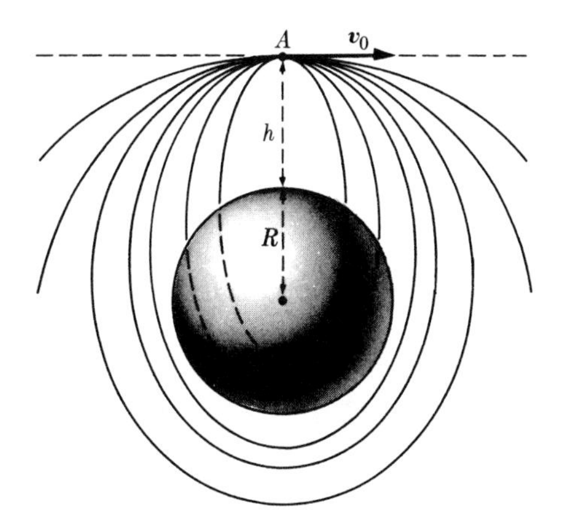

Figure 11.11 Paths of a particle thrown horizontally from a height h above the Earth's surface with a velocity v_0.

The orbit will be an ellipse, a parabola or a hyperbola depending on whether E is negative, zero or positive. In all cases the center of the Earth is at one focus of the path. Therefore the satellite will keep moving in an elliptical orbit, or will escape from the Earth, depending on whether

$$v_0^2 \lesseqqgtr \frac{2GM}{R+h}$$

If the energy is too low, the elliptical orbit will intersect the Earth and the satellite will fall back.

While the energy E alone determines the *size* or semi-major axis of the orbit, it can be shown that the *shape* or *eccentricity* of the orbit is determined by the total energy E and the angular momentum L. In the case of elliptical orbits, the larger the angular momentum, the less elongated is the orbit (Figure 11.12).

The same considerations apply to a natural satellite, such as the Moon. Clearly, for interplanetary space probes an orbit with positive energy relative to the Earth may be necessary. Generally, some guidance mechanism is required to adjust the path of the satellite after launching (see Chapter 12).

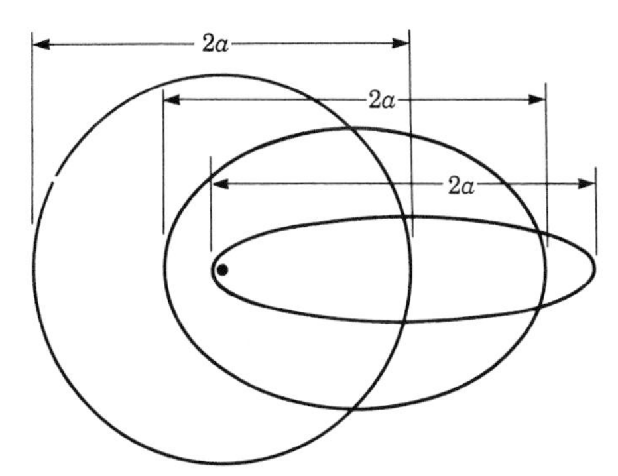

Figure 11.12 Elliptical orbits for different values of the angular momentum but the same energy. All orbits have the same focus and major axis, but differ in eccentricity.

We must note that a hyperbola has two branches and, under an inverse square attractive force, only the branch around the center of attraction is described (the right branch in Figure 11.13). If the force is repulsive, as may occur with electric forces, the orbit corresponds to the branch on the left in Figure 11.13. In this case (i.e. for a repulsive force), the potential energy is positive,

$$E_p = +\frac{C}{r}$$

Therefore the total energy $E = \frac{1}{2}mv^2 + C/r$ is always positive, and there are no bound orbits.

The preceding considerations would be sufficient to provide a complete analysis of planetary motion if the motion of a planet around the Sun was not affected by the other planets and heavenly bodies. In other words, the orbit of any planet would be a perfect ellipse if there were no forces other than the Sun's acting on it. However, the presence of other planets introduces *perturbations* in a planet's orbit. These perturbations can be calculated with great accuracy by means of special techniques called **celestial mechanics**. They can be analyzed, essentially, as two effects. One effect is that the elliptical path of a planet is not closed. Rather, the major axis of the ellipse rotates very slowly around the focus where the Sun is located, an effect called the **precession of the perihelion** (Figure 11.14(a)). The other effect is a periodic variation of the eccentricity of the ellipse about its average value, resulting in an oscillation of the shape of the orbit, as indicated in Figure 11.14(b). These changes occur very slowly. In the case of the Earth they have a period of the order of about 10^5 years (about 21′ of arc per century for the motion of the perihelion). Even so, they have produced noticeable effects, especially in slowly changing climatic conditions of the Earth. These changes have been identified by geophysicists who have studied the different layers of the Earth's crust. The orbit of the Moon around the Earth is also greatly perturbed by the Sun's attraction.

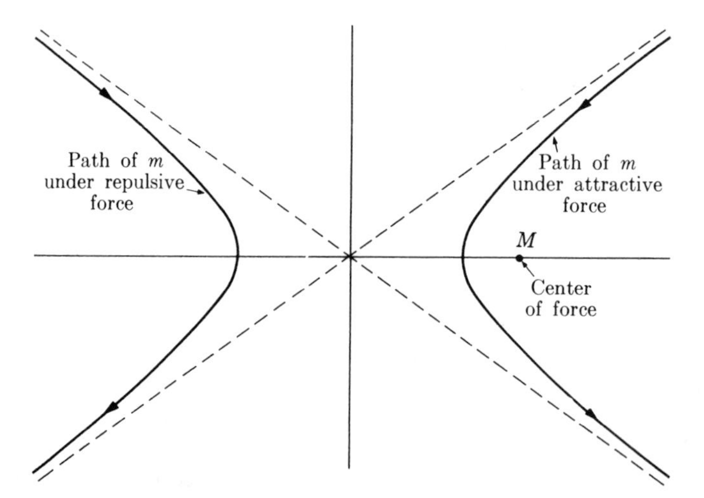

Figure 11.13 Hyperbolic paths under attractive and repulsive inverse square forces.

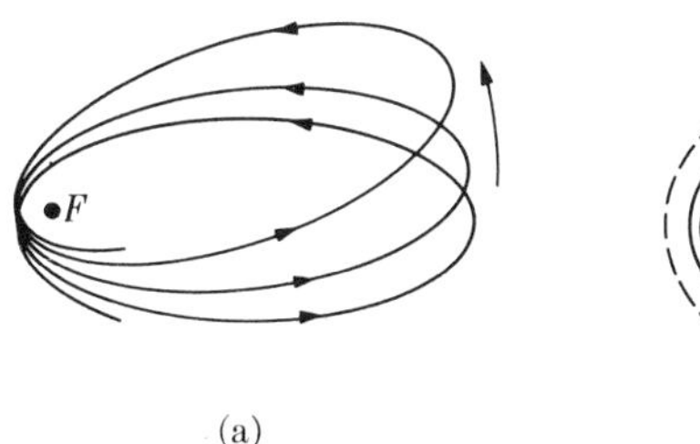

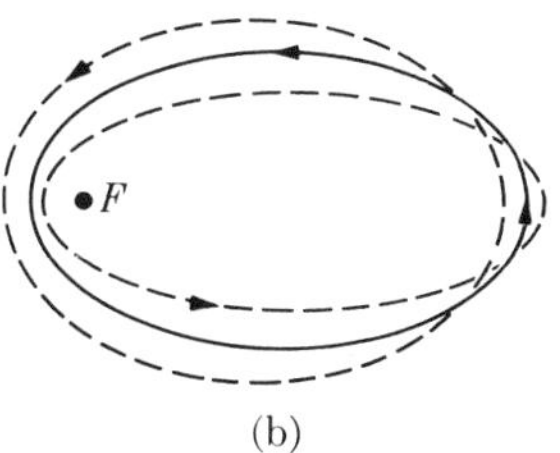

Figure 11.14 Perturbation effects on planetary motion. (a) Rotation of the axis of the ellipse. (b) Oscillation in the eccentricity of the ellipse. The two effects have been greatly exaggerated.

EXAMPLE 11.4

Computation of the **escape velocity** of a body from the Earth.

▷ The escape velocity is the minimum velocity with which a body must be fired from the Earth if it is to reach infinity. In order for a particle to reach infinity, the total energy must be zero or positive, and therefore the minimum velocity will correspond to zero total energy. Therefore, from Equation 11.8 with $E = 0$, and calling M the mass of the Earth, R its radius, and v_e the escape velocity of the projectile of mass m, we have

$$\tfrac{1}{2}mv_e^2 - \frac{GMm}{R} = 0 \quad \text{or} \quad \tfrac{1}{2}mv_e^2 = \frac{GMm}{R}$$

This equation gives the relation between v_e and R at the launching station. Thus the escape velocity from the Earth is

$$v_e = \left(\frac{2GM}{R}\right)^{1/2} = 1.12 \times 10^4 \text{ m s}^{-1} \qquad \textbf{(11.13)}$$

This is equal to 40 250 km h^{-1}. Note that the escape velocity is independent of the mass of the body. However, the **thrust** required to accelerate a body until it reaches the escape velocity does depend on the mass of the body. That is why heavier missiles and satellites require more powerful boosters.

A particle projected from the Earth with a velocity v_e given by Equation 11.13 will have zero velocity when it reaches infinity. If the velocity is greater than v_e, the particle will reach infinity with some velocity still left (recall Equation 11.11). If the launching velocity is less than v_e, the particle will fall back to the Earth, unless something is done to change its condition. For example, it may be placed into a bound orbit by successive stages of the propelling rocket as the direction of the velocity is adjusted. This was explained in connection with Figure 11.11.

The concept of escape velocity is also useful in determining the escape of gases from the Earth's atmosphere. If we assume that the gases composing the atmosphere are in thermal equilibrium, the average velocity of the molecules for gases found in the Earth's atmosphere at its average temperature are:

Hydrogen	1908 m s^{-1}
Helium	1350 m s^{-1}
Oxygen	477 m s^{-1}
Nitrogen	510 m s^{-1}
Carbon dioxide	407 m s^{-1}

In all cases the average velocity is much smaller than the escape velocity and thus we could conclude that no gas molecule in the atmosphere can overcome the gravitational attraction and escape from the Earth. But this would be a wrong conclusion. The notion of average velocity means that many molecules move with velocities either larger or smaller (see Chapter 17). Even if the average velocity is smaller than v_e, a certain number of molecules move with velocities that are equal to or larger than v_e. These molecules may escape the Earth, especially if they are in the upper layers of the atmosphere. From the above figures, we see that this effect is more important for the lighter gases than for the heavier. This is one of the reasons why hydrogen and helium are relatively scarce in our atmosphere. It has been estimated that, due to this gravitational effect, hydrogen is escaping from the Earth at the rate of about 600 kg per year. This does not represent the total loss of hydrogen from the Earth's atmosphere, however, and the net loss is different because of other processes.

For the planet Mercury, the escape velocity is much smaller than for the Earth; Mercury has probably lost its atmosphere completely. The same is true of the Moon, where there is no atmosphere. Venus has an escape velocity almost the same as the Earth's and still retains an appreciable atmosphere. In fact, the atmospheric pressure on Venus is about ninety times that of the Earth's atmospheric pressure. Mars has an escape velocity about one-sixth that of the Earth, and thus still retains some atmosphere, but it has lost a proportionately larger fraction of its atmosphere. Measurements by space probes have shown that the atmospheric pressure on Mars is much less than on the Earth. For the other planets, the escape velocity is greater than the Earth, and hence they still retain most of their original atmospheres. However, for other reasons, the composition of the atmospheres of those planets is different from the Earth's.

EXAMPLE 11.5

Black holes.

▷ According to the special theory of relativity (Chapter 19), the maximum velocity that can exist is the velocity of light, $c = 3 \times 10^8 \text{ m s}^{-1}$. We may then speculate on the mass and radius of a stellar body that has an escape velocity of c. If v is replaced by c in Equation 11.13, we have

$$c^2 = \frac{2GM}{R} \quad \text{or} \quad R = \frac{2GM}{c^2} \tag{11.14}$$

This quantity is called the **Schwarzschild radius** and is usually designated R_S. Introducing the numerical values for G and c, we have the relation

$$R_S = 1.485 \times 10^{-2} M \tag{11.15}$$

where M is in kg and R_S in m. For the mass of our Sun the value of the Schwarzschild radius is 2.96×10^3 m, which should be compared with the actual radius given in Table 11.1. For a body with the mass of Earth, R_S is 8.86×10^{-3} m or about the size of a pea.

A body with a mass M and radius R_S or smaller produces such a strong gravitational field at its surface that no particle on its surface can escape. This even applies to electromagnetic radiation (photons), including light. For that reason a body whose radius is equal to or smaller than its R_S is called a **black hole**, a name suggested by John A. Wheeler in 1969. Astrophysicists have identified several possible black holes in the universe, including one that might exist at the center of our galaxy with a radius about the same as that of the Sun but a mass many times larger.

The method we have used in this example to obtain the Schwarzschild radius was first suggested by J. Michell in 1784, at a time when the corpuscular theory of light was widely accepted and long before the theory of relativity was formulated. A correct analysis is presented in Section 19.10.

EXAMPLE 11.6

A body is released at a distance r from the center of the Earth. Find its velocity when it strikes the Earth's surface.

▷ The body's initial velocity is zero and its total energy, according to Equation 11.9, is therefore

$$E = -GMm/r$$

where m is the mass of the body and M the mass of the Earth. When it reaches the Earth's surface, its velocity is v and its distance from the center of the Earth is the Earth's radius R. Thus

$$E = \tfrac{1}{2}mv^2 - \frac{GMm}{R}$$

Equating both values of E, since the energy has remained constant (we neglect air friction), we have

$$\tfrac{1}{2}mv^2 - \frac{GMm}{R} = -\frac{GMm}{r}$$

Solving for v^2, we have

$$v^2 = 2GM\left(\frac{1}{R} - \frac{1}{r}\right)$$

Remembering from Equation 11.5 that $g = GM/R^2$, we obtain

$$v^2 = 2gR^2\left(\frac{1}{R} - \frac{1}{r}\right)$$

Because of the time reversibility of the laws of motion, this expression may also be used for finding the distance r reached by a body thrown vertically with a velocity v from the Earth's surface.

If the body is released at a great distance, so that $1/r$ is negligible compared with $1/R$, we get

$$v = (2gR)^{1/2} = \left(\frac{2GM}{R}\right)^{1/2} = 1.12 \times 10^4\ \mathrm{m\,s^{-1}}$$

in agreement with the result given in Equation 11.13 for the escape velocity from the Earth. This is not surprising, since this problem is just the reverse of the problem of Example 11.4. The above result gives, for example, an estimate of the velocity with which a meteorite strikes the upper atmosphere of the Earth.

EXAMPLE 11.7

Gravitational energy of a body that moves under a resultant force produced by two masses separated by a distance D.

▷ As a mass m moves along a line passing through a mass M, the gravitational potential energy $E_p = -GMm/r$ varies as shown in Figure 11.15. Therefore, as particle m approaches mass M and then recedes from it, its potential energy varies accordingly. For $E < 0$ the particle remains within a finite distance from M (bound orbit) as explained before, while if the particle has $E > 0$ it may approach from $-\infty$ and recede back to $+\infty$.

If we now consider two masses M_1 and M_2 separated by a distance D and combine the respective gravitational potential energies of a mass m, we obtain the result illustrated in Figure 11.16. The figure shows the variation of gravitational potential energy along the

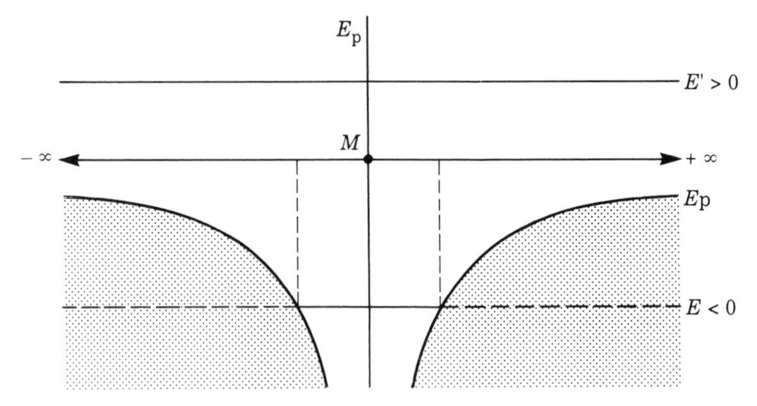

Figure 11.15

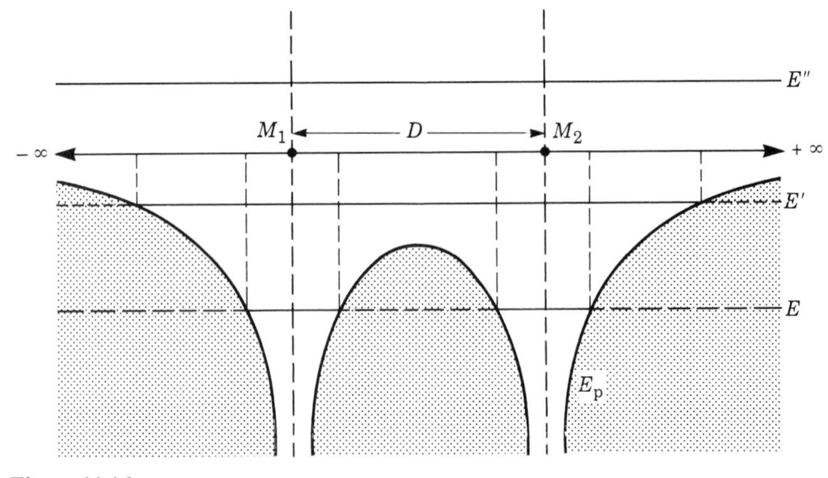

Figure 11.16

line joining the two masses. From the figure we note that if the energy of particle m is E, it moves in a bound orbit around either M_1 or M_2 (the inner solid lines in Figure 11.16), but neither is a true ellipse. For example, let M_1 be the Earth and M_2 be the Sun. Then the bound orbit around M_1 may correspond to the Moon, while the bound orbit around M_2 may correspond to the orbit of Mercury or Venus. For a larger energy, such as E', particle m is in a bound orbit but is not attached to either M_1 or M_2 and the orbits may become rather complex or even chaotic (see Note 11.5). For a positive energy, such as E'', particle m is unbound. This latter case may correspond, for example, to the orbit of an outer planet or to the path of a comet.

Note 11.1 General motion under gravitational attraction

In this note we shall analyze the possible orbits of a particle in a gravitational field in greater detail. First, recall Equation 9.39, which we write in the form

$$\left(\frac{\mathrm{d}r}{\mathrm{d}t}\right)^2 = \frac{2}{m}\left(E - E_p - \frac{L^2}{2mr^2}\right)$$

Next, we use the definition of angular momentum, $L = mr^2\omega$, with $\omega = \mathrm{d}\theta/\mathrm{d}t$, to write

$$\frac{\mathrm{d}\theta}{\mathrm{d}t} = \frac{L}{mr^2}$$

We can combine these two equations to obtain a relation between r and θ alone. The result is

$$\left(\frac{\mathrm{d}r}{\mathrm{d}\theta}\right)^2 = \frac{2mr^4}{L^2}\left(E - E_p - \frac{L^2}{2mr^2}\right)$$

This equation allows us to determine the relation between r and θ that describes the orbit when the potential energy $E_p(r)$ is known. Conversely, we can find $E_p(r)$ if we can determine $\mathrm{d}r/\mathrm{d}\theta$ from the equation of the orbit. We shall follow the second method, which is mathematically simpler.

From Appendix B (Equation B.37) the equation of a conic section in polar coordinates is given by

$$\frac{\varepsilon d}{r} = 1 + \varepsilon \cos\theta$$

where d is the distance from the focus to the directrix and ε is the eccentricity. The conic section is an ellipse, a parabola or a hyperbola depending on whether ε is smaller than, equal to or larger than one. Taking the derivative of the above equation with respect to θ, we obtain

$$-\frac{\varepsilon d}{r^2}\frac{\mathrm{d}r}{\mathrm{d}\theta} = -\varepsilon\sin\theta \qquad \text{or} \qquad \left(\frac{\mathrm{d}r}{\mathrm{d}\theta}\right)^2 = \frac{r^4\sin^2\theta}{d^2}$$

Equating the two equations for $(\mathrm{d}r/\mathrm{d}\theta)^2$ and canceling r^4 on both sides, we may write

$$\sin^2\theta = \frac{d^2 m}{L^2}\left(2(E - E_p) - \frac{L^2}{mr^2}\right)$$

Now, from the initial definition of the conic section above, we have $\cos\theta = d/r - 1/\varepsilon$. Then, using Equation B.4 we write

$$\sin^2\theta = 1 - \cos^2\theta = 1 - \left(\frac{d}{r} - \frac{1}{\varepsilon}\right)^2 = 1 - \frac{d^2}{r^2} + \frac{2d}{\varepsilon r} - \frac{1}{\varepsilon^2}$$

Substitution into the previous equation yields

$$1 - \frac{d^2}{r^2} + \frac{2d}{\varepsilon r} - \frac{1}{\varepsilon^2} = \frac{2d^2mE}{L^2} - \frac{2d^2mE_p(r)}{L^2} - \frac{d^2}{r^2}.$$

Canceling the d^2/r^2 terms on both sides and equating the two terms that are dependent on r we get

$$-\frac{2d^2mE_p}{L^2} = \frac{2d}{\varepsilon r} \quad \text{or} \quad E_p = -\frac{L^2}{md\varepsilon r}$$

This equation indicates that, to describe a conic section with the center of force at one focus, the potential energy E_p must vary with the distance as $1/r$; therefore, the force, which is $F = -\mathrm{d}E_p/\mathrm{d}r$, must vary as $1/r^2$. This generalizes Kepler's first law to include the hyperbola and the parabola in addition to the ellipse as possible orbits. When we equate those terms that are constant, we obtain

$$\frac{2d^2mE}{L^2} = 1 - \frac{1}{\varepsilon^2} \quad \text{or} \quad E = \frac{L^2}{2d^2m}\left(1 - \frac{1}{\varepsilon^2}\right)$$

which gives the total energy of the orbit in terms of fixed parameters. The total energy is negative, zero or positive depending on whether ε is smaller than, equal to or larger than one (that is, the orbit will be elliptical, parabolic or hyperbolic, respectively). For the case of elliptical orbits, the semi-major axis is given by Equation B.38

$$a = \frac{\varepsilon d}{1 - \varepsilon^2}$$

Therefore, the total energy can be written as

$$E = -\frac{L^2}{2\varepsilon dma}$$

Also, from Equation 11.6, $E_p = -GMm/r$, so that

$$-\frac{GMm}{r} = -\frac{L^2}{m\varepsilon dr} \quad \text{or} \quad \frac{L^2}{m\varepsilon d} = GMm$$

Making the corresponding substitution into the expression for the total energy, we see that

$$E = -G\frac{Mm}{2a}$$

Comparing this result with Equation 11.10, which was derived for circular orbits, we see that they are identical since $a = r$ for a circular orbit. This result also confirms that the total energy depends only on the semi-major axis a. So, all elliptical orbits having the same value for their semi-major axis have the same total energy, even though they have different eccentricities.

By using the expression $\varepsilon d = a(1 - \varepsilon^2)$, we may write another useful relation

$$L^2 = GMm^2\varepsilon d = GMm^2a(1 - \varepsilon^2)$$

Eliminating the semi-major axis a by using the previous expression for the total energy E, we can write the eccentricity of the orbit as

$$\varepsilon^2 = 1 + \frac{2E}{m}\left(\frac{L}{GMm}\right)^2$$

Thus we see that the eccentricity depends on the energy and the angular momentum.

In an inverse-square field, to a given total energy there may correspond many different angular momentum states and eccentricities.

This is of great importance in the discussion of atomic structure, because in an atom there may be several electrons that have the same energy but have different angular momenta.

We may summarize the preceding results by saying that

the 'size' of the orbit (as given by the semi-major axis) is determined by the energy and for a given energy, the 'shape' of the orbit (given by the eccentricity) is determined by the angular momentum.

Note 11.2 Gravitational energy of a spherical body

Consider a body of mass M and radius R. Suppose the body was formed by accretion; that is, by gradual addition of matter. As matter is added to the body, gravitational potential energy decreases (becomes more negative). For example, if a small mass m is added to the body, bringing it in from infinity until it is attached to the body, the additional gravitational potential energy of the system is (Figure 11.17)

$$E_p = -G\frac{Mm}{R}$$

This suggests that the gravitational energy of a body that started with a small amount of matter until it acquires the total mass M and a radius R is

$$E_p = -\alpha G\frac{M^2}{R}$$

where α is a constant, of order one, whose precise value depends on how matter was added to the body. For example, if the body is built maintaining a constant density, it can be shown that $\alpha = 3/5$ or 0.6. For the case of the Sun, using the values of Table 11.1, we obtain $E_p = -2.28 \times 10^{41}$ J.

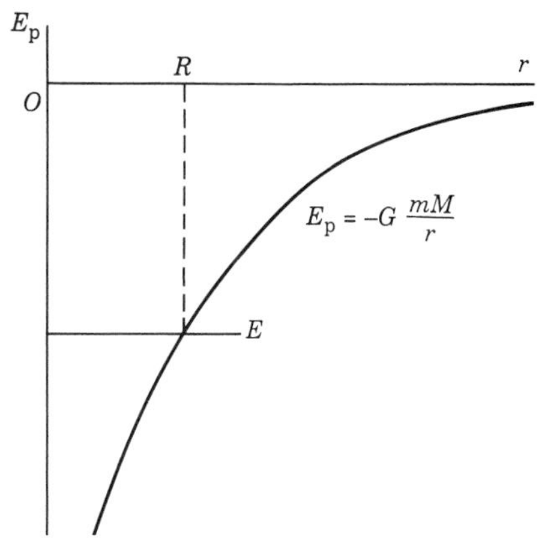

Figure 11.17

Note that if the radius of a body decreases while its mass is held constant, its gravitational potential energy decreases (becomes more negative). We can use the above equation to calculate the rate at which gravitational energy changes when a body of mass M contracts as a result of the gravitational attraction among its different components. This is a problem of great interest in analyzing stellar evolution. Assuming the mass M remains constant, the rate of change of gravitational energy of a star whose radius is changing is

$$\frac{dE_p}{dt} = \alpha G \frac{M^2}{R^2} \frac{dR}{dt}$$

If the body contracts, the radius diminishes and dR/dt is negative, making dE_p/dt also negative, which means that the gravitational potential energy decreases. In fact, what happens is that when a star suffers a gravitational contraction, part of the gravitational potential energy lost is released as gravitational radiation and part is transformed into kinetic energy of the nuclei composing the star. Therefore as the star contracts, its temperature increases and a series of nuclear reactions begins to take place that results in a further release of energy, as explained in Note 40.4. For example, the Sun radiates energy at a rate of 3.8×10^{26} W. If this radiated energy were all gravitational the Sun would be contracting at a rate of 36.7 m yr^{-1}, a quantity that would be easy to measure. No such rate of contraction has been observed. Therefore, we conclude that the source of the energy radiated by the Sun is not gravitational, but due to nuclear processes.

Note 11.3 Critical density of the universe

In Note 3.1 it was pointed out that the universe appears to be in a state of expansion. The relative velocity of separation of two galaxies is given by Hubble's law $V = HR$, where R is the separation of the two galaxies and H is Hubble's constant. Galaxies exert a gravitational attraction on each other, in the same way that the Sun attracts planets or the Earth attracts bodies near its surface. It seems reasonable, then, to think that gravitational expansion is at the expense of the kinetic energy in the universe.

Recall that when a body is thrown upward from the Earth, its kinetic energy decreases as its potential energy increases. The body will fall back or continue to move away forever, depending on its initial velocity (see Example 11.4 about escape velocity); that is, whether its total energy is negative or positive. For the same reason, the rate of expansion of the universe decreases continuously as its kinetic energy transforms into gravitational potential energy. Accordingly, the possibility exists that the universe will either continue to expand forever or stop expanding and begin to contract under the gravitational interaction. In such a case the gravitational potential energy will transform back into kinetic energy and all the matter will eventually collapse into a Big Crunch. The universe could then emerge with a new Big Bang and start all over again. This possibility corresponds to an **oscillating universe** (Figure 11.18) where similar phenomena repeat each cycle, although not necessarily in the same way each time. The first case, of continuous expansion, allows for new processes to occur, even though there is less energy available as time goes on.

We call an oscillating universe **closed**, while the expanding universe is **open**. Whether the universe is closed or open is a question of great interest to physicists and cosmologists. To estimate the necessary requirements for either case, consider a sphere of radius R, large enough so that it contains several galaxies (Figure 11.19). If ρ is the average density of matter in the universe, the mass inside the sphere is

$$M = \tfrac{4}{3}\pi R^3 \rho$$

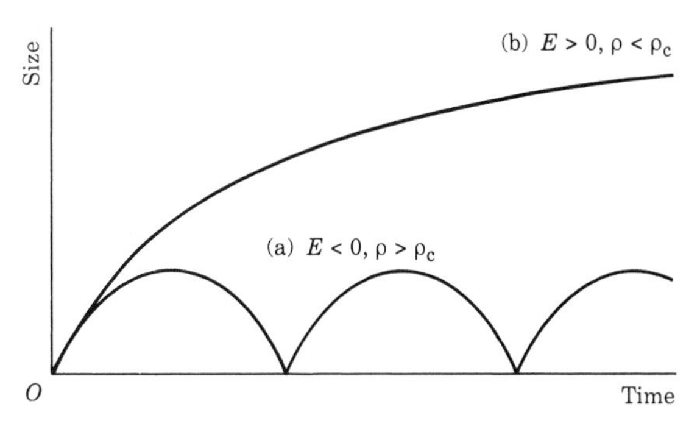

Figure 11.18

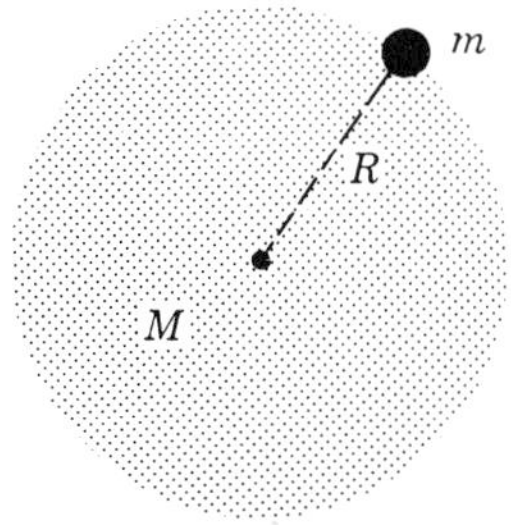

Figure 11.19

Now consider a system (perhaps a single galaxy) of mass m on the surface of the sphere. Its gravitational potential energy is

$$E_p = -G\frac{Mm}{R} = -G(\tfrac{4}{3}\pi R^2\rho)m$$

The kinetic energy of m, using Hubble's law, is

$$E_k = \tfrac{1}{2}mV^2 = \tfrac{1}{2}m(HR)^2 = \tfrac{1}{2}mH^2R^2$$

The total energy of the system of mass m is

$$E = E_k + E_p = \tfrac{1}{2}mH^2R^2 - (\tfrac{4}{3}\pi GR^2\rho)m = mR^2(\tfrac{1}{2}H^2 - \tfrac{4}{3}\pi G\rho)$$

According to Example 11.5, m will continuously move away from M or eventually start falling back, depending on whether $E > 0$ or $E < 0$. From the equation above the value of E will be positive, zero or negative depending on whether

$$\tfrac{1}{2}H^2 - \tfrac{4}{3}\pi G\rho \gtreqless 0$$

The average density of the universe for which $E = 0$ is called the *critical* density, designated ρ_c. Therefore, using accepted values for H and G, we have

$$\rho_c = \frac{3H^2}{8\pi G} \approx 2 \times 10^{-26}\,\mathrm{kg\,m^{-3}}$$

or about 12 protons per cubic meter. (Note that the best vacuum that can be produced at present is equivalent to a density of about $10^{-20}\,\mathrm{kg\,m^{-3}}$). Therefore, depending on whether the density ρ of the universe is greater or smaller than ρ_c, the universe should be closed or open. If $\rho = \rho_c$ we say that the universe is **flat**.

Current evidence, based on estimates of luminous matter in galaxies and other processes, indicates that the average density of the universe is of the order of $10^{-27}\,\mathrm{kg\,m^{-3}}$, supporting the idea that the universe is open. However, there is still great uncertainty about how much mass exists in the universe. Several considerations, which cannot be elaborated here, suggest the possible existence of a considerable amount of non-luminous or 'dark' matter that cannot be observed directly with current instrumentation but can be inferred by its gravitational effects on the internal motions of galaxies and clusters of galaxies. The added mass of the dark matter could make the universe flat or closed. However, the amount of dark matter that has been estimated is insufficient to make the universe closed, or even flat. One of the candidates for dark matter are the neutrinos that are emitted profusely in many nuclear processes that take place in stars. Another possibility for the missing mass is that intra-galactic dust may obscure some faintly luminous matter, which is then not observed on Earth. It has been claimed that one huge concentration of matter, called the **Great Attractor**, has been detected by its effect on the motion of many galaxies, including ours, which seem to be deflected towards a particular region of space.

On the other hand, strong theoretical considerations indicate that the universe should be flat. Calculations suggest that if the density had been slightly less than critical immediately after the Big Bang, the universe would have expanded so fast that neither stars nor galaxies could have been formed. And if the density at the start had been slightly larger than critical, the universe would have re-collapsed in less than a microsecond after the Big Bang. The problem still remains: what and where is the total mass in the universe?

11.7 Gravitational field

Suppose that we have a mass M and that we place another mass m at different positions $A, B, C, \dots$ around M (Figure 11.20). Due to its gravitational interaction with M, the mass m experiences a force at each position given by Equation 11.2. The force (recall Equation 11.9) can be written in vector form as

$$\boldsymbol{F} = -\frac{GMm}{r^2}\boldsymbol{u}$$

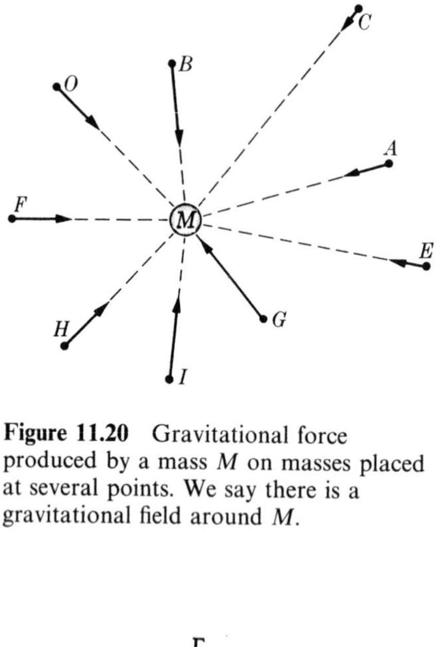

Figure 11.20 Gravitational force produced by a mass M on masses placed at several points. We say there is a gravitational field around M.

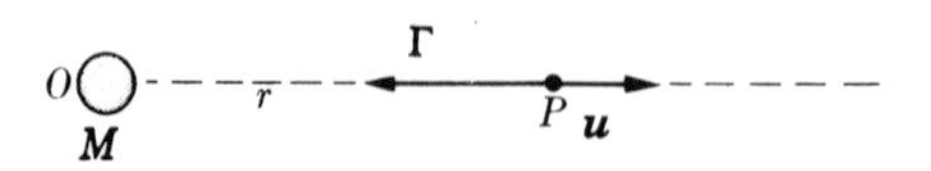

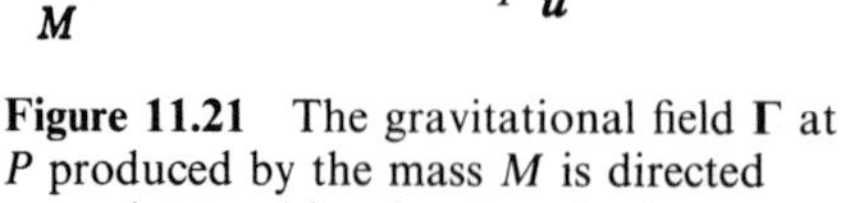

Figure 11.21 The gravitational field $\boldsymbol{\Gamma}$ at P produced by the mass M is directed toward M and is related to the force on mass m, which is used to probe the field, by $\boldsymbol{\Gamma} = \boldsymbol{F}/m$.

where $\boldsymbol{u}$ is a unit vector in the direction pointing from M to m (Figure 11.21). Of course, at each position of m the mass M experiences an equal but opposite force. However, we are interested only in what happens to m.

We may then conveniently say that the mass M produces, in the space around it, a physical situation that we call a **gravitational field**. This field is recognized by the force that M exerts on another mass, such as m, brought into that region. Whether something exists in the free space around M, even if we do not use a test mass m to probe the field, is something we can only speculate about. To a certain extent it is an irrelevant question, since we notice the gravitational field only when we bring in a second mass.

The **gravitational field strength $\boldsymbol{\Gamma}$** produced by the mass M at a point P is defined as the force exerted on the unit of mass placed at P. Then

$$\boldsymbol{\Gamma} = \frac{\boldsymbol{F}}{m} = -\frac{GM}{r^2}\boldsymbol{u} \tag{11.16}$$

Thus the gravitational field $\boldsymbol{\Gamma}$ has the direction opposite to that of the unit vector $\boldsymbol{u}$, which points along the line from the mass producing the field to the point where the field is computed (Figure 11.21). In other words,

the gravitational field always points toward the mass producing it.

Expression 11.16 gives the gravitational field at a distance r from a particle of mass M placed at O. We may then associate with each point in the space around M a vector given by Equation 11.16, such that the gravitational force exerted on any mass placed in that region is obtained by multiplying the mass of the particle by the corresponding field strength $\boldsymbol{\Gamma}$. That is,

$$\boldsymbol{F} = (\text{mass of particle}) \times \boldsymbol{\Gamma}$$

From its definition we see that the gravitational field strength is measured in N kg^{-1} or m s^{-2}, and it is dimensionally equivalent to an acceleration. Comparing Equation 11.16 with Equation 6.16,

$$\boldsymbol{g} = \frac{\boldsymbol{W}}{m}$$

we note that the acceleration of gravity may be considered as the gravitational field strength at the surface of the Earth.

Suppose now that we have several masses, $M_1, M_2, M_3, \ldots$ (Figure 11.22), each one producing its own gravitational field, $\boldsymbol{\Gamma}_1, \boldsymbol{\Gamma}_2, \boldsymbol{\Gamma}_3, \ldots$. The total force on a particle of mass m at P is

$$\begin{aligned}\boldsymbol{F} &= m\boldsymbol{\Gamma}_1 + m\boldsymbol{\Gamma}_2 + m\boldsymbol{\Gamma}_3 + \cdots \\ &= m(\boldsymbol{\Gamma}_1 + \boldsymbol{\Gamma}_2 + \boldsymbol{\Gamma}_3 + \cdots)\end{aligned} \tag{11.17}$$

We say that the resultant gravitational field at point P is the vector sum $\boldsymbol{\Gamma} = \boldsymbol{\Gamma}_1 + \boldsymbol{\Gamma}_2 + \boldsymbol{\Gamma}_3 + \cdots$. In other words, gravitational fields are linearly additive. Therefore, the relations

$$\boldsymbol{F} = m\boldsymbol{\Gamma} \quad \text{or} \quad \boldsymbol{\Gamma} = \frac{\boldsymbol{F}}{m} \tag{11.18}$$

are valid whether the gravitational field is produced by one mass or by many. This is one of the advantages of the notion of field. Gravitational fields can be calculated once we know the mass distribution.

A gravitational field can be represented by **lines of force**. A line of force is drawn in such a way that at each point the *direction* of the field is tangent to the line that passes through the point. By convention, the lines of force are drawn so

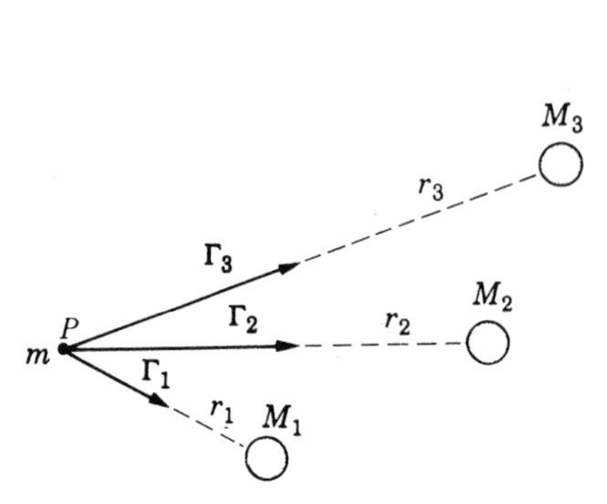

Figure 11.22 Resultant gravitational field of several masses.

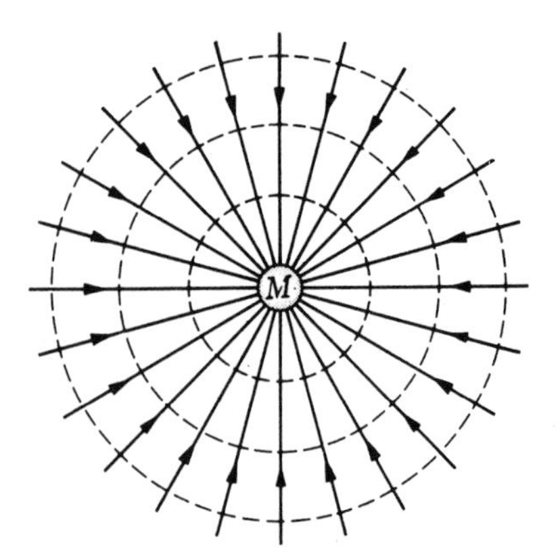

Figure 11.23 Lines of force and equipotential surfaces of the gravitational field of a point mass.

that their density is proportional to the *strength* of the field. Figure 11.23 depicts the field about a single mass; the lines of force are radial. Figure 11.24 shows the fields about two unequal masses, namely, the Earth and the Moon. The lines are not radial, and at point A the field strength is zero.

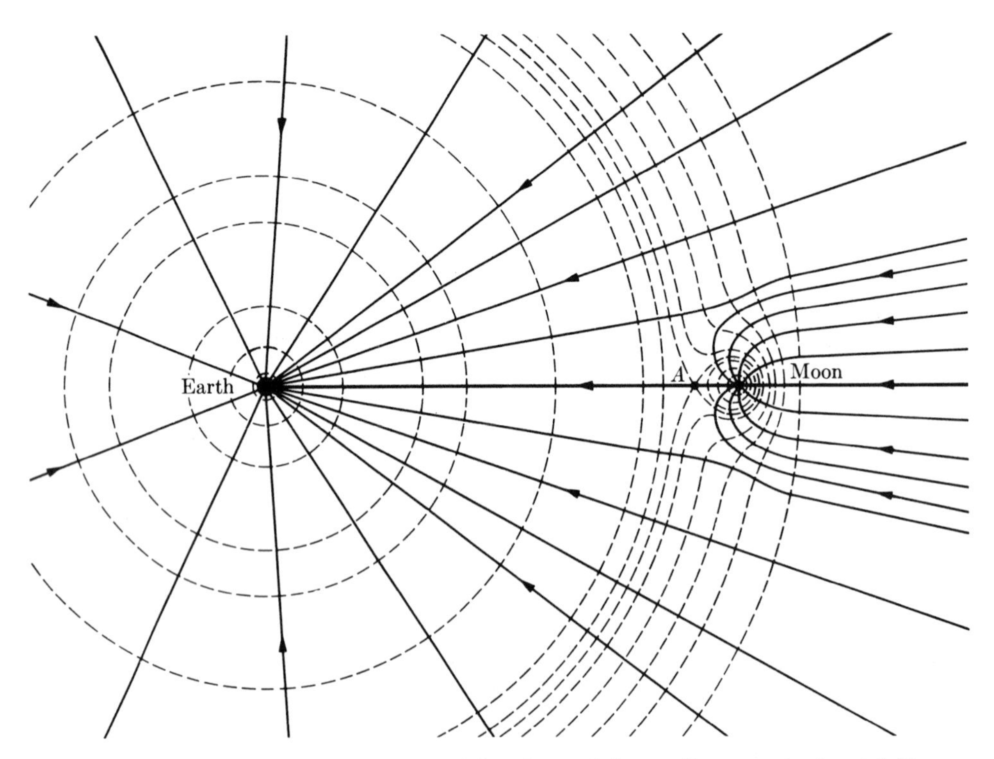

Figure 11.24 Lines of force and equipotential surfaces of the resultant gravitational field produced by the Earth and the Moon. At A the resultant gravitational field is zero. (After W. T. Scott, *Am. J. Phys.* **33**, 712 (1965).) The drawing contains only a few lines and surfaces.

11.8 Gravitational potential

The **gravitational potential** at a point is defined as the potential energy per unit mass placed at that point in a gravitational field. Thus if, at a certain point in a gravitational field, a mass m has a potential energy E_p, the gravitational potential at that point is

$$\Phi = \frac{E_p}{m} \quad \text{or} \quad E_p = m\Phi$$

The gravitational potential is expressed in the units J kg^{-1} or m^2 s^{-2}. For example, the potential energy of a body of mass m at a height y above the Earth's surface, with $y \ll R$, is $E_p = mgy$ (recall Equation 9.23). Therefore the gravitational potential at height y is

$$\Phi = \frac{E_p}{m} = \frac{mgy}{m} = gy$$

The energy of a particle in a gravitational field can then be written as

$$E = \tfrac{1}{2}mv^2 + m\Phi.$$

From Equation 11.8, the potential energy of masses M and m separated a distance r is $E_p = -GMm/r$. Dividing by m, we see that the gravitational potential at a distance r from a mass M is

$$\Phi = -\frac{GM}{r} \tag{11.19}$$

If, instead of one particle, we have several, as in Figure 11.22, the gravitational potential at P is the scalar sum

$$\Phi = \Phi_1 + \Phi_2 + \Phi_3 + \cdots$$

or

$$\Phi = -G\left(\frac{M_1}{r_1} + \frac{M_2}{r_2} + \frac{M_3}{r_3} + \cdots\right) = -G\sum_i \frac{M_i}{r_i} \tag{11.20}$$

We recall that the gravitational force $\boldsymbol{F}$ on a mass m is related to the gravitational potential energy E_p of the mass by the relation $F_s = -\mathrm{d}E_p/\mathrm{d}s$. But $\boldsymbol{F} = m\boldsymbol{\Gamma}$ and $E_p = m\Phi$. Hence we may write the following relation between the gravitational field Φ and the gravitational potential $\boldsymbol{\Gamma}$:

$$\Gamma_s = -\frac{\mathrm{d}\Phi}{\mathrm{d}s} \tag{11.21}$$

where Γ_s is the component of $\mathbf{\Gamma}$ in the direction of the displacement ds. Equation 11.21 shows that

> *the gravitational field is the negative of the gradient of the gravitational potential.*

If a particle of mass m moves from a point P_1 to a point P_2 along any path, the work done by the gravitational field is $W = -\Delta E_p$. Because $E_p = m\Phi$, we may write

$$W = -m\Delta\Phi \tag{11.22}$$

where $\Delta\Phi = \Phi_2 - \Phi_1$ is called the gravitational **potential difference**. By joining the points where the gravitational potential has the same value, we may obtain a series of surfaces called **equipotential surfaces**. For example, in the case of a single particle, when the potential is given by Equation 11.19, the equipotential surfaces correspond to the spheres $r = const.$, given by the dashed lines in Figure 11.23. In Figure 11.24 the equipotential surfaces have also been indicated by dashed lines. Note that in each case the equipotential surfaces are *perpendicular* to the lines of force. This can be verified in the following way.

Consider two points, very close to each other, on the same equipotential surface. When we move a particle from one of these points to the other, there is no change in potential energy because the two points have the same gravitational potential. Hence, the work done by the gravitational field acting on the particle is zero. The fact that the work is zero implies that the force is perpendicular to the displacement. Therefore

> *the direction of the gravitational field is perpendicular to the equipotential surfaces.*

11.9 Gravitational field of a spherical body

All the formulas stated so far in this chapter are strictly valid only for point masses. When we applied them in Section 11.5 to the motion of the planets around the Sun, it was under the assumption that their sizes are small compared with their separation. Even if this is true, their finite sizes may possibly introduce some geometrical factor in Equation 11.2. Similarly, when we were relating the acceleration of gravity g close to the Earth's surface to the mass and the radius of the Earth in Example 11.1, we used Equation 11.2, in spite of the fact that the above reasoning of relatively small size compared to the distance is not applicable. Newton himself was worried by this geometrical problem, and delayed the publication of his law until he found a correct explanation. We shall now examine this question in more detail.

Consider first a mass uniformly distributed over the surface of a sphere that is empty inside; that is, a spherical shell. It can be proved (see the calculation below) that the gravitational field and potential at points *outside* a mass uniformly distributed over a spherical shell is identical to the gravitational field and potential of a particle of the same mass located at the center of the sphere. At all points *inside* the spherical shell, the field is zero and the potential is constant. That is, if a is the radius of the shell and M its mass, the gravitational field and potential produced at the distance r from the center of the shell are

$$\mathbf{\Gamma} = -\frac{GM}{r^2}\mathbf{u} \qquad \Phi = -\frac{GM}{r} \qquad r > a \tag{11.23}$$

$$\mathbf{\Gamma} = 0 \qquad \Phi = -\frac{GM}{a} \qquad r < a \tag{11.24}$$

Figure 11.25 shows the variation of $\mathbf{\Gamma}$ and Φ with the distance from the center of the spherical shell. In moving from the center outward, the potential does not change in value when we cross the spherical shell. However, the gravitational field suffers a sudden change from 0 to $-GM/a^2$. Note also that the gravitational potential within the shell is constant and equal to the value at the surface.

Now suppose that the mass is uniformly distributed throughout the volume of the sphere, i.e. the sphere is solid. We may then consider the sphere to be built in an onion-like fashion, as the superposition of a series of thin spherical layers or shells. Each layer produces a field given by Equations 11.23 or 11.24. For a point *outside* the sphere (Figure 11.26), the distance r from the center to P is the same for all layers so that the masses add, again giving Equation 11.23. Therefore,

> *a solid homogeneous sphere produces, on outside points, a gravitational field and potential identical to those of a particle of the same mass located at the center of the sphere.*

However,

> *at points inside a homogeneous sphere the gravitational field is proportional to the distance from the center of the sphere.*

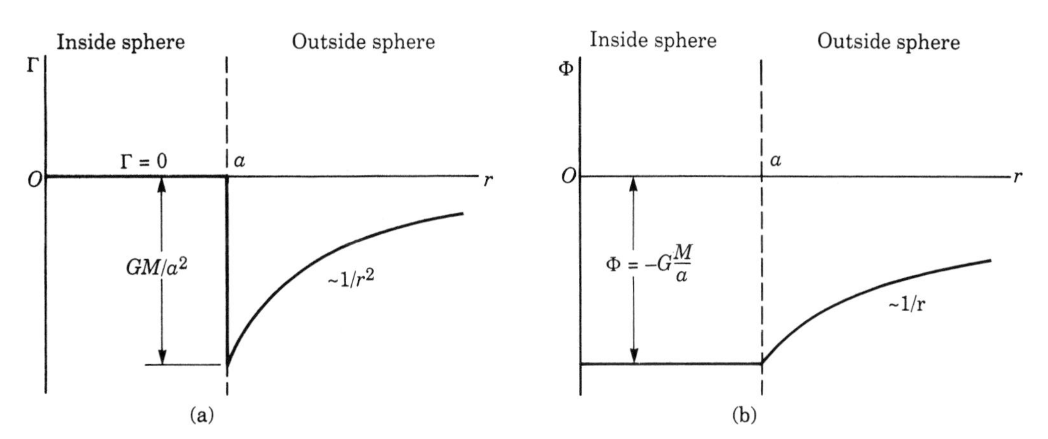

Figure 11.25 Variation of the gravitational field and potential, as a function of the distance from the center, for a mass distributed uniformly over a spherical shell.

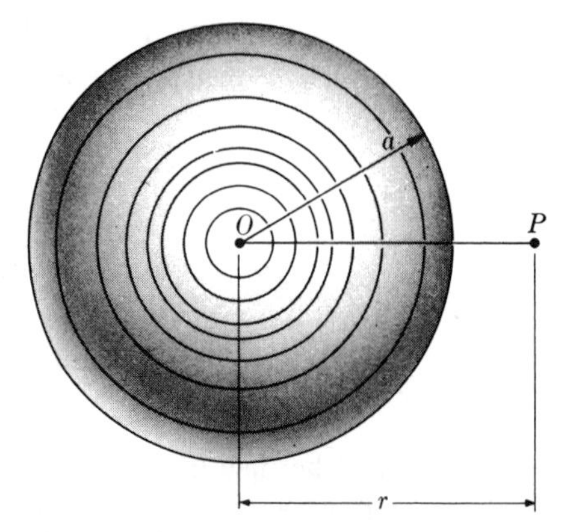

Figure 11.26 Calculation of the gravitational field at a point outside a solid sphere.

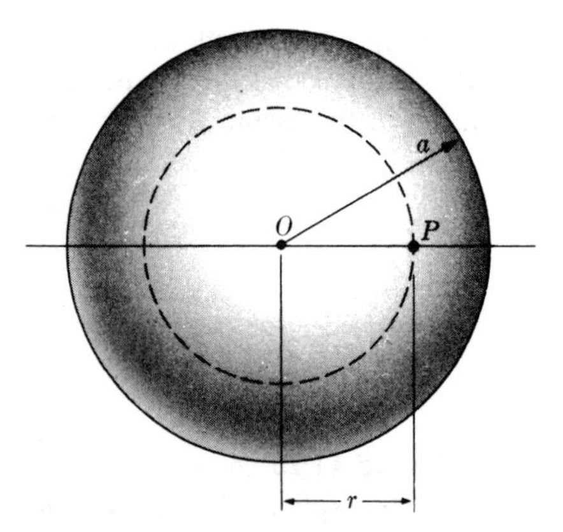

Figure 11.27 Calculation of the gravitational field at a point inside a solid sphere.

The reason why the field is proportional to the distance r inside the sphere is that, for an interior point such as P (Figure 11.27), only the mass within the sphere of radius $r = OP$ produces a field at P. As the point moves away from the center, the field decreases due to the inverse square law. However, the decrease is more than balanced by the increase in mass, which is proportional to the cube of the radius. Therefore, if a is the radius of the solid sphere, we have

$$\mathbf{\Gamma} = -\frac{GM}{r^2}\boldsymbol{u} \qquad r > a \tag{11.25}$$

$$\mathbf{\Gamma} = -\frac{GMr}{a^3}\boldsymbol{u} \qquad r < a \tag{11.26}$$

Figure 11.28 depicts the variation of $\mathbf{\Gamma}$ in terms of r for a solid homogeneous sphere. This figure gives, for example, the variation in the weight of a body as it moves from the center of the Earth to a point a great distance from it, if the Earth were homogeneous which is not the case (see Question 11.24).

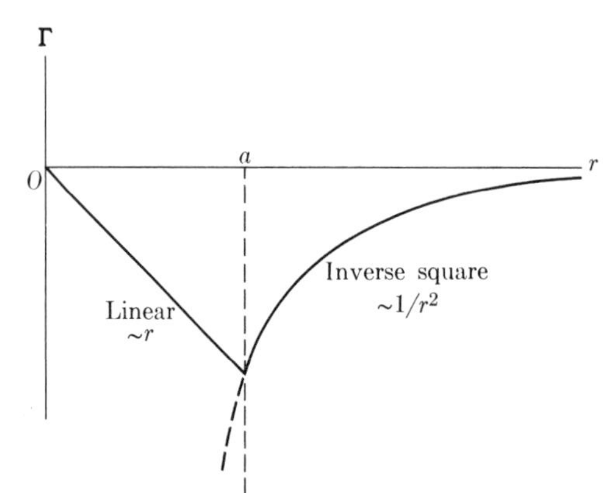

Figure 11.28 Variation of the gravitational field for a solid homogeneous sphere as a function of the distance from the center.

Note that in this spherical problem, the gravitational field at a point depends only on the distance from the point to the center of the sphere, not on the direction of the line joining the center to the point. This result was to be expected on the basis of the symmetry of the problem. If we were to consider a body with a different geometry or symmetry, or a non-homogeneous sphere (with the mass distributed without spherical symmetry), we should expect the angles or other factors to appear in the formulae for the gravitational field and potential. But for problems of spherical symmetry the properties depend on the distance from the point to the center only. The application of symmetry considerations greatly simplifies the solution of many problems in physics.

Calculation of the gravitational field of a spherical body

First consider the case of a spherical shell. Designate the radius of the shell as a and let r be the distance from any point P to the center C of the sphere. We are interested in obtaining the strength of the gravitational field at P. Consider first the case when P is outside the shell (Figure 11.29). We may divide the surface of the sphere into narrow circular strips, all with centers on the line AB. The radius of each strip is $a \sin\theta$ and the width is $a\,d\theta$. Therefore the area of the strip is

$$\begin{aligned}\text{Area} &= \text{length} \times \text{width}\\ &= (2\pi a \sin\theta)(a\,d\theta)\\ &= 2\pi a^2 \sin\theta\, d\theta.\end{aligned}$$

If M is the total mass, uniformly distributed over the surface of the sphere, the mass per unit area is $M/4\pi a^2$ and the mass of the circular strip is

$$\frac{M}{4\pi a^2}(2\pi a^2 \sin\theta\, d\theta) = \tfrac{1}{2}M \sin\theta\, d\theta$$

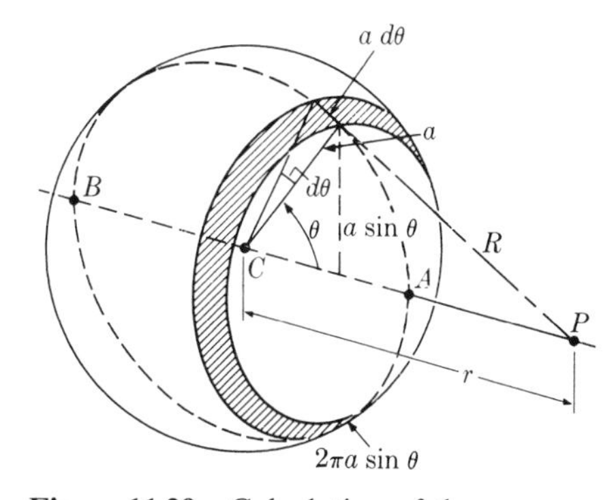

Figure 11.29 Calculation of the gravitational field at a point outside a mass distributed uniformly over a spherical shell.

All points of the strip are at the same distance R from P. Therefore, applying Equation 11.20, we find that the gravitational potential produced by the strip at P is

$$d\Phi = -G\frac{(\frac{1}{2}M \sin\theta\, d\theta)}{R} = -\frac{GM}{2R}\sin\theta\, d\theta$$

From Figure 11.30, using the law of cosines, Equation B.16, we note that

$$R^2 = a^2 + r^2 - 2ar\cos\theta$$

Differentiating, since a and r are constant, we obtain

$$2R\,dR = 2ar\sin\theta\, d\theta \qquad \text{or} \qquad \sin\theta\, d\theta = \frac{R\,dR}{ar}$$

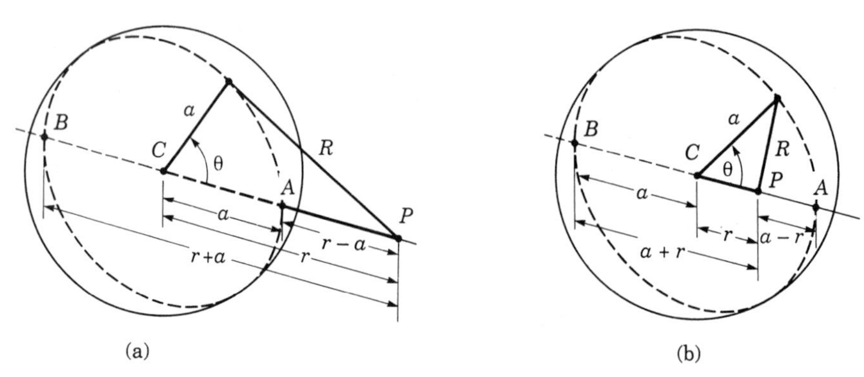

Figure 11.30 Calculation of the gravitational field at a point P (a) outside and (b) inside a mass distributed uniformly over a spherical shell.

Substituting in the general expression for $\mathrm{d}\Phi$, we get

$$\mathrm{d}\Phi = -G\frac{M}{2ar}\,\mathrm{d}R$$

To obtain the total gravitational potential we must integrate over all the surface of the sphere. The limits for R, when the point P is outside the sphere, are $r + a$ and $r - a$, corresponding to points B and A, respectively (Figure 11.30(a)). Therefore

$$\Phi = -\frac{GM}{2ar}\int_{r-a}^{r+a} \mathrm{d}R = -\frac{GM}{2ar}(2a) = -\frac{GM}{r}, \qquad r > a$$

is the potential at a point outside a homogeneous spherical shell. If the point P is inside the sphere (Figure 11.30(b)), the limits for R are $a + r$ and $a - r$. Integrating, we get

$$\Phi = -\frac{GM}{2ar}\int_{a-r}^{a+r} \mathrm{d}R = -\frac{GM}{2ar}(2r) = -\frac{GM}{a}, \qquad r < a$$

which yields a gravitational potential independent of the position of P; that is, the gravitational potential is constant.

Applying Equation 11.21, with s replaced by r, we find that the gravitational field at points outside and inside the homogeneous spherical shell is given by the expressions 11.23 and 11.24.

Now consider the case of a *solid* homogeneous sphere. We have already indicated that the sphere may be considered as composed of a series of thin spherical shells. Therefore, the gravitational field at points *outside* is the same as if all the mass were concentrated at the center.

To obtain the field *inside* the homogeneous sphere, consider a point P a distance r from the center, with $r < a$. We draw a sphere of radius r (Figure 11.27), and observe that those shells with radius larger than r do not contribute to the field at P, according to Equation 11.24, since P is inside them. All shells with a radius smaller than r produce a field similar to Equation 11.23. Let us call m' the mass inside the dashed sphere. By Equation 11.23, the field at P will be

$$\boldsymbol{\Gamma} = -\frac{Gm'}{r^2}\boldsymbol{u}$$

The volume of the whole sphere is $4\pi a^3/3$ and, since the sphere is homogeneous, the density (mass per unit volume) is $M/(4\pi a^3/3)$. The mass m' contained in the sphere of radius r is then

$$m' = \frac{M}{(4\pi a^3/3)}(4\pi r^3/3) = \frac{Mr^3}{a^3}$$

Substituting this result in the preceding equation, we finally obtain the field at a point inside the homogeneous sphere as

$$\mathbf{\Gamma} = -\frac{GMr}{a^3}\,\boldsymbol{u} \qquad (r < a)$$

If the sphere is non-homogeneous, then it is clear that the field inside obeys a different relation.

EXAMPLE 11.8

Gravitational force between two homogeneous spherical bodies.

▷ We are now in a position to verify that Equation 11.2 for the gravitational attraction between two point masses also holds for two homogeneous spherical bodies. Assume that we place a point mass m at a distance r from the center of a spherical mass (Figure 11.31). The field it experiences is $\Gamma = GM/r^2$, and the force on m is $F = m\Gamma = GMm/r^2$. By the law of action and reaction, m must exert an equal and opposite force on M. This force is interpreted as being due to the field created by m in the region occupied by M. Now, if we replace m by a homogeneous spherical body of the same mass, the field around M does not change, because of the theorem we have just proved, and therefore the force on M remains the same. Again we invoke the principle of action and reaction, and conclude that the force on the spherical mass m is still the same. Consequently, two homogeneous spherical masses attract each other according to the law 11.2, where r is the distance between their centers. If the masses are neither spherical nor homogeneous, some geometrical factors, including angles defining their relative orientation, will appear in the expression for their interaction.

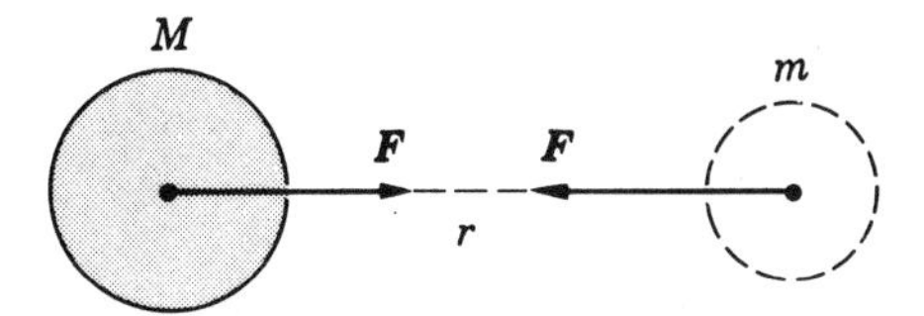

Figure 11.31 The gravitational interaction between two homogeneous spherical bodies depends only on the distance between their centers.

EXAMPLE 11.9

Variation of the acceleration of gravity at a small distance above or below the Earth's surface.

▷ Let h be the height of a body above the Earth's surface. If r is the Earth's radius, the distance of the body to the center is

$$r = R + h$$

The intensity of the gravitational field at height h, according to Equation 11.25, is

$$\Gamma = \frac{GM}{(R+h)^2} = \frac{GM}{R^2(1+h/R)^2}$$

where M is the Earth's mass. Considering that h is small compared with R and using the binomial approximation (B.21) and the result of Example 11.1, $g = GM/R^2$, we have

$$\Gamma = g\left(1 + \frac{h}{R}\right)^{-2} \approx g\left(1 - \frac{2h}{R}\right)$$

Introducing the values for g and R, we get

$$\Gamma = (9.81 - 3.06 \times 10^{-6}\, h)\,\mathrm{m\,s^{-2}}$$

This expression gives, approximately, the variation of the acceleration of gravity and of the weight as a body moves *up* from the Earth's surface a small distance h.

If instead, we move into the interior of the Earth a distance h, we have

$$r = R - h$$

Using Equation 11.26, with a replaced by R and r by $R - h$, we obtain

$$\Gamma = \frac{GM(R-h)}{R^3} = \frac{GM}{R^2}\left(1 - \frac{h}{R}\right) = g\left(1 - \frac{h}{R}\right)$$

Introducing the proper values of g and R gives

$$\Gamma = (9.81 - 1.53 \times 10^{-6}\, h)\,\mathrm{m\,s^{-2}}$$

So in both cases gravity decreases, but it decreases at a faster rate for points above the surface than below (recall Figure 11.28). To these results we must add the effects due to the rotation of the Earth (Chapter 5).

11.10 The principle of equivalence

A consequence of the equality of inertial and gravitational masses is that

> *all bodies at the same place in a gravitational field experience the same acceleration.*

An example of this is Galileo's discovery that all bodies fall to Earth with the same acceleration. This discovery, as we have already mentioned, is in turn an indirect proof of the identity of inertial and gravitational mass.

To prove the above statement, we note that in a place where the gravitational field is $\boldsymbol{\Gamma}$, the force on a body of mass m is $\boldsymbol{F} = m\boldsymbol{\Gamma}$ and its acceleration is $\boldsymbol{a} = \boldsymbol{F}/m = \boldsymbol{\Gamma}$. The acceleration is independent of the mass m of the body subject to the action of the gravitational field. Note that the acceleration of the body is equal to the field strength, which is consistent with our previous result that gravitational field is measured in $\mathrm{m\,s^{-2}}$.

If an experimenter's laboratory is placed in a gravitational field, all bodies that are subject to no other forces experience a common acceleration. The experimenter, by observing this common acceleration, *may* conclude that the laboratory is in a gravitational field.

However, this conclusion is not the *only* possible explanation for the observation of a common acceleration. In Section 3.9, we discussed relative motion. In that section we indicated that, when an observer O' has an acceleration $\boldsymbol{a}_0$ relative to an inertial observer O and $\boldsymbol{a}$ is the acceleration of a body as measured by the inertial observer O, the acceleration of A measured by the moving observer O' is expressed by

$$\boldsymbol{a}' = \boldsymbol{a} - \boldsymbol{a}_0$$

If the body is free, the acceleration $\boldsymbol{a}_B$ measured by the inertial observer is zero. Therefore, the acceleration measured by the accelerated observer is

$$\boldsymbol{a}' = 0 - \boldsymbol{a}_0 = -\boldsymbol{a}_0$$

Thus all free objects appear to the accelerated observer to have a common acceleration $-\boldsymbol{a}_0$, a situation identical to that found in a gravitational field of strength

$$\boldsymbol{\Gamma} = -\boldsymbol{a}_0$$

Thus we may conclude that:

an observer has no means of distinguishing whether his laboratory is in a uniform gravitational field or in an accelerated frame of reference.

This statement is known as the **principle of equivalence**, since it shows an equivalence, in so far as the description of motion is concerned, between a gravitational field and an accelerated frame of reference. Gravitation and inertia thus appear to be not two different properties of matter, but only two different aspects of a more fundamental and universal characteristic of all matter.

Suppose, for example, that an observer has a laboratory in a railroad car moving along a straight horizontal track with constant velocity. The windows of the laboratory are blackened so that the observer has no access to the outside world (Figure 11.32(a)). The experimenter drops some billiard balls and notes that all of them fall with the same acceleration. The observer may then conclude that the lab is surrounded by a vertical gravitational field in the downward direction, which is the normal interpretation. But the observer could equally well assume that the car is being lifted upward with a vertical acceleration, equal and opposite to that of the balls, and that the balls are free and not subject to a gravitational field.

Suppose now that the observer places the balls on a billiard table located in the car. When the observer notes that the balls on the table roll toward the rear of the car with a common acceleration, the experimenter may conclude that the laboratory either is acted on by a new horizontal gravitational field directed toward the rear of the car or that the laboratory is being accelerated horizontally in the forward direction. The second assumption is the usual one, associated with a decision by the train engineer to speed up the train (Figure 11.32(b)). However,

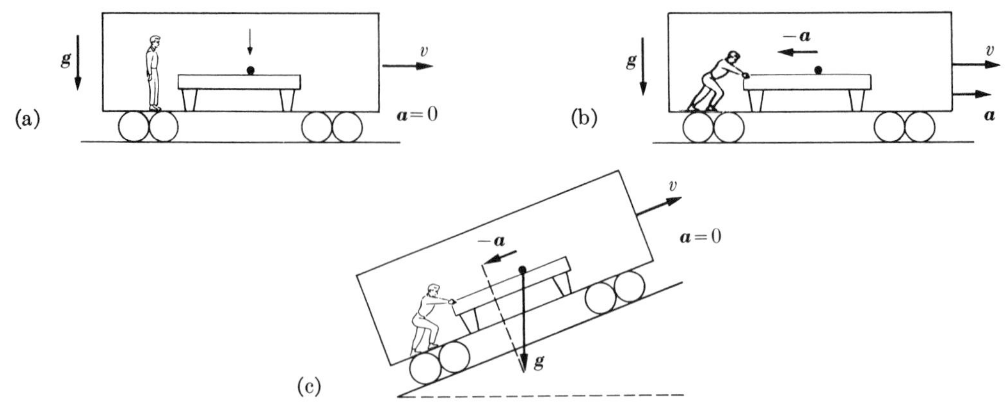

Figure 11.32 Equivalence between an accelerated frame of reference and a gravitational field.

the train could instead be going up a slope (Figure 11.32(c)); this is equivalent to producing a gravitational field parallel to the car's floor, with the same result to the motion of the billiard balls.

Because of the principle of equivalence,

the laws of nature must be written in such a way that it is impossible to distinguish between a uniform gravitational field and an accelerated frame of reference,

a statement that constitutes the basis of the **general principle of relativity**, proposed by Einstein in 1915 and discussed in Section 19.10. This principle requires that physical laws be written in a form independent of the state of motion of the frame of reference. The fundamental idea of the general principle of relativity is very simple; however, its mathematical formulation is rather complex.

Let us examine the case of an observer O' in a gravitational field $\mathbf{\Gamma}$ and having an acceleration $\boldsymbol{a}_0$ relative to an inertial observer O. The acceleration of bodies subject only to the gravitational field as measured by our accelerated observer is

$$\boldsymbol{a}' = \mathbf{\Gamma} - \boldsymbol{a}_0$$

As a concrete illustration, consider the case of a rocket accelerated upward from the Earth. We then have that

$$\mathbf{\Gamma} = \boldsymbol{g}$$

Let us write $\boldsymbol{a}_0 = -n\boldsymbol{g}$ for the rocket's upward acceleration relative to the Earth, where n gives the value of $\boldsymbol{a}_0$ in units of $\boldsymbol{g}$. (The minus sign is due to the fact that the rocket is accelerated in the upward direction or opposite to $\boldsymbol{g}$). Then

$$\boldsymbol{a}' = (n + 1)\boldsymbol{g}$$

is the acceleration, relative to the rocket, of a free body inside the rocket. For example, in a rocket accelerated upward with an acceleration four times that of gravity ($n = 4$), the apparent weight of all bodies inside the rocket is five times their normal weight. This apparent increase in weight is important at the launching stage, when the rocket's acceleration is largest.

Now consider, as another example, an orbiting satellite. Here $\boldsymbol{a}_0 = \boldsymbol{\Gamma}$, because the satellite is moving solely under the gravitational action of the Earth. In this case $\boldsymbol{a}' = 0$, and all bodies within the satellite *appear* to be weightless, since their acceleration relative to the satellite is zero. This is only a relative weightlessness because both the satellite and its contents are moving in the same gravitational field and have the same acceleration. Relative to the satellite, the bodies inside appear as free bodies unless other forces act on them; but, relative to a terrestrial observer, they are accelerated and subject to the gravitational field. Someone inside an elevator that is falling with the acceleration of gravity (due to a broken cable) would experience the same weightlessness relative to the elevator. In such a case (as in the satellite), $\boldsymbol{a}_0 = \boldsymbol{g}$, and again $\boldsymbol{a}' = 0$.

Weightlessness, we insist, does not mean that the gravitational force has ceased to act. It means that all bodies, including the one serving as frame of reference, are acted on by one and the same gravitational field, which produces a common acceleration. Therefore, there are no relative accelerations unless other forces act on the bodies. In other words, a gravitational field $\boldsymbol{\Gamma}$ can be 'washed off' if the observer moves through it with an acceleration $\boldsymbol{a}_0 = \boldsymbol{\Gamma}$ relative to an inertial frame.

11.11 Gravitation and intermolecular forces

In previous sections of this chapter we have seen how gravitational forces adequately describe planetary motion and the motion of bodies near the surface of the Earth. It is interesting to see if this interaction is responsible for keeping molecules together in a piece of matter or keeping atoms together in a molecule.

Consider the hydrogen molecule, composed of two hydrogen atoms separated the distance $r = 0.745 \times 10^{-10}$ m. The mass of each hydrogen atom is $m = 1.673 \times 10^{-27}$ kg. Therefore the gravitational interaction of the two atoms corresponds to a potential energy

$$E_p = -\frac{GMm}{r} = -2.22 \times 10^{-54}\ \text{J} = -1.39 \times 10^{-35}\ \text{eV}$$

However, the experimental value for the dissociation energy of a molecule of hydrogen is 7.18×10^{-19} J $= 4.48$ eV, or about 10^{35} times larger than the gravitational potential energy. Therefore we conclude that gravitational interaction *cannot* be responsible for the formation of a molecule of hydrogen. Similar results are obtained for more complex molecules.

The energy required to vaporize one mole of liquid water (18 g or 6.23×10^{23} molecules) is 4.06×10^3 J, corresponding to a separation energy per molecule of the order of 6.6×10^{-21} J, or 0.04 eV. The average separation of water molecules is about 3×10^{-10} m, and the mass of a molecule is 3×10^{-26} kg, corresponding to a gravitational potential energy of 2×10^{-52} J, again far too small to explain the existence of liquid water.

Therefore, we conclude that the forces giving rise to the association of atoms to form molecules or of molecules to form matter in bulk cannot be gravitational. But we know that there are other forces that seem to be responsible for the association of particles with electric charge: *electromagnetic interactions.*

However, gravitational interaction, being a mass effect, is very important in the presence of massive bodies that are electrically neutral, such as planets. For that reason gravitation is the strongest force we feel on the Earth's surface, in spite of the fact that it is the weakest of all forces known in nature. It is responsible for a large number of common phenomena affecting our daily lives. Tides, for example, are due to the gravitational interactions of the Moon and the Sun with the Earth. Gravitational interaction is also responsible for the large-scale structure of the universe (see Note 11.4).

Note 11.4 Gravitation and the large-scale structure of the universe

An important feature of the universe is that it shows some kind of structure or organization at all levels. At the fundamental or small-scale level, we find nuclei, atoms, molecules, solids and even living organisms. At a larger scale, we observe planets and stars. At a much larger scale, other kinds of structures are observed. One, and perhaps the better known, is our own planetary system where planets revolve around the Sun in a region about 10 light hours (10^{13} m) across. Newton's law of gravitation very nicely explains the dynamics of a planetary system, as we have seen earlier in this chapter. There are a large number of bright objects or stars, located at different distances from the Sun. The closest are the set of three stars called Alpha Centauri, at a distance of 4.3 Lyr (light years). The Sun forms part of a rotating conglomerate composed of about 10^{11} stars held together by their gravitational attraction and called the **Milky Way** or our **galaxy**. The Milky Way is a spiral galaxy with a bulge in the center. It has a radius of about 5×10^4 Lyr or 5×10^{20} m and is composed of about 10^{11} stars separated by an average distance of the order of 10 Lyr and has a total mass of the order of 10^{42} kg. The Sun is located in one of the arms of the spiral, about 3.1×10^4 Lyr or 3×10^{20} m from the center moving under the consolidated gravitational action of the huge number of stars in the Milky Way. Its orbital velocity about the center is approximately 250 km s^{-1}. The Milky Way is surrounded by several small conglomerates of stars, called **globular clusters**, each with about 10^5 to 10^6 stars. They resemble satellite mini-galaxies (Figure 11.33). The spiral arms are also filled with dust and gas, from which new stars keep forming.

Our galaxy is just one of more than 10^{10} galaxies that have been observed. These galaxies constitute the basic units of the large structure of the universe. Galaxies appear grouped in clusters of a few tens up to a few hundred. For example, our galaxy is part of the **Local Group**, which is a cluster of some 20 galaxies loosely bound by gravitational forces. The closest neighbors in the Local Group are two small galaxies (about 10^{10} stars each) known as the **Magellanic Clouds** because they were spotted in the southern sky by Magellan's crew. They are about 1.4×10^5 Lyr from the Sun. Andromeda is one galaxy in the Local Group that is very similar to ours. It contains about 3×10^{11} stars and is about 2×10^6 Lyr distant.

Clusters are prevalent throughout the universe. They seem to group in chain-like structures called **superclusters**, containing several thousands of galaxies in a region of the order of 10^9 Lyr in diameter (Figure 11.34). The Local Group is part of the **Local Supercluster**, which consists of thousands of galaxies grouped in a few hundred clusters, occupying a region 6×10^7 Lyr across. The largest cluster is Virgo, which is located near the center of our supercluster, while our Local Group is near an edge. One of the largest superclusters, discovered in 1989 and called the **Great Wall**, is a system of thousands of galaxies organized in a sheet-like structure with a width of about 5×10^8 Lyr.

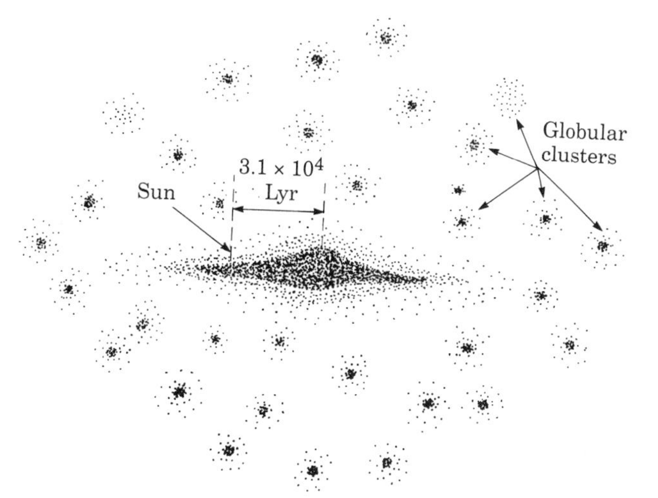

Figure 11.33 Side view of the Milky Way and the globular clusters. The position of the Sun is also shown.

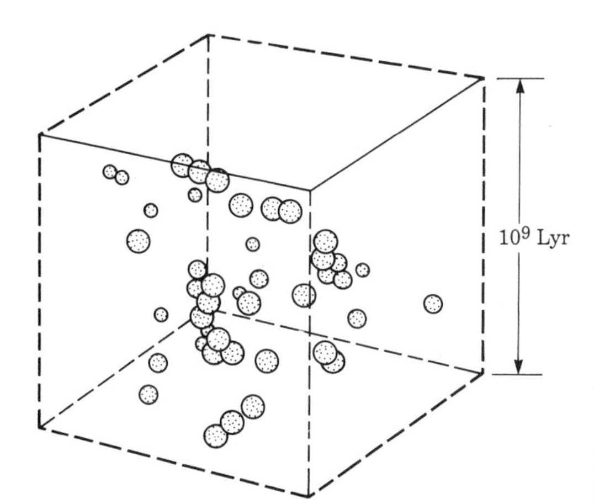

Figure 11.34 Supercluster in the direction of the Perseus and Pegasus constellations. The size is about 10^9 Lyr across.

In whatever direction we look into space, we observe clusters and superclusters, all moving away from us in accordance with Hubble's law. The superclusters are separated by regions called **voids** where the density of matter is less than 20% of the average. Some voids are as large as 10^8 Lyr in diameter.

The question that has preoccupied astrophysicists for a long time is how these large structures were formed after the Big Bang, which occurred about 1.5×10^{10} years ago. They are also interested in which forces played a role in their formation. From the discussion about fundamental forces in Note 6.1, we know that the weak and strong forces are of very short range and act only at the nuclear level. These forces were very important in shaping the events during the first few minutes after the Big Bang (see Chapter 41). They are still responsible for nuclear processes that occur in the stars, including our Sun (see Chapter 40). But, to explain the formation of large structures in the universe, long-range forces are needed. About 3×10^5 years after the Big Bang, gravitation, the weakest of all forces but always attractive, became the dominant factor at large scales in the universe.

The question is then how gravitation produced the large structures we have described. As a result of astronomical observations, astrophysicists have concluded that on a very large scale, matter in the universe is distributed uniformly and isotropically. This conclusion is called the **Cosmological Principle**. (However, the principle has been questioned, based on more recent data such as the Great Wall mentioned above.) The problem is then how, out of an initially homogeneous universe, did matter begin to clump into certain structures. One possible answer is that, although cosmic matter on the average is distributed uniformly, the distribution experiences local fluctuations. If, as a result of these fluctuations, matter concentrates slightly in a certain region, gravitational forces will tend to concentrate more matter in the region. That is, fluctuations act as seeds for the concentration of cosmic matter. But, for that to happen, the mass of a fluctuation must have a certain minimum value, called the **Jeans mass**, and cannot extend beyond a certain distance, called the **Jeans length**, after Sir James Jeans (1877–1946). Below the Jeans mass, cosmic matter cannot break into chunks. On the other hand, Jeans mass decreases as the kinetic energy or temperature decreases because it is then easier to hold matter together. Therefore, as the universe expands, Jeans mass decreases.

One way this process of aggregation could begin is with small fluctuations that result in galaxies. Then these galaxies, due to gravitational action, clump into clusters and then into superclusters. Another possibility is that fluctuations as large as superclusters are produced first. Then, as the universe expands and Jeans mass decreases, superclusters break down into clusters, which in turn eventually break into galaxies, until stars finally begin to be formed. Current thinking is that the second alternative might be the correct one, although some doubts have recently (1990) arisen. There is no doubt that gravitation plays a fundamental role in the process. To decide between the two alternatives in a definitive way, it is necessary to take into account many factors related to the composition and form of expansion of the universe. Some of these factors are not well understood. For example, an important question that needs to be clarified is whether, besides the 'luminous' cosmic matter (that is, matter that we observe because of the electromagnetic radiation it emits), there is also 'dark' matter that is affected by the gravitational force (recall Note 11.3).

Note 11.5 Gravitation and dynamical chaos

In this chapter we have considered the motion of two bodies (Sun–planet or Earth–Moon) under their mutual gravitational attraction. We saw that the two bodies describe elliptical orbits around their center of mass, in accordance with Kepler's laws. When three bodies move under their mutual gravitational field, such as the Sun–Earth–Moon system, the force on each one is just the vector sum of the gravitational forces exerted by the other two. Thus the respective equations of motion are easy to write down but difficult to solve because they are not linear.

However, since Newton's time many scientists have tried to find exact mathematical solutions of the 'three body problem' until, in this century, it has been proved to be impossible. Only under some circumstances are some simple solutions possible. For example, if the mass of one of the bodies is much smaller than that of the others (the so called 'restricted three body problem'), the system can be reduced (with good approximation) to a 'two body' plus 'one' problem. When the smaller body moves so that it remains mostly near one of the others, its motion can be very simple and predictable. This is why it is possible to calculate the motion of the Moon in the Sun–Earth–Moon system since the Moon is much closer to the Earth than to the Sun. However, the possible orbits, even in the 'restricted three body' problem, are so diverse and complicated, depending on the initial conditions, that no general formula can be written down to describe them, rendering the problem unsolvable by analytical methods. Motion in a 'many body' problem often gets even more complicated. The orbits may even become chaotic in the sense explained in Note 10.4 with regard to forced oscillatory motion.

For example, consider the case of two equal large masses M_1 and M_2 and a third, very small, mass M_3. For simplicity we shall assume that M_1 and M_2 move in a circular orbit around their center of mass. Choose a frame of reference centered at the circle that rotates in

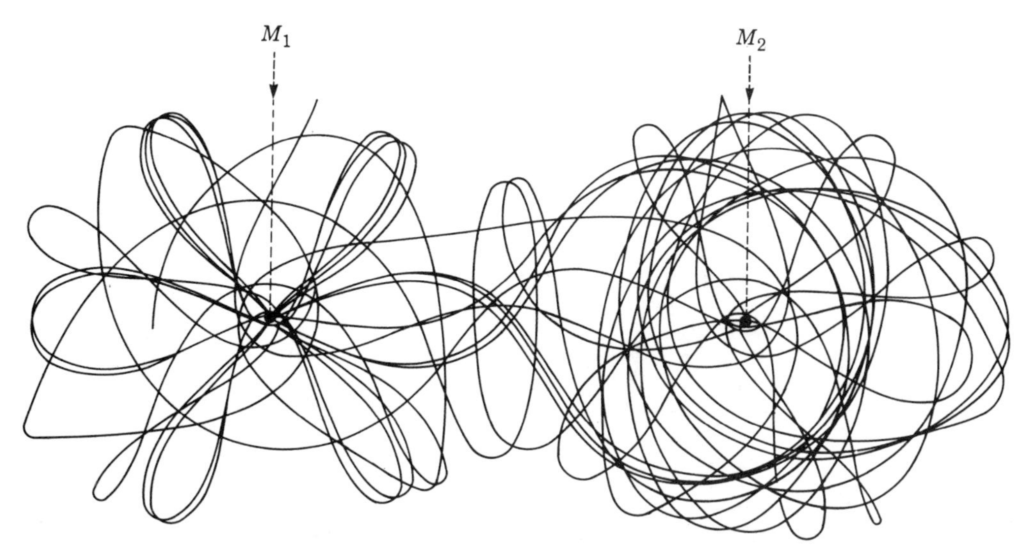

Figure 11.35 Chaotic gravitational motion of a mass M_3 in the restricted three-body problem. (From M. Hénon: 'Numerical Exploration of Hamiltonian Systems' in *Chaotic behavior of deterministic Systems*, Editors G. Iooss *et al.*, (North Holland, 1983).)

such a way that M_1 and M_2 appear at rest (Figure 11.35). Then, for the initial conditions chosen in this example, M_3 orbits around M_1 and M_2 in a rather complex fashion as shown in the figure. It is hard to predict when M_3 will be close to M_1 or M_2! If more bodies are present, the problem gets even more complicated and chaotic motion is more likely to occur. We do not observe such complications in planetary motion because, due to the strong dominant influence of the Sun, each planet, to a first approximation, moves as if the others were not present. Another reason is that the solar system is not old enough for most of the initial chaotic motions to have disappeared either because of collisions or as the result of damping caused by tidal actions. But very irregular chaotic motions *can* occur under the gravitational interaction. In fact, such motions do occur in the solar system. Also, we do not observe chaotic motion in the trajectory of a spaceship because its motion can be corrected whenever necessary either from a control center on the Earth or by an astronaut, both of which methods were used during the Apollo missions to the Moon.

One 'indirect' example of gravitational chaos is the uneven radial distribution of asteroids. Asteroids are a large number of small bodies, ranging in size from rocks about 1 km in diameter to planetoids, about 1000 km in diameter, much smaller than the Moon. They circulate around the Sun in a region called the asteroid belt, between the orbits of Mars and Jupiter, mostly at distances between 2.8×10^{11} m and 5×10^{11} m from the Sun. Presumably asteroids are remnants dating back to the early epoch of the solar system, when the planets began to be formed. Their periods of revolution vary with their distance from the Sun (recall Section 11.3), and most of them have between about $\frac{1}{4}$ and $\frac{1}{2}$ the period of Jupiter (12 years), corresponding to the distances indicated above. Although asteroids basically move under the attractive force of the Sun, the influence of Jupiter cannot be neglected since it is so massive and so close to the asteroid belt (the average radius of Jupiter's orbit is about 7.78×10^{11} m).

A feature discovered in 1867 by Daniel Kirkwood (1814–1895) is that the radial distribution of the asteroids is not uniform. It has been found that at certain distances, the number of asteroids drops sharply (Figure 11.36). These regions of asteroid depletion are called **Kirkwood gaps**. Since for each distance from the Sun there is a well-defined period of orbital motion, the gaps can be correlated with the ratios of the orbital period of Jupiter to that of the asteroids. This is indicated in Figure 11.36. Interestingly enough, it can be recognized that the regions of strong asteroid depletion occur at distances for which the ratio of the two periods is a simple rational number: 3/2, 2/1, 5/2, 3/1, 7/2, 4/1, ...

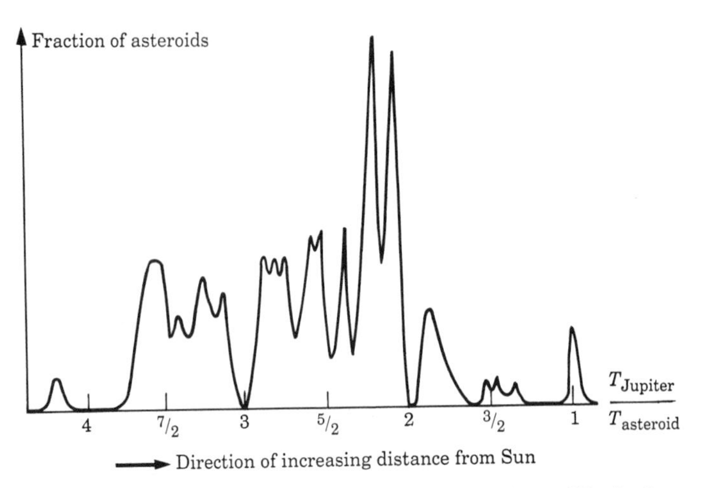

Figure 11.36 Distribution of asteroids as a function of Jupiter's orbital period to that of the asteroids' orbital periods, showing Kirkwood gaps.

Several explanations have been given, but the most reasonable one appears to be that the gaps are a resonance effect that results in chaos. When the ratio of Jupiter's period to that of an asteroid is a rational number, the Sun, Jupiter and the asteroid are successively in conjunction at regular intervals, resulting in a resonant perturbation of the asteroid's orbit. This perturbation adds or subtracts angular momentum (recall Example 6.4) and produces oscillations in the shape of the orbit. These oscillations bear a great resemblance to the forced oscillations of a pendulum and may result in chaotic behavior. When this perturbative effect is repeated over and over again, for thousands of millions of years, asteroids are eventually removed from the resonating orbits in an unpredictable way that has all the features of chaotic processes. Thus, gradually but relentlessly, the Kirkwood gaps appear as a consequence of a resonance effect associated with the gravitational forces due to Jupiter. In 1985 Jack L. Wisdom from MIT made a computer simulation of the motion of asteroids in resonating orbits. The simulation verifies that the motion indeed can become chaotic, resulting in asteroids being thrown far away from the asteroid belt; a few of them may even end as 'shooting stars' in the Earth's atmosphere. Therefore, asteroid dynamics constitutes a beautiful example of gravitational chaotic motion.

Incidentally it has been postulated by W. Alvarez, from the University of California, that about 65 million years ago a giant asteroid, which had been knocked from its orbit by the gravitational mechanism we have described, struck the Earth at a velocity of more than 10^4 m s^{-1}. The enormous energy released by the impact gave rise to a variety of events that adversely affected the biosphere. One of the consequences of this environmental turmoil was the extinction of more than half the plant and animal species, including the dinosaurs. A second theory, proposed by John O'Keefe at NASA, Goddard Space Flight Center, states that an asteroid struck the Moon, resulting in a ring of debris about the Earth. The debris would long ago have disappeared due to chaotic motion and slowly rained down on the Earth (as tektites). While in place, however, the ring would have shielded enough of the Sun's energy to result in enormous environmental changes. If either theory proves to be correct, it will provide a link between gravitational chaos and the way life has evolved on Earth.

Another interesting example of the occurrence of chaotic motion in the Solar System has been found by Wisdom, using information provided by Voyager 2. One of Saturn's moons, Hyperion, has a very oblong shape. It moves around Saturn with a quite regular period, but during its revolution it 'tumbles' around its center of mass in a chaotic way, i.e. it is very hard to predict the direction of its axis as a function of time.

QUESTIONS

11.1 What properties of the gravitational interaction are derived from Kepler's laws?

11.2 How was Newton, by comparing the acceleration of gravity with the acceleration of the Moon, able to conclude that the gravitational force is inversely proportional to the square of the distance?

11.3 Why do we say that gravitation is a 'universal' law?

11.4 Give an experimental fact that implies the equivalence of 'gravitational mass' and 'inertial mass'.

11.5 Is planetary motion in a plane? Explain.

11.6 What quantities are constant in planetary motion?

11.7 Why can there be no bound orbits if the gravitational force is repulsive?

11.8 Why is orbital motion under gravitational forces symmetric, as shown in Figure 11.10?

11.9 Why is the velocity of approach to a planet equal to the escape velocity from the planet?

11.10 Discuss the concept of 'weightlessness' in satellite motion and in a space probe.

11.11 Is 'gravitational field' a physical property that exists in space or is it merely a convenient invention for computational purposes?

11.12 How can we compute the component of the gravitational field in a certain direction if we know the gravitational potential as a function of coordinates?

11.13 Make a plot of the gravitational lines of force and equipotential surfaces corresponding to two equal masses separated by a certain distance. How would the plot change if one mass is twice the other?

11.14 Consider a very long, homogeneous, massive cylindrical body. Based on symmetry considerations alone, determine (a) the shape of the gravitational lines of force and equipotential surfaces of the gravitational field, (b) the variables that fix the magnitude of the gravitational field.

11.15 Repeat the preceding exercise for the case of an infinitely extended very thin plane sheet of material, and for two parallel sheets separated by a certain distance.

11.16 Assuming that the gravitational force is attractive at all distances, are stable configurations of matter possible if only this force exists in nature?

11.17 In Note 11.1 it was shown how the eccentricity of the orbit is related to the energy E and the angular momentum L of the body by

$$\varepsilon^2 = 1 + \frac{2E}{m}\left(\frac{L}{GMm}\right)^2$$

What happens to the eccentricity in elliptical orbits as the angular momentum increases, assuming the energy remains the same? What is the relation between E and L when the orbit is circular? What orbit results when $E = 0$? And when $L = 0$?

11.18 Using the results of Note 11.1, express the angular momentum of a body in terms of its energy and the eccentricity of its orbit.

11.19 Referring to Figure 10.11, what is the angular momentum of the satellite? What should be its value for (a) a circular orbit, (b) an orbit that barely goes around the Earth?

11.20 Verify that the gravitational potential difference between two points near the Earth's surface, separated by a vertical distance H, is $\Delta\Phi = gH$.

11.21 Verify that, when the field is radial, the relation $\mathbf{\Gamma} = -(\mathrm{d}\Phi/\mathrm{d}r)\boldsymbol{u}$ holds. Using Equation 11.19 for Φ, calculate $\mathbf{\Gamma}$ and compare with Equation 11.16.

11.22 What is the physical relevance of the principle of equivalence?

11.23 Why can gravitation not account for the formation of molecules?

11.24 Figure 11.37(a) gives the estimated variation of the Earth's density as a function of distance from the center. Justify that the corresponding variation of the acceleration due to gravity is indicated in Figure 11.37(b).

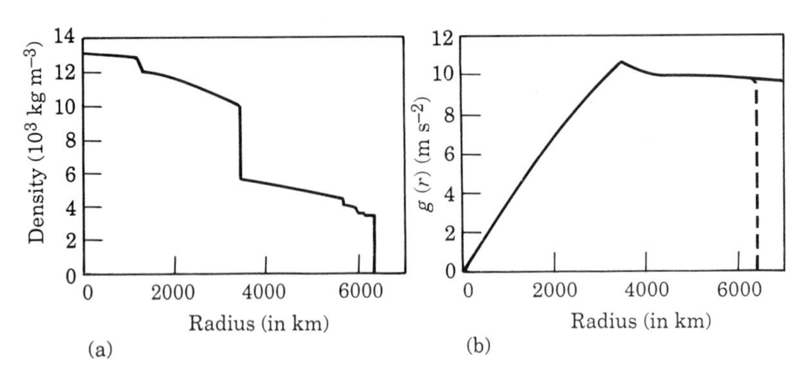

Figure 11.37

PROBLEMS

11.1 Calculate the gravitational attraction between the Earth and (a) the Moon, (b) the Sun. (c) Obtain the ratio between these two forces. How then do the two forces affect the motion of the Earth?

11.2 Compare the gravitational attraction produced on a body of mass m at the Earth's surface (a) by the Moon and (b) by the Sun, with the attraction of the Earth on the same body. (c) What do you conclude about the possibility of observing a change in the weight of a body during the daily rotation of the Earth?

11.3 Compare the forces of the Sun and the Earth on the Moon at (a) full moon and at (b) new moon. (c) Verify that the resultant force is always toward the Sun. (d) Show that the path of the Moon relative to the Sun is always concave. (e) Plot the path.

11.4 (a) Calculate the gravitational attraction between the two protons in a hydrogen molecule. Their separation is 0.74×10^{-10} m. (b) Compare with the electric repulsion between two protons at the same separation, which is 4.2×10^{-8} N. What do you conclude?

11.5 (a) Determine the gravitational attraction between the proton and the electron in a hydrogen atom, assuming that the electron describes a circular orbit with a radius of 0.53×10^{-10} m. (b) Compare it with the electric force between the electron and the proton, which is 8.2×10^{-8} N.

11.6 Two iron balls, each having a mass of 10 kg, are touching each other. (a) Find their gravitational attraction. (b) Compare it with the gravitational attraction of the Earth on each ball. (c) If you tried to separate the two balls, would you 'feel' the attraction between them? (Hint: the density of iron is listed in Table 2.2.)

11.7 A sphere of mass 5.0 kg is located in one pan of an equal-arm balance in equilibrium. A larger spherical mass (5.8×10^3 kg) is then rolled until it is directly underneath the first mass, the distance between the centers being 0.50 m. What mass must be placed in the other pan of the balance in order to restore equilibrium to the system? Assume $g = 9.80$ m s^{-2}. This method was used by G. von Jolly, in the 19th century, to determine the value of the gravitational constant.

11.8 A body has a mass of 70 kg. (a) What is the body's weight? Supposing that the radius of the Earth were doubled, how much would the body weigh (b) if the mass of the Earth remained constant, (c) if the average density of the Earth remained constant? (Ignore the rotation of the Earth.)

11.9 (a) Calculate the acceleration of gravity at the Sun's surface. Its radius is 110 times the radius of the Earth and its mass is 330 000 times the mass of the Earth. Repeat for (b) Venus, (c) Jupiter and (d) the Moon.

11.10 A body weighs 1100 N at the Earth's surface. Calculate how much the body weighs (a) at the surface of the Sun and (b) at the surface of the Moon. (c) What would be the body's mass at both places?

11.11 In a Cavendish experiment (Figure 11.5), the two small masses are equal to 10.0 g and the rod (of negligible mass) is 0.50 m long. The period of torsional oscillations of this system is 770 s. The two large masses are 10.0 kg each and are so placed that the distance between the centers of the large and small spheres is 0.10 m. Find the angular deflection of the rod.

11.12 (a) How high must one go above the Earth's surface for the acceleration of gravity to change by 1%? (b) How deep should one penetrate into the Earth to observe the same change? Assume a homogeneous Earth.

11.13 Suppose that a hole were drilled completely through the Earth along a diameter and a particle of mass m is released into the hole at the surface of the Earth. (a) Show that the force on the mass at a distance r from the center of the Earth is $F = -mgr/R$, if we assume that the density is uniform. (b) Show that the motion of the mass would be simple harmonic, with a period of about 85 min. (c) Write the equations for position, velocity and acceleration as functions of time, with numerical values for the constants.

11.14 A mass m is dropped from a great height h above a hole through the Earth. (a) Describe the overall motion of the particle. (b) With what velocity would m pass the center? (c) Is the motion simple harmonic? (d) Is the motion periodic? Give reasons for your answers. Consider both a homogeneous and an inhomogeneous Earth.

11.15 Write an equation that expresses the total energy of the system (a) Earth–Moon relative to a frame of reference attached to the Earth, (b) Sun–Earth–Moon, relative to a frame of reference attached to the Sun.

11.16 Estimate the kinetic energy, the potential energy, and the total energy of the Earth in its motion around the Sun. (Consider only the gravitational potential energy with the Sun.)

11.17 Verify that the energy required to build up a spherical body of radius R, by adding successive layers of matter in an onion-like fashion until the final radius is attained (keeping the density constant), is $E_p = -3GM^2/5R$ where M is the total mass.

11.18 Estimate the gravitational potential energy of the Milky Way, using the result of Problem 11.17. Assume that the Milky Way is a uniformly dense sphere with a radius of 10^{21} m and contains 10^{11} stars, each with a mass equal to our Sun.

11.19 Two bodies of masses m and $3m$ are separated a distance a. Find the point(s) where (a) the resultant gravitational field is zero, (b) the two masses produce gravitational fields that are the same in magnitude and direction, (c) the two masses produce identical gravitational potentials.

11.20 Two bodies of masses m and $2m$ are at two of the vertices of an equilateral triangle of side a. Find the gravitational field and the potential (a) at the midpoint between them, and (b) at the third vertex of the triangle. (c) Calculate the work required to move a mass m' from the vertex to the midpoint.

11.21 Three equal masses are located at the vertices of an equilateral triangle. Make a sketch of (a) the equipotential surfaces (actually their intersection with the plane of the triangle) and (b) of the lines of force of the gravitational field. (c) Is there any point where the gravitational field is zero? (d) What is the gravitational potential at that point? (e) What is the direction of the gravitational field along a line perpendicular to the plane of the triangle and passing through its center? (f) How does the magnitude of the gravitational field vary along that line?

11.22 Evaluate the period (a) for the Earth–Moon system and (b) for the solar system–Milky Way galaxy system.

11.23 Three bodies, each of mass m, are placed at the vertices of an equilateral triangle of side a. (a) Show that the particles, under their mutual gravitational interactions, can describe a circular orbit, whose center is at the center of the triangle. (b) Show also that the angular velocity of the motion is $\omega = (3GM/a^3)^{1/2}$.

11.24 (a) Using the data of Figure 6.2, verify that the force on the Sun is of the order of 6×10^{20} N. (b) Assuming that the galaxy is approximately spherical (which it is not), verify that the motion of the Sun can be considered due to the gravitational action of a mass of the order of 10^{41} kg. (c) The total mass of the galaxy is of the order of 10^{42} kg; verify whether Equation 11.26 applies.

11.25 (a) If the average separation of two stars in the galaxy is 10 Lyr and their average mass is 10^{31} kg, verify that the average gravitational force between stars is of the order of 10^{17} N. (b) Compare this result with the gravitational force on the Sun due to the galaxy (Problem 11.28) and with the gravitational force between the Sun and the Earth (Problem 11.1(b)). (c) What may you conclude from these comparisons?

12 Space exploration

An astronaut outside the cabin of the Space Shuttle Challenger uses a jet of nitrogen to maneuver. As the nitrogen escapes through the jet orifice, it exerts a reaction force like a small rocket engine. This force causes the acceleration of the astronaut needed for him to maneuver in space. (Courtesy of NASA.)

12.1 Introduction

The Earth is not isolated in space, but rather is subject to a multitude of external effects. The Sun is responsible for most of them. The Sun exerts a gravitational interaction that keeps the Earth in its orbit; the gravitational effect of the other planets and the more distant stars is negligible. The gravitational effect of the Moon, however, is noticeable in the generation of Earth's tides. The electromagnetic radiation from the Sun, mostly infrared, visible and ultraviolet (see Section 29.8), falls continuously on the Earth and produces a series of effects, such as weather, and is essential for photosynthesis. Perhaps the most important effect of solar radiation is the energy it provides, which is necessary to sustain life. The Sun also produces a strong magnetic field that extends beyond the Earth's orbit. Charged particles, many of which reach the Earth in what is called the **solar wind**, are also emitted by the Sun. The Earth is also bathed by electromagnetic radiation and charged particles that originate elsewhere in our galaxy. These radiations and particles are called the **cosmic radiation**.

Since the early days of civilization, about 10 000 years ago, people have always had a great curiosity about the world beyond the Earth's surface, a region we call 'space' for lack of a better term. Until very recently the only way humans could explore this exterior space was visually using light, first directly with a naked eye and then through telescopes, invented in the 16th century. However, visual observation from the surface of the Earth has two limitations. One is that light is only a small fraction of the radiation falling on the Earth. The other is that the radiation must travel through the Earth's atmosphere before reaching the surface, undergoing several changes, such as selective absorption, refraction and scattering. The first limitation has been reduced somewhat in the last few decades by using telescopes that are sensitive to other regions of the spectrum. Today we have instruments that receive information from the radio, microwave, infrared and ultraviolet regions. The second limitation can be reduced by building observatories on high mountains (such as Cerro Tololo in Chile, Chacaltaya in Bolivia and Mauna Kea in Hawaii). We also place probes in balloons that go several thousand meters high and, more recently (1990) place a telescope into an Earth orbit, as has been done with the Hubble telescope (Note 33.1).

Probably the best way to gain knowledge about space is to get out into it. We can do this by placing instruments and equipment at very large distances from the Earth's surface. We can position manned stations or laboratories in space. This is a formidable technological task. It requires the design of rockets, communication equipment, adequate protection for equipment and life support systems for any scientist involved. There are numerous other details that must be taken into account. Only since the 1960s has this effort been feasible.

Space research encompasses three major regions:

(a) The study of the atmosphere, up to heights above 100 km, using man-made Earth satellites. These satellites can also study the surface of the Earth. They help us understand atmospheric processes and evaluate the Earth's resources. Currently there are networks of Earth satellites that provide information on a 24-hour basis.

(b) The development of communications by radio and television between practically all points on the Earth, using **geostationary satellites**. These satellites stay above the same point on the equator all the time and receive and transmit signals.

(c) The exploration of the Moon, the planets and beyond, using space probes. These probes go beyond the gravitational influence of the Earth and orbit around a planet or the Moon or fall on their surface; some escape from the solar system.

In this chapter we briefly review how an understanding of the gravitational force helps in planning the exploration of space.

12.2 Earth's satellites

An Earth satellite is a body placed in a stable orbit about the Earth. These satellites are used for communications, meteorology, Earth measurements (gravitation and magnetic fields), resource evaluation (water, minerals), transmission of radio and TV signals, and as reference points for navigation. High altitude Earth satellites are also used for astronomical observations, since the effects of the atmosphere can be eliminated. Although most satellites are launched from ground-based stations, more recently some have been placed in orbit from one of NASA's space shuttles. A few satellites have been manned for small periods of time and a permanent space station is in the final planning stage and probably will be built within a few years.

The first Earth satellite, called *Sputnik*, was launched by the Soviet Union in October 1957. A month later another satellite was launched by the United States. Presently there are more than 100 satellites orbiting the Earth at different altitudes and orbits launched by the Soviet Union, the United States and the European Space Agency. The first manned satellite was launched by the Soviet Union in 1961 followed by the United States in 1962.

In principle a satellite can orbit forever. However, unless their altitude is sufficiently high (at least 160 km), the atmosphere gradually dissipates the energy of the satellites and eventually they fall back toward the Earth. They are then destroyed by the high temperatures developed due to friction with the upper atmosphere. On the other hand, a satellite that is very high has limited usefulness in terms of the information about the Earth it can detect.

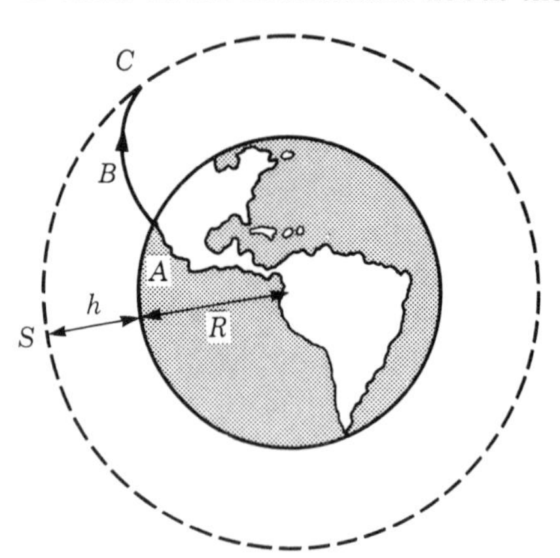

Figure 12.1 A satellite launched at point A and inserted into orbit at C.

There are two requirements needed to place a satellite in a stable orbit at an **insertion point** C (Figure 12.1). It is first necessary to bring the satellite to that altitude and then the satellite must be given the necessary orbiting velocity. The orbiting velocity for a circular orbit, also called the **insertion velocity**, is found by equating the centripetal force mv^2/r to the gravitational force GMm/r^2 and is given by $v^2 = GM/r$ (recall Section 11.6). Replacing r by $R + h$, where h is the height of the satellite above the Earth and R is the radius of the Earth, we get

$$v_{\text{ins}}^2 = \frac{GM}{R+h} = \frac{GM}{R}\frac{1}{(1+h/R)}$$

or

$$v_{\text{ins}} = \left(\frac{GM}{R+h}\right)^{1/2} = \left(\frac{GM}{R}\right)^{1/2}\left(1+\frac{h}{R}\right)^{-1/2}$$

Using the binomial expansion we may write then

$$v_{\text{ins}} = \left(\frac{GM}{R}\right)^{1/2}\left(1-\frac{h}{2R}\right) \quad \text{for } h \ll R \tag{12.1}$$

Note that the first factor is equal to the escape velocity from the Earth's surface divided by $\sqrt{2}$ and is equal to 7.9×10^3 m s^{-1}. In Table 12.1 the values of v_{ins} are

Table 12.1 Insertion velocity

h (km)	v_{ins} (km s^{-1})	v_{ins} (km h^{-1})	Period
10	7.894	28 418	1 h 25 min
160	7.809	28 112	1 h 28 min
320	7.715	27 774	1 h 31 min
1 600	7.068	25 446	1 h 58 min
35 880	3.070	11 052	24 h

given for several altitudes. The period, calculated using $P = 2\pi(R + h)/v_{ins}$, is also given. Note that a satellite at an altitude of 35 880 km has a period of 24 hours and therefore remains above the same point of the Earth's surface all the time if it is orbiting in an equatorial orbit; we say the satellite is **geostationary**. This type of satellite is very important for many purposes, particularly for a space station and for communications.

The orbit of a satellite depends on the value of the velocity at the insertion point C as explained in connection with Figure 11.11. The **escape velocity** of the satellite at the insertion height h (using Equation 11.13 with R replaced by $(R + h)$) is

$$v_{esc} = \left(\frac{2GM}{R + h}\right)^{1/2} = \sqrt{2}v_{ins} \tag{12.2}$$

If the velocity at insertion is less than v_{esc}, the satellite describes a closed elliptical orbit with the center of the Earth at one focus. If the velocity is equal to or larger than v_{esc}, the satellite will describe an open parabolic or hyperbolic orbit and move away from the Earth.

Another important factor is the minimum initial velocity that a satellite must have at launch after the rocket has burned all its fuel so that it has the proper velocity at the insertion point C. This initial velocity is called the **characteristic velocity** v_c. The value is easy to calculate for a circular orbit. The energy at the launch point A must be equal to the total energy of a circular orbit at height h. That is, using Equation 11.10 with r replaced by $R + h$,

$$\tfrac{1}{2}mv_c^2 - \frac{GMm}{R} = -\frac{GMm}{2(R + h)}$$

where the left-hand side gives the total energy at A and the right-hand side is the total energy in a circular orbit at C. Solving for v_c we obtain

$$v_c^2 = \frac{GM}{R}\left(2 - \frac{1}{(1 + h/R)}\right) = \frac{GM(1 + 2h/R)}{R(1 + h/R)} \approx \frac{GM}{R}\left(1 + \frac{h}{R}\right) \quad \text{if } h \ll R$$

or

$$v_c \approx \left(\frac{GM}{R}\right)^{1/2}\left(1 + \frac{h}{2R}\right) \quad \text{for } h \ll R \tag{12.3}$$

Table 12.2 Characteristic velocity

h (km)	v_c (km s^{-1})	v_c (km h^{-1})
10	7.906	28 463
160	8.003	28 810
320	8.093	29 136
1 600	8.664	31 190
35 880	10.752	38 706
∞	11.181	40 253

Table 12.2 gives the characteristic velocity for several altitudes. When the characteristic velocity is compared with the insertion velocity (Table 12.1), we see that as the altitude increases, v_c increases while v_{ins} decreases. Also, v_c approaches the escape velocity for very high altitudes. It is clear, then, that we need more initial kinetic energy to reach a higher altitude. It should be noted that once a satellite is placed in its orbit, it does not need any additional energy to keep moving in that orbit. However, energy is required to change the orbit of the satellite, as is done on some occasions.

EXAMPLE 12.1

Energy relations in launching a satellite.

▷ Placing an Earth satellite in orbit requires spending energy. Part of the energy is used in lifting the satellite to the desired orbit, and part in giving the satellite the necessary orbital velocity. We can look at the problem in two ways.

(a) If we launch the satellite with the characteristic velocity v_c, then the kinetic energy at launch provides for both parts and is given by

$$E_{k,c} = \tfrac{1}{2}mv_c^2 = \frac{GMm}{2R}\frac{1 + 2h/R}{1 + h/R}$$

The kinetic energy for orbiting at height h corresponds to the insertion velocity v_{ins}; that is,

$$E_{k,i} = \tfrac{1}{2}mv_{ins}^2 = \frac{GMm}{2R}\left(\frac{1}{1 + h/R}\right)$$

Thus the extra kinetic energy required for launching the satellite and to bring it to orbital altitude is

$$E_{k,h} = E_{k,c} - E_{k,i} = \frac{GMm}{2R}\left(\frac{2h/R}{1 + h/R}\right)$$

Then

$$\frac{E_{k,c}}{E_{k,i}} = 1 + \frac{2h}{R} \quad \text{and} \quad \frac{E_{k,h}}{E_{k,i}} = \frac{2h}{R}$$

This means that at low altitudes $E_{k,h} \ll E_{k,i}$ and practically all the initial kinetic energy provided by the burning fuel is used to insert the satellite in orbit. For example, for

$h = 300$ km, $E_{k,h}/E_{k,i} = 2h/R = 0.094$. This means that only 9.4% of the initial kinetic energy is used for lifting the satellite. The situation reverses for a high altitude satellite. For a geostationary satellite $h = 35\,880$ km and $2h/R \approx 11.2$, which means that the kinetic energy used in lifting is 11 times the kinetic energy for orbital motion.

(b) We could also proceed in a different way. First launch the satellite with enough initial kinetic energy just to reach the orbital height, and then add the required orbital velocity at the insertion point. A simple calculation, using the principle of conservation of energy, shows that the total kinetic energy is the same in both cases. In this second alternative, less initial kinetic energy is needed. However, a second rocket must be attached to the satellite to provide the orbital energy at insertion. This is the method used in the shuttle.

This analysis has been simplified by assuming that the mass of the whole system, satellite plus rocket engines, remains constant. In practice the mass of the fuel decreases continuously as it burns (recall Example 7.7) and multi-stage engines are used, with each stage dropped after it has completed its task. This diminishes the overall energy used. However the velocities, being independent of the mass of the satellite, remain the same as calculated in Equations 12.1 and 12.2. In carrying out these calculations we have not taken into account the effect due to the attraction of the Moon, which slightly affects the Earth's escape velocity.

EXAMPLE 12.2

Effect of the Earth's rotation and its orbital motion on the launching of a probe.

▷ In our calculations we have ignored the rotation of the Earth. However, all bodies on the Earth's surface have a velocity in the W–E direction given by (recall Example 5.3)

$$v = \omega R \cos \lambda = 463 \cos \lambda \quad \text{m s}^{-1} = 1700 \cos \lambda \quad \text{km h}^{-1}$$

where λ is the latitude of the body. That means that all probes launched from the Earth's surface have an eastward velocity in addition to their upward velocity. This has two important consequences. Whenever possible, launching sites are chosen on the East coast of continents, so that the initial eastward path is over the ocean. They are also chosen as close to the Equator as national boundaries allow, so that some initial orbital velocity is gained simply by the rotation of the Earth. This is why the United States uses Cape Canaveral, in Florida, at 30°N latitude, France uses French Guyana, at 8°N latitude and the former Soviet Union uses Baikonur, in Central Asia, at 45°N latitude.

For example, placing a satellite in an orbit at a height of 200 km requires an orbital velocity of 7.65×10^3 m s^{-1}. But for a satellite launched at the Equator, the Earth's rotation contributes 463 m s^{-1}, so that an insertion velocity of 7.19×10^3 m s^{-1} is all that is needed. A satellite launched from Cape Canaveral needs 7.29×10^3 m s^{-1} because the Earth's rotation contributes only 390 m s^{-1} in the eastward direction. Thus, for an equatorial launching, the Earth's rotation contributes 6% of the orbital velocity, and for Cape Canaveral the contribution is 5%. However the savings in kinetic energy compound to 12% and 10% respectively (because since $E = \frac{1}{2}mv^2$, $\Delta E/E = 2(\Delta v/v)$), which are important savings.

Since the insertion energy decreases with the height, the effect of the Earth's rotation is more important for higher altitude satellites.

The same effect occurs in launching a probe toward another planet. However, in this case the orbital velocity of the Earth around the Sun, which is about 3×10^4 m s^{-1}, is used. This situation is discussed in Example 12.3.

12.3 Voyage to the Moon

The Moon is the Earth's closest celestial body. It is at an average distance of 384 400 km or about 60 times the radius of the Earth. As seems natural, it was the first extraterrestrial body to be explored. When a probe is sent toward the Moon, it moves in the gravitational field shown in Figure 11.24. The velocity of the probe first decreases as it moves away from the Earth and its minimum initial velocity must be practically equal to the escape velocity, $1.13 \times 10^4 \text{ m s}^{-1}$ or about $40\,000 \text{ km h}^{-1}$. Eventually the probe reaches the equilibrium point A where the gravitational attraction of the Earth and the Moon are equal. This point is at a distance of 346 000 km from the Earth and 38 400 km from the Moon (recall Figure 11.24). Once the probe is beyond this point, the attraction of the Moon predominates over that of the Earth. The probe's velocity gradually increases, until it reaches the Moon's surface at $2.38 \times 10^3 \text{ m s}^{-1}$ (8480 km h^{-1}), the escape velocity of the Moon. The first probe to strike the Moon was Luna 2, launched by the Soviet Union in September 1959. The trip took 35 hours.

Since the Moon is in motion relative to the Earth (Figure 12.2), its position at launch time is C_1, while its position at arrival time is C_2. The path of the space probe must be a curved line, so that the actual calculations are a little more complex. It is possible that a probe, instead of falling on the Moon, passes near its surface. In this case, depending on its velocity and its distance of closest approach, the probe follows different paths. It may be captured into a stable orbit around the Moon. It may just go around the Moon and return to Earth. It may, as a final case, suffer a deviation and escape into the solar system. This latter was the case of Luna 1, launched by the Soviet Union in January 1959. After passing by the Moon at about 8000 km from its surface, Luna 1 entered an elliptical orbit around the Sun with a period of 443 days.

For a 'soft' landing on the Moon's surface, it is necessary to reduce the velocity of approach. Since the Moon has no atmosphere, the velocity is reduced by firing a rocket engine that produces a thrust in a direction opposite to the motion. Reducing the velocity from $2.38 \times 10^3 \text{ m s}^{-1}$ to almost zero requires the same energy and the same amount of fuel as would be needed to accelerate the probe up to that velocity. The first soft landing on the Moon was that of Luna 9, in February 1966. When Luna 9 was about 80 km above the Moon's surface its retro-rockets reduced its velocity to about 47 m s^{-1} or 170 km h^{-1}.

Returning from the Moon is easier in a sense because it is only necessary to provide an initial velocity equal to the lunar escape velocity, $2.38 \times 10^3 \text{ m s}^{-1}$. Once the probe passes the equilibrium point it continues 'falling' toward the Earth where it arrives with the Earth's escape velocity, $1.13 \times 10^4 \text{ m s}^{-1}$. A soft landing on Earth is also easier because the atmosphere slows the probe down and it is possible to use a parachute at the last moment.

A satellite can be placed into a lunar orbit (Figure 12.3) when it is near the Moon's surface, as at point C. It is only necessary to use the retro-rockets to adjust the velocity of the probe to that required for a stable orbit at the desired distance (recall Equation 12.1). The first Moon satellite, Luna 10, was placed in orbit in April 1966. It described an orbit with a radius of 2400 km, a velocity of $1.24 \times 10^3 \text{ m s}^{-1}$ and a period of 3 h.

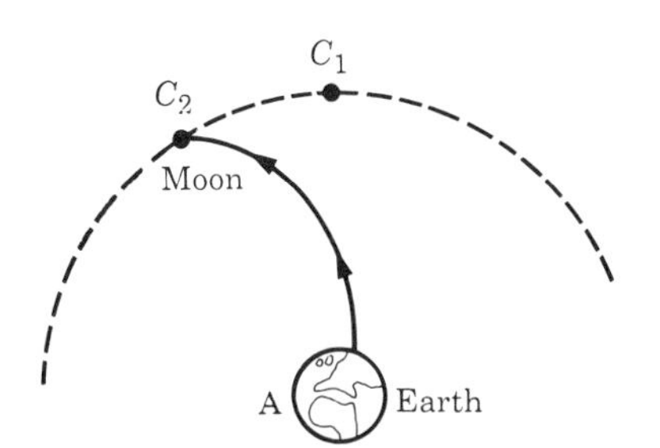

Figure 12.2 Launching a probe toward the Moon.

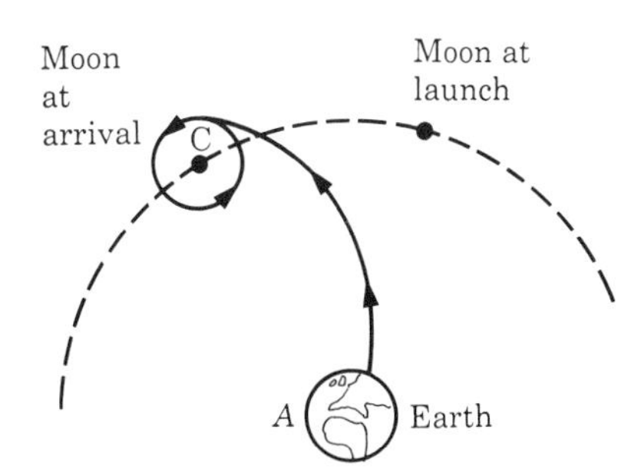

Figure 12.3 Insertion of a satellite into a lunar orbit.

The United States, Japan and the Soviet Union have sent several probes to the Moon, equipped with a variety of scientific instruments. These probes have obtained valuable information about the physical properties of the Moon's surface such as its composition, temperature, irradiation, seismicity etc.

A truly pioneering effort was the extraordinary exploration of the Moon in the Apollo series organized by the United States. The purpose was to send astronauts to the Moon's surface and bring them back. There have been nine manned trips to the Moon, of which six included a lunar landing. The first men to step on the Moon's surface were Neil A. Armstrong and Edward E. Aldrin, on 20 July 1969.

The system used to go to the Moon consisted of several components with a total initial mass of about 2.8×10^6 kg. The launching propulsion was produced by a three-stage Saturn V rocket, with an initial mass of about 2.6×10^6 kg; its function was to provide the velocity necessary to reach the Moon. The second component was the **service module** (SM) whose function was to place the vehicle in a lunar orbit; its initial mass was 23 000 kg. The third component was the **lunar module** (LM) in which the astronauts landed on the Moon; it also carried a rocket to bring the module down to the lunar surface and back into a lunar orbit; its initial mass was about 12 600 kg. The fourth component was the **command module** (CM) in which the three astronauts traveled to the Moon and returned to Earth; its mass was about 5 000 kg or 0.2% of the total initial mass.

Figure 12.4 shows the trajectory followed in the mission. The launching point is (1). Firing the first two stages of the Saturn V rocket placed the system in an Earth orbit with a velocity of 7.78×10^3 m s^{-1} at a height of 180 km. After circling for about three orbits, the third stage of the Saturn V rocket fired at (3) and propelled the vehicle until it reached escape velocity, 1.13×10^4 m s^{-1}. Then the Saturn V rocket separated from the rest of the vehicle and fell back to Earth. The rest of the system (service, command and lunar modules), with a mass of about 200 000 kg coasted toward the Moon along path (4) for about 60 hours without using fuel except for minor adjustments. Near the Moon the rocket in the service module decreased the velocity of the system to about 1.55×10^3 m s^{-1} and the vehicle entered into lunar orbit (5) about 110 km above the Moon's surface. After a few lunar orbits, the lunar module separated from the service and command modules (6) and carried two astronauts down to the Moon's surface. The service

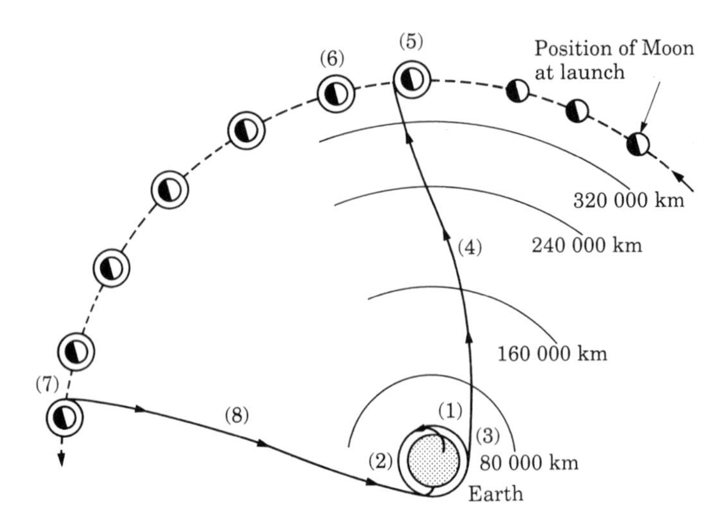

Figure 12.4 From Earth to the Moon and back.

and command modules remained in orbit with one astronaut aboard. After completion of the lunar mission, the rocket of the lunar module brought it back to rendezvous with the orbiting service and command modules (7). After the astronauts were transferred to the command module, the lunar module was jettisoned. Then the rockets in the service module fired to increase the velocity up to the Moon's escape velocity, $2.38 \times 10^3 \text{ m s}^{-1}$. The service and command modules returned (8) back to Earth, again taking about 60 hours. Close to the Earth the service module was separated from the command module, the only component that returned to the Earth's surface.

12.4 Exploration of the solar system

Several planets of the solar system have been reached from the Earth using unmanned vehicles or probes. The two planets that have been explored more thoroughly are Venus and Mars and several probes have landed on them.

The trip from the Earth to another planet requires several stages. First it is necessary that the vehicle acquire the Earth's escape velocity, $1.13 \times 10^4 \text{ m s}^{-1}$. Next, the vehicle must be placed in a solar orbit, under the Sun's gravitational attraction, which carries the vehicle to a rendezvous with the planet. When the vehicle is very close to the planet and subject to its gravitational attraction, it may fall on the planet, orbit around it, or suffer a deviation and continue its travel through the solar system. For example, in 1977 the probe Voyager 2 was launched for a 'tour' of the solar system, passing close to Jupiter (1979), Saturn (1981), Uranus (1986) and Neptune (1989), after which it entered a solar orbit. Those planets are relatively easy to visit in one 'trip' because their orbits are almost on the same plane (Figure 12.5). At each encounter the probe received an additional gravitational push to continue its voyage, as was explained in Example 6.4.

The program of planetary exploration can be carried out in different ways. One uses the shortest path, which results in a minimum time. Another is the path

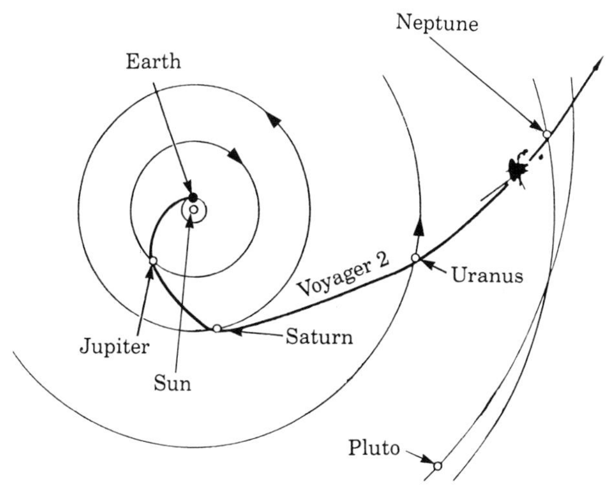

Figure 12.5 Voyager 2 and its exploration of the solar system.

that requires the least amount of energy. And of course there are many intermediate options.

The path where the least amount of energy is expended has the Earth, at launch, and the planet, at arrival, on a line that passes through the Sun (Figure 12.6). These paths are called **Hohmann orbits** and are ellipses, with the Sun at one focus. The major axis of the ellipse is equal to the sum of the radii of the Earth's and the planet's orbits. Recalling Equation 11.10, we see that the total energy of a probe of mass m in a Hohmann orbit is $E = -GMm/2a$, where M is the solar mass and $2a = r_E + r_P$ is the major axis of the orbit (Figure 12.7). In terms of its kinetic and potential energies, we can write $\frac{1}{2}mv^2 - GMm/r = -GMm/2a$. Solving for the velocity, we obtain

$$v^2 = GM\left(\frac{2a - r}{ar}\right) \tag{12.4}$$

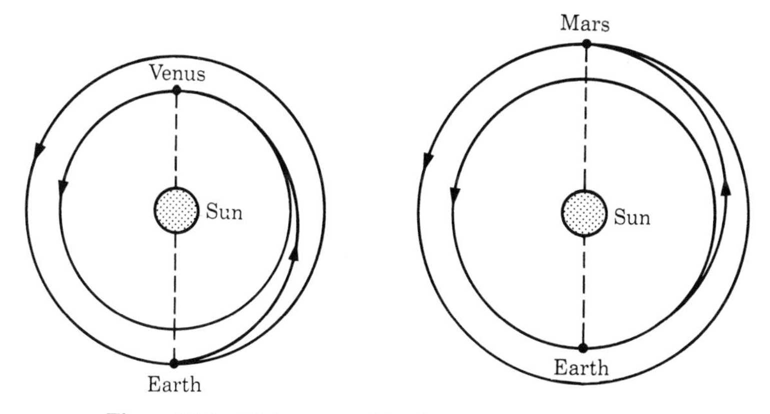

Figure 12.6 Hohmann orbits for Venus and Mars.

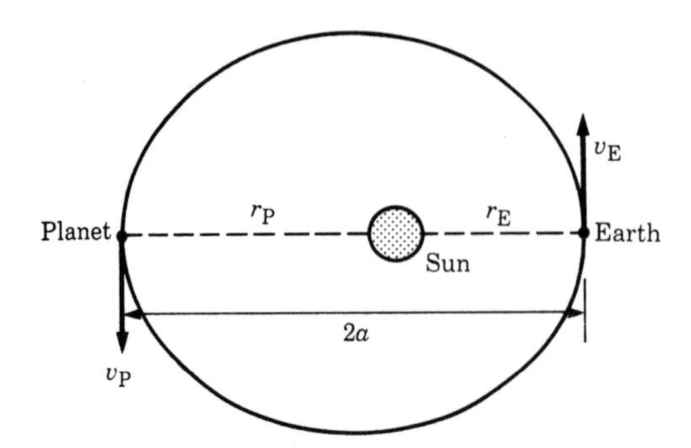

Figure 12.7 Velocities in a Hohmann orbit.

When the probe is launched from the Earth, $r = r_E$ and, noting that $2a - r_E = r_P$, we find the insertion velocity v_E relative to the Sun as

$$v_E = \left(\frac{GMr_P}{ar_E} \right)^{1/2}$$

The velocity v_P, also relative to the Sun, of the probe as it approaches the planet may be found by making $r = r_P$ in Equation 12.4. Then $2a - r_P = r_E$ and

$$v_P = \left(\frac{GMr_E}{ar_P} \right)^{1/2} \quad \textbf{(12.5)}$$

The first planet to be explored was Venus, using a probe named Mariner 2, launched by the United States in August 1962 (Figure 12.8). It arrived at Venus after 109 days. Afterwards Mariner 2 entered a solar orbit with a period of 346 days. The first exploration of Mars was with Mariner 4, launched by the United States in November 1964 (Figure 12.9). It took 227 days to reach Mars. After

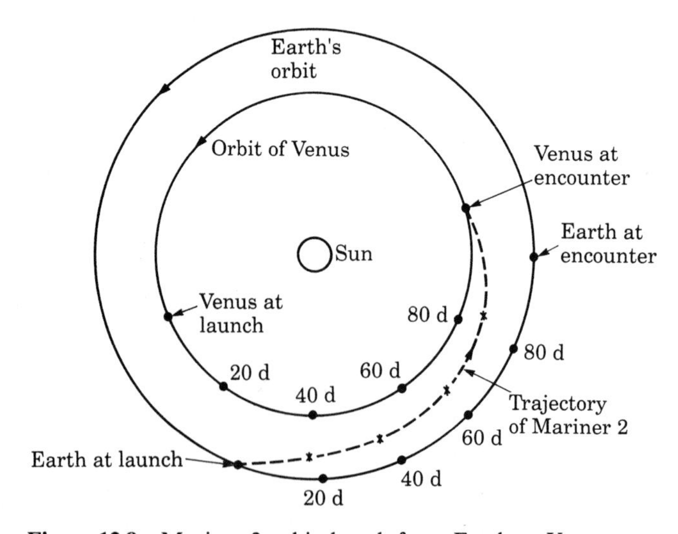

Figure 12.8 Mariner 2 orbital path from Earth to Venus.

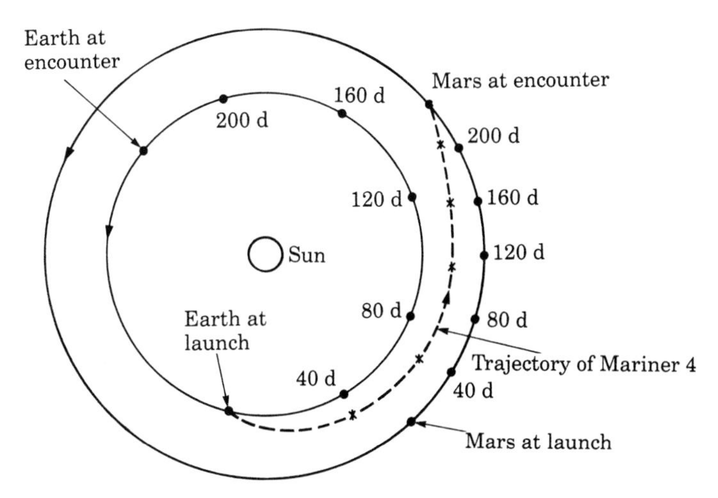

Figure 12.9 Mariner 4 orbital path from Earth to Mars.

passing at about 18 000 km from the planet, it too entered a solar orbit. Neither of these flights was along a true Hohmann orbit.

The three more recent space probes launched to explore the planets of the solar system have been the Magellan, Galileo and Ulysses missions. The Magellan probe was launched in April 1989 and arrived at Venus in September 1990 after going one and a half times around the Sun following a trajectory close to a Hohmann orbit. Once the probe reached Venus it entered into an orbit around the planet (Figure 12.10). The purpose of the Magellan mission was to explore the planet for about 243 Earth days or 1 Venus day. The Galileo probe was launched in October 1989 to explore Jupiter, arriving in December 1995. It follows the gravity assisted orbit, designated VEEGA, that was explained in Example 6.4. The orbit of Galileo is shown in Figure 12.11.

The same technique of gravity assisted propulsion of a space probe was used for the Ulysses solar probe, launched in October 1990. The path is shown in Figure 12.12. The probe was initially aimed at Jupiter. A close encounter

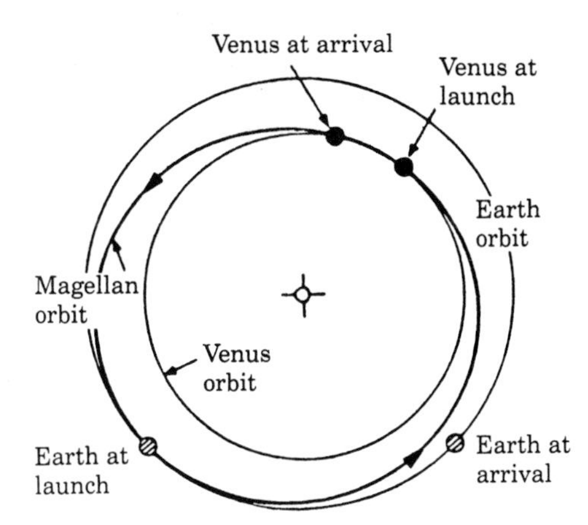

Figure 12.10 Earth to Venus Magellan trajectory.

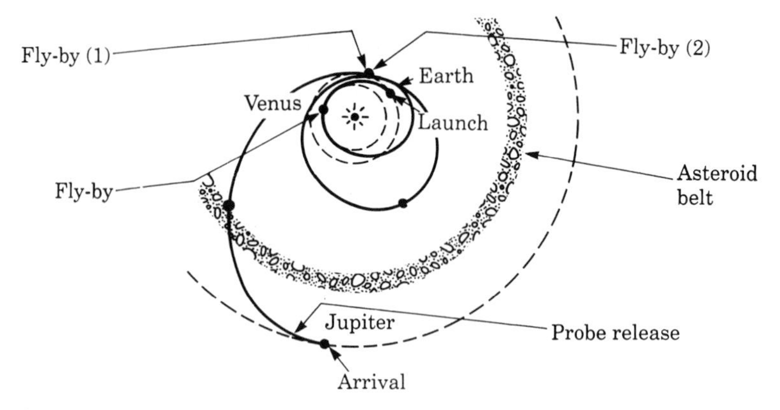

Figure 12.11 Galileo VEEGA flight path.

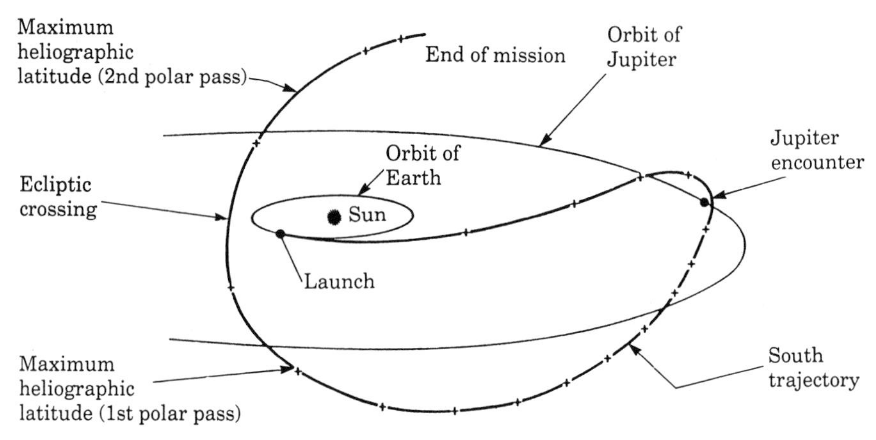

Figure 12.12 Ulysses Jupiter flight path.

with the planet in February 1992 used Jupiter's gravitational action to send Ulysses into a new orbit almost perpendicular to the solar equatorial plane. Ulysses will then enter a polar orbit around the Sun, allowing it to study the Sun from a previously unreachable angle. Ulysses passes over the Sun's south pole during May through September of 1994 and over the north pole a year later, after which it continues moving in this orbit. The closest distance (perihelion) to the Sun is about 1.9×10^{11} m, and the largest distance (aphelion) is about 8×10^{11} m.

EXAMPLE 12.3

Details of the Venus and Mars missions.

▷ In order to reach Venus we first accelerate the probe to the Earth's escape velocity 1.13×10^4 m s^{-1}. At a certain moment, when the velocity of the probe is very small relative to the Earth, the probe must be given the insertion velocity into a Hohmann orbit. Using Equation 12.4 and with the data from Table 11.1, the insertion velocity is found to be

27.3×10^3 m s^{-1}. Since at that time the velocity of the probe relative to the Sun is practically equal to the orbital velocity of the Earth, 29.8×10^3 m s^{-1}, it is necessary to *reduce* the probe velocity by 2.5×10^3 m s^{-1} so that it will 'fall in' toward Venus. The probe reaches Venus after 145 days with a velocity (using Equation 12.5) of 37.7×10^3 m s^{-1}. Since the orbital velocity of Venus is 35.05×10^3 m s^{-1}, the probe's velocity must be again *reduced* by 2.7×10^3 m s^{-1} so that it can be captured by Venus. The probe will fall on Venus's surface at the escape velocity, 10^4 m s^{-1}, unless other velocity adjustments are made.

In the case of Mars, the insertion velocity into a Hohmann orbit is 32.7×10^3 m s^{-1}, which is larger than the orbital velocity of the Earth, 29.8×10^3 m s^{-1}. Therefore, to reach insertion velocity, it is necessary to *increase* the velocity of the probe by 2.9×10^3 m s^{-1} so the probe will 'fall away' toward Mars. The vehicle arrives at Mars after 260 days with a velocity of 21.5×10^3 m s^{-1}. Mars' orbital velocity is 24.1×10^3 m s^{-1}. Therefore, it is necessary to *increase* the velocity of the probe by 2.6×10^3 m s^{-1} so that the probe can fall on Mars. The probe will fall on Mars with the escape velocity of 5×10^3 m s^{-1} unless other velocity adjustments are made.

QUESTIONS

12.1 Make a list of external factors affecting conditions in the Earth's surface and atmosphere.

12.2 Verify that for a very low altitude satellite launched with the characteristic velocity, the kinetic energy at launching corresponds to that of a circular orbit of radius R (see Example 12.1(a)).

12.3 Verify that for low altitudes the insertion and characteristic velocities differ by $(GM/R)^{1/2}(h/R)$.

12.4 Why is the escape velocity larger than the insertion velocity for the same height above the Earth's surface?

12.5 How does gravity assist in a planetary mission?

12.6 How does the Earth's orbital velocity assist in the orbital motion of a space probe?

12.7 Investigate what the advantages are for placing research satellites near the Earth's surface but above its atmosphere. Make a list of those satellites you have heard of and their purpose.

PROBLEMS

12.1 A rocket is fired vertically from the Earth toward the Moon, the fuel being consumed in a relatively short time after the firing. (a) At what point in its path toward the Moon is its acceleration zero? (b) Calculate the minimum initial velocity of the rocket in order to reach this point and fall on the Moon by the action of lunar attraction. (c) In the last case, calculate the velocity of the rocket when it hits the Moon. (Ignore the motion of the Moon relative to the Earth and assume the rocket moves in a straight line.)

12.2 Compute the escape velocity from the surface of (a) Mercury, (b) Venus, (c) Mars and (d) Jupiter. (Hint: to simplify the calculation, first compute the factor $(2G)^{1/2}$. Then you only have to multiply it by $(M/R)^{1/2}$ for each planet.)

12.3 (a) Calculate the insertion velocity of an Earth satellite in a circular orbit at 1000 km above the Earth's surface. Calculate also the characteristic velocity at launching.

12.4 A satellite, having a mass of 5000 kg, describes an equatorial circular path around the Earth of radius 8000 km. Find (a) its angular momentum and its (b) kinetic, (c) potential and (d) total energies.

12.5 The Apollo VIII spacecraft orbited the Moon in a circular orbit 113 km above the surface. Calculate: (a) the period of motion,

(b) the velocity and (c) the angular velocity of the craft.

12.6 Find the height and velocity of a satellite (in circular orbit in the equatorial plane) that remains over the same point on the Earth at all times. This is called a **geostationary** satellite.

12.7 An Earth satellite moves in a circular orbit concentric with the Earth at a height of 300 km above the Earth's surface. Find (a) its velocity, (b) its period of revolution and (c) its centripetal acceleration. (d) Compare the result of part (c) with the value of g at that height, as computed directly by the method of Example 11.1.

12.8 Calculate the launching velocity of the lunar module in the Apollo VIII mission from the Moon's surface so that it was able to join the command module smoothly, which was orbiting 113 km above the Moon's surface.

12.9 (a) Discuss in detail the path of a probe to Mercury, using a Hohmann orbit. (b) Calculate the velocities required at each stage. (c) Repeat the calculations for a mission to Jupiter.

12.10 (a) Calculate the velocity of a satellite when it is in an equatorial orbit 480 km above the Earth's surface. What additional velocity must be given to place the satellite in this orbit if it is launched at the equator (b) in the east–west direction or (c) in the west–east direction? (Hint: first calculate the initial velocity of the satellite due to Earth rotation.) (d) From your answers, what should be the preferred direction of launch for a satellite?

12.11 A meteorite is moving very slowly, relative to the Earth, when it passes perpendicularly at a distance from the center of the Earth equal to 20 Earth radii. If it is attracted toward the Earth, calculate what its velocity would be when it reaches a distance from the Earth's center equal to 2 Earth radii.

12.12 A probe is to be launched from the Earth with a velocity v_0, so that at a very large distance from the Earth its velocity, relative to the Earth, is v_E. (a) Show that if v_e is the escape velocity from the Earth (recall Equation 11.13), then $v_0 = (v_e^2 + v_E^2)^{1/2}$. (b) Calculate the velocity of the probe relative to the Sun.

12.13 A probe is to be launched in such a manner that its velocity relative to the Sun, when it is a great distance from the Earth, is 32.7 km s^{-1}. (This would be the velocity of a probe launched toward Mars in a Hohmann orbit.) Verify that the launching velocity must be 11.6 km s^{-1}. Recall that the orbital velocity of the Earth relative to the Sun is 30 km s^{-1}.

12.14 A probe is in a 'parking orbit' at a height above the Earth's surface of 2.26 R, where R is the Earth's radius. Verify that the velocity the probe needs to escape (a) from the Earth is 6.2 km s^{-1} and (b) from the solar system is 9.42 km s^{-1}. Recall that the orbital velocity of the Earth relative to the Sun is 30 km s^{-1}.

12.15 A probe is to be launched on an escape trajectory from a 'parking orbit' 180 km above the Earth's surface. Calculate the minimum escape velocity from that orbit. (b) Calculate the launching velocity to have the probe escape the Earth if launched from the Earth. (c) Discuss the advantage of a parking orbit.

13 Systems of particles I: Linear and angular momentum

After the diver's feet leave the board, his center of mass follows a parabolic path. (© Estate of Harold Edgerton. Courtesy of Palm Press, Inc.)

Notes

13.1 Introduction

In the theory of the dynamics of a particle, the rest of the universe is represented either by a force or a potential energy that depends only on the coordinates of the particle and certain parameters related to its environment. A more realistic and important problem is that of several interacting particles. In fact, it was with a system of two particles that we started our discussion of dynamics, when we stated the principle of conservation of momentum in Chapter 6. Two main results associated with a system of particles are discussed in this chapter: the motion of the center of mass and the conservation of angular momentum. The conservation of energy is discussed in the following chapter. We also assume that the observer is attached to an inertial frame $X_L Y_L Z_L$, which we call the **laboratory** or **L-frame**.

13.2 Motion of the center of mass of an isolated system of particles

Consider a system composed of particles of masses $m_1, m_2, \ldots$ and position vectors $\boldsymbol{r}_1, \boldsymbol{r}_2, \ldots$ relative to an inertial frame of reference $X_L Y_L Z_L$ (Figure 13.1). As we shall see, it is useful to define a point, more or less related to the average position of the particles, called the **center of mass** (CM) of the system of particles. The CM is defined as the point given by the position vector

$$\boldsymbol{r}_{CM} = \frac{m_1 \boldsymbol{r}_1 + m_2 \boldsymbol{r}_2 + \cdots}{m_1 + m_2 + \cdots} = \frac{\sum_i m_i \boldsymbol{r}_i}{M} \tag{13.1}$$

where $M = \sum_i m_i$ is the total mass of the system. In terms of rectangular coordinates, the center of mass is given by

$$x_{CM} = \frac{\sum_i m_i x_i}{M}, \qquad y_{CM} = \frac{\sum_i m_i y_i}{M}, \qquad z_{CM} = \frac{\sum_i m_i z_i}{M}$$

In Table 13.1 the position of the CM of bodies with several geometric shapes is illustrated.

When a system of particles is in motion relative to the observer, the CM of the system will also move with a velocity we shall designate $\boldsymbol{v}_{CM}$. Taking the time derivative of Equation 13.1, we obtain the velocity of the center of mass in terms of the velocities $\boldsymbol{v}_i$ of the particles relative to $X_L Y_L Z_L$ as

$$\boldsymbol{v}_{CM} = \frac{d\boldsymbol{r}_{CM}}{dt} = \frac{1}{M} \sum_i m_i \frac{d\boldsymbol{r}_i}{dt} = \frac{\sum_i m_i \boldsymbol{v}_i}{M} \tag{13.2}$$

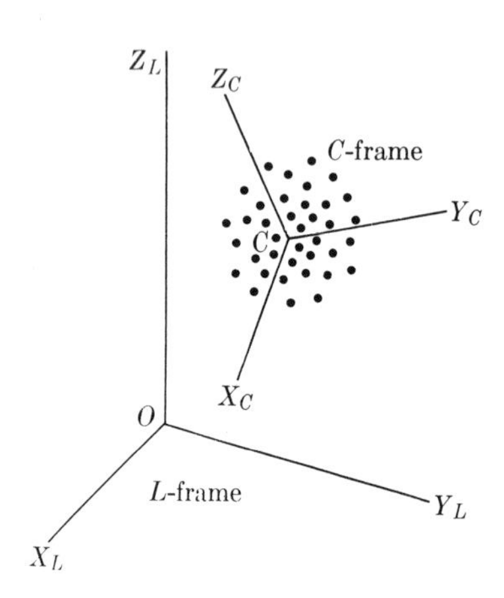

Figure 13.1 The c-frame of reference coincides with the center of mass of the system of particles. The l-frame of reference is fixed relative to the observer.

Noting that $\boldsymbol{p}_i = m_i \boldsymbol{v}_i$ is the momentum of the ith particle, we can also write Equation 13.2 as

$$\boldsymbol{v}_{CM} = \frac{1}{M} \sum_i \boldsymbol{p}_i = \frac{\boldsymbol{P}}{M} \quad \text{or} \quad \boldsymbol{P} = M\boldsymbol{v}_{CM} \tag{13.3}$$

where $\boldsymbol{P} = \sum_i \boldsymbol{p}_i$ is the total momentum of the system.

This suggests that the momentum of the system is the same as it would be if all the mass were concentrated at the center of mass, moving with velocity $\boldsymbol{v}_{CM}$. For that reason, $\boldsymbol{v}_{CM}$ is sometimes called the **system velocity**. Thus when we speak of the velocity of a moving body composed of many particles, such as an airplane or an automobile, the

Table 13.1 Centers of mass

Figure	Position of CM
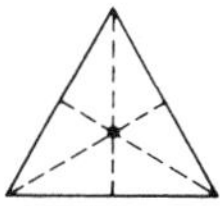	*Triangular plate* Point of intersection of the three medians
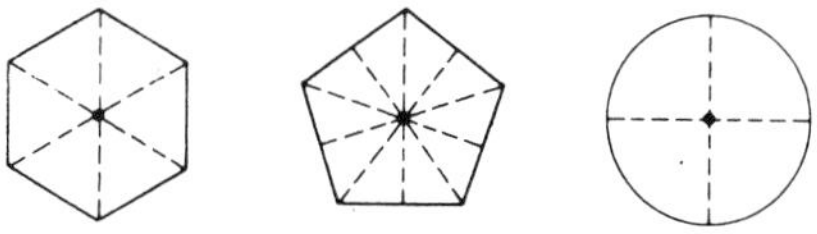	*Regular polygon and circular plate* At the geometrical center of the figure
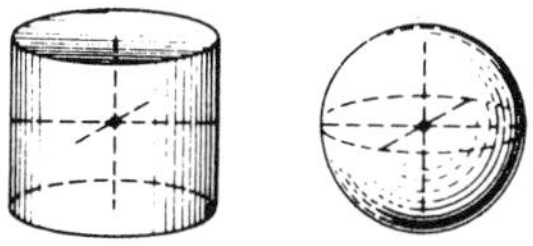	*Cylinder and sphere* At the geometrical center of the figure
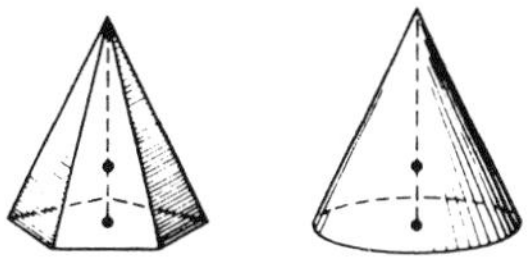	*Pyramid and cone* On line joining vertex with center of base and at $\frac{1}{4}$ of the length measured from the base
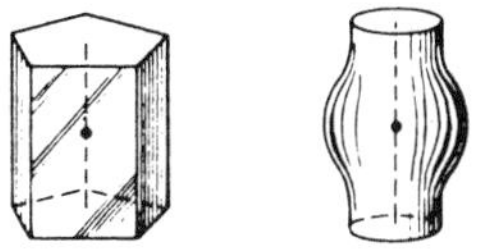	*Figure with axial symmetry* Some point on the axis of symmetry
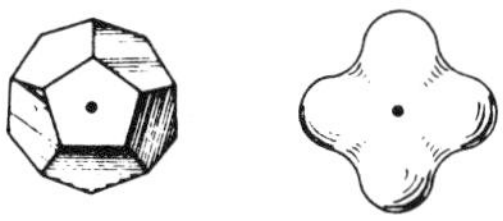	*Figure with center of symmetry* At the center of symmetry

Earth or the Moon, or even a molecule or a nucleus, we actually refer to its center of mass velocity $\boldsymbol{v}_{CM}$. For that reason we may sometimes treat a system of particles as if it were a single particle coincident with the CM. Therefore, the notion of CM is of great practical value.

If the system is isolated, that is, it is not subject to forces produced by particles external to the system, we know from the principle of conservation of momentum that $\boldsymbol{P}$ is constant. Therefore

> *the center of mass of an isolated system moves with constant velocity relative to any inertial frame.*

The collision of two sliding discs is shown in the multiflash photograph of Figure 13.2(a). The two discs form an isolated system and, from the graphical

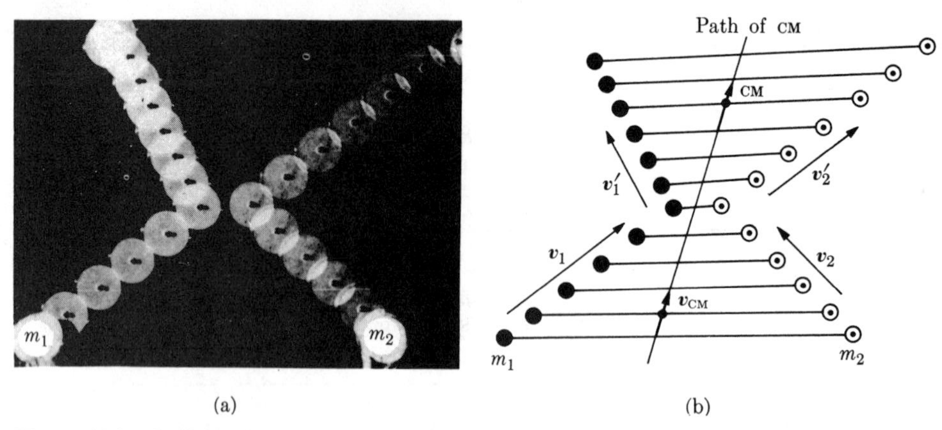

Figure 13.2 Collision between two bodies. (a) Multiflash photograph of collision. (b) Graphical analysis of the photograph, showing that the center of mass has moved in a straight line with constant velocity relative to the laboratory.

analysis of Figure 13.2(b), we see that the center of mass of the system moves in a straight line, with constant velocity.

We may attach a frame of reference, designated $X_C Y_C Z_C$, to the center of mass of a system (Figure 13.1). Relative to this frame, the center of mass is at rest ($\boldsymbol{v}_{CM} = 0$). This is called the **center of mass** or **C-frame of reference**. In view of Equation 13.3, the total momentum of a system of particles referred to the C-frame of reference is always zero:

$$\boldsymbol{P} = \sum_i \boldsymbol{p}_i = 0 \qquad \text{in the C-frame of reference}$$

For that reason the C-frame is sometimes called the **zero momentum frame**. The C-frame is important because many experiments we perform in our laboratory or L-frame can be more simply analyzed in the C-frame. It is clear that the C-frame moves with a velocity $\boldsymbol{v}_{CM}$ relative to the L-frame. When no external forces act on a system, the C-frame can be considered as inertial.

EXAMPLE 13.1

Center of mass of a two body system.

▷ Consider a system composed of two particles of masses m_1 and m_2 separated by a distance r (Figure 13.3). Take the origin of coordinates at the center of mass so that $\boldsymbol{r}_{CM} = 0$. Then Equation 13.1 gives

$$m_1\boldsymbol{r}_1 + m_2\boldsymbol{r}_2 = 0 \quad \text{or} \quad m_1\boldsymbol{r}_1 = -m_2\boldsymbol{r}_2$$

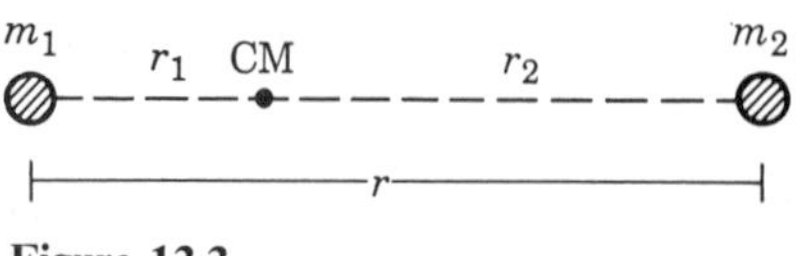

Figure 13.3

Using only magnitudes, we may write $m_1 r_1 = m_2 r_2$, so that the distances r_1 and r_2 are inversely proportional to the respective masses. Combining this relation with the fact that $r = r_1 + r_2$, we can write

$$r_1 = \frac{m_2}{m_1 + m_2} r, \qquad r_2 = \frac{m_1}{m_1 + m_2} r$$

We will apply these relations to some simple cases.

(a) Center of mass of the Earth–Moon system. For this case $m_E = 5.98 \times 10^{24}$ kg and $m_M = 7.34 \times 10^{22}$ kg. Therefore $m_E/m_M = 81.5$. The distance from the Earth to the Moon is 3.84×10^8 m. Therefore, the distance of the CM from the center of the Earth is

$$r_E = \frac{m_M}{m_E + m_M} r = \frac{r}{(m_E/m_M) + 1} = 1.21 \times 10^{-3} r = 4.665 \times 10^6 \text{ m}.$$

Recalling that the radius of the Earth is 6.37×10^6 m, the CM of the Earth–Moon system is about $\frac{2}{3}$ of the way inside the Earth. (The student can repeat the calculation for the Earth–Sun system.)

(b) Center of mass of the electron–proton system. In this case $m_p/m_e \simeq 1850$ and $r = 0.53 \times 10^{-10}$ m for the lowest energy state. Therefore, the distance of the CM from the proton is

$$r_p = \frac{m_e}{m_p + m_e} r = \frac{r}{(m_p/m_e) + 1} = 5.4 \times 10^{-4} r = 2.72 \times 10^{-14} \text{ m}$$

Therefore, the CM is practically coincident with the proton.

(c) Center of mass of the CO molecule. The atoms in a CO molecule are 1.13×10^{-10} m apart. The ratio $(m_O/m_C) = 16.0/12.0 = 1.33$ and the distance of the CM of the molecule from the O atom is

$$r_O = \frac{m_C}{m_O + m_C} r = \frac{r}{(m_O/m_C) + 1} = 0.4286\, r = 0.4843 \times 10^{-10} \text{ m}$$

while the distance from the C atom to the CM is

$$r_C = r - r_O = 0.5708\, r = 0.6450 \times 10^{-10} \text{ m}$$

EXAMPLE 13.2

Center of mass of the molecules (a) CO_2 and (b) H_2O (Figure 13.4).

▷ We will assume these molecules are formed from the isotopes ^{12}C, ^{16}O and ^{1}H.

(a) The CO_2 molecule is linear and symmetric (Figure 13.4(a)). Therefore, the CM coincides with the position of the carbon atom.

(b) The H_2O molecule is planar. We place the molecule on the XY-plane, with the oxygen atom at the origin and the X-axis bisecting the H–O–H angle. Due to the symmetry of the molecule, the CM falls along

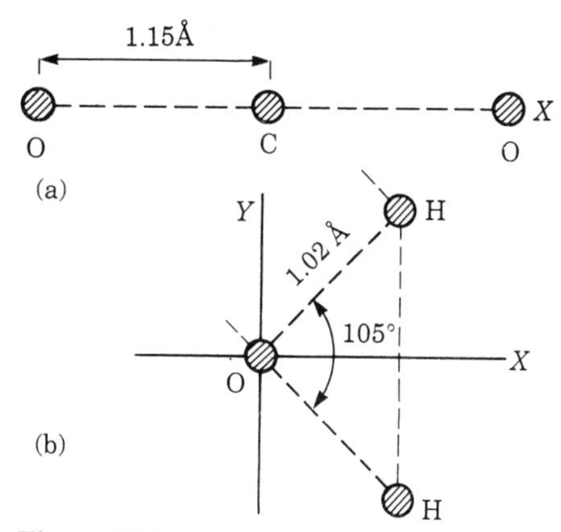

Figure 13.4

the X-axis at a distance from the oxygen atom given by

$$x_{CM} = \frac{2m(^1\mathrm{H}) \times (1.02\,\text{Å} \cos 52.5°)}{m(^{16}\mathrm{O}) + 2m(^1\mathrm{H})} = \frac{1.252\,\text{Å}}{18.001} = 0.070\,\text{Å}$$

Therefore, the center of mass of a water molecule is almost coincident with the oxygen atom. (Recall 1 Å = 10^{-10} m.)

EXAMPLE 13.3

The velocities of two particles of masses m_1 and m_2 relative to an inertial observer are $\boldsymbol{v}_1$ and $\boldsymbol{v}_2$. Determine the velocity of the center of mass relative to the observer and the velocity of each particle relative to the center of mass.

▷ From Equation 13.2 the velocity of the center of mass relative to the observer is

$$\boldsymbol{v}_{CM} = \frac{m_1\boldsymbol{v}_1 + m_2\boldsymbol{v}_2}{m_1 + m_2}$$

The velocity of each particle relative to the center of mass, using the Galilean transformation of velocities is

$$\boldsymbol{v}_1' = \boldsymbol{v}_1 - \boldsymbol{v}_{CM} = \boldsymbol{v}_1 - \frac{m_1\boldsymbol{v}_1 + m_2\boldsymbol{v}_2}{m_1 + m_2}$$

$$= \frac{m_2(\boldsymbol{v}_1 - \boldsymbol{v}_2)}{m_1 + m_2} = \frac{m_2\boldsymbol{v}_{12}}{m_1 + m_2}$$

and

$$\boldsymbol{v}_2' = \boldsymbol{v}_2 - \boldsymbol{v}_{CM} = \frac{m_1(\boldsymbol{v}_2 - \boldsymbol{v}_1)}{m_1 + m_2} = -\frac{m_1\boldsymbol{v}_{12}}{m_1 + m_2}$$

where $\boldsymbol{v}_{12} = \boldsymbol{v}_1 - \boldsymbol{v}_2$ is the relative velocity of the two particles. Thus, in the C-frame, the two particles appear to be moving in opposite directions (Figure 13.5) with velocities

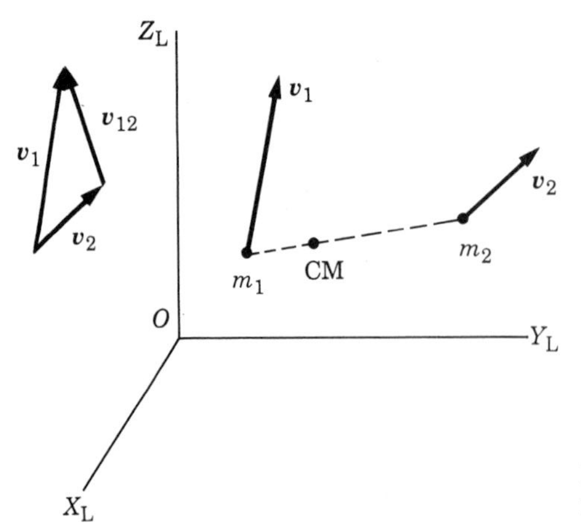

Figure 13.5 The velocity of m_2 relative to m_1 is $\boldsymbol{v}_{21} = \boldsymbol{v}_2 - \boldsymbol{v}_1$.

inversely proportional to their masses. If one particle is very massive compared with the other ($m_2 \gg m_1$) it remains practically at rest close to the CM of the system.

Also, relative to the CM, the two particles move with equal but opposite momenta, since

$$p'_1 = m_1 v'_1 = \frac{m_1 m_2 v_{12}}{(m_1 + m_2)} = -p'_2$$

The relations we have derived in this example are very important in scattering experiments. In these experiments the velocities of the particles are measured relative to the L-frame of reference attached to the laboratory. But the theoretical expressions for scattering are much simpler when they are related to the C-frame of reference.

13.3 Motion of the center of mass of a system of particles subject to external forces

Consider now a system of particles that is *not* isolated; in other words, the particles of the system are interacting with other particles that do not belong to the system. Suppose that our system, designated S, is composed of the particles within the dashed line in Figure 13.6. The particles in S may then interact among themselves as well as with those outside the dashed line, composing another system S'. To consider some concrete examples, the system S may be our galaxy and S' the rest of the universe. Or S may be the solar system and S' may be the rest of the galaxy. We may even consider a molecule or group of atoms and organize them into two systems S and S'.

We call the forces exerted between the particles composing the system S **internal forces**. These forces appear in pairs, having equal magnitude and opposite direction, because of the law of action and reaction. This is shown in Figure 13.7 for the case where system S is composed of only two particles. On the other hand, the **external forces** are those exerted on the particles of the system by the particles

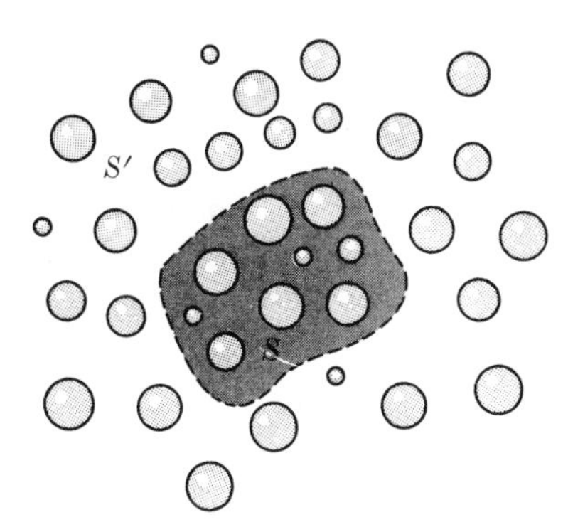

Figure 13.6 Interaction between a system S and its surroundings, or system S'.

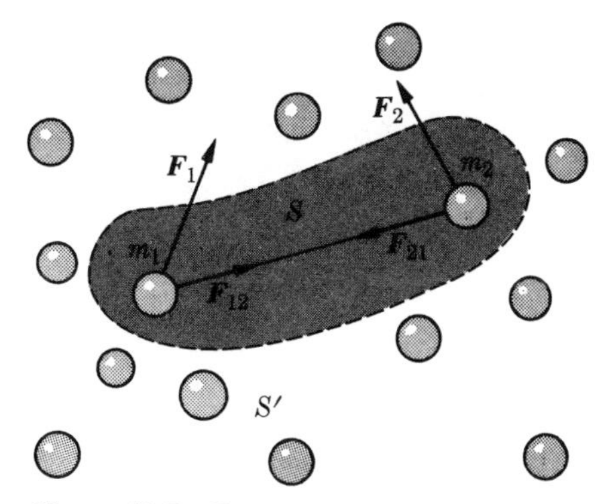

Figure 13.7 External and internal forces on a system S. F_{12} and F_{21} are internal. F_1 and F_2 are external.

in the surroundings, or system S'. Then we say that

the resultant external force on a system of particles is the sum of the external forces on each of its component particles.

That is,

$$F_{\text{ext}} = \sum_i F_i \tag{13.4}$$

where F_i is the **external** force on particle i. For example, in the case of a planetary system, the forces between the planets and the Sun are internal, while the forces exerted by the bodies outside the planetary system are external. Of course, because of the law of action and reaction, the particles in system S also exert forces on the particles in S'. However, we are not concerned now with what happens to system S'. Then, as shown at the end of the section,

the time rate of change of the total momentum of a system of particles is equal to the resultant external force applied to the system.

That is,

$$\frac{dP}{dt} = F_{\text{ext}} \quad \text{or} \quad \frac{d}{dt}\left(\sum_i p_i\right) = F_{\text{ext}} \tag{13.5}$$

Note that the rate of change of momentum of S is due only to the **external force** resulting from its interaction with S'. The **internal forces** in S due to the interactions among its component particles do not produce any net change in its total momentum. This occurs because, being in pairs, the internal forces produce equal and opposite changes in the momentum of the particles on which they act.

Since $P = Mv_{\text{CM}}$ (Equation 13.3), the velocity of the center of mass of S is $v_{\text{CM}} = P/M$, and we have from Equation 13.5 that

$$F_{\text{ext}} = M\frac{dv_{\text{CM}}}{dt} = Ma_{\text{CM}} \tag{13.6}$$

where a_{CM} is the acceleration of the center of mass of S. Comparing this result with Equation 6.15, $F = ma$, we see that

the center of mass of a system of particles moves as if it were a particle of mass equal to the total mass of the system and subject to the resultant external force applied to them.

Equations 13.5 and 13.6 clearly indicate that the interaction between two systems of particles can be described in terms identical to those found in Chapter 6 for two single particles. This justifies, *a posteriori*, the loose way we illustrated the application of the principles of dynamics in Chapter 6 where the interactions between the Earth and the Moon and between two molecules, and the motion of a rocket as well as that of an automobile, were all discussed as particles and not as extended bodies.

Consider some examples. Figure 13.8(a) shows the parabolic motion of the center of mass of an object thrown into the air, while the object rotates around its CM. Figure 13.8(b) shows the Earth in its motion around the Sun. The center of mass of the Earth moves in a way corresponding to a particle having a mass equal to that of the Earth and subject to a force equal to the sum of the forces exerted by the Sun (and the other heavenly bodies) on all the particles that compose the Earth. This was the assumption made implicitly in Chapter 11 when we discussed planetary motion. Figure 13.8(c) depicts a water molecule. Suppose, for example, that the molecule is subject to external electrical forces. Its center of mass moves as if it were a particle of mass equal to that of the molecule and subject to a force equal to the sum of the forces acting on all the charged particles composing the molecule. The three previous examples illustrate the case of bodies whose shape does not change during the motion. But the same applies to bodies with variable shape. Figure 13.8(d) illustrates the motion of a chain thrown into the air. The

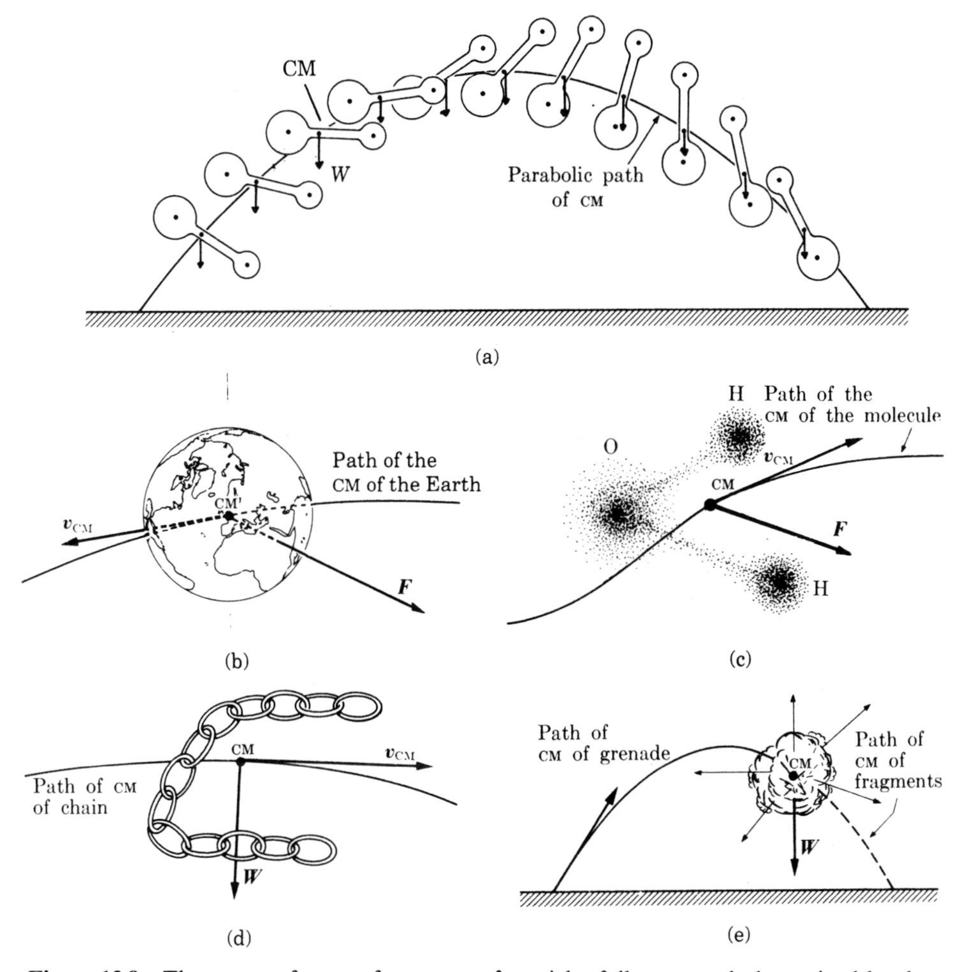

Figure 13.8 The center of mass of a system of particles follows a path determined by the external force on the system.

center of mass of the chain moves, describing a parabola, as if it were a particle of mass equal to that of the chain, subject to a force equal to the weight of the chain. The same applies to a swimmer jumping from a diving board, even if the body is rotated or the arms extended, or a cat jumping from a roof. Finally, in Figure 13.8(e), we have the case of a body exploding in flight, such as a grenade or a firework, when the explosion is due only to internal forces. The center of mass of the fragments will continue moving on the original parabola since the center of mass behaves like a particle of mass identical to the grenade and subject to the total weight of all fragments. The weight of the fragments does not change with the explosion because the force of gravity is practically independent of position at points near the surface of the Earth. However, if the weight were not constant but depended on the position, or the resistance of the air is taken into account, the fragments resulting from the explosion would be subject to external forces different from those along the original path. The path of the center of mass would not then continue to be the same as before the explosion because the sum of the external forces would be different. For example, if (due to some cosmic cataclysm) a planet in the solar system should break into fragments, the center of mass of the fragments would not follow the original elliptical path of the planet because the forces on the fragments would be different.

Relation between the external force and the time rate of change of momentum of a system of particles

We shall now justify Equation 13.5 in more detail. For simplicity, consider a system S composed of two particles (Figure 13.7). Call $\boldsymbol{F}_{12}$ the *internal* force on particle m_1 due to its interaction with m_2, and $\boldsymbol{F}_{21}$ the *internal* force on m_2 due to its interaction with m_1. The law of action and reaction requires that

$$\boldsymbol{F}_{12} = -\boldsymbol{F}_{21} \qquad \text{or} \qquad \boldsymbol{F}_{12} + \boldsymbol{F}_{21} = 0$$

Let $\boldsymbol{F}_1$ be the resultant *external* force on m_1 and $\boldsymbol{F}_2$ be the *external* force on m_2 due to their interaction with other particles outside the system. To obtain the equation of motion of each particle under the action of *all* the forces acting on the particle, we apply Equation 6.12:

$$\frac{\mathrm{d}\boldsymbol{p}_1}{\mathrm{d}t} = \boldsymbol{F}_1 + \boldsymbol{F}_{12} \qquad \frac{\mathrm{d}\boldsymbol{p}_2}{\mathrm{d}t} = \boldsymbol{F}_2 + \boldsymbol{F}_{21}$$

Adding these two equations and using $\boldsymbol{F}_{12} + \boldsymbol{F}_{21} = 0$, we find that

$$\frac{\mathrm{d}\boldsymbol{P}}{\mathrm{d}t} = \frac{\mathrm{d}}{\mathrm{d}t}(\boldsymbol{p}_1 + \boldsymbol{p}_2) = \boldsymbol{F}_1 + \boldsymbol{F}_2 = \boldsymbol{F}_{\mathrm{ext}}$$

Therefore the total rate of change of momentum of the system composed of m_1 and m_2 is equal to the sum of the *external* forces applied on m_1 and m_2. In general, for a system composed of an arbitrary number of particles subject to external forces $\boldsymbol{F}_i$,

$$\frac{\mathrm{d}\boldsymbol{P}}{\mathrm{d}t} = \frac{\mathrm{d}}{\mathrm{d}t}\left(\sum_i \boldsymbol{p}_i\right) = \sum_i \boldsymbol{F}_i = \boldsymbol{F}_{\mathrm{ext}}$$

EXAMPLE 13.4

A body, falling vertically, explodes into two equal fragments when it is at a height of 2000 m and has a downward velocity of $60\,\text{m s}^{-1}$. Immediately after the explosion one of the fragments is moving downward at $80\,\text{m s}^{-1}$. Find the position of the center of mass of the system 10 s after the explosion.

▷ We may follow one of two methods (see Figure 13.9). Since we know that the external forces do not change as a result of the explosion, we may assume that the center of mass continues moving as if there had not been any explosion. After the explosion, the center of mass will be at a height

$$y = y_0 + v_0 t + \tfrac{1}{2}gt^2$$

where $y_0 = 2000\,\text{m}$, $v_0 = -60\,\text{m s}^{-1}$, and $g = -9.8\,\text{m s}^{-2}$. Therefore at $t = 10\,\text{s}$, $y = 910\,\text{m}$.

As an alternative method, we directly compute the position of the center of mass from the positions of the fragments 10 s after the explosion. Since momentum is conserved in the explosion, we have that $mv_0 = m_1 v_1 + m_2 v_2$. But $m_1 = m_2 = \tfrac{1}{2}m$; thus

$$2v_0 = v_1 + v_2$$

Now $v_0 = -60\,\text{m s}^{-1}$ and $v_1 = -80\,\text{m s}^{-1}$. Therefore $v_2 = -40\,\text{m s}^{-1}$ and the second fragment moves downward also, but with a smaller velocity. After 10 s the position of the first fragment is

$$y_1 = y_0 + v_1 t + \tfrac{1}{2}gt^2 = 710\,\text{m}$$

and the second fragment has the position

$$y_2 = y_0 + v_2 t + \tfrac{1}{2}gt^2 = 1110\,\text{m}$$

The position of the center of mass is thus

$$\begin{aligned} y_{\text{CM}} &= [(\tfrac{1}{2}m)y_1 + (\tfrac{1}{2}m)y_2]/m \\ &= \tfrac{1}{2}(y_1 + y_2) = 910\,\text{m} \end{aligned}$$

which is in agreement with the previous result.

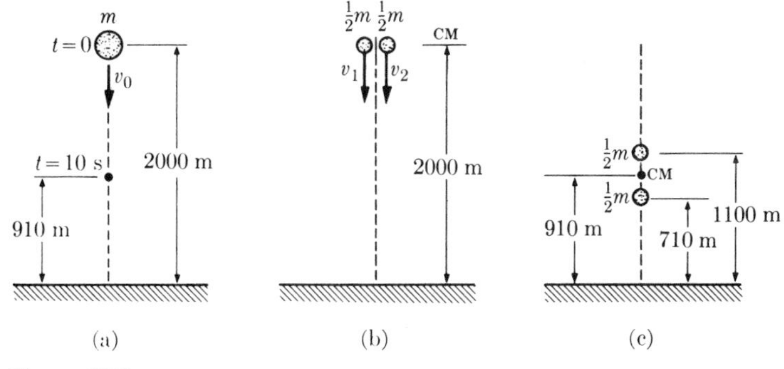

Figure 13.9

EXAMPLE 13.5

A nozzle throws a stream of gas against a wall with a velocity $\boldsymbol{v}$ much larger than the thermal agitation of the molecules. The wall deflects the molecules without changing the magnitude of their velocity. Also assume that the force exerted by the wall on the molecules is perpendicular to the wall. (This is not strictly true for a 'rough' wall.) Find the force exerted on the wall.

▷ When the molecules are moving toward the wall (Figure 13.10) at angle θ with the normal N, their velocity points toward the wall. After they strike the wall they move away with the same velocity. The force on the molecule is perpendicular to the wall. Therefore, the component of the velocity perpendicular to the wall is reversed, while the component of its velocity parallel to the wall does not change. The molecule bounces back moving at the same angle θ with the normal N. Each molecule, as a result of its impact on the wall, suffers a change in its velocity, $\Delta\boldsymbol{v}$, that is parallel to the normal N. The magnitude of the change is (Figure 13.10)

$$|\Delta\boldsymbol{v}| = 2v \cos \theta$$

The change in momentum of a molecule is

$$|\Delta\boldsymbol{p}| = m|\Delta\boldsymbol{v}| = 2mv \cos \theta$$

in the direction of the normal N. Let n be the number of molecules per unit volume. The number of molecules arriving at an area A of the wall per unit time is the number in a slanted cylinder whose length is equal to the velocity v and whose cross-section is $A \cos \theta$ (Figure 13.10(a)). Thus the number of molecules is $n(Av \cos \theta)$. Each molecule in this cylinder suffers a change of momentum equal to $2mv \cos \theta$. The change of momentum of the stream of gas per unit time in a direction perpendicular to the wall is then $(nAv \cos \theta) \times (2mv \cos \theta) = 2Anmv^2 \cos^2 \theta$ in a direction perpendicular to the wall. This, according to Equation 13.5, is equal to the force F exerted by the area A of the wall on the stream of gas, that is,

$$F = 2Anmv^2 \cos^2 \theta$$

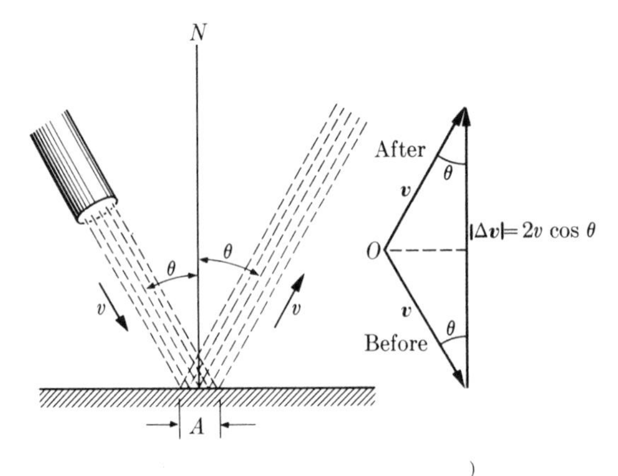

Figure 13.10 The *magnitude* of the change in the velocity of the molecules of a gas striking a wall is $|\Delta\boldsymbol{v}| = 2v \cos \theta$. The *direction* of the change in velocity is perpendicular to the wall.

Note that the force is proportional to the square of the velocity. By the law of action and reaction, the gas molecules exert an equal but opposite force on the wall.

Since the total force is not applied to a single particle of the wall, but rather over an area, we may introduce the useful concept of **pressure**. When forces act over an extended portion of the surface of a body, in a direction perpendicular to the surface, the pressure on the surface of the body is defined as the resultant normal force per unit area. Thus

$$\text{Pressure} = \frac{\text{Normal force}}{\text{Area}} = \frac{F}{A} \tag{13.7}$$

In the particular case of this example, the molecules in the gas stream exert a pressure on the wall equal to

$$\text{Pressure} = 2nmv^2 \cos^2 \theta \tag{13.8}$$

Note 13.1 Discussion of the interaction between two systems of particles

Consider two systems S and S' as illustrated in Figure 13.6. They can be, for example, two molecules. Assume that the systems together form an isolated system so that they experience only their mutual interactions.

The principle of conservation of momentum for the complete isolated system $S + S'$ is

$$\boldsymbol{P} = \underbrace{\textstyle\sum_i \boldsymbol{p}_i}_{\text{System } S} + \underbrace{\textstyle\sum_i \boldsymbol{p}'_i}_{\text{System } S'} = \textit{const.}$$

where the summations extend over all the particles of each system. Then we can write

$$\boldsymbol{P} = \boldsymbol{P}_S + \boldsymbol{P}_{S'} = \textit{const.}$$

Therefore any change in the momentum of S must be accompanied by an equal and opposite change in the momentum of S' so that the sum remains constant. That is,

$$\Delta \boldsymbol{P}_S = -\Delta \boldsymbol{P}_{S'}$$

Therefore the interaction between systems S and S′ can be described as an exchange of momentum. These results should be compared with Equations 6.8 and 6.5 for the case of two particles, and note the similarity.

Taking the time derivative of $\boldsymbol{P} = \boldsymbol{P}_S + \boldsymbol{P}_{S'}$, we have

$$\frac{\mathrm{d}\boldsymbol{P}_S}{\mathrm{d}t} = -\frac{\mathrm{d}\boldsymbol{P}_{S'}}{\mathrm{d}t}$$

since $\boldsymbol{P}$ is constant. If $\boldsymbol{F}_{\text{ext}}$ is the external force on each system the preceding result implies that

$$\boldsymbol{F}_{\text{ext}} = -\boldsymbol{F}'_{\text{ext}}$$

so that the forces the systems exert on each other are equal and opposite. This is the law of action and reaction for the interaction between systems S and S'. The above results show that in many instances two interacting systems can be treated as if they were single particles.

13.4 Reduced mass

Consider a system of two particles that are subject only to their mutual interaction; that is, there are no external forces acting on them (Figure 13.11). The two particles could be, for example, an electron and a proton in an isolated hydrogen atom. The mutual internal forces $\boldsymbol{F}_{12}$ and $\boldsymbol{F}_{21}$ satisfy the relation

$$\boldsymbol{F}_{12} = -\boldsymbol{F}_{21}$$

We have drawn these forces along the line $\boldsymbol{r}_{12}$, but there may be cases where this is not true. To discuss the *relative* motion of the two particles it is convenient to introduce a quantity called the **reduced mass** of the system. The reduced mass is designated by μ and defined as

$$\frac{1}{\mu} = \frac{1}{m_1} + \frac{1}{m_2} = \frac{m_1 + m_2}{m_1 m_2}$$

so that

$$\mu = \frac{m_1 m_2}{m_1 + m_2} \tag{13.9}$$

Let $\boldsymbol{F}_{12}$ be the force on particle 1 exerted by particle 2 and $\boldsymbol{a}_{12}$ the acceleration of particle 1 relative to particle 2. It may be shown (see proof below) that

$$\boldsymbol{F}_{12} = \mu \boldsymbol{a}_{12} \tag{13.10}$$

holds in the frame of reference attached to either particle even if it is not an inertial frame. That is:

> *the relative motion of two particles subject only to their mutual interaction is equivalent to the motion of a particle of mass equal to the reduced mass under a force equal to their interaction.*

For example, the motion of the Moon relative to the Earth can be reduced to a single particle problem, using the reduced mass of the Earth–Moon system and a force equal to the attraction of the Earth on the Moon. Similarly, when we talk about the motion of an electron around the nucleus, we may assume the system reduced to a particle with a mass equal to the reduced mass of the electron–nucleus system that moves under the force between the electron and the nucleus.

As a third example, consider two masses m_1 and m_2 attached by a spring that has an elastic constant k (Figure 13.12). When the spring is stretched, the particles are set in relative oscillatory motion. The internal force on particle 1 is $F_{12} = -kx$, where x is the stretch of the spring. Equation 13.10 then gives the relative acceleration as $\mu a_{12} = -kx$. By comparison with Equation 10.4, we conclude that the angular frequency of the oscillations is $\omega = (k/\mu)^{\frac{1}{2}}$ instead of $(k/m_1)^{\frac{1}{2}}$, which would have been the frequency if particle 2 had been held fixed in an

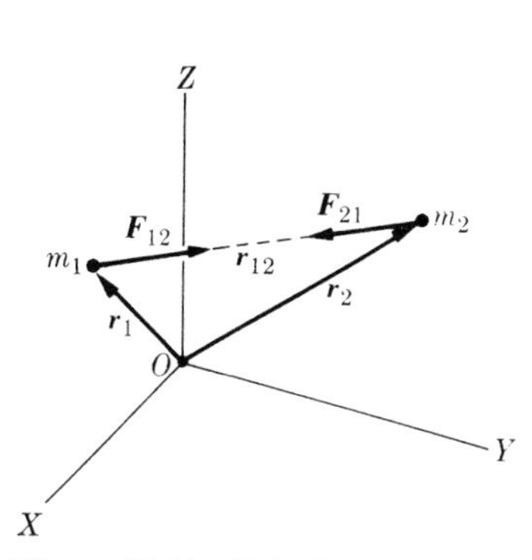

Figure 13.11 Relative motion of two particles.

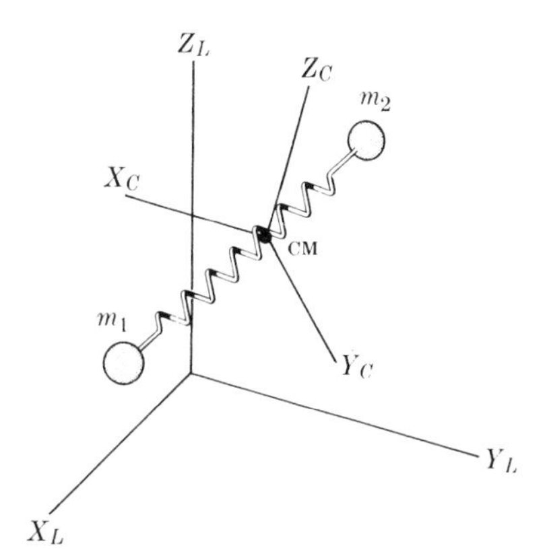

Figure 13.12 Relative motion of two particles coupled by a spring.

inertial frame. In Example 10.3 we used the reduced mass to analyze the vibration of atoms in a diatomic molecule.

In summary, to describe the motion of two particles under their mutual interaction, we may separate the motion of the system into two parts. One is the motion of the center of mass, whose velocity is constant, and the other is the relative motion of the two particles, given by Equation 13.10.

Note that Equation 13.9 can be written in the form

$$\mu = \frac{m_1}{1 + (m_1/m_2)}$$

If one of the particles, say m_2, has a much larger mass than the other, so that m_1/m_2 is negligible compared with unity, we get

$$\mu \approx m_1$$

That is, the reduced mass is approximately equal to the mass of the lighter particle. This is because the more massive particle can be considered approximately at rest in an inertial frame. For example, when we discuss the motion of an artificial satellite around the Earth we may use, with very good approximation, the mass of the satellite and not the reduced mass of the Earth–satellite system. On the other hand, if the two particles have the same mass ($m_1 = m_2$ or $m_1/m_2 = 1$), we then have

$$\mu = \tfrac{1}{2}m_1$$

which is the mass that must be used to discuss the system in the frame attached to particle 2. Such a situation exists when two protons interact, as in a hydrogen molecule, in proton–proton scattering or, approximately, in the proton–neutron system of a deuteron.

Relative motion of a system of two particles; reduced mass

The equation of motion for each particle relative to an inertial observer O (Fig. 13.11) using the L-frame is

$$m_1 \frac{d\boldsymbol{v}_1}{dt} = \boldsymbol{F}_{12} \quad \text{and} \quad m_2 \frac{d\boldsymbol{v}_2}{dt} = \boldsymbol{F}_{21}$$

where $\boldsymbol{v}_1$ and $\boldsymbol{v}_2$ are the velocities relative to the L-frame. Then

$$\frac{d\boldsymbol{v}_1}{dt} = \frac{\boldsymbol{F}_{12}}{m_1} \quad \text{and} \quad \frac{d\boldsymbol{v}_2}{dt} = \frac{\boldsymbol{F}_{21}}{m_2}$$

Subtracting these equations, we have

$$\frac{d\boldsymbol{v}_1}{dt} - \frac{d\boldsymbol{v}_2}{dt} = \frac{\boldsymbol{F}_{12}}{m_1} - \frac{\boldsymbol{F}_{21}}{m_2}$$

We use the equation, $\boldsymbol{F}_{12} = -\boldsymbol{F}_{21}$, and rewrite the preceding result as

$$\frac{d}{dt}(\boldsymbol{v}_1 - \boldsymbol{v}_2) = \left(\frac{1}{m_1} + \frac{1}{m_2}\right)\boldsymbol{F}_{12}$$

Note that $\boldsymbol{v}_1 - \boldsymbol{v}_2 = \boldsymbol{v}_{12}$ is the velocity of m_1 *relative* to m_2. Therefore, making

$$\frac{1}{\mu} = \frac{1}{m_1} + \frac{1}{m_2}$$

we may write

$$\frac{d\boldsymbol{v}_{12}}{dt} = \frac{1}{\mu}\boldsymbol{F}_{12}$$

Since $d\boldsymbol{v}_{12}/dt = \boldsymbol{a}_{12}$ is the acceleration of m_1 relative to m_2,

$$\boldsymbol{F}_{12} = \mu \boldsymbol{a}_{12}$$

which is Equation 13.10.

EXAMPLE 13.6

Reduced mass of the following systems: (a) electron–proton in a hydrogen atom, (b) proton–neutron in a deuteron nucleus and (c) CO molecule formed from ^{12}C and ^{16}O. In each case compare the result with the mass of the lighter particle.

▷ (a) For the electron–proton system, or a hydrogen atom, we have that $m_e = 9.109 \times 10^{-31}$ kg and $m_p = 1.6725 \times 10^{-27}$ kg, so that $m_e/m_p \approx 1/1850$. Thus,

$$\mu_{ep} = \frac{m_e m_p}{m_e + m_p} = \frac{m_e}{1 + m_e/m_p} = \frac{m_e}{1 + 1/1850} = 9.1031 \times 10^{-31}\ \text{kg}$$

and μ_{ep} differs from m_e by about 0.07%. In spite of the small difference, detectable results are observed in many atomic processes.

(b) For the proton–neutron system in the deuteron, we have that $m_n = 1.6748 \times 10^{-27}$ kg, which is almost the same as $m_p = 1.6725 \times 10^{-27}$ kg. Then

$$\mu_{np} = \frac{m_p m_n}{m_p + m_n} = 0.8368 \times 10^{-27}\ \text{kg}$$

which is approximately equal to one-half the mass of either particle.

(c) For the CO molecule, we have $m(^{12}\text{C}) = 12.000$ amu and $m(^{16}\text{O}) = 15.985$ amu. Therefore,

$$\mu_{CO} = \frac{m(^{12}\text{C}) \times m(^{16}\text{O})}{m(^{12}\text{C}) + m(^{16}\text{O})} = 6.854\ \text{amu} = 11.38 \times 10^{-27}\ \text{kg},$$

so that for the CO molecule, its reduced mass is considerably less than the mass of the carbon atom.

13.5 Angular momentum of a system of particles

In Equation 8.9 the angular momentum of a particle relative to a given point was defined as the vector quantity

$$\boldsymbol{L} = \boldsymbol{r} \times \boldsymbol{p} = m(\boldsymbol{r} \times \boldsymbol{v}) \tag{13.11}$$

Also, it was found that $\boldsymbol{L}$ is related to the torque $\boldsymbol{\tau} = \boldsymbol{r} \times \boldsymbol{F}$ of the applied force by Equation 8.12. That is,

$$\frac{\mathrm{d}\boldsymbol{L}}{\mathrm{d}t} = \boldsymbol{\tau} \tag{13.12}$$

The total angular momentum of a system of particles 1, 2, 3, ... relative to a given point is defined as

$$\boldsymbol{L} = \sum_i \boldsymbol{L}_i = \boldsymbol{L}_1 + \boldsymbol{L}_2 + \boldsymbol{L}_3 + \cdots$$

where each $\boldsymbol{L}_i$ has to be computed relative to the same point. We designate $\boldsymbol{\tau}_{ext}$ as the total torque exerted by the *external* forces acting on the particles of the system relative to the same point; that is,

$$\boldsymbol{\tau}_{ext} = \boldsymbol{\tau}_{1,ext} + \boldsymbol{\tau}_{2,ext} + \boldsymbol{\tau}_{3,ext} + \cdots$$

Then, as shown in the proof below, the following relation holds for a system of particles:

$$\frac{\mathrm{d}\boldsymbol{L}}{\mathrm{d}t} = \boldsymbol{\tau}_{ext} \tag{13.13}$$

This relation states that

> *the time rate of change of the total angular momentum of a system of particles relative to an arbitrary point is equal to the sum of the torques, relative to the same point, of the external forces acting on the particles of the system.*

This statement may be considered as the fundamental law of the dynamics of rotation. It is important to keep in mind that Equation 13.13 is valid only if the point relative to which $\boldsymbol{L}$ and $\boldsymbol{\tau}_{\text{ext}}$ are calculated is at rest in an inertial system or L-frame.

Also, Equation 13.13 shows that the change $\mathrm{d}\boldsymbol{L}$ of the angular momentum in a time $\mathrm{d}t$ has the same direction as the external torque $\boldsymbol{\tau}_{\text{ext}}$.

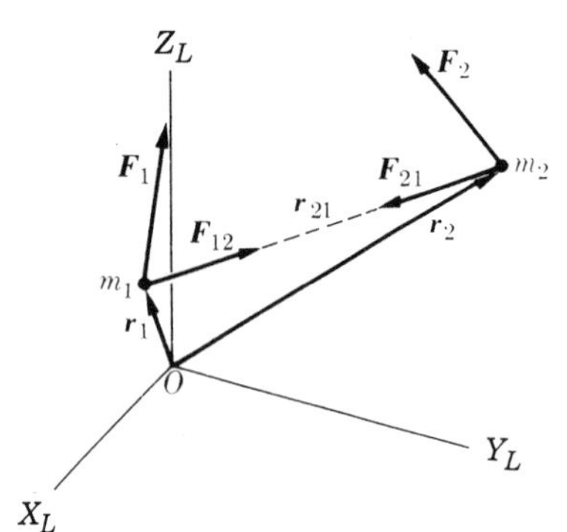

Figure 13.13 The torque relative to O due to internal forces $\boldsymbol{F}_{12}$ and $\boldsymbol{F}_{21}$ is zero.

It is not surprising that the torque of the internal forces does not appear in Equation 13.13. From Figure 13.13 we see that if the internal forces between pairs of particles are equal in magnitude and directly opposed, their torques about any point must add to zero.

If there are no external forces, or the sum of their torques relative to the chosen point is zero, $\boldsymbol{\tau}_{\text{ext}} = 0$. Then

$$\frac{\mathrm{d}\boldsymbol{L}}{\mathrm{d}t} = \frac{\mathrm{d}}{\mathrm{d}t}\left(\sum_i \boldsymbol{L}_i\right) = 0$$

which means that

$$\boldsymbol{L} = \sum_i \boldsymbol{L}_i = \boldsymbol{L}_1 + \boldsymbol{L}_2 + \boldsymbol{L}_3 + \cdots = \textit{const.} \tag{13.14}$$

Equation 13.14 shows that

> *the total angular momentum of an isolated system of particles, or a system with zero external torque, is constant in magnitude and direction.*

This statement constitutes the **law of conservation of angular momentum** for a system of particles. The law of conservation of angular momentum seems to be universally valid, applying to all processes so far observed. For example, the angular momentum of the electrons in an atom relative to the CM of the atom (which practically coincides with the nucleus) is constant when only the internal forces due to the electric repulsion of the electrons and the electric attraction of the nucleus are considered. Also, if we assume that the solar system is isolated (that is, if we neglect the forces due to the rest of the galaxy), the total angular momentum of all the planets relative to the center of mass of the solar system remains constant. This conclusion holds with a great degree of accuracy. Similarly, the Earth keeps rotating around its center of mass with a nearly constant angular momentum. The reason is that the external forces due to the Sun and the other planets pass close to the center of the Earth and therefore have approximately zero torque about the Earth's center of mass. However, due to tidal effects, as well as plastic deformations, the angular velocity of the Earth varies slightly over the centuries.

The law of conservation of angular momentum implies that if, in an isolated system, the angular momentum of some part of the system changes because of internal interactions, the rest of the system must experience an equal (but opposite) change of angular momentum.

For example, when a nucleus disintegrates, the emitted particles (in many cases an electron and a neutrino), possess some angular momentum. Since only internal forces act in the process, the angular momentum of the nucleus must change to compensate exactly for the angular momentum carried away by the emitted particles. Similarly, if an atom, molecule or nucleus emits electromagnetic radiation (or a photon), its angular momentum must change to compensate exactly for the angular momentum taken away by the radiation. Sometimes processes that would otherwise be possible in nature cannot occur because of some aspect, characteristic of the process, that makes it impossible for the process to satisfy the conservation of angular momentum.

Proof of the relation between torque and angular momentum

For simplicity consider first the case of only two particles. Equation 13.12 applied to particles 1 and 2 becomes

$$\frac{\mathrm{d}\boldsymbol{L}_1}{\mathrm{d}t} = \boldsymbol{\tau}_1 \qquad \text{and} \qquad \frac{\mathrm{d}\boldsymbol{L}_2}{\mathrm{d}t} = \boldsymbol{\tau}_2$$

where the angular momentum and the torque are computed relative to O (Figure 13.13). Adding the two equations, we obtain

$$\frac{\mathrm{d}}{\mathrm{d}t}(\boldsymbol{L}_1 + \boldsymbol{L}_2) = \boldsymbol{\tau}_1 + \boldsymbol{\tau}_2 \tag{13.15}$$

Let us assume that each particle, in addition to having a mutual interaction with the other particle, is acted on by an external force. The force on particle 1 is $\boldsymbol{F}_1 + \boldsymbol{F}_{12}$ and on particle 2 is $\boldsymbol{F}_2 + \boldsymbol{F}_{21}$. Then

$$\boldsymbol{\tau}_1 = \boldsymbol{r}_1 \times (\boldsymbol{F}_1 + \boldsymbol{F}_{12}) = \boldsymbol{r}_1 \times \boldsymbol{F}_1 + \boldsymbol{r}_1 \times \boldsymbol{F}_{12}$$

$$\boldsymbol{\tau}_2 = \boldsymbol{r}_2 \times (\boldsymbol{F}_2 + \boldsymbol{F}_{21}) = \boldsymbol{r}_2 \times \boldsymbol{F}_2 + \boldsymbol{r}_2 \times \boldsymbol{F}_{21}$$

Since $\boldsymbol{F}_{12} = -\boldsymbol{F}_{21}$, the total torque on the particles is

$$\boldsymbol{\tau}_1 + \boldsymbol{\tau}_2 = \boldsymbol{r}_1 \times \boldsymbol{F}_1 + \boldsymbol{r}_2 \times \boldsymbol{F}_2 + (\boldsymbol{r}_2 - \boldsymbol{r}_1) \times \boldsymbol{F}_{21}$$

Now the vector $\boldsymbol{r}_2 - \boldsymbol{r}_1$ has the direction of the line joining the two particles. If we make the *special assumption* that the internal forces $\boldsymbol{F}_{12}$ and $\boldsymbol{F}_{21}$ act along the line $\boldsymbol{r}_{21}$ joining the two particles, then the vectors $\boldsymbol{r}_2 - \boldsymbol{r}_1 = \boldsymbol{r}_{21}$ and $\boldsymbol{F}_{21}$ are parallel. Therefore

$$(\boldsymbol{r}_2 - \boldsymbol{r}_1) \times \boldsymbol{F}_{21} = 0$$

The last term in the above equation thus disappears, leaving only the torques due to the *external* forces. Thus, Equation 13.15 becomes

$$\frac{d}{dt}(\boldsymbol{L}_1 + \boldsymbol{L}_2) = \boldsymbol{r}_1 \times \boldsymbol{F}_1 + \boldsymbol{r}_2 \times \boldsymbol{F}_2 = \boldsymbol{\tau}_{1,\text{ext}} + \boldsymbol{\tau}_{2,\text{ext}}$$

Generalizing this result to any number of particles, we obtain Equation 13.13.

13.6 Internal and orbital angular momentum

The angular momentum of a system of particles depends on the point relative to which the angular momentum is computed. We define the **internal angular momentum** $\boldsymbol{L}_{\text{int}}$ of a many particle system as the total angular momentum computed relative to the center of mass, or origin of the C-frame (Figure 13.14). Internal angular momentum is thus a property of the system itself, and is independent of the observer. In the case of a rigid body or an elementary particle, the internal angular momentum is also called **spin**.

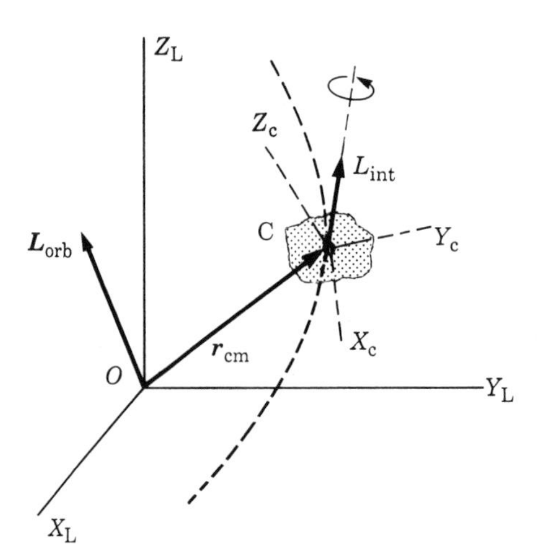

Figure 13.14 Internal and orbital angular momentum.

The **orbital angular momentum** of the system relative to the origin of the L-frame is defined by the expression

$$\boldsymbol{L}_{\text{orb}} = \boldsymbol{r}_{\text{CM}} \times \boldsymbol{P} = \boldsymbol{r}_{\text{CM}} \times (M\boldsymbol{v}_{\text{CM}})$$

where $\boldsymbol{P}$ is the total momentum of the system relative to the L-frame. Clearly $\boldsymbol{L}_{\text{orb}}$ is the angular momentum of a particle of mass $M = \sum_i m_i$ located at the CM of the system. Since the motion of a system can be considered as the superposition of the motion around the CM plus the motion of the CM itself we have that

the angular momentum of a system of particles can be expressed as the sum of the internal and orbital angular momenta of the system:

$$\boldsymbol{L} = \boldsymbol{L}_{\text{int}} + \boldsymbol{L}_{\text{orb}} \qquad \textbf{(13.16)}$$

The first term on the right gives the internal angular momentum relative to the C-frame of the system. The second term is the orbital angular momentum, relative to the L-frame, as if all the mass of the system were concentrated at the center of mass or origin of the C-frame.

For example, when a person throws a spinning ball, the angular momentum due to the spinning is given by $\boldsymbol{L}_{\text{int}}$, while the angular momentum relative to the person due to the orbital motion of the ball is $\boldsymbol{L}_{\text{orb}}$ (Figure 13.15(a)). The Earth

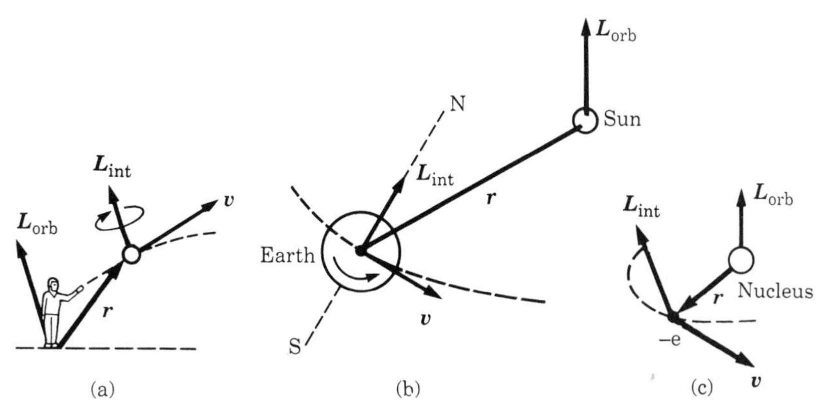

Figure 13.15 Internal and orbital angular momentum of (a) a ball, (b) the Earth, (c) an electron in an atom.

moves around the Sun and at the same time spins around its NS-axis. Then the Earth has an orbital angular momentum relative to the Sun and an internal angular momentum or spin relative to its center, as shown in Figure 13.15(b). A similar situation occurs for a spinning electron orbiting about a nucleus in an atom (Figure 13.15(c)).

The time rate of change of the orbital angular momentum is determined by the torque of the external forces as if they were all applied at the CM. Similarly,

the time rate of change of the internal angular momentum is equal to the external torque relative to the center of mass of the system:

$$\frac{\mathrm{d}\boldsymbol{L}_{\mathrm{int}}}{\mathrm{d}t} = \boldsymbol{\tau}_{\mathrm{CM}} \tag{13.17}$$

where $\boldsymbol{\tau}_{\mathrm{CM}}$ is the torque of the external forces relative to the CM. Relation 13.17 is formally identical to Equation 13.13, $\mathrm{d}\boldsymbol{L}/\mathrm{d}t = \boldsymbol{\tau}_{\mathrm{ext}}$, but there are some basic differences. Equation 13.13 is valid only when the angular momentum and torque are evaluated relative to a point fixed in an inertial frame of reference, usually the origin of coordinates. On the other hand, Equation 13.17 is valid for the center of mass even if it is not at rest in an inertial frame of reference.

13.7 Angular momentum of a rigid body

A system composed of many particles is a **rigid body** when the distances between all its component particles remain fixed under the application of a force or torque. A rigid body therefore conserves its shape during its motion. No body is absolutely rigid. However, depending on the forces applied to the body, many solids and also many molecules can be considered rigid, to a first approximation.

We may distinguish two types of motion of a rigid body. The motion is a **translation** when all the particles describe parallel paths so that the body always remains parallel to its initial position. The motion is a **rotation** around an axis when all the particles describe circular paths around a line called the axis of rotation. The axis may be fixed or it may be changing its direction relative to the body during the motion.

The most general motion of a rigid body can always be considered as a combination of a rotation and a translation. For example, the motion of the body in Figure 13.8(a), of the Earth in Figure 13.8(b) and of the water molecule in Figure 13.8(c) can be considered as the translational motion of the center of mass and a rotation around the center of mass.

The motion of the center of mass is determined by the external forces, in accordance with Equation 13.6:

$$M \frac{d\boldsymbol{v}_{CM}}{dt} = \boldsymbol{F}_{ext}$$

Therefore, the motion of the center of mass of a rigid body is identical to the motion of a single particle whose mass is equal to the mass of the body and is acted on by a force equal to the sum of all external forces applied to the body. This motion can be analyzed according to the methods explained in Chapters 6, 7 and 8 for the dynamics of particles. Therefore it does not involve special techniques.

The rotational motion of a rigid body around an axis that passes either through a point fixed in an inertial system or through the center of mass of the body, however, requires special treatment. Consider the case of a thin rigid plate rotating around an axis ZZ' perpendicular to the plate (Figure 13.16). To calculate the angular momentum relative to O, assume the plate to be composed of small volume element at A_i, each of mass m_i, at a distance R_i from O. Each volume element describes a circle of radius R_i as the plate rotates around ZZ'. In Section 8.4 we proved (Equation 8.11) that the angular momentum, with respect to O, of a particle of mass m_i that is at A_i and in circular motion is

$$\boldsymbol{L}_i = m_i R_i^2 \boldsymbol{\omega}$$

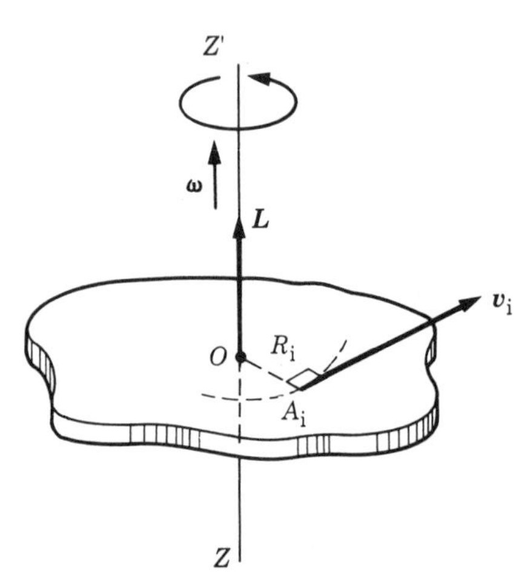

Figure 13.16 Angular velocity and angular momentum of a rotating plate.

where $\boldsymbol{\omega}$ is the angular velocity of the plate, represented by a vector in the ZZ' direction. Note that we write $\boldsymbol{\omega}$ and not $\boldsymbol{\omega}_i$ because all particles of the rigid plate move with the same angular velocity. The total angular momentum of the plate relative to O is then

$$\boldsymbol{L} = \boldsymbol{L}_1 + \boldsymbol{L}_2 + \boldsymbol{L}_3 + \cdots = \sum_i \boldsymbol{L}_i$$

or, using the expression for $\boldsymbol{L}_i$,

$$\boldsymbol{L} = \sum_i m_i R_i^2 \boldsymbol{\omega} = \left(\sum_i m_i R_i^2 \right) \boldsymbol{\omega}$$

This suggests that we define a quantity

$$I = \sum_i m_i R_i^2 = m_1 R_1^2 + m_2 R_2^2 + m_3 R_3^2 + \cdots \tag{13.18}$$

which is called the **moment of inertia** of the body relative to the axis of rotation ZZ'. Then

$$\boldsymbol{L} = I\boldsymbol{\omega} \tag{13.19}$$

This expression shows that in the case of a flat rigid body rotating around an axis perpendicular to the body, the angular momentum has the same direction as the angular velocity.

If, instead of a plate, we have a rigid body of arbitrary shape (Figure 13.17), relation 13.19 does not hold in general and the angular momentum may have a direction different from that of $\boldsymbol{\omega}$. For example, the angular momentum of a volume element at A_i relative to O is given by

$$\boldsymbol{L}_i = m_i \boldsymbol{r}_i \times \boldsymbol{v}_i$$

This is a vector perpendicular to the plane of $\boldsymbol{r}_i$ and $\boldsymbol{v}_i$, which means it is at an angle with the axis of rotation and with $\boldsymbol{\omega}$. When we add the vectors $\boldsymbol{L}_i$ for each particle, the resultant angular momentum $\boldsymbol{L} = \sum_i \boldsymbol{L}_i$ may have a direction different from $\boldsymbol{\omega}$. However, the *component* of the angular momentum $\boldsymbol{L}$ along the axis of rotation Z is always

$$L_z = I\omega \tag{13.20}$$

The difference between this equation and Equation 13.19 is that this is a scalar equation. It holds regardless of the shape of the body.

However, for each body, no matter what its shape, there are (at least) three mutually perpendicular directions for which the angular momentum is parallel to the axis of rotation. These are called the **principal axes of inertia**, and the corresponding moments of inertia are called the **principal moments of inertia**, designated by I_1, I_2 and I_3. Let us designate the principal axes by $X_0 Y_0 Z_0$; they constitute a frame of reference attached to the body that rotates with the body. When the body has some kind of symmetry, the principal axes coincide with the symmetry axes. For example, in a sphere, any axis passing through its center is a principal axis. For a cylinder, and in general for any body with cylindrical symmetry, the axis of

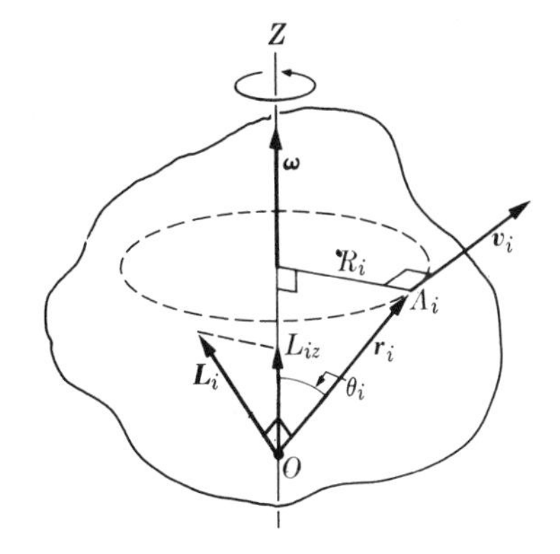

Figure 13.17 Angular momentum of a rotating rigid body.

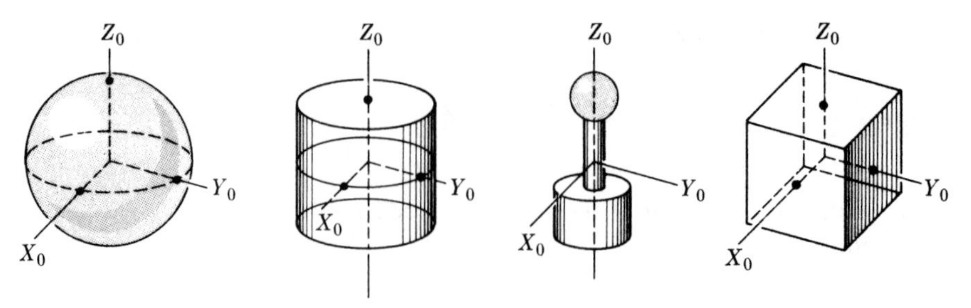

Figure 13.18 Principal axes of symmetrical bodies.

symmetry, as well as any axis perpendicular to it that passes through the CM, are principal axes. For a rectangular block the three principal axes are perpendicular to the surfaces and pass through the center of the block. These axes are illustrated in Figure 13.18.

When the body rotates around a principal axis of inertia, the total angular momentum $\boldsymbol{L}$ is parallel to the angular velocity $\boldsymbol{\omega}$, which is always along the rotation axis. We may therefore write the vector relation $\boldsymbol{L} = I\boldsymbol{\omega}$, where I is the corresponding principal moment of inertia. The notion of moment of inertia can be applied to any system of particles, but it is useful only for rigid bodies because only for rigid bodies is the moment of inertia about a fixed axis constant.

Table 13.2 Radii of gyration of some simple bodies, given by $K^2 = I/M$

K^2	Axis	K^2	Axis
$\frac{R^2}{2}$	Cylinder (L, R)	$\frac{L^2}{12}$	Thin rod (L)
$\frac{R^2}{4}+\frac{L^2}{12}$	Cylinder (L, R)	$\frac{R^2}{2}$	Disk (R)
$\frac{a^2+b^2}{12}$	Parallelepiped (a, b, c)	$\frac{R^2}{4}$	Disk (R)
$\frac{a^2+b^2}{12}$	Rectangular plate (a, b)	R^2	Ring (R)
$\frac{b^2}{12}$	Rectangular plate (a, b)	$\frac{2R^2}{5}$	Sphere (R)

Moment of inertia and radius of gyration

The moment of inertia of a body is a quantity I that depends on the mass of the body and its geometrical shape. In the SI system it is measured in m^2 kg. The calculation of the moment of inertia is a mathematical exercise, as shown in Appendix C. In Table 13.2 the moments of inertia of bodies with simple geometrical shapes are given in terms of the **radius of gyration** of the body, which is a quantity K defined by

$$K = \left(\frac{I}{M}\right)^{1/2} \quad \text{or} \quad I = MK^2$$

where I is the moment of inertia and M the mass of the body. Hence K represents the distance from the axis at which all the mass could be concentrated without changing the moment of inertia. It is a useful quantity because, for homogeneous bodies, it is determined entirely by their geometry. To obtain the moment of inertia from Table 13.2, just multiply K^2 by M.

If I_C is the moment of inertia relative to an axis passing through the center of mass of the body and I is the moment of inertia relative to a parallel axis at the distance a, the relation

$$I = I_C + Ma^2 \tag{13.21}$$

called **Steiner's theorem**, holds, as shown in Appendix C. For a thin rigid plate, it is also shown in Appendix C that

$$I_z = I_x + I_y$$

where the X- and Y-axes are on the plate and the Z-axis is perpendicular to it, with the origin of coordinates at the CM of the plate.

To obtain a relation between L_z and ω, we note that the angular momentum $\boldsymbol{L}_i$ of particle A_i makes an angle $\frac{1}{2}\pi - \theta_i$ with the axis of rotation Z (Figure 13.17). The magnitude of $\boldsymbol{L}_i$, according to Equation 11.1, is $m_i r_i v_i$. If we use

$$R_i = r_i \sin\theta_i \quad \text{and} \quad v_i = \omega R_i$$

the component of $\boldsymbol{L}_i$ parallel to the Z-axis is

$$\begin{aligned} L_{iz} &= (m_i r_i v_i)\cos(\tfrac{1}{2}\pi - \theta_i) \\ &= m_i(r_i \sin\theta_i)(\omega R_i) = m_i R_i^2 \omega \end{aligned}$$

The component of the total angular momentum of the rotating body along the rotation axis Z is therefore

$$\begin{aligned} L_z &= L_{1z} + L_{2z} + L_{3z} + \cdots \\ &= (m_1 R_1^2 + m_2 R_2^2 + m_3 R_3^2 + \cdots)\omega \\ &= \left(\sum_i m_i R_i^2\right)\omega = I\omega \end{aligned}$$

EXAMPLE 13.7

Angular momentum of the systems illustrated in Figure 13.19, which consist of two equal spheres of mass m mounted on arms connected to a bearing and rotating around the Z-axis. Neglect the masses of the arms.

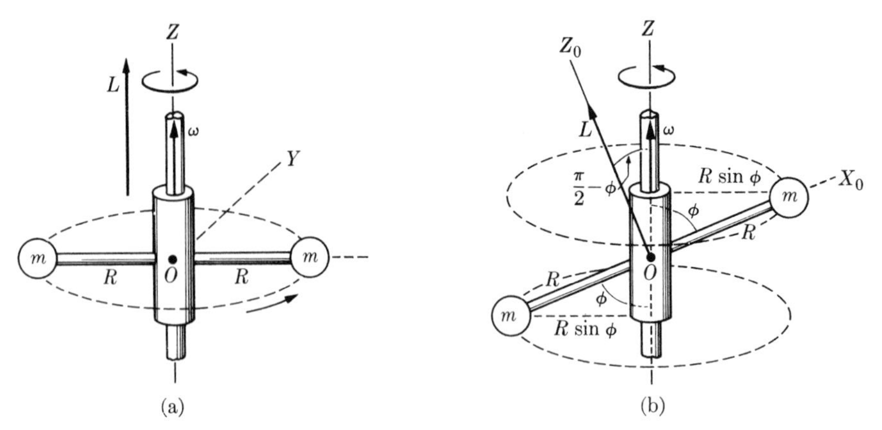

Figure 13.19

▷ In Figure 13.19(a) the two arms are perpendicular to the axis of rotation Z. Each sphere describes a circle of radius R with velocity $v = \omega R$. The angular momentum of each sphere relative to O is then $mRv = mR^2\omega$. Thus the total angular momentum of the system, directed along the Z-axis, is

$$L = 2mR^2\omega = I\omega$$

where $I = 2mR^2$ is the moment of inertia around the Z-axis. Note that in this case the system is rotating about a principal axis.

In Figure 13.19(b) the two arms make an angle ϕ with the axis of rotation Z, so that $\boldsymbol{\omega}$ is not parallel to a principal axis. The radius of the circle described by each sphere is $R \sin \phi$, so that their velocities are, in magnitude, $(R \sin \phi)\omega$. The angular momentum of each sphere relative to O is then $mRv = mR(R \sin \phi)\omega$ and is directed along the axis Z_0, perpendicular to the line that joins the two spheres and in the plane determined by the Z- and X_0-axes. The total angular momentum of the system is then

$$L = (2mR^2 \sin \phi)\omega$$

and makes an angle $(\frac{1}{2}\pi - \phi)$ with the rotation axis. Thus in this case the system is not rotating about a principal axis, as we may also see from the geometry of the system. Note that the vector $\boldsymbol{L}$ is rotating or, more correctly, **precessing** around the Z-axis with the same angular velocity as the system.

The component of $\boldsymbol{L}$ along the rotation axis is

$$L_z = L \cos(\tfrac{1}{2}\pi - \phi) = (2mR^2 \sin^2 \phi)\omega = I\omega$$

in agreement with Equation 13.20, since $I = 2m(R \sin \phi)^2$ is the moment of inertia of the system relative to the Z-axis.

EXAMPLE 13.8

Moment of inertia of a system of two particles, such as a diatomic molecule, relative to an axis passing through the CM.

▷ Consider a diatomic molecule composed of two atoms, with masses m_1 and m_2, separated a distance r. Let the molecule rotate about an axis that passes through the CM and is perpendicular to the molecule (Figure 13.20). The moment of inertia relative to this axis is

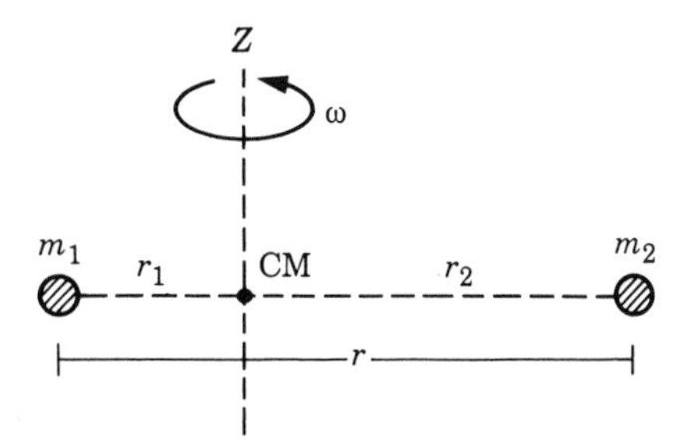

Figure 13.20

$$I = m_1 r_1^2 + m_2 r_2^2$$

However, according to Example 13.1, the distances of m_1 and m_2 to the CM are

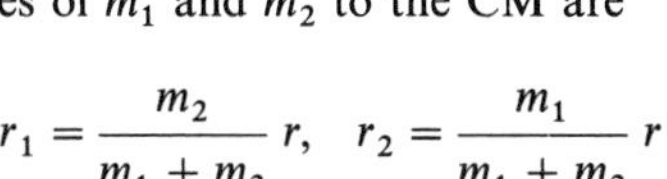

$$r_1 = \frac{m_2}{m_1 + m_2} r, \quad r_2 = \frac{m_1}{m_1 + m_2} r$$

Therefore,

$$I = \frac{m_1 m_2^2 r^2}{(m_1 + m_2)^2} + \frac{m_2 m_1^2 r^2}{(m_1 + m_2)^2} = \frac{m_1 m_2}{m_1 + m_2} r^2 = \mu r^2$$

where μ is the reduced mass of the molecule or of any two body system. Since atomic masses are of the order of 10^{-26} kg and interatomic distances are about 10^{-10} m, the moment of inertia of diatomic molecules is of the order of 10^{-46} kg m^2.

EXAMPLE 13.9

Moments of inertia of the molecules CO, CO_2 and H_2O relative to axes passing through their CM.

▷ (a) For the CO molecule we shall use an axis perpendicular to the axis of the molecule and passing through the CM. Using the results of Example 13.1, the distances from the CM to the C and O atoms are 0.645 Å and 0.485 Å, respectively. Therefore

$$\begin{aligned} I &= m(^{12}\mathrm{C})r_\mathrm{C}^2 + m(^{16}\mathrm{O})r_\mathrm{O}^2 \\ &= (12.000)(0.645 \times 10^{-10}\,\mathrm{m})^2 + (15.985)(0.485 \times 10^{-10}\,\mathrm{m})^2 \\ &= 8.752 \times 10^{-20}\ \mathrm{amu\ m^2} = 1.4531 \times 10^{-46}\ \mathrm{kg\ m^2} \end{aligned}$$

Instead, we could have used $I = \mu r^2$ with $\mu = 1.138 \times 10^{-26}$ kg and $r = 1.13 \times 10^{-10}$ m (Example 13.6), to get the same result.

(b) For the CO_2 molecule, where the bond length is 1.15 Å, the moment of inertia about an axis perpendicular to the axis and passing through the C atom (which coincides with the CM) is, using $m(^{16}\mathrm{O}) = 15.985$ amu,

$$\begin{aligned} I &= 2m(^{16}\mathrm{O})r^2 = 2(15.985)(1.15 \times 10^{-10}\,\mathrm{m})^2 \\ &= 42.28 \times 10^{-20}\ \mathrm{amu\ m^2} = 7.03 \times 10^{-46}\ \mathrm{kg\ m^2} \end{aligned}$$

(c) For the H_2O molecule, we shall calculate the moment of inertia about two axes. For the X-axis bisecting the H—O—H angle (recall Figure 13.4(b)) we have

$$\begin{aligned} I &= 2m(^{1}\mathrm{H})(1.02 \times 10^{-10} \sin 52.5°)^2 = 1.320 \times 10^{-20}\ \mathrm{amu\ m^2} \\ &= 2.19 \times 10^{-46}\ \mathrm{kg\ m^2} \end{aligned}$$

For the axis perpendicular to the plane of the molecule that passes through the CM (use the results of Example 13.2), we have

$$I = m(^{16}\text{O})r_\text{O}^2 + 2m(^1\text{H})r_\text{H}^2$$

where $r_\text{O} = 0.070$ Å and, by a straightforward calculation, $r_\text{H} = 0.979$ Å. Therefore,

$$\begin{aligned} I &= m(^{16}\text{O})(0.070 \times 10^{-10}\,\text{m})^2 + 2m(^1\text{H})(0.979 \times 10^{-10}\,\text{m})^2 \\ &= 2.011 \times 10^{-20}\ \text{amu m}^2 = 3.339 \times 10^{-46}\ \text{kg m}^2 \end{aligned}$$

13.8 Equation of motion for rotation of a rigid body

When the relation $\text{d}\boldsymbol{L}/\text{d}t = \boldsymbol{\tau}$ is applied to the rotation of a rigid body, it is important to distinguish whether the torque $\boldsymbol{\tau}$ and the angular momentum $\boldsymbol{L}$ are measured relative either to a point at rest in an inertial system or to the center of mass of the body. For example, consider the case when the only external force on a body is its weight, applied at the center of mass. Then we have $\boldsymbol{\tau} = 0$ and hence $\boldsymbol{L} = const.$ when $\boldsymbol{\tau}$ and $\boldsymbol{L}$ are both referred to the center of mass. That is, a body subject only to its own weight, such as that shown in Figure 13.8(a), rotates with a constant angular momentum relative to its center of mass.

Another important distinction is whether a rigid body rotates around a principal axis or not. For the case of a rigid body rotating around a principal axis, $\boldsymbol{L} = I\boldsymbol{\omega}$ and we may write

$$\frac{\text{d}(I\boldsymbol{\omega})}{\text{d}t} = \boldsymbol{\tau}$$

Since, for a rigid body, the principal moment of inertia is constant, this relation may also be written as

$$I\frac{\text{d}\boldsymbol{\omega}}{\text{d}t} = \boldsymbol{\tau} \quad \text{or} \quad I\boldsymbol{\alpha} = \boldsymbol{\tau} \quad \text{(for principal axes only)} \tag{13.22}$$

where $\boldsymbol{\alpha} = \text{d}\boldsymbol{\omega}/\text{d}t$ is the angular acceleration of the rigid body. It is a relation similar to $\boldsymbol{F} = m\boldsymbol{a}$ for a particle. This equation shows that the change of the angular velocity is parallel to the torque. We say, then, that the axis of rotation of a rigid body tends to precess in the direction of the applied torque.

When $\boldsymbol{\tau} = 0$, Equation 13.22 indicates that $\boldsymbol{\omega}$ is also constant. That is,

the angular velocity of a rigid body rotating around a principal axis is constant when no external torques are applied.

This may be considered as the law of inertia for rotational motion.

When a body is not rotating about a principal axis and the torque is zero, so that $\boldsymbol{L} = const$, $\boldsymbol{\omega}$ is not constant since the vector relation $\boldsymbol{L} = I\boldsymbol{\omega}$ does not hold in this case.

For example, the forces exerted by the Sun, the Moon and the planets on the Earth are, practically, applied at its center of mass, and thus the torque about the center of mass of the Earth is essentially zero (see Example 13.14). The Earth should then rotate with constant angular momentum. Since the Earth is practically a sphere and almost rotates about a principal axis, its angular velocity should also be constant, which is approximately correct. However, the Earth is slightly pear-shaped, and it is not at present rotating about a principal axis. Therefore the axis of rotation of the Earth is not fixed relative to the Earth. In addition, its angular velocity is slowing down at the rate of 1.3×10^{-14} rad s^{-1} per year.

The wobbling of a non-spherical ball (such as the one used in American football) after it has been kicked is due to the fact that the angular momentum imparted to the ball is not along one of its principal axes and the motion is not a pure spinning.

EXAMPLE 13.10

A disk of radius 0.5 m and mass 20 kg can rotate freely around a fixed horizontal axis passing through its center. A force of 9.8 N is applied by pulling a string wound around the edge of the disk. Find the angular acceleration of the disk and its angular velocity after 2 s.

▷ From Figure 13.21 we see that the only external forces on the disk are its weight Mg, the downward pull F, and the forces F' at the pivots. The axis ZZ' is a principal axis. Calculating the torques with respect to the center of mass C, we find that the torque of the weight is zero. The combined torque of the F' forces is also zero. Thus the net external torque is $\tau = FR$. Applying Equation 13.22, $I\alpha = \tau$, with $I = \frac{1}{2}MR^2$, we have

$$FR = (\tfrac{1}{2}MR^2)\alpha \quad \text{or} \quad F = \tfrac{1}{2}MR\alpha$$

giving an angular acceleration of

$$\alpha = \frac{2F}{MR} = \frac{2(9.8\ \text{N})}{(20\ \text{kg})(0.5\ \text{m})} = 1.96\ \text{rad s}^{-2}$$

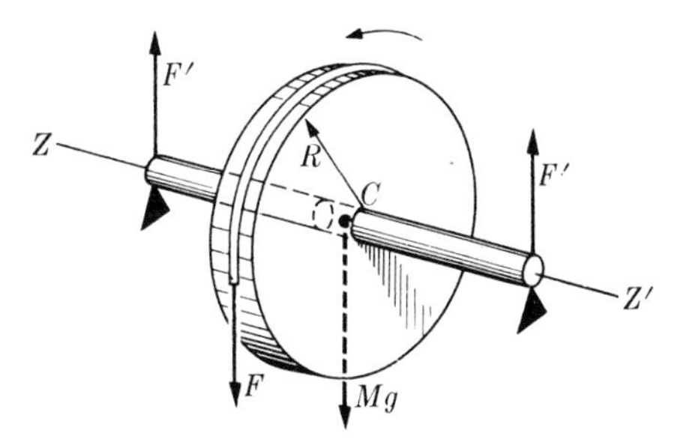

Figure 13.21

According to Equation 5.7, the angular velocity after 2 s if the disk started from rest is

$$\omega = \alpha t = (1.96\ \text{rad s}^{-2})(2\ \text{s}) = 3.92\ \text{rad s}^{-1}$$

Since the center of mass C is fixed, its acceleration is zero and we must have

$$2F' - Mg - F = 0 \quad \text{or} \quad F' = 102.9\ \text{N}$$

EXAMPLE 13.11

Angular acceleration of the system illustrated in Figure 13.22. The body has a mass of 1 kg, and the data for the disk are the same as in Example 13.10. The axis ZZ' is fixed and is a principal axis.

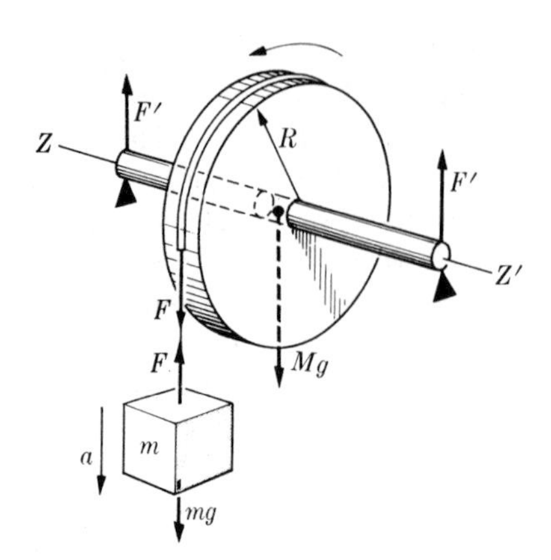

Figure 13.22

▷ Since the mass of the body is 1 kg, its weight is 9.8 N. This is the same value as the force F of Figure 13.21. Therefore there is a temptation to consider this case as identical to the previous one and assume the results are the same. This is not, however, the case. The mass m, when falling, exerts a downward pull F on the disk, and by the law of action and reaction, the disk exerts an equal but upward pull F on the mass m. Since the mass m is falling with accelerated motion, the net force on it cannot be zero. Thus F is not the same as mg, but smaller. Therefore the disk is subject to a smaller torque than in the previous problem.

The equation of motion of the mass m is

$$mg - F = ma \qquad \text{or} \qquad mg - F = mR\alpha$$

where the relation $a = R\alpha$ has been used (recall Equation 5.14). The equation for the rotational motion of the disk is $I\alpha = \tau$. With $\tau = FR$, $I\alpha = FR$. Then, since $I = \frac{1}{2}MR^2$,

$$F = \tfrac{1}{2}MR\alpha$$

Eliminating F between these two equations we find that the angular acceleration is

$$\alpha = \frac{mg}{(m + \frac{1}{2}M)R} = 1.80 \text{ rad s}^{-2}$$

which is smaller than our previous result. The downward acceleration of m is

$$a = R\alpha = \frac{mg}{m + \frac{1}{2}M} = 0.90 \text{ m s}^{-2}$$

which is much smaller than $g = 9.80 \text{ m s}^{-2}$, the value for free fall. The forces F' at the pivot can be found as in the previous example once the value of F has been calculated.

EXAMPLE 13.12

Angular acceleration of the disk of Figure 13.23. Assume the same data as for the disk of Example 13.10.

▷ The axis of rotation is the principal axis Z_0Z_0'. This problem differs from the previous examples, however, in that the center of mass of the disk is not fixed. The motion of the disk is similar to that of a yo-yo. The rotation of the disk about axis Z_0Z_0' is given by the equation $I\alpha = \tau$ or $I\alpha = FR$, since the torque of the weight Mg relative to C is zero. Thus, with $I = \frac{1}{2}MR^2$, we may write (after canceling a common factor R)

$$F = \tfrac{1}{2}MR\alpha$$

The downward motion of the center of mass has the acceleration $a = R\alpha$, and the resultant external force is $Mg - F$, so that, using Equation 13.6,

$$Mg - F = Ma = MR\alpha$$

Eliminating the force F and noting that the mass M cancels, we obtain

$$\alpha = \frac{2g}{3R} = 13.16 \text{ rad s}^{-2}$$

The downward acceleration of the center of mass is

$$a = R\alpha = \tfrac{2}{3}g = 6.53 \text{ m s}^{-2}$$

which is less than the acceleration of free fall. It is also independent of the size and mass of the disk.

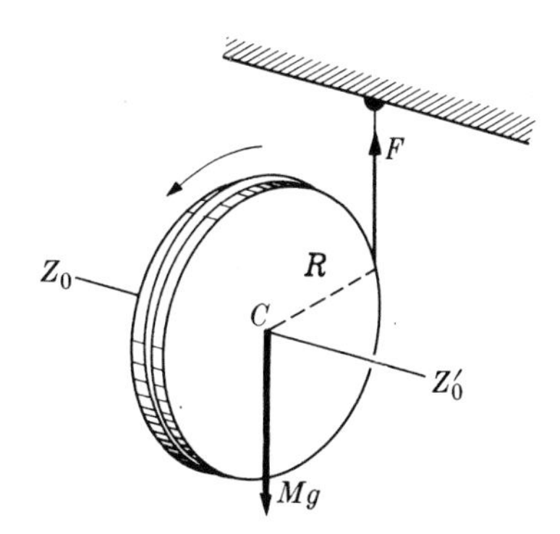

Figure 13.23

13.9 Oscillatory motion of a rigid body

We shall now illustrate the application of Equation 13.22 to two cases of oscillatory motion of a rigid body.

(i) Torsion pendulum

The torsion pendulum consists of a body suspended from a wire or fiber (Figure 13.24) such that line OC passes through the center of mass of the body. When the body is rotated an angle θ from its equilibrium position, the wire is twisted, exerting a torque on the body. The torque opposes the displacement θ and, if the twisting is not too large, has a magnitude proportional to θ, that is,

$$\tau = -\kappa\theta$$

where κ is the **torsion coefficient** of the wire. The torsion coefficient can be determined in terms of the physical and geometrical characteristics of the wire. If the body is released, the torque τ causes the body to oscillate around the line OC with SHM. Calling I the moment of inertia of the body around the axis OC, we find that the period of oscillation of

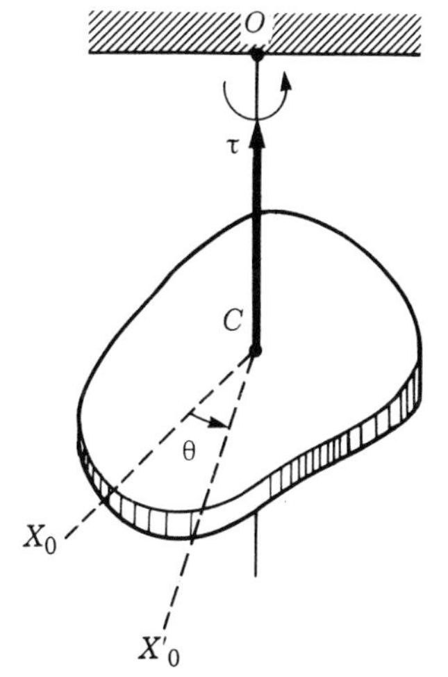

Figure 13.24 Torsion pendulum. The center of mass is at C.

the body is (see the calculation below)

$$P = 2\pi\left(\frac{I}{\kappa}\right)^{1/2} \tag{13.23}$$

Note that the larger the moment of inertia the larger the period, because it is more difficult for the torsional elastic torque to move the body. Also, the larger the torsion coefficient, which means a larger torsion torque, the smaller the period of the pendulum or the faster the pendulum oscillates. Both results were to be expected on physical grounds. Result 13.23 is interesting because we may use it to determine the moment of inertia of a body experimentally, by suspending the body from a wire whose torsion coefficient κ is known and measuring the period P of the oscillation. Conversely, this method may be used to measure κ if the moment of inertia I is known.

(ii) Physical pendulum

A physical (or compound) pendulum is any rigid body that can oscillate freely around a horizontal axis under the action of gravity. For oscillations of small amplitude, the pendulum moves with SHM. Let ZZ' (Figure 13.25) be the horizontal axis and C the center of mass of the body. The distance from the center of mass C to the axis of oscillation is labelled b and K is the radius of gyration of the body relative to the axis ZZ'. The period of oscillation when the amplitude is small, as shown in the calculation below, is

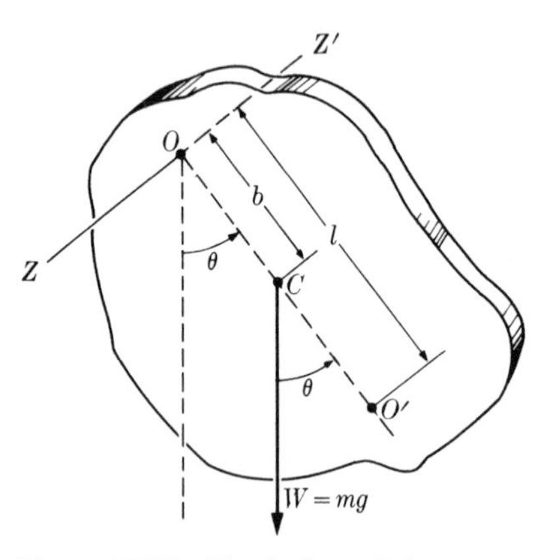

Figure 13.25 Physical pendulum.

$$P = 2\pi\left(\frac{K^2}{gb}\right)^{1/2} \tag{13.24}$$

The quantity $l = K^2/b$ is called the **length of the equivalent simple pendulum**, since a simple pendulum of that length has the same period as the body. The period of the physical pendulum is independent of its mass, as well as of its geometrical shape, as long as the ratio of K^2/b remains the same.

Calculation of the period of oscillation of a rigid body

(i) Torsion pendulum

In this case, the equation of motion of the body, using Equation 13.22 with $\alpha = \mathrm{d}^2\theta/\mathrm{d}t^2$ and $\tau = -\kappa\theta$, is

$$I\frac{\mathrm{d}^2\theta}{\mathrm{d}t^2} = -\kappa\theta \qquad \text{or} \qquad \frac{\mathrm{d}^2\theta}{\mathrm{d}t^2} + \frac{\kappa}{I}\theta = 0$$

where I is the moment of inertia relative to the axis of oscillation. Then we get Equation 10.6, where the angular motion is simple harmonic, with an angular frequency given by

$$\omega^2 = \frac{\kappa}{I}$$

from which we obtain

$$P = \frac{2\pi}{\omega} = 2\pi\left(\frac{I}{\kappa}\right)^{1/2}$$

(ii) Physical pendulum

Referring to Figure 13.25, we see that when the line OC makes an angle θ with the vertical, the torque acting on the body (which is due to its weight mg), is along the Z-axis and equal to

$$\tau = -mgb \sin \theta$$

If I is the moment of inertia of the body around the Z-axis, and $\alpha = d^2\theta/dt^2$ is the angular acceleration, then Equation 13.22, $I\alpha = \tau$, gives

$$I \frac{d^2\theta}{dt^2} = -mgb \sin \theta$$

Assuming that the oscillations are small, the approximation $\sin \theta \simeq \theta$ may be used. Then the equation of motion is

$$\frac{d^2\theta}{dt^2} = -\frac{mgb}{I}\theta \quad \text{or} \quad \frac{d^2\theta}{dt^2} + \frac{gb}{K^2}\theta = 0$$

where we have used $I = mK^2$. This equation of motion may be compared with Equation 10.6, showing that the oscillatory angular motion is simple harmonic, with an angular frequency given by $\omega^2 = gb/K^2$. Thus the period of the oscillation is

$$P = \frac{2\pi}{\omega} = 2\pi\left(\frac{K^2}{gb}\right)^{1/2}$$

EXAMPLE 13.13

A ring of radius 0.10 m is suspended from a peg, as illustrated in Figure 13.26. Determine its period of oscillation.

▷ We call the radius of the ring R. Its moment of inertia with respect to an axis passing through its center of mass C is $I_C = mR^2$ (see Table 13.2). Then, if we apply Steiner's theorem, Equation 13.21, with $a = R$, the moment of inertia relative to an axis passing through the point of suspension O is

$$I = I_C + mR^2 = mR^2 + mR^2 = 2mR^2$$

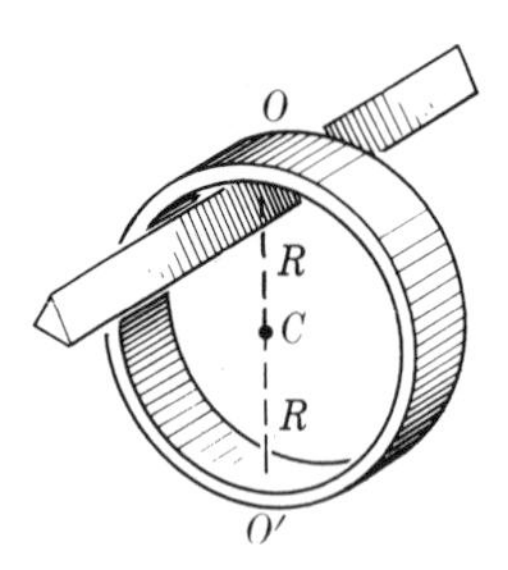

Figure 13.26

which yields a radius of gyration

$$K^2 = 2R^2$$

Also in our case $b = OC = R$. Therefore, using Equation 13.24, we obtain

$$P = 2\pi\left(\frac{2R^2}{gR}\right)^{1/2} = 2\pi\left(\frac{2R}{g}\right)^{1/2}$$

which indicates that the length of the equivalent simple pendulum is $OO' = 2R$, which is the diameter of the ring. Using the values $R = 0.10\,\text{m}$ and $g = 9.8\,\text{m}\,\text{s}^{-2}$, we obtain $P = 0.88\,\text{s}$.

13.10 Gyroscopic motion

The equation $\mathrm{d}\boldsymbol{L}/\mathrm{d}t = \boldsymbol{\tau}$ implies that in the absence of an external torque $\boldsymbol{\tau}$, the angular momentum $\boldsymbol{L}$ of the body remains constant. If the body is rotating about a principal axis, so that $\boldsymbol{L} = I\boldsymbol{\omega}$, the body will keep on rotating about that axis with constant angular velocity.

This fact can be illustrated by the **gyroscope** (Figure 13.27), which is a device for mounting a rotating wheel so that the axis can freely change its direction. The wheel G spins rapidly around the principal axis AB and is mounted so that the total torque around O is zero. Therefore, the angular momentum of the system is constant and parallel to AB or the Y-axis. If we neglect friction, the spin axis AB can move freely around both the horizontal X- and the vertical Z-axes. If we move the gyroscope around the room, AB always points in the same direction. Set the gyroscope axis so that AB is horizontal and points in the East–West direction (position 1 of Figure 13.28). Because of the Earth's rotation we observe that AB gradually tilts down (or up). After six hours it is in a vertical position (position 4 of Figure 13.28). This *apparent* rotation of AB is in fact due to the rotation of the Earth: while our laboratory moves from 1 to 4, the orientation of AB remains fixed in space. An experiment like this is sufficient to prove that the Earth rotates about its NS-axis.

If the torque on a gyroscope is not zero, then, according to the relation $\mathrm{d}\boldsymbol{L}/\mathrm{d}t = \boldsymbol{\tau}$, the angular momentum experiences a change in the time $\mathrm{d}t$ given by

$$\mathrm{d}\boldsymbol{L} = \boldsymbol{\tau}\,\mathrm{d}t \tag{13.25}$$

In other words, the change in angular momentum of the gyroscope is always in the direction of the torque (in the same way that the change of momentum of a particle is always in the direction of the force).

For the special case where the torque $\boldsymbol{\tau}$ is *perpendicular* to the angular momentum $\boldsymbol{L}$, the change $\mathrm{d}\boldsymbol{L}$ is also perpendicular to $\boldsymbol{L}$ and the angular momentum changes in direction but not in magnitude. That is, the axis of rotation changes in direction but the magnitude of the angular momentum remains constant. This

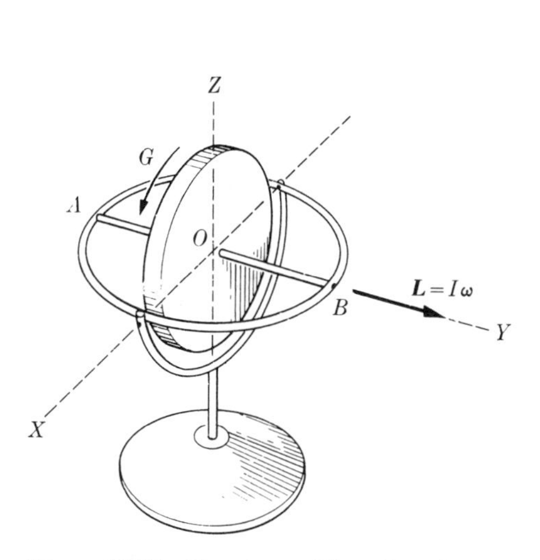

Figure 13.27 Gyroscope. The spin axis is AB.

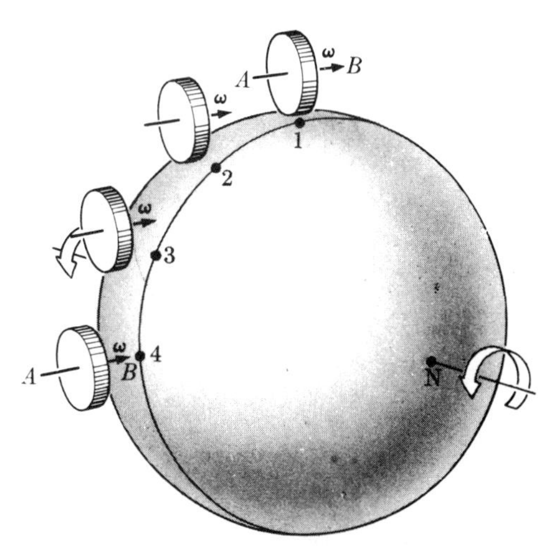

Figure 13.28 The axis of rotation of a gyroscope under no torque remains fixed in space, and therefore rotates relative to the Earth.

situation is similar to the case of circular motion under a centripetal force, where the force is perpendicular to the velocity and the velocity changes in direction but not in magnitude. The motion of the axis of rotation around a fixed axis due to an external torque is called **precession**.

This situation is found, for example, in the common top, a toy that is a type of gyroscope (Figure 13.29). The top spins around the principal axis Z_0. The axis X_0 has been chosen in the XY-plane, and thus Y_0 lies in the plane determined by Z and Z_0. Both $\boldsymbol{L}$ and $\boldsymbol{\tau}$ must be computed relative to the fixed point O. When the top rotates around its symmetry axis OZ_0 with angular velocity ω, its angular momentum $\boldsymbol{L}$ is also parallel to OZ_0. The external torque $\boldsymbol{\tau}$ is due to the weight Mg acting at the center of mass C and is equal to the vector product $(\boldsymbol{OC}) \times (M\boldsymbol{g})$ and is therefore perpendicular to the axes Z_0 and Z. Thus the torque $\boldsymbol{\tau}$ is along the axis X_0 and perpendicular to $\boldsymbol{L}$. Under the action of the torque $\boldsymbol{\tau}$, the axis Z_0 precesses around axis Z with an angular velocity

$$\Omega = \frac{Mgb}{I\omega} \tag{13.26}$$

where $b = OC$ (see Note 13.2). A more detailed discussion indicates that in general the angle ϕ does not remain constant, but oscillates between two fixed values. The end of $\boldsymbol{L}$, at the same time that it precesses around Z, oscillates between the two circles C and C' (Figure 13.30), describing the path indicated. This oscillatory motion of the Z_0 axis is called **nutation**.

Gyroscopic phenomena are of wide application. The tendency of a gyroscope to maintain its axis of rotation fixed in space is a principle used in stabilizers aboard ships and in automatic pilots on airplanes and space probes.

Most elementary particles have an internal angular momentum or spin. Because the particles are charged, they experience a torque in the presence of a

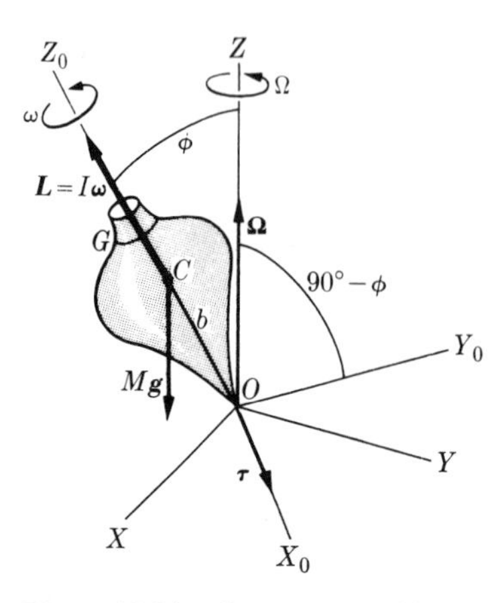

Figure 13.29 Gyroscope subject to an external torque.

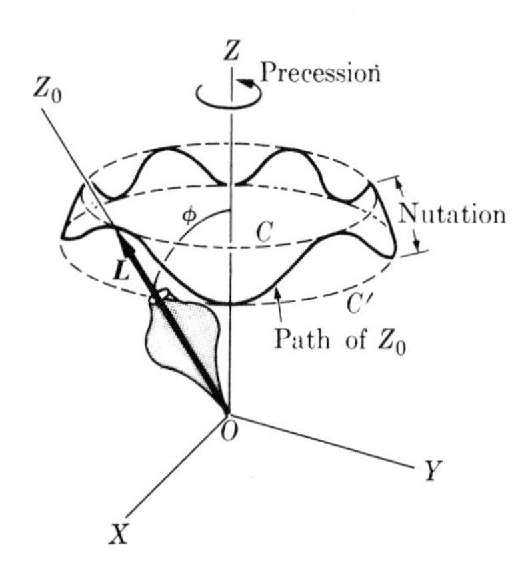

Figure 13.30 Precession and nutation of gyroscope axis.

magnetic field. This gives rise to a precession that manifests itself in several phenomena to be discussed later in the text.

Note 13.2 Precession of a gyroscope

As indicated in Figure 13.31, in a small time dt the angular momentum vector $\boldsymbol{L}$ changes from position OA to position OB. The change $d\boldsymbol{L} = \boldsymbol{AB}$ is parallel to the torque $\boldsymbol{\tau}$. The end of vector $\boldsymbol{L}$ describes a circle around Z of radius

$$AD = OA \sin \phi = L \sin \phi$$

where ϕ is the angle between axes Z_0 and Z. In time dt the radius AD moves through an angle $d\theta$ to the position BD. The angular velocity of precession Ω is defined as the rate at which the axis OZ_0 of the body rotates around the axis OZ fixed in the laboratory; that is,

$$\Omega = \frac{d\theta}{dt} \tag{13.27}$$

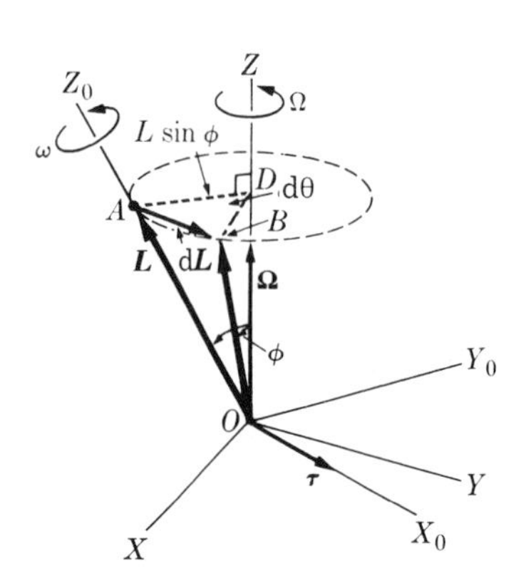

Figure 13.31 Precession of gyroscope axis.

and is represented by a vector parallel to OZ. The magnitude of $d\boldsymbol{L}$ is $|d\boldsymbol{L}| = (AD)\,d\theta = (L \sin \phi)(\Omega\, dt)$. But from Equation 8.12, we have that

$$|d\boldsymbol{L}| = \tau\, dt$$

Equating both results, we may write

$$\Omega L \sin \phi = \tau \tag{13.28}$$

From Figure 13.29 we see that the magnitude of the torque is

$$\tau = Mgb \sin \phi \tag{13.29}$$

where $b = OC$ is the distance from the CM to O. Then, comparing both relations $\Omega L = Mgb$. Making $L = I\omega$, we obtain

$$\Omega = \frac{Mgb}{I\omega}$$

which is Equation 13.26. This result is valid only if the spin angular velocity ω is very large compared with the precession angular velocity $\mathbf{\Omega}$. The reason is that if the body is precessing around OZ it also has an angular momentum about that axis. Therefore its total angular momentum is not $I\boldsymbol{\omega}$, as we have assumed, since the resultant angular velocity is $\boldsymbol{\omega} + \mathbf{\Omega}$. However, if the precession is very slow (that is, if $\mathbf{\Omega}$ is very small compared with $\boldsymbol{\omega}$), the angular momentum around OZ can be neglected.

EXAMPLE 13.14

Precession of the equinoxes.

▷ The plane of the equator makes an angle of 23°27′ with the plane of the Earth's orbit about the Sun. This plane is called the **ecliptic**. The intersection of the two planes is called the **line of equinoxes**. The Earth is a giant gyroscope whose axis passes through the North and South Poles. This axis is precessing around the normal to the plane of the ecliptic in the east–west direction, as indicated in Figure 13.32. The period is 27 725 years, giving a precessional angular velocity of 46.79″ of arc per year, or 7.19×10^{-12} rad s^{-1}. This precession of the Earth's axis results in an equal change in direction of the line of equinoxes, an effect discovered about 135 BC by Hipparchus. The precession of the Earth's axis is due to the torque exerted on the Earth by the Sun and the Moon. The Earth's axis also experiences a nutation with an amplitude of 9.2″ of arc and an oscillation period of 19 years. The net torque on the Earth is not zero because, among other reasons, the Earth is not a perfect homogeneous sphere.

It is simple to calculate the magnitude of the torque that must be exerted on the Earth in order to produce the observed rate of precession of the equinoxes. Using Equation 13.28, we have that $\tau = \Omega L \sin\phi$, where $\phi = 23°27'$ and $\Omega = 7.19 \times 10^{-11}$ rad s^{-1} is the precessional angular velocity of the Earth. We must first compute the angular momentum

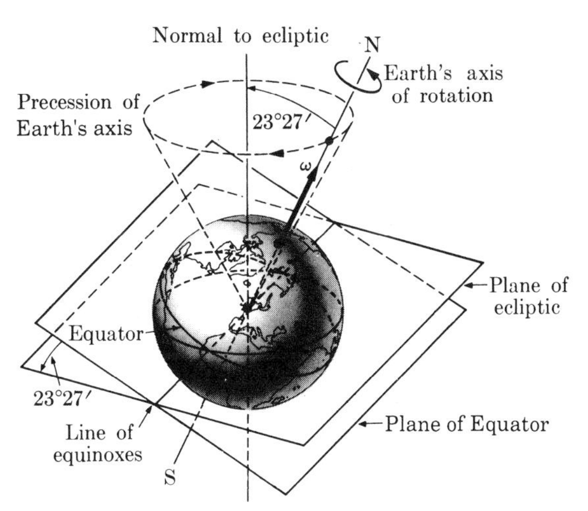

Figure 13.32 Precession of Earth's axis of rotation.

of the Earth. Since the Earth's axis of rotation deviates only slightly from a principal axis, we may use the relation

$$L = I\omega$$

The value of ω was given in Example 5.1 as 7.29×10^{-5} rad s^{-1}. The moment of inertia of the Earth, from Table 13.2, assuming that the Earth is spherical, is

$$I = \tfrac{2}{5}MR^2 = \tfrac{2}{5}(5.98 \times 10^{24}\,\text{kg})(6.38 \times 10^6\,\text{m})^2$$
$$= 9.72 \times 10^{37}\,\text{m}^2\,\text{kg}$$

Therefore $\tau = 2.76 \times 10^{27}$ N m is the torque about the CM exerted by the Sun and Moon on the Earth. This torque is rather small in planetary terms.

EXAMPLE 13.15
Gyroscopic compass.

▷ Suppose that we have a gyroscope in position G of Figure 13.33, where arrow 1 indicates the sense of rotation of the Earth. The gyroscope is arranged so that its axis must be kept in a horizontal plane. This can be done, for example, by floating the gyroscope in a liquid such as mercury. Assume that the gyroscope axis initially points in the W–E direction. When the Earth rotates, the horizontal plane and the W–E direction rotate in the same way. Therefore, if the axis of the gyroscope were maintained in the W–E direction, the axis would have to rotate as indicated by arrow 2. But that is equivalent to applying a torque around the South–North direction. Therefore the axis of the gyroscope, under the action of this torque, will turn around the vertical until it points North, as indicated by arrow 3. The gyroscopic compass has the special advantage of pointing toward the true North, since it is not subject to any local magnetic anomalies.

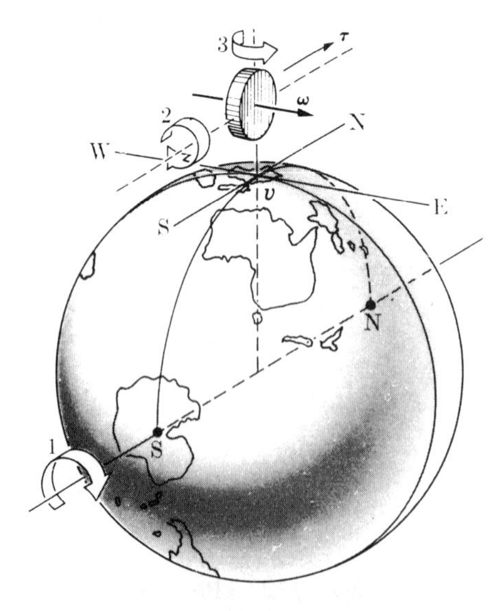

Figure 13.33 Gyroscopic compass. As a result of the Earth's rotation, indicated by arrow 1, the axis of rotation of the gyroscope rotates (precesses) as shown by arrow 3 until it points north.

13.11 Equilibrium of a body

In discussing the equilibrium of a rigid body, it is necessary to consider equilibrium relative to *both* translation and rotation. Therefore the two following conditions are required:

(i) *For translational equilibrium, the sum of all the* ***forces*** *must be zero*:

$$\sum_i \boldsymbol{F}_i = 0 \qquad \textbf{(13.30)}$$

which implies that the velocity of the CM is constant (or zero). The relation can be decomposed into the three relations

$$\sum_i \boldsymbol{F}_{ix} = 0, \qquad \sum_i \boldsymbol{F}_{iy} = 0, \qquad \sum_i \boldsymbol{F}_{iz} = 0$$

(ii) *For rotational equilibrium, the sum of all the* ***torques*** *relative to any point must be zero*:

$$\sum_i \boldsymbol{\tau}_i = 0, \qquad \textbf{(13.31)}$$

which implies that the angular momentum is constant (or zero). If the forces are all in one plane, the above conditions reduce to the following three algebraic equations:

$$\sum_i F_{ix} = 0, \qquad \sum_i F_{iy} = 0, \qquad \sum_i \tau_i = 0 \qquad \textbf{(13.32)}$$

Since these are three simultaneous equations, problems in a plane are determined by these equilibrium conditions only if there are no more than three unknown quantities. We now illustrate the technique of solving some simple problems of equilibrium of a body.

EXAMPLE 13.16

The bar of Figure 13.34 is resting in equilibrium on points A and B, under the action of the forces indicated. Find the forces exerted on the bar at points A and B. The bar weighs 40 N and its length is 4 m.

▷ First, applying the condition for translational equilibrium, Equation 13.30, where all the forces are only in the Y-direction and calling the upward forces positive (downward forces negative), we have

$$\sum F_i = F + F' - 200 - 500 - 40 - 100 - 300 = 0$$

or

$$F + F' = 1140\,\text{N}$$

Second, we apply condition 13.31 for rotational equilibrium. Since the sum of the torques is zero *about any point*, it is convenient to compute the torques relative to point A, because

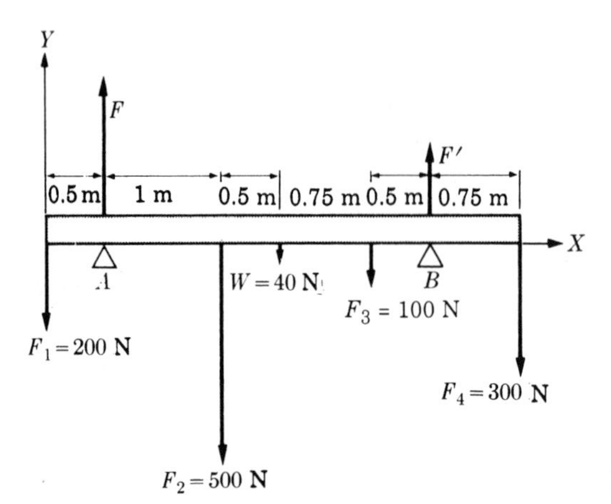

Figure 13.34

in this way we eliminate one of the unknowns. Thus

$$\begin{aligned}\sum_i \tau_i &= (-200)(-0.5) + F(0) + (-500)(1) + (-40)(1.5)\\ &\quad + (-100)(2.25) + F'(2.75) + (-300)(3.5) = 0\end{aligned}$$

or $F' = 630.9\,\text{N}$. Combining this result with the force equation, we obtain $F = 509.1\,\text{N}$, which solves the problem.

EXAMPLE 13.17

A ladder AB, weighing 40 N, rests against a vertical wall, making an angle of 60° with the floor. Find the forces on the ladder at A and B. The ladder is provided with rollers at A so that the friction with the vertical wall is negligible and any force exerted by the wall must only be perpendicular to the wall; that is, horizontal. However, the friction with the floor at B prevents the ladder from sliding down.

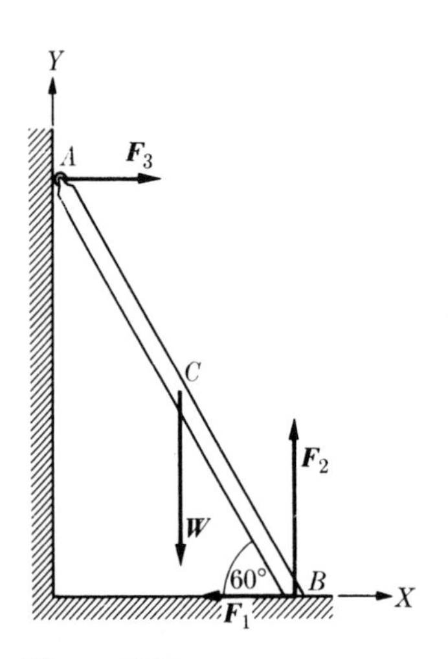

Figure 13.35

▷ The forces acting on the ladder are illustrated in Figure 13.35. The weight $\boldsymbol{W}$ is at the center C of the ladder because we assume the ladder is uniform. Force $\boldsymbol{F}_1$ is required to prevent the ladder from sliding and results from friction with the floor. Forces $\boldsymbol{F}_2$ and $\boldsymbol{F}_3$ are the normal (i.e. perpendicular) reactions at the floor and at the vertical wall. Using the three conditions of equilibrium, as stated in Equation 13.32, we have

$$\sum F_{ix} = -F_1 + F_3 = 0 \qquad \text{or} \qquad F_1 = F_3$$

$$\sum F_{iy} = -W + F_2 = 0 \qquad \text{or} \qquad F_2 = W$$

Calling L the length of the ladder and taking torques around B so that the torques of the unknown forces $\boldsymbol{F}_1$ and $\boldsymbol{F}_2$ are zero, we have for the third equation of equilibrium,

$$\sum \tau_i = W(\tfrac{1}{2}L \cos 60°) - F_3(L \sin 60°) = 0$$

or

$$F_3 = \frac{W \cos 60°}{2 \sin 60°} = 11.52 \text{ N}$$

Then the force equations above give

$$F_1 = F_3 = 11.52 \text{ N} \qquad \text{and} \qquad F_2 = \text{W} = 40 \text{ N}$$

Note that if the ladder has no roller at A, a frictional force parallel to the vertical wall would also be present at A. Thus we would have four unknown forces. Some additional assumption would be required to solve the problem since we have only three equations. The usual fourth condition is a relation between the friction force and the normal force.

QUESTIONS

13.1 State the properties of the C-frame of reference. Under what conditions is the C-frame an inertial frame of reference?

13.2 What is the path of the center of mass of a body (a) subject to no external forces? (b) Subject only to its own weight when thrown horizontally near the surface of the Earth? (c) Thrown horizontally at a very high altitude?

13.3 What is the path of the CM of a diver after jumping from a diving board? Make a diagram showing the path.

13.4 State Newton's law of action and reaction when referred to two interacting systems of particles. Apply it to two interacting atoms and two interacting galaxies.

13.5 Referring to Example 13.5, what happens to the pressure on the wall if the velocity of the molecules is doubled?

13.6 What is the usefulness of the concept of reduced mass?

13.7 Is the reduced mass of a system of two particles larger than, equal to or smaller than the mass of either particle?

13.8 Under what conditions is the reduced mass practically equal to the mass of one of the bodies? Give some examples of that situation. Under what conditions is the reduced mass equal to half the mass of one of the bodies?

13.9 Illustrate, with some examples, the principle of conservation of (i) momentum, (ii) angular momentum for (a) an isolated system and (b) two interacting systems.

13.10 Do you expect to find true rigid bodies in nature, or is this simply a convenient approximation valid under certain circumstances?

13.11 Is the angular momentum always parallel to the angular velocity of a rigid body? Is the change in angular momentum of a rigid body always parallel to the external torque?

13.12 Discuss the concept of a 'principal axis' of a rigid body. Justify why a symmetry axis should be a principal axis.

13.13 Under what conditions are the relations $L_z = I\omega$ and $\boldsymbol{L} = I\boldsymbol{\omega}$ valid?

13.14 Explain why the only possible motion of a rigid body relative to its center of mass is a rotation. In terms of your answer, justify that it is possible to split the angular momentum of a rigid body into two terms.

13.15 What is the origin of 'precessional' motion?

13.16 Consider a spring with an elastic constant k. When one end is fixed and a body of mass m is attached to the other end, the body oscillates with a certain frequency. Then the same spring, with the mass m at one end, has the other end free and attached to a mass M ($M > m$). If the spring is now stretched and released, is the frequency of oscillation larger than, equal to, or smaller than the frequency in the first case?

13.17 Consider a uniform stick of length L. Write the equation of motion for the center of mass of the stick when it is thrown in the air. The stick is released so that the bottom end is at rest in the L-frame and a height y_0 from the ground, the stick is vertical and the top end of the stick is given a velocity of $2v_0$ in the horizontal direction. Plot the curve followed by the CM.

13.18 Write an equation of motion for the bottom end of the stick, given the initial conditions in the previous question. Plot the curve followed by the bottom of the stick.

13.19 Discuss the orbital and internal angular momentum of the solar system in its motion around the center of the Milky Way (recall Figure 6.2).

13.20 Discuss the orbital and internal angular momentum of an electron orbiting around the nucleus of an atom.

13.21 Using Steiner's theorem (Equation 13.21), verify that the moment of inertia of a rigid body relative to an axis passing through its center of mass is always smaller than the moment of inertia relative to any other parallel axis.

13.22 Relative to which points should the angular momentum and the torque be computed so that the equation $d\boldsymbol{L}/dt = \boldsymbol{\tau}$ is valid?

13.23 Under what conditions are the relations $I(d\omega/dt) = \tau_z$ and $I(d\boldsymbol{\omega}/dt) = \boldsymbol{\tau}$ equivalent?

13.24 A rigid body can oscillate about a horizontal axis passing through the body. Does the period depend on (a) the mass of the body, (b) the shape of the body, (c) the size of the body, (d) the position of the axis relative to the center of mass of the body? What would be the period if the axis passes through the center of mass of the body?

13.25 Under what conditions do the internal forces in a system of particles *not* contribute to a change in the angular momentum?

13.26 Why does the equilibrium of a rigid body require more conditions than the equilibrium of a particle?

PROBLEMS

13.1 A system is composed of three particles with masses 3 kg, 2 kg and 5 kg. The first particle has a velocity of $\boldsymbol{i}(6)$ m s^{-1}. The second is moving with a velocity of 8 m s^{-1} in a direction making an angle of $-30°$ with the X-axis. Find the velocity of the third particle so that the CM appears at rest relative to the observer.

13.2 At a particular instant, three particles are moving as shown in Figure 13.36. They are subject only to their mutual interactions, so that no external forces act. After a certain time, they are observed again and it is found that m_1 is moving as shown, while m_2 is at rest. (a) Find the velocity of m_3. Assume that $m_1 = 2$ kg, $m_2 = 0.5$ kg, $m_3 = 1$ kg, $v_1 = 1$ m s^{-1}, $v_2 = 2$ m s^{-1}, $v_3 = 4$ m s^{-1} and $v_1' = 3$ m s^{-1}. (b) Find the velocity of the CM of the system at the two times mentioned in the problem. (c) At a given time the positions of the masses are $m_1(-0.8\text{ m}, -1.1\text{ m})$, $m_2(0.8\text{ m}, -1.1\text{ m})$ and $m_3(1.4\text{ m}, 0.8\text{ m})$. Draw a line showing the path of the CM of the system relative to the L-frame.

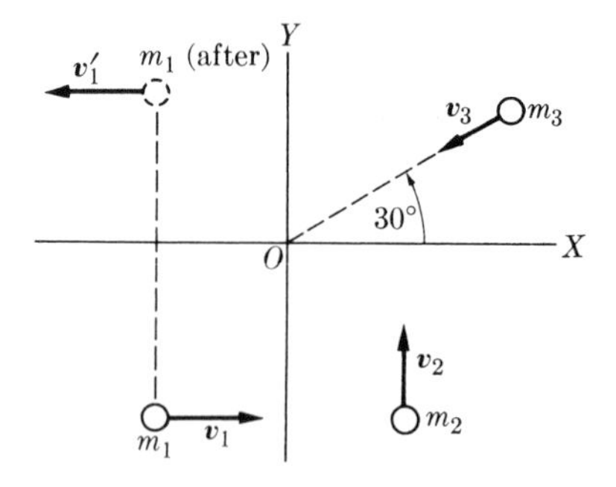

Figure 13.36

13.3 Two masses $m_1 = 10$ kg and $m_2 = 6$ kg are joined by a rigid bar of negligible mass (Figure 13.37). Being initially at rest, they are subject to forces $\boldsymbol{F}_1 = \boldsymbol{i}(8)$ N and $\boldsymbol{F}_2 = \boldsymbol{j}(6)$ N, as shown. (a) Find the coordinates of their CM as a function of time. (b) Express the total momentum as a function of time.

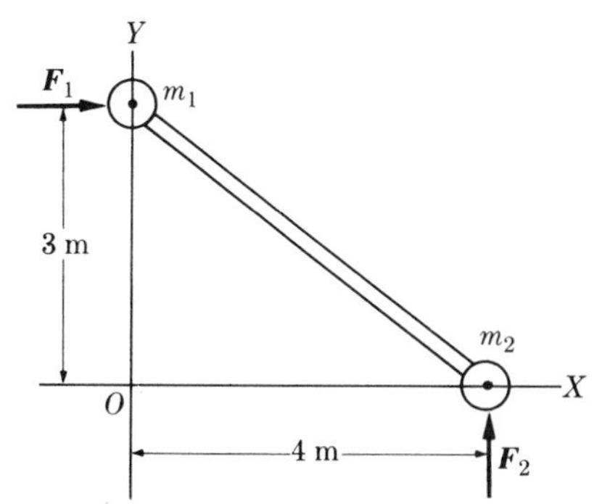

Figure 13.37

13.4 The two masses in Figure 13.38 are initially at rest. Assuming that $m_1 > m_2$, find (a) the velocity and (b) the acceleration of their CM at time t. Neglect the mass of the pulley.

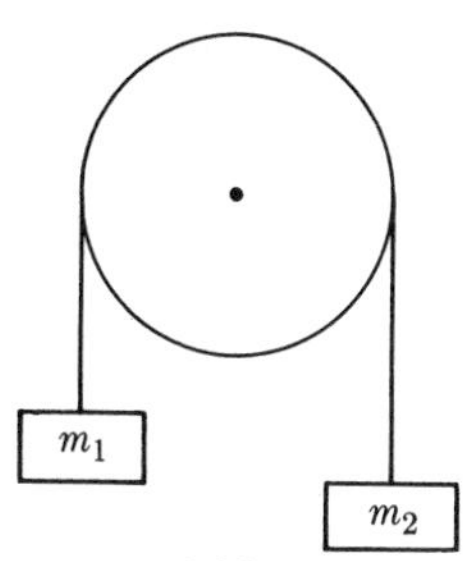

Figure 13.38

13.5 A nucleus, originally at rest, decays radioactively by emitting an electron of momentum 9.22×10^{-21} m kg s^{-1}, and, at right angles to the direction of the electron, a neutrino with momentum 5.33×10^{-21} m kg s^{-1}. (a) In what direction does the residual nucleus recoil? (b) What is its momentum? (c) Given that the mass of the residual nucleus is 3.90×10^{-25} kg, what is its velocity?

13.6 A ball, having a mass of 4 kg and a velocity of 1.2 m s^{-1}, collides head-on with another ball of mass 5 kg moving at 0.6 m s^{-1} in the same direction. Find (a) the velocities of the balls after the collision (assuming that it is elastic), (b) the change in momentum of each ball.

13.7 A particle having a mass of 0.2 kg while moving at 0.40 m s^{-1} collides with another particle of mass 0.3 kg, which is at rest. After the collision the first particle moves at 0.20 m s^{-1} in a direction making an angle of 40° with its initial direction. Find the velocity of the second particle.

13.8 Determine the position of the center of mass and the reduced mass for the LiH and HCl molecules. The bond lengths are 1.595×10^{-10} m and 1.27×10^{-10} m, respectively. (b) The NH_3 molecule (Figure 2.3) is a pyramid with the N atom at the vertex and the three H atoms at the base. The length of the N–H bonds is 1.01×10^{-10} m, and the angle between two such bonds is 108°. Find the moments of inertia relative to the Z_0 axes when in the frame where Z_0 is perpendicular to the base and X_0 and Y_0 are in the base of the pyramid.

13.9 At time t_0 two interacting particles with masses 2 kg and 3 kg are moving, relative to an observer, with velocities of 10 m s^{-1} along the $+X$-axis and 8 m s^{-1} at an angle of 120° with the $+X$-axis, respectively. (a) Express each velocity in vector form. (b) Find the velocity of their center of mass. (c) Express the velocity of each particle relative to the center of mass. (d) Find the momentum of each particle in the C-frame. (e) Find the relative velocity of the particles. (f) Calculate the reduced mass of the system. (g) Verify the relation given in Example 13.3. (h) Draw the path of the CM.

13.10 Assume that the particles of Problem 13.9 are initially at the points (0, 1, 1) and (−1, 0, 2), respectively. (a) Determine the system's angular momentum relative to their center of mass (b) Obtain the angular momentum relative to the origin. (c) Verify that the angular momentum found in part (b) is equal to the sum of the angular momentum found in part (a) plus an angular momentum $\boldsymbol{r}_C \times \boldsymbol{P}$, where $\boldsymbol{r}_C$ is the position of the center of mass and $\boldsymbol{P}$ is the linear momentum of the system.

13.11 Two masses connected by a light rod, as shown in Figure 13.39, are at rest on a horizontal frictionless surface. A third particle of mass 0.5 kg approaches the system with velocity $\boldsymbol{v}_0$ and strikes the 2 kg mass. Calculate the resulting motion of the center of mass of the two particles if the 0.5 kg mass bounces off with velocity $\boldsymbol{v}_f$ as shown.

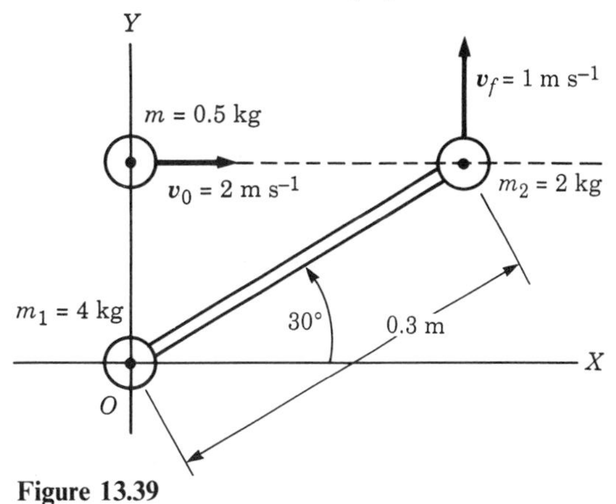

Figure 13.39

13.12 A thin rod 1 m long has a negligible mass. There are 5 bodies placed along it, each having a mass of 1.00 kg, and situated at 0 m, 0.25 m, 0.50 m, 0.75 m and 1.0 m from one end. Calculate the moment of inertia of the entire system with respect to an axis perpendicular to the rod that passes through (a) one end, (b) the second mass and (c) the center of mass. (d) Calculate the radius of gyration in each case.

13.13 Solve the previous problem again, this time for a rod whose mass is 0.20 kg.

13.14 Three masses, each of 2 kg, are situated at the vertices of an equilateral triangle whose sides measure 0.1 m each. Calculate (i) the moment of inertia of the system and (ii) its radius of gyration with respect to an axis perpendicular to the plane determined by the triangle and passing through (a) a vertex, (b) the middle point of one side, and (c) the center of mass.

13.15 Calculate the moment of inertia of (i) the CO molecule, where $\boldsymbol{r} = 1.13 \times 10^{-10}$ m, and (ii) the HCl molecule, where $\boldsymbol{r} = 1.27 \times 10^{-10}$ m relative to an axis (a) passing through the centers of mass and (b) perpendicular to the line joining the two atoms.

13.16 Determine the moments of inertia of the H_2O molecule relative to the X_0 axis shown in Figure 13.40, which passes through the center of mass. Use the data of Example 13.9.

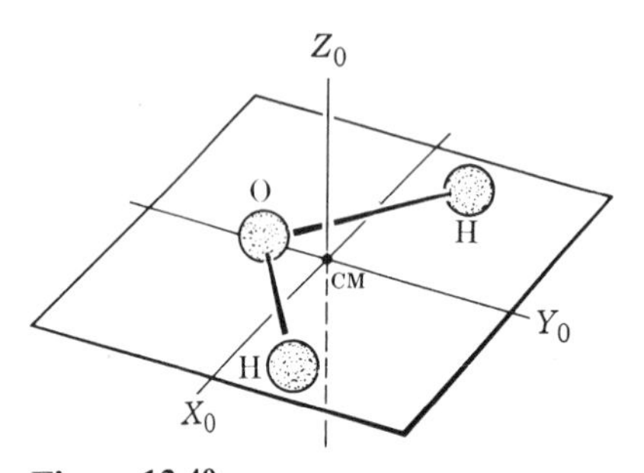

Figure 13.40

13.17 Two children, each with a mass of 25 kg, are sitting on the opposite ends of a horizontal plank which is 2.6 m long and has a mass of 10 kg. The beam is rotating at 5 rpm about a vertical axis passing through its center. (a) What will be the angular velocity if each child moves 60 cm toward the center of the beam without touching the floor?

13.18 The moment of inertia of a wheel is 1000 kg m^2. At a given instant its angular velocity is 10 rad s^{-1}. After it rotates through an angle of 100 radians, its angular velocity is 100 rad s^{-1}. Calculate the torque applied to the wheel.

13.19 A rotating wheel is subject to a frictional torque along its axis of 10 N m. The radius of the wheel is 0.6 m, its mass is 100 kg, and it is rotating at 175 rad s^{-1}. (a) How long will the wheel take to stop? (b) How many revolutions will it make before stopping?

13.20 The velocity of an automobile increases from 5 km h^{-1} to 50 km h^{-1} in 8 s. The radius of its wheels is 0.45 m. (a) What is their angular acceleration? The mass of each wheel is 30 kg and its radius of gyration is 0.3 m. (b) What is the initial and final angular momentum of each wheel?

13.21 The rotating parts of an engine have a mass of 15 kg and a radius of gyration of 15 cm. (a) Calculate the angular momentum when they are rotating at 1800 rpm. (b) What torque is necessary in order to reach this angular velocity in 5 s?

13.22 A uniform rod, hanging straight down, of length 1.0 m and mass 2.5 kg, is pivoted at its upper end. It is struck at the base by a horizontal force of 100 N that lasts only $\frac{1}{50}$ s. (a) Find the angular momentum acquired by the rod. (b) Will the rod reach a vertically upright position?

13.23 A rod of length L and mass M (Figure 13.41) can rotate freely around a pivot at A. A bullet of mass m and velocity v hits the rod at a distance a from A and becomes imbedded in it. (a) Find the angular momentum of the system around A immediately before and after the bullet hits the rod. (b) Determine the linear momentum of the system immediately before and after the collision. (c) Under what conditions will the linear momentum be conserved?

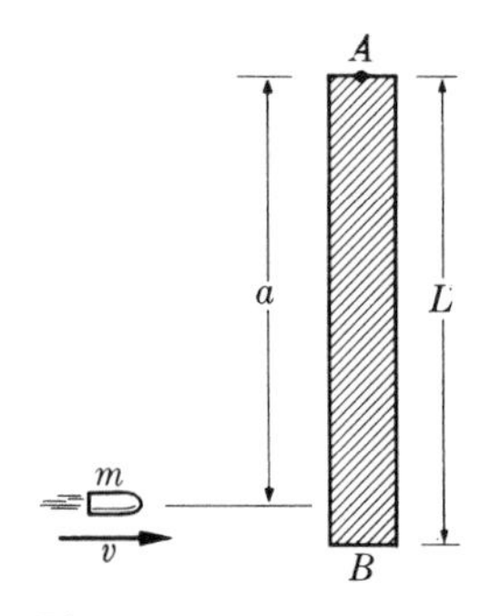

Figure 13.41

13.24 A bat of length L and mass m lies on a frictionless horizontal plane (Figure 13.42). During a very short interval Δt it is struck by a force F. The force acts at a point P which is a distance a from the center of mass. Find (a) the velocity of the center of mass, and (b) the angular velocity around the center of mass. (c) Determine the point R that initially remains at rest in the L-frame, showing that $b = K^2/a$, where K is the radius of gyration about the center of mass. The point R is called the **center of percussion**. (For example, a baseball player must hold the bat at the center of percussion in order to avoid feeling a stinging sensation when he hits the ball.) (d) Prove also that if the force strikes at R, the center of percussion will be at P.

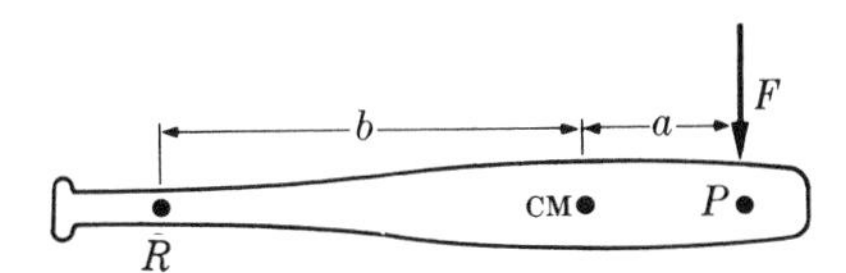

Figure 13.42

13.25 The wheel of Figure 13.22, which has a radius of 0.5 m and a mass of 25 kg, can rotate about its horizontal axis. A rope wrapped around the wheel has a mass of 10 kg hanging from its free end. Calculate (a) the angular acceleration of the wheel, (b) the linear acceleration of the body, and (c) the tension in the rope.

13.26 Calculate the acceleration of the system in Figure 13.43 if the radius of the pulley is R, its mass is M, and it rotates due to the friction on the rope. In this case $m_1 = 50$ kg, $m_2 = 200$ kg, $M = 15$ kg and $R = 0.10$ m. Ignore the effect of friction of m_1 with the table.

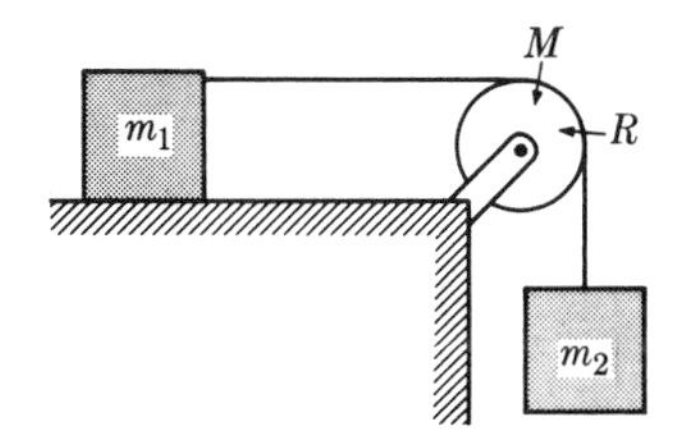

Figure 13.43

13.27 For the system in Figure 13.44, calculate (a) the acceleration of m and (b) the tension in the rope, assuming that the moment of inertia of the small disk of radius r is negligible. Let $r = 0.04$ m, $R = 0.12$ m, $M = 4$ kg, and $m = 2$ kg.

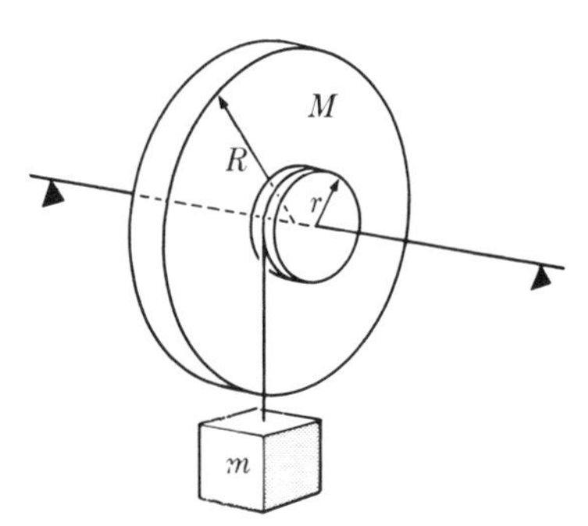

Figure 13.44

13.28 In the system represented in Figure 13.45, $M = 1.0$ kg, $m = 0.2$ kg, $r = 0.2$ m. Calculate (a) the linear acceleration of m, (b) the angular acceleration of the cylinder M, and (c) the tension in the rope. Neglect the mass of the pulley.

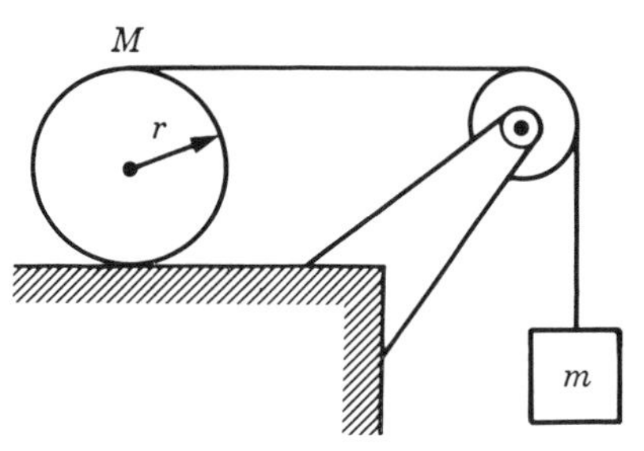

Figure 13.45

13.29 (a) Determine, for the system in Figure 7.5, the angular acceleration of the disk of mass M and radius R and the linear acceleration of m and m'. (b) Calculate the tension in each rope. Assume that $m = 0.60$ kg, $m' = 0.50$ kg, $M = 0.80$ kg, $R = 0.08$ m.

13.30 A rod 1 m long is suspended from one of its ends in such a way that it can oscillate freely. (a) Find the period and the length of the equivalent simple pendulum. (b) Find the period of oscillation if the rod is hung from an axis at a distance from one of its ends equal to the length of the equivalent pendulum found previously.

13.31 A solid disk of radius R can be hung from a horizontal axis a distance h from its center. (a) Find the length of the equivalent simple pendulum. (b) Find the position of the axis for which the period is a minimum. (c) Plot the period as a function of h.

13.32 A torsion pendulum consists of a rectangular wood block 8 cm × 12 cm × 3 cm with a mass of 0.3 kg suspended by means of a wire passing through its center in such a way that the shortest side is vertical. The period of the torsional oscillations is 2.4 s. What is the torsion constant κ of the wire?

13.33 A disk of mass m_1 is pivoted about a horizontal axis perpendicular to the disk and passes through its center. A particle of mass m_2 is attached at the rim of the disk. If the disk is released from rest with m_2 in a horizontal position, determine its angular velocity when m_2 reaches the bottom.

13.34 A string is wound around the outside of a uniform solid cylinder (mass M, radius R) and fastened to the ceiling as shown in Figure 13.23. The cylinder is held with the string vertical and then released from rest. As the cylinder descends, it unwinds the string without slipping. (a) Draw vectors showing the forces acting on the cylinder after it is released. (b) Find the acceleration of the center of mass of the cylinder as it unrolls from the string. (c) While descending, does the center of mass move toward the left, toward the right, or straight down?

13.35 The two disks in Figure 13.46 have equal masses m and radii R. The upper disk can rotate freely around a horizontal axis through its center. A rope is wrapped around both disks, and the lower one is permitted to fall. Find (a) the acceleration of the center of the lower disk, (b) the tension in the rope, and (c) the angular acceleration of each disk around its center of mass.

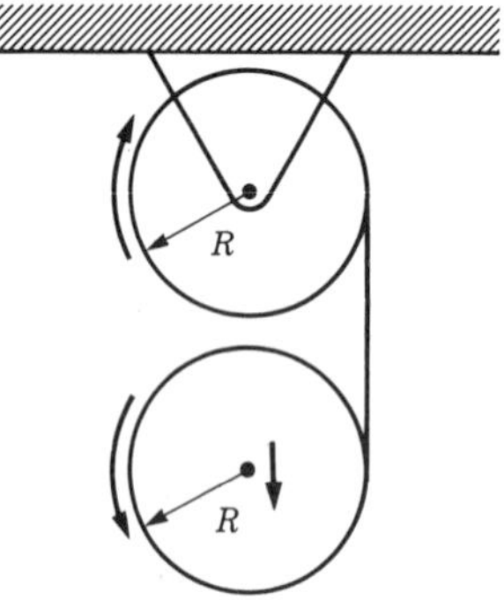

Figure 13.46

13.36 A ladder AB with a length of 3 m and a mass of 20 kg is resting against a frictionless wall (Figure 13.47). The floor, too, is frictionless; and to prevent the ladder from sliding, a rope OA has been attached. A man with a mass of 60 kg stands two-thirds of the way up the ladder. The rope suddenly breaks. Calculate (a) the initial acceleration of the center of mass of the system of ladder plus man, and (b) the initial angular acceleration around the center of mass. (Hint: note that the initial angular velocity of the ladder is zero.)

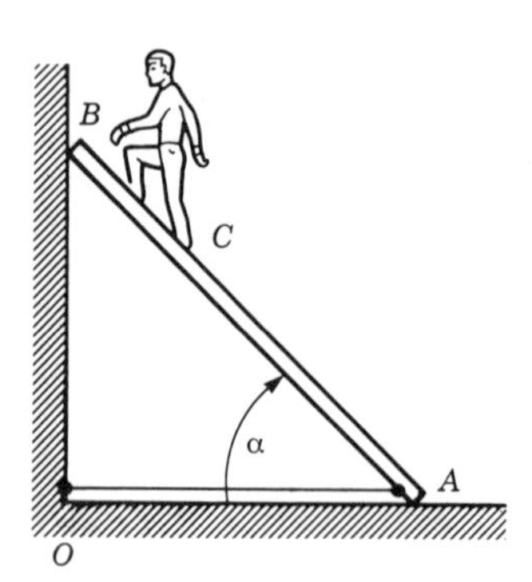

Figure 13.47

13.37 (a) Where must a block, weighing 10 N, hang along a uniform bar 2 m long and weighing 3 N if one end of the bar is supported on a pivot and a weight of 8 N is attached to the other end through the pulley arrangement shown in Figure 13.48? (b) Calculate the force exerted by the pivot.

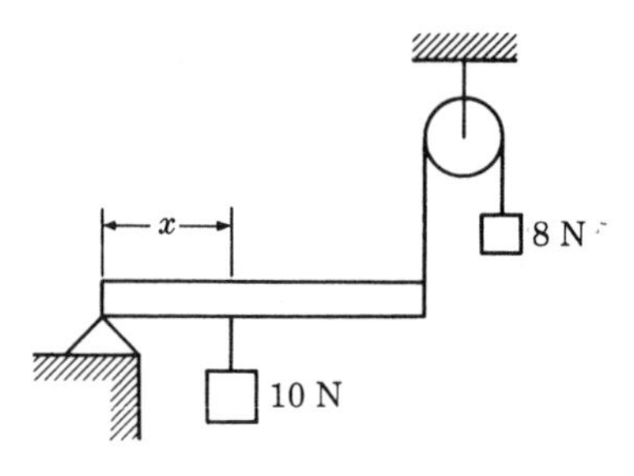

Figure 13.48

13.38 A stick is 2 m long and its weight is 5 N. There are forces of 30, 20 and 15 N acting downward at 0, 0.5 and 0.2 m from one end, and forces of 50 and 1, 30 N acting upward at 0.2 and 1.0 m from the same end. Determine the magnitude and position of the force required to balance the stick.

13.39 A beam AB is uniform and has a mass of 100 kg. It is resting on its ends A and B and is supporting the masses as shown in Figure 13.49. Calculate the reactions at the supports.

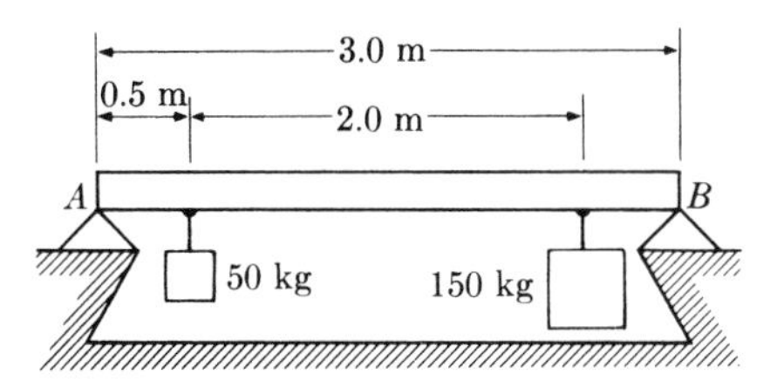

Figure 13.49

13.40 A 20 kg plank, 8.0 m long, rests on the banks of a narrow creek. A 100 kg man walks across the plank. Plot the reaction at each end of the plank as a function of the distance of the man from the first end.

13.41 A stick of mass 6 kg and length 0.8 m is placed resting on two smooth planes making a right angle, as shown in Figure 13.50. Determine the position of equilibrium and the reaction forces as a function of the angle α.

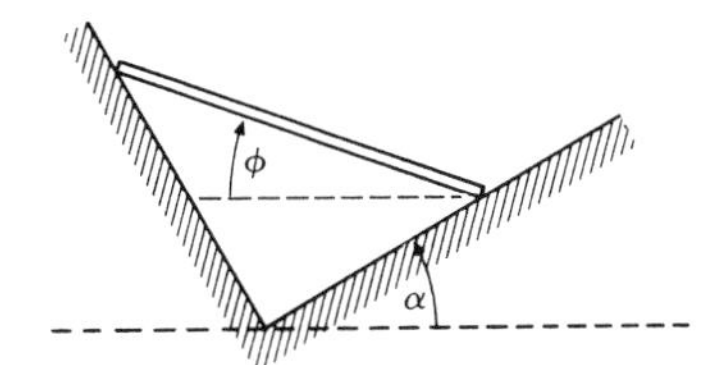

Figure 13.50

13.42 The uniform beam AB in Figure 13.51 is 4.0 m long and weighs 500 N. There is a fixed point C around which the beam can rotate. The beam is resting on point A. A man weighing 750 N is walking along the beam, starting at A. Calculate the maximum distance the man can go from A and still maintain equilibrium. Plot the reactions at A and C as a function of the distance x.

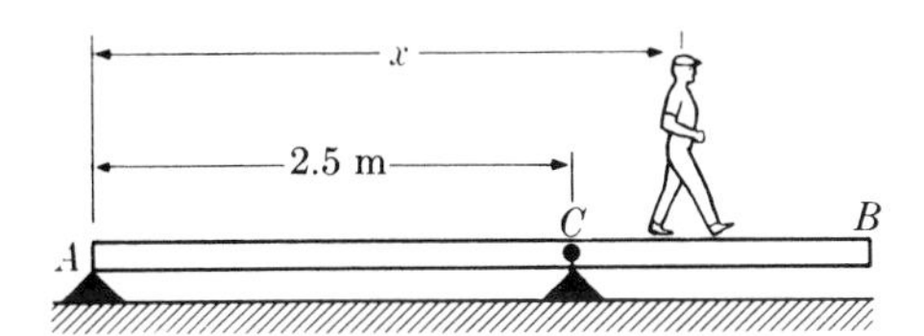

Figure 13.51

13.43 A bridge 100 m long and weighing 100 000 N is held up by two columns at its ends. What are the reactions on the columns when there are three cars on the bridge at 30 m, 60 m, and 80 m from one end, whose weights are, respectively, 15 000 N, 10 000 N, and 12 000 N?

13.44 Consider the three cars of Problem 13.43, all moving at the same speed, 10 m s^{-1}, and in the same direction. Plot the reactions of the columns as a function of time, with $t = 0$ at the position given in Problem 13.43. Extend your plot until all cars are off the bridge.

13.45 Repeat Example 13.17 of the text with a (vertical) frictional force at the wall that is 0.3 F_3. Everything else in the example remains the same.

13.46 Referring to Figure 7.26, assume the bar has a mass of 30 kg. Calculate the tension in the string AC and the forces at B.

13.47 The mass of the gyroscope in Figure 13.52 is 0.10 kg. The disk, which is located 10 cm from the ZZ'-axis, has a 5 cm radius. The disk is rotating about the YY'-axis with an angular velocity of 100 rad s^{-1}. What is the angular velocity of precession?

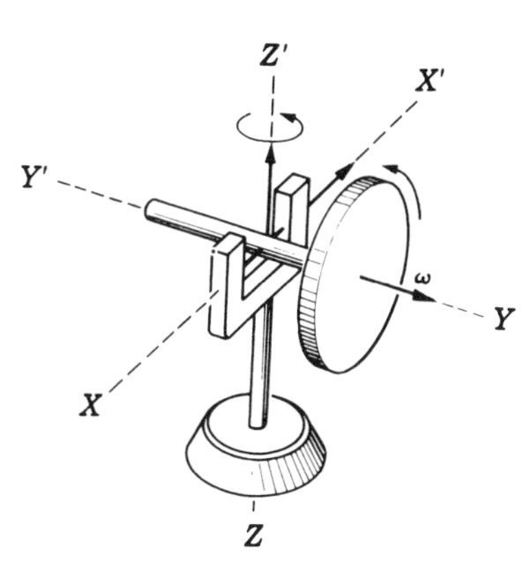

Figure 13.52

13.48 A gyroscope consists of a metal ring whose radius is 0.35 m, whose mass is 5 kg, and which is attached by spokes to an axis projecting 20 cm on each side. The axis is held in a horizontal position while the ring rotates at 300 rpm. Find the magnitude and direction of the force exerted at the axis in the following cases: (a) the axis is moved parallel to itself; (b) the axis is rotated about its center in a horizontal plane at 2 rpm; (c) the axis is rotated about its center in a vertical plane at 2 rpm. Also calculate what the angular velocity of the ring would have to be in order for its axis to remain horizontal if the gyroscope were to be supported by only one side of the axis.

14 Systems of particles II: Energy

As a skier slides down a slope his external gravitational potential energy, and part of his internal energy, is transformed into kinetic energy as well as into work to overcome the resistance of the air and the friction of the snow. (Courtesy of the British Columbia Government.)

Note

14.1 Introduction

In this chapter we analyze the energy of a system of particles and introduce a third conservation law, that of energy, which applies when the external work is zero.

The interaction of two systems can be interpreted as an exchange of linear momentum and/or of angular momentum. We will see that the interaction can also be expressed as an exchange of energy. In fact, most natural phenomena can be explained in terms of energy exchanges. Thus an understanding of the analysis of systems of particles in terms of energy is important not just for physics, but also for chemistry, biology and engineering.

14.2 Kinetic energy of a system of particles

The kinetic energy of a particle was defined, in Section 9.5, as $E_k = \frac{1}{2}mv^2$, where v is the velocity relative to the observer's frame. Section 9.5 also showed that the change in the kinetic energy is equal to the work done on the particle by all forces applied to it; that is, $\Delta E_k = W$. The same result also applies to a system of particles, if W refers to the total work done on the particles and ΔE_k refers to the change of the kinetic energy of the particles. The kinetic energy of a system of particles moving with velocities $\boldsymbol{v}_1, \boldsymbol{v}_2, \boldsymbol{v}_3, \ldots$ is

$$E_k = \sum_i \tfrac{1}{2}m_i v_i^2 = \tfrac{1}{2}m_1 v_1^2 + \tfrac{1}{2}m_2 v_2^2 + \tfrac{1}{2}m_3 v_3^2 + \cdots \tag{14.1}$$

In a system, both internal and external forces may act on the particles. Let W be the total work done on the particles by both the internal and external forces. We may separate the total work into the work W_{ext} done by the external forces and the work W_{int} done by the internal forces, so that $W = W_{ext} + W_{int}$. Hence for a system of particles, the relation $\Delta E_k = W$ becomes

$$\Delta E_k = W_{ext} + W_{int} \tag{14.2}$$

That is, the change in kinetic energy of a system of particles is equal to the work done on the system by the external and the internal forces. For example, when a gas is compressed, the external pressure applied does external work and the internal molecular forces do internal work. The two works combined result in a change in the kinetic energy of the gas molecules.

14.3 Conservation of energy of a system of particles

When the internal forces acting within a system of particles are conservative, there exists an **internal potential energy**, $E_{p,int}$, that depends on the nature of the internal forces. For example, if we consider a planetary system, the internal potential energy is the potential energy associated with the gravitational interaction between all the planets as well as each planet with the Sun. Similarly, in a hydrogen molecule, composed of two protons and two electrons, the internal potential energy is the sum of the terms corresponding to the electric interaction among the two protons, the two electrons, and each electron with the protons (Figure 14.1).

The internal potential energy depends on the relative position of the particles. Therefore $E_{p,int}$ changes as the relative positions of the particles vary during their motion. If the internal forces act along the lines joining the two particles, the internal potential energy depends only on the distances r_{ij} separating each pair of particles. The reason is that the potential energy due to a central force depends only on the distance r to the center of force (Section 9.8). The internal potential energy is then independent of the frame of reference because it contains only the distance between pairs of particles. This situation fairly well represents most of the interactions found in nature.

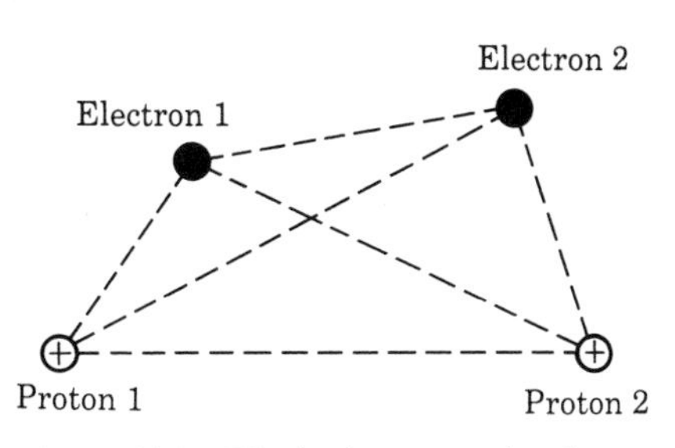

Figure 14.1 The hydrogen molecule.

Suppose the internal forces of the system do a work W_{int} during a certain time interval. This work is related to the change in the internal potential energy, $\Delta E_{\text{p, int}}$, by the same relation, Equation 9.21, that applies to a single particle; that is

$$W_{\text{int}} = -\Delta E_{\text{p, int}} \qquad \textbf{(14.3)}$$

This means that the work of the internal forces is the negative of the change in internal potential energy. When the internal forces do positive (negative) work the internal potential energy decreases (increases).

Substituting Equation 14.3 in Equation 14.2, we obtain

$$\Delta E_{\text{k}} = W_{\text{ext}} - \Delta E_{\text{p, int}}$$

We may rewrite this as

$$\Delta(E_{\text{k}} + E_{\text{p, int}}) = W_{\text{ext}} \qquad \textbf{(14.4)}$$

The quantity

$$U = E_{\text{k}} + E_{\text{p, int}} \qquad \textbf{(14.5)}$$

is called the **proper energy** of the system. It is equal to the sum of the kinetic energies of the particles *relative* to the inertial observer and their internal potential energy. As we indicated before, the internal potential energy is (under our assumption) independent of the frame of reference, while the kinetic energy is frame-dependent. The kinetic energy relative to the chosen frame of reference is given by Equation 14.5

$$E_{\text{k}} = \sum_{\text{all particles}} \tfrac{1}{2} m_i v_i^2 = \tfrac{1}{2} m_1 v_1^2 + \tfrac{1}{2} m_2 v_2^2 + \tfrac{1}{2} m_3 v_3^2 + \cdots$$

while the internal potential energy can be expressed in the form

$$E_{\text{p, int}} = \sum_{\text{all pairs}} E_{\text{p}, ij} = E_{\text{p},12} + E_{\text{p},13} + \cdots + E_{\text{p},23} + \cdots$$

The sum corresponding to the kinetic energy has one term for *each* particle, and therefore has n terms for a system of n particles. The sum corresponding to the internal potential energy has one term for *each pair* of particles because it refers to two-particle interactions and therefore has $\frac{1}{2}n(n-1)$ terms for an n-particle system. If there are no internal forces, the proper energy is all kinetic. This is a good approximation for many gases (Chapter 15).

Substituting the definition 14.5 of proper energy into Equation 14.4 gives

$$\Delta U = W_{\text{ext}} \qquad \textbf{(14.6)}$$

Equation 14.6 states that

the change in proper energy of a system of particles is equal to the work done on the system by the external forces.

Equation 13.5, $\boldsymbol{F}_{ext} = d\boldsymbol{P}/dt$, expresses a system's interaction with the outside world by its change in momentum. Equation 14.6 expresses the same interaction in terms of the system's change in energy.

When no external work can be done on the system ($W_{ext} = 0$), such as in an isolated system where there are no external forces, Equation 14.6 reads $\Delta U = 0$ or $U = U_0$. In a more explicit form we may write

$$E_k + E_{p,int} = const.$$

That is,

the sum of the kinetic energy and the internal potential energy, or the proper energy, of an isolated system of particles remains constant relative to an inertial observer.

This statement is called the **principle of conservation of energy**. So far this law is a consequence of the assumption that the internal forces are conservative. However, it seems to hold for all the processes we observe in our universe. Therefore the principle of energy conservation is considered of general validity, beyond the special assumptions under which we have stated it. In Chapter 16 we will examine this principle in a more general way.

If the kinetic energy of an isolated system increases, its internal potential energy must decrease by the same amount so that their sum remains the same. For example, in an isolated hydrogen molecule, the sum of the kinetic energy relative to some inertial frame of reference and the internal potential energy of the two protons and the two electrons remains constant. We may say that in an isolated system there is a continuous exchange of its kinetic energy into internal potential energy and conversely.

The principles of conservation of momentum, of angular momentum and of energy are fundamental laws. They seem to govern all the processes that may possibly occur in nature. They are unquestionably the three most important principles of physics.

As a simple example, consider a gas inside a balloon (Figure 14.2) drifting in air. The system is composed of the gas molecules inside the balloon. The *proper energy* of the gas is the sum of the kinetic energy of each molecule relative to the observer plus the internal potential energy of the gas molecules. The external forces on the gas have been

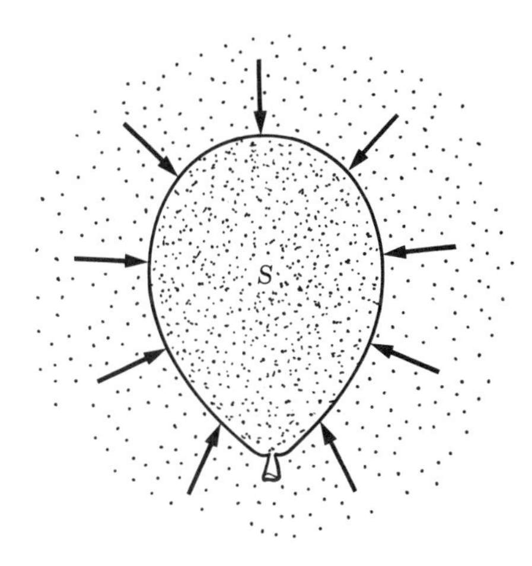

Figure 14.2

represented schematically by arrows pointing toward the surface of the balloon. As the balloon moves upward (downward), or its temperature increases (decreases), the balloon expands (contracts). In any and all of these processes external forces do work on the gas. The work done by the external forces shown in the figure corresponds to W_{ext} in Equation 14.6. But if the shape and size of the balloon do not vary, so that $W_{\text{ext}} = 0$, then the sum of the kinetic energy and internal potential energy of the molecules remains constant.

14.4 Total energy of a system of particles subject to external forces

It may happen that the external forces acting on a system are also conservative and are associated with a potential energy $E_{\text{p,ext}}$. In general $E_{\text{p,ext}}$ is a sum of one term for each particle on which external forces act. Then the external work, W_{ext}, done during a certain time interval can be written as

$$W_{\text{ext}} = -\Delta E_{\text{p,ext}}$$

where $\Delta E_{\text{p,ext}}$ is the change of the external potential energy. Equation 14.6 for the change in proper energy now becomes

$$\Delta U = -\Delta E_{\text{p,ext}}$$

which may be written as $\Delta(U + E_{\text{p,ext}}) = 0$ or $U + E_{\text{p,ext}} = const.$ The quantity

$$\begin{aligned} E &= U + E_{\text{p,ext}} \\ &= E_{\text{k}} + E_{\text{p,int}} + E_{\text{p,ext}} \end{aligned} \tag{14.7}$$

is called the **total energy** of the system subject to external conservative forces. It remains constant during the motion of the system when both internal and external forces are conservative. This result is similar to Equation 9.36 for a single particle.

For example, a hydrogen atom, composed of an electron and a proton, has a proper energy equal to the sum of the kinetic energies of the electron and the proton relative to the observer plus the internal potential energy of their electric interaction. If the atom is isolated, the proper energy is constant. However, when the atom is placed in an external electric field, its total energy must include, in addition, the potential energy due to the external electric field acting on the proton and the electron.

EXAMPLE 14.1

Energy of two masses m_1 and m_2 attached by a spring that has an elastic constant k (Figure 14.3).

▷ If the system is thrown into the air, the kinetic energy of the system relative to the observer is

$$E_{\text{k}} = \tfrac{1}{2}m_1v_1^2 + \tfrac{1}{2}m_2v_2^2$$

The internal potential energy is due to the extension or compression of the spring and is equal to

$$E_{p,int} = \tfrac{1}{2}kx_{12}^2$$

where x_{12} is the deformation of the spring. The external potential energy due to the Earth's gravitational attraction is

$$E_{p,ext} = m_1gy_1 + m_2gy_2$$

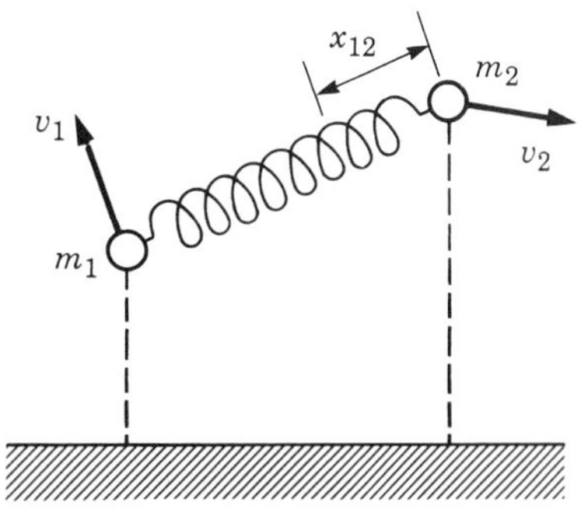

Figure 14.3

where y_1 and y_2 are the heights of the particles above the Earth's surface. The proper energy of the system is then

$$U = \tfrac{1}{2}m_1v_1^2 + \tfrac{1}{2}m_2v_2^2 + \tfrac{1}{2}kx_{12}^2$$

and, if no other forces act on the system besides gravity, the total energy is

$$E = \tfrac{1}{2}m_1v_1^2 + \tfrac{1}{2}m_2v_2^2 + \tfrac{1}{2}kx_{12}^2 + m_1gy_1 + m_2gy_2$$

and this energy must remain constant during the motion.

14.5 Internal energy of a system of particles

The kinetic energy of a particle depends on its velocity and the velocity, in turn, depends on the frame of reference used. Then, the value of the kinetic energy of a system of particles also depends on the frame of reference used to discuss the motion of the system. We call **internal kinetic energy** $E_{k,int}$ *the kinetic energy referred to the C-frame of reference* (recall Section 13.2). The internal potential energy, which depends only on the distance between the particles, has the same value in all frames of reference (as explained before). Thus we shall define the **internal energy** of a system as *the sum of its internal kinetic and potential energies.* That is,

$$U_{int} = E_{k,int} + E_{p,int} = (E_k + E_p)_{int} \qquad \textbf{(14.8)}$$

To obtain the kinetic energy relative to the observer's frame of reference, the internal kinetic energy of the system must be added to the kinetic energy associated with the motion of the CM; that is,

$$E_k = E_{k,int} + \tfrac{1}{2}Mv_{CM}^2 \qquad \textbf{(14.9)}$$

where $M = \sum m_i$ is the total mass of the system and v_{CM} is the velocity of the CM relative to the observer's frame. The term $\tfrac{1}{2}Mv_{CM}^2$ is the same as the kinetic energy of a particle of mass $M = \sum m_i$ moving with the velocity of the center of mass of the system. We call it the *orbital* kinetic energy of the system. That is,

$$E_{k,orb} = \tfrac{1}{2}Mv_{CM}^2 \qquad \textbf{(14.10)}$$

Hence,

> *the kinetic energy of a system can be expressed as the sum of the internal kinetic energy relative to the center of mass and the orbital kinetic energy associated with the motion of the center of mass.*

For example, the kinetic energy of the air inside the balloon (relative to the observer) in Figure 14.2, which is drifting in air, is the sum of the orbital kinetic energy, corresponding to the motion of the CM relative to the observer, and the kinetic energy of the motion of the molecules relative to the CM of the balloon.

Consider, as another example, a person throwing a spinning ball. The ball's total kinetic energy relative to the ground is the sum of its internal kinetic energy relative to the center of mass (corresponding to its spin and its molecular agitation), and the orbital kinetic energy of its center of mass relative to the ground. A similar situation is found in the case of a molecule.

Combining Equations 14.5, 14.8 and 14.9, we may write

$$U = U_{\text{int}} + \tfrac{1}{2}Mv_{\text{CM}}^2 \tag{14.11}$$

which shows that the proper and internal energies differ by the term $\frac{1}{2}Mv_{\text{CM}}^2$. Thus, when the center of mass of the system is at rest relative to the observer, the proper and internal energies coincide. The work–energy relation (Equation 14.6) now becomes

$$\Delta U_{\text{int}} + \Delta(\tfrac{1}{2}Mv_{\text{CM}}^2) = W_{\text{ext}} \tag{14.12}$$

which indicates that the work of the external forces is used to change the internal energy of the system and the kinetic energy of its orbital motion.

How W_{ext} is split into the two terms depends on each particular case. For example, if all external forces add to zero, then v_{CM} is constant. There is then no change in orbital kinetic energy so that

$$\Delta U_{\text{int}} = W_{\text{ext}} \tag{14.13}$$

Further, if $W_{\text{ext}} = 0$, then $\Delta U_{\text{int}} = 0$ (or $U_{\text{int}} = \textit{const.}$), which expresses the conservation of internal energy of an isolated system. On the other hand, if the internal energy does not change during the motion, $\Delta U_{\text{int}} = 0$, as usually is the case with a rigid body, then $\Delta(\frac{1}{2}Mv_{\text{CM}}^2) = W_{\text{ext}}$ and all the external work goes into changing the orbital kinetic energy. This is, for example, the case of a solid body, like a ball or a stone, thrown into the air.

When we are dealing with a system of particles and refer only to the internal energy, we shall omit the subscript 'int' unless otherwise specified.

14.6 Kinetic energy of rotation of a rigid body

In dealing with the motion of a 'rigid' or solid body, we may ignore its internal structure and assume its internal energy never changes. Consider a rigid body rotating around an axis with angular velocity $\boldsymbol{\omega}$. The velocity of each particle is $v_i = \omega R_i$, where R_i is the distance of the particle to the axis of rotation and ω is the same for all particles (Figure 14.4). Then the kinetic energy of rotation is

$$E_k = \sum_i \tfrac{1}{2} m_i v_i^2 = \sum_i \tfrac{1}{2} m_i R_i^2 \omega^2 = \frac{1}{2}\left(\sum_i m_i R_i^2\right)\omega^2$$

or, recalling definition 13.18 of the moment of inertia,

$$E_{k,rot} = \tfrac{1}{2} I \omega^2 \qquad \textbf{(14.14)}$$

Expression 14.14 is correct for any axis, even if it is not a principal axis. The magnitude of the velocity of each particle of the body is always $v_i = \omega R_i$, whether or not the rotation is about a principal axis. When the rotation is about a principal axis, we can use Equation 13.19, $\boldsymbol{L} = I\boldsymbol{\omega}$, and write

$$E_{k,rot} = \frac{L^2}{2I} \qquad \textbf{(14.15)}$$

Consider the situation where a rigid body rotates about an axis passing through its center of mass and, at the same time, has translational motion relative to the observer. As stated in Section 14.5, the kinetic energy of a body relative to an inertial frame of reference is

$$E_k = \tfrac{1}{2} M v_{CM}^2 + E_{k,int}$$

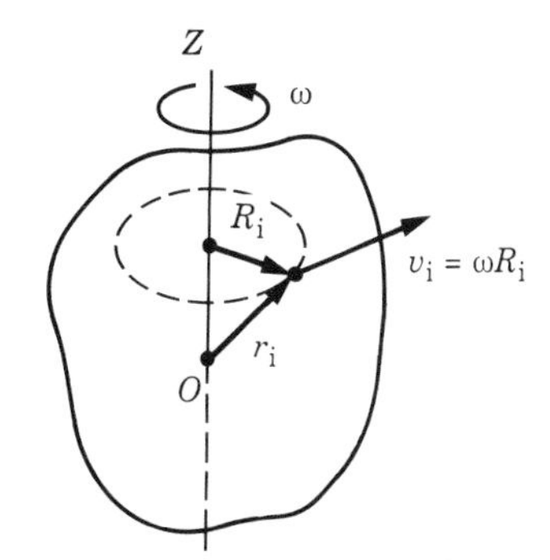

Figure 14.4 Rotation of a rigid body around the Z-axis.

For a rigid body, $\tfrac{1}{2} M v_{CM}^2$ is just its translational kinetic energy and $E_{k,int}$ is the rotational kinetic energy relative to the center of mass. This is true because the only motion a rigid body can have relative to the center of mass is rotation. The individual motion of the molecules relative to the center of mass can be ignored in a rigid body, although in all solid bodies the molecules are in vibratory motion and contribute to the internal kinetic energy. However, this contribution does not change during the motion. Therefore we may replace $E_{k,int}$ by $E_{k,rot}$ in the above expression and express the kinetic energy as the sum of the orbital and the rotational kinetic energies. Thus, using Equation 14.14,

$$E_k = E_{k,CM} + E_{k,rot} = \tfrac{1}{2} M v_{CM}^2 + \tfrac{1}{2} I \omega^2 \qquad \textbf{(14.16)}$$

where I is the moment of inertia relative to the axis of rotation passing through the center of mass.

For all practical purposes the distance between the particles in a rigid body does not change during the motion. So we may assume that its internal potential energy $E_{p,int}$ remains constant. Therefore we do not have to consider it when we are discussing the body's exchange of energy with its surroundings. Accordingly, the total energy, as expressed by Equation 14.7, reduces to

$$E = E_k + E_{p,ext} \tag{14.17}$$

where $E_{p,ext}$ is the potential energy associated with external forces. Equation 14.17 applies to a rigid body when the external forces are conservative. This result is similar to that for one particle as expressed in Equation 8.36. We call E, given above, the **total energy of a rigid body** relative to the observer. It remains constant as long as the external forces are conservative. In the future we will drop the subscript 'ext' for the external potential energy of a rigid body.

When we use Equation 14.16 for the kinetic energy, Equation 14.17 for the total energy of the body becomes

$$E = \tfrac{1}{2}Mv_{CM}^2 + \tfrac{1}{2}I\omega^2 + E_{p,ext} \tag{14.18}$$

For example, if the body is falling under the action of gravity, the external potential energy is $E_{p,ext} = Mgy$, where y refers to the height of the CM of the body relative to a horizontal reference plane. The total energy is then

$$E = \tfrac{1}{2}Mv_{CM}^2 + \tfrac{1}{2}I\omega^2 + Mgy \tag{14.19}$$

The first term is the energy of translation, the second that of rotation and the third the gravitational potential energy.

EXAMPLE 14.2

A sphere, a cylinder, and a ring, all with the same radius, roll down along an inclined plane starting at a height y_0. Find, in each case, the velocity at the base of the plane.

▷ Figure 14.5 shows the forces acting on the rolling body. They are the weight Mg, the reaction N of the plane, and the frictional force F at the point of contact with the plane. We could apply the methods used in Chapter 13 to solve this problem (and we recommend that the student try it). Instead we shall illustrate an application of the principle of conservation of energy, as expressed by Equation 14.19.

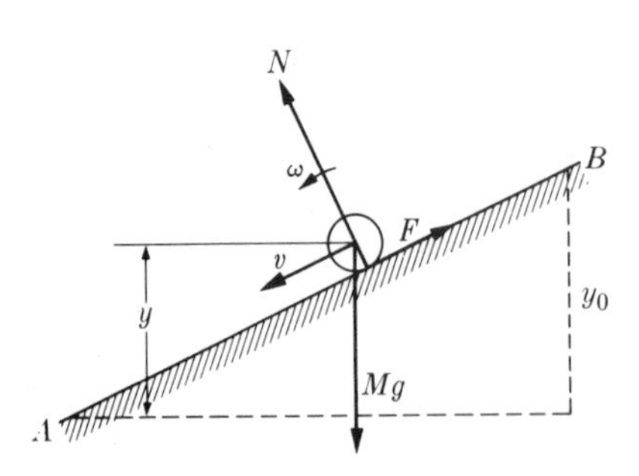

Figure 14.5 Rolling of a body along an inclined plane.

At the starting point B, when the body is at rest at an altitude y_0, its total energy is equal to the gravitational potential energy Mgy_0. At any intermediate position, the center of mass is moving with a translational velocity v and the body is rotating about the center of mass with angular velocity ω. As long as the body rolls without slipping, the relation $v = \omega R$ or $\omega = v/R$ is valid. The total energy is thus

$$E = \tfrac{1}{2}Mv^2 + \tfrac{1}{2}I\omega^2 + Mgy$$

$$= \tfrac{1}{2}Mv^2 + \tfrac{1}{2}I\left(\frac{v^2}{R^2}\right) + Mgy$$

Use Equation 13.23 to write the moment of inertia as $I = MK^2$, where K is the radius of gyration. Then we may express the total energy as

$$E = \tfrac{1}{2}M\left(1 + \frac{K^2}{R^2}\right)v^2 + Mgy$$

Since the total energy remains constant, this expression for the energy may be equated to the initial energy $E = Mgy_0$. Then the velocity at any height y is

$$v^2 = \frac{2g(y_0 - y)}{1 + (K^2/R^2)}$$

If, instead of a rolling rigid body, we had a body that freely slides down the plane, we would not include the rotational energy, and the result would be

$$v^2 = 2g(y_0 - y)$$

the same as for a freely falling particle. Thus we see that the rotational motion results in a slower translational motion because in a rolling body the initial potential energy must be used to produce both rotational and translational kinetic energy, while when the body slides down the plane without rotation, all the initial potential energy goes into translational kinetic energy.

From the above expression we see that the speed of a homogeneous body rolling down a slope does not depend on the mass or the actual dimensions of the body, but only on its *shape*, expressed by the ratio K/R. That is, all spheres translate identically regardless of their radius, but the ring will have less velocity than the cylinder, and this in turn less than the sphere.

14.7 Rotational energy of molecules

Although, as explained in Section 10.11, the atoms in a molecule are vibrating about their equilibrium positions, the amplitude of the vibration is small relative to their separation. Therefore, to a first approximation, it is possible to consider the rotating molecule as a rigid system. We shall assume that the molecule is rotating about a principal axis that passes through the CM with angular momentum $\boldsymbol{L} = I\boldsymbol{\omega}$, where I is the corresponding moment of inertia. For a diatomic molecule, $I = \mu r^2$, where μ is its reduced mass, as explained in Example 13.6. Then the kinetic energy of rotation of the molecule is

$$E_r = \frac{L^2}{2I}$$

In Example 8.9 we mentioned that the angular momentum of an electron in an atom is quantized. The same fact applies to the rotational motion of a molecule. Therefore, using Equation 8.25, $L^2 = \hbar^2 l(l+1)$, where l is a positive integer, $l = 0$, 1, 2, 3, . . . , the rotational kinetic energy of a diatomic molecule is

$$E_r = \frac{\hbar^2 l(l+1)}{2I} \qquad \textbf{(14.20)}$$

By giving successive values to l, the possible rotational kinetic energies of a molecule are obtained. The moment of inertia of a molecule is of the order of 10^{-46} kg m^2 and, using the value of $\hbar = 1.05 \times 10^{-34}$ kg m^2 s^{-1}, the term $\hbar^2/2I$ is of the order of 10^{-22} J or 10^{-3} eV. This value gives the order of magnitude of the separation of rotational energy levels for a diatomic molecule. The CO molecule has a moment of inertia $= 1.47 \times 10^{-46}$ kg m^2 (Example 13.6) so that $\hbar^2/2I = 0.38 \times 10^{-22}$ J or 2.38×10^{-4} eV.

When considering the energy of a molecule, we must include the kinetic energy of the motion of the CM, the internal rotational energy, E_r, the internal vibrational energy, E_v, and the internal electronic energy, E_e. That is,

$$U_{\text{molecule}} = E_{k,CM} + \underbrace{E_r + E_v + E_e}_{U_{\text{int}}}$$

where the last three terms correspond to the internal energy of the molecule. That is, $U_{\text{int}} = E_r + E_v + E_e$.

For polyatomic molecules, such as CO_2 and CH_3CH_3, which are linear, H_2O, which is planar and NH_3 and $ClCH_3$, which are three-dimensional molecules, the rotation can occur around any one (or all) of the three principal axes. Consequently, the structure of the energy levels is more complex.

14.8 Binding energy of a system of particles

When considering a system of particles, we might, for convenience, take the zero of internal energy when the particles are at relative rest and separated by very large distances. When the particles are brought together by some mechanism, so that a system is formed, the internal energy of the system is given by

$$U_{\text{int}} = (E_k + E_p)_{\text{int}}$$

Of course E_k is always positive but E_p may be positive or negative depending on the internal forces. Therefore U_{int} may be a positive or a negative energy (Figure 14.6) relative to the state where all particles are at rest at great distances from each other.

If U_{int} is positive, energy has to be supplied to the particles to arrange them in the system and therefore, without some external positive work it is impossible to arrange the system. The particles tend, if left to themselves, to separate by releasing the energy U_{int} and therefore a system with positive internal energy is unstable or unbound.

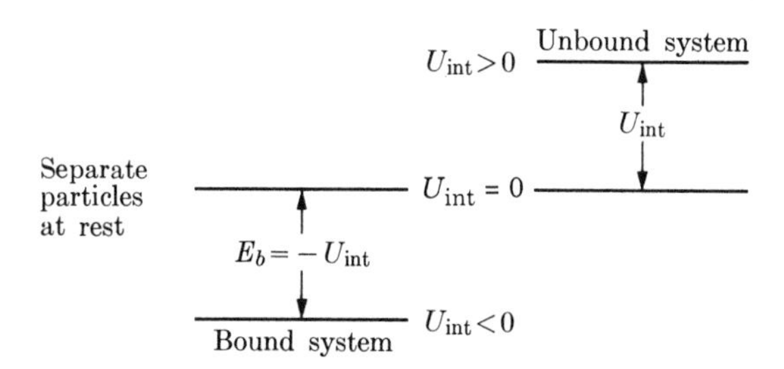

Figure 14.6 Binding energy of a system of particles.

On the other hand, if U_{int} is negative, the system has less energy than when the particles are separated. When the system is formed, energy, in the amount of $-U_{int}$, is released; this is what happens in a chemical reaction when new molecules are formed spontaneously. To separate the particles it is necessary to supply the same amount of energy (i.e. $-U_{int}$) to the system from an external source. Therefore a system with negative internal energy is stable or bound.

The **binding energy** of a system of particles is defined as

$$E_b = -U_{int} = -(E_k + E_p)_{int} \tag{14.21}$$

Therefore, the binding energy is the energy released when the system is formed, or the energy that must be supplied to the system in order to separate the particles.

Matter is composed of systems of particles, basically electrons, protons and neutrons. These particles are arranged in some stable configurations that we call nuclei, atoms, molecules and bodies (such as solids and liquids). A knowledge of the binding energies of such systems is of great interest. The binding energy of atoms and molecules is of the order of a few eV. For example, the binding energy of a hydrogen atom (Figure 14.7(a)), composed of a proton and an electron, is $E_b = 2.33 \times 10^{-18}$ J or 13.6 eV. This is the energy that must be supplied to separate the two particles, or the energy released when an electron is captured into the most stable orbit around a proton. Similarly, the binding energy of a hydrogen molecule (Figure 14.7(b)), composed of two hydrogen atoms, is 3.59×10^{-19} J or 2.24 eV. This is the energy that must be supplied to a hydrogen molecule to separate it into two hydrogen atoms (not into two protons and two electrons). It is also the energy released when two hydrogen atoms combine to form a hydrogen molecule. The binding energy of nuclei is of the order of several MeV, or a million times larger than binding energies in molecules. For instance, the deuteron (Figure 14.7(c)) is a nucleus composed of a proton and a neutron, and its binding energy is 3.56×10^{-13} J or 2.224 MeV. This large difference in binding energies suggests that the forces holding protons and neutrons in nuclei are much stronger (about

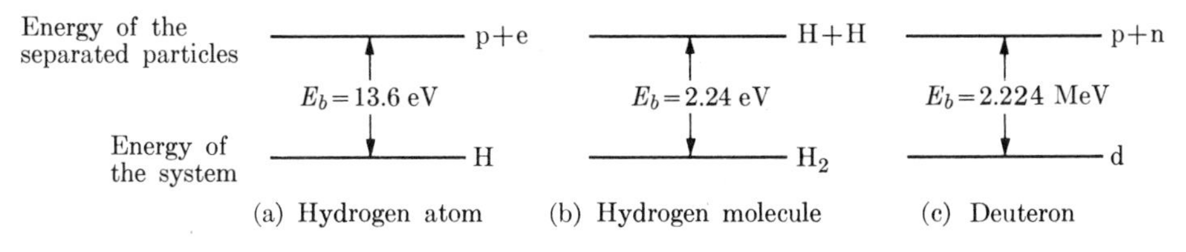

Figure 14.7 Binding energy of some simple systems.

10^6 times stronger) and of a different nature from those holding electrons and nuclei in atoms and molecules. On the other hand, the binding energy of molecules in liquids and solids is of the order of 10^{-1} eV. This relatively small value is due to the strength of intermolecular forces attributable to small residual electric forces. Since molecules are essentially electrically neutral systems, intermolecular forces are weaker than other forces.

14.9 Collisions

When two particles approach each other, their mutual interaction changes their motion. The interaction produces an exchange of momentum and of energy. We say that there has been a collision. We may say the same thing when we have two systems instead of two particles. Actually, in this section the terms 'particles' and 'systems' will be used as equivalent. A collision does not necessarily mean that the two particles or systems have been physically in contact, as happens in the case of the collision between two billiard balls. It means, in general, that an interaction entered into play when the two particles were close, such as in the shaded region of Figure 14.8. The interaction produced a measurable change in their motion in a relatively short time. For example, if an electron or a proton approaches an atom, electric forces come into effect, producing a pronounced perturbation in the motions of the particles. The bending of the path of a comet when it approaches the solar system is also a collision.

Sometimes the term **scattering** is used to refer to a collision when the final particles or systems are the same as the initial ones. In some collisions, however, the final particles or systems are not necessarily identical to the initial ones and the collision is sometimes called a **reaction**. For example, in a collision between an electron and a proton, the final product may be a hydrogen atom. In a collision between atom A and molecule B–C, the end result may be molecule A–B and atom C. In fact, this is the way that many chemical reactions take place. Similarly, a deuteron is a nucleus composed of a neutron and a proton. When a deuteron passes near another nucleus, the neutron may be captured by the second nucleus. Then the proton will continue separately and the final particles are the proton and a nucleus having an extra neutron. Hence, the particles of systems do not have to be the same before and after a collision.

In a collision experiment, the motions of the particles before the collision are usually known precisely, since their motion depends on how the experiment

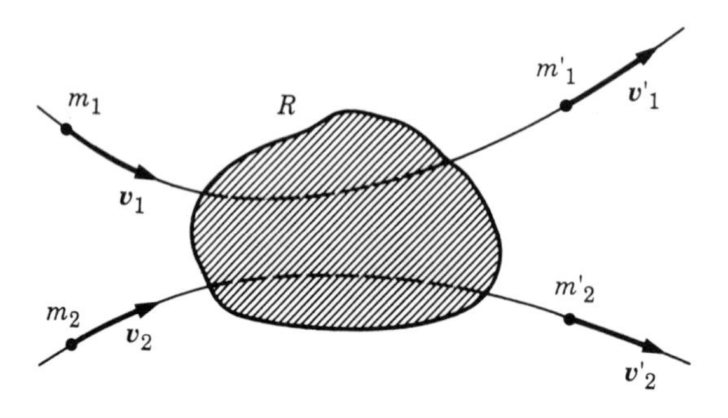

Figure 14.8 Conservation of energy and momentum in a collision. The collision region has been shaded.

has been prepared. For example, one particle may be a proton or an electron accelerated in an accelerator and the other particle may be an atom practically at rest in the laboratory. Then the final state is observed. That is, the motions of the two particles far away from the region where they collided are measured. If the forces between the particles are known, the final state may be calculated, so long as the initial state is well defined. Conversely, by comparing the final and initial states, it is possible to obtain information about the interaction. The analysis of such experiments thus provides valuable information about the interaction between the colliding particles. This is one of the reasons why collision experiments are so interesting to the physicist. On the other hand, collisions between two cars are more interesting to the police and insurance companies.

Since only internal forces enter into play in a collision, both the momentum and the total energy are conserved. Let $\boldsymbol{p}_1$ and $\boldsymbol{p}_2$ be the momenta of the particles relative to the L-frame before the collision and $\boldsymbol{p}_1'$ and $\boldsymbol{p}_2'$ be the momenta after the collision. The conservation of momentum requires that

$$\underbrace{\boldsymbol{p}_1' + \boldsymbol{p}_2'}_{\text{after}} = \underbrace{\boldsymbol{p}_1 + \boldsymbol{p}_2}_{\text{before}} \tag{14.22}$$

If we ignore any possible rotational motion, the kinetic energy of the particles relative to the L-frame before and after the collision are given by

$$E_k = \tfrac{1}{2}m_1 v_1^2 + \tfrac{1}{2}m_2 v_2^2 = \frac{p_1^2}{2m_1} + \frac{p_2^2}{2m_2} \qquad \text{(before)}$$

and **(14.23)**

$$E_k' = \tfrac{1}{2}m_1' v_1'^2 + \tfrac{1}{2}m_2' v_2'^2 = \frac{p_1'^2}{2m_1'} + \frac{p_2'^2}{2m_2'} \qquad \text{(after)}$$

The internal energy of the particles before the collision is U_{int}. After the collision, because there may be internal rearrangements, the internal energy may be different, with a value U_{int}'. Since no external forces act during the collision, the conservation of energy requires that

$$\underbrace{E_k' + U_{\text{int}}'}_{\text{after}} = \underbrace{E_k + U_{\text{int}}}_{\text{before}}$$

If the internal energy changes in the collision, the kinetic energy also changes. The Q of the reaction or collision, is defined by

$$Q = E_k' - E_k = (E_k)_{\text{after}} - (E_k)_{\text{before}} = \Delta E_k \tag{14.24}$$

Alternatively we may write

$$Q = U_{\text{int}} - U_{\text{int}}' = (U_{\text{int}})_{\text{before}} - (U_{\text{int}})_{\text{after}} = -\Delta U_{\text{int}}$$

When $Q = 0$, there is no change in kinetic energy or in internal potential energy and the collision is called **elastic**. Otherwise it is **inelastic**. When $Q < 0$, there is a decrease in kinetic energy with a corresponding increase in internal energy of the particles. We say that it is an *inelastic collision of the first kind* (or **endoergic**). When $Q > 0$, there is an increase of kinetic energy at the expense of the internal energy of the particles, which must decrease by the same amount. We have an *inelastic collision of the second kind* (or **exoergic**).

Combining Equations 14.23 and 14.24, we may write

$$\underbrace{\frac{p_1'^2}{2m_1'} + \frac{p_2'^2}{2m_2'}}_{E_k \text{ after}} = \underbrace{\frac{p_1^2}{2m_1} + \frac{p_2^2}{2m_2}}_{E_k \text{ before}} + Q \tag{14.25}$$

Equations 14.22 and 14.25 are sufficient to solve most collision problems.

Although we referred only to conservation of energy and momentum in a collision, it must be recalled that angular momentum is also conserved in collisions.

EXAMPLE 14.3

Calculation of Q in terms of the kinetic energy of the particles before and after they collide when one particle is initially at rest. This is the case in billiards and when a fast particle, such as a proton, is thrown against a target fixed in the laboratory.

▷ Assume that initially m_1 has a momentum $\boldsymbol{p}_1$ and that m_2 is at rest ($\boldsymbol{p}_2 = 0$) in the L-frame (see Figure 14.9). The masses of the particles after the collision are m_1' and m_2'.

The conservation of momentum gives

$$\boldsymbol{p}_1' + \boldsymbol{p}_2' = \boldsymbol{p}_1 \qquad \text{or} \qquad \boldsymbol{p}_2' = \boldsymbol{p}_1 - \boldsymbol{p}_1'$$

Squaring this equation,

$$\begin{aligned} p_2'^2 &= (\boldsymbol{p}_1 - \boldsymbol{p}_1')^2 = p_1^2 + p_1'^2 - 2\boldsymbol{p}_1 \cdot \boldsymbol{p}_1' \\ &= p_1^2 + p_1'^2 - 2p_1 p_1' \cos\theta \end{aligned}$$

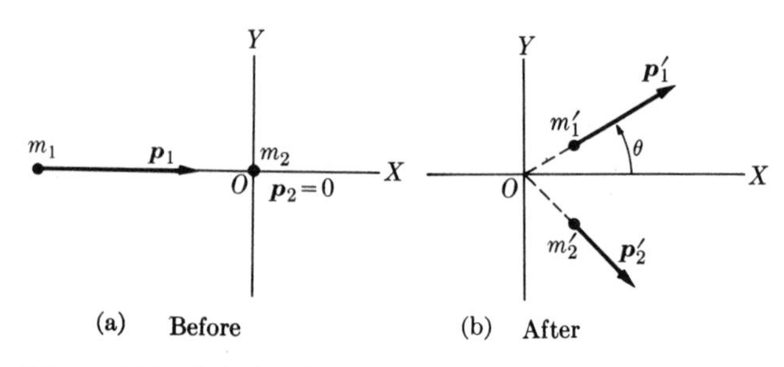

Figure 14.9 Relation between momenta relative to the L-frame before and after a collision when m_2 is initially at rest in the L-frame.

Using Equation 14.25, we have

$$Q = \frac{p_1'^2}{2m_1'} + \frac{p_2'^2}{2m_2'} - \frac{p_1^2}{2m_1} = \frac{p_1'^2}{2m_1'} + \frac{1}{2m_2'}(p_1^2 + p_1'^2 - 2p_1p_1' \cos\theta) - \frac{p_1^2}{2m_1}$$

or

$$Q = \frac{1}{2}\left(\frac{1}{m_1'} + \frac{1}{m_2'}\right)p_1'^2 + \frac{1}{2}\left(\frac{1}{m_2'} - \frac{1}{m_1}\right)p_1^2 - \frac{p_1p_1'}{m_2'}\cos\theta$$

Remembering that $E_k = p^2/2m$, we can express the above results as

$$Q = E_{k,1}'\left(1 + \frac{m_1'}{m_2'}\right) - E_{k,1}\left(1 - \frac{m_1}{m_2'}\right) - \frac{2(m_1m_1'E_{k,1}E_{k,1}')^{1/2}}{m_2'}\cos\theta \tag{14.26}$$

This result, known as the ***Q*-equation**, is frequently applied in analyzing nuclear collisions.

EXAMPLE 14.4

Elastic collisions of identical particles.

▷ Let one particle be initially at rest in the L-frame and assume the particles are identical so that $m_1 = m_1' = m_2 = m_2'$. When the collision is elastic ($Q = 0$) and particle 2 is initially at rest, the conservation of energy, Equation 14.25, gives

$$p_1'^2 + p_2'^2 = p_1^2$$

Squaring the equation $\boldsymbol{p}_1' + \boldsymbol{p}_2' = \boldsymbol{p}_1$ for conservation of momentum gives

$$p_1'^2 + p_2'^2 + 2\boldsymbol{p}_1'\cdot\boldsymbol{p}_2' = p_1^2$$

Comparing these results, we find that

$$\boldsymbol{p}_1'\cdot\boldsymbol{p}_2' = 0$$

This result implies that $\boldsymbol{p}_1'$ is perpendicular to $\boldsymbol{p}_2'$. Thus, in the L-frame, the two particles move at right angles after the collision. This may be seen in the photograph of Figure 14.10(a), which illustrates the collision of two identical sliding disks, one initially at rest. Figure 14.10(b) shows the collision of two He nuclei in a cloud chamber. The incoming He nucleus is an α-particle from a radioactive substance and the target He nucleus is from the gas in the chamber. In both cases, the two particles move at right angles after the collision. The condition $\boldsymbol{p}_1'\cdot\boldsymbol{p}_2' = 0$ is satisfied also if $\boldsymbol{p}_1' = 0$, in which case the conservation of momentum gives $\boldsymbol{p}_2' = \boldsymbol{p}_1$. Then particle 1 is at rest in the laboratory after the collision and its momentum is transferred to particle 2. This is observed in head-on collisions of billiard balls.

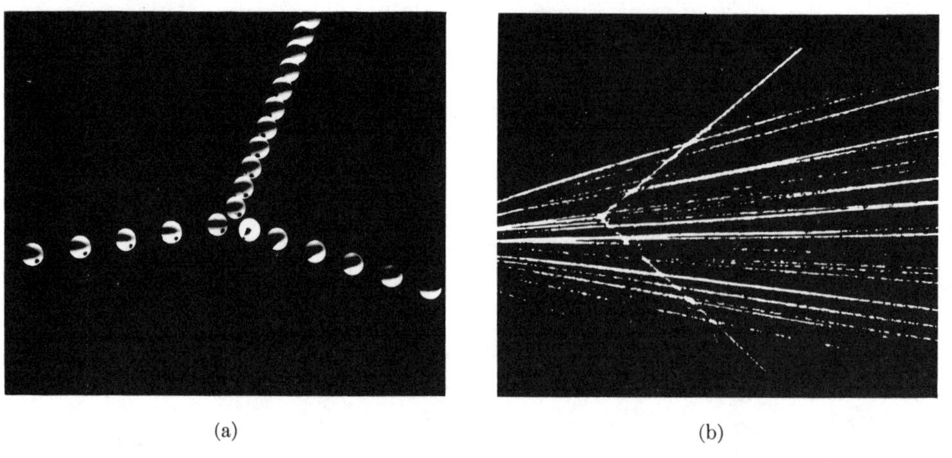

(a) (b)

Figure 14.10 (a) Collision of two identical sliding disks. (b) Collision between α-particles (He nuclei). In both cases, one of the particles was initially at rest in the L-frame and their momenta make angles of 90° in the L-frame after the collision. (Part (a) courtesy of Educational Services, Inc.)

EXAMPLE 14.5

Energies of the fragments of a body, initially at rest in the L-frame, which explodes into two fragments of masses m_1 and m_2.

▷ If the body is initially at rest, its total momentum is zero. After the explosion the two fragments separate in opposite directions with momenta $\boldsymbol{p}'_1$ and $\boldsymbol{p}'_2$ such that $\boldsymbol{p}'_1 + \boldsymbol{p}'_2 = 0$, or in magnitude, $p'_1 = p'_2$. Then, from Equation 14.25, with

$$E'_k = \frac{p'^2_1}{2m_1} + \frac{p'^2_2}{2m_2} = \frac{1}{2}\left(\frac{1}{m_1} + \frac{1}{m_2}\right)p'^2_1$$

and $E_k = 0$, we have

$$\frac{1}{2}\left(\frac{1}{m_1} + \frac{1}{m_2}\right)p'^2_1 = Q$$

We may rearrange this equation to read

$$p'_1 = \left(\frac{2m_1 m_2}{m_1 + m_2}\,Q\right)^{1/2} = (2\mu_{12}Q)^{1/2}$$

where μ_{12} is the reduced mass of the system. The kinetic energies of the fragments are (remember $p'_1 = p'_2$)

$$E'_{k,1} = \frac{p'^2_1}{2m_1} = \frac{m_2 Q}{m_1 + m_2} \quad \text{and} \quad E'_{k,2} = \frac{p'^2_2}{2m_2} = \frac{m_1 Q}{m_1 + m_2} \tag{14.27}$$

Note that the kinetic energies of the fragments are inversely proportional to their masses. This analysis applies equally well to the recoil of a firearm (remember Example 6.1), to the

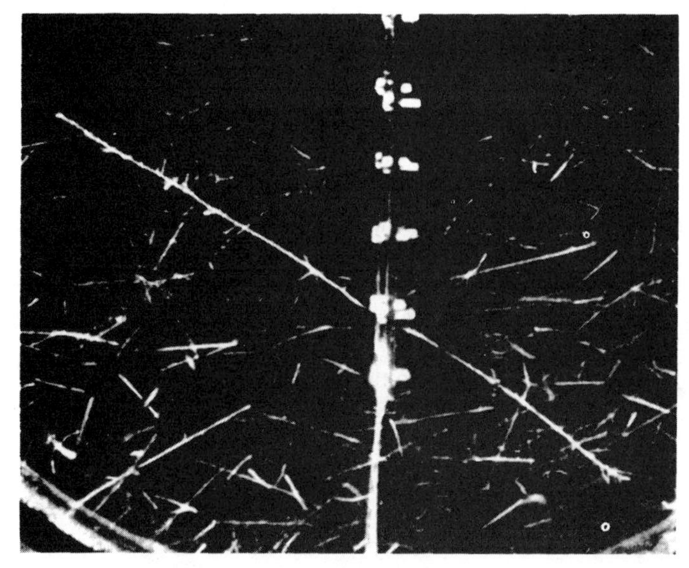

Figure 14.11 Cloud chamber photograph of the tracks of the two fragments resulting from the fission of a U nucleus (Bøggild, Brostrøm, and Lauritzen, *Phys. Rev.* **59**, 275 (1941)). Initially the U nucleus was at rest in the thin horizontal metal plate at the center of the photo. The two fragments move in opposite directions. From the analysis of the paths we can estimate the energies of the fragments which in turn (using the relation derived in Example 14.5) allow us to obtain the ratio of their masses.

fission of a nucleus into two fragments illustrated in Figure 14.11, to the dissociation of a diatomic molecule or to the disintegration of a nucleus.

If there are three fragments instead of two, several solutions are possible. There are three momenta involved, but only two physical conditions: conservation of energy and of momentum. For example, if two particles are observed in a reaction and the energy and momentum are not conserved, the physicist suspects the presence of a third, unobserved, particle. The particle may have been unobserved because it has no electric charge, or for some other reason. A missing momentum and energy are assigned to this hypothetical particle, in an effort to make the final total energy and momentum conform to the conservation laws, from which the mass of the particle can be guessed. Such an experiment led, for example, to the discovery of the neutron (see Example 14.7) and the neutrino (see Section 40.4). This procedure has so far always given results that are consistent with both theory and experiment.

EXAMPLE 14.6

Energy loss of a particle of mass m_1 (say a neutron) when it collides elastically ($Q = 0$) with a particle of mass m_2 (say a nucleus) at rest in the L-frame.

▷ This problem is of interest in the analysis of the slowing down of neutrons as they move through matter. The maximum energy loss occurs in a head-on collision, and this is the only case we consider. The conservation of momentum requires that (see Figure 14.12)

$$p_1' + p_2' = p_1 \tag{14.28}$$

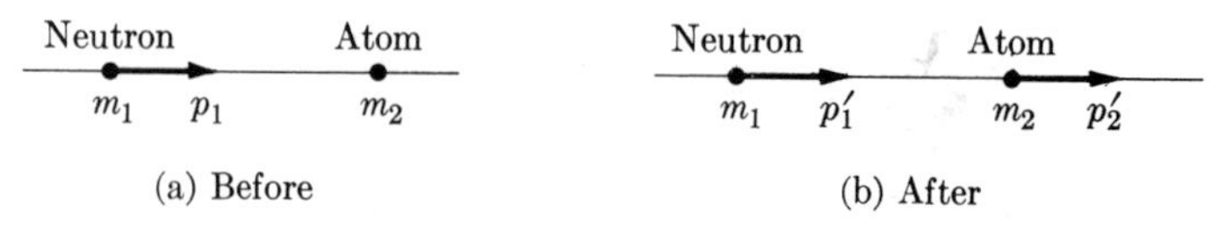

(a) Before (b) After

Figure 14.12 Collisions of a neutron with an atom.

and the conservation of energy, Equation 14.25, with $Q = 0$, may be written as

$$\frac{1}{2m_1} p_1'^2 + \frac{1}{2m_2} p_2'^2 = \frac{1}{2m_1} p_1^2$$

These two relations can be rewritten in the forms

$$p_1 - p_1' = p_2'$$

and

$$(p_1^2 - p_1'^2)/2m_1 = p_2'^2/2m^2$$

from which we obtain

$$p_2' = \left(\frac{m_2}{m_1}\right)(p_1 + p_1')$$

Combining this relation with Equation 14.28, we obtain

$$p_1' = \frac{m_1 - m_2}{m_1 + m_2} p_1 \quad \text{and} \quad p_2' = \frac{2m_2}{m_1 + m_2} p_1 \tag{14.29}$$

The kinetic energy of the neutron after the collision is

$$E_k' = \frac{1}{2m_1} p_1'^2 = \left(\frac{m_1 - m_2}{m_1 + m_2}\right)^2 E_k \tag{14.30}$$

where $E_k = p_1^2/2m_1$ is the kinetic energy of the neutron before the collision.

The energy E_k' becomes smaller the closer m_1 is to m_2 or the closer the ratio m_2/m_1 is to unity. This result is important when it comes to choosing the material to slow neutrons down quickly, as must be done in nuclear reactors. Atoms with the smallest value of m_2/m_1 relative to the neutrons are those of hydrogen ($m_2/m_1 \approx 1$). For that reason pure hydrogen would be the best moderator. However, even at room temperature, hydrogen is a gas, so that the number of hydrogen atoms per unit volume is relatively small. Therefore water is used instead. Water not only has the advantage of being abundant and inexpensive; in addition, it contains about 10^3 times more hydrogen atoms than hydrogen gas, per unit volume, at standard temperature and pressure. Unfortunately, hydrogen atoms tend to capture neutrons to form deuterium. On the other hand, deuterium atoms have a relatively small tendency to capture neutrons. Therefore, some nuclear reactors use **heavy water**, whose molecules are formed of deuterium (instead of hydrogen) and oxygen, as their moderator. In this case $m_2/m_1 = 2$ and $E_k' = E_k/9$. Another common moderator is carbon, used in the form of graphite; then $m_2/m_1 = 12$ and $E_k' = (11/13)^2 E_k$ (see Section 40.7).

EXAMPLE 14.7

The mass of the neutron.

▷ When beryllium is bombarded with α-particles (or helium nuclei, ${}^4_2He$), neutrons are produced, according to the process

$$ {}^9_4Be + {}^4_2He \rightarrow {}^{12}_6C + {}^1_0n $$

When the neutrons, in turn, pass through a material rich in hydrogen, protons are knocked out of the material. Also, nitrogen atoms are knocked out when neutrons pass through a substance rich in nitrogen. Applying the second relation of Equation 14.29 to hydrogen and nitrogen, where $p_1 = m_1 v_1$ refers to the neutron and p'_2 is replaced by $p'_H = m_H v_H$ and $p'_N = m_N v_N$, which correspond to the hydrogen and nitrogen nuclei, we obtain

$$ v'_H = \frac{2m_1}{m_1 + m_H} v_1 \quad \text{and} \quad v'_N = \frac{2m_1}{m_1 + m_N} v_1 $$

where v_1 is the same in both cases, since this is the velocity of the incoming neutron. From these two equations we can write

$$ \frac{v'_H}{v'_N} = \frac{m_1 + m_N}{m_1 + m_H} $$

This relation allows us to compute m_1 in terms of known and measurable quantities. In the original experiment, performed in 1932 by James Chadwick (1891–1974), the maximum velocities of the protons and nitrogen nuclei were estimated to be 3.3×10^7 m s^{-1} and 4.7×10^6 m s^{-1}, respectively. These results gave an estimated mass for the neutron of about 1.16 times the mass of the proton. The most recent measurements have yielded a value of 1.001 times the mass of the proton.

14.10 Fluid motion

The general principles we have discussed for many-particle systems can be applied to the study of fluids (i.e. liquids or gases) in motion. Consider a small volume of fluid, which we shall call a **volume element**. The volume element can be considered as a system of particles, since it contains many molecules. The velocity of the center of mass of the volume element is called the **fluid velocity** at the place occupied by the element.

The motion of a fluid is said to be **stationary** when the motion pattern does not change with time. This means that, although the velocity of a fluid element may change when the fluid element changes position, the velocity of the fluid at each point of space remains the same over time. To be more precise, let us follow a particular fluid element along its path of motion (Figure 14.13). When the element is at A its velocity is $\boldsymbol{v}$ and when it is at A' its velocity is $\boldsymbol{v}'$. But if the motion is stationary, *all* fluid elements have velocity $\boldsymbol{v}$ when they pass through A, and velocity $\boldsymbol{v}'$ when they pass through A'. Thus the velocity of the fluid may be considered as a function of position instead of a

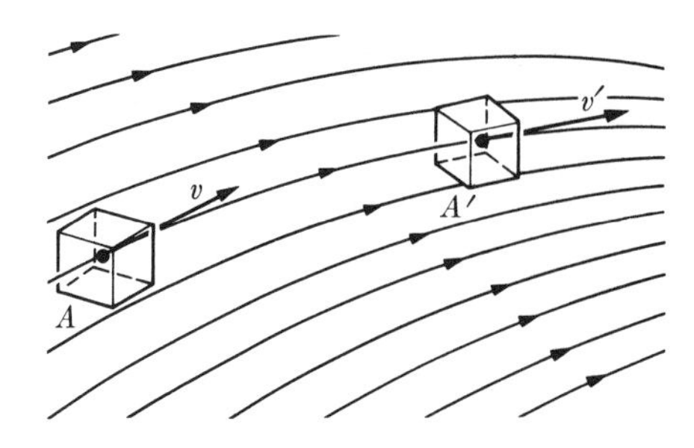

Figure 14.13 Stationary fluid flow. The lines shown are called **streamlines**.

function of time. When the motion is not stationary, the velocity at each position may change with time.

To discuss fluid motion it is useful to introduce the concept of **pressure**, defined as the normal force per unit area exerted on a surface (recall Example 13.5). If F is the normal force on an area A, the pressure is

$$p = \frac{F}{A} \tag{14.31}$$

Pressure must be expressed as a unit of force divided by a unit of area. Thus in SI units pressure is measured in *newtons per square meter*, or $\mathrm{N\,m^{-2}}$. To honor Blaise Pascal (1623–1662), this unit of pressure is called the **pascal** (abbreviated to Pa); i.e. $1\ \mathrm{N\,m^{-2}} = 1\ \mathrm{Pa}$. Another unit of pressure in common usage is the **atmosphere**, abbreviated atm and defined according to the equivalence

$$1\ \mathrm{atm} = 1.01325 \times 10^5\ \mathrm{Pa} = 101.325\ \mathrm{kPa}$$

One atmosphere is approximately the normal pressure exerted by the Earth's atmosphere on bodies at sea level (see Example 14.10).

The motion of a volume element is determined by the pressure produced by the rest of the fluid as well as the action of gravity and other forces (electric, magnetic etc.) applied externally. We first determine the resultant external force on the volume of fluid due to the pressure. Consider, for simplicity, a fluid moving along a cylinderical pipe of variable cross-section A (Figure 14.14). The X-axis is made coincident with the axis of the pipe. We shall concentrate on a volume element of thickness $\mathrm{d}x$ and volume $A\,\mathrm{d}x$.

Let p and p' be the values of the pressure at the left and the right of the volume element. Since pressure is force per unit area, the fluid at the left produces a force $F = pA$ on the volume element, directed toward the right, and the fluid at the right produces a force $F' = p'A$, directed toward the left. Thus the X-component

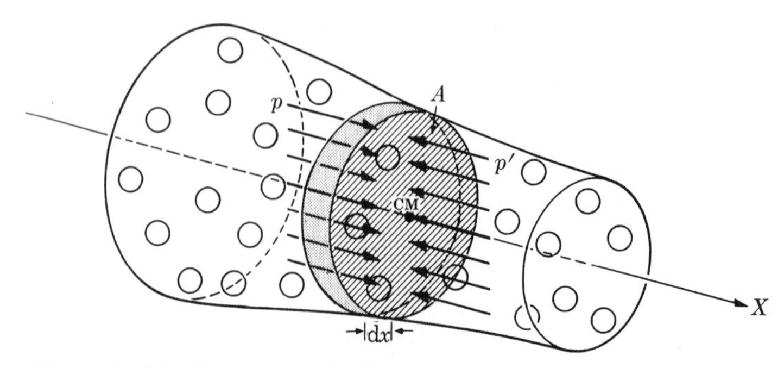

Figure 14.14 Forces on a fluid element due to the pressure.

of the resultant external force in the X-direction on the volume element due to pressure is

$$dF = F - F' = pA - p'A = -(p' - p)A$$

But $p' - p$ is the pressure difference between two points separated a distance dx. Therefore $p' - p = dp$. Thus

$$dF = -(dp)A = -\frac{dp}{dx}(A\,dx)$$

Since $A\,dx$ is the volume of the fluid, we conclude that the *force per unit volume* along the X-axis on the fluid due to pressure is

$$f = -\frac{dp}{dx} \tag{14.32}$$

which shows that the force acts in the direction in which the pressure decreases. Result 14.32, when compared with Equation 9.27, $F = -dE_p/dx$, suggests that *pressure may be considered as potential energy per unit volume.* This is dimensionally correct, since p is expressed in N m^{-2}, which is the same as (N m)m^{-3} or J m^{-3}.

In addition to the pressure, there may be other external forces (such as gravity or an external electric or magnetic field) acting on the fluid inside the volume element. Let E_p be the corresponding potential energy *per unit volume* due to other external forces. Then the total potential energy per unit volume due to pressure and other external forces is $p + E_p$.

We assume that the fluid is **incompressible**; that is, that the volume of the volume element does not change during the motion. For an incompressible fluid the internal energy does not change and we may ignore it when we consider the changes of energy of the volume element. Then the conservation of energy, applied to an incompressible fluid, states that

> *if all the forces acting on an incompressible fluid are conservative and we follow the motion of a small volume of the fluid, we find that the total energy per unit volume remains constant.*

Hence, using Equation 14.7, we may write

$$\tfrac{1}{2}\rho v^2 + p + E_p = const. \tag{14.33}$$

where ρ is the density of the fluid. This result, known as **Bernoulli's theorem**, expresses the conservation of energy in an incompressible fluid. The term $\frac{1}{2}\rho v^2$ is the kinetic energy per unit volume. The second term is interpreted as its potential energy per unit volume associated with the pressure. The third term is the potential energy per unit volume due to all other external forces.

In the particular case that the external force acting on the fluid is gravity, we have from Equation 9.23 that the potential energy per unit volume is $E_p = \rho g y$

and Equation 14.33 becomes

$$\tfrac{1}{2}\rho v^2 + p + \rho g y = const. \tag{14.34}$$

When the fluid is compressible we must take into account the internal potential energy because it might change during the motion. Also in this case the density does not remain constant.

EXAMPLE 14.8

The equation of continuity.

▷ One relation that is very important in discussing fluid motion is the **equation of continuity**, which expresses the conservation of mass of the fluid. Consider a fluid moving under steady conditions inside the pipe shown in Figure 14.15. Mass is not being added or lost at any point, so that mass is being conserved. Let A_1 and A_2 be two sections of the pipe. The volume of fluid that passes through A_1, per unit time, corresponds to a cylinder of base A_1 and length v_1. Thus the mass of fluid that passes through A_1 per unit time is $\rho_1 A_1 v_1$. Similarly, $\rho_2 A_2 v_2$ is the mass of the fluid that passes through A_2 per unit time. The conservation of mass, under the conditions stated, requires that

$$\rho_1 A_1 v_1 = \rho_2 A_2 v_2 \tag{14.35}$$

This is the equation of continuity for a fluid when the mass is conserved. If the fluid is incompressible, the density remains the same and Equation 14.35 reduces to

$$A_1 v_1 = A_2 v_2 \tag{14.36}$$

indicating that the velocity of an incompressible fluid is inversely proportional to the cross-section of the tube.

EXAMPLE 14.9

Motion in a horizontal pipe.

▷ When an incompressible fluid moves in the horizontal direction only, the term $\rho g y$ remains constant and Equation 14.30 reduces to

$$\tfrac{1}{2}\rho v^2 + p = const. \tag{14.37}$$

Thus, in a horizontal pipe, the greater the velocity, the lower the pressure, and conversely. Although air is not an incompressible fluid, this effect is used to produce the **lift** of an airplane (Figure 14.16). The profile of the wing is so designed that the air has a greater velocity above the wing surface than below it. Therefore, there is a larger pressure below than above. This results in a net resultant upward force. If A is the area of the wing, the upward force, or lift, is

$$F = A(p_2 - p_1) = \tfrac{1}{2}A\rho(v_1^2 - v_2^2)$$

where the subscripts 1 and 2 refer to the conditions above and below the wing, respectively.

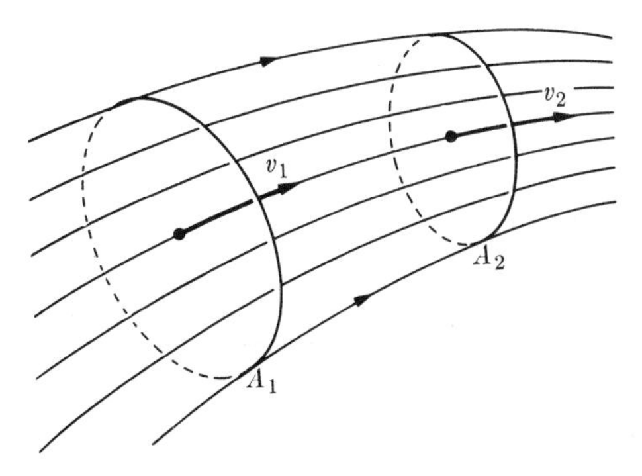

Figure 14.15

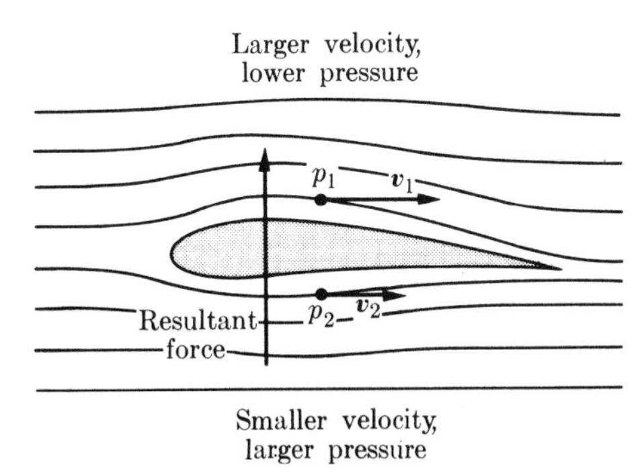

Figure 14.16 Lift on an airplane wing.

EXAMPLE 14.10

Motion of a fluid in a pipe of constant cross-section.

▷ In this case, according to the equation of continuity, the incompressible fluid moves with constant velocity along the pipe. Under such circumstances, the term $\frac{1}{2}\rho v^2$ may be dropped from Equation 14.33, which then reduces to

$$p + \rho g y = const.$$

Designating the constant by p_0, which is the pressure at $y = 0$, we have that the pressure in an incompressible fluid in steady motion with constant velocity is given by

$$p = p_0 - \rho g y \qquad \textbf{(14.38)}$$

and the difference in pressure between two points separated by a vertical distance y is $p_0 - p = \rho g y$. This result holds also for a fluid in equilibrium or at rest.

Therefore we can use Equation 14.38 to find the pressure difference between two points separated by a vertical distance y in a fluid at rest. It is the basis for the **barometer**, an instrument used to measure atmospheric pressure by balancing a column of mercury of length y. The **standard atmospheric pressure**, also called one atmosphere, corresponds to a column of mercury 76 cm long. Since the density of mercury is 13.6×10^3 kg m^{-3}, we have

$$1 \text{ atm} = (13.6 \times 10^3 \text{ kg m}^{-3})(9.8 \text{ m s}^{-2})(0.76 \text{ m}) = 1.01325 \times 10^5 \text{ Pa}$$

which is the value previously quoted. In that sense, pressure may be given in terms of so many centimeters of mercury, with 1 atm equal to 76 cm of Hg.

EXAMPLE 14.11

A method for determining the velocity of a fluid in a pipe: the Venturi meter (Figure 14.17).

▷ Two pressure gauges G_1 and G_2 measure the pressure in the pipe. One is placed in the pipe proper and the other at a constriction inserted in it. To obtain the expression for the velocity v_1, we note that if v_1 and v_2 are the velocities at A_1 and A_2, respectively, the equation of

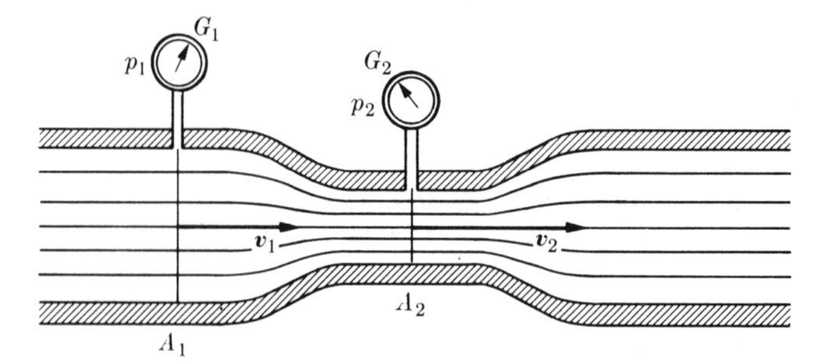

Figure 14.17 Venturi meter used for measuring the velocity of a fluid.

continuity gives

$$A_1 v_1 = A_2 v_2 \qquad \text{or} \qquad v_2 = \frac{A_1}{A_2} v_1$$

Then, if the pipe is horizontal, Bernoulli's theorem, given by Equation 14.37, allows us to write

$$\tfrac{1}{2}\rho v_1^2 + p_1 = \tfrac{1}{2}\rho v_2^2 + p_2$$

Inserting the value of v_2 just obtained above and solving for v_1, we finally obtain

$$v_1 = A_2 \left(\frac{2(p_1 - p_2)}{\rho(A_1^2 - A_2^2)} \right)^{1/2}$$

The amount of fluid passing through any section of the pipe per unit time is

$$V = A_1 v_1 = A_1 A_2 \left(\frac{2(p_1 - p_2)}{\rho(A_1^2 - A_2^2)} \right)^{1/2} = K[(p_1 - p_2)]^{1/2}$$

where K is a constant depending on the size of the pipe and the density of the fluid. Thus, by measuring p_1 and p_2 and calculating K, both the velocity of the fluid v and the volume per unit time are found.

Note 14.1 Invariance, symmetry and the conservation laws

The conservation laws for systems of particles that we have discussed (momentum, angular momentum and energy) are of universal validity and can be traced back to certain invariance and symmetry properties of physical systems. We shall now briefly consider some of these properties without entering into the mathematical details.

Space translation. We may assume that an isolated physical system placed in an otherwise empty space will behave the same no matter where it is located. Therefore, the properties of an isolated physical system are invariant with respect to a translation of the system relative to a frame of reference attached to the observer. For example, an isolated molecule, composed of several electrons and nuclei, must be described in exactly identical terms no matter where it is located relative to an observer. On the other hand, we know that the total momentum of an isolated system is constant. Thus it can be shown that

the conservation of momentum of an isolated system is a result of translational invariance of the laws describing the system.

In some cases the system may not be isolated, but the physical environment may exhibit a certain translational symmetry. For example, we find that the conservation of momentum in a given direction is a consequence of translational invariance of the physical conditions in that direction. Consider, for example, a particle moving over an extended (infinite) horizontal surface. The physical conditions of the particle do not change if the particle is displaced over the plane, since gravity acts only in the vertical direction. But we know that if the particle is set in motion by giving it a push, its momentum *parallel* to the plane remains constant. Similarly, consider an electron placed between two infinite parallel planes carrying equal but opposite charges. The physical conditions do not change if the electron is displaced parallel to the planes. We know then that when the electron is set in motion, its momentum parallel to the plane is constant. (This is usually stated by saying that the electric field produced by the charged planes is perpendicular to the planes.)

Space rotation. We may assume that the description of the properties of an isolated system are independent of its orientation in space relative to a given frame of reference. Consider an isolated molecule: the description of its properties is independent of the orientation of the molecule relative to the observer, since no external forces act on the molecule. On the other hand, we know that the total angular momentum of an isolated system is constant. It can thus be shown that

the conservation of angular momentum of an isolated system is a result of rotational invariance of the laws describing the system.

In some cases, even though a system may not be isolated, the physical system may exhibit a certain rotational symmetry. For example, the conservation of angular momentum about a given direction is a consequence of rotational symmetry of the physical conditions relative to that direction. A central field of force has spherical symmetry. We also know that the angular momentum (relative to the center of force) of a particle moving under central forces is constant and it can be shown that it is a result of the symmetry. If the field has cylindrical symmetry, as in the case of an electron in a diatomic molecule or a charged particle in a uniform magnetic field, then the component of the angular momentum relative to the symmetry axis remains constant.

Time translation. Invariance with respect to time translation means that if we prepare a physical system and leave it to evolve without external interference, the evolution of the system will be the same independent of the time at which it was prepared. In other words, the origin of time is not important. With a more elaborate logic, it can be shown that

the conservation of energy of an isolated system is a consequence of the invariance of the laws describing the system relative to the chosen origin of time;

that is, relative to a translation of time.

The correlation of the three conservation principles with the properties of space and time is a subject that is not yet fully understood. Later on we will identify other conservation laws that are consequences of more subtle symmetries in a physical system (Chapter 41).

QUESTIONS

14.1 Write the internal potential energy in the case of the system composed of (a) Earth and Moon, (b) Sun, Earth and Moon, (c) the two electrons and the nucleus in a helium atom.

14.2 Why does the internal potential energy of a system of n particles have $\frac{1}{2}n(n-1)$ terms?

14.3 Illustrate, with some examples, the principle of conservation of energy for a system of particles.

14.4 Why is the internal potential energy of a system of particles independent of the frame of reference (or of the observer)? Is the same true for the kinetic energy of the system?

14.5 What is the difference between proper energy and internal energy of a system?

14.6 Verify that the orbital kinetic energy can also be written as $E_{k,orb} = \boldsymbol{P}^2/2M$, where $\boldsymbol{P}$ is the total momentum of the system relative to the observer.

14.7 Referring to Example 14.2, compare the velocity of the sphere, cylinder, and ring at the bottom of the plane. Why are the velocities different?

14.8 Why can the internal energy of a rigid body be ignored when discussing its motion under external forces?

14.9 Analyze the relation between the binding energy of a system of particles and the stability of the system.

14.10 Are the expressions $E_r = \frac{1}{2}I\omega^2$ and $E_r = L^2/2I$ for the kinetic energy of rotation of a rigid body equivalent?

14.11 How do the rotational and vibrational energies of a molecule compare?

14.12 What quantities are conserved in elastic collisions? In inelastic collisions?

14.13 Which requires more energy: the separation of a hydrogen molecule into two hydrogen atoms or into two protons and two electrons?

14.14 Verify that the kinetic energy after the collision of the particle considered in Example 14.6 can be expressed in the form

$$E'_k = \left(\frac{(m_2/m_1) - 1}{(m_2/m_1) + 1}\right)^2 E_k$$

Under what conditions is E'_k larger or smaller than E_k? Discuss the variation of E'_k as a function of m_2/m_1.

14.15 What factors determine the value of Q in a collision?

14.16 Why is $\boldsymbol{p}'_2 = 0$ not another possibility in Example 14.4?

14.17 Discuss the conservation of energy per unit volume in the case of stationary motion of an incompressible fluid.

14.18 What conservation law is expressed by the equation of continuity?

14.19 Investigate how fluid pressure gauges work (look, for example, at an encyclopedia).

PROBLEMS

14.1 Verify that if the internal kinetic energy of a system of two particles is $E_{k,int}$, the magnitudes of the velocities of the particles relative to the CM are:

$$v_1 = \left(\frac{2m_2 E_{k,int}}{m_1(m_1 + m_2)}\right)^{1/2}$$

and

$$v_2 = \left(\frac{2m_1 E_{k,int}}{m_2(m_1 + m_2)}\right)^{1/2}$$

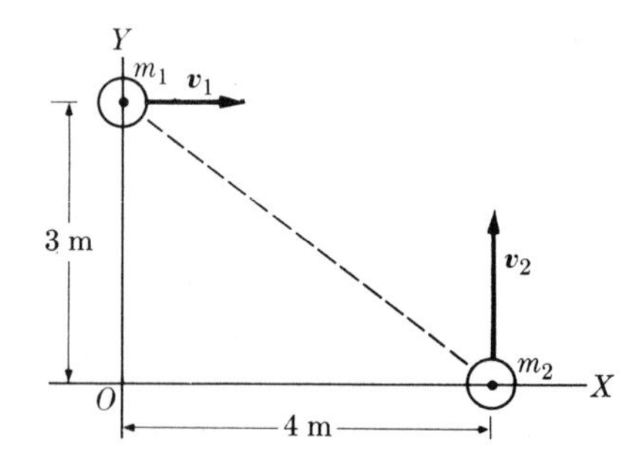

Figure 14.18

14.2 If, for the two particles in Figure 14.18, we know that $m_1 = 4$ kg, $m_2 = 6$ kg, $\boldsymbol{v}_1 = \boldsymbol{i}(2)$ m s^{-1} and $\boldsymbol{v}_2 = \boldsymbol{j}(3)$ m s^{-1}, (a) determine the total angular momentum of the system relative to O and relative to the CM and verify the relation between them. (b) Determine the total kinetic energy relative to O and relative to the CM and verify the relation between them.

14.3 Assume that the two particles of the preceding problem are joined by an elastic spring, whose constant is 2×10^{-3} N m^{-1} and is initially unstretched. (a) How will this affect the motion of the CM of the system? (b) What is the total internal energy of the system? (c) After a certain time, the spring is compressed by 0.4 m. Find the internal kinetic and potential energies of the particles. (d) Determine the magnitudes of the velocities relative to the CM (can you also determine their directions?). Also determine (e) the magnitude of their relative velocity, and (f) the angular momentum of the system relative to O and to the CM.

14.4 The arrangement in Figure 14.19 is called a **ballistic pendulum**. It is used to determine the velocity of a bullet by measuring the height h the block rises after the bullet is embedded in it. Verify that the velocity of the bullet is given by $v = (2gh)^{1/2}(m_1 + m_2)/m_1$, where m_1 is the mass of the bullet and m_2 the mass of the block.

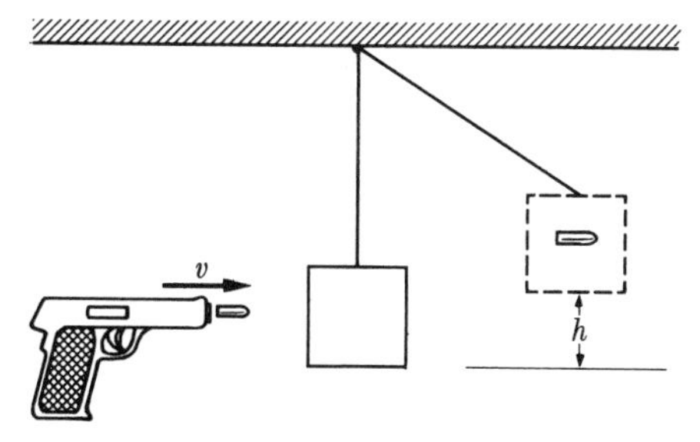

Figure 14.19

14.5 A bullet of mass m and velocity v passes through a pendulum bob of mass M and emerges with velocity $\frac{1}{2}v$ (Figure 14.20). The pendulum bob is at the end of a string of length l. Calculate the minimum value of v such that the pendulum bob will swing through a complete circle.

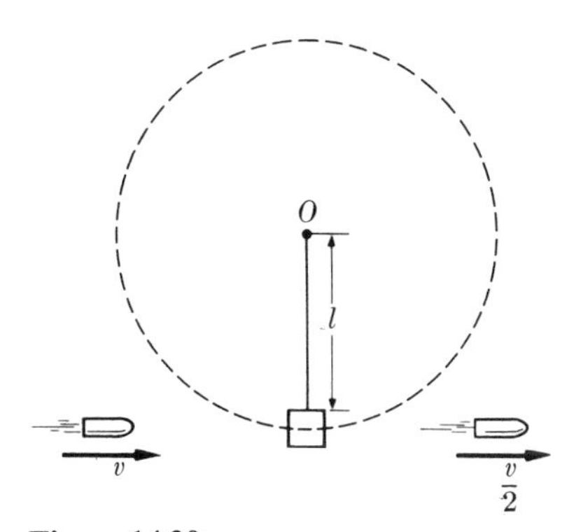

Figure 14.20

14.6 Consider a system composed of N identical particles, each of mass m (such as occurs in a gas). (a) Show that the kinetic energy of the system relative to an observer who sees the center of mass moving with velocity v_{CM} is equal to the kinetic energy of the particles relative to the C-frame of reference plus $\frac{1}{2}Nmv_{CM}^2$. (b) What is the relation among the average kinetic energy per particle between the C- and L-frames?

14.7 Two carts, each having a mass of 0.8 kg, are connected by a light cord C. A spring S is placed between them as shown in Figure 14.21. The cord prevents the spring S from expanding. Neglect the mass of the spring and all frictional effects.
(a) The carts are initially at rest. The cord C breaks, the spring S expands until it loses contact with the carts, and the carts move apart. If cart B has a speed of 0.3 m s^{-1}, find the speed of cart A. What is the Q of the process? How is it related to the potential energy of the spring? (b) The experiment is repeated with the sole exception that before cord C breaks, the carts have an initial velocity of 0.5 m s^{-1} to the right. Find the velocity of cart A and the velocity of cart B relative to the track after their separation.

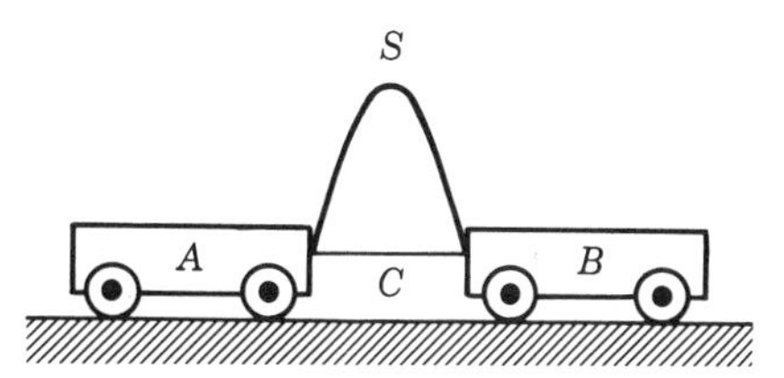

Figure 14.21

14.8 A particle of mass 5 kg, moving at 2 m s^{-1}, collides with a particle of mass 8 kg initially at rest. If the collision is elastic, find the velocity of each particle after the collision (a) if the collision is head-on, (b) if the first particle is deflected 50° from its original direction of motion. Express all directions relative to the direction of the incoming particle.

14.9 It is found experimentally that in the head-on collision of two solid spheres, such as two billiard balls, the velocities after the collision are related to those before by the expression $v_1' - v_2' = -e(v_1 - v_2)$ where e, called the **coefficient of restitution**, has a value between zero and one and depends on the materials of the balls. This result was discovered by Newton and has only approximate validity. Since momentum is conserved in the collision, verify that (a) the velocities after the collision are given by

$$v_1' = \frac{v_1(m_1 - m_2 e) + v_2 m_2(1 + e)}{m_1 + m_2}$$

and

$$v_2' = \frac{v_1 m_1(1 + e) + v_2(m_2 - m_1 e)}{m_1 + m_2}$$

(b) Verify also that the Q of the collision is

$$-\tfrac{1}{2}(1 - e^2)\frac{m_1 m_2}{m_1 + m_2}(v_1 - v_2)^2$$

(c) What value of e corresponds to an elastic collision?

14.10 In a **plastic collision** the two bodies move as one after the collision. (a) What is the value of the coefficient of restitution e? (b) Compute the Q of the reaction directly, and also by using the results of Problem 14.9 with the appropriate value of e.

14.11 If the masses of balls m_1 and m_2 in Figure 14.22 are 0.1 kg and 0.2 kg, respectively, and if m_1 is released when $d = 0.2$ m, find the heights to which they will return after colliding if the collision is (a) elastic ($e = 1$), (b) inelastic with a coefficient of restitution equal to 0.9, (c) plastic ($e = 0$).

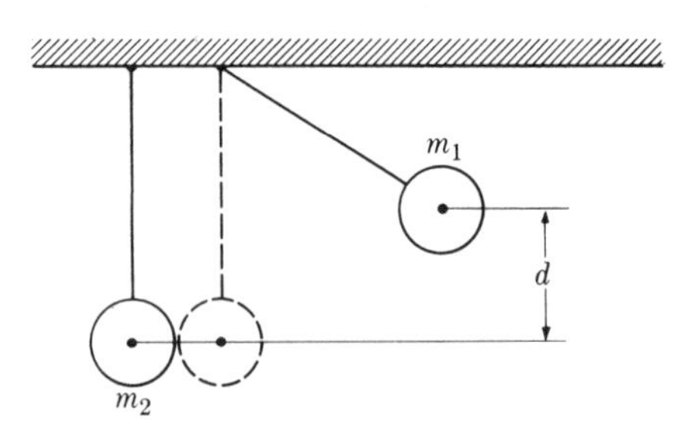

Figure 14.22

14.12 (a) Assuming that the second body in Problem 14.9 is at rest and that its mass is very large compared to that of the first, find the velocity of each body after the collision, and also find the value of Q. (b) Apply this result to determine how high a body, dropped from a height h, rebounds after hitting the floor. Do the experiment yourself with a marble and estimate from it the corresponding value of e. (c) Verify that the time required by the body to stop rebounding is

$$t = \left(\frac{2h}{g}\right)^{1/2}\left(\frac{1 + e}{1 - e}\right)$$

14.13 (a) Verify that if energy and momentum are conserved in an elastic collision, then $\boldsymbol{u}\cdot(\boldsymbol{v}_1' - \boldsymbol{v}_2') = -\boldsymbol{u}\cdot(\boldsymbol{v}_1 - \boldsymbol{v}_2)$, where $\boldsymbol{u}$ is a unit vector in the direction in which the momentum of either particle has changed. This result means that in the collision the component of the relative velocity along the direction of momentum exchange is reversed. (b) Apply this to the case of a head-on collision. (c) Compare this with the results of Problem 14.9 with $e = 1$. (Hint: write the two conservation laws, with all terms for each particle on each side of each equation.)

14.14 A grenade of mass M is falling with a velocity v_0, at height h, when it explodes into two equal fragments that initially move horizontally in the C-frame. The explosion has a Q value of Mv_0^2. Determine where the fragments will fall on the ground relative to the point directly below the grenade at the time of the explosion.

14.15 A ball, having a mass of 4 kg and a velocity of 1.2 m s^{-1}, collides head-on with another ball, of mass 5 kg, moving at 0.6 m s^{-1} in the same direction. Find (a) the velocities of the balls after the collision (assuming that it is elastic) and (b) the change in momentum of each ball.

14.16 A particle of mass m_1 moves with velocity $\boldsymbol{v}_1$ in the L-frame and collides with a particle of mass m_2 at rest in the L-frame. If, after the collision, the direction of the first particle makes angles θ and ϕ (relative to the L- and C-frames, respectively) with respect to its initial direction, verify that these two angles are related by:

$$\tan\theta = \frac{\sin\phi}{\cos\phi + 1/A}$$

where $A = m_2/m_1$.

14.17 For the particles of the previous problem, verify that if $m_1 = m_2$, then $\theta = \frac{1}{2}\phi$. What then is the maximum value of θ?

14.18 Referring to Problem 14.16, show that the maximum value of θ for arbitrary A when $A < 1$ is given by $\tan\theta = A/(1 - A^2)^{1/2}$. Discuss the situation when A is larger than one.

14.19 In analyzing the deflection of α-particles moving through hydrogen, physicists have found experimentally that the maximum deflection of an α-particle in the L-frame is about 15°. Using the results of Problem 14.18, estimate the mass of the α-particle relative to hydrogen. Compare your answer with the actual value obtained from Table A-1.

14.20 A particle having a mass of 0.2 kg while moving at $0.40\ \mathrm{m\,s^{-1}}$ collides with another particle of mass 0.3 kg, which is at rest. After the collision the first particle moves at $0.20\ \mathrm{m\,s^{-1}}$ in a direction making an angle of 40° with its initial direction. Find (a) the velocity of the second particle and (b) the Q of the process.

14.21 A particle of mass m moves with a velocity v and collides elastically and head-on with another particle of mass M. Compute in each case the velocity of the first particle after the collision if the second particle initially has (a) an equal but opposite momentum, (b) the same kinetic energy, but moving in the opposite direction. (c) Show that if M is initially at rest and much larger than m, the relative change in kinetic energy of m is $\Delta E_k/E_k \approx -(m/M)$.

14.22 A neutron, having an energy of 1 MeV, moves through (a) deuterium and (b) carbon. Estimate for each material how many head-on collisions are required to reduce the neutron's energy to a thermal value of about 0.025 eV.

14.23 A ^{236}U nucleus at rest splits into two fragments, having masses of 140 amu and 90 amu. The Q of the reaction is 190 MeV. Find the energies and the velocities of the two fragments.

14.24 A ^{238}U nucleus at rest disintegrates, emitting an α-particle ($m = 4$ amu) and leaving a residual nucleus of ^{234}Th ($M \approx 234$ amu). The total energy available is 4.18 MeV. Find (a) the kinetic energy of the α-particle and of the residual nucleus, (b) their momenta and (c) their velocities.

14.25 A nucleus of mass m breaks into several fragments. The explosion has a positive Q-value. (a) Show that if the nucleus explodes into two fragments, they move in the opposite directions in the C-frame of reference. (b) Also show that if the two fragments are equal, their momenta and velocities in the C-frame of reference are equal to $(mQ)^{1/2}$ and $(Q/m)^{1/2}$, respectively. (c) Show that if the nucleus explodes into three fragments, their momenta and velocities, all relative to the C-frame of reference, lie in one plane. (d) Also show that if the three fragments are equal and emitted symmetrically in the C-frame, their momenta and velocities in this frame are $(2mQ/3)^{1/2}$ and $(2Q/3m)^{1/2}$, respectively.

14.26 (a) Verify that if the body of Problem 14.12 strikes the ground at an angle α with the vertical, the body rebounds at an angle β, given by $\tan\beta = (1/e)\tan\alpha$, with a velocity $v' = v(e^2\cos^2\alpha + \sin^2\alpha)^{1/2}$. (b) Use these results to discuss the motion of a ball that has an initial horizontal velocity $\boldsymbol{v}_0$ and is dropped from a table. (c) Make a sketch of the ball's trajectory if the ball makes several collisions with the floor.

14.27 Assume that E and E' are the values of the total energy of a system of two interacting particles as measured by two inertial observers O and O' moving with relative velocity $\boldsymbol{v}$. Verify that $E = E' + (m_1 + m_2)(\boldsymbol{v}'_{CM}\cdot\boldsymbol{v} + \frac{1}{2}v^2)$.

14.28 A horizontal pipe has a cross-section of $10\ \mathrm{cm^2}$ in one region and of $5\ \mathrm{cm^2}$ in another. The water velocity at the first is $5\ \mathrm{m\,s^{-1}}$ and the pressure in the second is $2\times10^5\ \mathrm{N\,m^{-2}}$. Find (a) the velocity of the water in the second region and the pressure of the water in the first region, (b) the amount of water crossing a section in one minute, (c) the total energy per kilogram of water.

14.29 Repeat the previous problem for a case where the pipe is tilted and the second section is 2 m higher than the first.

14.30 Show that if there is a hole in the wall of a vessel and if the surface of the liquid inside the vessel is at a height h above the hole, the velocity of the liquid flowing through the hole is $v = (2gh)^{1/2}$.

14.31 Consider a cylindrical vessel having a diameter of 0.10 m and a height of 0.20 m. A hole $1\ \mathrm{cm^2}$ in cross-section is opened at its base. Water is flowing into the vessel at the rate of $1.4\times10^{-4}\ \mathrm{m^3\,s^{-1}}$. (a) Determine how high the water level will rise in the vessel. (b) After reaching that height the flow of water into the vessel is stopped. Find the time required for the vessel to empty.

14.32 A cylinder of height h and cross-section A stands vertically in a fluid of density ρ. The fluid pressure is given by Equation 14.38. Verify that the total upward force on the cylinder due to the fluid pressure is $B = V\rho_f g$, where V is the volume of the cylinder. Extend the result to a body of arbitrary shape by dividing it into thin vertical cylinders. (This result constitutes **Archimedes' principle**, and the force is known as the **buoyancy**.)

15 Gases

Robert Boyle is considered one of the founders of modern chemistry. He maintained that matter is ultimately composed of 'corpuscles' of various sorts and sizes that are capable of arranging themselves into groups to form a chemical substance. Boyle extensively studied the properties of gases, the propagation of sound, the thermal expansion of solids, the refraction of light, hydrostatics, combustion and many other scientific problems. In 1660 Boyle discovered the relation between the pressure and the volume of a gas at constant temperature, for which he is best known. Boyle was a co-founder of the Royal Society of London in 1662.

15.1 Introduction

The methods and results derived in Chapters 13 and 14 for systems of particles can be applied without great difficulty to systems composed of small numbers of particles for which it is possible to compute the individual terms that make up the momentum or the internal energy of the system. Two examples of such systems are our planetary system and an atom with a few electrons. However, when the number of particles is very large, as in a many-electron atom or in a gas composed of billions of molecules, calculations become mathematically unmanageable. In addition, in these complex systems we are usually not interested in the behavior of each individual component, which is generally not observable. Rather, we are interested in the behavior of the system as a whole. Therefore, certain methods of a statistical nature must be used for computing average values of the dynamical quantities rather than accurate individual values for each component of the system.

Perhaps the most important many-particle systems are those constituted by bodies or matter in bulk, containing large numbers of atoms or molecules. Our immediate contact with the universe is through our senses, which we use to observe the behavior of bodies around us. Consequently, the physics of matter in bulk was developed empirically long before our understanding of the behavior of atoms and molecules had evolved to any level of sophistication. Concepts such as those of temperature, pressure and heat, which we use in our daily life and are associated with sensorial experiences, were introduced early in the development of science. Today's physics must reconcile such sensorial concepts with our understanding of the structure of matter. Gases, being relatively simple systems of particles, make this task easier than liquids or solids. Therefore, in this chapter we will explore the application of statistical considerations to gases.

Note that, throughout this chapter and unless otherwise stated, we will ignore all bulk motion and refer all dynamical quantities relative to the C-frame of reference of the system.

15.2 Temperature

An important macroscopic statistical concept in many-particle systems, related to the sensations of hot and cold, is that of **temperature**. When we feel a body to be 'hot', we say it has a high temperature. Similarly, when we feel a body to be 'cold', we say its temperature is low. We learned to define and measure the temperature of a body long before we understood its physical nature. The instruments used to make the measurements are called **thermometers**, of which there are many different kinds. The temperature of a body is measured by observing some physical property of the thermometer, such as the length of a liquid column in a capillary tube. The volume of the liquid in the thermometer depends on the degree of 'hotness' or 'coldness' of the thermometer when the thermometer is placed in contact with the body and this results in a change in the length of the liquid column. Thermometers operate on the experimental fact that when two bodies are placed in contact they eventually reach the same temperature. It is said that they are then in **thermal equilibrium**.

In the **Celsius** scale the unit of temperature is a **degree centigrade**, denoted °C. The value of 0 °C is assigned to the temperature of melting ice; that is, the temperature at which ice and liquid water are in equilibrium. The value 100 °C is assigned to the temperature of boiling water. Both temperatures are determined at standard atmospheric pressure (1.01325×10^5 Pa; recall Section 14.10). These assignments are called the 'fixed points' for temperature measurement.

To determine other temperatures with a thermometer, we first measure the property X of the thermometer (the length of a liquid column, the electric resistance of the conductor, the emf of a cell etc.) at both fixed points, and obtain the values X_0 and X_{100}. The average change in value of the property per degree Celsius is

$$\frac{X_{100} - X_0}{100}$$

Let the property have a value X at a certain temperature. If the change in the property from the value at 0 °C is $X - X_0$, the temperature in °C is found by dividing the change $X - X_0$ by the average change per °C. Therefore

$$t(^\circ\text{C}) = 100\,\frac{X - X_0}{X_{100} - X_0}$$

Such an empirical method of measuring has its limitations. In particular, the method depends on the properties of the substance used in the thermometer and assumes a linear relation between the property considered and the temperature. Accordingly, thermometers using different properties agree only at the fixed points and show slight differences at all other temperatures. For that reason it is necessary to agree on a standard thermometer and calibrate all other thermometers against it. The standard thermometer is usually a **gas thermometer**.

EXAMPLE 15.1

Thermal coefficient of a physical property.

▷ Consider a physical system (a gas, a liquid, a solid) that has a property X (length, volume, density, pressure, electric resistance etc.) that varies with the temperature. We shall designate by X_0 the value of the property at a certain temperature t_0. When the temperature changes by the amount $\Delta t = t - t_0$, the property change is $\Delta X = X - X_0$. Then the **thermal coefficient** κ of the property is defined by

$$\kappa = \frac{1}{X_0}\frac{\Delta X}{\Delta t} \qquad \textbf{(15.1)}$$

For example, if X is the length L, κ is the coefficient of thermal linear expansion, if it is volume V, κ is the coefficient of thermal cubic expansion, and so on. Note that κ is always measured in $^\circ\text{C}^{-1}$, since the units of X cancel out.

The thermal coefficient κ in general is not constant and depends on the initial temperature t_0. For that reason κ is usually given for $X_0 = 0\,^\circ\text{C}$, as in Table 15.1, which lists the coefficient of thermal linear expansion for various solids. Finally, an important physical problem is to relate κ with the property X in terms of the atomic or molecular structure of the system. We will do that for gases in Example 15.2.

Table 15.1 Coefficients of linear thermal expansion ($^\circ\text{C}^{-1}$)*

Substance	Coefficient	Substance	Coefficient
Aluminum	23.8×10^{-6}	Iron	12.1×10^{-6}
Copper	16.8×10^{-6}	Quartz	0.6×10^{-6}
Glass	9×10^{-6}	Pyrex	3.1×10^{-6}
Bronze	17.1×10^{-6}	Nickel	13.1×10^{-6}
Platinum	9.0×10^{-6}	Steel	10×10^{-6}
Gold	14.2×10^{-6}	Zinc	26.3×10^{-6}
Invar	0.9×10^{-6}	Silver	19.4×10^{-6}
Iridium	6.6×10^{-6}	Tin	23.8×10^{-6}

* Average values for a range of temperature 0 °C to 100 °C.

15.3 The ideal gas temperature

Some properties of a gas are so sensitive to temperature changes that they may be used to measure the temperature of the gas. Consider a mass of gas that at a certain temperature occupies a volume V. The pressure of the gas, designated as p, is the average force per unit area the gas exerts on the walls of the container. As mentioned in Section 14.10, it is expressed in $\mathrm{N\,m^{-2}}$ or **pascal** (Pa). Another unit for expressing the pressure of a gas is the **atmosphere** (1 atm = $1.013\,25 \times 10^5$ Pa).

In the seventeenth century, Robert Boyle (1627–1691) and Edmé Mariotte (1620–1684) independently recognized that if the pressure of a gas is varied without changing the temperature of the gas, the volume the gas occupies varies in such a way that the product of pressure times volume remains practically constant as long as the density of the gas is small (Figure 15.1). This result, known as **Boyle's law**, may be expressed as

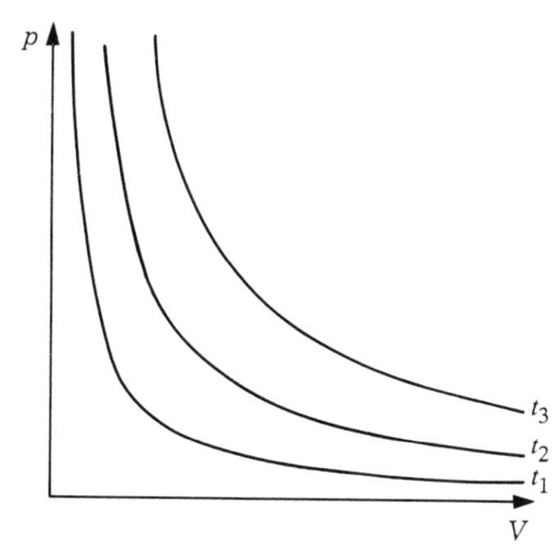

Figure 15.1 Isothermals of an ideal gas.

$$pV = const. \qquad \textbf{(15.2)}$$

The temperature of the gas can be maintained constant, for example, by immersing the container in a large volume of water kept at a certain fixed temperature. A process during which the temperature does not vary is called an **isothermal transformation**.

It is also an experimental result that, for a given mass of a gas, the value of the constant in Equation 15.2 depends on the temperature at which the isothermal transformation takes place. In other words, the constant for a given mass of gas, when maintained at 0 °C, differs from the constant for the same mass of gas when kept at another temperature. Therefore, the value of the product pV for a given mass of gas is an indicator of the temperature of the gas.

We define the **gas temperature** as a quantity proportional to the product pV. Designating the gas temperature by T, we may write

$$pV = CT \qquad \textbf{(15.3)}$$

where C is a constant proportional to the mass of gas present. Suppose that when the gas is at the normal freezing point of water (0 °C), the pressure and volume are p_0 and V_0. When the gas is at the normal boiling point of water (100 °C), these quantities are now p_1 and V_1. Then if T_0 and T_{100} are the corresponding gas temperatures,

$$p_0 V_0 = CT_0 \qquad \text{and} \qquad p_1 V_1 = CT_{100}$$

Next, we decide that the units for measuring the gas temperature are to be the same as those given by the Celsius scale so that $T_{100} = T_0 + 100$. Eliminating the constant C between these relations, we conclude that

$$T_0 = \frac{100 p_0 V_0}{p_1 V_1 - p_0 V_0}$$

is the temperature of the freezing point of water on the gas scale. It has been observed that the value of T_0 obtained in this way is essentially the same, regardless of the gas used, as long as the density of the gas is kept low. The numerical value, determined experimentally, is

$$T_0 = 273.15\,\mathrm{K}$$

The unit of the gas temperature scale is called **kelvin**, in honor of William Thomson, known also as Lord Kelvin (1824–1907), and designated by K. Accordingly, the values of the temperature of a substance when measured in the Kelvin and in the Celsius scales are related by

$$T = t\,^\circ\mathrm{C} + 273.15$$

In the SI the Kelvin scale is defined by assigning the value 273.16 K to the temperature of the triple point of water. This is the state at which water can exist in equilibrium in the solid, liquid and vapor phases. The triple point of water is 0.01 K above the normal freezing point of water.

Usually hydrogen and helium are used in **gas thermometers**. As indicated before, a gas thermometer agrees with other thermometers only at the fixed points because they are the calibration points; at other temperatures there are small numerical discrepancies.

Since for a given pressure and temperature the volume of the gas is proportional to the amount of gas, we shall designate the constant C in Equation 15.3 by R when we have only one mole of gas. Then if we have N moles of a gas we may write Equation 15.3 in the form

$$pV = \mathrm{N}RT \tag{15.4}$$

The constant R is essentially the same for all gases, and is called appropriately the **gas constant**. Its experimentally determined value is

$$R = 8.3144\ \mathrm{J\,K^{-1}\,mol^{-1}} \tag{15.5}$$

Equation 15.4 relates the pressure, volume and temperature of the gas and is called the **equation of state**. Gases follow Equation 15.4 only at high temperatures and low densities or large intermolecular separations, for which the effect of the intermolecular forces and the size of the molecules is negligible; gases increasingly depart from this equation as the temperature decreases and the density increases, or as the molecules move slower and get closer. A gas that follows Equation 15.4

at all temperatures and densities is called an **ideal gas**. Therefore T is called the *ideal gas temperature* or, more commonly, the **absolute temperature**, because it is independent of the nature of the gas.

We may write Equation 15.4 in another convenient form. If N is the number of molecules of the gas and N_A is Avogadro's number, the number of moles of the gas is

$$\mathrm{N} = \frac{N}{N_A}$$

Substituting this into Equation 15.4, we have

$$pV = N\left(\frac{R}{N_A}\right)T$$

We now introduce **Boltzmann's constant** (named in honor of Ludwig Boltzmann (1844–1906)), defined as

$$k = \frac{R}{N_A} = 1.3807 \times 10^{-23}\ \mathrm{J\,K^{-1}} = 8.6178 \times 10^{-5}\ \mathrm{eV\,K^{-1}} \qquad \textbf{(15.6)}$$

Thus we may write the ideal gas equation in the alternative form

$$pV = kNT \qquad \textbf{(15.7)}$$

This equation relates the four quantities pressure (p), volume (V), number of molecules (N), and temperature (T) that define the state of a gas in equilibrium.

Historically, the Boltzmann constant was first introduced in 1900 by Max Planck in his analysis of blackbody radiation (Section 31.7). By fitting his formula for blackbody radiation with the experimental results he obtained a value for k. In terms of it and the gas constant R, Planck obtained Avogadro's number N_A. However, at present there are several independent methods for measuring k and N_A, with consistent results.

EXAMPLE 15.2

Coefficient of thermal expansion of a gas at constant pressure.

▷ Consider a gas that occupies a volume V_0 at temperature T_0 and pressure p. Then, using the gas law Equation 15.3, $pV_0 = CT_0$. If we change the temperature to T without changing the pressure the new volume is such that $pV = CT$. Subtracting these two equations, we get

$$p(V - V_0) = C(T - T_0) \qquad \text{or} \qquad p\Delta V = C\Delta T$$

But $C = pV_0/T_0$. Therefore

$$\Delta V = (V_0/T_0)\Delta T \qquad \text{or} \qquad \Delta V/\Delta T = V_0/T_0$$

Then, using the definition of Equation 15.1, $\kappa = (1/X_0)(\Delta X/\Delta T)$, the thermal coefficient of volume expansion at constant pressure for an ideal gas is

$$\kappa_V = \frac{1}{V_0}\frac{\Delta V}{\Delta T} = \frac{1}{T_0}$$

If the initial temperature is 0 °C, then $T_0 = 273.15\,\text{K}$ and $\kappa_V = 0.003\,661\,\text{K}^{-1}$. That means that at 0 °C the volume of a gas increases by 0.3661% for each degree of increase of temperature at constant pressure.

Using the same procedure, it is easy to verify that the thermal coefficient of pressure change at constant volume for an ideal gas is

$$\kappa_p = \frac{1}{p_0}\frac{\Delta p}{\Delta T} = \frac{1}{T_0}$$

and therefore is equal to κ_V. The coefficients for most gases are very close to the values of κ_V and κ_p for an ideal gas.

15.4 Temperature and molecular energy

Historically, the concept of temperature was introduced without relating it to the molecular properties of the system. However, it is an experimental fact that in many cases when energy is supplied to the system (as, for example, when work is done *on* a system), the temperature increases. In the nineteenth century Benjamin Thompson, also known as Count Rumford (1753–1814), carefully studied the rapid increase in temperature of a cannon as its barrel was bored. During the process of boring, work is done on the cannon. The internal energy of the cannon material must necessarily increase, with a resulting increase in the energy of the molecules. Similarly, when a gas is compressed in a cylinder, for which work has to be done on the gas, the temperature of the gas increases if no other energy exchanges occur (Figure 15.2). As a result of these experiments (and many other similar experiments), we may conclude that the temperature of a system is a quantity related to the energy of its molecules.

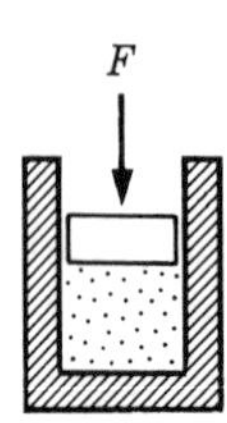

Figure 15.2 Compression of a gas.

Recall that, in any system of particles, we must distinguish between the energy associated with the motion of the system as a whole and the energy associated with the internal motion of its particles or molecules. The internal molecular motion is often called 'disordered' because molecules move in many different directions, while the motion of matter in bulk appears to be 'ordered'. For example, a solid body may have translational and rotational motion, which are 'ordered'; in addition, within each small volume of the solid, the molecules are all vibrating in different directions and with different energies; that is, their vibrations are 'disordered'. A stream of water or a blast of wind are systems containing both ordered and disordered motion. All adjacent small volumes in a stream of water or in a wind move in an apparently ordered way, determined by the current and wind velocities, respectively. However, in each case the molecules within a small volume of water or air have a disordered motion relative to a C-frame attached to the volume element.

From now on we shall exclusively consider the 'disordered' or internal motion of the molecules in a system. This motion is referred to a local C-frame of reference attached to each portion of the system so that the 'ordered' bulk motion is either properly subtracted or simply ignored.

Now, consider a system of particles of masses $m_1, m_2, m_3, \ldots$ and velocities $\boldsymbol{v}_1, \boldsymbol{v}_2, \boldsymbol{v}_3, \ldots$ in the system's C-frame of reference. The average kinetic energy of the particles of the system in the C-frame is

$$E_{\mathrm{k,ave}} = \frac{1}{N} \sum_i \tfrac{1}{2} m_i v_i^2 \tag{15.8}$$

where N is the total number of particles. If all the particles have the same mass, then

$$E_{\mathrm{k,ave}} = \frac{1}{N} \sum_i \tfrac{1}{2} m v_i^2 = \tfrac{1}{2} m \left(\frac{1}{N} \sum_i v_i^2 \right) = \tfrac{1}{2} m v_{\mathrm{rms}}^2 \tag{15.9}$$

In Equation 15.9 we have defined

$$v_{\mathrm{rms}}^2 = \frac{1}{N} \sum_i v_i^2 = \frac{1}{N} (v_1^2 + v_2^2 + v_3^2 + \cdots) = (v^2)_{\mathrm{ave}} \tag{15.10}$$

or $v_{\mathrm{rms}} = [(v^2)_{\mathrm{ave}}]^{1/2}$. The velocity v_{rms} is called the **root mean square velocity of the particles**. Therefore v_{rms} is the molecular velocity that corresponds to the average kinetic energy.

The **temperature** *of a system of particles is a quantity related to the average kinetic energy of the disordered motion of the particles in the C-frame of reference.* Thus temperature is a property of the system defined independently of the bulk motion of the system relative to the observer. In this sense we may speak of the temperature of a solid, of a gas, and even of a complex nucleus. However, the precise relation between temperature and average particle kinetic energy depends on the nature of the system. An important problem in physics is to determine that relation for each type of system. In Section 17.6 this is analyzed for the case of a gas. A system that has the same temperature throughout, so that the average kinetic energy of the molecules is the same in all regions of the system, is said to be in **thermal equilibrium**. Two bodies in contact at the same temperature are in thermal equilibrium, as indicated in Section 15.2.

This new definition of temperature can be reconciled with that associated with the sensations of 'hot' and 'cold' because our nervous system is affected by the average energy of the molecules, resulting in different sensorial feelings. Also, the average separation of the molecules of a liquid or solid increases with their average energy, resulting in a change of size. This justifies using the expansion of a substance to measure temperature.

The fact that we are referring the motions to local C-frames in order to define temperature is important. Suppose we have a 'hot' metal ball at rest in our laboratory and an identical 'cold' metal ball moving very fast relative to our laboratory. The hot ball has a high temperature, meaning a large molecular kinetic energy relative to its center of mass, which happens to be at rest in the laboratory. On the other hand, the cold ball has a low temperature. This means it has a small molecular kinetic energy relative to its CM, which in our case is in motion relative to the observer. The fast-moving cold ball may have a larger total kinetic energy relative to the laboratory than the stationary hot ball, but most of the cold ball's kinetic energy is translational kinetic energy associated with the ball's ordered motion and does not account for its temperature.

In an isolated system, whose total internal energy is constant, the temperature may change if the kinetic energy changes because of a change in internal potential energy. For example, a mass of gas in interstellar space may be condensing or a star may be contracting because of internal gravitational forces between its particles. The contraction results in a decrease of internal potential energy and a corresponding increase in 'disordered' molecular kinetic energy. As a result, the temperature of the gas will increase. If, on the other hand, the system is expanding, its internal potential energy increases (if the forces are attractive). The expansion then produces a decrease in molecular kinetic energy and a resultant drop in temperature. When the internal potential energy of an isolated system remains constant, as is the case for a gas contained in an insulated rigid box, the system's average molecular kinetic energy will also remain constant; i.e. its temperature will not change. However, if the system is not isolated, it may exchange energy with the rest of the universe, possibly resulting in a change of the system's internal molecular kinetic energy and thus of its temperature.

The new definition of temperature indicates that it should be expressed in joules/particle. However, for historical reasons it is customary to express temperature in kelvin (K) or in degrees Celsius (°C), as explained before.

15.5 Internal energy of an ideal gas

A gas may be pictured as a system composed of a large number of molecules moving freely in all directions with different velocities. Occasionally a molecule collides with another gas molecule or with the walls of the container and exchanges momentum and energy (Figure 15.3).

When a gas molecule collides with a molecule of the wall, there is an exchange of momentum between the two molecules. The exchange of momentum is accompanied by a force exerted by the gas molecule at the point of collision with the wall, and conversely. The individual forces exerted by the molecules fluctuate at each point. Because there is a large number of collisions over each small portion of the wall, the overall effect of the gas can be represented by an average force F acting on the whole area of the wall (recall Example 13.5). If A is the area of the wall, then the pressure p of the gas is

$$p = \frac{F}{A} \tag{15.11}$$

Gas pressure is a *statistical* concept valid only when a large number of molecules are involved.

Physical intuition, and Newton's second law, suggests that the larger the momentum mv of the molecules, the greater the force they exert when they hit the wall. Also, the larger the number n of molecules per unit volume and the larger the velocity v of the molecules the larger the number of molecules that hit the wall per unit time. Therefore, we may expect that the pressure exerted by the gas on the wall be proportional to $(n)(v)(mv)$ or nmv^2, where n is the number of molecules per unit volume, m their mass and v their velocity (again, recall Example 13.5). Since the molecules have different velocities the average pressure is determined by taking the average of v^2 or, using Equation 15.10, v_{rms}^2. So the gas pressure is proportional to nmv_{rms}^2. A precise calculation, which must also take into account the fact that molecules move in different directions (see the proof below), gives

$$p = \tfrac{1}{3}nmv_{\text{rms}}^2 \tag{15.12}$$

Equation 15.12 can also be written as $p = \frac{1}{3}(N/V)mv_{\text{rms}}^2$ because $n = N/V$, where N is the total number of molecules and V the volume of the container. Therefore

$$pV = \tfrac{1}{3}Nmv_{\text{rms}}^2 \tag{15.13}$$

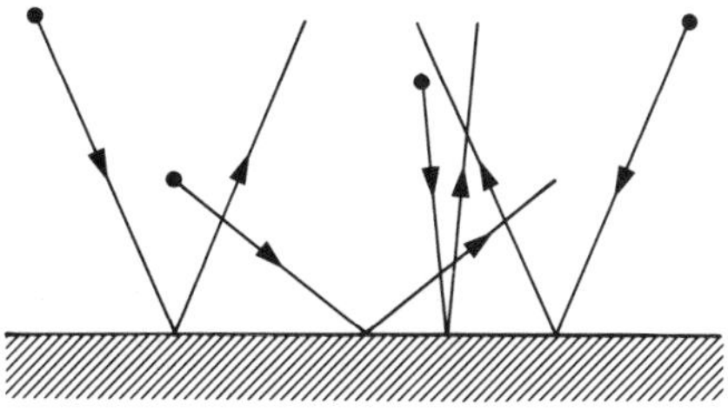

Figure 15.3 Molecular collisions with the walls of a container.

According to Equation 15.9, the average kinetic energy of a molecule in the gas is $E_{\text{k, ave}} = \frac{1}{2}mv_{\text{rms}}^2$. Therefore, we may rewrite Equation 15.13 as

$$pV = \tfrac{2}{3}NE_{\text{k, ave}} \tag{15.14}$$

Comparing this result with the ideal gas equation, $pV = kNT$ (Equation 15.7), we conclude that the average kinetic energy of a gas molecule is related to the absolute temperature of the gas by the expression

$$E_{\text{k, ave}} = \tfrac{3}{2}kT \tag{15.15}$$

This equation shows that the absolute temperature of an ideal gas is directly proportional to the average kinetic energy of its molecules. Equation 15.15 also indicates that one kelvin corresponds to 2.07×10^{-23} J or 1.29×10^{-4} eV (per particle).

When the average kinetic energy given in Equation 15.15 is combined with Equation 15.9, $E_{k,ave} = \frac{1}{2}mv_{rms}^2$, we obtain the rms velocity as

$$v_{rms} = \left(\frac{3kT}{m}\right)^{1/2} \tag{15.16}$$

This velocity corresponds to the average kinetic energy of a molecule of a gas at the temperature T. Multiplying both numerator and denominator by N_A, we may also write

$$v_{rms} = (3RT/M)^{1/2}$$

where M is the molar mass of the gas. For example, for hydrogen, $M = 2 \times 10^{-3}$ kg. Then at room temperature (300 K), $v_{rms} = 1.9 \times 10^3\ \mathrm{m\,s^{-1}}$ or $6.96 \times 10^3\ \mathrm{km\,h^{-1}}$.

In most gases, especially at high temperatures and low densities, the effect of the intermolecular forces is relatively weak. Thus the internal potential energy of the gas may be neglected in comparison with the kinetic energy of the molecules. We thus redefine an **ideal gas** as a gas whose intermolecular forces are negligible and whose molecules can be treated as point masses. Therefore, the internal energy of an ideal gas is exclusively molecular kinetic energy. That is, omitting the subscript 'int' for the internal energy,

$$U = \sum_i \tfrac{1}{2}mv_i^2 = N(\tfrac{1}{2}mv_{rms}^2) = NE_{k,av} \tag{15.17}$$

Comparing Equations 15.15 and 15.17, we conclude that the internal energy of an ideal gas composed of N molecules and at temperature T is then

$$U = \tfrac{3}{2}kNT \qquad \text{or} \qquad U = \tfrac{3}{2}\mathrm{N}RT \tag{15.18}$$

Therefore, the internal energy of an ideal gas depends only on the temperature of the gas. This relationship is not true for real gases, whose internal potential energy is not zero but rather depends on the intermolecular distances; that is, on the density of the gas. For low temperatures (when the molecular energy is relatively small) or high densities (when the molecules are close), the effect of molecular forces as well as the molecular sizes must be taken into account and Equation 15.8, for the equation of state, and Equation 15.17, for the internal energy, are not good approximations, as we will explain in Section 15.6.

The ideal gas law $pV = kNT$, indicates that the pressure of a gas increases with temperature (other factors remaining the same). This is to be expected, because the higher the temperature, the larger the molecular energies and velocities. Therefore, there is a larger change in molecular momentum when a gas molecule collides with the walls.

The ideal gas model fairly accurately describes real gases when the density of the gas is low enough so that the average intermolecular distance is large in comparison with molecular size and the temperature is high so that the average molecular interaction is small compared with the molecular kinetic energy.

Relation between the pressure of a gas and the average kinetic energy of its molecules

To obtain the pressure exerted by gas molecules when they collide with the walls of a vessel, we follow a method similar to that of Example 13.5. As explained in that example, when a molecule of mass m moving with velocity $\boldsymbol{v}$ hits a surface of area A at an angle θ with the normal OX to the surface (Figure 15.4), it rebounds with the same velocity, making the same angle with the normal. Let us designate the perpendicular components of the velocity by $v_x = v \cos \theta$ and the number of molecules of the gas per unit volume that move toward the wall with a perpendicular component of its velocity equal to v_x by n_x. Then, using Equation 13.15, the pressure exerted by these molecules on the surface is

$$p = 2n_x m v_x^2$$

In a gas the molecules move in different directions and with different velocities. Therefore, to obtain the pressure of the gas, v_x^2 must be replaced by the average $(v_x^2)_{\text{ave}}$ and n_x must now be the *total* number of molecules, per unit volume, moving toward the surface at any time, regardless of the value of v or of the direction of motion. That is, $p = 2n_x m(v_x^2)_{\text{ave}}$.

Statistically, at any time one-half the molecules per unit volume in the vicinity of the area A have a component of their velocity that points toward the wall; and the other half, away from the wall. Thus, if n is the total number of molecules per unit volume, we must replace n_x by $\frac{1}{2}n$ because only $\frac{1}{2}n$ are going to hit the wall at A. Making this change in the above expression for p, we obtain

$$p = 2(\tfrac{1}{2}n)m(v_x^2)_{\text{ave}}$$

The magnitude of the molecular velocity is related to its components by $v^2 = v_x^2 + v_y^2 + v_z^2$ so that

$$(v^2)_{\text{ave}} = (v_x^2)_{\text{ave}} + (v_y^2)_{\text{ave}} + (v_z^2)_{\text{ave}}$$

We may assume that the average molecular velocities are the same in any direction; that is, the molecular velocities are distributed **isotropically**. Thus $(v_x^2)_{\text{ave}} = (v_y^2)_{\text{ave}} = (v_z^2)_{\text{ave}}$ and therefore $(v_x^2)_{\text{ave}} = \frac{1}{3}(v^2)_{\text{ave}} = \frac{1}{3}v_{\text{rms}}^2$. Making this substitution in the expression for p, we obtain

$$p = 2(\tfrac{1}{2}n)m(\tfrac{1}{3}v_{\text{rms}}^2) = \tfrac{1}{3}nmv_{\text{rms}}^2$$

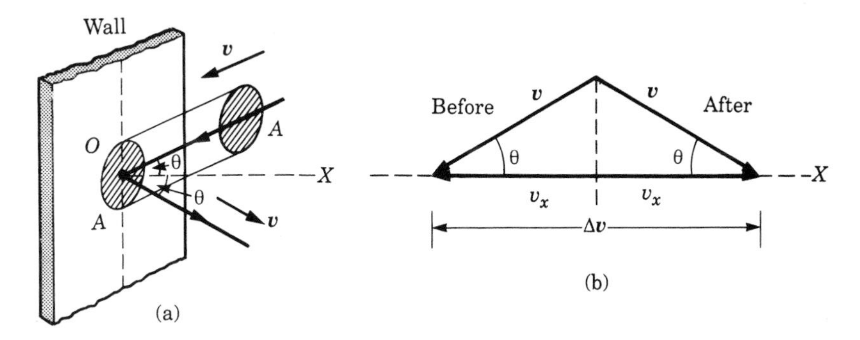

Figure 15.4 Change in velocity of a gas molecule after hitting a wall.

EXAMPLE 15.3

Free expansion of a gas: Joule's Experiment.

▷ Long before the development of the molecular theory of gases, there was some concern about the relation between the internal energy of a gas and its temperature. Many experiments have been carried out to clear up this issue. One of the classical experiments was performed by James P. Joule (1818–1889) in the middle of the 19th century. Two containers, 1 and 2, are joined through a stopcock (Figure 15.5). Also, the walls of the containers are properly insulated to eliminate any possible exchange of energy with the surroundings. With the stopcock closed, container 1 is filled with a gas up to a certain pressure and temperature while container 2 is evacuated. When the stopcock is opened some gas passes from container 1 to 2 until the pressure is the same throughout. In the process, the gas clearly has not done any external work ($W_{ext} = 0$) or exchanged any energy. Therefore in accordance with Equation 14.6, which is valid for any system of particles, $\Delta U = 0$. That is, the internal energy of the gas does not change during the **free expansion**. In the process it was observed that the pressure and the volume of the gas were changed. However, the temperature of the gas after the expansion remained the same, within the accuracy of Joule's measurements. We may conclude that the kinetic energy of the molecules of the gas did not change in the process. However, because the volume increases, the average separation of molecules is greater. An increase in separation implies that the internal potential energy of the gas should have increased, since negative internal work had to be done by the attractive intermolecular forces. Then, since U is the sum of the internal kinetic and potential energies, $\Delta U = 0$ would require a decrease in the internal kinetic energy with a corresponding decrease in temperature. The fact that the temperature did not change then implies that the internal potential energy is zero or too small for its effects to be measurable. Other more precise measurements, carried out more recently, have detected small changes in temperature. In any case, Joule's experiment was one of the first indications that, within some range of temperature values, the internal energy of most gases is, to a great approximation, only kinetic and thus depends only on the temperature, and the internal potential energy associated with molecular forces may be ignored.

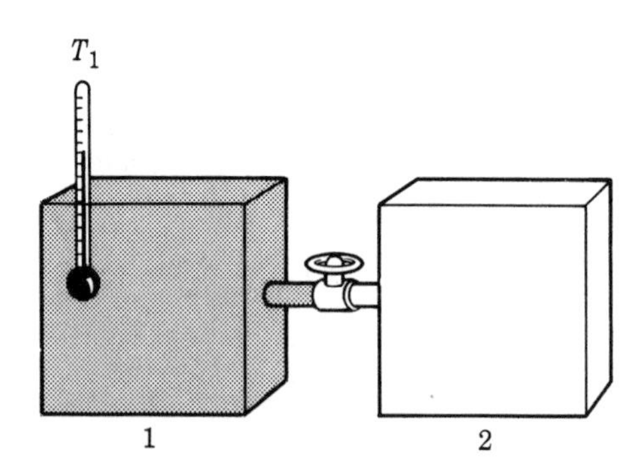

Figure 15.5 Free expansion of a gas.

15.6 Real gases

The *ideal* gas equation, $pV = kNT$, describes the behavior of a gas when there are no intermolecular forces and the molecules are considered as mass points. Therefore, the equation is a good approximation for real gases only in so far as the effects of intermolecular forces and of molecular sizes are negligible. Such conditions exist only at high temperatures (large molecular kinetic energies) or at low densities (large intermolecular separations) or both.

Many attempts have been made to empirically obtain an equation that describes the behavior of real gases at all temperatures and pressures. For example, in 1873 J. D. van der Waals (1837–1923) proposed to replace the ideal gas equation by an equation that took into account the effect of intermolecular forces and the size of the molecules in a very simple way. The first and most obvious correction to the ideal gas equation is to subtract the volume of the gas molecules from the volume V. This may be done by replacing V with $V - Nb$, where b is a constant proportional to the volume of a single molecule. Hence we may write $p(V - Nb) = kNT$ or

$$p = \frac{kNT}{V - Nb}$$

Secondly, a gas molecule in a gas is generally surrounded by other molecules, so that the average resultant of the attractive intermolecular forces on the molecule is zero. However, when a molecule approaches the wall of the container, it is surrounded by other molecules only on one side. This results in a net force that pulls the molecule away from the wall, reducing the velocity with which the molecule approaches the wall (Figure 15.6). Therefore, the attractive intermolecular forces, which become stronger as the molecules get closer, tend to reduce the pressure of the gas. We note that each molecule interacts with $N - 1$ molecules. Thus the effect of molecular forces on the pressure is of the order of $N(N - 1) \approx N^2$, if N is very large. However, the effect is reduced as the volume of the gas increases and the intermolecular distances increase. Following this reasoning, van der Waals added a negative term proportional to the square of the number of molecules per unit volume, that is $(N/V)^2$, to the right side of the above equation. This term results in a lowering of the pressure as the density of the gas increases. The modified equation is

$$p = \frac{kNT}{V - Nb} - a\left(\frac{N}{V}\right)^2 \qquad \textbf{(15.19)}$$

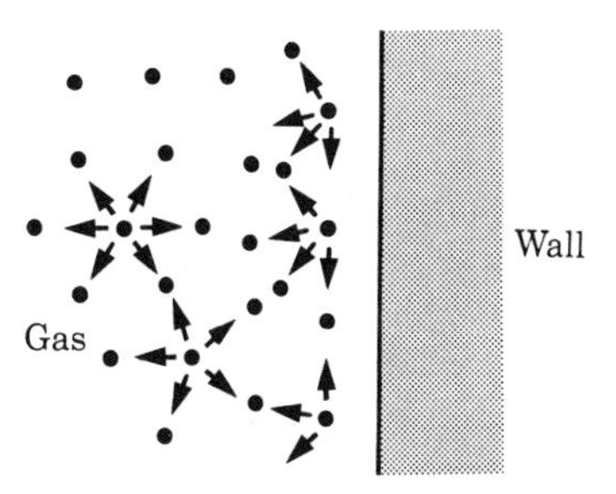

Figure 15.6 Effect of intermolecular forces on the pressure of a gas.

where a is a constant of proportionality related to the intermolecular forces. The above equation may be rewritten in the form

$$\left[p + a\left(\frac{N}{V}\right)^2\right](V - Nb) = kNT \qquad \textbf{(15.20)}$$

which is called the **van der Waals equation**. Equation 15.19 is an excellent description of the behavior of a gas for a large range of the variables p, V and T, given the values for a and b determined experimentally for each gas (Table 15.2). It can be seen that b is of the order of magnitude of molecular volumes.

Table 15.2 van der Waals coefficients and Boyle temperature (T_B)

Gas	a (Pa m^6 molecule^{-2})	b (m^3 molecule^{-1})	T_B (kelvin)
Monatomic			
He	0.0095×10^{-48}	3.936×10^{-29}	17.5
Ar	0.3729×10^{-48}	5.345×10^{-29}	505.4
Xe	1.1718×10^{-48}	8.477×10^{-29}	1001.3
Hg	2.2612×10^{-48}	2.816×10^{-29}	5817.1
Diatomic			
H_2	0.0683×10^{-48}	4.419×10^{-29}	112.0
O_2	0.3800×10^{-48}	5.286×10^{-29}	520.9
Cl_2	1.8142×10^{-48}	9.336×10^{-29}	1407.6
Triatomic			
N_2O	1.0567×10^{-48}	7.331×10^{-29}	1044.4
NO_2	1.4764×10^{-48}	7.346×10^{-29}	1456.3
H_2O	1.5267×10^{-48}	5.063×10^{-29}	2185.0
Polyatomic			
CH_4	0.6295×10^{-48}	7.104×10^{-29}	642.1
NH_3	1.1650×10^{-48}	6.156×10^{-29}	1371.4
CCl_4	5.6828×10^{-48}	22.966×10^{-29}	1793.1

For large volumes and low pressures, the isothermal transformation curve of a real gas practically coincides with that of an ideal gas (Figure 15.7). However, as the volume decreases, the $(N/V)^2$ term in Equation 15.19 shows that the pressure of a real gas falls below that of the ideal gas, due to the effect of the attractive intermolecular forces. But at even smaller volumes (that is as V approaches Nb), the first term in Equation 15.19 increases very rapidly and shoots the pressure up because the gas becomes less compressible due to the closeness of the molecules.

Below a certain temperature, called the **van der Waals temperature** or **critical temperature**, T_c, a new phenomenon occurs when a certain pressure is reached. For each temperature below T_c there is a certain pressure and corresponding volume at which a phase transition called **condensation** begins to occur. At that pressure and temperature, the gas can be in equilibrium with its liquid phase, with the proportion of each depending on the volume (Figure 15.8). For example, consider a gas that first follows the isothermal AB. At B the conditions are such that the molecules can either be in the gas or liquid phase, due to a balance between the thermal energy $\frac{3}{2}kT$ and the molecular potential energy. As more gas condenses into a liquid, the volume decreases without any change in pressure or temperature, as shown by line BC until point C is reached when all the gas has been condensed. As all the substance is in the liquid phase it becomes very difficult to compress, as shown by line CD. The pressure corresponding to line BC is the **equilibrium pressure** of the gas and liquid phases at the corresponding temperature. For example, for water at 100 °C, the equilibrium pressure is 1 atm, which defines the normal boiling point of water. At the critical temperature there is a minimum pressure, called the **critical pressure**, corresponding to point E in Figure 15.8 at

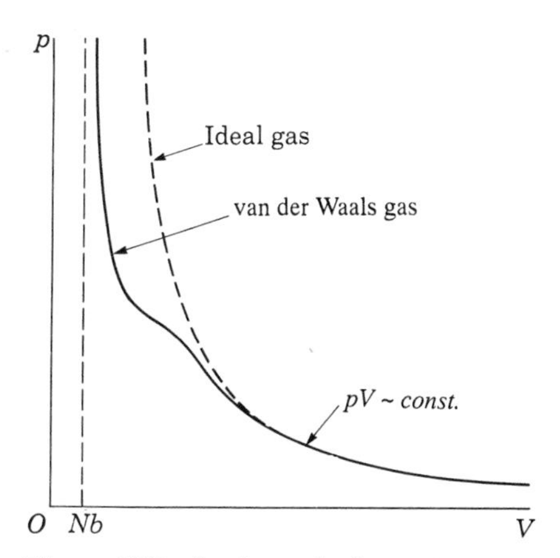

Figure 15.7 Isothermal of a real gas.

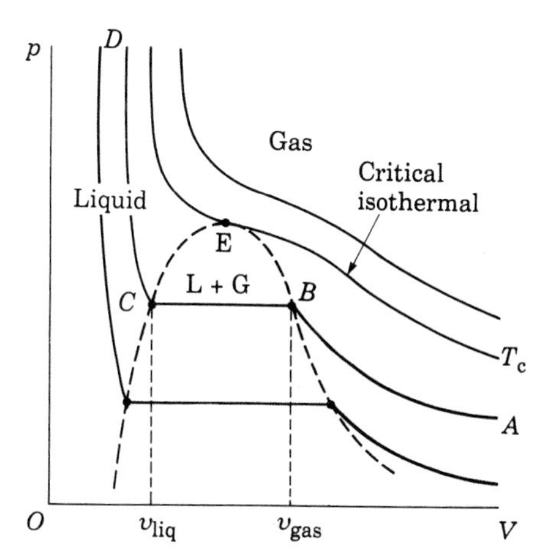

Figure 15.8 Isothermals of a gas–liquid system.

which the gas condenses into the liquid phase without a change in volume. At temperatures above T_c the gas cannot be liquified by compression alone.

EXAMPLE 15.4

The virial form of the real gas equation.

▷ There is a more convenient and general form of writing an approximate expression for the equation of state of a real gas. The new equation amounts to an expansion of the pressure in powers of the quantity N/V, and therefore allows for successive approximations as needed. This expression is called the **virial expansion** (after the virial theorem in dynamics first proposed by Rudolf Clausius (1822–1888)) and is given by

$$p = \frac{kNT}{V}\left[1 + A(T)\left(\frac{N}{V}\right) + B(T)\left(\frac{N}{V}\right)^2 + \cdots\right] \tag{15.21}$$

The quantities $A(T)$, $B(T)$, ... are called the **virial coefficients**. They depend only on the temperature and are characteristic for each gas. At the temperature where $A(T) = 0$ the real gas most resembles an ideal gas. This temperature is called **Boyle's temperature** and is designated by T_B.

For gases at very low density, all terms but the first in Equation 15.20 are negligible and we recover the ideal gas law, as expected for such conditions.

The van der Waals equation may be cast into the virial form in the following manner. Starting with Equation 15.19

$$p = \frac{kNT}{V - Nb} - a\left(\frac{N}{V}\right)^2$$

we write

$$p = \frac{kNT}{V(1 - Nb/V)} - a\left(\frac{N}{V}\right)^2 = \frac{kNT}{V}\left(1 - \frac{Nb}{V}\right)^{-1} - a\left(\frac{N}{V}\right)^2$$

$$= \frac{kNT}{V}\left(1 + \frac{Nb}{V} + \frac{N^2b^2}{V^2} + \cdots\right) - \frac{kNT}{V}\left(\frac{aN}{kTV}\right)$$

Collecting terms, we may write

$$p = \frac{kNT}{V}\left[1 + \left(b - \frac{a}{kT}\right)\left(\frac{N}{V}\right) + b^2\left(\frac{N}{V}\right)^2 + \cdots\right]$$

which shows that the first and second virial coefficients are

$$A(T) = b - \frac{a}{kT} \quad \text{and} \quad B(T) = b^2 \tag{15.22}$$

Setting $A(T) = 0$ in the above equation, we obtain the Boyle temperature as $T_B = a/kb$ (see Table 15.2).

15.7 Polyatomic gases

In our consideration of ideal gases, we have taken only the kinetic energy of the translational motion of the molecules with respect to the C-frame into account. For that reason we have written the internal energy of the gas (Equation 15.18) in the form

$$U = N(\tfrac{1}{2}mv_{\rm rms}^2) = NE_{\rm k,ave} = \tfrac{3}{2}kNT \tag{15.23}$$

so that the average energy per molecule is, recalling Equation 15.15,

$$E_{\rm k,ave} = \tfrac{3}{2}kT \tag{15.24}$$

This is correct as long as the molecules can be considered as point masses, or at least as very small balls. Monatomic gases can be treated satisfactorily in this way. However, this approximation does not work well with polyatomic gases; i.e. gases whose molecules are composed of two or more atoms and thus have some internal motion beside the motion of their centers of mass.

Therefore, in a polyatomic molecule we have to take into account the motion of its CM, the rotational motion of the molecule around its center of mass and the vibrational motion of the atoms relative to each other. Thus we must write the molecular energy as

$$\begin{aligned} E_{\rm molec.} &= (\text{trans. energy}) + (\text{rot. energy}) + (\text{vib. energy}) \\ &= E_{\rm k,ave} + E_{\rm r} + E_{\rm v} \end{aligned}$$

Strictly speaking we should include the electronic energy but at low temperatures this energy does not suffer any changes and we do not need to consider it.

As the temperature of a gas increases, the different components of the molecular energy also increase. But each component increases in a different manner, because each depends on different properties of the molecule. For example, the vibrational energy depends on the strength of the bonding of the atoms in the molecule. The stronger the atomic bonds, the more difficult it is to increase the vibrational energy. On the other hand, the rotational energy depends on the geometry (moment of inertia) of the molecule. In this case, the larger the moment of inertia, the less energy is required to increase the rotational energy by changing its angular momentum (recall Equation 14.20). However, it is relatively simple to estimate the contribution of each term to the total internal energy of a gas.

First we note that the translational motion of the molecule can be considered as the combination of three motions along the X-, Y- and Z-axes, each with the same average kinetic energy. Thus, since $E_{k,ave} = \frac{3}{2}kT$, we say that the molecule has three translational **degrees of freedom**, each with an average energy $\frac{1}{2}kT$. We may generalize this result and say that the average kinetic energy of each degree of freedom is $\frac{1}{2}kT$, a statement that is called the **principle of equipartition of energy**.

Consider, for simplicity, its application to a diatomic molecule. The molecule can rotate freely about an axis that is perpendicular to the molecular axis and passes through its center of mass. The rotational energy around the center of mass of the diatomic molecule has two degrees of freedom because we need two angles to fix the orientation of the molecular axis in space (Figure 15.9). Therefore, as the temperature increases, the average kinetic energy of rotation of a diatomic molecule approaches the value $E_{rot} = \frac{1}{2}kT + \frac{1}{2}kT = kT$. We recall, from Example 14.4, that the rotational energy of a molecule is quantized. Then, to excite the rotational energy of a molecule, kT must be at least of the order of $\hbar^2/2I$ or about 10^{-4} eV.

Also, the two atoms in the molecule are separated a certain distance that changes during its vibration. A diatomic molecule has one vibrational degree of freedom, since we only need to know that distance. Thus we could say that the average kinetic energy of vibration of the molecule is $\frac{1}{2}kT$. However, vibrational motion has kinetic and potential energy and their average values are equal (Section 10.4). Therefore the total average vibrational energy of the diatomic molecule is $E_{vib} = \frac{1}{2}kT + \frac{1}{2}kT = kT$. Actually, the relation of vibrational energy and temperature is more complex and only approaches the value kT at high temperatures. That is because the vibrational energies of a molecule are quantized and the energy of excitation is of the order of $\hbar\omega$, as explained in Section 10.11, or about 10^{-1} eV. Then, only for energies kT of the same order as $\hbar\omega$ or larger can the vibrations of a molecule be excited (see Example 17.3).

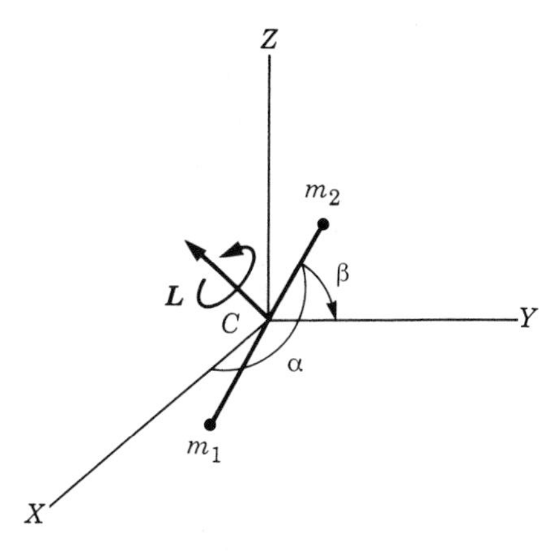

Figure 15.9 Rotation of a diatomic molecule.

Therefore, as the temperature of a diatomic gas increases, the rotational and vibrational energies also increase, with the rotational energy excited first. At high temperatures the average energy of a diatomic molecule tends asymptotically

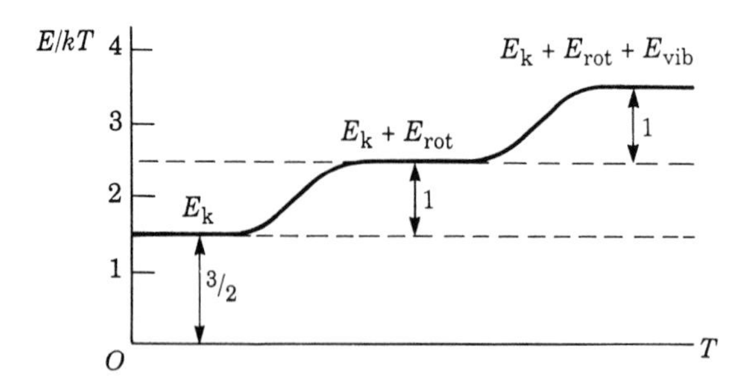

Figure 15.10 Molecular energy of a gas as a function of temperature.

to the value

$$E_{ave} = \underset{\text{(translation)}}{\tfrac{3}{2}kT} + \underset{\text{(rotation)}}{kT} + \underset{\text{(vibration)}}{kT}$$

Thus the average molecular energy in a diatomic gas varies with the temperature, as indicated schematically in Figure 15.10. The temperature at which the vibrations and rotations reach their asymptotic values depends on the energy levels of the molecule. However, as a rule, the rotational energy reaches its asymptotic values at temperatures lower than the vibrational energy because they need much less energy. The total internal energy of the diatomic gas varies in the same form.

In the case of molecules with more than two atoms, the analysis is similar. However, more internal degrees of freedom must be taken into account. Also, it is necessary to know whether the molecule is linear, planar or three-dimensional. Thus the analysis of the internal energy of a gas provides valuable information about the structure of its molecules.

If, in addition to the three degrees of freedom associated with the motion of the CM, the polyatomic molecule has f degrees of freedom for rotational and vibrational motion (with each vibration counting as two degrees), the average energy of the molecules at high temperatures is given by

$$E_{ave} = \frac{3+f}{2} kT \qquad \textbf{(15.25)}$$

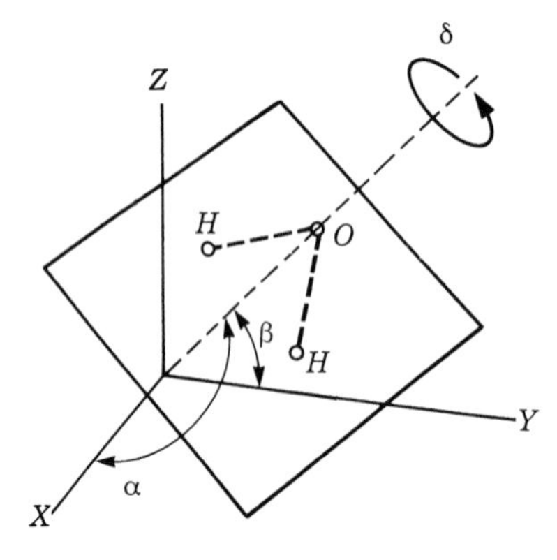

Figure 15.11 Rotation of a water molecule.

For temperatures below 400 K (energies of the order of 10^{-2} eV), only the rotational energy is excited. In that case, if the molecule is linear, like carbon dioxide (CO_2), there are only two rotational degrees of freedom and $f = 2$ (recall Figure 15.9). But if the molecule is planar, like water (H_2O), or three-dimensional, like ammonia (NH_3) there are three rotational degrees of freedom because, in addition to the angles α and β needed to fix the axis of rotation (Figure 15.11), we need the angle δ of rotation around the axis, so that $f = 3$.

EXAMPLE 15.5

Comparison of the thermal energy kT at room temperature, 300 K, with the energy necessary to excite the rotational and vibrational energy of the CO molecule.

▷ When $T = 300$ K, the thermal energy is $kT \simeq 0.026$ eV. According to Section 14.7, the rotational energy of the CO molecule is

$$E_r = \frac{\hbar^2}{2I}\,l(l+1)$$

where $I = \mu r^2 = 1.47 \times 10^{-46}$ kg m^2 (Example 13.6). Therefore, $\hbar^2/2I = 3.8 \times 10^{-21}$ J $=$ 0.00023 eV is the order of magnitude of excitation of rotational energy. Thus the rotational energy is much less than the thermal energy and many CO molecules are in excited rotational states at room temperature as a result of inelastic molecular collisions.

The vibrational frequency of CO is $\omega = 4 \times 10^{14}$ s^{-1} and the energy required to excite vibrations, according to Section 10.11, is $\hbar\omega = 4.2 \times 10^{-20}$ J $= 0.26$ eV, which is much larger than kT. Therefore, at room temperature very few CO molecules are in excited vibrational states as a result of inelastic molecular collisions.

QUESTIONS

15.1 Why is the pressure of a gas a macroscopic statistical concept? Can we speak of the pressure of a gas molecule? Explain your answer.

15.2 Can we speak of the temperature of a gas molecule, or must we refer to the temperature of the whole gas? Is temperature a statistical concept?

15.3 Write the defining expressions for the coefficients of thermal surface expansion, κ_S. Verify that if κ_L is the coefficient of linear expansion, then $\kappa_S = 2\kappa_L$, approximately.

15.4 What is meant by 'thermal equilibrium'?

15.5 Why must the pressure of a gas be proportional to the square of the velocity of the gas molecules? Why must the pressure of a gas be proportional to the number of molecules per unit volume?

15.6 State the difference between $(v_{ave})^2$ and $(v^2)_{ave}$. Are v_{ave} and v_{rms} the same?

15.7 Under what conditions can an ideal gas be an adequate model of real gases?

15.8 What is the effect on the pressure of a gas due to (a) the size of the molecules; (b) intermolecular forces?

15.9 Why should a real gas at low density behave very similarly to an ideal gas?

15.10 Joule's experiment shows that the effect of intermolecular forces is negligible in free expansion. What would happen to the temperature of a gas in free expansion if intermolecular forces were strong and attractive?

15.11 Discuss how the molecular energy of a polyatomic gas varies as the temperature increases.

15.12 Justify the term 'absolute' for the ideal gas temperature.

15.13 Is the effect of intermolecular forces more noticeable when a gas is at high temperature or at low temperature?

15.14 Referring to the process represented by isothermal *ABCD* in Figure 15.8, describe what happens as the gas proceeds (a) from *A* to *B*, (b) from *B* to *C* and (c) from *C* to *D*.

15.15 Suppose a gas, initially at a temperature and pressure higher than the critical values, is cooled at constant pressure. Draw the line representing the process on Figure 15.8. Explain what happens when the temperature reaches the value T_C.

PROBLEMS

Note: A substance at 273 K (0°C) and 1 atm is said to be at **standard temperature and pressure** (STP).

15.1 Verify that if κ is the thermal coefficient of linear expansion of a material at a certain temperature, the thermal coefficient of volume expansion is approximately 3κ.

15.2 A block of copper has a mass of 0.3 kg. Calculate its change in volume when it is heated from 27 °C to 100 °C.

15.3 The table below gives the pressure in $\mathrm{N\,m^{-2}}$, and the volume (in $\mathrm{m^3}$) for a gas at two different temperatures. (a) For each case calculate the product pV. (b) How does the product pV vary if the temperature does not change? (c) At which temperature is the product pV larger? What conclusion do you reach? (d) Represent on a graph the values of p against V at each temperature. These lines are called **isothermals**.

Temperature = 27 °C		Temperature = 130 °C	
$p(\mathrm{N/m^2})$	$V(\mathrm{m^3})$	$p(\mathrm{N/m^2})$	$V(\mathrm{m^3})$
0.81×10^6	2.43	1.85×10^6	1.41
1.30×10^6	1.51	2.19×10^6	1.20
1.74×10^6	1.13	2.46×10^6	1.07
2.05×10^6	0.96	2.94×10^6	0.89
2.56×10^6	0.77	3.14×10^6	0.83

15.4 The table below gives the values of the pressure and volume of a gas at different temperatures. (a) Calculate the value of the product pV for each temperature. (b) Plot the values of pV as a function of temperature. (c) Does pV vary linearly with the temperature? (d) Extend the graph until it intersects the temperature axis and estimate the value of that temperature. Is it possible to cool the gas below that temperature? (e) Write the equation that relates pV with T that satisfies this data.

Temperature (°C)	p(Pa)	$V(\mathrm{m^3})$
210	4.3×10^5	0.85
120	2.8×10^5	1.06
0	1.5×10^5	1.38
−80	1.0×10^5	1.45
−120	0.84×10^5	1.35

15.5 (a) From Equation 15.7, show that if the temperature of an ideal gas is constant, then $pV = const.$ or $p_1V_1 = p_2V_2$, a result known as **Boyle's law**. (b) Show also that if the pressure is constant, then $V/T = const.$ or $V_1/T_1 = V_2/T_2$, a result known as **Charles' law**. (c) Finally, show that if the volume is constant, then $p/T = const.$ or $p_1/T_1 = p_2/T_2$, a result known as **Gay-Lussac's law**. These laws were known experimentally long before they were synthesized into Equation 15.7.

15.6 Verify that Equation 15.4 can also be written in the form $p = \rho(RT/M)$, where ρ is the density of the gas and M is its molar mass (expressed in $\mathrm{kg/mol^{-1}}$).

15.7 An air bubble with a volume of $10\ \mathrm{cm^3}$ is formed in the water 40 m below the surface of a lake. If the temperature of the bubble remains constant as the bubble rises, determine the volume of the bubble just before it reaches the surface of the lake. Recall that one atmosphere = 1.0×10^5 Pa.

15.8 (a) Find the volume of one mole of an ideal gas at STP; that is, at a temperature of 0 °C and a pressure of one atmosphere. (b) Also show that the number of molecules of any gas per cubic centimeter at STP is 2.687×10^{19}. This is called the **Loschmidt number**.

15.9 A tank of 50 liters contains a gas at 20 °C and at a pressure of 13×10^5 Pa above atmospheric pressure. A valve is opened and gas is released until its pressure exceeds the atmospheric pressure by 6×10^5 Pa. What volume would the released gas occupy at STP?

15.10 A tank with a volume of 10 000 liters contains compressed air at 27 °C and a pressure of 5 atm. The tank is heated until the gas reaches a temperature of 100 °C. Then a valve is opened and gas is released until the pressure is again 5 atm. Finally it is cooled back to 27 °C. (a) What is the final pressure? (b) What mass of air was released?

15.11 The **bulk modulus** of a substance at constant temperature is defined as

$$\kappa_T = -\frac{1}{V}\left(\frac{\mathrm{d}V}{\mathrm{d}p}\right)_T$$

Find κ_T for (a) an ideal gas, and (b) a real gas following the van der Waals equation.

15.12 (a) Calculate the pressure at 27 °C of one mole of (i) Xe, (ii) O_2, (iii) NO_2, (iv) CH_4 using the van der Waals equation. (b) Compare with the value corresponding to the ideal gas approximation. Consult Table 15.2 for the values of the coefficients a and b.

15.13 Calculate the thermal coefficient of pressure of a real gas using the virial expansion (Example 15.4) with the coefficients a and b given by Equation 15.22.

15.14 (a) What is the average kinetic energy of a gas molecule at a temperature of 25 °C? Express it in joules and in eV. (b) What is the corresponding rms velocity if the gas is (i) hydrogen, (ii) oxygen, (iii) nitrogen, (iv) helium (monatomic) and (v) carbon dioxide? (Note that the first three molecules are diatomic.)

15.15 Find the internal energy of one mole of an ideal gas at 0 °C (273 K). Does it depend on the nature of the gas? Why?

15.16 Compute the rms velocity of (a) helium atoms at 20 K, (b) nitrogen molecules at 27 °C, and (c) mercury atoms at 100 °C.

15.17 (a) Show that atmospheric pressure varies with height h according to $\ln(p/p_0) = -Mgh/RT$, where M is the effective molar mass of air. Assume that the temperature T is independent of height, which is not correct. (b) Estimate the pressure at the summit of Mt McKinley, whose height is 6.19×10^3 m, if the air temperature is 0 °C.

15.18 Repeat Problem 15.17, assuming that the atmospheric temperature decreases linearly with height according to $T = T_0 - \alpha h$, where α is constant, and verify that atmospheric pressure is $\ln(p/p_0) = (Mg/R\alpha)\ln(1 - \alpha h/T_0)$. Find the limiting value when $\alpha \to 0$.

15.19 The **critical point** (p_C, V_C, T_C) of one mole of a real gas (Figure 15.8) is obtained by making $(dp/dV)_T = 0$ and $(d^2p/dV^2)_T = 0$ in the van der Waals equation. Show that $V_C = 3N_A b$, $p_C = a/27b^2$ and $T_C = 8a/27bk$. (b) Also verify that $a = 3(V_C/N_A)^2 p_C$, $b = V_C/3N_A$ and $k = 8p_C V_C/3N_A T_C$. (c) Substituting these values into the van der Waals equation, verify that it can be written in the form

$$\left(\frac{p}{p_C} + 3\frac{V_C^2}{V^2}\right)\left(\frac{V}{V_C} - \frac{1}{3}\right) = \frac{8}{3}\frac{T}{T_C}$$

where V is the volume of one mole. (d) Since this result is independent of a and b, what do you conclude?

15.20 Using the results of Problem 15.19 and the values of the constants a and b listed in Table 15.2, compute the critical values for (a) helium, (b) hydrogen, (c) oxygen and (d) water. Compare with the experimental values given below.

	T_C(K)	p_C(10^5 Pa)	V_C(10^{-6} m^3 mol^{-1})
Helium	5.3	2.29	57.8
Hydrogen	33.3	13.0	65.0
Oxygen	154.8	50.8	78
Water	647.4	221.2	56

15.21 An empirical equation of state for real gases, proposed by Dieterici, is

$$p(V - Nb)e^{Na/VRT} = NRT$$

Write the equation in virial form and compare it with Equation 15.20. (Hint: recall the expansion $e^{-x} = 1 - x + \frac{1}{2}x^2 \ldots$)

16 Thermodynamics

William Thomson, Lord Kelvin, is well known for extraordinary contributions to thermodynamics, telegraphy and precise electrical measurements, among his other achievements. In 1848 Kelvin proposed the absolute scale of temperature, which is independent of any particular thermometric substance. Based on the works of Carnot, Rumford and Joule, Kelvin elaborated a dynamical theory of heat that contributed to the formulation of the principle of conservation of energy and the first law of thermodynamics.

16.1 Introduction

A piece of matter, either solid, liquid or gas, is a system of particles composed of a very large number of interacting atoms or molecules. Many processes involving exchanges of energy between a piece of matter and its surroundings can be analyzed without explicitly considering the atomic or molecular structure of matter. The study of such processes is the subject of **thermodynamics**, which was developed during the eighteenth and nineteenth centuries as a rather formal and elegant empirical theory. Macroscopic concepts such as temperature, heat and pressure were introduced and gradually related to molecular properties.

Thermodynamics is extremely important for engineering purposes, since heat engines began to be developed at about the same time. Thermodynamics reached its climax at the end of the nineteenth century, with the work of N. L. Sadi Carnot (1796–1832), Rudolf Clausius, James P. Joule, William Thompson (Lord Kelvin), and others.

16.2 Internal energy and work

Recall from Section 14.5 that when no external forces act on the particles of a system (i.e. if the system is isolated from external actions), the internal energy of the system does not change. In other words,

the internal energy of an isolated system of particles remains constant.

When there are external forces acting on the particles of the system, the internal energy does not, in general, remain constant. In this chapter we shall designate the internal energy by U without the suffix 'int'. Suppose that the system is initially in a state with internal energy U_0. When the state of the system is modified by external forces, after a certain time the internal energy is U. We designate the total work done during the process by the external forces acting on the particles of the system by W_{ext}. The work W_{ext} is a sum of many terms, one for each particle subject to an external force, and represents the exchange of energy between the system and its surroundings. Recalling Equation 14.13, the conservation of energy requires that

$$U - U_0 = W_{ext} \quad \text{or} \quad \Delta U = W_{ext} \tag{16.1}$$

That is,

the change in the internal energy of a system of particles is equal to the work done on the system by the external forces.

If work is done on the system (W_{ext} positive), its internal energy increases ($U > U_0$), but if work is done by the system (W_{ext} negative), its internal energy decreases ($U < U_0$).

When we are dealing with thermal engines, it is often preferable to compute the external work done *by* the system, denoted by W_{syst}, instead of the external work done *on* the system, W_{ext}. Since both works correspond to the same energy exchange, but to forces equal and opposite, the two works are equal in magnitude but have opposite signs; that is,

$$W_{syst} = -W_{ext}$$

In the future we shall write W instead of W_{syst} for the work done *by* the system. The systems we consider in thermodynamics are composed of a very large number of particles. Therefore, calculating the external work as a sum of individual works on each particle poses serious difficulties. For that reason, it has been found convenient to express W_{ext} as a sum of two terms, both of a statistical nature, each of which can be calculated and measured in a relatively simple manner. One is still called work and the other is called **heat**.

16.3 Many-particle systems: work

When a system exchanges energy with the surroundings, the exchange can take place by various mechanisms. For example, consider a gas in a cylinder that has a piston P that can move as a result of the pressure of the gas (Figure 16.1). Also,

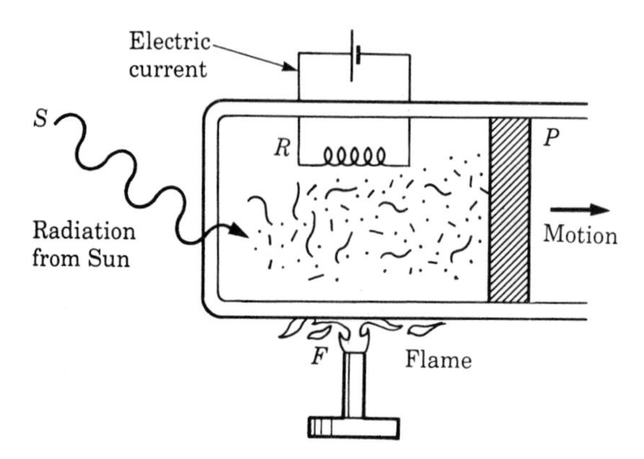

Figure 16.1 Energy exchanges in a system.

a flame F is applied to the cylinder, an electric current heats a resistor R inside the cylinder and the cylinder is exposed to radiation from the Sun S. Each one of these mechanisms produces an exchange of energy with the molecules of the gas. However, only one of the above energy exchanges corresponds to a measurable displacement under the action of a force: the motion of the piston P under the pressure exerted by the gas. It is the only energy exchange that can be calculated as force × displacement or work without reference to individual molecules. Thus we will adopt the convention of calling work the energy exchanges of a system of many particles that can be calculated by a macroscopic force and a collective displacement. It is sometimes called **mechanical work** because it is the type of work done by machines.

Suppose the pressure of the gas is p and the area of the piston is A. The force on the piston is then $F = pA$ (Figure 16.2). The work done by the gas when the piston is pushed the distance $\mathrm{d}x$ is $\mathrm{d}W = F\,\mathrm{d}x = pA\,\mathrm{d}x$ or, since $\mathrm{d}V = A\,\mathrm{d}x$ is the change in volume of the gas,

$$\mathrm{d}W = p\,\mathrm{d}V \tag{16.2}$$

When the volume changes from V_0 to V, the work done by the gas is

$$W = \int_{V_0}^{V} p\,\mathrm{d}V \tag{16.3}$$

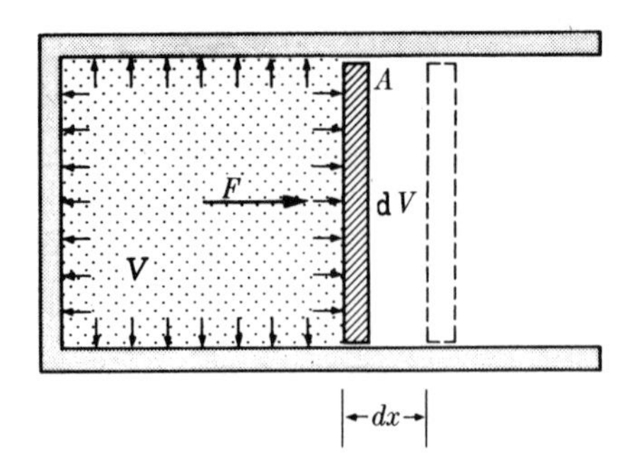

Figure 16.2 Work done in a gas expansion.

This expression is valid even if the gas is in a container whose volume can be changed by means other than a movable piston.

To compute the integral in Equation 16.3 we must know the relation between p and V at each state during the process. Assuming that this relation is known, we may represent the successive values of p and V on a diagram where the abscissa corresponds to the volume and the ordinate to the pressure. This is called a ***p*–*V* diagram**.

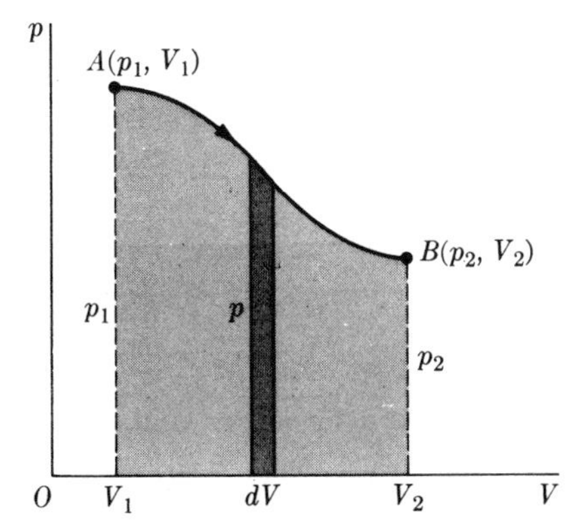

Figure 16.3 Diagram of a process in the $p-V$ plane. The work done by the system is indicated by the shaded area.

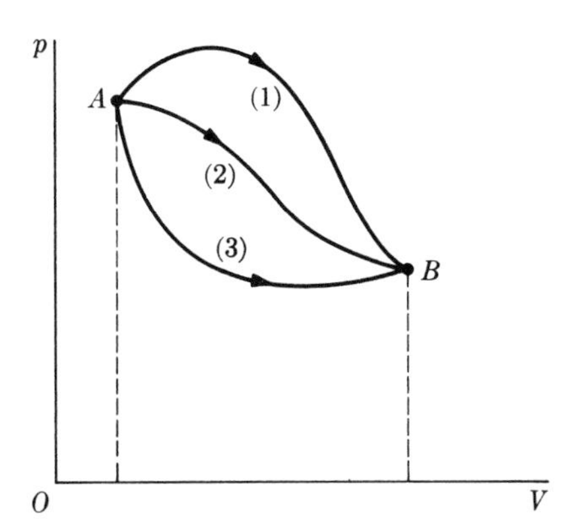

Figure 16.4 The work done in going from state A to state B depends on the process followed by the system.

In Figure 16.3 we have illustrated a process where a gas passes from state A to state B following a series of states indicated by the line joining the two points. The area of a strip of width $\mathrm{d}V$ and height p is $p\,\mathrm{d}V$. Therefore the work done by the system in going from A to B by the process is

$$W_{A\to B} = \int_{V_1}^{V_2} p\,\mathrm{d}V = \text{area under curve } AB$$

That is, the area under the curve AB in Figure 16.3 corresponds to the work done during the process or transformation. This result shows that *the work depends on the process* and not just on the change in volume.

Figure 16.4 indicates several processes, corresponding to curves (1), (2) and (3), all of which take a system from state A to state B. Since the area under each curve is different, the work done in each process is also different.

An interesting type of process is a **cycle**. This is a process where, at the end of the process, the system returns to the initial state. Therefore, in a $p-V$ diagram a cycle is represented by a closed line. One important example of a cyclical process is the operation of a thermal engine, where the cylinders move back and forth continuously.

Consider a cycle during which the system goes from A to B (Figure 16.5) along process (1) and returns from B to A along process (2). In going from A to B an expansion work equal to the area under curve (1) is done *by* the system. In returning from B to A the system does a negative work equal to the area under curve (2) because work is done *on* the system to compress it. The net work W done by the system during the cycle is thus the shaded area enclosed within the curve representing the cycle. That is,

$$W_{\text{cycle}} = \oint p\,\mathrm{d}V = \text{area under (1)} - \text{area under (2)} = \text{area within } A(1)B(2)A$$

Therefore, the work done by a thermal engine per cycle can be computed once we have the $p-V$ diagram of the cycle. If we know how many cycles the engine

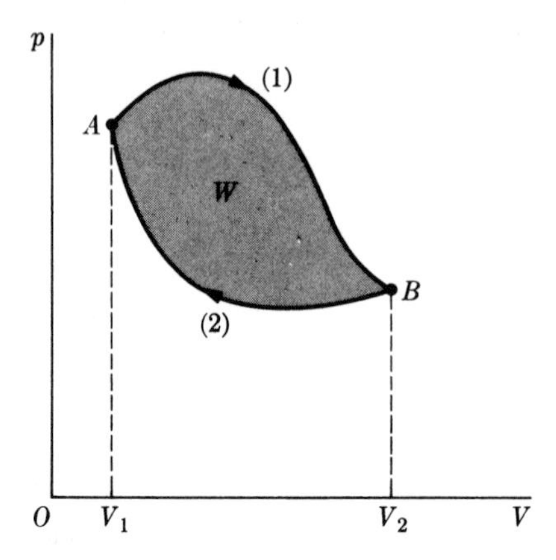

Figure 16.5 Cycle. The work done by the system in describing the cycle clockwise is equal to the area enclosed by the cycle in a $p-V$ diagram.

performs per unit time, we can obtain the **power** of the engine. If the cycle is reversed, then a net work is done on the system. This is the case, for example, for a refrigerator.

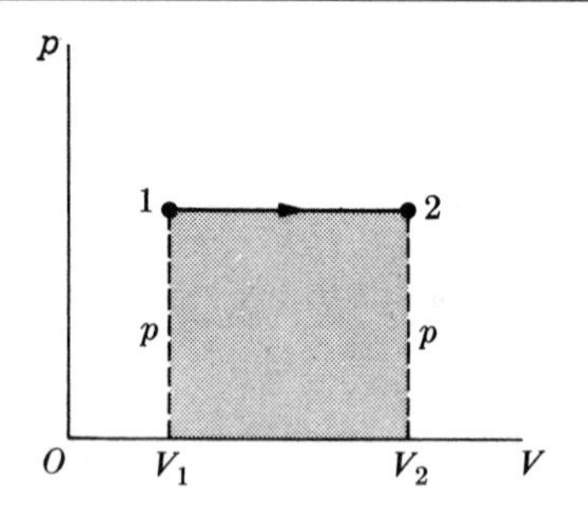

Figure 16.6 Isobaric expansion.

EXAMPLE 16.1

Work done in an expansion at constant pressure. When a process occurs at constant pressure, it is called **isobaric**. The process is illustrated in Figure 16.6. Apply the results to a gas with an initial volume of $0.30\ \mathrm{m}^3$ that exerts a pressure $p = 2 \times 10^5\ \mathrm{N\,m}^{-2}$. Find the work done by the gas when the volume expands to $0.45\ \mathrm{m}^3$ under constant pressure.

▷ When the pressure p remains constant, the work done by the system is

$$W = \int_{V_1}^{V_2} p\,\mathrm{d}V = p\int_{V_1}^{V_2} \mathrm{d}V = p(V_2 - V_1) \quad \text{or} \quad W = p\Delta V \tag{16.4}$$

This result is completely general and applies to any system whose volume changes under a constant pressure. Recalling that for an ideal gas $pV = kNT$, we may also write

$$W = kN(T_2 - T_1) \quad \text{or} \quad W = \mathrm{N}R(T_2 - T_1) = \mathrm{N}R\Delta T \tag{16.5}$$

which is valid only for ideal gases. Inserting the numerical values into Equation 16.4, we obtain $W = 3 \times 10^4\ \mathrm{J}$.

EXAMPLE 16.2

A gas expands in such a way that the temperature of the gas remains constant. This process is called **isothermal**. Find the work done when the volume expands from V_1 to V_2. The process is illustrated in Figure 16.7.

▷ Using Equation 15.7, $pV = kNT$, with T constant, we obtain

$$W = \int_{V_1}^{V_2} p\, \mathrm{d}V = kNT \int_{V_1}^{V_2} \frac{\mathrm{d}V}{V}$$

Integrating

$$W = kNT \ln \frac{V_2}{V_1} \quad \text{or} \quad W = \mathrm{N}RT \ln \frac{V_2}{V_1} \qquad \textbf{(16.6)}$$

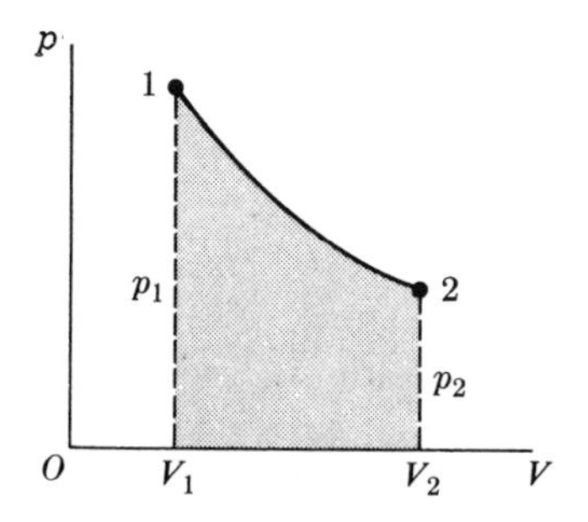

Figure 16.7 Isothermal expansion.

Therefore the work done in an isothermal expansion depends on the temperature and the ratio V_2/V_1 between the two volumes (called the **expansion ratio**). In the design of internal combustion engines, the compression (or expansion) ratio is one of the factors that determine the power of the engine.

16.4 Many-particle systems: heat

Many energy exchanges between a system and its surroundings do not result in a collective displacement. This is the case of the energy exchanges of the flame and the resistor with the gas molecules in the system of Figure 16.1, which result from collisions of high energy molecules in the flame or in the resistor with gas molecules or molecules in the walls of the container. In the case of the Sun, the energy transfer is through the absorption of **radiation**, which does not involve molecular collisions but a transfer of electromagnetic energy to the molecules of a gas. In fact, we use the first two methods every day in a stove and radiation in a microwave oven for cooking.

Those processes involve a large number of energy exchanges between the particles of the system and the surroundings. At each interaction, a small amount of energy is exchanged. But the total energy exchanged cannot be expressed macroscopically as force × displacement. If we could compute each one of these small amounts of energy and add all of them, we would have the total energy exchanged by the system. However this technique is impossible in practice because of the large number of terms involved. Accordingly, we shall define a new macroscopic or statistical concept called **heat**.

Heat refers to the aggregate of the individual exchanges of energy between a system and its surroundings that occur as a result of collisions between the molecules of the system and the molecules of the surroundings without any collective displacement. Therefore, heat is composed of a very large number of very small individual energy exchanges that cannot be expressed collectively as an average force times an average displacement.

By convention, the heat Q is considered positive when it corresponds to a net energy absorbed by the system and negative when it is equivalent to a net energy lost by the system. In the first case we say that heat is *absorbed* by the system and in the second, we say that heat is *given off* by the system. When a system suffers a transformation where heat is neither absorbed nor rejected, the transformation is said to be **adiabatic**.

Heat must not be considered as a new or different form of energy. It is just a name given to a special form of *energy transfer* where a very large number of particles participate through random collisions. Before the concept of interactions and the atomic structure of matter were clearly recognized, physicists had classified energy into two groups: *mechanical* energy, corresponding to kinetic and gravitational potential energy, and *non-mechanical* energy, divided into heat, chemical energy, electrical energy, radiation etc. This division is no longer justified. Today we recognize only kinetic and potential energy of particles. Potential energy is indicated by a different expression, depending on the nature of the corresponding physical interaction. Work, heat and radiation are simply mechanisms of energy transfer between particles. 'Chemical energy' is just a macroscopic term used to describe energy associated with electrical interactions in atoms and molecules that results in atomic rearrangements in molecules, and 'thermal' energy is associated with the random motion of the molecules of a system.

When two systems are in thermal equilibrium (that is, have the same temperature (Section 15.2)), there is no net exchange of energy between them. That is true only in a statistical sense, because individual molecules may exchange energy, but, on average, the same amount of energy is exchanged in one direction as in the other. Thermal equilibrium exists between two systems when the average molecular energies of the two interacting systems are the same. Then no net exchange of energy by molecular collision or heat transfer is possible. Therefore,

> *in two systems in thermal equilibrium (i.e. at the same temperature), there is no net energy exchanged as heat between the systems.*

We may conclude that energy is exchanged as heat only when the temperature of the two systems is different. Energy is transferred from the system at higher temperature (larger molecular energies) to the system at lower temperature (smaller molecular energies). This is the principle on which most heating systems operate.

Since heat corresponds to an energy exchange, it must be expressed in joules. However, heat is sometimes expressed in a unit called the **calorie**, whose definition was adopted in 1948 as

$$1 \text{ calorie} = 4.1840 \text{ J}$$

The calorie was first introduced as a unit of heat measurement when the nature of heat was unknown. It is practically equal to the amount of energy absorbed by one gram of water when its temperature increases by 1 °C. But the calorie is simply another unit for measuring energy, and not heat alone.

16.5 Many-particle systems: energy balance

We have seen that in systems composed of a very large number of particles, we may express the *total energy exchange* of a system with its surrounding as the sum of two parts:

$$\text{Total energy exchange} = Q + W_{\text{ext}}$$

where W_{ext} is the energy exchanged by the system with the surroundings expressed as external work *on* the system. It is computed as an average force times a distance, as discussed in Section 16.3. Q is the energy exchanged as heat, as discussed in Section 16.4. The conservation of energy, expressed by the equation $\Delta U = W_{ext}$, must now be written in the equivalent but more general and convenient form

$$\Delta U = Q + W_{ext} \tag{16.7}$$

where a clear distinction is made between the two types of energy exchanges. This equation may be expressed in words by saying that

the change of internal energy of a system is equal to the heat ***absorbed*** *plus the external work done* ***on*** *the system.*

Equation 16.7 can be seen pictorially in Figure 16.8(a): Heat Q is *absorbed by* the system and work W_{ext} is *done on* the system. Their sum $Q + W_{ext}$ is *stored* as internal energy, $\Delta U = U - U_0$ of the system.

Instead of writing the external work W_{ext} done *on* the system, the external work W done *by* the system is often used, especially in dealing with thermal engines. As explained before, W is the negative of the work done on the system. Making $W_{ext} = -W$ we have, instead of Equation 16.7,

$$\Delta U = Q - W \tag{16.8}$$

This relation states that

the change of internal energy of a system is equal to the heat ***absorbed*** *minus the external work done* ***by*** *the system.*

Equation 16.8 is illustrated in Figure 16.8(b): heat Q is *absorbed by* the system, work W is *done by* the system, and the difference $Q - W$ is *stored* as internal energy $\Delta U = U - U_0$ of the system.

The statements related to Equations 16.7 and 16.8 constitute what is called the **first law of thermodynamics**. This is simply the law of conservation of energy applied to systems having a very large number of particles. The external work has been conveniently split into two statistical terms, one still called work and the other called heat. A system can exchange energy with its surroundings by other mechanisms that do not correspond to our definitions of Q and W_{ext}. In fact, a very important process, as indicated in Figure 16.1, is the absorption (or emission) of radiation. For example, most of the energy exchanges of the Earth with its surroundings consist of the absorption of radiation from the Sun and the emission

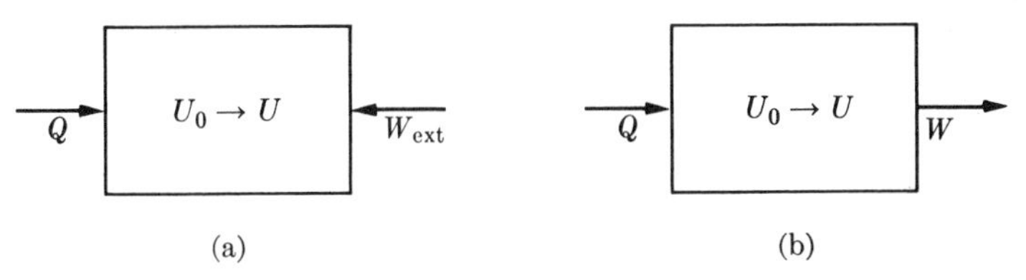

Figure 16.8 Relation between heat, work and internal energy.

of radiation back into space. Also, cooking in a microwave oven consists in absorption of radiation by the food.

If we designate the radiation energy absorbed by a system by R, we must then replace Equation 16.7 by

$$\Delta U = Q + W_{ext} + R \tag{16.9}$$

which is a more general formulation of the first law of thermodynamics. However, we shall defer until Chapters 30 and 31 any further consideration of radiation energy.

It is important to keep track of the signs when using Equations 16.8 or 16.9.

	Positive	Negative
Q	heat absorbed	heat released
W	work done *by* the system	work done *on* the system
R	radiation absorbed	radiation emitted
ΔU	internal energy increases	internal energy decreases

EXAMPLE 16.3

Expression of the first law for an infinitesimal transformation, that is, when the system experiences only a very small change.

▷ If the transformation suffered by a system is infinitesimal, we must replace ΔU by dU, Q by dQ, and W by dW in Equation 16.8, resulting in:

$$dU = dQ - dW \tag{16.10}$$

This equation expresses the first law of thermodynamics for infinitesimal transformations.

When the work dW is due to an expansion or change in volume, we have that $dW = p\,dV$, in accordance with Equation 16.2, and

$$dU = dQ - p\,dV$$

There may be other kinds of work in addition to expansion that will yield different expressions in terms of the variables of the problem.

EXAMPLE 16.4

Change of internal energy of one kilogram of H_2O when it passes (a) from ice (solid) to water (liquid) at 0 °C (273 K) and (b) from water to steam (gas) at 100 °C (373 K). In both cases assume the pressure is 1 atm or 1.01325×10^5 Pa.

▷ To solve this problem it is necessary to calculate how much heat is absorbed by a kilogram of H_2O when it changes **phase**, from ice to water and from water to steam, to increase the potential energy of the molecules without changing their kinetic energy. We also need to know the corresponding change in volume during these two phase changes.

(a) When ice melts at 0 °C, it is necessary to supply energy to break the molecular bonds that form the solid. The amount of heat required to melt one kilogram of ice is equal to

$$Q = 79.71 \times 10^3 \text{ cal} = 3.335 \times 10^5 \text{ J}$$

which is called the **heat of fusion**. The volumes of ice and water at a pressure of one atmosphere and 0 °C are $1.0908 \times 10^{-3}\ \text{m}^3\,\text{kg}^{-1}$ and $1.0001 \times 10^{-3}\ \text{m}^3\,\text{kg}^{-1}$, respectively. Therefore, the change in volume of one kilogram of H_2O during the process is

$$\Delta V = (1.0001 - 1.0908) \times 10^{-3}\ \text{m}^3 = -9.07 \times 10^{-5}\ \text{m}^3$$

Since the process takes place at constant pressure, the work done by the system is

$$W = p\Delta V = (1.01325 \times 10^5 \text{ Pa}) \times (-9.07 \times 10^{-5}\ \text{m}^3) = -9.19 \text{ J}$$

The work is quite small compared to the heat absorbed, and may be neglected. Therefore, the change in internal energy is

$$\Delta U = Q - W \approx 3.335 \times 10^5 \text{ J}$$

(b) To vaporize water at 100 °C and one atmosphere, it is necessary to supply energy to break the molecular bonds in the liquid phase. The amount of heat required to vaporize one kilogram of water is equal to

$$Q = 539.55 \times 10^3 \text{ cal} = 2.2575 \times 10^6 \text{ J}$$

which is called the **heat of vaporization**. The volumes of one kilogram of water and steam at 100 °C and 1 atmosphere are: $1.00 \times 10^{-3}\ \text{m}^3$ and $1.686\ \text{m}^3$. Accordingly, the change in volume during vaporization is

$$\Delta V = (1.686 - 0.001)\text{m}^3 = 1.685\ \text{m}^3$$

The work done during vaporization is then

$$W = p\Delta V = (1.01325 \times 10^5 \text{ Pa}) \times (1.685\ \text{m}^3) = 1.707 \times 10^5 \text{ J}$$

which is comparable with Q. Therefore, the change in internal energy of one kilogram of water at vaporization is

$$\Delta U = Q - W = 2.2575 \times 10^6 \text{ J} - 1.707 \times 10^5 \text{ J} = 2.0868 \times 10^6 \text{ J}$$

Note that in melting ice, the work due to the contraction is negligible, while in vaporizing water the work due to the expansion is appreciable.

In 1 kg of water there are 55.56 moles or 3.34×10^{25} molecules. The change in internal energy *per molecule* in the **fusion** process is then 9.985×10^{-21} J or 6.2×10^{-2} eV. Likewise, in the **vaporization** process the change in internal energy *per molecule* is 6.246×10^{-20} J or 0.390 eV. These energies are very small compared to the dissociation energy of water molecules (recall Section 14.8). Therefore, water molecules separate as whole units during melting and vaporization. This problem also shows that the intermolecular forces holding water molecules together in the solid or liquid state, although of electrical nature, are much weaker than the electric forces that hold the atoms of a water molecule together.

16.6 Special processes

The internal energy of a system depends only on the state of the system, since it is determined by the energy distribution of the molecules in the state considered. Therefore, when a system goes from a state where the internal energy is U_0 to a state with internal energy U, the change ΔU in internal energy is the same regardless of the process followed in going from the first state to the second. For example, referring to Figure 16.4, the change in internal energy of the system in going from state A to state B is the same for all paths shown in the figure. However, as explained before, the work done by the system does depend on the path.

Writing Equation 16.8 in the form

$$Q = \Delta U + W \tag{16.11}$$

we see that, although ΔU does not depend on the process, the heat Q also depends on the path since W does. Therefore, the heat absorbed along the three paths shown in Figure 16.4 is different in each case. We shall consider now some special transformations.

Cyclic transformation. When a system describes a cycle and returns to the initial state the change in internal energy is zero, that is, $U = U_0$ or $\Delta U_{\text{cycle}} = 0$. Then Equation 16.11 reduces to

$$Q = W \tag{16.12}$$

that is,

> *in a cyclic transformation, the work done by the system is equal to the net heat absorbed by the system and the change in internal energy is zero.*

This is the principle on which thermal engines operate; heat is supplied during each cycle to the engine, which in turn does work on the surroundings. During the cycle, the engine also releases some heat at a lower temperature (see Section 16.10). Refer now to Figure 16.5. Note that the area within the closed line corresponding to the cycle gives the work done by the system, as well as the heat it has absorbed.

Adiabatic transformation. When the transformation is **adiabatic**, the system neither absorbs nor rejects heat ($Q = 0$). Then Equation 16.8 becomes

$$\Delta U = -W \tag{16.13}$$

Equation 16.13 indicates that

> *in an adiabatic transformation the internal energy of the system decreases (increases) by an amount equal to the work done by (on) the system.*

In the case of an ideal gas the internal energy depends only on the temperature. Therefore the temperature of the gas decreases in an adiabatic expansion (W positive and ΔU negative) and increases during an adiabatic compression (W negative and ΔU positive).

The drop in temperature of a gas when it experiences an adiabatic expansion as a result of a sudden drop in pressure has many practical applications. One is in refrigeration using a **throttle valve**, which is a device that separates two regions at different pressures. In some refrigeration systems, a liquid is on the high pressure side of the throttle valve. The liquid is quickly evaporated when it passes to the low pressure side. This results in a drop in temperature because the kinetic energy of the molecules in the liquid is used to increase the potential energy of the molecules into a vapor (heat of vaporization).

Isochoric transformation. If no work is done by the system ($W = 0$), Equation 16.11 gives

$$\Delta U = Q \tag{16.14}$$

Equation 16.14 says that

> *when no work is done, the change in internal energy of the system is equal to the heat absorbed.*

This result applies, for example, to a process at constant volume, also called **isochoric transformation**. When a solid or a liquid is heated, the expansion work is negligible and all the heat absorbed goes into increasing the internal energy and thus the temperature.

Radiative process. If only radiation energy is absorbed (or emitted), such as when cooking with a microwave oven or when a body is exposed to the Sun's radiation, we must write

$$\Delta U = R$$

As indicated before, processes involving radiation energy will be considered in detail in Chapters 30 and 31.

EXAMPLE 16.5

Heat absorbed by an ideal gas during an isothermal expansion.

▷ During an isothermal expansion the temperature of the system remains constant. For an ideal gas the average kinetic energy of the molecules remains constant during such a process (recall Equation 15.15). Since in an ideal gas the intermolecular forces are assumed to be zero, the internal energy of the gas is entirely molecular kinetic energy and therefore depends only on the temperature (recall Equation 15.18). This was discussed in Example 15.3. Therefore, during an isothermal expansion of an ideal gas the internal energy does not change: $\Delta U = 0$. Then Equation 16.11 becomes $Q = W$. This means that, in an isothermal expansion of an ideal gas, all the heat absorbed is used to perform the expansion work. The reverse is true for an isothermal compression: the work done on the ideal gas is equal to the heat released by the gas. (We call attention to the fact that the relation $Q = W$ is valid only for an isothermal process in an ideal gas, whereas Equation 16.12 is valid for all substances but restricted to cyclic processes.)

Using Equation 16.6, we may write the heat absorbed by an ideal gas during an isothermal expansion as

$$Q = \text{N}RT \ln \frac{V_2}{V_1} \quad \text{or} \quad Q = kNT \ln \frac{V_2}{V_1} \tag{16.15}$$

For real gases these expressions are a good approximation to the extent that the intermolecular forces can be neglected.

16.7 Heat capacity

The **heat capacity** of a substance is defined as the heat absorbed by one mole of the substance per unit change of temperature. Thus if N moles absorb a heat dQ when the temperature increases dT, the heat capacity is

$$C = \frac{1}{\text{N}} \frac{dQ}{dT} \tag{16.16}$$

Heat capacity is expressed in J K^{-1} mole^{-1} in the SI system of units. However it is customary to use the alternative unit cal K^{-1} mole^{-1}.

Since the heat absorbed depends on the process, there is a heat capacity associated with each process. The two most widely used are the **heat capacity at constant pressure** (C_p) and the **heat capacity at constant volume** (C_V), which we may express as

$$C_p = \frac{1}{\text{N}} \left(\frac{dQ}{dT}\right)_{p=const.} \qquad C_V = \frac{1}{\text{N}} \left(\frac{dQ}{dT}\right)_{V=\text{const.}} \tag{16.17}$$

Note that when the volume is constant and therefore no work is done, we have that dQ = dU and

$$C_V = \frac{1}{\text{N}} \left(\frac{dU}{dT}\right)_{V=\text{const.}} \tag{16.18}$$

From the definition of the calorie, it follows that the average heat capacity of water at constant atmospheric pressure is 18.00 cal K^{-1} mole^{-1}.

The heat capacity of a substance depends on the temperature at which it is measured. This is the case because, as temperature increases, new modes for increasing the energy of the molecules may become available without necessarily increasing their kinetic energy. However, when the heat capacity may be considered as constant over a range of temperatures, the heat absorbed or rejected when the temperature changes from T_1 to T_2 is

$$Q = \text{N}C(T_2 - T_1)$$

Table 16.1 Heat capacities for various liquids and solids at room temperature

Liquid	C(cal K^{-1} $mole^{-1}$)	Solid	C(cal K^{-1} $mole^{-1}$)
Water (H_2O)	18.00	*Metals*	
Mercury	6.660	Aluminum	5.664
Methyl alcohol (CH_4O)	17.45	Copper	5.784
Ethyl ether $(C_2H_5)_2O$	39.21	Gold	5.916
Toluol (C_6H_8)	29.13	Silver	6.009
Carbon tetrachloride (CCl_4)	30.92		
		Non-metals	
		Ice (−20 °C)	8.370
		Zinc sulphide (ZnS)	11.20
		Quartz (SiO_2)	10.45

Otherwise an average heat capacity is used. It is important to know the heat capacity of substances (such as those fluids used for heating or cooling), since they give an indication of the amount of energy required to heat or cool the substance.

Table 16.1 gives the average heat capacity at constant pressure for several solids and liquids. For solids and liquids there is very little difference between C_p and C_V. Table 16.2 gives the heat capacities at constant pressure and at constant volume for several gases. Note that in all cases C_p is larger than C_V for gases. This is because at constant volume no work is done and all heat absorbed goes into increasing the internal energy. On the other hand, at constant pressure we must

Table 16.2 Heat capacities, at constant pressure and constant volume, for some gases at room temperature

Gas	C_p(cal K^{-1} $mole^{-1}$)	C_V(cal K^{-1} $mole^{-1}$)	γ
Monatomic			
He	5.004	3.014	1.660
Ar	4.990	2.993	1.667
Ne	4.966	3.024	1.642
Hg	4.983	2.991	1.666
Diatomic			
H_2	6.887	4.891	1.408
N_2	6.899	4.924	1.401
O_2	7.079	5.056	1.400
CO	6.919	5.335	1.297
NO	6.962	4.944	1.394
Triatomic			
CO_2	9.333	7.179	1.300
H_2O	7.946	5.956	1.334
N_2O	9.379	7.084	1.324
CS_2	12.19	9.838	1.239
Polyatomic			
Methyl alcohol (CH_4O)	13.30	10.59	1.256
Ethyl ether ($(C_2H_5)_2O$)	31.72	30.98	1.024
Chloroform ($CHCl_3$)	17.20	17.20	1.110

supply an additional amount of heat to account for the work done by the system. The quantity

$$\gamma = \frac{C_p}{C_V} \tag{16.19}$$

is also given in Table 16.2; this ratio of heat capacities appears in many calculations. For example, it can be shown that when an ideal gas suffers an adiabatic transformation, the pressure and volume are related by the expression (see the derivation below)

$$pV^{\gamma} = const. \qquad \text{or} \qquad p_1 V_1^{\gamma} = p_2 V_2^{\gamma} \tag{16.20}$$

Similarly, the work done by an ideal gas during an adiabatic transformation can be calculated by the expression

$$W = \frac{p_1 V_1 - p_2 V_2}{\gamma - 1} = \frac{\mathrm{N}R(T_1 - T_2)}{\gamma - 1} \tag{16.21}$$

Equation of state for the adiabatic transformation of an ideal gas

When a system undergoes an **adiabatic transformation** ($Q = 0$), we have $\Delta U = -W$. If the transformation is infinitesimal, we must rewrite this relation as $\mathrm{d}U = -\mathrm{d}W$. When the work is only due to a change in volume, $\mathrm{d}W = p\,\mathrm{d}V$, and therefore $\mathrm{d}U = -p\,\mathrm{d}V$. For an ideal gas, for which U depends only on the temperature, Equation 16.18 gives $\mathrm{d}U = \mathrm{N}C_V\,\mathrm{d}T$ and the preceding equation becomes

$$\mathrm{N}C_V\,\mathrm{d}T + p\,\mathrm{d}V = 0$$

By differentiating the equation of state of an ideal gas ($pV = \mathrm{N}RT$), we obtain

$$p\,\mathrm{d}V + V\,\mathrm{d}p = \mathrm{N}R\,\mathrm{d}T$$

The last two equations may be combined to eliminate $\mathrm{d}T$. We can then write

$$C_V V\,\mathrm{d}p + (C_V + R)p\,\mathrm{d}V = 0$$

As shown in Example 16.6, $C_p - C_V = R$ for an ideal gas and the above relation can be written as $C_V V\,\mathrm{d}p + C_p p\,\mathrm{d}V = 0$. Introducing the definition $\gamma = C_p/C_V$ and separating variables, we have

$$\frac{\mathrm{d}p}{p} + \gamma\frac{\mathrm{d}V}{V} = 0$$

Finally, integrating this equation,

$$\int \frac{\mathrm{d}p}{p} + \gamma \int \frac{\mathrm{d}V}{V} = const.$$

which gives $\ln p + \gamma \ln V = const.$, which is equivalent to

$$pV^{\gamma} = const.$$

This is expression 16.20, giving the relation between pressure and volume in an adiabatic transformation of a gas.

From $pV^{\gamma} = const.$, we have $p = const. V^{-\gamma}$ and the work done by an ideal gas during an adiabatic transformation is

$$W = \int_{V_1}^{V_2} p \, dV = const. \int_{V_1}^{V_2} V^{-\gamma} \, dV = const. \frac{(V_2^{1-\gamma} - V_1^{1-\gamma})}{1-\gamma} = \frac{p_2 V_2 - p_1 V_1}{1-\gamma}$$

Reversing the order of the terms, the work done by the system during an adiabatic transformation can be written as

$$W = \frac{p_1 V_1 - p_2 V_2}{\gamma - 1} = \frac{\text{N}R(T_1 - T_2)}{\gamma - 1}$$

which is expression 16.21.

A gas that experiences an adiabatic expansion has a final pressure, for the same change in volume, *less* than if the expansion is isothermal (Figure 16.9) because the temperature decreases during the adiabatic expansion, as discussed in Section 16.6.

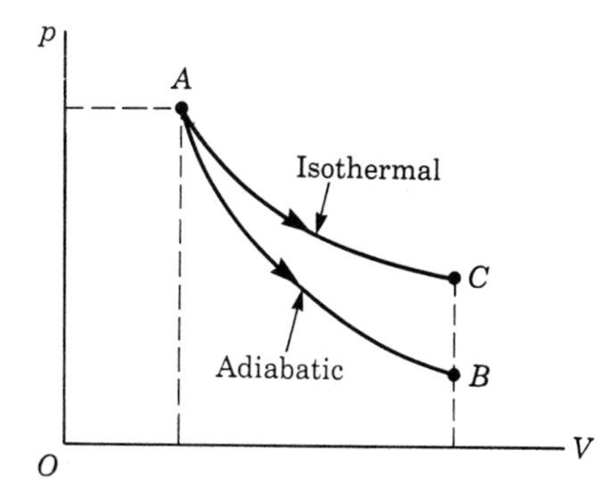

Figure 16.9

EXAMPLE 16.6

Heat capacities of an ideal monatomic gas.

▷ The internal energy of an ideal monatomic gas is given by Equation 15.18 as $U = \frac{3}{2}\text{N}RT$. When the temperature changes from T to $T + dT$, the change in internal energy is $dU = \frac{3}{2}\text{N}R \, dT$. If the process is at constant volume, so that no work is done, Equation 16.14 indicates that the heat absorbed is

$$(dQ)_{V=\text{const.}} = dU = \frac{3}{2}\text{N}R \, dT$$

Therefore Equation 16.17 gives $C_V = \frac{3}{2}R = 2.9807 \text{ cal K}^{-1} \text{mole}^{-1}$ so that we may write $U = \text{N}C_V T$.

If the change in temperature is at constant pressure, the work done by the gas, according to Equation 16.5, with $T_2 - T_1$ replaced by dT, is $dW = \text{N}R \, dT$. The heat absorbed, using Equation 16.10, is

$$(dQ)_{p=\text{const.}} = dU + dW = \frac{3}{2}\text{N}R \, dT + \text{N}R \, dT = \frac{5}{2}\text{N}R \, dT$$

Applying Equation 16.17 we obtain $C_p = \frac{5}{2}R = 4.9678 \text{ cal K}^{-1} \text{ mole}^{-1}$. Thus C_p and C_V are independent of the temperature for ideal gases. In real gases it is necessary to take into account their internal potential energy, in which case C_p and C_V depend slightly on the temperature.

From the values for C_p and C_V we obtain

$$\gamma = \frac{C_p}{C_V} = \frac{5}{3} = 1.67 \qquad \text{and} \qquad C_p - C_V = R \tag{16.22}$$

Comparing with the values given in Table 16.2 for γ, we see that there is a reasonable agreement for monatomic gases. The calculations do not apply to polyatomic gases because the rotational and vibrational energies of the molecules must be taken into account to obtain the internal energy of polyatomic gases. In general, if the molecules have f internal degrees of freedom (recall Section 15.6), $E_{ave} = [(3 + f)/2]kT$ (Equation 15.25), and

$$\gamma = \frac{5 + f}{3 + f}$$

Recalling Section 15.6, for the case of diatomic gases, where we took rotational energy into account, $f = 2$ and $\gamma = 7/5 = 1.4$, also in agreement with the table. If vibrational energy must also be considered, we must write $f = 2 + 2 = 4$, and $\gamma = 9/7 = 1.3$, which agrees with the values for CO and NO.

EXAMPLE 16.7

The pressure of one mole of helium at STP drops adiabatically to 0.4 atm. Calculate the final volume and temperature and the work done.

▷ Since helium is a monatomic gas we may assume $\gamma = 5/3$. Then initially $p_1 = 1$ atm, $V_1 = 22.4$ liters and $T_1 = 273$ K. Then we apply $pV^\gamma = const.$, which, in this case, is $p_1 V_1^{5/3} = p_2 V_2^{5/3}$, or $V_2 = V_1(p_1/p_2)^{3/5}$. Inserting values we get

$$V_2 = (22.41)(1 \text{ atm}/0.4 \text{ atm})^{3/5} = 22.41 \times 1.733 = 38.821$$

To calculate the final temperature we use $pV = \text{N}RT$, which we write in the form $p_1V_1/T_1 = p_2V_2/T_2$. Then

$$\frac{(1\,\text{atm})(22.41)}{273\,\text{K}} = \frac{(0.4\,\text{atm})(38.821)}{T_2}$$

from which we obtain $T_2 = 189.2$ K or -84 °C. Therefore, the temperature dropped 84 K during the adiabatic expansion.

To calculate the work done we use Equation 16.22; that is,

$$W = \frac{\text{N}R(T_1 - T_2)}{\gamma - 1} = \frac{(1\,\text{mol})(8.31\,\text{J K}^{-1}\,\text{mol}^{-1})(84\,\text{K})}{(2/3)} = 1045.8\text{ J}$$

16.8 Reversible and irreversible processes

When a process or transformation occurs very slowly, we may reasonably assume that the system at each instant is very close to equilibrium. That is, the transformation proceeds by infinitesimal steps, so that at each step the system is

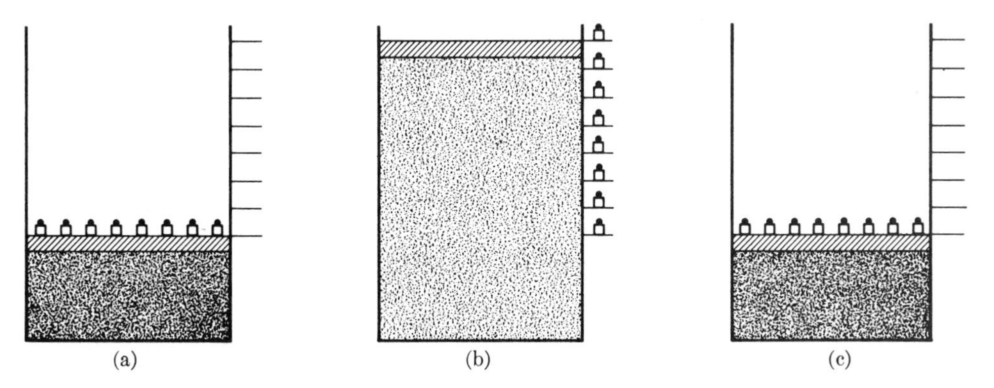

Figure 16.10 Reversible expansion and compression of a gas.

only slightly disturbed from its state of equilibrium. A process of this kind is said to be **reversible**.

We may use the expansion of a gas to illustrate a reversible process. Suppose that the piston in Figure 16.10 is held in position by many small weights, as indicated in (a). At equilibrium the pressure of the compressed gas is equal to the pressure due to the weights plus the atmospheric pressure. If we remove one of the weights, by sliding it to one side on to a platform, the external pressure decreases by a small amount and the equilibrium of the gas is slightly disturbed. The gas then undergoes a small expansion until equilibrium is (quickly) restored. When the process is repeated a number of times, the gas eventually expands up to the volume shown in (b) and the weights, which were previously on top of the piston, are stored as shown. Since the process has occurred very slowly, we may assume that the gas has remained continuously close to equilibrium, although it may have exchanged energy with the surroundings. We say that the expansion has been reversible.

To restore the gas to the original state, all we have to do is place back, in the reverse order, the same weights we took off. The energy exchanges with the surroundings are also reversed. At the end the gas is in its initial state, having completed a cycle, and no change has been produced in the surroundings. In other words,

in a cycle entirely composed of reversible transformations, no observable change is produced either in the system or in the surroundings.

On the other hand, an **irreversible** process occurs when the system deviates to a great extent from the equilibrium state. During the process, statistical quantities, such as pressure and temperature, defined only for equilibrium states, are undefined. Eventually, at the end of the process, the system returns to equilibrium in a new state, characterized by a certain pressure and temperature.

In general, irreversible processes occur at great speed. We may again use the expansion of a gas to illustrate an irreversible process. The gas in Figure 16.11(a) is as it was in Figure 16.10(a), but with all the weights consolidated into one, labeled A. Note that there is also a weight B at the upper level. If weight A is removed, the external pressure suddenly drops and the gas expands rapidly,

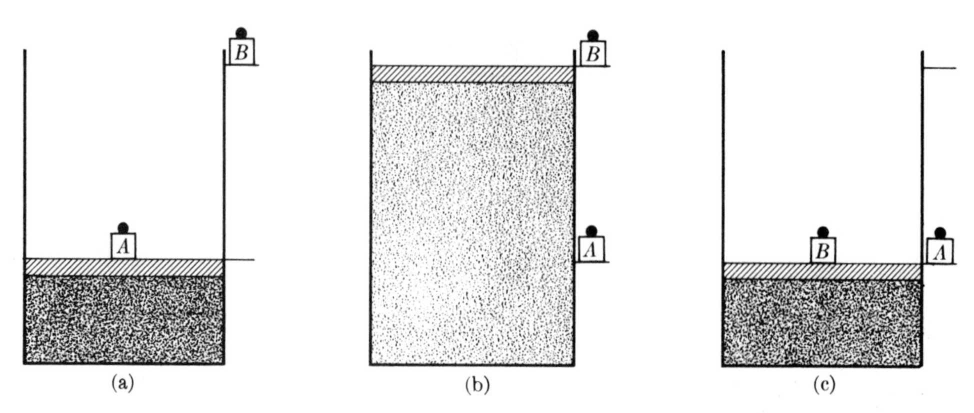

Figure 16.11 Irreversible expansion and compression of a gas.

with great turbulence in its molecular motion; i.e. the process is irreversible. Pressure and temperature are undefined for the system. Finally the piston comes to rest at a certain position. Eventually equilibrium is restored, with well-defined pressure and temperature throughout the system, as indicated in (b).

To take the gas back to its initial state, we may move weight B to the top of the piston. The piston then moves down in a process that may (but most probably will not) repeat, in reverse, the previous process. In the end, when equilibrium is once more restored, the gas is again in its initial state, as shown in (c); the gas has completed a cycle. However, a definite change has occurred in the surroundings, for weight B, which was initially at the top, is now at the bottom. Thus

> *in a cycle composed, in part or total, of irreversible transformations, the system returns to its initial state, but a permanent change is produced in the surroundings.*

A reversible transformation can be represented by a line on a p–V diagram joining the initial and final states because p, V and T are well defined at each step in the process. Thus diagrams such as those in Figures 16.3 through 16.7 correspond to reversible transformations. But an irreversible process cannot be represented in such a way.

An interesting aspect, easily recognizable from the above logic, is that a system does more work, and by the same reason absorbs more heat, when the process is reversible than when it is irreversible. For that reason engineers design thermal engines whose cycle of operation is as reversible as possible.

16.9 Entropy and heat

Thus far we have used four concepts to describe processes in many-particle systems when there are energy exchanges with other systems: internal energy, work, heat and temperature. Around 1854 Rudolf Clausius introduced another concept that is very useful in describing thermal processes. Later on (1865), he called this new concept **entropy**, a word taken from the Greek words '*trepein*', which means 'change' and '*en*', which means 'in' or 'content'. The letter S is used to designate entropy.

Suppose that a system, at temperature T, absorbs a small amount of heat, dQ, during an infinitesimal **reversible** transformation. Then the entropy S of a system is defined as a quantity whose **change** is calculated by the relation

$$dS = \frac{dQ}{T} \tag{16.23}$$

The change dS is positive if the system absorbs heat ($dQ > 0$) and negative if heat is rejected by the system ($dQ < 0$), since T is always positive, and zero if the reversible process is adiabatic ($dQ = 0$). Entropy is measured in $J\,K^{-1}$.

When the system passes from one state to another by a reversible transformation, the change in entropy of the system is

$$\Delta S = S_2 - S_1 = \int_1^2 \frac{dQ}{T} \tag{16.24}$$

Note that T is not necessarily a constant in Equation 16.24. An important property of entropy that is justified in Chapter 17 and illustrated in Example 16.9 for an ideal gas, is that

the change in entropy of a system when it goes from one state to another is independent of the transformation followed.

However, to compute the change in entropy, we must assume a reversible transformation unless the entropy is known as a function of the state variables. Thus, referring to Figure 16.12, the change in entropy, ΔS, is the same whether the system follows the reversible transformations (a) or (b) because in both cases the initial and final states are the same. A consequence of this property is that

when a system describes a cycle and returns to its initial state, the change in its entropy is zero;

that is,

$$\Delta S_{\text{cycle}} = \oint \frac{dQ}{T} = 0 \tag{16.25}$$

where the integral applies only to a reversible cycle.

These two properties show that the entropy of a system is a property of the state of the system, like internal energy, and is independent of how the system reached that state, whether by a reversible or an irreversible process. This is one of the underlying reasons for the importance of the concept of entropy. However, when the transformation is irreversible we may use a hypothetical reversible transformation that connects the initial and final states and use Equation 16.24 to calculate ΔS (Figure 16.13).

We have now identified several quantities that can be used to characterize the state of a system: its volume (V), its pressure (p), its temperature (T), its internal energy (U), and its entropy (S). It can be easily recognized that there is a formal parallelism between entropy and internal energy, although they are unrelated physical concepts: their changes are independent of the process followed

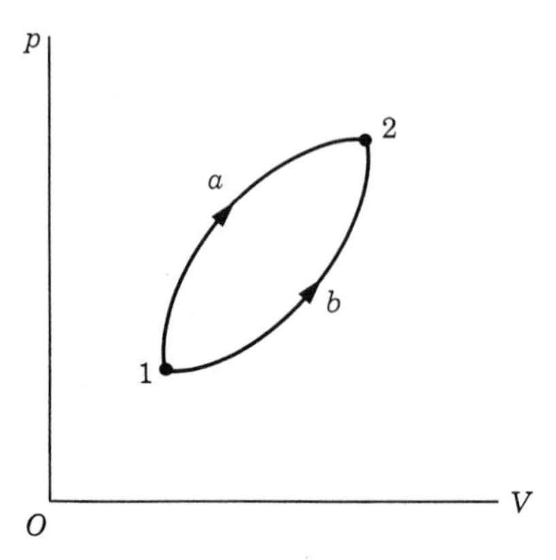

Figure 16.12 Reversible process.

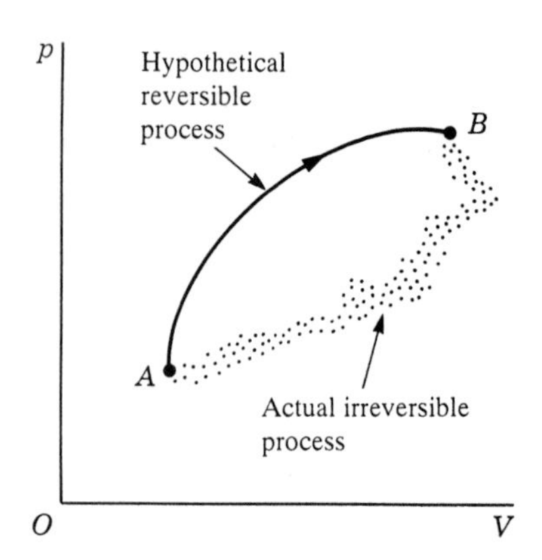

Figure 16.13 Irreversible process.

(being zero for a cyclic process) and, because entropy is defined as a difference, the zero of entropy can be chosen arbitrarily, as in the case for internal energy.

For an adiabatic reversible transformation, $dQ = 0$ and Equation 16.26 gives

$$\Delta S = 0 \qquad \text{or} \qquad S = const. \tag{16.26}$$

Thus adiabatic reversible transformations are at constant entropy, and for that reason they are also **isentropic**. Note, however, that an adiabatic *irreversible* transformation is not necessarily isentropic, because relation 16.24 does not hold for such a case.

From Equation 16.23 $dQ = T\,dS$ and therefore

$$Q = \int_1^2 T\,dS \tag{16.27}$$

gives the heat absorbed in going from state 1 to state 2 by a reversible transformation; this integral depends on the particular transformation.

The reversible transformation may be represented by a line in a diagram where the ordinate corresponds to the temperature T and the abscissa to the entropy S, as in Figure 16.14. Then $T\,dS$ is the area of a strip of width dS. Therefore Q is equal to the area under the curve from S_1 to S_2. For a cycle, such as $A(1)B(2)A$ (Figure 16.15) the net heat absorbed by the system in the cycle is, from Equation 16.27,

$$Q = \oint T\,dS = \text{area within the cycle in } (T, S) \text{ coordinates} \tag{16.28}$$

This is also the work done by the system during the cycle since $\Delta U = 0$ for any cycle. We wish to emphasize that entropy is a variable that may be used to describe a process in the same way as pressure, volume or temperature.

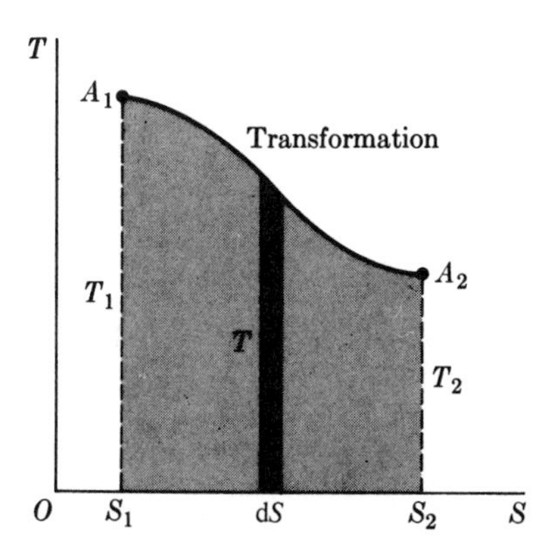

Figure 16.14 Diagram of a reversible process in the T–S plane. The heat absorbed during the process is given by the shaded area.

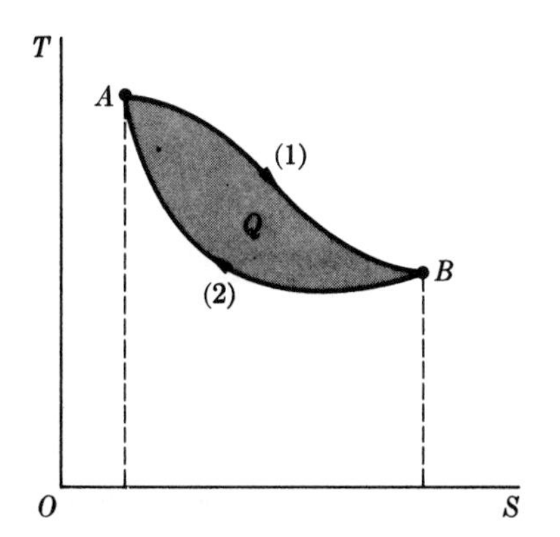

Figure 16.15 Cycle. The heat absorbed by the system in describing the cycle clockwise is equal to the area enclosed by the cycle in a T–S diagram.

EXAMPLE 16.8

Change in entropy during an isothermal reversible process.

▷ In the case of an **isothermal reversible** transformation, T is constant and Equation 16.24 becomes

$$\Delta S = \frac{1}{T}\int_1^2 \mathrm{d}Q = \frac{Q}{T}$$

or

$$Q = T\Delta S \qquad \textbf{(16.29)}$$

For example, ice melts at STP (0 °C or 273.1 K). During the process ice absorbs 79.71 kcal kg^{-1} or 1435 calories per mole. Therefore, the change in entropy of one mole of water at melting is

$$\Delta S = \frac{1435\,\text{cal}}{273.1\,\text{K}} = 5.25\ \text{cal}\,\text{K}^{-1}\ \text{mol}^{-1} = 1.255\ \text{J}\,\text{K}^{-1}\ \text{mol}^{-1}$$

For an ideal gas, using Equation 16.15, the change in entropy during an isothermal process is

$$\Delta S = \frac{Q}{T} = \text{N}R\ \ln\frac{V_2}{V_1} \qquad \text{or} \qquad \Delta S = kN\ \ln\frac{V_2}{V_1} \qquad \textbf{(16.30)}$$

since $\text{N}R = kN$. Therefore the entropy change depends only on the expansion ratio and is independent of the temperature. For example, if one mole of an ideal gas expands isothermally from a volume of 20 liters to a volume of 30 liters, the change in entropy of the gas is

$$\Delta S = (1\ \text{mol})(8.314\,\text{J}\,\text{K}^{-1}\ \text{mol}^{-1})\ln\frac{301}{201} = 3.37\ \text{J}\,\text{K}^{-1}\ \text{mol}^{-1}$$

EXAMPLE 16.9

Entropy of an ideal gas as a function of the volume and the temperature.

▷ An ideal gas is a sufficiently simple system that its entropy can be correlated with other state variables. Consider an ideal gas at temperature T occupying a volume V. Then, if N is the number of molecules of the gas, the pressure of the gas is $p = kNT/V$. Also, the energy of the gas is $U = \frac{3}{2}kNT$ (Equation 15.8). If the gas experiences an infinitesimal reversible transformation, the change in entropy of the gas is, using Equation 16.10,

$$\mathrm{d}S = \frac{\mathrm{d}Q}{T} = \frac{\mathrm{d}U + p\,\mathrm{d}V}{T} = \frac{\mathrm{d}U}{T} + \frac{p\,\mathrm{d}V}{T}$$

Introducing the values of U and p given above, we obtain

$$\mathrm{d}S = \tfrac{3}{2}\,kN\,\frac{\mathrm{d}T}{T} + kN\,\frac{\mathrm{d}V}{V} = kN\left(\tfrac{3}{2}\,\frac{\mathrm{d}T}{T} + \frac{\mathrm{d}V}{V}\right)$$

Integrating, we obtain

$$S = kN\left(\tfrac{3}{2}\int\frac{\mathrm{d}T}{T} + \int\frac{\mathrm{d}V}{V}\right) = kN\left(\tfrac{3}{2}\ln T + \ln V\right) + S_0$$

where S_0 is the entropy of some reference state. The above expression can also be written in the alternative form

$$S = kN\ln(VT^{3/2}) + S_0 \qquad \textbf{(16.31)}$$

This expression confirms that the entropy depends only on the state of the system, in this case determined by the volume V and the temperature T of the gas.

When the gas is subject to a transformation that takes it from the state (V_1, T_1) to the state (V_2, T_2), the change of entropy is

$$\Delta S = S_2 - S_1 = kN\ln(V_2/T_2^{3/2}) - kN\ln(V_1T_1^{3/2})$$

or

$$\Delta S = kN\left(\ln\frac{V_2}{V_1} + \tfrac{3}{2}\ln\frac{T_2}{T_1}\right) \qquad \textbf{(16.32)}$$

and is independent of the process followed.

Note that if the initial and final states have the same temperature, $T_1 = T_2$, the results stated in Equation 16.30 for a reversible isothermal transformation are obtained. But our result is more general, because it is independent of the process followed to go from the initial to the final state and directly relates the entropy to the state of the gas.

16.10 Efficiency of a thermal engine operating in a Carnot cycle

A thermal engine is a system that, operating cyclically, absorbs heat and performs work. We shall consider engines that operate in a Carnot cycle.

A **Carnot cycle** is a cycle composed of two isothermal and two adiabatic reversible transformations. It is represented by the rectangle $ABCD$ of Figure 16.16, where AB and CD are the isothermal transformations and BC and DA are the adiabatic (or isentropic) transformations. This diagram is valid for all Carnot cycles, whatever the specific mechanism used or the substances that describe the cycle. The cycle is described in a clockwise fashion, as indicated by the arrows. We designate the temperatures of the two isothermal processes by T_1 and T_2, with T_1 greater than T_2. During the isothermal process AB at the higher temperature T_1, the entropy increases and the system absorbs an amount of heat Q_1. During the isothermal process CD at the lower temperature T_2, the entropy decreases and an amount of heat Q_2 is rejected so that Q_2 is a negative quantity. During the two adiabatic transformations, no heat is exchanged with the surroundings and the entropy remains constant. The changes in entropy during each transformation, according to Equations 16.29 and 16.26, are

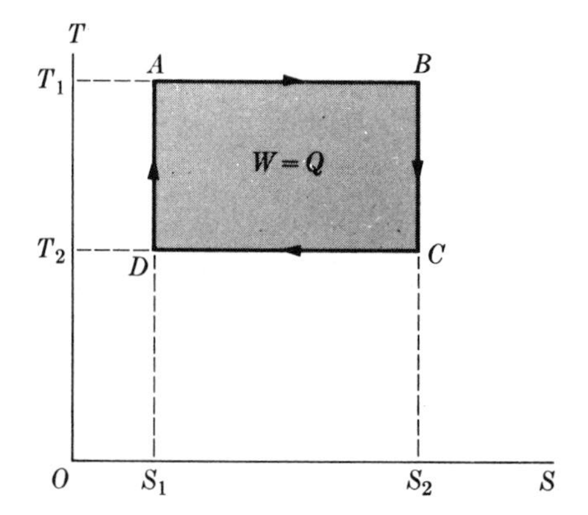

Figure 16.16 Carnot cycle in a T–S diagram. AB and CD are isothermals; BC and DA are adiabatics.

$\Delta S_{AB} = Q_1/T_1$	isothermal, heat absorbed, Q_1 positive
$\Delta S_{BC} = 0$	adiabatic, $Q = 0$
$\Delta S_{CD} = Q_2/T_2$	isothermal, heat rejected, Q_2 negative
$\Delta S_{DA} = 0$	adiabatic, $Q = 0$

Since the net change in entropy in the cycle is zero,

$$\Delta S_{\text{cycle}} = \frac{Q_1}{T_1} + \frac{Q_2}{T_2} = 0 \qquad \text{or} \qquad \frac{Q_1}{T_1} = -\frac{Q_2}{T_2} \tag{16.33}$$

This gives the relation between the heats absorbed and rejected and the corresponding temperatures. Since T_1 is larger than T_2, then Q_1 (heat absorbed) is larger than $-Q_2$ (heat rejected). Note that we have made no special assumption about the internal structure of the substance. Therefore, Equation 16.33 holds for any substance that undergoes a Carnot cycle, whether the substance is an ideal gas or not. In the case of a gas, the cycle is accomplished by a series of expansions and compressions (see Example 16.10).

The net heat absorbed by the system during the cycle is $Q = Q_1 + Q_2$ or $Q = Q_1 - |Q_2|$. This is also equal to the work W done by the system during the cycle. According to Equation 16.28, we may write

$$W = Q = \text{area of rectangle } ABCD = (T_1 - T_2)(S_2 - S_1)$$

Then, according to Equation 16.29, $Q = T\Delta S$ so that

$$Q_1 = T_1(S_2 - S_1)$$

The **efficiency** of a thermal engine operating according to a Carnot cycle is defined as the ratio of the total work done to the heat absorbed at the highest temperature, per cycle. That is,

$$\text{efficiency} = \frac{W}{Q_1} = \frac{(T_1 - T_2)(S_2 - S_1)}{T_1(S_2 - S_1)} = \frac{T_1 - T_2}{T_1} \tag{16.34}$$

Thus we see that

> *the efficiency of a thermal engine operating according to a reversible Carnot cycle is independent of the working substance and depends only on the two operating temperatures.*

This result is commonly known as **Carnot's theorem**. The theorem is important in the design of the **thermal engines** (Figure 16.17).

In practice, the source of heat at high temperature in a thermal engine may be, for example, a furnace where some fuel is burnt, or the core of a nuclear reactor. The heat sink at the low temperature may be air, a water current (river) or a large water reservoir (lake or ocean). In most cases, the working substance is a liquid (water) but it may also be a gas (air, CO_2, He). In internal combustion engines, the heat source is the combustion of a fuel (gasoline, alcohol, diesel oil) and the working substance is the gas mixture that results from the combustion.

A **heat pump** operates in the reverse cycle of a thermal engine (Figure 16.17(b)). Heat is extracted from a cold body, for which work has to be done on the system and a larger amount of heat is transferred to a hot body. Heat pumps are widely used for heating and for cooling, since they can either increase the temperature of the hotter body or reduce the temperature of the cooler body, depending on the direction in which the pump operates.

If the engine operates in an irreversible cycle, the work done per cycle is less than for a reversible cycle and therefore the efficiency of the engine is less than the value given by Equation 16.34, which is the maximum efficiency of an engine operating between the temperatures T_1 and T_2. It is for that reason that thermal engines must operate as reversibly as possible.

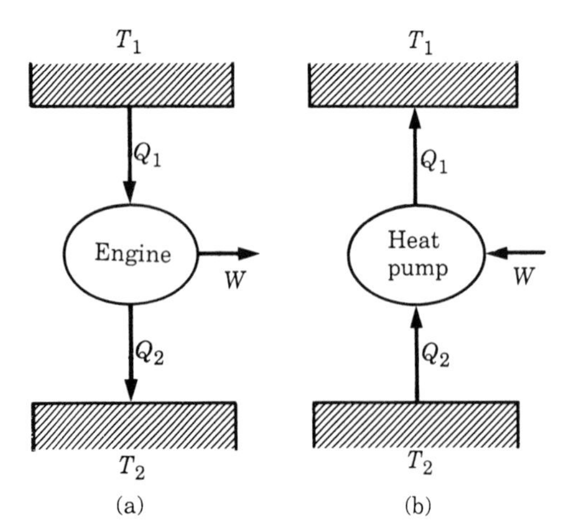

Figure 16.17 Thermal engine and heat pump. $T_1 > T_2$.

EXAMPLE 16.10

Carnot cycle of an ideal gas.

▷ In the preceding section we analyzed the Carnot cycle in a generic form, without reference to any particular substance. It is instructive to analyze the Carnot cycle for the specific case of an ideal gas. The cycle in this case consists of two expansions and two compressions.

In Figure 16.18 the Carnot cycle of an ideal gas has been represented on a p–V diagram. The processes $1 \to 2$ and $3 \to 4$ are isothermals at the temperatures T_1 and T_2. The adiabatic transformations are $2 \to 3$ and $4 \to 1$. Then for each of the processes the relations are $p_1 V_1 = p_2 V_2$, $p_2 V_2^\gamma = p_3 V_3^\gamma$, $p_3 V_3 = p_4 V_4$, $p_4 V_4^\gamma = p_1 V_1^\gamma$. Manipulating the four equations to eliminate the pressures, we find that $V_2/V_1 = V_3/V_4$. Therefore, to complete the cycle the two expansion ratios must be the same.

The total work done during the cycle is $W = W_{12} + W_{23} + W_{34} + W_{41}$. Using Equations 16.6 for the isothermal processes and 16.21 for the adiabatic and noting that $\ln V_4/V_3 = -\ln V_3/V_4 = -\ln V_2/V_1$, we have that

$$W_{12} = \text{N}RT_1 \ln \frac{V_2}{V_1} \qquad W_{23} = \frac{\text{N}R(T_1 - T_2)}{\gamma - 1}$$

$$W_{34} = -\text{N}RT_2 \ln \frac{V_2}{V_1} \qquad W_{41} = \frac{\text{N}R(T_2 - T_1)}{\gamma - 1}$$

Adding these four equations we get

$$W = \text{N}R(T_1 - T_2) \ln \frac{V_2}{V_1}$$

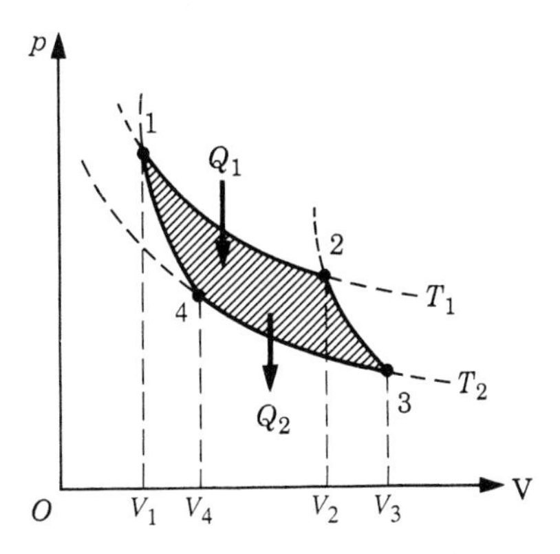

Figure 16.18 p–V diagram of an ideal gas Carnot cycle.

for the work done by the gas during the cycle. Recalling from Example 16.5 that during an isothermal process the heat absorbed by an ideal gas is equal to the work done by the gas, we have that $W_{12} = Q_1$ and $W_{34} = Q_2$, where Q_2 is negative because heat is rejected. The values of Q_1 and Q_2 are found from Equation 16.15. Then, using the above expressions for W_{12}, W_{34} and W, we obtain

$$\frac{Q_1}{T_1} = -\frac{Q_2}{T_2} \qquad \text{or} \qquad \frac{Q_1}{T_1} + \frac{Q_2}{T_2} = 0$$

and

$$\text{efficiency} = \frac{W}{Q_1} = \frac{T_1 - T_2}{T_1}$$

which are Equations 16.33 and 16.34.

Historically these relations were first derived in 1824 by Sadi Carnot prior to the introduction of the concept of entropy. In fact it was the relation $Q_1/T_1 = -Q_2/T_2$ that prompted Rudolf Clausius later on (upon some suggestions from Kelvin) to develop the concept of entropy and extend the theory beyond the specific case of an ideal gas. In particular, since $\Delta S_{12} = Q_1/T_1$, $\Delta S_{34} = Q_2/T_2$ and $\Delta S_{23} = \Delta S_{41} = 0$, we can see that

$$\Delta S_{\text{cycle}} = \Delta S_{12} + \Delta S_{23} + \Delta S_{34} + \Delta S_{41} = \frac{Q_1}{T_1} + \frac{Q_2}{T_2} = 0$$

so that the change of entropy is zero for the cycle. It is possible to prove that this result is true for any cycle described by the ideal gas and not just for a Carnot cycle.

16.11 The law of entropy

The law of conservation of energy (or the first law of thermodynamics) is not sufficient to determine which processes may occur in the universe. Some processes tend to develop in one direction but not in the opposite. In particular, irreversible processes always go in a certain direction: heat is transferred from a higher temperature to a lower temperature, a gas moves from higher to lower pressure, and so on. Furthermore, when all the systems that participate in a process are taken into account, so that they constitute a larger *isolated* system, it is found that when Q_i is the heat absorbed or rejected by component i at the temperature T_i, the quantity $\sum_i Q_i/T_i$ (where the summation extends over all the components of the system), is zero only if the process can be considered totally reversible. When the process, or even part of the process, is irreversible, $\sum_i Q_i/T_i$ is found to be larger than zero. That is, for an isolated system, $\sum_i Q_i/T_i \geqslant 0$, which in view of Equation 16.23 is equivalent to

$$\Delta S = \sum_i \Delta S_i \geqslant 0 \tag{16.35}$$

where ΔS_i is the change of entropy of component i in the process. That is:

> *for processes in an isolated system, where heat is exchanged among its different subsystems or components, the total entropy of the system remains constant if the processes are reversible or increases if the processes are irreversible.*

The above statement constitutes the **second law of thermodynamics**. It determines, together with the conservation of energy, which processes can occur in an isolated system and, by extension, in the universe. Processes occur only if energy is conserved and in the direction where the entropy of the whole system increases or does not change at all. In a process, the entropy of part of a system may decrease, but the entropy in some other part must increase by the same or a larger amount. This is why, as indicated in Section 16.8, when a component of a system undergoes a cycle that is at least partially irreversible, there is a permanent change in the system corresponding to an increase in entropy of the whole system even if the entropy of the component itself has not changed after completing the cycle.

EXAMPLE 16.11
Change of entropy in heat transfer.

▷ Suppose we have two bodies at temperatures T_1 and T_2, with $T_1 > T_2$. Heat can be transferred from one body to another; otherwise they constitute an isolated system. If the two bodies are placed in contact, an amount of energy in the form of heat is transferred from the hotter to the cooler body by an irreversible process without intervention of an external agent. If Q is the heat transferred, the change in entropy of each body is

$$\Delta S_1 = -\frac{Q}{T_1} \qquad \Delta S_2 = +\frac{Q}{T_2}$$

since the body at temperature T_1 rejects the heat Q and the body at temperature T_2 absorbs it. The change in entropy of the system is

$$\Delta S = \Delta S_1 + \Delta S_2 = -\frac{Q}{T_1} + \frac{Q}{T_2} = Q\left(\frac{T_1 - T_2}{T_1 T_2}\right) > 0$$

Note that if the energy were transferred from the cooler body to the hotter, the entropy of the system would have decreased, a situation not observed in nature except if an external agent intervenes. This is the case of a heat pump or a refrigerator, which use electric energy for their operation and therefore are not isolated systems.

EXAMPLE 16.12
Change of entropy during the free expansion of a gas.

▷ An interesting example of an irreversible process in an isolated system is the free expansion of a gas that was considered in Example 15.3 and referred to in Section 16.8. Consider a gas occupying a container of volume V (Figure 16.19) separated from an empty container, also of volume V, by a partition that can be removed without doing any work. We assume that the walls of the container are perfect insulators so that the gas cannot absorb or reject heat from the surroundings and therefore can be considered an isolated system.

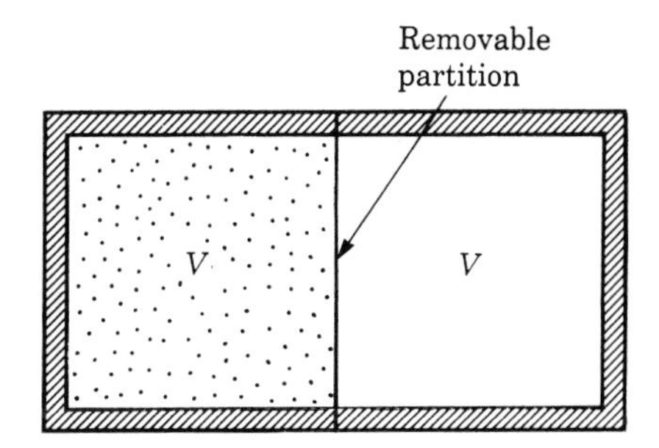

Figure 16.19 Free expansion of a gas.

Initially the gas is at temperature T. When we remove the partition, the gas rushes to fill the whole volume, $2V$. The process is irreversible and adiabatic ($Q = 0$). Also, the gas does no work ($W = 0$) and therefore the change in internal energy is also zero ($\Delta U = 0$). When the gas settles down and equilibrium is restored, we find that there is no change of temperature ($\Delta T = 0$) if it is an ideal gas.

We may follow either of two methods for calculating the change of entropy. For example, we may imagine a reversible transformation that connects the same initial and final states. Since the temperature has not changed during the expansion, we may consider a reversible isothermal expansion at temperature T, in which the gas pushes a piston very slowly until the volume is doubled. During the imaginary reversible process the gas does a work W and absorbs the heat Q. Then we apply Equation 16.30 with $V_1 = V$ and $V_2 = 2V$, so that

$$\Delta S = \mathrm{N}R \ln 2 > 0$$

Of course, during the imaginary reversible expansion, the gas is not isolated. However, its change of entropy is the same as for the irreversible adiabatic free expansion of the isolated gas because the initial and final states are the same in both cases. Therefore, the entropy of the isolated gas during the free expansion has increased as corresponds to an irreversible process in an isolated system.

Since we also know the entropy of an ideal gas as a function of the volume and the temperature, we could have used Equation 16.32 of Example 16.9, with $V_1 = V$, $V_2 = 2V$ and $T_1 = T_2$ to calculate ΔS.

Note that the gas after the irreversible expansion can be restored to its initial state by a reversible isothermal compression, decreasing its entropy, but that requires the intervention of an external agent. Thus the isolated gas cannot undergo a transformation by which its entropy decreases.

QUESTIONS

16.1 What is the difference between the work–energy relation, as stated by Equation 14.13, and the first law of thermodynamics?

16.2 What is the fundamental difference between work and heat in the thermodynamic sense?

16.3 What is meant by heat of fusion and heat of vaporization?

16.4 Referring to Figure 16.9, for a given change in volume which work is larger: that for an adiabatic expansion or for an isothermal expansion? Why? What is the meaning of the area between the two curves in Figure 16.9?

16.5 Why must the temperature of a gas decrease during an adiabatic expansion?

16.6 How is the temperature of a gas kept constant in an isothermal expansion?

16.7 Why is there no heat capacity at constant volume listed for solids and liquids?

16.8 The **specific heat** of a substance is defined as the amount of heat absorbed by the unit mass of a substance to increase its temperature by one degree. What is its relation with the heat capacity?

16.9 Analyze the possible changes in entropy of an isolated system that (a) is *not* in equilibrium and (b) is in equilibrium.

16.10 Verify that, for an infinitesimal reversible transformation, the first law can be written in the form $\mathrm{d}U = T\,\mathrm{d}S - p\,\mathrm{d}V$. Why does the transformation have to be reversible?

16.11 Referring to Equation 16.32, what is the change in entropy of a gas in (a) an isochoric transformation, (b) an isothermal transformation, (c) an isobaric transformation?

16.12 Obtain an expression for the change in entropy of an ideal gas when it expands at constant pressure from the temperature T_1 to T_2. Repeat for the case when the pressure is increased at constant volume. In which case is the change in entropy larger? Why?

16.13 Referring to Equation 16.31, verify that the entropy of an ideal gas can be expressed by $S = kN \ln(T^{5/2}/p) + S_0$, where S_0 is a new constant.

16.14 Referring to Equation 16.14, show that the entropy of an ideal gas in terms of the pressure and the volume is expressed by $S = kN \ln(V^{5/2}p^{3/2}) +$

S_0. Verify that the relation $pV^{5/3} = e^{2(S-S_0)/3kN}$ holds. Compare with Equation 16.19 for an adiabatic reversible transformation ($S = const.$) when γ for an ideal gas is used.

16.15 What is meant by a thermal engine? What is the best way of increasing the efficiency of a thermal engine? Is it possible to design a thermal engine that has 100% efficiency?

16.16 Is it possible for a cyclic reversible engine to absorb heat at a certain temperature and transform it completely into work without rejecting some heat at a lower temperature?

PROBLEMS

16.1 (a) Calculate the change in internal energy of one mole of an ideal gas when its temperature changes from 0 °C to 100 °C. (b) Do we also have to specify how the pressure and volume changed?

16.2 The process referred to in the previous problem occurs at constant volume. (a) What was the work done by the gas? (b) What was the heat absorbed?

16.3 Repeat Problem 16.1 when the process occurs at constant pressure.

16.4 A gas is maintained at a constant pressure of 20 atm while it expands from a volume of $5 \times 10^{-3}\ m^3$ to a volume of $9 \times 10^{-3}\ m^3$. Calculate the amount of energy that must be supplied as heat to the gas (a) to maintain its internal energy at a constant value, (b) to increase its internal energy by the same amount as the external work done. Express your result in calories and in joules.

16.5 A gas, initially at a pressure of 4 atm and having a volume of $4 \times 10^{-2}\ m^3$, expands at constant temperature. Calculate (a) the change of internal energy, (b) the work done and (c) the heat absorbed when the volume is doubled.

16.6 When a system is taken from state A to state B along the path ACB (see Figure 16.20), the system absorbs 80 J of heat and does 30 J of work.

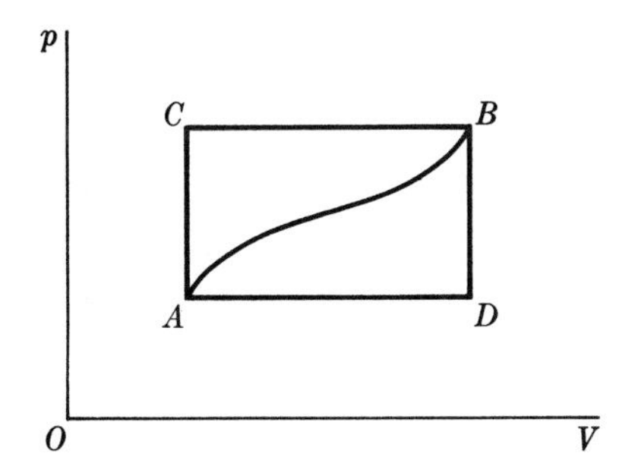

Figure 16.20

(a) Given that the work done is 10 J, how much heat is absorbed by the system along path ADB? (b) The system is returned from state B to state A along the curved path. The work done on the system is 20 J. Does the system absorb or liberate heat, and how much? (c) Given that $U_A = 20$ J and $U_D = 60$ J, determine the heat absorbed in the processes AD and AB.

16.7 Ten moles of a gas initially at STP are heated at constant volume until the pressure is doubled. Calculate (a) the work done, (b) the heat absorbed and (c) the change in internal energy of the gas.

16.8 A gas undergoes the cycle shown in Figure 16.21. The cycle is repeated 100 times per minute. Determine the power generated.

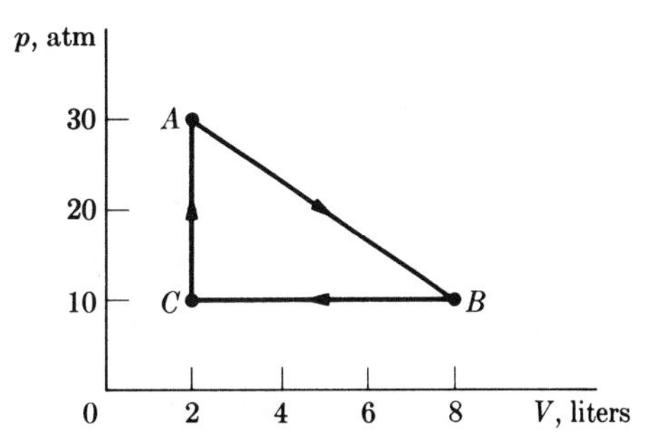

Figure 16.21

16.9 The heat capacity C_p of most substances (except at very low temperature) can be satisfactorily expressed by the empirical formula $C_p = a + 2bT - cT^{-2}$, where a, b and c are constants and T is the absolute temperature. (a) In terms of a, b and c, calculate the heat required to raise the temperature of one mole of the substance at constant pressure from T_1 to T_2. (b) Find the average heat capacity between temperatures T_1 and T_2. (c) If C_p is measured in $J\ K^{-1}\ mol^{-1}$, what are the units of the coefficients a, b and c?

16.10 When the formula of Problem 16.9 is applied to magnesium, the numerical values of the constants are $a = 25.7 \times 10^3\ \mathrm{J\,K^{-1}\,mol^{-1}}$, $b = 3.13\ \mathrm{J\,K^{-2}\,mol^{-1}}$ and $c = 3.27 \times 10^8\ \mathrm{J\,K\,mol^{-1}}$. Calculate (a) the heat capacity of magnesium at 300 K and (b) the average heat capacity between 200 K and 400 K.

16.11 Express the equation for an *adiabatic* process of an ideal gas in terms of: (a) pressure and temperature, (b) volume and temperature.

16.12 (a) Compare the slopes of an adiabatic and an isothermal transformation of an ideal gas at the same point in a p–V diagram. Conclude from the comparison that in an adiabatic expansion of an ideal gas the temperature decreases. (b) Explain why this is so.

16.13 (a) Using the first two terms of the virial expansion for the equation of state of a gas as given by Equation 15.22, calculate the work done by a gas when it expands isothermally from a volume V_1 to a volume V_2. (b) Compare with the value obtained using the ideal gas expression. (c) Apply the result to one mole of hydrogen at 300 K when it expands from a volume of $3 \times 10^{-2}\ \mathrm{m^3}$ to a volume of $5 \times 10^{-2}\ \mathrm{m^3}$.

16.14 An ideal gas at 300 K occupies a volume of $2\ \mathrm{m^3}$ at a pressure of 6 atm. It expands adiabatically until its volume is $4\ \mathrm{m^3}$. Next it expands isothermally until its volume is $6\ \mathrm{m^3}$. Calculate (a) the total work done by the gas, (b) the heat absorbed and (c) the change in internal energy. (d) Plot the process in a p–V diagram.

16.15 An ideal gas at 300 K occupies a volume of $0.5\ \mathrm{m^3}$ at a pressure of 2 atm. The gas expands adiabatically until its volume is $1.2\ \mathrm{m^3}$. Next the gas is compressed isobarically up to its original volume. Finally the pressure is increased isochorically until the gas returns to its initial state. (a) Plot the process in a p–V diagram. (b) Determine the temperature at the end of each transformation. (c) Find the work done during the cycle.

16.16 An ideal gas at 300 K occupies a volume of $0.5\ \mathrm{m^3}$ at a pressure of 2 atm. The gas expands adiabatically until its volume is $1.2\ \mathrm{m^3}$. Next the gas is compressed isothermally until the volume is the same as the original volume. Finally the pressure is increased isochorically until the gas returns to its initial state. (a) Plot the process in a p–V diagram. (b) Determine the temperature at the end of the adiabatic transformation. (c) Calculate the work done during the cycle.

16.17 A mixture of ice and water at 273 K (0 °C) is in an insulated foam container. An electric immersion heater with negligible heat capacity furnishes heat to the mixture at the rate of $40\ \mathrm{cal\,s^{-1}}$ for 20 minutes. The temperature of the contents is shown in Figure 16.22 as a function of time. The heat of fusion of ice is $79.7\ \mathrm{cal\,g^{-1}}$. (a) Explain what happens to the contents during the time interval 0 to 200 s, during the time interval 200 s to 1000 s, and during the time interval 1000 s to 1200 s. (b) How many grams of ice were in the container originally? (c) How many grams of water were in the container after all the ice melted?

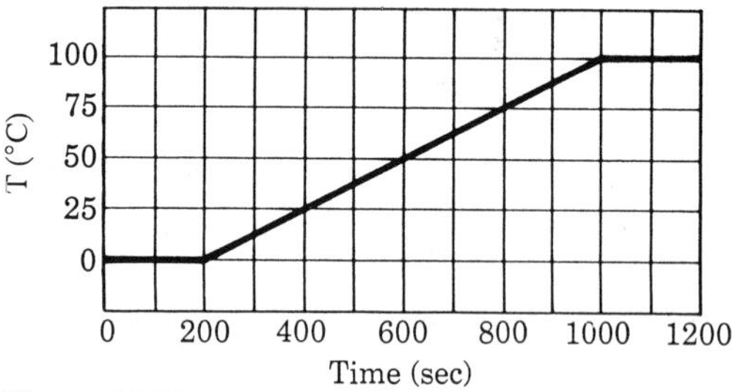

Figure 16.22

16.18 The normal boiling temperature of liquid helium is 4.2 K. The heat of vaporization of helium is $0.75\ \mathrm{cal\,g^{-1}}$, the atomic mass of helium is 4, and helium is monatomic in its gaseous state. Consider a 10 g sample of helium initially in the liquid state. (a) How much heat must the liquid helium at 4.2 K absorb to convert all the helium to gas at 4.2 K? (b) How much heat must be transferred to the helium gas to increase its temperature from 4.2 K to 77 K under a constant pressure of one atmosphere? (c) What is the increase in internal energy of this sample of helium gas in the process described in (b)?

16.19 Helium gas (monatomic) in a tank of constant volume absorbs 900 calories of heat so that its temperature changes from 300 K to 450 K. (a) How does the internal energy of the gas increase during this process? (b) What is the ratio of the average velocity of the helium atoms at the two temperatures?

16.20 A quantity of CO_2, with an initial volume of $100\ \mathrm{m^3}$ and an initial pressure of 30 MPa, is compressed isothermally along the path *ABCD* as shown in the p–V diagram of Figure 16.23. (a) Find the volume of CO_2 at point *B* if the CO_2 is treated as an ideal gas along the path *AB*. (b) Account for the change of slope of the curve at point *B*. (c) If the volume of CO_2 is $13\ \mathrm{m^3}$ at point *C*, calculate the work done on the CO_2 in the process *B*–*C*. (d) Account for the change in slope of the curve at point C. (e) Since the temperature remains constant, is heat added to or extracted from the CO_2 in the process B–C?

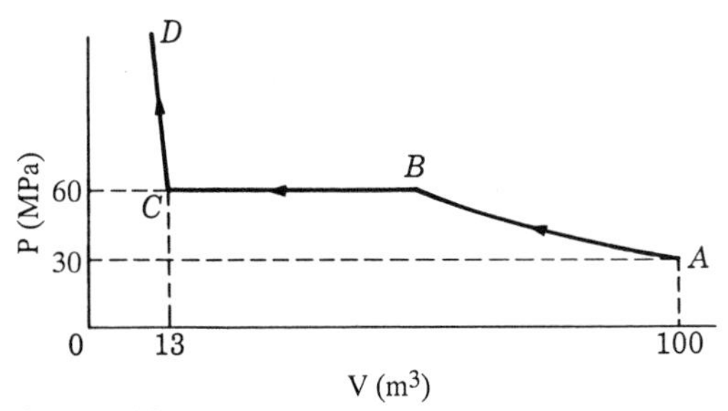

Figure 16.23

16.21 Consider 3 moles of an ideal monatomic gas initially in state 1 with volume $V_1 = 7.4 \times 10^3\ \text{cm}^3$, temperature $T_1 = 300\ \text{K}$, and pressure $p_1 = 1\ \text{MPa}$. (a) The gas is heated at constant volume from T_1 to $T_2 = 450\ \text{K}$. (b) From state 2 the gas expands at constant temperature to state 3. (c) From state 3 the gas is compressed at constant pressure back to its original state 1. Determine the heat absorbed, the work done and the change in internal energy of the gas in each process and during the whole cycle.

16.22 A 10 g sample of material, initially a liquid, is heated in a closed rigid container from 50 °C to 250 °C. The heat added to the sample and the corresponding temperature are shown in Figure 16.24. (a) Determine the specific heat of the sample between A and B and between C and D. (b) Discuss the state of the sample between points B and C on the graph. What does the heat absorbed during this stage accomplish? (c) Compare AB and CD and suggest a qualitative reason for the difference in slopes.

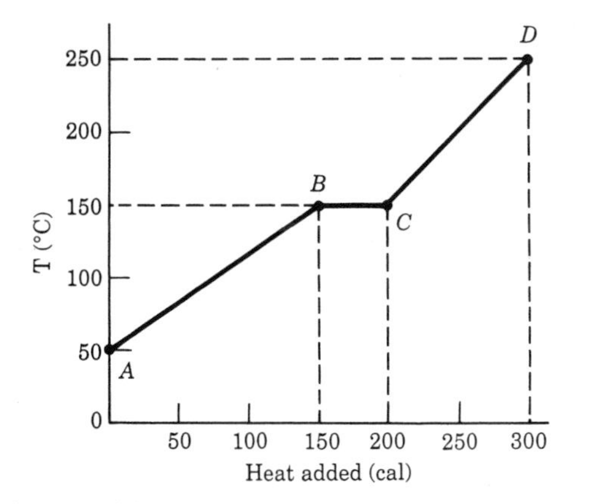

Figure 16.24

16.23 Consider 10 g of ice at an initial temperature of −20 °C. Heat is added at a constant rate of $5.0\ \text{cal s}^{-1}$. Plot a graph showing temperature as a function of time from the beginning until all that water vaporizes. Indicate numerical values for significant points on the graph.

16.24 One-tenth of a mole of an ideal monatomic gas undergoes a process described by the straight-line path AB shown in Figure 16.25. (a) How is the temperature of the gas related at points A and B? (b) How much heat must be absorbed by the gas, how much work is done by the gas and what is the change in internal energy during the process $A \rightarrow B$? (c) What is the highest temperature of the gas during the process?

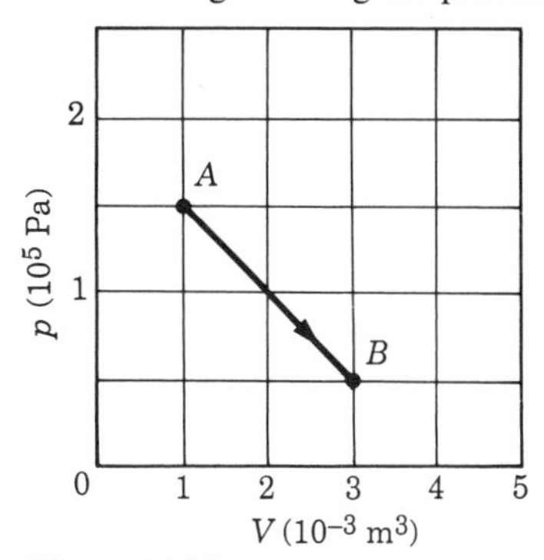

Figure 16.25

16.25 One mole of a monatomic ideal gas enclosed in a cylinder with a movable piston undergoes the process $ABCDA$ shown on the p–V diagram of Figure 16.26. (a) In terms of p_0 and V_0, calculate the work done and the net heat absorbed by the gas in the process. (b) At what points in the process are the temperatures equal? (c) In which of the segments is the amount of heat absorbed greater?

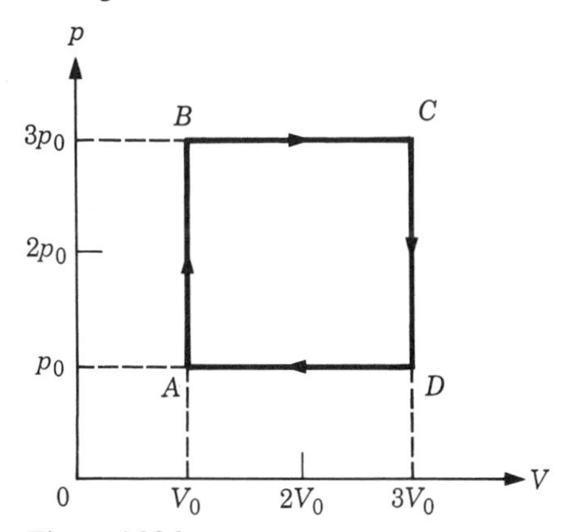

Figure 16.26

16.26 An ideal gas is initially at $T_1 = 300$ K, $p_1 = 3$ atm, and $V_1 = 4$ m^3. The gas expands isothermally to a volume of 16 m^3. This is followed by an isochoric process to such a pressure that an adiabatic compression returns the gas to the original state. Assume all the processes are reversible. (a) Draw the cycle on (i) a p–V diagram and (ii) a T–S diagram. Calculate (b) the work done and (c) the entropy change during each process. Also calculate (d) the work done and (e) the entropy change during the cycle.

16.27 (a) Show that the change in entropy of a substance heated reversibly at constant pressure (assuming that the heat capacity remains constant) is $\Delta S = \text{N}C_p \ln(T_2/T_1)$. (b) Apply your result to 1 kg of water heated from room temperature (298 K) up to its normal boiling point (373 K).

16.28 For water the heat of fusion is 1435 cal mol^{-1} and the heat of vaporization is 9712 cal mol^{-1}. Calculate the change in entropy of one mole of water, given that it is heated reversibly from -20 °C to 150 °C at a constant pressure of 1 atm. The heat capacity of ice is approximately 9.0 cal K^{-1} mol^{-1} and the heat capacity of steam at constant pressure is approximately 8.6 cal K^{-1} mol^{-1}.

16.29 (a) One kilogram of water at 0 °C is brought into contact with a large body at 100 °C. When the water has reached 100 °C, what has been the change in entropy of the water? Of the body? Of the universe? (b) If the water had been heated from 0 °C to 100 °C by first bringing it in contact with a large body at 50 °C and then with the large body at 100 °C, what would have been the change in entropy of the water and of the universe? (c) Explain how the water might be heated from 0 °C to 100 °C with no change in the entropy of the universe.

16.30 A body with heat capacity C_{p1} and containing N_1 moles at temperature T_1 is placed in thermal contact with another body of heat capacity C_{p2} and containing N_2 moles at temperature T_2. The only process that can occur is a heat exchange. (a) Is the process reversible or irreversible? (b) Show that when thermal equilibrium is reached the common temperature is

$$T = \frac{\text{N}_1 C_{p1} T_1 + \text{N}_2 C_{p2} T_2}{\text{N}_1 C_{p1} + \text{N}_2 C_{p2}}$$

Verify that T is a temperature that falls between T_1 and T_2. (c) Show that the total change in entropy is

$$\Delta S = \text{N}_1 C_{p1} \ln \frac{T}{T_1} + \text{N}_2 C_{p2} \ln \frac{T}{T_2}$$

(d) Also verify that ΔS is positive. (Hint: for part (d), assume that T_1 is smaller than T_2 so that the three temperatures are in the order $T_1 < T < T_2$. Add and subtract the quantity $\text{N}_2 C_{p2} \ln(T/T_1)$ from the expression for ΔS.)

16.31 Consider two samples of *different* gases, designated a and b, both at the same temperature T, composed of N_a and N_b molecules, respectively, and occupying adjoining containers of volumes V_1 and V_2, separated by a removable partition. When the partition is removed and both gases are mixed, we have N_a molecules of gas a and N_b molecules of gas b both occupying the volume $V_1 + V_2$. Show that (a) the temperature remains the same and (b) the change in entropy is $\Delta S = kN_a \ln(1 + V_2/V_1) + kN_b \ln(1 + V_1/V_2)$. (c) Verify that ΔS is positive.

16.32 One mole of nitrobenzene ($C_6H_5NO_2$) is vaporized at 483 K and a pressure of 10^5 Pa. The heat of vaporization is 9730 cal mol^{-1}. Compute (a) Q, (b) W, (c) ΔU and (d) ΔS.

16.33 One mole of an ideal gas at 27 °C and 1 atm is heated at constant pressure until its volume is tripled. Calculate (a) ΔU, (b) W, (c) Q and (d) ΔS.

16.34 Assume that 100 g of water at 363 K is poured into a 0.3 kg aluminum container initially at room temperature (298 K). Calculate the change in entropy of (a) the aluminum, (b) the water and (c) the entire system after thermal equilibrium has been reached. (d) Is the process reversible or irreversible? The heat capacity of aluminum is given in Table 16.1.

16.35 A liquid at a temperature T_1 is mixed with an equal amount of the same liquid at a temperature T_2. The system is thermally insulated. (a) Show that the entropy change of the system is $\Delta S = 2\text{N}C_p \ln{(T_1 + T_2)/2(T_1 T_2)^{1/2}}$. (b) Verify that ΔS is positive.

16.36 The adiabatic compressibility of a substance is defined as $\kappa_s = -(1/V)/(\mathrm{d}V/\mathrm{d}p)_s$. (a) Find κ_s for an ideal gas. (b) The propagation of elastic waves in a gas is an adiabatic process and the velocity of propagation is given by $v = (\kappa_s/\rho)^{1/2}$ where ρ is the density (see Section 28.6). Explain how, by measuring v, the value of γ may be calculated. (c) Express v in terms of the temperature T of the gas. (d) Assuming that air is composed of diatomic gases verify that $v = 20.55\,T^{1/2}$ m s^{-1}. (Recall that the average molar mass of air is 28.966.)

17 Statistical mechanics

Josiah W. Gibbs, first professor of mathematical physics at Yale University, was one of the founders of statistical mechanics, together with L. Boltzmann and J. C. Maxwell. His formulation of statistical mechanics was published in 1902. Gibbs established a rule for analyzing the equilibrium of heterogeneous substances, known as the 'phase rule'. He was one of the first to use the second law of thermodynamics to discuss exhaustively the relation between chemical, electrical and thermal energy and external work. Professor Gibbs developed vector analysis in its present form.

Notes

17.1 Introduction

Thermodynamics provides a general framework for describing exchanges of energy between a system of particles and its surroundings without explicitly considering the specific properties of the particles that compose the system. As such, several physical quantities of a macroscopic nature (such as heat, entropy, heat capacity etc.) that can be measured in the laboratory are introduced without relating them directly to the internal structure of the system. However, the properties of a system do depend critically on the properties of its constituents. For example, an assembly of electrons in a metal, of atoms in gaseous or liquid helium, and of molecules in vapor, liquid or solid water, behave in completely different ways. Therefore it is most important to relate the macroscopic behavior of a system to its microscopic structure. This is the task of **statistical mechanics**, a term coined in 1884 by J. W. Gibbs. Statistical mechanics bridges the gap between the empirical science of thermodynamics and the atomic structure of matter.

Statistical mechanics emphasizes methods to obtain collective or macroscopic properties of a system of particles, such as pressure or temperature, without considering the detailed motion of each particle. Rather, it focuses on their average behavior. In this chapter the term 'particle' is used in a rather broad sense. We mean either a fundamental particle, such as an electron, or a stable aggregate of fundamental particles, such as an atom or molecule. Thus a 'particle' will be each of the well-defined and stable units composing a given physical system.

A statistical approach is necessary to relate the macroscopic properties of matter with its atomic structure if we recognize that in one mole of any substance there are about 6×10^{23} molecules. It is not only practically impossible, but also unnecessary to take into account the detailed motions of each of these molecules to determine the bulk properties of a substance. This became clear in Chapter 15, where some simple statistical considerations were used to correlate the collective properties of gases with the properties of their molecules. Statistical mechanics was developed late in the nineteenth century and at the beginning of the twentieth century mainly as the result of the work of Ludwig Boltzmann (1844–1906), James C. Maxwell (1831–1879), and J. Willard Gibbs (1839–1903).

17.2 Statistical equilibrium

Consider an *isolated* system composed of a large number N of particles. Assume each particle has available to it several states with energies $E_1, E_2, E_3, \ldots$. The energy states may be quantized (as are the rotational and vibrational states in a molecule), or they may, instead, form a practically continuous spectrum (as for the translational kinetic energy of the molecules in a gas). At a particular time the particles are distributed in a certain way among the different states, so that n_1 particles have energy E_1, n_2 particles have energy E_2, and so on. The total number of particles is

$$N = n_1 + n_2 + n_3 + \cdots = \sum_i n_i \tag{17.1}$$

For the present we assume that N remains constant for all processes occurring in the system. The total energy of the system is

$$U = n_1 E_1 + n_2 E_2 + n_3 E_3 + \cdots = \sum_i n_i E_i \tag{17.2}$$

The set of numbers $(n_1, n_2, n_3, \ldots)$ constitutes a **partition**. A partition defines a *micro*-state of the system consistent with the *macro*-state or physical condition of the system, which is determined by the number of particles, the total energy, the structure of each particle, and some external parameters.

Expression 17.2 for the total energy of the system implicitly assumes that the particles are non-interacting (or that they interact only slightly, as in a gas), so that we can attribute to each particle a well-defined energy.

If we consider interactions, we must add to Equation 17.2 terms corresponding to the potential energy of interaction between pairs of particles. In such a case we cannot speak of the energy of each particle, but only of the system. (This is the situation in liquids and solids.) It may seem at first that our discussion is therefore unrealistic, since all particles that make up physical systems are interacting. However, under special conditions we can consider that each particle is subject to the *average* interaction of the others, and therefore has an average energy. Then we can still write U as in Equation 17.2. For cases where the interactions among the particles must be considered explicitly, other techniques must be used.

If the system is isolated, the total energy U must be constant. However, Equations 17.1 and 17.2 do not determine the partition $(n_1, n_2, n_3, \ldots)$ in a unique way and therefore there may be many micro-states compatible with the macro-state defined by N and U. So in a system, as a result of their mutual interactions and collisions, the distribution of the particles among the available energy states may be changing continuously. For example, in a gas a fast molecule may collide with a slow one; after the collision the fast molecule may have slowed down and the slow one may have sped up. Or an excited atom may collide inelastically with another atom, with a transfer of its excitation energy into kinetic energy of both atoms. Hence, in both examples, the particles after the collision are in different states. That is, the numbers $n_1, n_2, n_3, \ldots$, which give a partition or distribution of N particles among the available energy states, may be changing.

It is reasonable to assume that for each macroscopic state of a system of particles there is a partition that is more favored than any other. In other words, we may assume that

> *given the physical conditions of an isolated system of particles there is a most probable partition compatible with those conditions.*

When the most probable partition is achieved, the system is said to be in **statistical equilibrium**.

A system in statistical equilibrium will not depart from the most probable partition (except for small statistical fluctuations) unless it is disturbed by an external action. By this we mean that the partition numbers $n_1, n_2, n_3, \ldots$ may fluctuate around the values corresponding to the most probable partition without noticeable (or observable) macroscopic effects. For example, suppose that we have a gas in statistical equilibrium, and that a molecule of energy E_i collides with a molecule of energy E_j. After the collision their energies are E_r and E_s respectively. We may assume that within a short time another pair of molecules are moved into energy states E_i and E_j and two other molecules are removed from states E_r and E_s, so that, statistically, the partition has not changed and the macro-state is the same.

The key problem of statistical mechanics is to find the most probable partition (or distribution law) of an isolated system, given its composition. Once the most probable partition has been found, the next problem is to devise methods for deriving the macroscopically observed properties such as temperature, average energy and entropy from it. To obtain the distribution law, certain assumptions are required. Several plausible assumptions may be tried, until a distribution law in accordance with experimental results is obtained. Three distribution laws or statistics are presently used. One is called the **Maxwell–Boltzmann distribution law**, which is the basis of classical statistics. We shall study it in this chapter. Two other distribution laws, called **Fermi–Dirac** and **Bose–Einstein**, belong to quantum statistics and will not be considered.

EXAMPLE 17.1

A system is composed of 4000 particles distributed among three possible states of energies $E_1 = 0$, $E_2 = \varepsilon$ and $E_3 = 2\varepsilon$ (Figure 17.1). A particular partition corresponds to the occupation numbers $n_1 = 2000$, $n_2 = 1700$ and $n_3 = 300$. Calculate the total energy of the system and the average energy.

▷ The total energy, according to Equation 17.2, is

$$U = \sum_i n_i E_i = 2000 \times 0 + 1700 \times \varepsilon + 300 \times (2\varepsilon) = 2300\varepsilon$$

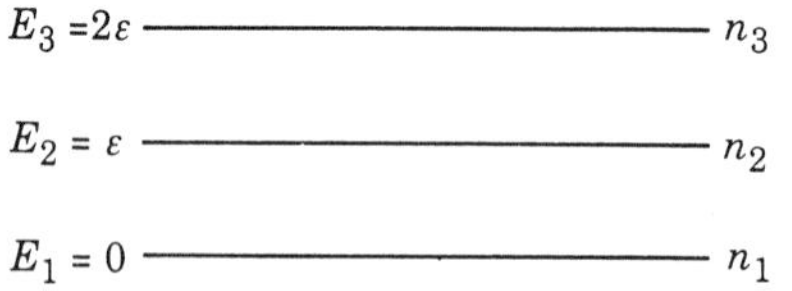

$E_3 = 2\varepsilon$ ——— n_3

$E_2 = \varepsilon$ ——— n_2

$E_1 = 0$ ——— n_1

Figure 17.1 System with three equally spread energy levels.

Then the average energy of the particles is

$$E_{\text{ave}} = \frac{U}{N} = \frac{2300\varepsilon}{4000} = 0.575\varepsilon$$

The average energy is smaller than ε because of the strong occupation of the level with $E_1 = 0$. How would the total and average energies be modified if 50 particles are shifted from level 1 to level 2 and the same number from level 3 to level 2 also?

17.3 Maxwell–Boltzmann distribution law

Consider again a system composed of a large number of identical particles; that is, particles that have the same structure and composition. This is the case, for example, of a gas composed of only one kind of molecule, or of free electrons in a metal. Suppose the particles can occupy the states of energies E_1, E_2, $E_3, \ldots$, and that there are $n_1, n_2, n_3, \ldots$ particles in each state (Figure 17.2). Then, the total number N of particles and the total energy U are given by Equations 17.1 and 17.2. If the system is isolated the quantities N and U are fixed.

The first task of statistical mechanics is to calculate the probability P of a partition (n_1, n_2, ..., n_i, ...), given certain assumptions about the particles. We shall assume that the probability of a particular partition is proportional to the number of different ways the particles

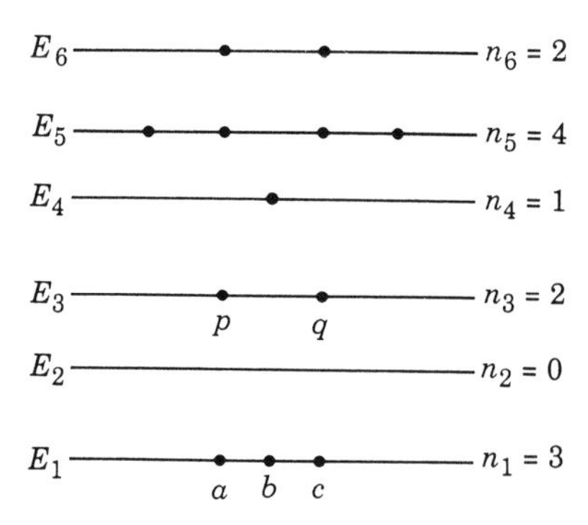

Figure 17.2 Distribution of particles among different energy states.

can be distributed among the available energy states compatible with the fixed values of N and U. The second step is to calculate the distribution that corresponds to the equilibrium microstate of the system. This implies finding the distribution for which the probability P has a maximum value (for the given values of N and U).

In carrying out this program it is assumed, in Maxwell–Boltzmann statistics, that there are no restrictions on the number of particles that can occupy a given energy state. However, it may happen that there are energy states that are more likely to be occupied than others. In other words, the energy states may have different **intrinsic probabilities**, designated by g_i, of being occupied. That is, the larger g_i, the larger the probability that the state will be occupied. For example, a certain energy state may be associated with more different angular momentum states of the particles than other states and therefore is more likely to be occupied. The values of the intrinsic probabilities g_i are determined in each case by the internal structure of the system.

We shall not enter into the mathematical procedure for calculating the partition probability P for Maxwell–Boltzmann statistics or for obtaining its maximum value. Instead, we will state the results and justify them in an intuitive way (see Note 17.1).

In the first place, it is reasonable to assume that, when statistical equilibrium is reached and the probability P is a maximum, the occupation numbers n_i should be proportional to the intrinsic probabilities g_i, since the larger g_i the more probable it is that a particle is in state E_i. Also we may assume that at equilibrium particles tend to favor the lower energy states. And so particles in higher energy states might tend to pass to states of lower energy. Or, in other terms, the larger the energy E_i, the less probable that a particle will be in that state when statistical equilibrium is reached. A negative exponential of the form $e^{-\beta E_i}$, where β is a positive parameter (that is shown later to be related to the temperature) satisfies this requirement. Therefore it is plausible to assume that the occupation numbers of the **most probable** or **equilibrium partition** should be of the form

$$n_i = \alpha g_i e^{-\beta E_i} \tag{17.3}$$

where α is a constant that depends on the structure of the system. For dimensional consistency, β must be expressed in reciprocal energy units; that is, J^{-1}, eV^{-1} etc.

Our next step is to relate the parameters α and β with the physical properties of the system. The quantity α may be expressed in terms of the number of particles of the system:

$$\begin{aligned} N &= n_1 + n_2 + n_3 + \cdots \\ &= \alpha g_1 e^{-\beta E_1} + \alpha g_2 e^{-\beta E_2} + \alpha g_3 e^{-\beta E_3} + \cdots \\ &= \alpha(g_1 e^{-\beta E_1} + g_2 e^{-\beta E_2} + g_3 e^{-\beta E_3} + \cdots) \\ &= \alpha\left(\sum_i g_i e^{-\beta E_i}\right) = \alpha Z \end{aligned}$$

where

$$Z = \sum_i g_i e^{-\beta E_i} \tag{17.4}$$

The quantity Z, called the **partition function**, is a very important expression which appears quite often in the calculation of the properties of the system. Note that Z is a function of the parameter β and depends on the values of g_i and E_i. We may thus write $\alpha = N/Z$, and Equation 17.3 becomes

$$n_i = \frac{N}{Z} g_i e^{-\beta E_i} \tag{17.5}$$

Expression 17.5 constitutes the **Maxwell–Boltzmann distribution law** for a system in statistical equilibrium. Note that n_i depends on the structure of the system through the g_is and E_is, and on its physical state through β.

The total energy of a system of particles in statistical equilibrium is

$$\begin{aligned} U &= n_1 E_1 + n_2 E_2 + n_3 E_3 + \cdots \\ &= \frac{N}{Z}(g_1 E_1 e^{-\beta E_1} + g_2 E_2 e^{-\beta E_2} + g_3 E_3 e^{-\beta E_3} + \cdots) \\ &= \frac{N}{Z}\left(\sum_i g_i E_i e^{-\beta E_i} \right) \end{aligned} \tag{17.6}$$

Using the definition of the partition function, and noticing that

$$\frac{d}{d\beta}(e^{-\beta E_i}) = -E_i e^{-\beta E_i}$$

we may write U in the alternative form

$$U = -\frac{N}{Z}\frac{d}{d\beta}\left(\sum_i g_i e^{-\beta E_i} \right) = -\frac{N}{Z}\frac{dZ}{d\beta}$$

This expression can also be written in the more compact form

$$U = -N\frac{d}{d\beta}(\ln Z) \tag{17.7}$$

The average energy of a particle is then

$$E_{ave} = \frac{U}{N} = -\frac{d}{d\beta}(\ln Z) \tag{17.8}$$

These expressions show the importance of the partition function for calculating properties of a system in statistical equilibrium in terms of its internal structure, which is determined by the values of g_i and E_i and the parameter β.

EXAMPLE 17.2

Determination of the most probable or equilibrium partition of the system considered in Example 17.1. Each particle can be in one of three energy states or levels, equally spaced, whose energies are 0, ε and 2ε. Assume each energy level has the same intrinsic probability g. This system, for example, could consist of atoms whose angular momentum has three possible orientations when placed in a magnetic field.

▷ We recall that the system is composed of 4000 particles and its total energy (see Figure 17.1) is 2300 ε. Using Equation 17.3 for the most probable partition, and setting $g_1 = g_2 = g_3 = g$, we have,

$$n_1 = \alpha g, \qquad n_2 = \alpha g e^{-\beta\varepsilon}, \qquad n_3 = \alpha g e^{-2\beta\varepsilon}$$

If we designate $e^{-\beta\varepsilon}$ by x, we can write $n_2 = n_1 x$ and $n_3 = n_1 x^2$. Thus conditions 17.1 and 17.2, which give the total number of particles and the total energy, respectively, become

$$n_1 + n_1 x + n_1 x^2 = 4000 \qquad \text{and} \qquad (n_1 x)\varepsilon + (n_1 x^2)(2\varepsilon) = 2300\varepsilon$$

Canceling the common factor ε in the second relation, we have

$$n_1(1 + x + x^2) = 4000 \qquad \text{and} \qquad n_1(x + 2x^2) = 2300$$

Dividing one equation by the other to eliminate n_1, we obtain an equation for x:

$$57x^2 + 17x - 23 = 0$$

or $x = 0.503\,37$. (Only the positive root is used. Why?) Returning to one of the two previous equations, and using this value for x, we find $n_1 = 2277$ (the figure has been rounded), $n_2 = 1146$ and $n_3 = 577$, which gives the equilibrium partition of the system. This distribution is quite different from that given in Example 17.1, and therefore the system of Example 17.1 was not in statistical equilibrium.

Note 17.1 Probability of a partition in Maxwell–Boltzmann statistics

It is relatively simple to estimate what should be the probability of a partition $(n_1, n_2, n_3, \ldots, n_i, \ldots)$ for a system of identical particles in Maxwell–Boltzmann statistics. First, we note that the intrinsic probability g_i gives the probability that the state of energy E_i be occupied by *one* particle. If there are no restrictions on the occupation of the state E_i by more than one particle, then the probability of occupation by *two* particles is $g_i \times g_i$ or g_i^2; for three particles, it is $g_i \times g_i \times g_i$ or g_i^3. Thus, the intrinsic probability of finding n_i particles in the state whose energy is E_i is $g_i^{n_i}$. This situation is similar to the calculation of the probability of obtaining the same face of a die for several successive throws: for one throw, it is $\frac{1}{6}$, for two throws it is $(\frac{1}{6})^2$, for three throws it is $(\frac{1}{6})^3$, and so on. Then we may reasonably assume that the probability P of the partition $(n_1, n_2, n_3, \ldots, n_i, \ldots)$ must be proportional to $g_1^{n_1} g_2^{n_2} g_3^{n_3} \ldots$.

Also, because the particles are identical, any permutation of particles in the same energy state does not give rise to a new partition. The number of permutations of n_i objects is $n_i!$. Therefore, we must expect that the probability P is inversely proportional to $n_1!n_2!n_3!\ldots$ Thus we write

$$P \approx \frac{g_1^{n_1} g_2^{n_2} g_3^{n_3} \ldots}{n_1!n_2!n_3!\ldots}$$

For example, for the partition illustrated in Figure 17.2, the probability is

$$P \approx \frac{g_1^3 g_2^0 g_3^2 g_4^1 g_5^4 g_6^2}{3!0!2!1!4!2!}$$

In this expression $g_2^0 = 1$ and $0! = 1$ because there is only one way of arranging that partition number when no particle is placed in that state. The above expression for P can be obtained using a more rigorous method, which we shall not develop.

To illustrate the meaning of the expression for P, consider the system of Examples 17.1 and 17.2 composed of 4000 particles in three energy levels, all with the same intrinsic probability. We first compare the relative probabilities of two non-equilibrium partitions. The partition in Example 17.1 has 2000 particles in the lower level, 1700 in the middle level, and 300 in the upper level. The second partition results from the transfer of one particle from the middle level to the lower level and another from the middle level to the upper level. Note that this process does not change the total energy.

According to the above expression for P, the probabilities for the first and second partitions are

$$P_1 \approx \frac{g^{4000}}{2000!1700!300!}, \qquad P_2 \approx \frac{g^{4000}}{2001!1698!301!}$$

Instead of computing the values of P_1 and P_2 (which we could do by using straightforward mathematical techniques), we shall simply find their ratios:

$$\frac{P_2}{P_1} = \frac{1700 \times 1699}{2001 \times 301} = \frac{2\,888\,300}{602\,301} = 4.8$$

Thus the mere transfer of two particles out of 4000 to other levels increases the probability by a factor of 4.8. This means that partitions P_1 and P_2 are both far from being the equilibrium partition; this situation is due to an excessive population of the middle level. Therefore the system will try to evolve to a state where the middle level is less populated. It is suggested that the student repeat the problem, considering other possible distributions of particles, all compatible with the same total energy. (Shift two more particles from the middle level or move one particle from the upper level and another from the lower level into the middle level and recompute the relative probabilities.)

In Example 17.2 we obtained the most probable or equilibrium partition for this system according to the Maxwell–Boltzmann distribution law and found the partition $n_1 = 2277$, $n_2 = 1146$, $n_3 = 577$. The corresponding partition probability is

$$P \approx \frac{g^{4000}}{2277!1146!577!}$$

Let us now compute the change in P when two particles are removed from the intermediate level and transferred to the upper and lower levels. The new partition probability is

$$P' \approx \frac{g^{4000}}{2278!1144!578!}$$

The ratio of the two probabilities is

$$\frac{P'}{P} = \frac{1146 \times 1145}{2278 \times 578} = \frac{1\,312\,170}{1\,316\,684} = 0.9966$$

Therefore the two probabilities are essentially the same since, if P is a maximum, the change in P must be very small for a small change in the distribution numbers. Thus the equilibrium microstate may fluctuate among partitions close to the most probable equilibrium partition without too much change. This confirms that the Maxwell–Boltzmann distribution law corresponds to the maximum of the probability P for a system satisfying both Equations 17.1 and 17.2.

17.4 Statistical definition of temperature

The parameter β is directly related to the physical quantity we have called temperature. As explained in Chapter 15 for gases, temperature is related to a statistical property, the average kinetic energy of the molecules of a gas in thermal equilibrium. Similarly, Equation 17.8 relates β to E_{ave}, in turn suggesting a relation between β and the temperature of the system.

Note first that the value of the exponential $e^{-\beta E_i}$ decreases (increases) as βE_i increases (decreases). We may conclude that as βE_i becomes larger (smaller), the occupation of the state with energy E_i becomes smaller (larger). Since the E_is are fixed by the composition of the system, we further conclude that a large β favors the smaller values of E_i and reduces the average energy of the particles. Likewise, a small β favors larger values of E_i and increases the average energy of the particles. On the other hand, a high (low) temperature implies large (small) average energy of the particles. In this way we see that the effect of a change in β is precisely opposite to that of a change in the temperature of the system. This suggests a general statistical definition of the temperature T of the system by relating it to the parameter β. Because the effects of β and T on the average molecular energy are opposite, we adopt the statistical definition of temperature by the inverse relation

$$\beta = \frac{1}{kT} \quad \text{or} \quad kT = \frac{1}{\beta} \tag{17.9}$$

where k is the Boltzmann constant introduced in Section 15.5, and has the value

$$k = 1.3805 \times 10^{-23}\ \text{J K}^{-1} = 8.6178 \times 10^{-5}\ \text{eV K}^{-1}$$

We need to introduce k for dimensional reasons because β is the reciprocal of an energy. Of course, we could measure temperature in energy units; however, that is not customary. By certain calculations which we shall omit, it can be verified that the temperature defined by Equation 17.9 is the same as the gas temperature defined by Equation 15.15 and used in Chapter 16 (see also Example 17.6).

Using Equation 17.9 we may now write the Maxwell–Boltzmann distribution law, Equation 17.6, in the form

$$n_i = \frac{N}{Z} g_i e^{-E_i/kT} \tag{17.10}$$

We emphasize that Equation 17.10 gives the *average* occupation numbers when the system is in statistical equilibrium. Similarly the partition function, Equation 17.4, becomes

$$Z = \sum_i g_i e^{-E_i/kT} \tag{17.11}$$

and is thus a function of the temperature of the system. From Equation 17.9, we have that $d\beta = -dT/kT^2$ and therefore Equations 17.7 and 17.8 can be written in terms of the temperature as

$$U = kNT^2 \frac{d}{dT}(\ln Z) \tag{17.12}$$

and

$$E_{ave} = kT^2 \frac{d}{dT}(\ln Z) \tag{17.13}$$

We conclude then that, for a system in statistical equilibrium, the total energy U and the average energy of the particles E_{ave} are both determined by the temperature of the system. The exact relation depends on the microscopic structure of the system, as expressed in the partition function Z. It is different for an ideal gas, a real gas, a liquid, or even a complex nucleus. Hence we may say that

> *the temperature of a system in statistical equilibrium is a parameter related to the average energy per particle of the system, the relation depending on the structure of the system.*

The statistical definition of temperature, as given by Equation 17.9, is valid only for a system of particles in statistical equilibrium, and hence does not apply to a single particle or to a system that is not in equilibrium. The reason for this is that the parameter β appears only in connection with the calculation of the most probable partition of a system, which corresponds, by definition, to the equilibrium state. If the system is not in equilibrium, we may still speak of an 'effective' temperature of a small portion of the system, assuming that the small portion is almost in equilibrium.

Since the exponential $e^{-E_i/kT}$ in Equation 17.10 is a decreasing function of E_i/kT, the larger the ratio E_i/kT, the smaller the value of the occupation number

n_i. Therefore, at a given temperature, the larger the energy E_i, the smaller the value of n_i. In other words, the occupation of states available to the particles decreases as their energy increases. At very low temperatures only the lowest energy levels are occupied, as shown in Figure 17.3. But at higher temperatures (corresponding to smaller values of E_i/kT for a given energy) the relative population of the higher energy levels increases (again as shown in Figure 17.3) by a transfer of particles from lower to higher energy levels. At the absolute zero of temperature, only the ground or lowest energy level is occupied. It is for that reason that it is customary to say that as the temperature of a system increases, the system becomes more disordered.

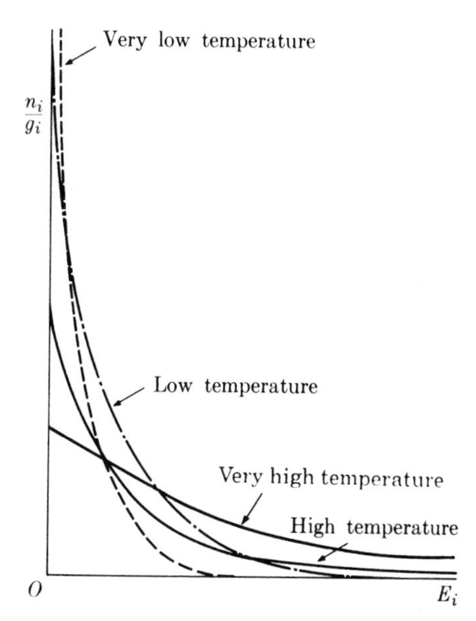

Figure 17.3 Maxwell–Boltzmann distribution at four different temperatures.

EXAMPLE 17.3

The particles of a system can be in two states of energy, $-\varepsilon$ and $+\varepsilon$. Calculate the average energy at the temperature T. Assume both states have the same intrinsic probability. This is the case, for example, of a spinning electron in a magnetic field.

▷ The particles can be in only two states of energy, $E_1 = -\varepsilon$ and $E_2 = +\varepsilon$, both with the same probability, $g_1 = g_2 = 1$ (Figure 17.4). Then, we can write the partition function as $Z = e^{-\beta E_1} + e^{-\beta E_2} = e^{\beta\varepsilon} + e^{-\beta\varepsilon}$ and the average energy of a particle as

$$E_{\text{ave}} = -\frac{d}{d\beta}(\ln Z) = -\frac{1}{Z}\frac{dZ}{d\beta} = -\varepsilon\frac{e^{\beta\varepsilon} - e^{-\beta\varepsilon}}{e^{\beta\varepsilon} + e^{-\beta\varepsilon}}$$

or

$$E_{\text{ave}} = -\varepsilon\frac{e^{\varepsilon/kT} - e^{-\varepsilon/kT}}{e^{\varepsilon/kT} + e^{-\varepsilon/kT}}$$

where β has been replaced by $1/kT$. The total energy is

$$U = NE_{\text{ave}} = -N\varepsilon\frac{e^{\varepsilon/kT} - e^{-\varepsilon/kT}}{e^{\varepsilon/kT} + e^{-\varepsilon/kT}}$$

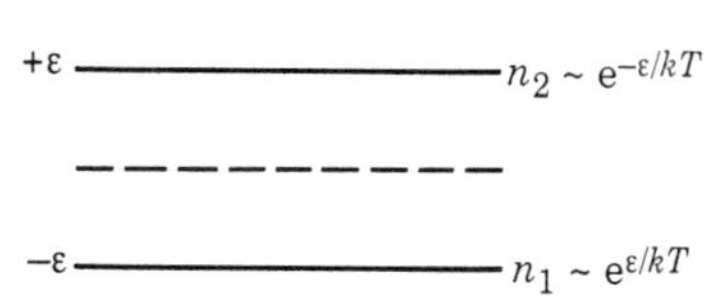

Figure 17.4 Two energy levels system.

which allows us to find U in terms of T. Note that for very low T the negative exponential becomes zero and we see that $E_{\text{ave}} \approx -\varepsilon$. This is the case because all particles tend to be at the lower level. For very large T, both exponentials become equal to 1 and we get $E_{\text{ave}} \approx 0$ because the particles tend to occupy both levels in the same proportion.

This example illustrates what happens to a paramagnetic substance when placed in a magnetic field and the temperature is changed (Section 26.4). At low temperature the substance is magnetized, but as the temperature increases, the substance loses its magnetization.

EXAMPLE 17.4

For temperatures of 100 K, 300 K (room temperature) and 1000 K, determine the ratio between the occupation numbers for two levels corresponding to an energy difference ΔE equal to: (a) 10^{-4} eV, a value equivalent to the spacing of rotational levels for many molecules, (b) 0.05 eV, corresponding to molecular vibrational levels, (c) 3.00 eV, of the order of magnitude of electronic excitation of atoms and molecules. Assume $g = 1$ for simplicity.

▷ The relation of the occupation numbers between two energy levels E_i and E_j is

$$\frac{n_j}{n_i} = \frac{g_j}{g_i} e^{-(E_j - E_i)/kT} = \frac{g_j}{g_i} e^{-\Delta E/kT}$$

where ΔE is the energy difference between the two levels. Thus n_i and n_j are comparable only if ΔE is much smaller than kT. The physical importance of this result is manifested in the rate of chemical reactions, the properties of semiconductors, the functioning of lasers etc.

From the value of k, $kT = 8.1678 \times 10^{-5}\ T$ eV. Then, if ΔE is expressed in electron volts,

$$\frac{\Delta E}{kT} = 1.1603 \times 10^4 \frac{\Delta E}{T}$$

For the specified values of ΔE, the values of n_2/n_1 at the three temperatures are given in the following table:

	n_2/n_1		
ΔE (eV)	100 K	300 K	1000 K
10^{-4}	0.9885	0.9962	0.9988
5×10^{-2}	3×10^{-3}	0.145	0.560
3.00	3×10^{-164}	8×10^{-49}	8×10^{-16}

From the table we see that, for $\Delta E = 10^{-4}$ eV, the two levels are practically equally populated at all temperatures considered; this means that at room temperature gas molecules are about equally distributed among the rotational levels. At $\Delta E = 5 \times 10^{-2}$ eV, the population of the upper level is almost negligible at the lower temperature but is appreciable at room temperature, where about 14% of the gas molecules are in an excited vibrational state. However, for $\Delta E = 3$ eV, the ratio n_2/n_1 is so small that it is plausible to consider the upper level as essentially empty at all temperatures considered. Thus most atoms and molecules (at room temperature) are in their ground electronic state. Only at extremely high temperatures (as in a flame, in the Sun's atmosphere, or in very hot stars), are there atoms and molecules in excited electronic states in any appreciable amount. Electronic excitations also may be produced by inelastic collisions of fast electrons with gas molecules in an electric discharge or by heating a gas.

Note 17.2 Heat capacity of a crystalline solid

A simple and useful application of statistical methods is to the properties of a crystalline solid. A crystalline solid is a regular arrangement or lattice of ions, atoms or molecules held in fixed positions. However, the ions are not completely at rest but can vibrate about their equilibrium positions. That means that the thermal internal energy of a solid is the vibrational energy of the units composing the lattice. Assume, for simplicity that all ions or atoms vibrate linearly with the same frequency ν. Then, since the energy of an oscillator is quantized, the possible energies are given by (recall Equation 10.27)

$$E_n = (n + \tfrac{1}{2})h\nu \qquad (n = 0, 1, 2, 3, \ldots)$$

The distribution of ions among the different energy levels depends on the temperature. The ions in state n have an energy $nh\nu$ relative to the ground state ($n = 0$). The relative occupation at temperature T is given by the Boltzmann factor, which, in this case, is $e^{-nh\nu/kT}$.

It is very simple to estimate how the internal vibrational energy of the solid varies with temperature. According to Equation 17.4, assuming that all vibrational states have the same relative probability of occupation, the partition function is

$$Z = \sum_n e^{-nh\nu/kT} = \sum_n x^n$$

where $x = e^{-h\nu/kT}$ and $x \leqq 1$. This is a decreasing geometrical series of the form $1 + x + x^2 + \ldots$, whose sum is $1/(1 - x)$, (see Appendix B, Equation B.30). The partition function may then be written as

$$Z = (1 - e^{-h\nu/kT})^{-1}$$

The total vibrational energy at the temperature T is, using Equation 17.12,

$$U = kNT^2 \frac{d}{dT} \ln Z$$

But $\ln Z = -\ln(1 - e^{-h\nu/kT})$ and therefore

$$\frac{d}{dT}(\ln Z) = \frac{(h\nu/kT^2)e^{-h\nu/kT}}{(1 - e^{-h\nu/kT})} = \frac{h\nu}{kT^2}\left(\frac{1}{e^{h\nu/kT} - 1}\right)$$

Therefore, we have

$$U = Nh\nu\left(\frac{1}{e^{h\nu/kT} - 1}\right)$$

If the temperature is very high, we may approximate the exponential in the denominator by $1 + (h\nu/kT)$ (recall that $e^x \approx 1 + x$ if $x \ll 1$; see Equation B.32), resulting in $U = kNT$. This

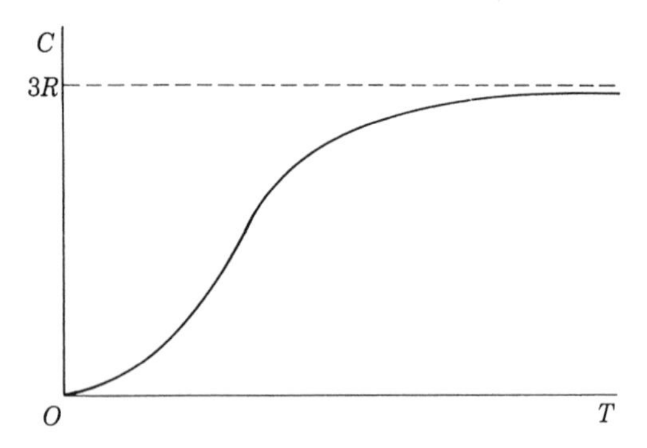

Figure 17.5 Variation of heat capacity with temperature.

would be the energy if the oscillators are only in one dimension. But the ions in the lattice can move along three directions. Therefore the internal vibrational energy in a solid is

$$U = 3kNT = 3\text{N}RT$$

where N is the number of moles. Note U is twice the value for an ideal monatomic gas. This is because in an oscillator we have to consider both kinetic and potential energies, and each has the same average value (recall Sections 10.4 and 15.6).

This expression can be used to estimate the heat capacity of a solid at high temperatures, resulting in

$$C_V = \frac{1}{\text{N}}\frac{\mathrm{d}U}{\mathrm{d}T} = 3R$$

in agreement with the value in Table 16.1 for metals.

At very low temperatures or large $h\nu/kT$ the expression for U can be approximated as $U = Nh\nu e^{-h\nu/kT}$, which indicates that the energy decreases exponentially. This is incorrect since, according to experiment, at very low temperatures the energy varies as T^4 and the heat capacity as T^3. Figure 17.5 shows how the heat capacity of a crystalline solid varies with the temperature. The theory can be improved by taking into account the fact that ions in a crystal lattice can vibrate with more than one frequency. In this way complete agreement with experiment is found.

This theory was developed by Einstein in 1906, one year after formulating his theory of special relativity, and was refined in 1912 by Peter J. W. Debye (1884–1966).

17.5 Energy and velocity distribution of the molecules in an ideal gas

In a gas the molecules move in all directions with different velocities and with rectilinear motion, except when they collide among themselves or with the walls of the container. At each collision an exchange of energy and momentum takes place. When equilibrium is reached there is a well-defined distribution of molecular energies and velocities.

In a gas all molecular energies (and velocities) are possible. Rather than speaking of how many molecules have a given energy (which would be rather difficult given the large number of molecules in the gas), from a practical point of view it is more reasonable to calculate the number $\mathrm{d}N$ of molecules in a small energy interval between the energies E and $E + \mathrm{d}E$. From the Maxwell–Boltzmann distribution law, it is possible to show by calculating the values of g_i for molecules within the energy interval $\mathrm{d}E$ that the number of molecules per unit energy interval (that is, dN/dE) of a gas composed of N molecules at the temperature T is given by the expression

$$\frac{dN}{dE} = \frac{2\pi N}{(\pi kT)^{3/2}} E^{1/2} e^{-E/kT} \tag{17.14}$$

This expression for the energy distribution of the molecules in an ideal gas was originally derived by Maxwell around 1857, before statistical mechanics was developed. Real gases follow this law rather closely at high temperatures and/or low densities. A plot of Equation 17.14 for two different temperatures is represented in Figure 17.6. Note that for each temperature there is an energy at which the number of molecules is the largest.

Sometimes we require the velocity distribution rather than the energy distribution. Recalling that $E = \frac{1}{2}mv^2$, the derivative of E with respect to v is $dE/dv = mv$. Therefore, we have

$$\frac{dN}{dv} = \frac{dN}{dE}\frac{dE}{dv} = mv\frac{dN}{dE}$$

Making the substitution $E = \frac{1}{2}mv^2$ in Equation 17.14, we get

$$\frac{dN}{dv} = 4\pi N\left(\frac{m}{2\pi kT}\right)^{3/2} v^2 e^{-mv^2/2kT} \tag{17.15}$$

which is Maxwell's formula for the velocity distribution at temperature T of the molecules in an ideal gas composed of N molecules, each of mass m. It gives the number dN of molecules moving with a velocity between v and $v + dv$, irrespective of the direction of motion. The velocity distribution in oxygen for two different temperatures is represented in Figure 17.7.

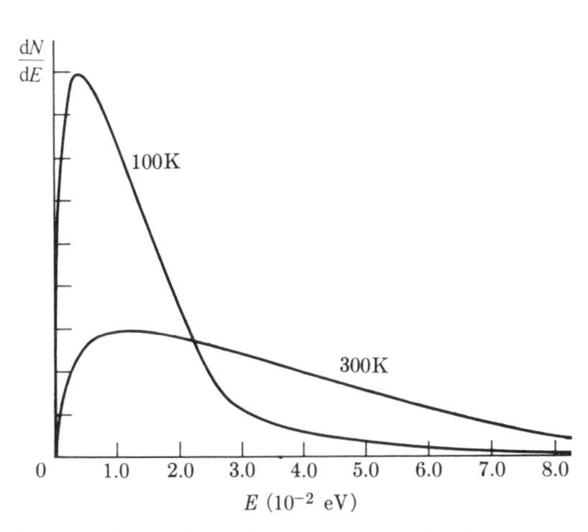

Figure 17.6 Molecular energy distribution at two temperatures (100 K and 300 K). Note, from Equation 17.14, that the energy distribution is independent of the molecular mass.

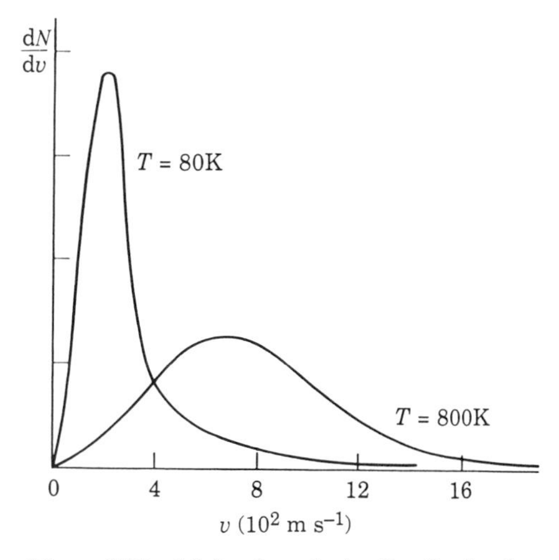

Figure 17.7 Molecular velocity distribution for oxygen at two temperatures (80 K and 800 K). Note, from Equation 17.15, that the velocity distribution depends on the molecular mass.

The energy distribution law 17.14 is the same for all gases at the same temperature, regardless of the mass of the molecules. However, the velocity distribution is different for each gas, due to the presence of the molecular mass m in Equation 17.15. Also, from the figures it can be seen that as the temperature increases, both energy and velocity distributions shift toward higher values, a fact that has important implications for such phenomena as the rate of chemical reactions.

EXAMPLE 17.5

Most probable energy and velocity of the gas molecules at a given temperature. These values correspond to the maxima of dN/dE and dN/dv, respectively.

▷ To obtain the maximum of dN/dE, given by Equation 17.14, at a certain temperature, it is necessary to compute only the maximum of $y = E^{1/2}\,e^{-E/kT}$. Thus

$$\frac{dy}{dE} = \left(\tfrac{1}{2}E^{-1/2} - \frac{E^{1/2}}{kT}\right)e^{-E/kT} = 0$$

from which we find that the most probable energy is $E = \tfrac{1}{2}kT$. Thus at room temperature, 300 K, for which $kT \approx 0.025$ eV, the most probable energy is $E \approx 0.012$ eV. On the other hand $E_{\text{ave}} = \tfrac{3}{2}kT \approx 0.036$ eV.

Similarly, to obtain the maximum of dN/dv, given by Equation 17.15, we must compute the maximum of $y = v^2\,e^{-mv^2/2kT}$. Then

$$\frac{dy}{dv} = \left(2v - \frac{mv^3}{kT}\right)e^{-mv^2/2kT} = 0$$

giving a most probable velocity of $v = (2kT/m)^{1/2}$. This velocity corresponds to an energy $E = kT$ and so is different from the most probable energy. Note that the smaller the mass of the molecules, the larger the most probable velocity. Thus, in air, nitrogen molecules move faster on the average than oxygen molecules and hydrogen molecules move even faster.

EXAMPLE 17.6

Average energy of the molecules in a gas.

▷ In a gas at temperature T, the total energy of the molecules that have energies between E and $E + dE$ is $E\,dN = E(dN/dE)\,dE$. Therefore the total energy of the gas is $U = \int_0^\infty E(dN/dE)\,dE$, and the average energy of the molecules is

$$E_{\text{ave}} = \frac{U}{N} = \frac{1}{N}\int_0^\infty E\left(\frac{dN}{dE}\right)dE$$

Inserting expression 17.14 we get

$$E_{\text{ave}} = \frac{2\pi}{(\pi kT)^{3/2}}\int_0^\infty E^{3/2}e^{-E/kT}\,dE$$

For calculating the integral it is simpler to make the substitution $x = E/kT$, so that $dE = kT\,dx$. Then

$$E_{ave} = \frac{2}{\pi^{1/2}} kT \int_0^\infty x^{3/2} e^{-x}\,dx$$

The value of the integral is $\frac{3}{4}\pi^{1/2}$ (see Appendix B, Section 6), so that we obtain $E_{ave} = \frac{3}{2}kT$, in agreement with Equation 15.15. This is further proof that the statistical definition of temperature associated with the Maxwell–Boltzmann temperature coincides with the ideal gas temperature.

17.6 Experimental verification of the Maxwell–Boltzmann distribution law

A test of the applicability of Maxwell–Boltzmann statistics to gases is to see if the energy and velocity distributions illustrated in Figures 17.6 and 17.7 actually occur. One way of doing this is to analyze the dependence of some molecular processes, such as the rate of chemical reactions, on the temperature. Suppose that a particular reaction occurs only if the colliding molecules have a certain energy equal to or larger than a certain energy E_a. The number of such particles, for two different temperatures, are given by the shaded areas under the low-temperature and high-temperature curves of Figure 17.8. We note that there are more molecules available at high than at low temperatures. By proper calculation, we can predict the effect of these additional molecules on the reaction rate, and the theoretical prediction can be compared with the experimental data. Experimental results are in excellent agreement with Equation 17.14, confirming the applicability of Maxwell–Boltzmann statistics to gases.

A more direct verification would consist in an actual 'count' of the number of molecules in each energy or velocity interval. Several experimental arrangements

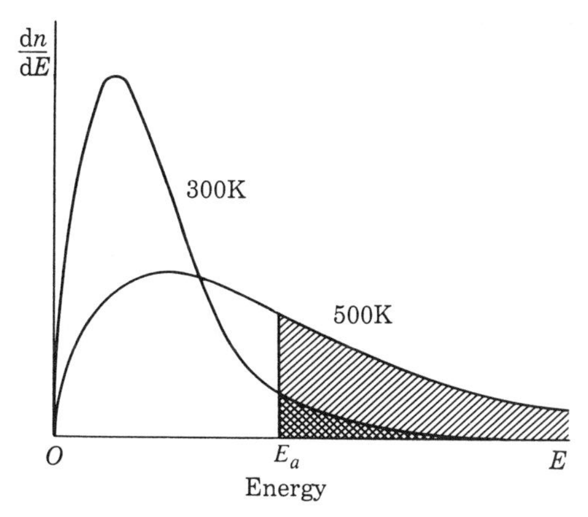

Figure 17.8 Number of molecules with energy larger than E_a at two different temperatures. The number in each case is indicated by the shaded area.

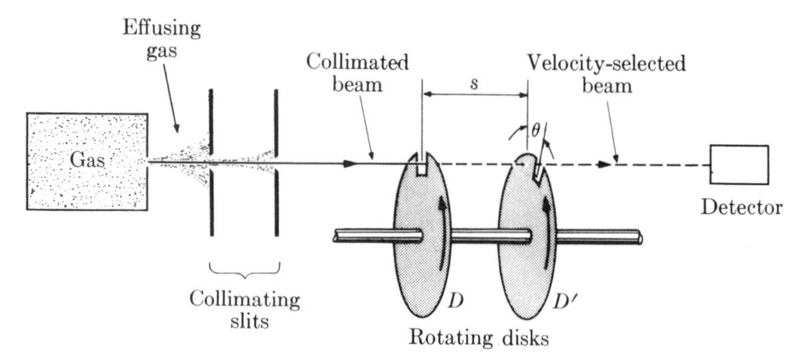

Figure 17.9 Molecular velocity selector.

have been used for this purpose. One method, using a velocity selector or 'chopper' is illustrated in Figure 17.9. The two slotted disks, D and D', rotate with an angular velocity ω and their slots are displaced by an angle θ. Gas molecules that escape from an oven at a certain temperature (a process called **effusion**), will pass through both slots and be received at the detector only if their velocity is $v = s\omega/\theta$, since $t = \theta/\omega$ is the time required for the disk to rotate through the angle θ. When either ω or θ is changed, the velocity of the molecules received at the detector can be changed. If several observations for different velocities v are made, velocity and energy distributions are obtained. The experimental results confirm the prediction of Maxwell–Boltzmann statistics, as expressed by Equations 17.14 and 17.15.

As a second example, neutrons produced in the fission process in a nuclear reactor are slowed down by means of a suitable moderator, such as water or graphite, until they are in thermal equilibrium at the temperature of the moderator. Neutrons in thermal equilibrium behave as an ideal gas and their energy distribution agrees with Maxwell's law, Equation 17.10; that is, thermal neutrons follow Maxwell–Boltzmann statistics. This fact is essential in nuclear reactor design. If the neutrons emerging from a porthole in a reactor are made to pass through an energy chopper similar to that of Figure 17.9, a beam of neutrons practically monoenergetic (depending on the opening in the chopper) is obtained.

17.7 Thermal equilibrium

Consider an isolated system composed of two different groups of particles, S_1 and S_2. We say that each group of particles constitutes a subsystem. Our two subsystems may, for example, consist of a liquid with a solid immersed in it, a mixture of two gases or liquids, two gases in containers with a common wall, or two solids in contact. By means of collisions and other interactions, energy may be exchanged between the particles composing the two subsystems, but the total energy of the whole system is assumed to be fixed. Let us designate the total numbers of particles in each subsystem by N and N' and the corresponding energy levels available to the particles by $E_1, E_2, E_3, \ldots$ and $E'_1, E'_2, E'_3, \ldots$. If there are no reactions between the particles in the two subsystems, the total number of particles in each subsystem is constant; also the total energy of the complete system is constant. But the energy of each subsystem is not conserved because, due to their interactions, energy may be exchanged between them. Therefore the following conditions for the two subsystems must be fulfilled by the occupation numbers $n_1, n_2, n_3, \ldots,$ and $n'_1, n'_2, n'_3, \ldots$ in a given partition:

$$N = \sum_i n_i = const. \qquad \text{(subsystem } S_1\text{)}$$

$$N' = \sum_j n'_j = const. \qquad \text{(subsystem } S_2\text{)}$$

$$U = \sum_i n_i E_i + \sum_j n'_j E'_j = const. \qquad \text{(both systems)}$$

The probability P of a given partition or distribution is given by the product of the separate probabilities P_1 and P_2 or $P = P_1 \times P_2$. We can obtain the equilibrium of the composite system by requiring that P be a maximum. The result is that the equilibrium distributions are

$$n_i = (N/Z)g_i e^{-\beta E_i}, \qquad n'_j = (N'/Z')g'_j e^{-\beta E'_j}$$

where Z and Z' are the respective partition functions of the two subsystems. We also note that the two subsystems in equilibrium have the same parameter β. In view of our definition of temperature, Equation 17.9, $kT = 1/\beta$, we may write

$$n_i = \frac{N}{Z} g_i e^{-E_i/kT}, \qquad n'_j = \frac{N'}{Z'} g'_j e^{-E'_j/kT}$$

where T is the common temperature of the two subsystems in thermal equilibrium. We conclude then that

> *two different and interacting systems of particles in statistical equilibrium must have the same temperature and therefore are in thermal equilibrium.*

The expressions for n_i and n'_j show that at thermal equilibrium each subsystem attains the same microstate as if it were isolated and at the common temperature T. These relations therefore express the fact that, in a statistical sense, after thermal equilibrium is attained the energy of each subsystem remains constant. This means that, although both systems may exchange energy at the microscopic level, the exchange takes place in both directions and, on the average, no net energy is exchanged in either direction. Therefore, statistically, the energy of each subsystem remains constant. It then follows that when two bodies at different temperatures are placed in contact they will exchange energy until both reach thermal equilibrium at the same temperature (Figure 17.10). At this point no further net exchange of energy takes place. This is also in agreement with our common understanding of the concept of temperature whereby, if a 'cold' and a 'hot' body are placed in

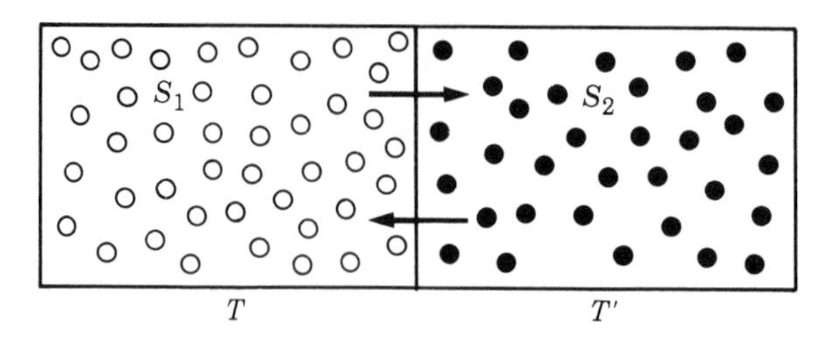

Figure 17.10 The systems S_1 and S_2 at temperatures T and T' exchange energy through the dividing wall. They reach equilibrium when their temperatures are the same.

contact, the cold warms up and the hot cools down. This goes on until both are 'felt' at the same temperature.

As mentioned in Chapter 15, this is the basis of the measurement of the temperature of a body. A body's temperature is determined by placing it in contact with a properly calibrated standard body or (thermometer) until thermal equilibrium is reached.

Note 17.3 Statistical analysis of work and heat

In this note we analyze the concepts of work and energy introduced in Chapter 16 from the statistical point of view. Consider a system where the particles can have energies E_1, E_2, $E_3, \ldots$. When the numbers of particles in each energy level are $n_1, n_2, n_3, \ldots$, the total energy of the system is

$$U = \sum_i n_i E_i$$

So far we have considered only isolated systems. If the system is not isolated, it may exchange energy with the surroundings. As a result of the energy exchange two effects may occur. One is that the particles will shift between energy levels, resulting in changes $dn_1, dn_2, dn_3, \ldots$ in the occupation numbers. The second effect may be changes $dE_1, dE_2, dE_3, \ldots$ in the energy levels because of changes in the structure or size of the system. For example, consider a gas in a box of side a. If the size of the box changes because its sides increase by an amount da, the energy levels of the molecules change as illustrated in Figure 17.11 (see Section 37.4).

The change in internal energy can then be expressed as

$$dU = \underbrace{\sum_i E_i \, dn_i}_{\text{redistribution of particles}} + \underbrace{\sum_i n_i \, dE_i}_{\text{change in structure or size of system}}$$

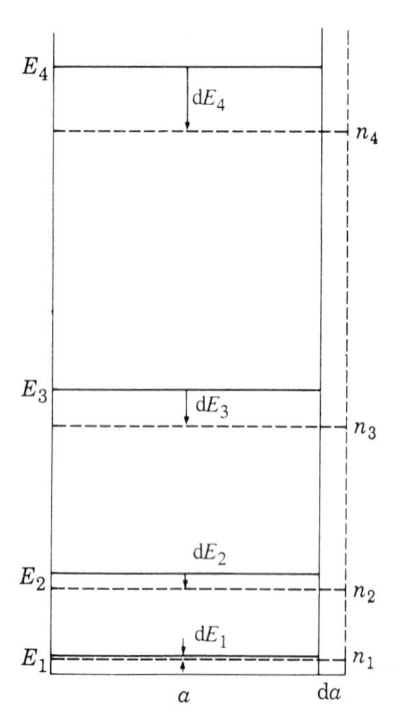

Figure 17.11 Change in energy levels of a gas in a box when the width of the box is changed.

The first sum, $\Sigma_i E_i\, dn_i$, corresponds to a change in internal energy due to a redistribution of the molecules among the available energy levels. The second sum, $\Sigma_i n_i\, dE_i$, corresponds to a change in internal energy due to a change in the energy levels. Let us examine the second term first. For the case of the expansion of a gas, the change in energy $\Sigma_i n_i\, dE_i$ is due to the change in the dimensions (i.e. volume) of the container. This energy change corresponds to what we have called work in Section 16.3. Thus, we conclude that the work done on the system is

$$dW_{ext} = \sum_i n_i\, dE_i$$

So we may also conclude that the first term is the heat absorbed by the system and write

$$dQ = \sum_i E_i\, dn_i$$

That is, the statistical quantity we have defined as heat in Section 16.4 corresponds to a change in energy of the system due to molecular jumps between energy levels. These jumps result from energy exchanges with the surroundings. When radiation energy is absorbed by the system, it also results in a redistribution of particles among energy levels; that is, it contributes to the term $\Sigma_i E_i\, dn_i$.

This analysis puts the statistical definitions of the work done and the heat absorbed by a system on a firmer basis than in our preliminary definition, given in Section 16.3.

17.8 Entropy

The equilibrium partition of a system depends on the properties of the components of the system. It corresponds to the most probable distribution of the molecules of the system among the different available energy states compatible with the macro-state of the system. Under such conditions the probability P of the partition is a maximum. If the system, although isolated, is not in equilibrium, we may assume that it is in a partition (or distribution) of lower probability than the maximum or equilibrium. In due time the system will evolve, under the interactions among its components or molecules, until it attains the partition of maximum probability. At this time the system reaches statistical equilibrium and no further increase in P is expected, unless the system is perturbed by an external action.

This natural trend toward statistical equilibrium, amounting to an evolution toward the partition of maximum probability, is similar to what happens to the entropy of an isolated system, as explained in Section 16.9. This similarity suggests that there is a relation between the entropy of a system and the probability of a partition. In fact, it was first shown by Boltzmann that entropy, as defined by Equation 16.24, that is, $S = \int dQ/T$, is related to the partition probability by

$$S = k \ln P \qquad \textbf{(17.16)}$$

where k is the Boltzmann constant, introduced to ensure that entropy is measured in J K^{-1}. Thus

the entropy of a system is proportional to the logarithm of the probability of the partition corresponding to the state of the system.

Such a statistical definition of entropy applies to any partition or state, either of equilibrium or non-equilibrium. This is in contrast to temperature, which is defined only for equilibrium partitions or states. The advantage of the statistical definition is that the entropy can be calculated for a variety of systems, both for equilibrium and non-equilibrium states, by calculating the partition probability P. For example, for any equilibrium state it can be shown that the entropy defined by Equation 17.16 is expressed by

$$S = \frac{U}{T} + kN \ln\left(\frac{Z}{N}\right)$$

The statistical definition of entropy confirms that it is a property of the state of the system. Therefore, the change of entropy of a system when it goes from one state to another is independent of the process followed, since it is determined by the probabilities of the initial and final partitions, a property that was also recognized in Section 16.9.

For two systems with respective probabilities P_1 and P_2, the total probability of the partition resulting from the combination of the two systems is $P = P_1 P_2$. Thus $\ln P = \ln P_1 + \ln P_2$ and

$$S = k \ln P = k \ln P_1 + k \ln P_2 = S_1 + S_2$$

We see then that entropy is an additive quantity.

17.9 Law of increase of entropy

The state of statistical equilibrium of a system corresponds to the most probable partition. Also the entropy of a system is directly related to the probability of its partition. We conclude then that the entropy of an isolated system in statistical equilibrium has the maximum value compatible with the physical conditions of the system. Hence

the only processes that may occur naturally in an isolated system after it reaches statistical equilibrium are those processes that are compatible with the requirement that the entropy does not change; that is, $\Delta S = 0$.

When an isolated system is not in equilibrium, the processes that are more likely to occur are those that carry the system toward the state of maximum probability or statistical equilibrium, for which the entropy is also maximum. Therefore,

an isolated system that is not in equilibrium will naturally evolve in the direction in which its entropy increases; that is, $\Delta S > 0$.

Thus the processes that are more likely to occur in an isolated system are those for which

$$\Delta S \geqq 0 \qquad \textbf{(17.17)}$$

The inequality holds when the isolated system is not initially in equilibrium and is equivalent to Equation 16.35. We may then state that

> *the most probable processes that may occur in an isolated system are those where the entropy either increases or remains constant.*

This statement is the statistical formulation of the **law of entropy** or second law of thermodynamics that was stated in Section 16.11 as an empirical law. However, there is an important difference and it is the use of the term 'most probable'. This implies that the law of entropy must be interpreted in a statistical sense. In a particular instance, due to the fluctuations in the molecular distribution, the entropy of an isolated system may decrease. But, the larger the decrease, the less probable it is to occur. The variation of entropy during the evolution of the system toward equilibrium may thus be represented by the irregular line in Figure 17.12.

The law of entropy expresses the well-known fact that in an isolated system there is a well-defined trend or direction of occurrence of processes. This trend is determined by the direction in which entropy increases. This is another reason why the concept of entropy is so important since it serves to characterize what processes are possible and what are not.

Transport phenomena, such as molecular diffusion and thermal conduction (Chapter 18), are good examples of processes that always take place in one direction. In both cases it may be verified that the entropy of the system increases. Diffusion takes place in the direction in which the concentration tends to be equalized, resulting in a homogeneous system. The reverse process, a spontaneous change of a homogeneous system into a non-homogeneous system, corresponding to a decrease in entropy, is never observed. For example, consider a drop of ink released at point A inside a vessel filled with water (Figure 17.13(a)). The ink molecules spread quickly throughout the water (Figure 17.13(b)), and after a relatively short time the water is uniformly colored (Figure 17.13(c)). In this process the entropy of the system has

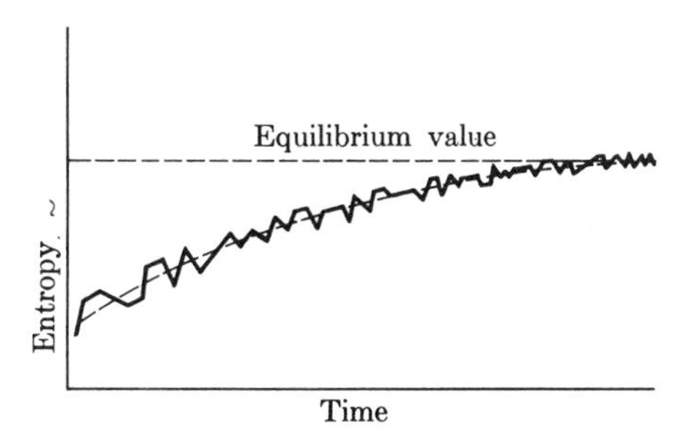

Figure 17.12 Variation of entropy of an isolated system while the system is evolving toward equilibrium.

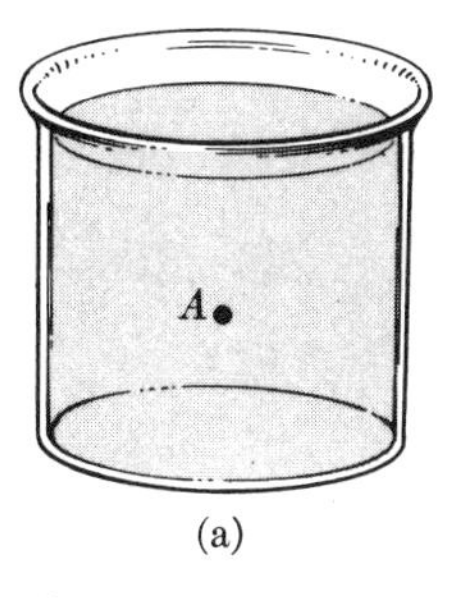

(a)

(b)

(c)

Figure 17.13 Irreversible diffusion of ink in water.

increased because of the increase in states available to the ink molecules. However, if at a given time the velocities of all the molecules were exactly reversed, all the ink would eventually collect back at A, resulting in a decrease in entropy. But this is surely an extremely improbable occurrence, and so far has never been observed. On the other hand, there may be small fluctuations in the concentration of ink molecules at different places, even after equilibrium has been reached. But these fluctuations, in most cases, are not noticeable. The same analysis can be made with the mixing of two substances at different temperatures.

If a system is not isolated, its entropy may decrease by interaction with other systems, whose entropy must then also change. But the total amount of all changes of entropy made by all systems involved in the process must be in agreement with Equation 17.17. That is, $\Delta S = 0$ will hold when all systems remain in equilibrium during the process and $\Delta S > 0$ when the systems are not in thermal equilibrium, either initially or during the process. For example, if a combination of two systems is isolated and the total entropy is $S = S_1 + S_2$, the processes occurring in the combined system must satisfy

$$\Delta S = \Delta S_1 + \Delta S_2 \geqq 0 \tag{17.18}$$

The entropy of one of the components may decrease during a process, but the net change of entropy for the whole system must be positive or zero.

The law of entropy indicates those processes that are more likely to occur in the universe as a whole. Therefore, many processes *could* occur because they comply with other laws, such as the conservation of energy. However, it is very improbable that they will occur, because they violate requirement 17.18. This prompted Clausius to make his famous statement: the energy of the universe is constant but its entropy grows constantly.

Clausius' statement has important cosmological implications. Since the origin of the universe, presumably at the Big Bang about 15×10^9 years ago, the universe has evolved continuously. This evolution is apparently accompanied by a gradual overall entropy increase. So, it appears that the law of entropy determines what can happen in the universe as time progresses. Of course, at certain places and times the local entropy may decrease due to the appearance of local organization, but this is offset by a larger increase somewhere else. Also it may be speculated that there must be some connection between the increase of entropy and the directionality or 'arrow' of time, a fascinating subject on which we cannot elaborate here.

EXAMPLE 17.7

Change of entropy of an ideal gas during a free expansion.

▷ Consider a container divided into two equal volumes 1 and 2 by a removable partition (Figure 17.14(a)). Initially, vessel 1 on the left contains the gas in thermal equilibrium, and vessel 2 on the right is empty. When the piston is removed, the gas undergoes a free expansion and in a short time the molecules are distributed over the whole container 1 + 2, with volume $2V$ (Figure 17.14(b)). During such an irreversible process, equilibrium is

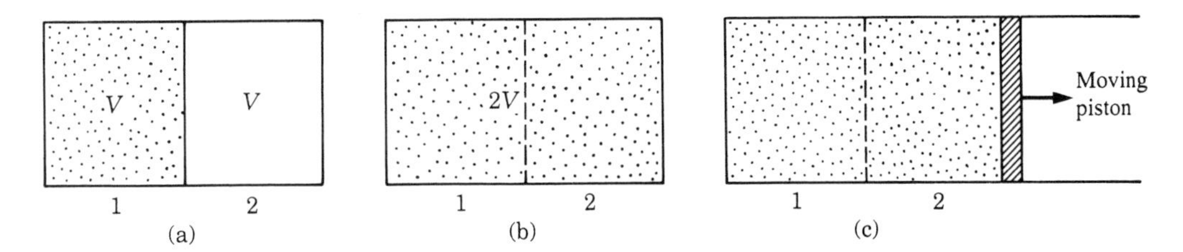

Figure 17.14 Free expansion of a gas.

destroyed for a certain time, but eventually statistical equilibrium is restored without any external action. As explained in Example 16.12, the temperature of the ideal gas does not change during the free expansion, and the same is true for the average energy of the molecules.

We may compute the change in entropy by analyzing the process from a simple probabilistic point of view. If P_1 and P_2 are equilibrium partition probabilities of the initial and final states we have, using Equation 17.16, that

$$\Delta S = S_2 - S_1 = k \ln P_2 - k \ln P_1 = k \ln \frac{P_2}{P_1}$$

In the case of an ideal gas, it is easy to calculate the ratio of the probabilities without calculating P_1 and P_2 separately. We note first that P_1 is the probability associated with the gas molecules all being in a volume V, while P_2 corresponds to the molecules occupying vessels 1 + 2 with volume $2V$. For just one molecule, the ratio P_1/P_2 of the two probabilities is the same as the ratio $V/2V$ of the two volumes or $\frac{1}{2}$, for two molecules it is $\frac{1}{2} \times \frac{1}{2}$ or $(\frac{1}{2})^2$ because molecules move independently from each other; for three molecules it is $\frac{1}{2} \times \frac{1}{2} \times \frac{1}{2} = (\frac{1}{2})^3$. In other words, if we take a series of pictures, we would find that for one molecule, in 50% of the pictures the molecule would be on the left side of the container; if we have two molecules, in 25% of the pictures the two molecules would simultaneously be on the left-hand side; and for three molecules, the probability is 12.5%. It is then possible in a short time to observe 'all' the molecules (one or two) on the left-hand side. But then, of course, statistical methods are unnecessary, and have no meaning. It may even be possible to calculate the exact times at which one of the particles, or two, or three, are on one side or the other if we know the initial conditions.

But when the number of molecules is large, the situation changes dramatically. For N molecules, the ratio is

$$\frac{P_1}{P_2} = \frac{1}{2^N} \quad \text{or} \quad \frac{P_2}{P_1} = 2^N$$

If N is a very large number, then P_2 is much larger than P_1. This accounts for the rapid rate at which the gas expands freely from volume V up to volume $2V$. For the same reason, the reverse process where all the molecules are simultaneously all in volume 1 on the left is highly improbable (although possible). No one expects that at a certain later time, as a result of molecular motions, all gas molecules will appear together on the left hand side of the container. This process is possible but very improbable and, in fact, is never observed. Therefore the free expansion of a gas is an example of the directionality of processes in the trend toward equilibrium and illustrates the probabilistic nature of the second law of thermodynamics.

Inserting the value $P_2/P_1 = 2^N$ in the expression for ΔS, we obtain

$$\Delta S = k \ln 2^N = kN \ln 2$$

which is precisely the value we obtained in Example 16.11. We must make clear that our analysis is valid only for the simple case of an ideal gas. For more complex systems, it is necessary to calculate P directly.

When vessel 2 in Figure 17.14 is not rigid but its volume gradually increases without the gas doing work (Figure 17.14(c)), the gas will never attain statistical equilibrium because it keeps expanding. The consequence of this continual expansion is that the entropy of the gas never reaches a maximum, but increases continuously. Since the universe is in a state of expansion, we may say that although the entropy of the universe is increasing continuously, the universe is not approaching a state of maximum entropy because it is not approaching equilibrium. In other words, cosmic expansion generates entropy.

Note 17.4 Systems far from equilibrium

In equilibrium statistical mechanics, it is assumed that an isolated or closed system that is not initially in equilibrium evolves inevitably and uniquely toward a well-defined state of statistical equilibrium. This is the state corresponding to the most probable partition, or rather maximum entropy, compatible with the constraints and parameters of the system. This assumption is generally correct, as long as the system initially is not far from equilibrium and its evolution can be described, to a good approximation, by linear equations.

A system not in equilibrium is **open** when it exchanges energy and/or matter with the surroundings. In this case, the system does not necessarily evolve toward statistical equilibrium. However, when the flow of energy and/or matter is regular, the system may reach a steady dynamical state so that certain properties, such as the gradients of temperature and matter distribution throughout the system, remain constant. This is the case for steady state transport processes in fluids (Chapter 18). The possibility exists that, in addition, the system evolves in complexity, developing certain structures that would not occur if the system were isolated, a process called **self-organization**. One example of self-organization in an open system is a fluid when the temperature gradient reaches a certain critical value (see Note 18.2). Another example of self-organization in an open system is that of living systems, from cells to complete organisms. The self-organization in an open system is very sensitive to the rates of flow of energy and/or matter as well as other parameters of the system. When the rates of flow, or the parameters, are altered, the self-organization may disappear, sometimes quite suddenly (in the case of a living system, death).

The complexity of most physical systems in nature makes them very sensitive to initial conditions. In addition, their description requires non-linear equations, making their mathematical analysis most difficult. In fact, it might not even be possible to write exact solutions, as we mentioned in Note 10.4 for a forced oscillator.

With the advent of computers, it has been possible to analyze complex physical systems in great detail. This has revealed an important feature of systems that obey non-linear equations. Because these systems are very sensitive to initial conditions and to external factors, even a small fluctuation in the parameters of the system, when it is far from statistical equilibrium, may be amplified in a quite unexpected manner, resulting in an unpredictable evolution of the system. Therefore, although the system is predictable in the short term it is unpredictable in the long term (Figure 17.15).

Complex systems far from equilibrium may exhibit chaotic behavior, in the sense described in Note 10.4 for simple dynamical systems. Air circulation in the atmosphere is a paramount example of a physical system exhibiting chaotic behavior. This is why, for example, long-term weather prediction is practically impossible. The dynamics of the atmosphere is very complex and is described by non-linear equations, with initial conditions and parameters known only within a certain degree of accuracy.

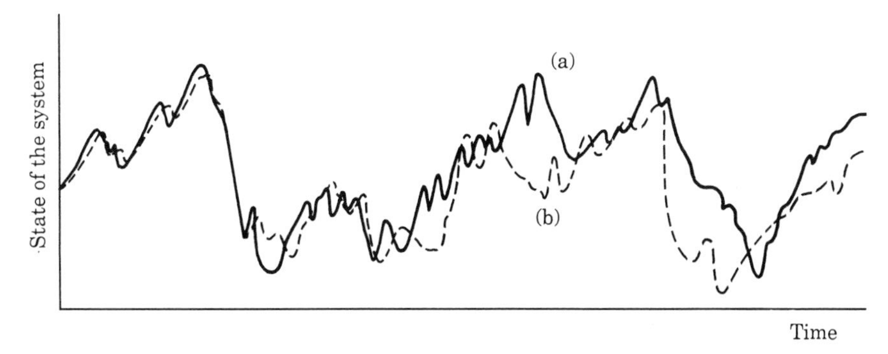

Figure 17.15 Evolution of a nonlinear system for slightly different initial conditions. After a while, the states become very different.

Examples of systems exhibiting chaotic behavior include coupled vibrating structures and electric circuits, turbulent fluid flow (Note 18.2), the orbits of asteroids (Note 11.9), oscillating chemical reactions, the motion of charged particles in accelerators or laser cavities, heartbeats, the electrical activity of the brain, and dripping faucets (taps), to mention a few.

QUESTIONS

17.1 Explain what is meant by a partition of a system of particles.

17.2 What is the relation between statistical equilibrium and the probability of a partition?

17.3 How does the occupation number in Maxwell–Boltzmann statistics for a given energy state E_i vary as the temperature (a) increases, (b) decreases? What is the physical reason for that kind of variation?

17.4 How does the partition function vary with the temperature? How is this reflected in the total energy of a system?

17.5 To which physical property of a system is the temperature related? Is the relation the same for all substances?

17.6 It can be shown that the partition function of an ideal gas is $Z = CVT^{3/2}$, where C is a constant. Using Equation 17.13, obtain the average energy of the gas molecules. What do you conclude?

17.7 Compare the notions of statistical equilibrium and of thermal equilibrium.

17.8 In Example 17.5, we found the most probable velocity (i.e. the maximum of the velocity distribution) of the molecules in a gas. Explain why this velocity does not correspond to the most probable energy for the distribution.

17.9 What formulation of the concept of entropy do you find has more physical insight: the one discussed in Section 16.9 or that in Section 17.7?

17.10 Using the partition function of an ideal gas given in Question 17.6 and the value $U = \frac{3}{2}kNT$ in Equation 17.17, verify that the entropy of an ideal gas is given by Equation 16.31. This shows the consistency of thermodynamics with statistical mechanics.

17.11 The deuteron is a nucleus composed of two particles (a proton and a neutron) while the uranium nucleus contains 238 particles (92 protons and 146 neutrons). To which of these nuclei is the concept of 'nuclear temperature' more appropriately applicable?

17.12 Referring to Example 17.3, draw a diagram that shows how the total energy U varies with the temperature T.

PROBLEMS

17.1 (a) Calculate the probability of distribution of 10 identical but distinguishable balls in 7 identical boxes so that one ball is placed in boxes 1 and 5, two in boxes 2 and 7, four in box 6 and none in boxes 3 and 4. (b) Repeat the calculation when there is one ball in boxes 1, 2, 6 and 7 and two each in boxes 3, 4 and 5.

17.2 Determine the temperature of the system of Example 17.2 when it is in statistical equilibrium. Assume $\varepsilon = 0.02$ eV.

17.3 (a) Referring to Example 17.3, find the ratio as a function of the temperature of the number of particles having an energy $+\varepsilon$ to those having an energy $-\varepsilon$. (b) Evaluate the ratio for temperatures of 10, 300, and 1000 K. Assume $\varepsilon = 0.02$ eV.

17.4 The possible particle energies of a system of particles are $0, \varepsilon, 2\varepsilon, \ldots, n\varepsilon, \ldots$. (a) Show that the partition function of the system (with $g_i = 1$) is $Z = (1 - e^{-\varepsilon/kT})^{-1}$. (b) Compute the average energy of the particles. (c) Find the limiting value of the average energy when ε is much smaller than kT.

17.5 Referring to the system of the previous problem, plot a graph showing the occupation numbers for (a) 100 K, (b) 300 K, (c) 800 K, given that the value of the energy ε is (i) 10^{-3} eV, (ii) 0.1 eV.

17.6 A reversible heat engine carries 1.0 mole of an ideal diatomic gas around the cycle shown in Figure 17.16. (a) Determine the pressure of the gas at point B. (b) For the process B–C, determine: the heat added, the change in internal energy and the work done on the gas. (c) Determine the root mean square velocity v_{rms} of the gas molecules at point A if the mass of a molecule is 5.0×10^{-26} kg. (d) Represent Maxwell's velocity distribution for the gas molecules at 300 K and at 600 K corresponding to points A and B of the figure.

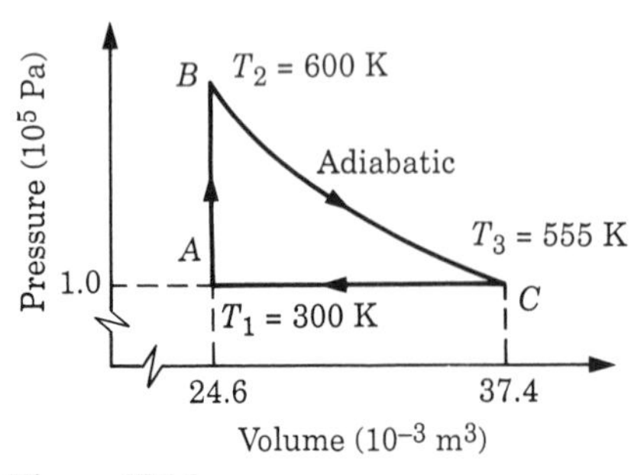

Figure 17.16

17.7 (a) Calculate the ratio of the average velocity of ^{14}C atoms to those of ^{12}C atoms at a temperature T where the atoms act as an ideal gas. (b) Repeat for ^{235}U and ^{238}U. (c) Consider the effectiveness of diffusion through a porous wall for these two cases as a means of isotopic separation.

17.8 Calculate the temperature when the number of molecules of an ideal gas per unit energy range at 2×10^{-2} eV is one-fourth of the number at 1×10^{-2} eV.

17.9 Find the ratio of the number of molecules of an ideal gas per unit energy range at energies of 0.2 eV to those at 0.02 eV, given that the gas temperature is (a) 100 K, (b) 300 K, (c) 600 K.

17.10 Compute enough points to construct graphs of the molecular energy distribution function in an ideal gas for one kilomole at (a) 200 K and (b) 600 K.

17.11 Calculate the (a) root mean square, (b) average and (c) most probable velocities of oxygen molecules at a temperature of 300 K. (d) Calculate the most probable velocity of oxygen molecules at the following temperatures: (i) 100 K, (ii) 500 K, (iii) 1000 K, (iv) 10 000 K.

17.12 (a) Compute the mean translational kinetic energy of an ideal gas molecule at 300 K. (b) Compute the root mean square velocity if the (ideal) gas is hydrogen (H_2), oxygen (O_2), or mercury vapor (Hg). (c) Compare your results for hydrogen and oxygen with the velocity of sound in those gases.

17.13 Compute the root mean square velocity of (a) helium atoms at 2 K, (b) nitrogen molecules at 27 °C, (c) mercury atoms at 100 °C.

17.14 Calculate the temperature of an ideal gas whose mean translational kinetic energy per molecule is equal to that of a single particle with energy (a) 1 eV, (b) 1000 eV, (c) 1 MeV.

17.15 Calculate the fractional number of molecules of an ideal gas with velocities between v_{ave} and 1.2 v_{ave} from Equation 17.15 making (a) $v = v_{ave}$ and $dv = 0.2\, v_{ave}$; (b) $v = 1.1\, v_{ave}$ and $dv = 0.1\, v_{ave}$.

17.16 It can be shown that the number of molecules of an ideal gas whose velocity has components in the range v_x and $v_x + dv_x$, v_y and $v_y + dv_y$, $v_z + dv_z$, is given by

$$dn = N(m/2\pi kT)^{3/2} e^{-mv^2/2kT}\, dv_x\, dv_y\, dv_z$$

Show that the number of molecules of an ideal gas

that have an x-component of velocity between v_x and $v_x + dv_x$ irrespective of the values of the v_y and v_z components is

$$dn = N(m/2\pi kT)^{1/2} e^{-mv_x^2/2kT} dv_x$$

(Hint: Recall that $v^2 = v_x^2 + v_y^2 + v_z^2$. The limits of integration in each case must be from $-\infty$ to $+\infty$. See Appendix B.6.)

17.17 Use the result of Problem 17.16 to obtain the average value of (a) v_x, (b) v_x^2, (c) $|v_x|$ for an ideal gas.

17.18 What fraction of the molecules of an ideal gas have positive X-components of velocity greater than $2v_{mp}$, where v_{mp} is the most probable velocity?

17.19 Compute the fraction of molecules of an ideal gas that have a velocity with a component along the X-axis (a) smaller than v_{mp}, (b) larger than v_{mp}, (c) smaller than v_{ave}, (d) larger than v_{ave}. (v_{mp} is the most probable velocity.)

17.20 Consider two samples of the *same* gas, both at the same temperature T, composed of N_1 and N_2 molecules, respectively, and occupying adjoining containers of volumes V_1 and V_2, separated by a removable partition. When the partition is removed, we have a sample of the gas composed of $N_1 + N_2$ molecules occupying the volume $V_1 + V_2$. Show that (a) the temperature remains the same and (b) the change in entropy is

$$\Delta S = kN_1 \ln\left(\frac{(V_1 + V_2)N_1}{(N_1 + N_2)V_1}\right) + kN_2 \ln\left(\frac{(V_1 + V_2)N_2}{(N_1 + N_2)V_2}\right)$$

(c) Show also that if the two gases were initially also at the same pressure, the change in entropy would be zero. Why?

17.21 A system is composed of particles whose internal degrees of freedom correspond to states of energy 0, ε, $2\varepsilon, \ldots, n\varepsilon, \ldots$, where n varies from zero to infinity. Find (a) the entropy and (b) the heat capacity due to the internal degrees of freedom of the system. (Hint: refer to Problem 17.4.)

17.22 A system is composed of particles that, due to their internal degrees of freedom, can exist only in either of two states, of energy $-\varepsilon$ and $+\varepsilon$, in addition to the translational kinetic energy of the particles. Calculate (a) the entropy and (b) the heat capacity (at constant volume) of the system due to the internal degrees of freedom of the particles, as a function of the temperature of the system. (c) Plot both quantities as a function of the absolute temperature of the system. (Hint: see Example 17.3.)

17.23 The number of particles in a system is N. The particles can only be in either of two states with energy $+\varepsilon$ or $-\varepsilon$, but the particles do not have any translational kinetic energy. (a) Given that the total energy of the system is U, show that the absolute temperature is given by

$$\frac{1}{T} = \frac{k}{2\varepsilon} \ln \frac{N - U/\varepsilon}{N + U/\varepsilon}$$

(b) Verify that the absolute temperature is positive (negative) if U is negative (positive).

18 Transport phenomena

Convection cells with characteristic polygonal geometry arise spontaneously when a thin layer of fluid is heated from below. When steady state is reached, the cells adopt a regular hexagonal form. In each cell the fluid rises in the center and sinks at the periphery. (Courtesy of Manuel G. Velarde, UNED, Madrid.)

18.1 Introduction

The fact that molecules in gases and liquids (and to a lesser extent in solids) are in continuous motion gives rise to a series of important physical processes of a statistical nature that have certain common features. All these processes are known under the general heading of transport phenomena. Transport phenomena are those processes where there is a net transfer (or transport) of either matter, energy or momentum, in a macroscopic amount. In this chapter we briefly discuss three kinds of transport phenomena: (a) molecular diffusion, (b) thermal conduction and (c) viscosity. Transport phenomena correspond to irreversible processes. They occur only in systems that are not in statistical equilibrium.

18.2 Molecular diffusion: Fick's law

When we open a bottle of perfume (or of any other liquid having a very distinct odor such as ammonia), we smell it very quickly in remote parts of a closed room or one without ventilation. The molecules of the liquid, after evaporating, diffuse

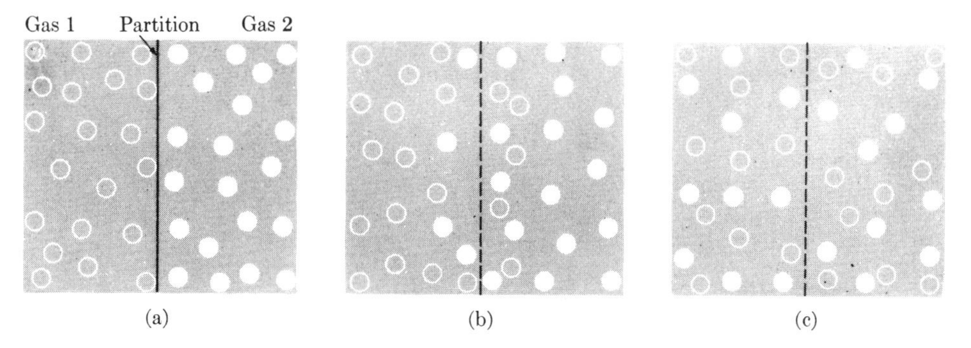

Figure 18.1 Gaseous diffusion. (a) The two gases are held separate by the partition. (b) Shortly after the partition is removed, a few molecules of each gas are found on the other's side. (c) After a certain time the mixture of the two gases is homogeneous, and no further diffusion takes place.

throughout the room; that is, they spread into all the surrounding space. The same thing happens if we place a sugar cube in a glass of water. The sugar cube dissolves gradually. At the same time, the dissolved sugar molecules diffuse and eventually are distributed throughout the water. As a final illustration, consider two gases in a vessel that keeps them separate by a partition, as shown in Figure 18.1. When the partition is removed, the two gases diffuse into each other until, after a short time, we have a homogeneous mixture. These and many other familiar examples illustrate a fundamental characteristic of the diffusion process; namely,

diffusion occurs whenever the space distribution of the molecules of the substance varies from place to place.

Call n the number of molecules of the diffusing substance per unit volume (this will be called the **concentration** of the substance). According to the above statement, this number must vary from place to place in order for diffusion to occur. A second characteristic is that

diffusion (molecular transport) occurs in the direction in which the concentration decreases.

Therefore, diffusion tends to equalize the molecular distribution of the diffusing substance over all space until statistical equilibrium is reached, unless some external action maintains a concentration distribution that is not uniform. Hence, there is a well-defined tendency for diffusion to occur, which is, in a sense, regulated by the law of entropy. This tendency, however, must be considered in a statistical or macroscopic way. Local fluctuations may occur that, for short intervals of time, produce a reversal of the molecular flow at certain spots.

Generally speaking, diffusion is the result of the fact that molecular agitation produces frequent collisions among the molecules. As a consequence, molecules are scattered throughout the region. Suppose that we have a gas occupying a region divided into two sections by a partition (Figure 18.2). Its density is different on each side, but the temperatures, and therefore also the molecular velocities, are the same on both sides. When the partition is removed, there are two currents of

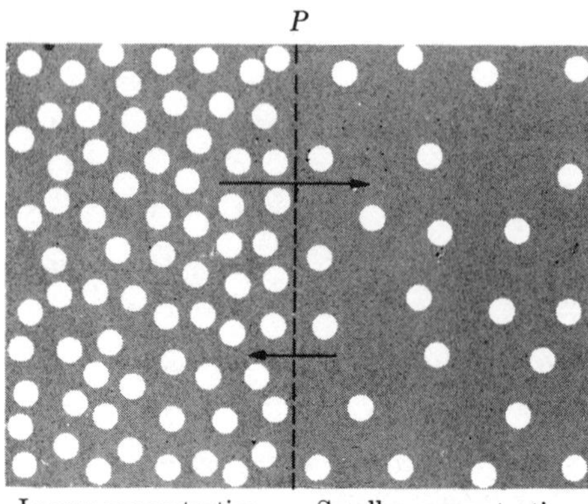

Figure 18.2 The diffusion current in the two directions is different.

molecules at the interface, indicated by the horizontal arrows. These currents are the result of collisions and scatterings that occur on both sides. However, the current from left to right is greater because there are more molecular collisions per second on the left, where the concentration is greater. Therefore, there is a net current to the right. This results in a diffusion from left to right, or from the region of greater concentration to that of lesser concentration.

It is easy to see that diffusion, being an irreversible process, is accompanied by an increase of entropy. Initially, the gases on both sides of the wall are in thermal equilibrium. As soon as we remove the wall, we make available to them a larger volume and therefore we disturb the state of equilibrium. After they diffuse into each other, with a corresponding increase in entropy, they reach a new state of equilibrium with a higher entropy (recall Example 17.7).

We consider only the diffusion of a substance into itself or through another medium that is homogeneous and whose molecules are essentially fixed. The case of two substances diffusing each into the other will not be discussed, since that requires a slightly different treatment. Finally, the effects of intermolecular forces will be ignored.

Suppose that the concentration of the substance varies in a certain direction, which will be designated as the X-axis (Figure 18.3). Otherwise, the concentration is assumed to be the same in all planes perpendicular to that direction. Then the number of atoms or molecules per unit volume (or the concentration) is a function of the x-coordinate alone; that is, $n(x)$. It is expressed in m^{-3}. Diffusion then occurs in the direction of the X-axis only.

We define the **particle current density**, designated by j, as the net number of particles that cross, per unit time, a unit area placed perpendicular to the direction of diffusion. This particle current density is expressed in $\text{m}^{-2}\,\text{s}^{-1}$. When the substance is homogeneous (that is, when n is constant), the current density j is zero because when n is constant the same number of particles passes in one direction as in the opposite, and no net mass transport exists. But when the substance is inhomogeneous and n varies from point to point, a net current, or mass transport, results. Physical intuition suggests that the current density is directly proportional to the variation of the concentration $n(x)$ per unit length. That is, the greater $\mathrm{d}n/\mathrm{d}x$, the larger the current density. Experiments also confirm

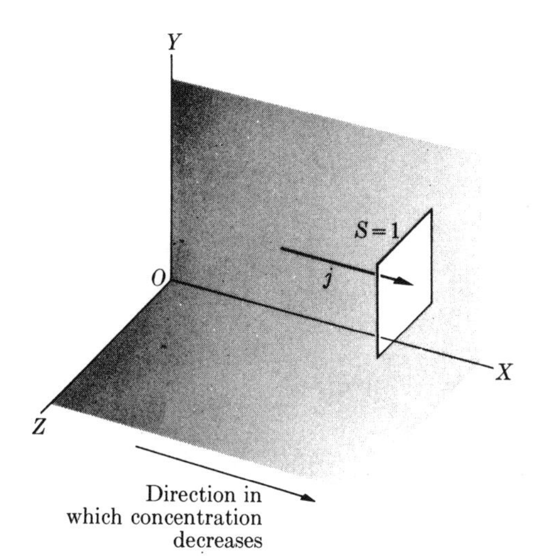

Figure 18.3 The net diffusion current is in the direction the concentration decreases.

our intuition. We may write this relation as

$$j = -D \frac{dn}{dx} \tag{18.1}$$

where D is a coefficient characteristic of the substance, called the **diffusion coefficient** and dn/dx is called the **concentration gradient**. The minus sign indicates that the net flow is in the direction in which n decreases. The diffusion coefficient D is expressed in $m^2 s^{-1}$, so that units are consistent. Equation 18.1 is called **Fick's law** and is of a statistical nature.

Most diffusion processes satisfy Fick's law quite well, except when the concentration n is extremely low, very high, or changes suddenly in a short distance. In such cases our assumptions are no longer valid.

18.3 Steady diffusion

Consider the special situation when the particle current density is the same across any cross-section, as it might be in the case of steady diffusion along a pipe. Then j is independent of x or $j = const.$ A constant current density means that, through any volume element, the number of diffusing particles entering per unit time through one end is the same as the number of particles leaving per unit time through the other end. That is, there is no accumulation or change in concentration at any region in the medium. We say then that the situation is **stationary** or steady.

When j is constant, Equation 18.1 may be rewritten as

$$j = D\left(\frac{n_0 - n}{x}\right) \quad \text{or} \quad n = n_0 - \frac{j}{D} x \tag{18.2}$$

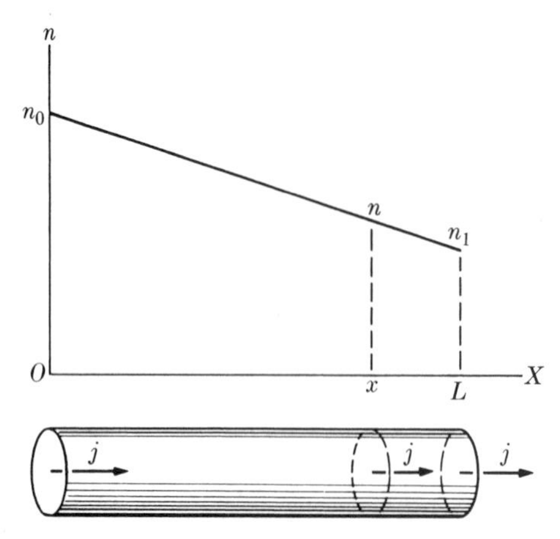

Figure 18.4 Change in concentration due to steady diffusion along a rod.

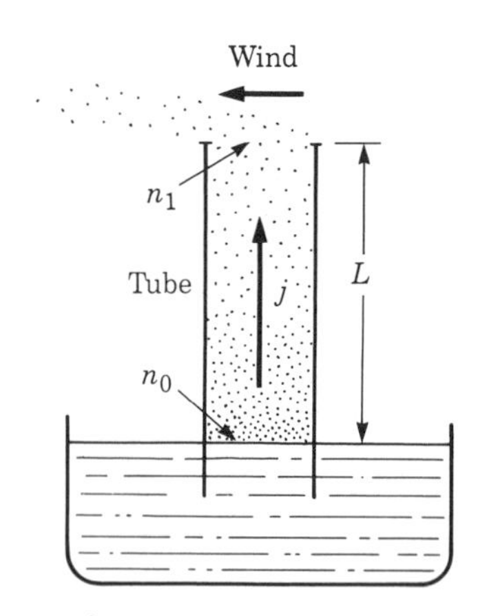

Figure 18.5 Diffusion of water vapor along a column.

where n_0 is the concentration at $x = 0$. Therefore, when j is constant, the concentration of particles decreases linearly with distance along the direction of diffusion, as shown in Figure 18.4. Note that, for diffusion along a pipe of length L (Figure 18.4) under stationary conditions, it is necessary to feed particles in at a constant rate at one end ($x = 0$) and remove them at the same rate at the other end.

Suppose, for example, that we have a vertical tube open at both ends. One end of the tube is in a liquid, which evaporates, so that the molecules of the liquid diffuse through the air in the tube (Figure 18.5). At the other end the molecules that arrive there are removed by some means. When a steady condition is reached, the number of molecules entering at the bottom because of evaporation is equal to the number of molecules removed at the top. This, in fact, is one method for measuring the diffusion coefficient D of a substance in air. We can find the quantity j by measuring the amount of liquid that evaporates in a given interval of time. We experimentally determine the concentration at the bottom and at the top. Then we apply Equation 18.2 to compute D.

The situation we have just considered assumes that a steady state has already been reached. Another important problem is to determine how this steady state is reached. Let us again consider the pipe shown in Figure 18.4. At a given time, $t = 0$, one end is connected to a source of the gas at a constant concentration n_0; the molecules are removed at a certain rate at the other end. If we measure the concentration of the molecules along the pipe at different times after the connection is made, we obtain the various curves shown in Figure 18.6. Only after a long time is the steady state reached. Then molecules are removed at $x = L$ at the same rate at which they are fed in at $x = 0$ and the concentration along the pipe is given by Equation 18.2. If the other end of the tube is closed, the change in concentration will be as shown in Figure 18.7, the steady state corresponding to a uniform concentration throughout.

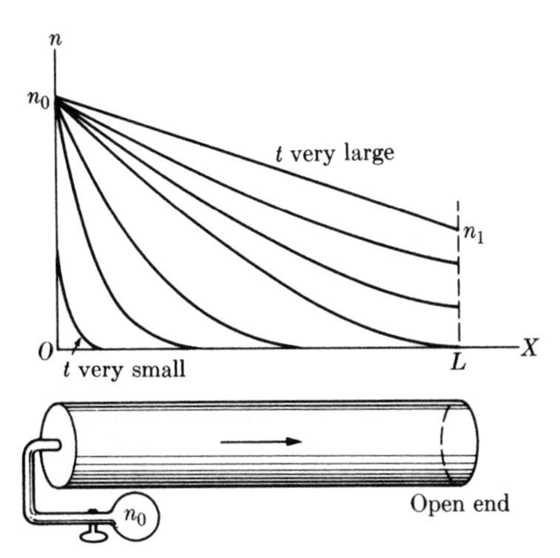

Figure 18.6 Change in concentration with time for diffusion along a pipe with an open end. The concentration changes until a steady state is reached.

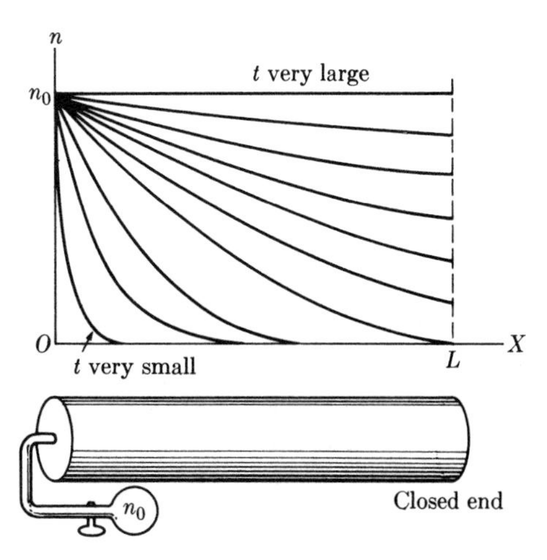

Figure 18.7 Change in concentration with time for diffusion along a pipe with a closed end. The steady state is reached when the concentration is uniform.

EXAMPLE 18.1

When water vapor diffuses through air, the diffusion coefficient is $2.19 \times 10^{-5}\,\mathrm{m^2\,s^{-1}}$ at standard pressure and at a temperature of 20 °C. In an experiment such as the one in Figure 18.5, the tube has a length of 1.0 m and a cross-section of $20\,\mathrm{cm^2}$. Find the amount of water evaporated per second through the tube.

▷ We must first determine the current density as given by Equation 18.2. Assume that the process is slow enough so that the region at the bottom of the tube can be considered as saturated at all times. Standard tables give the density of water vapor at 20 °C as $1.73 \times 10^{-2}\,\mathrm{kg\,m^{-3}}$. If m is the mass of one molecule, the number of molecules at the base of the tube per unit volume is $n_0 = (1.73 \times 10^{-2}/m)\ \mathrm{m^{-3}}$. If the concentration at the top is so small that we can assume it is zero, then, with $x = L$, we have $j = Dn_0/L$. Therefore

$$j = \frac{(2.19 \times 10^{-5}\,\mathrm{m^2\,s^{-1}})[(1.73 \times 10^{-2}/m)\ \mathrm{m^{-3}}]}{1.0\,\mathrm{m}} = \frac{3.78 \times 10^{-7}}{m}\ \mathrm{m^{-2}\,s^{-1}}$$

Let S be the area of the cross-section of the tube. Then, the mass evaporated per second through the tube is

$$M = jSm = (3.78 \times 10^{-7}\,\mathrm{kg\,m^{-2}\,s^{-1}})(2 \times 10^{-3}\,\mathrm{m^2}) = 7.56 \times 10^{-10}\,\mathrm{kg\,s^{-1}}$$

In one hour a mass of 2.73×10^{-6} kg, or about 2.73 mg will be evaporated. The mass of a water molecule is 2.98×10^{-26} kg and so 2.54×10^{16} molecules per second are evaporated.

EXAMPLE 18.2

Neutron diffusion in a non-multiplicative medium, such as water or graphite, that absorbs neutrons but does not produce them. This is the situation, for example, in the **thermal**

column of a nuclear reactor. (Nuclear reactors are very powerful sources of neutrons as a result of uranium fission. See Chapter 40.)

▷ The thermal column consists of a mass of water or graphite placed adjacent to one of the sides of the reactor. The neutrons produced inside the reactor are allowed to diffuse along the thermal column, and may be used for a number of experiments. The column is called thermal because the neutrons that emerge from it are in thermal equilibrium with the atoms of the material composing the column.

The physical arrangement is illustrated in Figure 18.8. Neutrons from the reactor core penetrate the thermal column from the left, diffuse along the column, and escape through S. During the diffusion process, the neutrons undergo collisions with the atoms of the material composing the column. In some cases the collision results in a capture of the neutron by the atoms of the material. If no neutrons were captured by the material of the thermal column, the neutron concentration, when a steady state is reached, would decrease linearly. This would be the same as the case of the open pipe illustrated in Figure 18.6. The situation has been indicated by the dashed line in Figure 18.8. The slope of the line would depend on the rate at which neutrons enter from the reactor and are removed through S. However, due to the capture of neutrons by the material in the thermal column, the concentration varies, as indicated by the solid line. The concentration falls more rapidly at the end close to the reactor because a larger neutron concentration results in a higher rate of capture by the material of the column.

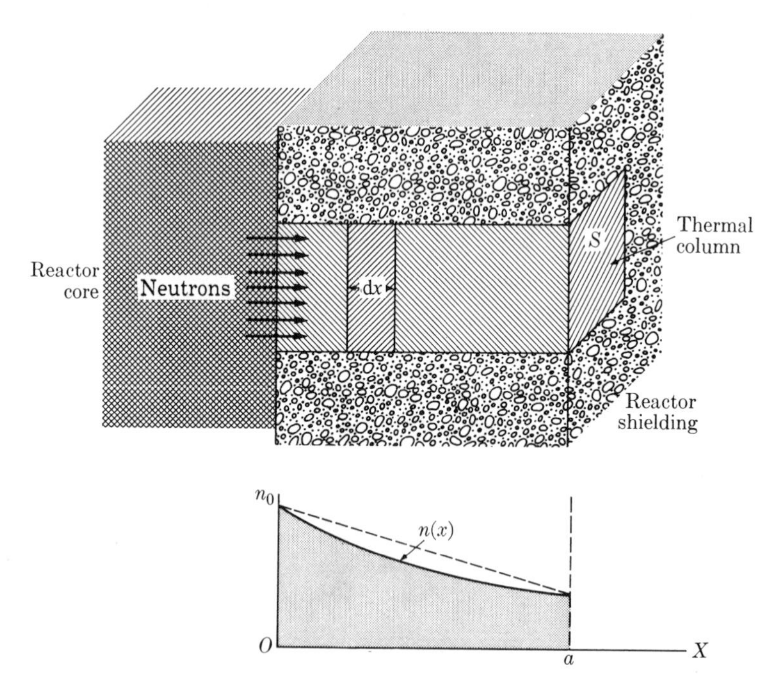

Figure 18.8 Neutron concentration along a thermal column in a nuclear reactor.

18.4 Thermal conduction: Fourier's law

Thermal conduction is a transport phenomenon where the energy due to molecular agitation is transferred from one place to another with a corresponding change in temperature. We find that

thermal conduction occurs whenever the temperature of the substance varies from place to place.

That is, there is thermal conduction when the average energy of the molecules is different in different parts of the substance. This temperature or average energy difference gives rise to a net energy flow (heat). Thus we may define thermal conduction as an energy transfer by temperature difference. This is an irreversible process, accompanied by an increase of entropy (recall Example 16.11).

Thermal conduction takes place from regions where the average energy of the molecules is large to regions where it is small or from where the temperature is high to where it is lower. That is,

thermal conduction is an energy transport that occurs in the direction in which the temperature decreases.

The mechanism of thermal conduction is different in solids, liquids and gases, due to differences in molecular mobility in these three states. In gases (and to a certain extent in liquids) thermal conduction is the result of collisions between fast molecules and slow molecules that result in a transfer of kinetic energy from the faster molecules to the slower.

Consider a chamber filled with a gas that has a temperature gradient throughout its volume. In places where the temperature is higher, the molecules have, on the average, higher velocities than in places where the temperature is lower. For example, in Figure 18.9, suppose that the gas at the left is hotter than the gas at the right. As a result of the collisions between the molecules at the boundary, and diffusion of the 'hot' molecules from left to right and of the 'cold' molecules from right to left, there is a net energy transfer from left to right.

In solids there is no such energy transfer by molecular displacements. The only motion of molecules in a solid is vibration about their equilibrium positions. Consequently, the process involved is, instead, a transport of this vibrational energy

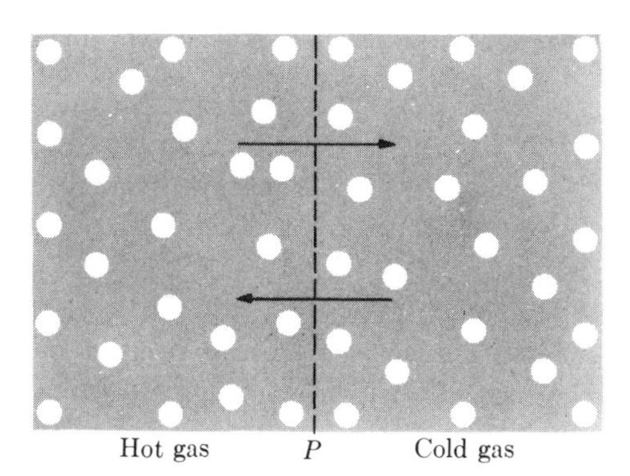

Figure 18.9 Thermal conduction in a gas.

across the crystal lattice of the solid. In metals, however, there is an additional effect, due to electrons (called **conduction electrons**) that are free to move through the volume of the metal. Conduction electrons behave in a manner similar to the molecules in a gas. They tend to diffuse through the metal from the hot to the cold region, transferring energy by collision with other electrons and with the lattice ions in the colder region.

In gases and liquids there may also be a bulk transfer of matter due to density differences created by temperature differences. This process, called **convection**, is not in the same category as those discussed here. The reason is that it is not due essentially to molecular agitation, but to a macroscopic condition of instability (see Note 18.1).

The **energy current density**, j_E, due to a temperature difference is the energy transferred per unit time through a unit area placed perpendicular to the direction in which the energy flow takes place. We shall assume that this direction is the X-axis. The energy flow takes place in a well-defined direction, namely, the direction in which the temperature decreases. The change of temperature per unit length (or the **temperature gradient**) of the material is $\mathrm{d}T/\mathrm{d}x$. It has been found experimentally that, unless the temperature changes very rapidly over a short distance, j_E is proportional to $\mathrm{d}T/\mathrm{d}x$. That is,

$$j_E = -\kappa \frac{\mathrm{d}T}{\mathrm{d}x} \tag{18.3}$$

where κ is a coefficient characteristic of each material, called **thermal conductivity**. The negative sign indicates that the energy flows in the direction in which the temperature decreases. Equation 18.3 is known as **Fourier's law**. This law is applicable only when the temperature changes slowly with the distance.

Note that j_E is expressed in $\mathrm{J\,m^{-2}\,s^{-1}}$ and that $\mathrm{d}T/\mathrm{d}x$ is expressed in $\mathrm{K\,m^{-1}}$. Therefore κ is expressed in $\mathrm{J\,m^{-1}\,s^{-1}\,K^{-1}}$ or $\mathrm{m\,kg\,s^{-3}\,K^{-1}}$. Sometimes κ is expressed in $\mathrm{cal\,m^{-1}\,s^{-1}\,K^{-1}}$.

Fourier's law for thermal conductivity is very similar to Fick's law for diffusion and is also of a statistical nature. In fact, there is a relationship between the thermal conductivity κ and the diffusion coefficient D. We shall talk about this in Section 18.8. Although the mechanism for heat conduction is different in gases, liquids and solids, Fourier's law applies to all three states of matter to a different degree of approximation in each case.

18.5 Steady thermal conduction

Consider the special case when the energy current density through any section of the substance is constant; that is, $j_E = const.$ Under such conditions there is no energy accumulation in any volume of the substance. We again say that we have a steady or stationary condition. For example, suppose that we have a rod surrounded by an insulating material so that no energy is lost through the lateral surfaces (Figure 18.10). The steady state requires that, for a given amount of energy

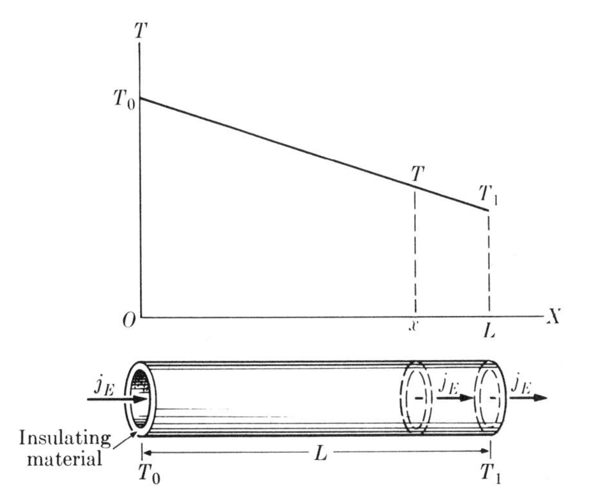

Figure 18.10 Steady temperature distribution along a thermally insulated rod with ends at fixed temperatures.

per unit time fed through the end at the higher temperature, the same amount of energy per unit time must be removed from the end at the lower temperature. This is necessary so that no energy is accumulating in the rod. When $j_E = const.$, Equation 18.3 may be written in the alternate form

$$j_E = \kappa\left(\frac{T_0 - T}{x}\right) \qquad \text{or} \qquad T = -\frac{j_E}{\kappa}x + T_0 \tag{18.4}$$

where T_0 is the temperature at $x = 0$. Therefore, under steady conditions the temperature decreases linearly along the rod. The variation of T with x is shown in the upper part of Figure 18.10. A different situation occurs if the surface of the rod is not thermally insulated, so that energy is lost in the form of heat or radiation across its surface into the surrounding medium. In such a case the rate of energy loss is proportional to the area of the surface, and to the temperature difference between the surface of the rod and the surroundings, so long as the difference is not very large. This statement is called **Newton's law of cooling.**

Even if the rod is surrounded by an insulating material, some heat is lost through its surface, since no insulation is perfect. According to Equation 18.4, the energy lost per unit time is proportional to the temperature difference between the inner and outer surfaces of the insulator. The result of this surface energy loss is that, in the steady state, the temperature decreases along the rod exponentially, as shown in Figure 18.11, instead of linearly, as in the case of the perfectly insulated bar.

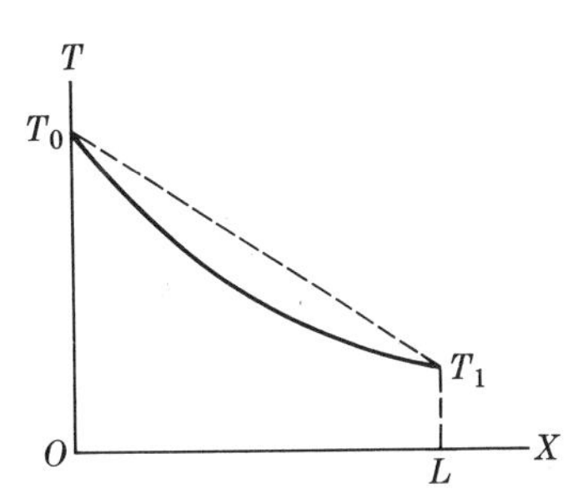

Figure 18.11

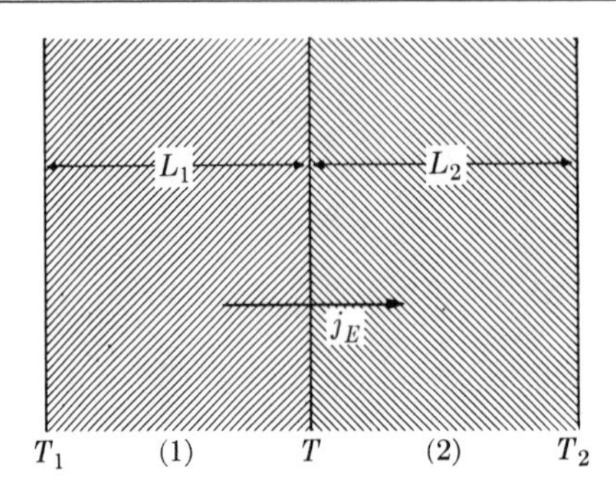

Figure 18.12 Heat flow through two slabs of different materials in contact.

EXAMPLE 18.3

Two slabs with thickness L_1 and L_2 and thermal conductivities κ_1 and κ_2 are in contact (Figure 18.12). The temperatures at the outer surfaces are T_1 and T_2. Compute the temperature at the common surface. Assume steady conditions.

▷ Since the energy flux is the same for both slabs, we have, from Equation 18.4, that

$$j_E = \kappa_1 \frac{T_1 - T}{L_1} \quad \text{and} \quad j_E = \kappa_2 \frac{T - T_2}{L_2}$$

Thus

$$\kappa_1 \frac{T_1 - T}{L_1} = \kappa_2 \frac{T - T_2}{L_2}$$

and then

$$T = \frac{\kappa_1 T_1 L_2 + \kappa_2 T_2 L_1}{\kappa_1 L_2 + \kappa_2 L_1} \tag{18.5}$$

is the temperature of the common wall.

18.6 Viscosity

A third transport phenomenon that holds for gases (and for fluids in general) is **viscosity** or **internal friction**. It is associated with collective currents that carry momentum from one region of the fluid to another (recall Section 14.10).

Consider a fluid where there is, in addition to the thermal agitation of the molecules, a collective movement or *current* of the whole fluid. Examples are an air current in a wind tunnel or water running in a canal or pipe under a pressure difference. Viscosity is manifest when the fluid current varies from place to place. Suppose that the fluid is moving as shown in Figure 18.13. The current velocity v is along the Y-axis, but its value varies as a function of the distance along the X-axis, as indicated at the right of Figure 18.13. Consider the plane P, perpendicular to the X-axis, parallel to the direction of the fluid motion. The molecules, however, are not restricted to motion parallel to the Y-axis, because, in addition to their collective motion, they have thermal motion and collide among themselves. As a result, molecules are continuously crossing the plane P, both from left to right and from right to left. Each molecule carries a momentum parallel to the Y-axis. In the situation illustrated in Figure 18.13, molecules crossing from left to right

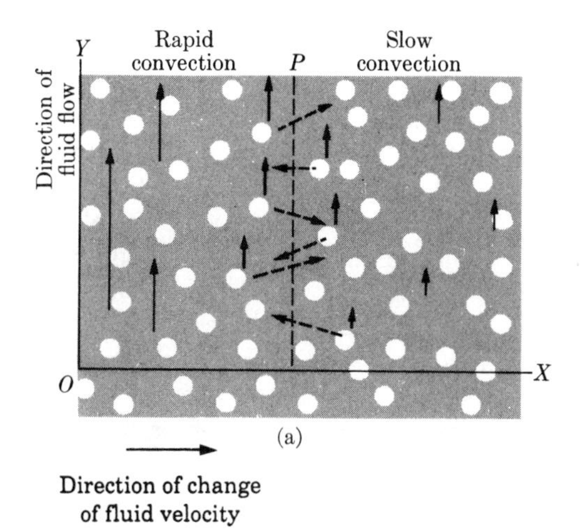

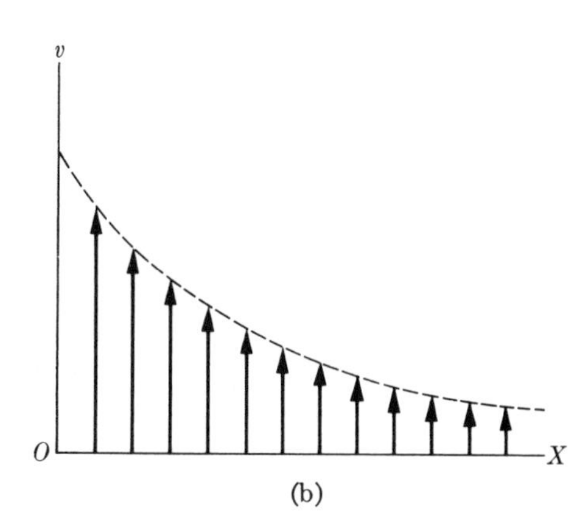

Figure 18.13 Fluid in which there is a change in velocity in a direction perpendicular to the direction of flow.

carry a larger collective momentum than those crossing from right to left. Therefore,

> *momentum transfer occurs in the direction in which the velocity of the fluid current decreases.*

The **momentum current density**, j_p, is the amount of collective momentum (parallel to the direction of the current or the Y-axis) transferred per unit time across a unit area perpendicular to the direction in which the fluid velocity changes. In our case, this direction is defined by the X-axis. Experiment shows that j_p is proportional to the variation of the fluid velocity v per unit length (along the X-axis), or the velocity gradient (that is, $\mathrm{d}v/\mathrm{d}x$). So we may write

$$j_p = -\eta \frac{\mathrm{d}v}{\mathrm{d}x} \tag{18.6}$$

Equation 18.6 is very similar to Fick's law for molecular diffusion and Fourier's law for thermal conduction and is therefore also a statistical law. The negative sign in Equation 18.6 is due to the fact that the momentum transfer takes place in the direction in which the fluid velocity decreases. The proportionality factor η is the **coefficient of viscosity** for the fluid (this coefficient was originally introduced in Section 7.5). Note that the units of j_p are expressed as $(\mathrm{m\,kg\,s^{-1}/s})\mathrm{m^{-2}}$ or $\mathrm{m^{-1}\,kg\,s^{-2}}$, and $\mathrm{d}v/\mathrm{d}x$ is expressed in $\mathrm{s^{-1}}$. Thus the coefficient of viscosity is expressed in $\mathrm{m^{-1}\,kg\,s^{-1}}$. One-tenth of this unit is called a **poise**, abbreviated P (see Table 7.2).

Equation 18.6 is applicable only when the fluid velocity is not very large and does not change very rapidly with distance, or when the pressure in the fluid does not change by a large amount in a short distance.

We may look at viscosity from a different point of view. The result of molecular transfer from left to right and from right to left is that the fluid to the right of plane P is gaining collective momentum (parallel to the Y-axis).

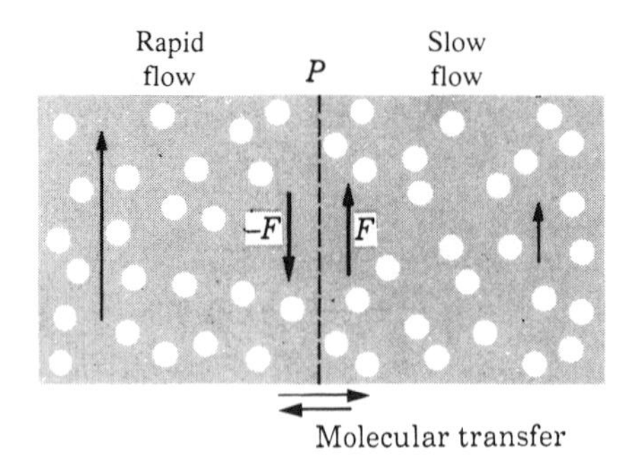

Figure 18.14 Shear stress in convective motion.

Similarly, the fluid to the left of P is losing collective momentum (also parallel to the Y-axis). Thus we may say that the fluid to the right of P is subject to a force parallel to the direction of the flow, while the fluid to the left of P is subject to an equal and opposite force. This force is called **internal friction** or **drag**. The value of this force, per unit area, is called **internal shear stress** (Figure 18.14) and is given by the rate of momentum transfer across P.

EXAMPLE 18.4

Force and power required to drag a floating barge with uniform motion in shallow water (Figure 18.15).

▷ Call F the force exerted by the tugboat on the barge, which moves with a velocity v relative to the bottom. The water in contact with the barge is also dragged with velocity v. On the other hand, the water at the bottom remains still. Hence as we move from the bottom ($x = 0$) to the surface ($x = h$), the convective velocity varies as shown in the figure. Therefore we may write $\mathrm{d}v/\mathrm{d}x = v/h$. Applying Equation 18.6 we have that $j_p = -\eta v/h$. The negative sign means that collective momentum per unit area is transferred in the $-X$ or downward direction. The water produces a dragging force on the barge equal to the rate of transfer of momentum to the water. If A is the area of the bottom of the barge, the total rate of momentum transfer is $j_p A$. Therefore the force F that must be exerted by the tugboat to move the barge with uniform motion is

$$F = -j_p A = \frac{\eta v A}{h}$$

and is proportional to the velocity of the barge. The power required from the tugboat is

$$P = Fv = \frac{\eta v^2 A}{h}$$

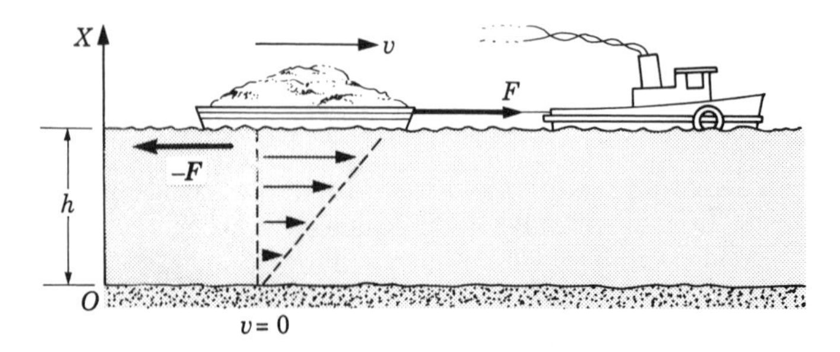

Figure 18.15

and hence increases as the square of the velocity. This example is somewhat unrealistic since a barge does not slide on the surface of the water, but is partially submerged. However, it is sufficient to illustrate the basic principles to be used when discussing the motion of a body in contact with a fluid.

18.7 Mean free path and collision frequency

Molecules in a gas collide frequently and, as a result, their paths zigzag (Figure 18.16). To describe the motion of these molecules, two concepts have been introduced: the **collision mean free path** and the **collision frequency**.

The *collision mean free path*, designated by l, is the average distance a molecule of a gas moves between collisions. It may be computed by following a molecule for a sufficiently long time and finding the average length of the paths between successive collisions. It may also be found by looking, at a particular instant of time, at a large number of molecules that have just suffered a collision, and finding the average distance moved by these molecules until their next collision. The two methods are statistically equivalent if the number of molecules is very large.

The *collision frequency*, designated by Γ, is the number of collisions a molecule suffers per unit time. The mean free path and the collision frequency are closely related. Given that v_{ave} is the average velocity of a molecule, the average time between two collisions is $t = l/v_{\text{ave}}$ and thus the number of collisions per unit time, or the collision frequency Γ, is

$$\Gamma = \frac{1}{t} = \frac{v_{\text{ave}}}{l} \qquad \textbf{(18.7)}$$

The collision frequency Γ is expressed in s^{-1}.

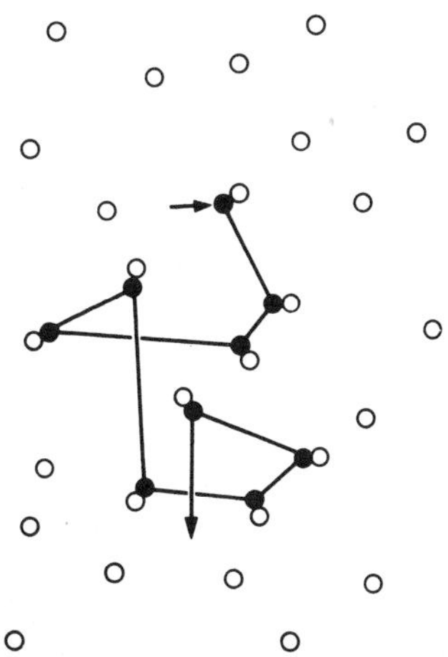

Figure 18.16 Molecular free paths.

Note 18.1 Relation between mean free path and molecular dimensions

Let us call n the number of molecules per unit volume and r the 'radius' of each molecule. When the word 'radius' is used, we do not mean to imply that the molecules are spheres. However, because of their rapid rotational motion, they effectively act as spheres. We shall neglect intermolecular forces and assume that molecules act like billiard balls. In order for two molecules to collide, the distance between their two centers, projected on a plane perpendicular to the direction of relative motion, must be less than $2r$ (Figure 18.17). Let molecule 1 move to the right. The effective region around the path of 1, within which the

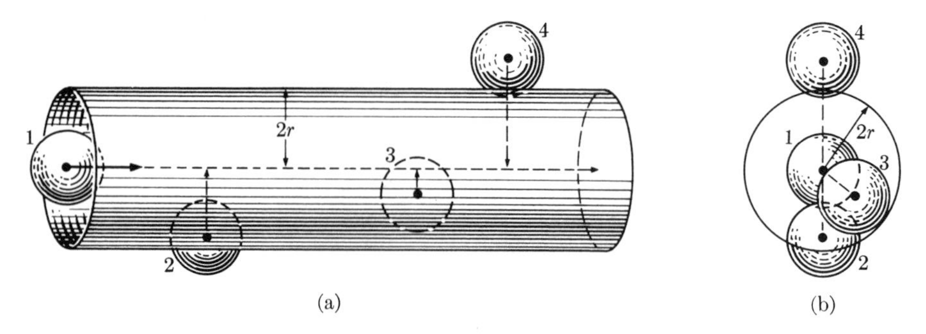

Figure 18.17 Collision cylinder swept out by a molecule. Molecules whose centers are within the cylinder would undergo collision with molecule 1. (a) Side view; (b) front view of excluded volume.

center of another molecule must be in order for a collision to occur, is a cylinder of radius $2r$. If the separation between the centers is less than $2r$, as it is for molecules 2 and 3, collision does occur. If it is greater than $2r$, as it is for molecule 4, there is no collision.

Consider a slab of thickness $\mathrm{d}x$ and area S (Figure 18.18). If n is the number of molecules per unit volume, the total number of gas molecules in the slab is $nS\,\mathrm{d}x$. Each molecule screens a certain area of radius $2r$, which prevents the free passage of molecule M through it. This area is

$$\sigma = \pi(2r)^2 = 4\pi r^2 \tag{18.8}$$

and is called the **microscopic collision cross-section**. The total area screened by all molecules in the slab (assuming that there is no superposition, which is correct if the molecular concentration is not very large and $\mathrm{d}x$ is very small) is $(nS\,\mathrm{d}x)\sigma$. Then the probability that M suffers a collision in passing through the slab is

$$\frac{(nS\,\mathrm{d}x)\sigma}{S} = n\sigma\,\mathrm{d}x$$

Thus the number of collisions of M per unit length is $n\sigma$. This is defined as the **macroscopic cross-section**,

$$\Sigma = n\sigma \tag{18.9}$$

and the collision mean free path of the molecule is

$$l = \frac{1}{\Sigma} = \frac{1}{n\sigma} = \frac{1}{4\pi r^2 n} \tag{18.10}$$

Equation 18.10 is correct as long as the molecules in the slab are considered fixed. However, if they are in motion, as they are in a gas, the probability of a collision may be greater, and thus the mean free path decreases. A more elaborate calculation is then needed to determine l.

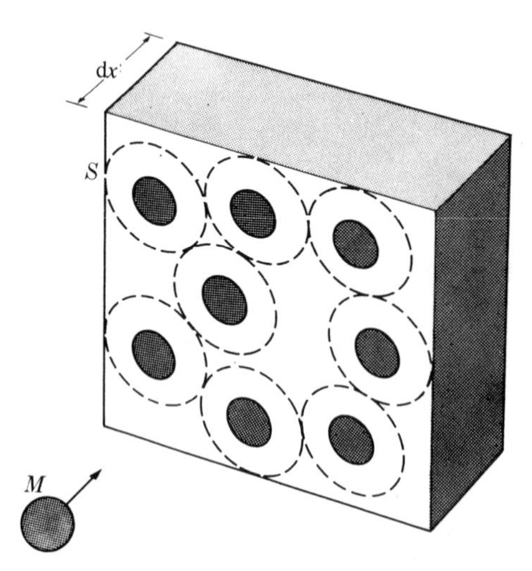

Figure 18.18 Each molecule in the slab screens an area whose radius is $2r$.

18.8 Molecular theory of transport phenomena

The discussion of transport phenomena has been based on Fick's law for diffusion (molecular transfer),

$$j = -D \frac{dn}{dx} \tag{18.11}$$

Fourier's law for thermal conduction (energy transfer),

$$j_E = -\kappa \frac{dT}{dx} \tag{18.12}$$

and the law of viscous flow (momentum transfer),

$$j_p = -\eta \frac{dv}{dx} \tag{18.13}$$

All these laws have an experimental basis, and are valid within the limits indicated in each case. In what follows, the quantities D, κ and η for gases are correlated with their molecular properties. Transport processes are due to molecular agitation. Molecular diffusion is due to the transfer of *molecules* from a region where they are more concentrated to a region of lower concentration. Thermal conduction is due to the transfer of *energy* from one region where molecules move fast, and thus the temperature is high, to another where they move slower and where the temperature is lower. Similarly, viscosity is due to the transfer of *momentum* associated with the collective motion, from one place where the fluid moves rapidly to another where it moves more slowly.

The larger the molecular velocity and the mean free path, the larger the diffusion coefficient should be, according to our physical intuition. In fact, the calculations show that the coefficient of molecular diffusion is related to molecular dynamics through the expression

$$D = \tfrac{1}{3} v_{\text{ave}} l \tag{18.14}$$

where v_{ave} is the average velocity and l is the collision mean free path of the molecules. For the coefficient of thermal conduction, we get the relation

$$\kappa = D(\tfrac{3}{2}kn) = \tfrac{1}{2} nkv_{\text{ave}} l \tag{18.15}$$

because each molecule carries an energy $\frac{3}{2}kT$. Finally, the coefficient of viscosity is expressed by

$$\eta = D(nm) = \tfrac{1}{3} nmv_{\text{ave}} l \tag{18.16}$$

because each molecule carries a momentum mv. Expressions 18.14, 18.15 and 18.16 show the close relationship among the three transport phenomena.

Table 18.1 Experimental values for the diffusion coefficient D, thermal conductivity κ and viscosity η, all at STP

Substance	$D(m^2 s^{-1})$	$\kappa(m\,kg\,s^{-3}\,K^{-1})$	$\eta(m^{-1}\,kg\,s^{-1})$	$r(m)$*
He	—	14.3×10^{-2}	1.86×10^{-5}	0.90×10^{-10}
Ne	4.52×10^{-5}	4.60×10^{-2}	2.97×10^{-5}	1.06×10^{-10}
Ar	1.57×10^{-5}	1.63×10^{-2}	2.10×10^{-5}	1.50×10^{-10}
Xe	0.58×10^{-5}	0.52×10^{-2}	2.10×10^{-5}	2.02×10^{-10}
H_2	12.8×10^{-5}	16.8×10^{-2}	0.84×10^{-5}	1.12×10^{-10}
O_2	1.81×10^{-5}	2.42×10^{-2}	1.89×10^{-5}	1.51×10^{-10}
N_2	1.78×10^{-5}	2.37×10^{-2}	1.66×10^{-5}	1.54×10^{-10}
CO_2	0.97×10^{-5}	1.49×10^{-2}	1.39×10^{-5}	1.89×10^{-10}
NH_3	2.12×10^{-5}	2.60×10^{-2}	0.92×10^{-5}	1.83×10^{-10}
CH_4	2.06×10^{-5}	3.04×10^{-2}	1.03×10^{-5}	1.70×10^{-10}

* Molecular radii have been computed from experimental data on viscosity.

We can estimate molecular dimensions by using the above relations. Measuring any one of these coefficients, we can compute the mean free path, and from it we can obtain the molecular radius. However, the calculated values of the mean free path and the molecular radius indicate only the orders of magnitude. Table 18.1 lists experimental values for some gases; it also gives their molecular radii, computed from viscosity data. The results are consistent with molecular dimensions calculated by other means.

The expressions for D, κ and η given above are obtained using the ideal gas approximation, so that the effects of intermolecular forces and molecular size are ignored. This is appropriate as long as the density of the gas is small.

EXAMPLE 18.5

Estimate the coefficients D and κ for hydrogen at STP. Compare with the experimental results.

▷ We must determine first the quantities v_{ave}, l, n and m for hydrogen under such conditions. The mass of a hydrogen molecule is 3.33×10^{-27} kg. The number of molecules per unit volume at $T = 273$ K and $p = 1.01 \times 10^5$ Pa is, using the ideal gas law $pV = NkT$ (Equation 15.11), $n = N/V = p/kT = 2.68 \times 10^{25}\ m^{-3}$. It can be shown that the average velocity of a gas molecule (see Examples 17.5 and 17.6) is given by $v_{ave} = (8kT/\pi m)^{1/2}$. In our case, then, $v_{ave} = 1.69 \times 10^3\ m\,s^{-1}$. Finally, the radius of a hydrogen molecule is $r = 1.12 \times 10^{-10}$ m. Using Equation 18.10, we obtain a mean free path of 2.37×10^{-7} m. With the above values, using Equations 18.15 and 18.16, we find that

$$D = 13.3 \times 10^{-5}\ m^2\,s^{-1} \qquad \text{and} \qquad \kappa = 7.40 \times 10^{-2}\ m\,kg\,s^{-3}\,K^{-1}$$

The corresponding experimental values are given in Table 18.1. Hence at least the correct orders of magnitude can be obtained by our simplified theory, which is based on an ideal monatomic gas. An extension of the theory that considers the energy transported by internal motion, such as rotation and vibration, improves the agreement between the theoretical and experimental values.

Note 18.2 Convective and turbulent transport

In our treatment of transport phenomena we have always assumed that the space rates of change (or gradients) of concentration, temperature or momentum were small. Fick's law 18.1 or Fourier's law 18.3 are linear approximations valid when dn/dx and dT/dx, respectively, are small. That is, the three types of transport phenomena we discussed were due mainly to molecular mobility rather than to bulk motion of matter and, to a certain extent, the fluid was assumed to be close to statistical equilibrium during the process. However, when the gradients of the physical parameters in a fluid become large, the motion of the fluid becomes rather complex. The fluid as a whole is far from equilibrium and each small volume of the fluid is subject to forces that set it in irregular motion. The transport process that results is due to the irregular bulk motion of matter rather than simply molecular mobility. Such a process is called **convective**. When the gradient is very large, the system is forced far from equilibrium, so that more violent motion results. The transport process may then change in a drastic way, which can become extremely difficult to analyze in detail. We say that the transport becomes **turbulent**. This applies to the motion of a fluid due to a gradient in pressure or height, which was discussed in Section 14.10. In such cases, when the pressure gradient is large, Bernoulli's Equation 14.32 must be modified, because some physical parameters, such as the fluid velocity at each point, fluctuate sharply. The turbulent transport of water through a pipe or in a mountain stream, or the transport of air past an airplane wing (together with the resulting drag forces as well as its 'lift') are problems of great practical interest that result from such strong variations of physical conditions with distance.

The onset of **thermal convection** in a liquid can be observed in a simple experiment. Consider a liquid placed between two parallel plates whose separation is very small compared with the size of the plates (Figure 18.19). Suppose the plates are kept at temperatures T_1 and T_2, with T_1 greater than T_2, so that thermal energy is transferred from the bottom to the top. For T_1 close to T_2, so that the gradient $(T_1 - T_2)/d$ is small, Fourier's law applies and thermal energy is transferred by molecular collisions without any mass transfer, resulting in a steady flow of energy. We then have thermal conduction (Section 18.4). As we increase the temperature gradient $(T_1 - T_2)/d$, there will be a value where the fluid becomes 'convectively unstable' and bulk movement of the fluid begins. This is because equilibrium is destroyed when the hotter fluid at the bottom expands and, due to the buoyancy effect, moves upward, pushing the fluid around it. However, when the fluid reaches the top plate and cools off, it tends to descend because it is denser than the surrounding fluid. Therefore, two bulk currents in opposite directions are set in motion, setting up a pattern, and we have thermal convection.

The precise pattern that convective motion forms depends sharply on the shape of the vessel and the conditions at the surface. Typically, the convective motion becomes organized into cells, called **Bénard cells**, as shown in Figure 18.20 and named after Henri Bénard who first observed them in 1900. The fluid in each cell constitutes a rolling convective current, with a well-defined sense of rotation which alternates from one cell to the next. Which cell is right-handed and which is left-handed cannot be predicted. Further, if the experiment is repeated several times, the cell pattern is not always the same.

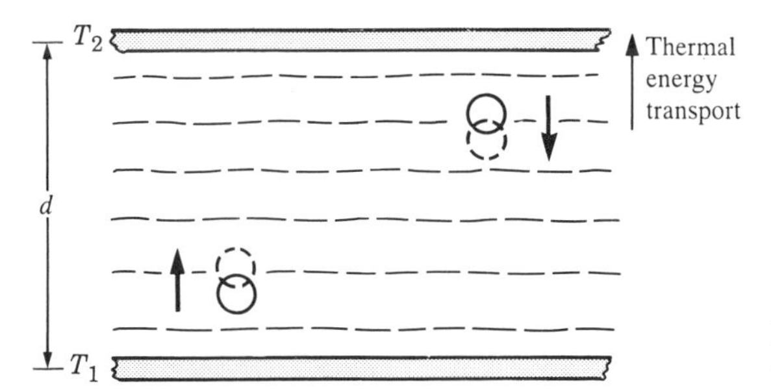

Figure 18.19 Thermal energy transport.

Figure 18.20 Bénard cells.

Thermal convection is the process by which thermal energy in the atmosphere, the oceans and the stars is transported. Convective currents constantly occur in the atmosphere, because air close to the Earth's surface is usually hotter than the air at high altitudes, and a pattern of hexagonal Bénard cells may sometimes be seen when flying over the clouds.

For very strong thermal gradients that push the system very far from statistical equilibrium, the convective motion becomes very irregular or **turbulent.** Turbulence is a manifestation of chaotic motion in fluids. The convection or turbulent transport of thermal energy from bottom to top can be many orders of magnitude greater than pure heat conduction.

QUESTIONS

18.1 Transport phenomena always occur in a well-defined direction. Relate this property to the second law of thermodynamics.

18.2 Would you say that transport phenomena are of a statistical nature? Explain your answer.

18.3 What transport phenomenon, due to molecular motion, is associated with (a) mass transport, (b) energy transport, (c) momentum transport?

18.4 In the text it is indicated that Fick's law is valid only under certain circumstances. State these limitations and justify them on physical grounds.

18.5 A fluid moves in the direction of the Y-axis. The velocity of the fluid varies in the X-direction. What is the direction of (a) the fluid momentum, (b) the transport of fluid momentum, (c) the transported fluid momentum, (d) the internal stress on a fluid layer parallel to the direction of motion?

18.6 What transport phenomena, if any, do you expect to occur in solids? Why?

18.7 A bar, covered with thermal insulation on its surface, has a small heating coil at its center while its ends are kept at room temperature. Show graphically the variation of temperature along the bar as time passes after the coil is turned on until a steady state is reached. What is the relation between the energy supplied by the heating coil and the energy removed at each end, per unit time, when the steady state is reached?

18.8 In the text, two operational definitions of collision mean free path are given. Under what conditions are both definitions statistically equivalent?

18.9 How do you expect the collision mean free path and the collision frequency to vary with (a) the concentration, (b) the temperature of a substance?

18.10 What is the difference between molecular diffusion, convection, and turbulence?

18.11 What is meant by steady diffusion? By steady thermal conduction?

18.12 Explain how by analyzing transport phenomena one may obtain an estimate of molecular sizes.

PROBLEMS

18.1 (a) Assuming stationary conditions, calculate the diffusion coefficient of carbon dioxide in air if the particle current density, through a tube 25 cm long, is $5.1 \times 10^{17}\ m^{-2}\ s^{-1}$. The CO_2 concentration changes from $1.41 \times 10^{22}\ m^{-3}$ to $8.6 \times 10^{21}\ m^{-3}$. (b) If the tube has a cross-section of $15\ cm^2$, calculate the number of molecules that pass per second through a given section of the tube.

18.2 The concentration of diffusing gas molecules in a 45 cm tube varies linearly and is given by the expression $n = n_0 - 6.45 \times 10^{23} x$, where n is in m^{-3} and x is in m. Determine the particle current density and mass of gas transferred per second if the gas is (a) neon, (b) argon, (c) ammonia. (See Table 18.1 for the diffusion coefficients for the gases.)

18.3 The temperature gradient along an insulated copper rod is $-2.5\ °C\ cm^{-1}$. (a) Compute the difference in temperature between two points separated 5 cm. (b) Determine the amount of heat crossing (per second) a unit area perpendicular to the rod. The thermal conductivity of copper is $3.84 \times 10^2\ m\ kg\ s^{-3}\ K^{-1}$.

18.4 A room has three windows with a total area of $3\ m^2$. The thickness of the glass is 0.4 cm. The indoor side is at 20 °C and the outdoor side is at 10 °C. Compute the amount of heat passing through the windows per second and per hour. The thermal conductivity of glass is $5.85 \times 10^{-1}\ m\ kg\ s^{-3}\ K^{-1}$.

18.5 Two rods, one of copper and the other of steel, both 1 m long and with cross-sections equal to $1\ cm^2$, are welded at a common end. The free end of the copper rod is kept at 100 °C and the free end of the steel rod is kept at 0 °C. Compute (a) the temperature at the common end, (b) the temperature gradient in the copper rod and in the steel rod, (c) the amount of heat crossing any section of the rod per unit time. (d) Make a plot of the temperature along the rod. The thermal conductivities of copper and steel are 3.84×10^2 and $0.46 \times 10^2\ m\ kg\ s^{-3}\ K^{-1}$, respectively.

18.6 Newton's law of cooling (see Section 18.5) may be expressed as $j_E = h(T - T_m)$, where j_E is the energy flow per unit time across the unit area of the surface of the body, T is the temperature of the body, T_m that of the surrounding medium, and h a constant called the **coefficient of thermal transfer**; the units of j_E are $J\ m^{-2}\ s^{-1}$ and of h are $J\ m^{-2}\ s^{-1}\ K^{-1}$. (a) Consider a small body of surface area S and heat capacity C initially at temperature T_0. Show that the temperature of the body, as a function of time, is given by $T = T_m + (T_0 - T_m)e^{-At}$ where $A = hS/C$. (b) Plot T versus t for T_0 (i) larger than, (ii) equal to, and (iii) smaller than T_m.

18.7 (a) Using the formula given in the previous problem, calculate the time needed for a body to cool from 100 °C to 30 °C if the surroundings are at 25 °C. The value of A may be taken as $10^{-3}\ s^{-1}$. (b) How long will the same cooling take if the surface area of the body is doubled?

18.8 When a fluid of viscosity η moves steadily through a horizontal pipe of radius R and length L, the volume of fluid crossing any section of the tube per unit time is given by $V = (\pi R^4/8\eta)(p_1 - p_2)/L$, where $p_1 - p_2$ is the pressure difference between the two ends of the tube. This result is known as **Poiseuille's formula**. (a) A water supply pipe between two reservoirs is 1 km long and has a diameter of 2 m. Calculate the pressure differential needed to transfer $55\ m^3\ s^{-1}$ of water through the pipe. (b) How much power is needed to maintain a steady flow? (c) What happens to the pressure differential if the walls of the pipe become encrusted with silt 15 cm thick? (d) How does the volume transfer change if the pressure differential is maintained but the pipe radius is decreased as indicated in (c)?

18.9 (a) The viscosity coefficient of human blood is 0.047 poise. What is the pressure differential per unit length in (a) capillary vessels whose cross-sectional area is $2.1 \times 10^{-6}\ cm^2$; (b) arteries whose cross-sectional area is $3.0 \times 10^{-1}\ cm^2$, if the blood flow is $4.2 \times 10^{-12}\ m^3\ s^{-1}$ and $1.2 \times 10^{-5}\ m^3\ s^{-1}$, respectively? (See Problem 18.8 for Poiseuille's formula.)

18.10 One end of a capillary tube 10 cm in length and 1 mm in diameter is connected to a water supply which has a pressure of 2 atm. The other end of the capillary tube is at a pressure of 1 atm. The viscosity coefficient of water is 0.01 poise. How much water does the tube deliver in 1 sec? (1 atm $\approx 10^5\ N\ m^{-2}$.) (Use Poiseuille's formula of Problem 18.8.)

18.11 At 25 °C, H_2 has a viscosity of 8.2×10^{-5} poise. If a tube 1 m long permits the flow of 1 liter min^{-1} under a pressure difference of 0.3 atm, what is the diameter of the tube? (Use Poiseuille's formula, given in Problem 18.8.)

18.12 The viscosity of acetone at certain temperatures is shown in the table:

Temp. (°C)	−60	−30	0	30
$\eta \times 10^2$ (poise)	0.932	0.575	0.339	0.295

By plotting $\ln \eta$ versus $1/T$, where T is the absolute temperature of the acetone, show that $\ln \eta = a + b/T$ represents the variation of η with T. Find constants a and b.

18.13 (a) Calculate the mean free path of the molecules of an ideal gas as a function of temperature and pressure. (b) Calculate the mean free path of hydrogen molecules at 100 °C and 10^{-6} atm. (c) Repeat for oxygen. (d) Repeat for both gases at STP. (See Example 18.5.)

18.14 Given that the molecular diameter of H_2 is 2.2×10^{-10} m, calculate the number of collisions made by a hydrogen molecule in 1 s if (a) $T = 300$ K and $p = 1$ atm, (b) $T = 500$ K and $p = 1$ atm, (c) $T = 300$ K and $p = 10^{-4}$ atm. (d) Compute the total number of collisions per second occurring in 1 cm^3 for each of the cases in (a), (b) and (c). (See Example 18.5.)

18.15 The effective diameter of an N_2 (nitrogen) molecule is about 3.0×10^{-10} m. (a) Compute the mean free path of N_2 at 300 K and 1 atm and at 0.01 atm. (b) A reasonably good vacuum system achieves a pressure of about 10^{-9} atm. What is the mean free path of N_2 at this pressure? (c) If the diameter of the evacuated tube at a pressure of 10^{-9} atm is 5 cm, how many times does a molecule of N_2 strike the walls between two successive collisions with other molecules of N_2 gas?

18.16 It can be shown that of the N_0 molecules in a gas that suffer a collision at a given instant, only the number $N = N_0 e^{-x/l}$ will have moved the distance x without suffering another collision (l is the mean free path). (a) Find the percentage of molecules that have yet to suffer their next collision at $x =$ (i) $0.5l$, (ii) l, (iii) $2l$, and (iv) $10l$. (b) Determine what value x must have if N is to equal 50% of N_0.

18.17 A group of oxygen molecules start their mean free paths at the same time at a temperature of 300 K and a pressure such that the mean free path is 2.0 cm. After a certain time interval, half of the molecules have undergone a second collision, and half have not. Determine this time interval. Assume that all molecules move at the average velocity v_{ave}. (See Example 18.5 for the value of v_{ave}.)

18.18 How do the coefficients D and κ for an ideal gas vary with (a) temperature and (b) pressure?

18.19 By inspection of Equation 18.16, verify that the coefficient of viscosity of an ideal gas depends only on the temperature. Determine the nature of this dependence.

18.20 The viscosity of carbon dioxide over a range of temperatures is given in the table below. (a) Make a plot of η versus $T^{1/2}$, where T is the absolute temperature. (b) Compute the ratio $\eta/T^{1/2}$. (c) In view of Problem 18.19, what do you conclude?

Temp. (°C)	−21	0	100	182	302
$\eta \times 10^6$ (poise)	12.9	13.9	18.6	22.2	26.8

18.21 Compute the coefficients D, κ and η for (a) helium and (b) oxygen at STP, and compare with the experimental results given in Table 18.1.

18.22 Obtain the theoretical values of the ratio κ/D and η/D for an ideal gas. Compare these ratios with the experimental results for real gases given in Table 18.1.

18.23 Two parallel plates 0.5 cm apart are maintained at 298 K and 301 K. The space between the two plates is filled with H_2. Calculate the flow of heat between the two plates, in J m^{-2} s^{-1}.

18.24 The coefficient of viscosity of ammonia at 0 °C is 9.2×10^{-5} poise. (a) Compute the mean free path and the collision frequency of an ammonia molecule at STP. (b) Also estimate its radius.

18.25 The coefficient of viscosity of methane at 0 °C is 10.3×10^{-5} poise. Calculate the molecular diameter of a methane molecule.

18.26 Compare the thermal conductivities of O_2 and H_2. Ignore the difference in molecular diameter.

19 The theory of relativity

Albert Einstein revolutionized our concepts about space, time, matter and energy with his special and general theories of relativity, formulated in 1905 and 1915, respectively. Einstein's explanation of the photoelectric effect (1905) was one of the foundations of the photon concept. Einstein also made many other important contributions. He was the first to apply quantum theory to calculate the heat capacity of solids (1906) and to explain blackbody radiation (1916), using a method that has been the basis for laser operation.

Notes

19.1 Introduction

A conceptual revolution was initiated in 1905 by Albert Einstein (1879–1955) with his formulation of the **theory of relativity**. This theory has changed our notions of space and time as well as of matter and energy. It has also provided a new basis for analyzing physical phenomena at high energies. About 10 years were needed to develop the theory fully, even though Einstein had covered all of its main aspects early in his work.

Einstein's work was preceded by efforts in the last decade of the 19th century by Hendrik Lorentz (1853–1928), Henri Poincaré (1854–1912) and others. They were trying to make the electromagnetic theory, developed by James C. Maxwell in the late 1800s (see Chapter 27), harmonize with the framework of mechanics based on Newton's principles. The name 'relativity' stems from the fact that the theory deals with the correlation of observations by observers in uniform relative motion. But the theory goes well beyond that aspect and has profound kinematical and dynamical implications.

19.2 The velocity of light

Until the end of the nineteenth century, it was assumed that space, empty of matter, was filled with 'ether'. Physicists had also assumed that vibrations of this hypothetical ether were related to light in the same way vibrations in air are related to sound. At that time, there was much discussion about how bodies moved through the ether and how this motion would affect the velocity of light as measured on the Earth.

Careful measurements have shown that light propagates with a velocity $c = 2.9979 \times 10^8$ m s^{-1}. Assuming the ether to be stationary, we may say that light propagates relative to the ether with a velocity c. If the Earth moves through the ether with a velocity v without disturbing it, then the velocity of light relative to the Earth should depend on the direction of light propagation. For example, according to the Galilean transformation of velocities (Section 4.6) the velocity of light relative to the Earth should be $c - v$ for a ray of light propagating in the same direction that the Earth is moving through the ether. The velocity of light should be $c + v$ for propagation in the opposite direction. If the path of the light ray, as observed from the Earth, is perpendicular to the Earth's motion the velocity of light relative to the Earth would be $(c^2 - v^2)^{1/2}$. (Remember Example 4.5 for a similar case pertaining to sound.)

Albert Michelson (1852–1931), and later with the assistance of Edward Morley (1838–1923), started a memorable series of experiments in 1881 to measure the velocity of light in different directions relative to the Earth, for which they used a special kind of interferometer for observing light interference (Figure 19.1).

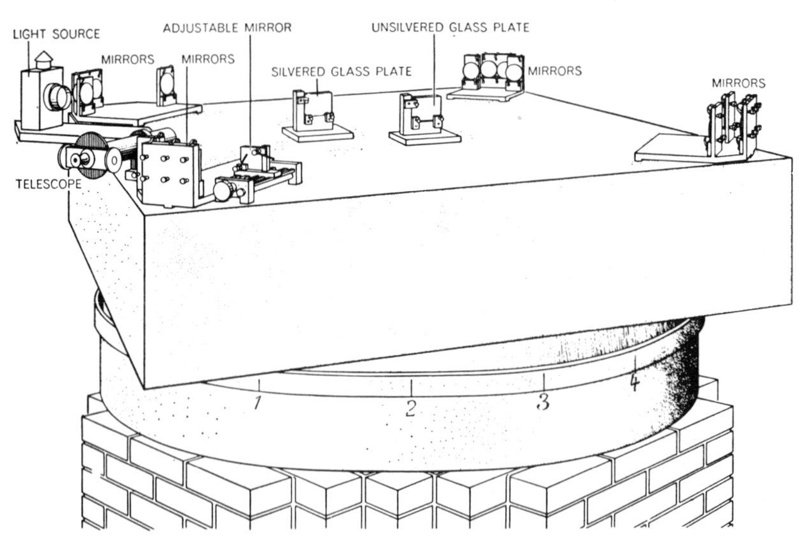

Figure 19.1 Interferometer used by Michelson and Morley in their measurements of the velocity of light. The sandstone table holding mirrors is fixed to a wooden ring which floats on mercury. The series of mirrors serves to lengthen the total path of light. The unsilvered plate is placed along one path to compensate for the fact that the other path must pass through the glass of the mirror. The telescope is used to observe the interference fringes. (Drawing courtesy of *Scientific American*.)

They also intended to determine the velocity of the Earth relative to the ether. Their experiments were carried out several times in different conditions over many years. To their great surprise, Michelson and Morley found that, within the high accuracy of their measurements, the velocity of light relative to the Earth was the same in all directions.

According to the Galilean transformation of velocities, no body may have the same velocity relative to two observers in uniform relative motion. In this case, one observer is supposed to be attached to the ether and the other to the Earth. One possible explanation of this experimental result would be that the Earth drags the ether with it, as it drags the atmosphere. Therefore, close to the Earth's surface the ether should be at rest relative to the Earth and therefore the velocity of light should be c in all directions. This is a rather improbable explanation, since the ether drag would manifest itself in other phenomena connected with light propagation, such as the change in the direction of light coming from stars as the Earth moves along its orbit. Such phenomena have never been observed.

The negative results of the Michelson and Morley experiment stimulated Einstein to discard the notion of the existence of an ether. In its place, he proposed, as a universal law of nature, that

> *the velocity of light is a physical invariant, having the same value for all observers in uniform relative motion.*

Therefore, the velocity of light should be independent of the relative motion of the emitting and receiving bodies.

Note 19.1 Analysis of the Michelson–Morley experiment

The arrangement for the Michelson–Morley experiment is shown schematically in Figure 19.2. In the figure, S is a monochromatic source of light and M_1 and M_2 are two mirrors placed at the same distance L' from the glass plate P as measured by a terrestrial observer; that is, $PM_1 = PM_2 = L'$. Light coming from S, when it reaches P, is partially transmitted toward M_1 and partially reflected toward M_2. Rays reflected at M_1 and M_2 retrace their paths and eventually reach the observer at O'. Note that the light path drawn in Figure 19.2 is relative to a frame $X'Y'Z'$ moving with the Earth and relative to which the instrument, called an **interferometer**, is at rest. (As an exercise, it is suggested that the student draw the path of light as seen by an observer at rest relative to the ether, and relative to whom the Earth is moving with a velocity $\mathbf{v}$.)

Assume the existence of an 'ether' and let c be the velocity of light as measured by an observer O stationary relative to the ether. Call the velocity of the Earth relative to the ether v and orient the interferometer so that the path PM_1 is parallel to the motion of the Earth. Using the Galilean transformation, we can find (recall Example 4.5)

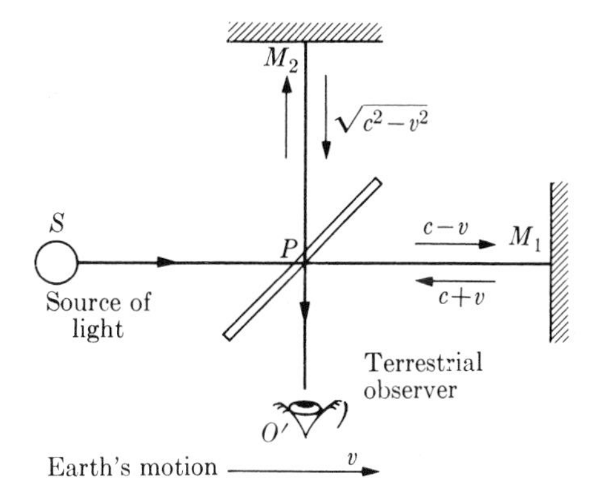

Figure 19.2 Paths of light in the Michelson–Morley experiment. $PM_1 = PM_2 = L'$.

the velocity of light, relative to the Earth, for various paths. When light goes from P to M_1, the relative velocity of light is $c - v$. From M_1 to P the relative velocity is $c + v$. Finally, from either P to M_2 or M_2 to P the relative velocity of light is $(c^2 - v^2)^{1/2}$. Thus, the time required by light to go along the parallel path from P to M_1 and back to P, as measured by the terrestrial observer O', is

$$t_1 = \frac{L'}{c - v} + \frac{L'}{c + v} = \frac{2L'c}{c^2 - v^2} = \frac{2L'/c}{1 - v^2/c^2}$$

However, the time required to go along the perpendicular path from P to M_2 and back to P, as measured by O', is

$$t_2 = \frac{2L'}{(c^2 - v^2)^{1/2}} = \frac{2L'/c}{(1 - v^2/c^2)^{1/2}}$$

Note that t_1 is larger than t_2. Therefore, the rays that reach observer O' have a certain path difference and (according to the theory that will be presented in Chapter 34) should result in a certain interference pattern. Surprisingly, *no* interference pattern is observed, suggesting that $t_1 = t_2$.

In the actual experiment performed by Michelson, it was very difficult to assure that the two arms of the interferometer or, to be more precise, the optical lengths of the two paths, were identical. Therefore, Michelson, to compensate for this difference and actually increase the precision of his measurement, rotated the instrument 90° and observed the *change* in the interference pattern. And although the theory, which was based on the Galilean transformation, predicted a shift in the interference pattern as a result of the rotation, no such shift was observed.

To explain this negative result, George F. Fitzgerald (1851–1901) in 1889 (and Lorentz independently in 1892), proposed that all objects moving through the ether suffer a 'real' contraction in the direction of motion. This contraction was just enough to make $t_1 = t_2$. For this to be true, the length of path PM_1 (as seen by an observer O at rest relative to the ether) must not be the same length as the length for PM_2. The first length is in the direction of the Earth's motion and therefore shortens to a distance which we will call L. The other length is perpendicular to it and so does not change. Writing L for L' in the expression above for t_1, we must have

$$t_1 = \frac{2L/c}{1 - v^2/c^2}$$

while t_2 is still given by the expression above with L'. Equating t_1 and t_2 we obtain, after simplifying,

$$L = (1 - v^2/c^2)^{1/2} L' \qquad \textbf{(19.1)}$$

This expression relates the lengths PM_1 and PM_2 as measured by an observer O at rest relative to the ether. Observer O', who is in motion relative to the ether, should not notice this contraction because the measuring stick used to measure the distance PM_1 is also contracted by the same factor when placed in the direction of the Earth's motion. Thus, to O', the lengths PM_1 and PM_2 are equal. But, according to the Lorentz–Fitzgerald hypothesis, the objects carried by O', including O''s measuring stick, are all shortened in the direction of motion. Therefore, O concludes that the 'real' length of PM_1 is L and that of PM_2 is L', both related by Equation 19.1. This 'real' difference in length is the source of the negative result obtained when the interference of the two light beams was examined.

An alternative explanation of the negative result of the Michelson–Morley experiment is to assume that the velocity of light is always the same in all directions, no matter the state of motion of the observer. Then observer O' uses c for all paths of Figure 19.2 and $t_1 = t_2 = 2L'/c$, as does observer O. This was the position adopted by Einstein when he formulated the postulate of the invariance of the velocity of light for all observers in uniform relative motion. But this implies giving up the Galilean transformation.

19.3 The Lorentz transformation

Under the assumption that the velocity of light is the same for all observers in uniform relative motion, the Galilean transformation cannot be the correct one. Therefore, we must replace it by another transformation, so that the speed of light is invariant or independent of the relative motion of the observers. In particular, the assumption we made in Chapter 4 that $t' = t$ can no longer be correct. Since velocity is distance divided by time, we have to adjust time as well as distance measurements if the velocity of light is to remain the same for observers in relative motion. In other words, the time interval between two events does not have to be the same when measured by observers in relative motion.

We assume that observers O and O' are moving with relative velocity v and that the X- and X'-axes point in the direction of their relative motion and the axes YZ and $Y'Z'$ are parallel (Figure 19.3). We may also assume that both observers set their clocks so that $t = t' = 0$ when O and O' coincide.

Under such conditions, it is shown below that the new transformation, compatible with the invariance of the velocity of light, is

$$\begin{aligned} x' &= \frac{x - vt}{(1 - v^2/c^2)^{1/2}} \\ y' &= y \\ z' &= z \\ t' &= \frac{t - vx/c^2}{(1 - v^2/c^2)^{1/2}} \end{aligned} \tag{19.2}$$

This set of relations, first obtained in this form by Einstein in 1905, was called the **Lorentz transformation** by him. He gave them this name because it was first obtained by Lorentz, in a somewhat different form, in connection with the problem of the electromagnetic field of a moving charge.

The velocity of light c is very large compared with the great majority of velocities that we normally encounter on Earth. Therefore, the ratio v/c is very

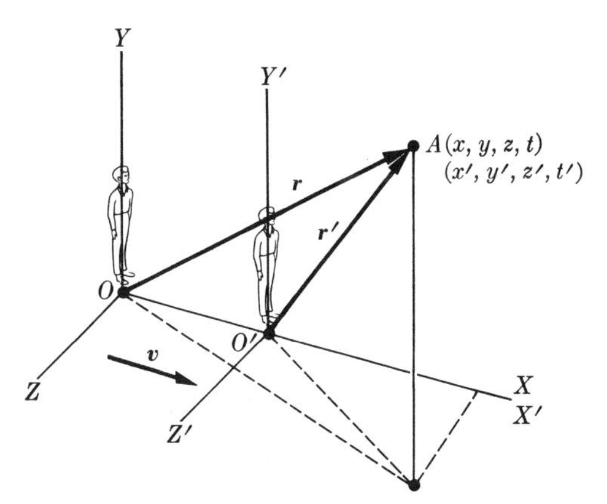

Figure 19.3 Frames of reference in uniform relative translational motion.

small and so the terms v^2/c^2 and vx/c^2 are, in general, negligible. Accordingly,

$$\gamma = \frac{1}{(1 - v^2/c^2)^{1/2}} \tag{19.3}$$

is practically equal to unity (see Figure 19.4) when $v \ll c$. From the practical point of view, then, there is no difference between the Lorentz and Galilean transformations for most measurements made on Earth. We may continue to use the latter in most of the problems we encounter. However, when we are dealing with very fast particles, such as particles in cosmic rays or those produced in high energy accelerators or even electrons in an atom, we must use the Lorentz (or relativistic) transformation.

Although it is true that in the majority of instances the numerical results of the Lorentz transformation do not differ to any great extent from those of the Galilean transformation, from a theoretical point of view the Lorentz transformation represents a most profound conceptual change, especially with regard to space and time, which now become closely related.

The **inverse** Lorentz transformation expresses the coordinates x, y, z and the time t measured by O in terms of the coordinates x', y', z' and the time t' measured by O'. Reversing the second and third relations in Equation 19.2 do not offer any difficulty. A straightforward method to handle the first and fourth equations is to look at them as a set of two simultaneous equations. Then, by a direct algebraic procedure, solve them for x and t in terms of x' and t'. We leave this method to the student as an exercise. We will, instead, proceed along a more intuitive line of reasoning. From the point of view of observer O', observer O recedes along the $-X'$ direction with a velocity $-v$. Therefore O' is entitled to use the same Lorentz transformation to obtain the values of x and t measured by O in terms of the values x' and t'. O' has only to replace v by $-v$ in Equation 19.2

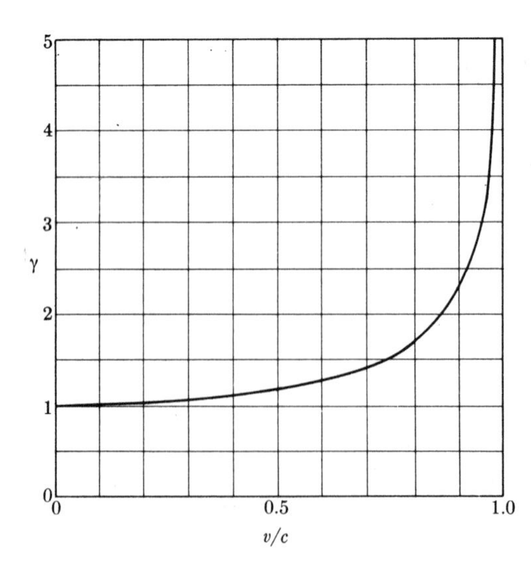

Figure 19.4 Change of $\gamma = 1/(1 - v^2/c^2)^{1/2}$ with v/c.

and exchange x by x' and t by t', resulting in

$$\begin{aligned} x &= \frac{x' + vt'}{(1 - v^2/c^2)^{1/2}} \\ y &= y' \\ z &= z' \\ t &= \frac{t' + vx'/c^2}{(1 - v^2/c^2)^{1/2}} \end{aligned} \qquad \textbf{(19.4)}$$

for the inverse Lorentz transformation.

Derivation of the Lorentz transformation

Referring to Figure 19.3, suppose that at $t = 0$ a flash of light is emitted at the common origin of the two observers in relative motion with velocity v. Both observers use parallel coordinate axes XYZ and $X'Y'Z'$ with the common X-axis in the direction of the relative motion. After a time t, observer O will note that the light has reached point A and will write $r = ct$, where c is the velocity of light. Since $x^2 + y^2 + z^2 = r^2$, we may also write

$$x^2 + y^2 + z^2 = c^2t^2 \qquad \textbf{(19.5)}$$

Similarly, observer O', whose position is no longer coincident with that of O, will note that the light arrives at the same point A in a time t', but also with velocity c. Therefore O' writes $r' = ct'$, or

$$x'^2 + y'^2 + z'^2 = c^2t'^2 \qquad \textbf{(19.6)}$$

Our next task is to obtain a transformation relating Equations 19.5 and 19.6. The symmetry of the problem suggests that $y' = y$ and $z' = z$. Also since $OO' = vt$ for observer O, it must be that $x = vt$ for $x' = 0$ (point O'). This suggests making $x' = \gamma(x - vt)$, where γ is a constant to be determined. We may also assume that $t' = a(t - bx)$, where a and b are constants to be determined (for the Galilean transformation $\gamma = a = 1$ and $b = 0$). Making all these substitutions in Equation 19.6, we have

$$\gamma^2(x^2 - 2vxt + v^2t^2) + y^2 + z^2 = c^2a^2(t^2 - 2bxt + b^2x^2)$$

or

$$(\gamma^2 - b^2a^2c^2)x^2 - 2(\gamma^2 v - ba^2c^2)xt + y^2 + z^2 = (a^2 - \gamma^2v^2/c^2)c^2t^2$$

This result must be identical to Equation 19.5. Therefore

$$\gamma^2 - b^2a^2c^2 = 1, \qquad \gamma^2 v - ba^2c^2 = 0, \qquad a^2 - \gamma^2v^2/c^2 = 1$$

Solving this set of equations for γ, a and b, we have

$$\gamma = a = \frac{1}{(1 - v^2/c^2)^{1/2}}, \qquad b = v/c^2$$

Inserting these values of γ, a and b in $x' = \gamma(x - vt)$ and $t' = a(t - bx)$, we obtain the Lorentz transformation 19.2.

19.4 Lorentz transformation of velocities and accelerations

Consider a particle A (refer to Figure 19.3 again) in motion relative to observers O and O' who are moving relative to each other with velocity $\boldsymbol{v}$. For simplicity, assume that A also moves in the same direction as $\boldsymbol{v}$, which we take as the X and X'-axes. Then the velocity V of A as measured by O is

$$V = \frac{dx}{dt} \tag{19.7}$$

Similarly, the velocity V' of A as measured by O', also along the X' direction, is

$$V' = \frac{dx'}{dt'} \tag{19.8}$$

Note that we now use dt' and not dt, because t and t' are no longer the same. A straightforward calculation shows that the velocities V and V' are related by the expression

$$V' = \frac{V - v}{1 - vV/c^2} \tag{19.9}$$

This equation gives the law for the Lorentz transformation of velocities. That is, Equation 19.9 is the rule for comparing the velocity of a body as measured by two observers in uniform relative translational motion. Equation 19.9 reduces to the Galilean transformation of velocities, $V' = V - v$, for relative velocities that are very small compared with the velocity of light.

To verify that Equation 19.9 is compatible with the assumption that the velocity of light is the same for both observers O and O', consider the case of a light signal propagating along the X direction. Then $V = c$ in Equation 19.9 and

$$V' = \frac{c - v}{1 - vc/c^2} = c$$

Therefore observer O' also measures a velocity c.

The inverse velocity transformation, obtained by solving Equation 19.9 for V is

$$V = \frac{V' + v}{1 + vV'/c^2} \tag{19.10}$$

Equation 19.10 gives the velocity, relative to O, of an object moving with velocity V' relative to O', which in turn is moving with velocity v relative to O.

Equations 19.9 and 19.10 relate the velocity of the *same* body as measured by two observers in relative motion. However, a given observer calculates relative velocities of *different* bodies in his own frame of reference according to the vector rules used in Chapter 4.

The relation between the accelerations of a particle as measured by two observers in uniform relative motion is somewhat more complex. For example, suppose the particle of Figure 19.2, that is moving with velocity V with respect to O, is also accelerated in the X direction. The accelerations of the particle measured by O and O' are related by the expression

$$a' = \frac{a(1 - v^2/c^2)^{3/2}}{(1 - vV/c^2)^3} \tag{19.11}$$

Therefore, the two observers measure different accelerations. This result differs from the Galilean transformation, which predicts that the acceleration is the same for observers in uniform relative motion. In other words, the requirement that the velocity of light be invariant in all frames of reference that are in uniform motion relative to each other destroys the invariance of acceleration relative to those frames.

Derivation of the law for transforming velocities and accelerations

Consider, for generality, that particle A moves relative to O and O' in an arbitrary direction. Then the components of the velocity of A as measured by O are

$$V_x = \frac{dx}{dt}, \qquad V_y = \frac{dy}{dt}, \qquad V_z = \frac{dz}{dt} \tag{19.12}$$

The components of the velocity of A as measured by O' are

$$V'_{x'} = \frac{dx'}{dt'}, \qquad V'_{y'} = \frac{dy'}{dt'}, \qquad V'_{z'} = \frac{dz'}{dt'} \tag{19.13}$$

Differentiating Equations 19.2 we have, noting that $dx = V_x\, dt$,

$$dx' = \frac{dx - v\, dt}{(1 - v^2/c^2)^{1/2}} = \frac{V_x - v}{(1 - v^2/c^2)^{1/2}}\, dt$$

$$dy' = dy$$

$$dz' = dz$$

$$dt' = \frac{dt - v\, dx/c^2}{(1 - v^2/c^2)^{1/2}} = \frac{1 - vV_x/c^2}{(1 - v^2/c^2)^{1/2}}\, dt$$

Dividing the first, second and third of these equations by the fourth gives

$$V'_{x'} = \frac{V_x - v}{1 - vV_x/c^2}$$

$$V'_{y'} = \frac{V_y(1 - v^2/c^2)^{1/2}}{1 - vV_x/c^2} \tag{19.14}$$

$$V'_{z'} = \frac{V_z(1 - v^2/c^2)^{1/2}}{1 - vV_x/c^2}$$

These equations constitute the general Lorentz transformation of velocities. When the motion is along the X-axis we have $V_x = V$, $V_y = V_z = 0$ and $V'_{x'} = V'$, $V'_{y'} = V'_{z'} = 0$ and so the first relation in 19.14 becomes

$$V' = \frac{V - v}{1 - vV/c^2}$$

which is the relation given in Equation 19.9. To obtain the expression for the acceleration, we first differentiate the preceding relation, obtaining

$$\mathrm{d}V' = \frac{(1 - v^2/c^2)}{(1 - vV/c^2)^2}\,\mathrm{d}V$$

Dividing this equation by the expression for $\mathrm{d}t'$ given before, with V_x replaced by V, we obtain

$$a' = \frac{\mathrm{d}V'}{\mathrm{d}t'} = \frac{(1 - v^2/c^2)^{3/2}}{(1 - vV/c^2)^3}\frac{\mathrm{d}V}{\mathrm{d}t}$$

Recalling that $a = \mathrm{d}V/\mathrm{d}t$ this is Equation 19.11.

19.5 Consequences of the Lorentz transformation

The scaling factor $\gamma = 1/(1 - v^2/c^2)^{1/2}$ that appears in the Lorentz transformation, Equation 19.2, presents an interesting implication. It suggests that, as a result of the constancy of the velocity of light, the lengths of bodies and the time intervals between events may not be the same when measured by observers in relative motion.

(i) Length contraction

The length of an object is defined as the distance between its two end-points. However, if the object is in motion relative to the observer who wishes to measure its length, the positions of the two end-points must be recorded **simultaneously**. Consider a bar at rest relative to O' and parallel to the X'-axis (Figure 19.5). Designating its two extremes by A and B, its length as measured by O' is $L' = x'_b - x'_a$. Simultaneity is not essential for O' because the bar is at rest in that frame. However, observer O, who sees the bar in motion, must measure the coordinates x_a and x_b of the endpoints at the *same* time t, obtaining $L = x_b - x_a$.

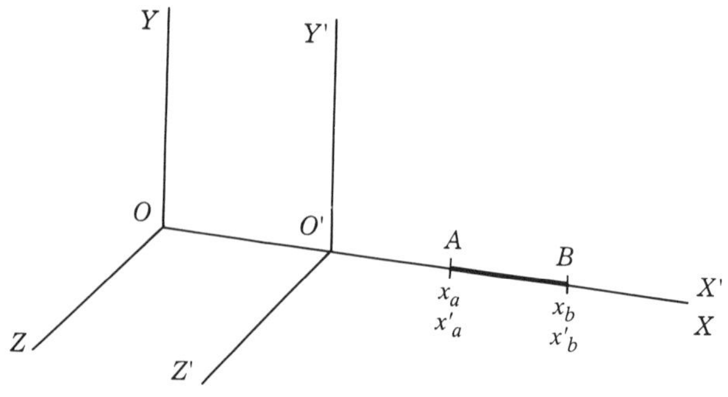

Figure 19.5

Applying the first relation in Equation 19.2, we find that

$$x'_a = \frac{x_a - vt}{(1 - v^2/c^2)^{1/2}} \qquad \text{and} \qquad x'_b = \frac{x_b - vt}{(1 - v^2/c^2)^{1/2}}$$

Note that we write the same t in both expressions. Now subtracting, we obtain

$$x'_b - x'_a = \frac{x_b - x_a}{(1 - v^2/c^2)^{1/2}}$$

or, since $x_b - x_a = L$ and $x'_b - x'_a = L'$,

$$L = (1 - v^2/c^2)^{1/2} L' \tag{19.15}$$

Since the factor $(1 - v^2/c^2)^{1/2}$ is smaller than unity, we have that L is smaller than L'. That is, observer O, who sees the object in motion, measures a **smaller** length than observer O', who sees the object at rest. Therefore

> *the length of a body is measured to be shorter when the body is in motion relative to the observer than when the body is at rest relative to the observer; that is,* $L_{\text{motion}} < L_{\text{rest}}$.

The length of a body at rest relative to an observer is called the **proper length** of the body.

The student should recognize that the 'real' contraction assumed by Lorentz to explain the negative result of the Michelson–Morley experiment, Equation 19.1, is exactly the same as the contraction given by Equation 19.15, found by using the Lorentz transformation and the principle of the invariance of the velocity of light. There is, however, a fundamental difference between the two underlying hypotheses used for obtaining these two apparently identical results:

1. The Lorentz–Fitzgerald contraction 19.1, obtained by means of the Galilean transformation, is assumed to be a *real* contraction suffered by all bodies moving through the ether. The v appearing in the formula is the velocity of the object relative to the ether.

2. Contraction 19.15 refers only to the *measured* valued of the length of the object in motion relative to the observer, and is a consequence of the invariance of the velocity of light. The v appearing in the formula is the velocity of the object relative to the observer. Therefore, the contraction of the length of an object is different for different observers in relative motion.

It was Einstein who was the first to realize that the idea of an ether was artificial and unnecessary, and that the logical explanation was the second one given above.

(ii) Time dilation

A time interval may be defined as the time elapsed between two events, as measured by an observer. An **event** is a specific occurrence that happens at a particular place and time. Thus, in terms of these definitions, when the bob of a pendulum reaches its lowest point during a swing, this constitutes an event. After a certain period of time it will return to this same position; this is a second event. The elapsed time between these two events is then a time interval. Thus a time interval is the time it takes to do something: for a pendulum to oscillate, for an electron to move around a nucleus, for a heart to beat etc.

Consider two events that occur at times t'_a and t'_b but at the same place x' relative to an observer O' in motion with respect to O. Relative to O the events occur at different places and at times t_a and t_b, respectively.

Applying the last relation of the inverse Lorentz transformation, given in Equation 19.4, to both events, we have

$$t_a = \frac{t'_a + vx'/c^2}{(1 - v^2/c^2)^{1/2}}, \qquad t_b = \frac{t'_b + vx'/c^2}{(1 - v^2/c^2)^{1/2}}$$

Note that x' is the same in both equations. Subtraction gives

$$t_b - t_a = \frac{t'_b - t'_a}{(1 - v^2/c^2)^{1/2}}$$

or, making $T = t_b - t_a$ and $T' = t'_b - t'_a$,

$$T = \frac{T'}{(1 - v^2/c^2)^{1/2}} \tag{19.16}$$

Now T' is the time interval between two events measured by an observer O' at *rest* with respect to the place x' where the events occurred. And T is the time interval measured by an observer O relative to whom the place where the events occurred is in *motion*. Since the factor $1/(1 - v^2/c^2)^{1/2}$ is larger than one, Equation 19.16 indicates that T is greater than T'. Therefore,

> *processes appear to take a longer time when they occur in a body in motion relative to the observer than when the body is at rest relative to the observer; that is,* $T_{\text{motion}} > T_{\text{rest}}$.

The time interval between two events occurring at points at rest relative to an observer is called the **proper time interval**. Thus in Equation 19.16 T' is the proper time interval.

EXAMPLE 19.1

Time dilation and muon decay.

▷ The *muon* is an unstable particle that has a mass about 207 times larger than the electron mass. These particles disintegrate or decay at a certain rate into other particles. The rate of decay is expressed in terms of the **half-life** τ, the time in which half of the particles initially present disintegrate. If we have a number of muons at rest in the laboratory, after a time τ, equal to one half-life, only one half, or $N/2$, of the muons remain. After a time 2τ (two half-lives) $N/4$ or $N/2^2$ muons are left. Clearly, after a time equal to n half-lives, only $(\frac{1}{2})^n$

of the original particles remain. Muons at rest in the laboratory have a half-life of about 1.5×10^{-6} s.

Collisions of cosmic rays with the atoms in the atmosphere at a height of about 60 km produce muons. These muons have a velocity relative to the Earth close to the velocity of light. Suppose that $v = 0.999c = 2.9949 \times 10^8 \text{ m s}^{-1}$. At that speed, the muons take a time, measured by a terrestrial observer

$$T = \frac{6 \times 10^4 \text{ m}}{2.9949 \times 10^8 \text{ m s}^{-1}} \approx 2 \times 10^{-4} \text{ s}$$

to reach the surface of the Earth when headed straight down for 60 km. This time is equivalent to 133 half-lives and, according to an observer on the Earth, we would have only a small fraction, $(\frac{1}{2})^{133}$ or 10^{-40}, of the original muons reaching the surface of the Earth. However, it is found that the number of muons at sea level is much larger. Due to time dilation the time T required by the muons to cross the atmosphere, as measured by a terrestrial observer, is much larger than the time T' measured by an observer at rest with the muons. Since in this case $v/c = 0.999$, we have that

$$(1 - v^2/c^2)^{1/2} = 4.5 \times 10^{-2}$$

Then, using Equation 19.16, we find that the time T' in the muon's frame of reference (that is, the muon proper time) needed to reach the Earth's surface is

$$T' = T(1 - v^2/c^2)^{1/2} \approx 9 \times 10^{-6} \text{ s}$$

which is equivalent to about six half-lives. In that time the number of muons still left when reaching the Earth is about $(\frac{1}{2})^6$ or 1/64 of the original number. This result is more consistent with the experimental evidence.

We may also analyze the problem differently. Relative to the muon frame of reference, the Earth moves toward the muon with a velocity $v = 0.999c$. Therefore, relative to the muon, the distance between the top of the atmosphere and the Earth's surface is shortened by a factor $(1 - v^2/c^2)^{1/2} = 4.5 \times 10^{-2}$. Accordingly, the time the Earth's surface takes to reach the muon is also shortened by the same factor, resulting again in a time interval equal to 9×10^{-6} s.

Similar results have been obtained with particles produced in the laboratory, using machines that accelerate the particles to very large velocities. Observing how the number of such particles decreases along the beam produces good confirmation of time dilation.

19.6 Special principle of relativity

Einstein's postulate that the velocity of light is the same for all observers in uniform relative motion, makes the introduction of an ether 'superfluous' (using Einstein's word). It also eliminates the need to consider a 'preferred' (again Einstein) observer at rest relative to the ether. The natural consequence is that all observers in uniform relative motion must be considered 'equivalent' (Einstein). This was the basis for Einstein's principle of relativity:

> *all laws of nature (not only of dynamics) must be the same for all inertial observers moving with constant velocity relative to each other.*

The **principle of relativity** requires that we must express all physical laws in a form that does not change when we transform from one inertial observer to another. We know that this holds for Newton's laws of dynamics when the Galilean transformation is used. Einstein's principle extends this invariance to all laws, not just those of dynamics. One consequence of the requirement is that it restricts the mathematical expressions of physical laws.

Among the laws that must remain invariant for all inertial observers are those describing electromagnetic phenomena; these will be discussed in detail in later chapters. We may state in advance that these laws, when expressed relative to an inertial observer, involve the velocity of light c. Therefore, a requirement of the principle of relativity, as formulated by Einstein, is that the velocity of light be the same for all observers in uniform relative motion.

We have seen that the velocity of light is the same to all inertial observers if their measurements are related by the Lorentz transformation for comparing measurements by two observers in relative motion. Accordingly, the principle of relativity may be restated:

> *observers in uniform relative motion must correlate their observations by means of the Lorentz transformation. All physical quantities must transform from one system to another in such a way that the expression of the physical laws is the same for all inertial observers.*

From the practical point of view, this new formulation of the principle of relativity is important only for velocities comparable to that of light. Therefore, it must be applied whenever bodies in motion have a very high velocity. For particles with low velocities, the Galilean transformation is still a very good approximation for relating physical quantities in two inertial frames. For such particles Newtonian mechanics provides a satisfactory formalism for describing these phenomena. The dynamical theory based on Einstein's postulates is usually called the *special* theory of relativity because it applies only to observers in uniform relative motion.

From the practical point of view, then, we may ignore the special theory of relativity in many instances. However, from the conceptual point of view, it has produced a profound modification in our theoretical approach to the analysis of physical phenomena. For example, the way the Lorentz transformation couples space and time suggests that they are not independent notions, and that rather we should think of **space–time** as a single entity, with the separation into space and time dependent on the relative motion of the observers.

19.7 Momentum

In Chapter 6 the momentum of a particle was defined as

$$\boldsymbol{p} = m\boldsymbol{v} \tag{19.17}$$

where the mass m is a constant characteristic of the particle. In Section 6.5 the conservation of momentum of two interacting particles was discussed. We found that the conservation of momentum holds for observers in uniform relative motion when they compare their measurements using the Galilean transformation of velocities. However, if we use this definition of momentum, together with the Lorentz transformation of velocities, the conservation of momentum does not hold. That means that the definition of momentum must be modified to comply with the principle of relativity. Recall that the force applied to a particle is related to the momentum by $\boldsymbol{F} = \mathrm{d}\boldsymbol{p}/\mathrm{d}t$. We then have that, by exerting forces on fast particles, it is possible to determine experimentally the corresponding expression for $\boldsymbol{p}$. We may, for example, observe the motion of fast electrons, protons, or other charged particles in electric and magnetic fields. Particles with different velocities may be produced in the laboratory by accelerators or found in cosmic rays. The result of these experiments and measurements indicates that the momentum of a particle moving with a velocity $\boldsymbol{v}$ relative to an observer must be expressed as:

$$\boldsymbol{p} = \frac{m\boldsymbol{v}}{(1 - v^2/c^2)^{1/2}} = \gamma m\boldsymbol{v} \qquad \textbf{(19.18)}$$

In Equation 19.18, m is the mass of the particle and $\gamma = 1/(1 - v^2/c^2)^{1/2}$ as defined in Equation 19.3. For small velocities ($v \ll c$), γ can be equated to one, and this new expression for $\boldsymbol{p}$ becomes identical to Equation 19.17. Therefore, the Newtonian definition of momentum is good for most motions observed on Earth where the velocity is small compared with c.

The variation of momentum with velocity according to Equation 19.18 is illustrated in Figure 19.6. This figure is essentially identical to Figure 19.4, since both give γ in terms of v/c. It can be seen that only at very large velocities is there any noticeable departure of the momentum of the particle from the Newtonian value $\boldsymbol{p} = m\boldsymbol{v}$. For example, even at $v = 0.5c$, $p/mv = 1.15$, or only a 15 per cent increase in the momentum over the Newtonian value. In addition, as the velocity of the particle approaches c the change in momentum for a given change in velocity

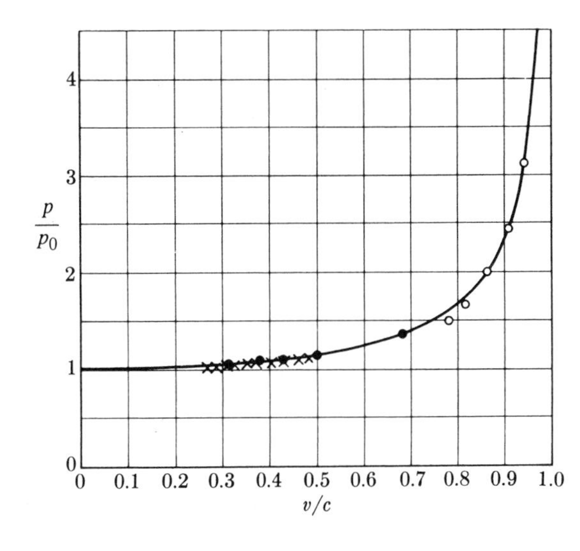

Figure 19.6 Experimental confirmation of the variation of momentum with velocity. The solid curve is a plot based on Equation 19.18. The experimental data of W. Kaufmann (1901) are plotted as open circles, those of A. Bucherer (1909) as solid circles, and those of C. Guye and C. Lavanchy (1915) as crosses. In the figure, $p_0 = mv$.

increases very rapidly. As seen in the next section, this means that it is more difficult to accelerate a particle as it moves faster relative to the observer.

Note 19.2 Relativistic momentum

It is possible, through an elaborate algebraic manipulation that we will not carry out, to verify that the momentum defined by Equation 19.18 satisfies the law of conservation of momentum when applied to the collision of two bodies, as well as the requirements of the principle of relativity. This provides further justification of definition 19.18.

The relativistic definition of momentum can also be justified in a simple way. Consider an observer O' attached to the moving body. If dt is the time it takes to move through a distance dx, as measured by observer O the velocity relative to O is $\boldsymbol{v} = d\boldsymbol{r}/dt$. Using Equation 19.16, the proper time interval of O' corresponding to dt is $d\tau = dt(1 - v^2/c^2)^{1/2}$. We now shall define the **proper velocity** as

$$\boldsymbol{v}_{\text{proper}} = \frac{d\boldsymbol{r}}{d\tau} = \frac{d\boldsymbol{r}}{dt(1 - v^2/c^2)^{1/2}} = \frac{\boldsymbol{v}}{(1 - v^2/c^2)^{1/2}}$$

and the momentum as

$$\boldsymbol{p} = m\boldsymbol{v}_{\text{proper}} = \frac{m\boldsymbol{v}}{(1 - v^2/c^2)}$$

This procedure has the advantage of using a time interval $d\tau$ that is independent of the observer who sees the body in motion. In this way, the definition is valid for all observers regardless of their relative motion and the momentum has the same vector property as the velocity since $d\tau$ is a true scalar or invariant property.

19.8 Force

The definition 19.18 of momentum clearly modifies the relation between force and acceleration, which in Newtonian mechanics is given by $\boldsymbol{F} = d\boldsymbol{p}/dt = m\, d\boldsymbol{v}/dt$. The relation $\boldsymbol{F} = d\boldsymbol{p}/dt$ is also maintained in relativistic mechanics, as indicated in the previous section, but with the new definition of momentum. Thus we restate force as

$$\boldsymbol{F} = \frac{d\boldsymbol{p}}{dt} = \frac{d}{dt}\left(\frac{m\boldsymbol{v}}{(1 - v^2/c^2)^{1/2}}\right) \tag{19.19}$$

When dealing with *rectilinear motion* we consider only the magnitudes, and thus we may write

$$F = \frac{d}{dt}\left(\frac{mv}{(1 - v^2/c^2)^{1/2}}\right)$$

or

$$F = \frac{m}{(1 - v^2/c^2)^{3/2}} \frac{dv}{dt} \tag{19.20}$$

On the other hand, in the case of *uniform circular motion*, the velocity remains constant in magnitude but not in direction, and Equation 19.19 becomes

$$\boldsymbol{F} = \frac{m}{(1 - v^2/c^2)^{1/2}} \frac{d\boldsymbol{v}}{dt}$$

But $d\boldsymbol{v}/dt$ is then the normal or centripetal acceleration whose magnitude is v^2/R, where R is the radius of the circle, according to Equation 5.13. Therefore, the magnitude of the normal or centripetal force becomes

$$F = \frac{m}{(1 - v^2/c^2)^{1/2}} \frac{v^2}{R} \tag{19.21}$$

In the general case of *curvilinear motion*, we recall that dv/dt is the tangential acceleration a_T and v^2/R the normal acceleration a_N. We conclude from Equations 19.20 and 19.21 that in curvilinear motion the components of the force along the tangent and the normal to the path are

$$F_T = \frac{m}{(1 - v^2/c^2)^{3/2}} a_T$$
$$F_N = \frac{m}{(1 - v^2/c^2)^{1/2}} a_N \tag{19.22}$$

An immediate consequence is that the force is *not* parallel to the acceleration (Figure 19.7) because the coefficients multiplying a_T and a_N are different. Thus a vector relation of the type $\boldsymbol{F} = m\boldsymbol{a}$ does not hold for particles that have high energy. However, the more fundamental relation $\boldsymbol{F} = d\boldsymbol{p}/dt$ still remains valid, because it is the law of motion. Equations 19.22 show that as the velocity of a particle approaches that of light the force required to produce a given acceleration increases considerably. These relations are important when analyzing the motion of particles in high energy accelerators.

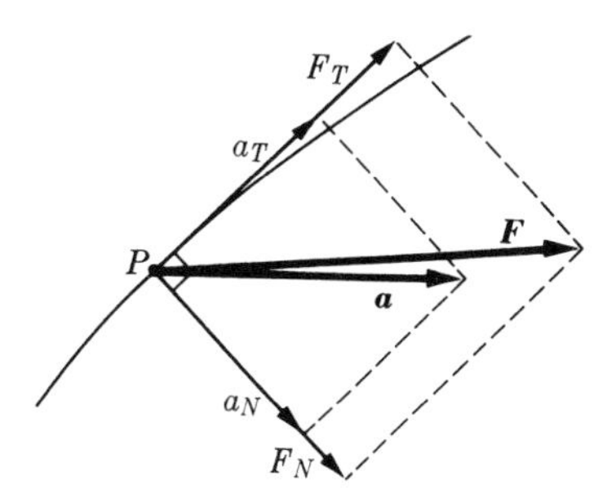

Figure 19.7 At high velocity, the force is not parallel to the acceleration.

EXAMPLE 19.2

Relation between force and acceleration for different particle velocities.

▷ According to Newtonian mechanics, the ratios F/a should be a constant and the same for both the tangential and centripetal components of the force. This is not the case, as shown in the table, calculated using Equations 19.22.

v/c	F_T/ma_T	F_N/ma_N
0.01	1.00015	1.00005
0.05	1.00376	1.00125
0.10	1.01519	1.00504
0.20	1.06315	1.02062
0.40	1.29892	1.09109
0.60	1.95312	1.25000
0.80	4.62963	1.66667
0.90	12.07451	2.29416
0.95	32.84680	3.20256
0.99	356.22171	7.08881

It is clear that as the velocity increases it becomes much more difficult to increase the magnitude of the velocity than to change its direction, and that as the velocity approaches that of light the tangential force becomes enormous. These results show how increasingly difficult it is to accelerate a particle up to velocities close to that of light and to keep them moving along a circular path.

19.9 Energy

In Chapter 9 a particle's kinetic energy was defined as

$$E_k = \tfrac{1}{2}mv^2 \tag{19.23}$$

However, when the expression for the momentum given in Equation 19.18 is used, the kinetic energy of a particle moving with velocity v relative to the observer is given by

$$E_k = \frac{mc^2}{(1 - v^2/c^2)^{1/2}} - mc^2 \tag{19.24}$$

which can be written alternatively as

$$E_k = mc^2(\gamma - 1)$$

where $\gamma = 1/(1 - v^2/c^2)^{1/2}$. This result may seem quite different from the Newtonian value. However, when v is small compared with c, we may expand the denominator in Equation 19.24, using the binomial theorem B.21:

$$(1 - v^2/c^2)^{-1/2} = 1 + v^2/2c^2 + 3v^4/8c^4 + \cdots$$

Substituting in Equation 19.24, we find that

$$E_k = \tfrac{1}{2}mv^2 + \tfrac{3}{8}mv^2(v/c)^2 + \cdots$$

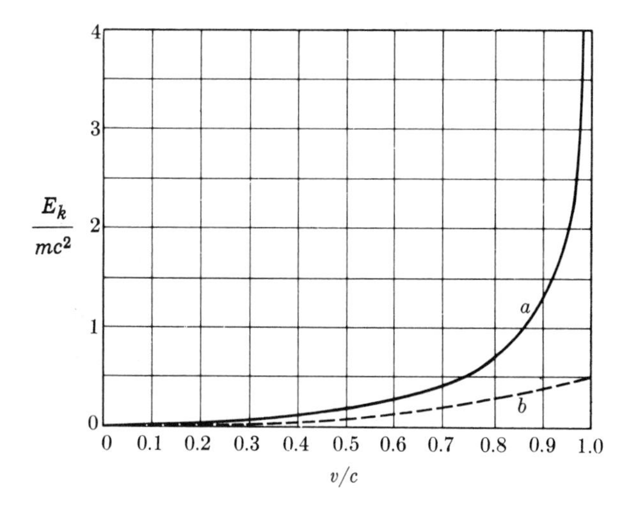

Figure 19.8 Variation of kinetic energy with velocity: (a) relativistic, Equation 19.24; (b) Newtonian, Equation 19.23.

The first term is the kinetic energy of Newtonian mechanics. The second and succeeding terms are negligible if $v \ll c$. In this way we again verify that Newtonian mechanics is an approximation of relativistic mechanics, valid for small velocities or energies.

In Figure 19.8, the variation of the kinetic energy E_k as given by Equation 19.24 has been indicated by curve *a*, and the Newtonian kinetic energy $E_k = \frac{1}{2}mv^2$ (divided by mc^2) by curve *b*. This figure clearly shows that, at equal velocities, the relativistic kinetic energy is larger than the Newtonian value. That means that it takes much more energy to accelerate a particle to a given velocity when the velocity is close to c.

We must note that the ratio E_k/mc^2 is the same for all particles having the same velocity. Thus, since the mass of the proton is about 1850 times the mass of the electron, relativistic effects in the motion of protons are noticeable only at kinetic energies 1850 times larger. For that reason the motion of protons and neutrons in atomic nuclei can often be treated without relativistic considerations, while the motion of atomic electrons usually requires a relativistic approach.

The quantity

$$E = E_k + mc^2 = \frac{mc^2}{(1 - v^2/c^2)^{1/2}} \qquad \textbf{(19.25)}$$

is called the **total energy** of the particle. It can also be written as $E_k = \gamma mc^2$. The quantity mc^2 is called the **rest energy** of the particle, since it is the value of E when $v = 0$. The total energy of the particle, as here defined, includes kinetic energy and rest energy, but not potential energy. Expression 19.25 is interesting because it suggests that to each mass m we may associate an energy

$$E = mc^2$$

and conversely, with an energy E we may associate a mass

$$m = E/c^2$$

This conclusion is very suggestive. It indicates that we may associate a change in mass Δm to any change in energy ΔE and conversely. Both changes are related by the expression

$$\Delta E = (\Delta m)c^2 \tag{19.26}$$

This relation is of extraordinary importance and we shall have occasion to refer to it many times. When Einstein arrived at this conclusion he was extremely cautious in his statements. He indicated the need to verify it experimentally and suggested some possible experiments. So far, it has been amply confirmed experimentally. Perhaps the most dramatic proof has been the calculation of the energy released in a nuclear bomb (see Chapter 40).

Calculation of the kinetic energy

To compute the kinetic energy of a particle using the new definition of momentum, Equation 19.18, we follow the same procedure as in Section 9.5. That is, recalling that $v = \mathrm{d}s/\mathrm{d}t$ we obtain

$$E_k = \int_0^v F_T\, \mathrm{d}s = \int_0^v \frac{\mathrm{d}p}{\mathrm{d}t}\,\mathrm{d}s = \int_0^v v\, \mathrm{d}p$$

Integrating by parts (Appendix B) and using the relativistic expression 12.18 for momentum, we have

$$E_k = vp - \int_0^v p\, \mathrm{d}v = \frac{mv^2}{(1 - v^2/c^2)^{1/2}} - \int_0^v \frac{mv\, \mathrm{d}v}{(1 - v^2/c^2)^{1/2}}$$

$$= \frac{mv^2}{(1 - v^2/c^2)^{1/2}} + mc^2(1 - v^2/c^2)^{1/2} - mc^2$$

Combining the first two terms of the right-hand side into one, we finally obtain the expression for the kinetic energy of a particle moving with velocity v relative to the observer,

$$E_k = \frac{mc^2}{(1 - v^2/c^2)^{1/2}} - mc^2 = mc^2\left(\frac{1}{(1 - v^2/c^2)^{1/2}} - 1\right) = mc^2(\gamma - 1)$$

EXAMPLE 19.3

Energy equivalence of one amu.

▷ Recall from Section 2.4 that 1 amu $= 1.6605 \times 10^{-27}$ kg. By using the relation $E = mc^2$ with $c = 2.9979 \times 10^8\ \mathrm{m\ s^{-1}}$ we have that

$$1\ \text{amu} = 1.4924 \times 10^{-10}\ \mathrm{J}$$

Since 1 J = 6.242×10^{12} MeV, we find that

$$1 \text{ amu} = 931.531 \text{ MeV} \tag{19.27}$$

Therefore, when the mass m of a particle is expressed in amu, the rest energy of the particle in MeV is

$$E = 931.531m \text{ MeV} \tag{19.28}$$

The values of the rest energy for the electron, proton and neutron are given in the following table.

Particle	Mass, m (amu)	Rest energy, mc^2 (MeV)
Electron	0.000 549	0.511
Proton	1.007 825	938.82
Neutron	1.008 665	939.60

EXAMPLE 19.4

Calculation of the energy required to accelerate (a) an electron, (b) a proton, up to a velocity half that of light. In each case, compare the relativistic and Newtonian values.

▷ When $v = 0.5c$, we have that $\gamma = 1/(1 - v^2/c^2)^{1/2} = 1.1547$. Then

$$E_k = mc^2(\gamma - 1) = 0.1547\, mc^2$$

for the relativistic value, and

$$E_k = \tfrac{1}{2}.mv^2 = mc^2\left(\frac{1}{2}\frac{v^2}{c^2}\right) = 0.125\, mc^2$$

Thus

$$\frac{E_k(\text{rel.})}{E_k(\text{Newt.})} = 1.2376$$

so that E_k(rel.) is about 24% higher than E_k (Newt.). Note that this ratio is independent of the mass of the particle. The value of E_k(rel.) for the electron and the proton, using the results above, are

electron: $E_k(\text{rel.}) = 0.079$ MeV and proton: $E_k(\text{rel.}) = 145.155$ MeV

This shows that to accelerate a proton to a certain velocity requires about 2000 times more energy than an electron, a result that is true both in Newtonian and in relativistic mechanics.

EXAMPLE 19.5

Calculation of the velocity of (a) an electron, (b) a proton that has kinetic energy of 10 MeV.

▷ In this case $E_k = 10$ MeV. Since the total energy is $E = E_k + mc^2$, we have,

(a) for the electron

$$E = 10 \text{ MeV} + 0.511 \text{ MeV} = 10.511 \text{ MeV}$$

so that $E/mc^2 = 20.569$. Then, using Equation 19.25, $E = \gamma mc^2$ where $\gamma = 1/(1 - v^2/c^2)^{1/2}$, we find $v/c = 0.9988$. That is, the electron moves at a velocity very close to that of light.

(b) For the proton we have

$$E = 10 \text{ MeV} + 938.3 \text{ MeV} = 948.3 \text{ MeV}$$

which gives $E/mc^2 = 1.01066$. Therefore, again using Equation 19.25, $v/c = 0.1448$. The proton moves at a velocity so small compared with that of light it could be treated non-relativistically. In fact, if we use the Newtonian relation $E_k = \frac{1}{2}mv^2$ for the proton, we get a value for v/c of 0.1460, which differs from the relativistic value by only 0.8%.

This example shows that for equal kinetic energy, protons move at much smaller velocities than electrons and often can be treated non-relativistically.

EXAMPLE 19.6

Velocity of a particle of mass m and energy E. The mass of a neutrino.

▷ From the expression $E = mc^2/(1 - v^2/c^2)^{1/2}$, we can solve for the velocity and get

$$v = c\left(1 - \frac{(mc^2)^2}{E^2}\right)^{1/2}$$

Note that as the energy increases, the velocity approaches the value c. Therefore, it is possible to increase the energy of the particle without appreciably changing its velocity. When $E \gg mc^2$, the above expression may be rewritten with the help of the binomial expansion as

$$v = c\left(1 - \frac{(mc^2)^2}{2E^2} + \cdots\right)$$

If we wish to calculate the time needed for a particle to travel a distance d, we get

$$t = \frac{d}{v} = \frac{d}{c}\left(1 - \frac{(mc^2)^2}{2E^2} + \cdots\right)^{-1} \approx \frac{d}{c}\left(1 + \frac{(mc^2)^2}{2E^2}\right)$$

This expression has been used to calculate the mass of the neutrinos produced in the explosion of the supernova SN1987a, which was observed in 1987. When the explosion occurred (approximately 250 000 years ago), two bursts of neutrinos of slightly different energy were emitted. They were received on Earth 10 seconds apart. Assuming that the two bursts were emitted simultaneously, but with different energies, their time of travel may be written as

$$t_1 = \frac{d}{c}\left(1 + \frac{(mc^2)^2}{2E_1^2}\right) \quad \text{and} \quad t_2 = \frac{d}{c}\left(1 + \frac{(mc^2)^2}{2E_2^2}\right)$$

Therefore,

$$\Delta t = t_2 - t_1 = \left(\frac{d}{c}\right)\frac{(mc^2)^2}{2}\left(\frac{1}{E_2^2} - \frac{1}{E_1^2}\right)$$

Assuming the energies are measured independently, the only unknown quantity in the above equation is the mass, which may be easily computed. If the neutrinos were emitted at different times, the expression above gives an upper limit to the mass of a neutrino. In this way, it has been estimated that $m_\nu c^2 \leqq 20$ eV (recall that the rest energy $m_e c^2$ of an electron is about 0.5 MeV or at least 25 000 times larger), although the matter is far from settled.

19.10 The general theory of relativity

The theory of relativity developed so far applies to observers in uniform relative motion. It incorporates electromagnetism through the invariance of the velocity of light (and of all electromagnetic waves) under a Lorentz transformation that couples space and time. However, this theory does not include gravitation. The way Newton formulated his theory implies instantaneous gravitational interaction between bodies. This is equivalent to saying that the gravitational interaction propagates with infinite velocity. This is not a satisfactory situation. Einstein, in 1916, formulated a more general theory of relativity that explicitly takes gravitation into account. The mathematical details of the general theory are too complex for discussion here. Rather, we will refer briefly to some of the fundamental ideas.

The first element that must be incorporated in the general theory of relativity is the principle of equivalence (Section 11.10). This principle establishes that it is impossible to determine if an observer's frame of reference is moving with accelerated motion relative to an inertial observer or is in a local gravitational field. Or, using Einstein's words,

the laws of physics must be of such a nature that they apply to reference systems in any kind of motion relative to the mass distribution of the universe.

The next step is to develop a formalism for relating events that are closely spaced and occur within a small interval of time in such a way that the above principle is taken into account. In the special theory of relativity, space and time for two observers are connected by the Lorentz transformation, which applies only to observers in uniform relative motion, without taking into account gravitational effects. Therefore, the above principle is not satisfied.

The gravitational field at any point in space is determined by the mass distribution in the whole universe. Therefore, the most important consequence of the general theory is that the *local* properties of space and time, or rather space–time, are determined by this mass distribution. To a first approximation, the relativistic correction to the predictions of Newtonian theory for the motion of a particle of mass m in a gravitational field equals the ratio of the gravitational potential energy E_p of m and its rest energy mc^2. That is,

$$\begin{pmatrix}\text{First-order}\\ \text{correction}\end{pmatrix} \sim \frac{E_p}{mc^2} = \frac{\Phi}{c^2} \tag{19.29}$$

where $\Phi = E_p/m$ is the gravitational potential relative to the place where the gravitational field is zero (recall Section 11.8). Therefore the correction is independent of the mass m. If the gravitational field is produced by a mass M at the distance r, the gravitational potential is $\Phi = -GM/r$, in accordance with Equation 11.19. Then

$$\begin{pmatrix}\text{First-order}\\ \text{correction}\end{pmatrix} \sim \frac{GM}{c^2 r} \tag{19.30}$$

which is a quantity of the order of $7.4 \times 10^{-28}(M/r)$ where M is measured in kg and r in m. Therefore the gravitational correction is negligible except near large masses. Note that the ratio of Equation 19.30 is dimensionless and thus independent of the units used.

To illustrate how gravitation affects space–time, suppose Δx and Δt are the space and time intervals between two events measured at a place where the gravitational potential is zero. If the same events occur at a place at a distance r from a mass M, the observed space and time intervals are given, to a first approximation, by

$$\Delta x' = \left(1 - \frac{GM}{c^2 r}\right)\Delta x, \qquad \Delta t' = \left(1 + \frac{GM}{c^2 r}\right)\Delta t \tag{19.31}$$

so that the space interval is contracted and the time interval is dilated. Equations 19.31 show that the gravitational effect of a body of mass M is to distort the measurements of distance and time in its vicinity. Although the local structure of space–time is determined by the mass distribution in the universe, the smallness of the ratio G/c^2 indicates that space–time is very ‘stiff’.

Gravitation also affects the measurements of velocity. If $\Delta x'$ and $\Delta t'$ correspond to the motion of a body, the velocity $v' = \Delta x'/\Delta t'$ in the gravitational field is related to the velocity in the absence of a gravitational field, $v = \Delta x/\Delta t$, by

$$v' = \frac{(1 - GM/c^2 r)}{(1 + GM/c^2 r)} v$$

Assuming $GM/c^2 r$ is a small quantity (at the Sun's surface it is about 2×10^{-6}), the binomial expansion $(1 + x)^{-1} \sim 1 - x$ for the denominator may be used, resulting in

$$v' = \left(1 - \frac{2GM}{c^2 r}\right) v \tag{19.32}$$

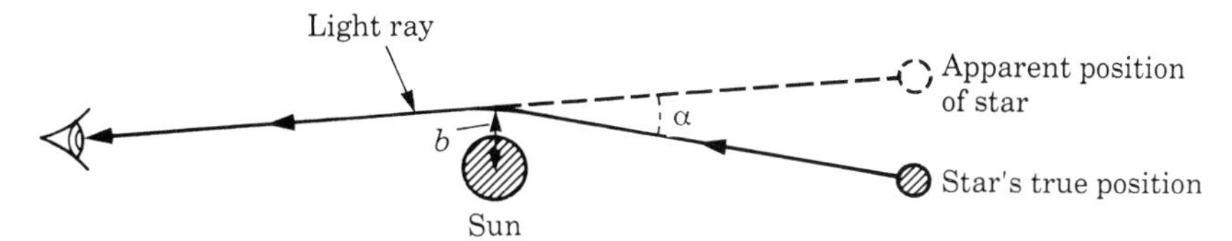

Figure 19.9 Deviation of a light ray from a star when passing close to the Sun. The deviation is given by $\alpha = 4GM/bc^2$, where M is the mass of the Sun and b is the distance of closest approach of the light ray.

so that v' is smaller than v and the body appears to move slower. This result applies as well to the velocity of light, which is reduced when light propagates through a gravitational field. It is even possible that near a massive compact star the velocity of light reduces to zero. Stars where this occurs are called **black holes** (recall Example 11.5 and see the discussion in Note 19.3 at the end of part ii).

There is no unique way of formulating general relativity in precise mathematical terms. Nevertheless, some consequences of the general theory of relativity are independent of any formulation. One consequence is that, since mass and energy are equivalent, the propagation of electromagnetic radiation (photons) is affected by a gravitational field. For example, a ray of light must be deviated by a gravitational field, such as that produced by the Sun (Figure 19.9), an effect that was first confirmed in a solar eclipse that took place in 1917 by observing the position of stars close to the edge of the Sun's shadow.

Another consequence is that since, as indicated above, the velocity of light is modified by a gravitational field, the time for an electromagnetic signal (light or radar) from a body in space to travel to Earth depends on the gravitational field through which it passes. Suppose radar tracking signals are sent from the Earth to a target in space. The target might be a planet such as Venus or Mars, or a spacecraft, such as one of the Mariners or Vikings, and the times for the reflected signals to return to Earth are measured. As the target moves along its orbit, the path of the radar signal gets closer to the Sun (Figure 19.10) and therefore passes through an increasing gravitational field. As a result, the round trip time for the signals increases due to the increased gravitational action of the Sun. According to the general theory of relativity, this gravitational time delay may amount to several hundred microseconds, depending on the distance of the path from the Sun. As the target recedes from the Sun, the time delays decrease.

Another effect is that the frequency of electromagnetic radiation depends on the gravitational potential at the place where it is measured, due to the change of time intervals. For example, light produced in the strong negative gravitational potential of a star has a lower frequency than the light emitted by the same source on Earth. We say that the star spectrum is shifted toward the red relative to the Earth.

All these effects are extremely small. However, they have been observed both in very precise laboratory experiments as well as with artificial satellites and space probes and in careful astronomical observations. The results agree very well with the predictions of the general theory of relativity.

The precession of the orbit of Mercury (Section 11.6), in excess of that due to the perturbation produced by the other planets, cannot be explained using

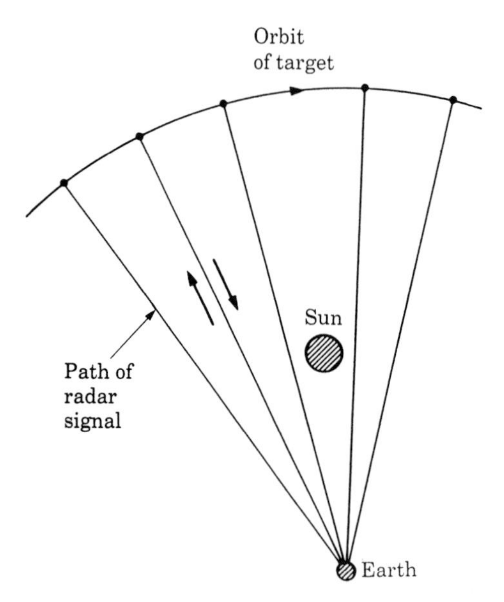

Figure 19.10 Time delay of a radar signal by the Sun as the path gets closer to the Sun.

Newton's theory. However, the general theory of relativity fully accounts for the precession. Similarly, the precession of the orbits in binary stars agrees very well with the predictions of general relativity.

The fact that the gravitational field of a body affects the structure of space–time around the body in proportion to its mass suggests a picture of the gravitational interaction different from that proposed by Newton. Instead of attractive forces between any two bodies (recall Fig. 11.4), we may assume that the deformation of space–time in the vicinity of a body induces all other bodies around to move toward it, following the 'easiest' or 'shortest' path in space–time, an effect manifested in space as a force. Since every body produces a deformation of space–time, the result is a universal attraction, the effect being larger close to massive bodies, where the deformation of space–time is larger, with the paths of all bodies in space–time determined by the mass distribution in the universe.

A simple analogy helps to understand the space–time gravitational effect. If we place several light balls on a horizontal stretched membrane, which is a two-dimensional space, they remain where they were placed because they practically do not affect the membrane. But if one of the balls is much more massive than the others, so that it pushes the membrane downward, all the other balls move toward the massive ball. An observer on the membrane would attribute the motions to the deformation in three-dimensional space. The analogy is not very satisfactory because the deformation of the membrane occurs in space, while the gravitational effect is in space–time, and cannot be visualized in three-dimensional geometric terms. Rather, it requires the use of certain mathematical techniques beyond the level of this book.

Another interesting idea is that as a body moves, the deformation of space–time it produces at a given point in space–time changes. This should result in a sort of ripple in space–time, or a **gravitational wave** propagating with a velocity that depends on the mass distribution (recall Equation 19.32) and is equal to c

in empty space. Because a gravitational wave is a 'traveling deformation' of space–time, it manifests itself in space as a tiny force on bodies through which it passes. However, because of the stiffness of space–time, the effect is so small that it requires special techniques such as the use of radar beams and laser interferometers to detect gravitational waves when they reach the Earth. To give an idea of the magnitude of a gravitational wave, the relative displacement of two masses separated by one meter should be no more than 1/100 of a nuclear diameter when hit by the wave! The gravitational waves that have better possibilities of being detected are those produced by massive bodies moving very fast and with large accelerations. This is the case, for example, of the slowing down of the binary pulsar 1913–16 in its orbital motion around its companion star. General relativity predicts that the rate of energy loss of the orbital motion of a pulsar must match the energy carried away by the gravitational waves produced. Another expected source of gravitational waves is the collapse of a supernova in our galaxy, a process that is thought to occur once every 30 years on the average. Active research is being carried out to improve the methods of detection of gravitational waves.

The general theory of relativity can be considered as well established. Despite its important cosmological implications, the theory has no major noticeable influence on most physical phenomena and will not be discussed further.

Note 19.3 Estimation of general relativistic effects

Making some approximations, the general relativistic corrections can be justified very easily using the Lorentz transformations for lengths and time intervals. Consider an observer O attached to a body of mass m coasting freely in a gravitational field of potential Φ, so that the gravitational effects are washed out relative to O. Then the total energy of the body measured by an observer O' at rest relative to the sources of the gravitational field is zero (recall the case of a body falling freely toward the Earth from infinity). Then $E_k + E_p = 0$. For E_k we use the relativistic expression given by Equation 19.24,

$$E_k = mc^2\left(\frac{1}{(1 - v^2/c^2)^{1/2}} - 1\right)$$

where v is the velocity of m measured by observer O and $E_p = m\Phi$. Therefore

$$mc^2\left(\frac{1}{(1 - v^2/c^2)^{1/2}} - 1\right) + m\Phi = 0$$

from which we obtain

$$\frac{1}{(1 - v^2/c^2)^{1/2}} = 1 - \frac{\Phi}{c^2}$$

Assume that that gravitational field is weak ($\Phi \ll c^2$) so that the acceleration is very small and special relativity can be applied in a first approximation to compare the measurements of O and O'. Let Δx be a distance along the direction of motion and Δt a time interval measured by observer O. Similarly, let $\Delta x'$ and $\Delta t'$ be the corresponding values measured by observer O' who is at rest relative to the masses producing the gravitational field. Then, using

Equation 19.15, we have

$$\Delta x' = (1 - v^2/c^2)^{1/2}\Delta x = \left(1 - \frac{\Phi}{c^2}\right)^{-1}\Delta x \approx \left(1 + \frac{\Phi}{c^2}\right)\Delta x$$

if $\Phi/c^2 \ll 1$. Similarly, using Equation 19.16,

$$\Delta t' = \frac{\Delta t}{(1 - v^2/c^2)^{1/2}} = \left(1 - \frac{\Phi}{c^2}\right)\Delta t$$

For the case of a body moving at the distance r from a body of mass M the gravitational potential is $\Phi = -GM/r$, so that

$$\Delta x' = \left(1 - \frac{GM}{c^2 r}\right)\Delta x, \qquad \Delta t' = \left(1 + \frac{GM}{c^2 r}\right)\Delta t$$

These are precisely Equations 19.31. This calculation should not be construed as an indication that general relativity can be derived from special relativity but rather as an indication of the order of magnitude of gravitational effects. The more elaborate theory developed by Einstein allows us to obtain these same results in a more rigorous way. We shall now illustrate how to calculate the three gravitational effects we have mentioned.

(i) Time delay of a signal bounced back from a planet, due to the gravitational effect of the Sun
The time delay of a light signal reflected back from a planet or an artificial satellite due to the gravitational action of the Sun can be easily calculated. We assume that the ray of light passes the Sun at a distance R. The time required by a light signal to go from the Earth E to a planet P and return (Figure 19.11) with velocity c, is

$$T = 2\int_P^E \mathrm{d}t = 2\int_P^E \frac{\mathrm{d}x}{c}$$

Since the actual velocity is $c' = (1 - 2GM/c^2 r)c$, according to Equation 19.32, where M is the mass of the Sun, the time for the round trip is

$$T' = 2\int_P^E \frac{\mathrm{d}x}{c'}$$

and the gravitational time delay is

$$\Delta T = T' - T = 2\int_P^E \left(\frac{1}{c'} - \frac{1}{c}\right)\mathrm{d}x$$

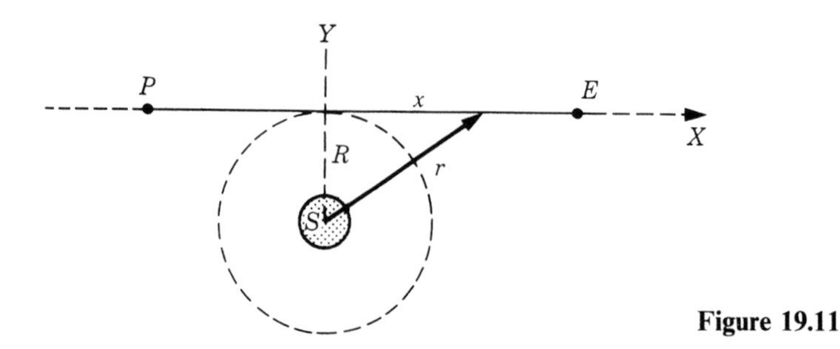

Figure 19.11

But, using Equation 19.32 to calculate c', we can write

$$\frac{1}{c'} = \frac{1}{c(1 - 2GM/c^2 r)} \simeq \frac{1}{c}\left(1 + \frac{2GM}{c^2 r}\right)$$

Therefore

$$\Delta T = \frac{2}{c}\int_P^E \frac{2GM}{c^2 r}\,\mathrm{d}x = \frac{4GM}{c^3} I$$

where, (see Appendix B.6)

$$I = \int_P^E \frac{\mathrm{d}x}{r} = \int_{-x_P}^{x_E} \frac{\mathrm{d}x}{(R^2 + x^2)^{1/2}} = \left[\ln[(R^2 + x^2)^{1/2} + x]\right]_{-x_P}^{x_E}$$

The case of greatest interest is when the ray passes very close to the surface of the Sun, so that R is almost the radius of the Sun. In this case R is very small compared to x_P and x_E, so that we can make the approximations

$$(R^2 + x_E^2)^{1/2} + x_E \simeq 2x_E \qquad \text{and} \qquad (R^2 + x_P^2)^{1/2} - x_P \simeq R^2/2x_P$$

Therefore

$$I \simeq \ln(2x_E) - \ln\left(\frac{R^2}{2x_P}\right) = \ln\frac{4x_P x_E}{R^2}$$

and

$$\Delta T = \frac{4GM}{c^3}\ln\frac{4x_E x_P}{R^2} \qquad \textbf{(19.33)}$$

gives the time delay of the light signal.

This expression has been tested by sending light signals to Mercury and Venus, and to artificial satellites, such as Mariners 6, 7 (both in 1970) and 9 (in 1972) and the Viking Mars orbiters (1976–78). The delay has been verified with a precision of 0.1%, thus supporting the predictions of the general theory of relativity.

(ii) Calculation of the deviation of light by a gravitational field

In Section 19.10 we indicated that one of the predictions of the general theory of relativity, verified by astronomical observation, is the bending of a ray of light as it moves through a gravitational field. Although the correct way to calculate this deviation is by solving the fundamental equations of general relativity, it is easy to obtain the same result (to a first order of approximation) by assuming that the dependence of the velocity of light with the position in a gravitational field results in a continuous refraction, as when light propagates in a non-homogeneous medium (see Chapter 34).

Consider two rays, 1 and 2, separated by a distance $\mathrm{d}y$ (Figure 19.12), passing close to a body of mass M, such as the Sun. The velocity of light is affected by how close it passes to M so that for ray 1 it is c' and for

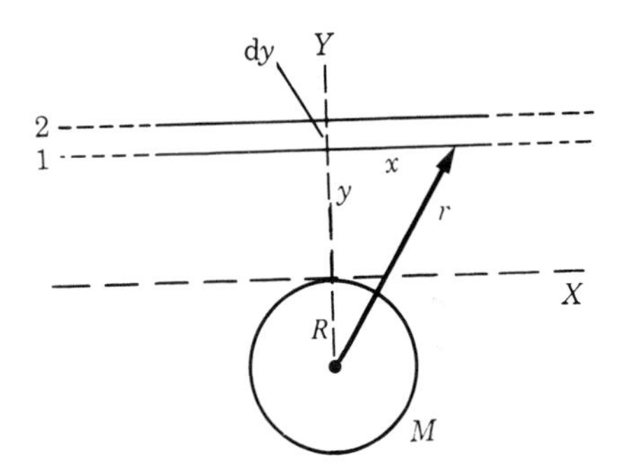

Figure 19.12

ray 2 it is $c' + (\mathrm{d}c'/\mathrm{d}y)\,\mathrm{d}y$. In a time $\mathrm{d}t$, light along ray 1 advances a distance $\mathrm{d}x = c'\,\mathrm{d}t$ while along ray 2, the light advances $\mathrm{d}x' = [c' + (\mathrm{d}c'/\mathrm{d}y)\,\mathrm{d}y]\,\mathrm{d}t$. Thus the wave front at ray 2 is ahead of ray 1 by the distance

$$\mathrm{d}x' - \mathrm{d}x = \left(\frac{\mathrm{d}c'}{\mathrm{d}y}\right)\mathrm{d}y\,\mathrm{d}t$$

Since this distance corresponds to two points of the wave front, separated a distance $\mathrm{d}y$, the result is a bending of the wave front by an angle given by

$$\mathrm{d}\alpha = \frac{\mathrm{d}x' - \mathrm{d}x}{\mathrm{d}y} = \left(\frac{\mathrm{d}c'}{\mathrm{d}y}\right)\mathrm{d}t = \frac{1}{c'}\left(\frac{\mathrm{d}c'}{\mathrm{d}y}\right)\mathrm{d}x$$

The total deviation or bending of the ray is then

$$\Delta\alpha = \int_{-\infty}^{\infty} \frac{1}{c'}\left(\frac{\mathrm{d}c'}{\mathrm{d}y}\right)\mathrm{d}x \simeq \frac{1}{c}\int_{-\infty}^{\infty}\left(\frac{\mathrm{d}c'}{\mathrm{d}y}\right)\mathrm{d}x$$

Using Equation 19.32, we have that

$$c' = c\left(1 - \frac{2GM}{c^2 r}\right) \tag{19.34}$$

whose derivative with respect to y gives

$$\frac{\mathrm{d}c'}{\mathrm{d}y} = \frac{2GM}{cr^2}\frac{\mathrm{d}r}{\mathrm{d}y}$$

From Figure 19.12 we see that $r^2 = (R + y)^2 + x^2$ and so

$$2r\frac{\mathrm{d}r}{\mathrm{d}y} = 2(R + y) \qquad \text{or} \qquad (\mathrm{d}r/\mathrm{d}y) = (R + y)/r$$

For a grazing ray, $y = 0$ and $\mathrm{d}r/\mathrm{d}y = R/r$, which gives

$$\frac{\mathrm{d}c'}{\mathrm{d}y} = \frac{2GMR}{cr^3}$$

Therefore,

$$\Delta\alpha = \frac{2GMR}{c^2}\int_{-\infty}^{\infty}\frac{\mathrm{d}x}{r^3}$$

But (see Appendix B.6)

$$\int_{-\infty}^{\infty}\frac{\mathrm{d}x}{r^3} = \int_{-\infty}^{\infty}\frac{\mathrm{d}x}{(R^2 + x^2)^{3/2}} = \left[\frac{x}{R^2(x^2 + R^2)^{1/2}}\right]_{-\infty}^{\infty} = \frac{2}{R^2}$$

so that

$$\Delta\alpha = \frac{4GM}{c^2 R} \tag{19.35}$$

is the deviation of light for a grazing ray as a result of the gravitational effect produced by the mass M. The deviation of light by a gravitational field gives rise to a **gravitational lens effect**: when a very distant galaxy is behind another galaxy 'nearer' to us, light from the distant galaxy is deviated by the gravitational field of the nearer galaxy, similar to the way light is deviated when it passes through a lens; the result is that one (or more) gravitational images of the distant galaxy are seen.

The paths of light rays are strongly affected near massive compact stars. Recalling Equation 11.24, the Schwarzschild radius of a star of mass M is $R_S = 2GM/c^2$ and Equation 19.34 becomes

$$c' = c\left(1 - \frac{R_S}{r}\right)$$

Suppose the star is shrinking, due to its own gravitational field. When the star contracts to a radius R_S, the velocity of light at its surface ($r = R_S$) becomes a zero. As the star shrinks further, a light ray passing at a distance from the star smaller than R_S is bent strongly toward the star and, in fact, is captured. Conversely, any light ray emitted by the star cannot go beyond a spherical surface of radius R_S and is eventually bent back toward the center of the star. The spherical surface of radius R_S is called the **event horizon** because an observer external to the surface cannot receive any signal produced inside the surface. By extension, no body inside the event horizon can escape from the star. Therefore, a star that shrinks to a radius smaller than R_S becomes invisible and is called a **black hole**.

(iii) Change in frequency of electromagnetic radiation in a gravitational field

Recall that the frequency of oscillatory phenomena is measured by the number of oscillations per unit time. So if there are N oscillations in the time interval Δt, the frequency is $\nu = N/\Delta t$. Suppose ν is the frequency of an electromagnetic wave emitted by a source at a place where the gravitational potential is zero. If the wave is instead produced by a source at a place where the gravitational potential is Φ, the frequency is $\nu' = N/\Delta t'$, where $\Delta t' = \Delta t(1 - \Phi/c^2)$. Therefore

$$\nu' = \frac{N}{\Delta t'} = \frac{N}{\Delta t(1 - \Phi/c^2)} \approx \nu\left(1 + \frac{\Phi}{c^2}\right) \tag{19.36}$$

where we have assumed that $|\Phi| \ll c^2$. Then the larger (smaller) the gravitational potential the greater (smaller) the measured frequency of the electromagnetic wave. If the wave is emitted at a distance r from the Sun (or a star of mass M), the gravitational potential is $\Phi = -GM/r$ and therefore

$$\nu' = \nu\left(1 - \frac{GM}{c^2 r}\right)$$

That means that electromagnetic radiation emitted near a star appears shifted to lower frequencies. This effect is called the **gravitational red shift**.

From Equation 19.36 we conclude that if electromagnetic radiation of frequency ν_1 is emitted at a place where the gravitational potential is Φ_1, the frequency measured at a place where the potential is Φ_2 is

$$\nu_2 = \nu_1\left(1 + \frac{\Delta\Phi}{c^2}\right)$$

where $\Delta\Phi = \Phi_2 - \Phi_1$ is the difference in gravitational potential. For example, a light

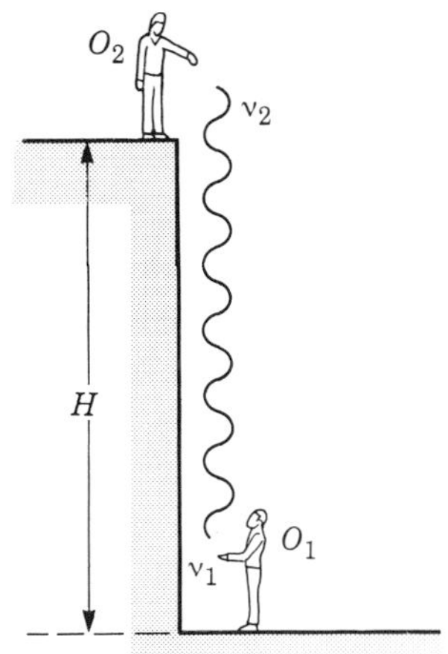

Figure 19.13

signal emitted at the ground level of a very tall building of height H has a frequency at the top of the building given by

$$\nu_2 = \nu_1\left(1 + \frac{gH}{c^2}\right)$$

since $\Delta\Phi = gH$ in this case (Figure 19.13).

QUESTIONS

19.1 What did the experiment of Michelson and Morley prove?

19.2 What physical quantity appears identical (invariant) to observers in relative motion, according to the Lorentz transformations?

19.3 State the similarities and differences between the classical and Einstein's special principles of relativity.

19.4 Explain why it is often, but incorrectly, said that the 'mass of a particle changes with velocity' according to the relation mass $= m/(1 - v^2/c^2)^{1/2}$.

19.5 Are relations $\boldsymbol{F} = \mathrm{d}\boldsymbol{p}/\mathrm{d}t$ and $\boldsymbol{F} = m\boldsymbol{a}$ valid in both the classical and Einstein's principles of relativity?

19.6 Is the acceleration of a very fast particle in the same direction as the applied force?

19.7 According to Equation 19.22, the force required to produce a given acceleration is larger if the force is parallel to the velocity than if it is perpendicular. Can you explain why this is the case?

19.8 What is the meaning of 'rest energy'? Which particle has a larger rest energy: an electron or a proton?

19.9 Make a plot of $E = mc^2/(1 - v^2/c^2)^{1/2}$ and verify that when v is very close to c a small change in v produces a relatively large change in E. Discuss the meaning of this result when we try to increase the energy of fast-moving particles in a particle accelerator.

19.10 Referring to Question 19.9, show that

$$\frac{\mathrm{d}E}{E} = \frac{v^2/c^2}{1 - v^2/c^2}\,\frac{\mathrm{d}v}{v}$$

Confirm the result obtained graphically in Question 19.9.

19.11 What is the difference between the special and the general principles of relativity?

19.12 Describe the relation between gravitation and the structure of space–time.

19.13 A clock is carried on an airplane traveling at an altitude of 10 000 m. Will it run faster than, slower than, or the same as an identical clock at rest at ground level?

PROBLEMS

19.1 A rocket ship heading toward the Moon passes the Earth with a relative velocity of $0.8c$. (a) How long does the trip from the Earth to the Moon take, according to an observer on the Earth? (b) What is the Earth–Moon distance, according to a passenger on the rocket? (c) How long does the trip take, according to the passenger?

19.2 Observers O and O' are in relative translational motion with $v = 0.6c$. They coincide at $t = t' = 0$. (a) When five years have passed, according to O, how long does O' calculate it will take a light signal to get from O to O'? (b) When the light signal arrives at O', how much time has elapsed according to O' since O and O' coincided? (c) A light placed at O is turned on for one year. How long is the light on, according to O'?

19.3 Answer the previous problem when the relative translational velocity is $0.9c$.

19.4 The average lifetime of a neutron, as a free particle at rest, is 15 min. It disintegrates spontaneously into an electron, a proton and a neutrino. What is the average minimum velocity with which a neutron must leave the Sun in order to reach the Earth before breaking up?

19.5 A muon, whose half-life is 2×10^{-6} s as measured by an observer at rest relative to the muon, is moving with a velocity of $0.9c$. If a large burst of such muons is produced at a certain point in the atmosphere, but only 1% reach the Earth's surface, estimate the height of the point where the burst originated.

19.6 A radioactive nucleus is moving with a velocity of $0.1c$ relative to the laboratory when the nucleus emits an electron with a velocity of $0.8c$ relative to the nucleus. What are the velocity and direction of the electron relative to the laboratory if, relative to the frame of reference attached to the decaying nucleus, the electron is emitted (a) in the direction of motion, (b) in the opposite direction, (c) in the perpendicular direction?

19.7 A rocket, whose proper length is 60 m, is moving directly away from the Earth. The ship is fitted with mirrors at each end. A light signal, sent from the Earth, is reflected back from the two mirrors. The first light signal is received after 200 s and the second 1.74 μs later. Find (a) the distance of the rocket from the Earth at the time of the measurement and (b) its velocity relative to the Earth.

19.8 An astronaut wishes to go to a star five light years away. (a) Calculate the velocity of the ship relative to the Earth so that the time, as measured by the astronaut's own clock, is one year. (b) What will be the time for this mission as recorded by a terrestrial observer?

19.9 An astronaut is given a task to be completed in one hour, as measured by a terrestrial clock. The astronaut moves at a velocity of $0.97c$ relative to the Earth and sends back a light signal when his clock reads one hour. How much time did the Earth observer record?

19.10 Observers O and O' are in relative translational motion with $v = 0.6c$. (a) Observer O sees that a stick, aligned parallel to the motion, is at rest relative to him and measures 2.0 m. How long is the stick according to O'? (b) If the same stick is at rest in O' and is aligned parallel to the motion, how long is the stick according to O and O'?

19.11 A meter stick is held at an angle of 45° with respect to the direction of motion of a coordinate system moving with a velocity of $0.8c$ relative to a laboratory system. What is the length of the stick and its orientation?

19.12 A particle with mass m and a velocity of $0.8c$ is subject to a force that is (a) parallel to the velocity, (b) perpendicular to the velocity. Determine the ratio of the force to the acceleration in each case.

19.13 Show that when v is almost equal to c, then $\gamma \approx 1/[2(1 - v/c)]^{1/2}$.

19.14 Show that when v is very small compared to c, then $\gamma \approx 1 + v^2/2c^2$.

19.15 Show that the relativistic force in rectilinear motion is given by $F = \gamma^3 m_0 a$ where $a = dv/dt$.

19.16 Verify that the momentum of a particle under the action of a *constant* force $\boldsymbol{F}$ is given by $p = (p_0^2 + F^2t^2)^{1/2}$ where $\boldsymbol{p}_0$ is the momentum of the particle at time $t = 0$. The particle is assumed to be moving initially in the X-direction, and the constant force is in the Y-direction.

19.17 Using the result of Problem 19.16, show that the total energy of a particle under the action of the constant force is $E = (E_0^2 + c^2F^2t^2)^{1/2}$ where $E_0 = c(m^2c^2 + p_0^2)^{1/2}$.

19.18 An electron is moving at a velocity $0.6c$ relative to an observer O. A force of 9.109×10^{-19} N, as measured in the frame of reference attached to the electron, is applied parallel to the relative velocity. Find the electron's acceleration relative to both frames of reference.

19.19 Solve Problem 19.18 for a case in which the force is applied perpendicular to the relative velocity.

19.20 (a) At what velocity is the momentum of a particle of mass m equal to mc? (b) What is the total energy and the kinetic energy in this case?

19.21 (a) Show that $v/c = [1 - (mc^2/E)^2]^{1/2}$. From this relation find the velocity of a particle when E is (b) equal to its rest energy, (c) twice its rest energy, (d) 10 times its rest energy and (e) one thousand times the rest energy. (f) Compute the corresponding energies in eV for an electron and a proton. Make a plot of v/c against E/mc^2.

19.22 Find (a) the exit velocity and (b) the momentum of a proton in an accelerator, given that the proton kinetic energy is 3×10^{10} eV.

19.23 The radius of the proton path in the accelerator of the preceding problem is 114 m.

Find the centripetal force required to hold the proton in orbit when the proton has reached its final kinetic energy.

19.24 Estimate the value of the corrective term $(3/8)mv^4/c^2$ relative to the term $\frac{1}{2}mv^2$ for (a) an electron in a hydrogen atom whose velocity is 2.2×10^6 m s^{-1}, (b) a proton coming from a cyclotron with a kinetic energy of 30 MeV, and (c) protons coming from an accelerator with a kinetic energy of 3×10^{10} eV.

19.25 Calculate, in eV, the energy required to accelerate an electron and a proton from (a) rest up to $0.500c$, (b) $0.500c$ to $0.900c$, (c) $0.900c$ to $0.950c$, (d) $0.950c$ to $0.990c$. What general conclusion do you reach?

19.26 Electrons are accelerated up to a kinetic energy of 10^9 eV. (a) Find (i) the ratio of their energy to the rest energy, (ii) the ratio of their velocity to the velocity of light, (iii) the ratio of their total energy to their rest energy. (b) Repeat the same problem for protons of the same kinetic energy.

19.27 Since energy/velocity has the same dimensions as momentum, the unit MeV/c has been introduced as a convenient unit for measuring the momentum of elementary particles. (a) Find, in terms of this unit, the momentum of an electron having a total energy of 5.0 MeV. (b) Repeat for a proton having a total energy of 2×10^3 MeV.

19.28 (a) Determine the total energy and the velocity of an electron having a momentum of 0.60 MeV/c. (b) Repeat for a proton.

19.29 *Discussion of simultaneity.* (a) Prove that if two events occur relative to observer O at times t_1 and t_2 and at places x_1 and x_2, and if $T = t_2 - t_1$, $L = x_2 - x_1$, the events appear to observer O' (moving relative to O with velocity v along the X-axis) at times t'_1 and t'_2 such that, if $T' = t'_2 = t'_1$, then $T' = \gamma(T - vL/c^2)$. (b) In general, are events that appear as simultaneous to O also simultaneous to O'? Under what conditions are events that appear simultaneous to O also simultaneous to all other observers in uniform relative motion? (c) Obtain the relation between L and T such that the order in which two events are observed by O' is reversed for O. (d) Suppose that events (x_1, t_1) and (x_2, t_2) observed by O are related causally. (That is, (x_2, t_2) is the result of some signal transmitted from (x_1, t_1) with velocity $V = L/T$, by necessity smaller than or equal to c.) Can the order of the events appear reversed to O'?

19.30 Prove that the general transformation for the velocity of a particle as measured by O and O' when the particle has a velocity V relative to O is

$$V'_x = \frac{V_x - v}{1 - vV_x/c^2}$$

$$V'_y = \frac{V_y(1 - v^2/c^2)^{1/2}}{1 - vV_x/c^2}$$

$$V'_z = \frac{V_z(1 - v^2/c^2)^{1/2}}{1 - vV_x/c^2}$$

Show that it reduces to Equation 19.9 when the velocity is parallel to the X-axis.

19.31 Prove that the general transformation for the acceleration of a particle as measured by O and O' when the particle moves with velocity V relative to O is

$$a'_x = \frac{a_x(1 - v^2/c^2)^{3/2}}{(1 - vV_x/c^2)^3},$$

$$a'_y = \frac{1 - v^2/c^2}{(1 - vV_x/c^2)^2}\left(a_y + a_x\frac{vV_y/c^2}{1 - vV_x/c^2}\right)$$

$$a'_z = \frac{1 - v^2/c^2}{(1 - vV_x/c^2)^2}\left(a_z + a_x\frac{vV_z/c^2}{1 - vV_x/c^2}\right)$$

Show that it reduces to Equation 19.11 when the particle moves parallel to the X-axis.

19.32 The general Lorentz transformation when the coordinate axes used by O and O' are not parallel to the relative velocity is

$$\boldsymbol{r}' = \boldsymbol{r} + (\gamma - 1)\frac{(\boldsymbol{r}\cdot\boldsymbol{v})\boldsymbol{v}}{v^2} - \gamma\boldsymbol{v}t$$

$$t' = \gamma(t - \boldsymbol{r}\cdot\boldsymbol{v}/c^2)$$

where $\boldsymbol{v}$ is the velocity of O' relative to O. Show that it reduces to Equation 19.2 when $\boldsymbol{v}$ is parallel to the X-axis.

19.33 Verify that the relative increase in momentum and in velocity of a particle are related by

$$\frac{\mathrm{d}p}{p} = \frac{1}{1 - v^2/c^2}\left(\frac{\mathrm{d}v}{v}\right)$$

How does $\mathrm{d}p/p$ vary as v approaches c?

19.34 Referring to Problem 19.33, find the relative change in momentum for $v/c = 0.50$, 0.80, 0.90 and 0.95 if the relative change in velocity is 1%. What do you conclude?

19.35 A particle of mass m at rest relative to the observer is subject to a constant force F. (a) Show that after a time t the velocity of the particle is

$$v = c\frac{(F/mc)t}{[1 + (F/mc)^2t^2]^{1/2}}$$

and that the displacement is given by

$$x = \frac{mc^2}{F}\left\{\left[1 + \left(\frac{F}{mc}\right)^2 t^2\right]^{1/2} - 1\right\}$$

(b) Find the limiting values of v and x for $Ft \ll mc$ and $Ft \gg mc$. (c) Represent x and v as a function of time. (d) Obtain the energy of the particle as a function of time.

19.36 When a ray of light passes at a distance R from the center of a mass M, it suffers a deviation given by Equation 19.35 according to the general theory of relativity. Calculate the deviation of a light ray from a star when it passes close to the surface of the Sun. How do you think this measurement can be carried out?

19.37 According to the general theory of relativity the time delay of a signal from the Earth to a planet and back is given by Equation 19.33. Calculate the maximum value of ΔT when a signal is sent from the Earth to Mercury and to Venus for a ray that passes within one, two and three Sun radii from the Sun. Use the data of Table 11.1.

19.38 General relativity predicts a precession of the perihelion of the orbit of a planet for each turn around the Sun given by $\Delta\theta = 6\pi GM/[c^2a(1 - \varepsilon^2)]$, where M is the mass of the Sun, a is the semi-major axis and ε the eccentricity of the planet's orbit. Using the data in Table 11.1, calculate $\Delta\theta$ for Mercury after 100 years.

19.39 Estimate the value of the general relativistic correction term, Equation 19.30, for a particle on the surface of (a) the Sun, (b) the Earth. (c) Repeat for an artificial satellite at a height of 30 000 km above the Earth's surface. What do you conclude?

19.40 It can be shown that if ν is the frequency of an electromagnetic wave in the absence of any gravitational field, the frequency when the wave propagates through a gravitational potential difference $\Delta\Phi$ is $\nu' = \nu(1 + \Delta\Phi/c^2)$. (See Note 19.3(iii).) Calculate the change in frequency $\Delta\nu = \nu' - \nu$ when a ray of blue light ($\nu = 6.5 \times 10^{14}$ Hz) emitted by a light source on the street reaches the top of the Empire State Building in New York ($h = 381$ m).

20 High energy processes

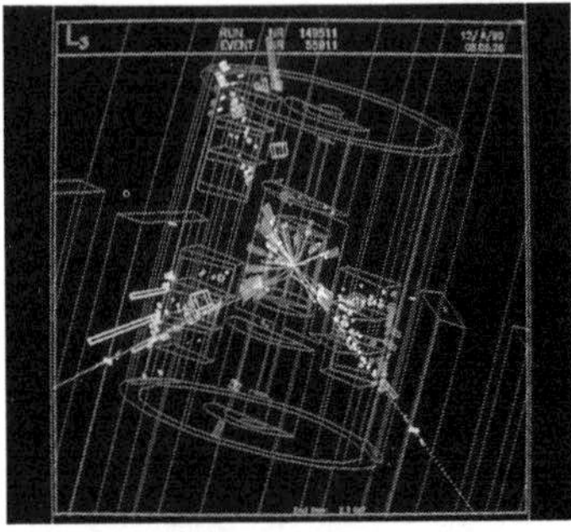

Computer generated representation of the trajectories of particles created when electrons and positrons collide in the L3 detector at CERN's LEP collider, at an energy of 10^{11} eV. At each collision, electronic sensors transmit about 500 000 electric signals that can be translated into information about the energy of a particle and its position, within 5×10^{-5} m. The computer system analyzes the event, discarding irrelevant information. In this way it has been possible to study elusive particles such as the Z^0. (Photo courtesy of CERN.)

Notes

20.1 Introduction

For most of our daily life experiences we do not need to use the results of Einstein's theory of relativity developed in the previous chapter. The velocities of bodies on Earth, even the fastest ones, such as rockets and missiles, are very small compared with the velocity of light. Also, their kinetic energies are usually very small compared with their rest energies mc^2. However, with the advent of high energy accelerators (see Section 20.4), it has been possible to accelerate particles to energies comparable to those found in cosmic rays and in processes occurring in stars and in other systems in the universe. These high energy processes require Einstein's theory of relativity for their analysis. This is, perhaps, its most important application.

We will not discuss the interactions among the fundamental particles and the processes that take place between them until Chapter 41. However, in this chapter we illustrate the procedures for analyzing such processes, in terms of energy and momentum, using the relativistic results of Chapter 19.

20.2 Energy and momentum

In order to analyze many high energy processes involving fundamental particles, it is not necessary to know the interactions that take place in any great detail. Rather, it is sufficient to apply the laws of conservation of energy and momentum to relate these quantities before and after the interaction. In Chapter 19 the momentum of a particle of mass m and velocity $\mathbf{v}$, relative to an observer, was given by the expression

$$\mathbf{p} = \frac{m\mathbf{v}}{(1 - v^2/c^2)^{1/2}} \tag{20.1}$$

Similarly, the energy was found to be

$$E = \frac{mc^2}{(1 - v^2/c^2)^{1/2}} \tag{20.2}$$

which includes the rest energy $E_0 = mc^2$ of the particle. Combining these two equations we see that

$$\mathbf{v} = \frac{c^2}{E}\mathbf{p} \tag{20.3}$$

Equation 20.3 gives the velocity of a particle in terms of its momentum and energy. Inserting Equation 20.3 in Equation 20.2 allows us to write

$$E^2 = (mc^2)^2 + (cp)^2 \qquad \text{or} \qquad E = c(m^2c^2 + p^2)^{1/2} \tag{20.4}$$

This is a very important relation that couples energy and momentum according to the theory of relativity. It is an easy relation to remember by referring to Figure 20.1.

If we write Equation 20.4 in the form $E = mc^2[1 + (p/mc)^2]^{1/2}$ and apply the binomial theorem, we can rewrite the equation as

$$E = mc^2\left[1 + \frac{1}{2}\left(\frac{p}{mc}\right)^2 - \frac{1}{8}\left(\frac{p}{mc}\right)^4 + \cdots\right] \tag{20.5}$$

Equation 20.5 reduces to the Newtonian expression for total energy $E = mc^2 + p^2/2m$ when $p \ll mc$.

At very large velocities, we may replace v by c in Equation 20.3, or neglect mc^2 in comparison with cp in Equation 20.4, so that

$$E \approx cp \tag{20.6}$$

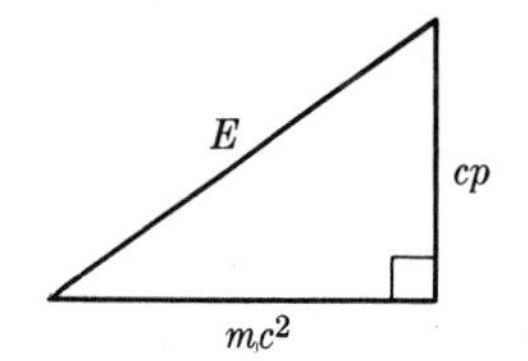

Figure 20.1
Graphical relation between energy, momentum and mass.

This relation is valid only at energies very high compared with the rest energy mc^2 of the particle. However, when the particle has zero mass ($m = 0$), Equation 20.4 reduces to $E = cp$ at all energies. From Equation 20.3, we then see that the velocity of a zero mass particle is $v = c$. Therefore, a particle with zero mass can move only with the velocity of light and can never be at rest in an inertial system. This is the case of the photon, which is the particle associated with electromagnetic radiation, and it also seems to be true for the neutrino.

In order to be consistent with the principle of relativity, the expression relating energy and momentum must remain invariant for all inertial observers. That is, energy and momentum must transform using the Lorentz transformation. To verify this statement, the energy–momentum relation 20.4 can be written in the alternative form

$$E^2 - (cp)^2 = (mc^2)^2 \tag{20.7}$$

Then an observer in another frame of reference, moving with constant velocity relative to the original frame, must be able to write

$$E'^2 - (cp')^2 = (mc^2)^2$$

The value of m remains the same because it corresponds to the mass of the same particle which is the same for all observers. Therefore the following relation holds for all observers in uniform relative motion,

$$E^2 - (cp)^2 = E'^2 - (cp')^2 \tag{20.8}$$

We may say that the expression $E^2 - (cp)^2$ is a **Lorentz invariant**. One consequence is that the quantities $c\boldsymbol{p}$ and E transform in the same way as $\boldsymbol{r}$ and ct under a Lorentz transformation (see Note 20.1). Another consequence is that the laws of conservation of momentum and energy are not independent. Rather, we must talk of the law of conservation of energy–momentum.

Note 20.1 Lorentz transformation of energy and momentum

Equation 20.4, which relates energy and momentum, must be the same for all inertial observers. It is therefore important to compare these quantities as measured by two observers in motion with relative velocity $\boldsymbol{v}$. We shall assume that both observers set their coordinate axes parallel with their X and X' axes parallel to the relative velocity (recall Figure 19.3). For observer O observing a particle of mass m and momentum p, Equation 20.4 may be written as

$$p^2 - \frac{E^2}{c^2} = -m^2c^2 \tag{20.9}$$

Recall that $\boldsymbol{p}$ is a vector with components p_x, p_y, p_z. Then $p^2 = p_x^2 + p_y^2 + p_z^2$ and the above equation becomes

$$p_x^2 + p_y^2 + p_z^2 - \frac{E^2}{c^2} = -m^2c^2 \tag{20.10}$$

In order to be consistent with the assumption that this expression must remain invariant for all inertial observers, when O' observes the same particle, O' must write Equation 20.4 as

$$p_{x'}'^2 + p_{y'}'^2 + p_{z'}'^2 - \frac{E'^2}{c^2} = -m^2c^2 \tag{20.11}$$

where m remains the same because it corresponds to the mass of the same particle. The structure of Equations 20.10 and 20.11 is similar to that of Equations 19.5 and 19.6, that is $x^2 + y^2 + z^2 - c^2t^2 = 0$ and the same for the primed coordinates. Therefore we may make the correspondence $p_x \to x$, $p_y \to y$, $p_z \to z$ and $ct \to E/c$ or $t \to E/c^2$. Therefore, the invariance of Equation 20.9 requires a transformation of p_x, p_y, p_z and E/c^2 similar to the Lorentz transformation that holds for x, y, z and t. This results in

$$\begin{aligned} p_{x'}' &= \frac{p_x - vE/c^2}{(1 - v^2/c^2)^{1/2}} \\ p_{y'}' &= p_y \\ p_{z'}' &= p_z \\ E' &= \frac{E - vp_x}{(1 - v^2/c^2)^{1/2}} \end{aligned} \tag{20.12}$$

In the fourth equation a common c^2 factor has been cancelled out. This result shows that both momentum and energy transform under a Lorentz transformation in the same way the position vector and time transform.

Note that we have found two sets of associated quantities (x, y, z, t and $p_x, p_y, p_z, E/c$) that appear to transform among themselves following the rules of the Lorentz transformation. Undoubtedly we may also expect other physical quantities to transform in a similar fashion. A common characteristic of these sets of quantities is that they have four 'components'; i.e. they are expressed by four numbers. The sets of quantities are therefore called **four-vectors** in a four-dimensional representative space. One method of adapting physical laws to the invariance requirements of the principle of relativity is by writing the laws as relations between scalars, four-vectors and other related quantities (tensors) in four-dimensional space. We shall not elaborate on this subject since it belongs with a more extensive discussion of the theory of relativity.

To express the quantities $p_x, p_y, p_z, E/c^2$ measured by O in terms of the corresponding quantities measured by O', we simply reverse the sign of v and exchange primed and unprimed quantities in Equation 20.12. This gives

$$\begin{aligned} p_x &= \frac{p_x' + vE'/c^2}{(1 - v^2/c^2)^{1/2}} \\ p_y &= p_{y'}' \\ p_z &= p_{z'}' \\ E &= \frac{E' + vp_{x'}'}{(1 - v^2/c^2)^{1/2}} \end{aligned} \tag{20.13}$$

Equations 20.13 are the inverse Lorentz transformation for energy and momentum. When the particle is at rest relative to O', and thus moves with velocity v relative to O, we have $p_{x'}' = p_{y'}' = p_{z'}' = 0$ and $E' = mc^2$. Using Equations 20.13, we get

$$p_x = \frac{mv}{(1 - v^2/c^2)^{1/2}}, \qquad p_y = 0, \qquad p_z = 0, \qquad E = \frac{mc^2}{(1 - v^2/c^2)^{1/2}} \tag{20.14}$$

The first three equations give the momentum and the last equation gives the energy of the particle as measured by O. Comparison with Equation 20.1 for the momentum and Equation 20.2 for the energy shows that they are precisely the momentum and energy of a particle moving along the X-axis with velocity v that we had obtained earlier in Chapter 19. This was to be expected since the particle, at rest relative to O', appears to be moving with velocity v relative to O. This calculation shows the consistency of the relativistic expressions for momentum and energy with the principle of relativity.

20.3 Systems of particles

Consider a system of particles of masses $m_1, m_2, \ldots$. Suppose that, relative to an inertial observer O, the momenta of the particles are $\boldsymbol{p}_1, \boldsymbol{p}_2, \ldots$, and the energies are $E_1, E_2, \ldots$. The total momentum and the total energy of the system relative to O are

$$\boldsymbol{P} = \sum_i \boldsymbol{p}_i \qquad \text{and} \qquad E = \sum_i E_i \tag{20.15}$$

Relative to another observer O', who is in motion relative to O, the total momentum and energy of the system are $\boldsymbol{P}'$ and E'. Although Equation 20.8 was derived for a single particle it applies also to a system of particles. Therefore, the momentum and energy of the system relative to observers O and O' are related by the expression

$$E^2 - (cP)^2 = E'^2 - (cP')^2 \tag{20.16}$$

In Chapter 13 we introduced the L- and C-frames of reference. In the C-frame the total momentum of the system is zero. Making $P' = 0$ in Equation 20.16, we have that the total energy E' in the C-frame is given by

$$\underbrace{E^2 - (cP)^2}_{\text{L-frame}} = \underbrace{E'^2}_{\text{C-frame}} \tag{20.17}$$

From Equation 20.17 we see that the energy of the system has its smallest value in the C-frame. We may interpret these results by saying that E' is the **internal energy** of the system, and henceforth we will write E_{int}. Therefore, $E = [E_{\text{int}}^2 + (cP)^2]^{1/2}$, that is, the total energy of the system is a combination of the internal energy and the translational energy of the system relative to the observer in the L-frame.

If all the particles are at rest in the C-frame and not interacting, the internal energy is

$$E_{\text{int}} = \sum_i m_i c^2 \tag{20.18}$$

However, if the particles are in motion relative to the C-frame and interacting, we must write instead, in accordance with Equation 14.8,

$$E_{\text{int}} = \sum_i m_i c^2 + (E_k + E_p)_{\text{int}}$$

Since the **binding energy** of the system was defined in Equation 14.21 as

$$E_b = -(E_k + E_p)_{\text{int}}$$

we may write

$$E_{\text{int}} = \sum_i m_i c^2 - E_b \tag{20.19}$$

But, according to the relation $E = mc^2$, we may associate a mass with any amount of energy, and conversely. Therefore, the mass M of the system is related to the internal energy by $E_{\text{int}} = Mc^2$. Equation 20.19 then becomes

$$Mc^2 = \sum_i m_i c^2 - E_b \qquad \text{or} \qquad M = \sum_i m_i - E_b/c^2 \tag{20.20}$$

This relation shows that the mass M of the system is different from the sum $\Sigma_i m_i$ of the rest masses of the particles by the amount E_b/c^2. If E_b is positive, as it is for all stable systems, then M is smaller than $\Sigma_i m_i$ and we may say, figuratively, that mass is 'lost' in assembling the system (see Figure 20.2).

Equation 20.20 may also be written in the form

$$E_b = \left(\sum_i m_i - M\right)c^2 \tag{20.21}$$

If we define $\Delta m = M - \Sigma_i m_i$ as the change in mass when the system is formed, we may write Equation 20.21 as

$$E_b = -(\Delta m)c^2$$

Recalling Equation 19.28, this may be written as

$$E_b = -931.531\Delta m \text{ MeV}$$

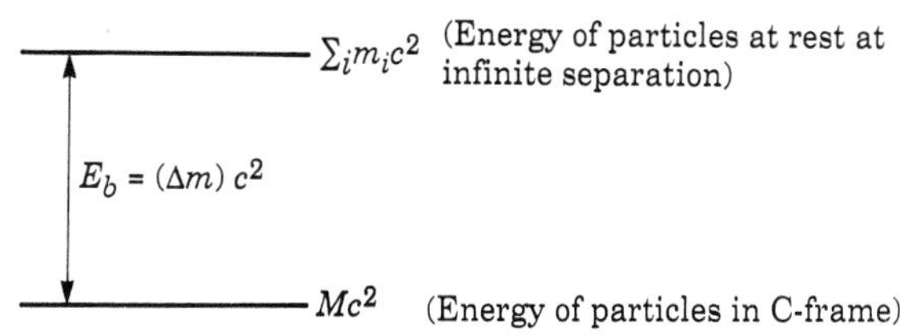

Figure 20.2 Binding energy of a system of particles.

where E_b is expressed in MeV and Δm in amu. Thus by measuring the change in mass of a system when it is formed, the binding energy of the system may be obtained. Conversely, a measurement of the binding energy of a system gives us the **mass defect** of a composite system.

There is ample evidence of a change in mass when a group of particles is assembled in a stable system. For example, in most nuclei the binding energy is about 6–8 MeV per nucleon (Section 39.4). This corresponds to a change in mass of about 7×10^{-3} amu per nucleon. Hence the mass of a nucleus is appreciably less than the sum of the masses of its protons and neutrons, a result that has been verified experimentally.

In the case of atoms and molecules, the binding energy is a few eV, which corresponds to a change in mass of the order of 10^{-9} amu. This is a very small change in mass and cannot be detected using present techniques. For that reason, chemists have always assumed that mass does not change in chemical reactions, a valid assumption from the practical point of view.

EXAMPLE 20.1

Binding energy of the deuteron.

▷ The deuteron is a nucleus composed of a proton and a neutron. The measured mass of the deuteron is $m_d = 2.014\,102$ amu. However, the sum of the masses of the proton and the neutron is

$$m_p\ (1.007\,825\ \text{amu}) + m_n\ (1.008\,665\ \text{amu}) = 2.016\,490\ \text{amu}$$

Therefore, in forming the deuteron there is a mass change or 'loss' $\Delta m = -0.002\,388$ amu, which corresponds to a binding energy

$$E_b = -931.48\Delta m\ \text{MeV} = 2.224\ \text{MeV}$$

This result is in complete agreement with the value of E_b measured directly by computing the energy released as γ-rays when a proton captures a neutron. If we want to separate the proton and the neutron it is necessary to supply the energy $E_b = 2.224$ MeV. This can be done when the deuteron absorbs a γ-ray or in a high energy inelastic collision.

EXAMPLE 20.2

Binding energy of a particle in uranium.

▷ Uranium has a number of different isotopes. Two are particularly important, ^{235}U and ^{238}U, in connection with the process called **fission** (Section 40.6). However, we shall consider the rarer isotope ^{232}U whose mass is $M(^{232}U) = 232.0372$ amu. We wish to determine whether spontaneous emission of particles such as neutrons, protons, deuterons, tritons and α-particles from such a nucleus is energetically possible. Call the emitted particle X and use Y for the resultant or **daughter nucleus**. The binding energy of ^{232}U relative to the system $X + Y$ is, according to Equation 20.21,

$$E_b = [m_X + m_Y - M(^{232}\text{U})]c^2 = 931.531[m_X + m_Y - M(^{232}\text{U})]\ \text{MeV}$$

The values of m_X, m_Y, and E_b foı the processes mentioned are given in Table 20.1.

Table 20.1 Binding energies of particles in ^{232}U

Particle	m_X (amu)	Daughter	m_Y (amu)	E_b (MeV)
n	1.0087	^{231}U	231.0363	7.26
p (^1H)	1.0073	^{231}Pa	231.0359	6.05
d (^2H)	2.0136	^{230}Pa	230.0344	10.5
t (^3H)	3.0160	^{229}Pa	229.0320	10.0
α (^4He)	4.0026	^{228}Th	228.0287	−5.50

All binding energies are positive except for the α-particle, which is negative. Hence we conclude that the nucleus ^{232}U can spontaneously emit α-particles but is stable with regard to the emission of all other particles considered in Table 20.1. In fact, it is known experimentally that ^{232}U is an α-particle emitter.

EXAMPLE 20.3

Energy production in stars.

▷ Nuclear reactions are the source of the energy in the Sun and other stars. One of the most important processes is equivalent to the **fusion** of four protons into a helium nucleus, with the emission of two positrons and two neutrinos. That is,

$$4\,{}^1_1H \rightarrow {}^4_2He + 2e^+ + 2\nu$$

where e^+ and ν refer to a positron and a neutrino, respectively. The net change in mass in the process may be easily calculated to be $\Delta m = 0.0283$ amu and the net energy liberated is 26.7 MeV, or about 6.6×10^{14} J per kg of ${}^1_1H$ consumed. (This value should be compared with the energy released when two hydrogen atoms combine to make H_2, which is 4.48 eV or 2.15×10^6 J per kg of hydrogen.) There are a number of ways the process may occur, but we will not discuss them at this time (see Note 40.4).

It is estimated that this process occurs in the Sun at a rate of 5.64×10^{11} kg per second of hydrogen fusing into helium, resulting in a loss of mass of about 4×10^9 kg s^{-1}. This results in a power output of 3.8×10^{26} W or about 1.2×10^{34} J yr^{-1}, corresponding to a loss of mass of 1.33×10^{17} kg yr^{-1}. Only about 1.8×10^{17} W falls on the Earth, mostly in the form of electromagnetic radiation. Still, this is about 10^5 times the total rate of energy generated on Earth.

Considering that the mass of the Sun is 1.98×10^{30} kg, the fraction of its mass lost per year due to nuclear fusion is about 10^{-13}. Therefore, we may expect the Sun to continue shining for millions of years without any appreciable change in mass.

20.4 High energy collisions

In Chapter 14 collisions involving low energy particles were considered. We shall now examine collisions involving fundamental particles, which only happen at high energies. Consequently, the concepts and techniques developed in this chapter must be used.

Consider two particles, whose masses are m_1 and m_2. Before the collision, when their interaction is negligible, they are moving with momenta $\boldsymbol{p}_1$ and $\boldsymbol{p}_2$ relative to a frame of reference L. The interaction between the particles is appreciable only during the small interval when the particles are very close, when it may result in the transformation of the initial particles into different ones. Suppose that after the collision, when the interaction is again negligible, the resulting particles have masses m'_1 and m'_2 and move with momenta $\boldsymbol{p}'_1$ and $\boldsymbol{p}'_2$ relative to the same frame of reference (Figure 20.3). The conservation of momentum and energy require that

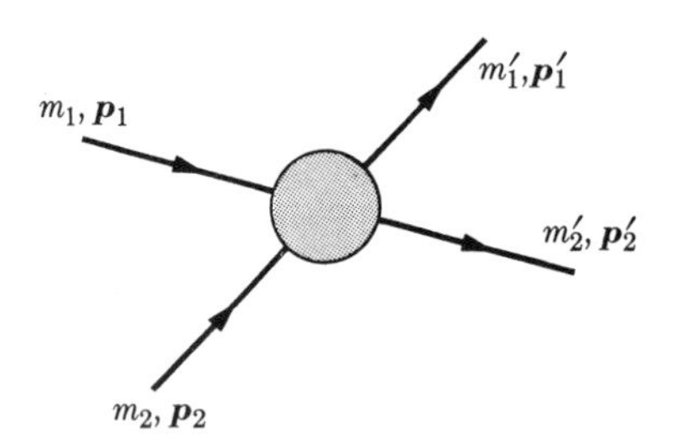

Figure 20.3 Energy and momentum are conserved in a collision.

$$\boldsymbol{p}_1 + \boldsymbol{p}_2 = \boldsymbol{p}'_1 + \boldsymbol{p}'_2$$

and **(20.22)**

$$E_1 + E_2 = E'_1 + E'_2$$

By properly manipulating Equations 20.22 we may find the final momenta of the particles in terms of the initial ones. The algebra, however, is generally rather complicated. Thus we shall consider only some simple cases.

Recalling that $E = E_k + mc^2$, where E_k is the kinetic energy, we may write the energy conservation equation as

$$E_{k1} + m_1c^2 + E_{k2} + m_2c^2 = E'_{k1} + m'_1c^2 + E'_{k2} + m'_2c^2$$

In Section 14.9, Equation 14.24, the quantity Q was defined as the *change in kinetic energy* during the collision,

$$Q = (E'_{k1} + E'_{k2}) - (E_{k1} + E_{k2})$$

Rearranging terms in the previous equation, we have that

$$Q = -[\underbrace{(m'_1 + m'_2)}_{\text{final mass}} - \underbrace{(m_1 + m_2)}_{\text{initial mass}}]c^2 = -(\Delta m)c^2 \qquad \textbf{(20.23)}$$

Therefore, Q is proportional to the change in mass during the collision. When there is *no* change of mass $Q = 0$, and the collision is called **elastic**. When $\Delta m > 0$, there is an *increase* in mass at the expense of a decrease in kinetic energy and $Q < 0$. Similarly, when $\Delta m < 0$, there is a *decrease* in mass and a corresponding increase in kinetic energy and $Q > 0$.

In many cases more than two particles are produced in a collision (Figure 20.4), and the above definition of Q becomes

$$Q = -[\underbrace{(\Sigma m')}_{\text{final mass}} - \underbrace{(m_1 + m_2)}_{\text{initial mass}}]c^2 = -(\Delta m)c^2 \qquad \textbf{(20.24)}$$

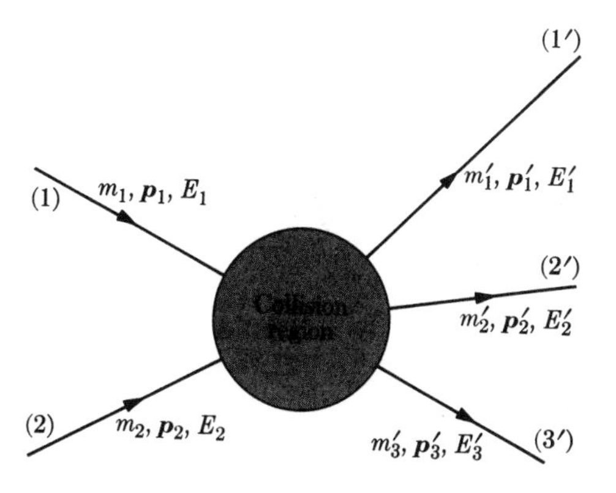

Figure 20.4 Collision where new particles are produced.

When $Q < 0$ or $\Delta m > 0$, in the L-frame there is a minimum or **threshold kinetic energy** of the incoming particles needed to supply the 'extra' mass needed and make the collision occur or 'go'. The threshold kinetic energy in the L-frame corresponds to the case when the resulting particles are all at rest in the C-frame. When particle 2 is initially at rest in the L-frame also, the value of the threshold kinetic energy is (see the calculation below)

$$E_k = -\frac{QM}{2m_2} \qquad \textbf{(20.25)}$$

which clearly implies that Q is negative ($Q < 0$). In Equation 20.25 M is the sum of the masses of all the particles involved before and after the collision; that is,

$$M = m_1 + m_2 + (\Sigma m_i')_{\text{final}}$$

However, if $\Delta m < 0$ so that Q is positive, no threshold kinetic energy exists. Then the rest energy of the initial particles is sufficient to produce the final particles and energy is released in the collision.

One of the chief uses of high-energy accelerators is to produce fast particles above the threshold kinetic energy in the L-frame. With such machines scientists can produce, in the laboratory and under controlled conditions, some processes that have been observed in cosmic rays or that occurred in the early history of the universe.

Calculation of the threshold kinetic energy

Let the target particle be at rest in the laboratory or L-frame. Designate the projectile and the target by P_1 and P_2 and the resulting particles by P_i'; we may write the process as

$$P_1 + P_2 \rightarrow \sum_i P_i'$$

If the momentum of P_1 is $\boldsymbol{p}$, in the L-frame, its total energy is $c(m_1^2c^2 + p^2)^{1/2}$. If P_2 is initially at rest in the L-frame, its momentum is zero and its total energy is m_2c^2. Thus the total initial energy in the L-frame is

$$E = c(m_1^2c^2 + p^2)^{1/2} + m_2c^2$$

and the total momentum is $\boldsymbol{p}$. Applying Equation 20.17 with $\boldsymbol{p}' = 0$, we have $E^2 - c^2p^2 = E'^2$, where E' is the total energy of the system in the C-frame before the collision.

The laws of conservation of energy and momentum require that, after the process has taken place, the total energy of the products in the C-frame still be E' and the total momentum be zero. Clearly, as mentioned above, the minimum energy required for the process corresponds to the situation where all the resulting particles are at rest in the C-frame, so that the total energy in such a frame is $E' = \Sigma_i m_i'c^2$. Substituting the values of E and E' into Equation 20.17 we get

$$m_1^2c^4 + m_2^2c^4 + 2m_2c^3(m_1^2c^2 + p^2)^{1/2} = \left(\sum_i m_i'\right)^2 c^4$$

But if E_k is the kinetic energy of the projectile in the L-frame, then $E = E_k + m_1c^2$. Using Equation 20.4, $c(m_1^2c^2 + p^2)^{1/2} = E_k + m_1c^2$. Making this substitution in the preceding equation and canceling the common c^2 factor, we get

$$(m_1 + m_2)^2c^2 + 2m_2E_k = \left(\sum_i m_i'\right)^2 c^2$$

Rearranging terms, we can write

$$\begin{aligned} 2m_2E_k &= -\left[(m_1 + m_2)^2 - \left(\sum_i m_i'\right)^2\right]c^2 \\ &= -\left(m_1 + m_2 - \sum_i m_i'\right)\left(m_1 + m_2 + \sum_i m_i'\right)c^2 \end{aligned}$$

Making $Q = (m_1 + m_2 - \Sigma m_i')c^2$ and $M = m_1 + m_2 + \Sigma m_i'$ we obtain Equation 20.25 $E_k = -QM/2m_2$

EXAMPLE 20.4

Proton–proton collisions.

▷ When two high energy particles collide (such as two protons or a proton and a pion), new particles may be produced. The nature and number of particles produced depend on the energy and nature of the colliding particles as well as on some other features. For example, electric charge and angular momentum must be conserved in a collision (see Chapter 41).

One common process is the production of pions in proton–proton collisions in particle accelerators when a beam of high energy protons hits a suitable target. For example, the following processes are observed,

$$p^+ + p^+ \rightarrow p^+ + p^+ + \pi^0 \qquad \text{and} \qquad p^+ + p^+ \rightarrow p^+ + n^0 + \pi^+$$

At very high energy several pions may be produced. We shall now calculate the threshold kinetic energy for the first process, assuming the target proton is at rest in the L-frame. For this case we have $m_2 = m_p$, $Q = -m_\pi c^2$ and $M = 4m_p + m_\pi$. Since the mass of a pion is 0.149 amu, $m_\pi c^2 = 140$ MeV and the threshold kinetic energy of the incoming proton must be

$$E_k = \frac{m_\pi c^2(4m_p + m_\pi)}{2m_p} = (2 + m_\pi/2m_p)m_\pi c^2 \approx 280 \text{ MeV}$$

Another interesting process is the production of an **antiproton** in a proton–proton collision. An antiproton is a particle whose mass is equal to that of a proton, but its electrical charge is equal in absolute value to that of a proton, but of negative sign. We designate the proton by p^+ and the antiproton by p^-, and write the process as

$$p^+ + p^+ \to p^+ + p^+ + p^+ + p^-$$

Note that a pair p^+, p^- is created, and not a single p^-. This is required by the principle of conservation of charge (Section 21.8). Antiprotons were first produced in the laboratory by this process in 1955 at the University of California.

Initially one of the protons is at rest (zero momentum) in the L-frame and the other is moving toward it with momentum $\boldsymbol{p}$. In this process all the particles have the same mass, m_p. Then $m_2 = m_p$, $Q = -2m_p c^2$ and $M = 6m_p$. The threshold energy of the incoming proton must then be

$$E_k = 6m_p c^2 = 5.64 \times 10^3 \text{ MeV}$$

or 5.64 GeV. Thus, in order to produce antiprotons in an accelerator by hitting a target of protons at rest in the laboratory, the incident protons must be accelerated until their kinetic energy in the L-frame is $6m_p c^2$, or almost 6 GeV. However, if the protons are moving toward each other with the same momentum, so that the center of mass of the system is at rest in the laboratory, the C- and L-frames coincide. Then only an energy of $2m_p c^2$ is required and each proton need only be accelerated to about 1 GeV. This can be accomplished in the laboratory by accelerating the protons in opposite directions, making them collide at a certain point (see Note 20.2).

20.5 Particle decay

An important fact of nature is that some fundamental particles are unstable; that is, they tend to transform spontaneously into other particles. One necessary precondition for spontaneous decay, imposed by energy conservation, is that the mass of the parent particle be larger than the sum of the masses of the daughter particles. In addition, as we have already mentioned, energy, momentum, angular momentum and charge are conserved. Other quantities, which we will consider later (Section 41.5), are also supposedly conserved, thus limiting the possible processes that might occur.

Consider the decay of a neutron,

$$n^0 \to p^+ + e^- + \bar{\nu} \qquad \textbf{(20.26)}$$

Here $\bar{\nu}$ means an antineutrino, a particle that supposedly has zero mass. The energy released in the decay is

$$Q = [m_n - (m_p + m_e + m_{\bar{\nu}})]c^2 = 0.7834 \text{ MeV}$$

The positive value of Q indicates that a free neutron is unstable and decays according to the process 20.26, a fact confirmed experimentally. It is assumed that, shortly after the Big Bang, about 15×10^9 years ago, there were about equal numbers of protons and neutrons in the universe. Then the neutrons decayed very rapidly and the few that were left were captured by protons to form deuterons (see Note 41.4). That explains why we do not find free neutrons and why there are many more protons than neutrons in the universe at present. Note that charge would be conserved by writing $n^0 \rightarrow p^+ + e^-$. However, if this were the actual process, the energy available would split in a fixed ratio between the proton and the electron, and all electrons emitted would have the same energy (recall Example 14.5). Experimentally, it is observed that the electrons are emitted with a wide range of energies. This fact suggests that a third particle, a neutrino, is produced, which carries away the rest of the energy. Careful measurements of this energy suggest that the mass of the neutrino is very close to zero (or actually zero).

A free proton cannot decay spontaneously; that is

$$p^+ \nrightarrow n^0 + e^+ + \nu \qquad (Q = -1.81 \text{ MeV}) \tag{20.27}$$

where ν is a neutrino, because the Q of the reaction is negative. Hence free protons are stable against decay into neutrons. Similarly, the process

$$p^+ + e^- \nrightarrow n^0 + \nu \qquad (Q = -0.58 \text{ MeV}) \tag{20.28}$$

called **electron capture**, is not energetically possible with free protons. This explains why hydrogen exists and is stable. If electron capture by free protons were possible, the proton would quickly capture the orbiting electron and the hydrogen atom would disappear. However, both of these processes may be produced with bound protons in nuclei if the excess energy required can be provided by the binding energy of the nucleus. We will discuss such cases in Chapter 40.

Another example of particle decay is that of the pion. The pion is found to decay by the process (Figure 20.5)

$$\pi^{\pm} \rightarrow \mu^{\pm} + \nu \qquad (Q = 33.3 \text{ MeV})$$

All muons are emitted with a single energy of about 4.1 MeV, indicating that this is a two-body process (recall, again, Example 14.5).

The muon also decays, as seen in Figure 20.5, according to the scheme

$$\mu^{\pm} \rightarrow e^{\pm} + \nu + \bar{\nu} \qquad (Q = 106 \text{ MeV})$$

The electrons and positrons are observed to have a wide range of energies, corresponding to a three-body process, even though the neutrinos are not directly observed.

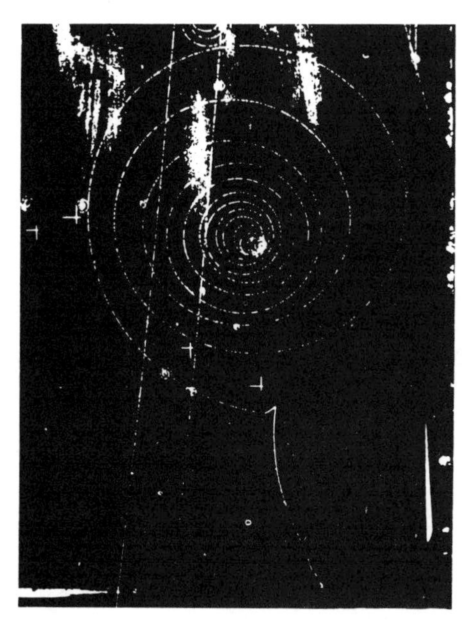

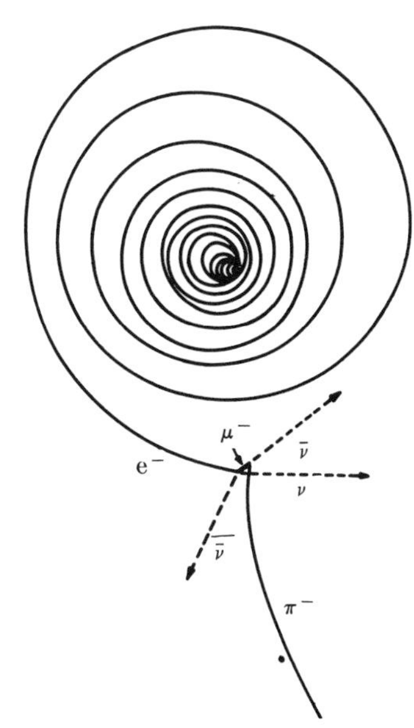

Figure 20.5 Decay of a π^--meson followed almost immediately by the decay of the muon. (Photograph courtesy of Brookhaven National Laboratory.)

EXAMPLE 20.5

The omega-minus experiment.

▷ An interesting example of the application of the conservation laws and the use of modern techniques for observing the decay of particles produced in high energy processes is the omega-minus (Ω^-) experiment, performed in 1963 at Brookhaven National Laboratory using the 33 GeV proton accelerator.

A theory formulated in 1961 by Murray Gell-Mann, to be discussed in Section 41.8, predicted the existence of a negatively charged particle called Ω^-, with a mass of about 3276 amu (or rest energy 1675 MeV). The theory also predicted that the Ω^- could be created by the process

$$K^- + p^+ \rightarrow \Omega^- + K^+ + K^0$$

where the Ks are K-mesons or kaons, with a mass of 967 amu or rest energy 494 MeV (see Table 41.1). The Q of the above process, using Equation 20.24, is $Q = -1229$ MeV. The process is produced by creating K^-s and using them as projectiles against protons in a liquid hydrogen bubble chamber. The threshold kinetic energy of the K^- particles needed for the reaction to take place is, using Equation 20.25,

$$E_k = -\frac{QM}{2m_2} = 2678 \text{ MeV} = 2.68 \text{ GeV}$$

The K^-s were produced in a collision of 33 GeV protons with a tungsten target. The experimental arrangement is shown in Figure 20.6. Only K^- particles with a kinetic energy

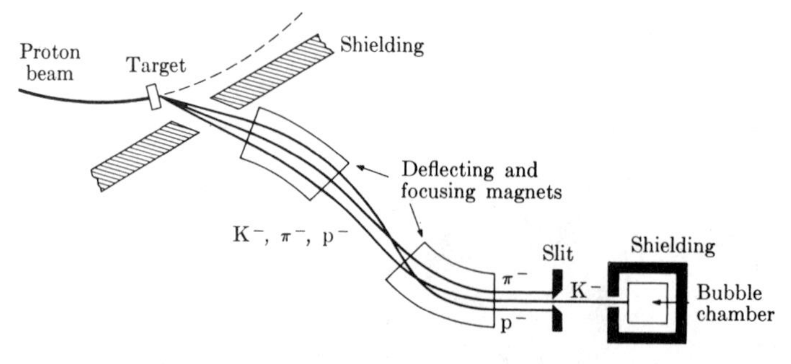

Figure 20.6 Schematic diagram of the Ω^--experiment. Before the separation there were approximately 800 pions and 10 antiprotons for every 10 K^- particles. After separation, there was only one pion and no antiprotons for every 10 K particles.

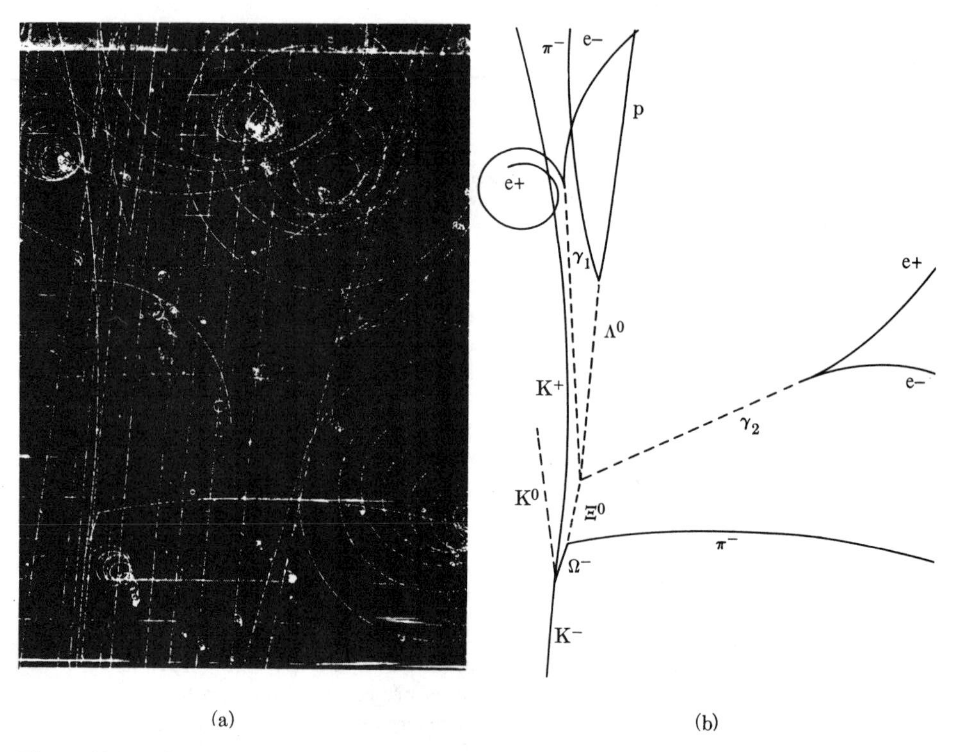

Figure 20.7 Photograph of Ω^- particle production. Particle tracks in a 2 m liquid hydrogen bubble chamber, placed in a strong magnetic field that forces the charged particles to follow curved paths. From analyses of the tracks, the properties of the different particles are derived. This photo, taken in 1964, provided the first evidence for the existence of the Ω^- particle, which had been previously predicted on a theoretical basis. (Photograph courtesy of Brookhaven National Laboratory.)

of 5.0 GeV (well above the threshold energy) were used for the experiment. After a run of several weeks, the experimenters had obtained about 50 000 photographs, of which one (shown in Figure 20.7(a)) corresponds to a process where an Ω^- had been produced and subsequently decayed according to the following sequence:

$$
\begin{aligned}
&K^- + p^+ \rightarrow \Omega^- + K^+ + K^0\\
&\qquad\quad \hookrightarrow \Xi^0 + \pi^-\\
&\qquad\qquad\quad \hookrightarrow \Lambda^0 + \gamma_1 + \gamma_2\\
&\qquad\qquad\qquad\qquad\qquad\quad \hookrightarrow e^+ + e^- \quad (\text{from } \gamma_2)\\
&\qquad\qquad\qquad\qquad\quad \hookrightarrow e^+ + e^- \quad (\text{from } \gamma_1)\\
&\qquad\qquad\qquad \hookrightarrow p^+ + \pi^- \quad (\text{from } \Lambda^0)
\end{aligned}
$$

where γ represents photons and the other particles are listed in Table 41.1. The neutral particles do not leave any track in the chamber, but are inferred from momentum considerations. The photograph is analyzed at the right in Figure 20.7(b). From an analysis of energy and momentum conservation at each decay vertex the experimenters could compute the energy and momentum of the Ω^-. The analysis gave a mass value between 1668 MeV and 1686 MeV. A second photograph, obtained a few weeks later, narrowed the mass to the range between 1671 and 1677 MeV, in excellent agreement with the theoretical prediction. The length of the Ω^- path in the chamber was about 2.5 cm, showing that the Ω^- lifetime is about $2.5 \times 10^{-2}\ \text{m}/(3 \times 10^8\ \text{m s}^{-1}) \sim 10^{-10}$ s. Because of this short life it is probable that without Gell-Mann's theoretical prediction it would have taken much longer for them to have been observed.

Note 20.2 Experimental techniques for producing high energy particle collisions

The study of high energy processes of elementary particles in the laboratory requires the use of highly sophisticated, complex and expensive experimental techniques. First, high-energy particles must be produced in machines called **accelerators**. Secondly, the particles must be made to collide with a **target** and the collision must occur at a suitable place. Finally, the particles produced by the collision must be observed so that their mass, charge, momentum and energy can be measured, for which special **detectors** are required.

Particles are accelerated using electric and magnetic forces. Although how electric and magnetic fields act on charged particles is not discussed until Chapters 21 and 22, it is interesting to have a general idea about the more advanced accelerators in use today. Accelerators are of two kinds: **linear** and **cyclic**.

In a *linear* accelerator (LINAC) (Figure 20.8), a bunch of charged particles (electrons, protons, positrons) moves through a series of tubes to which an alternating voltage is supplied. Successive tubes have opposite voltages, but the voltages alternate with the frequency of the applied voltage. If the lengths of the tubes are correctly chosen, the motion of the charged particles is synchronized with the alternation of voltage so that they cross the gap between successive tubes at the right time to receive a push that increases their velocity and energy. For that reason, the length of successive tubes must increase accordingly.

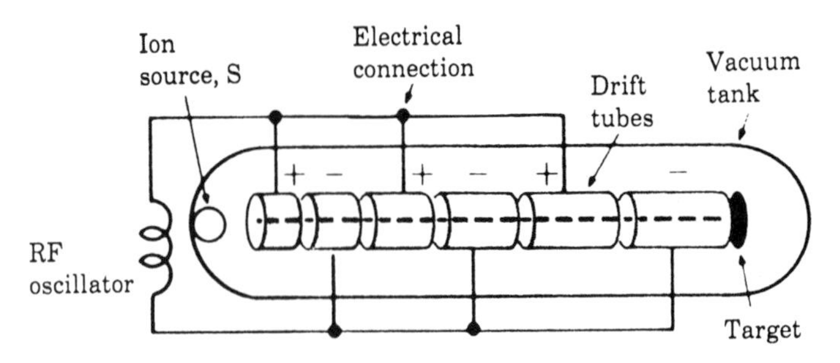

Figure 20.8 Linear accelerator.

The largest electron linear accelerator is at Stanford University and is called SLAC (Stanford Linear Accelerator Center). It is 3.2 km long and accelerates electrons to 25 GeV (or 25×10^9 eV) in bursts of 2μs duration 60 times per second.

In *cyclic* accelerators, the particles are forced by a magnetic field to describe a curved path along which they receive a push, or increase in energy, from electric fields at different points on its path. Cyclical accelerators can operate in a continuous form or in bursts. An example of the first type is the **cyclotron**, which was the first cyclical accelerator invented. The maximum energy that can be reached with a cyclotron is limited to several MeV (see Section 22.5). As demands for more energetic particles arose, cyclotrons were replaced by **synchrotrons** (see Example 22.5). The most powerful of these machines is the super proton synchrotron (SPS) at the European Centre for Nuclear Research (CERN), near Geneva, Switzerland, which can accelerate protons and antiprotons to an energy of 26 GeV.

The more advanced, and powerful, accelerators are the 'colliding beam' machines. We have explained that in a collision with a target at rest in the laboratory, only a fraction of the

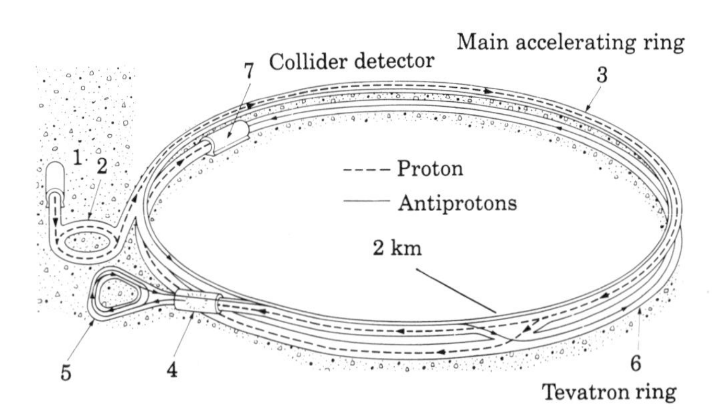

Figure 20.9 The Tevatron at the Fermi National Accelerator Laboratory (FNAL). A linear accelerator (1) produces protons that are energized up to 8 GeV in a proton synchrotron (2). Protons enter the main ring (3) where they are accelerated up to 150 GeV. Antiprotons are produced in proton–proton collisions (4) and stored in an accumulator ring (5). Protons and antiprotons, circulating in opposite directions are accelerated in the Tevatron ring (6) up to 900 GeV and collide inside the particle detector (7).

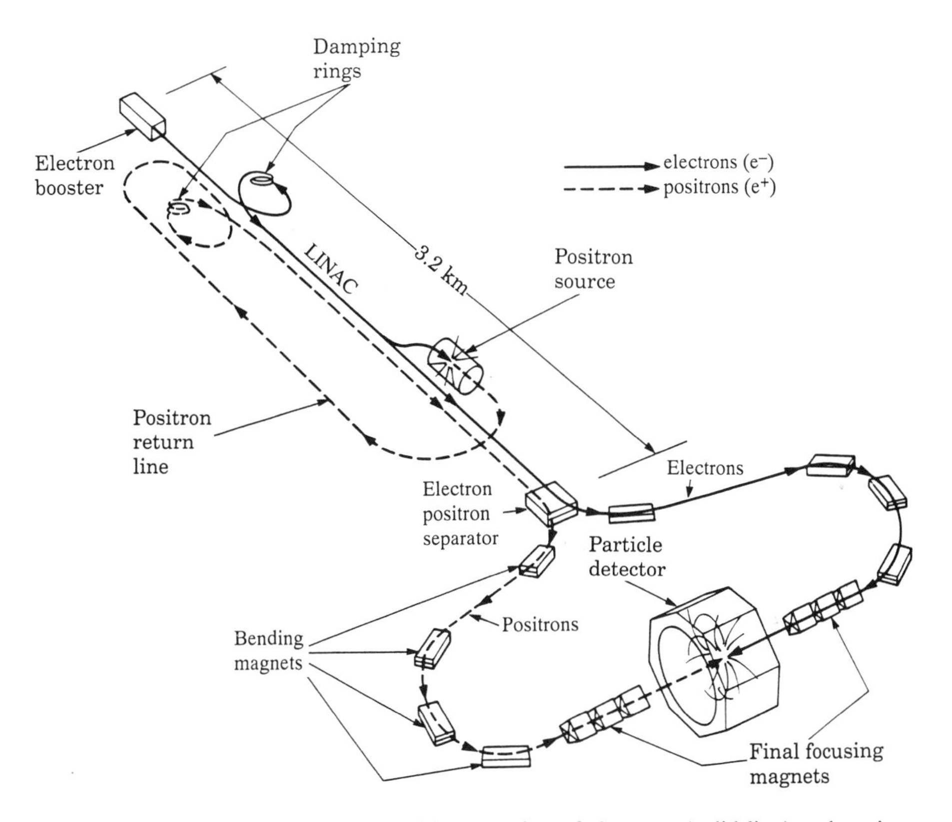

Figure 20.10 The Stanford Linear Collider. Bunches of electrons (solid line) and positrons (dashed line) are accelerated in a 3.2 km-long linear accelerator (LINAC) up to 50 GeV. Electrons and positrons are then separated and deflected into two large arcs, guided by magnets. The two opposite beams of particles collide head-on inside the particle detector at a combined energy of 100 GeV.

kinetic energy of the projectile can be used for the reaction because the momentum relative to the L-frame must be conserved. However, if we have two colliding beams, moving in opposite directions, with the same momentum, the total momentum is zero in the laboratory frame, which in this case is the C-frame. In this case all the kinetic energy is available for the reaction.

The three largest colliders in operation in 1992 are the Tevatron at Fermilab, Batavia, Illinois, the Stanford Linear Collider (SLC) at Stanford, California, and the Linear Electron–Positron Collider (LEPC) at CERN. The Tevatron (Figure 20.9) accelerates protons and antiprotons to an energy of 900 GeV in a circle of 1 km radius, so that the total energy available at collision is 1800 GeV or 1.8 TeV. A similar machine at CERN accelerates particles up to 300 GeV, resulting in 600 GeV on collision. Both machines use a synchrotron to provide the initial acceleration for the protons and antiprotons before they enter the main accelerator ring. The SLC (Figure 20.10) and the LEPC (Figure 20.11) accelerate electrons and positrons to 50 GeV, so that the collision energy is 100 GeV. The SLC uses a linear accelerator (LINAC) about 3 km long while the LEPC combines a linear accelerator with a synchrotron to provide the initial energy of the particles before they enter the main ring, which is 26.7 km in circumference and 100 m underground. More powerful machines are under construction.

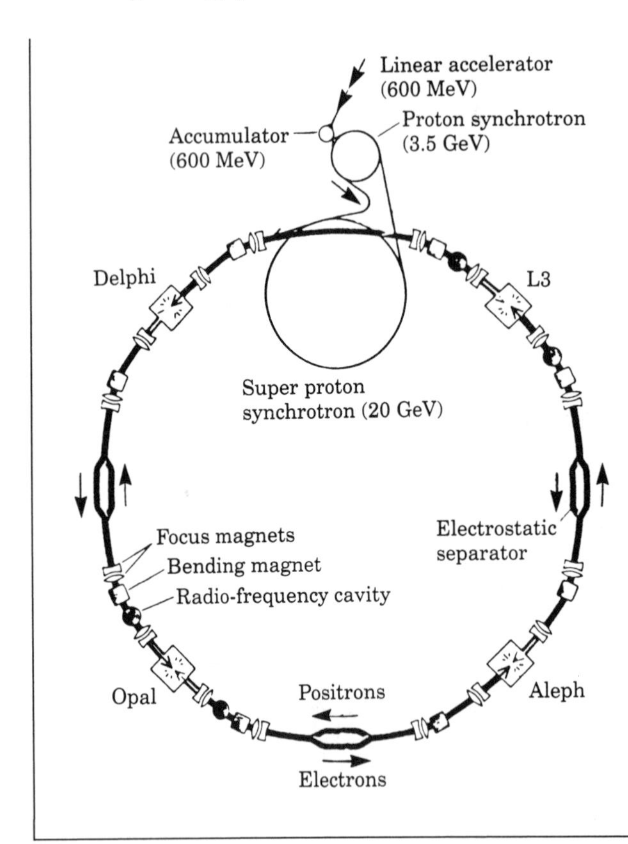

Figure 20.11 The LEPC collider at CERN. Electron and positron bunches are accelerated by a series of small machines. The storage rings have a radius of 4.25 km and are 100 m underground. Four bunches each of electrons and positrons circulate in opposite directions until they each attain an energy of 50 GeV. They then collide inside the four gigantic detectors: Aleph, Delphi, L3 and Opal. The bunches also cross paths at four intermediate sites, where they are prevented from colliding by electrostatic separator plates.

QUESTIONS

20.1 Why do we say that in the theory of relativity the laws of conservation of energy and momentum are not independent?

20.2 What is the definition of the C-frame in the theory of relativity? Is there any difference from the definition in Newtonian mechanics?

20.3 What is the 'mass defect' of a composite system? How is it related to the binding energy of a composite system?

20.4 When is there a threshold energy in a collision?

20.5 What are the main differences for processes with $Q > 0$ and $Q < 0$?

20.6 What is the advantage of 'colliding beams' in studying high energy processes?

20.7 Represent $E_k = c(m^2c^2 + p^2)^{1/2} - mc^2$ as a plot of E_k versus p, and compare with the non-relativistic plot of $E_k = p^2/2m$.

20.8 Under what conditions is the relation $E = cp$ valid (a) exactly, (b) approximately?

20.9 Is it possible to accelerate or decelerate a particle of zero mass?

20.10 Show that if a particle of mass m breaks into two particles of masses m_1 and m_2, their kinetic energies in the C-frame are $E_{k1} = Q^2/2mc^2 - Qm_2/m$ and $E_{k2} = Q^2/2mc^2 - Qm_1/m$ where $Q = [m - (m_1 + m_2)]c^2$. Apply this to the decay of a pion.

PROBLEMS

20.1 (a) Prove that the momentum of a particle can be written as $p = (E_k^2 + 2mc^2E_k)^{1/2}/c$. (b) Plot p/mc as a function of E_k/mc^2.

20.2 What maximum velocity must a particle have if its kinetic energy is to be written as $\frac{1}{2}mv^2$ with an error no greater than 1%? What is the kinetic energy, in eV, of (b) an electron and (c) a proton moving with that velocity?

20.3 Show, from Equation 20.4, that $v/c = [1 - (mc^2/E)^2]^{1/2}$. From this relation find the velocity of a particle when E is equal to (a) its rest energy, (b) twice its rest energy, (c) 10 times its rest energy, and (d) one thousand times the rest energy. Compute the corresponding energies in eV for (e) an electron and (f) a proton. Make a plot of v/c against E/mc^2.

20.4 Verify that the quantity E/c, where c is the velocity of light, has the dimensions of momentum. Then prove that $1\ \text{GeV}/c = 5.3 \times 10^{-19}\ \text{kg m s}^{-1}$.

20.5 (a) For what velocity may the total energy of a certain particle be written as pc with an error in the total energy not greater than 1%? (b) Calculate the kinetic energy of the particle.

20.6 Calculate the momentum, total energy and kinetic energy of a proton moving with a velocity $v = 0.99c$ relative to the laboratory (a) in the L-frame, (b) in the frame defined by the proton, (c) in the C-frame defined by the proton and a helium atom at rest in the laboratory.

20.7 A proton with a kinetic energy of 10^{10} eV collides with a proton at rest in the L-frame. Find (a) the total momentum and the total energy of the system in the L-frame, (b) the kinetic energy of the two particles in the C-frame.

20.8 An electron with a total energy E_e makes a head-on collision with a proton at rest. If the electron energy is very large compared with its rest energy, the electron must be treated relativistically, but if, in addition, it is small compared with the rest energy of the proton, the proton can be treated non-relativistically. Under those conditions, verify that (a) the proton recoils with a velocity approximately equal to $(2E_e/mc^2)c$, (b) the energy transferred from the electron to the proton is $2E_e^2/mc^2$. (c) Apply to a case where the electrons have a kinetic energy of 100 MeV. (Hint: for the electron, $E = cp$, while for the proton $E_k = p^2/2m_p$. Also note that if the proton moves forward, the electron bounces back, so that the *direction* of its momentum is reversed.)

20.9 One method for obtaining the energy needed for a nuclear reaction is to send two particles against each other. When the particles are identical and their energies are the same, the C-frame coincides with the laboratory (see Note 20.2). If the colliding particles are protons, of energy equal to 28 GeV, (a) calculate the total energy available for reaction and (b) calculate the kinetic energy of one of the protons in the frame of reference in which the other proton is at rest. This is the energy a proton would need to produce the same reaction when colliding with a target at rest in the laboratory. Do you see any advantage in using colliding beams of particles?

20.10 A particle of mass m_1 and momentum $\boldsymbol{p}_1$ collides with a particle of mass m_2 at rest in the laboratory. The two particles stick together, resulting in a particle of rest mass m_3. Find (a) the velocity of the resulting particle relative to the L-frame, (b) the Q of the collision.

20.11 A particle of mass m_1 and momentum $\boldsymbol{p}_1$ collides with a particle of mass m_2 at rest in the laboratory. The resulting products are a particle of mass m_3 and a particle of zero mass. Find the energy of the zero mass particle (a) in the C-frame, (b) in the L-frame.

20.12 A particle of mass m splits (or decays) into two particles of masses m_1 and m_2. (a) Verify that in the C-frame the energies of the resulting particles are $E_1' = (m^2 + m_1^2 - m_2^2)c^2/2m$ and $E_2' = (m^2 + m_2^2 - m_1^2)c^2/2m$. (b) Also find their momenta.

20.13 (a) Solve Problem 20.12 for the case of particles in the L-frame, given that the momentum of the particle of mass m in this frame is p. (b) Also verify that if p_1 and p_2 are the momenta of the resulting particles and θ the angle between them,

$$m^2c^4 = (m_1 + m_2)^2c^4 + 2E_1E_2 - 2m_1m_2c^4$$
$$-2p_1p_2c^2 \cos\theta.$$

20.14 In a collision between particles of masses m_1 and m_2, m_1 is moving with momentum p_1 and m_2 is at rest in the L-frame. After the collision, in addition to particles m_1 and m_2 there appear particles $m_3, m_4, \ldots$. (a) Verify that the threshold kinetic energy of m_1 in the L-frame for this process is $E_{k1} = Mc^2[1 + m_1/m_2 + M/2m_2]$, where $M = m_3 + m_4 + \dots$. (b) Apply this equation to the creation of a proton–antiproton pair, as discussed in Example 20.5.

20.15 A particle of mass m_1, moving relative to the L-frame with a total energy E_1 very large compared to its rest energy so that the relation $E_1 = p_1 c$ holds, collides with a particle of mass m_2 which is at rest. Show that the energy available in the C-frame is $(2E_1 m_2 c^2 + m_2^2 c^4)^{1/2}$.

20.16 Consider a reaction in which a particle of zero mass and energy E_1 collides with a particle of mass m_2 which is at rest in the laboratory. The final products of the reaction are two particles: one of mass m_2 and another of mass m_3. Show that the threshold energy E_1 for the reaction is $E_1 = m_3(1 + m_3/2m_2)c^2$.

20.17 Determine the Q-value and the threshold kinetic energy in the L-frame of the incident (the π^-) particle for both of the following reactions:

(a) $\pi^- + p^+ \to n + \pi^0$
(b) $\pi^- + p^+ \to \Sigma^- + K^+$

The masses of the particles are given in Table 41.1.

20.18 Calculate the threshold kinetic energy of a proton for the reaction $p^+ + p^+ \to p^+ + p^+ + x\pi^0$ to occur, where x is the number of π^0s produced.

20.19 A particle of mass m decays into other particles. The process has a non-zero Q-value. (a) Verify that if the particle divides into two equal fragments they must move in the C-frame in opposite directions with a momentum equal to $\frac{1}{2}(2mQ - Q^2/c^2)^{1/2}$. (b) Verify that if the particle distintegrates into three equal fragments, emitted symmetrically in the C-frame, the momentum of each particle is equal to $\frac{1}{3}(2mQ - Q^2/c^2)^{1/2}$.

20.20 Apply the result of part (b) of Problem 20.19 to the particle called a **tauon**, of mass $m_\tau = 8.8 \times 10^{-28}$ kg, which disintegrates into three identical **pions**, each of mass $m_\pi = 2.5 \times 10^{-28}$ kg. (a) Evaluate the Q of the process and (b) find the magnitude of the velocities of the fragments in the C-frame.

20.21 The transformation for energy and momentum can be written in the vector form as

$$\boldsymbol{p}' = \boldsymbol{p} - \frac{(\boldsymbol{p}\cdot\boldsymbol{v})\boldsymbol{v}}{v^2} + \gamma\left(\frac{(\boldsymbol{p}\cdot\boldsymbol{v})\boldsymbol{v}}{v^2} - \frac{\boldsymbol{v}E}{c^2}\right)$$

$$E' = \gamma(E - \boldsymbol{v}\cdot\boldsymbol{p})$$

Verify that it reduces to Equation 20.13 when the relative velocity $\boldsymbol{v}$ is parallel to the X-axis.

20.22 A particle of mass m_1 and moving with velocity $\boldsymbol{v}_1$ in the L-frame collides with a particle that has mass m_2 and is at rest in the L-frame. (a) Prove that the velocity of the C-frame of the system composed of the two particles is

$$\boldsymbol{v} = \frac{\boldsymbol{v}_1}{1 + A(1 - v_1^2/c^2)^{1/2}}$$

where $A = m_2/m_1$. (b) Show that in the C-frame the velocity of m_1 is

$$\boldsymbol{v}_1' = \frac{\boldsymbol{v}_1 A(1 - v_1^2/c^2)^{1/2}}{1 - v^2/c^2 + A(1 - v^2/c^2)^{1/2}}$$

and that the velocity of m_2 is $-\boldsymbol{v}$. (c) Compute the values of the preceding quantities when v_1 is small compared with c, and compare your result with those of Example 14.6.

20.23 Using the Lorentz transformation laws for energy and momentum, verify that if $\boldsymbol{v} = c^2\boldsymbol{P}/E$ is the system velocity relative to an observer O while the system velocity relative to another observer O', in motion relative to O with velocity V along the X-axis, is $\boldsymbol{v}' = c^2\boldsymbol{P}'/E'$, then $\boldsymbol{v}$, $\boldsymbol{v}'$, and $\boldsymbol{V}$ are related by Equations 19.14 for the transformation of velocities. (b) Also show that if $\boldsymbol{v}' = 0$ (or $\boldsymbol{P}' = 0$), then $\boldsymbol{v} = \boldsymbol{V}$. This was one of the basic assumptions in Section 20.2 for defining the system velocity.

20.24 A particle of mass m_1 and momentum p_1 collides elastically with a particle that has mass m_2, at rest in the L-frame. After the collision, m_1 is deviated through an angle θ. Obtain the momentum and energy of m_1 in the L-frame after the collision.

20.25 (a) Referring to Problem 20.24, verify that if particle 1 has zero mass ($E_1 = p_1 c$), then the relations for p_1' and E_1' reduce to the single relation

$$E_1' = E_1(m_2c^2)\frac{E_1(1 + \cos\theta) + m_2c^2}{(E_1 + m_2c^2)^2 - E_1^2\cos^2\theta}$$

(b) Verify that it can also be written as

$$\frac{1}{E_1'} - \frac{1}{E_1} = \frac{1}{mc^2}(1 - \cos\theta)$$

(This expression is important for the discussion of the Compton effect in Chapter 30).

21 Electric interaction

Charles A. de Coulomb was a military engineer who did extensive research on the elasticity of materials and the electric properties of conductors, showing that the electric charge of a conductor is confined to its surface. In 1777 Coulomb designed the torsion balance with which he was able to establish, in a quantitative way, the inverse square law of interaction between electric charges, now known as Coulomb's Law. (Photograph © Boyer–Viollet.)

Notes

21.1 Introduction

Historically, the notion of an electric force originated centuries ago in very simple experiments, such as rubbing two bodies together. Suppose that after we comb our hair on a very dry day, we bring the comb close to tiny pieces of paper. We note that they are swiftly attracted by the comb. Similar phenomena occur if we rub a glass rod or an amber rod with a cloth or with a piece of fur. It was thus concluded that, as a result of the rubbing, these materials acquire a new property called **electricity** (from the Greek word *elektron* meaning amber) and that this electrical property gives rise to an interaction much stronger than gravitation. There are, however, several fundamental differences between electrical and gravitational interaction.

In the first place there is only one kind of gravitational interaction, resulting in a universal attraction between any two masses. However, there are two kinds of electrical interactions. Suppose that we place an electrified glass rod near a small cork ball hanging from a string. We see that the rod attracts the ball (Figure

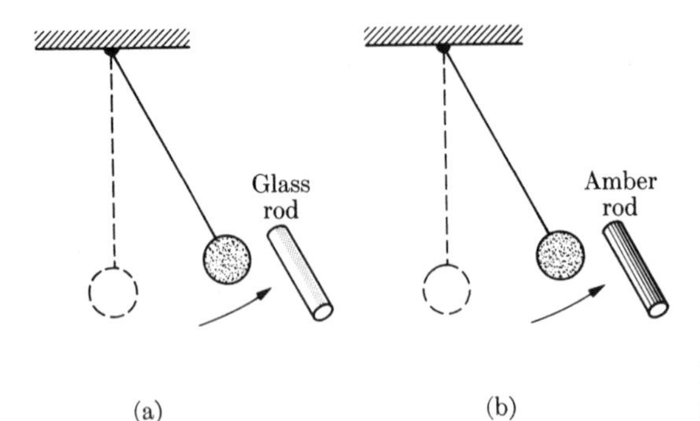

Figure 21.1 Experiments with electrified glass and amber rods.

21.1). If we repeat the experiment with an electrified amber rod, we observe the same attractive effect on the ball.

Suppose that we touch two cork balls with an electrified glass rod. We may assume that the two balls become electrified. If we bring the balls together we note that they repel each other. The same result occurs when we touch the balls with an electrified amber rod, so that they acquire the same electrification. However, if we touch one ball with the glass rod and the other ball with the amber rod, so that one ball has glass-like electrification and the other amber-like electrification, we observe that the balls attract each other. We conclude that there are two kinds of electrified states: one glass-like and the other amber-like. We may call the first **positive** and the other **negative** (we could, of course, make the opposite designation.) The fundamental law of electric interaction may then be stated as:

> *two bodies with the same kind of electrification (either positive or negative) repel each other, but if they have different kinds of electrification (one positive and the other negative), they attract each other.*

This statement is indicated schematically in Figure 21.2.

Therefore, while the gravitational interaction is always attractive, the electrical interaction may be either attractive or repulsive. If the electrical interaction had been only attractive (or only repulsive), we probably would never have noticed the existence of gravitation, since the electrical interaction is much stronger. However, most bodies seem to be composed of equal amounts of positive and negative electricity, so that the net electrical interaction between any two

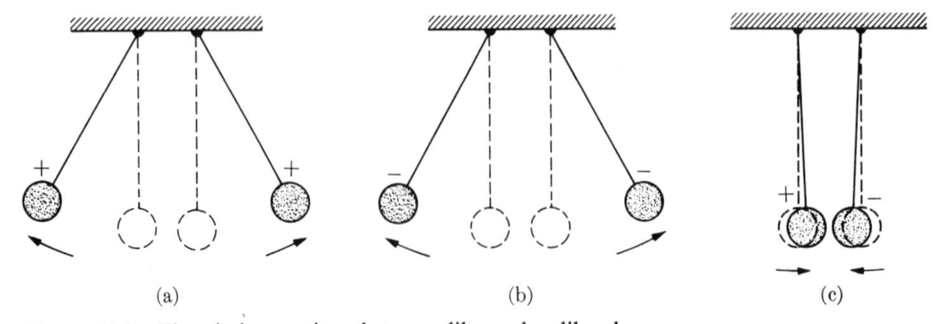

Figure 21.2 Electric interactions between like and unlike charges.

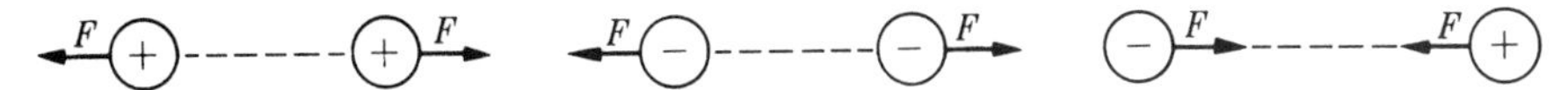

Figure 21.3 Forces between like and unlike charges.

macroscopic bodies is very small or zero. Thus, as a result of a cumulative mass effect, the dominant macroscopic interaction appears to be the much weaker gravitational interaction. However, it is the electric force that holds electrons in atoms, atoms together in molecules and molecules together in bulk matter. Thus bulk matter, and by extension also living beings, exists because of the electric interaction.

According to Newton's third law of motion, when two electric charges interact they exert equal and opposite forces on each other. The *direction* of the electric force on each charge depends on the signs of *both* charges. This is shown in Figure 21.3.

21.2 Electric charge

We characterize the strength of the gravitational interaction by attaching to each body a gravitational mass, which is identical to its inertial mass. In the same way, we characterize the state of electrification of a body by defining an **electric charge**, or simply charge, represented by the symbol q. Thus any piece of matter or any particle is characterized by two independent but fundamental properties: mass and charge.

Since there are two kinds of electrification, there are also two kinds of electric charge: positive and negative. A body exhibiting positive electrification has a positive electric charge. The net charge of a body is the algebraic sum of its positive and negative charges. A body having equal amounts of positive and negative charges (i.e. zero net charge) is called electrically **neutral**; this is the case for atoms and molecules. Since matter in bulk does not exhibit gross electrical forces, we may assume that atoms are composed of equal amounts of positive and negative charges.

To define the charge of an electrified body operationally, we adopt the following procedure. An arbitrary charged body Q (Figure 21.4) is chosen and, at a distance d from it, the charge q is placed and the force F on q is measured. Next, another charge q' is placed at the same distance d from Q and the force F' is measured. We define the values of the charges q and q' as proportional to the forces F and F'. That is,

$$\frac{q}{q'} = \frac{F}{F'} \tag{21.1}$$

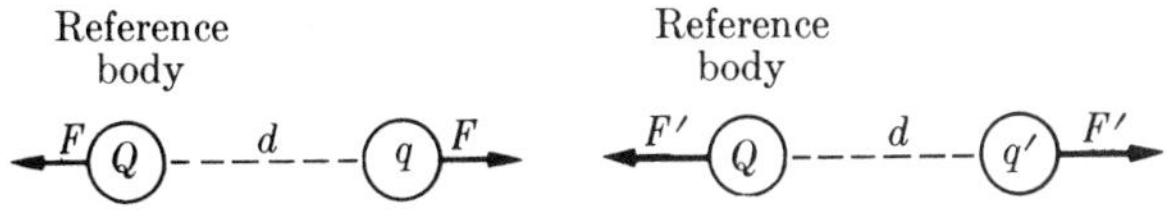

Reference body | Reference body

Figure 21.4 Comparison of electric charges q and q', showing their electric interactions with a third charge Q.

If we arbitrarily assign a value of unity to the charge q', we have a means of obtaining the value of q. This method of comparing charges is very similar to the one used in Section 11.3 for comparing the masses of two bodies. This definition of charge implies that, all geometrical factors being equal, the force of electrical interaction is proportional to the charges of the particles.

21.3 Coulomb's law

It is very important to determine how the electric interaction varies with the distance. Consider the electric interaction between two charged particles *at rest* in the observer's inertial frame of reference or, at most, moving with a very small velocity. The results of such an interaction constitute what is called **electrostatics**. The electrostatic interaction for two charged particles is given by **Coulomb's law**, named after Charles A. de Coulomb (1736–1806), which states that

> *the electric interaction between two charged particles at rest or in very slow relative motion is proportional to their charges and to the inverse of the square of the distance between them, and its direction is along the line joining the two charges.*

Coulomb's law may be expressed by

$$F = K_e \frac{qq'}{r^2} \tag{21.2}$$

where r is the distance between charges q and q', F is the force acting on either charge, and K_e is a constant to be determined by our choice of units. This law is very similar to the law for gravitational interaction (Equation 11.1):

$$F = G \frac{mm'}{r^2}$$

Thus we can apply many results that we proved in Chapter 11 simply by replacing Gmm' by $K_e qq'$.

For practical and computational reasons, it is more convenient to express K_e in the form

$$K_e = \frac{1}{4\pi\varepsilon_0} \tag{21.3}$$

where the new physical constant ε_0 is called the **vacuum permittivity**. Accordingly, we shall normally write Equation 21.2 in the form

$$F = \frac{qq'}{4\pi\varepsilon_0 r^2} \tag{21.4}$$

When using Equation 21.4, we must include the charges q and q' with their signs. A negative value of F corresponds to attraction and a positive value corresponds to repulsion.

We can experimentally verify the inverse square law 21.2 by measuring the force between two given charges placed at several distances. A possible experimental arrangement has been indicated in Figure 21.5, similar to the Cavendish torsion balance of Figure 11.3. The force F between the charge at B and the charge at D is found by measuring the angle θ through which the fiber OC is rotated to restore equilibrium.

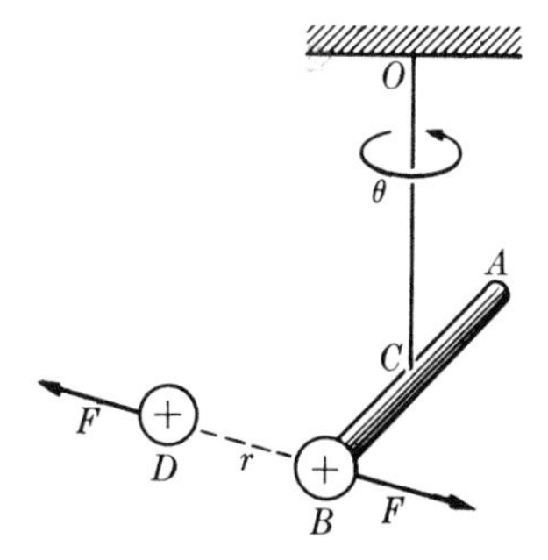

Figure 21.5 Cavendish torsion balance for verifying the law of electric interaction between two charges.

21.4 Units of charge

The constant K_e in Coulomb's law, Equation 21.2, is similar to the constant G in Newton's law of gravitation, Equation 11.1. But in Equation 11.1 the units of mass, distance and force were already defined and the value of G was determined experimentally in terms of known units. In the present case, however, although the units of force and distance have already been defined, the unit of charge is as yet undefined. If we make a definite statement about the unit of charge, then we must determine K_e experimentally. We may proceed, however, in the reverse order and assign to K_e a convenient value, in this way fixing the unit of charge. We shall adopt this second method and, using the *SI* system, set

$$K_e = 10^{-7}c^2 = 8.9874 \times 10^9$$

where $c = 2.9979 \times 10^8 \text{ m s}^{-1}$ is the velocity of light in vacuum. This choice makes it easier to write other expressions related to electromagnetic waves. For practical purposes, we may say that $K_e = 9 \times 10^9$. Once we have decided on the value of K_e, the unit of charge is fixed. This unit is called a **coulomb**, and is designated by the symbol C. Then, when the distance is measured in meters, the charge in coulombs and the force in newtons, Equation 22.2 becomes

$$F = (10^{-7}c^2)\frac{qq'}{r^2} \qquad \text{or} \qquad F = 9 \times 10^9 \frac{qq'}{r^2} \tag{21.5}$$

From Equation 21.5 we may state that:

the coulomb is that charge which, when placed one meter from an equal charge in vacuum, repels it with a force of $10^{-7}c^2$ or 8.9874×10^9 newtons.

As we shall see in Section 21.7, this definition of the coulomb is equivalent to the preliminary definition given in Section 2.3.

From Equation 21.2, K_e is expressed in units of $N\,m^2\,C^{-2}$ or $m^3\,kg\,s^{-2}\,C^{-2}$. Therefore, the vacuum permittivity ε_0 has the value

$$\varepsilon_0 = \frac{10^7}{4\pi c^2} = 8.854 \times 10^{-12}\ N^{-1}\,m^{-2}\,C^2 \tag{21.6}$$

For convenience, we shall usually omit the units of K_e or ε_0 when writing their numerical values.

EXAMPLE 21.1

Given the charge arrangement of Figure 21.6, where $q_1 = +1.5 \times 10^{-3}$ C, $q_2 = -0.5 \times 10^{-3}$ C, $q_3 = 0.2 \times 10^{-3}$ C and $AC = 1.2$ m, $BC = 0.5$ m, find the resultant force on charge q_3.

▷ The force F_1 between q_1 and q_3 is repulsive, while the force between q_2 and q_3 is attractive. Their respective values, when we use Equation 21.4, are

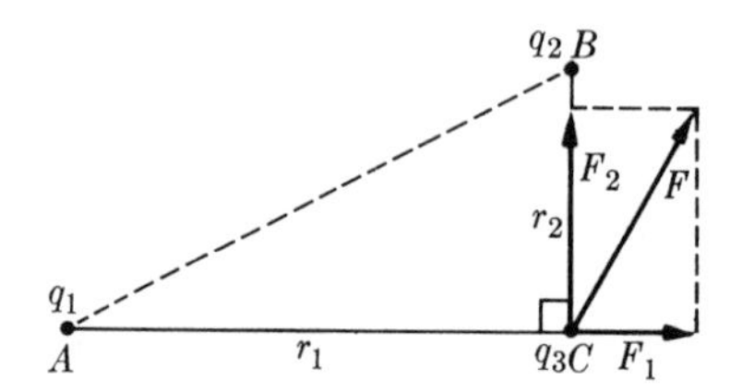

Figure 21.6 Resultant electric force on q_3 due to q_1 and q_2.

$$F_1 = \frac{q_1 q_3}{4\pi\varepsilon_0 r_1^2} = 1.88 \times 10^3\ N$$

$$F_2 = \frac{q_2 q_3}{4\pi\varepsilon_0 r_2^2} = -3.60 \times 10^3\ N$$

Therefore the resultant force is

$$F = (F_1^2 + F_2^2)^{1/2} = 4.06 \times 10^3\ N$$

EXAMPLE 21.2

Electric force on the electron and the proton in a hydrogen atom.

▷ In a hydrogen atom the electron and the proton are separated an average distance of 0.53×10^{-10} m approximately. The charge of both the electron and the proton is 1.60×10^{-19} C. Then the force on the electron due to the proton, or the force on the proton due to the electron, is

$$F = (9 \times 10^9) \times \frac{(1.60 \times 10^{-19}\ C)^2}{(0.53 \times 10^{-10}\ m)^2} = 8.2 \times 10^{-8}\ N$$

Since the mass of the proton is 1840 times the mass of the electron, its acceleration is 1840 times smaller and we may assume that the proton remains practically at rest in the inertial frame attached to the CM of the atom. The centripetal acceleration of the electron is $a = F/m = 0.90 \times 10^{23}\ m\,s^{-2}$. Using $a = \omega^2 r$ we find that the angular velocity of the electron is $\omega = 4.12 \times 10^{16}\ s^{-1}$, or a frequency 6.56×10^{15} Hz. The orbital velocity of the electron is $v = \omega r = 2.18 \times 10^6\ m\,s^{-1}$, which is about one hundredth the velocity of light. (Compare with Example 5.1.)

21.5 Electric field

We say there is an **electric field** in any region where an electric charge experiences a force. The force is due to the presence of other charges in that region. For example, a charge q placed in a region where there are other charges q_1, q_2, q_3 etc. (Figure 21.7) experiences a force $\boldsymbol{F} = \boldsymbol{F}_1 + \boldsymbol{F}_2 + \boldsymbol{F}_3 + \cdots$ and we say that in the region there is an electric field produced by the charges $q_1, q_2, q_3, \ldots$. (The charge q, of course, also exerts forces on $q_1, q_2, q_3, \ldots$, but we are not concerned with them now.) Since the force that each charge $q_1, q_2, q_3, \ldots$ produces on the charge q is proportional to q, the resultant force $\boldsymbol{F}$ is also proportional to q. Thus the force on a particle placed in an electric field is proportional to the charge of the particle.

The **intensity of the electric field** at a point is equal to the force per unit charge placed at that point. The symbol for electric field intensity is $\mathscr{E}$. Therefore

$$\mathscr{E} = \frac{\boldsymbol{F}}{q} \quad \text{or} \quad \boldsymbol{F} = q\mathscr{E} \tag{21.7}$$

The electric field intensity $\mathscr{E}$ is expressed in newtons/coulomb or $\mathrm{N\,C^{-1}}$, or, using the fundamental units, $\mathrm{m\,kg\,s^{-2}\,C^{-1}}$.

Note that, in view of the definition 21.7, if q is positive, the force $\boldsymbol{F}$ acting on the charge has the same direction as the field $\mathscr{E}$. However, if q is negative, the force $\boldsymbol{F}$ has the direction opposite to $\mathscr{E}$ (Figure 21.8). Therefore, if we apply an electric field to a region where positive and negative particles or ions are present, the field will tend to move the positively and negatively charged bodies in opposite directions. The result is a charge separation, an effect sometimes called **polarization**. This is what happens when atoms and molecules are placed in an electric field.

Just as in the case of a gravitational field, an electric field may be represented by **lines of force**. These lines are drawn so that, at each point, they are tangent to the direction of the electric field at the point. A *uniform* electric field has the same intensity and direction everywhere. Therefore, a uniform field is represented by parallel lines of force (Figure 21.9). One of the best ways to produce a

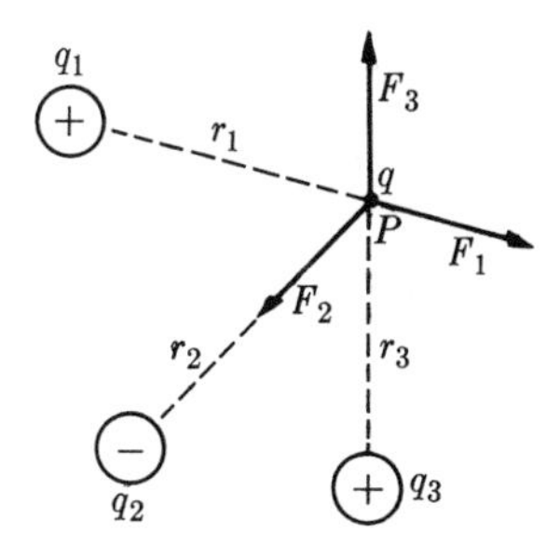

Figure 21.7 Resultant electric force on q produced by several charges.

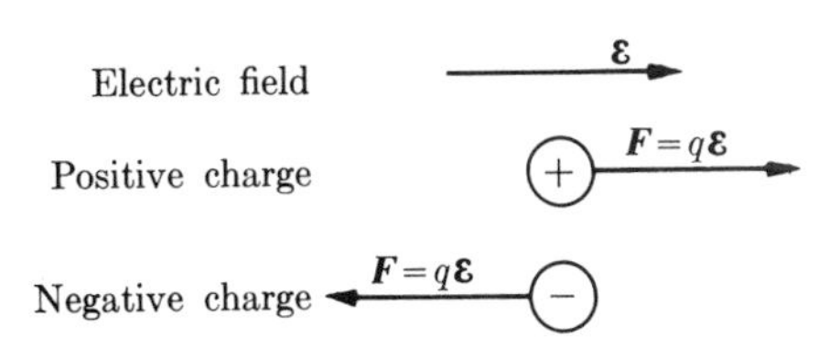

Figure 21.8 Direction of the force produced by an electric field on a positive and a negative charge.

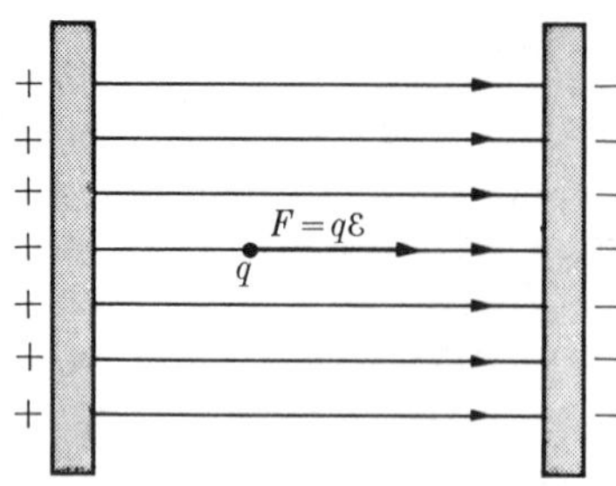

Figure 21.9 Uniform electric field.

uniform electric field is to charge two parallel metal plates with equal but opposite charges. Symmetry indicates that the electric field in the region between the plates is uniform, but later on (Example 25.1) we shall verify this assertion mathematically.

EXAMPLE 21.3

Motion of an electric charge in a uniform electric field.

▷ The equation of motion of an electric charge in an electric field is given by the equation

$$m\boldsymbol{a} = q\mathscr{E} \qquad \text{or} \qquad \boldsymbol{a} = \frac{q}{m}\mathscr{E}$$

The acceleration of a body in an electric field depends, therefore, on the ratio q/m. Since this ratio is in general different for different charged particles or ions, their accelerations in an electric field are also different. Thus there is a clear distinction between the acceleration in a gravitational field, which is the same for all bodies, and that in an electric field. If the field $\mathscr{E}$ is uniform, the acceleration $\boldsymbol{a}$ is constant. The path of the electric charge is then a straight line or a parabola, depending on the direction of the initial velocity relative to that of the electric field (recall Section 4.5).

An interesting case is that of a charged particle passing through an electric field that occupies a limited region in space (Figure 21.10). For simplicity, we assume that the initial velocity v_0 of the particle when it enters the field is perpendicular to the direction of the electric field. The X-axis is placed parallel to the initial velocity of the particle and the Y-axis is placed parallel to the field. The path AB followed by the particle when it moves through the field is a parabola. After crossing the field the particle resumes rectilinear motion, in a different direction and with a different velocity v. We then say that the electric field has produced a deflection measured by the angle α.

Using the results of Section 4.5, we find that the coordinates of the particle while it is moving through the field with an acceleration $a = (q/m)\mathscr{E}$ are given by

$$x = v_0 t, \qquad y = \frac{1}{2}\frac{q}{m}\mathscr{E}t^2$$

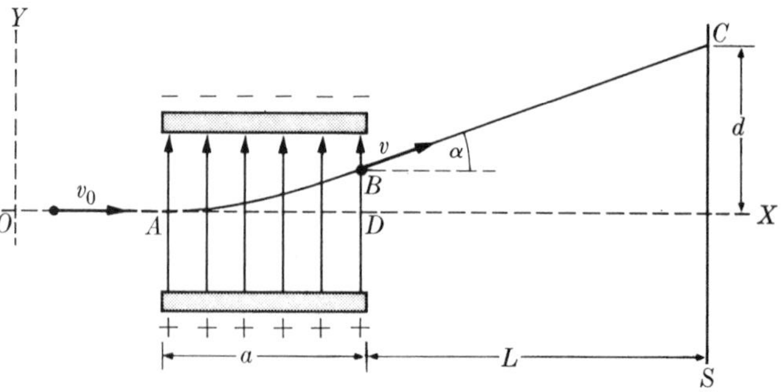

Figure 21.10 Deflection of a positive charge by a uniform electric field.

Eliminating the time t, we obtain the equation of the path as

$$y = \frac{1}{2}\frac{q}{m}\frac{\mathscr{E}}{v_0^2}x^2$$

thus verifying that it is a parabola. We obtain the deflection angle α by calculating the slope dy/dx of the path at $x = a$. The result is

$$\tan\alpha = \left(\frac{dy}{dx}\right)_{x=a} = \frac{q\mathscr{E}a}{mv_0^2}$$

Therefore, when a stream of particles, all having the same ratio q/m, passes through the electric field, they are deflected according to their velocities or kinetic energies. If we place a screen S at distance L, all particles with a given q/m ratio and velocity v_0 will reach a point C on the screen. Noting that $\tan\alpha$ is also approximately equal to d/L (because the vertical displacement BD is small compared with d if L is large), we have

$$\frac{q\mathscr{E}a}{mv_0^2} = \frac{d}{L} \tag{21.8}$$

By measuring d, L, a and $\mathscr{E}$, we may obtain the velocity v_0 (or the kinetic energy) if we know the ratio q/m; or, conversely, we may obtain q/m if we know v_0.

A device such as the one illustrated in Figure 21.10 may be used as an **energy analyzer**, which separates identical charged particles moving with different kinetic energies. For example, β-rays are electrons emitted by some radioactive materials; if we place a beta emitter at O, all the electrons emitted in the direction OA will concentrate at the same spot C on the screen if they have the same energy. But if they are emitted with different energies, they will be spread over a region of the screen. It is this second situation that is found experimentally. On the other hand, if the nucleus emits α-particles (positively charged helium nuclei) it is found that they have well-defined energies. Both are results of great importance from the point of view of nuclear structure (see Section 39.7).

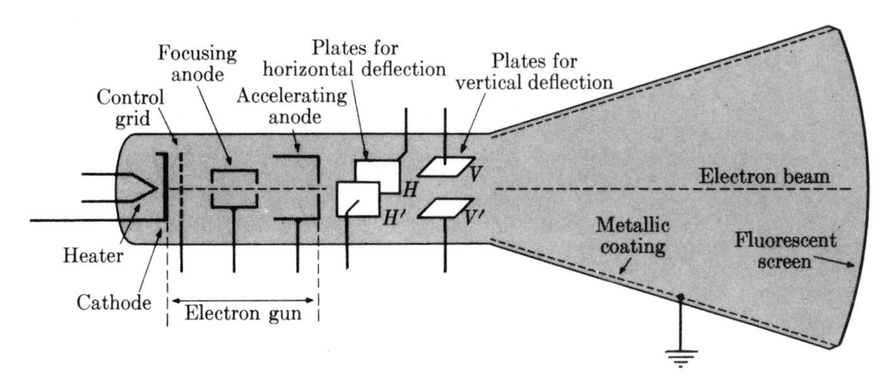

Figure 21.11 Motion of a charge under crossed electric fields. The electrons are emitted from the cathode and accelerated by an electric field. A hole in the accelerating anode allows the electrons to pass out of the electron gun and between the two sets of deflection plates. The metallic coating inside the tube shields the right end free of external electric fields and conducts away the electrons of the beam after they strike the fluorescent screen.

By using two sets of parallel charged plates, we can produce two mutually perpendicular fields, one horizontal along HH' and another vertical along VV', as shown in Figure 21.11. Suppose an electron beam is traveling along the space between the fields. By adjusting the relative intensity of the fields, we can obtain an arbitrary deflection of the electron beam to any spot on the screen. If the two fields are variable, the spot on the screen will describe a certain curve. Practical applications of this effect occur in television tubes and in oscilloscopes. In particular, if the electric fields vary in intensity with simple harmonic motion, the pattern traced out will be a Lissajous figure (Section 10.8).

21.6 Electric field of a point charge

Let us write Equation 21.4 in the form

$$F = q'\left(\frac{q}{4\pi\varepsilon_0 r^2}\right)$$

This gives the force produced by the charge q on the charge q' placed a distance r from q. We may also say, using Equation 21.7, that the electric field $\mathscr{E}$ at the point where q' is placed is such that $F = q'\mathscr{E}$. Therefore, by comparing both expressions for F, we conclude that the electric field at a distance r from a point charge q is

$$\mathscr{E} = \frac{q}{4\pi\varepsilon_0 r^2} \quad \text{or} \quad \mathscr{E} = 9 \times 10^9 \frac{q}{r^2} \tag{21.9}$$

In vector form,

$$\boldsymbol{\mathscr{E}} = \frac{q}{4\pi\varepsilon_0 r^2}\boldsymbol{u}_r \tag{21.10}$$

where $\boldsymbol{u}_r$ is the unit vector in the radial direction, away from the charge q, since $\boldsymbol{F}$ is along this direction. Expression 21.10 is valid for both positive and negative charges, with the direction of $\boldsymbol{\mathscr{E}}$ relative to $\boldsymbol{u}_r$ given by the sign of q. Thus $\boldsymbol{\mathscr{E}}$ is directed away from a positive charge and toward a negative charge.

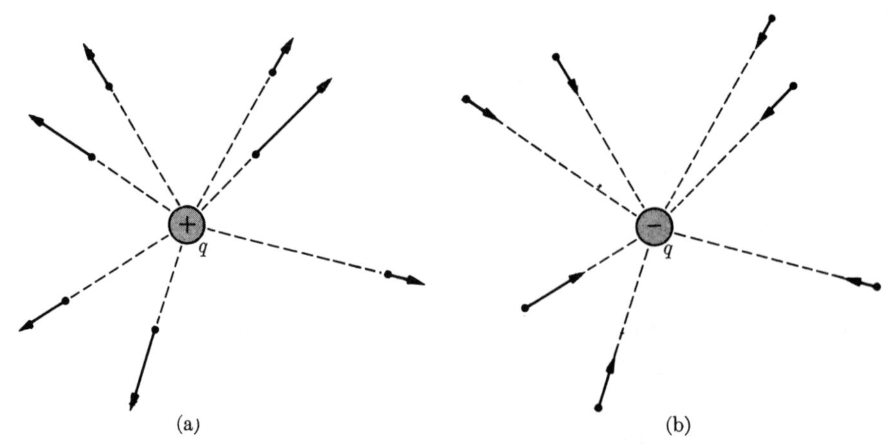

Figure 21.12 Electric field produced by a positive and negative charge.

Figure 21.12(a) indicates the electric field at points near a positive charge, and Figure 21.12(b) shows the electric field near a negative charge. The lines of force of the electric field of a positive and of a negative charge are shown in Figure 21.13. They are straight lines passing through the charge.

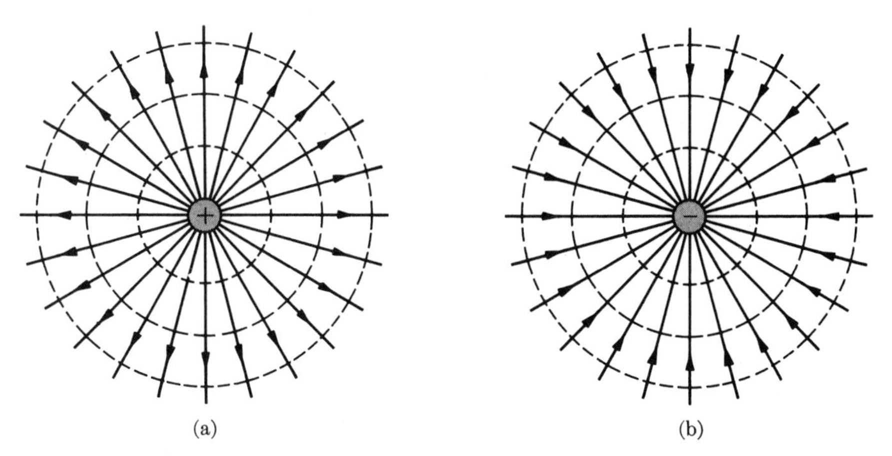

Figure 21.13 Lines of force and equipotential surfaces of the electric field of a positive and a negative charge.

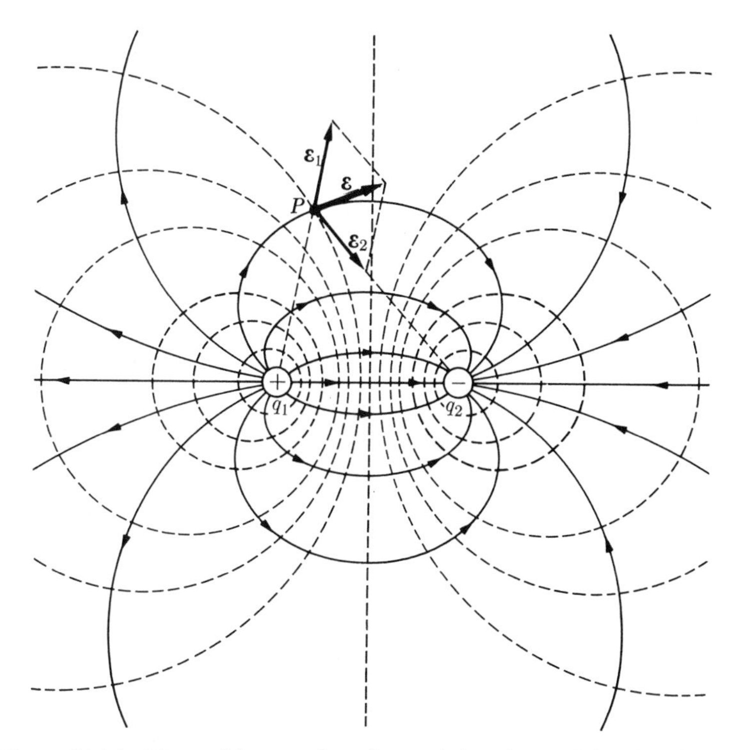

Figure 21.14 Lines of force and equipotential surfaces of the electric field of two equal but opposite charges, such as the electron and proton in a hydrogen atom.

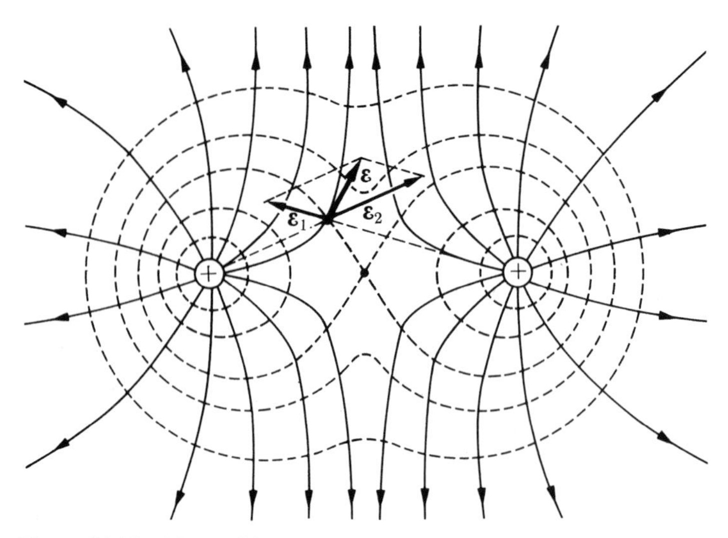

Figure 21.15 Lines of force and equipotential surfaces of the electric field of two identical charges, such as the field produced by the two protons in the H_2 and H_2^+ molecules. The electrons move in this field.

When several charges are present, as in Figure 21.7, the resultant electric field is the vector sum of the electric fields produced by each charge. That is, $\mathscr{E} = \mathscr{E}_1 + \mathscr{E}_2 + \mathscr{E}_3 + \cdots$. Figure 21.14 shows how to obtain the resultant electric field at a point P for the case of a positive and a negative charge of the same magnitude, such as a proton and an electron in a hydrogen atom. Figure 21.15 shows the lines of force for two equal positive charges, such as the two protons in a hydrogen molecule. In both figures the lines of force of the resultant electric field produced by the two charges have also been represented.

EXAMPLE 21.4

The electric field produced by charges q_1 and q_2 at C in Figure 21.16. The charges are given in Example 21.1.

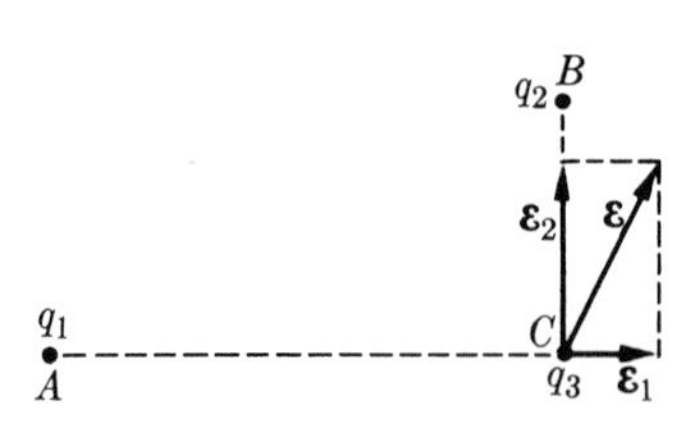

Figure 21.16 Resultant electric field at C produced by q_1 and q_2.

▷ We have two alternative solutions. Since, in Example 21.1, we found the force F on charge q_3 at C we have, using Equation 21.7, that $\mathscr{E} = F/q_3 = 2.03 \times 10^7\ \mathrm{N\,C^{-1}}$. Another procedure is to compute first the electric field produced at C (Figure 21.16) by each of the charges, using Equation 21.9. This gives $\mathscr{E}_1 = q_1/4\pi\varepsilon_0 r_1^2 = 0.94 \times 10^7\ \mathrm{N\,C^{-1}}$ and $\mathscr{E}_2 = q_2/4\pi\varepsilon_0 r_2^2 = 1.80 \times 10^7\ \mathrm{N\,C^{-1}}$. Therefore the resultant field is $\mathscr{E} = (\mathscr{E}_1^2 + \mathscr{E}_2^2)^{1/2} = 2.03 \times 10^7\ \mathrm{N\,C^{-1}}$. The two results clearly agree.

EXAMPLE 21.5

Electric field of a uniformly charged sphere.

▷ In the special case of a charged sphere, the electric field can be computed using the same technique explained in Section 11.7 for the gravitational case. The results are the same since both the gravitational and the electric field follow a $1/r^2$ law. It is only necessary to replace GM by $q/4\pi\varepsilon_0$ in Equations 11.26 and 11.27. Therefore, the field of a sphere of radius a with a charge Q uniformly distributed throughout all its volume is given at all exterior points ($r > a$) by

$$\mathscr{E} = \frac{Q}{4\pi\varepsilon_0 r^2}, \quad r > a \qquad \textbf{(21.11)}$$

This is the same result as if all the charge were concentrated at the center of the sphere. At all interior points ($r < a$) the electric field is given by

$$\mathscr{E} = \frac{Qr}{4\pi\varepsilon_0 a^3}, \quad r < a \qquad \textbf{(21.12)}$$

and is proportional to r. This field has been represented in Figure 21.17.

If the sphere is charged only on its surface, the field at points outside the surface is given by Equation 21.11 and the field at points inside the sphere is zero.

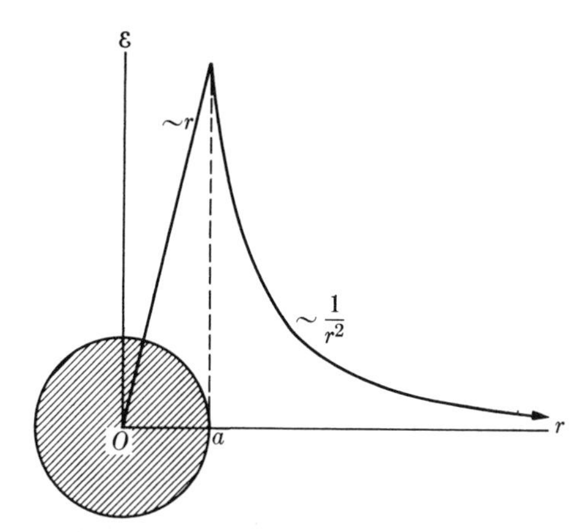

Figure 21.17 Electric field of a uniformly positively charged sphere of radius a.

EXAMPLE 21.6

Electric field produced by a very long, straight filament carrying a charge λ per unit length (Figure 21.18).

▷ Divide the filament of Figure 21.18 into small portions, each of length ds, and therefore carrying a small charge $\mathrm{d}q = \lambda\,\mathrm{d}s$. The magnitude of the electric field that the element ds produces at P is

$$\mathrm{d}\mathscr{E} = \frac{\lambda\,\mathrm{d}s}{4\pi\varepsilon_0 r^2}$$

and it is directed along the line AP. But, due to the symmetry of the problem, for every element ds at distance s above O, there is another element at the same distance below O. Therefore, when we add the electric fields produced by all elements, their components parallel to the filament give a total value of zero. Thus we have to consider only the components parallel to OP, given by $\mathrm{d}\mathscr{E}\cos\alpha$, and the resultant electric field is along OP. Therefore

$$\mathscr{E} = \int \mathrm{d}\mathscr{E}\cos\alpha = \frac{\lambda}{4\pi\varepsilon_0}\int \frac{\mathrm{d}s}{r^2}\cos\alpha$$

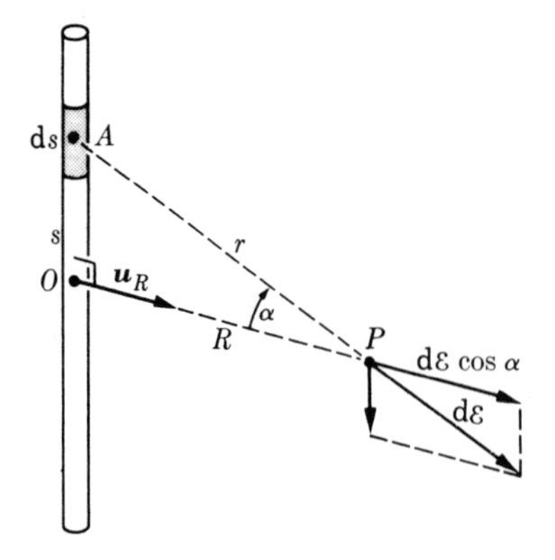

Figure 21.18 Electric field of a positively charged filament.

From the figure we note that $r = R \sec \alpha$ and $s = R \tan \alpha$. From the second relation we have $ds = R \sec^2 \alpha \, d\alpha$. Making these substitutions, integrating from $\alpha = 0$ to $\alpha = \pi/2$ and multiplying by two (since the two halves of the filament make the same contribution), we obtain

$$\mathscr{E} = \frac{2\lambda}{4\pi\varepsilon_0 R} \int_0^{\pi/2} \cos \alpha \, d\alpha = \frac{\lambda}{2\pi\varepsilon_0 R}$$

So the electric field of the filament varies as R^{-1}. In vector form,

$$\boldsymbol{\mathscr{E}} = \frac{\lambda}{2\pi\varepsilon_0 R} \boldsymbol{u}_R$$

21.7 The quantization of electric charge

Electric charge appears not just in any amount but as a multiple of a fundamental charge, or electric quantum. Of the many experiments devised to determine this fact, the classic one is that of Robert A. Millikan (1869–1953), who for several years during the early part of this century performed what is now known as the **oil-drop experiment**. Millikan set up, between two parallel horizontal plates A and B (Figure 21.19), a vertical electric field that could be switched on and off. The upper plate had at its center a few small perforations through which oil drops, produced by an atomizer, could pass. Most of the oil drops were charged by friction with the nozzle of the atomizer. When no electric field is applied, the drops fall by the action of gravity. The motion is essentially uniform because the viscosity of air quickly brings the drop to its terminal velocity (recall Section 7.10), which depends on the size of the drop. Consider a positively charged drop. When an electric field is applied in the upward direction, the motion of the drop is reversed

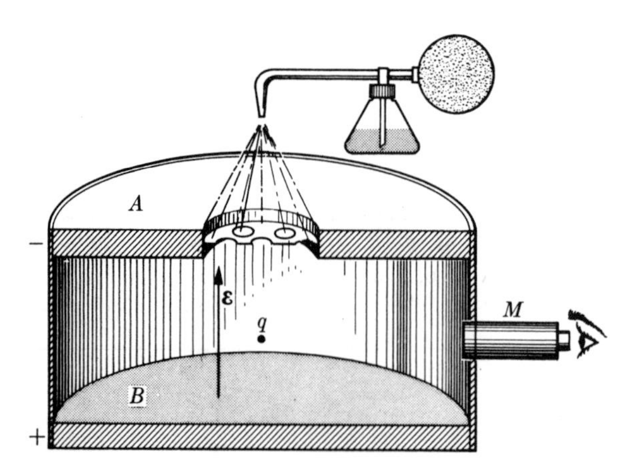

Figure 21.19 Millikan oil-drop experiment. The motion of the charged oil drop q is observed through the microscope M.

and moves upward, again with a constant velocity that depends on the size of the drop and the charge it carries.

Clearly if the charge on the drop changes, the upward velocity changes but the downward velocity remains the same. These changes in charge are due to the occasional ionization of the surrounding air by cosmic rays. The drop may pick up some of these ions while moving through the air. Changes in charge may also be induced by placing a source of X- or γ-rays, which also ionize the air, near the chamber.

By repeated observation of the rise and fall of the drop and by measuring the changes in the upward velocity, the changes Δq in the charge of the drop may be calculated. Sometimes Δq is positive and at other times negative, depending on the nature of the charge modification. Repeating the oil-drop experiment many times, with different drops and different electric fields, it has been found that the absolute values of the changes Δq are always multiples of a fundamental charge e called the **elementary charge**. The experimentally accepted value is

$$e = 1.602177 \times 10^{-19}\,\mathrm{C}$$

Hence, we may state that

> *all charges observed in nature are equal to, or are multiples of, the elementary charge e.*

It therefore seems to be a fundamental law of nature that electric charge is quantized. Until the present time, no one has found an explanation for this fact in terms of more fundamental concepts. However, particles called **quarks**, with charge $\frac{2}{3}e$ and $\frac{1}{3}e$ are assumed to be the basic components of nucleons and mesons. So far these particles have not been observed as free particles and probably never will be, even though evidence seems to point to their existence (see Section 41.8).

A second important aspect of electric charge is that the elementary charge is always associated with some fixed mass, giving rise to what we have called **fundamental particles** (recall Section 1.2). The charge of a proton is $+e$ and the charge of an electron is $-e$. The neutron is uncharged (Table 21.1).

At this point let us go back to our preliminary definition of the coulomb given in Section 2.3 and verify that the number of electrons or protons necessary to equal a negative or positive charge of one coulomb is

$$n = \frac{1\,\mathrm{C}}{1.602177 \times 10^{-19}\,\mathrm{C}} = 6.241508 \times 10^{18}$$

which is the value given in that section.

Table 21.1 Charge of fundamental particles

Particle	Mass, m (kg)	Charge, q	q/m ($\mathrm{C\,kg^{-1}}$)
Electron	$m_e = 9.1091 \times 10^{-31}$	$-e$	1.7588×10^{11}
Proton	$m_p = 1.6725 \times 10^{-27}$	$+e$	9.5792×10^{7}
Neutron	$m_n = 1.6748 \times 10^{-27}$	0	0

EXAMPLE 21.7

Comparison of the gravitational and electric interactions.

▷ In Section 11.10 we indicated that gravitational forces were not strong enough to produce the attraction necessary to bind atoms together in a molecule, or molecules together in a piece of matter. We will now compare the order of magnitude of the electrical and gravitational forces. Assuming that the distances are the same, the strength of the electrical interaction is determined by the coupling factor $q_1 q_2/4\pi\varepsilon_0$, and that of the gravitational interaction by Gm_1m_2. Therefore

$$\frac{\text{Electrical interaction}}{\text{Gravitational interaction}} = \frac{q_1 q_2}{4\pi\varepsilon_0 G m_1 m_2}$$

To obtain the order of magnitude, we set $q_1 = q_2 = e$ and $m_1 = m_2 = m_p$ so that for two protons or two hydrogen ions,

$$\frac{\text{Electrical interaction}}{\text{Gravitational interaction}} = \frac{e^2}{4\pi\varepsilon_0 G m_p^2} = 1.24 \times 10^{36}$$

This is just about the same as the factor by which the gravitational forces fell short of producing the required interaction (recall Chapter 16).

For the interaction between a proton and an electron, $m_1 = m_p$, $m_2 = m_e$, $q_1 = -q_2 = e$ and the above ratio is even larger: 2.27×10^{39}. We conclude, then, that the electrical interaction is of the order of magnitude needed to bind atoms to form molecules, or to bind electrons and nuclei to form atoms, and that gravitation plays no role in the structure of matter at the atomic level; it is important only for large 'lumps' of matter.

Note 21.1 Analysis of the Millikan oil-drop experiment

Let us call m the mass and r the radius of an oil drop. The terminal velocity v_1 for free fall with the electric field switched off is given by Equation 8.21, $v_1 = F/K\eta$, with $K = 6\pi r$ according to Equation 8.19 and $F = mg$; that is

$$v_1 = \frac{mg}{6\pi\eta r} \tag{21.13}$$

In order to be accurate, we should also take the buoyancy of the air into account, by writing $m - m_a$ instead of m, where m_a is the mass of air displaced by the drop, but we will ignore this modification. Assuming that the drop has a positive charge q, when we apply the electric field the resultant force in the upward direction is $F' = q\mathscr{E} - mg$ and the upward terminal velocity of the drop is given by

$$v_2 = \frac{F'}{K\eta} \quad \text{or} \quad v_2 = \frac{q\mathscr{E} - mq}{6\pi\eta r}$$

Solving for q, and using Equation 21.13 to eliminate mg, we find that the electric charge on the drop is

$$q = \frac{6\pi\eta r(v_1 + v_2)}{\mathscr{E}} \tag{21.14}$$

If the charge is negative, the upward motion is produced by applying a downward electric field.

Suppose now that the charge in the drop changes, with a corresponding change in the upward velocity. According to Equation 21.14, the changes Δq of charge and Δv_2 of upward velocity are related by

$$\Delta q = \frac{6\pi\eta r}{\mathscr{E}} \Delta v_2$$

This expression allows us to compute the changes Δq in terms of measurable quantities. In particular, we can find the radius of the drop by measuring v_1 and solving Equation 21.13 for r with $m = (\frac{4}{3}\pi r^3)\rho$, where ρ is the density of the oil.

21.8 Principle of conservation of charge

It has been found that, in all processes observed in nature, the net, or total, charge of an isolated system remains constant. In other words,

> *the net or total charge does not change for any processes occurring within an isolated system.*

No exception has been found to this rule, known as the **principle of conservation of charge**. We have already used it in Chapter 20, where the reaction

$$p^+ + p^+ \rightarrow p^+ + p^+ + p^+ + p^-$$

was discussed. On the left the total charge is twice the proton charge (that is, $+2e$) and on the right the three protons contribute three times the proton charge, while the antiproton contributes a negative proton charge. This gives a net charge of $+3e - e = +2e$, again equal to twice the proton charge.

Similarly, in the processes associated with the decay of fundamental particles discussed in Section 20.5, charge is conserved. For example, in β-decay $n \rightarrow p^+ + e^- + \bar{\nu}$ and $p^+ \rightarrow n + e^+ + \nu$ (where ν is a neutrino and $\bar{\nu}$ is an antineutrino), in muon decay $\mu^\pm \rightarrow e^\pm + \nu + \bar{\nu}$ and in pion decay $\pi^\pm \rightarrow \mu^\pm + \nu$, charge is conserved since the neutrino ν and the antineutrino $\bar{\nu}$ are uncharged particles.

21.9 Electric potential

A charged particle placed in an electric field has potential energy because the field does work when moving the particle from one place to another. The **electric potential** at a point in an electric field is defined as the potential energy per unit charge placed at the point. Designating the electric potential by V and the potential energy of a charge q by E_p, we have

$$V = \frac{E_p}{q} \quad \text{or} \quad E_p = qV \tag{21.15}$$

The electric potential is measured in joules/coulomb or J C^{-1}, a unit called a **volt**, abbreviated V, in honor of Alessandro Volta (1745–1827). In terms of the fundamental units we write $\text{V} = \text{m}^2\,\text{kg}\,\text{s}^{-2}\,\text{C}^{-1}$.

If a charge q moves from one point P_1 to another point P_2 along any path, the work done by the electric field is, using relation 9.21,

$$W = -\Delta E_p = E_{p1} - E_{p2} = q(V_1 - V_2)$$

And so, the difference in potential between points P_1 and P_2 is

$$V_1 - V_2 = \frac{W}{q} \quad \text{or} \quad \Delta V = -\frac{W}{q} \tag{21.16}$$

where $\Delta V = V_2 - V_1$ is the change in electric potential. Thus we may define the **electric potential difference** between two points as the work done by the electric field in moving a unit of positive charge from one point to the other. From the definition then, the electric potential *at a point* can be obtained by measuring the work done by the electric field in moving a unit positive charge from the point to a place where the electric potential is zero, usually at infinity.

Considering that the fundamental particles and nuclei have a charge that is equal to, or is a multiple of, the fundamental charge e, Equation 21.16 suggests that we define a new unit of energy, called the **electronvolt**, abbreviated eV, which was first introduced in Section 9.5. An electronvolt is equal to the work done on a particle of charge e when it is moved through a potential difference of one volt. Thus, using the value of e and Equation 21.16, we have

$$1\,\text{eV} = (1.6022 \times 10^{-19}\,\text{C})(1\,\text{V}) = 1.6022 \times 10^{-19}\,\text{J}$$

which is the equivalence given in Section 9.6. A particle of charge ve moving through a potential difference ΔV gains an energy of $v\Delta V$ electronvolts. Convenient multiples of the electronvolt are the *kiloelectronvolt* ($\text{keV} = 10^3\,\text{eV}$), the *megaelectronvolt* ($\text{MeV} = 10^6\,\text{eV}$) and the *gigaelectronvolt* ($\text{GeV} = 10^9\,\text{eV}$). Particle accelerators produce charged particles with energies of several MeV and larger. The rest energy of the fundamental particles in electronvolts was given in Example 19.3.

21.10 Relation between electric potential and electric field

Consider two points that are very close, separated the small distance $\text{d}s$. Then the potential difference between the two points is $\text{d}V$. If the electric field between the two points is $\mathscr{E}$, the force on a charge q is $\boldsymbol{F} = q\mathscr{E}$ and the work done in moving the charge from P_1 to P_2 is $dW = F_s\,ds = q\mathscr{E}_s\,ds$, where $\mathscr{E}_s$ is the component of the electric field along the line P_1P_2. In this case, Equation 21.16 becomes

$$dV = -\frac{dW}{q} = -\frac{q\mathscr{E}_s\,ds}{q} = -\mathscr{E}_s\,ds$$

Accordingly, the component $\mathscr{E}_s$ of the electric field along the direction corresponding to a displacement ds is given by

$$\mathscr{E}_s = -\frac{dV}{ds} \qquad \textbf{(21.17)}$$

The negative sign shows that the electric field points in the direction in which the electric potential decreases. Hence

> *the electric field is the negative of the directional derivative or gradient of the electric potential,*

a relation similar to that for the gravitational field, Equation 11.21.

Equation 21.17 indicates that the electric field can also be expressed in volts/meter, a unit equivalent to newton/coulomb given before. This can be seen in the following way:

$$\frac{\text{volt}}{\text{meter}} = \frac{\text{joule}}{\text{coulomb meter}} = \frac{\text{newton meter}}{\text{coulomb meter}} = \frac{\text{newton}}{\text{coulomb}}$$

By common usage, the term volt/meter, abbreviated V m^{-1}, is preferred to N C^{-1} for the electric field. Equation 21.17 is used to find the electric potential V when the field $\mathscr{E}$ is known, and conversely.

EXAMPLE 21.8

Electric potential in a uniform electric field.

▷ We shall illustrate the use of Equation 21.17 for the case of a uniform electric field. Placing the X-axis parallel to the field (Figure 21.20), we may write $\mathscr{E} = -dV/dx$. Since $\mathscr{E}$ is

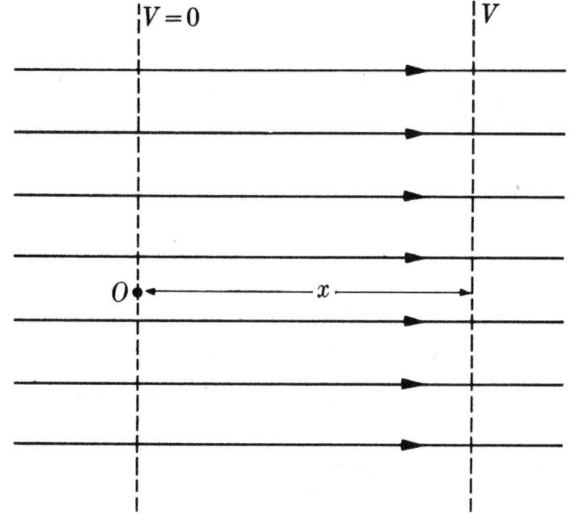

Figure 21.20 Uniform electric field.

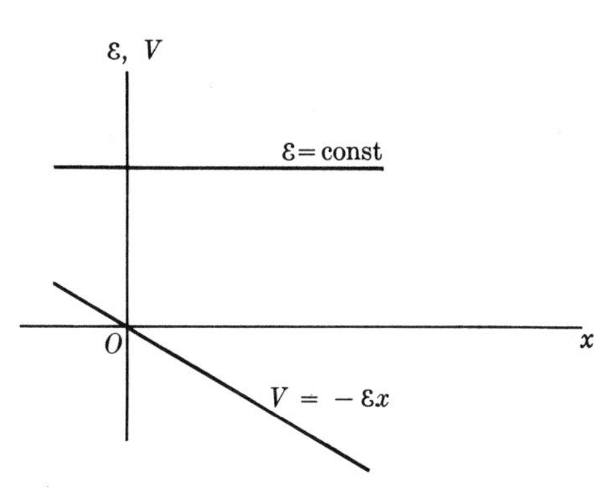

Figure 21.21 Variations of $\mathscr{E}$ and V for a uniform electric field.

constant, and we assume $V = 0$ at $x = 0$, by integration we have

$$\int_0^V dV = -\int_0^x \mathscr{E}\, dx = -\mathscr{E}\int_0^x dx \quad \text{or} \quad V = -\mathscr{E}x \tag{21.18}$$

This very useful relation has been represented in Figure 21.21. We may note that, because of the negative sign in Equation 21.18, the electric field points in the direction in which the electric potential decreases. When we consider two points x_1 and x_2, Equation 21.18 gives $V_1 = -\mathscr{E}x_1$ and $V_2 = -\mathscr{E}x_2$. Subtracting, we have $V_2 - V_1 = -\mathscr{E}(x_2 - x_1)$ or, calling $d = x_2 - x_1$, we obtain

$$\mathscr{E} = -\frac{V_2 - V_1}{d} \quad \text{or} \quad \mathscr{E} = \frac{V_1 - V_2}{d} \tag{21.19}$$

Although this relation is valid only for uniform electric fields, it can be used to estimate the average electric field between two points separated by a distance d, if the potential difference $V_1 - V_2$ between them is known. When the potential difference $V_1 - V_2$ is positive, the field points in the direction from x_1 to x_2, and if it is negative, the field points in the opposite direction.

21.11 Electric potential of a point charge

To obtain the electric potential due to a point charge, we use Equation 21.17, with s replaced by the distance r, since the electric field is along the radius, that is, $\mathscr{E} = -dV/dr$. Remembering Equation 21.9, $\mathscr{E} = q/4\pi\varepsilon_0 r^2$, we may write

$$\frac{1}{4\pi\varepsilon_0}\frac{q}{r^2} = -\frac{dV}{dr}$$

Integrating,

$$\int dV = -\frac{q}{4\pi\varepsilon_0}\int \frac{dr}{r^2} \quad \text{or} \quad V = \frac{q}{4\pi\varepsilon_0 r} + C$$

It is customary to assume $V = 0$ for $r = \infty$, as in the gravitational case, so that $C = 0$. Then

$$V = \frac{q}{4\pi\varepsilon_0 r} \tag{21.20}$$

which gives the electric potential at the distance r from the charge. Expression 21.20 could also have been obtained by replacing $-GM$ in Equation 11.19 by $q/4\pi\varepsilon_0$. In the corresponding formula for the gravitational field, the negative sign was written explicitly because the gravitational interaction is always attractive. Note that the electric potential is positive or negative depending on the sign of the charge q.

If we have several charges $q_1, q_2, q_3, \ldots$, the electric potential at a point P (Figure 21.7) is the scalar sum of their individual potentials. That is,

$$V = \frac{q_1}{4\pi\varepsilon_0 r_1} + \frac{q_2}{4\pi\varepsilon_0 r_2} + \frac{q_3}{4\pi\varepsilon_0 r_3} + \cdots = \frac{1}{4\pi\varepsilon_0}\sum_i \frac{q_i}{r_i}$$

If we place a charge q' at a distance r from a charge q, the potential energy of the system is $E_p = q'V$, or

$$E_p = \frac{qq'}{4\pi\varepsilon_0 r} \tag{21.21}$$

and the potential energy of a system of charges is

$$E_p = \sum_{\text{All pairs}} \frac{qq'}{4\pi\varepsilon_0 r} \tag{21.22}$$

This is important, for example, to calculate the energy of an atom or a molecule, since they are aggregates of electrons and positively charged nuclei. Note that, although the electric force is inversely proportional to the square of the distance, the electric potential energy varies only inversely proportional to the distance.

Surfaces having the same electric potential at all points, that is, $V = const.$, are called **equipotential surfaces**. At each point of an equipotential surface, the direction of the electric field is perpendicular to the surface because, in view of Equation 21.16, the work done by the electric force is zero when moving a charge over the equipotential surface. That is, the lines of force are orthogonal to the equipotential surfaces. (Recall the discussion in Section 11.7 for the gravitational field.) For a uniform field we see, from Equation 21.19, that $V = const.$ implies $x = const.$, and therefore the equipotential surfaces are planes, indicated by the dashed lines in Figure 21.20. For a point charge, Equation 21.20 indicates that the equipotential surfaces are the spheres $r = const.$, indicated by dashed lines in Figure 21.13(a) and (b). For two charges the equipotential surfaces have been indicated by dashed lines in Figures 21.14 and 21.15.

EXAMPLE 21.9

Electric potential energy of charge q_3 in Example 21.1.

▷ Refer back to Figure 21.6 and use Equation 21.20. The electric potentials produced at C by charges q_1 and q_2 at A and B are, respectively,

$$V_1 = \frac{q_1}{4\pi\varepsilon_0 r_1} = 11.2 \times 10^6 \text{ V} \quad \text{and} \quad V_2 = \frac{q_2}{4\pi\varepsilon_0 r_2} = -9.0 \times 10^6 \text{ V}$$

Thus the total electric potential at C is $V = V_1 + V_2 = 2.2 \times 10^6$ V. The potential energy of charge q_3 is then

$$E_p = q_3 V = (0.2 \times 10^{-3}\text{ C})(2.2 \times 10^6\text{ V}) = 4.4 \times 10^2\text{ J}$$

When we compare this example with Example 21.4, we see the difference between handling the electric field and the electric potential.

EXAMPLE 21.10

Energy of an electron in a hydrogen atom.

▷ The potential energy of an electron moving in a hydrogen atom at a distance of 0.53×10^{-10} m from the proton, keeping in mind that the electron and proton have opposite charges so that $q = e$ and $q' = -e$ is, using Equation 21.21,

$$E_p = \frac{-(8.987 \times 10^9)(1.6022 \times 10^{-19}\,\mathrm{C})^2}{0.53 \times 10^{-10}\,\mathrm{m}} = -4.353 \times 10^{-18}\,\mathrm{J} = -27.2\,\mathrm{eV}$$

For the total energy of the electron we must add its kinetic energy $E_k = \frac{1}{2}mv^2$, which, using the value of the velocity v found from Example 5.1 ($\omega = 4.13 \times 10^{16}$ rad s^{-1}, $v = \omega r = 2.189 \times 10^6$ m s^{-1}), is 2.177×10^{-18} J = 13.6 eV. Thus the total energy is

$$E = E_p + E_k = -2.177 \times 10^{-18}\,\mathrm{J} = -13.6\ \mathrm{eV}$$

EXAMPLE 21.11

Electric potential produced by a long straight filament with a charge λ per unit length.

▷ In Example 21.6 we proved that the electric field at a distance R from the filament (see Figure 21.18) is $\mathscr{E} = \lambda/4\pi\varepsilon_0 R$. To find the electric potential, we use the relation $\mathscr{E} = -\mathrm{d}V/\mathrm{d}R$, which gives

$$\frac{\mathrm{d}V}{\mathrm{d}R} = -\frac{\lambda}{2\pi\varepsilon_0 R}$$

Integration yields

$$\int \mathrm{d}V = -\frac{\lambda}{2\pi\varepsilon_0}\int \frac{\mathrm{d}R}{R} \qquad \text{or} \qquad V = -\frac{\lambda}{2\pi\varepsilon_0}\ln R + C$$

It is customary in this case to assign the zero of the potential at the point where $R = 1$, giving $C = 0$. Therefore we find that the electric potential is

$$V = -\frac{\lambda}{2\pi\varepsilon_0}\ln R$$

Note that in this case the potential is not zero at infinity.

EXAMPLE 21.12

The electric dipole.

▷ An **electric dipole** consists of two equal and opposite charges $+q$ and $-q$, separated by a very small distance a (Figure 21.22). Although in an electric dipole the two charges are equal and opposite, giving a zero net charge, the fact that they are slightly displaced is enough to produce a non-vanishing electric field. In many substances, like H_2O, the centers of the positive and negative charges of the molecule are slightly separated and thus resemble electric dipoles. We then say that the substance is composed of polar molecules.

The **electric dipole moment**, $\boldsymbol{P}$, is defined by

$$\boldsymbol{P} = q\boldsymbol{a} \tag{21.23}$$

where $\boldsymbol{a}$ is the vector displacement from the negative to the positive charge. The electric potential at a point R due to the electric dipole is, using Equation 21.20,

$$V = \frac{1}{4\pi\varepsilon_0}\left(\frac{q}{r_1} - \frac{q}{r_2}\right) = \frac{1}{4\pi\varepsilon_0}\frac{q(r_2 - r_1)}{r_1 r_2}$$

If the distance a is very small compared with r, we can set $r_2 - r_1 = a\cos\theta$ and $r_1 r_2 = r^2$, resulting in

$$V = \frac{qa\cos\theta}{4\pi\varepsilon_0 r^2}$$

Making $P = qa$ we finally obtain

$$V = \frac{P\cos\theta}{4\pi\varepsilon_0 r^2} \tag{21.24}$$

Therefore the electric potential of a dipole varies as r^{-2} instead of as r^{-1} for a point charge.

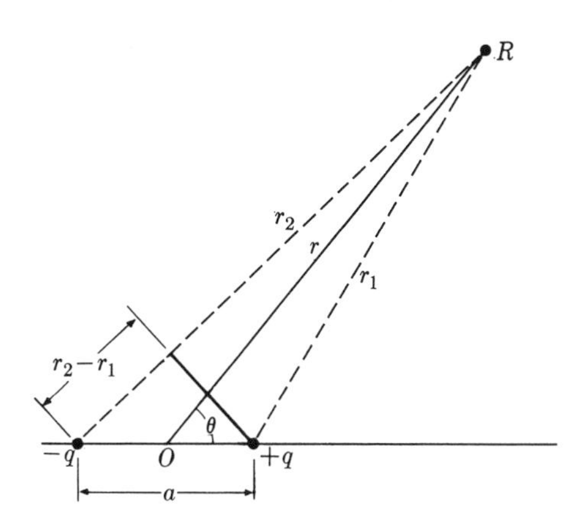

Figure 21.22 Electric dipole.

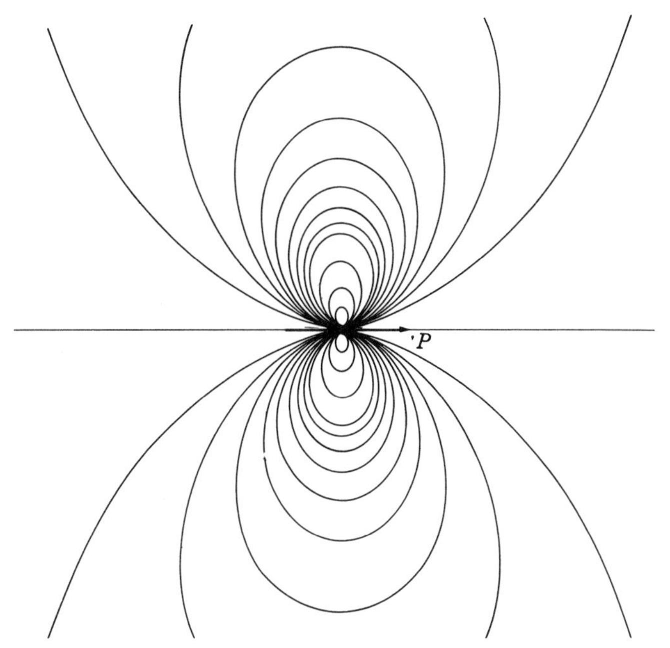

Figure 21.23 Lines of force of the electric field of an electric dipole. This is, for example, the electric field around a polar molecule such as CO.

We shall not compute the electric field of a dipole, which in this case depends on r and θ, but when Equation 21.24 is used, we find that the field varies as r^{-3} instead of the r^{-2} that corresponds to the electric field of a single charge. For example, along the axis of the dipole, $\theta = 0$, $V = P/4\pi\varepsilon_0 r^2$ and the electric field is

$$\mathscr{E} = -\frac{dV}{dr} = \frac{1}{4\pi\varepsilon_0}\frac{2P}{r^3} \tag{21.25}$$

The lines of force of the field produced by an electric dipole are indicated in Figure 21.23.

EXAMPLE 21.13

Force and torque of an electric field on an electric dipole.

▷ When an electric dipole is placed in an electric field, a force is produced on each charge of the dipole (Figure 21.24). The resultant force is $\boldsymbol{F} = q\boldsymbol{\mathscr{E}} + (-q\boldsymbol{\mathscr{E}}') = q(\boldsymbol{\mathscr{E}} - \boldsymbol{\mathscr{E}}')$, where $\boldsymbol{\mathscr{E}}$ and $\boldsymbol{\mathscr{E}}'$ are the values of the field at $+q$ and $-q$. Note that if the field is uniform, so that $\boldsymbol{\mathscr{E}} = \boldsymbol{\mathscr{E}}'$, we have $\boldsymbol{F} = 0$. Therefore, the resultant force produced by a uniform field on an electric dipole is zero.

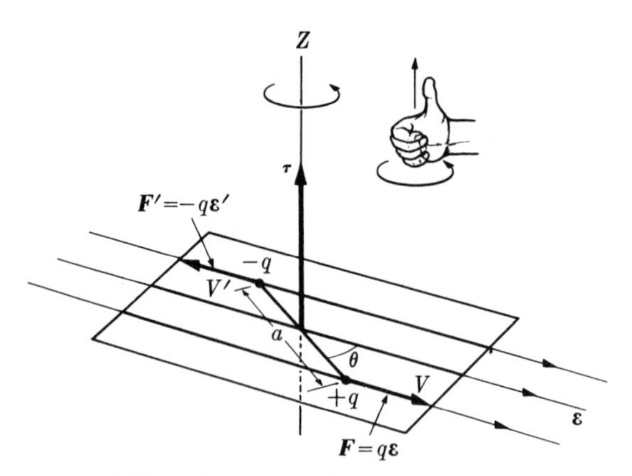

Figure 21.24 Forces on an electric dipole in an external field.

When the electric field is not uniform (Figure 21.25) and the dipole is oriented parallel to the field, then taking the X-axis along the direction of the field,

$$\mathscr{E} - \mathscr{E}' = \left(\frac{d\mathscr{E}}{dx}\right)a$$

and therefore

$$F = q(\mathscr{E} - \mathscr{E}') = P\left(\frac{d\mathscr{E}}{dx}\right) \tag{21.26}$$

This result shows that

> *an electric dipole oriented parallel to an electric field tends to move in the direction in which the field increases; the opposite result is obtained if the dipole is oriented antiparallel to the field.*

For example, when polar molecules such as CO and H_2O are placed in a non-uniform electric field, they tend to drift toward the region where the field is stronger.

An electric field produces a torque that tends to align the dipole along the field. As can be seen in Figure 21.24, the forces $q\mathscr{E}$ and $-q\mathscr{E}$ on the charges comprising the dipole form a couple whose torque, relative to O, is

$$\boldsymbol{\tau} = (\tfrac{1}{2}\boldsymbol{a}) \times (q\boldsymbol{\mathscr{E}}) + (-\tfrac{1}{2}\boldsymbol{a}) \times (-q\boldsymbol{\mathscr{E}}) = \boldsymbol{a} \times (q\boldsymbol{\mathscr{E}}) = (q\boldsymbol{a}) \times \boldsymbol{\mathscr{E}} = \boldsymbol{P} \times \boldsymbol{\mathscr{E}}$$

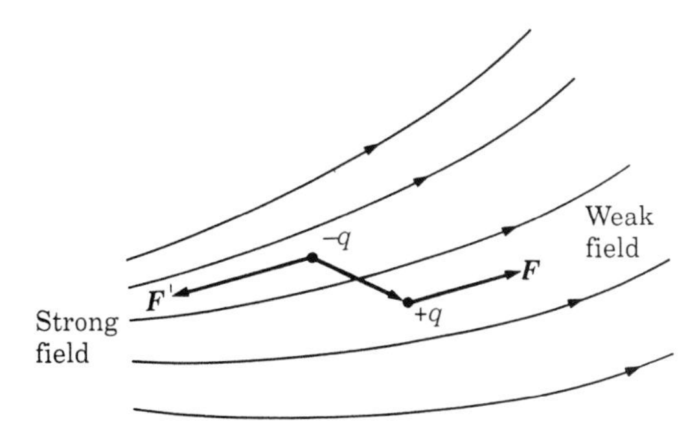

Figure 21.25 Electric dipole in a non-uniform external field, $F' > F$. The dipole moves toward where the field is stronger.

From the above expression, as well as from the figure, we see that

the torque of the electric field tends to align the dipole parallel to the field.

Therefore, when a substance whose molecules are polar is placed in an electric field, all molecules tend to orient their electric dipoles along the field. We say that the substance is **polarized**. The magnitude of the torque on the electric dipole is $\tau = P\mathscr{E} \sin\theta$, where θ is the angle between $\boldsymbol{P}$ and $\mathscr{E}$, and its direction is as indicated in Figure 21.24.

The potential energy of an electric dipole placed in an electric field is

$$E_p = qV - qV' = q(V - V') = -qa\left(-\frac{(V - V')}{a}\right)$$

and by using Equation 21.17, we find that the last factor is the component $\mathscr{E}_a$ of the field $\mathscr{E}$ parallel to $\boldsymbol{a}$. Therefore, if θ is the angle between the dipole and the electric field, $E_p = -qa\mathscr{E}_a = -qa\mathscr{E} \cos\theta$, or

$$E_p = -P\mathscr{E} \cos\theta = -\boldsymbol{P}\cdot\boldsymbol{\mathscr{E}} \qquad \textbf{(21.27)}$$

The potential energy is a minimum when $\theta = 0$, indicating that

the dipole is in equilibrium when it is oriented parallel to the field.

21.12 Energy relations in an electric field

The total energy of a charged particle or ion of mass m and charge q moving in an electric field is

$$E = E_k + E_p = \tfrac{1}{2}mv^2 + qV \qquad \textbf{(21.28)}$$

When the ion moves from position P_1 (where the electric potential is V_1) to position P_2 (where the potential is V_2), Equation 21.28, combined with the principle of conservation of energy, gives $\frac{1}{2}mv_1^2 + qV_1 = \frac{1}{2}mv_2^2 + qV_2$, which we may write as

$$\tfrac{1}{2}mv_2^2 - \tfrac{1}{2}mv_1^2 = q(V_1 - V_2) \qquad \textbf{(21.29)}$$

Note from Equation 21.29 that a positively charged particle ($q > 0$) gains kinetic energy when moving from a larger to a smaller potential ($V_1 > V_2$), while a negatively charged particle ($q < 0$), to gain energy, has to move from a lower to a higher potential ($V_1 < V_2$). Thus protons move in the direction in which the electric potential decreases, while electrons move in the direction in which the electric potential increases. There-fore, when there is a region where both positive and negative charges exist, such as in a plasma, and an electric field is applied, the charges tend to separate (Figure 21.26).

If we designate $\Delta V = V_2 - V_1$ and arrange our experiment so that at P_1 the ions have zero velocity ($v_1 = 0$), Equation 21.29 becomes (dropping the subscripts)

$$\tfrac{1}{2}mv^2 = q\Delta V \quad \text{or} \quad v^2 = 2\left(\frac{q}{m}\right)\Delta V \qquad \textbf{(21.30)}$$

which gives the kinetic energy and the velocity acquired by a charged particle when it moves through an electric potential difference ΔV. Equation 21.30 is similar to the relation $\frac{1}{2}mv^2 = mgh$ (or $v^2 = 2gh$), which gives the increase in kinetic energy of a body when it falls through a distance h in the Earth's gravitational field. This is, for example, the principle applied in **electrostatic accelerators** and in accelerating electrons in a television tube.

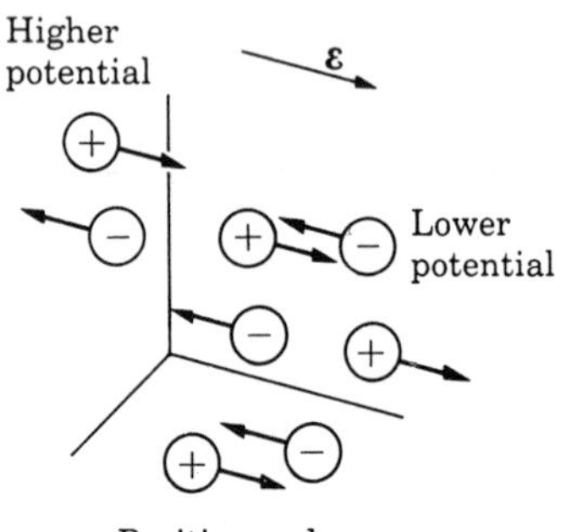

Figure 21.26 Motion of positive and negative ions in an electric field $\mathscr{E}$.

The widely used van de Graaff accelerator (Figure 21.27), in which charged particles are accelerated, consists of an evacuated tube through which an electric potential difference is applied between its extremes. At one end there is an ion source injecting charged particles into the tube. The particles arrive at the other end with a kinetic energy given by Equation 21.30. These fast ions impinge on a target made of a material chosen according to the nature of the experiment to be performed. The result of this collision is a scattering or some kind of nuclear reaction. The energy of the impinging ions is transferred to the target, which therefore must be constantly cooled, since otherwise it would melt or vaporize.

EXAMPLE 21.14

The accelerating potential in a van de Graaff accelerator (Figure 21.27) is 4.0×10^6 V. The particles are protons (or hydrogen ions). The ion source ionizes 1.00×10^{-9} g of atomic

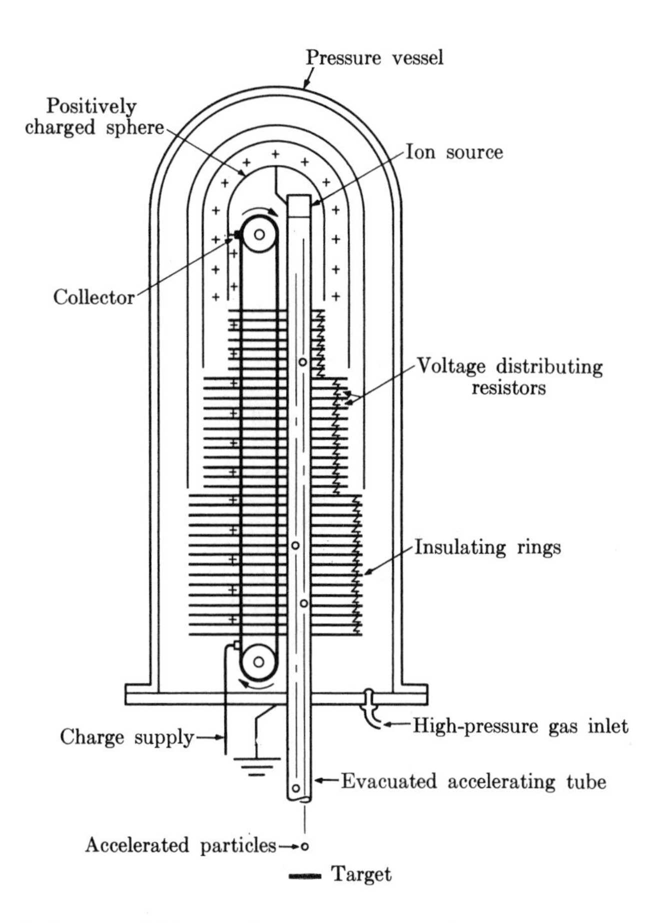

Figure 21.27 Simplified diagram of a van de Graaff accelerator. A high-speed motor runs a belt, made of an insulating material, over two pulleys. Electric charge from a voltage source is picked up by the belt at the lower end and conveyed upward. A collector draws off the charge to the metal sphere at the top, building up a high electric potential on it. Positive ions are produced at the high voltage end and are accelerated downward by the potential difference between the charged sphere and the ground potential at the other end.

hydrogen, which is equivalent to 5.98×10^{14} protons each second. Find the velocity of the protons when striking the target and the rate at which energy is transferred to the target.

▷ Equation 21.30, $\frac{1}{2}mv^2 = q\Delta V$, gives $v = (2q\Delta V/m)^{1/2}$ for the velocity of the protons when they strike the target. Setting $\Delta V = 4.0$ MeV and using the value of q/m for protons (see Table 21.1), we get $v = 2.77 \times 10^7$ m s^{-1}. The protons reach the target with a kinetic energy of 4 MeV, which is transferred to the target.

The power required to drive the accelerator depends on how many protons are accelerated per second. If the ion source produces n ions per second, then the power is given by $P = nq\Delta V$. Since the ion source produces 5.98×10^{14} protons each second, the power is

$$P = (5.98 \times 10^{14}\,\text{s}^{-1})(1.60 \times 10^{-19}\,\text{C})(4.0 \times 10^{6}\,\text{V}) = 383\,\text{W}$$

This is also the rate of energy transfer to the target.

QUESTIONS

21.1 State the main differences and similarities between the law of gravitational attraction and the law of electric interaction.

21.2 Suppose F is the force between two electric charges separated the distance r. Plot the points corresponding to the force when the separation is $\frac{1}{2}r$, $2r$, and $3r$. Join all the points by a smooth curve to show how the electric force varies with distance.

21.3 A positively charged particle enters a region where there is a uniform electric field. Describe the motion of the particle if the initial velocity is (a) in the direction of the field, (b) directly opposite to the field and (c) at an angle to the field.

21.4 Suppose E_p is the potential energy of two electric charges separated the distance r. Plot the points corresponding to the potential energy when the separation is $\frac{1}{2}r$, $2r$ and $3r$. Join all the points by a smooth curve to show how electric potential energy varies with distance.

21.5 Consider an electron revolving around a proton in a hydrogen atom. Is the electric force on the electron central? Which dynamical quantities are constants of motion of the electron?

21.6 If the Earth and the Sun are composed of electrically charged particles, why is it that the motion of the Earth around the Sun is described entirely in terms of the gravitational attraction?

21.7 Represent the variation of the electric field and potential of a long straight charged filament as a function of the distance R from the filament.

21.8 If an electric charge moves along a line of force in the direction of the electric field, how does the potential energy of the charge vary when the charge is (a) positive, (b) negative? In each case what is the sign of the work done by the electric field?

21.9 Represent the variation of the electric potential of an electric dipole (a) as a function of r for a fixed angle θ and (b) as a function of θ for a fixed distance r.

21.10 Why can we say that charge is quantized? Is mass also quantized?

21.11 How does an electric dipole move in (a) a uniform, (b) non-uniform electric field?

21.12 Explain how the resultant electric field of an electric dipole is not zero despite the fact that it is composed of two equal and opposite charges.

21.13 A beam of protons is accelerated in a linear accelerator. The ion source is at one end of the tube and the target is at the other end. Which end is at the higher potential? What is the direction of the electric field? Plot the kinetic energy and the velocity of the protons as a function of the distance along the tube.

21.14 Represent the lines of force of the resultant electric field of two identical electric dipoles: (a) when they are aligned along the line joining them and with their positive 'ends' pointing in the same direction and when they are oriented with positive 'ends' pointing toward each other; (b) when they are perpendicular to the line joining their centers and with their positive 'ends' both pointing 'up' and when they are oriented with positive 'ends' one 'up' and one 'down'.

PROBLEMS

21.1 (a) Find the electric force of repulsion between the two protons in a hydrogen molecule. Their separation is 0.74×10^{-10} m. (b) Compare the electric force with their gravitational attraction.

21.2 Compare the electric force of attraction between the proton and the electron in a hydrogen atom, assuming that the electron describes a circular orbit of 0.53×10^{-10} m with their gravitational attraction.

21.3 Two identical balls of mass m have equal charges q. They are attached to two strings of length l hanging from the same point. (a) Find the angle θ the strings make with the vertical when equilibrium is reached. (b) Draw the forces on each ball and their resultant.

21.4 A charge of 2.5×10^{-8} C is placed in an upward uniform electric field of 5×10^{4} N C^{-1}. Calculate the work of the electric field on the charge when the charge moves (a) 0.45 m to the right, (b) 0.80 m downward and (c) 2.60 m at an angle of 45° upward from the horizontal.

21.5 An electron is projected into an upward uniform electric field of 5000 N C^{-1}. The initial velocity of the electron is 10^{7} m s^{-1}, at an angle of 30° above the horizontal. Calculate (a) the time required for the electron to reach its maximum height, (b) the maximum distance the electron rises vertically above its initial position and (c) the horizontal distance traveled until the electron returns to its original elevation.
(d) Sketch the trajectory of the electron.

21.6 Two point charges, $5\ \mu C$ and $-10\ \mu C$, are spaced 1 meter apart. (a) Find the magnitude and direction of the electric field at a point 0.6 m from the first charge and 0.8 m from the second charge. (b) Where is the electric field zero due to these two charges? (c) Calculate the point along the line joining the charges where the electric field of each charge is identical both in magnitude and direction.

21.7 A small sphere, mass 2×10^{-3} kg, hangs by a thread between two parallel vertical plates 0.05 m apart. The charge on the sphere is 6×10^{-9} C. Calculate the potential difference between the plates if the thread makes an angle of $10°$ with the vertical.

21.8 The electric field between the deflecting plates of an oscilloscope is $3.0 \times 10^4\ N\ C^{-1}$. Calculate (a) the force on an electron in this region and (b) the acceleration of an electron when it is acted on by this force. Compare this value with the acceleration of gravity.

21.9 A uniform electric field exists between two oppositely charged parallel plates. An electron is released from rest at the surface of the negatively charged plate and strikes the surface of the opposite plate, 2.0×10^{-2} m from the first, in a time interval of 1.5×10^{-8} s. (a) Calculate the intensity of the electric field and (b) the velocity of the electron when it strikes the second plate. (c) What is the potential difference between the plates?

21.10 Three positive charges, 2×10^{-7} C, 1×10^{-7} C and 3×10^{-7} C, are in a straight line, with the second charge in the middle. The separation between adjacent charges is 0.10 m. Calculate (a) the resultant force on each charge due to the other two, (b) the potential energy of each charge due to the other two and (c) the internal energy of the system. (d) Compare the internal energy of the system with the sum of potential energies and explain any difference.

21.11 A charge of 25×10^{-9} C is placed at the origin of a coordinate system. A second charge of -25×10^{-9} C is placed at $x = 6$ m, $y = 0$ m. Calculate the electric field at (a) $x = 3$ m, $y = 0$ m and (b) $x = 3$ m, $y = 4$ m.

21.12 An oil droplet of mass 3×10^{-14} kg and radius 2×10^{-6} m carries 10 extra electrons. What is its terminal velocity when it falls (a) in a region where there is no electric field and (b) in an electric field whose intensity is $3 \times 10^5\ N\ C^{-1}$, directed downward? The viscosity of air is $1.80 \times 10^{-5}\ N\ s\ m^{-2}$. Neglect the buoyant force of air.

21.13 A charged oil drop, in a Millikan oil-drop apparatus, is observed to fall through a distance of 1 mm in 27.4 s in the absence of any electric field. The same drop is held stationary in an electric field of $2.37 \times 10^4\ N\ C^{-1}$. Calculate the number of excess electrons on the oil drop. The viscosity of air is given in Problem 21.12 and the density of the oil is $800\ kg\ m^{-3}$.

21.14 The electric potential at a certain distance from a point charge is 600 V and the electric field is $200\ N\ C^{-1}$. (a) Calculate the distance to the point charge. (b) Calculate the magnitude of the charge.

21.15 The maximum charge that can be retained by one of the spherical terminals of a large van de Graaff accelerator (see Figure 21.27) is about 10^{-3} C. Assume a positive charge of this amount, distributed uniformly over the surface of the sphere, 3.0 m in diameter, in otherwise empty space. (a) Calculate the magnitude of the electric field at a point outside the sphere, 5.0 m from its center. (b) If an electron were released at this point, calculate the magnitude and direction of its initial acceleration. (c) Calculate its acceleration just before it strikes the sphere. (d) Calculate the velocity with which it would strike the sphere if released from rest 5.0 m from the center of the sphere.

21.16 Two point charges, 2×10^{-7} C and 3×10^{-7} C, are separated by a distance of 0.1 m. Calculate the resultant electric field and potential at (a) the mid-point between them, (b) at a point 0.04 m from the first and on the line between them, (c) at a point 0.04 m from the first, on the line joining them, but outside them and (d) at a point 0.1 m from each. (e) Where is the electric field zero?

21.17 Referring to the previous problem, calculate the work required to move a charge of 4×10^{-7} C from the point in (c) to the point in (d). Is it necessary to specify the path?

21.18 Equal electric charges, of $1\ \mu C$ each, are placed at the vertices of an equilateral triangle whose sides are 0.1 m each. Calculate (a) the force and the potential energy of each charge as a result of interactions with the other two, (b) the resultant electric field and potential at the center of the triangle and (c) the internal potential energy of the system.

21.19 (a) Referring to the previous problem, make a plot of the lines of force of the electric field produced by the three charges. (b) Plot the equipotential surfaces [actually their intersection with the plane of the triangle].

21.20 There are an infinite number of alternating positive and negative charges, $\pm q$, along a straight line. All charges are separated the same distance r (Figure 21.28). Verify that the potential energy of any one charge is $E_p = (-q^2/2\pi\varepsilon_0 r)\ln 2$. (Hint: see Appendix B, Equation B.24)

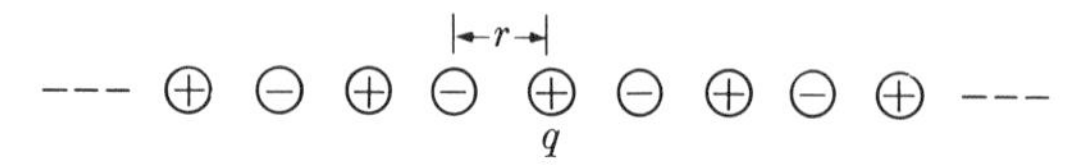

Figure 21.28

21.21 A ring of radius a carries a charge q uniformly distributed along its circumference. Calculate (a) the electric potential and (b) the electric field on points along the perpendicular axis through the center in terms of the distance to the center.

21.22 A disk of radius R carries a charge σ per unit area. Calculate the electric potential and electric field at points along the perpendicular axis through the center. (Hint: divide the disk into rings and add the contributions of all rings.)

21.23 Establish a numerical relationship that gives the velocity (in m s^{-1}) of (a) an electron and (b) a proton in terms of the potential difference (in V) through which they have moved, assuming that they were initially at rest. Assume the particles remain non-relativistic.

21.24 A potential difference of 1600 V is established between two parallel plates 0.04 m apart. An electron is released from the negative plate at the same instant that a proton is released from the positive plate. (a) How far will they be from the positive plate when they pass each other? (b) How do their velocities compare when they strike the opposite plates? (c) How do their energies compare when they strike the opposite plates?

21.25 An electron is accelerated in a certain X-ray tube from rest through a potential difference of 1.8×10^5 V. (Recognize that the electron is getting relativistic.) When it arrives at the anode calculate its (a) kinetic energy in eV and (b) its velocity.

21.26 Referring to the linear accelerator, illustrated in Figure 20.8, alternating sections of the tube are connected and an oscillating potential difference is applied between the two sets. (a) Show that, in order for the ion to be in phase with the maximum oscillating potential when it crosses from one tube to the next (consider the non-relativistic case), the lengths of successive tubes must be $L_1 n^{1/2}$, where L_1 is the length of the first tube. (b) Find L_1 if the maximum accelerating voltage is V_0 and its frequency is ν. (c) Calculate the energy of the ion emerging from the nth tube.

21.27 Suppose the potential difference between the terminal of a van de Graaff accelerator and the point at which charges are sprayed on to the upward moving belt is 2×10^6 V. If the belt delivers negative charge to the terminal at the rate of 2×10^{-3} C s^{-1} and removes positive charge at the same rate, calculate the power that must be expended to drive the belt against electric forces.

21.28 A linear accelerator, with a potential difference of 800 kV, produces a proton beam that carries a charge of 10^{-3} C s^{-1}. Calculate (a) the number of protons that strike the target per second, (b) the power required to maintain the proton beam and (c) the velocity of the protons when they hit the target. (d) Given that the protons lose 80% of their energy in the target, calculate the rate, expressed in cal s^{-1}, at which energy in the form of heat must be removed from the target to maintain its temperature constant.

22 Magnetic interaction

Sir Joseph John Thomson, while studying the properties of cathode rays, discovered (in 1897) that they consisted of negatively charged particles, with a mass more than 1000 times smaller than that of atoms, thereby establishing the existence of 'electrons'. He also measured their charge, verifying that it was the same as that of ions in electrolysis. Thomson extensively studied electric discharges in gases and the electric structure of matter.

Notes

22.1 Introduction

The ancient Greeks observed that certain iron ores, such as the lodestone, have the property of attracting small pieces of iron. The property is exhibited in the natural state by iron, cobalt and manganese and by many compounds of these metals. This property is unrelated to gravitation since, not only does it fail to be exhibited naturally by *all* bodies, but it appears to be concentrated at certain spots in the mineral ore. It is also apparently unrelated to the electrical interaction, because neither cork balls nor pieces of paper are attracted at all by these minerals. Therefore a new name, **magnetism**, was given to this physical property. The name magnetism is derived from the ancient city in Asia Minor called Magnesia, where, according to tradition, the phenomenon was first recognized. The regions of a body where the magnetism appears to be concentrated are called **magnetic poles**. A magnetized body is called a magnet.

The Earth itself is a huge magnet. For example, if we suspend a magnetized rod at any point on the Earth's surface and allow it to rotate freely about the vertical, the rod orients itself so that the same end always points toward the North

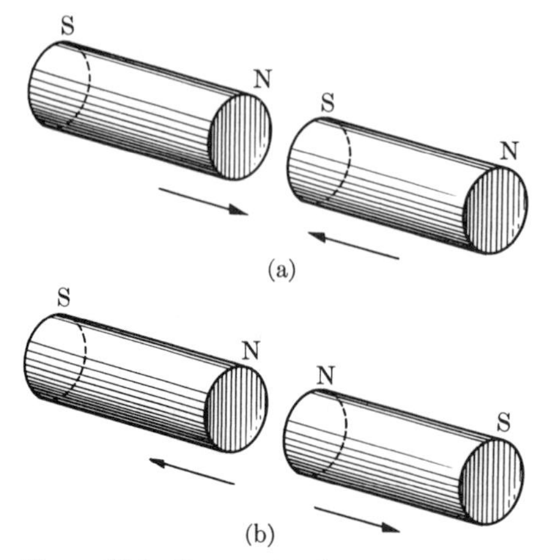

Figure 22.1 Interaction between two magnetized bars. (a) Unlike poles attract each other. (b) Like poles repel each other.

geographic Pole. This result shows that the Earth is exerting an additional force on the magnetized rod that it does not exert on unmagnetized rods.

This experiment also suggests that there are two kinds of magnetic poles. We may designate the two kinds by the signs + and −, or by the letters N and S, corresponding to the north-seeking and south-seeking poles, respectively. Experience shows that a magnetized bar has opposite magnetic poles at its ends. Two magnetized bars, placed as shown in Figure 22.1, will either repel or attract each other, depending on whether we place like or unlike poles facing each other. Thus we conclude that

the interaction between like magnetic poles is repulsive and the interaction between unlike magnetic poles is attractive.

We could measure the strength of a magnetic pole by defining a magnetic mass or charge, and investigating the dependence of the magnetic interaction on the distance between poles. Before physicists understood the nature of magnetism, this was the approach adopted. However, a fundamental difficulty appears when these measurements are attempted. Experiments have been able to isolate positive and negative electric charges and associate a definite amount of electric charge with the fundamental particles constituting matter. On the contrary, we have not been able to isolate a magnetic pole or identify a particle having only one kind of magnetism, either N or S. In addition, the notions of magnetic pole and magnetic mass have been found unnecessary for the description of magnetism. As we shall see, electric and magnetic interactions are very closely related, being only two different aspects of one property of matter, its electric charge. In fact, experience has shown that magnetism is a manifestation of electric charges in motion relative to the observer. For that reason, electric and magnetic interactions must be considered together under the more general name of **electromagnetic interaction**.

22.2 Magnetic force on a moving charge

Since we observe interactions between magnetized bodies even when they are separated, we may say, in analogy with the gravitational and electrical cases, that a magnetized body produces a **magnetic field** in the space around it. The direction of the magnetic field at a given point is determined by the direction of the force on the north pole of a small magnet placed at that point.

When we place an electric charge *at rest* in a magnetic field, no special force is observed on the charge. But when an electric charge *moves* in a region where there is a magnetic field, a force is observed on the charge in addition to those forces due to its gravitational and electric interactions.

By measuring, at the same point in a magnetic field, the force experienced by different charges moving in different ways, we may obtain a relation between the force, the charge and its velocity. First, it has been found experimentally that the magnetic force on a charged particle moving in the direction of the magnetic field is zero and is maximum when the particle moves perpendicular to the magnetic field. More precisely,

the force exerted by a magnetic field on a moving charge is proportional to the electric charge and to the component of the velocity of the charge in a direction perpendicular to the direction of the magnetic field.

If α is the angle between the velocity of the particle and the direction of the magnetic field, the component of the velocity perpendicular to the magnetic field is $v \sin \alpha$. Therefore, we may write the magnetic force as

$$F = q\mathscr{B}v \sin \alpha \qquad \textbf{(22.1)}$$

The quantity $\mathscr{B}$ is found at each point by comparing the observed value of F with those of q, v and α. Also it is found experimentally that $\mathscr{B}$ may vary from point to point, but at each point it is the same for all charges and velocities. This indicates that $\mathscr{B}$ is a property of the magnetic field, which we call the **magnetic field strength**. Note that the magnetic force is zero when $\alpha = 0$, or when $\boldsymbol{v}$ is parallel to $\mathscr{B}$, as indicated previously. On the other hand, the magnetic force is maximum when $\alpha = \pi/2$, or when $\boldsymbol{v}$ is perpendicular to $\mathscr{B}$, resulting in

$$F = qv\mathscr{B} \qquad \textbf{(22.2)}$$

A second experimental fact is that

the direction of the magnetic force is perpendicular to the plane determined by the velocity of the charge and the direction of the magnetic field,

as indicated in Figure 22.2. The direction is obtained using the right hand rule and is opposite for positive and negative charges.

We may combine these experimental requirements and, recalling the properties of the vector product, write the magnetic force as the vector product

$$\boldsymbol{F} = q\boldsymbol{v} \times \mathscr{B} \qquad \textbf{(22.3)}$$

When the particle moves in a region where there are both electric and magnetic fields, the total force is the vector sum of the electric force, $q\mathscr{E}$, and the magnetic force $q\boldsymbol{v} \times \mathscr{B}$. That is,

$$\boldsymbol{F} = q(\mathscr{E} + \boldsymbol{v} \times \mathscr{B}) \qquad \textbf{(22.4)}$$

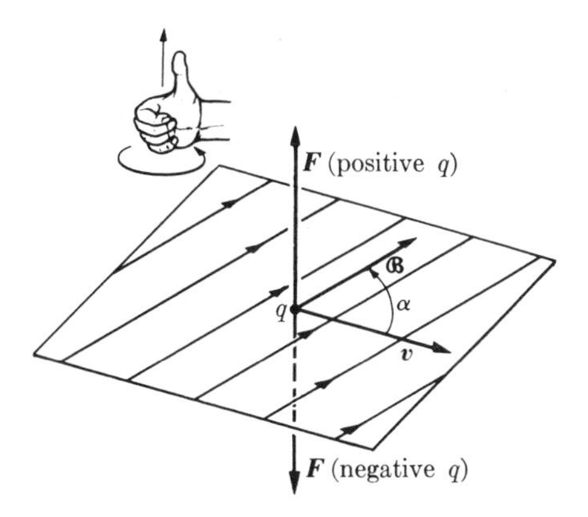

Figure 22.2 Vector relation between magnetic force, magnetic field, and charge velocity. The force is perpendicular to the plane containing $\mathscr{B}$ and $\boldsymbol{v}$.

This expression is called the **Lorentz force** because it was first identified in this form by Hendrik Lorentz.

Because the magnetic force is perpendicular to the velocity, its work is zero. Therefore, the magnetic force does not produce any change in either the magnitude of the velocity or the kinetic energy of the particle; it changes only the direction of the velocity.

From Equation 22.1, we may define the unit of magnetic field as $N/(C\ m\ s^{-1})$ or $kg\ s^{-1}\ C^{-1}$. This unit is called **tesla**, abbreviated T, in honor of Nikola Tesla (1856–1943). That is, $T = kg\ s^{-1}\ C^{-1}$. One tesla corresponds to the magnetic field that produces a force of one newton on a charge of one coulomb moving perpendicular to the field at one meter per second.

EXAMPLE 22.1

Force exerted on a cosmic ray proton that enters the magnetic field of the Earth.

▷ Let the proton initially move on the equatorial plane, perpendicular to the Earth's field with a velocity equal to $10^7\ m\ s^{-1}$. The intensity of the magnetic field near the Earth's surface at the equator is about $\mathscr{B} = 1.3 \times 10^{-5}$ T. The charge on the proton is $q = +e = 1.6 \times 10^{-19}$ C. Therefore the force on the proton is, using Equation 22.2, $qv\mathscr{B} = 2.1 \times 10^{-17}$ N, which is about 10^9 times larger than the force due to the Earth's gravitational attraction, $m_p g \approx 1.6 \times 10^{-26}$ N. The acceleration due to the magnetic force is $a = F/m_p = 1.2 \times 10^{10}\ m\ s^{-2}$. Thus the magnetic acceleration of the proton is also 10^9 times larger than the acceleration g of gravity. The magnetic force deviates the proton's path from that of the free fall on the Earth toward the west.

22.3 Motion of a charged particle in a uniform magnetic field

Consider a charged particle moving in a uniform magnetic field; i.e. a magnetic field that has the same intensity and direction at all its points. Suppose first that the particle moves in a direction perpendicular to the magnetic field (Figure 22.3). The force is then given by Equation 22.2. Since the force is perpendicular to the velocity, its effect is to change the direction of the velocity without changing its magnitude. As we saw in Chapter 8, this results in a uniform circular motion. We shall assume that the velocity of the particle is small compared with c so that we can ignore any relativistic effects. The acceleration is then centripetal, and using the equation of motion 8.2, we have $F = mv^2/r$, with F given by Equation 22.2. Hence we write $mv^2/r = qv\mathscr{B}$, from which we obtain

$$r = \frac{mv}{q\mathscr{B}} \tag{22.5}$$

giving the radius of the circle described by the particle. For example, using the data of Example 22.1, the protons would describe a circle of radius 8×10^3 m if the Earth's magnetic field were uniform.

Equation 22.5 also tells us that the curvature of the path of a charged particle in a magnetic field depends on the momentum $p = mv$ of the particle. The larger the momentum, the larger the radius of the path and the smaller the curvature, because it is more difficult for the magnetic field to change the direction of motion of the particle with greater momentum.

By writing $v = \omega r$ in Equation 22.5, where ω is the angular velocity of the particle, we have

$$\omega = \frac{q}{m}\mathscr{B} \tag{22.6}$$

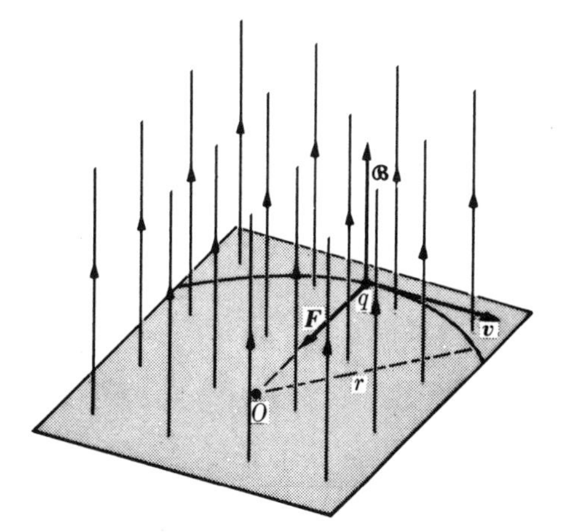

Figure 22.3 A charge moving perpendicular to a uniform magnetic field follows a circular path.

Therefore, the angular velocity is independent of the linear velocity v and depends only on the ratio q/m and the field $\mathscr{B}$. Expression 22.6 gives the magnitude of ω but not its direction. We recall that the acceleration in a uniform circular motion may be written in vector form as $\boldsymbol{a} = \boldsymbol{\omega} \times \boldsymbol{v}$ (Equation 5.11). And so, the equation of motion $\boldsymbol{F} = m\boldsymbol{a}$ becomes $m\boldsymbol{\omega} \times \boldsymbol{v} = q\boldsymbol{v} \times \mathscr{B}$. Reversing the vector product on the right-hand side and dividing by m, we get $\boldsymbol{\omega} \times \boldsymbol{v} = -(q/m)\mathscr{B} \times \boldsymbol{v}$, indicating that

$$\boldsymbol{\omega} = -\frac{q}{m}\mathscr{B} \tag{22.7}$$

which gives $\boldsymbol{\omega}$ both in magnitude and direction. The negative sign indicates that $\boldsymbol{\omega}$ is opposite to $\mathscr{B}$ for a positive charge and has the same direction for a negative charge. We call $\boldsymbol{\omega}$ the **cyclotron frequency**, for reasons to be explained in Section 22.5(iii).

It is customary to represent a field perpendicular to the paper by a dot (•) if it is directed toward the reader and by a cross (×) if it is directed into the page. Figure 22.4 represents the path of a positive (a) and a negative (b) charge moving perpendicular to a uniform magnetic field that is perpendicular to the page and directed toward the reader. In (a) $\boldsymbol{\omega}$ is directed into the page and in (b) toward the reader.

The bending of the path of an ion in a magnetic field therefore provides a means for determining whether its charge is positive or negative, if we know the direction of its motion. Figure 22.5 shows the paths of several charged particles made visible in a device called a cloud chamber placed in a strong magnetic field perpendicular to the page. Note that the paths are bent in either of two opposite senses, indicating that some particles are positive and others are negative. It may be observed that some of the particles describe a spiral of decreasing radius. This indicates that the particle is being slowed down by collisions with the gas molecules; this results in a decrease of the radius of the orbit because its velocity is decreasing. If we have a large number of charged particles moving through a magnetic field

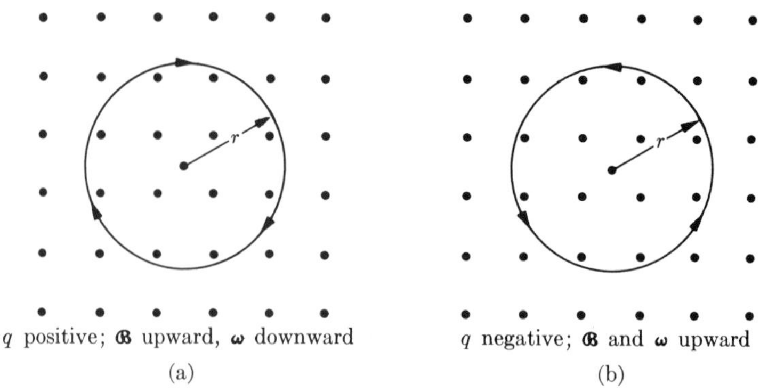

Figure 22.4 Circular path of positive and negative charges in a uniform magnetic field.

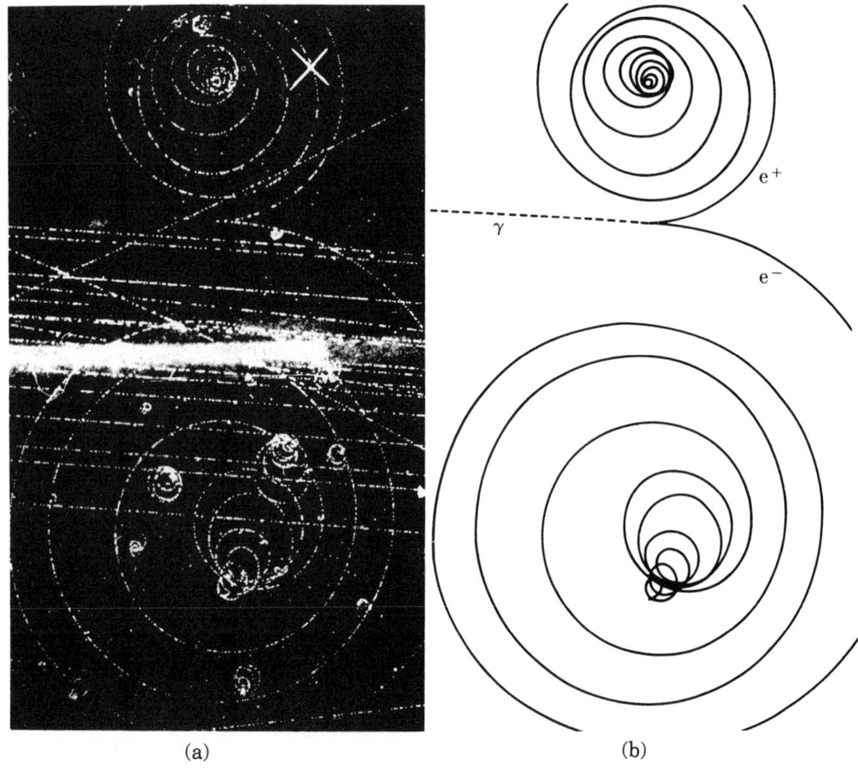

Figure 22.5 Cloud chamber photograph of the paths of an electron and a positron pair in a uniform magnetic field directed into the page. The pair was produced by a photon as shown in the diagram at right. Since the particles are oppositely charged they spiral in opposite senses. A cloud chamber contains a gas and vapor mixture where the path of a charged particle is made visible by condensing the vapor on ions of the gas. The ions are produced by the interaction of the charged particle and the gas molecules. Conditions for condensation are obtained by cooling the mixture by a rapid (adiabatic) expansion. The mixture may be air and water vapor. Cloud chambers, invented in 1911 by Charles T. R. Wilson (1869–1959), have been replaced by other devices that provide better definition of the path of a particle, such as the bubble chamber, invented in 1952 by Donald Glaser and which uses liquid hydrogen.

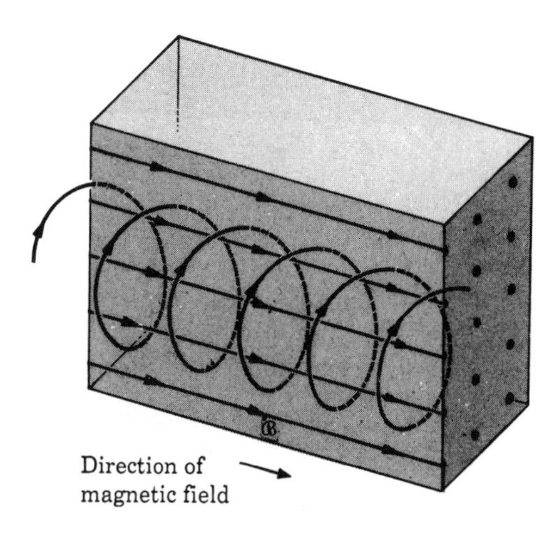

Figure 22.6 Helical path of a positive ion moving obliquely to a uniform magnetic field directed from left to right.

and the field strength increases in time but remains uniform, the particles tend to bunch together, because the radii decrease (recall Equation 22.5). Under such conditions the magnetic field acts as a compressor. This effect is used in fusion reactors to bring the plasma particles closer together (see Notes 26.1 and 40.3).

If a charged particle initially moves in a direction that is not perpendicular to the magnetic field, we may separate the velocity into components parallel and perpendicular to the field. The parallel component remains unaffected and the perpendicular component changes continuously in direction but not in magnitude. The motion is then the resultant of a uniform motion parallel to the field and a circular motion around the field, with angular velocity given by Equation 22.6. The path is a helix, as shown in Figure 22.6 for a positive ion (recall Example 8.3).

EXAMPLE 22.2

Discovery of the positron.

▷ The positron is a fundamental particle having the same mass m_e as the electron but a *positive* charge $+e$. They are found in cosmic rays, in the radioactive decay of some nuclei and can also be produced in particle accelerators. The existence of this particle had been predicted by Paul A. M. Dirac (1902–1984) a few years prior to its experimental discovery in 1932 by Carl D. Anderson (1905–1991), who obtained the cloud chamber photograph in Figure 22.7. The horizontal band seen in the figure is a lead slab 0.6 cm thick that has been inserted inside the cloud chamber, and through which the particle has passed. The lower part of the path of the particle is less curved than the upper part, indicating that the particle has less velocity and energy above the slab than below it. Therefore the particle is moving upward, since it must lose energy in passing through the slab. The curvature of the track of the particle and the sense of the motion relative to the magnetic field indicate that the particle is positive. The path looks very much like that of an electron — but a positive electron. From Equation 22.5 we may write that $p = mv = q\mathscr{B}r$. Therefore, if we measure r from the photograph and assume that $q = e$, we may compute p. By such a calculation, we find that p has an order of magnitude corresponding to a particle with the same mass as an electron. A more detailed analysis, involving the energy loss in passing through the slab, enables us to find the particle's velocity before and after its passage. We may then compute its mass m, thus obtaining full agreement with the electron mass.

Figure 22.7 Anderson's cloud chamber photograph of the path of a positron (positive electron) in a magnetic field directed into the page. This photograph presented the first experimental evidence of the existence of positrons, previously predicted theoretically by Dirac.

22.4 Motion of a charged particle in a non-uniform magnetic field

Consider now the case when a particle moves in a magnetic field that is not uniform. From Equation 22.5, $r = mv/q\mathscr{B}$, we can see that the larger the magnetic field, the smaller the radius of the path of the charged particle. Therefore, if the magnetic field is not uniform, the path is not circular. Figure 22.8 shows a magnetic

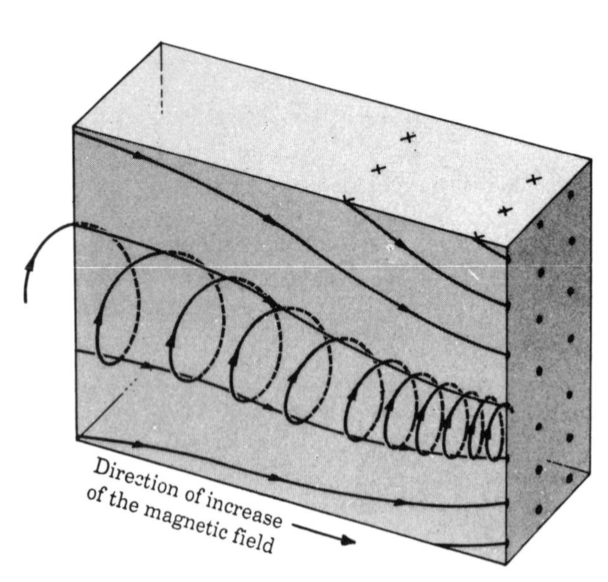

Figure 22.8 Path of a positive ion in a non-uniform magnetic field. The magnetic field increases toward the right.

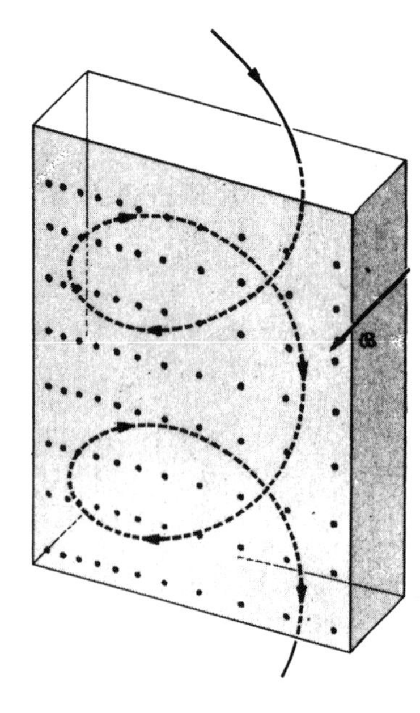

Figure 22.9 Plane motion of an ion drifting across a non-uniform magnetic field perpendicular to the page and toward the reader. The magnetic field decreases from left to right.

field directed from left to right with its strength increasing in that direction. Thus a charged particle injected at the left-hand side of the field describes a helix whose radius decreases continuously. In addition, the component of the velocity parallel to the field decreases (see Note 26.3) and therefore the pitch of the helix also decreases as the particle moves in the direction of increasing field strength. Eventually, given a strongly increasing magnetic field, the particle is forced to spiral back antiparallel to the magnetic field. Thus, as a magnetic field increases in strength, it begins to act as a reflector of charged particles, or, as it is frequently called, a **magnetic mirror**. This effect is used for containing ionized gases or plasmas (see Note 26.1) and in fusion experiments (see Note 40.4).

Another situation is depicted in Figure 22.9, where a magnetic field increases in intensity from right to left. The path of a positive ion injected perpendicular to the magnetic field has also been indicated, being more curved at the left, where the field is stronger, than at the right, where it is weaker. The path is not closed, and the particle drifts across the field perpendicular to the direction in which the magnetic field increases.

EXAMPLE 22.3

Motion of ions in the Earth's magnetic field.

▷ An example of the motion of ions in a non-uniform magnetic field is the case of charged particles falling on the Earth from outer space, constituting part of what are called **cosmic rays**. Figure 22.10 schematically shows the lines of force of the magnetic field of the Earth.

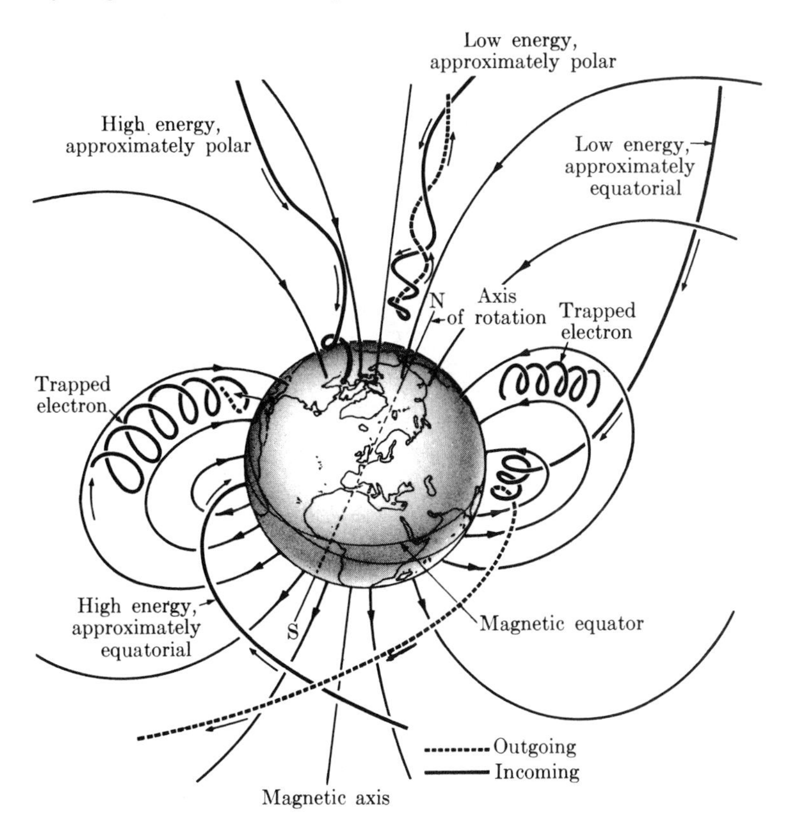

Figure 22.10 Motion of charged cosmic ray particles in the Earth's magnetic field.

Particles falling along the magnetic axis of the Earth do not suffer any deviation and reach the Earth even if they have very small energy. Particles falling at an angle with the magnetic axis of the Earth describe a helical path, and those moving very slowly may be bent so much that they do not reach the Earth's surface. Those arriving on the magnetic equator suffer the largest deflection because they are moving in a plane perpendicular to the magnetic field. Therefore, only the most energetic particles at the magnetic equator can reach the Earth's surface. In other words, the minimum energy that a charged cosmic particle must have to reach the Earth's surface increases from the Earth's magnetic axis to the Earth's magnetic equator.

Another effect due to the Earth's magnetic field is the *east–west asymmetry* of cosmic radiation. Particles of opposite sign are bent in opposite directions by the Earth's magnetic field. If cosmic rays are preponderantly positive, we should observe that the cosmic rays arriving at the Earth's surface in a direction east of the zenith will have a greater intensity than those arriving in a direction west of the zenith. Experimental results are highly in favor of a majority of positively charged particles.

22.5 Examples of motion of charged particles in a magnetic field

In this section we shall illustrate several practical situations where a charged particle or an ion moves in a magnetic field.

(i) Mass spectrometer

Consider the arrangement illustrated in Figure 22.11. Here I is an ion source (for electrons it may be just a heated filament) and S_1 and S_2 are two narrow slits through which the ions pass, being accelerated by the potential difference ΔV applied between them. The exit velocity of the ions is calculated from Equation 21.30, $\frac{1}{2}mv^2 = q\Delta V$, which gives

$$v^2 = 2\left(\frac{q}{m}\right)\Delta V \tag{22.8}$$

In the region below the slits there is a uniform magnetic field directed upward from the page. The ion will then describe a circular orbit, bent in one direction or the other depending on the sign of its charge q. After describing a semicircle the ions fall on a photographic plate P, leaving a mark on the emulsion. The

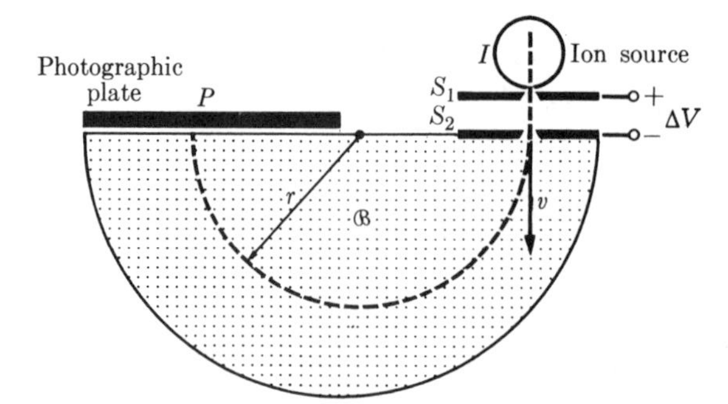

Figure 22.11 Dempster's mass spectrometer. I is an ion source. Slits S_1 and S_2 serve as collimators of the ion beam. ΔV is the accelerating potential difference applied between S_1 and S_2. P is a photographic plate which registers the arrival of the ions.

radius r of the orbit is given by Equation 22.5, $r = mv/q\mathscr{B}$, from which, solving for the velocity v, we obtain

$$v = \left(\frac{q}{m}\right)\mathscr{B}r \tag{22.9}$$

Combining Equations 22.8 and 22.9 to eliminate v, we obtain

$$\frac{q}{m} = \frac{2\Delta V}{\mathscr{B}^2 r^2} \tag{22.10}$$

which gives the ratio q/m in terms of three quantities, ΔV, $\mathscr{B}$ and r. We may apply this technique to electrons, protons and any other charged particle, atom or molecule. By measuring the charge q independently, we may obtain the mass of the particle.

The arrangement of Figure 22.11 constitutes a **mass spectrometer**, because it separates ions having the same charge q but different mass m since, according to Equation 22.10, the radius of the path of each ion will be different, depending on the ion's q/m value. Several other types of mass spectrometers, all based on the same principle, have been developed. Scientists using this technique discovered, in the 1920s, that atoms of the same chemical element do not necessarily have the same mass. As indicated in Section 1.3, the different varieties of atoms of one element which differ in mass are called **isotopes**.

The experimental arrangement of Figure 22.11 may also be used to obtain the variation of momentum with velocity for a particle moving with different velocities. Equation 22.5 may be written in the form $p = q\mathscr{B}r$. It has been found that, assuming q remains constant, p varies with the velocity as indicated by Equation 19.18; that is,

$$p = \frac{mv}{(1 - v^2/c^2)^{1/2}}$$

Therefore, we conclude that the electric charge is an invariant; that is, the same for all observers in uniform relative motion, but the momentum of the particle varies according to the predictions of the theory of relativity.

(ii) Thomson's experiments

During the latter part of the nineteenth century a large number of experiments were performed on electrical discharges. These experiments produced an electric discharge through a gas at low pressure. Two electrodes were placed within the gas and an electric potential difference of several thousand volts was applied. The negative electrode (C) is the **cathode** and the positive electrode (A) is the **anode**. Depending on the pressure of the gas in the tube, several luminous effects were observed. When the gas in the tube was kept at a pressure less than 10^{-3} atm or 10^2 Pa, a luminous spot was detected on the tube wall at O directly opposite the cathode C (Figure 22.12). Therefore it was considered that some radiation was

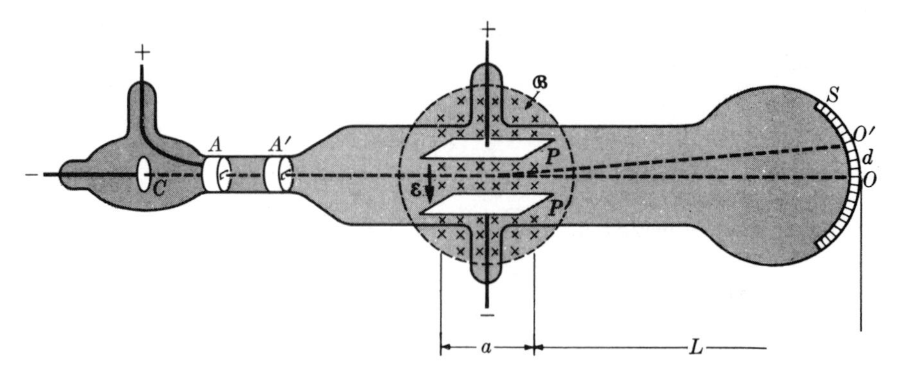

Figure 22.12 Thomson's experiment for measuring q/m. Cathode rays (electrons emitted by C and collimated by A and A') arrive at the screen S after passing through a region where electric and magnetic fields are applied.

emitted from the cathode that moved in a straight line toward O. Accordingly this radiation was called **cathode rays**.

When an electric field $\mathscr{E}$ was produced by applying a potential difference to the parallel plates P and P', it was observed that the luminous spot moved from O to O'. That is, the 'rays' were bent in the direction corresponding to a negative electric charge. This suggested that cathode rays were simply a stream of negatively charged particles. If q is the charge of each particle and v its velocity, the deviation $d = OO'$ can be computed by applying Equation 21.8

$$\frac{q\mathscr{E}a}{mv^2} = \frac{d}{L}$$

Solving for q/m we obtain

$$\frac{q}{m} = \frac{v^2 d}{\mathscr{E}La}$$

The electric force on the particle is $q\mathscr{E}$, and is directed upward. Next, suppose we apply, in the same region, a magnetic field directed into the paper. The magnetic force, according to Equation 22.4, is $qv\mathscr{B}$, and is directed downward because q is a negative charge. By properly adjusting $\mathscr{B}$, we can make the magnetic force equal to the electric force. This results in a zero net force, and the luminous spot returns from O' to O; that is, there is no deflection of the cathode rays. Then $q\mathscr{E} = qv\mathscr{B}$ or

$$v = \frac{\mathscr{E}}{\mathscr{B}}$$

This provides a measurement of the velocity of the charged particle. Substituting this value of v in the preceding expression, we obtain the ratio q/m of the particle constituting the cathode rays:

$$\frac{q}{m} = \frac{\mathscr{E}d}{\mathscr{B}^2 La}$$

This procedure provided one of the first reliable experiments for measuring q/m. It also gave indirect proof that cathode rays consist of negatively charged particles, since then called **electrons**.

These and similar experiments were performed in 1897 by Sir J. J. Thomson (1856–1940), who spent great effort and time trying to discover the nature of cathode rays. Today we know that free electrons present in the metal constituting the cathode C are pulled out from the cathode as a result of the strong electric field applied between C and A and are accelerated by the same field. Cathode rays have many applications, such as in oscilloscopes and TV tubes.

(iii) The cyclotron

The fact that the path of a charged particle in a magnetic field is circular has permitted the design of particle accelerators that operate cyclically. In electrostatic accelerators (described in Section 21.12), the acceleration depends on the total potential difference ΔV. To produce high energy particles, ΔV must be very large. However, in a cyclic accelerator an electric charge may receive a series of accelerations by passing many times through a relatively small potential difference. The first instrument working on this principle was the **cyclotron**, designed by E. O. Lawrence (1901–1958). The first practical cyclotron started operating in 1932. Since then many cyclotrons have been built all over the world, although by now they have been superseded by more powerful machines.

Essentially, a cyclotron (Figure 22.13) consists of a cylindrical cavity divided into two halves (each called a 'dee' because of its shape) and placed in a uniform magnetic field parallel to its axis. The two dees are electrically insulated from each other. An ion source S is placed at the center of the space between the dees. The system must be maintained within a high vacuum to prevent collisions between the accelerated particles and any gas molecules. An alternating potential difference of the order of 10^4 V is applied between the dees. When the ions are positive, they will be accelerated toward the negative dee. Once the ions get inside a dee, they experience no electrical force, since the electric field is zero in the

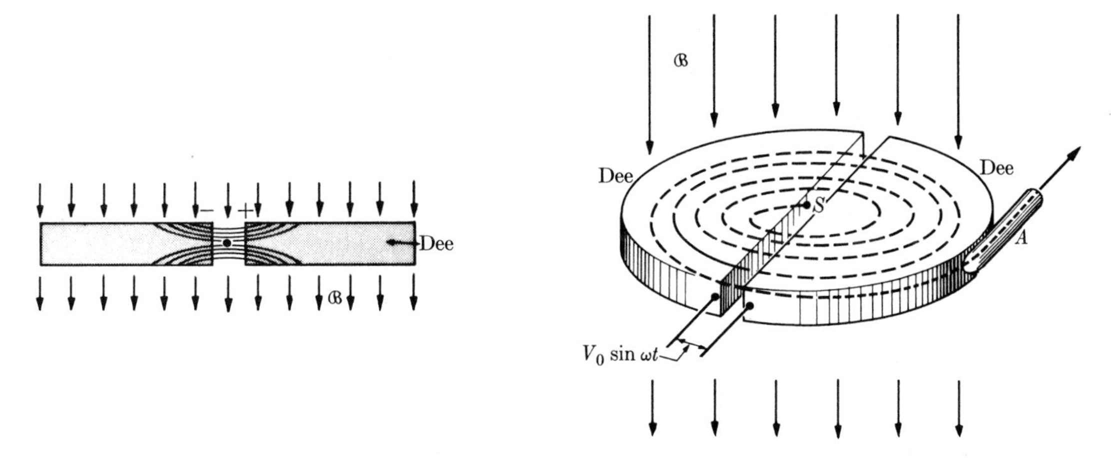

Figure 22.13 Basic components of a cyclotron. The path of an ion is shown by the broken lines.

interior of a conductor (recall Example 21.5 for the case of a spherical shell). However, the magnetic field makes the ions describe a circular orbit, with a radius given by Equation 22.5, $r = mv/q\mathscr{B}$, and an angular velocity given by Equation 22.6, $\omega = q\mathscr{B}/m$. The potential difference between the dees oscillates with a frequency equal to $\omega/2\pi$. In this way the potential difference between the two dees is in resonance with the circular motion of the ions.

As the ions describe half a revolution, the polarity of the dees is reversed. When the ions cross the gap between them, they receive another small acceleration. The next half-circle described then has a larger radius, but the same angular velocity. The process repeats itself several times, until the radius attains a maximum value R, which is practically equal to the radius of the dees. The poles of the magnet are designed so that the magnetic field at the edge of the dees decreases sharply and the ions move tangentially, escaping through a convenient opening. The maximum velocity v_{max} is related to the radius R by Equation 22.5, namely

$$R = \frac{mv_{max}}{q\mathscr{B}} \qquad \text{or} \qquad v_{max} = \left(\frac{q}{m}\right)\mathscr{B}R$$

The kinetic energy of the ions emerging from A is then

$$E_k = \tfrac{1}{2}mv_{max}^2 = \tfrac{1}{2}q\left(\frac{q}{m}\right)\mathscr{B}^2R^2 \tag{22.11}$$

and is determined by the charge and mass of the particle, the strength of the magnetic field and the radius of the cyclotron but is independent of the accelerating potential. When the potential difference is small, the ions have to make many turns before they pick up the final energy. But when it is large, only a few turns are required to attain the same energy.

The strength, of the magnetic field is limited by technological factors, such as the availability of materials with the required properties. But, by making magnets of a radius sufficiently large, we can, in principle, accelerate ions to any desired energy. However, the larger the magnet, the greater the weight and cost. There is also a physical limiting factor to the energy in a cyclotron. As the energy increases, the velocity of the ions also increases, requiring the use of the relativistic equation of motion Equation 19.21, with $F = qv\mathscr{B}$, and this results in a change in ω as the velocity increases.

When the energy is very large, the change in ω is enough to make the cyclotron frequency of the ions change noticeably. Therefore, unless the frequency of the electric potential is changed, the orbit of the ions will no longer be in phase with the oscillating potential, and further acceleration is difficult. Thus, in a cyclotron, the energy is limited by relativistic effects.

EXAMPLE 22.4

A cyclotron has an extraction radius of 0.92 m. The maximum magnetic field is 1.50 T and the maximum attainable oscillating frequency of the accelerating field is 15×10^6 Hz. Calculate the maximum energy of the protons and α-particles accelerated by the machine, and their cyclotron frequency.

▷ Using Equation 22.11 with the corresponding values for the charge and mass of protons ($q = e$, $m \approx 1$ amu) and α-particles ($q = 2e$, $m \approx 4$ amu), we find that the kinetic energies of both may be expressed as $E_k = 1.46 \times 10^{-11}$ J = 91 MeV. The cyclotron frequency for α-particles in this field is $\omega_\alpha = 7.2 \times 10^7\ \mathrm{s}^{-1}$ or a frequency $\nu_\alpha = \omega_\alpha/2\pi = 11.5 \times 10^6$ Hz, which is within the range of the maximum design frequency. For protons we find twice the frequency, or $\nu_p = 23 \times 10^6$ Hz. But the maximum design frequency of the cyclotron is 15×10^6 Hz, and therefore this machine cannot accelerate protons to the theoretical value of 91 MeV. Assuming the maximum available oscillatory frequency, we find $\omega_p = 9.42 \times 10^7\ \mathrm{s}^{-1}$. The corresponding magnetic field for cyclotron resonance is 0.948 T and we find the frequency-limited kinetic energy of protons to be $E_k = \frac{1}{2}mv^2 = \frac{1}{2}m\omega^2 R^2 = 0.63 \times 10^{-11}$ J = 39 MeV, which is very small compared with the rest energy of the proton (Example 19.3). Therefore, relativistic effects can be ignored.

Relativistic effects can be corrected by shaping the magnetic field, so that at each radius the value of ω remains constant in spite of the change in the velocity. The same effect may be accomplished by changing the frequency applied to the dees and keeping the magnetic field constant while the particle is spiraling, so that at each instant there is resonance between the particle motion and the applied potential. The first design is called a **synchrotron** and the second is called a **synchrocyclotron**. A synchrotron may operate continuously, but a synchrocyclotron operates in bursts because of the need to adjust the frequency. Sometimes, as in a **proton synchrotron**, both the frequency and the magnetic field are modified in order to keep the radius of the orbit constant.

22.6 Magnetic field of a moving charge

So far we have considered magnetic fields without going into how magnetic fields are produced, except by magnets. In Chapter 24 we will see that the best method to produce a magnetic field is by using electric currents. But an electric current is a stream of identically charged particles, all moving in the same direction within a conductor. An electric charge in motion, relative to the observer, produces a magnetic field in addition to its electric field. It has been found experimentally that the magnetic field at a distance r from a charge moving with a velocity $\boldsymbol{v}$ (small compared with the velocity of light) relative to an observer is (Figure 22.14)

$$\mathscr{B} = \frac{\mu_0}{4\pi}\frac{qv \sin\theta}{r^2} \qquad \textbf{(22.12)}$$

where μ_0 is a constant called the **vacuum permittivity** whose value in the SI is

$$\mu_0 = 4\pi \times 10^{-7}\ \mathrm{m\,kg\,C^{-2}}$$

The reasoning behind the choice of $4\pi \times 10^{-7}$ for this constant will be discussed in Section 24.12. Note that the magnitude of the magnetic field is zero along the line of motion, and has its

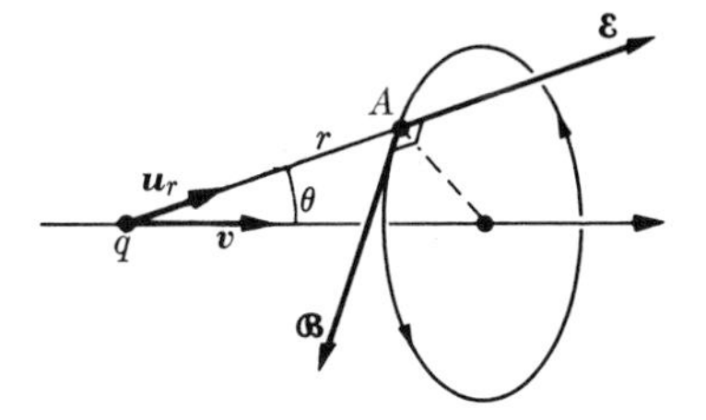

Figure 22.14 Electric and magnetic fields produced by a positive moving charge.

maximum value on the plane perpendicular to the line of motion and passing through the charge. The direction of the magnetic field is perpendicular to the vectors $\boldsymbol{r}$ and $\boldsymbol{v}$ as shown in the figure. Combining both properties of the magnetic field, we may express it in vector form as

$$\mathscr{B} = \frac{\mu_0}{4\pi} \frac{q\boldsymbol{v} \times \boldsymbol{u}_r}{r^2} \tag{22.13}$$

where $\boldsymbol{u}_r$ is a unit vector in the direction of $\boldsymbol{r}$. The magnetic lines of force are then circles, with their centers on the line of motion.

The electric field $\mathscr{E}$ produced by charge q at A, assuming that it is not affected by the motion of the charge (which is correct as long as $v \ll c$), is

$$\mathscr{E} = \frac{q}{4\pi\varepsilon_0 r^2} \boldsymbol{u}_r$$

Therefore we may write Equation 22.13 in the form

$$\mathscr{B} = \mu_0 \varepsilon_0 \boldsymbol{v} \times \mathscr{E} = \frac{1}{c^2} \boldsymbol{v} \times \mathscr{E} \tag{22.14}$$

which establishes a useful relationship between the electric and the magnetic fields produced by a moving charge. In the above expression we have set

$$c = \frac{1}{(\mu_0 \varepsilon_0)^{1/2}} = 2.9979 \times 10^8 \text{ m s}^{-1} \tag{22.15}$$

because, as we shall prove later (Section 29.1), this is the velocity of light or of any electromagnetic signal in vacuum. In round figures, $c = 3.0 \times 10^8 \text{ m s}^{-1}$.

Therefore, although a charge at rest produces only an electric field, a charge in motion relative to the observer produces both an electric field and a magnetic field. Moreover, the two fields are related by Equation 22.14. For example, ions moving along the axis of a linear accelerator produce a magnetic field and an electric field, relative to the laboratory, both related by Equation 22.14. Thus electric and magnetic fields are simply two aspects of one fundamental property of matter, and it is more appropriate to use the term **electromagnetic field** to describe the physical situation involving moving charges, as indicated before in Section 22.1. Another interesting property is that two observers in relative motion measure different velocities of the moving electric charge and therefore also measure different magnetic fields. In other words, magnetic fields depend on the relative motion of charge and observer.

The student must recognize that as the particle moves it carries its electric and magnetic fields with it. Thus, an observer who sees the particle in motion measures electric and magnetic fields that change with time as the particle approaches and recedes from the observer, while an observer at rest relative to the charge only measures an electric field, which is constant.

22.7 Magnetic dipoles

When a charged particle moves in a closed orbit, like an electron in an atom, it produces a magnetic field where the lines of force still make a loop with the orbit, but they are no longer circular (Figure 22.15). The lines of force follow the particle in its motion. However, if the particle moves very fast, we may think that the magnetic field is the statistical average of the field produced at each instant and, as such, it has a symmetry about the axis AB. It is possible to calculate the magnetic field at any point but it is mathematically complex. However, the field at the center O can be calculated very easily.

If the particle moves with uniform circular motion, the velocity of the particle at P is $v = \omega r$, where ω is the angular velocity, and it is perpendicular to the radius OP. Thus the magnetic field at O is, using Equation 22.12 with $\theta = \pi/2$,

$$\mathcal{B} = \frac{\mu_0}{4\pi}\frac{qv}{r^2}$$

The angular momentum of the particle is $L = mvr$. Thus the magnetic field at O is

$$\mathcal{B} = \left(\frac{\mu_0}{4\pi}\right)\left(\frac{q}{m}\right)\frac{L}{r^3}$$

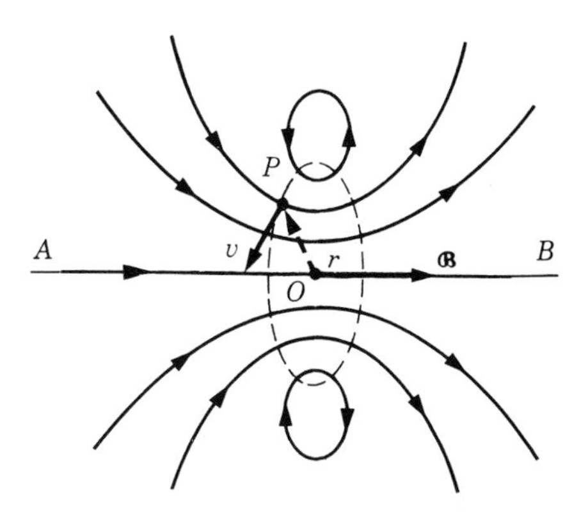

Figure 22.15 The magnetic field of an orbiting charge.

A charged particle, describing a small closed orbit, such as an atomic electron, constitutes a **magnetic dipole**. The vector quantity

$$\boldsymbol{M} = \left(\frac{q}{2m}\right)\boldsymbol{L} \qquad \textbf{(22.16)}$$

is called the **magnetic moment** of the orbiting particle. It is measured in $m^2\ s^{-1}\ C$. $\boldsymbol{M}$ has the same direction as $\boldsymbol{L}$ if q is positive and the opposite if q is negative (Figure 22.16). The magnetic field produced by the orbiting charge at its center, using the definition of magnetic moment, may be written as

$$\mathcal{B} = \left(\frac{\mu_0}{4\pi}\right)\frac{2M}{r^3} \qquad \textbf{(22.17)}$$

The magnetic field of a magnetic dipole along its axis will be calculated in Example 24.11. By comparison with the electric field of an electric dipole (Figure 21.23), it can be seen that the magnetic field of a magnetic dipole is very similar and resembles that of a small magnet.

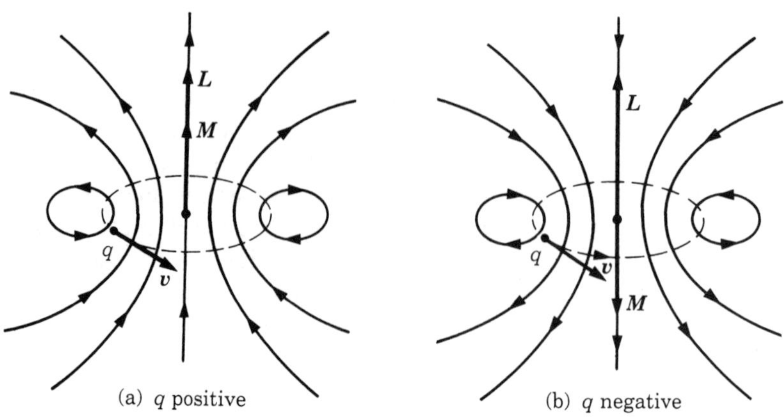

Figure 22.16 Vector relation between the magnetic moment and the angular momentum of an orbiting charge.

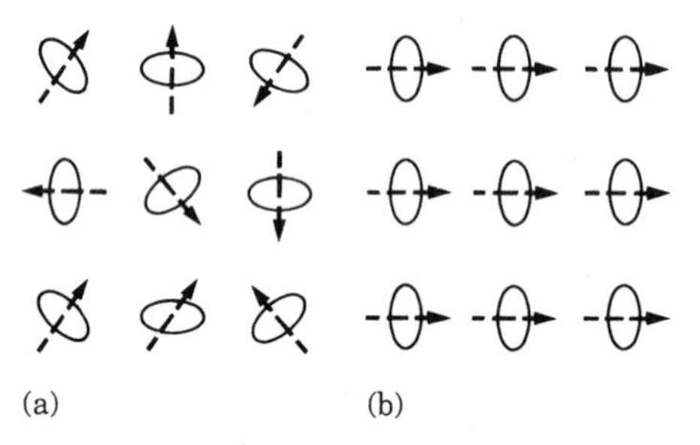

Figure 22.17 Magnetization of a substance.

Electrons in atoms are orbiting and spinning and thus each has a magnetic moment and produces a magnetic field oriented according to the magnetic moment. If their magnetic moments are oriented in different directions (Figure 22.17(a)), the net magnetic field is zero. But if, by some reason, the magnetic moments are aligned, the substance has a net magnetic moment and constitutes a magnet. This situation will be discussed in more detail in Chapter 26.

By similarity with the electric dipole case, an external magnetic field $\mathscr{B}$ produces a torque $\boldsymbol{\tau}$ on a magnetic dipole of moment $\boldsymbol{M}$ given by

$$\boldsymbol{\tau} = \boldsymbol{M} \times \boldsymbol{\mathscr{B}} \qquad \text{or} \qquad \tau = M\mathscr{B} \sin\theta \tag{22.18}$$

where θ is the angle between $\boldsymbol{\mathscr{B}}$ and $\boldsymbol{M}$. This torque tends to align $\boldsymbol{M}$ along $\boldsymbol{\mathscr{B}}$. The existence of the torque $\boldsymbol{\tau}$ can be easily understood if we analyze the forces produced by the magnetic field on an orbiting charge. Referring to Figure 22.18, we have drawn for simplicity the force $\boldsymbol{F} = q\boldsymbol{v} \times \boldsymbol{\mathscr{B}}$ on the positive charge q in the positions P_1 and P_2. It can be seen that the effect of the magnetic field is to distort the orbit, trying to place it in a plane perpendicular to $\boldsymbol{\mathscr{B}}$, aligning $\boldsymbol{M}$ with $\boldsymbol{\mathscr{B}}$. This is what happens to the orbiting electrons in atoms when a substance is placed in a magnetic field.

Again by similarity with the electric dipole case, the energy of an orbiting charged particle in a magnetic field is

$$E_{\mathrm{p}} = -\boldsymbol{M} \cdot \boldsymbol{\mathscr{B}} = -M\mathscr{B} \cos\theta \tag{22.19}$$

This can be verified by recalling that $\tau = -\mathrm{d}E_{\mathrm{p}}/\mathrm{d}\theta$ (Example 9.8). The minimum energy corresponds to $\theta = 0$; that is, when $\boldsymbol{M}$ is in the direction of $\boldsymbol{\mathscr{B}}$.

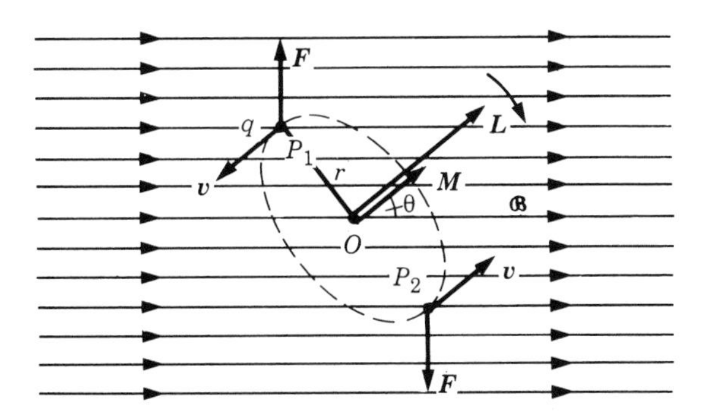

Figure 22.18 Magnetic torque on an orbiting positive electric charge. The forces $\boldsymbol{F}$ tend to align $\boldsymbol{M}$ and $\boldsymbol{\mathscr{B}}$.

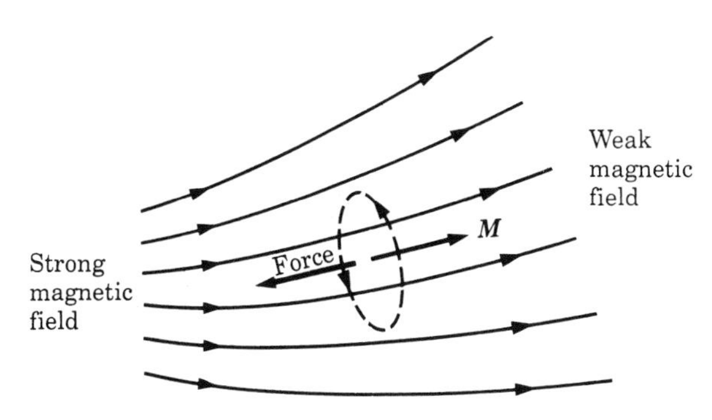

Figure 22.19 Magnetic dipole in non-uniform magnetic field. The dipole moves toward the stronger or weaker region of the field, depending on the orientation of the magnetic moment relative to the magnetic field.

Also, a magnetic dipole oriented parallel to a non-uniform magnetic field so that $\theta = 0$ and $E_p = -M\mathscr{B}$, experiences a force $F = -\mathrm{d}E_p/\mathrm{d}x$ given by

$$F = M\frac{\mathrm{d}\mathscr{B}}{\mathrm{d}x} \tag{22.20}$$

which is equivalent to Equation 21.26 for an electric dipole. This force tends to move the magnetic dipole in the direction in which the magnetic field increases. The opposite occurs if the dipole is oriented antiparallel to the magnetic field (Figure 22.19) moving in the direction of decreasing magnetic field. This is because $\theta = \pi$ and $E_p = +M\mathscr{B}$, so that $F = -\mathrm{d}E_p/\mathrm{d}x = -M\mathrm{d}\mathscr{B}/\mathrm{d}x$.

EXAMPLE 22.5

Comparison of the electric and magnetic interactions between two charges in relative motion.

▷ It is not easy to calculate the magnetic interaction between two charged particles in motion relative to an observer in a form similar to the electric interaction given by Coulomb's law, Equation 21.4. However, we may compare the order of magnitude of the magnetic interaction with the electric interaction. Consider two charges q and q' moving with velocities $\boldsymbol{v}$ and $\boldsymbol{v}'$ relative to an observer O. Since we want only orders of magnitude, we may simplify the writing of the formulas. Thus, we can say that the electric force produced by q' on q as measured by O is $q\mathscr{E}$. The magnetic field produced by q', if we use Equation 22.15, is of the order of magnitude of $v'\mathscr{E}/c^2$ and the magnetic force on q is of the order of

$$qv\mathscr{B} = qv\left(\frac{v'\mathscr{E}}{c^2}\right) = \left(\frac{vv'}{c^2}\right)q\mathscr{E}$$

Therefore, since $q\mathscr{E}$ is the electric force on q,

$$\frac{\text{Magnetic force}}{\text{Electric force}} \approx \frac{vv'}{c^2}$$

A similar result is obtained for the forces produced by q on q', as is clear from the symmetry of the expression. So, if the velocities of the charges are small compared with the velocity of light c, the magnetic force is negligible compared to the electric force and in many cases can be ignored. For example, if the charges have a velocity of the order of 10^6 m s^{-1}, corresponding to the orbital velocity of electrons in atoms, we have that

$$\frac{\text{Magnetic force}}{\text{Electric force}} \approx 10^{-4}$$

Magnetic forces, instead of electric forces, are used in electric motors and many other engineering devices because matter is normally electrically neutral, and consequently, the net electric force between two bodies is zero. But if charges are moving in the two bodies, the magnetic forces may not be zero. For example, when two wires are placed side by side, the net electric force between them is zero. When the wires are moved as a whole, the positive and negative charges move in the same direction with the same velocity so that the net relative motion is zero and thus the net magnetic force is also zero. But if by some method, such as applying a potential difference, we produce a motion of the negative charges relative to the positive, a net magnetic field results. The number of free electrons in a conductor is very large. Their cumulative effect, even if their velocities are small, produces a large magnetic field that results in an appreciable magnetic force between the wires (see Chapter 24).

Although the magnetic force is weak compared with the electric force, it is still very strong compared with the gravitational force.

Note 22.1 Van Allen radiation belts

The Van Allen radiation belts (Figure 22.20) are an example of cosmic charged particles interacting with the Earth's magnetic field. These belts are composed of rapidly moving charged particles, mainly electrons and protons, trapped in the Earth's magnetic field. The inner belt extends from about 800 km to about 4000 km above the Earth's surface, while the other belt extends up to about 60 000 km from the Earth. There is good evidence to show that the inner belt is composed of protons and electrons arising from the decay of neutrons that have been produced in the Earth's atmosphere by cosmic ray interactions. The outer belt mainly consists of charged particles that have been ejected by the Sun. An increase in the number of these particles is associated with solar activity. Their removal from the radiation belt is the cause for auroral activity and radio transmission blackouts.

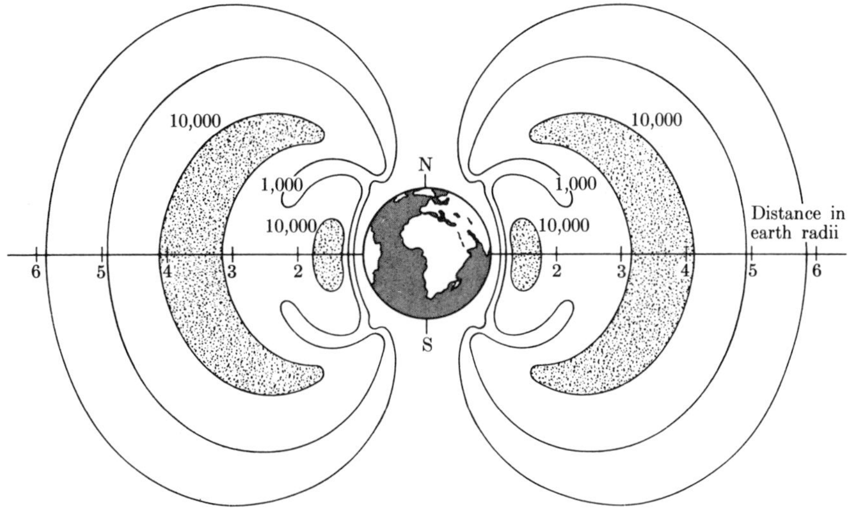

Figure 22.20 Van Allen radiation belts.

QUESTIONS

22.1 How do you recognize that a charged particle moves in a magnetic field?

22.2 How does the velocity of a charged particle vary when it moves in a uniform magnetic field? How does the energy vary? Compare your answers with the motion of a charged particle in a uniform electric field.

22.3 Under what conditions is the motion of a charged particle in a uniform magnetic field a straight line? A circle? A spiral? How is the particle accelerated in each case? Answer the same questions for the case of a uniform electric field.

22.4 What is the main difference between a cyclotron and (a) a linear accelerator; (b) a synchrotron?

22.5 What are the factors limiting the maximum energy attainable with a cyclotron?

22.6 A cyclotron is used to accelerate: protons, deuterons, tritons, ^{3}He and α-particles. Can the frequency of the voltage applied to the dees be the same in all cases? If the energy attained by the protons is E after acceleration, what is the energy of each of the other particles after they are accelerated?

22.7 Consider several identical observers in uniform rectilinear motion, all with different velocities relative to a certain charged particle. Do all observers measure the same electric and magnetic fields due to the particle? Which observer measures only an electric field? Is there an observer who measures only a magnetic field?

22.8 A positively charged, homogeneous sphere rotates with an angular velocity ω about a diameter. Does the sphere produce a magnetic field? Is the direction of the magnetic moment of the sphere related to the axis of rotation? Discuss the motion of the axis of rotation of a rotating charged sphere placed in a uniform magnetic field. (Hint: recall the discussion of gyroscopic motion in Chapter 13.)

22.9 Why may we say that the arrangement of the $\mathscr{E}$ and $\mathscr{B}$ fields between P and P' in Figure 22.12 acts as a velocity selector?

22.10 What are the differences between the electric and magnetic fields of a moving charge?

22.11 Verify that when a charged particle describes a circular or spiral orbit under the action of a magnetic field, its orbital magnetic moment is opposite to the magnetic field, regardless of whether the charge of the particle is positive or negative (recall Equations 22.7 and 22.16).

22.12 Using the result of the previous question and Equation 22.20, verify that a charged particle spiraling in a non-uniform magnetic field is subject to a force that acts in a direction opposite to that in which the magnitude of the magnetic field increases. From this fact, justify the magnetic mirror effect discussed in Section 22.4.

PROBLEMS

22.1 Electrons with a velocity of 10^6 m s^{-1} enter a region where a magnetic field exists. (a) Find the intensity of the magnetic field if the electron describes a circular path having a radius of 0.10 m. (b) Find the angular velocity of the electron.

22.2 Protons are accelerated, from rest, through a potential difference of 10^6 V. They are then shot into a uniform magnetic field of 2.0 T, with their velocities perpendicular to the field. Calculate (a) the trajectory radius and (b) the angular velocity of the protons.

22.3 A proton is in motion in a magnetic field at an angle of 30° with respect to the field. The velocity is 10^7 m s^{-1} and the field strength is 1.5 T. Compute (a) the radius of the helical motion, (b) the distance of advance per revolution (also called pitch) and (c) the frequency of the angular motion.

22.4 (a) A proton with a kinetic energy of 30 MeV moves transverse to a magnetic field of 1.50 T. Determine the radius of the path and the period of revolution. (b) Repeat the problem if the proton energy is 30 GeV. (Note that in (a) the proton can be treated classically, but in (b) it must be treated relativistically. Why?)

22.5 A uniform magnetic field lies in the OY-direction, as shown in Figure 22.21. Find the magnitude and direction of the force on a positive charge q whose instantaneous velocity is $\boldsymbol{v}$ for each of the directions shown in the figure. (The figure is a cube.)

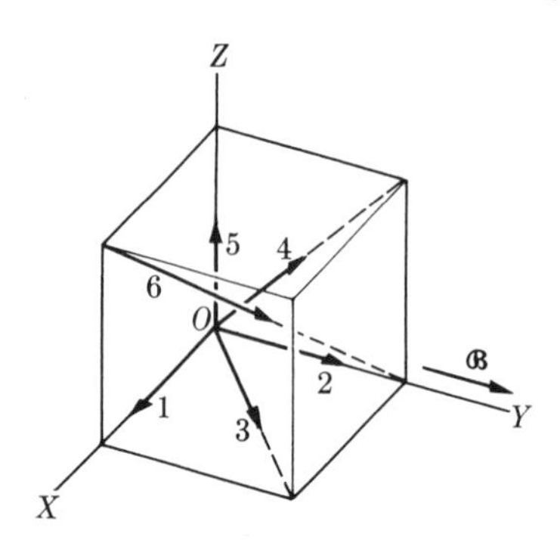

Figure 22.21

22.6 A particle has a charge of 4.0×10^{-9} C. When it moves with a velocity $\boldsymbol{v}_1 = 3.0 \times 10^4$ m s^{-1} at 45° above the Y-axis in the YZ-plane, a uniform magnetic field exerts a force $\boldsymbol{F}_1$ along the $-X$-axis. When the particle moves with a velocity $\boldsymbol{v}_2 = 2.0 \times 10^4$ m s^{-1} along the X-axis, there is a force $\boldsymbol{F}_2$ of 4.0×10^{-5} N exerted on it along the Y-axis. Calculate the magnitude and direction of the magnetic field. (See Figure 22.22.)

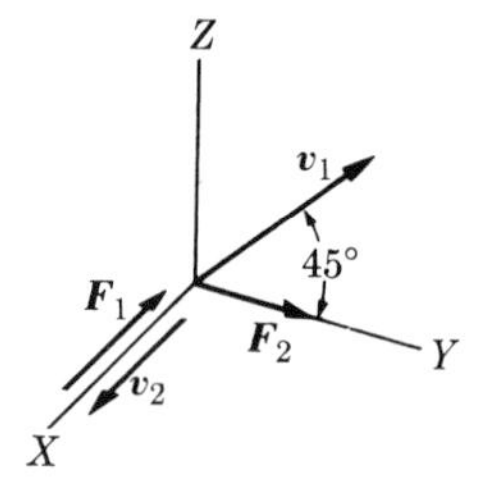

Figure 22.22

22.7 (a) Calculate the velocity of a beam of electrons when the simultaneous action of an electric field of intensity 3.4×10^5 V m^{-1} and a magnetic field of intensity 2.0×10^{-2} T, both fields perpendicular to the beam and to each other, produces no deflection of the electrons. (b) Show, in a diagram, the relative orientation of the vectors $\boldsymbol{v}$, $\mathscr{E}$ and $\mathscr{B}$.

22.8 In a mass spectrometer, such as that shown in Figure 22.11, a potential difference ΔV makes singly ionized ions of ^{24}Mg and ^{25}Mg describe a path of radius R and R'. (a) Calculate the ratio R/R'. (b) Calculate, in terms of ΔV and R/R', the separation of the points at which ^{24}Mg and ^{25}Mg ions hit the photographic plate. (c) Apply to the case where $\Delta V = 1000$ V. (Hint: assume that the masses, in amu, are the same as the mass numbers in the superscript of the chemical symbols.)

22.9 Dempster's mass spectrometer, illustrated in Figure 22.11, uses a magnetic field to separate ions having different masses but the *same energy*. Another arrangement is **Bainbridge's mass spectrometer** (Figure 22.23) that separates ions having the *same velocity*. The ions, after passing through the slits S_1 and S_2, pass through a velocity selector composed of crossed electric and magnetic fields $\mathscr{E}$ and $\mathscr{B}$. Those ions that pass undeviated through the crossed fields enter into a region where a second magnetic field, $\mathscr{B}'$, exists and are bent into circular orbits. A photographic plate registers the ions arrival. Show that $q/m = \mathscr{E}/r\mathscr{B}\mathscr{B}'$.

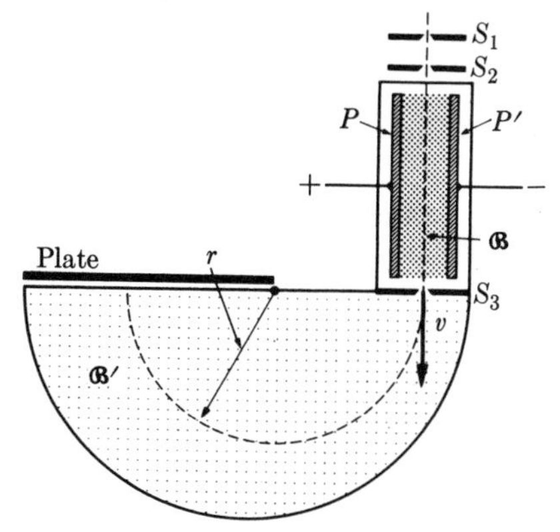

Figure 22.23

22.10 The electric field between the plates of the velocity selector in a Bainbridge mass spectrometer (Problem 22.9) is 1.20×10^5 V m^{-1} and both magnetic fields are 0.60 T. A stream of singly charged neon ions moves in a circular path of 7.28×10^{-2} m radius. Calculate (a) the velocity of the ions and (b) the mass of the neon isotope.

22.11 One of the processes for separating the isotopes ^{235}U and ^{238}U is based on the difference of radii of their paths in a magnetic field. Assume that singly ionized atoms of uranium start from a common source and move perpendicular to a uniform magnetic field. Find the maximum spatial separation of the beams when the radius of curvature of the ^{235}U beam is 0.5 m in a field of 1.5 T (a) if the energies of the two isotopes are the same and (b) if the velocities are the same. (For the purposes of this problem, the superscript of each isotope may be identified with the mass of the atom in amu.)

22.12 A Dempster mass spectrometer has an accelerating voltage of 5 keV and a magnetic field of 10^{-2} T. (a) Calculate the velocity of the zinc ions and determine if it will be necessary to use a relativistic correction. (b) Calculate the distance between the marks on a photographic plate corresponding to the two isotopes of zinc, ^{68}Zn and ^{70}Zn. (Hint: do not find the individual radii; rather, write an equation to find the separation directly.)

22.13 Protons in a cyclotron, just before they emerge, describe a circle of 0.40 m. The frequency of the alternating potential between the dees is 10^7 Hz. Neglecting relativistic effects, compute (a) the magnetic field, (b) the velocity of the protons, (c) the energy of the protons in J and MeV and (d) the minimum number of complete turns of the protons if the peak value of the electric potential between the dees is 20 000 V.

22.14 Repeat the previous problem for (a) a deuteron and (b) an α-particle (helium nucleus). Their respective masses are 2.014 amu and 4.003 amu.

22.15 The magnetic field in a cyclotron used to accelerate protons is 1.50 T. (a) Calculate the frequency of the alternating potential between the dees. (b) The maximum radius of the cyclotron is 0.35 m. Calculate the maximum velocity of the protons. (c) Through what electric potential difference would the protons have to be accelerated to attain the maximum cyclotron velocity?

22.16 Compute the magnetic dipole moment of the electron in a hydrogen atom orbiting in a circular path of radius 0.53×10^{-10} m about the proton.

22.17 (a) Use the results of Examples 21.12 and 21.13 to write the magnetic field along the axis of a magnetic dipole. (b) Obtain the force between two magnetic dipoles aligned along the Z-axis and separated a distance r.

22.18 Observer O' moves relative to observer O with a velocity $\boldsymbol{v}$ parallel to the common X-axis. Two charges q_1 and q_2 are at rest relative to O', separated by the distance r' and placed along the X-axis. (a) Find the forces on each charge as measured by O' and O. (b) Repeat the problem, assuming the charges are on the Y'-axis.

22.19 Consider two electrons moving in straight parallel paths separated by 1.0×10^{-3} m, relative to an observer O. (a) If they are moving side by side at the same velocity of 10^6 m s^{-1} as measured by O, find the electric and magnetic forces between them, as measured by observer O. (b) Calculate the force according to an observer O' moving with the electrons.

23 Electric structure of matter

Niels Bohr is considered the founder of the quantum theory of matter, introducing (in 1913) the notions of quantization of angular momentum and atomic energy levels. In this way he decoded the spectrum of hydrogen and explained the stability of the nuclear model of the atom, which had been proposed by E. Rutherford in 1911. He also proposed (in 1937) the liquid drop model of the atomic nucleus and used it to explain (1939), with J. A. Wheeler, the peculiarities of the fission of uranium isotopes.

Notes

23.1 Introduction

In this chapter the structure of atoms, which was briefly considered in Chapter 1, will be examined in more detail. Atoms consist of a small central region or nucleus, of the order of 10^{-14} m, composed of protons and neutrons. Electrons move about the nucleus in a region of the order of 10^{-10} m. We shall examine the experimental foundations of the nuclear model of the atom and analyze the motion of the electrons based on that model.

23.2 Electrolysis

The phenomenon of **electrolysis** is interesting not only because of its practical applications. It also provided one of the first clues about the electrical structure of matter and the quantization of electric charge. Suppose that an electric field $\mathscr{E}$ is applied (Figure 23.1) to a molten salt (such as KHF_2) or to a solution containing an acid (such as HCl), a base (such as NaOH) or a salt (such as NaCl). The field is produced by immersing two oppositely charged bars or plates, called **electrodes**, in the solution. We then observe that certain kinds of charged atoms move toward the positive electrode or **anode**, and others move to the negative electrode or **cathode**. This phenomenon suggests that the molecules of the dissolved substance have separated (or dissociated) into two different kinds of charged parts, or **ions**. Some ions are positively charged and move in the direction of the electric field; they are called *cations*. Other ions are negatively charged and move in the direction opposite to the electric field; they are called *anions*. For example, in the case of NaCl, Na ions move toward the cathode and therefore are positive ions, while the Cl ions go to the anode and are negative ions. We designate them Na^+ and Cl^- respectively. Then the dissociation of NaCl may be written as

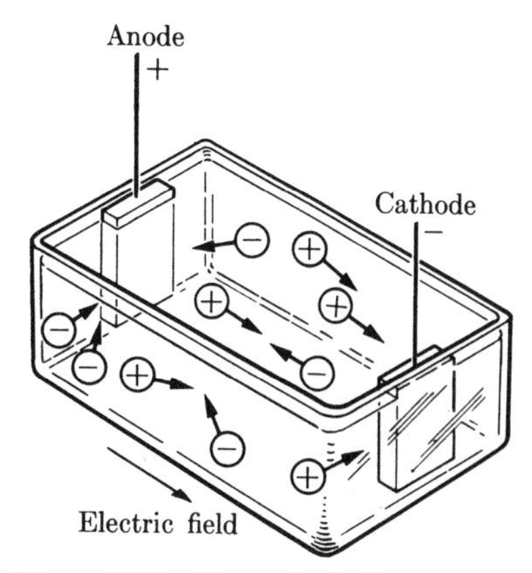

Figure 23.1 Electrolysis. Ions move under the action of the electric field produced by the charged electrodes.

$$NaCl \rightarrow Na^+ + Cl^-$$

Since normal molecules of NaCl do not exhibit any net electrical charge, we may assume that they are composed of equal amounts of positive and negative charges. When NaCl molecules dissociate, the charges are not split evenly. One part of the molecule carries an excess of negative electricity, and the other part an excess of positive electricity. Each of the two parts is thus an ion.

Suppose that the positive ions carry a charge $+ve$ and the negative ions a charge $-ve$. When the ions arrive at each electrode, they become neutralized by exchanging their charge with the charge available at the electrodes. Usually there follows a series of chemical reactions, which are of no concern to us here but which serve to identify the nature of the ions that move to each electrode.

After a certain time a number N of ions of each kind has gone to each electrode. The total charge Q transferred at each electrode is then, in absolute value,

$$Q = Nve \tag{23.1}$$

Therefore, the charge required to deposit one mole of ions on an electrode is

$$Q = N_A ve = Fv \tag{23.2}$$

where N_A is **Avogadro's constant** (the number of atoms or molecules in one mole of any substance) and

$$F = N_A e \tag{23.3}$$

is the **Faraday constant**, which is the charge of one mole of singly charged ions ($v = 1$). Its experimental value is

$$F = 9.6485 \times 10^4 \text{ C mole}^{-1} \tag{23.4}$$

From this value and the one previously found for e, we obtain Avogadro's constant as

$$N_A = 6.0221 \times 10^{23}\ \text{mole}^{-1} \tag{23.5}$$

in agreement with the values of N_A obtained by other methods. It has been found experimentally that v is an integer equal to the **chemical valence** of the ion concerned, suggesting that when two atoms bind together to make a molecule, they exchange a charge ve. One atom becomes positive and the other negative. The electrical interaction between the two charged atoms or ions holds them together.

The conclusion is that chemical processes, and in general the behavior of matter in bulk, are due to electrical interactions between atoms and molecules. A thorough understanding of the electrical structure of atoms and molecules is thus essential to explain the phenomena we observe, in both inert and living matter.

23.3 The nuclear model of the atom

Atoms are normally electrically neutral, since matter in bulk does not exhibit gross electrical forces. Therefore atoms must contain equal amounts of positive and negative electricity. In other words, any atom has equal numbers of protons and electrons, called its **atomic number** and designated by Z. Therefore, the atom consists of a positive charge $+Ze$ due to the protons and an equal negative charge due to the electrons.

The current model of an atom assumes that the protons, being more massive than the electrons, are clustered around the center of mass of the atom, forming a sort of **nucleus**. It also assumes that the electrons are moving around the nucleus, (Figure 23.2). In addition, the presence of neutrons in the nucleus is assumed to account for the total mass of an atom. However, there are difficulties with this model. For instance, we need to explain how the protons are held together in the nucleus in spite of their strong electrical repul-sion. We also have to explain how protons interact with neutrons, which have no charge. The fact that protons and neutrons stay together in a nucleus requires the existence of an interaction, called the **nuclear** or **strong** interaction, that affects both neutrons and protons and counterbalances the electrical repulsion. Neutrons contribute to the nuclear forces without adding electrical repulsion, thus producing a stabilizing effect.

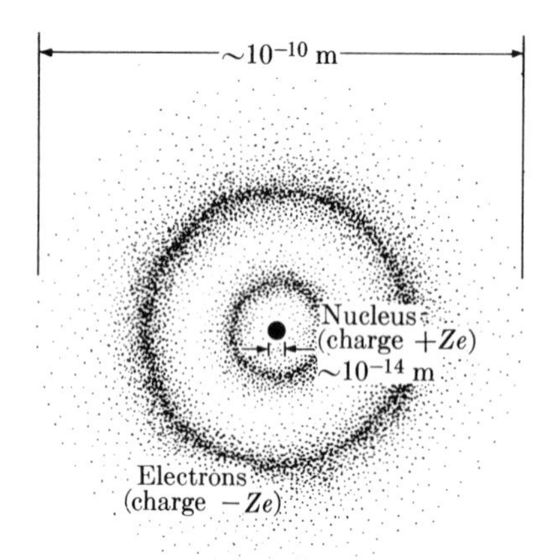

Figure 23.2 Electron distribution in an atom.

To discover the distribution of electrons and protons in an atom, we must probe its interior. One method is to send a stream of fast, charged particles such as hydrogen ions (that is, protons) or helium ions (called α-particles) against a target. We then observe how the particles are deviated or scattered by the electric interactions produced with the atoms in the target (recall Examples 8.8 and 9.10).

The deviation of a particle depends on its impact parameter b (Figure 23.3).

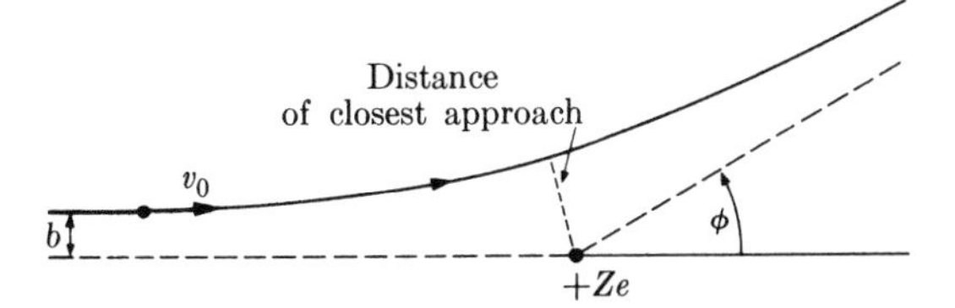

Figure 23.3 Deflection of a positive ion due to the Coulomb repulsion of the nucleus. The impact parameter is b and the deviation of the particle is given by ϕ.

The smaller the impact parameter, the closer the particle passes to the center of the atom and the larger the deviation ϕ if the positive electric charge is concentrated at the center (see Note 8.1). In the experimental arrangement, several particles are directed against a very thin foil and the deflected particles are observed. Since b cannot be controlled (because it is impossible to aim directly at a particular atom), we must make a statistical analysis to interpret the experimental results. It can be shown that the observed angular distribution of the scattered particles must be due to a repulsive central force proportional to $1/r^2$, indicating that the positive charge of an atom is concentrated near its center. Also, scattering experiments allow a determination of the charge of the nucleus. In this way it has been verified that the nuclear charge is a multiple of e, specifically $+Ze$, as assumed in our model.

Such experiments were performed for the first time during the period 1911–1913 by Hans Geiger (1882–1945) and Ernest Marsden (1889–1970), under the direction of Ernest Rutherford (1871–1937) and were the foundation for the nuclear model of the atom, which has been accepted since then as correct.

For each value of the impact parameter b, there is a distance of closest approach at which the bombarding particle is nearest to the center. The minimum distance occurs for a head-on collision ($b = 0$). From a number of different experiments, this distance is found to be of the order of 10^{-14} m for both protons and α-particles with energies of the order of 10^{-13} J (or one MeV). This distance gives an upper limit for the radius of the atomic nucleus. Therefore we conclude that protons and neutrons are concentrated in a region whose dimensions are of the order of 10^{-14} m. Further, we realize that most of the atomic volume is occupied by the moving electrons and is, in fact, empty.

For very small values of the impact parameter and high energy, when the incoming particle comes very close to the nucleus, experiments indicate the presence of the **nuclear forces** in addition to the electric forces.

EXAMPLE 23.1

Distance of closest approach for a head-on collision of a charged particle directed against an atomic nucleus.

▷ Let the charge of the nucleus be Ze and that of the projectile be ve. Assume that the mass of the nucleus is much greater than the mass m of the projectile (or that the nucleus is embedded in a crystal solid lattice) so that the motion of the nucleus can be ignored. The total energy of the system of projectile-plus-nucleus is then

$$E = \tfrac{1}{2}mv^2 + \frac{(ve)(Ze)}{4\pi\varepsilon_0 r}$$

where $\frac{1}{2}mv^2$ is the kinetic energy of the projectile. When the particle is very far away, all its energy is kinetic and equal to $\frac{1}{2}mv_0^2$. The conservation of energy requires that

$$\tfrac{1}{2}mv^2 + \frac{vZe^2}{4\pi\varepsilon_0 r} = \tfrac{1}{2}mv_0^2$$

At the point of closest approach R, the velocity v is zero if the collision is head-on. Therefore

$$\frac{vZe^2}{4\pi\varepsilon_0 R} = \tfrac{1}{2}mv_0^2 \quad \text{or} \quad R = \frac{vZe^2}{4\pi\varepsilon_0(\tfrac{1}{2}mv_0^2)} \tag{23.6}$$

This expression is valid as long as the particle does not get too close to the nucleus.

For example, suppose protons ($v = 1$), accelerated to an energy of 10 MeV, fall on a thin gold foil ($Z = 79$). The kinetic energy of the protons is $\tfrac{1}{2}mv_0^2 = 10$ MeV $= 1.6022 \times 10^{-12}$ J. Then, using Equation 23.6, the distance of closest approach of the protons that experience a head-on collision is $R = 1.139 \times 10^{-14}$ m. Since the radius of a gold nucleus is 8.15×10^{-15} m (Chapter 39) the protons are stopped and sent back before reaching the nucleus.

23.4 Bohr's theory of the atom

The simplest and lightest of all atoms are hydrogen atoms, which are composed of one electron revolving around a single proton. Then $Z = 1$, and the nucleus of the hydrogen atom is just one proton.

For atoms heavier than hydrogen, the nucleus is composed of protons and neutrons. The total number of particles in a nucleus is called the **mass number**, and is designated by A. Therefore an atom has Z electrons, Z protons and $A - Z$ neutrons.

The chemical behavior of an atom is an electrical effect, determined by the atomic number Z. Thus each chemical element is composed of atoms having the same atomic number Z. However, for a given Z there may be several values of the mass number A. In other words, to a given number of protons in a nucleus there may correspond different numbers of neutrons. Atoms having the same atomic number but different mass number are called **isotopes** of the same chemical element. Isotopes are designated by the symbol of the element (which also identifies the atomic number) along with a superscript to the left that indicates its mass number. For example, hydrogen ($Z = 1$) has three isotopes: ^1H, ^2H or deuterium, and ^3H or tritium. Similarly, two of the most important isotopes of carbon ($Z = 6$) are ^{12}C and ^{14}C. The isotope ^{12}C is the one used to define the atomic mass unit. Isotopes were first discovered in 1912 by J. J. Thomson using a simple mass spectrometer (recall Section 22.5).

An atomic electron is subject to an attractive electric $1/r^2$ force produced by the nucleus. Therefore, we may expect (for the same reasons given in Chapter 11 for planetary motion) that the orbits will be ellipses, with the nucleus at one focus. Consider an atom with a single electron and a nuclear charge $+Ze$; that is, hydrogen (H, $Z = 1$), single-ionized helium (He^+, $Z = 2$), double-ionized lithium (Li^{2+}, $Z = 3$) etc. Assume, as a first approximation, that the nucleus coincides with the center of mass of the atom and therefore is at rest in the C-frame. The potential energy of the system, using Equation 21.21, with $q = -e$ and $q' = +Ze$, is $E_p = -Ze^2/4\pi\varepsilon_0 r$. Therefore, the total electronic energy of the atom is

$$E = \tfrac{1}{2}m_e v^2 - \frac{Ze^2}{4\pi\varepsilon_0 r} \tag{23.7}$$

Within the context of classical mechanics there is no limitation to the possible values of the energy of an electron in an atom. However, there is ample experimental evidence based on the analysis of atomic and molecular spectra and on inelastic collisions, both of which will be considered later on, which suggests that

> *the energy of an electron in an atom can have only certain values E_1, E_2, $E_3, \ldots, E_n, \ldots$. That is, the energy of the electronic motion is quantized.*

The states corresponding to the allowed energies are called **stationary states**. The state having the lowest possible energy is the **ground state**.

In the case of hydrogen-like atoms it has been found experimentally, by analyzing the spectra of such atoms (Section 31.4), that the energy of the stationary states obeys an expression of the form

$$E = -\frac{2.177 \times 10^{-18} Z^2}{n^2}\ \text{J} = -\frac{13.598\, Z^2}{n^2}\ \text{eV} \qquad \textbf{(23.8)}$$

where n is an integer. The energy levels for H, He^+ and Li^{2+} are illustrated in Figure 23.4. The ground state corresponds to $n = 1$, since this is the minimum possible energy for the atom. The value of E_1 also gives the minimum energy required to ionize an atom that is in the ground state.

The existence of stationary states of one-electron atoms given by Equation 23.8 requires, as explained below, that the angular momentum $L = m_e vr$ of the electron be limited only to certain values. According to the Bohr theory, the possible values of the angular momentum are given by

$$L = n\hbar \qquad (n = 1, 2, 3, \ldots) \qquad \textbf{(23.9)}$$

where $\hbar = h/2\pi = 1.0545 \times 10^{-34}$ J s, which was mentioned in Example 8.6, and h is Planck's constant. This expression was derived by Niels Bohr (1885–1962) by assuming that the electron describes circular orbits (an assumption that is not entirely correct, for reasons given below). Therefore, we conclude that

> *the angular momentum of an electron in an atom can have only certain values L_1, L_2, L_3, ..., L_n, ... That is, the angular momentum of electronic motion is quantized.*

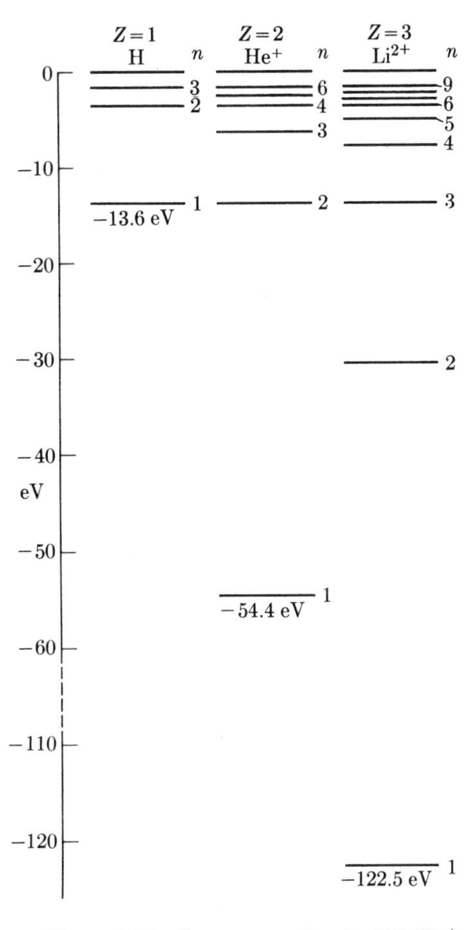

Figure 23.4 Some energy levels of H, He^+ and Li^{2+}.

The fact that energy and angular momentum are quantized is an indication that certain new principles must be taken into account to analyze electron motion. The corresponding theory is called **quantum mechanics**, which will be considered in Chapters 36 and 37. One of the main features of this theory is that electron orbits cannot be defined precisely, in the same way that we define planetary orbits. Rather, we may only speak of the regions where it is *more likely* to find the electron. Another important feature is that an atom can exchange energy only by amounts equal to the difference in the energy of two stationary states; that is

$$\Delta E = E_i - E_j$$

For example, the energy required to excite a hydrogen atom from its ground state $n = 1$ to its first excited state ($n = 2$) is

$$\Delta E = E_2 - E_1 = (-3.4\ \text{eV}) - (-13.6\ \text{eV}) = 10.2\ \text{eV}$$

The above theory was developed in 1913 by Bohr, who was working with Rutherford at the time the scattering experiments described in Section 23.3 were performed. Bohr developed his theory to provide a basis for the nuclear model of the atom and to explain the line spectra of atoms (see Section 31.4).

Details of the Bohr theory

Assume, for simplicity, that the electron moves around the nucleus in a circular orbit. Also assume that we can use Newtonian mechanics instead of quantum mechanics to analyze the motion. Then the equation of motion of the electron, according to Equation 8.2, is $m_e v^2/r = F$. where $F = Ze^2/4\pi\varepsilon_0 r^2$ is the force of attraction produced by the nucleus. Then

$$\frac{m_e v^2}{r} = \frac{Ze}{4\pi\varepsilon_0 r^2} \quad \text{or} \quad m_e v^2 = \frac{Ze^2}{4\pi\varepsilon_0 r}$$

When this relation is inserted in the expression for the energy $E = \frac{1}{2}mv^2 - Ze^2/4\pi\varepsilon_0 r$, we get

$$E = \frac{1}{2}\frac{Ze^2}{4\pi\varepsilon_0 r} - \frac{Ze^2}{4\pi\varepsilon_0 r} \quad \text{or} \quad E = -\frac{Ze^2}{4\pi\varepsilon_0 (2r)} \qquad \textbf{(23.10)}$$

Note that this expression agrees with Equation 11.6 for the gravitational case if we replace the gravitational coupling factor GMm by the electric coupling factor $Ze^2/4\pi\varepsilon_0$. We may express Equation 23.10 in terms of the angular momentum of the electron, which is a constant of motion since the force is central. We recall that $L = m_e vr$ and therefore

$$L^2 = m_e(m_e v^2)r^2 = \frac{m_e Ze^2 r}{4\pi\varepsilon_0} \qquad \textbf{(23.11)}$$

We use this relation to eliminate r from Equation 23.10 to obtain

$$E = -\frac{m_e e^4 Z^2}{2(4\pi\varepsilon_0)^2 L^2} = -\ const.\ \frac{Z^2}{L^2} \qquad \textbf{(23.12)}$$

This expression is valid only for circular orbits. Comparison with Equation 23.8 for the experimentally determined energy levels shows that L^2 must be proportional to n^2; that is, the angular momentum is quantized according to Equation 23.9, $L = n\hbar$ where $\hbar$ is a constant. Then Equation 23.12 for the allowed energy levels becomes

$$E = -\frac{m_e e^4 Z^2}{2(4\pi\varepsilon_0)^2 \hbar^2 n^2} \tag{23.13}$$

When we compare Equation 23.13 with the experimental result given by Equation 23.8, the value of $\hbar$ is obtained. Making $L = n\hbar$ in Equation 23.11, we find that the radius of an allowed orbit is

$$r = \frac{(4\pi\varepsilon_0)\hbar^2}{Ze^2 m_e} n^2 = \frac{a_0}{Z} n^2 \tag{23.14}$$

where

$$a_0 = \frac{(4\pi\varepsilon_0)\hbar^2}{e^2 m_e} = 5.292 \times 10^{-11}\ \text{m} \tag{23.15}$$

is called the **Bohr radius**. It corresponds to the radius of the hydrogen atom ($Z = 1$) in its ground state ($n = 1$). It must be emphasized, however, that the value of r obtained in Equation 23.14 must not be taken too literally. According to quantum mechanics it should be considered only as an indication of the order of magnitude of the region in which the electron is most likely to be found.

EXAMPLE 23.2

Inelastic collisions of electrons with atoms.

▷ The most important experimental evidence of stationary energy states in atoms is the existence of discrete atomic and molecular spectra. This was the basis on which Bohr established his theory. Atomic and molecular spectra are discussed in Chapter 30. However, the existence of stationary states is amply confirmed by many other experiments. The most characteristic experiment is that of inelastic collisions, where part of the kinetic energy of one particle is transferred to internal energy of another particle. These are called inelastic collisions of the **first kind**. Inelastic collisions of the **second kind** correspond to the reverse process. (Recall Section 14.9.)

Suppose that a fast particle q, which we will call the *projectile*, collides with a system A, which we call the *target* (which may be an atom, a molecule or a nucleus) in its ground state of energy E_1. As a result of the projectile–target interaction (which may be electromagnetic or nuclear), there is an exchange of energy. Let E_2 be the energy of the first excited state of the target. The collision will be elastic (i.e. the kinetic energy will be conserved) unless the projectile has enough kinetic energy to transfer the excitation energy $E_2 - E_1$ to the target. When this happens the collision is inelastic, and we may express it by

$$A + q_{\text{fast}} \rightarrow A^* + q_{\text{slow}}$$

where A^* represents the excited atom in the target. When the mass of the projectile q is very small compared with that of the target A (as happens for the case of an electron colliding with an atom), we may ignore all recoil effects. Then the condition for inelastic collisions is $E_k \geq E_2 - E_1$, where $E_k = \frac{1}{2}mv^2$ is the kinetic energy of the projectile before the collision. The kinetic energy of the projectile after the inelastic collision is

$$E'_k = E_k - (E_2 - E_1)$$

To give a concrete example, suppose that an electron of kinetic energy E_k moves through a substance, such as mercury vapor. Provided that E_k is smaller than the first excitation energy of mercury, the collisions are all elastic and the electron moves through the vapor, without exciting any atoms. However, if E_k is larger than $E_2 - E_1$ the electron may lose the energy $E_2 - E_1$ in a single inelastic collision. If the initial kinetic energy of the electron was not much larger than $E_2 - E_1$, the energy of the electron after the inelastic collision is insufficient to excite other atoms. Thereafter the successive collisions of the electron will be elastic. However, if the kinetic energy of the electron was initially very large compared with $E_2 - E_1$, it may still suffer a few more inelastic collisions. The electron loses kinetic energy $E_2 - E_1$ at each collision and produces more excited atoms. Finally, it will be slowed down below the threshold for inelastic collisions.

This process was observed for the first time in 1914 by James Franck (1882–1964) and Gustav Hertz (1887–1975). Their experimental arrangement is indicated schematically in Figure 23.5. A heated filament F emits electrons which are accelerated toward the grid G by a variable potential V. The space between F and G is filled with mercury vapor. A small retarding potential V', of approximately 0.5 V, is applied between the grid G and the collecting plate P. Then those electrons that are left with very little kinetic energy after one or more inelastic collisions cannot reach the plate and are not registered by the galvanometer. As the potential difference V is increased, the number of electrons arriving at the plate fluctuates as shown in Figure 23.6, the peaks occurring at a spacing of about 4.9 V. The first dip corresponds to electrons that lose all their kinetic energy after one inelastic collision with a mercury atom, which is then left in an excited state. The second dip corresponds to those electrons that suffered inelastic collisions with two mercury atoms, losing all their kinetic energy, and so on. Therefore we may conclude that the first excited state of mercury atoms is about 4.9 eV above the ground state. This conclusion is corroborated by the appearance of light emitted when the mercury atoms return to the ground state that corresponds to the spectrum of mercury (Sections 31.4 and 38.5).

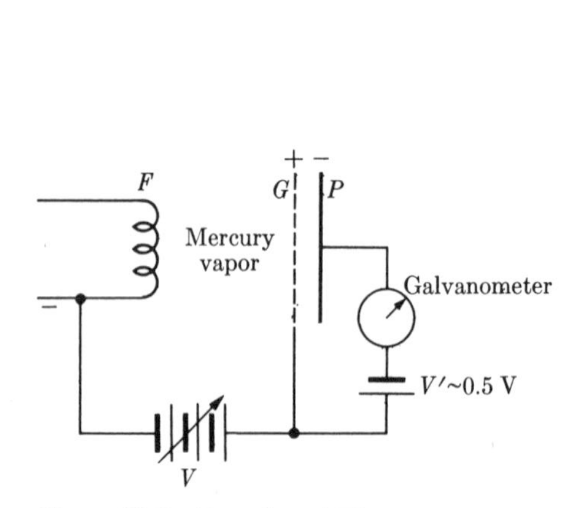

Figure 23.5 Franck and Hertz experimental arrangement for analyzing inelastic collisions of the second kind.

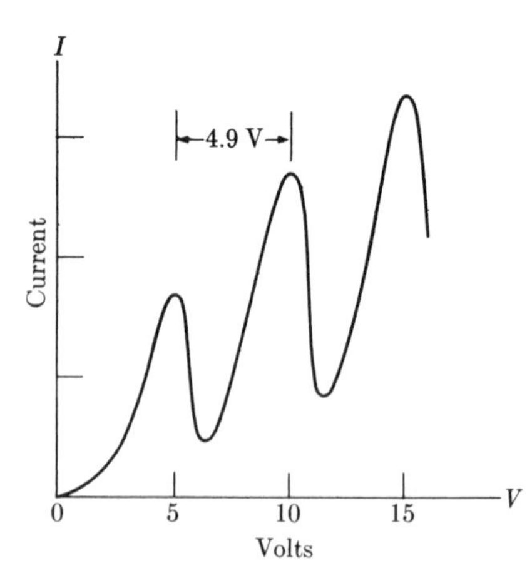

Figure 23.6 Electron current versus accelerating potential in the Franck–Hertz experiment.

23.5 Quantization of angular momentum

In the preceding section we showed that it is possible to justify the observed allowed energy levels of one-electron atoms by assuming that the angular momentum of the electron is quantized. The allowed values for the angular momentum according to Bohr's theory were given by Equation 23.9, $L = n\hbar$. This relation was derived by Bohr for the particular case of circular orbits, for which Equation 23.12 applies, using a classical description of the motion. However when the correct theory of quantum mechanics (Chapter 38) is used, it is found that the allowed values of the orbital angular momentum of an electron are given by

$$L = \sqrt{l(l+1)}\hbar \qquad \textbf{(23.16)}$$

where $l = 0, 1, 2, 3, \ldots$ is a positive integer (for a justification of Equation 23.16, see Note 23.1). Equation 23.16 agrees with Equation 23.9 when l is very large because in this case we may neglect 1 in comparison with l and write $L \approx l\hbar$. However, expression 23.13 for the energy levels remains valid when quantum mechanics is used, as a more elaborate calculation shows.

In classical mechanics, given the energy of the planetary orbit, the shape of the orbit varies according to the angular momentum of the particle. The same result applies in quantum mechanics: the region where the electron moves depends on the value of l. Accordingly, each energy level, determined by a value of n, is associated with several possible electron angular momentum states determined by the values of l. The values of l related with each value of n are given by quantum mechanics as

$$l = 0, 1, 2, \ldots, (n-1) \qquad (n \text{ values}) \qquad \textbf{(23.17)}$$

Angular momentum states are designated by letters according to the scheme shown in Table 23.1. This relation between n and l allows us to understand the electronic structure of atoms (Section 23.8).

Table 23.1 Angular momentum states

l	0	1	2	3	4
Symbol	s	p	d	f	g
$g = 2l + 1$	1	3	5	7	9

In addition to the quantization of the magnitude of the orbital angular momentum, there is experimental evidence, as indicated in Section 23.6, supported by quantum mechanics, that

the orientation of the angular momentum L relative to a given axis is limited to certain directions.

This result is called **space quantization**. Designating the given axis as Z (Figure 23.7), the allowed values of the Z-component of L are

$$L_z = m_l\hbar \qquad \textbf{(23.18)}$$

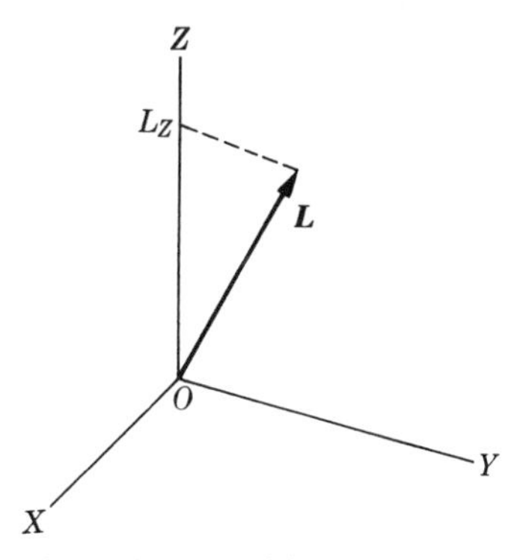

Figure 23.7 Orbital angular momentum $\boldsymbol{L}$ and its component L_z along the Z-axis.

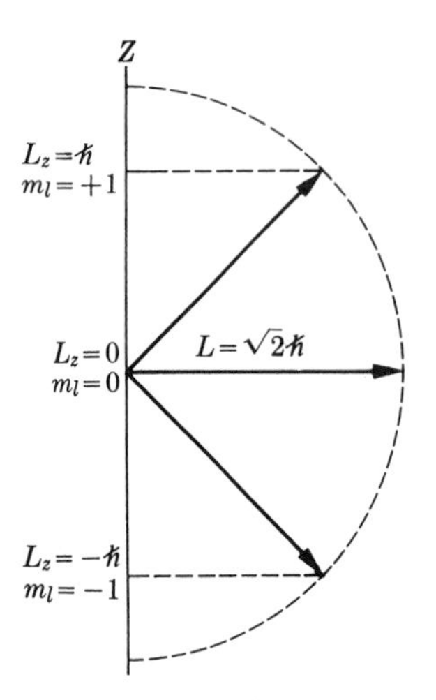

Figure 23.8 Possible orientation of angular momentum corresponding to $l = 1$ and $L = \sqrt{2}\hbar$.

where m_l is a positive or negative integer, having the values

$$m_l = 0, \pm 1, \pm 2, \ldots, \pm(l - 1), \pm l \tag{23.19}$$

The upper value of m_l is $\pm l$ because L_z cannot be larger than $|\boldsymbol{L}|$. For $l = 0$ (*s*-states) only $m_l = 0$ is possible. For $l = 1$ (*p*-states), we may have $m_l = 0, \pm 1$. In general for any l value there are $g = 2l + 1$ possible values of m_l; each value of m_l gives a possible orientation of L. Table 23.1 lists the g-values for various angular momentum states. In Figure 23.8 we have shown the three possible orientations of $\boldsymbol{L}$ for $l = 1$.

We conclude then that the state of an electron in a hydrogen-like atom is characterized by three quantum numbers: n, which gives the energy; l, which gives the angular momentum; and m_l, which gives the orientation of the angular momentum relative to the Z-axis.

An orbiting electron has a magnetic dipole moment which, according to Equation 22.16, is given by

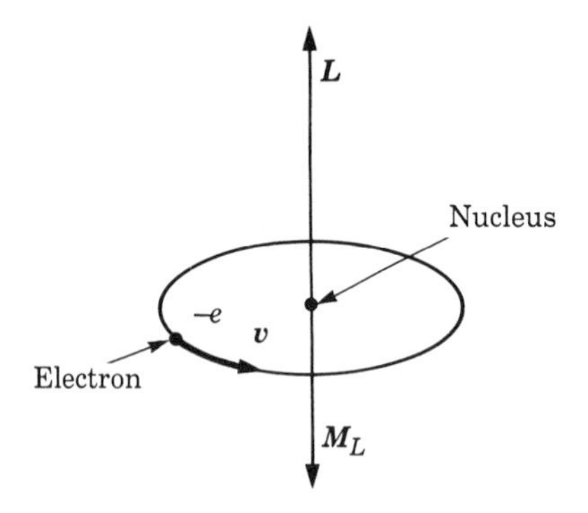

Figure 23.9 Vector relation between the magnetic moment and the angular momentum of an orbiting electron.

$$\boldsymbol{M}_L = -\frac{e}{2m_e}\boldsymbol{L} \tag{23.20}$$

when we replace q by $-e$ and m by m_e. The negative sign shows that $\boldsymbol{M}_L$ is opposed to $\boldsymbol{L}$ (Figure 23.9), corresponding to a negatively charged particle.

Because of the space quantization of $\boldsymbol{L}$, the orbital magnetic moment can have only $2l + 1$ orientations relative to the Z-axis. Also, the magnetic moment is zero for *s*-states ($l = 0$). Thus, the quantization of angular momentum determines the magnetic properties of many substances.

Note 23.1 Justification of relation $L^2 = l(l + 1)\hbar^2$

The expression $L^2 = l(l + 1)\hbar^2$ for the quantization of angular momentum can be justified in a simple manner. First note that for any orientation of $\boldsymbol{L}$,

$$L^2 = L_x^2 + L_y^2 + L_z^2$$

If the *direction* of $\boldsymbol{L}$ is not fixed, but $\boldsymbol{L}$ is precessing in some way around a certain direction without changing magnitude, the components of $\boldsymbol{L}$ are not constant and therefore

$$L^2 = (L_x^2)_{\text{ave}} + (L_y^2)_{\text{ave}} + (L_z^2)_{\text{ave}}$$

Also, in the absence of external fields all directions are equivalent so that the choice of axes is arbitrary. Then $(L_x^2)_{\text{ave}} = (L_y^2)_{\text{ave}} = (L_z^2)_{\text{ave}}$ and the above relation becomes

$$L^2 = 3(L_z^2)_{\text{ave}}$$

In Section 23.5 it was indicated that there is ample experimental evidence, such as the action of a magnetic field on an atom (Section 23.6), that $\boldsymbol{L}$ can have only $2l + 1$ orientations relative to the direction of the magnetic field, chosen as the Z-axis. Also, as indicated in Equation 23.18, each orientation of $\boldsymbol{L}$ corresponds to a Z-component $L_z = m_l\hbar$, with $m_l = 0, \pm 1, \pm 2, \ldots, \pm l$. Therefore,

$$(L_z^2)_{\text{ave}} = \frac{1}{2l+1}\left(\sum_{-l}^{l} m_l^2\hbar^2\right) = \frac{2\hbar^2}{2l+1}\left(\sum_{1}^{l} m_l^2\right),$$

where a factor of 2 has been introduced because the summation over negative values of m_l is the same as over positive values. Using relation B.29, we have

$$\sum_{1}^{l} m_l^2 = \tfrac{1}{6}l(l+1)(2l+1)$$

Therefore we conclude that

$$(L_z^2)_{\text{ave}} = \tfrac{1}{3}l(l+1)\hbar^2 \qquad \text{and} \qquad L^2 = 3(L_z^2)_{\text{ave}} = l(l+1)\hbar^2$$

which is Equation 23.16 for the angular momentum.

Since the summation given in Equation B.29 is valid for both integers or half-integers, as long as successive terms in the summation differ by one, we also conclude that Equation 23.16 is valid for both integer and half-integer values of l. Therefore, the summation also applies to the spin $\boldsymbol{S}$ (see Section 23.7) and to the total angular momentum $\boldsymbol{J} = \boldsymbol{L} + \boldsymbol{S}$, so that both obey Equation 23.16.

23.6 Effect of a magnetic field on electronic motion

There is ample experimental evidence for the existence of the quantization of angular momentum. Perhaps the most direct evidence is the effect of a magnetic field on atomic energy levels, called the **Zeeman effect**, named after Pieter Zeeman (1865–1943), who first observed this effect.

Suppose that an atom is placed in an external magnetic field. The magnetic energy of an orbiting electron is obtained by using Equation 22.19, which in this case is $E_{\text{mag}} = -\boldsymbol{M}_L \cdot \boldsymbol{\mathcal{B}}$ or, introducing the value of $\boldsymbol{M}_L$ given by Equation 23.20,

$$E_{\text{mag}} = (e/2m_e)\boldsymbol{\mathcal{B}} \cdot \boldsymbol{L} \qquad \textbf{(23.21)}$$

Placing the Z-axis in the direction of $\boldsymbol{\mathcal{B}}$, we may write Equation 23.21 as

$$E_{\text{mag}} = (e/2m_e)\mathcal{B}L_z \qquad \textbf{(23.22)}$$

If the angular momentum could have any orientation, the value of L_z could vary continuously between $+L$ and $-L$. Then the energy of the electron could be

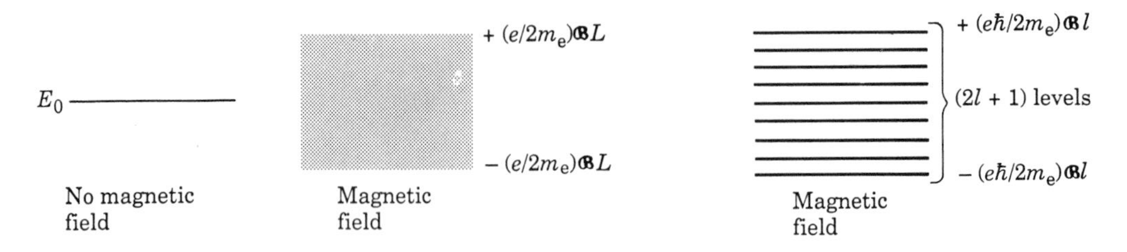

Figure 23.10 Splitting of an energy level by a magnetic field (Zeeman effect).

anywhere between $E_0 + (e/2m_e)\mathscr{B}L$ and $E_0 - (e/2m_e)\mathscr{B}L$, where E_0 is the energy in the absence of the magnetic field (Figure 23.10(a)). However if, according to space quantization, the values of L_z are limited to those given by Equation 23.18, the energy can attain only one of the $2l + 1$ values,

$$E = E_0 + \frac{e\hbar}{2m_e}\mathscr{B}m_l \tag{23.23}$$

as shown in Figure 23.10(b). In other words, each energy level is split into $2l + 1$ sublevels by the magnetic field. It is this second situation that is found experimentally, by observing the splitting of atomic spectral lines by a magnetic field, thus providing a proof of space quantization.

We shall call the quantity

$$\mu_B = \frac{e\hbar}{2m_e} = 9.2732 \times 10^{-24}\ \mathrm{J\,T^{-1}} = 5.6564 \times 10^{-4}\ \mathrm{eV\,T^{-1}} \tag{23.24}$$

a **Bohr magneton**. Then we may say that the energy levels in one-electron atoms are split by a magnetic field into $2l + 1$ levels, given by

$$E = E_0 + \mu_B \mathscr{B} m_l \tag{23.25}$$

The levels are equally spaced by the amount $\mu_B\mathscr{B}$.

There is another important action of a magnetic field on the electronic motion. In Section 22.7 it was shown that a magnetic field $\mathscr{B}$ exerts a torque on a magnetic dipole given by $\boldsymbol{\tau} = \boldsymbol{M} \times \mathscr{B}$, or using Equation 23.20, $\boldsymbol{\tau} = -(e/2m_e)\boldsymbol{L} \times \mathscr{B}$. This torque is perpendicular to $\boldsymbol{L}$ (Figure 23.11). Then, for the same reason as in a gyroscope (Section 13.11),

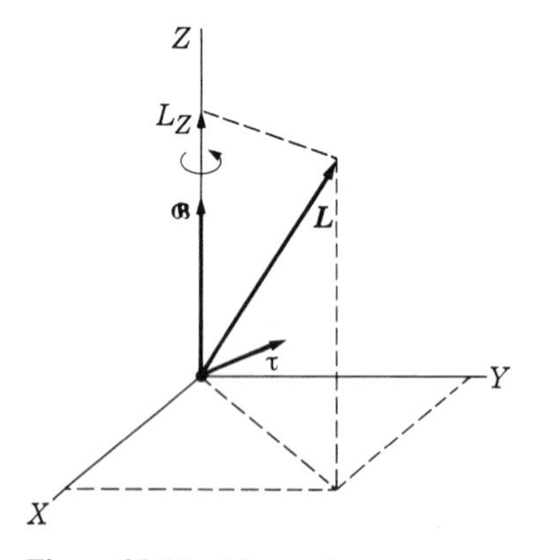

Figure 23.11 Magnetic precession of angular momentum due to the magnetic torque τ. The torque, since it is perpendicular to $\mathscr{B}$ and $\boldsymbol{L}$ rotates in the XY-plane as $\boldsymbol{L}$ precesses around $\mathscr{B}$.

the torque causes the angular momentum $\boldsymbol{L}$ to precess around the direction of $\mathscr{B}$ at a constant rate and at a constant angle with respect to $\mathscr{B}$.

23.7 Electron spin

Recall that the Earth, in addition to its orbital motion around the Sun, has a rotational or spinning motion about its own axis. Therefore, the total angular momentum of the Earth is the vector sum of its orbital angular momentum and its spin angular momentum (see Section 13.6). By analogy, we may suspect that an electron bound in an atom is also spinning. However, we may not describe the electron as a tiny spinning sphere because of our ignorance of its internal structure, if it has any at all. Thus the spin angular momentum of the electron cannot be computed in terms of its radius, mass and angular velocity in the same way that was used for the spin angular momentum of the Earth.

The existence of electron spin can be verified experimentally in several ways. For example, since the electron is a charged particle, electron spin should result in an intrinsic or spin magnetic moment $\boldsymbol{M}_S$ of the electron. By analogy with Equation 23.20, the relation between $\boldsymbol{M}_S$ and the spin angular momentum $\boldsymbol{S}$ is given by

$$\boldsymbol{M}_S = -g_S\left(\frac{e}{2m_e}\right)\boldsymbol{S} \tag{23.26}$$

where g_S, called the **gyromagnetic ratio** of the electron, accounts for the internal properties of the electron. The experimental value for g_S is 2.0024. For most practical purposes, we can make $g_S = 2$.

A direct way of verifying Equation 23.26 is the experiment performed by Otto Stern (1888–1969) and Walther Gerlach (1889–1979) in 1922. Suppose that a beam of hydrogen-like atoms is passed through a non-uniform magnetic field, as shown in Figure 23.12. The effect of such a magnetic field on a magnetic dipole is to exert a force whose direction and magnitude depend on the relative orientation of the magnetic field and the magnetic dipole. Recalling Equation 22.20, if the magnetic dipole is oriented parallel to the magnetic field, it tends to move in the direction in which the magnetic field increases. In contrast, if the magnetic dipole is oriented antiparallel to the magnetic field, it will move in the direction in which the magnetic field decreases.

In the Stern–Gerlach experiment, the non-uniform magnetic field is produced by shaping the pole faces as shown in the figure. The magnetic field increases in strength in the S–N direction. If the orbital angular momentum of the electron is zero (s-state or $l = 0$), the entire magnetic moment is due to the spin. Therefore, the atomic beam will be deviated by the magnetic field, depending on the orientation of $\boldsymbol{M}_S$, or, which is equivalent, the orientation of $\boldsymbol{S}$ relative to the magnetic field. The result of the experiment is that the atomic beam is split in two by the non-uniform magnetic field. This shows that

> *the electron spin may have only two orientations relative to a given direction: either 'parallel' or 'antiparallel'.*

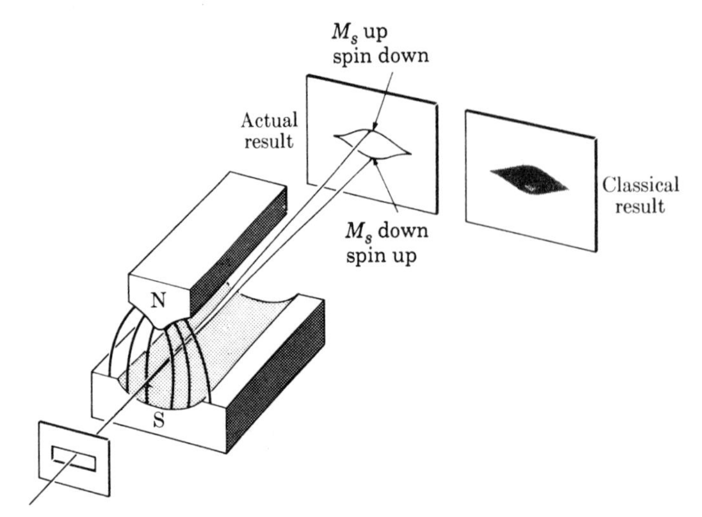

Figure 23.12 Stern–Gerlach experiment.

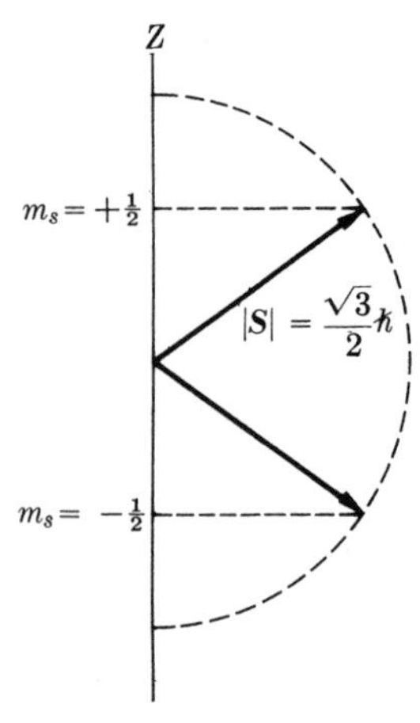

Figure 23.13 Possible orientations of the spin relative to the Z-axis.

According to our discussion in Section 23.6, the number of orientations of an angular momentum vector relative to a fixed axis is $g = 2l + 1$. Then, for the case of spin we have the value $g = 2$ or $l = \frac{1}{2}$. Designating the spin quantum number by s instead of l and the quantum number corresponding to the component S_z by m_s, we have that

$$S = \sqrt{s(s+1)}\hbar = \sqrt{\tfrac{3}{4}}\hbar \qquad s = \tfrac{1}{2} \tag{23.27}$$

$$S_z = m_s\hbar \qquad m_s = \pm\tfrac{1}{2}$$

The only permitted values of m_s (that is, $+\frac{1}{2}$ and $-\frac{1}{2}$), corresponding to the two possible orientations of $\boldsymbol{S}$, are shown in Figure 23.13. For brevity they are usually referred to as **spin up** (↑) and **spin down** (↓), although the spin is never actually directed along the Z-axis or opposite to it, but precesses around the Z-axis.

23.8 Spin–orbit interaction

The total angular momentum of an electron is $\boldsymbol{J} = \boldsymbol{L} + \boldsymbol{S}$, where $\boldsymbol{L}$ is the orbital angular momentum and $\boldsymbol{S}$ is the spin angular momentum. We shall designate by j the quantum number associated with $\boldsymbol{J}$, so that

$$|\boldsymbol{J}| = \sqrt{j(j+1)}\hbar \tag{23.28}$$

Since the electron spin may have only two possible orientations relative to the orbital angular momentum, we conclude that the possible values of j are

$$j = l + \tfrac{1}{2}, \qquad \text{when } \boldsymbol{S} \text{ is 'parallel' to } \boldsymbol{L}$$

$$j = l - \tfrac{1}{2}, \qquad \text{when } \boldsymbol{S} \text{ is 'antiparallel' to } \boldsymbol{L}$$

Table 23.2 Designation of electronic states

l	0	1		2		3	
j	$\frac{1}{2}$	$\frac{1}{2}$	$\frac{3}{2}$	$\frac{3}{2}$	$\frac{5}{2}$	$\frac{5}{2}$	$\frac{7}{2}$
Symbol	$s_{1/2}$	$p_{1/2}$	$p_{3/2}$	$d_{3/2}$	$d_{5/2}$	$f_{5/2}$	$f_{7/2}$

These two situations are illustrated in Figure 23.14 for $l = 2$, and thus $j = \frac{5}{2}$ or $\frac{3}{2}$. When $l = 0$ (or s-state), only $j = \frac{1}{2}$ is possible. The possible angular momentum states of an electron are designated as shown in Table 23.2.

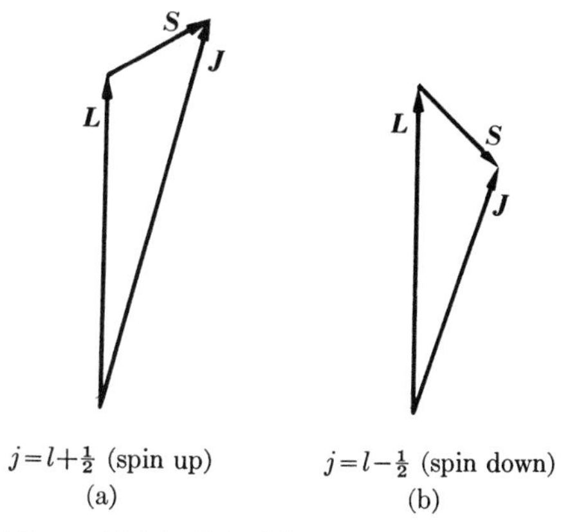

Figure 23.14 Possible relative orientations of $\boldsymbol{L}$ and $\boldsymbol{S}$, when $l = 2$.

The two possible orientations of the electron spin relative to the orbital momentum give rise to an important effect: the *doubling* of energy levels. All energy levels (except s-levels) of hydrogen-like atoms or, in general, of all atoms with one valence electron are split in two, depending on the relative orientation of $\boldsymbol{L}$ and $\boldsymbol{S}$. The doubling of the energy levels is a consequence of the so-called **spin–orbit interaction**. This is a magnetic effect, associated with the spin magnetic moment and the magnetic field of the orbital motion of the electron and is proportional to $\boldsymbol{S} \cdot \boldsymbol{L}$ (see Note 23.2).

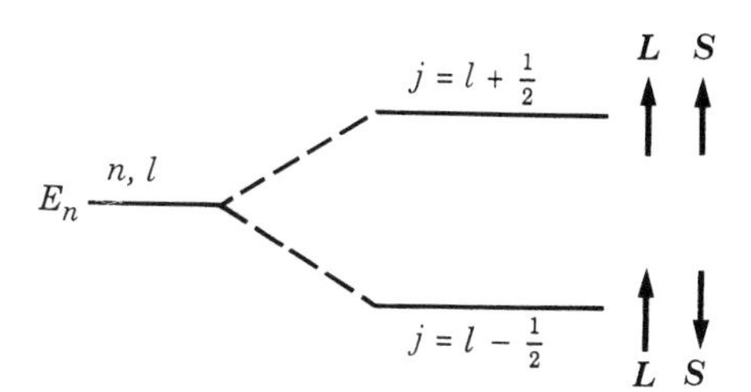

Figure 23.15 Splitting of energy levels due to the spin–orbit interaction.

Therefore, *the spin–orbit interaction splits each electron energy level with a given value of l into two closely spaced levels.* One level corresponds to $\boldsymbol{L}$ and $\boldsymbol{S}$ parallel or spin up (Figure 23.15) ($j = l + \frac{1}{2}$). The other level corresponds to $\boldsymbol{L}$ and $\boldsymbol{S}$ antiparallel or spin down ($j = l - \frac{1}{2}$). For s-levels ($l = 0$) there is no splitting. This doubling of electron energy levels was first observed in 1926 by George Uhlenbeck and Samuel Goudsmit (1902–1978). It led them to introduce the concept of electron spin and assign to it the value $s = \frac{1}{2}$.

Note 23.2 Origin of the spin–orbit interaction

Consider a frame of reference XYZ attached to the nucleus of an atom. In that frame the electron appears to revolve around the nucleus (Figure 21.16(a)) with angular momentum $\boldsymbol{L}$. But, in a frame of reference $X'Y'Z'$ attached to the electron, it is the nucleus that appears to revolve

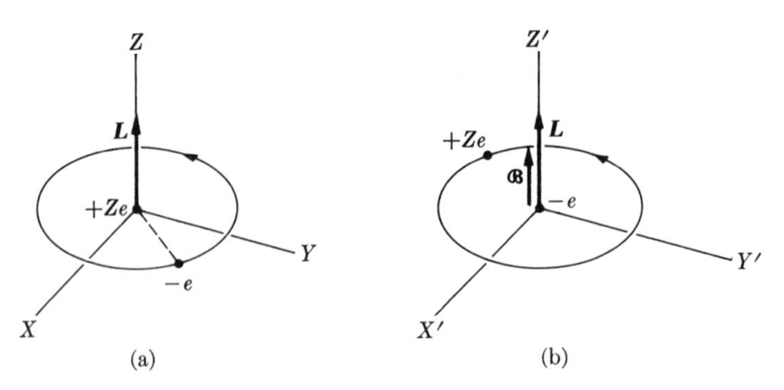

Figure 23.16 Origin of the spin–orbit interaction: (a) frame of reference attached to the proton; (b) frame of reference attached to the electron.

around the electron with angular momentum $\boldsymbol{L}$ (Figure 23.16(b)). The nucleus, since it has a positive charge, produces a magnetic field $\mathscr{B}$ in the $X'Y'Z'$ frame that is parallel to the angular momentum $\boldsymbol{L}$. Since the electron is at rest relative to $X'Y'Z'$, the only interaction of the nuclear magnetic field is with the spin magnetic moment $\boldsymbol{M}_S$ of the electron. This interaction is proportional to $\boldsymbol{M}_S \cdot \mathscr{B}$. But $\boldsymbol{M}_S$ is parallel to $\boldsymbol{S}$ and $\mathscr{B}$ is parallel to $\boldsymbol{L}$. Therefore the interaction is proportional to $\boldsymbol{S} \cdot \boldsymbol{L}$. This is why this effect is called **spin–orbit interaction**. We may then write, for the energy of the electron due to the spin–orbit interaction,

$$\Delta E_{SL} = a\boldsymbol{S} \cdot \boldsymbol{L}$$

The parameter a is a quantity whose value depends on the different variables that affect the electron's motion. We do not need to know its precise form at this time. It is sufficient to say that it is of the order of $|E_n|^2 Z^2 \alpha^2$, where

$$\alpha = \frac{e^2}{4\pi\varepsilon_0 \hbar c} \simeq \frac{1}{137}$$

is the **fine structure constant**. Therefore, ΔE_{SL} is of the order of $5 \times 10^{-5} |E_n|^2 Z^2$, which is very small compared to $|E_n|$, though it is measurable.

Given that E_n is the energy of the electronic motion, assuming only a central force, the total energy, when the spin–orbit interaction is added, is

$$E = E_n + \Delta E_{SL} = E_n + a\boldsymbol{S} \cdot \boldsymbol{L} \tag{23.29}$$

For a given value of $\boldsymbol{L}$ and $\boldsymbol{S}$, the spin–orbit interaction ΔE_{SL} depends on the relative orientation of these two vectors. But since $\boldsymbol{S}$ can have only two possible orientations relative to $\boldsymbol{L}$, we conclude that there are only two values of ΔE_{SL}.

23.9 Electron shells in atoms

When an atom has more than one electron, the calculation of the energies is more complex because the interactions between the electrons must also be included. However, from the discussion in the preceding sections we may conclude that to identify the state of an electron fully we need four quantum numbers n, l, m_l and m_s. The first three quantum numbers give the energy and angular momentum of the orbital motion and the fourth the orientation of the spin. In the absence of external electric or magnetic fields, the direction of the Z-axis is arbitrary. Therefore, the energy of the electrons depends only on n and l. When the spin–orbit effect is taken into account, the energy also depends on j. Thus each electronic state is identified by the symbol nl, that is, $1s$, $2s$, $2p$, $3s$, $3p$, $3d, \ldots$, to which a subscript for the values of j is sometimes added (Table 23.2).

To specify the state of an atom having many electrons, we must indicate the state of each electron. All electrons having the same nl quantum numbers are called *equivalent* because they have approximately the same energy. The complete state of the atom is specified by indicating the number of equivalent electrons in each nl state. This constitutes what is called a **configuration**. If there are x electrons in the state nl, this is indicated as $(nl)^x$. For example, in the ground state of helium the two electrons are in the $n = 1$, $l = 0$ or $1s$ state. Therefore the configuration is $(1s)^2$. But if one of the electrons is excited to the state $2s$ or $2p$, the configuration is written as $(1s)(2s)$ or $(1s)(2p)$, respectively.

The configuration of each chemical element in its ground and excited states is of great importance. It is well known that chemical elements exhibit certain regularities in their physical and chemical properties. These properties repeat themselves in different elements in a more or less cyclic form. Successive cycles or **periods** are completed at the atomic numbers $Z = 2$, 10, 18, 36, 54 and 86, corresponding to the (so-called) inert gases: helium, neon, argon, krypton, xenon and radon. Inert gases are characterized chemically by their very weak or almost non-existent capability for entering into combination with other elements. The periodicity in the properties of the elements is exemplified in a striking way by their *ionization energies*, as shown in Figure 23.17. We see that there is a trend that more or less repeats itself after each inert gas.

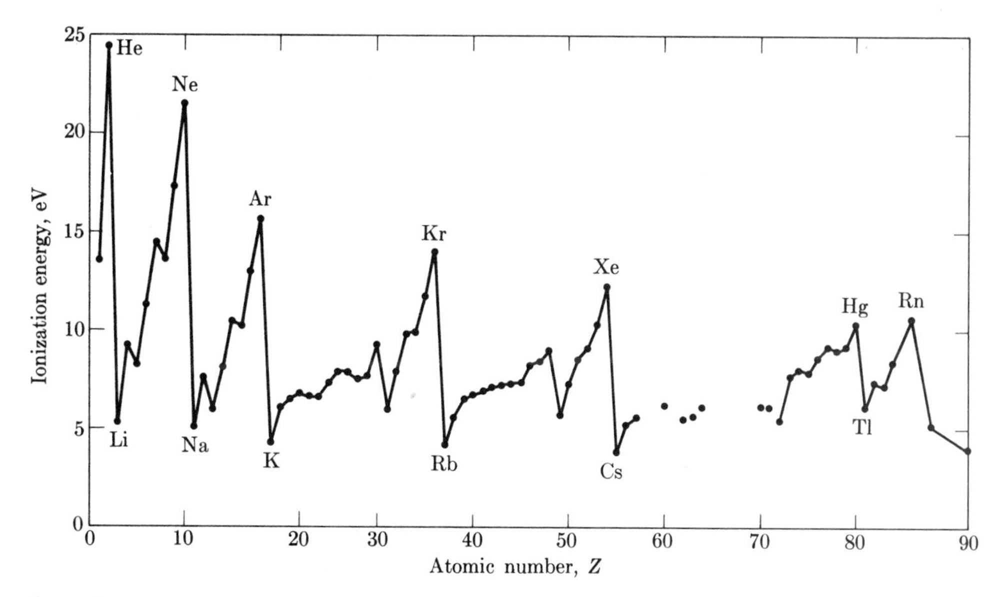

Figure 23.17 Ionization energy of the elements as a function of atomic number.

To explain this periodicity, in 1870 Dmitri Mendeleyev (1834–1907) proposed his celebrated periodic classification of the elements in cycles, or periods, composed of eight elements each. This classification, however, did not work well and some elements, such as the transition elements and the rare earths, did not fit easily into this simplified scheme.

The regularities in the atomic properties suggest certain regularities in the motion of the electrons in an atom. To explain these regularities, Wolfgang Pauli (1900–1958) proposed a new rule around 1925, since called the **exclusion principle**. All particles that have spin $\frac{1}{2}$, and not just electrons, obey the exclusion principle and are called **fermions**, after Enrico Fermi. Particles with spin 0 or 1 do not obey the exclusion principle and are called **bosons**, after Satyendranath Bose. This principle not only explains the periodic physical and chemical behavior of the elements in a beautiful and simple manner, but it also correlates many other important experimental facts of atomic structure. Pauli's exclusion principle states that *no two electrons in an atom may have the same set of quantum numbers, or which is equivalent, no two electrons in an atom may be in the same dynamical state.*

Pauli's exclusion principle can be used to determine the electronic configurations of atoms. First we calculate the number of combinations of quantum numbers m_l and m_s that are possible for each value of the angular momentum l. This gives the maximum number of electrons that can be accommodated in a state nl. We know that for each value l there are $2l + 1$ values of m_l, and for each set of (l, m_l) we may have the electron with spin up or down ($m_s = \pm\frac{1}{2}$). So, the maximum number of electrons that can be accommodated in a state nl without violating the exclusion principle is $2(2l + 1)$. This is indicated explicitly thus:

Angular momentum, l	0	1	2	3	4
Symbol	s	p	d	f	g
Occupation number, $2(2l + 1)$	2	6	10	14	18

Then, in the building-up process of the atoms from $Z = 1$ up to $Z = 92$ (and on), the ground state of an atom results when the electrons occupy the lowest-lying energy states available, each up to the maximum number allowed by the exclusion principle. Once an nl state has received its full quota of electrons, the next state begins to fill up.

The order in which the successive nl states fill up is indicated in Figure 23.18. There may be slight changes for some particular atoms, but in general the order is as shown. A more detailed analysis is given in Chapter 38. We observe in the first place certain 'energy gaps'; that is, regions where the energy difference between two stages or levels is much larger than it is between levels below the lower or above the upper. The energy gaps appear between $1s$ and $2s$, between $2p$ and $3s$, between $3p$ and $4s$, between $4p$ and $5s$, and so on. Energy levels grouped between two energy gaps constitute a **shell**. Traditionally the shells are designated by the letters K, L, M,.... Each nl state composing a shell is called a **subshell**. The maximum number of electrons in successive complete shells occurs precisely at

$Z = 2$, 10, 18, 36, 54 and 86, which are the inert gases, as required by experimental evidence. There may be another inert gas at $Z = 118$, but no atom with such a high number of electrons has been found in nature or yet made artificially.

The electrons filling the inner complete shells constitute the **core** or **kernel**. The remaining electrons in the outermost unfilled shells are called the *valence* electrons. The binding energy of the kernel electrons is much higher than that of the valence electrons. For that reason the kernel electrons are rather tightly bound and remain practically undisturbed in most of the processes in which the atom participates. It is the valence electrons that are mainly responsible for the chemical properties of the atom. These electrons are the ones that participate in chemical reactions and chemical binding.

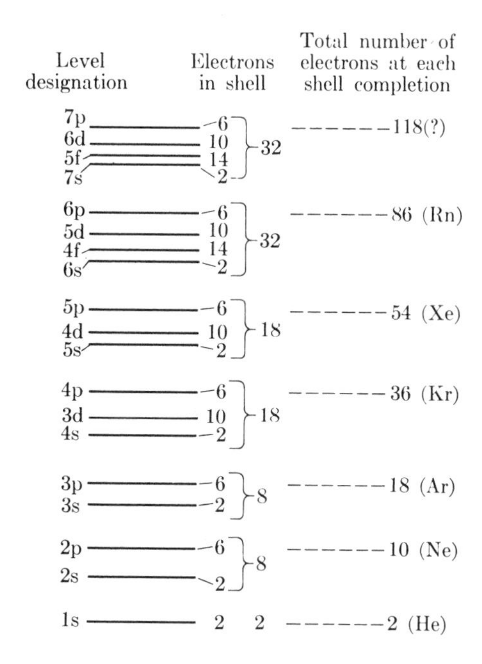

Figure 23.18 Shell structure of atomic energy levels.

The electron configuration of the ground state of the atoms of the different chemical elements is given in Table 23.3. The table also gives the orbital and total angular momentum of the atom in the ground state and the ionization energy. The block diagrams of Figure 23.19 show how the first 10 elements are built up. The *s*-subshells have one block accommodating two electrons with opposite spins, and the *p*-subshells three blocks ($m_l = +1, 0, -1$) accommodating six electrons.

From Figure 23.19, it can be seen that while hydrogen has a single 1*s*-electron, lithium has a single 2*s*-electron, which has the relatively close 2*p*-levels available and so it can be easily excited. Lithium also has a smaller ionization energy than hydrogen (5.5 eV compared to 13.6 eV). This may be attributed to the fact that the 2*s*-electron in lithium is, on the average, farther from the nucleus than the 1*s*-electron in hydrogen. These differences account for the distinct metallic behavior of lithium, not exhibited by hydrogen. Similarly, both helium and beryllium have completed *s*-shells, $1s^2$ for helium and $2s^2$ for beryllium. However, helium is an inert gas, while beryllium is not. This is because one of the 2*s*-electrons in beryllium can be excited to a nearby 2*p*-level, resulting in the excited state shown in Figure 23.20. This state corresponds to the valence 2 observed in beryllium compounds and gives beryllium its metallic character. The excitation energy is about 2.7 eV, while for the case of helium the minimum excitation energy is about 20 eV. The first excited state of boron may be obtained by moving a 2*s*-electron into a 2*p*-state, as shown in Figure 23.20, requiring an energy of 3.6 eV. This state explains the trivalence of boron. The situation for carbon, nitrogen and other atoms is more complicated. For them a number of excited states may be produced by re-orienting the spin and orbital angular momenta of the *p*-electrons rather

Table 23.3 Electronic configuration of atoms*

Z	Symbol	Ground state	Ground configuration	Ionization energy (eV)
1	H	^2S	$1s$	13.595
2	He	^1S	$1s^2$	24.581
3	Li	^2S	[He] $2s$	5.390
4	Be	^1S	$2s^2$	9.320
5	B	$^2P_{1/2}$	$2s^22p$	8.296
6	C	$^3P_0$	$2s^22p^2$	11.256
7	N	^4S	$2s^22p^3$	14.545
8	O	$^3P_2$	$2s^22p^4$	13.614
9	F	$^2P_{3/2}$	$2s^22p^5$	17.418
10	Ne	^1S	$2s^22p^6$	21.559
11	Na	^2S	[Ne] $3s$	5.138
12	Mg	^1S	$3s^2$	7.644
13	Al	$^2P_{1/2}$	$3s^23p$	5.984
14	Si	$^3P_0$	$3s^23p^2$	8.149
15	P	^4S	$3s^23p^3$	10.484
16	S	$^3P_2$	$3s^23p^4$	10.357
17	Cl	$^2P_{3/2}$	$3s^23p^5$	13.01
18	Ar	^1S	$3s^23p^6$	15.755
19	K	^2S	[A] $4s$	4.339
20	Ca	^1S	$4s^2$	6.111
21	Sc	$^2D_{3/2}$	$3d4s^2$	6.54
22	Ti	$^3F_2$	$3d^24s^2$	6.83
23	V	$^4F_{3/2}$	$3d^34s^2$	6.74
24	Cr	^7S	$3d^54s$	6.764
25	Mn	^6S	$3d^54s^2$	7.432
26	Fe	$^5D_4$	$3d^64s^2$	7.87
27	Co	$^4F_{9/2}$	$3d^74s^2$	7.86
28	Ni	$^3F_4$	$3d^84s^2$	7.633
29	Cu	^2S	$3d^{10}4s$	7.724
30	Zn	^1S	$3d^{10}4s^2$	9.391
31	Ga	$^2P_{1/2}$	$3d^{10}4s^24p$	6.00
32	Ge	$^3P_0$	$3d^{10}4s^24p^2$	7.88
33	As	^4S	$3d^{10}4s^24p^3$	9.81
34	Se	$^3P_2$	$3d^{10}4s^24p^4$	9.75
35	Br	$^2P_{3/2}$	$3d^{10}4s^24p^5$	11.84
36	Kr	^1S	$3d^{10}4s^24p^6$	13.996
37	Rb	^2S	[Kr] $5s$	4.176
38	Sr	^1S	$5s^2$	5.692
39	Y	$^2D_{3/2}$	$4d5s^2$	6.377
40	Zr	$^3F_2$	$4d^25s^2$	6.835
41	Nb	^7S	$4d^55s$	7.10
42	Mo	$^6D_{1/2}$	$4d^45s$	6.881
43	Tc	^6S	$4d^55s^2$	7.228
44	Ru	$^5F_5$	$4d^75s$	7.365
45	Rh	$^4F_{9/2}$	$4d^85s$	7.461
46	Pd	^1S	$4d^{10}$	8.33
47	Ag	^2S	$4d^{10}5s$	7.574
48	Cd	^1S	$4d^{10}5s^2$	8.991
49	In	$^2P_{1/2}$	$4d^{10}5s^25p$	5.785
50	Sn	$^3P_0$	$4d^{10}5s^25p^2$	7.342
51	Sb	^4S	$4d^{10}5s^25p^3$	8.639
52	Te	$^3P_2$	$4d^{10}5s^25p^4$	9.01
53	I	$^2P_{3/2}$	$4d^{10}5s^25p^5$	10.454
54	Xe	^1S	$4d^{10}5s^25p^6$	12.127
55	Cs	^2S	[Xe] $6s$	3.893
56	Ba	^1S	$6s^2$	5.210
57	La	$^3D_{3/2}$	$5d6s^2$	5.61
58	Ce	$^3H_5$	$4f5d6s^2$	6.54
59	Pr	$^4I_{9/2}$	$4f^36s^2$	5.48
60	Nd	$^5I_4$	$4f^46s^2$	5.51
61	Pm	$^6H_{5/2}$	$4f^56s^2$	
62	Sm	$^7S_0$	$4f^66s^2$	5.6
63	Eu	^8S	$4f^76s^2$	5.67
64	Gd	$^9D_2$	$4f^75d6s^2$	6.16
65	Tb	$^6H_{15/2}$	$4f^96s^2$	6.74
66	Dy	$^5I_8$	$4f^{10}6s^2$	6.82
67	Ho	$^4I_{15/2}$	$4f^{11}6s^2$	
68	Er	$^3H_6$	$4f^{12}6s^2$	
69	Tm	$^2F_{7/2}$	$4f^{13}6s^2$	
70	Yb	^1S	$4f^{14}6s^2$	6.22
71	Lu	$^2D_{3/2}$	$4f^{15}5d6s^2$	6.15
72	Hf	$^3F_2$	$4f^{14}5d^26s^2$	7.0
73	Ta	$^4F_{3/2}$	$4f^{14}5d^36s^2$	7.88
74	W	$^5D_0$	$4f^{14}5d^46s^2$	7.98
75	Re	^6S	$4f^{14}5d^56s^2$	7.87
76	Os	$^5D_4$	$4f^{14}5d^66s^2$	8.7
77	Ir	$^4F_{9/2}$	$4f^{14}5d^76s^2$	9.2
78	Pt	$^3D_3$	$4f^{14}5d^86s^2$	8.88
79	Au	^2S	[Xe, $4f^{14}5d^{10}$] $6s$	9.22
80	Hg	^1S	$6s^2$	10.434
81	Tl	$^2P_{1/2}$	$6s^26p$	6.106
82	Pb	$^3P_0$	$6s^26p^2$	7.415
83	Bi	^4S	$6s^26p^3$	7.287
84	Po	$^3P_2$	$6s^26p^4$	8.43
85	At	$^2P_{3/2}$	$6s^26p^5$	
86	Rn	^1S	$6s^26p^6$	10.745
87	Fr	^2S	[Rn] $7s$	
88	Ra	^1S	$7s^2$	5.277
89	Ac	$^2D_{3/2}$	$6d7s^2$	6.9
90	Th	$^3F_2$	$6d^27s^2$	
91	Pa	$^4K_{11/2}$	$5f^26d7s^2$	
92	U	$^5I_6$	$5f^36d7s^2$	4.0
93	Np	$^6L_{11/2}$	$5f^46d7s^2$	
94	Pu	$^7F_0$	$5f^67s^2$	
95	Am	^9S	$5f^77s^2$	
96	Cm	$^9D_2$	$5f^76d7s^2$	
97	Bk		$(5f^86d7s^2)$	
98	Cf		$(5f^96d7s^2)$	
99	E		$(5f^{10}6d7s^2)$	
100	Fm		$(5f^{11}6d7s^2)$	
101	Mv			
102	No			
103	Lw			
104				

*Chemical symbols in brackets indicate the equivalent configurations of the remaining electrons occupying filled shells. Configurations in parentheses are uncertain.

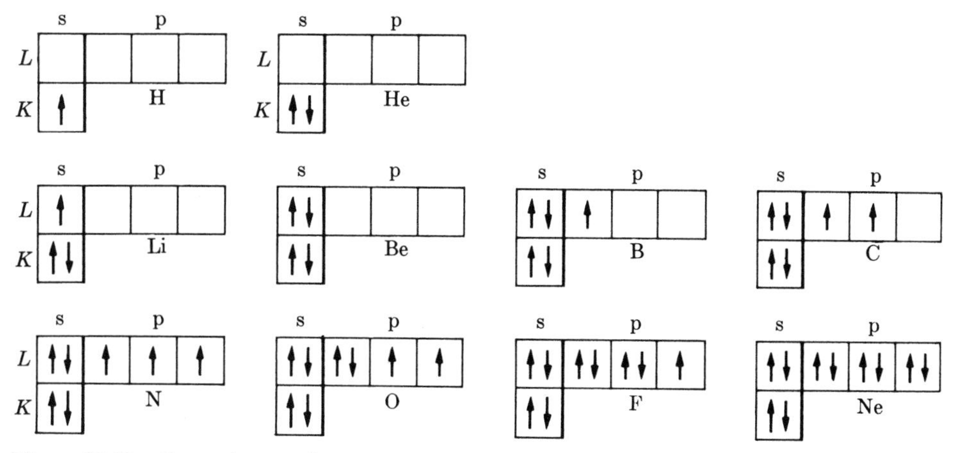

Figure 23.19 Ground state electronic configurations of the atoms of the first ten elements, showing how they are built up.

than promoting an *s*-electron to a *p*-level. A particularly important excited state of carbon is shown in Figure 23.20, which accounts for the tetravalence of carbon in many of its compounds.

The previous considerations help us to understand that the chemical and physical properties of an element depend on the ground-state electronic configuration as well as on the nearby excited levels. For example, what characterizes an inert gas (such as helium, neon or argon, which have $Z = 2$, 10 and 18, respectively) is that their atoms are composed of complete shells. To excite one of the electrons of an atom in an inert gas a comparatively large energy is needed, due to the large energy gap separating the last complete level from the first unoccupied level. On the other hand, atoms such as lithium, sodium, potassium etc., with $Z = 3, 11, 19, \ldots$, are composed of closed shells plus one electron in the first level above the energy gap. So lithium has a complete *K*-shell and a 2*s*-electron; sodium has complete *K*- and *L*-shells and a 3*s*-electron, and so on. This last electron is bound loosely and determines the metallic behavior of these elements. This is why Na ($Z = 11$), is so different from Ne ($Z = 10$), even if their atoms differ by only one electron. When the number of electrons beyond the last completed shells increases, the situation gets more and more complex and the energy levels follow a much more complicated pattern.

The quantization of energy and the existence of well-defined energy levels for the electronic motion in atoms is one of the reasons for the stability of matter and the uniform chemical behavior of each kind of atom and molecule. For example, not only are all hydrogen atoms composed of a proton and an orbiting electron, but the electron's energy in the ground state and its average distance from the

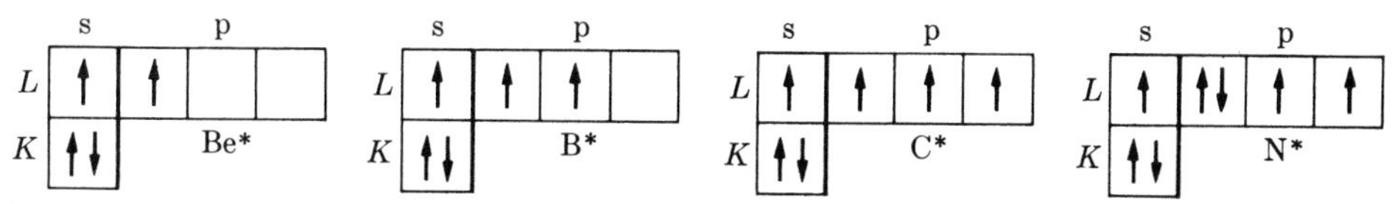

Figure 23.20 Excited electronic configurations of beryllium, boron, carbon and nitrogen.

proton, given by the Bohr radius, are always the same. That means that all hydrogen atoms are identical and have the same stable configuration as well as the same excitation energies. The same applies to all atoms with the same atomic number Z, because the Z electrons must be distributed in an identical way among the same available energy levels, in accordance with the exclusion principle, resulting in the stable configurations shown in Table 23.3. Further, all atoms with the same Z have the same size.

In addition, the temperatures in the Earth's crust and atmosphere are so low that they result in molecular kinetic energies well below 1 eV. This explains why atoms and molecules are stable under normal conditions, since electronic excitations are possible only by greatly raising the temperature or exposing the atoms or molecules to high energy electromagnetic radiation, such as ultraviolet and X-rays. On the other hand, the vibrational and rotational energy of molecules is easily modified (recall Example 17.3) without exciting the electronic motion.

23.10 Electrons in solids

When atoms are closely packed in a solid, the motion of their outer or valence electrons is profoundly disturbed, due to the electric field produced by the nearby atoms in the solid.

Consider, for example, atoms with one valence electron. We may look at these atoms as consisting of an electron plus a positive ion. Thus, Na atoms ($Z = 11$) may be considered as an electron (the valence electron) moving in the electric field of a Na^+ ion. The potential energy felt by the valence electron has been illustrated in Figure 23.21(a) where N represents the ion Na^+. If the energy of the valence electron is given by the horizontal line E, the electron classically can move between a and b. If we place two Na atoms very close to each other, each of the valence electrons is affected by the field of *both* Na^+ ions and they move in a potential shown in Figure 23.21(b). For the same energy as in Figure 23.21(a), the valence electrons can move between c and d, and hence are not attached to a particular sodium ion. If we now have a piece of solid sodium, composed of a very large number of sodium atoms regularly arranged in a lattice, the resultant potential energy felt by the valence electrons is as shown in Figure 23.21(c). We note that the valence electrons can move freely throughout the lattice of sodium ions and therefore are not localized at any atom in particular. When an electron moves through the lattice of sodium ions, the potential energy varies periodically with the period of the lattice. Also, it can be seen from the figure that the inner

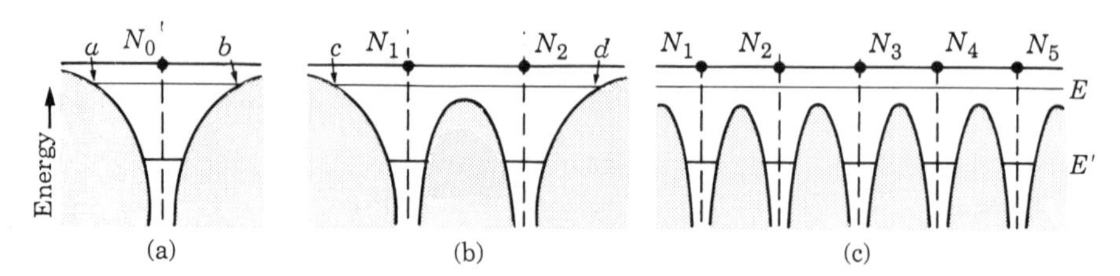

Figure 23.21 Coulomb potential energies due to (a) a single ion, (b) two ions, (c) several ions in a row. The allowed regions for energies E and E' are also shown.

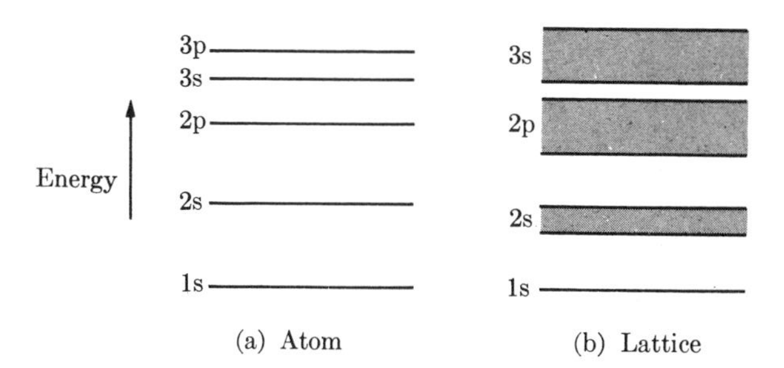

Figure 23.22 Comparison of atomic energy levels with energy bands in a lattice.

electrons, having a much greater binding energy E', remain essentially undisturbed and localized at their respective atoms.

Electrons in atoms can occupy only certain energy levels. In a lattice, however, the energy levels are modified, and we find instead that electrons can move within certain **energy bands** (Figure 23.22). Each band consists of closely spaced energy levels. The reasons for the bands are explained in Section 38.9.

Pauli's exclusion principle limits the number of electrons that can occupy a certain nl atomic energy level. In the same manner, the exclusion principle limits the maximum number of electrons that can be accommodated in an energy band of a lattice. For example, consider the band corresponding to the $3s$-atomic level in sodium. Each s-atomic level can accommodate up to two electrons. Therefore in a lattice of N sodium atoms, the $3s$ energy band can accommodate $2N$ electrons. However, sodium atoms have only one $3s$ electron each. Therefore in a lattice of N sodium atoms, there are only N electrons in the $3s$ energy band; that is, the band is half-filled and therefore it is easy to excite an electron within the band but not from one band to another.

In general, all energy levels in an atom are filled except the uppermost level, which may or may not be complete. We thus conclude that all low-lying energy bands in a solid are also filled. However, the uppermost (or valence) band, occupied by the valence electrons, may or may not be filled. The number of electrons in the valence band is very important in determining the electrical properties of a solid.

23.11 Conductors, semiconductors and insulators

An important property of solids is their **electrical conductivity**; that is, the ability for electrons to move freely through the solid under the action of an external electric field. Some materials, traditionally called *insulators*, are extremely poor conductors of electricity. Two examples are diamond and quartz. Other solids are exceedingly good *conductors* of electricity. In this group fall the metals, such as copper and silver. As a quantitative measure of their difference, the electrical conductivity of copper at room temperature is 10^{20} times greater than that of quartz. Intermediate between these two extremes is a third class of solids, called **semiconductors**. Although semiconductors are much poorer electrical conductors than the metals, it has been found experimentally that their conductivity increases with temperature, while that of metals decreases with temperature. Typical semiconductors are germanium and silicon. Whether a solid is a conductor, a semiconductor or an insulator depends on its band structure.

Consider a metal having the band structure shown in Figure 23.23, which might correspond, for example, to the energy levels of sodium ($Z = 11$). Bands corresponding to the $1s$, $2s$, and $2p$ atomic levels are completely filled because the respective atomic shells are also complete. But, as explained at the end of the previous section, the $3s$ band, which can accommodate up to two electrons per atom, is only half-filled, since the $3s$ level in each sodium atom has only one electron. Under the action of an external electric field, the uppermost electrons in the valence band may, without violating the exclusion principle, pick up additional small amounts of energy. This extra energy allows them to pass to any of the many nearby empty states within the band. In sharp distinction to disordered thermal excitation, the electrons excited by the electric field gain momentum in the direction opposite to the field. The result is a collective motion through the crystal, which constitutes an **electric current**. Therefore we conclude that a substance having a band structure such as that of Figure 23.23 should be a good conductor of electricity. It is the electrons in the uppermost, partially occupied band that are responsible for the process. In other words,

> *good conductors of electricity (also called metals) are those solids whose uppermost occupied band is not completely filled.*

Actually, the situation is slightly more complex because the uppermost bands may superpose due to the proximity of atomic levels. In fact, this overlapping of the uppermost bands is the common situation for most metals or conductors. For example, consider the case of magnesium ($Z = 12$). The magnesium atom has the configuration $1s^2 2s^2 2p^6 3s^2$, and therefore all the atomic levels are filled. However, the first excited level, $3p$, is rather close to $3s$, as shown in Figure 23.18. The corresponding $3s$ and $3p$ bands of solid magnesium are indicated schematically in Figure 23.24. Normally, with no overlapping, the $3s$ band should be filled and the $3p$ one empty, and magnesium should be an insulator. But because of the

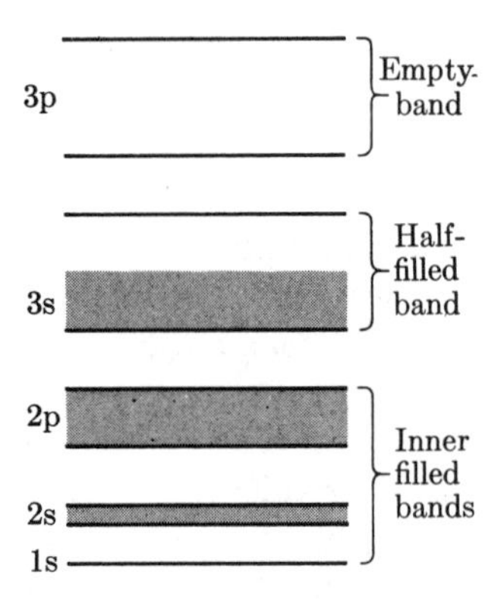

Figure 23.23 Energy bands in a conductor. The Fermi energy in this case corresponds to the maximum occupied energy level in the 3s band.

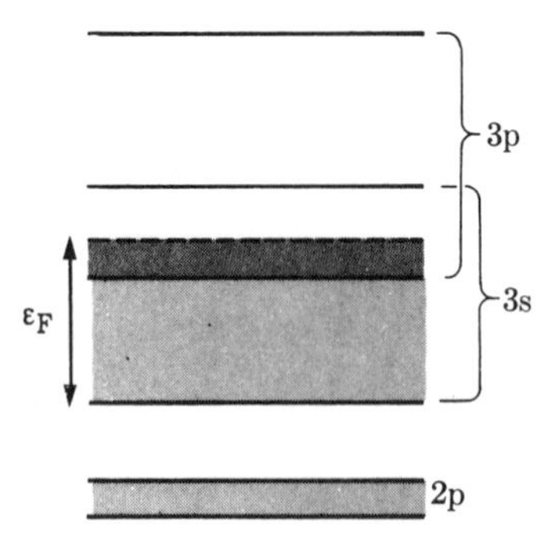

Figure 23.24 Overlapping of energy bands in a conductor.

overlapping, the uppermost electrons of the $3s$ band have the lowest energy states of the $3p$ band available to them. Thus some $3s$ electrons may move to occupy some low $3p$-levels until an equilibrium energy level for both bands is established. Since the total number of energy levels available from the $3s$ and $3p$ bands is $2N + 6N = 8N$ and we have only $2N$ electrons, there are $6N$ accessible empty states. Therefore magnesium should be a good conductor; this is in agreement with the experimental facts.

Substances whose atoms have complete shells but which, in the solid state, have their uppermost filled band overlapping an empty band, are also conductors.

Electrons fill the conduction band in accordance with the exclusion principle up to a maximum energy level called the **Fermi energy** ε_F. At very low temperatures all levels up to ε_F are occupied and those above are empty (Figure 23.25(a)). However, as the temperature increases some electrons are thermally excited above the Fermi energy, as shown in Figure 23.25(b), leaving some vacancies behind. The fraction of thermally excited electrons is of the order of kT/ε_F, where kT gives the average thermal energy. Only electrons in an energy range of the order of kT at the Fermi energy participate in electric conduction in metals because they can be thermally excited to nearby empty states.

When the temperature of the metal is sufficiently high, a new process takes place, called **thermal electronic emission**. Electrons in the upper or conduction band do not escape from the metal at normal temperatures because the electrons are held within the metal by a **potential barrier** at the surface. Unless an electron has enough energy to overcome this barrier, it cannot escape from the metal. One way to increase the energy of electrons is by heating the metal. The 'evaporated' electrons are then called **thermoelectrons**. This is the kind of electronic emission that exists in electron tubes.

A new situation occurs in a substance where the uppermost or valence band is completely filled and does not overlap the next band, which is totally empty (Figure 23.26). Since all states of the valence band are occupied, the electron motion is 'frozen'; that is, the electrons cannot change their state within the band without violating the exclusion principle. The only possibility for exciting an electron is to transfer it to the empty conduction band. But this may require an energy of a few electronvolts. Hence an applied electric field, in general, cannot

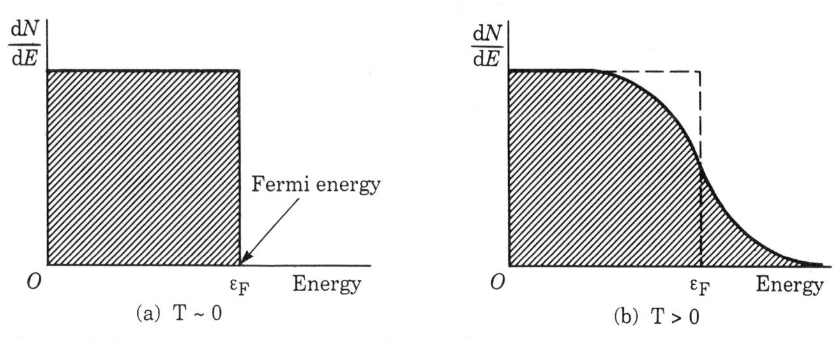

Figure 23.25 Occupation of the conduction band as a function of the temperature. The ordinate dN/dE gives the occupation number per unit energy.

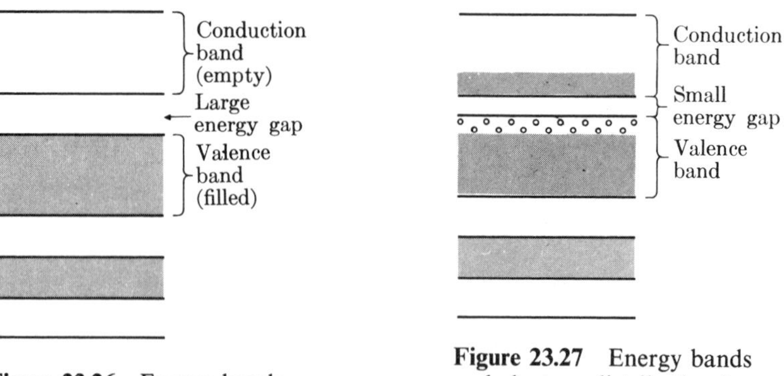

Figure 23.26 Energy bands of an insulator.

Figure 23.27 Energy bands and electron distribution in a semiconductor.

accelerate the electrons in the valence band and thus cannot produce a net electric current. Such a substance is therefore an **insulator**.

Of course, at sufficiently high temperatures or under very strong electric fields, some electrons in the valence band may be excited to the conduction band and an electric current is then possible, resulting in what is called an electric breakdown of the insulator. For example, at the equilibrium distance in diamond, about 1.5×10^{-10} m, the gap separating the valence band from the upper empty band is about 6 eV. This may be considered as a relatively large energy gap; it explains why diamond is such a good insulator. Hence,

> *insulators are substances whose uppermost valence band is completely filled and separated by an energy gap of a few eV from the next empty band.*

The same band scheme also applies to silicon and germanium. However, the gap between the valence and conduction bands at the equilibrium separation of the atoms is much smaller (1.1 eV in silicon and 0.7 eV in germanium). This smaller gap makes it much easier to excite the uppermost electrons of the valence band into the conduction band. The situation is illustrated in Figure 23.27. As the temperature increases, more electrons are able to jump into the next band. This results in two possibilities. First, the few electrons in the upper or conduction band act as they would in a metal. Secondly, the empty states, or **holes**, left in the lower or valence band act in a similar way, but as if they were positive electrons. Thus we have electric conduction from the excited electrons in the conduction band and from the holes in the valence band. Consequently, the conductivity increases rapidly with temperature because more electrons are thermally excited to the conduction band. For example, in silicon the number of excited electrons is increased by a factor of 10^6 when the temperature increases from 250 K to 450 K. Hence

> *semiconductors are substances in which the energy gap between the completely filled valence band and the conduction band is about 1 eV or less, so that it is relatively easy to excite electrons thermally from the valence to the conduction band.*

The energy gaps of some insulators and semiconductors are given in Table 23.4.

Table 23.4 Energy gaps (eV)

Insulators	eV	Semiconductors	eV
Diamond	5.33	Silicon	1.14
Zinc oxide	3.2	Germanium	0.67
Silver chloride	3.2	Tellurium	0.33
Cadmium sulfide	2.42	Indium antimonide	0.23

The temperature-dependent electrical conduction in semiconductors that we have described is called **intrinsic conductivity**. The conductivity of a semiconductor can also be enhanced by the addition of certain impurities, a technique called **doping**. Suppose that we replace some of the atoms of the semiconductor by atoms of a different substance (these atoms then constitute an impurity) that have *more* electrons than those of the semiconductor. For example, consider a crystal of silicon (or germanium), where each atom contributes *four* electrons to the valence band. If we add a few atoms of phosphorus (or arsenic), which contribute *five* electrons per atom to the valence band, we have an extra electron per impurity atom. These additional electrons (which cannot be accommodated in the valence band of the original lattice) occupy some discrete energy levels just below the conduction band, as shown in Figure 23.28(a). The separation between these energy levels and the conduction band may be a few tenths of an eV. These excess electrons are easily released by the impurity atoms and excited into the conduction band. The excited electrons then contribute to the electrical conductivity of the semiconductor. Such impurity atoms are called **donors**, and the semiconductor is called an **n-type** (or negative) semiconductor.

Conversely, the impurity may consist of atoms having *fewer* electrons than those of the semiconductor. For the cases where silicon and germanium are the host substances, the impurity atoms could be boron or aluminum, each of which contributes only three electrons. In this situation the impurity introduces vacant discrete energy levels, very close to the top of the valence band (Figure 23.28(b)). Therefore it is easy to excite some of the more energetic electrons in the valence band into the impurity levels. This process produces vacant states, or holes, in the valence band that act as 'positive' electrons. Such impurity atoms are called **acceptors** and the semiconductor is called a **p-type** (or positive) semiconductor.

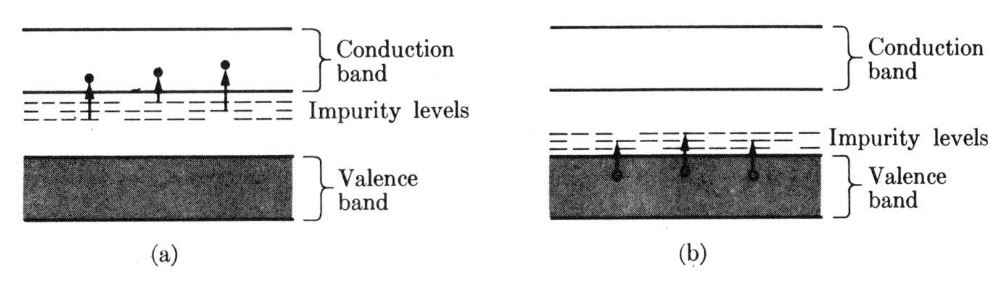

Figure 23.28 Impurities in a semiconductor: (a) donors, or n-type, (b) acceptors, or p-type.

To produce significant changes in the conductivity of a semiconductor, it is sufficient to have about one impurity atom per million semiconductor atoms. Semiconductors are used widely in industry as rectifiers, modulators, detectors, photocells, transistors and computer chips.

EXAMPLE 23.4

The p–n junction.

▷ One important application of semiconductors to modern electric circuitry is the **p–n junction**. Suppose that we have two samples of the same semiconductor—say germanium—one of p-type and the other of n-type (Figure 23.29(a)). If the two samples are placed in contact (Figure 23.29(b)), there is a diffusion or flow of holes from the p-type to the n-type and of

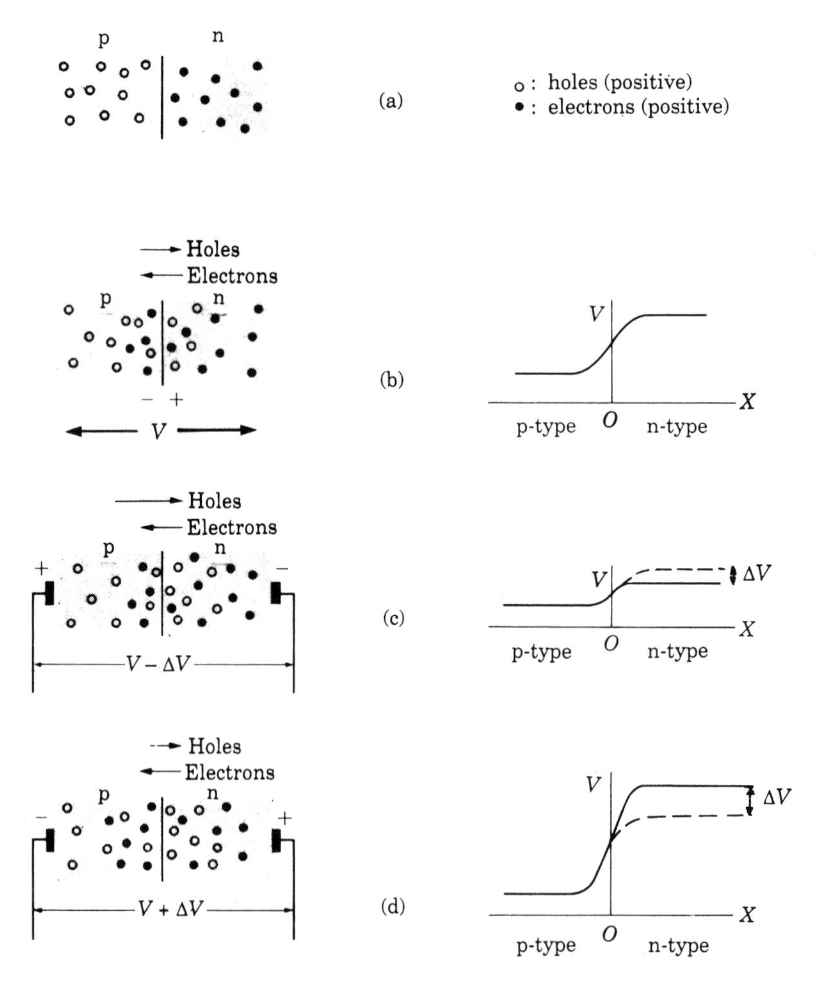

Figure 23.29 The p–n junction. The electric field ε across the junction is shown in the diagram on the right.

electrons in the opposite direction. This double flow produces a double layer of positive and negative charges on either side of the junction that produces an electric field from the n-type toward the p-type. An electric potential difference, V, is then set up across the junction, making the potential of the n-type higher than the potential of the p-type, as shown on the right in Figure 23.29(b). Equilibrium is reached when the potential difference reaches a value that opposes further flow of holes and electrons across the junction.

If a potential difference ΔV is applied that reduces the potential of the n-type relative to the p-type (Figure 23.29(c)), the electric field across the junction is reduced. Therefore, the flow of holes from the p-type to the n-type and the reverse flow of electrons is facilitated. However, if the potential difference ΔV increases the potential of the n-type relative to the p-type (Figure 23.29(d)) the electric field across the junction is increased. Accordingly, the flow of holes from the p-type and the reverse flow of electrons from the n-type is considerably reduced. Therefore, a p–n junction acts as a **rectifier**. When an external potential difference that decreases the potential of the n-side relative to the p-side is applied, the junction facilitates the flow of holes in the direction p → n, and of electrons in the n → p direction. When the external potential difference is applied in the opposite direction, the flow of charges across the junction is inhibited.

An important combination of p-type and n-type semiconductors is the n–p–n system, also called a **transistor**. It consists of a thin layer of p-type, called the base B, placed between two n-type semiconductors, called the emitter E and the collector C (Figure 23.30). The base B (p-type) is maintained at a small positive potential V with respect to the emitter E, and the collector C is, in turn, maintained at a positive potential V' relative to the base. This arrangement facilitates the flow of electrons from the emitter to the base where they tend to recombine with the holes in the base but not from the base to the collector. However, if the base B (p-type) is supplied with a sufficient number of electrons by lowering the potential of the emitter E, many can pass to the collector C (n-type) without recombining with the holes in the base. Therefore if an additional variable voltage V_1 is applied between the emitter and the base, the flow of electrons from emitter to base can be regulated. As the supply of electrons to the base varies, the number of electrons that can pass across the base and reach the collector also varies appreciably. This can produce a large variable potential difference V_2 between points c and d. In this way the n–p–n junction functions as an **amplifier**.

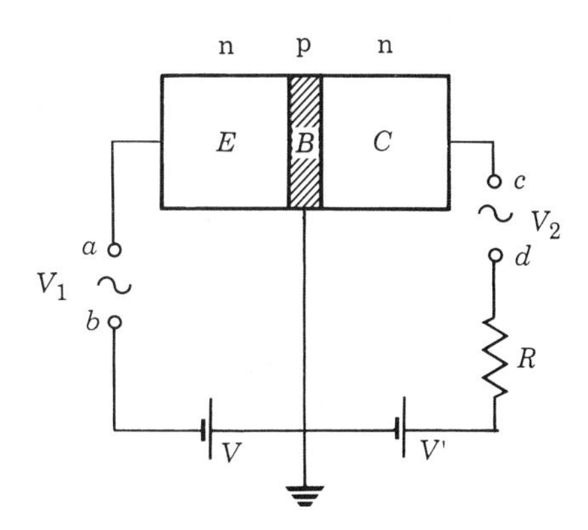

Figure 23.30 The n–p–n junction.

QUESTIONS

23.1 What is the difference between an atom and an ion?

23.2 What is the difference between a cation and an anion? Look at a chemistry book and make a list of cations and anions.

23.3 Explain how electrolysis provides a clue to the quantization of electric charge.

23.4 Explain the concept of ionic binding.

23.5 What factors determine the deviation of a charged particle in a scattering experiment?

23.6 Why does the scattering of α-particles over relatively large angles when they pass through a thin foil provide support for the nuclear model of the atom?

23.7 Explain what is meant by (a) energy quantization, (b) angular momentum quantization and (c) space quantization.

23.8 What experimental evidence supports space quantization?

23.9 Discuss the experimental evidence which indicates that electron spin may have only two orientations relative to a fixed direction.

23.10 Discuss the origin of the spin–orbit interaction in atoms.

23.11 Suppose a proton is thrown against another proton and is deviated (scattered) from its straight-line path by their coulomb interaction. Can you suggest any spin–orbit effect that may appear during the interaction?

23.12 How would electron energy levels be affected by the spin–orbit interaction if electrons had spin one instead of $\frac{1}{2}$?

23.13 Give some experimental evidence supporting the existence of electron shells in atoms.

23.14 What is the main difference between metals, semiconductors and insulators?

23.15 What is the difference between n-type and p-type semiconductors?

23.16 Are all planetary systems with the same number of planets identical in the same sense that all atoms with the same atomic number are identical? Explain your answer.

23.17 Why does the existence of energy levels and quantization explain the stability of matter?

23.18 How does the deflection of a particle in a scattering experiment vary as the energy of the particle is increased? As the angular momentum is increased?

PROBLEMS

23.1 (a) Calculate the mass of Cu (bivalent) that is deposited during one hour on an electrode when 2 C of electric charge passes through every second. (b) How many atoms have been deposited?

23.2 One mole of sodium is deposited on the cathode of an electrolytic cell in one day. (a) Calculate the charge that passed through the cell. (b) What mass was deposited and (c) how much charge passed every second through the cell?

23.3 A proton produced in a 1 MeV van de Graaff accelerator is sent against a gold foil. Calculate the distance of closest approach for a head-on collision and for collisions with impact parameters of 10^{-15} m and 10^{-14} m (recall Example 9.10).

23.4 An α-particle with a kinetic energy of 4.00 MeV is directed straight toward the nucleus of a mercury atom. The atomic number of mercury is 80. (a) Find the distance of closest approach of the α-particle to the nucleus. (b) Compare the result with the nuclear radius of mercury atoms, $\simeq 10^{-14}$ m.

23.5 When Geiger and Marsden investigated low atomic number nuclei with 4 MeV α-particles, they saw deviations from the predicted Coulomb scattering. Considering that nuclei have a radius of approximately 10^{-14} m, calculate the approximate atomic number at which deviations from Rutherford scattering for head-on collisions occur.

23.6 Calculate (a) the potential energy and (b) the kinetic energy of the electron in a hydrogen atom as a function of the quantum number n, assuming that the electron moves in circular orbits. (c) Evaluate the numerical coefficients. (d) Plot the calculated values as functions of n to determine their trend as the total energy of the electron increases.

23.7 (a) Calculate the velocity of the electron in the first three Bohr orbits ($n = 1, 2, 3$). (b) Also calculate the magnetic dipole moment of the electron for each case.

23.8 If the average lifetime of an excited state of hydrogen is of the order of 10^{-8} s, estimate how many orbits an electron makes (a) when it is in the state $n = 2$ and (b) when it is in the state $n = 15$, before it suffers a transition to state $n = 1$. (c) Compare these numbers with the number of orbits the Earth has made around the Sun in its 5×10^9 years of existence.

23.9 By inspection of Equation 23.8, explain why certain energy levels of H, He^+ and Li^{2+} coincide (see Figure 23.4). (b) Are the radii of any electron orbits in the three atoms the same?

23.10 (a) Calculate the energy released in the transition $n = 2$ to $n = 1$ for hydrogen. (b) Is the energy separation for the transition $n = 3$ to $n = 2$ greater than, less than, or the same as the energy found in part (a)?

23.11 The energy levels of helium-like ions, when one electron is in the ground state and the other in an excited state ($n > 1$), may be expressed approximately by

$$E = -13.6\left[Z^2 + \frac{(Z-1)^2}{n^2}\right]\text{eV} \quad (n > 1)$$

This expression assumes that the electron in the ground state fully screens one nuclear charge and the electron in the excited state does not affect the electron in the ground state. (a) Discuss the plausibility of this expression. (b) Compute the energy levels for helium when $n = 2$, 3 and 4 and compare with the experimental values of -58.08 eV, -56.04 eV and -55.37 eV, respectively. (c) Why does the accuracy of the above expression for E improve as n increases?

23.12 (a) Into how many levels does a magnetic field split the $n = 3$ energy level of the hydrogen atom? (b) What is the magnetic energy difference between these levels when the field is 0.4 T? (c) Compare with the energy difference between the electronic energies for $n = 2$ and $n = 3$.

23.13 (a) Given that the magnetic field gradient in a certain region is $1.5 \times 10^2\ \text{T m}^{-1}$, calculate the force exerted on an electron due to its spin magnetic dipole moment. (b) If a hydrogen atom moves 1.0 m in a direction perpendicular to such a field, calculate the transverse displacement when the velocity of the hydrogen atom is $10^5\ \text{m s}^{-1}$ and the electron spin is either parallel or antiparallel to the magnetic field. (Hint: use Equation 22.19.)

23.14 A beam of silver atoms with an average velocity of $7 \times 10^2\ \text{m s}^{-1}$ passes through an inhomogeneous magnetic field 0.1 m long that has a gradient of $3 \times 10^2\ \text{T m}^{-1}$ in a direction perpendicular to the motion of the atoms. Find the maximum separation of the two beams which emerge from the field region. Recall that each silver ion carries a charge of one electron and assume that the net magnetic moment of each atom is due to the spin of that electron.

23.15 An electron changes its value of m_s from $+\frac{1}{2}$ to $-\frac{1}{2}$ as a result of an interaction with a magnetic field. (a) Calculate the change in angular momentum of the electron. (b) If this happens in a magnetic field of 2 T, calculate the change in the electron's energy.

23.16 An empirical expression that fits the energy levels of the valence electron for one valence electron atoms is

$$E_n = -\frac{Rhc(Z-S)^2}{(n-\delta)^2}$$

where S is called the **screening constant** and δ is called the **quantum defect** (which depends on the n- and l-values of the particular valence electron). Values of δ for lithium and sodium are:

	s	p	d
Li ($Z = 3$)	0.40	0.04	0.00
Na ($Z = 11$)	1.35	0.85	0.01

For S, a value equal to the number of electrons in the kernel is a good estimate. Find (a) the energy of the ground state and (b) the first two excited states of the valence electron in (i) lithium and (ii) sodium. (c) Compare with the levels of hydrogen.

23.17 (a) Noting that $\boldsymbol{J} = \boldsymbol{L} + \boldsymbol{S}$, verify that the spin–orbit energy correction can be written as $\Delta E_{SL} = \frac{1}{2}a_{nl}[j(j+1) - l(l+1) - s(s+1)]$. According to quantum mechanics, $a_{nl} = |E_n|\, Z^2\alpha^2/nl(l+1)(l+\frac{1}{2})$.

(b) Obtain the values of ΔE_{SL} when $s = \frac{1}{2}$ and $j = l \pm \frac{1}{2}$.

23.18 According to quantum mechanics, the relativistic energy correction to the energy levels of a one electron atom is

$$\Delta E_r = |E_n|\frac{Z^2\alpha^2}{n}\left(\frac{3}{4n} - \frac{1}{l+\frac{1}{2}}\right)$$

Using the spin–orbit correction given in Problem 23.17 show that

$$\Delta E_r + \Delta E_{LS} = |E_n|\frac{Z^2\alpha^2}{n}\left(\frac{3}{4n} - \frac{1}{j+\frac{1}{2}}\right)$$

so that the total energy correction depends only on n and j.

24 Electric currents

Hans C. Oersted was a firm believer in the unity of forces in nature. During a demonstration at the University of Copenhagen in 1820, he discovered that an electric current produces a magnetic field, establishing the link between electricity and magnetism and the foundation of 'electromagnetism'. Oersted, also interested in chemistry, developed a method for the preparation of metallic aluminum (1825). He was an inspired teacher, concerned with making scientific knowledge accessible to the general public.

24.1 Introduction

Although matter is composed of electrically charged particles, and atoms are held together in molecules, liquids and solids by electric forces, we seldom find free electric charges and observe their individual electric and magnetic fields. However, the fact that in a conductor there are many charges that are free to move throughout the conductor under the action of external electric or magnetic fields gives rise to important collective macroscopic effects that can be detected rather easily and that have important practical applications. These effects will be discussed in this chapter.

Part A: Electric currents and electric fields

24.2 Electric current

An electric current consists of a stream of charged particles or ions. This applies to the charged particles in an accelerator, to ions in an electrolytic solution and in an ionized gas or plasma as well as to the electrons and holes in a conductor. In order to have an electric current, an electric field must be applied to move the charged particles in a well-defined direction.

The **intensity** of an electric current is defined as the electric charge passing per unit time through a section of the region where it flows, such as a section of an accelerator tube or of a metallic wire. Therefore if, in time Δt, N charged particles, each carrying a charge q, pass through a section of the conducting medium, the total charge passing is $\Delta Q = Nq$. Then the intensity of the current is

$$I = \frac{\Delta Q}{\Delta t} \tag{24.1}$$

The above expression gives the average current in time Δt; the instantaneous current at a given moment is

$$I = \frac{\mathrm{d}Q}{\mathrm{d}t} \tag{24.2}$$

The electric current is expressed in coulombs/second or $\mathrm{C\,s^{-1}}$, a unit called the **ampere** (abbreviated A) in honor of André M. Ampère (1775–1836). An ampere is the intensity of an electric current corresponding to a charge of one coulomb passing through a section of the material each second.

The *direction* of an electric current is assumed to be that of the motion of positively charged particles. It is the same direction as that of the applied electric field or of the potential *drop* that produces the motion of the charged particles (Figure 24.1(a)). Therefore, if a current is due to the motion of negatively charged

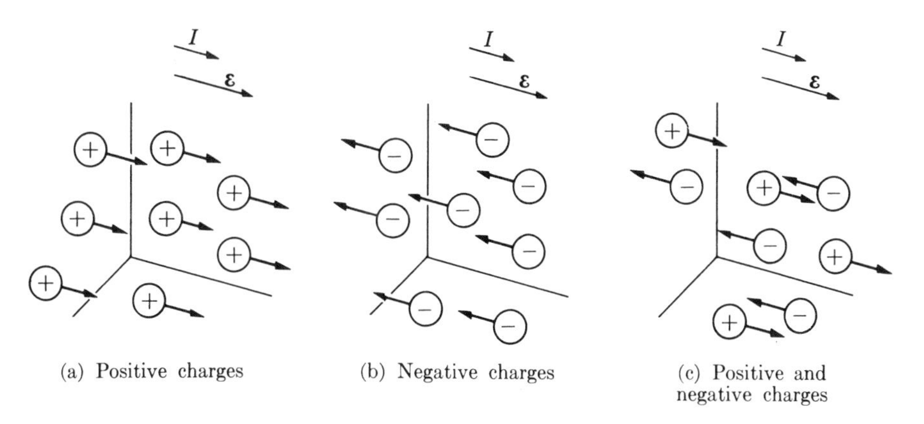

Figure 24.1 Motion of positive and negative ions resulting in an electric current I produced by an electric field $\mathscr{E}$. The electric potential decreases from left to right.

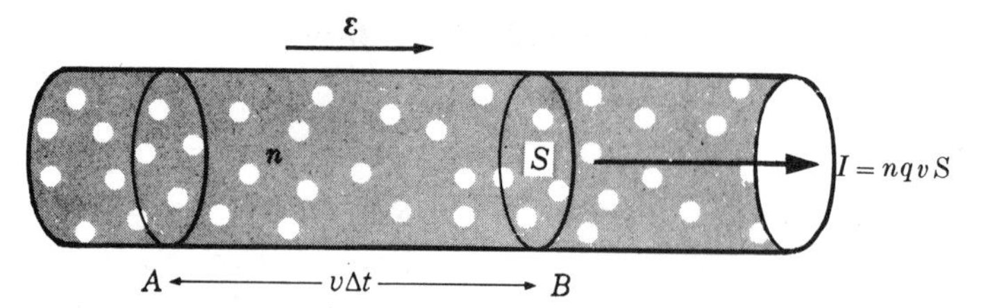

Figure 24.2 Electric current in a conductor.

particles, such as electrons, the conventional direction of the current is opposite to the actual motion of the electrons (Figure 24.1(b)). This convention may seem awkward, but it was adopted before it was known that in metallic conductors the current is due to negatively charged electrons; it then became difficult to change it. If there are oppositely charged particles, as in an electrolytic solution or a plasma, the electric current consists of positive and negative particles moving in opposite directions (Figure 24.1(c)).

A useful concept is that of **current density**. Suppose we have a conductor in the form of a filament (Figure 24.2) with a cross-section S. Suppose also that in the conductor there are n charged particles per unit volume, each with a charge q, all moving in the same direction with a velocity $\mathbf{v}$ under the action of an electric field. Then the charges that in time Δt pass through section B are those inside the volume limited by sections A and B, with $AB = v\Delta t$. This volume is $Sv\Delta t$. Then $\Delta Q = qnSv\Delta t$ and the current is

$$I = \frac{\Delta Q}{\Delta t} = qnSv$$

We define current density as the current per unit area of the conductor, or $j = I/S$. Therefore

$$j = qnv \tag{24.3}$$

Since the velocity of the particles is a vector, the current density is also a vector, so that

$$\mathbf{j} = qn\mathbf{v} \tag{24.4}$$

Therefore $\mathbf{j}$ has the same direction as $\mathbf{v}$, or the opposite direction, depending on the sign of q. The current density is measured in A m^{-2}.

EXAMPLE 24.1

In electroplating, the piece to be electroplated is immersed in a solution containing silver ions, Ag^+, and connected to the negative terminal of a battery, so that it becomes a cathode (recall Figure 23.1). A current of 10 amperes is passed through the solution for 10 minutes. Calculate the mass of silver deposited on the piece. How many atoms of silver are deposited?

▷ The intensity of the current is $I = 10$ A. It has passed during a time interval $\Delta t = 10$ min $=$ 600 s. Then, according to Equation 24.1, the charge that reached the cathode was $\Delta Q = I\Delta t = (10\text{ A})(600\text{ s}) = 6000$ C. But, according to Equation 23.4, the amount of charge required to deposit one mole of Ag on the cathode is $F = 9.6487 \times 10^4$ C. Therefore, the number of moles of silver deposited on the cathode is

$$\text{N} = \frac{\Delta Q}{F} = \frac{6 \times 10^3\text{ C}}{9.6487 \times 10^4\text{ C mole}^{-1}} = 6.22 \times 10^{-2}\text{ mole}$$

The mass of one mole of silver is 107.870 g; therefore, the mass of silver deposited on the cathode is $M = (6.22 \times 10^{-2}\text{ mole})(107.870\text{ g mole}^{-1}) = 6.71$ g. Also, recalling that the number of atoms in one mole is Avogadro's number, $N_A = 6.0225 \times 10^{23}\text{ mole}^{-1}$, the number of atoms of silver deposited is

$$N = \text{N}N_A = (6.22 \times 10^{-2}\text{ mole})(6.0225 \times 10^{23}\text{ mole}^{-1}) = 3.87 \times 10^{22}$$

24.3 Ohm's law

Metallic conductors are solids in which there are free electrons occupying the conduction band (Section 23.12). The uppermost electrons of the conduction band are easily set in collective motion when an external electric field is applied to the conductor. While the natural thermal motion of the conduction electrons is random and does not result in an electric current, the electronic motion due to the external electric field is organized in a direction opposite to that of the electric field. This motion results in an electric current through the conductor. It seems natural to assume that the intensity of the current must be related to the intensity of the applied electric field.

This relation, called **Ohm's law**, was first obtained by Georg Ohm (1787–1854). It states that

> *for a metallic conductor at constant temperature, the ratio of the potential difference between two points to the electric current is constant.*

This constant is called the **electrical resistance** R of the conductor between the two points. Thus if ΔV is the potential difference between the extremes of the conductor, also called the **voltage**, and I is the current through it, we may express this law by

$$\frac{\Delta V}{I} = R \quad \text{or} \quad \Delta V = RI \tag{24.5}$$

Ohm's law is obeyed with surprising accuracy by many conductors over a wide range of values of ΔV, I, and temperature of the conductor. When we plot the values of V against those of I, we should obtain a straight line. The slope of the line is the resistance of the conductor (Figure 24.3). It must be noted that there are many conductors that do not obey Ohm's law.

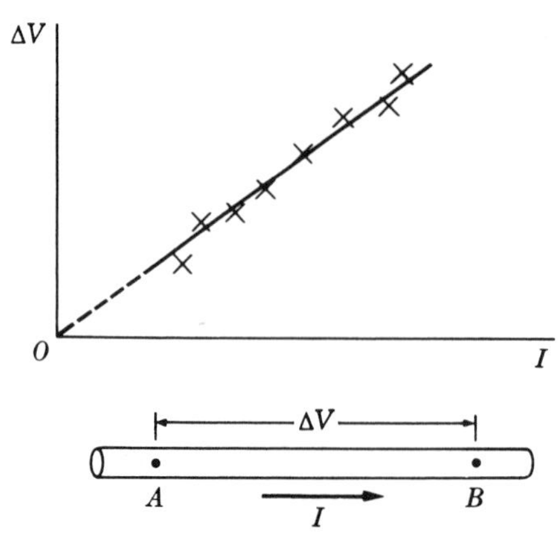

Figure 24.3 Relation between the potential difference and the current in a conductor.

From Equation 24.5, we see that R is expressed in volts/ampere or $m^2\,kg\,s^{-1}\,C^{-2}$, a unit called an **ohm**, and abbreviated Ω. Thus one ohm is the resistance of a conductor through which there is a current of one ampere when a potential difference of one volt is maintained across its ends. A conductor with a measurable resistance, also called a **resistor**, is represented diagrammatically in Figure 24.4.

Figure 24.4 Symbolic representation of a resistor.

24.4 Conductivity

Consider a cylindrical conductor of length l and cross-section S (Figure 24.5). The current may be expressed as $I = jS$, where j is the current density. The electric field along the conductor is $\mathscr{E} = \Delta V/l$ (recall Equation 21.28), so that $\Delta V = \mathscr{E}l$. Therefore we may write Ohm's law $\Delta V = RI$ in the form $\mathscr{E}l = RjS$, or

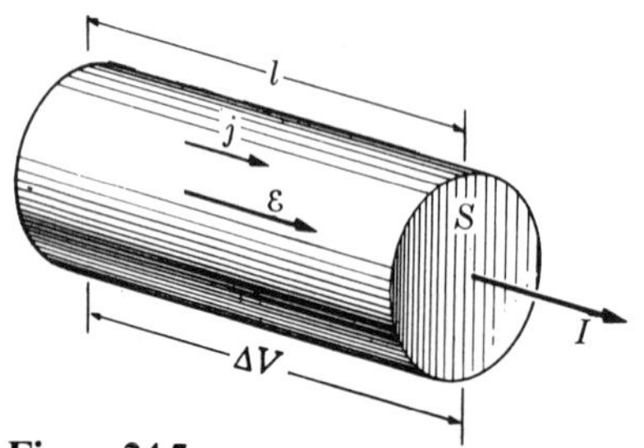

Figure 24.5

$$j = \left(\frac{l}{RS}\right)\mathscr{E} = \sigma\mathscr{E} \qquad \textbf{(24.6)}$$

where $\sigma = l/RS$ is called the **electrical conductivity** of the material. It is expressed in $\Omega^{-1}\,m^{-1}$ or $m^{-3}\,kg^{-1}\,s\,C^2$. Table 24.1 gives the electrical conductivity of several materials. The relation between σ and R is more conveniently written in the form

$$R = \frac{l}{\sigma S} \qquad \textbf{(24.7)}$$

Since $\mathscr{E} = -\,dV/dx$, Equation 24.6 may be rewritten as $j = -\,\sigma(dV/dx)$. Recalling Equation 18.1, we see that a current in a conductor may be considered a transport phenomenon where electric charge is the transported property. Equation 24.6 expresses a relation between the magnitudes of the vectors $\boldsymbol{j}$ and $\boldsymbol{\mathscr{E}}$. Assuming that they have the same direction, a situation found in most substances, we may replace

Table 24.1 Electrical conductivities at room temperature

Substance	$\sigma(\Omega^{-1}\,\mathrm{m}^{-1})$	Substance	$\sigma(\Omega^{-1}\,\mathrm{m}^{-1})$
Metals		*Semiconductors*	
Copper	5.81×10^{7}	Carbon	2.8×10^{4}
Silver	6.14×10^{7}	Germanium	2.2×10^{-2}
Aluminum	3.54×10^{7}	Silicon	1.6×10^{-5}
Iron	1.53×10^{7}		
Tungsten	1.82×10^{7}	*Insulators*	
		Glass	10^{-10} to 10^{-14}
Alloys		Lucite	$<10^{-13}$
Manganin	2.27×10^{6}	Mica	10^{-11} to 10^{-15}
Constantan	2.04×10^{6}	Quartz	1.33×10^{-18}
Nichrome	1.0×10^{6}	Teflon	$<10^{-13}$
		Paraffin	3.37×10^{-17}

Equation 24.5 by the vector equation

$$\boldsymbol{j} = \sigma \boldsymbol{\mathscr{E}} \tag{24.8}$$

which is merely another way of writing Ohm's law. Recalling Equation 24.4, $\boldsymbol{j} = qn\boldsymbol{v}$, with $q = -e$ for electrons, we may write $\boldsymbol{j} = -en\boldsymbol{v}$, where n is the number of electrons per unit volume in the conduction band near the Fermi energy level and $\boldsymbol{v}$ is the electrons' drift velocity due to the applied electric field $\boldsymbol{\mathscr{E}}$. Then Equation 24.8 may be written as

$$\boldsymbol{v} = -\frac{\sigma}{en}\boldsymbol{\mathscr{E}} \tag{24.9}$$

This equation shows that conduction electrons in a metal attain a *constant* drift velocity as a result of the external applied electric field. This is a conclusion quite different from our discussion of the motion of an ion along the evacuated tube of an accelerator (Example 21.14). There we found that the ions move with an acceleration $\boldsymbol{a} = -(e/m)\boldsymbol{\mathscr{E}}$, resulting in a velocity $\boldsymbol{v} = -(e/m)\,\boldsymbol{\mathscr{E}}t$ which increases continuously with time.

Note that this is not the first time we have encountered a situation like this. We know that a freely falling body, in vacuum, has a continually increasing velocity $\boldsymbol{v} = \boldsymbol{g}t$. But if the body falls through a viscous fluid, its motion becomes uniform with a constant limiting velocity, as we discussed in Section 7.5. By analogy, we may say that the effect of the crystal lattice may be represented by a 'viscous' force acting on the conduction electrons when their natural motion is disturbed by the applied electric field. The exact nature of this 'viscous' force depends on the dynamics of the electronic motion through the crystal lattice. We cannot enter into the details of how this electronic motion is hindered by the lattice. It is sufficient to say that the 'viscous' force is due to imperfections in the crystal lattice and to the thermal motion of the lattice ions, which hinder the free motion of the conduction electrons through the lattice and scatter the electrons continuously while they drift under the action of the electric field as shown in Figure 24.6

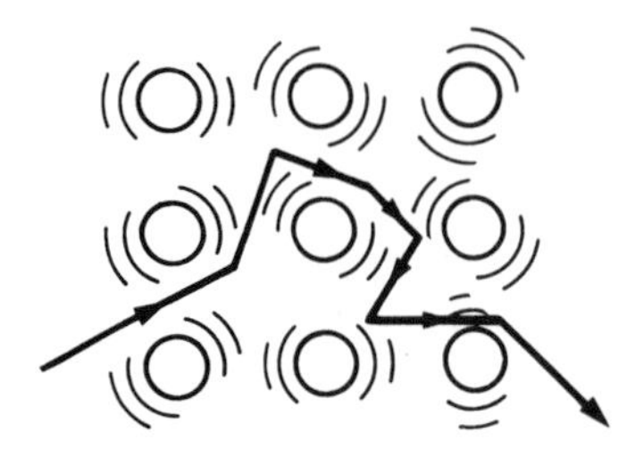

Figure 24.6

(see Note 24.1). In general, conductivity decreases as the temperature increases. This is because the vibrational energy of the ions forming the lattice increases with temperature, thereby increasing the scattering of the conduction electrons. Further, the higher the temperature, the higher the thermal kinetic energy of the conduction electrons, making it more difficult for them to move in a given direction under the action of the applied electric field.

There are several conducting substances, particularly some metals, alloys and metallic compounds, for which, when their temperature falls below a certain critical value T_C, characteristic for each substance (Table 24.2), several of their properties (electric, magnetic and thermal) change drastically. It is said that at T_C a *phase transition* occurs, similar to the phenomena of condensation and solidification. For example, the conductivity of these substances increases sharply below T_C. It is because of this property that these substances are called **superconductors**.

Superconductivity is a macroscopic quantum effect that results from a net attractive electron–electron pairing interaction, involving electrons in the conduction band near the Fermi level. Under certain conditions the lattice fosters the formation of what are called **Cooper pairs**, which consist of pairs of electrons with opposite spin and, consequently, zero net spin magnetic moment. An explanation of the cause for the pairing interaction is beyond the level of this book. One consequence of pairing is that Cooper pairs can move freely through the crystal lattice instead of being scattered, which happens to normal conduction electrons. The result is a large increase in conductivity. Since the pairing interaction is relatively weak, pairing can only occur below a critical temperature, when the thermal energy is less than the pairing energy; at higher temperatures thermal energy disrupts pair formation. Also, if the current is too large, the superconductivity disappears because the applied electric field disrupts the pairing.

Table 24.2 Critical temperatures of superconductors

Metals	T_C(K)	Alloys and compounds	T_C(K)	Oxides	T_C(K)	Fullerenes	T_C(K)
Al	1.18	Pb–In	7	$BaPb_{0.75}B_{0.25}O_3$	13	K_3C_{60}	18
In	3.41	Pb–Bi	8.3	$Ba_{0.6}K_{0.4}BiO_3$	30	Rb_3C_{60}	28
Sn	3.72	Nb–Ti	9.5	$La_{1.85}Ba_{0.15}CuO_4$	36	Cs_2RbC_{60}	33
Ta	4.47	Nb–Zr	10.7	$YBa_2Cu_3O_7$	94	$Rb_{2.7}Tl_{2.2}C_{60}$	42.5
V	5.40	Nb–N	16.0	$Y_2Ba_4Cu_8O_{16}$	81		
Pb	7.20	V_3Ge	15.3	$Bi_2Sr_2CaCu_2O_8$	84		
Nb	9.25	V_3Si	16.3	$Tl_2Ba_2CaCu_2O_8$	108		
Hg	4.12	Nb_3Sn	18.0	$Tl_2Ba_2Ca_2Cu_3O_{10}$	125		
Ga	1.07	Nb_3Ge	23.2				

Note: The value of T_C for alloys, compounds and oxides depends on the stoichiometry of the material. For the fullerenes, x is about 3. (These values are taken from the AIP *Vade Mecum* H. L. Anderson, ed. 1989.)

In the mid-1980s substances with higher critical temperatures, up to 125 K, were discovered. They are copper- and bismuth oxide-based superconductors, organic superconductors (that is, the lattice does not contain metal ions), and the more recently discovered K_xC_{60} and Rb_xC_{60} superconductors called **fullerenes** (see Section 38.8). The precise nature of the charge carriers and the mechanisms leading to their superconductivity are still the subject of intensive research.

EXAMPLE 24.2

A voltage of 0.10 V is applied to a copper wire 2 m long, whose diameter is 1 mm. Calculate the current in the wire. How many electrons cross a section of the wire per second?

▷ From Table 24.1 we find that the conductivity of copper is $\sigma = 5.81 \times 10^7\ \Omega^{-1}\,\mathrm{m}^{-1}$, and $S = \pi r^2$. Using Equation 24.7, the resistance of the wire is

$$R = \frac{l}{\sigma S} = \frac{2\,\mathrm{m}}{(5.81 \times 10^7\ \Omega^{-1}\,\mathrm{m}^{-1})\pi(5 \times 10^{-4}\,\mathrm{m})^2} = 4.383 \times 10^{-2}\ \Omega$$

Applying Ohm's law, we get

$$I = \frac{\Delta V}{R} = \frac{0.10\ \mathrm{V}}{4.383 \times 10^{-2}\ \Omega} = 2.282\ \mathrm{A}$$

The total charge that passes in one second through a section of the wire is

$$\Delta Q = I\Delta t = (2.282\ \mathrm{A}) \times (1\ \mathrm{s}) = 2.282\ \mathrm{C}$$

Then the number of electrons passing each second through a section of the conductor is

$$n = \frac{\Delta Q}{e} = \frac{2.282\ \mathrm{C}}{1.602 \times 10^{-19}} = 1.424 \times 10^{19}\ \text{electrons}$$

EXAMPLE 24.3

A potential difference of 100 volts is applied to the ends of a copper wire one meter long. Calculate the average drift velocity of the electrons. Compare this velocity with the thermal velocity of electrons at room temperature, 27 °C, assuming the electrons behave like an ideal gas.

▷ Since $\Delta V = 100\ \mathrm{V}$ and $l = 1\ \mathrm{m}$, the electric field in the conductor is $\mathscr{E} = \Delta V/l = 100\ \mathrm{V\,m^{-1}}$. The conductivity of Cu, from Table 24.1, is $\sigma = 5.81 \times 10^7\ \Omega^{-1}\,\mathrm{m}^{-1}$. Before using Equation 24.9, $v = -(\sigma/en)\mathscr{E}$, we must estimate the number of electrons per unit volume. The density of Cu is $8.92 \times 10^3\ \mathrm{kg\,m^{-3}}$. Since the atomic mass of Cu is 63.54 amu, we conclude that in 1 m^3 of Cu there are 1.4×10^5 moles or 8.45×10^{28} atoms. We may assume that each copper atom contributes a single conduction electron. Therefore, using magnitudes only,

$$v = \frac{\sigma}{en}\mathscr{E} = \frac{(5.81 \times 10^7\ \Omega^{-1}\,\mathrm{m}^{-1}) \times (100\ \mathrm{V\,m^{-1}})}{(1.602 \times 10^{-19}\ \mathrm{C}) \times (8.45 \times 10^{28}\ \mathrm{m}^{-3})} = 0.43\ \mathrm{m\,s^{-1}}$$

On the other hand, the rms thermal velocity of the electrons (assuming they behave like gas molecules) can be estimated using Equation 15.16; that is

$$v_{\text{rms}} = \left(\frac{3kT}{m}\right)^{1/2} = \left(\frac{3 \times (1.38 \times 10^{-23}\,\text{J K}^{-1}) \times (300\,\text{K})}{9.1 \times 10^{-31}\,\text{kg}}\right)^{1/2} = 1.17 \times 10^{5}\,\text{m s}^{-1}$$

This shows that the drift velocity under the action of an electric field is very small compared with the average thermal velocity but, because it corresponds to a collective motion, it is a measurable effect.

Note 24.1 Calculation of the electric conductivity

It is important to relate the electric conductivity, which is a macroscopic parameter, with the atomic properties of the material. Due to its motion, a conduction electron is scattered continuously by the imperfections and impurities in the lattice and the thermal motion of the lattice ions (Figure 24.6). However, the scattering is limited by the Pauli principle; that is, no electron can be scattered into a state already occupied by another electron. If λ is the average distance between two scatterings, also called **mean free path**, and v_{th} is the average thermal velocity, we have that the time between two scatterings is $t = \lambda / v_{\text{th}}$. When an electric field is applied to the metal, the electrons acquire an acceleration $a = (e/m)\mathscr{E}$. Assuming that at each scattering the electron loses its drift velocity and starts again, its average velocity during each mean free path is

$$v = \tfrac{1}{2}(0 + at) = \frac{1}{2}\left(\frac{e}{m}\right)\mathscr{E}t$$

Equating this velocity to the drift velocity $v = (\sigma/en)\mathscr{E}$, given in Equation 24.9, we obtain

$$\sigma = \frac{ne^2 t}{2m} = \frac{ne^2 \lambda}{2mv_{\text{th}}}$$

which is called **Drude's formula**. Because of our assumptions, we can expect that it only gives the order of magnitude. To verify its correctness we will apply it to Cu. In this case, as calculated in Example 24.3, $n = 8.45 \times 10^{28}\,\text{m}^{-3}$. To calculate λ we have that the average separation of the ions in a cubic lattice is of the order of $1/n^{1/3}$ or $2.3 \times 10^{-10}\,\text{m}$. The average thermal velocity, as calculated in Example 24.3, is $v_{\text{th}} = 1.17 \times 10^{5}\,\text{m s}^{-1}$. Thus $t \simeq \lambda / v_{\text{th}} = 2 \times 10^{-15}\,\text{s}$ and

$$\sigma = \frac{(8.45 \times 10^{28}\,\text{m}^{-3})(1.6 \times 10^{-19}\,\text{C})^2 (2 \times 10^{-15}\,\text{s})}{2 \times (9.11 \times 10^{-31}\,\text{kg})} = 2.37 \times 10^{6}\,\Omega^{-1}\,\text{m}^{-1}$$

which falls short of the experimental value by a factor of 20 due to our approximations.

24.5 Electric power

Maintaining an electric current requires energy because the moving charges are accelerated by the electric field. Suppose that in time Δt there are N ions, each with charge q, that move through a potential difference ΔV. Recalling Equation

21.29, each ion gains an energy $q\Delta V$ and the total energy they gain is $Nq\Delta V = Q\Delta V$. The energy per unit time, or the power required to maintain the current, is then

$$P = \frac{Q\Delta V}{\Delta t} \quad \text{or} \quad P = I\Delta V \qquad \textbf{(24.10)}$$

where $I = Q/\Delta t$. That is, power = (intensity of current) × (applied voltage). Expression 24.10 gives the power required to maintain an electric current through a potential difference applied to two points of any conducting media. This gives, for example, the power required to drive a particle accelerator. It also gives the rate at which energy is transferred to the accelerator's target, and therefore the rate at which energy must be removed by the target's coolant. Note from Equation 24.10 that

$$\text{volts} \times \text{amperes} = \frac{\text{joules}}{\text{coulomb}} \times \frac{\text{coulombs}}{\text{second}} = \frac{\text{joules}}{\text{second}} = \text{watts}$$

so that the units are consistent.

For conductors that follow Ohm's law, $\Delta V = RI$, Equation 24.10 may be written in the alternative form

$$P = RI^2 \qquad \textbf{(24.11)}$$

Many materials, however, do not follow Ohm's law, and for them Equation 24.11 is not correct, although Equation 24.10 remains valid.

The energy required to accelerate an ion in an accelerator or an electron tube is different from the energy required to maintain a current in a conductor. In the accelerator, all the energy supplied to the ions is spent in speeding them up. In a conductor, because of the interaction of the electrons and the crystal lattice, the energy supplied to the electrons is transferred to the lattice, increasing its vibrational energy. This leads to an increase in the internal energy of the material manifested by an increase in its temperature, which is the well-known heating effect of a current, called the **Joule effect**.

EXAMPLE 24.4

Rate of energy transfer to a crystal lattice.

▷ The rate at which energy is transferred to the crystal lattice may be easily estimated. Recalling Equation 9.9, the work done per unit time on an electron drifting with velocity $\boldsymbol{v}$ is $\boldsymbol{F}\cdot\boldsymbol{v} = -e\boldsymbol{\mathscr{E}}\cdot\boldsymbol{v}$ and if n is the number of electrons per unit volume, the work done per unit time and unit volume (or power per unit volume) is

$$p = n(-e\boldsymbol{\mathscr{E}}\cdot\boldsymbol{v}) = -en\boldsymbol{v}\cdot\boldsymbol{\mathscr{E}} = \boldsymbol{j}\cdot\boldsymbol{\mathscr{E}} = \sigma\mathscr{E}^2 \qquad \textbf{(24.12)}$$

Again, consider the cylindrical conductor of Figure 24.5, whose volume is Sl. The power required to maintain the current in it is $P = (Sl)p = (Sl)(j\mathscr{E}) = (jS)(\mathscr{E}l)$. But $jS = I$

is the current and $\mathscr{E}l = \Delta V$ is the potential difference. Therefore the power required to maintain the current in the conductor is $P = I\Delta V$, in agreement with Equation 24.10, which is independent of the nature of the conduction process.

24.6 Combination of resistors

Resistors can be combined in two kinds of arrangements: **series** and **parallel**. In the series combination (Figure 24.7), the resistors are connected in such a way that the same current I passes through all of them. The potential drop across each resistor, according to Ohm's law, is

$$\Delta V_1 = R_1 I, \quad \Delta V_2 = R_2 I, \ldots, \Delta V_n = R_n I$$

Thus the overall potential difference is

$$\Delta V = \Delta V_1 + \Delta V_2 + \cdots + \Delta V_n = (R_1 + R_2 + \cdots + R_n)I$$

The system can be reduced effectively to a single resistor R satisfying $\Delta V = RI$. Therefore

$$R = R_1 + R_2 + \cdots + R_n \qquad \text{(series)} \tag{24.13}$$

gives the resultant resistance for a series arrangement of resistors.

In the parallel combination (Figure 24.8), the resistors are connected in such a way that the potential difference ΔV is the same for all of them. The current through each resistor, according to Ohm's law, is

$$I_1 = \frac{\Delta V}{R_1}, \quad I_2 = \frac{\Delta V}{R_2}, \ldots, \quad I_n = \frac{\Delta V}{R_n}$$

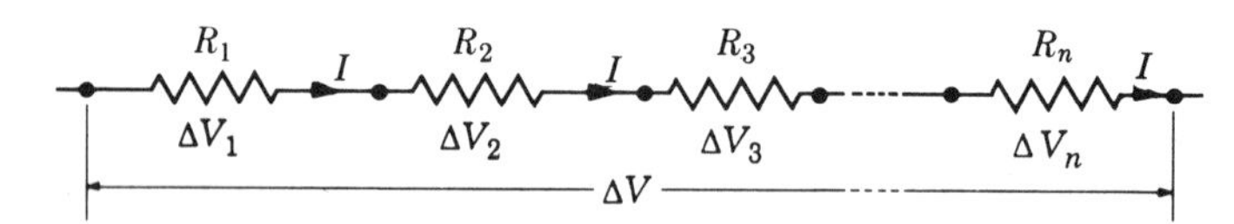

Figure 24.7 Series arrangement of resistors.

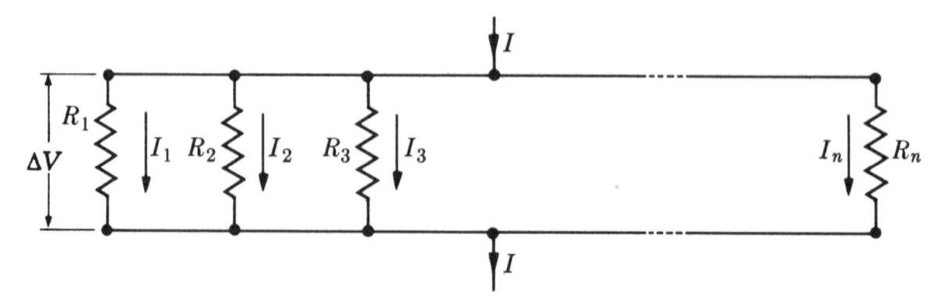

Figure 24.8 Parallel arrangement of resistors.

The total current I supplied to the system is

$$I = I_1 + I_2 + \cdots + I_n = \left(\frac{1}{R_1} + \frac{1}{R_2} + \cdots + \frac{1}{R_n}\right)\Delta V$$

The system can be reduced effectively to a single resistor R satisfying $I = \Delta V/R$. Therefore

$$\frac{1}{R} = \frac{1}{R_1} + \frac{1}{R_2} + \cdots + \frac{1}{R_n} \qquad \text{(parallel)} \tag{24.14}$$

is the resultant resistance for a parallel arrangement of resistors.

EXAMPLE 24.5

Find the effective resistance of the arrangement shown in Figure 24.9(a). Determine also the current in each conductor if the voltage applied between A and B is 30 V.

▷ We first compute the resistance of the 6 Ω and 12 Ω resistors in parallel, using Equation 24.14,

$$\frac{1}{R_1} = \frac{1}{6} + \frac{1}{12} = \frac{1}{4} \qquad \text{or} \qquad R_1 = 4\,\Omega$$

Then we may replace the arrangement of Figure 24.9(a) by that of Figure 24.9(b). The 1 Ω and 4 Ω resistors are equivalent to a resistor having a resistance $R_2 = 1\,\Omega + 4\,\Omega = 5\,\Omega$. Hence instead we have the arrangement of Figure 24.9 (c). Finally, the effective resistance of the system is

$$\frac{1}{R} = \frac{1}{5} + \frac{1}{20} = \frac{1}{4} \qquad \text{or} \qquad R = 4\,\Omega$$

If the voltage applied between A and B is 10 V, using Ohm's law the current through the 20 Ω resistor is

$$I = \frac{\Delta V}{R} = \frac{30\text{ V}}{20\,\Omega} = 1.5\text{ A}$$

The current through the 5 Ω resistor in Figure 24.9(c) is $30\text{ V}/5\,\Omega = 6\text{ A}$. This is also the current through the 1 Ω resistor in Figures 24.9(a) and (b). The voltage drop across the

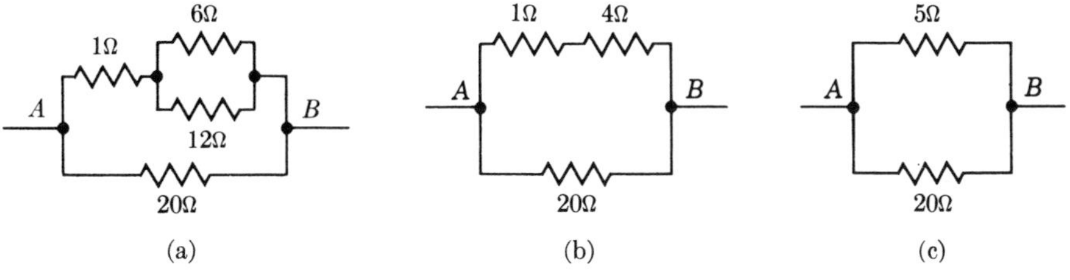

Figure 24.9

4 Ω resistor in Figure 24.9(b) is 4 Ω × 6 A = 24 V. Hence the currents across the 6 Ω and 12 Ω resistors in Figure 24.9(a) are 24 V/6 Ω = 4 A and 24 V/12 Ω = 2 A, respectively.

24.7 Direct current circuits

Ohm's law, as stated in Equation 24.5, relates the *potential difference* between two points of a conductor and the electric current in the conductor. Maintaining a current between two points in a conductor implies that energy must be supplied to the system by the source of the potential difference. Actually, the potential difference ΔV between the two points gives the work done by the source of the potential difference in moving one unit of charge from the first point to the second. Therefore the product RI, according to Ohm's law, is also equal to that work per unit charge.

Suppose now that we have a **closed circuit**; that is, a system of conductors forming a closed path. A charge moving inside a conductor is transferring energy to the crystal lattice and this process is irreversible; that is, the lattice does not give it back to the electrons. Therefore, unless a net amount of energy is supplied to the electrons, they cannot move steadily around a closed circuit. Accordingly, to maintain a current in a closed circuit, it is necessary to feed energy into the circuit at certain points A, A', A'',... (Figure 24.10). The suppliers of energy are called **electric generators** G, G', G'', ...

The total work done in moving a unit of charge around the circuit is called the **electromotive force**, emf, applied to the circuit. This is a poor name because an emf is not a force but an energy per unit charge, and is measured in volts. Therefore, since the rate of energy supplied to the charges moving in a circuit must match the rate of energy transfer to the crystal lattice of the conductor, we must have

$$\text{emf} = RI \tag{24.15}$$

where R is the *total* resistance of the circuit and I is the total current. The difference between this expression and Ohm's law for a simple conductor is that the emf is applied to a closed circuit, while ΔV is the potential difference between two points of a conductor. In general, we will use the symbol V to designate the emf applied to a circuit and ΔV for the potential difference across a conductor.

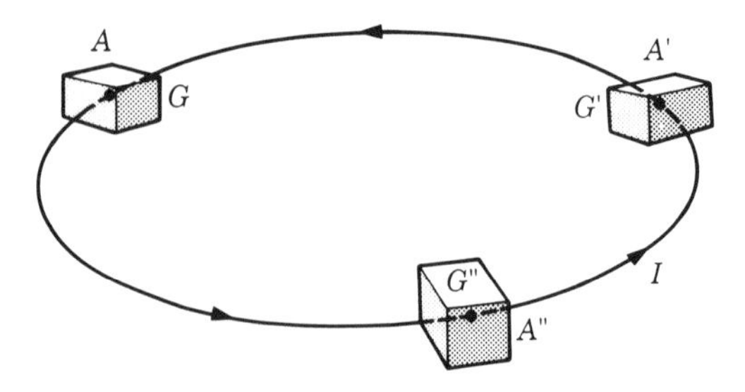

Figure 24.10 An electric current is maintained in a closed circuit by electric generators.

There are many ways to generate an electromotive force. A common method is by a chemical reaction, such as in a dry cell or a storage battery, where the internal energy released in the chemical reaction is transferred to the electrons. Another important method is by the phenomenon of electromagnetic induction, which will be discussed in Chapter 28.

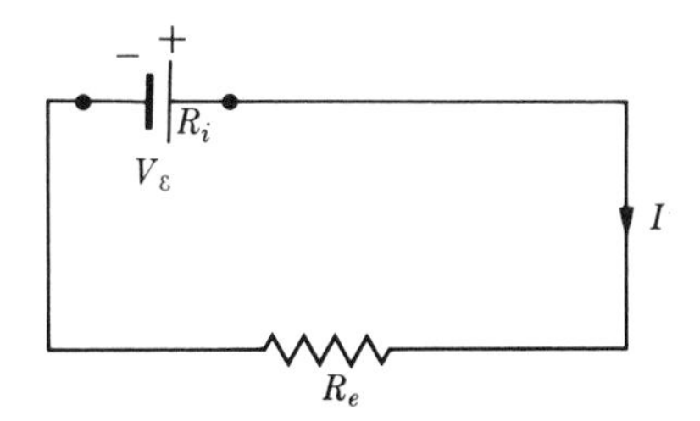

Figure 24.11 Representation of a circuit with an electromotive force.

A chemical generator, or battery, produces an electric potential difference between its two terminals. The terminal at the higher potential is the positive pole and the other is the negative. The terms positive and negative are conventional and have nothing to do with the signs of charges.

A source of emf is represented diagrammatically in Figure 24.11, where the sense of the current that is produced in the circuit *external to the source of emf* is from the long bar, or positive pole, to the short bar, or negative pole.

In a simple circuit such as that of Figure 24.11, we must recognize that the total resistance R is the sum of the **internal** resistance R_i of the source of emf and the **external** resistance R_e of the conductor connected to the generator (or battery). Thus $R = R_i + R_e$, and Ohm's law (Equation 24.15) for the circuit may be written in the form

$$\text{emf} = (R_e + R_i)I \qquad \textbf{(24.16)}$$

This may also be written as

$$\text{emf} - R_i I = R_e I$$

Each side of the equation gives the potential difference between the poles of the generator (or battery). Note that this potential difference is smaller than the emf. However, if the circuit is open, so that $I = 0$, then the potential difference across the poles of the generator will be equal to the emf.

EXAMPLE 24.6

Determine the conditions under which the current through R_5 in the circuit of Figure 24.12 is zero. This circuit is called a **Wheatstone bridge**.

▷ The current through R_5 will be zero when the potential difference between B and C is also zero. Under that condition, the current through R_1 and R_3 will be the same, and the current through R_2 and R_4 will also be the same. That is,

$$\Delta V_{AB} = R_1 I \qquad \Delta V_{BD} = R_3 I$$

$$\Delta V_{AC} = R_2 I' \qquad \Delta V_{CD} = R_4 I'$$

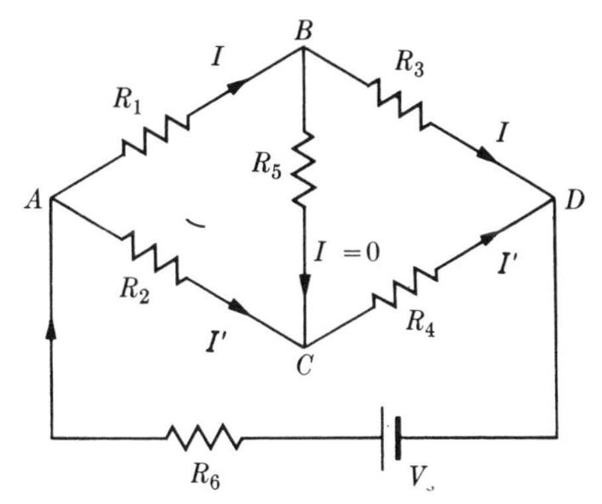

Figure 24.12 Wheatstone bridge. The current through R_5 is zero when the potential difference between B and C is zero.

The potential drop from A to B is equal to the potential drop from A to C and the potential drops from B to D and C to D are also equal, since B and C are at the same potential. That means that $\Delta V_{AB} = \Delta V_{AC}$ and $\Delta V_{BD} = \Delta V_{CD}$, or $R_1 I = R_2 I'$ and $R_3 I = R_4 I'$.

Combining these results to eliminate the currents I and I', we get

$$\frac{R_1}{R_2} = \frac{R_3}{R_4}$$

Thus, if we know R_2 and the ratio R_3/R_4, we can obtain the resistance R_1. For that reason this arrangement is used to measure resistances. Since bridge circuits rely on a null measurement, they are very sensitive measuring devices.

24.8 Methods for calculating the currents in an electric network

An electric network is a combination of conductors and emfs, such as the one illustrated in Figure 24.13. Usually the problem consists in finding the currents through the various conductors in terms of the emf's and their resistances. The rules to solve this kind of problem, known as **Kirchhoff's rules**, merely express the conservation of electric charge and energy. Kirchhoff's rules may be stated as follows:

1. *The sum of all currents at a junction in a network is zero.*
2. *The sum of all potential drops along any closed path in a network is zero.*

The first rule expresses the conservation of charge because, since charges are not accumulated at a junction, the number of charges that arrive at a junction in a certain time must leave it in the same time. The second rule expresses the conservation of energy, since the net change in the energy of a charge, after the charge completes a closed path, must be zero.

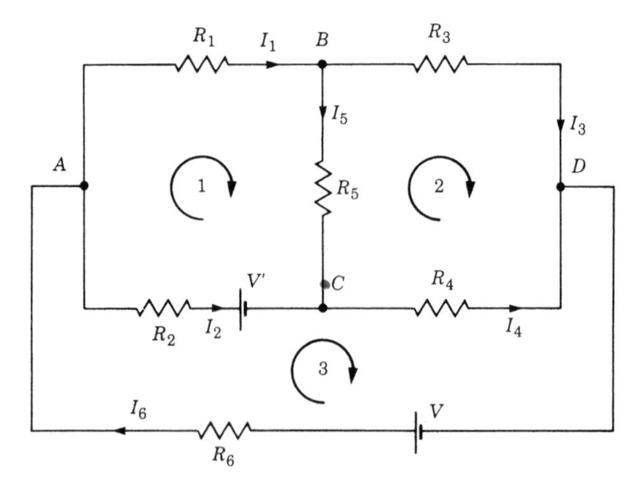

Figure 24.13 An electric network.

In writing the first rule, we consider those currents directed away from the junction as positive and those directed toward the junction as negative. In applying the second rule, we adopt the following conventions:

(a) A potential *drop* across a resistance is considered positive or negative depending on whether one moves in the same sense as the current or in the opposite sense.

(b) When we pass through an emf, we take the potential *drop* as negative or positive depending on whether we pass in the direction in which the emf acts (increase in potential) or in the opposite direction (drop in potential).

EXAMPLE 24.7

Calculation of the currents in the network illustrated in Figure 24.13.

▷ We shall now apply Kirchhoff's rules to the network of Figure 24.13. From the figure we see that there are six possible different currents, one through each resistor. Therefore, we will need six independent equations. There are four junctions in this network and so we can write only three equations that satisfy Kirchhoff's first rule. The first rule applied to junctions A, B and C gives:

$$\text{Junction A:} \quad -I_6 + I_1 + I_2 = 0$$

$$\text{Junction B:} \quad -I_1 + I_3 + I_5 = 0$$

$$\text{Junction C:} \quad -I_2 - I_5 + I_4 = 0$$

The second rule applied to the closed paths marked 1, 2 and 3 gives:

$$\text{Path 1:} \quad -R_2 I_2 + R_1 I_1 + R_5 I_5 - V' = 0$$

$$\text{Path 2:} \quad -R_5 I_5 + R_3 I_3 - R_4 I_4 = 0$$

$$\text{Path 3:} \quad R_6 I_6 + R_2 I_2 + R_4 I_4 - V + V' = 0$$

These six equations are enough to determine the six currents in the network.

To consider a numerical example, assume that $R_1 = R_5 = R_6 = 3\,\Omega$, $R_2 = 2\,\Omega$, $R_3 = 4\,\Omega$ and $R_4 = 1\,\Omega$, while $V = 10$ V and $V' = 5$ V. Substituting in the above equations and solving the system we find that $I_1 = 1.19$ A, $I_2 = 0.14$ A, $I_3 = 0.61$ A, $I_4 = 0.72$ A, $I_5 = 0.58$ A and $I_6 = 1.33$ A. If the value for any one of the currents had been found to be negative, this would only mean that the direction chosen was incorrect and is actually the reverse of that chosen.

Note 24.2 Electric currents in gases

When an electric field is applied to a gas, no electric current is produced because, in general, gas molecules are electrically neutral. However, under certain conditions it is possible to ionize gas molecules. One mechanism is by the passage of energetic charged particles, such as electrons, protons or helium nuclei (α-particles). If the kinetic energy of the particles is larger than the ionization energy of the molecule, the electric fields of these charged particles pull electrons out from molecules near their path, producing what is called an ion pair: an electron and the rest of the molecule, charged positively. Intense ultraviolet, X- and γ-radiation also produce ionization. The ionization produced by charged particles and radiation is not limited to gases and occurs in most materials, including the human body. Gases can also be ionized by increasing the temperature so that ionization results from inelastic collisions among the gas molecules (recall Example 23.2). The hot gas becomes a mixture of positive and negative ions, called a **plasma**. Plasmas exist, for example, in stellar atmospheres, where the temperatures are so high that atoms and molecules may not exist.

When gas molecules are ionized, they tend to recombine, returning to the neutral state. Therefore, if an ionization source is placed near a gas, both processes, ionization and recombination, occur. Eventually equilibrium is established when the rates of ionization and recombination are equal. Suppose now that an electric field $\mathscr{E}$ is produced in the ionized gas (Figure 24.14) by applying a voltage ΔV to two parallel plates, or electrodes, separated by a distance d. The electric field can be modified by changing either ΔV or d.

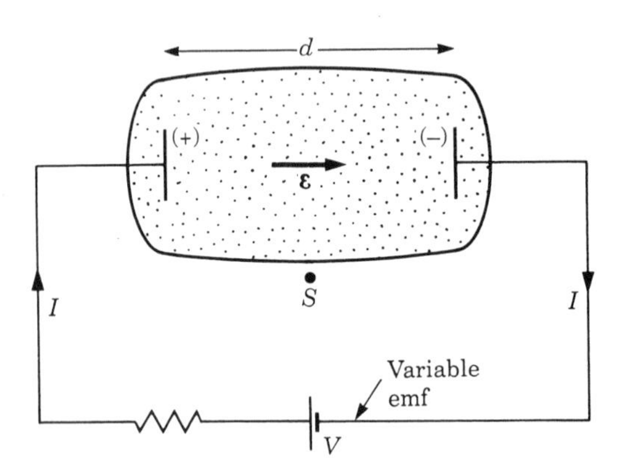

Figure 24.14 The electric field in the gas is $\mathscr{E} = V/d$. The ionization source is S.

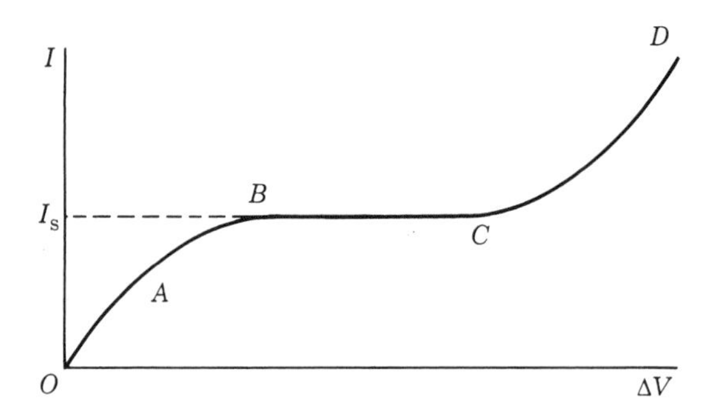

Figure 24.15 Ionization current vs. voltage. I_S is the saturation current. At D an electric discharge is produced.

Under the action of the electric field, some ions reach the electrodes before recombining, and an electric current I is produced. As the electric field increases, the current also increases because more ions reach the electrodes (Figure 24.15). Eventually, the electric field reaches a value where all the ions produced reach the electrodes before recombining. At this point no further increase in current is possible as the field increases. The current I_s is called the **saturation current**. The saturation current depends directly on the rate of ionization of the gas and can be used as a measure of the intensity of the ionizing source S. At a certain value of the electric field, which depends on the pressure and composition of the gas, a new phenomenon occurs. As the ions are accelerated toward the electrodes, they gain so much energy that they produce new ions in inelastic collisions with other gas molecules. This results in a multiplication of ions and an increase in the current. Eventually an avalanche of ions toward the electrodes is produced and there is an **electric discharge** in the gas. This phenomenon occurs in lightning, which is an electric discharge in the atmosphere, and is used in devices such as Geiger counters and dosimeters, for detecting and measuring radiations.

Part B: Electric currents and magnetic fields

24.9 Magnetic force on an electric current

Since an electric current is a stream of electric charges, when a conductor carrying an electric current is placed in a magnetic field, it experiences a force, perpendicular to the current, that is the resultant of the magnetic forces exerted on each of the moving charges (Figure 24.16). This is the principle on which electric motors operate.

Consider, for simplicity, the case of a rectilinear conductor of length L placed in a uniform magnetic field $\mathscr{B}$ (Figure 24.17). By placing the conductor in different positions with different currents and lengths, it has been concluded experimentally that the magnitude of the magnetic force is given by

$$F = IL\mathscr{B} \sin \theta \qquad (24.17)$$

where θ is the angle between the conductor and the magnetic field. The force is zero if the conductor is parallel to the field ($\theta = 0$) and maximum if it is perpendicular to it ($\theta = \pi/2$). Since all charged particles move within the conductor, the direction of the magnetic force on the electric current is perpendicular to the conductor. If $\boldsymbol{u}$ is a unit vector in the direction of the current we may combine both properties of the magnetic force and write the force as the vector product

$$\boldsymbol{F} = IL\boldsymbol{u} \times \mathscr{B} \qquad (24.18)$$

The direction of the force is found by applying the right-hand rule of the vector product. Both Equations 24.17 and 24.18 are simply extensions of the magnetic force on a moving charge (Equations 22.1 and 22.3; see the calculation below).

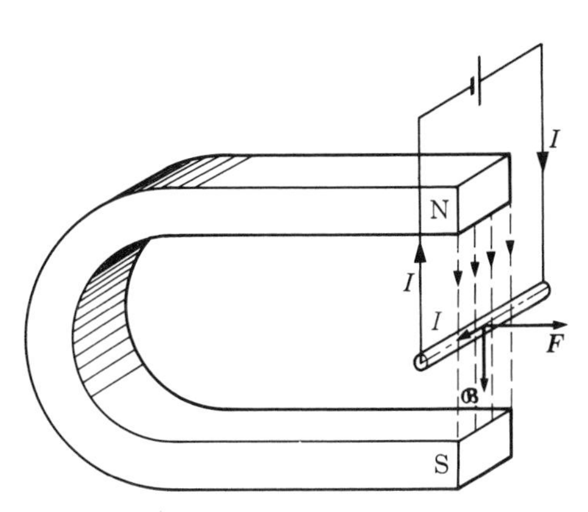

Figure 24.16 A current-carrying wire placed in a magnetic field experiences a force which is perpendicular to the current and to the magnetic field.

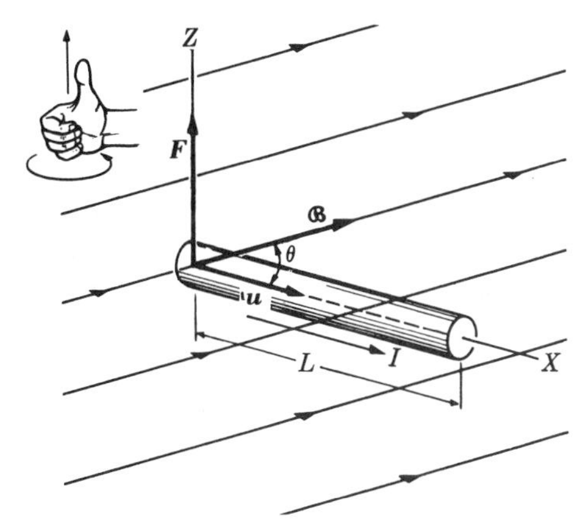

Figure 24.17 Vector relation between the magnetic force on a current-carrying conductor, the magnetic field, and the current. The force is perpendicular to the plane containing $\boldsymbol{u}$ and $\mathscr{B}$.

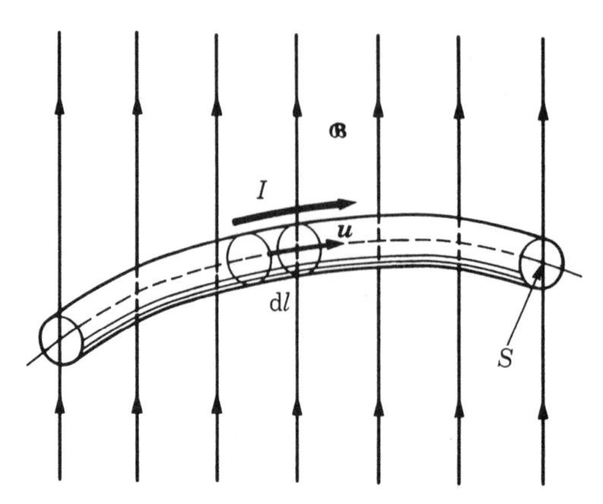

Figure 24.18 Current-carrying conductor in a magnetic field.

In the general case where the conductor has an arbitrary shape and the magnetic field is not uniform (Figure 24.18), we may divide the conductor in small segments, each of length dl, and apply Equation 24.18 to each segment. The result is that

$$\boldsymbol{F} = I\int_L \boldsymbol{u} \times \mathcal{B}\, dl \quad \text{or} \quad \boldsymbol{F} = I\int_L d\boldsymbol{l} \times \mathcal{B} \qquad \textbf{(24.19)}$$

where $d\boldsymbol{l} = \boldsymbol{u}\, dl$.

It is important to recognize that in the case of a current in a metallic conductor at rest relative to an observer, the magnetic force is exerted only on the moving electrons, because the positive lattice ions are practically at rest. However, the entire conductor experiences the magnetic force. The explanation is very simple. The magnetic force on the electrons is much weaker than the electric force coupling the electrons to the lattice. Then when the electrons experience the action of the magnetic force they drag the rest of the metal with them by their electric coupling with the lattice.

Calculation of the magnetic force on an electric current

Consider a cross-section of a conductor through which particles with charge q are moving with velocity $\boldsymbol{v}$. If there are n particles per unit volume, we have seen (Equation 24.4) that the current density is the vector $\boldsymbol{j} = nq\boldsymbol{v}$. Suppose now that the conductor is in a magnetic field. The force on each charge is given by Equation 22.3, $q\boldsymbol{v} \times \mathcal{B}$. Since there are n charged particles per unit volume, the *force per unit volume* $\boldsymbol{f}$ is $\boldsymbol{f} = nq\boldsymbol{v} \times \mathcal{B} = \boldsymbol{j} \times \mathcal{B}$. The total force on a small volume dV of the medium is $d\boldsymbol{F} = \boldsymbol{f}\, dV = (\boldsymbol{j} \times \mathcal{B})\, dV$. The total force on a finite volume, obtained by integrating this expression over all the volume, is

$$\boldsymbol{F} = \int_{\text{Vol}} (\boldsymbol{j} \times \mathcal{B})\, dV$$

Consider the case where the current is along a wire or filament of cross-section S. A volume element dV is given by $S\, dl$ (Figure 24.18), and therefore we may write

$$\boldsymbol{F} = \int_{\text{Filament}} (\boldsymbol{j} \times \mathcal{B}) S\, dl$$

Now $\boldsymbol{j} = j\boldsymbol{u}$, where $\boldsymbol{u}$ is a unit vector along the axis of the filament. Then

$$\boldsymbol{F} = \int (j\boldsymbol{u}) \times \mathscr{B} S \, \mathrm{d}l = \int (jS)(\boldsymbol{u} \times \mathscr{B}) \, \mathrm{d}l$$

which may be written as

$$\boldsymbol{F} = I \int (\boldsymbol{u} \times \mathscr{B}) \, \mathrm{d}l$$

where $jS = I$ is the intensity of the current along the wire, which is the same at all sections of a conductor, because of the law of conservation of electric charge. If the conductor is rectilinear and the magnetic field is uniform, both $\boldsymbol{u}$ and $\mathscr{B}$ are constant and we have

$$\boldsymbol{F} = I(\boldsymbol{u} \times \mathscr{B}) \int \mathrm{d}l = IL\boldsymbol{u} \times \mathscr{B}$$

which is Equation 24.18.

EXAMPLE 24.8

Magnetic force on a current-carrying conductor, of arbitrary shape, placed in a uniform magnetic field.

▷ Suppose a conductor L of arbitrary shape is placed in a uniform magnetic field $\mathscr{B}$ (Figure 24.19). Assume that the current I enters at A and leaves at B. Then the magnetic force on the conductor is obtained using Equation 24.19, $\boldsymbol{F} = I \int_L \mathrm{d}\boldsymbol{l} \times \mathscr{B}$. In this example, since the magnetic field is uniform it is the same at all points of L and it can be taken out of the integral, allowing us to write

$$\boldsymbol{F} = I\left(\int_L \mathrm{d}\boldsymbol{l}\right) \times \mathscr{B}$$

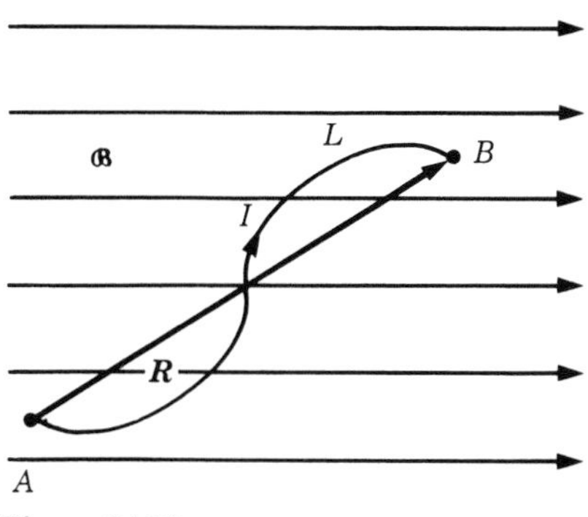

Figure 24.19

The sum of all the small displacements $\mathrm{d}\boldsymbol{l}$ along path L is equal to the vector $\boldsymbol{R}$ joining the two ends of the conductor. Therefore, $\boldsymbol{F} = I\boldsymbol{R} \times \mathscr{B}$, which shows that the resultant force on the conductor is independent of the shape of the conductor. Also, if the conductor is closed, then $\boldsymbol{R} = 0$ because points A and B coincide. That means that the force produced by a uniform magnetic field on a closed conductor carrying a current is zero. However, as will be seen later on, the magnetic field produces a torque on the circuit.

EXAMPLE 24.9

The Hall effect.

▷ In 1879, Edwin C. Hall (1855–1938) discovered that when a metal plate, along which an electric current is passing, is placed in a magnetic field perpendicular to the plate, a potential difference appears between opposite points on the edges of the plate. The **Hall effect** is an illustration of the magnetic force on the carriers of an electric current. This effect is also very useful in determining the sign of the charges composing an electric current.

Suppose that the carriers of the electric current in the metal plate are electrons, having a negative charge $q = -e$. Considering Figure 24.20(a), in which the Z-axis has been drawn parallel to the current I, we see that the actual motion of the electrons is in the negative Z-direction with the velocity $\boldsymbol{v}_-$. When a magnetic field $\mathscr{B}$ is applied perpendicular to the plate, or along the X-axis, the electrons are subject to the force $\boldsymbol{F} = (-e)\boldsymbol{v}_- \times \mathscr{B}$. The vector product $\boldsymbol{v}_- \times \mathscr{B}$ is along the negative Y-axis, but when it is multiplied by $-e$ the result is a vector $\boldsymbol{F}$ along the $+Y$-axis. Therefore the electrons drift to the right-hand side of the plate, which thus becomes negatively charged. The left-hand side, being deficient in the usual number of electrons, becomes positively charged. As a consequence, an electric field $\mathscr{E}$ parallel to the $+Y$-axis is produced. When the force $(-e)\mathscr{E}$ on the electrons, due to this electric field and directed to the left, balances the force to the right due to the magnetic field $\mathscr{B}$, equilibrium results. This in turn leads to a transverse potential difference between opposite sides of the conductor, with the left-hand side at the higher potential; the value of the potential difference is proportional to the magnetic field. This is the normal, or 'negative', Hall effect, exhibited by most metals, such as gold, silver, platinum, copper etc. But with some p-materials, such as semiconductors, an opposite, or 'positive', Hall effect is produced.

If the carriers of the current are positively charged particles, with $q = +e$, the situation is illustrated in Figure 24.20(b). The right-hand side of the plate becomes positively charged and the left-hand side negatively charged, producing a transverse electric field in the negative Y-direction. Therefore the potential difference is the reverse of that in the case of negative carriers, resulting in a positive Hall effect.

In many substances the carriers of the current are electron holes, which behave similarly to positive particles and move in the direction opposite to that of the negatively charged electrons. Thus the Hall effect provides a very useful method of determining the sign of the carriers of electric current in a conductor.

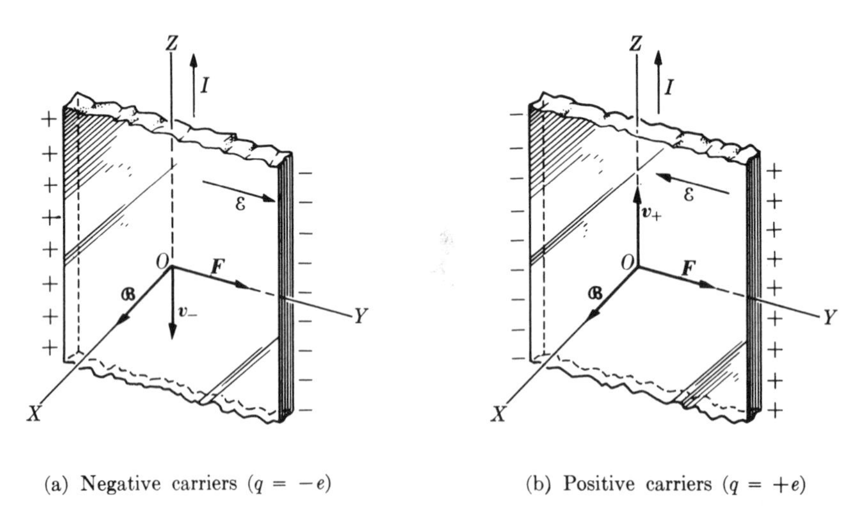

(a) Negative carriers ($q = -e$)

(b) Positive carriers ($q = +e$)

Figure 24.20 The Hall effect.

24.10 Magnetic torque on an electric current

The force exerted by a magnetic field on an electric circuit also gives rise to a magnetic torque. For simplicity, consider a rectangular circuit carrying a current I placed in a uniform magnetic field $\mathscr{B}$. The circuit is placed so that the normal $\boldsymbol{u}_N$ to its plane (oriented according to the right-hand rule) makes an angle θ with the field $\mathscr{B}$ and two sides of the circuit are perpendicular to the field (Figure 24.21). The forces $\boldsymbol{F}'$ acting on the sides L' have the same magnitude and opposite direction. They tend to deform the circuit, but produce no torque since they are in the plane of the circuit. The forces $\boldsymbol{F}$ on the sides L are, according to Equation 24.17, of magnitude $F = I\mathscr{B}L$ and constitute a couple whose lever arm is $L' \sin\theta$. Therefore they exert a torque on the circuit that tends to orient the circuit perpendicular to the magnetic field. The magnitude of the torque is $\tau = (I\mathscr{B}L)(L' \sin\theta)$. But $LL' = S$ is the area of the circuit. Then $\tau = (IS)\mathscr{B}\sin\theta$. The direction of the torque is perpendicular to the plane of the couple and is along the line PQ. If we define a vector

$$\boldsymbol{M} = IS\boldsymbol{u}_N \tag{24.20}$$

normal to the plane of the circuit, we may write the torque τ as

$$\tau = -M\mathscr{B}\sin\theta \tag{24.21}$$

or, in vector form,

$$\boldsymbol{\tau} = \boldsymbol{M} \times \mathscr{B} \tag{24.22}$$

Result 24.22 is mathematically similar to Equation 21.26, which gives the torque on an electric dipole due to an external electric field. Therefore the quantity $\boldsymbol{M}$ defined in Equation 24.20 is called the **magnetic moment** of the current. Note from

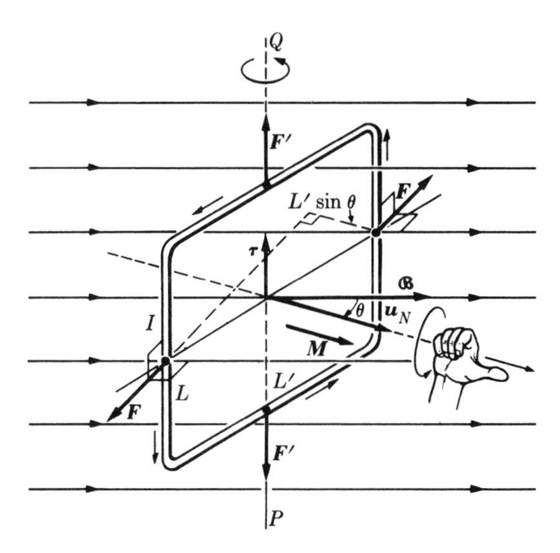

Figure 24.21 Magnetic torque on a rectangular electric circuit placed in a magnetic field. The torque is zero when the plane of the circuit is perpendicular to the magnetic field.

Equation 24.20 that the direction of $\boldsymbol{M}$ is that given by the right-hand rule as shown in Figure 24.22. Hence

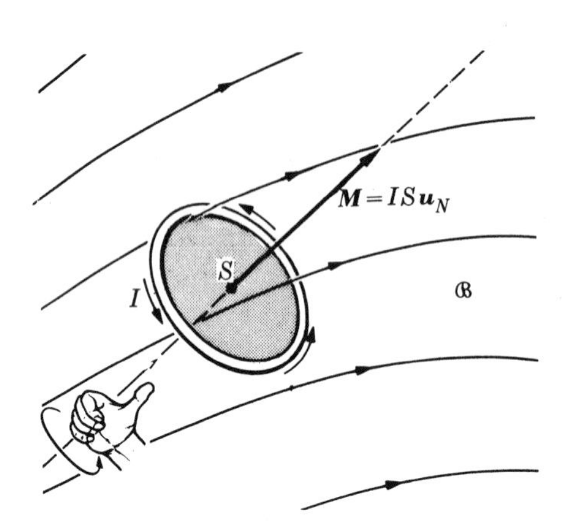

Figure 24.22 Relation between the magnetic moment of an electric current and the direction of the current, according to the right-hand rule.

a circuit in a magnetic field is subject to a torque, which tends to orient the magnetic moment of the circuit parallel to the field.

To obtain the energy of a circuit in a magnetic field, we apply Equation 9.32, $\tau = -\mathrm{d}E_p/\mathrm{d}\theta$, as we did in Example 22.7. We conclude that the potential energy of a circuit of magnetic moment $\boldsymbol{M}$ placed in a magnetic field $\mathscr{B}$ is

$$E_p = -M\mathscr{B}\cos\theta = -\boldsymbol{M}\cdot\mathscr{B} \qquad \textbf{(24.23)}$$

It can be shown that Equations 24.20, 24.22 and 24.23 are valid for a circuit of any shape.

EXAMPLE 24.10

Current-measuring devices: **galvanometers.**

▷ The fact that the magnetic torque on a circuit is proportional to the current can be used to measure the current. A simple design for a current-measuring device is illustrated in Figure 24.23(a). The current to be measured passes through a coil, of area S, suspended between the poles of a magnet. In some cases, the coil is wrapped around an iron cylinder C. The magnetic field exerts a torque given by Equation 24.21 on the coil when it carries

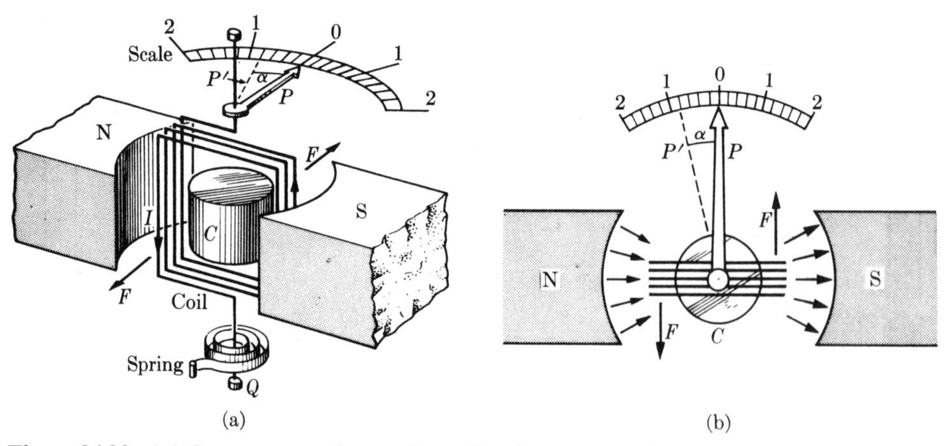

Figure 24.23 (a) Components of a moving-coil galvanometer. (b) Top view of galvanometer shown in (a).

a current, tending to place the coil perpendicular to the field and twisting the spring Q. The coil adopts an equilibrium position rotated through an angle α when the magnetic torque is balanced by the elastic torque $k\alpha$ produced by the spring, where k is the spring's elastic constant. The angle α is indicated by a pointer attached to the coil. The pole faces are shaped so that the magnetic field between the pole faces and the iron cylinder C is radial, as shown in the top view in Figure 24.23(b). In this case $\mathscr{B}$ is always in the plane of the circuit and θ in Equation 24.21 is $\pi/2$, so that $\sin\theta = 1$. Then the torque is given by $\tau = IS\mathscr{B}$, since $M = IS$. At equilibrium, when the torque due to the magnetic field is balanced by the torque due to the twisting of the spring, $IS\mathscr{B} = k\alpha$, and therefore $I = k\alpha/S\mathscr{B}$. If k, S and $\mathscr{B}$ are known, this equation gives the value of the current I in terms of the angle α. Usually the scale is calibrated so that the value of I can be read directly in some convenient units.

24.11 Magnetic field produced by a current

In 1820 Hans C. Oersted (1777–1851), by noting the deflection of a compass needle placed near a conductor through which a current passed, concluded that *an electric current produces a magnetic field* whose direction is perpendicular to the current. Because an electric current is a stream of electric charges all moving in the same direction, each charge produces a magnetic field. Therefore, the magnetic field of a current is the sum of the magnetic fields produced by each of the moving charges.

After many experiments made over a period of years by several physicists using circuits of different shapes A. M. Ampère and P. Laplace arrived empirically at a general expression for calculating the magnetic field produced by a current of any shape. This expression, called the **Ampère–Laplace law**, is

$$\mathscr{B} = K_m I \int \frac{\mathrm{d}\boldsymbol{l} \times \boldsymbol{u}_r}{r^2} \tag{24.24}$$

where the meaning of all symbols is indicated in Figure 24.24, and the integral is extended along the entire circuit. K_m is a constant that depends on the units chosen. In the SI system it has been determined (see Note 24.4) that

$$K_m = 10^{-7}\,\mathrm{T\,m/A} \quad \text{or} \quad \mathrm{m\,kg\,C^{-2}}$$

(We must note that the integral in Equation 24.24 is expressed in m^{-1} when r and l are given in meters.) Therefore

$$\mathscr{B} = 10^{-7} I \int \frac{\mathrm{d}\boldsymbol{l} \times \boldsymbol{u}_r}{r^2} \tag{24.25}$$

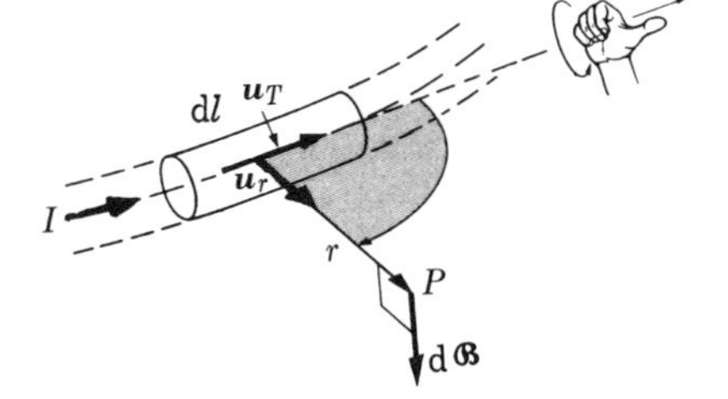

Figure 24.24 According to the Ampère–Laplace law, the small segment dl of the current contributes an amount d$\mathscr{B}$ to the magnetic field at point P. The field d$\mathscr{B}$ is perpendicular to the plane defined by $\boldsymbol{u}_T$ and $\boldsymbol{u}_R$.

It is customary to write $K_m = \mu_0/4\pi$, where μ_0 is a new constant called the **magnetic permeability of vacuum**. In the *SI* system of units

$$\mu_0 = 4\pi \times 10^{-7}\ \mathrm{m\,kg\,C^{-2}} = 1.2566 \times 10^{-6}\ \mathrm{m\,kg\,C^{-2}}$$

Thus Equation 24.24 for the Ampère–Laplace law may be written as

$$\mathscr{B} = \frac{\mu_0}{4\pi} I \int \frac{\mathrm{d}\boldsymbol{l} \times \boldsymbol{u}_r}{r^2} \tag{24.26}$$

Equation 24.26 can be interpreted as saying that the resultant magnetic field $\mathscr{B}$ at P produced by the current is the sum of a large number of very small or elementary contributions $\mathrm{d}\mathscr{B}$ by each of the segments or length elements $\mathrm{d}\boldsymbol{l}$ composing the circuit. Each elementary contribution is

$$\mathrm{d}\mathscr{B} = \frac{\mu_0}{4\pi} I \frac{\mathrm{d}\boldsymbol{l} \times \boldsymbol{u}_r}{r^2} \tag{24.27}$$

Note 24.3 Relation between the magnetic field of a current and the magnetic field of a moving charge

Since the electric current I is constant along the circuit we may write Equation 24.26 as

$$\mathscr{B} = \frac{\mu_0}{4\pi} \int \frac{(I\,\mathrm{d}\boldsymbol{l}) \times \boldsymbol{u}_r}{r^2}$$

But the volume of the section $\mathrm{d}l$ of the conductor is $\mathrm{d}V = S\,\mathrm{d}l$, where S is the cross-section, and recalling Equation 24.4, $j = nqv$, we have that $I\,\mathrm{d}l = (jS)\,\mathrm{d}l = j\,\mathrm{d}V = nqv\,\mathrm{d}V$ or in vector form, $I\,\mathrm{d}\boldsymbol{l} = nq\boldsymbol{v}\,\mathrm{d}V$. Therefore

$$\mathscr{B} = \frac{\mu_0}{4\pi} \int \frac{q\boldsymbol{v} \times \boldsymbol{u}_r}{r^2}\, n\,\mathrm{d}V$$

Since $n\,\mathrm{d}V$ is the number of particles in the volume $\mathrm{d}V$, we *may interpret* the above result by saying that each charged particle in the current produces a magnetic field at point P given by

$$\mathscr{B} = \frac{\mu_0}{4\pi} \frac{q\boldsymbol{v} \times \boldsymbol{u}_r}{r^2}$$

This is precisely the expression given in Equation 22.13 for the magnetic field of a moving charge, confirming that the magnetic field of a current is a collective effect of the charges that compose the current.

24.12 Magnetic field of a rectilinear current

Consider a very long and thin rectilinear current (Figure 24.25). Application of Equation 24.26 shows that the magnitude of the magnetic field at a perpendicular

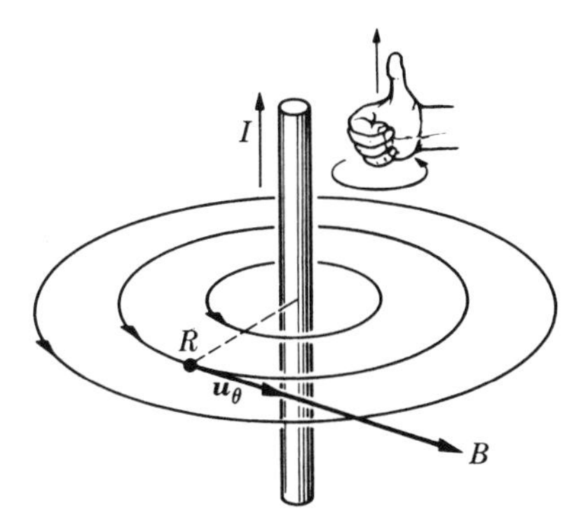

Figure 24.25 Magnetic field produced by a rectilinear current.

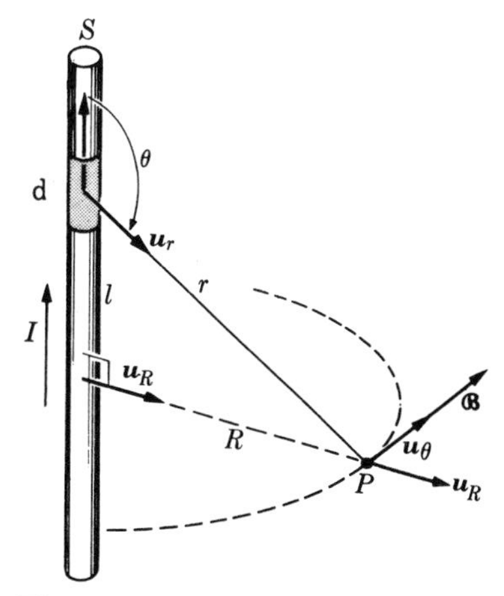

Figure 24.26 Computation of the magnetic field of a rectilinear current.

distance R from the current is (see the calculation below)

$$\mathscr{B} = \frac{\mu_0 I}{2\pi R} \quad \textbf{(24.28)}$$

a result that can be confirmed experimentally. The magnetic field is inversely proportional to the distance R, and the lines of force are circles concentric with the current and perpendicular to it, as indicated in the figure. The right-hand rule for determining the direction of the magnetic field relative to the direction of the current is also indicated in the figure. Result 24.28 is called the **Biot–Savart formula**. In vector form we may write Equation 24.28 as

$$\boldsymbol{\mathscr{B}} = \frac{\mu_0 I}{2\pi R}\,\boldsymbol{u}_\theta \quad \textbf{(24.29)}$$

where $\boldsymbol{u}_\theta$ is a unit vector tangent to the line of force.

In Chapter 21 we saw that the lines of force of an electric field go from the positive to the negative charges or perhaps, in some cases, from or to infinity. In Chapter 22 we noticed that the magnetic lines of force of a moving charge are loops concentric with the direction of motion of the charge. The same applies to an electric current (Figure 24.25) for which the lines of the magnetic field are closed lines, linked with the current. A field of this kind, which does not have point sources, such as magnetic poles, is called **solenoidal**.

Calculation of the magnetic field of a rectilinear current

Consider Figure 24.26. For any point P and any element dl of the current, the vector $\mathrm{d}\boldsymbol{l} \times \boldsymbol{u}_r$ is perpendicular to the plane determined by P and the current, and therefore its direction is that of the unit vector $\boldsymbol{u}_\theta$. The magnetic field at P, produced by dl, is then tangent to the circle of radius R that passes through P, is centered on the current, and is in a plane perpendicular to the current. Therefore, when we perform the integration in Equation 24.26, the contributions from all terms in the integral have the same direction $\boldsymbol{u}_\theta$, and the resultant magnetic field $\mathscr{B}$ is also tangent to the circle. Thus it is only necessary to find the magnitude of $\mathscr{B}$. The magnitude of $\mathrm{d}\boldsymbol{l} \times \boldsymbol{u}_r$ is $\mathrm{d}l \sin\theta$. Therefore, for a rectilinear current, we may write Equation 24.26 in magnitude as

$$\mathscr{B} = I\frac{\mu_0}{4\pi}\int_{-\infty}^{\infty}\frac{\sin\theta}{r^2}\,\mathrm{d}l$$

From the figure we see that $r = R\csc\theta$ and $l = -R\cot\theta$, so that $\mathrm{d}l = R\csc^2\theta\,\mathrm{d}\theta$. Substituting in the above equation and noting that $l = -\infty$ corresponds to $\theta = 0$ and $l = +\infty$ corresponds to $\theta = \pi$, we obtain

$$\mathscr{B} = \frac{\mu_0}{4\pi}I\int_0^{\pi}\frac{\sin\theta}{R^2\csc^2\theta}(R\csc^2\theta\,\mathrm{d}\theta) = \frac{\mu_0 I}{4\pi R}\int_0^{\pi}\sin\theta\,\mathrm{d}\theta = \frac{\mu_0 I}{4\pi R}(-\cos\theta)\Big|_0^{\pi} = \frac{\mu_0 I}{2\pi R}$$

24.13 Magnetic field of a circular current

The magnetic field of a circular current is illustrated in Figure 24.27. The lines of force form links with the current. They have axial symmetry relative to the line passing through the center of the current and are perpendicular to the plane. The orientation of the lines of force is obtained with the right-hand rule, as shown in the figure. Note that these are the kind of lines of force that result when a rectilinear conductor is bent into a circle.

The calculation of the magnetic field at an arbitrary point is a somewhat complicated mathematical problem, but at points *along the axis of the current* it is a fairly easy task. As shown below, the magnetic field at a point on the axis of a circular current of radius a, at a distance R from the center, is

$$\mathscr{B} = \frac{\mu_0 I a^2}{2(a^2 + R^2)^{3/2}} \tag{24.30}$$

The magnetic field at the center ($R = 0$) is $\mathscr{B} = \mu_0 I/2a$.

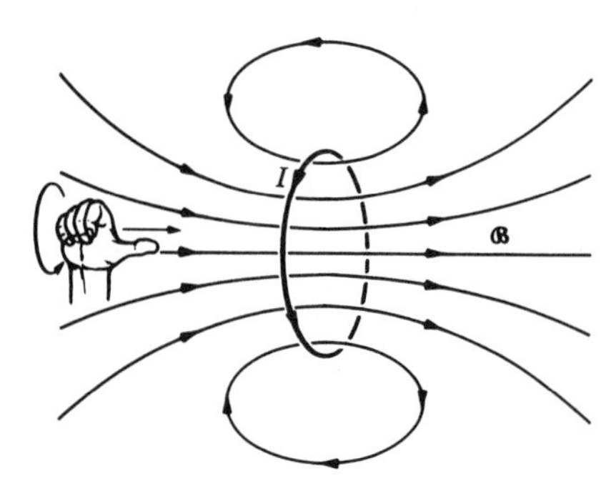

Figure 24.27 Magnetic field produced by a circular current.

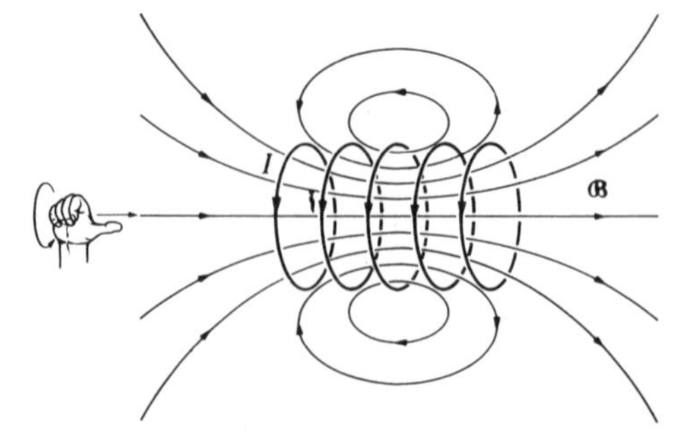

Figure 24.28 Magnetic field produced by a solenoidal current.

A **solenoidal current**, or simply a **solenoid**, is a current composed of several coaxial loops of the same radius, all carrying the same current (Figure 24.28). The magnetic field of a solenoid is found by adding the magnetic fields of each of the component circular currents. The resultant magnetic field is indicated by lines of force in the figure, where some fluctuations in the space between loops have been smoothed out. It can be shown (Section 26.3) that in the case of a very long solenoid of length L, with N total turns, the magnetic field at the center is

$$\mathscr{B} = \frac{\mu_0 IN}{L} \tag{24.31}$$

while at the ends the value is one-half the value at the center; that is,

$$\mathscr{B} = \frac{\mu_0 IN}{2L} \tag{24.32}$$

It is not surprising that the magnetic field at the ends is half of the value at the center. If a long solenoid is divided into two halves, the magnetic field at the common end is the sum of the fields produced by the two sections. Therefore each one must be half of the original value. Solenoids are used to produce fairly uniform magnetic fields in limited regions around their center.

Calculation of the magnetic field along the axis of a circular current

For a circular current, Figure 24.29, the vector product $\mathrm{d}\boldsymbol{l} \times \boldsymbol{u}_r$ in Equation 24.27 is perpendicular to the plane PAA' and has a magnitude $\mathrm{d}l$ because the two vectors are perpendicular. Therefore, the magnetic field produced by the length element $\mathrm{d}\boldsymbol{l}$ at P has the magnitude

$$\mathrm{d}\mathscr{B} = \frac{\mu_0}{4\pi} I \frac{\mathrm{d}l}{r^2}$$

It is perpendicular to the plane PAA' and thus oblique to the Z-axis. Decomposing $\mathrm{d}\mathscr{B}$ into a component $\mathrm{d}\mathscr{B}_\parallel$ parallel to the axis and a component $\mathrm{d}\mathscr{B}_\perp$ perpendicular to it, we see that, when we integrate along the circle, for each $\mathrm{d}\mathscr{B}_\perp$ there is another in the opposite direction from the length element directly opposed to $\mathrm{d}l$, and therefore all vectors $\mathrm{d}\mathscr{B}_\perp$ add to zero. The resultant $\mathscr{B}$ will be the sum of all the $\mathrm{d}\mathscr{B}_\parallel$, and therefore is parallel to the axis. Now, since $\cos\alpha = a/r$,

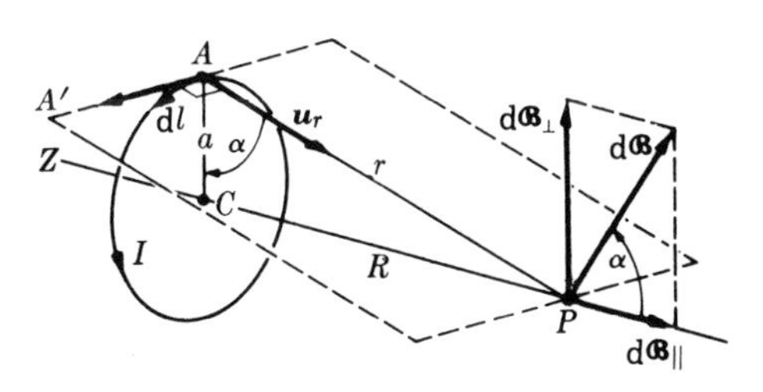

Figure 24.29 Computation of the magnetic field along the axis of a circular current.

$$\mathrm{d}\mathscr{B}_\parallel = (\mathrm{d}\mathscr{B})\cos\alpha = \frac{a}{r}\,\mathrm{d}\mathscr{B} = \frac{\mu_0 Ia}{4\pi r^3}\,\mathrm{d}l$$

The distance r remains constant when we integrate around the circle. Then, since for a circle $\int dl = 2\pi a$, the magnitude of the resultant magnetic field is

$$\mathscr{B} = \int d\mathscr{B} = \frac{\mu_0 I a}{4\pi r^3} \int dl = \frac{\mu_0 I a^2}{2r^3}$$

Noting that $r = (a^2 + R^2)^{1/2}$, the magnetic field for points on the axis of a circular current may be written as

$$\mathscr{B} = \frac{\mu_0 I a^2}{2(a^2 + R^2)^{3/2}}$$

EXAMPLE 24.11
Field of a magnetic dipole.

▷ A charged particle moving with uniform circular motion with a very small radius constitutes a magnetic dipole of moment $\boldsymbol{M} = (q/2m)\boldsymbol{L}$ (Equation 22.16), where $\boldsymbol{L}$ is the angular momentum of the particle. Also, the magnetic moment of a current is $M = IS$ (Equation 24.20). For the case of a circular current, $S = \pi a^2$ so that $M = \pi I a^2$.

The two definitions for magnetic moment are equivalent. First, a charge moving on a circle with angular velocity ω has an orbital frequency $v = \omega/2\pi$ and therefore is equivalent to a current $I = qv = q\omega/2\pi$. Therefore $M = \pi I a^2 = \frac{1}{2}q\omega a^2 = (q/2m)(m\omega a) = (q/2m)\,L$, since ωa is the velocity of the charge.

Replacing Ia^2 by M/π in Equation 24.30 we get for the magnetic field at points along the axis

$$\mathscr{B} = \left(\frac{\mu_0}{4\pi}\right) \frac{2M}{(a^2 + R^2)^{3/2}}$$

At the center ($R = 0$), we get

$$\mathscr{B} = \left(\frac{\mu_0}{4\pi}\right) \frac{2M}{a^3}$$

which is Equation 22.17 for the magnetic field of a charge moving in a circular orbit of radius a. Therefore, the magnetic field of a small circular current and of an electric charge describing a circular orbit are equivalent. This is not surprising, since an electric current is a flow of electric charges.

A circular current constitutes a magnetic dipole when $a \ll R$, in which case Expression 24.30 for the magnetic field along the axis, with Ia^2 replaced by M/π, becomes

$$\mathscr{B} = \left(\frac{\mu_0}{4\pi}\right) \frac{2M}{R^3}$$

and is equivalent to the expression for the electric field of an electric dipole given in Equation 21.25 if we replace r by R and $(1/4\pi\varepsilon_0)$ by $(\mu_0/4\pi)$. Because of the similarity of the two fields, we may use the results of Example 21.13 to calculate the force and energy of a magnetic dipole in an external magnetic field.

EXAMPLE 24.12

The tangent galvanometer.

▷ A **tangent galvanometer** consists of a circular coil (Figure 24.30) having N turns and carrying a current I. It is placed in a region where there is a magnetic field $\mathscr{B}$ so that a diameter of the coil is parallel to $\mathscr{B}$. The current I produces, at the center of the coil, a magnetic field $\mathscr{B}' = \mu_0 IN/2a$. Therefore, the resultant magnetic field $\mathscr{B}''$ at the center of the coil makes an angle θ with the axis of the coil given by

$$\tan\theta = \frac{\mathscr{B}}{\mathscr{B}'} = \frac{2a\mathscr{B}}{\mu_0 IN}$$

Thus if a small magnetic needle is placed at the center of the coil, it will turn and rest in equilibrium at an angle θ with the axis. This allows a measurement of the external field $\mathscr{B}$ if we know the current I, or conversely the current I when the field $\mathscr{B}$ is known. Usually $\mathscr{B}$ is the Earth's magnetic field. For precise measurements, the formula is corrected to take into account the finite length of the needle, since the field acting on it is not exactly the field at the center. The name 'tangent galvanometer' is derived from the trigonometric function appearing above.

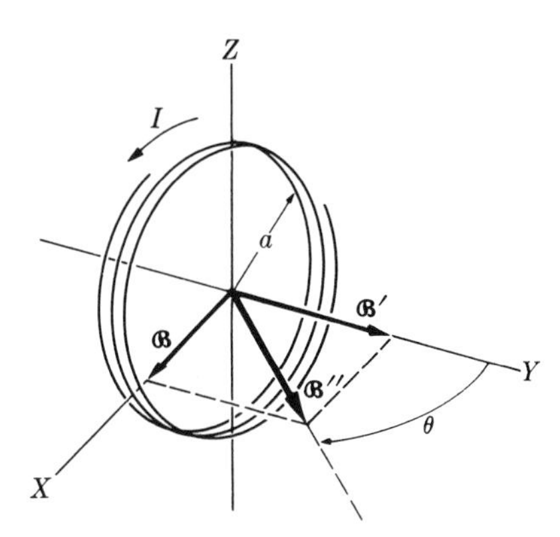

Figure 24.30 Tangent galvanometer.

24.14 Forces between electric currents

Since an electric current produces a magnetic field and, in turn, a magnetic field produces a force on an electric current, there is a magnetic interaction between electric currents. For simplicity, consider two parallel rectilinear currents I and I' (Figure 24.31), in the same direction and separated by a distance R. The magnetic field $\mathscr{B}$ due to I at any point of I' is given by Equation 24.28, and has the direction indicated. The force $\boldsymbol{F}'$ on I', using Equation 24.19, is

$$\boldsymbol{F}' = I' \int \mathrm{d}\boldsymbol{l}' \times \mathscr{B} \qquad \textbf{(24.32)}$$

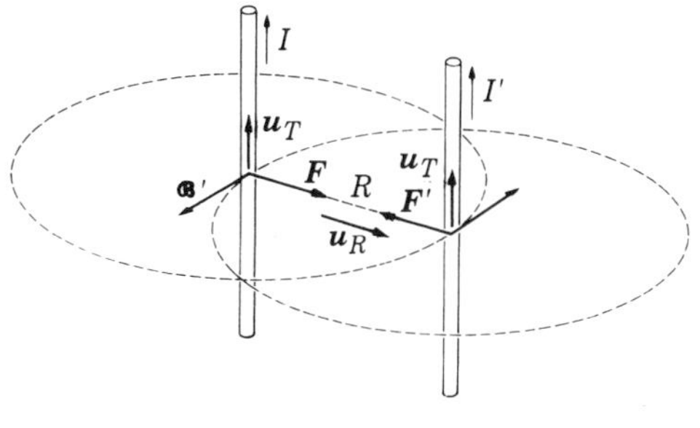

Figure 24.31 Magnetic forces between two rectilinear currents.

Now $d\boldsymbol{l}' \times \boldsymbol{\mathscr{B}} = -\boldsymbol{u}_R \mathscr{B}\, dl'$, where $\boldsymbol{u}_R$ is defined as the unit vector from I to I'. Therefore, using Equation 24.28 for $\mathscr{B}$, we have

$$\boldsymbol{F}' = I' \int \left(-\boldsymbol{u}_R \frac{\mu_0 I}{2\pi R}\right) dl' = -\boldsymbol{u}_R \left(\frac{\mu_0 I I'}{2\pi R}\right) \int dl'$$

The value of the integral is L', so that the force on I' is

$$\boldsymbol{F}' = -\boldsymbol{u}_R \left(\frac{\mu_0}{4\pi}\right) \frac{2II'}{R} L' \tag{24.33}$$

This result indicates that current I *attracts* current I' with a force per unit length equal to

$$f = \left(\frac{\mu_0}{4\pi}\right) \frac{2II'}{R}$$

A similar calculation of the force on I produced by I' gives a force of the same magnitude but opposite direction, and again represents an attraction. Therefore

> *two parallel currents in the same direction attract each other with a force, inversely proportional to their separation as a result of their magnetic interaction; if the parallel currents are in opposite directions, they repel each other.*

These interactions among currents are of great practical importance for electric motors, galvanometers, wattmeters and other applications.

Note 24.4 Note on electromagnetic units

We have two laws from which to choose the fourth basic unit besides those of distance, time and mass. They are: Coulomb's law for the electric interaction between two charges, given by Equation 21.2

$$F = \frac{1}{4\pi\varepsilon_0} \frac{qq'}{r^2}$$

and the law of interaction between two rectilinear currents, given by Equation 24.33,

$$F' = \frac{\mu_0}{4\pi} \frac{2II'}{R} L'$$

Although we have two constants, ε_0 and μ_0, corresponding to the electric and magnetic forces, there is only one degree of freedom because there is only one new physical quantity, the electric charge or the electric current, both related by the equation Current = Charge/Time. Therefore we can assign an arbitrary value to one of the constants only. The Eleventh General Conference on Weights and Measures, held in 1960, decided to define the ampere as the current which, circulating in two parallel conductors separated by a distance of one meter, results in a force on either conductor equal to 2×10^{-7} N per meter of length of each conductor (Figure 24.32). Once the ampere is so defined, the coulomb is the quantity of charge that flows across any cross-section of a conductor in one second when the current is one ampere.

An experimental arrangement called a **current balance** for measuring the force between two parallel conductors is shown in Figure 24.33. The balance is first set in equilibrium with no currents in the circuit. When the same current is sent through the two conductors, the attractive magnetic force exerted on the section L' is

$$F = 2 \times 10^{-7} \frac{I^2 L'}{R}$$

so that additional weights are required on the left pan to bring the balance back to equilibrium. From the known values of F, L' and R, the value of I is calculated. In practice two parallel circular coils are used. The expression for the force between the coils is then different.

Given the SI definition of the ampere, we have that

$$\mu_0 = 4\pi \times 10^{-7}\ \mathrm{N\,C^{-2}\,s^2}$$

As we shall see in Chapter 27, the velocity of light is related to ε_0 and μ_0 by $1/\varepsilon_0\mu_0 = c^2$. The constant c has been measured experimentally with very great accuracy. Actually, it has been used to define the unit of length (recall Section 2.4). Then we have that

$$\varepsilon_0 = \frac{1}{\mu_0 c^2} = \frac{10^7}{4\pi c^2}$$
$$\approx 8.5 \times 10^{-12}\ \mathrm{N\,m^2\,C^{-2}}$$

which is the value given in Equation 21.6. This explains our previous choice for K_e that may have appeared somewhat arbitrary.

One reason why the ampere was adopted as the fourth fundamental unit is that it is easier to measure the force between two currents, and therefore prepare a standard of current, than it is to set up a standard of charge and measure the force between two charges. However, from the physical point of view, the concept of charge is more fundamental than that of current. From the practical as well as the theoretical point of view, the MKSC and MKSA systems are entirely equivalent.

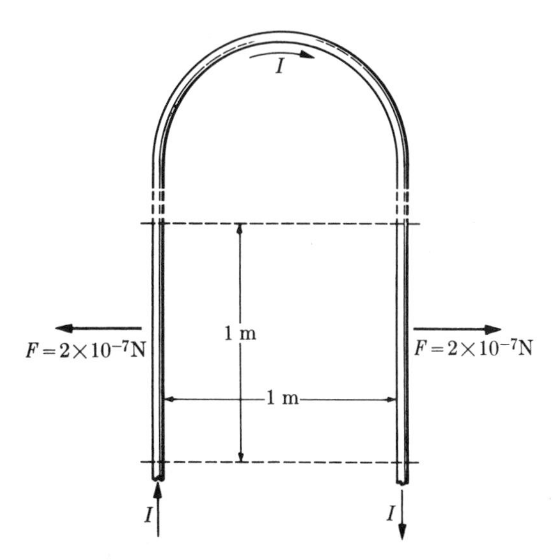

Figure 24.32 Experimental definition of the ampere.

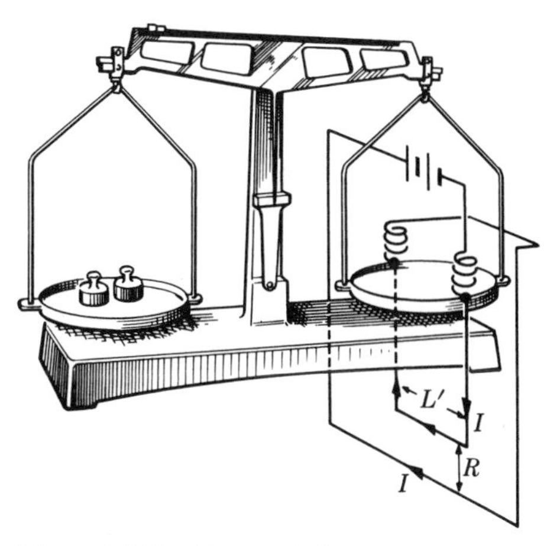

Figure 24.33 Current balance (simplified) for measuring a current in terms of the magnetic force between two parallel conductors.

QUESTIONS

Part A

24.1A Write Equation 24.4 for protons and for electrons.

24.2A The resistivity of a conductor is defined by $\rho = 1/\sigma$. What are the units of ρ? Write Equation 24.7 using ρ instead of σ.

24.3A Analyze how the resistance of a conductor should change with the temperature. Write an expression for the resistance in terms of the temperature, assuming a linear relation.

24.4A Referring to the Wheatstone bridge (Example 24.6), verify that if R_3 and R_4 are two pieces of the same wire but of lengths L and L' we may write $R_1/R_2 = L/L'$.

24.5A A uniform electric field is applied to a region where there are free electrons; describe the motion of the electrons. Repeat when the electrons are in a metal.

24.6A Is Ohm's law universally valid or of limited application?

24.7A Describe the origin of the Joule effect in a metallic conductor.

24.8A How should the potential difference in a conductor change to double the current? How is the power affected?

24.9A Express Equation 24.11 in terms of the potential difference and the resistance.

24.10A Given a collection of resistors, how should they be combined to obtain the maximum resistance? The minimum resistance?

24.11A Explain why a static electric field cannot maintain a current in a closed circuit.

24.12A Discuss the meaning of Kirchhoff's laws in terms of the conservation laws.

24.13A A solid cylindrical conductor has a radius R. Another conductor of the same material, length and radius is hollow along the center. The radius of the cavity is r. Compare the resistances of the two conductors.

24.14A What happens to the resistance of a cylindrical conductor if its length is doubled? If the radius is doubled? If both are doubled?

24.15A How do you determine the 'direction' of an electric current? Does the direction of an electric current always coincide with the direction of motion of the charges?

24.16A Why is an external power source needed to maintain a current across a potential difference in a conductor?

24.17A Verify that the current across a surface S placed in a region where there are n ions per unit volume, each with charge q and moving with velocity $\boldsymbol{v}$, is given by the surface integral $I = \int_S qn\boldsymbol{v} \cdot d\boldsymbol{S}$.

Part B

24.1B Verify that the unit $\mathrm{J\,T^{-1}}$ for the magnetic dipole moment is equivalent to $\mathrm{m^2\,s^{-1}\,C}$.

24.2B Show that the torque of a magnetic field on a circuit carrying a current is $\boldsymbol{\tau} = I \int_L \boldsymbol{r} \times (d\boldsymbol{l} \times \mathcal{B})$. Apply this to the case shown in Figure 24.21.

24.3B How does a magnetic dipole move in a non-uniform magnetic field? What is the direction of the force?

24.4B What relation exists between the magnetic force on a moving electric charge and on an electric current?

24.5B If we place a metallic wire in a magnetic field, why does a magnetic field exert a force on the wire only if there is an electric current on the wire?

24.6B Why does an electric current in a conductor produce a magnetic field but no electric field?

24.7B Investigate how a galvanometer must be designed and installed to measure a potential difference between two points of a conducting wire, in which case it is called a **voltmeter**, and for measuring an electric current, in which case it is called an **ammeter.**

24.8B What are the similarities and the differences between the fields of an electric dipole and a magnetic dipole?

24.9B A small electric circuit is placed in a uniform magnetic field. What is the direction of the torque exerted on the circuit? How does the

circuit tend to orient itself relative to the magnetic field?

24.10B Consider two parallel rectilinear currents of opposite directions. Show in a diagram similar to Figure 24.31 that they repel each other.

24.11B A battery has an emf V and an internal resistance r. Express the potential difference ΔV between the poles of the battery when the current in the circuit is I. Graphically represent the relation between ΔV and I.

PROBLEMS

Part A

24.1A (a) Find the resistance of 1 m of copper wire whose cross-section is 8 mm^2. If a current of 0.74 A is maintained in the wire, calculate (b) the potential drop between the ends of the wire and (c) the power dissipated. Repeat for a wire with double the radius.

24.2A A copper wire with a cross-section of 10^{-5} m^2 carries an electric current of 1.5 A. Assuming that there are 5×10^{28} conduction electrons per m^3, determine (a) the current density and (b) the electron drift velocity.

24.3A The maximum permissible current in the coil of a galvanometer is 2.4 A. Its resistance is 20 Ω. What must be done to the 'meter' so that it reads full scale when it is inserted (a) in an electric line carrying a current of 15 A, (b) between two points having a potential difference of 110 V?

24.4A The needle of a galvanometer suffers a full-scale (50 divisions) deviation when the current is 0.1 mA. The resistance of the galvanometer is 5 Ω. What must be done to change it into (a) an ammeter with each division corresponding to 0.2 A, (b) a voltmeter with each division corresponding to 0.5 V?

24.5A (a) Determine the total resistance in each of the arrangements shown in Figure 24.34.
(b) Also determine the current in, and the potential difference across, each resistor.

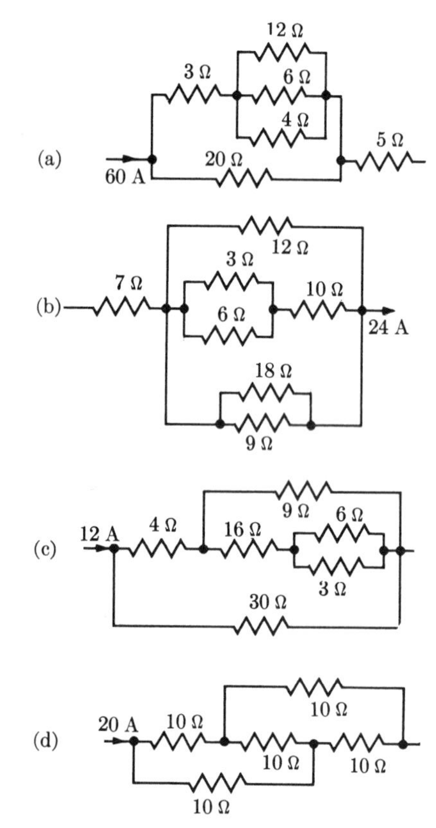

Figure 24.34

24.6A Each of the resistors in Figure 24.35 has a resistance of 4 Ω and can dissipate a maximum of 10 W without being destroyed. Calculate (a) the maximum power the circuit can dissipate,
(b) the current in each resistor and
(c) the maximum voltage that can be applied between the ends.

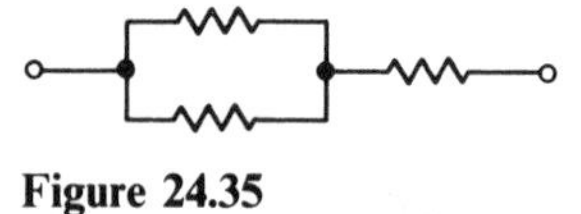

Figure 24.35

24.7A Three equal resistors are connected in parallel. When a certain potential difference is applied across the combination, the total power consumed is 7 W. What power would be consumed if the three resistors were connected in series across the same potential difference?

24.8A A cell is in series with a resistance of 5 Ω. When the switch is open, a voltmeter across the cell reads 2.06 V. When the switch is closed the voltmeter reads 1.72 V. Find (a) the emf and (b) the internal resistance of the cell.

24.9A The potential difference across the terminals of a rechargeable battery is 8.5 V when there is a current of 3 A in the battery from the negative to the positive terminal. When the current is 2 A in the reverse direction, the potential difference becomes 11 V. (a) What is the internal resistance of the battery? (b) What is its emf?

24.10A In the circuit of Figure 24.36, find the potential differences ΔV_{ab}, ΔV_{ac} and ΔV_{ad}.

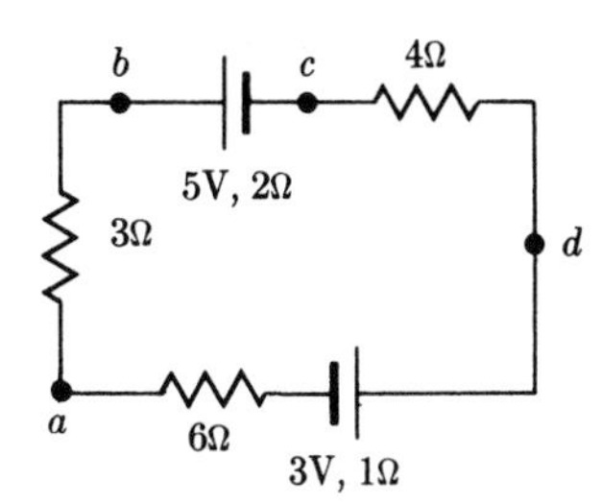

Figure 24.36

24.11A Determine the current in each conductor in the networks of Figure 24.37(a) and (b).

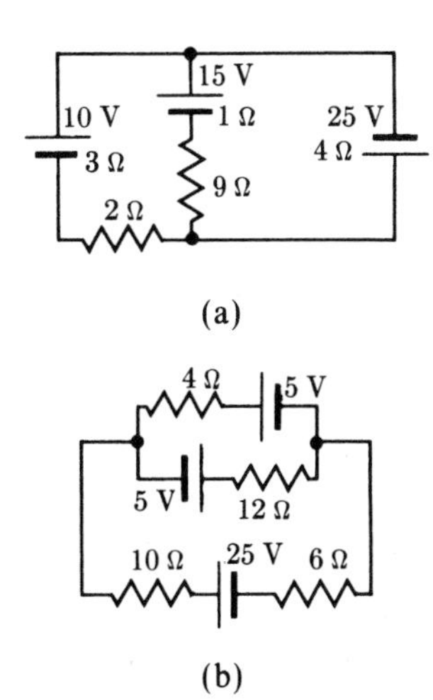

Figure 24.37

24.12A Figure 24.38 shows a **potentiometer** device, used to measure the emf V of cell N; B is a battery and St is a standard cell of emf V_{St}. When the switch is set at either 1 or 2, the tap b is moved until the galvanometer G reads zero. Show that if l_1 and l_2 are the corresponding distances from b to a, then $V = V_{St}(l_1/l_2)$.

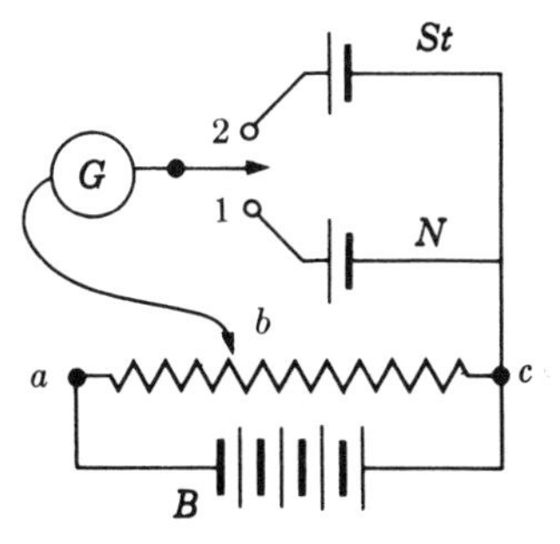

Figure 24.38

24.13A Referring to the potentiometer of Figure 24.38, the emf is approximately 3 V and its internal resistance is unknown. St is a standard cell of 1.0183 V. The switch is set at point 2, thus placing the standard cell in the galvanometer circuit. When the tap b is 0.64 of the distance from a to c, the galvanometer reads zero.
(a) What is the potential difference across the entire length of resistor ac? (b) The switch is set at point 1, and a new zero reading of the galvanometer is obtained when b is 0.53 of the distance from a to c. What is the emf of cell N?

Part B

24.1B Find the force on each of the wire segments shown in Figure 24.39 if the magnetic field, of magnitude 1.50 T, is parallel to OZ and the current in each segment is 20 A. An edge of the cube is 0.1 m.

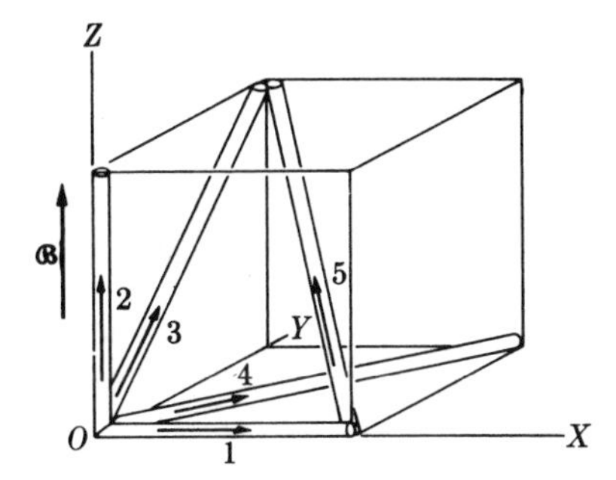

Figure 24.39

24.2B A rectangular loop of wire 5 cm × 8 cm is placed with the shorter sides parallel to a magnetic field of 0.15 T. (a) If the loop carries a current of 10 A, what torque acts on it? (b) What is the magnetic moment of the loop? (c) What is

the maximum torque that can be obtained with the same total length of wire carrying the same current in this magnetic field?

24.3B The rectangular loop in Figure 24.40 is pivoted about the Y-axis and carries a current of 10 A in the direction indicated. (a) If the loop is in a uniform magnetic field of 0.20 T parallel to the X-axis, calculate the force on each side of the loop. (b) Calculate the torque required to hold the loop in the position shown. (c) Repeat parts (a) and (b), but consider the loop in a uniform field 0.20 T parallel to the Z-axis. (d) What torque would be required in each case if the loop were pivoted about an axis through its center, parallel to the Y-axis?

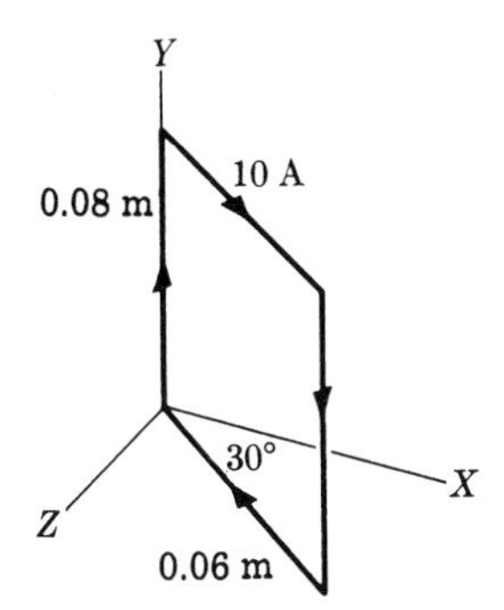

Figure 24.40

24.4B Find the current density (assumed uniform) required in a horizontal aluminum wire to make it 'float' in the Earth's magnetic field at the equator. The density of Al is 2.7×10^3 kg m^{-3}. Assume that the Earth's field is about 7×10^{-5} T and that the wire is oriented in the east–west direction.

24.5B The coil of a galvanometer has 50 turns and encloses an area of 6.0×10^{-4} m^2. The magnetic field in the region of the coil is 0.01 T and is radial. The torsional constant of the hairspring is 10^{-6} N m deg^{-1}. Find the angular deflection of the coil for a current of 1 mA.

24.6B A wire loop in the form of a square of side 0.1 m lies in the XY-plane, as shown in Figure 24.41. There is a current of 10 A in the loop as shown. If a magnetic field parallel to the Z-axis with an intensity of $0.10x$ T (where x is in meters) is applied, calculate (a) the resultant force on the loop and (b) the resultant torque relative to the point O.

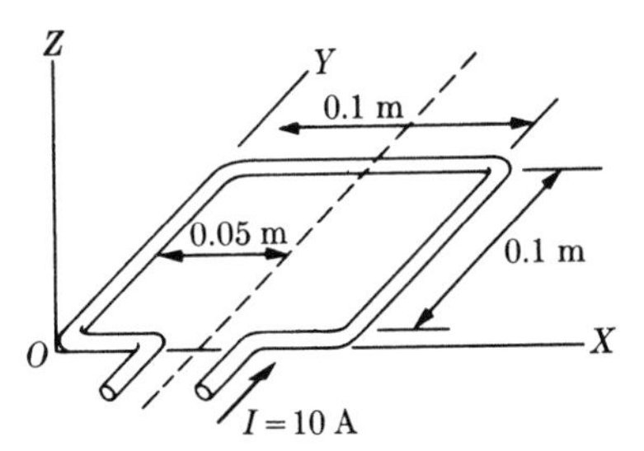

Figure 24.41

24.7B Consider a square coil, 6.0×10^{-2} m on a side, carrying a current of 0.10 A and placed in a uniform magnetic field of strength 10^{-4} T. (a) If the plane of the coil is initially parallel to the magnetic field, is there any torque on the coil? (b) If the plane of the coil is initially perpendicular to the field, is there now any torque on the coil? (c) Express the torque as a function of the angle the normal to the coil makes with the direction of the magnetic field. (d) Plot the torque as a function of angle, from 0 to 2π. (e) If, at the point where there is no torque on the coil, the coil has an angular velocity, what happens?

24.8B Compute the intensity of the magnetic field produced by an infinitely long wire 2 cm in diameter carrying an electric current of 1 A at a distance of (i) 0.50 cm and (ii) 1 m. Make a diagram showing how the magnetic field varies with the distance.

24.9B A long straight wire carries a current of 1.5 A. An electron travels with a velocity of 5.0×10^4 m s^{-1} parallel to the wire and in the same direction as the current. Calculate the force the magnetic field exerts on the moving electron.

24.10B Figure 24.42 is an end view of two long parallel wires perpendicular to the XY-plane, each carrying a current I, but in opposite directions.

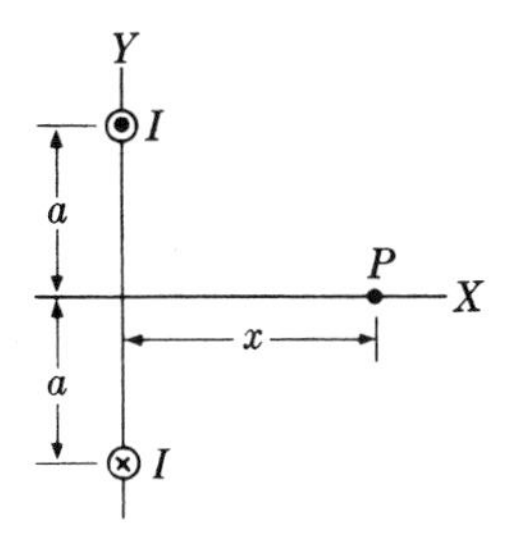

Figure 24.42

(a) Show the magnetic field of each wire, and the resultant magnetic field, at point *P*. (b) Derive the expression for the magnitude of the magnetic field at any point on the X-axis, in terms of the coordinate X of the point. (c) Construct a graph of the magnitude of the magnetic field at any point on the X-axis. (d) At what value of x is the magnetic field a maximum? (e) Repeat parts (b) and (c) for points along the Y-axis.

24.11B A solenoid is 0.30 m long and is wound with two layers of wire. The inner layer consists of 300 turns, the outer layer of 250 turns. The current in both layers is 3.0 A and in the same direction in both. Calculate the magnetic field at a point near the center of the solenoid.

24.12B A long horizontal wire rests on the surface of a table. Another wire, vertically above the first, is 1.00 m long and is free to slide up and down on two vertical metal guides. The two wires are connected through sliding contacts and carry a current of 50 A. The linear mass density of the moveable wire is 5.0×10^{-3} kg m^{-1}. To what equilibrium height will the wire rise?

24.13B A long straight wire and a rectangular wire frame lie on a table top. The side of the frame parallel to the wire is 0.30 m long, the side perpendicular to the wire is 0.50 m long. The currents are $I_1 = 10$ A and $I_2 = 20$ A. (a) Calculate the force on the loop. (b) Calculate the torque on the loop, relative to the wire as axis. (c) Calculate the torque on the loop relative to the dashed line as axis.

24.14B In Figure 24.42, a third long straight wire, parallel to the other two, passes through point *P*. Each wire carries a current of 20 A. Let $a = 0.30$ m and $x = 0.40$ m. Find the magnitude and direction of the force per unit length on the third wire (a) if the current in it is up. (b) Repeat part (a) if the current in the third wire is down into the plane of the paper.

24.15B Using the results obtained in Example 21.12 show that the magnetic field along the axis of a magnetic dipole is $(\mu_0/4\pi)(2M/r^3)$ and on a plane perpendicular to the axis and passing through the center of the dipole it is $(\mu_0/4\pi)(M/r^3)$.

25 The electric field

Carl F. Gauss is considered one of the greatest mathematicians of all time. His research included number theory, algebra, geometry and differential equations. Gauss applied his mathematical skills to geodesy, astronomy and physics. He devised (1801–1809) a method for calculating planetary orbits, using Newton's law of gravitation. Gauss did extensive research in electricity and magnetism, establishing (in 1831) a general property of the electric field, based on Coulomb's law, known as Gauss' law.

Notes

25.1 Introduction

In Chapter 21 we discussed the electric interaction and introduced the concept of the electric field. In this chapter we shall discuss some characteristics of the electric field, considering the field as an independent entity. We shall examine only the static or time-independent fields.

25.2 Electromotive force

Since the intensity of the electric field is equal to the force per unit charge, the work done by the electric field when moving one unit of charge along the path L is expressed by the line integral

$$\begin{pmatrix}\text{Work done along path}\\ L\text{ per unit charge}\end{pmatrix} = \int_L \mathscr{E} \cdot \mathrm{d}\boldsymbol{l} \qquad \textbf{(25.1)}$$

and therefore is measured in volts.

Recalling that for a static electric field the potential difference between two points is equal to the work done by the electric field in moving a unit charge from one point to the other, regardless of the path followed, as expressed by Equation 21.16, we may write

$$\int_L \mathscr{E} \cdot \mathrm{d}\boldsymbol{l} = V_A - V_B \tag{25.2}$$

where A and B are the two points joined by the path L. Thus the line integral of a static electric field between two points is equal to the potential difference between the points. Note that the path chosen to go from one point to the other is arbitrary.

In Section 24.6 the electromotive force along a closed path was defined as the work done when moving one unit of charge around the closed path. Therefore, using Equation 25.1, we may write

$$\begin{pmatrix}\text{Work done along a closed} \\ \text{path per unit charge}\end{pmatrix} = \text{emf} = \oint \mathscr{E} \cdot \mathrm{d}\boldsymbol{l} \tag{25.3}$$

where the circle on top of the integral sign is to indicate that the path is closed. Such an integral is called a **circulation** (see Appendix A.10).

If the electric field is static and the path is closed, points A and B coincide, and Equation 25.2 gives

$$\oint_L \mathscr{E} \cdot \mathrm{d}\boldsymbol{l} = 0 \tag{25.4}$$

that is,

the work done by a static electric field in moving a charge around a closed path is zero.

This is one of the fundamental properties of the static electric field. Recalling that energy is required to maintain a current in an electric circuit, we conclude that a static electric field cannot maintain a current in a closed circuit and that another source of energy is required.

25.3 Flux of the electric field

Consider a plane surface of area S in a uniform electric field $\mathscr{E}$. If the normal to the surface makes an angle θ with the direction of the electric field (Figure 25.1), we define the **electric flux** across the surface by the quantity

$$\text{Electric flux} = \mathscr{E} S \cos\theta \tag{25.5}$$

If the surface is perpendicular to the electric field, $\theta = 0$, $\cos\theta = 1$ and the flux is $\mathscr{E}S$, while if the surface is parallel to the field, $\theta = \pi/2$, $\cos\theta = 0$ and the flux is zero. The electric flux is positive if $\theta < \pi/2$ and negative if $\theta > \pi/2$.

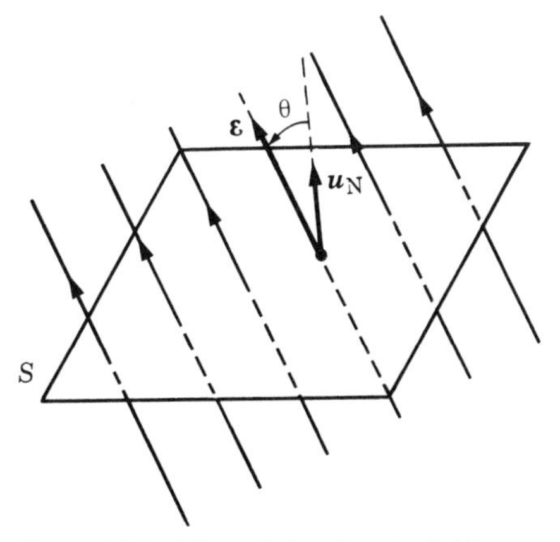

Figure 25.1 Flux of the electric field through a plane surface.

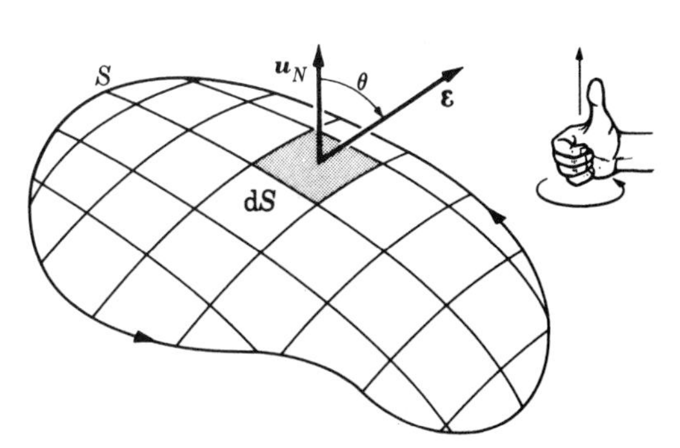

Figure 25.2 Flux of an electric field through an arbitrary surface.

If the surface is curved and/or the electric field is not uniform, we proceed as follows: first, we divide the surface into very small surfaces of areas dS_1, dS_2, dS_3,... (Figure 25.2). At each of them draw a unit vector $\boldsymbol{u}_1$, $\boldsymbol{u}_2$, $\boldsymbol{u}_3, \ldots,$ perpendicular to the surface at that point. Then $d\boldsymbol{S}_1 = \boldsymbol{u}_1\, dS_1$, $d\boldsymbol{S}_2 = \boldsymbol{u}_2\, dS_2$, $d\boldsymbol{S}_3 = \boldsymbol{u}_3\, dS_3, \ldots$ are the vectors representing each surface element. The unit vectors are oriented in the direction of the thumb when the fingers of the right hand point in the sense in which we decide to orient the rim of the surface. Let $\theta_1, \theta_2, \theta_3, \ldots$ be the angles between the normal vectors $\boldsymbol{u}_1, \boldsymbol{u}_2, \boldsymbol{u}_3, \ldots$ and the electric field vectors $\mathscr{E}_1$, $\mathscr{E}_2$, $\mathscr{E}_3, \ldots$ at each point on the surface. Then, the electric flux through the surface S is

$$\begin{aligned}\text{Electric flux} &= \mathscr{E}_1\, dS_1 \cos\theta_1 + \mathscr{E}_2\, dS_2 \cos\theta_2 + \mathscr{E}_3\, dS_3 \cos\theta_3 + \cdots \\ &= \mathscr{E}_1 \cdot d\boldsymbol{S}_1 + \mathscr{E}_2 \cdot d\boldsymbol{S}_2 + \mathscr{E}_3 \cdot d\boldsymbol{S}_3 + \cdots\end{aligned}$$

or, replacing the sum by an integral, we have

$$\text{Electric flux} = \int_S \mathscr{E} \cos\theta\, dS = \int_S \mathscr{E} \cdot d\boldsymbol{S} \qquad \textbf{(25.6)}$$

where the integral extends over all the surface, as indicated by the subscript S. For that reason an expression like Equation 25.6 is called a **surface integral** (see Appendix A.11). If the surface is closed (as it is for a sphere), the vectors $\boldsymbol{u}_N$ are oriented in the outward direction. Also, a circle is written on top of the integral sign, so that Equation 25.6 becomes

$$\text{Electric flux} = \oint_S \mathscr{E} \cos\theta\, dS = \oint_S \mathscr{E} \cdot d\boldsymbol{S} \qquad \textbf{(25.7)}$$

The total flux may be positive, negative or zero. When it is positive the flux is 'outgoing' and when it is negative the flux is 'incoming'.

25.4 Gauss' law for the electric field

Consider a point charge q (Figure 25.3). We will now compute the flux of its electric field $\mathscr{E}$ through a spherical surface concentric with the charge. Given that r is the radius of the sphere, the electric field produced by the charge at each point of the spherical surface is

$$\mathscr{E} = \frac{q}{4\pi\varepsilon_0 r^2}\,\boldsymbol{u}_r$$

The unit vector normal to a sphere coincides with the unit vector $\boldsymbol{u}_r$ along the radial direction. Therefore, the angle θ between the electric field $\mathscr{E}$ and the normal unit vector $\boldsymbol{u}_r$ is zero, and $\cos\theta = 1$. Noting that the electric field has the same magnitude at all points of the spherical surface and that the area of the sphere is $4\pi r^2$, Equation 25.6 gives the electric flux as

$$\oint_S \mathscr{E}\cdot \mathrm{d}\boldsymbol{S} = \mathscr{E}\oint_S \mathrm{d}S = \mathscr{E}S = \frac{q}{4\pi\varepsilon_0 r^2}(4\pi r^2) = \frac{q}{\varepsilon_0}$$

The electric flux through the sphere, then, is proportional to the charge and independent of the radius of the surface. Therefore, if we draw several concentric spherical surfaces $S_1, S_2, S_3, \ldots$ (Figure 25.4) around the charge q, the electric flux through all of them is the same and equal to q/ε_0. This result is due to the $1/r^2$ dependence of the field.

As shown below, the preceding result is quite general and applies to a charge q placed anywhere *inside any closed* surface. However, if the charge is *outside* the closed surface, the electric flux through the surface is zero.

If there are several charges $q_1, q_2, q_3, \ldots$ inside the arbitrary surface S (Figure 25.5), the total electric flux will be the sum of the fluxes produced by each charge. We may then state **Gauss' law for the electric field**:

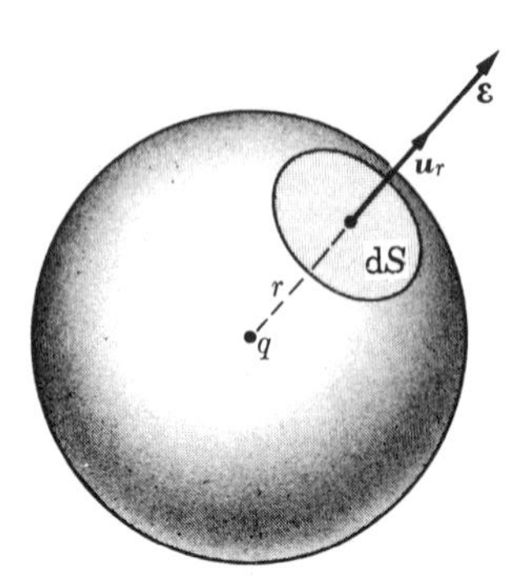

Figure 25.3 Electric flux of a point charge through a sphere which is concentric with the charge.

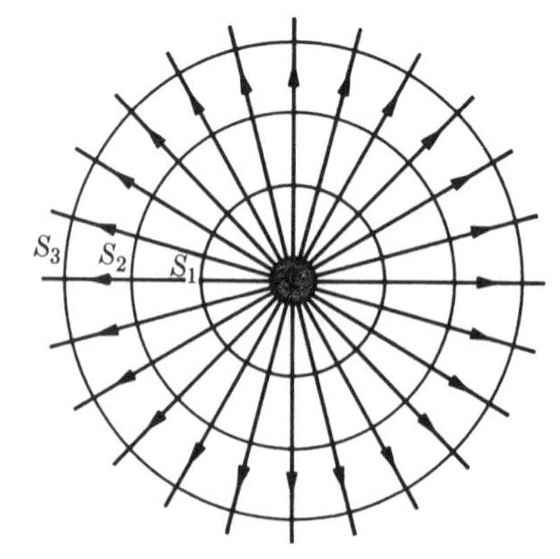

Figure 25.4 The electric flux through concentric spheres surrounding a charge is the same.

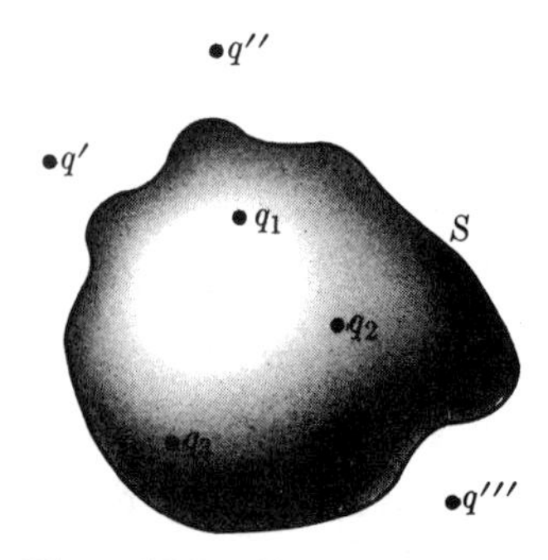

Figure 25.5 The electric flux through any closed surface is proportional to the net charge contained within the surface.

The electric flux through a closed surface surrounding charges $q_1, q_2, q_3, \ldots$ *is*

$$\oint_S \mathscr{E} \cdot \mathrm{d}\boldsymbol{S} = \frac{q}{\varepsilon_0} \tag{25.8}$$

where $q = q_1 + q_2 + q_3 + \ldots$ *is the total net charge inside the closed surface.*

The charges *outside* the closed surface (such as q', q'',... in Figure 25.5) do not contribute to the total flux because, for example, the positive flux at $\mathrm{d}S''$ is balanced by the negative flux at $\mathrm{d}S'$. Gauss' law as expressed by Equation 25.8 is another fundamental property of the electric field. Gauss' law is particularly useful when we wish to compute the electric field produced by charge distributions having certain geometric symmetries, as shown in the following examples.

Derivation of Gauss' law

Consider a charge q *inside* an arbitrary *closed* surface S (Figure 25.6). The total flux through S of the electric field produced by q is given by

$$\oint_S \mathscr{E} \cdot \mathrm{d}s = \oint_S \mathscr{E} \cos\theta \, \mathrm{d}S = \oint_S \frac{q}{4\pi\varepsilon_0 r^2} \cos\theta \, \mathrm{d}S = \frac{q}{4\pi\varepsilon_0} \oint_S \frac{\mathrm{d}S \cos\theta}{r^2}$$

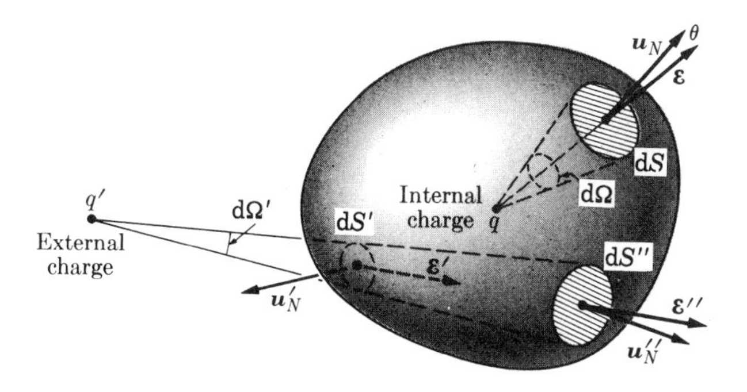

Figure 25.6 The electric flux through a closed surface surrounding a charge is independent of the shape of the surface and the position of the charge.

But $\mathrm{d}S \cos\theta / r^2$ is the solid angle $\mathrm{d}\Omega$ (see Appendix B.4) subtended by the surface element $\mathrm{d}S$ as viewed from the charge q. Since the total solid angle around a point is 4π, we then see that the above result becomes Equation 25.8. This result is valid for any closed surface, irrespective of the position of the charge within the surface. If a charge such as q' is *outside* a closed surface, the electric flux is zero, because the incoming flux is equal to the outgoing flux; hence a net zero flux results. For example, the electric flux of q' through $\mathrm{d}S'$ is equal in magnitude, but opposite in sign, to the electric flux through $\mathrm{d}S''$ because the solid angle is the same for both areas, and therefore the fluxes add to zero.

EXAMPLE 25.1

Electric field of (a) a charge uniformly distributed over a plane and (b) two parallel planes with equal but opposite charges.

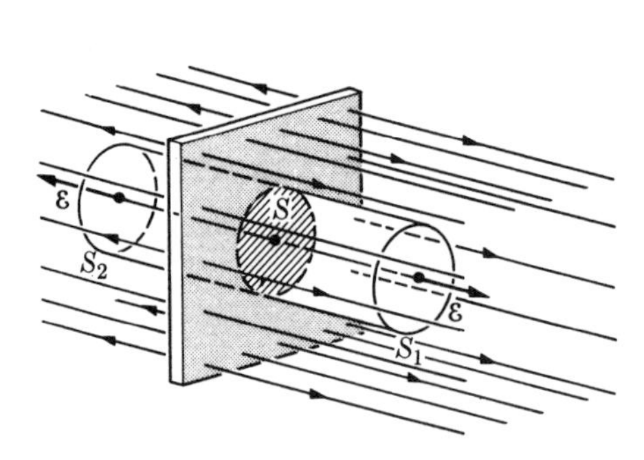

Figure 25.7 Electric field produced by a uniformly charged plane surface.

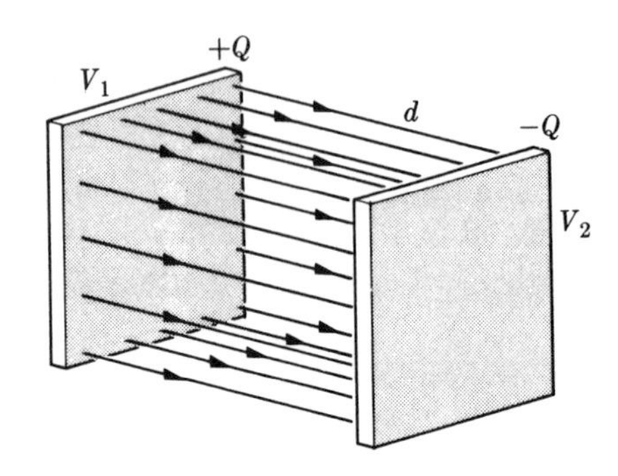

Figure 25.8 Electric field between a pair of plane parallel surfaces carrying equal but opposite charges ($V_1 > V_2$).

▷ (a) Consider the plane of Figure 25.7, which carries a charge σ per unit area. The symmetry of the problem indicates that the lines of force of the electric field are perpendicular to the plane and, if the charge is positive, the lines are oriented as indicated in the figure. Choosing the cylinder shown in the figure as the closed surface, the electric flux may be separated into three terms: (i) the flux through S_1, which is $+\mathscr{E}S$, where S is the area of the base of the cylinder; (ii) the flux through S_2, which is also $+\mathscr{E}S$ since, by symmetry, the electric field must be the same in magnitude and opposite in direction at points at the same distance on both sides of the plane; and (iii) the flux through the lateral surface of the cylinder, which is zero because the electric field is parallel to the surface. Therefore the total electric flux is $2\mathscr{E}S$. The charge inside the closed surface is in the shaded area and is $q = \sigma S$. Therefore, applying Gauss' law, Equation 25.8, we have

$$2\mathscr{E}S = \frac{\sigma S}{\varepsilon_0} \quad \text{or} \quad \mathscr{E} = \frac{\sigma}{2\varepsilon_0}$$

a result that indicates that the electric field is independent of the distance to the plane and is therefore uniform.

(b) Figure 25.8 shows two parallel planes with equal but opposite charges. We observe that in the region outside the two oppositely charged planes there are electric fields equal in magnitude but opposite in direction, giving a resultant field that is zero. But in the region between the planes, the fields are in the same direction, and the resultant field is twice as large as the field of a single plane, or $\mathscr{E} = \sigma/\varepsilon_0$. Thus the two parallel and oppositely charged planes produce a uniform field contained in the region between them. If d is the separation of the two planes, their potential difference is, using Equation 25.2

$$\Delta V = V_1 - V_2 = \mathscr{E}d = \frac{\sigma d}{\varepsilon_0} \tag{25.9}$$

EXAMPLE 25.2

Electric field of a spherical distribution of charge.

▷ Consider a sphere of radius a and charge Q (Figure 25.9(a)). The symmetry of the problem suggests that the field at each point must be radial and depends only on the distance r from the point to the center of the sphere. Therefore, drawing a spherical surface of radius r concentric with the charged sphere, we find that the electric flux through it is

$$\oint_S \mathscr{E} \cdot \mathrm{d}S = \mathscr{E} \oint_S \mathrm{d}S = \mathscr{E}(4\pi r^2)$$

First, considering $r > a$, we see that the charge inside the surface S is the total charge Q of the sphere. Thus, applying Gauss' law, Equation 25.8, we obtain $\mathscr{E}(4\pi r^2) = Q/\varepsilon_0$, which gives for the electric field outside the sphere

$$\mathscr{E} = \frac{Q}{4\pi\varepsilon_0 r^2} \qquad (r > a) \tag{25.10}$$

This is the same result as for the field of a point charge. Thus

the electric field at points outside a charged sphere is the same as if all charge were concentrated at its center.

For $r < a$ we have two possibilities. If all the charge is at the surface of the charged sphere, the total charge inside the spherical surface S' is zero, and Gauss' law gives

$$\mathscr{E}(4\pi r^2) = 0 \qquad \text{or} \qquad \mathscr{E} = 0 \quad (r < a)$$

Thus

the electric field at points inside a sphere that is charged only on its surface is zero.

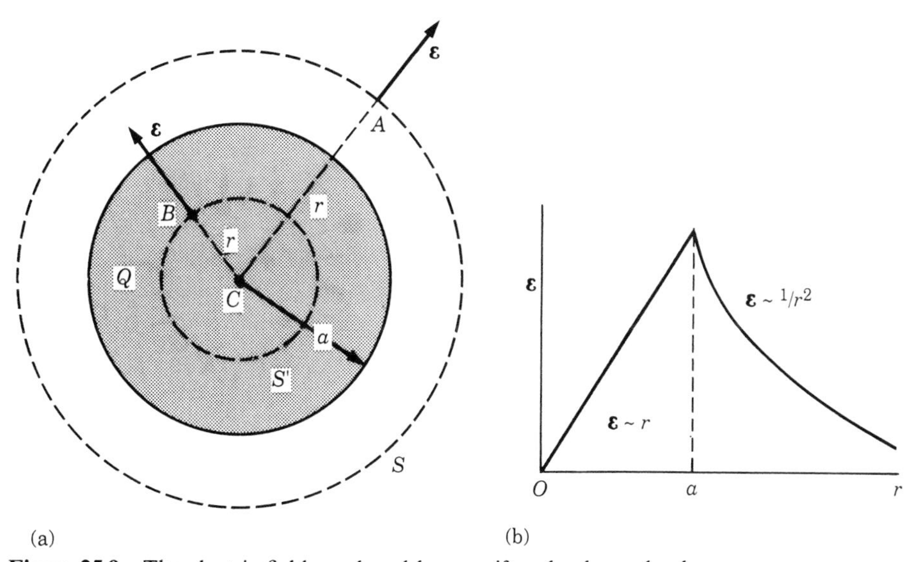

Figure 25.9 The electric field produced by a uniformly charged sphere.

But if the sphere is uniformly charged through its volume and Q' is the charge inside the surface S', we have

$$Q' = \frac{Q}{4\pi a^3/3}(4\pi r^3/3) = \frac{Qr^3}{a^3}$$

Therefore Gauss' law now gives

$$\mathscr{E}(4\pi r^2) = \frac{Q'}{\varepsilon_0} = \frac{Qr^3}{\varepsilon_0 a^3}$$

which gives the electric field as

$$\mathscr{E} = \frac{Qr}{4\pi\varepsilon_0 a^3} \qquad (r < a) \tag{25.11}$$

showing that (Figure 25.9(b))

the electric field at a point inside a uniformly charged sphere is directly proportional to the distance from the point to the center of the sphere.

These results have been mentioned already in Example 21.5. They were also obtained in Section 11.2 for a gravitational field using a direct calculation.

EXAMPLE 25.3

Electric field of a cylindrical charge distribution of infinite length.

▷ Consider a length L of cylinder C, whose radius is a (Figure 25.10(a)). If λ is the charge per unit length, the total charge in that portion of the cylinder is $q = \lambda L$. The symmetry of the problem indicates that the electric field at a point depends only on the distance from the point to the axis of the cylinder and is directed radially. We take as the closed surface

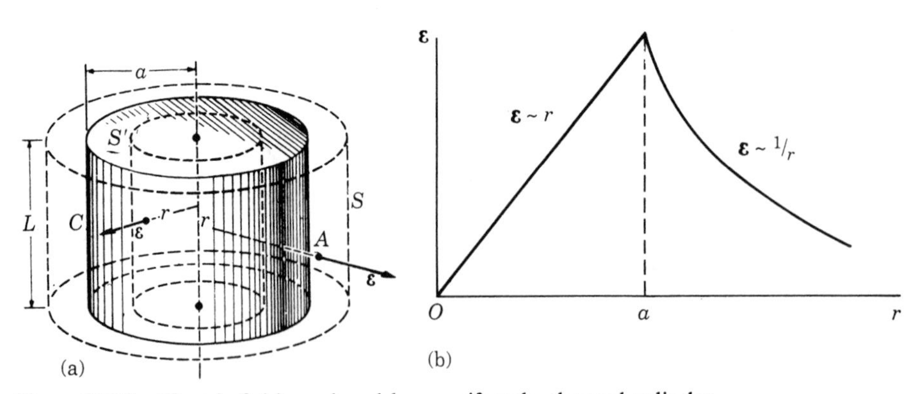

Figure 25.10 Electric field produced by a uniformly charged cylinder.

of integration a cylindrical surface of radius r, coaxial with the charge distribution. Then the electric flux through that surface has three terms. Two terms represent the flux through each base; but they are zero because the electric field is tangent to each base. Thus the flux through the lateral surface is all that remains, namely $\mathscr{E}S = \mathscr{E}(2\pi rL)$. For $r > a$ and using Gauss' law, Equation 25.8, with $q = \lambda L$, we get $2\pi rL\mathscr{E} = \lambda L/\varepsilon_0$, which gives for the electric field

$$\mathscr{E} = \frac{\lambda}{2\pi\varepsilon_0 r} \qquad (r > a) \tag{25.12}$$

Recalling Example 21.6, we conclude that

the electric field at points external to a cylindrical charge distribution of infinite length is the same as if all the charge were concentrated along the axis.

For $r < a$, we again have two possibilities. If all the charge is on the surface of the cylinder, there is no charge inside the surface S', and Gauss' law gives $2\pi rL\mathscr{E} = 0$ or $\mathscr{E} = 0$. Thus

the electric field, at points inside a cylinder charged only on its surface, is zero.

But if the charge is distributed uniformly over the volume of the cylinder C, we find the charge within the surface S' is

$$q' = \left(\frac{\lambda L}{\pi a^2 L}\right)(\pi r^2 L) = \frac{\lambda L r^2}{a^2}$$

and Gauss' law gives $2\pi rL\mathscr{E} = q'/\varepsilon_0 = \lambda L r^2/\varepsilon_0 a^2$, which gives the electric field as

$$\mathscr{E} = \frac{\lambda r}{2\pi\varepsilon_0 a^2} \qquad (r < a) \tag{25.13}$$

Thus (Figure 25.10(b))

the electric field at a point within a uniformly charged cylinder of infinite length is proportional to the distance of the point from the axis.

EXAMPLE 25.4

Electric potential and electric field in the region between two parallel planes charged at potentials V_1 and V_2 when there is a uniform charge distribution between the planes (Figure 25.11).

▷ Consider the small volume shown in Figure 25.11 of cross-section S and length $\mathrm{d}x$, and hence of volume $S\,\mathrm{d}x$. Since the electric field is parallel to the X-axis, the electric flux is zero across all sides of the volume except at those two that are parallel to the plates. Then we have

$$\text{Electric flux} = \mathscr{E}S - \mathscr{E}'S = (\mathscr{E} - \mathscr{E}')S$$

But $\mathscr{E} - \mathscr{E}' = \mathrm{d}\mathscr{E}$ because the two faces are rather close together. Therefore, the electric flux is $S\,\mathrm{d}\mathscr{E}$. On the other hand, if ρ is the charge density in the region occupied by the volume, the total charge within the volume is $\rho(S\,\mathrm{d}x)$. Applying Gauss' law,

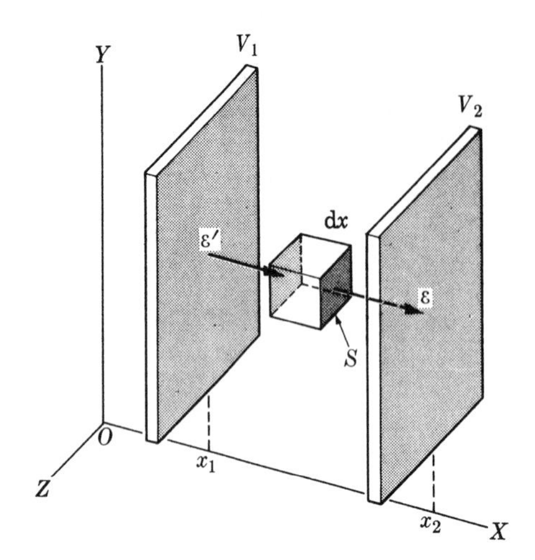

Figure 25.11 Parallel planes charged at potentials V_1 and V_2 with a uniform charge distribution between them ($V_1 > V_2$).

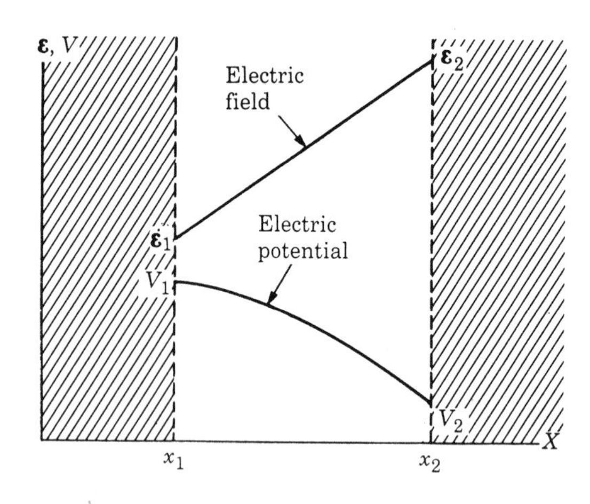

Figure 25.12 Electric field and potential in the region between the planes illustrated in Figure 25.11. The electric field corresponds to line $\varepsilon_1\varepsilon_2$ and the electric potential to line V_1V_2.

electric flux = (charge)/ε_0,

$$S\,\mathrm{d}\mathscr{E} = \frac{1}{\varepsilon_0}(\rho S\,\mathrm{d}x) \qquad \text{or} \qquad \frac{\mathrm{d}\mathscr{E}}{\mathrm{d}x} = \frac{\rho}{\varepsilon_0}$$

which relates the field at each point to the charge at the same point. Integrating this relation we have

$$\int_{\mathscr{E}_1}^{\mathscr{E}} \mathrm{d}\mathscr{E} = \frac{1}{\varepsilon_0}\int_{x_1}^{x} \rho\,\mathrm{d}x$$

where $\mathscr{E}_1$ is the electric field at $x = x_1$. When the charge is uniformly distributed ρ is constant and the field between the planes is

$$\mathscr{E} = \mathscr{E}_1 + \frac{\rho}{\varepsilon_0}(x - x_1) \tag{25.14}$$

showing that the electric field varies linearly with x, as illustrated in Figure 25.12.

25.5 Properties of a conductor placed in an electric field

Materials, such as metals, electrolytic solutions and ionized gases, contain charged particles that can move more or less freely through the medium; these materials are called **conductors**. In the presence of an electric field the free charged particles are set in motion, with the positively charged particles moving in the direction of the electric field and the negatively charged particles moving in the opposite direction. In the case of a metallic conductor the only moving particles are the negatively charged free electrons, as explained in Section 23.9. Unless properly removed, the mobile charges in a conductor, when placed in an electric field, accumulate on the surface until the field they produce within the conductor completely cancels the external applied field, thereby producing equilibrium (Figure 25.13). We conclude, then, that

> *in a conductor placed in a static electric field and which is in electrical equilibrium, the electric field at points inside is zero.*

For the same reason,

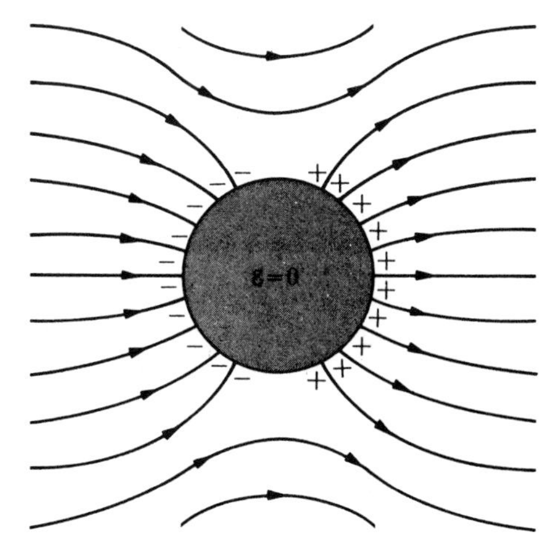

Figure 25.13 The electric field within a conductor is zero.

> *the electric field at the surface of a conductor in equilibrium is normal to the surface,*

since if there is a parallel component, the charges will move along the surface of the conductor. Furthermore, because the field inside the conductor is zero,

> *all points of a conductor that is in electrical equilibrium are at the same potential.*

If the electric field inside the conductor is zero, we have also that the electric flux across the surface of any small volume element within the conductor is zero. Therefore Gauss' law gives $q = 0$, and thus the net free charge at points within the conductor is zero. This means that

> *the entire electric charge of a conductor in equilibrium resides on its surface.*

By this statement we really mean that the net charge is distributed over the surface in a thin region having a thickness of several atomic layers, not a surface in the geometric sense.

The statements made in this section have to be considered in a statistical sense since, due to the internal structure of conducting solids, there are local fields that fluctuate from point to point, depending on the crystal lattice of the conductor (recall Section 23.11).

EXAMPLE 25.5

Electric field at the surface of a metallic conductor.

▷ Consider a conductor of arbitrary shape, as in Figure 25.14. To find the electric field at a point immediately outside the surface of the conductor, we construct a flat cylindrical surface similar to a pillbox, with one base immediately outside the surface of the conductor and the other base at a depth such that all the surface charge is within the cylinder and where the electric field is already zero. The electric flux through the pillbox is composed

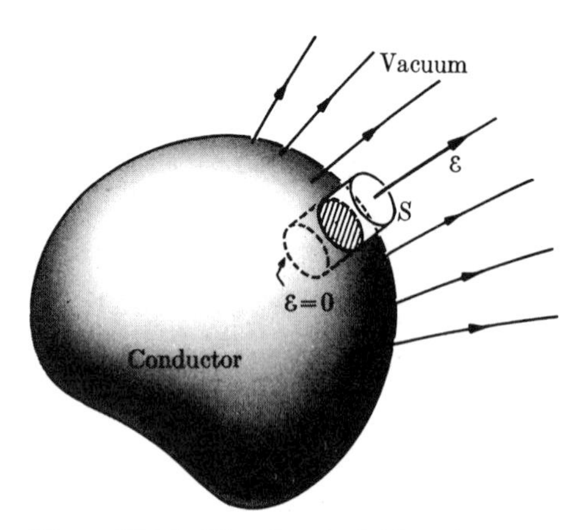

Figure 25.14 The electric field at the surface of a conductor is normal to the surface.

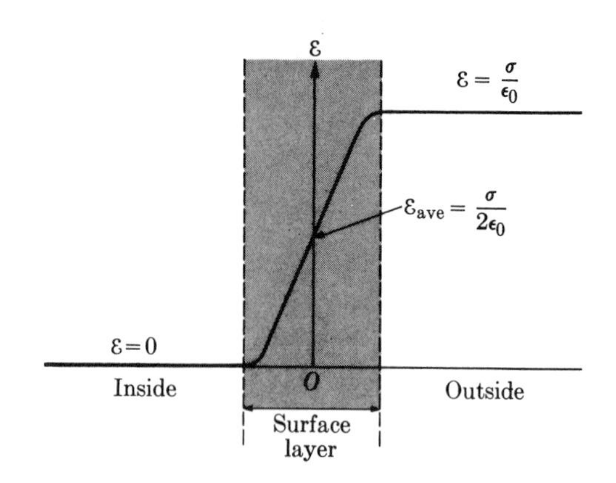

Figure 25.15 Variation of the electric field while crossing the surface of a conductor.

of three terms. The flux through the inner base is zero because the field is zero. The flux through the cylindrical side is zero because the field is tangent to this surface. Thus only the flux through the outer base remains. Given that the area of the base is S, the flux is $\mathscr{E}S$. On the other hand, if σ is the surface charge density of the conductor, the charge within the cylinder is $q = \sigma S$. Therefore, applying Gauss' law, $\mathscr{E}S = \sigma S/\varepsilon_0$, or

$$\mathscr{E} = \frac{\sigma}{\varepsilon_0} \tag{25.15}$$

This gives the electric field at a point immediately outside the surface of a charged conductor, while the field inside is zero. Therefore, as the surface of a charged conductor is crossed, the electric field varies in the way illustrated in Figure 25.15.

EXAMPLE 25.6

Force per unit area on the charges on the surface of a metallic conductor.

▷ The charges on the surface of a conductor are subject to a repulsive force due to the other charges in the conductor. The force per unit area, or electric stress, can be computed by multiplying the average electric field at the surface by the charge per unit area. The average field is, from Figure 25.15,

$$\mathscr{E}_{\text{ave}} = \tfrac{1}{2}(\mathscr{E}_{\text{inside}} + \mathscr{E}_{\text{outside}}) = \frac{1}{2}\left(0 + \frac{\sigma}{\varepsilon_0}\right) = \frac{\sigma}{2\varepsilon_0}$$

Therefore the **electric stress** is

$$F_s = \sigma\mathscr{E}_{\text{ave}} = \frac{\sigma^2}{2\varepsilon_0}$$

It is always positive, since it depends on σ^2, and therefore corresponds to a force pulling away from the conductor.

25.6 Electric polarization of matter

In isolated atoms, because of their spherical symmetry, the center of mass of the electrons coincides with the nucleus (Figure 25.16(a)). Therefore atoms do not have permanent electric dipole moments. However, when atoms are placed in an electric field, they become polarized, acquiring an *induced* electric dipole moment in the direction of the field. This dipole moment results from the perturbation of the motion of the electrons produced by the applied electric field (Figure 25.16(b)). Molecules may also acquire an induced electric dipole moment in the direction of the external field. Therefore if an insulator is placed in an electric field, its atoms or molecules become electric dipoles oriented in the direction of the applied field.

However, many molecules have a *permanent* electric dipole moment. Such molecules are called **polar**. For example, in the HCl molecule (Figure 25.17), the electron of the H atom spends more time moving around the Cl atom than around the proton of the H atom because the nucleus of Cl has a larger positive charge than that of H. Therefore the center of negative charges does not coincide with that of the positive charges, and the molecule has a dipole moment directed from the Cl atom to the H atom. That is, we may write H^+Cl^-. The electric dipole moment of the HCl molecule is 3.43×10^{-30} C m. In the CO molecule, the charge distribution is only slightly asymmetric because the nuclear charges of the C and O atoms are not very different and the electric dipole moment is relatively small, about 0.4×10^{-30} C m, with the carbon atom corresponding to the positive and the oxygen atom to the negative end of the molecule.

In a molecule such as H_2O, where the two H—O bonds are at an angle of 105° (Figure 25.18), the electrons try to crowd around the oxygen atom, which thereupon becomes slightly negative relative to the H atoms. Each H—O bond thus contributes to the electric dipole moment, whose resultant, because of symmetry, lies along the axis of the molecule and has a value equal to 6.2×10^{-30} C m. In the CO_2 molecule, all the atoms are in a straight line (Figure 25.19), and the resultant electric dipole moment is zero because of the symmetry. The values of the electric dipole moment of several polar molecules are given in Table 25.1.

In the absence of any external electric field, the dipole moments of polar molecules are in general oriented at random, and no macroscopic or collective

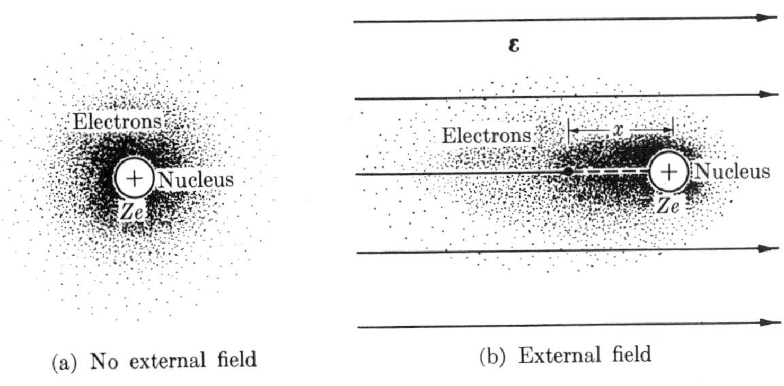

Figure 25.16 Polarization of an atom under an external electric field.

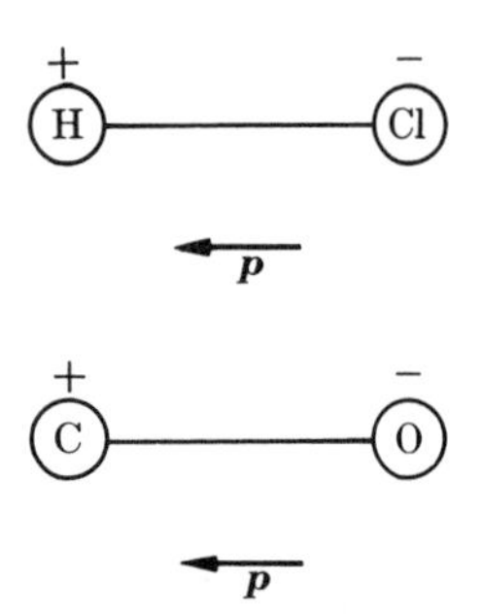

Figure 25.17 Polar diatomic molecules.

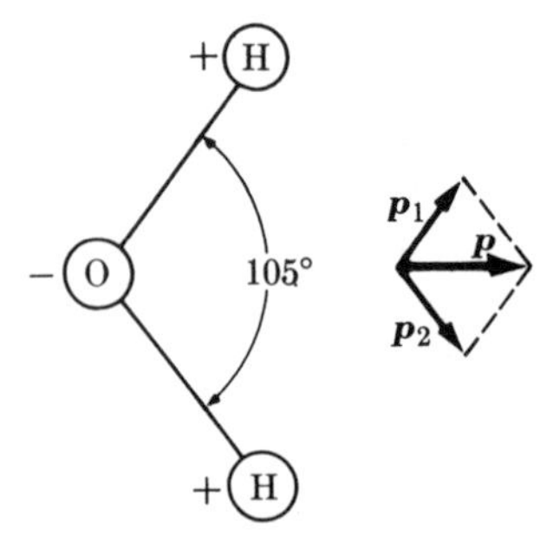

Figure 25.18 Electric dipole of H_2O molecule.

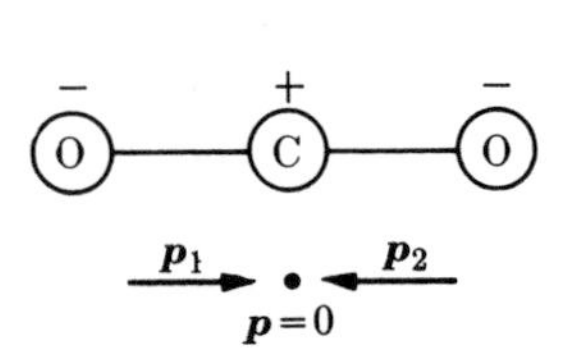

Figure 25.19 The CO_2 molecule has no electric dipole.

Table 25.1 Electric dipole moments for selected polar molecules*

Molecule	p(m C)	Molecule	p(m C)
HCl	3.43×10^{-30}	H_2S	5.3×10^{-30}
HBr	2.60×10^{-30}	SO_2	5.3×10^{-30}
HI	1.26×10^{-30}	NH_3	5.0×10^{-30}
CO	0.40×10^{-30}	C_2H_5OH	3.66×10^{-30}
H_2O	6.2×10^{-30}		

* Molecules with zero dipole moment include: CO_2, H_2, CH_4 (methane), C_2H_6 (ethane) and CCl_4 (carbon tetrachloride).

dipole moment is observed (Figure 25.20(a)). However, when a static electric field is applied, it tends to orient all the electric dipoles along the direction of the field. The alignment would be perfect in the absence of molecular interactions (Figure 25.20(b)); but the thermal molecular agitation tends to disarrange the orientation of the electric dipoles. The disarrangement is not complete because the applied electric field favors orientation along the field over orientation against it (Figure 25.20(c)). As a result, on the average, the molecules are oriented in the direction of the electric field.

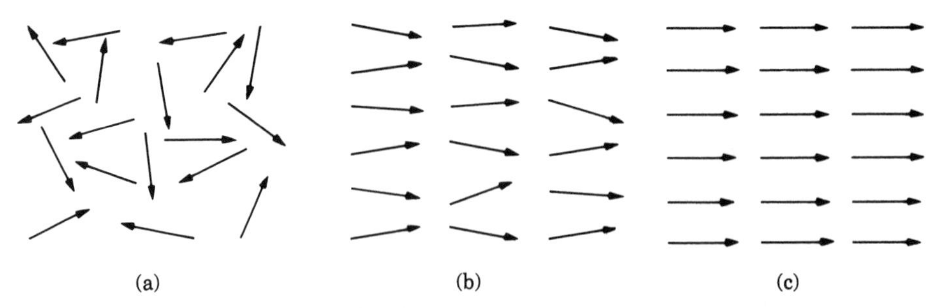

Figure 25.20 Orientation of electric dipoles in an electric field. (a) Thermal motion orients the dipoles at random. (b) An electric field produces a partial orientation. (c) At very low temperature the orientation is complete.

A special class of substances, called **ferroelectrics**, exhibits a permanent polarization, even in the absence of an external electric field; this suggests a natural tendency for the permanent dipoles of their molecules to align. The alignment results from the mutual interactions of polar molecules, which produces strong local fields that favor alignment. Among these substances we may mention $BaTiO_3$, $KNbO_3$, and $LiTaO_3$.

When the electric dipoles of a piece of matter become aligned either spontaneously or under the action of an external electric field, we say that the substance is **polarized**.

25.7 The polarization vector

We have seen that a piece of matter placed in an electric field may become electrically polarized. That is, its molecules (or atoms) become electric dipoles oriented in the direction of the local electric field, because of either the distortion of the electronic motion or the orientation of their permanent dipoles. A non-conducting medium that can be polarized by an electric field is called a **dielectric**. The polarization gives rise to a net positive charge on one side of the piece of matter and a net negative charge on the opposite side. The piece of matter then becomes a large electric dipole that *tends* to move in the direction where the field increases, as discussed in Example 21.7. This explains the phenomenon described in Section 21.1 in which an electrified glass rod or a comb attracts small pieces of paper or a cork ball.

The **polarization** $\mathscr{P}$ of a material is a vector quantity defined as the electric dipole moment of the medium per unit volume. Therefore, if $\boldsymbol{p}$ is the dipole moment induced in each atom or molecule and n is the number of atoms or molecules per unit volume, the polarization is $\mathscr{P} = n\boldsymbol{p}$. In general the polarization $\mathscr{P}$ has the same direction as the applied electric field.

Consider now a slab of material of thickness l and surface area S placed perpendicular to a uniform field $\mathscr{E}$ (Figure 25.21). The polariztion $\boldsymbol{\mathscr{P}}$, being parallel to $\boldsymbol{\mathscr{E}}$, is also perpendicular to $\boldsymbol{S}$. The volume of the slab is lS, and therefore its total electric dipole moment is $\mathscr{P}(lS) = (\mathscr{P}S)l$. But l is just the separation between the positive and negative charges that appear on the two surfaces. Since by definition the electric dipole moment is equal to charge × distance, we conclude that the total electric charge that appears on each of the surfaces is $\mathscr{P}S$, and the charge per unit area on the faces of the polarized slab is $\mathscr{P}$. Although this result has been obtained for a particular geometrical arrangement, it is of general validity, and

> *the charge per unit area on the surface of a polarized piece of matter is equal to the component of the polarization in the direction of the normal to the surface of the body.*

For example, in Figure 25.22 the charge density at A is given by $\sigma = \boldsymbol{\mathscr{P}} \cdot \boldsymbol{u}_N = \mathscr{P} \cos\theta$. Note that the polarization $\mathscr{P}$, since it is an electric dipole moment per unit volume, is measured in $(\mathrm{C\,m})\,\mathrm{m}^{-3}$ or $\mathrm{C\,m}^{-2}$, which corresponds to charge per unit area.

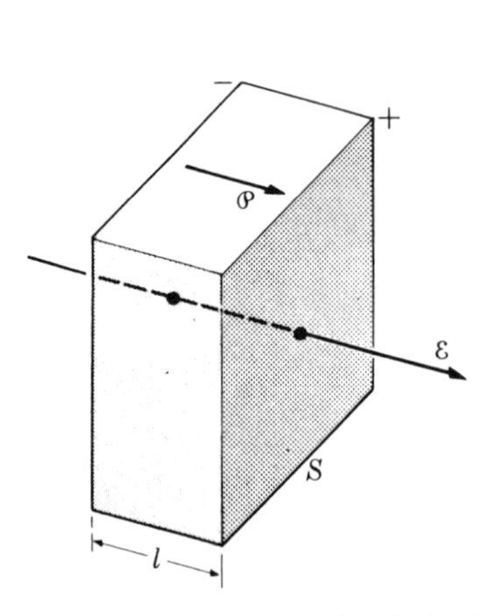

Figure 25.21 A slab of polarized material.

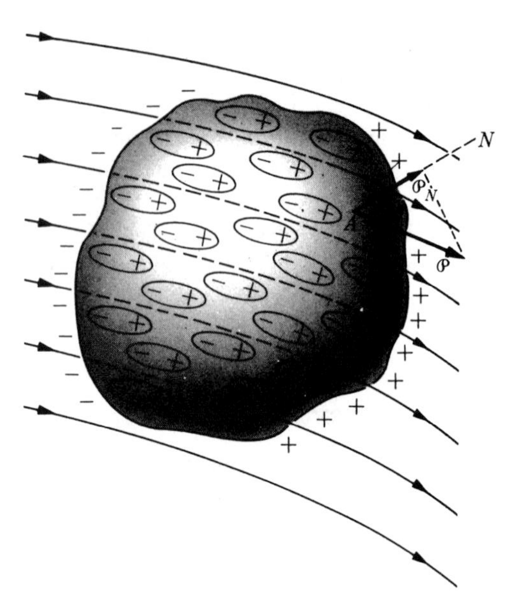

Figure 25.22 Polarization of matter by an electric field.

25.8 Electric displacement

A polarized dielectric has charges on its surface and also, unless the polarization is uniform, throughout its volume. These **polarization charges**, however, are 'frozen' in the sense that they are bound to specific atoms or molecules and are not free to move through the dielectric. In other materials, such as a metal or an ionized gas, there are electric charges capable of moving through the material, and which are therefore called **free charges**. In many instances it is desirable to make a clear distinction between free charges and polarization charges.

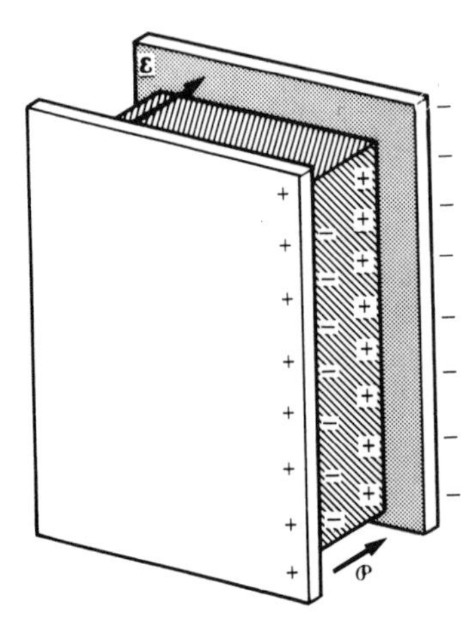

Figure 25.23 Dielectric placed between oppositely charged plates. Charges on the plates are free charges and charges on the dielectric surface are polarization charges.

Consider a slab of a dielectric material placed between two conducting parallel plates (Figure 25.23), carrying equal and opposite free charges. The surface charge density on the left-hand plate is $+\sigma_{\text{free}}$ and on the right-hand plate is $-\sigma_{\text{free}}$. These charges produce an electric field that polarizes the slab so that polarization charges appear on each surface of the slab. These polarization charges have a sign opposite to that on the plates. Therefore the polarization charges on the faces of the dielectric slab partially balance the free charges on the conducting plates. Given that $\mathscr{P}$ is the polarization in the slab, the surface charge density on the left face of the slab is $\sigma_{pol} = -\mathscr{P}$, while on the right face it is $\sigma_{pol} = +\mathscr{P}$. The effective, or net, surface charge density on the left is

$$\sigma = \sigma_{free} + \sigma_{pol} \qquad \text{or} \qquad \sigma = \sigma_{free} - \mathscr{P}$$

with a similar but opposite result on the right. These net surface charges give rise to a uniform electric field that, according to Equation 25.15, is given by $\mathscr{E} = \sigma/\varepsilon_0$. Thus, using the effective value of σ, we have

$$\mathscr{E} = \frac{1}{\varepsilon_0}(\sigma_{free} - \mathscr{P}) \qquad \text{or} \qquad \sigma_{free} = \varepsilon_0\mathscr{E} + \mathscr{P}$$

an expression that relates the free charges on the surface of a conductor surrounded by a dielectric to the electric field in the dielectric and the polarization of the dielectric. In the case we are discussing, $\boldsymbol{\mathscr{E}}$ and $\boldsymbol{\mathscr{P}}$ are vectors in the same direction, but in general their directions are different. The above result suggests the introduction of a new vector field, which is called the **electric displacement**, defined by

$$\boldsymbol{\mathscr{D}} = \varepsilon_0\boldsymbol{\mathscr{E}} + \boldsymbol{\mathscr{P}}. \tag{25.16}$$

Clearly, $\boldsymbol{\mathscr{D}}$ is expressed in $\mathrm{C\,m^{-2}}$, since those are the units of the two terms that appear on the right-hand side of Equation 25.16. In the special case we are considering, we find that $\sigma_{free} = \mathscr{D}$. That is, the free charges per unit area on the surface of the conductor are equal to the component of the electric displacement in the dielectric in a direction perpendicular to the surface. This result is of general validity and may be extended to conductors of any shape. Thus

the component of the electric displacement along the normal to the surface of a conductor embedded in a dielectric is equal to the surface charge density on the conductor.

That is,

$$\sigma_{free} = \boldsymbol{\mathscr{D}} \cdot \boldsymbol{u}_N = \mathscr{D}\cos\theta$$

The total free charge on a conductor is then

$$q_{free} = \oint_S \sigma_{free}\, dS = \oint_S \boldsymbol{\mathscr{D}} \cdot d\boldsymbol{S}$$

where the closed surface S surrounds the conductor. A more detailed analysis, which we shall omit, indicates that this equation is valid for any closed surface regardless of whether it surrounds a conductor or not. Therefore,

the flux of the electric displacement over a closed surface is equal to the total 'free' charge inside the surface:

$$\oint_S \boldsymbol{\mathscr{D}} \cdot d\mathbf{S} = q_{free} \tag{25.17}$$

In Equation 25.17 all charges due to the polarization of the medium are excluded. On the other hand, the flux of $\mathscr{E}$ is, from Equation 25.8, proportional to the *total* charge within the surface, including the polarization charges.

25.9 Electric susceptibility and permittivity

In general, the resulting polarization vector $\mathscr{P}$ is proportional to the applied electric field $\mathscr{E}$. Hence it is customary to write

$$\mathscr{P} = \varepsilon_0 \chi_e \mathscr{E} \tag{25.18}$$

The quantity χ_e is called the **electric susceptibility** of the material. It is a pure number since both $\mathscr{P}$ and $\varepsilon_0 \mathscr{E}$ are measured in C m^{-2}. For most substances, χ_e is a positive quantity. The electric susceptibility for a few materials is given in Table 25.2.

For cases where Equation 25.18 holds, we may write

$$\mathscr{D} = \varepsilon_0 \mathscr{E} + \varepsilon_0 \chi_e \mathscr{E} = (1 + \chi_e)\varepsilon_0 \mathscr{E} = \varepsilon \mathscr{E} \tag{25.19}$$

where the coefficient

$$\varepsilon = (1 + \chi_e)\varepsilon_0 \tag{25.20}$$

is called the **permittivity** of the medium, and is expressed in the same units as ε_0; that is, as $\text{m}^{-3}\,\text{kg}^{-1}\,\text{s}^2\,\text{C}^2$. The **relative permittivity** is defined as

$$\varepsilon_r = \frac{\varepsilon}{\varepsilon_0} = 1 + \chi_e \tag{25.21}$$

and is a pure number, independent of any system of units. The relative permittivity is also called the **dielectric constant**.

Table 25.2 Electric susceptibilities at room temperature

Substance	χ_e	Substance	χ_e
Solids		*Gases (at 1 atm. and 20°C)*	
Mica	5	Hydrogen	5.0×10^{-4}
Porcelain	6	Helium	0.6×10^{-4}
Glass	8	Nitrogen	5.5×10^{-4}
Bakelite	4.7	Oxygen	5.0×10^{-4}
		Argon	5.2×10^{-4}
Liquids		Carbon dioxide	9.2×10^{-4}
Oil	1.1	Water vapor	7.0×10^{-3}
Turpentine	1.2	Air	5.4×10^{-4}
Benzene	1.84	Air (100 atm)	5.5×10^{-2}
Alcohol (ethyl)	24		
Water	78		

When the relation $\mathscr{D} = \varepsilon\mathscr{E}$ holds for a medium, we may write Equation 25.17 as

$$q_{\text{free}} = \oint_S \varepsilon\mathscr{E} \cdot \mathrm{d}\boldsymbol{S}$$

and, if the medium is homogeneous so that ε is constant,

$$\oint_S \mathscr{E} \cdot \mathrm{d}\boldsymbol{S} = \frac{q_{\text{free}}}{\varepsilon} \qquad \textbf{(25.22)}$$

Comparing Equation 25.22 with Gauss' law, Equation 25.8, we see that the effect of the dielectric on the electric field $\mathscr{E}$ is to replace ε_0 by ε if only the free charges are taken into account. Therefore the electric field and potential produced by a point charge q embedded in a dielectric are

$$\mathscr{E} = \frac{q}{4\pi\varepsilon r^2}\boldsymbol{u}_{\mathrm{r}} \qquad \text{and} \qquad V = \frac{q}{4\pi\varepsilon r} \qquad \textbf{(25.23)}$$

The force and the potential energy of interaction between two point charges embedded in a dielectric are then

$$F = \frac{q_1 q_2}{4\pi\varepsilon r^2} \qquad \text{and} \qquad E_{\mathrm{p}} = \frac{q_1 q_2}{4\pi\varepsilon r} \qquad \textbf{(25.24)}$$

Since ε is in general larger than ε_0, the presence of the dielectric effectively reduces the interaction because of the screening due to the polarization of the molecules of the dielectric. This can be seen in Figure 25.24, which shows how the positive charge q surrounded by a dielectric orients the molecules of the dielectric in such a way that their negative ends point toward q, canceling it in part, and therefore its effective electric field is smaller.

The electric susceptibility χ_{e}, which describes the response of a medium to the action of an external electric field, is related to the properties of the atoms and molecules of the medium. For this reason, the electric susceptibility is different for static and oscillating electric fields. The induced electric susceptibility due to the distortion of electronic motion in atoms or molecules is essentially temperature-independent, since it is an effect related to the electronic structure of the atoms or molecules and not their thermal motion. On the other hand, the electric susceptibility due to the

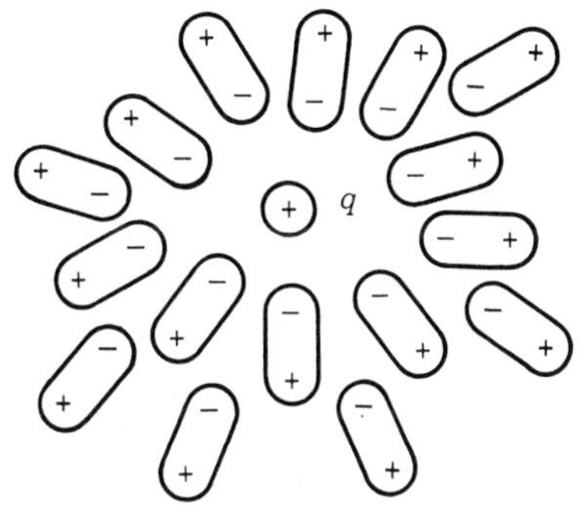

Figure 25.24 Orientation of fluid polar molecules by a charge placed in the fluid.

orientation of polar molecules varies inversely with temperature because, as the temperature increases, the thermal disorder more effectively offsets the ordering effect of the electric field, resulting in a smaller polarization of the substance.

It can be shown (and it has been confirmed experimentally) that the electric susceptibility of a substance varies with the temperature according to the expression

$$\chi_e = A + \frac{B}{T} \tag{25.25}$$

where the first term corresponds to the induced effect and the second to the orientation effect and the values of A and B are determined by the internal structure of the dielectric. Expression 25.25 constitutes **Curie's law.**

25.10 Electric capacitance: capacitors

The electric potential at the surface of a sphere containing a charge Q is $V = Q/4\pi\varepsilon_0 R$. If the sphere is surrounded by a dielectric, we have to replace ε_0 by ε, so that

$$V = \frac{Q}{4\pi\varepsilon R} \tag{25.26}$$

The relation Q/V for the sphere is then $4\pi\varepsilon R$, a constant quantity, independent of the charge Q. This is understandable because, if the potential is proportional to the charge producing it, the ratio of the two must be constant. This last statement is valid for all charged conductors of any geometrical shape. Accordingly, the electric **capacitance** of an isolated conductor is defined as the ratio of its charge to its potential,

$$C = \frac{Q}{V} \tag{25.27}$$

The capacitance of a spherical conductor of radius R, as we have just indicated, is

$$C = \frac{Q}{V} = 4\pi\varepsilon R \tag{25.28}$$

If the sphere is surrounded by vacuum instead of a dielectric, its capacitance is $C_0 = 4\pi\varepsilon_0 R$. Therefore surrounding a sphere, and in general any conductor, by a dielectric increases its electric capacitance by the factor $\varepsilon/\varepsilon_0$. This is due to the screening effect of the opposite charges that have been induced on the surface of the dielectric adjacent to the conductor. These charges reduce the effective charge of the conductor and decrease the potential of the conductor by the same factor.

The capacitance of a conductor is expressed in $\mathrm{C\,V^{-1}}$, a unit called the **farad** (abbreviated F) in honor of Michael Faraday (1791–1867). The farad is defined as the capacitance of an isolated conductor whose electric potential is one volt, after it receives a charge of one coulomb. In terms of the fundamental units, we have that $F = C\,V^{-1} = m^{-2}\,kg^{-1}\,s^2\,C^2$.

Other useful units are the microfarad (μF) $= 10^{-6}$ F, and the picofarad (pF) $= 10^{-12}$ F.

When a conductor is not isolated its capacitance is affected by the presence of the other conductors that modify its potential. Consider the case of two conductors having charges Q and $-Q$ (Figure 25.25). If ΔV is their potential *difference*, the capacitance of the system is defined as

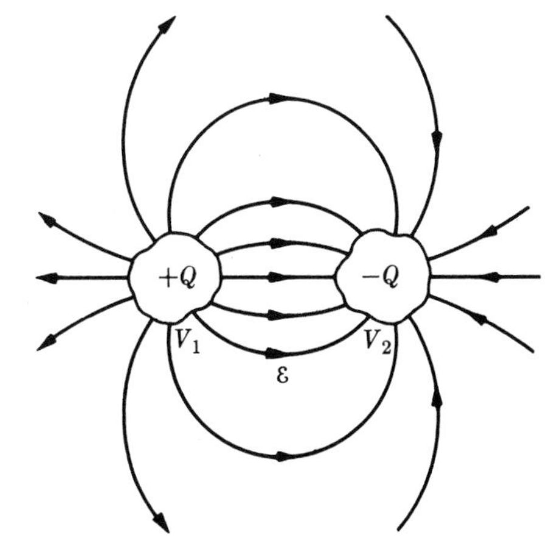

Figure 25.25 System of two conductors with equal but opposite charges.

$$C = \frac{Q}{\Delta V} \qquad \textbf{(25.29)}$$

This arrangement constitutes what is called a **capacitor**. Capacitors have wide application in electric circuits.

EXAMPLE 25.7

Plane capacitor.

▷ A typical capacitor is formed by two parallel plane conductors separated a distance d, with the space between them filled by a dielectric (Figure 25.26). If σ is the surface charge density, the potential difference between the plates, using Equation 25.9, is

$$\Delta V = V_1 - V_2 = \frac{\sigma d}{\varepsilon}$$

where ε_0 has been replaced by ε, due to the presence of a dielectric. On the other hand, if S is the area of the metal plates, we must have $Q = \sigma S$. Therefore, making the substitutions in Equation 25.29, we obtain the capacitance of the system as

$$C = \frac{\varepsilon S}{d} \qquad \textbf{(25.30)}$$

This suggests a practical means for measuring the permittivity or dielectric constant of a material. First we measure the capacitance of a capacitor with *no* material between the

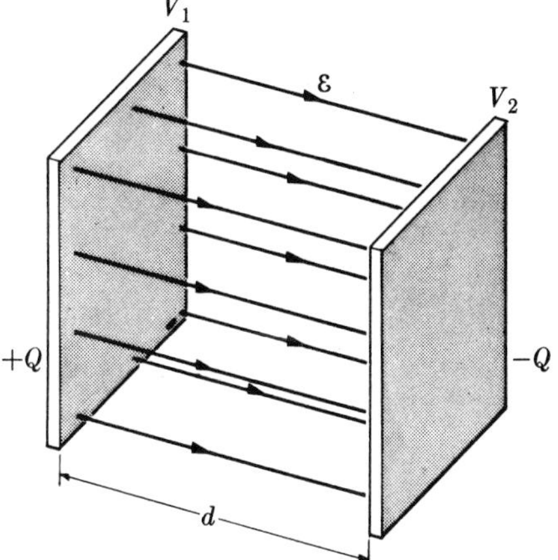

Figure 25.26 Parallel-plate capacitor. The space between the conducting plates is filled with a dielectric ($V_1 > V_2$).

plates, resulting in $C = \varepsilon_0 S/d$. Next we fill the space between the plates with the material being investigated, and measure the new capacitance, given by Equation 25.30. Then we have

$$\frac{C}{C_0} = \frac{\varepsilon}{\varepsilon_0} = \varepsilon_r$$

Therefore the ratio of the two capacitances gives the relative permittivity or dielectric constant of the material placed between the plates.

EXAMPLE 25.8

Combination of capacitors.

▷ Capacitors can be combined in two kinds of arrangement: **series** and **parallel**. In the series combination (see Figure 25.27(a)), the negative plate of one capacitor is connected to the positive of the next, and so on. As a result, all capacitors carry the same charge, positive or negative, on their plates. Call ΔV_1, $\Delta V_2, \ldots, \Delta V_n$ the potential differences across each capacitor. If C_1, C_2, ..., C_n are their respective capacitances we have that

$$\Delta V_1 = \frac{Q}{C_1}, \quad \Delta V_2 = \frac{Q}{C_2}, \ldots, \quad \Delta V_n = \frac{Q}{C_n}$$

Thus the overall potential difference is

$$\Delta V = \Delta V_1 + \Delta V_2 + \cdots + \Delta V_n = \left(\frac{1}{C_1} + \frac{1}{C_2} + \cdots + \frac{1}{C_n}\right) Q$$

The system can be equated to a single capacitor whose capacitance C satisfies the relation $\Delta V = Q/C$. Therefore

$$\frac{1}{C} = \frac{1}{C_1} + \frac{1}{C_2} + \cdots + \frac{1}{C_n} \tag{25.31}$$

which gives the resultant capacitance for a series arrangement of capacitors.

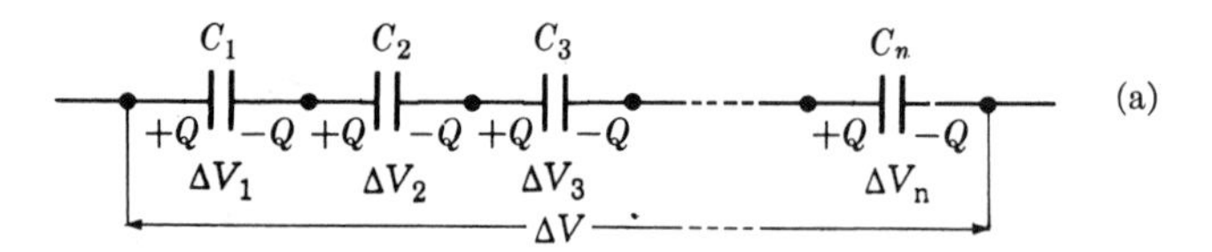

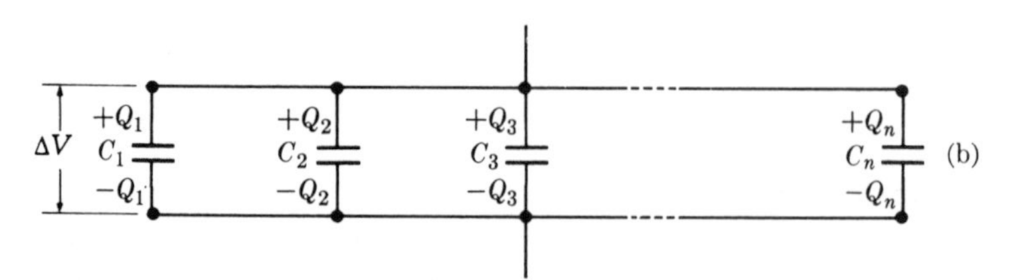

Figure 25.27 Series and parallel arrangements of capacitors.

In the parallel arrangement (Figure 25.27(b)), all positive plates are connected to a common point, and the negative plates are also connected to another common point, so that the potential difference ΔV is the same for all the capacitors. Thus, if their charges are $Q_1, Q_2, \ldots, Q_n$, we must have

$$Q_1 = C_1 \Delta V, \quad Q_2 = C_2 \Delta V, \ldots, \quad Q_n = C_n \Delta V$$

The total charge on the system is

$$Q = Q_1 + Q_2 + \cdots + Q_n = (C_1 + C_2 + \cdots + C_n)\Delta V$$

The system can be equated to a single capacitor whose capacitance C satisfies the relation $Q = C\Delta V$. Therefore

$$C = C_1 + C_2 + \cdots + C_n \tag{25.32}$$

gives the resultant capacitance for a parallel arrangement of capacitors.

Note 25.1 Charge and discharge of a capacitor

To charge a capacitor we need to apply to the plates a source of emf. Consider a circuit composed of a capacitor and a resistor in series with an applied emf V_0 (Figure 25.28). Suppose we close the circuit by pressing the switch. The source of the electromotive force produces a current I in the circuit in the direction indicated by the arrow. As a result, charges $+q$ and $-q$ appear on the plates of the capacitor. Since the rate of charge buildup on the capacitor is due to the current we have that

$$I = \frac{dq}{dt} \tag{25.33}$$

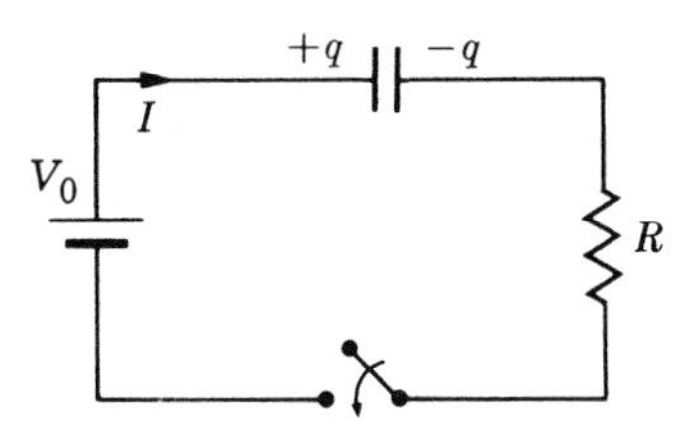

Figure 25.28 Changing a capacitator.

The charges in the capacitor produce a potential difference across the capacitor given by $\Delta V = q/C$.

To determine how the charge on the capacitor and the current vary with time when the switch is closed, we require that the combined potential drops across the resistor and the capacitor equal the applied emf so that

$$RI + \frac{q}{c} = V_0 \qquad \text{or} \qquad RI = V_0 - \frac{q}{C} \tag{25.34}$$

Using Equation 25.33 we may write Equation 25.34 as

$$R\frac{dq}{dt} = -\frac{1}{C}(q - V_0 C) \qquad \text{or} \qquad \frac{dq}{q - V_0 C} = -\frac{1}{RC}\,dt$$

Noting that at $t = 0$ the charge on the capacitor is zero ($q = 0$), we integrate and obtain

$$\int_0^q \frac{dq}{q - V_0C} = -\frac{1}{RC}\int_0^t dt$$

Evaluation of these integrals yields

$$\ln(q - V_0C) - \ln(-V_0C) = -t/RC, \qquad \text{or} \qquad \ln\left(\frac{q - V_0C}{-V_0C}\right) = -\frac{t}{RC}$$

Remembering that $\ln e^x = x$, we have

$$q - CV_0 = -CV_0e^{-t/RC} \qquad \text{or} \qquad q = V_0C(1 - e^{-t/RC}) \tag{25.35}$$

The second term in the parenthesis decreases with time and the charge builds up and asymptotically approaches V_0C (Figure 25.29(a)). The quantity $\tau = RC$ is called the **time constant** of the circuit. If τ is small, the charge reaches this value very fast, but if τ is large, it may take a long time before the capacitor is fully charged. The current in the circuit is

$$I = \frac{dq}{dt} = \frac{V_0}{R}e^{-t/RC}$$

At $t = 0$ the current has the value V_0/R given by Ohm's law, but it decreases exponentially with time until eventually it becomes zero when the potential difference across the capacitor equals the applied emf (Figure 25.29(b)).

Consider now a circuit composed of a charged capacitor and a resistor (Figure 25.30). When the switch is pressed the capacitor discharges through the resistor. Since there is no emf applied to the circuit, $V_0 = 0$ in Equation 25.34. Therefore

$$R\frac{dq}{dt} = -\frac{q}{C}$$

This expression may be written in the form

$$\frac{dq}{q} = -\frac{1}{RC}dt$$

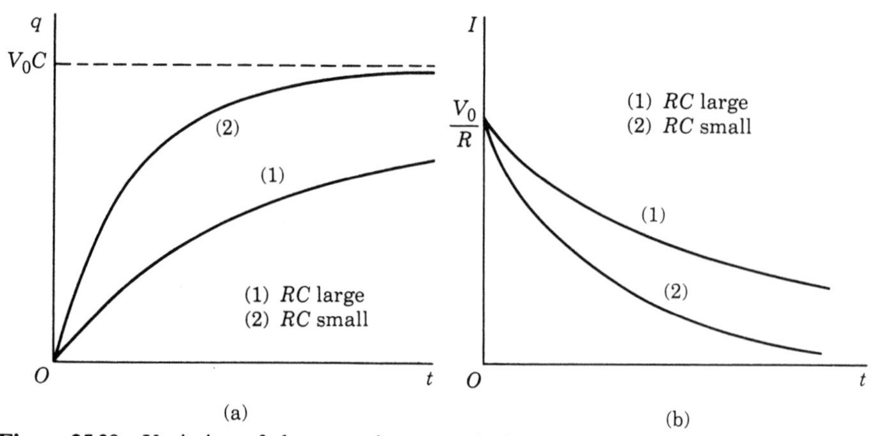

Figure 25.29 Variation of charge and current during the charging of a capacitator.

We shall denote the initial ($t = 0$) charge on the capacitor by q_0. Integrating, we have

$$\int_{q_0}^{q} \frac{dq}{dq} = -\frac{1}{RC}\int_0^t dt$$

which gives

$$\ln q - \ln q_0 = -\frac{t}{RC} \quad \text{or} \quad \ln\frac{q}{q_0} = -\frac{t}{RC}$$

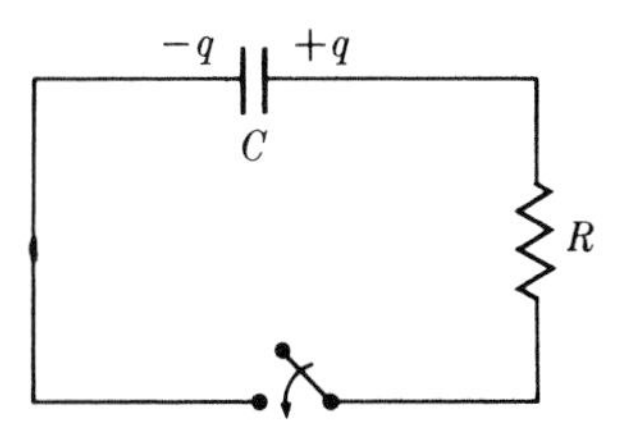

Figure 25.30 Discharge of a capacitor.

Removing logarithms, we have $q = q_0 e^{-t/RC}$. The current is given by

$$I = \frac{dq}{dt} = -\frac{q_0}{RC}e^{-t/RC} \qquad \textbf{(25.36)}$$

The charge and the current decrease exponentially, as shown in Figure 25.31.

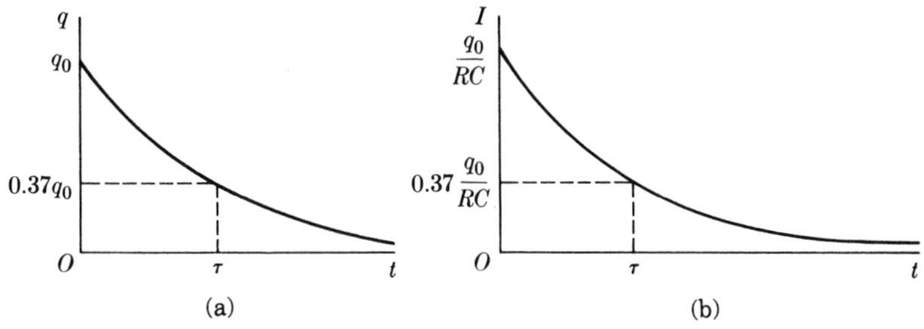

Figure 25.31 Variation of charge and current during the discharge of a capacitor.

25.11 Energy of the electric field

Charging a conductor requires expending energy because, to bring more charge to a conductor, work must be done to overcome the repulsion of the charge already present. This work results in an increase in the energy of the conductor. For example, consider a conductor of capacitance C having a charge q. Its potential is $V = q/C$. Thus if we add a charge dq to the conductor by bringing it from infinity, the work done is $dW = V\,dq$. This work is equal to the increase in energy dE of the conductor. Therefore, using the value of V, we have that

$$dE = \frac{q\,dq}{C}$$

The total increase in energy of the conductor when its charge is increased from zero to the value Q (which is equal to the work done during the process) is

$$E = \frac{1}{C}\int_0^Q q\,dq = \frac{Q^2}{2C} \qquad \textbf{(25.37)}$$

For the case of a spherical conductor charged on its surface, $C = 4\pi\varepsilon R$, and the energy is

$$E = \frac{1}{2}\left(\frac{Q^2}{4\pi\varepsilon R}\right) \tag{25.38}$$

Equation 25.37 is valid for calculating the energy of a charged conductor. However, in many instances we just need to assemble several charges, like the protons in a nucleus. It is possible to show (see Note 25.2) that the energy required to assemble a system of charges can be related to the electric field $\mathscr{E}$ produced by the charges by the expression

$$E = \tfrac{1}{2}\varepsilon \int \mathscr{E}^2 \, dV \tag{25.39}$$

where the integral extends over all the space. It can also be shown that Equation 25.39 is equivalent to Equation 25.37 for a charged conductor. This expression can be given an important physical interpretation. We may say that the energy spent in assembling the charges has been *stored* in the surrounding space, so that to the volume dV there corresponds an energy $\frac{1}{2}\varepsilon\mathscr{E}^2 \, dV$. Hence the energy per unit volume, or **energy density**, 'stored' in the electric field, is

$$\mathrm{E}_e = \tfrac{1}{2}\varepsilon\mathscr{E}^2 \tag{25.40}$$

This interpretation of the electric energy of a system of charged particles as distributed throughout all the space where the electric field is present is very useful in the discussion of many processes and can be extended to time-dependent fields.

EXAMPLE 25.9

The 'radius' of the electron.

▷ There is very little we know about the geometrical shape and internal structure, if any, of an electron. All we can say for certain is that an electron is a particle of charge $-e$ and mass m_e. It is interesting to estimate the size of the region where that charge is concentrated. To simplify our calculation, suppose that the electron is a sphere of radius R. To compute its electrical energy some assumptions must be made about how the charge is distributed over the volume of the electron. Assuming, for example, that it resembles a solid sphere of radius R and charge $-e$, charged only on its surface, we may use Equation 25.38 for the energy. Hence the electric energy of the electron, also called its **self-energy**, should be

$$E = \frac{1}{2}\left(\frac{e^2}{4\pi\varepsilon_0 R}\right)$$

We may equate this energy with the rest energy $m_e c^2$ of the electron, resulting in

$$m_e c^2 = \frac{1}{2}\left(\frac{e^2}{4\pi\varepsilon_0 R}\right) \quad \text{or} \quad R = \frac{1}{2}\left(\frac{1}{4\pi\varepsilon_0}\right)\frac{e^2}{m_e c^2} \tag{25.41}$$

However, if we adopt a different model for the charge distribution in the electron, we obtain a different numerical factor instead of $\frac{1}{2}$. For that reason, it is customary to adopt as the *definition* of the radius of the electron the quantity

$$r_e = \left(\frac{1}{4\pi\varepsilon_0}\right)\frac{e^2}{m_e c^2} = 2.8178 \times 10^{-15}\ \text{m} \tag{25.42}$$

This 'radius' is very small compared with the size of an atom, which is of the order of 10^{-10} m. Equation 25.42 can also be applied to obtain the 'radius' of a proton, substituting the proton mass for the electron mass, and a value of 1.530×10^{-18} m is obtained. We repeat that these 'radii' cannot be considered in a strictly geometrical sense, but mainly as an estimate of the size of the region where the electron or proton is 'concentrated'.

Recent evidence (Chapter 40) suggests that an electron is structureless and can be treated as a point charge. Thus, to calculate its self-energy, electromagnetic theory must be combined with quantum mechanics. On the other hand, the proton seems to be a system of three quarks and calculating its self-energy is an even more complex task.

Note 25.2 Relation between the electric field and the energy of the field

To establish the connection between Equation 25.37 and Equation 25.39, consider the simple case of a spherical conductor. The electric field produced by a charged spherical conductor at a distance r, larger than its radius, is $\mathscr{E} = Q/4\pi\varepsilon r^2$. To use Equation 25.39, we need to compute the integral of $\mathscr{E}^2$ over all the volume exterior to the sphere, but we do not need to consider the volume inside because the electric field there is zero. To obtain the volume element for the integration, we divide the outer space into thin spherical shells of radius r and thickness dr (Figure 25.32). The area of each shell is $4\pi r^2$, and therefore its volume is

$$\mathrm{d}V = \text{area} \times \text{thickness} = 4\pi r^2\ \mathrm{d}r$$

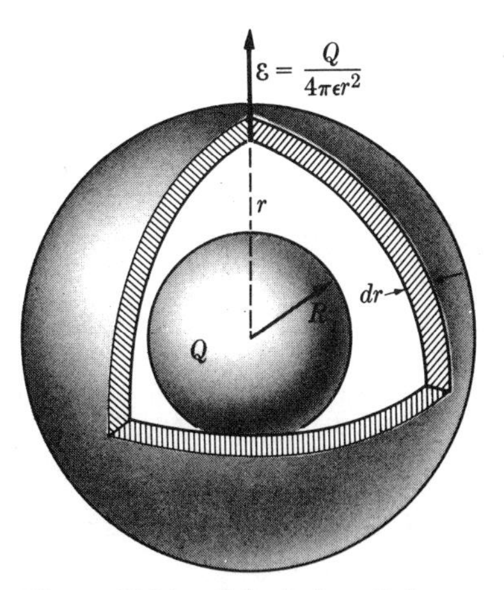

Figure 25.32 Calculation of the energy associated with the electric field of a spherical charged conductor.

Therefore we have

$$E = \tfrac{1}{2}\varepsilon\int_R^\infty \mathscr{E}^2\,\mathrm{d}V = \tfrac{1}{2}\varepsilon\int_R^\infty \left(\frac{Q}{4\pi\varepsilon r^2}\right)^2(4\pi r^2\ \mathrm{d}r) = \tfrac{1}{2}\frac{Q^2}{4\pi\varepsilon}\int_R^\infty \frac{\mathrm{d}r}{r^2} = \tfrac{1}{2}\left(\frac{Q^2}{4\pi\varepsilon R}\right)$$

Comparing this result with Equation 25.38, we see that the energy of a charged spherical conductor can be calculated using Equation 25.39. It is possible to show that Equation 25.39 is valid regardless of the geometry of the charge distribution.

QUESTIONS

25.1 Under what conditions is the flux of an electric field through a small plane surface (a) positive, (b) negative, (c) zero? Make a sketch illustrating each case.

25.2 What are the values of the electric flux through a closed surface and the circulation of an electric field along an arbitrary closed path?

25.3 A charged conductor contains an empty cavity. Show that the electric field inside the cavity is zero and that the charge on the inner surface is also zero.

25.4 Write the results of Example 25.4 if the charge density between the planes is zero. What kind of field is obtained? Compare with the results of Example 21.8.

25.5 What is the basic difference between free charges and polarization charges?

25.6 Can a substance have a negative electric susceptibility?

25.7 Why is CO a polar molecule while O_2 is not? What may be said of their structure if we are told that CH_4 (methane) has no dipole moment while CH_3Cl (methyl chloride) does?

25.8 A group of n identical capacitors are arranged in (a) series, (b) parallel. Write the capacitance of the system in each case.

25.9 Given a set of capacitors, how should they be arranged to obtain the maximum possible capacitance? The minimum possible capacitance?

25.10 Why is the capacitance of series and parallel combinations of capacitors opposite to that of resistors?

25.11 Verify that the time constant $\tau = RC$ gives the time necessary to charge a capacitor to 63% of its maximum value or to discharge it to 37% of its initial value.

25.12 How does the polarization of the dielectric surrounding a charged conductor affect (a) the effective surface charge, (b) the potential of the conductor?

25.13 What is the difference between electric field $\mathscr{E}$ and electric displacement $\mathscr{D}$?

25.14 What is the meaning of the 'radius' of an electron?

PROBLEMS

25.1 A non-conducting sphere of radius R_1 has a central cavity of radius R_2. A charge q is uniformly distributed over its volume. Find the electric field and the potential (a) outside the sphere (b) inside the sphere and (c) in the central cavity. (d) Plot the electric field and electric potential as functions of the distance from the center.

25.2 A conducting sphere of radius R_1 has a central cavity of radius R_2. At the center of the cavity there is a charge q. (a) Find the charge on the inner and outer surfaces of the conductor. (b) Compute the electric field and the potential outside the sphere, inside the sphere and in the cavity. (c) Plot the electric field and potential as functions of the distance from the center. (Hint: remember that the field inside a conductor in equilibrium is zero.)

25.3 Two conducting spheres of radii 1.0×10^{-3} m and 1.5×10^{-3} m have charges of 10^{-7} C and 2×10^{-7} C, respectively. They are placed in contact and separated. Calculate the charge on each sphere.

25.4 The permittivity of diamond is 1.46×10^{-10} $C^2\,m^{-2}\,N^{-1}$. (a) What is the dielectric constant of diamond? (b) What is the electric susceptibility of diamond?

25.5 An air capacitor consisting of two closely spaced parallel plates has a capacitance of 1000 pF. The charge on each plate is 1 μC. (a) Calculate the potential difference and the electric field between the plates. (b) Assuming that the charge is kept constant, calculate the potential difference and the electric field between the plates if the separation is doubled. (c) Calculate the work required to double the separation.

25.6 A capacitor can be made by sandwiching a sheet of paper 4.0×10^{-5} m thick between sheets of tin foil. The paper has a relative dielectric constant of 2.80 and will conduct electricity if it is in an electric field of strength $5.00 \times 10^7\ \text{V m}^{-1}$ or greater. That is, the **dielectric strength** of the paper is $50.0\ \text{MV m}^{-1}$. (a) Determine the plate area needed for a $0.30\ \mu\text{F}$ paper and foil capacitor. (b) Calculate the maximum electric potential difference that may be applied if the electric field in the paper is not to exceed one-half the dielectric strength.

25.7 Three capacitors of $1.5\ \mu\text{F}$, $2.0\ \mu\text{F}$ and $3.0\ \mu\text{F}$ are connected in (i) series, (ii) parallel, and a potential difference of 20 V is applied. For each case calculate (a) the capacitance of the system, (b) the charge on each capacitor, (c) the potential difference on each capacitor and (d) the energy of the system.

25.8 In the capacitor arrangement of Figure 25.33, the capacitors are $C_1 = 3.0\ \mu\text{F}$, $C_2 = 2.0\ \mu\text{F}$ and $C_3 = 4.0\ \mu\text{F}$. The voltage applied between points a and b is 300 V. Find (a) the charge on each capacitor, (b) the potential difference on each capacitor and (c) the energy of the system.

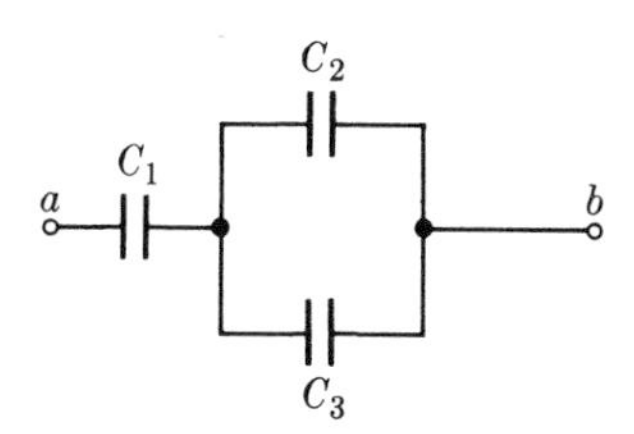

Figure 25.33

25.9 A dielectric slab is partially introduced between the two plates of a parallel plate capacitor, as shown in Figure 25.34. Calculate, as a function of x, (a) the capacitance of the system, (b) the energy of the system, and (c) the force on the slab. Assume that the potential applied to the capacitor is constant. (Hint: note that the systems may be considered as two capacitors in parallel.)

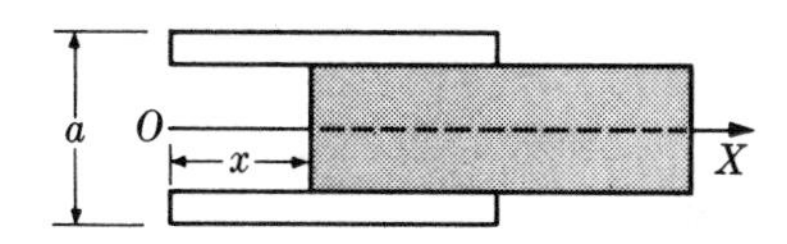

Figure 25.34

25.10 (a) Verify that the electric energy of an isolated charged conductor is $\frac{1}{2}CV^2$. (b) Also verify that the same result holds for a parallel plate capacitor and, in general, for any capacitor.

25.11 The plates of a parallel plate capacitor in vacuum have charges $+Q$ and $-Q$ and the distance between the plates is x. The plates are disconnected from the charging voltage and pulled apart a short distance dx. (a) Calculate the change in the capacity of the capacitor. (b) Calculate the change in the energy. (c) Equate the work to the increase in energy and find the force of attraction between the plates. (d) Explain why the force is not equal to $Q\mathscr{E}$, where $\mathscr{E}$ is the electric field strength between the plates.

25.12 A capacitor C_1 has an initial charge q_0. When the switch S is closed (Figure 25.35), the capacitor is connected in series with a resistor R and an uncharged capacitor C_2. (a) Show that the 'Ohm's law' equation for the circuit is given by

$$\frac{q}{C_1} + \frac{q_0 - q}{C_2} = RI$$

(b) Find q and I as functions of time.

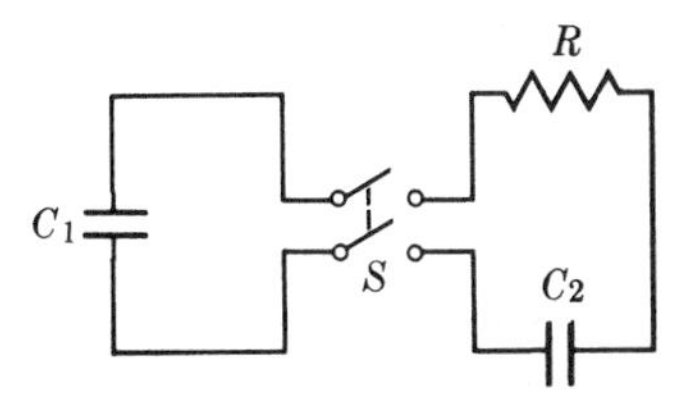

Figure 25.35

25.13 A spherical charged body of radius R has a charge Q distributed uniformly throughout its volume. What is the charge density? Suppose the body was built by aggregation, adding charge at a constant density until the radius R is reached. Show that energy of the system is $3/5\,(Q^2/4\pi\varepsilon_0 R)$. This result is useful to calculate the electric energy of an atomic nucleus (Chapter 38).

26 The magnetic field

André Marie Ampère, with Pierre Laplace, pioneered a mathematical theory of time-independent electromagnetic fields, which he called 'electrodynamics'. Shortly after H. C. Oersted's discovery, in 1820, of the magnetic field of an electric current, Ampère developed a general theory to explain complex magnetic phenomena involving electric currents in conductors of different shapes, including a method for calculating the magnetic interaction between two currents.

Notes

26.1 Introduction

Magnetic fields differ from electric fields in several respects. They are produced by charges moving relative to the observer, such as in electric currents, rather than charges at rest. Also, the lines of force of the magnetic field are closed; that is, they do not begin at one point and end at another and, in some way, they curve around the moving charges or electric current. In this chapter only static or time-independent magnetic fields will be considered.

26.2 Ampère's law for the magnetic field

Consider an infinite rectilinear current I (Figure 26.1). The magnetic field $\mathscr{B}$ at a point A at a distance r from the current is perpendicular to OA and is given by Equation 24.29,

$$\mathscr{B} = \frac{\mu_0 I}{2\pi r} \boldsymbol{u}_T$$

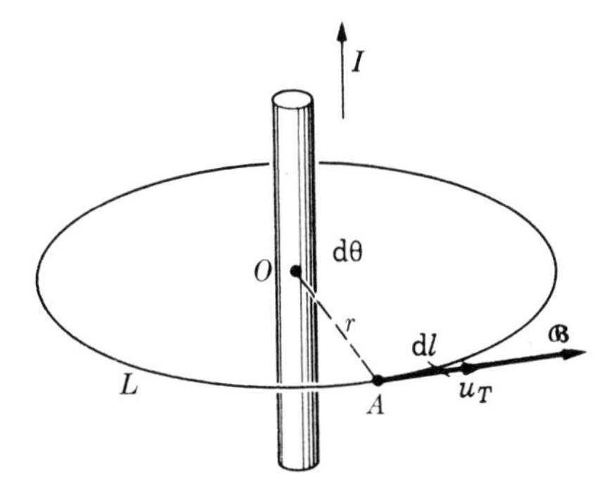

Figure 26.1 Magnetic field of a rectilinear current.

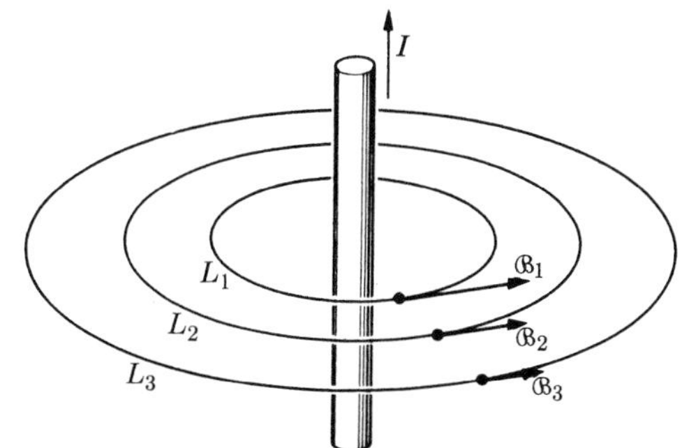

Figure 26.2 The magnetic circulation along all concentric circular paths around a rectilinear current is the same, and equal to $\mu_0 I$.

where $\boldsymbol{u}_T$ is a unit vector tangent to the circle of radius r. To compute the circulation $\oint \boldsymbol{\mathscr{B}} \cdot \mathrm{d}\boldsymbol{l}$, around the current, take a circular path of radius r that coincides with a magnetic line of force. Since the magnetic field $\boldsymbol{\mathscr{B}}$ is tangent to the path, we have that $\boldsymbol{\mathscr{B}} \cdot \mathrm{d}\boldsymbol{l} = \mathscr{B}\, \mathrm{d}l$, where $\mathrm{d}l$ is an arc element. Also, $\mathscr{B}$ is constant in magnitude along the circular line of force. Therefore the magnetic circulation, designated the **magnetomotive force** (mmf), although it is neither a force nor an energy, is

$$\text{mmf} = \oint_L \boldsymbol{\mathscr{B}} \cdot \mathrm{d}\boldsymbol{l} = \oint_L \mathscr{B}\, \mathrm{d}l = \mathscr{B} \oint_L \mathrm{d}l = \mathscr{B}L = \left(\frac{\mu_0 I}{2\pi r}\right)(2\pi r)$$

because $L = 2\pi r$. Thus

$$\oint_L \boldsymbol{\mathscr{B}} \cdot \mathrm{d}\boldsymbol{l} = \mu_0 I \tag{26.1}$$

The magnetic circulation is then proportional to the electric current I, and is independent of the radius of the path. Therefore, if we draw several circles L_1, L_2, L_3,... around the current I (Figure 26.2), the magnetic circulation around all of them is the same, and is equal to $\mu_0 I$.

A more elaborate analysis, which we shall omit here, indicates that Equation 26.1 is correct for *any* shape of the current, not necessarily only a rectilinear one, and for any shape of the closed path encircling or linking with the current. If we have several currents I_1, I_2, I_3,... linked by a closed line L (Figure 26.3), each current makes a contribution to the circulation of the magnetic field along L. Therefore we may state **Ampère's law** as

the circulation of the magnetic field or mmf along a closed line that links currents I_1, I_2, I_3,... is

$$\oint \boldsymbol{\mathscr{B}} \cdot \mathrm{d}\boldsymbol{l} = \mu_0 I \tag{26.2}$$

where $I = I_1 + I_2 + I_3 + \ldots$ stands for the total current linked by the path L.

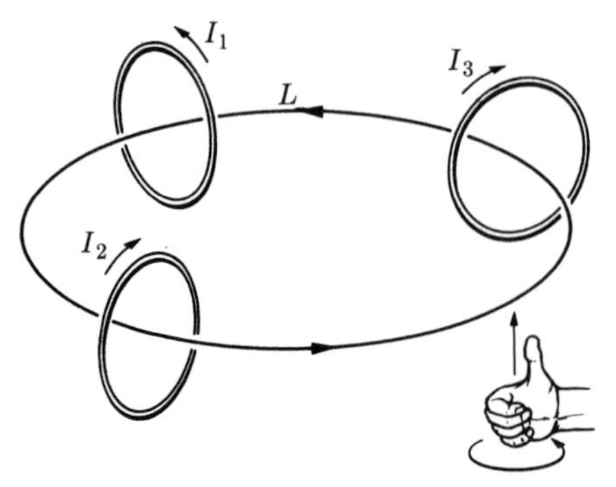

Figure 26.3 The magnetic circulation along any closed path is proportional to the net current through the path.

When we apply Equation 26.2, we take a current as positive if it passes through L in the sense indicated by the thumb when the right-hand rule is used to indicate how the path is oriented, and negative if it is in the opposite sense. Thus, in Figure 26.3, currents I_1 and I_3 are considered positive and I_2 negative. Ampère's law is particularly useful when we wish to compute the magnetic field produced by current distributions that have certain geometrical symmetries, as shown in the following examples.

EXAMPLE 26.1

Magnetic field produced by a current along a circular cylinder of infinite length.

▷ Consider a current I along a cylinder of radius a (Figure 26.4). The symmetry of the problem clearly suggests that the lines of force of the magnetic field are circles, with their centers along the axis of the cylinder, and that the magnetic field $\mathscr{B}$ at a point depends only on the distance from the point to the axis. Therefore, when we choose as our path L a circle of radius r concentric with the current, the magnetic circulation is

$$\oint_L \mathscr{B}\,\mathrm{d}l = \mathscr{B}\oint_L \mathrm{d}l = \mathscr{B}L = 2\pi r\mathscr{B}$$

If the radius r is larger than the current radius a, all the current I passes through the circle. Therefore, applying Equation 26.2, we have

$$2\pi r\mathscr{B} = \mu_0 I \qquad \text{or} \qquad \mathscr{B} = \frac{\mu_0 I}{2\pi r} \qquad (r > a) \tag{26.3}$$

This is just the result found in Chapter 24 for a current in a rectilinear filament. Therefore,

at points outside a cylindrical current, the magnetic field is the same as if all the current were concentrated along the axis.

But if r is smaller than a, we have two possibilities. If the current is only along the surface of the cylinder (as may occur if the conductor is a cylindrical sheath of metal), the current through L' is zero and Ampère's law gives $2\pi r\mathscr{B} = 0$ or $\mathscr{B} = 0$ $(r < a)$. Therefore,

the magnetic field at points inside a cylinder carrying a current on its surface is zero.

But if the current is uniformly distributed throughout the cross-section of the conductor, the current per unit area is $I/\pi a^2$, and the current through L' is

$$I' = \frac{I}{\pi a^2}(\pi r^2) = \frac{Ir^2}{a^2}$$

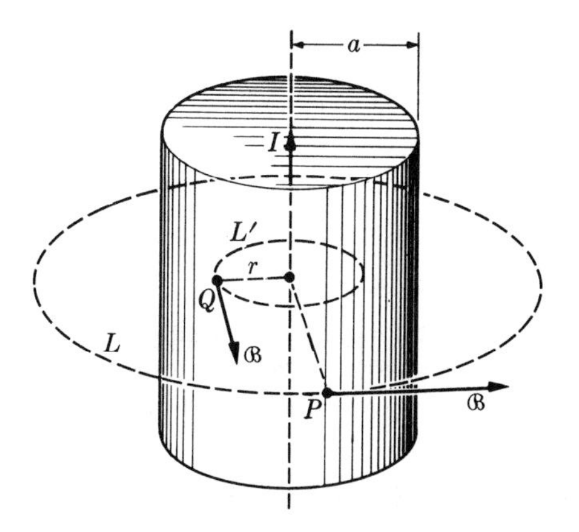

Figure 26.4 Magnetic field of a cylindrical conductor.

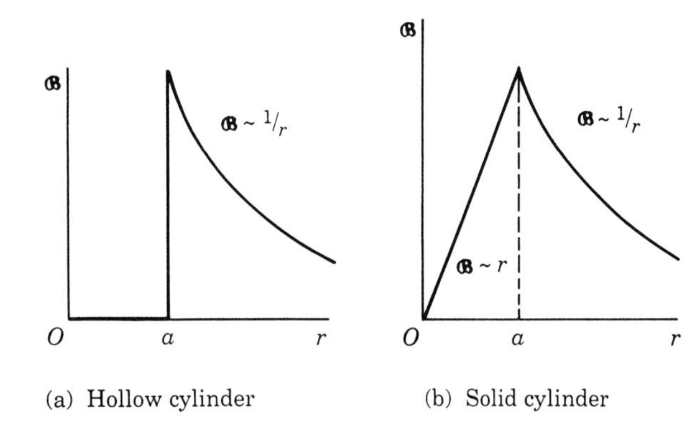

Figure 26.5 Variation of the magnetic field of current in a (a) hollow cylinder, (b) solid cylinder.

Therefore, applying Ampère's law, we get

$$2\pi r\mathscr{B} = \mu_0 I' = \frac{\mu_0 I r^2}{a^2} \quad \text{or} \quad \mathscr{B} = \frac{\mu_0 I r}{2\pi a^2} \quad (r < a) \tag{26.4}$$

Thus

> *the magnetic field at a point inside a cylinder carrying a current uniformly distributed throughout its cross-section is proportional to the distance from the point to the axis of the cylinder.*

The variation of the magnetic field with the distance for the two cases analyzed is shown in Figure 26.5.

EXAMPLE 26.2

Magnetic field produced by a toroidal coil.

▷ A toroidal coil consists of a wire uniformly wound on a torus, or doughnut-shaped surface, as in Figure 26.6. Let N be the number of turns, all equally spaced, and I be the electric current along them. The symmetry of the problem suggests that the lines of force of the magnetic field are circles concentric with the torus. Take a circle L *within* the torus as the first path of integration. The magnetic circulation is then $\mathscr{B}L$. Path L links once with all

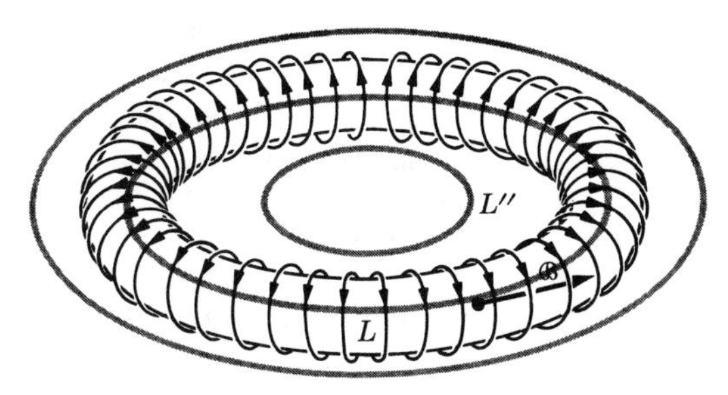

Figure 26.6 A toroidal coil.

the turns around the torus, and therefore the total current through it is NI. Thus, applying Ampère's law, we get $\mathscr{B}L = \mu_0 NI$ or $\mathscr{B} = \mu_0 NI/L$. If the cross-sectional radius of the torus is small compared with its radius, we may assume that L is practically the same for all interior paths. Given that $n = N/L$ is the number of turns per unit length, we conclude that the magnetic field inside the torus is uniform and has the constant value

$$\mathscr{B} = \mu_0 nI \tag{26.5}$$

For any path lying *outside* the torus, such as L' or L'', the total current linking with the path is zero. Therefore we get $\mathscr{B} = 0$. In other words, the magnetic field of a toroidal coil is entirely confined to its interior. This situation applies only when the turns of the wire are very closely spaced.

EXAMPLE 26.3

Magnetic field at the center of a very long solenoid.

▷ Consider the solenoid of Figure 26.7, which has n turns per unit length, carrying a current I. If the turns are very closely spaced and the solenoid is very long, the magnetic field is entirely confined to its interior and is uniform, as indicated by the lines of force in the figure, a fact confirmed experimentally. This, of course, is not true for a short solenoid.

We now apply Ampère's law to the path corresponding to the rectangle $PQRS$. The contribution of sides QR and SP to the magnetic circulation is zero because the field is perpendicular to them; also, the contribution of side RS is zero because there is no field there. Therefore, only side PQ contributes an amount $\mathscr{B}x$ to the circulation so that

$$\oint_{PQRS} \mathscr{B} \cdot \mathrm{d}\boldsymbol{l} = \mathscr{B}x$$

The total current linking with the integration path is nxI, since nx gives the number of turns in the length x. Therefore Ampère's law gives $\mathscr{B}x = \mu_0 nxI$ or $\mathscr{B} = \mu_0 nI$, which confirms the fact that the magnetic field within a long solenoid is uniform.

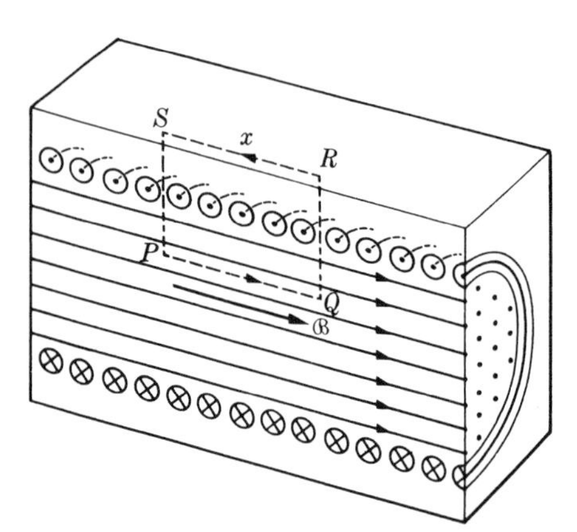

Figure 26.7 Magnetic field of a long solenoid. Only half of the solenoid is shown.

EXAMPLE 26.4

Magnetic field produced by a current along an infinite-plane conducting sheet.

▷ We will consider the sheet as composed of a very large number of filaments placed parallel to each other. Then, the resultant magnetic field may be found by adding the magnetic fields of each filament. When that is done, we find that the magnetic field must be parallel to the sheet in a direction perpendicular to the current, as shown in Figure 26.8(a), a fact that is confirmed experimentally.

However, to calculate the magnetic field, it is simpler to take advantage of the symmetry of the problem. Figure 26.8(b) shows a cross-section along line AB. The dots

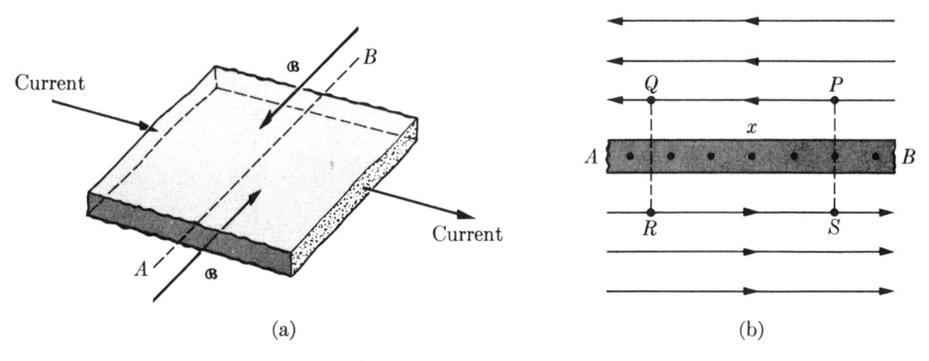

Figure 26.8 The magnetic field of a current along an infinite plane sheet is parallel to the sheet and is in opposite directions above and below.

indicate that the current is toward the reader. We consider the closed path $PQRS$ and apply Ampère's law. The contribution of sides QR and PS to the magnetic circulation is zero because the field is perpendicular to them; sides PQ and RS both contribute the same amount $\mathscr{B}x$, where $x = PQ = RS$, so that

$$\oint_{PQRS} \mathscr{B} \cdot d\boldsymbol{l} = 2\mathscr{B}x$$

If I is the current per unit width of the sheet, the total current linking with path $PQRS$ is Ix. Therefore, Ampère's law gives $2\mathscr{B}x = \mu_0 Ix$ or $\mathscr{B} = \frac{1}{2}\mu_0 I$. The magnetic field is therefore uniform since it does not depend on the distance to the plane and in opposite directions on both sides of the sheet.

Note 26.1 Magnetic confinement of a plasma

A **plasma** is a gaseous mixture of positive ions and electrons. A plasma is produced, for example, when a gas is heated to a very high temperature so that the atoms of the gas are ionized by collisions. It can also be produced by irradiating a gas with a strong source of X- or γ-rays. For the production of fusion reactions, as will be discussed in Section 38.7, it is necessary to heat plasmas up to at least 100 000 K. This poses two technical problems. One is to increase the temperature of the plasma and the other is to keep the ions away from the walls of the container. Plasma heating and containment can be achieved using magnetic fields. There are two arrangements: **linear** and **toroidal**.

In the linear arrangement (Figure 26.9), the plasma is contained in a tube through which a magnetic field is produced by a current that circulates around the tube. The coil carrying the current is designed such that the magnetic field increases sharply at the ends and is much stronger than at the center. The plasma ions describe spirals along the lines of force of the magnetic field, as shown in Figure 22.8. As the ions approach either end, the radii and pitch of the spirals decrease because of the magnetic mirror effect (see Section 22.4 and Note 26.2) and the ions are sent back toward the middle. If the current in the coils is increased, the magnetic field along the tube also increases, thus compressing the spirals of the ions. This magnetic compression increases the temperature of the plasma in the same way as the adiabatic compression of a gas.

Figure 26.9 Adiabatic magnetic compression of a plasma. Plasma is injected into the chamber while the field is weak. The magnetic field strength is then increased, compressing the plasma toward the center and raising its temperature. The magnetic mirrors may also be moved axially inward to provide additional compression and further increase in temperature.

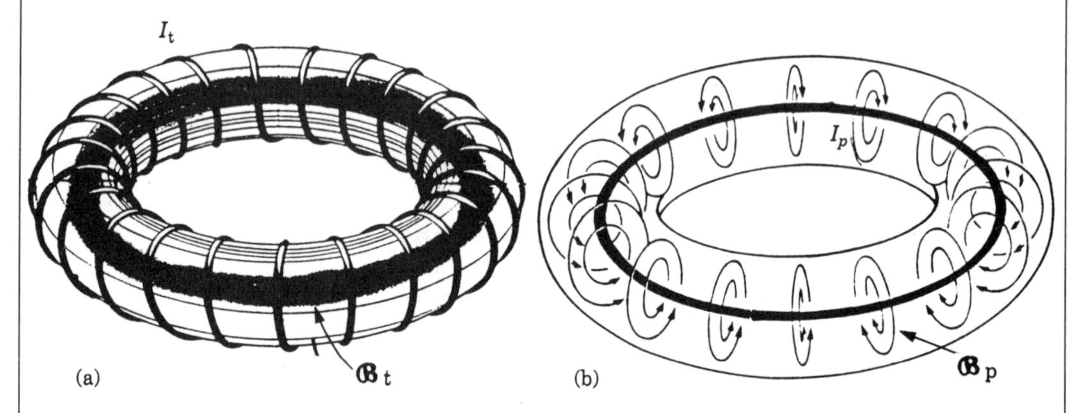

Figure 26.10 (a) Toroidal field $\mathscr{B}_t$ along the inside of the torus. (b) Poloidal or transverse magnetic field $\mathscr{B}_p$. The plasma current I_p is produced by using the electromagnetic induction effect (to be discussed in Chapter 27) and also contributes to heating the plasma. The cross section of the torus is shaped to optimize plasma containment and minimize losses.

In the toroidal arrangement two magnetic fields are used. The 'toroidal' magnetic field $\mathscr{B}_t$, produced by the current I_t, circulating around the torus (Figure 26.10(a)), is confined to the interior of the torus. A second magnetic field $\mathscr{B}_p$ (Figure 26.10(b)), called 'poloidal', is produced by a current I_p circulating in the plasma along the circumference of the torus and its closed lines of force are perpendicular to those of $\mathscr{B}_t$. The lines of force of the resultant magnetic field twist helically along the interior of the torus. When the magnetic fields increase, the plasma is compressed and its temperature increases. The most promising toroidal system is a design called **tokamak** (a Russian acronym for *toroidal magnetic chamber*), of which several are in operation in the United States, Canada, Japan, Europe and the former Soviet Union. The toroidal arrangement has the advantage of considerably reducing plasma losses that are difficult to avoid in linear systems.

26.3 Magnetic flux

The **magnetic flux** across any surface, closed or not, placed in a magnetic field is

$$\phi_m = \int_S \mathscr{B} \cdot dS \tag{26.6}$$

The concept of magnetic flux across an arbitrary surface is of great importance, especially when the surface is not closed (as we shall see in Chapter 27). The magnetic flux, being a magnetic field times an area, is expressed in T m^2, a unit called the **weber**, in honor of Wilhelm E. Weber (1804–1891). It is abbreviated Wb, so that Wb = T m^2 = m^2 kg s^{-1} C^{-1}, since T = kg s^{-1} C^{-1}.

Since there are no magnetic masses or poles (or at least they have not yet been observed), the lines of force of the magnetic field $\mathscr{B}$ are closed, as indicated in the examples discussed in Chapter 24. Therefore, if we consider a closed surface in a magnetic field, the inward magnetic flux is equal to the outward magnetic flux, since the same number of lines of force that enter the surface must come out. We then conclude that

the flux of the magnetic field through a closed surface is always zero.

$$\oint_S \mathscr{B} \cdot \mathrm{d}\boldsymbol{S} = 0 \tag{26.7}$$

This result can also be verified mathematically from the general expression for $\mathscr{B}$ given in Equation 22.13. The proof will be omitted. This result constitutes **Gauss' law for the magnetic field**. This law should be compared with Gauss' law for the electric field, Equation 25.7.

26.4 Magnetization of matter

Orbiting electrons in atoms can be treated as small magnetic dipoles, having a magnetic moment associated with both their orbital and spin angular momenta.

Atoms and molecules may or may not possess a net magnetic dipole moment, depending on their symmetry and on the relative orientation of their electronic orbits. Matter in bulk, with the exception of **ferromagnetic** substances, does not exhibit a net magnetic moment, because of the random orientation of the molecules, a situation similar to that found in the electric polarization of matter. However, the presence of an external magnetic field distorts the electronic motion, giving rise to a net magnetic polarization or **magnetization** of the material.

Substances can be grouped into several types, depending on the way in which they are magnetized by an external magnetic field. We shall briefly consider these types of magnetization.

(i) Diamagnetism. If an external magnetic field is applied to a substance, the electrons moving in the atoms or molecules are subject to an additional force due to the applied magnetic field. This results in a perturbation of the electronic motion that was discussed in Section 23.6. To evaluate this perturbation precisely, we have to use the methods of quantum mechanics, which are beyond the scope of this text (see Note 26.2). Thus we shall limit ourselves to stating the main results.

The effect of a magnetic field $\mathscr{B}$ on the electronic motion in an atom is equivalent to an additional current induced in the atom that produces a magnetic dipole moment oriented in a direction *opposite* to that of the magnetic field. This induced magnetic moment is rather small, of the order of 10^{-28} J T^{-1}. Since this effect is independent of the orientation of the atom and is the same for all atoms, we conclude that *the substance acquires a magnetization opposed to the magnetic field*, a result which is in contrast to that found in the electric field case. This behavior, called **diamagnetism**, is common to all substances, although in many cases it is masked by the paramagnetic effect described below. The possibility exists that the diamagnetic effect is so strong that the resultant magnetic field inside the substance is zero. This seems to be the case of **superconductors** (see Example 26.5).

(ii) Paramagnetism. Consider a substance whose atoms or molecules have a permanent magnetic dipole moment, which is associated with the angular momentum of its electrons and is of the order of the Bohr magneton, or 10^{-23} J T^{-1} (recall Section 23.6). In this case the presence of an external magnetic field produces a torque that tends to align the magnetic dipoles along the magnetic field, resulting in a magnetization called **paramagnetism**. In metals paramagnetism is also due to an alignment of the spin magnetic moments of the conduction electrons.

The orientation effect produced by the applied magnetic field is opposed by the disordering effect due to the thermal motion. The situation is entirely identical to that illustrated in Figure 25.20 in connection with the orientation effect produced by an electric field acting on polar molecules. The net result is that *paramagnetic substances acquire a magnetization that is in the same direction as the magnetic field.* Because permanent magnetic moments are about 10^3 times larger than the magnetic moments induced diamagnetically, diamagnetic effects are generally completely screened by the paramagnetic effects in most paramagnetic substances.

(iii) Ferromagnetism. A third class of magnetic substances is called **ferromagnetic**. The chief characteristic of ferromagnetic substances is that they exhibit a large permanent magnetization, which suggests a natural tendency for the atomic or molecular magnetic moments to align under their mutual interactions. Ferromagnetic substances are usually solids and good conductors. Ferromagnetism is thus similar to ferroelectricity in its overall behavior, although its origin is different. It is associated with a magnetic interaction between the spins of pairs of electrons. The result is a parallel orientation of electronic spins in microscopic regions called **domains** (Figure 26.11), which have dimensions of the order of 10^{-8} to 10^{-12} m^3 and contain from 10^{21} to 10^{17} atoms. The direction of magnetization of a domain depends on the crystal structure of the substance. For example, for iron, which crystallizes with a cubic structure, the directions of easy magnetization are along any one of the three axes of the cube.

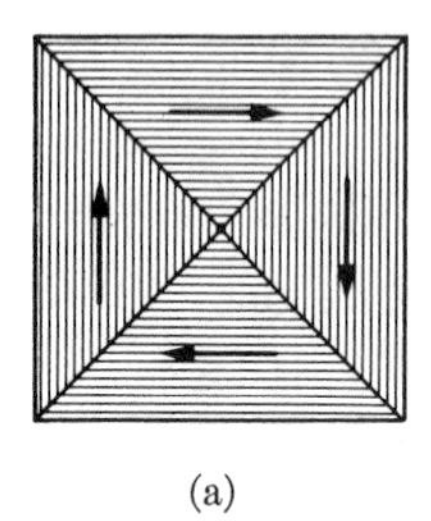

(a)

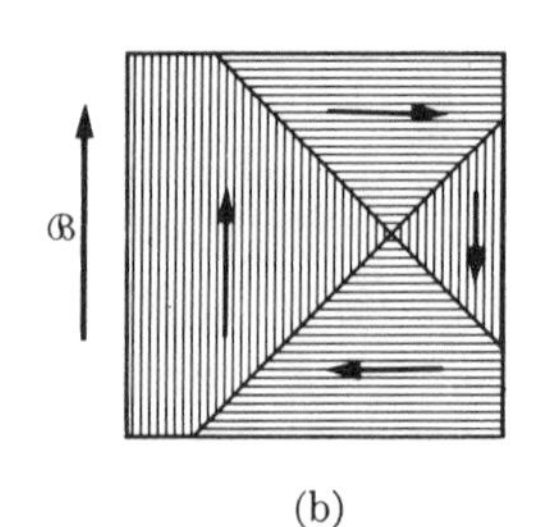

(b)

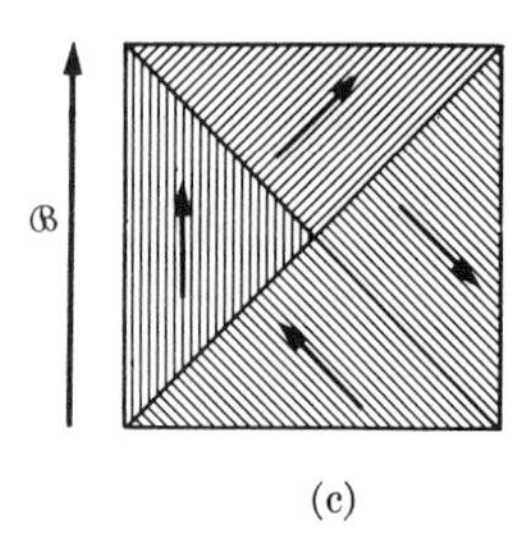

(c)

Figure 26.11 Magnetic domains. (a) Unmagnetized substances, (b) magnetization by domain growth, (c) magnetization by domain orientation.

In a piece of matter the domains themselves may be oriented in different directions, giving a net, or macroscopic, effect that may be zero or negligible. In the presence of an external magnetic field, the domains suffer two effects: those domains oriented favorably with respect to the magnetic field grow at the expense of those oriented less favorably (Figure 26.11(b)) due to a reorientation effect of the magnetic field; as the strength of the magnetic field increases, the magnetization of the domains tends to align in the direction of the field (Figure 26.11(c)), and the piece of matter becomes a **magnet**.

Ferromagnetism is a property that depends on temperature, and for each ferromagnetic substance there is a temperature, called the **Curie temperature**, above which the substance becomes paramagnetic. This transition occurs when thermal motion is great enough to offset the aligning forces due to the spin–spin interaction. Substances that are ferromagnetic at room temperature are: iron, nickel, cobalt and gadolinium. Their Curie temperatures are 770 °C, 365 °C, 1075 °C and 15 °C, respectively.

(iv) Antiferromagnetism. In some substances the electronic spins are naturally oriented in antiparallel directions, resulting in a zero net magnetization (Figure 26.12(b)). In this case the substance is called **antiferromagnetic**. Some antiferromagnetic substances are MnO, FeO, CoO and NiO.

(v) Ferrimagnetism. Another type of magnetization called **ferrimagnetism** is similar to antiferromagnetism, but the atomic or ionic magnetic moments in one direction are different from those oriented in the opposite direction, resulting in a net magnetization (Figure 26.12(c)) that is quite small. These substances are called **ferrites**, and can be generally represented by the chemical formula $XOFe_2O_3$, where X stands for Mn, Co, Ni, Cu, Mg, Zn, Cd etc. Note that if X is Fe, the compound Fe_3O_4, or magnetite, results, which is one of the most common natural magnetic materials.

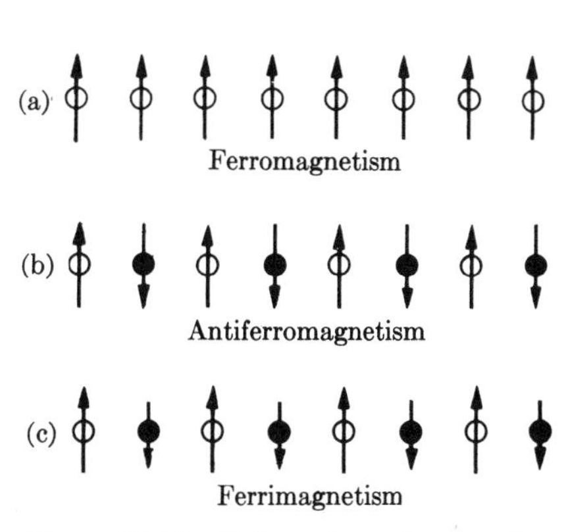

Figure 26.12 Orientation of magnetic dipole moments in various substances.

26.5 The magnetization vector

Consider, for simplicity, a substance in the form of a cylinder that is magnetized uniformly parallel to the axis of the cylinder (Figure 26.13). This means that the molecular magnetic dipoles are oriented parallel to the axis of the cylinder, and therefore the associated electronic currents are oriented perpendicular to the axis of the cylinder. We can see from Figure 26.13 (and in more detail in the front view of Figure 26.14), that the internal currents tend to cancel each other due to the contrary effects of adjacent currents. We shall assume that the cancellation is complete so that no net current results inside the substance. However, the magnetization gives rise to an equivalent net current I_{mag} on the surface of the material, which therefore behaves as a solenoid.

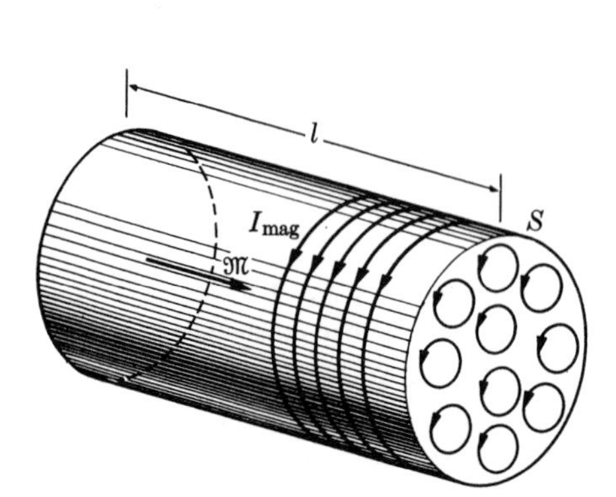

Figure 26.13 Magnetization surface current on a magnetized cylinder.

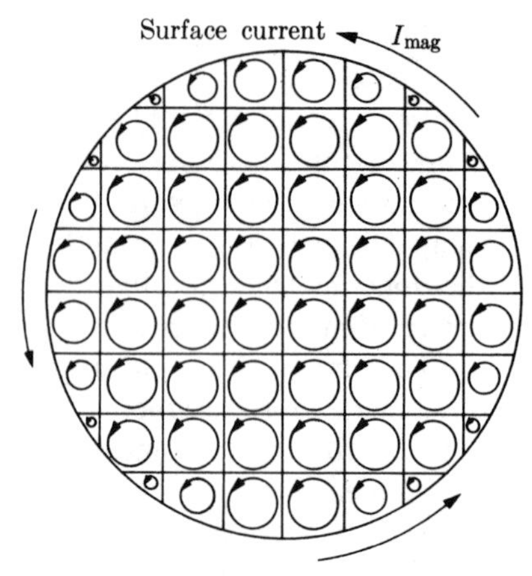

Figure 26.14 Elementary currents in a magnetized cylinder.

The **magnetization vector** $\mathcal{M}$ of a material is defined as the magnetic moment of the medium per unit volume. If $\boldsymbol{m}$ is the magnetic moment contributed by each atom or molecule and n is the number of atoms or molecules per unit volume, the magnetization is

$$\mathcal{M} = n\boldsymbol{m}$$

The magnetic moment of an elementary current is expressed in A m^2, and therefore the magnetization $\mathcal{M}$ is expressed in A m^2/m^3 = A m^{-1} or m^{-1} s^{-1} C, and is equivalent to current per unit length.

There is a relation between the equivalent surface current on the magnetized body and the magnetization $\mathcal{M}$. From Figure 26.13, note that I_{mag} has a direction perpendicular to $\mathcal{M}$. The cylinder itself behaves like a large magnet resulting from the superposition of all individual dipoles. If S is the area of the cross-section of the cylinder and l is its length, its volume is lS, and therefore its total magnetic moment is $\mathcal{M}(lS) = (\mathcal{M}l)S$. But S is just the cross-sectional area of the circuit formed by the surface current. Since magnetic moment = current × area, we may conclude that the magnetization is equivalent to an effective magnetization current $\mathcal{M}l$ on the surface of the cylinder, and therefore the effective current per unit length I_{mag} on the surface of the magnetized cylinder is $\mathcal{M}$, or

$$I_{mag} = \mathcal{M} \qquad \textbf{(26.8)}$$

Although this result has been obtained for a particular geometrical arrangement, it is of general validity. Thus we can say that

the effective magnetization current per unit length on the surface of a magnetized piece of matter is equal to the component of the magnetization vector $\mathcal{M}$ parallel to a plane tangent to the surface of the body, and it has a direction perpendicular to $\mathcal{M}$.

It must be emphasized that the magnetization current is not composed of electrons freely flowing on the surface of the substance that can be measured with a galvanometer. Rather, it is an effect due to the orientation of localized elementary currents associated with electronic motion in atoms, which together, from a magnetic point of view, amount to an effective current.

EXAMPLE 26.5

Magnetic properties of superconductors.

▷ Recall from Section 24.4 that when the temperature of a superconducting substance falls below the critical temperature T_C (Table 24.2), a phase transition occurs. One of the associated phenomena is that the substance's paramagnetism, which is associated with the orientation of the spin magnetic moment of the conduction electrons in a conductor under the action of an external magnetic field, suddenly disappears, and the substance becomes perfectly diamagnetic. This occurs because the Cooper pairs that are formed have no net magnetic moment, since the two electrons have opposite spins.

When a superconductor, initially at a temperature higher than T_C, is placed in a magnetic field, it appears paramagnetic, like any normal conductor. If the temperature is lowered below T_C, without changing the magnetic field, the paramagnetism will disappear as long as the magnetic field is sufficiently weak, so that the pairing interaction can overcome the orienting effect of the magnetic field and Cooper pairs can be formed. In the superconducting state, the applied magnetic field sets the Cooper pairs into orbital motion, similar to those currents shown in Figure 26.14. Since there is no lattice resistance to the motion of electron pairs, these microscopic induced currents increase very rapidly until a steady state is reached when the magnetic field they produce cancels the applied external magnetic field within the superconductor. The induced currents are generally known as **screening currents** or supercurrents. If the applied magnetic field is very strong, the Cooper pairs may not be formed and the material remains paramagnetic unless the temperature is lowered sufficiently to reduce the thermal energy, which also tends to disrupt Cooper pairs compensating for the orienting effect of the magnetic field.

The net average induced currents inside the superconductor also add to zero, as indicated in Figure 26.14, and only the surface screening current remains. This current, of course, also produces an external magnetic field such that the net or total magnetic field at the surface of the superconductor is tangent to the surface, as required by the continuity of magnetic lines of force. That is, no magnetic lines of force penetrate into the superconductor where the magnetic field is zero (actually, there is a small penetration for a very thin surface layer, of the order of 10^{-7} m, which may be ignored in our analysis). This phenomenon, discovered in 1933, is called the **Meissner effect** after one of its discoverers. Summarizing (Figure 26.15), when a superconductor, at a temperature lower than T_C, is in a sufficiently weak external magnetic field:

(1) the magnetic field inside the superconductor is zero;
(2) the magnetic field at the surface of the superconductor is tangent to the surface;
(3) the induced currents in the superconductor are all localized on a surface layer about 10^{-7} m thick.

Suppose now that the applied magnetic field is produced by a magnet placed in the vicinity of a superconductor in a direction perpendicular to the surface (Figure 26.16) and the temperature is lowered below T_C. Symmetry indicates that when the superconducting

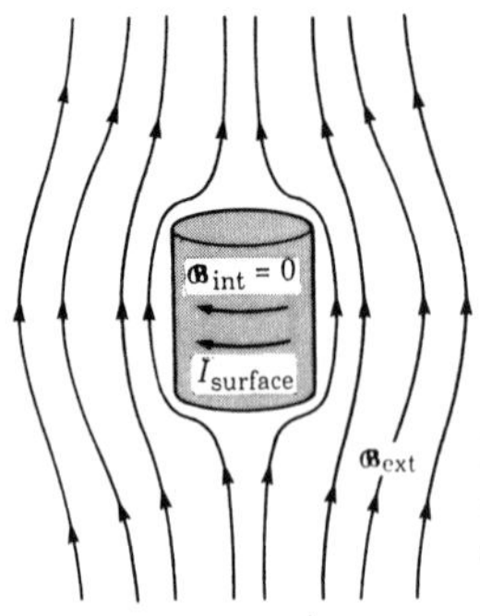

Figure 26.15 Magnetic field around a superconductor. Meissner effect.

state is reached, the screening currents in the superconductor are circular, centered at O. These currents cancel the magnetic field of the magnet inside the superconductor and produce a magnetic field outside, in the upward direction, that distorts the magnetic field of the magnet, as shown in Figure 26.17, and tends to push the magnet away. The magnet rests in the air at a position where its weight is balanced by the magnetic force from the magnetic field of the superconductor. This phenomenon, which is very easy to observe, is called **magnetic levitation**.

One way to obtain the resultant magnetic field *outside* the superconductor is to assume that, instead of the surface current, there is an imaginary magnet *inside*, located symmetrically relative to the surface (represented by dashed lines in Figure 26.17). Combining both magnetic fields outside results in a magnetic field tangent to the surface of the superconductor. This method also makes it easy to calculate the force of the superconductor on the small external magnet. When the magnet is small, it can be shown that the force is $-(\mu_0/4\pi)(3M^2/8z^4)$ (see Problem 26.13), where M is the magnetic dipole moment of the magnet and z is the distance from the surface (it has been assumed that the length of the magnet is very small compared with z). Equating this force to the weight mg of the magnet, the equilibrium position is found as

$$z = \left[\left(\frac{\mu_0}{4\pi}\right)\left(\frac{3M^2}{8mg}\right)\right]^{1/4}$$

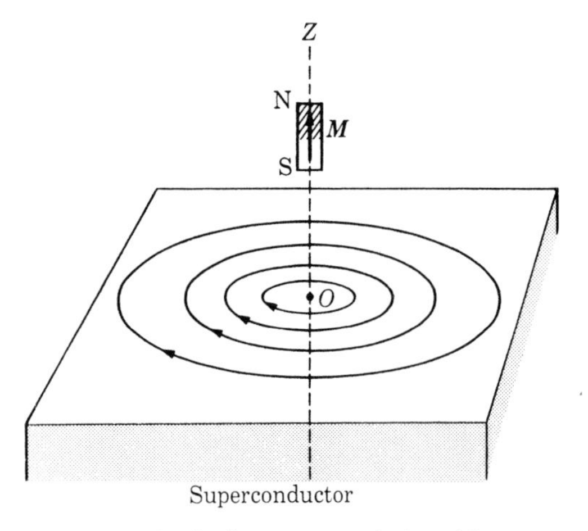

Figure 26.16 Surface currents induced by a static magnetic field in a superconductor.

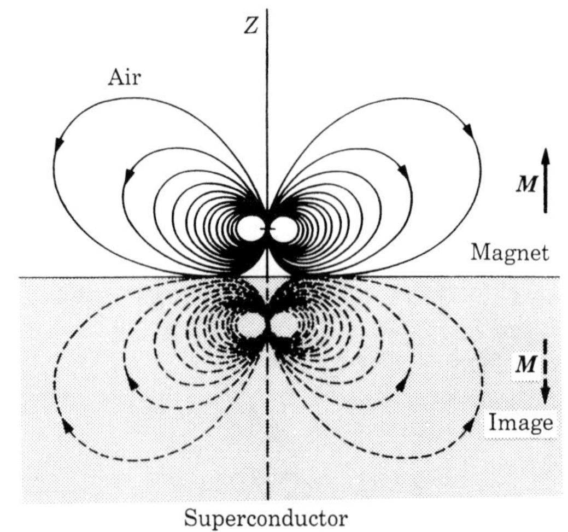

Figure 26.17 Magnetic field of a magnet above the surface of a superconductor.

Note 26.2 Magnetization of diamagnetic and paramagnetic substances

Diamagnetism. Consider a substance in which there are particles, each with charge q and mass m, in motion, such as electrons in an atom or ions in a plasma. When a magnetic field $\mathscr{B}$ is applied, the charges acquire an angular velocity around the direction of the magnetic field given by (recall Equation 22.7) $\boldsymbol{\omega} = -(q/m)\mathscr{B}$, which is identical for all charges. This is equivalent to superposing on the normal motion of the particles a circular motion in a plane perpendicular to the magnetic field. If r is the radius of this circular motion, the associated magnetic moment is (recall Equation 24.20)

$$\boldsymbol{M} = q\left(\frac{\omega}{2\pi}\right)(\pi r^2) = -\left(\frac{q^2 r^2}{2m}\right)\mathscr{B}$$

where $q(\omega/2\pi)$ is the current. To take into account the different orientations of the normal motions of the charges and that the radius is not the same for all the charges, we must replace r^2 by $\frac{1}{3}(r^2)_{\text{ave}}$, which gives

$$\boldsymbol{M} = -\left(\frac{q^2(r^2)_{\text{ave}}}{6m}\right)\mathscr{B} \tag{26.9}$$

This magnetic moment is independent of the sign of the charges. It is also antiparallel to the applied magnetic field. Therefore, the substance becomes magnetized in a direction opposed to the applied magnetic field. This is the **diamagnetic effect**, which occurs in all substances even though it may be masked by the stronger paramagnetic effect in substances having permanent magnetic moments.

For the case of atomic electrons, $q = -e$, r is of the order of the orbital radius or 10^{-10} m and m is the electron mass. Then the induced diamagnetic moment *per electron* is $M \approx 10^{-28}\,\mathscr{B}$ and the magnetization is $\mathscr{M} = nM = \approx 10^{-28}\,n\mathscr{B}$.

Similarly, the positive and negative ions in a plasma placed in a magnetic field acquire magnetic moments that are all opposite in direction to that of the magnetic field; their magnitudes are given by Equation 26.9. Therefore, the plasma becomes diamagnetic. When the magnetic field is not uniform, a new effect appears in the plasma. A magnetic dipole antiparallel to a non-uniform magnetic field is subject to a force in a direction opposite to that in which the field increases (recall Equation 22.20). Therefore, those ions whose thermal motion makes them drift in the direction in which the magnetic field increases are slowed down and eventually are sent back or 'reflected' if the gradient of the field is large enough. This is the magnetic mirror effect mentioned in Section 22.4 and Note 26.1.

Paramagnetism. Consider, for simplicity, a paramagnetic substance whose atoms have a magnetic moment that can be oriented only either parallel or antiparallel to the magnetic field. This is not unrealistic since the magnetic moment associated with the electron spin has this property. We designate the atomic magnetic moment by $\boldsymbol{M}$. Then according to Equation 24.23, the energy of the atom in a magnetic field $\mathscr{B}$ is $E = -\boldsymbol{M}\cdot\mathscr{B}$, and can have the values $E_1 = -M\mathscr{B}$ when $\boldsymbol{M}$ is oriented in the direction of $\mathscr{B}$ or $E_2 = +M\mathscr{B}$ when $\boldsymbol{M}$ is oriented in the opposite direction. This is the same situation we analyzed in Example 17.3. Therefore replacing ε by $M\mathscr{B}$ in the results of Example 17.3 we have that, if there are n atoms (or magnetic dipoles) per unit volume, the energy due to the magnetization is $E_{\text{mag}} = -nM\mathscr{B}\tanh(M\mathscr{B}/kT)$. Since $E_{\text{mag}} = -\mathscr{M}\mathscr{B}$, where $\mathscr{M}$ is the magnetization of the substance, we have that

$$\mathscr{M} = nM\tanh\frac{M\mathscr{B}}{kT} \tag{26.10}$$

Note that $M\mathscr{B}$ is the magnetic energy and kT is the thermal energy of the atoms. For $M\mathscr{B}/kT \ll 1$, we can make the approximation $\tanh M\mathscr{B}/kT \approx M\mathscr{B}/kT$ and we may write the magnetization as

$$\mathcal{M} = \frac{nM^2\mathcal{B}}{kT} \qquad \textbf{(26.11)}$$

which is **Curie's law for paramagnetic substances.** If the magnetic moment can have more orientations relative to the magnetic field, we obtain a similar result but with a different numerical coefficient. Thus Equation 26.11 expresses, in general terms, the paramagnetic effect.

For most substances M is of the order of a Bohr magneton (Equation 23.24), $\mu_B = 9.27 \times 10^{-24}\ \mathrm{J\,T^{-1}}$. Then, with $k = 1.28 \times 10^{-23}\ \mathrm{J\,K^{-1}}$, $M/k \sim 1\ \mathrm{K\,T^{-1}}$. Therefore, the condition $M\mathcal{B}/kT \ll 1$ implies $\mathcal{B}/T \ll 1\ \mathrm{T\,K^{-1}}$. This is the usual situation, since, for most magnetic fields in the laboratory, $\mathcal{B} \ll 1$ T and $T \sim 300$ K or room temperature.

In terms of order of magnitude at room temperature, Equation 26.11 gives a paramagnetic magnetization of $\mathcal{M} \sim 10^{-25}\, n\mathcal{B}$ while the magnetization of diamagnetic substances is $\mathcal{M} \sim 10^{-28}\, n\mathcal{B}$, thus confirming that the diamagnetic effect is about 10^3 times smaller than the paramagnetic for most substances.

For $M\mathcal{B}/kT \gg 1$, a situation that seldom occurs, except at extremely low temperatures, we can approximate $\tanh M\mathcal{B}/kT \approx 1$ and the magnetization becomes $\mathcal{M} = nM$, corresponding to complete orientation of the magnetic moments parallel to the magnetic field. That means the orientation effect dominates over thermal disordering.

26.6 The magnetizing field

Although a magnetized substance has certain effective magnetization currents on its surface (and through its volume if the magnetization is not uniform), these magnetization currents are 'frozen', in the sense that they are due to electrons bound to specific atoms or molecules and are not free to move through the substance. On the other hand, in certain substances such as metals, there are electric charges capable of moving through the substance. The electric currents due to these free charges, in order to make a distinction, are called **free currents.**

Consider again a cylindrical piece of matter placed inside a long solenoid that is carrying a current I (Figure 26.18). This current produces a magnetic field within the cylinder that magnetizes the cylinder and gives rise to a magnetization surface current on the cylinder in the same direction as I. According to Equation 26.8, the magnetization surface current per unit length is $I_{mag} = \mathcal{M}$. If the solenoid has n turns per unit length, the system of solenoid plus magnetized cylinder is equivalent to a single solenoid carrying a current per unit length equal to $nI + I_{mag}$ or $nI + \mathcal{M}$. This effective solenoidal current gives rise to a resultant magnetic field $\mathcal{B}$ parallel to the axis of the cylinder. The magnitude of this field is given by Equation 26.5, with nI re-

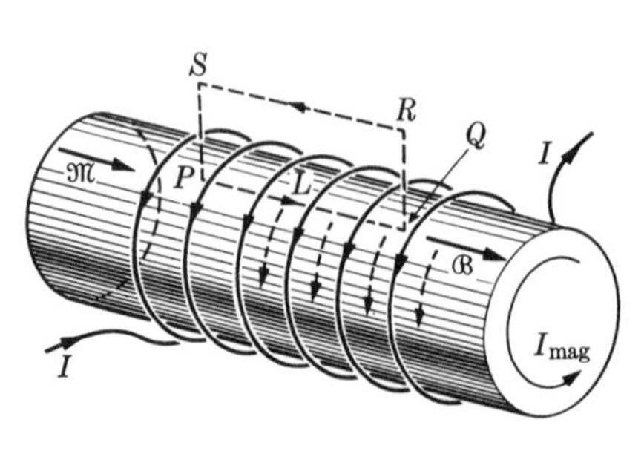

Figure 26.18 Magnetization currents in a cylinder.

placed by the total current per unit length. That is,

$$\mathscr{B} = \mu_0(nI + \mathscr{M}) \qquad \text{or} \qquad \frac{1}{\mu_0}\mathscr{B} - \mathscr{M} = nI$$

This expression relates the conduction or free current per unit length, nI, on the surface of the cylinder to the magnetic field $\mathscr{B}$ in the medium and the magnetization $\mathscr{M}$ of the medium. The above result suggests the introduction of a new field, called the **magnetizing field**, defined by

$$\mathscr{H} = \frac{1}{\mu_0}\mathscr{B} - \mathscr{M} \tag{26.12}$$

It is expressed in A m^{-1} or m^{-1} s^{-1} C, which are the units of the terms that appear on the right-hand side.

In our special example, we have that $\mathscr{H} = nI$, which relates $\mathscr{H}$ to the conduction or free current per unit length of the solenoid. When we consider a length $PQ = L$ parallel to the axis of the cylinder, we then have

$$\mathscr{H}L = LnI = I_{\text{free}} \tag{26.13}$$

where $I_{\text{free}} = LnI$ is the total free current of the solenoid corresponding to the length L. Computing the circulation around the rectangle $PQRS$, we have that

$$\oint_{PQRS} \mathscr{H} \cdot d\boldsymbol{l} = \mathscr{H}L$$

since $\mathscr{H}$ is zero outside the solenoid (both $\mathscr{B}$ and $\mathscr{M}$ are) and the sides QR and SP do not contribute to the circulation, since they are perpendicular to the magnetic field. Thus Equation 26.13 may be written in the form

$$\oint_{PQRS} \mathscr{H} \cdot d\boldsymbol{l} = I_{\text{free}}$$

where I_{free} is the total free current across the rectangle $PQRS$. This result is of more general validity than our simplified analysis may suggest. In fact, it can be verified that

> *the circulation of the magnetizing field along a closed line is equal to the total free current through the path.*

$$\oint_L \mathscr{H} \cdot d\boldsymbol{l} = I_{\text{free}} \tag{26.14}$$

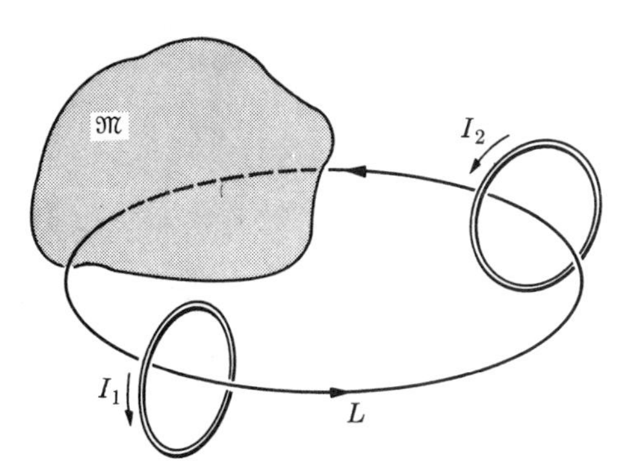

Figure 26.19

where I_{free} is the total current linking with the path L due to free-flowing charges in the medium or in an electric circuit, but excluding currents due to the magnetization of matter. For example, if the path L (Figure 26.19) links with circuits I_1 and I_2 and a body with magnetization $\mathscr{M}$, we must include only the free currents I_1 and I_2 in Equation 26.14, while in Ampère's law, Equation 26.2, for the magnetic field $\mathscr{B}$, we must include all currents; that is, I_1 and I_2, due to freely moving charges, as well as those due to the magnetization $\mathscr{M}$ of the body resulting from bound electrons.

26.7 Magnetic susceptibility and permeability

Equation 26.12 may be rewritten in the form

$$\mathscr{B} = \mu_0(\mathscr{H} + \mathscr{M}) \tag{26.15}$$

Since the magnetization $\mathscr{M}$ of the body is physically related to the resultant magnetic field $\mathscr{B}$, we could introduce a relation similar to the relation between $\mathscr{P}$ and $\mathscr{E}$, given in Equation 25.14. However, for historical reasons it is customary to proceed in a different manner, and express a relation between $\mathscr{M}$ and $\mathscr{H}$ instead, by writing

$$\mathscr{M} = \chi_m \mathscr{H} \tag{26.16}$$

The quantity χ_m is called the **magnetic susceptibility** of the material, and is a pure number independent of the units chosen for $\mathscr{M}$ and $\mathscr{H}$. The magnetic susceptibility χ_m, like the electric susceptibility χ_e, expresses the response of a medium to an external magnetic field and is related to the properties of the atoms and molecules of the medium.

Substituting Equation 26.16 into Equation 26.15, we may write

$$\mathscr{B} = \mu_0(\mathscr{H} + \chi_m \mathscr{H}) = \mu_0(1 + \chi_m)\mathscr{H} = \mu\mathscr{H} \tag{26.17}$$

where

$$\mu = \mu_0(1 + \chi_m) \tag{26.18}$$

is called the **permeability** of the medium, and is expressed in the same units as μ_0; that is, m kg^{-2} C. The **relative permeability** is defined by

$$\mu_r = \frac{\mu}{\mu_0} = 1 + \chi_m, \tag{26.19}$$

and is a pure number independent of the system and units. In Table 26.1 the magnetic susceptibility of several paramagnetic and diamagnetic substances is given. For both paramagnetic and diamagnetic substances, χ_m is very small compared with unity, and in many instances we may replace $\mu_r = 1 + \chi_m$ by one.

The magnetic susceptibility of diamagnetic substances is practically independent of temperature. But the magnetic susceptibility of paramagnetic substances is approximately inversely proportional to the absolute temperature of

Table 26.1 Magnetic susceptibilities at room temperature

Diamagnetic substances	χ_m	Paramagnetic substances	χ_m
Hydrogen (1 atm)	-2.1×10^{-9}	Oxygen (1 atm)	2.1×10^{-6}
Nitrogen (1 atm)	-5.0×10^{-9}	Magnesium	1.2×10^{-5}
Sodium	-2.4×10^{-6}	Aluminum	2.3×10^{-5}
Copper	-1.0×10^{-5}	Tungsten	6.8×10^{-5}
Bismuth	-1.7×10^{-5}	Titanium	7.1×10^{-5}
Diamond	-2.2×10^{-5}	Platinum	3.0×10^{-4}
Mercury	-3.2×10^{-5}	Gadolinium chloride ($GdCl_3$)	2.8×10^{-3}

the substance. This experimental result is a consequence of the increase in thermal disorder of the atoms of the substance as the temperature increases, which tends to offset the ordering effect of the applied magnetic field. Hence, ignoring the diamagnetic effect, we may write for a paramagnetic substance

$$\chi_m = \frac{C}{T} \tag{26.20}$$

a result similar to that for electric susceptibility (Equation 25.25), and also known as **Curie's law**.

When the relation $\mathscr{B} = \mu\mathscr{H}$ holds, we may write, instead of Equation 26.14,

$$\oint_L \frac{1}{\mu}\mathscr{B} \cdot \mathrm{d}\boldsymbol{l} = I_{\text{free}}$$

If the medium is homogeneous, so that μ is constant, the circulation of the magnetic field is then

$$\oint_L \mathscr{B} \cdot \mathrm{d}\boldsymbol{l} = \mu I_{\text{free}} \tag{26.21}$$

This result is similar to Ampère's law, Equation 26.2, but with the total current replaced by the free current and μ_0 by μ. We may then conclude that the effect of magnetized matter on the magnetic field $\mathscr{B}$ is to replace μ_0 by μ. For example, the magnetic field of a rectilinear current I embedded in a medium is

$$\mathscr{B} = \frac{\mu I}{2\pi r} \tag{26.22}$$

instead of the value given by Equation 24.29, which holds when the current is in a vacuum.

EXAMPLE 26.6
Magnetization of ferromagnetic materials.

▷ When the substance is ferromagnetic, the relation between $\mathscr{B}$ and $\mathscr{H}$ is more complex. Suppose, as a concrete example, that we have a sample of a ferromagnetic substance in the shape of a torus, with a coil wound around the torus. Assume we start with an unmagnetized substance and no current in the coil. As the current in the coil increases slowly, the magnetizing field $\mathscr{H}$ produced in the substance increases, according to Equation 26.14. The corresponding magnetic field $\mathscr{B} = \mu_0(\mathscr{H} + \mathscr{M})$ in the substance varies as indicated by the curve OA in Figure 26.20, so that the ratio $\mathscr{B}/\mathscr{H}$ is not constant. This result is explained as follows. When $\mathscr{H}$ is very small, the orienting effect on the domains is very small. But when $\mathscr{H}$ exceeds a certain value, the domains begin to change their random orientation, reorienting themselves along the magnetizing field $\mathscr{H}$. This results in a rapid increase of $\mathscr{B}$ due to an increase in the magnetization $\mathscr{M}$ of the substance. At a certain value of $\mathscr{H}$, all domains are practically reoriented and no further magnetization is possible. The substance is said to be **saturated** and $\mathscr{M}$ has reached its maximum value. Any further increase in $\mathscr{B}$ is solely due to an increase in $\mathscr{H}$ (or in the current of the coil).

If, at a point such as B, the field $\mathscr{H}$ is decreased (by decreasing the current in the coil), the substance does not retrace curve OB but instead follows curve BC because the magnetization does not decrease in the same proportion. When $\mathscr{H}$ becomes zero because the current is zero, the field $\mathscr{B}$ retains the value OC, indicative of a **residual magnetization** of the substance. This result can be explained by the fact that when the current is zero the thermal motion is not strong enough to offset completely the regular orientation of the domains that existed at B. However, if the substance is heated, the residual magnetization rapidly disappears.

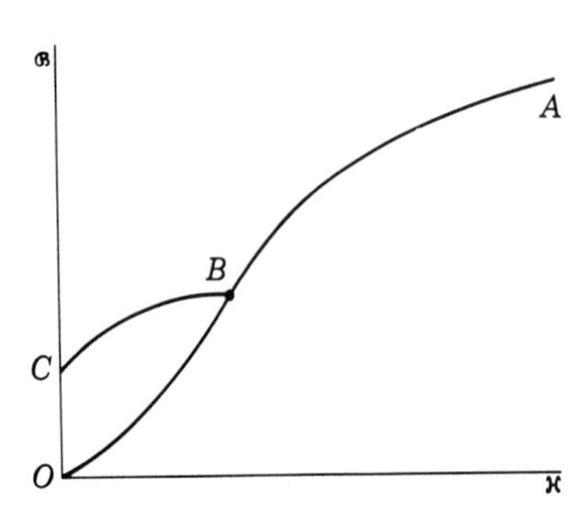

Figure 26.20 Magnetization curve of a ferromagnetic substance.

26.8 Energy of the magnetic field

To accelerate a particle of mass m and charge q from rest until it reaches a velocity $\boldsymbol{v}$ and a kinetic energy $E_k = \frac{1}{2}mv^2$ relative to an observer, assuming $v \ll c$, it is necessary to perform a work W equal to E_k. But at the same time, as the particle gains velocity, it builds up a magnetic field that (as long as $v \ll c$) is given by Equation 24.27, namely

$$\mathscr{B} = \frac{\mu_0}{4\pi} \frac{q\boldsymbol{v} \times \boldsymbol{u}_r}{r^2} \tag{26.23}$$

where the quantities are as shown in Figure 26.21. It seems natural to say that the energy of a moving charge is 'stored' in its magnetic field.

In fact it can be shown that the energy of a magnetic field is given by the expression

$$E_{\text{mag}} = \frac{1}{2\mu}\int \mathscr{B}^2 \, \mathrm{d}V \qquad \textbf{(26.24)}$$

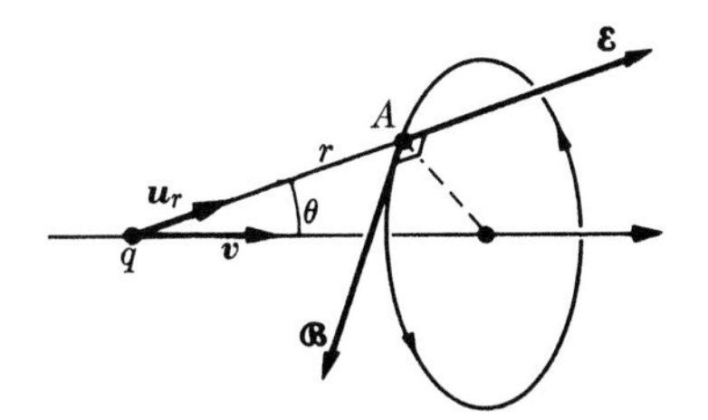

Figure 26.21 Electric and magnetic fields produced by a moving charge.

where the integral extends throughout all the volume in which the magnetic field exists, and $\mathrm{d}V$ is a volume element.

We may interpret expression 26.24 by saying that the energy spent in setting the charge in motion has been *stored* in the surrounding space so that there corresponds, to a volume $\mathrm{d}V$, an energy $(\mathscr{B}^2/2\mu)\,\mathrm{d}V$, and the energy per unit volume stored in the magnetic field is

$$\mathrm{E}_{\text{mag}} = \frac{1}{2\mu}\mathscr{B}^2 \qquad \textbf{(26.25)}$$

This is similar to the situation found in Section 25.11 for the energy of the electric field.

It may be shown that the energy of the magnetic field associated with a moving charge is (see Note 26.3)

$$E_{\text{mag}} = \frac{1}{2}\left(\frac{\mu_0}{4\pi}\right)\left(\frac{2q^2}{3R}\right)v^2 \qquad \textbf{(26.26)}$$

Equation 26.26 shows that the magnetic energy is proportional to the square of the velocity, v^2, of the electric charge. This result was to be expected since the magnetic field is proportional to the velocity of the charge. Therefore the magnetic energy is of the same type as the kinetic energy of a particle of mass

$$m = \left(\frac{\mu_0}{4\pi}\right)\left(\frac{2q^2}{3R}\right) \qquad \textbf{(26.27)}$$

For an electron, we replace q by $-e$ and use $c^2 = 1/\varepsilon_0\mu_0$ to eliminate μ_0. The result is

$$m_{\mathrm{e}} = \left(\frac{1}{4\pi\varepsilon_0}\right)\frac{2e^2}{3Rc^2} \qquad \text{or} \qquad R = \left(\frac{1}{4\pi\varepsilon_0}\right)\frac{2e^2}{3m_{\mathrm{e}}c^2}$$

which are of the same order of magnitude as the values obtained in Example 25.9 for the mass and radius of the electron in connection with the electric energy of the electron.

Therefore, it seems plausible to think that the rest energy of a charged particle is associated with the energy of its electric field, while the kinetic energy of the particle corresponds to the energy of the magnetic field. Further, it is logical to think that the fields associated with the other interactions existing in nature may also contribute to the rest and kinetic energies of a particle. However, our incomplete knowledge of those interactions makes it impossible to give a definite answer at this moment.

Note 26.3 Energy of the magnetic field of a slowly moving charge

A slowly moving charge produces a magnetic field whose lines of force are circles perpendicular to the direction of motion and whose magnitude is obtained from Equation 26.23 as

$$\mathscr{B} = \frac{\mu_0}{4\pi} q \frac{v \sin\theta}{r^2}$$

Substituting in Equation 26.24 with μ replaced by μ_0, the magnetic energy of the charge is

$$E_{\text{mag}} = \frac{1}{2}\frac{\mu_0}{(4\pi)^2} q^2 v^2 C$$

where

$$C = \int \frac{\sin^2\theta}{r^4} \, \mathrm{d}V$$

and the integral is extended over all the volume *external* to the charge, assuming that the magnetic field inside the charge is zero and the radius of the charge is R. To evaluate the integral, we shall take as our volume element the ring illustrated in Figure 26.22. It has a perimeter equal to $2\pi r \sin\theta$, and its cross-section has sides $\mathrm{d}r$ and $r\,\mathrm{d}\theta$, and therefore an area $r\,\mathrm{d}r\,\mathrm{d}\theta$. The volume is

$$\mathrm{d}V = \text{perimeter} \times \text{cross-section} = 2\pi r^2 \sin\theta \, \mathrm{d}r \, \mathrm{d}\theta.$$

Therefore, substituting this value for $\mathrm{d}V$ in the above equation, we have

$$C = \int_R^\infty \int_0^\pi \left(\frac{\sin^2\theta}{r^4}\right) 2\pi r^2 \sin\theta \, \mathrm{d}r \, \mathrm{d}\theta = \int_R^\infty \frac{\mathrm{d}r}{r^2} \int_0^\pi \sin^3\theta \, \mathrm{d}\theta = \frac{8\pi}{3R}$$

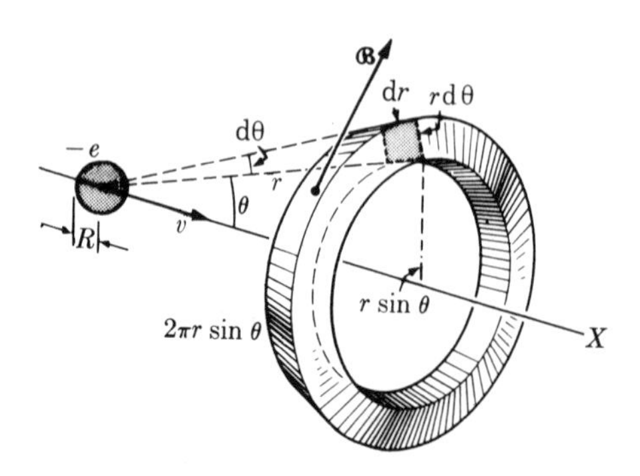

Figure 26.22

Inserting this result in the expression for E_{mag} we obtain

$$E_{\text{mag}} = \frac{1}{2}\left(\frac{\mu_0}{4\pi}\frac{2q^2}{3R}\right) v^2$$

for the energy of the magnetic field of the moving charge. This result depends on the model we have chosen for the charge distribution. To obtain the *total* magnetic energy, we would have to add the contribution due to the magnetic field *inside* the charged particle, which in turn requires that we know the charge distribution inside the particle. In any case, the above result gives an estimate of the order of magnitude. The most interesting feature is that E_{mag} depends on v^2, and therefore it resembles the kinetic energy of a particle.

26.9 Summary of the laws for static fields

In Chapter 25 and this chapter, we have discussed the static electric and magnetic fields as two separate entities. The sources of the electric field are electric charges at rest relative to the observer and the sources of the magnetic field are electric charges in motion or electric currents. Hence the static fields obey two separate sets of fundamental equations, which appear in Table 26.2. These equations allow us to compute the electric and magnetic fields if the charges and currents are known, and conversely. However, electric and magnetic fields are not independent fields and it can be shown that the electric and magnetic fields produced by the same set of charges, as measured by two observers in uniform relative motion, are related by a Lorentz transformation similar to that for space and time or energy and momentum. Thus the magnitude and direction of the electric and magnetic components of an electromagnetic field depend on the relative motion of the observer with the sources of the field.

Table 26.2 Equations of the static electromagnetic field

1. Gauss' law for the electric field Equation 25.6	$\oint_S \mathscr{E} \cdot dS = \frac{q}{\varepsilon_0}$	5. Energy density of the electric field	$E_{elec} = \frac{1}{2}\varepsilon\mathscr{E}^2$
2. Gauss' law for the magnetic field Equation 26.7	$\oint_S \mathscr{B} \cdot dS = 0$	6. Energy density of the magnetic field	$E_{mag} = \frac{1}{2\mu}\mathscr{B}^2$
3. Circulation of the electric field Equation 25.3	$\oint_L \mathscr{E} \cdot dl = 0$	7. Field relations	$\mathscr{D} = \varepsilon_0\mathscr{E} + \mathscr{P}$ $\mathscr{H} = \frac{1}{\mu_0}\mathscr{B} - \mathscr{M}$
4. Circulation of the magnetic field, (Ampère's law) Equation 26.2	$\oint_L \mathscr{B} \cdot dl = \mu_0 I$		

QUESTIONS

26.1 Make a comparison of the basic properties of static, or time-independent, electric and magnetic fields.

26.2 Why is the magnetic flux through a closed surface always zero?

26.3 State the main differences between
(a) diamagnetism and paramagnetism,
(b) ferromagnetism and antiferromagnetism,
(c) ferromagnetism and ferrimagnetism.

26.4 A hollow cylindrical conductor carries a current along its length. What is the magnetic field in the cavity, inside the conductor and outside?

26.5 How do the atomic and molecular properties of a substance determine its behavior in a magnetic field?

26.6 Why is the magnetic susceptibility of diamagnetic substances negative?

26.7 Why is it said that a superconductor is a perfect diamagnetic substance? What should its magnetic permeability be?

26.8 Why does the magnetization of paramagnetic and ferromagnetic substances decrease with the temperature? Do you expect the same effect in a diamagnetic substance? Explain.

26.9 Do you expect that a very strong magnetic field will affect a superconductor even if the temperature falls below the transition temperature?

26.10 What is the difference between the magnetic field $\mathscr{B}$ and the magnetizing field $\mathscr{H}$?

26.11 What is meant by the 'self-energy' of a charged particle? Is it the same if the particle is at rest or in motion relative to the observer?

26.12 A magnet is placed in front of a superconductor, as shown in Figure 26.23. Draw the lines of force of the field of the magnet. Show the position of the magnetic image. Can you suggest how to calculate the force between the magnet and the superconductor?

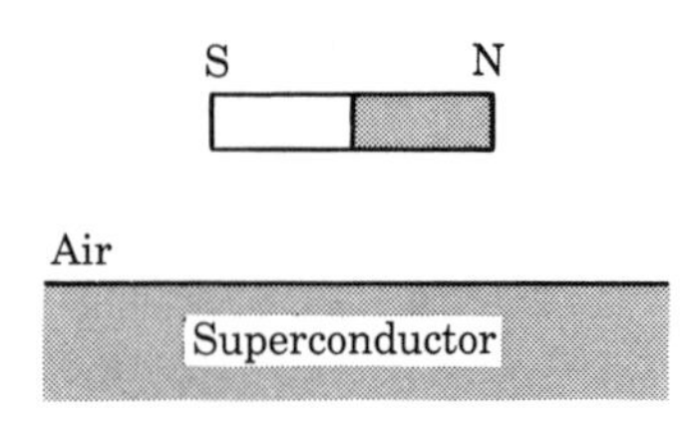

Figure 26.23

26.13 Explain why, when a superconductor is placed in a strong magnetic field, the critical temperature is lower than when it is in a weak magnetic field.

26.14 A strong magnet is placed on a superconducting metal plate (Figure 26.24) at a temperature high *above* T_C. Draw a diagram to show how the magnetic field varies within the plate. Explain why the superconducting state is not simultaneously established all over the plate as the temperature of the plate is lowered.

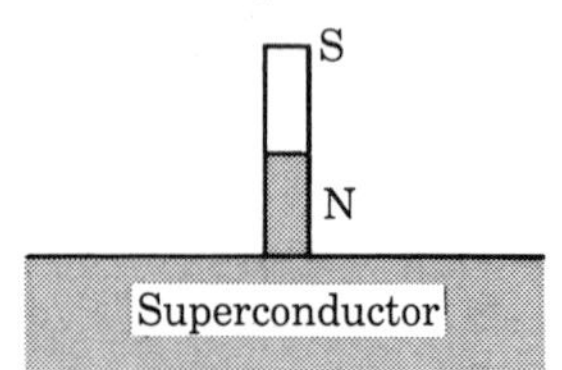

Figure 26.24

PROBLEMS

26.1 A toroid with a circumference 0.20 m is wrapped with 1500 turns of wire. Find the magnetic field within the toroid when the current is 1.50 A.

26.2 Find the magnetic field within a solenoid 0.20 m long wrapped with 1500 turns when the current through it is 1.50 A. Compare your answer with that of the previous problem.

26.3 Find the circulation of the magnetic field in a region where the magnetic field is uniform.

26.4 A very long rectilinear wire carries a current of 10 A. The radius of the wire is 4×10^{-2} m. (a) Calculate the magnetic field at the center of the wire, at 2×10^{-2} m from the center, and at the surface of the wire. (b) Calculate the magnetic field at the surface of the wire and at 8×10^{-2} m and 1.5 m from the center of the wire. (c) Calculate the point at which the field is 10^{-2} times as strong as that at the wire's surface.

26.5 A toroidal coil has a radius of 0.5 m and contains a constant current of 7 A. If the coil has been wound with 600 turns, calculate the magnitude of the magnetic field within the toroid.

26.6 A very long solenoid with 1400 turns per meter has a constant current of 7 A in its coils. Determine the magnetic field within the solenoid.

26.7 A solenoid is to be constructed with a magnetic field of 0.25 T in its interior. The radius of the solenoid is to be 0.1 m and the wire may carry a maximum current of 7 A. (a) How many turns per meter are needed? (b) If the solenoid is 1 m long, what length of wire is needed?

26.8 A hollow cylindrical conductor of radii R_1 and R_2 carries a current I uniformly distributed over its cross-section. Using Ampère's law, show (a) that the magnetic field at $r > R_2$ is $\mu_0 I/2\pi r$, (b) that the field for $R_1 < r < R_2$ is $\mu_0 I(r^2 - R_1^2)/[2\pi(R_2^2 - R_1^2)r]$ and (c) that the field is zero for $r < R_1$.

26.9 A coaxial cable is formed by surrounding a solid cylindrical conductor of radius R_1 with a concentric conducting shell of inner radius R_2 and outer radius R_3. In usual practice, a current I is sent down the inner wire and is returned via the outer shell. (a) Using Ampère's law, determine the magnetic field for all points about and within the cable. (b) Plot the magnetic field as a function of r. Assume uniform current density.

26.10 There is a uniform magnetic field $\mathscr{B} = \boldsymbol{i}(2\,\text{T})$ in a certain region (Figure 26.25). Calculate the magnetic flux (a) across the surface *abcd*, (b) across the surface *befc*, (c) across the surface *aefd* and (d) the flux through the complete closed surface.

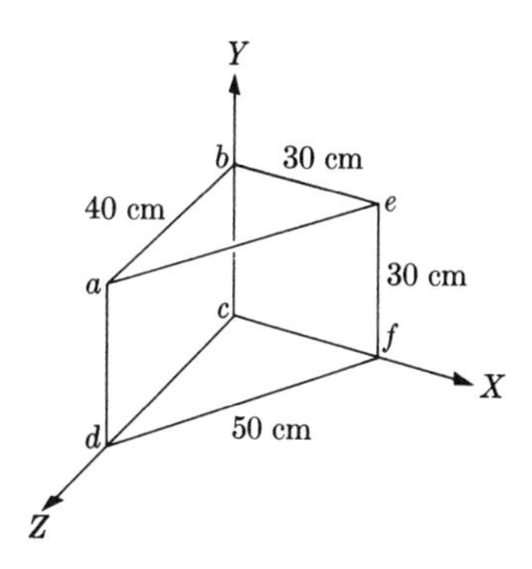

Figure 26.25

26.11 Determine the magnetic flux through the circuit of Figure 26.26 when there is a current I in the long straight wire.

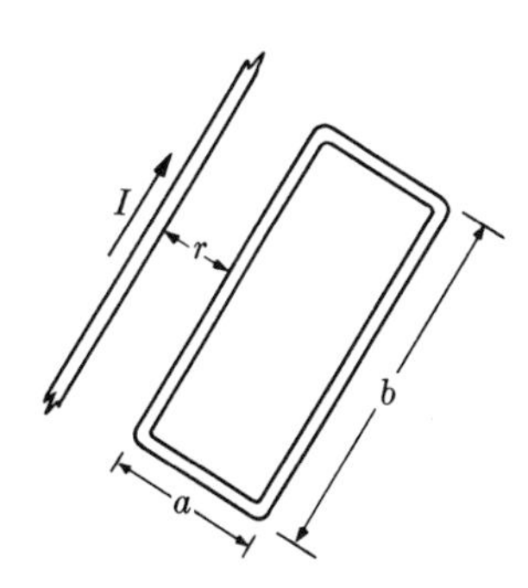

Figure 26.26

26.12 Using the result of Problem 24.15(b) for a magnetic dipole, show that the repulsive force between the magnetic dipole and its image in Figure 26.27 is $(\mu_0/4\pi)(3M^2/8z^4)$. (Hint: note that in this case $r = 2z$.)

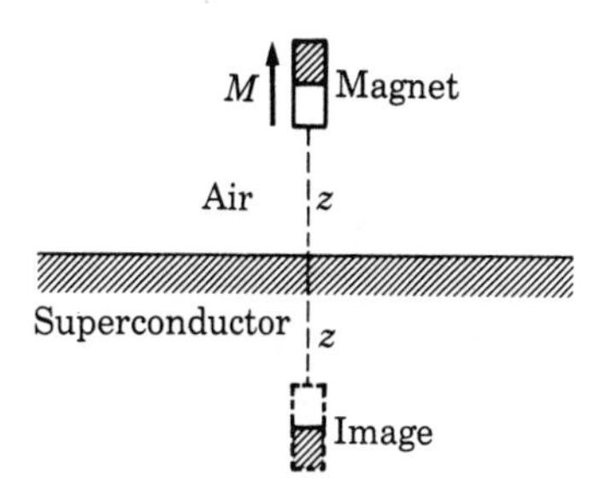

Figure 26.27

26.13 Consider a magnetic dipole of moment M placed in front of a superconductor as shown in Figure 26.28. Show that the force between the dipole and the superconductor is $(\mu_0/4\pi)(3M^2/16z^4)$. If the mass of the dipole is m, obtain the position of equilibrium. (Hint: compute the magnetic potential energy in the equatorial plane of the dipole using Equation 22.19; next compute the force on the image magnet, using Equation 22.20, and make $r = 2z$.)

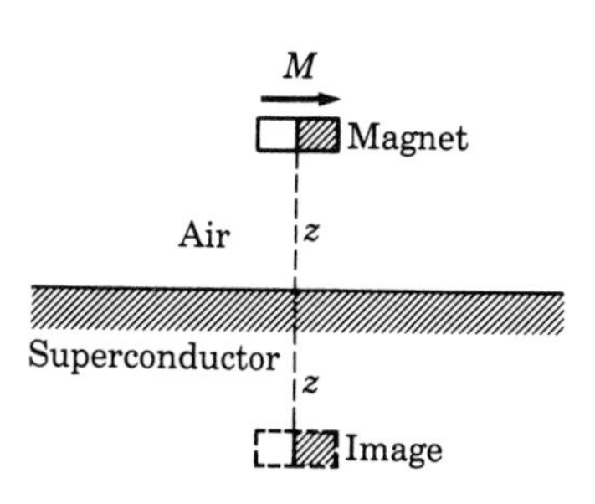

Figure 26.28

27 The electromagnetic field

James Clerk Maxwell made two of the most important contributions to 19th century physics. He established (1871) the mathematical foundations of the kinetic theory of gases and, by extension, of large assemblies of particles, thereby laying the foundations of statistical mechanics and transport phenomena. Maxwell's second and major contribution (1873) was the formulation of the theory of time-dependent electromagnetic fields, providing a unified view of electromagnetic phenomena. He introduced the notion of displacement current and showed that electromagnetic waves propagate with the same velocity as light, thus establishing the connection between the two phenomena.

27.1 Introduction

In this chapter we consider fields that are time-dependent; that is, they change with time. We shall see that

a varying magnetic field requires the presence of an electric field,

and conversely, that

a varying electric field requires a magnetic field.

The laws describing these two situations are called the **Faraday–Henry law** and the **Ampère–Maxwell law**. Finally, these laws will be applied to electric circuits carrying variable currents.

Part A: The laws of the electromagnetic field

27.2 The Faraday–Henry law

One of the phenomena that we often apply is **electromagnetic induction**, which was discovered, almost simultaneously, but independently, around 1830 by Michael Faraday and Joseph Henry (1797–1878). Electromagnetic induction is the working principle of the electric generator, the transformer, and many other devices in daily use. Suppose that an electric conductor, which forms a closed path or circuit, is placed in a region where a magnetic field exists. If the magnetic flux through the closed path *varies with time*, a current may be observed in the conductor while the flux is varying. The presence of an electric current indicates the existence of an electric field acting on the free charges in the conductor. This electric field produces an emf along the circuit, which is called *induced* emf. Measurement of this induced emf shows that it depends on the time *rate* of change of the magnetic flux. For example, if a magnet is placed near a closed conductor, an emf appears in the circuit when the magnet or the circuit is moved in such a way that the magnetic flux through the circuit changes. The magnitude of the induced emf depends on whether the relative motion of the magnet and the circuit is fast or slow. The greater the rate of change of the flux, the larger the induced emf. The direction in which the induced emf acts depends on whether the magnetic flux through the circuit is increasing or decreasing. One can use the right-hand rule to determine the direction in which the emf acts, as indicated in Figure 27.1. The thumb is placed in the direction of the magnetic field and the emf acts in the opposite (same) direction as the fingers when the flux increases (decreases).

In Figure 27.1, the path L has been oriented according to the right-hand rule. When the magnetic flux ϕ_m increases (that is, $d\phi_m/dt$ is positive), the induced emf acts in the negative sense, while when the magnetic flux decreases (that is, $d\phi_m/dt$ is negative), the emf acts in the positive sense. Thus the sign of the induced emf is always opposite to that of $d\phi_m/dt$. More detailed measurements reveal that the value of the induced emf is proportional to the time rate of change of the

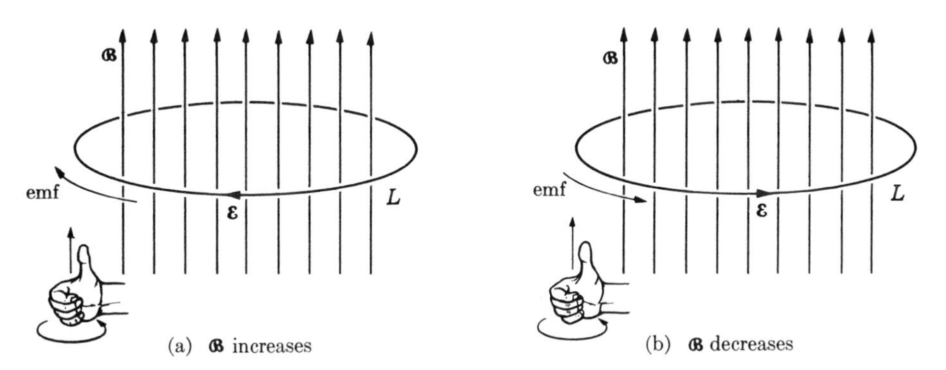

Figure 27.1 Electric field produced by a time-dependent magnetic field; (a) $d\phi/dt$ positive, V negative; (b) $d\phi/dt$ negative, V positive. The thumb of the right hand shows the direction of the magnetic field.

magnetic flux. Therefore we may write

$$\text{induced emf} = -\frac{\mathrm{d}\phi_{\mathrm{m}}}{\mathrm{d}t} \tag{27.1}$$

which expresses the **Faraday–Henry law** of electromagnetic induction:

in any closed circuit placed in a varying magnetic field there is an induced emf whose value is equal to the negative of the time rate of change of the magnetic flux through the circuit.

Equation 27.1 is consistent as far as units are concerned. We know that emf is expressed in V or $\mathrm{m^2\,kg\,s^{-2}\,C^{-1}}$. We recall from Section 26.3 that ϕ_{m} is expressed in Wb or $\mathrm{m^2\,kg\,s^{-1}\,C^{-1}}$ and $\mathrm{d}\phi_{\mathrm{m}}/\mathrm{d}t$ must be expressed in $\mathrm{Wb\,s^{-1}}$ or $\mathrm{m^2\,kg\,s^{-2}\,C^{-1}}$. Thus both sides of Equation 27.1 are expressed in the same units. Referring to Figure 27.2, the magnetic flux through L is, according to Equation 26.6,

$$\phi_{\mathrm{m}} = \int_S \mathscr{B}\cdot \mathrm{d}\boldsymbol{S}$$

Also the emf implies the existence of an electric field $\mathscr{E}$ such that, according to Equation 25.2,

$$\text{emf} = \oint_L \mathscr{E}\cdot \mathrm{d}\boldsymbol{l}$$

Thus we may write Equation 27.1 in the alternative form

$$\oint_L \mathscr{E}\cdot \mathrm{d}\boldsymbol{l} = -\frac{\mathrm{d}}{\mathrm{d}t}\int \mathscr{B}\cdot \mathrm{d}\boldsymbol{S} \tag{27.2}$$

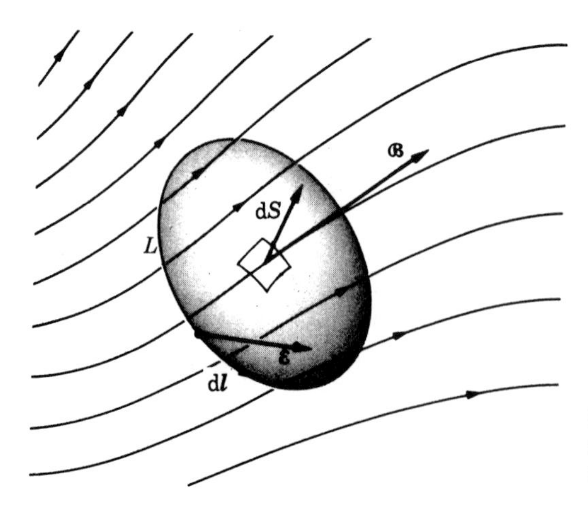

Figure 27.2 Relation between a time-varying magnetic flux and the electric circulation.

Equation 27.2 is valid for an arbitrary closed line L even if it does not coincide with an electric conductor. That is,

> *a time-dependent magnetic field implies the existence of an electric field such that the circulation of the electric field along an arbitrary closed path is equal to the negative of the time rate of change of the magnetic flux through a surface bounded by the path.*

This is another way of stating the Faraday–Henry law. Note that an electric current is produced by the induced emf only if a conductor is placed in a time-dependent magnetic field.

EXAMPLE 27.1

Emf induced in a plane circuit composed of N turns, each of area S, placed perpendicular to an alternating uniform magnetic field that varies sinusoidally with time.

▷ Let the magnetic field be expressed by $\mathscr{B} = \mathscr{B}_0 \sin \omega t$. The magnetic flux through one turn of the circuit is then $\phi_{\mathrm{m}} = S\mathscr{B} = S\mathscr{B}_0 \sin \omega t$ and the total magnetic flux through the N turns is $\phi_{\mathrm{m}} = NS\mathscr{B}_0 \sin \omega t$. Applying Equation 27.1, we obtain the induced emf as

$$\text{emf} = -\frac{\mathrm{d}\phi_{\mathrm{m}}}{\mathrm{d}t} = -NS\mathscr{B}_0\omega \cos \omega t \qquad \textbf{(27.3)}$$

which indicates that the induced emf is oscillatory and alternates with the same frequency as the magnetic field.

EXAMPLE 27.2

Consider a region of space where there is an axially symmetric time-dependent magnetic field parallel to the Z-axis; that is, the magnitude of the magnetic field varies with the distance r to the Z-axis only. Determine the electric field $\mathscr{B}$ at each point of space.

▷ Figure 27.3(a) shows a side view of the magnetic field and Figure 27.3(b) shows a cross-section. The symmetry of the problem suggests that the electric field $\mathscr{E}$ produced by the time-dependent magnetic field must depend on the distance r alone, and at each point be perpendicular to the magnetic field $\mathscr{B}$ and to the radius r. In other words, the lines of force of the electric field $\mathscr{E}$ are circles concentric with the Z-axis. Choosing our path L in Equation 27.2 as one of these circles, we have

$$\text{emf} = \oint_L \boldsymbol{\mathscr{E}} \cdot \mathrm{d}\boldsymbol{l} = \mathscr{E}(2\pi r)$$

Therefore, using Equation 27.1, we obtain

$$\mathscr{E}(2\pi r) = -\frac{\mathrm{d}\phi_{\mathrm{m}}}{\mathrm{d}t} \qquad \textbf{(27.4)}$$

The *average* magnetic field $\mathscr{B}_{\mathrm{ave}}$, in a region covering an area S is $\mathscr{B}_{\mathrm{ave}} = \phi_{\mathrm{m}}/S$, which gives $\phi_{\mathrm{m}} = \mathscr{B}_{\mathrm{ave}}S$. In our case then, $S = \pi r^2$, so that $\phi_{\mathrm{m}} = \mathscr{B}_{\mathrm{ave}}(\pi r^2)$. Then, using Equation 27.4,

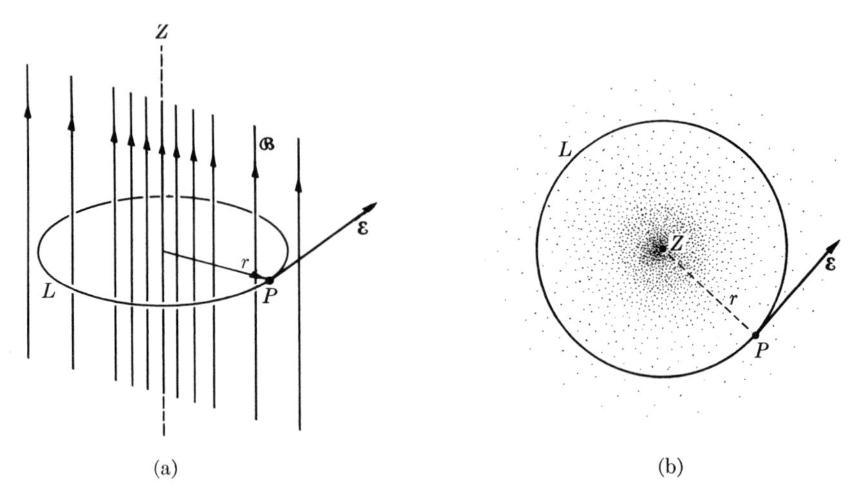

Figure 27.3 Electric field produced by a time-dependent magnetic field having cylindrical symmetry; (a) side view, (b) top view.

the electric field at a distance r from the axis is

$$\mathscr{E} = -\frac{1}{2}r\left(\frac{\mathrm{d}\mathscr{B}_{\text{ave}}}{\mathrm{d}t}\right) \tag{27.5}$$

For a uniform magnetic field (one independent of r), $\mathscr{B}_{\text{ave}} = \mathscr{B}$ and $\mathscr{E} = \frac{1}{2}r(\mathrm{d}\mathscr{B}/\mathrm{d}t)$.

EXAMPLE 27.3

Eddy currents.

▷ Consider a piece of conducting material placed in a magnetic field that is varying with time. The electric field associated with the time-dependent magnetic field acts on the electrons in the highest occupied states of the conduction band, setting them into organized motion. The result is the induction of electric currents throughout the whole volume of the conductor; these currents will persist as long as the magnetic field is time-dependent.

Let a conductor in the shape of a cylinder be placed in a uniform magnetic field (Figure 27.4). In (a) the magnetic field is *increasing* and in (b) it is *decreasing*. The associated electric field is transverse in both cases (recall Example 27.2). However, its direction depends on whether the magnetic field increases or decreases. The result is induced circular currents in the direction shown in Figure 27.4 for each case, which are called **eddy currents**. In either

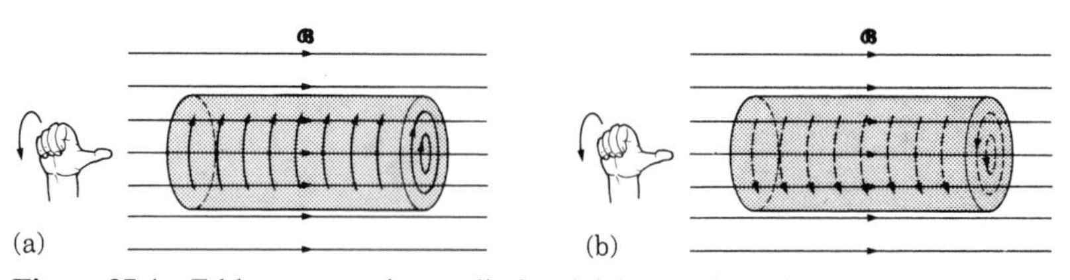

Figure 27.4 Eddy currents in a cylinder. (a) Increasing ($\mathrm{d}\mathscr{B}/\mathrm{d}t > 0$); (b) decreasing ($\mathrm{d}\mathscr{B}/\mathrm{d}t < 0$).

case, the direction of the eddy currents is such that the additional magnetic field they produce tends to compensate inside the material for the *change* in the applied external magnetic field.

Eddy currents are true conduction currents, unlike the localized magnetization currents in diamagnetic and paramagnetic substances placed in a static magnetic field (Section 26.5). Therefore, the conductor heats up because of the Joule effect due to the resistance of the conductor. This is the origin of the heating of the iron core in power transformers. To minimize heating due to eddy currents, the core of a transformer is made of metal plates separated by insulating material and placed in such a way that eddy currents have to be perpendicular to the plates. Another consequence of the Joule effect is that eddy currents rapidly disappear when the magnetic field stops changing, due to the dissipation of the energy of the electrons into the crystal lattice.

A different situation appears in superconductors (Section 24.4). In this case, due to a lack of resistance to the motion of the Cooper pairs through the crystal lattice, eddy currents may persist for a very long time after the magnetic field has ceased to change. In fact, it was the persistence of induced currents, a phenomenon first observed by Heike Kammerlingh Onnes (1853–1926) in 1911, that pointed out the existence of superconductivity. Currents produced by the electric field associated with a time-dependent magnetic field in a superconductor are a collective electronic motion. Therefore, these currents are quite different from the screening surface currents associated with the diamagnetic effect in superconductors analyzed in Example 26.5, which are due to the action of a static magnetic field on the normal motion of the Cooper pairs. When the applied magnetic field is time-dependent, both effects are present and, while the electric and magnetic fields inside the superconductor quickly drop to zero, the induced currents persist within a surface layer about 10^{-7} m thick.

Note 27.1 The betatron

The results of Example 27.2 were used to design an electron accelerator called a **betatron**, invented in 1941 by Donald Kerst. If an electron (or any kind of charged particle) is injected into a region where a varying magnetic field exists, and the magnetic field has axial symmetry, the electron will be accelerated by the associated electric field $\mathscr{E}$ as given by Equation 27.5. As the electron gains velocity, its path will be bent by the magnetic field $\mathscr{B}$. If the electric and magnetic fields are adjusted properly, the orbit of the electron is a circle. The electron gains energy with each revolution. Therefore, the greater the number of revolutions, the greater the energy of the electron.

The exact variation of $\mathscr{B}$ with r is determined by the fact that a certain stability of the orbital motion is required. That is, given the radius of the desired orbit, the forces on the electron must be such that if the motion of the electron is slightly disturbed (i.e. if it is pushed to one side or the other of the orbit), the electric and magnetic forces acting on the electron tend to place it back in the correct orbit.

In general, the magnetic field $\mathscr{B}$ is oscillatory with some angular frequency ω. Now, because of Equation 27.5, the electron is accelerated only while the magnetic field is increasing. On the other hand, since, in practice, electrons are injected with very small momentum, they must be injected when the magnetic field is zero. That means that only one quarter of the period of variation of the magnetic field is good for accelerating the electrons. The accelerating times have been indicated by the shaded areas in Figure 27.5(a).

Betatrons consist of a toroidal tube (Figure 27.5(b)) placed in the magnetic field produced by a magnet whose pole faces have been designed or shaped in such a way that

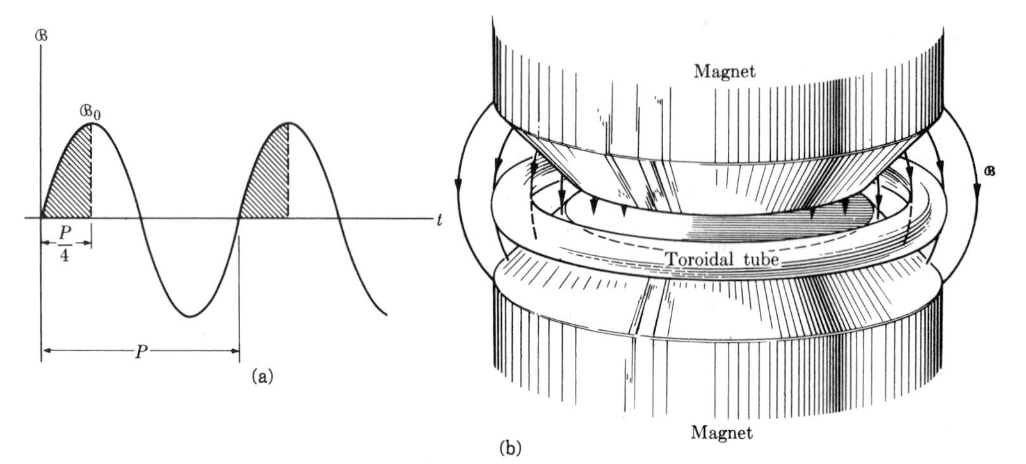

Figure 27.5 (a) Accelerating time in a betatron. (b) View of the accelerating tube and pole faces of a betatron.

the correct variation of the magnetic field $\mathscr{B}$ with r is produced and the stability conditions are fulfilled. The electrons are injected at the beginning of the accelerating period and slightly deflected at its end, so that they might hit a target properly located. The kinetic energy of the electrons is given off as electromagnetic radiation (Chapter 30) and/or as internal energy of the target, which is heated up. Betatrons have been built with energies up to 350 MeV. They are used for studying certain types of nuclear reactions and in medicine as a source of X-rays for the treatment of cancer.

27.3 Electromagnetic induction due to the relative motion of a conductor and a magnetic field

The law of electromagnetic induction, as expressed in Equation 27.2, implies the existence of an emf when the magnetic flux through the circuit changes with time. It is important to discover whether the same results occur when the change in flux is due to a motion or deformation of the path L without the magnetic field necessarily changing with time. We shall look at two simple cases.

(i) Moving conductor. Consider the arrangement of conductors illustrated in Figure 27.6, where the conductor PQ is moving parallel to itself with velocity $\boldsymbol{v}$ while maintaining contact with conductors RT and SU. The system $PQRS$ forms a closed circuit. Suppose also that there is a uniform magnetic field $\mathscr{B}$ perpendicular to the plane of the system.

According to Equation 22.1, each positive charge q in the moving conductor PQ is subject to a force $q\boldsymbol{v} \times \mathscr{B}$ acting along QP. Now the same force on the charge could be assumed to be due to an 'equivalent' electric field $\mathscr{E}_{\text{eq}}$ given by $q\mathscr{E}_{\text{eq}} = q\boldsymbol{v} \times \mathscr{B}$ or $\mathscr{E}_{\text{eq}} = \boldsymbol{v} \times \mathscr{B}$. Since $\boldsymbol{v}$ and $\mathscr{B}$ are perpendicular, the relation among the magnitudes is

$$\mathscr{E}_{\text{eq}} = v\mathscr{B} \qquad \textbf{(27.6)}$$

If $PQ = l$, there is a potential difference between P and Q given by $\mathscr{E}_{\text{eq}}l$ or $\mathscr{B}vl$. No forces are exerted on the section QR, RS, and SP, since they are stationary

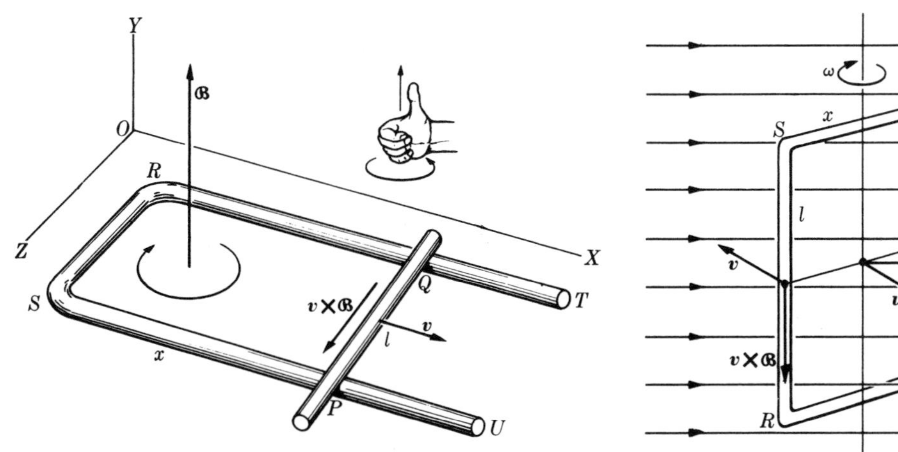

Figure 27.6 Emf induced in a conductor moving in a magnetic field. The fingers of the right hand show the positive sense of circulation around *PQRS*.

Figure 27.7 Emf induced in a rotating coil placed in a magnetic field.

relative to the magnetic field. Therefore, the circulation of $\mathscr{E}_{eq}$, or the emf, along circuit *PQRS* is just $\mathscr{B}vl$; that is, emf $= -\mathscr{B}vl$. The minus sign appears because $\mathscr{E}$ acts in the negative sense when the normal to the plane of the path is oriented in the direction of $\mathscr{B}$. If we designate the length *SP* by x, the area of *PQRS* is lx and the magnetic flux through *PQRS* is

$$\phi_m = \int_{PQRS} \mathscr{B} \cdot d\boldsymbol{S} = \mathscr{B}S = \mathscr{B}lx$$

The change of flux per unit time is then

$$\frac{d\phi_m}{dt} = \frac{d}{dt}(\mathscr{B}lx) = \mathscr{B}l\frac{dx}{dt} = \mathscr{B}lv \qquad \textbf{(27.7)}$$

since $dx/dt = v$. Therefore, recalling the value of the emf obtained above, we conclude that emf $= -\,d\phi_m/dt$, which is Equation 27.1.

(ii) Rotating coil. As a second example, consider a rectangular circuit rotating in a uniform magnetic field $\mathscr{B}$ with angular frequency ω (Figure 27.7). When the normal $\boldsymbol{u}_N$ to the circuit makes an angle $\theta = \omega t$ with the magnetic field $\mathscr{B}$, all points of *PQ* are moving with a velocity $\boldsymbol{v}$ such that the 'equivalent' electric field $\mathscr{E}_{eq} = \boldsymbol{v} \times \mathscr{B}$ points from *Q* to *P* and has a magnitude $\mathscr{E}_{eq} = v\mathscr{B}\sin\theta$. Similarly, for points on *RS*, the direction of $\boldsymbol{v} \times \mathscr{B}$ is from *S* to *R* and has the same magnitude. On the sides *RQ* and *PS*, we see that $\boldsymbol{v} \times \mathscr{B}$ is perpendicular to them and no potential difference exists between *S* and *P* and between *R* and *Q*. Therefore, if $PQ = RS = l$, the circulation of the equivalent electric field $\mathscr{E}_{eq}$ around *PQRS*, or the applied emf, is

$$\text{emf} = \oint_L \mathscr{E} \cdot d\boldsymbol{l} = \mathscr{E}_{eq}(PQ + SR) = 2lv\mathscr{B}\sin\theta$$

If $x = PS$, the radius of the circle described by the charges in PQ and SR is $\frac{1}{2}x$, and therefore $v = \omega(\frac{1}{2}x) = \frac{1}{2}\omega x$. Then, since $S = lx$ is the area of the circuit and $\theta = \omega t$, we may also write

$$\text{emf} = \omega \mathscr{B} S \sin \omega t$$

for the emf induced in the circuit as a result of its rotation in the magnetic field.

On the other hand, since the magnetic field is uniform, the magnetic flux through the circuit is $\phi_m = \mathscr{B} S \cos \theta = \mathscr{B} S \cos \omega t$. Then

$$-\frac{d\phi_m}{dt} = \omega \mathscr{B} S \sin \omega t = \text{emf}$$

Therefore we again verify that the induced emf resulting from the motion of the conductor can also be calculated by applying Equations 27.1 or 27.2 instead of Equations 22.1 and 25.2. The relative rotation of a circuit and a magnetic field is one of the methods used in practice to generate an alternating emf.

Although our discussion has dealt only with circuits of special shapes, a more detailed mathematical calculation indicates that for any circuit

> *the law of electromagnetic induction,* emf $= -\,d\phi_m/dt$, *can be applied when the change in magnetic flux is due to a change in the magnetic field or to a motion or a deformation relative to the magnetic field of the circuit along which the emf is calculated, or both.*

27.4 Electromagnetic induction and the principle of relativity

In spite of the fact that the law of electromagnetic induction is valid whether the change of magnetic flux is due to a change in the magnetic field or to the motion of a conductor, there is a profound difference in the physical situations in the two possibilities. When an observer, O, recognizes that the change of magnetic flux through a circuit stationary in the observer's frame of reference is due to a change in the magnetic field $\mathscr{B}$, an electric field $\mathscr{E}$ related to $\mathscr{B}$ as indicated by Equation 27.2 is measured by O. The presence of the electric field is recognized by measuring the force on a charge *at rest* in O's frame of reference. But when O recognizes that the change of magnetic flux is due to a motion of the conductor through a magnetic field relative to O's frame, no electric field is observed, but the emf measured is attributed to the force $q\boldsymbol{v} \times \mathscr{B}$ exerted by the magnetic field on the charges of the moving conductor, in accordance with Equation 22.1.

These two different and apparently unrelated situations are a consequence of the principle of relativity. We cannot go into a full mathematical analysis here; instead, we shall look at the situation from an intuitive point of view. Let us examine the case of the rotating circuit discussed in connection with Figure 27.7. In a frame of reference where the magnetic field $\mathscr{B}$ is constant (Figure 27.8(a)) and the circuit is rotating with an angular velocity $\boldsymbol{\omega}$, no electric field is observed and the forces on the electrons in the circuit are $\boldsymbol{F} = -e\boldsymbol{v} \times \mathscr{B}$. But an observer

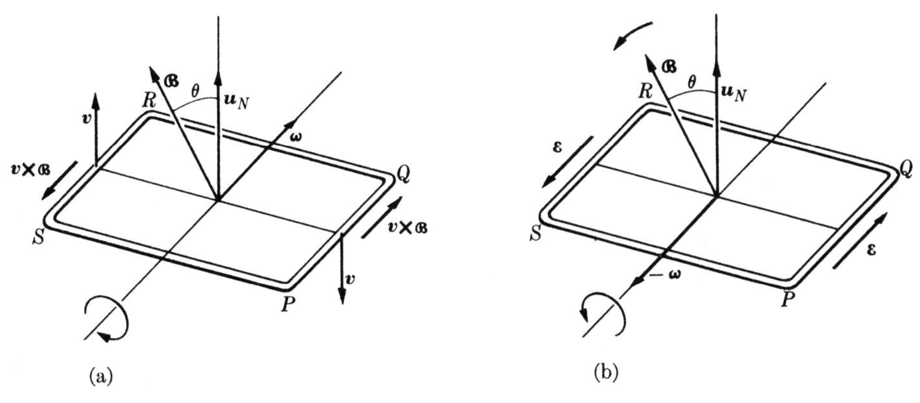

Figure 27.8 (a) Observer stationary relative to magnetic field. (b) Observer stationary relative to circuit; magnetic field appears to rotate.

attached to a frame moving with the circuit sees a stationary conductor and a magnetic field $\mathscr{B}$ whose direction rotates in space with angular velocity $-\boldsymbol{\omega}$ (Figure 27.8(b)). The observer then relates the forces on the electrons in the circuit to the electric field $\mathscr{E}$ associated with a changing magnetic field, in accordance with the law of electromagnetic induction. The results, in both cases, are the same, as required by the principle of relativity.

27.5 The principle of conservation of electric charge

The principle of conservation of electric charge states that

in all processes that occur in the universe, the net amount of electric charge must always remain the same.

This statement may be expressed in a quantitative way as follows. Consider a closed surface S (Figure 27.9), and designate, by q, the net charge inside it at a given time. Since our problem is dynamic, free charges (such as electrons in a metal or ions in a plasma) are moving through the medium, crossing the surface S. At some times there may be more outgoing charges than entering ones, resulting in a decrease in the net charge q within the surface S. At other times the situation may be reversed, and the incoming charges may exceed those leaving, resulting in an increase in the net charge q. Of course, if the outgoing and incoming charge fluxes through S are the same, the net charge q remains

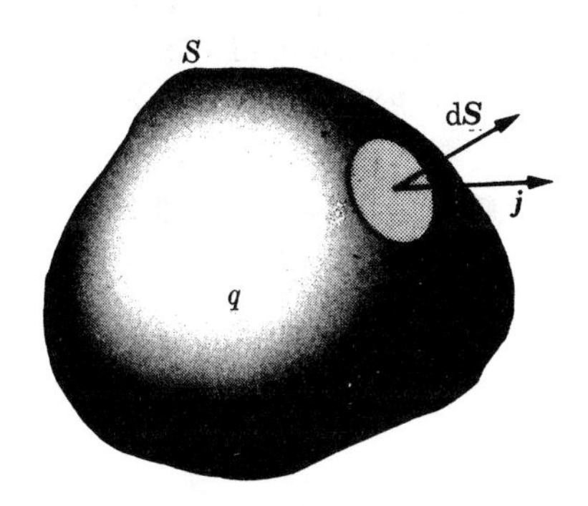

Figure 27.9 Current across a closed surface enclosing a charge q.

the same. The principle of conservation of charge requires that

$$\begin{pmatrix}\text{Rate of}\\ \text{charge loss}\end{pmatrix} = \begin{pmatrix}\text{outgoing}\\ \text{charge}\\ \text{flux}\end{pmatrix} - \begin{pmatrix}\text{incoming}\\ \text{charge}\\ \text{flux}\end{pmatrix} = \begin{pmatrix}\text{net outgoing}\\ \text{charge}\\ \text{flux}\end{pmatrix} \tag{27.8}$$

Now the net outgoing charge flux is the net current I passing through the surface. On the other hand, the loss of charge per unit time within the *closed* surface S is $-\mathrm{d}q/\mathrm{d}t$. Therefore, in mathematical terms

$$-\frac{\mathrm{d}q}{\mathrm{d}t} = I \qquad \text{or} \qquad \frac{\mathrm{d}q}{\mathrm{d}t} + I = 0 \tag{27.9}$$

The above equation expresses the principle of conservation of charge under the assumption that charge is neither created nor annihilated. Now, according to Gauss' law for the electric field, as given by Equation 25.7, the total charge within a closed surface is expressed in terms of the electric field at the surface by $q = \varepsilon_0 \oint_S \mathscr{E} \cdot \mathrm{d}\boldsymbol{S}$, so that

$$\frac{\mathrm{d}q}{\mathrm{d}t} = \varepsilon_0 \frac{\mathrm{d}}{\mathrm{d}t} \oint_S \mathscr{E} \cdot \mathrm{d}\boldsymbol{S} \tag{27.10}$$

Substituting this result in Equation 27.9, we obtain

$$I + \varepsilon_0 \frac{\mathrm{d}}{\mathrm{d}t} \oint_S \mathscr{E} \cdot \mathrm{d}\boldsymbol{S} = 0 \tag{27.11}$$

which expresses the principle of conservation of charge, in a way that incorporates Gauss' law. When the fields are static, the integral $\oint_S \mathscr{E} \cdot \mathrm{d}\boldsymbol{S}$ does not depend on time. Its time derivative is therefore zero, resulting in

$$I = 0 \tag{27.12}$$

This means that for static fields there is no accumulation or loss of charge in any region of space, and the net current across a closed surface is zero. In fact this idea was used in Section 24.6 in connection with direct current circuits.

27.6 Ampère–Maxwell law

The Faraday–Henry law, as expressed by

$$\oint_L \mathscr{E} \cdot \mathrm{d}\boldsymbol{l} = -\frac{\mathrm{d}}{\mathrm{d}t} \int_S \mathscr{B} \cdot \mathrm{d}\boldsymbol{S}$$

relates the circulation of the electric field to the time rate of change of the flux of the magnetic field through the surface defined by the closed path. We might expect that a similar expression must relate the circulation of the magnetic field to the time rate of change of the flux of the electric field. So far, we have found that the circulation of a static magnetic field is expressed by Ampère's law, Equation 26.37, as

$$\oint_L \mathscr{B} \cdot \mathrm{d}\boldsymbol{l} = \mu_0 I \tag{27.13}$$

where I is the net current across the surface bound by the path L. But this expression does not contain any time rate of change of the flux of the electric field. This is not surprising, since it was derived under static conditions. However, we may suspect that Ampère's law may need revising when it is applied to time-dependent fields.

Now Ampère's law in the form 27.13 applies to a surface S bounded by line L. The surface S is arbitrary so long as it is bounded by the line L. If the line L shrinks, the value of $\oint_L \mathscr{B} \cdot \mathrm{d}\boldsymbol{l}$ decreases (Figure 27.10). It eventually becomes zero when L shrinks to a point, and the surface S becomes a *closed* surface. Ampère's law, as expressed by Equation 27.13, then requires that $I = 0$. This result agrees with Equation 27.12 for the conservation of charge, so long as the field is static. However, when the field is not static but time-dependent, it is Equation 27.11, which incorporates Gauss' law, that is correct. This indicates that Ampère's law must be modified when dealing with time-dependent fields such that we must replace I in Equation 27.13 by

$$I + \varepsilon_0 \frac{\mathrm{d}}{\mathrm{d}t} \int_S \mathscr{E} \cdot \mathrm{d}\boldsymbol{S}$$

where I is the net current through the arbitrary surface S. This results in the expression

$$\oint_L \mathscr{B} \cdot \mathrm{d}\boldsymbol{l} = \mu_0 \left(I + \varepsilon_0 \frac{\mathrm{d}}{\mathrm{d}t} \int_S \mathscr{E} \cdot \mathrm{d}\boldsymbol{S} \right) \tag{27.14}$$

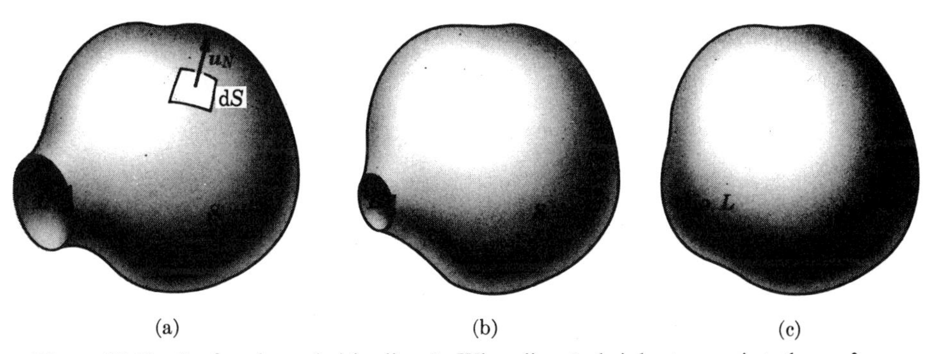

Figure 27.10 Surface bounded by line L. When line L shrinks to a point, the surface becomes closed.

Equation 27.14 reduces to Ampère's law for static fields, since then the last term is zero, and becomes Equation 27.11 when line L shrinks to a point and the surface S becomes closed so that the left-hand side is zero. Therefore, Equation 27.14 satisfies all the physical conditions previously found.

One necessary further step is to verify experimentally that Equation 27.14 is correct and that it describes the actual situation in nature. We may say in advance that it is so. The best proof is the existence of electromagnetic waves, a subject to be discussed in Chapter 29.

The modification of Ampère's law was proposed by James Clerk Maxwell (1831–1879) in 1873, and therefore Equation 27.14 is called the **Ampère–Maxwell law**.

We may observe in Equation 27.14 that the effect of a time-dependent electric field is to add to the current I a term $\varepsilon_0(\mathrm{d}/\mathrm{d}t)\int \mathscr{E}\cdot \mathrm{d}\boldsymbol{S}$. Maxwell interpreted this term as an additional current and called it the **displacement current**. However, the term 'displacement current' is misleading since there is no such current and Equation 27.14 simply expresses a correlation between $\mathscr{E}$, $\mathscr{B}$ and I at each point in space.

Ampère's law (Equation 27.13) relates a steady current to the magnetic field it produces. The Ampère–Maxwell law (Equation 27.14) goes a step further and indicates that a time-dependent electric field also contributes to the magnetic field. For example, in the absence of currents, so that $I = 0$, we have

$$\oint_L \mathscr{B}\cdot \mathrm{d}\boldsymbol{l} = \varepsilon_0\mu_0 \frac{\mathrm{d}}{\mathrm{d}t}\int_S \mathscr{E}\cdot \mathrm{d}\boldsymbol{S} \qquad \textbf{(27.15)}$$

which shows more clearly the relation between a time-dependent electric field and its associated magnetic field. In other words,

a time-dependent electric field implies the existence of a magnetic field at the same place, such that the circulation of the magnetic field along an arbitrary closed path is proportional to the time rate of change of the electric flux through a surface bounded by the path.

The relative orientation of the electric and magnetic fields is shown in Figure 27.11, corresponding to a time-dependent uniform electric field. If the electric

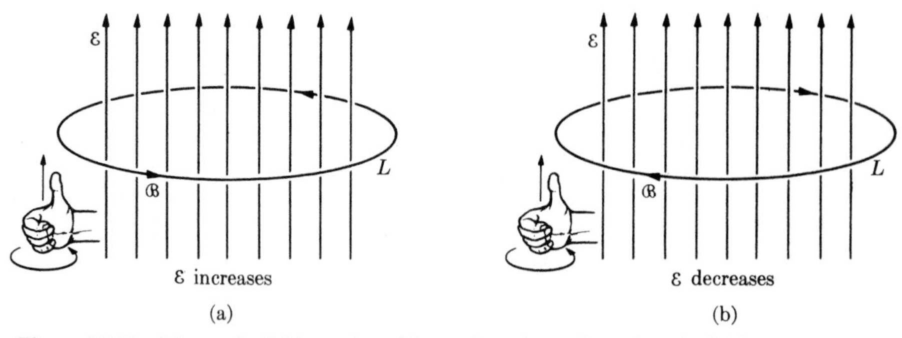

Figure 27.11 Magnetic field produced by a time-dependent electric field.

field increases (decreases), the orientation of the magnetic lines of force is the same as (opposite to) the sense indicated by the fingers of the right hand when the thumb points in the direction of the electric field. This result should be compared with Figure 27.1.

The Ampère–Maxwell law, as expressed by Equation 27.14, differs from the Faraday–Henry law, as expressed by Equation 27.2, in several respects. In the first place, in Equation 27.14 we have a term corresponding to an *electric* current, while in Equation 27.2 there is no term corresponding to a *magnetic* current. This is simply due to the fact that, apparently, there are no free magnetic poles in nature. In the second place, the time rate of change of the electric flux appears with a positive sign in Equation 27.14, while the time rate of change of the magnetic flux appears with a negative sign in Equation 27.2. It may also be verified that the factor $\varepsilon_0 \mu_0$ is consistent with the units.

27.7 Maxwell's equations

The **electromagnetic interaction** among the particles composing matter is associated with a property called **electric charge**. The electromagnetic interaction is described by an *electromagnetic field*, characterized by two vectors, the *electric field* $\mathscr{E}$ and the *magnetic field* $\mathscr{B}$, such that the force on an electric charge is given by the *Lorentz force*

$$\boldsymbol{F} = q(\mathscr{E} + \boldsymbol{v} \times \mathscr{B}) \quad \textbf{(27.16)}$$

The electric and magnetic fields $\mathscr{E}$ and $\mathscr{B}$ are, in turn, determined by the positions of the charges themselves and by their motions (or currents). The separation of the electromagnetic field into its electric and magnetic components depends on the relative motion of the observer and the charges producing the field. Also the fields $\mathscr{E}$ and $\mathscr{B}$ are directly correlated with each other by the Ampère–Maxwell and Faraday–Henry laws. All these relations are expressed by four laws, which we have analyzed in this and previous chapters, and which are written down in Table 27.1.

Table 27.1 Maxwell's equations for the electromagnetic field

1. Gauss' law for the electric field, Equation 25.7	$\oint_S \mathscr{E} \cdot d\boldsymbol{S} = \frac{q}{\varepsilon_0}$
2. Gauss' law for the magnetic field, Equation 26.7	$\oint_S \mathscr{B} \cdot d\boldsymbol{S} = 0$
3. Faraday–Henry law, Equation 27.2	$\oint_L \mathscr{E} \cdot d\boldsymbol{l} = -\frac{d}{dt}\int_S \mathscr{B} \cdot d\boldsymbol{S}$
4. Ampère–Maxwell law, Equation 27.15	$\oint_L \mathscr{B} \cdot d\boldsymbol{l} = \mu_0\left(I + \varepsilon_0 \frac{d}{dt}\int_S \mathscr{E} \cdot d\boldsymbol{S}\right)$

The theory of the electromagnetic field is condensed into these four laws. They are called **Maxwell's equations**, since it was Maxwell who, in addition to formulating the fourth law, recognized that they, together with Equation 27.16, constitute the basic framework of the theory of electromagnetic interactions. The electric charge q and the current I are called the **sources** of the electromagnetic field since, given q and I, Maxwell's equations allow us to compute the fields $\mathscr{E}$ and $\mathscr{B}$.

Even though Gauss' laws for the electric and magnetic fields, Equations 25.7 and 26.7, were derived for static fields, they are also valid for time-dependent fields. Also, the Faraday–Henry and Ampère–Maxwell laws provide the connection between the electric and magnetic fields that was absent in the equations for static fields (Table 26.2).

The synthesis of electromagnetic interactions as expressed by Maxwell's equations is one of the greatest achievements in physics. Much of modern civilization has been made possible by virtue of our understanding of electromagnetic interactions because they are responsible for most of the processes, natural and man-made, that affect our daily lives. Maxwell's equations work very well when dealing with electromagnetic interactions between large aggregates of charges, such as radiating antennas, electric circuits, and even beams of ionized atoms or molecules. But it has been found that electromagnetic interactions between fundamental particles (especially at high energies) must be treated in a somewhat different way, and according to the laws of quantum mechanics.

Part B: Application to electric circuits

27.8 Self-induction

Consider a circuit carrying a current I (Figure 27.12). According to Ampère's law the current produces a magnetic field which, at each point, is proportional to I. Therefore, the magnetic flux through the circuit due to its own magnetic field, called the **self-flux** and designated by ϕ_I, is proportional to the current I and we may write

$$\phi_I = LI \qquad \textbf{(27.17)}$$

The coefficient L depends on the geometrical shape of the conductor and is called the **self-inductance** of the circuit. It is expressed in Wb A^{-1}, a unit called the **henry**, in honor of Joseph Henry, and abbreviated H. That is, H = Wb A^{-1} = m^2 kg C^{-2}. If the current I

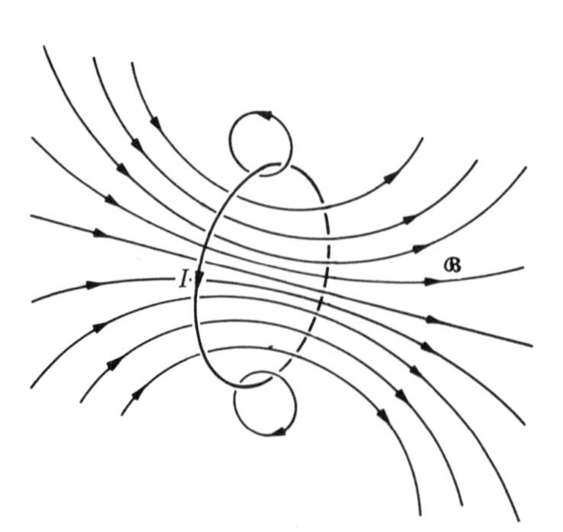

Figure 27.12 Self magnetic flux in a circuit.

changes with time, the magnetic flux ϕ_I through the circuit also changes and, according to the law of electromagnetic induction, an emf is induced in the circuit. This special case of electromagnetic induction is called **self-induction**.

Combining Equations 27.1 and 27.17, the self-induced emf is

$$V_L = -\frac{d\phi_I}{dt} \quad \text{or} \quad V_L = -L\frac{dI}{dt} \qquad \textbf{(27.18)}$$

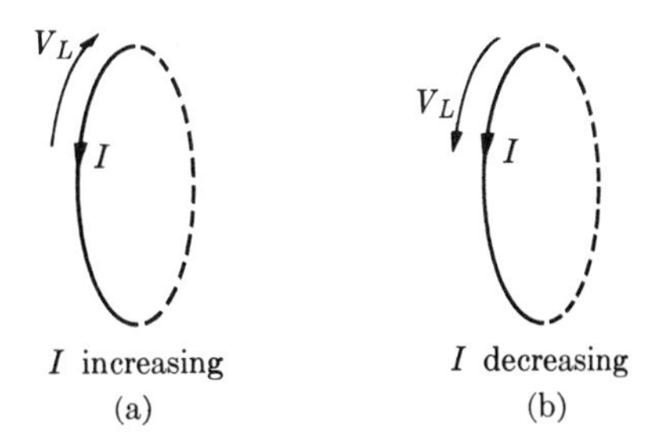

Figure 27.13 Direction of the self-induced emf in a circuit.

The minus sign indicates that the emf is opposed to the *change* in the current. So if the current increases, dI/dt is positive and V_L opposes the current (Figure 27.13(a)). If the current decreases, dI/dt is negative and V_L acts in the same direction as the current (Figure 27.13(b)). Therefore, the self-induced emf *acts in a direction that opposes the change in the current.*

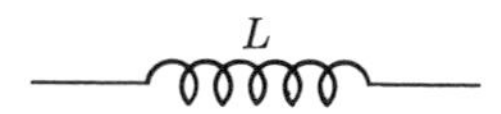

Figure 27.14 Representation of a self-inductance.

In diagrams drawn to indicate that a conductor has an appreciable inductance, the symbol of Figure 27.14 is used. However, we must note that the self-inductance of a conductor is not concentrated at a particular point, but is a property of the whole conductor that depends on its shape.

EXAMPLE 27.4

Self-inductance of a circuit composed of two coaxial cylindrical metallic sheets of radii a and b, each carrying a current I, but in opposite directions (Figure 27.15).

▷ Let the space between the cylinders be filled with a substance of permeability μ. Based on the calculations in Example 26.1 the magnetic field for this current arrangement was computed as $\mathscr{B} = \mu I/2\pi r$ in the region within the two cylinders, and zero elsewhere. Here we have replaced μ_0 used in Example 26.1 by μ, the permeability of the medium filling the space within the two cylinders, in accordance with Section 26.7. To obtain the self-inductance, we must compute the magnetic flux through any section of the conductor, such as $PQRS$, having a length l. If we divide this section into strips of width dr, the area of each strip is $l\,dr$. The magnetic field $\mathscr{B}$ is perpendicular to $PQRS$. Therefore

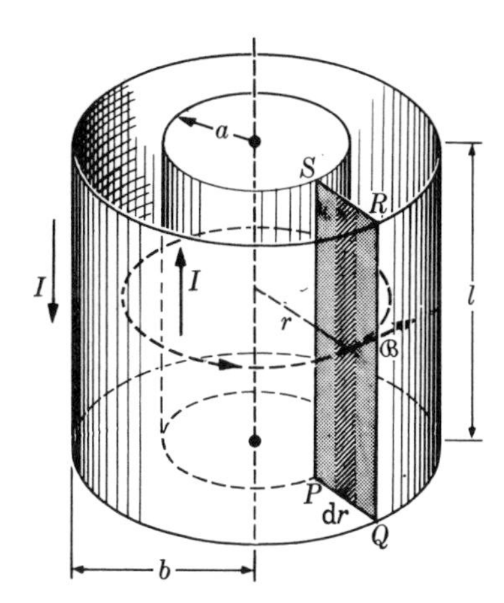

Figure 27.15

the magnetic flux is

$$\phi_I = \int_{PQRS} \mathscr{B}\, dS = \int_a^b \left(\frac{\mu I}{2\pi r}\right)(l\, dr) = \frac{\mu I l}{2\pi}\int_a^b \frac{dr}{r} = \frac{\mu I l}{2\pi}\ln\frac{b}{a}$$

Therefore the self-inductance of a portion of length l is

$$L = \frac{\phi_I}{I} = \frac{\mu l}{2\pi}\ln\frac{b}{a} \tag{27.19}$$

EXAMPLE 27.5

Establishment and decay of a current in a circuit with self-inductance.

▷ Consider a circuit having a resistance R and a self-inductance L. When an emf V_0 is applied to the circuit by closing a switch toward position 1 (Figure 27.16), the current does not instantaneously attain the value V_0/R corresponding to Ohm's law, but increases gradually, steadily approaching the value given by Ohm's law. This process is due to the self-induced emf $V_L = -L\,dI/dt$, which opposes the change in the current and is present while the current increases from zero up to its final constant value. The total emf applied to the circuit is then $V_0 + V_L$ so that

$$RI + V_L = V_0 \qquad \text{or} \qquad RI = V_0 - L\frac{dI}{dt} \tag{27.20}$$

Equation 27.21 may be written as

$$RI - V_0 = -L\frac{dI}{dt}$$

or, separating the variables I and t,

$$\frac{dI}{I - V_0/R} = -\frac{R}{L}\,dt$$

Noting that at $t = 0$ the current is also zero ($I = 0$), we can write the integral as

$$\int_0^I \frac{dI}{I - V_0/R} = -\frac{R}{L}\int_0^t dt$$

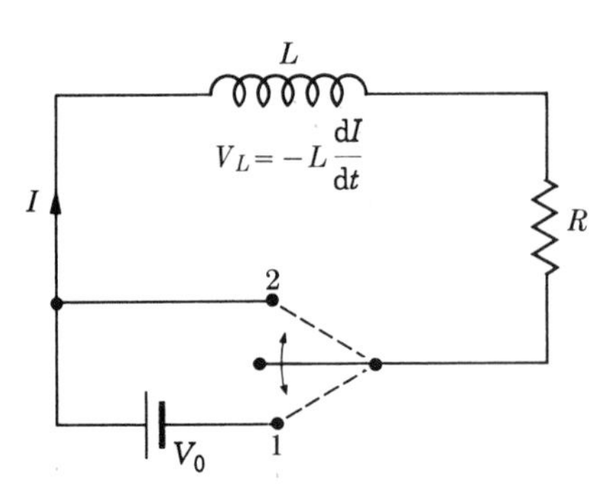

Figure 27.16 Circuit with a resistance and an inductance.

Integration gives

$$\ln\left(I - \frac{V_0}{RI}\right) - \ln\frac{-V_0}{R} = -\frac{R}{L}t$$

Remembering that $\ln e^x = x$, we have

$$\ln\left(I - \frac{V_0}{R}\right) = \ln\left(-\frac{V_0}{R}\right) + \ln e^{-Rt/L}$$

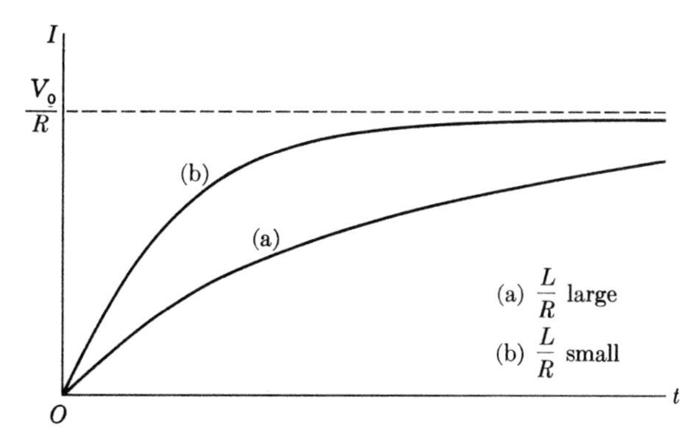

Figure 27.17 Establishment of a current in an RL-circuit.

which may be written as

$$I - \frac{V_0}{R} = -\frac{V_0}{R}\mathrm{e}^{-Rt/L} \qquad \text{or} \qquad I = \frac{V_0}{R}(1 - \mathrm{e}^{-Rt/L}) \tag{27.21}$$

The exponential term decreases with time and the current asymptotically approaches the value V_0/R, given by Ohm's law. The larger the resistance, or the smaller the inductance, the faster the current stabilizes at the value V_0/R. The quantity $\tau = L/R$ is called the **time constant** of the circuit. The larger (smaller) τ, the longer (shorter) time it takes the current to reach the limit value V_0/R. In Figure 27.17 we have shown the variation of the current with time for two values of the time constant.

Now consider the decay of the current in a circuit when the emf is removed without changing the resistance. Referring to Figure 27.16, if the switch has been in position 1 for a very long time, we may assume that the current in the circuit has achieved its limiting (or steady) value V_0/R. By moving the switch over to position 2, we remove the applied emf without actually opening the circuit. The only emf that remains is $V_L = -L\mathrm{d}I/\mathrm{d}t$ and Ohm's law becomes $V_L = RI$ or $RI = -L\mathrm{d}I/\mathrm{d}t$. If we measure time ($t = 0$) from the instant that V_0 is removed from the circuit, the initial current is V_0/R. Integrating, we have

$$\int_{V_0/R}^{I} \frac{\mathrm{d}I}{I} = -\frac{R}{L}\int_0^t \mathrm{d}t$$

Evaluating the integrals we obtain

$$\ln I - \ln \frac{V_0}{R} = -\frac{R}{L}t$$

or, using again $\ln \mathrm{e}^x = x$,

$$\ln I = \ln\left(\frac{V_0}{R}\right) + \ln \mathrm{e}^{-Rt/L} \qquad \text{or} \qquad I = \frac{V_0}{R}\mathrm{e}^{-Rt/L} \tag{27.22}$$

The current does not drop suddenly to zero but decreases exponentially, as shown in Figure 27.18. This is because the self-induction opposes the decrease in the current. The larger the resistance R, or the smaller the inductance L, the faster is the drop in the current. Incidentally,

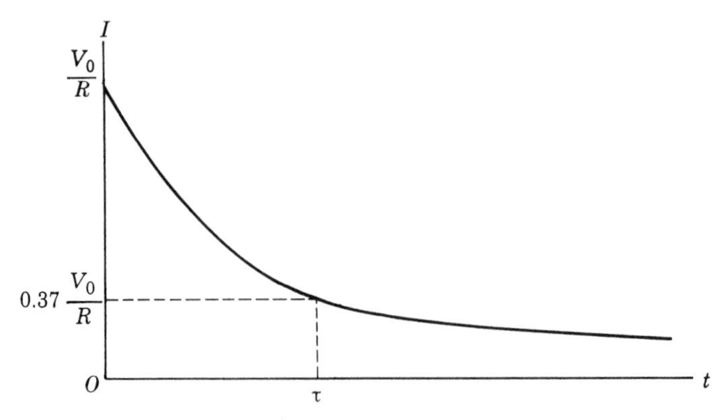

Figure 27.18 Decay of the current in a circuit after the emf has been removed.

in a superconductor, whose resistance is close to zero, a current may persist for extremely long times after the external emf is removed.

27.9 Free electrical oscillations

There are three parameters that characterize the flow of charges through an electric circuit: its capacitance C, its resistance R, and its self-inductance L. Previously we have considered the effect of each one separately. Now we shall analyze the way in which the three together determine the current produced by a given emf.

Consider first the situation where no external emf is applied to the circuit (Figure 27.19). The current in this case is started by charging the capacitor, or varying a magnetic flux through the inductance, or by inserting and later on removing an external emf. Equating the voltage drops across the resistor and the capacitor to the induced emf, we have

$$RI + \frac{q}{c} = V_L \qquad \text{or} \qquad RI = -L\frac{dI}{dt} - \frac{q}{C}$$

Taking the derivative of the whole equation with respect to t, we have

$$R\frac{dI}{dt} = -L\frac{d^2I}{dt^2} - \frac{1}{C}\frac{dq}{dt}$$

According to the direction of I we have chosen as positive, $I = dq/dt$ and placing all terms on the left-hand side of the equation, we get

$$L\frac{d^2I}{dt^2} + R\frac{dI}{dt} + \frac{1}{C}I = 0 \qquad \textbf{(27.23)}$$

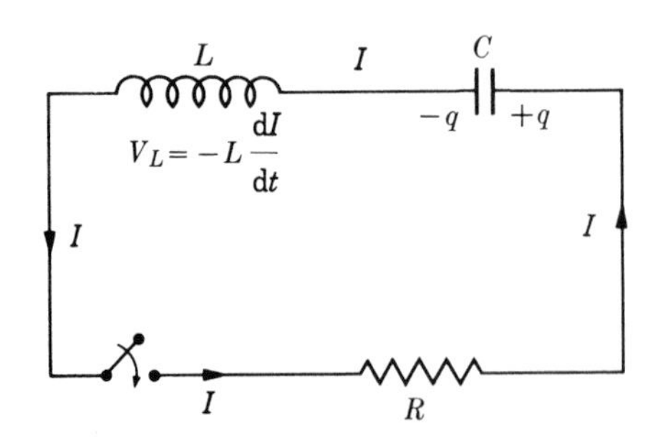

Figure 27.19 Series RCL circuit.

This is a differential equation whose solution gives the current I as a function of t and the parameters L, R, and C.

Suppose the resistance is negligible. Making $R = 0$ in Equation 27.23, we have

$$L\frac{d^2I}{dt^2} + \frac{1}{C}I = 0 \quad \text{or} \quad \frac{d^2I}{dt^2} + \frac{1}{LC}I = 0 \tag{27.24}$$

This equation is formally identical to Equation 10.12, $d^2x/dt^2 + \omega^2 x = 0$, for simple harmonic motion if we replace x by I and ω by

$$\omega_0 = \frac{1}{(LC)^{1/2}} \tag{27.25}$$

Accordingly, we conclude that the current in the circuit is oscillatory with angular frequency $\omega_0 = 1/(LC)^{1/2}$ and has constant amplitude I_0 (Figure 27.20). Therefore, the current can be expressed in the form $I = I_0 \sin \omega_0 t$.

Oscillation occurs because, as the capacitor discharges, the emf V_L of the self-inductance tends to maintain a current in the opposite direction that charges the capacitator. Once this is accomplished the process repeats in the opposite sense because the capacitor again tends to discharge. This back and forth flow of charge continues indefinitely as long as energy is not lost.

In the more general case where the resistance is not negligible, Equation 27.23 is formally identical to Equation 10.30, corresponding to the damped oscillations of a particle, if we establish the following correspondences:

$$L \leftrightarrow m, \qquad R \leftrightarrow \lambda, \qquad \frac{1}{C} \leftrightarrow k \tag{27.26}$$

The quantities γ and ω defined in Equation 10.31 are now

$$\gamma = \frac{R}{2L}, \qquad \omega = \sqrt{\frac{1}{LC} - \frac{R^2}{4L^2}} \tag{27.27}$$

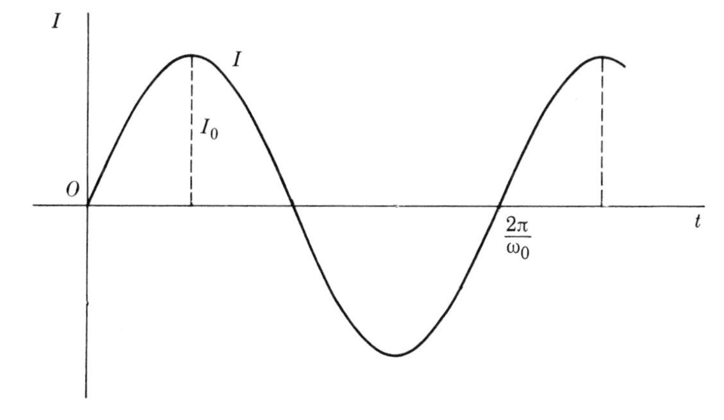

Figure 27.20 Current in an LC circuit ($R = 0$).

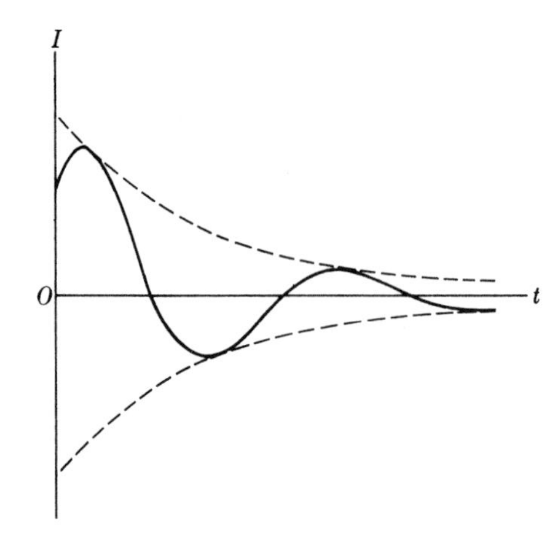

Figure 27.21 Variation of the current in an RCL circuit as a function of time when $R^2 < 4L/C$.

The resistance of the circuit makes the frequency of oscillation smaller than the characteristic frequency ω_0. The current is given as a function of time by expression 10.32, which we now write as

$$I = I_0 e^{-\gamma t} \sin(\omega t + \alpha) \qquad \textbf{(27.28)}$$

which the student may verify by direct substitution into Equation 27.23. The graph of the current versus time has been given in Figure 27.21. We see that an oscillatory or alternating current is established whose amplitude decreases with time. The damping in an electric circuit results from the dissipation of energy in the resistance. If the resistance is large enough so that $R^2/4L^2 > 1/LC$, the frequency ω becomes imaginary. In this case the current decreases gradually without oscillation because the circuit is 'overdamped', or too much energy is dissipated in the resistance.

27.10 Forced electrical oscillations: alternating current circuits

Forced electric oscillations are produced when we add an alternating emf of the form $V = V_0 \sin \omega t$, to an RCL circuit, as shown in Figure 27.22. This arrangement is called an alternating current (a.c.) series circuit. Equating the voltage drops across the resistor and the capacitor to the total emf, we have

$$RI + q/c = V_L + V_0 \sin \omega t$$

Inserting the value $V_L = -L(dI/dt)$, we obtain

$$RI = -L\frac{dI}{dt} - \frac{q}{C} + V_0 \sin \omega t \qquad \textbf{(27.29)}$$

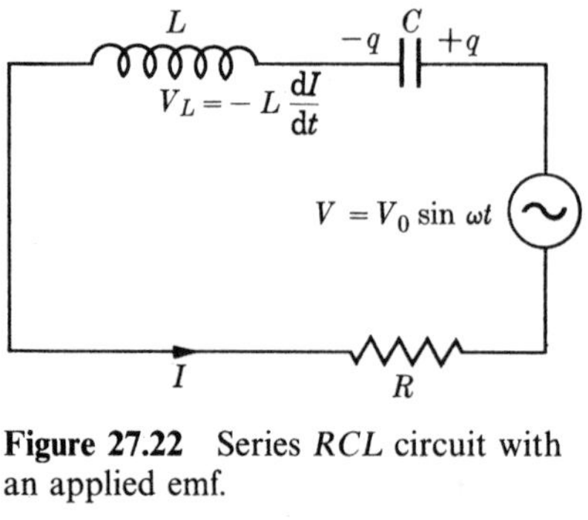

L
C
$-q$ $+q$
$V_L = -L\dfrac{dI}{dt}$
$V = V_0 \sin \omega t$
I
R

Figure 27.22 Series RCL circuit with an applied emf.

We differentiate the whole equation with respect to time and arrange the terms as

$$L\frac{d^2I}{dt^2} + R\frac{dI}{dt} + \frac{I}{C} = \omega V_0 \cos \omega t \qquad \textbf{(27.30)}$$

This equation is very similar to Equation 10.34 for forced oscillations of a particle, with this important difference: the frequency ω appears as a factor on the right-hand side of Equation 27.30. The reason for this is that, due to the relation

$I = dq/dt$, the current in an electric circuit corresponds to the velocity $v = dx/dt$ in the motion of a particle. We may now apply the formulas of Note 10.3, with the proper correspondence for the quantities, as given by Equation 27.26. We then have that the current is given by

$$I = I_0 \sin(\omega t - \alpha) \tag{27.31}$$

where the current amplitude is

$$I_0 = \frac{V_0}{[R^2 + (\omega L - 1/\omega C)^2]^{1/2}} \tag{27.32}$$

and

$$\tan \alpha = \frac{\omega L - 1/\omega C}{R} \tag{27.33}$$

gives the phase of the current relative to the applied emf. We can express the **impedance** of the electric circuit as

$$Z = [R^2 + (\omega L - 1/\omega C)^2]^{1/2} \tag{27.34}$$

The **reactance** of the circuit is

$$X = \omega L - \frac{1}{\omega C} \tag{27.35}$$

so that

$$Z = (R^2 + X^2)^{1/2} \tag{27.36}$$

and the phase difference α between the current and the applied emf is obtained from

$$\tan \alpha = \frac{X}{R} \tag{27.37}$$

The quantities Z, R, X and α are related as shown in Figure 27.23, which is a reproduction of Figure 10.31. Note that both the reactance and impedance are expressed in ohms. For example, the term ωL expressed in terms of the fundamental units gives $s^{-1}\,H = m^2\,kg\,s^{-1}\,C^{-2}$, which is the same expression obtained in Section 27.2 for the ohm. We can make the same verification for the term $1/\omega C$.

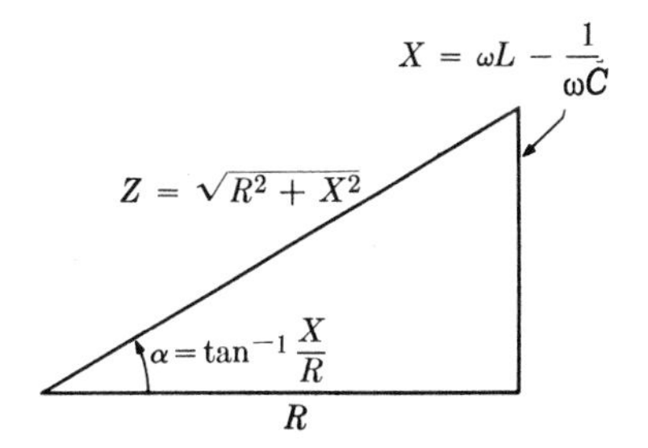

Figure 27.23 Relation between resistance, reactance, and impedance.

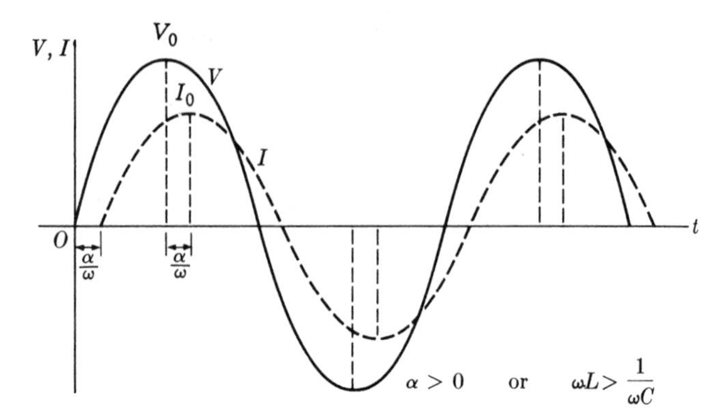

Figure 27.24 Variation of current and emf as a function of time in an a.c. circuit.

If R and X are expressed in ohms, then Z, in view of its definition 27.36, must also be expressed in ohms.

The current I lags (leads) the emf when α is positive (negative), which happens when ωL is larger (smaller) than $1/\omega C$. Figure 27.24 gives the plot of V and I versus time when I *lags* V by a time $t = \alpha/\omega$.

The **average power** required to maintain the current is obtained from Equation 10.44 with the proper correspondence. That is,

$$P_{ave} = \tfrac{1}{2}V_0 I_0 \cos\alpha = \tfrac{1}{2}RI_0^2 \tag{27.38}$$

Resonance, as discussed in Section 10.14, is obtained when P_{ave} is maximum, which occurs when $\alpha = 0$ or when $\omega L = 1/\omega C$, corresponding to a frequency $\omega = (1/LC)^{1/2}$, equal to the characteristic frequency ω_0 of the circuit (Equation 27.25). At resonance the current has maximum amplitude and is in phase with the emf, which results in maximum average power. In this case the applied emf V and the current I are in phase, as shown in Figure 27.25.

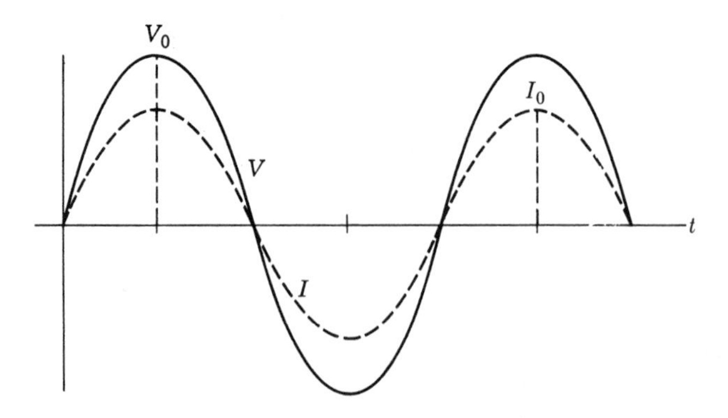

Figure 27.25 At resonance, V and I are in phase.

EXAMPLE 27.6

A circuit has a resistance of 40 Ω, a self-inductance of 0.1 H, and a capacity of 10^{-5} F. The applied emf has a frequency of 60 Hz. Calculate the impedance, the phase difference, and the resonant frequency.

▷ The angular frequency is $\omega = 2\pi\nu$, and since $\nu = 60$ Hz, we find that $\omega = 376.8\ s^{-1}$. Therefore, using Equation 27.35, we obtain $X = \omega L - 1/\omega C = -227.57\ \Omega$. The impedance is then $Z = (R^2 + X^2)^{1/2} = 231.2\ \Omega$. The phase shift, according to Equation 27.37, is $\tan\alpha = X/R = -5.680$ or $\alpha = -80°21'$. Therefore the current is leading the emf. For the resonant frequency, we find, using Equation 27.25, that

$$\omega_0 = \left(\frac{1}{LC}\right)^{1/2} = 10^3\ s^{-1} \qquad \text{or} \qquad \nu = \frac{\omega}{2\pi} = 159\ \text{Hz}$$

Note 27.2 Application of the method of rotating vectors to a.c. electric circuits

The results of Section 27.11 can be derived by means of the technique of rotating vectors or phasors, which was explained in Note 10.1. We shall start by considering some simple circuits to which an alternating emf $V = V_0 \sin \omega t$ is applied

(i) *R*-circuit (Figure 27.26). When a circuit is composed of just a resistor, Equation 27.29 becomes

$$RI = V_0 \sin \omega t \quad \text{or} \quad I = \frac{V_0}{R} \sin \omega t$$

Therefore the current in the circuit is in phase ($\alpha = 0$) with the emf and its amplitude I_0 is such that $I_0 = V_0/R$ or $V_0 = RI_0$. The vector diagram is shown in Figure 27.26. Note that $V_R = RI$ is the voltage drop across the resistor, which is in phase with the current.

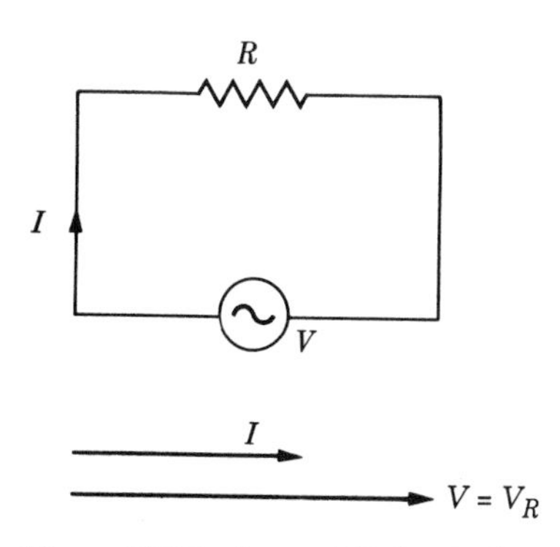

Figure 27.26 In an RV-circuit the current and voltage are in phase.

(ii) *L*-circuit (Figure 27.27). When the circuit consists of an inductor with negligible resistance, Equation 27.29

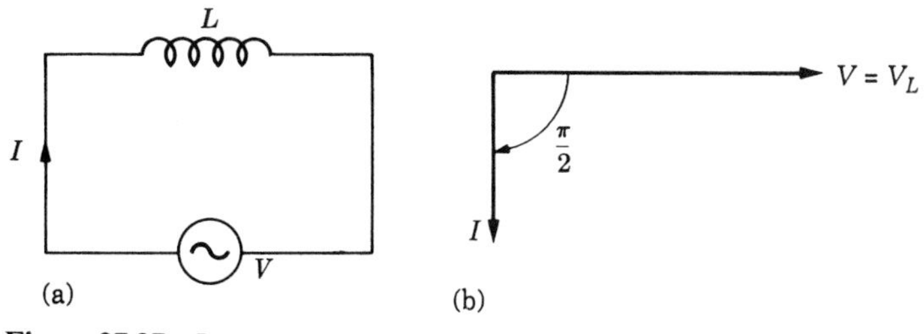

Figure 27.27 In an LV-circuit the current lags the voltage by $\pi/2$.

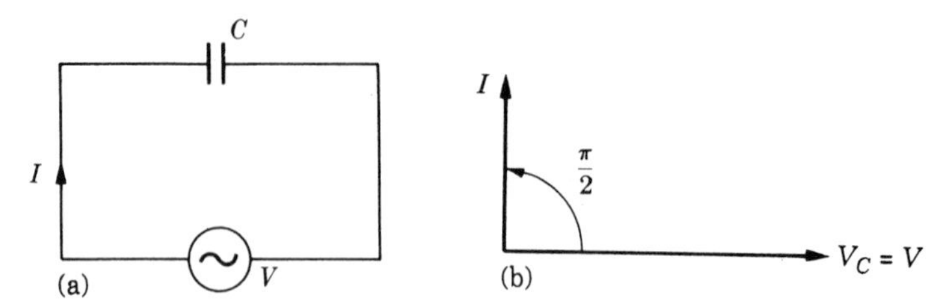

Figure 27.28 In a CV-circuit the current leads the voltage by $\pi/2$.

reduces to

$$L\frac{dI}{dt} = V_0 \sin \omega t$$

which shows that the voltage drop across the inductor is equal to the applied emf. Integrating,

$$I = -\frac{V_0}{\omega L}\cos \omega t = \frac{V_0}{\omega L}\sin\left(\omega t - \frac{\pi}{2}\right)$$

Therefore the current in the circuit *lags* the emf by an angle of $\pi/2$ and has an amplitude I_0 such that $I_0 = V_0/\omega L$ or $V_0 = \omega L I_0$. Conversely, the voltage across the inductance *leads* the current by $\pi/2$. The vector diagram is shown in Figure 27.27.

(iii) *C*-circuit (Figure 27.28). When the circuit consists of just one capacitor, Equation 27.29 gives

$$\frac{q}{C} = V_0 \sin \omega t$$

which implies that the voltage drop across the capacitor is equal to the applied emf. Taking the derivative of this equation (keeping in mind that $I = dq/dt$), we get

$$I = \omega C V_0 \cos \omega t = \omega C V_0 \sin\left(\omega t + \frac{\pi}{2}\right)$$

Therefore the current *leads* the emf by $\pi/2$ and has an amplitude I_0 such that $I_0 = \omega C V_0$ or $V_0 = (1/\omega C) I_0$. Conversely, the voltage drop across the capacitor *lags* the current by $\pi/2$. The vector diagram is shown in Figure 27.28.

(iv) *RCL*-series circuit (Figure 27.29). Consider a circuit composed of a resistor, an inductor and a capacitor in series. Then Equation 27.32 can be written in the form

$$RI + L\frac{dI}{dt} + \frac{1}{C}q = V_0 \sin \omega t$$

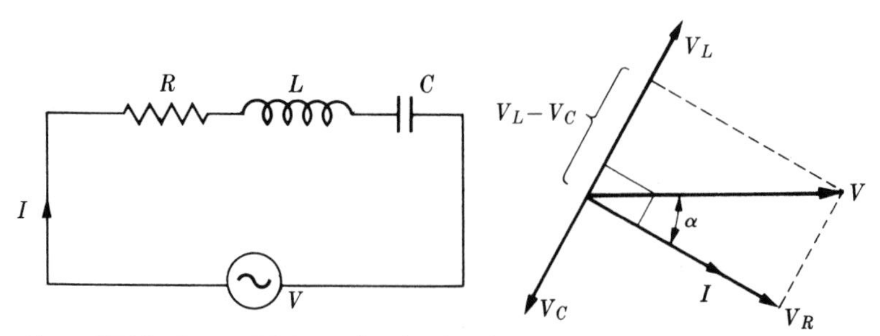

Figure 27.29 In an RLC–V circuit, the phase lag is α.

In the above equation we have $V_R = RI$, $V_L = L(\mathrm{d}I/\mathrm{d}t)$ and $V_C = q/C$ as the respective potential differences (or voltage drops) across the resistor, the inductor and the capacitor, so that $V_R + V_L + V_C = V$, where V is the applied emf. If we assume that $I = I_0 \sin(\omega t - \alpha)$, the rotating vector of the current lags that of the emf V by the angle α (Figure 27.29). We may now consider that the rotating vector of the emf is the sum of the rotating vectors corresponding to the three terms $V_R + V_L + V_C$ in the above equation. Therefore we may write, using the results of (i), (ii) and (iii) above, that

Potential drop	Amplitude	Phase
V_R	RI_0	*in phase* with I
V_L	ωLI_0	*leads* I by $\pi/2$
V_C	$\dfrac{1}{\omega C}I_0$	*lags* I by $\pi/2$

The three rotating vectors are shown in Figure 27.29, where the reference line is given by the rotating vector corresponding to V. Since the amplitude of their resultant must be equal to V_0, and V_L and V_C are in opposite directions, the relation between the amplitudes is

$$V_0 = I_0\sqrt{R^2 + \left(\omega L - \frac{1}{\omega C}\right)^2}$$

If we solve this equation for I_0, we obtain the result of Equation 27.32. From the figure we can compute the phase angle α, obtaining the value given in Equation 27.37, that is,

$$\tan\alpha = \frac{V_L - V_C}{V_R} = \frac{\omega L - 1/\omega C}{R}$$

To have resonance, the applied emf and the current must be in phase and therefore $\alpha = 0$, which requires $\omega L = 1/\omega C$. This happens when $V_L = V_C$; that is, when the inductance and capacitor voltages cancel each other. Recalling Note 10.1, the sharpness of the resonance in an a.c. circuit when the frequency of the applied emf is very close to the natural frequency $\omega_0 = 1/(LC)^{1/2}$, and using the correspondence $L \leftrightarrow m$ and $R \leftrightarrow \lambda$, is measured by the *Q-value* of the circuit given by

$$Q = \frac{L\omega_0}{R} \quad \text{or} \quad Q = \frac{1}{R}\left(\frac{L}{C}\right)^{1/2}$$

We recall that the larger Q, the sharper is the resonance. This is very important in tuning radio and TV sets. ∎

27.11 Energy of the electromagnetic field

In Section 24.4 we saw that to maintain a current I through a potential difference ΔV, a power or energy per unit time $(\Delta V)I$ is required. When the current is in a closed circuit, we must replace ΔV by the applied emf. Hence

$$\text{power in a closed circuit} = \text{emf} \times \text{current}$$

In the case of a self-induced current, the emf is $V_L = -L\,\mathrm{d}I/\mathrm{d}t$. This emf opposes the change in current. Therefore, to build up the current, energy must be supplied from some external source. Thus

$$\text{rate of energy supply} = -V_L I = \left(L\frac{\mathrm{d}I}{\mathrm{d}t} \right) I$$

where the negative sign is introduced because we are computing the energy supplied to the circuit and that requires us to apply an emf to oppose the induced emf V_L. We may interpret this expression as the rate at which energy must be supplied to build up the magnetic field associated with the current. Therefore, the rate of change of the magnetic energy of a current is

$$\frac{\mathrm{d}E}{\mathrm{d}t} = LI\frac{\mathrm{d}I}{\mathrm{d}t} \tag{27.39}$$

The magnetic energy required to increase a current from zero to the value I is

$$E = \int_0^E \mathrm{d}E = \int_0^I LI\,\mathrm{d}I \qquad \text{or} \qquad E = \tfrac{1}{2}LI^2 \tag{27.40}$$

For example, in the circuit of Example 27.4, the magnetic energy of a section of length l is, using Equation 27.19,

$$E = \frac{1}{2}\left(\frac{\mu l}{2\pi} \ln\frac{b}{a} \right) I^2 = \frac{\mu l I^2}{4\pi} \ln\frac{b}{a} \tag{27.41}$$

As indicated in Section 26.8, the magnetic energy corresponding to a magnetic field $\mathscr{B}$ can also be calculated by using the expression

$$E = \frac{1}{2\mu} \int \mathscr{B}^2 \,\mathrm{d}v \tag{27.42}$$

where the integral extends throughout all the volume where the magnetic field exists, and $\mathrm{d}v$ is a volume element. It can be shown that if we use in Equation 27.42 the value of $\mathscr{B}$ for the magnetic field corresponding to a given current, we recover Equation 27.40. The calculation for the current in Example 27.3 has been carried out below.

Using the same logic as in Section 26.8 for a moving electric charge, we may say that the energy spent in establishing the current has been *stored* in the magnetic field in the surrounding space so that the energy per unit volume or magnetic energy density is

$$\mathrm{E}_{\mathrm{mag}} = \frac{1}{2\mu}\mathscr{B}^2 \tag{27.43}$$

When both electric and magnetic fields are present, as required by the Faraday–Henry law, we must also consider the electric energy density given by Equation 25.39, and thus the total energy per unit volume in the electromagnetic field is

$$\mathrm{E} = \frac{1}{2}\varepsilon\mathscr{E}^2 + \frac{1}{2\mu}\mathscr{B}^2 \qquad \textbf{(27.44)}$$

EXAMPLE 27.7

Calculation of the magnetic energy density of the circuit of Figure 27.15.

▷ In Example 27.3, the magnetic field within the two cylinders was found to be $\mathscr{B} = \mu I/2\pi r$. When we take a cylindrical shell of radius r and thickness $\mathrm{d}r$ as the volume element, we find that its volume is $\mathrm{d}V = (2\pi r)l\,\mathrm{d}r$. Substituting in Equation 27.42, and remembering that the magnetic field extends only from $r = a$ to $r = b$, we find that the energy of the magnetic field in the region between the two cylinders is

$$E = \frac{1}{2\mu}\int_a^b \left(\frac{\mu I}{2\pi r}\right)^2 (2\pi l r\,\mathrm{d}r) = \frac{\mu l I^2}{4\pi}\int_a^b \frac{\mathrm{d}r}{r} = \frac{\mu l I^2}{4\pi}\ln\frac{b}{a}$$

We obtain, therefore, the same result as in Equation 27.41.

Although we have justified expression 27.42 for the magnetic energy by using a circuit of very special symmetry, a more detailed analysis, which will not be given here, would indicate that the result is completely general.

27.12 Coupled circuits

Consider two circuits such as (1) and (2) in Figure 27.30. When a current I_1 circulates in circuit (1), a magnetic field proportional to I_1 is established around it, and through circuit (2) there is a magnetic flux ϕ_2 that is also proportional to I_1. We may then write

$$\phi_2 = MI_1 \qquad \textbf{(27.45)}$$

where M is a coefficient representing the magnetic flux through circuit (2) per unit current in circuit (1). Similarly, if a current I_2 circulates in circuit (2), a magnetic field is produced, and it in turn produces a magnetic flux ϕ_1 through circuit (1),

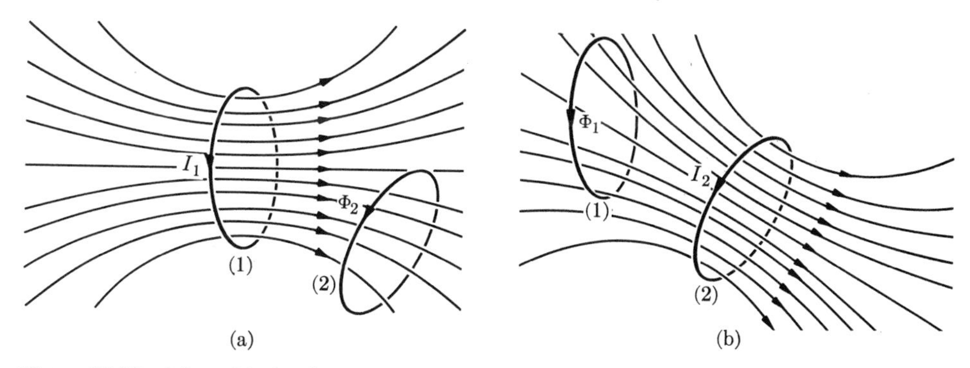

Figure 27.30 Mutual induction.

proportional to I_2. Hence we may write

$$\phi_1 = MI_2 \tag{27.46}$$

Note that in Equation 27.46 we have written the same coefficient M as in Equation 27.45. This means that the magnetic flux through circuit (1) due to a unit current in circuit (2) is the same as the magnetic flux through circuit (2) due to a unit current in (1), a statement that may be justified mathematically. The coefficient M is called the **mutual inductance** of the two circuits. The mutual inductance M depends on the shapes of the circuits and their relative orientation; it is also measured in henrys, since it corresponds to Wb A^{-1}.

If the current I_1 is variable, the flux ϕ_2 through circuit (2) changes and an emf V_{M2} is induced in this circuit, given by

$$V_{M2} = -M\frac{dI_1}{dt} \tag{27.47}$$

In writing this equation, we assume that the circuits are rigid and fixed in space, so that M is constant. Similarly, if the current I_2 is variable, an emf V_{M1} is induced in circuit (1), given by

$$V_{M1} = -M\frac{dI_2}{dt} \tag{27.48}$$

This is why M is called 'mutual inductance', since it describes the mutual effect or influence between the two circuits. In addition, if the circuits move relative to each other, resulting in a change in M, emfs are also induced in them.

The phenomenon of mutual induction indicates that there is an exchange of energy between two circuits when their currents vary with time. We then say that the circuits are coupled electromagnetically. A common and practical application of mutual induction is the **transformer** (see Example 27.9). Another application of mutual induction, in a broader sense, is the transmission of a signal from one place to another by producing a variable current in one circuit, called the **transmitter**. This circuit in turn acts on another circuit coupled to it, called the **receiver**. This is the case for the telegraph and for radio, television, radar etc. The discussion of these devices, however, requires a different technique, to be considered in Chapter 29.

The most important and fundamental aspect of mutual induction is that

energy can be exchanged between two circuits via the electromagnetic field.

We may say that the electromagnetic field produced by the currents in the circuits acts as a carrier of energy, transporting the energy through space from one circuit to the other. Mutual induction between two circuits is a macroscopic phenomenon, resulting from elementary interactions between the moving charges that constitute their respective currents. Thus we may conclude that

the electromagnetic interaction between two charged particles can be described as an exchange of energy via their electromagnetic field.

EXAMPLE 27.8

Mutual induction of a system composed of a coil with N turns wrapped around a toroidal solenoid having n turns per unit length and a cross-section of area S (Figure 27.31).

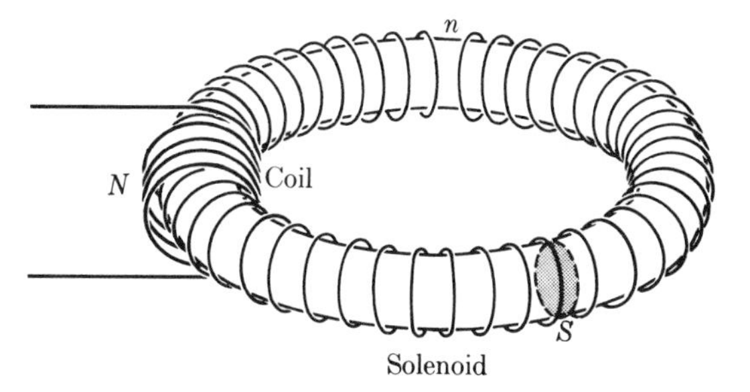

Figure 27.31

▷ We may solve the problem either by finding the magnetic flux through the solenoid when there is a current along the coil, or conversely, by finding the magnetic flux through the coil when there is a current along the solenoid. The second procedure, which is the easier of the two, will be followed. We may recall from Example 26.10 that, in the case of a toroidal solenoid, the magnetic field is confined to its interior and has a value given by Equation 26.40, $\mathscr{B} = \mu_0 nI$. The magnetic flux through any cross-section of the solenoid is $\phi_m = \mathscr{B}S = \mu_0 nSI$, where S is the cross-sectional area of the solenoid. This is the same as the flux through any turn of the coil, even if its cross section is larger. Therefore the magnetic flux through the coil is $\phi_{coil} = N\phi_m = \mu_0 nNSI$. Comparison with Equation 27.45 gives the mutual inductance of the system as $M = \mu_0 nNS$. This arrangement is widely used in the laboratory when a standard mutual inductance is required.

EXAMPLE 27.9

The transformer.

▷ A **transformer** consists of two coupled circuits, called the primary and the secondary, respectively. When a variable emf is applied to the primary, an emf, also variable, is produced in the secondary. Usually the primary and the secondary are wound around an iron core (Figure 27.32) with the purpose of concentrating the magnetic flux. When a variable emf V_1 is applied to the primary, consisting of N_1 turns of wire, a current is produced in the primary, which in turn produces a magnetic field contained almost entirely within the iron core. If ϕ is the magnetic flux through one turn of the primary, the total flux through the primary is $N_1\phi$ and we may write $V_1 = -N_1\,d\phi/dt$. The same flux ϕ passes through the N_2 turns of the secondary. Therefore the emf that appears in the secondary is $V_2 = -N_2\,d\phi/dt$. The ratio of the two emfs is $V_2/V_1 = N_2/N_1$ which means that the emf that appears in the secondary is larger or smaller than the emf applied to the primary in the proportion N_2/N_1. When N_2/N_1 is larger (smaller) than one, we talk of a step-up (-down) transformer.

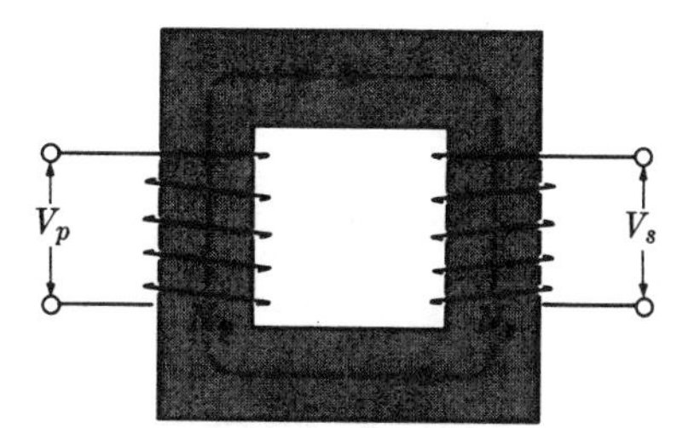

Figure 27.32 Transformer.

The analysis we have presented is highly simplified and has not taken into account several factors, such as flux losses, energy losses, and phase differences, as well as the effect of the load connected to the secondary.

QUESTIONS

Part A

27.1A What happens to the induced emf in a circuit if the rate of change of the magnetic flux through the circuit is (a) doubled, (b) reduced by one-half?

27.2A Does a static magnetic field require the existence of an electric field? What is the situation when the magnetic field is time-dependent?

27.3A Does a static electric field require the existence of a magnetic field? What is the situation when the electric field is time-dependent?

27.4A Explain why the principle of electromagnetic induction, as expressed in the Faraday–Henry law, is an illustration of the principle of relativity.

27.5A Compare the acceleration mechanism in a betatron with that in a cyclotron and state the main differences.

27.6A What are the 'sources' of an electromagnetic field?

27.7A Discuss the physical basis for introducing the concept of magnetic energy.

27.8A Is the principle of conservation of charge as universally valid as the principle of conservation of momentum?

Part B

27.1B Consider two coaxial identical circuits 1 and 2 separated a certain distance. There is a current I_1 in circuit 1. Determine the direction of the current in circuit 2 when (a) circuit 2 moves toward circuit 1, (b) circuit 2 moves away from circuit 1, (c) the current I_1 increases, (d) the current I_1 decreases.

27.2B Are two electric circuits coupled by their mutual inductance equivalent to a pair of mechanically coupled oscillators?

27.3B Why, in the circuit of Figure 27.15, is the magnetic field zero inside the inner cylinder and outside the outer cylinder?

27.4B Verify that the time constant $\tau = L/R$ gives the time required by the current in the circuit of Figure 27.16 to attain a value equal to 63% of its maximum value as well as for the current to be reduced to 37% of its maximum value when the circuit is opened.

27.5B Explain why the charge in the capacitance and the current in the inductance in an LC-circuit have a phase difference of $\pi/2$.

27.6B Consider an LC-circuit. Express the electric energy, the magnetic energy, and the total energy as functions of time. Plot these energies as functions of time during one period. Discuss the exchanges of energy that take place during the period.

27.7B State the conditions for the current to (a) lag, (b) lead, (c) be in phase with, the emf in an a.c. circuit. Plot the rotating vectors of the current and the emf in each case.

27.8B By looking at Figure 27.4, state a right-hand rule for determining the sense of eddy currents in a conducting material.

27.9B A solenoid is wound around a cylindrical conductor, as shown in Figure 27.33. Indicate the sense of the eddy currents in the cylinder when the current in the solenoid (a) increases, (b) decreases. What happens if the current alternately increases and decreases?

27.10B A magnet is moved toward a metal plate as shown in Figure 27.34. Show the direction of the eddy currents induced in the plate. Repeat if the magnet is moved away. Again repeat if the south pole points toward the plate. Analyze, in

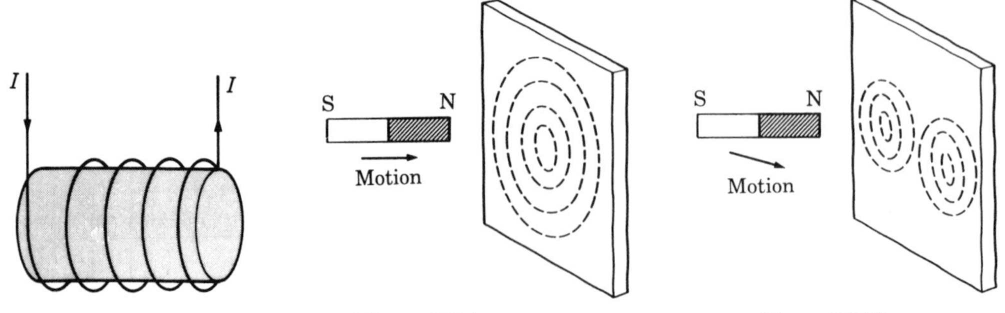

Figure 27.33 Figure 27.34 Figure 27.35

each case, the force between the magnet and the plate.

27.11B A magnet is moved parallel to a metal plate, as shown in Figure 27.35. Draw the direction of the eddy currents induced in the plate. Note that the magnetic field increases ahead of the magnet and decreases behind. Repeat if the south pole points toward the plate. In each case, analyze the force between the magnet and the plate.

PROBLEMS

27.1 A coil consisting of 200 turns and having a radius of 0.10 m is placed perpendicular to a uniform magnetic field of 0.2 T. Find the emf induced in the coil if, in 0.1 s, (a) the field is doubled, (b) the field is reduced to zero, (c) the field is reversed in direction, (d) the coil is rotated 90°, (e) the coil is rotated 180°. For each case draw a diagram, showing the direction of the emf.

27.2 Referring to the figure of Problem 26.11, if the current varies according to $I = I_0 \sin \omega t$, calculate the emf induced in the circuit.

27.3 The magnetic field $\mathscr{B}$ at all points within the dashed circle of Figure 27.36 equals 0.5 T. It is directed into the plane of the paper and is

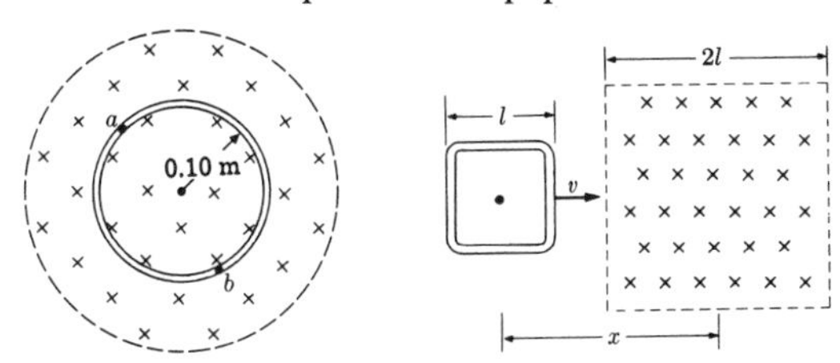

Figure 27.36 **Figure 27.37**

decreasing at the rate of $0.1\ \mathrm{T\,s^{-1}}$. (a) Draw the shape of the lines of force of the induced electric field within the dashed circle. (b) What are the magnitude and direction of this field at any point of the circular conducting ring, and what is the emf in the ring? (c) Find the potential difference between any two points of the ring separated by a distance s. (d) If the ring is cut at some point and the ends separated slightly, calculate the potential difference between the ends. Repeat if the magnetic field is increasing.

27.4 A square loop of wire is moved at constant velocity v across a uniform magnetic field confined to a square region whose sides are twice the length of those of the square loop (see Figure 27.37). Sketch a graph of the induced emf in the loop as a function of x, from $x = -2l$ to $x = +2l$, plotting clockwise emfs upward and counter-clockwise emfs downward.

27.5 A rectangular loop is moved through a region in which the magnetic field is given by

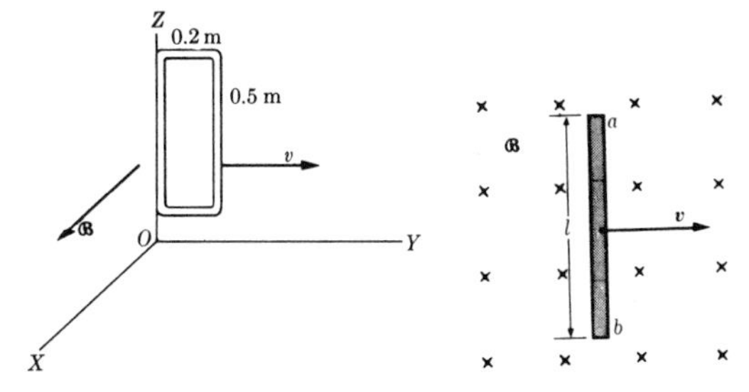

Figure 27.38 **Figure 27.39**

$\mathscr{B}_x = (6 - y)\mathrm{T}$, $\mathscr{B}_y = \mathscr{B}_z = 0$ (see Figure 27.38). Find the emf in the loop as a function of time, with $t = 0$ when the loop is in the position shown in the figure (a) if $v = 2\ \mathrm{m\,s^{-1}}$, (b) if the loop starts at rest and has an acceleration of $2\ \mathrm{m\,s^{-2}}$ (c) Repeat part (a) for motion parallel to OZ instead of OY.

27.6 Suppose that the loop in Problem 27.5 is pivoted about the OZ-axis. (a) Calculate the average emf during the first 90° of rotation if the period of rotation is 0.2 s. (b) Calculate the instantaneous emf as a function of time.

27.7 In Figure 27.39, let $l = 1.5$ m, $\mathscr{B} = 0.5$ T and $v = 4\ \mathrm{m\,s^{-1}}$. (a) Find the potential difference between the ends of the conductor. (b) Which end is at the higher potential?

27.8 The cube in Figure 27.40, one meter on a side, is in a uniform magnetic field of 0.2 T directed along the Y-axis. Wires A, C and D move in the directions indicated, each with a velocity of $0.5\ \mathrm{m\,s^{-1}}$. Calculate the potential difference between the ends of each wire.

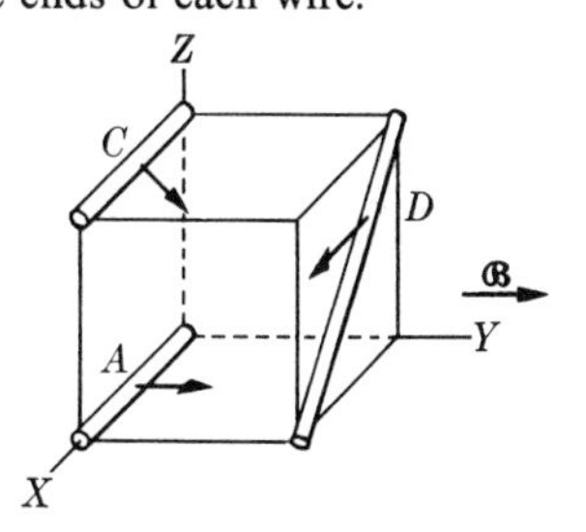

Figure 27.40

27.9 A metallic disk of radius a rotates with angular velocity ω in a plane where there is a uniform magnetic field $\mathscr{B}$ parallel to the disk axis. Verify that the potential difference between the center and the rim is $\frac{1}{2}\omega a^2 \mathscr{B}$.

27.10 Find the self-inductance of a toroidal solenoid of N turns. Assume that the radius of the coils is very small compared with the radius of the torus.

27.11 The magnetic flux through a circuit carrying a current of 2 A is 0.8 Wb. (a) Find its self-inductance. (b) Compute the emf induced in the circuit if, in 0.2 s, the current is (i) doubled, (ii) reduced to zero, (iii) reversed.

27.12 If the rectangular circuit of Figure 26.26 is moving away from the rectilinear current with velocity v, find the induced emf. (Hint: remember Problem 26.11 and note that $v = \mathrm{d}r/\mathrm{d}t$.)

27.13 Calculate the magnetic energy density (a) at a distance of 1.0 m from an infinitely long rectilinear wire carrying a current of 2.0 A; (b) within a toroid of radius 0.5 m with 2000 turns and a current of 1 A.

27.14 A solenoid has a length of 0.30 m and a cross-section of 1.2×10^{-3} m^2. Around its central section a coil of 300 turns is wound. Determine (a) their mutual inductance, (b) the emf in the coil if the initial current of 2 A in the solenoid is reversed in 0.2 s. The solenoid has 2000 turns.

27.15 Coils A and B have respectively 200 and 800 turns. A 2 A current in A produces a magnetic flux of 1.8×10^{-4} Wb in each turn of B. Compute (a) the coefficient of mutual inductance, (b) the magnetic flux through A when there is a current of 4 A in B, (c) the emf induced in B when the current in A changes from 3 A to 1 A in 0.3 s.

27.16 A coil having N turns is placed around a very long solenoid with cross-section S and n turns per unit length. Verify that the mutual inductance of the system is $\mu_0 nNS$.

27.17 A coil having 1000 turns is wound around a very long solenoid having 10^4 turns per meter and a cross-section of 2×10^{-3} m^2. The current in the solenoid is 10 A. In a short time the current in the solenoid is (a) doubled, (b) reduced to zero, (c) reversed. Find the emf induced in the coil for each case.

27.18 An alternating emf has a voltage amplitude of 100 V and an angular frequency of 120π rad s^{-1}. It is connected in series with a resistor of 1 Ω, a self-inductor of 3×10^{-3} H and a capacitor of 2×10^{-3} F. Determine (a) the amplitude and phase of the current and (b) the potential difference across the resistor, the capacitor and the inductor. (c) Make a diagram showing the rotating vectors corresponding to the applied emf, the current and the three potential differences. (d) Verify that the three potential difference vectors add to the emf vector.

27.19 A circuit is composed of a resistance and an inductance in series to which an alternating emf $V = V_0 \sin \omega t$ is applied. Show that the impedance of the circuit is $[R^2 + (\omega L)^2]^{1/2}$, and that the current lags the emf by an angle $\tan^{-1}(\omega L/R)$. (Hint: use the technique of rotating vectors.)

27.20 Repeat the preceding problem for a circuit composed of (a) a resistance and a capacitor and (b) an inductance and a capacitor.

27.21 A circuit consists of an alternating emf having a maximum value of 100 V, a resistance of 2 Ω, and a self-inductance of 10^{-3} F, all connected in series. Find the maximum value of the current for the following values of the angular frequency of the emf: (a) 0, (b) 10 s^{-1}, (c) 10^2 s^{-1}, (d) resonance, (e) 10^4 s^{-1}, and (f) 10^5 s^{-1}. Plot the current against the logarithm of the frequency.

27.22 A circuit is composed of a resistance and an inductance in parallel to which an emf $V = V_0 \sin \omega t$ is applied. Verify that the resultant impedance of the circuit is given by

$$\frac{1}{Z} = \left(\frac{1}{R^2} + \frac{1}{\omega^2 L^2}\right)^{1/2}$$

and the phase by $\tan^{-1}(R/\omega L)$. (Hint: note that the rotating vectors of the currents through R and L must add to the current through the emf.)

27.23 A circuit is composed of an inductance and a capacitor in parallel, connected in series with a resistance and an emf $V = V_0 \sin \omega t$. (a) Verify that the impedance of the circuit is

$$Z = \left(R^2 + \frac{\omega^2 L^2}{(1 - \omega^2 LC)^2}\right)^{1/2}$$

(b) Calculate the impedance when $\omega = 1/(LC)^{1/2}$. (In this case, there is said to be **antiresonance**.) (Hint: note that the rotating vectors of the currents through L and C must add to the current through R, while the rotating vectors corresponding to the potential difference must be identical for C and L.)

28 Wave motion

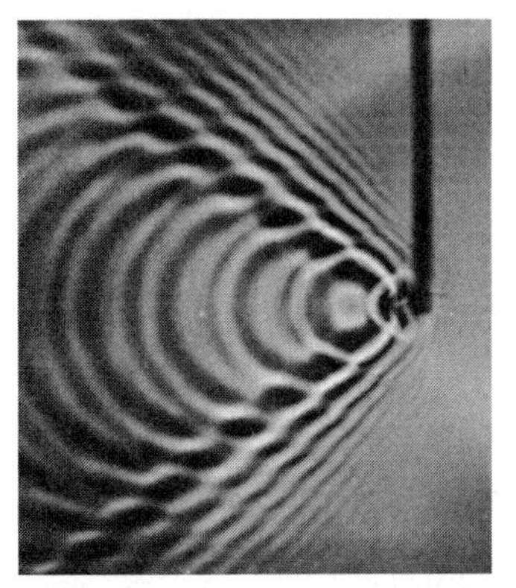

Shock waves (or Mach waves) are produced when the source of the waves moves faster than the velocity of propagation of the waves in the medium through which the source moves, resulting in a conical wave front. The aperture of the cone depends on the velocity of the source relative to the medium. The figure shows the shock waves produced when a rapidly vibrating reed moves, touching the water surface. Shock waves are also produced by supersonic planes. Particles moving in a medium at a very fast rate also produce electromagnetic shock waves called Cerenkov radiation.

Notes

28.1 Introduction

A common phenomenon is wave motion, such as waves on the surface of water, the transverse motion along a stretched string, or the vibration of a spring. Physicists have extended the concept of a wave to a large number of phenomena that do not resemble those on the surface of water, but correspond to physical situations described by a time-dependent field that propagates in both space and time.

There are several types of waves, such as elastic and electromagnetic waves. The most important aspects of waves are the velocity of their propagation and the modifications they suffer when the physical properties of the medium change (reflection, refraction, polarization), when different kinds of obstacles are interposed in their paths (diffraction, scattering), or when several waves coincide in the same region of space (interference).

28.2 Waves

When we strike a bell, play a musical instrument, turn on the radio or speak loudly, sound is heard at distant points. The sound is transmitted through the surrounding air. If we are at a beach and a speeding boat passes at a distance from the shore, we eventually are affected by the wake it has produced. When a light bulb is turned on, the room is filled with light. As a result of the physical relations between the electric and magnetic fields, it is possible to transmit an electric signal from one place to another; this is the foundation of modern telecommunications techniques. Although the mechanism may be different for each of the processes mentioned above, they all have a common feature. They are physical situations, produced at one point of space, propagated through space and felt later on at another point. These kinds of processes are all examples of **wave motion**.

Suppose that we have a physical property extending over a certain region of space. This property may be an electromagnetic field, a deformation in a spring, a strain in a solid, the pressure in a gas, a transverse displacement of a string, perhaps even a gravitational field. Suppose that the conditions at one place become time-dependent or dynamic, so that there is a perturbation of the physical state at that place. Depending on the physical nature of the system, the perturbation may propagate through space, disturbing the conditions at other places. Therefore, we speak of a **wave** associated with the particular physical property that is disturbed or time-dependent.

For example, consider the free surface of a liquid. The physical property in this case is the displacement of each point of the surface relative to the equilibrium shape. Under equilibrium conditions the free surface of a liquid is plane and horizontal. But if at one point the conditions at the surface are disturbed by dropping a stone into the liquid, this disturbance propagates in all directions along the surface of the liquid. To determine the mechanism of propagation and its

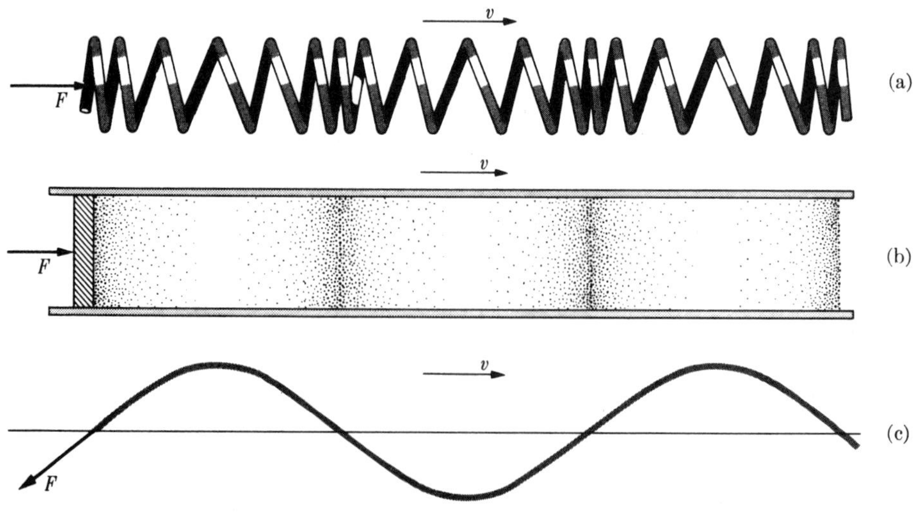

Figure 28.1 Elastic waves in (a) a spring, (b) a gas and (c) a string.

velocity, we must analyze how the displacement of a point at the surface of the liquid affects the rest of the surface. From this analysis, we set up the dynamical equations for the process. These equations then allow us to obtain quantitative information about the variation in space and time of the disturbance.

The various kinds of waves illustrated in Figure 28.1 are basically elastic waves resulting from a disturbance at some point. In such cases, we may ignore the molecular structure of matter and assume a continuous medium. This assumption is valid so long as the space fluctuation of the wave (determined by the wavelength) is large compared with the intermolecular separation.

28.3 Description of wave motion

Consider a function $\xi = f(x)$, represented graphically by the solid curve in Figure 28.2. If we replace x by $x - x_0$, we get the function

$$\xi = f(x - x_0)$$

It is clear that the shape of the curve has not changed; the same values of ξ occur for values of x increased by the amount x_0. In other words, assuming that x_0 is positive, we see that the curve has been displaced to the right an amount x_0, without deformation. Similarly we have that $\xi = f(x + x_0)$ corresponds to a rigid displacement of the curve to the left by the amount x_0.

Now if $x_0 = vt$, where t is the time, we get a 'traveling' curve; that is, $\xi = f(x - vt)$ represents a curve moving to the right with a velocity v, called the **phase velocity** (Figure 28.3(a)). Similarly $\xi = f(x + vt)$ represents a curve moving to the left with velocity v (Figure 28.3(b)). Therefore we conclude that a mathematical expression of the form

$$\xi(x, t) = f(x \pm vt) \qquad \textbf{(28.1)}$$

is adequate for describing a physical situation that 'travels' or 'propagates' without deformation along the positive or negative X-axis. The quantity $\xi(x, t)$ may represent a great diversity of physical quantities, such as the deformation in a solid, the pressure in a gas, an electric or magnetic field etc.

An especially interesting case is when $\xi(x, t)$ is a sinusoidal or harmonic function such as

$$\xi(x, t) = \xi_0 \sin k(x - vt) \qquad \textbf{(28.2)}$$

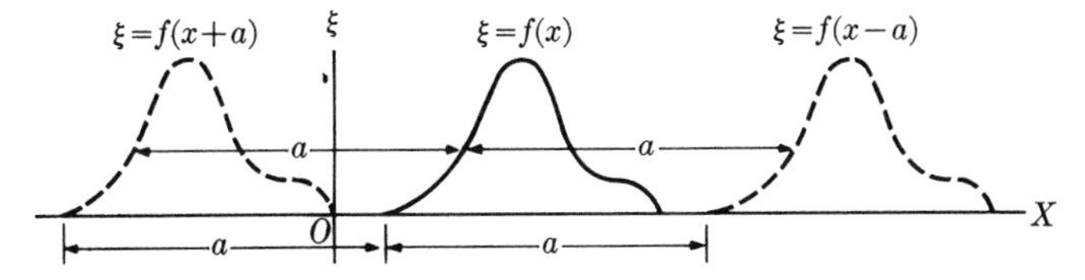

Figure 28.2 Undistorted translation of the function $\xi(x)$.

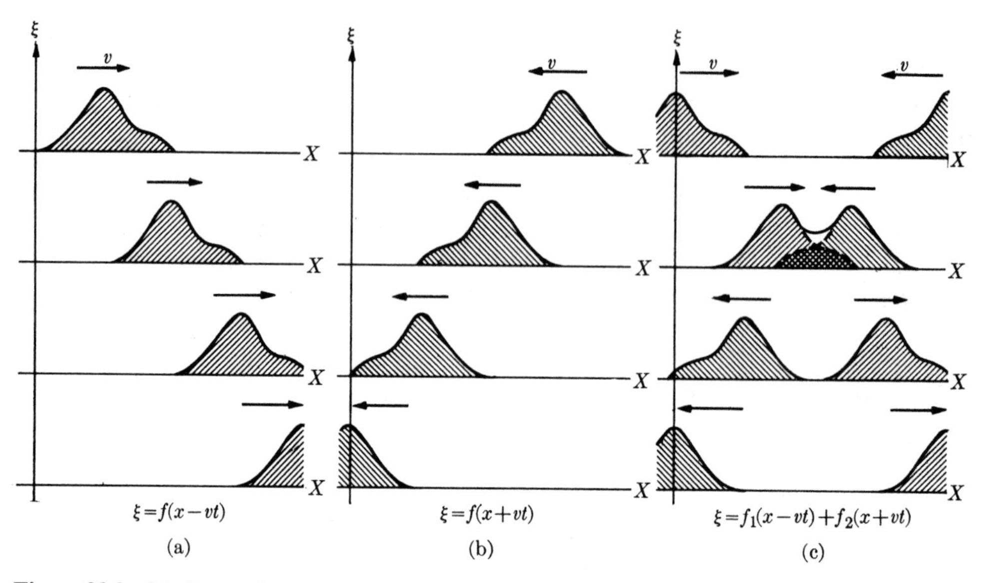

Figure 28.3 Undistorted propagation of a wave (a) to the right and (b) to the left. (c) Waves that propagate in opposite directions produce additive results where they interfere.

The quantity k has a special meaning. Replacing the value of x by $x + 2\pi/k$, we get the same value for $\xi(x, t)$; that is,

$$\xi\left(x + \frac{2\pi}{k}, t\right) = \xi_0 \sin k\left(x + \frac{2\pi}{k} - vt\right) = \xi_0 \sin[k(x - vt) + 2\pi] = \xi(x, t)$$

Then the quantity

$$\lambda = \frac{2\pi}{k} \tag{28.3}$$

designated the **wavelength**, is the 'space period' of the curve in Figure 28.4; that is, the curve repeats itself every length λ. The quantity $k = 2\pi/\lambda$ represents the number of wavelengths in the distance 2π and is called the **wave number**, although sometimes this name is reserved for $1/\lambda$ or $k/2\pi$, corresponding to the number of

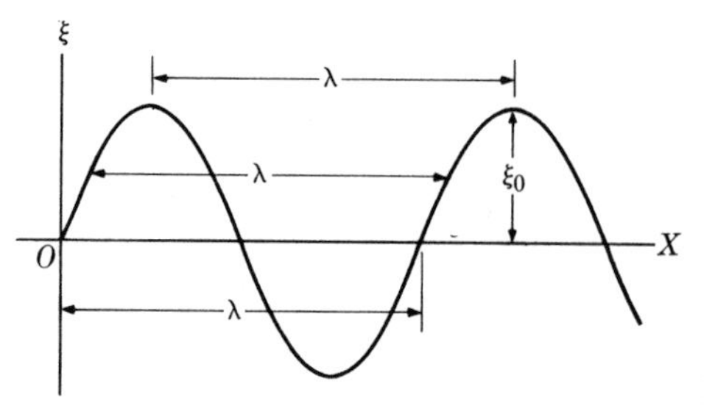

Figure 28.4 Harmonic wave.

wavelengths in one unit of length. Therefore

$$\xi(x, t) = \xi_0 \sin k(x - vt) = \xi_0 \sin \frac{2\pi}{\lambda}(x - vt) \qquad \textbf{(28.4)}$$

represents a sinusoidal or harmonic wave of wavelength λ propagating to the right along the X-axis with velocity v. Equation 28.4 can also be written in the form

$$\xi(x, t) = \xi_0 \sin(kx - \omega t) \qquad \textbf{(28.5)}$$

where

$$\omega = kv = \frac{2\pi v}{\lambda} \qquad \textbf{(28.6)}$$

gives the **angular frequency** of the wave. Since, according to Equation 10.2, $\omega = 2\pi\nu$, where ν is the **frequency** with which the physical situation varies at every point x, we have the important relation

$$\lambda\nu = v \qquad \textbf{(28.7)}$$

between the wavelength, the frequency, and the velocity of propagation. Also, if P is the **period** of oscillation at each point, given by $P = 2\pi/\omega = 1/\nu$, according to Equation 10.2, we may also write Equation 28.4 as

$$\xi = \xi_0 \sin 2\pi\left(\frac{x}{\lambda} - \frac{t}{P}\right) \qquad \textbf{(28.8)}$$

Similarly, the expressions

$$\xi = \xi_0 \sin k(x + vt) = \xi_0 \sin(kx + \omega t) = \xi_0 \sin 2\pi\left(\frac{x}{\lambda} + \frac{t}{P}\right) \qquad \textbf{(28.9)}$$

represent a sinusoidal or harmonic wave moving in the negative X-direction.

The function $\xi(x, t)$ has been represented in Figure 28.5 at times t_0, $t_0 + \frac{1}{4}P$, $t_0 + \frac{1}{2}P$, $t_0 + \frac{3}{4}P$, and $t_0 + P$. Note that as the physical situation propagates to the right, it repeats itself in *space* after one period. The reason for this is that, combining $\lambda\nu = v$ with $P = 1/\nu$, we get

$$\lambda = \frac{v}{\nu} = vP \qquad \textbf{(28.10)}$$

which shows that

the wavelength is the distance advanced by the wave motion in one period.

Therefore in sinusoidal wave motion we have two periodicities: one in time, given by the period P, and one in space, given by the wavelength λ, with the two related by $\lambda = vP$.

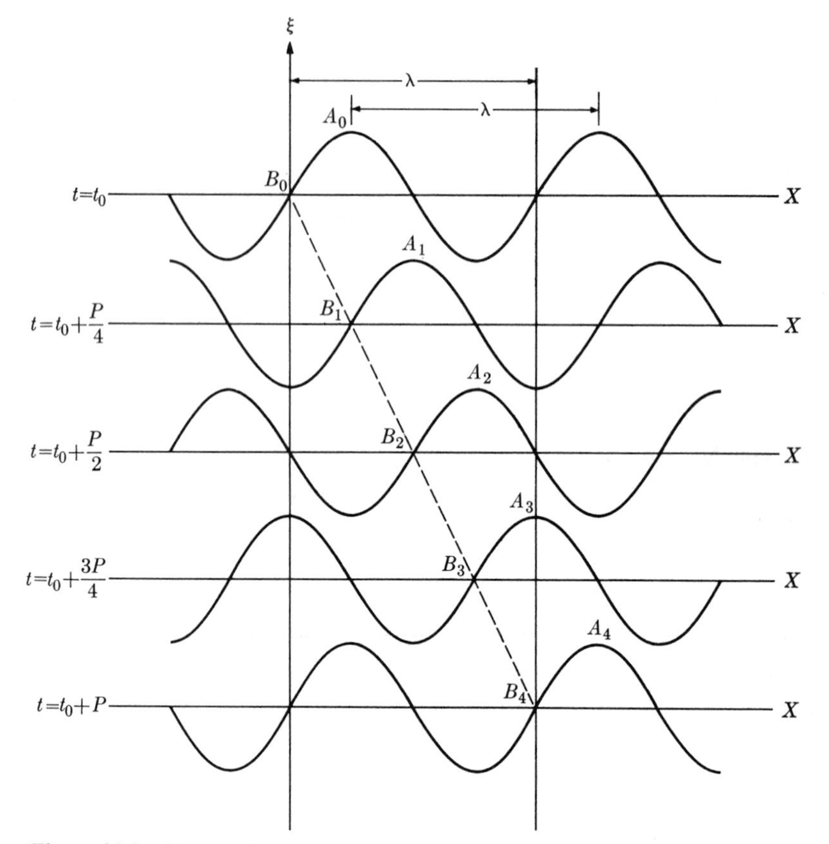

Figure 28.5 Harmonic wave propagating to the right. It advances the distance λ in the time P.

EXAMPLE 28.1

A tuning fork oscillates with a frequency of 440 Hz. If the velocity of sound in air is 340 m s^{-1}, find the wavelength and wave number of the sound. Write the expressions that represent the wave.

▷ Using Equation 28.7, we have

$$\lambda = \frac{v}{\nu} = \frac{340\,\text{m s}^{-1}}{440\,\text{Hz}} = 0.772\,\text{m} \qquad \text{and} \qquad k = \frac{2\pi}{\lambda} = \frac{2\pi}{0.772\,\text{m}} = 8.14\,\text{m}^{-1}$$

Accordingly, the sound wave can be represented by either of the following expressions:

$$\xi = \xi_0 \sin 8.14(x - 340t) \qquad \text{or} \qquad \xi = \xi_0 \sin(8.14x - 2.77 \times 10^3\, t)$$

where x is measured in meters and t in seconds.

EXAMPLE 28.2

Light propagates in vacuum with a velocity of $3 \times 10^8\ \text{m s}^{-1}$. Find the wavelength and wave number corresponding to a frequency of 5×10^{14} Hz, which is that of light in the red region of the visible spectrum.

▷ Again using Equation 28.7, we obtain

$$\lambda = \frac{v}{\nu} = \frac{3 \times 10^8\ \text{m s}^{-1}}{5 \times 10^{14}\ \text{Hz}} = 6 \times 10^{-7}\ \text{m} \quad \text{and} \quad k = \frac{2\pi}{\lambda} = \frac{2\pi}{6 \times 10^{-7}} = 1.05 \times 10^7\ \text{m}^{-1}$$

The light wave can be represented by either of the following expressions:

$$\xi = \xi_0 \sin 1.05 \times 10^7(x - 3 \times 10^8 t) \quad \text{or} \quad \xi = \xi_0 \sin(1.05 \times 10^7 x - 3.15 \times 10^{15} t)$$

Comparing these two examples will give the student a feeling for the difference in the orders of magnitude involved in sound and in light waves.

28.4 The general equation of wave motion

Although the fields associated with each physical process in a wave are governed by dynamical laws characteristic of each process, there is an equation applicable to *all* kinds of wave motion in the same way that Newton's law $\boldsymbol{F} = \mathrm{d}\boldsymbol{p}/\mathrm{d}t$ applies to all particle motions. Then, every time we recognize that a particular field, as a result of its physical properties, satisfies such an equation, we may be sure that the field propagates through space with a definite velocity and without distortion. Conversely, if we experimentally observe that a field propagates in space with a definite velocity and without distortion, we are prepared to describe the field by a set of equations compatible with the wave equation.

The equation that describes a wave motion propagated with a definite velocity v and without distortion along either the $+X$- or $-X$-direction is

$$\frac{\mathrm{d}^2\xi}{\mathrm{d}t^2} = v^2 \frac{\mathrm{d}^2\xi}{\mathrm{d}x^2} \tag{28.11}$$

This is called the **equation of wave motion**. (Strictly speaking, we should use the notation of partial derivatives in writing the wave equation. However, for our purpose it is enough to write Equation 28.11 in the form we have used.) The general solution of this equation is of the form of Equation 28.1. That is,

$$\xi(x, t) = f_1(x - vt) + f_2(x + vt) \tag{28.12}$$

Thus the general solution of Equation 28.1 can be expressed as the superposition of two wave motions propagating in opposite directions. Of course, for a wave propagating in one direction, only one of the two functions appearing in Equation 28.12 is required. However, when (for example) we have an incoming wave, along the $+X$-direction, and a reflected wave, along the negative X-direction, the general form of Equation 28.12 must be used.

To prove that an expression of the form of Equation 28.12 is a solution of the wave equation, we would have to use some mathematical techniques that are somewhat involved. Instead we shall verify, as a concrete example, that the wave equation is satisfied by the sinusoidal wave, $\xi = \xi_0 \sin k(x - vt)$. The space and time derivatives give

$$\frac{d\xi}{dx} = k\xi_0 \cos k(x - vt) \qquad \frac{d^2\xi}{dx^2} = -k^2\xi_0 \sin k(x - vt)$$

$$\frac{d\xi}{dt} = -kv\xi_0 \cos k(x - vt) \qquad \frac{d^2\xi}{dt^2} = -k^2v^2\xi_0 \sin k(x - vt)$$

Therefore $d^2\xi/dt^2 = v^2\, d^2\xi/dx^2$, in agreement with Equation 28.11. The same results are obtained with $\xi = \xi_0 \sin k(x + vt)$.

For the waves discussed in this chapter, the wave equation results from the dynamical laws of each process, under certain approximations such as small amplitude or long wavelength, etc. Therefore, the theory related to Equation 28.11 is applicable only under the stated approximations.

28.5 Elastic waves

Two important examples of elastic waves are longitudinal waves along a rod and along a spring.

(i) Elastic waves in a rod. When a disturbance is produced at one end of a solid rod, say by hitting it with a hammer, the disturbance propagates along the rod and eventually is felt at the other end. We say that an elastic wave has propagated along the rod with a velocity determined by the physical properties of the rod.

Consider a rod of uniform cross-section A (Figure 28.6) subject to a stress along its axis indicated by the force F. The force F is not necessarily the same at all sections, and may vary along the axis of the rod. At each cross-section (as

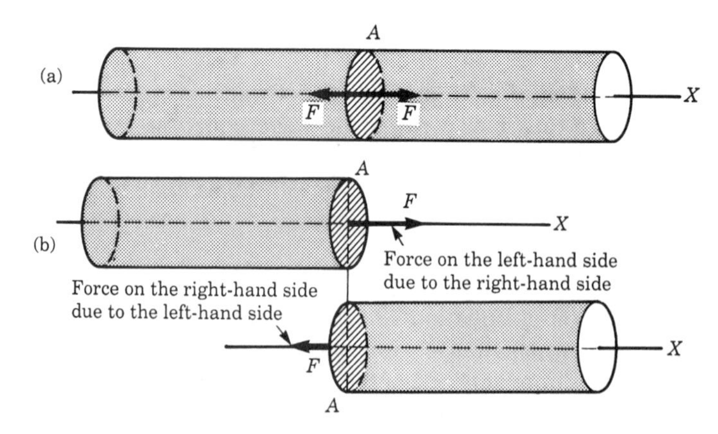

Figure 28.6 The forces on any section of a rod under stress are equal and opposite.

shown in Figure 28.6) there are two equal and opposite forces. One is the pull on the left part of the rod due to the right part and the other is the pull on the right part of the rod due to the left part. The **normal stress** $\mathscr{S}$ at a section of the rod is defined as the force per unit area acting perpendicular to the cross-section in either direction. Then

$$\mathscr{S} = \frac{F}{A} \tag{28.13}$$

The stress is expressed in N m^{-2} or Pa, which is the same as the unit of pressure.

Under the action of such forces, each section of the rod suffers a displacement parallel to the axis. We are interested in the case when there is deformation, so that the displacement varies along the rod, for which the force F must also vary along the rod. Suppose for simplicity that the rod is clamped at the left end and the other end is pulled away, stretching the rod. Designating the displacement of a section A at distance x from end O by ξ, we have that ξ is a function of x. Consider another section A' separated from A a distance $\mathrm{d}x$ in the undisturbed situation (Figure 28.7). When the rod is stretched, section A is displaced a distance ξ and section A' is displaced a distance $\xi + \mathrm{d}\xi$. Then the separation between A and A' in the deformed state is $\mathrm{d}x + \mathrm{d}\xi$. The deformation of the rod in that region has therefore been $\mathrm{d}\xi$. The **linear strain** ε in the rod is defined as the deformation along the axis per unit length. Since the deformation $\mathrm{d}\xi$ corresponds to a length $\mathrm{d}x$, we see that the strain in the rod is

$$\varepsilon = \frac{\mathrm{d}\xi}{\mathrm{d}x} \tag{28.14}$$

When there is no deformation, ξ is the same at all cross-sections and $\varepsilon = 0$. That is, there is no linear strain. The strain, since it is a quotient of two lengths, is a pure number or dimensionless quantity.

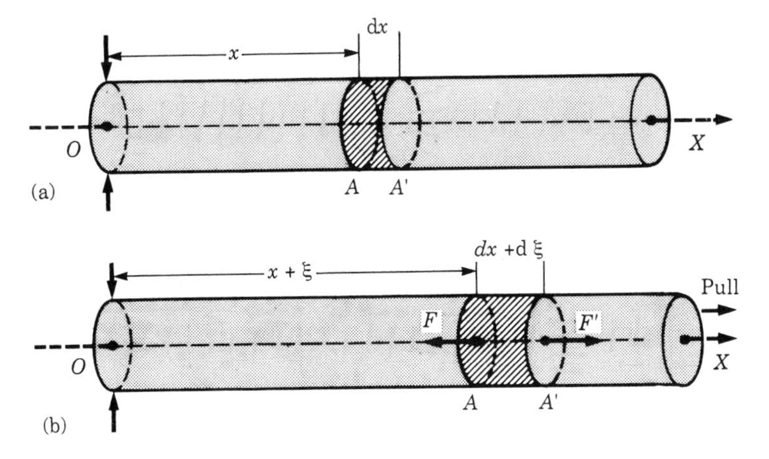

Figure 28.7 (a) Undeformed rod. (b) Deformed rod.

Between the normal stress $\mathscr{S}$ and the linear strain ε of the rod there exists an approximate relation called **Hooke's law**, which states that

the normal stress is proportional to the linear strain;

$$\mathscr{S} = Y \frac{d\xi}{dx} \tag{28.15}$$

where the proportionality constant Y is called **Young's modulus of elasticity** (Table 28.1). It is expressed in $N\,m^{-2}$, since $d\xi/dx$ is a dimensionless factor and $\mathscr{S}$ is expressed in $N\,m^{-2}$. Hooke's law is a good approximation of the elastic behavior of a substance as long as the deformations are small. For large stresses and deformations, Equation 28.15 no longer holds, and the description of the physical situation becomes much more complicated. The stress at which Hooke's law is no longer obeyed is called the **elastic limit** of the substance.

Introducing Equation 28.15 into Equation 28.13 and solving for F, we get

$$F = YA \frac{d\xi}{dx} \tag{28.16}$$

When the rod is not in equilibrium, the force on each section is not the same along the rod. As a result, a section of the rod of thickness dx is subject to a net or resultant force. For example, in Figure 28.7, the side A' of the section of thickness dx is subject to a force F' toward the right, due to the pull of the right part of the rod, while the side A is subject to a force F pointing to the left, due to the pull of the left part of the rod. The net force to the right on the section is $F' - F$. This results in an accelerated motion of the section of the rod. Application of the laws of motion to the section shows (see the derivation below) that the displacement ξ satisfies the wave equation 28.11 with a velocity of propagation

$$v = \sqrt{\frac{Y}{\rho}} \tag{28.17}$$

where ρ is the density of the material of the rod, as long as the deformation ξ is not very large. This result is confirmed experimentally by independently measuring the three quantities. Therefore, we may conclude that the deformation field ξ propagates along the rod as a wave.

We may note that Equation 28.17 checks dimensionally since Y is expressed in $N\,m^{-2}$ and ρ in $kg\,m^{-3}$. Therefore their ratio is $(N\,m^{-2})(kg\,m^{-3})^{-1} = m^2\,s^{-2}$, which is the square of a velocity.

Table 28.1 Elastic constants ($10^{11}\,N\,m^{-2}$)

Material	Young's modulus, Y	Bulk modulus, κ	Shear modulus, G
Aluminum	0.70	0.61	0.24
Copper	1.25	1.31	0.46
Iron	2.06	1.13	0.82
Lead	0.16	0.33	0.054
Nickel	2.1	1.64	0.72
Steel	2.0	1.13	0.80

It is important to note that the wave we are describing corresponds to a physical property, the deformation ξ, oriented along the direction of propagation of the wave, i.e. the X-axis. This kind of wave motion is called **longitudinal**.

(ii) Elastic waves in a spring. Suppose a disturbance is produced along a stretched spring (Figure 28.1(a)). The resulting wave consists in a series of compressions and expansions that propagate along the spring. Let ξ be the longitudinal displacement suffered by a section of the spring. It can be shown, using a logic similar to that explained for a solid rod, that the longitudinal displacement ξ satisfies the wave equation 28.11 with a velocity of propagation of the longitudinal wave along the spring

$$v = \sqrt{\frac{kL}{m}} \qquad \textbf{(28.18)}$$

where k is the elastic constant of the spring, introduced in Example 9.3, L is the length of the unstretched spring, and m is its mass per unit length or linear density. In the above equation k is measured in $\mathrm{N\,m^{-1}}$ or $\mathrm{kg\,s^{-2}}$, L in m and m in $\mathrm{kg\,m^{-1}}$. Thus the quotient is then $\mathrm{m^2\,s^{-2}}$, which is the square of a velocity.

Derivation of the equation for longitudinal waves along a rod

The net force on a section AA' of the rod (Figure 28.7) is $F' - F = \mathrm{d}F$. If ρ is the density of the material of the rod, the mass of the section is $\mathrm{d}m = \rho\,\mathrm{d}V = \rho A\,\mathrm{d}x$, where $A\,\mathrm{d}x$ is the volume of the section. The acceleration of this mass is $\mathrm{d}^2\xi/\mathrm{d}t^2$. Therefore, applying the dynamical relation force = mass × acceleration, we may write the equation of motion of the section as

$$\mathrm{d}F = (\rho A\,\mathrm{d}x)\frac{\mathrm{d}^2\xi}{\mathrm{d}t^2} \quad \text{or} \quad \frac{\mathrm{d}F}{\mathrm{d}x} = \rho A\frac{\mathrm{d}^2\xi}{\mathrm{d}t^2} \qquad \textbf{(28.19)}$$

In this problem we have *two* fields; one is the displacement ξ of each section of the rod, where ξ is a function of position and time, and the other is the force F at each section, where F is also a function of position and time. These two fields are related by Equations 28.16 and 28.19, which describe the physical conditions of the problem. Now we shall combine Equations 28.16 and 28.19. Taking the derivative of Equation 28.16, $F = YA(\mathrm{d}\xi/\mathrm{d}x)$, with respect to x, we have

$$\frac{\mathrm{d}F}{\mathrm{d}x} = YA\frac{\mathrm{d}^2\xi}{\mathrm{d}x^2}$$

Substituting this result in Equation 28.19 and canceling the common factor A, we get

$$\frac{\mathrm{d}^2\xi}{\mathrm{d}t^2} = \frac{Y}{\rho}\frac{\mathrm{d}^2\xi}{\mathrm{d}x^2} \qquad \textbf{(28.20)}$$

which is of the form of the equation of wave motion 28.11 with

$$v = \left(\frac{Y}{\rho}\right)^{1/2} \qquad \textbf{(28.21)}$$

which is Equation 28.17.

EXAMPLE 28.3

Velocity of propagation of longitudinal elastic waves in a steel bar.

▷ Using the values of Table 28.1 and a value of $7.8 \times 10^3\ \mathrm{kg\,m^{-3}}$ for the density of steel, we have from Equation 28.17 that

$$v = \left(\frac{Y}{\rho}\right)^{1/2} = \left(\frac{2.0 \times 10^{11}\ \mathrm{N\,m^{-2}}}{7.8 \times 10^3\ \mathrm{kg\,m^{-3}}}\right)^{1/2} = 5.06 \times 10^3\ \mathrm{m\,s^{-1}}$$

The experimental value is $5.10 \times 10^3\ \mathrm{m\,s^{-1}}$ at 0 °C. This value should be compared with the velocity of sound in air, which is about $340\ \mathrm{m\,s^{-1}}$, showing that elastic waves propagate much faster in metals than in air.

EXAMPLE 28.4

An unstretched spring has a length of 1.0 m and a mass of 0.2 kg. When a body of mass equal to 2 kg is hung from the spring it stretches 3 cm. What is the velocity of longitudinal waves along the spring?

▷ In this problem $L = 1.0$ m. The stretch of the spring is $\Delta L = 3$ cm $= 0.03$ m under a force $F = Mg = (0.2\ \mathrm{kg}) \times (9.8\ \mathrm{m\,s^{-2}}) = 1.96\ \mathrm{N}$. Then $k = F/\Delta L = 65.3\ \mathrm{N\,m^{-1}}$. Also the mass per unit length is $m = (0.2\ \mathrm{kg})/(1.0\ \mathrm{m}) = 0.2\ \mathrm{kg\,m^{-1}}$. Therefore

$$v = \left(\frac{kL}{m}\right)^{1/2} = \left[\frac{(65.3\ \mathrm{N\,m^{-1}})(1.0\ \mathrm{m})}{0.2\ \mathrm{kg\,m^{-1}}}\right]^{1/2} = 18.07\ \mathrm{m\,s^{-1}}$$

28.6 Pressure waves in a gas

Elastic waves also result from pressure variations in a gas. These waves consist of a series of compressions and expansions propagating along the gas. Sound is the most important example of this type of wave.

There is an important difference between elastic waves in a gas and elastic waves in a solid rod. Gases are very compressible, and when pressure fluctuations are set up in a gas, the density of the gas will suffer the same kind of fluctuations as the pressure, while in a rod the density remains practically constant.

Consider a gas column within a pipe or tube (Figure 28.8). We call p and ρ the pressure and density in the gas. Under equilibrium conditions, p and ρ are the same throughout the volume of the gas; that is, they are independent of x. If the pressure of the gas is disturbed, a volume element such as $A\,\mathrm{d}x$ is set in motion because the pressures p and p' on both sides are different, giving rise to a net force. As a result, section A is displaced an amount ξ and section A' an amount $\xi + \mathrm{d}\xi$, so that the thickness of the volume element after the deformation is $\mathrm{d}x + \mathrm{d}\xi$. So far, all seems identical to the case of the solid rod. However, because of the relatively large change in volume, due to the greater compressibility of the gas, there is now a change in density as well. To describe this change in density, it is convenient to

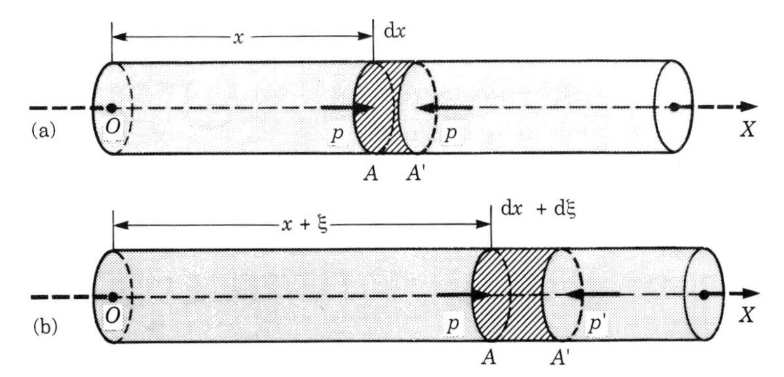

Figure 28.8 (a) Gas in equilibrium. (b) Disturbance in a gas.

define the quantity

$$\kappa = \rho\left(\frac{\mathrm{d}p}{\mathrm{d}\rho}\right) \tag{28.22}$$

called the **bulk modulus of elasticity**, which gives the change in pressure ($\mathrm{d}p$) per unit change in density ($\mathrm{d}\rho/\rho$). It is expressed in $\mathrm{N\,m^{-2}}$, the same units used to express pressure, since the units of density cancel out.

The gas at the left of the volume element limited by the sections A and A' pushes to the right with a force pA and the gas at the right pushes to the left with a force $p'A'$. Therefore, when $A = A'$ the resultant force on the volume element is $(p - p')A$. The motion of the volume element results in a wave propagating in the gas. We shall not discuss the equation of motion of the volume element, which is mathematically involved due to the compressibility of the gas, but shall simply state that, if the fluctuations in pressure are not very large, the displacement ξ satisfies the wave equation 28.11 with a velocity of propagation

$$v = \sqrt{\frac{\kappa}{\rho}} \tag{28.23}$$

where ρ is the equilibrium or average density. Therefore, the displacement due to a pressure disturbance in a gas propagates as a longitudinal wave.

The pressure also obeys the same wave equation, so that the variations of pressure produced by the wave propagate with the velocity given by Equation 28.23. This is why we call elastic waves in a gas **pressure waves**. Sound is simply a pressure wave in air. Similarly, the gas density obeys an equation of the same form, with ξ replaced by the variation in the density of the gas. Therefore, when referring to a gas, we may speak of a displacement wave, a pressure wave and a density wave. For the case of a harmonic pressure wave, we have $p - p_0 = P_0 \sin(kx - \omega t)$ where p_0 is the average pressure and P_0 is the amplitude of the wave.

In a gas the displacement is a vector field and the wave associated with the displacement motion corresponds to a longitudinal **vector** wave parallel to the direction of propagation. However, neither the pressure nor the density are vectors (and no direction is associated with them) and the wave motion corresponding to the pressure and the density is a **scalar** wave.

EXAMPLE 28.5

Wave velocity in a gas as a function of the pressure.

▷ Wave motion in gases is generally an adiabatic process, since the variations are so rapid that there is no time for heat transfer. We recall that under adiabatic conditions the equation of state of a gas is Equation 17.32, $pV^\gamma = const.$ or $p = const. \times (1/V)^\gamma$. But the density is inversely proportional to the volume, $\rho \sim 1/V$. Therefore we may write $p = C\rho^\gamma$. Differentiating this equation we have

$$\frac{\mathrm{d}p}{\mathrm{d}\rho} = \gamma C \rho^{\gamma - 1}$$

and from the definition of the bulk modulus

$$\kappa = \rho\left(\frac{\mathrm{d}p}{\mathrm{d}\rho}\right) = \gamma C \rho^\gamma = \gamma p$$

Then substituting into Equation 28.23, we find that the velocity of sound in a gas is

$$v = \sqrt{\frac{\gamma p}{\rho}} \tag{28.24}$$

Thus the higher the pressure and the lower the density, the faster the wave propagation is.

EXAMPLE 28.6

Relation between the velocity of a pressure wave (or sound) in a gas and the temperature of the gas.

▷ As we showed in Section 15.5, the relation between pressure and volume in an ideal gas is $pV = \text{N}RT$, where N is the number of moles in the volume V. But since $\rho = m/V$, we have that

$$\frac{p}{\rho} = \frac{\text{N}RT}{m} = \frac{RT}{M}$$

where $M = m/\text{N}$ is the mass of one mole of the gas, expressed in kilograms. Then we may write

$$v = \sqrt{\frac{\gamma p}{\rho}} = \sqrt{\frac{\gamma RT}{M}} = \alpha\sqrt{T} \tag{28.25}$$

where $\alpha = (\gamma R/M)^{1/2}$. Therefore, the velocity of sound in a gas varies as $(T)^{1/2}$, and the proportionality constant α depends on the gas. For example, we know from experimental measurements that at $T = 273.15$ K (or 0 °C), the velocity of sound in air is 331.45 m s^{-1}. Therefore the velocity of sound in air at any temperature is $v = 20.055(T)^{1/2}$ m s^{-1}, a result that agrees with experiment over a fairly large temperature range.

28.7 Transverse waves on a string

Under equilibrium conditions a string subject to a tension T is straight. Suppose that we now displace the string sidewise, or perpendicular to its length, by a small amount, as shown in Figure 28.9. Consider a small section AB of the string, of length $\mathrm{d}x$, that has been displaced a distance ξ from the equilibrium position. On each end a tangential force T is acting, the one at B produced by the pull of the string on the right and the one at A due to the pull of the string on the left. Due to the curvature of the string, the two forces are not directly opposed but make angles α and α' with the X-axis. The upward components of each force are $T'_y = T \sin \alpha'$ and $T_y = T \sin \alpha$. The resultant upward force on the section AB of the string is $F_y = T'_y - T_y$. Under the action of this force, the section AB of the string moves up and down. Applying the laws of motion (see the derivation below), the displacement ξ satisfies the wave equation 28.11 with a velocity of propagation

$$v = \sqrt{\frac{T}{m}} \qquad \textbf{(28.26)}$$

where m is the **linear density** of the string (or mass per unit length) expressed in $\mathrm{kg\,m^{-1}}$, provided that the amplitude is small. Therefore, a transverse disturbance propagates with a velocity v along the string subject to the tension T.

The wave motion is **transverse**; that is, the physical property, the displacement ξ, is perpendicular to the direction of propagation of the wave, which is along the X-axis. But there are many directions of displacement perpendicular to the X-axis. If we choose two mutually perpendicular directions Y and Z as references, we can express the transverse displacement ξ, a vector, in terms of its components along the Y- and Z-axes. While the disturbance propagates, the direction of ξ may change from point to point, resulting in a twisting of the string (Figure 28.10). However, if all the displacements are in the same direction, let us say along the Y-axis, the string is always in the XY-plane, and we say that the wave motion is **linearly polarized** (Figure 28.11). It is clear that a transverse wave can always be considered as the combination of two linearly polarized waves in perpendicular directions. If ξ has a constant length but changes in direction, so that the string lies over a circular cylindrical surface (Figure 28.12), the wave is **circularly polarized.** In this case, each portion of the string moves in a circle around the X-axis. The polarization of transverse waves is a very important subject, which we shall discuss in more detail in Chapter 32.

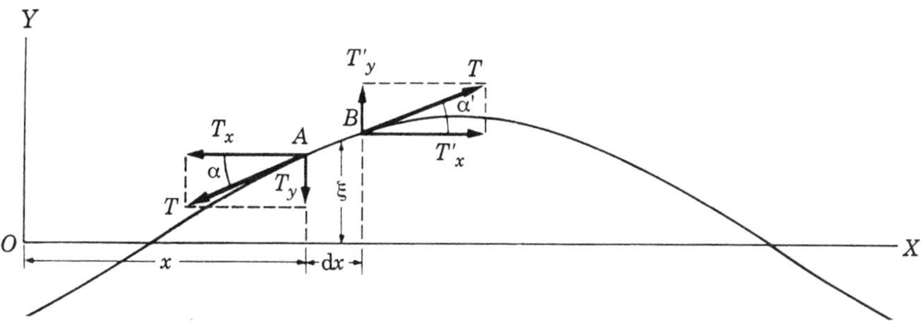

Figure 28.9 Forces on a section of a transversely displaced string.

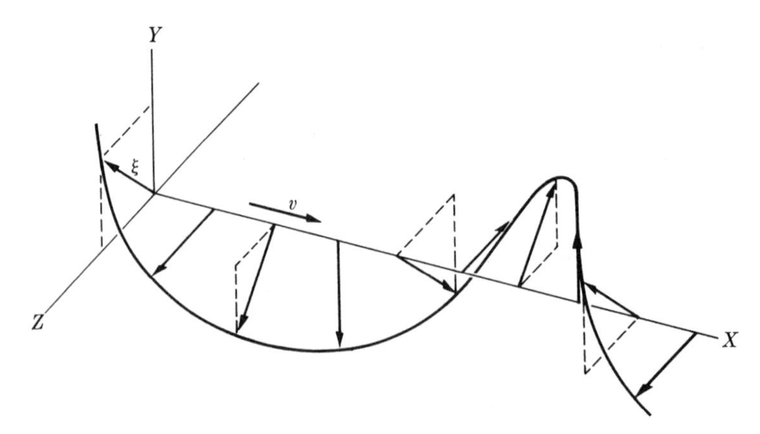

Figure 28.10 Non-polarized transverse waves in a string.

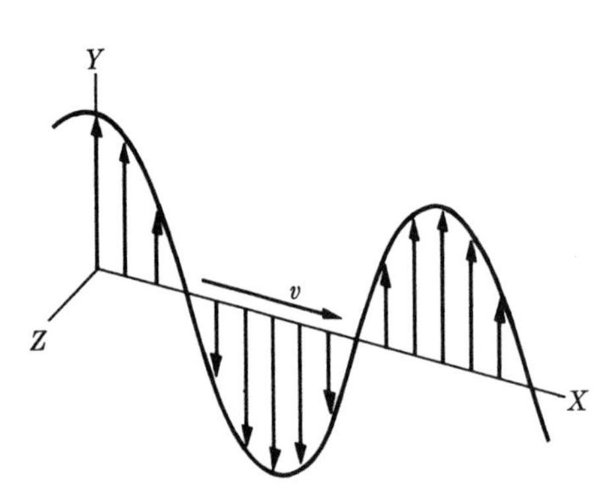

Figure 28.11 Linearly polarized transverse wave in a string.

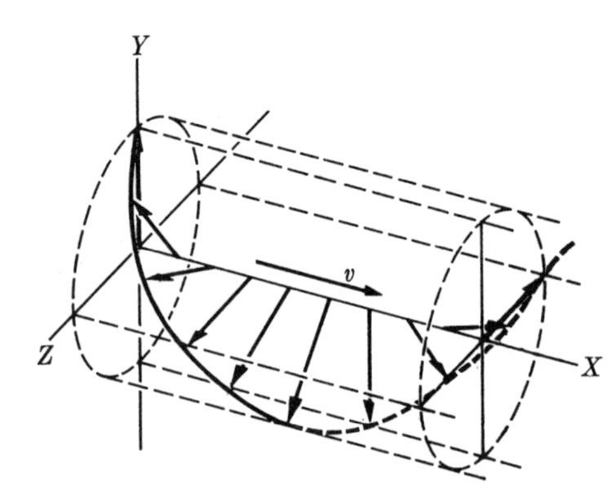

Figure 28.12 Circularly polarized transverse wave in a string.

Derivation of the velocity of propagation of waves in a string

The net upward force on the section AB of the string (Figure 28.9) is

$$F_y = T'_y - T_y = T(\sin \alpha' - \sin \alpha)$$

But, since α and α' are almost equal, we may write $F_y = T \, \mathrm{d}(\sin \alpha)$. If the curvature of the string is not very large, the angles α and α' are small, and the sines can be replaced by their tangents. So the upward force is

$$F_y = T \, \mathrm{d}(\tan \alpha) = T \frac{\mathrm{d}}{\mathrm{d}x} (\tan \alpha) \, \mathrm{d}x$$

But $\tan \alpha$ is the slope of the curve adopted by the string, which is equal to $\mathrm{d}\xi/\mathrm{d}x$. Then

$$F_y = T \frac{\mathrm{d}}{\mathrm{d}x} \left(\frac{\mathrm{d}\xi}{\mathrm{d}x} \right) \mathrm{d}x = T \frac{\mathrm{d}^2 \xi}{\mathrm{d}x^2} \, \mathrm{d}x$$

This force must be equal to the mass of the section AB multiplied by its upward acceleration, $d^2\xi/dt^2$. Since m is the linear density of the string, the mass of the section AB is $m\,dx$. We use the relation force = mass × acceleration and write the equation of motion of this section of the string as

$$(m\,dx)\frac{d^2\xi}{dt^2} = T\frac{d^2\xi}{dx^2}\,dx$$

from which, after cancellation of the common factor dx and placing m on the right-hand side of the equation, we obtain

$$\frac{d^2\xi}{dt^2} = \frac{T}{m}\frac{d^2\xi}{dx^2} \quad \textbf{(28.27)}$$

This is of the form of the wave equation 28.11 with

$$v = \sqrt{\frac{T}{m}}$$

28.8 Transverse elastic waves in a rod

Elastic waves in a solid rod can also be transverse. Consider a rod that, in its undistorted state, is represented by the straight dashed lines in Figure 28.13. If we start the rod vibrating by hitting it transversely, at a particular instant it adopts the shape of the curved lines, and we may assume that each section of the rod moves up and down, with no horizontal motion. Then, the transverse displacement of a section at a particular time will be ξ. This displacement must also be a function of position because, if it were constant, it would correspond to a parallel displacement of the rod. The quantity $d\xi/dx$, which is the change of *transverse* displacement per unit length along the rod, is called the **shearing strain**. Like the longitudinal strain, it is a pure number.

As a result of the deformation, each section of thickness dx is subject to the opposed forces F and F', which are transverse to the rod and therefore tangent to the section (compare with the situation in Figure 28.7) and are produced by those portions of the bar on each side of the section. The tangential or transverse force per unit area, $\mathscr{S} = F/A$, is called the **shearing stress**.

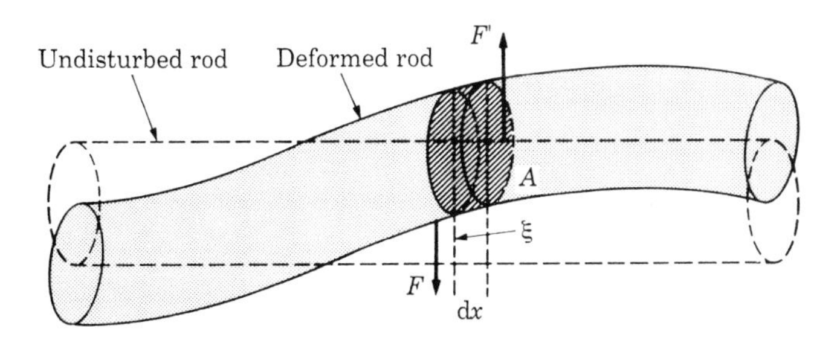

Figure 28.13 Tranverse or shear wave in a rod.

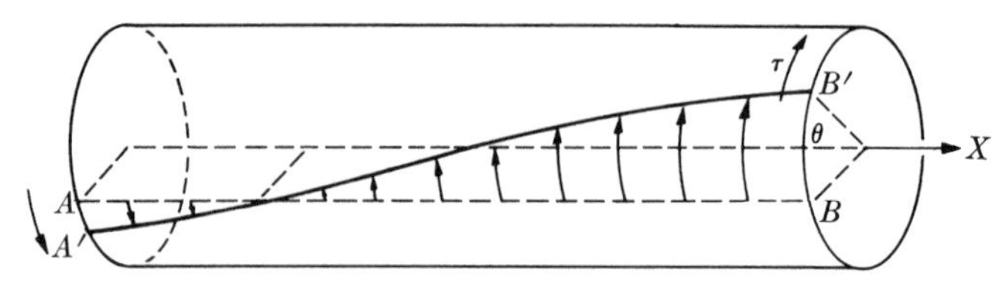

Figure 28.14 Torsional wave in a rod.

Recall Equation 28.15 for the relationship between normal stress and linear strain. The shearing stress and the shearing strain also have a relation similar to Hooke's law; that is,

$$\mathscr{S} = G \frac{\mathrm{d}\xi}{\mathrm{d}x} \tag{28.28}$$

where G is a coefficient characteristic of the material and called the **shear modulus** (Table 28.1). Replacing $\mathscr{S}$ by F/A, we obtain

$$F = A\mathscr{S} = AG \frac{\mathrm{d}\xi}{\mathrm{d}x} \tag{28.29}$$

The resultant transverse force on the section of thickness $\mathrm{d}x$ is $F' - F$. Again applying the laws of motion to the section, we find that the transverse displacement when the displacements are small obeys the wave equation 28.11 with a velocity of propagation given by

$$v = \sqrt{\frac{G}{\rho}} \tag{28.30}$$

indicating that a transverse deformation propagates along the bar. More properly, the wave should be called a **shear wave**.

Another example of a shear wave is a **torsional wave**. Suppose that, at the free end of a rod clamped at one end, a variable torque is applied. This produces a twisting of the rod (Figure 28.14). If the torque is time-dependent, the angle of the twist, which is called **torsion**, changes with time, resulting in a torsional wave propagated along the rod. A mathematical analysis of the problem shows that, irrespective of the shape of the cross-section of the rod, the velocity of propagation of the torsional wave is also given by Equation 28.30 because both processes are essentially due to the shearing stress in the rod. An interesting aspect of torsional waves is that they correspond not to displacements either parallel or transverse to the axis of the rod, but to angular displacements around the axis without changing its shape.

28.9 Surface waves in a liquid

Waves on the surface of a liquid are the most familiar kinds of waves; they are the waves we observe on the oceans and lakes, or simply when we drop a stone into a pond. The mathematical aspect, however, is more complicated than in the previous examples, and will be omitted. Instead we shall present only a descriptive discussion.

The undisturbed surface of a liquid is plane and horizontal. A disturbance of the surface produces a displacement of all molecules directly underneath the surface as a result of intermolecular forces (Figure 28.15). The amplitude of the horizontal and the vertical displacements of a volume element of a fluid varies, in general, with depth. Of course, the molecules at the bottom do not suffer any vertical displacement, since they cannot separate from the bottom. At the surface of the liquid other forces enter into play. One force is the surface tension of the liquid, which gives an upward or downward force on an element of the surface similar to the case of a string. Another force is the weight of the liquid above the undisturbed level of the liquid's surface. The resulting equation for the surface displacement is not exactly of type 28.11, but slightly more complicated. However, it is satisfied by harmonic waves of wavelength λ, and the velocity of propagation of the surface wave is given by

$$v = \sqrt{\frac{g\lambda}{2\pi} + \frac{2\pi\mathscr{T}}{\rho\lambda}} \qquad \textbf{(28.31)}$$

where ρ is the density of the liquid, $\mathscr{T}$ is the surface tension (measured in N m^{-1}), and g is the acceleration due to gravity. This expression is valid only when the depth is very great compared with the wavelength λ.

The most interesting aspect of Equation 28.31 is that

the velocity of propagation depends on the wavelength,

a situation not encountered previously. Since the wavelength is related to the frequency and the velocity of propagation through $\nu = v/\lambda$, we conclude that

the velocity of propagation depends on the frequency of the waves.

When the wavelength λ is large enough so that the second term in Equation 28.31 can be neglected, we have

$$v = \sqrt{\frac{g\lambda}{2\pi}} \qquad \textbf{(28.32)}$$

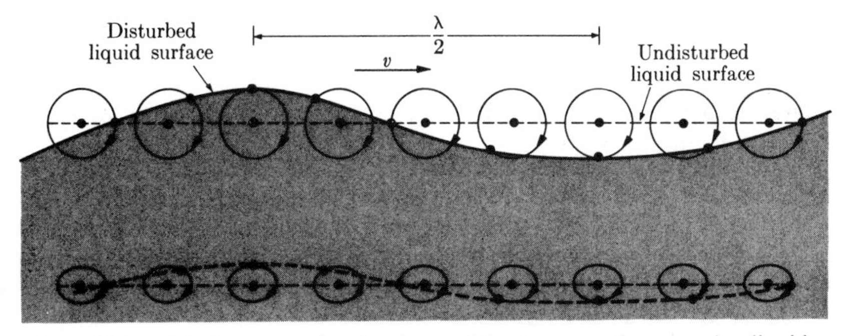

Figure 28.15 Displacement of molecules resulting from a surface wave in a liquid.

The waves in this case are called **gravity waves**. In this approximation the velocity of propagation is independent of the nature of the liquid, since no factor related to the liquid (such as density or its surface tension) appears in Equation 28.32. We see that in this case the velocity of propagation is proportional to the square root of the wavelength, and that the longer the wavelength, the faster the propagation. It is for this reason that a strong steady wind produces waves of longer wavelength than a swift but gusty wind. If a disturbance is produced in a liquid, Equation 28.32 implies that the initial disturbance is distorted in such a way that the components of longer wavelength 'escape' from the disturbance by moving faster than the components of shorter wavelength.

When the wavelength λ is very small, the dominant term is the second in Equation 28.31, which then gives the velocity of propagation as

$$v = \sqrt{\frac{2\pi\mathcal{T}}{\rho\lambda}} \tag{28.33}$$

These waves are called **ripples** or **capillary waves** because of their dependence on the surface tension. These are the waves observed when a very gentle wind blows, or when a container is subject to vibrations of high frequency. In this case, the longer the wavelength, the slower the propagation.

When the velocity of propagation of a wave motion depends on the wavelength or the frequency, we say that there is **dispersion**. If a wave motion resulting from the superposition of several harmonic waves of different frequencies impinges on a **dispersive** medium, the wave is distorted, since each of its component waves propagates with a different velocity. Dispersion is a very important phenomenon present in several types of wave propagation. In particular, it appears when electromagnetic waves propagate through matter, as we shall see in the next chapter.

28.10 What propagates in wave motion?

In wave motion, what propagates is a physical condition generated at some place and transmitted to other regions. All the waves discussed in the previous sections correspond to certain kinds of motion of atoms or molecules of the medium through which the wave propagates, but the atoms remain, on the average, at their equilibrium positions (Figure 28.16). Then it is not matter that propagates, but the *state of motion* or dynamic condition of matter that is transferred from one region to another. Since the dynamic condition of a system is described in terms of momentum and energy, we may say that,

in wave motion, energy and momentum are transferred or propagated.

The **intensity** of a wave is defined as the energy that flows per unit time across a unit area perpendicular to the direction of propagation. Let us designate by E the average energy density of the wave; that is, the average energy per unit volume in the medium through which the wave propagates. Clearly E is expressed in $\mathrm{J\,m^{-3}}$. If v is the velocity of propagation the intensity of the wave is

$$I = v\mathrm{E} \quad (28.34)$$

The intensity of the wave is expressed in $(\mathrm{m\,s^{-1}})\,(\mathrm{J\,m^{-3}}) = \mathrm{J\,s^{-1}\,m^{-2}} = \mathrm{W\,m^{-2}}$, or power per unit area.

If the wave propagates through a bounded medium, such as a rod or a pipe, of cross-section A, the total average energy crossing per unit time across a section of the medium or rate of energy flow is

$$\left(\frac{\mathrm{d}E}{\mathrm{d}t}\right)_{\mathrm{ave}} = IA = v\mathrm{E}A \quad (28.35)$$

This expression also gives the power required to maintain the wave; that is, the rate at which energy must be fed continuously at one end to maintain a continuous train of waves along the medium.

In the case of a harmonic elastic wave expressed by $\xi = \xi_0 \sin(kx - \omega t)$, we may use the results of Chapter 10 to obtain the average energy density for an elastic wave. Recalling Equation 10.11, which gives the energy of an oscillator, and using ξ_0 for the amplitude and ρ for the mass since we want the energy per unit volume, therefore

$$\mathrm{E} = \tfrac{1}{2}\rho\omega^2\xi_0^2 \quad (28.36)$$

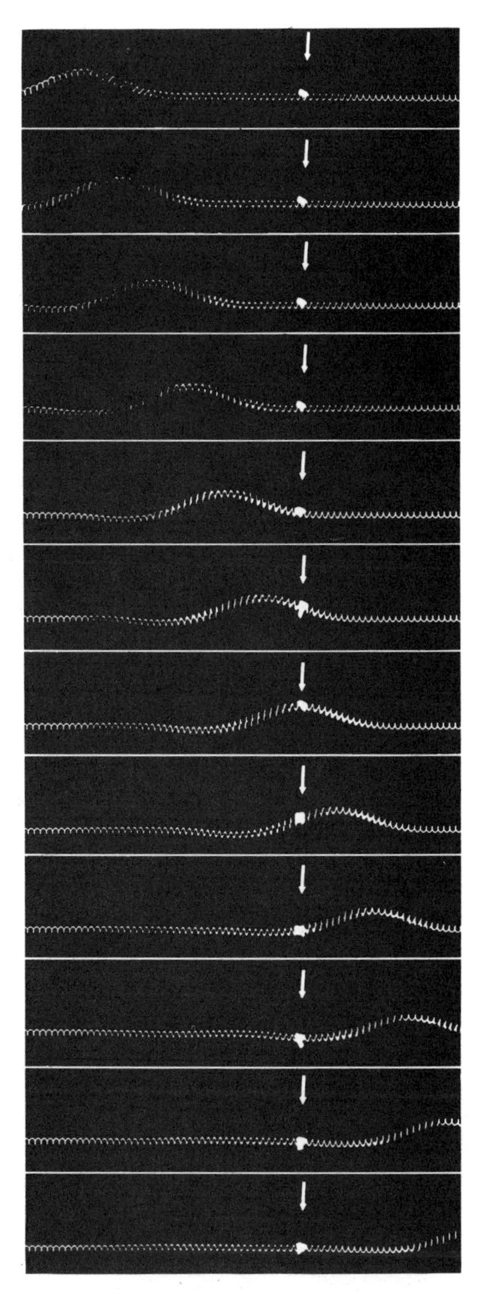

Figure 28.16 Propagation of a pulse on a spring. The sections of the spring move up and down as the pulse moves from left to right. (Courtesy of D. C. Heath, *PSSC Physics*.)

EXAMPLE 28.7

Intensity of the waves in a gas column expressed in terms of the amplitude of the pressure wave.

▷ It can be shown that the amplitudes of pressure, P_0, and displacement, ξ_0, harmonic waves in a gas are related by $P_0 = v\rho\omega\xi_0$. For example, at 400 Hz, the faintest sound that can be heard corresponds to a pressure amplitude of about $8 \times 10^{-5}\,\mathrm{N\,m^{-2}}$. The corresponding displacement amplitude, assuming an air density of $1.29\,\mathrm{kg\,m^{-3}}$ and a velocity of sound of $345\,\mathrm{m\,s^{-1}}$ and recalling that $\omega = 2\pi\nu$, is

$$\xi_0 = \frac{P_0}{v\rho\omega} = 7.15 \times 10^{-11}\,\mathrm{m}$$

This amplitude is of the order of molecular dimensions but is much smaller than the average molecular separation in a gas.

The energy density of the wave is (see Equation 28.36),

$$E = \tfrac{1}{2}\rho\omega^2\xi_0^2 = \frac{P_0^2}{2v^2\rho} \qquad \textbf{(28.37)}$$

and the intensity of the wave, according to Equation 28.34, is $I = vE = P_0^2/2v\rho$. In our example I is $7.19 \times 10^{-12}\,\mathrm{W\,m^{-2}}$. The sensitivity of the human ear is such that for each frequency there is a minimum intensity or **threshold of hearing**, below which sound is not audible, and a maximum intensity or **threshold of feeling**, above which sound produces discomfort or pain. This is illustrated for each frequency by the two curves of Figure 28.17, which also indicates the intensity and pressure amplitudes.

Intensity is also expressed by another unit called the **decibel**, after Alexander G. Bell (1847–1922) who invented the telephone. The **intensity level** of any sound (or of any wave motion) is expressed in decibels, abbreviated dB, according to the definition

$$B = 10 \log \frac{I}{I_0}$$

where I_0 is a reference intensity. For the case of sound in air the reference level I_0 has been arbitrarily chosen as $10^{-12}\,\mathrm{W\,m^{-2}}$. For example, the pressure amplitude given above for the faintest sound that can be heard at 400 Hz has an intensity of $7.19 \times 10^{-12}\,\mathrm{W\,m^{-2}}$; the intensity level is 8.57 dB.

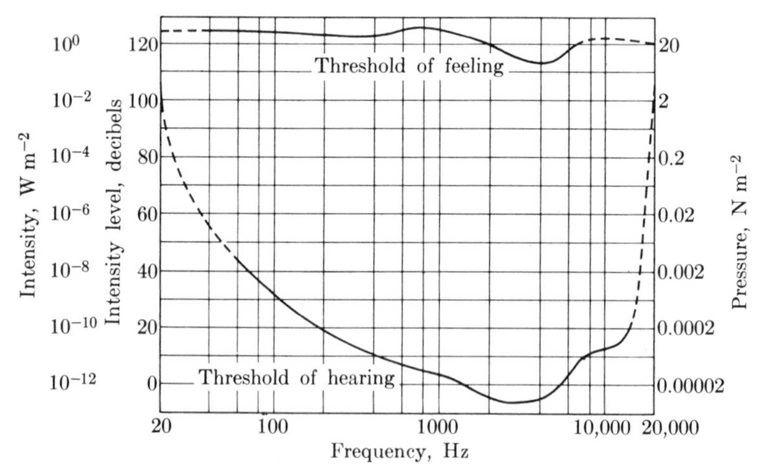

Figure 28.17 Average auditory range for the human ear. The lower curve is the threshold of hearing and the upper curve is the threshold of feeling for each frequency.

28.11 Waves in two and three dimensions

Although $\xi = f(x - vt)$ represents a wave motion propagating along the X-axis, we do not necessarily have to interpret it as meaning a wave concentrated *on* the X-axis. If the physical disturbance described by ξ is extended over all space, we have that at a given time t, the function $\xi = f(x - vt)$ takes the same value at all points having the same x. But $x = const.$ represents a plane perpendicular to the X-axis (Figure 28.18), called a **wave surface**. Therefore in three dimensions the expression $\xi = f(x - vt)$ describes a **plane wave** propagating parallel to the X-axis.

We must note that what is relevant in a plane wave is the direction of propagation, indicated by a unit vector $\boldsymbol{u}$ perpendicular to the plane of the wave, and that the orientation of the coordinate axes is a more or less arbitrary matter. It is therefore convenient to express the plane wave $\xi = f(x - vt)$ in a form that is independent of the orientation of the axes. In the case of Figure 28.19, the unit vector $\boldsymbol{u}$ is parallel to the X-axis. If $\boldsymbol{r}$ is the position vector of *any* point in the wave front, we have that $x = \boldsymbol{u} \cdot \boldsymbol{r}$, and therefore we may write

$$\xi = f(\boldsymbol{u} \cdot \boldsymbol{r} - vt) \tag{28.38}$$

If $\boldsymbol{u}$ is pointing in an arbitrary direction (Figure 28.19), the quantity $\boldsymbol{u} \cdot \boldsymbol{r}$ is still a distance measured from an origin O along the direction of the unit vector $\boldsymbol{u}$. Therefore Equation 28.38 represents a plane wave propagating in the direction $\boldsymbol{u}$. In the case of a harmonic or sinusoidal plane wave propagating in the direction $\boldsymbol{u}$, we write

$$\xi = \xi_0 \sin[k(\boldsymbol{u} \cdot \boldsymbol{r} - vt)]$$

It is convenient to define a vector $\boldsymbol{k} = k\boldsymbol{u}$. This vector has a length given by Equation 28.3, $k = 2\pi/\lambda = \omega/v$, and points in the direction of propagation, and thus is perpendicular to the wave surface. It is usually called the **propagation vector** or **wave number vector**. Then, since $\omega = kv$, a plane harmonic wave is expressed as

$$\xi = \xi_0 \sin(\boldsymbol{k} \cdot \boldsymbol{r} - \omega t) = \xi_0 \sin(k_x x + k_y y + k_z z - \omega t) \tag{28.39}$$

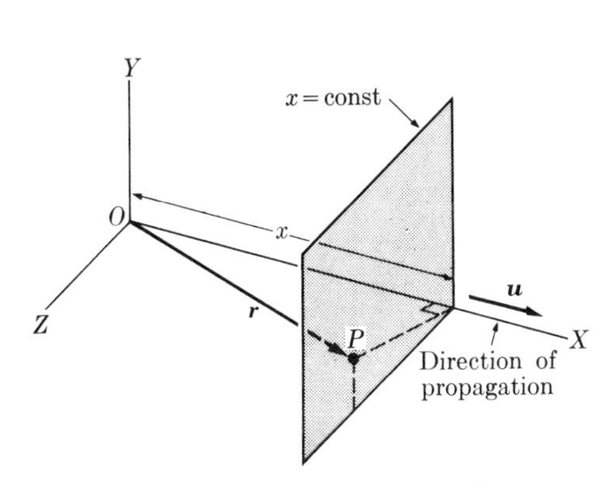

Figure 28.18 Plane wave propagating along the X-axis.

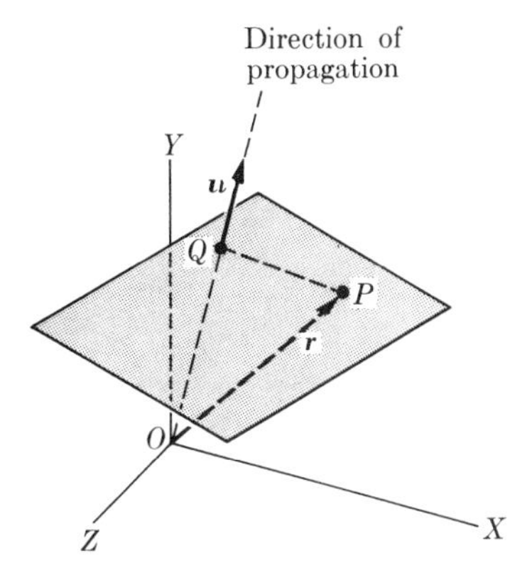

Figure 28.19 Plane wave propagating in arbitrary direction.

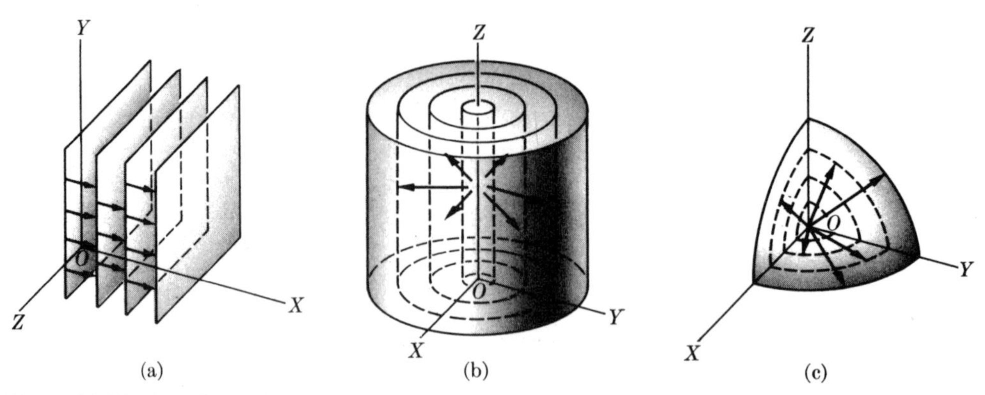

Figure 28.20 (a) Plane, (b) cylindrical, and (c) spherical waves.

where k_x, k_y and k_z are the components of $\boldsymbol{k}$ satisfying the relation

$$k_x^2 + k_y^2 + k_z^2 = k^2 = \omega^2/v^2 \qquad \textbf{(28.40)}$$

The plane waves 28.38 or 28.39, although they contain the three coordinates x, y, z, are really one-dimensional problems, since the propagation is along one particular direction and the physical situation is the same in all planes perpendicular to the direction of propagation (Figure 28.20(a)). But in nature there are other kinds of waves that propagate in several directions. The two most interesting cases are **cylindrical** and **spherical** waves. In the case of cylindrical waves, the wave surfaces are coaxial cylinders parallel to a given line, say the Z-axis, and thus perpendicular to the XY-plane (Figure 28.20(b)). The disturbance propagates in all directions perpendicular to the Z-axis. This type of wave is produced, for example, when we have a series of sources uniformly distributed along an axis, all oscillating in phase, or when a pressure wave in air is produced by a long vibrating string.

If a disturbance that originates at a certain point propagates with the same velocity in all directions, i.e. the medium is **isotropic** (*isos*: the same, *tropos*: direction), spherical waves result. The wave fronts are spheres concentric with the point where the disturbance originated (Figure 28.20(c)). Such waves are produced, for example, when there is a sudden change of pressure at a point in a gas, like an explosion.

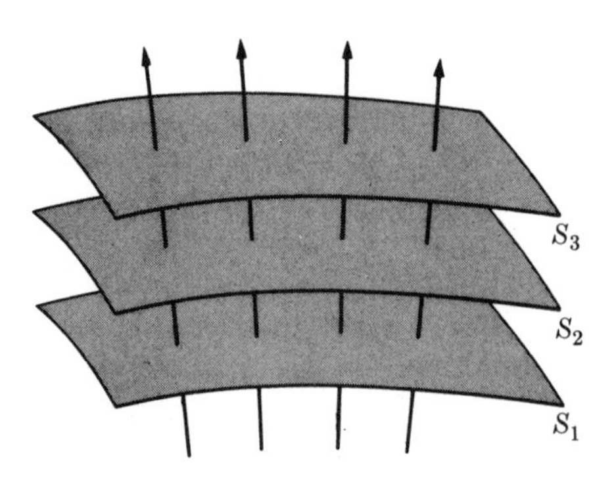

Figure 28.21

Sometimes the velocity of propagation is not the same in all directions, in which case the medium is called **anisotropic**. For example, in a gas where there is a temperature gradient, in a solid under certain strains, or in a large crystal having different elastic properties in different directions, there is a different velocity of propagation for each direction. In these media, the waves are not spherical (Figure 28.21).

28.12 Spherical waves in a fluid

One important case of spherical waves is pressure waves generated by a point source in a homogeneous, isotropic fluid. The velocity of propagation is given by the same expression obtained for plane waves in a fluid, Equation 28.23; that is,

$$v = \sqrt{\frac{\kappa}{\rho}} \tag{28.41}$$

To maintain a wave motion it is necessary to supply energy to the medium through which the wave propagates. As spherical waves propagate, the energy of the wave motion spreads over larger surfaces, resulting in a decrease in the intensity of the wave motion as the distance from the source increases. If energy is not absorbed or scattered in the medium, the average energy that passes through any surface per unit time is also $(\mathrm{d}E/\mathrm{d}t)_{\mathrm{ave}}$. Therefore, according to Equation 28.35, the intensity of the wave at the distance r is related to the surface area, $A = 4\pi r^2$, by the expression (Figure 28.22)

$$\left(\frac{\mathrm{d}E}{\mathrm{d}t}\right)_{\mathrm{ave}} = IA = I(4\pi r^2) = const.$$

Therefore $I = const./r^2$. If we designate the constant by I_0, we finally obtain the intensity of the spherical wave at a distance r from the source as

$$I = \frac{I_0}{r^2} \tag{28.42}$$

We conclude then that

> *in a spherical wave that propagates without absorption or scattering, the intensity decreases inversely with the square of the distance from the source.*

a result that is frequently applied in both acoustics and optics.

Since the intensity of a wave is proportional to the square of the amplitude, we conclude that Equation 28.42 requires the amplitude of a spherical wave to be inversely proportional to r. This means that the amplitude of the pressure wave drops as the distance from the source increases, a result confirmed experimentally. If the fluid is isotropic the wave has the same amplitude in all directions at the same distance from the source. If, in addition, the wave propagates without

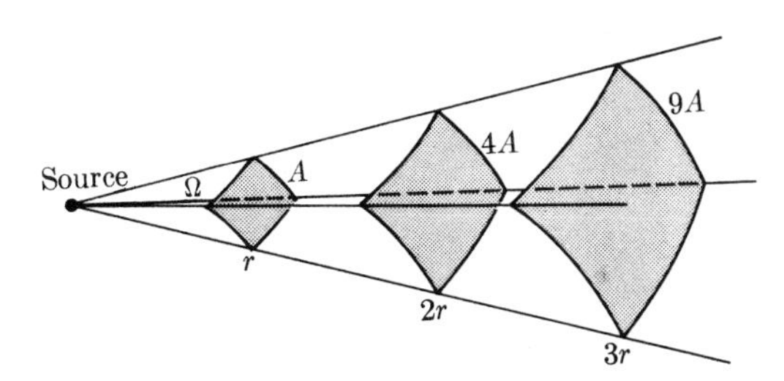

Figure 28.22

absorption or scattering, the pressure wave is given by the expression

$$p - p_0 = \frac{1}{r} f(r - vt) \tag{28.43}$$

where p_0 is the average or equilibrium pressure. Therefore, we now have a geometrical factor $1/r$ that was not present in a plane wave, which accounts for a decrease in pressure variations with distance from the source. A particularly interesting case is that of a spherical harmonic pressure wave expressed by

$$p - p_0 = \frac{P_0}{r} \sin(kr - \omega t) \tag{28.44}$$

The amplitude of the pressure wave at the distance r from the source is P_0/r. The displacement corresponding to this pressure wave is given by a more complicated expression. But at large distances from the source, the amplitude of the displacement wave also decreases inversely proportional to the distance from the source. Then, the displacement may be expressed by

$$\xi = \frac{\xi_0}{r} \cos(kr - \omega t) \tag{28.45}$$

where $\xi_0 = P_0/v\rho\omega$, a relation identical to that for plane waves (Example 28.8).

EXAMPLE 28.8

Intensity of a spherical wave in a gas.

▷ For a harmonic spherical wave, the displacement is given by Equation 28.45. We note that the amplitude is now ξ_0/r instead of ξ_0. The energy per unit volume at large distances is now given, according to Equation 28.37, by

$$\mathrm{E} = \frac{1}{2} \frac{\rho\omega^2\xi_0^2}{r^2} = \frac{P_0}{2v^2\rho r^2}$$

Hence E decreases with the distance from the source as $1/r^2$, or inversely proportional to the square of the distance from the source.

According to Equation 28.34, the intensity of the wave at the distance r from the source, or the average energy crossing the unit area per unit time, is $I = v\mathrm{E} = P_0/2v\rho r^2 = I_0/r^2$, with $I_0 = P_0^2/2v\rho$, which is the same expression obtained in Example 28.7 for plane waves.

28.13 Group velocity

The velocity of propagation $v = \omega/k$, as given by Equation 28.6 for a harmonic wave of angular frequency ω and wavelength $\lambda = 2\pi/k$, is called the **phase velocity**. However, this is not necessarily the velocity we observe when we analyze a non-harmonic wave motion. If we have a continuous harmonic wave (or, as it is sometimes called, a **wave train**) of infinite length, the wave has a single wavelength and a single frequency. But a wave of this nature is not adequate for transmitting a signal, because a signal implies something that begins at a certain time and ends at a certain later time; i.e. the wave must have a shape similar to that indicated in Figure 28.23. A wave with such a shape is called a **pulse**. A signal is a codified set of pulses. Therefore, if we measure the velocity with which the signal is transmitted, we are essentially implying the velocity with which the pulses travel.

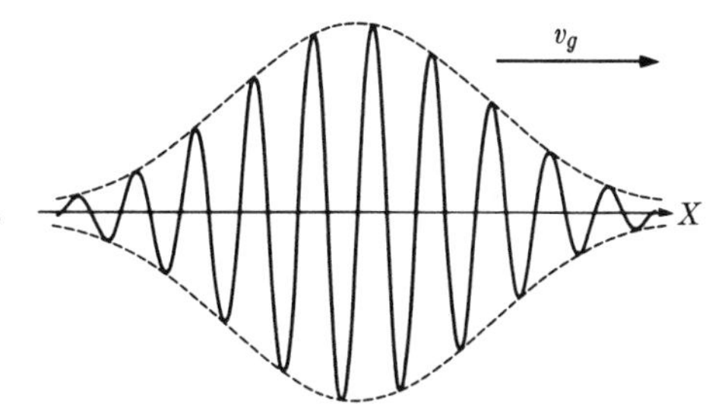

Figure 28.23 Wave pulse.

However, the pulse depicted in Figure 28.23 is *not* a harmonic wave, since its amplitude is not constant along the direction of propagation. The pulse actually contains several frequencies and wavelengths. Of course, if the velocity of propagation is independent of the frequency (i.e. if there is no dispersion), then all components of the pulse travel with the same velocity, and the velocity of the pulse is the same as the phase velocity of the wave. However, in a dispersive medium each component of the pulse has its own velocity of propagation, and the velocity of the pulse is not the same as the phase velocity.

For simplicity, consider a case where the wave motion may be broken down into two frequencies, ω and ω', which are almost equal, so that $\omega' - \omega$ is very small. We shall also assume that their amplitudes are the same. Then we have

$$\begin{aligned}
\xi &= \xi_0 \sin(kx - \omega t) + \xi_0 \sin(k'x - \omega' t) \\
&= \xi_0[\sin(kx - \omega t) + \sin(k'x - \omega' t)] \\
&= 2\xi_0 \cos \tfrac{1}{2}[(k' - k)x - (\omega' - \omega)t] \times \sin \tfrac{1}{2}[(k' + k)x - (\omega' + \omega)t]
\end{aligned}$$

Since ω and ω' as well as k and k' are almost equal, we may replace $\frac{1}{2}(\omega' + \omega)$ by ω and $\frac{1}{2}(k' + k)$ by k, so that

$$\xi = 2\xi_0 \cos \tfrac{1}{2}[(k' - k)x - (\omega' - \omega)t] \sin(kx - \omega t) \qquad \textbf{(28.46)}$$

Equation 28.46 represents a wave motion determined by the factor $\sin(kx - \omega t)$, whose amplitude is modulated according to

$$2\xi_0 \cos \tfrac{1}{2}[(k' - k)x - (\omega' - \omega)t]$$

(see Figure 28.24) so that the wave motion resembles a series of pulses. The modulated amplitude itself corresponds to a wave motion propagated with a velocity

$$v_g = \frac{\omega' - \omega}{k' - k} = \frac{d\omega}{dk} \qquad \textbf{(28.47)}$$

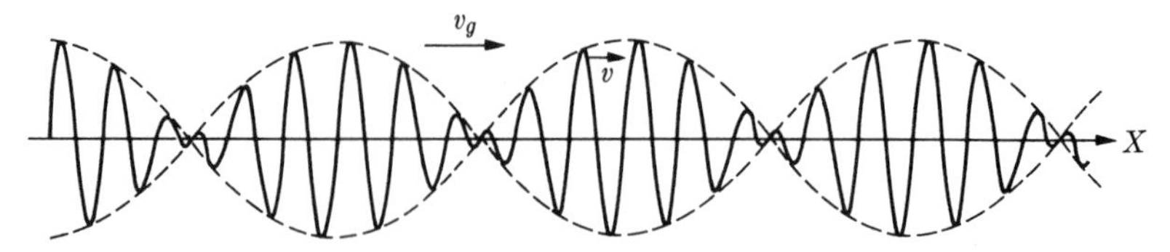

Figure 28.24 Phase and group velocity.

called the **group velocity**. This is the velocity with which the amplitude wave, represented by the dashed line in Figure 28.24, propagates. That is, each maximum of the wave in Figure 28.24 propagates with the group velocity v_g. Therefore, in a dispersive medium the group velocity is the velocity with which a signal is transmitted.

If we recall that $\omega = kv$, Equation 28.47 becomes

$$v_g = v + k\frac{dv}{dk} \tag{28.48}$$

If the phase velocity is independent of the wavelength, as in non-dispersive media, $dv/dk = 0$ and the phase and group velocities are equal; that is, $v_g = v$. Therefore, in non-dispersive media there is no difference between phase velocity and group velocity. But in a dispersive medium, the group velocity may be larger or smaller than the phase velocity.

EXAMPLE 28.9

Group velocity of surface waves in a liquid.

▷ In the case of surface waves in a liquid in the long wave approximation, the phase velocity is given by Equation 28.32, and since $k = 2\pi/\lambda$,

$$v = \sqrt{\frac{g\lambda}{2\pi}} = \sqrt{\frac{g}{k}}$$

Then

$$\frac{dv}{dk} = -\frac{1}{2k}\sqrt{\frac{g}{k}} = -\frac{v}{2k}$$

and Equation 28.48 gives $v_g = \frac{1}{2}v$, so that the group velocity is just half the phase velocity. This means that if a disturbance is produced in a liquid, the velocity of the peak of the disturbance is half the velocity of propagation of each harmonic component. For the short wave approximation, we use Equation 28.33,

$$v = \sqrt{\frac{2\pi\mathcal{T}}{\rho\lambda}} = \sqrt{\frac{k\mathcal{T}}{\rho}}$$

Then

$$\frac{dv}{dk} = \frac{1}{2}\sqrt{\frac{\mathcal{T}}{\rho k}} = \frac{v}{2k}$$

so that $v_g = \frac{3}{2}v$ and the group velocity is larger than the phase velocity.

28.14 The Doppler effect

When the source of a wave and the observer are in relative motion with respect to a medium in which waves propagate, the frequency of the waves observed is different from the frequency of the source. This phenomenon is called the **Doppler effect**, after Christian J. Doppler (1803–1853), who first noticed it in sound waves.

Suppose we have a source of waves, such as a vibrating body, moving to the right (Figure 28.25) through a still medium such as air or water. Observing the source at several positions 1, 2, 3, 4, ..., we note that after a certain time counted from the time when the source was at position 1, the waves emitted at the several positions occupy the spheres 1, 2, 3, 4, ..., which are not concentric. The waves are more closely spaced on the side toward which the body is moving and are more widely separated on the opposite side. To an observer at rest on either side, this corresponds, respectively, to a shorter and a longer effective wavelength or a larger and a smaller effective frequency. The waves will reach an observer in motion at a different rate. For example, if the observer is approaching the source from the right an even shorter wavelength (or higher frequency) will be observed, since the motion is into the waves. The opposite happens if the observer is moving away from the source and thus moving away from the waves.

We shall designate v_s and v_o as the velocities of the source and the observer relative to the medium through which the waves propagate with phase velocity v. Then the relation between the frequency ν of the source and the frequency ν' noticed by the observer when both are moving along the direction of propagation (Figure 28.26) is given by the expression (see the derivation below)

$$\nu' = \frac{v - v_o}{v - v_s}\,\nu \tag{28.49}$$

In this expression v_o and v_s are considered positive if they have the same direction as the vector ***SO*** that goes from the source to the observer and negative if they have the opposite direction.

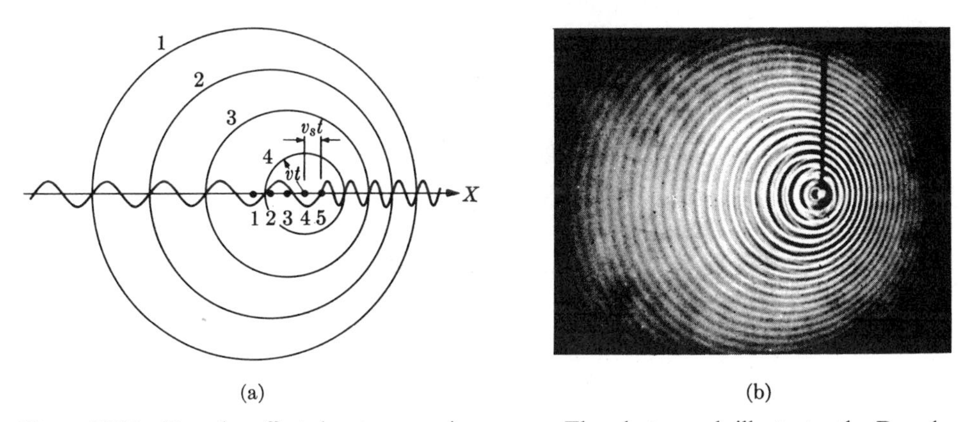

Figure 28.25 Doppler effect due to a moving source. The photograph illustrates the Doppler effect on a liquid surface.

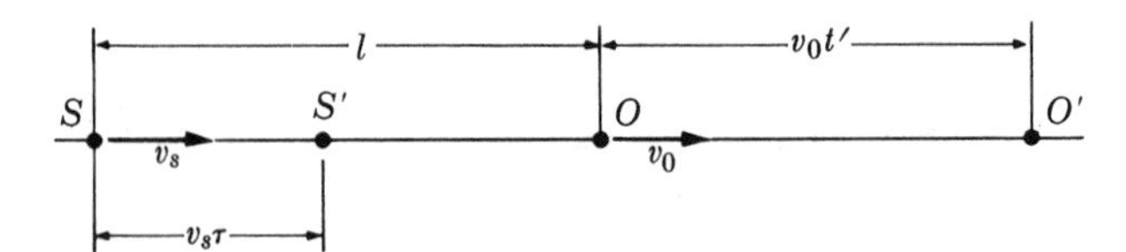

Figure 28.26 The positive direction is always taken in the direction of SO, regardless of whether O is to the right or left of S.

When both v_o and v_s are very small compared with v, the expression relating ν and ν' can be written in the form

$$\nu' = \left(1 - \frac{v_{os}}{v}\right)\nu \tag{28.50}$$

where $v_{os} = v_o - v_s$ is the velocity of the observer relative to the source. If v_{os} is positive, the observer is receding from the source and the frequency observed is lower. But if v_{os} is negative, the observer and the source are approaching each other and the frequency observed is higher.

A special situation occurs when the observer is at rest but the source moves with a velocity greater than the phase velocity of the waves; that is, v_s is larger than v. Then in a given time the source advances further than the wave front. For example, if in time t the source moves from A to B (Figure 28.27), the wave emitted at A has traveled only from A to A'. The surface tangent to all successive waves is a cone, whose axis is the line of motion of the source and whose aperture α is given by

$$\sin\alpha = \frac{v}{v_s} \tag{28.51}$$

The resultant wave motion is then a **conical** wave that propagates as indicated by the arrows in Figure 28.27. This wave is sometimes called a **Mach wave** or a **shock wave**, and is the sudden and violent sound we hear when a supersonic plane passes nearby. These waves are also observed in the wakes of boats moving faster than the velocity of surface waves on the water.

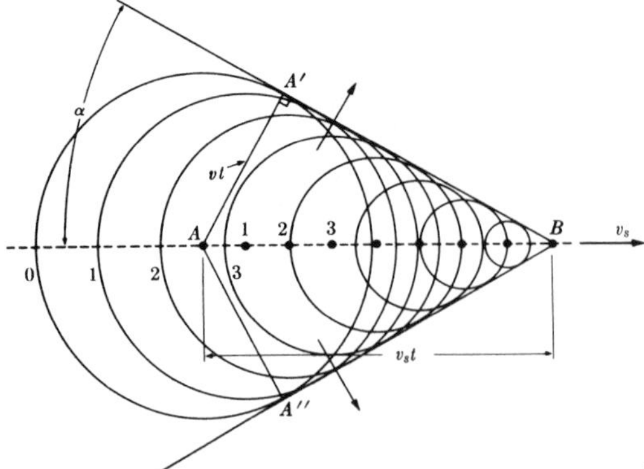

Figure 28.27 Mach or shock wave.

Derivation of Doppler's relation

To obtain the relation between the frequency ν of the waves produced by the source and the frequency ν' recorded by the observer, we use the following reasoning. For simplicity, assume that the source and the observer are moving along the same line (see Figure 28.26) and that the observer O is to the right of the source S. In any case the positive direction is always taken as that of the vector $\boldsymbol{SO}$.

Suppose that at time $t = 0$, when the source and the observer are separated by the distance $SO = l$, the source emits a wave that reaches the observer at a later time t. In that time the observer has moved the distance $v_o t$ and the total distance traversed by the wave in the time t has been $l + v_o t$. If v is the velocity of propagation of the wave, this distance is also vt. Then

$$vt = l + v_o t \qquad \text{or} \qquad t = \frac{l}{v - v_o}$$

At time $t = \tau$ the source is at S', and the wave emitted at that instant will reach the observer at a time t', measured from the same time origin as before. The total distance traveled by the wave from the time it is emitted at S' until it is received by the observer is $(l - v_s\tau) + v_o t'$. The actual travel time of the wave has been $t' - \tau$ and the distance traveled is $v(t' - \tau)$. Therefore,

$$v(t' - \tau) = l - v_s\tau + v_o t' \qquad \text{or} \qquad t' = \frac{l + (v - v_s)\tau}{v - v_o}$$

The time interval reckoned by the observer between the two waves emitted by the source at S and at S' is

$$\tau' = t' - t = \frac{v - v_s}{v - v_o}\,\tau$$

Now if ν is the frequency of the source, the number of waves emitted in the time interval τ is $\nu\tau$. Since the waves are received by the observer in the time interval τ', the frequency observed is

$$\nu' = \frac{\nu\tau}{\tau'} \qquad \text{or} \qquad \nu' = \frac{v - v_o}{v - v_s}\,\nu$$

which is the relation given in Equation 28.49.

Dividing the numerator and denominator by v, we have

$$\nu' = \left(\frac{1 - v_o/v}{1 - v_s/v}\right)\nu = \left(1 - \frac{v_o}{v}\right)\left(1 - \frac{v_s}{v}\right)^{-1}\nu$$

Using the first two terms of the binomial expansion (Equation B.22), we have $(1 - v_s/v)^{-1} \approx (1 + v_s/v)$ when v_s/v is much smaller than one. Then we may write

$$\nu' = \left(1 - \frac{v_o}{v}\right)\left(1 + \frac{v_s}{v}\right)\nu = \left(1 - \frac{v_o}{v} + \frac{v_s}{v} - \frac{v_o v_s}{v^2}\right)\nu$$

When we multiply both parentheses we must be consistent with our approximation, and keep only first-order terms. Then, neglecting the term $v_o v_s/v^2$, we have that the frequency measured by the observer reduces to

$$\nu' = \left(1 - \frac{v_o - v_s}{v}\right)\nu = \left(1 - \frac{v_{os}}{v}\right)\nu$$

which is the approximate Equation 28.50.

EXAMPLE 28.10

The frequency of the sound produced by a bell is 500 Hz. The velocity of the source relative to still air is $60\ \mathrm{m\,s^{-1}}$. An observer moves at $30\ \mathrm{m\,s^{-1}}$ along the same line as the source. Calculate the sound frequency measured by the observer. Consider all the possible cases.

▷ Assume the velocity of sound to be $340\ \mathrm{m\,s^{-1}}$ and that the relative position of the source and the observer is as shown in Figure 28.26.

(i) The source and the observer move to the right. Since $v_s > v_o$, the source is approaching the observer. Also v_s and v_o are positive. Then, omitting some units,

$$\nu' = \frac{v - v_o}{v - v_s}\nu = \left(\frac{340 - 30}{340 - 60}\right)(500\ \mathrm{Hz}) = 553\ \mathrm{Hz}$$

(ii) The source and the observer move to the left. Then the source recedes from the observer and v_s and v_o are negative. Then

$$\nu' = \left(\frac{340 - (-30)}{340 - (-60)}\right)(500\ \mathrm{Hz}) = 462\ \mathrm{Hz}$$

(iii) The source moves to the right and the observer to the left so that they approach each other. Then v_s is positive and v_o is negative, and

$$\nu' = \left(\frac{340 - (-30)}{340 - 60}\right)(500\ \mathrm{Hz}) = 660\ \mathrm{Hz}$$

(iv) The source moves to the left and the observer to the right so that they recede from each other. Then v_s is negative and v_o is positive and

$$\nu' = \left(\frac{340 - 30}{340 - (-60)}\right)(500\ \mathrm{Hz}) = 387\ \mathrm{Hz}$$

The student should repeat the example assuming that initially the observer is to the left of the source. What conclusions can be derived from this example?

Note 28.1 Acoustics

With the exception of waves on the surface of a liquid, all other waves discussed in this chapter fall in the category of **elastic waves**, where the disturbance—whether a strain, a pressure, or a bulk displacement involving many atoms—propagates with a velocity depending on the elastic properties of the medium, or in other words on the intermolecular forces.

These elastic waves are also called **sound**. In everyday language the concept of sound is related to the sensation of **hearing**. Whenever an elastic wave propagates through a gas, a liquid, or a solid and reaches our ear, it produces vibrations in the ear's membrane. These vibrations provoke a nervous response constituting the process known as hearing. Our nervous system produces a hearing sensation only for frequencies from about 16 Hz up to about 20 000 Hz. The frequency range may vary from one person to another and is different for other animals. Outside these limits, sounds are not audible, but the elastic waves are still called sound. The physics of elastic waves with frequencies above 20 000 Hz is called **ultrasonics**. Ultrasonics finds many applications in industry and medicine.

Table 28.2 Velocity of sound (m s^{-1})

Solids (20 °C)		Liquids (25 °C)		Gases (0 °C, 1 atm)	
Granite	6000	Fresh water	1493.2	Air	331.45
Iron	5130	Sea water	1532.8	Air (100 atm)	350.6
Copper	3750	(3.6% salinity)		Hydrogen	1269.5
Aluminum	5100	Kerosene	1315	Oxygen	317.2
Lead	1230	Mercury	1450	Nitrogen	339.3
Lucite	1840	Alcohol	1210	Steam (100 °C)	404.8

The science dealing with the methods of generation, reception and propagation of sound is called **acoustics**. It really covers many fields and is closely related to several branches of engineering. Among the fields of acoustics are the design of acoustical instruments, which includes **electroacoustics**, dealing with electrical production and recording of sound (microphones, loudspeakers, amplifiers etc.). **Architectural acoustics** deals with the design and construction of rooms and buildings and the behavior of sound waves in closed spaces. **Musical acoustics** deals more directly with sound in relation to music.

Sound involves the displacement of atoms and molecules of the medium through which it propagates. But this displacement is an ordered collective motion where all atoms in a small volume suffer essentially the same displacement. The ordered collective motion is then superposed on the random or disordered molecular agitation of liquids and gases. As the wave propagates, part of its energy is transferred from the ordered wave motion to disordered thermal motion, a process called **dissipation**. The net result is that the intensity of sound decreases or is **attenuated** while the sound wave propagates and some of the wave's energy is taken away by molecules of the medium through collisions. This results in an increase in the molecular internal energy, mainly rotational molecular motion or translational kinetic energy. In liquids, the viscosity, which in essence is an effect of the intermolecular forces on the molecular motion (recall Section 18.6), also plays an important role in sound attenuation.

The velocity of sound in a gas is practically independent of frequency for a very large range, extending up to more than 10^8 Hz. The value of this velocity for different substances is given in Table 28.2. The velocity of sound is, however, rather sensitive to temperature and pressure changes because of the dependence of velocity on density.

QUESTIONS

28.1 How far does a harmonic wave advance in one period? How long does it take to move through a distance of one wavelength?

28.2 Verify that $\xi = \xi_0 \sin k(x + vt)$ satisfies the wave equation.

28.3 State the difference between normal stress and shear stress. Can a fluid withstand a shear stress?

28.4 How is the velocity of propagation of a transverse wave along a string modified if the tension is (a) doubled, (b) halved? How must the tension be changed to (c) double, (d) halve, the velocity of propagation?

28.5 A string has a cross-sectional radius r and a density ρ. Express the velocity of propagation of transverse waves on the string in terms of r and ρ.

28.6 Why is wave motion in gases considered an adiabatic process?

28.7 Distinguish phase velocity from group velocity. In what case are both velocities equal? Which of the two is a signal velocity?

28.8 Why is a harmonic wave not adequate to transmit a message?

28.9 What is meant by a dispersive medium? How does dispersion affect (a) a harmonic wave, (b) a non-harmonic wave?

28.10 Consider several spherical surfaces concentric with a point source. How do the area of each surface, the energy flowing per unit time through each surface and the intensity of the wave motion at each surface vary with distance from the source?

28.11 Under what condition is the Doppler effect the same when the source is stationary relative to the medium and the observer approaches the source with a certain speed as when the observer is stationary and the source approaches the observer with the same relative speed?

28.12 Write the wave equation for (a) longitudinal waves in a spring, (b) density and pressure waves in a gas, indicating explicitly in each case the velocity of propagation.

28.13 Show that, when a force F is applied at one end of a rod of length L that has the other end fixed, the deformation at the distance x is $\xi = Fx/YA$. Graphically represent how ξ varies with x. Analyze how this relation can be used to measure Young's modulus of a rod or wire by measuring the deformation ΔL.

28.14 Express a harmonic wave in terms of (a) k and v, (b) k and ω and (c) λ and P.

28.15 Show that the expression $\xi = \xi_0 \sin \omega(t \pm x/v)$ also satisfies the wave equation and is equivalent to Equations 28.5 and 28.9.

PROBLEMS

28.1 By rocking a boat, surface waves are produced on a quiet lake. The boat performs 12 oscillations in 20 seconds, each oscillation producing a wave crest. It takes 6 s for a given crest to reach the shore 12 m away. (a) Calculate the wavelength of the surface waves. (b) Write the expression of the surface waves.

28.2 The equation of a certain wave is $\xi = 10 \sin 2\pi(2x - 100t)$, where x is in meters and t is in seconds. Find (a) the amplitude, (b) the wavelength, (c) the frequency and (d) the velocity of propagation of the wave. (e) Make a sketch of the wave, showing the amplitude and the wavelength.

28.3 Given the wave $\xi = 2 \sin 2\pi(0.5x - 10t)$, where t is in seconds and x is in meters, (a) plot ξ over several wavelengths for $t = 0$ and $t = 0.025$ s. (b) Repeat the problem for $\xi = 2 \sin 2\pi(0.5x + 10t)$ and compare results.

28.4 Assuming that in the previous problem the wave corresponds to an elastic transverse wave, plot (a) the velocity $d\xi/dt$ and (b) the acceleration $d^2\xi/dt^2$ at $t = 0$ and $t = \frac{1}{4}P$.

28.5 Given the equation for a wave on a string $\xi = 0.03 \sin(3x - 2t)$, where ξ and x are in meters and t is in seconds, (a) at $t = 0$, what are the values of the displacement at $x = 0$, 0.1 m, 0.2 m and 0.3 m? (b) At $x = 0.1$ m, what are the values of the displacement at $t = 0$, 0.1 s, and 0.2 s? (c) What is the equation for the velocity of oscillation of the particles of the string? (d) What is the maximum velocity of oscillation? (e) What is the velocity of propagation of the wave?

28.6 Consider longitudinal waves along a rod (Section 28.5) and assume that the deformation at each point is $\xi = \xi_0 \sin 2\pi(x/\lambda - t/P)$. (a) Using relation 28.16, obtain the expression for the force along the rod. (b) Show that the ξ and F waves have a phase difference of one-quarter wavelength. (c) Plot ξ and F against x, on the same set of axes, at a given time.

28.7 A spring having a normal length of 1 m and a mass of 0.2 kg is stretched 0.04 m by a force of 10 N. Calculate the velocity of propagation of longitudinal waves along the spring.

28.8 Compute the velocity of propagation of sound in (a) hydrogen, (b) nitrogen and (c) oxygen at 0 °C. Compare with the experimental results. For the three gases, assume $\gamma = 1.40$.

28.9 Find the change of sound velocity in air per unit change in temperature at 27 °C.

28.10 From the value given in Example 28.6 for the coefficient $\alpha = (\gamma R/M)^{1/2}$ for air, obtain the effective molecular mass of air and compare with the result obtained by other means (recall Problem 2.4). Assume that for air $\gamma = 1.40$.

28.11 A string is fastened at one end to a fixed support. The other end passes over a pulley at 5 m from the fixed end and carries a load of 2 kg. The mass of the string between the fixed end and the pulley is 0.6 kg. (a) Find the velocity of propagation of transverse waves along the string. (b) Suppose that a harmonic wave of amplitude 10^{-3} m and wavelength 0.3 m propagates along the string; find the maximum transverse velocity of any point of the string. (c) Write the equation of the wave. (d) Determine the average rate of energy flow across any section of the string.

28.12 A steel wire with a diameter of 2×10^{-4} m is subject to a tension of 200 N. (a) Calculate the velocity of propagation of transverse waves along the wire. (b) Write the expression of a transverse wave having a frequency of 400 Hz.

28.13 A string of length 2 m and mass 4×10^{-3} kg is held horizontally, with one end fixed. The other end passes over a pulley and supports a mass of 2 kg. Calculate the velocity of transverse waves in the string.

28.14 The end of a stretched string is forced to vibrate with a displacement given by the equation $\xi = 0.1 \sin 6t$, where ξ is in meters and t is in seconds. The tension in the string is 4 N and the mass per unit length is 0.010 kg m^{-1}. Calculate (a) the wave velocity in the string, (b) the frequency of the wave, (c) the wavelength, (d) the equation of the displacement at a point 1 m from the source and at a point 3 m from the source.

28.15 A steel wire having a length of 2 m and a radius of 5×10^{-4} m hangs from the ceiling. (a) If a body having a mass of 100 kg is hung from the free end, calculate the elongation of the wire. (b) Also determine the displacement and the downward pull at a point at the middle of the wire. (c) Determine the velocity of longitudinal and transverse waves along the wire when the mass is attached.

28.16 (a) Verify that a transverse elastic wave propagating along the X-axis and corresponding to a displacement ξ having components $\xi_y = \xi_0 \sin(kx - \omega t)$ and $\xi_z = \xi_0 \cos(kx - \omega t)$ is circularly polarized. (b) Determine the sense of rotation of ξ as seen by an observer on the X-axis. (c) Write the expressions for ξ_y and ξ_z for a wave having an opposite polarization.

28.17 (a) Obtain the velocity of shear waves in steel. (b) Compare with the result for longitudinal waves given in Example 28.3.

28.18 Consider a canal of rectangular cross-section having a depth of 4 m. Determine the velocity of propagation of waves having a wavelength of (a) 0.01 m, (b) 1 m, (c) 10 m and (d) 100 m. In each case use the expression that corresponds best to the order of magnitude of the quantities involved. The water in the canal has a surface tension of 7×10^{-2} N m^{-1}.

28.19 (a) Compare the relative importance of the two terms in the velocity of surface waves in deep water (Equation 28.33) for the following wavelengths: (i) 10^{-3} m, (ii) 10^{-2} m, (iii) 1 m. (b) At what wavelength are the two terms equal? For water the surface tension is about 7×10^{-2} N m^{-1}.

28.20 A thin steel rod is forced to transmit longitudinal waves by means of an oscillator coupled to one end. The rod has a diameter of 4×10^{-3} m. The amplitude of the oscillations is 10^{-4} m and the frequency is 10 oscillations per second. Find (a) the equation of the waves along the rod, (b) the energy per unit volume of the wave, (c) the average energy flow per unit time across any section of the rod and (d) the power required to drive the oscillator.

28.21 The faintest sound that can be heard has a pressure amplitude of about 2×10^{-5} N m^{-2}, and the loudest that can be heard without pain has a pressure amplitude of about 28 N m^{-2}. Determine, in each case, (a) the intensity of the sound both in W m^{-2} and in dB and (b) the amplitude of the oscillations if the frequency is 500 Hz. Assume an air density of 1.29 kg m^{-3} and a velocity of sound of 345 m s^{-1}.

28.22 Two sound waves have intensity levels differing by (a) 10 dB, (b) 20 dB. Find the ratio of their intensities and pressure amplitudes.

28.23 (a) How is the intensity of a sound wave changed when the pressure amplitude is doubled? (b) How must the pressure amplitude change to increase the intensity by a factor of 10?

28.24 Find the group velocity of ripples in water. Use Equation 28.33 for the phase velocity.

28.25 The pitch of the whistle of a locomotive is 500 Hz. Determine the frequency of the sound heard by a person standing at the station if the train is moving with a velocity of 72 km h^{-1} (a) toward and (b) away from the station. Assume the velocity of sound in still air is 340 m s^{-1}.

28.26 A sound source has a frequency of 10^3 Hz and moves at 30 m s^{-1} relative to the air. Assuming that the velocity of sound relative to still air is 340 m s^{-1}, find the effective wavelength and the frequency recorded by an observer at rest relative to the air who sees the source (a) moving away and (b) moving toward the observer.

28.27 Repeat Problem 28.26, assuming that the source is at rest relative to the air but that the observer moves at 30 m s^{-1}. From your results do you conclude that it is immaterial whether it is the source or the observer that is in motion?

29 Electromagnetic waves

Note

During 1885–1889 Heinrich Hertz was able to produce and detect electromagnetic waves in his laboratory, providing a definitive experimental confirmation of Maxwell's electromagnetic theory that predicted their existence. Hertz carried out a long series of experiments in which he studied the reflection, refraction, polarization and interference of electromagnetic waves. He also measured their velocity of propagation, which is the same as that of light, establishing the relation between electromagnetic waves and light. Hertz' discoveries opened the way to radio communication.

29.1 Introduction

The phenomenon of electromagnetic induction indicates the possibility of transmitting a signal from one place to another using a time-dependent electromagnetic field. Near the end of the nineteenth century, Heinrich Hertz (1857–1894) proved beyond any doubt that the electromagnetic field does propagate in vacuum with a velocity equal to that of light. Before Hertz performed his experiments, the existence of electromagnetic waves had been predicted by Maxwell, as a result of analysis of the equations of the electromagnetic field (Table 27.1) that showed that the electric and magnetic fields satisfy a wave equation of the form of Equation 28.11. Understanding about the production, propagation and absorption of electromagnetic waves has opened the door to modern methods of communication.

29.2 Plane electromagnetic waves

Maxwell's equations for the electromagnetic field in vacuum (that is, in a region where there are no free charges or currents) admit, as a special solution, an electric field $\mathscr{E}$ and a magnetic field $\mathscr{B}$ that are perpendicular to each other and vary only along a direction perpendicular to both fields (Figure 29.1). If we place the Y-axis parallel to the $\mathscr{E}$-field and orient the Z-axis parallel to the $\mathscr{B}$-field, we have, for this special case, $\mathscr{E}_x = 0$, $\mathscr{E}_y = \mathscr{E}$, $\mathscr{E}_z = 0$ and $\mathscr{B}_x = 0$, $\mathscr{B}_y = 0$, $\mathscr{B}_z = \mathscr{B}$. Under these conditions, as shown at the end of this section, the electric and magnetic fields satisfy the wave equation; that is,

$$\frac{d^2\mathscr{E}}{dt^2} = c^2 \frac{d^2\mathscr{E}}{dx^2} \quad \text{and} \quad \frac{d^2\mathscr{B}}{dt^2} = c^2 \frac{d^2\mathscr{B}}{dx^2} \tag{29.1}$$

indicating that the electric field $\mathscr{E}$ and the magnetic field $\mathscr{B}$ propagate along the X-axis with a velocity c given by

$$c = \frac{1}{(\varepsilon_0 \mu_0)^{1/2}} \approx 3 \times 10^8 \text{ m s}^{-1} \tag{29.2}$$

Hence the electric and magnetic fields can be expressed as

$$\mathscr{E} = \mathscr{E}(x - ct), \quad \mathscr{B} = \mathscr{B}(x - ct) \tag{29.3}$$

and correspond to a plane electromagnetic wave. Let us, in particular, consider the case of harmonic waves of frequency $\nu = \omega/2\pi$ and wavelength $\lambda = 2\pi/k$. In such a case

$$\mathscr{E} = \mathscr{E}_0 \sin k(x - ct) \quad \text{and} \quad \mathscr{B} = \mathscr{B}_0 \sin k(x - ct) \tag{29.4}$$

The amplitudes $\mathscr{E}_0$ and $\mathscr{B}_0$ are not independent but, as required by Maxwell's equations, are related by

$$\mathscr{E}_0 = c\mathscr{B}_0 \quad \text{or} \quad \mathscr{B}_0 = \frac{1}{c}\mathscr{E}_0 \tag{29.5}$$

Relation 29.5 between the amplitudes means that the instantaneous values, given by Equation 29.4, are also related by

$$\mathscr{E} = c\mathscr{B} \quad \text{or} \quad \mathscr{B} = \frac{1}{c}\mathscr{E} \tag{29.6}$$

It may be verified, through the use of Maxwell's equations, that the same relations hold true when applied to fields defined in the more general form of Equation 29.3.

From Equation 29.6, we see that the $\mathscr{E}$ and $\mathscr{B}$ fields are in phase, reaching their zero and maximum values at the same time. The electromagnetic wave

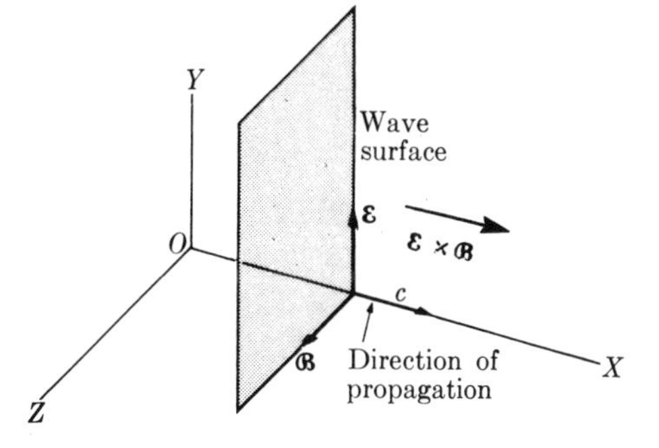

Figure 29.1 Orientation of the electric and magnetic fields relative to the direction of propagation of a plane electromagnetic wave. The vector $\varepsilon \times \mathscr{B}$ is in the direction of propagation.

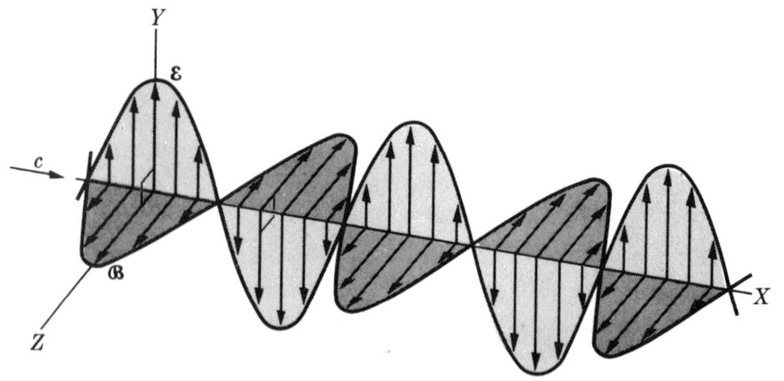

Figure 29.2 Electric and magnetic fields in a harmonic plane electromagnetic wave propagating in the positive X-direction. The plane of polarization is XY.

described by Equation 29.4 is represented in Figure 29.2. The electric field oscillates in the XY-plane and the magnetic field in the XZ-plane. This corresponds to a **plane linearly polarized** wave. The plane of polarization is defined as the plane in which the electric field oscillates, in this case the XY-plane. Thus an electromagnetic wave actually consists of two coupled waves in phase: the electric wave and the magnetic wave.

Another plane wave solution of Maxwell's equations is one where the electric and magnetic fields remain constant in magnitude but rotate around the direction of propagation, resulting in a **circularly polarized** wave (Figure 29.3). Circular polarization can be obtained by combining two perpendicular linearly polarized solutions with equal amplitudes for each component and with a phase difference of $\pm\pi/2$ (recall Section 10.8). The circular polarization may be right- or left-handed according to the phase difference and the sense of rotation of the fields with respect to the direction of propagation.

If the amplitudes of the two rectangular components of each field are different, **elliptical polarization** results. In addition, other plane wave solutions of Maxwell's equations are possible that do not correspond to any particular state of polarization.

Since the choice of the XYZ-axes is a matter of convenience, we may conclude that the plane wave solutions of Maxwell's equations we have discussed are completely general, and that

plane electromagnetic waves are transverse, with the $\mathscr{E}$ and $\mathscr{B}$ fields perpendicular both to each other and to the direction of propagation of the waves.

This theoretical prediction of Maxwell's equations has been amply confirmed by experiment, and it results in several phenomena that will be considered in subsequent chapters.

Beside plane wave solutions of Maxwell's equations, there are also cylindrical and spherical electromagnetic waves. At a large distance from the source a limited portion of a cylindrical or a spherical wave can be considered as practically plane, and in this case the electric and magnetic fields are also perpendicular to each other and to the direction of propagation (i.e. radial) as indicated in Figure 29.4. Note that the direction of propagation coincides with the direction of the vector $\mathscr{E} \times \mathscr{B}$.

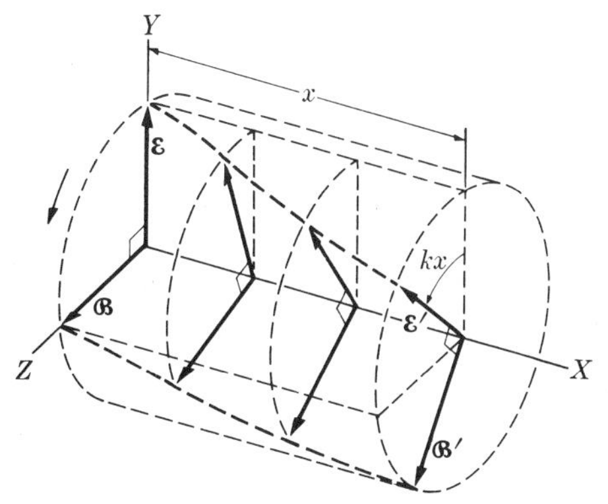

Figure 29.3 Right-handed or dextro-circularly polarized electromagnetic wave. The $\mathscr{E} \times \mathscr{B}$ fields rotate around the direction of propagation but maintain the same magnitude.

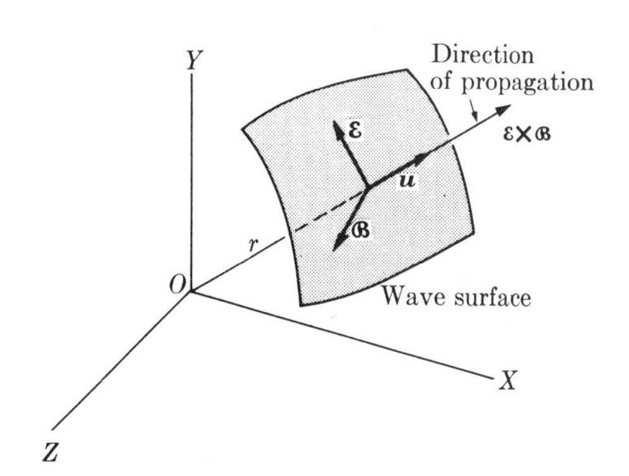

Figure 29.4 Spherical electromagnetic wave at a large distance from the source.

Derivation of the wave equations for electric and magnetic fields

In Figure 29.5 we have represented the lines of force of an electric field parallel to the Y-axis and a magnetic field parallel to the Z-axis. Although we have drawn only the lines of force in the XY- and the XZ-planes, the electric and magnetic fields extend over the whole space. Consider first the small rectangle $PQRS$ located as shown in Figure 29.5(a). The sides of the rectangle are $QR = PS = \mathrm{d}x$ and $PQ = SR = \mathrm{d}y$. We shall apply the Faraday–Henry law,

$$\oint \mathscr{E} \cdot \mathrm{d}\boldsymbol{l} = -\frac{\mathrm{d}}{\mathrm{d}t} \int \mathscr{B} \cdot \mathrm{d}\boldsymbol{S} \tag{29.7}$$

to this rectangle. In computing the circulation of the electric field around $PQRS$, note that the electric field is perpendicular to sides QR and SP. Therefore these sides do not contribute to the circulation of $\mathscr{E}$. Calling $\mathscr{E}$ and $\mathscr{E}'$ the values of the electric field along sides PQ and RS,

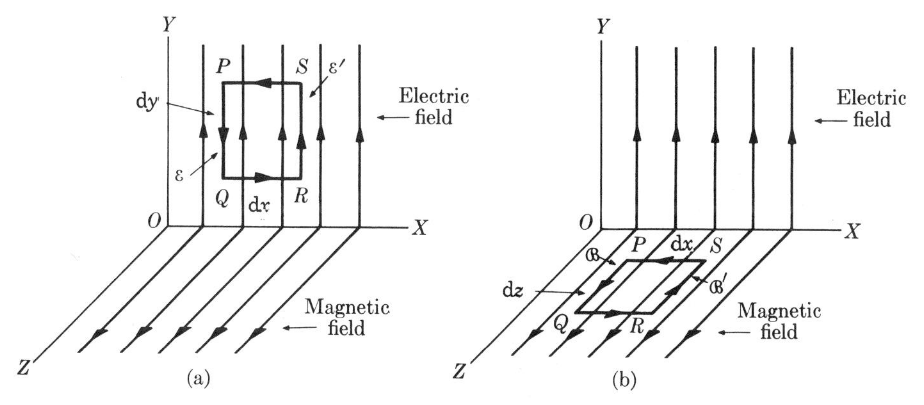

Figure 29.5 Application of Maxwell's equations to the rectangles $PQRS$ yields the wave equation for the electromagnetic fields. The wave propagates in the positive X-direction.

we may then write

$$\oint_{PQRS} \mathscr{E}\cdot \mathrm{d}\boldsymbol{l} = \mathscr{E}'(SR) - \mathscr{E}(PQ) = (\mathscr{E}' - \mathscr{E})\,\mathrm{d}y$$

The negative sign in the second term is due to the fact that the electric field has a direction opposite to that in which PQ is traversed. But $\mathscr{E}'$ and $\mathscr{E}$ are the values of the electric field along two lines separated by a distance $\mathrm{d}x$. Therefore, $\mathscr{E}' - \mathscr{E} = \mathrm{d}\mathscr{E} = (\mathrm{d}\mathscr{E}/\mathrm{d}x)\,\mathrm{d}x$. Hence, the circulation of the electric field (the left-hand side of Equation 29.7) may be written as

$$\oint_{PQRS} \boldsymbol{\mathscr{E}}\cdot \mathrm{d}\boldsymbol{l} = \frac{\mathrm{d}\mathscr{E}}{\mathrm{d}x}\,\mathrm{d}x\,\mathrm{d}y$$

Next we note that $\mathrm{d}S$ is the area of the rectangle $PQRS$ and is equal to $\mathrm{d}x\,\mathrm{d}y$. Also $\mathrm{d}\boldsymbol{S}$ is a vector perpendicular to the XY-plane and therefore it is parallel to $\boldsymbol{\mathscr{B}}$. Then we may compute the right-hand side of Equation 29.7 as

$$\int_{PQRS} \boldsymbol{\mathscr{B}}\cdot \mathrm{d}\boldsymbol{S} = \mathscr{B}\,\mathrm{d}x\,\mathrm{d}y$$

Substituting in Equation 29.7, after canceling the common factor $\mathrm{d}x\,\mathrm{d}y$, we get

$$\frac{\mathrm{d}\mathscr{E}}{\mathrm{d}x} = -\frac{\mathrm{d}\mathscr{B}}{\mathrm{d}t} \tag{29.8}$$

Next we apply the Ampère–Maxwell law when there are no currents present,

$$\oint \boldsymbol{\mathscr{B}}\cdot \mathrm{d}\boldsymbol{l} = \varepsilon_0\mu_0 \frac{\mathrm{d}}{\mathrm{d}t}\int \boldsymbol{\mathscr{E}}\cdot \mathrm{d}\boldsymbol{S} \tag{29.9}$$

to the rectangle $PQRS$ in Figure 29.5(b). The sides of the rectangle are $\mathrm{d}x$ and $\mathrm{d}z$. Designating $\mathscr{B}$ and $\mathscr{B}'$ as the values of the magnetic field along the sides PQ and RS, we may write

$$\oint_{PQRS} \boldsymbol{\mathscr{B}}\cdot \mathrm{d}\boldsymbol{l} = -\mathscr{B}'(RS) + \mathscr{B}(PQ) = -(\mathscr{B}' - \mathscr{B})\,\mathrm{d}z = -\frac{\mathrm{d}\mathscr{B}}{\mathrm{d}x}\,\mathrm{d}x\,\mathrm{d}z$$

where $\mathrm{d}\mathscr{B} = \mathscr{B}' - \mathscr{B}$. Again, since $\mathrm{d}\boldsymbol{S}$ (with magnitude $\mathrm{d}x\,\mathrm{d}z$) is a vector perpendicular to the ZX-plane and thus parallel to $\boldsymbol{\mathscr{E}}$, we have

$$\oint_{PQRS} \boldsymbol{\mathscr{E}}\cdot \mathrm{d}\boldsymbol{S} = \mathscr{E}\,\mathrm{d}x\,\mathrm{d}z$$

Substituting these results in Equation 29.9 and canceling the common factor $\mathrm{d}x\,\mathrm{d}z$, we obtain

$$-\frac{\mathrm{d}\mathscr{B}}{\mathrm{d}x} = \varepsilon_0\mu_0 \frac{\mathrm{d}\mathscr{E}}{\mathrm{d}t} \tag{29.10}$$

Equations 29.8 and 29.10 couple the electric and magnetic fields at each point of space. To obtain an equation where only the electric field appears, we take the derivative of Equation 29.8 with respect to x and of Equation 29.10 with respect to t. The results are

$$\frac{\mathrm{d}^2\mathscr{E}}{\mathrm{d}x^2} = -\frac{\mathrm{d}^2\mathscr{B}}{\mathrm{d}x\,\mathrm{d}t}, \qquad -\frac{\mathrm{d}^2\mathscr{B}}{\mathrm{d}x\,\mathrm{d}t} = \varepsilon_0\mu_0 \frac{\mathrm{d}^2\mathscr{E}}{\mathrm{d}t^2}$$

Eliminating $d^2\mathscr{B}/dx\,dt$ between these two equations, we obtain

$$\frac{d^2\mathscr{E}}{dx^2} = \varepsilon_0\mu_0 \frac{d^2\mathscr{E}}{dt^2}$$

Making $c^2 = 1/\varepsilon_0\mu_0$, the first of Equations 29.1 results. By the same method we obtain a similar equation for the magnetic field.

To obtain relation 29.5, we first see from Equation 29.4 that

$$\frac{d\mathscr{E}}{dx} = k\mathscr{E}_0 \cos k(x - ct)$$

and

$$\frac{d\mathscr{B}}{dt} = -kc\mathscr{B}_0 \cos k(x - ct)$$

Substituting these results in Equation 29.8 and canceling common factors, we get $\mathscr{E}_0 = c\mathscr{B}_0$.

29.3 Energy and momentum of an electromagnetic wave

Recalling Equations 25.34 and 26.17, the energy densities associated with the electric and magnetic fields of an electromagnetic wave are

$$E_e = \tfrac{1}{2}\varepsilon_0\mathscr{E}^2 \qquad \text{and} \qquad E_m = \frac{1}{2\mu_0}\mathscr{B}^2$$

When we use Equations 29.6, $\mathscr{B} = \mathscr{E}/c$, and 29.2, $c = 1/(\varepsilon_0\mu_0)^{1/2}$, the magnetic energy density becomes

$$E_m = \frac{1}{2\mu_0}\mathscr{B}^2 = \frac{1}{2\mu_0 c^2}\mathscr{E}^2 = \tfrac{1}{2}\varepsilon_0\mathscr{E}^2 = E_e$$

That is, the electric and magnetic energy densities of an electromagnetic wave are equal. The total energy density of the electromagnetic wave is

$$E = E_e + E_m = \varepsilon_0\mathscr{E}^2 \tag{29.11}$$

The **intensity** of the electromagnetic wave (that is, the energy passing in the unit time through the unit area perpendicular to the direction of propagation is, according to Equation 28.34,

$$I = Ec = c\varepsilon_0\mathscr{E}^2 \tag{29.12}$$

Energy and momentum are closely related, so that we may expect that an electromagnetic wave carries, in addition to its energy, a certain momentum. We may use the relation between energy and momentum given by Equation 20.3,

$p = vE/c^2$, to obtain the momentum per unit volume associated with an electromagnetic wave. Since electromagnetic radiation propagates with velocity c, we make $v = c$ and obtain

$$\mathrm{P} = \frac{\mathrm{E}}{c} \tag{29.13}$$

We may write the above equation in vector form as

$$\mathbf{P} = \frac{\mathrm{E}}{c}\,\boldsymbol{u} \tag{29.14}$$

where $\boldsymbol{u}$ is the unit vector in the direction of propagation, which is the direction of the vector $\mathscr{E} \times \mathscr{B}$. Although the relation given in Equation 29.13 has been established per unit volume, to an arbitrary amount of energy E in a plane wave there corresponds a momentum $p = E/c$ in the direction of propagation.

An electromagnetic wave also possesses an intrinsic angular momentum, or spin, similar to the spin of particles such as electrons and protons (remember Section 23.6). For circularly polarized plane waves it can be shown that the spin has a component along the direction of propagation $S = \pm E/\omega$, depending on whether the polarization is right- or left-handed. For a linearly polarized wave, the average value of the component of the spin along the direction of propagation is zero. This is because a plane polarized wave can be considered as the superposition of the two opposite circularly polarized waves.

In summary, we conclude that

an electromagnetic wave carries momentum and angular momentum as well as energy.

Therefore, when a charged particle absorbs or emits electromagnetic radiation, not only do its energy and momentum change but also its angular momentum changes accordingly, a result that has been verified experimentally, both directly and indirectly.

This supports the concept that the electromagnetic interaction between two electric charges means an exchange of energy, momentum and angular momentum between the charges, by means of the electromagnetic field, which is the carrier of the energy, momentum and angular momentum exchanged.

EXAMPLE 29.1

Radiation pressure.

▷ Since electromagnetic waves carry momentum, they give rise to a certain pressure when they are reflected or absorbed at the surface of a body. The basic principle is the same as in the case of the pressure exerted by a gas on the walls of a container, as explained in Example 13.5.

First consider some simple cases. Suppose that a plane electromagnetic wave falls perpendicularly on a perfectly absorbing surface (Figure 29.6). The incident momentum

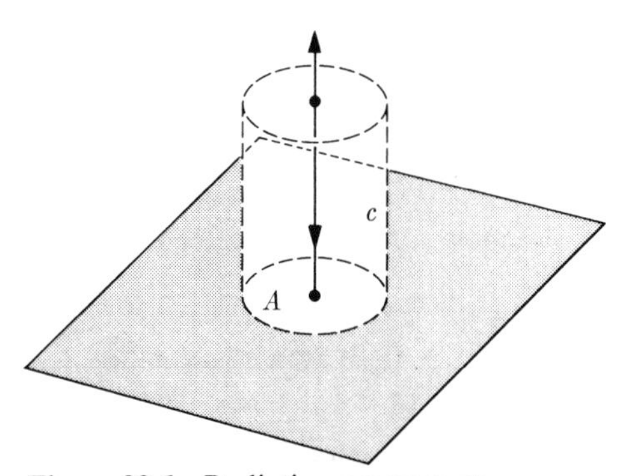

Figure 29.6 Radiation pressure at normal incidence.

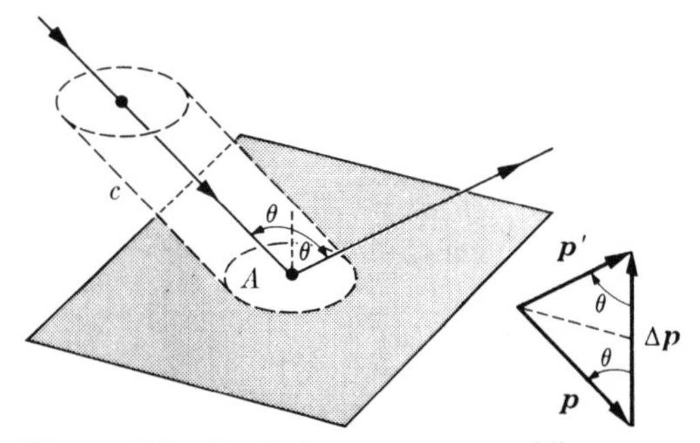

Figure 29.7 Radiation pressure at oblique incidence. The momentum diagram is shown at the right. The incident momentum is p and the reflected momentum is p'. (The diagrams are drawn in perspective.)

per unit volume is P and the amount of momentum in the radiation falling, per unit time, on the surface A is obtained by multiplying P by the volume cA; that is, $\mathrm{P}cA$. If the radiation is completely absorbed by the surface, this is also the momentum absorbed per unit time by the surface A; that is, the force on A. Dividing by A, we get the pressure due to the radiation

$$P_{\text{rad}} = c\mathrm{P} = \mathrm{E} = \varepsilon_0 \mathscr{E}^2$$

Thus for normal incidence the radiation pressure on a perfect *absorber* is equal to the energy density in the electromagnetic wave.

On the other hand, if the surface is a perfect *reflector*, the radiation after reflection has a momentum equal in magnitude but opposite in direction to the incident radiation. For perpendicular incidence, the change in momentum per unit volume is thus $2\mathrm{P}$, and the radiation pressure is accordingly

$$P_{\text{rad}} = 2c\mathrm{P} = 2\mathrm{E} = 2\varepsilon_0 \mathscr{E}^2$$

These results can be generalized to the case of oblique incidence (Figure 29.7), where the change in momentum of the radiation per unit volume at the perfectly reflecting surface is $2\mathrm{P} \cos \theta$, and the corresponding radiation pressure is (recall Example 13.5)

$$P_{\text{rad}} = 2c\mathrm{P} \cos^2 \theta = 2\mathrm{E} \cos^2 \theta$$

The student may verify that this is identical to the result of Equation 13.8, if $c\mathrm{P}$ is replaced by nmv^2. If the radiation propagates in all directions, we must sum over all directions. The result obtained is

$$P_{\text{rad}} = \tfrac{2}{3}c\mathrm{P} = \tfrac{2}{3}\mathrm{E}$$

If the surface is a perfect absorber, no momentum is reflected and the radiation pressure is then $\frac{1}{3}\mathrm{E}$. The existence of radiation pressure, which has been verified experimentally and is responsible for several important phenomena, provides an indirect verification of Equation 29.13. For example, the curvature of a comet tail can be explained in terms of radiation pressure, due to the electromagnetic radiation from the Sun.

To estimate the pressure on the Earth's surface due to the Sun's radiation, we must consider that the rate of incident energy is approximately $1.4 \times 10^3\ \mathrm{W\,m^{-2}}$, corresponding to an energy density (when we divide by c) equal to $4.7 \times 10^{-6}\ \mathrm{J\,m^{-3}}$. Assuming that the Earth is a perfect absorber and that the radiation comes from all directions, the radiation pressure is

$$P_{rad} = \tfrac{1}{3}E = 1.6 \times 10^{-6}\,\mathrm{N\,m^{-2}}$$

Comparing this with atmospheric pressure, which is about $10^5\,\mathrm{N\,m^{-2}}$, the radiation pressure from the Sun is so small it does not affect the orbital motion of the Earth.

29.4 Radiation from oscillating dipoles

In our discussion so far, we have considered electromagnetic waves without mentioning how they are produced; in other words, without explaining what are the *sources* of electromagnetic waves. In dealing with elastic waves (such as sound) we say that the source of the waves is some vibrating body, such as the membrane of a drum, or a violin string. In the case of electromagnetic waves the sources of the waves are the same as the sources of electromagnetic fields; that is, moving electric charges. Maxwell's equations predict that accelerated charges radiate electromagnetic waves. We shall consider two important cases. In one case the moving charges constitute an oscillating electric dipole and in the other they correspond to an oscillating magnetic dipole.

(i) Electric dipole radiation. An oscillating electric dipole arises when the relative motion of the positive and negative charges can be described collectively by an electric dipole whose moment changes with time according to $\mathscr{P} = \mathscr{P}_0 \sin \omega t$. This could be the case, for example, of an electron within an atom, when the normal motion of the electron is perturbed, or of an oscillating current in a linear antenna of a broadcasting station. When the electric dipole moment is constant, the only field produced is electric. But when the electric dipole moment is oscillating, the electric field is also oscillating, and is therefore time-dependent. This means that a magnetic field is also present, as stated in the Ampère–Maxwell law. This can also be seen from the fact that an oscillating electric dipole is equivalent to a linearly oscillating current, and an oscillating electric current always produces an oscillating magnetic field as well.

We may use our physical intuition to determine the main characteristics of the field of an oscillating electric dipole. At points very close to the electric dipole, the effect of retardation due to the finite velocity of propagation of the electromagnetic waves is negligible because the distance r is very small. The oscillating electric field is then similar to the field created by a static electric dipole, as computed in Example 21.12. At large distances, however, the finite propagation of the waves produces a modification in the field. The solution of the wave equation shows that the electric and magnetic fields are generally oriented as indicated in Figure 29.8, so that $\mathscr{E}$ and $\mathscr{B}$ are perpendicular. The direction of the vector $\mathscr{E} \times \mathscr{B}$, which gives the direction of energy and momentum flow, is outward.

The amplitude and direction of the fields $\mathscr{E}$ and $\mathscr{B}$ depend on the angle θ that $\boldsymbol{r}$ makes with the dipole axis. Both $\mathscr{E}$ and $\mathscr{B}$ are zero for points along the Z-axis. This means that the intensity of the electromagnetic wave of an oscillating electric dipole is zero along the direction of oscillation. On the other hand, $\mathscr{E}$ and $\mathscr{B}$ have maximum amplitude for points on the XY-plane. Therefore, the electro-

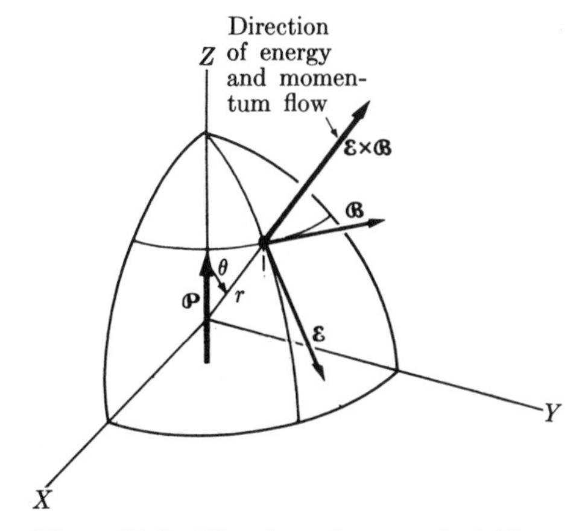

Figure 29.8 Electric and magnetic fields produced by an electric dipole oscillating parallel to the Z-axis.

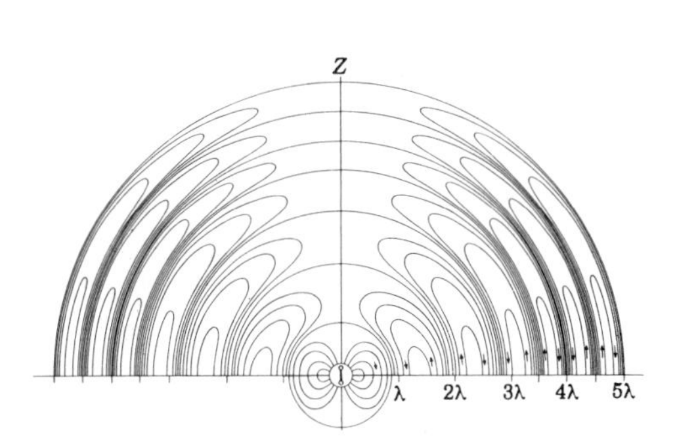

Figure 29.9 Electric field lines produced by an oscillating electric dipole $\mathscr{P}$.

magnetic wave of an oscillating electric dipole has its maximum intensity in the equatorial plane. The waves are linearly polarized, with the electric field oscillating in a meridian plane and the magnetic field oscillating parallel to the XY-plane. Figure 29.9 shows the electric lines of force in a meridian plane at a given time. Each loop corresponds to one complete oscillation. Successive loops are separated one wavelength. The magnetic lines of force are circles parallel to the XY-plane with their centers on the Z-axis.

(ii) Magnetic dipole radiation. In Section 22.6 we defined a magnetic dipole as an electric charge describing a small closed path, or in the case of many charges, as a small current loop. Suppose that the loop is in the XY-plane with its center at the origin (Figure 29.10). If the current oscillates according to $I = I_0 \sin \omega t$, it constitutes an oscillating magnetic dipole with a magnetic dipole moment $\mathscr{M} = \mathscr{M}_0 \sin \omega t$, where $\mathscr{M}_0 = I_0 A$, A is the area of the loop and $\mathscr{M}$ is oriented in a direction perpendicular to the loop. A static magnetic dipole produces only a constant magnetic field. When the magnetic dipole oscillates, its magnetic field at each point of space is also oscillating or time-dependent. This means that an electric field is also present, in accordance with the Faraday–Henry law.

The radiation field of an oscillating magnetic dipole has the roles of the electric and magnetic fields reversed from those in the electric dipole case. That is, at large distances from the magnetic dipole, the magnetic field is in a meridian plane, while the electric field

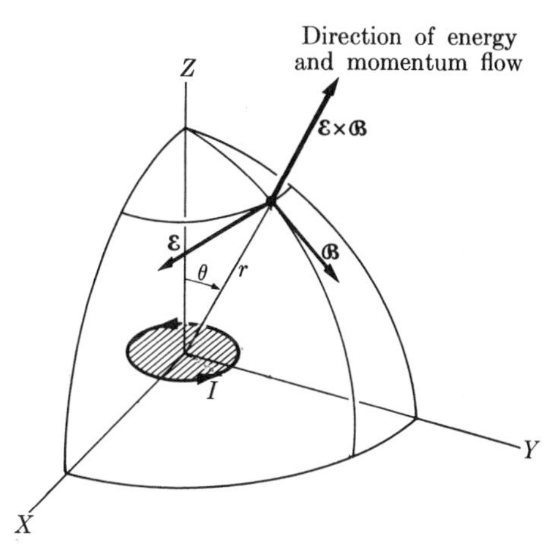

Figure 29.10 Electric and magnetic fields produced by an oscillating magnetic dipole $\mathscr{M}$.

is in a transverse direction, so that the electric lines of force are circles concentric with the Z-axis and parallel to the XY-plane. In other words, the plane of polarization is rotated 90° with respect to electric dipole waves. This affords a means of distinguishing electric dipole radiation from magnetic dipole radiation.

The relative orientation of the $\mathscr{E}$ and $\mathscr{B}$ fields for an oscillating magnetic dipole is illustrated in Figure 29.10. Note that the vector $\mathscr{E} \times \mathscr{B}$, which gives the direction of energy and momentum flow, is in the radial direction at large distances. As in the case of the electric dipole, the magnetic dipole radiation has its maximum intensity at the equatorial plane and is zero along the axis of the dipole.

Since energy flows away from the oscillating electric and magnetic dipoles, it is necessary to supply energy to them at a steady rate to maintain their oscillation. Electric dipole radiation and magnetic dipole radiation are two of the most effective ways for producing electromagnetic waves and constitute the most important mechanisms by which atoms, molecules and nuclei emit (and absorb) electromagnetic radiation.

Note 29.1 Comparison of electric and magnetic dipole radiation

The calculation of the energy radiated by a system of oscillating charges is a very important practical problem because of its many applications in the transmission and reception of electromagnetic signals, such as in broadcasting stations, radio-telephones, radar and microwave signaling, etc. However, as indicated before, it is a rather elaborate mathematical procedure, beyond the level of this text. Therefore we shall limit ourselves to quote the results.

In the case of an oscillating electric dipole of amplitude $\mathscr{P}_0$ and frequency $\nu = \omega/2\pi$, the average energy radiated per unit time by the electric dipole is given by

$$\left(\frac{\mathrm{d}E}{\mathrm{d}t}\right)_{\text{elec.}} = \frac{\mathscr{P}_0^2\omega^4}{12\pi\varepsilon_0 c^3} \qquad \textbf{(29.15)}$$

The average energy radiated per unit time by an oscillating magnetic dipole of amplitude $\mathscr{M}_0$ and frequency $\nu = \omega/2\pi$ is

$$\left(\frac{\mathrm{d}E}{\mathrm{d}t}\right)_{\text{mag.}} = \frac{\mu_0\mathscr{M}_0^2\omega^4}{12\pi c^3} \qquad \textbf{(29.16)}$$

Therefore, since $\varepsilon_0\mu_0 = 1/c^2$,

$$\frac{(\mathrm{d}E/\mathrm{d}t)_{\text{mag.}}}{(\mathrm{d}E/\mathrm{d}t)_{\text{elec.}}} = \frac{\mathscr{M}_0^2}{c^2\mathscr{P}_0^2}$$

In the case of an electron in an atom $\mathscr{P}_0$ is of the order of magnitude of ez_0, where z_0 is of the order of atomic dimensions or about 10^{-10} m. Therefore $\mathscr{P}_0 \sim 10^{-29}$ m C. Also, in the case of an electron in an atom we have, from Equation 23.23, that $\mathscr{M}_0 = (e/2m_e)L$, where L is the orbital angular momentum of the electron. The quantity $e/2m_e$ is 1.759×10^{11} C kg^{-1} and the angular momentum L is of the order of Planck's constant, 10^{-34} J s^{-1}, so that $\mathscr{M}_0 \sim 10^{-23}$ C m^2 s^{-1}, and

$$\frac{(\mathrm{d}E/\mathrm{d}t)_{\text{mag.}}}{(\mathrm{d}E/\mathrm{d}t)_{\text{elec.}}} \sim 10^{-6}$$

Therefore, we conclude that for atoms (and also molecules) the ratio of the intensity of the magnetic dipole radiation to the electric dipole radiation is of the order of 10^{-6}. This indicates that, for the same frequency, magnetic dipole radiation from atoms is negligible compared with electric dipole radiation and must be taken into consideration only when electric dipole radiation is absent for some reason. However, in nuclei, magnetic dipole radiation is relatively more intense than in atoms and molecules because z_0, being of the order of nuclear sizes, is about 10^4 times smaller, while the magnetic moment is only 10^3 times smaller, because of the larger proton mass. Then

$$\frac{(\mathrm{d}E/\mathrm{d}t)_{\text{mag.}}}{(\mathrm{d}E/\mathrm{d}t)_{\text{elec.}}} \sim 10^{-4}$$

In both the atomic and nuclear cases there is a dependence on the frequency of the radiation that has not been taken into account in our approximate expressions.

The case of an antenna from a broadcasting station is different because there is a large number of electrons involved in the magnetic dipole radiation. A calculation, which we omit, shows that for broadcasting antennas the magnetic mode of radiation is not much weaker than the electric mode.

29.5 Radiation from an accelerated charge

The electric and magnetic fields of a charge in uniform motion, that is, moving with constant velocity, were discussed in Section 27.10. The electric field is radial and the magnetic field is transverse, with circular lines of force concentric with the line of motion. Figure 29.11 shows the electric field $\mathscr{E}$ and the magnetic field $\mathscr{B}$ at four symmetric points P_1, P_2, P_3 and P_4. The moving particle carries the field (and therefore the field's energy and momentum as well) with itself. At points fixed in our laboratory frame of reference and behind the moving charge, the electromagnetic field is decreasing. At points ahead of the charge the field is being built up at the same rate. This requires a transfer of energy in the direction of motion of the charge relative to the L-frame of reference. However, the total energy of the field, as measured in the L-frame of reference, remains constant. We conclude then that

a charge in uniform rectilinear motion does not radiate electromagnetic energy.

When a charge is in accelerated motion, the electric field is no longer radial and does not have the fore–aft symmetry it has when the motion is uniform. The lines of force have a pattern similar to that shown in Figure 29.12. The reason for the asymmetry is that as the particle moves, the field behind the charge and the field ahead of the charge change as in the previous case. But, because of the acceleration, the new field

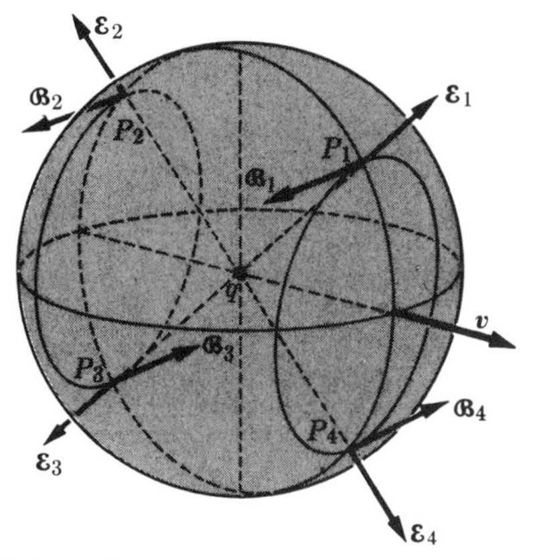

Figure 29.11 Electric and magnetic fields of a uniformly moving charge.

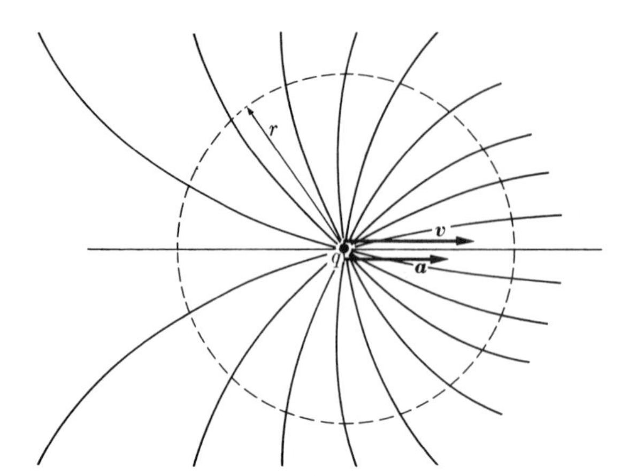

Figure 29.12 Electric lines of force of the field produced by an accelerated charge.

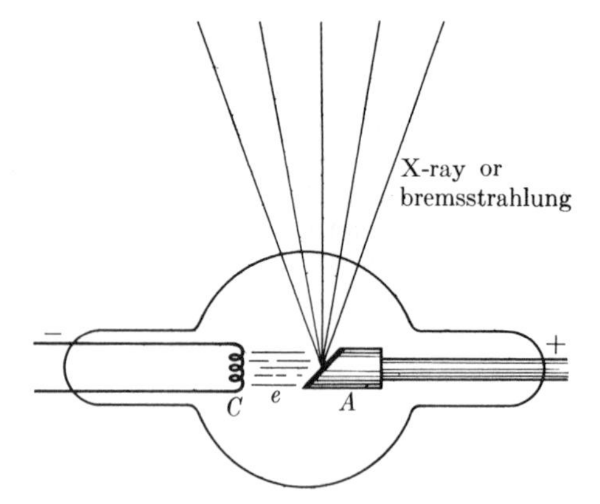

Figure 29.13 Radiation emitted by a charge decelerated when hitting the target A in an X-ray tube. The target must be constructed from material with a large atomic number and a high melting point and it must be continuously cooled.

(which corresponds to the new velocity) is different from the field that existed previously (corresponding to an earlier velocity). For this reason, energy must be transferred to all space to adjust the field. Therefore

an accelerated charge radiates electromagnetic energy.

This means that, when attempting to determine the motion of a charged particle under an applied force, we must take into account the radiation of energy. When the acceleration is small, the radiation is negligible. At small velocities, the energy radiated by the charge has maximum intensity along directions perpendicular to the acceleration. Also the energy radiated per unit time by an accelerated charge is proportional to the square of the acceleration of the charge. In particular, the rate of energy radiation by an accelerated charge which is momentarily at rest or is moving slowly relative to the observer, is given by

$$\frac{\mathrm{d}E}{\mathrm{d}t} = \frac{q^2 a^2}{6\pi\varepsilon_0 c^3} \tag{29.17}$$

where a is the acceleration of the charge. This result is called **Larmor's formula**.

One important conclusion is that, to maintain a charge in accelerated motion, we must supply energy to compensate for the energy lost by radiation. This means that when an ion is accelerated in a linear accelerator, a fraction of the energy supplied to the ion is lost as electromagnetic radiation. The loss, however, is negligible except at very high accelerations. This may be seen as follows. Assume a charged particle of mass m, moving along an accelerator of length s under a constant force F. Using Equations 9.1 and 9.15, we have $a = F/m = E_k/ms$. Therefore, Equation 29.11 gives

$$\frac{\mathrm{d}E}{\mathrm{d}t} = \frac{q^2 c}{6\pi\varepsilon_0 s^2}\left(\frac{E_k}{mc^2}\right)^2 \tag{29.18}$$

As long as E_k is small compared with the rest energy of the particle, the rate of energy radiation is very small. Further, for a given final energy, radiation losses can be reduced by increasing the length of the linear accelerator, thereby reducing the acceleration.

When a particle is decelerated, it also radiates energy. This is what happens, for example, when a fast charged particle, such as an electron or a proton, hits a target and is quickly decelerated. A substantial part of its total energy goes off as radiation, called **deceleration radiation**, or more commonly **bremsstrahlung** (from the German *Bremsung*, deceleration, and *Strahlung*, radiation) (Figure 29.13). This is the main mechanism by which radiation is produced in X-ray tubes used for physical, medical and industrial applications.

Although Figure 29.12 shows the case where the acceleration is in the same direction as the motion, our discussion holds true for any kind of motion where there is acceleration. For example, a charged particle moving in a circular path has a centripetal acceleration, and hence emits radiation. Therefore when an ion is accelerated in a cyclical accelerator, such as a cyclotron, a betatron or a high-energy collider (Note 20.2), a fraction of the energy supplied to the charged particle is lost as electromagnetic radiation, called **synchrotron radiation**. As can be seen from Equation 29.18, this radiation is more important the smaller the mass of the particle and therefore is more serious in electron accelerators than in proton accelerators.

The intensity of synchrotron radiation increases very rapidly with the energy of the particles. Therefore, in low energy accelerators, such as cyclotrons and betatrons, synchrotron radiation is not a problem. However, when particles reach very high energies, as they do in synchrotrons and colliders, the radiation becomes a serious limiting factor. Since synchrotron radiation decreases as the radius of the orbit increases (recall that centripetal acceleration is v^2/r), one way to minimize synchrotron radiation in high energy accelerators is by increasing the radius of the path of the particles (recall Figures 29.9, 29.10 and 29.11 where the radius is one kilometer or more).

These properties of synchrotron radiation can be verified by a simple calculation. Since, for a circular orbit, $a = v^2/r = p^2/m^2r$, we may rewrite Equation 29.18 as

$$\frac{\mathrm{d}E}{\mathrm{d}t} = \frac{q^2}{6\pi\varepsilon_0 c^3 r^2}\left(\frac{p}{m}\right)^4$$

At low energies, $E_k = p^2/2m$ and we may write

$$\frac{\mathrm{d}E}{\mathrm{d}t} = \frac{4q^2 c}{6\pi\varepsilon_0 r^2}\left(\frac{E_k}{mc^2}\right)^2 \qquad (E_k \ll mc^2)$$

However, at high energies, much greater than mc^2, we may use Equation 20.6, $E_k = cp$, so that

$$\frac{\mathrm{d}E}{\mathrm{d}t} = \frac{q^2 c}{6\pi\varepsilon_0 r^2}\left(\frac{E_k}{mc^2}\right)^4 \qquad (E_k \gg mc^2)$$

which shows the critical dependence of synchrotron radiation on the energy and the mass of the particle.

A particle trapped in a magnetic field moves along a spiral, as discussed in Section 22.3. Therefore it emits synchrotron radiation. Radiation coming from the charged particles trapped in the Earth's magnetic field, from sunspots, or from some more distant bodies (such as certain galaxies) is basically of this kind.

EXAMPLE 29.2

A proton is accelerated in a van de Graaff accelerator through a potential difference of 5×10^5 V. The length of the tube is 2 m. Determine the energy radiated and compare it with the energy gained.

▷ If t is the time for the proton to travel along the length s of the accelerator, we have $t = 2s/v$, with $v = (2E_k/m)^{1/2}$ (where E_k is the kinetic energy of the particle). The total energy radiated is $E_{rad} = (dE/dt)t$. Using the result obtained above for the rate of energy radiation in linear accelerators, we have

$$E_{rad} = \left(\frac{dE}{dt}\right)t = \frac{2^{1/2}q^2}{6\pi\varepsilon_0 s}\left(\frac{E_k}{mc^2}\right)^{3/2}$$

and the ratio of the energy radiated to that gained is

$$\frac{E_{rad}}{E_k} = \frac{2^{1/2}q^2}{6\pi\varepsilon_0 mc^2 s}\left(\frac{E_k}{mc^2}\right)^{1/2}$$

Introducing numerical values, with $E_k = 5 \times 10^5$ *eV*, we have $E_{rad}/E_k = 1.67 \times 10^{-20}$. Radiation losses are then negligible in this accelerator.

EXAMPLE 29.3

A proton is accelerated in a cyclotron having a radius of 0.92 m. The frequency of the potential applied to the dees is 1.5×10^7 Hz, and the peak value of the potential difference is 20 000 V (see Example 22.4). Compare the energy lost by radiation in one revolution with the kinetic energy gained.

▷ The maximum kinetic energy gained by the proton in each revolution is $E_k = 2e\Delta V_{max}$, since it crosses between the dees' gap twice. The acceleration of the proton is $a = \omega^2 r = 4\pi^2\nu^2 r$, and we may neglect relativistic effects. Thus Equation 29.18, with $q = +e$, yields

$$\frac{dE}{dt} = \frac{e^2(4\pi^2\nu^2 r)^2}{6\pi\varepsilon_0 c^3} = \frac{8\pi^3 e^2\nu^4 r^2}{3\varepsilon_0 c^3}$$

and the energy radiated in one revolution (time $= 1/\nu$) is

$$E_{rad} = \left(\frac{dE}{dt}\right)\frac{1}{\nu} = \frac{8\pi^3 e^2\nu^3 r^2}{3\varepsilon_0 c^3}$$

Introducing numerical values yields $E_{rad}/E_k = 4.0 \times 10^{-15}$. Here E_{rad} is still much smaller than E_k, but it is greater than in our previous example of the linear accelerator. As indicated previously, radiation losses are relatively more important in circular than in linear accelerators.

29.6 Propagation of electromagnetic waves in matter; dispersion

Experiments reveal that the velocity of propagation of an electromagnetic wave through matter is different from its velocity of propagation in vacuum. The reason is that when an electromagnetic wave propagates through matter, even if there are no free charges and currents, it induces oscillations in the charged particles of the atoms or molecules, which then emit secondary or 'scattered' waves. These scattered waves are superposed on the original wave, giving a resultant wave. The phases of the secondary waves are, in general, different from those of the original wave, since a forced oscillator is not always in phase with the driving force (remember Section 10.12). A detailed analysis, which we shall omit, indicates that this phase difference affects the resultant wave in such a way that the resultant wave appears to have a velocity different from the wave velocity in vacuum.

If the substance is homogeneous and isotropic, it can be proved that the net effect of the polarization and magnetization of the medium by the electromagnetic wave is to replace the constants ε_0 and μ_0 in Maxwell's equations by the electric permittivity ε and the magnetic permeability μ characteristic of the material. Everything in the discussion of Section 29.2 remains the same, except that the velocity of the wave now becomes

$$v = \frac{1}{\sqrt{\varepsilon\mu}} \tag{29.19}$$

The ratio between the velocity of electromagnetic waves in vacuum, c, and in a substance, v, is called the **index of refraction** of the substance, designated by n. It is a useful concept for describing the properties of materials in relation to electromagnetic waves. Thus

$$n = \frac{c}{v} = \frac{1}{\sqrt{\varepsilon_0\mu_0}}\sqrt{\varepsilon\mu} = \sqrt{\frac{\varepsilon\mu}{\varepsilon_0\mu_0}} = \sqrt{\varepsilon_r\mu_r} \tag{29.20}$$

where ε_r and μ_r are the relative permittivity and permeability of the medium, which were defined in Sections 25.9 and 26.7. In general μ_r differs very little from 1 for the majority of substances that transmit electromagnetic waves (see Table 26.1), and we can write, as a satisfactory approximation,

$$n \approx \sqrt{\varepsilon_r} \tag{29.21}$$

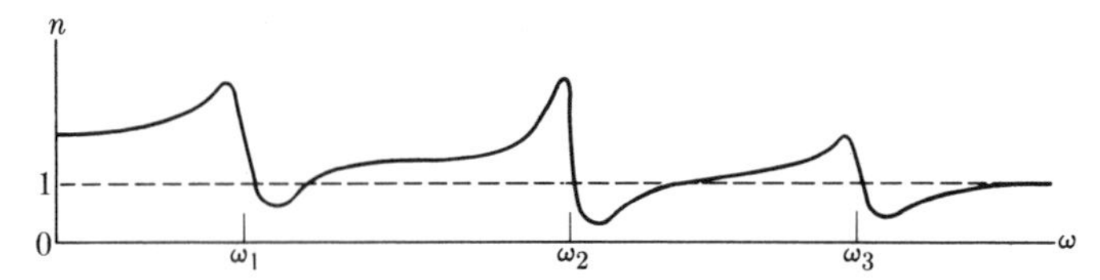

Figure 29.14 Variation of the index of refraction with frequency and wavelength.

This relation affords a simple experimental method for determining the relative permittivity of the substance if the index of refraction is obtained independently (as may be done; see Chapter 33). The consistency of the values of ε_r obtained by this method with those from other kinds of measurement gives a satisfactory experimental foundation to the theory.

In Section 25.9 it was indicated that ε_r is a quantity that depends on the frequency of the electromagnetic field. Therefore the index of refraction depends on the wave frequency, and hence also on the wavelength, in a manner shown in Figure 29.14, where $\omega_1, \omega_2, \ldots$ are certain characteristic frequencies of the substance. (The significance of these characteristic frequencies will be explained in Section 30.11.) Consequently, the velocity $v = c/n$ of the electromagnetic wave in matter also depends on the frequency of the radiation. Therefore electromagnetic waves suffer dispersion when propagating in matter. That is, an electromagnetic pulse containing several frequencies will be distorted because each component will travel with a different velocity.

There is the possibility that a charged particle, q, emitting electromagnetic waves, moves in the medium with a velocity V larger than the velocity v of the electromagnetic waves. This corresponds to the situation depicted in Figure 28.27 for Mach waves in a fluid. Then the electromagnetic waves propagate along conical surfaces, making an angle α with the direction of propagation given by $\sin \alpha = V/v$, in accordance with Equation 28.51. These electromagnetic waves are called **Cerenkov radiation**. Because the effective direction of propagation of the wave front is related to the velocity of the charged particle, Cerenkov radiation may be used to measure the particle velocity.

29.7 The Doppler effect in electromagnetic waves

The Dopper effect for electromagnetic waves is different from the Doppler effect for elastic waves, which consist of matter in motion; this was discussed in Section 28.13. In the first place, electromagnetic waves do not consist of matter in motion and, therefore, the velocity of the source relative to the medium does not enter into the discussion, only its velocity v relative to the observer. Secondly, the velocity of propagation in vacuum is c, and is the same to all observers irrespective of their relative velocity v. For these reasons, the Doppler effect for electromagnetic waves must necessarily be computed by means of the theory of relativity.

As shown in the derivation below, the frequency ν' measured by an observer O' moving with a velocity v relative to a source that has a frequency ν when measured by an observer O at rest relative to the source is

$$\nu' = \nu\left(\frac{1 - v/c}{1 + v/c}\right)^{1/2} \tag{29.22}$$

In this expression, v is positive if the observer and the source are receding from each other and negative if they are approaching each other. Also, the Doppler effect for electromagnetic waves depends only on the velocity of the observer *relative* to the source.

For velocities small compared to c, that is, $v \ll c$, Equation 29.22 becomes

$$\nu' \approx \nu\left(1 - \frac{v}{c}\right) \tag{29.23}$$

which is the same as Equation 28.52 for motion of the observer relative to the source along the line of propagation. Note that v_{os} in Equation 28.52 is now denoted by v, and v by c.

Assuming that O is at rest relative to the source of the electromagnetic wave, we see that if observers O and O' are receding from each other (v positive), O' observes a lower frequency, or correspondingly longer wavelength. This is observed in the spectrum of galaxies, nebulae and many stars and is called the **red shift**, since the visible spectrum of the light from receding stars is shifted toward the red (or longer) wavelength limit. This factor allows us to estimate the velocity with which these systems are receding.

Figure 29.15 shows the red shift of the calcium H and K lines observed in the spectra of several galaxies. The shift is indicated by the horizontal line. Note that the larger the shift, and therefore the larger the relative velocity, the greater the distance of the nebula. Therefore, the red shift of radiation from distant galaxies and nebulae supports the theory of an expanding universe, according to **Hubble's law** (recall Note 3.1). The relative velocity of recession of two galaxies is $V = HR$, where $H = 2.32 \times 10^{-18}\,\mathrm{s}^{-1}$ is Hubble's constant and R is their separation.

It is interesting to note that light from the Andromeda galaxy shows a shift toward shorter wavelengths, or a blue shift. This seems to indicate that the present motion of our solar system within our rotating galaxy is in a direction toward this galaxy.

Figure 29.16 shows the shift of the spectrum from the star Arcturus, which is about 36 light years from the Sun. The two spectra were recorded six months apart, and we see that the shift of one is toward the red, while the shift of the other is toward the blue. This shift is due to the reversal in the direction of the motion of the Earth relative to Arcturus in the six-month period.

In Chapter 31 it will be mentioned that the universe is filled with a low intensity cosmic radiation that is essentially homogeneous and isotropic. However, relative to the Earth this radiation shows a slight anisotropy that has been attributed to a Doppler effect, as a result of the motion of our galaxy toward the center of the Virgo cluster of galaxies, with a velocity of about $3 \times 10^5\,\mathrm{m\,s^{-1}}$. We may speculate then that there is a frame of reference where the background cosmic radiation is exactly isotropic and consider that frame as a preferred standard reference frame.

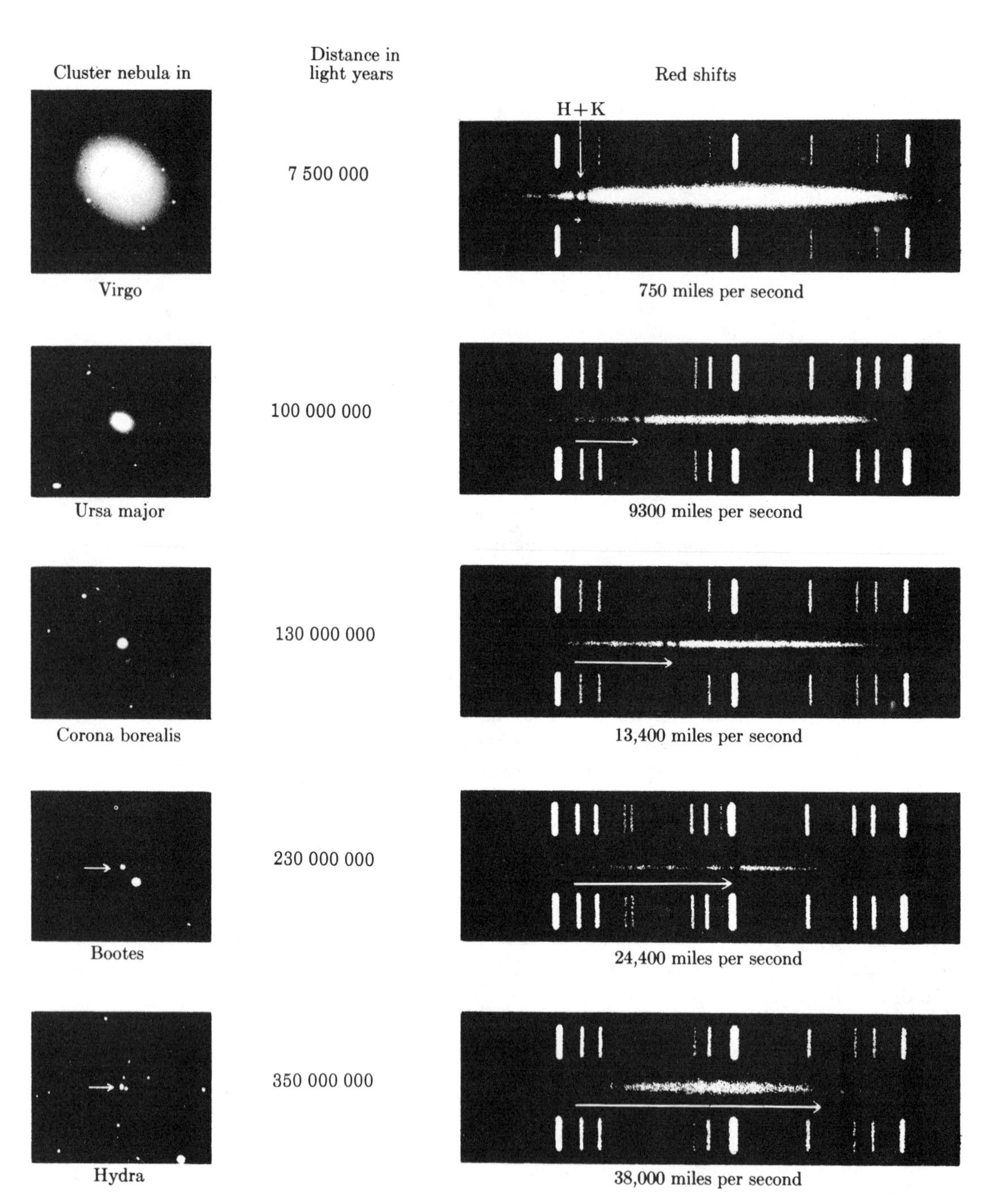

Figure 29.15 Doppler effect in nebulas. The red shift of the spectral H and K calcium lines (indicated by the arrow) increases with the distance of the nebula, suggesting greater recessional velocities. (Photograph courtesy of Mt Wilson and Palomar Observatories.)

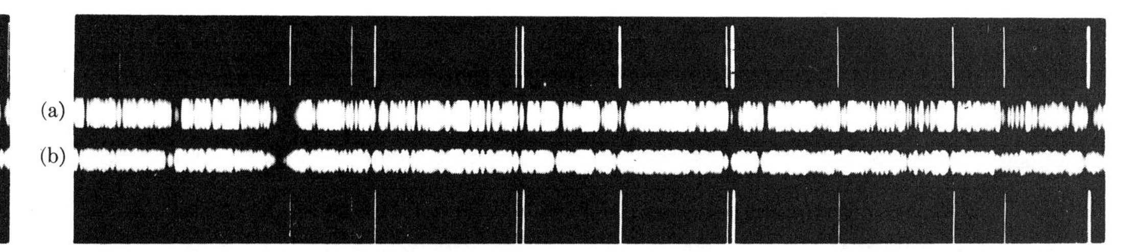

Figure 29.16 Spectra ($\lambda = 4200$ Å to $\lambda = 4300$ Å) of the constant-velocity star Arcturus taken about 6 months apart. (a) 1 July 1939; measured velocity $+18\ \text{km s}^{-1}$ relative to Earth. (b) 19 January 1940; measured velocity $-32\ \text{km s}^{-1}$. The velocity difference of $50\ \text{km s}^{-1}$ is entirely due to the change in orbital velocity of the Earth. The shift in the spectral lines may be seen when compared with the two reference spectra. (Photograph courtesy of Mt Wilson and Palomar Observatories.)

Derivation of the Doppler effect for electromagnetic waves

Consider two observers O and O' in relative motion, and let v be the velocity of O' relative to O. To observer O, a plane harmonic electromagnetic wave can be described by a function of the form $\sin k(x - ct)$ multiplied by an appropriate amplitude factor. To observer O', attached to a different inertial frame of reference, the coordinates x and t must be replaced by x' and t', as given by the Lorentz transformation 19.2. Therefore O' will write the function as $\sin k'(x' - ct')$, where k' does not necessarily have to be the same as for the other observer. On the other hand, the principle of relativity requires that $k(x - ct)$ must remain invariant for the two observers. Therefore we must have

$$k(x - ct) = k'(x' - ct')$$

Using the first and fourth equations of the inverse Lorentz transformation given in Equation 19.4, we have

$$k\left(\frac{x' + vt'}{(1 - v^2/c^2)^{1/2}} - c\,\frac{t' + vx'/c^2}{(1 - v^2/c^2)^{1/2}}\right) = k'(x' - ct')$$

Therefore

$$k' = k\,\frac{1 + v/c}{(1 - v^2/c^2)^{1/2}} = k\left(\frac{1 - v/c}{1 + v/c}\right)^{1/2}$$

Remembering that $\omega = ck$ and $\nu = \omega/2\pi$, we obtain

$$\nu' = \nu\left(\frac{1 - v/c}{1 + v/c}\right)^{1/2}$$

For $v \ll c$, we may approximate the numerator and denominator using the binomial theorem, which gives

$$\nu' \approx \nu\,\frac{1 - \frac{1}{2}(v/c)}{1 + \frac{1}{2}(v/c)} \approx \nu\left(1 - \frac{v}{c}\right)$$

This approximation agrees with the non-relativistic expression for the Doppler effect.

EXAMPLE 29.4

Line broadening by Doppler effect.

▷ In sources of electromagnetic radiation, such as a hot gas, the atoms and molecules are moving relative to the observer. This results in a change of the frequency of the radiation as measured by the observer due to a Doppler effect. Since the velocity of the atoms is very small compared with c, we may use the non-relativistic Equation 29.23, that is $\nu = \nu_0(1 - v/c)$, where ν_0 is the frequency of the radiation measured in the frame of reference of the emitting source, ν is the frequency measured by the observer, and v is the velocity of the source relative to the observer measured along the line joining the source and observer, or X-axis. It is positive if the source moves away from the observer and negative if it moves toward the observer.

For atoms whose velocity is v we have that $\Delta\nu = \nu_0 - \nu = (v/c)\nu_0$. Recalling Maxwell's law of distribution of molecular velocities in a gas (Equation 17.14), we may write the number of atoms moving with velocity between v and $v + \Delta v$ as $N(v) = N_0 e^{-(mv^2/kT)}$. Replacing the velocity v by $(c/\nu_0)\Delta\nu$ in this expression and noting that the intensity of the radiation emitted is proportional to the number of atoms, the intensity of the radiation displaced a frequency $\Delta\nu$ with respect to ν_0 is

$$I(\nu) = I_0 e^{(mc^2/2kT\nu_0^2)(\Delta\nu)^2}$$

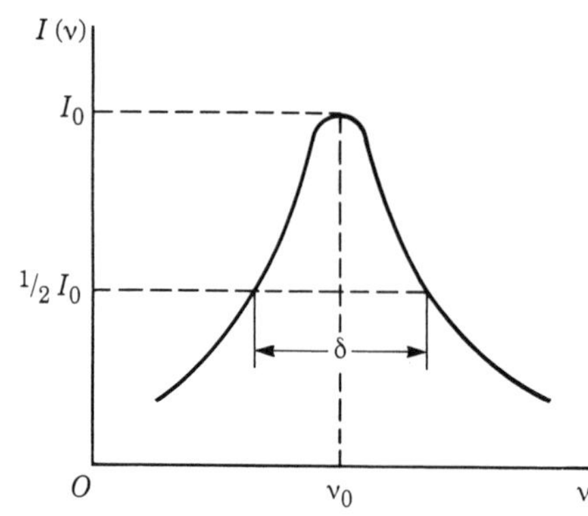

Figure 29.17 Doppler width of a spectral line.

which can be represented by the curve of Figure 29.17. The **line width** is defined as the frequency difference corresponding to the intensity $\frac{1}{2}I_0$. Making $I(\nu) = \frac{1}{2}I_0$ and taking logarithms, we obtain the value of the line width as

$$\delta = 2\Delta\nu = \frac{\nu_0}{c}\left(\frac{2kT\ln 2}{m}\right)^{1/2}$$

For example, the Na yellow lines have $\nu_0 \approx 5.10 \times 10^{15}$ Hz. For a temperature of 500 K, we obtain $\delta = 1.73 \times 10^{11}$ Hz, so that $\Delta\nu/\nu_0 = 3.39 \times 10^{-5}$, a quantity that can be measured in high resolution spectroscopes.

29.8 The spectrum of electromagnetic radiation

Electromagnetic waves cover a wide range of frequencies or wavelengths and, for convenience, may be classified according to their main source and their main effect when interacting with matter. The classification does not have very sharp boundaries, since different sources may produce waves in overlapping ranges of frequencies. We summarize here the usual classification of the electromagnetic spectrum.

(i) Radio-frequency waves. These have wavelengths ranging from a few kilometers down to 0.3 m. The frequency range is from a few Hz up to 10^9 Hz. These waves, which are used in television and radio broadcasting systems, are generated by electronic devices, mainly oscillating circuits. They are also used in techniques such as nuclear magnetic resonance imaging (NMRI).

(ii) Microwaves. The wavelengths of microwaves range from 0.3 m down to 10^{-3} m. The frequency range is from 10^9 Hz up to 3×10^{11} Hz. These waves are used in radar and other communication systems, as well as in the analysis of very fine details of atomic and molecular structure, and are also generated by electronic devices. The microwave region is also designated as UHF (ultra-high frequency relative to radio-frequency).

(iii) Infrared spectrum. This covers wavelengths from 10^{-3} m down to 7.8×10^{-7} m (780 nm or 7800 Å). The frequency range is from 3×10^{11} Hz up to 4×10^{14} Hz. The region is subdivided into three: the **far infrared**, from 10^{-3} m to 3×10^{-5} m; the **middle infrared**, from 3×10^{-5} m to 3×10^{-6} m; and the **near infrared**, extending to about 7.8×10^{-7} m. These waves are produced by molecules and hot bodies whose atoms have been excited by thermal energy. They have many applications in industry, medicine, astronomy etc.

(iv) Light or visible spectrum. This is a narrow band of wavelengths to which our retinas are sensitive. It extends from a wavelength of 7.8×10^{-7} m to 3.8×10^{-7} m and frequencies from 4×10^{14} Hz to 8×10^{14} Hz. Light is produced by atoms and molecules as a result of internal adjustments in the motion of their components, principally that of the electrons. It is not necessary to emphasize the importance of light in our world.

Light is so important that a special branch of applied physics has evolved, called **optics**. Optics deals with light phenomena as well as vision, and includes the design of optical instruments. The different sensations that light produces on the eye, called *colors*, depend on the frequency (or wavelength) of the electromagnetic wave, and correspond to the following ranges for the average person:

Color	λ(m)	ν(Hz)
Violet	$3.90–4.55 \times 10^{-7}$	$7.69–6.59 \times 10^{14}$
Blue	4.55–4.92	6.59–6.10
Green	4.92–5.77	6.10–5.20
Yellow	5.77–5.97	5.20–5.03
Orange	5.97–6.22	5.03–4.82
Red	6.22–7.80	4.82–3.84

The sensitivity of the eye also depends on the wavelength of light; this sensitivity is a maximum for wavelengths of approximately 5.6×10^{-7} m. Because of the relation between color and wavelength or frequency, an electromagnetic wave of well-defined wavelength or frequency is also called a **monochromatic wave** (*monos*: one; *chromos*: color).

Vision is the result of the signals transmitted to the brain by two elements present in a membrane called the **retina**, lying in the back of the eye. These elements are the **cones** and the **rods**. Cones are those elements that are active in the presence of intense light, such as during the daylight hours. Cones are sensitive to wavelength or color. Rods, on the other hand, are elements that are able to act under very

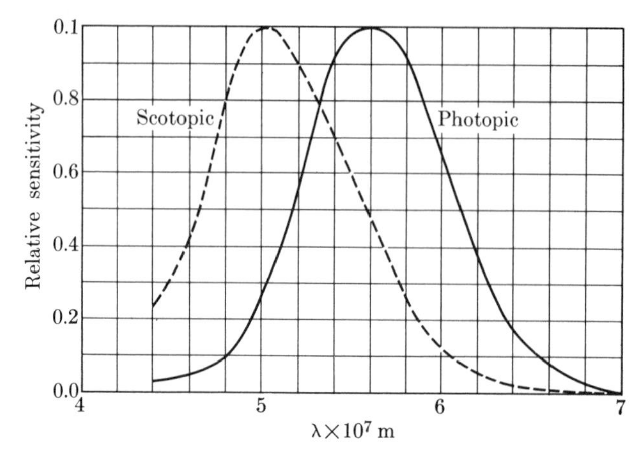

Figure 29.18 Sensitivity of the eye for scotopic and photopic vision.

dim illumination, such as in a darkened room; they are less sensitive to color. The vision due to cones is called **photopic**; that due to rods is called **scotopic**. The sensitivity of the eye for different wavelengths for both kinds of vision is shown in Figure 29.18.

(v) Ultraviolet rays. This region covers from 3.8×10^{-7} m down to about 6×10^{-10} m, with frequencies from 8×10^{14} Hz to about 3×10^{17} Hz. These waves are produced by excited atoms and molecules as well as in electric discharges. Their energy is of the order of magnitude of the energy involved in atomic ionization and molecular dissociation, which accounts for many of the chemical effects of ultraviolet radiation. The Sun is a very powerful source of ultraviolet radiation, which is mainly responsible for suntans. The Sun's ultraviolet radiation also interacts with the atoms in the upper atmosphere, producing a large number of ions. This explains why the upper atmosphere, at a height greater than about 90 km is highly ionized. For this reason it is called the **ionosphere**. When some micro-organisms absorb ultraviolet radiation, they can be destroyed as a result of the chemical reactions produced by the ionization and dissociation of molecules. For that reason ultraviolet rays are used in some medical applications and also in sterilization processes.

(vi) X-rays. This part of the electromagnetic spectrum extends from wavelengths of about 10^{-9} m down to wavelengths of about 6×10^{-12} m, or frequencies between 3×10^{17} Hz and 5×10^{19} Hz. X-rays, discovered in 1895 by Wilhelm Röntgen (1845–1923) when he was studying cathode rays, are produced by the inner, or more tightly bound, electrons in atoms. Another source of X-rays is the bremsstrahlung or decelerating radiation mentioned in Section 29.5. In fact, this is the most common way of producing X-rays.

X-rays produce profound effects on the atoms and molecules of the substances through which they propagate, such as dissociation or ionization. They are used in medical diagnosis because the different absorption of X-radiation by bone and different tissues allows for a fairly well-defined pattern on a photographic film. They also, as a result of the molecular processes they induce, cause serious damage to living tissues and organisms. It is for this reason that X-rays are used for treatment of cancer, to destroy diseased tissue. It should be emphasized that even a small amount of X-radiation may destroy some good tissue as well; for that reason, an exposure to a large dose of X-rays may cause enough destruction to produce sickness or death.

(vii) γ-rays. These electromagnetic waves are of nuclear origin. They overlap the upper limit of the X-ray spectrum. Their wavelength runs from about 10^{-10} m to well below 10^{-14} m, with a corresponding frequency range from 3×10^{18} Hz to more than 3×10^{22} Hz. The energies of these waves are of the same order of magnitude as those involved in nuclear processes and therefore the absorption of γ-rays may produce some nuclear changes. γ-rays are produced by many radioactive substances, and are present in large quantities in nuclear reactors and in cosmic radiation. They are not easily absorbed by most substances, but when they are absorbed by living organisms they produce serious effects. Even so, γ-rays are used to treat some forms or cancer.

Figure 29.19 relates the various sections of the electromagnetic spectrum in terms of frequency and wavelength. (The energy of the corresponding photons, Section 30.6, is also given.)

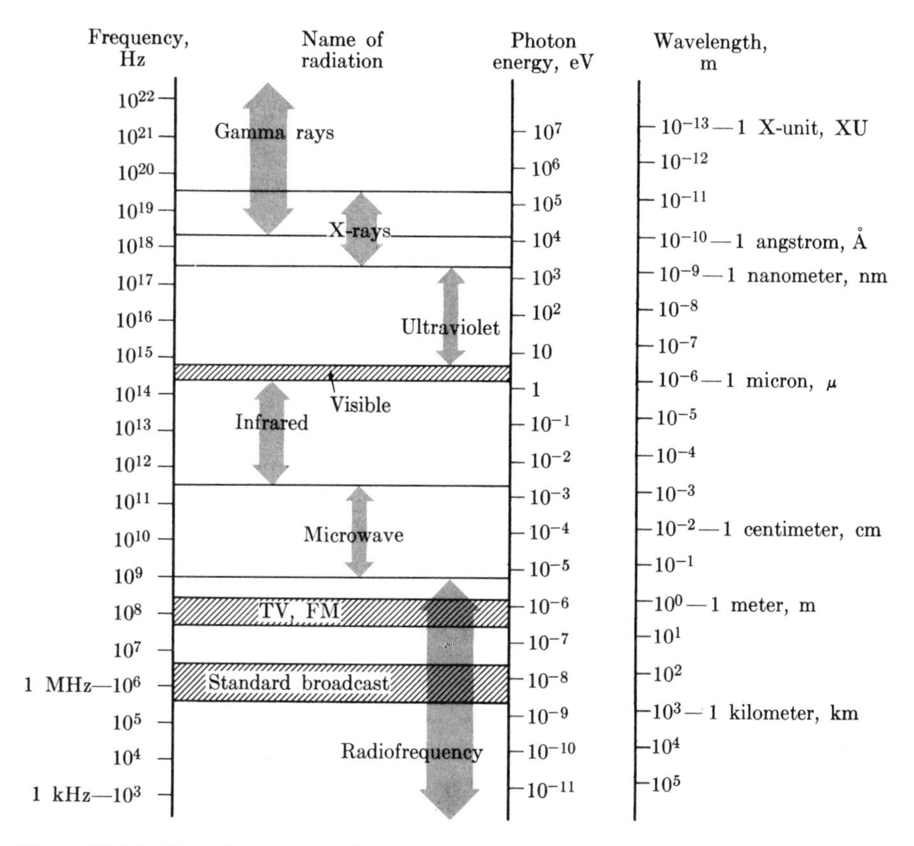

Figure 29.19 The electromagnetic spectrum. (The scale labeled 'photon energy' will be considered in Chapter 30.)

QUESTIONS

29.1 Are plane electromagnetic waves (a) longitudinal or transverse, (b) scalar or vector?

29.2 Consider a plane linearly polarized electromagnetic wave. What is the angle between the directions of the electric and magnetic fields? What is the phase difference between the electric and magnetic fields? How do the directions of the electric and magnetic fields change with time? How do their magnitudes change?

29.3 Repeat Question 29.2 for the case of plane circularly polarized electromagnetic waves.

29.4 Explain why the radiation pressure on a perfect absorber is half as large as on a perfect reflector.

29.5 State the main differences between electric and magnetic dipole radiation.

29.6 Explain why a charge in uniform rectilinear motion does not radiate electromagnetic energy.

29.7 Is a harmonic electromagnetic wave distorted when propagating in a dispersive medium? What about a non-harmonic wave?

29.8 Discuss how the Doppler effect is related to the relativistic time dilation discussed in Section 19.5.

29.9 Explain why different regions of the electromagnetic spectrum show different effects when propagating through a substance.

29.10 Express the momentum P per unit volume of a plane electromagnetic wave in terms of the vector $\mathscr{E} \times \mathscr{B}$.

29.11 Write Equation 29.22 in terms of the wavelengths λ and λ'.

29.12 Radiation emitted by an electric dipole is polarized. Assuming we have a gas in which many atoms are radiating, do you expect the radiation to be polarized? What if the gas is in an electric field?

29.13 Verify, using Equation 29.18, that synchrotron radiation in a cyclic accelerator is proportional to $(E/mr)^2$ when the energy E of the particle is small.

29.14 Consider a charge e oscillating with angular frequency ω and an amplitude z_0. Verify that Equation 29.18 reduces to $dE/dt = e^2 z_0^2 \omega^4 / 12\pi\varepsilon_0 c^3$. Compare this result with that given in Note 29.1 for the electric radiation of a linear antenna, with $\mathscr{P}_0 = ez_0$.

PROBLEMS

29.1 The electric field of a plane electromagnetic wave in vacuum is represented by $\mathscr{E}_x = 0$, $\mathscr{E}_y = 0.5 \cos[2\pi \times 10^8(t - x/c)]$ and $\mathscr{E}_z = 0$.
(a) Determine (i) the wavelength, (ii) the state of polarization and (iii) the direction of propagation.
(b) Calculate the magnetic field of the wave.
(c) Calculate the average intensity, or energy flux per unit area and per unit time, of the wave.

29.2 Repeat the previous problem for the wave $\mathscr{E}_x = 0, \mathscr{E}_y = 0.5 \cos[4\pi \times 10^7(t - x/c)]$ and $\mathscr{E}_z = 0.5 \sin[4\pi \times 10^7(t - x/c)]$. What is the state of polarization of the wave?

29.3 Write the equations of the $\mathscr{E}$- and $\mathscr{B}$-fields describing the following electromagnetic waves that propagate along the X-axis: (a) a linearly polarized wave whose plane of oscillation of the electric field lies at an angle of 45° with the XY-plane; (b) a linearly polarized wave whose plane of oscillation of the electric field lies at an angle of 120° with the XY-plane; (c) a wave with right-handed circular polarization; (d) a wave with left-handed circular polarization.

29.4 A plane sinusoidal linearly polarized electromagnetic wave of wavelength $\lambda = 5.0 \times 10^{-7}$ m travels in vacuum in the direction of the X-axis. The average intensity of the wave per unit area is 0.1 W m^{-2} and the plane of oscillation of the electric field is parallel to the Y-axis. Write the equations describing the (a) electric and (b) magnetic fields of this wave.

29.5 The electric field of a plane electromagnetic wave has an amplitude of 10^{-2} V m^{-1}. Find (a) the magnitude of the magnetic field and (b) the energy per unit volume of the wave. (c) If the wave is completely absorbed when it falls on a body, determine the radiation pressure. (d) What is the radiation pressure if the body is a perfect reflector?

29.6 Electromagnetic radiation from the Sun falls on the Earth's surface at the rate of 1.4×10^3 W m^{-2}. Assuming that this radiation can be considered as plane waves, estimate the magnitude of the (a) electric and (b) magnetic field amplitudes of the wave.

29.7 The average power of a broadcasting station is 10^5 W. Assume that the power is radiated uniformly over any hemisphere concentric with the station. For a point 10 km from the source, determine the amplitudes of the (a) electric and (b) magnetic fields. Note that at that distance the wave can be considered plane.

29.8 A radar transmitter emits its energy within a cone having a solid angle of 10^{-2} sterad. At a distance of 10^3 m from the transmitter the electric field has an amplitude of 10 V m^{-1}. Calculate (a) the amplitude of the magnetic field and (b) the power of the transmitter.

29.9 A system of oscillating charges concentrated around a point radiates energy at the rate of 10^4 W. Assuming that the energy is radiated isotropically, find, for a point at a distance of 1 m (a) the amplitude of the electric and magnetic fields, and (b) the energy and momentum densities. (Hint: at 1 m from the source, a small portion of the wave front can be considered as plane.)

29.10 A gaseous source emits light of wavelength 5×10^{-7} m. Assume that each molecule acts as an electric dipole of charge e and amplitude 10^{-10} m. (a) Compute the average rate of energy radiation per molecule. (b) If the total rate of energy radiation of the source is 1 W, calculate the number of molecules emitting simultaneously.

29.11 Estimate the value of $(\mathrm{d}E/\mathrm{d}t)_{\text{ave}}$ as given by Equation 29.15 for a proton in a nucleus. Assume z_0 of the order of 10^{-15} m and ω about 5×10^{20} Hz, for low-energy γ-rays.

29.12 (a) Calculate the power radiated by a 50 keV electron moving along a 1 m circular path. (b) Repeat for a 50 keV proton moving along a 1 m circular path.

29.13 Consider a glass plate of index of refraction n and thickness Δx interposed between a monochromatic source S and an observer O, as shown in Figure 29.20. (a) Verify that if absorption by the glass plate is neglected, the effect of the glass plate on the wave received by O

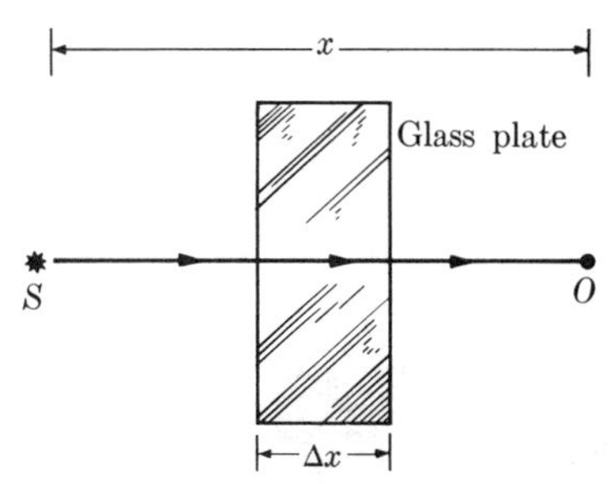

Figure 29.20

is to add a phase difference equal to $-\omega(n-1)\Delta x/c$ without changing the amplitude $\mathscr{E}_0$ of the wave. (b) If the phase difference is small, either because Δx is very small or because n is very close to one, verify that the wave received at O can be considered as a superposition on the original wave of amplitude $\mathscr{E}_0$, with no plate present, of a wave of amplitude $\mathscr{E}_0\,\omega(n-1)\Delta x/c$, having a phase shift of $-\pi/2$. (This problem shows the effect of a material medium on an electromagnetic wave.)

29.14 Astronomers measure the Doppler red shift by a quantity z defined as $1 + z = (v/v')$. (a) Express z in terms of v/c. (b) Verify that $z = \Delta\lambda/\lambda$. (c) Estimate the velocity of galaxies showing red shifts of $z = 0.5, 1, 10, 10^2, 10^3$ and 10^4. (d) Plot v/c against z.

29.15 Verify that recessional velocities may be obtained by measuring the relative wavelength shift $(\Delta\lambda/\lambda)$ and that its value is approximately

$$\frac{\Delta\lambda}{\lambda} = \frac{v/c}{(1 - v^2/c^2)^{1/2}}$$

29.16 When the new celestial bodies called **quasars** were first observed, the relative wavelength shift $(\Delta\lambda/\lambda)$ was measured as 0.158 for the body 3C273. (a) Using the results of Problem 29.15, calculate the recessional velocity associated with 3C273. (b) Using Hubble's law (Note 3.1), calculate the distance of 3C273 from Earth.

29.17 The relative wavelength shift for a distant quasar has been measured at $\Delta\lambda/\lambda = 2.223$. (a) Calculate the recessional velocity of this body and its distance. (b) The calcium H line is in the violet region of the spectrum for a stationary source. In what region of the spectrum is the shifted line? Use the approximation of Problem 29.15. (c) Calculate the values without approximating.

30 Interaction of electromagnetic radiation with matter: photons

Arthur H. Compton observed, in 1922, that when X-rays are scattered by electrons, the scattered radiation has a longer wavelength than that of the incident radiation and that the amount of increase is dependent on the direction of scattering. Compton was able to deduce that X-rays, when interacting with electrons, obey the laws of conservation of energy and momentum, behaving like particles with zero mass, now called photons. This provided the definitive experimental confirmation of the quantization of electromagnetic waves when interacting with matter. The corresponding theory is called quantum electrodynamics.

Notes

30.1 Introduction

When electromagnetic radiation interacts with systems of electric charges, such as atoms, molecules and nuclei, energy from the wave is absorbed as well as emitted by the system. The absorption and emission of electromagnetic waves by atoms is a complicated problem that requires extensive mathematical calculations and the use of quantum mechanics, but the fundamental ideas are easy to understand. The most important aspect is that it gives rise to the concept of photons or quanta of radiation.

30.2 Emission of radiation by atoms, molecules and nuclei

Atoms, molecules and nuclei are normally in the state of lowest energy, or the ground state. An atom, molecule or nucleus may be excited to a state of energy greater than the ground state by several means. One method is by inelastic collisions in which, for example, a fast particle (say an electron or a proton) collides with an atom, a molecule or a nucleus, or when a neutron is captured by a nucleus. In either case the fast projectile transfers part of its kinetic energy to the target (the atom or the nucleus). In Example 23.2, we discussed this kind of excitation of Hg atoms by collisions with electrons. When the process takes place as the result of the electric interaction of both colliding systems, it is called **Coulomb excitation**. Also, by raising the temperature of a solid or a gas, the kinetic energy of the atoms or molecules can be sufficiently increased, resulting in inelastic collisions between atoms or molecules.

An important experimental result is that the excited atoms, molecules, and nuclei release their excess energy in the form of electromagnetic radiation. For example, if in a tube containing hydrogen at low pressure we produce an electric discharge by applying a large potential difference to two suitably located electrodes (Figure 30.1), we observe that the tube glows. We explain this result by saying that the electrons emitted by the negative electrode and accelerated by the electric field collide inelastically with some of the hydrogen molecules (or atoms), leaving them in an excited state. The observed radiation is emitted by the excited atoms as they return to a lower energy state or to their ground state.

The radiation emitted by each substance is composed of well-defined frequencies, ν_1, ν_2, ν_3,..., which are characteristic of the substance.

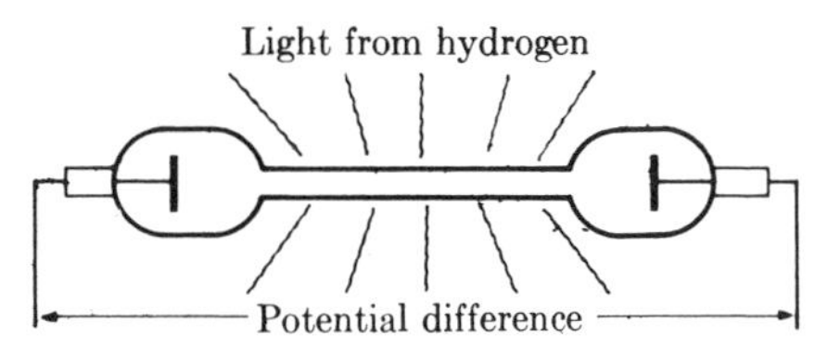

Figure 30.1 Gas under low pressure in the tube emits light when an electric discharge is maintained in the gas by applying a large potential difference.

The set of all characteristic frequencies is designated the **emission spectrum** of the substance.

The emission spectra of atoms, molecules, solids and nuclei have marked differences. Atomic spectra lie mostly in the visible and ultraviolet regions and are composed of frequencies sufficiently spaced as to appear as separate lines in a spectroscope (Section 32.7); for that reason atomic spectra are also designated as **line spectra**. Figure 30.2 shows the lines in the visible portion of the spectrum of some atoms.

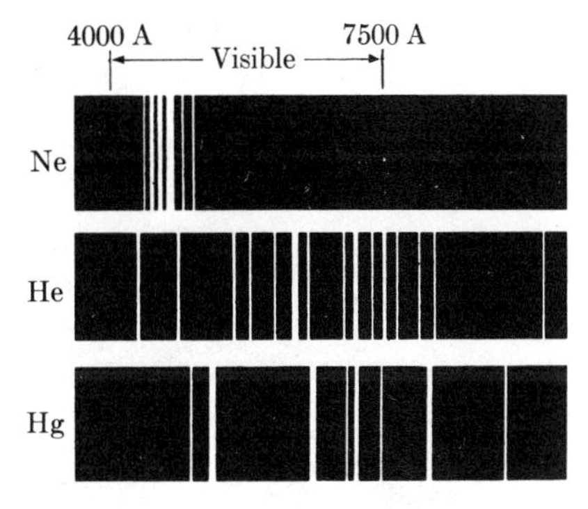

Figure 30.2 Part of the line emission spectrum of mercury vapor, of helium, and of neon. (Redrawn from photographic records.) (Courtesy of A. B. Arons, *Development of concepts of physics*. Addison-Wesley, 1965.)

Molecular spectra extend from the far infrared up to the ultraviolet and are composed of groups of frequencies that have very similar values and thus appear as bright bands in a spectroscope with low resolving power. For that reason molecular spectra are also called **band spectra**. Figure 30.3 shows the band grouping in the spectrum of nitrous oxide, NO. Solid spectra are somewhat

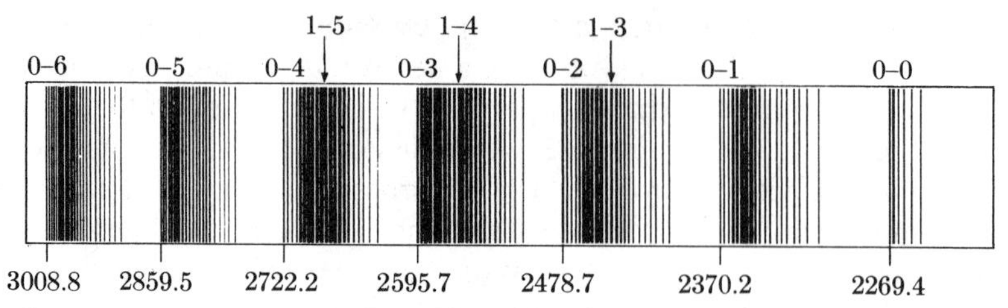

Figure 30.3 Simplified molecular spectrum of nitrous oxide (NO) from 3100 Å to 2100 Å. The lower numbers are the wavelengths of the head of each band and the upper numbers are the assigned values of the vibrational levels responsible for the transition. (Freely adapted from a photograph in G. Herzberg *Spectra of diatomic molecules*. New York, D. Van Nostrand Co., 1950.)

more complex. They will be analyzed in Section 31.6. Nuclear spectra fall in the X- and γ-regions and the frequencies are arranged in a rather complex way.

30.3 Absorption of electromagnetic radiation by atoms, molecules and nuclei

When an electromagnetic wave interacts with a system of charges, such as an atom, a molecule or a nucleus, the electric and magnetic fields of the wave disturb the motion of the charges. In a simplified form, we could say that the wave impresses a forced oscillation on the natural motion of the charges. This results in absorption of energy by the system of charges. An oscillator responds most easily when the frequency of the forced oscillations is the same as its natural frequency, in which case there is **resonance**. At resonance, the rate at which the oscillator absorbs energy is maximum (recall Section 10.12).

It has been found experimentally that atoms, molecules and nuclei—in general, any assembly of charged particles—have a series of resonating frequencies at which the absorption of electromagnetic radiation is appreciable. At all other frequencies the absorption is negligible. The resonating frequencies constitute the **absorption spectrum** of the substance. When a system absorbs electromagnetic radiation, it passes to some other state of higher energy, or a more excited state.

The frequencies observed in the absorption spectrum of a system of charges are also observed in the emission spectrum of the system.

For example, in Figure 30.4 we have compared the absorption and emission spectra of sodium. It can be seen that all lines appearing in the absorption spectrum are

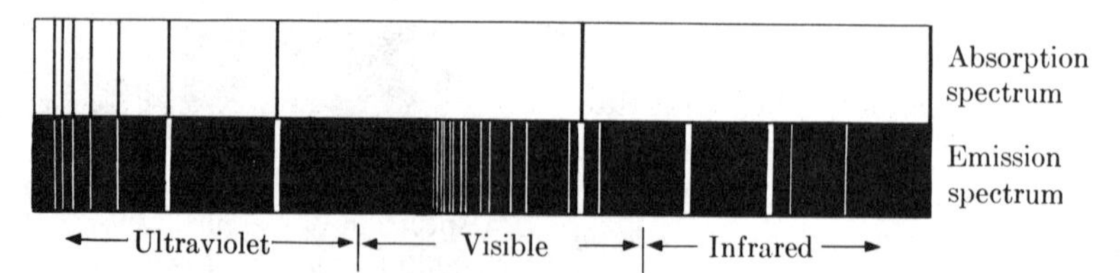

Figure 30.4 Comparison of the line absorption spectrum and the line emission spectrum of sodium vapor. (Coutesty of A. B. Arons, *Development of concepts of physics*. Addison-Wesley, 1965.)

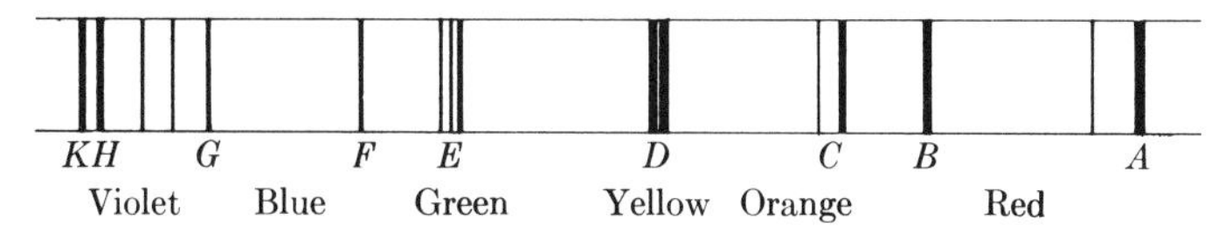

Figure 30.5 The Fraunhofer dark lines in the visible part of the solar spectrum. Only a few of the most prominent lines are represented here.

present also in the emission spectrum (although the converse is not true; see Section 31.5).

Historically absorption spectra were discovered in 1802 by William Wollaston (1766–1828), as he analyzed light from the Sun with a spectroscope. Wollaston observed seven dark lines against a continuous colored background. A few years later Joseph von Fraunhofer (1787–1826), with a better instrument, identified many more dark lines, which he labelled A, B, C,.... Today more than 15000 dark lines, called **Fraunhofer lines**, have been recognized in the spectrum of the radiation from the Sun extending from the infrared to the ultraviolet regions. Some lines in the visible region are shown in Figure 30.5.

As first suggested by Gustav Kirchhoff, the Fraunhofer lines result from the absorption of the corresponding frequencies by the atoms and molecules in the atmospheres of the Sun and the Earth through which this radiation must pass. Similar absorption lines have been observed in the radiation from other stars, thus providing a clue to the composition of their atmospheres. It is interesting to note that the existence of helium was suspected, before it was even found on Earth, from the corresponding Fraunhofer lines in the solar spectrum that could not be ascribed to any known terrestrial element.

30.4 Scattering of electromagnetic waves by bound electrons

Another process that occurs when an electromagnetic wave passes through an atom (or molecule) is **scattering**. The atoms (or molecules) whose electrons are disturbed by the electromagnetic wave absorb energy from the electromagnetic radiation. By the reciprocal process, the excited atom subsequently emits the electromagnetic radiation that had been absorbed from the incident wave by the atom's bound electrons. The radiation emitted is the **scattered wave** (Figure 30.6).

Scattering decreases the intensity of the primary or incident wave because the energy absorbed from the wave is re-emitted in all directions, resulting in an effective removal of energy from the primary radiation.

The intensity of the scattered waves depends on the frequency of the primary wave and on the angle of scattering.

The scattered waves are more intense when the frequency of the incident radiation is equal to one of the frequencies of the spectrum of the atom (or molecule).

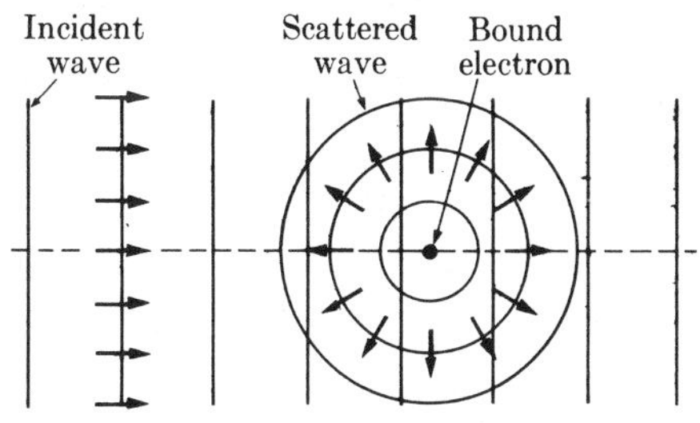

Figure 30.6 Scattering of electromagnetic radiation by a bound electron.

This occurs because scattering is more intense at those frequencies at which the energy absorption from the wave is greater, and these are the same frequencies as the emission spectrum of the atom. However, at frequencies different from those of the emission spectrum, scattering may still be appreciable.

For gases whose molecules have an emission spectrum in the ultraviolet region, the scattering of electromagnetic waves falling in the visible region increases with their frequency because the higher the frequency in the visible region, the closer it is to the ultraviolet resonant frequencies of the molecules. As an illustration, the brightness and blue color of the sky are attributed to the scattering of sunlight by air molecules. In particular the blue color is the result of the more intense scattering of the higher frequencies (or shorter wavelengths). The same process accounts for the bright red colors observed at sunrise and sunset, when the Sun's direct rays traverse a very large thickness of air before reaching the Earth's surface, resulting in a strong attenuation of the high frequencies (or short wavelengths) on account of scattering.

Scattering can also be produced by small particles (such as smoke or dust) or water droplets (such as clouds) suspended in the air. Haze in many industrial cities is the result of light scattering by particles in the air. Liquids carrying a suspension of particles, as in a colloid, show a strong scattering; this is called the **Tyndall effect**. The scattering in these cases is due to the reflection of light by the particles instead of the mechanism indicated above for molecular scattering.

When the primary radiation is linearly polarized, the electric dipole oscillations of the atoms such as S (Figure 30.7) are in the fixed direction of the electric field of the wave and the scattered radiation, such as SA, SB, SC etc., has the polarization characteristic of the electric dipole radiation (Figure 30.7(a)). However, even if the incident radiation is not polarized, the scattered radiation is always partially polarized. When an incident wave is unpolarized (Figure 30.7(b)), the electric dipole oscillations induced in an atom S are parallel to the electric field of the wave. Therefore they are all in a plane P perpendicular to the direction of propagation IA of the incident wave. The polarization of the scattered radiation

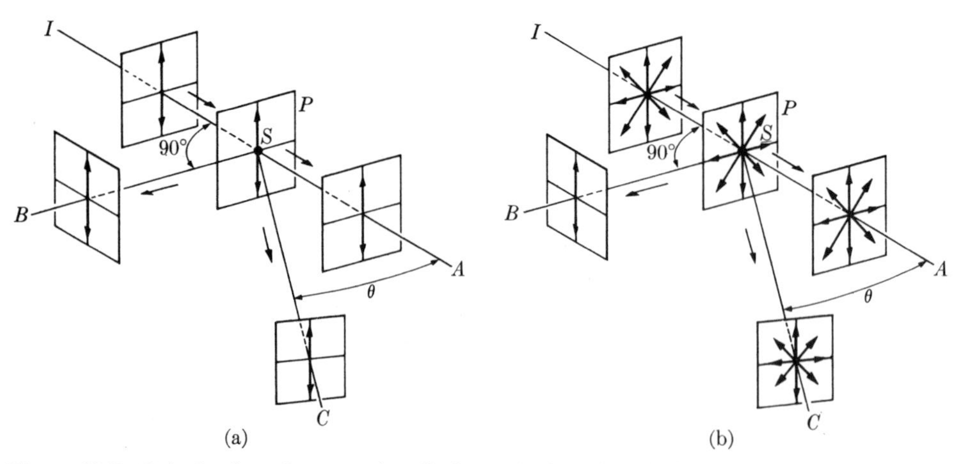

Figure 30.7 Polarization of scattered radiation. (a) Linearly polarized and (b) unpolarized incident radiation.

in each direction depends on the direction of the dipole oscillations, and therefore is not always the same when the incident wave is unpolarized. But for any direction SB perpendicular to IA, the scattered radiation is linearly polarized parallel to the plane P, perpendicular to IA, since for these directions the dipoles always oscillate in such a plane. For other directions the degree of polarization of the scattered radiation depends on the angle the direction of scattering makes with IA. Along IA, if the incident radiation is unpolarized, the scattered radiation is completely unpolarized.

30.5 Scattering of electromagnetic radiation by a free electron: the Compton effect

The scattering of electromagnetic radiation by a free electron requires that the electron absorbs energy from an electromagnetic wave and re-radiates it as scattered radiation. An electromagnetic wave carries energy and momentum, which are related by $\boldsymbol{p} = (E/c)\boldsymbol{u}$, as explained in Section 29.3, which is the same relation that holds for a particle of zero mass (recall Section 20.2). Therefore, if some energy E is removed from the wave, a corresponding amount of momentum $p = E/c$ must also be removed from the wave.

Now a free electron cannot absorb an amount of energy E and at the same time increase its momentum in the amount $p = E/c$ because the relation between kinetic energy and momentum for an electron

$$E_k = c\sqrt{m_e^2 c^2 + p_e^2} - m_e c^2 \approx \frac{p_e^2}{2m_e}$$

is incompatible with the relation $p = E/c$ if $E = E_k$, as required by the conservation of energy, assuming the electron was initially at rest relative to the observer. We should conclude then that a free electron cannot absorb electromagnetic energy without violating either the principle of conservation of momentum or of conservation of energy.

In scattering and absorption of electromagnetic waves by electrons bound in atoms or molecules, the conservation of momentum and energy is possible because the energy and momentum absorbed are shared by both the electron and the ion corresponding to the remaining part of the atom, and it is always possible to split both energy and momentum in the correct proportions. However, the ion, having a much larger mass, carries (along with some momentum) only a small fraction of the energy available, and so it is not usually considered. In the case of a free electron, there is no other particle with which the electron can share the energy and the momentum absorbed from the radiation and no absorption or scattering should be possible.

However, when electromagnetic radiation passes through a region where free electrons are present, in addition to the incident radiation, another radiation of *different* frequency is observed. This new radiation is interpreted as the radiation scattered by the free electrons. The frequency of the scattered radiation is *smaller* than the incident frequency, and accordingly its wavelength is *longer* than the incident wavelength (Figure 30.8). The wavelength of the scattered radiation is

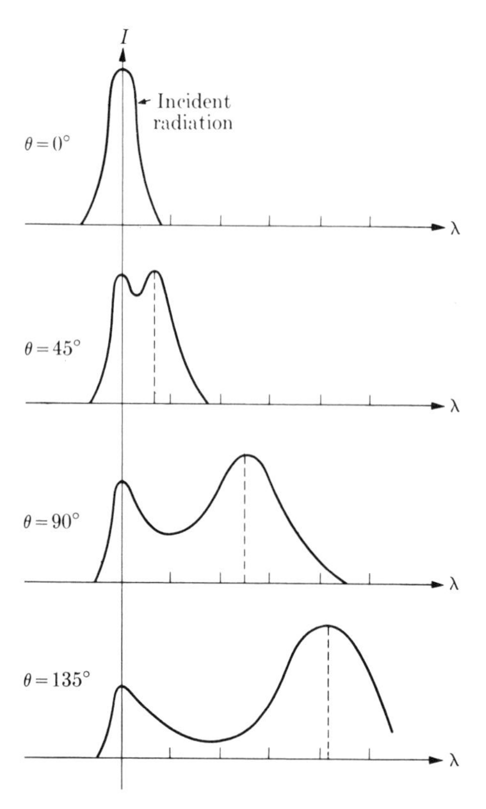

Figure 30.8 Intensity distribution of the radiation scattered by a free electron at different scattering angles.

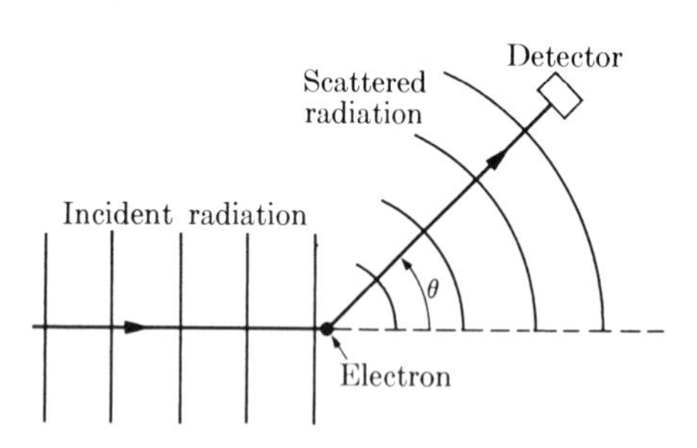

Figure 30.9 Compton scattering.

also different for each direction of scattering. This phenomenon is called the **Compton effect**, after Arthur H. Compton (1892–1962), who first observed and analyzed it in 1922.

Given that λ is the wavelength of the incident radiation and λ' that of the scattered radiation, Compton found that λ' is determined solely by the direction of scattering. That is, if θ is the angle between the incident waves and the direction in which the scattered waves are observed (Figure 30.9), the wavelength of the scattered radiation λ' is solely determined by the angle θ. The experimental relation is

$$\lambda' - \lambda = \lambda_C(1 - \cos\theta) \qquad \textbf{(30.1)}$$

where λ_C is a constant, whose value, if λ and λ' are measured in meters, is $\lambda_C = 2.4262 \times 10^{-12}$ m. It is called the **Compton wavelength for electrons**.

Remembering that $\lambda = c/\nu$, where ν is the frequency of the wave, we may write Equation 30.1 in the form

$$\frac{1}{\nu'} - \frac{1}{\nu} = \frac{\lambda_C}{c}(1 - \cos\theta) \qquad \textbf{(30.2)}$$

The scattering of an electromagnetic wave by an electron may be visualized as a 'collision' between the wave and the electron since it comprises an exchange of energy and momentum. Further, since the wave propagates with the velocity c and its energy–momentum relationship, $E = cp$, is similar to that for a particle of zero mass, this scattering must resemble a collision in which one of the particles has a zero mass and is moving with velocity c. It can be shown (see Note 30.1) that if E and E' are the energies of the particle of zero mass before and after the collision,

$$\frac{1}{E'} - \frac{1}{E} = \frac{1}{m_e c^2}(1 - \cos\theta) \qquad \textbf{(30.3)}$$

where m_e is the mass of the other particle involved in the collision, in this case an electron, initially at rest relative to the observer. The similarity between Equations 30.2 and 30.3 is striking, and goes beyond a simple algebraic similarity. Both equations apply to a collision process and the energy–momentum relationship $E = cp$ for an electromagnetic wave is of the same type as that corresponding to a particle of zero mass. A reasonable conclusion is to link the frequency ν and the energy E by writing

$$E = h\nu, \tag{30.4}$$

where h is a universal constant that describes the proportionality between the frequency of an electromagnetic wave and the energy associated with it in the 'collision' process. Then Equation 30.3 becomes

$$\frac{1}{h\nu'} - \frac{1}{h\nu} = \frac{1}{m_e c^2}(1 - \cos\theta)$$

which may be written as

$$\frac{1}{\nu'} - \frac{1}{\nu} = \frac{h}{m_e c^2}(1 - \cos\theta) \tag{30.5}$$

which is similar to Equation 30.2. Then, by comparison, the Compton wavelength for an electron is related to the mass of the scattering electron by

$$\lambda_C = \frac{h}{m_e c} \tag{30.6}$$

From the known values of λ_C, m_e and c, we may obtain for the constant h the value $h = 6.6256 \times 10^{-34}$ J s or m^2 kg s^{-1}. This is precisely the value of **Planck's constant** that was introduced in Section 23.4 in connection with the quantization of angular momentum of an electron and with the oscillation and rotation of molecules (Sections 10.11 and 14.7).

A proton, which has a mass different from that of an electron, has a Compton wavelength (using the above value of h) of

$$\lambda_{C,p} = \frac{h}{m_p c} = 1.3214 \times 10^{-15}\ \mathrm{m} \tag{30.7}$$

This result has been experimentally confirmed by analyzing the scattering of γ-rays by free protons, which assures the general validity of our assumption 30.4. However, because the Compton wavelength of the proton is much smaller than that of the electron, the Compton effect is much less noticeable unless the radiation has a very short wavelength, as in the case of γ-rays.

We may then conclude that we can 'explain' the scattering of electromagnetic radiation by a free electron by identifying the process with the collision of the electron and a particle of zero mass, which we call a **photon**, a name proposed by G. N. Lewis in 1926.

Note 30.1 Collisions involving a zero mass particle

Let us call E and E' the energy of the particle of zero mass before and after the collision; then $p = E/c$ and $p' = E'/c$ are the corresponding values of the momentum. We assume that the other particle (electron) is initially at rest relative to the observer. Given that p_e is the momentum of the electron after the collision, the principles of conservation of energy and momentum yield

$$\boldsymbol{p} = \boldsymbol{p}' + \boldsymbol{p}_e \tag{30.8}$$

$$E + m_e c^2 = E' + c(m_e^2 c^2 + p_e^2)^{1/2} \tag{30.9}$$

From Equation 30.8 we get $\boldsymbol{p}_e = \boldsymbol{p} - \boldsymbol{p}'$. Squaring, we obtain

$$p_e^2 = p^2 + p'^2 - 2\boldsymbol{p}\cdot\boldsymbol{p}' = \frac{1}{c^2}(E^2 + E'^2 - 2EE'\cos\theta)$$

where θ is the angle through which the particle of zero mass has been deviated or scattered. Solving Equation 30.9 for p_e^2 gives us

$$p_e^2 = \frac{1}{c^2}(E + m_e c^2 - E')^2 - m_e^2 c^2 = \frac{1}{c^2}[E^2 + E'^2 + 2(E - E')m_e c^2 - 2EE']$$

Equating the two results for p_e^2 and canceling common terms, we obtain

$$E - E' = \frac{EE'}{m_e c^2}(1 - \cos\theta)$$

Dividing both sides by EE' yields Equation 30.3,

$$\frac{1}{E'} - \frac{1}{E} = \frac{1}{m_e c^2}(1 - \cos\theta)$$

30.6 Photons

Our 'explanation' of the Compton effect implies the following assumptions: (a) the scattering of electromagnetic radiation by a free electron can be considered as a collision between the electron and a particle of zero mass; (b) electromagnetic radiation plays the role of the particle of zero mass, which for brevity will be called, from now on, a *photon*; and (c) the energy and momentum of the particle of zero mass (or photon) are related to the frequency and wavelength of the electromagnetic radiation by

$$E = h\nu \qquad \text{and} \qquad p = \frac{h}{\lambda} \tag{30.10}$$

The second relation is due to the fact that $p = E/c = h\nu/c$ and $\nu/c = 1/\lambda$. Thus we may 'visualize' the Compton effect as the collision illustrated in Figure 30.10, in which a photon of frequency ν collides with an electron at rest, transferring to it certain energy and momentum. As a result of the interaction, the energy of

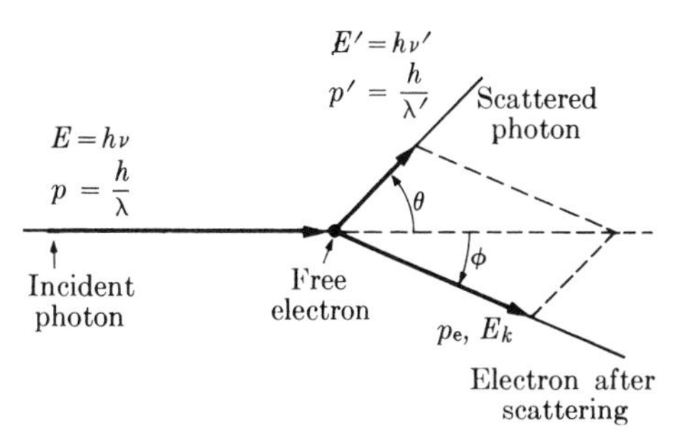

Figure 30.10 Momentum and energy relations in Compton scattering.

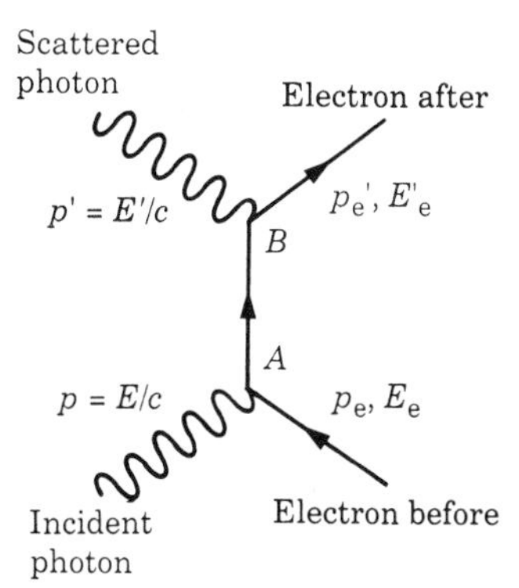

Figure 30.11 Scattering of a photon by an electron is a two-step process with $E_e + E = E_e' + E'$ and $\mathbf{p}_e + \mathbf{p} = \mathbf{p}_e' + \mathbf{p}'$.

the scattered photon is smaller, with a correspondingly smaller frequency ν'. A further test is to check to see whether the electron, after the scattering, has a momentum equal to the difference between the momentum of the incident photon and that of the scattered photon. It is a difficult experiment, but it has been performed and the results check very well.

Instead of the simple collision of Figure 30.10, the Compton effect is rather a two-step process represented in Figure 30.11. We may interpret the photon energy $E = h\nu$ and the momentum p, with $p = h/\lambda$, as the energy and momentum absorbed by the free electron from the incident electromagnetic wave. This momentarily violates the conservation of energy and momentum. The photon energy $E' = h\nu'$ and momentum $p' = h/\lambda'$ corresponds to the energy and momentum of a photon re-emitted by the electron into the scattered radiation. The electron acquires an energy $E - E'$ and gains a momentum $\mathbf{p} - \mathbf{p}'$, as required by the conservation of energy and momentum.

Similarly, in the scattering of electromagnetic radiation discussed in Section 30.4, photons from the incident wave are absorbed by the atom, molecule or nucleus, which subsequently emits another photon of the same or different energy (or frequency) in the direction of incidence or any other direction. Using the photon concept, scattering can be expressed as $A + h\nu \rightarrow A^* + h\nu'$, where A and A^* refer to the atom before and after the scattering.

The physical meaning of the photon concept and of the defining relations 30.10 is not necessarily that electromagnetic radiation is a stream of photons. Rather, the photon is the 'quantum' of electromagnetic energy and momentum absorbed or emitted in a single process by a charged particle. It is entirely determined by the frequency of the radiation and the direction of propagation. The concept of a photon is applicable to interactions between electromagnetic radiation and charged particles, either free or bound, and not just with free electrons. Therefore we may state that

when an electromagnetic wave interacts with an electron (or any other charged particle), the amounts of energy and momentum that can be exchanged in the process are those corresponding to a photon.

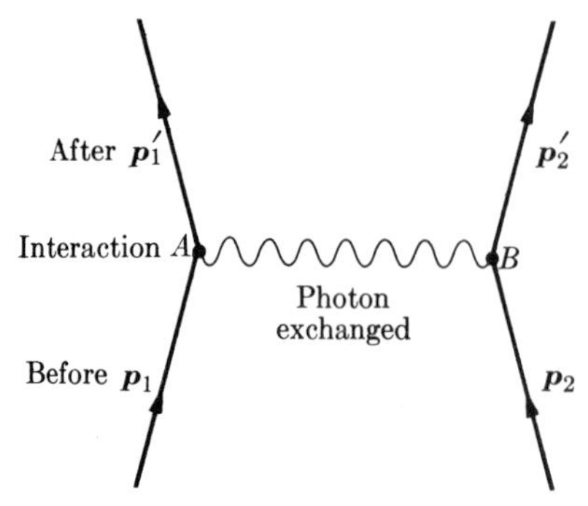

Figure 30.12 Electromagnetic interaction considered as an exchange of photons. The photons transfer energy and momentum from one charged particle to the other.

The principle stated above is one of the fundamental laws of physics. It is applicable to all radiative processes involving charged particles and electromagnetic fields.

The concept of the photon suggests a simple pictorial representation of the electromagnetic interaction between two charged particles, as shown in Figure 30.12. The interaction corresponds to an exchange of momentum and energy between the two particles. The initial momenta $\boldsymbol{p}_1$ and $\boldsymbol{p}_2$ of the particles become $\boldsymbol{p}_1'$ and $\boldsymbol{p}_2'$ after the interaction. Although the interaction is not localized at a particular instant, for simplicity we have indicated it at a particular time, and at positions A and B. Particle 1 interacts with particle 2 via its electromagnetic field, with the result that particle 2 takes a certain amount of energy and momentum from the field, equivalent to a photon, with a corresponding change in its motion. The motion of particle 1 must then be adjusted to correspond to the new field, which is the original field minus one photon. Of course, the reverse process is also possible, and particle 1 may absorb a photon from the field of particle 2. We may say then that what has happened is that between particles 1 and 2 there has been an exchange of photons. Note that energy and momentum are not conserved at A and at B but they are conserved when both A and B are taken together since the violation at A is cancelled by the violation at B.

In Section 29.3 we indicated that circularly polarized radiation has angular momentum $S_z = \pm E/\omega$ along the direction of propagation. Recalling that $\omega = 2\pi\nu$ and $E = h\nu$ for a photon, we have $S_z = \pm h/2\pi = \pm\hbar$. Thus a photon can also be assumed to carry spin along the direction of propagation with $m_s = \pm 1$. So we may say that photons carry one unit of spin in the same direction as the photon momentum (Figure 30.13) or the opposite. Therefore,

> *electromagnetic interactions can be pictured as the result of the exchange of photons between interacting charged particles; the photons are the carriers of energy, momentum and angular momentum between the particles.*

In Figure 29.20 the energy of the photons associated with the various regions of the electromagnetic spectrum was indicated. The behavior of photons when

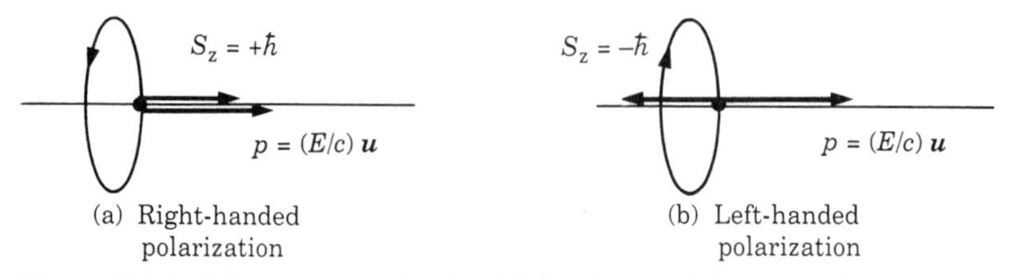

Figure 30.13 Momentum and spin of (a) right-handed polarization, (b) left-handed polarization.

interacting with matter is closely related to their energy. When we look at the very broad spectrum of electromagnetic radiation, we can easily understand why the different parts of it behave in a different way when propagating through matter. For example, those electromagnetic waves having photons with an energy comparable to the excitation energies of atoms and molecules will interact more strongly with matter. This is the case for infrared, visible and ultraviolet radiation. In particular, 'visible' light corresponds to frequencies whose photon energies are adequate for exciting molecules in the cells of the retina and produce a nervous response. Radiation with longer wavelength, carrying photons with less energy, in general interacts weakly with matter. This is the case for radio-frequency waves. However, these waves may interact with electrons in solids that have smaller excitation energies. Waves that have photons of high energy or very short wavelength, such as X- and γ-rays, are weakly absorbed by atoms and molecules but they can excite nuclei; however, their effects on matter are more profound, producing not only atomic and molecular ionization or dissociation, but also in many cases nuclear break-up.

EXAMPLE 30.1

Energy of a photon in electronvolts when the wavelength is given in meters. Apply the result to obtain the wavelength of X-rays in terms of the accelerating voltage applied to an X-ray tube.

▷ From $E = h\nu$ and $\nu\lambda = c$, we may write $E = hc/\lambda$. But

$$hc = (6.6256 \times 10^{-34}\ \text{J s})(2.9979 \times 10^{8}\ \text{m s}^{-1}) = 1.9863 \times 10^{-25}\ \text{J m}$$

Remembering that $1\ \text{eV} = 1.6021 \times 10^{-19}\ \text{J}$, we have that $hc = 1.2397 \times 10^{-6}\ \text{eV m}$. Therefore, when E is expressed in electronvolts and λ in meters,

$$E = \frac{1.2397 \times 10^{-6}}{\lambda}$$

As we explained in connection with Figure 29.13, X-rays are produced by the impact of fast electrons against the anode material of an X-ray tube. The energy of an electron may be radiated as a result of successive collisions, giving rise to several photons, or it may all be radiated in just one collision. Therefore, the most energetic photons coming out of the X-ray tube would be those emitted in the latter process, and they correspond to the shortest wavelength. In other words, given that V is the accelerating voltage, the wavelengths of the X-rays produced are equal to or larger than the threshold wavelength, satisfying the relation

$$\lambda = \frac{1.2397 \times 10^{-6}}{V} \approx \frac{1.24 \times 10^{-6}}{V}$$

since in this case the energy E of the photon is equal to the kinetic energy of the electron, which in turn is equal to V if expressed in electronvolts. For example, consider a television tube in which electrons are accelerated by a potential difference of 10 000 V. When the electrons reach the screen of the tube, they are abruptly stopped, emitting X-rays for the same reason as in the X-ray tube. The intensity, however, is quite low because the electron current is small. The minimum wavelength of the X-rays produced when the electrons are stopped at the screen is then $\lambda = 1.24 \times 10^{-10}$ m.

30.7 More about photons: the photoelectric effect in metals

The **photoelectric effect** is the emission of electrons from metals and other substances when they absorb energy from an electromagnetic wave. In 1887 Heinrich Hertz observed that by illuminating the electrodes between which an electric discharge takes place with ultraviolet light, he could increase the intensity of the discharge. This suggested the availability of more charged particles, later on identified as electrons. A year later Wilhelm Hallwachs (1859–1922) observed an electronic emission when the surfaces of certain metals, such as Zn, Rb, K, Na etc., were illuminated. These electrons are called **photoelectrons** in view of the method of their production.

Electronic emission increases with the intensity of the radiation falling on the metal surface, since more energy is available to release electrons. But a characteristic frequency dependence is also observed. For each substance there is a minimum frequency ν_0 of electromagnetic radiation such that, for radiation of frequency less than ν_0, no photoelectrons are produced, regardless of the intensity of the radiation. Also, the maximum kinetic energy of the photoelectrons varies linearly with the frequency of the electromagnetic radiation and is independent of the intensity of the radiation.

The photoelectric effect has a simple explanation using the concept of photons. We have seen (Section 23.9) that in a metal the electrons in the upper or conduction band are more or less free to move throughout the crystal lattice. These electrons are held within the metal by a **potential barrier** at the surface. Unless the electrons have enough energy to overcome this barrier they cannot escape from the metal (Figure 30.14). In Section 23.9 it was explained that one way to increase the energy of the electrons is by heating the metal. But another way to release electrons from a metal is by absorption of energy from electromagnetic radiation.

We will designate the energy required by a certain electron to escape from a given metal by ϕ. Then, if the electron absorbs an energy E larger than ϕ, the difference $E - \phi$ appears as the kinetic energy of the electron, and we may write

$$E_k = E - \phi \tag{30.11}$$

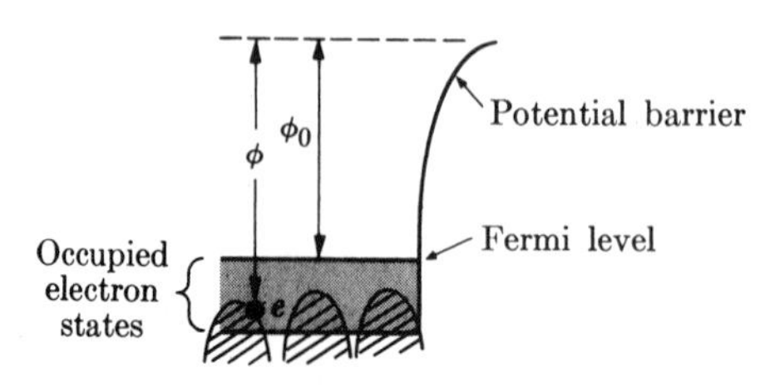

Figure 30.14 Work function of a metal.

According to the photon concept, the energy absorbed by an electron from the electromagnetic radiation must be that of a photon. Therefore, if ν is the frequency of the radiation, then $E = h\nu$, according to Equation 30.10, and we may rewrite Equation 30.11 as

$$E_k = h\nu - \phi \tag{30.12}$$

This equation was first proposed by Albert Einstein in 1905, prior to the development of the photon concept, in order to fit the experimental data. The great contribution of Einstein was to propose that electrons absorb electromagnetic energy in bundles or quanta equal to $h\nu$, following a previous theory developed by Max Planck to explain blackbody radiation (Section 31.7). However, Einstein resisted the photon concept for many years.

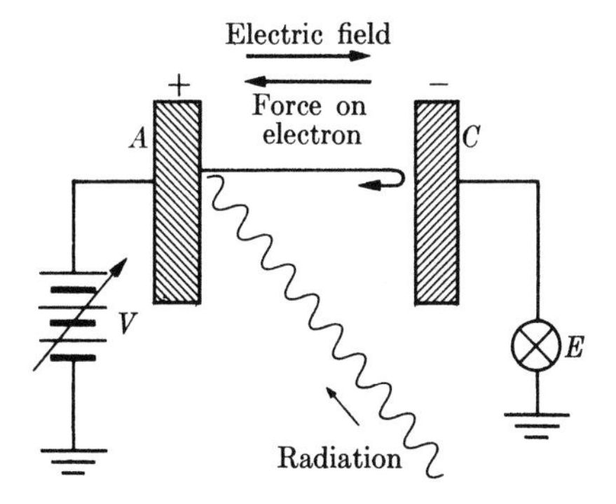

Figure 30.15 Experimental arrangement for observing the photoelectric effect.

Not all electrons require the same energy to escape from a metal. Those electrons occupying the uppermost states in the conduction band, close to the Fermi level, require less energy. We call the minimum energy ϕ_0 the **work function** of the metal (Figure 30.14). Then the maximum kinetic energy of the electrons is

$$E_{\mathrm{k,max}} = h\nu - \phi_0 \tag{30.13}$$

From this equation, we see that the kinetic energy of the electrons is zero for the frequency $\nu_0 = \phi_0/h$. Therefore, ν_0 is the minimum or **threshold frequency** at which there is photoelectric emission. For frequencies smaller than ν_0, there is no emission at all. The work function ϕ_0 can be calculated in terms of ν_0 and h, and the value obtained agrees with that calculated by other methods, confirming the validity of Equation 30.13.

We do not have to consider the conservation of momentum in the photoelectric effect because the electron absorbing the electromagnetic radiation is bound to the crystal lattice of the metal and the momentum of the absorbed photon is shared by the electron and the lattice. Because of the relatively large mass of the lattice, its kinetic energy is negligible and we may assume (without noticeable error) that all the energy of the photon goes to the electron.

The photoelectric effect is important not only because it is one of the experimental foundations of quantum theory but also because of its many practical applications, such as photocells and photometers used in several devices such as photographic and video cameras to monitor the intensity of illumination.

Note 30.2 Experimental verification of Einstein's equation

The maximum kinetic energy $E_{\mathrm{k,max}}$ of the photoelectrons can be measured by the method indicated in Figure 30.15. By applying a potential difference V between the plates A and C, we can retard the motion of the photoelectrons emitted by A when electromagnetic radiation of frequency ν falls on it. At a particular voltage V_0 the current indicated by the electrometer E drops suddenly to zero, indicating that no electrons, even the fastest, are reaching plate C.

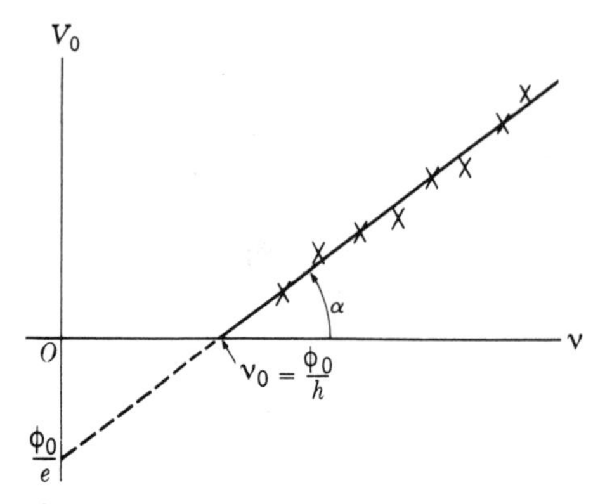

Figure 30.16 Relation between stopping potential and frequency in the photoelectric effect. The crosses (X) indicate experimental points. Note that $\tan \alpha = h/e$.

Then, using Equation 21.30, we have $E_{k,max} = eV_0$, and Equation 30.13 becomes

$$eV_0 = h\nu - \phi_0 \tag{30.14}$$

By changing the frequency ν, we can obtain a series of values for the stopping potential V_0. If Equation 30.14 is correct, the result of plotting the values of V_0 against ν should be a straight line, as shown in Figure 30.16. The slope of the straight line is $\tan \alpha = h/e$. Measuring α, and using the known value of e, we may recalculate Planck's constant h and obtain the same result found for the Compton effect, or conversely. This agreement can be considered as a further justification of the photon concept.

QUESTIONS

30.1 Justify, in terms of energy considerations, why in the Compton effect the scattered radiation has a longer wavelength than the incident radiation.

30.2 Discuss the main differences in the scattering of electromagnetic radiation by bound and by free electrons.

30.3 Can we change the momentum of a photon without changing its energy? Can we change the energy of a photon without changing its momentum? In each case illustrate your answer with an example. (Hint: note that momentum is a vector quantity.)

30.4 Make a list of devices familiar to you that operate using the photoelectric effect.

30.5 Explain, using the photon concept, the scattering of photons by bound electrons.

30.6 Which phenomenon fully reveals the properties of a photon: the Compton effect or the photoelectric effect?

30.7 Why does the concept of photons allow us to explain the existence of a threshold frequency in photoelectric emission?

30.8 Using the photon concept, explain what happens to the energy of an atom when the frequency of the scattered radiation is (a) the same, (b) smaller, (c) larger than the frequency of the incident radiation.

30.9 Describe how to explain the electromagnetic interaction using photons.

PROBLEMS

30.1 Find the energy and wavelength of a photon having the same momentum as a 40 MeV (a) proton, (b) electron. Identify the region of the spectrum in which it falls. (Hint: note that the proton can be treated non-relativistically, but for the electron relativistic mechanics is required.)

30.2 A photon having an energy of 10^4 eV collides with a free electron at rest and is scattered through an angle of 60°. Find (a) the changes in energy, frequency and wavelength of the photon and (b) the kinetic energy, momentum and direction of the recoiling electron.

30.3 Radiation having a wavelength of 10^{-10} m (or 1 Å) undergoes Compton scattering. The scattered radiation is observed at a direction perpendicular to that of incidence. Find (a) the wavelength of the scattered radiation and (b) the kinetic energy and direction of motion of the recoil electrons.

30.4 Referring to the preceding problem, if the electrons recoil at an angle of 60° relative to the incident radiation, find (a) the (i) wavelength and (ii) direction of the scattered radiation and (b) the kinetic energy of the electron.

30.5 When a certain metal surface is illuminated with light of different wavelengths, the stopping potentials of the photoelectrons are measured as shown in the table:

$\lambda(\times 10^{-7}$ m)	V_0(V)
3.66	1.48
4.05	1.15
4.36	0.93
4.92	0.62
5.46	0.36
5.79	0.24

Plot the stopping potential as ordinate against the frequency of light as abscissa. Determine from the graph: (a) the threshold frequency, (b) the photoelectric work function of the metal and (c) the ratio h/e.

30.6 The photoelectric work function of potassium is 2.0 eV. When light with wavelength of 3.6×10^{-7} m falls on potassium, determine (a) the stopping potential of the photoelectrons and (b) the kinetic energy and velocity of the fastest electrons ejected.

30.7 Electromagnetic radiation of wavelength 10^{-5} m falls normally on a metal sample of mass 10^{-1} kg and an electron is ejected in a direction opposite to the incident radiation. Using the laws of conservation of energy and momentum, obtain (a) the energy of the electron and (b) the recoil energy of the metal sample. Assume that the work function is negligible. (c) Does the result justify not considering momentum conservation in the calculation of the photoelectric effect?

30.8 Often the reciprocal of the wavelength, $1/\lambda$, is used to describe the energy associated with that wavelength. Verify that for a photon we may use the equivalences 1 eV = 8065.8 cm^{-1} and 1 cm^{-1} = 1.2398×10^{-4} eV.

30.9 The minimum light intensity that can be perceived by the eye is about 10^{-10} W m^{-2}. Calculate the number of photons per second (wavelength 5.6×10^{-7} m) that enter the pupil of the eye at this intensity. The pupil area may be considered as 4×10^{-5} m^2.

30.10 (a) Compare the wavelength of a photon of 2 eV in the visible region of the spectrum with the dimensions of an atom. (b) Repeat for a γ-ray photon of 1 MeV and nuclear dimensions.

30.11 A monochromatic beam of electromagnetic radiation has an intensity of 1 W m^{-2}. Calculate the average number of photons per m^3 for (a) 1 kHz radio waves and (b) 10 MeV γ-rays.

30.12 A uniform monochromatic beam of light, of wavelength 4.0×10^{-7} m, falls on a material having a work function of 2.0 eV. The beam has an intensity of 3.0×10^{-9} W m^{-2}. Calculate (a) the number of electrons emitted per m^2 and per second, (b) the energy absorbed per m^2 and per second and (c) the kinetic energy of the photoelectrons.

30.13 Find (a) the energy and (b) the wavelength of a photon that can impart a maximum kinetic energy of 60 keV to a free electron.

30.14 Verify that when a free electron is scattered in a direction making an angle ϕ with the incident photon in a Compton scattering, the kinetic energy of the electron is $E_k = h\nu(2\alpha \cos^2 \phi)/[(1 + \alpha)^2 - \alpha^2 \cos^2 \phi]$, where $\alpha = h\nu/m_e c^2$.

30.15 Verify that, in Compton scattering, the relation between the angles defining the directions of the scattered photon and the recoil electron is $\cot \phi = (1 + \alpha) \tan \frac{1}{2}\theta$, where $\alpha = h\nu/m_e c^2$.

31 Radiative transitions

Max K. Planck became the founder of quantum theory with the publication, in 1900, of an expression that faithfully reproduced the energy distribution of blackbody radiation. In his derivation Planck introduced the revolutionary idea of the quantization of the energy levels of an oscillator, which was used later on by Einstein (1905) to explain the photoelectric effect and by Bohr (1913) to explain atomic structure. Planck's theory was the first step in the great conceptual revolution of the 20th century, quantum mechanics. Planck also conducted extensive studies in thermodynamics, clarifying the concept of entropy.

Notes

31.1 Introduction

We have seen that when electromagnetic radiation interacts with matter, photons play a critical role. We saw that in scattering, either by free or bound electrons, a photon is absorbed and another is re-emitted with either the same or a different energy. In other instances a photon is completely absorbed, as in the photoelectric effect. As a result, the system passes to a state of higher energy or an excited state. Conversely, a system in an excited state may release the excess energy as a photon and pass to a state of lower energy. In this chapter we are going to see how to explain the emission and absorption spectra considered in Sections 30.2 and 30.3 using the concept of a photon.

31.2 Stationary states

The existence of emission and absorption spectra composed of well-defined or characteristic frequencies was a problem that puzzled physicists at the end of the 19th century and the beginning of the 20th. To solve this problem, a new and revolutionary idea was advanced in 1913 by Niels Bohr (1885–1962). Suppose that an atom in a state of energy E absorbs radiation of frequency ν and thus passes to another state of higher energy E'. The change in energy of the atom is $E' - E$. On the other hand, the energy absorbed from the radiation in a single process must be that of a photon of energy $h\nu$. Conservation of energy requires that both quantities be equal. Therefore

$$E' - E = h\nu \qquad \text{or} \qquad \Delta E = h\nu \tag{31.1}$$

an expression called **Bohr's formula**. Similarly, if the atom passes from a state of energy E' to another state of lower energy, E, the frequency of the emitted radiation must be given by Equation 31.1.

The fact that only certain frequencies $\nu_1, \nu_2, \nu_3, \ldots$ are observed in emission and absorption can be explained if we assume that the energy of the atom can have only certain values $E_1, E_2, E_3, \ldots$, a hypothesis we introduced in Section 23.4 and called **energy quantization**. It will be recalled that each allowed energy value is called an **energy level**. Then the only possible frequencies that result in emission or absorption of radiation are those corresponding to transitions between two allowed energy levels; that is,

$$\nu = \frac{E' - E}{h} \qquad \text{or} \qquad \nu = \frac{\Delta E}{h} \tag{31.2}$$

where E and E' are the energies of the two states involved. We may therefore say that atomic spectra are an indirect proof of the quantization of energy and the existence of energy levels. Bohr's assumption may now be stated as:

> *The energy of a bound system of charges—either an atom, a molecule or a nucleus—can have only certain values $E_1, E_2, E_3, \ldots$; that is, the energy is quantized. The states corresponding to these energies are called stationary states and the possible values of the energy are called energy levels.*

Absorption of electromagnetic radiation (or of any other energy) results in a transition of the atom (or molecule or nucleus) from one **stationary state** to another of higher energy; emission of electromagnetic radiation results in the reverse process. The frequency of the radiation involved in either process is given by Equation 31.2. Some emission and absorption transitions are shown schematically in Figure 31.1. A process in which an atom in its ground state, represented by A, absorbs a photon and passes to an excited state, represented by A^*, is designated by

$$A + h\nu \rightarrow A^*$$

The reverse process, photon emission, by an atom in an excited state, can be expressed by

$$A^* \rightarrow A + h\nu$$

An atom can also be excited to a stationary state of higher energy by inelastic collisions (recall Example 23.2). For example, if fast electrons pass through a gas,

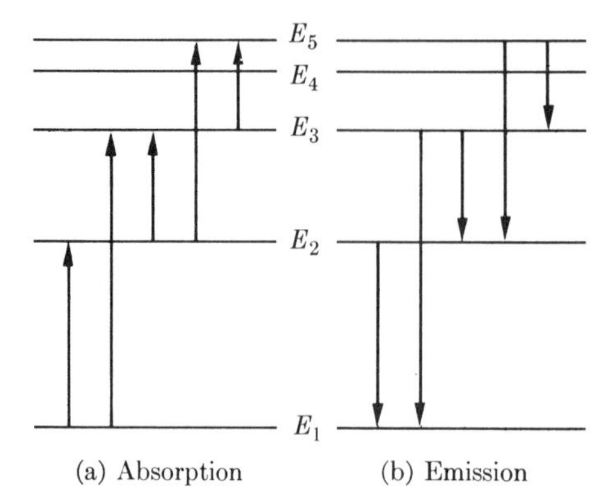

Figure 31.1 Transitions between stationary states. The relative spacing of the energy levels and the possible transitions depend on the nature of the system. In each transition, a photon of frequency given by Equation 31.2 is either absorbed or emitted.

many atoms in the gas are excited; that is,

$$A + \text{e(fast)} \rightarrow A^* + \text{e(slow)}$$

The excited atoms return to the ground state with emission of a photon. This is the principle on which mercury lamps operate. This is also why an electric discharge in a gas is accompanied by light emission, as in lightning. Therefore, we may state that

> *when a bound system of charges, either an atom, a molecule, a solid, or a nucleus, absorbs or emits energy as a photon or in another form, it passes from one stationary state to another.*

The idea of stationary states poses a serious difficulty to Maxwell's electromagnetic theory. When an electron revolves around a nucleus in an atom, its motion has both tangential and centripetal acceleration; i.e. its motion is accelerated. Therefore one would think that the electron would be radiating energy continuously (recall Section 29.5). As a result, the electron's energy would be decreasing continuously and its orbit would be shrinking. This would make the existence of stationary states impossible. However, neither this contraction of matter nor the continuous radiation of energy associated with it have been observed. Therefore, since the predictions of electrodynamics as formulated by Maxwell are not followed, we may conclude that an electron (or a charged particle) moving in a stationary state is governed by some additional principles that we have not yet considered. These principles constitute the branch of physics called **quantum mechanics**, which we shall consider in Chapters 36 through 38.

In many instances all values of the energy in a certain energy range are allowed, and a **continuous energy spectrum** results. Consider, for example, the case of an electron and a proton, and take the zero of energy when both the electron and proton are at rest and separated by a very large distance. Then, as explained in Section 23.4, all discrete stationary states have negative energy and correspond to bound states where the electron moves around the proton to form a hydrogen atom. The energy of such states can have only certain values $E_K, E_L, E_M, \ldots$ (Figure 31.2) given by Equation 23.12. We say that their energy is quantized. On the other hand, the states of positive energy are unbound and their energy is not quantized. They correspond to the situation in which an electron is thrown from a very large distance, and with a certain initial kinetic energy, against a proton; the electron, after passing near the proton, is deflected from its original direction of motion and recedes to an infinite separation without the formation of a bound system, resulting in a scattering. The energy of the system in this case is determined by the initial kinetic energy of the electron, which may be arbitrarily chosen and therefore is not quantized. It is possible that in the process the electron emits a photon, decreasing its energy. If the energy of the photon is not very large, the energy of the electron remains positive. However, if the photon energy is very large, the electron may be captured into one of the discrete negative energy levels corresponding to a hydrogen atom.

Therefore, transitions may occur between two states of the discrete energy spectrum, such as *ab* and *cd* in Figure 31.2, or between a state of the discrete spectrum and a state of the continuous spectrum, such as *ef*, or between two states of the continuous spectrum, such as *gh*.

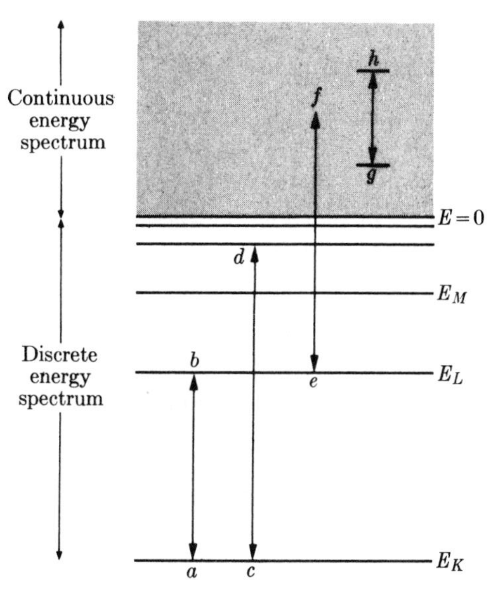

Figure 31.2 Discrete and continuous energy spectrum of an electron–proton system. The discrete spectrum corresponds to the stationary states of the hydrogen atom.

When an atom that is in an excited state E' (Figure 31.3(a)) and at rest in the L-frame emits a photon of energy $h\nu$, the photon also carries a momentum $p = h\nu/c$. Conservation of momentum requires that the atom recoil with a momentum that is also p, and thus with a kinetic energy $p^2/2m_{\text{atom}}$. Hence in emission we must write

$$\Delta E = h\nu + (\text{recoil kinetic energy of atom})$$

On the other hand, when an atom, in its ground state E and at rest in the L-frame (Figure 31.3(b)), absorbs a photon of energy $h\nu$ and momentum $p = h\nu/c$, the

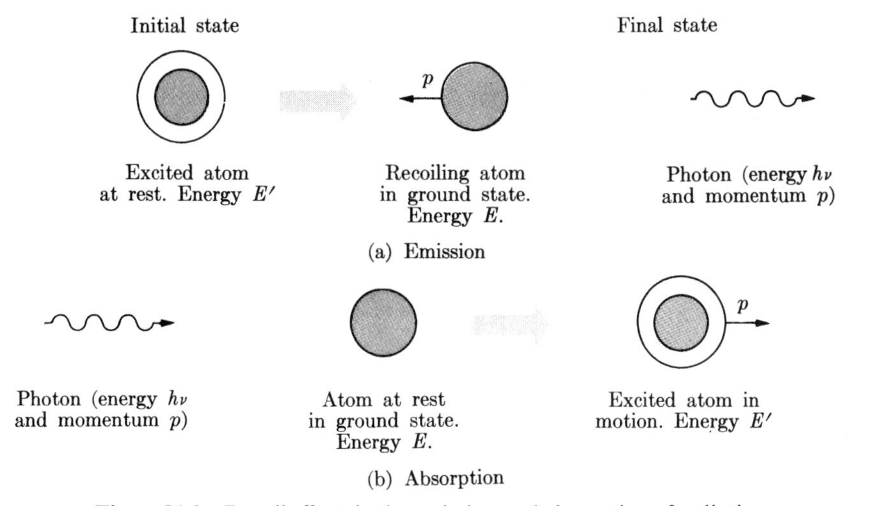

Figure 31.3 Recoil effects in the emission and absorption of radiation.

atom is not only excited to a state of energy E', but is also set in motion with momentum p. Therefore, in absorption we must write

$$h\nu = \Delta E + (\text{recoil kinetic energy of atom})$$

We therefore conclude that Bohr's Equation 31.1 must be corrected because

in emission, the energy of the photon is smaller (and in absorption must be larger) than the difference of energies of the two stationary states involved, by an amount equal to the recoil energy of the system.

The same result also applies to molecules and nuclei. In general the recoil energy is negligible and can be ignored in many instances, especially in atomic and molecular transitions. In nuclear transitions, however, the recoil effects are relatively more important because of the larger energy of the photons. One way to minimize the recoil effect is to embed the atom in a crystal lattice. The recoil energy is then $p^2/2m_{\text{crystal}}$ and since m_{crystal} is very large, the recoil energy is very small.

Another factor that affects Equation 31.2 is a quantum effect called **broadening** of the energy level. This effect will be discussed in Example 36.4 but we might mention here that it is related to what is called the **transition probability** from one energy level to another of less energy. The consequence is that the energy of the stationary states is not precisely defined but is extended over the range $E - \varepsilon$ to $E + \varepsilon$, where ε is very small compared with E and is of the order of 10^{-7} eV for atomic energy levels. The result is that the frequency of the transition between two energy levels is not precisely defined and spectral lines appear a little fuzzy in a high resolution spectroscope. It is said that the spectral line has a certain **natural width**, to be added to the Doppler width, which was discussed in Example 29.4.

EXAMPLE 31.1

Recoil energy of a mercury atom when emitting a photon of energy 4.86 eV (visible region) and of a ^{60}Ni nucleus when emitting a γ-ray photon of 1.33 MeV.

▷ The momentum of a photon of 4.86 eV is 2.59×10^{-27} kg m s^{-1}. The mass of a mercury atom is 3.34×10^{-25} kg. Therefore the recoil energy of the mercury atom is

$$\frac{p^2}{2m_{\text{Hg}}} = 1.008 \times 10^{-29}\ \text{J} = 6.3 \times 10^{-11}\ \text{eV}$$

which is negligible in comparison with the energy of the photon and is very difficult to detect. Hence we may ignore the recoil effects and apply Equation 31.1.

The momentum of a photon with an energy of 1.33 MeV is 7.10×10^{-22} kg m s^{-1}. The mass of a Ni atom is 9.7×10^{-26} kg. Therefore the recoil energy is

$$\frac{p^2}{2m_{\text{Ni}}} = 2.60 \times 10^{-18}\ \text{J} = 16.2\ \text{eV}$$

a quantity which, although still small compared with the energy of the photon, is relatively more important than in the atomic case and is large enough to be detected. Incidentally, the recoil energy in γ-emission or absorption is of the order of the binding energy of an atom in a molecule and may result in an atom dissociating from the molecule, a process that has been observed experimentally. This is one of the reasons why prolonged exposure to γ-rays may result in material damage.

31.3 Interaction of radiation with matter

The interaction of radiation with matter is one of the fundamental processes responsible for many phenomena occurring in the universe. For example, the Earth is subject to a continuous flow of electromagnetic radiation from the Sun, which makes life on Earth possible through the process of **photosynthesis** (that is, the formation of new compounds, mainly carbohydrates, out of the synthesis of carbon dioxide and water as a result of the absorption of photons; a compound called chlorophyll plays an important role in the reaction). The process can be written in a simplified form as

$$6CO_2 + 6H_2O + nh\nu \rightarrow C_6H_{12}O_6 + 6O_2$$

The number n of photons involved is not fixed; their energy falls mostly in the visible region of the spectrum. The process is much more complicated than the above equation would suggest and active research on it is still going on. Photosynthesis is important not only because it produces carbohydrates, which are the ultimate source of food (and thus of energy) for most living organisms, but also because it controls the amount of oxygen in the atmosphere by liberating oxygen, which, on the other hand, is quickly consumed in the many oxidizing processes occurring on the Earth, such as combustion and respiration. Photosynthesis is just one example of many reactions initiated by the absorption of photons. The study of such reactions is called **photochemistry**. Each photochemical reaction requires the intervention of photons of a certain energy.

Another process due to absorption of radiation is the dissociation of a molecule by the absorption of a photon, a process called **photodissociation**. That is,

$$AB + h\nu \rightarrow A + B$$

One such reaction, of great geophysical and biological importance, is the dissociation of oxygen in the atmosphere by the absorption of ultraviolet radiation of wavelength in the range of 1600 Å to 2400 Å (that is, photons with an energy between 7.8 eV and 5.2 eV). We may express this process by the equation

$$O_2 + h\nu \rightarrow O + O$$

The atomic oxygen produced combines with molecular oxygen to form ozone, O_3, which in turn undergoes photochemical dissociation by absorption of ultraviolet radiation of wavelength between 2400 Å and 3600 Å (that is, photons of energy between 5.2 eV and 3.4 eV). The reaction is

$$O_3 + h\nu \rightarrow O + O_2$$

These two reactions absorb ultraviolet radiation so strongly that they remove practically all the ultraviolet radiation coming from the Sun before it reaches the Earth's surface. If this ultraviolet radiation were able to reach the Earth's surface, it would destroy many organisms by means of photochemical reactions with cell components, enzymes etc. This is why it is very important not to irreversibly disturb the ozone layer in the upper atmosphere.

The photographic process is a photochemical reaction. Under the action of radiation, molecules of silver bromide undergo decomposition, with the silver atoms forming a so-called latent image on a sensitized film. Later, in the developing process, the film is treated in such a way that the silver ions are fixed on the film and a permanent image is formed.

When a photon has enough energy, its absorption by an atom or a molecule may result in the ejection of an electron, leaving behind an ionized atom or molecule. We may write the process as

$$A + h\nu \rightarrow A^+ + e^-$$

This process, called **photoionization**, is the equivalent of the photoelectric effect in metals discussed in Section 30.7. For that reason it is also called the **atomic photoelectric effect**. As a result of photoionization, when a beam of ultraviolet, X-, or γ-radiation passes through matter, it produces ionization along its path. Measuring that ionization is one of the methods of detecting X- and γ-radiation. Photoionization occurs, for example, near X-ray machines.

The energy required to extract an electron from an atom or molecule, designated by I, is called the **ionization potential**. Then the kinetic energy of the photoelectrons is given by

$$E_k = h\nu - I$$

an expression analogous to Equation 30.12. (In writing this equation, we have neglected the recoil energy of the ion.) In order to produce photoionization, the energy of the photon must be equal to or larger than I. The value of I depends on the stationary state initially occupied by the ejected electrons. It is equal in value to the binding energy of the electron corresponding to the energy level. For example, if an electron is to be ejected from the ground state in a hydrogen atom, the minimum energy of the photon must be 13.6 eV. But if the electron is in the first excited state, only 3.4 eV are required.

In the region of the upper atmosphere called the **ionosphere**, the large concentration of ions and free electrons (about 10^{11} per m^3) is due mostly to the photoelectric effect in atoms and molecules produced by ultraviolet and X-radiation from the Sun. Some of the reactions that occur more frequently are

$$NO + h\nu \rightarrow NO^+ + e^- \qquad (5.3\text{ eV})$$

$$N_2 + h\nu \rightarrow N_2^+ + e^- \qquad (7.4\text{ eV})$$

$$O_2 + h\nu \rightarrow O_2^+ + e^- \qquad (5.1\text{ eV})$$

$$He + h\nu \rightarrow He^+ + e^- \qquad (24.6\text{ eV})$$

The ionization energies are indicated in parentheses. Many other secondary reactions take place in the atmosphere as a result of these ionizations.

Photoionization and photodissociation have some important biological applications. For example, X-rays and γ-rays are used in the treatment of cancer and for sterilization of food and beverages because of the damage they produce in cells and micro-organisms by the ionization they produce. Consequently, exposure to high-energy photons (more than 5 eV or 5.4×10^{16} Hz or wavelength less than $\lambda = 5.5 \times 10^{-7}$ m) should be avoided whenever possible.

A process that is the reverse of photoionization is **radiative capture**. In radiative capture, a free electron with kinetic energy E_k is captured into a bound state by an ion with the emission of a photon. That is,

$$A^+ + e^- \rightarrow A + h\nu$$

If recoil effects are neglected, the energy of the photon is $h\nu = E_k + I$.

If the energy of the photons is sufficiently high, the photons may interact with the atomic nuclei, resulting in **photonuclear reactions**. For example, the photodisintegration of the deuteron, which is a system composed of a neutron bound to a proton (Section 39.9), occurs according to the equation $h\nu + d \rightarrow n + p$. For this process to occur, the photon must have an energy of at least 2.224 MeV, which is the binding energy of the deuteron.

Other processes involving the interaction of electromagnetic radiation with matter include scattering (such as Compton scattering, Section 30.5) and pair production (see Figure 22.5), which can be written as $h\nu \rightarrow e^+ + e^-$.

31.4 Atomic spectra

In discussing atomic spectra it is convenient to examine the cases of atoms with one electron separate from the case of atoms with many electrons.

(i) One-electron atoms. The energy of the stationary states of hydrogen-like atoms is given by Equation 23.17:

$$E_n = -\frac{m_e e^4 Z^2}{2(4\pi\varepsilon_0)^2 \hbar^2 n^2}$$

which for practical reasons is written in the form

$$E_n = -\frac{RhcZ^2}{n^2} \qquad \textbf{(31.3)}$$

where, recalling that $h = 2\pi\hbar$,

$$R = \frac{m_e e^4}{4\pi(4\pi\varepsilon_0)^2 \hbar^3 c} = 1.0974 \times 10^7 \text{ m}^{-1}$$

is called **Rydberg's constant**. When the values of the constants are inserted Equation 23.12, $E_n = -13.598\, Z^2/n^2$ eV, results. Since the energy of the stationary states is negative, it increases with the quantum number n. The difference in energy between the states corresponding to n_1 and n_2 ($n_2 > n_1$) for a hydrogen-like ion is

$$E_2 - E_1 = \left(-\frac{RhcZ^2}{n_2^2}\right) - \left(-\frac{RhcZ^2}{n_1^2}\right)$$

which reduces to

$$E_2 - E_1 = RhcZ^2\left(\frac{1}{n_1^2} - \frac{1}{n_2^2}\right) \tag{31.4}$$

When we apply Bohr's condition, $\nu = (E_2 - E_1)/h$, and neglect recoil effects, the frequency of the electromagnetic radiation emitted or absorbed by the atom in a transition between stationary states corresponding to n_1 and n_2 is

$$\nu = RcZ^2\left(\frac{1}{n_1^2} - \frac{1}{n_2^2}\right) \tag{31.5}$$

or, using the numerical values of R and c,

$$\nu = 3.2898 \times 10^{15} Z^2\left(\frac{1}{n_1^2} - \frac{1}{n_2^2}\right) \text{ Hz} \tag{31.6}$$

This expression is called **Balmer's formula**, and is applicable only to hydrogen-like atoms or ions.

Since in a spectroscope (either prism or grating; see Sections 33.6 and 35.6), radiation of a given frequency appears as a line (which is the image of the slit), the spectrum is called a **line spectrum**.

The hydrogen spectrum ($Z = 1$) (and similarly for the spectra of other atoms) is classified in terms of series, each series formed by transitions that have the lowest energy level in common. Figure 31.4 represents the following hydrogen series:

1. Lyman series: $n_1 = 1$, $n_2 = 2, 3, 4, \ldots$
2. Balmer series: $n_1 = 2$, $n_2 = 3, 4, \ldots.$
3. Paschen series: $n_1 = 3$, $n_2 = 4, 5, \ldots$
4. Brackett series: $n_1 = 4$, $n_2 = 5, 6, \ldots$
5. Pfund series: $n_1 = 5$, $n_2 = 6, 7, \ldots$

The Balmer series, which is mostly in the visible region, is easily observed with a common spectroscope. The Lyman series falls in the ultraviolet region and the others in the infrared. The transitions indicated in Figure 31.4 correspond to the emission spectrum; the reverse transitions take place in the absorption spectrum.

(ii) Many-electron atoms. The energy levels of an atom with several electrons are much more complex than those of atoms with one electron. The resulting spectra are therefore rather complex and contain many more lines.

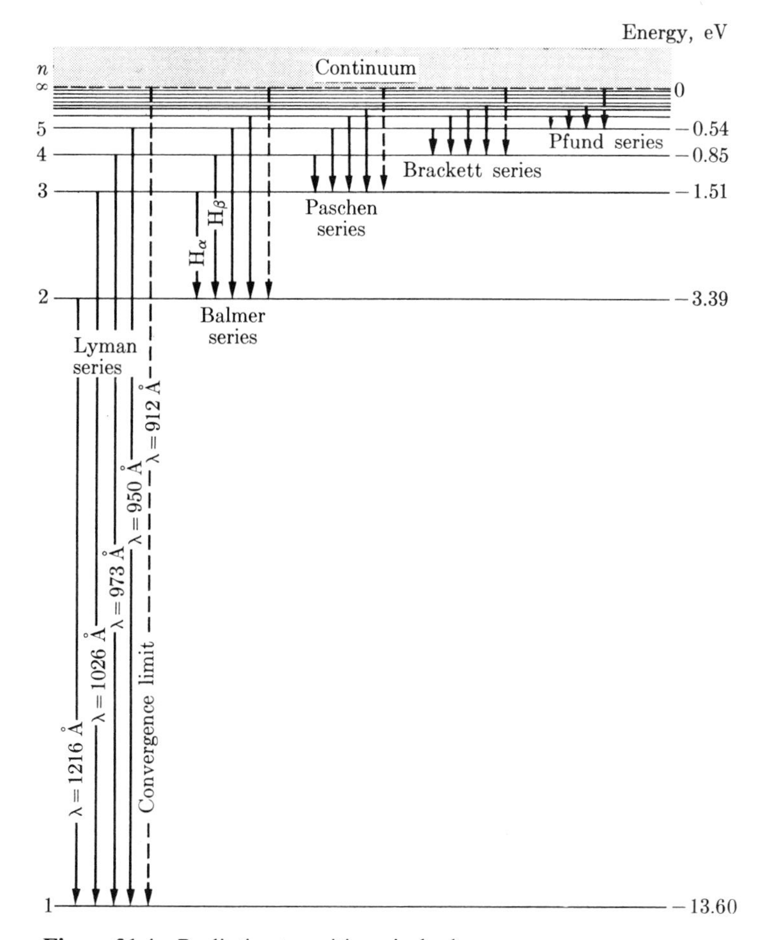

Figure 31.4 Radiative transitions in hydrogen.

Therefore we shall mention only some of the most important features. In the first place, in many-electron atoms it is necessary to distinguish between transitions involving the outer or valence electrons and those in the inner shells or **kernel** electrons.

Transitions involving the valence electrons result in what is usually called the *optical* spectrum because most of the lines fall in the visible region, although many fall in the ultraviolet region due to the energies involved. The spectrum of atoms with only one valence electron, such as Li, Na, K etc., is rather similar to that of hydrogen-like atoms; but when the number of valence electrons is larger than one, the spectrum is rather complex. This matter will be considered in more detail in Sections 38.4 and 38.5.

(iii) X-ray spectrum. Transitions in which the electrons in inner shells participate give rise to the so-called X-ray **characteristic spectrum**. Suppose that a vacant state (or hole) is produced in an inner shell of an atom, say the K-shell. Another electron in a higher energy level (or even a free electron) may fall into the vacant state in the K-shell. Since the amount of energy involved is fairly large,

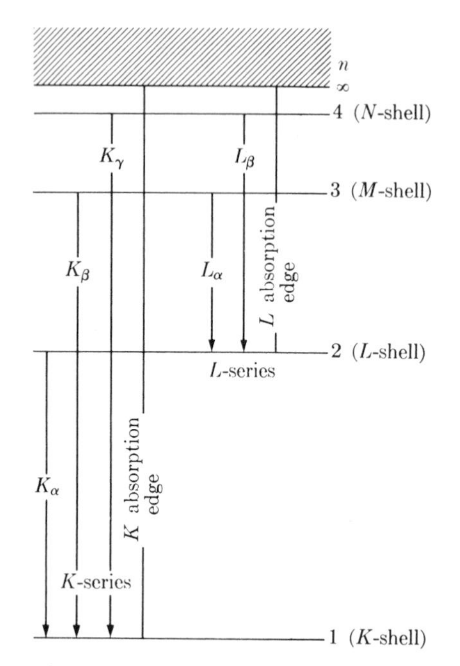

Figure 31.5 X-ray transitions in an atom with an atomic number $Z \sim 36$.

the radiation emitted by the electron falling into the vacant state has a rather large frequency or small wavelength and lies in the X-ray region of the spectrum. The electron falling into the vacant state of the K-shell may proceed from the L, M, N etc. shells, and therefore we have a series of X-ray lines designated K_α, K_β, K_γ etc. If the vacant state is in the L-shell, the electrons may fall from the M, N etc. shells, giving rise to the L_α, L_β etc. X-ray lines, and so on. The X-ray spectrum of an atom with $Z \sim 36$ is illustrated schematically in Figure 31.5. The actual spectrum is more complex because, as may be recalled from Section 23.9, each inner shell is composed of several close energy levels.

Note that X-ray emission is possible only if a vacancy is produced in an inner shell, since the exclusion principle prevents a transition from a higher level into a lower, but fully occupied, shell. The vacant space in an inner shell may be produced by absorption of radiation, resulting in a photoelectric effect for electrons of the K, L, M etc. shells. Another way of producing a vacant state in an inner shell is by high-energy electron impact. For example, when the electrons in an X-ray tube strike the anode, their energy may be sufficient to knock out one of the electrons of the target atoms. Then in an X-ray tube, in addition to the X-ray bremsstrahlung (see Section 29.5), the characteristic X-ray spectrum of the material composing the anode is observed.

Historically, the problem of explaining the line spectra of hydrogen and other elements resulted in the first application of the quantum theory to the atom. The mathematician Johann Balmer (1825–1898), long before the advent of the quantum theory, empirically obtained formula 31.5 in 1885, without any theoretical explanation related to atomic structure. Some years later, in 1913, Bohr derived Equation 31.5 by introducing, for the first time, the concept of stationary states and the quantization of energy and angular momentum, as explained in Section 23.4.

EXAMPLE 31.2

First excitation potential and ionization energy of hydrogen.

▷ The **first excitation potential** is the energy required to take an atom from its ground state to its first (or lowest-lying) excited state. These states in hydrogen-like atoms correspond, respectively, to $n_1 = 1$ (ground state) and $n_2 = 2$ (first excited state). Setting $n_1 = 1$ and

$n_2 = 2$ in Equation 31.4 with $Z = 1$, the energy required to excite the atom from the ground state to the first excited state is $E_2 - E_1 = 10.2$ eV. If a hydrogen atom is carried to its first excited state by an inelastic collision, as happens in a gas discharge tube, it returns to the ground state by emitting radiation of frequency $\nu = (E_2 - E_1)/h = 2.47 \times 10^{15}$ Hz and wavelength $\lambda = c/\nu = 1.216 \times 10^{-7}$ m, which in this case falls in the ultraviolet region.

The **ionization energy** is the energy required to take the electron from the ground state ($n_1 = 1$) to the state of zero energy ($n_2 = \infty$), and thus is equal to $I = -E_1 = 13.6$ eV. Ionization may result from either an inelastic collision of the hydrogen atom with an electron or another charged particle or with another atom, or from the atom absorbing a photon having a frequency equal to or larger than $I/h = 3.29 \times 10^{15}$ Hz or a wavelength equal to or shorter than 9.12×10^{-8} m.

31.5 Molecular spectra

In discussing molecular spectra, we must consider three separate effects associated respectively with the rotation of the molecule as a whole, the vibration of the nuclei about their equilibrium positions, and electronic transitions. The vibration and rotation energies of a molecule were discussed in Chapters 10 and 14 and will be reviewed here in the context of molecular processes.

(i) Rotational spectra. In a first approximation it is possible to consider a molecule as a rigid body, whose shape is determined by the equilibrium position of the nuclei. The simplest nuclear motion is that of rotation of the molecule around its center of mass. We can study the rotational motion of a molecule most easily by considering the rotation about the principal axes of inertia. In the case of diatomic molecules (Figure 31.6), the principal axes are the line joining the two nuclei, N_1 and N_2, or the Z_0 axis, and any line perpendicular to it through the center of mass. Because the mass of the electrons is so small, we may assume that the angular momentum of the molecule relative to the Z_0 axis is zero. Thus the angular momentum of the molecule is perpendicular to the molecular axis. Given that r_0 is the equilibrium separation of the nuclei and μ the reduced mass of the molecule, the moment of inertia about an axis perpendicular to Z_0 and passing through the center of mass of the molecule is (see Example 13.6) $I = \mu r_0^2$. We may then write the kinetic energy of rotation of the molecule as (recall Example 14.2)

$$E_r = \frac{L^2}{2I} = \frac{L^2}{2\mu r_0^2}$$

By virtue of the quantization of the angular momentum we have, according to Equation 23.20, $L^2 = \hbar^2 l(l+1)$, where $l = 0, 1, 2, 3, \ldots$ (that is, a positive integer). Therefore the kinetic energy of rotation of the molecule is

$$E_r = \frac{\hbar^2}{2I} l(l+1) \qquad (31.7)$$

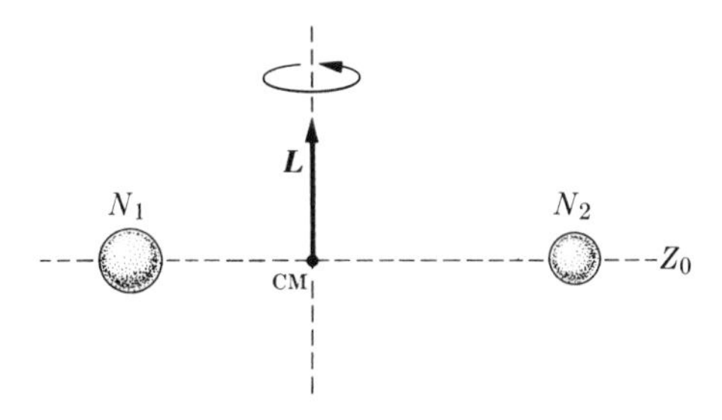

Figure 31.6 Rotation of a diatomic molecule about its center of mass.

This expression was mentioned previously in Example 14.2. The quantity $\hbar^2/2I$ (in eV) for several molecules is given in Table 31.1. The rotational energy levels of the molecule are found by giving successive values to l (Figure 31.7). Succcessive energy levels, corresponding to l and $l' = l + 1$, are separated by the amount

$$\Delta E_r = \frac{\hbar^2}{I}(l + 1) \tag{31.8}$$

The frequencies associated with transitions between those adjacent states are

$$\nu = \frac{\Delta E_r}{h} = \frac{\hbar}{2\pi I}(l + 1)$$

an expression that closely describes the rotational spectra of diatomic molecules. The allowed rotational transitions are those for which $\Delta l = \pm 1$ (see Section 38.3). This restriction is called a **selection rule**. Thus the only rotational transitions possible are those between adjacent levels. Some of the rotational transitions, corresponding to absorption, are indicated in Figure 31.7. Pure rotational spectra fall in the microwave or far-infrared regions of the spectrum with frequencies of the order of 10^{12} Hz or energies of 10^{-4} eV.

(ii) Vibrational spectra. The nuclei in a molecule are not fixed relative to each other, but rather are in relative oscillatory motion, as explained in Section 10.11. If the potential energy were that of a simple harmonic oscillator, the relative oscillatory motion of the two nuclei would be simple harmonic with a frequency ν_0. Then it can be shown using the methods of quantum mechanics (Section 37.8) that the vibrational energy levels for a diatomic molecule are expressed by

$$E_v = (n + \tfrac{1}{2})h\nu_0 \tag{31.9}$$

where $n = 1, 2, 3, \ldots$ (positive integer), corresponding to the quantized energy levels of a harmonic oscillator. This expression was mentioned in Section 10.11 and will be analyzed in more detail in Section 37.8. If $\omega_0 = 2\pi\nu_0$ is the corresponding angular frequency, we have that $h\nu = \hbar\omega$ since $\hbar = h/2\pi$. Equation 31.9 can then

Table 31.1 Rotational and vibrational constants of some diatomic molecules

Molecule	$\hbar^2/2I$ (eV)	$h\nu_0$ (eV)
H_2	8.0×10^{-3}	0.543
Cl_2	3.1×10^{-5}	0.0698
N_2	2.48×10^{-4}	0.292
Li_2	8.3×10^{-4}	0.0434
O_2	1.78×10^{-4}	0.194
CO	2.38×10^{-4}	0.268
HF	2.48×10^{-3}	0.510
HCl	1.31×10^{-3}	0.369
HBr	1.05×10^{-3}	0.326
BeF	1.84×10^{-4}	0.151

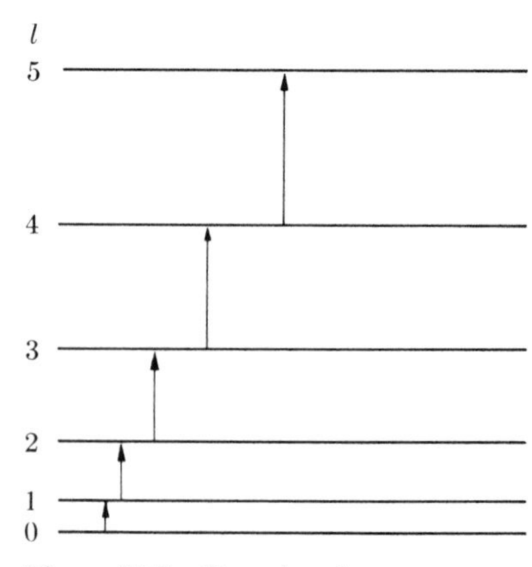

Figure 31.7 Rotational energy levels of a diatomic molecule.

be written in the alternative form

$$E_v = (n + \tfrac{1}{2})\hbar\omega_0$$

Therefore the vibrational energy levels of the molecule are equally spaced an amount $\hbar\omega_0$ (Figure 31.8) and the molecule has a **zero-point vibrational energy** equal to $\frac{1}{2}\hbar\omega_0$. Table 31.1 gives some values of $h\nu_0$.

The energy difference between adjacent levels n and $n + 1$ is $\Delta E_v = h\nu_0$, and therefore the frequency of the radiation emitted or absorbed in a transition between these two vibrational states is $\nu_0 = \Delta E_v/h$. This is the only vibrational frequency observed experimentally. That means that the only allowed vibrational transitions are between neighboring energy levels and that the selection rule for transitions among vibrational levels is $\Delta n = \pm 1$. Therefore, the only frequency absorbed or emitted in a vibrational transition is equal to the natural classical frequency ν_0, which is of the order of 10^{14} Hz, corresponding to an energy about 0.1 eV.

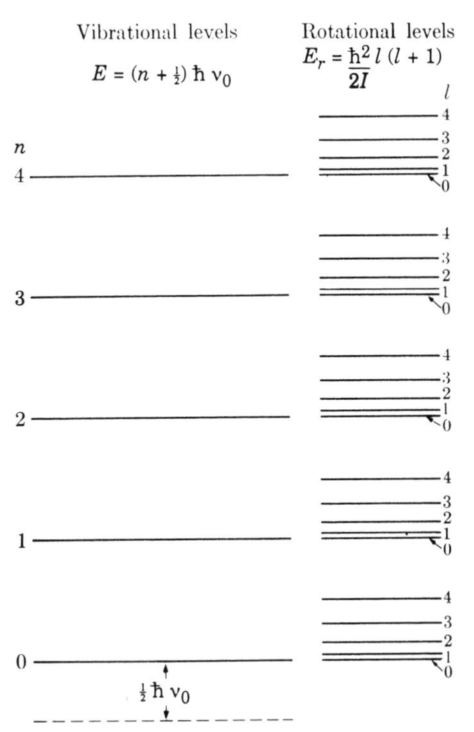

Figure 31.8 Vibrational and rotational energy levels of a diatomic molecule.

The vibrational frequencies of most diatomic molecules fall in the infrared region of the spectrum. For a vibrational transition to occur, either in emission or absorption, the diatomic molecule must have a permanent electric dipole moment so that it behaves like an oscillating electric dipole. Thus homonuclear molecules such as H_2 or N_2, which have no permanent electric dipole moments, do not show pure vibrational transitions. But polar molecules, such as HCl, show strong vibrational transitions.

The molecular energy due to both rotation and vibration is obtained by combining Equations 31.7 and 31.9. In general, the rotational energy is much smaller ($\sim 10^{-4}$ eV) than the vibrational ($\sim 10^{-1}$ eV), and we may say that, to each vibrational level, there correspond several rotational levels, as shown in Figure 31.9. When the selection rules

$$\Delta n = \pm 1 \qquad \text{and} \qquad \Delta l = \pm 1 \tag{31.10}$$

are taken into account for a transition between two rotational levels belonging to two adjacent vibrational levels, the rotation–vibration spectrum results. Some of the transitions are shown in Figure 31.9. For polyatomic molecules the rotational and vibrational spectra become much more complex.

(iii) Electronic spectra. Finally, we consider electronic transitions in molecules. A given molecule may have several electronic configurations or stationary

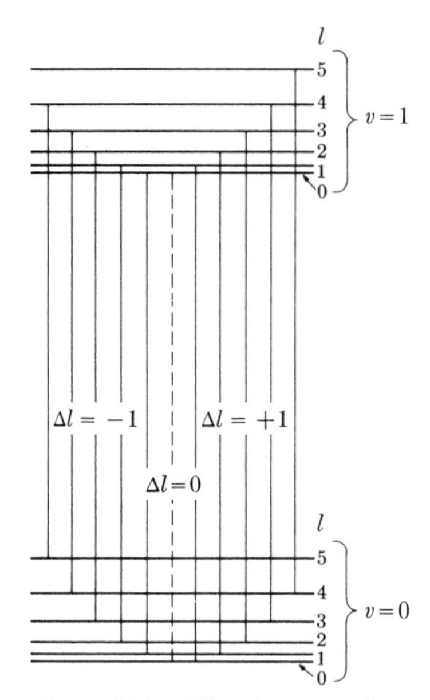

Figure 31.9 Vibration–rotation transitions in a diatomic molecule.

states, each with its own energy. The energy required to excite the electronic motion in molecules is of the same order as in atoms, that is, 1 to 10 eV. Thus when a molecule experiences an electronic transition, jumping from one electronic configuration to another, the radiation involved falls in the visible or the ultra-violet regions of the spectrum.

To a given electronic state there correspond many vibrational states and to each vibrational state there correspond several rotational states. As a first approximation, we may write the energy of the molecule in the form

$$E = E_e + E_v + E_r \tag{31.11}$$

where E_e refers to the electronic energy. We already used this expression in Section 15.7 in discussing the molecular energy of polyatomic gases. In an electronic transition all three energies may change. Therefore we must write the energy change in the electronic transition as $\Delta E = \Delta E_e + \Delta E_v + \Delta E_r$, where ΔE_e is the change in electronic energy, ΔE_v is the vibrational energy change, and ΔE_r is the rotational energy change. The frequency of the radiation emitted or absorbed in an electronic transition is therefore the sum of three terms,

$$\nu = \frac{\Delta E}{h} = \nu_e + \nu_v + \nu_r$$

where $\nu_e = \Delta E_e/h$ is due to the change in electronic energy, and $\nu_v = \Delta E_v/h$ and $\nu_r = \Delta E_r/h$ correspond to the changes in vibrational and rotational energies, respectively. The frequency ν_e is the largest. For a given electronic transition the spectra consist of a series of **bands**; each band corresponds to a given value of ν_v and all the possible values of ν_r (see Figure 30.3).

The spectrum of a molecule is one of the most important sources of information about its properties and structure.

31.6 Radiative transitions in solids

The band theory of solids, discussed in Chapter 23, accounts for most properties of solids in a consistent fashion. But is there any direct evidence for the existence of energy bands and energy gaps? Is it possible to measure the Fermi energy? The answer to these questions is yes. Perhaps the most direct evidence comes from radiative transitions in solids. We shall first discuss X-ray emission and then some absorption processes.

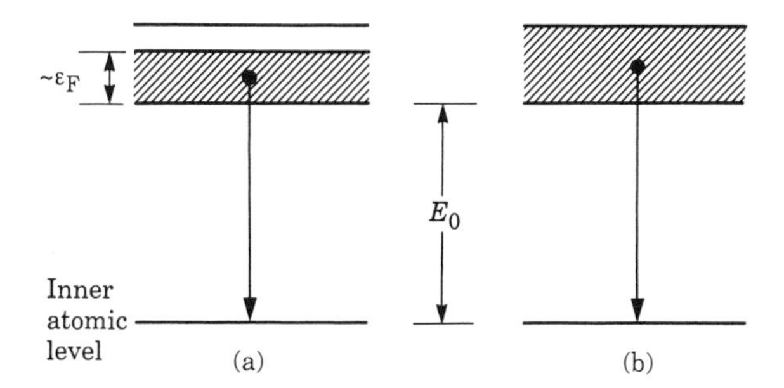

Figure 31.10 Electronic transition from the uppermost band into a vacant inner atomic state; (a) conductor, (b) insulator.

X-ray emission in atoms results when an electron occupying one of the outer shells falls into a vacant state left in one of the inner or low-lying inner shells. According to the band theory, we may say that X-ray emission in solids takes place when, for example, an electron in the uppermost band undergoes a transition into a lower-level band or into an inner atomic energy level where a vacant state exists, as indicated in Figure 31.10. These vacant states are produced by electron bombardment or by absorption of radiation.

In the atomic case a given transition corresponds to initial and final energy levels of well-defined energy, resulting in a photon of a certain energy or a sharp line of a given wavelength. But the band theory suggests a different situation in solids. The electron making the transition can start from any of the possible energy levels in the uppermost band and therefore the energy of the emitted photons has a spread of the order of the Fermi energy, ε_F (that is, the photons have energies between E_0 and $E_0 + \varepsilon_F$). In the case of an insulator, the energy spread is equal to the width of the band. Therefore, instead of a single X-ray line, the spectrum consists of a band of wavelengths. This, in fact, is what is observed experimentally.

Figure 31.11 indicates some typical situations of absorption. In (a) we have a conductor with the uppermost band B partially filled and separated by an energy gap from the empty band B'. Electrons in B can be excited into nearby states in the *same* band when they absorb photons with energies from zero up to the energy required to reach the top of the band. The other photons that can be absorbed are those that take an electron from B into the empty band B'. Therefore, the absorption spectrum consists of two groups of photons, with a gap for the energy region at which no photons can be absorbed. If bands B and B' overlap, as in Figure 31.11(b), no energy gap exists and a continuous absorption spectrum results. In the case of an insulator (Figure 31.11(c)), only transitions from the valence band B into the conduction band B' are possible. Therefore, to induce electronic transitions, the photons must have a minimum energy of a few eV. The resulting absorption spectrum is shown in Figure 31.11(c). This would also be the case for semiconductors, except that due to the smallness of their energy gap, some electrons occupy band B'; therefore, the absorption spectrum resembles the situation illustrated for case (b). Lattice defects, or impurities, such as in p- and n-type semiconductors, have important consequences. They introduce new energy levels, which may fall in the energy gap (Figure 31.11(d)). Electron transitions into these energy levels allow absorption of photons of much lower energies than are needed to go from B to B', and the absorption is as shown.

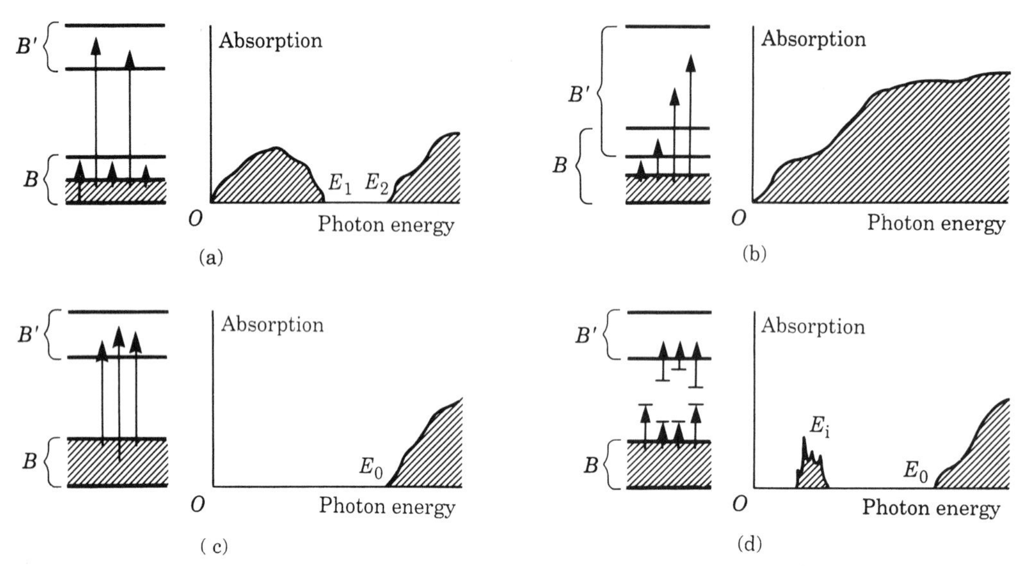

Figure 31.11 Absorption transitions in a solid. (a) and (b) are conductors, (c) is an insulator while (d) is an insulator with impurities.

Photons in the visible region of the electromagnetic spectrum have energies in the range of 1.6 eV up to 3.2 eV. A solid is transparent or opaque depending on its absorption properties in that energy range. Conductors and semiconductors are all opaque, and have absorption curves similar to those given in Figures 31.11(a) and (b). Pure insulators are transparent if E_0 in Figure 31.11(c) is larger than about 3.2 eV. But if, as a result of impurities, the peaks E_i in Figure 31.11(d) fall in the visible region, the insulator is colored (or even opaque). For this reason these impurities are called **color centers** or **F-centers** (from the German *farben*: color). For example, pure corundum (Al_2O_3) should be transparent, but ruby (which is Al_2O_3 with a small chromium impurity) shows a strong red color. This is because the chromium atoms induce a strong absorption in the green region of the spectrum, resulting in a red color for the solid when it is illuminated with white light.

Another important radiative property of solids is **luminescence**. In general, when the electrons in atoms, molecules or solids are excited by some means (e.g. absorption of radiation or electronic bombardment), there are several processes that compete to bring about de-excitation (e.g. radiative transitions and inelastic collisions). In some instances the favored process is radiative transition, and the substance glows when it is illuminated with radiation of the proper wavelength or is excited by some other means. Substances having this property are called luminescent.

Luminescence in solids is closely related to impurities and lattice defects. Figure 31.12 illustrates some processes that take place in luminescent solids. When an electron is removed from the valence band into the conduction band, a hole is left in the valence band (Figure 31.12(a)). In a perfectly pure and regular lattice, the electron usually returns to the valence band, although it may take some time to do so, since both electron and hole have great mobility and they may wander

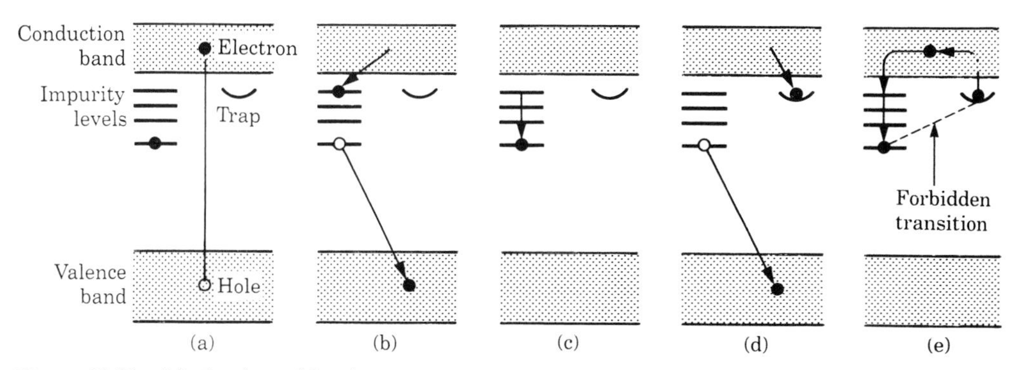

Figure 31.12 Mechanism of luminescence.

in different directions. However, if the lattice has some impurity that introduces energy levels in the forbidden region, an electron in a low-lying impurity level may fill the hole in the valence band, while the electron in the conduction band may fall into one of the (normally empty) high impurity energy levels, as shown in Figure 31.12(b). These transitions generally involve photons of small energy that do not fall in the visible region. Finally, an electron may fall from the high-energy impurity level to the empty low-energy one, emitting radiation of lower energy (or longer wavelength) than the incident radiation; this constitutes the luminescence (Figure 31.12(c)).

In some instances, instead of going through the process shown in Figure 31.12(b), the electron in the conduction band may fall into an energy level called a **trap**, from which a radiative transition to the ground state impurity energy level is forbidden (Figure 31.12(d)). In such a case, the electron is in a state similar to an atom or molecule in a metastable state; the trapped electron may wait until, by some mechanism, it is returned to the conduction band, after which it follows steps (b) and (c), as shown in Figure 31.12(e). Due to the time delay involved, which may amount to many seconds, the process is called **phosphorescence**. These substances are therefore called **phosphors**. One such substance is zinc sulfide. Phosphorescent materials are used in the screen of cathode-ray and TV tubes. In scintillation detectors that are used to detect γ-rays, the phosphor is NaI activated with thallium.

31.7 Spontaneous and stimulated radiative transitions

A bound system of electric charges, such as an atom, a molecule, a solid or a nucleus, in an excited energy level may fall to a lower energy level *spontaneously* (Figure 31.13(a)). Also, if radiation of the proper frequency is present, the system may be *induced* or *stimulated* to jump into the lower level (Figure 31.13(b)). In either case the system emits radiation (a photon) of frequency given by Bohr's condition, $\Delta E = h\nu$, where $\Delta E = E_2 - E_1$ is the energy difference between the two levels.

The transition of a system from a lower to a higher energy level with absorption of energy (Figure 31.13(c)) is also an induced transition in the sense that it can occur only in the presence of electromagnetic radiation of the proper

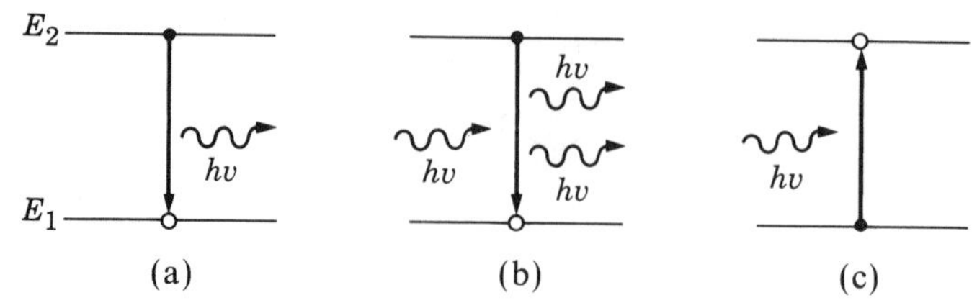

Figure 31.13 Radiative transitions: (a) Spontaneous emission; (b) induced emission; (c) induced absorption. In all cases $h\nu = E_2 - E_1$.

frequency, so that the system can absorb from the radiation a photon of energy given by Bohr's condition. Absorption transitions can also occur as a result of inelastic collisions, as explained in Example 23.2 and Section 31.2. This form of excitation occurs, for example, when charged particles, such as fast electrons, pass through a gas, as in an electric discharge or in cosmic rays falling on the Earth's atmosphere.

The probability of spontaneous emission of radiation depends only on the internal structure of the system. On the other hand, induced emission or absorption of radiation is the result of the action of the electromagnetic field of the incoming radiation on the system. Therefore the probability of an induced transition depends on the intensity of the radiation as well as on the internal structure of the system. Some transitions can be restricted or forbidden because of *selection rules*, which will be analyzed in Chapter 38. Excited states whose transitions to lower energy states are restricted are called **metastable** because the system may stay in the excited state for some time.

To a certain extent, induced transitions can be considered as forced oscillations of the system, and therefore bear a constant phase difference relative to the incoming radiation. That means that in the case of induced emission all systems radiate in phase with the incident radiation, resulting in the induced emission of monochromatic **coherent** radiation. On the other hand, spontaneous emission transitions occur at random, with no correlation between the times at which individual systems undergo transitions. Therefore the phases of the spontaneous emission transitions are distributed randomly, resulting in radiation that, though monochromatic, is **incoherent**.

In general, when several sources radiate in phase emitting coherent radiation, the intensity of the radiation is much larger than if the phases are distributed randomly and the radiation is incoherent. This is because in the case of coherent radiation the resultant amplitude of the electromagnetic field is the sum of the individual amplitudes; that is,

$$\text{Resultant coherent amplitude} = N \times \text{source amplitude}$$

where N is the number of sources. Since the intensity of the radiation is proportional to the square of the amplitude, we have that:

$$\text{Resultant coherent intensity} = N^2 \times \text{source intensity}$$

On the other hand, if the sources emit radiation incoherently it is the individual intensities that add linearly; that is:

$$\text{Resultant incoherent intensity} = N \times \text{source intensity}$$

Therefore, if the number of sources is large, as happens with atoms and molecules in a gas, the coherent stimulated radiation may be much more intense than the incoherent spontaneous radiation. This is the principle on which lasers and masers operate, as well as several interference arrays that will be discussed in Chapters 34 and 35.

31.8 Lasers and masers

When matter is in the presence of radiation of the proper frequency, all the processes described in the preceding section take place as atoms in the ground state are excited and atoms in excited states return to the ground state. Which processes are more important depend on the intensity of the radiation and on whether matter is in thermal equilibrium or not. If matter is in thermal equilibrium, the number N_2 of atoms in an excited state of energy E_2 is much smaller than the number N_1 of atoms in the ground state or in a lower energy state of energy E_1, that is $N_2 \ll N_1$ (recall Example 17.3). Therefore absorption dominates over emission of radiation.

But if, by some means, the occupation of an excited level in a piece of matter is increased appreciably at the expense of the population of the ground level, so that N_2 becomes much larger than the thermal equilibrium value, a situation called **population inversion**, the total emission rate may become larger than the absorption rate. In other words, if radiation passes through the material, the radiation that comes out as a result of induced emission has more photons of frequency $\nu = (E_2 - E_1)/h$ than the incident radiation, resulting in a **coherent amplification** of radiation of that frequency.

Since more atoms are de-excited by emission transitions than excited by absorption transitions, the upper energy level is quickly depleted, so that the radiation amplification decreases until the thermal equilibrium distribution is re-established. Thus, to sustain a constant amplification, it is necessary to continuously replenish the atoms in the upper level or remove atoms from the lower level by some means. The devices in which this is done are called **masers** and **lasers**. These are coined words or acronyms for *M*icrowave *A*mplification by *S*timulated *E*mission of *R*adiation and *L*ight *A*mplification by *S*timulated *E*mission of *R*adiation, depending on the region of the electromagnetic spectrum in which they operate. Lasers and masers use a resonating cavity to trap part of the radiation and use it to produce more stimulated emission transitions.

Laser and maser beams have several unique features: (i) they produce intense coherent monochromatic induced radiation; (ii) the incoherent spontaneous radiation component in the beams, called **noise**, is negligible; (iii) the beams are highly directional and collimated, experiencing only a very small angular dispersion as they propagate; (iv) the beams can be focused very sharply. These features make laser and maser beams very useful for several applications.

Maser and laser beams are used whenever very low noise and high directionality are of prime importance in signal processing, as in microwave spectrometry, satellite communication and radio astronomy. For example, the distance from the Earth to the Moon has been measured very precisely by sending a laser pulse to the Moon and measuring the time interval for receiving the reflected pulse. For the same reasons, lasers are useful in reproducing encoded digital information, as in audio and video discs.

Laser beams are also used whenever a strong and well focused beam of coherent radiation is required. This is why lasers are used in metallurgy for precision etching, drilling and cutting metals, and in medicine for minimal invasive procedures, as in eye surgery. By properly selecting beam width, pulse duration, frequency, intensity and other parameters, a laser beam can be used to weld, cut, seal, ablate and coagulate tissues. Biomedical applications of lasers include unclogging obstructed arteries, breaking up of kidney stones, clearing cataracts and even altering genetic material. By matching the frequency of the laser radiation with the absorption band of the target structure, an intense pulse of laser radiation can be absorbed selectively by the target, which is destroyed with minimal damage to surrounding tissues. This technique is called *selective photothermolysis.* In combination with optical fibers, laser beams can be directed to almost any part in the human body or used for long distance transmission of signals for communication purposes. Other applications of laser beams are to the accurate measurement of distances and time intervals. In this way it has been possible to analyze in great detail chemical processes that take less than 10^{-15} s, and that involve molecular collisions, the formation of an intermediate state and the emergence of new molecules. Last, but not least, laser technology has been applied to the restoration of artistic works.

Several means have been devised to overpopulate an upper or excited energy level in a steady fashion. All methods require expenditure of energy, and the efficiency of a maser or a laser is the ratio between the energy output and the energy input. Two frequently used methods are optical pumping and collision energy transfer.

EXAMPLE 31.4

Population inversion by optical pumping. The ruby laser.

▷ In **optical pumping**, radiation energy is supplied, either continuously or in bursts, to excite atoms from the ground state E_1 (Figure 31.14) to a state of higher energy E_3. Some excited atoms decay to an intermediate state of energy E_2, which is metastable; that is, the probability for a transition to the ground state is very low. Therefore these intermediate states may become highly populated relative to the ground state. Radiation of frequency $\nu = (E_2 - E_1)/h$ will stimulate transitions from E_2 to E_1, producing the desired coherent amplification.

A simple laser that operates on the principle of optical pumping is the **ruby laser**. As indicated in Section 31.6, ruby is a variety of a very hard mineral called corundum, which is a crystalline form of aluminum oxide (Al_2O_3), containing a small amount of chromium. The structure of the electronic energy levels of ruby is shown in simplified form in Figure 31.15. When ruby is illuminated with white light, it absorbs photons of energy in the green and blue regions of the spectrum. Electrons are thus excited to the energy bands E_3 and E_4. That is why ruby appears a bright red color. Some of the excited electrons release part of the energy to the crystal lattice and fall to the energy level E_2 which is relatively sharp and metastable. Therefore, level E_2 becomes overpopulated. When ruby is exposed to radiation of frequency $\nu = (E_2 - E_1)/h = 4.326 \times 10^{14}$ Hz or wavelength of 6.934×10^{-7} m, stimulated transitions from E_2 to the ground state occur.

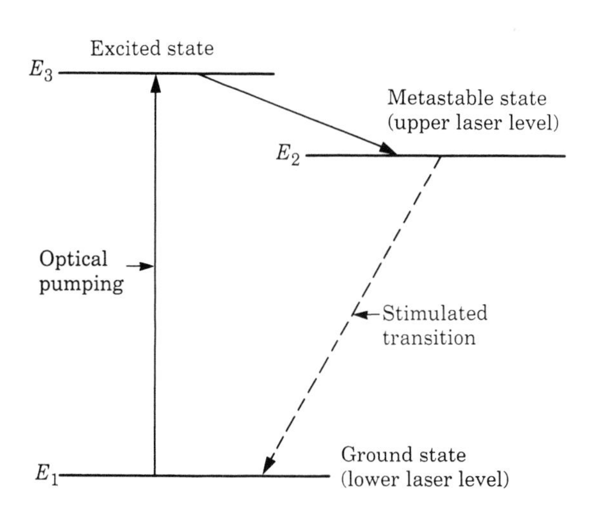

Figure 31.14 Three-level laser.

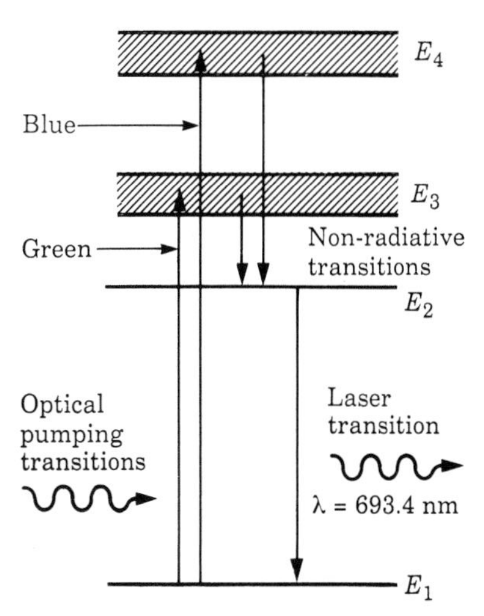

Figure 31.15 Energy level diagram for chromium ions in ruby. Absorption bands E_3 and E_4 are excited by optical pumping from the ground state E_1.

Figure 31.16 illustrates schematically the design of a pulsed ruby laser. A high intensity pulsed lamp is coiled around a ruby rod. The ends of the rod are polished and silvered, forming a pair of parallel mirrors. One end is a perfect reflector while the other is only partially reflecting. When the lamp is activated, emitting light pulses, some electrons are excited to bands E_3 and E_4 by absorption of photons from the pulse. Many of the excited electrons experience transitions to level E_2, increasing its population. In turn, electrons in level E_2 suffer induced transitions to the ground level E_1, resulting in a pulse of coherent light. If the length of the rod is adjusted properly so that it constitutes an optical resonating cavity, similar to an organ pipe for sound (see Section 34.5), tuned to the laser frequency or wavelength, the light pulse travels back and forth along the rod inducing many more emission transitions from E_2 to E_1. The result is a fast amplification of coherent light of frequency 4.326×10^{14} Hz. When the rate of de-excitation by induced emission matches the rate of excitation, determined by the driving power, the laser reaches maximum amplification and power output. Some of the laser light escapes through the partially transmitting end of the rod, constituting an intense and highly collimated laser beam.

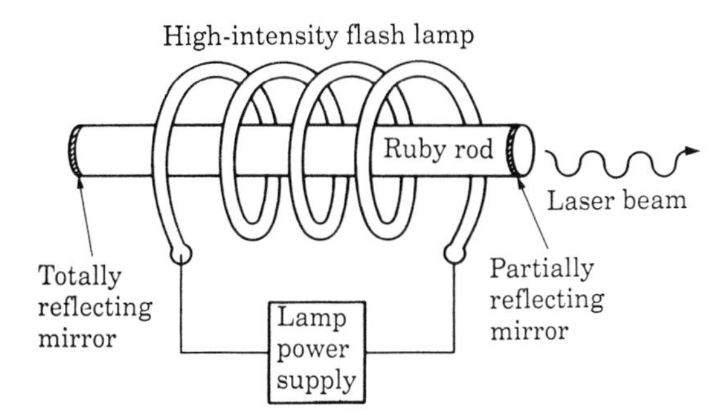

Figure 31.16 Schematic diagram of a ruby laser.

EXAMPLE 31.5

Population inversion by collision energy transfer. The helium–neon laser.

▷ **Collision energy transfer** consists in overpopulating an excited energy level by inelastic collisions. It is used to produce population inversion in the helium–neon laser. In Figure 31.17, some energy levels of He and Ne atoms are shown in a simplified form. Recalling Table 23.3, an He atom in its ground state has two electrons in the 1*s* level. To transfer one of the electrons to the 2*s* level, which is metastable, requires absorbing 20.61 eV. This is possible, for example, by inelastic collisions of fast electrons with the He electrons in the 1*s* level. An Ne atom has six electrons in the 2*p* level, which can be excited to nearby levels. In particular, the energy required to excite an electron from the 2*p* to the 5*s* level is 20.66 eV, which is practically the same as the excitation energy of He. Therefore if an excited He atom collides with an Ne atom, it may transfer its excitation energy to a 2*p* electron in Ne and excite it to the 5*s* level. The 5*s* electron cannot jump to the 3*s* and 4*s* levels because of the selection rules referred to in Section 31.7. However, the transition to the 3*p* level is possible with emission of a photon of 1.96 eV or wavelength of 6.328×10^{-7} m, which can be used to operate a laser.

Figure 31.18 shows schematically the components of an He–Ne laser. A mixture of He and Ne at a pressure of the order of 10^2 Pa (or 0.001 atm) is placed inside a tube with one end reflecting perfectly and the other only partially, as in the case of the ruby laser. The distance between mirrors is chosen so that it is a resonating cavity tuned to the laser wavelength. When an electric discharge is produced along the tube, the high-energy electrons in the discharge excite the He atoms in inelastic collisions. The He atoms quickly transfer their excitation energy to Ne atoms by collisions. Subsequently, the excited Ne atoms pass

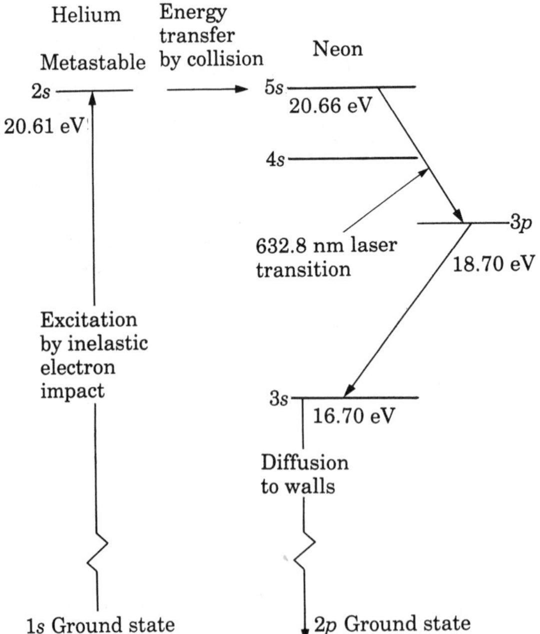

Figure 31.17 Energy level diagram for a helium–neon laser. Only the relevant energy levels are shown. The state of the excited electron is also indicated.

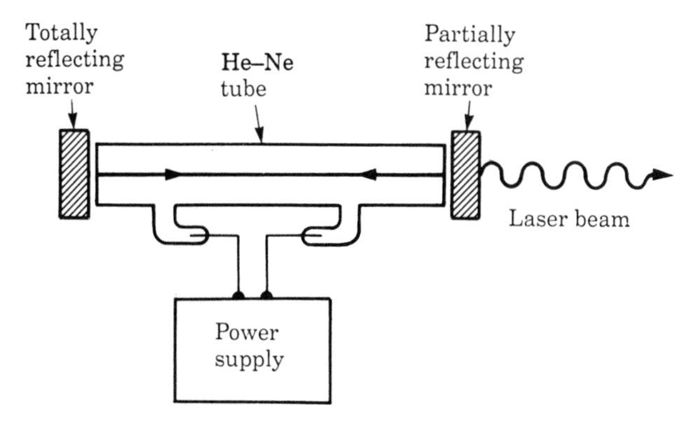

Figure 31.18 Schematic diagram of a helium–neon laser.

from the 5s to the 3p level emitting light of wavelength 6.328×10^{-7} m. This induces more transitions of excited Ne atoms from the 5s to the 3p level, resulting in a coherent amplification of light. Part of this light is transmitted out of the tube as a laser beam.

Note 31.1 Tuning of a laser

We have indicated that the active medium of a laser must be enclosed in an optical resonating cavity of the right size in order to enhance radiation of frequency equal to the laser frequency. This contributes to the production of a strong beam of highly monochromatic light. Resonating cavities will be analyzed in Section 34.4, but we may state that the frequencies of the standing waves propagating with velocity c along a tube of length L much larger than its cross section and with perfectly reflecting ends are given by

$$\nu = \frac{nc}{2L}$$

where n is an integer. In our case, c is the velocity of light, 3×10^8 m s^{-1}, and L is of the order of a few centimeters, so that n is of the order of 10^6 for optical frequencies ($\simeq 10^{15}$ Hz).

For the proper functioning of a laser, the length L must be adjusted so that one of the resonating frequencies ν coincides or is very close to the laser frequency ν_0, an operation called *tuning* the laser. It is similar to the tuning of musical instruments. It is important to determine the precision with which the length L must be adjusted to properly tune a laser. If the length L is known with an error ΔL, then differentiating the above expression for ν, the error in the resonating frequency is

$$\Delta\nu \simeq \frac{nc}{2L^2}\,\Delta L \qquad \text{or} \qquad \frac{\Delta\nu}{\nu} \simeq \frac{\Delta L}{L}$$

It must be recalled that spectral lines are not very sharp but have an intensity distribution as shown in Figure 31.19. The line width δ is the range of frequencies at half the intensity at the center of the line. It is due to several factors, such as the magnitude of the transition probability (Example 36.5) and the Doppler effect (Section 29.7). In the optical region, for

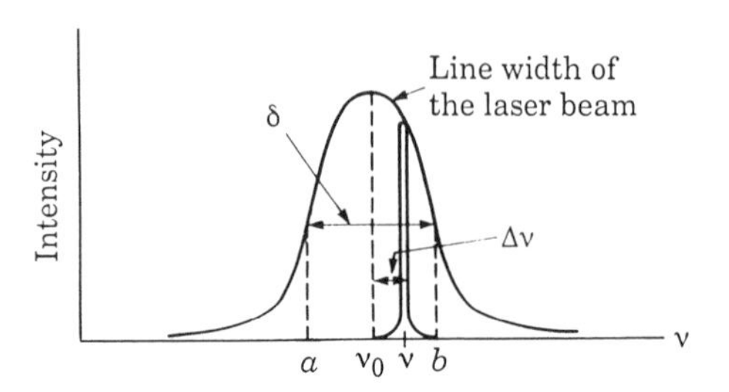

Figure 31.19 The laser frequency is ν_0, the laser beam width is δ and the resonating frequency of the laser cavity is ν. For laser operation $\Delta\nu < \frac{1}{2}\delta$.

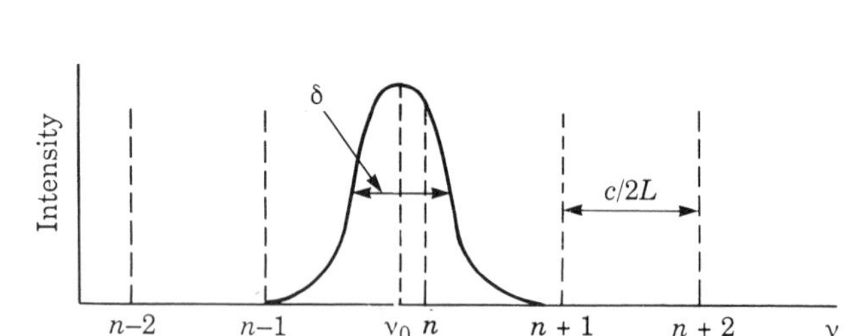

Figure 31.20 When the separation $c/2L$ is larger than δ, only one resonating frequency falls within the linewidth of the laser frequency. When $c/2L$ is smaller than δ more than one resonating frequency falls within the bandwidth.

which the laser frequency is of the order of 10^{15} Hz, the line width δ is of the order of 10^8 to 10^9 Hz. Therefore, since for laser operation the resonating frequency ν must be as close as possible to the laser frequency ν_0, the difference $\Delta\nu = \nu - \nu_0$ must be smaller than $\frac{1}{2}\delta$ or about 10^8 Hz. Then $\Delta\nu/\nu < 10^{-7}$ or $\Delta L < 10^{-7} L$, which gives the precision with which the cavity must be constructed.

A second requirement needed to obtain a highly monochromatic laser beam is that only one resonating frequency ν fall close enough to the laser frequency ν_0. This implies that the separation $\Delta\nu$ of consecutive resonating frequencies must be larger than the natural width δ. The difference between consecutive resonating frequencies is

$$\Delta\nu = \frac{(n+1)c}{2L} - \frac{nc}{2L} = \frac{c}{2L}$$

and therefore $c/2L > \delta$ or $L < c/2\delta$ (Figure 31.20). For example, for δ about 10^8–10^9 Hz, L must be smaller than one meter. In general, L is about a few centimeters.

31.9 Blackbody radiation

Consider a cavity whose walls are at a certain temperature. The atoms composing the walls are emitting electromagnetic radiation and at the same time they absorb radiation emitted by other atoms of the walls. The electromagnetic radiation field occupies the whole cavity. When the radiation trapped within the cavity reaches equilibrium with the atoms of the walls, the amount of energy emitted by the atoms per unit time is equal to the amount absorbed by them. Hence, when the radiation in the cavity is at equilibrium with the walls, the energy density of the electromagnetic field is constant.

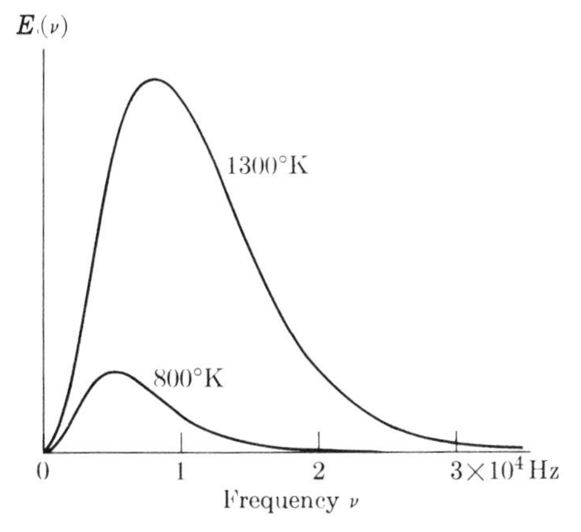

Figure 31.21 Monochromatic energy density of blackbody radiation at different temperatures as a function of the frequency.

Experiment has shown that, at equilibrium, the trapped electromagnetic radiation has a well-defined energy distribution; that is, to each frequency there corresponds an energy density that depends solely on the temperature of the walls and is independent of their material. The energy density corresponding to radiation with frequency between ν and $\nu + d\nu$ is written $E(\nu)\, d\nu$, where $E(\nu)$ is the energy density per unit frequency range, sometimes called **monochromatic energy density**. The observed variation of $E(\nu)$ with the frequency ν is illustrated in Figure 31.21 for two temperatures. It may be seen from the curves that for each temperature the energy density shows a pronounced maximum at a certain frequency or wavelength. Note that the frequency at which the energy density is maximum increases (or the wavelength decreases) as the temperature increases. This explains the change in color of a radiating body as its temperature varies.

If a small hole is opened in one of the walls of the cavity, some of the radiation escapes and can be analyzed without disturbing the thermal equilibrium in the cavity. The hole appears very bright when the body is at high temperatures and the intensity of the equilibrium radiation within the cavity is high, but it appears completely black at low temperatures, when the intensity of the equilibrium radiation is negligible in the visible region of the spectrum. For that reason the radiation coming out of the cavity was called **blackbody radiation** by those who analysed it in the 19th century.

The problem of finding what mechanism causes radiating atoms to produce the observed energy distribution of blackbody radiation led to the birth of the quantum theory. By the end of the 19th century all attempts to explain blackbody energy distribution using the concepts available at that time had failed completely. Max Planck (1858–1947) suggested that if the radiation in the cavity was in equilibrium with the atoms of the walls, there should be a correspondence between the energy distribution in the radiation and the energies of the atoms in the cavity. As a model for the radiating atoms, Planck assumed that

atoms behave like harmonic oscillators that absorbed or emitted radiation energy only in an amount proportional to their frequency ν.

If ΔE is the energy absorbed or emitted in a single process of interaction of an oscillator with electromagnetic radiation, then according to Planck's assumption $\Delta E = h\nu$, where h is a proportionality constant, the same for all oscillators regardless of their frequency. Hence, when an oscillator absorbs or emits electromagnetic radiation, its energy increases or decreases by an amount $h\nu$. Planck's assumption implies that

the energy of atomic oscillators is quantized.

That is, the energy of an oscillator of frequency ν can attain only certain values, which are (assuming that the minimum energy of the oscillator is zero), 0, $h\nu$, $2h\nu$, $3h\nu$, ... Thus, in general, the possible values of the energy of an oscillator of frequency ν are

$$E_n = nh\nu \tag{31.12}$$

where n is a positive integer. As we know, the energy of an oscillator is proportional to the square of its amplitude (Section 10.4) and, *a priori*, by properly adjusting the amplitude of the oscillations, we can make an oscillator of a given frequency have any arbitrarily chosen energy. Therefore Planck's idea was an *ad hoc* assumption that could be justified only because it 'worked', and because at the time (1900) there was no better explanation. Now it is recognized that the quantization of some physical quantities is a fundamental fact of nature.

By applying some considerations of a statistical nature, using the Maxwell–Boltzmann distribution law $N_E = Ae^{-E/kT}$ (Equation 17.6), together with $E_n = nh\nu$, Planck obtained the energy density in blackbody radiation as

$$E(\nu) = \frac{8\pi h\nu^3}{c^3} \frac{1}{e^{h\nu/kT} - 1} \tag{31.13}$$

This expression, which agrees surprisingly well with the experimental values of $E(\nu)$ at different temperatures, has been accepted as the correct expression for blackbody radiation. It is called **Planck's radiation law**.

In Equation 31.13 the arbitrary constant h was introduced by Planck, which for that reason is called **Planck's constant**. The value, obtained by Planck by making Equation 31.13 fit the experimental results for $E(\nu)$, was $h = 6.6256 \times 10^{-34}$ J s, which agrees with the values obtained years later from the photoelectric and Compton effects and from atomic and molecular spectra.

Actually Planck first obtained Equation 31.13 empirically as the mathematical expression that best fit the experimental energy distribution shown in Figure 31.21, and after that sought a theoretical explanation based on energy quantization. Later on, Einstein and others refined Planck's derivation (Note 31.2).

EXAMPLE 31.6

Cosmic background radiation.

▷ An interesting application of Planck's blackbody radiation equation has been to the analysis of what is called the **cosmic background radiation**. In 1961 two physicists, Arno Penzias and Robert Wilson, who were testing a very sensitive microwave detector at the Bell Telephone Laboratories in the United States, made an unexpected discovery. They found that, no matter what direction they pointed their detector, there was a background radiation that always had the same intensity and spectral distribution. The radiation was even the same whether the measurements were made during the day or at night. Later on it was verified that the radiation was the same all year round. This showed that the radiation was independent of the rotation and the orbital motion of the Earth and that it must therefore come from outside the solar system or even outside the galaxy. The spectral distribution very closely resembles blackbody radiation that corresponds to a temperature of about

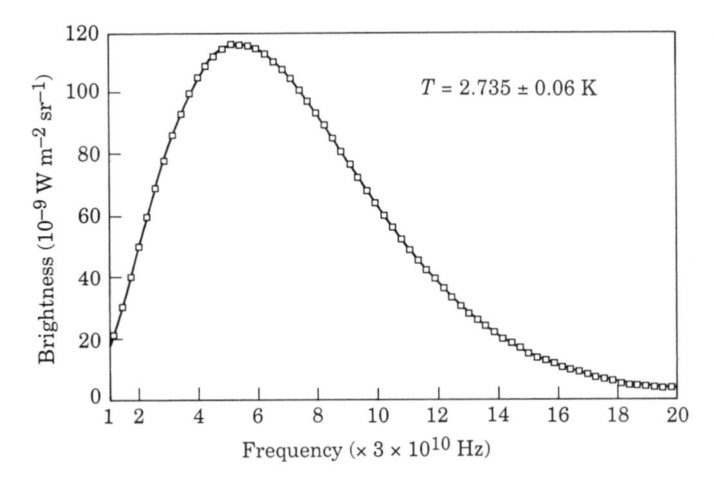

Figure 31.22 Cosmic background radiation spectrum measured by the Far Infrared Absolute Spectrometer (FIRAS) aboard the Cosmic Background Explorer (COBE) Satellite. The data has been fitted to a Planck blackbody curve at a temperature of 2.735 K.

2.7 K. Thanks to the pioneering work of Robert Dicke, James Peebles and many others, it is now accepted that the 2.7 K radiation pervades the whole universe and is the remnant of the electromagnetic radiation that existed shortly after the Big Bang, when the universe was extremely hot (see Note 41.4). Due to the expansion of the universe and accompanying cooling, the original electromagnetic radiation has shifted toward the red, or longer, wavelengths or, equivalently, to lower temperatures.

Until recently, it has been difficult to measure, with any precision, the full spectrum of the cosmic background radiation, particularly at short wavelengths (i.e. high frequencies), because of the scattering of radiation by the Earth's atmosphere. However, using instruments aboard Earth satellites, it is possible to eliminate the effects of the atmosphere. Figure 31.22 shows the spectrum of the cosmic background radiation as measured by the Far Infrared Absolute Spectrometer (FIRAS) aboard the Cosmic Background Explorer (COBE) satellite that was launched in November 1989. It took FIRAS only 9 minutes of observation time, in a direction near the galactic north pole, to obtain these data. The data have been fitted to a Planck blackbody spectrum corresponding to 2.735 ± 0.06 K. The fit is excellent at all wavelengths (or frequencies). This is one example of the importance of space research.

Actually, the cosmic background radiation is not exactly isotropic but shows a slight asymmetry in the direction of the Virgo cluster of galaxies, which is attributed to a Doppler effect due to the motion of our galaxy in that direction with a velocity of 3×10^5 m s^{-1}. In this respect, it is possible to define a 'preferred' frame of reference: one in which the cosmic background radiation is exactly isotropic.

Note 31.2 Analysis of spontaneous and stimulated transitions

Consider a substance whose atoms or molecules can be in states of energies E_1 and E_2. For example, E_1 may correspond to the ground state and E_2 to an excited state. Suppose that radiation of frequency $\nu = (E_2 - E_1)/h$ falls on the substance. As explained in Section 31.7, atoms in state E_1 may absorb photons of energy $h\nu$ and pass to the higher energy state E_2. Also, atoms in the state E_2 may emit photons of energy $h\nu$ and pass to state E_1 either spontaneously or stimulated by the radiation. If N_1 and N_2 are the number of atoms in states E_1 and E_2 at a given time, we may write

$$\begin{pmatrix}\text{Rate of energy}\\ \text{absorption}\end{pmatrix} \simeq N_1 \times \begin{pmatrix}\text{stimulated absorption}\\ \text{probability}\end{pmatrix}$$

$$\begin{pmatrix}\text{Rate of energy}\\ \text{emission}\end{pmatrix} \simeq N_2 \times \left[\begin{pmatrix}\text{spontaneous emission}\\ \text{probability}\end{pmatrix} + \begin{pmatrix}\text{stimulated emission}\\ \text{probability}\end{pmatrix}\right] \tag{31.14}$$

The **stimulated** absorption and emission probabilities are proportional to the intensity of the radiation. On the other hand, the **spontaneous** emission probability is independent of whether or not radiation is present.

If the substance is in thermal equilibrium, the occupation number N_2 is much smaller than N_1. Therefore, unless the intensity of the radiation is very high or the population of the higher energy state is increased by some means, as explained in Section 31.8, absorption predominates over emission of radiation. There is a particular intensity of the radiation for which it is possible that absorption matches emission without disturbing thermal equilibrium.

When matter and radiation are in thermal equilibrium, no net absorption or emission of radiation occurs because on the average the total number of emission and absorption transitions in a given time are the same. That is, if a photon of the incident radiation is absorbed by an atom, passing to a higher energy state, another atom in the same excited state soon emits a photon of the same energy, as a result of a spontaneous or a stimulated emission transition. That means that on the average the occupation numbers N_1 and N_2 remain constant. Combining expression 31.14 for the rates of absorption and emission, the thermal equilibrium between matter and radiation can be expressed by the following relation:

$$N_2 \times \left[\begin{pmatrix}\text{spontaneous emission}\\ \text{probability}\end{pmatrix} + \begin{pmatrix}\text{stimulated emission}\\ \text{probability}\end{pmatrix}\right] = N_1 \times \begin{pmatrix}\text{stimulated absorption}\\ \text{probability}\end{pmatrix} \tag{31.15}$$

If the substance is in thermal equilibrium, the occupation numbers N_1 and N_2 are determined by the Maxwell–Boltzmann distribution law, Equation 17.10. Since the frequency of the radiation emitted or absorbed is given by $h\nu = (E_2 - E_1)$, we have

$$\frac{N_2}{N_1} = e^{-(E_2 - E_1)/kT} = e^{-h\nu/kT}$$

Inserting this relation in Equation 31.15, we obtain

$$e^{-h\nu/kT} \times \left[\begin{pmatrix}\text{spontaneous emission}\\ \text{probability}\end{pmatrix} + \begin{pmatrix}\text{stimulated emission}\\ \text{probability}\end{pmatrix}\right] = \begin{pmatrix}\text{stimulated absorption}\\ \text{probability}\end{pmatrix}$$

which expresses the condition for equilibrium between matter and radiation of frequency ν at the temperature T. Using the techniques of quantum mechanics (Chapter 36), it is possible to show that the stimulated emission and absorption probabilities are equal, a fact that was first recognized by Einstein in 1916. Then the above equation can be written in the alternate form

$$\frac{\text{Spontaneous emission probability}}{\text{Stimulated emission probability}} = e^{h\nu/kT} - 1 \tag{31.16}$$

This relation shows that at the high frequency end of the spectrum of blackbody radiation, for which the photon energy $h\nu$ is much larger than the thermal energy kT ($h\nu/kT \gg 1$), spontaneous emission is more important than stimulated emission, while the reverse occurs for the low frequency end of the spectrum, for which $h\nu/kT \ll 1$.

Relation 31.16 was used by Einstein to derive the energy distribution of radiation in thermal equilibrium with matter at temperature T, without any specific assumption about atomic oscillators, as Planck originally did. To simplify the writing we shall designate by A_{12} the spontaneous emission probability between states 1 and 2. The stimulated emission probability is proportional to the energy density $E(\nu)$ of the radiation corresponding to frequencies between ν and $\nu + \mathrm{d}\nu$. Then we may write it as $B_{12}E(\nu)$ and Equation 31.16 becomes

$$\frac{A_{12}}{B_{12}E(\nu)} = \mathrm{e}^{h\nu/kT} - 1 \quad \text{or} \quad E(\nu) = \frac{A_{12}/B_{12}}{\mathrm{e}^{h\nu/kT} - 1} \tag{31.17}$$

By calculating the ratio A_{12}/B_{12}, which is a bit too complex to be done here, Einstein obtained Planck's radiation law, Equation 31.13, for the energy distribution in blackbody radiation; that is,

$$E(\nu) = \frac{8\pi h\nu^3}{c^3} \frac{1}{\mathrm{e}^{h\nu/kT} - 1} \tag{31.18}$$

From Equation 31.18 it is possible to verify that the maximum energy density occurs at a wavelength such that

$$\lambda_m T = 2.8978 \times 10^{-3}\ \mathrm{m\,K}. \tag{31.19}$$

This result is known as **Wien's law**. Thus by measuring λ_m the temperature of the blackbody can be estimated.

Incidentally, the total energy in the blackbody radiation is given by

$$U = \int_0^\infty E(\nu)\,\mathrm{d}\nu = \int_0^\infty \frac{8\pi h}{c^3} \frac{\nu^3\,\mathrm{d}\nu}{\mathrm{e}^{h\nu/kT} - 1}$$

Making $x = h\nu/kT$ we may write

$$U = aT^4 \tag{31.20}$$

where

$$a = \frac{8\pi hk^4}{c^3} \int_0^\infty \frac{x^3\,\mathrm{d}x}{\mathrm{e}^x - 1} = 7.5643 \times 10^{-16}\ \mathrm{J\,m^{-3}\,K^{-4}}$$

Equation 31.20 shows that the energy of the blackbody radiation increases very rapidly with the temperature. Expression 31.20 constitutes the **Stefan–Boltzmann law**, which has been confirmed extensively by measurements of the radiation emitted by hot bodies. This law also allows us to estimate the temperature of hot bodies such as a furnace, the Sun or a star by approximating them to a blackbody.

QUESTIONS

31.1 What conservation laws must be fulfilled in a transition in which an atom, a molecule or a nucleus absorbs or emits electromagnetic radiation?

31.2 Explain why the energy of a photon emitted or absorbed by a system (atom, molecule or nucleus) is not exactly equal to the difference in energy between the two stationary states involved.

31.3 A beam of electrons is passed through a monatomic gas. Is there a threshold energy for the inelastic electron–atom collisions?

31.4 Discuss the possible processes that may occur, as a function of photon energy, when (a) an atom, (b) a molecule, (c) a nucleus interacts with a photon.

31.5 What processes require the notion of stationary states for their explanation?

31.6 How do impurities affect the color of a solid?

31.7 Discuss possible mechanisms for the transition of a system to a state of higher energy. What condition must be satisfied?

31.8 Explain the difference between spontaneous and stimulated emission of radiation.

31.9 Why is radiation by stimulated emission coherent while spontaneous radiation is incoherent?

31.10 What is the purpose of the resonating cavity in a laser?

31.11 Discuss the main features of lasers that make them useful. Investigate some of the applications in medicine.

31.12 Explain how the color of a hot body should change as the temperature increases.

31.13 Why may the background cosmic radiation serve for defining a 'preferred' inertial frame of reference?

31.14 Repeat the diagram of Figure 31.20 when $c/2L < \nu_0$. How does this affect the quality of the laser beam?

31.15 What special assumptions were made by Planck to explain the energy distribution of blackbody radiation?

PROBLEMS

31.1 A minimum energy of 11 eV is required to separate the carbon and oxygen atoms in the CO (carbon monoxide) molecule. Calculate the minimum frequency and maximum wavelength of the electromagnetic radiation required to dissociate the molecule.

31.2 Determine the frequency and the wavelength of the photons absorbed by the following systems: (a) a nucleus absorbing 10^3 eV, (b) an atom absorbing 1 eV, (c) a molecule absorbing 10^{-2} eV.

31.3 Sodium atoms absorb or emit electromagnetic radiation of 5.9×10^{-7} m, corresponding to the yellow region of the visible spectrum. Determine the energy and momentum of the photons that are absorbed or emitted.

31.4 A photon having an energy of 10^4 eV is absorbed by a hydrogen atom at rest. As a result, the electron is ejected in the same direction as the incident radiation. Neglecting the energy required to separate the electron (about 13.6 eV), find the momentum and the energy of (a) the electron and of (b) the proton.

31.5 Compute the wavelength difference between the H_α (that is, $n = 3 \rightarrow n = 2$) lines of hydrogen, deuterium and tritium which results from the mass difference of these atoms.

31.6 (a) Which of the lines of the spectrum of hydrogen fall in the visible region of the spectrum (between 4000 Å and 7000 Å)? (b) Which lines of He^+ fall in the same region? (c) How could you distinguish whether there was hydrogen mixed in with a helium sample?

31.7 In the upper atmosphere molecular oxygen is dissociated into two oxygen atoms by photons from the Sun. The maximum photon wavelength that causes this process is 1.75×10^{-7} m. Calculate the binding energy of O_2.

31.8 The binding energy of an inner electron in lead is 9×10^4 eV. When lead is irradiated with a certain electromagnetic radiation and the photoelectrons enter a magnetic field of 10^{-2} T, they describe a circle of radius 0.25 m. Compute (a) the momentum and energy of the electrons and (b) the energy of the photons absorbed.

31.9 The following K_α lines have been measured:

magnesium:	9.87 Å	cobalt:	1.79 Å
sulfur:	5.36 Å	copper:	1.54 Å
calcium:	3.35 Å	rubidium:	0.93 Å
chromium:	2.29 Å	tungsten:	0.21 Å

(a) Plot the square root of the K_α frequency against the atomic number of the element. H. G. Moseley, found, in 1912, an empirical relationship of the form $(\nu)^{1/2} = A(Z - c)$. (b) From your plot, verify this relation and estimate the values of A and c.

31.10 Calculate the wavelengths and energies for the K_α X-ray lines of aluminum, potassium, iron, nickel, zinc, molybdenum and silver. Use the Moseley relation of the previous problem. Compare with observed values.

31.11 The K_α line for cobalt is 1.785 Å. (a) Calculate the energy difference between the 1*s*- and 2*p*-orbits in cobalt. (b) Compare with the energy difference between the 1*s*- and 2*p*-orbits in hydrogen (i.e. the first Lyman line). (c) Why is the difference much larger for cobalt than for hydrogen?

31.12 (a) Calculate the energy and wavelength of the photon absorbed when a $^{200}Hg^{35}Cl$ molecule ($r_0 = 2.23$ Å) makes the rotational transitions $l = 0 \rightarrow l = 1$ and $l = 1 \rightarrow l = 2$. (b) In what region of the electromagnetic spectrum are these lines found?

31.13 The adjacent lines in the pure rotational spectrum of $^{35}Cl^{19}F$ are separated by a frequency of 1.12×10^{10} Hz. Calculate the interatomic distance of the molecule. (Hint: recall Example 13.9.)

31.14 (a) Calculate the energy of the three lowest vibrational levels in HF, given that the force constant is 9.7×10^2 N m^{-1}. (b) Calculate the wave number of the radiation absorbed in the transition $n = 0 \rightarrow n = 1$.

31.15 The infrared spectrum of CO, at low resolution, shows an absorption band centered at a wavelength of 4.608×10^{-6} m. (a) Find the force constant in CO. (b) Plot, to scale, the potential energy curve.

31.16 What is the force constant for the HCl molecule, given that the vibrational frequency is 9×10^{13} Hz? Also find the zero-point energy.

31.17 The three vibrational frequencies of CO_2 are 2.002×10^{13} Hz, 4.165×10^{13} Hz and 7.048×10^{13} Hz. Make a sketch of the first few vibrational energy levels of this molecule.

31.18 Find the recoil energy and recoil velocity of a hydrogen atom when it suffers a transition from the state $n = 4$ to the state $n = 1$, emitting a photon. From the result, justify the validity of the italicized statement made near the end of Section 31.2 about recoil energy.

31.19 Estimate the order of magnitude of the standing waves in a laser when the length of the resonating cavity is about 1 m and the wavelength is (a) 3.30×10^{-6} m (infrared; Ne laser), (b) 3.371×10^{-7} m (ultraviolet; N_2 laser), (c) 6.328×10^{-7} m (visible; He laser).

31.20 A laser has mirrors 20 cm apart. The natural width of the laser emission line is 10^8 Hz. Calculate how many resonating frequencies respond to the laser action.

31.21 The natural width of the Ne line, with a wavelength 6.328×10^{-7} m, is 6×10^{-11} m. The resonating cavity of an He–Ne laser has a length of 30 cm. (a) Determine the resonating frequencies of the cavity that will respond to the laser action. (b) Calculate the optimum value of *n* for this laser. (Recall that $\Delta\nu/\nu = -\Delta\lambda/\lambda$.)

31.22 Assuming that the Sun is a spherical blackbody with a radius of 7×10^8 m, calculate the Sun's temperature and the radiation density within it. The intensity of the Sun's radiation at the surface of the Earth (which is 1.5×10^{11} m distant from the Sun) is 1.4×10^3 W m^{-2}. Are the numbers realistic? Explain.

31.23 Write the asymptotic form of Planck's radiation law (Equation 31.13), for a case (a) where the frequency is very high ($h\nu \gg kT$) and (b) where it is very low ($h\nu \ll kT$). The first relation is called **Wien's radiation law** and the second is called the **Rayleigh–Jeans radiation law**.

31.24 Verify that when the recoil energy of the atom is taken into account, Equation 31.1 must be replaced by the following relation: $E_i - E_f = h\nu(1 \pm h\nu/2Mc^2)$, where *M* is the mass of the atom and the + (−) sign applies in emission (absorption) of a photon of energy $h\nu$. Verify also that when $h\nu \ll 2Mc^2$ and $\Delta E = E_i - E_f$ the above relation can be written as $h\nu = \Delta E(1 \mp \Delta E/2Mc^2)$.

31.25 Calculate the shortest 'bremsstrahlung' wavelength observed when an electron accelerated through a potential difference of 40 kV is suddenly stopped at the anode of an X-ray tube.

32 Reflection, refraction and polarization

When light enters a thin transparent filament or fiber, it suffers many internal total reflections at the surface of the fiber while propagating along the fiber until it emerges at the other end. Optical fibers allow the transmission of light over long distances, following curved paths and with little absorption of its energy. By combining optical fibers in bundles, it is possible to transmit an image. (Photograph courtesy of ISA/SPEX Industries, Inc.)

32.1 Introduction

The velocity of propagation of waves depends on some physical properties of the medium through which the waves propagate. For example, the velocity of elastic waves depends on an elastic modulus and the density of the medium. The velocity of electromagnetic waves depends on the permittivity and permeability of the substance through which they propagate.

The dependence of the velocity of propagation of a wave on the properties of the medium gives rise to the phenomena of **reflection** and **refraction**, which occur when a wave crosses a surface separating two media where the wave propagates with different velocities. The **reflected wave** is a new wave that propagates back into the medium through which the initial wave was propagating. The **refracted wave** is the wave transmitted into the second medium. The energy of the incident wave is divided between the reflected and the refracted waves. In many instances it is the reflected wave that receives more energy, as is true of reflection by mirrors. In other cases, the refracted wave carries most of the energy.

32.2 Rays and wave surfaces

Referring to Figure 32.1, we note that we may draw a series of lines L, L', L'',... perpendicular to the wave surfaces S, S', S'',.... These lines are called **rays** and correspond to the lines of propagation of the energy and momentum of the wave. We must observe that the geometric relation between rays and wave surfaces is similar to the relation between lines of force and equipotential surfaces. Points on different wave surfaces joined by a given ray, such as a, a', a'' or b, b', b'' are called **corresponding points**. The time required by the wave to go from surface S to surface S'' must be the same measured along any ray. From this statement we conclude that the distances aa'', bb'', cc'' etc. must depend on the velocity of the wave motion at each point.

In a homogeneous isotropic medium, where the velocity is the same at all points and in all directions, the separation between two wave surfaces must be the same for all corresponding points. Another important fact is that

in a homogeneous isotropic medium, the rays must be straight lines,

because symmetry suggests that there is no reason for them to bend to one side or another. We have illustrated this fact for plane and for spherical waves in parts (a) and (b) of Figure 32.2.

Consider now the case where a wave propagates through a succession of different homogeneous isotropic media. At the crossing of each interface separating two adjacent media, the direction of propagation may change (that is, the rays may change direction) as a result of a modification of the wave surfaces; but while the waves are going through a given medium the rays will still be straight lines perpendicular to the wave surfaces. Let S (Figure 32.3) be one wave

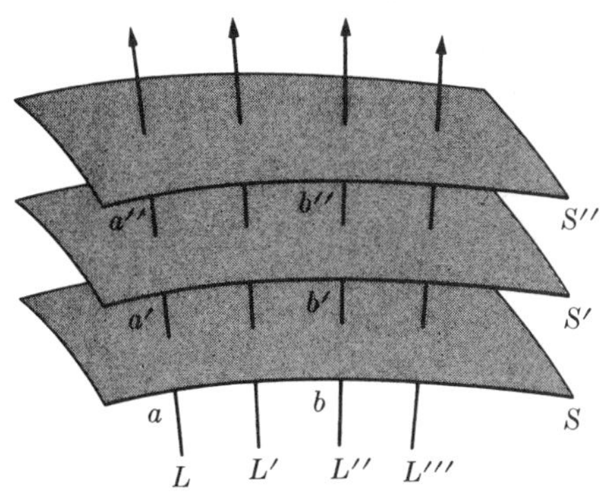

Figure 32.1 Rays and wave surfaces.

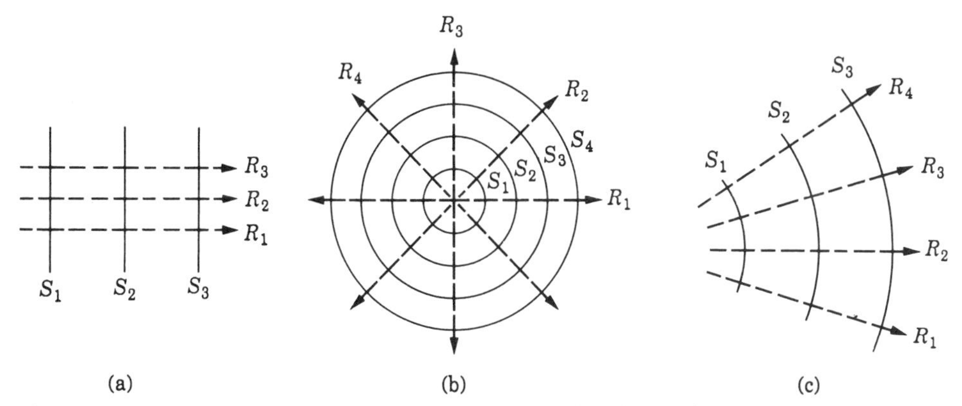

Figure 32.2 Plane waves, spherical waves, and waves of arbitrary shape, with some of their rays. S: wave surfaces. R: rays.

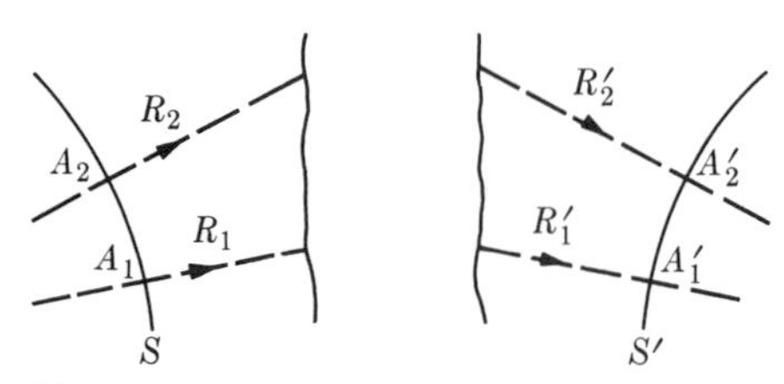

Figure 32.3 Corresponding rays in incident and outgoing waves.

surface in the first medium. Then we may trace two rays, R_1 and R_2, which are perpendicular to S. Successive wave surfaces in that medium must be perpendicular to R_1 and R_2. If, after the wave motion passes through all different media, we observe a new wave surface S', we find that the rays R'_1 and R'_2, which are the continuation of R_1 and R_2, are also perpendicular to S'. In other words, the relation of orthogonality between rays and wave surfaces is conserved throughout all the process of wave propagation.

32.3 Reflection and refraction of plane waves

Consider a plane wave propagating in medium (1) in the direction of the unit vector $\boldsymbol{u}_i$ (Figure 32.4). Experience indicates that, when the wave reaches the plane surface AB that separates medium (1) from medium (2), a wave is transmitted into the second medium and another plane wave is reflected back into medium (1). These are the *refracted* and *reflected* waves, respectively. When the direction of incidence is oblique to the surface AB, observation shows that the refracted waves propagate in a direction indicated by unit vector $\boldsymbol{u}_r$, which is different from $\boldsymbol{u}_i$, and the reflected waves propagate in a direction, indicated by unit vector $\boldsymbol{u}'_r$, which is symmetric to $\boldsymbol{u}_i$ with respect to the surface AB. Figure 32.5 indicates the corresponding situation for the incident, reflected and refracted rays. The angles θ_i, θ_r and θ'_r that the rays make with the normal N to the surface are related by the following experimentally verified laws:

1. *The directions of incidence, refraction, and reflection are all in one plane, which is normal to the surface of separation and therefore contains the normal N to the surface.*
2. *The angle of incidence is equal to the angle of reflection. That is,*

$$\theta_i = \theta'_r \tag{32.1}$$

3. *The ratio between the sine of the angle of incidence and the sine of the angle of refraction is constant.* This is called ***Snell's law***, and is expressed by

$$\frac{\sin\theta_i}{\sin\theta_r} = n_{21} \tag{32.2}$$

The constant n_{21} is called the **index of refraction** of medium (2) *relative* to medium (1). Its numerical value depends on the nature of the wave and on the properties of the two media.

These laws remain valid when neither the wave surface nor the interface are plane because, at each point, there is a limited section of either surface that can be considered as plane and the rays at that point behave according to Equations 32.1 and 32.2.

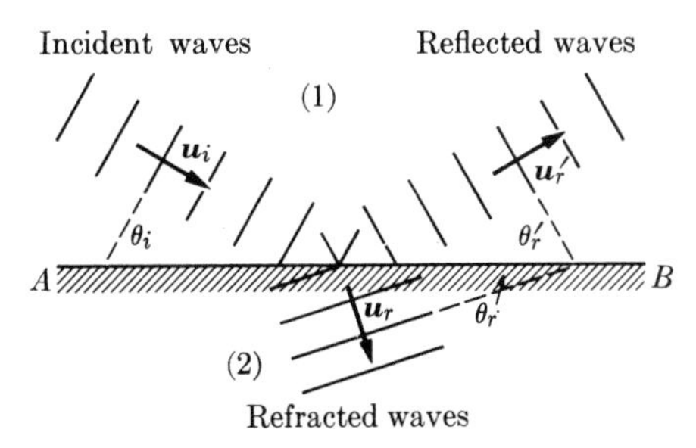

Figure 32.4 Incident, reflected, and refracted plane waves.

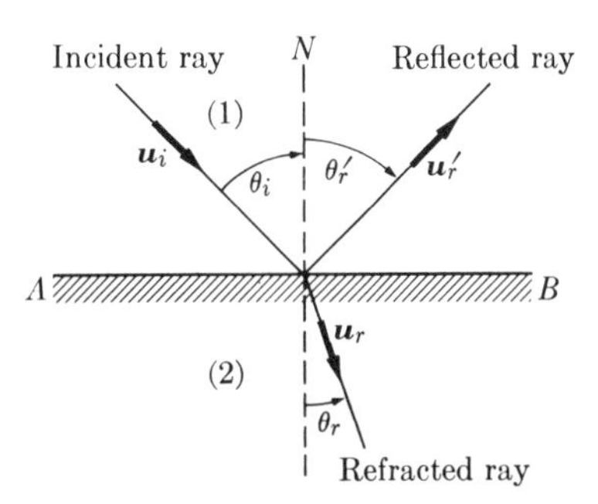

Figure 32.5 Incident, reflected and refracted rays.

Note that when $n_{21} > 1$, then $\sin \theta_i > \sin \theta_r$ and therefore $\theta_i > \theta_r$, indicating that the refracted ray is closer to the normal than the incident ray (Figure 32.6(a)) while if $n_{21} < 1$, then $\theta_i < \theta_r$ and the refracted ray is farther from the normal than the incident ray (Figure 32.6(b)).

When $n_{21} < 1$, there is an angle of incidence θ_i for which $\sin \theta_i = n_{21}$. Then Equation 32.2 gives $\sin \theta_r = 1$ or $\theta = \pi/2$, indicating that the refracted ray is parallel to the surface (Figure 32.7). The angle θ_i is then called the **critical angle** and is designated by λ. Therefore

$$\sin \lambda = n_{21} \qquad (n_{21} < 1) \tag{32.3}$$

If $\theta_i > \lambda$ or $\sin \theta_i > n_{21}$, then $\sin \theta_r > 1$, which is impossible. Therefore when $n_{21} < 1$ and the angle of incidence is larger than the critical angle there is no refracted ray, only a reflected ray. We then say that there is **total reflection**. This situation may come about, for example, when light passes from glass or water to air. Strictly speaking, there is a wave propagating in the second medium parallel to the surface, but the amplitude of the wave decreases very rapidly with distance and the wave is confined to a very thin layer along the surface.

Total reflection has several applications in the design of optical instruments, but perhaps the most important is to the transmission of signals in **optical fibers**, to be discussed in Section 34.8.

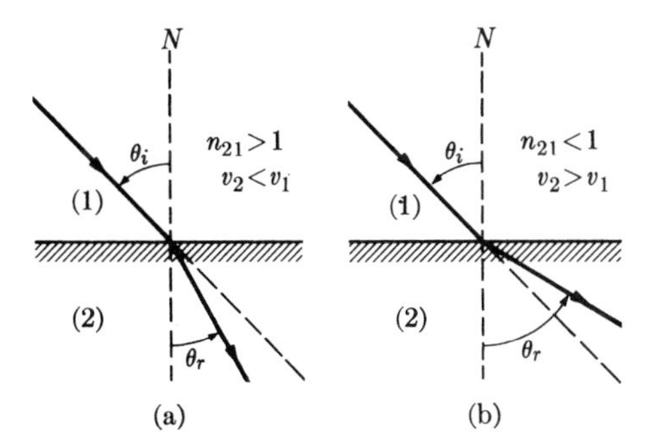

Figure 32.6 Refracted rays for $n_{21} > 1$ and $n_{21} < 1$.

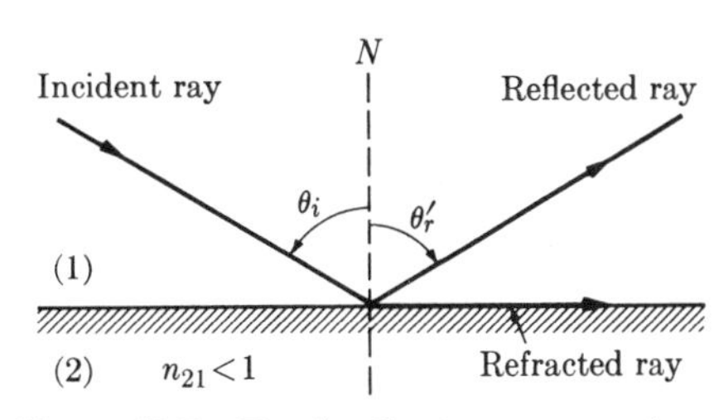

Figure 32.7 Total reflection occurs when $n_{21} < 1$ and Θ_i is larger than the critical angle λ.

If v_1 and v_2 are the velocities of propagation of the wave in the substances, it is shown below that the relative index of refraction is given by

$$n_{21} = \frac{v_1}{v_2} \tag{32.4}$$

Let us take a particular medium as reference or standard and designate by c the velocity of propagation of the wave in that medium. The *standard* index of refraction of any other medium is defined by

$$n = \frac{c}{v} \tag{32.5}$$

For electromagnetic waves, the reference medium is vacuum, and thus $c \approx 3 \times 10^8$ m s^{-1}, as explained in Section 29.6, so that $n = (\varepsilon_r \mu_r)^{1/2}$.

Considering any two substances, we then have that

$$\frac{n_2}{n_1} = \frac{c}{v_2} \times \frac{v_1}{c} = \frac{v_1}{v_2} = n_{21} \tag{32.6}$$

so that the *relative* index of refraction of two substances is equal to the ratio of their standard indexes of refraction. Therefore, it is enough to measure the standard indexes of refraction of relevant substances since, by combination, we can obtain all possible indexes of refraction.

Using relation 32.6, we can write Snell's law, Equation 32.2, in the more symmetric and convenient form,

$$n_1 \sin \theta_i = n_2 \sin \theta_r \tag{32.7}$$

which requires only the standard indexes of refraction.

Proof of the laws of reflection and refraction

The three laws of reflection and refraction can also be justified theoretically, using the concepts of wave propagation and rays. Since the incident ray and the normal determine a plane, and there is no *a priori* reason for the refracted and reflected rays to be deflected from this plane, we conclude that all rays must be in the same plane that includes the normal, as required by the first law.

To prove the second and third laws, consider two incident rays R_1 and R_2 (Figure 32.8) that are parallel, since the incident waves S are plane. Ray R_1 hits the interface at A and R_2 hits it at B'. Because the geometrical situation is the same at A and B', we conclude that the refracted rays R_1' and R_2', as well as the reflected rays R_1'' and R_2'', are also parallel. Since rays R_1 and R_2 were chosen arbitrarily, we then have that the refracted and reflected waves S' and S'' are also plane because they must be

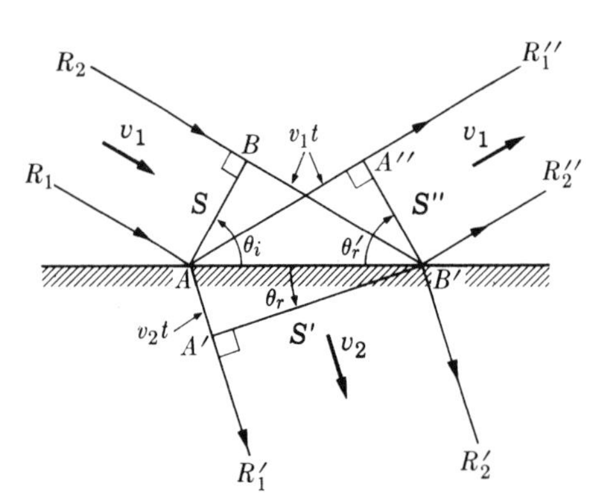

Figure 32.8 Verification of the laws of reflection and refraction.

perpendicular to a corresponding set of parallel rays, as demanded by orthogonality requirements.

Consider the following wave surfaces: AB in the incident wave, $A'B'$ in the refracted wave, and $A''B'$ in the reflected wave. The rays between corresponding points in the wave surfaces must be traversed in equal times. Let t be the time taken by the incident wave S to go from B to B' along ray R_2 with velocity v_1. In the same time the refracted wave S' moved along ray R_1' from A to A' with velocity v_2 and the reflected wave S'' moved along ray R_1'' from A to A'' with velocity v_1. Then $BB' = v_1 t$, $AA' = v_2 t$ and $AA'' = v_1 t$, and, from the geometry of the figure,

$$\sin\theta_i = \frac{BB'}{AB'} = \frac{v_1 t}{AB'}, \qquad \sin\theta_r = \frac{AA'}{AB'} = \frac{v_2 t}{AB'}, \qquad \sin\theta_r' = \frac{AA''}{AB'} = \frac{v_1 t}{AB'}$$

Comparing the first and third relations, $\sin\theta_i = \sin\theta_r'$ or $\theta_i = \theta_r'$, which is the law for reflection, Equation 32.1. Dividing the first relation into the second yields

$$\frac{\sin\theta_i}{\sin\theta_r} = \frac{v_1}{v_2} = n_{21}$$

which expresses Snell's law, Equation 32.2, in accordance with Equation 32.4.

EXAMPLE 32.1

Transmission of a wave through a medium limited by plane parallel sides.

▷ Consider a plate of thickness a and a ray AB (Figure 32.9) whose angle of incidence is θ_i. Here we shall disregard the reflected ray. The angle of refraction is θ_r, corresponding to the refracted ray BC. Using relation 32.7, we have that $n_1 \sin\theta_i = n_2 \sin\theta_r$. At C the refraction is from medium (2) into medium (1) with an angle of incidence which is also θ_r, so that Equation 32.7 gives $n_2 \sin\theta_r = n_1 \sin\theta_i'$. Comparing both relations, we get $\sin\theta_i = \sin\theta_i'$ or $\theta_i = \theta_i'$, which proves that the emergent ray CD is parallel to the incident ray AB but the ray suffers a lateral displacement. From Figure 32.9 it can also be verified that the **lateral displacement** of the ray is

$$d = a\frac{\sin(\theta_i - \theta_r)}{\cos\theta_r}$$

It may also be easily verified that if, instead of one, there are several parallel plates of different materials, the emergent and the incident rays are still parallel.

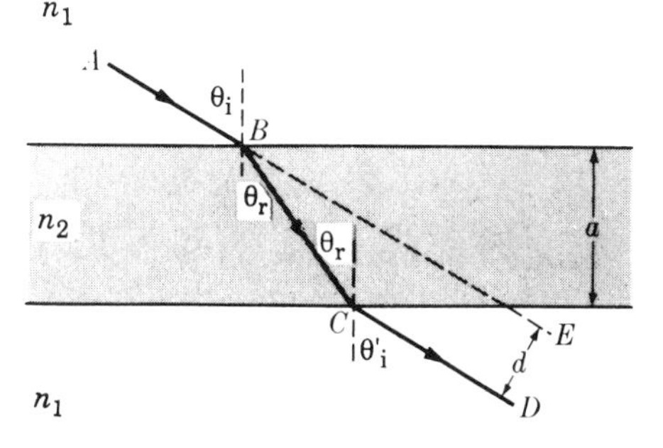

Figure 32.9 Propagation of a ray through a parallel plate.

32.4 Reflection and refraction of spherical waves

Consider spherical waves generated at a point source O and incident on a plane surface S. Two new sets of waves are then produced: the reflected and the refracted or transmitted, as shown in Figure 32.10. In order to trace the shape of the reflected and refracted wave fronts, it would be necessary to draw many reflected and refracted rays. The corresponding reflected and refracted wave surfaces are perpendicular to the rays. In Figure 32.10, one set of these rays has been drawn at B, assuming $n_{21} > 1$. According to laws (2) and (3) for reflection and refraction, we have

$$\theta_i = \theta_r' \qquad \frac{\sin \theta_i}{\sin \theta_r} = n_{21}$$

The reflected ray BD, when it is extended back into medium (2), intersects the extended normal AO at point I'. Because OAB and $I'AB$ are right triangles and the angles at O and I' are the same, we have that $AO = AI'$. Since B was an arbitrary point, we conclude that

all reflected rays pass through a point I', symmetric from O relative to the plane surface.

The point I' is called the **image** of O due to reflection.

Therefore, when spherical waves fall on a plane surface, the reflected waves are spherical and symmetrical with respect to the incident waves. This symmetry was to be expected in view of the fact that the reflected waves propagate backward with the same velocity as the incident waves.

Concerning the refracted ray BC, we see that when we extend it into medium (1) it intersects the normal OA at a point I such that $\tan \theta_r = AB/AI$ and $\tan \theta_i = AB/AO$. Therefore

$$\frac{\tan \theta_i}{\tan \theta_r} = \frac{AI}{AO} \qquad \text{or} \qquad AI = AO\frac{\tan \theta_i}{\tan \theta_r} = AO\left[n_{21}\frac{\cos \theta_r}{\cos \theta_i}\right] \tag{32.8}$$

where Snell's law of refraction has been used in the last equation. The ratio $\cos \theta_r/\cos \theta_i$ is not constant, so that the distance AI depends on the angle of

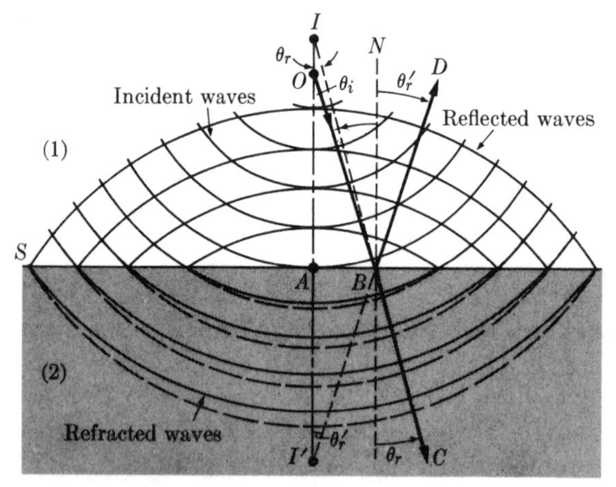

Figure 32.10 Incident, reflected and refracted spherical waves.

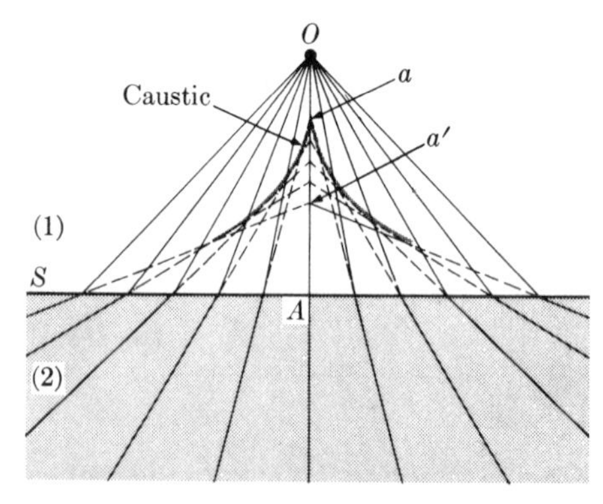

Figure 32.11 Refraction of rays that proceed from a point source. The refracted rays, when extended back, do not intersect at a single point.

incidence θ_i. Therefore the refracted rays do not all pass through the same point. We conclude then that when spherical waves fall on a plane surface, the refracted waves are not spherical.

The refracted rays, since they do not pass through a single point, do not form a point image of O as the reflected rays do. The refracted rays intersect at several points along the normal OA, as well as on a conical surface called the **caustic**, shown in Figure 32.11. The point a, formed by the intersection of the least-inclined rays, can be found very easily, because then the angles θ_i and θ_r of Figure 32.10 are very small and we may replace the cosines by unity in Equation 32.8, resulting in

$$AI \approx n_{21}(AO) \tag{32.9}$$

which is valid for small angles only.

32.5 Reflection and transmission of transverse waves on a string

When a wave reaches the surface separating two different media, the energy of the wave is split between the reflected and the refracted wave. The analysis of the general case is rather complex. Therefore, to illustrate the main points, we shall analyze the simple case of waves propagating along two different strings having a common end and subject to a tension T.

Consider two strings (1) and (2) (Figure 32.12), attached at one point; this common point will be chosen as our origin of coordinates. Suppose an incident wave propagates from the left, having the form

$$\xi_i = A_i \sin \omega\left(t - \frac{x}{v_1}\right)$$

which is equivalent to the wave expression 28.5 (see Question 28.15). The transmitted or refracted wave, propagating along string (2), is

$$\xi_t = A_t \sin \omega\left(t - \frac{x}{v_2}\right)$$

The reflected wave, which propagates back along string (1), is

$$\xi_r = A_r \sin \omega\left(t + \frac{x}{v_1}\right)$$

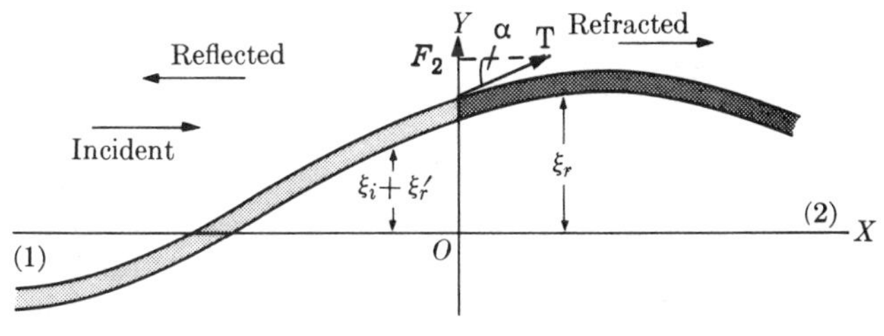

Figure 32.12 Transverse waves in two attached strings of different linear densities.

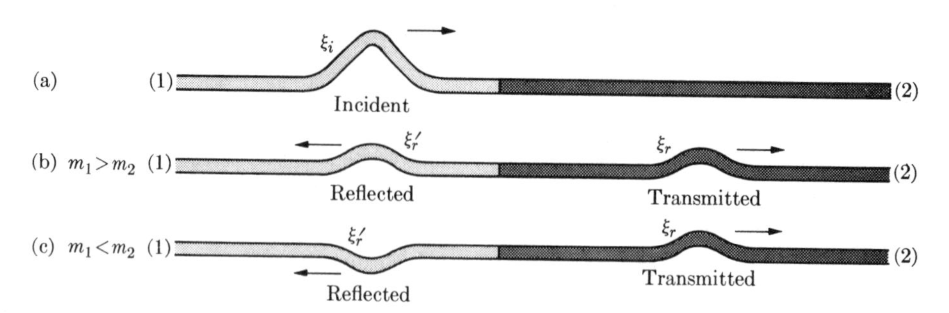

Figure 32.13 Incident, reflected and transmitted waves in two attached strings of different linear densities. In (b) the string carrying the incident wave is heavier; in (c) the string carrying the incident wave is lighter.

Note that we use v_1 for the reflected and incident waves because they propagate in the same medium: string (1).

By analyzing the motion of the point at which the two strings are joined, it is possible to show (see the calculations at the end of the section) that the amplitudes of the reflected and transmitted waves are given by

$$A_t = \frac{2v_2}{v_2 + v_1} A_i \quad \text{and} \quad A_r = \frac{v_2 - v_1}{v_2 + v_1} A_i \tag{32.10}$$

Since, in the case of transverse waves in a string, the velocity of propagation is $v = (T/m)^{1/2}$, according to Equation 28.26, where m is the mass per unit length, we may also write

$$A_t = \frac{2(m_1)^{1/2}}{(m_1)^{1/2} + (m_2)^{1/2}} A_i \quad \text{and} \quad A_r = \frac{(m_1)^{1/2} - (m_2)^{1/2}}{(m_1)^{1/2} + (m_2)^{1/2}} A_i \tag{32.11}$$

The ratios A_t/A_i and A_r/A_i are called, respectively, the **coefficients of transmission** and **reflection**, designated by **T** and **R**, respectively. Thus

$$\mathbf{T} = \frac{2(m_1)^{1/2}}{(m_1)^{1/2} + (m_2)^{1/2}} \quad \text{and} \quad \mathbf{R} = \frac{(m_1)^{1/2} - (m_2)^{1/2}}{(m_1)^{1/2} + (m_2)^{1/2}} \tag{32.12}$$

We note that **T** is always positive, so that A_t always has the same sign as A_i and the transmitted wave is always in phase with the incident wave. But **R** is positive or negative depending on whether m_1 is greater or smaller than m_2, so that the reflected wave may be in phase with or in opposition to the incident wave. In the second case it amounts to adding a phase shift of π to the reflected wave. The two situations are illustrated in Figure 32.13 when the incident wave is a pulse.

Reflection and transmission of waves along two attached strings

Referring to Figure 32.12, we note that at the left of O the displacement is the result of combining the incident and reflected waves or $\xi_i + \xi_r$. To the right of O the displacement is due only to the transmitted wave, or ξ_t. At O, or $x = 0$, where the two strings are attached, the displacement can be computed either way. Therefore $\xi_i + \xi_r = \xi_t$ at $x = 0$. Using the expressions given before for ξ_i, ξ_r and ξ_t, we get

$$A_i + A_r = A_t$$

Also, as indicated in Section 28.7, if T is the tension of the strings, the upward force on any point is given by $T \sin \alpha$. But if α is small we can replace $\sin \alpha$ by $\tan \alpha$, which is the same as $d\xi/dx$. Thus the upward force is $T\, d\xi/dx$.

Accordingly, at the left of O the vertical force is $T\,(d\xi_i/dx + d\xi_r/dx)$ while to the right of O it is $T\, d\xi_t/dx$. Again at O the upward force must be the same regardless of the method of calculation. Therefore, since the factor T is common, we have

$$\frac{d\xi_i}{dx} + \frac{d\xi_r}{dx} = \frac{d\xi_t}{dx} \qquad \text{at } x = 0$$

which results in

$$\frac{1}{v_1}(A_i - A_r) = \frac{1}{v_2} A_t$$

Combining the two relations among the amplitudes we obtain

$$A_t = \frac{2v_2}{v_2 + v_1} A_i \qquad \text{and} \qquad A_r = \frac{v_2 - v_1}{v_2 + v_1} A_i \tag{32.13}$$

which are the relations given in Equation 32.10.

It should be stressed that the relations between the three amplitudes have been obtained by imposing certain physical conditions at the point where there is a sudden change or discontinuity in the properties of the string, in this case its mass per unit length. This is a simple illustration of a general procedure applied whenever there is a discontinuity in the properties of the medium through which any wave propagates.

32.6 Reflection and refraction of electromagnetic waves

The reflection and refraction of electromagnetic waves requires special attention because it involves the electric and the magnetic components of the wave. Both electric and magnetic fields are perpendicular to the direction of propagation of each wave, but otherwise they may have any orientation around it. We shall designate the component *parallel* to the plane of incidence by the subscript π and the component *perpendicular* to the plane of incidence by the subscript σ. Because of the perpendicularity of $\mathscr{E}$ and $\mathscr{B}$, we have a component $\mathscr{E}_\pi$ associated with $\mathscr{B}_\sigma$ and a component $\mathscr{E}_\sigma$ associated with $\mathscr{B}_\pi$. Since, as we indicated in Chapter 29, the polarization of an electromagnetic wave is conventionally determined by the direction of the electric field, we have indicated the $\mathscr{E}$ and $\mathscr{B}$ components in Figure 32.14 for polarization in the plane of incidence, and in Figure 32.15 for polarization

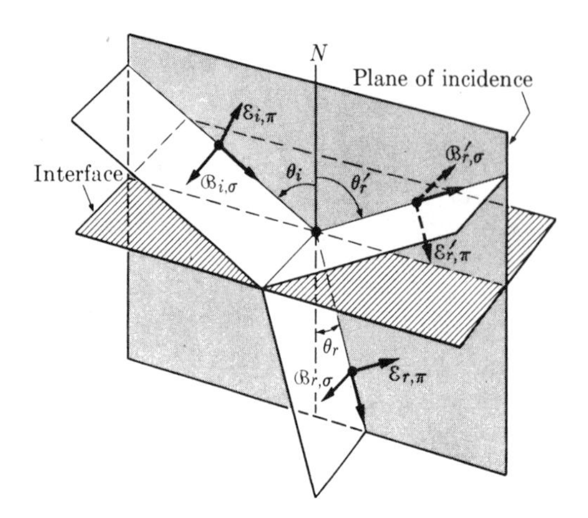

Figure 32.14 Electric and magnetic fields in the incident, reflected, and refracted electromagnetic waves for polarization parallel to the plane of incidence.

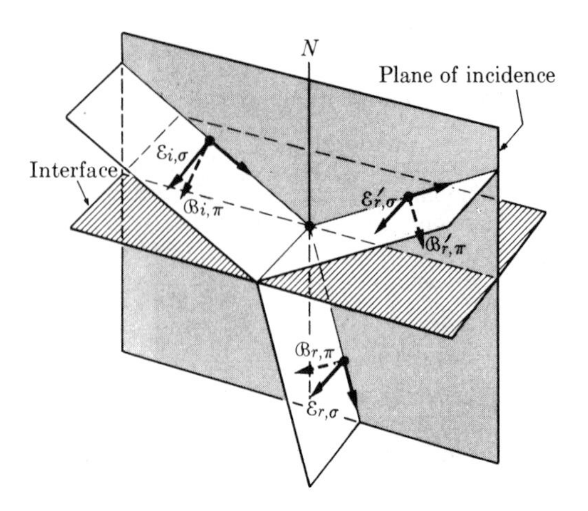

Figure 32.15 Electric and magnetic fields in the incident, reflected and refracted electromagnetic waves for polarization perpendicular to the plane of incidence.

perpendicular to the plane of incidence. The arrows in each case indicate the directions considered positive for the $\mathscr{E}$ components. The general case is a combination of both polarizations.

Maxwell's equations provide certain relations among the parallel and perpendicular components of the electric and magnetic fields on both sides of the surface separating two media that result in relations between the components of the electric field in the incident, transmitted and reflected waves in a way similar to what was done for a string. From them it is possible to calculate the coefficients of reflection and transmission, which depend on whether the incident wave is polarized parallel or perpendicular to the plane of incidence.

There is an important case corresponding to the situation when, regardless of the polarization of the incident wave, the reflected wave is totally polarized with the electric field perpendicular to the plane of incidence, a phenomenon called **polarization by reflection**. This situation is illustrated in Figure 32.16, where only the components of the electric field are indicated; for simplicity, we have omitted the corresponding components of the magnetic field.

It can be shown that polarization by reflection occurs when the reflected and transmitted (or refracted) rays are perpendicular. This statement is known as **Brewster's law**. The proper angle of incidence is called the **polarizing angle**.

From Figure 32.16 we see that in this case $\theta_r + \pi/2 + \theta_r' = \pi$ or $\theta_r + \theta_r' = \pi/2$. Thus, $\sin\theta_r = \cos\theta_r' = \cos\theta_i$ since $\theta_r' = \theta_i$. Accordingly, Snell's law, $\sin\theta_i/\sin\theta_r = n_{21}$, becomes $\sin\theta_i/\cos\theta_i = n_{21}$, or

$$\tan\theta_i = n_{21} \tag{32.14}$$

which expresses the polarizing angle in terms of the relative index of refraction, a result confirmed experimentally.

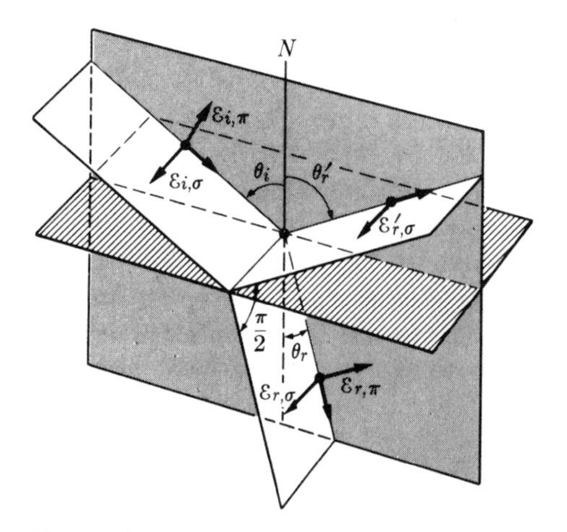

Figure 32.16 Polarization of an electromagnetic wave by reflection.

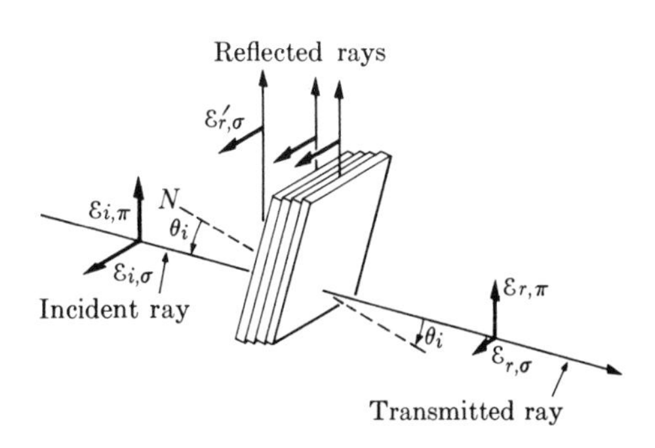

Figure 32.17 Polarization of an electromagnetic wave by successive refractions.

Also it can be shown that the refracted wave can never be completely polarized. However, if an electromagnetic wave is transmitted through a stack of thin parallel plates (Figure 32.17), with an incident angle equal to the polarizing angle, the perpendicular component of the electric field tends to go with the reflected wave each time the wave passes from one plate to the next. Therefore the transmitted wave is partially polarized, with the electric field oscillating in the plane of incidence.

In addition, the transmitted wave is always in phase with the incident wave. However, the reflected wave may be in phase with the incident wave or have a phase shift of π relative to it.

32.7 Propagation of electromagnetic waves in an anisotropic medium

When a transverse wave propagates through an anisotropic medium, the velocity of propagation of the wave may depend both on the directions of polarization and of propagation of the wave. This is particularly true in the case of electromagnetic waves (which are the only ones we shall consider in this section). The polarizability of most molecules is not the same in all directions. Since the molecules in gases and liquids are oriented at random, this directional dependence of the polarizability does not give rise to any particular effect. Therefore, the medium effectively behaves macroscopically as an isotropic substance unless it is subjected to strong static electric or magnetic fields that tend to orient the molecules in a fixed direction. In a crystalline solid, the molecules are already more or less oriented by the local fields in the crystal lattice and their orientation is 'frozen'; that is, they are not free to rotate around their equilibrium positions within the crystal lattice. Thus the properties of the crystal in general depend on the direction along which they are measured. Depending on what their molecular structure and arrangement is, crystalline solids may behave optically as either isotropic or anisotropic media.

The fact that the polarizability of the medium is not the same in all directions means that, in general, the polarization field $\mathscr{P}$ does not have the same direction as the electric field $\mathscr{E}$. However, there is at least one set of three perpendicular directions, called **principal axes**, characteristic of each substance, along which $\mathscr{P}$ and $\mathscr{E}$ are parallel. Orienting the coordinate axes XYZ parallel to these principal axes, we find that to each of the principal axes there corresponds a different permittivity and therefore a different index of refraction, which are called principal indexes of refraction and are designated n_1, n_2 and n_3.

Both experiment and theory (based on Maxwell's equations) show that, when an electromagnetic wave penetrates an anisotropic substance, it splits into two waves, polarized at right angles to each other and propagating with different phase velocities and therefore having different indices of refraction. This situation gives rise to the phenomenon of **double refraction** or **birefringence**. When a beam of light falls on an anisotropic material the incident ray splits into two refracted rays, each refracted along a different direction and with different polarizations. If the transmitted light falls on a screen, two images instead of one are observed (Figure 32.18).

In isotropic media all three principal indexes of refraction are equal: $n_1 = n_2 = n_3$. The index of refraction is the same in all directions and no double refraction occurs. **Cubic crystals**, as well as most non-crystalline solids, behave this way.

When two principal indexes of refraction are the same, say $n_2 = n_3$, the direction corresponding to the unequal index n_1 is called the **optical axis**; it is an axis of symmetry of the crystal. For that reason these substances are called **uniaxial crystals**. To this class belong the **trigonal, hexagonal** and **tetragonal** crystal systems.

In this case we may define two waves: the ordinary and the extraordinary.

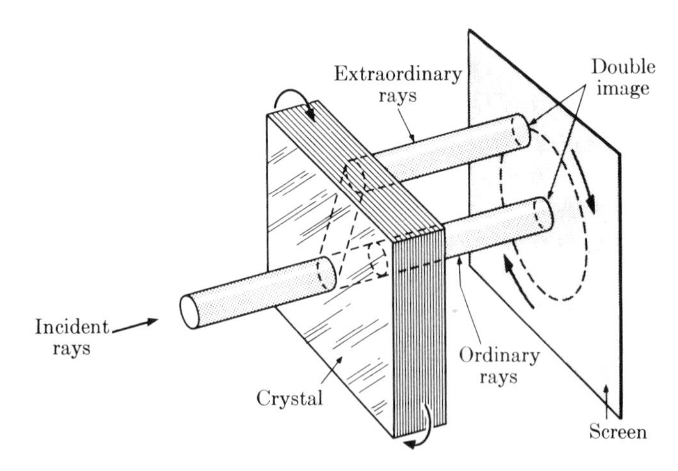

Figure 32.18 A narrow beam of unpolarized light can be split into two beams by a doubly refracting crystal. If the crystal is rotated, the extraordinary beam rotates around the ordinary ray. The two beams are linearly polarized at right angles with respect to each other.

The **ordinary wave** is linearly polarized perpendicular to the plane determined by the direction of propagation and the optical axis. The ordinary wave propagates in all directions with the same velocity $v_2 = c/n_2$. It therefore behaves like a wave in an isotropic medium, and this is the reason why it is called ordinary. The **extraordinary wave** is linearly polarized in the plane determined by the direction of propagation and the optical axis; but its velocity depends on the direction of propagation, varying from v_2 to v_1 (corresponding to an index of refraction between n_2 and n_1). For propagation along the optical axis, the ordinary and extraordinary waves propagate with the same velocity v_2. This may be considered as another definition of the optical axis: the optical axis is that direction along which there is only one velocity of propagation regardless of the polarization of the wave.

In the general case of three different indexes of refraction, there are two directions for which the velocities of propagation of the two polarized waves are equal. These directions are also called optical axes. These substances are called **biaxial**, and they belong to the **orthorhombic**, **monoclinic** and **triclinic** crystal systems. Table 32.1 lists the indexes of refraction for selected uniaxial and biaxial materials.

Many substances that are normally isotropic become anisotropic when subject to mechanical stresses or to strong static electric or magnetic fields. In all cases the anisotropy of the substance is due to the partial orientation of the molecules that results from the stresses or the fields. Induced anisotropy is an important tool used to analyze materials under stress.

Table 32.1 Principal indexes of refraction of several crystals*

Substance	n_1	n_2	n_3
Uniaxial:			
Apatite	1.6417	1.6461	
Calcite	1.4864	1.6583	
Quartz	1.5533	1.5422	
Zircon	1.9682	1.9239	
Biaxial:			
Aragonite	1.5301	1.6816	1.6859
Gypsum	1.5206	1.5227	1.5297
Mica	1.5692	1.6049	1.6117
Topaz	1.6155	1.6181	1.6250

* For sodium light, $\lambda = 5.893 \times 10^{-7}$ m.

EXAMPLE 32.2

Calculation of the phase difference between the ordinary and extraordinary waves and the state of polarization of the emergent wave when a linearly polarized wave falls on a thin plate of a uniaxial material (for example, quartz), cut with the faces parallel to the optical axis.

▷ Figure 32.19 shows the experimental arrangement. The crystal plate has been placed with its optical axis horizontal. The direction of the optical axis has been designated by Y. The perpendicular Z-direction corresponds to the polarization of the ordinary ray. Suppose that a linearly polarized wave, with the electric field making an angle α with the Y-axis, falls on the plate. Then we may write $\mathscr{E} = \mathscr{E}_0 \sin(\omega t - kx)$ for the electric field in the incident wave.

When the linearly polarized wave propagates through the crystal, it is separated into two waves with their electric fields along the Y- and Z-axes, respectively. These component waves correspond to the extraordinary and ordinary waves. Since the velocity of propagation of each wave is $v_1 = c/n_1$ and $v_2 = c/n_2$, we have that

$$k_1 = \frac{\omega}{v_1} = \frac{\omega n_1}{c} = kn_1 \qquad \text{and} \qquad k_2 = \frac{\omega}{v_2} = \frac{\omega n_2}{c} = kn_2$$

where $k = \omega/c = 2\pi/\lambda$ and λ is the wavelength in vacuum. Therefore, after the waves have traversed the thickness d, the respective electric fields are represented by the expressions

$$\mathscr{E}_y = \mathscr{E}_{0y} \sin(\omega t - k_1 d) \qquad \text{and} \qquad \mathscr{E}_z = \mathscr{E}_{0z} \sin(\omega t - k_2 d)$$

resulting in a phase difference between the two waves of

$$\delta = (k_1 - k_2)d = k(n_1 - n_2)d = \frac{2\pi(n_1 - n_2)d}{\lambda}$$

After traversing the anisotropic plate the two waves recombine into a single one. According to our discussion in Section 10.9, we conclude that, because of the phase difference, the transmitted wave will in general be elliptically polarized. The axes of the ellipse will be parallel to the Y- and Z-axes if δ is an odd multiple of $\pi/2$, or if

$$(n_1 - n_2)d = \text{odd integer} \times \frac{\lambda}{4}$$

The transmitted wave will be linearly polarized if δ is a multiple of π, or if

$$(n_1 - n_2)d = \text{integer} \times \frac{\lambda}{2}$$

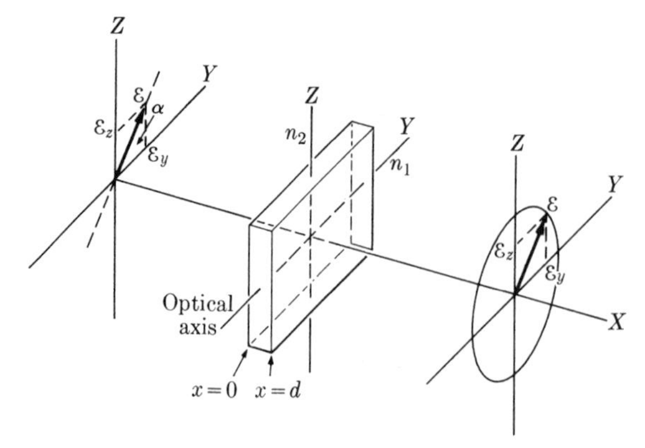

Figure 32.19 Change of polarization of an electromagnetic wave after traversing a parallel plate cut from a uniaxial crystal.

In this case, if the integer is even, the transmitted wave is linearly polarized in the same plane as the incident wave, but if the integer is odd, it is polarized in a plane symmetric with the XZ-plane. The plates corresponding to the two conditions given above are called a **quarter-wave plate** and a **half-wave plate**, respectively. These types of plates are widely used in the analysis of polarized light.

EXAMPLE 32.3

A ray of light falls on a calcite crystal cut so that its surface is parallel to the optical axis. Calculate the angular separation between the ordinary and the extraordinary rays, assuming that the plane of incidence is perpendicular to the optical axis and that the angle of incidence is 50°.

▷ When the propagation of the wave is in a direction perpendicular to the optical axis, the ordinary rays propagate with velocity v_2 corresponding to the refractive index n_2, and the extraordinary waves propagate with velocity v_1 corresponding to the refractive index n_1. Therefore, using Snell's law and the principal indexes of refraction as given in Table 32.1, we have

$$\frac{\sin\theta_i}{\sin\theta_0} = n_2 = 1.6583 \qquad \text{and} \qquad \frac{\sin\theta_i}{\sin\theta_e} = n_1 = 1.4864$$

Given that $\theta_i = 50°$, we obtain $\theta_0 = 27°30'$ and $\theta_e = 31°5'$. The angular separation of the two rays is thus $\theta_e - \theta_0 = 3°35'$.

EXAMPLE 32.4

Variation in the intensity when linearly polarized electromagnetic waves are observed through a polarizing device called an **analyzer**.

▷ Consider Figure 32.20. The analyzer is a device that transmits a wave whose electric field is parallel to its axis AA'. When the axis AA' of the analyzer makes an angle θ with the electric field of the incident linearly polarized wave, it transmits only the component $\mathscr{E}_A = \mathscr{E}\cos\theta$. Therefore, since the intensity of the wave is proportional to the square of the electric field, we have the relation $I = I_0\cos^2\theta$, where I_0 is the intensity of the incident wave and I that of the transmitted wave. When $\theta = 0$ or π, the intensity of the transmitted light is maximum; when $\theta = \pi/2$ or $3\pi/2$, it is zero. Therefore, when the analyzer is rotated, the intensity of the transmitted wave fluctuates between 0 and I_0.

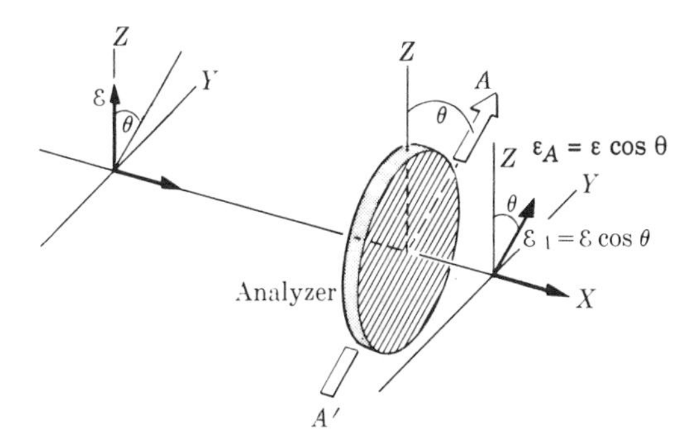

Figure 32.20 Change in the intensity of linearly polarized light with change in orientation of analyzer.

This fluctuation in intensity affords a means of determining whether an electromagnetic wave is polarized or not. For unpolarized or circularly polarized waves, no fluctuation in intensity is observed. For elliptically polarized waves, the transmitted wave fluctuates between a maximum and a minimum value. These two extremes are obtained when the analyzer is parallel either to the larger or the smaller axis of the ellipse.

32.8 Reflection and refraction at metallic surfaces

When an electromagnetic wave penetrates into a conductor, such as a metal or an ionized gas, it is rapidly attenuated and becomes practically zero after penetrating a very short distance. The attenuation is the result of the dissipation of the wave energy in setting the electrons in the conduction band into motion.

This situation explains two important facts concerning conductors. One fact is their opacity, resulting from the strong absorption of electromagnetic waves, so that no wave is transmitted through the conductor unless it is a very thin sheet. Conductors are therefore excellent for shielding a region from electromagnetic waves. (This is done, for example, by surrounding the region by a metal grid.) The other is the great reflectivity of conductors, which results from the fact that only a small fraction of the energy of the incident wave penetrates the conductor, and most of the energy goes into the reflected wave.

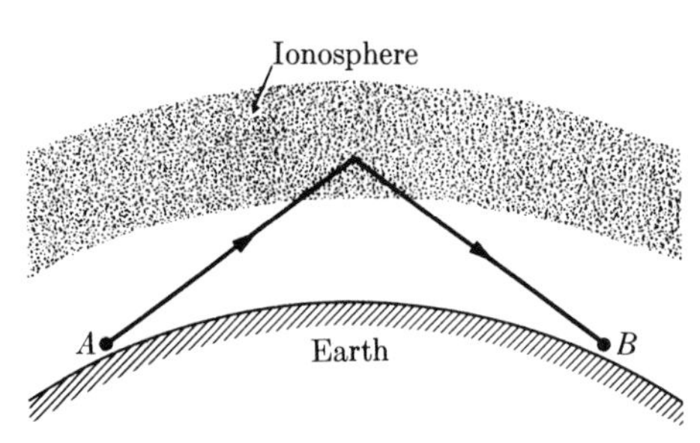

Figure 32.21 Reflection of radio waves by the ionosphere.

An ionized layer of gas can also act as a conductor, reflecting electromagnetic waves falling on it. This principle is used, for example, in radio communication, to transmit a radio signal around the Earth. The signal is reflected back to the Earth when it reaches a highly ionized layer in the atmosphere, called the **ionosphere**, which is about 100 km above the Earth's surface. In this way communication is possible between points A and B, something which cannot be achieved by a wave propagating in a straight line between the points (Figure 32.21). However, modern communications systems rely more on a network of geostationary satellites that receive and transmit waves from the ground, as mentioned in Chapter 12.

QUESTIONS

32.1 Can a medium be (a) homogeneous and isotropic, (b) homogeneous and anisotropic, (c) heterogeneous and isotropic, (d) heterogeneous and anisotropic? Give examples for your answers.

32.2 What is the condition for total reflection at an interface between two media? Define critical angle.

32.3 What is the condition for total polarization by reflection? Define polarization by reflection.

32.4 Given an anisotropic medium, why may the polarization vector not necessarily be parallel to the electric field?

32.5 Which of the following properties of a wave vary and which remain unchanged in refraction: (a) frequency, (b) wavelength, (c) phase velocity, (d) state of polarization, (e) direction of propagation? Explain in each case.

32.6 Make a plot of the fluctuations of the intensity of linearly polarized light transmitted through an analyzer, as a function of the angle of rotation of the analyzer.

32.7 How do you distinguish whether a beam of light is (a) linearly polarized, (b) circularly polarized, (c) elliptically polarized, (d) unpolarized?

32.8 A beam of light is circularly polarized. If it is observed through an analyzer, how does the intensity of the transmitted light vary with the angle of rotation? What kind of polarization does the transmitted light have?

32.9 Answer Question 32.8 for light that is elliptically polarized.

32.10 Two polarizing devices are set with their axes at right angles. Is there any light transmitted? A third polarizing device is inserted between the first two. Describe the variation of intensity in the transmitted light as a function of the angular orientation of the third device when the system is illuminated with unpolarized light.

32.11 Investigate the use of photoelasticity in stress analysis of materials of different shapes.

32.12 Investigate some practical applications of polarized light.

32.13 Some substances, called **optically active**, rotate the plane of polarization of plane polarized light. Some rotate the plane to the right (dextro) and others to the left (levo). Make a list of dextro- and levo-optically active substances. The angle of rotation is proportional to the distance traveled by light through the substance. Investigate how, using a polarizer and an analyzer, the angle of rotation can be determined. The device is called a **polarimeter**.

32.14 Referring to the previous question, a solution containing sugar is optically active. The angle of rotation for a given distance is proportional to the sugar concentration. Investigate how, measuring the angle of rotation, the sugar concentration may be determined.

PROBLEMS

32.1 The following rule has been proposed to construct the refracted ray (Figure 32.22): at the point of incidence, two circles of radii 1 and n are drawn (using arbitrary units). The incident ray is extended until it intersects the circle of radius 1. A perpendicular to the surface is drawn through that point, and its intersection with the circle of radius n is found. The refracted ray passes through this point. (a) Justify this rule. (b) Draw the case when $n = 1.5$ and the angle of incidence i is 60°. (c) Repeat for $n = 0.80$ with an angle of incidence of 30°, and also for 60°. Verify your results by the use of Snell's law.

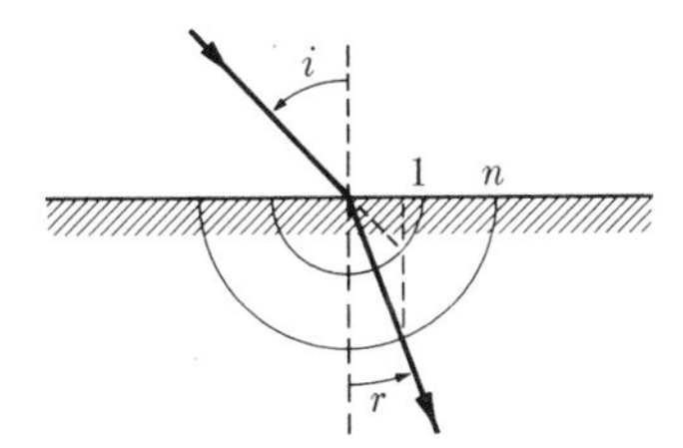

Figure 32.22

32.2 A plate of glass ($n = 1.6$) with parallel sides is 8 cm thick. (a) Calculate the lateral displacement of a ray of light whose angle of incidence is 45°. (b) Using the method of Problem 32.1, plot the path of the ray.

32.3 A ray of light makes an angle of incidence of 35° on a plate of glass whose index of refraction is 1.3 and which is 6 cm thick. Directly below the first plate is another plate whose index of refraction is 1.5. (a) What is the angle of incidence and angle of refraction at the boundary between the plates and at the exit from the second plate? (b) If the second plate is 5 cm thick, determine the total amount of lateral displacement of the ray after it emerges.

32.4 A ray of light falls on a wedge of glass with $n = 1.6$. (See Figure 32.23.) Calculate the minimum angle of α so there is total reflection at the second surface.

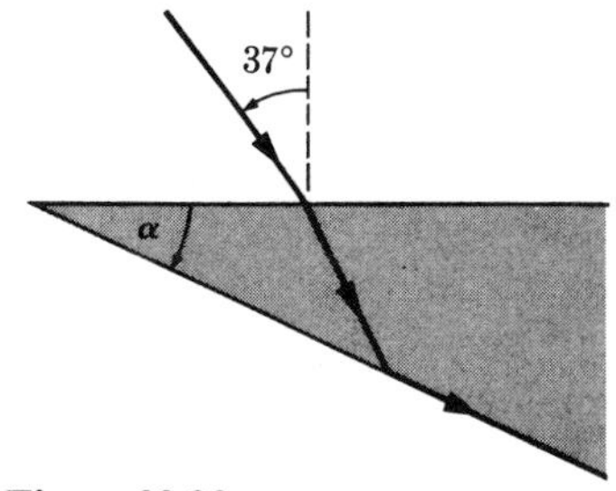

Figure 32.23

32.5 For the situation discussed in Section 32.5, verify that the sum of the intensity of the transmitted wave and the intensity of the reflected wave equals the intensity of the incident wave. What is the physical meaning of this result?

32.6 Copper and steel wires of the same radii are joined, making a long string. Find **T** and **R** at the junction for waves propagated along the string. Let the common radius be 1 mm. Assuming that the incident wave has a frequency of 10 Hz, that the amplitude is 2 cm, and that the tension is 50 N, write the equations for the incident, reflected, and transmitted waves. (The density of copper is 8.89×10^3 kg m^{-3}, of steel 7.80×10^3 kg m^{-3}.)

32.7 The index of refraction of a piece of glass is 1.50. Compute the angles of incidence and of refraction when the light reflected from a glass interface is completely polarized.

32.8 The critical angle of light in a certain substance is 45°. What is the polarizing angle?

32.9 (a) Calculate the angle the Sun must be above the horizontal so that sunlight reflected from the surface of a calm lake is completely polarized. (b) What is the plane of the vector in the reflected light?

32.10 A parallel beam of linearly polarized light of wavelength 5.90×10^{-7} m (in vacuum) is incident on a calcite crystal as in Figure 32.19. (a) Calculate the wavelengths of the ordinary and extraordinary waves in the crystal. (b) Also calculate the frequency of each ray.

32.11 Find the thickness of a calcite plate needed to produce a phase difference of (a) $\lambda/4$, (b) $\lambda/2$, (c) λ between the ordinary and the extraordinary rays for a wavelength of 6×10^{-7} m.

32.12 Determine the state of polarization of the light transmitted by a quarter-wave plate when the electric vector of the incident linearly polarized light makes an angle of 30° with the optical axis.

32.13 A polarizer and an analyzer are so oriented that the maximum amount of light is transmitted. To what fraction of its maximum value is the intensity of the transmitted light reduced when the analyzer is rotated through (a) 30°, (b) 45°, (c) 60°, (d) 90°, (e) 120°, (f) 135°, (g) 150°, (h) 180°? Plot I/I_{max} for a complete turn of the analyzer.

32.14 A beam of white linearly polarized light falls with normal incidence on a plate of quartz 0.865 mm thick, cut so the surfaces are parallel to the optical axis. The arrangement is similar to that in Figure 32.19. The plane of the electric field is at an angle of 45° to the *Y*-axis. The principal indexes of refraction of quartz for sodium light are listed in Table 32.1. Disregard the variation of $n_1 - n_2$ with wavelength. (a) Which wavelengths between 6.0×10^{-7} m and 7.0×10^{-7} m emerge from the plate linearly polarized? (b) Which wavelengths emerge circularly polarized? (c) Suppose that the beam emerging from the plate passes through an analyzer whose transmission axis is perpendicular to the plane of the incident electric field (i.e. at 135° with the $+Y$-axis). Which wavelengths are missing in the transmitted beam?

32.15 A beam of light, after passing through a polarizer P_1, traverses a cell containing a scattering medium. The cell is observed at right angles through another polarizer, P_2. Originally, the polarizers are oriented so that the brightness of the field seen by the observer is a maximum. (a) Polarizer 2 is rotated through 90°. Is extinction produced? (b) Polarizer P_1 is now rotated through 90°. Is the field through P_2 bright or dark? (c) Polarizer P_2 is then restored to its original position. Is the field through P_2 bright or dark?

32.16 It is known experimentally that for each gram of sugar dissolved in one cm^3 of water, the rotation of the plane of polarization of a linearly polarized electromagnetic wave is +66.5° per cm of path (see Question 32.13). A tube 30 cm long contains a sugar solution with 15 g of sugar per 100 cm^3 of solution. Find the angle of rotation of polarized light.

32.17 Find the amount of sugar in a cylindrical tube 30 cm long and 2 cm^2 in cross-section if the plane of polarization is rotated 39.7°. (Hint: see preceding problem.)

32.18 The optical activity of sugar may be used to determine the amount of sugar in a specimen of urine. If a typical specimen is 100 cm^3, will a urinalysis tube 30 cm long be sensitive enough to detect a difference in sugar concentration of 1 milligram per cm^3? (You may assume that the instrument can be read accurately to 0.1°.)

32.19 When a plane electromagnetic wave falls perpendicularly on a plane surface separating a medium of index n_1 from a medium of index n_2, it can be shown that the coefficients of reflection and refraction are $\mathsf{R} = (n_1 - n_2)/(n_1 + n_2)$ and $\mathsf{T} = 2n_1/(n_1 + n_2)$. Discuss what may be said about the intensity and phase of the reflected and transmitted (refracted) waves for the cases $n_1 > n_2$, $n_1 < n_2$ and $n_1 = n_2$.

32.20 (a) Light falls perpendicularly on a glass plate ($n = 1.5$). Find the coefficients of reflection and transmission. (b) Repeat the calculation if the light is passing from the glass into the air. (c) Discuss in each case the changes of phase. (Hint: see the previous problem.)

33 Wave geometry

Reflecting radiotelescope at Parkes, New South Wales, Australia. The reflector is 64 m in diameter. The reflector can rotate about the vertical as well as change in zenith angle and is steerable over most of the visible sky. The telescope is designed for optimum performance at the 21 cm hydrogen line wavelength, although it is still sensitive down to wavelengths of a few cm. The location, 330 km from Sydney, was chosen so that there would be minimum electrical interference. (Photograph courtesy of the Australian News and Info. Bureau; © CSIRO Division of Radiophysics.)

Notes

33.1 Introduction

The reflection and refraction of waves that occur at surfaces of discontinuity can be analyzed geometrically using the ray concept when no other changes happen at the surface. This method is what is called **wave geometry**, or **ray tracing**. In particular, for electromagnetic waves in the visible and near-visible regions, it constitutes **geometrical optics**, which is a very important branch of applied physics.

This geometrical treatment is adequate so long as the surfaces and other discontinuities encountered by the wave during its propagation are very large compared with the wavelength. So long as this condition is fulfilled, the treatment applies equally well to light waves, acoustic waves (especially ultrasonic waves), earthquake waves etc. However, in most of our examples we shall consider light waves, since they are perhaps the more familiar and important from this point of view, and for this reason this is a chapter on geometrical optics.

A characteristic example of the use of rays is in the image produced by a pinhole camera (Figure 33.1). Such a camera consists of a box with a very small

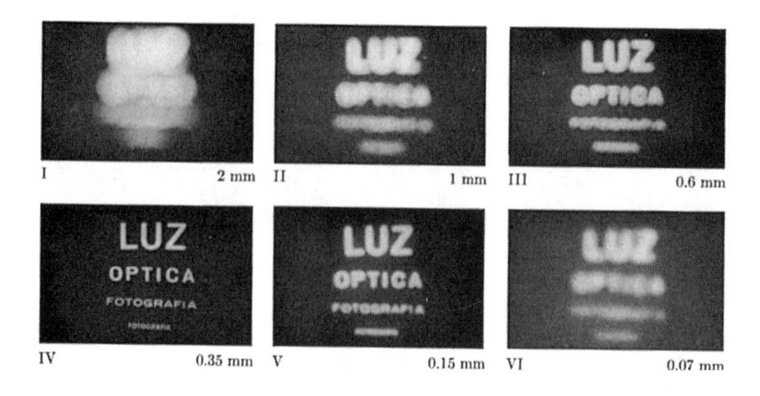

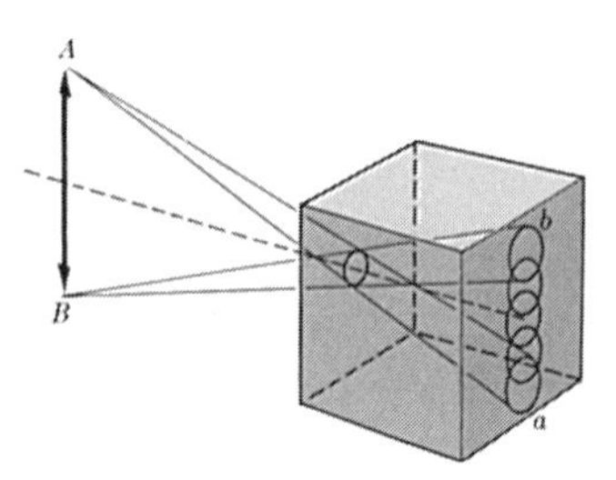

Figure 33.1 Image formation by a pinhole camera. The line drawing illustrates the ray paths. The series of photos shows the change in the clarity of the image as the diameter of the hole is decreased. Note that there is an optimum diameter for image clarity. (Photograph courtesy of Dr N. Joel, UNESCO Pilot Project for the teaching of physics.)

hole in one side. If an object AB emitting waves is placed in front of the camera, the rays Bb and Aa will form an image ab on the opposite side. This image is well defined when the hole is very small, so that only a small fraction of the wave fronts pass through it, and therefore, for each point of the object, there is a corresponding point in the image. If the hole is too large, the image appears blurred because, to each point of the object, there corresponds a spot in the image. Further, the hole must not be so small that its radius is comparable with the wavelength, because then diffraction effects begin to appear and the image ab again appears blurred (diffraction is discussed in Chapter 35).

33.2 Reflection at a spherical surface

To consider the reflection of waves at a spherical surface, we must establish certain definitions and sign conventions. The center of curvature C is the center of the spherical surface (Figure 33.2) and the point O is the pole of the spherical cap. The line passing through O and C is called the **principal axis**. If we take our origin of coordinates at O, all quantities measured to the right of O are positive and all those to the left are negative. Table 33.1 lists the sign conventions used in this text.

(i) Descartes' formula. Suppose that point P in Figure 33.2 is a source of spherical waves. The ray PA is reflected as the ray AQ, which intersects the principal axis at Q. Calling $OP = p$ and $OQ = q$, then as shown below, if the angles α_1 and

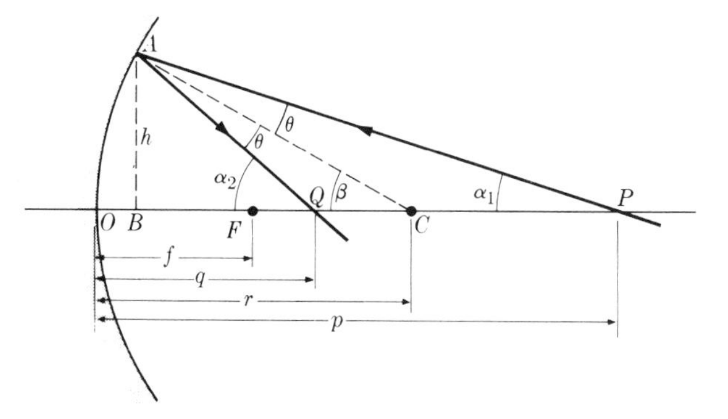

Figure 33.2 Path of a ray reflected at a spherical surface.

Table 33.1 Sign conventions in spherical mirrors

	+	−
Radius r	Concave	Convex
Focus f	Convergent	Divergent
Object p	Real	Virtual
Image q	Real	Virtual

α_2 are small (that is, the rays are paraxial), these two quantities are related by the expression

$$\frac{1}{p} + \frac{1}{q} = \frac{2}{r} \tag{33.1}$$

which is **Descartes' formula for reflection at a spherical surface**. Since there is no reference to point A in this expression, we conclude that, under the approximations made, all incident rays passing through P will go through Q after reflection at the surface. We say that Q is the **image** of the **object** P.

(ii) Focal point. For the special case where the incident ray is parallel to the principal axis, which is equivalent to placing the object at a very large distance from the mirror, we have $p = \infty$. Designating the distance of the image by f, Equation 33.1 becomes

$$\frac{1}{f} = \frac{2}{r} \quad \text{or} \quad f = \frac{r}{2}$$

and the image falls at point F, a distance from the mirror given by $f = r/2$. Point F is called the **focus** of the spherical mirror, and its distance from the mirror is called the **focal length**. Then Equation 33.1 can be written in the form

$$\frac{1}{p} + \frac{1}{q} = \frac{1}{f} \tag{33.2}$$

This equation relates the position of the object, given by p, to the position of the image, given by q, and the focal length f of the mirror. Since f can be determined experimentally by observing the point of convergence of rays that are parallel to the principal axis, it is not necessary to know the radius r in order to apply Equation 33.3. Note that if $q = \infty$, then $p = f$, so that all incident rays passing through the focus F are reflected parallel to the principal axis.

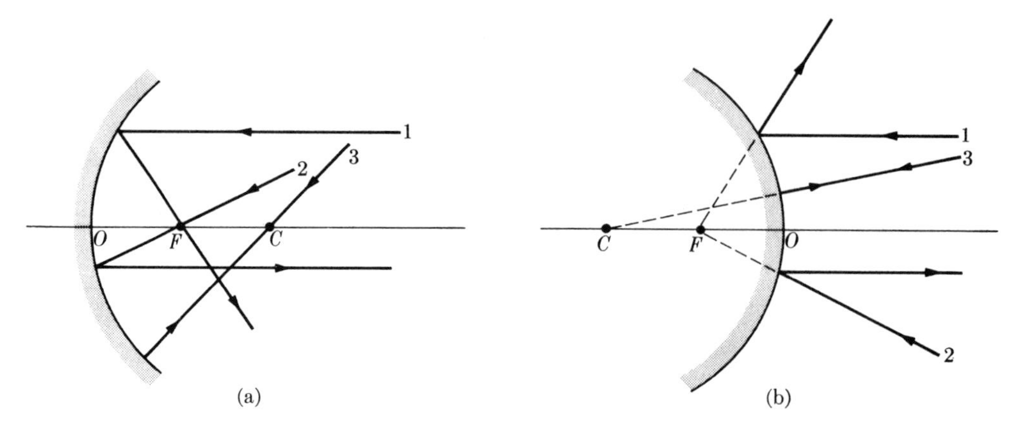

Figure 33.3 Principal rays in spherical mirrors. (a) Concave and (b) convex.

Due to our sign convention, concave surfaces have a positive radius, while convex surfaces have a negative radius. Therefore the signs of the corresponding focal lengths are positive and negative, respectively.

(iii) Image construction. Figure 33.3 shows the so-called **principal rays** for a concave and a convex surface. Ray 1 is a parallel ray, 2 is a focal ray, and 3 is a central ray, which is normal to the mirror. In Figure 33.4 these rays are used to illustrate the formation of an image by a spherical reflecting surface. The object is AB and the image is ab. In Figure 33.4(a) the image is **real** (since the reflected rays *do* cross), and in Figure 33.4(b) it is **virtual** (because the rays only seem to have crossed behind the mirror).

(iv) Spherical aberration. When the aperture of a mirror is large, so that it accepts rays of large inclination, Equation 33.2 is no longer a good approximation. In such a case, there is not a well-defined point image corresponding to a point object, but an infinite number of them; hence the image of an extended object appears blurred. Figure 33.5 shows the rays coming from the point P and reflected at the mirror. We see that the rays do not all intersect at the same point, but on a segment QQ', along the axis, an effect called **spherical aberration**. The point Q,

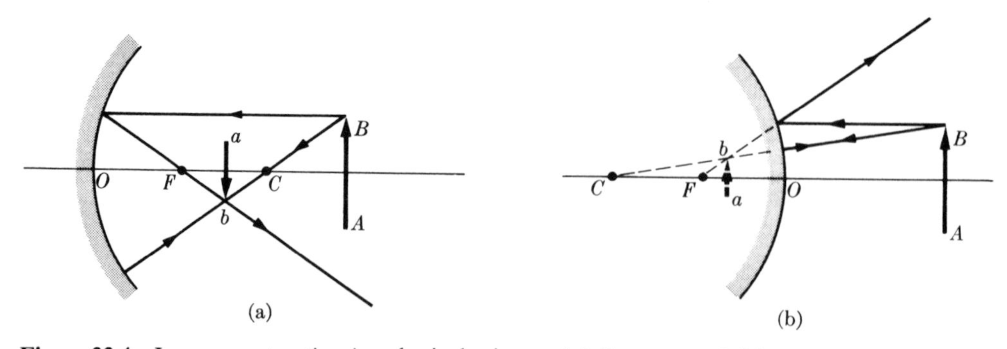

Figure 33.4 Image construction in spherical mirrors. (a) Concave and (b) convex.

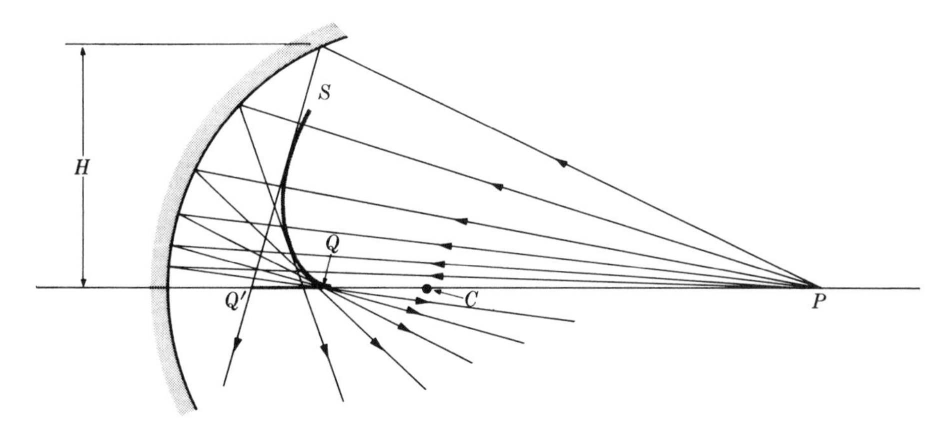

Figure 33.5 Spherical aberration and caustic in a concave mirror.

which corresponds to the rays that make a very small angle with the axis, is determined by Equation 33.2; Q' corresponds to the rays that make a maximum inclination. The reflected rays also intersect along a conical surface, which is indicated by the heavy line QS, and is called the **reflection caustic**.

Spherical aberration cannot be completely eliminated, but by proper design of the surface it can be suppressed for certain positions, called **anastigmatic**. Anastigmatic positions can be modified by changing the shape of the surface. For example, an elliptical mirror is anastigmatic for an object placed at one focus of the ellipse whose image falls exactly at the other focus (Figure 33.6(a)). Similarly, a parabolic mirror produces no aberration for rays that are parallel to the principal axis; they must all pass through the focus of the parabola (Figure 33.6(b)), which is the anastigmatic point. This is why parabolic mirrors are used in telescopes, not only for receiving rays in the visible region but also for receiving rays in the radio-frequency region, as in radio telescopes.

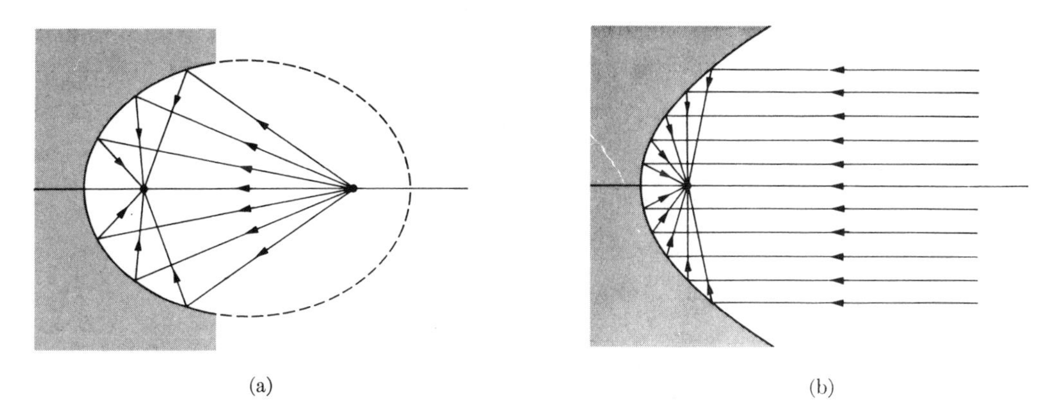

Figure 33.6 (a) Elliptical mirror. (b) Parabolic mirror.

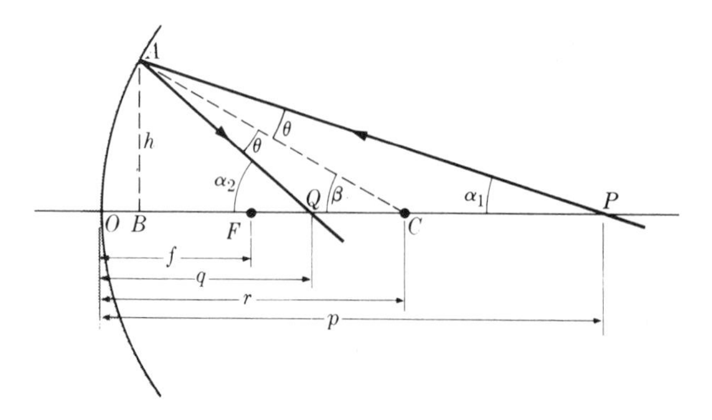

Figure 33.7 Path of a ray reflected at a spherical surface.

There are other defects besides spherical aberration that we observe in the images produced by reflection (or refraction) at spherical surfaces. However, we shall not discuss them, since they belong to rather specialized branches of optics.

Derivation of Descartes' formula for reflection at a spherical surface

Referring to Figure 33.7, let P be a point object on the principal axis. A ray like PA passing through P is reflected as the ray AQ and, since the angles of incidence and reflection are equal, we have from the figure that $\beta = \theta + \alpha_1$ and $\alpha_2 = \beta + \theta$, resulting in

$$\alpha_1 + \alpha_2 = 2\beta \tag{33.3}$$

Assuming that the angles α_1, α_2 and β are very small (i.e. the rays are **paraxial**) and defining the distance AB from A to the principal axis by the symbol h, we may write with good approximation

$$\alpha_1 \approx \tan \alpha_1 = \frac{AB}{BP} \approx \frac{h}{p}, \qquad \alpha_2 \approx \tan \alpha_2 = \frac{AB}{BQ} \approx \frac{h}{q} \qquad \text{and} \qquad \beta \approx \tan \beta = \frac{AB}{BC} \approx \frac{h}{r}$$

Substituting in Equation 33.3 and canceling the common factor h, we get

$$\frac{1}{p} + \frac{1}{q} = \frac{2}{r}$$

It must be emphasized that this relation is valid only for small values of the angles α_1, α_2 and β. For larger values the reflected rays pass through different points of the axis. A first correction to the equation is

$$\frac{1}{p} + \frac{1}{q} = \frac{2}{r} + \frac{h^2}{r}\left(\frac{1}{r} - \frac{1}{p}\right)^2$$

which shows that q depends on the distance h and therefore there is no unique point image.

EXAMPLE 33.1

Magnification produced by a spherical mirror.

▷ The **magnification** M of an optical system is defined as the ratio of the size of the image to that of the object. That is (Figure 33.8),

$$M = \frac{\text{image}}{\text{object}} = \frac{ab}{AB}$$

The magnification may be positive or negative, depending on whether the image is erect or inverted with respect to the object. From Figure 33.8 we see that

$$\tan\theta = \frac{AB}{OA} = \frac{AB}{p}$$

and

$$\tan\theta' = -\frac{ab}{Oa} = -\frac{ab}{q}$$

where the minus sign is due to the fact that ab is negative, since the image is inverted. Thus, considering that $\theta = \theta'$, we have then that

$$M = -\frac{q}{p} \tag{33.4}$$

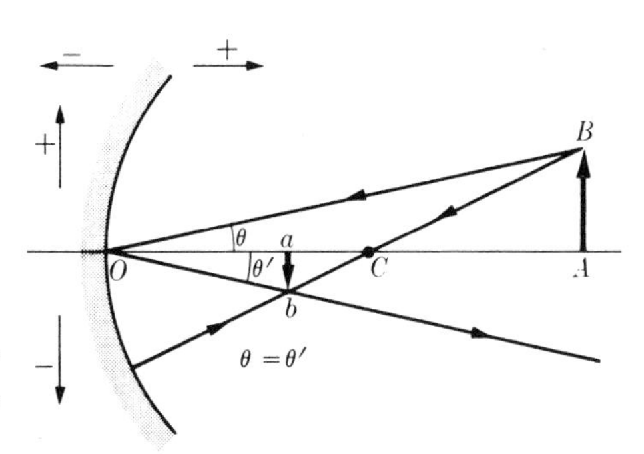

Figure 33.8 Calculation of magnification by a spherical mirror.

EXAMPLE 33.2

A spherical mirror has a radius of 0.40 m. An object is placed in front of the mirror at a distance of 0.30 m. Find the position of the image and the magnification if the mirror is (i) concave, (ii) convex.

▷ **(i) Concave mirror.** Following the sign convention of Table 33.1, the radius $r = +0.40$ m and the focal length $f = r/2 = +0.20$ m are positive quantities. Also $p = +0.30$ m. Applying Equation 33.2,

$$\frac{1}{0.30} + \frac{1}{q} = \frac{1}{0.20} \quad \text{or} \quad q = +0.60\text{ m}$$

The positive sign indicates a real image. The magnification is

$$M = -\frac{q}{p} = -\frac{0.60}{0.30} = -2.0$$

which indicates that the image is twice as large as the object and inverted.

(ii) Convex mirror. In this case the focal length is $f = -0.20$ m, but p is still $+0.30$ m. Therefore Equation 33.2 gives

$$\frac{1}{0.30} + \frac{1}{q} = \frac{1}{-0.20} \quad \text{or} \quad q = -0.12\text{ m}$$

The negative sign indicates a virtual image. The magnification is

$$M = -\frac{q}{p} = -\frac{-0.12}{0.30} = +0.40$$

which indicates an image smaller than the object and erect.

33.3 Refraction at a spherical surface

Consider refraction at a spherical surface that separates two media whose indexes of refraction are n_1 and n_2 (Figure 33.9). The fundamental geometric elements are the same as defined in the previous section. Table 33.2 lists the sign conventions used in the text. The sign conventions are the same as for spherical mirrors in Table 33.1 except that a real image falls within medium 2 or to the left of O, giving a negative sign for q, and the reverse if the image is virtual, as in the case of Figure 33.9.

(i) Descartes' formula. An incident ray such as PA is refracted along AD which, when it is extended back into the first medium, intersects the principal axis at Q. If all refracted rays pass through Q we say that Q is the **image** of P. Making $OP = p$ and $OQ = q$, it is shown below that when the incident rays are paraxial these two quantities are related by the expression

$$\frac{n_1}{p} - \frac{n_2}{q} = \frac{n_1 - n_2}{r} \tag{33.5}$$

which is **Descartes' formula for refraction at a spherical surface**. For incident rays that make large angles with the principal axis, the expression relating p and q is more complex.

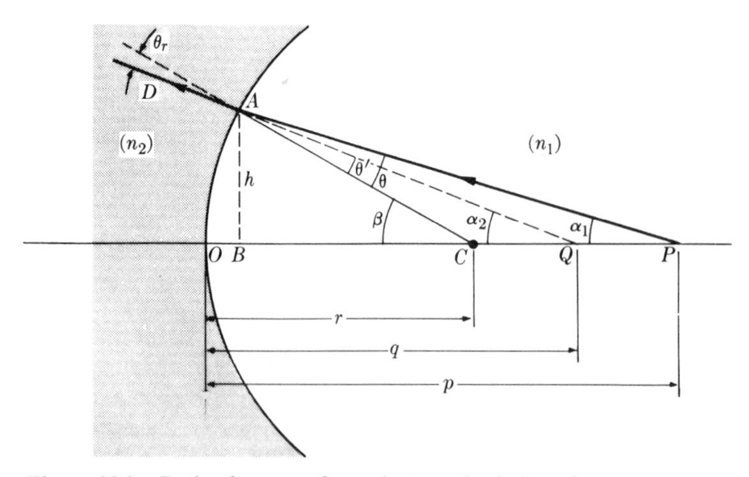

Figure 33.9 Path of a ray refracted at a spherical surface.

Table 33.2 Sign conventions for a spherical refracting surface

	+	−
Radius r	Concave	Convex
Focus f	Convergent	Divergent
Object p	Real	Virtual
Image q	Virtual	Real

(ii) Focal points. The **object focus** F_o, also called the **first focal point**, is the position of a point object on the principal axis such that the refracted rays are parallel to the principal axis, which amounts to having the image of the point at infinity, or $q = \infty$. The distance of the object from the spherical surface is called the **object focal length**, designated by f. Setting $p = f$ and $q = \infty$ in Equation 33.5, we have

$$\frac{n_1}{f} = \frac{n_1 - n_2}{r} \quad \text{or} \quad f = \frac{n_1}{n_1 - n_2} r \tag{33.6}$$

Similarly, when the incident rays are parallel to the principal axis, which is equivalent to having the object at a very large distance from the spherical surface ($p = \infty$), the refracted rays pass through a point F_i on the principal axis called the **image focus** or **second focal point**. In this case the distance of the image from the spherical surface is called the **image focal length**, designated by f'. Making $p = \infty$ and $q = f'$ in Equation 33.5, we have

$$-\frac{n_2}{f'} = \frac{n_1 - n_2}{r} \quad \text{or} \quad f' = -\frac{n_2}{n_1 - n_2} r \tag{33.7}$$

Note that $f + f' = r$. By combining Equations 33.5 and 33.6, we may write Descartes' formula as

$$\frac{n_1}{p} - \frac{n_2}{q} = \frac{n_1}{f} \tag{33.8}$$

When f is positive the system is called **convergent**, and when f is negative it is called **divergent**.

(iii) Image construction. Figure 33.10 shows the construction of the principal rays for the case where $r > 0$ and $n_1 > n_2$, which corresponds to a concave convergent refracting surface. The construction of the image of an object, given those same conditions, is shown in Figure 33.11. The student may draw similar figures for the remaining cases; that is, $r > 0$ and $n_1 < n_2$, and $r < 0$ and $n_1 \gtrless n_2$.

Equation 33.5 also indicates that, for each point object, there is a unique point image. This is correct so long as the spherical surface is of small aperture,

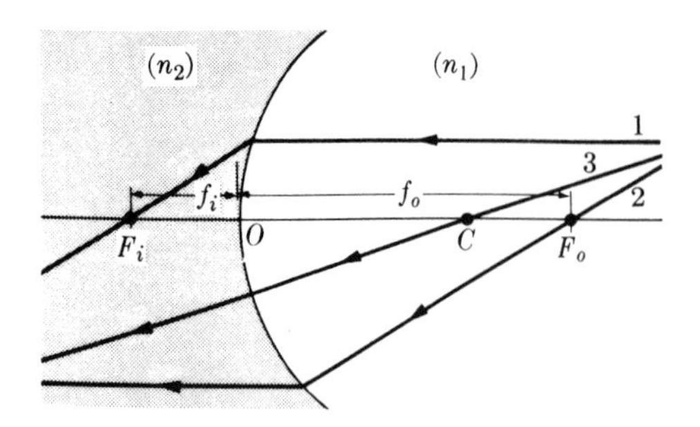

Figure 33.10 Principal rays for a refracting spherical surface. We assume $r > 0$ and $n_1 > n_2$.

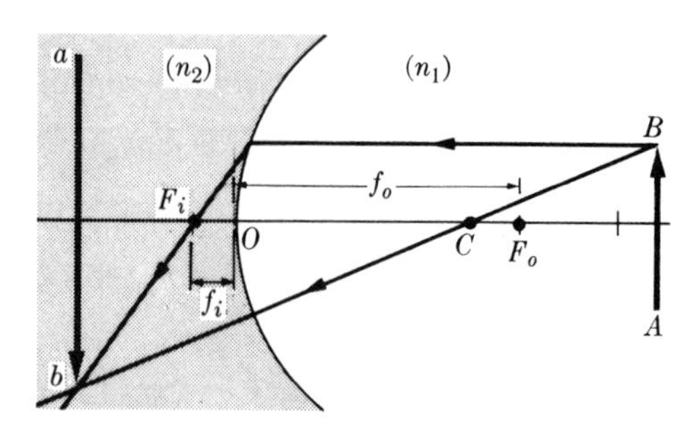

Figure 33.11 Image formation by refraction at a concave spherical surface. We assume $r > 0$ and $n_1 > n_2$.

admitting only rays of very small inclination, so that our approximations are valid. For refracting surfaces of larger aperture, the situation is similar to that encountered in Figure 33.6 for a spherical mirror, and results in the same phenomenon of spherical aberration.

Derivation of Descartes' formula for refraction at a spherical surface

Referring to Figure 33.9, consider ray PA that has AD as the refracted ray. From the triangles PAC and QAC, we observe that $\beta = \theta + \alpha_1$ and $\beta = \theta' + \alpha_2$. Now from Snell's law, $n_1 \sin\theta = n_2 \sin\theta'$. We assume, as we did in the preceding section, that the rays have a very small inclination. Hence the angles θ, θ', α_1, α_2 and β are all very small, and we may write $\sin\theta \approx \theta$ and $\sin\theta' \approx \theta'$, so that Snell's law becomes $n_1\theta = n_2\theta'$, or

$$n_1(\beta - \alpha_1) = n_2(\beta - \alpha_2) \qquad \textbf{(33.9)}$$

Now from Figure 33.9 we find that, as in the case of a spherical mirror, $\alpha_1 \approx h/p$, $\alpha_2 \approx h/q$ and $\beta \approx h/r$, so that, when we substitute in Equation 33.6, cancel the common factor h, and rearrange terms, we get

$$\frac{n_1}{p} - \frac{n_2}{q} = \frac{n_1 - n_2}{r}$$

which is Descartes' formula 33.5. When the aperture of the refracting surface is large the expression relating p and q is more complicated and depends on the distance h.

EXAMPLE 33.3

Magnification produced by a spherical refracting surface.

▷ This problem is similar to that of Example 33.1. Considering Figure 33.12, in which AB is an object and ab is its (virtual) image, we have that $M = ab/AB$. We also have

$$\tan\theta = \frac{AB}{OA} = \frac{AB}{p}, \qquad \tan\theta' = \frac{ab}{Oa} = \frac{ab}{q}$$

so that $AB = p \tan \theta$ and $ab = q \tan \theta'$, and therefore

$$M = \frac{q \tan \theta'}{p \tan \theta} \approx \frac{q \sin \theta'}{p \sin \theta}$$

where the last approximation is valid whenever the angles are small and we can replace the tangents by sines. Then, using Snell's law, $n_1 \sin \theta = n_2 \sin \theta'$, we have

$$M = \frac{n_1 q}{n_2 p} \tag{33.10}$$

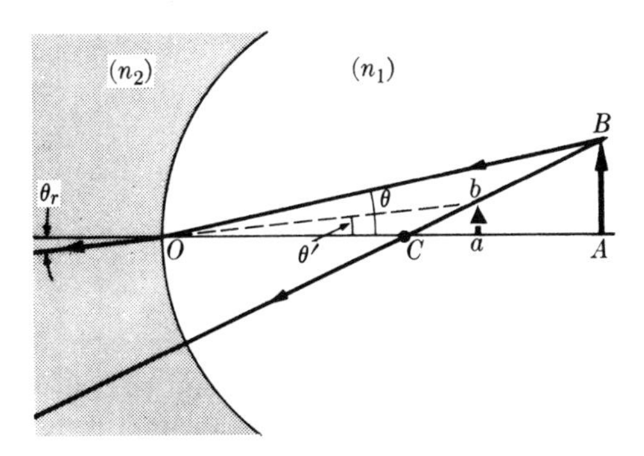

Figure 33.12 Magnification by refraction at a spherical surface.

EXAMPLE 33.4

Determination of the focal lengths, the position of the image, and the magnification for a concave surface, whose radius is 0.50 m, separating a medium whose index of refraction is 1.20 from another whose index is 1.60. An object is placed in the first medium 0.80 m from the surface.

▷ In this case $r = +0.50$ m, $n_1 = 1.20$, and $n_2 = 1.60$. Therefore, using Equations 33.6 and 33.7, we obtain

$$f = \frac{n_1 r}{n_1 - n_2} = -1.50 \text{ m} \quad \text{and} \quad f' = -\frac{n_1 q}{n_1 - n_2} = +2.00 \text{ m}$$

The system is therefore divergent. Using Equation 33.5, we find that

$$\frac{1.20}{0.80} - \frac{1.60}{q} = \frac{1.20 - 1.60}{0.50} \quad \text{or} \quad q = +0.69 \text{ m}$$

The positive sign indicates that the image is virtual. For the magnification we use Equation 33.10, $M = (1.20 \times 0.69)/(1.60 \times 0.80) = 0.65$. Since M is positive, the image is erect; that is, it is in the same direction as the object.

33.4 Lenses

A **lens** is a transparent medium bounded by two curved (usually spherical or cylindrical) surfaces, although one of the faces of the lens may be plane. An incident wave therefore suffers two refractions in going through the lens. For simplicity assume that the medium on both sides of the lens is the same and has index of refraction one (which is approximately true for air), while the index of refraction of the lens is n. We shall also consider only thin lenses, i.e. lenses where the thickness is very small compared with the radii. The sign conventions for lenses are the same as those given in Table 33.2 for a spherical refracting surface.

(i) Descartes' formula. The **principal axis** of a lens is the line determined by the two centers C_1 and C_2 (Figure 33.13). Consider the incident ray PA that passes through P. At the first surface the incident ray is refracted along ray AB. If extended, the ray AB would pass through Q', which is, therefore, the image of P produced by the first refracting surface. The distance q' of Q' from O_1 is obtained by applying Equation 32.5, with n_1 replaced by 1 and n_2 by n; that is

$$\frac{1}{p} - \frac{n}{q'} = \frac{1-n}{r_1} \tag{33.11}$$

At B the ray suffers a second refraction and becomes ray BQ. Then we say that Q is the final image of P produced by the system of two refracting surfaces that constitute the lens. Now, regarding the refraction at B, the object (virtual) is Q' and the image is Q at a distance q from the lens. Therefore, again applying Equation 33.5, with p replaced by q', n_1 by n and n_2 by 1, we have

$$\frac{n}{q'} - \frac{1}{q} = \frac{n-1}{r_2} \tag{33.12}$$

Note that the order of the indexes of refraction has been reversed because the ray goes from the lens into the air. Strictly speaking, the distances appearing in Equations 33.11 and 33.12 must be measured from O_1 and O_2 in each case, so that in Equation 33.12 we should write $q' - t$ instead of q', where $t = O_2O_1$ is the thickness of the lens. But since the lens is very thin we may neglect t, which is equivalent to measuring all distances from a common origin O. Combining Equations 33.11 and 33.12 to eliminate q', we find that

$$\frac{1}{p} - \frac{1}{q} = (n-1)\left(\frac{1}{r_2} - \frac{1}{r_1}\right) \tag{33.13}$$

which is **Descartes' formula for a thin lens**. In this equation q is negative if the image is real because it is to the left of the lens, and the opposite if the image is virtual. Also, for a lens of large diameter admitting rays making large angles with the axis, the expression relating p and q is more complex.

(ii) Focal points. As in the case of a single refracting surface, the **object focus** F_o, or first focal point, of a lens is the position of the object for which the rays emerge parallel to the principal axis ($q = \infty$) after traversing the lens. The distance of F_o from the lens is called the **object focal length**, designated by f. Then, setting $p = f$ and $q = \infty$ in Equation 33.13, we obtain the object focal length as

$$\frac{1}{f} = (n-1)\left(\frac{1}{r_2} - \frac{1}{r_1}\right) \tag{33.14}$$

which is sometimes called the **lensmaker's equation**. Combining Equations 33.13 and 33.14, we have

$$\frac{1}{p} - \frac{1}{q} = \frac{1}{f} \tag{33.15}$$

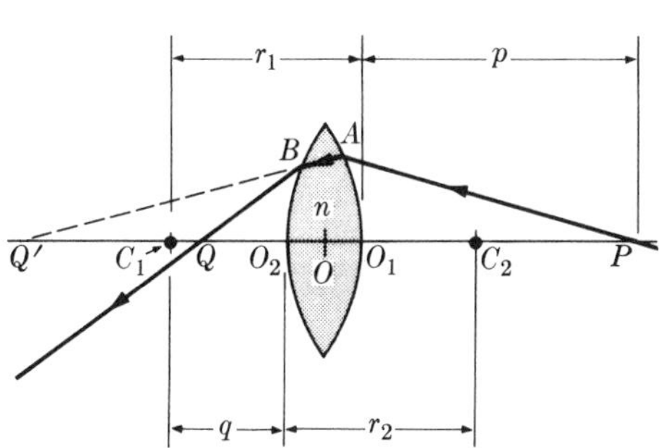

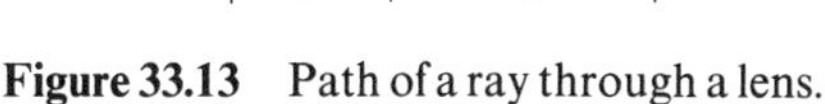
Figure 33.13 Path of a ray through a lens.

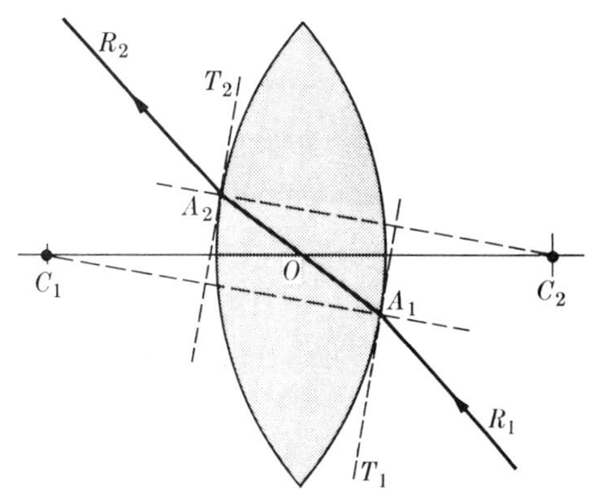

Figure 33.14 Optical center of a lens.

If we determine f experimentally, we may use Equation 33.15 to relate the position of the object and its image without necessarily knowing the index of refraction or the radii of the lens.

For an incident ray that is parallel to the principal axis ($p = \infty$), the emergent ray passes through a point F_i having $q = -f$ and called the **image focus**, or second focal point. Therefore, in a thin lens, the two foci are symmetrically located on both sides. If f is positive the lens is called **convergent**, and if it is negative, **divergent**.

(iii) Optical center. The point O in Figure 33.13 is chosen so that it coincides with the **optical center** of the lens. The optical center is a point defined so that any ray passing through it emerges in a direction parallel to the incident ray. To see that such a point exists, consider, in the lens of Figure 33.14, two parallel radii C_1A_1 and C_2A_2. The corresponding tangent planes T_1 and T_2 are also parallel. For ray R_1A_1, whose direction is such that the refracted ray is A_1A_2, the emergent ray A_2R_2 is parallel to A_1R_1. From the similarity of triangles C_1A_1O and C_2A_2O, we see that the point O is so positioned that

$$\frac{C_1O}{C_2O} = \frac{C_1A_1}{C_2A_2} = \frac{r_1}{r_2}$$

and therefore its position is independent of the particular ray chosen. Therefore all incident rays whose internal ray passes through point O emerge without angular deviation.

(iv) Image construction. For purposes of ray tracing, we may represent a thin lens by a plane perpendicular to the principal axis passing through O. Figure 33.15 shows the construction of the principal rays for a convergent and divergent lens, and in Figure 33.16 these rays have been used for the construction of the image of an object in two cases.

(v) Spherical aberration. The theory we have developed is correct so long as the rays have a very small inclination, so that spherical aberration is negligible. For lenses having a large diameter, the image of a point is not a point, but a line segment. In particular, incident rays that are parallel to the principal axis intersect at different points, depending on their distance from that axis. Spherical aberration is measured then by the difference $f' - f$ between the focal lengths for a marginal

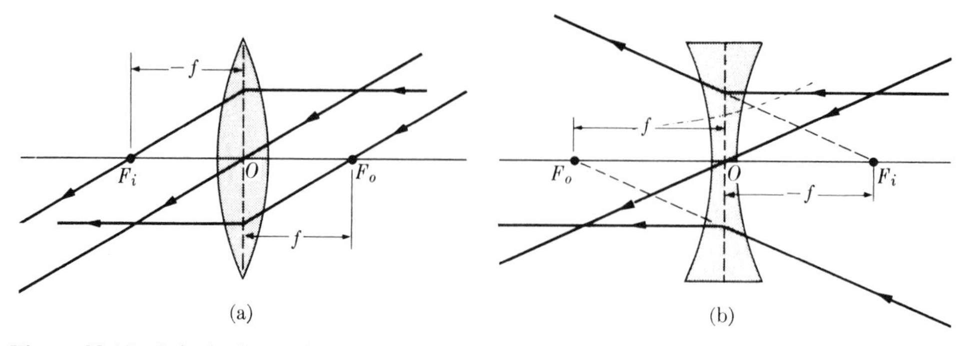

Figure 33.15 Principal rays for (a) convergent and (b) divergent lenses.

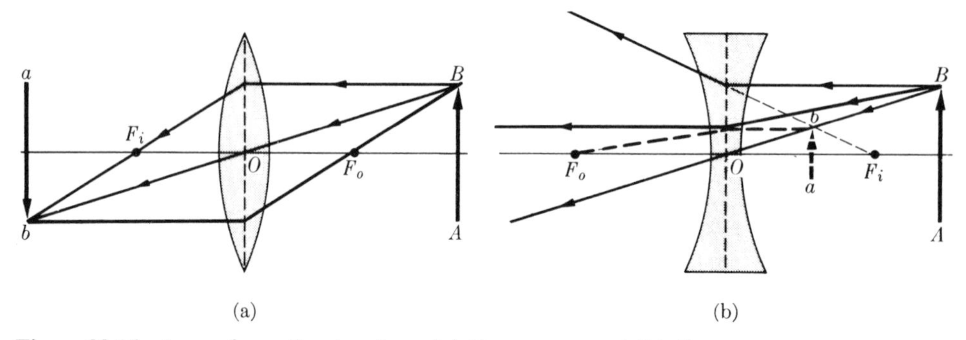

Figure 33.16 Image formation in a lens. (a) Convergent and (b) divergent.

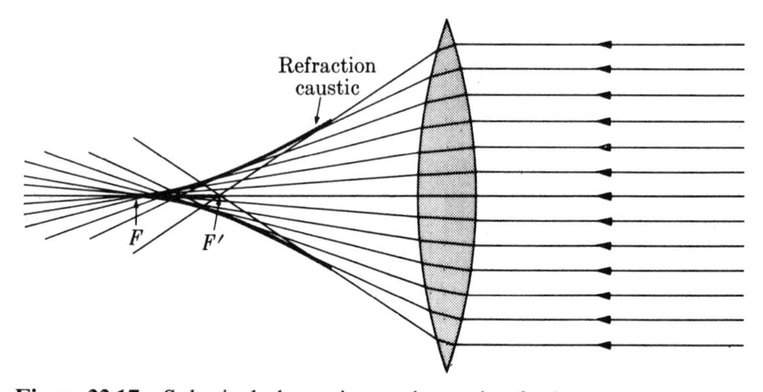

Figure 33.17 Spherical aberration and caustic of a lens.

ray and for an axial ray (Figure 33.17). The refracted rays intersect over a conical surface called the **refraction caustic**.

EXAMPLE 33.5

Magnification produced by a lens.

▷ As before, the magnification is defined as (Figure 33.18) $M = ab/AB$. If O is the optical center of the lens, we have that $\tan \alpha = AB/OA$ and $\tan \alpha = ab/Oa$. Both relations are

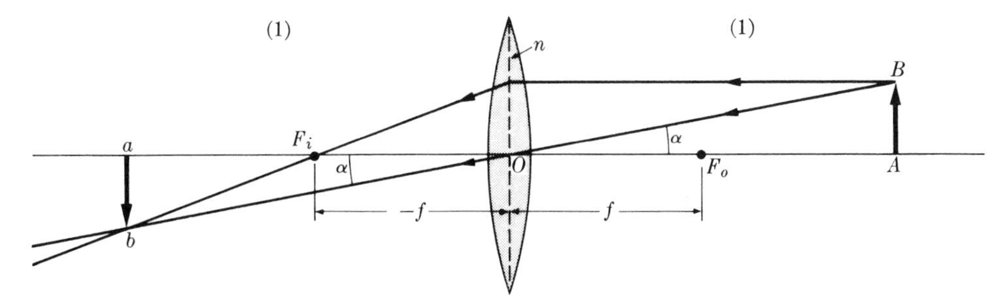

Figure 33.18 Magnification produced by a lens.

correct algebraically, i.e. in magnitude and sign. Therefore

$$\frac{ab}{AB} = \frac{Oa}{OA} \quad \text{or} \quad M = \frac{q}{p} \tag{33.16}$$

which gives the magnification in terms of the object and image distances.

EXAMPLE 33.6

A spherical lens has two convex surfaces of radii 0.80 m and 1.20 m. Its index of refraction is $n = 1.50$. Find its focal length and the position of the image of a point 2.00 m from the lens.

▷ According to the sign conventions of Table 33.2, since the first surface appears convex and the second concave as seen from the side of the object which is placed on the right (see Figure 33.13), we must write $r_1 = O_1C_1 = -0.80$ m, $r_2 = O_2C_2 = +1.20$ m. Therefore, using Equation 33.14, we have

$$\frac{1}{f} = (1.50 - 1)\left(\frac{1}{1.20} - \frac{1}{-0.80}\right) \quad \text{or} \quad f = +0.96 \text{ m}$$

The fact that f is positive indicates that this lens is convergent. In order to find the position of the image we use Equation 33.15 with $p = 2.00$ m and the above value of f, which yields

$$\frac{1}{2.00} - \frac{1}{q} = \frac{1}{0.96} \quad \text{or} \quad q = -1.81 \text{ m}$$

The negative sign of q indicates that the image falls on the left side of the lens and thus it is real. Finally the magnification is $M = q/p = -0.905$. In view of the negative sign, the image must be inverted, and, since M is less than one, the image will be slightly smaller than the object.

EXAMPLE 33.7

Positions of the foci of a system of two thin lenses separated by a distance t.

▷ A system of thin lenses is illustrated in Figure 33.19, which shows, in (a), the path of a ray passing through point P. The image of P produced by the first lens is Q'. Let us call p

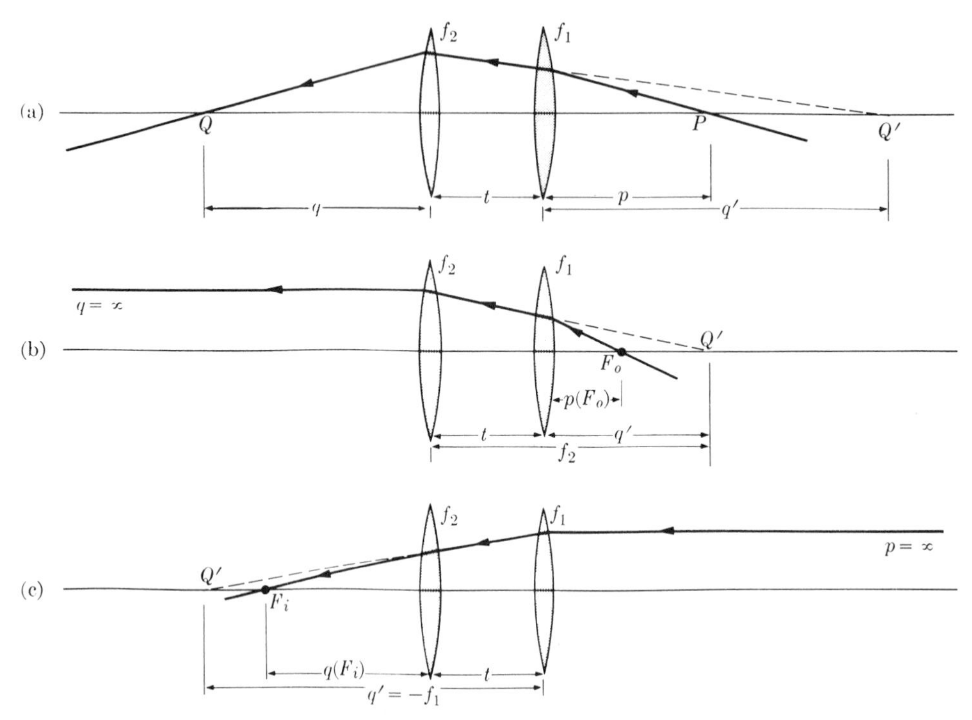

Figure 33.19 System of two thin lenses.

the distance of the object from the first lens. Then the position of Q' is determined by

$$\frac{1}{p} - \frac{1}{q'} = \frac{1}{f_1}$$

Point Q' acts as an object with respect to the second lens, which produces a final image at Q. Since the distance of Q' from the second lens is $q' + t$, we have that

$$\frac{1}{q' + t} - \frac{1}{q} = \frac{1}{f_2}$$

where q is the distance of the final image from the second lens. The above set of equations allows us to obtain the position of the image corresponding to any position of the object.

The object focus F_o (Figure 33.19(b)) of the lens system is the position of the object for which the image Q is at infinity. Similarly, the image focus F_i corresponds to the point on the principal axis through which the final ray passes when the incident ray is parallel to the axis (Figure 33.19(c)).

An important situation occurs when the two lenses are in contact, so that t can be neglected. Then the equation relating q' and q becomes

$$\frac{1}{q'} - \frac{1}{q} = \frac{1}{f_2}$$

which, combined with the first equation, becomes

$$\frac{1}{p} - \frac{1}{q} = \frac{1}{f_1} + \frac{1}{f_2}$$

This shows that a set of thin lenses in contact is equivalent to a single lens of focal length F given by

$$\frac{1}{F} = \frac{1}{f_1} + \frac{1}{f_2}$$

This relation is very important in the design of optical instruments, which usually employ several systems of lenses.

33.5 Optical instruments

Ray tracing in lens and mirror systems is especially important for the design of optical instruments. We shall discuss two: the microscope and the telescope.

(i) The microscope. A microscope is a lens system that produces an enlarged virtual image of a small object. The simplest microscope is a convergent lens, commonly called a **magnifying glass**. The object AB (Figure 33.20) is placed between the lens and the focus F_0, so that the image is virtual and falls at a distance q equal to the minimal distance of distinct vision, δ, which for a normal person is about 25 cm. Since p is almost equal to f, we may write for the magnification,

$$M = \frac{q}{p} \approx \frac{\delta}{f} \tag{33.17}$$

The **compound microscope** (Figure 33.21) consists of two convergent lenses, called the **objective** and the **eyepiece**. The focal length f of the objective is much smaller than the focal length f' of the eyepiece. Both f and f' are much smaller than the distance between the objective and the eyepiece. The object AB is placed at a distance from the objective slightly greater than f. The objective forms a real first image $a'b'$ at a distance L from the objective. This image acts as an object

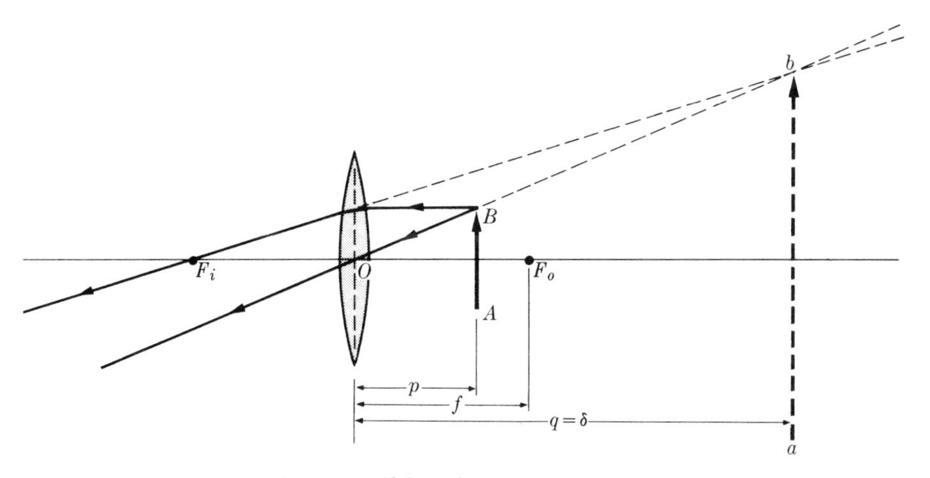

Figure 33.20 Ray tracing in a magnifying glass.

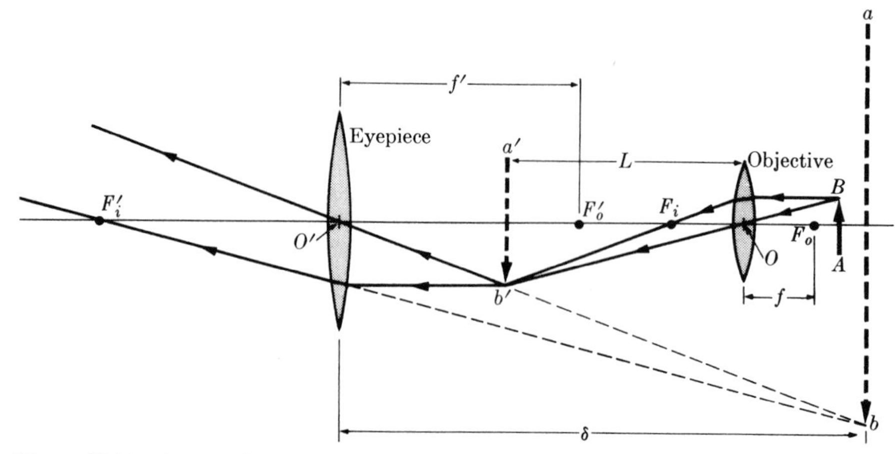

Figure 33.21 Ray tracing in a compound microscope.

for the eyepiece. The image $a'b'$ must be at a distance from the eyepiece slightly less than f'. The final image, ab, is virtual, inverted and much larger than the object. The object AB is so placed that ab is at a distance from the eyepiece equal to the minimal distance of distinct vision, δ (about 25 cm). This condition is attained by the operation called **focusing**, which consists in moving the whole microscope relative to the object. The respective magnifications are

$$M_O = \frac{a'b'}{AB} \approx \frac{L}{f} \quad \text{and} \quad M_E = \frac{ab}{a'b'} \approx \frac{\delta}{f'}$$

Therefore the total magnification is

$$M = M_O M_E = \frac{ab}{AB} = \frac{\delta L}{ff'} \tag{33.18}$$

In an actual microscope, L is practically the same as the distance between the objective and the eyepiece. Since δ and L are fixed, the magnification depends solely on the focal lengths of the objective and the eyepiece.

The useful magnification in a microscope is limited by its **resolving power**; that is, the minimum distance between two points in the object that can be seen as distinct in the image. This resolving power is, in turn, determined by the diffraction at the objective. A detailed calculation gives the resolving power as

$$R = \frac{\lambda}{2n \sin \theta} \tag{33.19}$$

where λ is the wavelength of the light used, n the index of refraction of the medium in which the object is immersed, and θ the angle a marginal ray makes with the axis of the microscope. For most glass lenses $2n$ is about three, so that $R \approx \frac{1}{3}\lambda$. On the other hand, the resolving power of the eye, determined by the structure

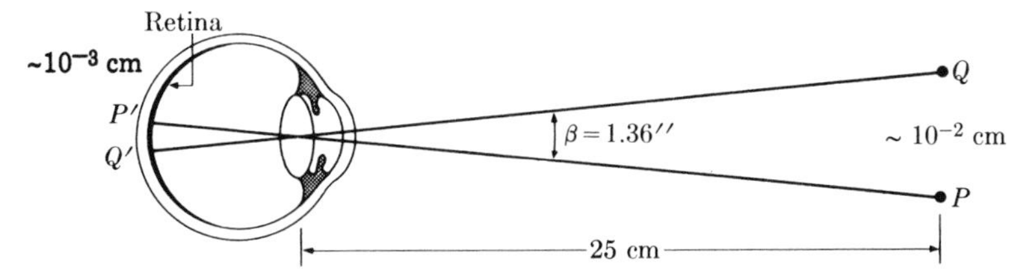

Figure 33.22 Resolving power of the eye.

of the retina, is about 10^{-2} cm for an object at about 25 cm (Figure 33.22). Therefore the maximum useful magnification is

$$M = \frac{10^{-2}\text{ cm}}{\lambda/3} \approx \frac{3 \times 10^{-2}\text{ cm}}{\lambda}$$

For example, for light with $\lambda = 5 \times 10^{-7}$ m, at about the center of the visible spectrum, M is about 600. The resolving power and resolution of optical systems is discussed further in Chapter 35. To minimize spherical and chromatic aberration in a microscope, the eyepieces and the objectives are very complex lens systems composed of several lenses of different materials.

(ii) The telescope. A telescope is an optical system designed to produce a near image of a distant object. In the **refracting** telescope the objective (Figure 33.23) is a convergent lens having a very large focal length f, sometimes of several meters. Since the object AB is very distant, its image $a'b'$, produced by the objective, falls at its focus F_0. We have indicated only the central rays Bb' and Aa', since that is all that is necessary because we know the position of the image. The eyepiece is also a convergent lens, but of a much smaller focal length f'. It is placed so that the intermediate image $a'b'$ falls between O' and F'_0 and the final image ab falls at the minimum distance of distinct vision. Focusing is performed by moving the eyepiece relative to the image $a'b'$.

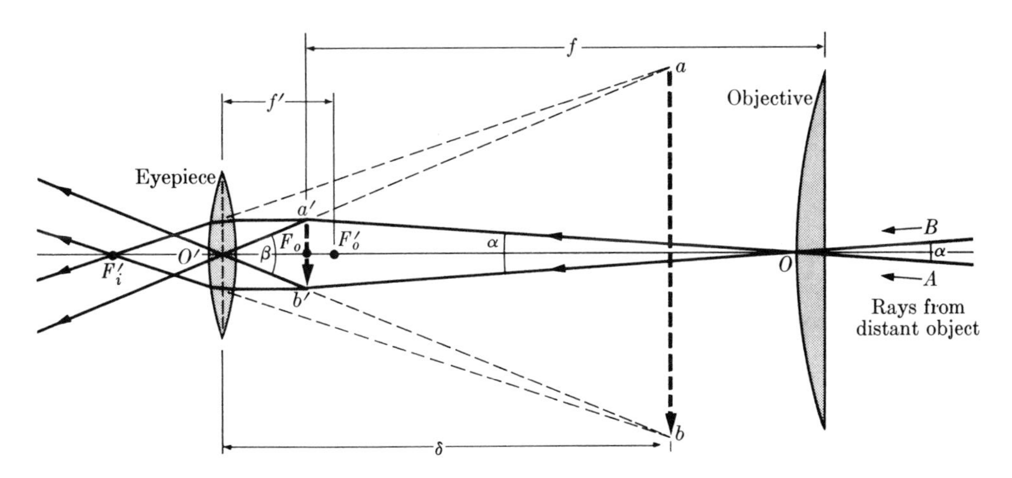

Figure 33.23 Ray tracing in a refracting telescope.

The largest refracting telescope is at the University of Chicago's Yerkes Observatory. The objective has a focal length of 19.8 m, while the eyepiece focal length is 2.8 cm. The diameter of the objective is 1.02 m. This is about the largest lens that can be built both free of aberration at the focus as well as having reasonable elastic stresses. For that reason reflecting telescopes are preferred for astronomical work.

In the **reflecting** telescope, the objective is a concave mirror designed so that it forms an image at its focus that is free from spherical aberration. In reflecting telescopes two methods are used for observing an image. In the **Newtonian mounting** (Figure 33.24(a)), a small mirror mounted on the axis deflects the light toward the eyepiece, mounted on the side. In the **Cassegrainian mounting** (Figure 33.24(b)), a small convex mirror on the axis sends the reflected light back toward the objective, which has a hole in the center. The eyepiece is located in front of this hole. The two largest single mirror reflecting telescopes are the Hale at Mount Palomar, California, whose diameter is about 5 m, and on Mount Pastrekhov, in the Caucasus, the former Soviet Union, that has a diameter of 6 m.

There are three parameters that determine the performance of a telescope. They are the magnification, the resolving power and the aperture of the objective. The magnification produced by a telescope is not linear, but angular, because the image is always smaller than the object. The **angular magnification** is defined as the ratio between the angle β subtended by the image and the angle α subtended by the object; that is,

$$M = \frac{\beta}{\alpha} \tag{33.20}$$

Because of the proximity of the image, the angle β is much larger than α, and this is what creates the sensation of magnification. From Figure 33.23, considering that angles α and β are small, we may write $\alpha \approx \tan \alpha = a'b'/f$ and $\beta \approx \tan \beta \approx a'b'/f'$, because the distance from $a'b'$ to O' is practically f'. Substituting into Equation 33.20, we get

$$M = \frac{f}{f'} \tag{33.21}$$

The same relation applies to a reflecting telescope. Therefore, to obtain a large amplification, the focal length of the objective must be very large and that of the eyepiece very small. Practically, the length of the instrument is determined by the focal length f of the objective.

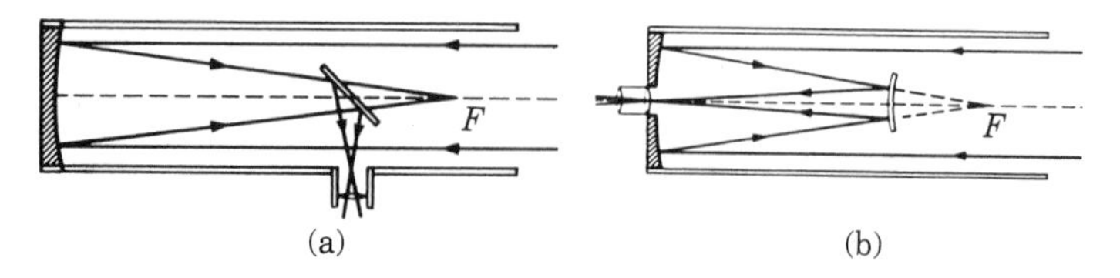

Figure 33.24 Ray tracing in a reflecting telescope. (a) Newtonian mounting; (b) Cassegrainian mounting.

The magnification of a telescope is limited by the **resolving power** of the objective and of the eye of the observer. For an objective lens whose diameter is D, the resolving power (i.e. the minimum angle subtended by two points of the object that appear as distinct or different in the image) is, as will be explained in Chapter 35,

$$\alpha \approx 1.22 \frac{\lambda}{D} \qquad \text{radians} \tag{33.22}$$

On the other hand, as we discussed in connection with the microscope, the resolving power of the eye (Figure 33.22), expressed in terms of an angle, is

$$\beta = \frac{10^{-2}\ \text{cm}}{25\ \text{cm}} = 4 \times 10^{-4}\ \text{radians} = 1.36''$$

Therefore the maximum useful magnification of a telescope is

$$M = \frac{\beta}{\alpha} = \frac{4 \times 10^{-4} D}{1.22\lambda} \approx 3.3 \times 10^{-4} \frac{D}{\lambda} \tag{33.23}$$

Nothing is gained by increasing the ratio f/f' to obtain a magnification larger than Equation 33.23, without improving the resolving power, that is, by increasing D. A larger magnification obtained by increasing f results in a larger intermediate image $a'b'$ without revealing more details. And decreasing f' to produce a larger final image does not add any details that were not present in the intermediate image. For example, for light of $\lambda = 5 \times 10^{-7}$ m, we have $M \approx 660D$, where D is in meters. In the case of the Yerkes telescope, D is about 1 m, resulting in a magnification of about 660 diameters and a resolving power of $1.22(5 \times 10^{-7})$rad or about 10^{-1} second of arc. For the Hale telescope the magnification is about five times larger, or 3600.

The magnification of a telescope is limited by the size of the objective lens or mirror that can be built free of imperfection. Another factor in refracting telescopes is chromatic aberration, to be discussed in Section 33.8. This is another reason why the largest telescopes are not refracting but reflecting.

Note 33.1 New telescope technologies

The construction of reflecting telescopes with mirrors larger than 6 m poses severe problems. The Hale's 5 m mirror is 0.5 m thick and weighs 1.3×10^5 N. It requires a massive support to keep it from sagging under its own weight, which would distort its curvature and change the mirror's focus. Further, the bigger and heavier a mirror, the more susceptible it is to warping due to temperature fluctuations between day and night.

Several new technologies have been developed to improve reflecting telescopes. One technique is to build a large mirror from smaller, lighter pieces. This is called a **multiple mirror telescope** (MMT) (Figure 33.25(a)). The Keck telescope, atop Mauna Kea in Hawaii (scheduled to be completed in 1992), has a mirror with a diameter of 10 meters, but it is composed of 36 concave hexagonal pieces, each 1.8 m wide and 7.6 cm thick, that fit together like tiles on a floor to make a single concave reflecting surface. The hexagons are inserted in a metal frame

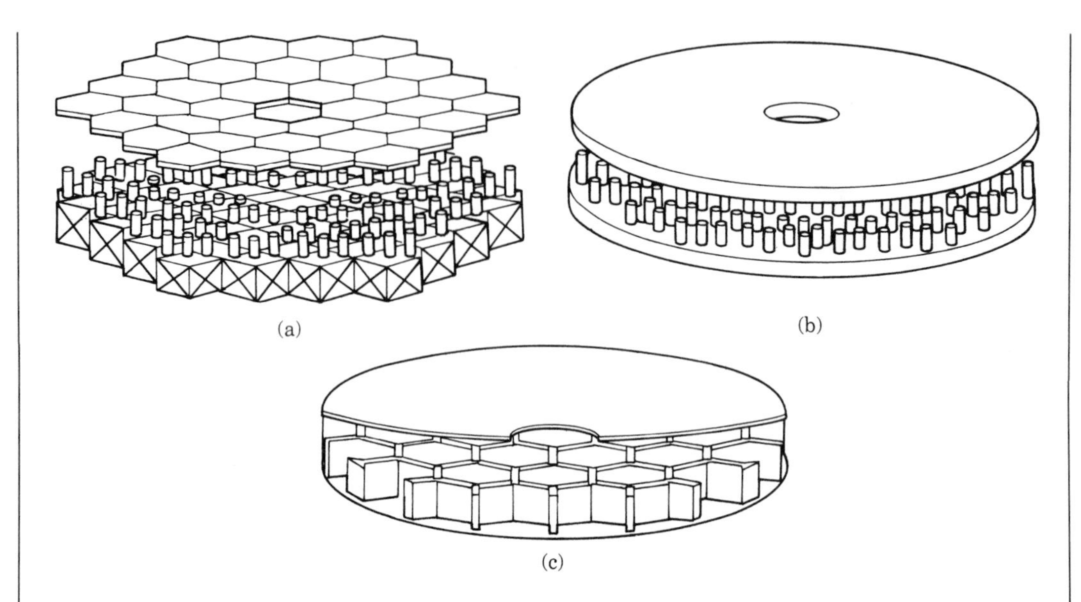

Figure 33.25 New telescope mirrors. (a) Multiple mirror; (b) meniscus; (c) honeycomb. From 'Mirroring the Cosmos', by Corey S. Powell. Copyright © 1991 by Scientific American, Inc. All rights reserved.

and each one is held in place by three movable pistons. This allows adjustment of each hexagon to improve overall focusing once the telescope has been pointed in a given direction.

Another new technology is to use a thin mirror, called a **meniscus**, with multiple support (Figure 33.25(b)). This technology has been used in the New Technology Telescope (NTT) at La Silla, Chile (completed in 1990). It uses a single, relatively small mirror, 3.56 m in diameter. The mirror is only 24 cm thick, making it relatively flexible, and is attached to 75 computer controlled movable supports, which allows changes in its shape to compensate for sagging and temperature distortion. Once the telescope is pointed in a given direction, the shape is adjusted to obtain as sharp an image as possible. Some of the sharpest pictures ever produced with a telescope have been taken at La Silla. Two larger similar telescopes, with diameters about 8 m, are under construction (1992) at La Silla and in Japan.

A third mirror technology is the 'honeycomb' design (Figure 33.25(c)), which consists of two thin curved pieces of glass held together in one piece by a hollow rib structure. This technology has been chosen for the two 8.4 m mirrors of the Columbus double telescope, under construction (1992) at Mt Graham, Arizona, with a focal length of 9.7 m.

Even if the mirror of a telescope is free from aberration, other corrections might be necessary to produce a sharp image. For example, the rapid fluctuations in the temperature of the atmosphere, due to atmospheric turbulence, continuously distorts the wave front of the light from a star. The result is a fuzzy image. To correct the wave front a technique called **adaptive optics**, which combines auxiliary deformable mirrors and computers, has been developed. One technology uses a rubber-like mirror made of a piezoelectric material that changes its shape when an electric current passes through it. A computer controlled light sensor instructs the mirror to change its shape as much as 100 times a second to continuously correct the distortions of the wave front.

In addition to new mirror designs, new materials are used for the mirrors, such as glass ceramics and fused quartz doped with titanium oxide, that reduce the thermal expansion coefficient. Also, rather than visual observation of the image produced by a telescope, photographic plates and, more recently, charge coupled devices (CCD) such as those in video cameras are used to record the image.

As may be seen from Equations 33.22 and 33.23, the magnification and the resolving power of a telescope can be increased by increasing the diameter of the objective. Also, the larger the diameter, the more light that reaches the focus, allowing observation of faint distant objects. This light is contained within a cone of aperture given approximately by D/f. (The same applies to microscopes and photographic and video cameras.) However, in a reflecting telescope, the larger D is relative to f, the more the mirror must depart from the spherical shape or the more aspheric the mirror must be, making it more expensive to build. For that reason, old reflecting telescopes had diameters between 2.2 and 5 times smaller than the focal length (or ***f*-number** of $f/2.2$–$f/5$). This makes the telescope rather long, resulting in severe structural problems and requiring large buildings. The new mirror technologies allow the construction of large aspherical mirrors with apertures close to one. This results in shorter telescopes that are cheaper to build and house and are easier to manage. The aperture of the Keck telescope is $f/1.75$, that of the NTT is $f/1.8$, while the aperture of the Columbus telescope will be $f/1.14$. These new telescope technologies allow us to observe faint light sources as far away as 12×10^9 light years.

33.6 The prism

A prism is a medium bounded by two plane surfaces making an angle A (Figure 33.26). We assume that the medium has an index of refraction n surrounded by a medium having unit index, such as air. An incident ray such as PQ suffers two refractions and emerges deviated an angle δ relative to the incident direction. From the figure it is easily seen that the following relations hold:

$$\sin i = n \sin r \qquad \textbf{(33.24)}$$

$$\sin i' = n \sin r' \qquad \textbf{(33.25)}$$

$$r + r' = A \qquad \textbf{(33.26)}$$

$$\delta = i + i' - A \qquad \textbf{(33.27)}$$

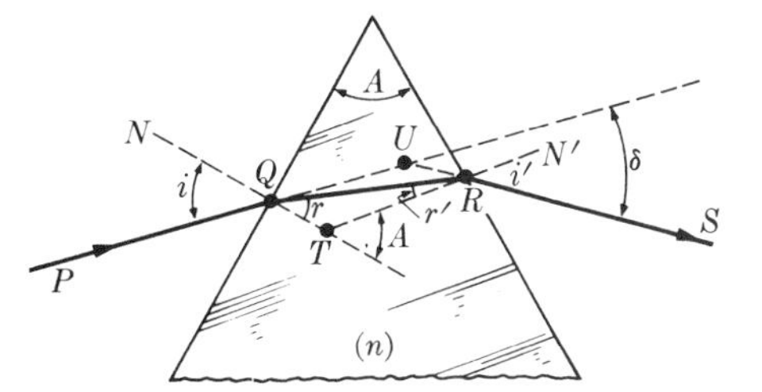

Figure 33.26 Path of a ray through a prism.

The first and second equations are simply Snell's law applied to the refractions at Q and R. The third follows when we use triangle QTR, and the fourth when we use triangle QRU. The first three equations serve to trace the path of the ray and the last allows us to find the deviation.

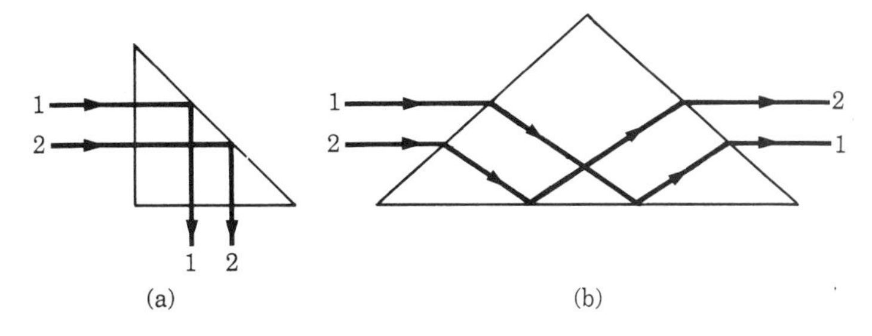

Figure 33.27 Total reflection prisms.

As shown below, the deviation has a minimum value, δ_{min}, when $i = i'$, and therefore $r = r'$. According to Equations 33.26 and 33.27, this requires that

$$i = \tfrac{1}{2}(\delta_{min} + A), \qquad r = \tfrac{1}{2}A \tag{33.28}$$

Note that in this case the path of the ray is symmetric with respect to the two faces of the prism. Introducing Equation 33.28 in Equation 33.24, we obtain

$$n = \frac{\sin \tfrac{1}{2}(\delta_{min} + A)}{\sin \tfrac{1}{2}A} \tag{33.29}$$

which is a convenient formula for measuring the index of refraction of a substance by finding δ_{min} experimentally in a prism of known angle A.

In certain instances it is possible that the inner ray suffers total reflection at the second surface and is sent back into the prism. That occurs when angle r' is larger than the critical angle λ. Figure 33.27 illustrates two such cases. In (a) the rays are bent 90° and in (b) the relative position of the rays is reversed. These kinds of prism have several important applications.

Derivation of the condition for minimum deviation in a prism

Minimum deviation is obtained by making $d\delta/di = 0$. From Equation 33.27, we have $d\delta/di = 1 + di'/di$, and for $d\delta/di = 0$ we must have

$$\frac{di'}{di} = -1 \tag{33.30}$$

From Equations 33.24, 33.25 and 33.26, we have $\cos i \, di = n \cos r \, dr$, $\cos i' \, di' = n \cos r' \, dr'$ and $dr = -dr'$. Therefore

$$\frac{di'}{di} = -\frac{\cos i \cos r'}{\cos i' \cos r} \tag{33.31}$$

Since the four angles i, r, i', and r' are smaller than $\tfrac{1}{2}\pi$ and satisfy the symmetric conditions 33.24 and 33.25, Equations 33.30 and 33.31 can be satisfied simultaneously only if $i = i'$ and $r = r'$.

33.7 Dispersion

When a wave is refracted into a dispersive medium whose index of refraction depends on the frequency (or wavelength), the angle of refraction will also depend on the frequency or wavelength. If the incident wave, instead of being monochromatic, is composed of several frequencies or wavelengths, each component wavelength will be refracted through a different angle, a phenomenon called **dispersion**.

In the particular case of electromagnetic waves, the indices of refraction of some materials that vary with the wavelength in the visible region are shown in

Figure 33.28. We remind the student that colors are associated with wavelength intervals. Therefore white light is decomposed into colors when refracted from air into another substance such as water or glass. If a piece of glass is in the form of a plate with parallel sides, the rays which emerge are parallel, the different colors are superposed again (Figure 33.29(a)), and no dispersion is observed except at the very edges of the image. Even so, this effect is not normally noticeable.

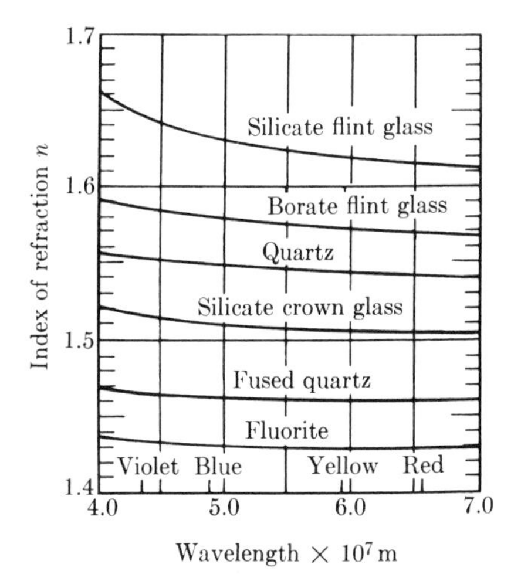

Figure 33.28 Variation of refractive index with wavelength, in the visible region, for some materials.

But if the light passes through a prism (Figure 33.29(b)), the emerging rays are not parallel for the different colors and the dispersion is clearly noticeable, especially at the edges of the image. For that reason prisms are widely used for analyzing light in instruments called **spectroscopes**. A simple type of spectroscope is illustrated in Figure 33.30. Light emitted by a source S, and limited by a slit, is transformed into parallel rays by the lens L. After being dispersed by the prism the rays of different colors pass through another lens L'. Since all rays of the same color (or wavelength) are parallel, they are focused on the same point of the screen. But rays that differ in color (or wavelength) are not parallel; therefore different colors are focused at different points on the screen. The different colors or wavelengths emitted by the source S appear displayed on the screen in what is called the **spectrum** of the light coming from S. If the deviation varies rapidly with the wavelength λ, the colors appear widely spaced on the screen. For each wavelength a line appears on the screen, which is the image of the slit for that wavelength. This is the origin of the term 'line spectra'.

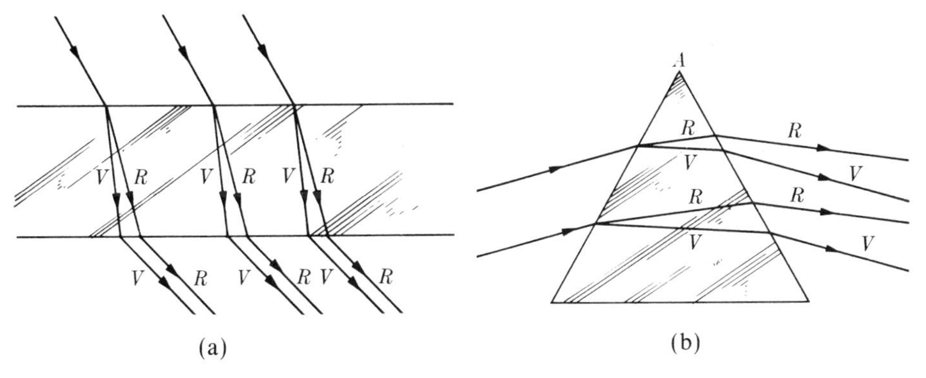

Figure 33.29 Dispersion when light passes through (a) a plate with parallel sides. Rays of different wavelength emerge parallel and no colors are observed; (b) a prism. Rays of different wavelength emerge in different directions and colors are separated.

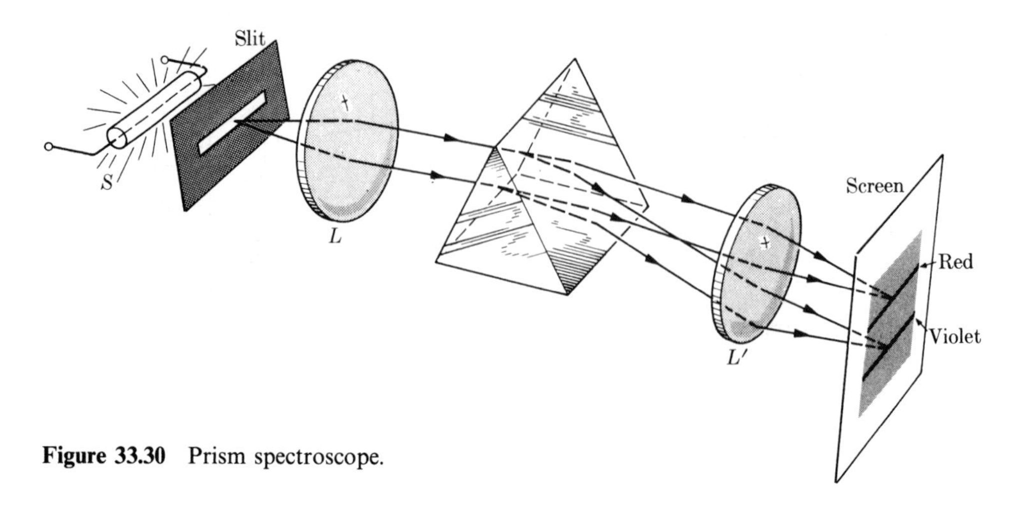

Figure 33.30 Prism spectroscope.

33.8 Chromatic aberration

When light composed of several frequencies or wavelengths (such as white light) passes through a lens, it suffers dispersion, and the edges of the image produced by the lens appear colored. This effect is called **chromatic aberration.** It is easy to understand the reason for this effect if we recognize that a lens can be compared to two prisms attached at their bases (for a convergent lens) or their vertexes (for a divergent lens).

From Equation 33.14,

$$\frac{1}{f} = (n - 1)\left(\frac{1}{r_2} - \frac{1}{r_1}\right)$$

we see that f is determined by the index of refraction n, which in turn depends on the wavelength. Therefore, a lens has a focus for each color or wavelength. For transparent substances whose index of refraction decreases with increasing wavelength in the visible region (see Figure 33.28), violet has a shorter focal length than red. In Figure 33.31, the chromatic aberration of a convergent and a divergent lens is shown for such materials.

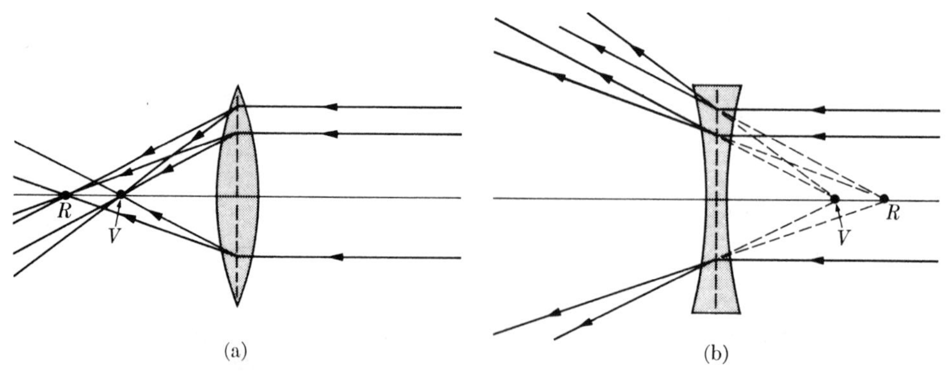

Figure 33.31 Chromatic aberration in lenses. (a) Convergent and (b) divergent lens.

Table 33.3 Indexes of refraction and dispersive power

Fraunhofer line	C	D	F	
Wavelength × 10^7 m	6.563	5.890	4.862	Dispersive power, ω
Crown glass	1.514	1.517	1.524	0.0193
Flint glass	1.622	1.627	1.639	0.0271
Alcohol	1.361	1.363	1.367	0.0165
Benzene	1.497	1.503	1.514	0.0338
Water	1.332	1.334	1.338	0.0180

The chromatic aberration of a lens is measured by the difference $f_C - f_F$ between the focal distances corresponding to the wavelengths 6.563×10^{-7} m and 4.962×10^{-7} m, emitted by hydrogen and designated as the C- and F-Fraunhofer lines. It can easily be shown that the longitudinal chromatic aberration of the lens is

$$A = f_C - f_F = \frac{n_F - n_C}{n_D - 1} f_D \tag{33.32}$$

where f_D is the focal length corresponding to the D-Fraunhofer line. The quantity

$$\omega = \frac{f_C - f_F}{f_D} = \frac{n_F - n_C}{n_D - 1} \tag{33.33}$$

is called the **dispersive power** of the material of the lens. Table 33.3 gives the indices of refraction of some transparent materials at the C-, D-, and F-Fraunhofer lines, as well as their dispersive power.

The kind of chromatic aberration we have discussed for lenses is called **longitudinal** because it is measured along the principal axis. There is also a **transverse** chromatic aberration. Consider an object AB in front of a lens L (Figure 33.32). Unless the light from the object is monochromatic, it will be dispersed as it goes through the lens and, instead of one image, a series of images differing in size will be formed, one for each wavelength or color. The figure shows only the extreme images corresponding to red and violet, and their separation has

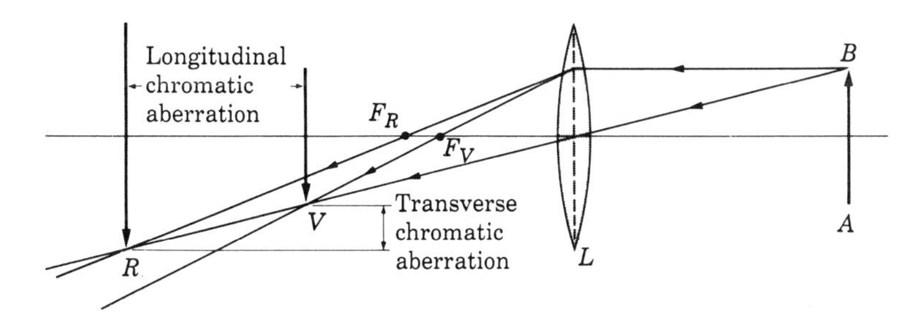

Figure 33.32 Longitudinal and transverse chromatic aberration in a lens.

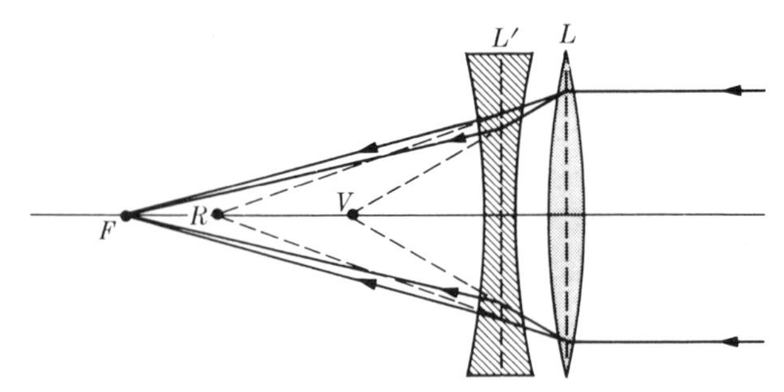

Figure 33.33 Achromatic system of lenses.

been greatly exaggerated. Because of this lateral dispersion, the edges of the images will appear colored.

Chromatic aberration can be reduced or even eliminated by combining lenses of different materials, resulting in what is called an **achromatic system**. To see how this can be done, suppose we have the lens system in Figure 33.33, where, for example, lens L is made of crown glass and lens L' of flint glass. Lens L would have the chromatic aberration indicated by the segment VR. But if the divergent lens L', which has a negative chromatic aberration, is properly designed, rays of all wavelengths should focus at F.

Note 33.2 The Hubble Space Telescope (HST)

When light from outer space reaches the Earth, it must pass through the atmosphere before reaching the Earth's surface. In traversing the atmosphere the light, or in general any electromagnetic radiation, is modified through absorption and scattering by air molecules and dust particles. Fluctuations in atmospheric properties, temperature and pressure, as well as other weather factors diminish the brightness and sharpness of the images produced by Earth-bound telescopes. To reduce these effects, most astronomical observatories are located on mountain tops and far from large urban centers. However, the most effective way of eliminating these distortion effects is to place a telescope in orbit around the Earth, well above the atmosphere. This is the concept of the HST, which was launched in 1990 from Cape Canaveral, Florida in an orbit 613 km above the Earth's surface, with a period of about 90 min and an orbital velocity relative to the Earth of about 28 000 km h^{-1}.

The HST consists of an **Optical Telescope Assembly** (OTA), the **Scientific Instruments** (SI), and the **Support Systems Module** (SSM). The OTA is a reflecting Cassegrain telescope consisting of a 2.4 m primary concave mirror and a 0.3 m secondary convex mirror, separated 4.8 m. The light reflected by the secondary mirror passes through a 0.6 m hole in the primary mirror converging on the focal plane, 1.5 m behind (Figure 33.34). It is most important to maintain the mirrors at a constant temperature of 21 °C to reduce distortions in the image. To increase the sharpness of the image, the telescope has several baffles that reduce the effect of diffuse light. The light at the focal plane is diverted to the scientific instruments, where the image is processed electronically.

One scientific instrument is the **Wide-Field and Planetary Camera**, designed to obtain information about galaxies, stars, planets and comets, which is sensitive to wavelengths from the infrared to the ultraviolet between 1.30×10^{-7} m and 11.00×10^{-7} m. A second instrument is the **Faint Object Camera**, built by the European Space Agency, which is sensitive to wavelengths in the UV and visible regions between 1.15×10^{-7} m and 6.50×10^{-7} m. The camera is designed to observe very faint objects, such as globular clusters, binary star systems and extrasolar planets, with exposures of up to 10 hours.

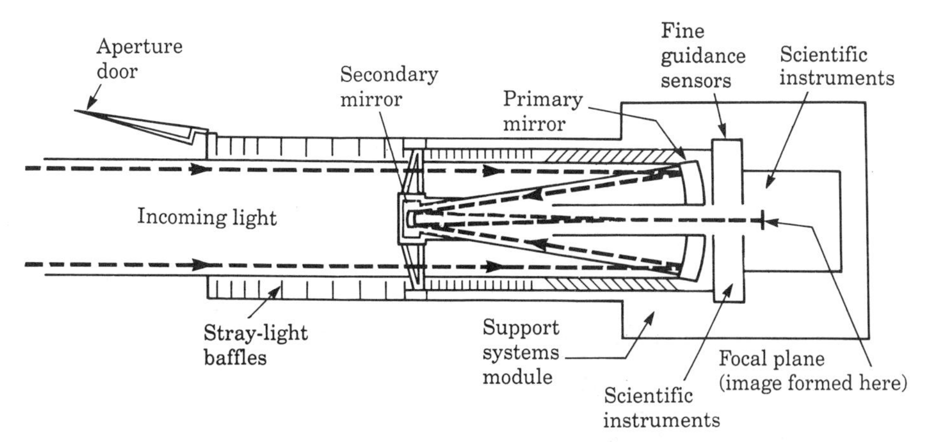

Figure 33.34 Light from distant celestial objects enters the Hubble Space Telescope's front aperture, strikes the 94 inch primary mirror and is reflected back to the 12 inch secondary mirror. The smaller mirror redirects the concentrated light back through a 24 inch hole in the primary mirror to the focal plane and scientific instruments behind this mirror. (Courtesy of NASA.)

The HST carries two spectrographs. The **Faint Object Spectrograph** is designed to obtain the spectra of very faint objects in the region between 1.05×10^{-7} m and 8.50×10^{-7} m corresponding to the visible and ultraviolet regions. It can collect information about the abundance of elements in galaxies, quasars and comets, among others.

The **High Resolution Spectrograph** operates only in the ultraviolet region between wavelengths of 1.05×10^{-7} m and 3.20×10^{-7} m. However, it has a high resolution. The HRS detects radiation from extremely faint sources that does not reach the Earth's surface and permits investigation, for example, of the exchange of mass in binary stars, planetary atmospheres and cosmic gas clouds.

Finally, the **High Speed Photometer** is designed to provide accurate observation of the total electromagnetic spectrum emitted by an object in space in the wavelength range from 1.20×10^{-7} m to 7.00×10^{-7} m, corresponding to the visible and ultraviolet regions. For example, it can make precise measurements of rapidly pulsating objects and binary systems, as well as measuring zodiacal light from the Sun and diffuse galactic light.

The Support System Module (SSM) provides the HST instruments with the power, communications, pointing control and other electronic support required for operation. Because of the high resolution of the Optical Telescope Assembly, it is necessary to point the telescope precisely and then lock its orientation during exposure while the HST orbits. This is the function of the **Fine Guidance Sensors**. Four massive variable speed rotors or gyroscopes, which operate on the principle of conservation of angular momentum (recall Section 13.10), serve to change the orientation of the spacecraft, which can then be locked in with an accuracy better than 10^{-2} seconds of arc. The system also allows precise measurement of the positions of planets, stars and galaxies.

The **Data Management System** (DMS) processes information gathered by the scientific instruments and relays it to Earth via high gain antennas at the rate of 10^6 bits per second. The DMS also receives instructions from the Earth, relaying them to the instruments via the **Scientific Instrument Control System**. The communication is carried out as indicated in Figure 33.35. Instructions from the Space Telescope Science Institute, at the Johns Hopkins University in Maryland, are translated into commands for the HST at the nearby Goddard Space Flight Center and from there they are relayed to the White Sands station in New Mexico, through a commercial satellite. The White Sands station then sends a signal to NASA's geosynchronous Tracking and Data Relay Satellite (TDRS). The TDRS forwards the commands to the HST. Information from the HST is returned to Earth along the same route, but in reverse.

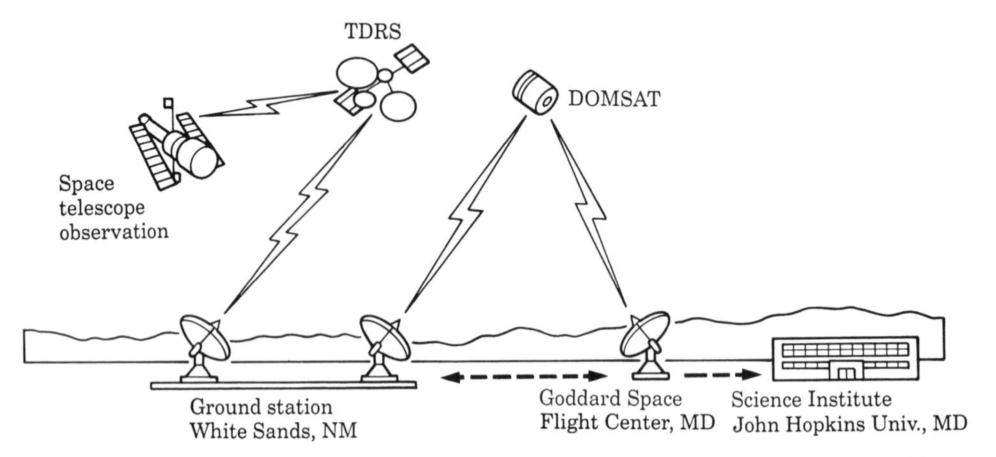

Figure 33.35 Telemetry from the Hubble Space Telescope is first sent to NASA's Tracking and Data Relay Satellites, which then relays the data to White Sands, NM. This installation boosts the HST signal to a domestic communications satellite (DOMSAT). The second satellite then sends the electronic information to the Space Telescope Operations Control Center at Goddard Space Flight Center. Goddard forwards scientific data to the Space Telescope Science Institute at Johns Hopkins University. (Courtesy of NASA.)

Electrical power for the telescope is provided by two panels containing 48 000 solar cells. The construction of the HST is modular so that its parts can be changed or serviced as needed, using the space shuttle. The HST has a mass of 12 000 kg, it is 13 m long and its diameter at the widest point is 4.3 m. It has been designed to last for about 15 years.

Although the primary mirror was constructed with a precision of about 3×10^{-8} mm, a small error was made at some points on the edge of the mirror. The resulting aberration has limited the sharpness of the images produced by the OTA. To correct this defect, astronauts in a shuttle mission planned for 1994 will install corrective lenses in the path of the reflected light before it reaches the scientific instruments. In spite of this error, the HST has provided unique information about distant objects, particularly galaxies.

Note 33.3 Non-imaging optics

Although mirrors and lenses are normally used to form a sharp image of an object, for which aberrations must be corrected, it is also possible to use them to channel light rays in certain directions or to certain places, without formation of an image, in which case aberrations do not matter. This application is called **non-imaging optics**. For example, concave mirrors can be used to concentrate light, or electromagnetic radiation, in a small region. The concentration is enhanced by using a conical crystal of high index of refraction, such as sapphire. This device is called a **non-imaging concentrator**. It is placed at the mirror's focal region with the larger base toward the mirror. Light reflected from the mirror enters the concentrator and is funneled toward the smaller base, assisted by total reflections at the lateral surfaces. In this way the intensity of light can be increased several thousand times relative to the intensity of the light coming from the source. Figure 33.36 illustrates the use of a concentrator in a solar energized laser. Non-imaging techniques have also been used to detect very faint sources of radiation, such as the cosmic background radiation measured by COBE (recall Example 31.5), in experiments with high energy particles to detect the radiation emitted by fast moving particles, and in certain illumination systems using fluorescent lamps to concentrate the light in a specific direction.

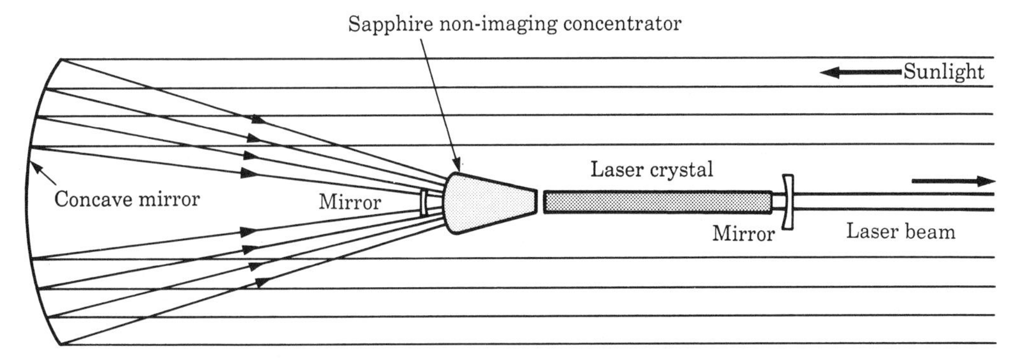

Figure 33.36 In a solar energized laser, light from the Sun is reflected by a concave mirror into a sapphire non-imaging concentrator. The concentrated sunlight enters a laser crystal at one end and induces atoms in the crystal to emit light of a certain frequency or frequencies. The emitted light bounces back and forth between two partially reflecting mirrors positioned outside the crystal. Some of the light 'leaks' through the mirror on the right, forming a laser beam.

QUESTIONS

33.1 Distinguish between (a) a real and a virtual image (b) a real and a virtual object.

33.2 Indicate which images are real and which are virtual: (a) our own image as seen in a plane mirror; (b) the image of an object seen through a magnifying glass; (c) the image projected on the screen by a movie projector or a slide projector; (d) the image of an object as projected by the eye's lens on the retina; (e) the image produced on a photographic film by the objective lens of a camera; (f) the image of an object seen through a microscope or a telescope.

33.3 Distinguish the relation of the image to the object when the magnification is (a) greater than one, (b) one, (c) less than one, (d) positive and (e) negative. Consider the case of a mirror as well as a lens.

33.4 Consider a refracting surface separating two media, with $n_1 > n_2$. Give the sign of the radius and indicate whether the surface is concave or convex when the refracting surface is (a) convergent or (b) divergent. Repeat with $n_1 < n_2$.

33.5 State what is meant by (a) spherical aberration, (b) chromatic aberration.

33.6 Which of the following systems exhibit spherical aberration: (a) a spherical mirror; (b) a spherical refracting surface; (c) a spherical lens? Which of these systems exhibit chromatic aberration?

33.7 Discuss means of reducing or eliminating (a) spherical aberration, (b) chromatic aberration.

33.8 What is meant by the term 'focusing' in an optical instrument? Discuss how focusing is done with (a) a microscope, (b) a telescope, (c) a slide projector, (d) a photographic or video camera.

33.9 Sketch the path of rays and image formation in (a) a photographic camera, (b) a slide projector.

33.10 What is the principle on which a prism spectroscope operates?

33.11 Is dispersion observed when monochromatic light passes through a prism?

33.12 Can a substance whose index of refraction does not vary appreciably with wavelength be used for the prism of a spectroscope?

33.13 Investigate the use of total reflection prisms in optical instruments.

33.14 Explain the difference between the magnification of a telescope and a microscope.

33.15 Explain how the resolving power of the eye limits the magnification of a (a) telescope, (b) microscope.

33.16 Explain the meaning of f-number. How does one change the f-number of a camera as the intensity of light decreases?

PROBLEMS

33.1 Verify that when a plane mirror is rotated by an angle α, the reflected ray rotates through an angle twice as large; that is, $\beta = 2\alpha$ in Figure 33.37.

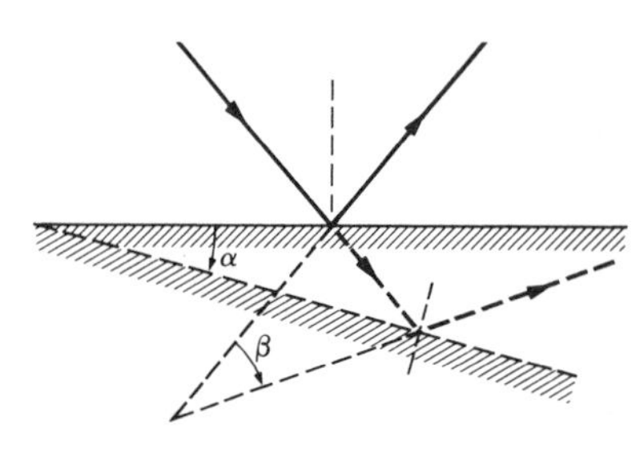

Figure 33.37

33.2 Verify that if a plane mirror is displaced parallel to itself a distance x in the direction of the normal, the image moves a distance $2x$.

33.3 A concave mirror has a radius of 1.00 m. Calculate the position of the image of an object and its magnification if the object is at a distance from the mirror equal to (a) 1.40 m, (b) 1.00 m, (c) 0.80 m, (d) 0.50 m and (e) 0.30 m.

33.4 A convex mirror has a radius of 1.00 m. (a) Calculate the position of the image of an object and the magnification if the distance from the object to the mirror is 0.60 m. Also consider a virtual object at a distance of (b) 0.30 m, and (c) 0.80 m.

33.5 A concave shaving mirror has a focal length of 15 cm. (a) Find the optimum distance of a person from the mirror if the distance of distinct vision is 25 cm. (b) Calculate the magnification.

33.6 A concave mirror produces a real, inverted image three times larger than the object and at a distance of 28 cm from the object. Calculate the focal length of the mirror.

33.7 When an object, initially 60 cm from a concave mirror, moves 10 cm closer to it, the separation between the object and its image becomes 2.5 times larger. Determine the focal length of the mirror.

33.8 The **spherical aberration** of a (spherical) mirror is defined as the difference between the focal length f for a ray close to the mirror's axis and the focal length f' for a ray at the edge. Verify that $f - f' = f(\sec \alpha - 1)$, where α is the angle of incidence for a ray parallel to the axis that strikes the lens at its edge.

33.9 Apply the result of Problem 33.8 to a concave mirror that has a radius of 10 cm. The base of the mirror has a radius of 8 cm. Find the spherical aberration of the mirror and compare with its focal length.

33.10 If x_1 and x_2 are the distances of the object and its image, measured from the focus of a spherical mirror, verify that Equation 33.2 gives $x_1 x_2 = f^2$. This is called **Newton's equation** for a mirror. Can you conclude then that the object and its image are always on the same side of the focus? (Hint: note that $x_1 = FP = OP - OF$, and similarly for x_2.)

33.11 Given the focal length f of a spherical mirror and the magnification M of a given object, verify that the positions of the object and the image may be written as $p = f(M - 1)/M$ and $q = -f(M - 1)$.

33.12 A transparent substance is limited by a concave spherical surface of radius equal to 0.60 m. Its index of refraction is 1.5. (a) Calculate the focal lengths. (b) Determine the position of the image and the magnification of an object placed (a) 2.40 m, (b) 1.60 m and (c) 0.60 m from the surface. Repeat the problem for a convex surface.

33.13 A transparent rod 40 cm long is cut flat at one end and rounded to a convex hemispherical surface of 12 cm radius at the other end. An object is placed on the axis of the rod, 10 cm from the hemispherical end. (a) What is the position of the final image? (b) What is its magnification? Assume the refractive index to be 1.50.

33.14 Sketch the various possible thin lenses that may be obtained by combining two surfaces whose radii of curvature are 10 cm and 20 cm. Which are converging and which are diverging? Find the focal length in each case. Assume $n = 1.5$.

33.15 A double convex lens has an index of refraction of 1.5 and its radii are 0.20 m and 0.30 m. (a) Calculate the focal length. (b) Determine the position of the image and the magnification of an object which is at a distance from the lens equal to (i) 0.80 m, (ii) 0.48 m, (iii) 0.40 m, (iv) 0.24 m and (v) 0.20 m.

33.16 Using the lens of the previous problem, consider the case of a virtual object that is 0.20 m behind the lens.

33.17 A double concave lens has an index of refraction of 1.5 and its radii are 0.20 m and 0.30 m. (a) Find the focal length. (b) Determine the position of the image and the magnification of an object which is 0.20 m from the lens.

33.18 Using the lens of the previous problem, consider a virtual object at a distance of (a) 0.40 m and (b) 0.20 m.

33.19 A lens system is composed of two convergent lenses in contact with focal lengths 30 cm and 60 cm. (a) Calculate the position of the image and the magnification of an object placed 0.20 m from the system. (b) Consider also a virtual object placed at a distance of 0.40 m from the system.

33.20 A convergent lens has a focal length of 0.60 m. Calculate the position of the object to produce an image that is (a) real and three times larger, (b) real and one-third as large and (c) virtual and three times larger.

33.21 An object is placed 1.20 m from a lens. Determine the focal length and the nature of the lens that produces an image (a) real and 0.80 m from the lens, (b) virtual and 3.20 m from the lens, (c) virtual and 0.60 m from the lens and (d) real and twice as large.

33.22 The **ocular** (or eyepiece) of an optical instrument is composed of two identical convergent lenses of focal length 5 cm each, separated by 2.5 cm. Find the position of the foci of the system as measured from the closer lens.

33.23 The objective of a microscope has a focal length of 4 mm. The image formed by this objective is 180 mm from its second focal point. The eyepiece has a focal length of 31.25 mm. (a) Calculate the magnification of the microscope. (b) The unaided eye can distinguish two points as being separate if they are about 0.1 mm apart. What is the minimum separation that can be perceived with the aid of this microscope?

33.24 The diameter of the Moon is 3.5×10^3 km and its distance from the Earth is 3.8×10^5 km. Find the angular diameter of the image of the Moon formed by a telescope if the focal length of the objective is 4 m and that of the eyepiece 10 cm.

33.25 A prism has an index of refraction of 1.5 and an angle of 60°. (a) Determine the deviation of a ray incident at an angle of 40°. (b) Find the minimum deviation and the corresponding angle of incidence.

33.26 The minimum deviation of a prism is 30°. The angle of the prism is 50°. Find (a) its index of refraction and (b) the angle of incidence for minimum deviation.

33.27 Show that if the angle of a prism is very small and the incident rays fall almost perpendicular to one of the faces, the deviation is $\delta = (n - 1)A$.

33.28 If a ray reaches the second surface of a prism at an angle larger than the critical angle, total reflection occurs and the ray is reflected back instead of passing out of the prism. This principle is used in many optical instruments. (a) Show that if $n > 1$, the condition that at least one ray will emerge is that $A \leqslant 2\lambda$, where λ is the critical angle. (b) Discuss the range of values of the angle of incidence if the ray is to emerge on the other side. This range is given by the angle α shown in Figure 33.38. Verify that it is given by $\cos \alpha = n \sin(A - \lambda)$. (c) Discuss the variation of α with A.

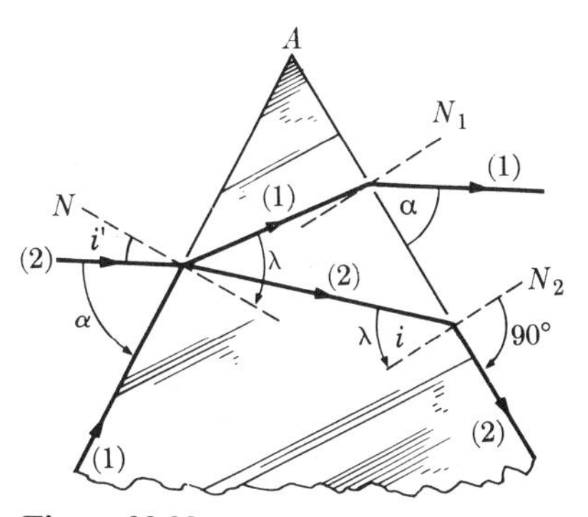

Figure 33.38

33.29 Apply the discussion of the preceding problem to the case of a prism having a refracting angle of 45° and an index of refraction of 1.5. (a) Obtain the value of α. (b) Discuss the path of a ray that falls perpendicular to one of the faces. (c) Consider the case when the prism's angle is 35°.

33.30 A lens system is composed of two lenses in contact, one plano-concave of flint glass and another double convex of crown glass. The radius of the common face is 0.20 m and the radius of the other face of the crown lens is 0.12 m. Calculate (a) the focal length of the system and (b) the chromatic aberration.

34 Interference

Albert A. Michelson designed an interferometer with which he performed over the years, beginning in 1887, a series of crucial and precise experiments to measure how the orbital motion of the Earth affected the velocity of light. At that time it was believed that a medium called ether permeated all space. Michelson was not able to detect any change in the velocity of light regardless of its direction of propagation relative to the Earth. The negative results of Michelson's experiments provided Einstein with one of the justifications on which to base his theory of special relativity. Michelson's interferometer has many other applications such as in spectroscopy and in the measurement of slight changes in length.

Notes

34.1 Introduction

When two or more waves coincide in space and time, there is **interference**. In Chapter 10 we discussed the superposition of two simple harmonic motions; the theory developed there can be applied directly to this problem for the case of harmonic or monochromatic waves. Interference occurs, for example, in the region where reflected and incident waves coincide. In fact, this is one of the most common methods of producing interference. Interference also occurs when a wave motion is confined to a limited region of space, such as a string with its two ends fixed, or a liquid in a channel, or an electromagnetic wave in a metallic cavity, giving rise to **standing waves**.

In order to apply the formulas developed in Chapter 10, we shall write for a harmonic wave moving in the $+X$-direction

$$\xi = A_0 \sin(\omega t - kx) \quad \textbf{(34.1)}$$

and for one moving in the $-X$-direction

$$\xi = A_0 \sin(\omega t + kx) \quad \textbf{(34.2)}$$

(recall Section 28.3). The theory developed here is applicable to any kind of wave motion, but our examples and applications will refer mostly to acoustic and electromagnetic waves.

34.2 Interference of waves produced by two synchronous sources

Consider two point sources S_1 and S_2 (Figure 34.1), which oscillate in phase with the same angular frequency ω and amplitudes A_1 and A_2. Their respective spherical waves are

$$\xi_1 = A_1 \sin(\omega t - kr_1) \quad \textbf{(34.3)}$$

and

$$\xi_2 = A_2 \sin(\omega t - kr_2) \quad \textbf{(34.4)}$$

where r_1 and r_2 are the distances from any point in the medium to S_1 and S_2, respectively. Note that, although the two sources may be identical, they do not produce the same amplitude at P if r_1 and r_2 are different because, as we know from Section 28.11, the amplitude of a spherical wave falls according to a $1/r$ law. However, in our analysis we shall not consider explicitly the variation of the amplitude of the spherical waves with the distance from the source. This is a valid approximation for points for which r_1 and r_2 are not too different.

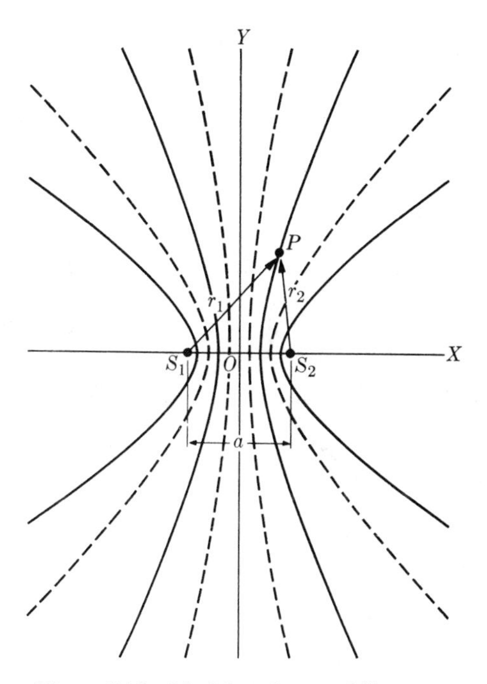

Figure 34.1 Nodal and ventral lines resulting from interference of waves produced by two identical sources.

(i) Phase condition. Suppose that ξ is a scalar property, such as a pressure disturbance. If ξ corresponds to a vector quantity, such as a displacement or an electric field, we assume that ξ_1 and ξ_2 are in the same direction so that the combination of the two waves can be treated in a scalar manner. When we compare Equations 34.3 and 34.4 with Equation 10.1, that is, $\xi = A \sin(\omega t + \alpha)$, the quantities kr_1 and kr_2 play the same role as the initial phases, except for a change in sign, which is irrelevant for our analysis. Then the phase difference between the two wave motions at any point P (if we remember that $k = 2\pi/\lambda$) is

$$\delta = kr_1 - kr_2 = \frac{2\pi}{\lambda}(r_1 - r_2) \tag{34.5}$$

When we use the technique of rotating vectors, which was explained in Section 10.3, the two interfering wave motions can be represented by rotating vectors, of length A_1 and A_2 respectively, that make angles $\alpha_1 = kr_1$ and $\alpha_2 = kr_2$ with a reference axis designated OX in Figure 34.2. The amplitude A and phase α of the resulting wave motion are given by their vector resultant. Therefore we may express the amplitude of the resulting disturbance at P by

$$A = \sqrt{A_1^2 + A_2^2 + 2A_1A_2 \cos \delta} \tag{34.6}$$

We see that A falls between the values $(A_1 + A_2)$ and $(A_1 - A_2)$, depending on whether $\cos \delta = +1$ or -1 or $\delta = 2n\pi$ or $(2n + 1)\pi$, where n is either a positive or a negative integer. In the first case we have maximum reinforcement of the two wave motions, or **constructive interference**, and in the second case maximum attenuation, or **destructive interference**. That is,

$$\delta = \begin{cases} 2n\pi & \text{constructive interference} \\ (2n+1)\pi & \text{destructive interference} \end{cases}$$

Using Equation 34.5, we have

$$\frac{2\pi}{\lambda}(r_1 - r_2) = \begin{cases} 2n\pi & \text{constructive interference} \\ (2n+1)\pi & \text{destructive interference} \end{cases} \tag{34.7}$$

which may be written as

$$r_1 - r_2 = \begin{cases} n\lambda & \text{constructive interference} \\ (2n+1)\dfrac{\lambda}{2} & \text{destructive interference} \end{cases} \tag{34.8}$$

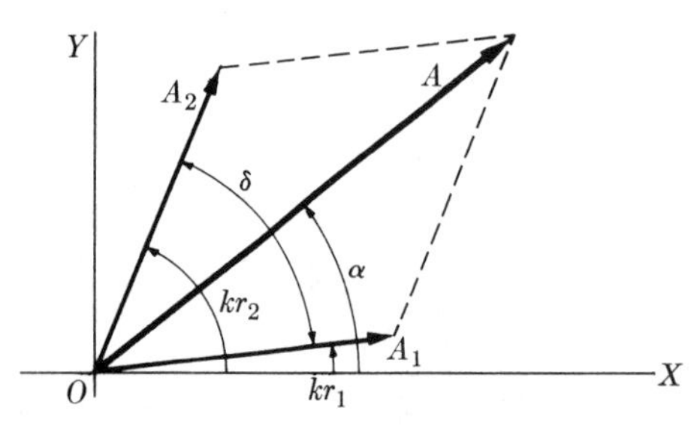

Figure 34.2 Resultant amplitude of two interfering waves. The line OX has been taken as the reference line.

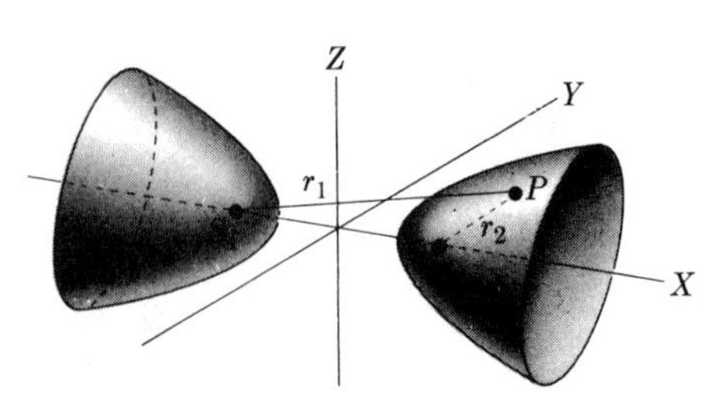

Figure 34.3 Surfaces of constant phase difference for spherical waves produced by two coherent point sources S_1 and S_2.

But $r_1 - r_2 = const.$ defines a hyperbola whose foci are S_1 and S_2, or, since the problem is actually in space, this equation defines hyperbolic surfaces of revolution, such as those in Figure 34.3. Therefore we conclude from Equation 34.8 that at hyperbolic surfaces whose equations are $r_1 - r_2 = 0, \pm\lambda, \pm 2\lambda, \pm 3\lambda, \ldots$ the two wave motions interfere with reinforcement. These surfaces are called **ventral** or **antinodal surfaces**. At the hyperbolic surfaces whose equations are $r_1 - r_2 = \pm\frac{1}{2}\lambda, \pm\frac{3}{2}\lambda, \ldots$ the two wave motions interfere destructively. These surfaces are called **nodal surfaces**. The overall pattern is thus a succession of alternate ventral and nodal surfaces. The intersections of these surfaces with a plane passing through the line joining the two sources are the hyperbolas illustrated in Figure 34.1, where the common line is the X-axis.

(ii) Coherence. The situation described is such that, at each point of space, the resulting wave motion has a characteristic amplitude, given by Equation 34.6, so that $\xi = A \sin(\omega t - \alpha)$, where α is as indicated in Figure 34.2. Therefore the result of the interference does not have the appearance of a progressive wave motion but a **stationary** situation where, at each point of space, the oscillatory motion has a fixed amplitude and phase. The reason for this is that the two sources oscillate with the same frequency and maintain a constant phase difference, and hence are said to be **coherent**.

If the phase difference of the sources changes erratically with time, even if they have the same frequency, no stationary interference pattern is observed, and the sources are said to be **incoherent**. This is what happens with light sources composed of a large number of the same kind of atoms, which emit light of the same frequency. Since there are many atoms involved in each source and they do not oscillate in phase, no definite interference pattern is observed. For that reason we do not observe interference from two light bulbs.

To circumvent this difficulty and produce two coherent beams of light, several devices have been designed. One effective way is by splitting a laser beam in two and recombining the beams at a point. One way of splitting a light beam is by use of **Fresnel's biprism**. It is composed of two prisms, P_1 and P_2, joined at their base. Light coming from the source S is refracted in each prism and separated into two coherent beams that apparently proceed from two coherent sources, S_1 and S_2. These are the images of S produced by each prism. Coherence

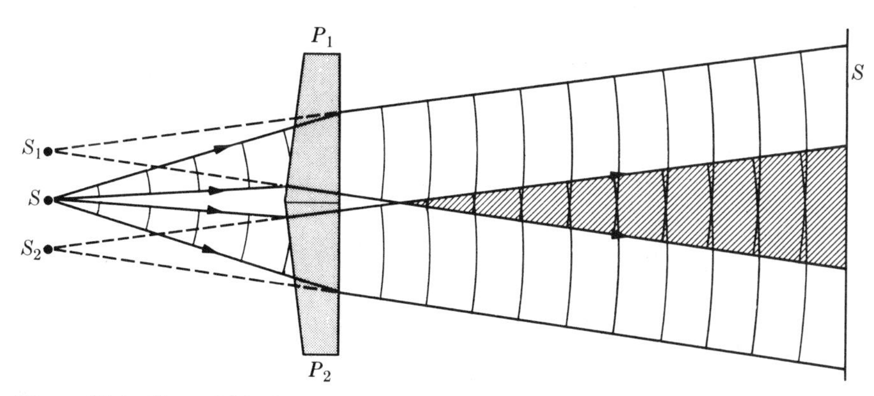

Figure 34.4 Fresnel biprism.

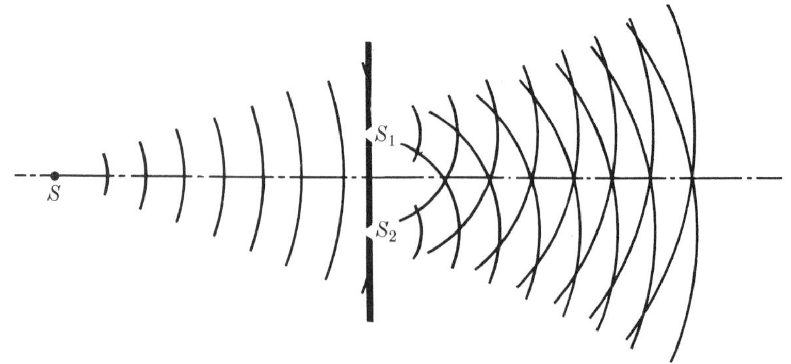

Figure 34.5 Interference of two coherent sources. Young's double-slit experiment.

is assured in this case because the two beams proceed from the same source. The beams interfere in the shaded region. For large phase differences the coherence is destroyed, because the interfering beams are produced by the source at two widely separated times so that, microscopically speaking, the source is not the same at both times and the phases are not constant.

Another method for producing interference of light is the Michelson interferometer that was discussed in Section 19.2. An even simpler device is the one used by Thomas Young (1773–1829), who, in his early experiments on light interference, proved conclusively that light was a wave phenomenon. His arrangement (Figure 34.5) consists of two small closely spaced holes or slits, S_1 and S_2, in a screen with a light source S placed behind it. The slits S_1 and S_2 behave as a pair of coherent sources whose waves interfere on the right-hand side of the screen.

In the case of light, the interference pattern is observed on a second screen placed parallel to the two sources S_1 and S_2. A series of alternate bright and dark fringes appears on the screen, as shown in Figure 34.6, due to the intersection of the screen with the hyperbolic ventral and nodal surfaces. For other regions of the electromagnetic spectrum, different kinds of detectors are used to observe the interference pattern.

(iii) Interference patterns. From the geometry of Figure 34.7, and considering that θ is a small angle, we have $r_1 - r_2 \approx S_1B = a \sin\theta$. Hence

$$\delta = \frac{2\pi}{\lambda}(r_1 - r_2) = \frac{2\pi}{\lambda} a \sin\theta \qquad \textbf{(34.9)}$$

Figure 34.6 Photograph of the interference fringes produced on a screen by a pair of slits illuminated by a point monochromatic light source. Note that there is a fading of the fringes near the edges, due to the loss of coherence.

Therefore, according to Equation 34.7, the points of a maximum intensity correspond to $\delta = 2n\pi$, or

$$\sin\theta = \frac{n\lambda}{a} \tag{34.10}$$

where n is a positive or negative integer. The intensity distribution in the interference pattern is shown in Figure 34.8 as a function of the angle θ. The maxima correspond to $(a \sin\theta)/\lambda = n$ (integer). When θ is small $\sin\theta \approx \tan\theta = x/D$ and Equation 34.10 becomes

$$x = \frac{nD}{a}\lambda \tag{34.11}$$

which gives the position of the bright fringes on the screen. The separation between two successive bright fringes is $\Delta x = (D/a)\lambda$. Therefore, by measuring Δx, D and a, we may obtain the wavelength λ. This is, in fact, one of the standard methods of measuring wavelengths.

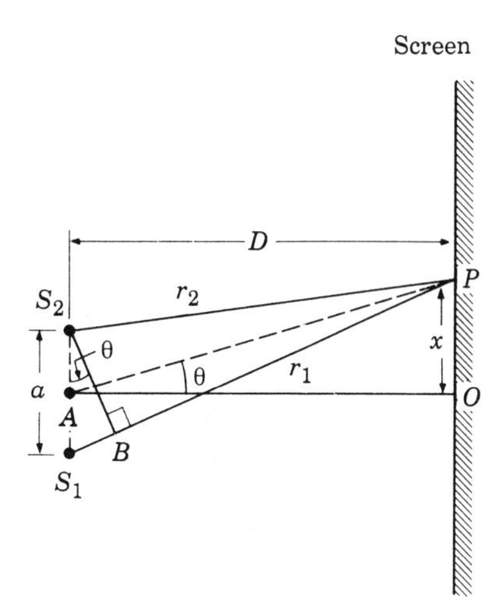

Figure 34.7 Schematic diagram for determining the intensity of the resultant wave motion on a screen due to the interference of two coherent sources.

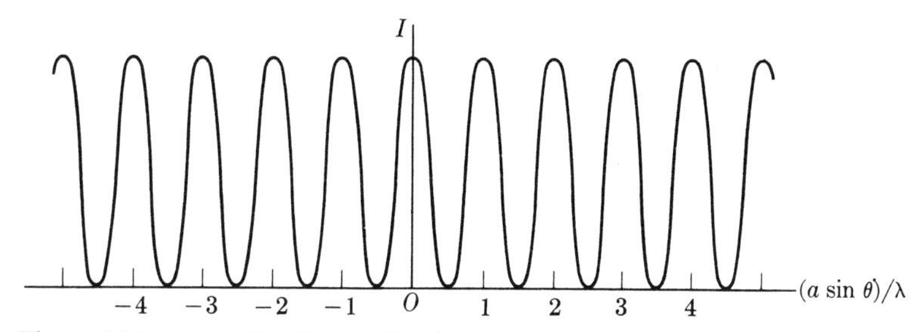

Figure 34.8 Intensity distribution in the interference pattern produced by two coherent sources.

EXAMPLE 34.1

In an experiment similar to Young's, the two slits are separated by a distance of 0.8 mm. The slits are illuminated with monochromatic light of wavelength 5.9×10^{-7} m, and the interference pattern is observed on a second screen, which is at a distance of 0.50 m from the slits. Determine the separation between two successive bright or dark fringes.

▷ The quantities appearing in Equation 34.11 are, in this case, $a = 0.8 \text{ mm} = 8 \times 10^{-4}$ m, $D = 5 \times 10^{-1}$ m and $\lambda = 5.9 \times 10^{-7}$ m. Therefore the positions of the bright fringes are

$$x = n\left(\frac{D\lambda}{a}\right) = 3.7 \times 10^{-4} n \text{ m} = 0.37n \text{ mm}$$

In general, the fringes have to be observed with a magnifying glass or a microscope. The separation between successive bright fringes is 0.37 mm, which is the same as the separation between two dark fringes.

EXAMPLE 34.2
Surface analysis by optical interferometry.

▷ There are many devices, such as videotapes, photographic films, computer disks and ball bearings, that require an extremely smooth and carefully shaped surface in order to function properly. Optical interferometry using laser beams offers the possibility of high precision non-contacting and non-destructive analysis of surface textures. There are several interferometric systems designed for that purpose, of which we will describe one using the Michelson interferometer. A laser beam (1) (Figure 34.9) is split in two by a half-silvered mirror. One of the split beams (2) is sent to a reference mirror where it is reflected. The other split beam (3) is sent toward the surface to be studied, where it is also reflected. The two split beams recombine and interfere (4). A video camera records the intensity of the recombined beam.

The intensity of the recombined beam (4) depends on the phase difference δ between beams (2) and (3), in accordance with Equation 34.8. As the surface being analyzed is moved sidewise, the path of beam (3) changes depending on whether it hits a bump or a dip in the surface. This in turn affects the phase difference between beams (2) and (3), and therefore the intensity of beam (4).

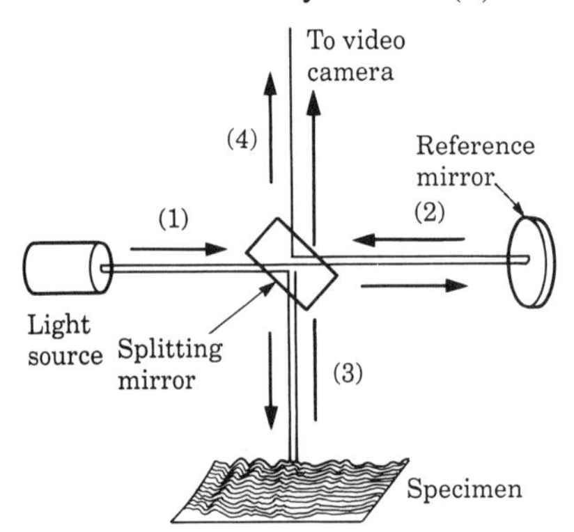

Figure 34.9 Interferometric method for detecting relative smoothness of a surface.

The scanning of the surface is performed three times, each time slightly varying the distance of the reference mirror to the splitting mirror. In this way three recordings of beam (4) (or **interferograms**) with different sets of phase differences are obtained. The data of the three interferograms are fed into a computer where, with the proper software, a three-dimensional representation of the surface is obtained. In this way, variations in the surface texture as small as a few angstroms (10^{-10} m) can be detected.

This technique is called *computer-analyzed optical interferometry*. Since the laser beams can be very narrow, the resolution of the surface is very high.

34.3 Interference of several synchronous sources

Consider now the case of several (N) synchronous and identical sources arranged linearly, as illustrated in Figure 34.10. To simplify our discussion, assume that the resulting wave motion is observed at a distance very large compared with the separation of the sources, so that the interfering rays are effectively parallel. Between successive rays there is a constant phase difference given by Equation 34.9,

$$\delta = \frac{2\pi}{\lambda} a \sin \theta \tag{34.12}$$

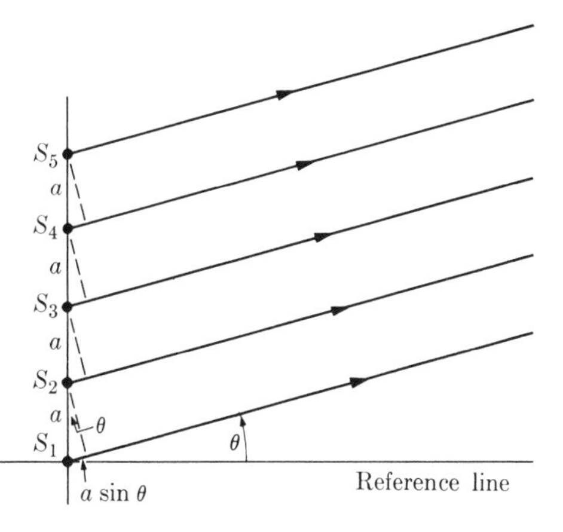

Figure 34.10 Linear series of five equally spaced coherent sources.

To obtain the resulting amplitude for the direction of observation, given by the angle θ, we must evaluate the vector sum of the corresponding rotating vectors for each source. If all the sources are alike, their rotating vectors all have the same length A_1, and successive vectors are deviated by the same angle δ, as indicated in Figure 34.11(a). It is clear from the figure that the maximum resultant amplitude occurs when all rotating vectors are collinear, as shown in Figure 34.11(b). This requires that $\delta = 2n\pi$ or

$$\sin \theta = \frac{n\lambda}{a} \tag{34.13}$$

where n is a positive or negative integer. This result can be easily understood because when Equation 34.13 holds, rays coming from any two adjacent sources are in phase and interfere by reinforcement. The resultant amplitude in this case is $A = NA_1$, where N is the number of sources; therefore the total intensity is $I = N^2 I_1$. The total intensity may be rather large if the number of sources is large.

The intensity is zero when the vector sum (Figure 34.12) is a closed polygon. This requires that $N\delta = 2n'\pi$ or

$$\sin \theta = \frac{n'\lambda}{Na} \tag{34.14}$$

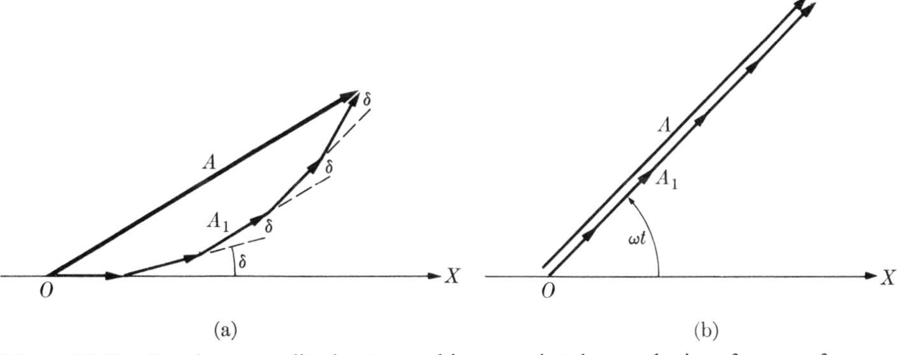

Figure 34.11 Resultant amplitude at an arbitrary point due to the interference of waves generated by five equally spaced linear coherent sources. (a) Arbitrary phase difference; (b) zero phase difference. OX is the reference line for phase measurements.

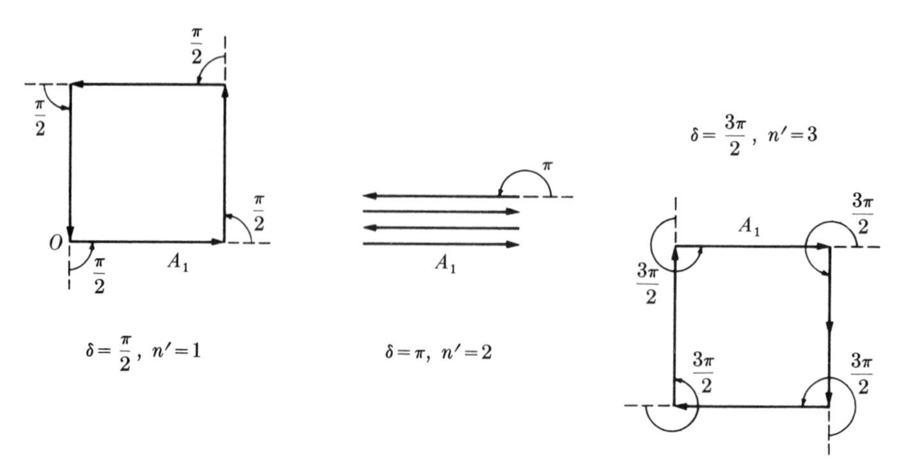

Figure 34.12 Zeros of amplitude for four synchronous sources of amplitude A_1.

where n' is an integer that goes from 1 to $(N-1)$, from $(N+1)$ to $(2N-1)$ etc. since $n' = 0, N, 2N, \ldots$ are excluded, because then Equation 34.14 would transform into Equation 34.13, which gives maximum intensity. The situation for $N = 4$ has been shown in Figure 34.12.

Equation 34.14 implies that between every two principal maxima given by Equation 34.13 there are $(N-1)$ zeros. Between two minima there must always be a maximum. Therefore we conclude that there are also $(N-2)$ additional maxima between the principal maxima given by Equation 34.13. Their amplitudes are, however, relatively much smaller, especially if N is large.

The graph of I/I_0 in terms of $a \sin \theta/\lambda$ is shown in Figure 34.13 for $N = 4$, 8, and very large. For comparison, the case when $N = 2$ has been added. We see that when we increase N the system becomes highly directional, since the resulting wave motion is important only for narrow bands of values of δ, or for narrow bands of values of the angle θ.

These results are widely used in broadcasting or receiving stations when a directional effect is desired. In this case several antennas are arranged in such a form that the intensity of the radiation emitted (or received) is maximum only for certain directions, given by Equation 34.13. For example, for four linear antennas separated by $a = \lambda/2$, Equation 34.13 gives $\sin \theta = 2n$. Then only $n = 0$ is possible for the principal maxima, giving only $\theta = 0$ and π. For the zeros, or nodal planes, Equation 34.14 gives $\sin \theta = \frac{1}{2}n'$, allowing for $n' = \pm 1$ and ± 2 or $\theta = \pm\pi/6$ and $\pm\pi/2$. The situation is illustrated in the **polar diagram** of Figure 34.14, where the intensity is plotted in terms of the angle. This antenna arrangement then transmits and receives preferentially in a direction perpendicular to the line joining the sources, and is called a **broadside array**. The same directional effect is used in radio telescopes. Several parabolic antennas are placed at equal distances along a straight line, with their axes parallel. For a given spacing and orientation of the axes, the wavelength of the radio waves received is determined by Equation 34.13. In this case a much higher resolution is also obtained.

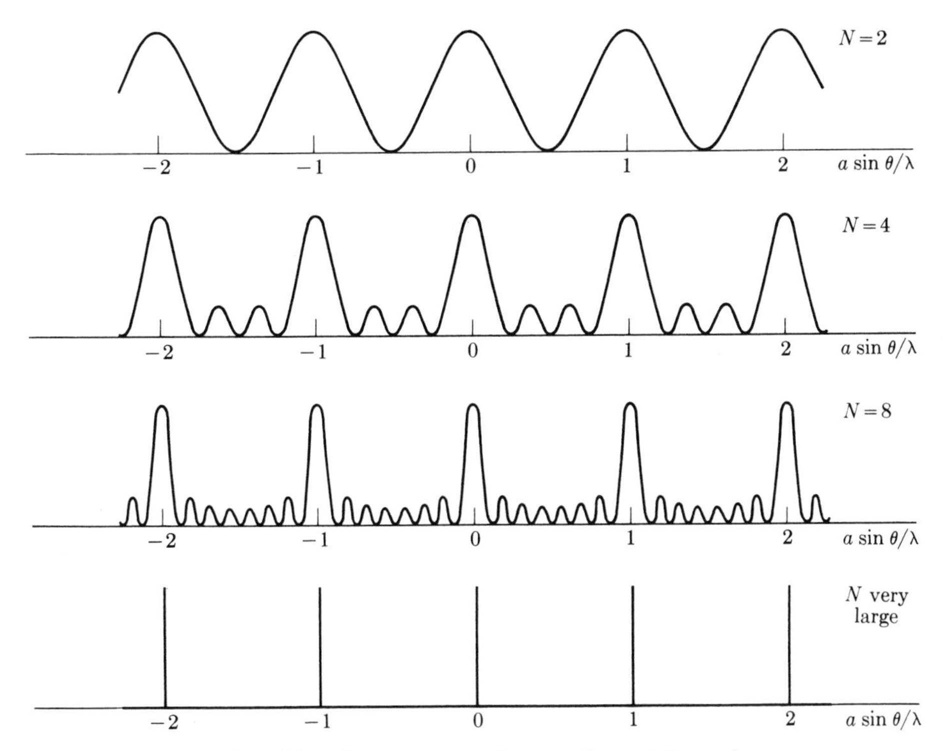

Figure 34.13 Intensity of interference pattern for two, four, eight, and very many sources each of intensity I_0. The source spacing is kept constant. The ordinates give the ratio of I/N^2I_0.

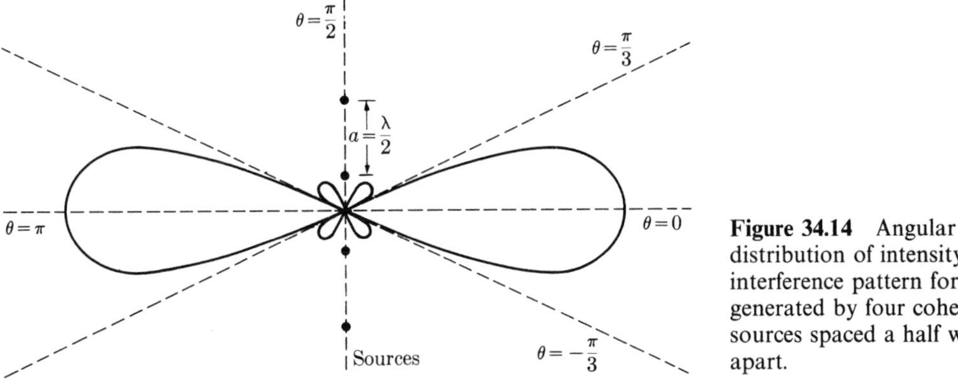

Figure 34.14 Angular distribution of intensity in the interference pattern for waves generated by four coherent linear sources spaced a half wavelength apart.

EXAMPLE 34.3

Interference by reflection from or transmission through thin films.

▷ Consider (Figure 34.15) a thin film of thickness a with plane waves falling on it at an angle of incidence θ_i. Part of a ray such as AB is reflected along BG, and part of it is transmitted

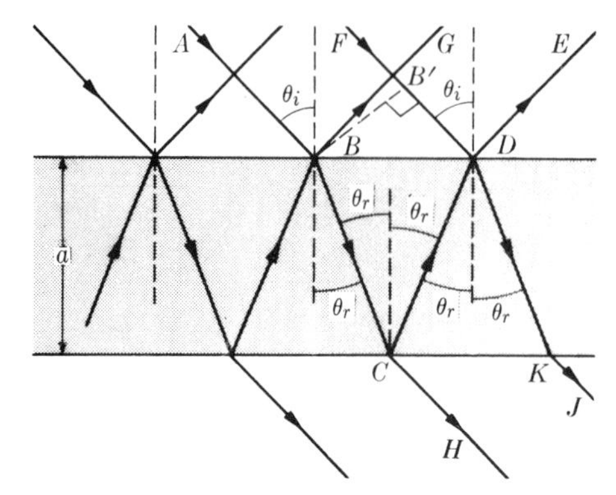

Figure 34.15 Interference by reflection or refraction through a thin film.

along BC. Ray BC in turn is partly reflected at C along CD and partly transmitted along CH. Ray CD again is partly reflected at D along DK, being superposed on the refracted ray of FD and partly transmitted along DE, superposed with the reflected ray of FD. Similarly, the reflected ray BG also contains contributions from several rays to the left. Therefore, interference phenomena occur along the reflected and transmitted rays. The situation is then similar to the case illustrated in Figure 34.10, with N very large, but with an important difference: interfering rays do not all have the same intensity because each successive reflection or refraction decreases the intensity.

If we ignore this change in intensity, which is justified if we consider only a few reflected and transmitted rays, the maxima for interference by reflection or transmission depends on the phase difference δ between interfering rays such as $ABCDE$ and FDE. For a given θ_i, the value of δ is different depending on whether we consider the reflected or transmitted rays. A calculation, which we omit, shows that, if n is the index of refraction of the film, the condition for maximum intensity of the reflected wave is

$$2an \cos \theta_r = \tfrac{1}{2}(2N - 1)\lambda \text{ (Maximum reflection, minimum transmission)} \quad \mathbf{(34.15)}$$

with N an integer. The condition for maximum intensity of the transmitted wave is

$$2an \cos \theta_r = N\lambda \qquad \text{(Maximum transmission, minimum reflection)} \quad \mathbf{(34.16)}$$

which differs from Equation 34.15 by a half wavelength. The difference is due to the phase change in the reflected waves. The color we observe on reflection is not the same as the color we observe by transmission. These are determined in each case by the wavelengths satisfying Equations 34.15 or 34.16.

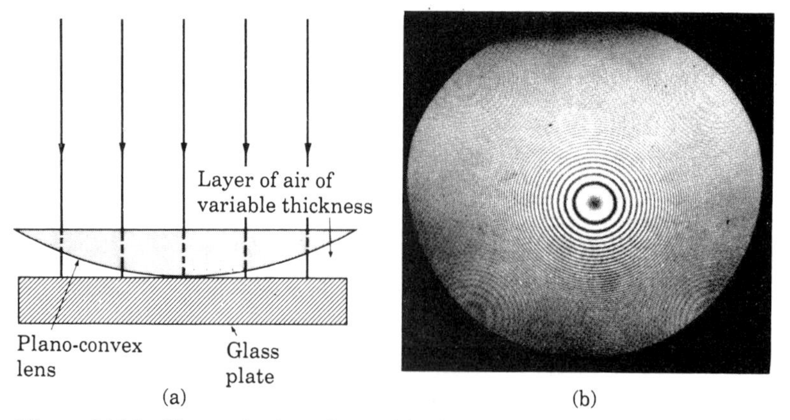

Figure 34.16 Newton's rings, formed by interference in the air film between a convex and a plane surface. (a) Schematic diagram. (b) Photograph of rings. (Courtesy Bausch & Lomb Optical Co.)

If the incident light is not monochromatic, Equations 34.15 and 34.16 give different values of θ_r, and then θ_i, for each λ. Also, if the film is of variable thickness, conditions 34.15 and 34.16 are not fulfilled at all points for a given wavelength; this results, in the case of monochromatic light, in a succession of dark and bright bands, and in the case of white light, it results in a succession of colored bands. This explains the colors we observe in thin oil films on water surfaces. The interference pattern can also easily be seen by placing a planoconvex lens on a plane glass plate, as shown in Figure 34.16(a). The space between the lens and the glass plate is a layer of air, of varying thickness. The resulting interference pattern consists of a series of concentric colored rings, known as **Newton's rings**, shown in Figure 34.16(b).

34.4 Standing waves in one dimension

What happens when a wave is restricted to travel in a limited region is a problem of great practical application. For simplicity we shall consider only waves in one dimension.

(i) Standing waves on a string. Consider a string OX with one end O fixed, as indicated in Figure 34.17. An incident transverse wave moving to the left, expressed by $\xi = A\sin(\omega t + kx)$, is reflected at O, producing a new wave propagating to the right and expressed as $\xi = A'\sin(\omega t - kx)$. Note that we write a different amplitude for the incident and reflected waves to take into account a possible change of amplitude at reflection.

The displacement at any point of the string is the result of the interference or superposition of the two waves, that is,

$$\xi = A\sin(\omega t + kx) + A'\sin(\omega t - kx) \tag{34.17}$$

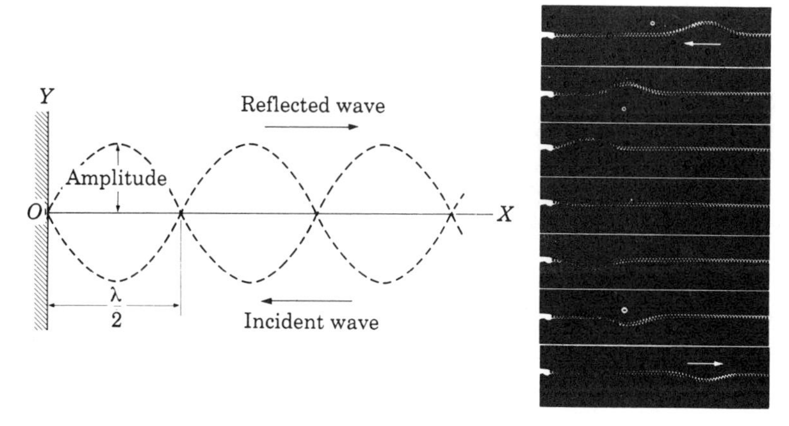

Figure 34.17 Change in phase of a reflected wave on a string having one end fixed. (From *Physics*, Boston, D. C. Heath, 1960.)

At O we have $x = 0$, so that

$$\xi_{(x=0)} = (A + A') \sin \omega t$$

But O is fixed, which means that $\xi_{(x=0)} = 0$ at all times. This requires that $A' = -A$. In other words, the wave undergoes a phase change of π but no change in amplitude when it is reflected at the fixed end. The phase change may be seen in the series of photographs of Figure 34.17, which show an incident and a reflected pulse rather than a continuous wave. Then Equation 34.17 becomes

$$\xi = A[\sin(\omega t + kx) - \sin(\omega t - kx)]$$

Using the trigonometric relation (see Appendix B, Equation B.7)

$$\sin \alpha - \sin \beta = 2 \sin \tfrac{1}{2}(\alpha - \beta) \cos \tfrac{1}{2}(\alpha + \beta)$$

we obtain

$$\xi = 2A \sin kx \cos \omega t \tag{34.18}$$

The expressions $\omega t \pm kx$ no longer appear and Equation 34.18 does not represent a traveling wave. We say that it corresponds to a **standing wave**. It effectively represents a simple harmonic motion whose amplitude varies from point to point, and is given by

$$\xi_o = 2A \sin kx \tag{34.19}$$

This amplitude has been indicated by the dashed lines in Figure 34.17. The amplitude is zero for $kx = n\pi$, where n is an integer. Since $k = 2\pi/\lambda$, this result may also be written as

$$x = \tfrac{1}{2}n\lambda \tag{34.20}$$

These points are called **nodes**. Successive nodes are separated by the distance $\frac{1}{2}\lambda$. The wavelength λ, however, is determined by the frequency and the velocity of propagation, according to $\lambda = v/\nu$.

Suppose now that we impose a *second* condition: that the point $x = L$, which may be the other end of the string, is also fixed. That means that $x = L$ is a node and must satisfy the condition $\xi_{(x=L)} = 0$, which requires that

$$\sin kL = 0 \qquad \text{or} \qquad kL = n\pi \tag{34.21}$$

If we use Equation 34.20, we have that

$$L = \tfrac{1}{2}n\lambda \tag{34.22}$$

which indicates that the length of the string must be a multiple of $\frac{1}{2}\lambda$. Therefore, the second condition automatically limits the wavelengths of the waves that can be sustained on this string to the values given by Equation 34.22; that is,

$$\lambda = \frac{2L}{n} = 2L, \frac{2L}{2}, \frac{2L}{3}, \cdots \tag{34.23}$$

and in turn, the frequencies of oscillations are also limited to the values

$$\nu = \frac{v}{\lambda} = \frac{nv}{2L} \tag{34.24}$$

Recalling expression 28.26, $v = (T/m)^{1/2}$, for the velocity of propagation of waves along a string subject to a tension T and having a mass m per unit length, the possible frequencies are determined by

$$\nu_n = \frac{n}{2L}\sqrt{\frac{T}{m}} = \nu_1, 2\nu_1, 3\nu_1, \ldots \tag{34.25}$$

where

$$\nu_1 = \frac{1}{2L}\sqrt{\frac{T}{m}}$$

is called the **fundamental frequency**. Thus the possible frequencies of oscillation (called **harmonics**) are all multiples of the fundamental. We may say that the frequencies and wavelengths are **quantized**, and that the quantization is the result of the boundary conditions imposed at *both* ends of the string. This is a situation that appears in many physical problems.

Figure 34.18 indicates the amplitude distribution for the first three modes of vibration ($n = 1, 2, 3$). The nodes or points of zero amplitude are determined by means of Equation 34.20. The points of maximum amplitude are the antinodes. The distance between successive antinodes is also $\lambda/2$. Of course, the separation between a node and an antinode is $\lambda/4$. Observe that while $\xi = 0$ at the nodes, $d\xi/dx = 0$ at the antinodes, since the amplitude is maximum.

If, instead of imposing the condition that $\xi = 0$ at the ends of the string, we impose other conditions

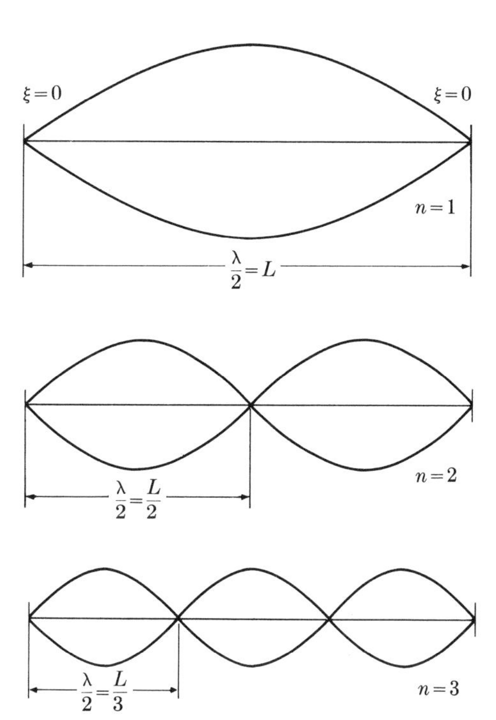

Figure 34.18 Standing transverse waves on a string with both ends fixed.

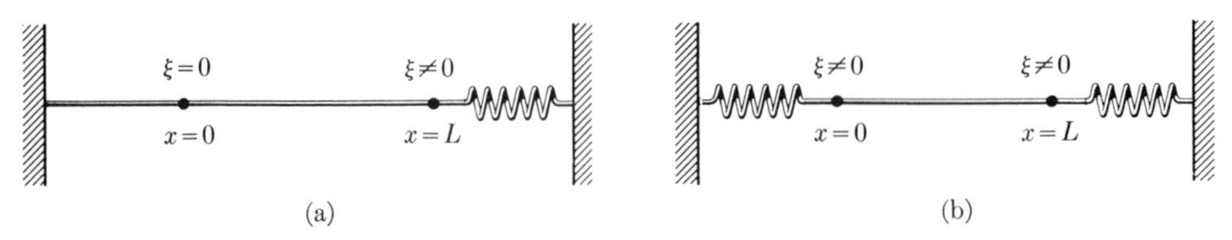

Figure 34.19 (a) String with a 'soft' end. (b) String with two 'soft' ends.

because the physical situation at the ends is different, as in the strings with 'soft' ends illustrated in Figure 34.19, we end up with a set of allowed wavelengths and frequencies different from those given by Equations 34.22 and 34.24.

(ii) Standing waves in a pipe. It is instructive to examine two other simple examples, related to standing waves in the air inside a pipe, such as an organ pipe. Consider first a tube open at both ends (Figure 34.20). Air is blown in at one end through the mouthpiece, and standing waves are produced because of the reflection occurring at the other end. The fundamental difference between this case and the string with fixed ends considered previously is that in the pipe both ends are free, and therefore ξ has a maximum value at these ends; in other words, there is an antinode at each end. Therefore, in an open tube the nodes and antinodes are interchanged in comparison with their positions in a string. This has been illustrated by the broken lines in Figure 34.20. We may conclude then that, as in the case of a string with fixed ends, the length of the tube must be a multiple of $\frac{1}{2}\lambda$; that is,

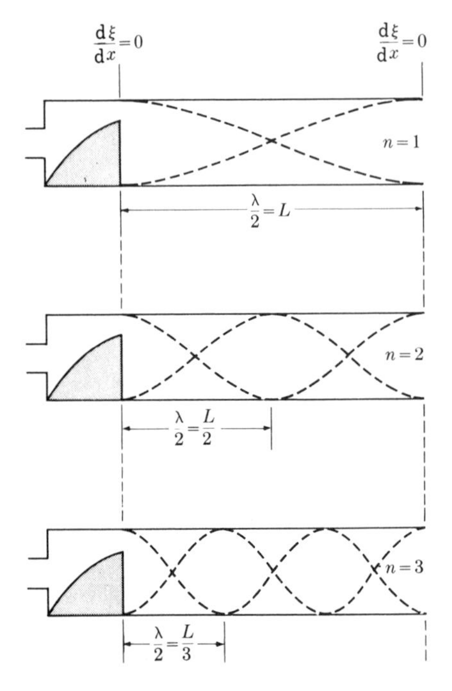

Figure 34.20 Standing pressure wave in an air column with both ends open.

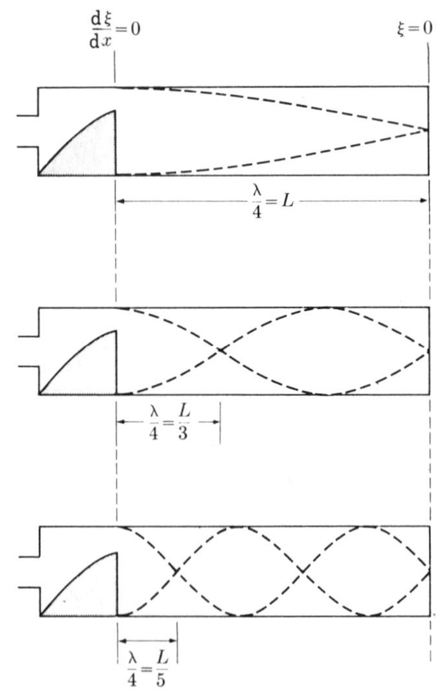

Figure 34.21 Standing pressure wave in an air column with one end closed.

$L = \frac{1}{2}n\lambda$ or

$$\lambda = \frac{2L}{n} = 2L, \frac{2L}{2}, \frac{2L}{3}, \cdots \tag{34.26}$$

The allowed frequencies of the standing waves are

$$\nu_n = \frac{v}{\lambda} = n\left(\frac{v}{2L}\right) = \nu_1, 2\nu_1, 3\nu_1, \ldots \tag{34.27}$$

with $n = 1, 2, 3, \ldots$, and therefore the allowed frequencies comprise all the harmonics corresponding to a fundamental tone of frequency $\nu_1 = v/2L$.

Next consider a tube closed at the end opposite to the mouthpiece (Figure 34.21). The physical conditions at the closed end have changed, while at the mouthpiece they remain the same as in the preceding case. Therefore at the mouthpiece we must again have an antinode, but at the closed end ($x = L$) we must have a node. The distribution of nodes and antinodes in a closed tube has been shown by the broken lines in Figure 34.21. It is clear from the figure that the length of the tube is related to the wavelength by the expression $L = \frac{1}{2}n\lambda + \frac{1}{4}\lambda$, or

$$\lambda = \frac{4L}{2n+1} = 4L, \frac{4L}{3}, \frac{4L}{5}, \cdots \tag{34.28}$$

The allowed frequencies of the standing waves are

$$\nu = \frac{v}{\lambda} = (2n+1)\frac{v}{4L} = \nu_1, 3\nu_1, 5\nu_1, \ldots \tag{34.29}$$

The modes of vibration are now different from those corresponding to a tube open at both ends. The most important feature is that a tube closed at one end can vibrate only with *odd* harmonics of the fundamental $\nu_1 = v/4L$. For equal lengths, the fundamental frequency of a closed tube is one-half that of an open tube. A pipe is an example of what is called a **resonating cavity**.

EXAMPLE 34.4

A steel string has a length of 40 cm and a diameter of 1 mm. Given that its fundamental vibration is 440 Hz, corresponding to the musical tone A (or lah) in the diatonic scale, key of C, find its tension. Assume that the density of the string is $\rho = 7.86 \times 10^3\ \text{kg m}^{-3}$.

▷ The mass per unit length is $m = \pi r^2 \rho$. With $r = 5 \times 10^{-4}$ m and $\rho = 7.86 \times 10^3\ \text{kg m}^{-3}$, we get $m = 6.15 \times 10^{-3}\ \text{kg m}^{-1}$. Solving Equation 34.25 for the tension T, setting $n = 1$ since we want the fundamental tone, we obtain $T = 4L^2 m\nu_1^2$. Setting $L = 40$ cm $= 0.40$ m, $\nu_1 = 440$ Hz, and introducing the value of m that we calculated, we finally obtain $T = 762.0$ N. As the student may now realize, stringed musical instruments are tuned by adjusting the tensions or the lengths of their strings. For the case of a pipe organ, the tones are selected by adjusting the length of the pipes.

34.5 Standing electromagnetic waves

Interference phenomena are so characteristic of waves that their presence has always been accepted as conclusive proof that a process can be interpreted as a wave phenomenon. For that reason, when Young, Fresnel, Fraunhofer and others in the nineteenth century observed light interference (and diffraction phenomena), the wave theory of light became generally accepted. At that time electromagnetic waves were not known, and light was assumed to be an elastic wave in a subtle medium, called ether, that pervaded all matter. It was not until the end of the nineteenth century that Maxwell predicted the existence of electromagnetic waves (Chapter 29), and Hertz, by means of interference experiments that gave rise to standing electromagnetic waves, verified in 1888 the existence of electromagnetic waves in the radio-frequency range. Later their velocity was measured and found to be equal to that of light. The reflection, refraction and polarization of electromagnetic waves were also found to be similar to those of light. The conclusion was to identify light with electromagnetic waves of certain frequencies.

Consider a plane polarized electromagnetic wave falling with perpendicular incidence on the plane surface of a good conductor located at the YZ-plane (Figure 34.22). Take the X-axis as the direction of propagation and the Y- and Z-axes as parallel to the electric and the magnetic fields, respectively. The oscillating incident electric field is then parallel to the surface of the conductor. But at the surface of a conductor the electric field must be perpendicular to the conductor; that is, the electric field cannot have a tangential component. The only way to make this condition compatible with the orientation of the electric field in the incident wave is to require that the resultant electric field always be zero at the surface of the conductor. This means that the electric field of the reflected wave at the surface must at all times be equal and opposite to that of the incident wave, thus giving $\mathscr{E} = \mathscr{E}_i + \mathscr{E}_r = 0$ for $x = 0$. This condition is mathematically equivalent to the condition for the reflection of waves in a string with one end fixed, discussed in the previous section. Since the mathematics is the same, we may use Equation

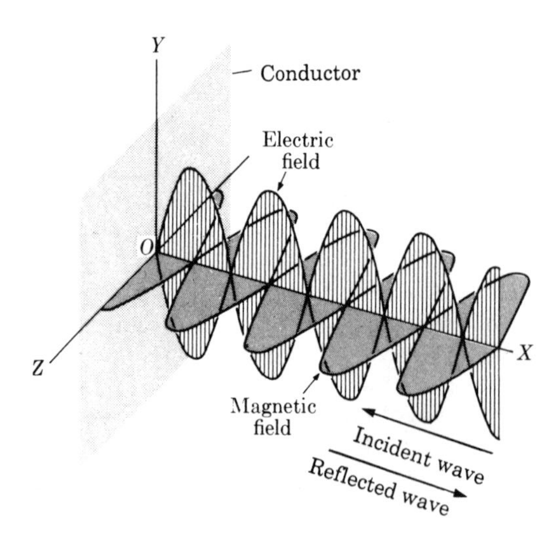

Figure 34.22 Standing electromagnetic waves produced by reflection from a conducting surface coincident with the YZ plane. The incident wave propagates from X to O and the reflected wave from O to X.

34.18 to write an expression for the resultant electric field,

$$\mathscr{E} = 2\mathscr{E}_0 \sin kx \cos \omega t$$

The magnetic field oscillates in the XZ-plane. Using Equation 29.8, we find that the magnetic field is expressed by

$$\mathscr{B} = 2\mathscr{B}_0 \cos kx \sin \omega t$$

with $\mathscr{B}_0 = \mathscr{E}_0/c$. Therefore, there is a phase difference of $\frac{1}{2}\lambda$ in the space variations and of $\frac{1}{2}P$ in the time variations of the two fields. From the mathematical expression for $\mathscr{B}$, note that the magnetic field has maximum amplitude at the surface ($x = 0$). Thus, although the electric fields interfere destructively at the surface, the magnetic fields interfere constructively there.

The amplitudes of the electric and magnetic fields of the resulting wave at a distance x from the surface are $2\mathscr{E}_0 \sin kx$ and $2\mathscr{B}_0 \cos kx$. They are indicated by the shaded lines in Figure 34.22. At the points where $kx = n\pi$ or $x = \frac{1}{2}n\lambda$, the electric field is zero and the magnetic field is maximum. At the points where $kx = (n + \frac{1}{2})\pi$ or $x = (2n + 1)\lambda/4$ the electric field has a maximum value but the magnetic field is zero.

Note 34.1 Hertz's experiment

Hertz's simple arrangement is shown on the left in Figure 34.23. The transformer T charges the metallic plates C and C'. These plates discharge through the gap P, which becomes a dipole oscillator. Along the line PX, the direction of the electric field is parallel to the Y-axis and that of the magnetic field along the Z-axis. To observe the waves, Hertz used a short wire, bent in circular shape, but with a small gap. This simple device is called a **resonator**. The

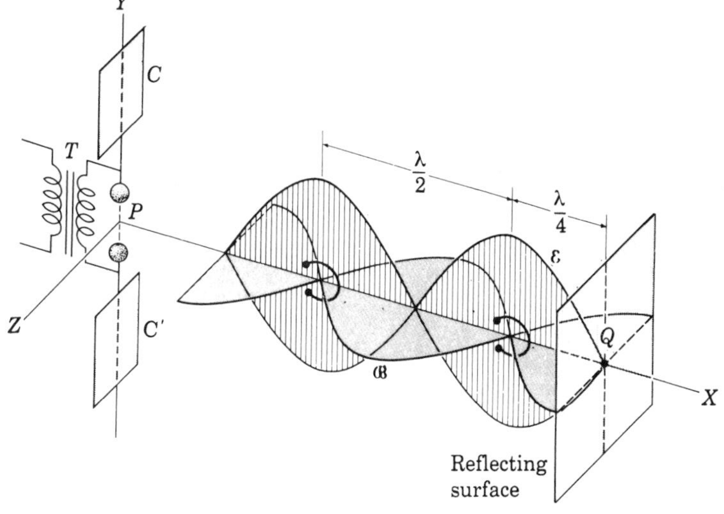

Figure 34.23 Hertz' experiment on interference of electromagnetic waves.

diameter of the resonator used in this kind of experiment must be very small compared with the wavelength of the waves. If the resonator is placed with its plane perpendicular to the magnetic field of the wave, the varying magnetic field induces an emf in the resonator, resulting in sparks at its gap. On the other hand, if the plane of the resonator is parallel to the magnetic field, no emf is induced and no sparks are observed at the gap.

To produce standing electromagnetic waves, Hertz placed a reflecting surface (made of a good conductor) at Q. In such a case, when the resonator is at a node of the magnetic field, no matter what its orientation, it will show no induced emf (or sparks). At an antinode of the magnetic field, however, the sparking is greatest when the resonator is oriented perpendicular to the magnetic field. By moving the resonator along the line PQ, Hertz found the position of the nodes and antinodes and the direction of the magnetic field. The results obtained by Hertz coincided with the theoretical analysis we have given. By measuring the distance between two successive nodes, Hertz could calculate the wavelength λ, and since he knew the frequency ν of the oscillator, he could calculate the velocity c of the electromagnetic waves by using the equation $c = \lambda\nu$. It was by this means that Hertz obtained the first experimental value for the velocity of propagation of electromagnetic waves.

34.6 Standing waves in two dimensions

Consider a rectangular membrane stretched over a frame so that its edges are fixed. If the surface of the membrane is disturbed, waves are produced that propagate in all directions and are reflected at the edges, resulting in interference. Consider the special case when plane waves of only one frequency are generated in the membrane. Assume further that these waves propagate parallel to either side, as indicated in Figure 34.24. Instead of nodes and antinodes we get nodal lines and antinodal or ventral lines, designated by continuous and dashed lines in Figure 34.24. In Figure 34.24(a), the membrane is fixed at the left ($x = 0$) and the right ($x = a$), but the other two sides are free. The waves propagate, both to the left and to the right, resulting in a system of nodal and antinodal lines parallel to the Y-axis. At $x = 0$ and $x = a$ we must have nodal lines. Therefore the condition

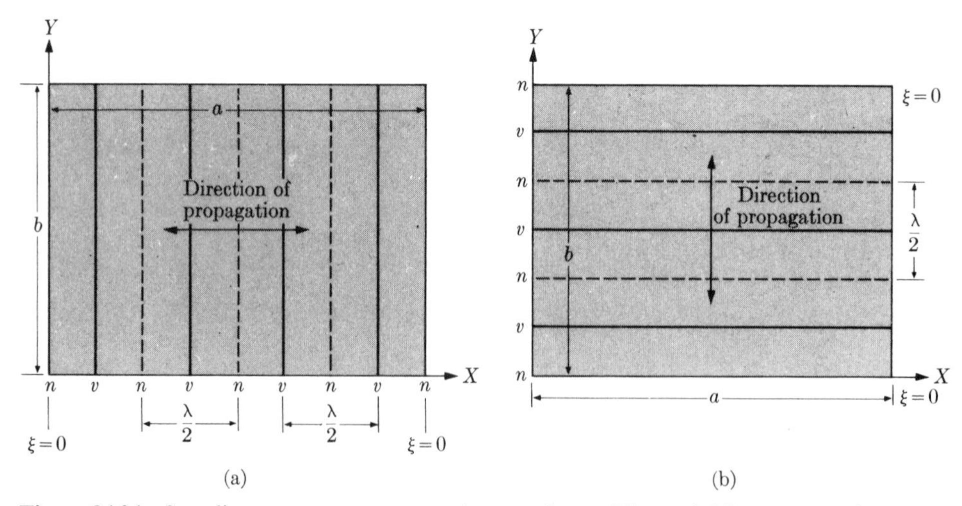

Figure 34.24 Standing waves on a rectangular membrane. The nodal lines are continuous and the ventral lines are dashed.

for standing waves is similar to Equation 34.21; that is,

$$ka = n\pi \qquad \text{or} \qquad k = \frac{n\pi}{a} \tag{34.31}$$

The corresponding frequencies are given by

$$\nu = \frac{v}{\lambda} = \frac{kv}{2\pi} \qquad \text{or} \qquad \nu = n\left(\frac{v}{2a}\right) \tag{34.32}$$

where v is the velocity of propagation of the waves along the surface of the membrane and $k = 2\pi/\lambda$.

In Figure 34.24(b) the membrane is fixed at the bottom ($y = 0$) and at the top ($y = b$). For waves propagating up and down, the nodal and antinodal lines are parallel to the X-axis. The condition for standing waves is similar to Equation 34.31, with a replaced by b, yielding

$$kb = n\pi \qquad \text{or} \qquad k = \frac{n\pi}{b} \tag{34.33}$$

with frequencies

$$\nu = n\left(\frac{v}{2b}\right) \tag{34.34}$$

Next consider a membrane with all four sides fixed and plane waves traveling in an arbitrary direction along its surface. First we recall that a plane wave in two dimensions, making $z = 0$ in Equation 28.39, is expressed by

$$\xi = A \sin[\omega t - (k_1 x + k_2 y)] \tag{34.35}$$

where we have followed our present convention of writing the time factor first. The quantities k_1 and k_2 are the components of a vector $\boldsymbol{k}$ parallel to the direction of propagation on the membrane or in the XY-plane and a magnitude

$$k = \sqrt{k_1^2 + k_2^2} \tag{34.36}$$

For an initial ray PQ (Figure 34.25), characterized by the components k_1, k_2, there is a reflected ray QR characterized by k_1, $-k_2$. From R to S the ray is characterized by $-k_1$, $-k_2$. And from S toward the Y-axis, the ray is characterized

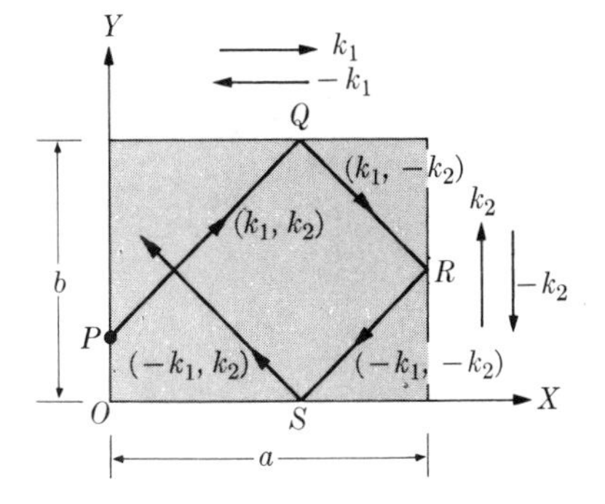

Figure 34.25 Successive reflections of a wave in a rectangular membrane.

by the components $-k_1$, k_2. In successive reflections of this ray, no new combinations of k_1 and k_2 appear. The path of the ray is similar to that of a ball in billiards. We conclude then that along the membrane there is a system of four waves, due to reflection at the four sides. These four waves must interfere in such a way that at $x = 0$ and a, and at $y = 0$ and b the resultant value of ξ is zero. Then instead of conditions 34.31 and 34.33 we have that k_1 and k_2 must satisfy the conditions

$$k_1 a = n_1 \pi \quad \text{or} \quad k_1 = \frac{n_1 \pi}{a} \qquad k_2 b = n_2 \pi \quad \text{or} \quad k_2 = \frac{n_2 \pi}{b} \tag{34.37}$$

where n_1 and n_2 are integers. Then, using Equation 34.36, we have

$$k = \pi \sqrt{\frac{n_1^2}{a^2} + \frac{n_2^2}{b^2}} \tag{34.38}$$

Since $k = 2\pi/\lambda$ and $\lambda\nu = v$, we have $\nu = kv/2\pi$ and the possible frequencies are

$$\nu = \frac{v}{2}\sqrt{\frac{n_1^2}{a^2} + \frac{n_2^2}{b^2}} \tag{34.39}$$

We may note that the possible frequencies are no longer integers of a fundamental frequency, but follow a different sequence. The nodal lines form the rectangular patterns shown in Figure 34.26.

For the case of a circular membrane, symmetry suggests that the nodal lines are now circles and radii, as indicated in Figure 34.27.

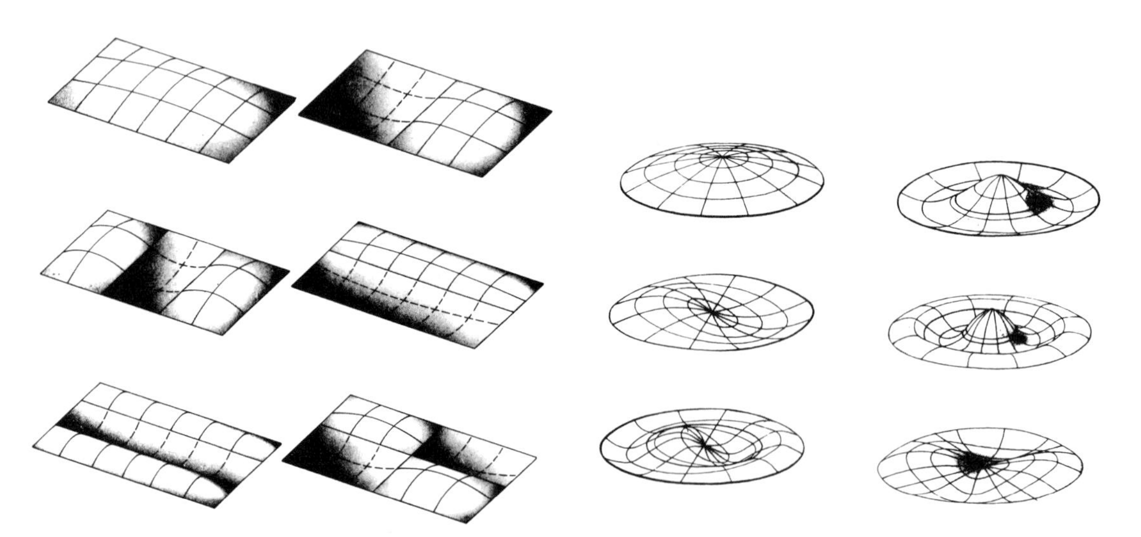

Figure 34.26 First few modes of vibration of a rectangular membrane, showing nodal lines.

Figure 34.27 Some possible modes of vibration of a circular membrane. (From *Vibration and sound*, by Philip M. Morse. McGraw-Hill Book Co., 1948.)

34.7 Standing waves in three dimensions; resonating cavities

The problem of standing waves in three dimensions is a simple extension of the case for two dimensions. Consider a rectangular cavity of sides a, b and c, which has perfectly reflecting walls (Figure 34.28) so that $\xi = 0$ at all six faces. A plane wave in a space is characterized by a vector $\boldsymbol{k}$, perpendicular to the plane of the wave, with three components, k_1, k_2 and k_3, along the three axes. When a wave is produced inside the cavity, it is reflected successively at all faces, and a set of eight waves, resulting from all the different combinations possible among $\pm k_1$, $\pm k_2$, $\pm k_3$, is established. The interference or superposition of these eight waves gives rise to standing waves if k_1, k_2 and k_3 have the appropriate values. By analogy with Equation 34.37, these values are

$$k_1 = \frac{n_1 \pi}{a}, \qquad k_2 = \frac{n_2 \pi}{b}, \qquad k_3 = \frac{n_3 \pi}{c} \tag{34.40}$$

where n_1, n_2 and n_3 are integers. Since $k = \sqrt{k_1^2 + k_2^2 + k_3^2}$, we may write

$$k = \pi \sqrt{\frac{n_1^2}{a^2} + \frac{n_2^2}{b^2} + \frac{n_3^2}{c^2}} \tag{34.41}$$

and the possible frequencies of the standing waves in the cavity are

$$\nu = \frac{v}{2} \sqrt{\frac{n_1^2}{a^2} + \frac{n_2^2}{b^2} + \frac{n_3^2}{c^2}} \tag{34.42}$$

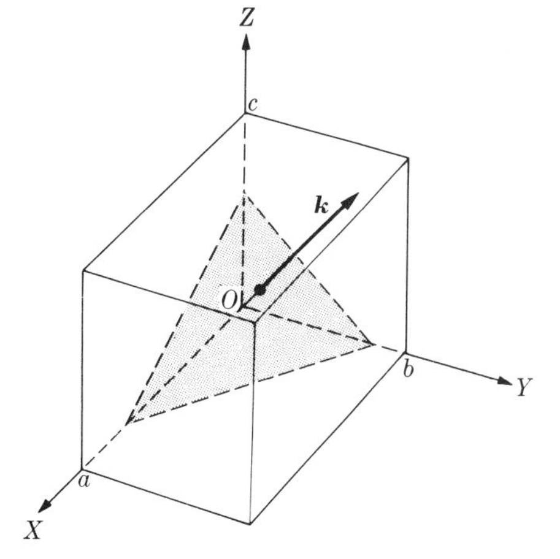

Figure 34.28 A rectangular cavity.

A cavity such as the one shown in Figure 34.28 will therefore resonate, sustaining waves for the frequencies given by Equation 34.42.

In the case of a spherical or cylindrical cavity, the mathematical treatment is more complex, but again we find that only certain frequencies are allowed.

The results we have obtained for standing waves in cavities find many applications. In acoustics, for example, resonating cavities are used for sound analysis. Resonating cavities for electromagnetic waves have walls made of materials that are good electrical conductors, so that the walls are the best possible reflectors. These cavities can sustain standing electromagnetic waves of definite frequencies with very little attenuation of the waves through energy loss by reflection. This means that the cavities serve as storage spaces for electromagnetic energy. The detailed theory of electromagnetic standing waves in cavities is slightly more complicated than our discussion here indicates, because of the transverse character of the waves. But results such as Equation 34.42 remain the same. Such cavities are used for frequency measurement or analysis (in the same manner as acoustic resonators), for frequency control in oscillating circuits, and for measuring the properties of the material filling the cavity. They are also used in lasers to enhance the laser beam. In this case the cavity is long and narrow. The standing waves are mostly along the longest dimension so that n_2 and n_3 are practically zero, and Equation 34.42 becomes $\nu = n_1 v/2a$, which was the expression used in the discussion of lasers in Section 31.8.

34.8 Waveguides

The cavities discussed above only allow standing waves. Traveling waves may also be produced in certain enclosures called **waveguides**, which are long cavities open at both ends. Waves are fed in at one end and received at the other. We shall discuss one simple type of guide, consisting of two parallel planes separated by a distance a (Figure 34.29). A simple example of a parallel waveguide in the optical region is two parallel mirrors, such as those found in some barber shops.

Consider a wave inside the cavity at an angle with the planes, as determined by the vector $\boldsymbol{k}$, with components k_1 and k_2 parallel and perpendicular, respectively, to the planes. The wave will suffer successive reflections at both limiting surfaces, bouncing back and forth between them. Since the space is not limited in the direction parallel to the planes (as was the case for the cavities), the wave will continue to progress to the right.

Let us choose the X-axis parallel, and the Y-axis perpendicular, to the planes so that the vector $\boldsymbol{k}$ is in the XY-plane. In Figure 34.29 the path of a particular ray has been shown. Along PQ the ray is characterized by the components k_1, k_2; from Q to R it is characterized by the components k_1, $-k_2$. From R on, it is again characterized by k_1, k_2, and so on. We conclude then that in the space between the reflecting planes we have two sets of waves, one corresponding to k_1, k_2 and the other to k_1, $-k_2$. (Remember that in the case of two-dimensional standing waves, such as those in a membrane, we had four waves, because of the additional waves generated by reflections at the right and left ends.) The two waves interfere as they progress along the space between the planes.

In the X-direction we have *one* traveling wave with wave number k_1 and therefore represented by the factor $\cos(\omega t - k_1 x)$. In the Y-direction we have *two* waves of wave numbers k_2 and $-k_2$, giving rise to **standing** waves with $\xi = 0$ at $y = 0$ and $y = a$. Thus, recalling Equation 34.19 with x replaced by y, the standing waves have an amplitude given by $2A \sin k_2 y$. Therefore, the resultant wave is

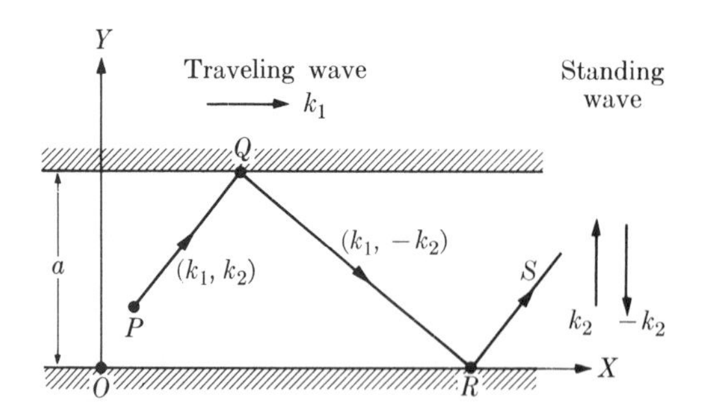

Figure 34.29 Ray propagating between two parallel reflecting planes.

expressed by

$$\xi = 2A \sin k_2 y \cos(\omega t - k_1 x) \tag{34.43}$$

where

$$k_2 = \frac{n\pi}{a} \tag{34.44}$$

to satisfy the boundary condition $\xi = 0$ at $y = a$. The term $\cos(\omega t - k_1 x)$ corresponds to a wave traveling in the X-direction with a phase velocity

$$v_p = \frac{\omega}{k_1} = \left(\frac{k}{k_1}\right)v \tag{34.45}$$

since $k = \omega/v$. Noting that $k_1 \leqslant k$ because k_1 is a component of $\boldsymbol{k}$, Equation 34.45 indicates that the phase velocity of the wave traveling along the cavity is larger than the phase velocity $v = \omega/k$ of the wave in free space. Now from $k^2 = k_1^2 + k_2^2$ and Equation 34.44, we have that

$$k^2 = k_1^2 + \frac{n^2\pi^2}{a^2}$$

or, making $k = \omega/v$,

$$\frac{\omega^2}{v^2} = k_1^2 + \frac{n^2\pi^2}{a^2} \tag{34.46}$$

The group velocity associated with the phase velocity given by Equation 34.45 is, by Equations 28.49 and 34.46,

$$v_g = \frac{d\omega}{dk_1} = \frac{k_1}{\omega}v^2 = \left(\frac{k_1}{k}\right)v \tag{34.47}$$

which is smaller than the free space velocity v since $k_1 \leqslant k$. Multiplying Equation 34.45 by Equation 34.47, we get $v_p v_g = v^2$. For electromagnetic waves in vacuum ($v = c$), $v_p v_g = c^2$. We see then that, even if it is empty, an electromagnetic waveguide acts as a dispersive medium with an index of refraction less than one, and thus a phase velocity larger than c, but a group velocity smaller than c. Only the group velocity is involved in transmitting a signal along the waveguide.

In order for a wave to propagate along the waveguide, it is necessary, according to Equation 34.46, that $\omega^2/v^2 > n^2\pi^2/a^2$, which yields

$$\omega > \frac{n\pi v}{a} \quad \text{or} \quad \nu > \frac{nv}{2a} \tag{34.48}$$

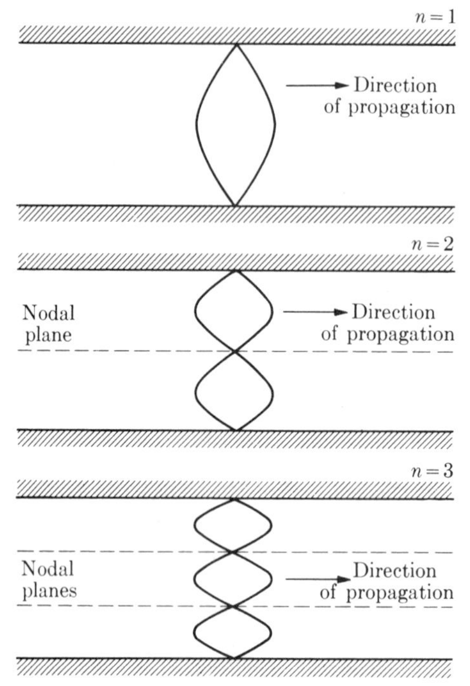

Figure 34.30 First three modes of propagation of a wave between two parallel reflecting planes.

In other words, only those waves with frequencies satisfying Equation 34.48 are propagated along the guide. Each mode is determined by the value of n, and for each mode there is a cut-off frequency, equal to $nv/2a$, below which propagation is impossible. Waveguides then act as frequency filters.

Although the wave propagates within the waveguide in the X-direction, the amplitude is modulated transversely in the Y-direction by the factor $\sin k_2 y$ in Equation 34.43. The transverse variation of the amplitude is indicated in Figure 34.30 for $n = 1$, 2 and 3. In practice, waveguides have either a rectangular or a circular cross-section. Waveguides are used to transmit electromagnetic signals, mainly in the microwave and optical regions.

An important type of waveguide in the near infrared and optical regions consists of transparent fibers with a diameter of a few microns (10^{-6} m), called **optical fibers**, coated with a material of lower refractive index to enhance total reflection at the walls of the fibers. These fibers are made of glass or quartz, although other materials, such as nylon, are also used. A ray entering at one end propagates along the fiber as a result of successive reflections, emerging at the other end whether or not the fiber is straight or bent (Figure 34.31). When the fibers are arranged in bundles, an image can be transmitted from one point to another and around corners. This technique is used in industry and medicine to explore regions not directly accessible. Optical fibers are used in medicine to carry laser beams to different parts of the body for diagnostic and therapeutic purposes. For example, in the procedure called *laser angioplasty*, a flexible optical fiber, 4×10^{-4} m in diameter, is inserted in the peripheral or coronary system until an internal blockage is reached. A laser pulse is sent along the optical fiber, destroying the blockage, unclogging the artery and restoring normal circulation.

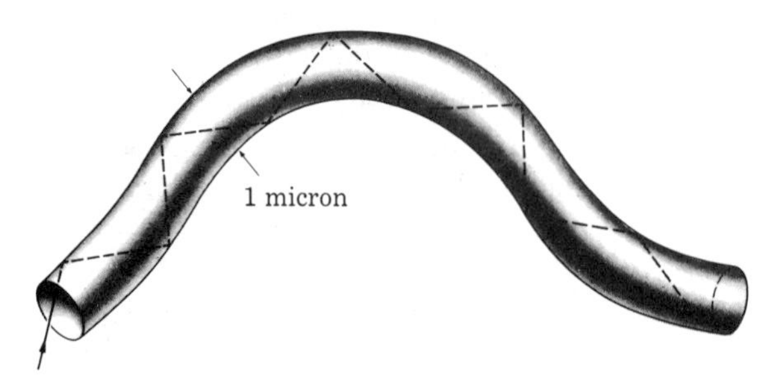

Figure 34.31 Optical fibers act as waveguides for light.

More recently, optical fibers are used for telecommunications (telephone, television, computer data). A single fiber can simultaneously carry a large number of different signals. To decrease the absorption of the electromagnetic waves in a fiber, a wavelength in the near infrared of the order of 10^{-6} m is used. Also, the radius of the fiber is of the order of 10^{-6} m to minimize the number of reflections inside the fiber, since they tend to decrease the intensity of the signal.

Acoustic waveguides are also very common. The air ducts in the air conditioning system of a house, for example, act as acoustic waveguides that are capable of transmitting the noises from the furnace or sounds from one room to another. The inner ear is essentially an acoustic waveguide.

QUESTIONS

34.1 Why do we say that interference phenomena are an indication of the wave nature of a process?

34.2 Why is coherence essential for the observation of interference?

34.3 Is it possible to observe an interference pattern when the two sources have (a) different frequencies, (b) a fixed phase difference, (c) a randomly varying phase difference?

34.4 The two slits in a Young's experiment are illuminated with light of wavelengths λ_1 and λ_2. On the same diagram plot the intensity distribution for each wavelength and describe the interference pattern observed. Assume $\lambda_1 > \lambda_2$. What is the requirement for the two interference patterns to appear distinguishable?

34.5 In a Young's experiment white light is used. What kind of interference pattern do you expect?

34.6 If polarized light falls on the two slits in a Young's experiment, does an interference pattern appear? What happens if the light is unpolarized?

34.7 Is coherence a property that affects (a) reflection, (b) refraction, (c) polarization, (d) interference?

34.8 Why is the device used by Michelson (described in Chapter 19 and in Example 34.2) called an interferometer?

34.9 In Example 34.2 interference by thin films was discussed. Explain why the films must be thin.

34.10 In what respects can we consider the interference phenomena produced by sound waves and light waves as identical? In what respects are they different?

34.11 How does a change of temperature affect the frequency of standing waves in a tube?

34.12 How does a change in tension affect the frequency of standing waves in a string?

34.13 Is the wavelength of the traveling wave in a parallel plane waveguide larger, smaller or equal to the wavelength of the wave in free space?

34.14 Explain why there is a cut-off frequency or a maximum wavelength of waves that can be transmitted in a waveguide.

34.15 Show that the amplitude of standing waves on a string, given by Equation 34.19, satisfies the equation

$$\frac{d^2\psi}{dx^2} + k^2\psi = 0$$

Verify that this equation is also satisfied by any solution of the wave equation 28.11 corresponding to standing waves of the form $\xi(x, t) = \psi(x) \sin \omega t$.

34.16 What properties of an electromagnetic wave are demonstrated in Hertz's experiment?

PROBLEMS

34.1 Two slits, separated a distance of 1 mm, are illuminated with red light of wavelength 6.5×10^{-7} m. The interference fringes are observed on a screen placed 1 m from the slits. (a) Find the distance between two bright fringes and between two dark fringes. (b) Determine the distance of the third dark fringe and the fifth bright fringe from the central fringe.

34.2 By means of a Fresnel biprism (Figure 34.4), interference fringes are produced on a screen 0.80 m away from the biprism, using light of wavelength equal to 6.0×10^{-7} m. Find the distance between the two images produced by the biprism if 21 fringes cover a distance of 2.44 mm on the screen.

34.3 (a) Verify that if a source is placed at a distance d from a Fresnel biprism having an index n and a very small angle A, the distance between the two images is $a = 2(n - 1)Ad$, where A is expressed in radians. (b) Calculate the spacing of the fringes of green light of wavelength 5×10^{-7} m produced by a source placed 5 cm from a biprism having an index equal to 1.5 and an angle of 2°. The screen is 1 m from the biprism (recall Problem 33.27).

34.4 Discuss the interference pattern on a screen when the sources S_1 and S_2, separated by the small distance a, are placed along a line perpendicular to the screen (Figure 34.32).

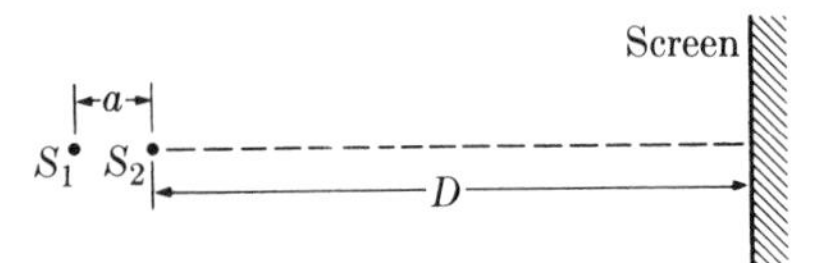

Figure 34.32

34.5 Two synchronized sound sources send out waves of equal intensity at a frequency of 680 Hz. The sources are 0.75 m apart. The velocity of sound in air is 340 m s^{-1}. Find the positions of minimum intensity: (a) on a line that passes through the sources, (b) in a plane that is the perpendicular bisector of the line between the sources and (c) in a plane that contains the two sources. (d) Is the intensity zero at any of the minima?

34.6 One technique for observing an interference pattern produced by two slits is to illuminate them with parallel rays of light, place a convergent lens behind the plane of the slits, and observe the interference pattern on a screen placed at the focal plane of the lens (Figure 34.33). Verify that the position of the bright fringes, relative to the central fringe, is given by $x = n(f\lambda/a)$, while the dark fringes correspond to $x = (2n + 1)(f\lambda/2a)$ where n is an integer, f the focal length of the lens, and a is the separation of the slits.

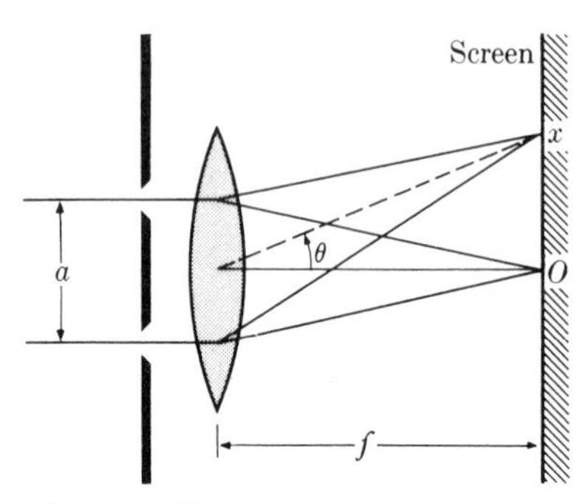

Figure 34.33

34.7 An interferometric arrangement used in radio astronomy consists of two radio telescopes separated a certain distance. The antennas can be oriented in different directions, but are always kept parallel (Figure 34.34). (a) Verify that the directions of incidence for which the resultant signal is maximum are $\sin\theta = n\lambda/a$. What advantages does this arrangement have over a single antenna? (b) Make a polar plot of the intensity of the signal as a function of the angle θ. (c) In such an interferometer, which operates at a wavelength of 11 cm, the distance a between the two radio telescopes can be adjusted up to 2700 m. Find the angle subtended by the central intensity maximum at the largest separation of the two telescopes.

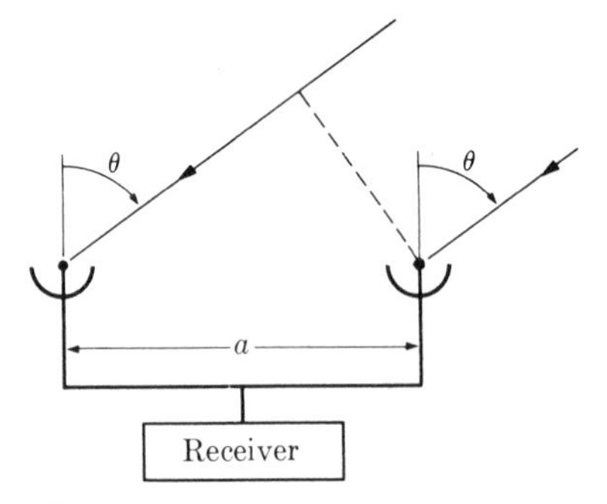

Figure 34.34

34.8 Suppose that, instead of two parallel slits as in Young's experiment, there are three parallel slits equally spaced a distance a. Make a plot of the intensity distribution of the interference pattern observed on a distant screen.

34.9 Discuss the angular distribution of intensity for (a) three and (b) five identical sources of waves equally spaced the distance a along a straight line. Assume that $a = \lambda/2$.

34.10 The first multiple radio interferometer, built in 1951, consists of 32 antennas, 7 m apart. The system is tuned to a wavelength of 21 cm. The system is thus equivalent to 32 equally spaced sources. Find (a) the angular separation between successive principal maxima and (b) the angular width of the central maximum. Compare the intensity of the maximum with that of one reflector. What are the advantages of this interferometric array?

34.11 Two rectangular pieces of plate glass are laid one upon the other. A thin strip of paper is placed between them at one edge so that a very thin wedge of air is formed. The plates are illuminated by a beam of sodium light, $\lambda = 5.9 \times 10^{-7}$ m, at normal incidence. Ten interference fringes are formed per centimeter length of wedge. Find the angle of the wedge.

34.12 A thin film having a thickness of 2.4×10^{-6} m and an index of refraction of 1.4 is illuminated with monochromatic light of wavelength 6.2×10^{-7} m. Find the smallest angles of incidence for which there is maximum (a) constructive and (b) destructive interference by reflection. (c) Repeat the problem for transmitted light.

34.13 Two glass plates, having a length of 5 cm, are placed in contact at one end and separated at the other by a thin paper sheet, thus forming an air prism. When the prism is illuminated by light of wavelength 5.9×10^{-7} m at normal incidence, 42 dark fringes are observed. Find the thickness of the paper sheet.

34.14 Verify that if R is the radius of the convex side of a plano-convex lens used to produce the interference patterns called Newton's rings (Figure 34.16), the radii of the bright rings are given by $r^2 = N\lambda R$ and the radii of the dark rings by $r^2 = (2N + 1)(\lambda R/2)$, where N is a positive integer. The index of refraction of air has been taken as one.

34.15 The radius of curvature of the convex surface of a plano-convex lens is 1.20 m. The lens is placed on a plane glass plate with the convex side down, and illuminated from above with red light of wavelength 6.5×10^{-7} m. Find the diameter of the third bright ring of the interference pattern (see Problem 34.14).

34.16 How is the fundamental frequency of a string changed if (a) its tension is doubled, (b) its mass per unit length is doubled, (c) its radius is doubled, (d) its length is doubled? Repeat the problem if the quantities listed are halved.

34.17 A tube whose length is 0.60 m is (a) open at both ends, and (b) closed at one end and open at the other. Find its fundamental frequency and the first overtone if the temperature of the air is 27 °C. Plot the amplitude distribution along the tube corresponding to the fundamental frequency and the first overtone for each case.

34.18 Estimate the percentage change in the fundamental frequency of an air column open at both ends per degree change in temperature at a temperature of 27 °C (see Example 28.6).

34.19 Two surface waves, $A \sin k(x - vt)$ and $A \sin k(y - vt)$, propagate in perpendicular directions along a membrane. (a) Calculate the resulting motion, showing that these waves are equivalent to a modulated wave propagating in a direction making an angle of 45° with the X-axis and with a phase velocity equal to $2^{1/2}v$. (b) Verify that the wavelength is reduced by the factor $2^{1/2}$. (c) Verify that the amplitude is zero along the lines $x - y = (2n + 1)\pi/k$.

34.20 (a) Verify that for a square membrane of side a, if $\nu_0 = v/2a$ is the fundamental frequency, the successive frequencies are $\nu = 2^{1/2}\nu_0, 2\nu_0, 5^{1/2}\nu_0, 2(2)^{1/2}\nu_0, 3\nu_0, 10^{1/2}\nu_0, (13)^{1/2}\nu_0, \ldots$. (b) Determine the number of different combinations of n_1 and n_2 needed to obtain the fundamental and each successive mode of vibration. The number of different combinations gives the **degeneracy** of the vibrating mode.

34.21 Repeat Problem 34.20 for a cubical cavity of side a.

34.22 A wave with a wavelength of 2.0 m and a frequency of 1000 Hz is propagating between two plane parallel reflectors separated by a distance of 1.2 m. (a) Determine the phase velocity of the guided wave for $n = 1$. (b) What is the angle of incidence with the planes?

34.23 (a) Calculate the cut-off frequency for the waveguide defined in the previous problem. (b) Consider a waveguide with two sets of parallel sides, each 1.2 m apart. What is the cut-off frequency of such a device?

34.24 A waveguide consists of a long tube with a rectangular cross-section of sides a and b. Verify that (a) the resultant wave is described by $\xi = 4\xi_0 \sin k_2 y \sin k_3 z \cos(\omega t - k_1 x)$, (b) the only frequencies transmitted along the waveguide are those satisfying $\nu \geqslant \frac{1}{2}v(n_1^2/a + n_2^2/b^2)^{1/2}$ where n_1 and n_2 are integers. (c) Discuss the nodal planes in the waveguide for $n_1 = 2$ and $n_2 = 3$.

35 Diffraction

William H. Bragg and his son William L. Bragg (pictured here) are well known for their outstanding research on the structure of crystals using X-rays, determining the spacing between the atomic planes and the position of the atoms in a crystal. For that purpose they designed an X-ray spectrometer, with which they were the first to measure the wavelength of X-rays, supporting the idea that X-rays were a kind of electromagnetic wave.

35.1 Introduction

Diffraction is a phenomenon characteristic of wave motion that occurs when a wave is distorted by an obstacle. The obstacle may be a screen with a small opening or a slit that allows only a small portion of the incident wave front to pass, or a small object, such as a wire or a disk, that blocks the passage of a small portion of the wave front. For example, we know from common experience, especially in the case of sound waves and surfaces waves in water, that waves extend around the obstacles interposed in their path, as illustrated in Figure 35.1. This effect becomes more and more noticeable as the dimensions of the slits or the size of the obstacles approach the wavelength of the waves.

We shall consider only **Fraunhofer diffraction**, which occurs when the incident waves are plane, so that the rays are parallel, and the pattern is observed at a distance sufficiently large so that only parallel diffracted rays are received. This diffraction is named after Joseph von Fraunhofer (1787–1826) who was one of the first to study the phenomenon.

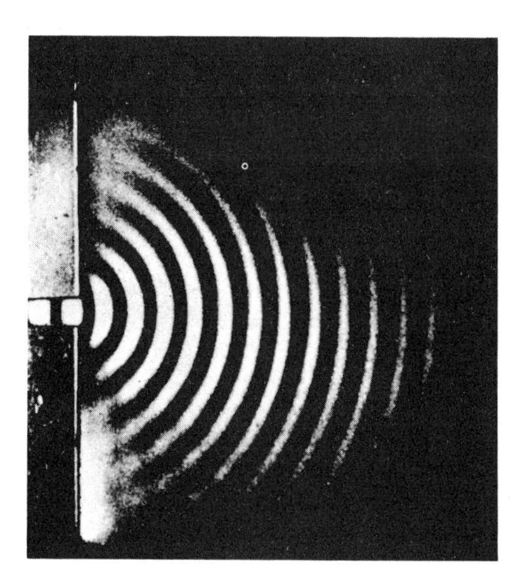

Figure 35.1 Behavior of a wave on the surface of a liquid impinging from the left on a wall with a small opening.

35.2 Huygens' principle

Around 1680 Christiaan Huygens (1629–1695) proposed a geometric procedure for tracing the propagation of elastic or mechanical waves in a material medium. A **wave surface** or a **wave front** is a surface composed of all points of the medium reached by the wave motion at the same time. Therefore the displacement at all points of a wave surface has the same phase. For example, for a plane wave propagating in the direction of the unit vector $\boldsymbol{u}$ the disturbance is expressed by $f(\boldsymbol{u}\cdot\boldsymbol{r} - vt)$, and a wave surface is composed of all points at which the phase $\boldsymbol{u}\cdot\boldsymbol{r} - vt$ has the same value at a given time. Therefore the wave surface is given by the equation $\boldsymbol{u}\cdot\boldsymbol{r} - vt = const.$, which, for a given t, corresponds to a plane perpendicular to the unit vector $\boldsymbol{u}$. Similarly, for spherical waves, the wave surfaces are given by $r - vt = const.$, where r is the distance from the source. For a given t, the surface corresponds to a sphere concentric with the source.

Consider a wave surface S (Figure 35.2). According to Huygens, when the wave motion reaches S, each particle $a, b, c, \ldots$ on the surface becomes a secondary source of waves, emitting **secondary waves** (indicated by the small semi-circles), which reach the next layer of particles in the medium. These particles are then set in motion, forming the next wave surface S', which is tangent to all secondary waves. The process keeps repeating itself, resulting in the propagation of a wave through the medium.

This pictorial representation of the propagation of a wave looks very reasonable for an elastic wave resulting from vibrations of atoms or molecules in a body. However, Huygens' construction has no physical meaning in such cases as, for example, an electromagnetic wave propagating in vacuum, where there are no vibrating particles. Therefore, it required a revision when it was recognized that other waves, of a different kind, also exist in nature. This revision was accomplished at the end of the 19th century by Kirchhoff, who replaced Huygens' intuitive construction by a more mathematical treatment.

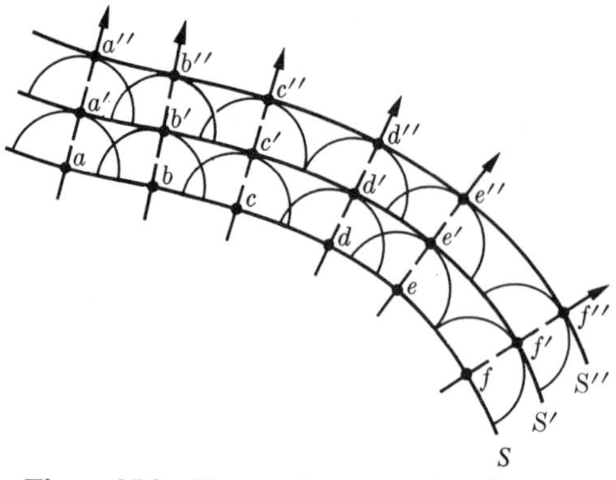

Figure 35.2 Huygens' construction for a progressing wave.

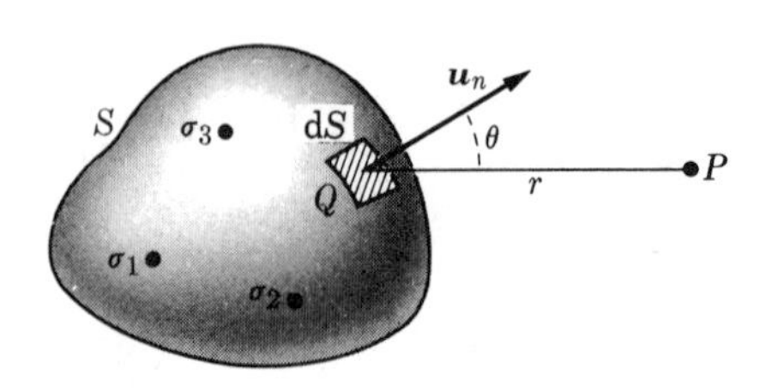

Figure 35.3 The wave at P can be computed if the wave at the points on the closed surface S is known.

Wave motion is regulated by the wave equation 28.11, which in three dimensions is

$$\frac{d^2\xi}{dt^2} = v^2\left[\frac{d^2\xi}{dx^2} + \frac{d^2\xi}{dy^2} + \frac{d^2\xi}{dz^2}\right] \tag{35.1}$$

where ξ may be the displacement of the atoms of a substance in the case of an elastic wave, the electric or magnetic field in the case of an electromagnetic wave, and so forth. Analyzing wave propagation in any given medium consists in obtaining a solution to the wave equation that satisfies the physical conditions of the problem; that is, the position and nature of the sources, physical surfaces of discontinuity etc. These conditions are called the **boundary conditions**. The theory of differential equations states that, under special conditions, we can find a solution of an equation such as Equation 35.1 without reference to the sources if we know the values of ξ over a surface S (Figure 35.3). Suppose that we want to evaluate the wave motion at a point P. If we know the sources $\sigma_1, \sigma_2, \sigma_3, \ldots$, we may add all their contributions at P to obtain the resultant wave motion. Now suppose that, instead, we only know the value of ξ at all points of the arbitrary but closed surface S enclosing the sources. In this case we may also obtain the wave at P, even if we ignore the distribution of the sources. Then Huygens' principle, as modified by Kirchhoff, amounts to the statement that

we can obtain the perturbation at a point P at time t if we know the perturbation at each surface element dS on a surface S and we assume that the surface elements act as secondary sources of waves. The wave motion at any point is obtained by adding the wave motions due to these secondary sources.

EXAMPLE 35.1

Diffraction pattern of a razor's edge.

▷ Suppose a beam of parallel monochromatic light falls on a razor's edge (Figure 35.4). Half of the wave surface S is interrupted by the razor. Each point of the wave front to the right of the edge acts as a secondary source according to Huygens' principle. Therefore the light

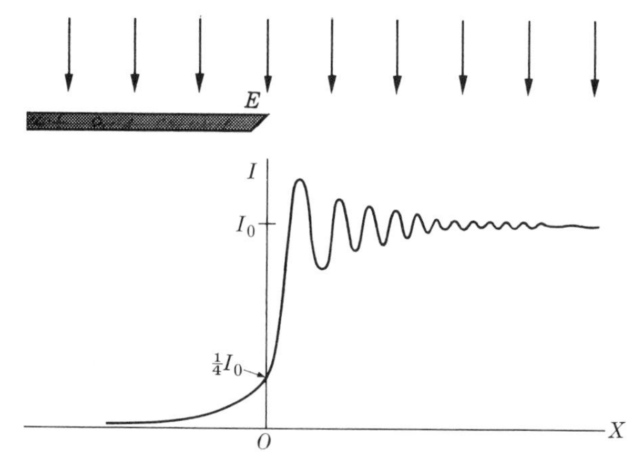

Figure 35.4 Intensity distribution for diffraction by a straight edge.

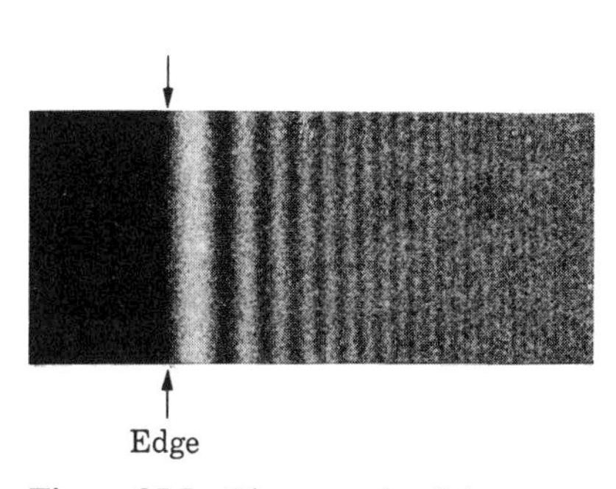

Figure 35.5 Photograph of the diffraction pattern of a straight edge. The arrows indicate the position of the edge.

falling on a screen behind the razor is the result of the interference of the secondary waves produced by half the original wave, resulting in the intensity distribution shown in Figure 35.5. The intensity falls off gradually to zero within the geometrical shadow, and fluctuates during the first few wavelengths within the geometrical region of illumination. Note that at the points directly behind the edge the intensity is one fourth the intensity without the screen. This is because the screen stops the passage of half of the wave front, resulting in an amplitude reduced to one half and hence an intensity reduced to one fourth.

35.3 Fraunhofer diffraction by a rectangular slit

Consider waves incident normally to a very narrow and a very long rectangular slit. Only the part of the wave front passing through the slit contributes to the transmitted or *diffracted* waves. According to Huygens' principle, the intensity distribution of the diffracted waves is obtained by adding the waves emitted by each point between A and E (Figure 35.6), which are considered as secondary sources of waves. Observing the diffracted waves at different angles θ with respect to the direction of incidence, we find that for certain directions their intensity is zero. These directions are given by the relation

$$\sin\theta = \frac{n\lambda}{b} \qquad (n \neq 0) \qquad (35.2)$$

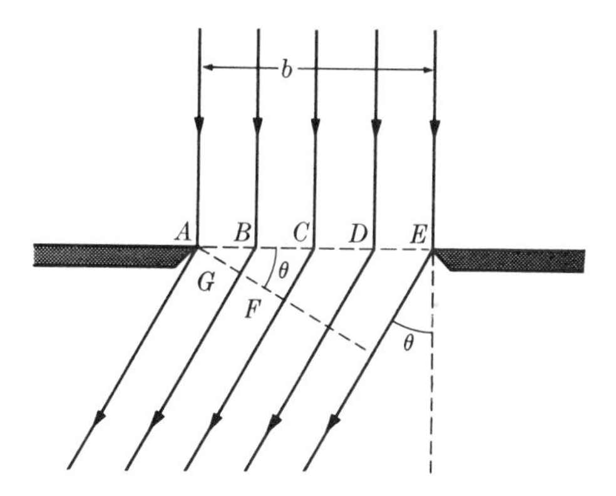

Figure 35.6 Diffraction by a long narrow slit of width b.

where n is a positive or a negative integer, b is the width of the slit, and λ the

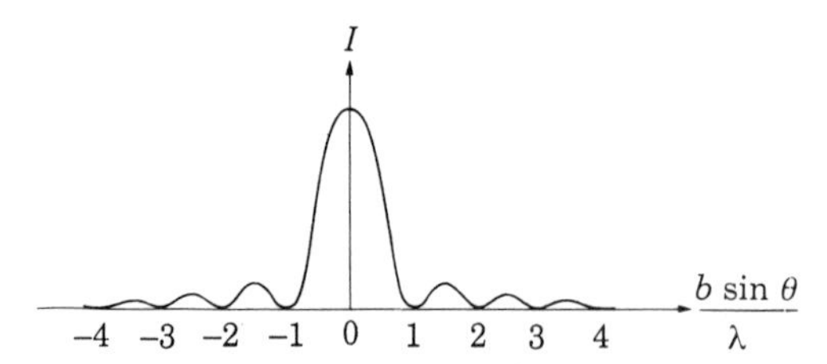

Figure 35.7 Intensity distribution of the diffraction pattern of a long narrow slit.

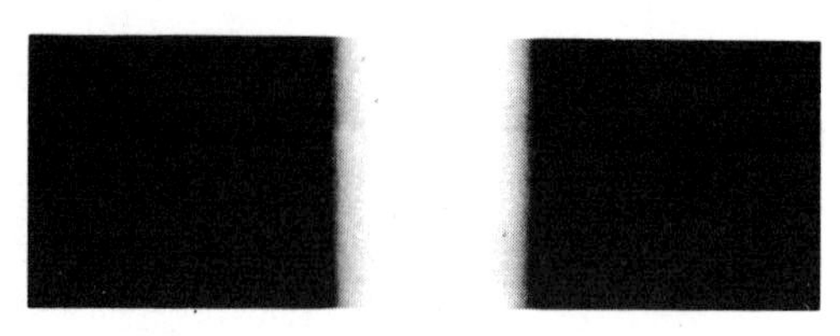

Figure 35.8 Fraunhofer diffraction pattern produced by a long narrow slit.

wavelength of the incident waves. The value $n = 0$ is excluded, because it corresponds to observation along the direction of incidence, which obviously implies a maximum of illumination.

From Equation 35.2, the intensity is zero for angles given by

$$\sin\theta = \pm\frac{\lambda}{b}, \pm\frac{2\lambda}{b}, \pm\frac{3\lambda}{b}, \cdots \tag{35.3}$$

Between each two zeros of intensity given by Equation 35.3 there is a maximum; but these maxima gradually decrease in intensity because they correspond to points farther away from the slit. This is a situation different from that for interference. The intensity of the diffracted waves as a function of θ is represented in Figure 35.7. Note that the central maximum has twice the width of the others. Figure 35.8 shows the actual diffraction pattern of a rectangular slit. For λ very small compared with b, the first zeros of intensity on either side of the central maximum correspond to an angle

$$\theta \approx \sin\theta = \pm\frac{\lambda}{b} \tag{35.4}$$

obtained by setting $n = \pm 1$ in Equation 35.2. This is shown in Figure 35.9.

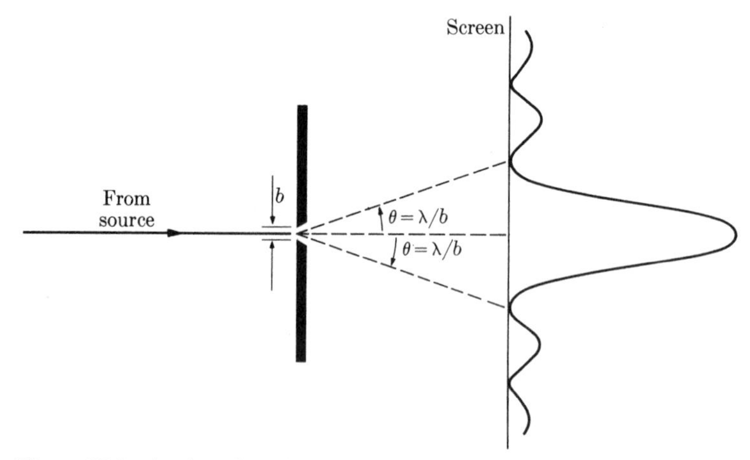

Figure 35.9 Angle subtended by the central intensity peak of the diffraction pattern of a single slit.

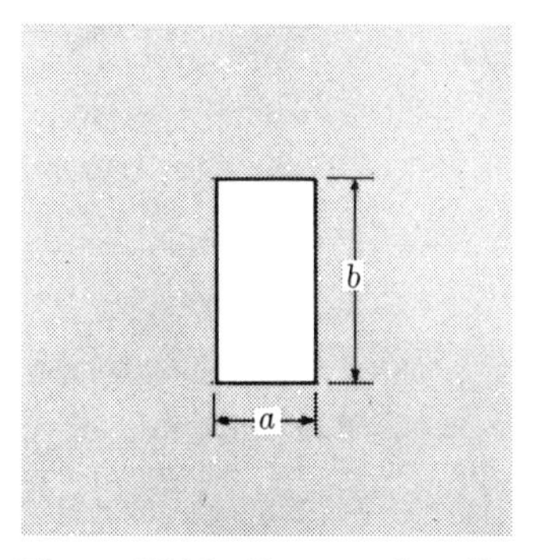

Figure 35.10 Rectangular slit.

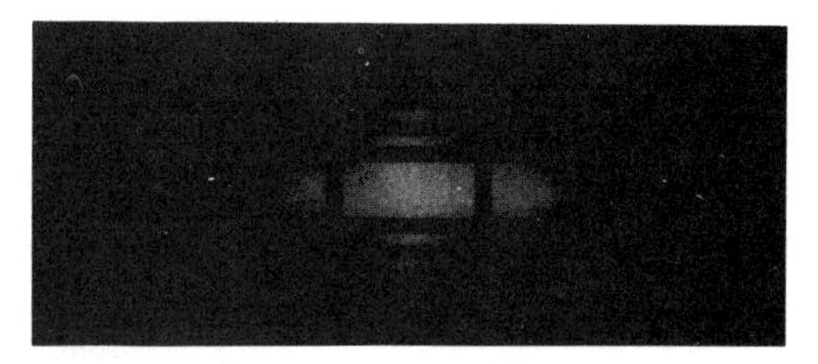

Figure 35.11 Fraunhofer diffraction pattern of a rectangular slit whose length is twice its width.

If the slit is rectangular, with sides a and b that are of comparable size (Figure 35.10), the diffraction pattern is the combination of the two patterns due to each pair of sides. Instead of the series of bands, we get a series of rectangles arranged in a crosswise form, as in the photograph of Figure 35.11.

Proof of the condition for zero intensity

To justify Equation 35.2, we recall from Equation 34.8 that when the difference in path length for two rays is $r_1 - r_2 =$ odd integer $\times (\frac{1}{2}\lambda)$, destructive interference results. From Figure 35.6 we see that for rays coming from A and the midpoint C we have

$$r_1 - r_2 = CF = \tfrac{1}{2}b \sin\theta = n(\tfrac{1}{2}\lambda)$$

Thus for $n = 1, 3, 5, \ldots$, these two rays, as well as all other pairs of rays originating at points between A and E separated by $\frac{1}{2}b$, interfere destructively, and no wave is observed in the direction θ. For even n, consider points A and B separated by $b/4$. Then

$$r_1 - r_2 = BG = \tfrac{1}{4}b \sin\theta = (\tfrac{1}{2}n)(\tfrac{1}{2}\lambda)$$

So when $n/2$ is an odd integer, or $n = 2, 6, 10, \ldots$, these two rays, as well as all other pairs of rays originating at points separated by $b/4$ interfere destructively. Therefore, no wave is observed in the directions corresponding to these angles. The procedure can be extended until all integers are included, so that the zeros of intensity occur when $b \sin\theta = n\lambda$, from which Equation 35.2 follows. For $\theta = 0$, which corresponds to the direction of incidence, however, there is no phase difference for the rays coming from different points, and the interference is constructive, resulting in a pronounced maximum.

EXAMPLE 35.2

Resolving power of a slit.

▷ The **resolving power** of a slit, as defined by Lord Rayleigh, is the minimum angle subtended by two incident waves, coming in different directions from two distant point sources, that will permit their respective diffraction patterns to be distinguished. When waves coming from two distant sources S_1 and S_2 pass through the same slit in two directions, making

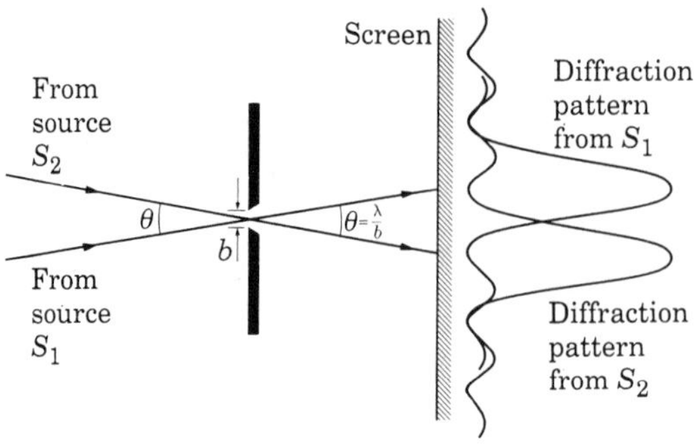

Figure 35.12 Rayleigh's rule for resolving power of a rectangular slit.

an angle θ (Figure 35.12), the diffraction patterns of the two sets of waves are superposed. They begin to be distinguishable when the central maximum of one falls on the first zero on either side of the central maximum of the other, as indicated on the right in Figure 35.12. But then, in view of Equation 35.4 and Figure 35.9, the angle θ must be

$$\theta = \frac{\lambda}{b} \tag{35.5}$$

which gives the resolving power of the slit according to Rayleigh's definition.

Assuming that S_1 and S_2 are two points on a distant object, Equation 35.5 gives the minimum angular separation between them in order for the two points to be recognizable as different when the object is observed through the slit. If the light passing through the slit forms an image on a screen, and that image is observed with a microscope, for example, it is not possible, no matter what the magnification of the microscope, to observe more detail in the image than that allowed by the resolving power of the slit. These considerations must be taken into account in the design of optical instruments.

35.4 Fraunhofer diffraction by a circular aperture

When plane waves fall perpendicularly on a screen that has a circular aperture, the diffraction pattern consists of a bright disk surrounded by alternate dark and bright rings, as shown in Figure 35.13. The radii of the central disk and successive rings do not follow a simple sequence. We shall omit the mathematical analysis of the problem, which is much more involved than in the case of the rectangular slit because of the geometrical arrangement. Assuming that R is the radius of the aperture (Figure 35.14), the angle corresponding to the first dark ring is given by

$$\sin\theta = 1.22\frac{\lambda}{2R} \tag{35.6}$$

When λ is much smaller than R we may write

$$\theta = 1.22\frac{\lambda}{2R} = 1.22\frac{\lambda}{D} \tag{35.7}$$

where $D = 2R$ is the diameter of the aperture and θ is expressed in radians. A lens is actually a circular aperture. Therefore, the image of a point, which in

Figure 35.13 Fraunhofer diffraction pattern of a circular slit.

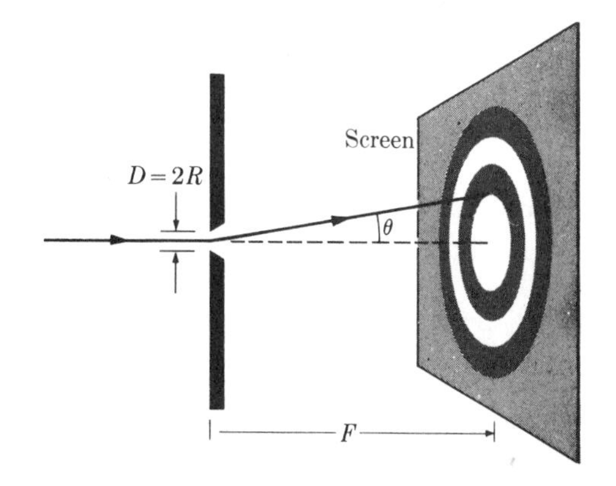

Figure 35.14 Diffraction by a circular slit.

Chapter 33 was assumed to be another point, is in fact a diffraction pattern. However, the radius of a lens is, in general, so large compared with the wavelength of light that, for most practical purposes, diffraction effects may be ignored.

EXAMPLE 35.3

Resolving power of a circular aperture.

▷ Expression 35.7 also gives the resolving power for a circular aperture, defined, again according to Lord Rayleigh, as the minimum angle between the directions of incidence of two plane waves coming from two distant point sources that will allow their respective diffraction patterns to be distinguished. This takes place when the center of the bright disk of the diffraction pattern of one source falls on the first dark ring of the diffraction pattern of the second (Figure 35.15). The angular separation is given by $\theta = 1.22\,\lambda/D$. This expression was mentioned in Section 33.5 when we discussed the magnification of a microscope and a telescope.

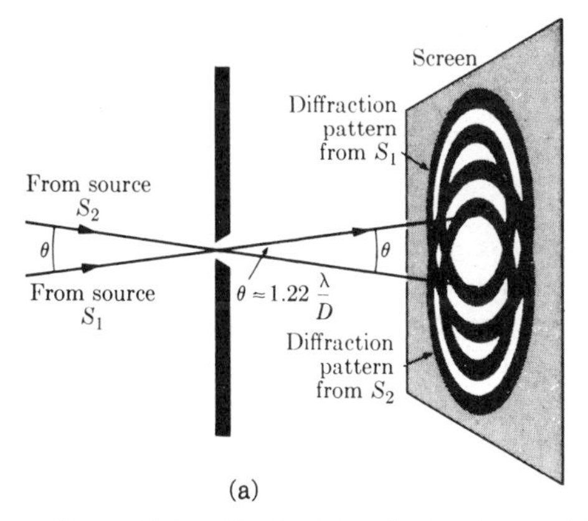

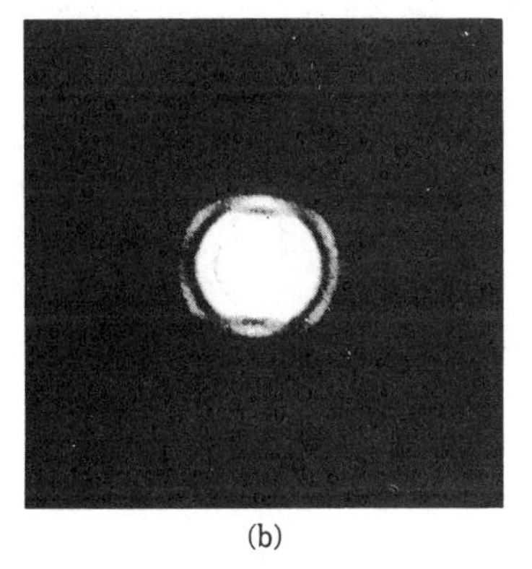

Figure 35.15 Rayleigh's rule for the resolving power of a circular slit. Part (b) shows two distant point sources imaged through a lens and just resolved.

EXAMPLE 35.4

Compute the radius of the central disk of the diffraction pattern observed in a plane at the focus of a lens with a diameter of 2 cm and a focal length of 40 cm. Assume the lens is illuminated with a beam of parallel monochromatic light of wavelength 5.9×10^{-7} m.

▷ When we use Equation 35.7, the angle subtended by the central disk in the diffraction pattern is

$$\theta = 1.22 \times \frac{5.9 \times 10^{-7}\ \text{m}}{2 \times 10^{-2}\ \text{m}} = 3.60 \times 10^{-5}\ \text{rad} = 7.42''$$

This is also the resolving power of the lens. The radius of the central disk is $r = f(\frac{1}{2}\theta) = 40\ \text{cm} \times 1.80 \times 10^{-5}\ \text{rad} = 7.2 \times 10^{-4}\ \text{cm}$, and thus, for practical purposes, we may say that the image at the focal plane is a point.

35.5 Fraunhofer diffraction by two equal parallel slits

Consider two slits, each of width b, displaced by a distance a (Figure 35.16). For a direction corresponding to the angle θ, we now have two sets of diffracted waves coming from each slit, and what we actually observe is the result of the interference of these diffracted waves. In other words, we now have a combination of diffraction and interference.

If the two slits are identical, the interference pattern is that of two synchronous sources, with maxima at the directions given by Equation 34.10; that is,

$$\sin\theta = \frac{n\lambda}{a} \tag{35.8}$$

The intensity distribution of the interference pattern is modulated by the intensity distribution for the diffraction pattern of a single slit.

The zeros of the diffraction pattern are given by Equation 35.2,

$$\sin\theta = \frac{n'\lambda}{b}$$

where $n' = \pm 1, \pm 2, \ldots$. Since a is larger than b, the zeros of the diffraction pattern are more widely spaced than the maxima of the interference pattern. Therefore, the bright fringes of two slits are much narrower and more closely spaced than those produced by a single slit. The resulting intensity distribution is shown in Figure 35.17 and in the photograph of Figure 35.18.

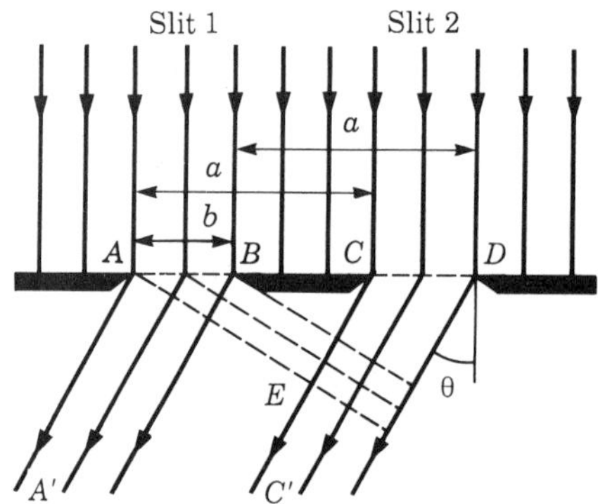

Figure 35.16 Front view and cross section of two parallel long narrow slits.

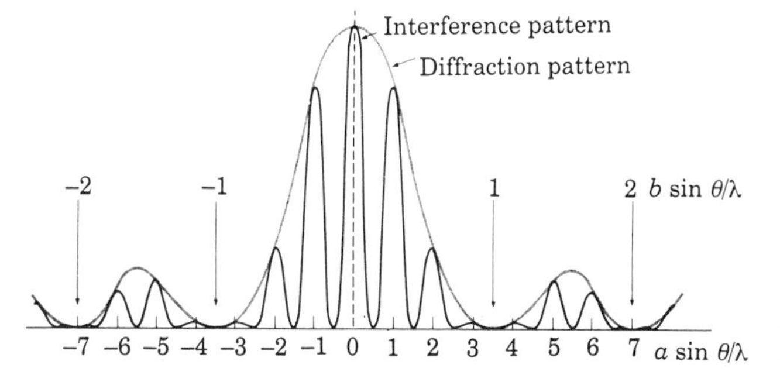

Figure 35.17 Intensity distribution (along a plane set normal to the incident light) resulting from two parallel long narrow slits.

Figure 35.18 Fraunhofer diffraction pattern due to two parallel long narrow slits.

35.6 Diffraction gratings

A **diffraction grating** is a set of several parallel slits of equal width b, equally spaced the distance a. Let N be the number of slits and consider plane waves falling normally on the grating (Figure 35.19). By similarity with the problem of the double slit, in the direction corresponding to the angle θ we observe the interference due to N synchronous sources (one per slit) modulated by the diffraction pattern of one slit.

If the number of slits is large, the pattern will consist of a series of narrow bright fringes corresponding to the main maxima of the interference pattern, which for normal incidence according to Equation 34.10, are given by

$$\sin\theta = \frac{n\lambda}{a} \tag{35.9}$$

where $n = 0, \pm 1, \pm 2, \ldots,$ but their intensities are modulated by the diffraction pattern whose zeros are given by

$$\sin\theta = \frac{n'\lambda}{b}$$

where $n' = \pm 1, \pm 2, \ldots$. Figure 35.20 shows the case for eight slits ($N = 8$). According to the value of n, the principal maxima are called the first, second, third, etc. **order of diffraction**.

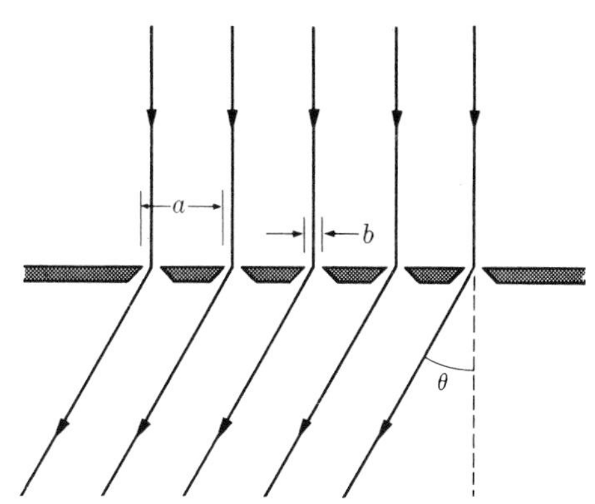

Figure 35.19 A transmission diffraction grating.

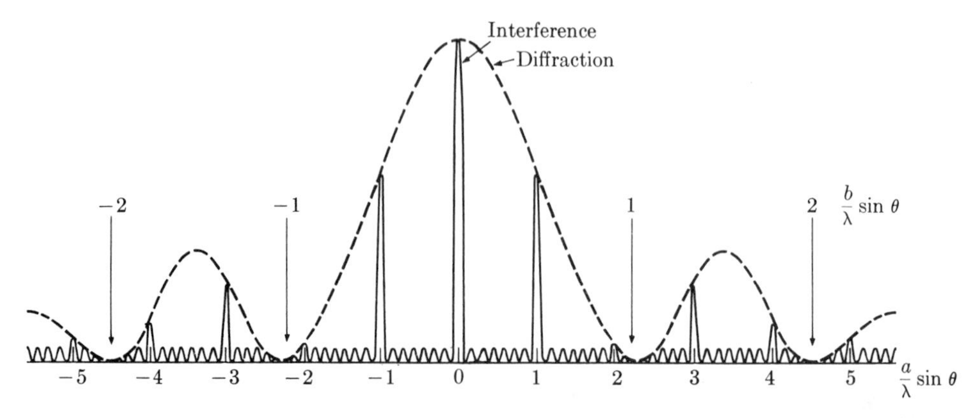

Figure 35.20 Intensity distribution produced by a diffraction grating on a plane placed normal to the incident light and parallel to the grating.

The system we have just discussed is called a **transmission diffraction grating**. For analyzing near-infrared, visible or ultraviolet light, transmission diffraction gratings consist of several thousands of closely spaced slits, obtained by etching a series of parallel lines on a transparent film. The lines then act as the opaque spaces between the slits. A diffraction grating can also work by reflection. A series of parallel lines are etched on a metallic surface. The narrow strips between the etchings reflect light, producing a diffraction pattern. Sometimes the surface is made concave to improve focusing.

When light of several wavelengths falls on a diffraction grating, the different wavelengths produce diffraction maxima at different angles, except for the zero order, which is the same for all. The set of maxima of a given order for all wavelengths constitutes a **spectrum**. So we have spectra of first, second, third etc. orders and the longer the wavelength, the larger the deviation for any order of spectrum. Therefore, red is deviated more than violet, which is the opposite of what happens when light is dispersed by a prism.

Diffraction gratings can be used to analyze a wide range of regions of the electromagnetic spectrum, and have several distinct advantages over prisms. One advantage is that diffraction gratings do not depend on the dispersive properties of the material, but only on the geometry of the grating. Figure 35.21 shows the basic elements of a diffraction grating spectroscope.

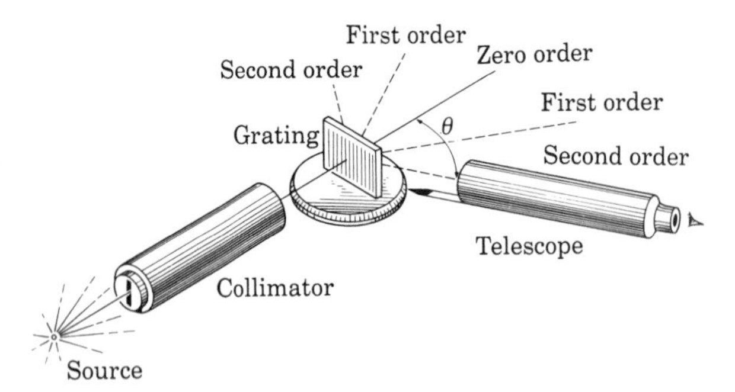

Figure 35.21 Grating spectroscope. The source is placed in front of the slit on the collimator. The diffraction grating is placed perpendicular to the collimator's axis and the spectra of different orders are investigated by moving the telescope. The slit is placed at the focal plane of the collimator's lens so the light on the grating is parallel.

EXAMPLE 35.5

Angular separation of the whole visible spectrum for the first order and the second order, for a grating with 20 000 lines and a length of 4 cm. Assume that the visible spectrum goes from 3.90×10^{-7} m, up to 7.70×10^{-7} m, as mentioned in Section 24.8.

▷ We have that $a = 4 \times 10^{-2}\ \text{m}/20\,000 = 2 \times 10^{-6}$ m. Therefore, using Equation 35.9, we have for $n = 1$,

$$\sin \theta_{\text{red}} = \frac{7.70 \times 10^{-7}}{2 \times 10^{-6}} = 0.335, \quad \text{or} \quad \theta_{\text{red}} = 19°34'$$

$$\sin \theta_{\text{violet}} = \frac{3.90 \times 10^{-7}}{2 \times 10^{-6}} = 0.195, \quad \text{or} \quad \theta_{\text{violet}} = 11°15'$$

Therefore the first-order spectrum covers an angle of 8°10′. Similarly, for the second-order spectrum, the angle is 22°27′. Is a full third-order spectrum possible?

EXAMPLE 35.6

Resolving power of a diffraction grating.

▷ When plane waves of slightly different wavelengths fall on a diffraction grating, the principal maxima of the same order for each wavelength may fall so close to each other that it is impossible to distinguish whether the original beam was monochromatic or not. In order that two wavelengths may be distinguished (or resolved) in a given order, it is necessary that the principal maximum for one of the wavelengths fall on the first zero on either side of the principal maximum of the other wavelength. Given that $\Delta\lambda$ is the minimum wavelength difference for which the above condition is met at a wavelength λ, the **spectral resolving power** of the grating is $R = \lambda/\Delta\lambda$. It can be verified, using the results of Section 34.3, that the resolving power is given by

$$R = \frac{\lambda}{\Delta\lambda} = Nn \qquad \textbf{(35.10)}$$

Therefore, the greater the total number of lines of the grating and the higher the order of the spectrum, the smaller $\Delta\lambda$ is, and so the greater the resolving power of the grating. On the other hand, the resolving power is independent of the size and spacing of the ruling in the grating. The spectral resolving power of a diffraction grating is important for spectroscopic analysis to assure that lines of the spectrum with close wavelengths appear separated or 'resolved'.

EXAMPLE 35.7

Determine whether the grating of Example 35.5 can resolve the two yellow lines of sodium, whose wavelengths are 5.890×10^{-7} m and 5.896×10^{-7} m.

▷ The average wavelength of the two lines is 5.893×10^{-7} m, and their separation is 6×10^{-10} m. From Equation 35.10 we have that the resolving power of the grating is $R = Nn = 2 \times 10^4\, n$. At the given wavelength, the minimum wavelength separation in the first-order spectrum is

$$\Delta\lambda = \frac{\lambda}{Nn} = \frac{5.893 \times 10^{-7}}{2 \times 10^4 \times 1} = 2.947 \times 10^{-11} \text{ m}$$

which is one-twentieth of the separation of the two sodium lines. The two D-lines in the first-order spectrum produced by the grating could therefore be quite clearly distinguished.

35.7 X-ray scattering by crystals

Electromagnetic waves with wavelengths shorter than the ultraviolet, such as X-rays and γ-rays, are not noticeably affected by slits, gratings and objects of the dimensions used for the optical region. However, in a crystal lattice, the atoms or molecules are regularly spaced at distances of the order of 10^{-10} m (Figure 35.22). Thus the atoms in a crystal can serve as scattering centers for electromagnetic waves with wavelengths of the same order of magnitude as (or smaller than) interatomic distances, which is the case for X- or γ-rays.

When X- or γ-rays pass through a crystal, the intensity of the scattered rays shows a pattern that is the result of the interference, along the direction of observation, of the waves scattered by each atom in that direction, modulated by a scattering factor characteristic of the atoms. In this respect, the crystal is similar to a three-dimensional grating with the diffraction of each slit replaced by the atomic scatterer. When the crystal is composed of more than one class of atoms, each kind of atom contributes in a different way to the scattering of the X-rays. Thus, to simplify our calculation, we shall assume that we have only one class of atoms, and only one atom per unit cell in the crystals.

To analyze X-ray scattering by crystals, it is convenient to imagine a series of parallel planes, all equally spaced, passing through layers of atoms of the crystal. Figure 35.23 shows several possible groups of parallel planes in a cubic crystal. Groups of parallel planes differ in their spacing and in the density of scattering centers as the figure shows for planes a, b and c.

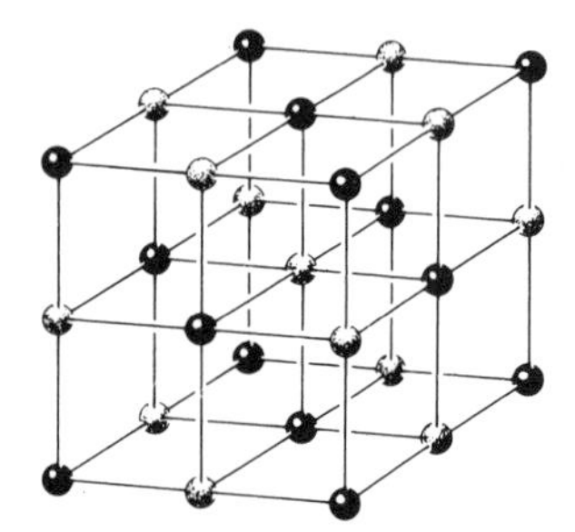

Figure 35.22 Representation of a NaCl crystal showing the regular arrangement of atoms forming a cubic lattice.

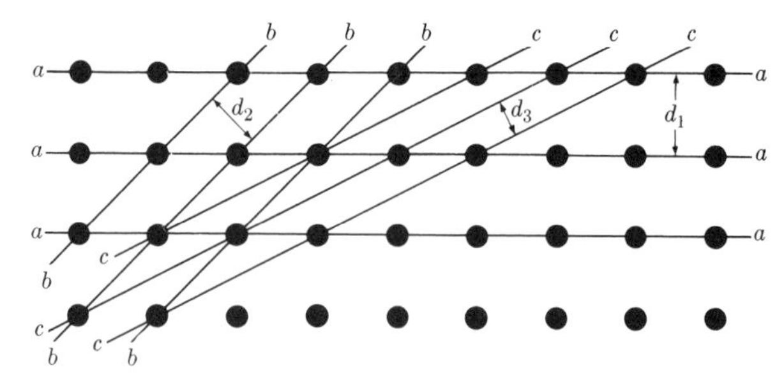

Figure 35.23 Several possible parallel scattering planes in a cubic crystal: $d_2 = d_1/2^{1/2}$; $d_3 = d_1/5^{1/2}$.

Consider rays falling on a crystal making an angle θ with a certain family of planes (Figure 35.24) and investigate the scattered rays along a *symmetric* direction, which also makes an angle θ with the family of planes. As shown below, if d is the separation of consecutive planes, maximum intensity of the scattered waves, resulting from constructive interference, occurs when

$$\sin\theta = \frac{n\lambda}{2d} \qquad \textbf{(35.11)}$$

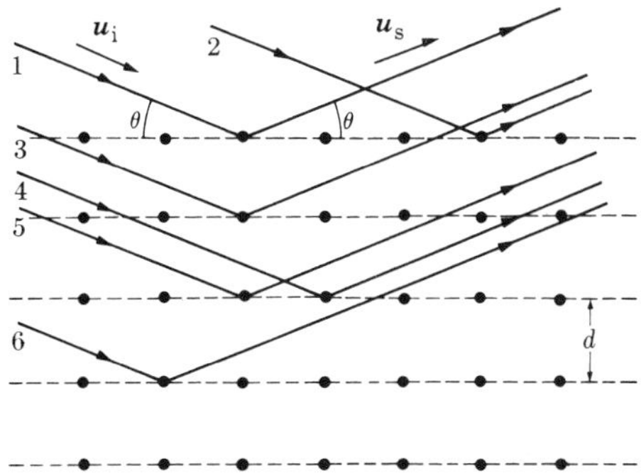

Figure 35.24 Parallel scattering planes in a crystal.

where n is an integer. This expression is known as **Bragg's equation**, to honor William H. Bragg (1862–1942) and his son William L. Bragg (1890–1971), who together studied X-ray scattering by crystals. The values of n are limited by the condition that $\sin\theta$ must be smaller than one. For rays such as 1 and 2 in Figure 35.24, which are scattered by atoms in the same plane, the phase difference is zero ($n = 0$) and they interfere constructively. This, however, happens for any angle of incidence. The importance of Bragg's condition is that rays such as 3, 4, 5,... coming from successive planes also interfere constructively, giving rise to a very intense maximum. Therefore Bragg's condition expresses a sort of collective effect, where the rays scattered by all atoms in certain parallel planes interfere constructively. For fixed planes (or fixed d) and wavelength λ, changing the angle θ alternately produces positions of maximum and minimum intensity, corresponding to constructive (as given by Equation 35.11), or destructive interference. Note that Equation 35.11 can be used to measure the plane separation d if the wavelength λ is known, and conversely. A schematic drawing of the experimental arrangement for observing Bragg's scattering of X-rays, a device called a **crystal spectrometer**, is shown in Figure 35.25.

For a given direction of incidence relative to the whole crystal, we have a series of maxima corresponding to scattering from all families of planes for which Bragg's equation holds. The maxima are in different directions because of the different orientations of the families of planes.

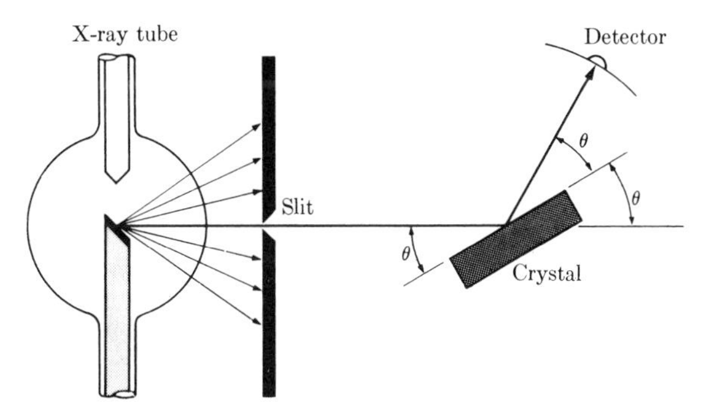

Figure 35.25 Crystal spectrometer for analysis of X-ray diffraction. *X*-rays, generated by the tube at left and collimated by a slit in a lead screen, are diffracted by the crystal. The diffracted X-rays are observed by a movable detector, usually an ion chamber.

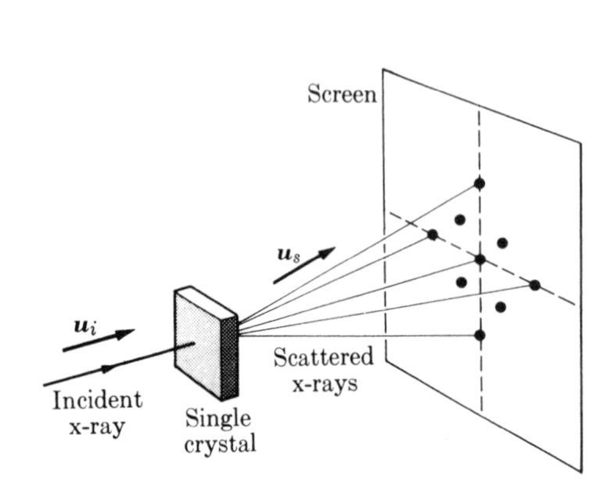

Figure 35.26 *X*-ray scattering by a single crystal.

Figure 35.27 Laue X-ray pattern for a quartz crystal.

The intensity depends on the number and spacing of atoms in each family of planes. If a photographic plate or fluorescent screen is interposed in the path of the scattered rays from a single crystal (Figure 35.26), a regular pattern, which is characteristic of the crystal structure, appears. It is called a **Laue pattern**, after Max von Laue (1879–1960) who performed original work on crystal structure. Each dot in the pattern corresponds to the direction of scattering from a family of planes. The photograph of Figure 35.27 shows one such Laue pattern.

If the scatterer, instead of being a single crystal, is a powder containing a large number of small crystals, all randomly oriented, the corresponding directions of scattering satisfying Bragg's condition are distributed on conical surfaces about the direction of incidence, as shown in Figure 35.28. On a photographic film, each conical surface produces a bright ring (Figure 35.29), resulting in the so-called **Debye–Scherrer patterns**, named for Peter Debye (1884–1966) and his student Paul Scherrer (1890–1979). By analyzing patterns such as those of Figures 35.28 and 35.29, the internal structure of a crystal can be derived or, conversely, the wavelength of the X-rays can be found.

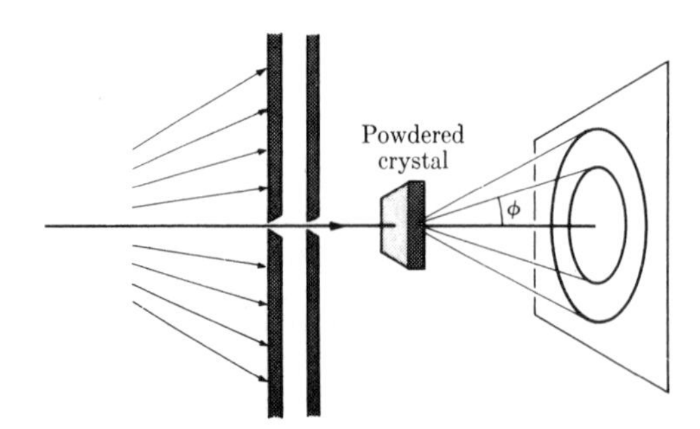

Figure 35.28 Powder X-ray scattering.

Figure 35.29 Debye–Scherrer X-ray pattern for powdered aluminum.

When Wilhelm Röntgen observed X-rays for the first time at the end of the nineteenth century, there was doubt about whether they were waves or particles. To answer this question, interference and scattering experiments were performed, using equipment similar to that used for experiments with light; the results were either negative or unconvincing. It was not until Laue, Bragg, and others studied the passage of X-rays through crystals that the wave character of X-radiation was confirmed.

Derivation of Bragg's equation

Consider two atoms A and B, separated the distance r (Figure 35.30). Let $\boldsymbol{u}_i$ be a unit vector along the direction of propagation of the incident waves and $\boldsymbol{u}_s$ a similar unit vector along the direction of the scattered waves. The path difference for the incident and scattered waves for those two atoms is $AD - BC$. The phase shift is then given by $\delta = 2\pi/\lambda(AD - BC)$. But, using the vector $\boldsymbol{r}$, we see that $AD = \boldsymbol{u}_s \cdot \boldsymbol{r}$ and $BC = \boldsymbol{u}_i \cdot \boldsymbol{r}$. Therefore, we can rewrite the phase shift as

$$\delta = \frac{2\pi}{\lambda}(\boldsymbol{u}_s - \boldsymbol{u}_i) \cdot \boldsymbol{r} = \frac{2\pi}{\lambda}\boldsymbol{v} \cdot \boldsymbol{r} \quad (35.12)$$

where $\boldsymbol{v} \equiv \boldsymbol{u}_s - \boldsymbol{u}_i$ (see the inset in Figure 35.30). Designating the angle between $\boldsymbol{u}_s$ and $\boldsymbol{u}_i$ by 2θ, the magnitude of $\boldsymbol{v}$ can be written as

$$v = 2 \sin \theta \quad (35.13)$$

The condition for constructive interference in the direction $\boldsymbol{u}_s$ is $\delta = 2n\pi$ or, in view of Equation 35.12,

$$\boldsymbol{v} \cdot \boldsymbol{r} = n\lambda \quad (35.14)$$

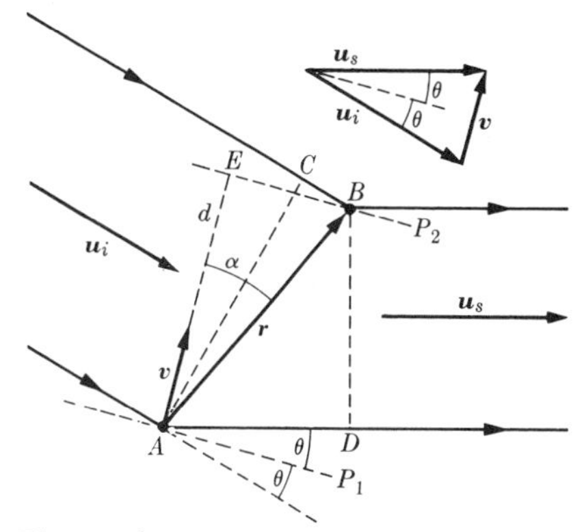

Figure 35.30 X-ray scattering by two atoms A and B.

where, as before, n is a positive or negative integer. Equation 35.14 represents a plane perpendicular to the vector $\boldsymbol{v}$. Therefore, for a given wavelength λ and a given direction of incidence, Equation 35.14 gives a series of parallel planes, one for each value of n. Figure 35.30 shows two such planes, P_1 and P_2. For all atoms located on these planes, condition 35.14 holds and they all contribute to a maximum of intensity in the direction $\boldsymbol{u}_s$. From Figure 35.30, and using Equation 35.13, we see that $d = AE = r \cos \alpha$ is the distance between planes P_1 and P_2. Then $\boldsymbol{v} \cdot \boldsymbol{r} = vr \cos \alpha = vd = 2d \sin \theta$, and Equation 35.14 becomes $2d \sin \theta = n\lambda$ or $\sin \theta = n\lambda/2d$.

EXAMPLE 35.8

The first-order spectrum of a beam of X-rays scattered by a rock-salt crystal corresponds to an angle of 6°50′ and the distance between the planes is 2.81×10^{-10} m. Determine the wavelength of the X-rays and the position of the second-order spectrum.

▷ Using Bragg's relation 35.11 with $d = 2.81 \times 10^{-10}$ m, $\theta = 6°50'$, and $n = 1$, we find that $\lambda = 2d \sin \theta = 6.69 \times 10^{-11}$ m. To find the position of the second-order spectrum, we set

$n = 2$. Thus

$$\sin\theta = \frac{n\lambda}{2d} = 0.238 \qquad \text{or} \qquad \theta = 13°46'$$

Note that the maximum scattering order is limited by the condition $n\lambda/2d < 1$, which in our case amounts to $n < 8.4$ or $n_{max} = 8$.

Note 35.1 Holography

The virtual image of an object produced by a plane mirror displays the three-dimensional appearance of the object. This is because the reflected light seems to come from points at different distances behind the mirror. On the other hand, when an image is formed on a photographic film, it is a two-dimensional, or flat, reproduction of the object. However, it is possible, using a photographic film combined with interference and diffraction, to produce a three-dimensional virtual image of an object. The method is called **holography** (*holos*: whole in Greek) and was originally proposed by Dennis Gabor in 1947. However, it was not fully developed until fine-grained photographic films, and highly coherent laser beams of light (Section 31.8) became available a few years later.

Suppose a monochromatic laser beam (Figure 35.31) is split in two by a semi-transparent mirror. One beam falls directly on the photographic film. The other falls on the object and is reflected or scattered toward the photographic film. The two beams interfere and produce an interference pattern, called a **hologram**, on the photographic film. The interference pattern depends on the shape of the object. Once the photographic film is developed, the hologram does not show an image but appears as a distribution of dots with different degrees of exposure. Each object produces a different hologram.

If the hologram of an object is illuminated with laser light of the same wavelength, the hologram functions like a diffraction grating except that, instead of regularly spaced lines, it consists of a series of dots that transmit light with different intensity and that are associated with the shape of the original object (Figure 35.32). If the pattern of interference of the diffracted light is observed at an angle with the film, a virtual image that is a precise three-dimensional copy, or wave front reconstruction, of the original waves reflected by the object is perceived. The image shows a different perspective of the object for each position of the observer. In addition, there is a real image that can be observed by placing a screen in the proper position.

The method we have described is a *transmission* hologram; it is also possible to have a reflection hologram as well as using white instead of monochromatic light.

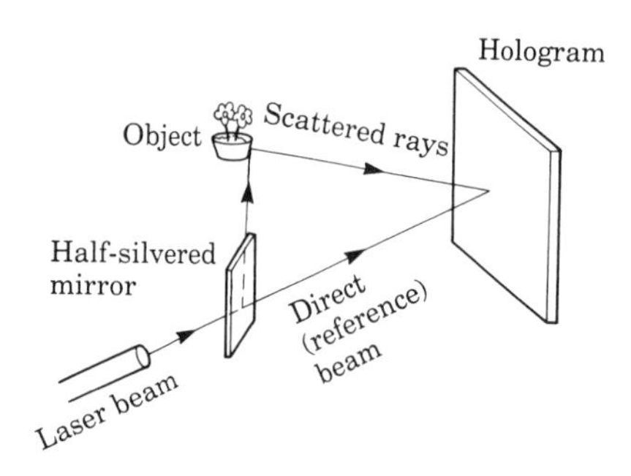

Figure 35.31 Construction of a hologram.

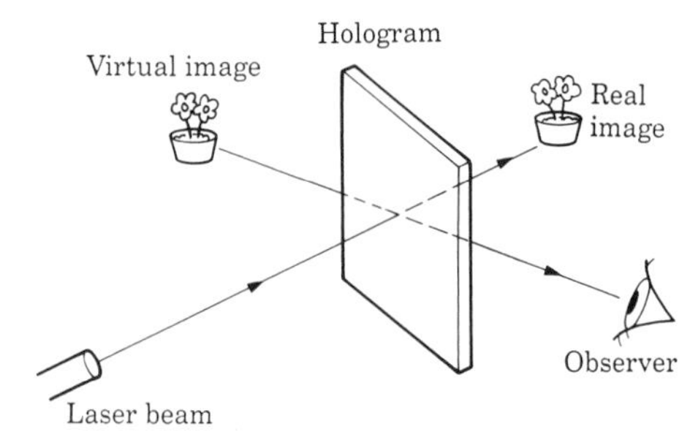

Figure 35.32 Reconstruction of a hologram.

QUESTIONS

35.1 How does Huygens' principle help to explain how a wave propagates around an obstacle?

35.2 Why is the width of the central diffraction maximum of a slit twice that of the other maxima?

35.3 Why should the intensity of a diffraction pattern decrease as n (or θ) increases?

35.4 Determine the size of the rectangles bounded by the zero intensity line in Figure 35.11.

35.5 Mention some instances where diffraction of sound waves is observed.

35.6 Describe the diffraction pattern observed when a slit is illuminated by white light.

35.7 Must we take diffraction effects into account when discussing image formation by a lens?

35.8 Is Young's double-slit experiment (Section 34.2) a pure interference effect or a mixture of interference and diffraction?

35.9 How does increasing the ratio a/b in a double-slit experiment affect the diffraction–interference pattern?

35.10 What is understood by the resolving power of: (a) a single slit; (b) a pair of slits; (c) a diffraction grating?

35.11 Derive Equation 35.10 for the spectral resolving power of a diffraction grating.

35.12 Why do we say that when X-rays pass through a crystal we observe a scattering–interference effect?

35.13 Why is the intensity at the edge of the slit $\frac{1}{4}$ of the maximum intensity (Figure 35.4)?

35.14 Explain why we observe regular (or Laue) patterns when X-rays pass through a crystal and not when they pass through amorphous matter.

35.15 Can we use X-ray scattering to detect impurities and irregularities in crystals?

35.16 Is it possible to observe X-ray scattering by crystals if the wavelength, λ, is larger than $2d$?

PROBLEMS

35.1 Parallel rays of green light, of wavelength 5.6×10^{-7} m, pass through a slit of width 0.4 mm covering a lens of focal length 40 cm. What is the distance from the central maximum to the first minimum on a screen at the focal plane of the lens?

35.2 The Fraunhofer diffraction pattern of a single slit, reproduced at twice its size in Figure 35.8, was formed on a photographic film in the focal plane of a lens of focal length 0.60 m. The wavelength of the light used was 5.9×10^{-7} m. Compute the width of the slit. (Hint: measure the distance between corresponding minima on the right and the left of the central maximum.)

35.3 A telescope is used to observe two distant point sources 1 m apart. The objective of the telescope is covered with a screen in which there is a slit 1 mm wide. Calculate the maximum distance, in meters, at which the two sources will be resolved. Assume $\lambda = 5.0 \times 10^{-7}$ m.

35.4 The Fraunhofer diffraction pattern of a single slit is observed in the focal plane of a lens with a focal length of 1 m. The width of the slit is 0.4 mm. The incident light contains two wavelengths, λ_1 and λ_2. The fourth minimum corresponding to λ_1 and the fifth minimum corresponding to λ_2 occur at the same point, 5 mm from the central maximum. Calculate λ_1 and λ_2.

35.5 Discuss the intensity distribution of the Fraunhofer diffraction by three identical slits equally spaced. Assume normal incidence on the slits.

35.6 A plane monochromatic wave of wavelength 6.0×10^{-7} m is incident perpendicularly on a screen that has a rectangular aperture of 0.5 mm × 1.0 mm. (a) Describe the diffraction pattern observed in the focal plane of a converging lens of focal length 2 m placed directly behind the aperture. (b) Calculate the sides of the rectangle formed by the dark lines surrounding the central maximum.

35.7 In a double-slit diffraction pattern, the third principal maximum is missing because that interference maximum coincides with the first diffraction zero. (a) Find the ratio a/b. (b) Plot the intensity distribution over several maxima on either side of the central maximum. (c) Make a sketch of the fringes as they would appear on a screen.

35.8 Calculate the radius of the central disk of the Fraunhofer diffraction pattern of the image of a star formed by (a) a camera lens 2.5 cm in diameter and focal length 7.5 cm, (b) a telescopic objective 0.15 m in diameter, with a 1.5 m focal length. Assume light of wavelength 5.6×10^{-7} m.

35.9 The headlights of an approaching automobile are 1.30 m apart. Estimate the distance at which the two headlights can be resolved by the naked eye if the resolution of the eye is determined by diffraction alone. Assume a mean wavelength of 5.5×10^{-7} m and assume that the diameter of the pupil of the eye is 5 mm. Compare with the result obtained by the resolving power of the eye as given in Section 33.5.

35.10 Using the method of Example 35.2, obtain the resolving power of a double slit. Compare with that of a single slit. What do you conclude?

35.11 Plane monochromatic light of wavelength 6.0×10^{-7} m is incident normally on a plane transmission grating having 500 lines per mm. Determine the angles of deviation for the first-, second-, and third-order spectra.

35.12 A plane transmission grating is ruled with 4000 lines per cm. Compute the angular separation in degrees, in the second-order spectrum, between the lines of atomic hydrogen, whose wavelengths are 6.56×10^{-7} m and 4.10×10^{-7} m. Assume normal incidence.

35.13 A transmission grating 4 cm long has 4000 lines per cm. (a) Calculate the resolving power of the grating in the first-order spectrum. (b) Will the grating separate the two lines of wavelength 5.890×10^{-7} m and 5.896×10^{-7} m which constitute the sodium yellow doublet?

35.14 Verify that, no matter what the grating spacing, the violet of the third-order spectrum overlaps the red of the second-order spectrum. Assume normal incidence.

35.15 A **reflection grating** is made by etching fine lines on a polished metal surface (Figure 35.33). The polished spaces left between adjacent rulings are the equivalent of the slits in a transmission grating. Verify that the principal maxima are obtained by the condition $a(\sin i - \sin \theta) = n\lambda$, where a is the separation between consecutive lines.

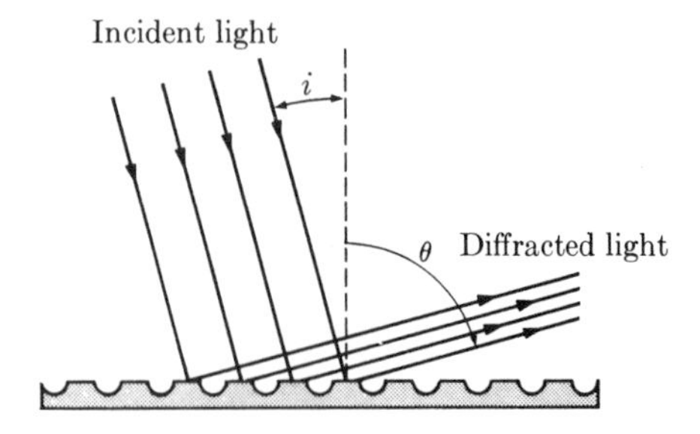

Figure 35.33

35.16 The spacing between the principal planes in an NaCl crystal is 2.82×10^{-10} m. It is found that a first-order Bragg scattering of a monochromatic X-ray beam occurs at an angle of 10°. (a) Compute the wavelength of the X-rays. (b) What angle corresponds to the second-order spectrum?

35.17 Potassium iodide, KI, is a cubic crystal having a density of 3.13 g cm^{-3}. (a) Calculate the smallest interplanar distance; i.e. the length of a unit cell. (b) Determine the angles corresponding to the first two Bragg scatterings for X-rays of wavelength 3.0×10^{-10} m.

35.18 An X-ray tube accelerates electrons through a potential difference of 10^5 V. The X-rays produced are examined by means of the crystal described in Problem 35.16. Calculate the angle at which the first-order spectrum of the shortest wavelength produced by the tube occurs.

35.19 A beam of X-rays, of wavelength 5×10^{-11} m, falls on a powder composed of microscopic crystals of KCl oriented at random. The lattice spacing in the crystal is 3.14×10^{-10} m. A photographic film is placed 0.1 m from the powder target. (a) Calculate the radii of the circles corresponding to the first- and second-order spectra from planes having the same spacing as the lattice spacing. (b) Determine the radii for the circles resulting from planes making an angle of 45° with those in (a).

36 Quantum mechanics: fundamentals

Werner Heisenberg was one of the main contributors to the early development of quantum mechanics in the 1920s. He first published his ideas about quantum mechanics in 1925, which he initially called matrix mechanics. In 1927 Heisenberg formulated the **uncertainty principle**, which established a limitation, inherent to nature, on the precision with which simultaneous measurement of the position and momentum of a particle is possible. The uncertainty principle showed that the classical view of atomic particles had to be replaced by a quantum-mechanical approach. With W. Pauli, he laid the foundations of quantum field theory in 1929. Heisenberg also contributed to the formulation of the theory of nuclear forces (1932) and of positrons (1934) and to an explanation of nuclear fission.

36.1 Introduction

The motion of the bodies we observe around us can be described in terms of general rules based on experimental evidence. These rules or principles are: (1) the conservation of momentum, (2) the conservation of angular momentum, and (3) the conservation of energy. Based on these conservation laws, the formalism called *classical mechanics* was developed in the nineteenth century, under the assumptions that particles are localized in space and that we can observe them without appreciably disturbing their motion. However, when applied to the motion of the basic constituents of matter, such as electrons and atoms, classical mechanics gives only approximate results, and in some instances it is entirely inadequate.

As a result of experimental evidence, several new and revolutionary concepts and methods have been introduced to describe the behavior of matter at the atomic and subatomic levels. Although the laws of conservation of momentum, angular momentum and energy remain valid, the detailed description of the motion of atomic particles in the sense of Newtonian mechanics is no longer applicable. The quantization of energy and other physical quantities is another idea that does not appear in classical mechanics and a satisfactory theory must contain information about the allowed values of such quantities. The interaction of radiation and matter by means of absorption or emission of photons is yet another concept to be incorporated.

The new formalism, called **quantum mechanics**, is the result of the original work of Louis de Broglie (1892–1987), Erwin Schrödinger (1887–1961), Werner Heisenberg (1901–1976), Paul Dirac (1902–1984), Max Born (1882–1970), Albert Einstein (1879–1955) and others who developed it in the 1920s. Quantum mechanics is essential for understanding the behavior of the fundamental constituents of matter. The term **quantics** is used to designate the study of systems that behave according to the laws of quantum mechanics.

36.2 Particles and fields

Our sensory experience tells us that the objects we touch and see have a well-defined shape and size and therefore are localized in space. We thus tend to extrapolate and think of the fundamental particles (i.e. electrons, protons, neutrons etc.) as having shape and size, and imagine them as being somewhat like small spheres, with a characteristic radius, as well as mass and charge. This, however, is an extrapolation beyond our direct sensory experience and we must analyze this picture carefully before we accept it.

Experiments have shown that our extrapolated sensory picture of the basic constituents of matter is erroneous. The dynamical behavior of atomic and subatomic particles requires that we associate with each particle a field—a **matter field**—in the same way that, in the reverse manner, we associate a photon (which can be considered as equivalent to a particle) with an electromagnetic field. The matter field describes the dynamical condition of a particle in the same sense that the electromagnetic field corresponds to photons that have precise momentum and energy. In discussing the connection between the matter field and the dynamical properties of the particle (i.e. momentum and energy), we may be guided by the relations $E = h\nu$, $p = h/\lambda$, previously found for the photon. Writing these relations in reverse, we may assume that the wavelength λ and the frequency ν of the field associated with a particle of momentum p and energy E are given by

$$\lambda = \frac{h}{p}, \qquad \nu = \frac{E}{h} \tag{36.1}$$

where h, as before, is Planck's constant. These relations were first proposed in 1924 by de Broglie, and for that reason $\lambda = h/p$ is called the **de Broglie wavelength** of a particle. Introducing the wave number $k = 2\pi/\lambda$ and the angular frequency $\omega = 2\pi\nu$, and, recalling that $\hbar = h/2\pi = 1.0544 \times 10^{-34}$ J s, we may write the above relations in the more symmetric form

$$p = \hbar k, \qquad E = \hbar\omega \tag{36.2}$$

If our assumption, as expressed by Equations 36.1 or 36.2, is correct, we may expect that whenever the motion of a particle is disturbed in such a way that the field associated with the particle cannot propagate freely, interference, diffraction and scattering phenomena should be observed, as is the case for elastic and electromagnetic waves. This is indeed what happens, as discussed in the next section.

Although we will continue using the term particle when referring to electrons, protons, neutrons and other fundamental components of matter, we must refrain from considering them as little balls. Their behavior is quite different and sometimes against our sensory perceptions. The 'particle' picture applies only when the motion is in a region large compared with atomic or nuclear dimensions, such as electrons in a TV tube or protons in an accelerator. This is why our discussion of the motion of charged particles in electric and magnetic fields (Chapters 21 and 22) is valid. However, we must refine our analysis of the motion of electrons within an atom, a molecule or a solid, which we presented in Chapter 23, because it requires the introduction of the matter field.

EXAMPLE 36.1

The de Broglie wavelength of electrons accelerated through an electric potential difference. Apply the result to electrons accelerated by a voltage of 10 000 V, which is of the order of voltages in a TV tube.

▷ Electrons accelerated by an electric potential difference ΔV gain an energy $e\Delta V$; their kinetic energy is $p^2/2m_e = e\Delta V$ so that $p = (2m_e e\Delta V)^{1/2}$. Therefore, introducing the values of e, m_e and h, we obtain the de Broglie wavelength of such electrons as

$$\lambda = \frac{h}{(2m_e e\Delta V)^{1/2}} = \frac{1.23 \times 10^{-9}}{(\Delta V)^{1/2}} \text{ m} \qquad \textbf{(36.3)}$$

where ΔV is expressed in volts. This formula can also be used when the kinetic energy of the electron is expressed in electronvolts.

When $\Delta V = 10\,000$ V, the wavelength is 1.23×10^{-11} m. This is of the order of magnitude of atomic dimensions. Therefore it is sufficient to produce a sharp image on a TV screen. A longer wavelength or smaller accelerating voltage may produce a less well-defined image.

36.3 Scattering of particles by crystals

Consider electrons with energy equivalent to that acquired by moving through a potential difference of the order of 10^4 V. In Example 36.1 it was shown that the wavelength of these electrons is about 10^{-11} m, comparable to the wavelength of X-rays. This means that if we send a beam of sufficiently fast electrons through a crystal, we should obtain patterns that result from scattering of the matter field, similar to those observed for X-rays, discussed in Section 35.7.

In 1927 G. P. Thomson (1882–1975) began a series of experiments to study the passage of a beam of electrons through a thin film of crystalline material. After the electrons passed through the film, they struck a photographic plate, as shown in Figure 36.1. If the electrons had behaved as particles in the macroscopic sense,

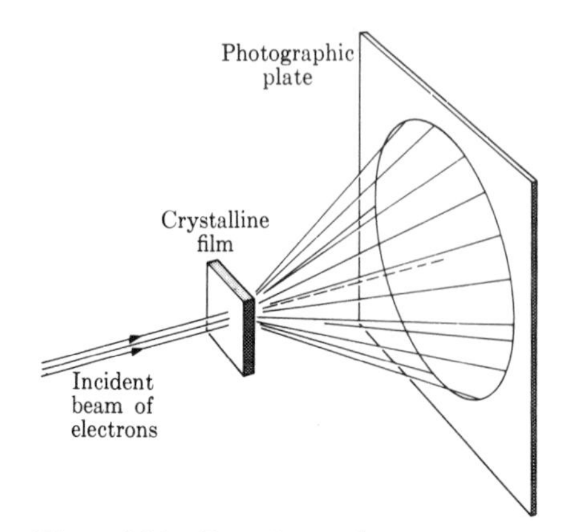

Figure 36.1 Experimental arrangement for observing electron scattering through crystalline material.

a blurred image would have been observed because each electron would undergo, in general, a different scattering by the atoms in the crystal. However, the result obtained was identical to the Debye–Scherrer patterns for X-ray scattering by a polycrystalline substance, as indicated in the photograph of Figure 36.2. Similarly, when an electron beam passes through a single crystal, Laue spot patterns are produced, also observed with X-rays, as seen in the photograph of Figure 36.3. From the structure of these patterns we can compute the de Broglie wavelength λ if the spacing between the crystal planes is known and the formulas derived for X-rays are applied. The resulting values of λ can be compared with those obtained from Equation 36.3. The result is complete agreement, within the limits of experimental error.

In the experiments by C. Davisson and L. Germer (made at about the same time as those of Thomson), a beam of electrons was sent at an angle to the face of the crystal. The scattered electrons were observed by means of a detector symmetrically located, as indicated in Figure 36.4. This is similar to the Bragg arrangement for observing X-ray scattering (Section 35.7). It was found that the electron current registered by the detector was a maximum every time the Bragg condition, Equation 35.11 (originally derived for X-rays),

$$\sin\theta = \frac{n\lambda}{2d} \tag{36.4}$$

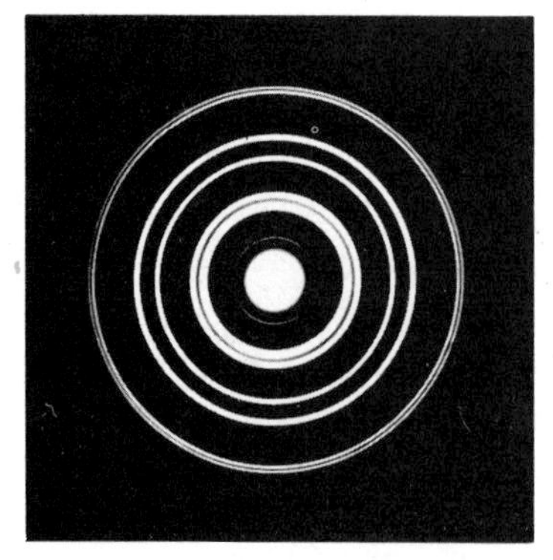

Figure 36.2 Scattering of electrons by crystal powder (courtesy Dr Lester Germer).

Figure 36.3 Scattering of electrons by a single carbon (graphite) crystal (courtesy of R. Heidenreich, Bell Telephone Laboratories).

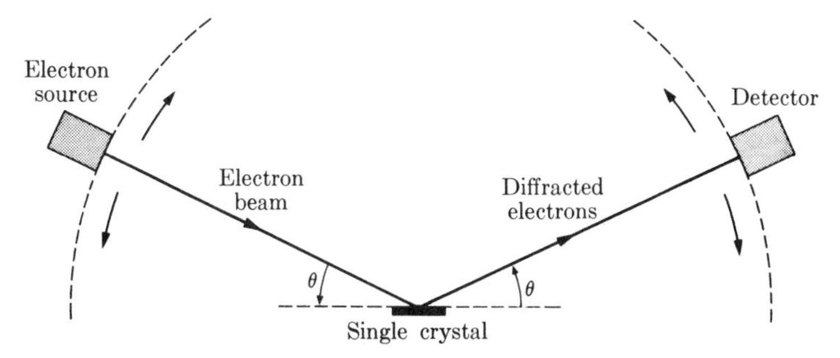

Figure 36.4 Davisson and Germer arrangement for observing Bragg scattering of electrons.

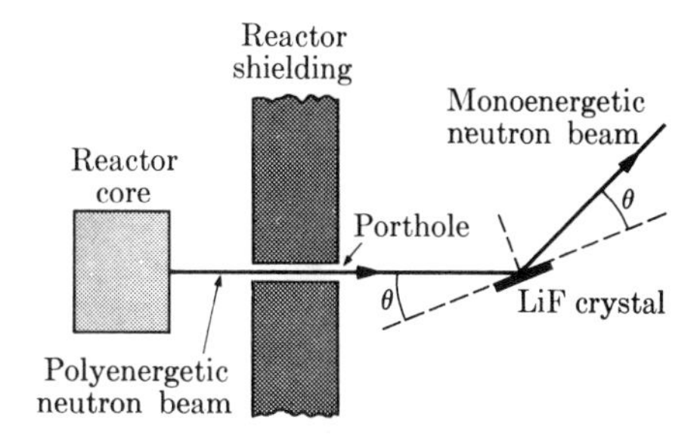

Figure 36.5 Neutron crystal spectrometer.

was fulfilled. In Equation 36.4 d is the separation of successive atomic layers in the crystal, and λ is given by Equation 36.3.

The same phenomenon of Bragg scattering has been observed in experiments with protons and neutrons. Neutron scattering is especially useful, since it is one of the most powerful means of studying crystal structure. For example, neutrons emerging from a nuclear reactor through a porthole (Figure 36.5) have a wide range of energies and thus they vary widely in momentum. Therefore, the neutron beam contains a spectrum composed of many de Broglie wavelengths. When the neutron beam from the reactor falls on a crystal, of LiF for example, the neutrons observed in the symmetric direction correspond only to the wavelength λ as given by Bragg's condition 36.4. Therefore they have a well-defined energy and momentum. The crystal then acts as an energy filter or **monochromator**. The monoenergetic neutron beam is in turn used to study other materials or to analyze nuclear reactions involving neutrons.

EXAMPLE 36.2

The de Broglie wavelength of thermal neutrons at a temperature of 25 °C.

▷ By **thermal neutrons** we mean neutrons that are in thermal equilibrium with matter at a given temperature. Thus the neutrons have an average kinetic energy identical to that of the molecules of an ideal gas at the same temperature. Therefore the average kinetic energy of thermal neutrons is $E_{\text{ave}} = \frac{3}{2}kT$, where T is the absolute temperature and k is Boltzmann's constant (see Equation 15.6). Given a temperature of 25 °C, or $T = 298$ K, we have

$$E_{\text{ave}} = \tfrac{3}{2}kT = 6.17 \times 10^{-21}\ \text{J} = 3.85 \times 10^{-2}\ \text{eV}$$

The corresponding momentum is

$$p = (2m_n E_{\text{ave}})^{1/2} = 4.55 \times 10^{-24}\ \text{kg m s}^{-1}$$

Then, using Equation 36.1, we find that the average de Broglie wavelength of thermal neutrons is $\lambda = 1.85 \times 10^{-10}$ m. Noting that the separation of the planes in a NaCl crystal is $d = 2.82 \times 10^{-10}$ m, we see that the first Bragg maximum for neutrons of this wavelength diffracted by the crystal occurs at an angle $\theta = 19°$.

36.4 Particles and wave packets

Using relations 36.1, we may represent the field corresponding to a free particle moving with a well-defined momentum p and non-relativistic kinetic energy $E = p^2/2m$ by a harmonic wave of constant amplitude, as shown in Figure 36.6(a). Symmetry demands that the amplitude of the wave be the same throughout all space, since there are no forces acting on the particle that could alter the momentum of the particle and distort the associated matter field more in some regions of space than in others. The phase velocity of the field of the free particle is

$$v_{\text{phase}} = \lambda \nu = \frac{h}{p}\frac{E}{h} = \frac{E}{p} = \frac{p}{2m} = \frac{1}{2}v$$

That is, the phase velocity of the matter field is one-half the particle velocity. This has no experimental consequence, however, because the fact that the amplitude of the harmonic matter field is the same throughout all space suggests that the matter field of a free particle does not give any information about the localization in space and time of a free particle of well-defined momentum and direction of motion; therefore it is impossible to measure the velocity of such a particle by timing its passage through two points separated a given distance.

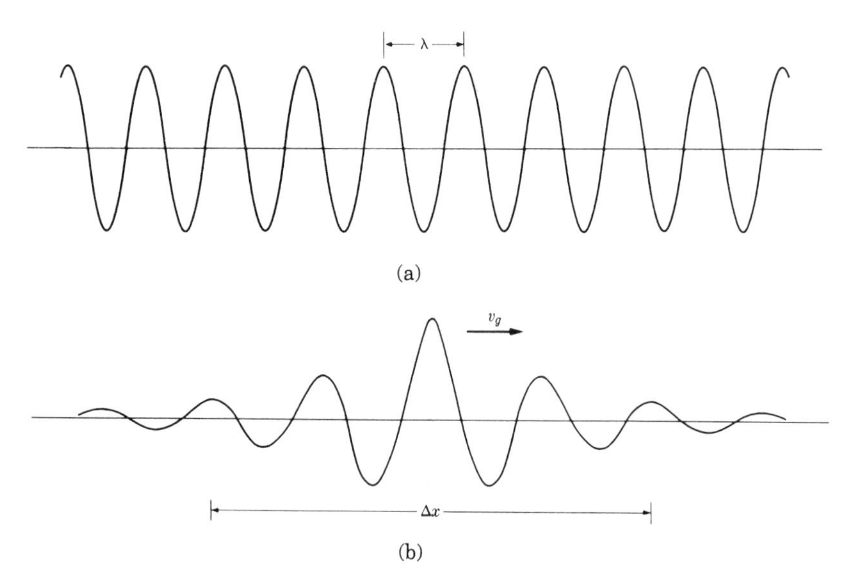

Figure 36.6 (a) Continuous wave train corresponding to an unlocalized particle. (b) Wave packet corresponding to a particle localized within the distance Δx.

The matter field that describes a particle localized within a certain region Δx of space must have an amplitude or intensity that is large in that region and very small outside it. A field may be built up in a certain region and attenuated outside that region by superposing waves of different wavelengths. The result is a **wave packet** or pulse, as shown in Figure 36.6(b). The velocity with which the wave packet propagates is the group velocity v_{group}, which is given by Equation 28.48,

$$v_{\text{group}} = \frac{d\omega}{dk}$$

Multiplying numerator and denominator by $\hbar$ and using relations 36.2 with $E = p^2/2m$, we may rewrite the group velocity of the matter field wave packet corresponding to a free particle as

$$v_{\text{group}} = \frac{dE}{dp} = \frac{p}{m} = v$$

Thus,

> *the group velocity of the matter field, which is the velocity of propagation of the packet, is equal to the velocity of the particle.*

We conclude then that a moving particle localized in a certain region of space at a given time is associated with a traveling wave packet whose amplitude is important only in the region occupied by the moving particle; the group velocity of the wave packet is the velocity of the particle.

36.5 Heisenberg's uncertainty principle for position and momentum

A wave packet localized in space is the result of superposing several fields of different wavelengths λ (or with different values of the wave number k). If the wave packet extends over a region Δx, the values of the wave numbers of the interfering waves that compose the wave packet fall within a range Δk such that, according to a mathematical analysis that we omit, $\Delta x \Delta k \approx 1$. This relation shows that Δx and Δk are inversely proportional and therefore the smaller the size Δx of the wave packet the larger the spread Δk of wave numbers or wavelengths needed to built up the wave packet (Figure 36.7).

But, according to Equations 36.1 or 36.2, different wavelengths λ or wave numbers k mean that there are several values of p such that $\Delta p = \hbar \Delta k$. Therefore, we see that the expression $\Delta x \Delta k \approx 1$ becomes

$$\Delta x \Delta p \approx \hbar \qquad \textbf{(36.5)}$$

The physical meaning of relation 36.5 is as follows: if a particle is within the region $x - \frac{1}{2}\Delta x$ and $x + \frac{1}{2}\Delta x$ (that is, Δx is the uncertainty in the position of the particle), its associated matter field is obtained by superposing matter fields corresponding to momenta between $p - \frac{1}{2}\Delta p$ and $p + \frac{1}{2}\Delta p$, where Δp is related to Δx by relation

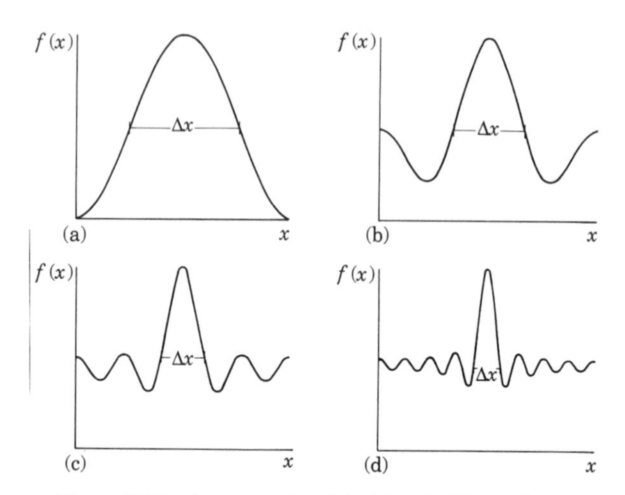

Figure 36.7 An example of $\Delta x \Delta k \approx 1$. Plots of $f(x) = (1/n)\Sigma \cos kx$ for $k = 1$ to n. (a) $n = 1$; (b) $n = 2$; (c) $n = 4$; (d) $n = 8$. As the spread in values for k increases, the width of the main peak becomes smaller.

36.5. We say that Δp is the uncertainty in the momentum of the particle. Relation 36.5 implies that the larger Δx is, the smaller is Δp, and conversely.

In other words, information about the localization of a particle in space is obtained at the expense of losing knowledge about its momentum. The more precise our knowledge of the position of the particle, the more imprecise is our information about its momentum, because to localize a wave packet we have to combine more wavelengths. Conversely, the more precise the momentum the fewer wavelengths we need in the wave packet and the more extended is the packet. This is why a particle of well-known momentum ($\Delta p = 0$) is represented by a wave of constant amplitude extending over all space ($\Delta x \approx \infty$), so that our knowledge of its position is nil. Conversely, if the localization of a particle is very precise ($\Delta x = 0$) our knowledge of the momentum of the particle is nil ($\Delta p = \infty$) and we have no idea about the momentum of the particle. (Of course these two extreme situations never occur in practice.) We cannot accurately determine both the position and the momentum of a particle simultaneously so that $\Delta x = 0$ and $\Delta p = 0$ at the same time because that violates relation 36.5, which relates the maximum precision with which we can know both quantities.

The result expressed by relation 36.5 is called **Heisenberg's uncertainty principle** for position and momentum, which may be stated as follows:

it is impossible to know simultaneously and with exactness both the position and the momentum of the fundamental particles constituting matter. The optimum precision in the knowledge of position and momentum is determined by Equation 36.5.

This principle expresses one of the fundamental facts of nature. It has been suggested that the uncertainty principle should rather be called the *limiting* principle, since it expresses a fundamental limitation in nature that also limits the precision of our measurements.

The uncertainty principle implies that we can never define the path of an atomic particle with the absolute precision postulated in classical mechanics. Classical mechanics still holds true when the uncertainty implied by relation 36.5 is much smaller than the experimental errors in the measured values of x and p. For example, for describing the motion of atomic particles under certain conditions, such as electrons in a TV tube and electrons and protons in an accelerator, we can use classical mechanics when Δx and Δp as permitted by Equation 36.5 are much smaller than the experimental error in the momentum and the position of the particles. However, for particles bound to move in regions of atomic dimensions, the concept of trajectory has no meaning, since it cannot be defined precisely; therefore a picture of the motion different from that of classical physics is required. For the same reason, concepts such as velocity, acceleration and force are of limited use in quantum mechanics. On the other hand, the concept of energy is of primary importance, since it is related to the 'state' of the system rather than to its 'path'.

36.6 Illustrations of Heisenberg's principle

To illustrate Heisenberg's principle consider some simple situations or 'thought' experiments. Suppose, for example, that we want to determine the X-coordinate of a particle moving along the Y-axis (Figure 36.8) by observing whether or not the particle passes through a slit (of width b) in a screen perpendicular to the direction of motion. The precision of the particle's position is limited by the size of the slit; that is, $\Delta x \approx b$. But the slit disturbs the field associated with the particle, and this results in a corresponding change in the motion of the particle after going through the slit, as seen by the diffraction pattern produced. The uncertainty in the particle's momentum parallel to the X-axis is determined by the

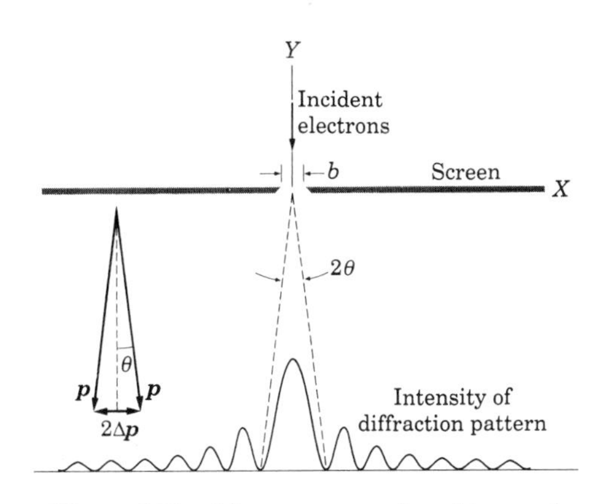

Figure 36.8 Measurement of position and momentum of a particle passing through a slit.

angle θ, corresponding to the central maximum of the diffraction pattern, since the particle, after traversing the slit, is most probably moving within the angle 2θ. According to the theory of the diffraction produced by a rectangular slit (Section 35.3), the angle θ is given by $\sin\theta = \lambda/b$. Then

$$\Delta p \approx p \sin\theta = \frac{h}{\lambda}\frac{\lambda}{b} = \frac{h}{b} \approx \frac{h}{\Delta x}$$

is the uncertainty in the momentum of the particle parallel to the X-axis. Therefore $\Delta x \Delta p \approx h$, which agrees with relation 36.5 (the missing 2π is irrelevant for this analysis). Note that to improve our ability to determine the exact position of the particle along the X-axis, we must use a very narrow slit. But a very narrow slit produces a very wide central maximum in the diffraction pattern, which means a large uncertainty in our knowledge of the X-component of the momentum of the particle after passing the slit. Conversely, in order to reduce the uncertainty in our knowledge of the X-component of the momentum, the central maximum in the diffraction pattern must be very narrow. This requires a very wide slit which, in turn, results in a large uncertainty in the X-coordinate of the particle.

Another 'thought' situation that illustrates Heisenberg's principle is the case where we try to determine the position of an electron by means of a microscope (Figure 36.9). To observe the electron, we must illuminate it with light of some wavelength λ. The only light that passes through the microscope is what has been scattered by the electron under observation. The momentum of the scattered photons is $p_{\text{photon}} = h/\lambda$ and, to penetrate into the objective, the photons must move within the cone of angle α, so that the X-component of their momenta has an uncertainty

$$\Delta p \approx p_{\text{photon}} \sin\alpha \approx \left(\frac{h}{\lambda}\right)\left(\frac{d}{2y}\right) = \frac{hd}{2\lambda y}$$

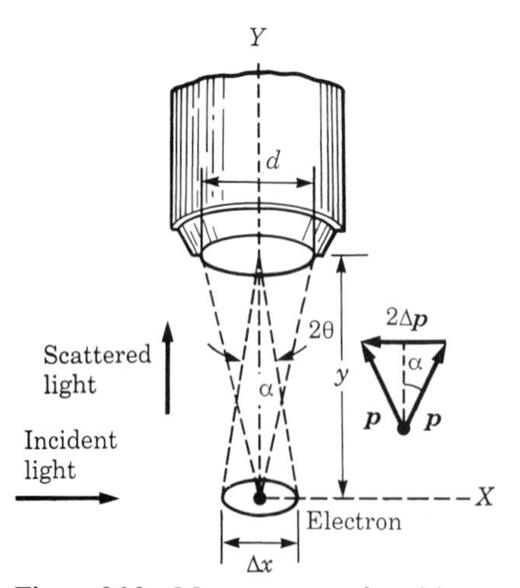

Figure 36.9 Measurement of position and momentum of a particle by means of a microscope.

since $\sin\alpha \approx d/2y$. This is also the uncertainty in the X-component of the electron momentum after the scattering of light, since in the scattering process some momentum is exchanged between the electron and the photon. On the other hand, the exact position of the electron is uncertain because of the diffraction of light when it passes through the objective of the microscope. The uncertainty in the position of the electron is thus equal to the diameter of the central disk in the diffraction pattern. This diameter is given by $2y \sin\theta$, with $\sin\theta \approx \lambda/d$. (We have disregarded the factor 1.22 which appears in the theory of diffraction of a plane wave by a circular aperture (Section 35.4).) Hence

$$\Delta x \approx 2y \sin\theta \approx \frac{2y\lambda}{d}$$

Therefore again $\Delta x \Delta p \approx h$. Note that to improve the accuracy of our knowledge of the position of the electron we must use a radiation of very small wavelength, but this produces a large disturbance in the momentum. Conversely, in order to produce a small disturbance in the momentum, we must use radiation of very long wavelength, which in turn gives rise to a great uncertainty in the position.

These two 'thought' experiments show how the uncertainty principle is related to the process of measurement. At the atomic level, measurement inevitably introduces a significant perturbation in the system, due to the interaction between the measuring device and the measured quantity, which limits the precision of the measurement.

EXAMPLE 36.3

Minimum energy of an electron in a hydrogen-like atom.

▷ Heisenberg's principle allows us to estimate the minimum energy of an electron in an atom. If r is the radius of the region where the electron can be found, we can assume that $\Delta x \approx r$, and if p is the momentum of the electron we may also assume that $\Delta p \approx p$. In that case the uncertainty relation becomes $rp \approx \hbar$, which, when inserted in the expression for the total energy of the electron,

$$E = \frac{p^2}{2m_e} - \frac{Ze^2}{(4\pi\varepsilon_0)r}$$

gives

$$E \approx \frac{\hbar^2}{2m_e r^2} - \frac{Ze^2}{(4\pi\varepsilon_0)r}$$

To calculate the minimum energy compatible with the uncertainty principle, we set $dE/dr = 0$ and solve for r. This gives

$$\frac{1}{r} = \frac{m_e Ze^2}{(4\pi\varepsilon_0)\hbar^2}$$

When this value is inserted in the expression for E, we get

$$E \approx -\frac{m_e Z^2 e^4}{2(4\pi\varepsilon_0)^2\hbar^2}$$

This is precisely the energy obtained with $n = 1$ in Equation 23.13, derived using Bohr's theory. The result is not surprising because the uncertainty relation $rp \approx \hbar$ is equivalent to Bohr's quantization condition $L = nh/2\pi$ when $n = 1$. Thus the uncertainty principle makes it very difficult to push the energy of the electron below a certain minimum value. Also, the condition $rp = \hbar$ is equivalent to $r \approx \hbar/p = \lambda$, so that the minimum energy corresponds to the wavelength that can be accommodated in the region where the electron can move.

EXAMPLE 36.4

The uncertainty principle and the mass of stars.

▷ Consider a star composed of N particles, each of mass m, of the order of the proton mass 1.7×10^{-27} kg, so that the total mass is $M = Nm$. As the star contracts under gravitational action, the kinetic energy of the particles increases (recall Note 11.4). At the relativistic limit, the kinetic energy is of the order of cp, where p is the momentum of the particles (recall Equation 20.6). As the particles get very close, at an average separation r, the uncertainty principle requires that $rp \sim \hbar$. Therefore, the average kinetic energy of each particle is

$$E_{\mathrm{k,ave}} \approx \frac{\hbar c}{r}$$

The volume of a star when it has contracted considerably is given approximately by $N(\frac{4}{3}\pi r^3)$ so that its radius is $R = N^{1/3}r$. Then, the average gravitational potential energy of each particle is of the order of $-GMm/R$, within a numerical coefficient of order one, which depends on the position of the particle within the star. Therefore

$$E_{\mathrm{p,ave}} \approx -\frac{GN^{2/3}m^2}{r}.$$

The electric potential energy of each particle is of the order of $e^2/4\pi\varepsilon_0 r = \hbar c/137r$ (see Note 23.2) and therefore can be ignored when compared with the average kinetic energy. Thus the average total energy of a particle in the star is

$$E_{\mathrm{ave}} \approx \frac{\hbar c}{r} - \frac{GN^{2/3}m^2}{r} = \frac{\hbar c - GN^{2/3}m^2}{r}$$

The collapse of a star is prevented as long as its internal kinetic energy dominates its gravitational potential energy. But the larger the number of particles (i.e., the mass of the star) the greater the gravitational effect. Gravitational collapse is inevitable when $E_{\mathrm{ave}} = 0$, or

$$N = \left(\frac{\hbar c}{Gm^2}\right)^{3/2} \approx 2 \times 10^{57}$$

The corresponding mass of the star is $M = Nm \approx 3 \times 10^{30}$ kg, which is about 1.4 times the mass of the Sun. The number N above is called the **Chandrasekhar limit** and M is the **Chandrasekhar mass** (usually denoted N_{C} and M_{C}, respectively) after S. Chandrasekhar, who, in 1931, determined their values using a more rigorous method. Because of the approximations made in the calculations, both N_C and M_C only indicate the order of magnitude for the size of stars bound to suffer gravitational collapse (see the end of Note 40.4 for further discussion).

Our simplified analysis shows the connection between the quantum mechanical properties of matter, represented by $\hbar$, relativity, represented by c, gravitation, represented by G and the size of the large structures in the universe.

36.7 The uncertainty relation for time and energy

In addition to the uncertainty relation $\Delta x \Delta p \approx \hbar$ between a coordinate and the corresponding momentum of a moving particle, there is an uncertainty relation between time and energy. Suppose that we want to measure not only the energy of a particle but also the time at which the particle has such energy. If Δt and ΔE are the uncertainties in the values of these quantities, the following relation holds:

$$\Delta t \Delta E \approx \hbar \tag{36.6}$$

We can understand this relation in the following way. If we want to define the time when a particle passes through a given point, we must represent the particle by a pulse or wave packet having a very short duration Δt. But to build such a pulse it is necessary to superpose fields that have different frequencies, with an amplitude appreciable only in a frequency range $\Delta\omega$. The mathematical theory of wave pulses, which we cannot reproduce here, requires that

$$\Delta t \Delta \omega \approx 1 \tag{36.7}$$

Multiplying by $\hbar$ and recalling from Equation 36.2 that $E = \hbar\omega$, we obtain relation 36.6. We conclude then that it is impossible to know simultaneously and with exactness the energy of a particle and the time at which it has that energy.

EXAMPLE 36.5

Width of energy levels and lifetime of an excited state.

▷ Consider an electron in an excited stationary state in an atom. The electron, after a certain time, will suffer a radiative transition into another stationary state of less energy. However, we have no means of predicting with certainty how long the electron will remain in the stationary state before making the transition. The most we can talk about is the **probability** that the electron will jump into a lower energy state. Therefore, the average time the electron is in the stationary state, which is called the **lifetime** of the state, is known within an uncertainty Δt that is inversely proportional to the **transition probability** P, that is, $\Delta t \simeq 1/P$.

Hence the energy of the stationary state of the electron is not known precisely but has an uncertainty ΔE, such that relation 36.6 holds. Often ΔE is designated as the **energy width** of the state whose energy is most probably between $E - \frac{1}{2}\Delta E$ and $E + \frac{1}{2}\Delta E$ (Figure 36.10). Since Δt is of the order of magnitude of the lifetime of the excited state, the shorter

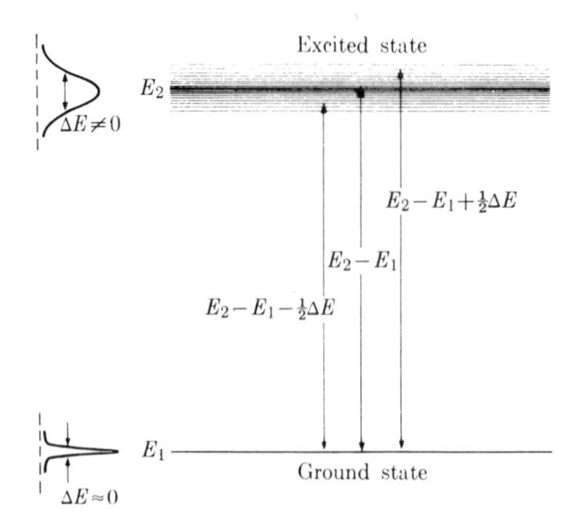

Figure 36.10 Natural width of energy levels.

the lifetime of an excited state or the higher the transition probability, the larger the uncertainty in the energy of the state, that is, $\Delta E \simeq \hbar/\Delta t \simeq \hbar P$. The lifetime of excited atomic states is of the order of 10^{-8} s, giving an energy width of the order of 10^{-26} J or 10^{-7} eV. States with much longer lifetimes are called **metastable**. For a ground state, whose lifetime is infinite because a system that is in its ground state cannot suffer a transition to a stationary state of lower energy, we have $\Delta t \approx \infty$. This yields $\Delta E = 0$ and the energy of the ground state can be determined precisely.

Due to the energy width of stationary states, the energy released or absorbed in a transition is not well defined. Therefore, in a transition between states of energy E_1 and E_2, the photons emitted or absorbed fall in the energy range $E_2 - E_1 \pm \frac{1}{2}\Delta E$, where ΔE is the total energy width of both states, resulting in a broadening of the spectral lines with a frequency range of the order of $\Delta E/h$. In most atomic and molecular transitions the Doppler effect due to thermal motion (Section 29.7 and Example 29.4) is much greater than the broadening due to the uncertainty principle, which can be neglected. Line broadening is, however, important in laser tuning (Note 31.1).

36.8 Stationary states and the matter field

We are now in a position to give a theoretical justification to the idea of stationary states. When a particle is in a bound state and confined to move within a limited region of space, such as an electron in an atom or a proton in a nucleus, the associated matter field must also be confined to that region. The situation is similar to that of waves on a string with fixed ends or within a cavity. But we know that in such cases only certain wavelengths are possible and the allowed waves are called standing waves. Therefore, we may expect that in the case of a bound particle only the states corresponding to the allowed wavelengths of the matter field are possible.

A simple example is a particle constrained to move inside a certain region, such as a gas molecule in a box. The molecule moves freely until it hits the walls, which forces the molecule to bounce back. A similar situation exists for an electron in the conduction band in a metal if the height of the potential barrier at the metal surface is much larger than the electron's kinetic energy. The electron can move freely through the metal but cannot escape from it.

We may represent each of these physical situations by the rectangular potential energy of Figure 36.11, which is an oversimplification of the potential energies that actually occur in nature. This simplified potential energy diagram is called a **one-dimensional potential box** which constrains the particle to move along the X-axis between $x = 0$ and $x = a$. We have $E_p(x) = 0$ for $0 < x < a$, since the particle moves freely in that region. But the potential energy increases sharply to

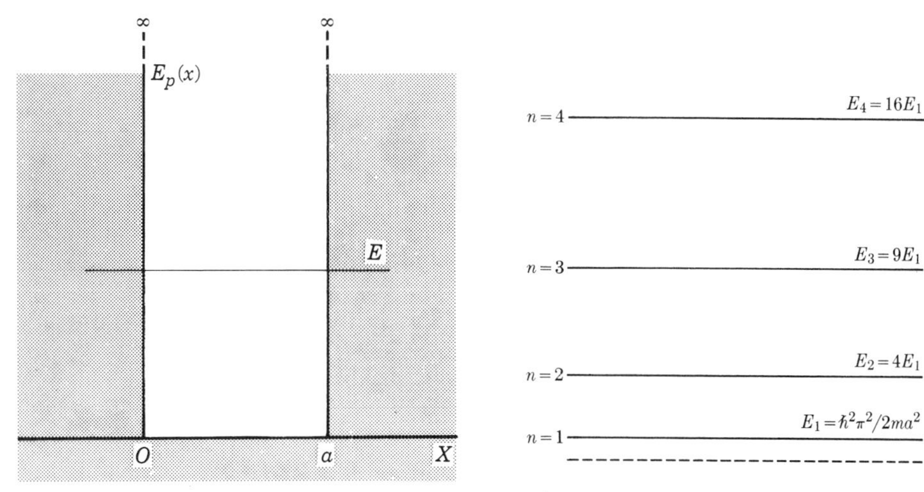

Figure 36.11 One-dimensional potential box of width a.

Figure 36.12 Energy levels for a one-dimensional potential box.

infinity at $x = 0$ and $x = a$. This means that very strong forces act on the particle at those two points, forcing the particle to reverse its motion. Then, no matter what the value of the energy E, the particle cannot be to the left of $x = 0$ or to the right of $x = a$. The situation is formally identical to that corresponding to standing waves on a string with fixed ends or in a cavity closed at both ends. It may be recalled that in order to have standing waves on a string with fixed ends or a cavity with closed ends a distance a apart, the wavelength λ must have the values $\frac{1}{2}\lambda = a, a/2, a/3, \ldots, a/n$, or

$$\lambda = \frac{2a}{n} \tag{36.8}$$

We may thus assume that the same relation applies to the wavelength of the matter field of a particle within a potential box of width a. Hence, according to Equation 36.1, the only possible values of the momentum of the particle are

$$p = \frac{h}{\lambda} = \frac{nh}{2a} = \frac{n\pi\hbar}{a} \tag{36.9}$$

The energy of the particle corresponding to the value given by Equation 36.9 is

$$E = \frac{p^2}{2m} = \frac{n^2\pi^2\hbar^2}{2ma^2} \tag{36.10}$$

or, if $E_1 = \hbar^2\pi^2/2ma^2$ is the energy for $n = 1$, then $E = E_1, 4E_1, 9E_1, \ldots, n^2E_1$. We conclude then that the particle cannot have any arbitrary energy, but only those values given by Equation 36.10 and shown in Figure 36.12; that is, the energy of the particle is **quantized**.

As a second example of energy quantization, consider an electron in a hydrogen-like atom. Suppose that the electron describes a circular orbit, as shown in Figure 36.13. Its momentum p is constant for a circular orbit. In order that the orbit correspond to a stationary state, it seems logical that it must be able to sustain standing waves of wavelength $\lambda = h/p$. We can see from Figure 36.13 that this requires that the length of the orbit be equal to an integral multiple of the wavelength λ; that is,

$$2\pi r = n\lambda = \frac{nh}{p} \quad \text{or} \quad rp = \frac{nh}{2\pi} \tag{36.11}$$

Noting that rp is the angular momentum of the electron, we see that the stationary states are those for which the angular momentum is an integral multiple of $\hbar = h/2\pi$. Since $p = mv$, we may also write Equation 36.11 as

$$L = mvr = n\hbar \tag{36.12}$$

which expresses the quantization of angular momentum. Equation 36.12 was used in Section 23.4 to obtain the energy of the stationary states of hydrogen.

However, the preceding analysis is not completely correct because, as we pointed out when discussing Heisenberg's uncertainty principle, it is impossible to define clearly the orbit of an electron in a hydrogen atom. Instead, as examined in the next section, we can talk only of the region where it is more likely that the electron will be found. Therefore Equations 36.11, or their equivalent, Equation 36.12, cannot be rigorously valid. Instead, as indicated in Section 23.5 and shown in Note 23.4, the allowed values of the orbital angular momentum are given by

$$L = \sqrt{l(l+1)}\hbar \tag{36.13}$$

where $l = 0, 1, 2, \ldots$. From this analysis we may consider that *the quantization of energy and angular momentum is a consequence of the need to accommodate the matter field within a limited region.*

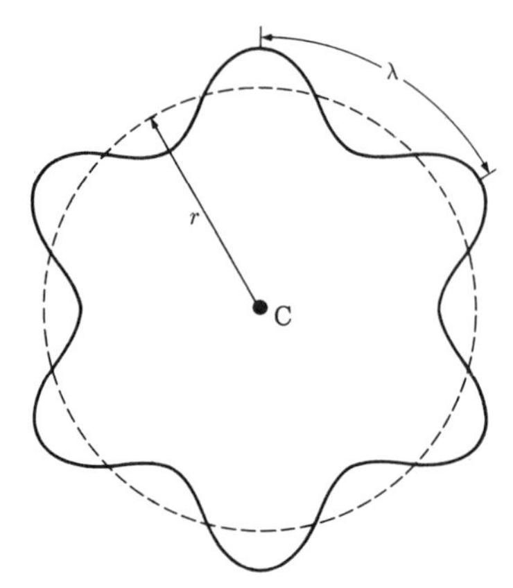

Figure 36.13 Fitting a particle's matter field of wavelength λ to a circle.

EXAMPLE 36.6

Zero point energy of a particle in a potential box.

▷ It is interesting to note that the minimum energy of a particle in a potential box is not zero, as one would suspect, but $E_1 = \hbar^2\pi^2/2ma^2$. This minimum energy is related to the uncertainty principle. The uncertainty in the position of the particle is, obviously, $\Delta x \approx a$. The particle is moving back and forth with a momentum p; the uncertainty in the momentum is then $\Delta p \approx 2p$. The uncertainty principle requires that, at least, $\Delta x \Delta p \approx \hbar$. Therefore

$a(2p) \approx \hbar$ or $p \approx \hbar/2a$, giving $E = p^2/2m \approx \hbar^2/8ma^2$, which is of the order of magnitude of E_1.

The existence of a **zero-point energy**, as E_1 is sometimes called, is typical of all problems where a particle is confined to move in a limited region. We already saw such a case in Example 36.3 for an electron in an atom and in Section 10.11 for an oscillator.

EXAMPLE 36.7

The wavelength of the matter field of an electron in the ground state ($n = 1$) of a hydrogen atom.

▷ Making $n = 1$ in Equation 36.11 gives a wavelength $\lambda = 2\pi r$. For the radius we will take the Bohr radius $a_0 = 5.3 \times 10^{-11}$ m, which was given in Equation 23.19. Therefore, $\lambda = 3.33 \times 10^{-10}$ m. This wavelength must be taken only as an indication of the order of magnitude because of the lack of definition of the electronic orbit.

36.9 Wave function and probability density

We have seen that, according to quantum mechanics, we cannot talk about the trajectory of an atomic particle in the sense of classical mechanics. We cannot, for example, ask whether or not the electrons move in elliptical orbits around the nucleus in an atom. This question would be meaningless even if the forces acting on the particles could produce such orbits. But if we cannot talk about the trajectory of an electron or of any other atomic particle, how may we describe its motion?

The information to answer this question is provided by the matter field, to which we assign the symbol Ψ. To obtain such information, we are guided by our knowledge of waves even if the matter field is not a wave in the same sense as a sound wave or an electromagnetic wave. First we note that, in general, the matter field is a function of both position and time. In the case of motion in one direction (which we designate as along the X-axis), the matter field has the functional form $\Psi(x, t)$.

We know that, in general, the intensity of a wave motion is proportional to the square of the field. Therefore the **intensity of the matter field** is given by $|\Psi(x, t)|^2$. The wave function $\Psi(x, t)$ is sometimes expressed by a complex function; that is, a function containing $i = \sqrt{-1}$. The complex conjugate of a complex function is obtained by replacing each i by $-i$. The complex conjugate of a function Ψ is designated by Ψ^*. Then $|\Psi(x, t)|^2 \equiv \Psi^*(x, t)\Psi(x, t)$. For a real function $\Psi \equiv \Psi^*$. Since the matter field describes the motion of a particle, we may say that

the regions of space where the particle is more likely to be found at time t are those where the intensity of the field, given by $|\Psi(x, t)|^2$, is large.

To be more quantitative, we say that

the probability of finding the particle described by the wave function $\Psi(x, t)$ in the interval dx *around the point x at time t is $|\Psi(x, t)|^2$* dx.

In other words, the probability per unit length of finding the particle at x and time t is

$$P(x, t) = |\Psi(x, t)|^2 \tag{36.14}$$

The probability of finding the particle between $x = a$ and $x = b$ is then $\int_a^b |\Psi|^2 \, dx$. Therefore, since we are sure that the particle must be somewhere along the X-axis we must have

$$\int_{-\infty}^{\infty} |\Psi(x, t)|^2 \, dx = 1 \tag{36.15}$$

This expression is called the **normalization condition**. It severely restricts the mathematical form of the matter field and therefore the possible matter field states of the particle.

A special situation occurs when the particle is in a state of well-defined energy, E. In this case, recalling Equation 36.2, the frequency of the matter field at all points in space is $\omega = E/\hbar$ and the time dependence of the matter field can be expressed by $e^{-i\omega t}$ or $e^{-iEt/\hbar}$ (see Section 37.2). Then the expression of the matter field becomes

$$\Psi(x, t) = e^{-iEt/\hbar}\psi(x) \tag{36.16}$$

where $\psi(x)$ is called the **wave function** for historical reasons, although the name is misleading. Perhaps it would be better just to call it the **matter field amplitude**. Then

$$|\Psi(x, t)|^2 = \Psi^*\Psi = [e^{+iEt/\hbar}\psi^*(x)][e^{-iEt/\hbar}\psi(x)] = |\psi(x)|^2 \tag{36.17}$$

Therefore, in a state of well-defined energy, the probability distribution of the particle is independent of time and given solely by the matter field amplitude (or wave function) $\psi(x)$. That is

$$P(x) = |\psi(x)|^2 \tag{36.18}$$

We say the matter field is **stationary** (recall standing waves, Chapter 34) and that the state of the particle is also stationary. The normalization condition is now

$$\int |\psi(x)|^2 \, dx = 1 \tag{36.19}$$

which also restricts the possible forms of the wave function $\psi(x)$. For example, the wave function $\psi(x)$ for a particle confined mainly to the region between A and B is shown in Figure 36.14(a). Note that $\psi(x)$ decreases very rapidly outside the region AB where the particle is likely to be found while the wave function is oscillating within such a region. The intensity of the matter field, given by $|\psi(x)|^2$, is indicated in Figure 36.14(b).

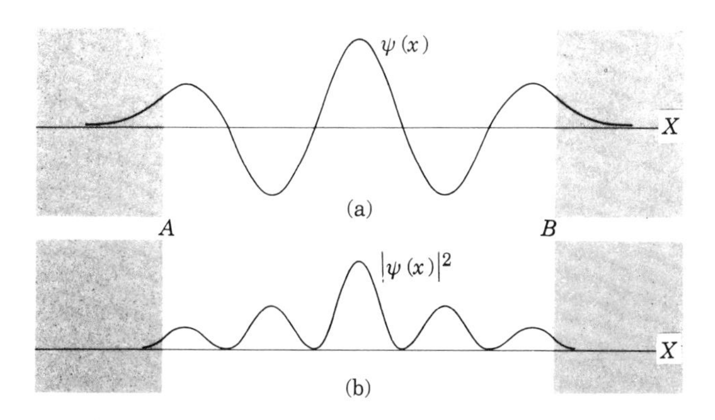

Figure 36.14 (a) Wave function of a particle moving between A and B. (b) Probability distribution corresponding to the wave function shown in (a).

In the general case of motion in space, the matter field depends on the three coordinates x, y, z as well as time; that is, $\Psi(x, y, z, t)$. Then $|\Psi(x, y, z, t)|^2 \, dx \, dy \, dz$ is the probability of finding the particle in the volume $dx \, dy \, dz$ around the point having coordinates x, y, z at time t, or

$$P = |\Psi(x, y, z, t)|^2 \tag{36.20}$$

is the probability per unit volume, or the **probability density**, of finding the particle, at time t, at x, y, z.

For motion in space the normalization condition is

$$\int_{\text{all space}} |\Psi|^2 \, dx \, dy \, dz = 1 \tag{36.21}$$

Consider, for example, the case of an atomic particle, such as an electron in an atom in a stationary state. The electron never moves too far away from the nucleus; it is essentially confined to a small region of space with dimensions of

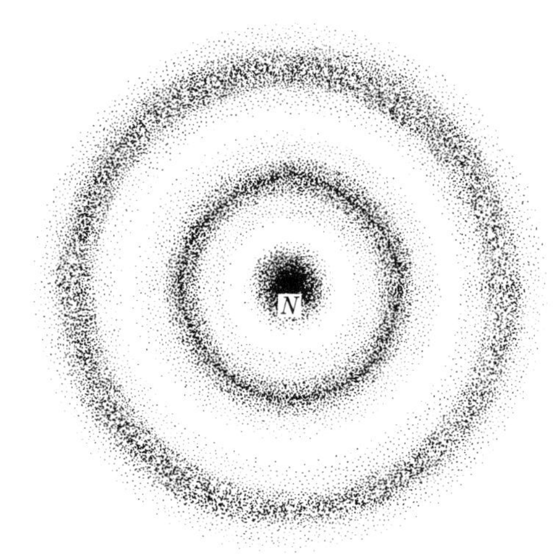

Figure 36.15 Probability distribution for an electron in an atom.

the order of 10^{-10} m. Thus its associated matter field may be expressed in terms of standing waves localized in this region and corresponding to the wave function $\Psi(x, y, z, t) = e^{-iEt/\hbar}\psi(x, y, z)$. We can compute ψ and plot $|\psi|^2$ as in Figure 36.15, where N is the nucleus and the degree of darkness is proportional to the value of $|\psi|^2$. Thus the darker zones represent the regions where the probability of finding the electron is greatest. This statement is the most we can say about the localization of the electron in an atom and it is impossible to talk about the precise orbit of the electron.

EXAMPLE 36.8

Motion of wave packets.

▷ For particles that are not confined to a limited region of space, such as a proton in an accelerator or an electron in a TV tube, the matter field corresponds to a traveling wave packet with a pronounced maximum in the region where the particle is most probably to be found. To describe a localized wave packet, we have to combine fields of several frequencies or energies, as explained in connection with Section 36.4. That means that the matter field is described by an expression of the form

$$\Psi(x, t) = \sum_E a_E e^{-iEt/\hbar}\psi_E(x) \tag{36.22}$$

obtained by having a linear combination of functions of the form given by Equation 36.16 for each energy and $\psi(x)$ is the wave function corresponding to the energy E. The coefficients

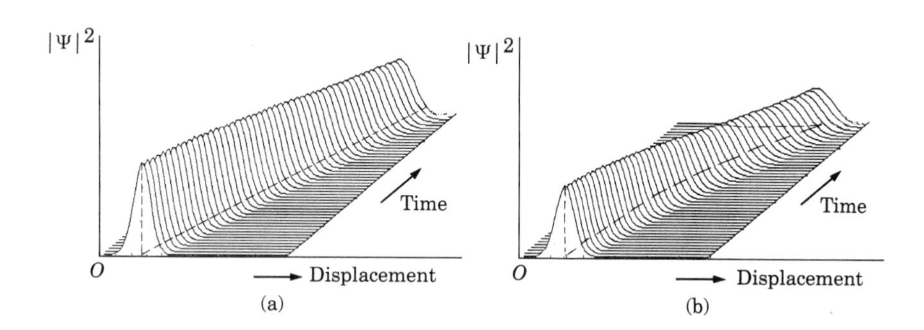

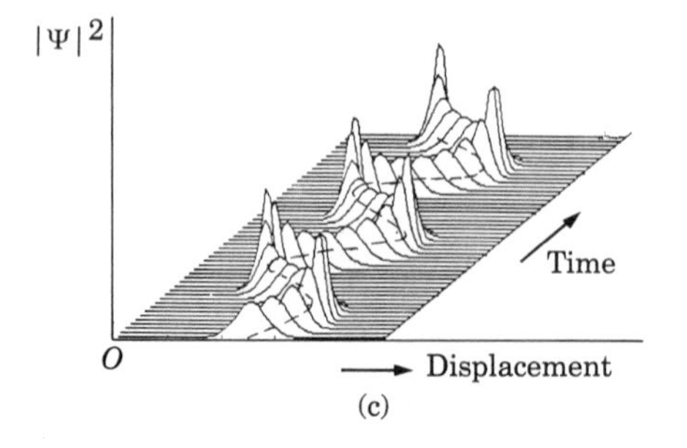

Figure 36.16 Evolution of a particle wave packet in space and time: (a) free particle, (b) particle under a constant force, (c) harmonic oscillator (By permission from Prof. A. T. Rakhimov, Institute of Nuclear Physics, Moscow State University, Russia). The dotted line represents the path of the maximum of the wave packet.

a_E are chosen to obtain the proper initial shape ($t = 0$) of the wave packet. As time evolves, the wave packet advances with the particle, as illustrated in Figure 36.16 for a free particle, for a particle under the action of a constant force and for a particle with harmonic oscillatory motion.

One of the objectives of quantum mechanics is to obtain, for each particular situation, the precise form of the wave packet of a particle that is compatible with the uncertainty principle and to determine the motion of the wave packet, given the forces acting on the particle. This is equivalent to solving the equation of motion in classical mechanics, given the forces on the particle.

QUESTIONS

36.1 Explain why a matter field expressed by a harmonic wave of constant amplitude does not provide information about the location of a particle. What information does it exactly provide?

36.2 To improve the information about the location of a particle, the matter field must correspond to a narrow wave packet. What physical quantity (or quantities) associated with the particle are necessarily less exactly known?

36.3 Are the uncertainty relations a result of a deficiency of the measuring apparatus, a consequence of the measuring process itself, or a fundamental property of matter?

36.4 How do the wavelength and frequency of the matter field associated with a particle vary as the energy of the particle increases?

36.5 Analyze the concept of 'particle' from the point of view of quantum mechanics.

36.6 Do a photon and an electron of the same momentum have the same wavelength and if the two have the same energy?

36.7 What is meant by the 'lifetime' of a state? Explain why the ground state has an infinite lifetime.

36.8 Discuss the notion of the width of an energy level. How is the width of an energy level related to transition probability?

36.9 Discuss the factors that determine line broadening.

36.10 What is the origin of the zero point energy in a bound system?

36.11 Is it justified to talk about the 'wavelength' of a stone or a ball?

36.12 Since $p = mv$, verify that the uncertainty principle can also be written as $\Delta x \Delta v \approx \hbar/m$. In view of this relation, if the positions of an electron and a proton are known with the same precision, which particle's velocity can be measured with greater precision?

36.13 What kind of information is provided by the matter field $\Psi(x, t)$?

36.14 Discuss the meaning of $|\Psi(x, t)|^2$ and of $|\psi(x)|^2$.

36.15 Under what conditions is $\Psi(x, t) = e^{-iEt/\hbar}\psi(x)$ the correct expression of the matter field?

36.16 Discuss why, if matter at the atomic and subatomic levels is composed of 'particles' described by a localized matter field (wave packet), large aggregates of matter can be considered as balls or sand grains.

PROBLEMS

36.1 (a) Calculate the de Broglie wavelength of an electron when its energy is 1 eV, 100 eV, 1 keV and 1 MeV. (b) Which of these wavelengths would be significantly scattered in a nickel crystal, in which the atomic separation is about 2.15 Å? (c) Calculate the energy of those electrons that are Bragg-scattered by the crystal at an angle of 30°.

36.2 A narrow beam of thermal neutrons produced by a nuclear reactor falls on a crystal with lattice spacing of 1.60 Å. Determine the Bragg angle such that 2 eV neutrons are strongly scattered.

36.3 Express Bragg's condition for the scattering of particles by a crystal lattice in terms of the accelerating voltage and the mass of the particles.

36.4 Repeat Example 36.2 for a neutron beam at a temperature of 100 °C.

36.5 A beam of electrons with a de Broglie wavelength of 10^{-5} m passes through a slit 10^{-4} m wide. Calculate the angular spread introduced because of diffraction by the slit.

36.6 A beam of particles used to probe an atomic structure must have wavelengths smaller than the object being studied. Calculate the minimum particle energy if (a) photons, (b) electrons and (c) neutrons are used to probe a nucleus whose diameter is 10^{-14} m, and (d) a nucleon (diameter 10^{-15} m).

36.7 The velocity of a nucleon in the X-direction is measured to an accuracy of 10^{-7} m s^{-1}. Determine the limit of accuracy with which the particle can be located simultaneously (a) along the X-axis and (b) along the Y-axis. Repeat for an electron.

36.8 The position of an electron is determined with an uncertainty of 0.1 Å. (a) Calculate the uncertainty in its momentum. (b) If the electron's energy is of the order of 1 keV, estimate the uncertainty in its energy. Repeat for a proton confined to a nuclear diameter ($\approx 10^{-14}$ m) with an energy of the order of 2 MeV.

36.9 Calculate the line width and frequency spread for a 1 nanosecond (10^{-9} s) pulse from a ruby laser ($\lambda = 6.3 \times 10^{-7}$ m).

36.10 A particle is represented by the wave function

$$\psi(x) = e^{-(x-x_0)^2/2\alpha} \sin kx$$

where α is a constant $\gg \lambda^2$ and $\lambda = 2\pi/k$. (a) Plot the wave function $\psi(x)$ and the probability distribution $|\psi(x)|^2$. (b) Estimate the uncertainty in the position and in the momentum of the particle.

36.11 In Example 25.9, the 'radius' of the electron was estimated as 2.82×10^{-15} m. This can be considered as the maximum precision with which the position of an electron can be measured. (a) What is the maximum precision with which the velocity of an electron can be measured? (b) Repeat the calculation for a proton, whose 'radius' is at least 1000 times smaller than that of an electron. What do you conclude? (c) Which of the two particles' motions can be better described using classical mechanics?

36.12 Electrons in a TV tube are accelerated through a potential difference of 10^4 V. Assuming that they start from rest, what is the ratio of their uncertainty in velocity (see Problem 36.7) to their final velocity?

36.13 The velocity, the distance traveled along the tube and the time taken to reach the screen for the electrons of the previous problem are related by $x = \frac{1}{2}vt$, since $v_{ave} = \frac{1}{2}v$ (recall Section 3.5). Assuming that the uncertainties in x and v are those given in Problem 36.7, find the uncertainty in the time to reach the screen.

37 Quantum mechanics: applications

Erwin Schrödinger was one of the founders of quantum mechanics. In 1926 he began publishing a series of papers that developed a formalism, initially designated as wave mechanics, in which he extended the ideas of Louis de Broglie (1923). He applied his theory to atomic structure, partially replacing the methods proposed by Bohr in 1913. Schrödinger introduced the concept of wave function or matter field, which describes, in space and time, the dynamical behavior of an electron and formulated an equation for obtaining the wave function, from which he could derive many atomic properties.

37.1 Introduction

We have seen in the preceding chapter that when dealing with the fundamental particles that compose matter, we must use a formalism called quantum mechanics and associate a 'matter field' or 'wave function' $\Psi(x, y, z, t)$ with each particle. The wave function provides information about the dynamics of the particle, including the probability of the space localization and time evolution of the particle, its energy, its momentum, its angular momentum, and other dynamical quantities. Further, this wave function is different for each state of the particle, whether an electron in an atom, a molecule or a solid, or a proton or neutron in a nucleus. Finding the wave function for a given system is one of the most important problems of quantum mechanics. For simplicity, we shall consider in this chapter only one-dimensional problems, so that the wave function is $\Psi(x, t)$.

37.2 Schrödinger's equation

The wave function $\Psi(x, t)$ for each dynamical state of a particle is determined by the forces acting on the particle. The rule for finding $\Psi(x, t)$ is expressed in a form called **Schrödinger's equation**, which was formulated in 1926 by Erwin Schrödinger, and is based on the expression for the total energy of the particle, $E = p^2/2m + E_p(x)$, where $p^2/2m$ is the kinetic energy of the particle and $E_p(x)$ is its potential energy. His method, later on refined by P. A. M. Dirac, is too complex to be considered here and we must limit ourselves to quoting the results. Schrödinger's equation (for one-dimensional problems) is

$$-\frac{\hbar^2}{2m}\frac{d^2\Psi}{dx^2} + E_p(x)\Psi = i\hbar\frac{d\Psi}{dt} \tag{37.1}$$

This equation is as fundamental to quantum mechanics as Newton's equation $\boldsymbol{F} = d\boldsymbol{p}/dt$ is to classical mechanics, or Maxwell's equations are to electromagnetism.

Equation 37.1 can be simplified when the particle is in a state of well-defined energy, E, such that the matter field can be expressed by Equation 36.16,

$$\Psi(x, t) = e^{-iEt/\hbar}\psi(x) \tag{37.2}$$

where $\psi(x)$ is the matter field amplitude. Then

$$\frac{d}{dt}[\Psi(x, t)] = -\frac{iE}{\hbar}e^{-iEt/\hbar}\psi(x)$$

Inserting this expression in Equation 37.1 and canceling the factor $e^{-iEt/\hbar}$ that appears in all terms, we obtain the following equation for the field amplitude $\psi(x)$:

$$-\frac{\hbar^2}{2m}\frac{d^2\psi}{dx^2} + E_p(x)\psi = E\psi \tag{37.3}$$

which is **Schrödinger's time-independent equation**. Recall that in Equation 37.3 E is the total energy of the particle. This suggests that the first term on the left corresponds to the kinetic energy, while the second is related to the potential energy of the particle. The solutions ψ of Equation 37.3 depend on the form of the potential energy $E_p(x)$. Also, only certain values of the energy E give solutions that can be normalized, i.e. that can satisfy the condition $\int_{-\infty}^{\infty} |\psi(x)|^2\, dx = 1$. They are the energies of the stationary states. Thus when we solve Schrödinger's equation, we obtain not only the wave function $\psi(x)$ but also the energy E of the stationary states of the system. Schrödinger's time-independent equation can be written in the alternative form

$$\frac{d^2\psi}{dx^2} + \frac{2m}{\hbar^2}[E - E_p(x)]\psi = 0 \tag{37.4}$$

In the more general case of a three-dimensional problem, where the wave function depends on the three coordinates x, y, z, Schrödinger's time-independent equation becomes

$$-\frac{\hbar^2}{2m}\left(\frac{d^2\psi}{dx^2}+\frac{d^2\psi}{dy^2}+\frac{d^2\psi}{dz^2}\right)+E_p(x, y, z)\psi = E\psi \tag{37.5}$$

Note that because Schrödinger's Equation 37.1 is of first order in time, it is not a wave equation, such as Equation 28.11. This is why its solutions are different from those for elastic and electromagnetic waves found in Chapters 28 and 29.

37.3 Free particle

For the case of a free particle, the potential energy is zero (that is, $E_p(x) \equiv 0$), and Schrödinger's time-independent equation 37.4 becomes

$$\frac{d^2\psi}{dx^2}+\left(\frac{2mE}{\hbar^2}\right)\psi = 0$$

But for a free particle, $E = p^2/2m$. Setting $p = \hbar k$, according to Equation 36.2, where k is the wave number, we have

$$E = \frac{\hbar^2 k^2}{2m} \qquad \text{or} \qquad k^2 = \frac{2mE}{\hbar^2}$$

Therefore, Schrödinger's equation for a free particle reduces to

$$\frac{d^2\psi}{dx^2}+k^2\psi = 0 \tag{37.6}$$

This is of the same mathematical form as the equation for the amplitude of a standing wave on a string.

Recalling that $i = \sqrt{-1}$ and $i^2 = -1$, we see, by direct substitution, that Equation 37.6 admits as solutions the wave functions

$$\psi(x) = e^{ikx} \qquad \text{and} \qquad \psi(x) = e^{-ikx} \tag{37.7}$$

Inserting Equation 37.7 in the general expression 37.2 for the wave function $\Psi(x, t)$, with $\omega = E/\hbar$, results in

$$\Psi(x, t) = e^{-i\omega t}e^{\pm ikx} = e^{-i(\omega t \mp kx)}$$

which corresponds to a 'complex' harmonic wave propagating in the $\pm X$-direction with a velocity $v = \omega/k = E/p$ (recall Section 36.4). Therefore, the wave function $\psi = e^{ikx}$ represents a free particle of momentum $p = \hbar k$ moving in the $+X$-direction, and the wave function $\psi = e^{-ikx}$ represents a free particle of the same momentum but moving in the opposite, or $-X$, direction.

Note that either solution yields

$$|\psi(x)|^2 = \psi^*(x)\psi(x) = e^{-ikx}e^{ikx} = 1$$

The fact that $|\psi(x)|^2 = 1$, or a constant, means that the probability of finding the particle is the same at all points. In other words, $\psi = e^{\pm ikx}$ describes a situation where we have complete uncertainty about position. This is in agreement with the uncertainty principle, because $\psi = e^{\pm ikx}$ describes a particle whose momentum, $p = \hbar k$, is known precisely; that is, $\Delta p = 0$, which requires that $\Delta x \to \infty$.

37.4 Potential wall

Consider a particle that is free to move in the region $0 < x < \infty$, but at $x = 0$ there is a very strong force to the right so that the particle can never cross to the region $x < 0$ or to the left of O (Figure 37.1(a)). This is equivalent to a sudden increase in the potential energy at $x = 0$. We say that at $x = 0$ there is a **potential wall**, which acts as a very strong force at $x = 0$ directed to the right. In this case a particle moving toward O from the right is bounced back when it reaches $x = 0$ and is forced to reverse its motion. We may say that the matter field $\psi(x)$ is reflected at $x = 0$, which is similar to the reflection of a wave along a string with a fixed end we discussed in Chapter 34. The situation depicted in Figure 37.1 approximately corresponds to the case of an electron in a metal that finds a very large potential wall when it reaches the surface.

If the particle has an energy $E = p^2/2m = \hbar^2 k^2/2m$, the wave function in this case must incorporate motion in both directions along the X-axis and therefore must be a superposition of the two solutions given in Equation 37.7, that is

$$\psi(x) = Ae^{ikx} + Be^{-ikx}$$

However we must impose the condition that $\psi(x) = 0$ at $x = 0$ since the particle cannot be in the region $x < 0$. This is the same condition as with a string with a fixed end and requires that $\psi(0) = A + B = 0$ or $B = -A$, which means that the motions in both directions have the same amplitude but opposite phase. Therefore

$$\psi(x) = A(e^{ikx} - e^{-ikx}) = 2iA \sin kx = C \sin kx \tag{37.8}$$

where $C = 2iA$ (Figure 37.1(b)). Note that the particle can have any energy given by $E = \hbar^2 k^2/2m$, and momentum $p = \hbar k$, corresponding to a wavelength $\lambda = 2\pi/k$. This is understandable because we may throw the particle against the wall with any energy, and it is reflected with the same energy.

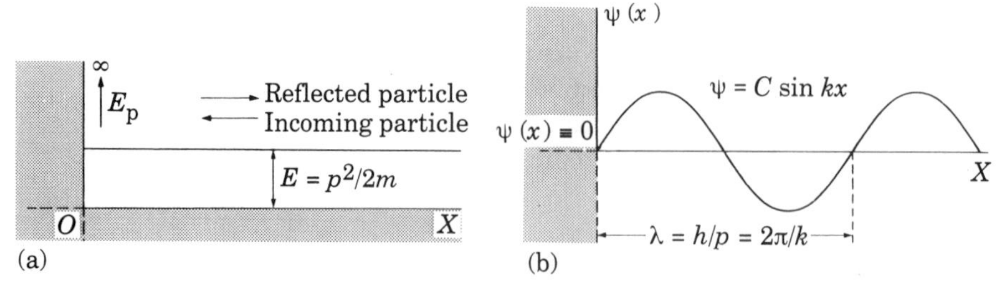

Figure 37.1 (a) Potential wall. (b) Wave function of a free particle near a potential wall.

37.5 Potential box

Consider a particle confined to the region $0 < x < a$ where it can move freely ($E_p \equiv 0$) but subject to strong forces at $x = 0$ and $x = a$, which means that $E_p \to \infty$ at those two points. As indicated in Section 36.8, we call this situation a one-dimensional potential box. Since the particle moves freely in the region $0 < x < a$ (Figure 37.2), Schrödinger's equation for a particle in a potential box is

$$\frac{d^2\psi}{dx^2} + k^2\psi = 0, \qquad k^2 = \frac{2mE}{\hbar^2} \qquad (0 < x < a)$$

Because the particle can move back and forth freely between $x = 0$ and $x = a$, the wave function is again of the form

$$\psi(x) = Ae^{ikx} + Be^{-ikx}$$

which contains motion in both directions. The boundary conditions require that $\psi(x) = 0$ at $x = 0$ and $x = a$ since the particle cannot be outside the walls of the box. Then $\psi_{(x=0)} = A + B = 0$ or $B = -A$. So, again,

$$\psi(x) = A(e^{ikx} - e^{-ikx}) = 2iA \sin kx = C \sin kx$$

where $C = 2iA$. This is the same situation we found in the previous section for a particle restricted only at $x = 0$. The boundary condition at $x = a$ gives $\psi_{(x=a)} = C \sin ka = 0$. Since C cannot be zero, because we would then have no wave function, we conclude that $\sin ka = 0$ or $ka = n\pi$, where n is an integer. Solving for k, we have

$$k = \frac{n\pi}{a} \qquad \text{or} \qquad p = \hbar k = \frac{n\pi\hbar}{a} \qquad \textbf{(37.9)}$$

which gives the possible values of the momentum $p = \hbar k$ of the particle. The possible values of the energy are

$$E = \frac{p^2}{2m} = \frac{\pi^2\hbar^2}{2ma^2}n^2 \qquad \textbf{(37.10)}$$

These results are identical to Equations 36.9 and 36.10 obtained using a more intuitive method.

Note that because we now have two boundary conditions (because the particle is confined to a certain region), the energy can only have certain values. In other words, **energy quantization** is a consequence of confining a particle to a certain region.

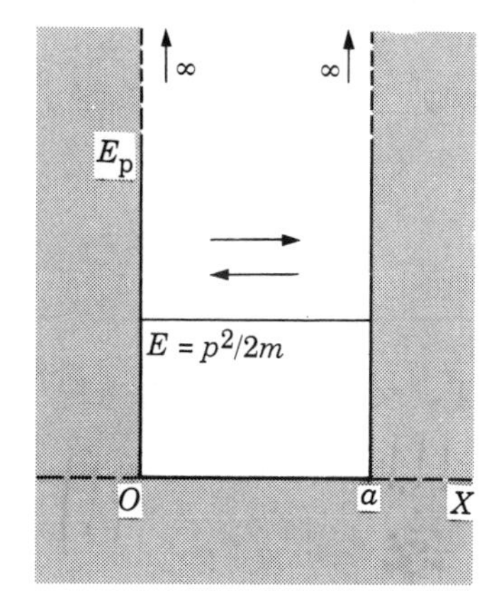

Figure 37.2 Potential box.

The wave functions corresponding to $n = 1$, 2 and 3 have been illustrated in Figure 37.3(a). Note that as the energy of the particle increases, the greater the number of zeros (or nodes) in the wave function. In Figure 37.3(b) the corresponding probability densities for a particle in a potential box are shown.

The constant C can be obtained using the normalization condition, which in this case becomes

$$C^2 \int_0^a \sin^2 kx \, dx = 1$$

Recalling that $\sin^2 kx = \frac{1}{2}(1 + \cos 2kx) = \frac{1}{2}[1 + (\cos 2n\pi x/a)]$, we have that

$$\int_0^a \sin^2 kx \, dx = \frac{1}{2} \int_0^a \left(1 + \cos \frac{2n\pi x}{a}\right) dx = \frac{a}{2}$$

Therefore $(\frac{1}{2}a)C^2 = 1$ or $C = (2/a)^{1/2}$. Thus the normalized functions of a particle in a box are

$$\psi_n(x) = \left(\frac{2}{a}\right)^{1/2} \sin \frac{n\pi x}{a} \tag{37.11}$$

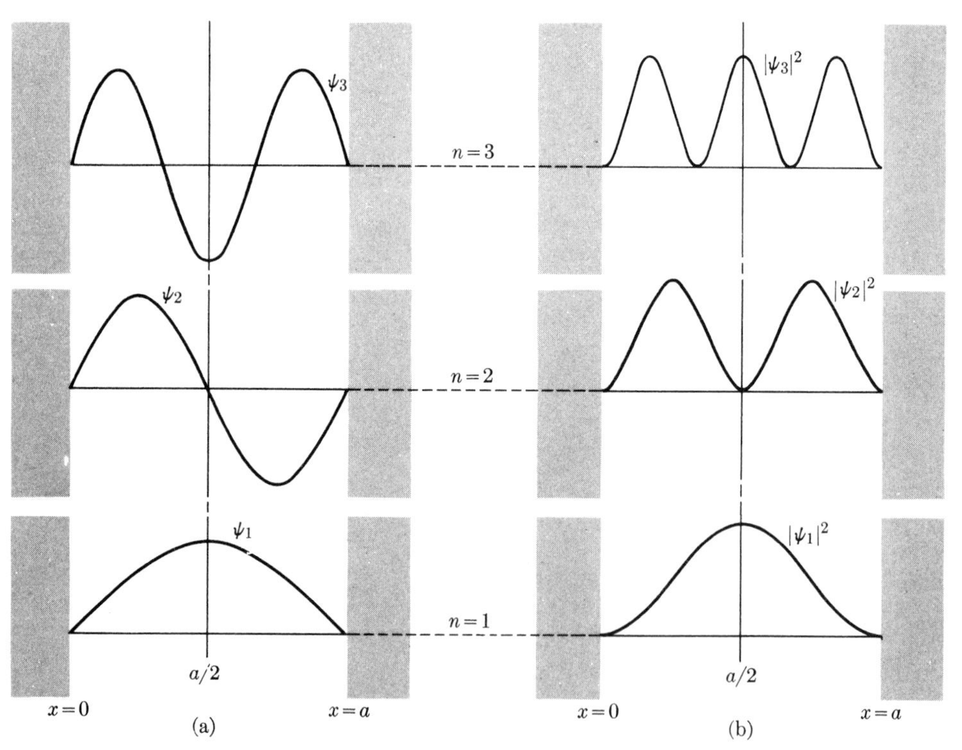

Figure 37.3 (a) First three wave functions for a particle in a potential box. (b) Corresponding probability densities.

Table 37.1 Energy levels and degeneracies in a cubical potential box ($E_1 = \pi^2\hbar^2/2ma^2$).

Energy	Combinations of n_1, n_2, n_3	Degeneracy
$3E_1$	(1, 1, 1)	1
$6E_1$	(2, 1, 1)(1, 2, 1)(1, 1, 2)	3
$9E_1$	(2, 2, 1)(2, 1, 2)(1, 2, 2)	3
$11E_1$	(3, 1, 1)(1, 3, 1)(1, 1, 3)	3
$12E_1$	(2, 2, 2)	1
$14E_1$	(1, 2, 3)(3, 2, 1)(2, 3, 1)	6
	(1, 3, 2)(2, 1, 3)(3, 1, 2)	

If instead of a one-dimensional potential box, we have a three-dimensional box or cavity of side a, it can be easily verified that the wave function has the form

$$\psi(x, y, z) = \left(\frac{2}{a}\right)^{3/2} \sin\frac{n_1 \pi x}{a} \sin\frac{n_2 \pi y}{a} \sin\frac{n_3 \pi z}{a} \tag{37.12}$$

where n_1, n_2 and n_3 are positive integers. Note that $\psi(x, y, z)$ is zero for x, y, z equal to zero or a, since the particle must be confined to the region inside the box.

The energy of the stationary state is now given by

$$E = \frac{\pi^2\hbar^2}{2ma^2}(n_1^2 + n_2^2 + n_3^2) \tag{37.13}$$

which is the generalization of Equation 37.10. The possible energy values are obtained by giving n_1, n_2 and n_3 the values 1, 2, 3, ... They are given in Table 37.1 and are represented in Figure 37.4. The zero point energy for the three-dimensional box is $E_1 = 3\pi^2\hbar^2/2ma^2$.

An energy value in Equation 37.13 can be obtained with different combinations of n_1, n_2, n_3, as shown in Table 37.1. For each combination of n_1, n_2, n_3 there are different wave functions $\psi(x, y, z)$. Whenever different wave functions are associated with the same energy, it is said that there is a **degeneracy**. This is an important quantum feature that normally is associated with some symmetry of the physical system. In this example it is a consequence of the fact that the potential box is cubical and therefore we can rotate the box through an angle of $\pi/2$ about any coordinate axis and it remains the same. This is not the case if the box has different dimensions a, b, c.

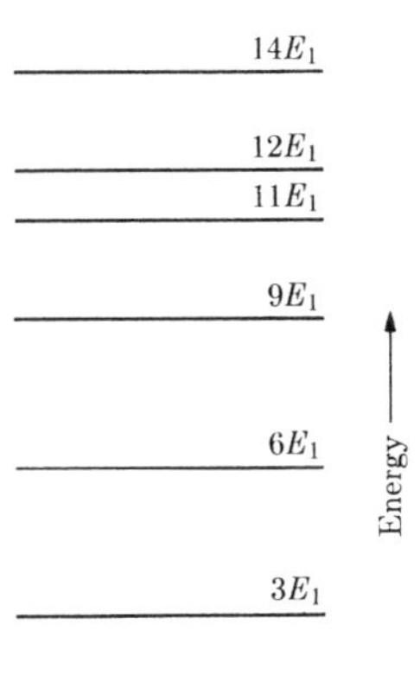

Figure 37.4 Energy levels for a cubical potential box.

No physical problem corresponds to a potential box, which is an oversimplification. Nevertheless, the potential box is a useful approximation for analyzing the situation where a particle is confined to a finite region of space, such as an electron in an atom or a nucleon in a nucleus.

EXAMPLE 37.1

Comparison of the excitation energy of a proton in a nucleus and of an electron in an atom.

▷ From Equation 37.10 the ground state energy of a particle in a one-dimensional potential box, obtained by making $n = 1$, is $E_1 = \pi^2\hbar^2/2ma^2$. The first excited energy state, corresponding to $n = 2$, is $E_2 = 4\pi^2\hbar^2/2ma^2$. Therefore, the excitation energy of a particle in a box of size a is

$$E_{\text{exc}} = E_2 - E_1 = \frac{3\pi^2\hbar^2}{2ma^2}$$

The most interesting feature of this energy–size relation is that the excitation energy is inversely proportional to ma^2.

Electrons in atoms and nucleons in nuclei are confined to move in limited regions of space under electric and nuclear forces, respectively. Although the potential energies are not equivalent to potential boxes, we may assume that in both cases the excitation energies are inversely proportional to ma^2, where a gives the order of magnitude of the region where the particle moves. Therefore we may write

$$\frac{(E_{\text{exc}})_{\text{nucl}}}{(E_{\text{exc}})_{\text{elec}}} \simeq \frac{(ma^2)_{\text{elec}}}{(ma^2)_{\text{nucl}}}$$

For an electron in an atom $m_e = 9.109 \times 10^{-31}$ kg and $a \simeq 10^{-10}$ m, giving $ma^2 \simeq 10^{-50}$ m^2 kg. For a nucleon in a nucleus, $m_n \simeq 1.67 \times 10^{-27}$ kg and $a \simeq 10^{-15}$ m, giving $ma^2 \simeq 1.7 \times 10^{-57}$ m^2 kg. Therefore

$$\frac{(E_{\text{exc}})_{\text{nucl}}}{(E_{\text{exc}})_{\text{elec}}} \simeq \frac{10^{-50}}{1.7 \times 10^{-57}} \simeq 10^6$$

so that, in order of magnitude, the energy required to excite a nucleon in a nucleus is about one million times larger than the energy required to excite an electron in an atom. In fact, the excitation energies of nucleons in a nucleus are of the order of millions of eV, while the excitation energies of electrons in an atom are only a few eV.

The Earth is surrounded by electromagnetic radiation from the Sun, which contains mostly photons with energies well below 1 MeV. This is one of the reasons why matter on Earth can exist in stable atomic and molecular configurations and why most of the phenomena we observe on Earth involve excitation and transfer of electrons, while the nuclei remain highly stable. Nuclear excitations occur naturally only in the Sun and other stars, where the energies available are much higher. On Earth, nuclear excitations and reactions are possible only by the use of high energy accelerators, except for a few naturally occurring radioactive substances.

37.6 Potential well

Another situation of great practical interest is that of a particle moving in a potential well (Figure 37.5). At $x = 0$ the potential energy goes to infinity, equivalent to a very strong force to the right. For $0 < x < a$ the potential energy has the constant value $-E_0$, which is equivalent to no force. At $x = a$ the potential energy jumps suddenly to the value $E_p = 0$, which is equivalent to a strong force to the left. For $x > a$ the potential energy remains zero which again means no force. The quantity a is called the **range** of the potential. This potential represents, in a simplified form, the potential energy of a neutron–proton system, as in the deuteron, or of the binding of two atoms in a molecule, or even of an electron in an atom.

Consider, first, a particle moving to the left (toward O) from a great distance with a *positive* energy E (Figure 37.5(a)) that is all kinetic because, for $x > a$, the potential energy is zero. When the particle reaches $x = a$ it suffers a sudden acceleration and its kinetic energy becomes $E + E_0$. The particle is reflected at O and sent back to the right. We say that the particle has been **scattered** by the potential.

While the particle is in the external region $x > a$, which we call 'e', for external, it has a momentum $p_e = (2mE)^{1/2}$ and the matter field has a wave number $k_e = p_e/\hbar$. When the particle is in the inner region $0 < x < a$, which we call 'i', for internal, the momentum is larger, becoming $p_i = [2m(E + E_0)]^{1/2}$ and the matter field has a wave number $k_i = p_i/\hbar$. The wavelength, being inversely proportional to the momentum, is smaller in the inner region $x < a$ than in the external region $x > a$. Thus, recalling the results of Section 37.2, the wave function for the region $x < a$ is given by Equation 37.8, that is

$$\psi_i(x) = C \sin k_i x \tag{37.14}$$

The wave function for $x > a$ must be written as

$$\psi_e(x) = C' \sin(k_e x + \delta) \tag{37.15}$$

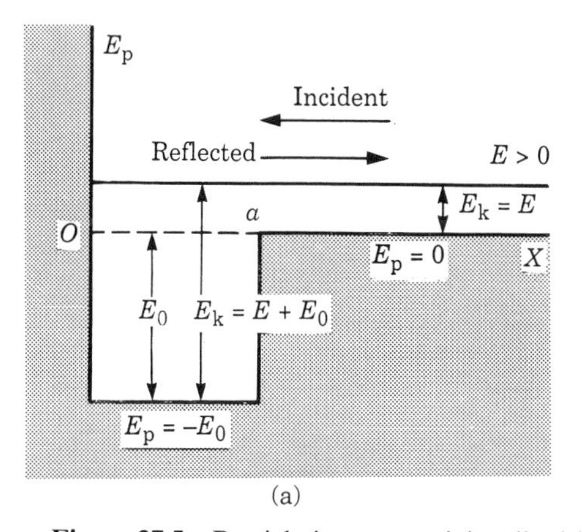

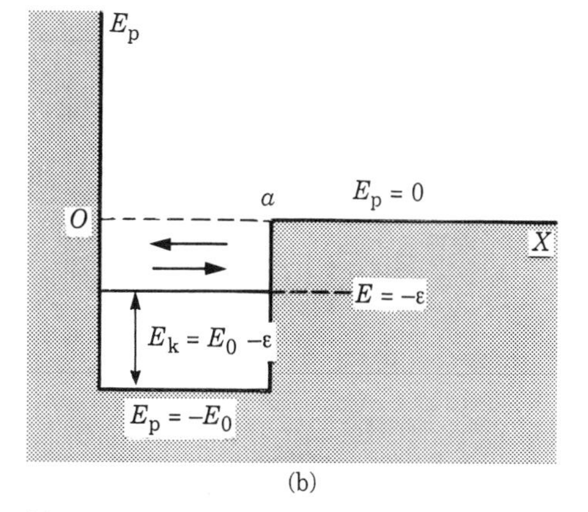

Figure 37.5 Particle in a potential well with (a) positive and (b) negative energy.

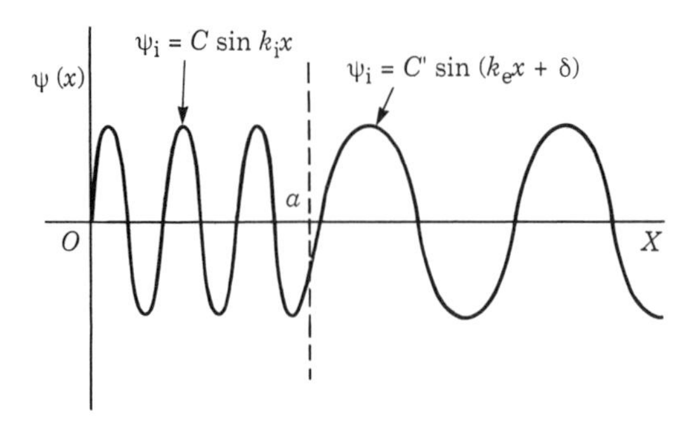

Figure 37.6 Wave function of a particle in a potential well with positive energy ($E > 0$).

where δ is a phase shift that must be introduced because now we do not require that $\psi(x)$ be zero at $x = a$. The phase shift is adjusted by requiring that the two functions join smoothly at $x = a$. The wave functions $\psi_i(x)$ and $\psi_e(x)$ have been represented in Figure 37.6.

The phase shift is a consequence of the distortion of the free particle wave function in the region $0 < x < a$ by the potential well. The phase shift can be measured experimentally and, from it, information about the range a and the depth E_0 of the potential well is obtained. This is how many scattering experiments using particle accelerators are analyzed.

Now consider the case when the energy of the particle is negative, $E < 0$, which corresponds to bound states (Figure 37.5(b)). In such a case the particle is bound, in the Newtonian sense, to move in the region between $x = 0$ and $x = a$, or region 'i'. In this case region 'e' or $x > a$ is called 'forbidden'. We shall make $E = -\varepsilon$, and we say that ε is the **binding energy** of the particle, i.e. the *positive* energy required to pull the particle out of the potential well. Then the kinetic energy of the particle is $E_k = E_0 - \varepsilon$ and the momentum is $p_i = [2m(E_0 - \varepsilon)]^{1/2}$.

When Schrödinger's equation is solved (see Note 37.1), we find that the wave function for $0 < x < a$ is again

$$\psi_i(x) = C \sin k_i x \tag{37.16}$$

with

$$k_i^2 = \frac{p_i^2}{\hbar^2} = \frac{2m(E_0 - \varepsilon)}{\hbar^2} \tag{37.17}$$

For the 'forbidden' region, $x > a$, the wave function is not zero but given by

$$\psi_e(x) = De^{-\alpha x} \tag{37.18}$$

where

$$\alpha^2 = \frac{2m\varepsilon}{\hbar^2} \tag{37.19}$$

Thus the wave function $\psi_e(x)$ decreases very rapidly as x increases. This implies that the probability of finding the particle in the forbidden region, $x > a$, though finite, is very small. The fact that the wave function is not zero in the forbidden region is one of the most striking results of quantum mechanics.

Because the wave functions $\psi_i(x)$ and $\psi_e(x)$ must join smoothly at $x = a$, the following relation between k_i and α must be satisfied:

$$k_i \cot k_i a = -\alpha \qquad \textbf{(37.20)}$$

The binding energy ε can be obtained from Equation 37.20 since both k_i and α depend on ε. In this way the possible energy levels can be found in terms of E_0 and a.

Depending on the depth E_0 and the range a of the well, there may be none, one, two etc. possible energy levels. The wave functions for bound states have been represented in Figure 37.7 for the case where there are three energy levels.

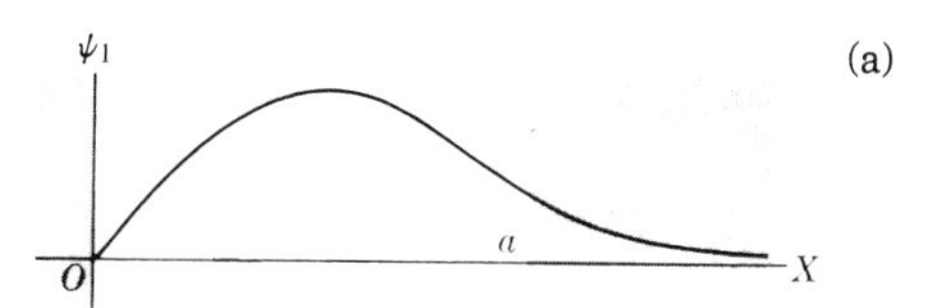

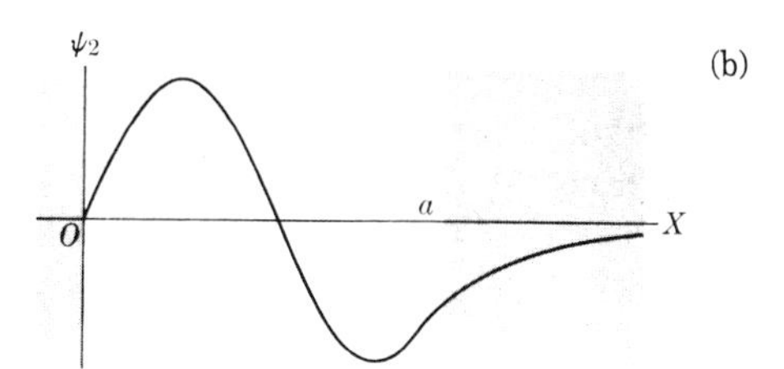

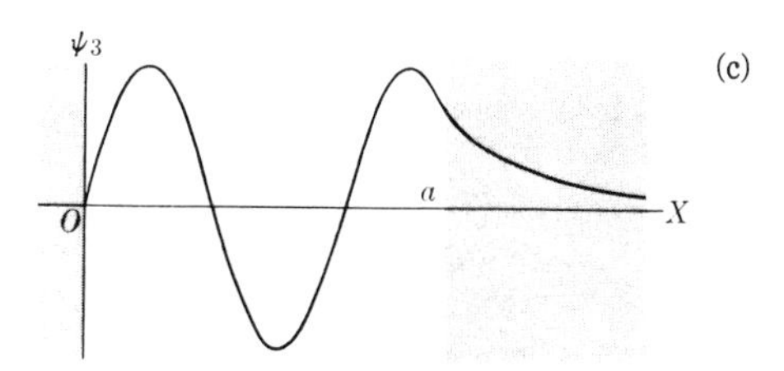

Figure 37.7 Wave functions of a particle in a potential well with negative energy ($E < 0$). (a) Ground state, (b) and (c) excited states.

Note 37.1 Energy states in a potential well

Schrödinger's equation 37.4 for region 'i', $x < a$, with $E_p(x) = -E_0$ and $E = -\varepsilon$, becomes

$$\frac{d^2\psi}{dx^2} + \frac{2m}{\hbar^2}(E_0 - \varepsilon)\psi = 0 \qquad \text{or} \qquad \frac{d^2\psi}{dx^2} + k_i^2\psi = 0$$

where $k_i^2 = 2m(E_0 - \varepsilon)/\hbar^2$, and $p_i = \hbar k_i$ is the momentum inside the well. The solution of this equation is the same as for a free particle. Since the particle can move back and forth within the well,

$$\psi_i(x) = Ae^{ik_i x} + Be^{-ik_i x}$$

Recalling the case of the potential wall, we must require that $\psi_i = 0$ at $x = 0$ because the potential energy increases to ∞ very steeply; this means that $A + B = 0$ or $B = -A$. Thus

$$\psi_i(x) = A(e^{ik_i x} - e^{-ik_i x}) = C \sin k_i x$$

where $C = 2iA$. In the external region 'e', corresponding to $x > a$, Schrödinger's equation with $E_p(x) = 0$ and $E = -\varepsilon$ becomes

$$\frac{d^2\psi}{dx^2} - \frac{2m\varepsilon}{\hbar^2} = 0 \qquad \text{or} \qquad \frac{d^2\psi}{dx^2} - \alpha^2\psi = 0 \qquad (x > a)$$

where $\alpha^2 = 2m\varepsilon/\hbar^2$. The wave function in the external region $x > 0$ must decrease very rapidly as x increases since the probability of finding the particle in the 'forbidden' region must be very small. Therefore the solution of this equation satisfying that condition is

$$\psi_e(x) = De^{-\alpha x},$$

as is verified by direct substitution. Next we must make sure that ψ_i and ψ_e join smoothly at $x = a$; i.e. $\psi_i = \psi_e$ and $d\psi_i/dx = d\psi_e/dx$ for $x = a$. This yields

$$C \sin k_i a = De^{-\alpha a} \quad \text{and} \quad k_i C \cos k_i a = -\alpha De^{-\alpha a}$$

Dividing these two equations to eliminate the constants C and D, we obtain Equation 37.20, $k_i \cot k_i a = -\alpha$. The possible energy levels may now be found in terms of E_0 and a, by inserting the values of k_i and α, in terms of ε, into the equation. The number of possible energy levels depends on the depth E_0 and the range a of the well, and may be found as follows.

The sine wave function in the internal region must approach $x = a$ with a negative slope so that it can join smoothly with the decreasing exponential wave function in the external region. Thus, as can be seen from Figure 37.7(a), the range a must be larger than $\lambda_i/4$ or $a > 2\pi/4k_i$, which can be written as $k_i a > \pi/2$. This can also be seen by noting that $\cot k_i a$ is negative and therefore we must have at least $k_i a > \pi/2$. Using the value $k_i^2 = 2m(E_0 - \varepsilon)/\hbar^2$, we have

$$2m(E_0 - \varepsilon)/\hbar^2 a^2 > \pi^2/4 \quad \text{or} \quad a^2(E_0 - \varepsilon) > \pi^2\hbar^2/8m$$

Thus the condition for having at least one energy level is $E_0 a^2 > \pi^2\hbar^2/8m$. We conclude that for $E_0 a^2 < \pi^2\hbar^2/8m$ there is no bound state. Thus a very shallow potential well (E_0 small) or a very short range force (a small) may result in the impossibility of having a bound state. Similarly if $\pi^2\hbar^2/8m < E_0 a^2 < 9\pi^2\hbar^2/8m$ there is only one bound state; if $9\pi^2\hbar^2/8m < E_0 a^2 < 25\pi^2\hbar^2/8m$ there are two bound states, and so on. Thus the number of energy levels depends on the value of the product $E_0 a^2$ or energy × (range)^2.

37.7 Particles in a general potential

The potential box and potential well are extreme simplifications of the potential energies found in atoms, molecules, solids and nuclei. A more realistic potential leading to bound and unbound states is illustrated in Figure 37.8.

At large x the potential energy $E_p(x)$ becomes essentially constant with a value usually chosen as the zero of energy. The potential energy drops as the distance increases from zero to x_0, indicating a force directed to the right, or

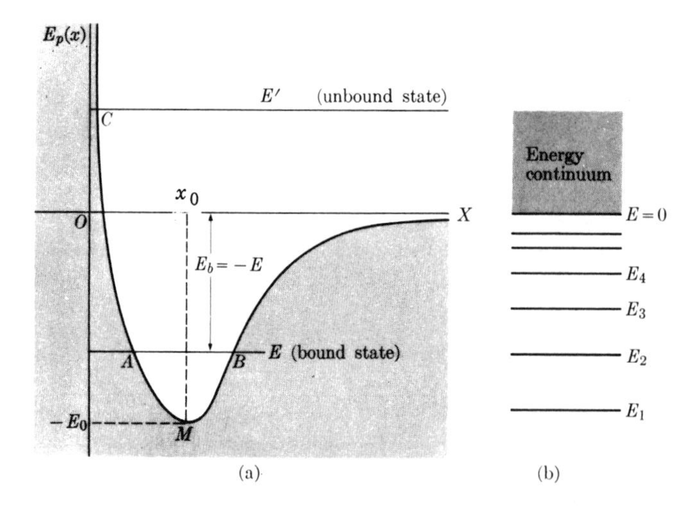

Figure 37.8 (a) General potential energy curve corresponding to strong repulsion at small values of x and negligible interaction at large values of x. (b) Energy levels corresponding to the potential energy shown in (a).

repulsive. At $x = x_0$, the potential energy is a minimum, and is equal to $-E_0$; the distance from O is the *equilibrium separation.* As the distance increases further, the potential energy increases, indicating a force to the left, or attractive.

For negative energies the motion is bounded. For example, when the particle has an energy E, the classical motion is oscillatory between A and B. These two points are called the **classical limits of oscillation** (recall the discussion in Section 9.11). For positive energies ($E > 0$) such as E', a particle coming from the right is stopped when it reaches C and bounces back, receding to infinity. This is why C is called the **classical turning point**.

In quantum mechanics the description is basically the same, but certain new features appear, as seen in the examples we have worked out before.

(i) Bound states occur only for *certain* negative energies, so that for $E < 0$ there is a **discrete spectrum** of energy levels or states. This is because, for $E < 0$, the motion is limited at both classical limits of oscillation, thus imposing two boundary conditions on the wave functions. Hence appropriate solutions of Schrödinger's equation exist only for certain energies; that is, the energy is quantized. A characteristic feature of the bound states is that there is a state of minimum energy or **ground state** that is dictated by the requirements of the uncertainty principle (recall Example 36.3), or alternatively because of the need to accommodate a wave function in a limited region of space.

Even if the potential energy is attractive at all distances and does not show a minimum, there is always a stationary state of minimum energy. The reason is that as the total (negative) energy is lowered, the size of the region where the particle can move, determined by the turning points, decreases (Δx decreases). This results in an increase of the momentum (Δp increases) and therefore of the kinetic energy. Eventually an energy, and thus a size of the system, is reached such that, at lower energies, the kinetic energy required by the uncertainty principle would overcompensate the attractive potential energy. Thus once more we see that the uncertainty principle and the existence of a matter field provide an explanation for the stability of matter. This is the same situation we found, for example, in the case of the potential box and potential well.

(ii) Unbound states exist for all positive energies; we then say that, for $E > 0$, there is a **continuous spectrum** of energy levels or states. The reason for this is that, when $E > 0$, the motion is limited at only one point, the classical turning point, so that only one boundary condition is required. Enough flexibility is then left to allow a solution of Schrödinger's equation with one arbitrary constant: the energy. This situation has already arisen in the case of the potential well (Figure 37.5). Physically, the situation arises because we can always arbitrarily fix the energy of the particle when it is at a great distance from O, as we do, for example, when a charged particle is accelerated in a machine and thrown against a target nucleus; or when one atom collides with another in a gas. The discrete and continuous energy spectra are shown schematically in Figure 37.8(b).

When the particle is in a bound state of negative energy E, the minimum energy that must be supplied to the particle to remove it to a very large distance is called the **binding energy** of the particle in that state. Sometimes, when we are referring to the ground state of a diatomic molecule, we call it the **dissociation energy**, because it is the minimum energy required to separate the atoms when the molecule is initially in the ground state. For an electron in an atom or molecule, we call it the **ionization energy**, because it is the minimum energy required to remove the electron from the ground state.

For $E < 0$, the wave functions corresponding to the potential energy of Figure 37.8 resemble those of the potential well, except that their exact shape depends on the form of the potential energy $E_p(x)$. Figure 37.9 shows the general shape of the wave functions for a negative energy level ($E < 0$) of the discrete spectrum (a bound state) and the wave function for a positive energy level ($E > 0$) of the continuous spectrum (an unbound state). In both cases the wave functions extend beyond the classical limits of motion, but decrease very rapidly outside those limits. Therefore, there is a small probability for finding the particle outside the classical limits of oscillation. For large x, the wave function of a positive energy or unbound state resembles that of a free particle, but the wavelength is not constant because the momentum of the particle depends on the distance from O.

However, if the potential energy $E_p(x)$ never levels off, as in Figure 37.10, then the spectrum of possible energies obtained by solving the Schrödinger equation is always discrete, as we found in the case of a potential box. The spacing of the energy levels depends on the specific form of $E_p(x)$. The mathematical expressions for the wave functions also depend on the form of $E_p(x)$. They have the same oscillatory form we have found before but, of course, they are not sine functions and do not exhibit any symmetry unless $E_p(x)$ is symmetric. The general form of $\psi(x)$ is also illustrated in Figure 37.10 for three energy levels. One interesting feature of these wave functions is that they again extend beyond the classical limits of oscillation although they decrease very rapidly. Also, as the energy increases, the number of nodes (or roots) increases.

The conclusion is then that it is very difficult to confine a particle precisely within a certain region unless the potential energy increases very sharply (or, which is equivalent, the forces are very strong) at the boundaries of the region.

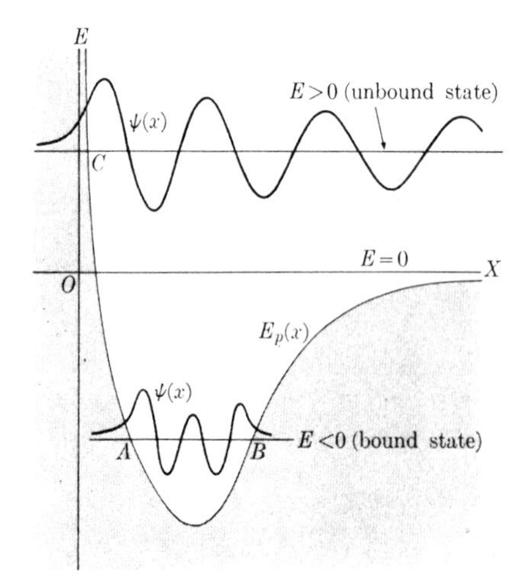

Figure 37.9 Shape of wave functions for bound and unbound states corresponding to the potential energy shown in Figure 37.8(a).

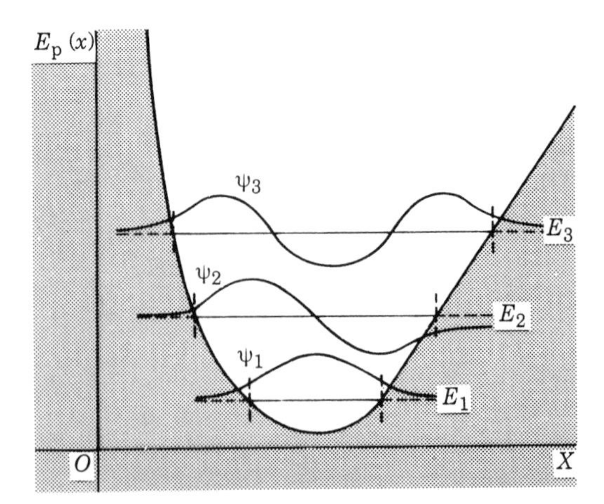

Figure 37.10 Potential energy curve and wave functions for bound states.

37.8 The simple harmonic oscillator

An interesting and important physical system is the simple harmonic oscillator, corresponding to the potential energy $E_p(x) = \frac{1}{2}kx^2$ (Figure 37.11), which is similar to that in Figure 37.10 but is symmetric about $x = 0$. The Schrödinger equation is

$$-\frac{\hbar^2}{2m}\frac{d^2\psi(x)}{dx^2} + \frac{1}{2}kx^2\psi(x) = E\psi(x)$$

Finding the solution of this equation is too elaborate a mathematical problem to be discussed here. However, when it is solved, the possible values of the energy are found to be

$$E = (n + \tfrac{1}{2})\hbar\omega \tag{37.21}$$

where $n = 0, 1, 2, 3, \ldots$, integer, and $\omega = (k/m)^{1/2}$ is the angular frequency of the oscillator. The energy levels, represented in Figure 37.11, are equally spaced an amount $\hbar\omega$. Since $\omega = 2\pi\nu$ and $\hbar = h/2\pi$, expression 37.21 can also be written in the alternate form

$$E = (n + \tfrac{1}{2})h\nu \tag{37.22}$$

Expressions 37.21 and 37.22 for the energy levels have already been mentioned in Sections 10.11 and 31.5 in connection with molecular vibrations.

We shall not give the expression for the wave functions of the simple harmonic oscillator. However, in Figure 37.12 the wave functions corresponding to $n = 0$, 1 and 2 have been represented. They again extend beyond the classical limits of oscillation. The corresponding probability densities are shown in Figure 37.13.

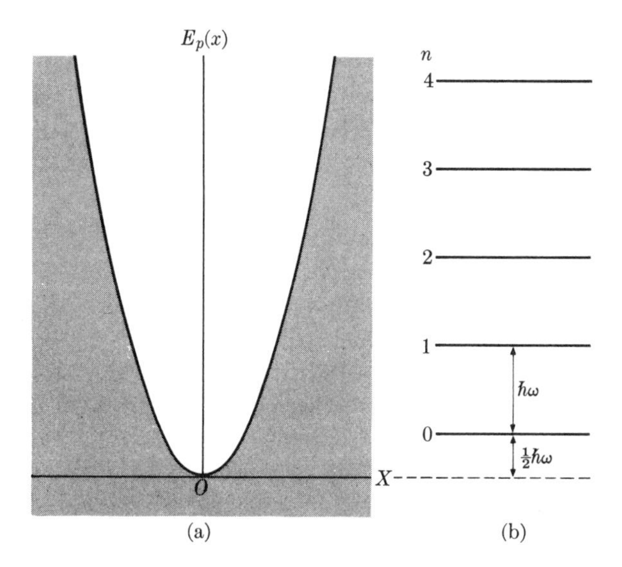

Figure 37.11 (a) Potential energy of a simple harmonic oscillator. (b) Energy levels.

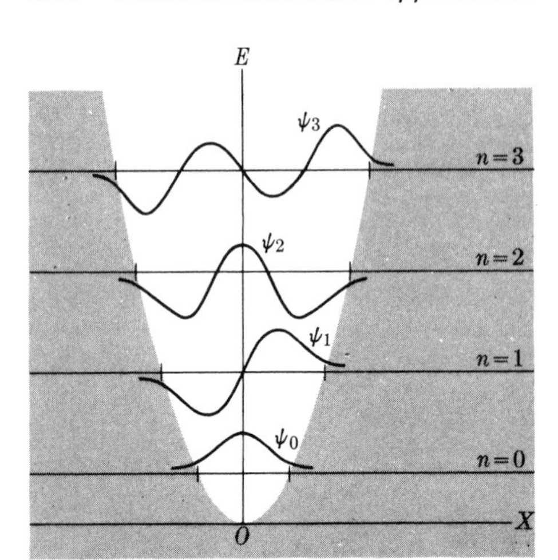

Figure 37.12 Wave functions corresponding to the first four energy levels of a harmonic oscillator.

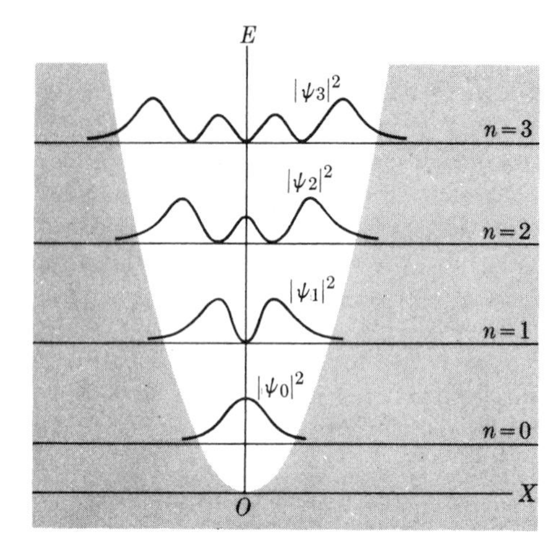

Figure 37.13 Probability densities corresponding to the first four energy levels of a harmonic oscillator.

Note that because the potential energy is symmetric with respect to $x = 0$, the probability distributions are also symmetric. However, the wave functions are either *symmetric* (for $n = 0, 2, 4, \ldots$) or *antisymmetric* (for $n = 1, 3, \ldots$). We say that the wave functions for $n = 0, 2, 4, \ldots$ are *even* and for $n = 1, 3, \ldots$ are *odd*. This is a situation that is encountered in many other problems and which determines several properties of the system, such as the probability of transition from one state to another.

EXAMPLE 37.2

Zero-point energy of a simple harmonic oscillator.

▷ The **zero-point energy** of the oscillator is $\frac{1}{2}\hbar\omega$, and is again a consequence of the uncertainty principle. This can be seen very easily as follows. For the lowest, or ground, energy state the amplitude of the oscillations is very small and we can make $\Delta x \approx [(x^2)_{\text{ave}}]^{1/2}$ and $\Delta p \approx [(p^2)_{\text{ave}}]^{1/2}$, where $(x^2)_{\text{ave}}$ and $(p^2)_{\text{ave}}$ are the average values of x^2 and p^2. Applying the uncertainty relation $\Delta x \Delta p \approx \hbar$ we get $(p^2)_{\text{ave}} \approx (\Delta p)^2 \approx \hbar^2/(\Delta x)^2$, which is valid only in order of magnitude and only for the ground state. Then the total energy of the oscillator $E = (p^2)_{\text{ave}}/2m + \frac{1}{2}k(x^2)_{\text{ave}}$, can be written in the form

$$E \simeq \frac{\hbar^2}{2m}\frac{1}{(\Delta x)^2} + \frac{1}{2}k(\Delta x)^2$$

To obtain the minimum energy compatible with the uncertainty relation, we solve the equation $\mathrm{d}E/\mathrm{d}(\Delta x) = 0$. The result is $(\Delta x)^2 = \hbar/(mk)^{1/2}$, which, when inserted in the expression for the energy, gives $E \sim \hbar(k/m)^{1/2} = \hbar\omega$ for the zero-point energy. There is a discrepancy of a factor $\frac{1}{2}$ because of the approximate nature of our calculation.

37.9 Potential barrier penetration

The fact that a wave function may extend beyond the classical limits of motion gives rise to an important phenomenon called **potential barrier penetration.** Consider the potential represented in Figure 37.14. The potential energy is zero for $x < 0$ and $x > a$ and has the value E_0 for $0 < x < a$. It is called a **potential barrier**. The cases $E < E_0$ and $E > E_0$ will be considered separately. Classical mechanics requires that a particle coming from the left with an energy $E < E_0$ should be reflected back at $x = 0$. However, when we consider the problem according to quantum mechanics by obtaining the solution of Schrödinger's equation for regions (I), (II) and (III), we find that the wave function has, in general, the form illustrated in Figure 37.15. The wave function ψ_1 corresponds to free particles with momentum $p = (2mE)^{1/2}$. The wave function ψ_2 in the forbidden region decays exponentially. Because ψ_2 is not yet zero at $x = a$, the wave function continues into region (III) with the free particle wave function ψ_3. This wave function represents *transmitted* free particles that have the same energy and momentum as the incident particles, but an amplitude which, in general, is smaller. Since ψ_3 is not zero, there is a finite probability of finding the particle in region (III). In other words,

it is possible for a particle to go through a potential barrier even if its kinetic energy is less than the height of the potential barrier.

When $E > E_0$, all particles should cross the potential barrier and reach the right-hand side. However, according to quantum mechanics, some particles are reflected at $x = 0$ and at $x = a$. Hence the wave functions in the three regions are

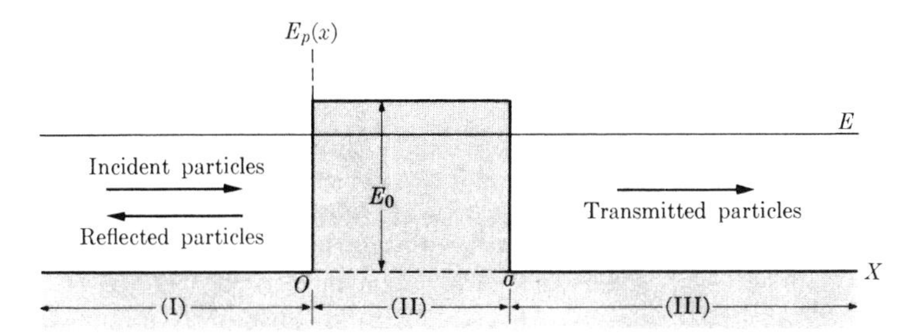

Figure 37.14 Rectangular potential barrier of width a and height E_0.

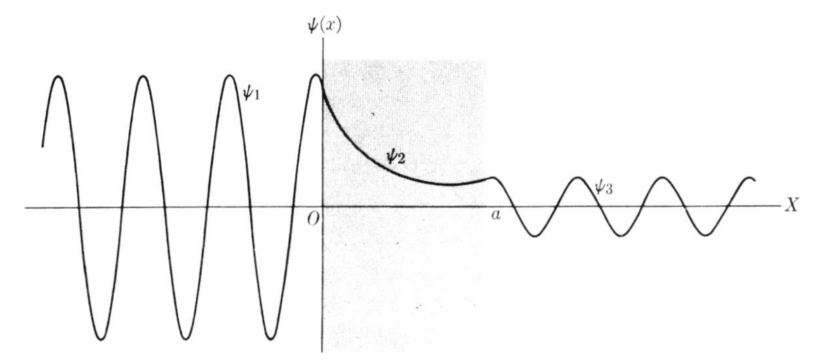

Figure 37.15 Wave function corresponding to the potential barrier of Figure 37.14 for an energy less than the height of the barrier ($E < E_0$).

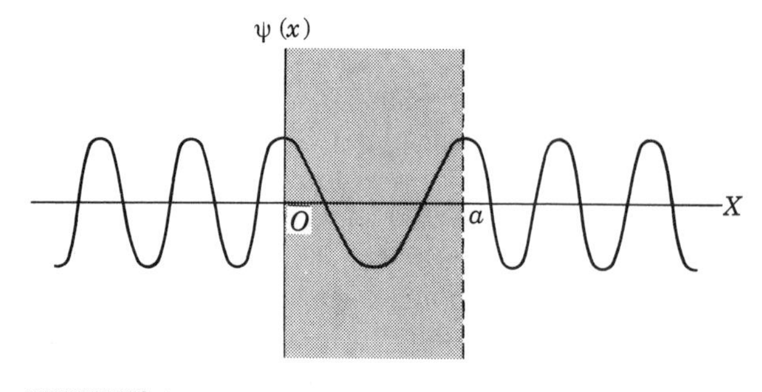

Figure 37.16 Wave function for a potential barrier when the energy of the particle is higher than the height of the barrier ($E > 0$).

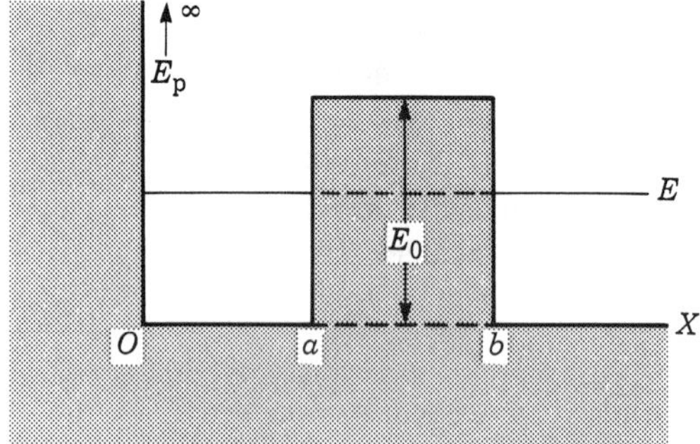

Figure 37.17 Potential wall with barrier of height E_0.

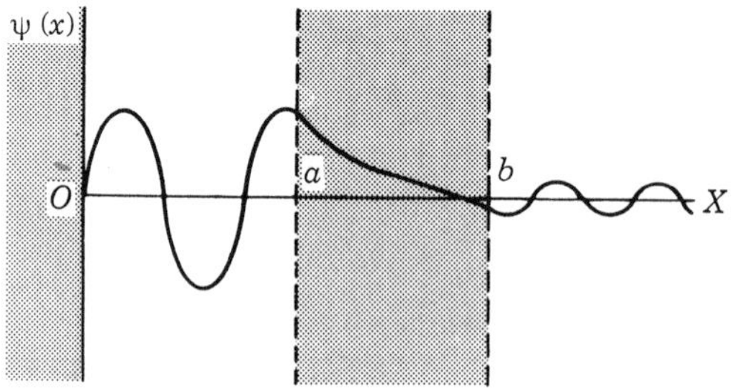

Figure 37.18 Wave function corresponding to the potential of Figure 37.17.

now free particle wave functions, but in region II the momentum of the particles while they are crossing the potential barrier is smaller than at either side and the wavelength is larger (Figure 37.16) because the particles are slowed down while going through the barrier.

An interesting case is that of a particle in a potential of the form illustrated in Figure 37.17 with a potential wall at $x = 0$ and a potential barrier in the region $a < x < b$. Consider the case where the energy E is smaller than the potential barrier height E_0. Classically, if the particle is initially in the region $0 < x < a$, it cannot escape from the well or, if it is initially in the region $x > b$, it cannot fall into the well. However, in accordance with our previous discussion, the wave function resulting from the solution of Schrödinger's equation should be as shown in Figure 37.18. That means that for a particle initially in the well ($x < a$ and $E < E_0$), there is a finite probability of escaping (or 'sneaking out') from the well and reaching the region $x > b$ without having to surmount the potential barrier. In Chapters 39 and 40 we will encounter some examples of potential barrier penetration such as the emission of α-particles by a nucleus. Barrier penetration shows once more that our sensorial perception of a macroscopic 'particle' is not applicable to atomic and subatomic particles.

EXAMPLE 37.3

Inversion motion of the nitrogen atom in an NH_3 molecule.

▷ The ammonia molecule, NH_3, is a pyramid with the nitrogen atom at the vertex and the three hydrogen atoms at the base, as shown in Figure 37.19(a). The nitrogen atom may

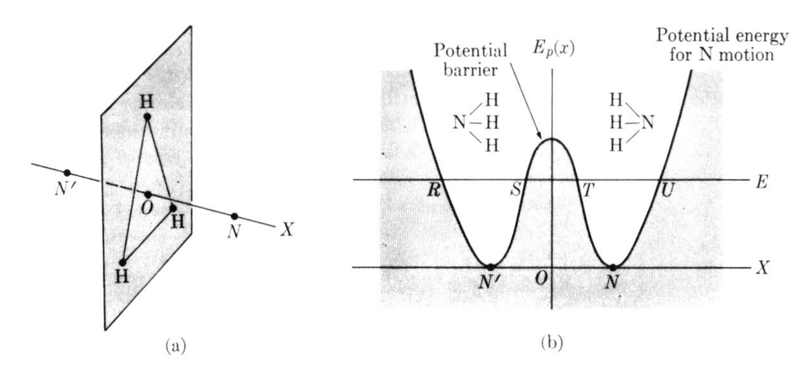

Figure 37.19 (a) Inversion motion of the nitrogen atom in the ammonia molecule. (b) Potential energy for the inversion motion.

be at *either* of the two symmetric equilibrium positions, N and N', on opposite sides of the base of the pyramid. Since both N and N' must be equilibrium positions, the potential energy for the motion of the nitrogen atom along the axis of the pyramid must also be symmetric and have two minima as indicated in Figure 37.19(b), with a potential barrier between N and N'. If the nitrogen atom is initially at N, it may eventually leak through the potential barrier and appear at N'. Thus when the energy is less than the height of the potential barrier, such as the energy level E in the figure, the motion of the nitrogen atom is composed of an oscillatory motion between R and S or between T and U, depending on which side of the plane it happens to be, plus a much slower oscillatory motion between the two classical regions passing through the potential barrier. The frequency of this second motion is 2.3786×10^{10} Hz for the ground state of NH_3. It is this second type of motion that has been used in atomic clocks (see Section 2.3).

From Figure 37.19(a) it can be seen that the potential energy for the motion of the N atom along the X-axis can be considered as two oscillators separated by the potential barrier. If the nitrogen atom were restricted to stay on either side of the potential barrier its wave functions would be similar to those shown in Figure 37.12 for the harmonic oscillator. Therefore in the ground state, $n = 1$, the wave function would be like ψ_R (Figure 37.20(a)) if the nitrogen atom is on the right and as ψ_L if it is on the left. However, due to the potential barrier penetration the wave functions must join smoothly across the barrier. Thus the actual wave functions resemble more the combinations $\psi_R + \psi_L$ and $\psi_R - \psi_L$, which are symmetric and antisymmetric relative to the plane of hydrogen atoms, as indicated in Figure 37.21. The corresponding probability distributions are shown in Figure 37.22. The two probability distributions are almost identical but $|\psi_s|^2$ is finite at $x = 0$ while $|\psi_a|^2$ is zero. Therefore, they correspond to two different energy levels separated a small amount ΔE, which is related to the frequency of inversion ν by $\Delta E = h\nu$. Since $\nu = 2.3786 \times 10^{10}$ Hz, we then have that $\Delta E = 9.84 \times 10^{-5}$ eV, a result that can be verified experimentally.

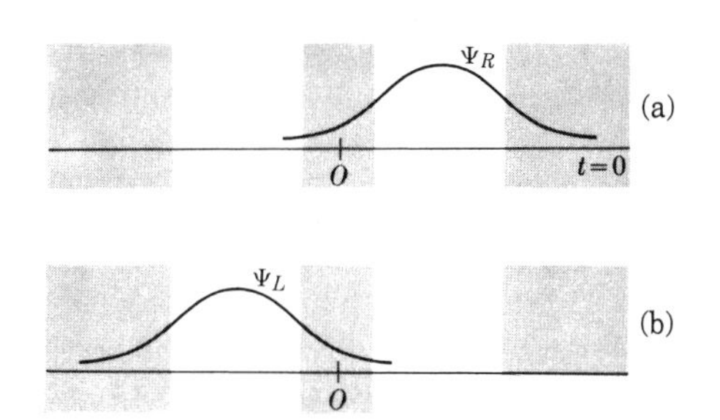

Figure 37.20 Description of the inversion motion in NH_3 by means of time-dependent wave functions.

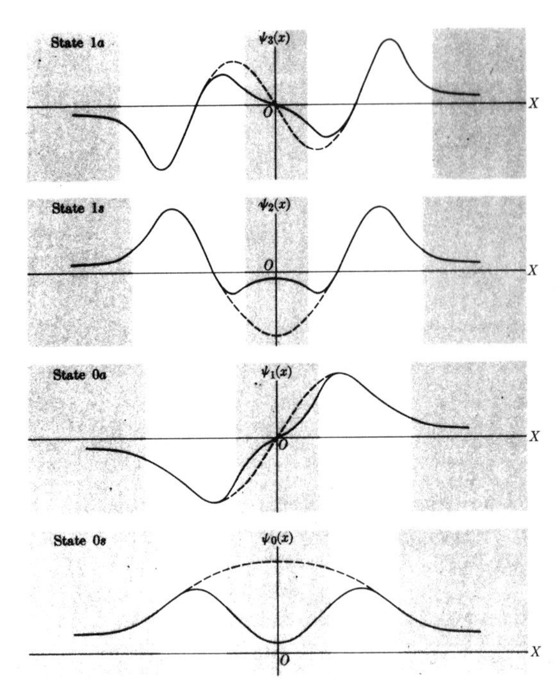

Figure 37.21 Symmetrical and antisymmetrical wave functions corresponding to the ground energy levels of the inversion motion in NH_3.

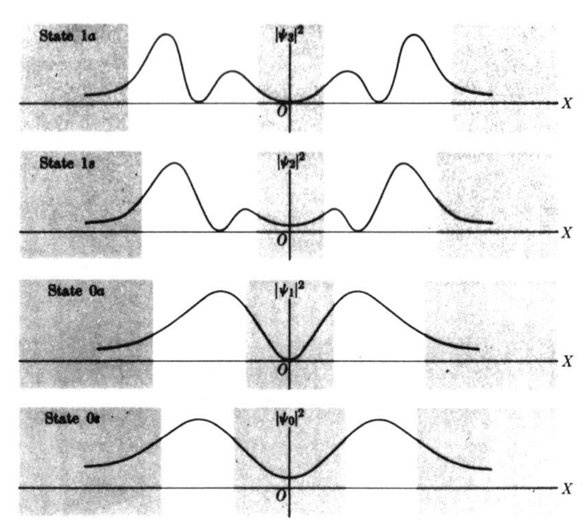

Figure 37.22 Probability densities corresponding to the wave functions shown in Figure 37.21.

QUESTIONS

37.1 What are the differences between Schrödinger's time-dependent equation 37.1 and the wave equation 28.11?

37.2 Which of the following are solutions of Schrödinger's time-dependent equation: (a) $e^{i(kx-\omega t)}$, (b) $e^{-i\omega t}\sin kx$, (c) $\sin(kx \pm \omega t)$ and (d) $\cos(kx \pm \omega t)$.

37.3 What is the difference between the states of a particle described by the wave functions $e^{i(kx-\omega t)}$ and $e^{i\omega t}\sin kx$?

37.4 What is a degenerate state?

37.5 Why must the wave function be zero at a potential wall?

37.6 How does the excitation energy of a particle in a bound state vary with the mass of the particle and the size of the region where it is confined?

37.7 Can a particle confined inside a potential well also be outside the well?

37.8 Describe the kind of potential energy where a particle can be in (a) only discrete energy states, (b) discrete and continuous energy states.

37.9 Why, if the potential energy is symmetric, may the wave functions be symmetric or antisymmetric?

37.10 Explain why potential barrier penetration is possible. Analyze how the height and the width of the potential barrier affect the penetration of the barrier.

37.11 How does the probability of penetrating a potential barrier vary as (a) the width, (b) the height of the barrier increases?

37.12 Does potential barrier penetration by a subatomic 'particle' show that its motion must be considered in different terms from that of a macroscopic 'particle'?

37.13 Consider the **potential step** of Figure 37.23. Discuss the wave functions of a particle moving from the extreme right toward the left with energy (a) greater than and (b) less than E_0. Repeat for a particle moving from the extreme left toward the right with an energy greater than E_0.

37.14 Consider the potential energy of Figure 37.24. Discuss the wave functions of a

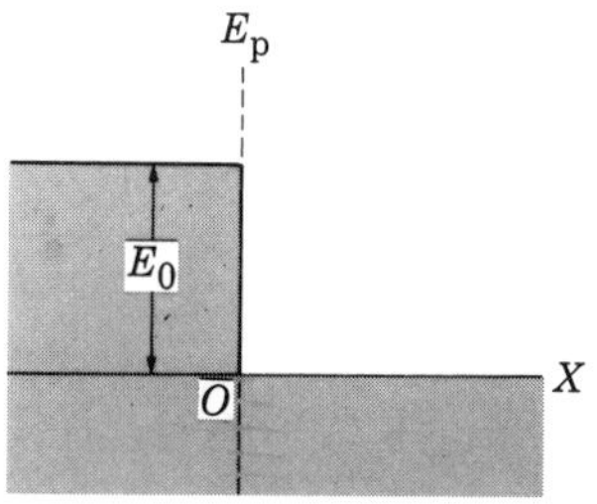

Figure 37.23

particle moving from the extreme right toward the left with energy (a) greater than and (b) less than E_0.

37.15 Write Equation 37.20 explicitly in terms of E, using the values of k_i and α. What kind of equation results?

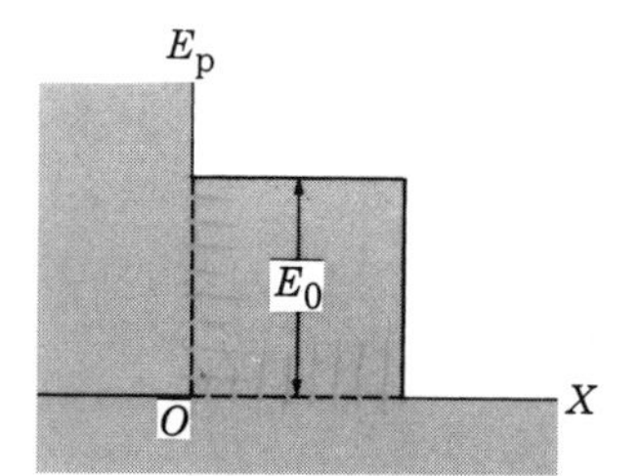

Figure 37.24

37.16 Write the time-dependent wave function corresponding to Equation 37.8.

37.17 Why is the zero-point energy of a particle in a cubical box three times that of a particle in a one-dimensional box?

PROBLEMS

37.1 Consider an electron in a one-dimensional potential box of width 2.0 Å. (a) Calculate the zero-point energy. (b) Using the uncertainty principle, discuss the effect of incident radiation designed to locate the electron with a 1% accuracy (that is, $\Delta x = 0.02$ Å).

37.2 Estimate the zero-point energy of (a) an electron, (b) a proton and (c) a neutron confined inside a region of size 10^{-14} m, which is the order of magnitude of nuclear dimensions. (d) Estimate the respective wavelengths of the electron. On the basis of this comparison, discuss the possibility that an electron, a proton or a neutron can exist within a nucleus.

37.3 (a) Verify that the energy levels and wave functions of a particle moving in the XY-plane within a two-dimensional potential box of sides a and b are

$$E = \left(\frac{\hbar^2\pi^2}{2m}\right)\left(\frac{n_1^2}{a^2} + \frac{n_2^2}{b^2}\right),$$

$$\psi = C \sin\left(\frac{n_1 \pi x}{a}\right) \sin\left(\frac{n_2 \pi y}{b}\right)$$

(b) Discuss the degeneracy of energy levels when $a = b$. (Hint: refer to Sections 34.6 and 37.5.)

37.4 Calculate the energy levels and wave functions for a particle moving within a three-dimensional box of sides a, b and c. Note that this is an extension of the previous problem.

37.5 Calculate the zero-point energy and the spacing for the energy levels of (a) a one-dimensional harmonic oscillator, with a frequency of 400 Hz, (b) a three-dimensional harmonic oscillator with the same frequency.

37.6 A three-dimensional harmonic oscillator, oscillating with a frequency ω, may be considered as a linear combination of three oscillators, each vibrating alone one of the X-, Y- and Z-axes. (a) Justify that the energy of the oscillator is $E = (n + 3/2)\hbar\omega$, where $n = n_1 + n_2 + n_3$ and n_1, n_2 and n_3 correspond to oscillations along the three axes. (b) Verify that the degenercy of each state is $g = \frac{1}{2}(n + 1)(n + 2)$. (c) Explain why the zero point energy of such an oscillator is $\frac{3}{2}\hbar\omega$.

37.7 A particle moves along the X-axis in a region of a potential energy $E_p(x) = -E_0 e^{-\alpha x^2}$. (a) Plot $E_p(x)$. (b) Make a sketch of the wave functions when the total energy is negative and positive. (c) Do you expect to have quantized energy levels? Repeat for the potential energy $E_p = E_0 e^{\alpha x^2}$.

37.8 What is the effect on the energy level of a one-dimensional potential well as the depth of the well (a) decreases and (b) increases? Repeat the analysis for the case where the width changes but the depth remains constant.

37.9 Discuss the effect on the energy levels and wave functions of a potential well of range a when a hard core of range b is added, as shown in Figure 37.25.

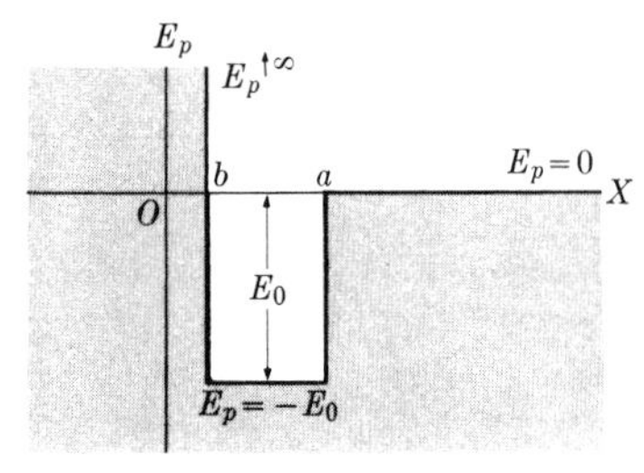

Figure 37.25

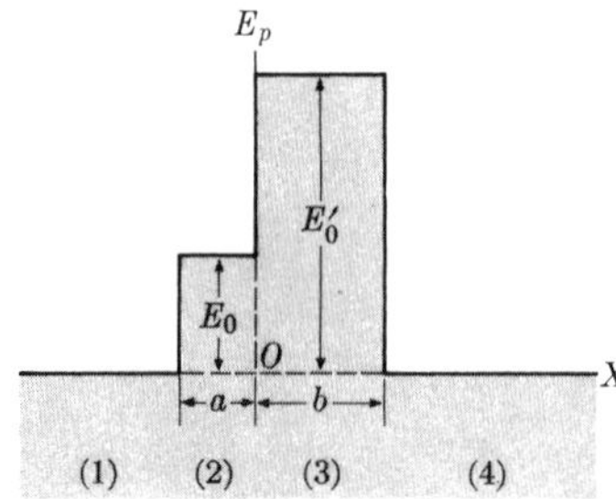

Figure 37.26

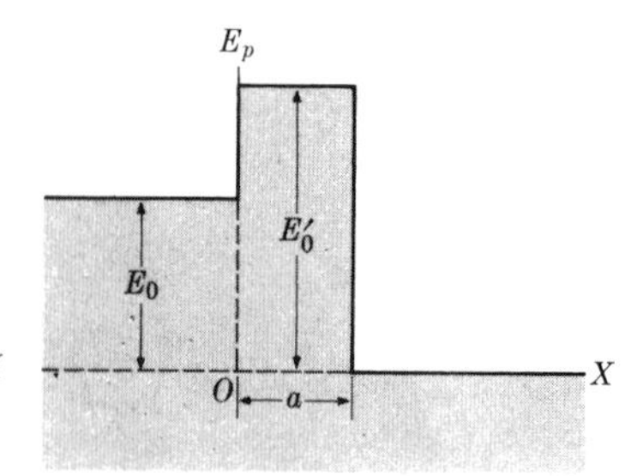

Figure 37.27

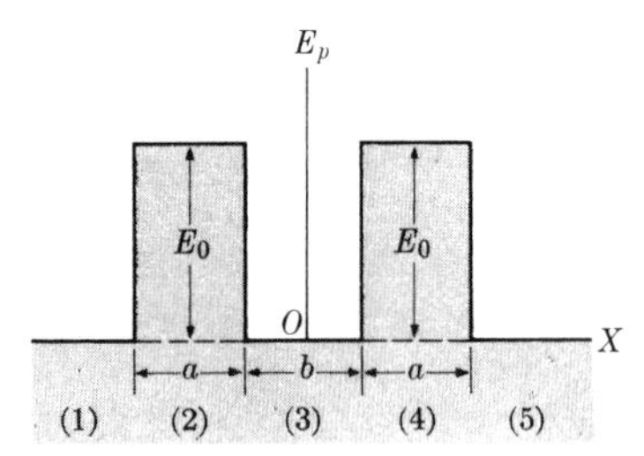

Figure 37.28

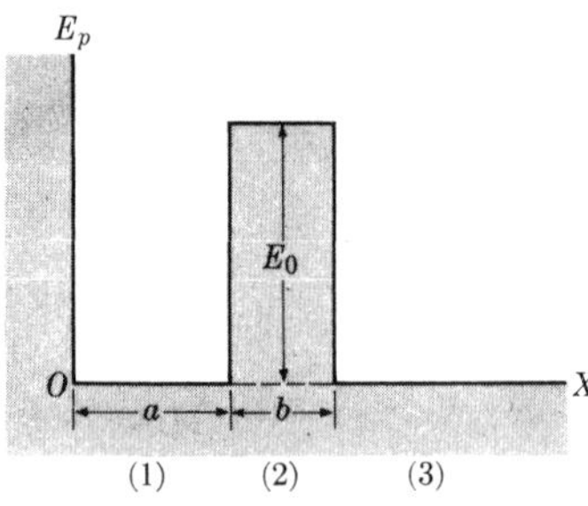

Figure 37.29

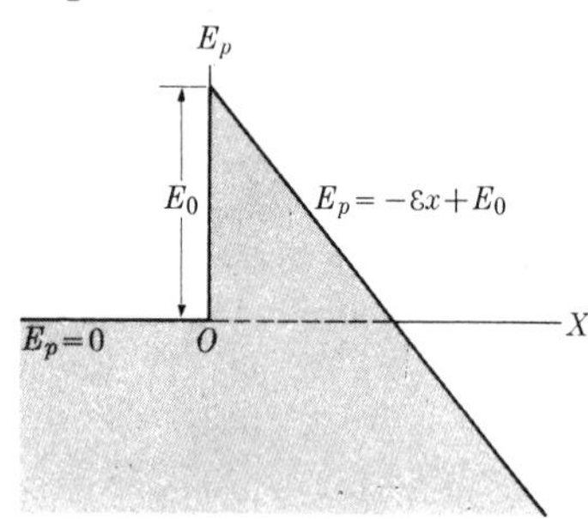

Figure 37.30

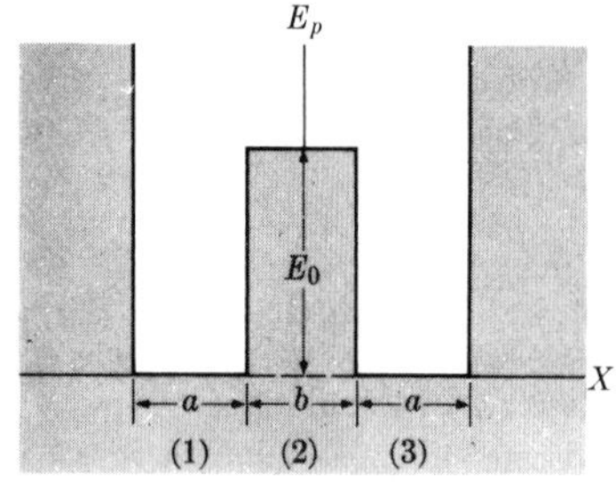

Figure 37.31

37.10 Estimate the well depth E_0 for a neutron in a one-dimensional rectangular well of width 3×10^{-15} m, given that its binding energy E_b is 2.0 MeV, and assuming that only one energy level is possible.

37.11 Sketch the wave functions in each of the regions of the potential energy shown in Figure 37.14 for (a) $E < E_0$ and (b) $E > E_0$. This requires imposing the continuity of ψ on the boundaries of the barrier.

37.12 Sketch the wave functions in each of the regions of the potential energy shown in Figure 37.26. Consider that the incident particles come from the left and discuss the three distinct cases of $E < E_0$, $E_0 < E < E_0'$, and $E > E_0'$.

37.13 Repeat the preceding problem for the case where the particles are initially incident from the right.

37.14 Consider the potential energy shown in Figure 37.27. Discuss the general shape of the wave function for a particle incident from the right when its energy is (a) $E < E_0$, (b) $E_0 < E < E_0'$, and (c) $E > E_0'$. Repeat the last two cases for a particle incident from the left.

37.15 Sketch the wave function in each of the regions of the potential energy shown in Figure 37.28. Consider a particle incident from the left, first with $E < E_0$, and then with $E > E_0$. Is there a possible stationary state for a particle initially in region (3)?

37.16 Sketch the wave function in each of the regions of the potential energy shown in Figure 37.29 for different energies of the particle.

37.17 Given the potential energy shown in Figure 37.30, sketch the wave functions for a particle coming from the right and having a total energy that is (a) negative, (b) between zero and E_0 and (c) larger than E_0. How does the wavelength of the particle change as the particle moves in the region $x > 0$ and $x < 0$, if the total energy is larger than E_0? Repeat for a case in which the particle comes from the left. This situation approximately corresponds to the potential energy of an electron in a metal when an external electric field $\mathscr{E}$ is applied.

37.18 Given the potential energy shown in Figure 37.31, sketch the wave function for each region and indicate the energy levels for $E < E_0$ and $E > E_0$. Note that the wave functions are symmetric (even) and antisymmetric (odd). Discuss the effect on adjacent energy levels as either E_0 or b increases.

38 Atoms, molecules and solids

Wolfgang Pauli was one of the brightest original contributors to the development of quantum mechanics. In 1925 he formulated the **exclusion principle**, which states that no two electrons in an atom can exist in the same quantum state. Pauli also made other important contributions, such as the elaboration of the concept of electron spin (1924), the initial formulation of field theory (1927), the postulate that in β-decay a particle of zero mass (neutrino) must be emitted (1929) to conserve energy and momentum, the connection between spin and statistics and many others.

38.1 Introduction

One of the most important and far-reaching applications of quantum mechanics is to the analysis of the properties of atoms, molecules and solids. Recall that atoms consist of a number of electrons moving in a region with a radius of the order of 10^{-10} m around a positively charged nucleus composed of protons and neutrons occupying a region whose radius is of the order of 10^{-14} m. The motion of the electrons is due to their electric interaction with the nucleus. In Section 23.4 we used Newtonian mechanics, combined with Bohr's quantization of angular momentum, to obtain the energy levels of electrons in atoms. However, to analyze the electronic structure of atoms, molecules and solids properly, it is necessary to use the methods of quantum mechanics.

38.2 Angular wave function under a central force

The simplest atom is one composed of one electron, charge $-e$, moving in the electric field produced by a nucleus composed of Z protons, charge $+Ze$, with

$Z = 1$ for H, $Z = 2$ for He^+, $Z = 3$ for Li^{++} and so on. These atoms are called *hydrogen-like*. The electronic potential energy is then $E_p(r) = -Ze^2/4\pi\varepsilon_0 r$, corresponding to a central force. As for any central force problem, the energy and the angular momentum of the electron are constants of motion.

An atom is a three-dimensional system and the electronic motion is not in a plane, as Bohr's model might suggest because of its similarity to planetary motion. To locate an electron in an atom we need three coordinates x, y, z, (Figure 38.1). Then, to discuss its motion, we have to solve the three-dimensional Schrödinger equation

$$-\frac{\hbar^2}{2m}\left(\frac{d^2\psi}{dx^2} + \frac{d^2\psi}{dy^2} + \frac{d^2\psi}{dz^2}\right) + E_p(r)\psi = E\psi \tag{38.1}$$

This is a rather complex mathematical problem. However we can learn a lot without attempting to solve this equation.

Because the electronic potential energy depends only on the distance r, an isolated atom is spherically symmetric and the choice of axes XYZ is arbitrary. (The symmetry might be destroyed, however, when an atom is placed in an external electric or magnetic field or is surrounded by other atoms.) Therefore, to discuss an electron in an atom it is better to replace the coordinates x, y, z by the distance r and the angles θ and ϕ, as shown in Figure 38.1. The set r, θ, ϕ are called the *spherical coordinates* of the electron. Then instead of writing the wave function as $\psi(x, y, z)$ we shall write $\psi(r, \theta, \phi)$.

The wave function $\psi(r, \theta, \phi)$ for a single electron in a central field with potential energy $E_p(r)$ can be written as a product of two factors, one that depends on the distance of the electron from the origin and another that depends on the orientation of the position vector r, given by the angles θ and ϕ. Thus we may write the wave function as

$$\psi(r, \theta, \phi) = R(r)Y(\theta, \phi) \tag{38.2}$$

The radial part $R(r)$ is determined by the potential energy $E_p(r)$ corresponding to the force acting on the electron. However the angular part $Y(\theta, \phi)$, as a consequence of the spherical symmetry of the central force, is independent of the potential energy. In other words, the angular functions $Y(\theta, \phi)$ are the same for all central force problems and are determined entirely by the angular motion corresponding to the magnitude and the Z-component of the angular momentum of the electron.

As explained in Chapter 23, the magnitude of the angular momentum is determined by the quantum number l and the Z-component is determined by m_l so that $L^2 = l(l+1)\hbar^2$ and $L_z = m_l\hbar$ with $m_l = 0, \pm 1, \pm 2, \ldots, \pm l$ or a total of $2l + 1$ values. For that reason the angular functions corresponding to specific values of L^2 and L_z will be designated as $Y_{lm}(\theta, \phi)$. The angular functions for $l = 0$, 1 and 2 or *s*, *p* and *d* states have been represented in Figures 38.2, 38.3 and 38.4 in a form suitable to the discussion of molecular binding. It can be seen that, for $l = 0$ or *s*-states (Figure 38.2), the angular wave function is spherically symmetric; that is, there is no preferred direction for the motion of the electron. This is

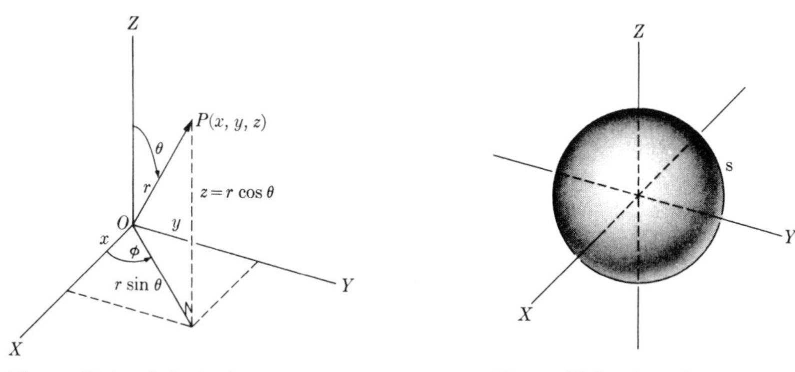

Figure 38.1 Spherical coordinates.

Figure 38.2 Angular wave function for *s*-states ($l = 0$).

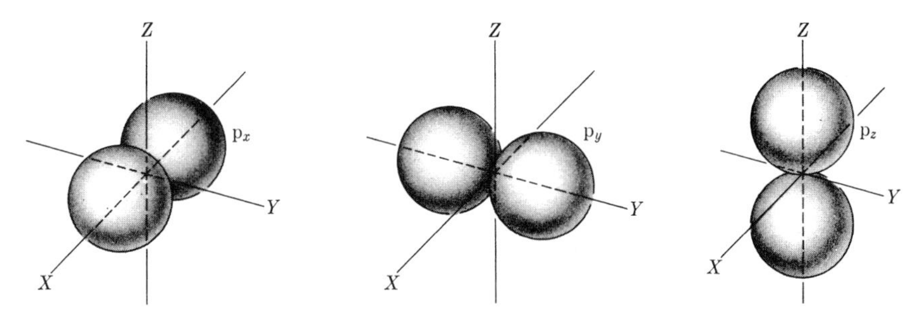

Figure 38.3 Angular wave functions for *p*-states ($l = 1$).

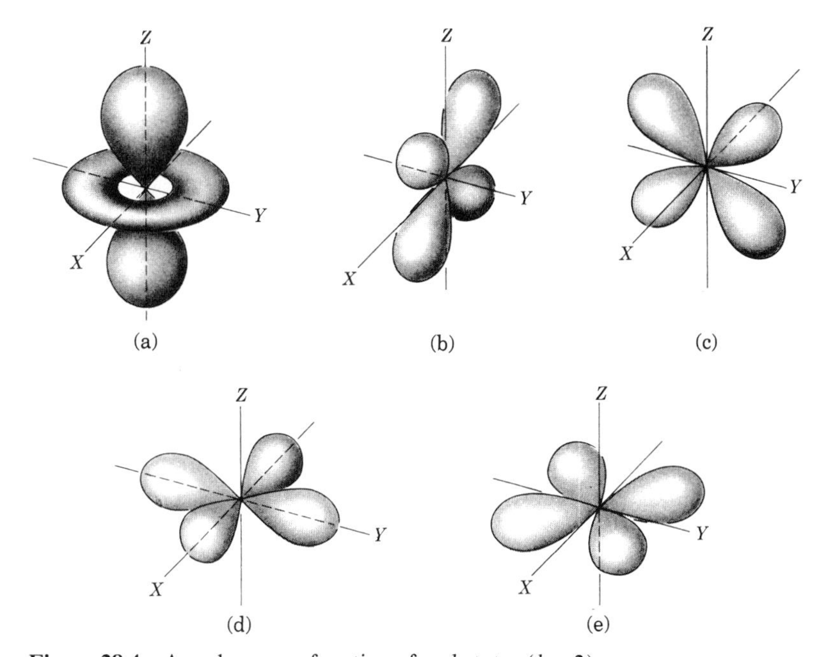

Figure 38.4 Angular wave functions for *d*-states ($l = 2$).

understandable because, if the angular momentum is zero, there is no preferred orientation of the orbit of the electrons.

For $l = 1$ or p-states there are three angular wave functions corresponding to the three possible orientations of the angular momentum or the three values $m_l = 0, \pm 1$. However, since the energy cannot depend on the orientation of the angular momentum or on the arbitrary choice of the axes XYZ, *any* linear combination of the angular wave functions is also an angular wave function for $l = 1$. The three p-wave functions represented in Figure 38.3, designated by p_x, p_y and p_z, correspond to preferred motion along each of the chosen coordinate axes with p_z corresponding to $m_l = 0$ while p_x and p_y are linear combinations of $m_l = \pm 1$.

For $l = 2$ or d-states (Figure 38.4) there are five different angular functions corresponding to the five possible orientations of the angular momentum or $m_l = 0, \pm 1, \pm 2$. As can be seen from the figure, the angular distribution of these states is more complex than for $l = 1$. For larger values of l, the situation becomes even more complex.

The form of the angular momentum wave functions is very important for describing chemical binding, as we shall see in Sections 38.4 and 38.5.

38.3 Atoms with one electron

Consider again a one-electron, or hydrogen-like, atom. The Coulomb potential energy of the electron is $E_p = -Ze^2/4\pi\varepsilon_0 r$, and has been represented in Figure 38.5. We recall from Note 9.2 that, for motion under a central force, we have to consider an additional term to the potential energy given in Equation 9.39 or $L^2/2mr^2$. Thus, using $L^2 = l(l+1)\hbar^2$, the *effective* potential energy of the electron is

$$E_{\text{eff}}(r) = -\frac{Ze^2}{4\pi\varepsilon_0 r} + \frac{l(l+1)\hbar^2}{2mr^2} \tag{38.3}$$

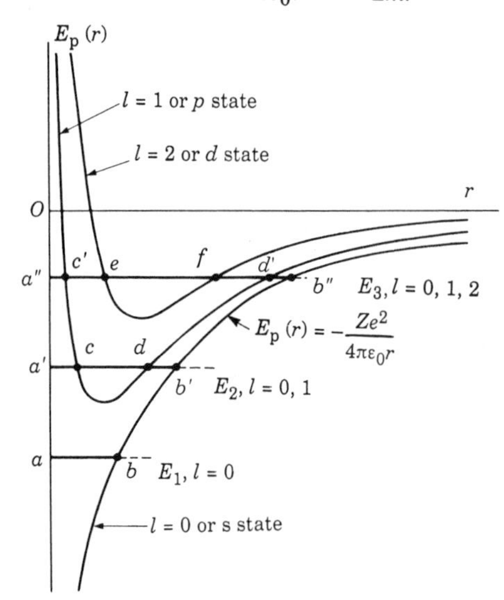

Figure 38.5 Energy levels in a Coulomb potential.

For $l = 0$ the effective potential energy reduces to the Coulomb term $-Ze^2/(4\pi\varepsilon_0)r$, which has no minimum. However for $l = 1, 2, \ldots$ the effective potential energy increases very rapidly as $r \to 0$ and has a minimum, which is different for each value of l, as shown in Figure 38.5 for $l = 1$ and 2. We have already described this situation in Figure 9.27.

Designate the energy of the electron corresponding to $n = 1$ by E_1. From Figure 38.5 we conclude that the classical turning points of the electron motion are a and b, which means that the electron has zero angular momentum and can move between $r = 0$ and some value of r. That is, an electron with $n = 1$ can be only in an s-state.

For the energy E_2, or $n = 2$, the electron can move between a' and b' if $l = 0$ or between c and d if $l = 1$. Thus the electron can be in an s- or a p-state. Similarly, for an energy like E_3 or $n = 3$, the electron can move between a'' and b'' if $l = 0$ (s-state), between c' and d' if $l = 1$ (p-state), and between e and f if $l = 2$ (d-state), and so on.

From this analysis we can derive several important conclusions: (1) for all $l = 0$ or s-states, the electron can move between $r = 0$ and some other value of r, which depends on the energy; thus in the s-states the electron can be very close to the nucleus and we say that the orbit is **penetrating**; (2) as the energy increases, more angular momentum states are possible, conforming to the rule $l = 0, 1, 2, \ldots, n - 1$, that we have mentioned before; and (3) as the energy and angular momentum increase, the region in which it is more probable to find the electron gets farther away from the nucleus ($r = 0$). These characteristics of electronic motion are reflected in many important atomic properties. For example, s-electrons are much more sensitive to the size, shape and internal structure of the nucleus than electrons with higher values of angular momentum.

Because the region in which the electron can move depends on the energy and the angular momentum but not its direction, the radial wave functions depend on n and l, but not on m_l, and are designated $R_{nl}(r)$. In Figure 38.6 some radial functions are represented for a few values of n and l for the Coulomb potential. In Figure 38.7 the corresponding radial probability distributions of the electron are given (see Example 38.1). For comparison, the Bohr radius of the orbit is indicated in each case. It can be seen that for $l = 0$ there is a larger probability of finding the electron near the nucleus than for larger values of l, and that the most probable region is farther from O as n and l increase.

By solving Schrödinger's equation with the Coulomb potential it can be verified that the energy of the stationary states depends only on n. (This might not be true for other forms of the potential energy.) Also the possible energies are

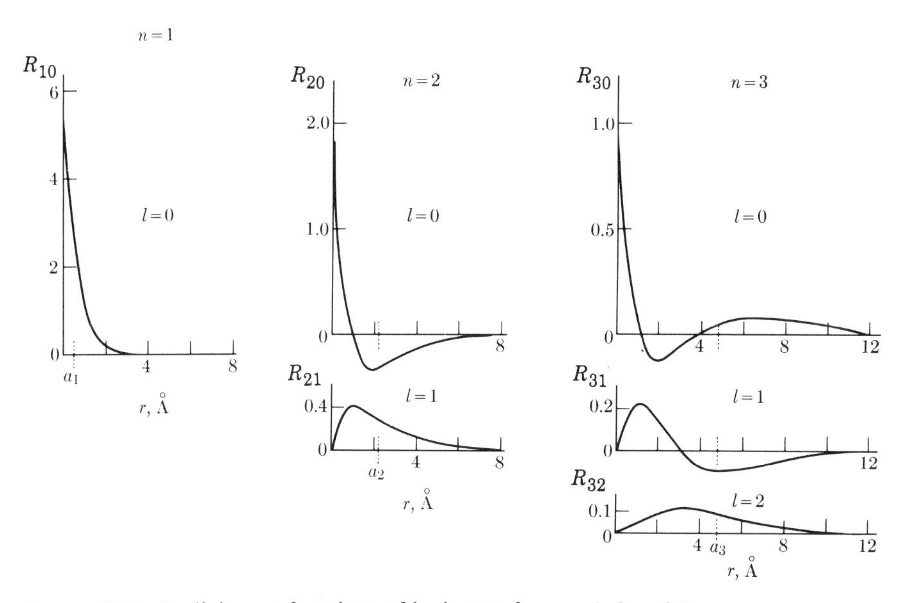

Figure 38.6 Radial wave functions of hydrogen for $n = 1$, 2 and 3.

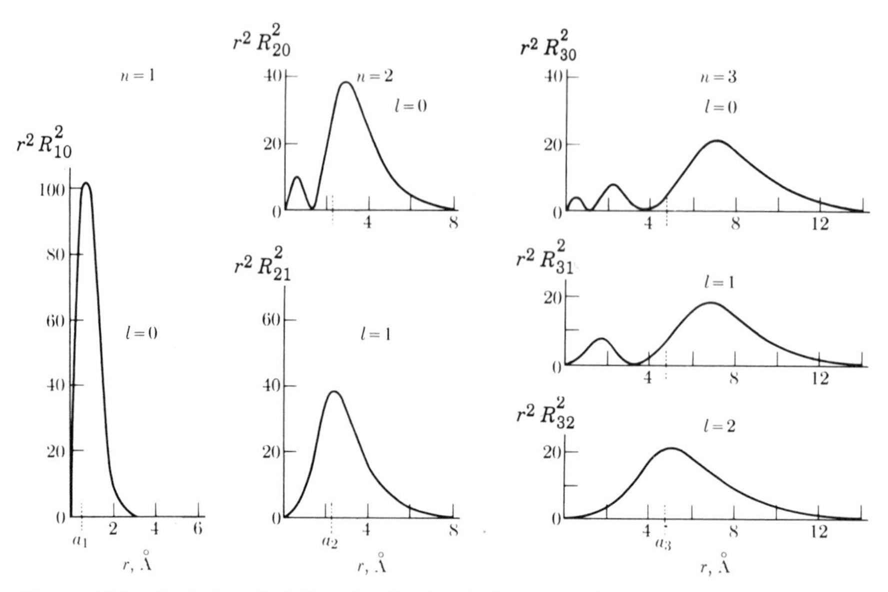

Figure 38.7 Radial probability distribution in hydrogen for $n = 1$, 2 and 3.

given by the same expression, Equation 23.13, obtained using Bohr's theory,

$$E_n = \frac{m_e e^4 Z^2}{2(4\pi\varepsilon_0)^2\hbar^2 n^2} = -\frac{RhcZ^2}{n^2} \qquad (n = 1, 2, 3, \ldots) \tag{38.4}$$

where $R = 1.0974 \times 10^7\ \text{m}^{-1}$ is *Rydberg's constant* as defined in Section 31.4. To be more precise, the electron mass m_e must be replaced by the reduced mass (recall Section 13.4). In this respect, Bohr's theory and quantum mechanics agree completely. That there is a minimum energy E_1 reflects the need (already encountered in the potential box and the potential well) to join the wave function in the classical region of motion smoothly with the decreasing function in the forbidden region. Such a need requires a minimum wavelength and therefore a certain momentum for the electron. This minimum energy is also required by the uncertainty principle (see Example 36.5).

Thus, in a Coulomb field each energy level, which corresponds to a given n, contains n different angular momentum states, all with the same energy and with l ranging from 0 to $n - 1$, as shown in Figure 38.8. These levels are indicated by *ns*, *np*, *nd* etc. In a more refined theory of one-electron atoms, which takes into account other effects (such as relativistic and spin–orbit corrections discussed in Chapter 23), the different angular momentum states corresponding to the same n appear with different energies.

If the force is not inverse square, those levels that have the same value of n but different angular momenta (e.g. levels 4*s*, 4*p*, 4*d* etc.) do not necessarily all have the same energy. Thus under central forces, the energy depends in general on n and l, but not on m_l, since in a central field of force the orientation of the orbit or of the Z-axis is irrelevant. Also, in general, the more penetrating the orbit or the lower the value of l, the lower also is the value of the energy, as shown schematically in Figure 38.9, because the particle gets closer to the center of force.

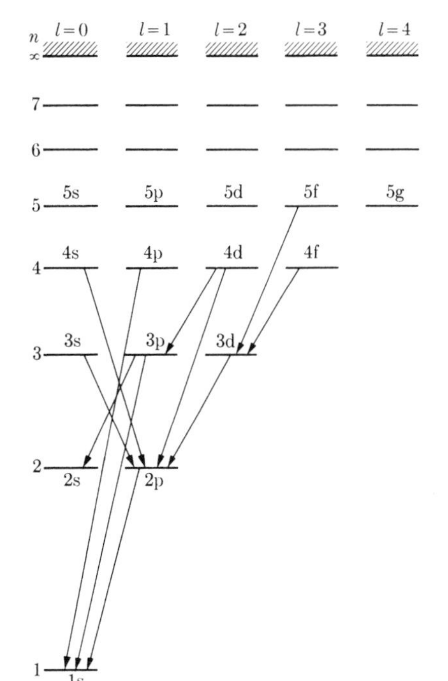

Figure 38.8 Transitions among different angular momentum states.

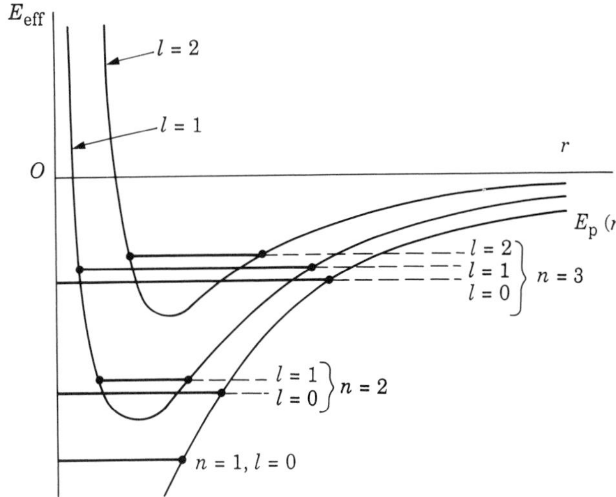

Figure 38.9 Energy levels in a non-Coulombic central force. The curves represent the effective potential energy for different values of *l*.

The fact that each level in a hydrogen-like atom is composed of several angular momentum states is, however, important from the point of view of radiative transitions, which are restricted by *selection rules*. For motion in a central force potential, the selection rules for electric dipole transitions, as shown in Figure 38.8, are

$$\Delta l = \pm 1, \quad \Delta m_l = 0, \pm 1 \tag{38.5}$$

These selection rules are imposed by the law of conservation of angular momentum since the emitted or absorbed photon carries a spin angular momentum of one (recall Section 30.6). Therefore the angular momentum of the atom must change by one unit to compensate for the angular momentum carried by the emitted or absorbed photon.

EXAMPLE 38.1

Radial probability distribution of an electron in a hydrogen atom.

▷ We may recall from Section 36.8 that the probability per *unit volume* of finding the electron at a given point is given by $|\psi(\mathbf{r})|^2$, which in general depends on the direction of $\mathbf{r}$. However, in central force problems it is more important to determine the probability of finding the electron at a certain distance r from the nucleus, regardless of the direction, since this is equivalent to determining the *size* of the electron's orbit. If we consider a shell of radius r and thickness $\mathrm{d}r$, its volume is $4\pi r^2\ \mathrm{d}r$. The probability of finding the electron on some point of the shell is then $(4\pi r^2\,\mathrm{d}r)|R_{nl}(r)|^2$, where we have to use only the radial part of the wave function. Then the probability per *unit length* of finding the electron between the distances r and $r + \mathrm{d}r$, regardless of direction, is $P_{nl}(r) = 4\pi r^2|R_{nl}|^2$. This probability was shown in Figure 38.7. The distance at which it is more probable to find the electron is obtained by making $\mathrm{d}P_{nl}(r)/\mathrm{d}r = 0$. For example, for the state $n = 1$, $l = 0$, it can be shown that the distance is a_1/Z, which is the same as the radius in Bohr's model (Section 23.4).

38.4 Atoms with two electrons

Except for hydrogen-like atoms and ions, all atoms contain several electrons. Of all many electron atoms, the simplest are those with two electrons, such as the negative hydrogen ion H^- ($Z = 1$), the helium atom He ($Z = 2$), the positive lithium atom Li^+ ($Z = 3$) and so on (Figure 38.10). We call these systems *helium-like* atoms or ions. The potential energy of the system comprises the interaction of each electron with the nucleus, which is attractive, and the interaction among the electrons, which is repulsive; that is

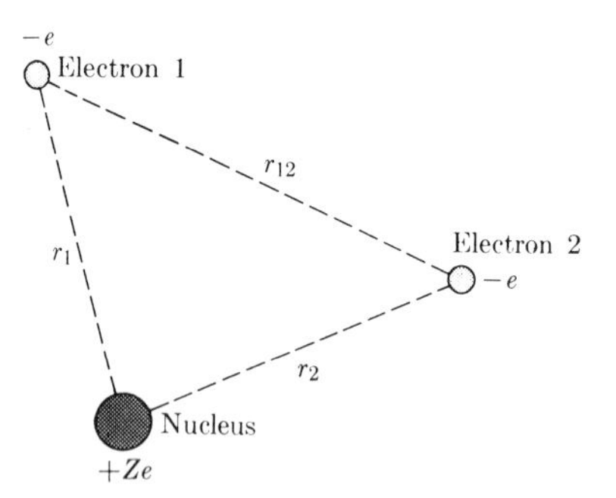

Figure 38.10 A helium atom ($Z = 2$) or helium-like ion ($Z > 2$).

$$E_p = -\frac{Ze^2}{4\pi\varepsilon_0 r_1} - \frac{Ze^2}{4\pi\varepsilon_0 r_2} + \frac{e^2}{4\pi\varepsilon_0 r_{12}} \qquad \textbf{(38.6)}$$

A new situation appears in this case, which is the coupling of the motion of the two electrons through their mutual interaction. Suppose for a time that we ignore the mutual interaction. We can then assume that the electrons move independently of each other. The energy of each electron may then be computed using Equation 38.4 for hydrogen-like atoms; that is,

$$E_n = -\frac{RhcZ^2}{n^2} = -Z^2E_H = -\frac{13.6}{n^2}Z^2 \text{ eV}$$

Thus, if each electron is in the $n = 1$ state, the energy of the helium atom ($Z = 2$) in the ground state would be

$$E_{He} = 2Z^2E_H = 2 \times 4 \times (-13.6) \text{ eV} = -108.8 \text{ eV}$$

The experimental value, however, is $E_{He} = -78.98$ eV. Thus, our approximation gives an energy that is too negative because we have ignored the repulsion of the two electrons, which tends to raise the energy of the atom.

One way of improving our calculation is to consider that each electron screens the charge of the nucleus by a certain amount $\mathscr{S}$, which results in reducing the effective atomic number of the nucleus. The energy of the atom may then be written as

$$E = 2(Z - \mathscr{S})^2 E_H \quad \textbf{(38.7)}$$

Comparing the theoretical value with the experimental value for the ground state of He, we obtain $\mathscr{S} = 0.32$. That is, the **nuclear screening effect** of each electron on the other is about one third of the electronic charge.

The calculation of the wave functions of helium-like atoms is a complex mathematical procedure, but we can get valuable information under the approximation of ignoring the electronic interaction. Suppose that electron 1 is in state a and electron 2 is in state b. Then the wave function of the system is (Figure 38.11(a)) $\psi_{ab}(1, 2) = \psi_a(1)\psi_b(2)$.

However, electrons are identical and indistinguishable and the wave function $\psi_{ab}(2, 1) = \psi_a(2)\psi_b(1)$ corresponds to exactly the same state of the atom with exactly the same energy (Figure 38.11(b)). Therefore, the most we can say is that one electron is in state a and other in state b. In other words, the wave function must be invariant for an exchange of electrons 1 and 2, except perhaps for a change of sign. This is called **exchange degeneracy**. This means that the wave function of the atom must be one of the two linear combinations (Figure 38.11(c))

$$\psi_{ab} = \psi_a(1)\psi_b(2) \pm \psi_a(2)\psi_b(1) \quad \textbf{(38.8)}$$

When the positive sign is used, the function ψ_{ab} is *symmetric* in the two electrons, and if the negative sign is used the function is *antisymmetric*. Each combination corresponds to a different space distribution of the electrons, and therefore also a different energy of the atom, when the electronic interaction is included.

Next, consider the spin of the electrons, which is $s = \frac{1}{2}$. The spin of the electrons can be combined symmetrically (↑↑), giving a total spin of $S = 1$ or antisymmetrically (↑↓), giving a total spin of $S = 0$. The total wave function of the system of two electrons must be the product of the orbital wave function ψ_{ab}, given by Equation 38.8, and the spin wave function. That is

$$\psi_{\text{total}} = (\text{orbital wave function}) \times (\text{spin wave function})$$

An examination of the energy levels of He reveals that states with symmetric orbital wave function always have $S = 0$ or an antisymmetric spin wave function, while those with antisymmetric orbital wave function always have $S = 1$ or a

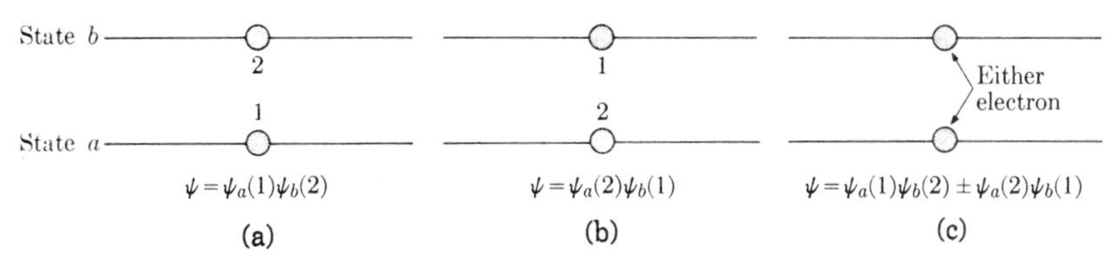

Figure 38.11 Symmetric and antisymmetric wave functions.

symmetric spin wave function. Thus it seems that the only states allowed in nature are

$$\psi_{\text{total}} = \begin{pmatrix} \text{symmetric orbital} \\ \text{wave function} \end{pmatrix} \times \begin{pmatrix} \text{antisymmetric spin wave} \\ \text{function } (S = 0) \end{pmatrix} \tag{38.9}$$

or

$$\psi_{\text{total}} = \begin{pmatrix} \text{antisymmetric orbital} \\ \text{wave function} \end{pmatrix} \times \begin{pmatrix} \text{symmetric spin wave} \\ \text{function } (S = 1) \end{pmatrix} \tag{38.10}$$

Since the product of a symmetric and antisymmetric wave function is antisymmetric, we reach the general conclusion that

the total wave function of a system of electrons must be antisymmetric.

This statement is a more precise way of stating Pauli's exclusion principle that was used in Section 23.9 to analyze the distribution of electrons in an atom, because if two electrons have the same set of quantum numbers or are in the same dynamical state, the antisymmetric wave function is identically zero. Incidentally, all particles with spin $\frac{1}{2}$, and not only electrons, are described by antisymmetric wave functions and obey the exclusion principle. They are called **fermions**, after Enrico Fermi (1901–1954), who studied their properties in great detail.

In view of the two forms of ψ_{total}, the energy levels of helium-like atoms can be grouped in two sets: those corresponding to $S = 0$ and those corresponding to $S = 1$. Figure 38.12 shows the energy levels when one of the electrons is in the $1s$ ground state and the other is in an nl state. Several features are clear from this diagram. The ground state of He corresponds to both electrons in state $1s$, with the same set of orbital quantum numbers $n = 1$, $l = 0$, $m_l = 0$, a configuration called $1s^2$. The orbital wave function is symmetric with $S = 0$. There is no configuration $1s^2$ with $S = 1$ because the orbital wave function would have to be antisymmetric, and if the two electrons have the same set of orbital quantum numbers we get

$$\psi_{100}(1)\psi_{100}(2) - \psi_{100}(2)\psi_{100}(1) \equiv 0$$

Also, s-states have lower energy than p-states, and these in turn have less than d-states and so on. This is because the screening of the nuclear charge is less for s-electrons than for other states, as explained before.

For a given nl, the states with $S = 1$ or $(\uparrow\uparrow)$ have lower energy than the states with $S = 0$ (Figure 38.12). This is because it is less probable for the electrons to be close to each other when $S = 1$ (i.e. when they are described by the antisymmetric orbital wave function) than when $S = 0$ (associated with the symmetric orbital wave function). Therefore, the average repulsive energy of the electrons is less for $S = 1$ or parallel spins than for $S = 0$ or antiparallel spins.

In Figure 38.12 some of the possible transitions between stationary states are shown. They must conform to the selection rules given in Equation 38.5; that is, $\Delta l = \pm 1$, $\Delta m_l = 0, \pm 1$. For that reason the first radiative excitation corresponds to the transition $1s^2 \to 1s2p$, requiring a little more than 21 eV. An He atom can be excited to the $1s2s$ configuration only by inelastic collisions. Also if the atom

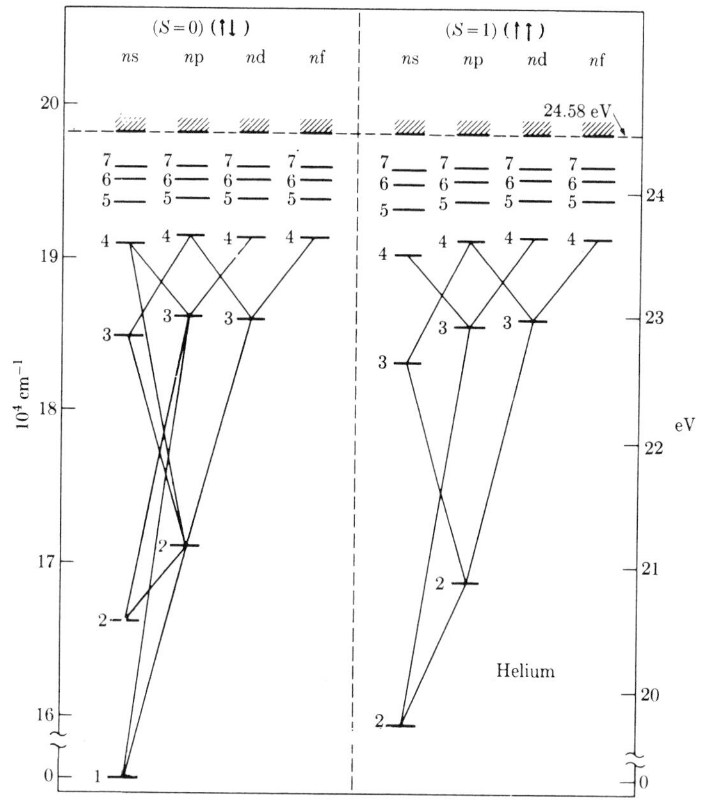

Figure 38.12 Energy levels of helium, showing some transitions. $S = 0$ corresponds to antiparallel spins (↑↓) and $S = 1$ to parallel spins (↑↑).

is in the configuration $1s2s$, it is very difficult to make a transition to the lower energy level configuration $1s^2$ because that would violate the selection rule $\Delta l = \pm 1$. For that reason the $1s2s$ configuration is called a **metastable state** (recall the discussion of the He–Ne laser in Section 31.8).

The probability of transitions between $S = 0$ and $S = 1$ states is extremely small because it implies a reversal of the relative orientation of the spins of the electrons. Since the electronic spins are coupled with their magnetic moments, spin reversal implies a magnetic force. But in Chapter 22 we saw that the magnetic interaction is much weaker than the electric interaction, resulting in a very improbable spin reversal. Thus we may say that helium-like atoms are of two kinds, those with $S = 0$, called **parahelium**, and those with $S = 1$, called **orthohelium**.

38.5 Atoms with many electrons

In Section 23.9 electron shells in atoms were discussed in terms of the exclusion principle. The possible configurations were given in Table 23.3, and the general configuration of energy levels was shown in Figure 23.18. Quantum mechanics allows us to explain these features.

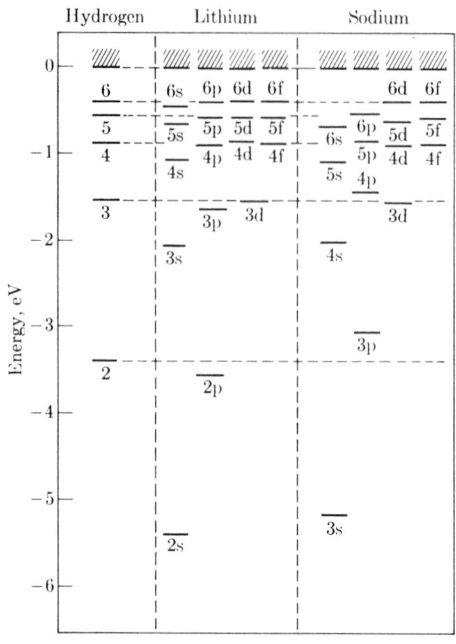

Figure 38.13 Comparison of energy levels of hydrogen and single valence electron lithium and sodium.

The electrons in an atom do not move independently, but are subject to their mutual interactions in addition to the attraction of the nucleus. Thus the description of electron states and configurations using one-electron wave functions with given n, l, m_l and m_s, as well as the organization of the energy levels, is only approximate and valid mostly for the uppermost or more energetic valence electrons. To explain the general ordering of the energy levels, we note that the lower the angular momentum, the more penetrating the electron orbits, thus exposing the electrons to a stronger attractive nuclear Coulomb field. Therefore, s-electrons have a lower energy than p- or d-electrons, which are screened by the electrons in between.

For atoms with one or two valence electrons the energy levels are very similar to those of hydrogen and helium,

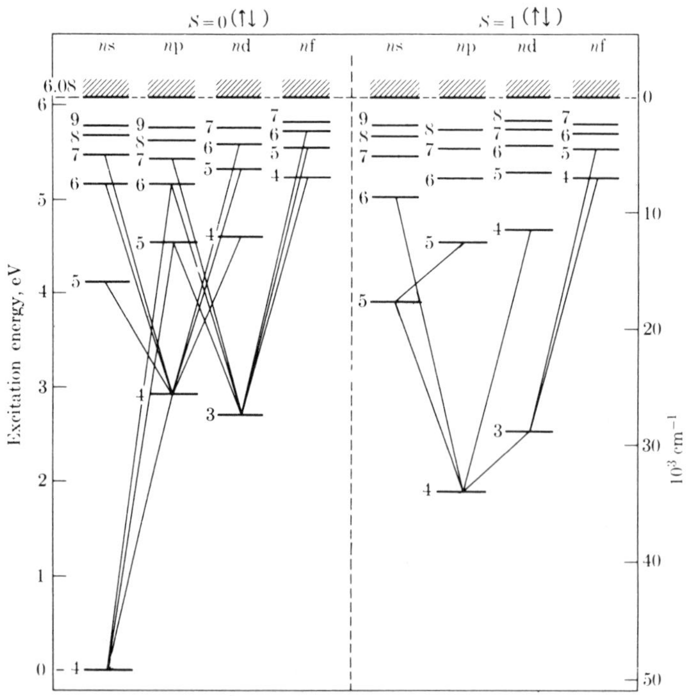

Figure 38.14 Energy levels of calcium, having two valence electrons, showing some transitions.

as shown in Figures 38.13 and 38.14, because these electrons move mostly outside the core constituted by the nucleus and the intermediate electrons. Atoms with more than two valence electrons have an energy level structure that is much more complex.

There is a special situation in the case of atoms like carbon and silicon, which is important for explaining molecular binding. Recalling Figure 23.19, the electrons in the L-shell of the carbon atom can be in the excited configuration sp^3. The angular momentum wave functions for s- and p-states were shown in Figures 38.2 and 38.3 and they are superposed in Figure 38.15. If the s- and p-states have approximately the same energy, any linear combination of the angular wave functions s, p_x, p_y, and p_z is also a good angular wave function for the same energy. For example, if the s and p_z wave functions are combined as $s + p_z$ and $s - p_z$, the two wave functions shown in Figure 38.16 result, which correspond to electrons moving around the Z-axis predominantly above or below the XY-plane. Another possibility is to combine the s, p_x, and p_y functions, resulting in three angular wave functions ψ_1, ψ_2, ψ_3 (Figure 38.17), which correspond to electrons moving predominantly in directions 120° apart in the XY-plane. Finally, the s, p_x, p_y and p_z wave functions can be combined so that the four resulting angular wave functions point towards the vertex of a regular tetrahedron (Figure 38.18). This linear combination of s- and p-functions is called **hybridization** and implies that the angular momentum of the electron is no longer a constant of motion. This technique can be used when the spherical symmetry of central forces has been eliminated by external forces, as in molecular binding. The three types of hybrid wave functions are designated sp, sp^2 and sp^3 respectively.

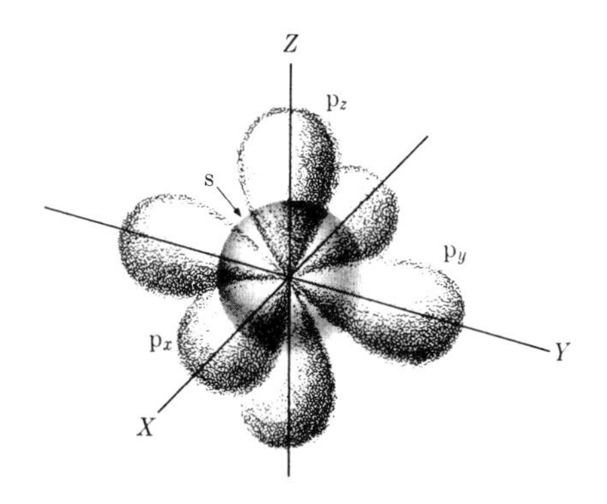

Figure 38.15 s, p_x, p_y and p_z wave functions.

For atoms with valence electrons in d-states, even more complex energy levels and angular wave functions result that very nicely explain the physical and chemical properties of such atoms.

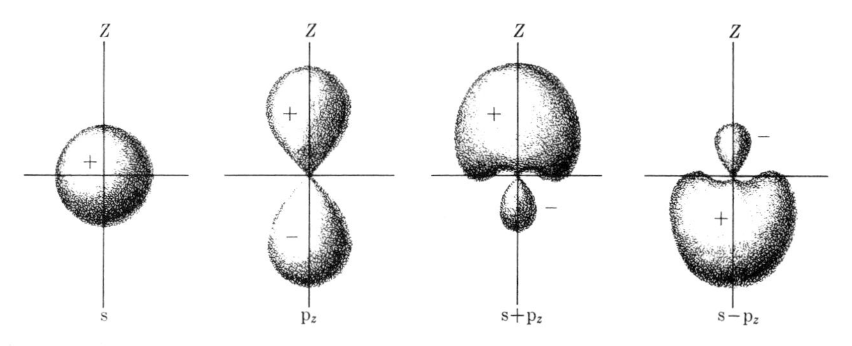

Figure 38.16 Wave functions resulting from sp hybridization.

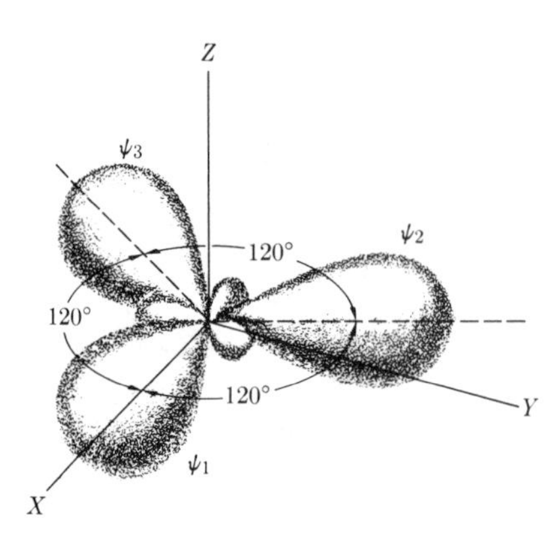

Figure 38.17 Wave functions resulting from sp^2 hybridization.

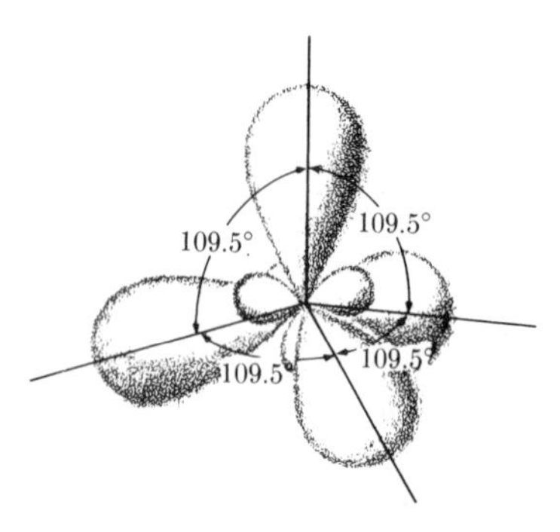

Figure 38.18 Wave functions resulting from sp^3 hybridization.

EXAMPLE 38.2

Hybridization of angular wave functions.

▷ The simplest hybridization is the combination of s and p_z angular wave functions. The normalized s wave function ($l = 0$, $m_l = 0$) is $s = 1/(4\pi)^{1/2}$. There is no angular dependence because of the spherical symmetry of the s-state. The normalized p_z wave function ($l = 1$, $m_l = 0$) is $p_z = (3/4\pi)^{1/2} \cos\theta$. Then the hybridized wave functions are

$$\psi_+ = \frac{1}{(2)^{1/2}}(s + p_z) = \frac{1}{(8\pi)^{1/2}}[1 + (3)^{1/2}\cos\theta]$$

$$\psi_- = \frac{1}{(2)^{1/2}}(s - p_z) = \frac{1}{(8\pi)^{1/2}}[1 - (3)^{1/2}\cos\theta]$$

The wave function ψ_+ shows a maximum for $\theta = 0$ and a minimum for $\theta = \pi$, while the maximum of ψ_- corresponds to $\theta = \pi$ and the minimum to $\theta = 0$. Therefore, ψ_+ is concentrated along the $+Z$-axis and ψ_- along the $-Z$-axis, as shown in Figure 38.16.

38.6 Diatomic molecules

The simplest molecules are those composed of two atoms and the simplest of all diatomic molecules is the H_2^+ ion, composed of one electron and two protons, separated by a distance $r = 1.06 \times 10^{-10}$ m $= 1.06$ Å (Figure 38.19). The potential energy of the system consists of two terms for the attractive interactions of the electron with each proton plus a term for the electric repulsion of the two protons; that is

$$E_p = -\frac{e^2}{(4\pi\varepsilon_0)r_1} - \frac{e^2}{(4\pi\varepsilon_0)r_2} + \frac{e^2}{(4\pi\varepsilon_0)r}$$

where r_1 and r_2 are the distances of the electron to the two protons. The electron does not move in a central field but rather in an electric field like the one represented

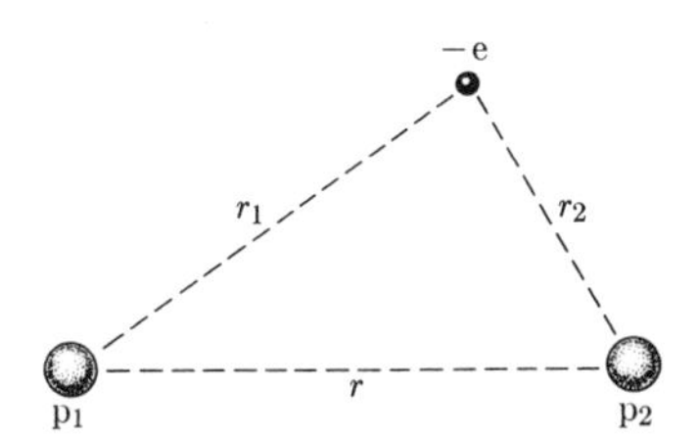

Figure 38.19 The H_2^+ molecule ion.

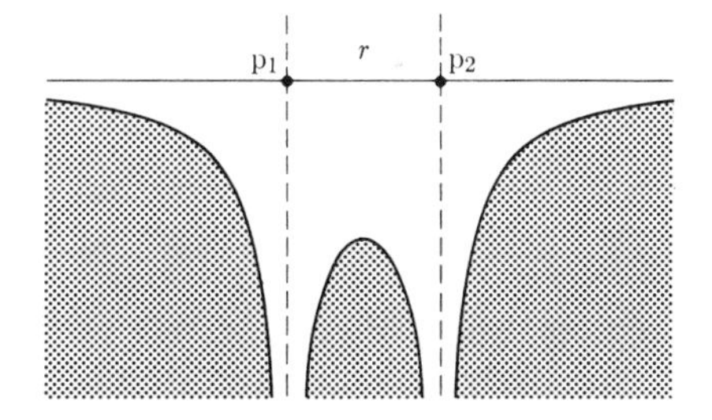

Figure 38.20 Potential energy along the line joining the two nuclei in H_2^+.

in Figure 21.15, which has axial symmetry along the line joining the two protons. Therefore, the angular momentum of the electron is no longer a constant of the motion. However, the component L_z along the proton–proton line is still a constant.

Another feature is that the protons repel each other and thus tend to fly apart. In order to produce a stable system, the electron's attraction on each proton must compensate for the repulsion between protons. This can be achieved if the electron stays for an appreciable time in the region between the two protons. To see how this can be achieved, let us look at the electron's potential energy along the line joining the two protons, which has been represented in Figure 38.20.

If the two protons were quite far apart, the electron in the ground state would only move around one of the protons in a 1*s* orbit and its wave function would be like the hydrogen 1*s* function represented in Figure 38.7. The two possibilities are represented by ψ_1 and ψ_2 in Figure 38.21(a) and (b). When the two protons approach each other, the motion of the electron is perturbed and the wave function is deformed in such a way that, because of the symmetry, the electron can be around either proton with equal probability. A satisfactory approximate wave function meeting this requirement is obtained by linearly combining the wave functions ψ_1 and ψ_2, since both correspond to the same 1*s* atomic energy level. Thus we obtain the wave functions $\psi_S = \psi_1 + \psi_2$ and $\psi_A = \psi_1 - \psi_2$, which are

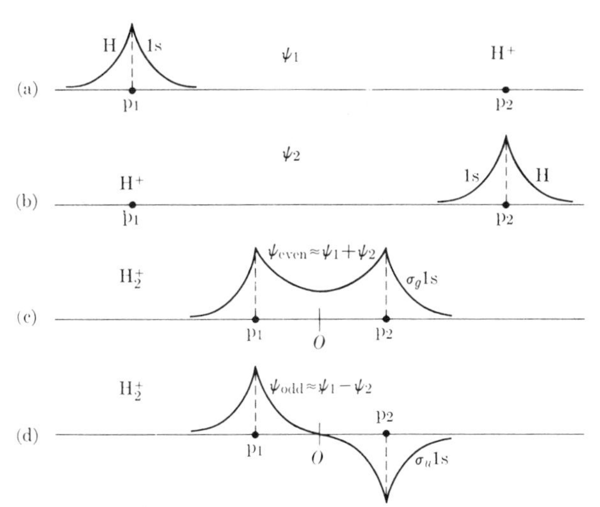

Figure 38.21 Symmetric and antisymmetric molecular orbitals in H_2^+.

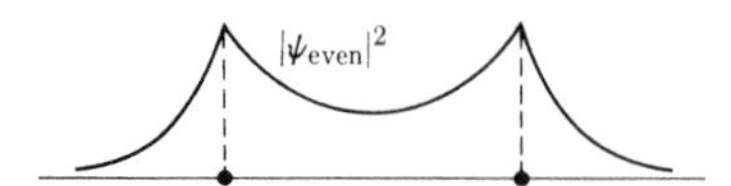

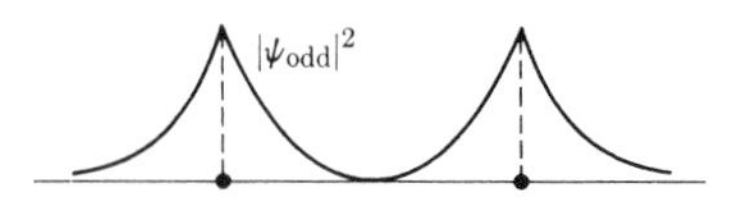

Figure 38.22 Probability density for symmetric and antisymmetric molecular orbitals in H_2^+ along the line joining the protons.

represented in Figure 38.21(c) and (d), and are symmetric and antisymmetric respectively with respect to the midpoint O. The corresponding probability distributions are shown in Figure 38.22. In both cases the probability distribution is symmetric with respect to O but there is a big difference: $|\psi_S|^2$ is finite while $|\psi_A|^2$ is zero in the region between the two protons. Thus, ψ_S results in a stable system but ψ_A does not. We say that ψ_S is a **bonding** wave function and ψ_A is **antibonding**.

If we calculate the energy of the system as a function of the distance between the protons, we obtain the two curves E_S and E_A shown in Figure 38.23. For the bonding state E_S, corresponding to ψ_S, the curve has a minimum energy of -2.648 eV at a proton equilibrium separation of 1.06 Å. For the antibonding state E_A, given by ψ_A, there is no minimum and therefore there is a repulsion at all distances and no molecule can be formed.

The next diatomic molecule in order of complexity is that of hydrogen, H_2, which consists of two protons and two electrons (Figure 38.24). The potential energy now has six terms, corresponding to the attractive interaction of each electron with each proton, the repulsion between the two electrons and the repulsion between the two protons. The system still has axial symmetry around the line P_1P_2.

As a first approximation, we may use for each electron the wave functions ψ_S and ψ_A found for H_2^+. The result is that if *both* electrons are described by ψ_S, so that both electrons tend to be in the region between the protons with equal probability, a stable molecule results and the energy of the system has a minimum of -4.476 eV at a proton separation of 0.74×10^{-10} m, as shown by curve E_S in Figure 38.25. Otherwise, the system is unstable with repulsion at all distances, as shown by curve E_A in Figure 38.25. Note that the equilibrium distance for H_2 is smaller that for H_2^+; this is due to the stabilizing effect of the second electron. Also, at distances smaller than r_0 the proton repulsion predominates, while at distances larger than r_0 the electron attraction predominates.

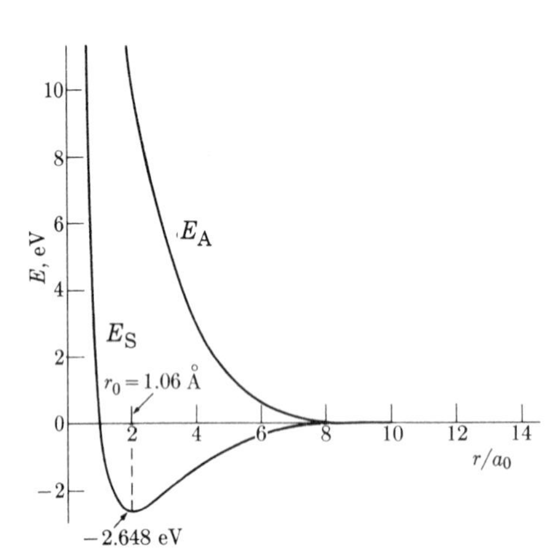

Figure 38.23 Potential energy as a function of the internuclear distance in H_2^+ for symmetric and antisymmetric states.

In order to accommodate the two electrons in the ψ_S orbital state in a symmetric way and at the same time satisfy the exclusion principle, it is necessary that the spin state be antisymmetric, that is, the two electrons have opposite spins,

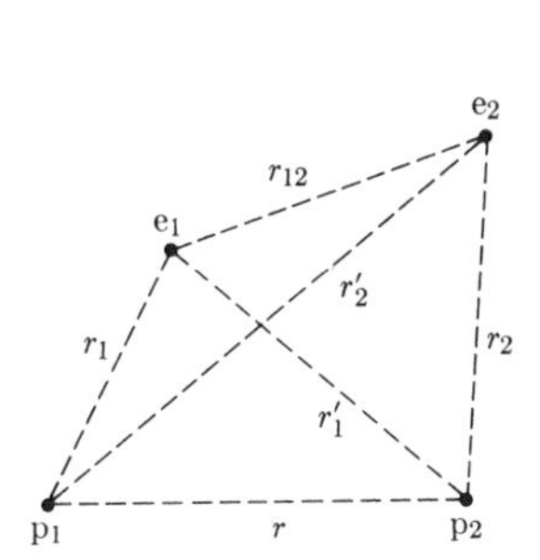

Figure 38.24 The H_2 molecule.

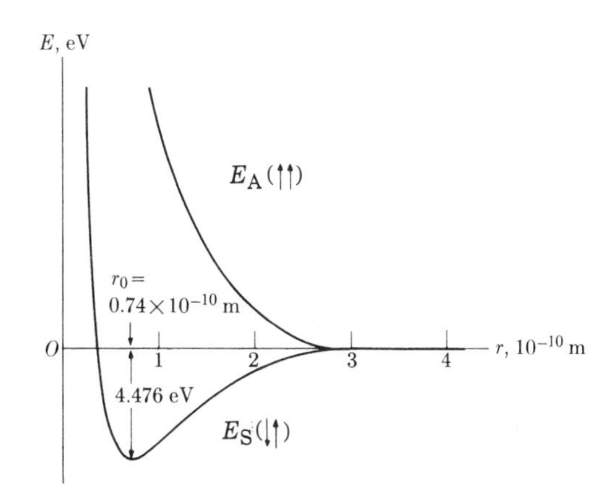

Figure 38.25 Potential energy as a function of internuclear separation for bonding and antibonding states in H_2.

as shown in the figure, a result that is confirmed experimentally (the magnetic dipole moment of H_2 is zero). By the same token, when the electrons are accommodated antisymmetrically in the orbital state, their spins must be in the same direction. This allows us to state the following rule, which is of general validity:

bonding between atoms results when electrons can be located in the region between the nuclei with a symmetrical orbital wave function and opposite spins.

Although our discussion has been limited to diatomic molecules composed of two identical atoms, the analysis can be extended to all kinds of diatomic molecules.

38.7 Linear molecules

Linear molecules are molecules whose atoms are arranged in a linear configuration, like beads on a string. Suppose, for example, a molecule composed of four nuclei N_1, N_2, N_3, N_4 arranged linearly (Figure 38.26). If the electrons are localized around each nucleus, their wave functions would be as shown at the bottom in Figure 38.26. On the other hand, if the bonding electrons were free to move along the molecule, with their motion limited at both ends, it would be equivalent to motion in a potential box and their wave functions would be similar to those in Figure 36.16; these wave functions are shown by dashed lines in Figure 38.26. Each one corresponds to a different energy. Thus, when we combine both kinds of wave functions, we obtain the electron wave functions ψ_1, ψ_2, ψ_3 and ψ_4 indicated by the solid lines in Figure 38.26. These functions are alternately symmetric and antisymmetric, and each corresponds to a different energy and different degree of bonding between contiguous nuclei. Note that each wave function can accommodate two electrons with antiparallel spins, in accordance with the exclusion principle.

The energy of the system for each electron wave function depends on the interatomic distances, as shown in Figure 38.27. Therefore, we can conclude that

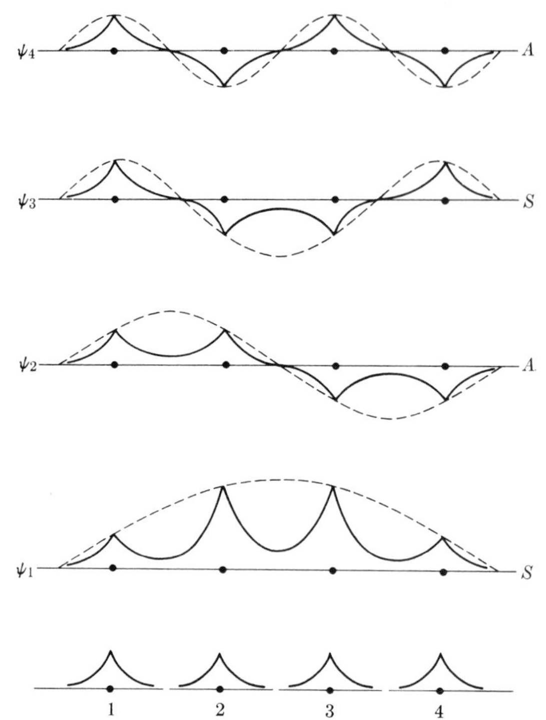

Figure 38.26 Molecular orbitals for p_z electrons in butadiene, C_4H_6.

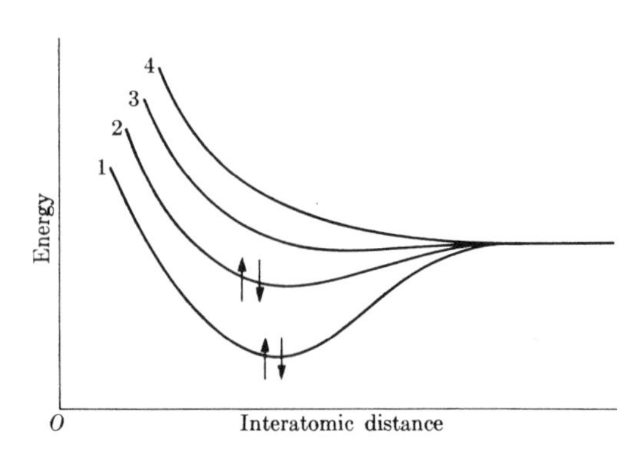

Figure 38.27 Electronic potential energy of p_z electrons in butadiene as a function of internuclear separation.

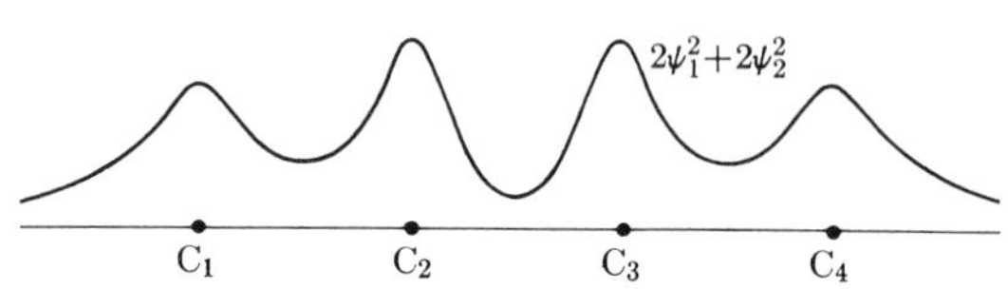

Figure 38.28 Total probability distribution of p_z electrons in butadiene.

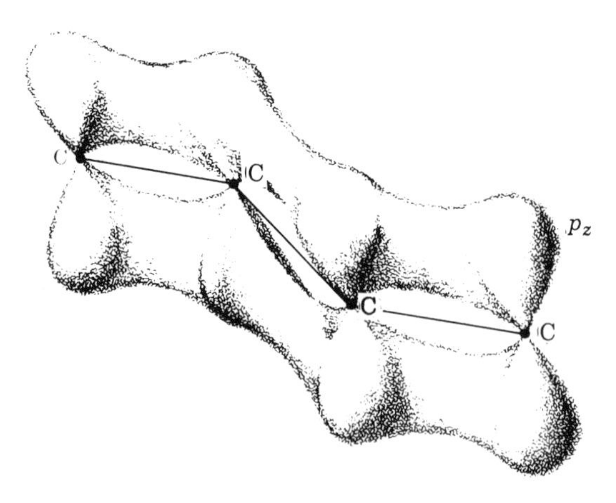

Figure 38.29 Electron distribution in butadiene corresponding to non-localized p_z electrons.

in a linear molecule composed of n atoms, each of the uppermost atomic energy states splits into n closely spaced molecular energy states, each one accommodating two electrons. As the interatomic distances decrease, the energy splitting becomes more pronounced. For example, in the molecule of butadiene, composed of four carbon atoms, each of the four electrons originally in a $2p_z$-state of each carbon atom can move freely along the molecule, occupying the states ψ_1 and ψ_2, resulting in the probability distribution shown in Figure 38.28 and in three dimensions in Figure 38.29. An interesting feature is that these p_z electrons are not localized, but are free to move along the molecule from one end to the other and back, like in a potential box. The same situation occurs in several other molecules.

38.8 The geometry of molecules

Quantum mechanics not only allows us to explain how atoms bind together to form stable molecules, but also the geometrical shape of molecules. We shall consider a few examples, starting with the water molecule, H_2O. The ground state configuration of the oxygen atom (Table 23.3) is $1s^2 2s^2 2p^4$, but two of the p-electrons have their spins in the same direction (recall Figure 23.19), for the same reason that they do in the C atom. It is possible that the two electrons occupy the p_x and p_y angular momentum states respectively (Figure 38.3), so that they tend to move along the chosen X and Y axes. Thus, in the H_2O molecule (Figure 38.30), two hydrogen atoms must join the oxygen atom so that their $1s$ electrons can be optimally coupled with the p_x and p_y electrons of oxygen. The result is that in the H_2O molecule the two bonds are at right angles. However, because the two protons repel each other, the actual angle is 104.5°.

The H_2O molecule has two other important features. The oxygen valence electrons are attracted to the two protons, so that the p_x and p_y angular wave functions of the oxygen atom are deformed slightly toward the hydrogen atoms, as shown in the figure. Because the nuclear charge of the oxygen atom is $+8e$, the hydrogen electrons tend to be closer to the nucleus of the oxygen atom than to the protons of the hydrogen atoms. The result is an asymmetry of the charge distribution, tending to make the oxygen atom negative and the hydrogen atoms positive. This creates a polarized molecule, with an electric dipole moment of 6.2×10^{-30} m C along the line bisecting the bond angle. This was already mentioned in Section 25.6. The polarization of water molecules results in very important physical and biological properties; in particular, it is the reason why water is a solvent of many other substances.

A similar situation occurs with the ammonia molecule, NH_3. From Figure 23.19 we see that the N atom has three p-electrons in its ground state, with their spins all in the same direction. We may assume that they occupy the p_x, p_y and p_z states of Figure 38.3, and therefore are concentrated along the X-, Y- and Z-axes. Therefore, the NH_3 molecule has a pyramidal structure, with the N atom at the vertex and the three H atoms at the base (Figure 38.31). The angles of the bonds at the N vertex are 107.3° instead of 90°, again because of the repulsion between the protons of the H atoms. The pyramidal structure gives rise to an electric dipole moment of 5.0×10^{-30} m C, directed along the axis of the pyramid.

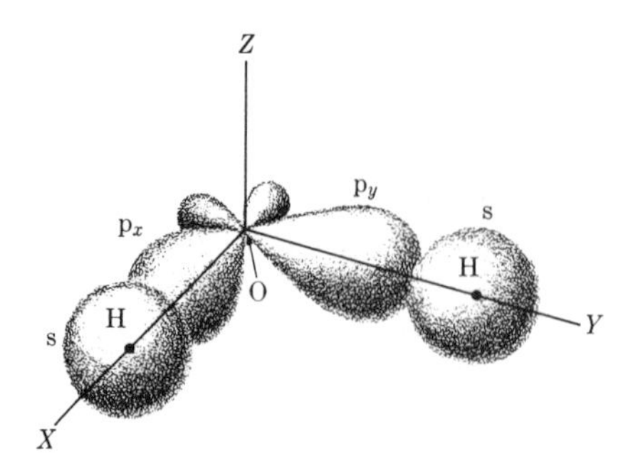

Figure 38.30 Electronic distribution in the H_2O molecule.

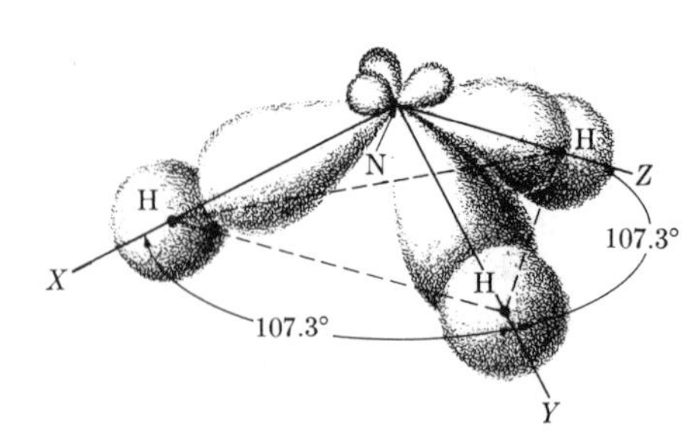

Figure 38.31 Electronic distribution in the NH_3 molecule.

The CH_4 molecule has the tetrahedral structure shown in Figure 38.32. This geometry occurs because, by the procedure of sp^3 hybridization (Figure 38.18), the electrons in the $2s2p^3$ configuration of the carbon atom can be arranged to move in directions pointing towards the vertices of a tetrahedron. The symmetry of the CH_4 molecule means that it has no net dipole moment even if each bond is polarized.

In the ethane molecule $H_3C—CH_3$ (Figure 38.33), three of the sp^3-hybridized wave functions are used by each carbon atom to combine with the hydrogen atoms, and the fourth wave function is used to join the carbon atoms together, resulting in a particularly strong bond. This points to the possibility of forming long chains of carbon atoms:

$$
\begin{array}{c}
\;|\quad\;|\quad\;|\qquad\qquad\;|\quad\;| \\
—C—C—C—\cdots—C—C— \\
\;|\quad\;|\quad\;|\qquad\qquad\;|\quad\;|
\end{array}
$$

using the sp^3 functions. This gives rise to the existence of **polymers** that play an important role in modern technology of materials. The structure of $H_3C—CH_3$ allows for the rotation of a CH_3 group relative to the others, a fact that has been verified experimentally.

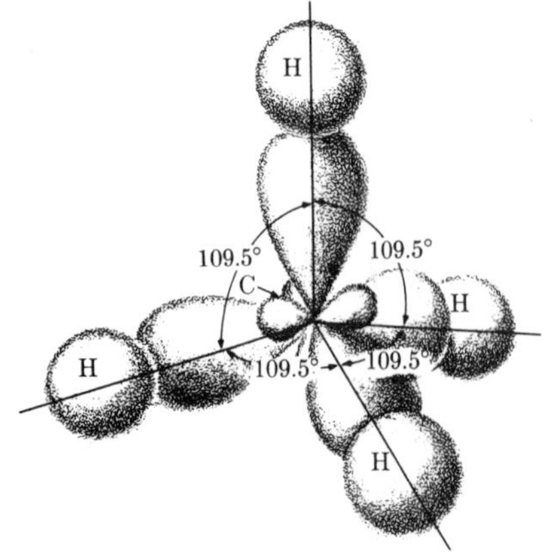

Figure 38.32 Electronic distribution in the CH_4 molecule, using sp^3 orbitals.

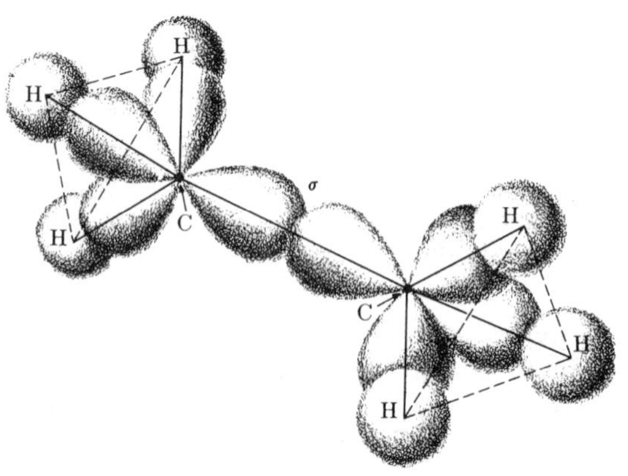

Figure 38.33 Electronic distribution in the C_2H_6 molecule, using sp^3 orbitals.

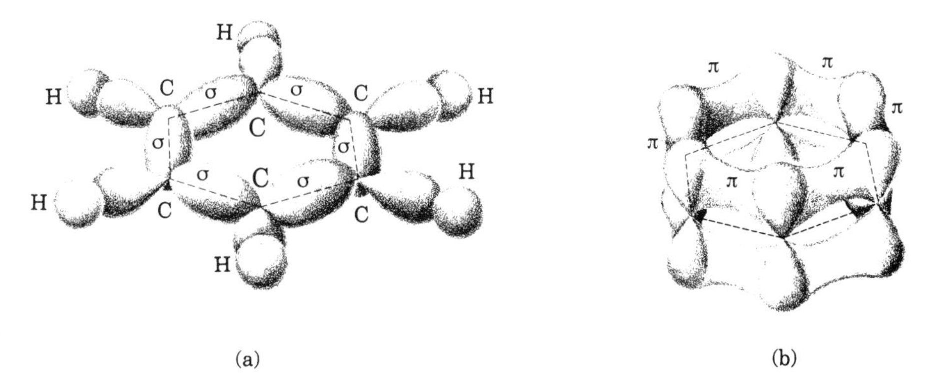

Figure 38.34 Benzene molecular orbitals. (a) Localized sp^2 bonds; (b) unlocalized p_z bonds.

As another example, consider the benzene molecule, C_6H_6. It has been verified that this molecule has the form of a hexagon, as shown in Figure 38.34. Each carbon atom supplies four valence electrons occupying the p_z and the three sp^2 orbitals shown in Figure 38.17. One of the sp^2 functions of each carbon atom is used for binding with a H atom. The other two sp^2 functions are used to bind the C atom to the other two adjacent carbon atoms. There remain the p_z electrons of each atom. They behave like the p_z electrons in butadiene that we represented in Figure 38.29, and they are free to circulate around the hexagon without being bound to any specific carbon atom. However, because the motion is cyclic, the wave functions must be periodic and that determines their energy.

Since 1985 a new type of carbon molecule, with the generic composition C_n, where n can go from 32 to several hundred, has been identified. In these molecules the carbon atoms are bound by sp^2 hybrid wave function and occupy the vertices of a hollow polyhedral structure composed of a certain number of hexagons but always with 12 pentagons in between. These structures resemble the geodesic domes designed by R. Buckminster Fuller and for that reason, they have been called **fullerenes** by R. E. Smalley. The most important fullerene is C_{60} (Figure 38.35(a)), where the carbon atoms are arranged forming 20 hexagons, each similar

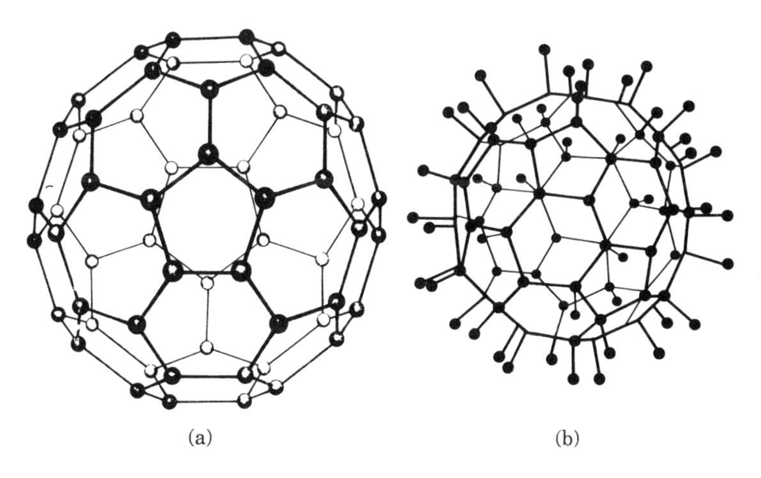

Figure 38.35 (a) Hexagonal arrangement of the carbon atoms in C_{60}. Note that in some places the spaces between the hexagons are pentagons. (b) Arrangement of carbon and hydrogen atoms in $C_{60}H_{60}$. The hydrogen atoms are at the end of the lines attached like pins to the carbon atoms.

to the benzene ring, along with the 12 required pentagons. As in benzene, the p_3 electrons are not localized and can move about the molecule. C_{60} has the most perfect spherical symmetry of any molecule. It can form several compounds, such as $C_{60}H_{60}$, which is a sort of spherical benzene (Figure 38.35(b)), $C_{60}(OsO_4)$, as well as even more complex compounds using the p_3 electrons. What makes C_{60} particularly important is that in the solid state, where the molecules form a face-centered cubic lattice, the material is a semiconductor. However, when doped with certain elements (for example, K, Rb or Tl) and the doping atoms are placed either *inside* each molecule or at interstitial spaces in the lattice, the material functions as an insulator, a conductor, a semiconductor or a superconductor, depending on the amount of doping. For example, a solid with the stoichiometric composition K_3C_{60} (or three potassium atoms for each C_{60} molecule) is a conductor that, when cooled below 18 K becomes a superconductor (recall Table 24.2). If, instead, too much potassium is added, the solid is an insulator. Fullerenes can also have tubular, helical and thin-film shapes, each with different properties.

38.9 Structure of solids

In the solid state the atoms (or molecules) are closely packed and held in more or less fixed positions by electromagnetic forces that are of the same order of magnitude as those involved in molecular binding. In most solids the atoms (or molecules) do not exist as isolated entities; rather their properties are modified by the nearby atoms. This distorts the energy levels of the outermost electrons. There is a class of solids whose structure exhibits a regularity or periodicity constituting what is called a **crystal lattice**. Many solids do not exhibit any regular structure; these are called **amorphous**.

Determining the structure of a solid does not differ fundamentally from determining that of a molecule. It consists in finding a stable configuration of nuclei and electrons with the proper matching of the wave functions. The two main differences between the structure of a solid and that of a molecule are the large number of atoms involved in the solid and the regularity in their arrangement.

The theory of solids must correlate their structure with macroscopic properties such as elasticity and hardness, thermal and electric conductivity, reflectivity and refractivity, and so on. Although it is impossible to classify solids in a precise way, most solids fall into certain types according to the predominant type of bonding. Some of these types are the following.

(i) Covalent solids. In a covalent solid, atoms are bound together by localized directional bonds such as those we encountered in the H_2, CH_4 and $H_3C—CH_3$ molecules. A typical case is diamond, where the four bonding electrons of each carbon atom are oriented according to the sp^3 hybrid wave functions. Each carbon atom is bound to four other carbon atoms by means of the sp^3 functions, resulting in the crystal lattice structure indicated in Figure 1.12. Each ball represents a carbon nucleus and each bar a pair of localized bonding electrons. The separation between two carbon atoms is 1.54×10^{-10} m. In a sense covalent solids are huge molecules.

Covalent solids, because of their rigid electronic structure, exhibit several common macroscopic features. They are extremely hard and difficult to deform.

All are poor conductors of heat and electricity because there are no free electrons to carry energy or charge from one place to another. Also, a relatively high energy is required to excite whole-crystal vibrations in a covalent solid due to the rigidity of the bonds. Whole-crystal vibrations therefore have a high frequency. Similarly, electronic excitation energies of covalent solids are of the order of a few electron volts (for example, the first electronic excitation energy of diamond is about 6 eV). This electronic excitation energy is relatively large compared with the average vibrational thermal energy (of the order of kT), which at room temperature (298 K) is about 2.4×10^{-2} eV; hence covalent solids are normally in their electronic ground state. Many covalent solids are transparent or rather, colorless, especially diamond, because their first electronic state is higher than the photon energies in the visible spectrum, which lie between 1.8 and 3.1 eV.

(ii) Ionic crystals. At the other extreme are ionic crystals, which consist of a regular array of positive and negative ions resulting from the transfer of one electron (or more) from one kind of atom to another. This is, for example, the case for NaCl and CsCl, whose crystal structures are shown in Figure 1.10. The separation between the Na and Cl ions is about 2.81×10^{-10} m, while the shortest distance between identical atoms is 3.97×10^{-10} m. The ions are so arranged that a stable configuration is produced under their mutual electronic interactions.

These solids, because they have no free electrons, are also poor conductors of heat and electricity. However, at high temperatures the ions may gain some mobility, resulting in better electrical conductivity. Ionic crystals are usually hard, brittle, and have a high melting point due to the relatively strong electrostatic forces between the ions. Some ionic crystals strongly absorb electromagnetic radiation in the far infrared region of the spectrum. This property is associated with the energy needed for exciting lattice vibrations, which is less than 1 eV. This energy is generally lower for ionic than for covalent crystals, due to their relatively weaker binding energy.

(iii) Hydrogen bond solids. Closely related to ionic crystals are the hydrogen bond solids, which are characterized by strongly polar molecules having one or more hydrogen atoms, such as water, H_2O, and hydrofluoric acid, HF. The positive hydrogen ions, since they are relatively small, may attract the negative end of other molecules, forming chains such as $(H^+F^-)(H^+F^-)(H^+F^-)$ .... This is particularly interesting in the case of ice, where the water molecules have the tetrahedral arrangement shown in Figure 1.9 and are held together by a bond between the oxygen atom of a molecule and the hydrogen atom of another molecule. The relatively open structure of ice accounts for the larger volume of ice with respect to water in its liquid phase. Hydrogen bonding is very important in many biomolecules such as DNA (Figure 1.8).

(iv) Molecular solids. These solids are mostly made of substances whose molecules are not polar. All electrons in these molecules are paired, so that no covalent bonds between atoms of two different molecules may be formed. Molecules in this type of solid retain their individuality. They are bound by the same intermolecular forces that exist between molecules in a gas or a liquid: **van der Waals forces**, which are very weak, and which correspond roughly to the forces between electric dipoles. This may be explained as follows: although, on the average, these molecules do not have a permanent electric dipole moment, their electronic configuration at each instant may give rise to an instantaneous electric dipole, resulting in dipole–dipole interactions whose average, over time, is not zero. For the above reasons molecular solids are not conductors of heat and electricity, have a very low melting point, and are very compressible and deformable. Examples of molecular solids are CH_4, Cl_2, I_2, CO_2, C_6H_6 etc. in their solid state. Inert gases, whose outer shells are complete, solidify as molecular solids.

(v) Metals. As explained in Section 23.10, metals are elements that have relatively small ionization energies and whose atoms have only a few weakly bound electrons in their outermost incomplete shells. These outermost weakly bound electrons are easily set free using the energy released when the crystal is formed. A metal thus has a regular lattice of spherically symmetric positive ions pervaded by an electron 'gas' formed by the released electrons, which can move through the crystal lattice and therefore are not localized. These electrons are responsible for metallic bonding and occupy the valence and conduction bands.

Metallic solids exhibit good thermal and electrical conductivity, for which the free electrons are mainly responsible, the reason being that the free electrons, particularly those close to the Fermi level in the conduction band, easily absorb any energy from electromagnetic radiation or lattice vibrations—no matter how small—and thus increase their kinetic energy and their mobility. For this same reason metals are also opaque, since the free electrons can absorb photons in the visible region and be excited to one of the many close-by quantum states available to them. The free electrons are also largely responsible for another characteristic exhibited by metals: their high reflection coefficient for electromagnetic waves, due to the surface effects induced by the waves.

Some solids are a mixture of more than one type. An example is graphite (Figure 1.11), which consists of layers of carbon atoms arranged in a hexagonal lattice. Carbon atoms on each layer are bound by sp^2 wave functions. Layers are held together by the p_z electrons. The atoms in a layer are bonded by localized covalent bonds which use sp^2 hybrid wave functions and non-localized p_z-bonds, as in benzene (see the previous section). The non-localized electrons are free to move parallel to the layers but not perpendicular to them. Successive layers of atoms act as macromolecules. Layers are held together by weak van der Waals forces, just as molecular crystals are, which accounts for the flaky, slippery nature of graphite. In fact, it is because of this structure that graphite is used as a lubricant. The length of the sp^2 bonds in graphite is 1.42×10^{-10} m and the separation between the layers is 3.35×10^{-10} m. Amorphous carbon is characterized by an irregular arrangement of benzene groups loosely held together by p_z bonds. This is why carbon is so soft and can be broken very easily.

The structure of each of the types of solid that we have described is determined by the electronic structure of the component atoms, which determines the type and number of electrons available for bonding, as well as the energy required to adjust their motion to the conditions prevailing in the lattice (conditions that are different from those in the isolated atoms).

Crystal lattices are not perfect, and the imperfections may be due to several causes. Figure 38.36 illustrates some of the most common imperfections. In (a) we have a **vacancy** left by a missing atom, while in (b) we have a **substitutional impurity** replacing an atom of the lattice. Imperfections due to an **interstitial atom**—either

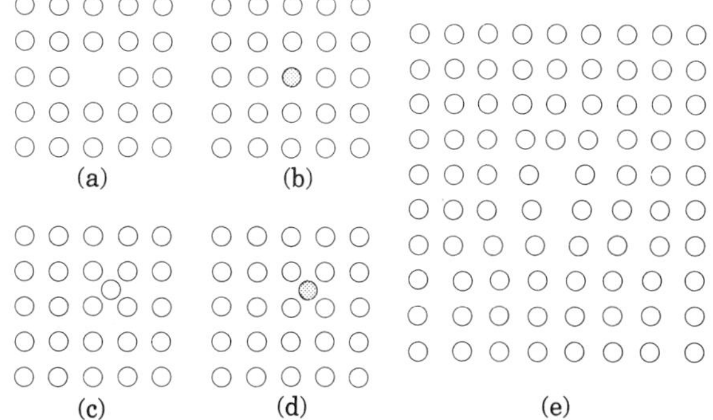

Figure 38.36 Imperfections in a crystal lattice. (a) Vacancy; (b) impurity; (c) and (d) interstitial atom; (e) dislocation.

of the same type or of an impurity—are shown in (c) and (d). Finally, (e) shows an **edge dislocation**, which may be looked upon as if an extra layer of atoms had been pushed halfway through the top of the lattice. Lattice imperfections have a most important effect on many electric, elastic and optical properties of solids.

38.10 Electrons in metals

Quantum mechanics can explain the existence of energy bands in metals, a topic considered in Section 23.10. The positive ions in a metal are arranged in a regular manner, forming a three-dimensional crystal lattice. Conduction electrons therefore move through a periodic potential that exhibits the same regularity, or periodicity, as the lattice. This periodicity in the electric potential allows us to estimate the possible allowed electron energies without having to solve Schrödinger's equation. For simplicity, consider a one-dimensional lattice (Figure 38.37). An electron having an energy such as E_1 cannot move freely through the lattice, but is confined mainly to one of the classically allowed regions AB, CD etc. around specific ions. It is true that it can go from AB to CD by leaking through the potential barrier interposed between the two allowed regions, but the barrier is relatively so high and wide that its penetration by the electron is highly improbable. This is why the innermost electrons in a crystal are essentially localized and their energies and wave functions may be considered the same as in the isolated atoms. An electron with energy E_2 is not bound so strongly to a particular ion and, by leaking through the potential barrier, it can move about in the lattice. Finally, an electron with energy E_3 is not bound to any atom in particular; it has great freedom of movement throughout the lattice. These quasi-free electrons, which are characteristic of metals, are not only responsible for most of the collective properties of the lattice (such

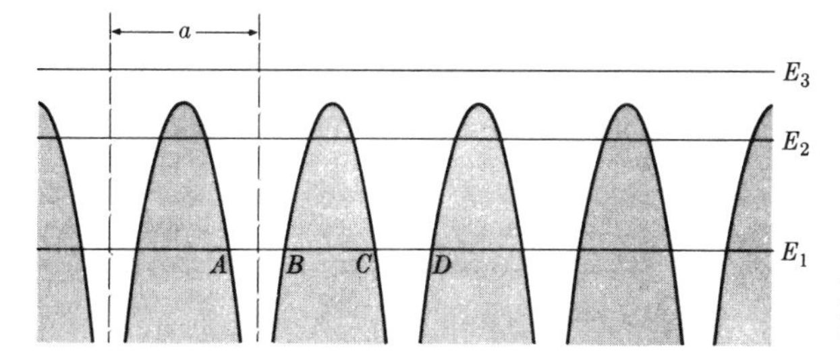

Figure 38.37 Types of energy levels in a linear crystal lattice.

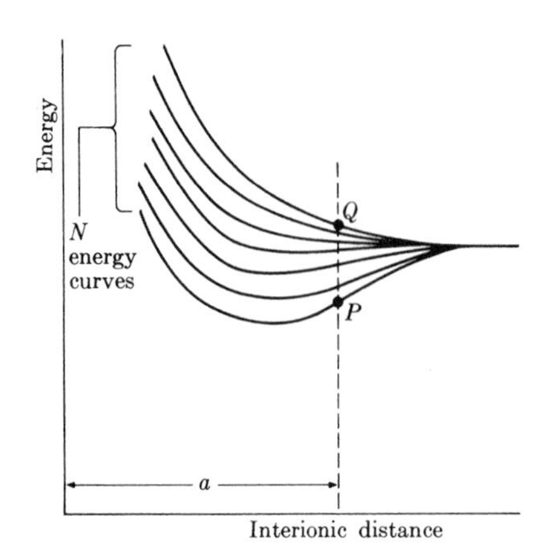

Figure 38.38 Energy levels in a linear crystal lattice composed of N atoms as a function of the interionic distance.

as the electric and thermal conductivities), but they also provide for the binding of the atomic ions which form the crystal structure. The electronic potential energy in diatomic molecules is due to two ions and, as a result, the atomic energy levels split into two as the interionic distance decreases. This was illustrated in Figures 38.23 and 38.25 for the H_2^+ and H_2 molecules. Similarly, in linear molecules the bonding electrons move in a periodic potential and each atomic energy level splits into a number of levels equal to the number of atoms. This was illustrated in Figure 38.29 for butadiene. Therefore, in a linear lattice composed of N atoms, each atomic energy level splits into N closely spaced levels. Their spacing and position depend on the interionic separation, as shown qualitatively in Figure 38.38. For example, for an interionic distance a, the possible energy levels fall between P and Q. When N is very large, the different energy levels are closely spaced and one may say they form an energy band. In the real case of a three-dimensional lattice the situation is similar but more complex. Figure 38.39 shows the energy bands corresponding to several energy levels at an interionic distance a. Note that as the interionic distance decreases some bands overlap.

Since, according to Pauli's exclusion principle, each energy level can accommodate two electrons, one with spin up and the other with spin down, an energy band corresponding to a given atomic state can accommodate a maximum of $2N$ electrons, or two electrons per ion. The bands are full or are incompletely

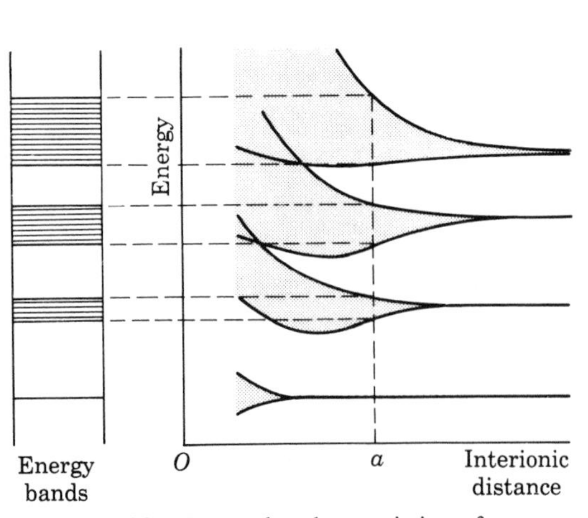

Figure 38.39 Energy bands, consisting of closely spaced energy levels. The fine structure of the band is not shown.

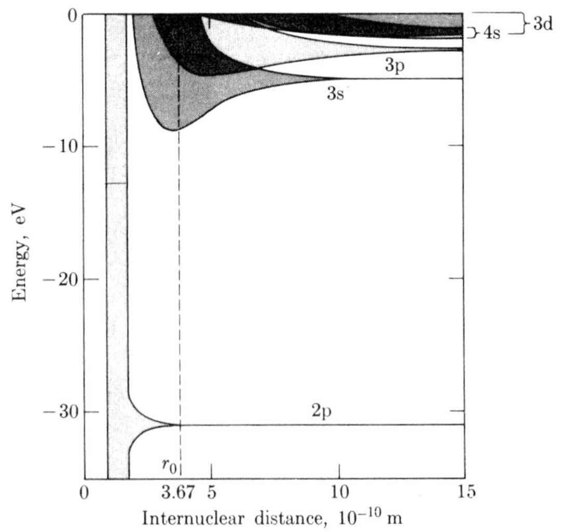

Figure 38.40 Energy bands of sodium.

filled depending on the number of electrons provided by each atom. The bands are designated as *s*-, *p*-, *d*- etc. according to the value of the angular momentum of the atomic state to which they are related. The actual band structure for sodium is shown in Figure 38.40.

QUESTIONS

38.1 What part of the wave function in a central field depends on the potential energy and what part is a result of spherical symmetry?

38.2 Why does the angular wave function in a central field not depend on n?

38.3 Why does the radial wave function in a central field not depend on m_l?

38.4 Explain why an electron's *s*-orbits are more penetrating than the other orbits. Why is their energy lower than for *p*-orbits?

38.5 Explain why valence electrons in atoms generally have parallel spins while bonding electron pairs in molecules have opposite spins.

38.6 The probability of finding the electron in a hydrogen atom in the ground state at some distance between r and $r + \mathrm{d}r$ is given by $P(r) = (32\pi/a_0)(r/a_0)^2 e^{-2r/a_0}$, where a_0 is the Bohr radius defined in Equation 23.15. Make a plot of $P(r)$ in terms of r/a_0 and estimate the radius of the shell in which the electron is more likely to be found. Compare with the radius of the first Bohr orbit.

38.7 Which property of the wave function of two electrons depends on the identity of the two electrons and which property depends on Pauli's principle?

38.8 Designate the spin wave function by $\sigma(m_s)$ with $m_s = \pm\frac{1}{2}$. Write the possible spin wave function of two electrons, 1 and 2, when $S = 0$ and when $S = 1$.

38.9 When is hybridization of *s*- and *p*-wave functions possible?

38.10 In Example 38.2 the angles for the maxima of ψ_+ and ψ_- were found. Find the angles when ψ_+ and ψ_- are zero.

38.11 What property must the wave function in H_2^+ and H_2 have in order to produce a bonding system?

38.12 Analyze the possible excited states of the p_z electrons in C_6H_6, using the same method as for butadiene. Note that in this case the wave function must have a period of 2π.

38.13 How do the wave functions determine the geometry of molecules?

38.14 Schematically draw the potential energy of rotation of the CH_3 groups with respect to the C—C line in C_2H_6 as a function of the rotation angle. What relative motions of the two CH_3 groups are possible in that potential energy? Recall Example 9.9.

38.15 Why do electrons in the uppermost atomic energy levels determine the properties of a solid?

38.16 Why are there electron energy bands in metals?

PROBLEMS

38.1 The angular wave functions corresponding to $l = 1$ are $Y_{l0} = (3/4\pi)^{1/2} \cos\theta$, $Y_{l\pm1} = \mp(3/8\pi)^{1/2} \sin\theta\, e^{\pm i\phi}$. Show that it is possible to transform the Y_{l0} and $Y_{l\pm1}$ functions, by combining them linearly, into the functions $p_x = (\frac{3}{4}\pi)^{1/2} \sin\theta \cos\theta$, $p_y = (\frac{3}{4}\pi)^{1/2} \sin\theta \cos\phi$ and $p_z = (\frac{3}{4}\pi)^{1/2} \cos\theta$ represented in Figure 38.3.

38.2 The average value of r in hydrogen-like atoms is

$$r_{\text{ave}} = \frac{n^2 a_0}{Z}\left[1 + \frac{1}{2}\left(1 - \frac{l(l+1)}{n^2}\right)\right]$$

(a) Compute r_{ave} for all states with $n = 1$, 2 and 3. Compare these values with the corresponding Bohr radii (Figure 38.6). (b) Using r_{ave} as a measure of the size of the orbit, arrange the nl states according to increasing average distances from the nucleus for $n = 1$, 2, 3 and 4.

38.3 The energy levels of helium-like atoms, when one electron is in the ground state ($n = 1$) and the other in an excited state ($n > 1$), can be expressed approximately by

$$E = -RhcZ^2 - \frac{Rhc(Z-1)^2}{n^2}$$

This expression assumes that the electron in the ground state fully screens one nuclear charge. Discuss the plausibility of this expression. Compute the energy levels for helium when $n = 2$, 3 and 4, and compare with the experimental result. Why does the accuracy of the above expression for E increase when n increases?

38.4 The sp^2 hybrid wave functions are, within a normalization constant, $s + \sqrt{2}p_x$, $s - (1/\sqrt{2})p_x + \sqrt{3/2}p_y$ and $s - (1/\sqrt{2})p_x - \sqrt{3/2}p_y$, where the wavefunctions $s = 1/(4\pi)^{1/2}$ and p_x, p_y and p_z are given in Problem 38.1. Determine the directions for which they have maximum values and the angles between them.

38.5 The sp^3 hybrid wave functions are, within a normalization constant, $s + p_x + p_y + p_z$, $s + p_x - p_y - p_z$, $s - p_x + p_y - p_z$ and $s - p_x - p_y + p_z$. Determine the directions for which they have maximum values and the angles between them.

38.6 The ground state energy of H_2^+ relative to the system composed of hydrogen in its ground state and H^+ at infinite separation is -2.65 eV. Compute: (a) the energy of H_2^+ relative to the system $H^+ + H^+ + e$ at infinite separation; (b) the energy of the system $H_2^+ + e$ at infinite separation relative to the system H + H, again at infinite separation, and with both atoms in their ground states; (c) the ionization energy of H_2 if the dissociation energy of this molecule into two hydrogen atoms in their ground state is 4.48 eV. Compare the result in (c) with the ionization energy of atomic hydrogen.

38.7 Dissociation and ionization energies are frequently expressed in kcal mol^{-1}. (a) Show that one kcal mol^{-1} is equal to 4.338×10^{-2} eV. (b) Express the dissociation energy of H_2 in kcal mol^{-1}.

38.8 (a) Explain why the bond length of H_2^+ is 1.06 Å, while H_2 has a shorter bond length equal to 0.74 Å. (b) Also explain why the dissociation energy of H_2 (103.2 kcal mol^{-1}) is less than the dissociation energy of H_2^+ (61.06 kcal mol^{-1}).

38.9 (a) Explain why the H_2^- ion is less stable than the He_2^+ ion if both have the same electronic configuration. (b) Which ion should have the larger internuclear separation?

38.10 Discuss the bond structure in (a) C_2H_8, (b) C_3H_6 and (c) C_3H_4.

38.11 Analyze the bond structure of (a) BH_3 and (b) BH_4^-, indicating whether the molecules are planar or pyramidal.

38.12 Molecules of the form CH_3–(CH=CH–)$_n$$CH_3$ are called **conjugate**. Analyze the binding of a conjugate molecule, using sp^2 wave functions and p_z electrons. The effective total length of the conjugate molecule CH_3–(CH=CH–)$_4$$CH_3$ is 9.8 Å. (a) Plot the energy levels occupied by the p_z electrons. (b) Estimate the energy and wavelength of the photons absorbed when one of the uppermost p_z electrons is excited.

38.13 Suppose that the equilibrium separation in the $H\,^{35}Cl$ and $H\,^{37}Cl$ molecules is the same and equal to 1.27 Å. Compute for each molecule (a) the energy of the first two excited rotational levels, (b) the frequencies and wavelengths corresponding to the transitions $l = 0 \rightarrow l = 1$ and $l = 1 \rightarrow l = 2$, and (c) the frequency difference for successive lines.

38.14 (a) Calculate the energy of the first three excited rotational states in CO and CO_2. (b) Determine the wavelength of the photons absorbed in transitions among such energy levels.

38.15 A diatomic molecule is not rigorously rigid and, due to a centrifugal effect, the internuclear distance increases as the angular momentum of the molecules increases. (a) How should this molecular stretching affect the energy levels? (b) An empirical expression for the rotational energy is $E_{rot} = (\hbar^2/2I)\{l(l+1) - \delta[l(l+1)]^2\}$, where δ is the stretching constant. Obtain an expression for the frequencies due to transitions among successive rotational levels.

38.16 (a) Show that the number of carbon–carbon bonds in diamond is twice the number of carbon atoms (see Figure 1.12). (b) The energy required to dissociate one mole of diamond is 170 kcal. Determine the energy per bond and express it in eV.

38.17 Estimate the radius of the 1s-electron orbit in sodium ($Z = 11$). Compare this value with the equilibrium distance between sodium ions in the solid state and decide whether such electrons are affected in the metal.

39 Nuclear structure

Ernest Rutherford first observed the scattering of α-particles (helium nuclei) by thin gold foils (1906) and was surprised to find that some particles were scattered backward. As a result, he formulated the nuclear model of the atom (1911). He assumed that the scattering was due to Coulomb repulsion between the α-particles and nuclei. Further experimentation indicated some deviations and Rutherford suggested (1919) the existence of a strong or nuclear force.

39.1 Introduction

Gravitational and electromagnetic interactions explain most of observed physical phenomena. Gravitation seems to explain satisfactorily phenomena that involve large masses, such as planetary motion and motion on the Earth's surface, and electromagnetism together with quantum mechanics explains atomic, molecular and solid structure and therefore most of the physical and chemical phenomena involving interactions between atoms and molecules. There are, however, several processes that do not directly affect our daily life in an obvious way that indicate the existence of other interactions, the **strong** and the **weak**, which were briefly analyzed in Note 6.1. The strong interaction is responsible for most nuclear properties while the weak interaction is more directly related to some properties of the elementary particles.

39.2 The nucleus

Recall that the atomic nucleus is a cluster of protons and neutrons occupying a small region at the center of the atom, with a diameter of the order of 10^{-14} m;

that is, about one ten-thousandth the diameter of the atom. Protons and neutrons are designated by the common name of **nucleons**. Both particles have spin $\frac{1}{2}$ and obey the Pauli exclusion principle. The number of protons is the atomic number Z, and the total number of nucleons (protons and neutrons) is the mass number A. Nuclei with the same Z are **isotopes** while those with the same A are **isobars**.

When we compare the structure of the nucleus with that of the atom, several new features strike us. First, all the particles that make up the nucleus have practically the same mass, while in atoms the electrons are very light compared with the nucleus, which remains essentially fixed at the center of mass of the atom. Therefore, we cannot speak of a central dominant force that acts on the particles of the nucleus and is produced by a body at the center of the nucleus; rather we must imagine all particles moving under their mutual interactions.

Second, in an atom it is possible to explain electronic motion in terms of electromagnetic interactions between the negatively charged electrons and the positively charged nucleus. But a nucleus is composed of protons, with positive charge, and neutrons, with no charge at all. Therefore, we cannot attribute the stability of the nucleus to electric attraction. On the contrary, it seems that electric repulsion between the protons would send the nucleus flying apart. The mere fact that nuclei, composed of protons and neutrons, exist is a clear indication of an interaction between them that is not directly related to electric charges and which is *much* stronger than the electromagnetic interaction. This interaction is called the **nuclear interaction**. Our knowledge of the nuclear interaction is still incomplete, but at least we do know some of its more important characteristics.

39.3 Properties of the nucleus

To obtain a clue about the nature of the forces holding the protons and neutrons together in a nucleus, it is necessary to analyze several physical properties of nuclei besides charge and mass.

(ii) Size. If we assume that a nucleus is spherical, we may express its size in terms of its radius R. We must, however, refrain from picturing the nucleus as a solid ball with a well-defined surface. Rather, the radius R gives only the order of magnitude of the region where the nucleons are concentrated most of the time. We can obtain the radius R by measuring several nuclear properties, such as the lifetime of α-particle emitters and the isotope shift in line spectra, and nuclear processes such as scattering of fast electrons and nuclear reactions induced by charged particles. The experimental results indicate that the nuclear radius is proportional to $A^{1/3}$, where A is the mass number of the nucleus. That is,

$$R = r_0 A^{1/3} \qquad \textbf{(39.1)}$$

where r_0 is an empirical coefficient, approximately the same for all nuclei. Its accepted value is $r_0 = 1.4 \times 10^{-15}$ m. Some nuclei depart substantially from the spherical shape and must be assumed to be ellipsoidal or even pear-shaped.

Since the volume of a sphere is $\frac{4}{3}\pi R^3$, we conclude from Equation 39.1 that the nuclear volume is

$$V = \tfrac{4}{3}\pi r_0^3 A = 1.12 \times 10^{-45} A \text{ m}^3$$

That is, the volume of a nucleus is proportional to the number of nucleons, A. This suggests that nucleons are closely packed at fixed average distances, independent of the number of particles, so that the volume per nucleon, $V_0 = \frac{4}{3}\pi r_0^3$, is a constant quantity, the same for all nuclei. Another conclusion is that the density of nuclear matter is about the same for all nuclei. This may be seen as follows: since 1 amu $= 1.66 \times 10^{-27}$ kg, the mass of a nucleus of mass number A is approximately $M = 1.66 \times 10^{-27} A$ kg. Therefore the average density of nuclear matter is

$$\rho = \frac{M}{V} = \frac{1.66 \times 10^{-27} A \,\mathrm{kg}}{1.12 \times 10^{-45} A \,\mathrm{m}^3} = 1.49 \times 10^{18} \mathrm{~kg~m}^{-3},$$

which is independent of A. This density is about 10^{15} times greater than the density of matter in bulk as we know it on Earth, and gives us an idea of the degree of compactness of the nucleons in a nucleus. It also shows that matter in bulk is essentially empty since most of the mass is concentrated in the nuclei, which occupy a small fraction of the atomic volume.

(ii) Angular momentum. The resultant angular momentum of a nucleus is called (for historical reasons) the **nuclear spin** but it does not imply that the nucleus is rotating like a solid body. Both protons and neutrons, like electrons, have spin $\frac{1}{2}$. In addition, protons and neutrons possess orbital angular momentum associated with their motion in the nucleus. The resultant nuclear angular momentum (or spin) is obtained by combining, in a proper way, the orbital angular momenta and the spins of the nucleons composing the nucleus. The nuclear spin is designated by a quantum number I such that the magnitude of the nuclear spin is $\hbar[I(I+1)]^{1/2}$. The component of the nuclear spin in a given direction such as the Z-axis (Figure 39.1) is given by $I_Z = m_I \hbar$, where

$$m_I = \pm I, \pm(I-1), \ldots, \pm\tfrac{1}{2} \text{ or } 0$$

depending on whether I is a half-integer or an integer. Therefore there are $2I + 1$ possible orientations of the nuclear spin. Since the spin of nucleons is $\frac{1}{2}$, the values of I are integers (if A is even) or half-integers (if A is odd) ranging from zero, as in $^4_2\mathrm{He}$ and $^{12}_6\mathrm{C}$, up to 7, as in $^{176}_{71}\mathrm{Lu}$.

Practically all even–even nuclei (i.e. nuclei that have an even number of both neutrons and protons) have $I = 0$, which indicates that identical nucleons tend to pair their angular momenta in opposite directions. Even–odd nuclei (i.e. nuclei that have an odd number of either protons or neutrons) all have half-integral spin, and it is reasonable to assume that the nuclear spin coincides with the angular momentum of the last unpaired nucleon, a result which seems to hold in many cases. Odd–odd nuclei have two unpaired nucleons (one neutron

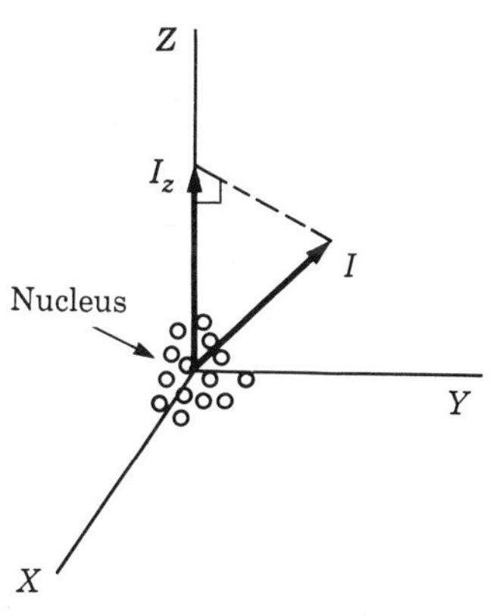

Figure 39.1 Nuclear spin.

and one proton) and the experimental results are a little more difficult to predict. Nevertheless, their angular momenta are integers, since there is an even number of particles.

(iii) Magnetic moment. We may recall that a moving charge possesses an orbital magnetic moment $\boldsymbol{M}_L$ which is proportional to its orbital angular momentum $\boldsymbol{L}$ and which, according to Equation 23.33, is given by $\boldsymbol{M}_L = (q/2m)\boldsymbol{L}$. For the case of protons, the charge is $q = +e$ and $\boldsymbol{M}_L$ has the same direction as $\boldsymbol{L}$. Since neutrons have no charge, they do not have an orbital magnetic moment. The component of the orbital magnetic moment of a proton along the Z-axis is

$$M_{L,z} = \frac{e}{2m_\mathrm{p}} L_z = \frac{e\hbar}{2m_\mathrm{p}} m_l = \mu_\mathrm{N} m_l \qquad \text{(proton)}$$

where the constant

$$\mu_\mathrm{N} = \frac{e\hbar}{2m_\mathrm{p}} = 5.0504 \times 10^{-27}\ \mathrm{J\ T^{-1}}$$

is called a **nuclear magneton** (recall Section 23.6).

If the particle has spin $\boldsymbol{S}$, it may have an additional spin magnetic moment. The spin magnetic moment of the proton is parallel to its spin, as corresponds to a positive charge. It is interesting to note that the neutron, although it has no electric charge, has a spin magnetic moment that is antiparallel to its spin.

The spin magnetic moment for the proton and the neutron is expressed as

$$\boldsymbol{M}_S = g_S\left(\frac{e}{2m_\mathrm{p}}\right)\boldsymbol{S} \qquad \text{and} \qquad M_{S,z} = g_S \mu_\mathrm{N} m_s \tag{39.2}$$

where g_S is a constant characteristic of each particle, called the **spin gyromagnetic ratio**. The value for the proton is $g_{S,\mathrm{p}} = +5.5855$, and for the neutron $g_{S,\mathrm{n}} = -3.8263$. (Recall that for the electron $g_{S,\mathrm{e}} \approx -2.0$, Section 23.7.) Since m_s is $\pm\frac{1}{2}$, the resulting magnetic moments, expressed in nuclear magnetons, are 2.7927 for the proton and -1.9131 for the neutron. These results suggest that the proton and the neutron have a complex structure, which will be analyzed in Section 41.8.

The resultant magnetic moment of a nucleus is obtained by combining, in a proper way, the magnetic moments of all the nucleons. The resultant magnetic moment $\boldsymbol{M}$ is directly proportional to the nuclear spin $\boldsymbol{I}$ and the relation may be written as $\boldsymbol{M}_\text{nucleus} = \mu_\mathrm{N} g \boldsymbol{I}$, where g is a constant characteristic of each nucleus and is called the gyromagnetic ratio of the nucleus.

39.4 Nuclear binding energy

The **binding energy** of a system is the energy released when the system is formed, or the energy that must be supplied to the system to separate it into its components; that is, into the individual nucleons of a nucleus. The binding energy of a system of mass M, composed of particles of masses m_i, is given by (recall Equation 20.21),

$$E_\mathrm{b} = \left(\sum_i m_i - M\right)c^2 \tag{39.3}$$

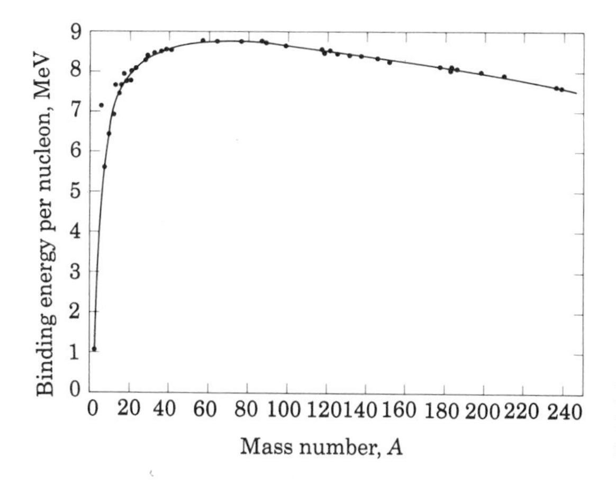

Figure 39.2 Binding energy per nucleon as a function of mass number.

Because the strength of the nuclear interaction is much larger than that for atoms or molecules, the binding energy of nuclei is very large, resulting in nuclear masses that are appreciably smaller than the sum of the masses of their nucleons.

The binding energy of a nucleus of mass M composed of A nucleons, of which Z are protons and $A - Z$ are neutrons may be written as

$$\begin{aligned} E_b &= [Zm_p + (A - Z)m_n - M]c^2 \text{ J} \\ &= 931.48[Zm_p + (A - Z)m_n - M] \text{ MeV} \end{aligned} \qquad \textbf{(39.4)}$$

In the first expression, the masses must be expressed in kilograms while in the second expression all masses must be expressed in amu.

An indication of the stability of a nucleus is the average binding energy per nucleon, E_b/A. Its value for various nuclei is shown in Figure 39.2. It may be seen that the binding energy per nucleon is a maximum for nuclei in the region of mass number $A = 60$. Therefore, if two light nuclei are joined together to form a medium mass nucleus (a process called **fusion**), energy is liberated, and if a heavy nucleus is divided into two medium mass fragments (a process called **fission**), energy is also liberated.

The fact that the binding energy per nucleon varies by less than 10% above $A = 10$ suggests that each nucleon in the nucleus interacts only with its immediate neighbors, independently of the total number of nucleons present in the nucleus. The small decrease after $A = 60$ is due to the destablizing effect of the long-range repulsive Coulomb force between the protons in a nucleus.

39.5 Nuclear forces

We now summarize the main properties of nuclear forces that have been determined experimentally.

(i) The nuclear force is of short range. Short range means that the nuclear force is appreciable only when the interacting particles are very close, at a separation of the order of 10^{-15} m or less. At greater distances the nuclear force is negligible. We may infer that the nuclear force is of short range because at distances greater than 10^{-14} m, corresponding to nuclear dimensions, the interaction regulating the scattering of nucleons and the grouping of atoms into molecules is electromagnetic. If the nuclear force were of long range, the nuclear interaction between the atomic nuclei would be fundamental in determining molecular formation, dominating the weaker electromagnetic forces, in the same way that the electromagnetic interaction dominates the even weaker gravitational interaction in the formation of atoms and molecules.

The range of nuclear forces may be determined directly by performing scattering experiments. Suppose, for example, that we send a proton against a nucleus. The proton, on approaching the nucleus, is subject to both the electric repulsion and the nuclear force. If the nuclear force were of a range comparable to that of the electric force, the motion of the proton, no matter how close or how distant it is when it passes the nucleus, would be affected by both types of forces, and the angular distribution of the scattered protons would differ appreciably from the results for pure electric (or Coulomb) scattering.

If, however, the range of the nuclear force is small, those protons passing at a distance from the nucleus greater than the range of the nuclear force essentially experience only an electric force. Only those protons with enough kinetic energy to overcome the Coulomb repulsion and pass close to the nucleus are affected by the nuclear force, and their scattering is different from Coulomb scattering, as discussed in Section 23.3. This is the situation observed experimentally, confirming the short range of the nuclear force.

Accordingly, in a nucleus each nucleon interacts only with its nearest neighbors, as indicated in Figure 39.3. Since the Coulomb repulsion is long range, each proton interacts with all the other protons, like the electrons in an atom.

(ii) The nuclear force seems to be independent of electric charge. This means that nuclear interactions between two protons, two neutrons, or one proton and one neutron are basically the same. For example, from the analysis of proton–proton and neutron–proton scattering, scientists have concluded that the nuclear part is essentially the same in both cases. This is also supported by the fact that the binding energy per nucleon is the same irrespective of the mix of neutrons and protons in the nucleus. Because of this property, protons and neutrons are considered equivalent in so far as the nuclear force is concerned and they are designated by the common name of **nucleons**.

(iii) The nuclear force depends on the relative orientation of the spins of the interacting nucleons. This fact has been confirmed by scattering experiments and by analysis of the nuclear energy levels. It has been found that the energy of a two-nucleon system where the two nucleons have their spins parallel is different from the energy of such a system where one has spin up and the other down. In fact, the neutron–proton system has a bound state, the **deuteron**, where the two nucleons have their spins parallel ($S = 1$), but no bound state exists if the spins are antiparallel ($S = 0$).

(iv) The nuclear force is not completely central; it depends on the orientation of the spins relative to the line joining the two nucleons. This property has been deduced by noting that even in the simplest nucleus (the deuteron), the orbital angular momentum of the two nucleons relative to their center of mass is not constant, contrary to the situation when forces are central.

(v) The nuclear force, at very short distances much smaller than the range, becomes repulsive. This assumption has been introduced to explain the constant average separation of nucleons, resulting in a nuclear volume proportional to the total number of nucleons, as well as to account for certain features of nucleon–nucleon scattering.

In spite of all this information about nuclear forces, the correct expression for the potential energy for the nuclear interaction between two nucleons is not yet well known. Several expressions have been proposed. About 1935 Hideki Yukawa (1907–1981) suggested the expression

$$E_p(r) = -\frac{E_0 a}{r} e^{-r/a} \qquad \textbf{(39.5)}$$

where E_0 and a are two empirical constants. The constant a is related to the range of the nuclear force and E_0 gives the strength of the interaction. The decreasing exponential factor $e^{-r/a}$ drops the Yukawa potential to zero faster than the electric potential energy, which varies as $1/r$. However, there are even some doubts that the nuclear interaction can be described in terms of a potential energy function in the same way that we have been able to explain the gravitational and electromagnetic interactions. The reason is that the nuclear force is a residual effect of the strong interaction among the quarks that compose the protons and neutrons, which, in turn make up a nucleus. (This will be analyzed in Section 41.8.)

In any case, for most problems the nuclear interaction at low energies may be represented schematically by a potential energy, as shown in Figure 39.4(a). Beyond a certain distance the potential energy is practically constant, or (which is equivalent) the force is zero. At very short distances a repulsive core might be added. This is the type of potential to be used when discussing n–n and n–p interaction. For p–p interactions, however, we must also include the Coulomb repulsion $e^2/4\pi\varepsilon_0 r$, which is important only outside the range of the nuclear forces (Figure 39.4(b)).

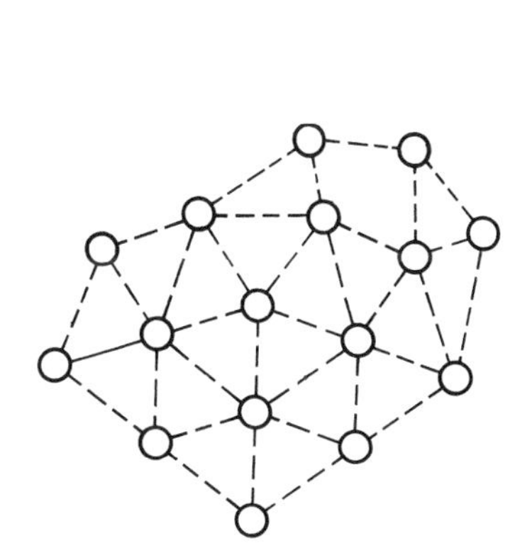

Figure 39.3 Packing of nucleons in a nucleus.

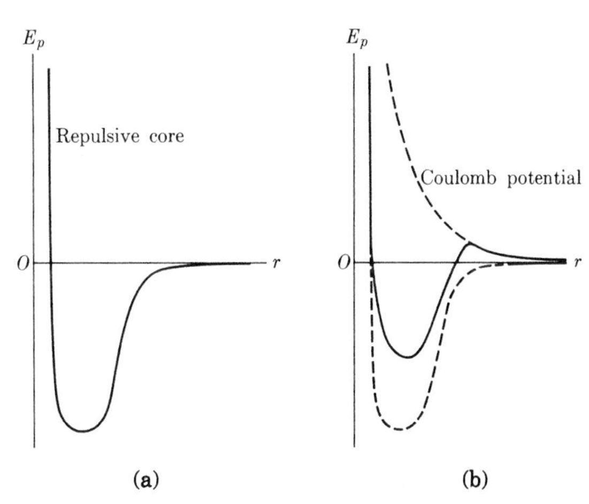

Figure 39.4 Empirical shapes for the nuclear potential energy.

39.6 The deuteron

The deuteron, composed of a proton and a neutron, is the simplest of all nuclei (if we exclude the trivial case of the nucleus of hydrogen, which is a single proton). Basically, only the nuclear interaction is operative in binding a neutron and a proton together (we may neglect, in a first approximation, the small magnetic interaction resulting from their magnetic moments). Therefore, from a detailed analysis of the properties of the deuteron, we can obtain valuable information about the nature of nuclear forces.

The deuteron has only one stationary state, with an energy $E = -E_b = -2.224$ MeV. This is equal to the energy required to separate the neutron and the proton and is the binding energy of the deuteron. Also, the spin of the deuteron is $I = 1$. We may then assume that the proton and neutron have their spins parallel (that is, $S = 1$) and that the orbital angular momentum of their relative motion around their center of mass is zero (that is, $L = 0$). In a state with $L = 0$ there is no orbital magnetic moment, and the magnetic moment of the deuteron should be equal to the spin magnetic moment or, since $S = 1$, $\mu_p + \mu_n = 2.7927 - 1.9131 = 0.8796$ nuclear magnetons, which is very close to the experimental value $\mu_d = 0.8574$ nuclear magnetons. Therefore our assumption about the ground state seems to be a good approximation.

To obtain the properties of the deuteron, we must proceed as we did in Chapter 37 when discussing atomic structure. First, we must write Schrödinger's equation with a proper potential energy, as indicated in Figure 39.5. Since the exact form of the nuclear potential energy is not known, we may, for simplicity, represent the nuclear potential by a rectangular spherical potential well, as shown in Figure 39.5; that is, $E_p(r) = -E_0$ for $0 < r < a$ and 0 for $r > a$, where a is the range of the nuclear forces and E_0 is the depth of the potential well. This potential energy is similar to the potential well discussed in Section 37.6 with $E_0 = \varepsilon$, but with an important difference. The potential well in Section 37.6 was linear or one-dimensional, while in the deuteron the potential well is spherical or three-dimensional. This introduces some changes in Schrödinger's equation and in the wave functions because the relative orbital angular momentum of the proton and the neutron must be taken into account. However for $L = 0$, or the s-state, the wave function is spherically symmetric and we may use the results of Section 37.5. That means that the radial wave function is similar

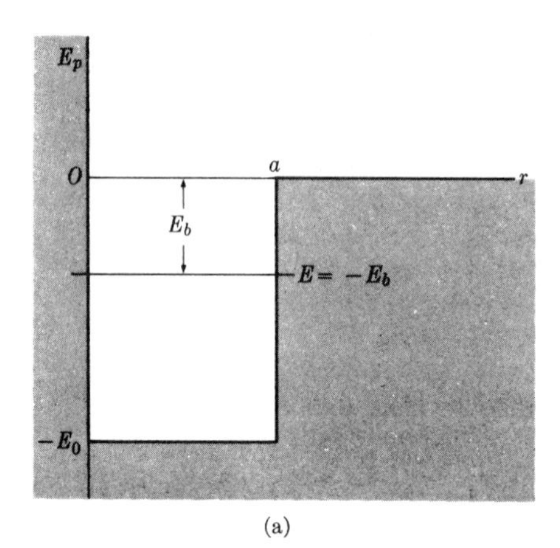

(a)

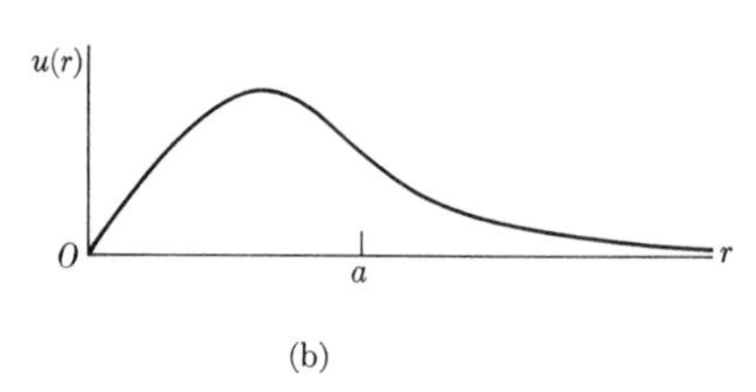

(b)

Figure 39.5 (a) Deuteron spherical rectangular potential well. (b) Ground state radial wave function.

(except for a $1/r$ factor) to that represented in Figure 37.7(a). This wave function has been represented in Figure 39.5(b).

The energy is determined by applying the method explained in Section 37.5 for a potential well. Using the known value of E_b for the energy of the ground state of the deuteron and following the technique outlined in Note 37.1, we get $E_0 a^2 \simeq 1.48 \times 10^{-28}$ MeV m^2. To obtain E_0 and a separately, another relation is necessary. From scattering experiments we can estimate the range a (see the next section). Thus if a is of the order of 2×10^{-15} m, we find that E_0 is about 37 MeV. If we use some other type of short-range potential instead of a square well, we obtain different values for E_0 and a. Therefore, this value of E_0 must be taken only as an indication of the order of magnitude of the nuclear potential.

For a more precise description of the neutron and proton system in the deuteron, it is necessary to include an orbital angular momentum $L = 2$ in addition to $L = 0$. This allows for the computation of some properties of the deuteron, such as its magnetic moment, with great precision. The mixing of $L = 0$ and $L = 2$ states is a confirmation that the nuclear force is not strictly central. Note that $L = 2$ combined with $S = 1$ still gives $I = 1$ if the two vectors are in opposite directions.

39.7 Neutron–proton scattering

Another source of information about the nuclear force between two nucleons is scattering experiments. To perform proton–proton scattering experiments, a beam of protons from an accelerator is made to impinge on a target containing hydrogen atoms, and the scattered protons are analyzed. The deviation from pure Coulomb scattering gives information about the nuclear force. In neutron–proton scattering, a beam of neutrons from a nuclear reactor or other neutron source is projected onto a hydrogen target and the scattered neutrons are observed. The scattering in this case is only due to the nuclear force.

Consider a beam of neutrons moving along the Z-axis toward a target composed of hydrogen atoms (Figure 39.6). Only neutrons that pass very close to the target are subject to the short range nuclear forces resulting from their interaction with the protons and are deviated from their initial direction of motion.

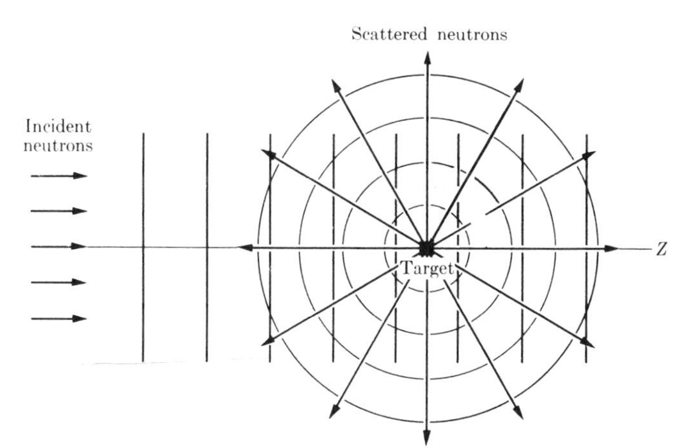

Figure 39.6 Scattering of particles.

The scattered neutrons move in all directions. However, the intensity of the scattering is not necessarily the same in all directions. From the observed angular distribution of the scattered neutrons, it is possible to estimate the nuclear forces that give the observed distribution. These experiments confirm that the nuclear force is of short range because only those neutrons that pass close to the protons are affected by the nuclear force.

Another experimental result is that the scattering is not the same when the neutron and the proton have their spins pointing in the same direction (↑↑) as when they are in the opposite directions (↑↓). This confirms that the nuclear force is spin-dependent. For example, assuming a range $a = 2 \times 10^{-15}$ m it has been found that $E_0 = 33$ MeV for parallel spins ($S = 1$) and $E_0 = 21$ MeV for antiparallel spins ($S = 0$). The first result agrees with that obtained for the deuteron. Since E_0 is too small for antiparallel spins to make it possible to have a bound n–p state, the deuteron exists only with $S = 1$.

Neutron–neutron scattering experiments are more difficult, since it is impossible to have a target composed only of neutrons; therefore, some indirect methods are necessary, which we will not discuss.

39.8 The shell model

A basic problem in nuclear physics is to determine the motion of the nucleons in a nucleus, and from the motion to derive nuclear properties, in both the ground and the excited states. The problem is more complex than in the atomic case, due to the lack of a dominant central force and the existence of two classes of particles – neutrons and protons – each class separately obeying the exclusion principle, which restricts their motion inside the nucleus. However, it seems reasonable to assume that each nucleon moves in an average field of force produced by the other nucleons that, in a first approximation, may be considered central. Thus, using an independent-particle model as we did for the electrons in an atom, we may characterize the nucleon energy states by quantum numbers n and l, giving the energy level and the orbital angular momentum. Quantum number n is chosen to label the order of increasing energy in which successive levels with the same l appear. Thus $1s$, $2s$, $3s, \ldots$, etc. mean the first, second, third, ... levels with $l = 0$ in order of increasing energy.

In the same way that atoms exhibit a shell structure, and have complete shells when all occupied energy levels have their full quota of electrons, nuclear levels also exhibit a shell-like structure. Since there are two classes of particles in a nucleus, there is a double-shell arrangement, one for protons and another for neutrons. For values of Z or N corresponding to complete shells, nuclei result that are particularly stable, in the same way that inert gases consist of atoms with certain complete electronic shells. These values of Z or N (commonly called '*magic numbers*') are 2, 8, 20, 28, 50, 82 and 126. Magic-numbered nuclei show an abnormally high first excitation energy (Figure 39.7), suggesting that at the magic numbers there is a large energy gap between the last filled energy level or shell and the next empty state. This is also the case for the atoms of inert gases.

To explain the values of the magic numbers, it is assumed that, in addition to the average central force, there is a strong **spin–orbit interaction** in nuclei (not

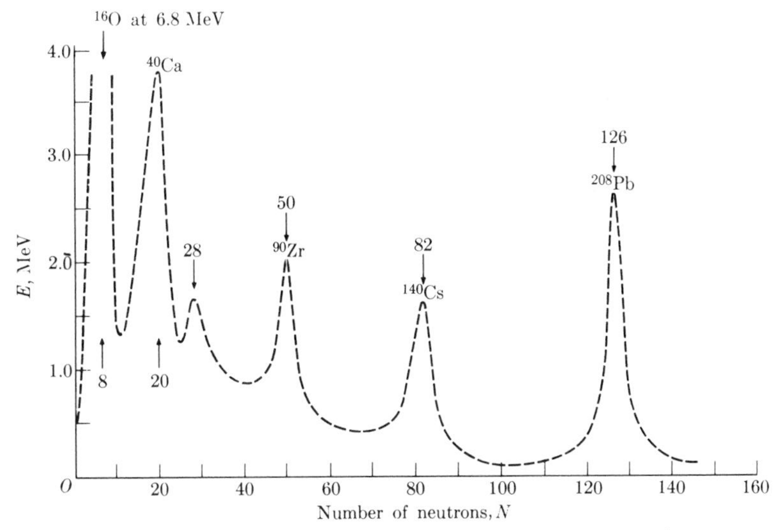

Figure 39.7 Energy of the first excited state for even-even nuclei.

necessarily of electromagnetic origin, as in the case of the atomic electrons), which acts on each nucleon and is proportional to $\boldsymbol{S}\cdot\boldsymbol{L}$. The existence of a spin–orbit interaction is amply supported by experimental evidence.

The total angular momentum of a nucleon is the sum of the orbital and spin angular momenta $\boldsymbol{J} = \boldsymbol{L} + \boldsymbol{S}$, with $|\boldsymbol{J}| = \sqrt{j(j+1)}\hbar$. Since $\boldsymbol{S}$ can be either parallel or antiparallel to $\boldsymbol{L}$, giving two different values of the total angular momentum $\boldsymbol{J}$, each (n, l) level is separated into two parts by the spin–orbit interaction, with the lower energy corresponding to $\boldsymbol{L}$ and $\boldsymbol{S}$ parallel. Thus a nucleon energy state is characterized by the quantum numbers n, l, j, with $j = l \pm \frac{1}{2}$. For a given l, the state with $j = l + \frac{1}{2}$ has lower energy than the state with $j = l - \frac{1}{2}$. This is contrary to the electronic case in atoms, and shows that the nuclear spin–orbit interaction is not of electromagnetic origin.

To each n, l, j, state there are $2j + 1$ values of m corresponding to the $2j + 1$ possible orientations of $\boldsymbol{J}$ relative to a given axis corresponding to $J_z = m\hbar$ with

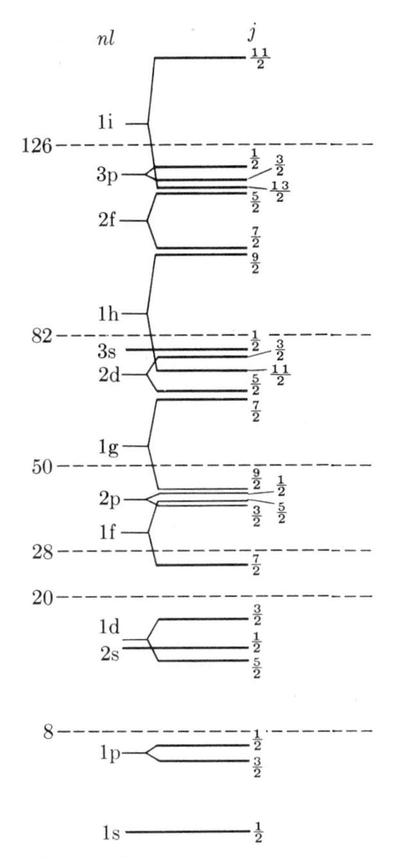

Figure 39.8 Single-particle energy levels in the shell model.

$m = \pm j, \pm(j-1), \ldots \pm\frac{1}{2}$. Thus according to the exclusion principle, the maximum number of neutrons or protons in a given n, l, j level or shell is $2j+1$; that is,

j	$\frac{1}{2}$	$\frac{3}{2}$	$\frac{5}{2}$	$\frac{7}{2}$	$\frac{9}{2}$	$\frac{11}{2}$	$\frac{13}{2}$	...
Maximum number of protons or neutrons	2	4	6	8	10	12	14	...

As for atomic electrons, each nucleon state is designated by the letters *s*, *p*, *d* etc., corresponding to the value of l, and a subscript giving the value of j. Thus for $l = 0$ we have an $s_{1/2}$ state, for $l = 1$ we have states $p_{1/2}$ and $p_{3/2}$, for $l = 2$ the states are $d_{3/2}$ and $d_{5/2}$, and so on.

The arrangement of single-particle energy levels, shown schematically in Figure 39.8, indicates how each (n, l) level, due to the average central force, is split by the spin–orbit interaction. The ordering follows the requirements of the experimental evidence. The energy gaps at the magic numbers occur every time a new high value of l appears, producing a large spin–orbit splitting.

39.9 Nuclear radiative transitions

A nucleus, like an atom or a molecule, may have several excited states. These states may be classified into two groups: **particle excitations** and **collective excitations**. In a particle excitation, one or more nucleons are shifted to another level of higher energy without essentially modifying the motion of the other nucleons. This is similar to electronic excitations in atoms and molecules. Particle excitation energies are of the order of one MeV.

Collective excitations are a result of the tight binding of the nucleons, so that if the motion of one nucleon is slightly distorted, the disturbance extends to the rest of the nucleons in the nucleus. Nucleons occupying unfilled shells in nuclei far from magic numbers exert a sort of polarizing action on the core and tend to give the nucleus an equilibrium nonspherical shape closely resembling an ellipsoid. This nuclear deformation is particularly large for nuclei in the regions $90 < N < 114$ and $Z > 88$, where the number of nucleons outside closed shells is relatively large. Deformed nuclei, like molecules, can rotate in space in what is called **collective rotation**. The rotational energy is quantized, and therefore a deformed nucleus has several rotational energy levels. For even–even nuclei, the rotational energy levels are given by

$$E_{\text{rot}} = \frac{\hbar^2}{2\mathscr{I}} I(I+1) \tag{39.7}$$

where $I(I+1)\hbar^2$ is the square of the angular momentum of rotation of the whole nucleus and $\mathscr{I}$ is the effective moment of inertia. For symmetry reasons, I can take only even values; that is, $I = 0, 2, 4, \ldots$. The effective moment of inertia of the nucleus $\mathscr{I}$ is always less than the moment of inertia obtained if the nucleus were considered a rigid ellipsoid. This shows that not all nucleons participate in the same way in the rotational motion. Figure 39.9 shows the rotational energy levels for ^{180}Hf and ^{238}Pu. These energy levels clearly follow the pattern given by Equation 39.7. For nuclei other than even–even ones, the arrangement of rotational levels is more complex.

Some nuclei are capable of sustaining vibrations in shape around an equilibrium shape without a change in volume, resulting in **collective vibrations**.

Each nucleus may have several modes of vibration. A nucleus may be in several vibrational excited states that are separated by the same energy $\hbar\omega$, where ω is the vibrational angular frequency. Figure 39.10 shows some vibrational levels of ^{82}Kr, ^{126}Te, ^{136}Xe and ^{192}Pt.

Both collective rotational and vibrational excitation energies are in general much less than particle excitation energies, and amount to a few keV. This accounts for the low-lying first excited levels of highly deformed nuclei that lie between magic-number nuclei. Nuclei with complete shells are spherical and do not exhibit low-lying rotational or vibrational levels. Nuclei with almost-complete shells are only slightly deformed and do not exhibit rotational excited levels, but only vibrational levels. From our discussion we see that the energy levels of a nucleus are as complex as those of a molecule. However, to predict theoretically the energy levels of a nucleus is more difficult than to predict those of a molecule because of our incomplete knowledge of the nuclear interaction.

A nucleus may be carried to an excited energy level when it absorbs a photon of the proper energy or when it undergoes an inelastic collision with a nearby passing fast particle, such as a proton or a neutron. An excited nucleus may give off its excess energy and undergo a transition to the ground level with emission of electromagnetic radiation or **γ-rays**. Thus the nuclear γ-ray spectrum is similar in origin to atomic and molecular spectra. The γ-spectrum results from the readjustment of the motion of the nucleons in a transition between two nuclear stationary states. Some γ-transitions of ^{111}Cd, ^{131}Xe and ^{130}Xe are shown in Figure 39.11. The angular momentum of each level is given on the left and the energy of each level relative to the ground state appears on the right.

Because γ-ray photons are very energetic, they produce ionization, molecular dissociation, atomic displacements in solids, and even excitations in nuclei and emission of nucleons when propagating through matter. For that reason, appropriate protection when handling sources of γ-rays is a necessity. But, for the same reasons, γ-emitters are widely used in medicine and industry.

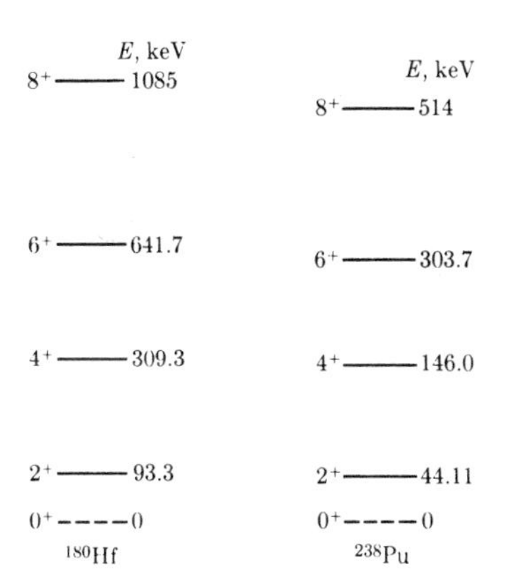

Figure 39.9 Rotational energy levels of ^{180}Hf and ^{238}Pu. The value of I is given on the left of each energy level.

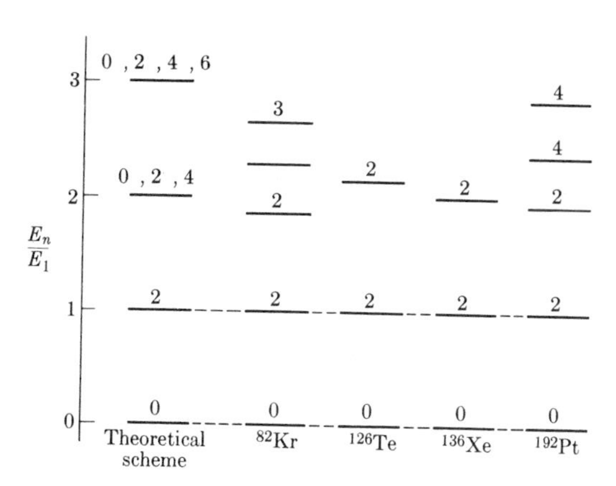

Figure 39.10 Vibrational energy levels of some nuclei.

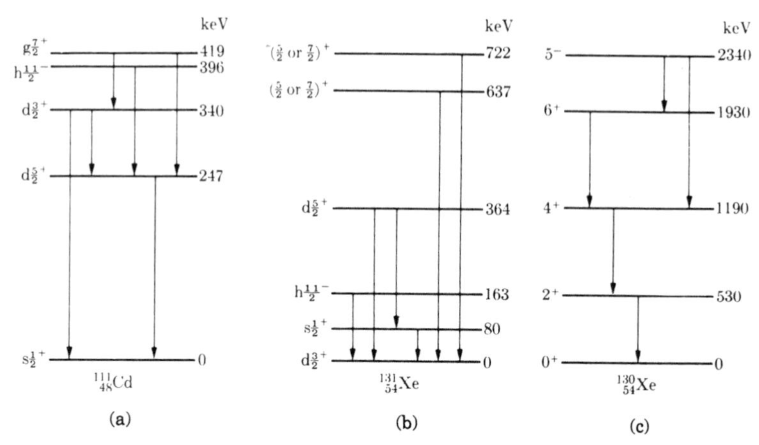

Figure 39.11 Radiative γ-transitions of $^{111}_{48}$Cd, $^{131}_{54}$Xe and $^{130}_{54}$Xe.

QUESTIONS

39.1 What is the difference between isotopes and isobars?

39.2 How do the 'volume' and the 'surface' of a nucleus vary with the mass number A?

39.3 Why must we invoke the existence of the nuclear interaction to explain nuclear stability?

39.4 What is the evidence suggesting the 'pairing of nucleons' in a nucleus?

39.5 Can you estimate how the electric component of the binding energy of a nucleus must vary with Z and A?

39.6 Is the mass of a nucleus equal to, larger than or smaller than the sum of the masses of the nucleons? How is it related to the strength of the nuclear force?

39.7 What is the evidence suggesting the nuclear force is of short 'range'?

39.8 What is the experimental evidence that indicates that the nuclear force is not central?

39.9 What information about the nuclear force is obtained from neutron–proton scattering?

39.10 Why is a strong spin–orbit interaction required to explain the shell structure of nuclei?

39.11 Discuss the concept of 'magic numbers' in nuclei. How are they related to the nuclear spin–orbit interaction?

39.12 Why does Figure 39.7 point toward magic numbers?

39.13 What are the similarities and differences between Figure 39.7 for nuclei and Figure 23.17 for atoms?

39.14 Prepare a table comparing the properties of the nuclear force with the electromagnetic interaction.

39.15 Look at a table of nuclides and identify some 'doubly magic' nuclei. Also compare the number of isotopes of magic nuclei with those of other nuclei. What do you conclude?

39.16 Explain the difference between particle excitations and collective excitations in nuclei.

PROBLEMS

Note: A chart of nuclides is necessary to solve some of the problems for this chapter.

39.1 Ordinary boron is a mixture of the ^{10}B and ^{11}B isotopes. The composite (or chemical) atomic mass is 10.811 amu. What percentage of each isotope is present in natural boron (a) by number and (b) by mass?

39.2 (a) Calculate the nuclear radius of ^{16}O, ^{120}Sn and ^{208}Pb. (b) Calculate the radius of the K-electron orbits for these nuclei. What do you conclude?

39.3 Estimate the kinetic energy of a nucleon inside a nucleus, by (a) using the quantum mechanical picture of a particle in a potential box of width 10^{-15} m and (b) considering the de Broglie wavelength λ of the nucleon, which is of the order of $2\pi r$, where r is 10^{-15} m.

39.4 (a) Estimate the Coulomb repulsion energy of the two protons in ^{3}He (assume that they are 1.7×10^{-15} m apart). (b) Compare this energy with the difference in the binding energies of ^{3}H and ^{3}He. Is the result compatible with the assumption that nuclear forces are charge independent?

39.5 Calculate the total binding energy and the binding energy per nucleon for ^{7}Li, ^{16}O, ^{57}Fe and ^{176}Lu.

39.6 (a) Calculate the binding energies for ^{14}O, ^{15}O, ^{16}O, ^{17}O, ^{18}O and ^{19}O. (b) Repeat for ^{14}C, ^{15}N, ^{16}O, ^{17}F and ^{18}Ne. (c) From the results of (a) and (b), explain the variation in binding energy as a neutron or a proton is added to a nucleus.

39.7 The following mass differences have been found experimentally:

(a) $^1H_2 - {}^2H = 1.5434 \times 10^{-3}$ amu
(b) $3\,{}^2H - \frac{1}{2}\,{}^{12}C = 4.2300 \times 10^{-2}$ amu
(c) $^{12}C\,{}^1H_4 - {}^{16}O = 3.6364 \times 10^{-2}$ amu

On the basis of $^{12}C = 12.000$ amu, calculate the atomic masses of ^{1}H, ^{2}H and ^{16}O. Compare with experimental values.

39.8 An aluminum foil scatters 10^3 α-particles per second in a given direction. If the aluminum foil is replaced by a gold foil of identical thickness, calculate the number of α-particles that will be scattered per second in the same direction. (It may be shown that the probability of scattering is directly proportional to the atomic number of the scatterer.)

39.9 A beam of 12.75 MeV α-particles is scattered by an aluminum foil. It is found that the number of particles scattered in a given direction begins to deviate from the predicted coulomb scattering value at approximately 54°. If the α-particles are assumed to have a radius of 2×10^{-15} m, estimate the radius of the aluminum nucleus. (Hint: use the equation for the distance of closest approach given in Example 9.10, where k is replaced by $zZe^2/4\pi\varepsilon_0$ and z is the charge of the α-particle and Z is the atomic number of aluminum.)

39.10 The moment of inertia of a nucleus of mass M and average radius R, if considered a solid sphere, is $I = \frac{2}{5}MR^2$. With a mass number A equal to (a) 50, (b) 100 and (c) 150, estimate the energy (in MeV) of the γ-ray emitted in a transition from a rotational energy level with $l = 2$ to one with $l = 0$. Compare with the energy of γ-rays emitted by even–even nuclei in those regions. What do you conclude?

39.11 An ellipsoidal nucleus of axial symmetry has semi-major axes a and b. The mean nuclear radius R is the radius of the sphere that has the same volume as the ellipsoid, so that $R^3 = ab^2$. (a) Verify that if $a = R + \Delta R$ then $b = R - \frac{1}{2}\Delta R$, to a first approximation. The ratio $\delta = \Delta R/R$ is called the *deformation* of the nucleus. (b) The moment of inertia of an ellipsoid of revolution around an axis perpendicular to the symmetry axis is $I = \frac{1}{5}M(a^2 + b^2)$. Express the moment of inertia in terms of δ, neglecting ΔR^2 terms. (c) Use your result to calculate the rotational energy levels of a deformed nucleus.

39.12 The **electric quadrupole moment** of a deformed nucleus is $Q = \frac{6}{5}Z(a^2 - b^2)$. Express it in terms of δ (see Problem 39.11).

39.13 (a) From the rotational energy levels shown in Figure 39.9, estimate the moment of inertia of ^{180}Hf. (b) Estimate the deformation $\Delta R/R$ of ^{180}Hf, and calculate the value obtained from the electric quadrupole moment. (Hint: recall Problem 39.11.)

39.14 Refer to Figure 39.11(a) and (b), and assume that each of the levels shown is due to a single particle transition; with the aid of the shell-model level scheme of Figure 39.8, write the configuration for the ground state and each of the excited states.

39.15 (a) Using the result of Problem 25.13, show that the electric energy of a nucleus of atomic number Z and mass number A is $(1/4\pi\varepsilon_0)\frac{3}{5}(Ze^2/r_0A^{1/3})$ or $0.62Z^2A^{-1/3}$ MeV. (b) Calculate its value for the nucleus $^{238}_{92}$U. Compare with the binding energy of the nucleus. (c) Repeat the calculation for $^{56}_{26}$Fe. What is your conclusion?

39.16 Compare the nuclear density with the density of a black hole having the same mass as the Sun (recall Example 11.5). What do you conclude?

40 Nuclear processes

Enrico Fermi was one of the leaders in the early development of nuclear physics in the 1930s. He studied β-decay, formulating a theory of the process (1933), and the production of radioisotopes by neutron bombardment, obtaining the first transuranium elements (1934). Fermi's research in neutron physics was instrumental in the study of nuclear fission. In 1942 he and his colleagues at the University of Chicago achieved the first successful nuclear chain reaction. Fermi was also interested in thermodynamics, statistical mechanics and field theory. He was a great science communicator and educator.

Notes

40.1 Introduction

In this chapter, we will examine nuclear processes, such as radioactive decays and nuclear reactions, where there is a rearrangement in the configuration of the nucleons along with absorption or emission of energy. Many nuclear processes occur naturally, but others are produced artificially in the laboratory using different types of accelerating machines or nuclear reactors.

40.2 Radioactive decay

Most nuclei are stable combinations of nucleons (Figure 40.1). However some combinations of protons and neutrons do not lead to a stable nuclear configuration. These nuclei are therefore unstable or **radioactive**. Unstable nuclei tend to approach a stable configuration by releasing certain particles as well as energy. These particles, when they were first observed at the end of the last century by Antoine Becquerel

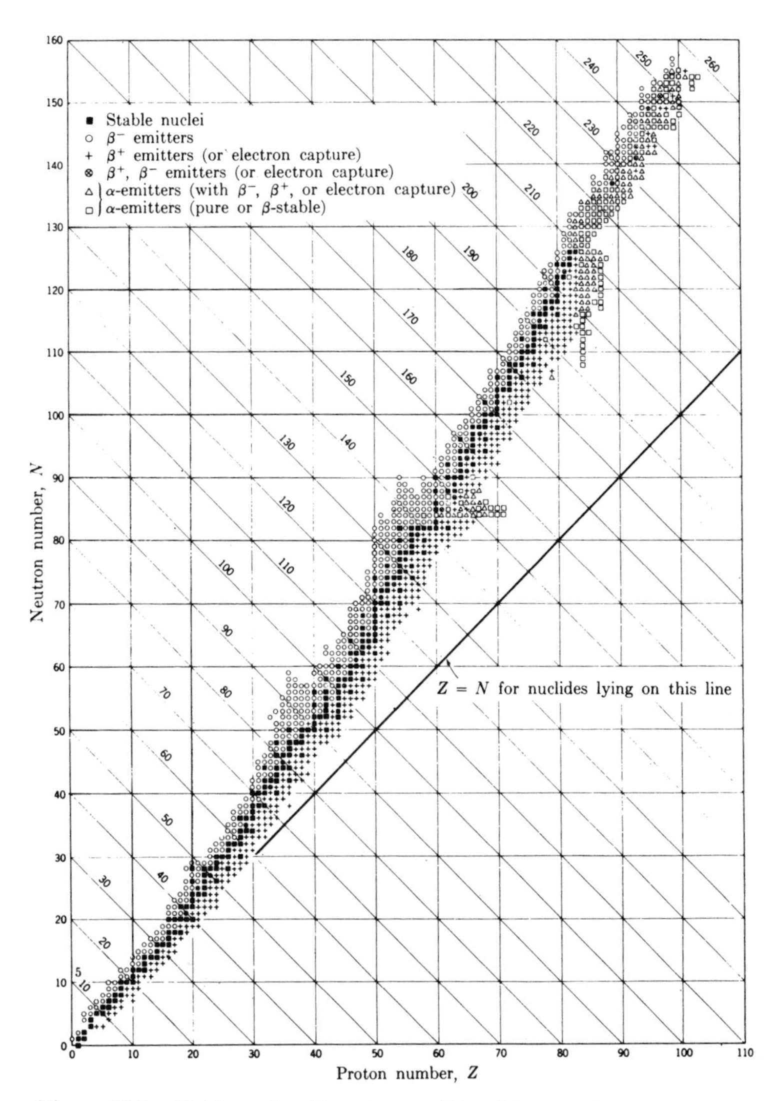

Figure 40.1 Stable and radioactive nuclides. The 45° lines correspond to nuclides of equal mass number A as shown.

(1852–1908). Pierre (1859–1906) and Marie Curie (1867–1934), and others, were designated as α- and β-particles.

α-particles are helium nuclei, ${}^4_2He$, composed of two protons and two neutrons. **β-particles** are electrons, which carry a negative charge $-e$, or positrons, with a charge of $+e$. The two types of β-decay are designated β^- and β^+, respectively. In β-decay, a neutrino is also emitted. In both α- and β-decay a new nucleus results. The residual or **daughter nucleus** is sometimes left in an excited state and, in the transition to its ground state, emits γ-rays.

Most isotopes of elements with $Z > 81$ (or $A > 206$) are naturally radioactive. A few other naturally existing lighter nuclei, such as ${}^{14}C$ and ${}^{40}K$, are also radioactive. Many more radioactive nuclei have been produced in the laboratory

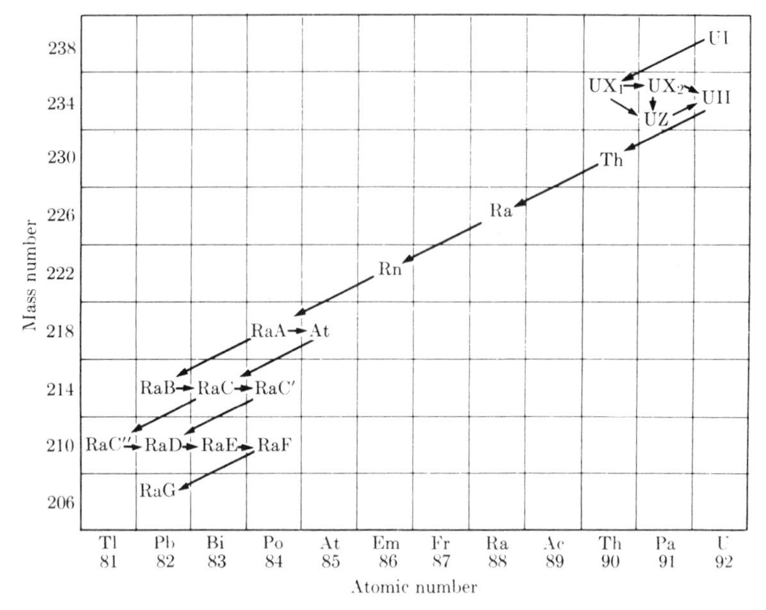

Figure 40.2 The naturally radioactive uranium series.

using nuclear reactors and particle accelerators. Figure 40.2 shows one of the three natural radioactive chains, the so-called **uranium series**, which begins with $^{238}_{92}U$ and ends when the stable nucleus $^{206}_{82}Pb$, or RaG, is produced. The two other natural radioactive chains are the **actinium series** and the **thorium series**. The heavier nuclides in these two series are $^{235}_{92}U$ and $^{232}_{90}Th$, respectively.

It has been observed that all radioactive processes follow an exponential law of decay. Therefore, if N_0 is the initial number of unstable nuclei, the number of nuclei remaining after a time t is given by

$$N = N_0 e^{-\lambda t} \tag{40.1}$$

where λ is a constant characteristic of each nuclide, called the **distintegration constant**. It is expressed in s^{-1} (or in the reciprocal of any other time unit). Equation 40.1 is represented in Figure 40.3. For each radioactive nuclide there is a fixed time interval T, called the **half-life**, during which the number of nuclei at the beginning of the interval is reduced, by the end of the interval, to one-half. So if we initially had N_0 nuclei, after time T only $N_0/2$ are left, after time $2T$ only $N_0/4$ remain, and so on. To find this time T, we set $N = \frac{1}{2}N_0$ and $t = T$ in Equation 40.1. Then $\frac{1}{2}N_0 = N_0 e^{-\lambda T}$ or $e^{\lambda T} = 2$. Taking logarithms, we have $\lambda T = \ln 2 = 0.693$ or

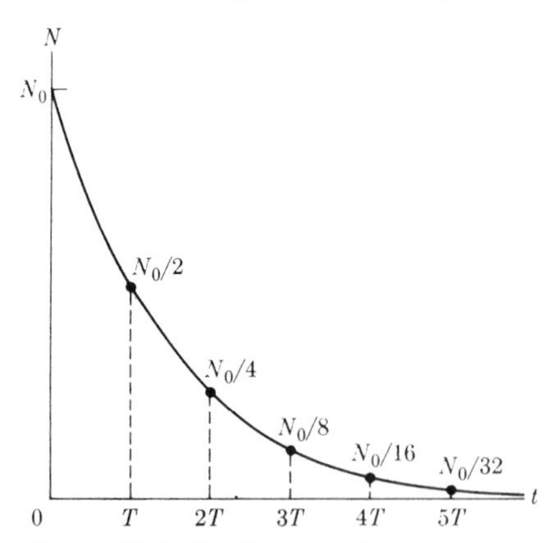

Figure 40.3 Radioactive decay as a function of time.

$$T = 0.693/\lambda \tag{40.2}$$

which relates T and λ. The recorded half-lives go from a great many years – such as the half-life of the α-decay of ^{209}Bi, which is about 2×10^{18} years, and the β^--decay of ^{115}In, which is about 6×10^{14} years – down to fractions of a second, such as ^{8}Be, which has an α-decay half-life of the order of 10^{-16} s.

The distintegration rate of radioactive nuclei is called the **activity** of the substance. From Equation 40.1 we can find the rate at which nuclei disintegrate:

$$\frac{dN}{dt} = -\lambda N_0 e^{-\lambda t} = -\lambda N \tag{40.3}$$

This indicates that the disintegration rate dN/dt is proportional to the number of nuclei present. Therefore the disintegration rate or activity decreases in the same proportion and with the same half-life as the number of nuclei N.

Disintegration rates are usually expressed in **curies**, abbreviated Ci, in honor of Pierre and Marie Curie, the discoverers of polonium and radium. The curie is defined as the activity of a substance when 3.7000×10^{10} nuclei disintegrate per second. This rate is approximately equal to the activity of 1 g of Ra. Submultiples of the basic unit are the millicurie ($1\ \text{mCi} = 10^{-3}\ \text{Ci}$) and the microcurie ($1\ \mu\text{Ci} = 10^{-6}\ \text{Ci}$).

Equations 40.1 and 40.3 are both statistical laws, which are valid only when the number of nuclei is very large. Therefore, we cannot speak of the half-life of a single nucleus or predict with certainty when a given nucleus will disintegrate. On the other hand, λ gives the probability per unit time for a nucleus to disintegrate. It can be calculated based on theoretical considerations.

EXAMPLE 40.1

Calculation of the mass of 1.00 Ci of ^{14}C. The half-life of ^{14}C is 5570 years.

▷ Since $T = 5570\ \text{yr} = (5.570 \times 10^3\ \text{yr}) \times (3.1536 \times 10^7\ \text{s yr}^{-1}) = 1.757 \times 10^{11}\ \text{s}$, the disintegration constant is $\lambda = 0.693/T = 3.94 \times 10^{-12}\ \text{s}^{-1}$. Also $|dN/dt| = 1\ \text{Ci} = 3.70 \times 10^{10}\ \text{s}^{-1}$. Therefore, using Equation 40.3 with absolute values, we find that

$$N = \frac{1}{\lambda}\left|\frac{dN}{dt}\right| = 9.39 \times 10^{21} \text{ nuclei of C}$$

which is also the number of carbon atoms present. The atomic mass of ^{14}C is 14.0077 amu. Thus the mass of the above number of carbon atoms is

$$M = (14.0077 \times 1.6604 \times 10^{-27}\ \text{kg atom}^{-1}) \times (9.39 \times 10^{21}\ \text{atoms}) = 2.18 \times 10^{-4}\ \text{kg}$$

40.3 α-decay

α-decay consists in the emission of an α-particle or helium nucleus, $^4_2$He, composed of two protons and two neutrons. When a nucleus decays and emits an α-particle, the daughter nucleus has an atomic number two units less and a mass number four units less than its parent (Figure 40.4). Thus, if we denote the parent and

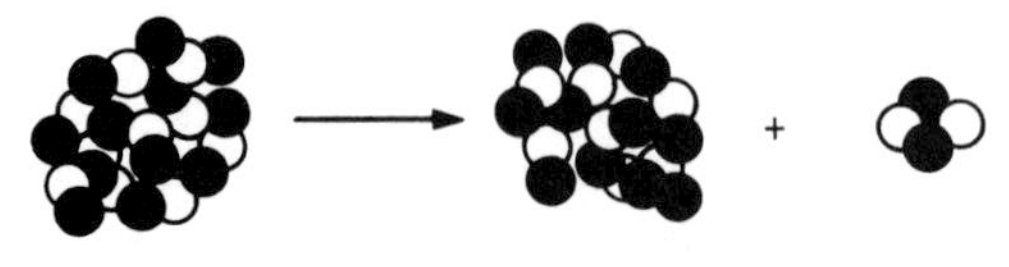

Figure 40.4 α-decay. The daughter nucleus has two protons less and two neutrons less.

daughter nuclei by X and Y, respectively, we may write the process of α-decay as

$$^{A}_{Z}\mathrm{X} \rightarrow {}^{A-4}_{Z-2}\mathrm{Y} + {}^{4}_{2}\mathrm{He} \qquad \textbf{(40.4)}$$

For example, $^{238}_{92}\mathrm{U}$ is an α-emitter and disintegrates according to the scheme $^{238}_{92}\mathrm{U} \rightarrow {}^{234}_{90}\mathrm{Th} + {}^{4}_{2}\mathrm{He}$. Most α-emitters are heavy nuclei, corresponding to nuclides at the end of the periodic table.

α-particles, being double-magic nuclei, have an extraordinary stability, and therefore behave in many instances as a single unit or particle, similar to protons and neutrons, and for that reason they are also called **helions**. Ernest Rutherford used α-particles as projectiles to probe the interior of the atom and establish the nuclear model (recall Section 23.3). We must not think, however, that α-particles exist as such inside the nucleus. Supposedly, there are certain correlations in the motion of the nucleons that occasionally cause some of them to group themselves into an α-particle-like configuration, which acts as a dynamical unit for a short time. When such a unit is near the surface of the nucleus, there is a certain probability that the group of nucleons escapes as an α-particle, if it is energetically possible.

The potential energy of interaction of an α-particle with the rest of the nucleus, which is similar to that of a proton (recall Figures 37.17 and 39.4), is indicated in Figure 40.5. The energy of the α-particles (about 4 to 9 MeV) is less than the height of the Coulomb barrier (about 40 MeV for most α-emitters) at the surface of the nucleus, and the α-particle can escape only by penetrating the potential barrier. The disintegration probability per unit time, λ, may be calculated in terms of the probability P of penetrating the barrier (Section 37.8). The quantity P is found using quantum mechanical methods. The results agree fairly well with experimental values for λ.

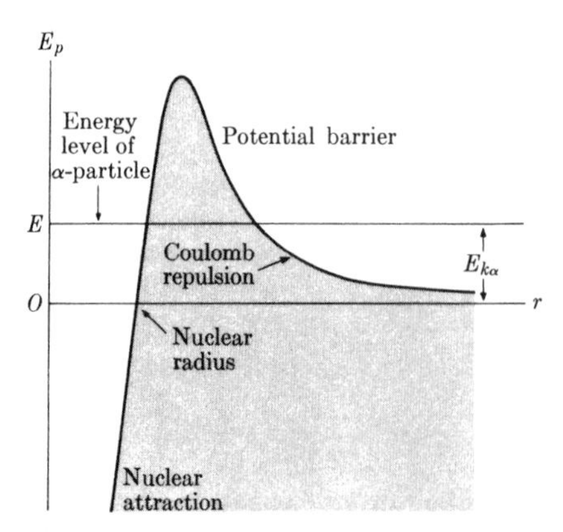

Figure 40.5 Potential energy of an α-particle and a nucleus.

The energy released in α-decay, designated Q, is obtained from the mass change in the process (recall Section 20.4); that is,

$$Q = (m_{\mathrm{X}} - m_{\mathrm{Y}} - m_{\alpha})c^2 \qquad \textbf{(40.5)}$$

For decay to occur naturally, it is necessary that $Q > 0$. When the masses are expressed in amu and Q is expressed in MeV, Equation 40.5 becomes

$$Q = 931.48(m_{\mathrm{X}} - m_{\mathrm{Y}} - m_{\alpha}) \qquad \textbf{(40.6)}$$

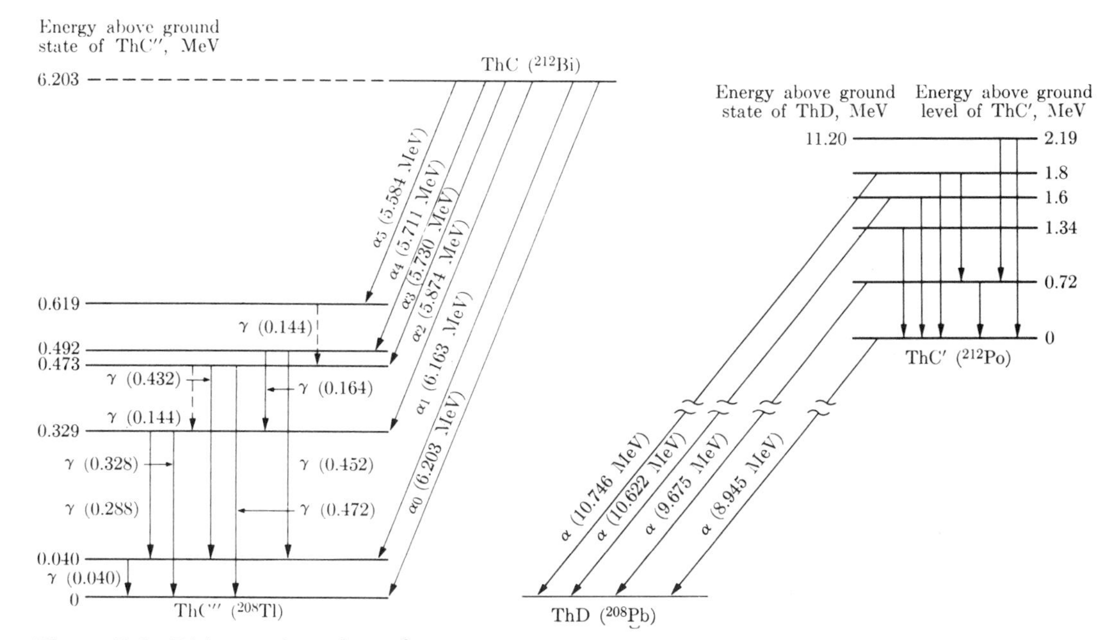

Figure 40.6 Disintegration scheme for α-decay of ${}^{212}_{83}\text{Bi}$ to ${}^{208}_{81}\text{Tl}$.

Figure 40.7 Disintegration scheme for α-decay of ${}^{212}_{84}\text{Po}$ to ${}^{208}_{82}\text{Pb}$.

Note that α-decay, as expressed by Equation 40.4, is a two-body process equivalent to the explosion of a grenade into two fragments. Therefore, conservation of energy and momentum require that, for each α-decay, the α-particles have a well-defined energy, a fact confirmed by experiment. The energy of α-particles is slightly less than Q because part of the energy is carried away by the recoiling daughter nucleus.

In many cases α-particles from a given nuclide do not all have the same energy. For example, the α-particles from ${}^{238}\text{U}$ have energies of 4.18 MeV and 4.13 MeV. This occurs because, although the parent nucleus may be in its ground state, the daughter nucleus may be formed in its ground state or in an excited state. This situation is illustrated in Figure 40.6, which shows the decay scheme of ${}^{212}_{83}\text{Bi}$; the six α-particle transitions are indicated by the arrows. In this case the α-particles are accompanied by γ-rays, as shown by the vertical lines. In other cases, the reverse situation occurs: the parent nucleus may be in the ground state or in one of several excited states, and the daughter nucleus be in the ground state, as shown in Figure 40.7 for the disintegration scheme of ${}^{212}_{84}\text{Po}$. Some γ-transitions are also shown.

EXAMPLE 40.2

Calculation of the kinetic energy of α-particles emitted from ${}^{232}_{92}\text{U}$.

▷ The decay process is ${}^{232}_{92}\text{U} \rightarrow {}^{228}_{90}\text{Th} + {}^{4}_{2}\text{He}$. The masses that are involved are $m_X = 232.1095$ amu, $m_Y = 228.0998$ amu and $m_\alpha = 4.0039$ amu. Applying Equation 40.6, we obtain $Q = 5.40$ MeV. A positive Q means that the process may occur spontaneously. The energy Q is distributed between the α-particle and the daughter nucleus in inverse proportion to their masses (recall Example 14.5), which in this case is $4/228$. Therefore the kinetic energies are $E_{k(Th)} = 0.10$ MeV and $E_{k(\alpha)} = 5.30$ MeV. We obtain this value of $E_{k(\alpha)}$ assuming that the $^{238}_{90}Th$ is in its ground state. But if $^{238}_{90}Th$ is left in an excited state, then the value of $E_{k(\alpha)}$ is less. The experimental value of the most energetic α-particles from $^{232}_{92}U$ is $E_k = 5.32$ MeV, so that our interpretation seems to be correct.

EXAMPLE 40.3

Stability of $^{232}_{92}U$ relative to the emission of other kinds of particles.

▷ Other than α-decay, no nuclei exhibit proton, neutron, deuteron or some other type of decay, because the Q-values for these processes are negative. Therefore these decays cannot occur unless energy is supplied and the parent nucleus is raised to an excited state, which may happen in certain nuclear reactions. As an illustration, we may compute the Q-values for the emission of several kinds of particles from the $^{232}_{92}U$ nucleus, using Equation 40.6 with m_α and m_Y replaced by the corresponding masses of the emitted particle and the daughter nucleus. The results are listed in Table 40.1. All Qs are negative, so that $^{232}_{92}U$ is stable against decay into such products. The reason α-emission by $^{232}_{92}U$ is possible lies in the relatively small mass of the α-particle due to its relatively large binding energy, which is equivalent to appreciably reducing the 'effective' masses of each of the two protons and two neutrons that make it.

Table 40.1 Q-values for emission of different nuclear particles from ^{232}U

Particle	Mass	Daughter	Mass	Q (MeV)
n	1.0090	^{231}U	231.1082	−7.16
p(^1H)	1.0081	^{231}Pa	231.1078	−6.05
d(^2H)	2.0147	^{230}Pa	230.1060	−10.4
t(^3H)	3.0170	^{229}Pa	229.1033	−10.1

40.4 β-decay

Nuclei that have too many neutrons compared with the number of protons may be unstable and emit electrons (charge $-e$), a process called β^--decay. The daughter nucleus has the same mass number, A, but has an atomic number larger by one than the parent nucleus. That is, in β^--decay a neutron is replaced by a proton (Figure 40.8). The process can thus be expressed by

$$^A_ZX \rightarrow {}_{Z+1}^{\;\;A}Y + e^- \qquad \textbf{(40.7)}$$

Note that total charge is conserved, since the charge on the left of Equation 40.7 is Ze and on the right it is $(Z+1)e - e = Ze$. Also the total number of nucleons is conserved, since A remains the same. For example, ^{14}C is a β^--emitter and transforms according to the scheme $^{14}_6C \rightarrow {}^{14}_7N + e^-$.

Nuclei that have a relatively large number of protons compared to neutrons may also be unstable and suffer β^+-decay, a process that consists in the emission of **positrons** (charge $+e$). Recall that positrons are particles with the same mass and spin as electrons, but their charge is positive instead of negative. In β^+-decay, the

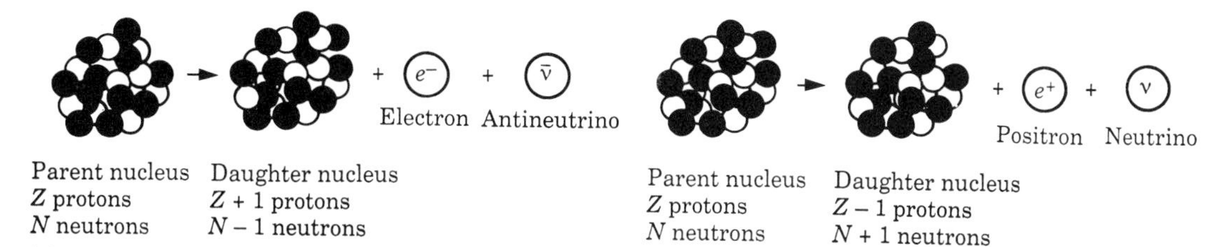

Figure 40.8 (a) In β^--decay, the daughter nucleus gains one proton and loses one neutron. (b) In β^+-decay, the daughter nucleus loses a proton and gains a neutron.

atomic number of the daughter nucleus is smaller by one unit, in agreement with the law of conservation of charge, but its mass number is the same as that of the parent nucleus, in agreement with the conservation of nucleons. Thus in β^+-decay a proton is replaced by a neutron. Therefore the process may be expressed by

$$ {}^{A}_{Z}X \rightarrow {}_{Z-1}{}^{A}Y + e^+ \tag{40.8}$$

As an example, ${}^{11}_{6}C$ is a β^+-emitter and transforms according to the scheme ${}^{11}_{6}C \rightarrow {}^{11}_{5}B + e^+$.

The daughter nucleus resulting from β-decay may be left in its ground state or in an excited state; in the latter case the process is followed by γ-emission. Figure 40.9 shows the disintegration schemes of some β-emitters.

An interesting feature of β-decay is that the electrons and positrons are emitted with a wide range of kinetic energies (and momenta), from zero to a maximum compatible with the total energy available. In other words, the electrons and positrons have a continuous energy spectrum. But Equations 40.8 and 40.9

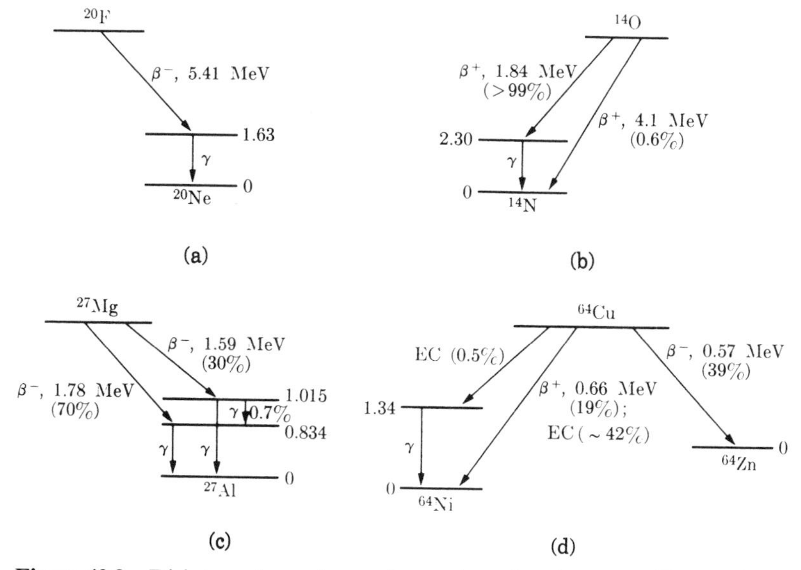

Figure 40.9 Disintegration schemes for various β-emitters. (a) ${}^{20}_{9}F$, (b) ${}^{14}_{8}O$, (c) ${}^{27}_{12}Mg$, and (d) ${}^{64}_{29}Cu$.

are two-body processes, similar to α-decay, and the laws of conservation of energy and momentum require that in the center of mass frame of reference, where the parent nucleus is at rest, the energy released must be split in a fixed ratio among the daughter nucleus and the electron or positron. This is in contradiction to experimental results.

To settle this new difficulty, Wolfgang Pauli suggested, in 1930, that another particle must also be involved in β-decay, so that three particles result. The third particle must be neutral, to comply with the law of conservation of charge, and of very small mass, since the total mass is essentially accounted for by the other observed particles. For these two reasons the new particle was called a **neutrino.** (This name was proposed by Enrico Fermi, and means 'the little neutron'.) It is designated by the symbol ν. It has been found that there are two kinds of neutral particles, almost identical, associated with β-decay. One, the neutrino, is emitted in β^+-decay, while the particle emitted in β^--decay is an **antineutrino**, and is designated by $\bar{\nu}$. However, in this chapter we shall refer in most cases to both particles by the name 'neutrino'. Therefore processes 40.7 and 40.8 must be rewritten in the following form:

$$\beta^-\text{-decay:} \quad {}^A_Z\text{X} \rightarrow {}_{Z+1}^{\ \ A}\text{Y} + \text{e}^- + \bar{\nu} \tag{40.9}$$

$$\beta^+\text{-decay:} \quad {}^A_Z\text{X} \rightarrow {}_{Z-1}^{\ \ A}\text{Y} + \text{e}^+ + \nu \tag{40.10}$$

The neutrino is assumed to carry away the energy and momentum needed to restore the conservation of both quantities. In addition, the neutrino must have spin $\frac{1}{2}$ to compensate for the spin of the electron and assure the conservation of angular momentum. In the center of mass frame of reference, the momenta of the three resulting particles must add to zero (Figure 40.10). But there is an infinite number of ways in which the total energy released can be split among the three products, and this very nicely explains the continuous energy distribution of the electrons or positrons.

In some instances, a nucleus may capture an electron from one of the innermost atomic shells, such as a K-electron. These electrons have very penetrating orbits that reach very close to the nucleus; therefore, their probability of being captured by a proton is relatively large. This process, called **electron capture** (EC), results in the replacement of a proton by a neutron in the daughter nucleus. It may be expressed by

$${}^A_Z\text{X} + \text{e}^- \rightarrow {}_{Z-1}^{\ \ A}\text{Y} + \nu$$

Note that the electric charge and the nucleon number are conserved in the process. Also, for electron capture to occur, energy must be available. Electron capture is followed by X-ray emission by the daughter nucleus when an outer electron falls into the vacant state left in the K-shell. These X-rays are the same as the characteristic rays of the daughter atom, as discussed in Section 39.9. Sometimes the X-rays knock out an electron in an upper shell, in a sort of internal photoelectric effect.

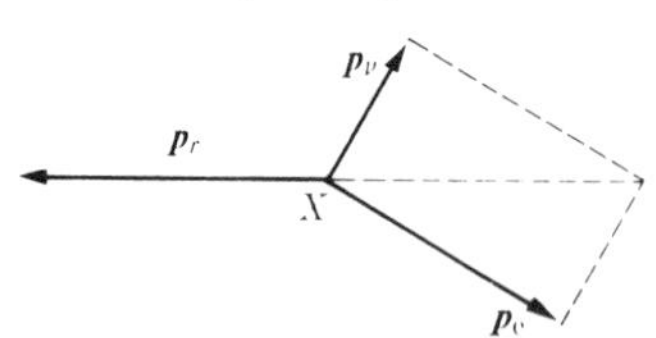

Figure 40.10 Momentum conservation in β-decay.

So far we may say that the neutrino is an interesting invention to save two conservation laws, but there is ample experimental evidence of its existence. On the theoretical side, a statistical analysis can be made to determine how the energy released in the decay process is shared by the electron (or the positron) and the neutrino. When this analysis is compared with the experimental energy distribution of the electrons (or the positrons) and the neutrino, excellent agreement is obtained, provided that the mass of the neutrino is very small (less than $10^{-3}m_e$). For most purposes the neutrino's mass is considered to be zero ($m_\nu = 0$). On the experimental side, since the neutrino is insensible to the action of electric and magnetic fields, its very small or zero mass does not allow us to use mass measurements to observe its emission or capture. In fact the neutrino eluded direct observation until 1956, when Clyde Cowan and Francis Reines observed processes that could only be triggered by absorption of neutrons. Since then neutrinos have been observed directly in many experiments. In fact, it is speculated that free neutrinos are very abundant in the universe and the Earth is subject to a flux of neutrinos from the Sun and from processes in outer space, such as supernova explosions.

Since the daughter nucleus ${}_{Z-1}^{\ A}\mathrm{Y}$ in process 40.9 has one neutron less and one proton more than the parent nucleus ${}_{Z}^{A}\mathrm{X}$, and the reverse occurs in process 40.10, we may explain these decay processes by assuming that, *in β^--decay, a neutron transforms into a proton* according to the scheme

$$\mathrm{n} \to \mathrm{p} + \mathrm{e}^- + \bar{\nu} \tag{40.11}$$

while *in β^+-decay, a proton transforms into a neutron* according to the schemes

$$\mathrm{p} \to \mathrm{n} + \mathrm{e}^+ + \nu, \qquad \mathrm{p} + \mathrm{e}^- \to \mathrm{n} + \nu \tag{40.12}$$

In any of these three ways a nucleus may get rid of its excess neutrons or protons without emitting either of these particles. The assumption expressed by relations 40.11 and 40.12 has far-reaching consequences, since it amounts to a statement that the nucleons, although they have well-defined properties, have an internal structure, a property to be discussed in Chapter 41, and may transform into other particles within the limitations imposed by the conservation laws. In fact, since the mass of the neutron exceeds the combined mass of the proton and the electron by 0.728 MeV, process 40.11 can take place with free neutrons. It has been observed that free neutrons disintegrate according to Equation 40.11 with a half-life of about 12 minutes. On the other hand, processes 40.12 cannot occur with free protons. It occurs only in nuclei where the protons can use part of the binding energy of the nucleus for the decay. This explains why hydrogen is abundant in the universe but there are no free neutrons.

From the experimental analysis of many β-decay processes and the need to explain the transformation of protons and neutrons into each other, it has been concluded that this process must be due to a special interaction different from the nuclear force and called the **weak interaction**. The strength of the weak interaction is of the order of 10^{-14} when compared with that of the strong or nuclear interaction, or about 10^{-12} when compared with the electromagnetic interaction.

EXAMPLE 40.4
Energy released in β-decay.

▷ As an illustration of the energy balance in β-decay, let us analyze the process ${}^{20}_{9}F \rightarrow {}^{20}_{10}Ne^* + e^- + \bar{\nu}$, which is followed by ${}^{20}_{10}Ne^* \rightarrow {}^{20}_{10}Ne + \gamma$ (Figure 40.9(a)). The asterisk is used to indicate an excited state. It has been determined experimentally that the maximum energy of the electrons in the β-decay is 5.41 MeV and the energy of the γ-ray is 1.63 MeV. This gives a combined energy of 7.04 MeV, to which we must add the mass of the electron (0.511 MeV), so that the total energy is 7.55 MeV. This energy is equivalent to a mass change of 7.56×10^{-3} amu. This is the difference in the masses of ${}^{20}_{9}F$ and ${}^{20}_{10}Ne$.

As a second example, consider the process ${}^{14}_{8}O \rightarrow {}^{14}_{7}N^* + e^+ + \nu$, which is followed by ${}^{14}_{7}N^* \rightarrow {}^{14}_{7}N + \gamma$ (Figure 40.9(b)). The maximum energy of the positron is 1.84 MeV and the energy of the γ-ray is 2.30 MeV, so that the total energy released is 4.14 MeV, to which we add the mass of the positron (0.511 MeV), giving a total of 4.65 MeV. It is possible also for ${}^{14}_{8}O$ to decay directly into ${}^{14}_{7}N$ by emitting positrons with a maximum energy of 4.1 MeV. Therefore the total energy is the same in both cases, as it should be, since the initial and final states are the same. In this analysis we have ignored all recoil effects.

40.5 Nuclear reactions

When two nuclei, overcoming their Coulomb repulsion, come within the range of the nuclear force, a rearrangement of nucleons may occur. This may result in a **nuclear reaction**, similar to the rearrangement of atoms in reacting molecules during a chemical reaction. Nuclear reactions are usually produced by bombarding a target nucleus (M_i) with a nuclear projectile (m_i), in most cases a nucleon (neutron or proton) or a light nucleus such as a deuteron or an α-particle. Heavier nuclei are not generally used because the electric repulsion between heavy nuclei requires a projectile with a large kinetic energy and the products of the collision might be very diverse. Sometimes photons are used as projectiles. Most of the reactions result in the same particle, or another particle (m_f) being ejected and a residual or final nucleus (M_f) that is left in its ground state or in an excited state.

In general, when the energies of the particles involved are not too high, a nuclear reaction supposedly occurs in two steps. First, the incoming particle or projectile is captured, resulting in the formation of an **intermediate or compound nucleus**, which is in a highly excited state. In the second step, the compound nucleus may be de-excited, either by emission of a particle, which might be the same as the incoming particle, or by some other means. For example, the bombardment of ${}^{14}_{7}N$ by α-particles can be written in the form

$${}^{14}_{7}N + {}^{4}_{2}He \rightarrow [{}^{18}_{9}F]^* \rightarrow {}^{1}_{1}H + {}^{17}_{8}O$$

where ${}^{18}_{9}F$ is the intermediate or compound nucleus.

Generally speaking, to a given first step in a nuclear reaction, there might be several modes of de-excitation for the compound nucleus. For example, when ${}^{27}_{13}Al$ is bombarded with protons, several products result, some of which are listed below:

$$^{27}_{13}\text{Al} + ^{1}_{1}\text{H} \rightarrow [^{28}_{14}\text{Si}]^* \rightarrow \begin{cases} ^{24}_{12}\text{Mg} + ^{4}_{2}\text{He} \\ ^{27}_{14}\text{Si} + \text{n} \\ ^{28}_{14}\text{Si} + \gamma \\ ^{24}_{11}\text{Na} + 3\,^{1}_{1}\text{H} + \text{n} \end{cases}$$

Nuclear reactions are essentially collision processes in which energy, momentum, angular momentum, number of nucleons and charge must be conserved; the methods of Newtonian and relativistic mechanics are used to compute some of these quantities.

If the incoming and outgoing particles are the same, the process is called **scattering**. Scattering is *elastic* if the nucleus is left in the same state, so that kinetic energy is conserved, and *inelastic* if the nucleus is left in a different state. In the inelastic case, the kinetic energy of the outgoing particle differs from the kinetic energy of the incoming particle (recall Section 14.9).

In many cases the nucleus that results from a nuclear reaction is unstable or radioactive. In fact, it is through nuclear reactions that the artificially radioactive nuclei are formed. Artificial radioactivity was discovered by Frédéric Joliot-Curie (1900–1958) and his wife Irene Joliot-Curie (1897–1956) in 1934, while they were studying nuclear reactions produced by bombarding light elements with α-particles. One of the reactions they observed was $^{10}_{5}\text{B} + ^{4}_{2}\text{He} \rightarrow [^{14}_{7}\text{N}]^* \rightarrow ^{13}_{7}\text{N} + \text{n}$. The nucleus $^{13}_{7}\text{N}$ is unstable and decays according to the scheme $^{13}_{7}\text{N} \rightarrow ^{13}_{6}\text{C} + \text{e}^+ + \nu$.

A simple way of producing β^- radioactive nuclei is by neutron capture, for which a sample of the material is exposed to a strong neutron flux. For example, when $^{59}_{27}\text{Co}$ is bombarded by neutrons, $^{60}_{27}\text{Co}$ is produced, which is β^- radioactive and decays into $^{60}_{28}\text{Ni}$ with a half-life of 5.27 years and the emission of an electron and two γ-rays of energies 1.17 MeV and 1.33 MeV according to

$$^{59}_{27}\text{Co} + \text{n} \rightarrow ^{60}_{27}\text{Co} \rightarrow ^{60}_{28}\text{Ni} + \text{e}^- + \bar{\nu} + \gamma_1 + \gamma_2$$

The radionuclide $^{60}_{27}\text{Co}$ is widely used in radiotherapy and for the analysis of defects in metal structures.

An interesting series of reactions are those that result from neutron capture and subsequent β^--decay of the uranium isotopes, producing new nuclei with $Z = 93$ (neptunium), $Z = 94$ (plutonium), $Z = 95$ (americium), up to $Z = 109$ called **transuranic**. Elements beyond plutonium have been produced in the laboratory in extremely small amounts by neutron capture.

Note 40.1 Discovery of the neutron

An illustration of the application of the conservation laws in nuclear processes is the story of the discovery of the neutron. In 1930 Walther Bothe (1881–1957) and his student Herbert Becker observed that when they bombarded boron and beryllium with α-particles, a highly penetrating radiation was produced. This radiation was not composed of charged particles because it was not affected by electric or magnetic fields. For that reason they thought that the radiation consisted of high energy γ-rays, and they wrote the reaction as

$$^{9}_{4}Be + ^{4}_{2}He \rightarrow [^{13}_{6}C]^* \rightarrow ^{13}_{6}C + \gamma$$

The Q of this reaction is 10.4 MeV. Since the kinetic energy of the α-particle was about 5 MeV, the total energy available is about 15 MeV. This energy must be shared by the $^{13}_{6}C$ atom and the γ-photon. Thus the γ-rays should have an energy less than 15 MeV. Two years later (1932) the Joliot-Curies observed that when the radiation from the above reaction was passed through a hydrogenous material, highly energetic protons were produced, with a maximum energy of about 7.5 MeV. The natural interpretation was to assume that the protons had been knocked out by collision with γ-photons produced in the above reaction. The most energetic protons should result from a head-on collision in which the photons recoil or are deflected 180°. Given that E_γ and $p_\gamma = E_\gamma/c$ are the energy and momentum of the incident photon, E'_γ and $p'_\gamma = E'_\gamma/c$ those of the recoil photon and E_k the energy of the protons, the conservation of energy and momentum give $E_\gamma = E'_\gamma + E_k$ and $E_\gamma/c = -E'_\gamma/c + (2m_p E_k)^{1/2}$, from which we get

$$E_\gamma = \tfrac{1}{2}\{E_k + [2(m_p c^2)E_k]^{1/2}\}$$

Inserting the maximum value of E_k, 7.5 MeV, and recalling that $m_p c^2$ is about 938 MeV, we then get $E_\gamma \approx 64$ MeV. This value of the γ-ray photons is much larger than the energy available from the reaction, which is 15 MeV. What is worse, considering the effect of the products of the Be–He reaction on other substances, other values of E_γ are measured, in some cases as high as 90 MeV. Therefore no consistent results for the energy of the γ-rays, compatible with energy and momentum conservation are found.

In 1932 James Chadwick (1891–1974) showed that all these difficulties disappeared and the conservation laws were restored if, instead of γ-rays, neutral particles were emitted, having a mass close to that of protons. Those neutral particles were called neutrons, designated $^{1}_{0}n$, and the process written instead as

$$^{9}_{4}Be + ^{4}_{2}He \rightarrow [^{13}_{6}C]^* \rightarrow ^{12}_{6}C + ^{1}_{0}n$$

Chadwick made careful measurements of the kinetic energy of protons and nitrogen atoms knocked out when neutrons passed through a substance containing hydrogen and nitrogen, respectively. This allowed the mass of the neutron to be calculated, resulting in a value close to that of a proton, as explained in Example 14.7.

40.6 Nuclear fission

Nuclear **fission** consists in the division of a massive nucleus, such as uranium or thorium, into two fragments of comparable size. Fission as a natural process is very rare ($^{238}_{92}U$ is believed to fission spontaneously with a half-life of approximately 10^{16} years). One method for producing fission artificially is to excite the nucleus. The threshold or minimum activation energy required for fission of a heavy nucleus is from 4 to 6 MeV. Another means of inducing fission is by neutron capture. The binding energy of the captured neutron is, in some cases, enough to excite the nucleus above the threshold energy, so that the division into two fragments takes place. This is the case of the nucleus $^{235}_{92}U$, which undergoes fission after capturing a neutron even if it is very slow. The process may be expressed by the equation

$$^{235}_{92}U + ^{1}_{0}n \rightarrow [^{236}_{92}U]^* \rightarrow X + Y$$

where $[^{236}_{92}U]^*$ is the excited nucleus formed when $^{235}_{92}U$ captures a neutron.

For other nuclei, the binding energy of the captured neutron is not sufficient for fission to take place and the neutrons must have some kinetic energy as well. This is what occurs with $^{238}_{92}\text{U}$, which fissions only after capturing a *fast* neutron that has a kinetic energy of the order of 1 MeV. The capture of slow neutrons by $^{238}_{92}\text{U}$ results instead in the production of neptunium and plutonium according to the process

$$^{238}_{92}\text{U} + ^{1}_{0}\text{n} \rightarrow [^{239}_{92}\text{U}]^* \rightarrow {}^{239}_{93}\text{Np} + \text{e}^- + \bar{\nu} + \gamma$$
$$\hookrightarrow {}^{239}_{94}\text{Pu} + \text{e}^- + \bar{\nu} + \gamma$$

For that reason, large quantities of plutonium are produced in nuclear reactors. The reason for this different behavior lies in some details of the structure of the different nuclei. The nucleus $^{235}_{92}\text{U}$ is even–odd, with 143 neutrons, and when a neutron is captured, an even–even nucleus, $^{236}_{92}\text{U}$, is formed. The captured neutron is coupled or paired with the last odd neutron of $^{235}_{92}\text{U}$, releasing an additional pairing energy of about 0.57 MeV. On the other hand, $^{238}_{92}\text{U}$ is an even–even nucleus, with 146 neutrons, all paired, and when a neutron is captured, an even–odd nucleus, $^{239}_{92}\text{U}$, results, with no extra pairing energy available. For the same reason $^{239}_{94}\text{Pu}$, with 145 neutrons, undergoes fission by slow neutron capture.

A nucleus, if properly excited, may experience collective vibrations around its equilibrium shape. When the excitation energy is low, the oscillations are small. Eventually, the excitation energy is released in the form of radiation and the nucleus returns to its equilibrium shape. The process has therefore been a radiative neutron capture and may be written as

$$^{A}_{Z}\text{X} + \text{n} \rightarrow {}^{A+1}_{\ Z}\text{Y}^* \rightarrow {}^{A+1}_{\ Z}\text{Y} + \gamma$$

But if the excitation energy is large enough, the nucleus may depart so much from its equilibrium shape that the long-range electrical repulsion between the protons becomes larger than the short-range nuclear interaction, and there is a probability that the nucleus, instead of returning to its equilibrium shape, deforms more and more until it divides into two fragments.

EXAMPLE 40.5

Energy released in the fission of $^{235}_{92}\text{U}$ by slow neutrons.

▷ Consider the particular case where the fission products are $^{95}_{42}\text{Mo}$ and $^{139}_{57}\text{La}$. We may write the process as

$$\text{n} + {}^{235}_{92}\text{U} \rightarrow {}^{236}_{92}\text{U}^* \rightarrow {}^{95}_{42}\text{Mo} + {}^{139}_{57}\text{La} + 2\text{n}$$

Since the incoming neutrons are slow, we may ignore their kinetic energy in the energy balance and consider only mass. The initial masses are

$$m(\text{n}) + m(^{235}_{92}\text{U}) = 1.0090 \text{ amu} + 235.0439 \text{ amu} = 236.0529 \text{ amu}$$

This is also the mass of $^{236}_{92}\text{U}^*$. Since the mass of $^{236}_{92}\text{U}$ is 236.0456, the excess mass is 0.0073 amu, corresponding to an excitation energy of 6.796 MeV. The fission threshold of $^{236}_{92}\text{U}$ is about 5.3 MeV, so that the excitation energy is sufficient to result in fission.

The masses of the final products are

$$m({}^{139}_{57}\text{La}) + m({}^{95}_{42}\text{Mo}) + 2m(\text{n}) = 138.9061 \text{ amu} + 94.9058 \text{ amu}$$
$$+ 2 \times 1.0090 \text{ amu} = 235.8299 \text{ amu}$$

The excess mass of the initial nuclei relative to the final nuclei is 0.223 amu. Therefore, the energy released in fission is 207.16 MeV, which appears as kinetic energy of the final products.

Actually, the total energy released is larger because the ${}^{95}_{42}$Mo and ${}^{139}_{57}$La nuclei are unstable and emit electrons, neutrinos and γ-rays until stable nuclei are reached.

40.7 Fission chain reactions

Two properties of fission make it a very important process for practical applications: one is that *neutrons are released in fission* and the other is that *energy is released in fission.*

For the heaviest nuclei, such as uranium, the ratio of neutrons to protons is $N/Z \simeq 1.55$. This should also be the ratio for the resulting fragments. However, for medium mass stable nuclei the ratio is $N/Z \simeq 1.30$. This means that the resulting fragments have too many neutrons and some are released at the time of fission. The average number of neutrons released per fission is about 2.5. For the same reason, the fragments are β^--radioactive.

Energy is released in nuclear fission because the binding energy per nucleon (recall Figure 39.2) is less in heavy nuclei than in medium mass nuclei. For a heavy nucleus, the binding energy is about 7.5 MeV per nucleon, but for medium mass nuclei, corresponding to the two fission fragments, it is about 8.4 MeV per nucleon, resulting in an increase of binding energy per nucleon of about 0.9 MeV, or a total of about 200 MeV for all nucleons in a uranium nucleus. This is the order of magnitude of the energy liberated in the fission of a uranium atom, which appears as kinetic energy of the two fragments, of the released neutrons and of the disintegration products (that is, electrons and neutrinos) resulting from the β-decay of the radioactive fragments, as well as electromagnetic radiation. Since the neutrinos emitted in β-decay normally escape from the material in which fission takes place, only about 185 MeV per atom can be retained, an energy still considerably larger than the energy liberated in a chemical reaction (which is of the order of 3 to 10 eV per atom).

The fact that for each neutron absorbed in order to produce one fission, more than two neutrons are emitted (on average) makes a **chain reaction** possible. That is, if after each fission, at least one of the new neutrons produces another fission, and of the neutrons released in this fashion, again at least one produces a fission, and so on, a self-sustaining process results (Figure 40.11). Such *chain reactions* are very common in chemistry. Combustion is a chain reaction. Burning requires that a molecule have a certain excitation energy so that it can combine with an oxygen molecule. But once the first molecules are excited and combine with oxygen, the energy liberated is enough to excite more molecules of the fuel, and burning results.

If, in each stage of a fission chain process, more than one neutron per fission produces a new fission, the number of fissions increases exponentially and a divergent chain reaction results. This is what happens in an **atomic bomb**. But if

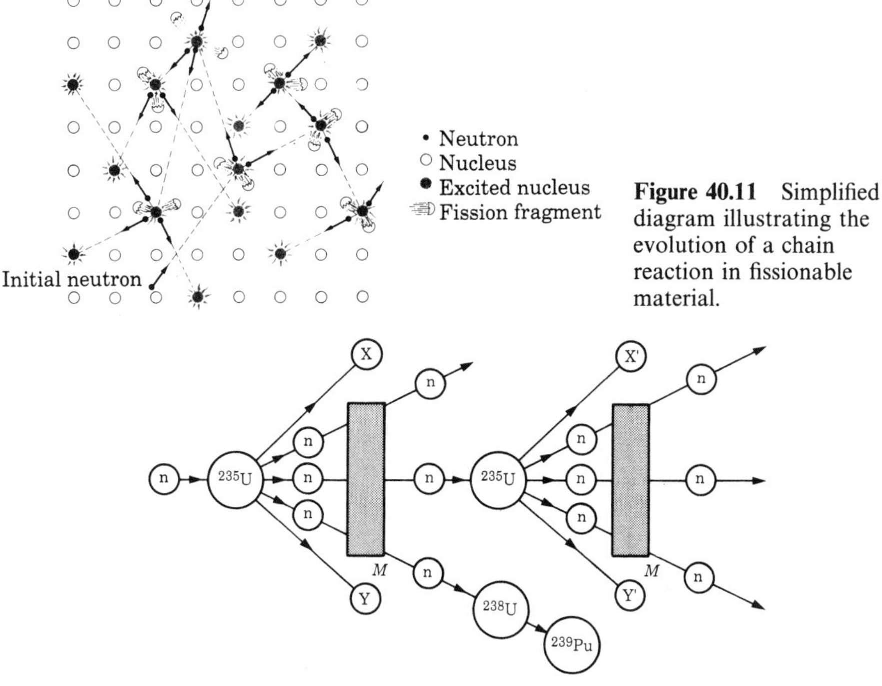

Figure 40.11 Simplified diagram illustrating the evolution of a chain reaction in fissionable material.

Figure 40.12 Moderation of a fission chain reaction. M is a moderating substance.

things are arranged so that on the average, only one neutron of each fission produces a new fission, a steady chain reaction is maintained under controlled conditions. This is what happens in a **nuclear reactor** (see Note 40.2).

In *fast* nuclear reactors the neutrons are used at the same energies (1 to 2 MeV) at which they are released in the fission process. But in *thermal* nuclear reactors the neutrons are first slowed down by allowing them to collide with the atoms of some other substance, called a *moderator*, until they come to thermal equilibrium with the substance (Figure 40.12). The neutrons are then called *thermal*. The moderator must be a substance with a small mass number (recall Example 14.6) and does not tend to capture neutrons. Water, heavy water and graphite are the substances most commonly used as moderators.

Note 40.2 Nuclear fission reactors

The systems in which a nuclear fission chain reaction is produced and maintained under control are called nuclear reactors. There are various types of nuclear reactors. In the case of *thermal* reactors all have the same basic components (Figure 40.13).

1. **Nuclear fuel** is a fissionable material such as U and Pu. The uranium may be natural, which contains only 0.7% of $^{235}_{92}U$, or **enriched**, where it contains a larger proportion of $^{235}_{92}U$, which in some cases may reach up to 90%. The fuel, usually in the form of uranium oxide, is placed inside metallic tubes that constitute the **fuel elements**. A set of fuel elements make up the reactor **core**. As the fuel is used in the reactor the $^{235}_{92}U$ content decreases and part of the $^{238}_{92}U$ is converted into $^{239}_{94}Pu$ and other Pu isotopes. The fission products remain inside the fuel elements. The fuel elements must be replaced periodically, as the fuel is used.

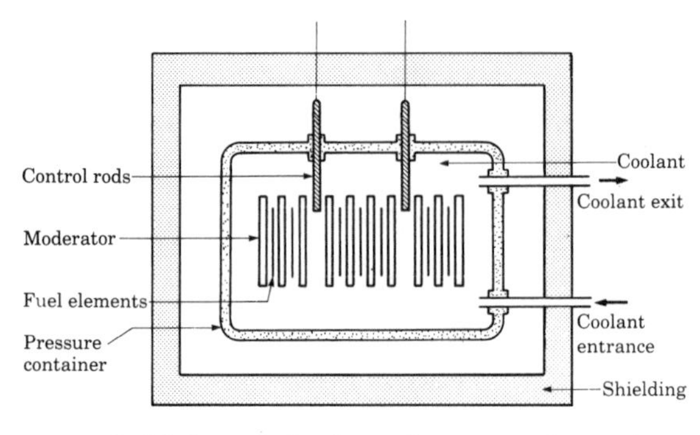

Figure 40.13 Components of a thermal nuclear reactor.

2. **Moderator.** In thermal reactors the moderator, which surrounds the fuel elements, must be a substance of low nuclear mass that does not absorb neutrons or absorbs them slightly. The moderator can be water if the uranium is enriched, and heavy water or graphite if the uranium is natural. Since the moderator decreases the energy of the neutrons from about one MeV down to less than 0.1 eV, when they become 'thermalized', the thermal energy of the moderator increases and its temperature rises.

3. **Reflector.** To diminish the loss of neutrons from the reactor, the core is surrounded by a substance, such as water or graphite, that scatters, back into the core, some of the neutrons that were not used in fissions and that otherwise would escape from the reactor.

4. **Control rods.** To regulate the availability of neutrons for fissions and maintain control over their number, a series of rods made of a neutron absorbing material (B, Hf, Cd) are used. The rods are moved in and out of the core as needed to regulate the neutron flux.

5. **Coolant.** To extract the energy generated in the reactor core and prevent its temperature increasing excessively, a fluid is circulated through the core and the moderator. The coolant must be a substance that absorbs neutrons very slightly. The fluid may be water, heavy water, air or a gas like He or CO_2, depending on the reactor design.

In **research reactors** the γ-rays and the neutrons produced in the fissions are used to carry out research or produce new radioactive substances. In **power reactors** the energy extracted by the coolant is used to produce steam that feeds a turbine and generates electric power like in any conventional thermal power station. In **boiling water reactors** (Figure 40.14), the coolant itself is transformed into steam within the reactor vessel. In **pressurized water reactors** (Figure 40.15), the coolant passes through a steam generator or heat exchanger. Thus the steam feeding the turbine flows in a loop external to the reactor. In 1992 there were about 430 reactors in operation in the world generating 300 GW of electric power or about 20% of the electric power generated worldwide.

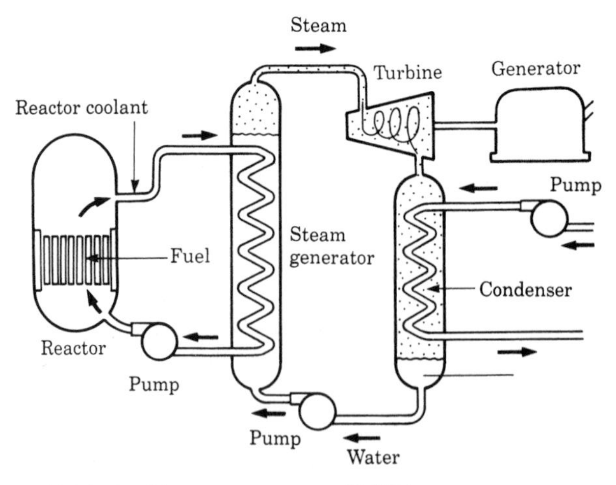

Figure 40.14 Boiling water reactor (BWR).

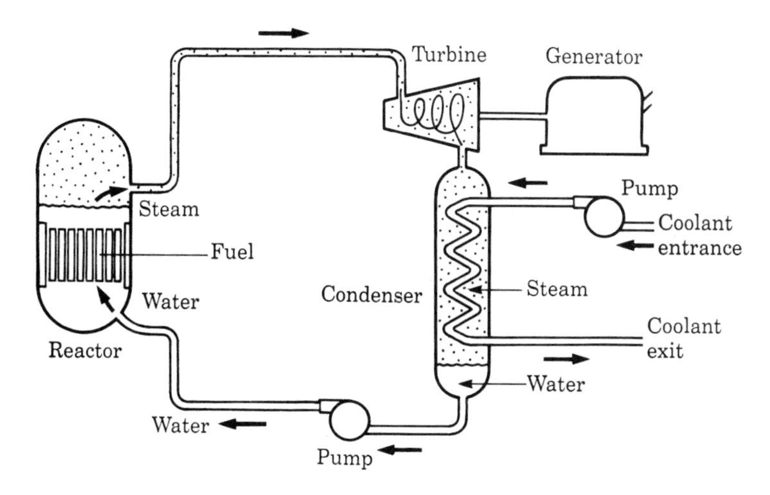

Figure 40.15 Pressurized water reactor (PWR).

40.8 Nuclear fusion

Nuclear **fusion** consists in the coalescence of two colliding nuclei into a larger nucleus. Because of the Coulomb repulsion, the colliding nuclei must have a minimum kinetic energy to overcome the Coulomb barrier and get close enough so that nuclear forces produce the necessary consolidating action (Figure 40.16). This problem does not appear in nuclear fission because the neutrons do not have an electric charge, and thus can approach a nucleus even if their kinetic energy is very small or practically zero. Since the Coulomb barrier increases with atomic number, nuclear fusion occurs at reasonable kinetic energies only for very light nuclei with low atomic number or nuclear charge.

When two nuclei of atomic numbers Z_1 and Z_2 are in contact, the electric potential energy of two nuclei is $E_p = Z_1 Z_2 e^2/4\pi\varepsilon_0 r$, where r is equal to the sum of the nuclear radii, or about 10^{-14} m, so that we obtain

$$E_p \simeq 2.4 \times 10^{-14} Z_1 Z_2 \text{ J} = 1.5 \times 10^5 Z_1 Z_2 \text{ eV} = 0.15 Z_1 Z_2 \text{ MeV}$$

This gives the height of the barrier and thus the minimum kinetic energy the nuclei must have for fusion to occur. At lower energies, there is still a certain probability of penetrating the potential barrier, but it is very small.

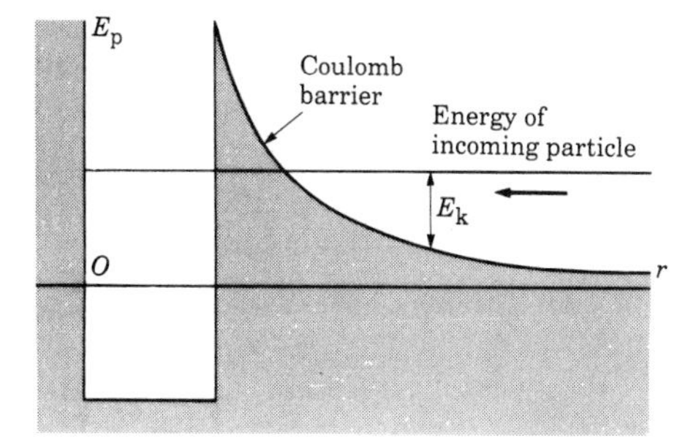

Figure 40.16 Potential barrier for fusion.

The average kinetic energy of a system of particles having temperature T is of the order of kT, or about $8.6 \times 10^{-5}T$ eV, where T is in kelvin. Thus an energy of 10^5 eV corresponds to a temperature of about 10^9 K, which is much higher than the temperature believed to exist at the center of the Sun. Even so, fusion of light nuclei is one of the most important processes occurring in the Sun and is its main source of energy (see Note 40.3).

For nuclear fusion of a large number of nuclei to take place, it is necessary for the reacting nuclei to be at temperatures much higher than those generated by even the most exoergic chemical reaction, creating a problem of containment of the reacting particles, since no known material can sustain such temperatures. Also, at these extreme temperatures, the nuclei are stripped of all their surrounding electrons (because of collisions) and the substance consists of a neutral mixture of positively charged nuclei and negative electrons called a **plasma**. Containment and heating can be done by means of magnetic fields and by laser beams (see Notes 26.2 and 40.3).

Energy is liberated in nuclear fusion of light nuclei ($A < 20$) because when two light nuclei coalesce into a heavier one, the binding energy of the product nucleus is greater than the sum of the binding energies of the two lighter nuclei (recall Figure 39.2). In fact, it is thought that all elements up to Fe have been produced by fusion in stars. Beyond Fe, nuclear binding energy decreases and fusion does not occur, so other processes are required to produce the elements beyond.

If conditions are appropriate, the energy liberated in fusion is enough to excite other nuclei, and a chain reaction results. If the chain reaction takes place rapidly, so that a large amount of energy is liberated in a very short time, it becomes a nuclear explosion. The chain reaction may also occur under controlled conditions, although no fully satisfactory fusion reactor has yet been built.

The simplest fusion reaction is the capture of a neutron by a proton (or hydrogen nucleus) to form a deuteron:

$$^1_1\mathrm{H} + \mathrm{n} \rightarrow {}^2_1\mathrm{H} + 2.224\ \mathrm{MeV} \tag{40.13}$$

The great advantage of this reaction is that there is no electrical repulsion to be overcome. Reaction 40.13 occurs, for example, when neutrons from a nuclear reactor diffuse through a hydrogenous substance, such as water or paraffin.

Another simple fusion reaction is that between two protons. Since a diproton nucleus does not exist, the process is accompanied by the conversion of one of the protons into a neutron and the emission of a positron and a neutrino. That is,

$$^1_1\mathrm{H} + {}^1_1\mathrm{H} \rightarrow {}^2_1\mathrm{H} + \mathrm{e}^+ + \nu + 1.35\ \mathrm{MeV} \tag{40.14}$$

A third important fusion reaction is that between hydrogen and deuterium, resulting in a tritium nucleus:

$$^1_1\mathrm{H} + {}^2_1\mathrm{H} \rightarrow {}^3_1\mathrm{H} + \mathrm{e}^+ + \nu + 4.6\ \mathrm{MeV} \tag{40.15}$$

Another reaction, of practical importance, is the fusion of two deuterons. Two possible results occur with about the same probability:

$$^2_1\mathrm{H} + {}^2_1\mathrm{H} \rightarrow \begin{cases} {}^3_2\mathrm{He} + \mathrm{n} + 3.2\ \mathrm{MeV} \\ {}^3_1\mathrm{H} + {}^1_1\mathrm{H} + 4.2\ \mathrm{MeV} \end{cases} \tag{40.16}$$

Two fusion reactions that liberate a great amount of energy per unit mass are between deuterium and tritium,

$$^2_1H + ^3_1H \rightarrow ^4_2He + n + 17.6\ \text{MeV} \quad \textbf{(40.17)}$$

and between deuterium and helium-3,

$$^2_1H + ^3_2He \rightarrow ^4_2He + ^1_1H + 18.3\ \text{MeV} \quad \textbf{(40.18)}$$

However, tritium and ^3He are not readily available and have to be manufactured. On the other hand, the fusion of two deuterons has the advantage of using only one class of nuclei.

Although the energy released in a single fusion reaction is much less than that released in a single fission reaction, the energy per unit mass is larger (because deuterium is a very light fuel). For the deuterium–deuterium fusion reaction, the energy is about 2×10^{14} J per kilogram of fuel. This is more than double the value for uranium fission. The deuterium–tritium reaction is four times larger. Due to the relative abundance of deuterium (approximately one deuterium atom for every 7000 hydrogen atoms), and the relatively low cost of extracting deuterium from water (about 30 cents per gram), once fusion-controlled devices become practical, the fusion process will provide an almost unlimited source of energy.

Fusion reactions are also the main source of the energy released in stars, including our Sun. The most common fusion process is the fusion of four protons (or hydrogen nuclei) into a helium nucleus, that in simplified form is,

$$4\ ^1_1H \rightarrow ^4_2He + 2e^+ + 2\nu + 26.7\ \text{MeV} \quad \textbf{(40.19)}$$

It is estimated that this fusion process is occurring in the Sun at the rate of 5.64×10^{11} kg of hydrogen per second, with a release of 3.7×10^{25} W. Of this, only about 1.8×10^{16} W falls on the Earth, mostly in the form of electromagnetic radiation; however, this is still 10^4 times greater than all the industrial power generated on the Earth. Fusion is the mechanism by which the light elements are synthesized in stars (see Note 40.4).

Note 40.3 Nuclear fusion reactors

From the point of view of energy per unit mass of fuel, the two most interesting fusion processes are the deuteron–tritium reaction, Equation 40.17 and the deuteron–helium-3 reaction, Equation 40.18. The second reaction is more attractive because no neutrons are produced and therefore no radioisotopes are produced by neutron capture, so that it is a 'clean' reaction. Deuterium can be easily extracted from sea water. However, there is not enough ^3_2He on Earth to feed fusion reactors and therefore this reaction is not practical for power generation. There is plenty of cosmic ^3_2He trapped in the Moon's surface and if it can be brought back to Earth, it might contribute to the solution to fusion power.

The deuterium–tritium reaction is attractive but it has two problems. One is that tritium is scarce. However it can be produced easily by bombarding lithium with neutrons according to the reaction

$$^6_3Li + n \rightarrow ^4_2He + ^3_1H$$

This reaction can be carried out, for example, in fission reactors by providing a lithium blanket around the core. Thus fission reactors can provide the tritium needed by fusion reactors. The second problem is that the deuterium tritium reaction produces neutrons, which in turn may

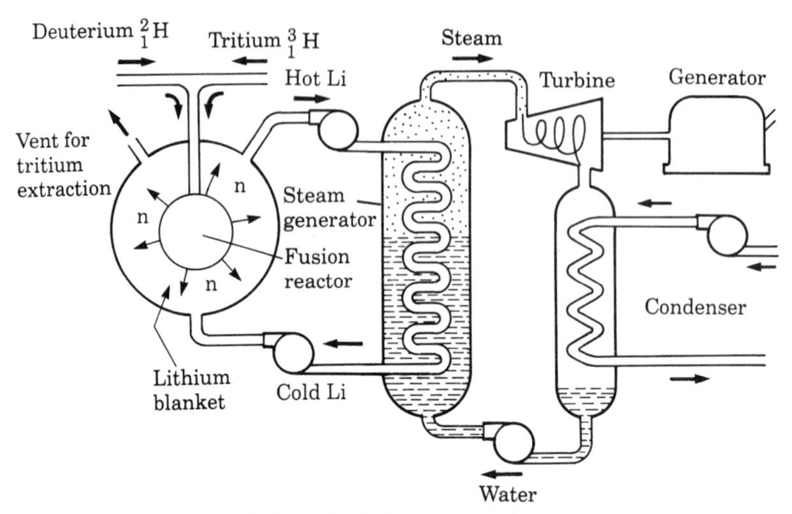

Figure 40.17 Possible design of a fusion power plant.

produce radioisotopes. This can be simplified by also using a lithium blanket around the fusion reactor so that the fusion neutrons are captured and more tritium is produced.

The main obstacle to practical fusion reactors is to reach the temperatures, pressures and densities required to produce enough fusions in a deuterium–tritium plasma for enough time so that appropriate amounts of energy are generated larger than the input energy required to operate the reactor. One method is by **magnetic confinement** in a tokamak, as discussed in Note 26.1. When the intensity of the magnetic field in a tokamak is rapidly increased, the plasma is compressed adiabatically and its temperature increases until fusion begins.

Another promising method is by **inertial confinement**, using laser beams. A pellet with diameter less than 1 mm containing deuterium and tritium is subject to very intense laser beams from many directions. The momentum and energy of the photons in the beams heat and compress the pellet until it reaches a density about 10 000 times the normal density, a process called 'implosion'. At such densities and temperatures fusion occurs. In both cases the fusion reactor operates in pulses and requires an appreciable input of energy to produce the magnetic fields or generate the laser beams. Even so, the energy balance is expected to be positive.

Figure 40.17 shows how a fusion power plant might look using deuterium–tritium plasma and a lithium blanket.

Note 40.4 The formation of the elements

One of the clues we use for our speculation on how the elements were formed from primeval matter is the relative abundance in the universe of the different chemical elements, as shown in Figure 40.18, and another is their isotopic composition. From the figure we see that hydrogen is the most abundant element, followed by helium. In fact, these two elements comprise about 98% of all nuclei in the universe. After the sudden drop in abundance corresponding to lithium, beryllium and boron, the abundance follows a regularly decreasing trend, but with some pronounced maxima, especially for iron and neighboring nuclides. The elements stop at $Z = 92$, since the amount of naturally existing nuclei with $Z > 92$ is essentially zero, although they have been produced artificially in the laboratory, up to $Z = 109$. As to isotopic composition, the lighter elements are richer in isotopes with about equal numbers of protons and neutrons, while the heavier nuclei are richer in isotopes with higher neutron content to counterbalance the Coulomb repulsion among protons. Also, even-A nuclei are more abundant than odd-A nuclei, which is attributed to the pairing energy. Another interesting feature is that no nuclei with $A = 5$ or 8 are found in nature.

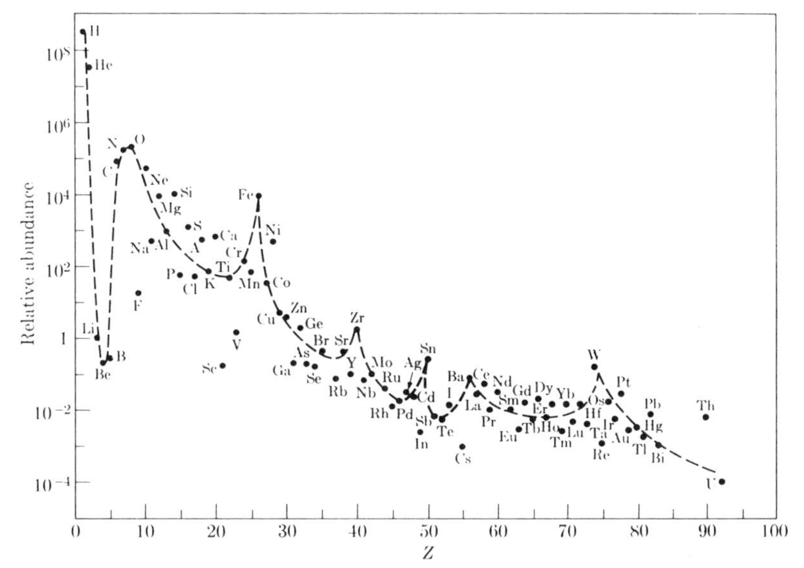

Figure 40.18 Cosmic abundance of the elements. (H. Urey and H. Brown, *Phys. Rev.* **88**, 248 (1952).)

The relative abundance and isotopic composition have been found to be the same not only in samples taken from various parts and depths of the Earth's crust but also in samples taken from meteorites that have fallen on the Earth from outer space and from spectroscopic measurements of light from stars. This constant composition suggests that, at least in our galaxy, the elements have all been formed by approximately the same process. At the very early stages of the universe, about 10^{10} years ago, some of the protons and neutrons that were formed shortly after the Big Bang combined and formed small amounts of deuterium, tritium and helium nuclei on a cosmic scale according to fusion processes discussed in Section 40.8. Afterwards the heavier elements have been synthesized in stars under varying conditions, a process that still continues.

The sequence of events *might* have been as follows. Initially, due to statistical fluctuations and gravitational interaction, some of the very large number of protons (hydrogen nuclei), together with the much smaller number of deuterium, tritium, helium, and neutrons, condensed into clusters, or stars. Eventually these clusters reached a density of the order of 10^6 kg m^{-3}, which is one thousand times more dense than ordinary matter. In the process of condensation there is a transformation of gravitational potential energy into kinetic energy, resulting in an increase of temperature (to about 10^7 K). At such temperatures the fusion of hydrogen into helium is accelerated by a series of fusion reactions designated as the **proton–proton cycle** (Figure 40.19), which consists of the following steps:

$$^1_1\text{H} + {}^1_1\text{H} \rightarrow {}^2_1\text{H} + e^+ + \nu$$

$$^1_1\text{H} + {}^2_1\text{H} \rightarrow {}^3_2\text{He}$$

$$^3_2\text{He} + {}^3_2\text{He} \rightarrow {}^4_2\text{He} + 2\,{}^1_1\text{H}$$

Combining all the reactions and canceling common terms, we have

$$4\,{}^1_1\text{H} \rightarrow {}^4_2\text{He} + 2e^+ + 2\nu + 26.7\ \text{MeV}$$

which was the process mentioned in Equation 40.19. The net energy liberated in the process is about 6.6×10^{14} J per kg of $^1_1$H consumed. The time required for the proton–proton cycle to be completed in the Sun is about 3×10^9 years.

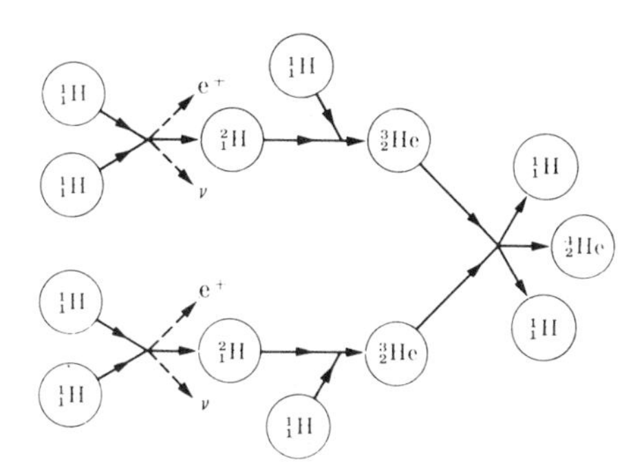

Figure 40.19 The proton–proton cycle.

At these high temperatures and densities, it is also possible that, as by-product reactions, relatively small quantities of nuclei with higher mass may be formed in stars. For example, ^7_3Li is produced by the processes

$$^3_1H + ^4_2He \rightarrow ^7_3Li + \gamma$$

or

$$^3_2He + ^4_2He \rightarrow ^7_4Be + \gamma, \qquad ^7_4Be + e^- \rightarrow ^7_3Li + \nu$$

Part of the lithium reverts back to helium through the reaction

$$^7_3Li + ^1_1H \rightarrow ^4_2He + ^4_2He$$

Other fusion reactions, such as deuterium–deuterium and deuterium–tritium, might also take place, but in much lesser amount.

Since helium is more massive than hydrogen, the helium nuclei produced in the fusion process are carried to the center (or core) of the star by gravitational action, gaining kinetic energy in the process. The density at the core may then become as high as 10^8 kg m^{-3}. The resultant gain in kinetic energy of the helium nuclei at the core increases further core temperature (up to values of 10^8 K). The increase in temperature and density of helium nuclei at the core allows for the production of ^8_4Be by the helium fusion reaction

$$^4_2He + ^4_2He \rightarrow ^8_4Be,$$

which is followed, in about 10^{-16} s, by the reverse decay process

$$^8_4Be \rightarrow ^4_2He + ^4_2He$$

When the helium concentration is large enough, it is possible for another helium nucleus to be captured before this decay occurs, resulting in the reaction

$$^8_4Be + ^4_2He \rightarrow ^{12}_6C + \gamma$$

Alternatively, the intermediate products in the fusion process ^2_1H and ^3_2He may be captured by the beryllium, forming $^{10}_5B$ and $^{11}_6C$, respectively, although in much smaller amounts than $^{12}_6C$. Also many other less probable reactions can take place, which result in other light elements.

As the amount of $^{12}_6C$ increases, a new fusion process called the **carbon cycle** becomes important at these high temperatures. The cycle occurs in the following steps (Figure 40.20):

$$^1_1H + ^{12}_6C \rightarrow ^{13}_7N \rightarrow ^{13}_6C + e^+ + \nu$$

$$^1_1H + ^{13}_6C \rightarrow ^{14}_7N$$

$$^1_1H + ^{14}_7N \rightarrow ^{15}_8O \rightarrow ^{15}_7N + e^+ + \nu$$

$$^1_1H + ^{15}_7N \rightarrow ^{12}_6C + ^4_2He$$

Combining these reactions, we recover Equation 40.19. Therefore, the carbon cycle is also equivalent to the fusion of hydrogen into helium. Note that the carbon atom is a sort of catalyzer, since it is regenerated at the end of the cycle. The time required for a carbon atom to go through this cycle in the Sun is about 6×10^6 years.

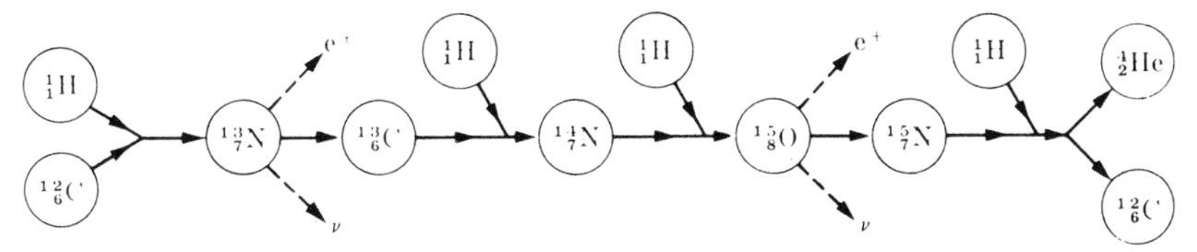

Figure 40.20 The carbon cycle.

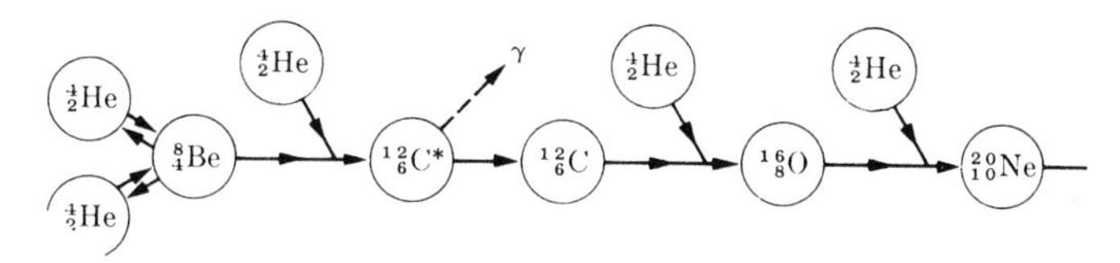

Figure 40.21 The helium burning process.

The chain of events described explains how the gaps at $A = 5$ and 8 can be by-passed. This mechanism also makes the scarcity of lithium, beryllium and boron understandable, as well as explaining the relatively great abundance of $^{12}_{6}$C. By the process of hydrogen and helium capture, the production of successively more massive nuclei such as $^{16}_{8}$O, $^{20}_{10}$Ne, and others is possible (Figure 40.21). The only limitation to helium capture is the amount of energy required by the $^{4}_{2}$He nuclei to overcome the Coulomb repulsion of the heavier nuclei.

With the production of heavier nuclei, a further gravitational contraction of the star takes place, leading to core densities of the order of 10^9 kg m^{-3} (with a corresponding increase in the kinetic energy of the nuclei and a core temperature that approaches 10^9 K). Under such conditions of very high density and extreme temperature, other nuclear reactions are possible that may produce nuclei of higher mass number, up to the iron group (about $A = 60$), but not heavier, because the heavier nuclei have lower binding energies per nucleon.

Neutrons may be produced in some instances through reactions such as

$$^{13}_{6}\text{C} + ^{4}_{2}\text{He} \rightarrow ^{16}_{8}\text{O} + \text{n}, \qquad ^{17}_{8}\text{O} + ^{4}_{2}\text{He} \rightarrow ^{20}_{10}\text{Ne} + \text{n}, \qquad ^{21}_{11}\text{Ne} + ^{4}_{2}\text{He} \rightarrow ^{24}_{12}\text{Mg} + \text{n}$$

These neutrons are available for extending the production of nuclei to higher mass numbers – that is, nuclei beyond the iron group – by neutron capture rather than by the capture of charged particles, followed by β^--decay. The resultant nuclei have a higher atomic number and the chain can thus advance toward higher Z-values. As time passes, the number of free neutrons decreases, which explains why the heavier elements are much less abundant, since the production of the heavier elements is dependent on neutron capture.

Stars in the universe do not all follow the same sequence at the same rate so that stars are presently in different stages of evolution. The Sun itself is still in its first stage of evolution; its composition is assumed to be 81.76% hydrogen, 18.17% helium and 0.07% for the rest of the elements. Stars where hydrogen burning presently appears to be the dominant process are called **main sequence stars**. Those stars in which, at present, the most important process is helium-burning are called **red giants** because of their color. In many stars the three stages of nucleosynthesis might be taking place simultaneously, with hydrogen-burning occurring at the surface, helium reactions prevailing at an intermediate (and hotter) layer, and heavy elements being produced in the much hotter core of the star. Stars that have evolved in this way are called **first generation stars**.

Instabilities that arise during the evolution and ageing of a star may result in the ejection of some of its material into interstellar space. This is what happens, for example, in the explosion of a **supernova**. The ejected material mixes with uncondensed hydrogen and other particles in outer space. Condensation of some of this mixture at a later time results in **second-** (and later) **generation stars** where some elements are already present but hydrogen fusion may still continue when the appropriate temperature is reached. This is the case of the Sun.

The Earth, including ourselves, and the rest of the planets are also the result of the condensation of cosmic material or of debris thrown out by exploding stars. However, the temperature of the planets is too low for any fusion reaction to take place.

Once the nuclear fuel of a star has been used, its further evolution is governed primarily by gravitational forces among its components and depends, very critically, on their masses. Stars with masses between $10^{-3} M_C$ and less than $10 M_C$, where M_C is the Chandrasekhar mass defined in Example 36.4, eventually contract to a radius of a few thousand kilometers and have a density between 10^4 kg m^{-3} and 10^{13} kg m^{-3} so that electrons and nuclei are packed as closely as allowed by the exclusion principle. These stars are called **white dwarfs**. The Sun is expected to become a white dwarf in a few billion years. Stars with masses close to $10 M_C$ contract even further, crushing electrons into protons so they are transformed into neutrons by the process of electron capture, which was described in Section 40.4. Such stars are called **neutron stars** and their collapse is limited by the exclusion principle as it applies to neutrons. Neutron stars have a radius of a few kilometers and a density between 10^{13} kg m^{-3} and 10^{20} kg m^{-3}, which is of the order of the density of nuclear matter (see Section 39.3). Stars whose mass is larger than $10 M_C$ contract even further, becoming **black holes**, with sizes smaller than their Schwarzschild radii (see Example 11.3) and therefore have extremely high densities. Planets, with masses smaller than 10^{27} kg (Table 11.1) or $10^{-3} M_C$ are gravitationally stable, giving electric binding forces a more important role in holding the system together. For even smaller masses, electric forces become the dominant factor. There is, then, a connection between all fundamental forces at all levels of the large structures in the universe.

QUESTIONS

40.1 Show that the activity of a substance obeys the law $A = A_0 e^{-\lambda t}$.

40.2 Is the concept of 'half-life' of a radioactive nuclide directly related to exponential decay?

40.3 How do we know that α-decay is a two-body process while β-decay must be a three-body process?

40.4 Referring to Figure 40.2, mark those processes due to α-decay and those due to β-decay.

40.5 Why must an α-particle penetrate a potential barrier before escaping from a nucleus?

40.6 An unstable nucleus is characterized by having too many (a) neutrons, (b) protons. What kind of β-decay is to be expected in each case? Check with a table of nuclides.

40.7 How can one determine if a nucleus is stable relative to emission of a certain kind of particle?

40.8 By using an expression similar to Equation 40.6 investigate the stability of $^{238}_{92}U$ with respect to the emission of protons and neutrons.

40.9 By looking at a nuclear physics book make drawings similar to Figure 40.4 for the actinium and thorium series.

40.10 Why is it necessary to introduce a weak interaction?

40.11 Why is an intermediate nucleus formed in nuclear reactions? If the energy of the projectile is very high is it expected that an intermediate nucleus will be formed?

40.12 Why does the binding energy per nucleon begin to decrease for $A > 60$?

40.13 Why is energy liberated in the fission of a heavy nucleus or in the fusion of two light nuclei?

40.14 Why does fission occur with heavy nuclei while fusion occurs with light nuclei?

40.15 What is the fundamental difference between a 'thermal' and a 'fast' nuclear reactor?

40.16 What are the requirements that must be met to make a fusion reactor practical?

PROBLEMS

Note: A chart of nuclides is necessary to solve some of the problems for this chapter.

40.1 The half-life of ^{90}Sr is 28 years. Calculate: (a) the disintegration constant for ^{90}Sr, (b) the activity of 1 mg of ^{90}Sr in curies and as nuclei per second, (c) the time for the 1 mg to reduce to 250 μg, (d) the activity at this later time.

40.2 A freshly prepared sample of a radioactive material, which decays to a stable nuclide, has its activity measured every 20 seconds. The following activity (in μCi) is measured starting at $t = 0$ s: 410; 190; 90; 43; 20; 9.6; 4.5; 2.15; 1.00; 0.48; 0.23. (a) Plot the natural logarithm of the activity against the time. (b) Find the disintegration constant and the half-life of the sample. (c) How many radioactive nuclei were present in the sample at $t = 0$ s?

40.3 A material is composed of two different radioactive substances having half-lives equal to 2 h and 20 min, respectively. Initially there is one mCi of the first substance and 9 mCi of the second. (a) Using semi-logarithmic paper, plot, as functions of time, the activity of each substance and also of the whole material. (b) At what time is the total activity one mCi? (c) At what time is the activity of the short-lived substance 1% of that of the long-lived substance?

40.4 The activity of a material is measured with a Geiger counter every 30 s and the following values (in counts per minute) are found: 1167; 264; 111; 67; 48.3; 37.1; 30.0; 24.6; 20.9; 18.1; 15.7; 13.9; 12.3; 11.1; 9.84; 8.85; 7.83; 7.02; 6.26; 5.60; 5.00. (a) Plot the logarithm of the activity versus time. (b) Determine how many radioactive substances are present and compute the half-life and disintegration constant of each. (Hint: first subtract the activity of the longest-lived substance by extending the tail of the curve in a straight line back to zero time. Plot the remaining activity and repeat the process until a straight line, corresponding to the shortest-lived substance, remains.)

40.5 A steel piston ring with a mass of 25 g is irradiated in a nuclear reactor until its activity is 9.0 mCi (due to the nuclide ^{59}Fe which has a half-life of 3.90×10^6 s). Two days later the piston ring is installed in a test engine. After a ten-day test run, the crankcase oil is drained and studied for activity due to ^{59}Fe. Investigation shows an average activity of 9.8×10^2 distintegrations per second for a 200 cm^3 sample of the oil. Calculate the mass of iron worn off the piston ring, given that the crankcase has a capacity of 7.6 liters.

40.6 The activity of carbon found in living specimens is 0.007 μCi per kilogram, due to the ^{14}C present. The charcoal taken from the fire pit of an old campsite has an activity of 0.0048 μCi kg^{-1}. The half-life of ^{14}C is 5760 years. Calculate how long ago the campsite was last used.

40.7 Calculate the energy of the α-particle released by ^{144}Nd when it decays to ^{140}Ce. Also calculate the recoil energy of the daughter nucleus. The masses are 143.9100 amu and 139.9054 amu, respectively.

40.8 An elastic collision between an α-particle and a nucleus of unknown mass is observed in a cloud chamber. The α-particle is deflected 55° from its original direction, while the nucleus leaves a track that makes an angle of 35° with the incident direction. Calculate the mass of the nucleus.

40.9 The α-decay spectrum from ^{226}Ra has a triplet structure, with α-particle energies of 4.777, 4.593 and 4.342 MeV. Assuming that the daughter nuclide ^{222}Rn is produced in the ground or one of the two excited states, draw the energy level diagram and show the γ-ray emission associated with the transition.

40.10 (a) Determine the possible modes of decay for ^{40}K, which has a mass of 39.9640 amu. (b) Compute the energy available for each possible process.

40.11 Obtain the expressions that give the energy released in (a) β^--decay, (b) β^+-decay and (c) electron capture. (Hint: you have to keep track of how many electrons were there before and after the decay.)

40.12 (a) Verify that ^{64}Cu may decay by β^+- and β^--emission, or electron capture (EC). Experimentally, we know that ^{64}Cu has a half-life of 12.8 hours with 39% β^-, 19% β^+, and 42% EC. (b) Compute the energy available for each of the three processes. The mass of ^{64}Cu is 63.929 77 amu while ^{64}Zn = 63.929 14 amu and ^{64}Ni = 63.927 96 amu.

40.13 (a) Verify that ^{7}Be decays by electron capture. Its mass is 7.016 929 amu. (b) Compute the energy and the momentum of the neutrino and the daughter ^{7}Li nucleus of mass 7.016 00 amu.

40.14 Calculate the maximum energy for the electron in the β^--decay of ^{3}H.

40.15 When ^{14}O decays by β^+-emission, the daughter ^{14}N nuclide is almost always in an excited state (>99%). From the experimental data which are given in Figure 40.9(b) and from the fact that ^{14}N has a mass of 14.003 074 amu, calculate the mass of ^{14}O.

40.16 Complete the following nuclear reaction equations, substituting the correct nuclide or particle wherever an X appears:

(a) $^{27}_{13}$Al(n, α)X (b) $^{31}_{15}$P(γ, n)X
(c) $^{31}_{15}$P(d, p)X (d) $^{12}_{6}$C(X, α)$^8_4$Be
(e) $^{11}_{5}$B(γ, X)$^8_4$Be (f) $^{115}_{49}$In(n, γ)X
(g) $^{58}_{28}$Ni(p, n)X (h) $^{59}_{27}$Co(n, X)$^{60}_{27}$Co

40.17 For the photonuclear reaction ^{24}Mg(γ, n)^{23}Mg, determine the threshold energy of the photon. The masses of the parent and product nuclides are 23.985 04 and 22.994 12 amu, respectively.

40.18 A certain accelerating machine can accelerate singly charged particles to an energy of 2 MeV and doubly charged particles to 4 MeV. What reactions may be observed when ^{12}C is bombarded by protons, deuterons and α-particles from this machine?

40.19 When ^{7}Li is bombarded by 0.70 MeV protons, two α-particles are produced, each with 9.0 MeV kinetic energy. (a) Calculate the Q of the reaction. (b) Calculate the difference between the total kinetic energy of the α-particles and the kinetic energy of the initial proton in the L-frame.

40.20 The mass of $^{27}_{13}$Al is 26.981 54 amu. Find the mass of the product nuclei for the following reactions:

(a) ^{27}Al(n, γ)^{28}Al $Q = 7.722$ MeV
(b) ^{27}Al(p, α)^{24}Mg $Q = 1.594$ MeV
(c) ^{27}Al(d, p)^{28}Al $Q = 5.497$ MeV
(d) ^{27}Al(d, α)^{25}Mg $Q = 6.693$ MeV

40.21 A sample of natural silicon is bombarded by a 2.0 MeV beam of deuterons. (a) Write all the possible reactions in which either a proton or an α-particle is the ejected particle. (b) In each case, find the kinetic energy of the ejected particle in the C-frame and its maximum and minimum kinetic energy in the L-frame.

40.22 A tantalum foil 0.02 cm thick whose density is 1.66×10^4 kg m^{-3} is irradiated for 2 h in a beam of thermal neutrons of flux 10^{16} N m^{-2} s^{-1}. The nucleus ^{182}Ta, with a half-life of 114 days, is formed as a result of the reaction ^{181}Ta(n, γ)^{182}Ta. Immediately after irradiation, the foil has an activity of 1.23×10^7 disintegrations per second per cm^2. Calculate the number of ^{182}Ta nuclei formed.

40.23 In the photofission of ^{235}U into ^{90}Kr, ^{142}Ba and three neutrons, calculate, from the mass differences, the total energy released. Compare this energy with the initial Coulomb repulsion energy of the two charged fragments, assuming that they are just touching when fission occurs.

40.24 Calculate the energy required to split a ^{4}He nucleus into (a) ^{3}H and p, and (b) ^{3}He and n. Explain the difference between these energies in terms of the properties of nuclear forces.

40.25 (a) Verify that the energy released in the fission of uranium (185 MeV per atom) is equivalent to 8.3×10^{13} J kg^{-1}. (b) At what rate should uranium fission so that 1 MW of power is generated? (c) How long would it take for 1 kg of uranium to be used up, given that it is continuously generating 1 MW of power?

40.26 (a) What must be the average temperature of a deuterium plasma in order for fusion to take place? (Hint: we may estimate this by calculating the Coulomb repulsion energy between deuterons when they are within the range of the nuclear force, $\simeq 10^{-15}$ m.) (b) Calculate the energy released in the fusion of two deuterium nuclei into an α-particle.

40.27 (a) Calculate the energy released in the fusion process $3\,^4\text{He} \rightarrow {}^{12}\text{C}$. This process occurs in the second stage of nucleosynthesis in stars. (b) Determine the power generated by this process in a star in which 5×10^9 kg of ^{4}He are fused in ^{12}C per second.

40.28 Show that, if the energy produced by the Sun is to be accounted for, its mass must decrease at the rate of 4.6×10^9 kg s^{-1}. How much time must pass for the mass of the Sun to decrease by 1% (the mass of the Sun is given in Table 11.1)?

40.29 How long would it take for the length of the year to increase by one second due to the loss of mass of the Sun by radiation? (See preceding problem.) Recall that the square of the period of motion of a planet around the Sun is inversely proportional to the mass of the Sun.

41 The ultimate structure of matter

Richard P. Feynman (1918–1988), one of the brightest scientists of this century and in his own words a 'curious character', was a theoretical physicist at the California Institute of Technology, in Pasadena. Among his achievements was the development of a technique for calculating the processes that occur between elementary particles, for which he received the 1965 Nobel Prize in Physics. Feynman was also an enthusiastic and challenging communicator of science, as exemplified by his many lectures and books. He was the first to point out the main cause for the disastrous failure of the space shuttle *Challenger*.

Notes

41.1 Introduction

The ultimate building blocks of matter are called **fundamental** or **elementary particles**. We have already referred to protons, neutrons, electrons, positrons, neutrinos and photons. The first three particles are the only ones needed to explain the structure of atoms and nuclei. The positron is similar to an electron but with positive charge and is produced in certain nuclear processes. The neutrino is required to satisfy three basic laws in the process of β-decay: conservation of energy, momentum and angular momentum. Finally, the photon is the carrier of the electromagnetic interaction between charged particles (Figure 41.1(a)).

We have also referred to other particles, such as muons (μ) and pions (π). The nuclear interaction between protons and neutrons can be considered as a result of an exchange of **pions** (Figure 41.1(b)). The existence of pions was predicted in 1935 by Hideki Yukawa to explain the short range of nuclear forces. Pions

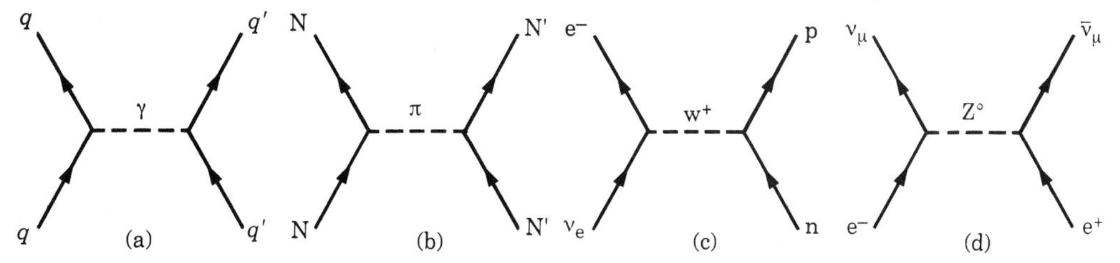

Figure 41.1 Interaction resulting from the exchange of 'particles'.

have been observed in cosmic rays and are produced profusely in high-energy accelerators and in proton–proton collisions. The number of pions produced depends on the kinetic energy of the protons. Some of the reactions are:

$$p^+ + p^+ \to \begin{cases} p^+ + p^+ + \pi^0 \\ p^+ + n + \pi^+ \\ p^+ + p^+ + \pi^+ + \pi^- \end{cases}$$

The weak interaction is associated with three carrier particles, called the $W^\pm$ and Z^0 bosons that have been observed experimentally. By extension, the gravitational interaction between two masses should require that there be another carrier particle, called a **graviton**, that has yet to be observed.

In the last 30 years many other particles have been observed. More than 30 relatively stable particles (mean lives longer than 10^{-10} s) have been identified. More than 100 extremely short-lived particles (mean lives of the order of 10^{-21} s), sometimes called **resonances**, have also been identified.

These discoveries have been made possible by the construction of very high energy accelerators (cyclotrons, synchrotrons, linacs, colliding beam machines etc.) that produce particles with energies up to several GeV, by the refinement in the techniques for observing particles (cloud, bubble and spark chambers and photographic emulsions, in all of which visible tracks are produced by the passage of charged particles), by various kinds of detectors (Geiger–Müller, scintillation and Cerenkov counters) and by sophisticated electronic circuitry (recall Note 20.3).

Our current understanding of the properties of fundamental particles constitutes a theory called the 'standard model' (Section 41.8), formulated during the 1980s, through the efforts of many physicists and as a result of the pioneering work, in 1964, of Murray Gell-Mann and George Zweig. The study of the fundamental particles is important because it displays the underlying structures of matter that are not apparent in the study of atoms, molecules and matter in bulk.

41.2 The 'fundamental' particles

Table 41.1 presents a list of stable, or long-lived ($>10^{-10}$ s) particles that have been observed and also gives some of their properties. The three basic quantities used to identify the particles are mass, charge and spin; some other identifying properties will be given later. According to their dominant interaction, the fundamental particles fall into four groups or families: (a) **carrier bosons**, associated

Table 41.1 'Fundamental' particles

Particle	Symbol	Mass (MeV)
Carrier bosons (spin 1)		
Photon	γ	0
Weak bosons	$W^{\pm}$ Z^0	$80–90 \times 10^3$
Leptons (fermions) (spin $\frac{1}{2}$)		
Neutrino	ν_e, ν_μ, ν_τ	0
Electron	e^-	0.511
Muon	μ^-	106.1
Tauon	τ^-	1780
Mesons (bosons) (spin 0)		
Pion	π^+	140
	π^0	135
Kaon	K^+	494
	K^0	498
Eta	η^0	549
Baryons (fermions) (spin $\frac{1}{2}$)		
Nucleons		
proton	p^+	938.3
neutron	n^0	939.6
Hyperons		
Lambda	Λ^0	1116
Sigma	Σ^+	1189
	Σ^0	1192
	Σ^-	1197
Xi	Ξ^0	1315
	Ξ^-	1321
Omega (spin $\frac{3}{2}$)	Ω^-	1674

with the different interactions, (b) **leptons**, or light fermions, (c) **mesons**, or intermediate mass bosons, and (d) **baryons**, or massive fermions. Also recall that fermions obey the exclusion principle and bosons do not. Baryons and mesons are subject to all four interactions; i.e. strong, electromagnetic, weak and gravitational, and for that reason are collectively called **hadrons** (from the Greek for heavy). Leptons are *not* sensitive to strong interactions. The carrier bosons are related to specific interactions, and correspond to what are called *force fields* rather than the *matter fields* related to the other particles.

The masses of fundamental particles do not seem to show any kind of regularity. On the other hand, the particles are either uncharged or have charges $\pm e$, a fact related to the law of conservation of charge. Fermions have spin $\frac{1}{2}$, except Ω^-, which has spin $\frac{3}{2}$, and obey the Pauli exclusion principle. Bosons have zero or integral spin and do not obey the exclusion principle. Most of the particles also have a magnetic dipole moment. There are three kinds of neutrinos, designated ν_e, ν_μ and ν_τ, associated with the electron, muon and tauon, respectively.

EXAMPLE 41.1

Masses of particles that mediate an interaction.

▷ Particles interact through the exchange of bosons, which are the carriers of momentum, angular momentum and energy and, in some cases, also of electric charge. The bosons exchanged are called **virtual** because they exist only briefly during the exchange process, although when sufficient energy is available, they can exist as free particles.

The range of *nuclear* forces may be explained with this model for interactions. It is assumed that a proton is not a static system but is continuously ejecting and reabsorbing virtual pions. When a virtual pion is emitted, the proton energy changes by an amount at least equal to $\Delta E \sim m_\pi c^2 \sim 140$ MeV, with a corresponding change in momentum. According to Heisenberg's uncertainty principle, Equation 36.5, this pion can exist, without any violation of the conservation laws during a time

$$\Delta t \sim \hbar/\Delta E \sim \hbar/m_\pi c^2 \sim 10^{-22}\ \text{s}$$

After this time the virtual pion must be reabsorbed by the nucleon or exchanged with another nucleon to restore energy and momentum conservation. In this time, if we assume that the pion travels with a velocity close to that of light, the maximum distance it can go is about 10^{-14} m. This then gives the maximum distance at which a second nucleon must be located to absorb the virtual pion and, therefore, gives the order of magnitude of the range of the nuclear interaction.

In the case of *electromagnetic* interactions, we assume that a charged particle is continuously emitting and absorbing virtual photons. The virtual photons have zero mass and therefore, the energy fluctuation ΔE of a charged particle may have any arbitrary value, depending on the frequency associated with the photon. Consequently, the length of time that the virtual photon can exist before it is reabsorbed or exchanged with another charged particle is also arbitrary. Therefore, electromagnetic forces are of long range. A calculation shows that the force between two slowly moving charged particles, resulting from the exchange of zero-mass photons, must vary as the inverse square of the distance of the two charges, in agreement with Coulomb's law.

The particles that are assumed to mediate the *weak* interaction, $W^{\pm}$ and Z^0, have masses of 81 GeV and 91 GeV, respectively. They then have a range of about 10^{-17} m, or 10^3 times smaller than the range of the nuclear interaction. This is why the weak interaction is not involved in most processes we observe on Earth except radioactive decay. The weak interaction can be represented by processes illustrated in Figure 41.1(c) and (d).

41.3 Particles and antiparticles

It has been found that each particle has an associated **antiparticle**. The antiparticles are listed in the last column of Table 41.1. An antiparticle is designated by the same symbol as the particle, but with a bar over it. (Sometimes, for simplicity, the bar is omitted.) Antiparticles have the same mass and spin as the particles but opposite electromagnetic properties, such as charge and magnetic moment, but other properties, to be mentioned later on, are also opposite. The π^0 and η^0 mesons are their own antiparticles. In Figure 41.2 the particles and antiparticles are arranged in a symmetric way. The first antiparticle to be observed was the positron, in 1933 (recall Figure 22.7). Antiprotons were first observed in 1955; antineutrons were observed shortly afterward.

The neutrino is distinguished from its antiparticle by the fact that a neutrino always has its spin pointing in a direction *opposite* to that of its momentum (which is in the direction of motion) while, for an antineutrino, the momentum and the spin are in the *same* direction (Figure 41.3). A neutrino is said to have **negative helicity**, equal to -1, and an antineutrino is said to have **positive helicity**, or $+1$. This property is a consequence of the *zero* mass of the neutrino. For other particles of spin $\frac{1}{2}$ that have *non-zero* mass (such as electrons), both particle and antiparticle may have either positive or negative helicity.

A particle and its antiparticle may combine, disappearing or **annihilating** each other. Their total energy, including their rest energy, reappears in the form of other particles. For example, the electron e^- and its antiparticle, the positron e^+, may annihilate, emitting photons, a process called **electron–positron annihilation,**

$$e^- + e^+ \rightarrow 2\gamma \tag{41.1}$$

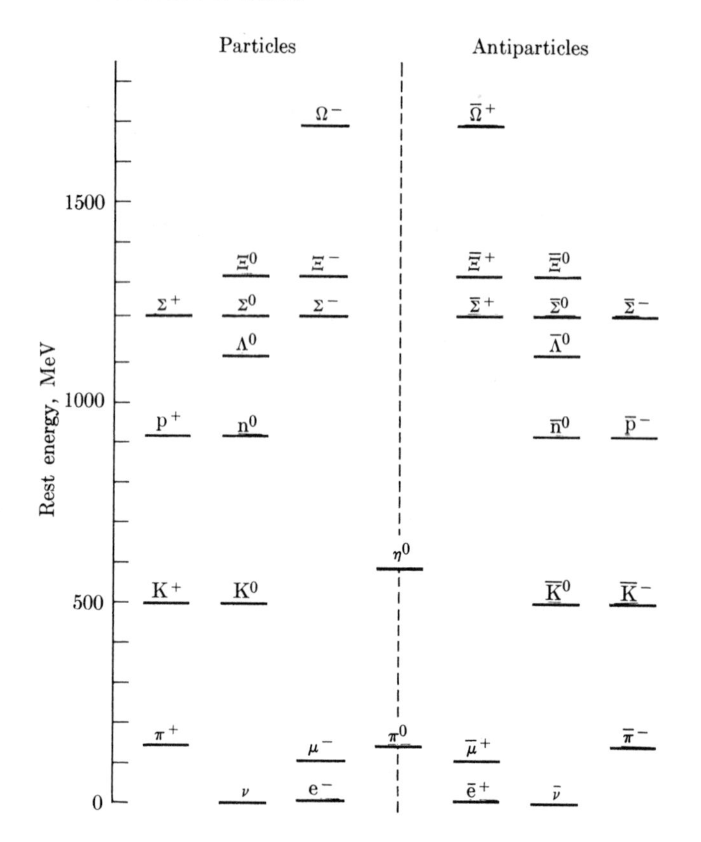

Figure 41.2 Particles and antiparticles arranged according to their rest energy and charge.

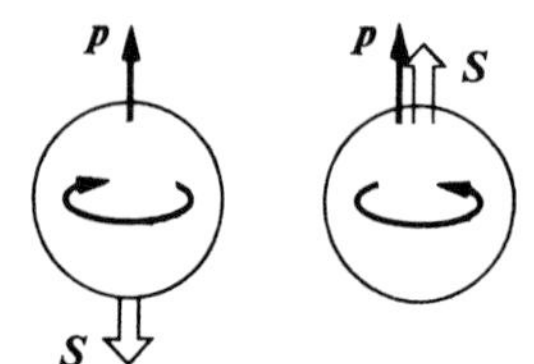

Figure 41.3 The neutrino and antineutrino, both with momentum *p*, have opposite helicities or spin *S*.

The reason that two photons, rather than one, are emitted is that energy and momentum must be conserved. If the electron and positron are at rest in the laboratory (which then coincides with the C-frame), the total energy available is $2m_ec^2 = 1.022$ MeV and the total momentum is zero. If a photon of energy $E_\gamma = 2m_ec^2$ were emitted, it would have a momentum $p_\gamma = E_\gamma/c = 2m_ec$ and the principle of conservation of momentum would be violated, since the initial momentum was zero. To conserve energy and momentum, two equal photons must be emitted in opposite directions. Thus each photon must have an energy equal to $m_ec^2 = 0.511$ MeV. Photons of this energy are observed when positrons pass through matter. If the electron's energy is sufficiently large, other processes might occur, as will be explained in Section 41.8. Also, charge is conserved in process 41.1.

Proton–antiproton annihilation is a more complex process, involving the production of several particles, most of them pions. One such annihilation is

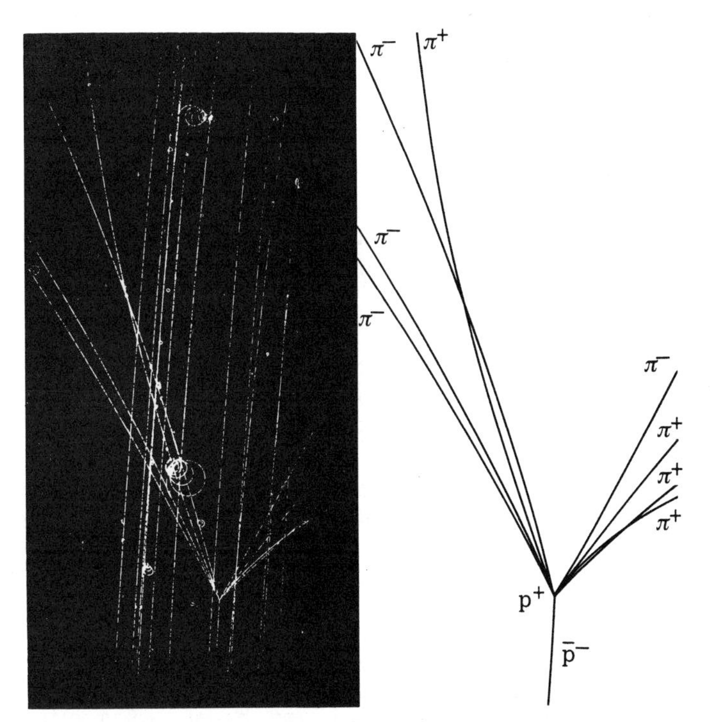

Figure 41.4 Proton–proton annihilation in a bubble chamber. The incoming antiproton annihilates with one of the protons in the hydrogen gas chamber. (Photograph courtesy of Brookhaven National Laboratory.)

shown in Figure 41.4, corresponding to the process

$$p^+ + p^- \rightarrow 4\pi^+ + 4\pi^- + x\pi^0 \tag{41.2}$$

If charge is to be conserved, the number of positive and negative pions must be the same. The number of π^0-mesons produced is difficult to ascertain, since they do not leave a track in the chamber. The total number of particles produced depends on the energy available.

Conversely, a particle and its antiparticle may be produced simultaneously. In Figure 22.5 the production of an electron–positron pair by a photon was illustrated. The process is

$$\gamma \rightarrow e^- + e^+ \tag{41.3}$$

The track of the photon is not visible. For an electron pair to be produced, the photon must have an energy equal to at least $2m_e c^2 = 1.022$ MeV. In order for energy and momentum to be conserved in Equation 41.3, the process must occur near a nucleus. Then, as a result of the electromagnetic coupling between the nucleus and the electron–positron pair, the nucleus will take up the energy and momentum required for the conservation of both quantities. For this reason, electron pair production is more intense in materials that have high atomic numbers (such as lead), since these materials provide a stronger electromagnetic coupling with the electron–positron pair. Pair production is one of the main processes that accounts for the absorption of high-energy photons by various materials. At low energy the photoelectric effect (Section 30.7) is the more important process, and at energies between 0.1 MeV and 1 MeV it is the Compton effect (Section 30.5). However, if the energy of the photons is sufficiently large, other processes might occur, such as μ^+, μ^- creation.

Similarly, a **proton–antiproton pair** may be produced in a high-energy proton–proton collision according to the scheme

$$p^+ + p^+ \rightarrow p^+ + p^+ + p^+ + p^- \qquad \textbf{(41.4)}$$

For this process to occur, the threshold kinetic energy of the incoming proton (if the target proton is at rest) has to be at least 5.64 GeV (see Example 20.4 for the calculation of this value).

The universe seems to be composed mainly of particles (rather than a uniform mixture of particles and antiparticles), making it stable against annihilation.

Note 41.1 The antiproton experiment

The purpose of the experiment that led to the discovery of the antiproton (in 1955) was to detect particles having a charge $-e$ and a mass m_p. Protons, accelerated up to 6.2 GeV by the University of California bevatron, hit a suitable target where several reactions produced many particles, including some antiprotons, p^-. A deflecting magnet M_1 selected only negative particles, which were allowed to pass through an opening in the shielding (Figure 41.5). Scintillation detectors and Cerenkov detectors, labeled S and C respectively, allow measurement of the energy and the velocity of the particle, respectively. These instruments are designed to be sensitive only to particles in a certain energy range. A second deflecting magnet M_2 was placed between S_1 and S_2. This magnet then acted as a momentum selector and only particles with momentum $p = eBr$ were properly deflected toward S_2. The magnetic field was chosen to correspond to the momentum at which most antiprotons should have been produced.

Many negatively charged particles (K^-, π^-, μ^-) and a few antiprotons (it was estimated that the proportion was 40 000 other particles to one antiproton) passed through magnet M_2. To discriminate the antiprotons, the scintillation detectors S_1 and S_2 were placed in such a way that they would give a signal only for antiprotons of momentum p. Hence the observation of such signals was an indication of the passage of an antiproton. This experiment illustrates the variety of techniques required in experimental particle physics.

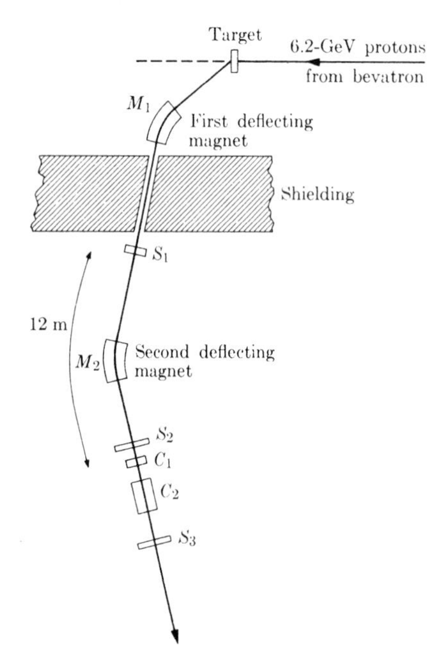

Figure 41.5 Experimental arrangement for observation of the antiproton.

41.4 Particle instability

The creation and annihilation of particles, as well as the process of β-decay, indicate that *fundamental particles can transform into others as a result of their interactions.* Some particles undergo **spontaneous decay**, with a well-defined half-life. Table 41.2 shows the decay modes and half-lives of several unstable particles. In some cases more than one decay mode is possible. Only three particles (and their antiparticles) are stable against spontaneous decay: the neutrino, the electron and the proton, although there are reasons to believe that protons might decay (with a half-life of more than 10^{32} yr!). Of all the unstable particles, the one having the longest half-life is the neutron. This may explain why matter is composed of electrons, protons, and neutrons. Note that mesons have a half-life of the order of 10^{-8} s (with the exception of the π^0 and η^0), while baryon half-lives are of the order of 10^{-10} s.

A condition necessary for spontaneous decay, imposed by energy conservation, is that the mass of the parent particle be larger than the sum of the masses of the daughter particles. In addition, energy, charge, momentum and angular momentum must be conserved. These criteria are used as guiding principles when analyzing decays of fundamental particles.

The negative muon, with spin $\frac{1}{2}$, discovered in cosmic rays in 1937, was the first unstable particle observed. The decay of muons always results in electrons that have a continuous kinetic energy spectrum, similar to that found in β-decay.

Table 41.2 Some particle decay modes*

Particle	Decay mode	Half-life (s)
Leptons		
Neutrino	Stable	
Electron	Stable	
Muon	$\mu^- \to e^- + \nu_\mu + \bar{\nu}_e$	1.52×10^{-6}
Tauon	$\tau^- \to e^- + \nu_\tau + \bar{\nu}_e$ $\to \mu^- + \nu_\tau + \bar{\nu}_\mu$	3.5×10^{-13}
Mesons		
Pion	$\pi^\pm \to \mu^\pm + \nu_\mu$ or $\bar{\nu}_\mu$ $\to e^\pm + \nu_e$ or $\bar{\nu}_e$	1.80×10^{-8}
	$\pi^0 \to \gamma + \gamma$ $\to \gamma + e^+ + e^-$	6×10^{-17}
Kaon	$K^+ \to \mu^+ + \nu$ $\to \pi^+ + \pi^0$	8.56×10^{-9}
	$K^0 \to \pi^\pm + e^\mp + \nu$ $\to \pi^\pm + \mu^\mp + \nu$	4×10^{-8}
	$K^0 \to \pi^+ + \pi^-$ $\to 2\pi^0$	6.0×10^{-11}
Eta	$\eta^0 \to \gamma + \gamma$ $\to \pi^0 + \gamma + \gamma$	$< 10^{-16}$
Baryons		
Proton	Stable	(10^{31} yr?)
Neutron	$n^0 \to p^+ + e^- + \bar{\nu}$	7.0×10^2
Lambda	$\Lambda^0 \to p^+ + \pi^-$ $\to n^0 + \pi^0$	1.76×10^{-10}
Sigma	$\Sigma^+ \to p^+ + \pi^0$ $\to n^0 + \pi^+$	5.6×10^{-11}
	$\Sigma^0 \to \Lambda^0 + \gamma$	$< 7 \times 10^{-15}$
	$\Sigma^- \to n^0 + \pi^-$	1.1×10^{-10}
Xi	$\Xi^0 \to \Lambda^0 + \pi^0$	2.0×10^{-10}
	$\Xi^- \to \Lambda^0 + \pi^-$	1.2×10^{-10}
Omega	$\Omega^- \to \Lambda^0 + K^-$ $\to \Xi^0 + \pi^-$	10^{-10}

* To obtain the decay of antiparticles, change all particles into antiparticles on both sides of the equations.

Therefore, energy and momentum conservation requires that *two* neutrinos, ν_e and ν_μ, be emitted with the electron (see Table 41.2). Charged pions, with spin 0, decay into either a muon or an electron with a well-defined kinetic energy, and therefore only a neutrino (ν_μ or ν_e) is emitted (see Table 41.2 and Figure 20.5).

Figure 41.6 shows proton–antiproton annihilation, which gives rise to new particles that undergo subsequent decays:

$$
\begin{aligned}
&p^+ + p^- \to \Xi^+ + \Xi^- \\
&\qquad\qquad\qquad\quad \hookrightarrow \pi^- + \Lambda^0 \\
&\qquad\qquad\quad\ \hookrightarrow \pi^+ + \bar{\Lambda}^0 \\
&\qquad\qquad\qquad\qquad\quad \hookrightarrow \pi^+ + p^-
\end{aligned}
\tag{41.5}
$$

Each of the decays, when checked for conservation of energy and momentum, enables us to identify the neutral particles that do not leave any track.

Figure 41.7 illustrates a more complex process by which a K^- and a p^+ collide:

$$
\begin{aligned}
&K^- + p^+ \to \Xi^- + K^0 + \pi^+ \\
&\qquad\qquad\qquad\qquad\quad \hookrightarrow \pi^+ + \pi^- \\
&\qquad\qquad\quad\ \hookrightarrow \pi^- + \Lambda^0 \\
&\qquad\qquad\qquad\qquad\quad \hookrightarrow \pi^- + p^+
\end{aligned}
\tag{41.6}
$$

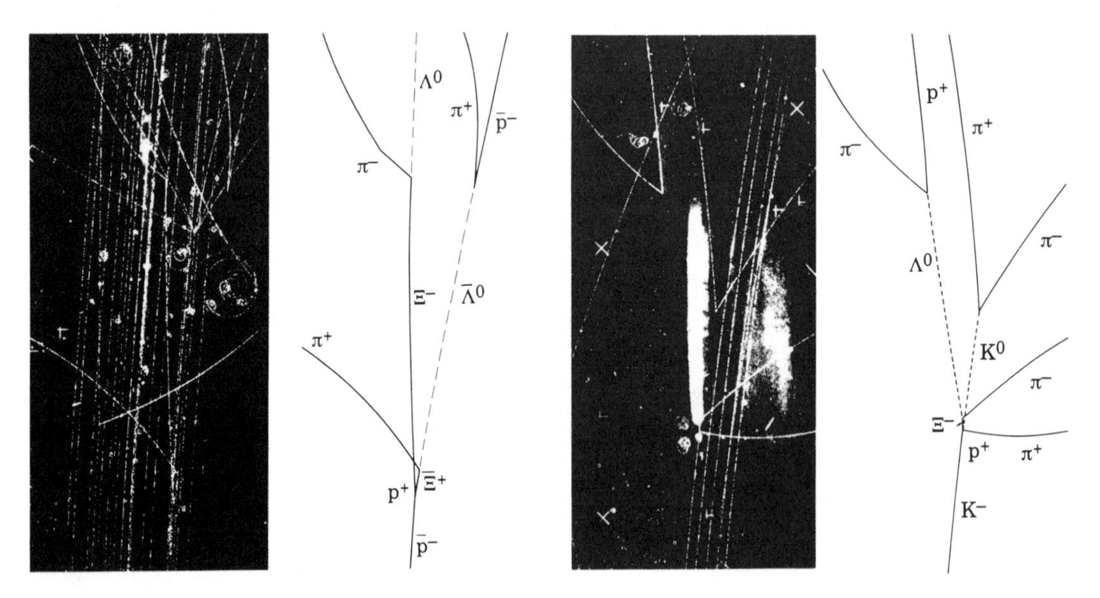

Figure 41.6 Events triggered by a proton-antiproton annihilation. (Photograph courtesy of Brookhaven National Laboratory.)

Figure 41.7 Events triggered by a K^-–p collision. (Photograph courtesy of Brookhaven National Laboratory.)

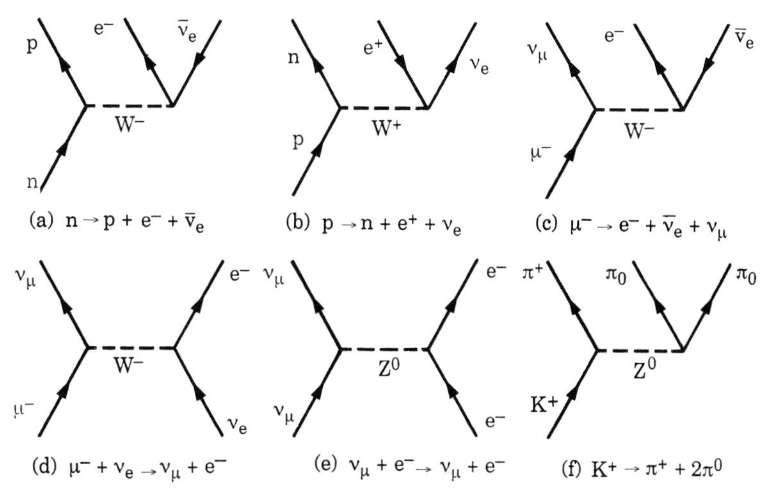

Figure 41.8 Graphical representation of processes involving W^+ and Z^0 bosons.

These particle decays are basically the result of the weak interaction, carried by $W^\pm$ and Z^0 bosons. In Figure 41.8(a), (b) and (c), the mechanism of neutron, proton and muon decay is described graphically in terms of the exchange of $W^\pm$ particles. The first two correspond to β-decay discussed in Section 40.4. The other processes are described in Figure 41.8(d), (e) and (f).

Many of the processes involving fundamental particles have been observed in **cosmic rays**, which consist of energetic particles (mostly protons) falling on the Earth as a result of nuclear processes occurring in the Sun and other parts of the

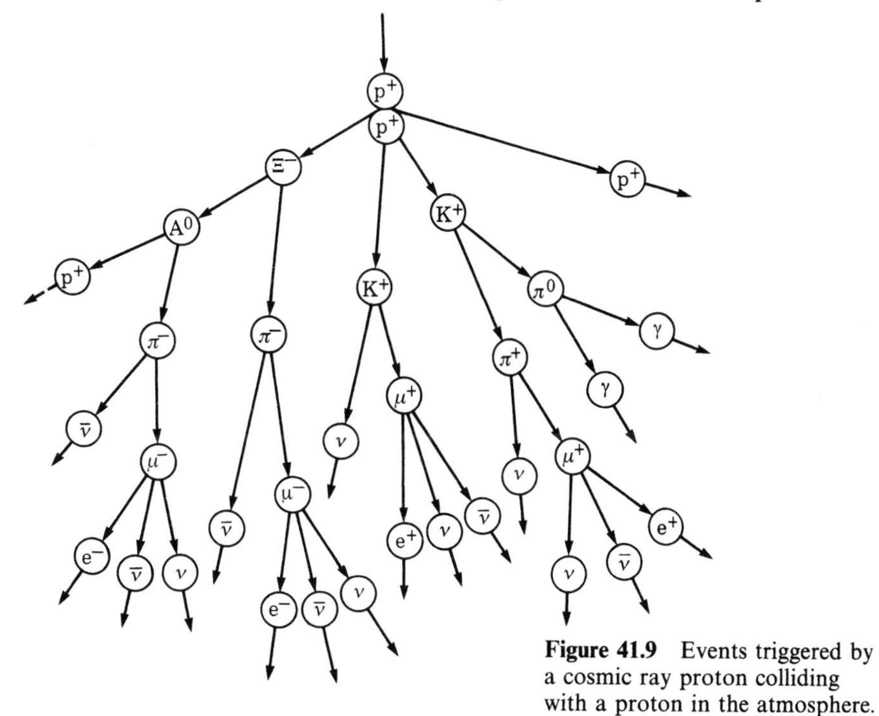

Figure 41.9 Events triggered by a cosmic ray proton colliding with a proton in the atmosphere.

universe. Cosmic ray particles trigger a chain of reactions when they interact with nuclei in the upper atmosphere. One such chain reaction is represented in Figure 41.9. Such a process continues until the energy of the particles is no longer great enough to produce new particles and only stable particles remain.

EXAMPLE 41.2

Energy analysis of muon and pion decay.

▷ Electrons emitted in muon decay have a continuous kinetic energy spectrum of up to 53 MeV. Since the maximum energy of the electron is much larger than its rest energy (0.5 MeV), the maximum momentum of the electron is $p_{max} \approx E_{max}/c = 53$ MeV/c. This maximum momentum occurs when the two neutrinos are both emitted in the opposite direction to that of the electron, and both carry a total momentum of 53 MeV/c and their total energy is 53 MeV. Thus, the total energy released in muon decay is about 106 MeV, which gives the rest energy of the muon (Table 41.1). The spin of the muon has been measured independently and found to be $\frac{1}{2}$; thus, the law of conservation of angular momentum forbids the decay of a muon into an electron and a neutrino, each having spin $\frac{1}{2}$, but it allows for a muon decay into an electron and two neutrinos. Because the two neutrinos behave in slightly different ways, it has been concluded that they are different: one is a muon-neutrino (ν_μ) and the other an electron-neutrino (ν_e).

The muons produced in pion decay have a fixed kinetic energy of about 4.1 MeV and a rest energy of 105.7 MeV. Since the momentum of a muon with that kinetic energy is 29.5 MeV/c, this must also be the momentum of the neutrino, which is emitted in the opposite direction. The neutrino energy is then 29.5 MeV. Thus, the total energy released in pion decay is 105.7 MeV + 4.1 MeV + 29.5 MeV = 139.3 MeV, which again approximately agrees with the rest energy of the pion, 140 MeV. From the decay $\pi^+ \rightarrow \mu + \nu_\mu$, we conclude that the spin of the pion is either 0 or 1. From an analysis of other processes involving pions, it has been determined that the spin of a pion is zero.

41.5 The conservation laws

In all the processes involving fundamental particles, it has been found that the following conservation laws hold without exception: (1) conservation of momentum, (2) conservation of angular momentum, (3) conservation of energy, and (4) conservation of charge. The limitations imposed by these four conservation laws restrict many processes involving fundamental particles. For example, the electron and the positron are stable because there are no other lighter charged particles into which they might decay without violating charge conservation. But there are several processes that comply with the four conservation laws but do not occur in nature. For example, $p^+ \rightarrow \pi^+ + \nu$, $\Lambda^0 \rightarrow p^- + \pi^+$, or $\pi^+ + p^+ \rightarrow \Sigma^+ + \pi^+$ are not observed. To explain this situation, new conservation laws, which resemble the law of conservation of charge rather than the first three conservation laws, have been formulated. We shall consider only two of them: (5) conservation of leptons and (6) conservation of baryons. Experimental data has led to the conclusion that leptons and baryons are restricted with regard to the number that may be produced or annihilated in a single process. For example, the process of electron pair production $\gamma \rightarrow e^- + e^+$, where two leptons – an electron and its antiparticle, a positron – are created out of a photon, is possible. But the process of electron–proton production $\gamma \to e^- + p^+$, in which one lepton and one baryon are produced, is not observed. It is experimental facts of this kind that form the basis for the fifth and sixth conservation laws.

To express the **conservation of leptons**, a **lepton quantum number** $\mathscr{L} = 1$ is assigned to lepton particles, $\mathscr{L} = -1$ for lepton antiparticles and $\mathscr{L} = 0$ for non-lepton particles. Then

in any process, the total lepton number must remain unchanged.

For example, in the process $\gamma \to e^- + e^+$, the lepton number on the left is $\mathscr{L} = 0$ and on the right $\mathscr{L} = +1 - 1 = 0$. However, in the non-observed process $\gamma \to e^- + p^+$, we have $\mathscr{L} = 0$ on the left and $\mathscr{L} = +1$ on the right, so that the conservation of leptons is violated. The law of conservation of leptons requires that neutron decay be written in the form

$$n \to p^+ + e^- + \bar{\nu}_e \qquad \textbf{(41.7)}$$

to ensure a total lepton number of zero on both sides. Thus conservation of leptons requires that, in β-decay, an antineutrino $\bar{\nu}_e$ ($\mathscr{L} = -1$) and *not* a neutrino ($\mathscr{L} = +1$) be emitted together with the electron ($\mathscr{L} = +1$). By examining the decay schemes of Table 41.2, the law of conservation of leptons can be verified in each case and why an antineutrino is indicated in some decays.

To express the **conservation of baryons**, a **baryon quantum number** $B = +1$ is assigned to all baryon particles, with $B = -1$ for all baryon antiparticles and $B = 0$ for all non-baryon particles. then

in any process, the total baryon number must remain unchanged.

This law is satisfied in all processes previously mentioned involving baryons, as well as in the decay schemes of Table 41.2. Since the proton is the lightest of all baryons, its decay into lighter particles (which cannot be baryons) would violate the law of conservation of baryons, and this explains why the proton is a stable particle. Thus the world in which we live is, to a certain extent, the result of the laws of conservation of leptons and baryons.

41.6 Symmetry and interactions

The behavior of physical systems under certain symmetry operations provides some indication of the properties of the fundamental interactions. We shall consider three symmetry operations.

(i) Parity. This concept refers to the operation of space reflection, either in a plane or through a point, such as the origin of coordinates. Let us first see the behavior of some dynamical quantities regarding reflection in a plane. Consider a particle A (Figure 41.10(a)) moving with momentum $\boldsymbol{p}$. The mirror image of A is another particle A' moving with momentum $\boldsymbol{p}'$ such that $p_\parallel = p'_\parallel$, $p_\perp = -p'_\perp$, where $\parallel$ and $\perp$ refer to directions parallel and perpendicular to the plane, respectively. Next consider a particle moving as shown in Figure 41.10(b), with angular momentum $\boldsymbol{L}$. The mirror image of A is another particle A', revolving as shown in the figure and thus having an angular momentum $\boldsymbol{L}'$ such that $L_\parallel = -L'_\parallel$, $L_\perp = L'_\perp$. We see then that in reflection in a plane, $\boldsymbol{p}$ and $\boldsymbol{L}$ behave in different ways. For that reason $\boldsymbol{p}$ is called a *polar* vector and $\boldsymbol{L}$ an *axial* vector. All vectors that play a part in physical laws are either polar or axial. By examining Figure

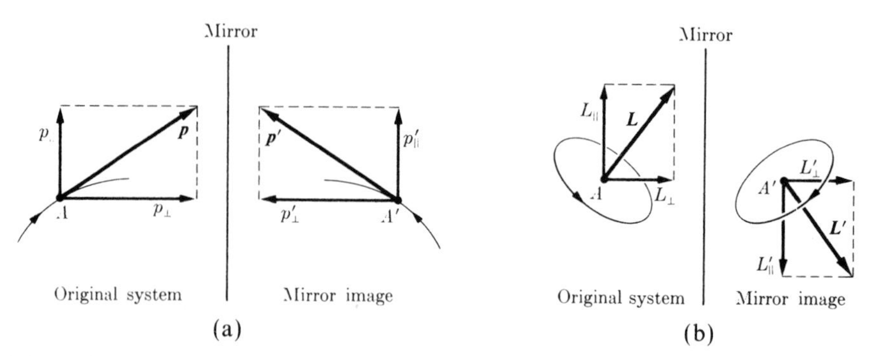

Figure 41.10 Reflection in a plane of (a) momentum, (b) angular momentum.

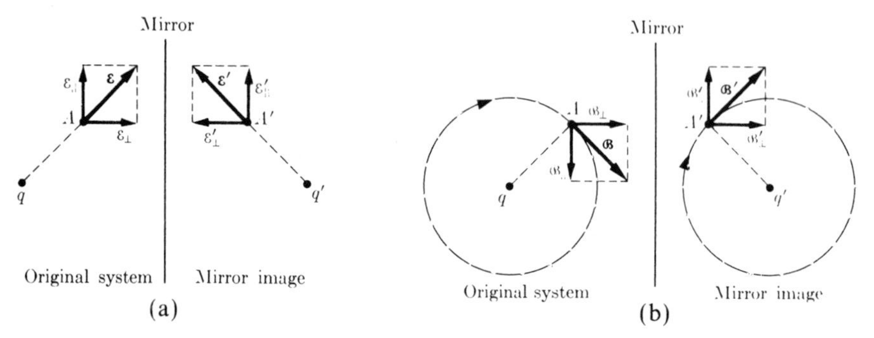

Figure 41.11 Reflection in a plane of (a) electric field of a positive charge q, (b) the magnetic field of a positive charge q moving into the page.

41.11, it can be verified that the electric field is polar and that the magnetic field is axial. Force is also a polar vector.

It has always been assumed that the laws of physics are invariant relative to the parity operation. In other words, if a given physical system satisfies the laws of motion, its mirror image also satisfies the laws of motion, and it may also occur in nature. This means that in the mathematical statements of physical laws, polar and axial quantities must appear in such a way that the relation among them is not changed by space reflection. We may then state the **law of conservation of parity**:

> *parity is conserved in a process if the mirror image of the process is also a process that can occur in nature; that is the interaction involved in the process must be invariant with respect to the parity operation.*

It can be shown that Maxwell's equations for the electromagnetic field are invariant to space reflection, and that, therefore, parity is conserved in electromagnetic interactions. In 1956, Tsung D. Lee and Chen N. Yang questioned the validity of invariance under space reflection for processes due to the weak interaction. Soon after, it was found that parity is violated in such processes. As an example, consider the pion decay $\pi^+ \rightarrow \mu^+ + \nu_\mu$, shown in Figure 41.12(a). The pion has zero spin and, in order to conserve angular momentum, the spins of the muon and the neutrino must be in opposite directions. Also, to conform with the helicity of the neutrino, their spins must be as indicated in the figure (Recall Figure

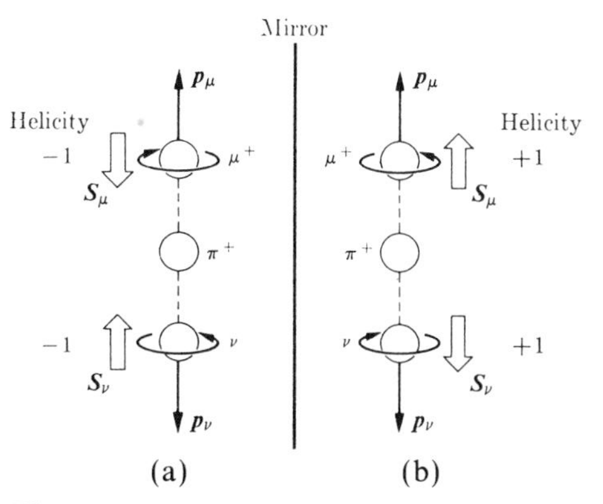

Figure 41.12 The nonconservation of parity in π-decay. (a) Original decay. (b) Space-reflected decay.

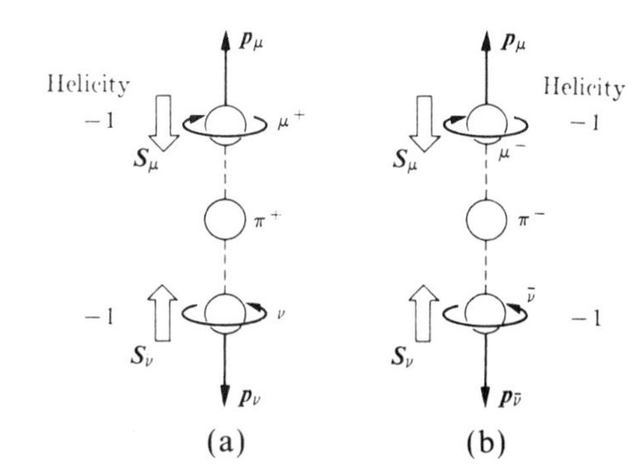

Figure 41.13 Application of charge conjugation to the π-decay. (a) Original decay. (b) Charge conjugate decay.

41.3). However, the mirror system, shown in Figure 41.12(b), does not occur in nature because the neutrino would have the wrong helicity. Therefore, parity is violated in pion decay, which is due to the weak interaction.

(ii) Charge conjugation. This operation consists in replacing all particles by their antiparticles, and conversely, without changing any other physical property, such as momentum or spin.

An interaction is invariant with regard to charge conjugation when, if a process due to the interaction is possible in a system, the corresponding process is also possible in the charge conjugate system.

It seems that processes due to strong and electromagnetic interactions are invariant relative to charge conjugation. However, processes involving weak interactions are not invariant relative to charge conjugation. To use pion decay as an example (Figure 41.13(a)), the charge conjugate system (Figure 41.13(b)) cannot occur because the antineutrino would have the wrong helicity.

Nevertheless, a comparison of Figures 41.12 and 41.13 suggests a new possibility. Suppose that we first perform a space reflection or parity operation P on the system as shown in Figure 41.14(a) and (b) and then perform a subsequent charge conjugation C, as shown in Figure 41.14(c); the system resulting from the combined operation (CP) also occurs in nature; that is, $\pi^- \to \mu^- + \bar{\nu}$. Thus:

weak interactions are invariant under the combined CP operation.

(iii) Time reversal. This operation consists in changing t into $-t$. Under time reversal, velocity, momentum and angular momentum are reversed and in a collision the initial and final states are exchanged (Figure 41.15). Our intuition, supported by strong experimental evidence both at the macroscopic and the atomic levels, points toward invariance of physical laws relative to time reversal. It can be shown that the laws of electromagnetism are invariant under time reversal. Also, it is assumed that the strong interaction is considered invariant under time reversal. An important theorem relating the three symmetry operations states that

all physical laws are considered to be invariant under the CPT operation;

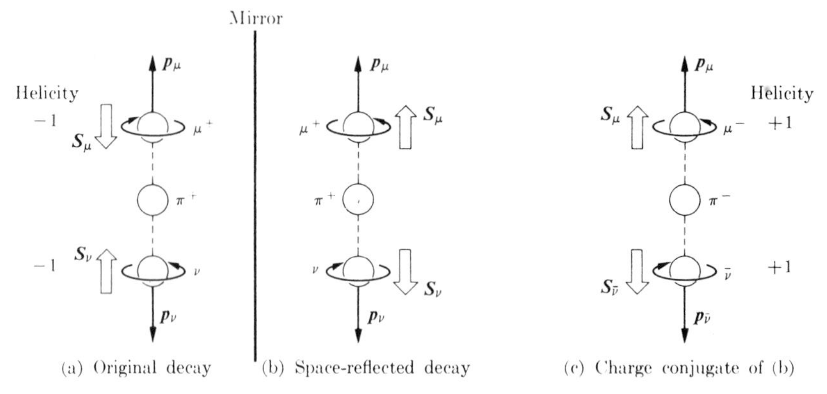

Figure 41.14 The *CP* operation applied to π-decay. (a) Original decay. (b) Space-reflected decay. (c) Charge conjugate of (b).

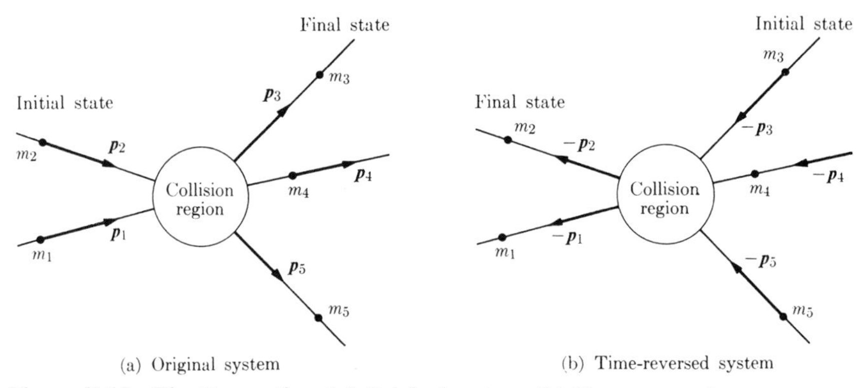

Figure 41.15 The *T* operation. (a) Original system. (b) Time-reversed system.

that is, under the combined operations of time reversal T, parity P and charge conjugation C. This theorem is true even in cases where the laws may not be invariant under the individual operations; therefore, it constitutes a fundamental property of the universe and a requirement of all physical laws. One consequence of the CPT theorem is that particles and their antiparticles should have exactly the same masses and lifetimes, a fact that has been observed in all cases.

Recent experiments regarding a special form of decay of the neutral kaon, which is due to the weak interaction and occurs in only 0.3% of the cases, seem to indicate CP violation. Such a violation implies violation of invariance under time reversal in order to preserve CPT invariance.

Note 41.2 Parity violations in β-decay

To test the conservation of parity in processes due to weak interactions, an experiment was performed by Chien S. Wu and her collaborators in 1957. A sample of $^{60}_{27}Co$ was polarized so that the nuclei had their spins aligned. Then they found (Figure 41.16) that the electrons resulting from the β-decay of the $^{60}_{27}Co$ nuclei were emitted in greater quantities in the direction of the $^{60}_{27}Co$ spin (or the direction of polarization) than in the opposite direction. This proved that there is a larger probability that a $^{60}_{27}Co$ nucleus will decay, emitting an electron in the

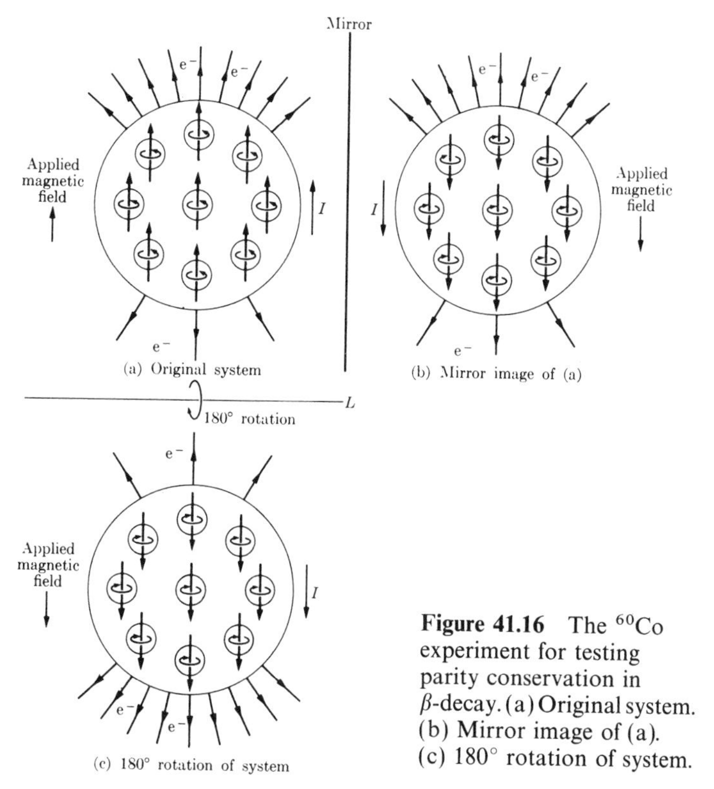

Figure 41.16 The ^{60}Co experiment for testing parity conservation in β-decay. (a) Original system. (b) Mirror image of (a). (c) 180° rotation of system.

direction of its spin, than in the opposite direction. When they reversed the whole system by rotating it through 180° around the line L, the polarization, the spins, and the magnetic field were reversed (Figure 41.16(c)), and the direction of maximum intensity of electron emission also rotated 180°. The mirror image of (a) is shown in (b), where the spins and the applied magnetic field have been reversed, because they are axial vectors, but the direction of maximum intensity of electron emission remains the same. Comparison of (b) and (c) clearly indicates that (b) does not correspond to the situation found in nature. Therefore, Wu's experiment provides direct experimental evidence that parity is not conserved in weak interactions, which are responsible for β-decay. After the ^{60}Co experiment, many other experimental proofs were obtained for the nonconservation of parity in weak interactions. However, in processes due to strong and to electromagnetic interactions, parity is conserved.

41.7 Resonances

Experimental evidence has established the existence of particles, called **resonances**, with lifetimes so short, of the order of 10^{-20} s or less, that they do not leave any recognizable track in bubble or spark chambers but are recognized by their decay products. Resonances are hadrons and decay by means of the strong interaction, which accounts for their extremely short lives. Figure 41.17 shows some meson resonances and Figure 41.18 gives some of the known baryon resonances.

Resonances correspond to excited states of mesons and baryons. Resonances have the same quantum numbers as the mesons and baryons to which they are related, and are designated by the same symbol but with a subscript 1, 2, 3,... according to increasing mass.

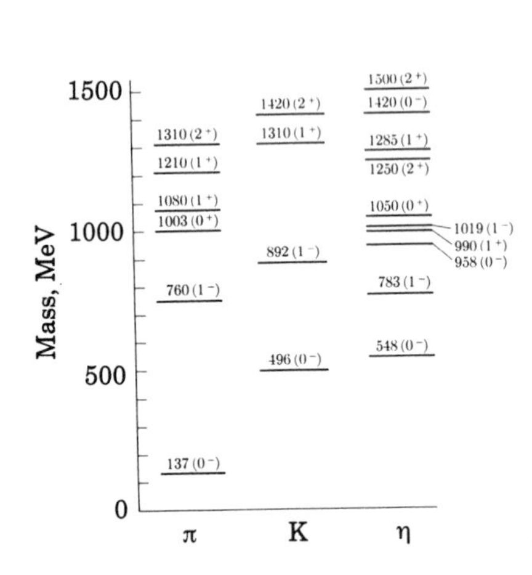

Figure 41.17 Meson resonances. The spin and parity of each resonance is given in parentheses.

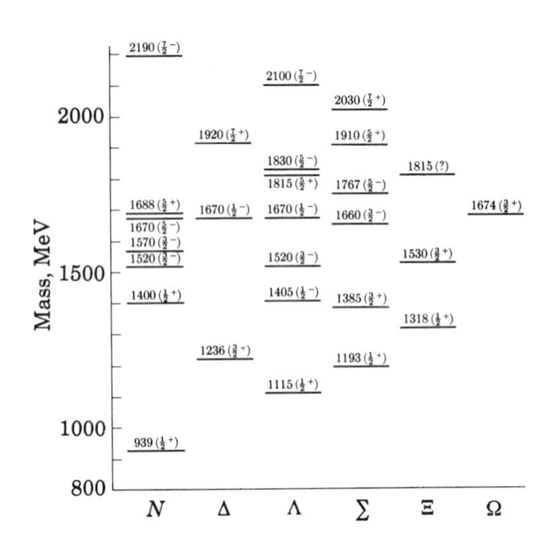

Figure 41.18 Baryon resonances. The spin and parity of each resonance is given in parentheses.

The fact that mesons and baryons can have excited states is an indication that these particles must have some kind of internal structure. In Figure 41.19 the excitation energies of molecules, atoms, nuclei and hadrons have been compared, only to orders of magnitude. The large differences explain the different physical behavior of the systems and of the forces involved in forming them. In Example 36.5, the reason for these great differences in energy was given.

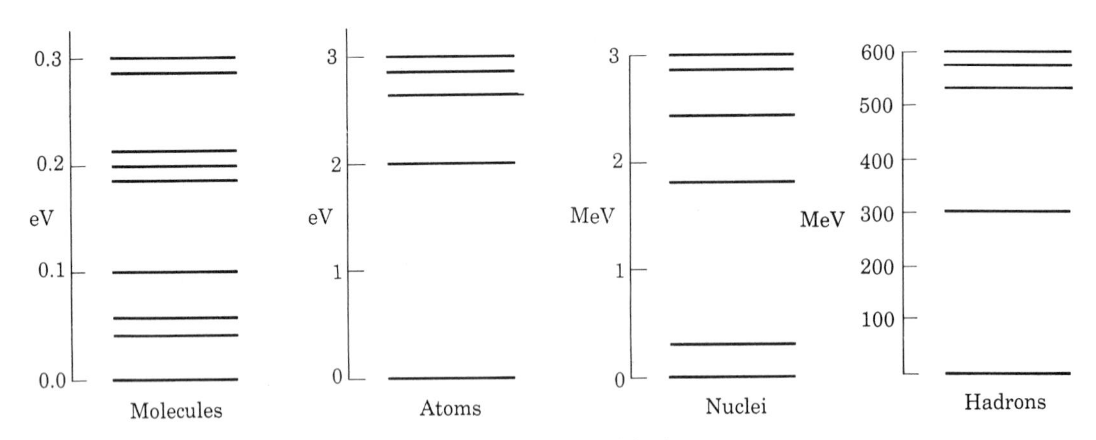

Figure 41.19 Energy levels in atoms, molecules, nuclei and hadrons.

41.8 The standard model

In the same way that atoms are composed of certain basic ingredients (electrons, protons and neutrons) and nuclei are composed of protons and neutrons, it appears that all hadrons are composed of certain building blocks or fundamental particles. Murray Gell-Mann has proposed the name **quark** for these fundamental particles. Baryons and mesons can be in the ground state or in one of several excited states (or resonances), each state having its own attributes, such as energy, angular momentum and parity. On the other hand, no excited states of leptons have been observed.

The theory that describes the properties and structure of the fundamental particles, as well as their interactions, is called the **standard model**. Our discussion will be sufficient to understand how the model explains the particle properties and processes we have discussed so far, without entering into theoretical considerations and the experimental foundations. The standard model assumes that matter is made up of two kinds of particles, *leptons* and *quarks*, to which we add the *intermediate bosons* that carry the interactions between particles. Both leptons and quarks are assumed to be structureless or point-like.

According to the theory, there are six leptons and six quarks, all with spin $\frac{1}{2}$ and therefore all are fermions obeying Pauli's exclusion principle. The six leptons are: electrons (e), muons (μ), and tauons (τ), all with charge $-e$, and the three corresponding neutrinos, ν_e, ν_μ, ν_τ, although the experimental evidence for the ν_τ is not yet conclusive. The upper limits for the neutrino masses are 10 eV, 250 eV and 70 MeV respectively. The electric charge of the leptons is either $-e$ or zero. Thus we have three groups of leptons:

$$\begin{matrix}\text{charge } 0: \\ \text{charge } -e:\end{matrix} \quad \begin{pmatrix}\nu_e \\ e\end{pmatrix} \quad \begin{pmatrix}\nu_\mu \\ \mu\end{pmatrix} \quad \begin{pmatrix}\nu_\tau \\ \tau\end{pmatrix} \qquad \textbf{(41.8)}$$

Similarly, the theory assumes six kinds or 'flavors' of quarks called *up* (u), *down* (d), *charm* (c), *strange* (s), *top* (t) and *bottom* (b), although the experimental evidence for the top quark is not yet definitive. (These names have been chosen arbtrarily and carry no special meaning.) The electric charge of the quarks is either $\frac{2}{3}e$ or $-\frac{1}{3}e$ according to the following scheme that groups quarks in three categories:

$$\begin{matrix}\text{charge } \frac{2}{3}e: \\ \text{charge } -\frac{1}{3}e:\end{matrix} \quad \begin{pmatrix}u \\ d\end{pmatrix} \quad \begin{pmatrix}c \\ s\end{pmatrix} \quad \begin{pmatrix}t \\ b\end{pmatrix} \qquad \textbf{(41.9)}$$

The masses of the quarks are not well known because they have never been observed as free particles but only as bound states in the form of hadrons. However, they probably range between 300 MeV up to 5 GeV. To each of the leptons and quarks corresponds an antiparticle, with opposite charge but the same spin and mass.

In addition to electric charge, quarks possess another attribute called **color charge**, which plays the same role as the electric charge in the electromagnetic interaction. (The word color has no relation to the colors we see and is used only for the sake of convenience.) To explain why only certain stable combinations of quarks occur, it has been assumed that there are three kinds of color charges arbitrarily called *red* (r), *green* (g) and *blue* (b), to which we add the **anti**-color charges $\bar{r}$, $\bar{g}$, $\bar{b}$. A combination of three quarks with three different colors or of a quark with a color charge and antiquark with the anticolor charge gives a color neutral or colorless (*white*) system. A colorless combination of quarks and antiquarks is the equivalent of a neutral combination of negative and positive charges, as in an atom. Hadrons are assumed to be colorless combinations of quarks.

Quarks are held together by the strong or 'color' force. It is assumed that quarks with different colors attract each other, forming stable colorless combinations such as (q_r, q'_g, q''_b) or $(q_r, \bar{q}_{\bar{r}})$. For that reason the stable quark arrangements are colorless combinations of three quarks or of a quark and an antiquark. (Note that you cannot have any other colorless system of quarks.)

Two colorless hadrons feel practically no color force when they are far apart (similar to the way two electrically neutral atoms feel practically no electric force when they are distant). But if two hadrons come close to each other, the colored quarks in each may feel the color force from the quarks in the other. This force is called a 'residual interaction', which results in a variety of nuclear phenomena (similar to when two atoms or molecules, which are electrically neutral, get too close and interact and chemical reactions may result). This is the origin of the nuclear force between protons and neutrons; in other words, nuclei exist because of a residual effect of the color force.

Baryons are composed of three quarks (u, d or s), while mesons are combinations of a quark and antiquark (Table 41.3). For that reason baryons have spin $\frac{1}{2}$ (↑↑↓) or $\frac{3}{2}$ (↑↑↑) and are fermions, while mesons have spin 0 (↑↓) or 1 (↑↑) and are bosons. Resonances are then excited states of these quark systems. This model also explains many other properties of the hadrons. For example, neutrons, which are (ddu), have zero electric charge but have a magnetic moment that results from a combination of the magnetic moments of the three quarks. Only u, d and s quarks are involved in ordinary matter; the other quarks manifest themselves in very high energy processes. The strong binding energy due to the color force makes the masses of hadrons much smaller than the sum of the masses of the quarks composing them, a situation similar to that found in nuclei with respect to nucleons.

In the same way that the electromagnetic forces are carried by photons or γ-bosons, it is assumed that the strong interaction or color force between quarks is mediated by bosons called **gluons** (from glue), which are supposed to be massless, have spin 1, and carry color charge but no electric charge. There are eight gluons,

Table 41.3 Quark structure of some hadrons

Baryons with $S = \frac{1}{2}$		Mesons with $S = 0$	
p^+	(uud)	π^+	$(u\bar{d})$
n^0	(ddu)	π^0	$(u\bar{u} + d\bar{d})$
Σ^+	(uus)	π^-	$(\bar{u}d)$
Σ^0	(uds)	K^+	$(u\bar{s})$
Σ^-	(dds)	K^0	$(d\bar{s})$
Λ^0	(uds)	η^0	$(u\bar{u} - d\bar{d})$

corresponding to the color combinations $r\bar{r}$, $r\bar{g}$, $r\bar{b}$, $g\bar{r}$,... (only eight different gluons exist because of the invariance of the colorless combination $\bar{r}r + \bar{g}g + \bar{b}b$, which gives white). Because of their color, gluons can interact among themselves as well as with quarks, providing a large variety of processes. Also, because gluons carry color charge, the strong interaction mediated by them is of short range (less than 10^{-15} m) even if the gluon is massless. The theory dealing with color-carrying particles is called **quantum chromodynamics**.

Leptons, which possess no color charge, interact among themselves and with quarks only through the electromagnetic and weak interactions (exchange of γ, $W^{\pm}$, Z^0 bosons). Both interactions can be combined together in what is called the **electroweak theory**. Accordingly, electron–positron annihilation is not limited to the production of two photons, as indicated in Equation 41.1, by $e^+ + e^- \rightarrow \gamma + \gamma$, which is a purely electromagnetic process. If the energy of the electron–positron pair is sufficiently high, the production of several other particle–antiparticle pairs is possible through the interaction of the $W^{\pm}$, Z^0 weak bosons, so that we may have $e^+ + e^- \rightarrow \mu^+ + \mu^-, \pi^+ + \pi^-, K^0 + \overline{K^0}, K^+ + K^-$ etc. Due to the extremely short range of the weak interaction, weak processes can occur only if the particles come very close to each other, less than 10^{-15} m. Quarks, however, because of their color charge, also interact among themselves via the strong or color interaction (exchange of gluons). In processes due to the strong or color force, quarks can change color but not flavor; however, in processes due to the electroweak interaction, quarks can change flavor but not color, through the action of weak bosons.

Quark–antiquark pairs can be produced and annihilated through the intervention of photons and $W^{\pm}$, Z^0 bosons. Thus in weak processes, such as β-decay, a quark may be changed into another with the emission of a $W^{\pm}$ or a Z^0 boson, which in turn produces a pair of leptons or a quark–antiquark pair. The nuclear interaction between two nucleons is due to an exchange of a quark and an antiquark via the strong interaction, which effectively amounts to the exchange of a pion.

Once quarks or antiquarks combine to form a hadron, they cannot be separated, a property called **confinement**. The nature of gluons is such that it produces a force between quarks that increases very rapidly with distance. This explains why there are no more free quarks in the universe. They were all confined within hadrons shortly after the Big Bang.

Note 41.3 Experimental evidence of the internal structure of protons

High energy proton–proton collisions show the existence of nucleon excited states and therefore that a nucleon may have an internal structure. The experimental setup appears in Figure 41.20(a). Protons from an accelerator hit a liquid hydrogen target. Some of the protons are scattered through a fixed angle θ and pass through a magnetic field $\mathscr{B}$. The field then deflects some of these protons through an arc of a circle of fixed radius r, determined by the initial direction of motion and the position of the detectors D_1 and D_2. The momentum of these protons is then given by $p = e\mathscr{B}r$. By varying the intensity of the magnetic field, we can change the momentum of the protons passing through the fixed detectors. The momentum distribution

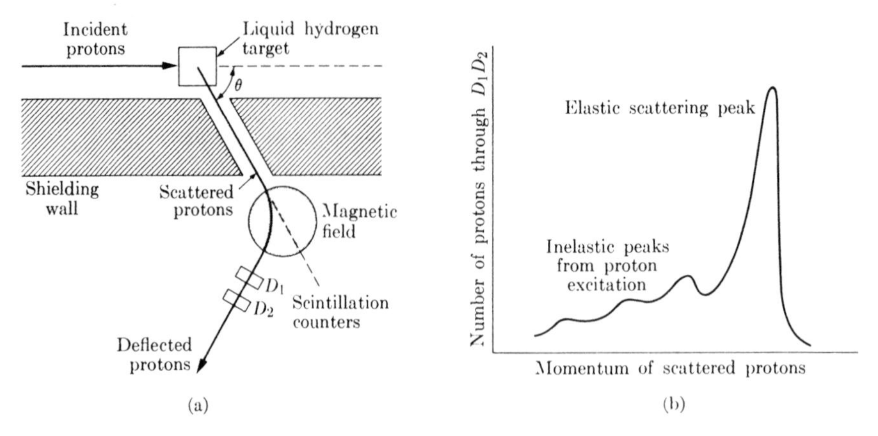

Figure 41.20 (a) Schematic diagram of proton–proton inelastic scattering experiment. (b) Experimental results showing inelastic scattering peaks.

of the scattered protons may therefore be determined. The experimental result is shown in Figure 41.20(b). In addition to a main peak in the momentum distribution of the protons (corresponding to elastic scattering by the target), there are several secondary peaks, corresponding to protons of lesser momentum or kinetic energy. These secondary peaks correspond to inelastic scattering, where the incoming protons lose some of their kinetic energy to a target proton, which is raised to an excited state of well-defined energy. This experiment is similar to the Franck and Hertz experiment (Example 23.2) of inelastic electron–atom collision, which showed the existence of excited stationary atomic states of well-defined energy.

Another experiment that provides a clue to the internal structure of protons is high-energy electron scattering. Rutherford's experiment (Section 23.3) on the scattering of α-particles provided direct evidence of internal structure of atoms, with most of the mass and all of the positive charge concentrated in the center, or nucleus, and the negative electrons moving around it. In the same way, the scattering of high energy electrons or neutrinos by protons serves to determine if protons have an internal structure (Figure 41.21).

To probe the interior of a proton, the wavelength of an electron must be smaller than the size of the proton, or about 10^{-15} m. Using relation 36.1, $\lambda = h/p$, the momentum of the electrons must be about 6.6×10^{-19} kg m s^{-1}. At this high momentum, the electron energy can be equated to $pc \sim 2 \times 10^{-10}$ J or about 3 GeV. Probing of protons using electrons became possible only after accelerators producing electrons of that energy or higher were available. This occurred in the 1970s. The results of the experiments clearly point to an internal structure where the charge is localized at some places inside the proton in a form compatible with the three-quark model. The theoretical analysis is more complex because quarks are strongly coupled and carry large momenta.

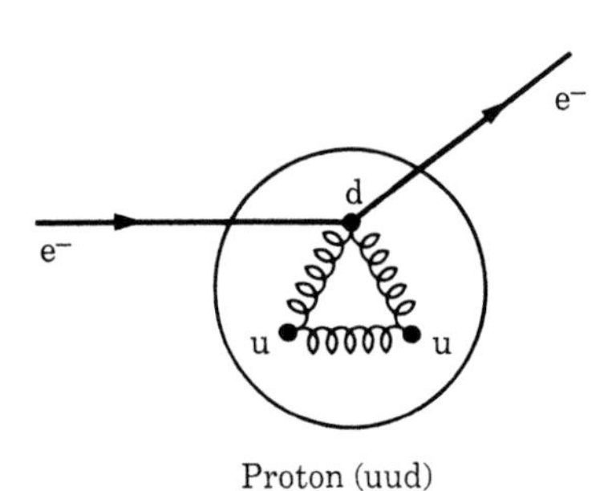

Figure 41.21 Electron scattering by a quark inside a proton.

Experiments using high-energy neutrinos and pions as projectiles have similar results. Therefore it seems that there is sound experimental evidence in support of the quark model of hadrons, as consisting of quarks surrounded by a cloud of gluons. The excited states of protons found in these experiments then correspond to quark excitations inside the proton, in the same way that electrons may be excited in an atom.

41.9 The evolution of the universe

Progress in the understanding of the fundamental components of matter and their interactions has helped develop a theory, or model, of the evolution of the universe that assumes that the universe had a beginning (although it does not say anything about what might have existed before). Since then the universe has been in a continuous state of evolution toward more complex structures, although its final fate is difficult to predict (see Note 41.4). According to an idea first proposed in 1948 by Ralph Alpher, George Gamow (1904–1968) and Robert Herman, the universe began in space and time about 15×10^9 years ago in what is loosely designated as the **Big Bang**. This term was coined by the astronomer Fred Hoyle to describe graphically the magnitude and speed of the early events. There are three basic reasons in support of the Big Bang theory. One is the isotropic expansion of the universe (Note 3.1), another is the presently cold isotropic background radiation (Example 31.5) and the third reason is the relative cosmic abundance of hydrogen, deuterium, helium and lithium (Note 40.4).

The Big Bang is supposed to have been a sort of explosion of a universe in an unstable state of extremely large energy density. Right after this initial event, the universe began an expansion process. As mentioned in Note 3.1, the current rate of expansion is estimated to be 22 km s^{-1} per million light years of separation (the Hubble constant). The expansion has been accompanied by a gradual decrease in the average energy per particle (which is a sort of Doppler effect) and a corresponding decrease in the 'temperature' of the universe.

As the average energy of the particles decreased, several phenomena occurred, called **phase transitions** or **symmetry breakings**, when certain energies were reached. These transitions resulted in important changes in the composition and structure of the universe. The transitions continued until about 10^6 years after the Big Bang, when the universe finally reached a structure and composition not very different from its current form.

Any theory about the very early stages of the universe is not subject to direct experimental verification with present techniques because of the high energies involved. However, some assumptions can be verified on a much smaller scale in the laboratory by using high-energy accelerators now in existence or planned. Also, the concepts of time and energy must be extrapolated back to extremely small and large values, respectively; whether that extrapolation is completely valid or not is not really relevant so long as it provides an appropriate frame of reference. In addition, the conditions of the universe before the Big Bang are impossible to ascertain, and probably will remain unknown to us.

It is hard to guess what the universe looked like or how it was composed at the Big Bang and immediately after, except that the average particle energy must have been extremely large, of the order of 10^{20} GeV or higher. Under those conditions all interactions were probably indistinguishable and all particles looked alike. It is assumed on theoretical grounds that, shortly after or about 10^{-42} s, the universe went through a rapid inflationary process during which its size probably increased by a factor as large as 10^{30} and conditions were established for the further 'normal' expansion that allowed it to reach its present state. One result of the inflation was that the energy of the particles dropped considerably.

Another result was that gravitation became completely separated from the other interactions. However, because of its weakness, gravitation did not play a role in the universe until a much later time, about 10^6 yr as explained in step (vi) below. After the initial inflation, the 'normal' expansion proceeded as follows:

(i) Up to about 10^{-32} s, or for particle energies above 10^{15} GeV (which is extremely large compared with the rest energies of all known particles), all particles appeared as massless and there was no difference between quarks and leptons, as well as among the strong and electroweak interactions. During this era we may visualize the universe as a mixture of fermions and bosons, subject to two fundamental interactions, gravitation and strong–electroweak, described by what is called the **Grand Unification Theory** (GUT). The GUT requires the intervention of super massive colored bosons, designated X, with rest energy of about 10^{15} GeV, to carry the interaction responsible for transitions between fermions (quarks and leptons). One interesting feature of the theory is the possibility of proton decay via X-bosons: $p \rightarrow \pi^0 + e^+$ or $p \rightarrow \pi^+ + \bar{\nu}$, with a half-life of about 10^{31} yr, a prediction that has not yet been verified experimentally. When the particle energy dropped to about 10^{15} GeV (temperature about 10^{28} K), a phase transition (or symmetry breaking) occurred: X-bosons could no longer be created in free states because the colliding particles did not have enough energy. Then quarks (colored fermions) and leptons (non-colored fermions) became different kinds of fermions with no possibility of transitions among them. As a consequence, the strong (color sensitive) force separated from the electroweak (color insensitive) force.

(ii) During the time up to 10^{-10} s, particle energies gradually dropped down to 10^2 GeV and the universe was a mixture of quarks and leptons in interaction through the exchange of gluons and electroweak bosons ($W^{\pm}$, Z^0, γ). Therefore, the only forces in operation were gravitation, strong (or color) and electroweak. Early during this era the predominance of matter over antimatter was established, although it is not yet well understood how this happened. When the particle energies reached 10^2 GeV (temperature 10^{15} K), a new phase transition or symmetry breaking occurred. Below this energy the more massive electroweak bosons ($W^{\pm}$, Z^0) cannot be created in free states in electroweak processes and they began to disappear from the universe. The result was a separation of the electric and weak interactions. The universe then reduced to a mixture of quarks, leptons, gluons and photons.

(iii) Between 10^{-10} s and 1 s, particle energies fell from 10^2 GeV down to 1 MeV. When the particle energy was down to about 100 MeV (temperature 10^{12} K), the strong (color) forces were able to cluster quarks and gluons into colorless hadrons. Due to the nature of the strong interaction, all quarks and gluons disappeared because they coalesced and were confined into hadrons. As a result the color force ceased to be a dominant factor in the overall evolution of the universe except as a residual force between nucleons. Since most hadrons have a very short life, they quickly decayed into nucleons (neutrons and protons) or leptons. The most heavy leptons also decayed into electrons and neutrinos and by $t \approx 1$ s the universe became essentially a mixture or plasma of nucleons, electrons, neutrinos and photons, interacting among themselves. The two predominant processes became $p + e^- \rightarrow n + \nu_e$ and $n + e^+ \rightarrow p + \nu_e$. Free neutrons are unstable and decay, with a half-life of about 900 s, according to $n \rightarrow p^+ + e^- + \bar{\nu}_e$. Also, because neutrons are more massive than protons, the second process above is more probable than the first, favoring proton production. Thus, by the end of this period, the number of neutrons decreased down to only about 12% compared to the number of protons.

(iv) At about 1 s, or energies of 1 MeV (temperature 10^{10} K), a new process became possible: **nucleosynthesis**, which can be considered as another phase transition. At that energy the residual strong or color interaction between nucleons was sufficient to bind them into stable structures (recall that the binding energy of the deuteron is 2.2 MeV) and the first light nuclei began to form (^2_1H, ^3_1H, ^3_2He, ^4_2He). At that stage the nuclear force ceased to play a critical role in the evolutionary process. Nucleosynthesis froze the number of neutrons in the universe and it became a mixture of protons (75%), light nuclei up to ^4_2He (25%), electrons, neutrinos and photons held by electromagnetic interactions.

(v) Only after 10^6 yr, when particle energies dropped to about 10 eV (temperature 10^5 K), was it possible for the electric forces exerted by nuclei to bind electrons into stable structures or atoms, mostly hydrogen and helium. This is also a sort of phase transition. At that stage most of the charged particles disappeared as free particles. The universe became mostly a mixture of hydrogen and helium atoms with traces of deuterium, neutrinos and photons. The interaction of photons with matter was then greatly reduced. It is said that the photons, whose average energy was then below the ionization energy of the atoms, decoupled from matter, and that the Universe became transparent to radiation. This might have stabilized the residual background electromagnetic radiation that currently corresponds to a temperature of 2.7 K (or 10^{-3} eV) (recall Example 31.5).

(vi) From 10^6 yr up to the present (a period of about 1.5×10^{10} yr), the large structures (clusters, galaxies, stars etc.) appeared under the action of gravitation (recall Note 11.8). Gravity became, by default, the dominant long range interaction involving all matter, in spite of being the weakest of all forces. Nuclear processes continue to occur in stars, including our Sun, with the fusion of hydrogen into helium and the synthesis of nuclei with mass number larger than 4 (recall Section 40.9), but they occur at a much smaller scale than in earlier phases of the universe.

A relatively recent development in the evolution of the universe has been the emergence of life; that is, self-replicating systems whose functioning depends critically on energy exchanges, of the order of 1 eV, between rather complex molecules (proteins, enzymes, nucleic acids etc.). This process began on Earth about 3.5×10^9 years ago. How life originated and evolved on Earth is still a subject of intensive research. Also, whether life exists in other parts of the Universe, in the same form as on Earth or in a different form, is a subject open to speculation. In any case, many molecules that are precursors to 'life molecules' have been observed in the intergalactic gas. The relation between these molecules and life on Earth is not yet well understood.

Note 41.4 The cosmological fate

In Section 41.9 we explained how, based on energy considerations, we can arrive at a reasonable answer to the problem of the origin and evolution of the universe, from its earliest moments up to the present. Although the physical laws that define nature and our understanding of energy exchanges may provide clues about the future evolution of the universe, such predictions are highly speculative because of the lack of total information, the extremely long period of time involved and, perhaps, the attendant impossibility of experimental verification of our conclusions. The issue centers around three questions. 1. Is the universe closed, flat or open? 2. If the universe is not closed, how will it evolve? 3. Can intelligent life affect the course of cosmological evolution?

The answer to the first question depends critically on the mass-energy density throughout the universe, since the bulk of the energy is presently in the form of mass and the force dominating large-scale dynamics is gravitation. As indicated in Note 11.4, if mass is distributed in the universe with a density greater than 2×10^{-26} kg m^{-3} (the *critical density*), or about 12 nucleons per cubic meter, the universe is closed. That means that the kinetic energy of expansion will eventually be transformed into gravitational energy. After that stage, the universe will begin to contract, with a reversal of events that occurred during expansion. The gravitational potential energy will go back into kinetic energy; as the mass-energy density increases, all other forces will begin to enter into play. After some time, perhaps at $t \sim 10^{32}$ yr, the universe ends in a **Big Crunch** followed perhaps by a new Big Bang, not necessarily identical to the preceding one. We then have an **oscillating universe** (Figure 11.30).

If the mass-energy density in the universe is equal to or less than the critical density, the universe is either flat or open. It will then continue to expand forever, with new events occurring as conditions change. Current evidence, based on estimates of luminous matter (i.e. matter that interacts with electromagnetic fields) in galaxies as well as on the process of nucleosynthesis, indicates that the present average mass-energy density in the universe is of the order of 10^{-27} kg m^{-3}, supporting the idea that the universe is indeed open. However, there is still great uncertainty about how much mass-energy exists in the universe and several considerations, which we cannot elaborate on here, suggest the existence of a considerable amount of 'dark' matter (i.e. matter not coupled to electromagnetic fields). Unresolved questions are how much dark matter exists, how is it distributed and what is it made of (recall Notes 11.3 and 11.4)? None of these questions has received a definite answer yet. Therefore, we may conclude, depending on the amount of dark matter, that the estimated mass-energy density corresponds to an open or, at most, a flat ever-expanding universe.

The future of an open universe is basically dominated by the gravitational energy, with an appreciable contribution from nuclear energy. It has been estimated that, after 10^{14} years, most of the stars will have exhausted the hydrogen process and, depending on their size and other factors, will have become white dwarfs, neutron stars or black holes. By that time all life will have disappeared from the universe due to a lack of photons of adequate energy to sustain life. This means that life can be possible in the universe during a span of 10^{10} years to 10^{14} years at the most.

A process that might occur shortly afterwards, at about 10^{15} years, is the disturbance of planetary systems by collisions and gravitational action as they move through their respective galaxies. The next process in time should be the loss of stars by galactic or intergalactic encounters, with time of the order of 10^{19} years. Collapse of planetary orbits into their suns or of stars toward the center of their galaxies, accompanied by gravitational radiation, is a much slower process, with a time-scale of 10^{20} years to 10^{24} years. After about 10^{31} years proton decay will have contributed appreciably to the disappearance of nucleons. During all this time a substantial amount of matter will have concentrated in black holes, releasing considerable amounts of gravitational radiation. However, at about 10^{64} years to 10^{100} years, depending on their size, black holes will collapse and decay, emitting electromagnetic radiation. At the end there will be a large burst of radiation, with a power that might reach 10^{24} W. Other processes may occur at still later times, but it might seem a bit futile for our purpose to look that far into the future.

One fact seems to be certain. As the universe expands, the mass-energy density will continue to decrease and the universe will cool steadily. The universe will gradually be reduced to a cold mixture of electrons, positrons, neutrinos and photons. Perhaps it will eventually reach a stable cold configuration.

This brings us to the third question: can intelligent structures, capable of handling information (i.e. codified energy) alter the course of the universe? Intelligent systems are a relatively recent event in the universe and require special environmental conditions that probably are found only in a relatively small number of places in the universe. On the other hand, intelligence has evolved on Earth in a dramatic manner. Only about 10^6 years ago the highest level of intelligence known to us, the hominids (Genus Homo) emerged, and *Homo Sapiens* has existed for only about 10^5 years. And even the intelligence of *Homo Sapiens* has evolved dramatically in the last 10^4 years. There is no reason why intelligence will not continue to grow beyond what we see today. The major limiting factor will probably be how future intelligent societies can use energy and other resources under the conditions available to them.

It is conceivable that in other parts of the universe intelligence has already developed well beyond the levels found on Earth. Unfortunately, we only know how to transmit information by using electromagnetic energy and this energy is transmitted with a finite velocity, that of light. Given interstellar distances, it appears very difficult to establish useful communication with other intelligent societies because of the time that would be involved. Intelligence on Earth will remain isolated and the same may apply to intelligence in other parts of the universe. The conclusion then might be that intelligence, being a sparsely localized phenomenon in the universe, will not be able to alter the course of evolution of the universe. But intelligence can change the local conditions where it exists, as humans are doing with the planet Earth, and not necessarily for the better, especially because of misuse of its energy and other resources. It is our collective responsibility to see that we do not do any irreparable harm to our planet. Physicists, in particular, bear a greater responsibility because of their special understanding of the physical universe.

QUESTIONS

41.1 What is understood by the term 'fundamental particle'?

41.2 What is the role of the carrier bosons in the interaction between fermions?

41.3 Can you see any relation between the fact that fermions have fractional spin and carrier bosons have integral spin and the law of conservation of angular momentum?

41.4 State the similarities and differences between particles and antiparticles.

41.5 What is meant by particle instability? State the general laws that restrict the possible decays of a particle.

41.6 Is the instability of a particle an indication that the particle is a composite system?

41.7 State the conservation laws that must be followed in *all* processes involving fundamental particles. Are there conservation laws of more restricted validity?

41.8 How may an investigator determine whether a particle decays into two, three or more particles, if some of the decay particles are neutral and cannot be detected directly?

41.9 What role does symmetry play in the fundamental interactions?

41.10 Why are resonances an indication of an internal structure of hadrons?

41.11 Which are the basic elements of the standard model?

41.12 List the fundamental particles that are subject to (a) the color force, (b) the weak force, (c) the electromagnetic force, (d) gravitation.

41.13 What are the differences between the color or strong force and the nuclear force?

41.14 Compare the role of gluons and photons in their respective interactions.

41.15 Why are only three quarks or a quark and an antiquark the only possible arrangements to produce a colorless system?

41.16 What is meant by phase transitions in the evolutionary process of the universe? Why do they occur at specific values of the energy of the particles?

41.17 Is the universe still in a state of evolution?

41.18 Describe why (a) the expansion of the universe (Note 3.1), (b) the isotropy of the 3K background radiation (Example 31.5) and (c) the cosmic abundance of light elements (hydrogen, deuterium and lithium) are important arguments in favor of the Big Bang theory.

PROBLEMS

41.1 A neutral kaon with a kinetic energy of 100 MeV decays in flight into two oppositely charged pions. The kinetic energy of one of the pions is 200 MeV. Compute the momentum of each pion and the angles their paths make in the *L*-frame.

41.2 A neutral K^0-meson is observed to decay into a pair of oppositely charged pions. Initially the pion tracks are perpendicular in the L-frame, but are bent by a magnetic field of 8.5×10^{-1} T, so that they have radii of 0.8 m and 1.6 m, respectively. Calculate the rest mass of the K^0-meson and its kinetic energy.

41.3 A beam of negative pions (π^-) entering a bubble chamber triggers the reaction $\pi^- + p^+ \rightarrow \Lambda^0 + K^0$. The successive decay processes of Λ^0 and K^0 also take place within the chamber. A magnetic field is applied in the region occupied by the chamber. Make a diagram showing the whole process. Neutral particles should be represented by dashed lines and charged particles by continuous lines with the proper curvature.

41.4 A 1 MeV positron collides with an electron that is at rest in the L-frame and the two annihilate. (a) What is the energy, in the C-frame, of the two photons emitted? (b) Given that one photon is emitted in the direction of motion of the positron and the other in the opposite direction, find their energies in the L-frame.

41.5 The annihilation probability per unit time of positrons is $\lambda = 7.49 \times 10^{-15} n\,\mathrm{s}^{-1}$, where n is the number of electrons per unit volume. Show that the half-life of positrons moving through argon ($Z = 18$) is $T = 2.67 \times 10^{-7}/p$, where p is the pressure, in atmospheres, of argon.

41.6 A negative pion may be captured into a stable orbit around the nucleus. (a) Calculate the energy and the radius of the pion orbit. (b) What energy is released when a free pion at rest is captured into the ground state around a proton? (c) Estimate the nucleus for which the radius of the ground state orbit of the pion is equal to the radius of the nucleus.

41.7 Determine the Q value and the projectile threshold energy in the L-frame for the following reactions:

(a) $\pi^- + p^+ \rightarrow \Lambda^0 + K^0$
(b) $p^+ + p^+ \rightarrow p^+ + p^+ + \pi^0$
(c) $p^+ + p^+ \rightarrow p^+ + n + \pi^+$
(d) $p^+ + p^+ \rightarrow p^+ + \Lambda^0 + K^+$

41.8 Calculate the minimum kinetic energy of the incoming proton necessary to trigger the process shown in Figure 41.9.

41.9 Determine the energy of the photon in the process $\pi^- + p^+ \rightarrow n + \gamma$ when both colliding particles have negligible kinetic energy.

41.10 Calculate the threshold kinetic energy of the incoming particle for the processes illustrated in Figures 41.6 and 41.7.

41.11 Find the magnetic field for the chosen value of the $\bar{p}^-$ momentum in the antiproton experiment (Note 41.1).

41.12 A π^0-meson decays into two photons of energies E_1 and E_2. Show that $E_1E_2 = \frac{1}{2}(m_\pi c^2)^2/(1 - \cos\theta)$, where θ is the angle between the photon directions as measured in the L-frame.

41.13 Calculate the energy of the photon emitted when a Σ^0 particle, at rest in the L-frame, decays to a Λ^0 particle.

41.14 Show that the processes (a) $\pi^- + p^+ \rightarrow n + \gamma$, (b) $\pi^- + d \rightarrow 2n + \gamma$, and (c) $\pi^+ + d \rightarrow 2p^+$ all imply that the spin of the pion is either 0 or 1.

41.15 ^{152}Eu decays by electron capture according to $^{152}\mathrm{Eu} + e^- \rightarrow {}^{152}\mathrm{Sm}^* + \nu$. In turn, the excited samarium decays by γ-emission according to $^{152}\mathrm{Sm}^* \rightarrow {}^{152}\mathrm{Sm} + \gamma$. Both ^{152}Eu and ^{152}Sm nuclei have zero spin. By observing the γ-rays emitted in a direction opposite to that of emission of the neutrinos, the photons are found to be predominantly right-handed polarized (that is, negative helicity). Verify that this implies that the neutrino also has negative helicity. (Hint: analyze the conservation of angular momentum for the overall process.)

41.16 Using relativity theory and statistical mechanics, it can be shown that the time t since the Big Bang (measured in s), the average particle energy E (measured in eV), and the cosmic temperature T (measured in K) should be related by $tE^2 \approx 2.5 \times 10^{11}$, $tT^2 \approx 10^{20}$ and $T \approx 10^4 E$. Make a plot of these relations on logarithmic paper. At what time was E equal to (a) the threshold energy for producing the $W^\pm$ and Z^0 bosons, (b) the binding energy of the deuteron and (c) the binding energy of the electron in the hydrogen atom? Calculate the temperature in each case.

Appendices

A: Vectors

Because so many physical quantities are vectors, it is most important to learn the algebra of vectors, which is briefly explained in this Appendix. The relations obtained here are used frequently throughout the text. A more extensive treatment is found in any standard calculus or analytical geometry text.

A.1 Concept of direction

Given a straight line, we can move along it in two opposite senses; these are distinguished by assigning to each a sign, plus or minus. Once the positive sense has been determined, we say that the line is oriented and call it an **axis**. The positive sense is usually indicated by an arrow. The coordinate axes X and Y, used to draw graphs on paper, are shown in Figure A.1. An oriented line or axis defines a **direction**. Parallel lines oriented in the same sense define the same direction (Figure A.2(a)), but if they have opposite orientations they define opposite directions (Figure A.2(b)).

Directions in a plane are determined by an angle, which is the angle between a **reference** direction (or axis) and the direction we want to indicate, measured counterclockwise (Figure A.3). Opposite directions are determined by the angles θ and $\pi + \theta$ (or $180^\circ + \theta$).

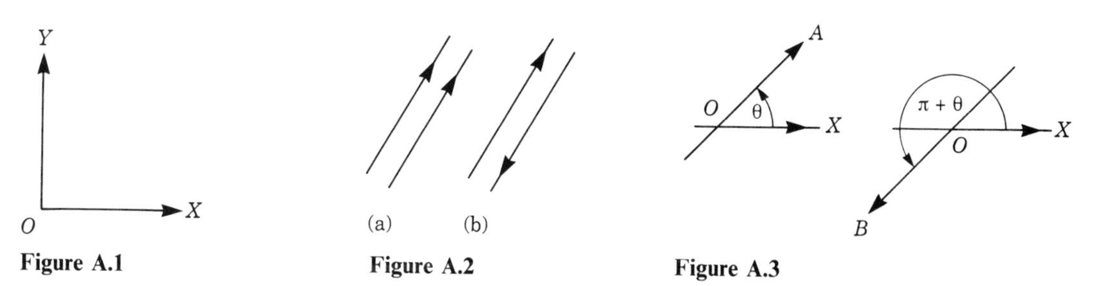

Figure A.1 **Figure A.2** **Figure A.3**

A.2 Scalars and vectors

Many physical quantities are completely determined by their magnitude, expressed in some convenient unit. These quantities are called **scalars**. Volume, temperature, time, mass, charge and energy are scalar quantities. Other physical quantities require, for their complete determination, a direction in addition to their magnitude. Such quantities we call **vectors**. The most obvious case is **displacement**. The displacement of a body is determined by the **distance** it has moved and the **direction** in which it moved; that is, by the vector $\boldsymbol{AB}$ (Figure A.4). Velocity, acceleration and force are vector quantities.

Vectors are represented graphically by line segments having the same direction as the vector (indicated by an arrow) and a length proportional to the magnitude. A symbol in bold italic type, such as $\boldsymbol{V}$, indicates a vector (i.e., magnitude plus direction), while V refers to the magnitude only. A **unit vector** is a vector whose magnitude is one. A vector $\boldsymbol{V}$ parallel to the unit vector $\boldsymbol{u}$ can be expressed in the form

$$\boldsymbol{V} = \boldsymbol{u}V \tag{A.1}$$

The negative of a vector is another vector that has the same magnitude but opposite direction.

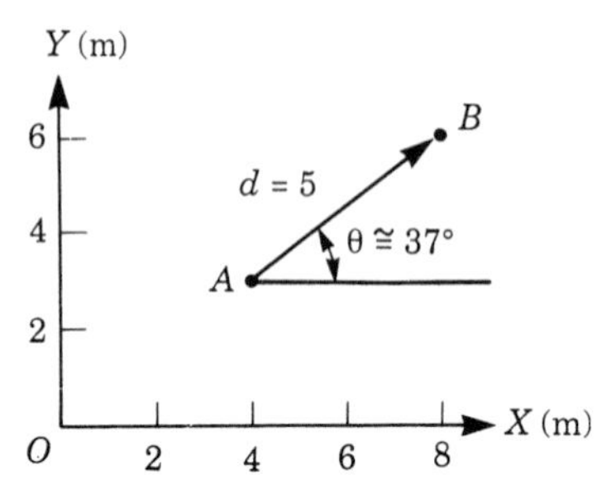

Figure A.4

A.3 Addition and subtraction of vectors

To understand the rule for addition of vectors consider first the case of displacements. If a particle is displaced first from A to B (Figure A.5), as indicated by vector $\boldsymbol{d}_1 = \boldsymbol{AB}$ and then from B to C, as indicated by the vector $\boldsymbol{d}_2 = \boldsymbol{BC}$, the result is equivalent to a single displacement from A to C, or $\boldsymbol{d} = \boldsymbol{AC}$. We say that $\boldsymbol{d} = \boldsymbol{AC}$ is the vector sum of vectors $\boldsymbol{d}_1 = \boldsymbol{AB}$ and $\boldsymbol{d}_2 = \boldsymbol{BC}$, which we write symbolically as

$$\boldsymbol{AC} = \boldsymbol{AB} + \boldsymbol{BC} \qquad \text{or} \qquad \boldsymbol{d} = \boldsymbol{d}_1 + \boldsymbol{d}_2 \tag{A.2}$$

This expression must not be confused with $d = d_1 + d_2$, which refers only to the magnitudes and does not hold in this case. The procedure can be generalized to any kind of vector and we write $\boldsymbol{V} = \boldsymbol{V}_1 + \boldsymbol{V}_2$ (Figure A.6). The vector sum is commutative, the result being the same if the order in which the vectors are added is reversed; that is, $\boldsymbol{V}_1 + \boldsymbol{V}_2$ yields the same result as $\boldsymbol{V}_2 + \boldsymbol{V}_1$. To compute the magnitude of $\boldsymbol{V}$ we see from Figure A.7 that if θ is the angle formed by vectors $\boldsymbol{V}_1$ and $\boldsymbol{V}_2$, $(AC)^2 = (AD)^2 + (DC)^2$. But $AD = AB + BD = V_1 + V_2 \cos\theta$ and

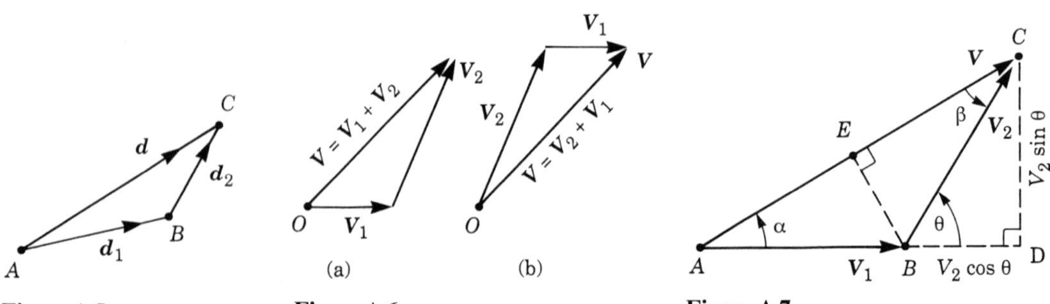

Figure A.5

Figure A.6

Figure A.7

$DC = V_2 \sin\theta$. Since $AC = V$ and $\sin^2\theta + \cos^2\theta = 1$, we write

$$V^2 = (V_1 + V_2\cos\theta)^2 + (V_2\sin\theta)^2 = V_1^2 + V_2^2 + 2V_1V_2\cos\theta$$

or

$$V = (V_1^2 + V_2^2 + 2V_1V_2\cos\theta)^{1/2} \tag{A.3}$$

To determine the direction of $\boldsymbol{V}$, we need to know the angle α or β. In triangle ACD, $CD = AC\sin\alpha$, and in triangle BDC, $CD = BC\sin\theta$. Therefore $V\sin\alpha = V_2\sin\theta$ or

$$\frac{V}{\sin\theta} = \frac{V_2}{\sin\alpha}$$

Similarly, $BE = V_1\sin\alpha = V_2\sin\beta$ or

$$\frac{V_2}{\sin\alpha} = \frac{V_1}{\sin\beta}$$

Combining both results, we have the symmetrical relation

$$\frac{V}{\sin\theta} = \frac{V_1}{\sin\beta} = \frac{V_2}{\sin\alpha} \tag{A.4}$$

Equations A.3 and A.4, are designated as the **Law of cosines** and the **Law of sines**, respectively. In the special case when $\boldsymbol{V}_1$ and $\boldsymbol{V}_2$ are perpendicular (Figure A.8), $\theta = \frac{1}{2}\pi$, $\cos\theta = 0$, $\sin\beta = \cos\alpha$ and the following relations hold:

$$V = (V_1^2 + V_2^2)^{1/2} \qquad \tan\alpha = V_2/V_1 \tag{A.5}$$

The **difference** between two vectors is obtained by adding to the first vector the negative (or opposite) of the second (Figure A.9); that is, $\boldsymbol{D} = \boldsymbol{V}_1 - \boldsymbol{V}_2 = \boldsymbol{V}_1 + (-\boldsymbol{V}_2)$. The magnitude of the difference is

$$D = [V_1^2 + V_2^2 + 2V_1V_2\cos(\pi - \theta)]^{1/2}$$

or, since $\cos(\pi - \theta) = -\cos\theta$,

$$D = (V_1^2 + V_2^2 - 2V_1V_2\cos\theta)^{1/2} \tag{A.6}$$

Note (Figure A.10) that $\boldsymbol{V}_2 - \boldsymbol{V}_1 = -\boldsymbol{D}$; that is, if the vectors are subtracted in the reverse order, the opposite vector results; the magnitude of the difference is not changed but the direction is reversed. We say that the vector difference is anticommutative.

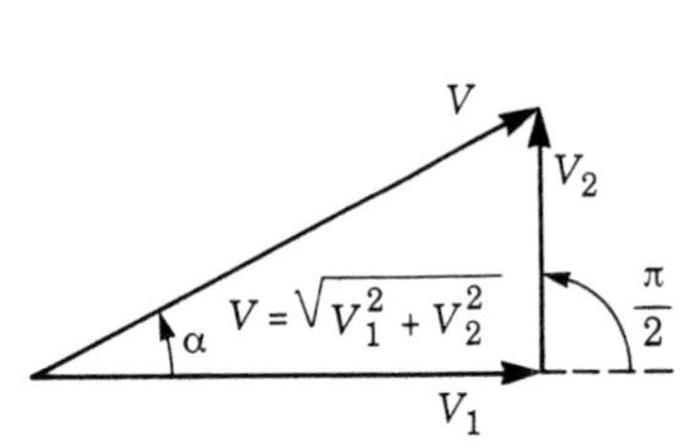

Figure A.8

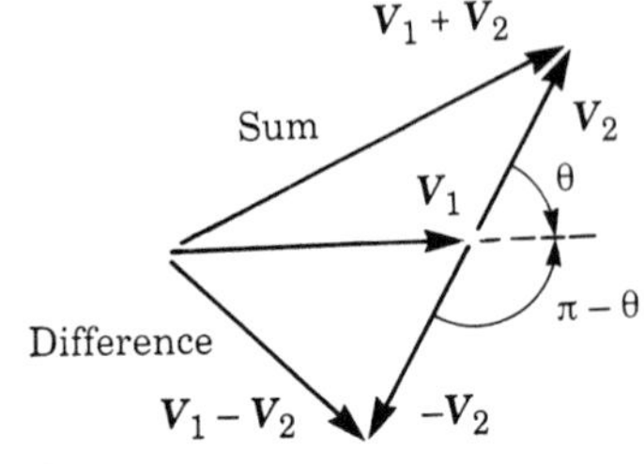

Figure A.9

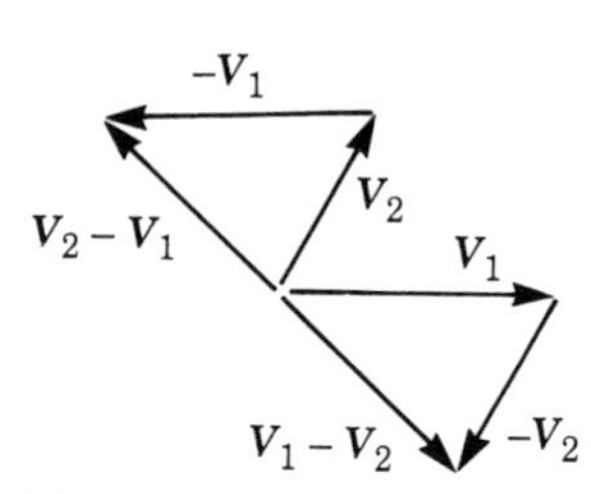

Figure A.10

A.4 Components of a vector

Any vector $\boldsymbol{V}$ can always be considered as the sum of two vectors. The vectors which, when added, give $\boldsymbol{V}$, are called the **components** of $\boldsymbol{V}$. The most commonly used components are the **rectangular components**, where the vector is expressed as the sum of two mutually perpendicular vectors (Figure A.11); that is, $\boldsymbol{V}=\boldsymbol{V}_x+\boldsymbol{V}_y$, with

$$V_x = V\cos\alpha \qquad \text{and} \qquad V_y = V\sin\alpha \tag{A.7}$$

Defining unit vectors $\boldsymbol{i}$ and $\boldsymbol{j}$ in the directions of the X- and Y-axes respectively, we note that $\boldsymbol{V}_x=\boldsymbol{i}V_x$ and $\boldsymbol{V}_y=\boldsymbol{j}V_y$. Therefore we have

$$\boldsymbol{V}=\boldsymbol{i}V_x+\boldsymbol{j}V_y \qquad V^2=V_x^2+V_y^2 \qquad \text{and} \qquad \tan\alpha = V_y/V_x \tag{A.8}$$

The components of a **unit vector** ($V=1$), which makes an angle α with the $+X$-axis, are $\cos\alpha$ and $\sin\alpha$ and the unit vector can be written as

$$\boldsymbol{u}=\boldsymbol{i}\cos\alpha+\boldsymbol{j}\sin\alpha \tag{A.9}$$

We may use a set of three orthogonal axes, X, Y, Z, to describe objects in space. Accordingly, there are three rectangular components in space, V_x, V_y, V_z (Figure A.12). Defining three unit vectors $\boldsymbol{i}$, $\boldsymbol{j}$ and $\boldsymbol{k}$, parallel to the X-, Y- and Z-axes, respectively, we have

$$\boldsymbol{V}=\boldsymbol{i}V_x+\boldsymbol{j}V_y+\boldsymbol{k}V_z \tag{A.10}$$

From Figure A.12 we see that $V^2=OD^2=OE^2+ED^2=OE^2+V_z^2$. But $OE^2=OA^2+AE^2=V_x^2+V_y^2$. Substituting OE^2 into the above expression, the magnitude of the vector is

$$V^2=V_x^2+V_y^2+V_z^2 \tag{A.11}$$

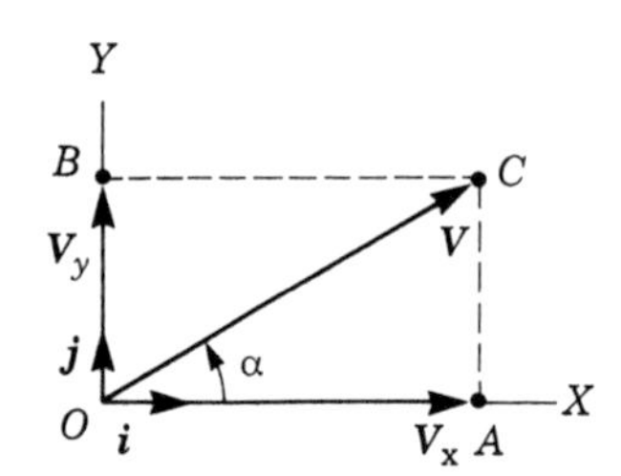

Figure A.11

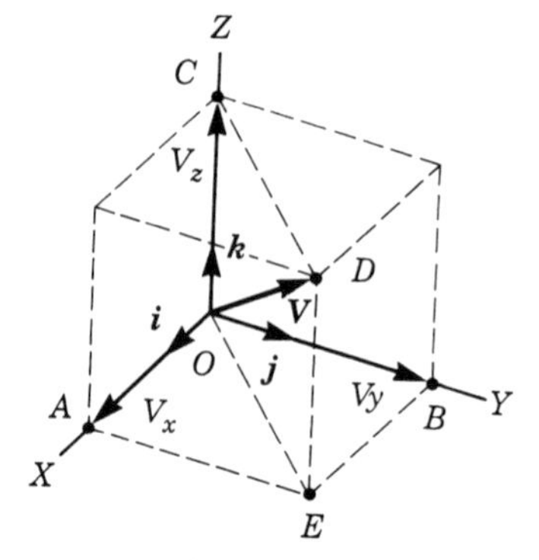

Figure A.12

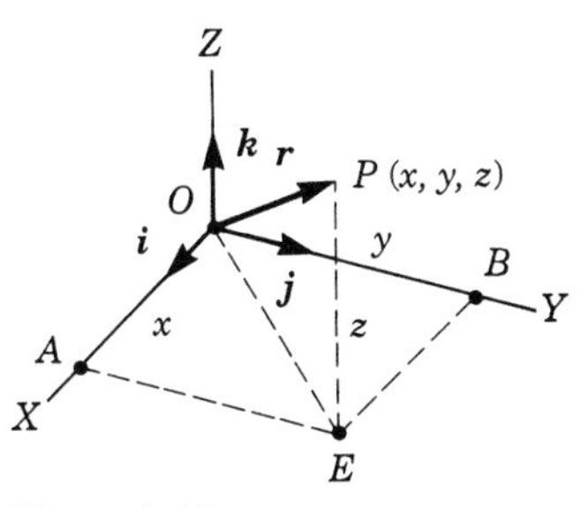

Figure A.13

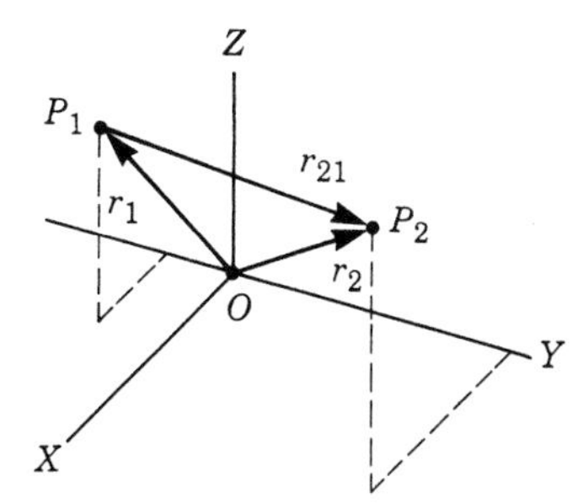

Figure A.14

The **position vector** $\boldsymbol{r} = \boldsymbol{OP}$ of a point P having coordinates x, y, z (Figure A.13) is

$$\boldsymbol{r} = \boldsymbol{OP} = \boldsymbol{i}x + \boldsymbol{j}y + \boldsymbol{k}z \tag{A.12}$$

The **relative position vector** of point P_2 with respect to point P_1 is $\boldsymbol{r}_{21} = \boldsymbol{P_1P_2}$ (Figure A.14). From the figure we have $\boldsymbol{OP}_2 = \boldsymbol{OP}_1 + \boldsymbol{P_1P_2}$, so that $\boldsymbol{P_1P_2} = \boldsymbol{OP}_2 - \boldsymbol{OP}_1$

$$\boldsymbol{r}_{21} = \boldsymbol{r}_2 - \boldsymbol{r}_1 = \boldsymbol{i}(x_2 - x_1) + \boldsymbol{j}(y_2 - y_1) + \boldsymbol{k}(z_2 - z_1) \tag{A.13}$$

Note that $\boldsymbol{P_2P_1} = -\boldsymbol{P_1P_2}$. Applying Equation A.11 to Equation A.13, the distance between two points is:

$$r_{21} = [(x_2 - x_1)^2 + (y_2 - y_1)^2 + (z_2 - z_1)^2]^{1/2} \tag{A.14}$$

A.5 Addition of several vectors

To add several vectors $\boldsymbol{V}_1, \boldsymbol{V}_2, \boldsymbol{V}_3, \ldots$, we extend the procedure indicated in Figure A.6 for the case of two vectors. The method for three vectors is shown in Figure A.15. That is, we draw one vector after another, the vector sum being indicated by the line going from the origin of the first to the end of the last, and we write $\boldsymbol{V} = \boldsymbol{V}_1 + \boldsymbol{V}_2 + \boldsymbol{V}_3$. There is no simple formula to express $\boldsymbol{V}$ in terms of $\boldsymbol{V}_1$, $\boldsymbol{V}_2$ and $\boldsymbol{V}_3$ and it is better to use the method of components. In this case, the rectangular components of $\boldsymbol{V}$ are

$$V_x = \Sigma_i V_{ix}, \quad V_y = \Sigma_i V_{iy}, \quad V_z = \Sigma_i V_{iz} \tag{A.15}$$

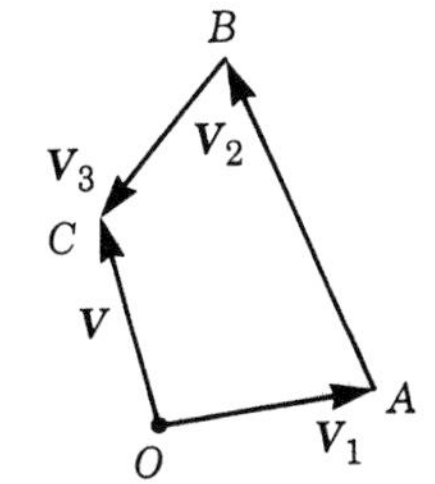

Figure A.15

A.6 The scalar product

The **scalar product** of two vectors $\boldsymbol{A}$ and $\boldsymbol{B}$, represented by the symbol $\boldsymbol{A} \cdot \boldsymbol{B}$ (read: '$\boldsymbol{A}$ dot $\boldsymbol{B}$'), is defined as the **scalar quantity** obtained by taking the product of the magnitudes of $\boldsymbol{A}$ and $\boldsymbol{B}$ and multiplying by the cosine of the angle (less than 180°) between the two vectors,

$$\boldsymbol{A} \cdot \boldsymbol{B} = AB \cos \theta \tag{A.16}$$

The scalar product is commutative; that is, $\boldsymbol{A} \cdot \boldsymbol{B} = \boldsymbol{B} \cdot \boldsymbol{A}$, since $\cos \Theta$ is the same in both cases.

Clearly, $\boldsymbol{A} \cdot \boldsymbol{A} = A^2$, since the angle in this case is zero. If the two vectors are perpendicular ($\theta = \pi/2$), the scalar product is zero. Therefore the condition of perpendicularity is expressed by $\boldsymbol{A} \cdot \boldsymbol{B} = 0$. The scalar products among the unit vectors $\boldsymbol{i}$, $\boldsymbol{j}$ and $\boldsymbol{k}$ are

$$\boldsymbol{i} \cdot \boldsymbol{i} = \boldsymbol{j} \cdot \boldsymbol{j} = \boldsymbol{k} \cdot \boldsymbol{k} = 1, \boldsymbol{i} \cdot \boldsymbol{j} = \boldsymbol{j} \cdot \boldsymbol{i} = \boldsymbol{i} \cdot \boldsymbol{k} = \boldsymbol{k} \cdot \boldsymbol{i} = \boldsymbol{j} \cdot \boldsymbol{k} = \boldsymbol{k} \cdot \boldsymbol{j} = 0 \quad \textbf{(A.17)}$$

It can be shown that the scalar product is distributive with respect to the sum; that is,

$$\boldsymbol{C} \cdot (\boldsymbol{A} + \boldsymbol{B}) = \boldsymbol{C} \cdot \boldsymbol{A} + \boldsymbol{C} \cdot \boldsymbol{B} \quad \textbf{(A.18)}$$

To express the scalar product of two vectors in terms of their components, we write the vectors in terms of their rectangular components, and apply the distributive law A.19. Then

$$\begin{aligned} \boldsymbol{A} \cdot \boldsymbol{B} = (\boldsymbol{i}A_x + \boldsymbol{j}A_y + \boldsymbol{k}A_z) \cdot (\boldsymbol{i}B_x + \boldsymbol{j}B_y + \boldsymbol{k}B_z) &= (\boldsymbol{i} \cdot \boldsymbol{i})A_xB_x + (\boldsymbol{i} \cdot \boldsymbol{j})A_xB_y + (\boldsymbol{i} \cdot \boldsymbol{k})A_xB_z \\ &+ (\boldsymbol{j} \cdot \boldsymbol{i})A_yB_x + (\boldsymbol{j} \cdot \boldsymbol{j})A_yB_y + (\boldsymbol{j} \cdot \boldsymbol{k})A_yB_z \\ &+ (\boldsymbol{k} \cdot \boldsymbol{i})A_zB_x + (\boldsymbol{k} \cdot \boldsymbol{j})A_zB_y + (\boldsymbol{k} \cdot \boldsymbol{k})A_zB_z \end{aligned} \quad \textbf{(A.19)}$$

Applying relations A.17, we finally obtain

$$\boldsymbol{A} \cdot \boldsymbol{B} = A_xB_x + A_yB_y + A_zB_z \quad \textbf{(A.20)}$$

Note that $A^2 = \boldsymbol{A} \cdot \boldsymbol{A} = A_x^2 + A_y^2 + A_z^2$, in agreement with Equation A.5.

We can apply the properties of the scalar product to derive Equation A.3 for the magnitude of the sum of two vectors. From $\boldsymbol{V} = \boldsymbol{V}_1 + \boldsymbol{V}_2$, we have

$$V^2 = \boldsymbol{V} \cdot \boldsymbol{V} = (\boldsymbol{V}_1 + \boldsymbol{V}_2) \cdot (\boldsymbol{V}_1 + \boldsymbol{V}_2) = V_1^2 + V_2^2 + 2\boldsymbol{V}_1 \cdot \boldsymbol{V}_2 = V_1^2 + V_2^2 + 2V_1V_2 \cos\theta$$

A.7 The vector product

The vector product of two vectors $\boldsymbol{A}$ and $\boldsymbol{B}$, represented by the symbol $\boldsymbol{A} \times \boldsymbol{B}$ (read '**A** cross **B**'), is defined as the **vector** whose **magnitude** is given by

$$|\boldsymbol{A} \times \boldsymbol{B}| = AB \sin\theta \quad \textbf{(A.21)}$$

where θ is the angle (less than 180°) between $\boldsymbol{A}$ and $\boldsymbol{B}$. The direction of $\boldsymbol{A} \times \boldsymbol{B}$ is perpendicular to the plane determined by $\boldsymbol{A}$ and $\boldsymbol{B}$ and in the direction of the thumb when one's right hand is placed as shown in Figure A.16, with the fingers pointing in the direction of rotation from $\boldsymbol{A}$ to $\boldsymbol{B}$ through the smaller angle.

From the definition of the vector product, we conclude that $\boldsymbol{A} \times \boldsymbol{B} = -\boldsymbol{B} \times \boldsymbol{A}$, so that the vector product is anticommutative. If two vectors are parallel, $\theta = 0°$, $\sin\theta = 0$ and the vector product is zero. Therefore the condition of parallelism is expressed by $\boldsymbol{A} \times \boldsymbol{B} = 0$. It is then clear that $\boldsymbol{A} \times \boldsymbol{A} = 0$.

The vector products among the unit vectors $\boldsymbol{i}$, $\boldsymbol{j}$ and $\boldsymbol{k}$ are

$$\boldsymbol{i} \times \boldsymbol{j} = -\boldsymbol{j} \times \boldsymbol{i} = \boldsymbol{k}, \boldsymbol{j} \times \boldsymbol{k} = -\boldsymbol{k} \times \boldsymbol{j} = \boldsymbol{i}, \boldsymbol{k} \times \boldsymbol{i} = -\boldsymbol{i} \times \boldsymbol{k} = \boldsymbol{j}, \boldsymbol{i} \times \boldsymbol{i} = \boldsymbol{j} \times \boldsymbol{j} = \boldsymbol{k} \times \boldsymbol{k} = 0 \quad \textbf{(A.22)}$$

It can be shown that the vector product is distributive relative to the sum; that is,

$$C \times (A + B) = C \times A + C \times B \quad (A.23)$$

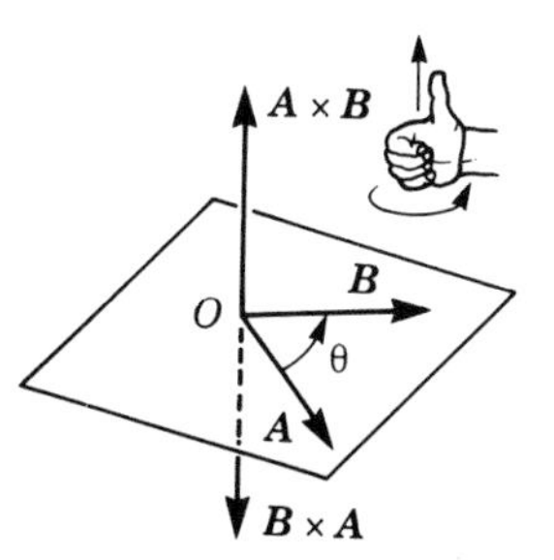

Figure A.16

The vector product of two vectors lying in a plane can be expressed in terms of its components in the following manner. Writing A and B in terms of their rectangular components, we have $A = iA_x + jA_y + kA_z$ and $B = iB_x + jB_y + kB_z$. Applying the distributive law,

$$A \times B = (i \times i)A_xB_x + (i \times j)A_xB_y + (i \times k)A_xB_z + (j \times i)A_yB_x + (j \times j)A_yB_y + (j \times k)A_yB_z + (k \times i)A_zB_x + (k \times j)A_zB_y + (k \times k)A_zB_z$$

Using the relations of Equation A.22, this expression may be reduced to three terms, one in each of the coordinate directions:

$$A \times B = i(A_yB_z - A_zB_y) + j(A_zB_x - A_xB_z) + k(A_xB_y - A_yB_x) \quad (A.24)$$

Equation A.24 may be written in the more compact determinant form

$$A \times B = \begin{vmatrix} i & j & k \\ A_x & A_y & A_z \\ B_x & B_y & B_z \end{vmatrix} \quad (A.25)$$

If the vectors A and B are in the XY-plane, then $A_z = B_z = 0$ and the vector product is

$$A \times B = k(A_xB_y - A_yB_x) \quad (A.26)$$

so that it is parallel to the Z-axis.

A.8 Vector representaton of an area

Consider a parallelogram whose sides are the vectors A and B (Figure A.17). The area of the parallelogram is

$$\text{Area} = \text{base} \times \text{height} = Ah = A(B \sin \theta) = |A \times B|$$

Therefore, the area of the parallelogram can be represented by the vector product $S = A \times B$, where S is perpendicular to the plane of A and B. Since any plane area can be considered as the sum of many small parallelograms, plane areas can be represented by a vector perpendicular to the plane, with a magnitude equal to the area (Figure A.18). The orientation of the vector is given by the thumb of the right hand when the fingers of the right hand point in the sense in which the boundary of the area is oriented.

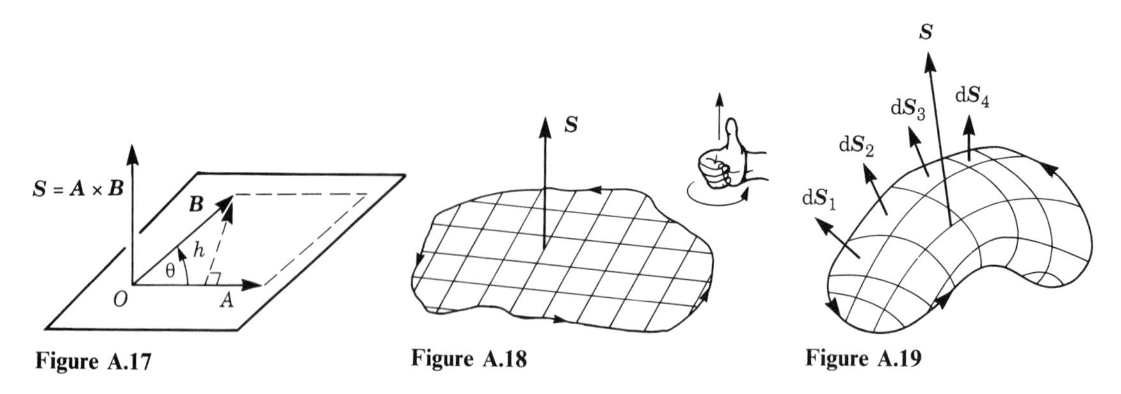

Figure A.17 **Figure A.18** **Figure A.19**

When the surface is curved, it still can be represented by a vector. In this case the area is divided into small rectangules (Figure A.19), each considered as a plane and represented by vectors $\boldsymbol{S}_1$, $\boldsymbol{S}_2$, $\boldsymbol{S}_3, \ldots$, perpendicular to the rectangles. Then $\boldsymbol{S} = \boldsymbol{S}_1 + \boldsymbol{S}_2 + \boldsymbol{S}_3 + \ldots = \Sigma_i \boldsymbol{S}_i$ is the vector representing the curved surface. However, in this case the magnitude of $\boldsymbol{S}$ is not equal to the area of the surface.

A.9 Gradient of a scalar function

Consider a scalar function $V(x, y, z)$ that depends on the three coordinates of a point. Draw the surfaces (Figure A.20) $V(x, y, z) = V_1$ and $V(x, y, z) = V_2$. In moving from a point A on V_1 to any point B on V_2, the function $V(x, y, z)$ always experiences the same change $V_2 - V_1$. If V_1 and V_2 differ by an infinitesimal amount, we may write $\mathrm{d}V = V_2 - V_1$. The change in $V(x, y, z)$ per unit length, or the **directional derivative** is

$$\frac{\mathrm{d}V}{\mathrm{d}s} = \frac{(V_2 - V_1)}{\mathrm{d}s}$$

Consider the case when A and B are along a normal N common to the two surfaces. The directional derivative along the normal AN is $\mathrm{d}V/\mathrm{d}n$. But, $\mathrm{d}n = \mathrm{d}s \cos\theta$. Thus

$$\frac{\mathrm{d}V}{\mathrm{d}s} = \frac{\mathrm{d}V}{\mathrm{d}n}\frac{\mathrm{d}n}{\mathrm{d}s} = \frac{\mathrm{d}V}{\mathrm{d}n}\cos\theta$$

which relates the directional derivative along the normal with the directional derivative along any other direction. Since $\cos\theta$ has its maximum value for $\theta = 0$, we conclude that $\mathrm{d}V/\mathrm{d}n$ gives the maximum directional derivative of $V(x, y, z)$. Introducing the unit vector $\boldsymbol{u}_n$, perpendicular to the surface at A, we define the gradient of $V(x, y, z)$ by

$$\text{grad } V = \boldsymbol{u}_n \frac{\mathrm{d}V}{\mathrm{d}n} \qquad \textbf{(A.27)}$$

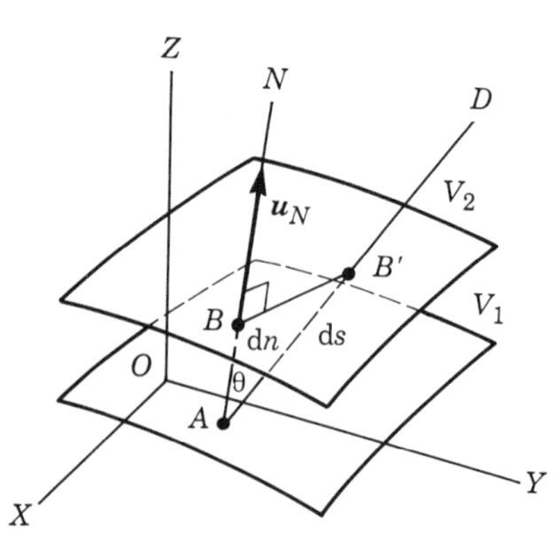

Figure A.20

Thus the gradient is a vector perpendicular at each point to the surface $V(x, y, z) = const.$ and is equal to the maximum directional derivative of $V(x, y, z)$.

A.10 The line integral of a vector: Circulation

Consider a line joining points A and B (Figure A.21). We can divide the line into small segments $d\boldsymbol{r}_1, d\boldsymbol{r}_2, d\boldsymbol{r}_3, \ldots$. Then for any vector $\boldsymbol{V}$ that is a function of position, the quantity

$$\boldsymbol{V}_1 \cdot d\boldsymbol{r}_1 + \boldsymbol{V}_2 \cdot d\boldsymbol{r}_2 + \boldsymbol{V}_3 \cdot d\boldsymbol{r}_3, \ldots = \int_A^B \boldsymbol{V} \cdot d\boldsymbol{r} \tag{A.28}$$

along some path joining points A and B is designated the **line integral** of $\boldsymbol{V}$. In general, its value depends on the path followed from A to B.

When the path is closed and points A and B coincide (Figure A.22), the line integral is called the **circulation** of the vector function $\boldsymbol{V}$ and is designated by

$$\text{circulation} = \oint \boldsymbol{V} \cdot d\boldsymbol{r} \tag{A.29}$$

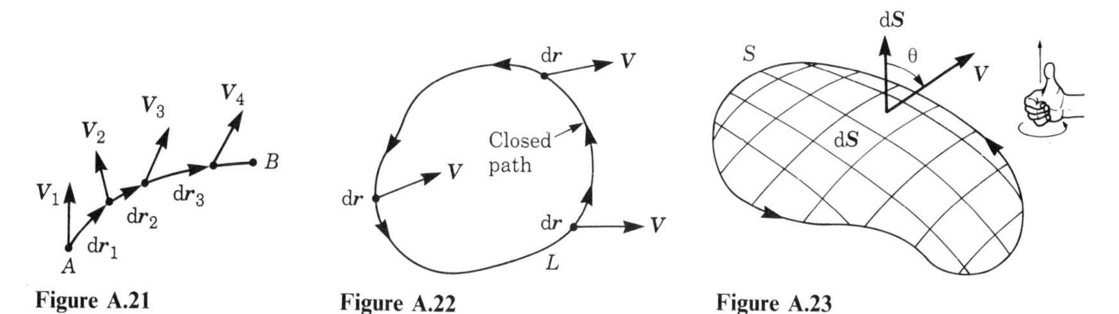

Figure A.21 **Figure A.22** **Figure A.23**

A.11 The surface integral of a vector: Flux

Consider a surface S placed in a region and a vector function $\boldsymbol{V}(x, y, z)$ (Figure A.23). Divide the surface into very small surfaces of areas $dS_1, dS_2, dS_3, \ldots$. At each of them draw the vectors $d\boldsymbol{S}_1, d\boldsymbol{S}_2, d\boldsymbol{S}_3, \ldots$, perpendicular to the surface at that point. These vectors are oriented in the direction of the thumb when the fingers of the right hand point in the sense in which we decide to orient the rim of the surface. If the surface is closed, the vectors $d\boldsymbol{S}$ are oriented in the outward direction. Let $\theta_1, \theta_2, \theta_3, \ldots$ be the angles between the normal vectors $d\boldsymbol{S}_1, d\boldsymbol{S}_2, d\boldsymbol{S}_3, \ldots$ and the vectors $\boldsymbol{V}_1, \boldsymbol{V}_2, \boldsymbol{V}_3, \ldots$ at each point on the surface. Then, by definition, the **flux** of the vector function $\boldsymbol{V}(x, y, z)$ through the surface S is

$$\begin{aligned}\text{Flux of } \boldsymbol{V} &= V_1 dS_1 \cos\theta_1 + V_2 dS_2 \cos\theta_2 + V_3 dS_3 \cos\theta_3 + \ldots \\ &= \boldsymbol{V}_1 \cdot d\boldsymbol{S}_1 + \boldsymbol{V}_2 \cdot d\boldsymbol{S}_2 + \boldsymbol{V}_3 \cdot d\boldsymbol{S}_3 + \ldots \\ &= \int_S V \cos\theta \, dS = \int_S \boldsymbol{V} \cdot d\boldsymbol{S}\end{aligned} \tag{A.30}$$

where the integral extends over all the surface, as indicated by the subscript S. An expression like Equation A.30 is also called a **surface integral**. Because of the $\cos\theta$ factor, the flux through the surface element may be positive or negative, depending on whether θ is smaller or larger than $\pi/2$. If the field $\mathbf{V}$ is tangent or parallel to the surface element dS, the angle θ is $\pi/2$ and $\cos\theta = 0$, resulting in zero flux through dS. The total flux may also be positive, negative or zero. When it is positive the flux is 'outgoing' and when it is negative the flux is 'incoming'. If the surface is closed (as it is for a sphere), a circle is written on top of the integral sign, so that Equation A.30 becomes

$$\oint_S V\cos\theta\, dS = \oint_S V\cdot dS \tag{A.31}$$

B: Mathematical relations

This appendix is a quick reference of mathematical formulas frequently used in the text. Proofs and a discussion of most of the formulas may be found in any standard calculus text.

B.1 Trigonometric functions

(i) Definition of the trigonometric functions (Figure B.1):

$$\sin\alpha = y/r, \qquad \cos\alpha = x/r, \qquad \tan\alpha = y/x \tag{B.1}$$

$$\csc\alpha = r/y, \qquad \sec\alpha = r/x, \qquad \cot\alpha = x/y \tag{B.2}$$

(ii) Trigonometric relations:

$$\tan\alpha = \sin\alpha/\cos\alpha \tag{B.3}$$

$$\sin^2\alpha + \cos^2\alpha = 1, \qquad \sec^2\alpha - 1 = \tan^2\alpha, \qquad \csc^2\alpha - 1 = \cot^2\alpha \tag{B.4}$$

$$\sin(\alpha \pm \beta) = \sin\alpha\cos\beta \pm \cos\alpha\sin\beta \tag{B.5}$$

$$\cos(\alpha \pm \beta) = \cos\alpha\cos\beta \mp \sin\alpha\sin\beta \tag{B.6}$$

$$\sin\alpha \pm \sin\beta = 2\sin\tfrac{1}{2}(\alpha \pm \beta)\cos\tfrac{1}{2}(\alpha \mp \beta) \tag{B.7}$$

$$\cos\alpha + \cos\beta = 2\cos\tfrac{1}{2}(\alpha + \beta)\cos\tfrac{1}{2}(\alpha - \beta) \tag{B.8}$$

$$\cos\alpha - \cos\beta = -2\sin\tfrac{1}{2}(\alpha + \beta)\sin\tfrac{1}{2}(\alpha - \beta) \tag{B.9}$$

$$\sin\alpha\sin\beta = \tfrac{1}{2}[\cos(\alpha - \beta) - \cos(\alpha + \beta)] \tag{B.10}$$

$$\cos\alpha\cos\beta = \tfrac{1}{2}[\cos(\alpha - \beta) + \cos(\alpha + \beta)] \tag{B.11}$$

$$\sin\alpha\cos\beta = \tfrac{1}{2}[\sin(\alpha - \beta) + \sin(\alpha + \beta)] \tag{B.12}$$

$$\sin 2\alpha = 2\sin\alpha\cos\alpha, \qquad \cos 2\alpha = \cos^2\alpha - \sin^2\alpha \tag{B.13}$$

$$\sin^2\tfrac{1}{2}\alpha = \tfrac{1}{2}(1 - \cos\alpha), \qquad \cos^2\tfrac{1}{2}\alpha = \tfrac{1}{2}(1 + \cos\alpha) \tag{B.14}$$

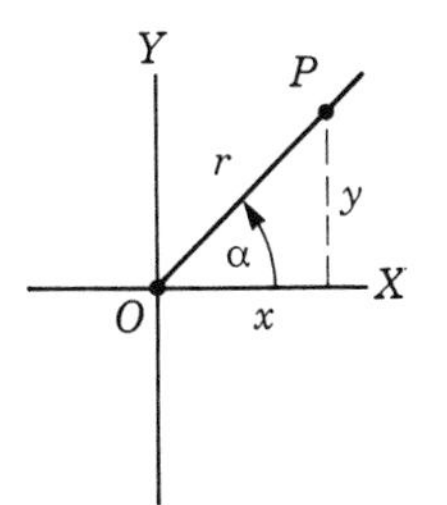

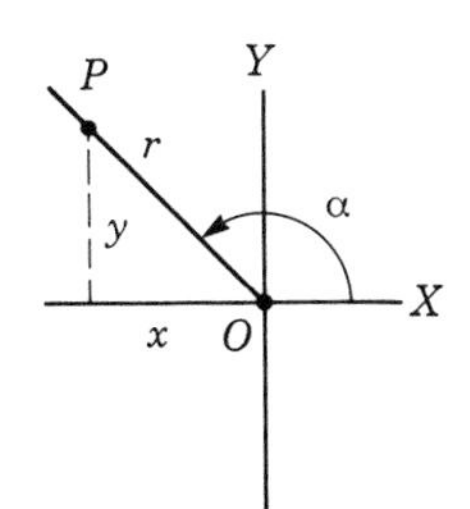

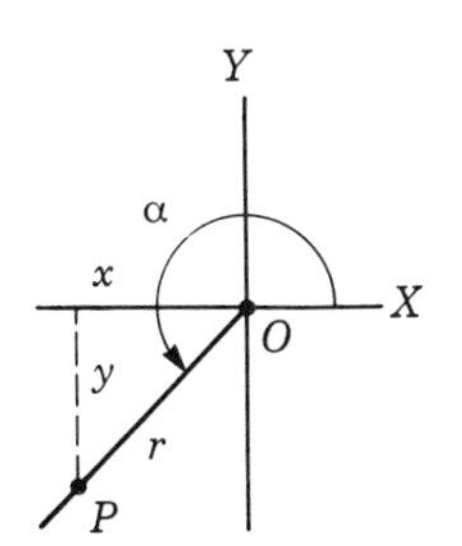

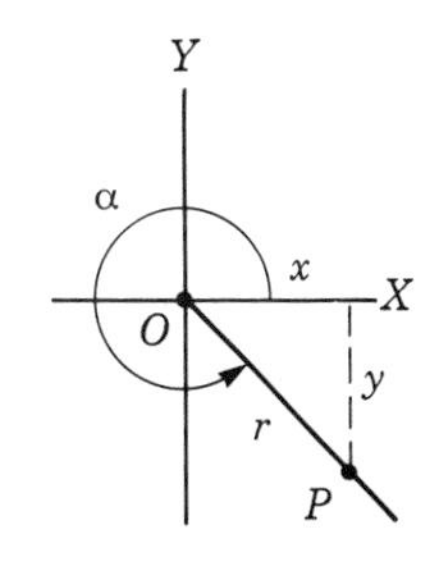

Figure B.1

(iii) Laws of sines and cosines (Figure B.2).

Law of sines: $\dfrac{a}{\sin A} = \dfrac{b}{\sin B} = \dfrac{c}{\sin C}$ **(B.15)**

Law of cosines: $a^2 = b^2 + c^2 - 2bc \cos A$ **(B.16)**

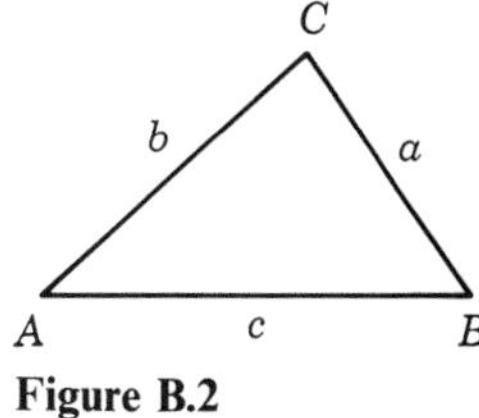

Figure B.2

B.2 Logarithms

(i) Definition of e:

$$\mathrm{e} = \lim_{n \to \infty} \left[1 + \frac{1}{n} \right]^n = 2.7182818\ldots \quad \textbf{(B.17)}$$

The exponential functions $y = \mathrm{e}^x$ and $y = \mathrm{e}^{-x}$ are plotted in Figure B.3.

(ii) Natural logarithm (Figure B.4):

$$y = \ln x \qquad \text{if } x = \mathrm{e}^y \quad \textbf{(B.18)}$$

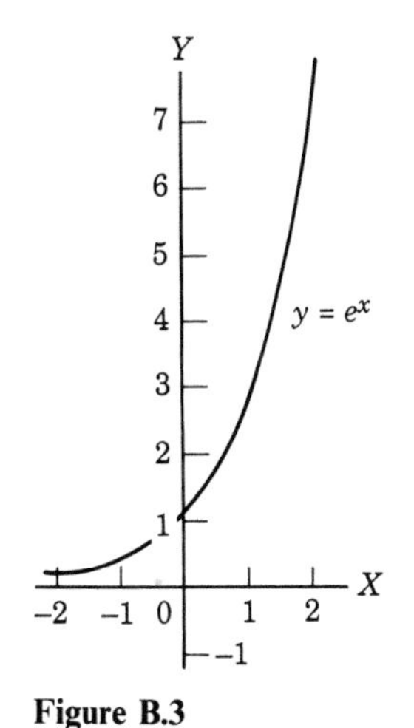

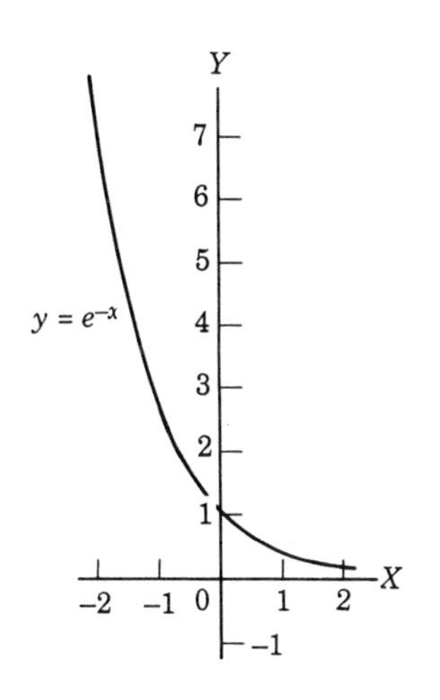

Figure B.3

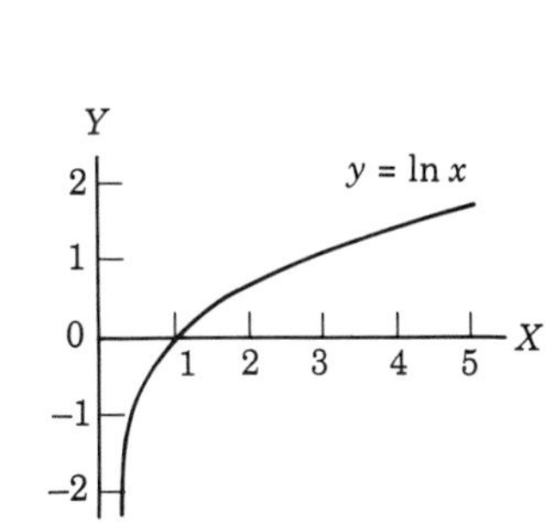

Figure B.4

(iii) Common logarithm, base 10:

$$y = \log x \qquad \text{if } x = 10^y \tag{B.19}$$

$$\ln x = 2.303 \log x, \qquad \log x = 0.434 \ln x \tag{B.20}$$

B.3 Power expansions

(i) The binomial expansion:

$$(a+b)^n = a^n + na^{n-1}b + \frac{n(n-1)}{2!}a^{n-2}b^2 + \frac{n(n-1)(n-2)}{3!}a^{n-3}b^3 + \ldots \tag{B.21}$$

When n is a positive integer, the expansion has $n + 1$ terms. In all other cases, the expansion has an infinite number of terms. When $a = 1$ and $b = x$, the binomial expansion is written

$$(1+x)^n = 1 + nx + \frac{n(n-1)}{2!}x^2 + \frac{n(n-1)(n-2)}{3!}x^3 + \ldots \tag{B.22}$$

(ii) Other useful expansions:

$$e^x = 1 + x + \frac{1}{2!}x^2 + \frac{1}{3!}x^3 + \ldots \tag{B.23}$$

$$\ln(1+x) = x - \frac{x^2}{2} + \frac{x^3}{3} + \ldots \tag{B.24}$$

$$\sin x = x - \frac{1}{3!}x^3 + \frac{1}{5!}x^5 - \ldots \tag{B.25}$$

$$\cos x = 1 - \frac{1}{2!}x^2 + \frac{1}{4!}x^4 - \ldots \tag{B.26}$$

$$\tan x = x + \frac{1}{3}x^3 + \frac{1}{15}x^5 + \ldots \tag{B.27}$$

$$\sum_1^n x_i = 1 + 2 + 3 + \ldots + n = \frac{n(n+1)}{2} \tag{B.28}$$

$$\sum_i^n x_1^2 = 1 + 2^2 + 3^2 + \ldots + n^2 = \frac{n(n+1)(2n+1)}{6} \tag{B.29}$$

$$\sum_{n=0}^{\infty} x^n = 1/(1-x) \qquad \text{for } x < 1 \tag{B.30}$$

For $x \ll 1$, the following approximations are satisfactory:

$$(1+x)^n \approx 1+nx \tag{B.31}$$

$$e^x \approx 1+x, \qquad \ln(1+x) \approx x \tag{B.32}$$

$$\sin x \approx x, \qquad \cos x \approx 1, \qquad \tan x \approx x \tag{B.33}$$

In Equations B.25, B.26, B.27 and B.33, x must be expressed in radians.

B.4 Plane and solid angles

There are two systems for measuring plane angles: **degrees** and **radians**. The circumference of a circle is divided into 360 degrees (°). Each degree is divided into 60 minutes (′) and each minute into 60 seconds (″).

To express a plane angle in radians (see Figure 2.4), one uses the relation $\theta(\text{rad}) = l/R$ where l is the length of the arc subtended by the angle θ. The equivalence between degrees and radians is given by

$$1^\circ = \pi/180 \text{ rad} = 0.017\,453 \text{ rad}, \qquad 1 \text{ rad} = 180^\circ/\pi = 57^\circ 17' 44.9'' \tag{B.34}$$

A **solid angle** is the space included inside a conical (or pyramidal) surface, as in Figure B.5. Its value, expressed in **steradians** (abbreviated sterad), is obtained by drawing, with arbitrary radius R and center at the vertex O, a spherical surface and applying the relation

$$\Omega = \frac{S}{R^2} \tag{B.35}$$

where S is the area of the spherical cap intercepted by the solid angle. Since the surface area of a sphere is $4\pi R^2$, the complete solid angle around a point is 4π steradians.

When the solid angle is small (Figure B.6), the surface area S becomes dS, and is not necessarily a spherical cap, but may be a small plane surface perpendicular to OP so that

$$d\Omega = \frac{dS}{R^2} \tag{B.36}$$

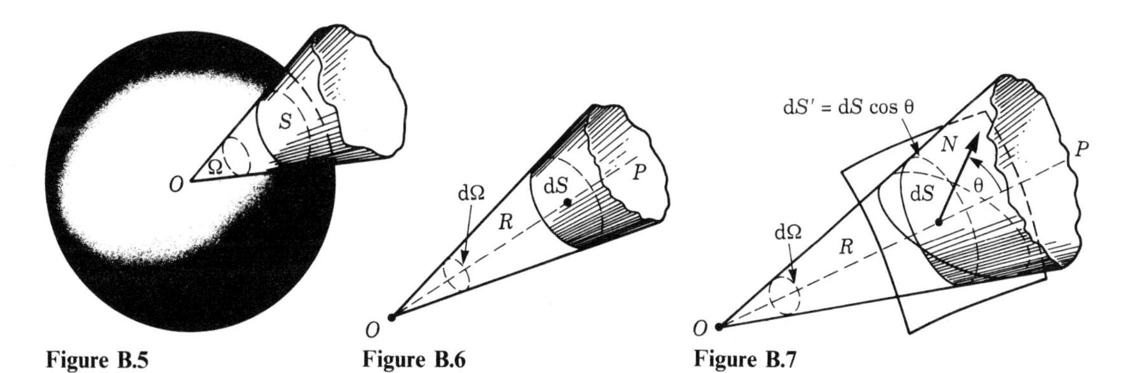

Figure B.5 **Figure B.6** **Figure B.7**

In some instances the surface dS is not perpendicular to OP, but its normal N makes an angle θ with OP (Figure B.7). Then it is necessary to project dS on a plane perpendicular to OP, which gives us the area $dS' = dS \cos\theta$. Thus $d\Omega = dS \cos\theta / R^2$.

B.5 Basic derivatives and integrals

$f(u)$	df/du	$\int f(u)\,du$
u^n	$nu^{n-1}\,du/dx$	$[u^{n+1}/(n+1)] + C \quad (n \neq -1)$
u^{-1}	$-(1/u^2)\,du/dx$	$\ln u + C$
$\ln u$	$(1/u)\,du/dx$	$u \ln u - u + C$
e^u	$e^u\,du/dx$	$e^u + C$
$\sin u$	$\cos u\,du/dx$	$-\cos u + C$
$\cos u$	$-\sin u\,du/dx$	$\sin u + C$
$\tan u$	$\sec^2 u\,du/dx$	$-\ln \cos u + C$
$\cot u$	$-\csc^2 u\,du/dx$	$\ln \sin u + C$

B.6 Special integrals

$$\int \frac{dx}{(R^2 + x^2)^{1/2}} = \ln[(R^2 + x^2)^{1/2} + x] + C$$

$$\int_{-\infty}^{\infty} \frac{dx}{(R^2 + x^2)^{3/2}} = \left.\frac{x}{R^2(x^2 + R^2)^{1/2}}\right]_{-\infty}^{\infty} = \frac{2}{R^2}$$

$$\int_0^{\infty} x^n e^{-x}\,dx = \begin{cases} n! & \text{if } n \text{ is an integer} \\ \dfrac{1 \times 3 \times 5 \times \cdots \times (2k+1)}{2^{k+1}} \sqrt{\pi} & \text{if } n = \tfrac{1}{2}(2k+1)\text{, a half-integer} \end{cases}$$

B.7 Average value of a function

The **mean** or **average value** of a function $y = f(x)$ in the interval (a, b) is defined by

$$y_{\text{ave}} = \frac{1}{b-a} \int_a^b y\,dx$$

The average value of y^2 is defined by

$$(y^2)_{\text{ave}} = \frac{1}{b-a} \int_a^b y^2\,dx$$

The quantity $[(y^2)_{\text{ave}}]^{1/2}$ is called the **root mean square** value of $y = f(x)$ in the interval (a, b) and is designated y_{rms}.

B.8 Conic sections

A conic section is defined as a curve generated by a point moving in such a way that the ratio of its distance to a point, called the **focus**, and to a line, called the **directrix**, is constant. There are three kinds of conic section, called ellipse, parabola and hyperbola, depending on whether this ratio (called the **eccentricity**) is smaller than, equal to or larger than one. Designating the eccentricity by ε, the focus by F and the directrix by the line HDQ (Figure B.8(a)), we see that $\varepsilon = PF/PQ$. Now, $PF = r$ and, if we say that $FD = d$, then $PQ = FD - FB = d - r\cos\theta$. Therefore, $\varepsilon = r/(d - r\cos\theta)$ or,

$$\frac{\varepsilon d}{r} = 1 + \varepsilon \cos\theta \tag{B.37}$$

In the case of an ellipse ($\varepsilon < 1$), point A corresponds to $\theta = 0$ and point A' to $\theta = \pi$. Thus, according to the polar equation B.37, we see that

$$r_1 = \frac{\varepsilon d}{1+\varepsilon} \quad \text{and} \quad r_2 = \frac{\varepsilon d}{1-\varepsilon}$$

The **semi-major axis** of the ellipse is a. Then, since $r_1 + r_2 = 2a$,

$$a = \tfrac{1}{2}(r_1 + r_2) = \frac{\varepsilon d}{1-\varepsilon^2} \tag{B.38}$$

Similarly, the **semi-minor axis**, b, is

$$b = a(1-\varepsilon^2)^{1/2} \tag{B.39}$$

The area of an ellipse is

$$S = \pi ab = \pi a^2 (1-\varepsilon^2)^{1/2} \tag{B.40}$$

A circle is a special case of an ellipse, where $\varepsilon = 0$.

In a hyperbola ($\varepsilon > 1$), there are two branches and the axes are a and b, such that $b = a(\varepsilon^2 - 1)^{1/2}$. Also, the **foci** F and F' are located (Figure B.8(b)) so that $OF = OC$. Therefore, $OF = (a^2 + b^2)^{1/2} = a\varepsilon$. In a parabola ($\varepsilon = 1$) there is only one branch.

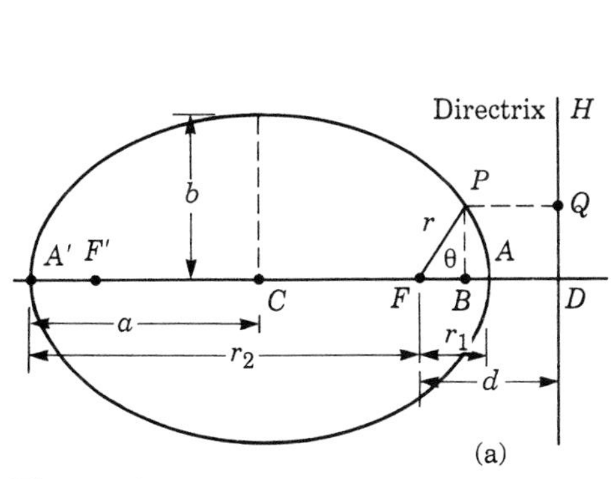

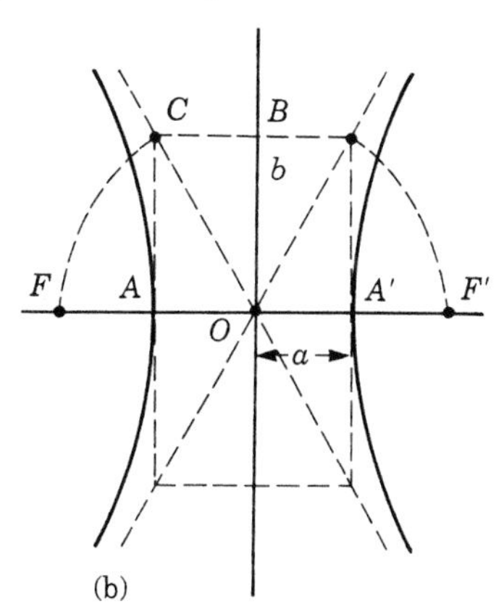

Figure B.8

C: Calculation of the moment of inertia

A rigid body is composed of a very large number of closely packed particles, so that the sum in Equation 13.2 may be replaced by an integral,

$$I = \sum_i m_i R_i^2 = \int R^2 \, dm$$

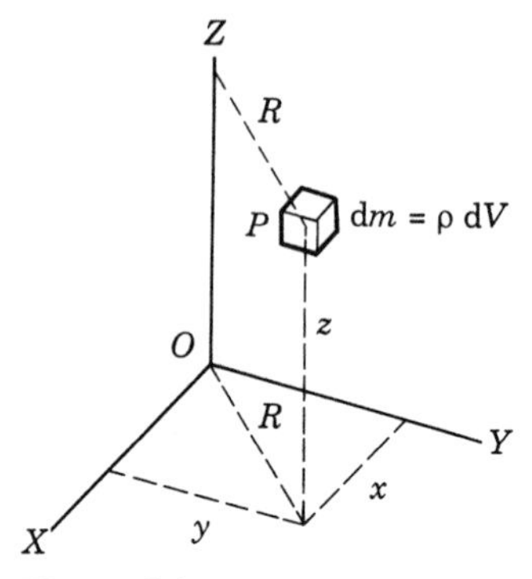

Figure C.1

If ρ is the density of the body, $dm = \rho \, dV$ and

$$I = \int \rho \, R^2 \, dV$$

When the body is homogeneous, its density is constant, and we may write

$$I = \rho \int R^2 \, dV \tag{C.1}$$

The integral thus reduces to a geometrical factor, the same for all bodies with the same shape and size. We note from Figure C.1 that $R^2 = x^2 + y^2$, and therefore the moment of inertia around the Z-axis is

$$I_z = \int \rho (x^2 + y^2) \, dV \tag{C.2}$$

with similar relations for I_x and I_y.

If the body is a thin plate, as indicated in Figure C.2, we note that the moments of inertia relative to the X- and Y-axes may be written as

$$I_x = \int \rho y^2 \, dV \quad \text{and} \quad I_y = \int \rho x^2 \, dV$$

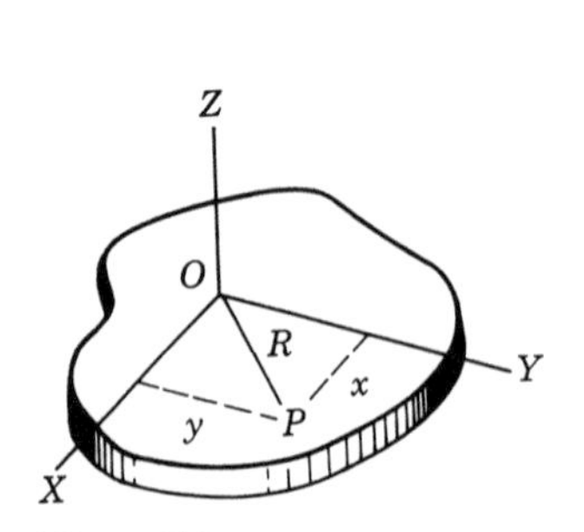

Figure C.2

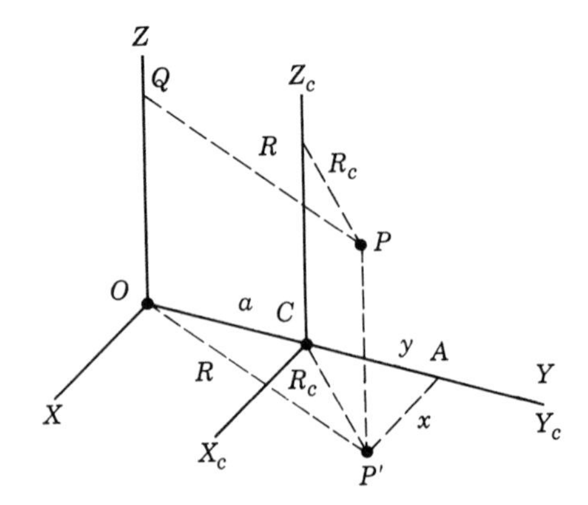

Figure C.3

because the Z-coordinate is essentially zero. Comparison with the expression for I_z shows that for thin plates

$$I_z = I_x + I_y \tag{C.3}$$

Let Z be an arbitrary axis and Z_C a parallel axis passing through the center of mass of the body (Figure C.3). Choose the axes $X_C Y_C Z_C$ so that their origin is at the center of mass C and the Y_C axis is in the plane determined by Z and Z_C. The axes XYZ are chosen so that Y coincides with Y_C. The point P is any arbitrary point in the body. Noting that $P'A$ is perpendicular to Y_C and $P'A = x$, $CA = y$ and $OC = a$, we have $R_C^2 = x^2 + y^2$, and thus

$$R^2 = x^2 + (y + a)^2 = x^2 + y^2 + 2ya + a^2 = R_C^2 + 2ya + a^2$$

Now the moment of inertia relative to the Z-axis is

$$I = \Sigma m R^2 = \Sigma m(R_C^2 + 2ya + a^2) = \Sigma m R_C^2 + 2a(\Sigma my) + a^2(\Sigma m)$$

The first term is just the moment of inertia I_C relative to the Z_C-axis, and in the last term, $\Sigma m = M$ is the total mass of the body. Therefore

$$I = I_C + 2a\Sigma my + Ma^2$$

To evaluate the middle term, we recall from Equation 13.1 that the position of the center of mass is given by $y_{CM} = \Sigma my/\Sigma m$. But in our case $y_{CM} = 0$ because the center of mass coincides with the origin C of the frame $X_C Y_C Z_C$. Then $\Sigma my = 0$ and

$$I = I_C + Ma^2 \tag{C.4}$$

This result is known as **Steiner's theorem**.

The **radius of gyration** of a body is a quantity K defined by

$$K \equiv (I/M)^{1/2} \qquad \text{or} \qquad I = MK^2 \tag{C.5}$$

where I is the moment of inertia and M the mass of the body. Hence K represents the distance from the axis at which all the mass could be concentrated without changing the moment of inertia. It is a useful quantity because it can be determined, for homogeneous bodies, entirely by their geometry. It can be tabulated and thus used to compute the moments of inertia of homogeneous, symmetrical bodies. The calculation of the moment of inertia is straightforward and is explained in a standard calculus text.

Answers to selected problems

CHAPTER 2

2.1 (a) 1.674×10^{-27} kg; (b) 2.657×10^{-26} kg; (c)(i) 5.975×10^{26} atoms; (ii) 3.764×10^{25} atoms.

2.3 (a) 2.685×10^{25} molecules of H_2, 2.689×10^{25} molecules of O_2, 2.688×10^{25} molecules of N_2; (b) all gases contain about 2.7×10^{25} molecules per m^3 at STP.

2.5 (a) 0.6 atoms cm^{-3}.

2.7 (a) 3.340×10^{-9} m; (b) 3.104×10^{-10} m; (c) 2.276×10^{-10} m.

2.9 Earth: 5.52×10^3 kg m^{-3} and Sun: 1.40×10^3 kg m^{-3}. Other planets and the Moon are in the same range.

2.11 6.35×10^4 AU.

2.13 (a) 3.07×10^{16} m; (b) 3.25 lyr; (c) 2.06×10^5 AU.

2.15 For small angles and θ in radians $\theta \approx \sin\theta \approx \tan\theta$.

2.17 2.01×10^8 m.

2.19 (a) 9.062×10^{-3} radians or 31.15′; (b) 9.280×10^{-3} radians or 31.90′.

CHAPTER 3

3.1 (a) 31 m s^{-1}, 119 m; (b) -25 m s^{-1}, -77 m; (c) $v = 3 \pm 4t$, $x = 3t \pm 2t^2$.

3.3 (a) 4.00×10^3 m min^{-2}, 125 m; (b) 5.0 s; (c) 222.2 m.

3.9 (a) 10 m; (b) 2.7 s and 0 s; (c) 4 m s^{-1}; (d) $16 - 12t_0 - 6\Delta t$ m s^{-1}; (e) $16 - 12t$ m s^{-1}; (f) 16 m s^{-1}; (g) 1.33 s and 10.7 m only; (h) -12 m s^{-2}; (i) $-$ 12 m s^{-2}; (j) never; (k) retarded for $0 < t < 1.33$ s, accelerated (in negative X-direction) for $t > 1.33$ s.

3.11 (a) $v = 4(2 - e^{4t})$; (b) $x = 8t + 1 - e^{4t}$; (c) $x = \frac{1}{4}v - 1 - 2\ln(\frac{1}{4}v - 2)$.

3.13 110 m s^{-1} downward, 610 m; (b) 86 m s^{-1} downward, 370 m below initial point.

3.15 (a) 44.1 m above ground; (b) 36.4 m; (c) 29.4 m s^{-1}.

3.17 18 s.

3.19 175.4 m.

3.21 81.8 km h^{-1} at 132°; 138.9 km h^{-1} at 214°.

3.23 (a) 14.1 km h^{-1}; (b) 81.2 km h^{-1}.

3.25 2.52 km h^{-1} at S37.5°W.

3.26 (b) S41°18′E; (c) 1h 34 min.

3.29 (a) 5 km h^{-1}; (b) 15 min; (c) 0.75 km.

3.31 (a) Parabolic path, with a horizontal velocity of 30 m s^{-1} backwards; (b) parabolic path, with a horizontal velocity of 460 m s^{-1} backwards; (c) up or down at 28.65°.

CHAPTER 4

4.2 (a) $v_x = 2t + \Delta t$ m s^{-1}, $v_y = 2(t-1) + \Delta t$ m s^{-1}, $a_x = 2$ m s^{-2}, $a_y = 2$ m s^{-2}; (b) $v_x = 5$ m s^{-1}; $v_y =$ 3 m s^{-1}; (c) $v_x = 4$ m s^{-1}, $v_y = 2$ m s^{-1}.

4.5 (a) $\boldsymbol{v} = \boldsymbol{i}(2t + \Delta t) + \boldsymbol{j}(2[t-1] + \Delta t)$ m s^{-1}, $\boldsymbol{a} = 2(\boldsymbol{i} + \boldsymbol{j})$ m s^{-2}; (b) $\boldsymbol{v}_{ar} = \boldsymbol{i}5 + \boldsymbol{j}3$ m s^{-1}; (c) $\boldsymbol{v} = \boldsymbol{i}4 + \boldsymbol{j}2$ m s^{-1}.

4.6 (a) 883.7 m; (b) 382.6 m; (c) 17.67 s; (d) 51.28 m s^{-1}, 376 m.

4.8 (a) 15.65 s; (b) 183.1 m s^{-1}; (c) 1565 m.

4.13 (a) $\boldsymbol{v}_{rel} = \boldsymbol{i}7$ m s^{-1}.

4.15 5 m s^{-1}, arctan(4/3); 18 m s^{-2} in $+Y$-direction.

CHAPTER 5

5.1 (a) 2.2 rad s^{-1}; (b) 2.9 s, 0.35 Hz; (d) 6.2 s; (e) 34.3 s.

5.3 (a) 2.60×10^{-6} rad s^{-1}; (b) 9.97×10^2 m s^{-1}; (c) 2.60×10^{-3} m s^{-2}.

5.5 2.4×10^5 m s^{-1}; (b) 2.4×10^{-10} m s^{-2}.

5.7 (a) 2 Hz; (b) 0.5 s; (c) 4π rad s^{-1}; (d) 12π m s^{-1}.

5.9 (a) $\theta = 10t^4 - 8t^3 + 8t^2$ rad;
(b) $\omega = 40t^3 - 24t^2 + 16t$ rad s^{-1};
(c) $156t^2 - 62.4t + 20.8$ m s^{-2};
(d) $83.2t^2(5t^2 - 3t + 2)^2$ m s^{-2}.

5.11 (b) 18.1 min; (c) 1.95×10^4 rad.

5.13 (b) $v_x = R\omega(1 - \cos \omega t)$, $v_y = R\omega \sin \omega t$, $a_x = R\omega^2 \sin \omega t$, $a_y = R\omega^2 \cos \omega t$.

5.15 (a) 2.55×10^{-4} m s^{-2} for North and South flow, 3.61×10^{-4} m s^{-2} for East flow and always the right bank; (b) same accelerations but water now pushes against left bank.

CHAPTER 6

6.1 (a) 0.472.

6.3 $m_A = 1$ kg, $m_B = 2$ kg.

6.5 (a) 0.186 m s^{-1}, 27.5° below the X-axis;
(b) $\Delta \boldsymbol{p}_1 = \boldsymbol{i}(-0.049) + \boldsymbol{j}(0.026)$ kg m s^{-1},
$\Delta \boldsymbol{v}_1 = \boldsymbol{i}(-0.247) + \boldsymbol{j}(0.129)$ m s^{-1}, $\Delta \boldsymbol{p}_2 = -\Delta \boldsymbol{p}$,
$\Delta \boldsymbol{v}_2 = \boldsymbol{i}(0.165) + \boldsymbol{j}(-0.086)$ kg m s^{-1}.

6.6 (a) 15 m s^{-1}, W44.9′S; (b) $\boldsymbol{W}$19.2 + $\boldsymbol{N}$8.0 kg m s^{-1}.

6.7 3.33×10^4 m s^{-1}, $\theta = 82.6°$.

6.10 (a) -0.3 kg m s^{-1}, -3.0 N;
(b) -0.45 kg m s^{-1}, -4.5 N.

6.12 -347 N.

6.14 -9.8 N.

6.16 (a) $a = (F_0 - kt)/m$; (b) $v = (F_0 t - \frac{1}{2}kt^2)/m$, $x = (\frac{1}{2}F_0 t^2 - \frac{1}{6}kt^3)/m$.

6.18 (a) 250 kg m s^{-1}; (b) 25 N.

CHAPTER 7

7.1 (a) 7.3° forward; (b) 21° backward.

7.3 (a) 588 N; (b) 588 N; (c) 768 N; (d) 408 N; (e) 0.

7.5 (a) 1.11 m s^{-2}; (b) $T_1 = 11.11$ N,
$T_2 = 27.78$ N; (c) 0.769 m s^{-2}, $T_1 = 23.08$ N,
$T_2 = 34.62$ N.

7.7 (a) $a = (F - m_2 g)/(m_1 + m_2) = 1.66$ m s^{-2},
$T = m_2(a + g) = 917$ N; (b)
$a = (F + m_1 g - m_2 g)/(m_1 + m_2) = 5.43$ m s^{-2},
$T = m_2(a + g) = 1218$ N.

7.8 (a) $a = g(m_2 - m_1 \sin \alpha)/(m_1 + m_2)$
$= 2.06$ m s^{-2}, $T = m_1 m_2 g(1 + \sin \alpha)/(m_1 + m_2)$
$= 139.3$ N;
(b) $a = g(m_2 \sin \beta - m_1 \sin \alpha)/(m_1 + m_2)$
$= 1.44$ m s^{-2},
$T = m_1 m_2 g(\sin \beta + \sin \alpha)/(m_1 + m_2) = 126.8$ N.

7.11 (a) 15 kg; (b) $\frac{1}{5}g$.

7.13 (a) 11.5 m; 9.03 m s^{-1}.

7.16 (a) 4.9×10^3 m s^{-1}; (b) 0.99 m s^{-1}.

7.18 (a) $a = m_0(F_0 + kv_0)/(m_0 - kt)^2$, where m_0, v_0 and F_0 are the initial values and $k = |dm/dt|$;
(b) $F = F_0 - [2k(F_0 + kv_0)/m_0]t$.

7.21 2867 N, 4096 N.

7.24 (a) 255.8 N each; (b) 392 N each;
(c) $T_{AC} = 196.0$ N, $T_{BC} = 339.5$ N;
(d) $T_{AC} = 392$ N, $T_{BC} = 554.4$ N.

7.27 (a) 220.5 N; (b) 441 N.

7.29 358.7 N, 253.6 N.

7.31 574.3 N.

CHAPTER 8

8.1 (a) 22.5 N; (b) 39.5 rad s^{-1}; (c) yes.

8.3 8.81×10^{-8} N.

8.5 7.10 m.

8.9 (a) $\boldsymbol{F} = \boldsymbol{i}36 + \boldsymbol{j}(-144t)$ N;
(b) $\boldsymbol{\tau} = \boldsymbol{k}(-288t^3 + 863t^2)$ N m;
(c) $\boldsymbol{p} = \boldsymbol{i}(36t - 36) + \boldsymbol{j}(-72t^2)$ kg m s^{-1},
$\boldsymbol{L} = \boldsymbol{k}(-72t^4 + 288t^3)$ kg m^2 s^{-1}.

8.11 (a) 3.03×10^4 m s^{-1}; (b) aphelion: 1.92×10^{-7} rad s^{-1}, perihelion: 2.06×10^{-7} rad s^{-1}.

8.13 (a) zero X and Y components,
Z component is $-\frac{1}{2}(mg/v_0)t^2 \cos \alpha$;
(b) $d\boldsymbol{L}/dt = -\boldsymbol{k}(mg/v_0)t \cos \alpha$; (c) same as (b).

CHAPTER 9

9.2 2978 J, 74.4 W.

9.5 227.3 W.

9.7 (a) 1875 W; (b) 1.5×10^4 kg m s^{-1}, 7.5×10^4 J; (c) 7.5×10^4 J; (d) 9.38 kW.

9.9 $E'_k = E_k - \frac{1}{2}mv^2 - m\boldsymbol{V} \cdot \boldsymbol{v}$.

9.11 (a) 9.40×10^7 m s^{-1}; (b) 7.56×10^6 m s^{-1}.

9.13 (a) 45J; (b) 47J.

9.17 (a) 2.5×10^4 J, 0, 2.5×10^4 J; (b) 6.02×10^3 J, 1.90×10^4 J; (c) 5.4×10^3 J, 1.96×10^4 J; (d) 25.5 m.

9.19 7.25×10^{-2} m.

9.23 26.5 J.

9.25 $\mu m g^2 \cos\alpha(\sin\alpha - \mu\cos\alpha)t$.

9.27 (a) $-(V_0/r)(r_0/r + 1)e^{-r/r_0}$; (b) $-(2V_0/r_0)e^{-1}$; (c) let $x = r/r_0$, then solve $(1 + x)e^{-x} - 0.02e^{-1}x^2 = 0$ to get $x = 3.806$.

CHAPTER 10

10.1 (a) 2 s; (b) $\frac{1}{2}$ Hz; (c) 0.30 m; (d) $x = 0.30 \cos \pi t$ m.

10.4 (a) 318.3 Hz; $\pi \times 10^{-3}$ s; (b) $x = 10^{-3}\cos 2000t$ m, $v = -2\sin 2000t$ m s^{-1}.

10.6 (a) at middle: $a = 0$, $v = 0.188$ m s^{-1}, at extremes: $a = 1.18 \times 10^2$ m s^{-2}, $v = 0$; (b) $x = 3 \times 10^{-4}\cos(200\pi t)$ m.

10.9 (a) 8.88×10^{-2} m s^{-1}, -0.140 m s^{-2}; (b) 0.126 m s^{-1}, 0.

10.11 $-20\pi^2$ m s^{-2}, $-10\pi^2$ N, $\frac{1}{4}\pi^2$ J, $\frac{3}{4}\pi^2$ J.

10.12 (a) $x = x_0\cos(\frac{1}{8}\pi t + \frac{1}{4}\pi)$.

10.15 (c) $\boldsymbol{F} = -\boldsymbol{j} m\omega^2 y$; (d) $v = (v_0^2 + \omega^2 y_0^2 \cos^2 \omega t)^{1/2}$, $a = \omega^2 y_0 \sin\omega t$.

10.16 0, $\frac{1}{2}x_0^2$.

10.19 0.102% ($\Delta l/l = \Delta g/g$).

10.21 (a) 0.2838 rad; (b) $F_T = 5.88(2h - h^2)^{1/2}$ N, where h is the height, in meters, above minimum point; (c) $a_T = F_T/m$; (d) $v = [2g(h_0 - h)]^{1/2}$ m s^{-1}; (e) $\theta = \arccos(1 - h)$ rad; (f) 1.65 N, 2.74 m s^{-2}, 0, 16.3°; (g) 0, 0, 0.885 m s^{-1}, 0.

10.23 (a) $14\cos 2t$, $10\cos(2t + 0.93)$, $-2\cos 2t$.

10.26 $\boldsymbol{r} = \boldsymbol{i}x_0\cos\omega_1 t + \boldsymbol{j}y_0\cos(2\omega_1 t + \alpha)$ from which the general solution is $y^2 = (x_0^2 - x^2) - x_0^2\sin^2\omega_1 t + y_0^2(\cos 2\omega_1 t\cos 2\alpha - \sin 2\omega_1 t \sin 2\alpha)$.

10.28 (b) $0.61A_0$; (c) 1.39τ; (d) $\frac{1}{4}A_0$, $\frac{1}{8}A_0$, ... etc.

10.31 (a) $F_0/m\omega_0^2$, $-\frac{1}{2}\pi$; (b) $(F_0/m\omega_f^2)(\omega_f^2 - 4\gamma^2)^{-1/2}$, $\arctan(\omega_f/2\gamma)$.

10.34 (a) $4(2h/g)^{1/2}$; (b) oscillatory, not SHM.

10.37 (a) No; (b) in $+X$-direction, no; (c) $F = -kx + ax^2$.

CHAPTER 11

11.2 Moon: 3.39×10^{-4}%, Sun: 6.01×10^{-2}% force of the Earth.

11.5 (a) 3.62×10^{-47} N.

11.7 3.95×10^{-6} kg.

11.9 (a) 267.3 m s^{-2}; (b) 8.86 m s^{-2}; (c) 26.0 m s^{-2}; (d) 1.62 m s^{-2}.

11.12 (a) 32.1 km; (b) 64.1 km.

11.14 (a) Particle will oscillate through the Earth to a height h on both sides; (b) $[2GM/(h + R)]^{1/2}$; (c) no; (d) yes.

11.16 2.64×10^{33} J, -5.28×10^{33} J, -2.64×10^{33} J.

11.19 (a) 0.366 a from m on the line between m and $3m$; (b) $1.366a$ from m on the same line, but outside the bodies; (c) defining $\theta = 0$ for the line from m to $3m$ and using m as origin: $r = (\frac{1}{8}a\cos\theta)[-1 \pm (1 + 8\sec^2\theta)^{1/2}]$.

11.21 (a) At the geometric center of the triangle; (d) $-3\sqrt{3}GM/a$, where a is a side of the triangle; (e) always along that line and pointing towards the plane; (f) $-3Ma/(\sqrt{3}r^3)$, where $r = (z^2 + \frac{1}{3}a^2)^{1/2}$ and z is the distance of the point from the plane.

CHAPTER 12

12.2 (a) 4.25×10^3 m s^{-1}; (b) 1.04×10^4 m s^{-1}; (c) 5.03×10^3 m s^{-1}; (d) 6.02×10^4 m s^{-1}.

12.4 (a) 2.822×10^{14} kg m s^{-1}; (b) 1.244×10^{11} J; (c) -2.489×10^{11} J; (d) -1.244×10^{11} J.

12.7 (a) 7.73×10^3 m s^{-1}; (b) 5.42×10^3 s; (c) 8.96 m s^{-2}.

12.11 7.50 km s^{-1}.

12.15 (a) 7.80×10^3 m s^{-1}; (b) 11.18×10^3 m s^{-1}.

CHAPTER 13

13.1 $\boldsymbol{i}(-6.37) + \boldsymbol{j}1.6$ m s^{-1}.

13.3 (a) $\boldsymbol{r}_C = \boldsymbol{i}(1.50 + 0.25t^2) + \boldsymbol{j}(1.88 + 0.188t^2)$ m; (b) $\boldsymbol{P} = \boldsymbol{i}8t + \boldsymbol{j}6t$ kg m s^{-1}.

13.5 (a) 120° or 210° with respect to the direction of the electron; (b) 1.06×10^{-20} kg m s^{-1}; (c) 2.73×10^4 m s^{-1}.

13.7 (a) LiH: 1.393×10^{-10} m from H atom, HCl: 1.235×10^{-10} m from H atom; (b) 2.34×10^{-12} m from N atom.

13.9 (a) $\boldsymbol{i}(-8.32) + \boldsymbol{j}(-16.8) + \boldsymbol{k}(-25.1)$ kg m^2 s^{-1}; (b) $\boldsymbol{i}(-41.6) + \boldsymbol{j}(-4.0) + \boldsymbol{k}(-40.8)$ kg m^2 s^{-1}.

13.11 (a) 1.942 kg m^2; (b) 0.967 kg m^2; (c) 0.642 kg m^2.

13.13 2.04×10^{-47} kg m^2.

13.16 (a) 10.5 min; (b) 8773 rad.

13.18 (a) 63.6 kg m^2 s^{-1}; (b) 12.72 Nm.

13.20 (a) mva, $I\omega$, where $I = ma^2 + \frac{1}{3}ML^2$; (b) mv, $(ma + \frac{1}{2}ML)\omega$; (c) only if the rod is unpinned at A and allowed to move freely.

13.22 (a) 8.71 rad s^{-2}; (b) 4.36 m s^{-2}; (c) 54.4 N.

13.24 (a) 0.98 m s^{-2}; (b) 17.64 N.

13.27 (a) 1.64 s; (b) 0.667 m; (c) 1.53 s.

13.29 3.56×10^{-3} Nm.

13.31 (b) 6.53 m s^{-2}.

13.34 (a) 1.3 m from the pivot; (b) 5 N.

13.35 110 N down at 1.11 m.

13.38 $\phi = \arctan(\cot 2\alpha)$, $F_{\text{left}} = mg \sin \alpha$, $F_{\text{right}} = mg \cos \alpha$.

13.39 2.33 m.

13.42 $F_1 = F_3 = 9.84$ N, $F_2 = 37.05$ N.

13.43 (a) $T = 762.3$ N, $B_{\text{H}} = 539.0$ N, $B_{\text{V}} = 147.0$ N; (b) $T = 933.6$ N; $B_{\text{H}} = 466.8$ N, $B_{\text{V}} = 122.5$ N (down); (c) $T = 419.3$ N, $B_{\text{H}} = 321.2$ N, $B_{\text{V}} = 416.5$ N.

13.45 (a) 0; (b) 20 N in each hand, one up, one down; (c) 20 N in each hand, one forward, one backward.

CHAPTER 14

14.2 (a) $\boldsymbol{L} = \boldsymbol{k}$ 48 kg m^2 s^{-1}, $\boldsymbol{L}_{\text{C}} = \boldsymbol{k}$ 14.4 kg m^2 s^{-1}, $\boldsymbol{r}_{\text{C}} \times \boldsymbol{P} = \boldsymbol{k}$ 33.6 kg m^2 s^{-1}; (b) 35J, 15.6J, $\frac{1}{2}Mv_{\text{C}}^2 = 19.4$J.

14.5 $(4M/m)\sqrt{gl}$.

14.6 (b) $(E_{k,\text{C}} - E_{k,\text{L}})_{\text{arc}} = \frac{1}{2}m(v_{\text{CM}}^2 - v_{\text{arc}}^2)$.

14.8 (a) $v_5' = -0.462$ m s^{-1}, $v_8' = 1.538$ m s^{-1}; (b) $v_5' = 1.57$ m s^{-1} at 50°, $v_8' = 0.974$ m s^{-1} at $-50.7°$.

14.10 (a) 0; (b) $-\frac{1}{2}\,\mu v_{12}^2$.

14.12 (a) $v_1' = -ev_1$, $v_2' = 0$, $-\frac{1}{2}(1 - e^2)m_1 v_1^2$; (b) $h' = e^2 h$.

14.15 (a) $v_4' = 0.53$ m s^{-1}, $v_5' = 1.13$ m s^{-1}; (b) $\Delta p_4 = -2.67$ kg m s^{-1}, $\Delta p_5 = -\Delta p_4$.

14.19 0.2588 (compared to 0.2528).

14.21 (a) $v' = -v$; (b) $v' = -v[a + \sqrt{a} - 1]/(1 + a)$, where $a = M/m$.

14.23 90 amu particle: 115.6 MeV, 1.57×10^4 m s^{-1}, 140 amu particle: 74.4 MeV, 1.01×10^4 m s^{-1}.

14.28 (a) 10 m s^{-1}, 2.37×10^5 Nm^{-2}; (b) 3×10^5 cm^3 min^{-1}; (c) 250 J kg^{-1}.

14.31 (a) 0.1 m; (b) 11.2 s.

CHAPTER 15

15.2 1.24×10^{-7} m^3.

15.4 (a) 3.655×10^5 Pa m^3, 2.968×10^5 Pa m^3, 2.07×10^5 Pa m^3, 1.45×10^5Pa m^3, 1.134×10^5 Pa m^3; (b) yes; (c) -270°C; (d) $pV = 761.94(t + 270)$ Pa m^3.

15.8 (a) 2.24×10^{-2} m^3 (22.4 liters).

15.10 (a) 4.02 atm; (b) 11.4 kg of air.

15.11 (a) $1/p$; (b) $[1 - b(N/V)]/[p - a(N/V)^2 + 2a(N/V)^3]$.

15.14 (a) 6.19×10^{-21} J, 3.86×10^{-2} eV; (b) (i) 1.92 km s^{-1}, (ii) 482 m s^{-1}, (iii) 514 m s^{-1} (iv) 1.36 km s^{-1}, (v) 411 m s^{-1}.

15.16 (a) 1.11 km s^{-1}; (b) 5.16 km s^{-1}; (c) 2.16 km s^{-1}.

15.20 (a) 5.2 K, 2.27×10^5 Pa, 71.1×10^{-6} m^3 mol^{-1}; (b) 33.2 K, 12.95×10^5 Pa, 79.8×10^{-6} m^3 mol^{-1}; (c) 154.3 K, 50.37×10^5 Pa, 95.5×10^{-6} m^3 mol^{-1}; (d) 647.1 K, 220.6×10^5 Pa, 91.5×10^{-6} m^3 mol^{-1}.

CHAPTER 16

16.1 (a) 1.295 kJ; (b) no.

16.2 (a) 0; (b) 1.295 kJ.

16.4 (a) 1.94 kcal, 8.10 kJ; (b) 3.87 kcal, 16.2 kJ.

16.6 (a) 60 J; (b) 70 J liberated; (c) $Q_{ad} = 50$ J, $Q_{ab} = 70$ J along the curved path.

16.9 (a) $Q = a\Delta T + 2b\Delta T(T_1 + T_2) - c\Delta T/T_1 T_2$, where $\Delta T = T_2 - T_1$; (b) $C_{p,\text{ave}} = a + 2b(T_1 + T_2) - c/T_1 T_2$.

16.11 (a) $p^{1-\gamma}T = \text{const.}$; (b) $TV^{\gamma-1} = \text{const.}$

16.13 (a) $kNT[\ln(V_2/V_1) + AN\Delta V/V_1 V_2 + \frac{1}{2}BN^2\Delta V(V_1 + V_2)/V_1^2 V_2^2$, where $\Delta V = V_2 - V_1$; (c) 1193.9 J vs 1274.3 for an ideal gas.

16.15 (b) $T_2 = 167.4$ K, $T_3 = 69.7$ K; (c) 33.8 kJ.

16.17 (b) 100.4 g; (c) 320 g.

16.19 (a) 3.77 kJ, 1.225.

16.21 (a) 5.55 kJ, 0, 5.55 kJ; (b) 11.1 $\ln(V_3/V_1)$ kJ, 11.1 $\ln(V_3/V_1)$ kJ, 0; (c) -9.25 kJ $-$ 3.70 kJ, -5.55 kJ; overall, $\Delta Q = \Delta W = 11.1 \ln(V_3/V_1) - 3.7$ kJ and $\Delta U = 0$.

16.24 (a) $T_B = (p_B V_B/p_A V_A)T_A$; (b) 200 J, 200J, 0; (c) 240.5 K.

16.26 (b) $W_{12} = 1.69$ MJ, $W_{23} = 0$, $W_{31} = -1.30$ MJ; (c) $\Delta S_{12} = 5.39$ kJ K^{-1}, $\Delta S_{23} = -5.39$ kJ K^{-1}, $\Delta S_{31} = 0$; (d) $W_T = 0.388$ MJ; (e) $\Delta S_{\text{cycle}} = 0$.

16.29 (a) 312 cal K^{-1}, -268.1 cal K^{-1}, $+43.9$ cal K^{-1}; (b) 312 cal K^{-1}, $+23.2$ cal K^{-1}.

16.32 (a) 9730 cal; (b) 9730 cal; (c) 0; (d) $+20.14$ cal K^{-1}.

16.34 (a) $+7.91$ cal K^{-1}; (b) -7.17 cal K^{-1}; (c) $+0.74$ cal K^{-1}.

CHAPTER 17

17.1 (a) 3.69×10^{-11}; (b) 4.42×10^{-10} (12 times as great).

17.3 (a) $e^{-2\varepsilon/kT}$; (b) 6.95×10^{-21}, 0.213, 0.629.

17.6 (a) 2×10^5 Pa; (b) 0, -553.5 J, $+553.5$ J; (c) 407 m s^{-1}.

17.8 83.7 K.

17.12 (a) $450k = 6.213 \times 10^{-21}$ J $= 0.0388$ eV; (b) 1.93 km s^{-1}, 481 m s^{-1}, 193 m s^{-1}.

17.14 (a) 7736 K; (b) 7.74×10^6 K; (c) 7.74×10^9 K.

17.18 $\frac{1}{2}[1 - \text{erf}(2)] = 2.35 \times 10^{-3}$, where $\text{erf}(x) = \frac{2}{\sqrt{\pi}}\int_0^x e^{-x}\,dx$ and $x = (m/2kT)^{1/2}v_x$.

17.21 (a) $S = -kN(\varepsilon U/kT) - \ln z$; (b) $C_v = 4kN_A(\varepsilon/kT)^2[e^{\varepsilon/kT} + e^{-\varepsilon/kT}]^{-2}$.

CHAPTER 18

18.1 (a) 2.31×10^{-5} m^2 s^{-1}; (b) 7.65×10^{14} particles s^{-1}.

18.3 (a) 12.5 K; (b) 9.6×10^4 J m^{-2} s^{-1}.

18.5 (a) 89.3°C; (b) 10.7 K m^{-1}, 89.3 K m^{-1}; (c) 0.411 J s^{-1}.

18.7 (a) 42.1 min; (b) 21 min.

18.9 (a) 1.12×10^7 N m^{-3}; (b) 1.57×10^{-1} N m^{-3}.

18.11 0.32 mm.

18.12 $a = -8.52$, $b = 830$.

18.4 (a) 6×10^9; (b) 4.65×10^9; (c) 6.0×10^5; (d) 1.46×10^{29}, 6.80×10^{28}, 1.46×10^{21}.

18.16 (a) 60.6%, 36.8%, 13.5%, 4.5×10^{-3}%; (b) $x = 0.693l$.

18.18 (a) $D \propto T/p$; (b) $k \propto \sqrt{T}$.

18.20 (b) 1.72×10^{-6} P K^{-1}.

18.22 $\kappa/D = \frac{3}{2}p/T = 555$ at STP, $\eta/D = pm/kT = 4.45 \times 10^{-2}\,m$ at STP (m in amu).

18.24 (a) 6.25×10^{-8} m, 9.33×10^9 s^{-1}; (b) 1.83×10^{-11} m.

CHAPTER 19

19.2 (a) 7.5 yr; (b) 6.25 yr; (c) 1.25 yr.

19.4 $0.483c$.

19.5 8.37 km.

19.7 (a) 3.01×10^9 m; (b) $0.897c$.

19.8 $0.98c$; (b) 5.1 yr.

19.10 (a) 1.6 m; (b) O': 2.0 m, O: 1.6 m.

19.12 (a) 4.63 m; (b) 1.67 m.

19.18 $a_{\text{elec}} = 1.0 \times 10^{12}$ m s^{-1}, $a_L = 5.12 \times 10^{11}$ m s^{-2}.

19.20 (a) $c/\sqrt{2}$; (b) $\sqrt{2}mc^2$, $0.414mc^2$.

19.22 (a) $(1 - 4.59 \times 10^{-4})c$; (b) 1.65×10^{-17} kg m s^{-1}.

19.25 (a) 0.075 MeV; (b) 0.582 MeV; (c) 0.464 MeV; (d) 1.986 MeV.

19.27 (a) 4.97 MeV/c; (b) 1766 MeV/c.

19.36 (b) for $Ft \ll mc$: $v \approx at$ and $x \approx \frac{1}{2}at^2$, for $Ft \gg mc$: $v \approx c$ and $x \approx ct$;
(c) $E = mc^2[1 + \frac{3}{2}x^2]/[1 + x^2]$, where $x = Ft/mc$.

19.38 Mercury: 1.77×10^{-4} s, 1.49×10^{-4} s, 1.34×10^{-4} s; Venus: 2.34×10^{-4} s, 2.07×10^{-4} s, 1.91×10^{-4} s.

19.40 (a) 2.24×10^{-6}; (b) 6.93×10^{-10}; (c) 1.22×10^{-10}.

CHAPTER 20

20.3 (a) 0; (b) $0.866c$; (c) $0.995c$; (d) $(1 - 5 \times 10^{-6})c$; (b) 0.511 MeV, 1.022 MeV, 5.11 MeV, 511 MeV; (f) 938.3 MeV, 1.876 GeV, 9.38 GeV, 938 GeV.

20.5 (a) $0.9901c$; (b) $7.12\ mc^2$.

20.7 (a) 10.898 GeV/c, 11.876 GeV; (b) 2.175 GeV.

20.9 (a) 57.9 GeV; (b) 1780 GeV.

20.10 (a) $c^2 p_1/E$, where $E = \sqrt{(m_1c^2)^2 + (p_1c)^2} + m_2c^2$;
(b) $-[m_3 - (m_1 + m_2)]c^2$.

20.12 (b) $p_1' = (c/2m)[(m^2 + m_1^2 - m_2^2)^2 - 4m^2m_1^2]^{1/2}$, $p_2' = -p_1'$.

20.17 (a) +4.82 MeV, none; (b) −519.4 MeV, 644.1 MeV.

20.20 (a) 73.0 MeV; (b) $0.866c$.

20.24 $p' = p(A \cos\theta + BD^{1/2})/([B/c]^2 - p_1^2\cos^2\theta)$ and
$E' = (AB + D^{1/2}p_1^2c^2\cos\theta)/([B/c]^2 - p_1^2\cos^2\theta)$
where $A = m_1^2c^2 + m_2E_1$, $B = E_1 + m_2c^2$ and $D = m_2^2 - m_1^2\sin^{-2}\theta$.

CHAPTER 21

21.1 (a) 4.21×10^{-8} N; (b) 1.16×10^{36} times stronger.

21.3 (a) $\tan\theta \sin^2\theta = K_e q^2/4mgl^2$.

21.6 (a) 1.87×10^5 NC^{-1} at 11.5°, with respect to the line parallel to the line of the charges, toward the negative charge; (b) 2.414 m from the positive charge, along the line joining the charges; (c) 0.414 m from the positive charge and 1.414 m from the negative charge.

21.7 28.8 kV.

21.9 (a) 1.01×10^3 NC^{-1}; (b) 2.66×10^6 m s^{-1}; (c) 20.2 V.

21.11 (a) $\boldsymbol{i}$ 50 NC^{-1}; (b) $\boldsymbol{i}$ 10.8 NC^{-1}.

21.13 2.

21.14 (a) 3 m; (b) 0.2 μC.

21.17 7.7×10^{-3} J.

21.21 (a) $V = K_e q/(a^2 + x^2)^{1/2}$;
(b) $\mathcal{E} = \boldsymbol{i}K_e qx/(a^2 + x^2)^{3/2}$, where x is the distance along the axis.

21.23 (a) $v_e = 5.93 \times 10^5 (V)^{1/2}$ m s^{-1};
(b) $v_p = 1.38 \times 10^4 (V)^{1/2}$.

21.26 (b) $(2eV_0/mv^2)^{1/2}$; (c) neV_0.

21.28 (a) 6.25×10^{15} protons s^{-1}; (b) 800 W; (c) 1.24×10^7 m s^{-1}; (d) 153 cal s^{-1}.

CHAPTER 22

22.1 (a) 5.5×10^{-5} T; (b) 9.6×10^6 rad s^{-1}.

22.3 (a) 0.138 m; (b) 0.106 m; (c) 7.1×10^7 rad s^{-1}.

22.6 $\frac{1}{2}$T, in negative Z direction.

22.7 (a) 1.75×10^7 m s^{-1}.

22.10 (a) 4×10^6 m s^{-1}; (b) 3.50×10^{-26} kg.

22.12 (a) Classical velocities, around 10^5 m s^{-1}; (b) 0.122 m.

22.14 (a) 1.31 T, 2.51×10^7 m s^{-1}, 1.06×10^{-12} J = 6.59 MeV, 165 turns;
(b) 1.30 T, 2.51×10^7 m s^{-1}, 2.01×10^{-12} J = 13.1 MeV, 164 turns.

22.16 9.29×10^{-24} A m^2.

22.19 (a) 2.3×10^{-20} N electric repulsion, 2.56×10^{-25} N magnetic attraction;
(b) 2.3×10^{-20} N electric repulsion only.

CHAPTER 23

23.1 (a) 2.73 g; (b) 2.32×10^{22} atoms of Cu.

23.3 1.1391×10^{-13} m, 1.1392×10^{-13} m, 1.148×10^{-13} m.

23.5 About 14.

23.7 (a) 2.18×10^6 m s^{-1}, 1.09×10^6 m s^{-1}, 7.28×10^5 m s^{-1}; (b) 9.27×10^{-24} A m^2, 1.85×10^{-23} A m^2, 2.78×10^{-23} A m^2.

23.8 (a) 8.20×10^6; (b) 1.94×10^4; (c) 610:1, 2.6×10^5:1.

23.10 (a) 10.19 eV; (b) much less, 1.88 eV.

23.12 (a) 9, if spin is not considered; (b) 2.31×10^{-4} eV.

23.14 1.58×10^{-4} m.

23.17 For $j = l + \frac{1}{2}$, $\Delta E_{SL} = \frac{1}{2}a_{nl}l$; for $j = l - \frac{1}{2}$, $\Delta E_{SL} = -\frac{1}{2}a_{nl}l(l+1)$.

CHAPTER 24

24.1A (a) 2.15 mΩ; (b) 1.59 mV; (c) 1.18 mW.

24.3A (a) Place a 3.8 Ω resistor in parallel with the meter; (b) place a 21.6 Ω resistor in series with the meter.

24.5A (a) 9 Ω, 10 Ω, 7.5 Ω, 10 Ω; (b) for case (*a*): 3 Ω: 48 A, 144 V; 12 Ω: 8 A, 96 V; 6 Ω: 16 A, 96 V; 4 Ω: 24 A, 96 V; 20 Ω: 12 A, 240 V; 5 Ω: 60 A, 300 V. For case (*c*): 4 Ω: 9 A, 36 V; 9 Ω: 6 A, 54 V; 16 Ω: 3 A, 48 V; 3 Ω: 2 A, 6 V; 30 Ω: 3 A, 90 V.

24.6A (a) 15 W; (b) 1.58 A in the single resistor and 0.791 A in each of the parallel resistors; (b) 9.49 V.

24.8A (a) 2.06 V; (b) 0.99 Ω.

24.10A −1.5 V, 2.5 V, −0.5 V.

24.13A (a) 1.591 V; (b) 0.843 V.

24.1B 1: $-\boldsymbol{j}0.3$ N; 2: 0; 3: $\boldsymbol{i}0.296$ N; 4: $(\boldsymbol{i}-\boldsymbol{j})\,0.296$ N; 5: $(\boldsymbol{i}+\boldsymbol{j})0.253$ N.

24.3B (a) On the 2 sides parallel to the Y-direction, equal and opposite forces of 0.16 N in the Z-direction; on the 2 horizontal sides, equal and opposite forces in the Y-direction of 0.06 N; (b) $\boldsymbol{j}8.31 \times 10^{-3}$ Nm.

24.6B (a) $\boldsymbol{i}10^{-2}$ N; (b) $-\boldsymbol{k}5 \times 10^{-4}$ Nm.

24.8B (i) 1.0×10^{-5} T; (ii) 2.0×10^{-7} T.

24.11B 6.9×10^{-3} T.

24.13B (a) 3.2×10^{-5} N, repulsive; (b) 0; (c) 0.

CHAPTER 25

25.1 (a) $\mathscr{E} = A/r^2$, $V = A/r$; (b) $\mathscr{E} = (A/r^2)(r^3 - R_2^3)/(R_1^3 - R_2^3)$, $V = (A/R_1) + A(R_1 - r)[\frac{1}{2}(R_1 + r) - R_2^3/rR_1]/(R_1^3 - R_2^3)$; (c) $\mathscr{E} = 0$, $V =$ a constant $= (A/R_1) + A(R_1 - R_2)[\frac{1}{2}(R_1 + R_2) - R_2^2/R_1]/(R_1^3 - R_2^3)$.

25.3 0.12 μC on the smaller and 0.18 μC on the larger.

25.5 (a) 10^3 V, $10^3/d$, where d is the distance between the plates; (b) 2×10^9 V, electric field is the same; (c) 5×10^{-4} J.

25.8 (a) C_1: 600 μC, C_2: 200 μC, C_3: 400 μC; (b) C_1: 200 V, C_2: 100 V, C_3: 100 V; (c) 9.0×10^{-2} J.

25.11 (a) $-(\varepsilon_0 S/x^2)\,dx$; (b) $\frac{1}{2}(Q^2/\varepsilon_0 S)\,dx$; (c) $\frac{1}{2}(Q^2/\varepsilon_0 S)$.

25.12 (b) $q = (q_0/A)[A + C_1(e^{At/RC_1C_2} - 1)]$, $I = (q_0/RC_2)e^{At/RC_1C_2}$, where $A = C_2 - C_1$.

CHAPTER 26

26.2 1.4×10^{-2} T.

26.4 (a) 0, 2.5×10^{-5} T, 5×10^{-5} T; (b) 5×10^{-5} T, 2.5×10^{-5} T, 1.33×10^{-6} T; (c) 4×10^{-4} m and 4.0 m.

26.6 4.40×10^{-2} T.

26.9 (a) For $r < R_1$: $2K_m I/R_1^2$, for $R_1 < r < R_2$: $2K_m I/r$, for $R_2 < r < R_3$: $2K_m(I/r)(R_3^2 - r^2)/(R_3^2 - R_2^2)$, for $r > R_3$: 0.

26.11 $\mu_0 Ib \ln(r + a/r)$.

CHAPTER 27

27.1 (a) -4πV; (b) $+4\pi$V; (c) $+8\pi$V; (d) $+4\pi$V; (e) $+8\pi$V.

27.3 (b) 5×10^{-3} NC^{-1}, clockwise; (c) 5×10^{-3} s V; (d) $\pi \times 10^{-3}$ V.

27.6 (a) 11.8 V; (b) $V(t) = \mathscr{B}S \sin \omega t$.

27.8 A: 0; C: 0.707 V; D: 1.414 V.

27.10 $\frac{1}{2}\mu_0 NR$, where R is the radius of the torus.

27.12 $2K_m Iabv/[(r + vt)(r + a + vt)]$.

27.14 (a) 3.01 mH; (b) 6.03×10^{-2} V.

27.17 (a) −0.251 V; (b) 0.251 V; (b) 0.502 V.

27.20 (a) $[R^2 + (1/\omega C)^2]^{1/2}$ current leads voltage by $\arctan(1/\omega RC)$; (b) $[\omega L - (1/\omega C)]$, angle is $\pm\pi/2$.

CHAPTER 28

28.1 (a) 3.33 m; (b) $\xi = \xi_0 \sin 2\pi(0.3x - 0.6t)$.

28.5 (a) 0, 8.86×10^{-3} m, 1.69×10^{-2} m, 2.35×10^{-2} m; (b) 8.86×10^{-3} m, 2.99×10^{-3} m, -2.99×10^{-3} m; (c) $-6 \times 10^{-2} \cos(3x - 2t)$ m s^{-1}; (d) 6×10^{-2} m s^{-1}; (e) 0.667 m s^{-1}.

28.7 35.1 m s^{-1}.

28.10 28.8 amu.

28.12 (a) 9.03 m s^{-1}; (b) $\xi = \xi_0 \sin 2\pi(44.3x - 400t)$.

28.14 (a) 5×10^{-2} m s^{-1}; (b) 0.955 Hz; (c) 5.24×10^{-2} m; (d) $\xi = 0.1 \sin(6t - 120)$ m, $\xi = 0.1 \sin(6t - 360)$ m.

28.16 (b) counterclockwise, standing on $+X$-axis, looking in toward the origin; (c) $\xi_y = \xi_0 \sin(kx - \omega t)$, $\xi_z = -\cos(kx - \omega t)$.

28.18 (a) 0.244 m s^{-1}; (b) 1.25 m s^{-1}; (c) 3.95 m s^{-1}; (d) 12.5 m s^{-1}.

28.20 (a) $\xi = 10^{-4} \sin 2\pi(1.98 \times 10^{-3}x - 10t)$; (b) $1.56\pi \times 10^{-2}$ J m^{-3}; (c) $1.26\pi^3 \times 10^{-3}$ W; (d) same as (c).

28.21 Faintest: (a) 4.49×10^{-13} W m^{-2}; -3.45 db; (b) 1.43×10^{-11} m; Loudest: (a) 0.881 W m^{-2}, $+119$ db; (b) 2.01×10^{-5} m.

28.23 (a) Increased $4\times$; (b) increase by $(10)^{1/2}$.

28.25 (a) 531.2 Hz; (b) 472.2 Hz.

28.27 (a) 0.373 m, 911.8 Hz; (b) 0.312 m, 1088.2 Hz. Who is moving is important only if motion is large relative to the velocity of sound.

CHAPTER 29

29.1 (a) (i) 3 m, (ii) linearly polarized in the XY-plane, (iii) propagates in the $+X$ direction; (b) $\mathscr{B}_x = \mathscr{B}_y = 0$, $\mathscr{B}_z = \frac{1}{6} \times 10^{-8} \cos[2\pi \times 10^8(t - x/c)]$T; (c) 3.32×10^{-4} W m^{-2}.

29.3 (a) $\mathscr{E}_x = 0$, $\mathscr{E}_y = \mathscr{E}_z = \varepsilon_0 \sin\theta$ and $\mathscr{B}_x = 0$, $\mathscr{B}_y = -(\varepsilon_0/c)\sin\theta$, $\mathscr{B}_z = (\varepsilon_0/c)\sin\theta$; (c) $\mathscr{E}_x = 0$, $\mathscr{E}_y = \varepsilon_0 \cos\theta$, $\mathscr{E}_z = \varepsilon_0 \sin\theta$ and $\mathscr{B}_x = 0$, $\mathscr{B}_y = -(\varepsilon_0/c)\sin\theta$, $\mathscr{B}_z = (\varepsilon_0/c)\cos\theta$. In all cases $\theta = kx - \omega t$.

29.5 (a) $\frac{1}{3} \times 10^{-10}$ T; (b) 4.42×10^{-16} J m^{-3}; (c) 8.85×10^{-16} Pa; (b) 4.42×10^{-16} Pa.

29.6 (a) 1.03×10^3 NC^{-1}; (b) 3.42×10^{-6} T.

29.8 (a) $\frac{1}{3} \times 10^{-7}$ T; (b) 1.33 kW.

29.10 (a) 5.76×10^{-12} W; (b) 1.74×10^{11} (2.8×10^{-13} moles).

29.12 (a) 1.77×10^{-21} W; (b) 5.26×10^{-28} W.

29.14 (a) $z = [(1 + v/c)/(1 - v/c)]^{1/2} - 1$; (c).

29.16 (a) $0.024c$; (b) 3.15×10^{24} m = 333 Mlyr.

CHAPTER 30

30.1 (a) 274 MeV, 4.53×10^{-15} m; (b) 40 MeV, 3.10×10^{-14} m.

30.4 (a) (i) 1.01×10^{-10} m, (ii) 58.8°; (b) 143.4 eV.

30.6 (a) 1.45 V; (b) 1.45 eV, 7.13×10^5 m s^{-1}.

30.9 1.13×10^4 photons s^{-1}.

30.12 (a) 6.04×10^9 electrons s^{-1} m^{-2}; (b) 1.93×10^{-9} J s^{-1} m^{-2}; (c) 1.10 eV maximum.

30.13 (a) 157.4 keV; (b) 7.88×10^{-12} m.

CHAPTER 31

31.1 2.66×10^{16} Hz, 1.13×10^{-8} m.

31.3 3.37×10^{-19} J = 2.10 eV, 2.10 eV$/c = 1.12 \times 10^{-27}$ kg m s^{-1}.

31.5 $\Delta\lambda_{H-D} = 1.78$ Å, $\Delta\lambda_{H-T} = 2.38$ Å.

31.7 7.09 eV.

31.10 Al: 8.40 Å, 1470 eV; K: 3.59 Å, 3452 eV; Fe: 1.74 Å, 7121 eV; Ni: 1.55 Å, 7965 eV; Zn: 1.34 Å, 9221 eV; Mo: 0.664 Å, 18,700 eV; Ag: 0.525 Å, 23,600 eV.

31.12 (a) 2.8×10^{-4} eV, 4.44×10^{-3} m; 5.6×10^{-4} eV, 2.22×10^{-3} m; (b) far infrared.

31.15 (a) 1.9×10^3 N m^{-1}.

31.19 (a) 6.06×10^5; (b) 5.93×10^6; (c) 3.16×10^6.

31.21 (a) One; (b) about 950,000.

31.23 (a) $E(\nu) = (8\pi h\nu^3/c^3)e^{-h\nu/kT}$; (b) $E(\nu) = (8\pi\nu^2/c^3)kT$.

31.25 3.5×10^{-11} m.

CHAPTER 32

32.3 (a) 26°11′, 22°29′; (b) 1.37 cm.

32.4 16.6°.

32.7 56.3°, 33.7°.

32.9 (a) 53.1° for water with $n = 1.33$; (b) horizontal, parallel to the water's surface.

32.10 (a) 3.97×10^{-7} m, 3.55×10^{-7} m; (b) both 5.096×10^{14} Hz.

32.12 Elliptic polarization, semimajor axis along Y-axis.

32.14 (a) 6055 Å, 6297 Å, 6559 Å, 6844 Å; (b) 6173 Å, 6425 Å, 6714 Å, 6996 Å; (c) 6055 Å and 6559 Å.

32.16 299.25°.

32.18 Yes.

32.20 (a) -0.2, 0.8; (b) 0.2, 1.2; (c) only phase change takes place on reflection from glass to air.

CHAPTER 33

33.3 (a) 0.778 m, −0.556; (b) 1.0 m, −1.0; (c) 1.33 m, −1.67; (d), ∞, ∞; (e) −0.75 m, −2.5.

33.5 (a) 7.97 cm; (b) 2.1.

33.7 Either 40 cm or 37.5 cm.

33.9 3.2 cm, compared with a focal length of 5 cm.

33.13 (a) 43.8 cm, to the right of the flat surface; (b) 1.71.

33.15 (a) 0.24 m; (b) (i) −0.343 m, −0.43; (ii) −0.48 m, −1; (iii) −0.6 m, −1.5; (iv) ∞, ∞; (v) 1.2 m, 6.

33.19 (a) ∞, ∞; (b) 60 cm, 1.5.

33.20 (a) 0.8 m; (b) 2.4 m; (c) 0.4 m.

33.22 $F_i = 4.17$ cm, $F_0 = -1.67$ cm.

33.24 21.1°.

33.26 (a) 1.521; (b) 40°.

33.30 (a) 0.266 m; 0.34 cm.

CHAPTER 34

34.1 (a) 0.65 mm; (b) 1.62 mm, 3.25 mm.

34.3 (b) 0.29 mm.

34.5 (a) With respect to the midpoint, there are two minima between the sources, $\frac{1}{8}$ m on either side of the midpoint; outside either source, the condition is always minimum ($\delta = 3\pi$); (b) none, the plane is a plane of constructive interference ($\delta = 0$).

34.7 (c) 4″ of arc.

34.10 (a) 0.11° = 1.9 × 10^{-3} rad; (b) 1.72°.

34.13 1.18 × 10^{-5} m.

34.16 (a) Increases by $\sqrt{2}$; decreases by $1/\sqrt{2}$; (c) halved; (d) halved.

34.18 2.88%.

34.21 (a) With $v_0 = v/2a$, $v = v_0\sqrt{n_1^2 + n_2^2 + n_3^2}$; (b) v_0: triply degenerate with integers 1, 0 and 0; $v_1 = \sqrt{2}v_0$, triply on integers 1, 1 and 0; $v_2 = \sqrt{3}v_0$, nondegenerate on integers 1, 1 and 1; $v_3 = 2v_0$, triply degenerate on integers 2, 0 and 0; $v_4 = \sqrt{5}v_0$, sixtuple degenerate on integers 2, 1 and 0; $v_5 = \sqrt{6}v_0$, $v_6 = 3v_0$.

34.23 (a) 1.25 n × 10^8 Hz; (b) $1.25(n_1^2 + n_2^2)^{1/2} \times 10^8$ Hz.

CHAPTER 35

35.1 0.56 mm.

35.4 5 × 10^{-7} m and 4 × 10^{-7} m.

35.6 (b) 0.24 mm × 0.12 mm.

35.8 (a) 2.05 × 10^{-6} m; (b) 7.03 × 10^{-6} m.

35.9 9.69 km.

35.12 12.5°.

35.16 (a) 9.79 × 10^{-11} m; (b) 20.3°.

35.17 (a) 2.13 × 10^{-9} m; (b) 4.03°, 8.10°.

35.18 1.25°.

CHAPTER 36

36.1 (a) 1.29 × 10^{-9} m, 1.23 × 10^{-10} m, 3.88 × 10^{-12} m, 8.58 × 10^{-13} m; (b) the 1 eV and 100 eV electrons are most scattered; (c) 32.7 eV.

36.3 $\sin\theta = (nh/2d)[2me\Delta V]^{-1/2}$.

36.5 5.7°.

36.7 (a) ±0.63 m; (b) 0 m, totally determinable.

36.9 2.12 × 10^{-13} m, 1.6 × 10^8 Hz.

36.11 (a) $\Delta p = 3.72 \times 10^{-20}$ kg m s^{-1}, which makes the electron's motion highly relativistic; (b) $\Delta p \sim 3.72 \times 10^{-7}$ kg m s^{-1} and the proton would also be relativistic.

36.12 5 × 10^{-5} ΔV, where ΔV is the uncertainty in the accelerating voltage.

CHAPTER 37

37.1 (a) 29.4 eV.

37.5 (a) 8.27 × 10^{-13} eV; (b) 1.24 × 10^{-12} eV.

37.7 (c) Yes, when $E < 0$.

37.10 2.7 MeV.

CHAPTER 38

38.2 (a) $n = 1$: $3a_0/2Z$; $n = 2$: $l = 0$, $6a_0/Z$, $l = 1$, $5a_0/Z$; $n = 3$: $l = 0$, $27a_0/2Z$, $l = 1$, $25a_0/2Z$, $l = 2$, $21a_0/2Z$.

38.3 −57.8 eV, −55.9 eV, −55.2 eV.

38.5 109.5°.

38.6 (a) −16.25 eV; (b) 10.95 eV; (c) 15.43 eV.

38.12 (b) 4.89 × 10^{-3} eV.

38.13 For $^{1}H^{35}Cl$: (a) 2.645×10^{-3} eV, 7.936×10^{-3} eV; (b) 6.396×10^{11} Hz, 4.691×10^{-4} m and 1.280×10^{12} Hz, 2.345×10^{-4} m.

38.16 3.686 eV.

CHAPTER 39

39.1 (a) 18.9% ^{10}B and 81.1% ^{11}B; (b) 19.78% ^{10}B and 80.22% ^{11}B.

39.3 (a) 64.0 MeV; (b) 20.76 MeV.

39.5 ^{7}Li: 39.24 MeV, 5.61 MeV/nucleon; ^{16}O: 127.62 MeV; ^{57}Fe: 499.90 MeV; ^{176}Lu: 1417.97 MeV.

39.7 ^{1}H: 1.0078 amu; ^{2}H: 2.0141 amu; ^{16}O: 15.9949 amu.

39.10 (a) 222 keV; (b) 74 keV; (c) 38.2 keV.

39.13 (a) $I_{av} = 2.24 \times 10^{-54}$ kg m^2.

39.15 (b) 847 MeV; (c) 109.7 MeV.

CHAPTER 40

40.2 (b) 18.02 s; (c) 1.517×10^{7} atoms.

40.5 28 mg.

40.7 1.811 MeV, 51.8 keV.

40.9 γ-ray emissions of 0.184 MeV, 0.251 MeV and 0.255 MeV.

40.11 (a) $Q(\beta^-) = (M_z - M_{z+1})c^2$; (b) $Q(\beta^+) = (M_z - M_{z-1} - 2m_e)c^2$; (c) $Q(\text{EC}) = (M_z - M_{z-1})c^2$, where the M's are atomic masses.

40.12 (b) $Q(\beta^+) = 0.64$ MeV, $Q(\beta^-) = 0.565$ MeV, $Q(\text{EC}) = 1.66$ MeV.

40.14 18.63 keV.

40.16 (a) $^{24}_{11}Na$; (b) $^{30}_{15}P$; (c) $^{32}_{15}P$; (d) γ; (e) $^{3}_{1}H$; (f) $^{116}_{46}In$; (g) $^{58}_{29}Cu$; (h) γ.

40.18 Protons: $^{12}C(p, \gamma)\,^{13}N$; deuterons: $^{12}C(d, \gamma)^{14}N$, $^{12}C(d, p)^{13}C$, $^{12}C(d, n)^{13}N$, $^{12}C(d, \alpha)^{10}B$; alphas: $^{12}C(\alpha, \gamma)^{16}O$.

40.20 (a) 27.9824 amu; (b) 23.9867 amu; (c) 27.9819 amu; (d) 24.5859 amu.

40.22 1.74×10^{14} atoms.

40.23 188 MeV.

40.25 (b) 1.2×10^{-8} kg s^{-1}; (c) 2.6 yr.

40.27 (a) 7.28 MeV; (c) 2.92×10^{23} W.

40.28 1.28×10^{11} yr.

CHAPTER 41

41.2 502 MeV, 680 MeV.

41.4 (a) 0.511 MeV; forward electron: 1.44 MeV, other: 0.58 MeV.

41.6 (a) 2.53 keV, 2.85 × [illegible]^{13} m; (b) 2.53 keV; (c) $^{110}_{47}Ag$.

41.7 (a) $Q = -495.7$ MeV, 4.836 GeV; (b) -140 MeV, 290.45 MeV.

41.9 148.7 MeV.

41.13 143.8 MeV.

Index

Periodic Table of the Elements

The atomic masses, based on the exact number 12.00000 as the assigned atomic mass of the principal isotope of carbon, ^{12}C, are the most recent (1961) values adopted by the International Union of Pure and Applied Chemistry. The unit of mass used in this table is called *atomic mass unit* (amu): 1 amu $= 1.6604 \times 10^{-27}$ kg. The atomic mass of carbon is 12.01115 on this scale because it is the average of the different isotopes naturally present in carbon. (For artificially produced elements, the approximate atomic mass of the most stable isotope is given in brackets.)

Group→		I	II	III	IV	V	VI	VII	VIII			0
Period	Series											
1	1	1 H 1.00797										2 He 4.0026
2	2	3 Li 6.939	4 Be 9.0122	5 B 10.811	6 C 12.01115	7 N 14.0067	8 O 15.9994	9 F 18.9984				10 Ne 20.183
3	3	11 Na 22.9898	12 Mg 24.312	13 Al 26.9815	14 Si 28.086	15 P 30.9738	16 S 32.064	17 Cl 35.453				18 Ar 39.948
4	4	19 K 39.102	20 Ca 40.08	21 Sc 44.956	22 Ti 47.90	23 V 50.942	24 Cr 51.996	25 Mn 54.9380	26 Fe 55.847	27 Co 58.9332	28 Ni 58.71	
	5	29 Cu 63.54	30 Zn 65.37	31 Ga 69.72	32 Ge 72.59	33 As 74.9216	34 Se 78.96	35 Br 79.909				36 Kr 83.80
5	6	37 Rb 85.47	38 Sr 87.62	39 Y 88.905	40 Zr 91.22	41 Nb 92.906	42 Mo 95.94	43 Tc [99]	44 Ru 101.07	45 Rh 102.905	46 Pd 106.4	
	7	47 Ag 107.870	48 Cd 112.40	49 In 114.82	50 Sn 118.69	51 Sb 121.75	52 Te 127.60	53 I 126.9044				54 Xe 131.30
6	8	55 Cs 132.905	56 Ba 137.34	57–71 Lanthanide series*	72 Hf 178.49	73 Ta 180.948	74 W 183.85	75 Re 186.2	76 Os 190.2	77 Ir 192.2	78 Pt 195.09	
	9	79 Au 196.967	80 Hg 200.59	81 Tl 204.37	82 Pb 207.19	83 Bi 208.980	84 Po [210]	85 At [210]				86 Rn [222]
7	10	87 Fr [223]	88 Ra [226.05]	89–Actinide series**								

* Lanthanide series:	57 La 138.91	58 Ce 140.12	59 Pr 140.907	60 Nd 144.24	61 Pm [147]	62 Sm 150.35	63 Eu 151.96	64 Gd 157.25	65 Tb 158.924	66 Dy 162.50	67 Ho 164.930	68 Er 167.26	69 Tm 168.934	70 Yb 173.04	71 Lu 174.97
** Actinide series:	89 Ac [227]	90 Th 232.038	91 Pa [231]	92 U 238.03	93 Np [237]	94 Pu [242]	95 Am [243]	96 Cm [245]	97 Bk [249]	98 Cf [249]	99 Es [253]	100 Fm [255]	101 Md [256]	102 No	103

Fundamental Constants

Constant	Symbol	Value
Velocity of light	c	2.9979×10^{8} m s^{-1}
Elementary charge	e	1.6021×10^{-19} C
Electron rest mass	m_e	9.1091×10^{-31} kg
Proton rest mass	m_p	1.6725×10^{-27} kg
Neutron rest mass	m_n	1.6748×10^{-27} kg
Planck constant	h	6.6256×10^{-34} J s
	$\hbar = h/2\pi$	1.0545×10^{-34} J s
Charge-to-mass ratio for electron	e/m_e	1.7588×10^{11} kg^{-1} C
Quantum charge ratio	h/e	4.1356×10^{-15} J s C^{-1}
Bohr radius	a_0	5.2917×10^{-11} m
Compton wavelength:		
of electron	$\lambda_{C,e}$	2.4262×10^{-12} m
of proton	$\lambda_{C,p}$	1.3214×10^{-15} m
Rydberg constant	R	1.0974×10^{7} m^{-1}
Bohr magneton	μ_B	9.2732×10^{-24} J T^{-1}
Avogadro constant	N_A	6.0225×10^{23} mol^{-1}
Boltzmann constant	k	1.3805×10^{-23} J K^{-1}
Gas constant	R	8.3143 J K^{-1} mol^{-1}
Ideal gas normal volume (STP)	V_0	2.2414×10^{-2} m^{3} mol^{-1}
Faraday constant	F	9.6487×10^{4} C mol^{-1}
Coulomb constant	K_e	8.9874×10^{9} N m^{2} C^{-2}
Vacuum permittivity	ϵ_0	8.8544×10^{-12} N^{-1} m^{-2} C^{2}
Magnetic constant	K_m	1.0000×10^{-7} m kg C^{-2}
Vacuum permeability	μ_0	1.2566×10^{-6} m kg C^{-2}
Gravitational constant	γ	6.670×10^{-11} N m^{2} kg^{-2}
Acceleration of gravity at sea level and at equator	g	9.7805 m s^{-2}

Numerical constants: $\pi = 3.1416$; $e = 2.7183$; $\sqrt{2} = 1.4142$; $\sqrt{3} = 1.7320$